AN INTRODUCTION TO
GENETIC ANALYSIS

AN INTRODUCTION TO
GENETIC ANALYSIS

Seventh Edition

Anthony J. F. Griffiths
University of British Columbia

Jeffrey H. Miller
University of California, Los Angeles

David T. Suzuki
University of British Columbia

Richard C. Lewontin
Harvard University

William M. Gelbart
Harvard University

W. H. FREEMAN AND COMPANY
New York

Acquistions Editor:	Nicole Folchetti
Development Editor:	Randi Rossignol
Project Editor:	Mary Louise Byrd
Cover and Text Design:	Cambraia (Magalhaes) Fernandes
Illustration Coordinator:	Lou Capaldo
Illustrations:	Hudson River Studio
Photo Researcher:	Jennifer MacMillan
Production Coordinator:	Susan Wein
New Media and Supplements Editor:	Charles Van Wagner
Composition:	Progressive Information Technologies
Manufacturing:	Von Hoffman Press
Marketing Director:	John Britch

Library of Congress Cataloging-in-Publication Data

An introduction to genetic analysis/Anthony J. F. Griffiths . . . [et
al.].—7th ed.
p.   cm.
Includes bibliographical references and index.
ISBN 0-7167-3520-2
1. Genetics.   I. Griffiths, Anthony J. F.
QH430.I62 1999
576.5—dc21   99-43254
CIP

Printed in the United States of America.
Third Printing, 2002

Contents in Brief

Contents

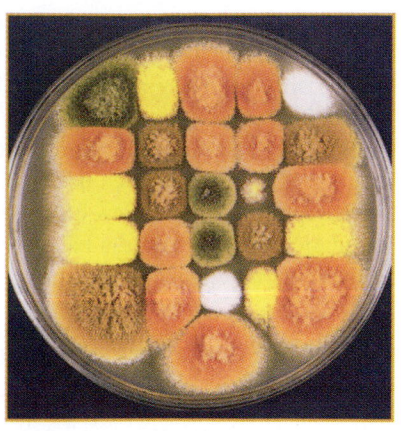

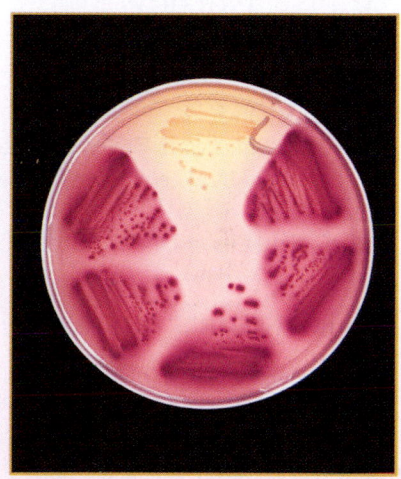

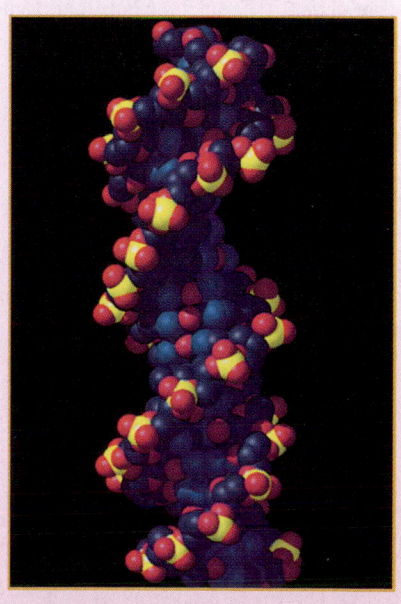

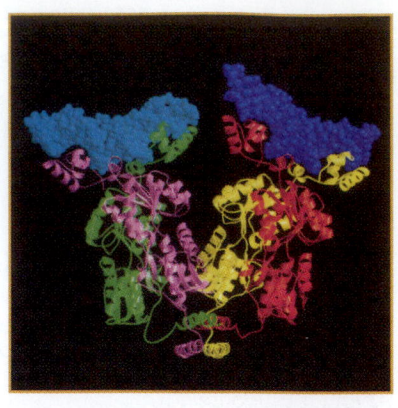

GENERATION OF GENETIC VARIATION

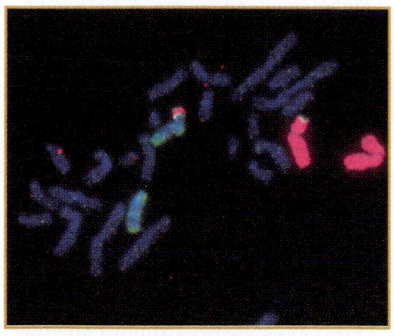

DEVELOPMENT

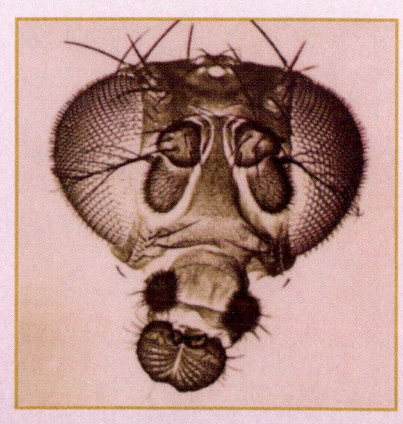

Contents

GENES AT THE POPULATION LEVEL

Preface

Genetics and Education

Genetics has become an indispensable component of almost all research in modern biology and medicine. Research publications investigating any biological process, from the molecular level all the way to the population level, use the "genetic approach" to gain understanding of that process. Thus, no student of the life sciences can afford to be ignorant of the science of genetics.

Genetics has also risen to a position of prominence in human affairs. Special types of plants, animals, and microbes have been developed for human foods, drugs, and myriad other uses. Molecular genetics is the central foundation of the burgeoning biotechnology industry. At the philosophical level, genetics has presented humans with a large number of ethical dilemmas, which regularly surface in the media. Some examples are genetically modified foods, eugenics, privacy of genetic information about individuals, and loss of genetic diversity in nature. Students must be knowledgeable about genetics in order to understand these issues and make informed decisions about them. Lastly, genetic insight has radically affected the human worldview — the way we see ourselves in relation to other organisms.

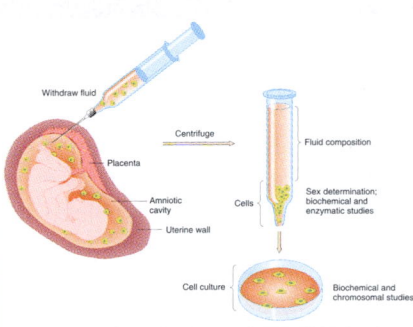

Figure 13-28

Figure 13-18

The Balanced Approach to Teaching Genetics

Genetics has risen to such prominence through the powerful merger of classical and molecular approaches. Each analytical approach has its unique strengths: classical (organismal) genetics is unparalleled in its ability to explore uncharted biological terrain; molecular genetics is equally unparalleled in its ability to unravel cellular mechanisms. It would be unthinkable to teach one without the other, and each is given due prominence in this book. Armed with both approaches, students are able to form an integrated view of genetic principles.

The partnership of classical and molecular genetics has always presented a teaching dilemma: which of the two partners should the student be introduced to first, the classical or the molecular? We believe that students begin much as biologists did at the turn of the century, asking general questions about the laws governing heredity. Therefore, the first half of the book introduces the intellectual framework of classical eukaryotic genetics in more or less historical sequence. However, molecular information is provided where appropriate. Our students' knowledge base and our years of teaching have together caused us to rethink the traditional organization. In this new edition we have integrated a good deal of molecular genetics into the early chapters. Thus, we reinforce the students' knowledge of DNA structure and function and avoid presenting the gene as an abstraction. The second half of the book pursues the details of molecular genetics.

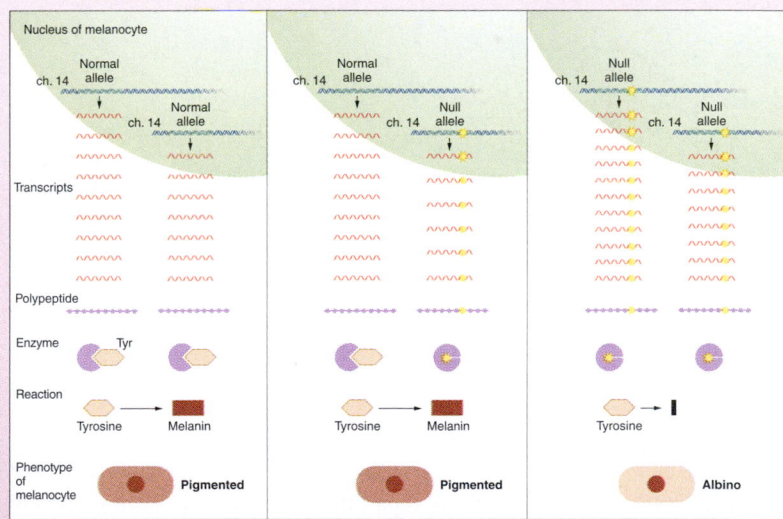

Figure 1-15

Focus on Genetic Analysis

True to its title, the theme of this book is genetic analysis. This theme emphasizes our belief that the best way to understand genetics is by learning how genetic inference is made. On almost every page, we recreate the landmark experiments in genetics and have the students analyze the data and draw conclusions as if they had done the research themselves. This proactive process teaches students how to think like scientists. The modes of inference and the techniques of analysis are the keys to future exploration.

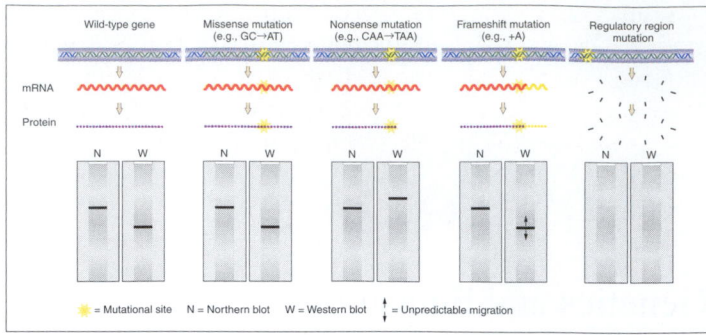

Figure 15-2

Similarly, quantitative analysis is central to the book because many of the new ideas in genetics, from the original conception of the gene to such modern techniques as SSLP mapping, are based on quantitative analysis. The problems at the end of each chapter provide students with the opportunity to test their understanding in quantitative analyses that effectively simulate the act of doing genetics.

Fostering Analytical Skills

A great deal of effort has been put into encouraging students to practice and hone their analytical and problem-solving skills. We provide a great variety of solved and unsolved problems and a wide range of study aids.

The **Problems** section continues to be one of the strengths of the book. Problems are generally arranged to start from simple and proceed to the more difficult. Particularly challenging problems are marked with an asterisk. All problems have been classroom tested. Answers to selected problems are found at the back of the book, and the full set of solutions is in the Student Companion, all prepared by Bill Fixsen (Harvard University).

54. To understand the genetic basis of locomotion in the diploid nematode *Caenorhabditis elegans*, recessive mutations were obtained, all making the worm "wiggle" ineffectually instead of moving with its usual smooth gliding motion. These mutations presumably affect the nervous or muscle systems. Twelve homozygous mutants were intercrossed, and the F₁ hybrids were examined to see if they wiggled. The results were as follows, where a plus sign means that the F₁ hybrid was wild type (gliding) and "w" means that the hybrid wiggled.

	1	2	3	4	5	6	7	8	9	10	11	12
1	w	+	+	+	w	+	+	+	+	+	+	+
2		w	+	+	+	w	+	w	+	w	+	+
3			w	w	+	+	+	+	+	+	+	+
4				w	+	+	+	+	+	+	+	+
5					w	+	+	+	+	+	+	+
6						w	+	w	+	w	+	+
7							w	+	+	+	w	w
8								w	+	w	+	+
9									w	+	+	+
10										w	+	+
11											w	w
12												w

a. Explain what this experiment was designed to test.

b. Use this reasoning to assign genotypes to all 12 mutants.

20. In *Drosophila*, a cross (cross 1) is made between two mutant flies, one homozygous for the recessive mutation bent wing (*b*) and the other homozygous for the recessive mutation eyeless (*e*). The mutations *e* and *b* are alleles of two different genes that are known to be very closely linked on the tiny autosomal chromosome 4. All the progeny were wild-type phenotype. One of the female progeny was crossed to a male of genotype *b e/b e*; call this cross 2. The progeny of cross 2 were mostly of the expected types, but there was also one rare female of wild-type phenotype.

a. Explain what the common progeny are expected to be from cross 2.

b. Could the rare wild-type female have arisen by

   (1) crossing-over?

   (2) nondisjunction?

Explain.

c. The rare wild-type female was testcrossed to a male of genotype *b e/b e* (cross 3). The progeny were

   $\frac{1}{6}$ wild type

   $\frac{1}{6}$ bent, eyeless

   $\frac{1}{3}$ bent

   $\frac{1}{3}$ eyeless

Which of the explanations in part b are compatible with this result? Explain the genotypes and phenotypes of progeny of cross 3 and their proportions.

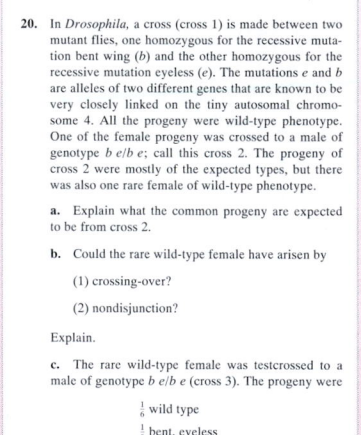 **Unpacking the Problem**

1. Define *homozygous, mutation, allele, closely linked, recessive, wild type, crossing-over, nondisjunction, testcross, phenotype,* and *genotype.*

2. Does this problem concern sex linkage? Explain.

3. How many chromosomes does *Drosophila* have?

4. Draw a clear pedigree summarizing the results of crosses 1, 2, and 3.

Most chapters have an exercise in problem solving called **Unpacking the Problem.** This exercise grew from the idea that a genetics problem represents only the tip of a vast iceberg of knowledge (we originally considered calling them "iceberg problems"). It is only when the underlying levels of knowledge are exposed that the problem can be solved in a constructive manner. The unpacking activities access this underlying knowledge without actually solving the problem. Some of the component questions in the unpacking exercise might sound trivial, but often these address the kind of fundamental levels of misunderstanding that prevent students from successfully solving problems.

The problems at the end of each chapter are prefaced by **Solved Problems** that illustrate the ways that geneticists apply principles to experimental data. Research in science education has shown that this application of principles is a process that professionals find second nature, whereas students find it a major stumbling block. The Solved Problems demonstrate this process and prepare the students for solving problems on their own.

The **Chapter Integration Problems** are solved problems that emphasize concept integration both within and between chapters. These chapter integration problems help to show how one set of learned skills builds on and interacts with previous ones. They also enable students to develop a holistic perspective as they begin to organize diverse concepts into a coherent body of knowledge.

CHAPTER INTEGRATION PROBLEM

1. The human pedigree below concerns a rare visual abnormality in which the person affected loses central vision while retaining peripheral vision.

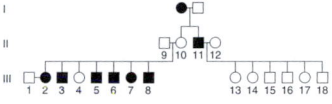

a. What inheritance pattern is shown? Can it be explained by nuclear inheritance? Mitochondrial inheritance? Molecular geneticists studied the mitochondrial DNA of the 18 members of generations II and III. A restriction fragment 212 base pairs long of the mtDNA from each person was digested with another restriction enzyme, *Sfa*N1, with the following results:

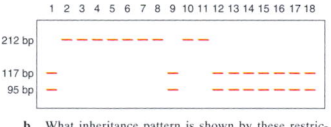

b. What inheritance pattern is shown by these restriction fragments?

c. How does the restriction-pattern inheritance relate to the inheritance of the disease?

d. How can you explain individuals 4 and 10?

e. What is the likely nature of the mutation?

f. How would this analysis be useful in counseling this family?

♦ Solution ♦

a. On the basis of the pedigree alone, it is possible, but unlikely, that the disease is caused by a dominant nuclear allele. But we would have to invoke lack of penetrance in individual 10, who would have to carry the allele because it is passed on to her children. In addition, we have to explain the ratios in generation III. The matings 9 × 10 and 11 × 12 would have to be $A/a × a/a$, and the phenotypic ratio of affected to normal then expected among the children in each family is 1:1. So overall this model is not an attractive one for explaining the results.

SOLVED PROBLEMS

2. A yeast plasmid carrying the yeast *leu2*$^+$ gene is used to transform nonrevertible haploid *leu2*$^-$ yeast cells. Several *leu*$^+$-transformed colonies appear on a medium lacking leucine. Thus, *leu2*$^+$ DNA presumably has entered the recipient cells, but now we have to decide what has happened to it inside these cells. Crosses of transformants to *leu2*$^-$ testers reveal that there are three types of transformants, A, B, and C, representing three different fates of the *leu2*$^+$ gene in the transformation. The results are:

Type A × *leu2*$^-$ ⟶ $\frac{1}{2}$ *leu*$^-$

$\frac{1}{2}$ *leu*$^+$, × standard *leu2*$^+$

⟶ $\frac{3}{4}$ *leu*$^+$

$\frac{1}{4}$ *leu*$^-$

Type B × *leu2*$^-$ ⟶ $\frac{1}{2}$ *leu*$^-$

$\frac{1}{2}$ *leu*$^+$, × standard *leu2*$^+$

⟶ 100% *leu*$^+$

0% *leu*$^-$

Type C × *leu2*$^-$ ⟶ 100% *leu*$^+$

What three different fates of the *leu2*$^+$ DNA do these results suggest? Be sure to explain *all* the results according to your hypotheses. Use diagrams if possible.

♦ Solution ♦

If the yeast plasmid does not integrate, then it replicates independently of the chromosomes. In meiosis, the daughter plasmids would be distributed to the daughter cells, resulting in 100 percent transmission. This percentage was observed in transformant type C.

If one copy of the plasmid is inserted, in a cross with a *leu2*$^-$ line, the resulting offspring would have a ratio of 1 *leu*$^+$:1 *leu*$^-$. This ratio is seen in type A and type B.

When the resulting *leu*$^+$ cells are crossed with standard *leu2*$^-$ lines, the data from type A cells suggest that the inserted gene is segregating independently of the standard *leu2*$^+$ gene; so the *leu2*$^+$ transgene has inserted ectopically into another chromosome.

Key Concepts

When two haploid genomes containing different recessive mutations are combined in one cell and the phenotype is mutant, the mutations must be in the same gene (alleles).

When two haploid genomes containing different recessive mutations are combined in one cell and the phenotype is wild type, the mutations must be in different genes.

The phenotypes of some heterozygotes reveal types of dominance other than full dominance.

Some mutant alleles can kill the organism.

Most characters are determined by sets of genes that interact with one another and with the environment.

Modified monohybrid ratios reveal allelic interactions.

Modified dihybrid ratios reveal gene interactions.

The **Key Concepts** at the chapter openings give an overview of the main principles to be covered in the chapter, stated in simple prose without genetic terminology. These provide a strong pedagogic direction for the reader.

MESSAGE

The linear sequence of a protein folds up to yield a unique three-dimensional configuration. This configuration creates specific sites to which substrates bind and at which catalytic reactions take place. The three-dimensional structure of a protein, which is crucial for its function, is determined solely by the primary structure (linear sequence) of amino acids. Therefore, genes can control enzyme function by controlling the primary structure of proteins.

Throughout the chapters, boxed **Messages** provide convenient milestones at which the reader can pause and contemplate the material just presented.

SUMMARY

After making dihybrid crosses of sweet pea plants, William Bateson and R. C. Punnett discovered deviations from the 9:3:3:1 ratio of phenotypes expected in the F_2 generation. The parental gametic types outnumbered the other two classes. Later, in his studies of two different autosomal genes in *Drosophila*, Thomas Hunt Morgan found a similar deviation from Mendel's law of independent assortment. Morgan postulated that the two genes were located on the same pair of homologous chromosomes. This relation is called linkage.

Linkage explains why the parental gene combinations stay together but not how the nonparental combinations arise. Morgan postulated that in meiosis there may be a physical exchange of chromosome parts by a process now called crossing-over. Thus, there are two types of meiotic recombination. Recombination by Mendelian independent assortment results in a recombinant frequency of 50 percent.

Chapter **Summaries** provide a short distillation of the chapter material and an immediate reinforcement of the concepts. All these items are useful in text review, especially for exam study.

> Now draw a concept map for this chapter, interrelating as many of the following terms as possible. Note that the terms are listed in no particular order.
>
> genotype / phenotype / norm of reaction / environment / development / organism

A **Concept Map** exercise appears at the end of every chapter. Concept maps grew out of the constructivist movement in education, which asserts that student learning is most effective when new information is brought into direct conflict with previous understanding. Concept mapping can be a powerful method for visualizing and resolving such conflicts, while aiding concept integration.

New Features of the Seventh Edition

New emphasis on the integration of classical and molecular genetics

The linearity of the teaching process means that concepts have to be introduced one at a time to avoid bewilderment. Nevertheless, genetics is an integrated subject in which organismal and molecular manipulations go hand in hand. Therefore, integration is a key issue in teaching and a key goal in learning. This edition integrates organismal and molecular aspects of genetics wherever possible, starting in the early chapters. Chapter 1 presents the basics of DNA structure and function and uses albinism in humans as an example to establish the relationship between DNA, genes, proteins, and phenotype. In Chapter 2 we examine patterns of inheritance (both Mendelian and non-Mendelian) and explain them at the molecular level, using examples such as PKU. The complementarity of the two genetic approaches continues with the exploration of gene interactions at the molecular level, as exemplified by flower color in Blue-eyed Mary and sickle cell anemia in humans (Chapter 4).

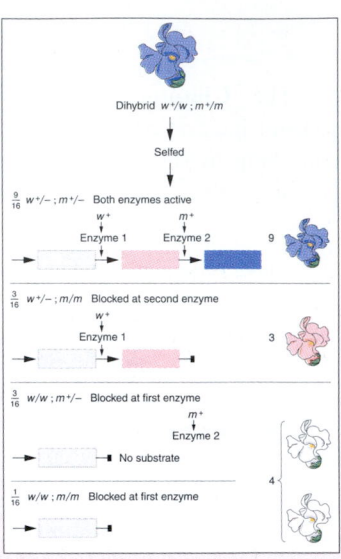

Figure 4-13

New chapter sequence for greater topic cohesiveness

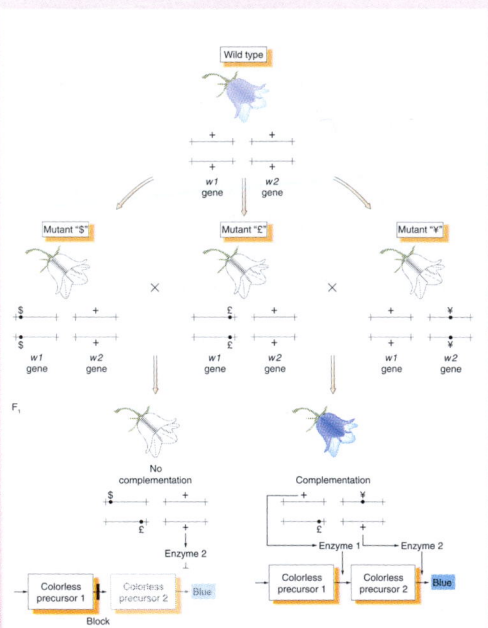

Figure 4-1

Our goal to better emphasize overarching principles has resulted in the relocation of several chapters. The book now consists of six major blocks that group topics related by common underlying principles: general aspects of inheritance are covered in Chapters 1–4; recombination and mapping of genes in Chapters 5–7; molecular genetics in Chapters 8–14; genetic change and variation in Chapters 15–20; developmental genetics in Chapters 21–23; and population genetic analysis in Chapters 24–26.

The benefits of this reorganization can be found throughout the text. Chapter 2, for instance, now covers both Mendelian genetics and sex-linked inheritance in order to emphasize general patterns of inheritance. Similarly, the incorporation of eukaryotic genome structure with the chromosomal basis of heredity in Chapter 3 provides mutual reinforcement of principles. Chapter 4's new title, Gene Interaction, reflects its inclusion of new material on complementation and its new emphasis on overarching principles of how genes interact. Chapter 11 now juxtaposes prokaryotic and eukaryotic gene regulation so that principles common to both processes emerge. Chapter 23, Developmental Genetics, now includes coverage of gene regulation during development, a topic previously covered in a separate chapter.

Figure 3-28

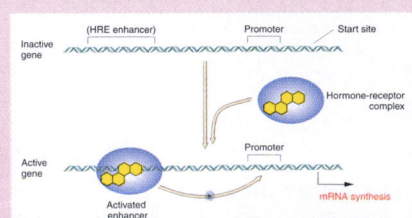

Figure 11-37

New streamlined and simplified explanations

In an effort to focus on overarching principles, we have examined the level of detail in many chapters and have chosen to simplify and streamline the coverage of several topics. This is particularly noticeable in Chapter 7 (Gene Transfer in Bacteria and Their Viruses), Chapter 9 (Genetics of DNA Function), Chapter 16 (Mechanisms of Gene Mutation), Chapter 19 (Mechanisms of Recombination), and Chapter 21 (Extranuclear Genes).

Two new chapters and updates throughout

Two new chapters have been introduced: Chapter 22, on Cancer as a Genetic Disease, and Chapter 26, on Evolutionary Genetics. Chapter 22 presents the integrated control mechanisms of cell proliferation and cell death and what happens when these mechanisms are disrupted. Chapter 26 discusses the evolutionary process in terms of both natural selection and random factors and includes sections on speciation and the origin of new genes. Notable updates in other chapters are: lod scores (Chapter 5); rolling circle replication and synteny (Chapter 8); functional genomics including yeast 2-hybrid analysis, micro arrays/DNA chips, and global regulation (Chapter 14); the molecular basis of chromosome rearrangements (Chapter 17); mitochondrial DNA, aging and human disease (Chapter 21); and programmed cell death (Chapter 24).

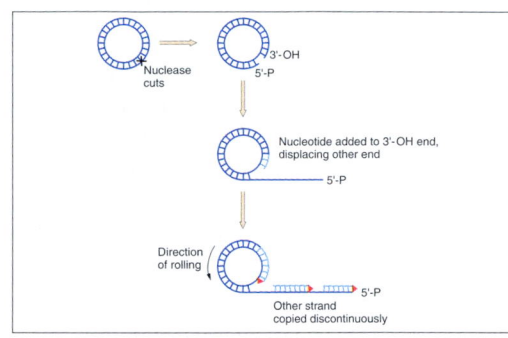

Figure 8-19

New problems

More than 50 new problems have been added, including many problems involving molecular analysis.

55. In corn, synthesis of purple pigment is controlled by two genes acting sequentially through colorless (white) intermediates:

$$\text{white 1} \xrightarrow{\text{gene } A} \text{white 2} \xrightarrow{\text{gene } B} \text{purple}$$

Recessive nonsense mutations (a^n and b^n) were obtained in genes A and B. Each of these mutations gave a white phenotype, and each could be suppressed by the nonsense-suppressor mutation T^S (wild-type allele T^+)

a. Would you expect T^S to be dominant to T^+? Explain.

b. A trihybrid A/a^n ; B/b^n ; T^S/T^+ is selfed. If all the genes are unlinked, what phenotypic ratio do you expect in the progeny? Explain, preferably with a diagram.

56. A plant believed to be heterozygous for a pair of alleles B/b (where B encodes yellow and b encodes bronze) was selfed and in the progeny there were 280 yellow and 120 bronze individuals. Do these results support the hypothesis that the plant is B/b?

57. A plant thought to be heterozygous for two independently assorting genes (P/p ; Q/q) was s   d th   ny w

Supplements

The following supplementary materials are available to accompany *Introduction to Genetic Analysis*.

Solutions Manual
William Fixsen, Harvard University, 0-7167-3525-3
The *Solutions Manual* contains worked-out solutions to all the problems in the textbook.

Introduction to Genetic Analysis CD-ROM
(hybrid format for Windows and Macintosh)
Packaged with every copy of the textbook, this CD has over 30 original animations that are available in two formats. Topics such as transcription, complementation, and DNA replication bring the textbook figures to life. Students can view each animation as a series of steps that make up the process or watch the animation in its entirety.

Text and media come together by denoting figures of genetic processes that come to life as animations on the Freeman Genetics CD-ROM.

Introduction to Genetic Analysis Web Site
W. H. Freeman and Sumanas, Inc., with contributions from William Sofer, The State University of New Jersey at Rutgers

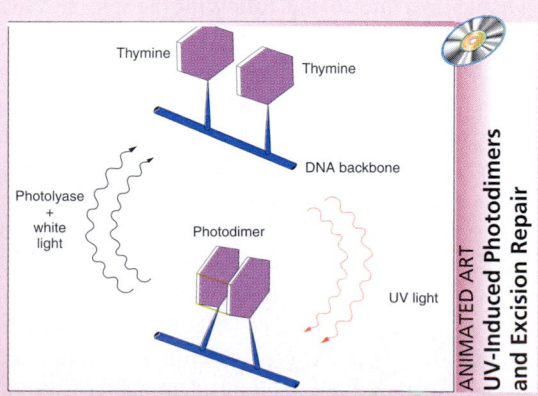

ANIMATED ART
UV-Induced Photodimers
and Excision Repair

Figure 16-25

This multimedia learning tool complements and enriches the textbook. Practice tools such as interactive quizzes, tutorials, flashcards, key concepts, and Web links in every chapter help students review for exams. All text images will be available for downloading. The Introduction to Genetic Analysis Web site, at *www.whfreeman.com/iga/* will be updated regularly.

Essays on Genetics Education

Understanding Genetics: Ideas for Teachers

Anthony J. F. Griffiths and Jolie A. Mayer-Smith, both of the University of British Columbia

This biweekly electronic publication explores the problems faced by instructors in genetics education and offers creative and thoughtful strategies for promoting student understanding of genetics and learning in general. A complete collection of the essays will be available for purchase early in 2000.

Instructor Resource Manual and Test Bank

Sally Allen, University of Michigan, and Ewan Harrison

Printed: 0-7167-3530X; CD ROM: 0-7167-3528-8

The *Instructor Resource Manual* contains over 700 test questions in multiple-choice, true-false, and matching formats. It also contains complete sample exams and teaching hints. The electronic version (both Windows and Mac formats on one CD) allows instructors to download, edit, add, and re-sequence questions to suit their particular needs.

Online Testing

With Diploma, the computerized test bank package from Brownstone Research Group, instructors can create and administer exams on paper, over a network, and now, over the Internet as well. Instructors can include multimedia, graphics, movies, and sound in their questions. Security features allow instructors to restrict tests to specific computers or time blocks. The package also includes an impressive suite of grade book and question-analysis features.

Instructor's Resource CD-ROM

0-7167-3952-6

Our instructor's CD offers all the images from the textbook and the animations in two formats — as part of our Presentation Manager Pro software and in JPEG files. Presentation Manager lets instructors quickly prepare play lists of images for display during lectures. The JPEG files are for instructors who use commercially available presentation software.

Transparency Set

0-7167-3526-1

A full-color overhead transparency set of 150 key illustrations from the textbook is available free of charge to qualified adopters.

Course Syllabi

For a two-semester course, the entire text provides an appropriate course structure and syllabus that reflects the range of modern genetics. A syllabus for a one-semester course can be designed around selected chapters. One possible selection of chapters for a one-semester course is Chapters 1, 2, 3, 4, 5, 8, 11, 12, 15, 17, 18, 23, and 24. A one-semester course in molecular genetics could be based on Chapters 8 through 23.

Acknowledgments

Thanks are due to the following people at W. H. Freeman and Company for their considerable support throughout the preparation of this edition: Randi Rossignol, Senior Developmental Editor; Philip McCaffrey, Managing Editor; Mary Louise Byrd, Project Editor; Cambraia Magalhaes, Designer; Bill Page, Illustration Coordinator; Lou Capaldo, Assistant Illustration Coordinator; Kathy Bendo, Senior Photo Editor; Jennifer MacMillan, Assistant Photo Editor; Susan Wein, Production Coordinator; Nicole Folchetti, Acquisitions Editor; John Britch, Executive Marketing Manager; Patrick Shriner, Media and Supplements Director; Charles Van Wagner, Media and Supplements Editor; Ellen Cash, Vice President, Director of Production; and Shawn Churchman and Melanie Mays, Editorial Assistants. We also thank the copy editor, Patricia Zimmerman; the layout artist, Marsha Cohen; the indexer, Chris Hunt; and the proofreader, Elaine Rosenberg.

Jeffrey H. Miller would like to thank Anh Miller for her constant support.

Finally, we extend our thanks and gratitude to our colleagues who reviewed this edition and whose insights and advice were most helpful:

Colleen Belk	*University of Minnesota, Duluth*
Ralph Bertrand	*Colorado College*
John Bowman	*University of California, Davis*
Glen Collier	*University of Tulsa*
David S. Durica	*University of Oklahoma*

Deborah Eastman	*Southwestern University*	Michael H. Perlin	*University of Louisville*
Ronald Ellis	*University of Michigan*	Dennis T. Ray	*The University of Arizona*
John Ellison	*Texas A&M — College Station*	Sue Jinks-Robertson	*Emory University*
Robert Fowler	*San Jose State University*	Laura Runyen-Janecky	*Southwestern University*
Dan Garza	*Florida State University*	Mark Sanders	*University of California, Davis*
Elliott S. Goldstein	*Arizona State University*	David E. Sheppard	*University of Delaware*
Muriel Herrington	*Concordia University*	Laurence Von Kalm	*University of Central Florida*
Robert Holmgren	*Northwestern University*	Bruce Walsh	*University of Arizona*
Andrew Hoyt	*The Johns Hopkins University*	Edmund J. Zimmerer	*Murray State University*
Lynne A. Hunter	*University of Pittsburgh*		
David Hyde	*University of Notre Dame*		
Bob Ivarie	*University of Georgia*		
Fordyce G. Lux III	*Lander University*		
William McGinnis	*University of California, San Diego*		
Bruce McKee	*University of Texas, Knoxville*		
Gregg Orloff	*Emory University*		

We believe this edition to be a true celebration of genetics. As authors, we hope that our love of the subject comes through and that the book will stimulate the reader to do some firsthand genetics, whether as professional scientist, student, amateur breeder, or naturalist. Failing this, we hope to impart some lasting impression of the incisiveness, elegance, and power of genetic analysis.

AN INTRODUCTION TO
GENETIC ANALYSIS

1

GENETICS AND THE ORGANISM

Genetic variation in the color of corn kernels.
Each kernel represents a separate individual with a distinct
genetic makeup. The photograph symbolizes the history of
humanity's interest in heredity. Humans were breeding corn
thousands of years before the advent of the modern discipline of
genetics. Extending this heritage, corn today is one of the main
research organisms in classical and molecular genetics.

(William Sheridan, University of North Dakota; photograph by Travis Amos.)

Key Concepts

The hereditary material is DNA.

DNA is a double helix composed of two intertwined
nucleotide chains oriented in opposite directions.

In the copying of DNA, the chains separate and serve as
molds for making two identical daughter DNA molecules.

The functional units of DNA are genes.

A gene is a segment of DNA that can be copied to make
RNA.

The nucleotide sequence in RNA is translated into the
amino acid sequence of a protein.

Proteins are the main determinants of the basic structural
and physiological properties of an organism.

The characteristics of a species are encoded by its genes.

Variation within a species may be from hereditary
variation, environmental variation, or both.

Hereditary variation is caused by variant forms of genes
(alleles).

Why study genetics? There are two basic reasons. First, genetics occupies a pivotal position in the entire subject of biology. Therefore, for any serious student of plant, animal, or microbial life, an understanding of genetics is essential. Second, genetics, like no other scientific discipline, is central to numerous aspects of human affairs. It touches our humanity in many different ways. Indeed, genetic issues seem to surface daily in our lives, and no thinking person can afford to be ignorant of its discoveries. In this chapter, we take an overview of the science of genetics, showing how it has come to occupy its crucial position. In addition, we provide a perspective from which to view the subsequent chapters.

First, we need to define what genetics is. Some define it as the study of heredity, but hereditary phenomena were of interest to humans long before biology or genetics existed as the scientific disciplines that we know today. Ancient peoples were improving plant crops and domesticated animals by selecting desirable individuals for breeding. They also must have puzzled about the inheritance of individuality in humans and asked such questions as, "Why do children resemble their parents?" and "How can various diseases run in families?" But these people could not be called geneticists. Genetics as a set of principles and analytical procedures did

not begin until the 1860s, when an Augustinian monk named Gregor Mendel (Figure 1-1) performed a set of experiments that pointed to the existence of biological elements that we now call genes. The word *genetics* comes from the word "gene," and genes are the focus of the subject. Whether geneticists study at the molecular, cellular, organismal, family, population, or evolutionary level, genes are always central in their studies. Simply stated, genetics is the study of genes.

What is a gene? A gene is a section of a threadlike double helical molecule called **deoxyribonucleic acid,** abbreviated **DNA.** The discovery of genes and understanding their molecular structure and function have been sources of profound insight into two of the biggest mysteries of biology:

1. What makes a species what it is? We know that cats always have kittens and people always have babies. This common-sense observation naturally leads to questions about the determination of the properties of a species. The determination must be hereditary because, for example, the ability to have kittens is inherited by every generation of cats.

2. What causes variation within a species? We can distinguish each other as well as our own pet cat from other cats. Such differences within a species require explanation. Some of these distinguishing features are clearly familial; for example, animals of a certain unique color often have offspring with the same color, and, in human families, certain features such as the shape of the nose definitely "run in the family." Hence we might suspect that a hereditary component explains at least some of the variation within a species.

The answer to the first question is that genes dictate the inherent properties of a species. The products of most genes are specific **proteins.** Proteins are the main macromolecules of an organism. When you look at an organism, what you see is either protein or something that has been made by a protein. The amino acid sequence of a protein is encoded in a gene. The timing and rate of production of proteins and other cellular components are a function both of the genes within the cells and of the environment in which the organism is developing and functioning.

The answer to the second question is that any one gene can exist in several forms that differ from each other, generally in small ways. These forms of a gene are called **alleles.** Allelic variation causes hereditary variation within a species. At the protein level, allelic variation becomes protein variation.

The next two sections show how genes influence the inherent properties of a species and how allelic variation contributes to variation within a species. These sections are an overview; most of the details will be presented in later chapters.

Genes as determinants of the inherent properties of species

What is the nature of genes, and how do they perform their biological roles? Three fundamental properties are required of genes and the DNA of which they are composed.

1. *Replication.* Hereditary molecules must be capable of being copied at two key stages of the life cycle (Figure 1-2). The first stage is the production of the cell type that will ensure the continuation of a species from one generation to the next. In plants and animals, these cells are the gametes: egg and sperm. The other stage is when the first cell of a new organism undergoes multiple rounds of division to produce a multicellular organism. In plants and animals, this is the stage at which the fertilized egg, the **zygote,** divides repeatedly to produce the complex organismal appearance that we recognize.

2. *Generation of form.* The working structures that make up an organism can be thought of as form or substance. Looked at in this way, DNA has essential "information"; in other words, "that which is needed to give form."

3. *Mutation.* A gene that has changed from one allelic form into another has undergone mutation—an event that happens rarely but regularly. Mutation is not only a basis for variation within a species, but, over the long term, also the raw material for evolution.

We will examine replication and the generation of form in this section and mutation in the next.

DNA and its replication

An organism's basic complement of DNA is called its **genome.** The body cells of most plants and animals contain two genomes (Figure 1-3). These organisms are **diploid.** The cells of most fungi, algae, and bacteria contain just one genome. These organisms are **haploid.** The genome itself is made up of one or more extremely long molecules of DNA that are organized into **chromosomes.** For instance, human body cells contain two sets of 23 chromosomes, for a total of 46. Genes are simply the functional regions of chromosomal DNA. Each chromosome in the genome carries a different array of genes. In diploid cells, each chromosome and

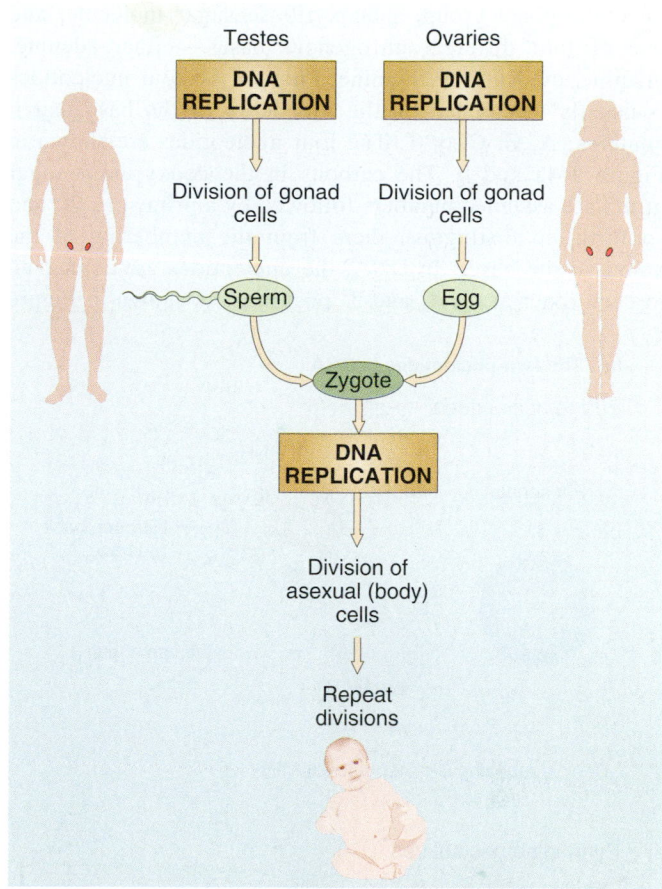

Figure 1-2   DNA replication is the basis of the perpetuation of life through time.

its component genes are present twice. Two chromosomes with the same gene array are said to be **homologous.** When a cell divides, all the chromosomes (one or two genomes) are replicated; so each daughter cell contains the full complement. Therefore, the unit of replication is the chromosome; when a chromosome is replicated, all the genes of that chromosome are automatically replicated along with it.

To understand replication, we need to understand the basic nature of DNA. DNA is a linear, double-helical structure looking rather like a molecular spiral staircase. The double helix is composed of two intertwined chains of building blocks called **nucleotides.** Each nucleotide consists

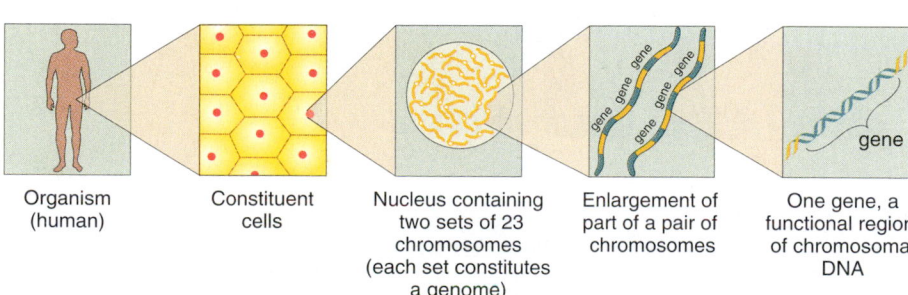

| Organism (human) | Constituent cells | Nucleus containing two sets of 23 chromosomes (each set constitutes a genome) | Enlargement of part of a pair of chromosomes | One gene, a functional region of chromosomal DNA |

Figure 1-3   Successive enlargements bringing the genetic material of an organism into sharper focus.

of a phosphate group, a deoxyribose sugar molecule, and one of four different nitrogenous bases — either adenine, guanine, cytosine, or thymine. Each of the four nucleotides is usually designated by the first letter of the base that it contains: A, G, C, or T. The four nucleotides are shown in Figure 1-4a and b. The carbons in the deoxyribose sugar group are assigned numbers followed by a prime (1′, 2′, and so forth) to distinguish them from the numbering of the atoms in the bases. In DNA, the nucleotides are connected to each other at the 3′ and 5′ positions, as shown in Figure

1-4(c); hence each chain is said to have **polarity,** with one end having a 5′ phosphate group and the other a 3′ OH group. The connecting bonds between the repeating sugar and phosphate groups are called **phosphodiester bonds.**

The polarities of the two intertwined nucleotide chains are in opposite directions; the chains are said to be **antiparallel.** The two nucleotide chains are held together by weak bonds called **hydrogen bonds,** between bases. Hydrogen bonding is very specific because of a "lock-and-key" fit between the shape and atomic charge of the bases. Adenine

(a) The four nucleotides in DNA

Purine nucleotides

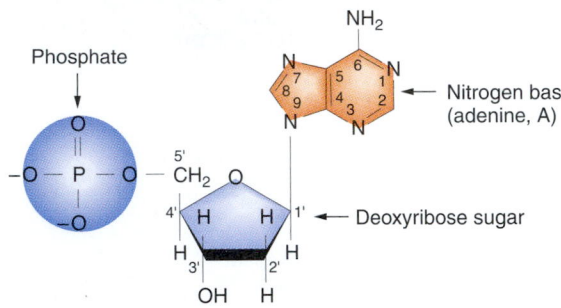 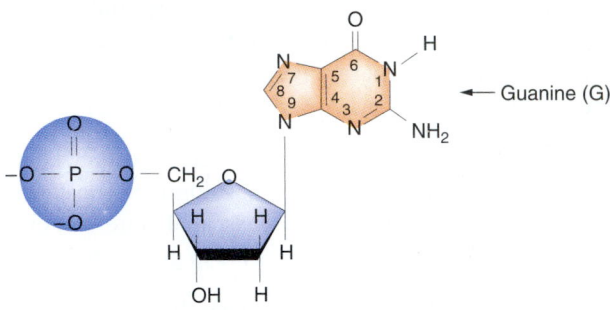

Pyrimidine nucleotides

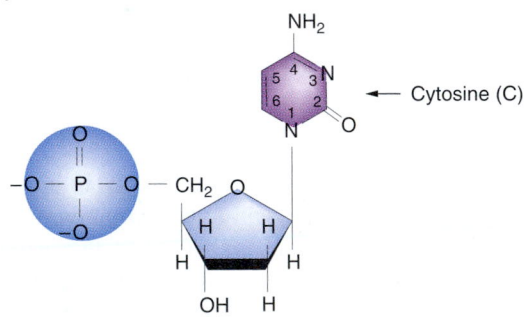 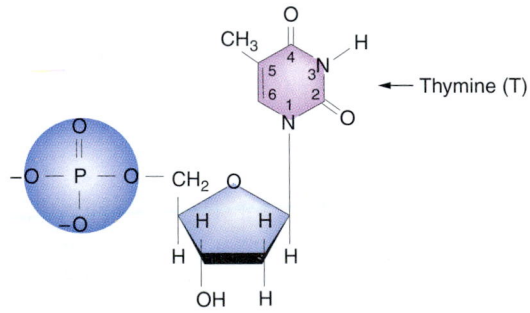

(b) Schematic diagram of the four nucleotides

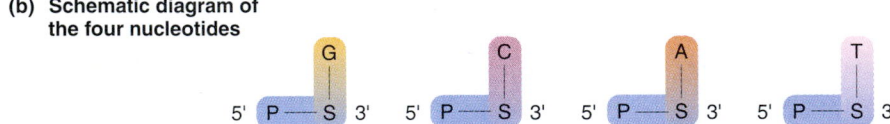

(c) Polymerized nucleotides

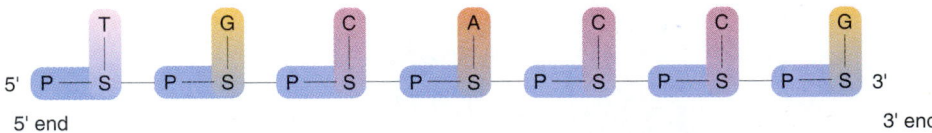

Figure 1-4   The fundamental building blocks of DNA. (a) Chemical structures of the four nucleotides (two with purine bases and two with pyrimidine bases). The sugar is called deoxyribose because it is a variation of a common sugar, ribose, which has one more oxygen atom. (b) Diagrammatic representation of nucleotides in DNA. (c) DNA is a polymer of nucleotides. Abbreviations: P, phosphate; S, deoxyribose sugar; A, T, G, and C, the bases adenine, thymine, guanine, and cytosine, respectively.

pairs only with thymine, and guanine pairs only with cytosine. The bases that form base pairs are said to be **complementary.** Hence a short segment of DNA drawn with arbitrary nucleotide sequence might be

$$5'\text{-CAGT-}3'$$
$$3'\text{-GTCA-}5'$$

The same structure is depicted in more detail in Figure 1-5. Although hydrogen bonds are weak individually, their combined bonding holds the two chains together in a stable manner. Furthermore, it is important that the bonds between the bases be relatively weak because, as we shall see, the two chains have to be pulled apart to allow the replication process to work. The base pairs, which run down the center of the double helix, are flat hydrophobic structures that have a tendency to stack owing to the exclusion of water molecules. This stacking draws the two intertwined strands of DNA into its helical structure (Figure 1-6).

For replication to take place, the two strands of the double helix must unwind in one direction, rather like the opening of a zipper. The two exposed nucleotide chains then act as alignment guides, or **templates,** for the deposition of free nucleotides. These nucleotides have been synthesized inside the cell and arrive in the nucleus by diffusion. Their polymerization into a new strand is catalyzed by the enzyme **DNA polymerase.** This enzyme initially binds to double-helical DNA at a specific nucleotide sequence called the **origin of replication** and then moves along the DNA, polymerizing new chains as shown in Figure 1-7. The crucial point illustrated in Figure 1-7 is that, because of base complementarity, the two **daughter DNA** molecules are identical with each other and with the original molecule. However, note that each daughter molecule is half old and half newly polymerized. This method of replication is therefore called **semiconservative.**

> **MESSAGE** ·······················
> DNA is composed of two antiparallel nucleotide strands held together by complementary hydrogen bonding of A with T and G with C.

In the foregoing description of DNA and its replication, we see two principles that are the bases of most genetic transactions:

1. Complementary bases bind to each other.

2. Certain proteins (for example, DNA polymerase) act by binding to a specific base sequence in a nucleic acid.

We will encounter these two ideas numerous times throughout the book.

> **MESSAGE** ·······················
> Base complementarity and the binding of proteins to specific base sequences in a nucleic acid are two recurring mechanisms in genetics.

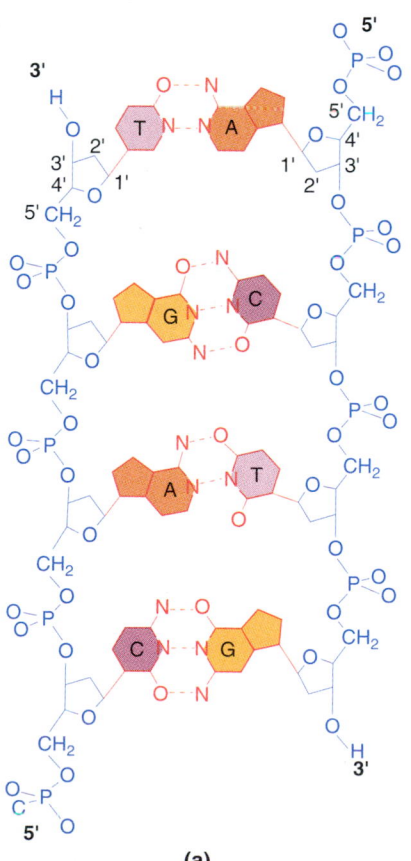

(a)

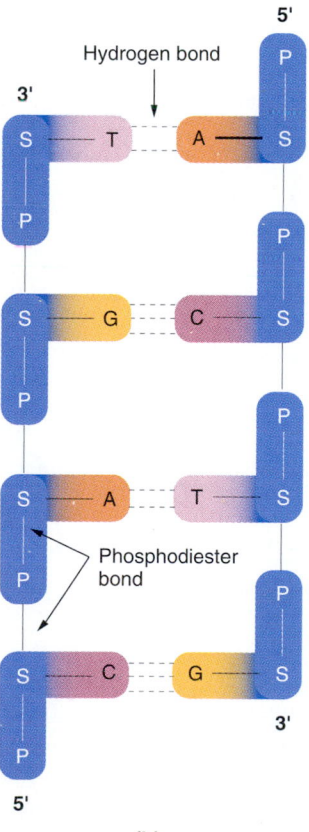

(b)

Figure 1-5 The arrangement of the components of DNA in which a segment of the double helix has been unwound to show the structures more clearly. (a) An accurate chemical diagram showing the sugar-phosphate backbone in blue and the hydrogen bonding of bases in the center of the molecule. (b) A simplified version of the same segment emphasizing the antiparallel arrangement of the nucleotides, which are represented as sock-shaped structures with 5′ phosphate "toes" and 3′ "heels."

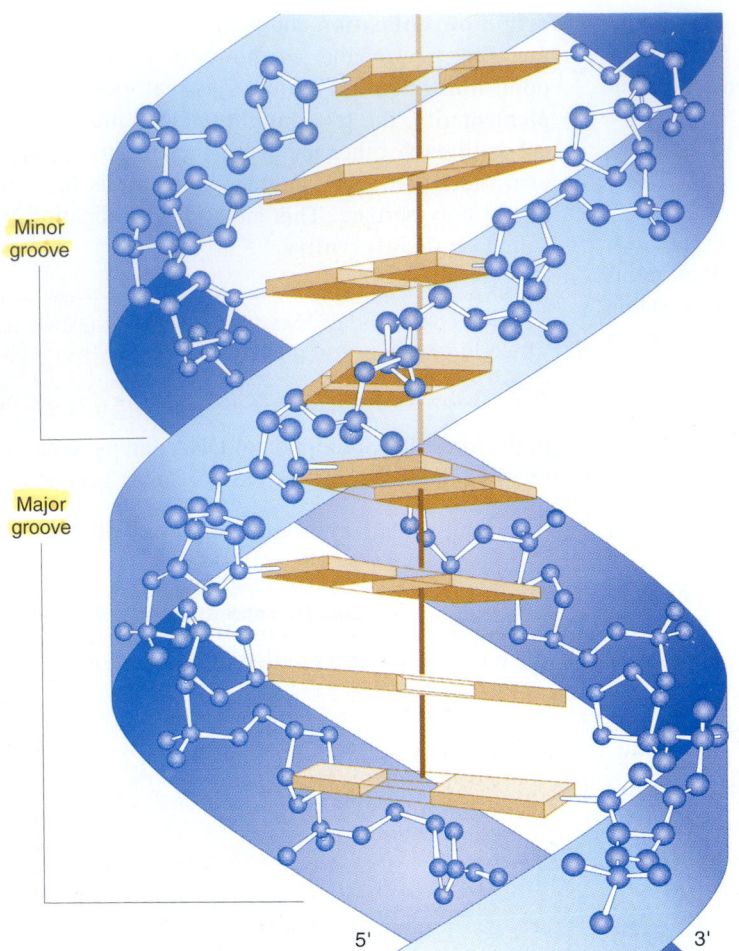

Minor groove

Major groove

5'   3'

Figure 1-6 Ribbon representation of the DNA double helix. Blue = sugar-phosphate backbone; brown = base pairs.

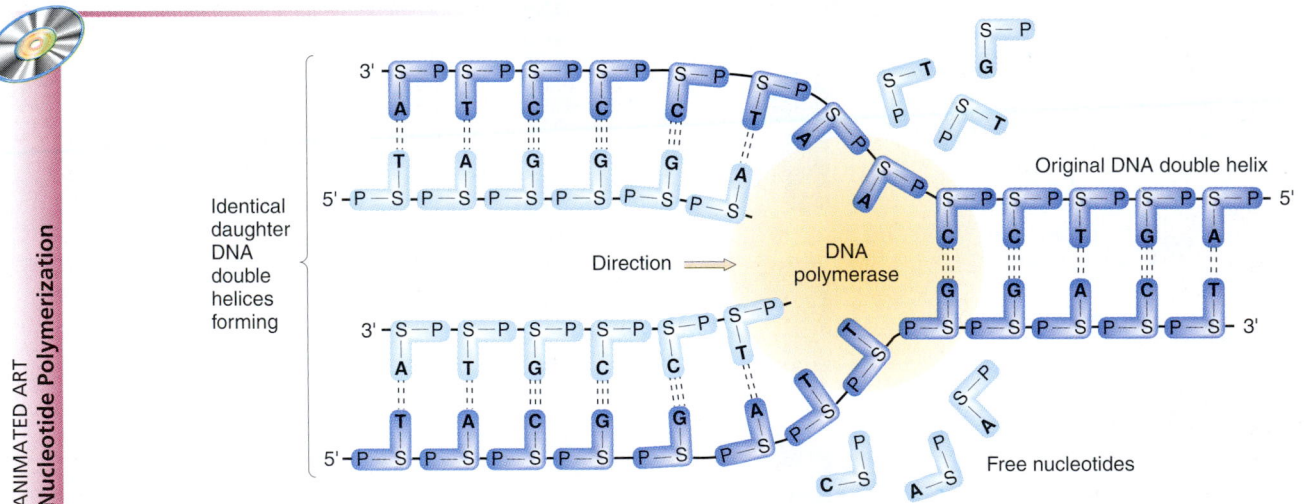

ANIMATED ART
Nucleotide Polymerization

Figure 1-7 Semiconservative replication in process.

Generation of form

If DNA represents information, what constitutes form at the cellular level? The simple answer is "protein" because the great majority of structures in a cell are protein or have been made by protein. In this section, we trace the steps through which information becomes form.

Every functional gene is read by the cellular machinery to produce the product of that gene. For most genes, the product is a specific protein. The first step taken by the cell to make the protein is to copy, or **transcribe,** the information encoded in the DNA of the gene as a related but single-stranded molecule called **ribonucleic acid (RNA).** This

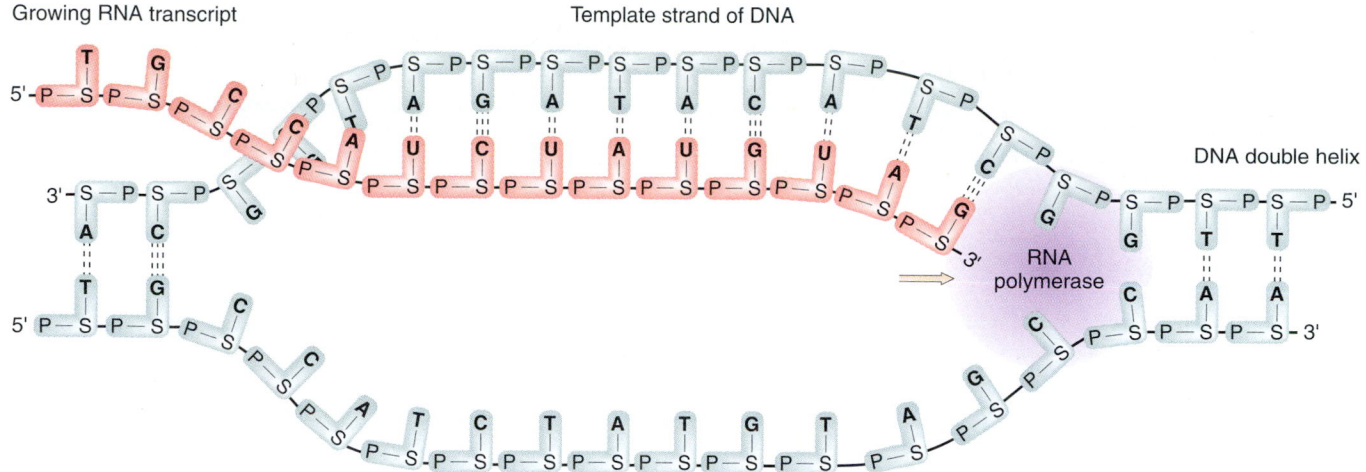

Figure 1-8   Transcription process.

RNA molecule represents a "working copy" of the gene. Indeed, a gene can be defined as a segment of DNA that specifies a functional RNA. Like DNA, RNA is composed of nucleotides, but these nucleotides contain the sugar ribose instead of deoxyribose. Furthermore, instead of thymine, RNA contains uracil (U), a base that has hydrogen-bonding properties identical with those of thymine. Hence the RNA bases are A, G, C, and U. The RNA copy made from the gene is called a **transcript.**

The polymerization of ribonucleotides to form RNA is catalyzed by the enzyme **RNA polymerase.** This enzyme binds to a specific sequence, the transcriptional start site, at one end of a gene. It separates the two strands of DNA. It moves along the gene, maintaining the separated "bubble," and, as it proceeds, it uses only *one* of the separated strands as a template, synthesizing an ever-growing tail of polymerized nucleotides that eventually become the full-length transcript. The addition of ribonucleotides by RNA polymerase is always at the 3′ end of the growing chain. Transcription is represented diagrammatically in Figure 1-8. Note again the two powerful principles of macromolecular interactions: RNA polymerase binds to a specific initiation sequence on DNA and moves along the DNA, illustrating protein–nucleic acid binding; the action of the DNA template strand in aligning ribonucleotides is based on the principle of base complementarity, this time between DNA bases and RNA bases. Note that the RNA transcript has the same sequence as the *nontemplate* strand of the DNA (Figure 1-8).

MESSAGE ·
During transcription one of the DNA strands of a gene acts as a template; ribonucleotides are added at a 3′ growing tip, catalyzed by the enzyme RNA polymerase.
· ·

The biological role of most genes is to carry or encode information on the composition of proteins. This composition, together with the timing and amount of production of each protein, is an extremely important determinant of the structure and physiology of organisms. The primary structure of a protein is a linear chain of building blocks called **amino acids.** This primary chain is coiled and folded — and, in some cases, associated with other chains — to form a functional protein. Proteins are important either as structural components — such as the proteins that constitute hair, skin, and muscle — or as active agents in cellular processes — for example, enzymes and active-transport proteins. Each gene is responsible for coding for one specific protein or a part of a protein. Let's look at how a gene is organized. For the purpose of this overview, we will focus on the protein-coding genes of **eukaryotes.** Eukaryotes are those organisms whose cells have a membrane-bound nucleus. Outside the nucleus of each cell are a complex array of membranous structures, including the endoplasmic reticulum and Golgi apparatus, and organelles such as mitochondria and chloroplasts. Animals, plants, algae, and fungi are eukaryotes.

Figure 1-9 shows the general structure of a eukaryotic gene. At one end, there is a regulatory region to which various

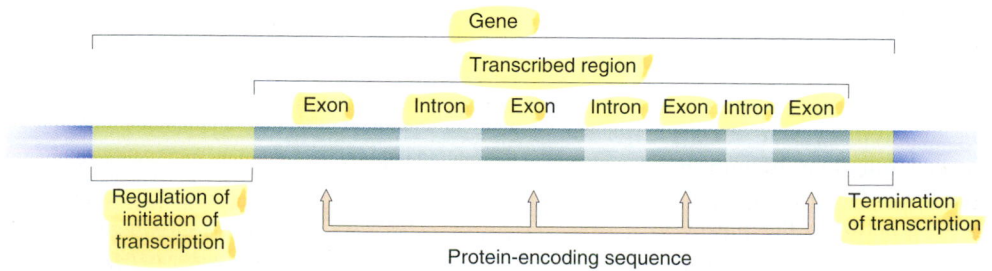

Figure 1-9   Generalized structure of a eukaryotic gene. This example has three introns and four exons.

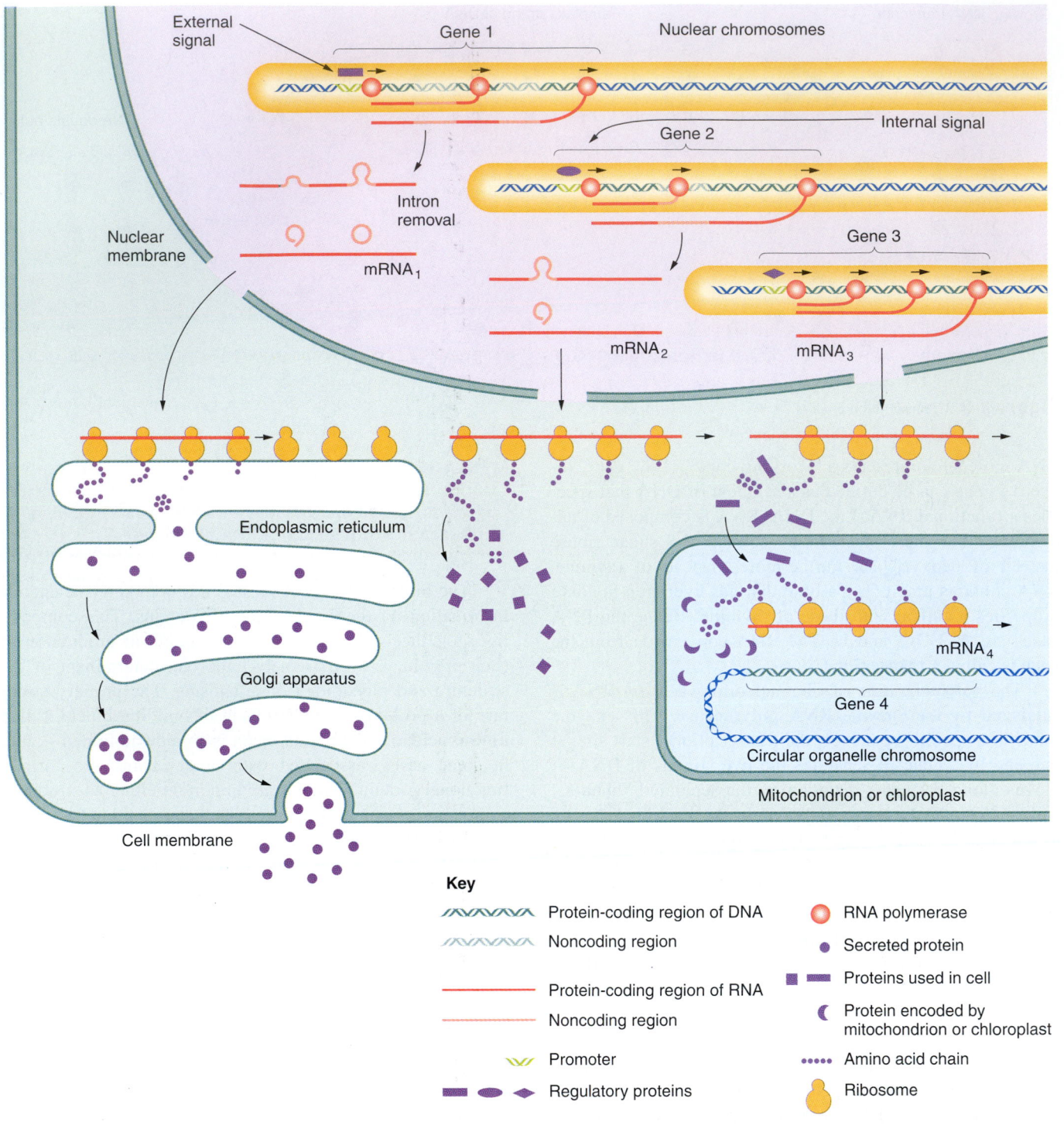

Figure 1-10 Simplified view of gene action in a eukaryotic cell. The basic flow of genetic information is from DNA to RNA to protein. Four types of genes are shown. Gene 1 responds to external regulatory signals and makes a protein for export; gene 2 responds to internal signals and makes a protein for use in the cytoplasm; gene 3 makes a protein to be transported into an organelle; gene 4 is part of the organelle DNA and makes a protein for use inside its own organelle. Note that many organelle genes have introns and that an RNA-synthesizing enzyme is needed for organelle mRNA synthesis. These details have been omitted from the diagram of the organelle for clarity. (Introns will be explained in detail in subsequent chapters.)

proteins (such as RNA polymerase) bind, causing the gene to be transcribed at the right time and in the right amount. A region at the other end of the gene contains sequences encoding the termination of transcription. In the genes of many eukaryotes, the protein-encoding sequence is interrupted by segments (ranging in number from one to many) called **introns.** The origin and functions of introns are still unclear. They are excised from the primary transcript. The split-up coding sequences between the introns are called **exons.**

Figure 1-10 illustrates the essentials of gene action in a generalized eukaryotic cell. The nucleus contains most of the DNA, but note that mitochondria and chloroplasts also contain small chromosomes.

Some protein-encoding genes are transcribed more or less constantly; they are the "housekeeping" genes that are always needed for basic reactions. Other genes may be rendered unreadable or readable to suit the functions of the organism at particular moments and under particular external conditions. The signal that masks or unmasks a gene may come from outside the cell; for example, from a steroid hormone or a nutrient. Or the signal may come from within the cell itself as the result of the reading of other genes. In either case, special *regulatory* sequences in the DNA are directly affected by the signal, and they in turn affect the transcription of the protein-encoding gene. The regulatory substances that are the signals bind to the regulatory DNA of the target genes to control the synthesis of transcripts.

When the introns have been cut out of the primary transcript, the remaining RNA sequence is called **messenger RNA (mRNA).** The mRNA molecules exit the nucleus through nuclear pores and enter the cytoplasm. The cytoplasm is where protein synthesis takes place. The nucleotide sequence of an mRNA molecule is "read" from the 5′ end to the 3′ end, in groups of three. These groups of three are called **codons.**

$$5'\ldots AUU\quad CCG\quad UAC\quad GUA\quad AAU\quad UUG\ldots 3'$$
Codon Codon Codon Codon Codon Codon

Because there are four nucleotides, there are $4 \times 4 \times 4 = 64$ different codons, each one standing for an amino acid or a signal to terminate translation. For instance, AUU stands for isoleucine, CCG for proline, and UAG is a translation-termination ("stop") codon.

Protein synthesis takes place on cytoplasmic organelles called ribosomes. A ribosome attaches to the 5′ end of an mRNA molecule and moves along the mRNA, catalyzing the assembly of the string of amino acids that will constitute the primary structure of the protein. This primary chain is called a **polypeptide.** Each amino acid is brought to the ribosome by a specific **transfer RNA (tRNA)** molecule that docks at a codon of the mRNA. Docking is by base pairing between a three-nucleotide tRNA segment called an **anticodon** and the codon:

| Anticodon in tRNA | 3′-GGA-5′ |
| Codon in mRNA | 5′-CCU-3′ |

Because this process of reading the mRNA sequence and converting it into an amino acid sequence is rather like converting one language into another, the process of protein synthesis is called **translation.** The process of translation is shown in Figure 1-11.

Trains of ribosomes pass along an mRNA molecule, each member of a train making the same type of polypeptide. At the end of the mRNA, a termination codon causes the ribosome to detach and recycle to another mRNA.

Let's pause to reconsider this amazing process in relation to the principles of macromolecular association. The alignment of tRNA anticodon to mRNA codon is by base complementarity. The attachment of the ribosome to the mRNA, its movement along the mRNA, and the binding of the tRNA to the ribosome are all examples of protein–nucleic acid bonding.

MESSAGE ···
The information in protein-coding genes is used by the cell in two steps of information transfer:

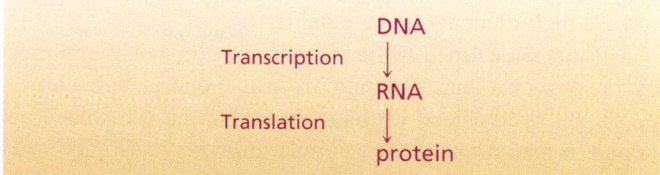

$$\text{DNA}$$
Transcription $\downarrow$
$$\text{RNA}$$
Translation $\downarrow$
$$\text{protein}$$

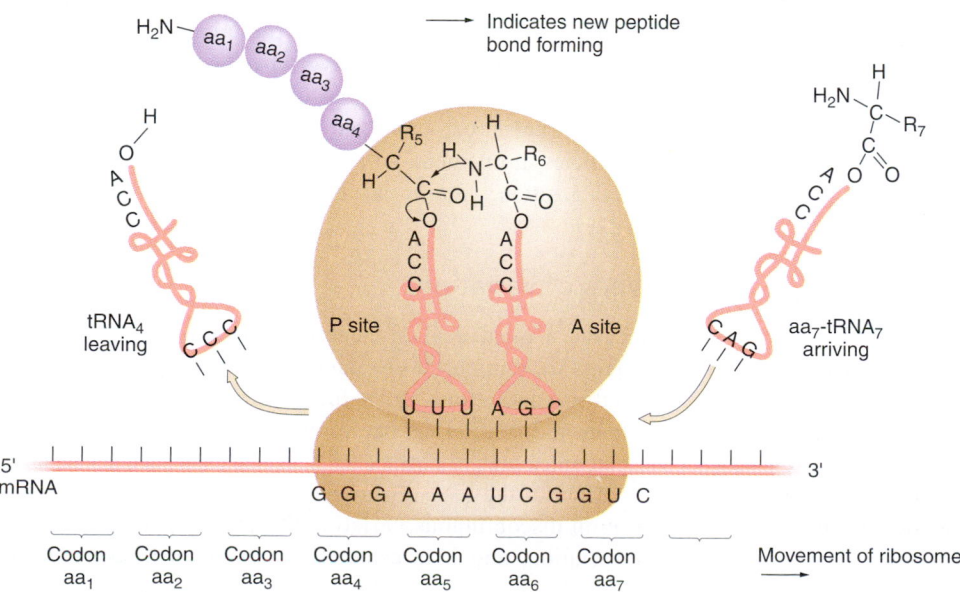

Figure 1-11   The addition of a single amino acid to the growing polypeptide chain in the translation of mRNA.

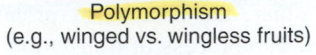

Polymorphism
(e.g., winged vs. wingless fruits)

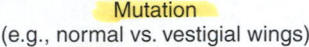

Mutation
(e.g., normal vs. vestigial wings)

Discontinuous variation

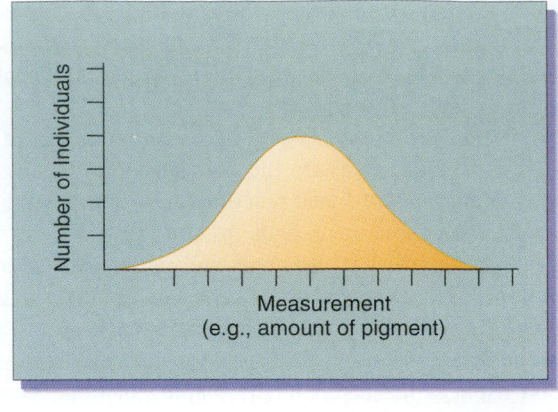

Continuous variation

Figure 1-12   Discontinuous and continuous variation in natural populations. In populations showing discontinuous variation for a particular character, each member possesses one of several discrete alternatives. For example, in the left-hand panel, a population of plants manifests two distinct, common fruit types: winged and wingless (see Figure 1-14a). Such variation is called a *polymorphism*. Sometimes most of the population is of one kind, with mutation providing only the occasional alternative, as in the vestigial wing type in the fruit fly, *Drosophila*. Discontinuous variants are often determined by the alleles of a single gene. Continuous variation, on the other hand, does not show such discrete alternatives; a character may be found in phenotypic gradations in a population. There may be no genetic basis — that is, all variation may be environmentally caused — or genes may play a role, often several or many genes.

Each gene encodes a separate protein, each with specific functions either within the cell (for example, the purple-rectangle proteins in Figure 1-10) or for export to other parts of the organism (the purple-circle proteins). The synthesis of proteins for export (secretory proteins) takes place on ribosomes that are located on the surface of the rough endoplasmic reticulum, a system of large flattened vesicles. The completed amino acid chains are passed into the lumen of the endoplasmic reticulum, where they fold up spontaneously to take on their protein shape. The proteins may be modified at this stage, but they eventually enter the chambers of the Golgi apparatus and from there, the secretory vessels, which eventually fuse with the cell membrane and release their contents to the outside.

Proteins destined to function in the cytoplasm and most of the proteins that function in mitochondria and chloroplasts are synthesized in the cytosoplasm on ribosomes not bound to membranes. For example, proteins that function as enzymes in the glycolysis pathway follow this route. The proteins destined for organelles are specially tagged to target their insertion into specific organelles. In addition, mitochondria and chloroplasts have their own small circular DNA molecules. The synthesis of proteins encoded by genes on mitochondrial or chloroplast DNA takes place on ribosomes inside the organelles themselves. Therefore the proteins in mitochondria and chloroplasts are of two different origins: either encoded in the nucleus and imported into the organelle or encoded in the organelle and synthesized within the organelle compartment.

Prokaryotes are one-celled organisms such as bacteria, whose cellular structure is simpler than that of eukaryotes; there is no nucleus separated from the cytoplasm by a nuclear membrane. Protein synthesis in prokaryotes is generally similar to that in eukaryotes mRNA, tRNA, and ribosomes are used), but there are some important differences. For example, prokaryotic genes have no introns. Furthermore, there are no compartments bounded by membranes through which the RNA or protein must pass.

MESSAGE ....
The flow of information from DNA to RNA to protein is a central focus of modern biology.

Genetic variation

If all members of a species have the same set of genes, how can there be genetic variation? As indicated earlier, the answer is that genes come in different forms called alleles. In a population, for any given gene there can be from one to many different alleles; however, because most organisms carry only one or two chromosome sets per cell, any individual organism can carry only one or two alleles per gene. The alleles of one gene will always be found in one chromosomal position. Allelic variation is the basis for hereditary variation.

Types of variation

Because a great deal of genetics concerns the analysis of variants, it is important to understand the types of variation found in populations. A useful classification is into discontinuous and continuous variation (Figure 1-12). Allelic variation contributes to both.

Most of the research in genetics in the past century has been on discontinuous variation because it is a simpler type of variation, and it is easier to analyze. In **discontinuous**

variation, a character is found in a population in two or more distinct and separate forms called **phenotypes.** Such alternative phenotypes are often found to be encoded by the alleles of one gene. A good example is albinism in humans, which concerns phenotypes of the character of skin pigmentation. In most people, the cells of the skin can make a dark brown or black pigment called melanin, the substance that gives our skin its color ranging from tan color in people of European ancestry to brown or black in those of tropical and subtropical ancestry. Although always rare, albinos are found in all races; they have a totally pigmentless skin and hair (Figure 1-13). The difference between pigmented and unpigmented is caused by two alleles of a gene taking part in melanin synthesis. The alleles of a gene are conventionally designated by letters. The allele that codes for the ability to make melanin is called *A* and the allele that codes for the inability to make melanin (resulting in albinism) is designated *a* to show that they are related. The allelic constitution of an organism is its **genotype,** which is the hereditary underpinning of the phenotype. Because humans have two sets of chromosomes in each cell, genotypes can be either *A/A, A/a,* or *a/a* (the slash shows that they are a pair). The phenotype of *A/A* is pigmented, *a/a* is albino, and *A/a* is pigmented. The *ability* to make pigment is expressed over *inability* (*A* is said to be dominant, as we shall see in Chapter 2).

Although allelic differences cause phenotypic differences such as pigmented and albino, this does not mean that only one gene affects skin color. It is known that there are several. However, the *difference* between pigmented, of

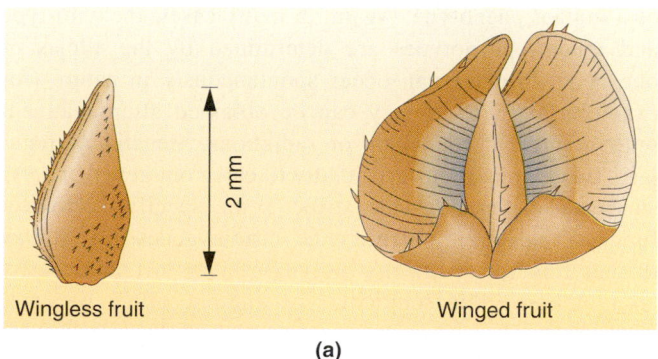

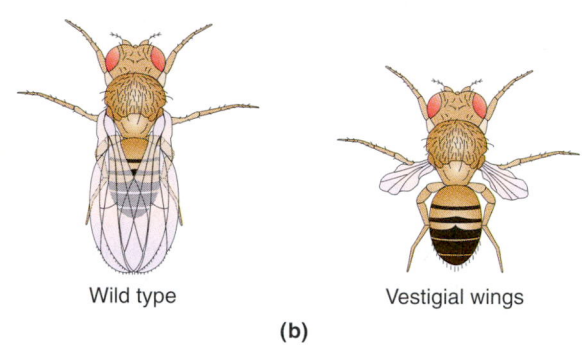

Figure 1-14   A dimorphism. (a) The fruits of two different forms of *Plectritis congesta,* the sea blush. Any one plant has either all wingless or all winged fruits. In every other way, the plants are identical. (b) A *Drosophila* mutant with abnormal wings and a normal fly (wild type) for comparison. In both cases, the two phenotypes are caused by the alleles of one gene.

Figure 1-13   An albino. The phenotype is caused by two doses of a recessive allele — *a / a.* The dominant allele *A* determines one step in the chemical synthesis of the dark pigment melanin in the cells of skin, hair, and eye retinas. In *a / a* individuals, this step is nonfunctional, and the synthesis of melanin is blocked. (© Yves Gellie/Icone.)

whatever shade, and albino is caused by the *difference* at one gene; the state of all the other pigment genes is irrelevant.

In discontinuous variation, there is a predictable one-to-one relation between genotype and phenotype under most conditions. In other words, the two phenotypes (and their underlying genotypes) can almost always be distinguished. In the albinism example, the *A* allele always allows some pigment formation, whereas the white allele always results in albinism when homozygous. For this reason, discontinuous variation has been successfully used by geneticists to identify the underlying alleles and their role in cellular functions.

Geneticists distinguish two categories of discontinuous variation on the basis of simple allelic differences. In a natural population, the existence of two or more *common* discontinuous variants is called **polymorphism** (Greek; many forms), and an example is shown in Figure 1-14a. The various forms are called **morphs.** It is often found that morphs are determined by the alleles of a single gene. Why do populations show genetic polymorphism? Special types of natural selection can explain a few cases, but, in other cases, the morphs seem to be selectively neutral.

Rare, exceptional discontinuous variants are called **mutants,** whereas the more common "normal" companion phenotype is called the **wild type.** Figure 1-14b shows an example of a mutant phenotype. Again, in many cases, the wild-type and mutant phenotypes are determined by the alleles of one gene. Mutants can occur spontaneously in nature (for example, albinos) or they can be obtained after treatment with mutagenic chemicals or radiations. Geneticists regularly induce mutations artificially to carry out genetic analysis because mutations that affect some specific biological function under study identify the various genes that interact in that function. Note that polymorphisms originally arise as mutations, but somehow the mutant allele becomes common.

MESSAGE ·······························
In many cases, an allelic difference at a single gene may result in discrete phenotypic forms that make it easy to study the gene and its associated biological function.

Continuous variation of a character shows an unbroken range of phenotypes in the population (see Figure 1-12). Measurable characters such as height, weight, and color intensity are good examples of such variation. Intermediate phenotypes are generally more common than extreme phenotypes and, when phenotypic frequencies are plotted as a graph, a bell-shaped distribution is observed. In some such distributions, all the variation is environmental and has no genetic basis at all. In other cases, there is a genetic component caused by allelic variation of one or many genes. In most cases, there is both genetic and environmental variation. In continuous distributions, there is no one-to-one correspondence of genotype and phenotype. For this reason, little is known about the types of genes underlying continuous variation, and only recently have techniques become available for identifying and characterizing them.

Continuous variation is encountered more commonly than discontinuous variation in everyday life. We can all identify examples of continuous variation in plant or animal populations that we have observed — many examples exist in human populations. One area of genetics in which continuous variation is important is in plant and animal breeding. Many of the characters that are under selection in breeding programs, such as seed weight or milk production, have complex determination, and the phenotypes show continuous variation in populations. Animals or plants from one extreme end of the range are chosen and selectively bred. Before such selection is undertaken, the sizes of the genetic and environmental components of the variation must be known. We shall return to these specialized techniques in Chapter 20, but, for the greater part of the book, we shall be dealing with the genes underlying discontinuous variation.

Molecular basis of allelic variation

Consider the difference between the pigmented and the albino phenotypes in humans. The dark pigment melanin has a complex structure that is the end product of a biochemical synthetic pathway. Each step in the pathway is a conversion of one molecule into another, with the progressive formation of melanin in a step-by-step manner. Each step is catalyzed by a separate enzyme protein encoded by a specific gene. Most cases of albinism result from changes in one of these enzymes — tyrosinase. The enzyme tyrosinase catalyzes the last step of the pathway, the conversion of tyrosine into melanin.

$$\longrightarrow \underset{\text{(not a pigment)}}{\text{tyrosine}} \xrightarrow{\text{tyrosinase}} \underset{\text{(pigment)}}{\text{melanin}}$$

To perform this task, tyrosinase binds to its substrate, a molecule of tyrosine, and facilitates the molecular changes necessary to produce the pigment melanin. There is a specific "lock-and-key" fit between tyrosine and the active site of the enzyme. The **active site** is a pocket formed by several crucial amino acids in the polypeptide. If the DNA of the tyrosinase-encoding gene changes in such a way that one of these crucial amino acids is replaced by another amino acid or lost, then there are several possible consequences. First, the enzyme might still be able to perform its functions but in a less efficient manner. Such a change may have only a small effect at the phenotypic level, so small as to be difficult to observe, but it might lead to a reduction in the amount of melanin formed and, consequently, a lighter skin coloration. Note that the protein is still present more or less intact, but its ability to convert tyrosine into melanin has been compromised. Second, the enzyme might be incapable of any function, in which case the mutational event in the DNA of the gene would have produced an albinism allele, referred to earlier as an *a* allele. Hence a person of genotype *a/a* is an albino. The genotype *A/a* is interesting. It results in normal pigmentation because transcription of one copy of the wild-type allele (*A*) can provide enough tyrosinase for synthesis of normal amounts of melanin. Alleles are termed *haplosufficient* if roughly normal function is obtained when there is only a single copy of the normal gene. Alleles commonly appear to be haplosufficient, in part because small reductions in function are not vital to the organism. Alleles that fail to code for a functional protein are called **null** ("nothing") **alleles** and are generally not expressed in combination with functional alleles (in individuals of genotype *A/a*). The molecular basis of albinism is represented in Figure 1-15.

The mutational site in the DNA can be of a number of types. The simplest and most common type is **nucleotide-pair substitution,** which can lead to amino acid substitution or to premature stop codons. Small **deletions and duplication** also are common. Even a single base deletion or insertion produces widespread damage at the protein level; because mRNA is read from one end "in frame" in groups of three, a loss or gain of one nucleotide pair shifts the reading frame, and all the amino acids translationally downstream will be incorrect. Such mutations are called **frameshift mutations.**

At the protein level, mutation changes the amino acid composition of the protein. The most important outcomes are change in shape and size. Such change in shape or size

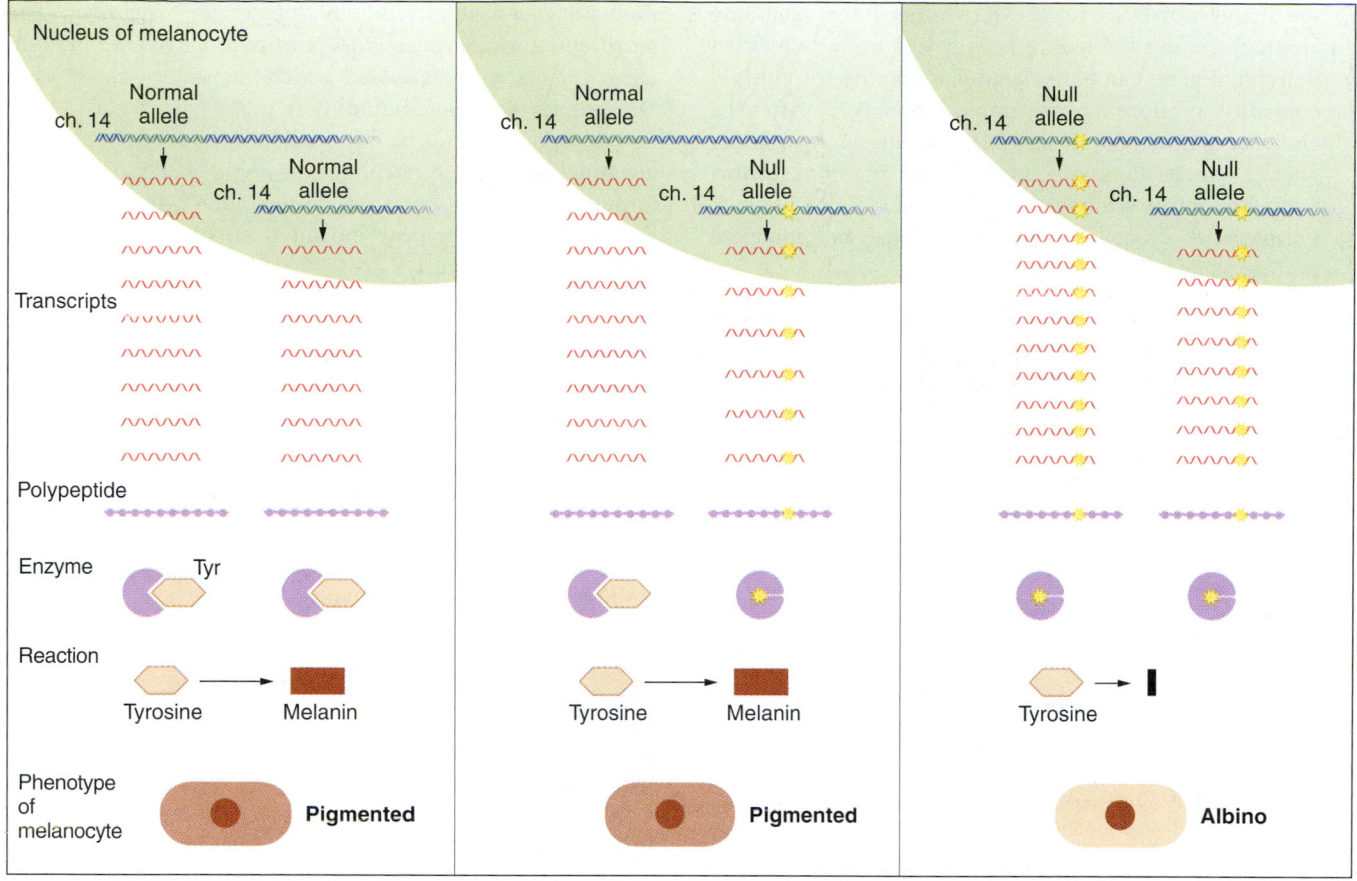

Figure 1-15   Molecular basis of albinism. Expression in cells containing 2, 1, and 0 copies of the normal tyrosinase allele on chromosome 14. Melanocytes are specialized melanin-producing cells.

can result in no biological function (which would be the basis of a null allele), or reduced function. More rarely, mutation can lead to new function of the protein product.

MESSAGE ···

New alleles formed by mutation can result in no function, less function, or new function at the protein level.

Methodologies used in genetics

The study of genes has proved to be a powerful approach to understanding biological systems. Because genes affect virtually every aspect of the structure and function of an organism, being able to identify and determine the role of genes and the proteins that they encode is an important step in charting the various processes that underly a particular character under investigation. It is interesting that geneticists study not only hereditary mechanisms, but *all* biological mechanisms. Many different methodologies are used to study genes and gene activities, and these methodologies can be summarized briefly as follows:

1. *Isolation of mutants affecting the process under study.* Each mutant gene reveals a genetic component of the process, and together they show the range of proteins that interact in that specific process.

2. *Analysis of progeny of controlled matings ("crosses") between mutants and other discontinuous variants.* This type of analysis identifies genes and their alleles, their chromosomal locations, and their inheritance patterns. These methods will be introduced in Chapter 2.

3. *Biochemical analysis of cellular processes controlled by genes.* Life is basically a complex set of chemical reactions; so studying the ways in which genes are relevant to these reactions is an important way of dissecting this complex chemistry. Mutant alleles underlying defective function (see method 1) are invaluable in this type of analysis. The basic approach is to find out how the cellular chemistry is disturbed in the mutant individual and, from this information, deduce the role of the gene. The deductions from many genes are assembled to reveal the larger picture.

4. *Microscopic analysis.* Chromosome structure and movement have long been an integral part of genetics, but new technologies have provided ways of labeling genes and gene products so that their locations can be easily visualized under the microscope.

5. *Analysis of DNA directly.* Because the genetic material is composed of DNA, the ultimate characterization is

the analysis of DNA itself. Many procedures, including cloning, are used. Cloning is a procedure by which an individual gene can be isolated and amplified (multiply copied) to produce a pure sample for analysis. After the clone of a gene has been obtained, its nucleotide sequence can be determined and hence important information about its structure and function can be obtained. Furthermore cloned genes can be used as biological probes, as described next.

Cloning will be fully described in Chapters 12, 13, and 14, but a brief overview is necessary here so that some concepts can be applied in earlier chapters. Cloning works by inserting the gene to be amplified into a small accessory chromosome and letting this chromosome do the job of

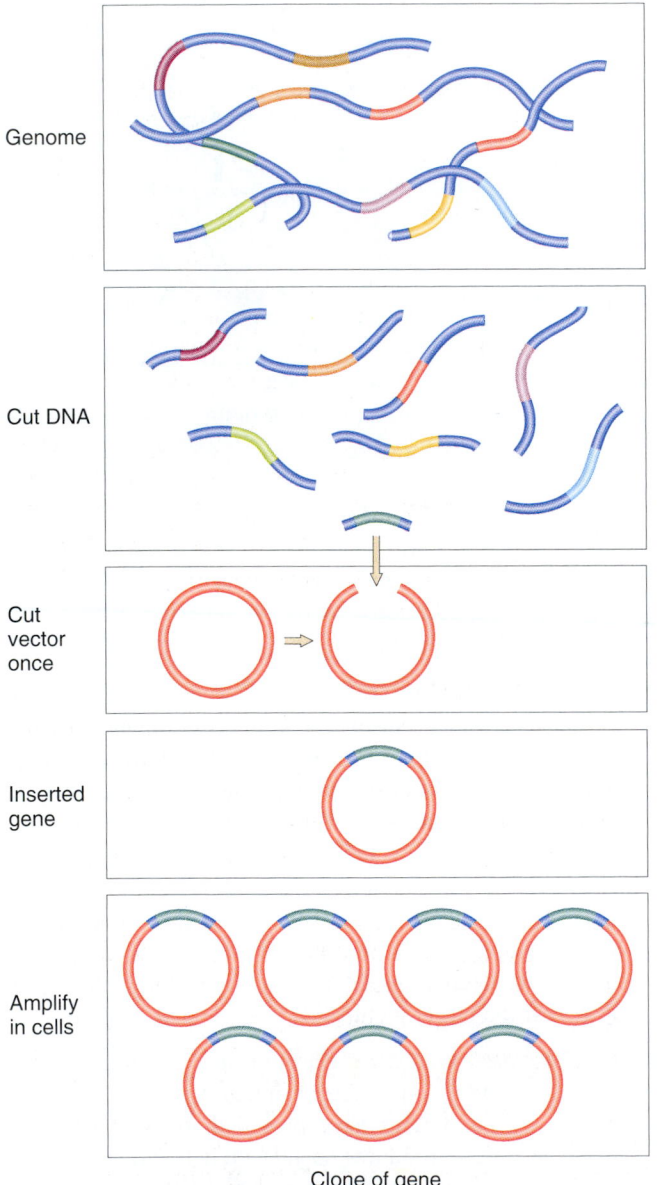

Genome

Cut DNA

Cut
vector
once

Inserted
gene

Amplify
in cells

Clone of gene

Figure 1-16   Basic cloning methodology.

replicating and amplifying its "passenger" fragment. This small chromosome is called a **vector** or carrier. Commonly used vectors are plasmids, which are nonessential extra DNA molecules found naturally in many bacteria. The DNA of any organism can be inserted into a vector; the hybrid construct is then introduced into a single bacterial cell and a large culture grows from this cell (Figure 1-16). The vector and insert can then be removed from disrupted cells and studied as appropriate.

Detecting specific molecules of DNA, RNA, and protein

Because the main macromolecules of genetics are DNA, RNA and protein, genetic analysis often requires the detection of specific molecules of each of these three types. How can specific molecules be identified among the thousands of types in the cell? The most extensively used method for detecting specific macromolecules in a mixture is **probing**. This method makes use of the specificity of intermolecular binding, which we have already encountered several times. The probe is labeled in some way, either by a radioactive atom or by a flourescent compound, so that the site of binding can easily be detected. Let's look at probes for DNA, RNA, and protein.

Probing for a specific DNA.   A cloned gene can act as a probe for finding segments of DNA that have the same or a very similar sequence. For example, if a gene *G* from a fungus has been cloned, it might be of interest to determine whether plants have the same gene. The use of a cloned gene as a probe takes us back to the principle of base complementarity. The probe works through the principle that, in solution, the random motion of probe molecules enables them to find and bind to complementary sequences. The experiment must be done with separated DNA strands, because then the bonding sites of the bases are unoccupied. DNA from the plant is extracted and cut with one of the many available types of **restriction enzymes,** which cut DNA at specific target sequences of four or more bases. The target sequences are at the same positions in all the plant cells used, so the enzyme cuts the genome into defined populations of segments of specific sizes. The fragments can be fractionated by using electrophoresis.

Electrophoresis fractionates a population of nucleic acid fragments on the basis of size. The cut mixture is placed in a small well in a gelatinous slab (a gel), and the gel is placed in a powerful electrical field. The electricity causes the molecules to move through the gel at speeds inversely proportional to their size. After fractionation, the separated fragments are blotted onto a piece of porous membrane, where they maintain the same relative positions. This procedure is called a **Southern blot.** After having been heated to separate the DNA strands and hold the DNA in position, the membrane is placed in a solution of the probe.

The single-stranded probe will find and bind to its complementary DNA sequence. For example,

TAGGCATCG   Probe
ACTAATCCATAGCTTA   Genomic fragment

On the blot, this binding concentrates the label in one spot, as shown in Figure 1-17a.

Probing for a specific RNA.   It is often necessary to determine whether a gene is being transcribed in some particular tissue. This determination can be accomplished by using a modification of the Southern analysis. Total mRNA is extracted from the tissue, fractionated electrophoretically, and blotted onto a membrane (this procedure is called a **Northern blot**). The cloned gene is used as a probe and its label will highlight the mRNA in question if it is present (Figure 1-17b).

Probing for a specific protein.   Probing for proteins is generally performed with antibodies because an antibody has a specific lock-and-key fit with its antigen. The protein mixture is electrophoresed and blotted onto a membrane (this procedure is a **Western blot**). The position of a specific protein on the membrane is revealed by bathing the membrane in a solution of antibody obtained from a rabbit into which the protein has been injected. The position of the protein is revealed by the position of the label that the antibody carries (Figure 1-17c).

Genes, the environment, and the organism

The net outcome of the reading of a gene is that a protein is made that generally has one of two basic functions, depending on the gene. First, the protein may be a **structural protein,** contributing to the physical properties of cells or organisms. Examples are microtubule, muscle, and hair proteins. Second, the protein may be an **enzyme** that catalyzes one of the chemical reactions of the cell. Therefore, by coding for proteins, genes determine two important facets of biological structure and function. However, genes cannot dictate the structure of an organism by themselves. The other crucial component in the formula is the environment. The environment influences gene action in many ways, about which we shall learn in subsequent chapters. In the present context, however, it is relevant to note that the environment provides the raw materials for the synthetic processes controlled by genes. For example, animals obtain several of the amino acids for their proteins as part of their diet. Most of the chemical syntheses in plant cells use carbon atoms taken from the air as carbon dioxide. Bacteria and fungi absorb from their surroundings many substances that are simply treated as carbon and nitrogen skeletons, and their enzymes convert these components into the compounds that constitute the living cell. Thus, through genes, an organism builds the orderly process that we call life out of disorderly environmental materials.

Genetic determination

From our brief look at the role of genes, we can see that living organisms mobilize the components of the world around them and convert these components into their own living material. An acorn becomes an oak tree, by using in the process only water, oxygen, carbon dioxide, some inorganic materials from the soil, and light energy.

An acorn develops into an oak, whereas the spore of a moss develops into a moss, although both are growing side by side in the same forest. The two plants that result from these developmental processes resemble their parents and differ from each other, even though they have access to the same narrow range of inorganic materials from the environment. The parents pass to their offspring the specifications for building living cells from environmental materials. These specifications are in the form of genes in the fertilized egg. Because of the information in the genes, the acorn develops into an oak and the moss spore becomes a moss.

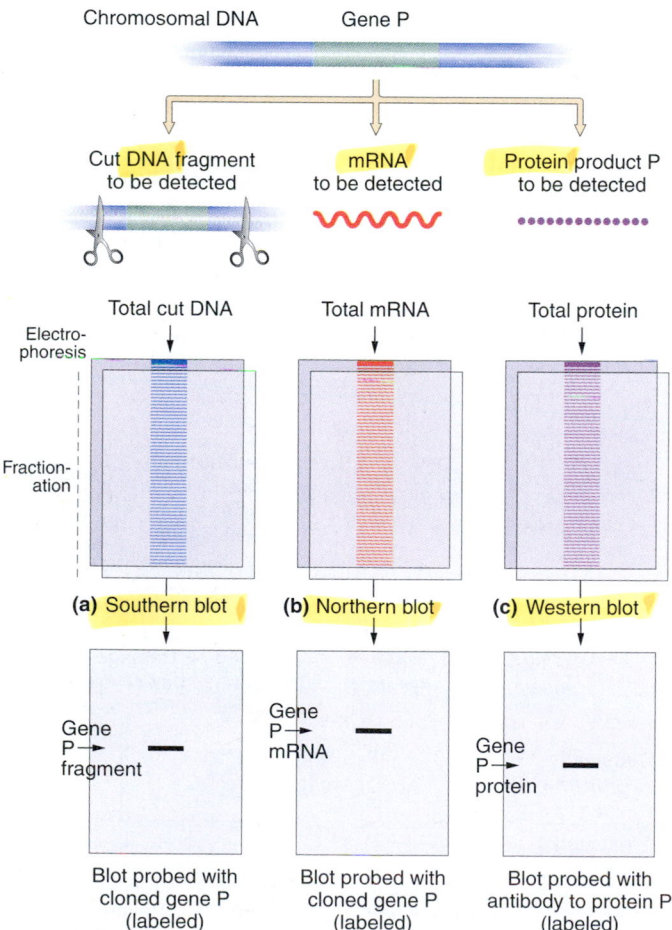

Figure 1-17   Probing DNA, RNA, and protein mixtures.

Just as genes maintain differences between species such as the oak and moss, they also maintain differences within species. Consider plants of the species *Plectritis congesta,* the sea blush. Two forms of this species are found wherever the plants grow in nature: one form has wingless fruits, and the other has winged fruits (see Figure 1-14a). These plants will self-pollinate, and we can observe the offspring that result from such "selfs" when they are grown in a greenhouse under uniform conditions; we commonly observe that all the selfed progeny of a winged-fruited plant are winged fruited and that all the selfed progeny from a wingless-fruited plant have wingless fruits. Because all the progeny were grown in the same environment, we can rule out the possibility that environmental differences cause some plants to bear wingless fruits and others to bear winged. We can safely conclude that the fruit-shape difference between the original plants, which each passed on to its selfed progeny, results from the different genotypes. It can be shown that the winged and wingless morphs are determined by alleles of one gene.

Plectritis shows two inherited forms that are both perfectly normal. The determinative power of genes is probably more often demonstrated by differences in which one form is normal and the other abnormal. The human inherited disease sickle-cell anemia is a good example. The underlying cause of the disease is a variant of hemoglobin, the oxygen-transporting protein molecule found in red blood cells. Normal people have a type of hemoglobin called *hemoglobin A,* the information for which is encoded in a gene. A minute chemical change at the molecular level in the DNA of this gene results in the production of a slightly changed hemoglobin, termed *hemoglobin S.* In people possessing only hemoglobin S, the ultimate effect of this small change is severe ill health and usually death. The gene works its effect on the organism through a complex "cascade effect," as summarized in Figure 1-18.

Such observations lead to the model of how genes and the environment interact shown in Figure 1-19. In this view, the genes act as a set of instructions for turning more or less undifferentiated environmental materials into a specific organism, much as blueprints specify what form of house is to be built from basic materials. The same bricks, mortar, wood, and nails can be made into an A-frame or a flat-roofed house, according to different plans. Such a model implies that the genes are really the dominant elements in the determination of organisms; the environment simply supplies the undifferentiated raw materials.

Environmental determination

Consider two monozygotic ("identical") twins, the products of a single fertilized egg that divided and produced two complete babies with identical genes. Suppose that the twins are born in England but are separated at birth and taken to different countries. If one is reared in China by Chinese-speaking foster parents, she will speak Chinese, whereas her sister reared in Budapest will speak Hungarian. Each will absorb the cultural values and customs of her environment. Although the twins begin life with identical genetic properties, the different cultural environments in which they live will produce differences between them (and differences from their parents). Obviously, the differences in this case are due to the environment, and genetic effects are of little importance in determining the differences.

This example suggests the model of Figure 1-20, which is the converse of that shown in Figure 1-19. In the model in Figure 1-20, the genes impinge on the system, giving certain general signals for development, but the environment determines the actual course of action. Imagine a set of specifications for a house that simply calls for "a floor that will sup-

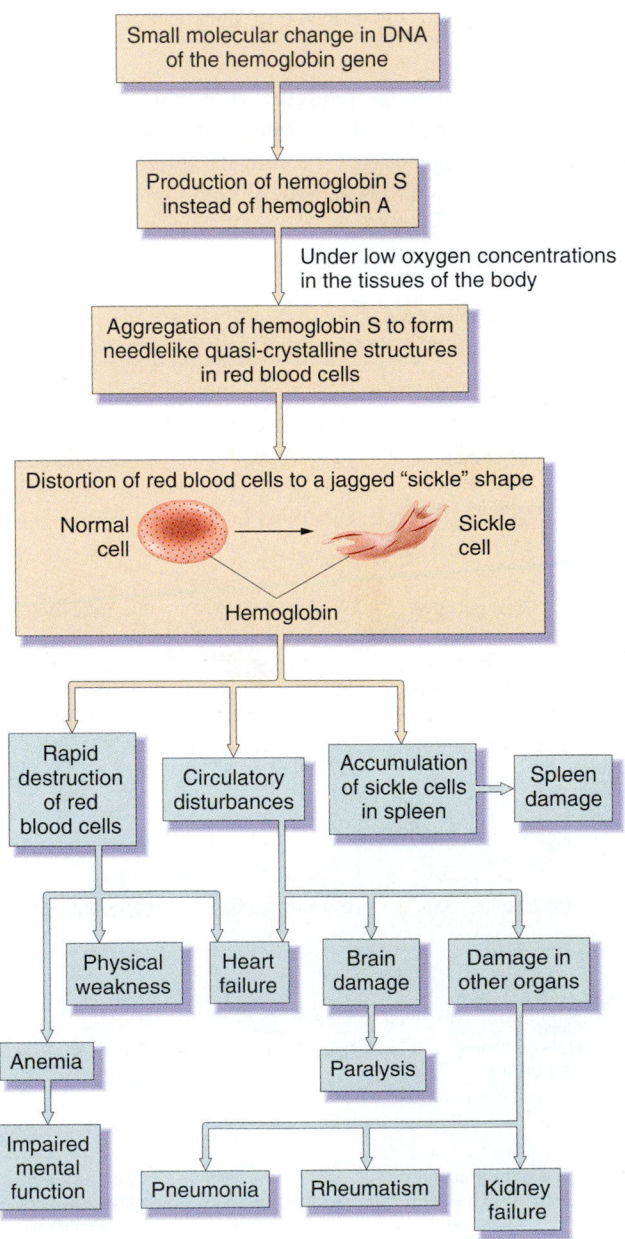

Figure 1-18   Chain of events in human sickle-cell anemia.

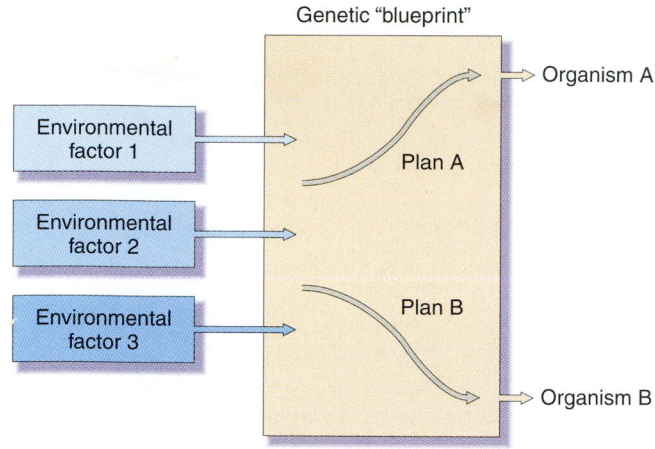

Genetic "blueprint"

Figure 1-19   A model of determination that emphasizes the role of genes.

port 300 pounds per square foot" or "walls with an insulation factor of 15 inches"; the actual appearance and other characteristics of the structure would be determined by the available building materials.

Our different types of example — of purely genetic effect versus the effect of the environment — lead to two very different models. First, consider the seed example: given a pair of seeds and a uniform growth environment, we would be unable to predict future growth patterns solely from a knowledge of the environment. In any environment that we can imagine, if the organisms develop, the acorn becomes an oak and the spore becomes a moss. The model of Figure 1-19 applies here. Second, consider the twins: no information about the set of genes that they inherit could possibly enable us to predict their ultimate languages and cultures. Two individuals that are *genetically different* may develop differently in the *same environment,* but two *genetically identical* individuals may develop differently in *different environments.* The model of Figure 1-20 applies here.

In general, we deal with organisms that differ in both genes and environment. If we wish to predict how a living organism will develop, we must first know the genetic constitution that it inherits from its parents. Then we must know the historical sequence of environments to which the devel-

oping organism is exposed. Every organism has a developmental history from birth to death. What an organism will become in the next moment depends critically both on the environment that it encounters during that moment and on its present state. It makes a difference to an organism not only what environments it encounters, but in what sequence it encounters them. A fruit fly *(Drosophila melanogaster),* for example, develops normally at 20°C. If the temperature is briefly raised to 37°C early in its pupal stage of development, the adult fly will be missing part of the normal vein pattern on its wings. However, if this "temperature shock" is administered just 24 hours later, the vein pattern develops normally. A general model in which genes and the environment jointly determine (by some rules of development) the actual characteristics of an organism is depicted in Figure 1-21.

MESSAGE ⋯⋯⋯⋯⋯⋯⋯⋯⋯⋯⋯⋯⋯⋯⋯⋯⋯

As an organism transforms developmentally from one stage to another, its genes interact with its environment at each moment of its life history. The interaction of genes and environment determines what organisms are.

The use of genotype and phenotype

In studying how genes and the environment interact to produce an organism, geneticists have developed some useful terms that are introduced in this section.

A typical organism resembles its parents more than it resembles unrelated individuals. Thus, we often speak as if the individual characteristics themselves are inherited: "He gets his brains from his mother," or "She inherited diabetes from her father." Yet the preceding section shows that such statements are inaccurate. "His brains" and "her diabetes" develop through long sequences of events in the life histories of the affected people, and both genes and environment play roles in those sequences. In the biological sense, individuals inherit only the molecular structures of the fertilized eggs from which they develop. Individuals inherit their genes, not the end products of their individual developmental histories.

To prevent such confusion between genes (which are inherited) and developmental outcomes (which are not), geneticists make the fundamental distinction between the genotype and the phenotype of an organism. Organisms

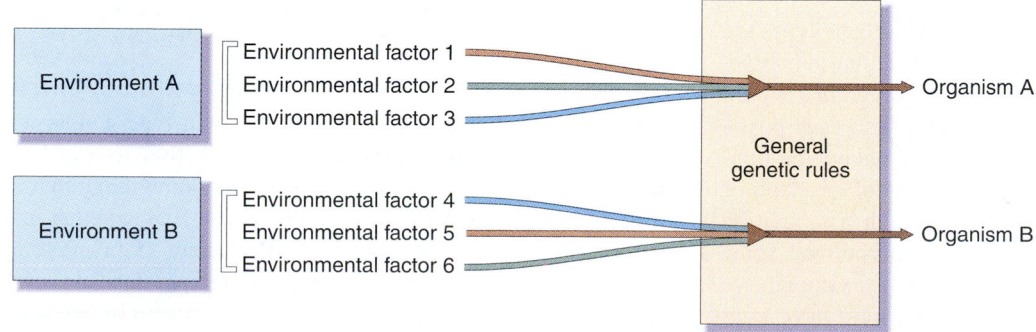

Figure 1-20   A model of determination that emphasizes the role of the environment.

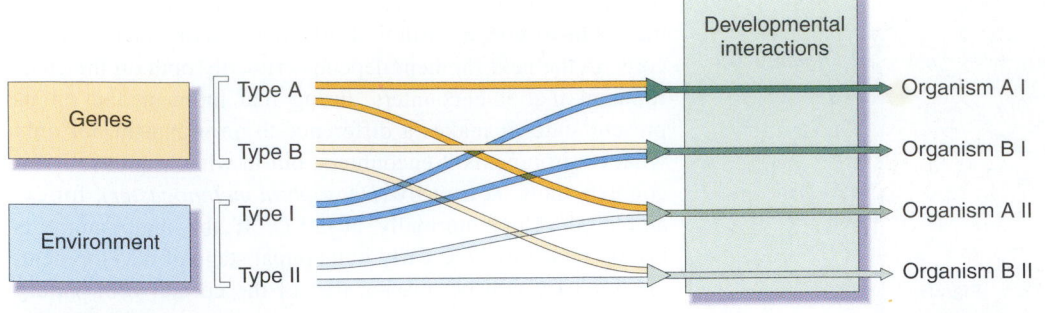

Figure 1-21   A model of determination that emphasizes the interaction of genes and environment.

have the same genotype in common if they have the same set of genes. Organisms have the same phenotype if they look or function alike.

Strictly speaking, the genotype describes the complete set of genes inherited by an individual, and the phenotype describes all aspects of the individual's morphology, physiology, behavior, and ecological relations. In this sense, no two individuals ever belong to the same phenotype, because there is always some difference (however slight) between them in morphology or physiology. Additionally, except for individuals produced from another organism by asexual reproduction, any two organisms differ at least a little in genotype. In practice, we use the terms *genotype* and *phenotype* in a more restricted sense. We deal with some partial phenotypic description (say, eye color) and with some subset of the genotype (say, the genes that affect eye pigmentation).

MESSAGE ···

When we use the terms *phenotype* and *genotype*, we generally mean "partial phenotype" and "partial genotype," and we specify one or a few traits and genes that are the subsets of interest.

··

Note one very important difference between genotype and phenotype: The genotype is essentially a fixed character of an individual organism; the genotype remains constant throughout life and is essentially unchanged by environmental effects. Most phenotypes change continually throughout the life of an organism as its genes interact with a sequence of environments. Fixity of genotype does not imply fixity of phenotype.

Norm of reaction

How can we quantify the relation between the genotype, the environment, and the phenotype? For a particular genotype, we could prepare a table showing the phenotype that would result from the development of that genotype in each possible environment. Such a set of environment-phenotype relations for a given genotype is called the **norm of reaction** of the genotype. In practice, we can make such a tabulation only for a partial genotype, a partial phenotype, and some particular aspects of the environment. For example, we might specify the eye sizes that fruit flies would have after developing at various constant temperatures; we could do

this for several different eye-size genotypes to get the norms of reaction of the species.

Figure 1-22 represents just such norms of reaction for three eye-size genotypes in the fruit fly *Drosophila melanogaster*. The graph is a convenient summary of more extensive tabulated data. The size of the fly eye is measured by counting its individual facets, or cells. The vertical axis of the graph shows the number of facets (on a logarithmic scale); the horizontal axis shows the constant temperature at which the flies develop.

Three norms of reaction are shown on the graph. When flies of the wild-type genotype that is characteristic of flies in natural populations are raised at higher temperatures, they develop eyes that are somewhat smaller than those of wild-type flies raised at cooler temperatures. The graph shows that wild-type phenotypes range from more than 700 to 1000 facets — the wild-type norm of reaction. A fly that has the *ultrabar* genotype has smaller eyes than those of wild-type flies regardless of temperature during development. Temperatures have a stronger effect on development of *ultrabar* genotypes than on wild-type genotypes, as we see by noticing that the *ultrabar* norm of reaction slopes more steeply than the wild-type norm of reaction. Any fly of the *infrabar* genotype also has smaller eyes than those of any wild-type fly, but temperatures have the opposite effect on flies of this genotype; *infrabar* flies raised at higher temperatures tend to have larger eyes than those of flys raised at lower temperatures. These norms of reaction indicate that the relation between genotype and phenotype is complex rather than simple.

MESSAGE ···

A single genotype may produce different phenotypes, depending on the environment in which organisms develop. The same phenotype may be produced by different genotypes, depending on the environment.

··

If we know that a fruit fly has the wild-type genotype, this information alone does not tell us whether its eye has 800 or 1000 facets. On the other hand, the knowledge that a fruit fly's eye has 170 facets does not tell us whether its genotype is *ultrabar* or *infrabar*. We cannot even make a general statement about the effect of temperature on eye size in *Drosophila*, because the effect is opposite in two different genotypes. We see from Figure 1-22 that some geno-

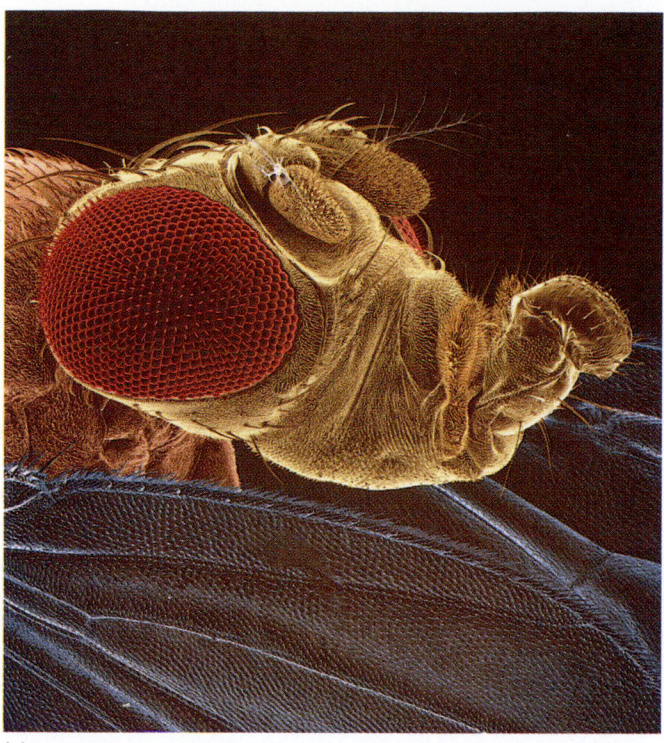

(a)

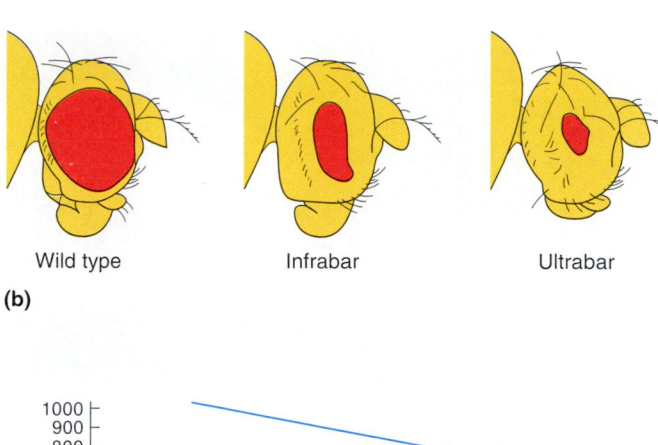

Wild type   Infrabar   Ultrabar

(b)

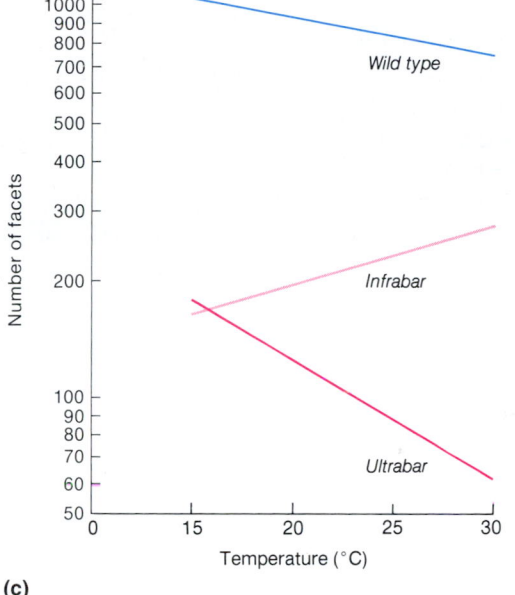

(c)

Figure 1-22   Norms of reaction to temperature for three different eye-size genotypes in *Drosophila*. (a) Closeup showing how the normal eye comprises hundreds of units called *facets*. The number of facets determines eye size. (b) Relative eye sizes of wild-type, *infrabar*, and *ultrabar* flies raised at the higher end of the temperature range. (c) Norm-of-reaction curves for the three genotypes. (Part a, Don Rio and Sima Misra, University of California, Berkeley.)

types do differ unambiguously in phenotype, no matter what the environment: any wild-type fly has larger eyes than any *ultrabar* or *infrabar* fly. But other genotypes overlap in phenotypic expression: the eyes of an *ultrabar* fly may be larger or smaller than those of an *infrabar* fly, depending on the temperatures at which the individuals developed.

To obtain a norm of reaction such as the norms of reaction in Figure 1-22, we must allow different individuals of identical genotype to develop in many different environments. To carry out such an experiment, we must be able to obtain or produce many fertilized eggs with identical genotypes. For example, to test a human genotype in 10 environments, we would have to obtain genetically identical sibs and raise each individual in a different milieu. However, that is possible neither biologically nor ethically. At the present time, we do not know the norm of reaction of any human genotype for any character in any set of environments. Nor is it clear how we can ever acquire such information without the unacceptable manipulation of human individuals.

For a few experimental organisms, special genetic methods make it possible to replicate genotypes and thus to determine norms of reaction. Such studies are particularly easy in plants that can be propagated vegetatively (that is, by cuttings). The pieces cut from a single plant all have the same genotype; so all offspring produced in this way have identical genotypes. Such a study has been done on the yarrow plant, *Achillea millefolium* (Figure 1-23a). The experimental results are shown in Figure 1-23b. Many plants were collected, and three cuttings were taken from each plant. One cutting from each plant was planted at low elevation (30 meters above sea level), one at medium elevation (1400 meters), and one at high elevation (3050 meters). Figure 1-23b shows the mature individuals that developed from the cuttings of seven plants; each set of three plants of identical genotype is aligned vertically in the figure for comparison.

First, we note an average effect of environment: in general, the plants grew poorly at the medium elevation. This is not true for every genotype, however; the cutting of plant 4 grew best at the medium elevation. Second, we note that no genotype is unconditionally superior in growth to all others. Plant 1 showed the best growth at low and high elevations but showed the poorest growth at the medium elevation. Plant 6 showed the second-worst growth at low elevation and the second-best at high elevation. Once again, we see

(a)

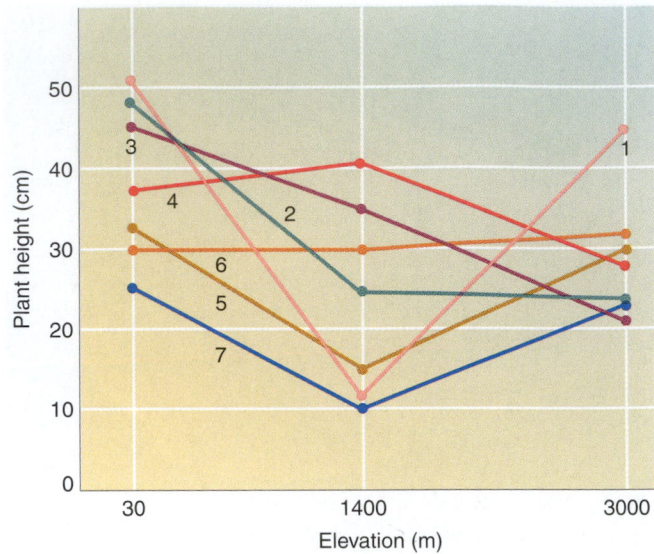

Figure 1-24   Graphic representation of the complete set of results of the type shown in Figure 1-23. Each line represents the norm of reaction of one plant.

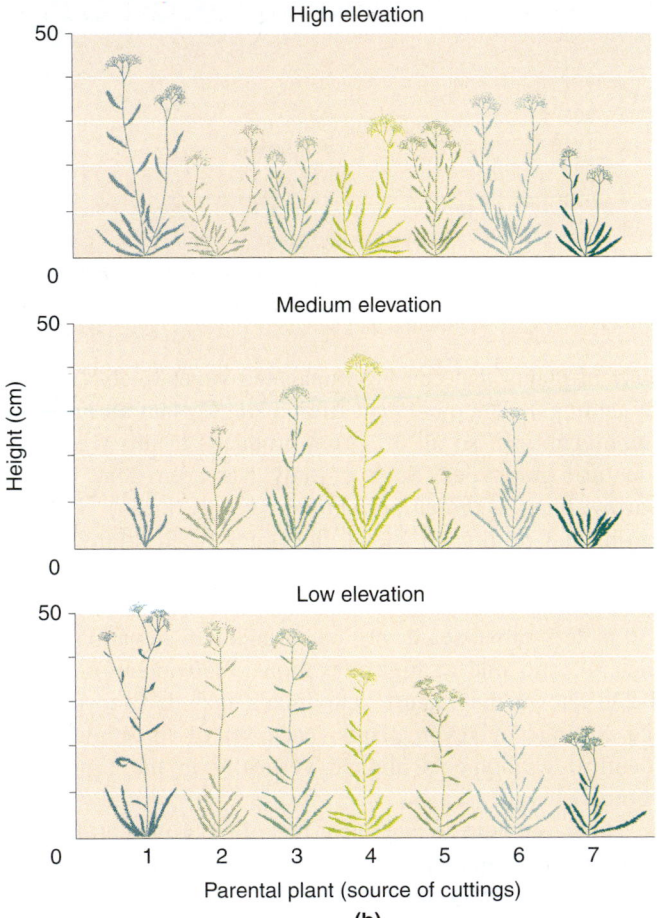

(b)

Figure 1-23   (a) *Achillea millefolium.* (b) Norms of reaction to elevation for seven different *Achillea* plants (seven different genotypes). A cutting from each plant was grown at low, medium, and high elevations. (Part a, Harper Horticultural Slide Library; part b, Carnegie Institution of Washington.)

the complex relation between genotype and phenotype. Figure 1-24 graphs the norms of reaction derived from the results shown in Figure 1-23b. Each genotype has a different norm of reaction, and the norms cross one another; so we cannot identify either a "best" genotype or a "best" environment for *Achillea* growth.

We have seen two different patterns of reaction norms. The difference between the wild-type and the other eye-size genotypes in *Drosophila* is such that the corresponding phenotypes show a consistent difference, regardless of the environment. Any fly of wild-type genotype has larger eyes than any fly of the other genotypes; so we could (imprecisely) speak of "large eye" and "small eye" genotypes. In this case, the differences in phenotype between genotypes are much greater than the variation within a genotype caused by development in different environments. But the variation for a single *Achillea* genotype in different environments is so great that the norms of reaction cross one another and form no consistent pattern. In this case, it makes no sense to identify a genotype with a particular phenotype except in regard to response to particular environments.

Developmental noise

Thus far, we have assumed that a phenotype is uniquely determined by the interaction of a specific genotype and a specific environment. But a closer look shows some further unexplained variation. According to Figure 1-22, a *Drosophila* of wild-type genotype raised at 16°C has 1000 facets in each eye. In fact, this is only an average value; one fly raised at 16°C may have 980 facets and another may have 1020. Perhaps these variations are due to slight fluctuations in the local environment or slight differences in genotypes.

However, a typical count may show that a fly has, say, 1017 facets in the left eye and 982 in the right eye. In another fly, the left eye has slightly fewer facets than the right eye. Yet the left and right eyes of the same fly are genetically identical. Furthermore, under typical experimental conditions, the fly develops as a larva (a few millimeters long) burrowing in homogeneous artificial food in a laboratory bottle and then completes its development as a pupa (also a few millimeters long) glued vertically to the inside of the glass high above the food surface. Surely the environment does not differ significantly from one side of the fly to the other. But if the two eyes experience the same sequence of environments and are identical genetically, then why is there any phenotypic difference between the left and right eyes?

Differences in shape and size are partly dependent on the process of cell division that turns the zygote into a multicellular organism. Cell division, in turn, is sensitive to molecular events within the cell, and these events may have a relatively large random component. For example, the vitamin biotin is essential for *Drosophila* growth, but its average concentration is only one molecule per cell. Obviously, the rate of any process that depends on the presence of this molecule will fluctuate as the concentration of biotin varies. But cells can divide to produce differentiated eye cells only within the relatively short developmental period during which the eye is being formed. Thus, we would expect random variation in such phenotypic characters as the number of eye cells, the number of hairs, the exact shape of small features, and the variations of neurons in a very complex central nervous system — even when the genotype and the environment are precisely fixed. Even such structures as the very simple nervous systems of nematodes vary at random for this reason. Random events in development lead to variation in phenotype called **developmental noise.**

MESSAGE ···

In some characteristics, such as eye cells in *Drosophila*, developmental noise is a major source of the observed variations in phenotype.

Like noise in a verbal communication, developmental noise adds small random variations to the predictable development governed by norms of reaction. Adding developmental noise to our model of phenotypic development, we obtain something like Figure 1-25. With a given genotype and environment, there is a range of possible outcomes for each developmental step. The developmental process does contain feedback systems that tend to hold the deviations within certain bounds so that the range of deviation does not increase indefinitely through the many steps of development. However, this feedback is not perfect. For any given genotype developing in any given sequence of environments, there remains some uncertainty regarding the exact phenotype that will result.

Three levels of development

Chapters 22 and 23 of this book are concerned with developmental genetics. Those chapters are entirely concerned with the way in which genes mediate development, and nowhere in those chapters do we consider the role of the environment or the influence of developmental noise. How can we, at the beginning of the book, emphasize the joint role of genes, environment, and noise in influencing phenotype, yet, in our later consideration of the genetics of development, ignore the environment? The answer is that modern developmental genetics is concerned with very basic processes of differentiation that are common to all individual members of a species and, indeed, are common to animals as different as fruit flies and mammals. How does the front end of an animal become differentiated from the back end, the ventral from the dorsal side? How does the body become segmented, and why do limbs form on some segments and not on others? Why do eyes form on the head and not in the middle of the abdomen? Why do the antennae, wings, and legs of a fly look so different even though they are derived in evolution from appendages that looked alike in the earliest ancestors of insects? At this level of development, constant across individuals and species, normal

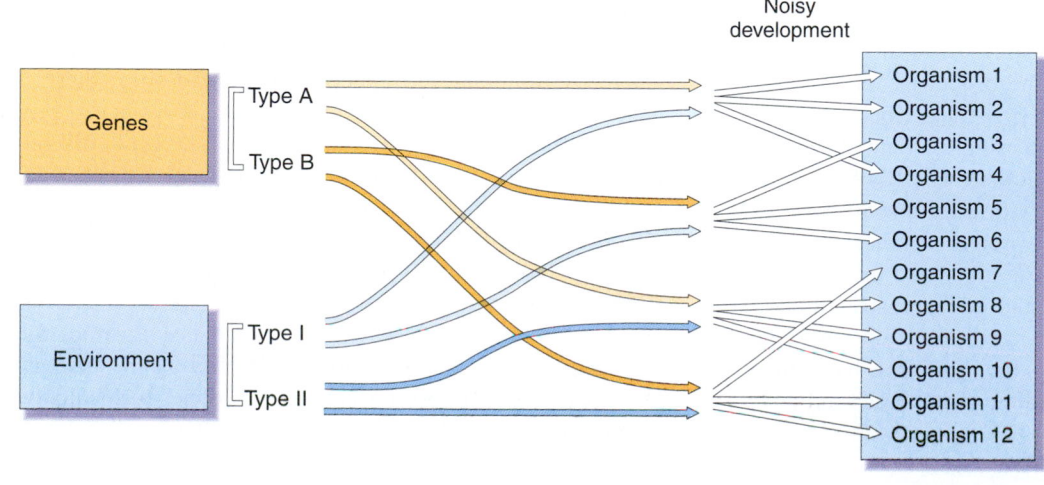

Noisy development

Genes — Type A — Type B

Environment — Type I — Type II

Organism 1
Organism 2
Organism 3
Organism 4
Organism 5
Organism 6
Organism 7
Organism 8
Organism 9
Organism 10
Organism 11
Organism 12

Figure 1-25 A model of phenotypic determination that shows how genes, environment, and developmental noise interact to produce a phenotype.

environmental variation plays no role, and we can speak correctly of genes "determining" the phenotype. Precisely because the effect of genes can be isolated at this level of development and because the processes seem to be general across a wide variety of organisms, they are easier to study than are characteristics for which environmental variation is important, and developmental genetics has concentrated on understanding them.

At a second level of development, there are variations on the basic developmental themes that are different between species but are constant within species, and these too could be understood by concentrating on genes, although at the moment they are not part of the study of developmental genetics. So, although both lions and lambs have four legs, one at each corner, lions always give birth to lions and lambs to lambs, and we have no difficulty in distinguishing between them in any environment. Again, we are entitled to say that genes "determine" the difference between the two species, although we must be more cautious here. Two species may differ in some characteristic because they live in quite different environments, and, until we can raise them in the same environment, we cannot always be sure whether environmental influence plays a role. For example, two species of baboons in Africa, one living in the very dry plains of Ethiopia and the other in the more productive areas of Uganda, have very different food-gathering behavior and social structure. Without actually transplanting colonies of the two species between the two environments, we cannot know how much of the difference is a direct response of these primates to different food conditions.

It is at the third level, the differences in morphology, physiology, and behavior between individuals within species, that genetic, environmental, and developmental noise factors become intertwined, as discussed in this chapter. One of the most serious errors in the understanding of genetics by nongeneticists has been a confusion between variation at this level and variation at the higher levels. The experiments and discoveries to be discussed in Chapters 22 and 23 are not, and are not meant to be, models for the causation of individual variation. They apply directly only to those characteristics, deliberately chosen, that are general features of development and for which environment appears to be irrelevant. There are, as yet, no experiments, or any ideas about how to perform them, that will bring together the explanations for the different levels of development.

SUMMARY

Genetics is the study of genes at all levels from molecules to populations. As a modern discipline, it began in the 1860s with the work of Gregor Mendel, who first formulated the idea that genes exist. We now know that a gene is a functional region of the long DNA molecule that constitutes the fundamental structure of a chromosome. DNA is composed of four nucleotides, each containing deoxyribose sugar, phosphate, and one of four bases — adenine (A), thymine (T), guanine (G), and cytosine (C). DNA is two antiparallel nucleotide chains held together by bonding A with T and G with C. In replication, the two chains separate and their exposed bases are used as templates for the synthesis of two identical daughter DNA molecules.

Most genes encode the structure of a protein (proteins are the main determinants of the properties of an organism). To make protein, DNA is first transcribed by the enzyme RNA polymerase into a single-stranded working copy called messenger RNA (mRNA). The RNA is synthesized in the $5' \rightarrow 3'$ direction with only one of the DNA strands used as a template. In protein synthesis, the nucleotide sequence in the mRNA is translated into an amino acid sequence that constitutes the primary structure of a protein. Amino acid chains are synthesized on ribosomes. Each amino acid is brought to the ribosome by a tRNA molecule that docks by the binding of its triplet anticodon to a triplet codon in mRNA.

Variation is of two types: discontinuous, showing two or more distinct phenotypes, and continuous, showing phenotypes with a wide range of quantitative values. Discontinuous variants are often determined by alternative forms (alleles) of one gene. For example, people with normal skin pigmentation have the functional allele coding for the enzyme tyrosinase, which converts tyrosine into the dark pigment melanin, whereas albinos have a mutated form of the gene that codes for a protein that can no longer make the conversion.

The relation of genotype to phenotype across an environmental range is called the norm of reaction. In the laboratory, geneticists study discontinuous variants under conditions where there is a one-to-one correspondence between genotype and phenotype. However, in natural populations, where environment and genetic background vary, there is generally a more complex relation, and genotypes can produce overlapping ranges of phenotypes. As a result, discontinuous variants have been the starting point for most experiments in genetic analysis.

The main tools of genetics are breeding analysis of variants, biochemistry, microscopy, and DNA cloning. DNA cloning is achieved by inserting the gene of interest into a small accessory chromosome, such as a plasmid, and allowing bacteria to do the job of amplifying the inserted DNA. Cloned DNA can provide useful probes for detecting the presence of related DNA and RNA.

CONCEPT MAP

Each chapter contains an exercise on drawing concept maps. As you draw these maps, you will be organizing the knowledge that you have just acquired. As we learn, we must place new knowledge into the overall conceptual framework of the discipline. An essential part of doing so is to discern the interrelations between new knowledge and previously acquired knowledge. It is not as easy as it sounds, and the mind can sometimes trick us into believing that we have the overall picture. Concept maps are a good way of proving to ourselves that we really do know how the various structures, processes, and ideas of genetics interrelate. The maps can also help in identifying errors in our understanding.

A selection of terms is given, and the challenge is to draw lines or arrows between the ones that you think are related, with a description on the arrow showing just what the relation is. Draw as many relations as possible. Force yourself to make connections, no matter how remote they might first seem. The arrow description must be clear enough that another person reading the map can understand what you are thinking. Sometimes the maps reveal a simple series of consecutive steps, sometimes a complex network of interactions. There is no one correct answer; there are many correct variations. But you may make some incorrect connections, and finding these misunderstandings is part of the purpose of the exercise.

A sample concept map interrelating terms from several different chapters is shown in Figure 1-26.

> base-pair substitution / recessive / 3 : 1 ratio / albino / environment / melanin / metabolic pathway / active site / enzyme / gene

Now draw a concept map for this chapter, interrelating as many of the following terms as possible. Note that the terms are listed in no particular order.

> genotype / phenotype / norm of reaction / environment / development / organism

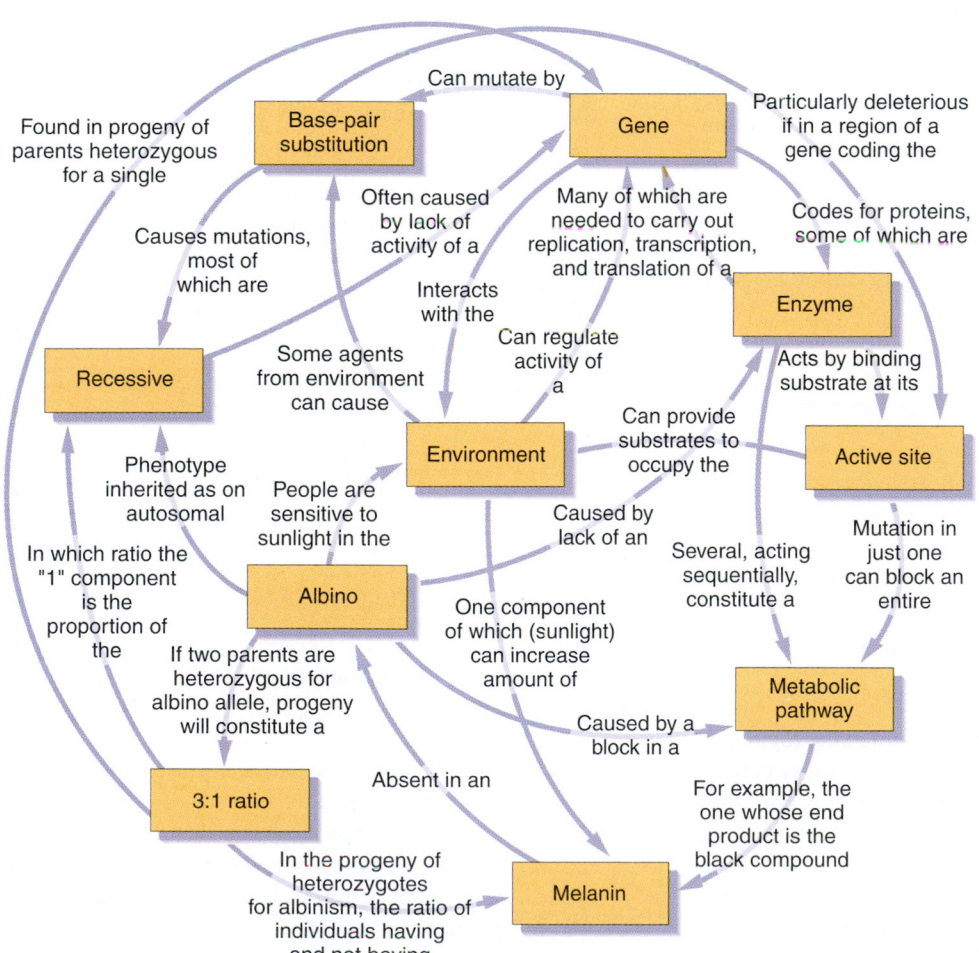

Figure 1-26 Sample concept map.

PROBLEMS

1. Define genetics. Do you think the ancient Egyptian racehorse breeders were geneticists? How might their approaches have differed from those of modern geneticists?

2. How does DNA dictate the general properties of a species?

3. What are the two features of DNA that suit it for its role as a hereditary molecule? Can you think of alternative types of hereditary molecules that might be found in extraterrestrial life forms?

4. How many different DNA molecules 10 nucleotide pairs long are possible?

5. If thymine makes up 15% of the bases in a certain DNA sample, what percentage of the bases must be cytosine?

6. If the G + C content of a DNA sample is 48%, what are the proportions of the four different nucleotides?

7. Draw a simple diagram of DNA that makes clear

   a. what 5′ and 3′ ends are.

   b. what a sugar-phosphate backbone is.

8. Each cell of the human body contains 46 chromosomes.

   a. How many DNA molecules does this statement represent?

   b. How many different types of DNA molecules does it represent?

9. A certain segment of DNA has the following nucleotide sequence in one strand:

   5′-ATTGGTGCATTACTTCAGGCTCT-3′

   What must the sequence in the other strand be? (Label its 5′ and 3′ ends.)

10. In a single strand of DNA, is it ever possible for the number of adenines to be greater than the number of thymines?

11. In normal double-helical DNA, is it true that

    a. A plus C will always equal G plus T?

    b. A plus G will always equal C plus T?

12. Suppose that the following DNA molecule replicates to produce two daughter molecules. Draw these daughter molecules by using black for previously polymerized nucleotides and red for newly polymerized nucleotides.

    3′-TTGGCACGTCGTAAT-5′
    5′-AACCGTGCAGCATTA-3′

13. In the DNA molecule in Problem 12, assume that the bottom strand is the template strand and draw the RNA transcript, labeling the 5′ and 3′ ends.

14. Draw Northern and Western blots of the three genotypes in Figure 1-15. (Assume the probe used in the Northern blot to be a clone of the tyrosinase gene.)

15. The most common elements in living organisms are carbon, hydrogen, oxygen, nitrogen, phosphorus, and sulfur. Which of them is not found in DNA?

16. What is a gene? What are some of the problems with your definition?

17. The gene for the human protein albumin spans a chromosomal region 25,000 nucleotide pairs (25 kilobases, or kb) long from the beginning of the protein-coding sequence to the end of the protein-coding sequence, but the messenger RNA for this protein is only 2.1 kb long. What do you think accounts for this huge difference?

18. DNA is extracted from cells of *Neurospora*, a fungus that has one set of seven chromosomes; pea, a plant that has two sets of seven chromosomes; and housefly, an animal that has two sets of six chromosomes. If powerful electrophoresis is used to separate the DNA on a gel, how many bands will each of these three species produce?

19. Devise a formula that relates size of RNA to gene size, number of introns, and average size of introns.

20. If a codon in mRNA is 5′-UUA-3′, draw the tRNA anticodon that would bind to this codon (label the 5′ and 3′ ends).

21. Consider the following segment of DNA:

    5′ GCTTCCCAA 3′
    3′ CGAAGGGTT 5′

    Suppose that the top strand is the template strand used by RNA polymerase.

    a. Will it be transcribed from left to right as drawn or from right to left?

    b. Draw the RNA transcribed.

    c. Label its 5′ and 3′ ends.

22. Two mutations arise in separate cultures of a normally red fungus (which has only one set of chromosomes). The mutations are found to be in different genes. Mutation 1 gives an orange color and mutation 2 gives a yellow color. Biochemists working on the synthesis of the red pigment in this species have already described the following pathway:

colorless precursor $\xrightarrow[\text{enzyme A}]{}$ yellow pigment $\xrightarrow[\text{enzyme B}]{}$

orange pigment $\xrightarrow[\text{enzyme C}]{}$ red pigment

a.   Which enzyme is defective in mutant 1?

b.   Which enzyme is defective in mutant 2?

c.   What would be the color of a double mutant (1 and 2)?

23.   In sweet peas, the purple color of the petals is controlled by two genes, *B* and *D*. The pathway is:

colorless precursor $\xrightarrow[]{\text{gene }B}$ red anthocyanin pigment
colorless precursor $\xrightarrow[\text{gene }D]{}$ blue anthocyanin pigment
$\Big\}$ purple

a.   What color petals would you expect in a plant that carries two copies of a null mutation for *B*?

b.   What color petals would you expect in a plant that carries two copies of a null mutation for *D*?

c.   What color petals would you expect in a plant that is a double mutant; that is, it carries two copies of a null mutation for both *B* and *D*?

 Unpacking the Problem
....

In many chapters, one of the problems will ask "unpacking" questions. These questions are designed to assist in working the problem by showing the often quite extensive content that is inherent in the problem. The same approach can be applied to other questions by the solver. If you have trouble with Problem 23, try answering the following unpacking questions. If necessary, look up material that you cannot remember. Then try to solve the problem again by using the information emerging from the unpacking.

1. What are sweet peas, and how do they differ from edible peas?

2. What is a pathway in the sense used here?

3. How many pathways are evident in this system?

4. Are the pathways independent?

5. Define the term *pigment*.

6. What does colorless mean in this problem? Think of an example of any solute that is colorless.

7. What would a petal that contained only colorless substances look like?

8. Is color in sweet peas anything like mixing paint?

9. What is a mutation?

10. What is a null mutation?

11. What might be the cause of a null mutation at the DNA level?

12. What does "two copies" mean? (How many copies of genes do sweet peas normally have?)

13. What is the relevance of proteins to this problem?

14. Does it matter whether genes *B* and *D* are on the same chromosome?

15. Draw a representation of the wild-type allele of *B* and a null mutant at the DNA level.

16. Repeat for gene *D*.

17. Repeat for the double mutant.

18. How would you explain the genetic determination of petal color in sweet peas to a gardener with no scientific training?

24.   Twelve null alleles of an intron-less *Neurospora* gene are examined, and all the mutant sites are found to cluster in a region occupying the central third of the gene. What might be the explanation for this finding?

25.   An albino mouse mutant is obtained whose pigment lacks melanin, normally made by an enzyme T. Indeed, the tissue of the mutant lacks all detectable activity for enzyme T. However, a Western blot clearly shows that a protein with immunological properties identical with those of enzyme T is present in the cells of the mutant. How is this possible?

26.   In Norway in 1934, a mother with two mentally retarded children consulted the physician Asbjørn Følling. In the course of the interview, Følling learned that the urine of the children had a curious odor. He later tested their urine with ferric chloride and found that, whereas normal urine gives a brownish color, the children's urine stained green. He deduced that the chemical responsible must be phenylpyruvic acid. Because of chemical similarity to phenylalanine, it seemed likely that this substance had been formed from phenylalanine in the blood, but there was no assay for phenylalanine. However, a certain bacterium could convert phenylalanine into phenylpyruvic acid; so the level of phenylalanine could be measured by using the ferric chloride test. The children were indeed found to have high levels of phenylalanine in their blood, and the phenylalanine was probably the source of the phenylpyruvic acid. This disease, which came to be known as phenylketonuria (PKU), was shown to be inherited and caused by a recessive allele.

It became clear that phenylalanine was the culprit and that this chemical accumulated in PKU patients and was converted into high levels of phenylpyruvic acid, which then interfered with the normal development of nervous tissue. This finding led to the formulation of a special diet low in phenylalanine, which could be fed to newborn babies diagnosed with PKU and which allowed normal development to continue without retardation. Indeed, it was found that, after the child's nervous system had developed, the patient could be taken off the special diet. However, tragically, many PKU women who had developed normally with the special diet were found to have babies who were born mentally retarded, and the special diet had no effect on these children.

a. Why do you think the babies of the PKU mothers were born retarded?

b. Why did the special diet have no effect on them?

c. Explain the reason for the difference in the results between the PKU babies and the babies of PKU mothers.

d. Propose a treatment that might allow PKU mothers to have unaffected children.

e. Write a short essay on PKU, integrating concepts at the genetic, diagnostic, enzymatic, physiological, pedigree, and population levels.

27. Normally the thyroid growth hormone thyroxine is made in the body by an enzyme as follows:

$$\text{tyrosine} \xrightarrow{\text{enzyme}} \text{thyroxine}$$

If the enzyme is deficient, the symptoms are called genetic goiterous cretinism (GGC), a rare syndrome consisting of slow growth, enlarged thyroid (called a goiter), and mental retardation.

a. If the normal allele is haplosufficient would you expect GGC to be inherited as a dominant or a recessive phenotype? Explain.

b. Speculate on the nature of the GGC-causing allele, comparing its molecular sequence with that of the normal allele. Show why it results in an inactive enzyme.

c. How might the symptoms of GGC be alleviated?

d. At birth, infants with GGC are perfectly normal and only develop symptoms later. Why do you think this is so?

28. Compare and contrast the processes by which information becomes form in an organism and in house building.

29. Try to think of exceptions to the statement made in this chapter that, "When you look at an organism, what you see is either a protein or something that has been made by a protein."

30. What is the relevance of norms of reaction to phenotypic variation within a species?

31. What are the types and the significance of phenotypic variation within a species?

32. Is the formula "genotype + environment = phenotype" accurate?

2

PATTERNS OF INHERITANCE

The small monastery garden that the monk Gregor Mendel used for his experiments with peas.
A statue of Mendel is visible in the background. Today, this part of Mendel's monastery is a museum, and the curators have planted red and white begonias in an array that graphically represents some of Mendel's results. (Anthony Griffiths.)

Key Concepts

The existence of genes can be inferred by observing standard progeny ratios in the descendants of matings between different phenotypes.

Discrete phenotypic difference in a character is often determined by a difference in a single gene.

In plants and animals, each type of gene is represented twice in each cell, once on each member of a chromosome pair.

Inheritance patterns are based on chromosome behavior at meiosis.

In gamete formation, each member of a gene pair separates into half the gametes.

In gamete formation, gene pairs on different chromosome pairs behave independently of one another.

Genes on the sex chromosomes show unique inheritance patterns.

The gene, the basic functional unit of heredity, is the focal point of the discipline of modern genetics. In all lines of genetic research, the gene is the common unifying thread of a great diversity of experimentation. In this chapter, we analyze the patterns in which phenotypes are inherited in plants and animals. We shall see that these patterns are regular and predictable. It was these regular patterns of inheritance that first led to the concept of the gene, and that is where we will begin the story.

The concept of the gene (but not the word) was first proposed in 1865 by Gregor Mendel. Until then, little progress had been made in understanding the mechanisms of heredity. The prevailing notion was that the spermatozoon and egg contained a sampling of essences from the various parts of the parental body; at conception, these essences somehow blended to influence the development of the new offspring. This idea of **blending inheritance** evolved to account for the fact that offspring typically show characteristics that are similar to those of both parents. However, some obvious problems are associated with this idea, one of which is that offspring are not always an intermediate blend of their parents' characteristics. Attempts to expand and improve this theory, originally conceived by Aristotle, led to no better understanding of heredity.

As a result of his research with pea plants, Mendel proposed instead a theory of particulate inheritance. According to Mendel's theory, characters are determined by discrete units that are inherited intact down through the generations. This model explained many observations that could not be explained by the idea of blending inheritance. It also served well as a framework for the later, more detailed understanding of the mechanism of heredity.

The importance of Mendel's ideas was not recognized until about 1900 (after his death). His written work was then rediscovered by three scientists, after each had independently obtained the same kind of results. Mendel's work constitutes the prototype for genetic analysis. He laid down an experimental and logical approach to heredity that is still used today. Therefore, although the following description is historical, the experimental sequence is the one still used by geneticists.

Mendel's experiments

Gregor Mendel was born in the district of Moravia, then part of the Austro-Hungarian Empire. At the end of high school, he entered the Augustinian monastery of St. Thomas in the city of Brünn, now Brno of the Czech Republic. His monastery was dedicated to teaching science and to scientific research, so Mendel was sent to a university in Vienna to obtain his teaching credentials. However, he failed his examinations and returned to the monastery at Brünn. There he embarked on the research program of plant hybridization that was posthumously to earn him the title of founder of the science of genetics.

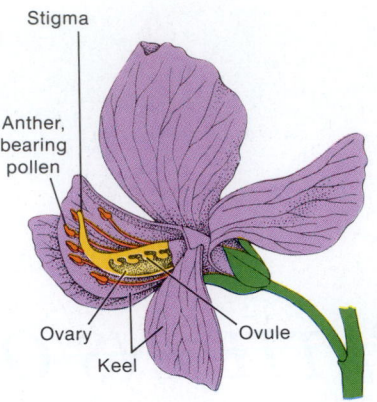

Figure 2-1   A pea flower with the keel cut and opened to expose the reproductive parts. The ovary is shown in a cutaway view. (After J. B. Hill, H. W. Popp, and A. R. Grove, Jr., *Botany*. Copyright © 1967 by McGraw-Hill.)

Mendel's studies constitute an outstanding example of good scientific technique. He chose research material well suited to the study of the problem at hand, designed his experiments carefully, collected large amounts of data, and used mathematical analysis to show that the results were consistent with his explanatory hypothesis. The predictions of the hypothesis were then tested in a new round of experimentation.

Mendel studied the garden pea *(Pisum sativum)* for two main reasons. First, peas were available from seed merchants in a wide array of distinct shapes and colors that could be easily identified and analyzed. Second, peas can either **self** (self-pollinate) or be cross-pollinated. The peas self because the male parts (anthers) and female parts (ovaries) of the flower—which produce the pollen containing the sperm and the ovules containing eggs, respectively—are enclosed by two petals fused to form a compartment called a keel (Figure 2-1). The gardener or experimenter can **cross** (cross-pollinate) any two pea plants at will. The anthers from one plant are removed before they have opened to shed their pollen, an operation called emasculation that is done to prevent selfing. Pollen from the other plant is then transferred to the receptive stigma with a paintbrush or on anthers themselves (Figure 2-2). Thus, the experimenter can choose to self or to cross the pea plants.

Other practical reasons for Mendel's choice of peas were that they are inexpensive and easy to obtain, take up little space, have a short generation time, and produce many offspring. Such considerations enter into the choice of organism for any piece of genetic research.

Plants differing in one character

Mendel chose seven different *characters* to study. The word **character** in this regard means a specific property of an organism; geneticists use this term as a synonym for characteristic or trait.

Figure 2-2 One of the techniques of artificial cross-pollination, demonstrated with the *Mimulus guttatus,* the yellow monkey flower. To transfer pollen, the experimenter touches anthers from the male parent to the stigma of an emasculated flower, which acts as the female parent. (Anthony Griffiths.)

For each of the characters that he chose, Mendel obtained lines of plants, which he grew for two years to make sure that they were pure. A **pure line** is a population that breeds true for (shows no variation in) the particular character being studied; that is, all offspring produced by selfing or crossing within the population are identical for this character. By making sure that his lines bred true, Mendel had made a clever beginning: he had established a fixed baseline for his future studies so that any changes observed subsequent to deliberate manipulation in his research would be scientifically meaningful; in effect, he had set up a control experiment.

Two of the pea lines studied by Mendel bred true for the character of flower color. One line bred true for purple flowers; the other, for white flowers. Any plant in the purple-flowered line — when selfed or when crossed with others from the same line — produced seeds that all grew into plants with purple flowers. When these plants in turn were selfed or crossed within the line, their progeny also had purple flowers, and so forth. The white-flowered line similarly produced only white flowers through all generations. Mendel obtained seven pairs of pure lines for seven characters, with each pair differing in only one character (Figure 2-3).

Each pair of Mendel's plant lines can be said to show a **character difference** — a contrasting difference between two lines of organisms (or between two organisms) in one particular character. Contrasting phenotypes for a particular character are the starting point for any genetic analysis. The differing lines (or individuals) represent different forms that the character may take: they can be called *character forms,*

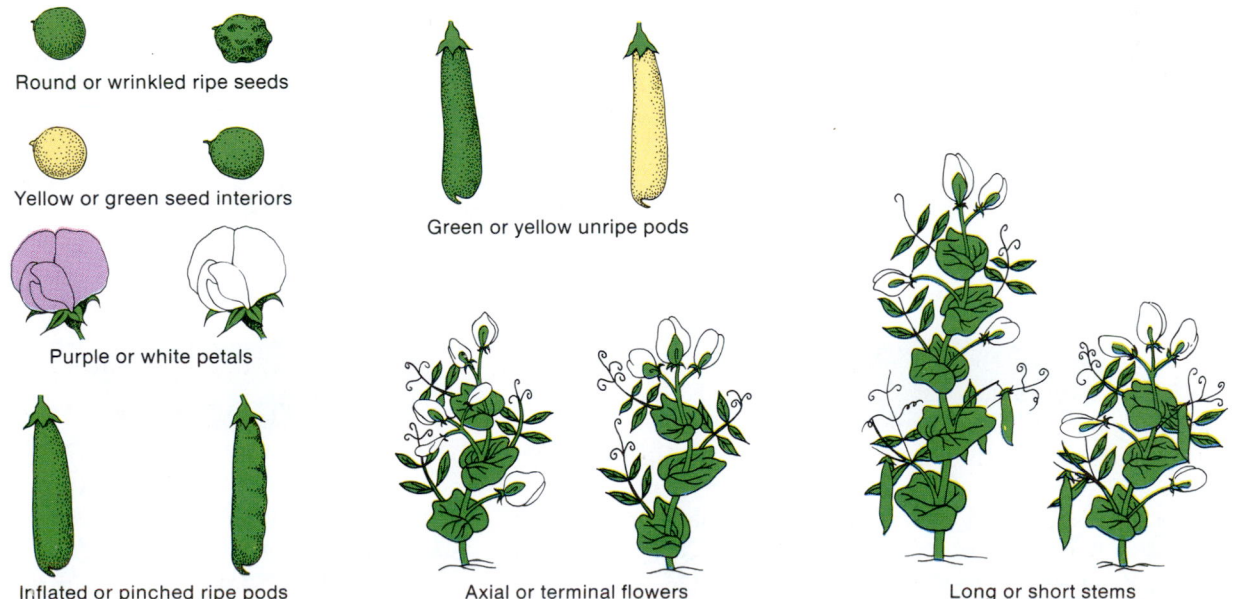

Round or wrinkled ripe seeds

Yellow or green seed interiors

Purple or white petals

Inflated or pinched ripe pods

Green or yellow unripe pods

Axial or terminal flowers

Long or short stems

Figure 2-3 The seven character differences studied by Mendel. (After S. Singer and H. Hilgard, *The Biology of People.* Copyright © 1978 by W. H. Freeman and Company.)

character variants, or **phenotypes.** The term *phenotype* (derived from Greek) literally means "the form that is shown"; it is the term used by geneticists today. Even though such words as gene and phenotype were not coined or used by Mendel, we shall use them in describing Mendel's results and hypotheses.

Figure 2-3 shows the seven pea characters, each represented by two contrasting phenotypes. The description of characters is somewhat arbitrary. For example, we can state the color-character difference in at least three ways:

Character	Phenotype
flower color	purple versus white
flower purpleness	presence versus absence
flower whiteness	absence versus presence

Fortunately, the description does not alter the final conclusions of the analysis, except in the words used.

We turn now to Mendel's analysis of the lines breeding true for flower color. In one of his early experiments, Mendel pollinated a purple-flowered plant with pollen from a white-flowered plant. We call the plants from the pure lines the **parental generation** (P). All the plants resulting from this cross had purple flowers (Figure 2-4). This progeny generation is called the **first filial generation** (F_1). (The subsequent generations produced by selfing are symbolized F_2, F_3, and so forth.)

Mendel made **reciprocal crosses.** In most plants, any cross can be made in two ways, depending on which pheno-type is used as male (♂) or female (♀). For example, the following two crosses

phenotype A ♀ × phenotype B ♂
phenotype B ♀ × phenotype A ♂

are reciprocal crosses. Mendel's reciprocal cross in which he pollinated a white flower with pollen from a purple-flowered plant produced the same result (all purple flowers) in the F_1 (Figure 2-5). He concluded that it makes no difference which way the cross is made. If one pure-breeding parent is purple flowered and the other is white flowered, all plants in the F_1 have purple flowers. The purple flower color in the F_1 generation is identical with that in the purple-flowered parental plants. In this case, the inheritance is *not* a simple blending of purple and white colors to produce some intermediate color. To maintain a theory of blending inheritance, we would have to assume that the purple color is somehow "stronger" than the white color and completely overwhelms any trace of the white phenotype in the blend.

Next, Mendel selfed the F_1 plants, allowing the pollen of each flower to fall on its own stigma. He obtained 929 pea seeds from this selfing (the F_2 individuals) and planted them. Interestingly, some of the resulting plants were white flowered; the white phenotype had reappeared. Mendel then did something that, more than anything else, marks the birth of modern genetics: he *counted* the numbers of plants with each phenotype. This procedure had seldom, if ever, been used in studies on inheritance before Mendel's work. Indeed, others had obtained remarkably similar results in

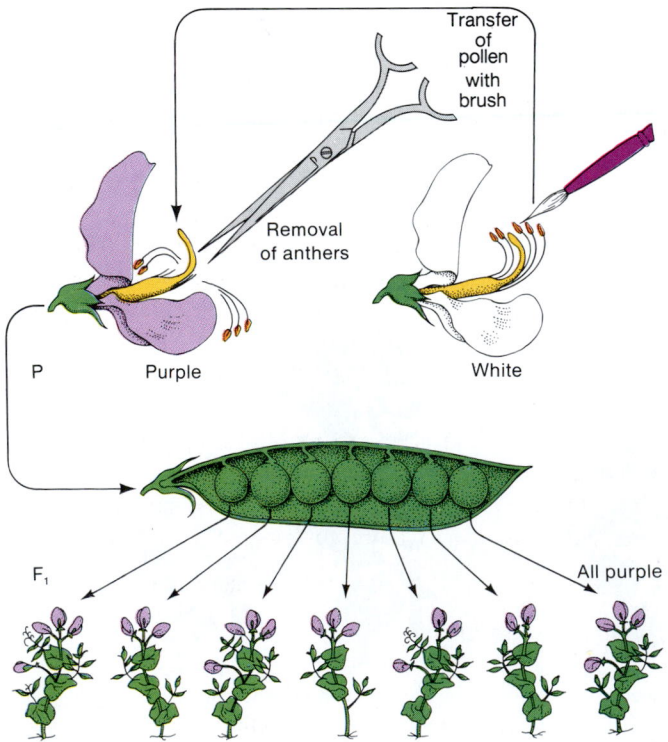

Figure 2-4   Mendel's cross of purple-flowered ♀ × white-flowered ♂.

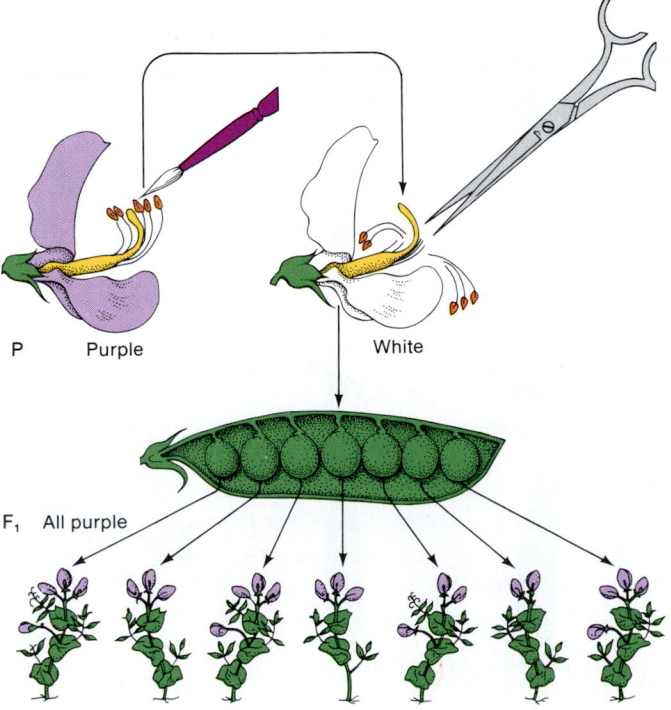

Figure 2-5   Mendel's cross of white-flowered ♀ × purple-flowered ♂.

2-1 ▷ TABLE	Results of All Mendel's Crosses in Which Parents Differed in One Character		
Parental phenotype	F_1	F_2	F_2 ratio
1. Round × wrinkled seeds	All round	5474 round; 1850 wrinkled	2.96:1
2. Yellow × green seeds	All yellow	6022 yellow; 2001 green	3.01:1
3. Purple × white petals	All purple	705 purple; 224 white	3.15:1
4. Inflated × pinched pods	All inflated	882 inflated; 299 pinched	2.95:1
5. Green × yellow pods	All green	428 green; 152 yellow	2.82:1
6. Axial × terminal flowers	All axial	651 axial; 207 terminal	3.14:1
7. Long × short stems	All long	787 long; 277 short	2.84:1

breeding studies but had failed to count the numbers in each class. Mendel counted 705 purple-flowered plants and 224 white-flowered plants. He noted that the ratio of 705:224 is almost exactly a 3:1 ratio (in fact, it is 3.1:1).

Mendel repeated the crossing procedures for the six other pairs of pea character differences. He found the same 3:1 ratio in the F_2 generation for each pair (Table 2-1). By this time, he was undoubtedly beginning to believe in the significance of the 3:1 ratio and to seek an explanation for it. In all cases, one parental phenotype disappeared in the F_1 and reappeared in one-fourth of the F_2. The white phenotype, for example, was completely absent from the F_1 generation but reappeared (in its full original form) in one-fourth of the F_2 plants.

It is very difficult to apply the theory of blending inheritance to devise an explanation of this result. Even though the F_1 flowers were purple, the plants evidently still carried the *potential* to produce progeny with white flowers. Mendel inferred that the F_1 plants receive from their parents the abilities to produce both the purple phenotype and the white phenotype and that these abilities are retained and passed on to future generations rather than blended. Why is the white phenotype not expressed in the F_1 plants? Mendel used the terms **dominant** and **recessive** to describe this phenomenon without explaining the mechanism. The purple phenotype is dominant to the white phenotype and the white phenotype is recessive to purple. Thus the operational definition of dominance is provided by the phenotype of an F_1 established by intercrossing two pure lines. The parental phenotype that is expressed in such F_1 individuals is by definition the dominant phenotype.

Mendel went on to show that, in the class of F_2 individuals showing the dominant phenotype, there were in fact two genetically distinct subclasses. In this case, he was working with seed color. In peas, the color of the seed is determined by the genetic constitution of the seed itself, not by the maternal parent as in some plant species. This autonomy is convenient because the investigator can treat each pea as an individual and can observe its phenotype directly without having to grow a plant from it, as must be done for flower color. It also means that much larger numbers can be examined, and studies can be extended into subsequent generations. The seed colors that Mendel used were yellow and

green. He crossed a pure yellow line with a pure green line and observed that the F_1 peas that appeared were all yellow. Symbolically,

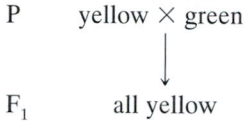

P   yellow × green

F_1 all yellow

Therefore, by definition, yellow is the dominant phenotype and green is recessive.

Mendel grew F_1 plants from these F_1 peas and then selfed the plants. The peas that developed on the F_1 plants constituted the F_2 generation. He observed that, in the pods of the F_1 plants, three-fourths of the F_2 peas were yellow and one-fourth were green:

F_1 all yellow (self)

F_2 $\frac{3}{4}$ yellow
         $\frac{1}{4}$ green

Here, again, in the F_2 we see a 3:1 phenotypic ratio. Mendel took a sample consisting of 519 yellow F_2 peas and grew plants from them. These yellow F_2 plants were selfed individually, and the peas that developed were noted. Mendel found that 166 of the plants bore only yellow peas, and each of the remaining 353 plants bore a mixture of yellow and green peas in a 3:1 ratio. Plants from green F_2 peas were then grown and selfed and were found to bear only green peas. In summary, all the F_2 greens were evidently pure breeding, like the green parental line; but, of the F_2 yellows, two-thirds were like the F_1 yellows (producing yellow and green seeds in a 3:1 ratio) and one-third were like the pure-breeding yellow parent. Thus the study of the individual selfings revealed that underlying the 3:1 phenotypic ratio in the F_2 generation was a more fundamental 1:2:1 ratio:

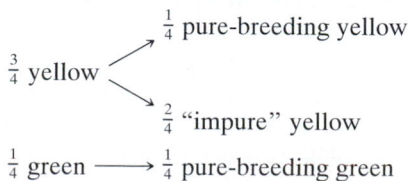

$\frac{3}{4}$ yellow → $\frac{1}{4}$ pure-breeding yellow

→ $\frac{2}{4}$ "impure" yellow

$\frac{1}{4}$ green → $\frac{1}{4}$ pure-breeding green

Further studies showed that such 1:2:1 ratios underlie all the phenotypic ratios that Mendel had observed. Thus, the problem really was to explain the 1:2:1 ratio. Mendel's explanation is a classic example of a creative model or hypothesis derived from observation and suitable for testing by further experimentation. He deduced the following explanation:

1. *The existence of genes.* There are hereditary determinants of a particulate nature. We now call these determinants *genes.*

2. *Genes are in pairs.* Alternative phenotypes of a character are determined by different forms of a single type of gene. The different forms of one type of gene are called **alleles.** In adult pea plants, each type of gene is present twice in each cell, constituting a **gene pair.** In different plants, the gene pair can be of the same alleles or of different alleles of that gene. Mendel's reasoning here was obvious: the F_1 plants, for example, must have had one allele that was responsible for the dominant phenotype and another allele that was responsible for the recessive phenotype, which showed up only in later generations.

3. *The principle of segregation.* The members of the gene pairs segregate (separate) equally into the gametes, or eggs and sperm.

4. *Gametic content.* Consequently, each gamete carries only one member of each gene pair.

5. *Random fertilization.* The union of one gamete from each parent to form the first cell (**zygote**) of a new progeny individual is random — that is, gametes combine without regard to which member of a gene pair is carried.

These points can be illustrated diagrammatically for a general case by using A to represent the allele that determines the dominant phenotype and a to represent the gene for the recessive phenotype (as Mendel did). The use of A and a is similar to the way in which a mathematician uses symbols to represent abstract entities of various kinds. In Figure 2-6, these symbols are used to illustrate how the preceding five points explain the 1:2:1 ratio. As mentioned in Chapter 1, the members of a gene pair are separated by a slash (/). This slash is used to show us that they are indeed a pair; the slash also serves as a symbolic chromosome to remind us that the gene pair is found at one location on a chromosome pair.

The whole model made logical sense of the data. However, many beautiful models have been knocked down under test. Mendel's next job was to test his model. He did so in the seed-color crosses by taking an F_1 plant that grew from a yellow seed and crossing it with a plant grown from a green seed. A 1:1 ratio of yellow to green seeds could be predicted in the next generation. If we let Y stand for the allele that determines the dominant phenotype (yellow seeds) and y stand for the allele that determines the recessive phenotype (green seeds), we can diagram Mendel's predictions, as shown in Figure 2-7. In this experiment, Mendel obtained 58 yellow (Y/y) and 52 green (y/y), a very close approxima-

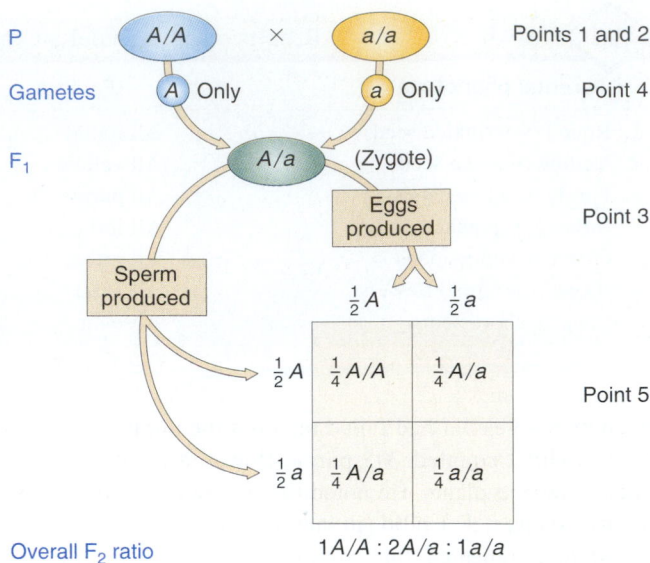

Figure 2-6   Mendel's model of the hereditary determinants of a character difference in the P, F_1, and F_2 generations. The five points are those listed in the text.

tion to the predicted 1:1 ratio and confirmation of the equal segregation of Y and y in the F_1 individual. This concept of **equal segregation** has been given formal recognition as **Mendel's first law:** *The two members of a gene pair segregate from each other into the gametes; so half the gametes carry one member of the pair and the other half of the gametes carry the other member of the pair.*

Now we need to introduce some more terms. The individuals represented by A/a are called **heterozygotes** or, sometimes, **hybrids,** whereas the individuals in pure lines are called **homozygotes.** In such words, *hetero-* means "different" and *homo-* means "identical." Thus, an A/A plant is

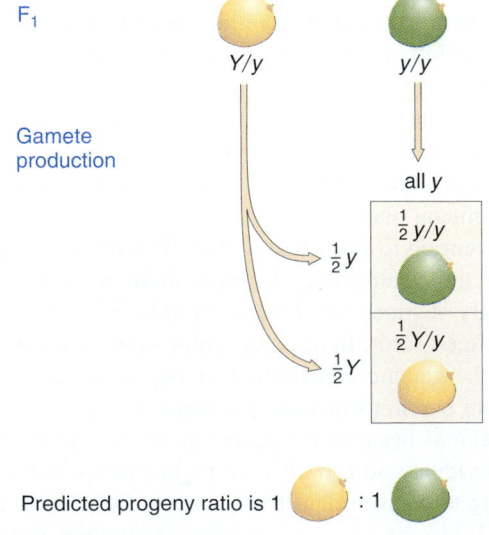

Figure 2-7   Using pure-breeding lines to deduce genotypes and dominance and recessiveness.

said to be **homozygous dominant;** an a/a plant is homozygous for the recessive allele, or **homozygous recessive.** As stated in Chapter 1, the designated genetic constitution of the character or characters under study is called the **genotype.** Thus, Y/Y and Y/y, for example, are different genotypes even though the seeds of both types are of the same phenotype (that is, yellow). In such a situation, the phenotype is viewed simply as the outward manifestation of the underlying genotype. Note that, underlying the 3 : 1 phenotypic ratio in the F_2, there is a 1 : 2 : 1 genotypic ratio of $Y/Y : Y/y : y/y$.

Note that, strictly speaking, the expressions *dominant* and *recessive* are properties of the phenotype. The dominant phenotype is established in analysis by the appearance of the F_1. However, a phenotype (which is merely a description) cannot really exert dominance. Mendel showed that the dominance of one phenotype over another is in fact due to the dominance of one member of a gene pair over the other.

Let's pause to let the significance of this work sink in. What Mendel did was to develop an analytic scheme for the identification of genes regulating any biological character or function. Let's take petal color as an example. Starting with two different phenotypes (purple and white) of one character (petal color), Mendel was able to show that the difference was caused by one gene pair. Modern geneticists would say that Mendel's analysis had identified a gene for petal color. What does this mean? It means that, in these organisms, there is a gene that greatly affects the color of the petals. This gene can exist in different forms: a dominant form of the gene (represented by C) causes purple petals, and a recessive form of the gene (represented by c) causes white petals. The forms C and c are *alleles* (alternative forms) of that gene for petal color. The same letter designation is used to show that the alleles are forms of one gene. We can express this idea in another way by saying that there is a gene, called phonetically a "see" gene, with alleles C and c. Any individual pea plant will always have two "see" genes, forming a gene pair, and the actual members of the

gene pair can be C/C, C/c, or c/c. Notice that, although the members of a gene pair can produce different effects, they both affect the same character. The basic route of Mendelian analysis for a single character is summarized in Table 2-2.

Molecular basis of Mendelian genetics

Let us consider some of Mendel's terms in the context of the cell. First, what is the molecular nature of alleles? When alleles such as A and a are examined at the DNA level by using modern technology, they are generally found to be identical in most of their sequences and differ only at one or a few nucleotides of the thousands of nucleotides that make up the gene. Therefore, we see that the alleles are truly different versions of the same basic gene. Looked at another way, gene is the generic term and allele is specific. (The pea-color gene has two alleles coding for yellow and green.) The following diagram represents the DNA of two alleles of one gene; the letter "x" represents a difference in the nucleotide sequence:

What about dominance? We have seen that, although the terms dominant and recessive are defined at the level of phenotype, the phenotypes are clearly manifestations of the different actions of alleles. Therefore we can legitimately use the phrases *dominant allele* and *recessive allele* as the determinants of dominant and recessive phenotypes. Several different molecular factors can make an allele either dominant or recessive. One commonly found situation is that the dominant allele encodes a functional protein, and the recessive allele encodes the lack of the protein or a nonfunctional

2-2 TABLE	Summary of the Modus Operandi for Establishing Simple Mendelian Inheritance	

Experimental procedure	1. Choose pure lines showing a character difference (purple versus white flowers).
	2. Cross the lines.
	3. Self the F_1 individuals.

Results: F_1 is all purple; F_2 is $\frac{3}{4}$ purple and $\frac{1}{4}$ white.

| Inferences: | 1. The character difference is controlled by a major gene for flower color. |
| | 2. The dominant allele of this gene causes purple petals; the recessive allele causes white petals. |

Symbolic interpretation:

Character	Phenotype	Genotype	Allele	Gene
Flower color	Purple (dominant)	C/C (homozygous dominant)	C (dominant)	Flower-color gene
		C/c (heterozygous)	c (recessive)	
	White (recessive)	c/c (homozygous recessive)		

form of it. In the heterozygote, the protein produced by the functional allele is enough for the normal needs of the cell; so the functional allele acts as a dominant allele. An example of the dominance of the functional allele in a heterozygote was presented in the discussion of albinism in Chapter 1. The general idea can be stated as a formula as follows:

$$A \quad\quad \text{plus} \quad\quad a \quad = \quad A/a$$
(functional protein) plus (nonfunctional protein) = function

What is the cellular basis of Mendel's first law, the equal segregation of alleles at gamete formation? In a diploid organism such as peas, all the cells of the organism contain two chromosome sets. Gametes, however, are haploid, containing one chromosome set. Gametes are produced by specialized cell divisions in the diploid cells in the germinal tissue (ovaries and anthers). These specialized cell divisions are accompanied by nuclear divisions called **meiosis.** The highly programmed chromosomal movements in meiosis cause the equal segregation of alleles into the gametes. In meiosis in a heterozygote A/a, the chromosome carrying A is pulled in the opposite direction from the chromosome carrying a; so half the resulting gametes carry A and the other half carry a. The situation can be summarized in a simplified form as follows (meiosis will be revisited in detail in Chapter 3):

$$
\begin{array}{ccc}
\underline{\quad\quad A \quad\quad} & \nearrow \tfrac{1}{2} & \underline{\quad\quad A \quad\quad} \\[2ex]
\underline{\quad\quad a \quad\quad} & \searrow \tfrac{1}{2} & \underline{\quad\quad a \quad\quad}
\end{array}
$$

The force pulling the chromosomes to cell poles is generated by the nuclear spindle, a series of microtubules made of the protein tubulin. Microtubules attach to the centromeres of chromosomes by interacting with another specific set of proteins located in that area. The orchestration of these molecular interactions is complex, yet constitutes the basis of the laws of hereditary transmission in eukaryotes.

Plants differing in two characters

Mendel's experiments described so far stemmed from two pure-breeding parental lines that differed in one character. As we have seen, such lines produce F_1 progeny that are heterozygous for one gene (genotype A/a). Such heterozygotes are sometimes called **monohybrids.** The selfing or intercross of identical heterozygous F_1 individuals (symbolically $A/a \times A/a$) is called a **monohybrid cross,** and it was this type of cross that provided the interesting 3:1 progeny ratios that suggested the principle of equal segregation. Mendel went on to analyze the descendants of pure lines that differed in *two* characters. Here we need a general symbolism to represent genotypes including two genes. If two genes are on different chromosomes, the gene pairs are separated by a semicolon — for example, A/a ; B/b. If they are on the same chromosome, the alleles on one chromosome are written adjacently and are separated from those on the other chromosome by a slash — for example, $A\ B/a\ b$ or $A\ b/a\ B$. An accepted symbolism does not exist for situations in which it is not known whether the genes are on the same chromosome or on different chromosomes. For this situation, we will separate the genes with a dot — for example, $A/a \cdot B/b$. A double heterozygote, $A/a \cdot B/b$, is also known as a **dihybrid.** From studying **dihybrid crosses** ($A/a \cdot B/b \times A/a \cdot B/b$), Mendel came up with another important principle of heredity.

The two specific characters that he began working with were seed shape and seed color. We have already followed the monohybrid cross for seed color ($Y/y \times Y/y$), which gave a progeny ratio of 3 yellow:1 green. The seed-shape phenotypes were round (determined by allele R) and wrinkled (determined by allele r). The monohybrid cross $R/r \times R/r$ gave a progeny ratio of 3 round:1 wrinkled (Table 2-1 and Figure 2-8). To perform a dihybrid cross, Mendel started with two parental pure lines. One line had yellow, wrinkled seeds; because Mendel had no concept of the chromosomal location of genes, we must use the dot representation to write this genotype as $Y/Y \cdot r/r$. The other line had green, round seeds, the genotype being $y/y \cdot R/R$. The cross

Figure 2-8   Round (R/R or R/r) and wrinkled (r/r) peas in a pod of a selfed heterozygous plant (R/r). The phenotypic ratio in this pod happens to be precisely the 3:1 ratio expected on average in the progeny of this selfing. (Molecular studies have shown that the wrinkled allele used by Mendel is produced by insertion into the gene of a segment of mobile DNA of the type to be discussed in Chapter 20.) (Madan K. Bhattacharyya.)

between these two lines produced dihybrid F_1 seeds of genotype $R/r \cdot Y/y$, which he discovered were round and yellow. This result showed that the dominance of R over r and of Y over y was unaffected by the presence of heterozygosity for either gene pair in the $R/r \cdot Y/y$ dihybrid. Next Mendel made the dihybrid cross by selfing the dihybrid F_1 to obtain the F_2 generation. The F_2 seeds were of four different types in the following proportions:

$\frac{9}{16}$ round yellow,

$\frac{3}{16}$ round green,

$\frac{3}{16}$ wrinkled yellow, and

$\frac{1}{16}$ wrinkled green,

as shown in Figure 2-9. This rather unexpected $9:3:3:1$ ratio seems a lot more complex than the simple $3:1$ ratios of the monohybrid crosses. What could be the explanation? Before attempting to explain the ratio, Mendel made dihybrid crosses that included several other combinations of characters and found that *all* of the dihybrid F_1 individuals produced $9:3:3:1$ progeny ratios similar to that obtained for seed shape and color. The $9:3:3:1$ ratio was another consistent hereditary pattern that needed to be converted into an idea.

Mendel added up the numbers of individuals in certain F_2 phenotypic classes (the numbers are shown in Figure 2-9) to determine if the monohybrid $3:1$ F_2 ratios were still present. He noted that, in regard to seed shape, there were 423

round seeds ($315 + 108$) and 133 wrinkled seeds ($101 + 32$). This result is close to a $3:1$ ratio. Next, in regard to seed color, there were 416 yellow seeds ($315 + 101$) and 140 green ($108 + 32$), also very close to a $3:1$ ratio. The presence of these two $3:1$ ratios hidden in the $9:3:3:1$ ratio was undoubtedly a source of the insight that Mendel needed to explain the $9:3:3:1$ ratio, because he realized that it was nothing more than two independent $3:1$ ratios combined at random. One way of visualizing the random combination of these two ratios is with a branch diagram, as follows:

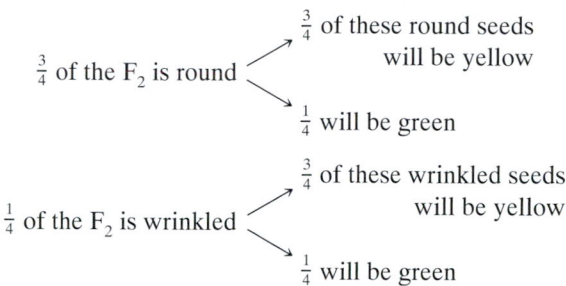

The combined proportions are calculated by multiplying along the branches in the diagram because, for example, $\frac{3}{4}$ of $\frac{3}{4}$ is calculated as $\frac{3}{4} \times \frac{3}{4}$, which equals $\frac{9}{16}$. These multiplications give us the following four proportions:

$\frac{3}{4} \times \frac{3}{4} = \frac{9}{16}$ round yellow

$\frac{3}{4} \times \frac{1}{4} = \frac{3}{16}$ round green

$\frac{1}{4} \times \frac{3}{4} = \frac{3}{16}$ wrinkled yellow

$\frac{1}{4} \times \frac{1}{4} = \frac{1}{16}$ wrinkled green

These proportions constitute the $9:3:3:1$ ratio that we are trying to explain. However, is this not merely number juggling? What could the combination of the two $3:1$ ratios mean biologically? The way that Mendel phrased his explanation does in fact amount to a biological mechanism. In what is now known as **Mendel's second law,** he concluded that *different gene pairs assort independently in gamete formation.* With hindsight about the chromosomal location of genes, we now know that this "law" is true only in some cases. Most cases of independence are observed for genes on different chromosome. Genes on the same chromosome generally do not assort independently, because they are held together on the chromosome. Hence the modern version of Mendel's second law is stated as the following message.

MESSAGE ···
Gene pairs on separate chromosome pairs assort independently at meiosis.
···

We have explained the $9:3:3:1$ phenotypic ratio as two combined $3:1$ phenotypic ratios. But the second law pertains to packing alleles into gametes. Can the $9:3:3:1$ ratio be explained on the basis of gametic genotypes? Let us consider the gametes produced by the F_1 dihybrid R/r ; Y/y (the semicolon shows that we are now assuming the genes to be on different chromosomes). Again, we will use the branch diagram to get us started because it illustrates

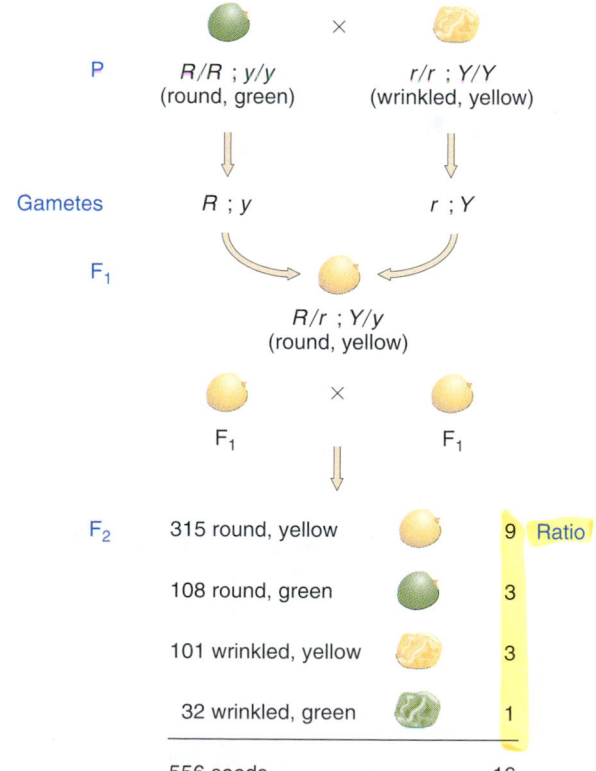

Figure 2-9 The F_2 generation resulting from a dihybrid cross.

independence visually. Combining Mendel's laws of equal segregation and independent assortment, we can predict that

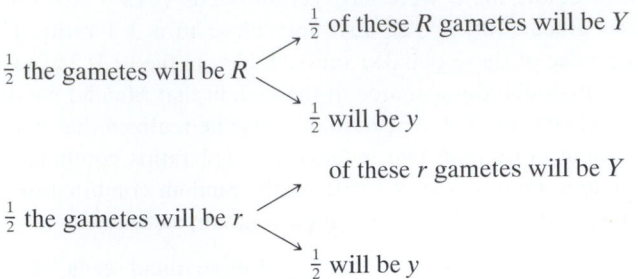

Multiplication along the branches gives us the gamete proportions:

$$\frac{1}{4} R \; ; \; Y$$
$$\frac{1}{4} R \; ; \; y$$
$$\frac{1}{4} r \; ; \; Y$$
$$\frac{1}{4} r \; ; \; y$$

These proportions are a direct result of the application of the two Mendelian laws. However, we still have not arrived at the $9:3:3:1$ ratio. The next step is to recognize that both the male and the female gametes will show the same proportions just given, because Mendel did not specify different rules for male and female gamete formation. The four female gametic types will be fertilized randomly by the four male gametic types to obtain the F_2, and the best way of showing this graphically is to use a 4×4 grid called a *Punnett square,* which is depicted in Figure 2-10. Grids are useful in genetics because their proportions can be drawn according to genetic proportions or ratios being considered, and thereby a visual data representation is obtained. In the Punnett square in Figure 2-10, for example, we see that the areas of the 16 boxes representing the various gametic fusions are each one-sixteenth of the total area of the grid, simply because the rows and columns were drawn to correspond to the gametic proportions of each. As the Punnett square shows, the F_2 contains a variety of genotypes, but there are only four phenotypes and their proportions are in the $9:3:3:1$ ratio. So we see that, when we work at the biological level of gamete formation, Mendel's laws explain not only the F_2 phenotypes, but also the genotypes underlying them.

Mendel was a thorough scientist; he went on to test his principle of independent assortment in a number of ways. The most direct way zeroed in on the $1:1:1:1$ gametic ratio hypothesized to be produced by the F_1 dihybrid $R/r \; ; \; Y/y$, because this ratio sprang from his principle of independent assortment and was the biological basis of the $9:3:3:1$ ratio in the F_2, as we have just demonstrated by using the Punnett square. He reasoned that, if there were in fact a $1:1:1:1$ ratio of $R \; ; \; Y$, $R \; ; \; y$, $r \; ; \; Y$, and $r \; ; \; y$ gametes, then, if he crossed the F_1 dihybrid with a plant of genotype $r/r \; ; \; y/y$, which produces only gametes with recessive alleles (genotype $r \; ; \; y$), the progeny proportions of this cross should be a

direct manifestation of the gametic proportions of the dihybrid; in other words,

$$\frac{1}{4} R/r \; ; \; Y/y$$
$$\frac{1}{4} R/r \; ; \; y/y$$
$$\frac{1}{4} r/r \; ; \; Y/y$$
$$\frac{1}{4} r/r \; ; \; y/y$$

These proportions were the result that he obtained, perfectly consistent with his expectations. Similar results were obtained for all the other dihybrid crosses that he made, and these and other types of tests all showed that he had in fact

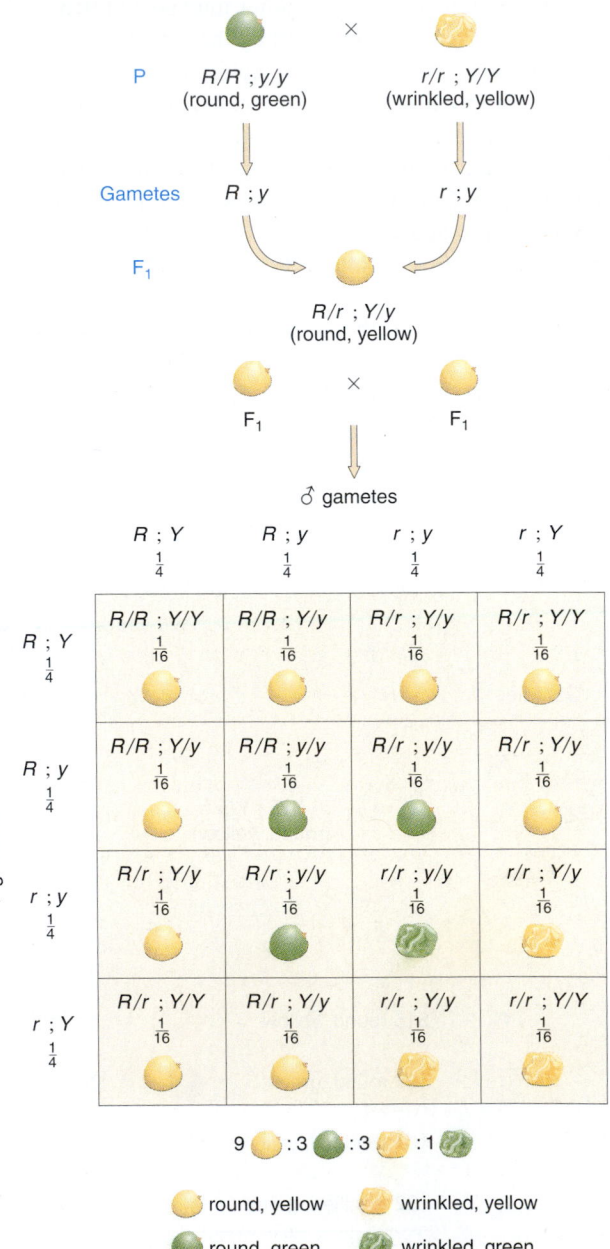

Figure 2-10   Punnett square showing predicted genotypic and phenotypic constitution of the F_2 generation from a dihybrid cross.

devised a robust model to explain the inheritance patterns observed in his various pea crosses.

The type of cross just considered, of an individual of unknown genotype with a fully recessive homozygote, is now called a **testcross.** The recessive individual is called a **tester.** Because the tester contributes only recessive alleles, the gametes of the unknown individual can be deduced from progeny phenotypes.

When Mendel's results were rediscovered in 1900, his principles were tested in a wide spectrum of eukaryotic organisms (organisms with cells that contain nuclei). The results of these tests showed that Mendelian principles were generally applicable. Mendelian ratios (such as $3:1$, $1:1$, $9:3:3:1$, and $1:1:1:1$) were extensively reported, suggesting that equal segregation and independent assortment are fundamental hereditary processes found throughout nature. Mendel's laws are not merely laws about peas, but laws about the genetics of eukaryotic organisms in general. The experimental approach used by Mendel can be extensively applied in plants. However, in some plants and in most animals, the technique of selfing is impossible. This problem can be circumvented by crossing identical genotypes. For example, an F_1 animal resulting from the mating of parents from differing pure lines can be mated to its F_1 siblings (brothers or sisters) to produce an F_2. The F_1 individuals are identical for the genes is question, so the F_1 cross is equivalent to a selfing.

Using genetic ratios

An important part of genetics today is concerned with predicting the types of progeny that emerge from a cross and calculating their expected frequency — in other words, their probability. We have already examined two methods for doing so — Punnett squares and branch diagrams. Punnett squares can be used to show hereditary patterns based on one gene pair, two gene pairs (as in Figure 2-10), or more. Such squares are a good graphic device for representing progeny, but making them is time consuming. Even the 16-compartment Punnett square in Figure 2-10 takes a long time to write out, but, for a trihybrid, there are 2^3, or 8, different gamete types, and the Punnett square has 64 compartments. The branch diagram (top right) is easier and is adaptable for phenotypic, genotypic, or gametic proportions, as illustrated for the dihybrid A/a ; B/b. (The dash means that the allele can be present in either form; that is, dominant or recessive.)

Note that the "tree" of branches for genotypes is quite unwieldy even in this case, which uses two gene pairs, because there are $3^2 = 9$ genotypes. For three gene pairs, there are 3^3, or 27, possible genotypes.

The application of simple statistical rules is the third method for calculating the probabilities (expected frequencies) of specific phenotypes or genotypes coming from a cross. The two probability rules needed are the **product rule** and the **sum rule,** which we will consider in that order.

Progeny genotypes from a self | Progeny phenotypes from a self | Gametes

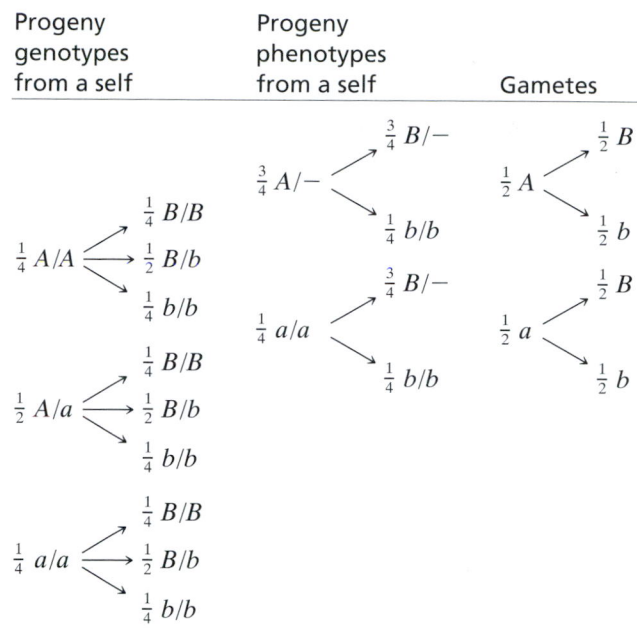

MESSAGE

The product rule states that the probability of independent events occurring together is the product of the probabilities of the individual events.

The possible outcomes of rolling dice follow the product rule because the outcome on each separate die is independent of the others. As an example, let us consider two dice and calculate the probability of rolling a pair of 4s. The probability of a 4 on one die is $\frac{1}{6}$ because the die has six sides and only one side carries the 4. This probability is written as follows:

$$p \text{ (of a 4)} = \tfrac{1}{6}$$

Therefore, with the use of the product rule, the probability of a 4 appearing on both dice is $\frac{1}{6} \times \frac{1}{6} = \frac{1}{36}$, which is written

$$p \text{ (of two 4s)} = \tfrac{1}{6} \times \tfrac{1}{6} = \tfrac{1}{36}$$

MESSAGE

The sum rule states that the probability of either of two mutually exclusive events occurring is the sum of their individual probabilities.

In the product rule, the focus is on outcomes A *and* B. In the sum rule, the focus is on the concept of outcome A *or* B. Dice can also be used to illustrate the sum rule. We have already calculated that the probability of two 4s is $\frac{1}{36}$, and, with the use of the same type of calculation, it is clear that the probability of two 5s will be the same, or $\frac{1}{36}$. Now we can calculate the probability of either two 4s or two 5s. Because these outcomes are mutually exclusive, the sum rule can be used to tell us that the answer is $\frac{1}{36} + \frac{1}{36}$, which is $\frac{1}{18}$. This probability can be written as follows:

$$p \text{ (two 4s or two 5s)} = \tfrac{1}{36} + \tfrac{1}{36} = \tfrac{1}{18}$$

Now we can consider a genetic example. Assume that we have two plants of genotypes *A/a* ; *b/b* ; *C/c* ; *D/d* ; *E/e* and *A/a* ; *B/b* ; *C/c* ; *d/d* ; *E/e*, and that, from a cross between these plants, we want to recover a progeny plant of genotype *a/a* ; *b/b* ; *c/c* ; *d/d* ; *e/e* (perhaps for the purpose of acting as the tester strain in a testcross). To estimate how many progeny plants need to be grown to stand a reasonable chance of obtaining the desired genotype, we need to calculate the proportion of the progeny that is expected to be of that genotype. If we assume that all the gene pairs assort independently, then we can do this calculation easily by using the product rule. The five different gene pairs are considered individually, as if five separate crosses, and then the appropriate probabilities are multiplied together to arrive at the answer.

From *A/a* × *A/a*, one-fourth of the progeny will be *a/a* (see Mendel's crosses); from *b/b* × *B/b*, one-half of the progeny will be *b/b*; from *C/c* × *C/c*, one-fourth of the progeny will be *c/c*; from *D/d* × *d/d*, one-half of the progeny will be *d/d*; from *E/e* × *E/e*, one-fourth of the progeny will be *e/e*. Therefore, the overall probability (or expected frequency) of progeny of genotype *a/a* ; *b/b* ; *c/c* ; *d/d* ; *e/e* will be $\frac{1}{4} \times \frac{1}{2} \times \frac{1}{4} \times \frac{1}{2} \times \frac{1}{4} = \frac{1}{256}$. So we learn that hundreds of progeny will need to be isolated to stand a chance of obtaining at least one of the desired genotype. This probability calculation can be extended to predict phenotypic frequencies or gametic frequencies. Indeed, there are thousands of other uses of this method in genetic analysis, and we will encounter many in later chapters.

Sex chromosomes and sex-linked inheritance

Most animals and many plants show sexual dimorphism; in other words, an individual can be either male or female. In most of these cases, sex is determined by special sex chromosomes. In these organisms, there are two categories of chromosomes, **sex chromosomes** and **autosomes** (the chromosomes other than the sex chromosomes). The rules of inheritance considered so far, with the use of Mendel's analysis as an example, are the rules of autosomes. Most of the chromosomes in a genome are autosomes. The sex chromosomes are fewer in number, and, generally in diploid organisms, there is just one pair.

Let us look at the human situation as an example. Human body cells have 46 chromosomes: 22 homologous pairs of autosomes plus 2 sex chromosomes. In females, there is a pair of identical sex chromosomes called the **X chromosomes**. In males, there is a nonidentical pair, consisting of one X and one Y. The **Y chromosome** is considerably shorter than the X. At meiosis in females, the two X chromosomes pair and segregate like autosomes so that each egg receives one X chromosome. Hence the female is said to be the **homogametic sex.** At meiosis in males, the X and the Y pair over a short region, which ensures that the X and Y separate so that half the sperm cells receive X and the other

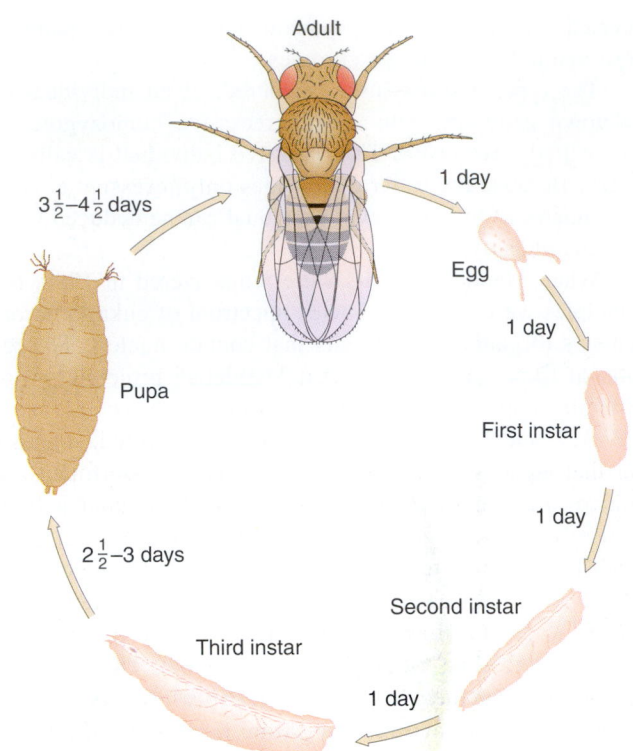

Figure 2-11   Life cycle of *Drosophila melanogaster*, the common fruit fly.

half receive Y. Therefore the male is called the **heterogametic sex.**

The fruit fly *Drosophila melanogaster* has been one of the most important research organisms in genetics; its short, simple life cycle contributes to its usefulness in this regard (Figure 2-11). Fruit flies also have XX females and XY males. However, the mechanism of sex determination in *Drosophila* differs from that in mammals. In *Drosophila*, the number of X chromosomes determines sex: two X's result in a female and one X results in a male. In mammals, the presence of the Y determines maleness and the absence of a Y determines femaleness. This difference is demonstrated by the sexes of the abnormal chromosome types XXY and XO, as shown in Table 2-3. However, we postpone a full discussion of this topic until Chapter 23.

Vascular plants show a variety of sexual arrangements. **Dioecious** species are the ones showing animal-like sexual dimorphism, with female plants bearing flowers containing only ovaries and male plants bearing flowers containing

2-3 TABLE	Chromosomal Determination of Sex in *Drosophila* and Humans			
	SEX CHROMOSOMES			
Species	XX	XY	XXY	XO
Drosophila	♀	♂	♀	♂
Human	♀	♂	♂	♀

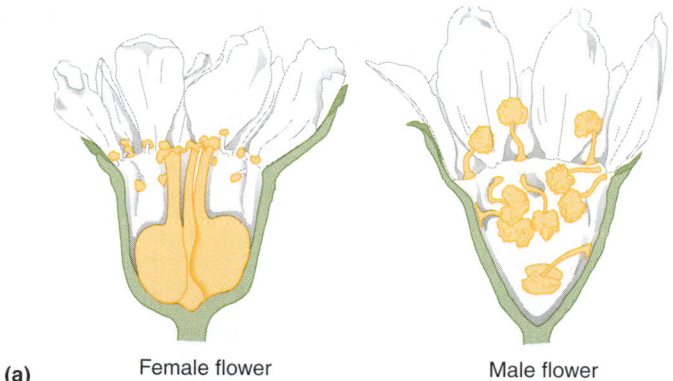

(a)   Female flower   Male flower

(b)

Figure 2-12   Two dioecious plant species: (a) *Osmaronia dioica;* (b) *Aruncus dioicus.* (Part a, Leslie Bohm; part b, Anthony Griffiths.)

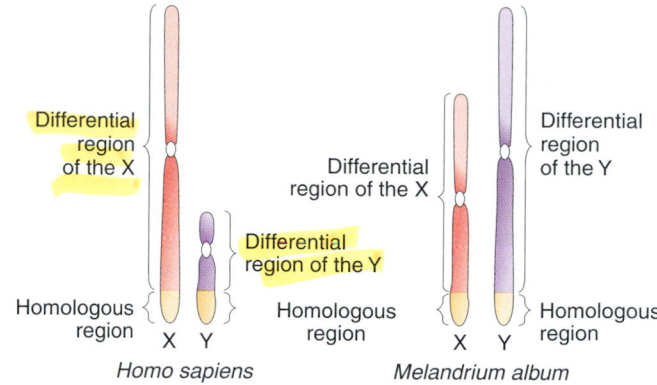

Homo sapiens   *Melandrium album*

Figure 2-13   Differential and pairing regions of sex chromosomes of humans and of the plant *Melandrium album.* The regions were located by observing where the chromosomes paired up in meiosis and where they did not.

only anthers (Figure 2-12). Some, but not all, dioecious plants have a nonidentical pair of chromosomes associated with (and almost certainly determining) the sex of the plant. Of the species with nonidentical sex chromosomes, a large proportion have an XY system. For example, the dioecious plant *Melandrium album* has 22 chromosomes per cell: 20 autosomes plus 2 sex chromosomes, with XX females and XY males. Other dioecious plants have no visibly different pair of chromosomes; they may still have sex chromosomes but not visibly distinguishable types.

Cytogeneticists have divided the X and Y chromosomes of some species into homologous and nonhomologous regions. The latter are called *differential* regions (Figure 2-13). These differential regions contain genes that have no counterparts on the other sex chromosome. Genes in the differential regions are said to be **hemizygous** ("half zygous") in males. Genes in the differential region of the X show an inheritance pattern called **X linkage**; those in the differential region of the Y show **Y linkage**. Genes in the homologous region show what might be called **X-and-Y linkage**. In general, genes on sex chromosomes are said to show **sex linkage.**

The genes on the differential regions of the sex chromosomes show patterns of inheritance related to sex. The in-

heritance patterns of genes on the autosomes produce male and female progeny in the same phenotypic proportions, as typified by Mendel's data (for example, both sexes might show a 3:1 ratio). However, crosses following the inheritance of genes on the sex chromosomes often show male and female progeny with different phenotypic ratios. In fact, for studies of genes of unknown chromosomal location, this pattern is a diagnostic of location on the sex chromosomes. Let's look at an example from *Drosophila.* The wild-type eye color of *Drosophila* is dull red, but pure lines with white eyes are available (Figure 2-14). This phenotypic difference is determined by two alleles of a gene located on the differential region of the X chromosome. When white-eyed males are crossed with red-eyed females, all the F_1 progeny have red eyes, showing that the allele for white is recessive. Crossing the red-eyed F_1 males and females produces a 3:1 F_2 ratio of red-eyed to white-eyed flies, but all the white-eyed flies are males. This inheritance pattern is explained by the alleles being located on the differential region of the X chromosome; in other words, by X-linkage. The genotypes

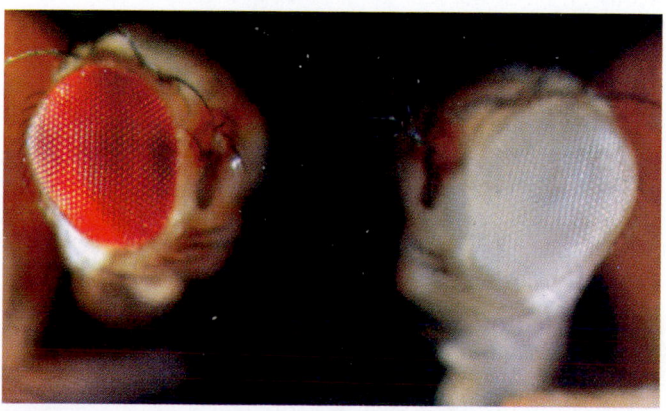

Figure 2-14   Red-eyed and white-eyed *Drosophila.* (Carolina Biological Supply.)

are shown in Figure 2-15. The reciprocal cross gives a different result. A reciprocal cross between white-eyed females and red-eyed males gives an F_1 in which all the females are red eyed, but all the males are white eyed. The F_2 consists of one-half red-eyed and one-half white-eyed flies of both sexes. Hence in sex linkage, we see examples not only of different ratios in different sexes, but also of differences between reciprocal crosses.

In *Drosophila,* eye color has nothing to do with sex determination, so we see that genes on the sex chromosomes are not necessarily related to sexual function. The same is true in humans, for whom pedigree analysis has revealed many X-linked genes, of which few could be construed as being connected to sexual function.

MESSAGE ···
Sex-linked inheritance regularly shows different phenotypic ratios in the two sexes of progeny, as well as different ratios in reciprocal crosses.
··

Human genetics

Human matings, like those of experimental organisms, show inheritance patterns both of the type discovered by Mendel (autosomal inheritance) and of sex linkage. Because con-

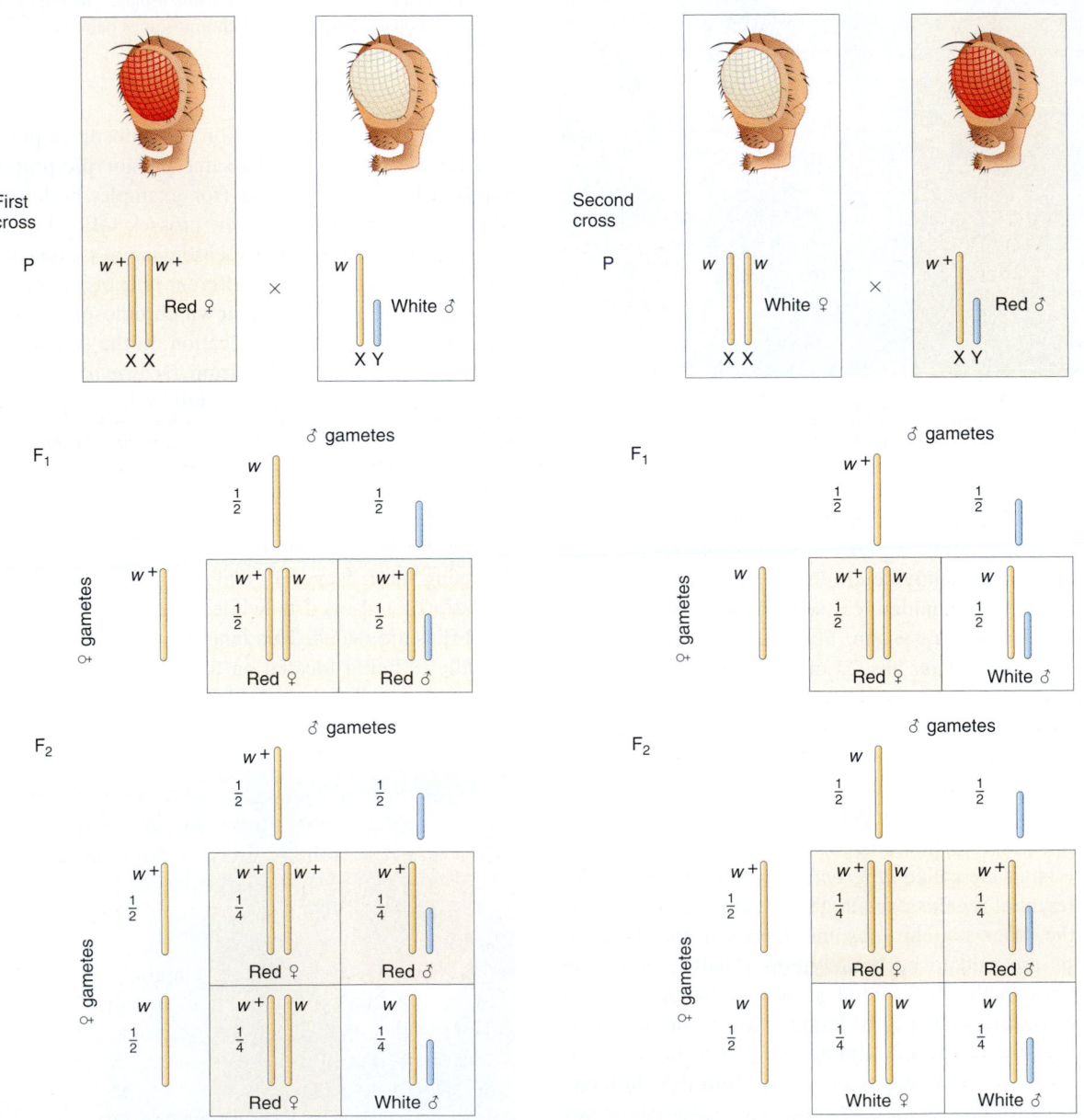

Figure 2-15   Explanation of the different results from reciprocal crosses between red-eyed (red) and white-eyed (white) *Drosophila.* (In *Drosophila* and many other experimental systems, a superscript plus sign is used to designate the normal, or wild-type allele. Here w^+ = red and w = white.)

trolled experimental crosses cannot be made with humans, geneticists must resort to scrutinizing records in the hope that informative matings have been made by chance. Such a scrutiny of records of matings is called **pedigree analysis.** A member of a family who first comes to the attention of a geneticist is called the **propositus.** Usually the phenotype of the propositus is exceptional in some way (for example, the propositus might be a dwarf). The investigator then traces the history of the phenotype in the propositus back through the history of the family and draws a family tree, or pedigree, by using the standard symbols given in Figure 2-16.

Many pairs of contrasting human phenotypes are determined by pairs of alleles. Inheritance patterns in pedigree analysis can reveal such allelic determination, but the clues in the pedigree have to be interpreted differently, depending on whether one of the contrasting phenotypes is a rare disorder or whether both phenotypes of a pair are morphs of a polymorphism. Rare inherited disorders are the domain of medical genetics.

Medical genetics

In the study of rare disorders, four general patterns of inheritance are distinguishable by pedigree analysis: autosomal recessive, autosomal dominant, X-linked recessive, and X-linked dominant.

Autosomal recessive disorders. The affected phenotype of an autosomal recessive disorder is determined by a recessive allele, and the corresponding unaffected phenotype is determined by a dominant allele. For example, the human disease phenylketonuria is inherited in a simple Mendelian manner as a recessive phenotype, with PKU determined by the allele p and the normal condition by P. Therefore, sufferers from this disease are of genotype p/p, and people who do not have the disease are either P/P or P/p. What patterns in a pedigree would reveal such an inheritance? The two key points are that (1) generally the disease appears in the progeny of unaffected parents and (2) the affected progeny include both males and females. When we know that both male and female progeny are affected, we can assume that we are dealing with simple Mendelian inheritance, not sex-linked inheritance. The following typical pedigree illustrates the key point that affected children are born to unaffected parents:

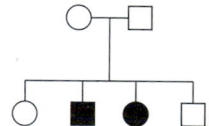

From this pattern, we can immediately deduce simple Mendelian inheritance of the recessive allele responsible for the exceptional phenotype (indicated in black). Furthermore, we can deduce that the parents are both heterozygotes, say A/a; both must have an a allele because each contributed an a allele to each affected child, and both must have an A allele because they are phenotypically normal.

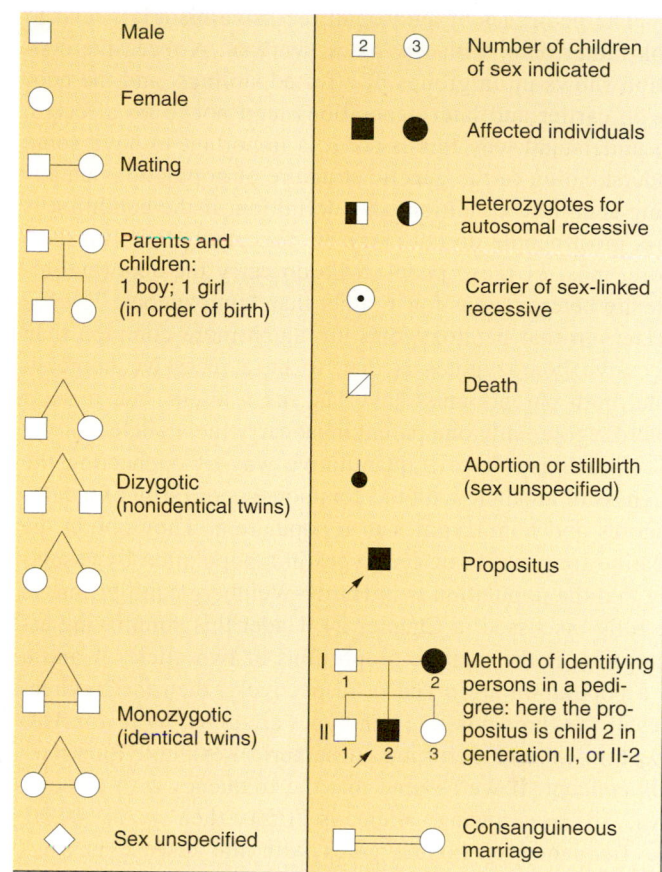

Figure 2-16 Symbols used in human pedigree analysis. (After W. F. Bodmer and L. L. Cavalli-Sforza, *Genetics, Evolution, and Man.* Copyright © 1976 by W. H. Freeman and Company.)

We can identify the genotypes of the children (in the order shown) as $A/-$, a/a, a/a, and $A/-$. Hence, the pedigree can be rewritten as follows:

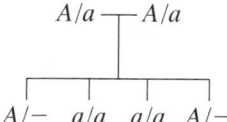

Note that this pedigree does not support the hypothesis of X-linked recessive inheritance, because, under that hypothesis, an affected daughter must have a heterozygous mother (possible) and a hemizygous father, which is clearly impossible, because he would have expressed the phenotype of the disorder.

Notice another interesting feature of pedigree analysis: even though Mendelian rules are at work, Mendelian ratios are rarely observed in families, because the sample size is too small. In the preceding example, we see a 1:1 phenotypic ratio in the progeny of a monohybrid cross. If the couple were to have, say, 20 children, the ratio would be something like 15 unaffected children and 5 with PKU (a 3:1 ratio); but, in a sample of 4 children, any ratio is possible, and all ratios are commonly found.

The pedigrees of autosomal recessive disorders tend to look rather bare, with few black symbols. A recessive condition shows up in groups of affected siblings, and the people in earlier and later generations tend not to be affected. To understand why this is so, it is important to have some understanding of the genetic structure of populations underlying such rare conditions. By definition, if the condition is rare, most people do not carry the abnormal allele. Furthermore, most of those people who do carry the abnormal allele are heterozygous for it rather than homozygous. The basic reason that heterozygotes are much more common than recessive homozygotes is that, to be a recessive homozygote, both parents must have had the *a* allele, but, to be a heterozygote, only one parent must carry the *a* allele.

Geneticists have a quantitative way of connecting the rareness of an allele with the commonness or rarity of heterozygotes and homozygotes in a population. They obtain the relative frequencies of genotypes in a population by assuming that the population is in Hardy-Weinberg equilibrium, to be fully discussed in Chapter 24. Under this simplifying assumption, if the relative proportions of two alleles *A* and *a* in a population are *p* and *q*, respectively, then the frequencies of the three possible genotypes are given by p^2 for *A/A*, $2pq$ for *A/a*, and q^2 for *a/a*. A numerical example illustrates this concept. If we assume that the frequency *q* of a recessive, disease-causing allele is 1/50, then *p* is 49/50, the frequency of homozygotes with the disease is $q^2 = (1/50)^2 = 1/250$, and the frequency of heterozygotes is $2pq = 2 \times 49/50 \times 1/50$, or approximately 1/25. Hence, for this example, we see that heterozygotes are 100 times as frequent as disease sufferers, and, as this ratio increases, the rarer the allele becomes. The relation between heterozygotes and homozygotes recessive for a rare allele is shown in the following illustration. Note that the allele frequencies *p* and *q* can be used as the gamete frequencies in both sexes.

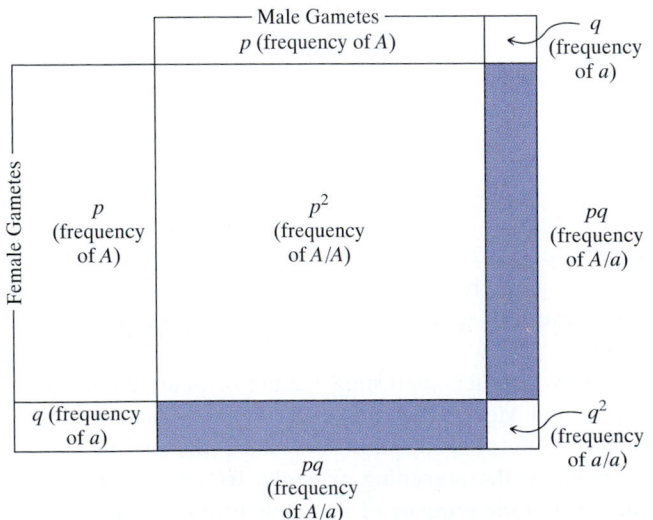

The formation of an affected person usually depends on the chance union of unrelated heterozygotes. However, inbreeding (mating between relatives) increases the chance

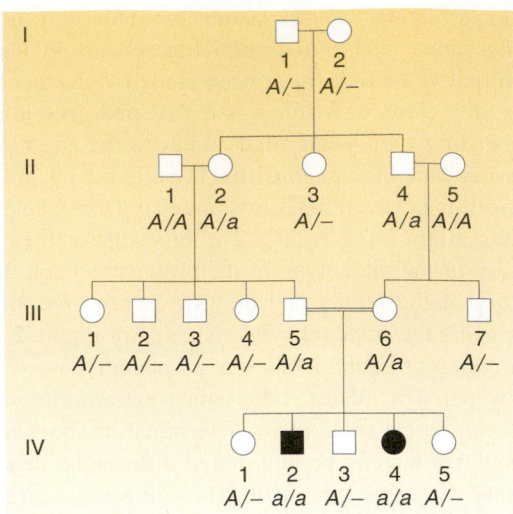

Figure 2-17   Pedigree of a rare recessive phenotype determined by a recessive allele *a*. Gene symbols are normally not included in pedigree charts, but genotypes are inserted here for reference. Note that individuals II-1 and II-5 marry into the family; they are assumed to be normal because the heritable condition under scrutiny is rare. Note also that it is not possible to be certain of the genotype in some persons with normal phenotype; such persons are indicated by A/–.

that a mating will be between two heterozygotes. An example of a marriage between cousins is shown in Figure 2-17. Individuals III-5 and III-6 are first cousins and produce two homozygotes for the rare allele. You can see from Figure 2-17 that an ancestor who is a heterozygote may produce many descendants who also are heterozygotes. Hence two cousins can carry the *same* rare recessive allele inherited from a common ancestor. For two *unrelated* persons to be heterozygous, they would have to inherit the rare allele from *both* their families. Thus matings between relatives generally run a higher risk of producing abnormal phenotypes caused by homozygosity for recessive alleles than do matings between nonrelatives. For this reason, first-cousin marriages contribute a large proportion of the sufferers of recessive diseases in the population.

What are some examples of human recessive disorders? PKU has already served as an example of pedigree analysis, but what kind of phenotype is it? PKU is a disease of processing of the amino acid phenylalanine, a component of all proteins in the food that we eat. Phenylalanine is normally converted into tyrosine by the enzyme phenylalanine hydroxylase:

$$\text{phenylalanine} \xrightarrow{\text{phenylalanine hydroxylase}} \text{tyrosine}$$

However, if a mutation in the gene encoding this enzyme alters the amino acid sequence in the vicinity of the enzyme's active site, the enzyme cannot bind or convert phenylalanine (its substrate). Therefore phenylalanine builds up in the body and is converted instead into phenylpyruvic acid, a compound that interferes with the development of the nervous system, leading to mental retardation.

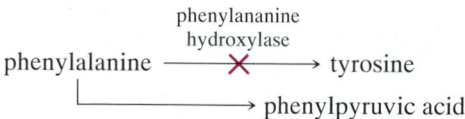

phenylalanine $\xrightarrow{\text{phenylalanine hydroxylase}}$ tyrosine

phenylpyruvic acid

Babies are now routinely tested for this processing deficiency at birth. If the deficiency is detected, phenylalanine can be withheld by use of a special diet, and the development of the disease can be arrested.

Cystic fibrosis is another disease inherited according to Mendelian rules as a recessive phenotype. The allele that causes cystic fibrosis was isolated in 1989, and the sequence of its DNA was determined. This has led to an understanding of gene function in affected and unaffected persons, giving hope for more effective treatment. Cystic fibrosis is a disease whose most important symptom is the secretion of large amounts of mucus into the lungs, resulting in death from a combination of effects but usually precipitated by upper respiratory infection. The mucus can be dislodged by mechanical chest thumpers, and pulmonary infection can be prevented by antibiotics; so, with treatment, cystic fibrosis patients can live to adulthood. The disorder is caused by a defective protein that transports chloride ions across the cell membrane. The resultant alteration of the salt balance changes the constitution of the lung mucus.

Albinism, which served as a model of allelic determination of contrasting phenotypes in Chapter 1, also is inherited in the standard autosomal recessive manner. The molecular nature of an albino allele and its inheritance are diagrammed in Figure 2-18. This diagram shows a simple autosomal recessive inheritance in a pedigree and shows the molecular nature of the alleles involved. In this example, the recessive allele *a* is caused by a base pair change that introduces a stop codon into the middle of the gene, resulting in a truncated polypeptide. The mutation, by chance, also introduces a new target site for a restriction enzyme. Hence, a probe for the gene detects two fragments in the case of *a* and only one in *A*. (Other types of mutations would produce different effects at the level detected by Southern, Northern, and Western analyses.)

In all the examples heretofore considered, the disorder is caused by an allele for a defective protein. In heterozygotes, the single functional allele provides enough active protein for the cell's needs. This situation is called haplosufficiency.

MESSAGE ·····················

In human pedigrees, an autosomal recessive disorder is revealed by the appearance of the disorder in the male and female progeny of unaffected persons.

Autosomal dominant disorders. Here the normal allele is recessive, and the abnormal allele is dominant. It may seem paradoxical that a rare disorder can be dominant, but remember that dominance and recessiveness are simply properties of how alleles act and are not defined in terms of how common they are in the population. A good example of a rare dominant phenotype with Mendelian inheritance is

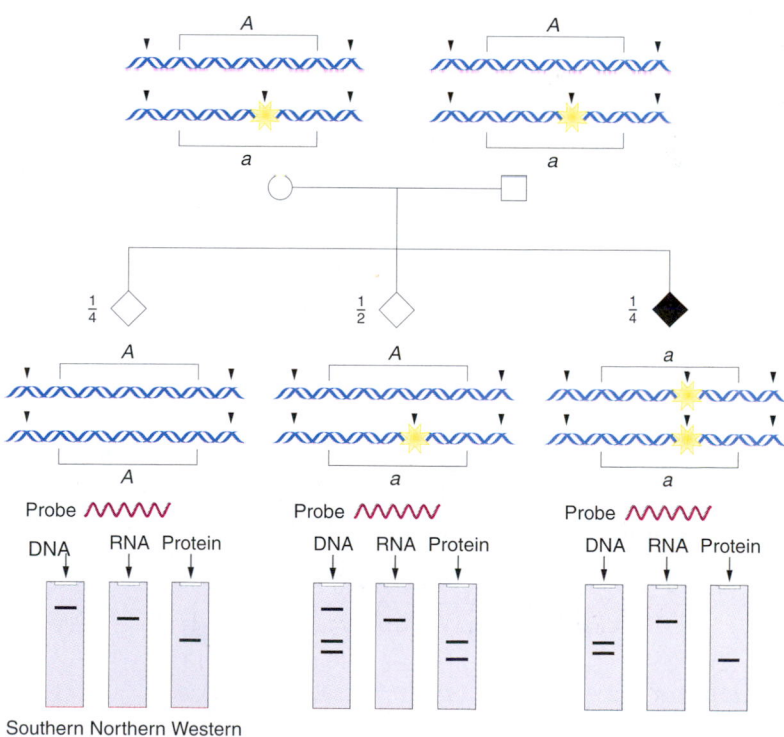

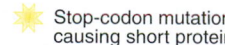

Figure 2-18 The molecular basis of Mendelian inheritance in a pedigree.

▼ Cutting site for restriction enzyme

✦ Stop-codon mutation causing short protein

pseudo-achondroplasia, a type of dwarfism (Figure 2-19). In regard to this gene, people with normal stature are genotypically *d/d*, and the dwarf phenotype in principle could be *D/d* or *D/D*. However, it is believed that the two "doses" of the *D* allele in the *D/D* genotype produce such a severe effect that this is a lethal genotype. If this is true, all the dwarf individuals are heterozygotes.

In pedigree analysis, the main clues for identifying a dominant disorder with Mendelian inheritance are that the phenotype tends to appear in every generation of the pedigree and that affected fathers and mothers transmit the phenotype to both sons and daughters. Again, the equal representation of both sexes among the affected offspring rules out sex-linked inheritance. The phenotype appears in every generation because generally the abnormal allele carried by a person must have come from a parent in the preceding generation. Abnormal alleles can arise de novo by the process of mutation. This event is relatively rare but must be kept in mind as a possibility. A typical pedigree for a dominant disorder is shown in Figure 2-20. Once again, notice that Mendelian ratios are not necessarily observed in families. As with recessive disorders, persons bearing one copy of the rare *A* allele (*A/a*) are much more common than those bearing two copies (*A/A*), so most affected people are heterozygotes, and virtually all matings concerning dominant disorders are *A/a* × *a/a*. Therefore, when the progeny of

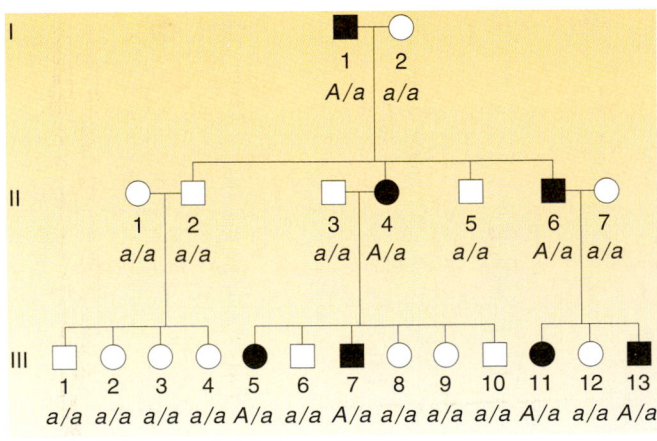

Figure 2-20 Pedigree of a dominant phenotype determined by a dominant allele *A*. In this pedigree, all the genotypes have been deduced.

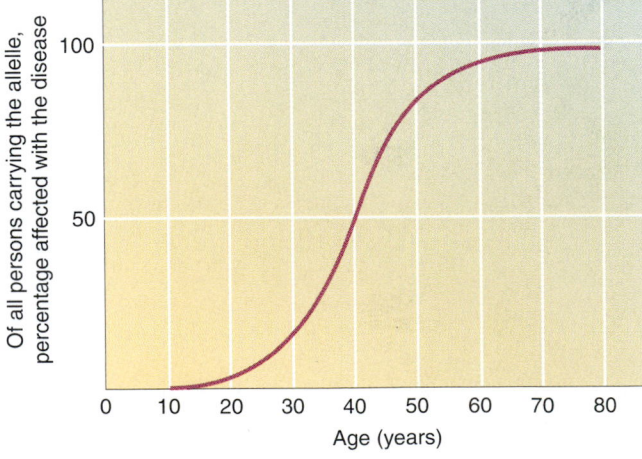

Figure 2-21 The age of onset of Huntington disease. The graph shows that people carrying the allele generally do not express the disease until after child-bearing age.

such matings are totaled, a 1:1 ratio is expected of unaffected (*a/a*) to affected (*A/a*) persons.

Huntington disease is an example of a disease inherited as a dominant phenotype determined by an allele of a single gene. The phenotype is one of neural degeneration, leading to convulsions and premature death. However, it is a late-onset disease, the symptoms generally not appearing until after the person has begun to have children (Figure 2-21). Each child of a carrier of the abnormal allele stands a 50

percent chance of inheriting the allele and the associated disease. This tragic pattern has led to a great effort to find ways of identifying people who carry the abnormal allele before they experience the onset of the disease. The application of molecular techniques has resulted in a promising screening procedure.

Some other rare dominant conditions are polydactyly (extra digits) and brachydactyly (short digits), shown in Figure 2-22, and piebald spotting, shown in Figure 2-23.

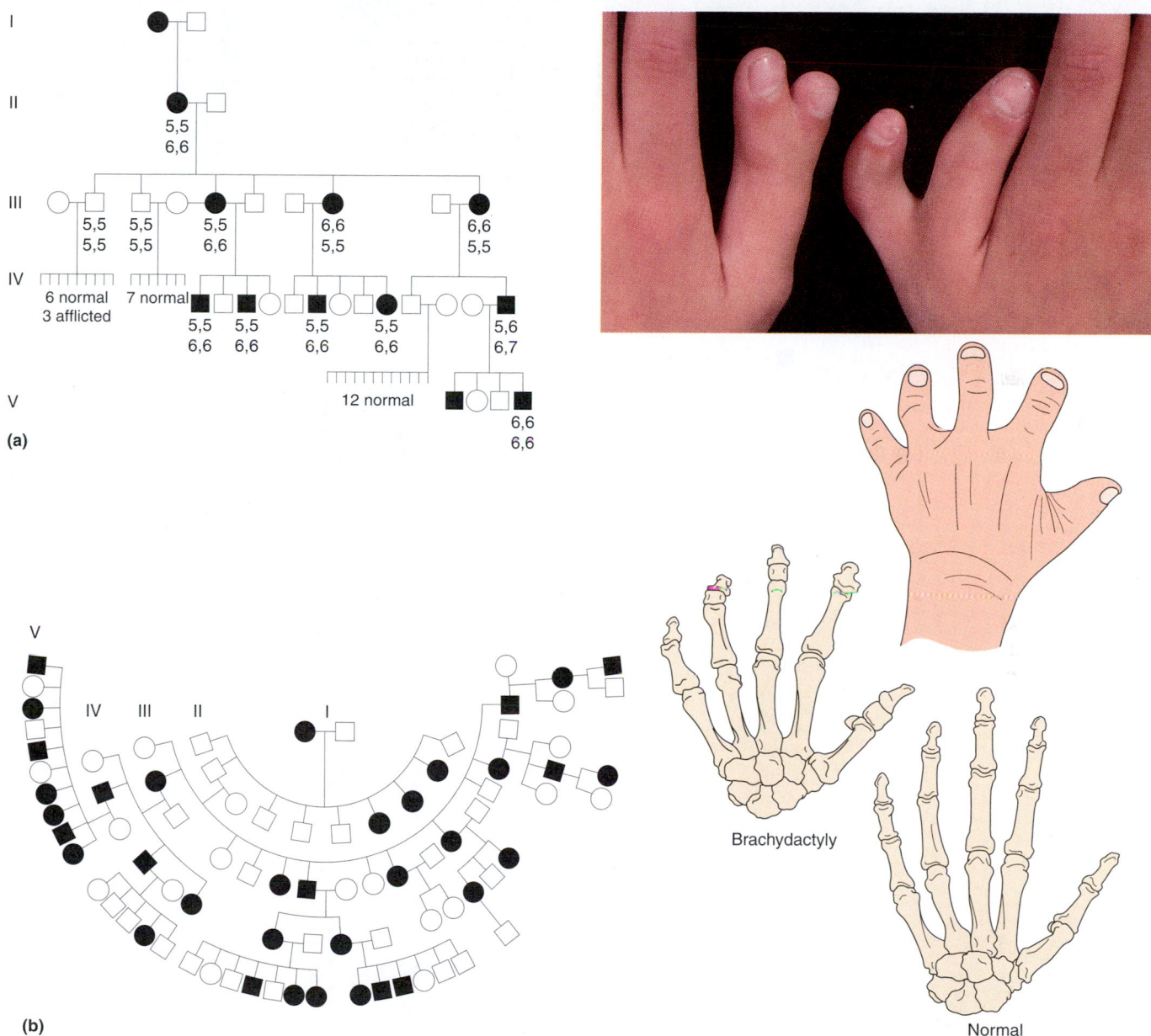

Figure 2-22   Some rare dominant phenotypes of the human hand. (a) *(right)* Polydactyly, a dominant phenotype characterized by extra fingers, toes, or both, determined by an allele *P*. The numbers in the accompanying pedigree *(left)* give the number of fingers in the upper lines and the number of toes in the lower. (Note the variation in expression of the *P* allele, a topic covered specifically in Chapter 4.) (b) *(right)* Brachydactyly, a dominant phenotype of short fingers, determined by an allele *B*. Note the very short terminal bones in the fingers compared with those in a normal hand. The pedigree for a family with brachydactyly shows a typical inheritance pattern for a rare dominant condition. All affected persons are *B/b* and unaffected persons are *b/b*. (Part a, photograph © Biophoto Associates/Science Source; part b, after C. Stern, *Principles of Human Genetics*, 3d ed. Copyright © 1973 by W. H. Freeman and Company.)

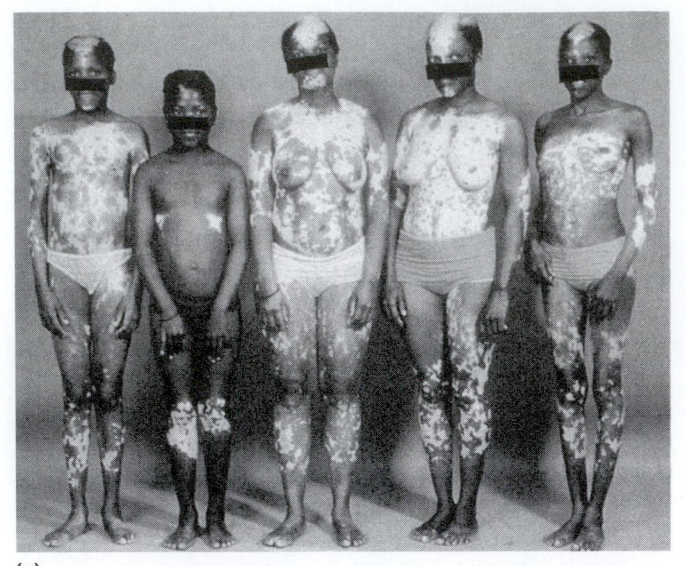
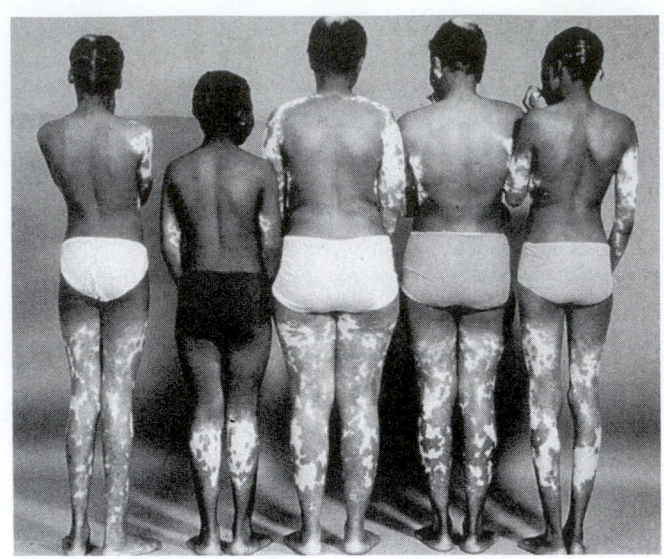

(a)

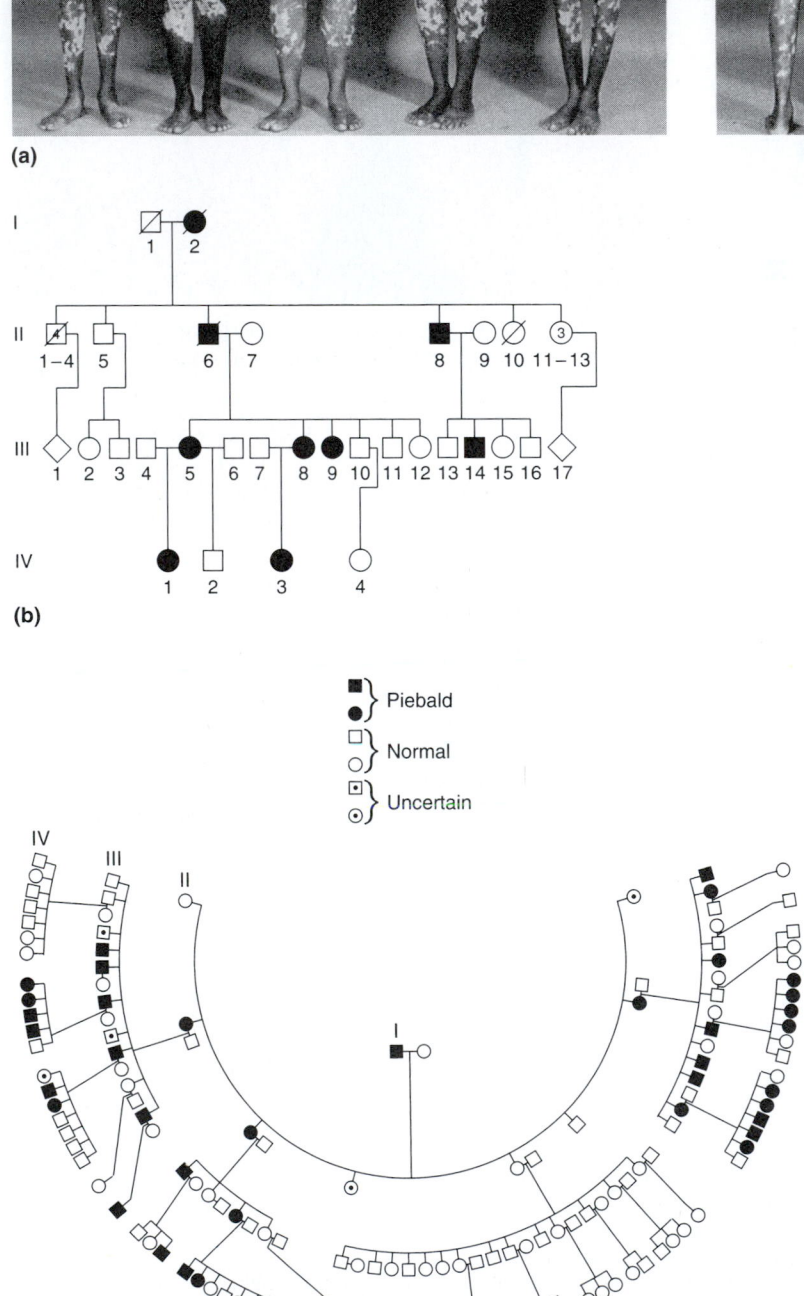

(b)

(d)

■ } Piebald
● }

□ } Normal
○ }

⊡ } Uncertain
⊙ }

(c)

Figure 2-23   Piebald spotting, a rare dominant human phenotype. Although the phenotype is encountered sporadically in all races, the patterns show up best in those with dark skin. (a) The photographs show front and back views of affected persons IV-1, IV-3, III-5, III-8, and III-9 from (b) the family pedigree. Notice the variation between family members in expression of the piebald gene. (c) A larger pedigree of a Norwegian family. It is believed that the patterns are caused by the dominant allele interfering with the migration of melanocytes (melanin-producing cells) from the dorsal to the ventral surface in the course of development. The white forehead blaze is particularly characteristic and is often accompanied by a white forelock in the hair. The same basic condition is known in mice (d), and, again, the melanocytes fail to cover the top of the head and the ventral surface.

Piebaldism is not a form of albinism; the cells in the light patches have the genetic potential to make melanin; but, because they are not melanocytes, they are not developmentally programmed to do so. In true albinism, the cells lack the potential to make melanin. (The DNA of the piebald allele has recently been characterized as an allele of c-*kit,* a type of gene called a protooncogene, to be discussed in Chapter 7. (Parts a and b from I. Winship, K. Young, R. Martell, R. Ramesar, D. Curtis, and P. Beighton, "Piebaldism: An Autonomous Autosomal Dominant Entity," *Clinical Genetics* 39, 1991, 330; part c from C. Stern, *Principles of Human Genetics,* 3d ed. Copyright © 1973 by W. H. Freeman and Company; part d provided by R. A. Fleishman, University of Texas, Southwestern Medical Center, Dallas — also see R. A. Fleishman, D. L. Saltman, V. Stastny, and S. Znemier, "Deletion of the c-*kit* Protooncogene in the Human Development Defect Piebald Trait," *Proceedings of the National Academy of Sciences of the United States of America* 88, 1991, 10885.)

> **MESSAGE** ·································
> Pedigrees of Mendelian autosomal dominant disorders show affected males and females in each generation; they also show that affected men and women transmit the condition to equal proportions of their sons and daughters.

X-linked recessive disorders.   Phenotypes with X-linked recessive inheritance typically show the following patterns in pedigrees:

1. **Many more males than females show the phenotype under study.** This is because a female showing the phenotype can result only from a mating in which both the mother and the father bear the allele (for example, $X^A\,X^a \times X^a\,Y$), whereas a male with the phenotype can be produced when only the mother carries the allele. If the recessive allele is very rare, almost all persons showing the phenotype are male.

2. None of the offspring of an affected male are affected, but all his daughters are "carriers," bearing the recessive allele masked in the heterozygous condition. Half of the sons of these carrier daughters are affected (Figure 2-24). Note that, in *common* X-linked phenotypes, this pattern might be obscured by inheritance of the recessive allele from a heterozygous mother as well as the father.

3. None of the sons of an affected male show the phenotype under study, nor will they pass the condition to their offspring. The reason behind this lack of male-to-male transmission is that a son obtains his Y chromosome from his father, so he cannot normally inherit the father's X chromosome too.

In the pedigree analysis of rare X-linked recessives, a normal female of unknown genotype is assumed to be homozygous unless there is evidence to the contrary.

Perhaps the most familiar example of X-linked recessive inheritance is red-green colorblindness. People with this condition are unable to distinguish red from green and see them as the same. The genes for color vision have been characterized at the molecular level. Color vision is based on three different kinds of cone cells in the retina, each sensitive to red, green, or blue wavelengths. The genetic determinants for the red and green cone cells are on the X chromosome. As with any X-linked recessive, there are many more males with the phenotype than females.

Another familiar example is *hemophilia,* the failure of blood to clot. Many proteins must interact in sequence to make blood clot. The most common type of hemophilia is

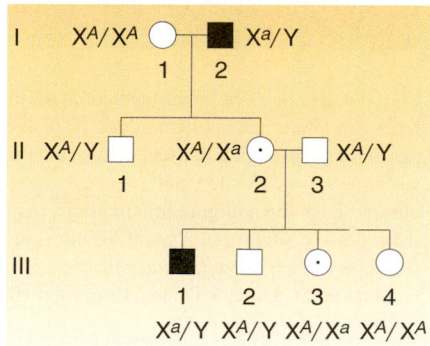

Figure 2-24   Pedigree showing that X-linked recessive alleles expressed in males are then carried unexpressed by their daughters in the next generation, to be expressed again in their sons. Note that III-3 and III-4 cannot be distinguished phenotypically.

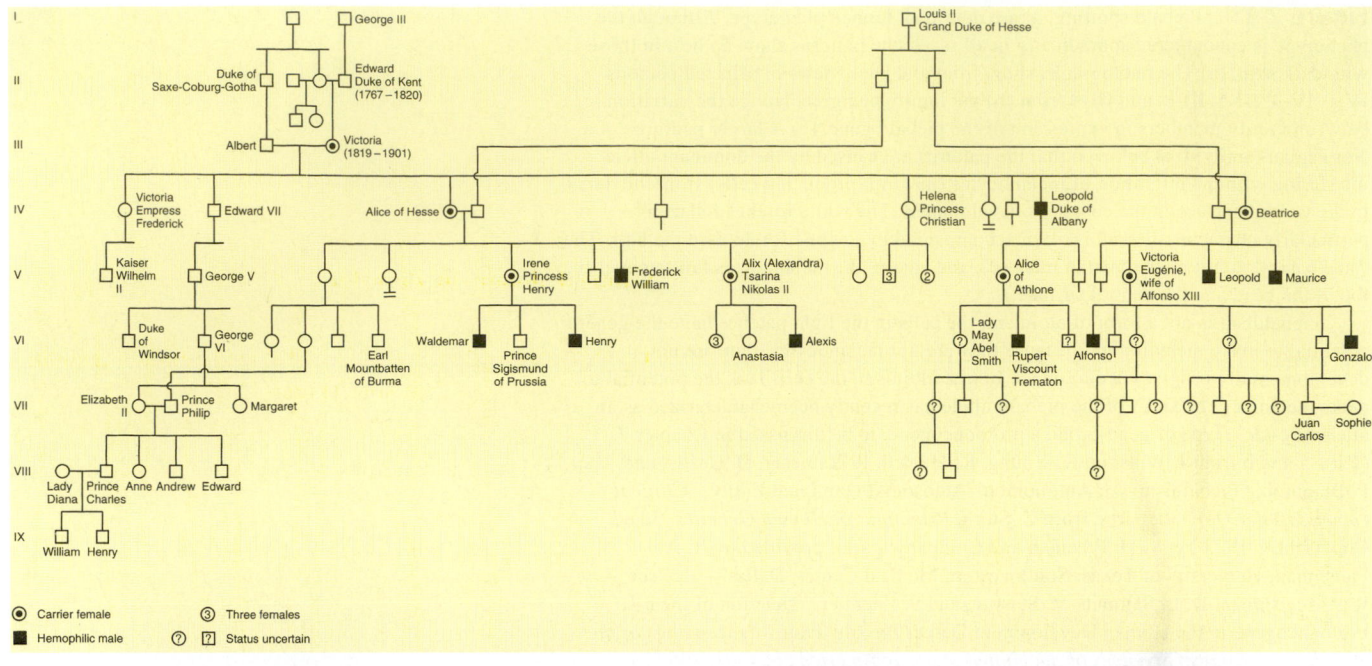

Figure 2-25   The inheritance of the X-linked recessive condition hemophilia in the royal families of Europe. A recessive allele causing hemophilia (failure of blood clotting) arose in the reproductive cells of Queen Victoria, or one of her parents, through mutation. This hemophilia allele spread into other royal families by intermarriage. (a) This partial pedigree shows affected males and carrier females (heterozygotes). Most spouses marrying into the families have been omitted from the pedigree for simplicity. Can you deduce the likelihood of the present British royal family's harboring the recessive allele? (b) A painting showing Queen Victoria surrounded by her numerous descendants. (Part a, after C. Stern, *Principles of Human Genetics*, 3d ed. Copyright © 1973 by W. H. Freeman and Company; part b, Royal Collection, St. James's Palace. Copyright Her Majesty Queen Elizabeth II.)

caused by the absence or malfunction of one of these proteins, called *Factor VIII*. The most well known cases of hemophilia are found in the pedigree of interrelated royal families in Europe (Figure 2-25). The original hemophilia allele in the pedigree arose spontaneously (as a mutation) either in

the reproductive cells of Queen Victoria's parents or of Queen Victoria herself. The son of the last czar of Russia, Alexis, inherited the allele ultimately from Queen Victoria, who was the grandmother of his mother Alexandra. Nowadays, hemophilia can be treated medically, but it was for-

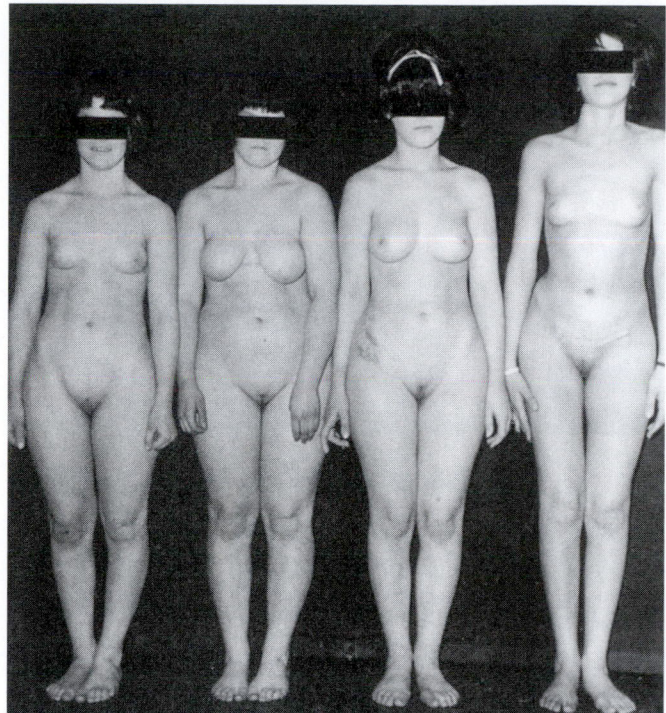

Figure 2-26 Four siblings with testicular feminization syndrome (congenital insensitivity to androgens). All four subjects in this photograph have 44 autosomes plus an X and a Y chromosome, but they have inherited the recessive X-linked allele conferring insensitivity to androgens (male hormones). One of their sisters (not shown), who was genetically XX, was a carrier and bore a child who also showed testicular feminization syndrome. (Leonard Pinsky, McGill University.)

merly a potentially fatal condition. It is interesting to note that, in the Jewish Talmud, there are rules about exemptions to male circumcision that show clearly that the mode of transmission of the disease through unaffected carrier females was well understood in ancient times. For example, one exemption was for the sons of women whose sisters' sons had bled profusely when they were circumcised.

Duchenne muscular dystrophy is a fatal X-linked recessive disease. The phenotype is a wasting and atrophy of muscles. Generally the onset is before the age of 6, with confinement to a wheelchair by 12, and death by 20. The gene for Duchenne muscular dystrophy has now been isolated and shown to encode the muscle protein dystrophin. This discovery holds out hope for a better understanding of the physiology of this condition and, ultimately, a therapy.

A rare X-linked recessive phenotype that is interesting from the point of view of sexual differentiation is a condition called *testicular feminization syndrome,* which has a frequency of about 1 in 65,000 male births. People afflicted with this syndrome are chromosomally males, having 44 autosomes plus an X and a Y, but they develop as females (Figure 2-26). They have female external genitalia, a blind vagina, and no uterus. Testes may be present either in the

labia or in the abdomen. Although many such persons marry, they are sterile. The condition is not reversed by treatment with the male hormone androgen, so it is sometimes called *androgen insensitivity syndrome.* The reason for the insensitivity is that the androgen receptor malfunctions, so the male hormone can have no effect on the target organs that contribute to maleness. In humans, femaleness results when the male-determining system is not functional.

X-linked dominant disorders. These disorders have the following characteristics:

1. Affected males pass the condition to all their daughters but to none of their sons (Figure 2-27).

2. Affected heterozygous females married to unaffected males pass the condition to half their sons and daughters (Figure 2-28).

There are few examples of X-linked dominant phenotypes in humans. One example is *hypophosphatemia,* a type of vitamin D-resistant rickets.

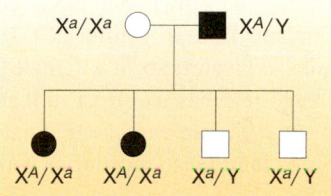

Figure 2-27 Pedigree showing that all the daughters of a male expressing an X-linked dominant phenotype will show the phenotype.

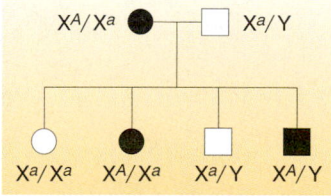

Figure 2-28 Pedigree showing that females affected by an X-linked dominant condition are usually heterozygous and pass the condition to half their sons and daughters.

X-chromosome inactivation

Early in the development of female mammals, one of the X chromosomes in each cell becomes inactivated. The inactivated X chromosome becomes highly condensed and is visible as a darkly staining spot called a **Barr body** (Figure 2-29). Surprisingly, this chromosomal inactivation persists through all the subsequent mitotic divisions that produce the mature body of the animal. The inactivation process is random, affecting either of the X chromosomes. As a result of this inactivation, the adult female body is a mixture, or **mosaic,** of cells with either of the two different X chromosome genotypes

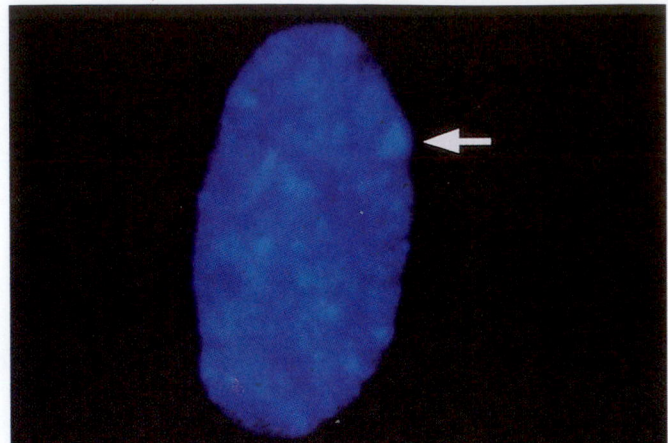

Figure 2-29   A Barr body, a condensed inactivated X chromosome, in the nucleus of a cell of a normal woman. Men have no Barr bodies. The number of Barr bodies in a cell is always equal to the total number of X chromosomes minus one. (Karen Dyer Montgomery.)

Figure 2-31   A calico cat. Both calico and tortoiseshell cats are females heterozygous for two alleles of an X-linked coat-color gene, O (orange) and o (black). The orange and black sectors are caused by X-chromosome inactivation. The white areas are caused by a separate genetic determinant present in calicos, but not in tortoiseshell cats. (Anthony Griffiths.)

(Figure 2-30). During the growth and development of tissues, the mitotic descendants of a progenitor cell often stay next to each other, forming a cluster; so, if a female is heterozygous for an X-linked gene that has its effect in that tissue, the two alleles of the heterozygote are expressed in patches, or sectors. A mosaic phenotype familiar to most of us is the coat pigmen-

tation pattern of tortoiseshell and calico cats (Figure 2-31). Such cats are females heterozygous for the alleles O (which causes fur to be orange) and o (which causes it to be black). Inactivation of the O-bearing X chromosome produces a black patch expressing o, and inactivation of the o-bearing X chromosome produces an orange patch expressing O.

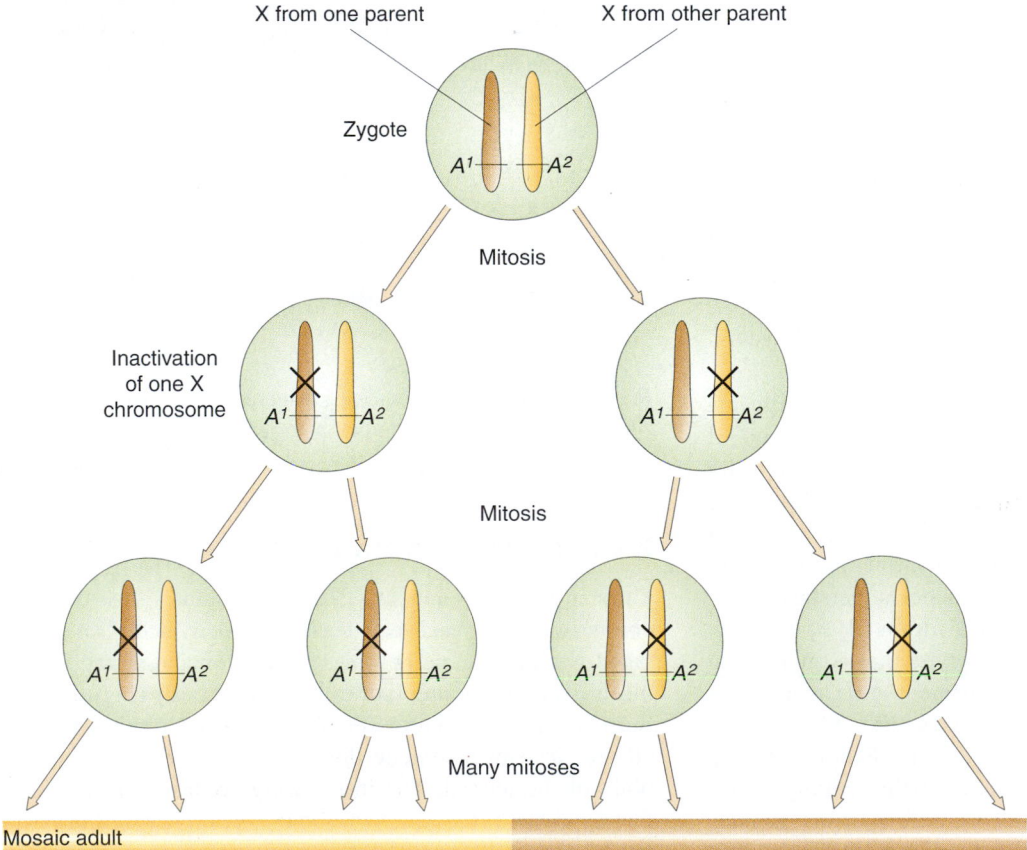

Figure 2-30   X-chromosome inactivation in mammals. The zygote of a female mammal heterozygous for an X-linked gene becomes a mosaic adult composed of two cell lines expressing one or the other of the alleles of the heterozygous gene because one or the other X chromosome is inactivated in all cell lines. For simplicity, inactivation is shown at the two-cell stage, but it can take place at other low cell numbers, too.

Figure 2-32 Somatic mosaicism in three generations of females heterozygous for sex-linked anhidrotic ectodermal dysplasia (absence of sweat glands). Areas without sweat glands are shown in green. The extent and location of the different tissues is determined by chance, but each female exhibits the characteristic mosaic pattern.

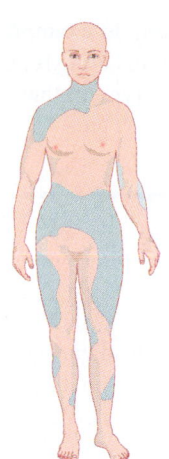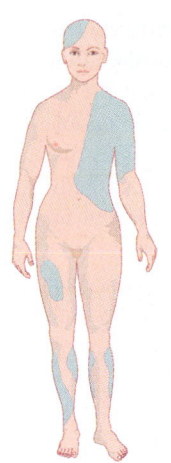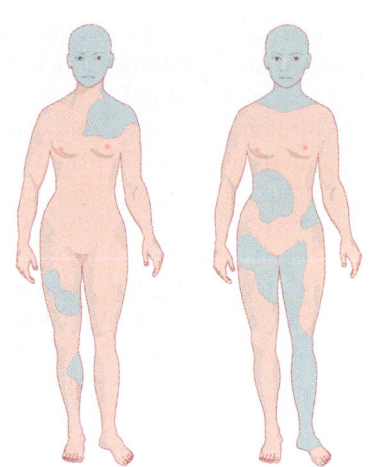

Although all human females have one of their X chromosomes inactivated in every cell, this inactivation is detectable only when a female is heterozygous for an X-linked gene. This is particularly striking when, as in tortoiseshell cats, the phenotype is expressed on the exterior of the body. Such a condition is *anhidrotic ectodermal dysplasia.* Males carrying the responsible allele (let us call it *d*) in its hemizygous condition have no sweat glands. A heterozygous (*D/d*) female has a mosaic of D and d sectors across her body, as shown in Figure 2-32.

Interestingly, the X-chromosome location of the gene causing testicular feminization was confirmed when it was shown microscopically that, in females heterozygous for the gene, half their fibroblast cells bind androgen but the other half do not. It should be noted that X inactivation is canceled in the female germinal tissue, so both X chromosomes are passed into the eggs.

MESSAGE ··
Inheritance patterns with an unequal representation of phenotypes in males and females can locate the genes concerned to one or both of the sex chromosomes.
··

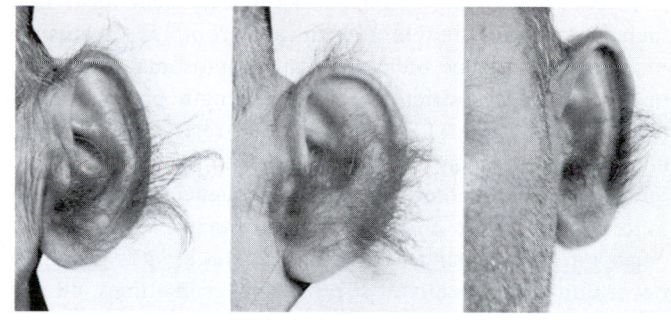

Figure 2-33 Hairy ear rims. This phenotype has been proposed to be caused by an allele of a Y-linked gene. (From C. Stern, W. R. Centerwall, and S. S. Sarkar, *The American Journal of Human Genetics* 16, 1964, 467. By permission of Grune & Stratton, Inc.)

Y-linked inheritance

Genes on the differential region of the human Y chromosome are inherited only by males, with fathers transmitting the region to their sons. The gene that plays a primary role in maleness is the **TDF** gene, which codes for **testis-determining factor.** The *TDF* gene has been located and mapped on the differential region of the Y chromosome (see Chapter 23). However, other than maleness itself, no human phenotype has been conclusively proved to be Y linked. Hairy ear rims (Figure 2-33) has been proposed as a possibility. The phenotype is extremely rare among the populations of most countries but more common among the populations of India. An Indian geneticist, K. Dronamraju, studied the phenotype in his own family. Every male in the family descended from a certain male ancestor showed the phenotype. In other Indian families, however, males seem to transmit the phenotype to only some of their sons, which is part of the reason that the evidence for Y-linked inheritance is considered to be inconclusive.

Human autosomal polymorphisms

Recall from Chapter 1 that a polymorphism is the coexistence of two to several *common* phenotypes of a character in a population. The alternative phenotypes of polymorphisms are often inherited as alleles of a single gene. In humans, there are many examples; consider, for example, the dimorphisms brown versus blue eyes, dark versus blonde hair, chin dimples versus none, widow's peak versus none, and attached versus free earlobes.

The interpretation of pedigrees for polymorphisms is somewhat different from that of rare disorders, because, by definition, the morphs are common. Let's look at a pedigree for an interesting human dimorphism. Most human populations are dimorphic for the ability to taste the chemical phenylthiocarbamide (PTC). That is, people can either detect it as a foul, bitter taste, or — to the great surprise and disbelief of tasters — cannot taste it at all. From the pedigree in Figure 2-34, we can see that two tasters sometimes produce nontaster children, which makes it clear that the allele that confers the ability to taste is dominant and that the allele for nontasting is recessive. Notice that almost all people who marry into this family carry the recessive allele either in heterozygous or in homozygous condition. Such a pedigree thus differs from those of rare recessive disorders for which it is conventional to assume that all who marry into a family are homozygous normal. Because both PTC alleles are common, it is not surprising that all but one of the family members in this pedigree married persons with at least one copy of the recessive allele.

Polymorphism is an interesting genetic phenomenon. Population geneticists have been surprised at how much polymorphism there is in natural populations of plants and animals generally. Furthermore, even though the genetics of polymorphisms is straightforward, there are very few polymorphisms for which there is satisfactory explanation for the coexistence of the morphs. But polymorphism is rampant at every level of genetic analysis, even at the DNA level; indeed, polymorphisms observed at the DNA level have been invaluable as landmarks to help geneticists find their way around the chromosomes of complex organisms.

One useful type of molecular chromosomal landmark, or marker, is a restriction fragment length polymorphism (RFLP). In Chapter 1, we learned that restriction enzymes are bacterial enzymes that cut DNA at specific base sequences in the genome. The target sequences have no biological significance in organisms other than bacteria — they occur purely by chance. Although the target sites generally occur quite consistently at specific sites, sometimes, on any one chromosome, a specific site is missing or there is an extra site. If such restriction-site presence or absence flanks the sequence hybridized by a probe, then a Southern hybridization will reveal a length polymorphism, or RFLP.

Consider this simple example in which one chromosome of one parent contains an extra site not found in the other chromosomes of that type in that cross:

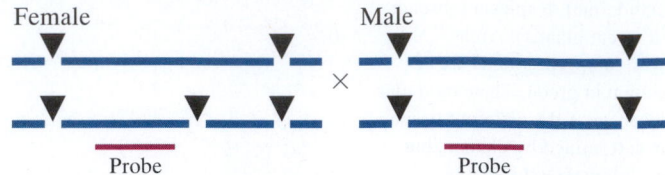

The Southern hybridizations will show two bands in the female and only one in the male. The "heterozygous" fragments will be inherited in exactly the same way as a gene. The preceding cross could be written as follows:

long/short   ×   long/long
and the progeny will be $\frac{1}{2}$ long/short
                           $\frac{1}{2}$ long/long

according to the law of equal segregation.

MESSAGE ···
Populations of plants and animals (including humans) are highly polymorphic. Contrasting morphs are generally determined by alleles inherited in a simple Mendelian manner.
··

Mendel's work has withstood the test of time and has provided us with the basic groundwork for all modern genetic study. He was the first person to draw attention to the mathematical regularity of inheritance patterns. From these patterns, he was able to make deductions about the fundamental nature of inheritance. Mendel's approach is still used by geneticists today. Yet his work went unrecognized and neglected for 35 years after its publication. Why? There are many possible reasons, but here we shall consider just one. Perhaps it was because biological science at that time could not provide evidence for any real physical units within cells that might correspond to Mendel's genetic particles. Chromosomes had certainly not yet been studied, meiosis had not yet been described, and even the full details of plant life cycles had not been worked out. Without this basic knowledge, it may have seemed that Mendel's ideas were mere numerology.

Above the doorway into the Mendel museum in Brno there is a wistful quip by Mendel inscribed in Czech, "MÁ DOBA PŘIJDE," meaning "My time will come." Mendel's time did come; in the twentieth century, research and the understanding of heredity flowered, all stemming from Mendel's seminal studies in the tiny monastery garden. His hypothetical "factors" (genes, as we now call them) are a well-understood molecular reality, and even whole genomes are becoming characterized. It is possible to take the latest dramatic research on cloning, gene therapy, transgenics, the human genome project, and so forth, and trace it all back through the research literature to that single paper entitled "Experiments on Plant Hybridization," presented in 1865 to the Brünn Natural History Society.

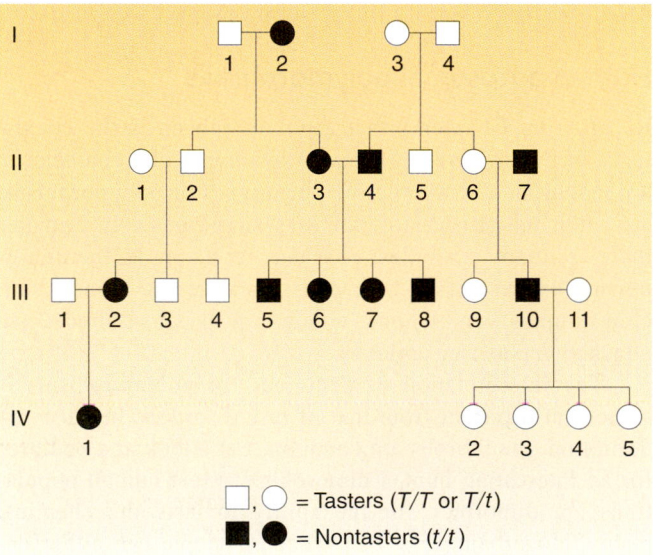

Figure 2-34   Pedigree for the ability to taste the chemical PTC.

SUMMARY

Modern genetics is based on the concept of the gene, the fundamental unit of heredity. As a result of his experiments with the garden pea, Mendel was the first to recognize the existence of genes. For example, by crossing a pure line of purple-flowered pea plants with a pure line of white-flowered pea plants and then selfing the F_1 generation, which was entirely purple, Mendel produced an F_2 generation of purple plants and white plants in a 3:1 ratio. In crosses such as those of pea plants bearing yellow seeds and pea plants bearing green seeds, he discovered that a 1:2:1 ratio underlies all 3:1 ratios. From these precise mathematical ratios, Mendel concluded that there are hereditary determinants of a particulate nature, now known as genes. In higher plant and animal cells, genes exist in pairs. Variant forms of a gene are called alleles. Gene pairs can be identical (homozygous) or carry two different alleles (heterozygous. An allele can be either dominant, for example Y (yellow), or recessive, for example y (green). Dominance is defined as the phenotype expressed in a heterozygote.

In a cross of heterozygous yellow (Y/y) plants with homozygous green (y/y) plants, a 1:1 ratio of yellow to green plants was produced. From this ratio, Mendel confirmed his so-called first law, which states that two members of a gene pair segregate from each other in gamete formation into equal numbers of gametes. Thus, each gamete carries only one member of each gene pair. The union of gametes to form a zygote is random in regard to which allele the gametes carry. The law of equal segregation is based on the segregation of homologous chromo-

somes at the first division of meiosis. This law has been found to be applicable in all organisms that undergo meiotic division.

The foregoing conclusions came from Mendel's work with monohybrid crosses. In dihybrid crosses, Mendel found 9:3:3:1 ratios in the F_2, which are really two 3:1 ratios combined at random. From these ratios, Mendel inferred that alleles of the two genes in a dihybrid cross behave independently. This concept is Mendel's second law. This law is generally applicable to genes on separate chromosomes. The basis for the law is the independent segregation of different chromosomes at meiosis.

In many organisms, sex is determined by special chromosomes called sex chromosomes. Examples are *Drosophila,* human beings, and certain dioecious plants. The genes on the sex chromosomes show a pattern of inheritance different from that of the autosomes, which show strictly Mendelian patterns. In certain crosses, different phenotypic ratios are found in the two sexes of progeny. Furthermore, in certain reciprocal crosses of the same two contrasting phenotypes, the progeny ratios differ.

Mendelian genetics has great significance for humans. Many human disorders are determined by abnormal recessive or dominant alleles of genes on the autosomal chromosomes. These alleles are inherited in a strict Mendelian manner. Other conditions clearly show sex-linked inheritance. These inheritance patterns can all be deduced from pedigree analysis by using certain standard rules. The morphs of human dimorphisms also are inherited in a Mendelian manner.

CONCEPT MAP

Draw a concept map interrelating as many of the following terms as possible. Note that the terms are listed in no particular order.

independent assortment/genotype/phenotype/dihybrid/(testcross/9:3:3:1 ratio/self/gametes/fertilization/1:1:1:1 ratio

CHAPTER INTEGRATION PROBLEM

Each chapter has one solved problem in which we stress the integration of concepts from different chapters. Learning in any discipline is a linear process, moving from topic to topic in some kind of appropriate sequence. The discipline itself is not linear; rather it is a set of integrated parts that the professional sees as one whole. We hope that focusing on integration will clarify the overall structure of genetics and that the reader will not see the contents of each chapter in isolation. As we pass from chapter to chapter, the levels of understanding build on the preceding ones, and the subject is assembled like the layers of an onion.

◆ Problem ◆

Crosses were made between two pure lines of rabbits that we can call A and B. A male from line A was mated with a female from line B, and the F_1 rabbits were subsequently intercrossed to produce an F_2. It was discovered that $\frac{3}{4}$ of the F_2 animals had white subcutaneous fat, and $\frac{1}{4}$ had yellow subcutaneous fat. Later, the F_1 was examined and was found to have white fat. Several years later, an attempt was made to repeat the experiment by using the same male from line A and the same female from line B. This time, the F_1 and all the F$_2$ (22 animals) had white fat. The only difference between the original experiment and the repeat that seemed relevant was that, in the original, all the animals were fed fresh vegetables, and, in the repeat, they were fed commercial rabbit chow. Provide an explanation for the difference and a test of your idea.

◆ Solution ◆

The first time that the experiment was done, the breeders would have been perfectly justified in proposing that a pair of alleles determine white versus yellow body fat, because the data clearly resemble Mendel's results in peas. White must be dominant, so we can represent the white allele as *W* and the yellow allele as *w*. The results can then be expressed as follows:

$$
\begin{array}{ll}
\text{P} & W/W \times w/w \\
\text{F}_1 & W/w \\
\text{F}_2 & \frac{1}{4}\ W/W \\
& \frac{1}{2}\ W/w \\
& \frac{1}{4}\ w/w
\end{array}
$$

No doubt, if the parental rabbits had been sacrificed, it would have been predicted that one (we cannot tell which) would have white fat and the other yellow. Luckily, this was not done, and the same animals were bred again, leading to a very interesting, different result. Often in science, an unexpected observation can lead to a novel principle, and, rather than moving on to something else, it is useful to try to explain the inconsistency. So why did the 3:1 ratio disappear? Here are some possible explanations.

First, perhaps the genotypes of the parental animals had changed. This type of spontaneous change affecting the whole animal, or at least its gonads, is very unlikely, because even common experience tells us that organisms tend to be stable to their type.

Second, in the repeat, the sample of 22 F$_2$ animals did not contain any yellow simply by chance ("bad luck"). This again seems unlikely, because the sample was quite large, but it is a definite possibility.

A third explanation draws on the principle covered in Chapter 1 that genes do not act in a vacuum; they depend on the environment for their effects. Hence, the useful catchphrase arises, "Genotype plus environment equals phenotype." A corollary of this catchphrase is that genes can act differently in different environments; so

Genotype 1 plus environment 1 equals phenotype 1

and

Genotype 1 plus environment 2 equals phenotype 2

In the present question, the different diets constituted different environments, so a possible explanation of the results is that the recessive allele *w* produces yellow fat only when the diet contains fresh vegetables. This explanation is testable. One way to test it is to repeat the experiment again and use vegetables as food, but the parents might be dead by this time. A more convincing way is to interbreed several of the white-fatted F$_2$ rabbits from the second experiment. According to the original interpretation, about $\frac{3}{4}$ would bear at least one recessive *w* allele for yellow fat, and, if their progeny are raised on vegetables, yellow should appear in Mendelian proportions. For example, if we choose two rabbits, *W/w* and *w/w*, the progeny would be $\frac{1}{2}$ white and $\frac{1}{2}$ yellow.

If this outcome did not happen and no yellow progeny appeared in any of the F$_2$ matings, one would be forced back to explanations 1 or 2. Explanation 2 can be tested by using larger numbers, and, if this explanation doesn't work, we are left with number 1, which is difficult to test directly.

As you might have guessed, in reality the diet was the culprit. The specific details illustrate environmental effects beautifully. Fresh vegetables contain yellow substances called xanthophylls, and the dominant allele *W* gives rabbits the ability to break down these substances to a colorless ("white") form. However, *w/w* animals lack this ability, and the xanthophylls are deposited in the fat, making it yellow. When no xanthophylls have been ingested, both *W/−* and *w/w* animals end up with white fat.

SOLVED PROBLEMS

This section in each chapter contains a few solved problems that show how to approach the problem sets that follow. The purpose of the problem sets is to challenge your understanding of the genetic principles learned in the chapter. The best way to demonstrate an understanding of a subject is to be able to use that knowledge in a real or simulated situation. Be forewarned that there is no machinelike way of solving these problems. The three main resources at your disposal are the genetic principles just learned, common sense, and trial and error.

Here is some general advice before beginning. First, it is absolutely essential to read and understand all of the question. Find out exactly what facts are provided, what assumptions have to be made, what clues are given in the question, and what inferences can be made from the available information. Second, be methodical. Staring at the question rarely helps. Restate the information in the question in your own way, preferably using a diagrammatic representation or flowchart to help you think out the problem. Good luck.

1. Consider three yellow, round peas, labeled A, B, and C. Each was grown into a plant and crossed to a plant grown from a green, wrinkled pea. Exactly 100 peas issuing from each cross were sorted into phenotypic classes as follows:

   A: 51 yellow, round
   49 green, round

B: 100 yellow, round

C: 24 yellow, round
       26 yellow, wrinkled
       25 green, round
       25 green, wrinkled

What were the genotypes of A, B, and C? (Use gene symbols of your own choosing; be sure to define each one.)

◆ Solution ◆

Notice that each of the crosses is

yellow, round × green, wrinkled

↓

progeny

Because A, B, and C were all crossed to the same plant, all the differences among the three progeny populations must be attributable to differences in the underlying genotypes of A, B, and C.

You might remember a lot about these analyses from the chapter, which is fine, but let's see how much we can deduce from the data. What about dominance? The key cross for deducing dominance is B. Here, the inheritance pattern is

yellow, round × green, wrinkled

↓

all yellow, round

So yellow and round must be dominant phenotypes, because dominance is literally defined in terms of the phenotype of a hybrid. Now we know that the green, wrinkled parent used in each cross must be fully recessive; we have a very convenient situation because it means that each cross is a testcross, which is generally the most informative type of cross.

Turning to the progeny of A, we see a 1:1 ratio for yellow to green. This ratio is a demonstration of Mendel's first law (equal segregation) and shows that, for the character of color, the cross must have been heterozygote × homozygous recessive. Letting Y = yellow and y = green, we have

$$Y/y \times y/y$$

↓

$\frac{1}{2} Y/y$ (yellow)
$\frac{1}{2} y/y$ (green)

For the character of shape, because all the progeny are round, the cross must have been homozygous dominant × homozygous recessive. Letting R = round and r = wrinkled, we have

$$R/R \times r/r$$

↓

R/r (round)

Combining the two characters, we have

$$Y/y \; ; R/R \times y/y \; ; r/r$$

↓

$\frac{1}{2} Y/y \; ; R/r$
$\frac{1}{2} y/y \; ; R/r$

Now, cross B becomes crystal clear and must have been

$$Y/Y \; ; R/R \times y/y \; ; r/r$$

↓

$Y/y \; ; R/r$

because any heterozygosity in pea B would have given rise to several progeny phenotypes, not just one.

What about C? Here, we see a ratio of 50 yellow:50 green (1:1) and a ratio of 49 round:51 wrinkled (also 1:1). So both genes in pea C must have been heterozygous, and cross C was

$$Y/y \; ; R/r \times y/y \; ; r/r$$

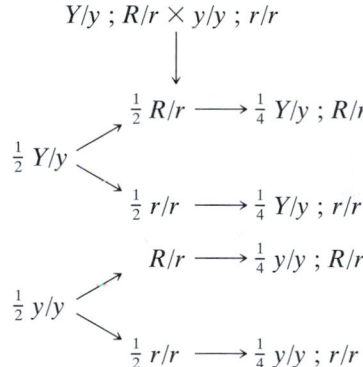

which is a good demonstration of Mendel's second law (independent behavior of different genes).

How would a geneticist have analyzed these crosses? Basically, the same way that we just did but with fewer intervening steps. Possibly something like this: "yellow and round dominant; single-gene segregation in A; B homozygous dominant; independent two-gene segregation in C."

2. Phenylketonuria (PKU) is a human hereditary disease resulting from the inability of the body to process the chemical phenylalanine, which is contained in the protein that we eat. PKU is manifested in early infancy and, if it remains untreated, generally leads to mental retardation. PKU is caused by a recessive allele with simple Mendelian inheritance.

A couple intends to have children but consults a genetic counselor because the man has a sister with PKU and the woman has a brother with PKU. There are no other known cases in their families. They ask the genetic counselor to determine the probability that their first child will have PKU. What is this probability?

◆ Solution ◆

What can we deduce? If we let the allele causing the PKU phenotype be p and the respective normal allele be P, then the sister and brother of the man and woman, respectively, must have been *p/p*. To produce these affected persons, all four grandparents must have been heterozygous normal. The pedigree can be summarized as follows:

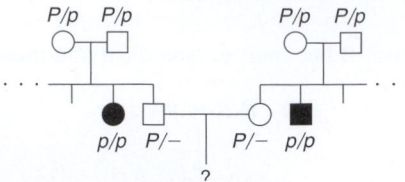

When these inferences have been made, the problem is reduced to an application of the product rule. The only way in which the man and woman can have a PKU child is if both of them are heterozygotes (it is obvious that they themselves do not have the disease). Both the grandparental matings are simple Mendelian monohybrid crosses expected to produce progeny in the following proportions:

$$\left. \begin{array}{l} \frac{1}{4}P/P \\ \frac{1}{2}P/p \end{array} \right\} \quad \text{Normal } (\frac{3}{4})$$

$$\frac{1}{4}p/p \quad \text{PKU } (\frac{1}{4})$$

We know that the man and the woman are normal, so the probability of them being a heterozygote is $\frac{2}{3}$, because within the $P/-$ class, $\frac{2}{3}$ are P/p and $\frac{1}{3}$ are P/P.

The probability of *both* the man and the woman being heterozygotes is $\frac{2}{3} \times \frac{2}{3} = \frac{4}{9}$. If they are both heterozygous, then one-quarter of their children would have PKU, so the probability that their first child will have PKU is $\frac{1}{4}$, and the probability of their being heterozygous *and* of their first child having PKU is $\frac{4}{9} \times \frac{1}{4} = \frac{4}{36} = \frac{1}{9}$, which is the answer.

3. A rare human disease afflicted a family as shown in the accompanying pedigree.

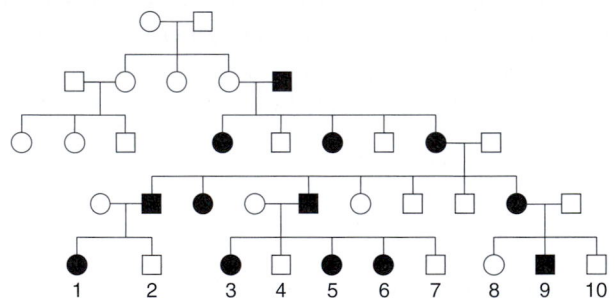

a. Deduce the most likely mode of inheritance.

b. What would be the outcomes of the cousin marriages 1×9, 1×4, 2×3, and 2×8?

◆ Solution ◆

a. The most likely mode of inheritance is X-linked dominant. We assume that the disease phenotype is dominant because, after it has been introduced into the pedigree by the male in generation II, it appears in every generation. We assume that the phenotype is X linked because fathers do not transmit it to their sons. If it were autosomal dominant, father-to-son transmission would be common.

In theory, autosomal recessive could work, but it is improbable. In particular, note the marriages between affected members of the family and unaffected outsiders. If the condition were autosomal recessive, the only way in which these marriages could have affected offspring is if each person marrying into the family were a heterozygote; then the matings would be *a/a* (affected) × *A/a* (unaffected). However, we are told that the disease is rare; in such a case, it is highly unlikely that heterozygotes would be so common. X-linked recessive inheritance is impossible, because a mating of an affected woman with a normal man could not produce affected daughters. So we can let *A* represent the disease-causing allele and *a* represent the normal allele.

b. 1×9: Number 1 must be heterozygous *A/a* because she must have obtained *a* from her normal mother. Number 9 must be *A/Y*. Hence, the cross is *A/a* ♀ × *A/Y* ♂.

Female gametes	Male gametes	Progeny
$\frac{1}{2}A$	$\frac{1}{2}A$	$\frac{1}{4}A/A$ ♀
	$\frac{1}{2}Y$	$\frac{1}{4}A/Y$ ♂
$\frac{1}{2}a$	$\frac{1}{2}A$	$\frac{1}{4}A/a$ ♀
	$\frac{1}{2}Y$	$\frac{1}{4}a/Y$ ♂

1×4: Must be *A/a* ♀ × *a/Y* ♂.

Female gametes	Male gametes	Progeny
$\frac{1}{2}A$	$\frac{1}{2}a$	$\frac{1}{4}A/a$ ♀
	$\frac{1}{2}Y$	$\frac{1}{4}A/Y$ ♂
$\frac{1}{2}a$	$\frac{1}{2}a$	$\frac{1}{4}a/a$ ♀
	$\frac{1}{2}Y$	$\frac{1}{4}a/Y$ ♂

2×3: Must be *a/Y* ♂ × *A/a* ♀ (same as 1×4).
2×8: Must be *a/Y* ♂ × *a/a* ♀ (all progeny normal).

PROBLEMS

Generally the straightforward problems are at the beginning of a set. Particularly challenging problems are marked with an asterisk.

1. What are Mendel's laws?

2. If you had a fruit fly (*Drosophila melanogaster*) that was of phenotype A, what test would you make to determine if it was *A/A* or *A/a*?

3. Two black guinea pigs were mated and over several years produced 29 black and 9 white offspring. Explain these results, giving the genotypes of parents and progeny.

4. Look at the Punnett square in Figure 2-10.

   a. How many genotypes are there in the 16 squares of the grid?

   b. What is the genotypic ratio underlying the 9:3:3:1 phenotypic ratio?

   c. Can you devise a simple formula for the calculation of the number of progeny genotypes in dihybrid, trihybrid, and so forth, crosses? Repeat for phenotypes.

   d. Mendel predicted that, within all but one of the phenotypic classes in the Punnett square, there should be several different genotypes. In particular, he performed many crosses to identify the underlying genotypes of the round, yellow phenotype. Show two different ways that could be used to identify the various genotypes underlying the round, yellow phenotype. (Remember, all the round, yellow peas look identical.)

5. You have three dice: one red (R), one green (G), and one blue (B). When all three dice are rolled at the same time, calculate the probability of the following outcomes:

   a. 6(R) 6(G) 6(B)

   b. 6(R) 5(G) 6(B)

   c. 6(R) 5(G) 4(B)

   d. no sixes at all

   e. 2 sixes and 1 five on any dice

   f. 3 sixes or 3 fives

   g. the same number on all dice

   h. a different number on all dice

6. You have three jars containing marbles, as follows:

jar 1	600 red	and	400 white
jar 2	900 blue	and	100 white
jar 3	10 green	and	990 white

   a. If you blindly select one marble from each jar, calculate the probability of obtaining

   (1) a red, a blue, and a green
   (2) three whites
   (3) a red, a green, and a white
   (4) a red and two whites
   (5) a color and two whites
   (6) at least one white

   ***b.** In a certain plant, R = red and r = white. You self a red R/r heterozygote with the express purpose of obtaining a white plant for an experiment. What mini-

mum number of seeds do you have to grow to be at least 95 percent certain of obtaining at least one white individual? [**Hint:** consider your answer to part a(6).]

   c. When a woman is injected with an egg fertilized in vitro, the probability of it implanting successfully is 20%. If a woman is injected with five eggs simultaneously, what is the probability that she will become pregnant? (Part c from Margaret Holm.)

7. **a.** The ability to taste the chemical phenylthiocarbamide is an autosomal dominant phenotype, and the inability to taste it is recessive. If a taster woman with a nontaster father marries a taster man who, in a previous marriage, had a nontaster daughter, what is the probability that their first child will be

   (1) a nontaster girl
   (2) a taster girl
   (3) a taster boy

   b. What is the probability that their first two children will be tasters of any sex?

8. John and Martha are contemplating having children, but John's brother has galactosemia (an autosomal recessive disease) and Martha's great-grandmother also had galactosemia. Martha has a sister who has three children, none of whom has galactosemia. What is the probability that John and Martha's first child will have galactosemia?

 Unpacking the Problem

In some chapters, we expand one specific problem with a list of exercises that help mentally process the principles and other knowledge surrounding the subject area of the problem. You can make up similar exercises yourself for other problems.

Before attempting a solution to problem 8, consider some questions such as the following, which are meant only as examples.

1. Can the problem be restated as a pedigree? If so, write one.

2. Can parts of the problem be restated by using Punnett squares?

3. Can parts of the problem be restated by using branch diagrams?

4. In the pedigree, identify a mating that illustrates Mendel's first law.

5. Define all the scientific terms in the problem, and look up any other terms that you are uncertain about.

6. What assumptions need to be made in answering this problem?

7. Which unmentioned family members must be considered? Why?

8. What statistical rules might be relevant, and in what situations can they be applied? Do such situations exist in this problem?

9. What are two generalities about autosomal recessive diseases in human populations?

10. What is the relevance of the rareness of the phenotype under study in pedigree analysis generally, and what can be inferred in this problem?

11. In this family, whose genotypes are certain and whose are uncertain?

12. In what way is John's side of the pedigree different from Martha's side? How does this difference affect your calculations?

13. Is there any irrelevant information in the problem as stated?

14. In what way is solving this kind of problem similar to or different from solving problems that you have already successfully solved?

15. Can you make up a short story based on the human dilemma in this problem?

Now try to solve the problem. If you are unable to do so, try to identify the obstacle and write a sentence or two describing your difficulty. Then go back to the expansion questions and see if any of them relate to your difficulty.

9. Holstein cattle are normally black and white. A superb black and white bull, Charlie, was purchased by a farmer for $100,000. All the progeny sired by Charlie were normal in appearance. However, certain pairs of his progeny, when interbred, produced red and white progeny at a frequency of about 25 percent. Charlie was soon removed from the stud lists of the Holstein breeders. Use symbols to explain precisely why.

10. Suppose that a husband and wife are both heterozygous for a recessive allele for albinism. If they have dizygotic (two-egg) twins, what is the probability that both of the twins will have the same phenotype for pigmentation?

11. The plant blue-eyed Mary grows on Vancouver Island and on the lower mainland of British Columbia. The populations are dimorphic for purple blotches on the leaves — some plants have blotches, and others don't. Near Nanaimo, one plant in nature had blotched leaves. This plant, which had not yet flowered, was dug up and taken to a laboratory, where it was allowed to self. Seeds were collected and grown into progeny. One randomly selected (but typical) leaf from each of the progeny is shown in the illustration at the upper right.

    a. Formulate a concise genetic hypothesis to explain these results. Explain all symbols and show all genotypic classes (and the genotype of the original plant).

    b. How would you test your hypothesis? Be specific.

12. Can it ever be proved that an animal is *not* a carrier of a recessive allele (that is, not a heterozygote for a given gene)? Explain.

13. In nature, the plant *Plectritis congesta* is dimorphic for fruit shape; that is, individual plants bear either wingless or winged fruits, as shown below. Plants were

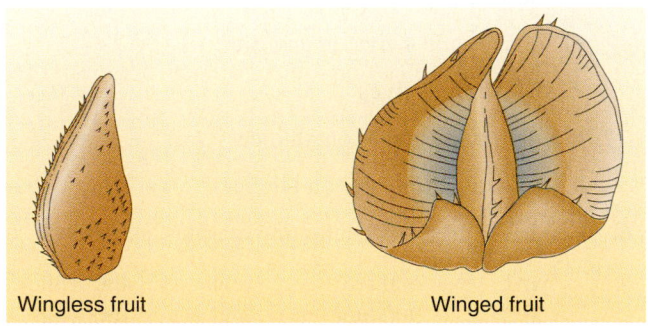

Wingless fruit   Winged fruit

collected from nature before flowering and were crossed or selfed with the following results:

Pollination	NUMBER OF PROGENY	
	Winged	Wingless
winged (selfed)	91	1*
winged (selfed)	90	30
wingless (selfed)	4*	80
winged × wingless	161	0
winged × wingless	29	31
winged × wingless	46	0
winged × winged	44	0
winged × winged	24	0

Interpret these results, and derive the mode of inheritance of these fruit-shaped phenotypes. Use symbols. (**Note:** The phenotypes of progeny marked by asterisks probably have a nongenetic explanation. What do you think it is?)

14. The accompanying pedigree is for a rare, but relatively mild, hereditary disorder of the skin.

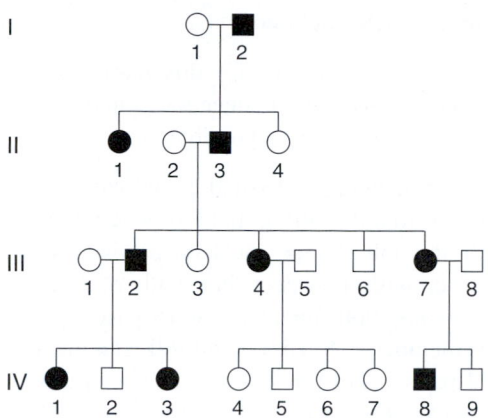

a. Is the disorder inherited as a recessive or a dominant phenotype? State reasons for your answer.

b. Give genotypes for as many individuals in the pedigree as possible. (Invent your own defined allele symbols.)

c. Consider the four unaffected children of parents III-4 and III-5. In all four-child progenies from parents of these genotypes, what proportion is expected to contain all unaffected children?

15. Here are four human pedigrees. The black symbols represent an abnormal phenotype inherited in a simple Mendelian manner.

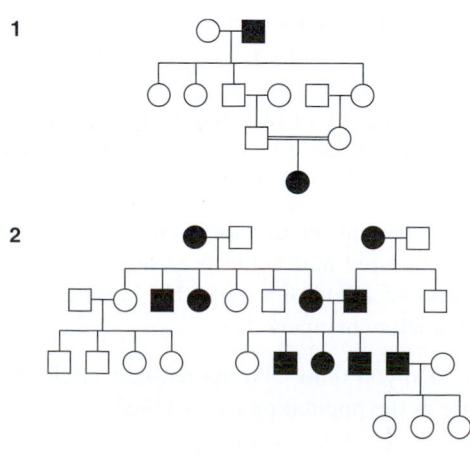

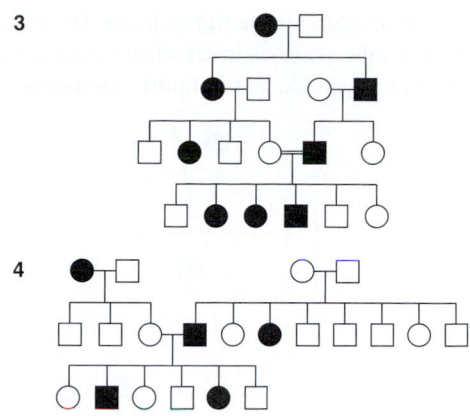

a. For each pedigree, state whether the abnormal condition is dominant or recessive. Try to state the logic behind your answer.

b. For each pedigree, describe the genotypes of as many persons as possible.

16. Tay-Sachs disease ("infantile amaurotic idiocy") is a rare human disease in which toxic substances accumulate in nerve cells. The recessive allele responsible for the disease is inherited in a simple Mendelian manner. For unknown reasons, the allele is more common in populations of Ashkenazi Jews of eastern Europe. A woman is planning to marry her first cousin, but the couple discovers that their shared grandfather's sister died in infancy of Tay-Sachs disease.

a. Draw the relevant parts of the pedigree, and show all the genotypes as completely as possible.

b. What is the probability that the cousins' first child will have Tay-Sachs disease, assuming that all people who marry into the family are homozygous normal?

17. The accompanying pedigree was obtained for a rare kidney disease.

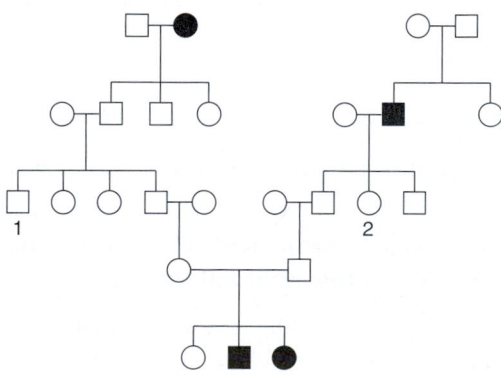

a. Deduce the inheritance of this condition, stating your reasons.

b. If individuals 1 and 2 marry, what is the probability that their first child will have the kidney disease?

18. The accompanying pedigree is for Huntington disease (HD), a late-onset disorder of the nervous system. The slashes indicate deceased family members.

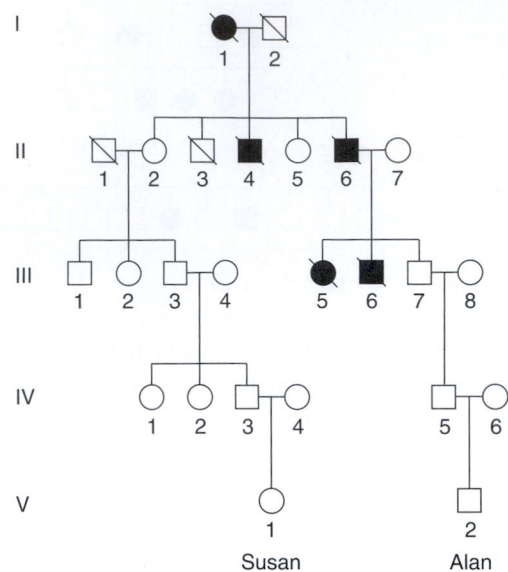

a. Is this pedigree compatible with the mode of inheritance for HD mentioned in the chapter?

b. Consider two newborn children in the two arms of the pedigree, Susan in the left arm and Alan in the right arm. Study the graph in Figure 2-21 and come up with an opinion on the likelihood that they will develop HD. Assume for the sake of the discussion that parents have children at age 25.

19. Consider the accompanying pedigree of a rare autosomal recessive disease, PKU.

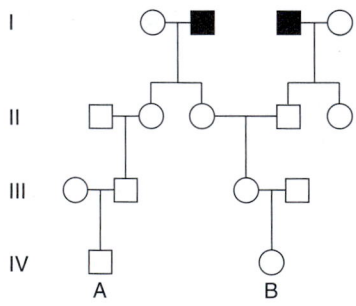

a. List genotypes of as many of the family members as possible.

b. If individuals A and B marry, what is the probability that their first child will have PKU?

c. If their first child is normal, what is the probability that their second child will have PKU?

d. If their first child has the disease, what is the probability that their second child will be unaffected?

(Assume that all people marrying into the pedigree do not have the abnormal allele.)

20. A curious polymorphism in human populations has to do with the ability to curl up the sides of the tongue to make a trough ("tongue rolling"). Some people can do this trick, and others simply cannot. Hence it is an example of a dimorphism. Its significance is a complete mystery. In one family, a boy was unable to roll his tongue, but, to his great chagrin, his sister could. Furthermore, both his parents were rollers, and so were both grandfathers, one paternal uncle, and one paternal aunt. One paternal aunt, one paternal uncle, and one maternal uncle could not.

a. Draw the pedigree for this family, defining your symbols clearly, and deduce the genotypes of as many individual members as possible.

b. The pedigree that you drew is typical of the inheritance of tongue rolling and led geneticists to come up with the inheritance mechanism that no doubt you came up with. However, in a study of 33 pairs of identical twins, both members of 18 pairs could roll, neither member of 8 pairs could roll, and one of the twins in 7 pairs could roll but the other could not. Because identical twins are derived from the splitting of one fertilized egg into two embryos, the members of a pair must be genetically identical. How can the existence of the seven discordant pairs be reconciled with your genetic explanation of the pedigree?

21. A rare, recessive allele inherited in a Mendelian manner causes the disease cystic fibrosis. A phenotypically normal man whose father had cystic fibrosis marries a phenotypically normal woman from outside the family, and the couple consider having a child.

a. Draw the pedigree as far as described.

b. If the frequency in the population of heterozygotes for cystic fibrosis is 1 in 50, what is the chance that the couple's first child will have cystic fibrosis?

c. If the first child does have cystic fibrosis, what is the probability that the second child will be normal?

22. In human hair, the black and brown colors are produced by various amounts and combinations of chemicals called *melanins*. However, red hair is produced by a different type of chemical substance about which little is known. Red hair runs in families, and the illustration (at the top of the facing page) shows a large pedigree for red hair. (Pedigree from W. R. Singleton and B. Ellis, *Journal of Heredity* 55, 1964, 261.)

a. Does the inheritance pattern in this pedigree suggest that red hair could be caused by a dominant or a recessive allele of a gene that is inherited in a simple Mendelian manner?

b. Do you think that the red-hair allele is common or rare in the population as a whole?

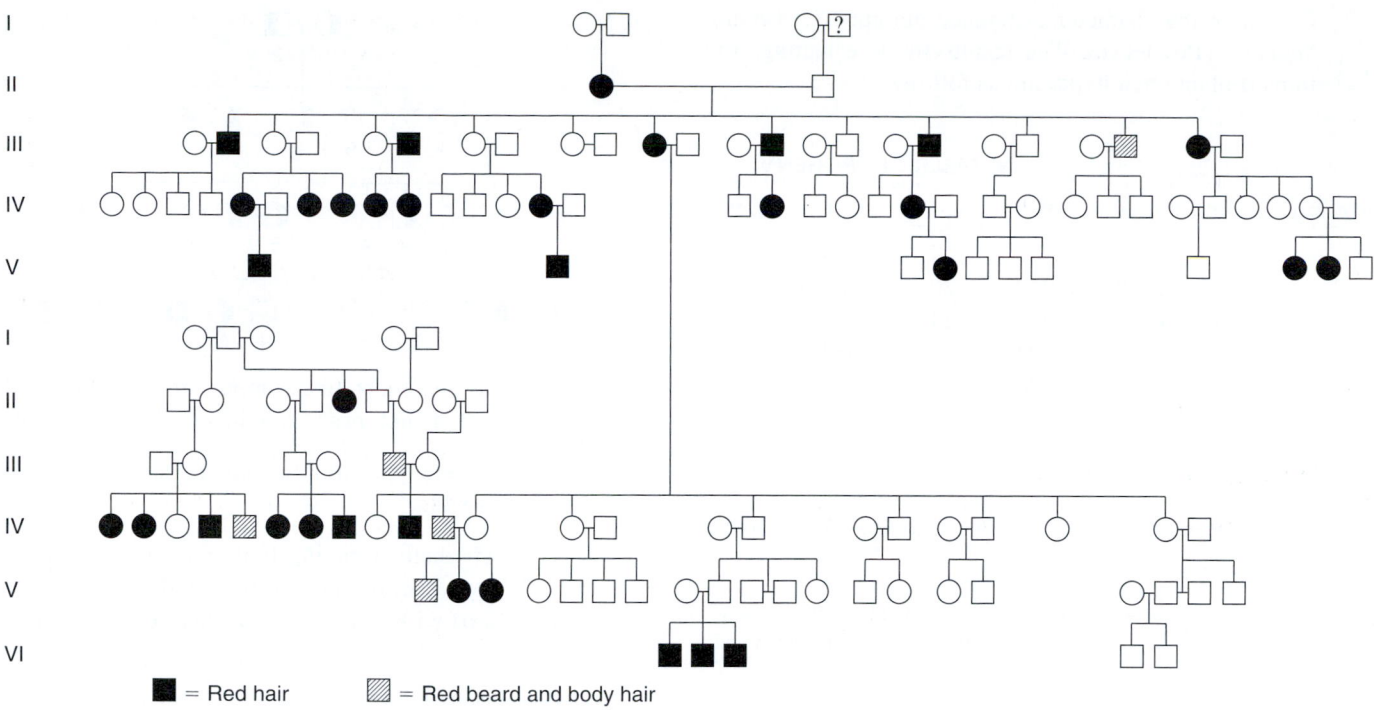

■ = Red hair   ▨ = Red beard and body hair

23. When many families were tested for the ability to taste the chemical PTC, the matings were grouped into three types and the progeny totaled, with the results shown below:

Parents	Number of families	CHILDREN Tasters	CHILDREN Non-tasters
taster × taster	425	929	130
taster × nontaster	289	483	278
nontaster × nontaster	86	5	218

Assuming that PTC tasting is dominant (*P*) and nontasting is recessive (*p*), how can the progeny ratios in each of the three types of mating be accounted for?

24. In tomatoes, red fruit is dominant to yellow, two-loculed fruit is dominant to many-loculed fruit, and tall vine is dominant to dwarf. A breeder has two pure lines: red, two-loculed, dwarf and yellow, many-loculed, tall. From these two lines, he wants to produce a new pure line for trade that is yellow, two-loculed, and tall. How exactly should he go about doing this? Show not only which crosses to make, but also how many progeny should be sampled in each case.

25. In humans, achondroplastic dwarfism and neurofibromatosis are both extremely rare dominant conditions. If a woman with achondroplasia marries a man with neurofibromatosis, what phenotypes could be produced in their children and in what proportions? (Be sure to define any symbols that you use.)

26. In dogs, dark coat color is dominant over albino and short hair is dominant over long hair. Assume that these effects are caused by two independently assorting genes and write the genotypes of the parents in each of the crosses shown here, in which D and A stand for the dark and albino phenotypes, respectively, and S and L stand for the short-hair and long-hair phenotypes.

Parental phenotypes	NUMBER OF PROGENY D, S	D, L	A, S	A, L
a. D, S × D, S	89	31	29	11
b. D, S × D, L	18	19	0	0
c. D, S × A, S	20	0	21	0
d. A, S × A, S	0	0	28	9
e. D, L × D, L	0	32	0	10
f. D, S × D, S	46	16	0	0
g. D, S × D, L	30	31	9	11

Use the symbols *C* and *c* for the dark and albino coat-color alleles and the symbols *S* and *s* for the short-hair and long-hair alleles, respectively. Assume homozygosity unless there is evidence otherwise. (Problem 26 reprinted by permission of Macmillan Publishing Co., Inc., from *Genetics* by M. Strickberger. Copyright © 1968 by Monroe W. Strickberger.)

27. In tomatoes, two alleles of one gene determine the character difference of purple (P) versus green (G) stems, and two alleles of a separate, independent gene determine the character difference of "cut" (C) versus "potato" (Po) leaves. The results for five matings of tomato-plant phenotypes are as follows:

Mating	Parental phenotypes	NUMBER OF PROGENY			
		P, C	P, Po	G, C	G, Po
1	P, C × G, C	321	101	310	107
2	P, C × P, Po	219	207	64	71
3	P, C × G, C	722	231	0	0
4	P, C × G, Po	404	0	387	0
5	P, Po × G, C	70	91	86	77

a. Determine which alleles are dominant.

b. What are the most probable genotypes for the parents in each cross?

(Problem 27 from A. M. Srb, R. D. Owen, and R. S. Edgar, *General Genetics,* 2d ed. Copyright © 1965 by W. H. Freeman and Company.)

28. We have dealt mainly with only two genes, but the same principles hold for more than two genes. Consider the following cross:

A/a ; B/b ; C/c ; D/d ; E/e × a/a ; B/b ; c/c ; D/d ; e/e

a. What proportion of progeny will *phenotypically* resemble (1) the first parent, (2) the second parent, (3) either parent, and (4) neither parent?

b. What proportion of progeny will be *genotypically* the same as (1) the first parent, (2) the second parent, (3) either parent, and (4) neither parent?

Assume independent assortment.

29. Most *Drosophila melanogaster* have brown bodies, but members of the species that are homozygous for the recessive allele *y* have yellow bodies. However, if larvae of pure *Y/Y* lines are reared on food containing silver salts, the resulting adults are yellow. These flies are called *phenocopies,* environmentally induced copies of genetically determined phenotypes. If you were presented with a single yellow fly, how would you determine if it was a yellow genotype or a yellow phenocopy? (Can you think of examples of phenocopies in humans?)

30. The accompanying pedigree shows the pattern of transmission of two rare human phenotypes: cataract and pituitary dwarfism. Family members with cataract are shown with a solid *left* half of the symbol; those with pituitary dwarfism are indicated by a solid *right* half.

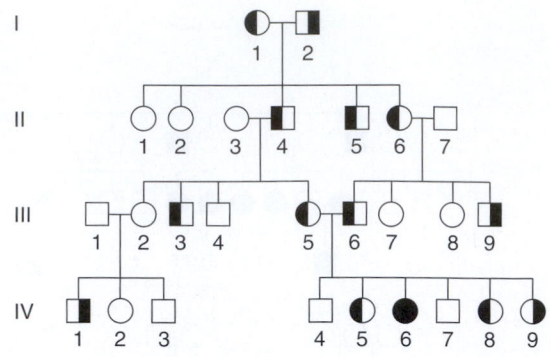

a. What is the most likely mode of inheritance of each of these phenotypes? Explain.

b. List the genotypes of all members in generation III as far as possible.

c. If a hypothetical mating took place between IV-1 and IV-5, what is the probability of the first child's being a dwarf with cataracts? A phenotypically normal child?

(Problem 30 after J. Kuspira and R. Bhambhani, *Compendium of Problems in Genetics.* Copyright © 1994 by Wm. C. Brown.)

31. A corn geneticist has three pure lines of genotypes a/a ; B/B ; C/C, A/A ; b/b ; C/C, and A/A ; B/B ; c/c. All the phenotypes determined by *a, b,* and *c* will increase the market value of the corn, so naturally he wants to combine them all in one pure line of genotype a/a ; b/b ; c/c.

a. Outline an effective crossing program that can be used to obtain the a/a ; b/b ; c/c pure line.

b. At each stage, state exactly which phenotypes will be selected and give their expected frequencies.

c. Is there more than one way to obtain the desired genotype? Which is the best way?

(Assume independent assortment of the three gene pairs. **Note:** Corn will self or cross-pollinate easily.)

32. **a.** A population of annual plants was composed exclusively of individuals of genotype *a/a*. One year, a flood introduced many seeds of genotype *A/A* and *A/a* into the population. Immediately after the introduction, there were 55% *A/A*, 40% *A/a*, and 5% *a/a* individuals. Because the region had no insects capable of cross-pollinating this species of plant, all plants routinely self-pollinated. After three generations of selfing, what were the proportions of *A/A*, *A/a*, and *a/a* genotypes?

b. Imagine a plant that has 40% of all its gene pairs heterozygous (*A/a, B/b, C/c,* and so forth). After three generations of selfing, what proportion of gene pairs will still be heterozygous?

c. How does the answer to part b relate to the answer to part a?

33. The recessive allele s causes *Drosophila* to have small wings and the s^+ allele causes normal wings. This gene is known to be X linked. If a small-winged male is crossed with a homozygous wild-type female, what ratio of normal to small-winged flies can be expected in each sex in the F_1? If F_1 flies are intercrossed, what F_2 progeny ratios are expected? What progeny ratios are predicted if F_1 females are backcrossed with their father?

34. Assuming the sex chromosomes to be identical, name the proportion of all alleles that you have in common with

   a. your mother

   b. your brother

35. A wild-type female schmoo who is graceful (G) is mated to a non-wild-type male who is gruesome (g). Their progeny consist solely of graceful males and gruesome females. Interpret these results and give genotypes. (Problem 35 from E. H. Simon and H. Grossfield, *The Challenge of Genetics.* Copyright 1971 by Addison-Wesley.)

36. A man with a certain disease marries a normal woman. They have eight children (four boys and four girls); all of the girls have their father's disease, but none of the boys do. What inheritance is suggested?

   a. autosomal recessive

   b. autosomal dominant

   c. Y linked

   d. X-linked dominant

   e. X-linked recessive

37. An X-linked dominant allele causes hypophosphatemia in humans. A man with hypophosphatemia marries a normal woman. What proportion of their sons will have hypophosphatemia?

38. A condition known as icthyosis hystrix gravior appeared in a boy in the early eighteenth century. His skin became very thick and formed loose spines that were sloughed off at intervals. When he grew up, this "porcupine man" married and had six sons, all of whom had this condition, and several daughters, all of whom were normal. For four generations, this condition was passed from father to son. From this evidence, what can you postulate about the location of the gene?

39. Duchenne's muscular dystrophy is sex linked and usually affects only males. Victims of the disease become progressively weaker, starting early in life.

   a. What is the probability that a woman whose brother has Duchenne's disease will have an affected child?

   b. If your mother's brother (your uncle) had Duchenne's disease, what is the probability that you have received the allele?

   c. If your father's brother had the disease, what is the probability that you have received the allele?

40. The following pedigree is concerned with an inherited dental abnormality, amelogenesis imperfecta.

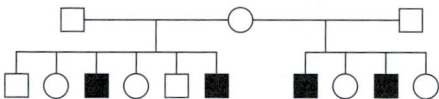

   a. What mode of inheritance *best* accounts for the transmission of this trait?

   b. Write the genotypes of all family members according to your hypothesis.

41. A sex-linked recessive allele c produces a red-green colorblindness in humans. A normal woman whose father was colorblind marries a colorblind man.

   a. What genotypes are possible for the mother of the colorblind man?

   b. What are the chances that the first child from this marriage will be a colorblind boy?

   c. Of the girls produced by these parents, what proportion can be expected to be colorblind?

   d. Of all the children (sex unspecified) of these parents, what proportion can be expected to have normal color vision?

42. Male house cats are either black or orange; females are black, orange, or calico.

   a. If these coat-color phenotypes are governed by a sex-linked gene, how can these observations be explained?

   b. Using appropriate symbols, determine the phenotypes expected in the progeny of a cross between an orange female and a black male.

   c. Repeat part b for the reciprocal of the cross described there.

   d. Half the females produced by a certain kind of mating are calico, and half are black; half the males are orange, and half are black. What colors are the parental males and females in this kind of mating?

   e. Another kind of mating produces progeny in the following proportions: $\frac{1}{4}$ orange males, $\frac{1}{4}$ orange females,

$\frac{1}{4}$ black males, and $\frac{1}{4}$ calico females. What colors are the parental males and females in this kind of mating?

43. A man is heterozygous B/b for one autosomal gene, and he carries a recessive X-linked allele d. What proportion of his sperm will be $b\,d$?

44. The accompanying pedigree concerns a certain rare disease that is incapacitating but not fatal.

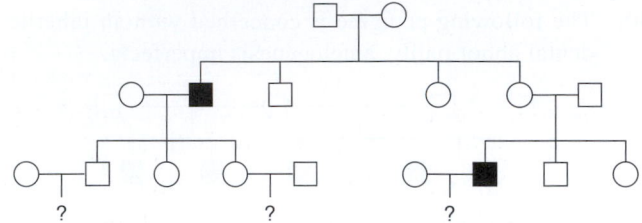

a. Determine the most likely mode of inheritance of this disease.

b. Write the genotype of each family member according to your proposed mode of inheritance.

c. If you were this family's doctor, how would you advise the three couples in the third generation about the likelihood of having an affected child?

45. Assume that this pedigree is straightforward, with no complications such as illegitimacy.

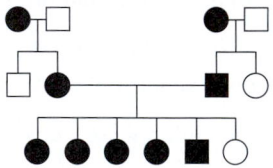

Phenotype W, found in the individuals represented by the shaded symbols, is rare in the general population. Which of the following patterns of transmission for W are consistent with this pedigree? Which are excluded?

a. autosomal recessive

b. autosomal dominant

c. X-linked recessive

d. X-linked dominant

e. Y linked

(Problem 45 from A. M. Srb, R. D. Owen, and R. S. Edgar, *General Genetics,* 2d ed. Copyright © 1965 by W. H. Freeman and Company.)

46. A mutant allele in mice causes a bent tail. Six pairs of mice were crossed. Their phenotypes and those of their progeny are given in the following table. N is normal phenotype; B is bent phenotype. Deduce the mode of inheritance of this phenotype.

| | PARENTS | | PROGENY | |
Cross	♀	♂	♀	♂
1	N	B	All B	All N
2	B	N	$\frac{1}{2}$ B, $\frac{1}{2}$ N	$\frac{1}{2}$ B, $\frac{1}{2}$ N
3	B	N	All B	All B
4	N	N	All N	All N
5	B	B	All B	All B
6	B	B	All B	$\frac{1}{2}$ B, $\frac{1}{2}$ N

a. Is it recessive or dominant?

b. Is it autosomal or sex linked?

c. What are the genotypes of all parents and progeny?

47. The normal eye color of *Drosophila* is red, but strains in which all flies have brown eyes are available. Similarly, wings are normally long, but there are strains with short wings. A female from a pure line with brown eyes and short wings is crossed with a male from a normal pure line. The F_1 consists of normal females and short-winged males. An F_2 is then produced by intercrossing the F_1. *Both* sexes of F_2 flies show phenotypes as follows:

$\frac{3}{8}$ red eyes, long wings

$\frac{3}{8}$ red eyes, short wings

$\frac{1}{8}$ brown eyes, long wings

$\frac{1}{8}$ brown eyes, short wings

Deduce the inheritance of these phenotypes, using clearly defined genetic symbols of your own invention. State the genotypes of all three generations and the genotypic proportions of the F_1 and F_2.

 Unpacking the Problem

Before attempting a solution to this problem, try answering the following questions.

1. What does the word "normal" mean in this problem?

2. The words "line" and "strain" are used in this problem. What do they mean and are they interchangeable?

3. Draw a simple sketch of the two parental flies showing their eyes, wings, and sexual differences.

4. How many different characters are there in this problem?

5. How many phenotypes are there in this problem, and which phenotypes go with which characters?

6. What is the full phenotype of the F_1 females called "normal"?

7. What is the full phenotype of the F_1 males called "short winged"?

8. List the F_2 phenotypic ratios for each character that you came up with in question d.

9. What do the F_2 phenotypic ratios tell you?

10. What major inheritance pattern distinguishes sex-linked inheritance from autosomal inheritance?

11. Do the F_2 data show such a distinguishing criterion?

12. Do the F_1 data show such a distinguishing criterion?

13. What can you learn about dominance in the F_1? the F_2?

14. What rules about wild-type symbolism can you use in deciding which allelic symbols to invent for these crosses?

15. What does "deduce the inheritance of these phenotypes" mean?

Now try to solve the problem. If you are unable to do so, make a list of questions about the things that you do not understand. Inspect the key concepts at the beginning of the chapter and ask yourself which are relevant to your questions. If this doesn't work, inspect the messages of this chapter and ask yourself which might be relevant to your questions.

48. The wild-type (W) *Abraxas* moth has large spots on its wings, but the lacticolor (L) form of this species has very small spots. Crosses were made between strains differing in this character, with the following results:

Cross	PARENTS		PROGENY	
	♀	♂	F_1	F_2
1	L	W	♀ W	♀ $\frac{1}{2}$ L, $\frac{1}{2}$ W
			♂ W	♂ *W*
2	W	L	♀ L	♀ $\frac{1}{2}$ W, $\frac{1}{2}$ L
			♂ W	♂ $\frac{1}{2}$ W, $\frac{1}{2}$ L

Provide a clear genetic explanation of the results in these two crosses, showing the genotypes of all individuals.

49. A certain gene that governs the activity of the enzyme glucose 6-phosphate dehydrogenase (G6PD) has two common alleles in Mediterranean and African populations. One allele stands for normal G6PD activity, and the other allele, which stands for reduced G6PD activity, confers resistance to malaria.

When the red blood cells of a certain African woman were examined under the microscope, precisely half the cells were found to contain the malarial parasite, whereas the other half appeared normal. Provide a genetic explanation for this finding.

50. Medical literature records the interesting case of a woman whose right breast was larger than the left, who had no pubic hair to the right of the midline, and who suffered from menstrual irregularities. On investigation of her family, it was discovered that a brother, a son, and a grandson showed testicular feminization syndrome. One of her daughters had three normal sons. Draw the pedigree from this information, determine if it fits the inheritance mode described in this chapter, and speculate on the cause of the symptoms in the propositus.

51. The pedigree below is for a rare human disease called spastic paraplegia, a nervous disorder in which there is an inability to coordinate voluntary movements.

a. What mode of inheritance is suggested by this pedigree?

b. Which family members must be heterozygous according to your model?

(Pedigree from V. A. McKusick, *On the Chromosomes of Man.* Copyright © 1964 by American Institute of Biological Science, Washington, D.C.)

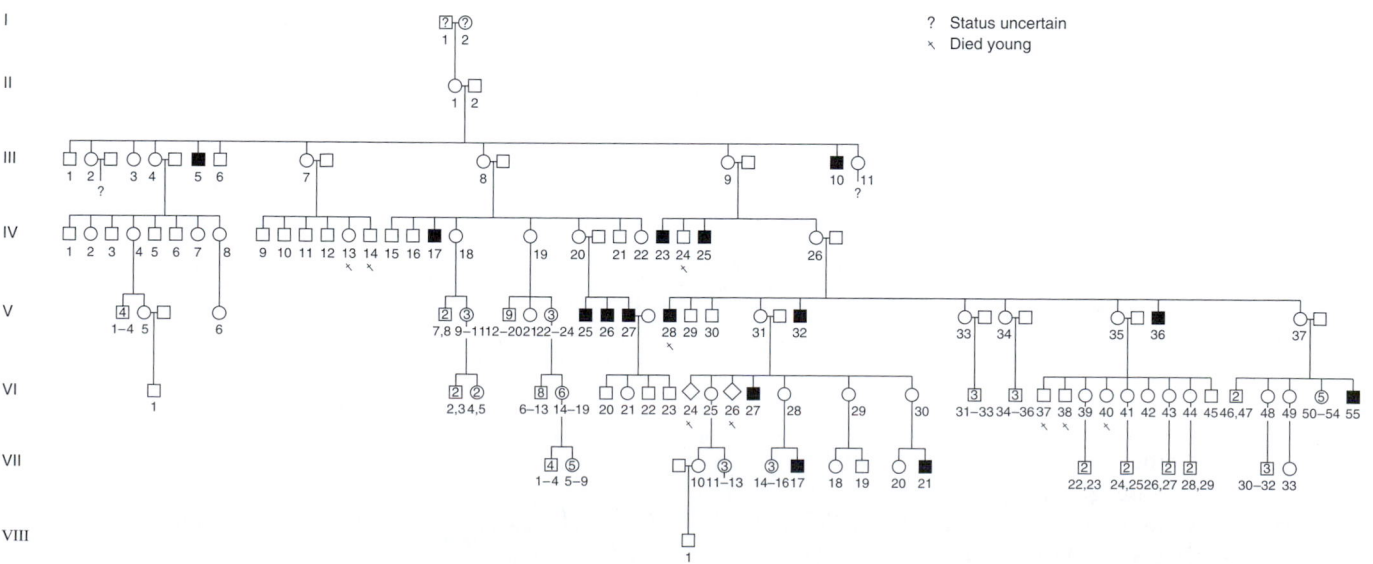

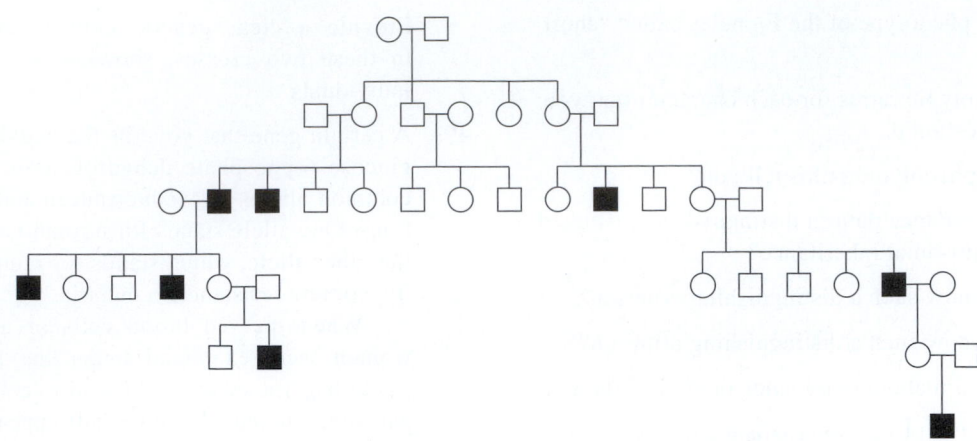

52. The pedigree above shows the inheritance of a rare human disease. Is the pattern best explained as being caused by an X-linked recessive allele or by an autosomal dominant allele with expression limited to males? (Pedigree modified from J. F. Crow, *Genetics Notes,* 6th ed. Copyright © 1967 by Burgers Publishing Company, Minneapolis.)

53. In humans, color vision depends on genes encoding three pigments. The *R* (red pigment) and *G* (green pigment) genes are on the X chromosome, whereas the *B* (blue pigment) gene is autosomal. A mutation in any one of these genes can cause colorblindness. Suppose that a colorblind man married a woman with normal color vision. All their sons were colorblind, and all their daughters were normal. Specify the genotypes of both parents and all possible children, explaining your reasoning. (A pedigree drawing will probably be helpful.) (Problem by Rosemary Redfield.)

54. A certain type of deafness in humans is inherited as an X-linked recessive. A man who suffers from this type of deafness marries a normal woman, and they are expecting a child. They find out that they are distantly related. Part of the family tree follows.

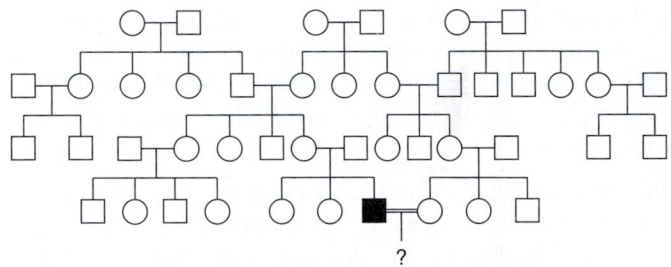

How would you advise the parents about the probability of their child being a deaf boy, a deaf girl, a normal boy, or a normal girl? Be sure to state any assumptions that you make.

3

CHROMOSOMAL BASIS OF HEREDITY

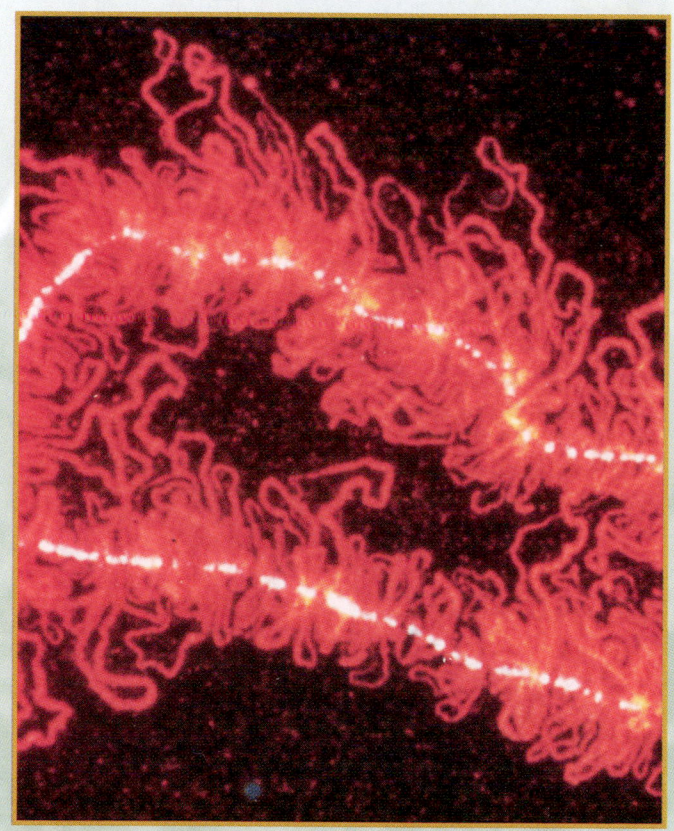

Lampbrush chromosomes.
The chromosomes of some animals take on this lampbrush appearance during meiotic diplotene in females. The lampbrush structure is thought to be reminiscent of the underlying organization of all chromosomes: a central scaffold (here stained brightly) and projecting lateral loops (stained red) formed by a folded continuous strand of DNA with associated histone proteins.
(M. Roth and J. Gall.)

Key Concepts

Genes are parts of chromosomes.

Mitosis is the nuclear division that results in two daughter nuclei whose genetic material is identical with that of the original nucleus.

Meiosis is the nuclear division by which a reproductive cell with two equivalent chromosome sets divides twice to produce four meiotic products, each of which has only one set of chromosomes.

Mendel's laws of equal segregation and independent assortment are based on the separation in meiosis of members of each chromosome pair and on the independent meiotic behavior of different chromosome pairs.

Chromosomes can be identified microscopically by using various visible landmarks.

A chromosome contains a single, long DNA molecule.

DNA winds around protein spools, and the spooled unit then coils, loops, and supercoils, forming a chromosome.

A large proportion of eukaryotic DNA is present in multiple copies.

Most of the multiple-copy DNA has no known function.

The beauty of Mendel's analysis is that it is not necessary to know what genes are or how they control phenotypes to analyze the results of crosses and to predict the outcomes of future crosses. All that is necessary is to apply the simple laws of equal segregation and independent assortment. We can do so by simply representing abstract, hypothetical factors of inheritance (genes) by symbols — without any concern about their molecular structures or their locations in a cell. Nevertheless, our interest naturally turns to the next questions: Where are genes located in the cell? and what is the precise way in which segregation and independent assortment are achieved at the cellular level?

Genetics took a major step forward with the notion that the genes, as characterized by Mendel, are parts of specific cellular structures, the chromosomes. This simple concept has become known as the **chromosome theory of heredity.** Although simple, the idea has had enormous implications, providing a means of correlating the results of breeding experiments with the behavior of structures that can be seen under the microscope. This fusion between genetics and cytology is still an essential part of genetic analysis today and has important applications in medical genetics, agricultural genetics, and evolutionary genetics. First, we shall consider the history of the idea.

Historical development of the chromosome theory

How did the chromosome theory take shape? Evidence accumulated from a variety of sources. One of the first lines of evidence came from observations of how chromosomes behave during the division of a cell's nucleus. In the interval between Mendel's research and its rediscovery, many biologists were interested in heredity even though they were unaware of Mendel's results, and they approached the problem in a completely different way. These investigators wanted to locate the hereditary material in the cell. An obvious place to look was in the gametes, because they are the only connecting link between generations. Egg and sperm were believed to contribute equally to the genetic endowment of offspring, even though they differ greatly in size. Because an egg has a great volume of cytoplasm and a sperm has very little, the cytoplasm of gametes seemed an unlikely seat of the hereditary structures. The nuclei of egg and sperm, however, were known to be approximately equal in size, so the nuclei were considered good candidates for harboring hereditary structures.

Evidence from nuclear division

What was known about the contents of cell nuclei? It became clear that the most prominent components were the chromosomes, which proved to possess unique properties that set them apart from all other cellular structures. A property that especially intrigued biologists was the constancy of the number of chromosomes from cell to cell within an or-

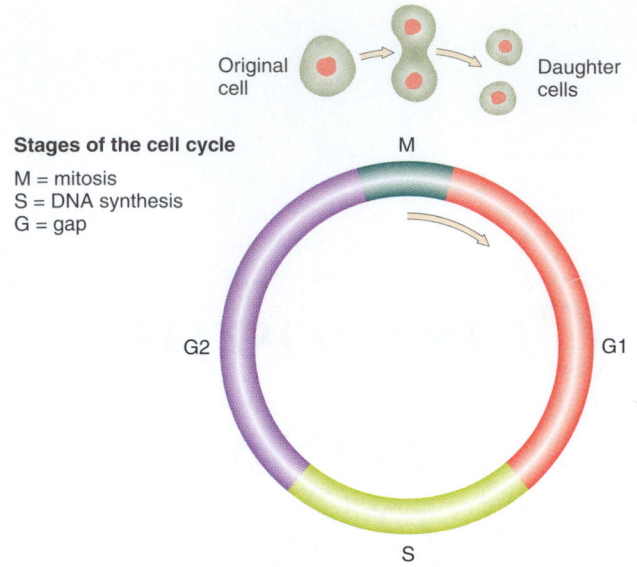

Stages of the cell cycle

M = mitosis
S = DNA synthesis
G = gap

Figure 3-1   Stages of the cell cycle.

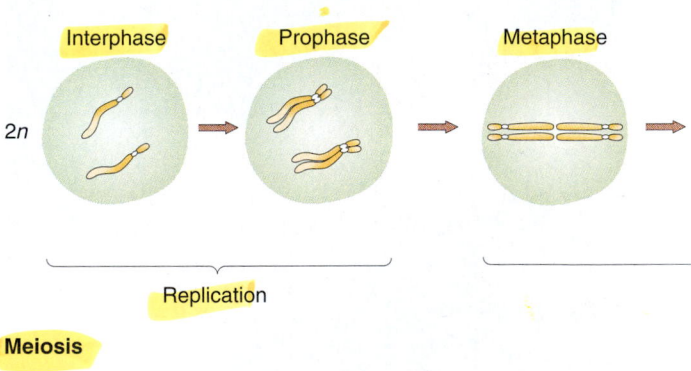

Mitosis

Interphase   Prophase   Metaphase

2n

Replication

Meiosis

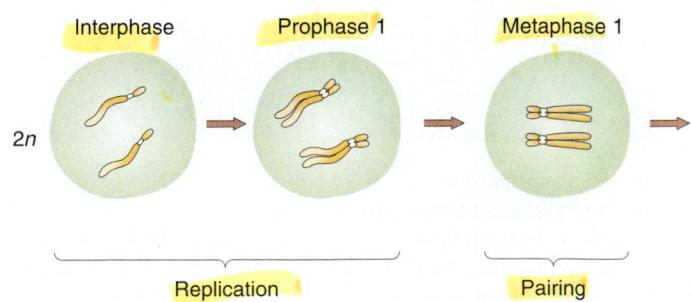

Interphase   Prophase 1   Metaphase 1

2n

Replication   Pairing

ganism, from organism to organism within any one species, and from generation to generation within that species. The question therefore arose: How is the chromosome number maintained? The question was answered by observing under the microscope the behavior of chromosomes during cell division. From those observations arose the hypothesis that chromosomes are the structures that contain the genes.

Mitosis is the nuclear division associated with the division of somatic cells — cells of the eukaryotic body that are not destined to become sex cells. The stages of the cell-division cycle (Figure 3-1) are similar in most organisms. The two basic parts of the cycle are **interphase** (comprising gap 1, synthesis, and gap 2) and mitosis. The key event in interphase takes place in the **S phase** (synthesis phase) in which the DNA of each chromosome replicates. As a result of DNA synthesis, each chromosome becomes two **sister chromatids,** which lie side by side. These sister chromatids cannot be seen during interphase but do become visible during **prophase,** an early stage of mitosis in which the chromosomes contract into a series of coils that are more easily

moved around. A simplified version of the main events of mitosis is shown in Figure 3-2. The next stage is **metaphase,** at which the sister chromatid pairs come to lie in the equatorial plane of the cell. In **anaphase,** the sister chromatids are pulled to opposite ends of the cell by microtubules that attach to the centromeres. The microtubules are part of the **nuclear spindle,** a set of parallel fibers running from one pole of the cell to the other. The pulling-apart process is complete at **telophase,** during which a nuclear membrane reforms around each nucleus, and the cell divides into two **daughter cells.** Each daughter cell inherits one of each pair of sister chromatids, which now become chromosomes in their own right. Thus this kind of division produces two genetically identical cells from a single progenitor cell. Successive cell divisions and the accompanying mitotic divisions result in a population of genetically identical cells. For example, mitosis is the type of division that allows a multicellular organism to be constructed from a single fertilized egg cell. From this simplified description, we see that the two fundamental processes of mitosis are *replication*

Figure 3-2   Simplified representation of mitosis and meiosis in diploid cells (2*n*, diploid; *n*, haploid). Detailed versions are shown in Figures 3-3 and 3-4).

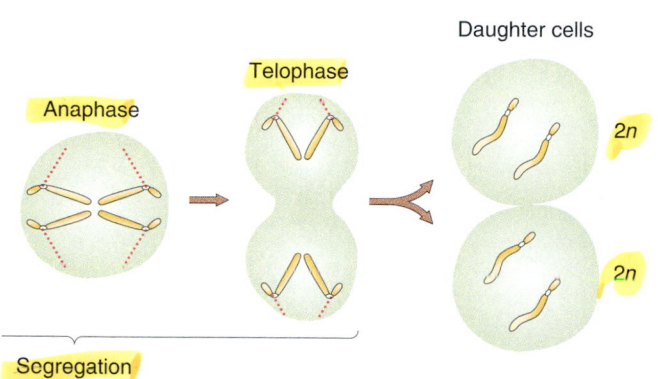

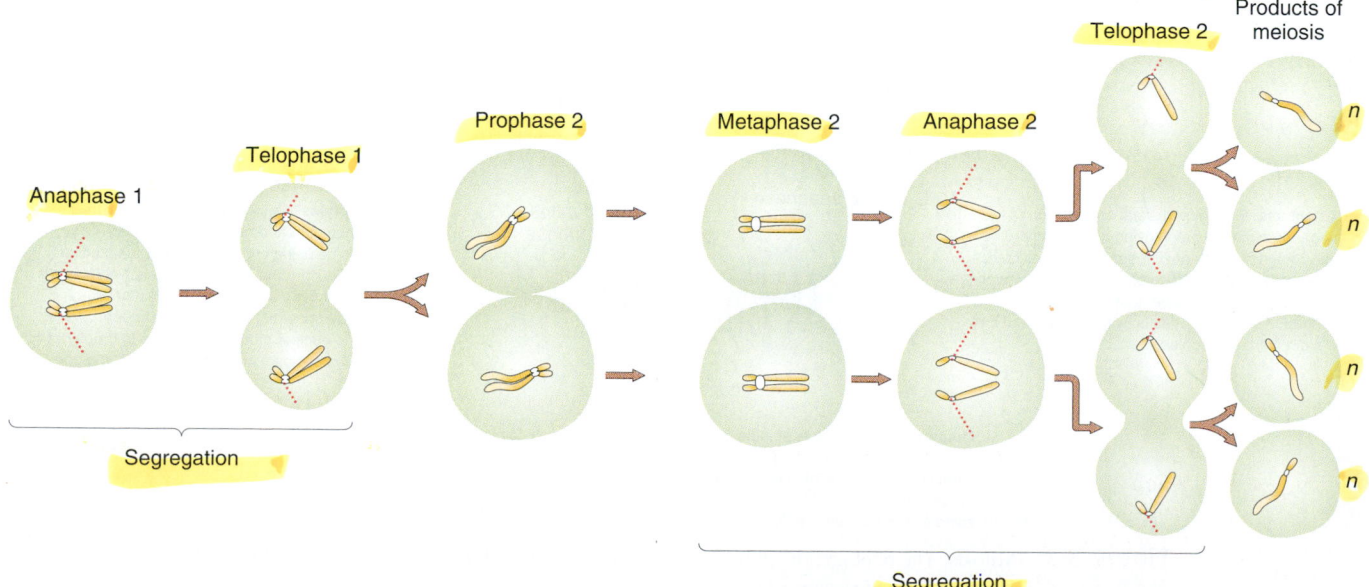

followed by *segregation.* Segregation is the name given to the separation of two homologous chromosomes or chromatids. (A complete description of mitosis in a plant is shown in Figure 3-3.)

Even though the early investigators did not know about DNA or that it is replicated during interphase, it was still evident that mitosis is the way in which the chromosome number is maintained during cell division. Thus the chromosomes seemed to be the natural candidates for the carriers of the genes. But there was still a puzzle concerning the joining of two gametes in the fertilization event. They knew that, in this process, two nuclei fuse but that the chromosome number nevertheless remains constant. What prevented the doubling of the chromosome number at each generation? This puzzle was resolved by the prediction of a special kind of nuclear division that *halved* the chromosome number. This special division, which was eventually discovered in the gamete-producing tissues of plants and animals, is called *meiosis.* A simplified representation of meiosis is shown in Figure 3-2.

Meiosis is the name given to the two successive nuclear divisions called meiosis I and meiosis II in special cells called **meiocytes.** The two meiotic divisions and their accompanying cell divisions give rise to a group of four cells that are called **products of meiosis.** In animals and plants, the products of meiosis become the haploid **gametes.** In humans and other animals, meiosis takes place in the gonads, and the products of meiosis are the gametes — sperm (more properly, spermatozoa) and eggs (ova). In flowering plants, meiosis takes place in the anthers and ovaries, and the products of meiosis are **meiospores,** which eventually give rise to gametes. Before meiosis, an S phase duplicates each chromosome's DNA to form sister chromatids, just as in mitosis. As in mitosis, the sister chromatids become visible in prophase I. However, in contrast with mitosis, the homologous chromosomes then pair (in metaphase I) to form groups of four chromatids called **tetrads.** Nonsister chromatids engage in a breakage and reunion process called **crossing over,** to be discussed in detail in Chapter 5. At anaphase I, each of the two pairs of sister chromatids is pulled into a different daughter nucleus. At anaphase II, the sister chromatids themselves are pulled into the daughter nuclei resulting from that division. We see therefore that the fundamental events of meiosis are DNA *replication,* followed by homologous *pairing,* by *segregation,* and then by another *segregation.* Hence, for each chromosomal type, the number of DNA molecules goes from $2 \rightarrow 4 \rightarrow 2 \rightarrow 1$, and each *product of meiosis* must therefore contain one chromosome of each type, half the number of the original meiocyte. (A detailed description of meiosis is given in Figure 3-4.)

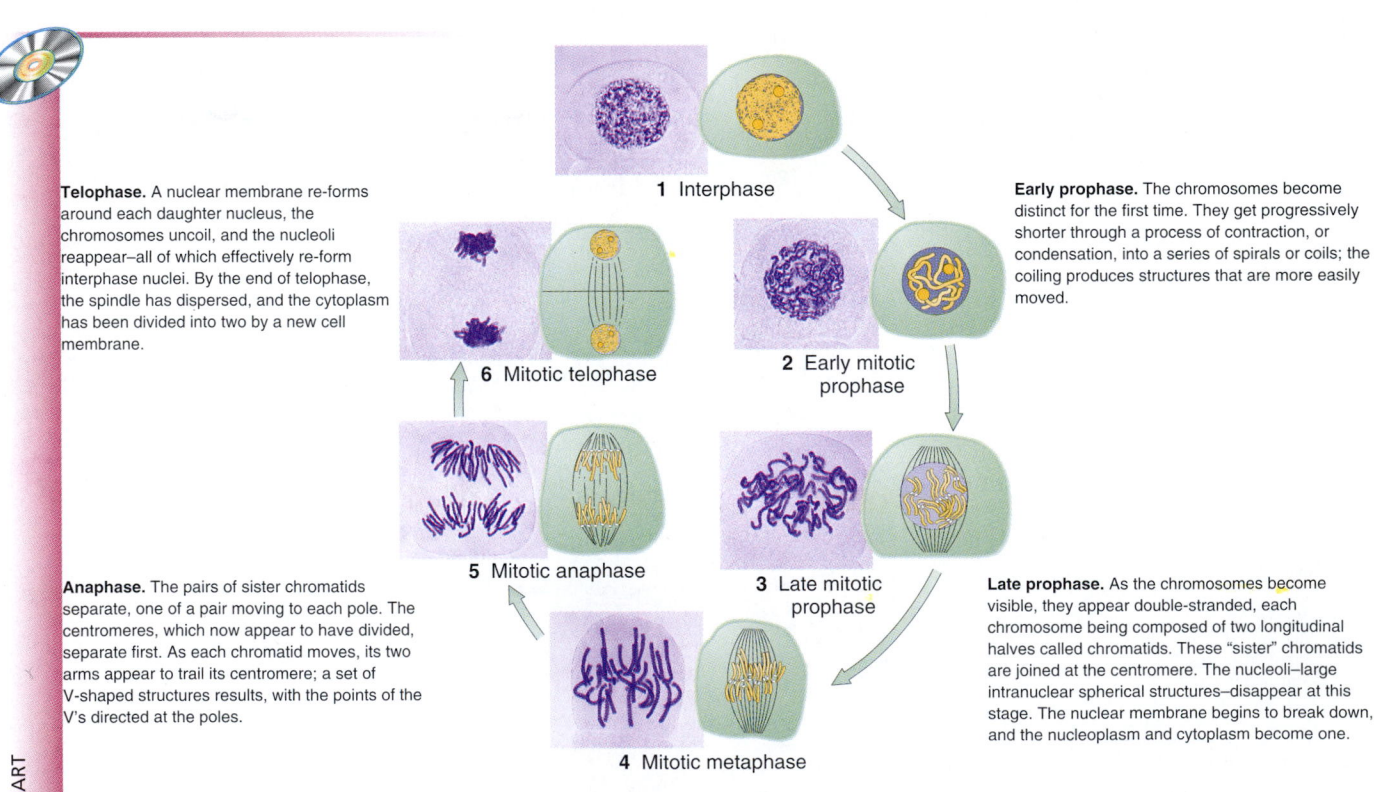

Telophase. A nuclear membrane re-forms around each daughter nucleus, the chromosomes uncoil, and the nucleoli reappear–all of which effectively re-form interphase nuclei. By the end of telophase, the spindle has dispersed, and the cytoplasm has been divided into two by a new cell membrane.

6 Mitotic telophase

1 Interphase

2 Early mitotic prophase

Early prophase. The chromosomes become distinct for the first time. They get progressively shorter through a process of contraction, or condensation, into a series of spirals or coils; the coiling produces structures that are more easily moved.

5 Mitotic anaphase

Anaphase. The pairs of sister chromatids separate, one of a pair moving to each pole. The centromeres, which now appear to have divided, separate first. As each chromatid moves, its two arms appear to trail its centromere; a set of V-shaped structures results, with the points of the V's directed at the poles.

3 Late mitotic prophase

Late prophase. As the chromosomes become visible, they appear double-stranded, each chromosome being composed of two longitudinal halves called chromatids. These "sister" chromatids are joined at the centromere. The nucleoli–large intranuclear spherical structures–disappear at this stage. The nuclear membrane begins to break down, and the nucleoplasm and cytoplasm become one.

4 Mitotic metaphase

Metaphase. The nuclear spindle becomes prominent. The spindle is a birdcage-like structure that forms in the nuclear area; it consists of a series of parallel fibers that point to each of two cell poles. The chromosomes move to the equatorial plane of the cell, where the centromeres become attached to a spindle fiber from each pole.

ANIMATED ART
Mitosis

Figure 3-3 Mitosis. The photographs show nuclei of root-tip cells of *Lilium regale.* (After J. McLeish and B. Snoad, *Looking at Chromosomes.* Copyright © 1958, St. Martin's, Macmillan.)

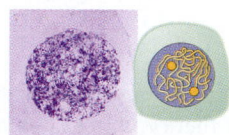

1 Leptotene

Prophase I: Leptotene. The chromosomes become visible as long, thin single threads. The process of chromosome contraction continues in leptotene and throughout the entire prophase. Small areas of thickening (chromomeres) develop along each chromosome, which give it the appearance of a necklace of beads.

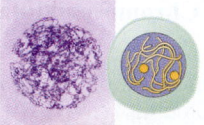

2 Zygotene

Prophase I: Zygotene. Active pairing of the threads makes it apparent that the chromosome complement of the meiocyte is in fact two complete chromosome sets. Thus, each chromosome has a pairing partner, and the two become progressively paired, or synapsed, side by side as if by a zipper.

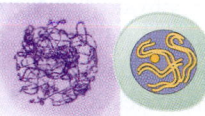

3 Pachytene

Prophase I: Pachytene. This stage is characterized by thick, fully synapsed chromosomes. Thus, the number of homologous pairs of chromosomes in the nucleus is equal to the number *n*. Nucleoli are often pronounced during pachytene. The beadlike chromomeres align precisely in the paired homologs, producing a distinctive pattern for each pair.

4 Diplotene

Prophase I: Diplotene. Although each homolog appeared to be a single thread in leptotene, the DNA had already replicated during the premeiotic S phase. This fact becomes manifest in diplotene as a longitudinal doubleness of each paired homolog. Hence, because each member of a homologous pair produces two sister chromatids, the synapsed structure now consists of a bundle of four homologous chromatids. At diplotene, the pairing between homologs becomes less tight; in fact, they appear to repel each other, and, as they separate slightly, cross-shaped structures called chiasmata (singular, chiasma) appear between nonsister chromatids. Each chromosome pair generally has one or more chiasmata.

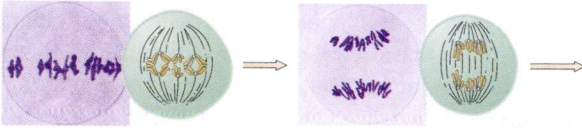

5 Diakinesis   **6 Metaphase I**

Prophase I. Diakinesis. This stage differs only slightly from diplotene, except for further chromosome contraction. By the end of diakinesis, the long, filamentous chromosome threads of interphase have been replaced by compact units that are far more maneuverable in the movements of the meiotic division.

Metaphase I. The nuclear membrane and nucleoli have disappeared by metaphase I, and each pair of homologs takes up a position in the equatorial plane. At this stage of meiosis, the centromeres do not divide; this lack of division is a major difference from mitosis. The two centromeres of a homologous chromosome pair attach to spindle fibers from opposite poles.

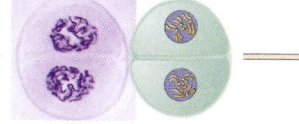   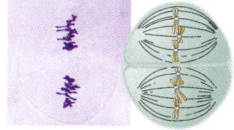   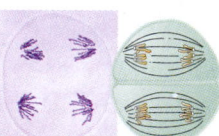

7 Early anaphase I   **8 Later anaphase I**   **9 Telophase I**   **10 Interphase**

Anaphase I. Anaphase begins when chromosomes move directionally to the poles. The members of a homologous pair move to opposite poles.

Telophase I. Telophase and the ensuing interphase, called interkinesis, are not universal. In many organisms, these stages do not exist, no nuclear membrane re-forms, and the cells proceed directly to meiosis II. In other organisms, telophase I and the interkinesis are brief in duration; the chromosomes elongate and become diffuse, and the nuclear membrane re-forms. In any case, there is never DNA synthesis at this time, and the genetic state of the chromosomes does not change.

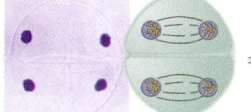   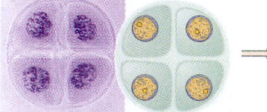   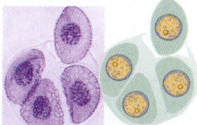

11 Prophase II   **12 Metaphase II**   **13 Anaphase II**

Prophase II. The presence of the haploid number of chromosomes in the contracted state characterizes prophase II.

Metaphase II. The chromosomes arrange themselves on the equatorial plane in metaphase II. Here the chromatids often partly dissociate from each other instead of being closely appressed as they are in mitosis.

Anaphase II. Centromeres split and sister chromatids are pulled to opposite poles by the spindle fibers.

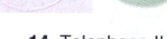

14 Telophase II   **15 The tetrad   16 Young pollen grains**

Telophase II. The nuclei re-form around the chromosomes at the poles.

In the anthers of a flower, the four products of meiosis develop into pollen grains. In other organisms, differentiation produces other kinds of structures from the products of meiosis, such as sperm cells in animals.

Figure 3-4   Meiosis and pollen formation. The photographs are of *Lilium regale.* **Note:** For simplicity, multiple chiasmata are drawn between only two chromatids; in reality, all four chromatids can take part. (After J. McLeish and B. Snoad, *Looking at Chromosomes.* Copyright © 1958, St. Martin's, Macmillan.)

MESSAGE ·

In mitosis, each chromosome replicates to form sister chromatids, which segregate into the daughter cells. *In meiosis,* each chromosome replicates to form sister chromatids. Homologous chromosomes physically pair and segregate at the first division. Sister chromatids segregate at the second division.

Credit for the chromosome theory of heredity — the concept that the invisible and hypothetical entities called genes are parts of the visible structures called chromosomes — is usually given to both Walter Sutton (an American who at the time was a graduate student) and Theodor Boveri (a German biologist). In 1902, these investigators recognized independently that the behavior of Mendel's particles during the production of gametes in peas precisely parallels the behavior of chromosomes at meiosis: genes are in pairs (so are chromosomes); the alleles of a gene segregate equally into gametes (so do the members of a pair of homologous chromosomes); different genes act independently (so do different chromosome pairs). After recognizing this parallel behavior (which is summarized in Figure 3-5), both investigators reached the same conclusion that the parallel behavior of genes and chromosomes suggests that genes are located on chromosomes.

Thus, observations on both mitosis and meiosis seemed to point to the same conclusion. To modern biology students, the chromosome theory may not seem very earthshaking. However, early in the twentieth century, the Sutton-Boveri hypothesis (which potentially united cytology and the infant field of genetics) was a bombshell. The first response to publication of the hypothesis was to try to pick holes in it. For years after, there was a raging controversy over the validity of the chromosome theory of heredity.

It is worth considering some of the objections raised to the Sutton-Boveri theory. For example, at the time, chromosomes could not be detected at interphase (the stage between cell divisions). Boveri had to make some very detailed studies of chromosome position before and after interphase before he could argue persuasively that chromosomes retain their physical integrity through interphase, even though they are cytologically invisible at that time. It was also pointed out that, in some organisms, several different pairs of chromosomes look alike, making it impossible to say from visual observation that they are not all pairing randomly, whereas Mendel's laws absolutely require the orderly pairing and segregation of alleles. However, in species in which chromosomes do differ in size and shape, it was verified that chromosomes come in pairs and that these physically pair and segregate in meiosis.

In 1913, Elinor Carothers found an unusual chromosomal situation in a certain species of grasshopper — a situation that permitted a direct test of whether different chromosome pairs do indeed segregate independently. Studying grasshopper testes, she found one case in which there was a chromosome pair that had nonidentical members; such a pair is called a heteromorphic pair, and the chromosomes presumably show only partial homology. In addition, she found that another chromosome, unrelated to the heteromorphic pair, had no pairing partner at all. Carothers was able to use these unusual chromosomes as visible cytological markers of the behavior of chromosomes during assortment. By looking at anaphase nuclei, she could count the number of times that each dissimilar chromosome of the heteromorphic pair migrated to the same pole as the chromosome with no pairing partner (Figure 3-6). She observed the two patterns of chromosome behavior with equal frequency. Although these unusual chromosomes are not typical, the re-

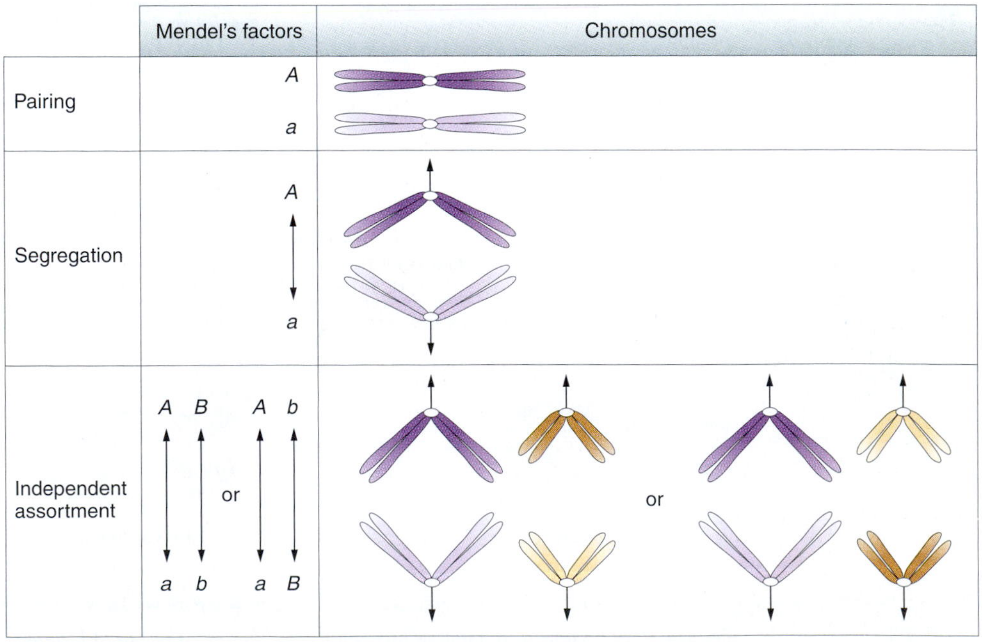

Figure 3-5 Parallels in the behavior of Mendel's genes and in that of chromosomes during meiosis. A dark shade represents one member of a homologous pair; a light shade represents the other member.

	Mendel's factors	Chromosomes
Pairing	*A* *a*	
Segregation	*A* ↕ *a*	
Independent assortment	*A B A b* ↑ ↑ or ↑ ↑ ↓ ↓ ↓ ↓ *a b a B*	or

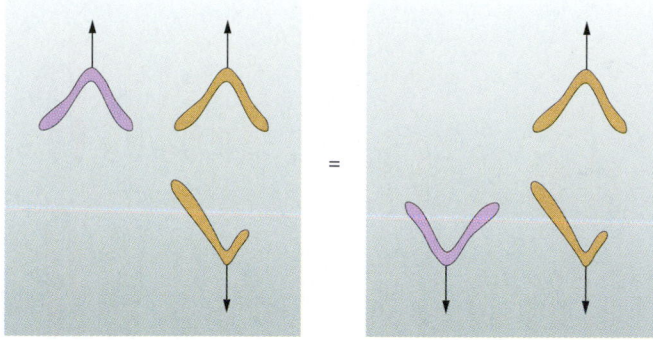

Figure 3-6 Two equally frequent patterns by which a heteromorphic pair and an unpaired chromosome move into gametes, as observed by Carothers.

Normal

Rolled   Glossy   Buckling   Elongate

Echinus   Cocklebur   Microcarpic   Reduced

Poinsettia   Spinach   Glove   Ilex

Figure 3-7 Fruits from *Datura* plants, each having one different extra chromosome. Their characteristic appearances suggest that each chromosome is different. (From E. W. Sinnott, L. C. Dunn, and T. Dobzhansky, *Principles of Genetics,* 5th ed. McGraw-Hill Book Company, New York.)

sults do suggest that nonhomologous chromosomes assort independently.

Other investigators argued that, because all chromosomes appear as stringy structures, qualitative differences between them are of no significance. It was suggested that perhaps all chromosomes were just more or less made of the same stuff. It is worth introducing a study out of historical sequence that effectively counters this objection. In 1922, Alfred Blakeslee performed a study on the chromosomes of jimsonweed *(Datura stramonium),* which has 12 chromosome pairs. He obtained 12 different strains, each of which had the normal 12 chromosome pairs plus an extra representative of one pair. Blakeslee showed that each strain was phenotypically distinct from the others (Figure 3-7). This result would not be expected if there were no genetic differences between the chromosomes.

All these results indicated that the behavior of chromosomes closely parallels that of genes. This indication made the Sutton-Boveri theory attractive, but there was as yet no real proof that genes are located on chromosomes. The argument was just based on correlation. Further observations, however, did provide such proof, and they began with the discovery of sex linkage.

Evidence from sex linkage

Most of the early crosses analyzed gave the same results in reciprocal crosses, as shown by Mendel. The first exception to this pattern was discovered in 1906 by L. Doncaster and G. H. Raynor. They were studying wing color in the magpie moth *(Abraxas),* by using two different lines: one with light wings and the other with dark wings. If light-winged females are crossed with dark-winged males, all the progeny have dark wings, showing that the allele for light wings is recessive. However, in the reciprocal cross (dark female × light male), all the female progeny have light wings and all the male progeny have dark wings. Thus, this pair of reciprocal crosses does not give similar results, and the wing phenotypes in the second cross are associated with the sex of the moths. Note that the female progeny of this second cross are phenotypically similar to their fathers, as the males are to their mothers. Let's consider another similar example.

William Bateson had been studying the inheritance of feather pattern in chickens. One line had feathers with alternating stripes of dark and light coloring, a phenotype called barred. Another line, nonbarred, had feathers of uniform coloring. In the cross barred male × nonbarred female, all the progeny were barred, showing that the allele for non-barred is recessive. However, the reciprocal cross (barred female × nonbarred male) gave barred males and nonbarred females. Again, the result is that female progeny have their father's phenotype and male progeny have their mother's.

The explanation for such results came from the laboratory of Thomas Hunt Morgan, who in 1909 began studying inheritance in a fruit fly *(Drosophila melanogaster).* The life cycle of *Drosophila* is typical of the life cycles of many insects. The flies grow vigorously in the laboratory. The choice of *Drosophila* as a research organism was a very fortunate one for geneticists — and especially for Morgan, whose work earned him a Nobel Prize in 1934.

The normal eye color of *Drosophila* is dull red. Early in his studies, Morgan discovered a male with completely white eyes (see Figure 2-14). He found a difference between reciprocal crosses and an association of different phenotypic ratios in different sexes of progeny, as discussed in Chapter 2. This result was very similar to the outcomes in the examples of chickens and moths, but there was a difference: in chickens and moths, the progeny are like the parent of opposite sex when the parental males carry the recessive alleles; in the *Drosophila* cross, this outcome is seen when the female parents carry the recessive alleles.

Before turning to Morgan's explanation of the *Drosophila* results, we should look at some of the cytological information that he was able to use in his interpretations. In 1891, working with males of a species of Hemiptera (the true bugs), H. Henking observed that meiotic nuclei contain 11 pairs of chromosomes and an unpaired element that moved to one of the poles in the first meiotic division. Henking called this unpaired element an "X body"; he interpreted it as a nucleolus, but later studies showed it to be a chromosome. Similar unpaired elements were later found in other species. In 1905, Edmond Wilson noted that females of *Protenor* (another Hemipteran) have six pairs of chromosomes, whereas males have five pairs and an unpaired chromosome, which Wilson called (by analogy) the X chromosome. The females, in fact, have a pair of X chromosomes.

Also in 1905, Nettie Stevens found that males and females of the beetle *Tenebrio* have the same number of chromosomes, but one of the chromosome pairs in males is heteromorphic. One member of the heteromorphic pair appears identical with the members of a pair in the female; Stevens called this the *X* chromosome. The other member of the heteromorphic pair is never found in females; Stevens called this the *Y* chromosome (Figure 3-8). She found a similar situation in *Drosophila melanogaster,* which has four pairs of chromosomes, with one of the pairs being heteromorphic in males. Figure 3-9 summarizes the two basic situations: the

extra unpaired chromosome and the heteromorphic pair. (You may be wondering about the male grasshoppers studied by Carothers that had both a heteromorphic chromosome pair and an unpaired chromosome. This situation is very unusual, and we needn't worry about it at this point.)

With this background information, Morgan constructed an interpretation of his genetic data. First, it appeared that the X and Y chromosomes determine the sex of the fly. *Drosophila* females have four chromosome pairs, whereas males have three matching pairs plus a heteromorphic pair. Thus, meiosis in the female produces eggs that each bear one X chromosome. Although the X and Y chromosomes in males are heteromorphic, as mentioned in Chapter 2, they seem to pair and segregate like homologs (Figure 3-10). Thus, meiosis in the male produces two types of sperm, one type bearing an X chromosome and the other bearing a Y chromosome.

Morgan next turned to the problem of eye color. He postulated that the alleles for red or white eye color are present on the X chromosome but that there is no counterpart for this gene on the Y chromosome. Thus, females would have two alleles for this gene, whereas males would have only one. The genetic results were completely consistent with the known meiotic behavior of the X and Y chromosomes (as described in Chapter 2). This experiment strongly supports the notion that genes are located on chro-

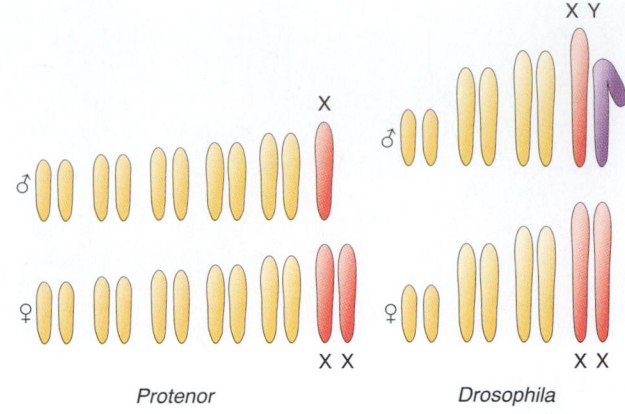

Figure 3-9   The chromosomal constitutions of males and females in two insect species.

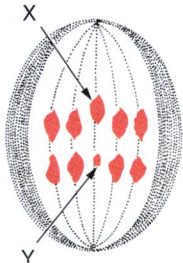

Figure 3-8   Segregation of the heteromorphic chromosome pair (X and Y) in meiosis in a *Tenebrio* male. The X and Y chromosomes are being pulled to opposite poles during anaphase I. (From A. M. Srb, R. D. Owen, and R. S. Edgar, *General Genetics,* 2d ed. Copyright © 1965 by W. H. Freeman and Company.)

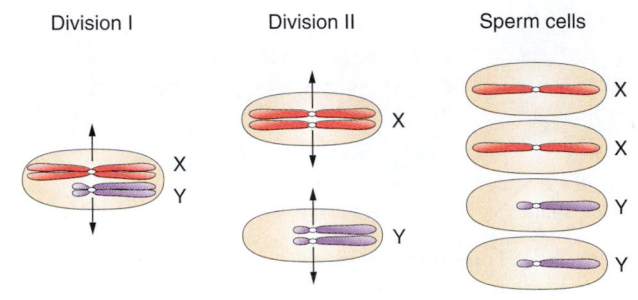

Figure 3-10   Meiotic pairing and the segregation of the X and Y chromosomes into equal numbers of sperm.

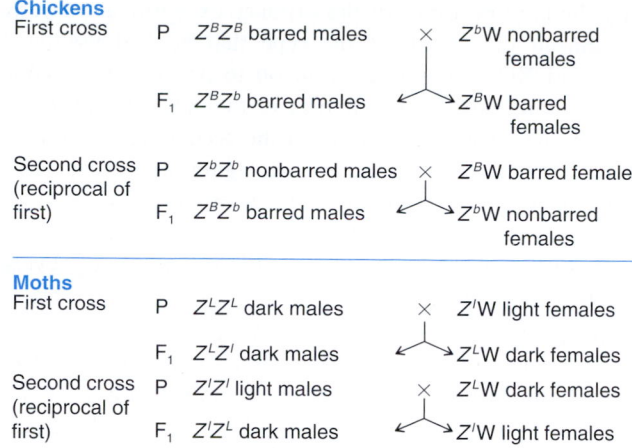

Figure 3-11 Inheritance pattern of genes on the sex chromosomes of two species having the WZ mechanism of sex determination.

mosomes. However, again it is only a correlation; it does not constitute a definitive proof of the Sutton-Boveri theory.

Can the same XX and XY chromosome theory be applied to the results of the earlier crosses made with chickens and moths? You will find that it cannot. However, Richard Goldschmidt recognized immediately that these results can be explained with a similar hypothesis, making the simple assumption that, in this case, the *males* have pairs of identical chromosomes, whereas the females have the different pair. To distinguish this situation from the XY situation in *Drosophila,* Morgan suggested that the sex chromosomes in chickens and moths be called W and Z, with males being ZZ and females being WZ. Thus, if the genes in the chicken and moth crosses are on the Z chromosome, the crosses can be diagrammed as shown in Figure 3-11. The interpretation is consistent with the genetic data. In this case, cytological data provided a confirmation of the genetic hypothesis. In 1914, J. Seiler verified that both chromosomes are identical in all pairs in male moths, whereas females have one different pair.

MESSAGE ··································

> The special inheritance pattern of some genes makes it likely that they are borne on the sex chromosomes, which show a parallel pattern of inheritance.

An aside on genetic symbols

In *Drosophila,* a special symbolism for allele designation was introduced to define variant alleles in relation to a "normal" allele. This system is now used by many geneticists and is especially useful in genetic dissection. For a given *Drosophila* character, the allele that is found most frequently in natural populations (or, alternatively, the allele that is found in standard laboratory stocks) is designated as the standard, or **wild type.** All other alleles are then mutant alleles. The symbol for a gene comes from the first mutant allele found. In Morgan's *Drosophila* experiment, this allele was that for white eyes, symbolized by w. The wild-type counterpart allele is conventionally represented by adding a

superscript plus sign; so the normal red-eye-determining allele is written w^+.

In a polymorphism, several alleles might be common in nature and all might be regarded as wild type. In this case, the alleles can be distinguished by using superscripts. For example, populations of *Drosophila* show a dimorphism for forms of the enzyme alcohol dehydrogenase. These two forms move at different speeds on an electrophoretic gel. The alleles coding for these forms are designated Adh^F (fast) and Adh^S (slow).

The wild-type allele *can be dominant or recessive* to a mutant allele. For the two alleles w^+ and w, the use of the lowercase letter indicates that the wild-type allele is dominant over the one for white eyes (that is, w is recessive to w^+). As another example, this one concerning the character wing shape, the wild-type phenotype of a fly's wing is straight and flat, and a mutant allele causes the wing to be curled. Because the latter allele is dominant over the wild-type allele, it is written Cy (short for *Curly*), whereas the wild-type allele is written Cy^+. Here note that the capital letter indicates that Cy is dominant over Cy^+. (Also note from these examples that the symbol for a single gene may consist of more than one letter.)

A critical test of the chromosome theory

The correlations between the behavior of genes and the behavior of chromosomes made it very likely that genes are parts of chromosomes. But these correlations were not a critical test of the chromosome theory, and debate continued. The critical analysis came from one of Morgan's students, Calvin Bridges. Bridges's work began with a fruit fly cross that we have considered before. Letting the capital letters X and Y represent the X and Y chromosomes, we can write the parental genotypes by using our new symbolism as X^wX^w (white eyed) $\times$ $X^{w^+}Y$ (red eyed). We know that the expected progeny are $X^{w^+}X^w$ (red-eyed females) and X^wY (white-eyed males). When Bridges made the cross on a large scale, he observed a few exceptions among the progeny. About 1 of every 2000 F_1 progeny was a white-eyed female or a red-eyed male. Collectively, these individuals were called *primary exceptional progeny*. All the primary exceptional males proved to be sterile. However, when Bridges crossed the primary exceptional white-eyed females with normal red-eyed males, in addition to the expected red-eyed female and white-eyed male progeny, a higher proportion of exceptional progeny was observed, 4 percent white-eyed females and red-eyed males that were fertile. These exceptional progeny of primary exceptional mothers were called *secondary exceptional progeny* (Figure 3-12). How did Bridges explain these types of exceptional progeny?

The exceptional females—which, like all females, have two X chromosomes—must get both of these chromosomes from their mothers because the exceptional females are homozygous for w. Similarly, exceptional males must get their X chromosomes from their fathers because these chromosomes

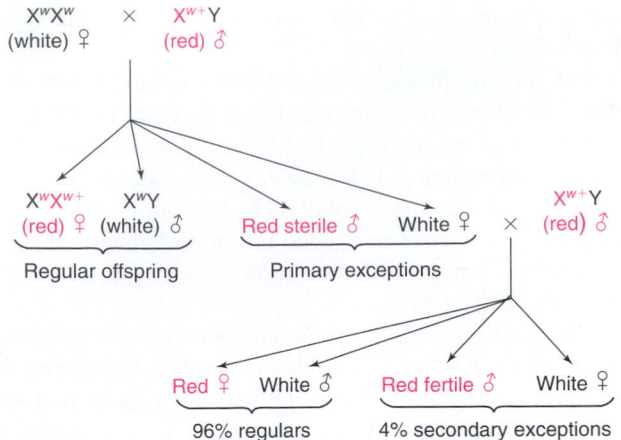

Figure 3-12 *Drosophila* crosses from which primary and secondary exceptional progeny were originally obtained. Red, red-eyed; white, white-eyed *Drosophila*.

carry w^+. Bridges hypothesized rare mishaps in the course of meiosis in the female whereby the paired X chromosomes failed to separate in either the first or the second division. This failure would result in meiotic nuclei containing either two X chromosomes or no X at all. Such a failure to separate is called **nondisjunction.** Fertilization of eggs having these types of nuclei by sperm from a wild-type male produces four zygotic classes (Figure 3-13). It is important to note that the representation of a chromosome in these diagrams is in each case not a single chromosome but a pair of daughter chromatids.

Bridges assumed that XXX and YO zygotes die before development is complete, so the two types of viable exceptional progeny are expected to be $X^w X^w Y$ (white-eyed female) and $X^{w+}O$ (red-eyed sterile male). What about the sterility of the primary exceptional males? This sterility makes sense if we assume that a male must have a Y chromosome to be fertile.

To summarize, Bridges explained the primary exceptional progeny by postulating rare abnormal meioses that gave rise to viable XXY females and XO males. He tested this model in several ways. First, he examined microscopi-

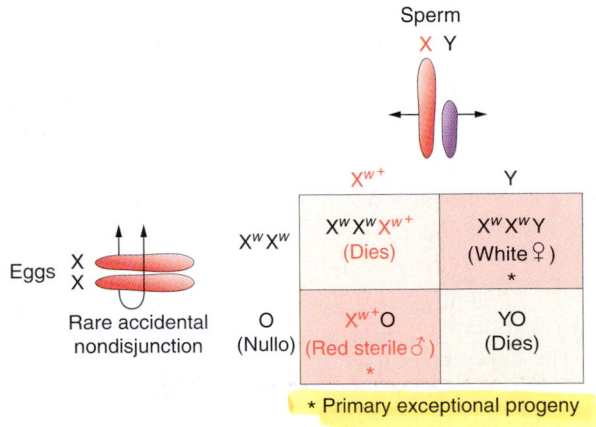

Figure 3-13 Proposed explanation of primary exceptional progeny through nondisjunction of the X chromosomes in the maternal parent. Red, red-eyed; white, white-eyed *Drosophila*.

cally the chromosomes of the primary exceptional progeny, and indeed they were of the type that he had predicted, XXY and XO. Second, he went on to predict the possible chromosome pairings in meiosis in the XXY females and, from these pairings, the nature of the secondary exceptional progeny that arose from them. Once again, microscopy confirmed that all his predictions were correct. Therefore, by assuming the chromosomal location of the eye-color gene, Bridges was able to accurately predict the nature of several unusual genetic processes.

MESSAGE ....
When Bridges used the chromosome theory to successfully predict the nature of certain genetic oddities, the chromosome location of genes was established beyond reasonable doubt.
....

Mendelian genetics in eukaryotic life cycles

So far, we have mainly been considering diploid organisms — organisms with two homologous chromosome sets in each cell. As we have seen, the diploid condition is designated $2n$, where n stands for the number of chromosomes in one chromosome set. For example, the pea cell contains two sets of seven chromosomes, so $2n = 14$. The organisms that we encounter most often in our daily lives (animals and flowering plants) are diploid in most of their tissues.

However, a large part of the biomass on the earth comprises organisms that spend most of their life cycles in a haploid condition, in which each cell has only one set of chromosomes. Important examples are most fungi and algae. Bacteria could be considered haploid, but they form a special case because they do not have chromosomes of the type that we have been considering. (Bacterial cycles are discussed in Chapter 10.) Also important are organisms that are haploid for part of their life cycles and diploid for another part. Such organisms are said to show **alternation of generations,** referring to the alternation of the $2n$ and n stages. All types of plants show alternation of generations; however, the haploid stage of flowering plants and conifers is an inconspicuous specialized structure dependent on the diploid part of the plant. Other types of plants, such as mosses and ferns, have independent haploid stages.

Do all these various life cycles show Mendelian genetics? The answer is that Mendelian inheritance patterns characterize any species that has meiosis as part of its life cycle, because Mendelian laws are based on the process of meiosis. All the groups of organisms mentioned, except bacteria, undergo meiosis as part of their cycles.

Diploids

Figure 3-14 summarizes the diploid cycle, the cycle of most animals (including humans). The adult body is composed of diploid cells, and meiosis takes place in specialized diploid cells, the meiocytes, that are found in the gonads (testes and ovaries). The products of meiosis are the gametes (eggs or

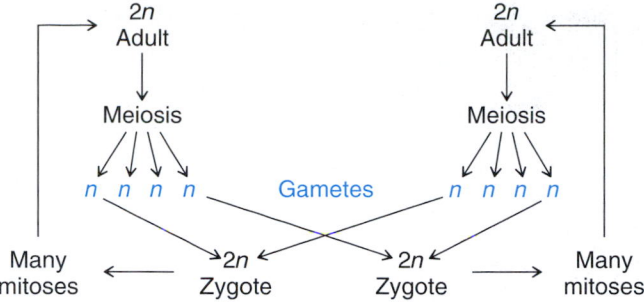

Figure 3-14 The diploid life cycle.

quency of each of the four genotypes is $\frac{1}{4}$. This gametic distribution is that postulated by Mendel for a dihybrid, and it is the one that we insert along one edge of the Punnett square. The random fusion of these gametes results in the $9 : 3 : 3 : 1$ F_2 phenotypic ratio.

The diagram in Figure 3.15 shows exactly why chromosomal behaviors produce the Mendelian ratios. Notice that Mendel's first law (equal segregation) describes what happens to a pair of alleles when the pair of homologs that carry them separate into opposite cells at the first meiotic division. Notice also that Mendel's second law (independent assortment) results from the independent segregation of

sperm). Fusion of haploid gametes forms a diploid zygote, which, through mitosis, produces a multicellular organism.

Meiosis in a dihybrid is shown in Figure 3-15. The genotype of the cell is A/a ; B/b, and the two allelic pairs, A/a and B/b, are shown on two different chromosome pairs. The hypothetical cell has four chromosomes: a pair of homologous long chromosomes and a pair of homologous short ones. Such size differences between pairs are common.

Parts 4 and 4′ of Figure 3-15 show that two equally frequent spindle attachments to the centromeres in the first anaphase result in two different allelic segregation patterns. Meiosis then produces four cells of the genotypes shown from each of these segregation patterns. Because segregation patterns 4 and 4′ are equally common, the meiotic product cells of genotypes A ; B, a ; b, A ; b, and a ; B are produced in equal frequencies. In other words, the frequency of each of the four genotypes is

Figure 3-15 Meiosis in a diploid cell of genotype A/a ; B/b, showing how the segregation and assortment of different chromosome pairs give rise to the $1 : 1 : 1 : 1$ Mendelian gametic ratio.

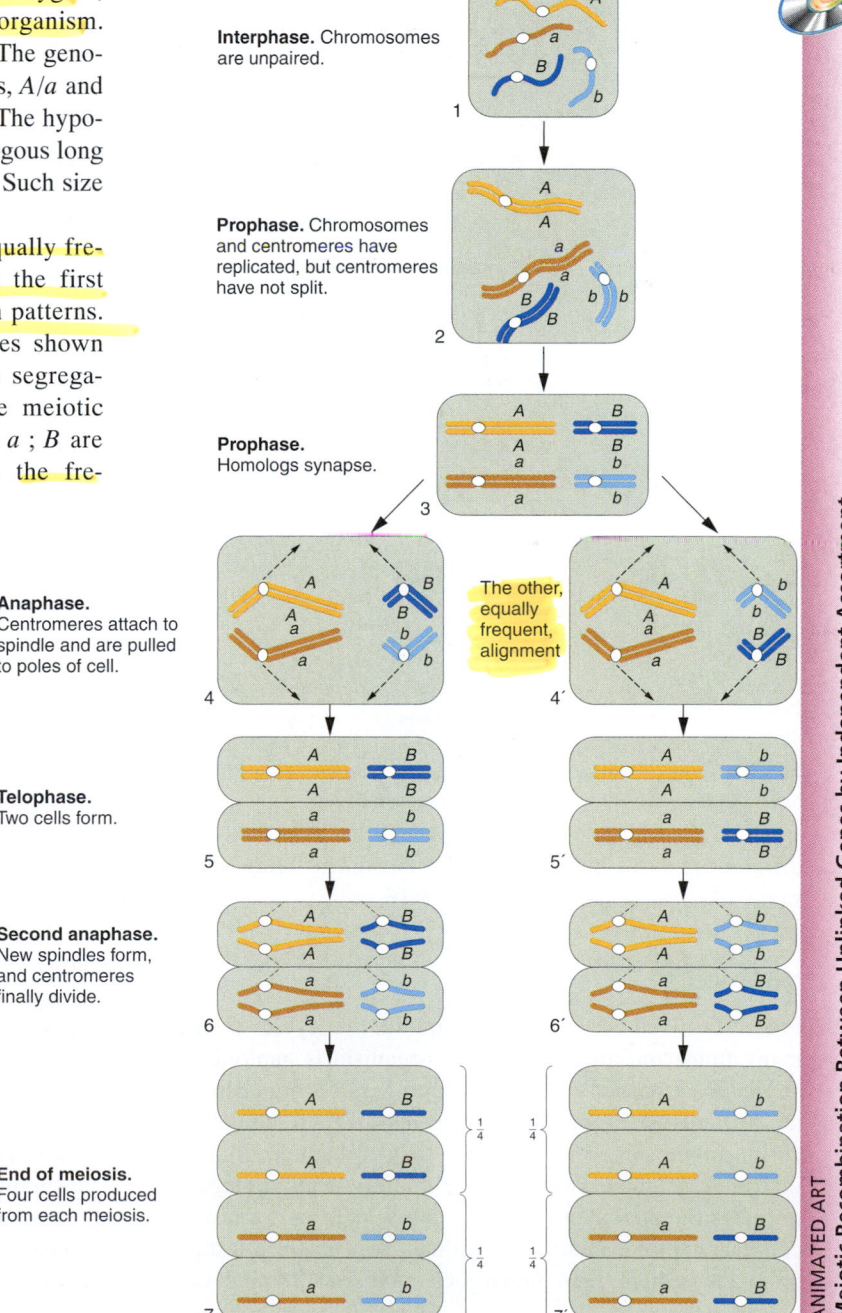

ANIMATED ART
Meiotic Recombination Between Unlinked Genes by Independent Assortment

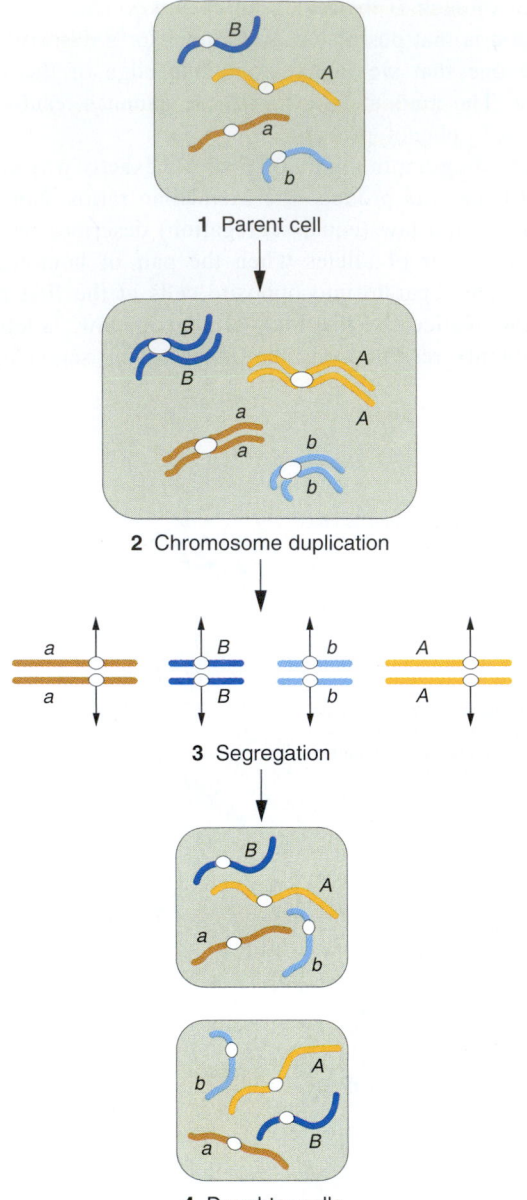

1 Parent cell

2 Chromosome duplication

3 Segregation

4 Daughter cells

Figure 3-16   Mitosis in a diploid cell of genotype *A/a* ; *B/b*. The heterozygous genes are on separate chromosome pairs.

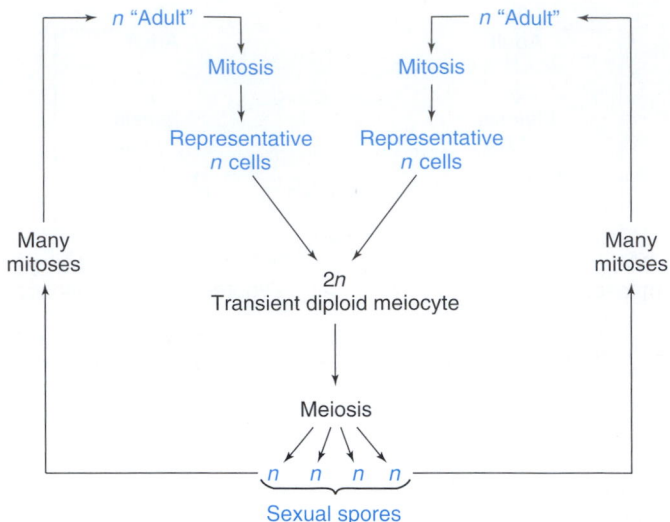

Figure 3-17   The haploid life cycle.

different pairs of homologous chromosomes. Mitosis in the same dihybrid is shown in Figure 3-16.

Haploids

Figure 3-17 shows the basic haploid life cycle, found in many fungi and algae. Here, the organism is haploid. How can meiosis possibly take place in a haploid organism? After all, meiosis requires the pairing of two homologous chromosome sets. The answer is that all haploid organisms that undergo meiosis create a temporary diploid stage that provides the meiocytes. In some cases, such as in yeast, unicellular, haploid, mature individuals fuse to form a diploid meiocyte, which then undergoes meiosis. In other cases,

specialized cells from different parents fuse to give rise to the meiocytes. Note that these fusing cells are properly called *gametes,* so we see that, in these cases, gametes arise from mitosis. Meiosis, as usual, produces haploid products, which are called **sexual spores.** The sexual spores in some species become new unicellular adults; in other species, they each develop through mitosis into a multicellular haploid individual. A cross between two haploid organisms includes only one meiosis, whereas a cross between two diploid organisms includes a meiosis in each diploid organism. As we shall see, this simplicity makes haploids very attractive for genetic analysis. In haploids, mitosis proceeds as shown in Figure 3-18.

Let's consider a cross in a specific haploid. A convenient organism for demonstration is the orange-colored bread mold *Neurospora crassa.* This fungus is a multicellular haploid in which the cells are joined end to end to form **hyphae,** or threads of cells. The hyphae grow through the substrate and send up aerial branches that bud off haploid cells known as **conidia** (asexual spores). Conidia can detach and disperse to form new colonies or, alternatively, they can act as paternal gametes and fuse with a maternal structure of a different individual (Figure 3-19). However, that different individual must be of the opposite **mating type.** In fungi there are no true sexes, and all haploid cultures develop similarly. However, populations contain distinct genetically determined mating types. In *Neurospora,* there are two mating types, called *A* and *a*, and the meiotic (sexual) part of the life cycle can take place only if two haploids of different mating type unite. Mating types can be thought of as "physiological sexes," and, although this definition is inadequate, it is a useful phrase that stresses the unseen difference between mating types.

A maternal gamete waits inside a specialized knot of hyphae, and eventually a haploid maternal and a haploid paternal nucleus pair up and divide mitotically to produce nu-

1 Parent cell

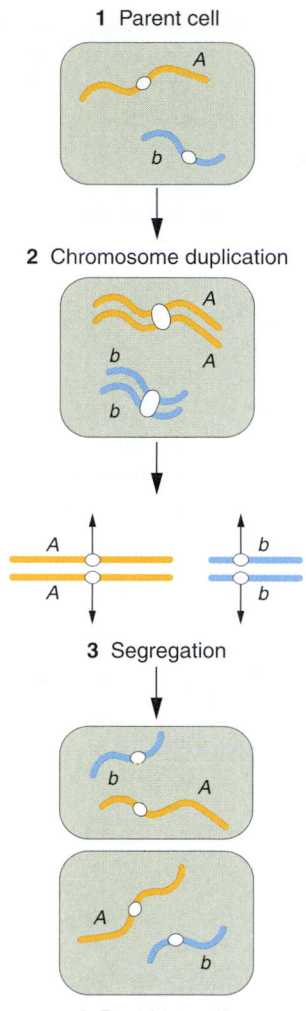

2 Chromosome duplication

3 Segregation

4 Daughter cells

Figure 3-18   Mitosis in a haploid cell of genotype *A* ; *b*. The genes are on separate chromosomes.

merous pairs. The pairs eventually fuse to form diploid meiocytes. Meiosis takes place and, in each meiocyte, four haploid nuclei, which represent the four products of meiosis, are produced. For an unknown reason, these four nuclei divide mitotically, resulting in eight nuclei, which develop into eight football-shaped sexual spores called **ascospores.** The ascospores are shot out of a flask-shaped fruiting body that has developed from the knot of hyphae that originally contained the maternal gametic cell. The ascospores can be isolated, each into a culture tube, where each ascospore will grow into a new culture by mitosis (Figure 3-20).

What characters can be studied in such an organism? One character is the color of the conidia. Variants of the normal orange color can be found. Figure 3-21 shows a normal culture and some color variants, including an albino. Another possible character to study is the compactness of the culture, and one pair of contrasting phenotypes are normal *spreading* growth and a densely branching variant called *colonial.*

Let's cross a spreading, orange culture with a colonial, albino culture of opposite mating type. We isolate ascospores (progeny) and grow each one into a culture. We would find the following progeny phenotypes and proportions:

25% spreading, orange

25% colonial, albino

25% spreading, albino

25% colonial, orange

In total, half the progeny are spreading and half are colonial. Thus, this phenotypic difference must be determined

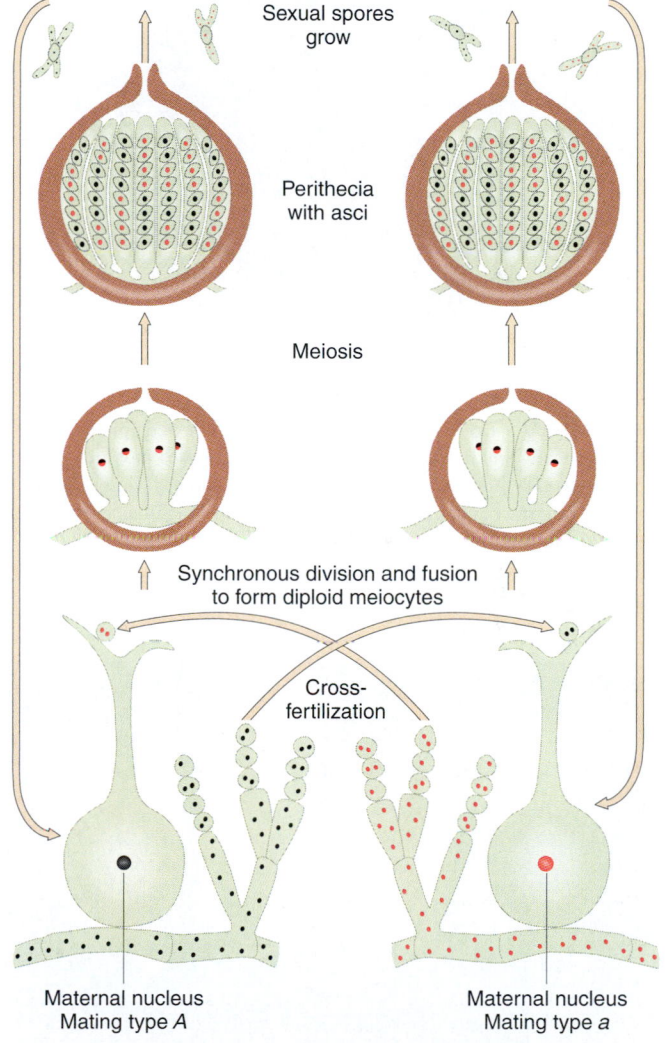

Figure 3-19   The life cycle of *Neurospora crassa,* the orange bread mold. Self-fertilization is not possible in this species: there are two mating types, determined by the alleles *A* and *a* of one gene. A cross will succeed only if it is *A* × *a*. An asexual spore from the opposite mating type fuses with a receptive hair, and a nucleus travels down the hair to pair with a nucleus in the knot of cells. The *A* and *a* pair then undergo synchronous mitoses, finally fusing to form diploid meiocytes.

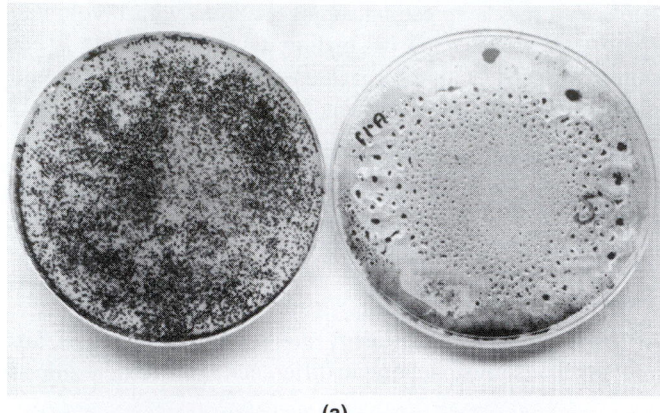

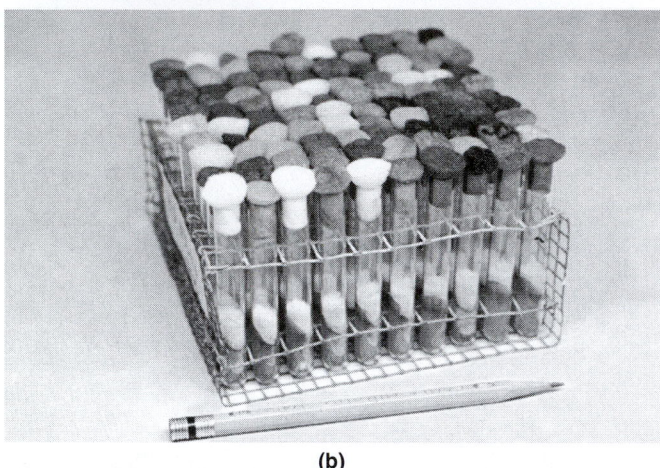

Figure 3-20 (a) A *Neurospora* cross made in a petri plate (at the left). The many small black spheres are fruiting bodies in which meiosis has taken place; the ascospores (sexual spores) were shot as a fine dust into the condensed moisture on the lid (which has been removed and is to the right of the plate). (b) A rack of progeny cultures, each resulting from one isolated ascospore. (Anthony Griffiths.)

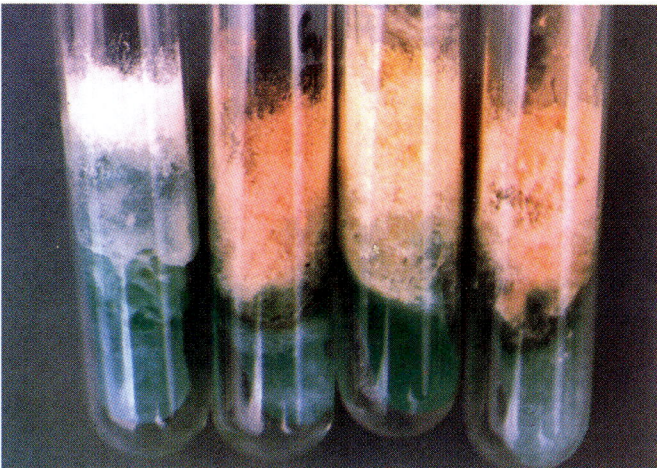

Figure 3-21 Genetically determined color variants of the fungus *Neurospora*. The orange wild-type color is shown at right; the variants are, starting at left, albino, yellow, and brown. Their genotypes are wild type ($al^+ \cdot ylo^+ \cdot ad^+$), albino ($al \cdot ylo^+ \cdot ad^+$), yellow ($al^+ \cdot ylo \cdot ad^+$), and brown ($al^+ \cdot ylo^+ \cdot ad$). (Anthony Griffiths.)

by the alleles of one gene that have segregated equally at meiosis. We can call these alleles col^+ (spreading) and col (colonial). The same logic can be applied to the other character: half the progeny are orange and half are albino, so the phenotypic difference in color also is determined by a pair of alleles, which we can name al^+ (orange) and al (albino). We can represent the parents and the four progeny types as follows:

Parents

$col^+ ; al^+$ $\times$ $col ; al$

(spreading, orange)   (colonial, albino)

Progeny

$col^+ ; al^+$	(spreading, orange)
$col ; al$	(colonial, albino)
$col^+ ; al$	(spreading, albino)
$col ; al^+$	(colonial, orange)

The 1:1:1:1 ratio is a result of equal segregation and independent assortment, as illustrated in the following branch diagram:

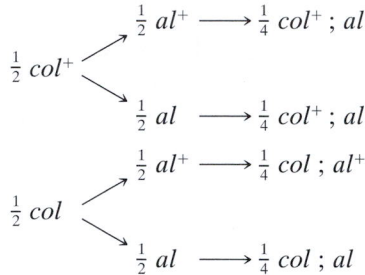

So we see that, even in such a lowly organism, Mendel's laws are still in operation.

Alternating haploid and diploid

In an organism with alternation of generations, the life cycle comprises two stages: one diploid and one haploid. One stage is usually more prominent than the other. For example, what we recognize as a fern plant is the diploid **sporophyte** stage. This stage is the one that undergoes meiosis, producing sexual spores. However, the organism also has a small, independent, photosynthetic haploid stage that is usually much more difficult to spot on the forest floor. This haploid stage is the **gametophyte** stage, which produces gametes by mitosis. In contrast, the green moss plant is the haploid gametophyte stage, and the brownish stalk that grows up out of this plant is a dependent, diploid sporophyte that is effectively parasitic on the gametophyte.

In flowering plants, the main green stage is the diploid sporophyte. The haploid gametophytes of flowering plants are extremely reduced and dependent on the diploid. These gametophytes are found in the flower. In the anther and the ovary, meiocytes undergo meiosis, and the resulting haploid products of meiosis are called **spores.** A spore undergoes a few mitotic divisions to produce a small, multicellular gametophyte. The male gametophyte of seed plants is known

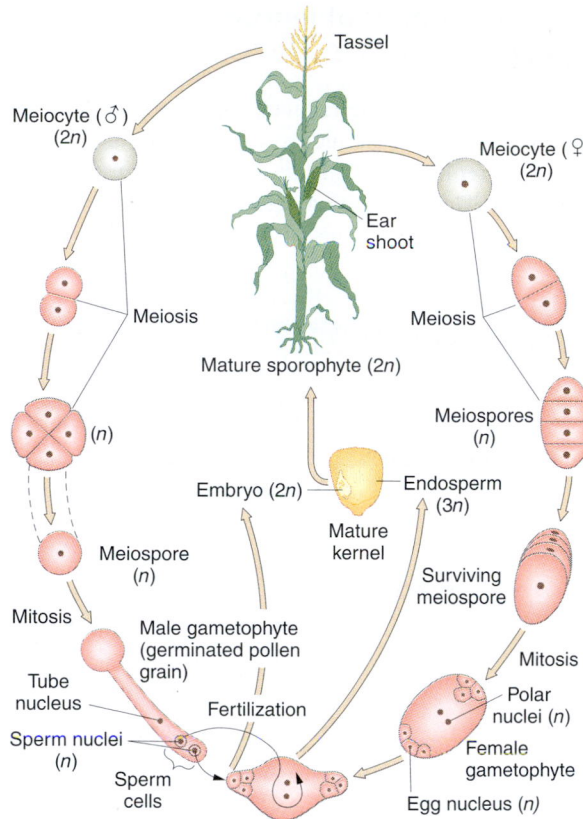

Figure 3-22 Alternation of generations of corn. The male gametophyte arises from a meiocyte in the tassel. The female gametophyte arises from a meiocyte in the ear shoot. One sperm cell from the male gametophyte fuses with an egg nucleus of the female gametophyte, and the diploid zygote thus formed develops into the embryo. The other sperm cell fuses with the two polar nuclei in the center of the female gametophyte, forming a triploid ($3n$) cell that generates the endosperm tissue surrounding the embryo. The endosperm provides nutrition to the embryo during seed germination. Which parts of the diagram represent the haploid stage? Which parts represent the diploid stage?

as a pollen grain. Figure 3-22 shows that, in flowering plants, cells of the gametophytes act as eggs or sperm in fertilization. The generalized cycle of alternation of generations is shown in Figure 3-23.

In mosses and ferns, the sperm cells are motile and travel from one gametophyte to another in a film of water to effect fertilization. Let us consider a cross that we might make in a moss. The character to be studied can pertain to the gametophyte or the sporophyte. Assume that we have a gene whose alleles affect the "leaves" of the gametophyte, with w causing wavy edges and w^+ causing smooth edges. Also assume that a separate gene affects the color of the sporophyte, with r causing reddish coloration and r^+ causing the normal brown coloration. We fertilize a smooth-leaved w^+ gametophyte that also bears the unexpressed allele r by transferring onto it male gametes from a wrinkly-leaved w gametophyte, also carrying r^+ (Figure 3-24). Hence, the cross is $w^+ ; r \times w ; r^+$.

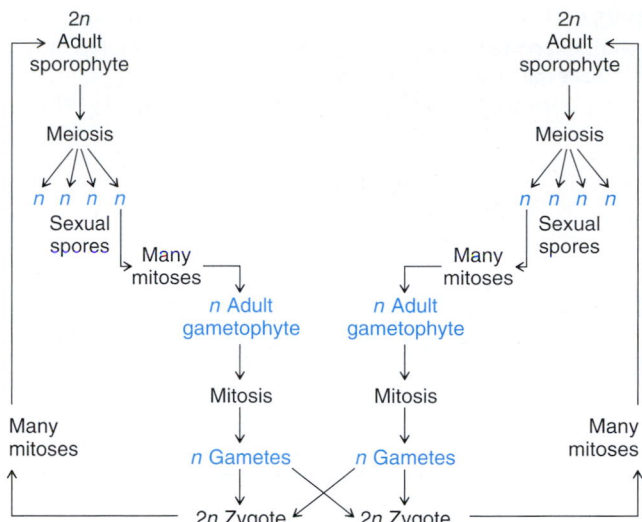

Figure 3-23 The alternation of diploid and haploid stages in the life cycle of plants.

A diploid sporophyte of genotype $w^+/w ; r^+/r$ develops on the gametophyte, and it is brown because reddish is recessive. Cells of this sporophyte act as meiocytes, and sexual spores (products of meiosis) are produced in the following proportions:

$25\% \; w^+ \; ; \; r^+$

$25\% \; w^+ \; ; \; r$

$25\% \; w \; ; \; r^+$

$25\% \; w \; ; \; r$

We can directly classify only the leaf character in these gametophytes, and we would have to make the appropriate intercrosses to determine whether each individual is r^+ or r.

Once again, Mendel's laws dictate the inheritance patterns. These patterns may be deduced simply by keeping track of the ploidy in each part of the cycle and applying the simple Mendelian ratios.

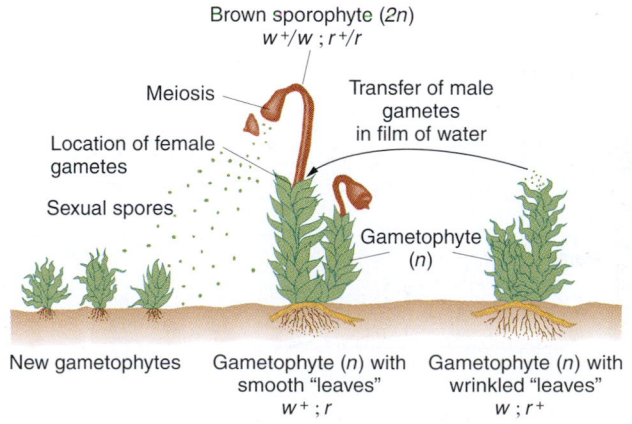

Figure 3-24 Mendelian genetics in a hypothetical cross in a moss. Only the haploid gametophyte expresses the w^+ or w, and only the diploid sporophyte expresses the r^+ or r alternatives.

MESSAGE ·····························

Mendelian laws apply to meiosis in any organism and may be generally stated as follows:

1. At meiosis, the alleles of a gene segregate equally into the haploid products of meiosis.
2. At meiosis, the alleles of one gene segregate independently of the alleles of genes on other chromosome pairs.

···

The molecular basis of mitosis and meiosis

We know that, at the genetic level, a chromosome is a single DNA molecule. Cell and nuclear division are made possible by the DNA replication that takes place during S phase before division. This replication creates two sister chromatids. The formation of chromatids at the DNA level is shown in

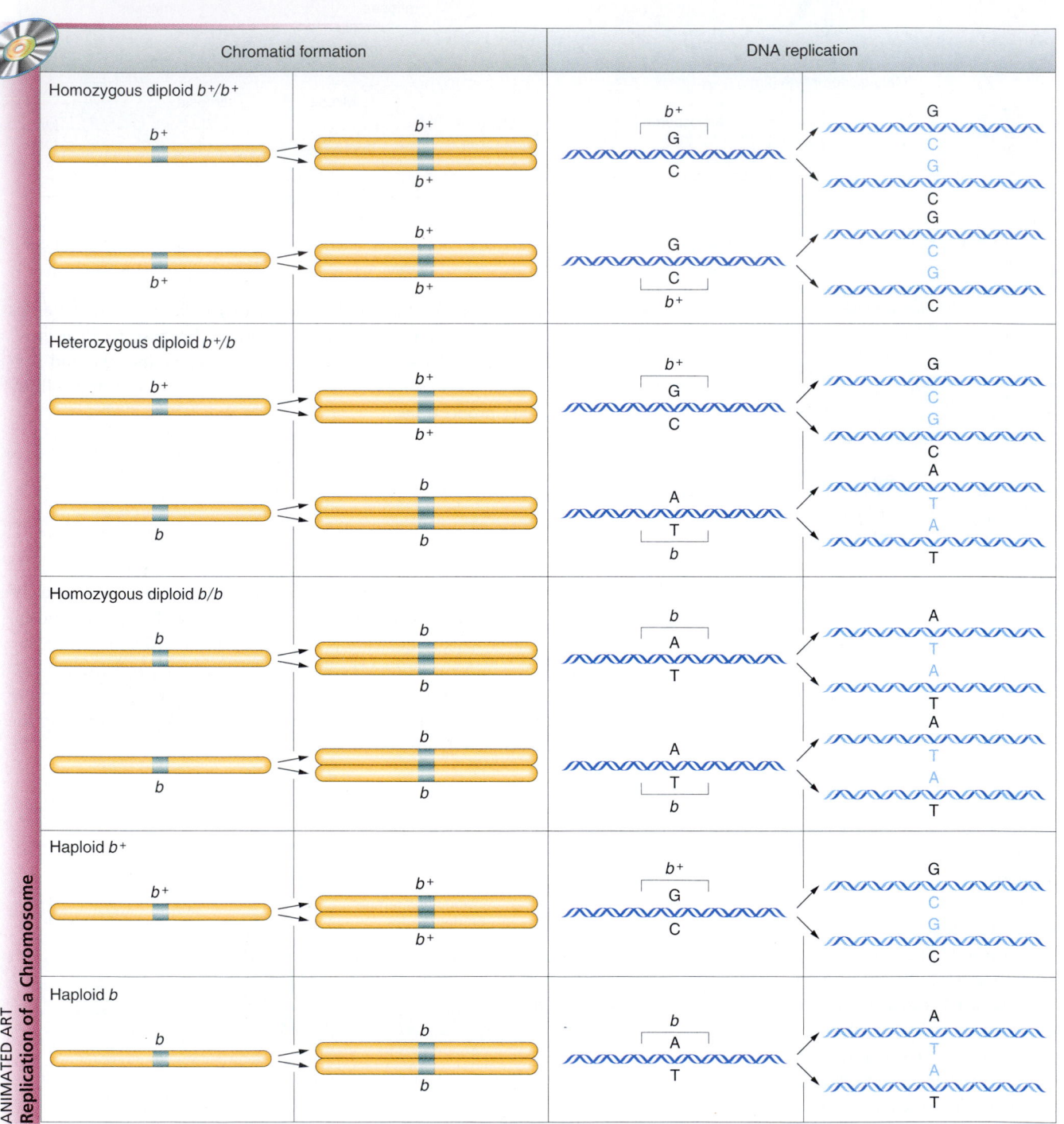

ANIMATED ART
Replication of a Chromosome

Figure 3-25   Chromatid formation and its underlying DNA replication. Three different diploid and two different haploid genotypes are represented. The wild-type allele is called b^+ and a mutant allele is b. At the mutant site, a single base pair changes from GC in the wild type to AT in the mutant.

Figure 3-25. The events of mitosis and meiosis can be repre- sented at the DNA level, as diagrammed for a diploid het- erozygote in Figure 3-26.

Adhesion and pairing also are key molecular features. At both mitosis and meiosis, sister chromatids remain at- tached along their length until pulled apart at anaphase of mitosis or anaphase II of meiosis. This **sister chromatid**

adhesion is due to special adhesive proteins. Pairing of ho- mologous chromosomes at meiosis is accomplished by molecular assemblages called **synaptonemal complexes** along the middle of paired sister chromatids (Figure 3-27). Although the existence of synaptonemal complexes has been known for some time, the precise working of these structures is still a topic of research.

Mitosis in a^+/a diploid

Meiosis in a^+/a diploid

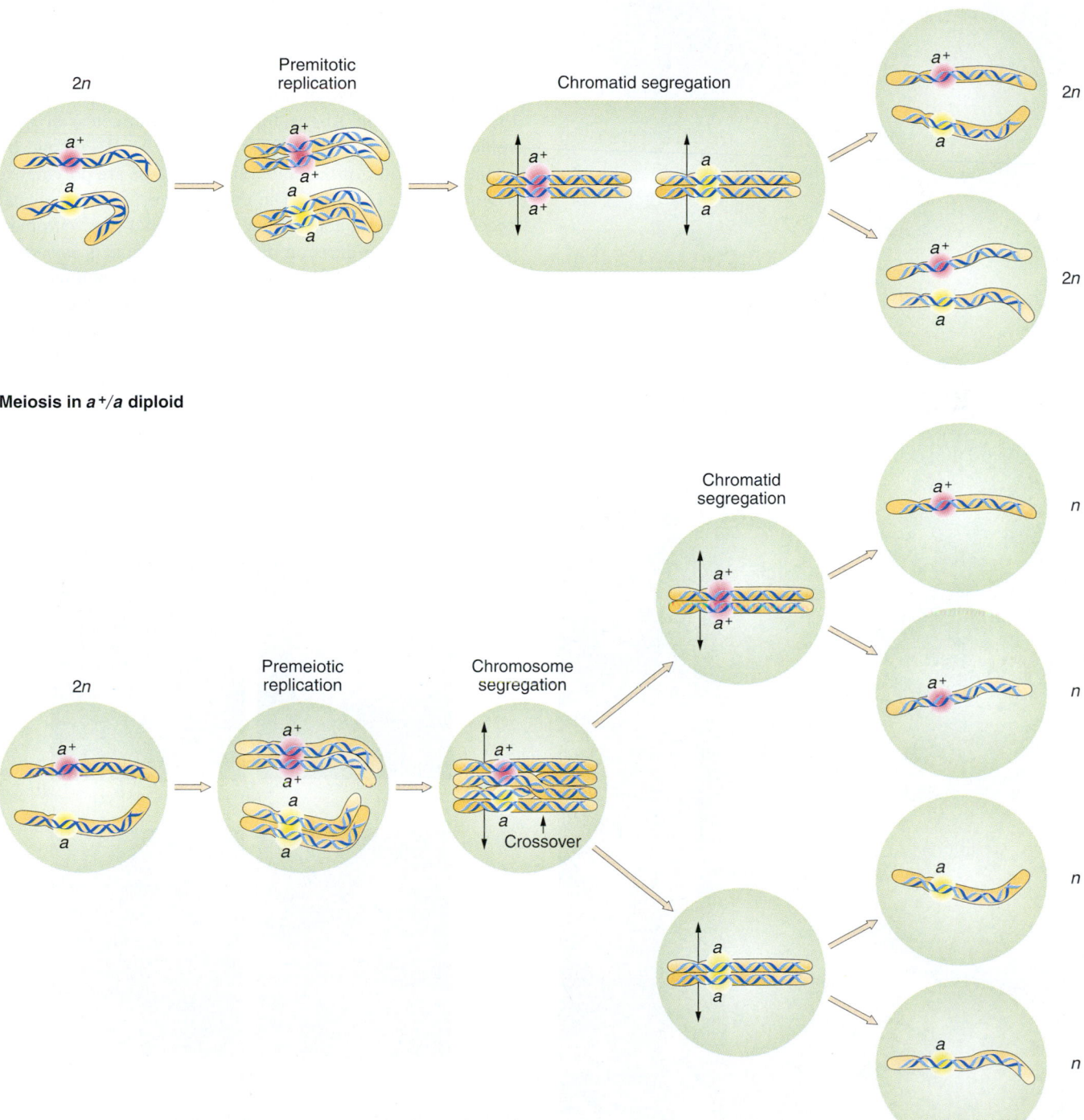

Figure 3-26   Simplified representation of mitosis and meiosis at the DNA level. Both are shown starting with a diploid cell of genotype a^+/a.

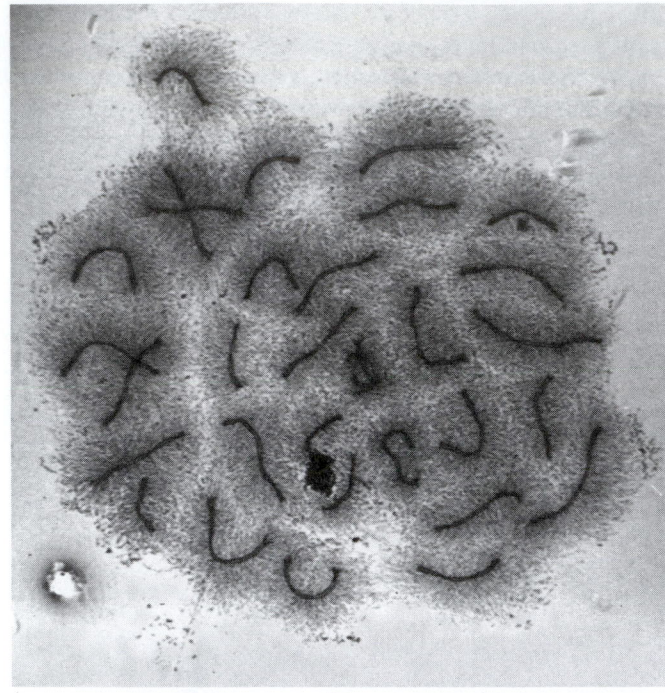

(a)

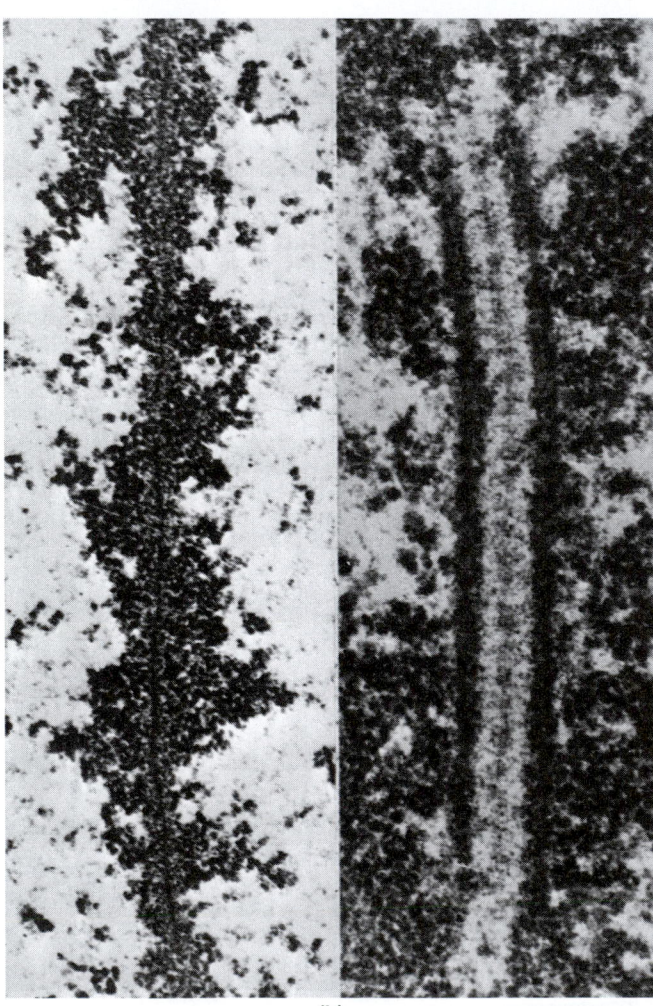

(b)

Figure 3-27 Synaptonemal complexes. (a) In *Hyalophora cecropia*, a silk moth, the normal male chromosome number is 62, giving 31 synaptonemal complexes. In the individual shown here, one chromosome *(center)* is represented three times; such a chromosome is termed *trivalent*. The DNA is arranged in regular loops around the synaptonemal complex. The black, dense structure is the nucleolus. (b) Regular synaptonemal complex in *Lilium tyrinum*. Note *(right)* the two lateral elements of the synaptonemal complex and *(left)* an unpaired chromosome, showing a central core corresponding to one of the lateral elements. (Courtesy of Peter Moens.)

The motive force of mitosis and meiosis is produced by the nuclear spindle fibers (Figure 3-28). In nuclear division, spindle fibers form that are parallel to the cell axis and that connect the poles of the cells. Spindle fibers are polymers of a protein called tubulin. The chromosomal centromere is a specific DNA sequence that is essential for chromatid movement during division. Each centromere acts as a site at which a multiprotein complex called the **kinetochore** binds. The kinetochores in turn act as the sites for attachment to microtubules. From one to many microtubules from one pole attach to one kinetochore, and a similar number from the opposite pole attach to the other kinetochore. The spindle fibers then pull the chromosomes to opposite poles. The spindle apparatus and the complex of kinetochores and centromeres determine the fidelity of nuclear division.

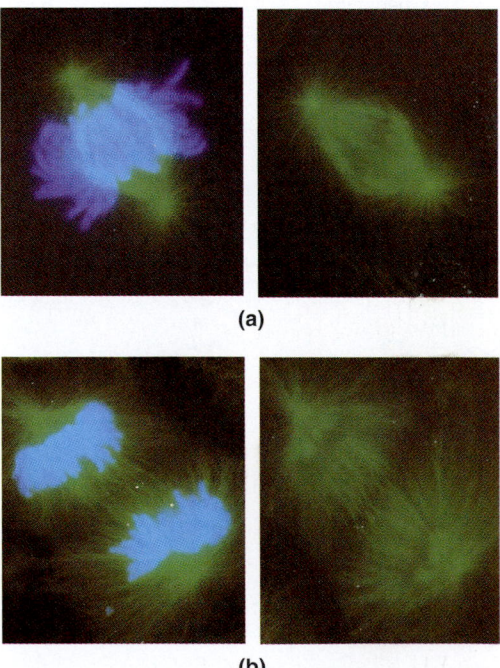

(a)

(b)

Figure 3-28 Fluorescent label of the nuclear spindle (green) and chromosomes (blue) in mitosis: (a) before the chromatids are pulled apart; (b) during the pulling apart. (From J. C. Waters, R. W. Cole, and C. L. Rieder, *J. Cell Biol.* 122, 1993, 361; courtesy of C. L. Rieder.)

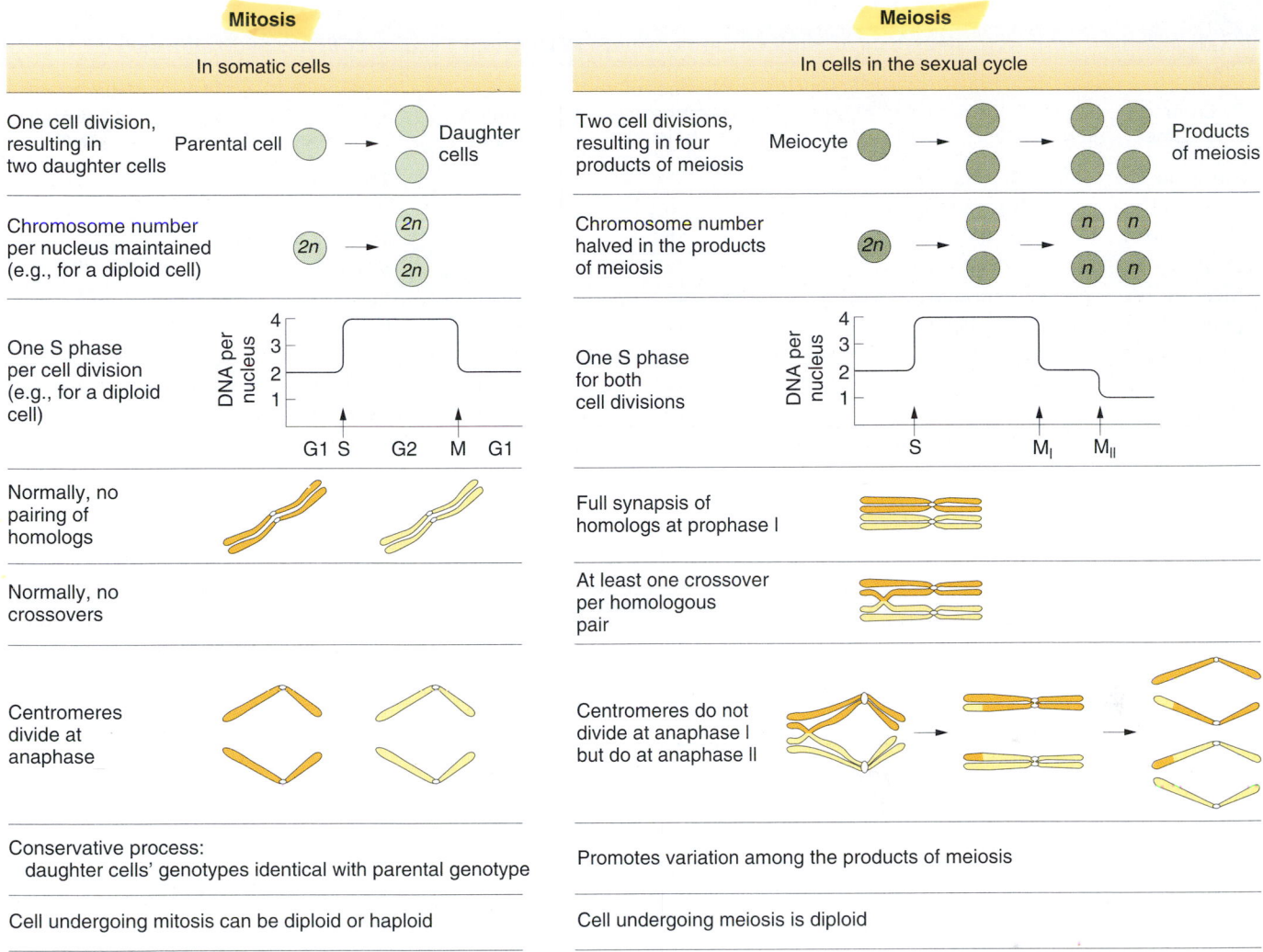

Figure 3-29   Comparison of the main features of mitosis and meiosis.

The key features of mitosis and meiosis are compared in Figure 3-29. In the next sections, we probe deeper into the way that a chromosome is structured.

Topography of the chromosome set

So far, we have treated chromosomes as wormlike structures that contain DNA (genes). In fact, chromosomes are highly varied in size and shape, with characteristics that allow cytogeneticists to identify specific chromosomes in many cases. In this section, we shall consider the features that allow cytogeneticists to distinguish one chromosome set from another, as well as one chromosome from another.

Chromosome number

Different species have highly characteristic chromosome numbers, and examples are shown in Table 3-1. The range is immense, from two in some flowering plants to many hundreds in certain ferns.

Chromosome size

The chromosomes of a single genome may differ considerably in size. In the human genome, for example, there is about a three- to fourfold range in size from chromosome 1 (the biggest) to chromosome 21 (the smallest), as shown in Table 3-2. In studying the chromosomes of some species, a cytogeneticist may have difficulty identifying individual chromosomes by size alone but may be able to group chromosomes of similar size. A change may then be detected in, for example, "one of the chromosomes in size group A."

Centromeres

The centromere is the region of the chromosome to which spindle fibers attach. The centromere region usually appears

3-1 TABLE — Numbers of Pairs of Chromosomes in Different Species of Plants and Animals

Common name	Species	Number of chromosome pairs	Common name	Species	Number of chromosome pairs
Mosquito	*Culex pipiens*	3	Wheat	*Triticum aestivum*	21
Housefly	*Musca domestica*	6	Human	*Homo sapiens*	23
Garden onion	*Allium cepa*	8	Potato	*Solanum tuberosum*	24
Toad	*Bufo americanus*	11	Cattle	*Bos taurus*	30
Rice	*Oryza sativa*	12	Donkey	*Equus asinus*	31
Frog	*Rana pipiens*	13	Horse	*Equus caballus*	32
Alligator	*Alligator mississipiensis*	16	Dog	*Canis familiaris*	39
Cat	*Felis domesticus*	19	Chicken	*Gallus domesticus*	39
House mouse	*Mus musculus*	20	Carp	*Cyprinus carpio*	52
Rhesus monkey	*Macaca mulatta*	21			

3-2 TABLE — Human Chromosomes

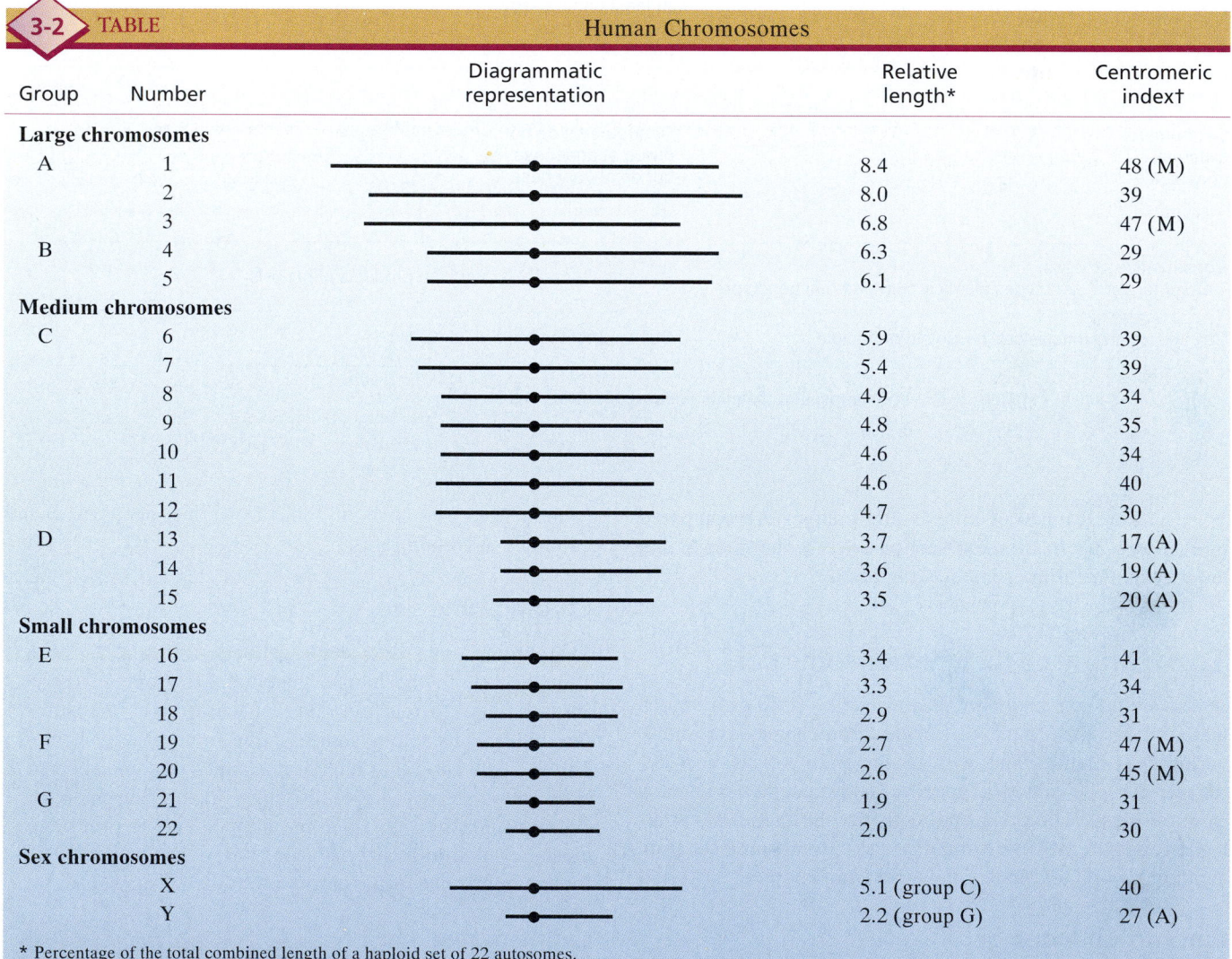

Group	Number	Diagrammatic representation	Relative length*	Centromeric index†
Large chromosomes				
A	1		8.4	48 (M)
	2		8.0	39
	3		6.8	47 (M)
B	4		6.3	29
	5		6.1	29
Medium chromosomes				
C	6		5.9	39
	7		5.4	39
	8		4.9	34
	9		4.8	35
	10		4.6	34
	11		4.6	40
	12		4.7	30
D	13		3.7	17 (A)
	14		3.6	19 (A)
	15		3.5	20 (A)
Small chromosomes				
E	16		3.4	41
	17		3.3	34
	18		2.9	31
F	19		2.7	47 (M)
	20		2.6	45 (M)
G	21		1.9	31
	22		2.0	30
Sex chromosomes				
	X		5.1 (group C)	40
	Y		2.2 (group G)	27 (A)

* Percentage of the total combined length of a haploid set of 22 autosomes.
† Percentage of a chromosome's length spanned by its short arm. The four most metacentric chromosomes are indicated by an (M); the four most acrocentric by an (A).

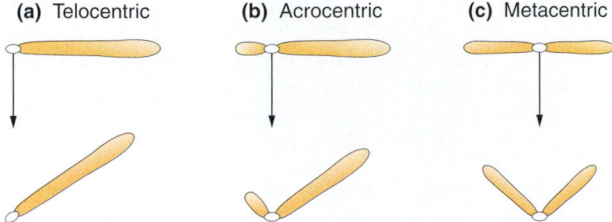

(a) Telocentric **(b)** Acrocentric **(c)** Metacentric

Figure 3-30 The classification of chromosomes by the position of the centromere. A telocentric chromosome has its centromere at one end; when the chromosome moves toward one pole of the cell during the anaphase of cellular division, it appears as a simple rod. An acrocentric chromosome has its centromere somewhere between the end and the middle of the chromosome; during anaphase movement, the chromosome appears as a J. A metacentric chromosome has its centromere in the middle and appears as a V during anaphase.

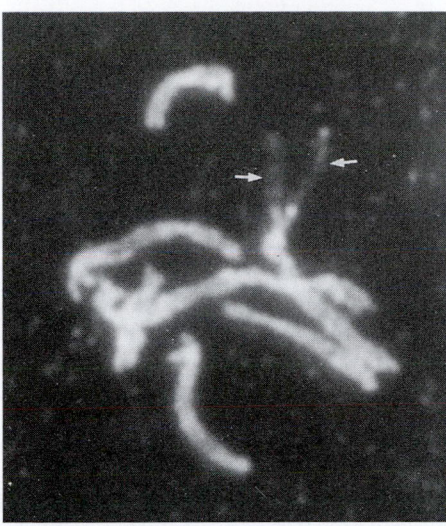

Figure 3-31 Nucleolus and nucleolar organizer of the haploid fungus *Neurospora*. (Namboori Raju.)

to be constricted, and the position of this constriction defines the ratio between the lengths of the two chromosome arms; this ratio is a useful characteristic (Table 3-2). Centromere positions can be categorized as **telocentric** (at one end), **acrocentric** (off center), or **metacentric** (in the middle).

The centromere position determines not only arm ratio, but also the shapes of chromosomes as they migrate to opposite poles during anaphase. These anaphase shapes range from a rod to a J to a V (Figure 3-30). In some organisms, such as the lepidoptera, centromeres are "diffuse," so spindle fibers attach all along the chromosome.

Position of nucleolar organizers

Nucleoli are intranuclear organelles that contain ribosomal RNA, an important component of ribosomes. Different organisms are differently endowed with nucleoli, which range in number from one to many per chromosome set. The diploid cells of many species have two nucleoli. The nucleoli reside next to secondary constrictions of the chromosomes, called **nucleolar organizers** (Figure 3-31), which have highly specific positions in the chromosome set. Nucleolar organizers contain the genes that code for ribosomal RNA. Their positions, like those of centromeres, are landmarks for cytogenetic analysis.

Chromomere patterns

Chromomeres are beadlike, localized thickenings found along the chromosome during prophase of mitosis and meiosis. Homologous chromosomes tend to have homologous sets of chromomeres. Although they can be useful as markers, their molecular nature is not known.

Heterochromatin patterns

When chromosomes are treated with chemicals that react with DNA, such as Feulgen stain, distinct regions with different staining characteristics are visually revealed. Densely staining regions are called **heterochromatin;** poorly staining regions are said to be **euchromatin.** The distinction refers to the degree of compactness, or coiling, of the DNA in the chromosome. Heterochromatin can be either *constitutive* or *facultative.* The constitutive type is a permanent feature of a specific chromosome location and is, in this sense, a hereditary feature; see Figure 5-15 for good examples of constitutive heterochromatin in the tomato. The facultative type of heterochromatin, as its name suggests, is sometimes but not always found at a particular chromosomal location. The patterns of heterochromatin and euchromatin along a chromosome are good cytogenetic markers.

Banding patterns

Special chromosome-staining procedures have revealed specific sets of intricate bands (transverse stripes) in many different organisms. The positions and sizes of the bands are highly chromosome specific. One of the basic chromosomal banding patterns is that produced by Giemsa reagent, a DNA stain applied after mild proteolytic digestion of the chromosomes. This reagent produces patterns of light-staining (G-light) regions and dark-staining (G-dark) regions. An example of G bands in human chromosomes is shown in Figure 3-32. The patterns are consistent within species. In the complete set of 23 human chromosomes, there are approximately 850 G-dark bands visible at metaphase of mitosis. These bands have provided a useful way of subdividing the various regions of chromosomes, and each band has been assigned a specific number.

The difference between dark- and light-staining regions was believed to be caused by differences in the relative pro-

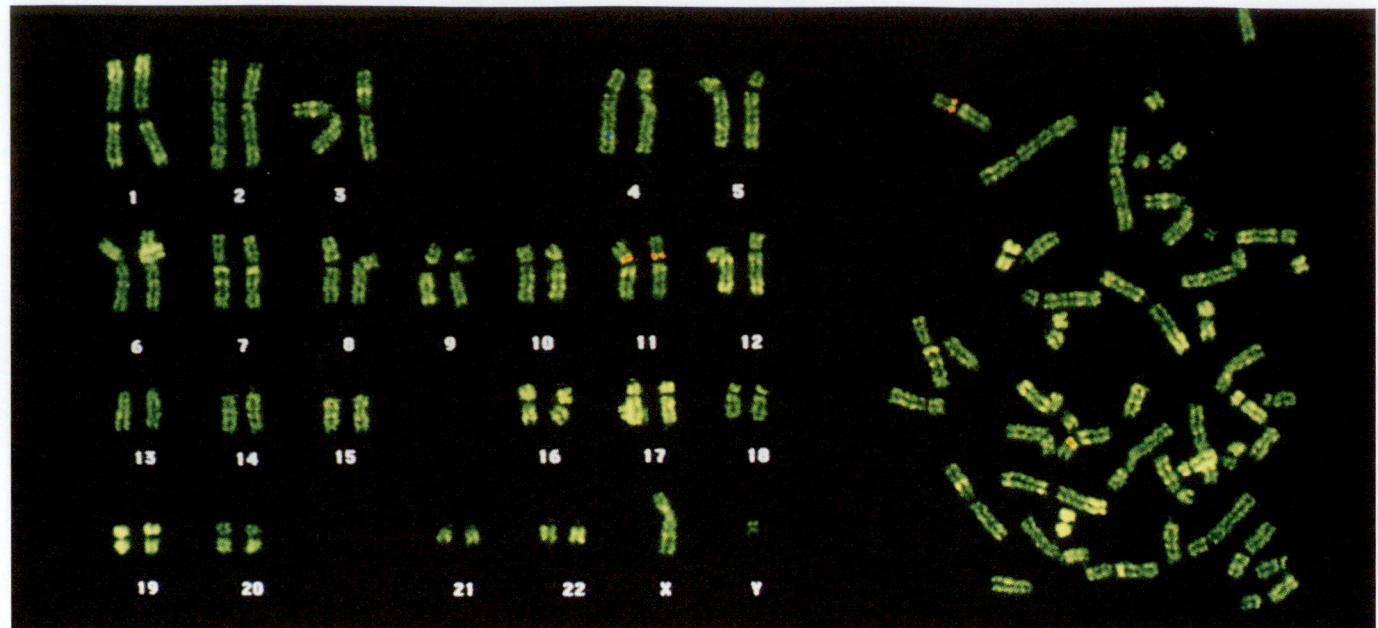

Figure 3-32   Chromosomes of a human male stained with Giemsa to reveal bands.

portions of bases: the G-light bands being relatively GC-rich, and the G-dark bands AT-rich. However, it is now thought that the differences are too small to account for banding patterns. The crucial factor appears to be chromatin packing density: the G-dark regions are packed more densely, with tighter coils, which results in a higher density of DNA to take up the stain.

In addition, various other correlations have been made. For example, deoxynucleotide-labeling studies showed that G-light bands are early replicating. Furthermore, if polysomal (poly*ribo*somal) mRNA (representing genes being actively transcribed) is used to label chromosomes in situ, then most label binds to the G-light regions, suggesting that these regions contain most of the active genes. From such an analysis, it was presumed that the density of active genes is higher in the G-light bands.

Our view of chromosome banding is based largely on how chromosomes stain when they are in mitotic metaphase. Nevertheless, the domains revealed by metaphase banding must still be in the same relative position in interphase, whether banded or not.

A rather specialized kind of banding, which has been used extensively by cytogeneticists for many years, is characteristic of the so-called **polytene** chromosomes in certain organs of the dipteran insects. Polytene chromosomes develop in the following way. In secretory tissues, such as the Malpighian tubules, rectum, gut, footpads, and salivary glands of the diptera, the chromosomes replicate their DNA many times without actually separating into distinct chromatids. As a chromosome increases its number of replicas, it elongates and thickens. This bundle of replicas becomes the polytene chromosome. We can look at *Drosophila* as an

example. This insect has a 2*n* number of 8, but the special organs contain only four polytene chromosomes. There are four and not eight, because, in the replication process, the homologs unexpectedly become tightly paired. Furthermore, all four polytene chromosomes become joined at a structured called the **chromocenter,** which is a coalescence of the heterochromatic areas around the centromeres of all four chromosome pairs. The chromocenter of *Drosophila* salivary gland chromosomes is shown in Figure 3-33, in which the letters L and R stand for arbitrarily assigned left and right arms.

Along the length of a polytene chromosome are transverse stripes called **bands.** Polytene bands are much more numerous than Q, G, or R bands, numbering in the hundreds on each chromosome. The bands differ in width and morphology, so the banding pattern of each chromosome is unique and characteristic of that chromosome. In addition, there are regions that may at times appear swollen (**puffs**) or greatly distended (**Balbiani rings),** and these regions correspond to regions of RNA synthesis. As we shall see in Chapter 16, recent molecular studies have shown that, in any chromosomal region of *Drosophila,* there are more genes than there are polytene bands, so there is not a one-to-one correspondence of bands and genes as was once believed. Similarly, the significance of the banding patterns of the chromosomes of vertebrates is not clear. However, it is known that the chromosome bands relate to DNA nucleotide composition and that most of the active genes reside in the light bands.

By using all the available chromosomal landmarks together, cytogeneticists can distinguish each of the chromosomes in many species. As an example, Figure 3-34 is a map of the chromosomal landmarks of the genome of corn.

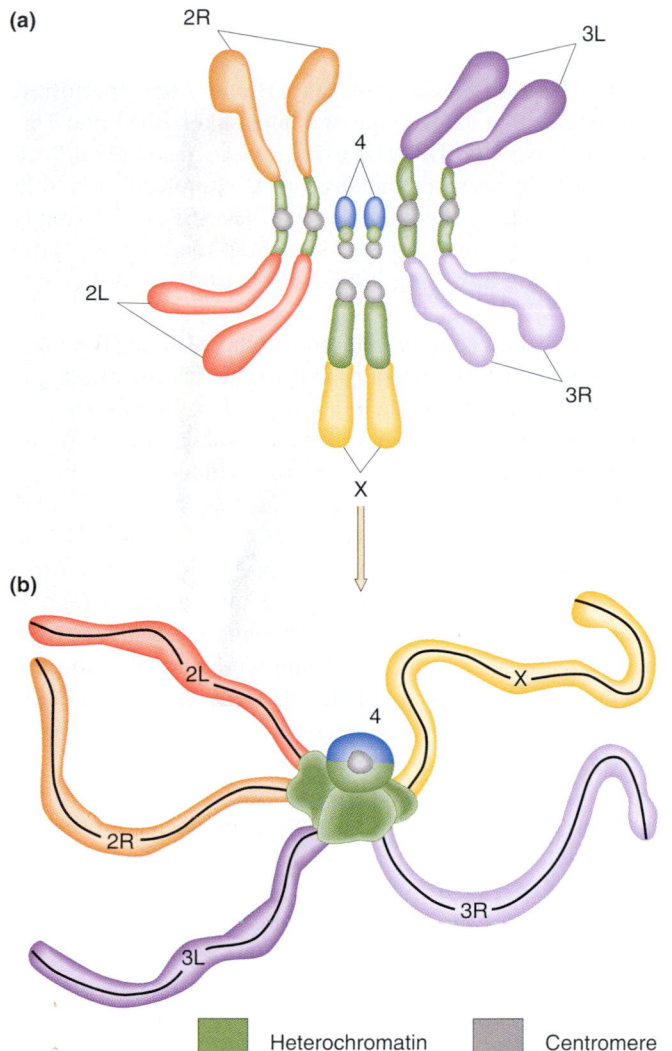

(a)

(b)

Heterochromatin   Centromere

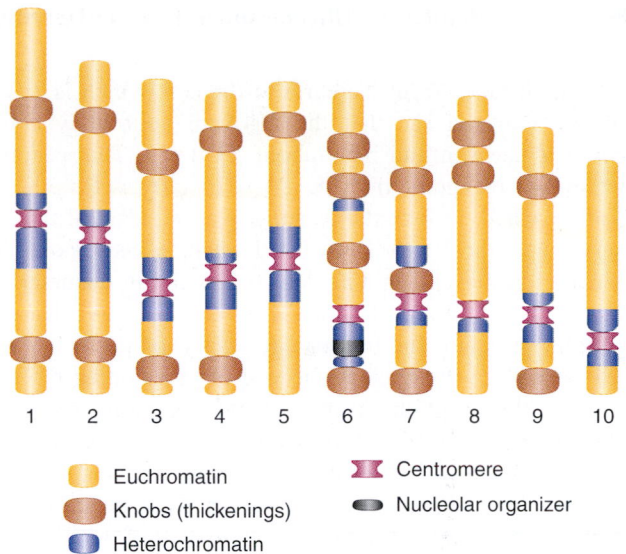

Euchromatin

Knobs (thickenings)

Heterochromatin

Centromere

Nucleolar organizer

Figure 3-34   Landmarks that distinguish the chromosomes of corn.

Notice how the landmarks enable each of the 10 chromosomes to be distinguished under the microscope.

MESSAGE ···································

Features as size, arm ratio, heterochromatin, number and position of thickenings, number and location of nucleolar organizers, and banding pattern identify the individual chromosomes within the set that characterizes a species.

···································

Three-dimensional structure of chromosomes

How much DNA is there in a chromosome set? The single chromosome of the prokaryote *Escherichia coli* is about 1.3 mm of DNA. In stark contrast, a human cell contains about 2 m of DNA (1 m per chromosome set). The human body consists of approximately 10^{13} cells and therefore contains a total of about 2×10^{13} m of DNA. Some idea of the extreme length of this DNA can be obtained by comparing it

(c)

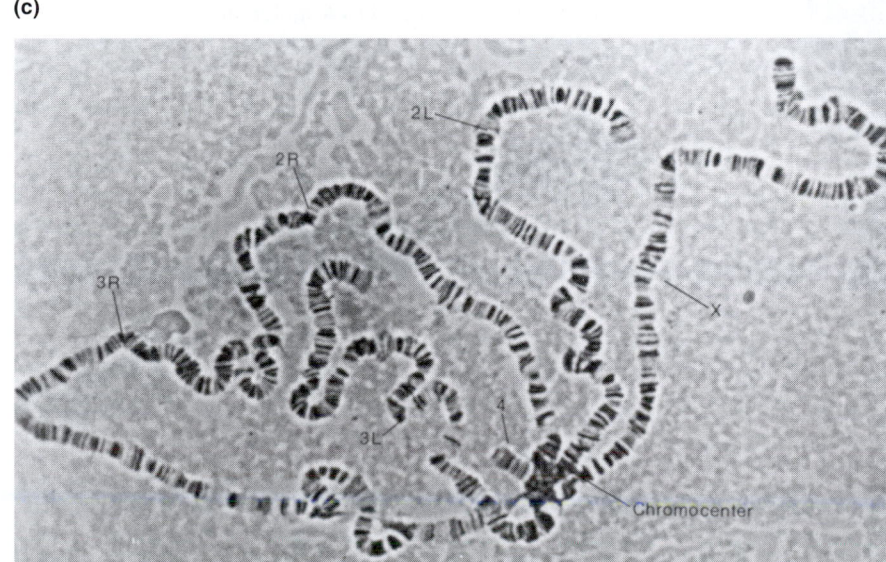

Figure 3-33   Polytene chromosomes form a chromocenter in a *Drosophila* salivary gland. (a) Mitotic metaphase chromosomes, with arms represented by different shades. (b) Heterochromatin coalesces to form the chromocenter. (c) Photograph of polytene chromosomes. (Tom Kaufman.)

with the distance from the earth to the sun, which is 1.5×10^{11} m. You can see that the DNA in your body could stretch to the sun and back about 50 times. This peculiar fact makes the point that the DNA of eukaryotes is efficiently packed. In fact, the packing is at the level of the nucleus, where the 2 m of DNA in a human cell is packed into 46 chromosomes, all in a nucleus 0.006 mm in diameter. In this section, we have to translate what we have learned about the structure and function of eukaryotic genes into the "real world" of the nucleus. Instead of envisioning replication and transcription machinery moving along the airy-looking straight lines that we have used to represent genes in Chapters 1 and 2, we must now come to grips with the fact that these processes take place in what must be very much like the inside of a densely wound ball of wool.

One DNA molecule per chromosome

If eukaryotic cells are broken and the contents of their nuclei are examined under the electron microscope, the chromosomes appear as masses of spaghetti-like fibers with diameters of about 30 nm. Some examples are shown in the electron micrograph in Figure 3-35. In the 1960s, Ernest DuPraw studied such chromosomes carefully and found that there are no ends protruding from the fibrillar mass. This finding suggests that each chromosome is one, long, fine fiber folded up in some way. If the fiber somehow corresponds to a DNA molecule, then we arrive at the idea that each chromosome is one densely folded DNA molecule.

In 1973, Ruth Kavenoff and Bruno Zimm performed experiments that showed this was most likely the case. They studied *Drosophila* DNA by using a viscoelastic recoil technique, which measures the size of DNA molecules in solution by measuring their elastic recoil properties. Put simply, the procedure is analogous to stretching out a coiled spring and measuring how long it takes to return to its fully coiled state. DNA is stretched by spinning a paddle in the solution and is then allowed to recoil into its relaxed state. The recoil time is known to be proportional to the size of the largest molecules present. In their study of *Drosophila melanogaster*, which has four pairs of chromosomes, Kavenoff and Zimm obtained a value of 41×10^9 daltons for the largest DNA molecule in the wild-type genome. Then they studied two chromosomal rearrangements of differing size and showed that viscoelasticity was proportional to size. It looked as if the chromosome was indeed one strand of DNA, continuous from one end, through the centromere, to the other end. Kavenoff and Zimm were also able to piece together electron micrographs of DNA molecules about 1.5 cm long, each presumably corresponding to a *Drosophila* chromosome (Figure 3-36).

Today, geneticists can demonstrate directly that certain chromosomes contain single DNA molecules by using pulsed field gel electrophoresis, a technique for separating very long DNA molecules by size. If the DNA of an organism with relatively small chromosomes, such as *Neurospora*, is subjected to electrophoresis for long periods of time in this apparatus, the number of bands that separate on the gel is equal to the number of chromosomes (seven, for *Neurospora*). If each chromosome contained more than one DNA molecule, the number of bands would be expected to be greater than the number of chromosomes. Such separations cannot be made for organisms with large chromosomes (such as humans and *Drosophila*), because the DNA molecules are too large to move through the gel; nevertheless, all the evidence points to the general principle that a chromosome contains one DNA molecule.

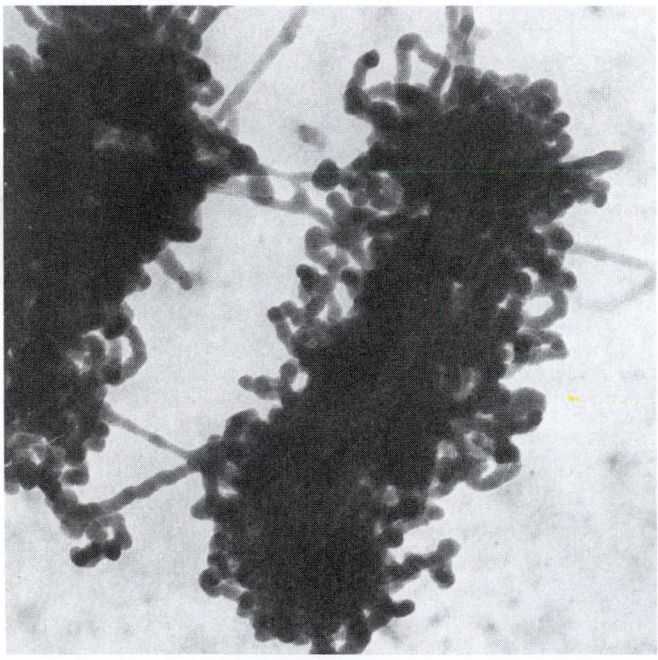

Figure 3-35 Electron micrograph of metaphase chromosomes from a honeybee. The chromosomes each appear to be composed of one continuous fiber 30 nm wide. (From E. J. DuPraw, *Cell and Molecular Biology.* Copyright © 1968 by Academic Press.)

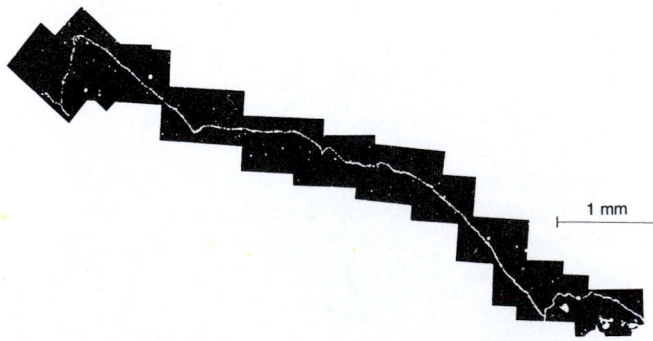

1 mm

Figure 3-36 Composite electron micrograph of a DNA molecule from *Drosophila.* The overall length is 1.5 cm and is thought to correspond to one chromosome. (From R. Kavenoff, L. C. Klotz, and B. H. Zimm, *Cold Spring Harbor Symposia on Quantitative Biology,* 38, 1974, 4.)

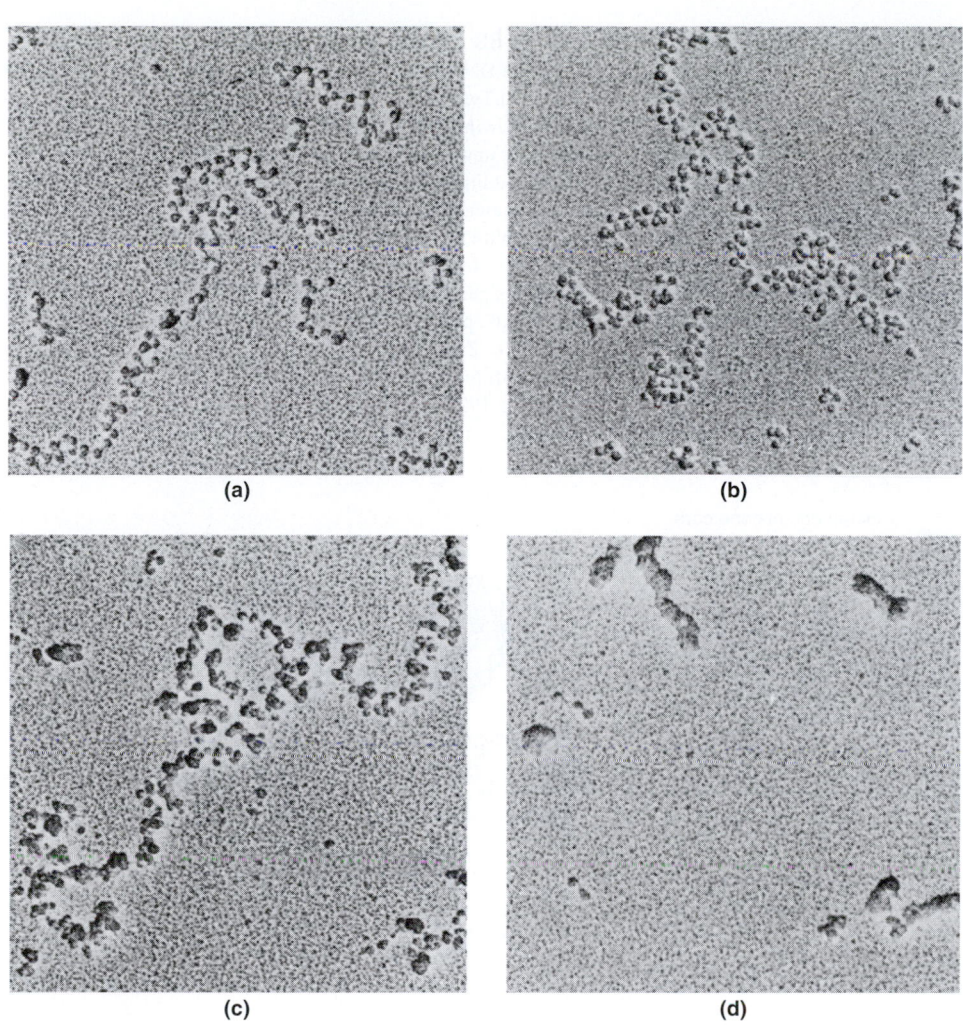

Figure 3-37 Condensation of chromatin with increasing salt concentration is demonstrated in electron micrographs made by Fritz Thomas and Theo Koller. At a very low salt concentration, as in part a, chromatin forms a loose fiber about 10 nm thick; nucleosomes are connected by short stretches of DNA. At a concentration with an ionic strength closer to that of normal physiological conditions, as in part d, chromatin forms a thick fiber some 30-nm thick. The origin of this solenoid can be deduced by an examination of chromatin at increasing intermediate ionic strengths, as in parts b and c. It arises from a shallow coiling of the nucleosome filament. The chromatin is enlarged here about 80,000 diameters.

> **MESSAGE**
>
> Each eukaryotic chromosome contains a single, long, folded DNA molecule.

Role of histone proteins in packaging DNA

We have seen that, because the length of a chromosomal DNA molecule is much greater than the length of a chromosome, there must be an efficient packaging system. What are the mechanisms that pack DNA into chromosomes? How is the very long DNA thread converted into the relatively thick, dense rod that is a chromosome? The overall mixture of material that chromosomes are composed of is given the general name **chromatin.** It is DNA and protein. If chromatin is extracted and treated with differing concentrations of salt, different degrees of compaction, or condensation, are observed under the electron microscope (Figure 3-37). With low salt concentrations, a structure about 10 nm in diameter that resembles a bead necklace is seen. The string between the beads of the necklace can be digested away with the enzyme DNase, so the string can be inferred

to be DNA. The beads on the necklace are called **nucleosomes,** which consist of special chromosomal proteins, called **histones,** and DNA. Histone structure is remarkably conserved across the gamut of eukaryotic organisms, and nucleosomes are always found to contain an octamer of two units each of histones H2A, H2B, H3, and H4. The DNA is wrapped twice around the octamer, as shown in Figure 3-38a. When salt concentrations are higher, the nucleosome bead necklace gradually assumes a coiled form called a **solenoid** (Figure 3-38b). This solenoid produced in vitro is 30 nm in diameter and probably corresponds to the in vivo spaghetti-like structures that we first encountered in Figure 3-35. The solenoid is thought to be stabilized by another histone, H1, that runs down the center of the structure, as Figure 3-38b shows.

We see, then, that to achieve its first level of packaging, DNA winds onto histones, which act somewhat like spools. Further coiling results in the solenoid conformation. However, it takes at least one more level of packaging to convert the solenoids into the three-dimensional structure that we call the *chromosome*.

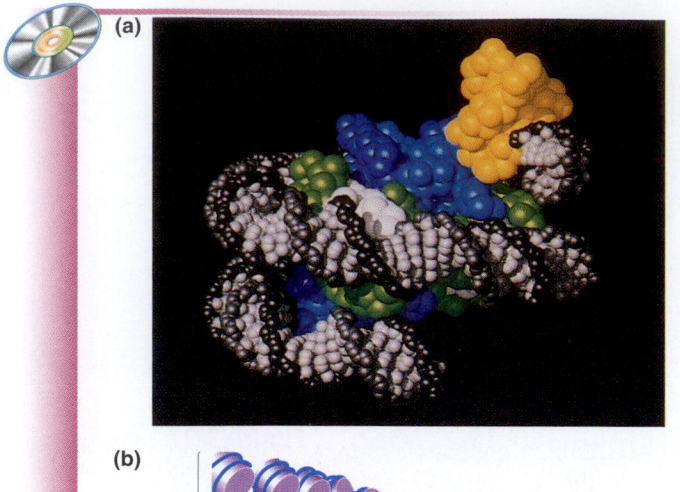

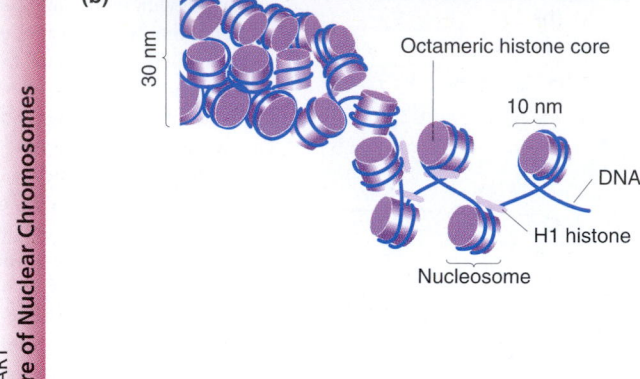

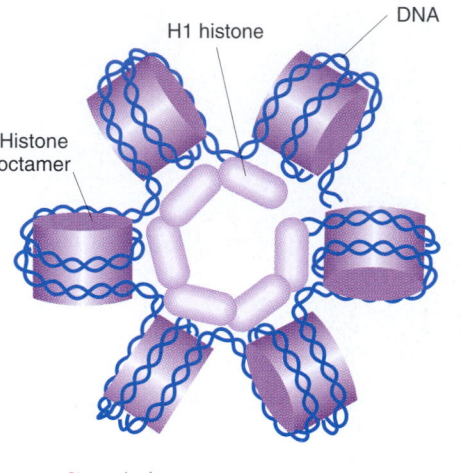

Figure 3-38   (a) Model of a nucleosome showing the DNA wrapped twice around a histone octamer. (b) Two views of a model of the 30-nm solenoid showing histone octamers as purple disks. *(Left)* Partly unwound lateral view. *(Right)* End view. The additional histone H1 is shown running down the center of the coil, probably acting as a stabilizer. With increasing salt concentrations, the nucleosomes close up to form a solenoid with six nucleosomes per turn. (Part a, Alan Wolffe and Van Moudrianakis; part b from H. Lodish, D. Baltimore, A. Berk, S. L. Zipursky, P. Matsudaira, and J. Darnell, *Molecular Cell Biology,* 3d ed. Copyright © 1995 by Scientific American Books.)

High-order coiling

Many cytogenetic studies show that chromosomes appear to be coiled, and Figure 3-39 shows a good example from the nucleus of a protozoan. Whereas the diameter of the solenoids is 30 nm, the diameter of these coils is the same as the diameter of the chromosome during cell division, often about 700 nm. What produces these supercoils? One clue comes from observing mitotic metaphase chromosomes from which the histone proteins have been removed chemically. After such treatment, the chromosomes have a densely staining central core of nonhistone protein called a **scaffold,** as shown in Figure 3-40 and in the electron micrograph on the first page of this chapter. Projecting laterally

Figure 3-39   Drawings of chromosomes in meiotic prophase in a protozoan, demonstrating different degrees of coiling and supercoiling visible with the light microscope. Two large chromosomes are shown: one yellow and the other orange; from part a to part d is a progression. (a) Coiling is seen, though duplication becomes apparent. (b) Duplication is well advanced. (c) Supercoiling is beginning. (d) Supercoiling is well advanced. (From L. R. Cleveland, "The Whole Life Cycle of Chromosomes and Their Coiling Systems," *Transactions of the American Philosophical Society* 39, 1941, 1.)

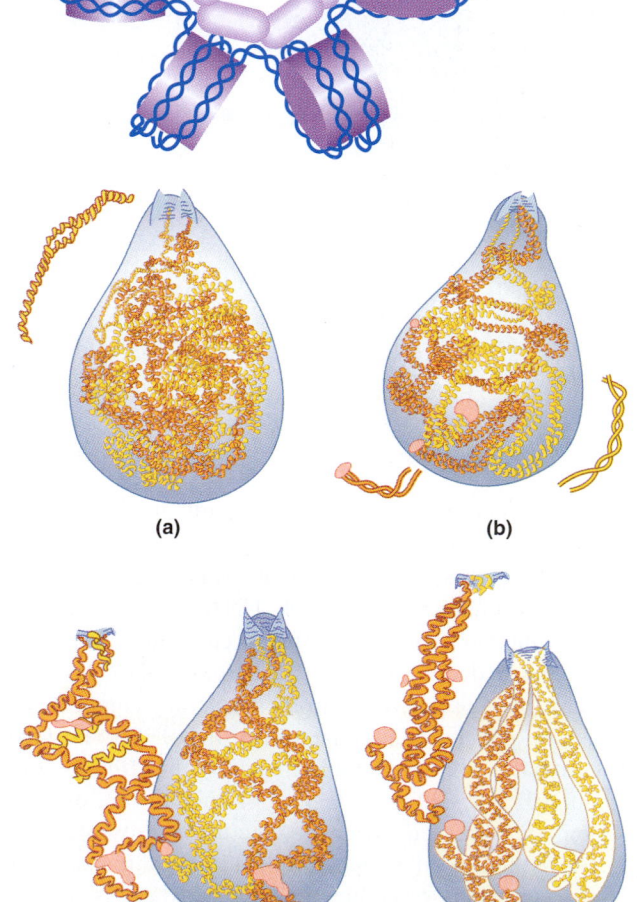

Figure 3-40 Electron micrograph of a metaphase chromosome from a cultured human cell. Note the central core, or scaffold, from which the DNA strands extend outward. No free ends are visible at the outer edge. At even higher magnification, it is clear that each loop begins and ends near the same region of the scaffold. (From W. R. Baumbach and K. W. Adolph, *Cold Spring Harbor Symposium on Quantitative Biology,* Cold Spring Harbor Laboratory, Cold Spring Harbor, N.Y., 1977.)

from this protein scaffold are loops of DNA. At high magnifications, it is clear from electron micrographs that each DNA loop begins and ends at the scaffold. The central scaffold in metaphase chromosomes is largely composed of the enzyme topoisomerase II. This enzyme has the ability to pass a strand of DNA through another cut strand. Evidently, this central scaffold manipulates the vast skein of DNA during replication, preventing many possible problems of unwinding DNA strands at this crucial stage. In any case, it is well established that there is a scaffold in eukaryotic chromosomes, and it seems to be a major organizing device for these chromosomes.

Now let us return to the question of how the supercoiling of the chromosome is produced. The best evidence suggests that the solenoids arrange in loops emanating from the central scaffold matrix, which itself is in the form of a spiral. We see the general idea in Figure 3-41, which shows a

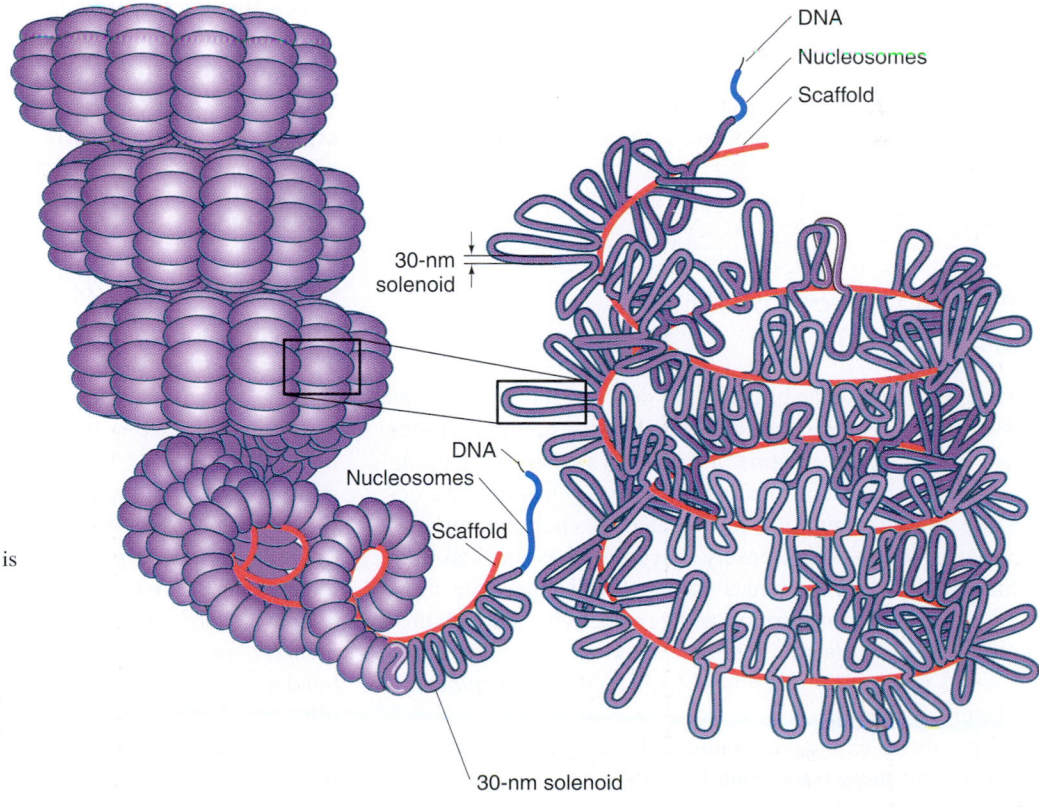

Figure 3-41 Model for chromosome structure. On the left is shown tight coiling, representing metaphase: here the loops are so densely packed that only their tips are visible. At the free ends, a solenoid is shown uncoiled to give an approximation of relative scale. On the right is shown a relaxed supercoil, as at interphase.

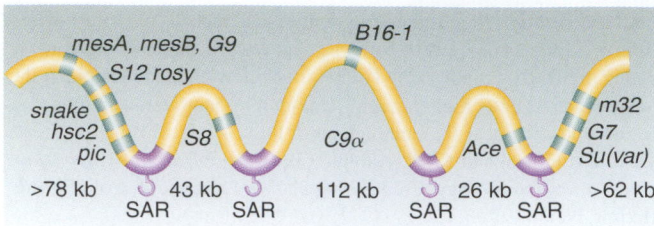

Figure 3-42   Loop domains that have been mapped in *Drosophila*. Gene loci and scaffold attachment regions (SARs) are shown.

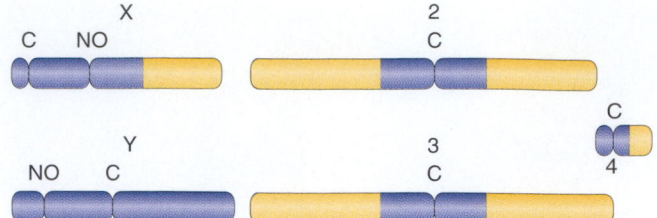

Figure 3-43   Position of heterochromatin in *Drosophila*. C represents the centromeres, and NO the nucleolar organizers. A secondary chromosomal constriction is located on chromosome 2. Heterochromatin is more darkly colored (shown in blue here). (From A. Hilliker and C. B. Sharp, *Chromosome Structure and Function*, J. P. Gustafson and R. Appels, eds. Plenum, 1988, pp. 91–95.)

representation of loosely coiled interphase chromosomes and the more tightly coiled metaphase chromosomes. How do the loops attach to the scaffold? There appear to be special regions along the DNA called **scaffold attachment regions (SARs)**. The evidence for these regions is as follows. When histoneless chromatin is treated with restriction enzymes, the DNA loops are cut off the scaffold, but special regions of DNA remain attached to the scaffold. These regions are resistant to exonuclease digestion and have been shown to have protein bound to them. When the protein is digested away, the remaining DNA regions can be analyzed and have been shown in *Drosophila* to contain sequences that are known to be specific for topoisomerase binding. This finding makes it likely that these regions are the SARs that glue the loops onto the scaffold. Some specific loops that have been mapped in *Drosophila* are shown in Figure 3-42. The size of the loops ranges between 4.5 and 112 kb. The SARs are only in nontranscribed regions of the DNA.

MESSAGE ··
In the progressive levels of chromosome packing,
1. DNA winds onto nucleosome spools.
2. The nucleosome chain coils into a solenoid.
3. The solenoid forms loops, and the loops attach to a central scaffold.
4. The scaffold plus loops arrange themselves into a giant supercoil.

Nature of heterochromatin and euchromatin

What is the molecular basis of the densely staining regions called *heterochromatin* and the less densely staining regions called *euchromatin* (Figure 3-43)? Chromosome mapping experiments show that most of the active genes that are detectable through mutation are located in euchromatin. Euchromatin stains less densely because it is packed less tightly, and the general idea is that this is the state most compatible with transcription and gene activity. Heterochromatin in most organisms is found flanking the centromeres, but some whole chromosomes, such as the *Drosophila* Y, are heterochromatic.

The most extensive studies of the genetics of heterochromatin have been done in *Drosophila*. In this organism, some essential genes are found in heterochromatin, but

the function of most of these genes is unknown. The proportion of *Drosophila* heterochromatin that maps as genes is only 1/100 of the proportion of *Drosophila* euchromatin that maps as genes; thus there are long stretches of DNA between the functional genes. There is clearly a major distinction in the gene activity of the two kinds of chromatin.

In thinking about the architecture of chromosomes, key questions are how heterochromatin differs from euchromatin and, more important, how this difference is maintained. First, the difference, as we have seen, is the degree of compaction. Whether this difference is similar to the difference between interphase and metaphase chromosomes — in other words, a difference in the tightness of coiling as shown in Figure 3-41 — or whether it resembles other structural differences is not known. The question of how this difference between euchromatin and heterochromatin is maintained also is difficult to answer. In fact, the answer is not known at present but is under intensive investigation.

MESSAGE ··
Euchromatin contains most of the active genes. Heterochromatin is more condensed, and densely staining.

Sequence organization

How are genes arranged on chromosomes? What proportion of chromosomal DNA comprises active or potentially active genes? What is the nature of the DNA between the genes? In this section, we will answer some of these questions. As in most aspects of research, the situation turns out to be more complex than expected and reveals some surprises.

Some of the first informative results came from studies in which DNA samples from eukaryotic nuclei were heated to separate the two strands of the double helix and then allowed to cool. Whenever this procedure is done, complementary sequences, such as those that are initially hydrogen bonded together in a double helix, eventually find each other through random movements of the molecules in solution. This process is called **reannealing.** But, surprisingly, in these studies, the DNA started coming together much faster than was expected on the basis of separated strands of

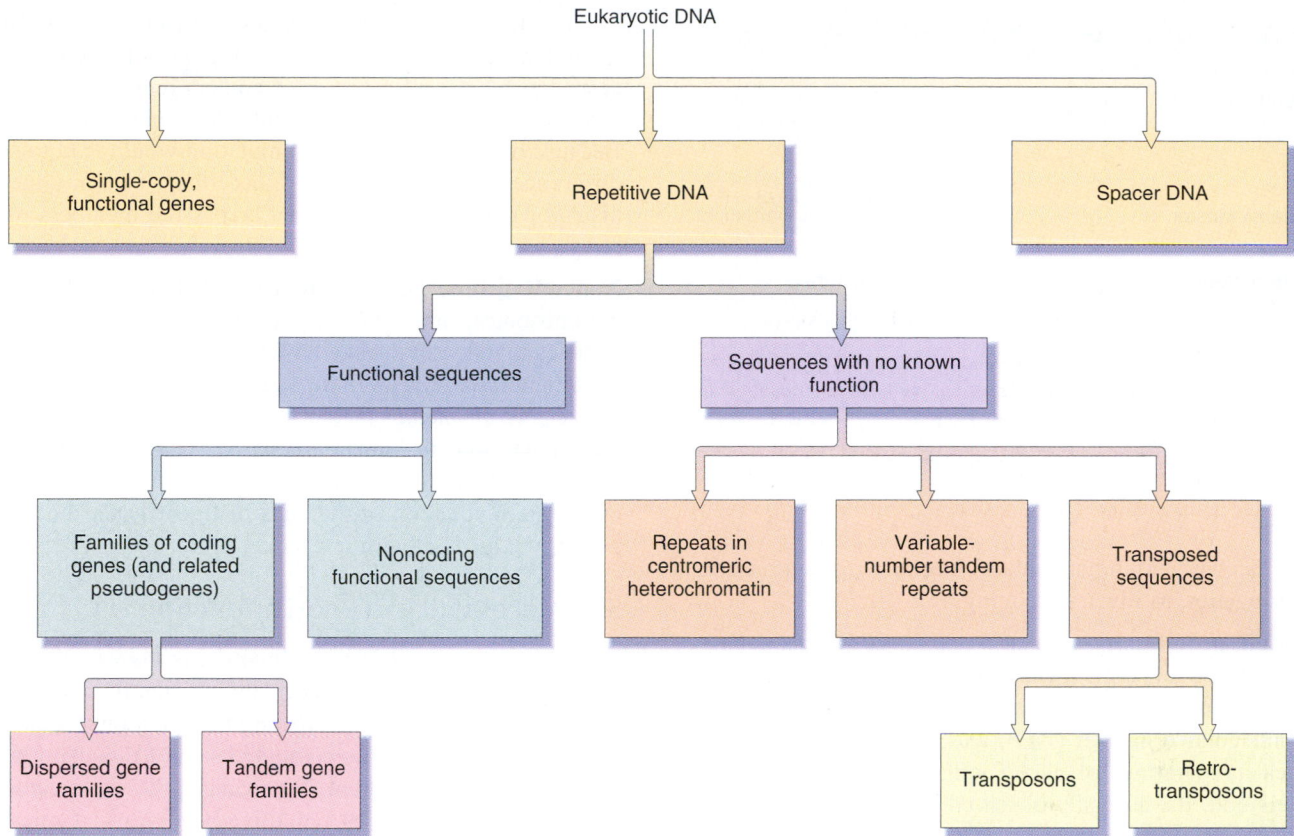

Figure 3-44 Classification of eukaryotic DNA.

single genes finding each other. The kinetics of reannealing required postulation of a class of DNA that was present in many copies in the genome. This class of DNA became known as **repetitive DNA.** Although the single-copy genes reanneal eventually, they take much longer. Now it is known that there are several classes of repetitive DNA. We can classify eukaryotic DNA as shown in Figure 3-44. The kind of genes that we have spent most time on so far in this book constitutes the unique single-copy genes (Figure 3-45), which are embedded in a diverse assembly of repetitive DNAs.

Functional repetitive sequences

Dispersed gene families. Several types of proteins are encoded by families of homologous genes spread throughout the genome. Such families may comprise only a few genes or very many, as some examples illustrate: actins, from 5 to 30; keratins more than 20; myosin heavy chain, from 5 to 10; tubulins, from 3 to 15; insect eggshell proteins, 50; globins, as many as 5; immunoglobulin variable region, 500; ovalbumin, 3; and histones, from 100 to 1000. The exact DNA sequences of the genes within a family may diverge, and the different homologous genes may come to have slightly different functions. Some genes within families

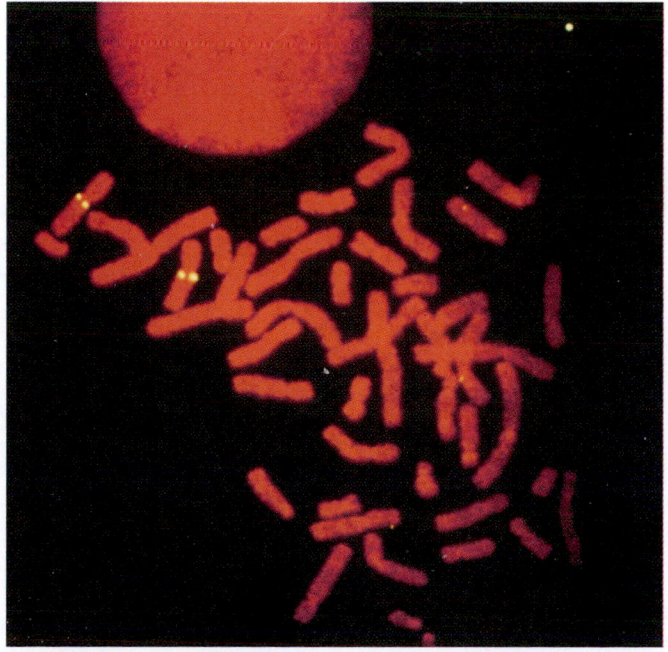

Figure 3-45 Chromosomes probed in situ with a fluorescent probe specific for a gene present in a single copy in each chromosome set — in this case, a muscle protein. Only one locus shows a fluorescent spot, corresponding to the probe bound to the muscle protein gene. (From Peter Lichter et al., *Science* 247, 1990, 64.)

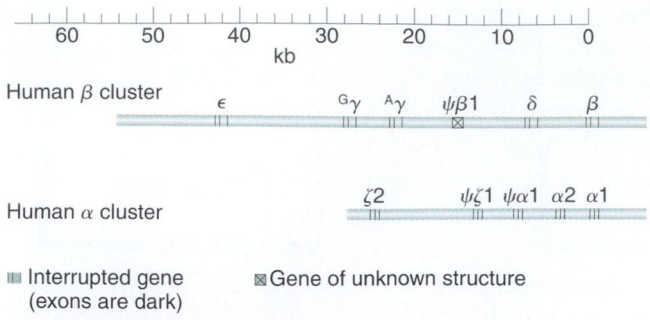

Figure 3-46 The human α-like and β-like globin gene families are each organized into a single cluster that includes functional genes and pseudogenes; the latter are denoted here by ψ. (After B. Lewin, *Genes*. Wiley, 1983.)

have become nonfunctional, giving rise to untranscribed **pseudogenes,** as illustrated in Figure 3-46.

Tandem gene family arrays. Cells need large amounts of the products of some genes, and families of these genes have evolved into tandem arrays. A good example is seen in the nucleolar organizer (NO), which was observed in cytological preparations of nuclei long before its function was understood. It was easily observed because it does not stain with normal chromatin stains. The role of the NO was revealed by a variety of genetic and molecular studies. We now know that the NOs on the *Drosophila* X and Y chromosomes contain 250 and 150 tandem copies of rRNA genes, respectively. One human NO has about 250 copies. Such redundancy is one way of ensuring a large amount of rRNA per cell.

MESSAGE ··
The nucleolar organizer, which is cytologically distinct, is a tandem array of genes that encode ribosomal RNA.
···

Another example of a tandem array consists of the genes for tRNA. In humans, about 50 chromosomal sites correspond to the different tRNA types, and each site contains between 10 and 100 copies.

Finally, the genes for histones themselves are arranged in tandem arrays in some species (Figure 3-47). For histone arrays, as for the other arrays of tandem repeats, sequencing analysis has shown that the multiple copies are identical. Because one might expect that mutation would lead to some differences at noncrucial sites, there must be some mechanism for maintaining constancy across the members of the array.

Noncoding functional sequences. Telomeres, the tips of chromosomes, have tandem arrays of simple DNA sequences that do not encode an RNA or a protein product but nevertheless have a definite function. For example, in the ciliate *Tetrahymena,* there is repetition of the sequence TTGGGG, and, in humans, it is TTAGGG. The telomeric repeats are there to solve a functional problem that is inherent in the replication of the ends of linear DNA molecules, which will be dealt with in Chapter 8.

Sequences with no known function

For some repetitive DNA, no function is known. This category contains DNA sequences that generally have many more copies in the genome than do the functional sequences heretofore discussed. The combined size of this class is surprising; for example, about 20 percent of the human genome consists of nonfunctional repetitive sequences of one kind or another. We will consider three types.

Highly repetitive centromeric DNA. When genomic DNA is spun for a long time in a cesium chloride density gradient in an ultracentrifuge, the DNA settles into one prominent visible band. However, satellite bands are often visible, distinct from the main DNA band. Such **satellite DNA** consists of multiple tandem repeats of short nucleotide sequences, stretching to as much as hundreds of kilobases in length. When probes are prepared from such simple-sequence DNA and used in chromosomal in situ labeling experiments, the great bulk of the satellite DNA is found to reside in the heterochromatic regions flanking the centromeres. There can be either one or several basic repeating units, but usually they are less than 10 bases long. For ex-

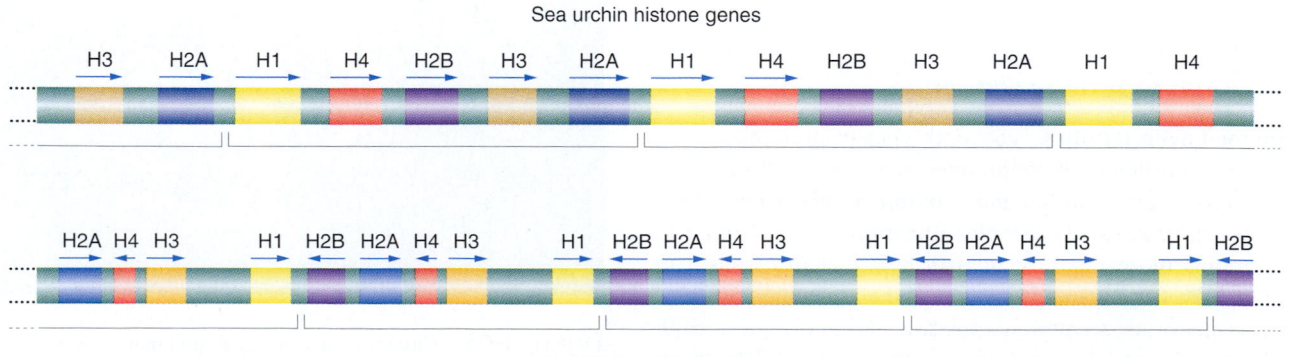

Figure 3-47 Tandem repeats of histone genes in sea urchin and fruit fly. Only a small fraction of the repeats is shown. Arrows indicate direction of transcription.

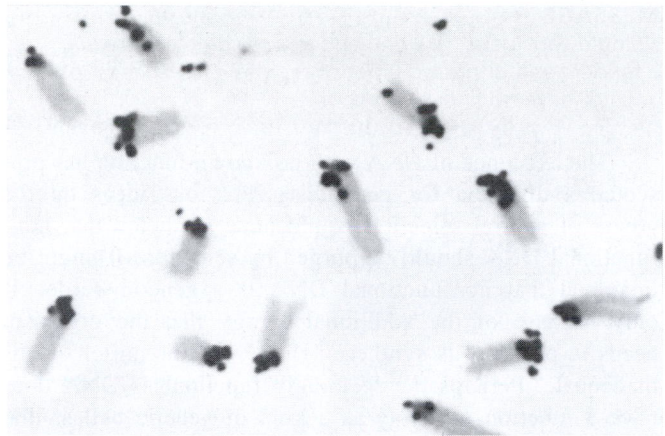

Figure 3-48   Autoradiographic localization of simple-sequence mouse DNA to the centromeres. A radioactive probe for simple-sequence DNA was added to chromosomes whose DNA had been denatured. (Note that all mouse chromosomes have their centromeres at one end.) (From M. L. Pardue and J. G. Gall, *Science* 168, 1970, 1356.)

ample, in *Drosophila melanogaster,* the sequence AATAA-CATAG is found in tandem arrays around all centromeres. Similarly, in the guinea pig, the shorter sequence CCCTAA is arrayed flanking the centromeres. In situ labeling of a mouse satellite DNA is shown in Figure 3-48.

Because the centromeric repeats are a nonrepresentative sample of the genomic DNA, the G + C content can be significantly different from the rest of the DNA. For this reason, the DNA forms a separate satellite band in an ultracentrifuge density gradient. There is no demonstrable function for centromeric repetitive DNA, nor is there any understanding of its relation to heterochromatin or to genes in the heterochromatin. Some organisms have staggering amounts of this DNA; for example, as much as 50 percent of kangaroo DNA can be centromeric satellite DNA.

VNTRs.   A special class of tandem repeats shows variable number at different chromosomal positions and in different individual members of a species. This type of repeat is called a **variable number tandem repeat (VNTR),** sometimes called **minisatellite DNA.** The VNTR loci in humans are 1- to 5-kb sequences consisting of variable numbers of a repeating unit from 15 to 100 nucleotides long. If a VNTR probe is available and the total genomic DNA is cut with a restriction enzyme that has no target sites within the VNTR arrays, then a Southern blot reveals a large number of different-sized fragments that are bound to the probe. Because of the variability in the number of tandem repeats from individual to individual, the set of fragments that shows up on the Southern autoradiogram is highly individualistic. In fact, these patterns are called **DNA fingerprints.** They are used extensively in forensic medicine (Chapter 14).

Another class of dispersed repetitive DNA is composed of dinucleotide repeats. This class of repetitive DNA is called **microsatellite DNA.** Although not normally included in the VNTR class, microsatellites are in fact dispersed regions composed of variable numbers of dinucleotides repeated in tandem. Because the number of repeats varies between individuals, this type of DNA has been most useful in providing a dense array of molecular markers for mapping the huge expanses of the human genome, as we shall see in Chapter 14.

Transposed sequences.   A large proportion of a eukaryotic genome is composed of repetitive elements that have propagated within the genome by making copies of themselves, which can move into new locations. These elements are collectively called transposed sequences, and they are described in detail in Chapter 21. Those elements that move as DNA are called **transposons.** Many genomes have multiple copies of such elements or truncated versions of them dispersed throughout the genome.

Another general class of transposed sequences comprises the **retrotransposons** (Figure 3-49), DNA sequences that propagate themselves through the action of reverse transcriptase, an enzyme that can make a DNA strand from RNA. One type within this category consists of repetitive sequences whose structures are related to retroviruses. They move by reverse transcription of their RNA transcripts into DNA, which then inserts throughout the genome. Examples of retroviral-like retrotransposons are the copia elements of *Drosophila* (5-kb sequences present at about 50 sites per genome), and the Ty elements of yeast (6-kb elements with about 30 full-length copies per genome). The **long interspersed elements (LINES)** of mammals are 1- to 5-kb nonviral retroelements of which there are from 20,000 to 40,000 copies per human genome.

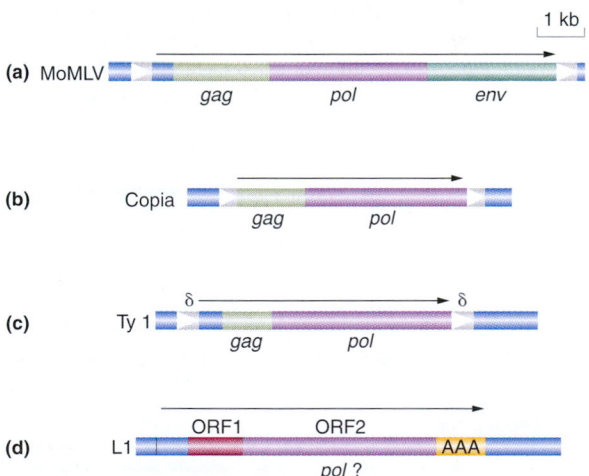

Figure 3-49   Structures of four retroelements found in eukaryote genomes: *gag, pol,* and *env* are viral genes; ORFs are genes of unknown function; AAA is a poly(A) tail (found in mRNAs); triangles are direct repeats. (a) A retrovirus, Moloney murine leukemia virus. (b) A retrotransposon of *Drosophila,* Copia. (c) A retrotransposon of yeast, Ty1. (d) A human LINE, L1. (After J. R. S. Fincham, *Genetic Analysis: Principles, Scope and Objectives.* Blackwell, 1994.)

The Alu repetitive sequence in the human genome, so named because it contains an *Alu* restriction enzyme target site, is an example of a class of retrotransposons that are not related to retroviruses. The human genome contains hundreds of thousands of whole and partial Alu sequences, scattered between genes and within introns and making up about 5 percent of human DNA. The full Alu sequence is about 200 nucleotides long and bears remarkable resemblance to 7SL RNA, an RNA that is part of a complex whose role is to secrete newly synthesized polypeptides through the endoplasmic reticulum. Presumably, the Alu sequences originated as reverse transcripts of these RNA molecules. Short interspersed repeats such as Alu sequences are collectively called **SINEs,** for **short interspersed elements.**

Other examples of this class of moderately repetitive elements are the many scattered *pseudogenes* that have clearly been created by the reverse transcription process inasmuch as they do not contain the introns that are found in the original functional gene.

Spacer DNA

The final category of DNA is **spacer DNA,** which is basically what is left after all the recognizable units have been identified. Needless to say, little is known about spacer DNA. Possibly its only function is to space, but no studies have yet been performed to delete such DNA and observe the consequences.

MESSAGE ··
Single-copy genes are embedded in a complex array of tandem and dispersed types of repetitive DNA, most of which have no known function.
··

The existence of DNA with no known function has presented a dilemma for geneticists. Previous ideas on the power of natural selection would have predicted that nonfunctional DNA should be purged by selection. It might be imagined that nonfunctional DNA is a genetic burden if only because of the additional energy that the organism needs to put into its synthesis. However, this notion seems inadequate. Perhaps the seemingly functionless DNA does have a function, possibly as a kind of genetic ballast that gives the chromosome bulk to permit efficient partitioning at cell division or perhaps to separate the functional elements (genes) for their efficient regulation. Alternatively, the repetitive elements that make up most of the nonfunctional DNA might have hit upon a way of avoiding detection by the forces of natural selection. Such DNA has been termed **selfish DNA,** a type of DNA that exists only for the purpose of existing and is never exposed to the rigors of phenotype.

Figure 3-50 is a stylized diagram of the overall organization of a hypothetical eukaryotic chromosome, summarizing much of the preceding discussion. Some actual spacings known from sequencing are shown in Figure 3-51.

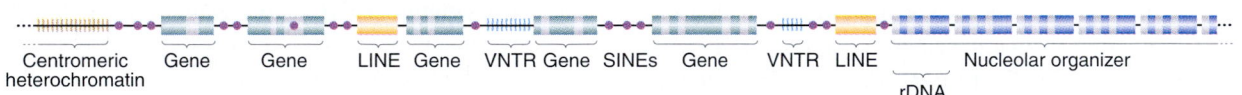

Centromeric heterochromatin   Gene   Gene   LINE   Gene   VNTR   Gene   SINEs   Gene   VNTR   LINE   Nucleolar organizer   rDNA

Figure 3-50   General depiction of a eukaryotic chromosomal landscape. This small region of a chromosome happens to have five protein-coding genes, one end of a nucleolar organizer, and one end of centromeric heterochromatin. Various kinds of repetitive DNAs are shown. (Each chromosome would normally have several thousand genes.)

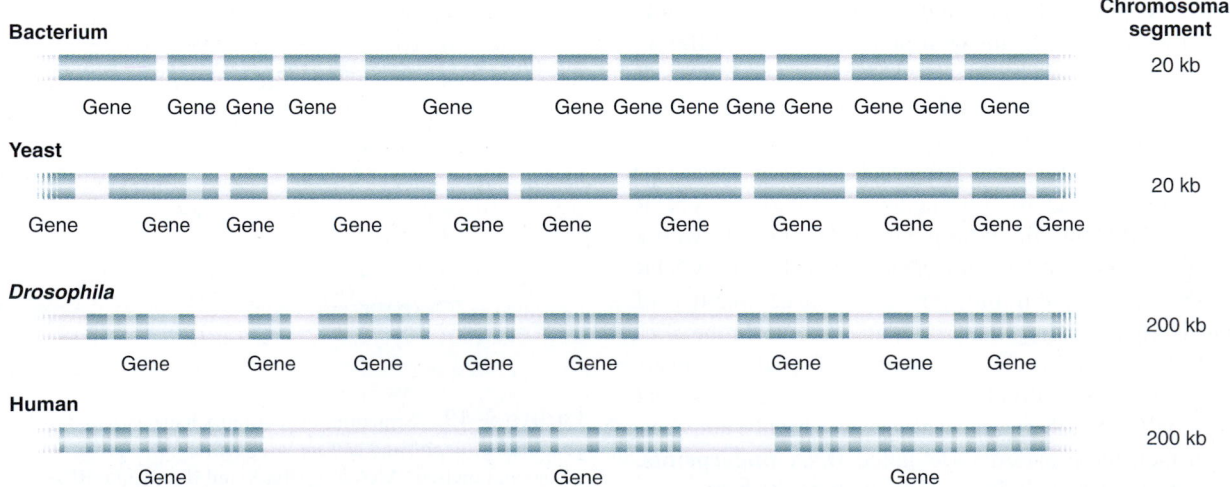

Figure 3-51   Approximate sizes (in kilobases) and spacing of genes in several representative organisms. Coding sequences are in green; introns are in yellow. Note that the *Drosophila* and human segments shown are ten times larger than the bacterium and yeast segments.

SUMMARY

After the rediscovery of Mendelian principles in 1900, scientists set out to discover what structures within cells correspond to Mendel's hypothetical units of heredity, which we now call genes. Recognizing that the behavior of chromosomes during meiosis parallels the behavior of genes, Sutton and Boveri suggested that genes were located in or on the chromosomes. Morgan's evidence from sex-linked inheritance strengthened the hypothesis. Proof that genes are on chromosomes came from Bridges's use of aberrant chromosome behavior to explain inheritance anomalies.

We now know which chromosome behaviors produce Mendelian ratios. Mendel's first law (equal segregation) results from the separation of a pair of homologous chromosomes into opposite cells at the first division. Mendel's second law (independent assortment) results from independent behavior of separate pairs of homologous chromosomes.

Because Mendelian laws are based on meiosis, Mendelian inheritance characterizes any organism with a meiotic stage in its life cycle, including diploid organisms, haploid organisms, and organisms with alternating haploid and diploid generations.

Chromosomes are distinguished by a number of topological features such as centromere and nucleolar position, size, and banding pattern. The stuff that chromosomes are made of is chromatin, composed of DNA and protein. Each chromosome is one DNA molecule wrapped around octamers of histone proteins. Between cell divisions, the chromosomes are in a relatively extended state, although still associated with histones. At cell division, the chromosomes condense by a tightening of the coiling. This condensed state permits easy manipulation by the nuclear spindle.

The DNA in eukaryotic genomes consists partly of unique sequences (genes) and partly of repetitive DNA. Functional gene families, either tandem or dispersed, account for some repetitive DNA. Several types of highly repetitive DNA consist of short-sequence repeats of unknown function. The number of repeats of a given type at one chromosome position can vary among individuals, giving a unique DNA fingerprint.

CONCEPT MAP

Draw a concept map interrelating as many of the following terms as possible. Note that the terms are listed in no particular order.

Genes / chromosomes / meiosis / equal segregation / independent assortment / haploid / DNA / replication / chromatid / histones / repetitive DNA

CHAPTER INTEGRATION PROBLEM

Two *Drosophila* flies that had normal (transparent, long) wings were mated. In the progeny, two new phenotypes appeared, dusky wings (having a semiopaque appearance) and clipped wings (with squared ends). The progeny were as follows:

Females	179 transparent, long
	58 transparent, clipped
Males	92 transparent, long
	89 dusky, long
	28 transparent, clipped
	31 dusky, clipped

a. Provide a genetic explanation for these results, showing genotypes of parents and of all progeny classes under your model.

b. Design a test for your model.

◆ Solution ◆

a. The first step is to state any interesting features of the data. The first striking feature is the appearance of two new phenotypes. We encountered the phenomenon in Chapter 2, where it was explained as recessive alleles masked by their dominant counterparts. So first we might suppose that one or both parental flies have recessive alleles of two different genes. This inference is strengthened by the observation that some progeny express only one of the new phenotypes. If the new phenotypes always appeared together, we might suppose that the same recessive allele determines both.

However, the other striking aspect of the data, which we cannot explain by using the Mendelian principles from Chapter 2, is the obvious difference between the sexes; although there are approximately equal numbers of males and females, the males fall into four phenotypic classes, but the females constitute only two. This fact should immediately suggest some kind of sex-linked inheritance. When we study the data, we see that the long and clipped phenotypes are segregating in both males and females, but only males have the dusky phenotype. This observation suggests that the inheritance of wing transparency differs from the inheritance of wing shape. First, long and clipped are found in a 3:1 ratio in both males and females. This ratio can be explained if the parents are both heterozygous for an autosomal gene; we can represent them as L/l, where L stands for long and l stands for clipped.

Having done this partial analysis, we see that it is only the wing transparency inheritance that is associated with sex. The most obvious possibility is that the alleles for transparent (D) and dusky (d) are on the X chromosome, because we have seen in Chapter 2 that gene location on this chromosome gives inheritance patterns correlated with sex. If this suggestion is true, then the parental female must be the one sheltering the d allele, because, if the male had the d, he would have been dusky, whereas we were told that he had transparent wings. Therefore, the female parent would be D/d and the male D. Let's see if this suggestion works: if it is true, all female progeny would inherit the D allele from their father, so all would be transparent winged. This was observed. Half the sons would be D (transparent) and half d (dusky), which was also observed.

So, overall, we can represent the female parent as D/d ; L/l and the male parent as D ; L/l. Then the progeny would be

Females

Males

b. Generally, a good way to test such a model is to make a cross and predict the outcome. But which cross? We have to predict some kind of ratio in the progeny, so it is important to make a cross from which a unique phenotypic ratio can be expected. Notice that using one of the female progeny as a parent would not serve our needs: we cannot say from observing the phenotype of any one of these females what her genotype is. A female with transparent wings could be D/D or D/d, and one with long wings could be L/L or L/l. It would be good to cross the parental female of the original cross with a dusky, clipped son, because the full genotypes of both are specified under the model that we have created. According to our model, this cross is:

$$D/d ; L/l \times d ; l/l$$

From this cross, we predict

Females

Males

SOLVED PROBLEMS

Two corn plants are studied; one is A/a, and the other is a/a. These two plants are intercrossed in two ways: using A/a as female and a/a as male; using a/a as female and A/a as male. Recall from Figure 3-22 that the endosperm is $3n$ and is formed by the union of a sperm cell with the two polar nuclei of the female gametophyte.

a.   What endosperm genotypes does each cross produce? In what proportions?

b.   In an experiment to study the effects of "doses" of alleles, you wish to establish endosperms with genotypes $a/a/a$, $A/a/a$, $A/A/a$, and $A/A/A$ (carrying 0, 1, 2, and 3 "doses" of A, respectively). What crosses would you make to obtain these endosperm genotypes?

◆ Solution ◆

a.   In such a question, we have to think about meiosis and mitosis at the same time. The meiospores are produced by meiosis; the nuclei of the male and female gametophytes in higher plants are produced by the mitotic division of the meiospore nucleus. We also need to study the corn life cycle to know what nuclei fuse to form the endosperm.

First cross: A/a ♀ $\times$ a/a ♂

Here, the female meiosis will result in spores of which half will be A and half will be a. Therefore, similar proportions of haploid female gametophytes will be produced. Their nuclei will be either all A or all a, because mitosis reproduces genetically identical genotypes. Likewise, all nuclei in every male gametophyte will be a. In the corn life cycle, the endosperm is formed from two female nuclei plus one male nucleus, so two endosperm types will be formed as follows:

♀ Spore	♀ Polar Nuclei	♂ Sperm	3n Endosperm
$\frac{1}{2} A$	A and A	a	$\frac{1}{2} A/A/a$
$\frac{1}{2} a$	a and a	a	$\frac{1}{2} a/a/a$

Second cross: a/a ♀ $\times$ A/a ♂

♀ Spore	♀ Polar Nuclei	♂ Sperm	3n Endosperm
all *a*	all *a* and *a*	$\frac{1}{2} A$	$\frac{1}{2} A/a/a$
		$\frac{1}{2} a$	$\frac{1}{2} a/a/a$

The phenotypic ratio of endosperm characters would still be Mendelian, even though the underlying endosperm genotypes are slightly different. (However, none of these problems arise in embryo characters, because embryos are diploid.)

b. This kind of experiment has been very useful in studying plant genetics and molecular biology. In answering the question, all we need to realize is that the two polar nuclei contributing to the endosperm are genetically identical. To obtain endosperms, all of which will be *a/a/a*, any *a/a* × *a/a* cross will work. To obtain endosperms, all of which will be *A/a/a*, the cross must be *a/a* ♀ × *A/A* ♂. To obtain embryos, all of which will be *A/A/a*, the cross must be *A/A* ♀ × *a/a* ♂. For *A/A/A*, any *A/A* × *A/A* cross will work. These endosperm genotypes can be obtained in other crosses but only in combination with other endosperm genotypes.

PROBLEMS

1. Name the key function of mitosis.

2. Name two key functions of meiosis.

3. Can you design a different nuclear division system that would achieve the same outcome as meiosis?

4. In a possible future scenario, male fertility drops to zero, but, luckily, scientists develop a way for women to produce babies by virgin birth. What were meiocytes are converted directly into zygotes, which implant in the usual way. What would be the short- and long-term effects in such a society?

5. In what ways does the second division of meiosis differ from mitosis?

6. Normal mitosis takes place in a diploid cell of genotype *A/a* ; *B/b*. Which of the following genotypes represent possible daughter cells? *A* ; *B, a* ; *b, A* ; *b, a* ; *B, A/A* ; *B/B, A/a* ; *B/b, a/a* ; *b/b*.

7. In an attempt to simplify meiosis for the benefit of students, mad scientists develop a way of preventing premeiotic S phase and making do with just having one division, including pairing, crossing-over, and segregation. Would this system work and would the products of such a system differ from those of the present system?

8. In a diploid organism of $2n = 5$, assume that you can label all the centromeres derived from its female parent and all the centromeres derived from its male parent. When this organism produces gametes, how many male/female labeled centromeric combinations are possible in the gametes?

9. In corn, DNA measurements are made on several nuclei in light-absorption units. The measurements were:

    0.7, 1.4, 2.1, 2.8, and 4.2

    Which cells could have been used for these measurements?

10. A certain species of seaweed shows alternation of generations, but its diploid phase looks just like its haploid phase. A red mutant of the normally brown species was obtained. The red mutant was placed in a container with a brown selected at random. Soon, new plants started appearing, half of which were red and half brown. Suggest an explanation for this result.

11. Draw a haploid mitosis of the genotype a^+ ; b in the style of Figure 3-18.

12. The plant *Haplopappus gracilis* has a $2n$ of 4. A diploid cell culture was established and, at premitotic S phase, a radioactive nucleotide was added and was incorporated into newly synthesized DNA. The cells were then removed from the radioactivity, washed, and allowed to proceed through mitosis. Radioactive chromosomes or chromatids can be detected by placing photographic emulsion on the cells; radioactive chromosomes or chromatids appeared covered with spots of silver from the emulsion. (The chromosomes "take their own photograph".) Draw the chromosomes at prophase and telophase of the first and second mitotic divisions after the radioactive treatment. If they are radioactive, show it in your diagram. If there are several possibilities, show them, too.

13. In the species of Problem 12, you can introduce radioactivity by injection into the anthers at the S phase before meiosis. Draw the four products of meiosis with their chromosomes, and show which are radioactive.

14. Peas are diploid and $2n = 14$. Neurospora is haploid and $n = 7$. If it were possible to fractionate genomic DNA from both by using pulsed field electrophoresis, how many distinct bands would be visible in each species?

15. The broad bean *(Vicia faba)* is diploid and $2n = 18$. Each haploid chromosome set contains approximately 4 m of DNA. The average size of each chromosome during the metaphase of mitosis is 13 μm. What is the

average packing ratio of DNA at metaphase? (Packing ratio = length of chromosome to length of DNA molecule therein.) How is this packing achieved?

16. Boveri said, "The nucleus doesn't divide; it is divided." What was he getting at?

17. Galton, a geneticist of the pre-Mendelian era, devised the principle that half of our genetic makeup is derived from each parent, one-quarter from each grandparent, one-eighth from each great grandparent, and so forth. Was he right? Explain.

18. If children obtain half of their genes from one parent and half from the other parent, why aren't siblings identical?

19. The DNA double helices of chromosomes can be partly unwound in situ by special treatments.

    a. If such a preparation is bathed in a radioactive probe specific for one unique gene, what pattern of radioactivity is expected?

    b. If such a preparation is bathed in a radioactive probe specific for dispersed repetitive DNA, what pattern of radioactivity is expected?

    c. If such a preparation is bathed in a radioactive probe specific for ribosomal DNA, what pattern of radioactivity is expected?

20. If genomic DNA is cut with a restriction enzyme and fractionated by size by electrophoresis, what pattern of Southern hybridization is expected for the three probes shown in Problem 19?

21. In corn, the allele *s* causes sugary endosperm, whereas *S* causes starchy. What endosperm genotypes result from the following crosses?

    s/s female $\times$ S/S male
    S/S female $\times$ s/s male
    S/s female $\times$ S/s male

22. In corn, the allele *f'* causes floury endosperm, and *f"* causes flinty endosperm. In the cross f'/f' ♀ $\times$ $f"/f"$ ♂, all the progeny endosperms are floury, but, in the reciprocal cross, all the progeny endosperms are flinty. What is a possible explanation? (Check the corn life cycle.)

23. In moss, the genes *A* and *B* are expressed only in the gametophyte. A sporophyte of genotype A/a ; B/b is allowed to produce gametophytes.

    a. What proportion of the gametophytes will be A ; B?

    b. If fertilization is random, what proportion of sporophytes in the next generation will be A/a ; B/b?

24. When a cell of genotype A/a ; B/b ; C/c having all the genes on separate chromosome pairs divides mitotically, what are the genotypes of the daughter cells?

25. In working with a haploid yeast, you cross a purple (ad^-) strain of mating type *a* and a white (ad^+) strain of mating type α. If ad^- and ad^+ are alleles of one gene and *a* and α are alleles of an independently inherited gene on a separate chromosome pair, what progeny do you expect to obtain? In what proportions?

26. State where cells divide mitotically and where they divide meiotically in a fern, a moss, a flowering plant, a pine tree, a mushroom, a frog, a butterfly, and a snail.

27. Human cells normally have 46 chromosomes. For each of the following stages, state the number of chromosomes present in a human cell:

    a. metaphase of mitosis

    b. metaphase I of meiosis

    c. telophase of mitosis

    d. telophase I of meiosis

    e. telophase II of meiosis

    (In your answers, count chromatids as chromosomes.)

28. Four of the following events are part of both meiosis and mitosis, but one is only meiotic. Which one? (1) Chromatid formation, (2) spindle formation, (3) chromosome condensation, (4) chromosome movement to poles, (5) chromosome pairing.

29. Suppose that you discover two interesting *rare* cytological abnormalities in the karyotype of a human male. (A karyotype is the total visible chromosome complement.) There is an extra piece (or satellite) on *one* of the chromosomes of pair 4, and there is an abnormal pattern of staining on one of the chromosomes of pair 7. Assuming that all the gametes of this male are equally viable, what proportion of his children will have the same karyotype as his?

30. Suppose that meiosis occurs in the transient diploid stage of the cycle of a haploid organism of chromosome number *n*. What is the probability that an individual haploid cell resulting from the meiotic division will have a complete parental set of centromeres (that is, a set all from one parent or all from the other parent)?

31. Pretend that the year is 1868. You are a skilled young lens maker working in Vienna. With your superior new lenses, you have just built a microscope that has better resolution than any others available. In your testing of this microscope, you have been observing the cells in the testes of grasshoppers and have been fascinated by the behavior of strange elongated structures that you have seen within the dividing cells. One day, in the library, you read a recent journal paper by G. Mendel on hypothetical "factors" that he claims explain the results of certain crosses in peas. In a flash of revelation, you are struck by the parallels between your grasshopper

studies and Mendel's, and you resolve to write him a letter. What do you write? (Based on an idea by Ernest Kroeker.)

32. The plant *Haplopappus gracilis* is diploid and $2n = 4$. There are one long pair and one short pair of chromosomes. The diagrams below represent anaphases ("pulling apart" stages) of individual cells in meiosis or mitosis in a plant that is genetically a dihybrid (A/a ; B/b) for genes on different chromosomes. The lines represent chromosomes or chromatids, and the points of the V's represent centromeres. In each case, say if the diagram represents a cell in meiosis I, meosis II, or mitosis. If a diagram shows an impossible situation, say so.

4

GENE INTERACTION

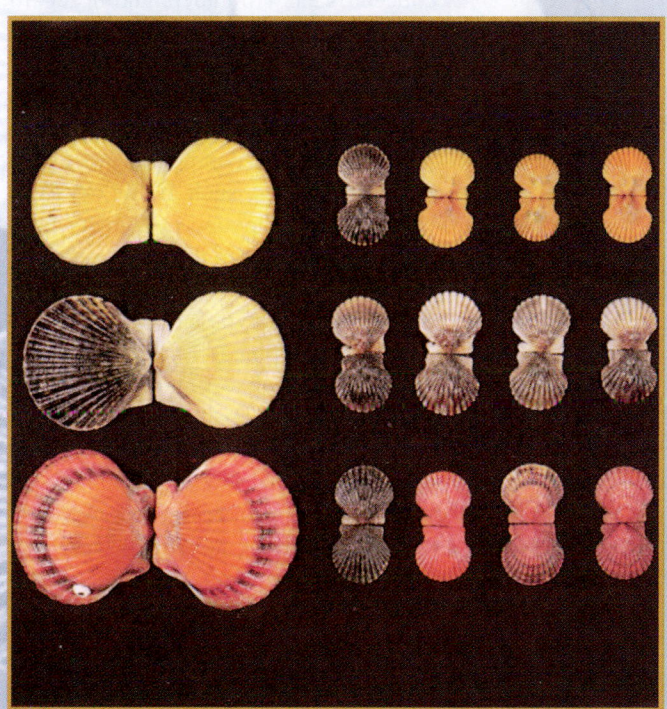

Variation in shell coloration of the bay scallop _(Argopecten irradians)_ caused by three alleles of one gene.
Yellow, black, and orange are determined by the alleles p^y, p^b, and p^o, respectively. The groups of small shells represent the proportions obtained upon selfing individuals of genotype p^y / p^b (top row), p^b / p^b (middle row), and p^o / p^b (bottom row). These results demonstrate the allelic relationship and also show that p^y and p^o are both dominant to p^b. _(From L. Adamkewicz and M. Castagna, Journal of Heredity 79, 1988, 15/BPS.)_

Key Concepts

When two haploid genomes containing different recessive mutations are combined in one cell and the phenotype is mutant, the mutations must be in the same gene (alleles).

When two haploid genomes containing different recessive mutations are combined in one cell and the phenotype is wild type, the mutations must be in different genes.

The phenotypes of some heterozygotes reveal types of dominance other than full dominance.

Some mutant alleles can kill the organism.

Most characters are determined by sets of genes that interact with one another and with the environment.

Modified monohybrid ratios reveal allelic interactions.

Modified dihybrid ratios reveal gene interactions.

So far in this book the emphasis has been on individual genes. We have examined the inheritance patterns of individual genes and deduced the laws governing their inheritance. We have also begun the process of examining the cellular processes by which individual genes can affect the phenotype. But it is now time to expand the discussion to recognize that genes interact in many ways. An individual gene cannot achieve anything by itself; it must act in a cellular setting that is determined by many other genes and by the environment.

From genes to phenotypes

At one level, geneticists tend to think of genes in isolation. Every time we make a cross between a red-eyed (wild-type) strain of *Drosophila* and a white-eyed mutant strain, each parent and each offspring contains 20,000 or so genes. Nonetheless, in such a cross, we would take note only of the one segregating genetic difference that we care about — the eye color of each fly. In reality, genes do not act in isolation. The proteins and RNAs that they encode contribute to specific cellular pathways that also receive input from the products of many other genes. Furthermore, the expression of a single gene depends on many factors, including the specific genetic backgrounds of the flies and a range of environmental conditions — temperature, nutritional conditions, population density, and so forth. (Such dependence of gene expression on environmental factors was considered in Chapter 1 in the section on the norm of reaction.)

Gene action is a term that covers a very complex set of events, and there is probably no case for which we understand all of the events that transpire from the level of expression of a single gene to the level of an organism's phenotype. In this chapter, we shall consider some of the methods that geneticists use to unravel the mechanistic strands that connect gene to final phenotype. We can start by making a couple of generalizations about the complexity of gene action.

1. *There is a one-to-many relation of genes to phenotypes.* This relation is called *pleiotropy*, which is inferred from the observation that mutations selected for their effect on one specific character are often found to affect other characters of the organism. This effect might mean that related physiological pathways contribute to a similar phenotype in several tissues. For example, the white eye-color mutation in *Drosophila* results in lack of pigmentation not only in compound eyes, but also in ocelli (simple eyes), in sheaths of tissue surrounding the male gonad, and in the Malpighian tubules (the fly's kidneys). In all of these tissues, pigment formation requires the uptake of pigment precursors by the cells. The white allele causes a defect in this uptake, thereby blocking pigment formation in all of these tissues.

   Often, pleiotropy includes multiple events that are not obviously physiologically related. For example, the dominant *Drosophila* mutation *Dichaete* not only causes the wings to be held out laterally, but also removes certain hairs on the back of the fly; furthermore the mutation is lethal when homozygous. This example illustrates a real limitation in the way that dominant and recessive mutations are named. The reality is that a single mutation can be both dominant and recessive, depending on which aspect of its pleiotropic phenotype is under consideration. In general, genetic terminology is not up to the task of representing this level of pleiotropy and complexity in one symbol, and, as will be discussed later, how we name alleles has a certain arbitrary or historical aspect. In this chapter, we consider some of the variations on the general theme of dominance/recessiveness and some of the traditional ways in which alleles are named.

2. *There is a one-to-many relation of phenotypes to genes.* This idea, the converse of the first, is based on the observation that many different genes can affect a single phenotype. This concept is easy to understand in regard to a character such as eye color, for which there are complex metabolic pathways with numerous enzymatic steps, each controlled by one or more gene products. Thus, in *Drosophila,* an estimated 100 or more genes contribute to pigmentation of the compound eye of the fly. For more complex processes, such as the development of the structure of the compound eye, many hundreds of genes probably contribute to the process.

   A goal of genetic analysis is to identify all the genes that affect a specific phenotype and to understand their genetic, cellular, developmental, and molecular roles. To do so, we need ways of sorting mutations and genes. In the first section of this chapter, we shall consider how to use genetic analysis to determine if two mutants are caused by mutational hits in the same gene (that is, they are alleles) or in different genes. Later in the chapter, we shall consider how genetic analysis can be used to make inferences about how genes interact in developmental and biochemical pathways.

A diagnostic test for alleles

Many genetic research programs are undertaken to attempt to understand the genes that contribute to one particular biological process. Such an analysis begins with a collection of related mutant phenotypes centered on that particular process. For example, if a geneticist were interested in the genes determining locomotion in a nematode worm, the genetic dissection would begin by isolating a set of different mutants with defective locomotion. An important task is to determine how many different genes are represented by the mutations that determine the related phenotypes, because this number defines the set of genes that affect the process under study. Hence it is necessary to have a test to find out if the mutations are alleles of one gene or of different genes. The allelism test having the widest application is the **complementation test,** which is illustrated in the following example.

Consider a species of harebell *(Campanula)* in which the wild-type flower color is blue. Let's assume that, by applying mutagenic radiation, we have induced three white-petalled mutants and that they are available as homozygous pure-breeding strains. We can call the mutant strains $, £, and ¥, using currency symbols so that we do not prejudice our thinking concerning dominance. When crossed with wild type, each mutant gives the same results in the F_1 and F_2, as follows:

white $ × blue ⟶ F_1 all blue ⟶ F_2 $\frac{3}{4}$ blue, $\frac{1}{4}$ white
white £ × blue ⟶ F_1 all blue ⟶ F_2 $\frac{3}{4}$ blue, $\frac{1}{4}$ white
white ¥ × blue ⟶ F_1 all blue ⟶ F_2 $\frac{3}{4}$ blue, $\frac{1}{4}$ white

In each case, the results show that the mutant condition is determined by the recessive allele of a single gene. However are they three alleles of one gene? or of two or three genes? The question can be answered by asking if the mutants complement each other. Complementation is defined as follows:

MESSAGE ···
Complementation is the production of a wild-type phenotype when two haploid genomes bearing different recessive mutations are united in the same cell.
··

A harebell plant (*Campanula* species). (Gregory G. Dimijian/ Photo Researchers.)

(The demonstration of the recessive nature of individual mutants is a crucial result that allows us to proceed with a complementation test. Dominant mutations cannot be used in a complementation test.)

In a diploid organism, the complementation test is performed by intercrossing homozygous recessive mutants two at a time. The next step is to observe whether the progeny have the wild-type phenotype.

This unites the two mutations as haploid gametes to form a diploid nucleus in one cell (the zygote). If recessive mutations represent alleles of the same gene, then they will not complement, because both mutations represent lost gene function. Such alleles can be thought of generally as a' and a'', by using primes to distinguish between two different mutant alleles of a gene whose wild-type allele is a^+. These alleles could have different mutant sites, but they would be functionally identical (that is, both nonfunctional). The heterozygote a'/a'' would be:

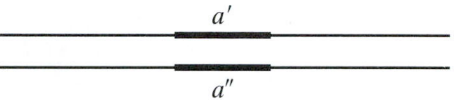

However, two recessive mutations in different genes would have wild-type function provided by the respective wild-type alleles. Here we can name the genes *a1* and *a2*, after their mutant alleles. We can represent the heterozygotes as follows, depending on whether the genes are on the same or different chromosomes:

Different chromosomes:

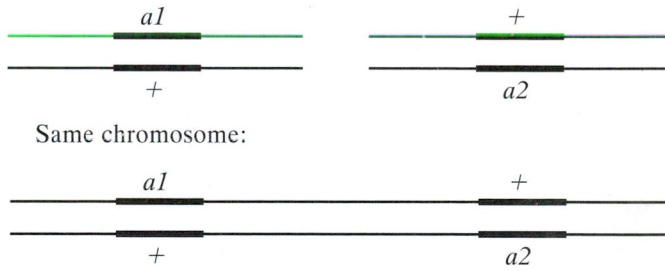

Same chromosome:

Let us return to the harebell example and intercross the mutants to unite the mutant alleles to test for complementation. We can assume that the results of intercrossing mutants $, £, and ¥ are as follows:

white $ × white £ ⟶ F_1 all white
white $ × white ¥ ⟶ F_1 all blue
white £ × white ¥ ⟶ F_1 all blue

From this set of results, we can conclude that mutants $ and £ must be caused by alleles of one gene (say, *w1*) because they do not complement; but ¥ must be caused by a mutant allele of another gene *(w2)*.

The molecular explanation of such results is often in relation to biochemical pathways in the cell. How does complementation work at the molecular level? Although the convention is to say that it is the *mutants* that complement, in fact, the active agents in complementation are the proteins produced by the *wild-type* alleles. The normal blue color of the flower is caused by a blue pigment called anthocyanin. Pigments are chemicals that absorb certain parts of the visible spectrum; in the harebell, the anthocyanin absorbs all wavelengths except blue, which is reflected into the eye of the observer. However, this anthocyanin is made from chemical precursors that are not pigments; that is, they do not absorb light of any specific wavelength and simply reflect back the white light of the sun to the observer, giving a white appearance. The blue pigment is the end product of a series of biochemical conversions of nonpigments. Each step is catalyzed by a specific enzyme encoded by a specific

gene. We can accommodate the results with a pathway as follows:

$$\text{gene } w1^+ \qquad \text{gene } w2^+$$
$$\downarrow \qquad \qquad \downarrow$$
$$\text{enzyme 1} \qquad \text{enzyme 2}$$

$$\longrightarrow \text{precursor 1} \longrightarrow \text{precursor 2} \longrightarrow \text{blue anthocyanin}$$

A mutation in either of the genes in homozygous condition will lead to the accumulation of a precursor that will simply make the plant white. Now the mutant designations could be written as follows:

$$\$ \qquad w1_\$/w1_\$ \cdot w2^+/w2^+$$
$$£ \qquad w1_£/w1_£ \cdot w2^+/w2^+$$
$$¥ \qquad w1^+/w1^+ \cdot w2_¥/w2_¥$$

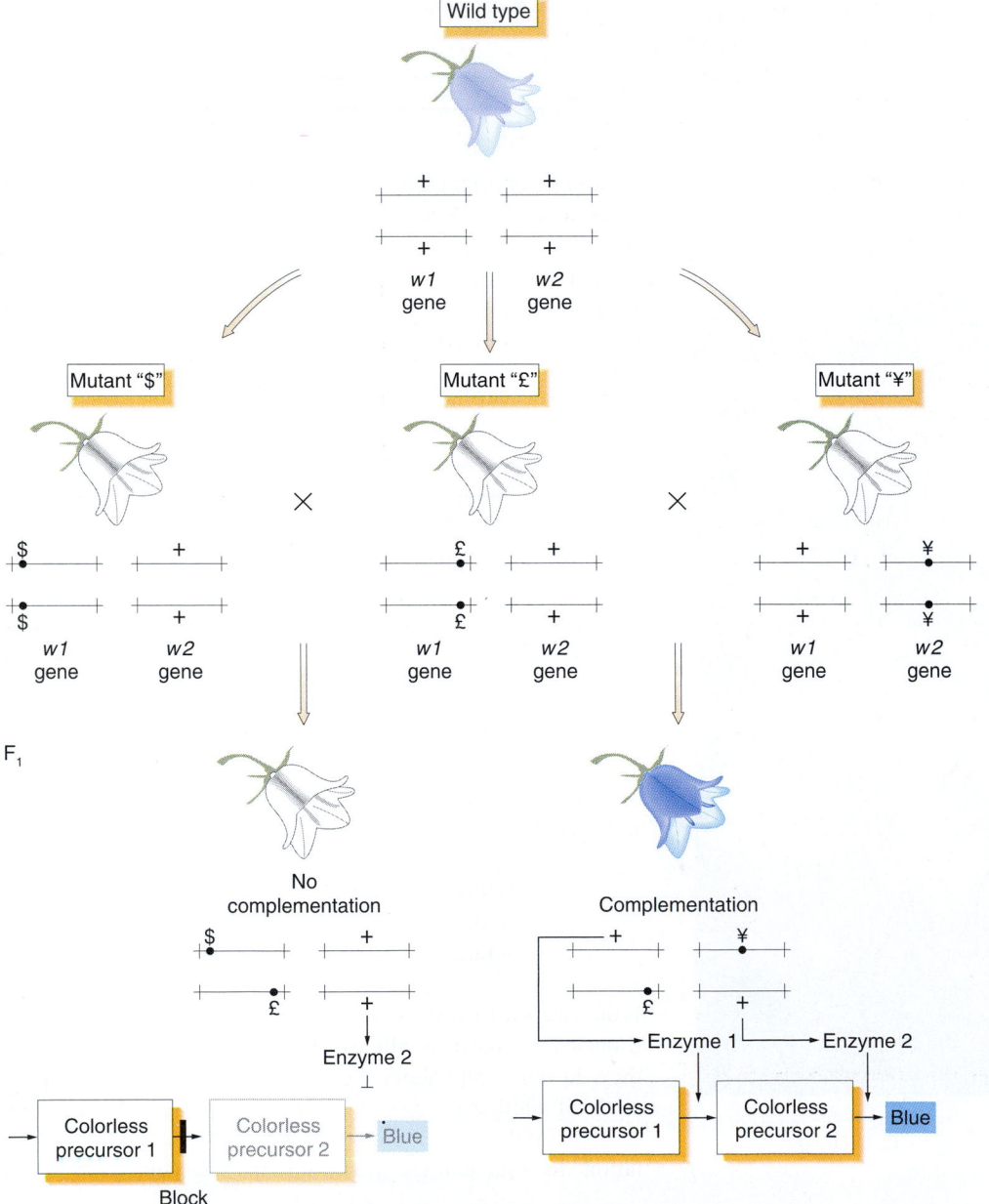

Wild type

Mutant "$"   Mutant "£"   Mutant "¥"

$\times$ $\times$

$w1$ gene $w2$ gene $w1$ gene $w2$ gene $w1$ gene $w2$ gene

F₁

No complementation   Complementation

Enzyme 2

Colorless precursor 1   Colorless precursor 2   Blue

Block

Enzyme 1   Enzyme 2

Colorless precursor 1   Colorless precursor 2   Blue

Figure 4-1 The molecular basis of genetic complementation. Three phenotypically identical white mutants — $, £, and ¥ — are intercrossed to form heterozygotes whose phenotypes reveal whether the mutations complement each other. (Only two of the three possible crosses are shown here.) If two mutations are in different genes (such as £ and ¥), then complementation results in the completion of the biochemical pathway (the end product is a blue pigment in this example). If mutations are in the same gene (such as $ and £), no complementation occurs, because the biochemical pathway is blocked at the step controlled by that gene, and the intermediates in the pathway are colorless (white). (What would you predict to be the result of crossing $ and ¥?)

However, in practice, the subscript symbols would be dropped and the genotypes written as follows:

$$\$ \quad w1/w1 \cdot w2^+/w2^+$$
$$\pounds \quad w1/w1 \cdot w2^+/w2^+$$
$$\yen \quad w1^+/w1^+ \cdot w2/w2$$

Hence an F_1 from $\$ \times \pounds$ will be:

$$w1/w1 \cdot w2^+/w2^+$$

which will have two defective alleles for *w1* and will therefore be blocked at step 1. Even though enzyme 2 is fully functional, it has no substrate on which to act, so no blue pigment will be produced and the phenotype will be white.

The F_1s from the other crosses, however, will have the wild-type alleles for both the enzymes needed to take the interconversions to the final blue product. Their genotypes will be:

$$w1^+/wl \cdot w2^+/w2$$

Hence we see the reason why complementation is actually a result of the cooperative interaction of the wild-type alleles of the two genes. Figure 4-1 is a summary diagram of the interaction of the complementing and noncomplementing white mutants.

In a haploid organism, the complementation test cannot be performed by intercrossing. In fungi, an alternative way to test complementation is to make a **heterokaryon** (Figure 4-2). Fungal cells fuse readily and, when two different strains fuse, the haploid nuclei from the different strains occupy one cell, which is called a heterokaryon (Greek; different kernels). The nuclei in a heterokaryon generally do not fuse. In one sense, this condition is a "mimic" diploid. Assume that, in different strains, there are mutations in two different genes conferring the same mutant phenotype — for example, arginine requirement. We can call these genes *arg-1* and *arg-2*. The two strains, whose genotypes can be represented as $arg\text{-}1 \cdot arg\text{-}2^+$ and $arg\text{-}1^+ \cdot arg\text{-}2,$ can be

fused to form a heterokaryon with the two nuclei in a common cytoplasm:

$$\text{Nucleus 1 is } arg\text{-}1 \cdot arg\text{-}2^+$$
$$\text{Nucleus 2 is } arg\text{-}1^+ \cdot arg\text{-}2$$

Because gene expression takes place in a common cytoplasm, the two wild-type alleles can exert their dominant effect and cooperate to produce a heterokaryon of wild-type phenotype. In other words, the two mutations complement, just as they would in a diploid. If the mutations had been alleles of the same gene, there would have been no complementation.

MESSAGE ··

When two independently derived recessive mutant alleles producing similar recessive phenotypes fail to complement, the alleles must be of the same gene.

··

Interactions between the alleles of one gene

The alleles of one gene can interact in several different ways at the functional level, resulting in variations in the type of dominance and in markedly different phenotypic effects in different allelic combinations.

Incomplete dominance

Four-o'clocks are plants native to tropical America. Their name comes from the fact that their flowers open in the late afternoon. When a wild-type four-o'clock plant with red petals is crossed with a pure line with white petals, the F_1 has pink petals. If an F_2 is produced by selfing the F_1, the result is:

$\frac{1}{4}$ of the plants have red petals
$\frac{1}{2}$ of the plants have pink petals
$\frac{1}{4}$ of the plants have white petals

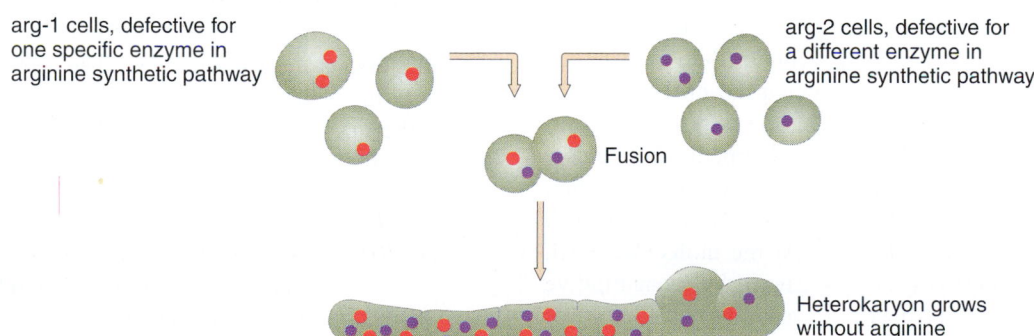

arg-1 cells, defective for one specific enzyme in arginine synthetic pathway

arg-2 cells, defective for a different enzyme in arginine synthetic pathway

Fusion

Heterokaryon grows without arginine

Figure 4-2   Formation of a heterokaryon of *Neurospora,* demonstrating both complementation and recessiveness. Vegetative cells of this normally haploid fungus can fuse, allowing the nuclei from the two strains to intermingle within the same cytoplasm. If each strain is blocked at a different point in a metabolic pathway, as are *arg-1* and *arg-2* mutants, all functions are present in the heterokaryon and the *Neurospora* will grow; in other words, complementation takes place. The growth must be due to the presence of the wild-type gene; hence, the mutant genes *arg-1* and *arg-2* must be recessive.

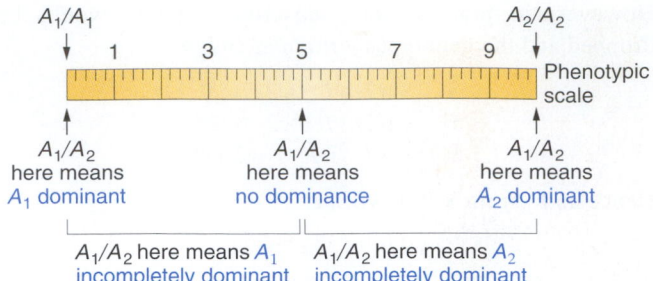

A_1/A_1 A_2/A_2

Figure 4-3   Summary of dominance relations. The ruler represents some sort of phenotypic measurement, such as amount of pigment.

Red (c^+/c^+), pink (c^+/c), and white (c/c) phenotypes of four o'clock plants. The pink heterozygote demonstrates incomplete dominance. (R. Calentine/Visuals Unlimited.)

Because of the 1:2:1 ratio in the F_2, we can deduce an inheritance pattern based on two alleles of a single gene. However, the heterozygotes (the F_1 and half the F_2) are intermediate in phenotype, suggesting an *incomplete* type of dominance. By inventing allele symbols, we can list the genotypes of the four-o'clocks in this experiment as c^+/c^+ (red), c/c (white), and c^+/c (pink). **Incomplete dominance** describes the general situation in which the phenotype of a heterozygote is intermediate between the two homozygotes on some quantitative scale of measurement. Figure 4-3 gives terms for all the theoretical positions on the scale, but, in practice, it is difficult to determine exactly where on such a scale the heterozygote is located. At the molecular level, incomplete dominance is generally caused by a quantitative effect of the number of "doses" of a wild-type allele; two doses produces most functional transcript and therefore most functional protein product; one dose produces less transcript and product, whereas zero dose has no functional transcript or product. In cases of full dominance, in the wild-type/mutant heterozygote either half the normal amount of transcript and product is adequate for normal cell function (the gene is said to be **haplosufficient**), or the

wild-type allele is "upregulated" to bring the concentration of transcript up to normal levels.

Codominance

The human ABO blood groups are determined by three alleles of one gene that show several types of interaction to produce the four blood types of the ABO system. The allelic series includes three major alleles i, I^A, and I^B, but one person can have only two of the three alleles or two copies of one of them. There are six different genotypes: the three homozygotes and three different types of heterozygotes.

Genotype	Blood type
I^A/I^A, I^A/i	A
I^B/I^B, I^B/i	B
I^A/I^B	AB
i/i	O

In this allelic series, the alleles I^A and I^B each determine a unique antigen, which is deposited on the surface of the red blood cells. These antigens are two different forms of a single protein. However, the allele i results in no antigenic protein of this type. In the genotypes I^A/i and I^B/i, the alleles I^A and I^B are fully dominant to i. However, in the genotype I^A/I^B, each of the alleles produces its own antigen, so they are said to be **codominant.**

The human disease sickle-cell anemia is a source of interesting insight into dominance. The gene concerned affects the molecule hemoglobin, which transports oxygen and is the major constituent of red blood cells. The three genotypes have different phenotypes, as follows:

Hb^A/Hb^A: Normal; red blood cells never sickle.
Hb^S/Hb^S: Severe, often fatal anemia; abnormal hemoglobin causes red blood cells to have sickle shape.
HB^A/Hb^S: No anemia; red blood cells sickle only under low oxygen concentrations.

Figure 4-4 shows an electron micrograph of sickle cells. In regard to the presence or absence of anemia, the Hb^A allele is dominant. In regard to blood-cell shape, however, there is incomplete dominance. Finally, as we shall now see,

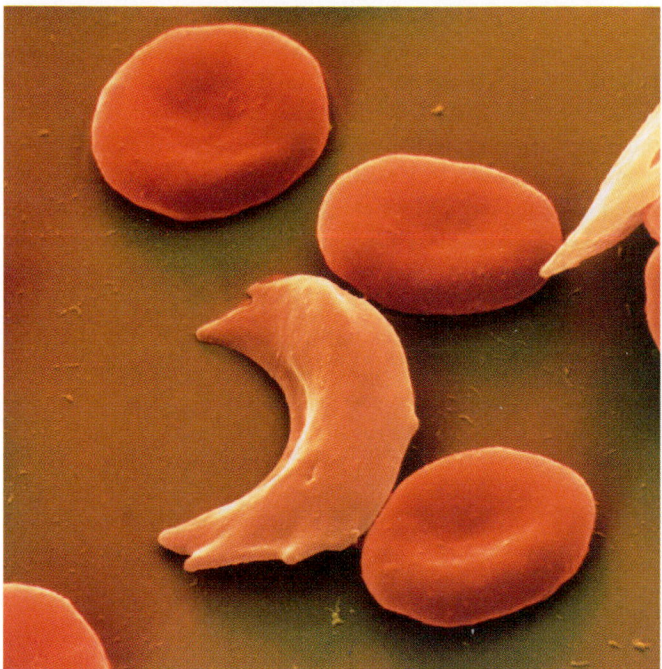

Figure 4-4   An electron micrograph of a sickle-shaped red blood cell. Other, more rounded cells appear almost normal. (Meckes/Ottawa/Photo Researchers.)

mal people have one type of hemoglobin (A) and anemics have type S, which moves more slowly in the electric field. The heterozygotes have both types, A and S. In other words, there is codominance at the molecular level.

Sickle-cell anemia illustrates that the terms *dominance, incomplete dominance,* and *codominance* are somewhat arbitrary. The type of dominance inferred depends on the phenotypic level at which the observations are being made — organismal, cellular, or molecular. Indeed the same caution can be applied to many of the categories that scientists use to classify structures and processes; these categories are devised by humans for convenience of analysis.

The population analysis of the Hb^A and Hb^S alleles will be considered in Chapter 24.

MESSAGE ·
The type of dominance is determined by the molecular functions of the alleles of a gene and by the investigative level of analysis.

The leaves of clover plants show several variations on the dominance theme. Clover is the common name for plants of the genus *Trifolium*. There are many species. Some are native to North America, whereas others grow there as introduced weeds. Much genetic research has been done with white clover, which shows considerable variation among individuals in the curious V, or chevron, pattern on the leaves. Figure 4-7 shows that the different chevron forms (and the absence of chevrons) are determined by multiple alleles. In this example, we are dealing with a genetic polymorphism, so the wild-type/mutant allele symbolism is not used. Study the photographs to determine the type of dominance of each allele in various combinations. List the alleles in a way that expresses how they relate to one another in dominance. Are there uncertainties? Does the photographic evidence permit us to say anything about the dominance or recessiveness of allele *v*?

in regard to hemoglobin itself, there is codominance. The alleles Hb^A and Hb^S code for two different forms of hemoglobin differing by a single amino acid, and both these forms are synthesized in the heterozygote. The different hemoglobin forms can be visualized by **electrophoresis,** a technique that separates macromolecules with different charge or size (Figure 4-5). It so happens that the A and S forms of hemoglobin have different charges, so they can be separated by electrophoresis (Figure 4-6). We see that homozygous nor-

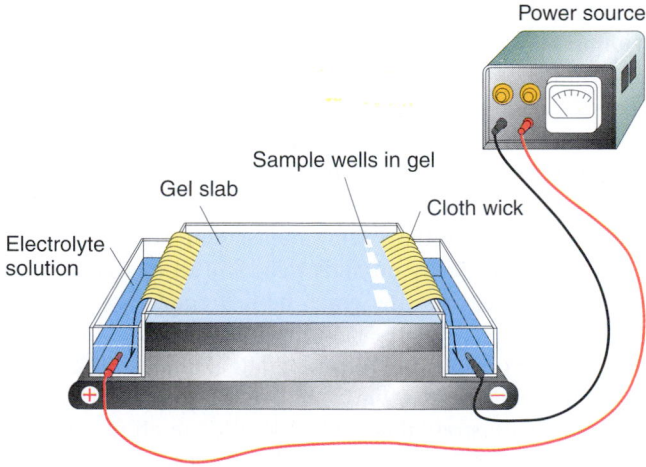

Figure 4-5   Apparatus for electrophoresis. Each sample is placed in a well in a gelatinous slab (a gel). The molecules in the samples migrate different distances on the gel owing to their different electric charges. Several samples are tested at the same time (one in each well). The positions to which the molecules have migrated are later revealed by staining.

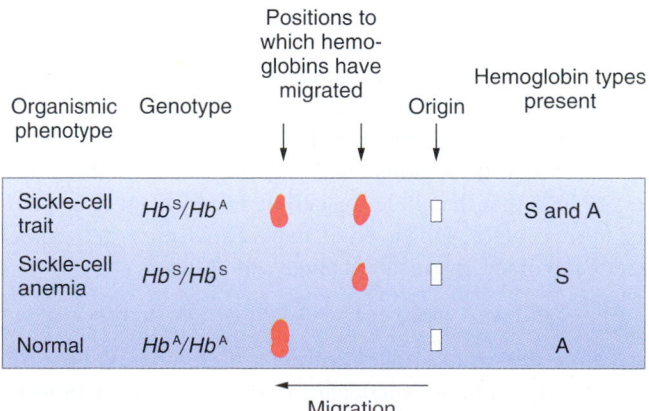

Figure 4-6   Electrophoresis of hemoglobin from a person with sickle-cell anemia, a heterozygote (called sickle-cell trait), and a normal person. The smudges show the positions to which the hemoglobins migrate on the starch gel.

v/v

V^l/V^l

V^h/V^h V^l/V^h

V^f/V^f V^l/V^f V^h/V^f

V^{ba}/V^{ba} V^l/V^{ba} V^h/V^{ba} V^f/V^{ba}

V^b/V^b V^l/V^b V^h/V^b V^f/V^b V^{ba}/V^b

V^{by}/V^{by} V^l/V^{by} V^h/V^{by} V^f/V^{by} V^{ba}/V^{by} V^b/V^{by}

Figure 4-7 Multiple alleles determine the chevron pattern on the leaves of white clover. The genotype of each plant is shown below it. (After photograph by W. Ellis Davies.)

Lethal alleles

Normal wild-type mice have coats with a rather dark overall pigmentation. A mutation called *yellow* (a lighter coat color) illustrates an interesting allelic interaction. If a yellow mouse is mated to a homozygous wild-type mouse, a 1 : 1 ratio of yellow to wild-type mice is always observed in the progeny. This observation suggests (1) that a single gene with two alleles determines these phenotypic alternatives, (2) that the yellow mouse was heterozygous for these alleles, and (3) that the allele for yellow is dominant to an allele for normal color. However, if two yellow mice are crossed with each other, the result is always as follows:

$$\text{yellow} \times \text{yellow} \longrightarrow \tfrac{2}{3} \text{ yellow}, \tfrac{1}{3} \text{ wild type}$$

Note two interesting features in these results. First, the 2 : 1 phenotypic ratio is a departure from the expectations for a monohybrid self-cross. Second, because no cross of yellow × yellow ever produced all yellow progeny, as there would be if either parent were a homozygote, it appears that it is impossible to obtain homozygous yellow mice.

The explanation for such results is that all yellow mice are heterozygous for one special allele. A cross between two heterozygotes would be expected to yield a monohybrid genotypic ratio of 1 : 2 : 1. However, if all the mice in one of the homozygous classes died before birth, the live births would then show a 2 : 1 ratio of heterozygotes to the surviving homozygotes. The allele A^Y for yellow is dominant to the wild-type allele A with respect to its effect on color, but A^Y acts as a recessive **lethal** allele with respect to a character that we call *viability*. Thus, a mouse with the homozygous genotype A^Y/A^Y dies before birth and is not observed among the progeny. All surviving yellow mice must be heterozygous A^Y/A, so a cross between yellow mice will always yield the following results:

$$A^Y/A \times A^Y/A$$

Progeny $\tfrac{1}{4} A^Y/A^Y$ lethal
$\tfrac{1}{2} A^Y/A$ yellow
$\tfrac{1}{4} A/A$ wild type

The expected monohybrid ratio of 1 : 2 : 1 would be found among the zygotes, but it is altered to a 2 : 1 ratio in the progeny born because zygotes with a lethal A^Y/A^Y genotype do not survive to be counted. This hypothesis is supported by the removal of uteri from pregnant females of the yellow × yellow cross; one-fourth of the embryos are found to be dead. Figure 4-8 shows a typical litter from a cross between yellow mice.

The A^Y allele produces effects on two characters: coat color and survival. The A^Y allele is *pleiotropic*. It is entirely possible, however, that both effects of the A^Y pleiotropic allele result from the same basic cause, which promotes yellowness of coat in a single dose and death in a double dose.

The tailless Manx phenotype in cats (Figure 4-9) also is produced by an allele that is lethal in the homozygous state. A single dose of the Manx allele, M^L, severely interferes with normal spinal development, resulting in the absence of a tail in the M^L/M heterozygote. But in the M^L/M^L homozygote, the double dose of the gene produces such an extreme developmental abnormality that the embryo does not survive.

There are indeed many different types of lethal alleles. Some lethal alleles produce a recognizable phenotype in the heterozygote, as in the yellow mouse and Manx cat. Some lethal alleles are fully dominant and kill in one dose in the heterozygote. Others (the much more frequent case) confer no detectable effect in the heterozygote at all, and the lethality is fully recessive. Furthermore, lethal alleles differ in the developmental stage at which they express their effects. Human lethals illustrate this very well. We are all estimated to be heterozygous for a small number of recessive lethals in our genomes. The lethal effect is expressed in the homozygous progeny of a mating between two people who by chance carry the same recessive lethal in the heterozygous condition. Some lethals are expressed as deaths in utero, where they either go unnoticed or are noticed as sponta-

Figure 4-8 A mouse litter from two parents heterozygous for the yellow coat-color allele, which is lethal in a double dose. Not all progeny are visible. (Anthony Griffiths.)

neous abortions. Other lethals, such as those responsible for Duchenne muscular dystrophy, cystic fibrosis, or Tay-Sachs disease, exert their effects in childhood. The time of death can even be in adulthood, as in Huntington disease. The total of all the deleterious and lethal genes that are present in individual members of a population is called *genetic load,* a kind of genetic burden that the population has to carry.

Exactly what goes wrong in lethal mutations? In many cases, it is possible to trace the cascade of events that leads to death. A common situation is that the allele causes a deficiency in some essential chemical reaction. The human diseases PKU and cystic fibrosis are good examples of this kind of deficiency. In other cases, there is a structural defect. For example, a lethal allele of rats determines abnormal cartilage protein, and the effect of this abnormality is expressed phenotypically in several different organs, resulting in lethal symptoms, as shown in Figure 4-10. Sickle-cell anemia, discussed earlier, is another example.

Whether an allele is lethal or not often depends on the environment in which the organism develops. Whereas certain alleles are lethal in virtually any environment, others are viable in one environment but lethal in another. For example, the human hereditary disease cystic fibrosis is a disease that would be lethal without treatment. Furthermore,

Figure 4-9 Manx cat. All such cats are heterozygous for a dominant allele that causes no tail to form. The allele is lethal in the homozygous condition. The dissimilar eyes are unrelated to taillessness. (Gerard Lacz/NHPA.)

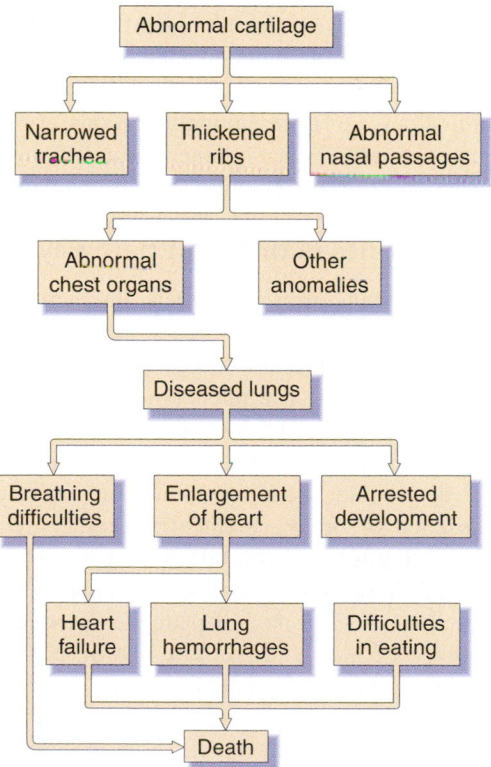

Figure 4-10 Diagram showing how one specific lethal allele causes death in rats. (From I. M. Lerner and W. J. Libby, *Heredity, Evolution, and Society,* 2d ed. W. H. Freeman and Company, 1976; after H. Gruneberg.)

many of the alleles favored and selected by animal and plant breeders would almost certainly be eliminated in nature as a result of competition with the members of the natural population. Modern grain varieties provide good examples; only careful nurturing by farmers has maintained such alleles for our benefit.

Geneticists commonly encounter situations in which expected phenotypic ratios are consistently skewed in one direction by reduced viability caused by one allele. For example, in the cross $A/a \times a/a$, we predict a progeny ratio of 50 percent A/a and 50 percent a/a, but we might consistently observe a ratio such as 55 percent:45 percent or 60 percent:40 percent. In such a case, the recessive phenotype is said to be *subvital,* or *semilethal,* because the lethality is expressed in only some individuals. Thus, lethality may range from 0 to 100 percent, depending on the gene itself, the rest of the genome, and the environment.

> **MESSAGE** ···
>
> A gene can have several different states or forms—called multiple alleles. The alleles are said to constitute an allelic series, and the members of a series can show various degrees of dominance to one another.

Gene interaction and modified dihybrid ratios

Genetic analysis can identify the genes that interact in the determination of a particular biological property. The key diagnostic is that two interacting genes produce *modified dihybrid ratios.* There are various types of interactions, and they lead to a range of different modifications. An important distinction is between genes interacting in different biological pathways and those interacting in the same pathway.

Interacting genes in different pathways

A simple, yet striking, example of gene interaction is the inheritance of skin coloration in corn snakes. The natural color is a repeating black-and-orange camouflage pattern, as shown in Figure 4-11a. The phenotype is produced by two separate pigments, both of which are under genetic control. One gene determines the orange pigment, and the alleles that we shall consider are o^+ (presence of orange pigment) and o (absence of orange pigment). Another gene determines the black pigment, with alleles b^+ (presence of black pigment) and b (absence of black pigment). These two genes are unlinked. The natural pattern is produced by the genotype $o^+/-$; $b^+/-$. A snake that is o/o ; $b^+/-$ is black because it lacks orange pigment (Figure 4-11b), and a snake that is $o^+/-$; b/b is orange because it lacks the black pigment (Figure 4-11c). The double homozygous recessive o/o ; b/b is albino, as shown in Figure 4-11d. Notice, however, that the faint pink color of the albino is from yet another pigment, the hemoglobin of the blood that is visible through this snake's skin when the other pigments are absent. The albino snake also clearly shows that there is another element to the skin-pigmentation pattern in addition to

pigment, and this element is the repeating motif in and around which pigment is deposited.

Because there are two genes in this system, we obtain a typical dihybrid inheritance pattern, and the four unique phenotypes form a 9:3:3:1 ratio in the F_2. A typical analysis might be as follows:

female o^+/o^+ ; b/b (orange) $\times$ male o/o ; b^+/b^+ (black)

$\downarrow$

F_1 o^+/o ; b^+/b (camouflaged)

female o^+/o ; b^+/b (camouflaged)
$\times$ male o^+/o ; b^+/b (camouflaged)

$\downarrow$

F_2 9 $o^+/-$; $b^+/-$ (camouflaged)
            3 $o^+/-$; b/b (orange)
            3 o/o ; $b^+/-$ (black)
            1 o/o ; b/b (albino)

In summary, the 9:3:3:1 dihybrid ratio is produced because the mutations are in two parallel biochemical pathways.

$$\left[\begin{array}{l} \text{precursors} \xrightarrow{\;b^+\;} \text{black pigment} \\ \text{precursors} \xrightarrow{\;\;o^+\;\;} \text{orange pigment} \end{array} \right\} \text{(camouflaged)}$$

Generally, interacting genes in two different pathways produce an F_2 with four phenotypes corresponding to the four possible genotypic classes, as in the snake example. However, when the mutations are in one biological pathway, different ratios are seen. Usually there are only two or three phenotypes resulting from various combinations of the genotypic classes. Likewise, the F_2 ratio is a modification of the 9:3:3:1 ratio, produced by grouping various components of the ratio. Examples are shown in the next section.

Interacting genes in the same pathway

For an example of a modified ratio produced by genes in the same pathways, we need only to return to the example of petal color in harebells.

Mutations with the same phenotype. The harebell anthocyanin pathway terminated in a blue pigment and the inter-

Figure 4-11 Analysis of the genes for skin pigment in the corn snake. The wild type (a) has a skin pigmentation pattern made up of a black and an orange pigment. The gene O determines an enzyme in the synthetic pathway for orange pigment; when this enzyme is deficient (o/o), no orange pigment is made and the snake is black (b). Another gene, B, determines an enzyme for black pigment; when this enzyme is deficient (b/b), the snake is orange (c). When both enzymes are deficient, the snake is albino (d). Hence the four homozygous genotypes are O/O ; B/B (a), o/o ; B/B (b), O/O ; b/b (c), and o/o ; b/b (d). A cross of wild type (a) $\times$ albino (d) or black (b) $\times$ orange (c) would give a dihybrid wild-type F_1 and a 9:3:3:1 ratio of the four phenotypes in the F_2. (Anthony Griffiths.)

(a)

(b)

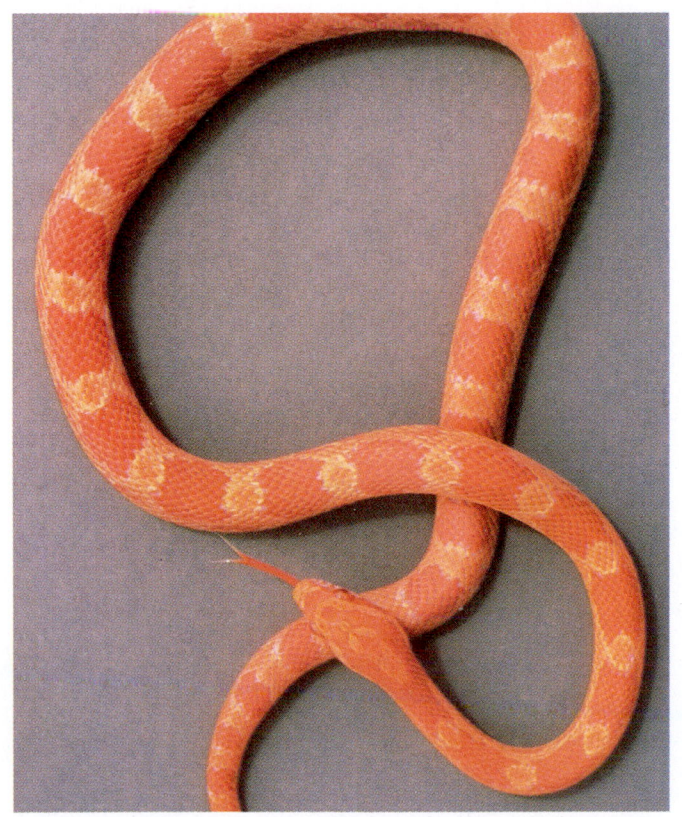

(c)

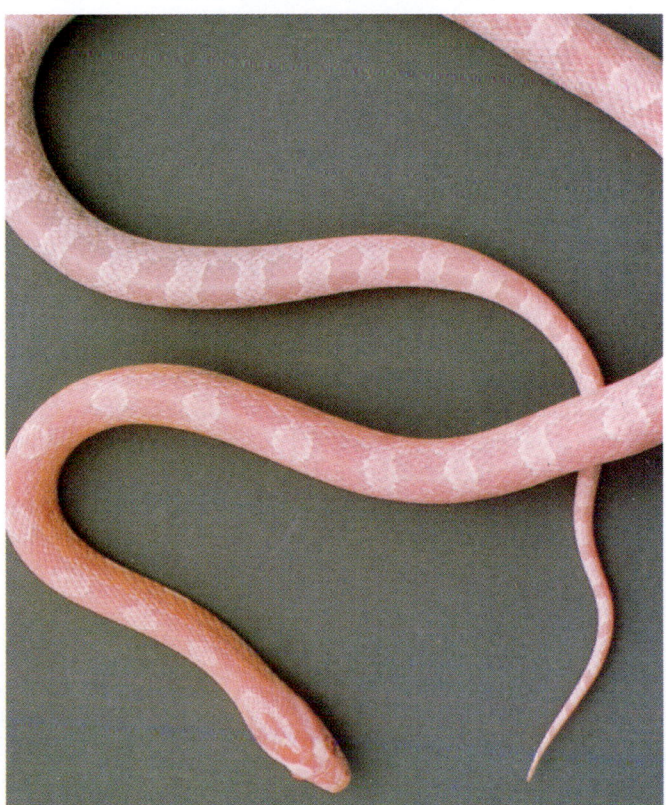

(d)

mediates were all colorless. Two different white-petalled homozygous lines of harebells were crossed and the F_1 was blue flowered, showing complementation. What will be the F_2 resulting from crossing the F_1 plants? The F_2 shows both blue and white plants in a ratio of 9:7. How can these results be explained? The 9:7 ratio is clearly a modification of the dihybrid 9:3:3:1 ratio with the 3:3:1 combined to make 7. The cross of the two white lines and subsequent generations can be represented as follows:

white $w1/w1$; $w2^+/w2^+$ $\times$ white $w1^+/w1^+$; $w2/w2$

$\downarrow$

F_1 $w1^+/w1$; $w2^+/w2$ (blue)

$w1^+/w1$; $w2^+/w2$ $\times$ $w1^+/w1$; $w2^+/w2$

$\downarrow$

F_2 $9\ w1^+/-$; $w2^+/-$ (blue) $\Big\}$ 9
$3\ w1^+/-$; $w2/w2$ (white)
$3\ w1/w1$; $w2^+/-$ (white) $\Big\}$ 7
$1\ w1/w1$; $w2/w2$ (white)

The results show that homozygosity for the recessive mutant allele of *either* gene or *both* genes causes a plant to have white petals. To have the blue phenotype, a plant must have at least one dominant allele of both genes.

An important type of gene interaction at the molecular level is the interaction between a regulatory gene and the gene that it regulates (Figure 4-12). Such genes also show a type of complementation. A common situation is that the regulatory gene produces a regulatory protein that binds to the upstream regulatory site of the target gene, possibly facilitating the action of RNA polymerase (Figure 4-12a). In the absence of the regulatory protein, the target gene would be transcribed at very low levels, inadequate for cellular needs. This type of gene interaction can be followed in a situation in which a dihybrid is heterozygous for a null mutation of the regulatory gene (r^+/r) and heterozygous for a null mutation — say, a translation-termination mutation (a^+/a). Normally translation-termination codons are at the 3' end of every mRNA, but mutation can introduce a termination codon within the coding sequence. In this position, it results in a short polypeptide. Assume that the mutation in the present example is close to the 5' end of the protein-coding sequence. The possible genotypes are shown in Figure 4-12. The stop codon will lead to premature transcriptional termination — in this case, a protein of such small size that it is negligible. The r^+/r ; a^+/a dihybrid will give the following progeny:

Proportion	Genotype	Functional a^+ protein	Ratio
$\frac{9}{16}$	$r^+/-$; $a^+/-$	yes	9
$\frac{3}{16}$	$r^+/-$; a/a	no	
$\frac{3}{16}$	r/r ; $a^+/-$	no	7
$\frac{1}{16}$	r/r ; a/a	no	

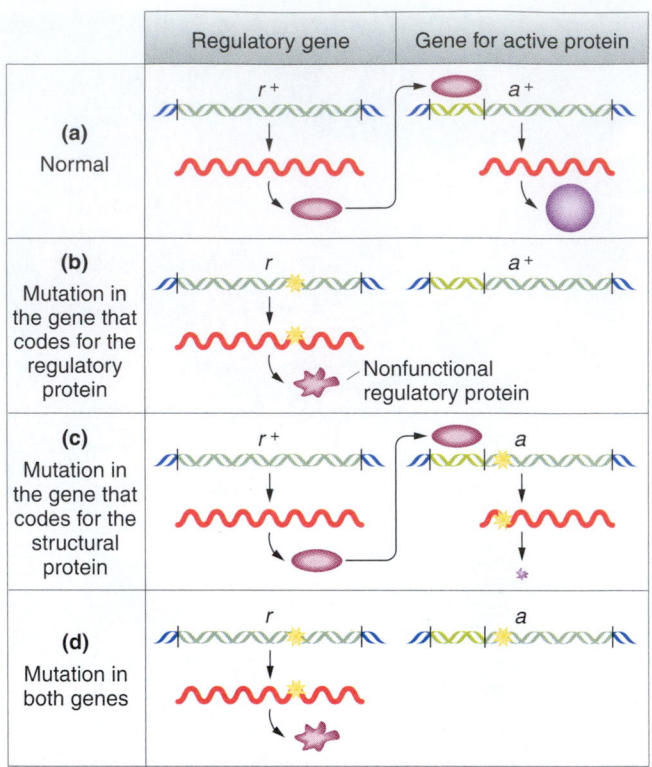

Figure 4-12   Interaction between a regulating gene and its target.

Here then we see another mechanism for complementation, the cooperation of a regulatory and a regulated gene.

Mutations with different phenotypes.   If one or more intermediates in a biochemical pathway is colored, then a different F_2 ratio is produced. In this example, taken from the plant blue-eyed Mary *(Collinsia parviflora),* the pathway is as follows:

$$colorless \xrightarrow{\text{gene } w^+} magenta \xrightarrow{\text{gene } m^+} blue$$

The w and m genes are not linked. If homozygous white and magenta plants are crossed, the F_1 and F_2 are as follows:

(white) w/w ; m^+/m^+ $\times$ (magenta) w^+/w^+ ; m/m

F_1 w^+/w ; m^+/m (blue)

$\downarrow$

w^+/w ; m^+/m $\times$ w^+/w ; m^+/m

$\downarrow$

F_2 $9\ w^+/-$; $m^+/-$ (blue)   9
$3\ w^+/-$; m/m (magenta)   3
$3\ w/w$; $m^+/-$ (white) $\Big\}$ 4
$1\ w/w$; m/m (white)

Complemention results in a wild-type F_1. However, in the F_2, a 9:3:4 phenotypic ratio is produced. This kind of interaction is called **epistasis,** which literally means "standing on"; in other words, an allele of one gene masks the expres-

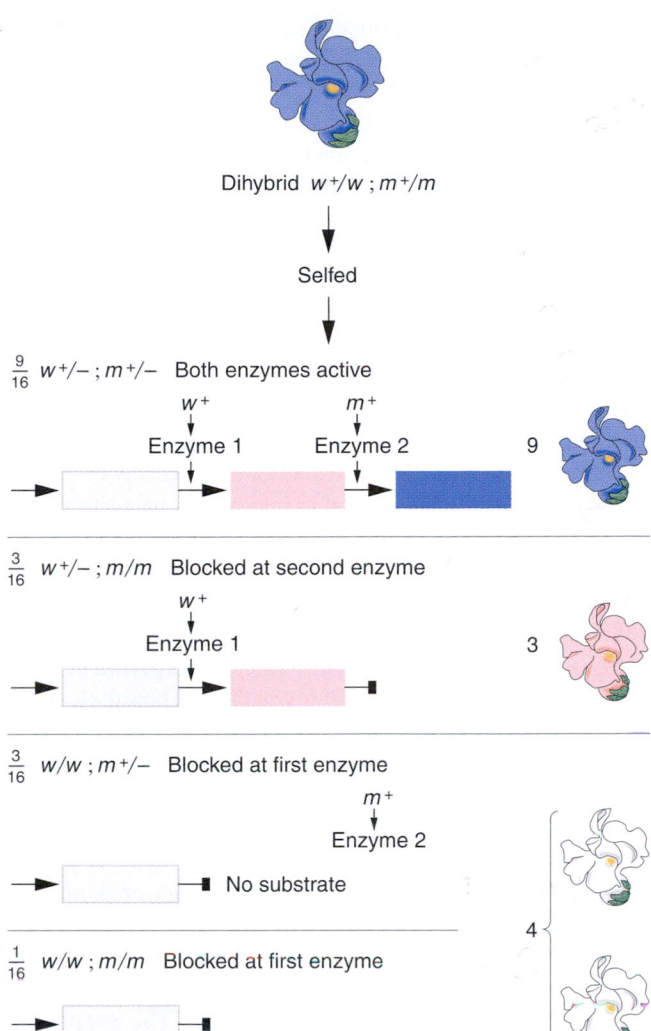

Dihybrid w^+/w ; m^+/m

↓

Selfed

↓

$\frac{9}{16}$ $w^+/-$; $m^+/-$ Both enzymes active

w^+ m^+
↓   ↓
Enzyme 1   Enzyme 2   9

$\frac{3}{16}$ $w^+/-$; m/m Blocked at second enzyme

w^+
↓
Enzyme 1   3

$\frac{3}{16}$ w/w ; $m^+/-$ Blocked at first enzyme

m^+
↓
Enzyme 2

No substrate

4

$\frac{1}{16}$ w/w ; m/m Blocked at first enzyme

Figure 4-13   A molecular mechanism for recessive epistasis. Two representative genes encode enzymes catalyzing successive steps in the synthesis of a blue petal pigment. The substrates for these enzymes are colorless and pink, respectively, so null alleles of the genes will result in colorless (white) or pink petals. The epistasis is revealed in the double mutant because it shows the phenotype of the earlier of the two blocks in the pathway (that is, white). Hence the mutation in the earlier gene precludes expression of any alleles of a gene acting at the later step.

sion of the alleles of another gene. In this example, the w allele is epistatic on m^+ and m. Conversely, m^+ and m can be expressed only in the presence of w^+. Because a recessive allele is epistatic, this is a case of **recessive epistasis.** The state of the synthetic pathway in the various genotypes is illustrated in Figure 4-13.

MESSAGE ·································
Epistasis is inferred when an allele of one gene masks expression of alleles of another gene and expresses its own phenotype instead.

In general, every time one gene is higher, or upstream, in some biochemical pathway, we would expect there to be an epistatic effect of a defective allele on alleles of genes later in the sequence. Therefore, finding a case of epistasis (for example, by the 9:4:3 modified dihybrid ratio) can be a source of insight about the sequence in which genes act. This principle can be useful in piecing together biochemical pathways.

Another case of recessive epistasis well known to most people is the yellow coat color of Labrador retriever dogs. Two alleles, B and b, stand for black and brown coats, respectively, but the allele e of another gene is epistatic on these alleles, giving a yellow coat (Figure 4-14). Therefore the genotypes $B/-$; e/e and b/b ; e/e are both of yellow phenotype, whereas $B/-$; $E/-$ and b/b ; $E/-$ are black and brown, respectively. This case of epistasis is *not* caused by an upstream block in a pathway leading to dark pigment. Yellow dogs can make black or brown pigment, as can be seen in their noses and lips. The action of the allele e is to prevent deposition of the pigment in hairs. In this case, the epistatic gene is *developmentally downstream;* it represents

Figure 4-14   Coat-color inheritance in Labrador retrievers. Two alleles B and b of a pigment gene determine (a) black and (b) brown, respectively. At a separate gene, E allows color deposition in the coat, and e/e prevents deposition, resulting in (c) the gold phenotype. This is a case of recessive epistasis. Thus the three homozygous genotypes are (a) B/B ; E/E, (b) b/b ; E/E, and (c) B/B ; e/e or b/b ; e/e. The dog in part c is most likely B/B ; e/e — the animal still has the ability to make black pigment (as indicated by the black nose and lips) but not to deposit this pigment in the hairs. The progeny of a dihybrid cross would produce a 9:3:4 ratio of black:brown:golden. (Anthony Griffiths.)

(a)

(b)

(c)

a kind of developmental target that has to be of E genotype before pigment can be deposited.

> **MESSAGE** ·······················
> Epistasis points to interaction of genes in some biochemical or developmental sequence.

Suppressors. Another important type of gene interaction is **suppression.** A suppressor is an allele that reverses the effect of a mutation of another gene, resulting in the normal (wild-type) phenotype. For example, assume that an allele a^+ produces the normal phenotype, whereas a recessive mutant allele a results in abnormality. A recessive mutant allele s at another gene suppresses the effect of a so that the genotype $a/a \cdot s/s$ will have wild-type (a^+-like) phenotype. The phenotype of a suppressor alone (genotype $a^+/a^+ \cdot s/s$) is sometimes wild type or near wild type; in other cases, the suppressor produces its own abnormal phenotype.

Suppressors also result in modified dihybrid ratios. Let's look at a real example from *Drosophila* and consider a recessive suppressor su of the unlinked recessive purple eye color allele pd. A homozygous purple-eyed fly is crossed with a homozygous red-eyed stock carrying the suppressor.

$$pd/pd \; ; \; su^+/su^+ \text{ (purple)} \times pd^+/pd^+ \; ; \; su/su \text{ (red)}$$

$$\downarrow$$

F$_1$ all $pd^+/pd \; ; \; su^+/su$ (red)

$$pd^+/pd \; ; \; su^+/su \text{ (red)} \times pd^+/pd \; ; \; su^+/su \text{ (red)}$$

$$\downarrow$$

F$_2$
$$\left.\begin{array}{l} 9 \; pd^+/- \; ; \; su^+/- \quad \text{(red)} \\ 3 \; pd^+/- \; ; \; su/su \quad \text{(red)} \\ 1 \; pd/pd \; ; \; su/su \quad \text{(red)} \end{array}\right\} \; 13$$
$$3 \; pd/pd \; ; \; su^+/- \quad \text{(purple)} \qquad 3$$

The overall ratio in the F$_2$ is 13 red : 3 purple. This ratio is characteristic of a recessive suppressor acting on a recessive mutation. Both recessive and dominant suppressors are found, and they can act on recessive or dominant mutations. These possibilities result in a variety of different phenotypic ratios.

Suppression is sometimes confused with epistasis. However, the key difference is that a suppressor cancels the expression of a mutant allele and restores the corresponding wild-type phenotype. The modified ratio is an indicator of this type of interaction. Furthermore, often only two phenotypes segregate (as in the preceding example), not three, as in epistasis.

How do suppressors work at the molecular level? There are many possible mechanisms. A well-researched type is the **nonsense suppressor,** which acts on a mutation caused by a translation-termination (nonsense) codon within a coding sequence. Nonsense mutants produce premature amino acid chain termination. However, a mutation in a tRNA anticodon that allows the tRNA to insert an amino acid at a nonsense codon will suppress the effect of the mutation by allowing protein synthesis to proceed past the site of the mutation in the mRNA. Because tRNA genes are often present in several copies, a suppressor mutation in one of them will be perfectly viable. Another type of suppression is possible in protein–protein interactions. If two proteins normally fit together to provide some type of cellular function, when a mutation causes a shape change in one protein, no bonding occurs and hence no function (Figure 4-15). However, a compensatory shape change by mutation in the second protein can act as a suppressor to restore normal binding. Finally, in situations in which a mutation causes a block in a metabolic pathway, the suppressor finds some way of circumventing the block—for example, by channeling in substances beyond the block from related pathways.

Because of the demonstrable interaction of a suppressor with its target gene, geneticists deliberately seek suppressors as another way of piecing together a set of interacting genes that affect one biological process or structure. The approach is relatively easy because all that is necessary is to perform a large-scale mutation-induction experiment starting with a mutant line (say, genotype m) and simply look for rare individuals that are wild type. Most of these wild types will be m^+ reverse mutations, but some will be suppressed ($m \cdot su$) and distinguishable by the dihybrid ratios produced on crossing. This procedure can be very easily applied in haploid organisms. For example, if large numbers of cells of an arginine-requiring mutant (*arg*) are spread on a plate of growth medium lacking arginine, most cells will not grow, but reverse mutations to the true wild-type allele (*arg$^+$*) and suppressed mutations (*arg · su*) will grow and announce their presence by forming visible colonies. The suppressed

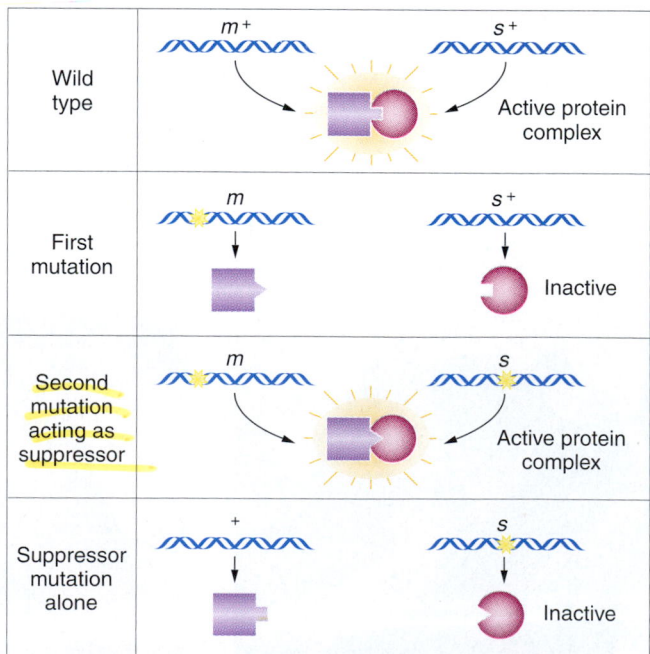

Figure 4-15   A molecular mechanism for suppression.

colonies can be detected by crossing to wild type because arginine-requiring progeny will be produced:

$$arg \cdot su \qquad \times \qquad arg^+ \cdot su^+$$

$$\left.\begin{array}{l} arg^+ \cdot su^+ \\ arg^+ \cdot su \\ arg \cdot su^+ \end{array}\right\} \quad \text{do not require arginine}$$

$$arg \cdot su^+ \qquad \text{arginine requiring}$$

MESSAGE ·······································

Suppressors cancel the expression of a mutant allele of another gene, resulting in normal wild-type phenotype.

···

Duplicate genes. Our final example of gene interaction in the same pathway is based on the idea that some genes may be present more than once in the genome. The example concerns the genes that control fruit shape in the plant called shepherd's purse, *Capsella bursa-pastoris.* Two different lines have fruits of different shapes: one is "heart shaped"; the other, "narrow." Are these two phenotypes determined by two alleles of a single gene? A cross between the two lines produces an F_1 with heart-shaped fruit; this result is consistent with the hypothesis of determination by a pair of alleles. However, the F_2 shows a 15:1 ratio of heart-shaped to narrow, and this ratio suggests a specific modification of the dihybrid 9:3:3:1 Mendelian ratio in which the 9, 3, and 3 are grouped. The genetic control of fruit shape can be explained by means of duplicate genes (Figure 4-16). Apparently, heart-shaped fruits result from the presence of at least one dominant allele of *either* gene. The two genes appear to be identical in function. (Contrast this 15:1 ratio with the 9:7 ratio where *both* dominant genes are necessary to produce a specific phenotype.)

MESSAGE ·······································

Duplicate genes provide alternative genetic determination of a specific phenotype.

···

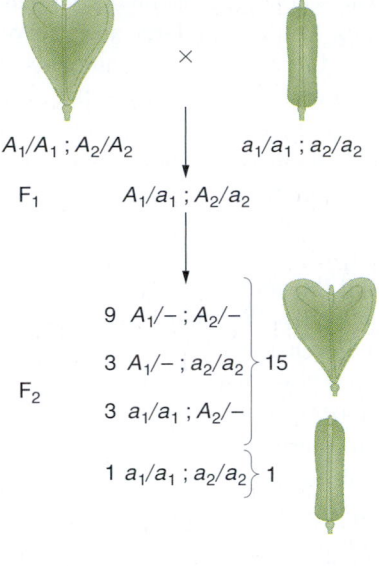

$$A_1/A_1 \; ; A_2/A_2 \qquad\qquad a_1/a_1 \; ; a_2/a_2$$

F_1 $\qquad A_1/a_1 \; ; A_2/a_2$

F_2

$$\left.\begin{array}{l} 9 \; A_1/- \; ; A_2/- \\ 3 \; A_1/- \; ; a_2/a_2 \\ 3 \; a_1/a_1 \; ; A_2/- \end{array}\right\} 15$$

$$\left.1 \; a_1/a_1 \; ; a_2/a_2 \right\} 1$$

Figure 4-16 Inheritance pattern of duplicate genes controlling fruit shape in shepherd's purse. Either A_1 or A_2 can cause a heart-shaped fruit.

The next two sections present examples of how more than two genes interact in determining some specific character. One example is from a plant and one from an animal.

Gene interaction in petal color of foxgloves

Genetic variants of foxgloves (*Digitalis purpurea*) are excellent examples of gene interaction in the determination of the overall appearance of an organism. Three important genes interact to determine petal coloration. The first gene determines the ability of the plant to synthesize purple pigment (a type of anthocyanin). The *M* allele of this gene stands for ability to synthesize anthocyanin, whereas *m* stands for inability to synthesize this pigment, resulting in white petals. Figure 4-17 shows the phenotypes to be dis-

Figure 4-17 Pigment phenotypes in foxgloves, determined by three separate genes. *M* codes for an enzyme that synthesizes anthocyanin, the purple pigment seen in these petals; *m/m* produces no pigment and produces the phenotype albino with yellowish spots. *D* is an enhancer of anthocyanin, resulting in a darker pigment; *d/d* does not enhance. At the third locus, *w/w* allows pigment deposition in petals, but *W* prevents pigment deposition except in the spots and so results in the white, spotted phenotype. Genotypes (and phenotypes) from left to right are *M/−* ; *W/−* ; *−/−* (white with purple spots), *m/m* ; *−/−* ; *−/* (white with yellowish spots), *M/−* ; *w/w* ; *d/d* (light purple), and *M/−* ; *w/w* ; *D/−* (dark purple). (Anthony Griffiths.)

cussed here. The second gene is a modifier gene. One allele, D, determines the synthesis of large amounts of anthocyanin (dark purple), and d stands for low amounts (light purple). Possibly the D and d alleles regulate the synthesis of pigment by M. The third gene affects pigment deposition. The allele W prevents pigment deposition in all parts of the petal except in the throat spots, whereas the recessive allele w allows deposition of pigment all over the petal. Thus these three genes control the ability to synthesize, the amount synthesized, and the ability for the pigment to be deposited in specific petal cells. We shall consider a variety of dihybrid crosses and even a trihybrid cross.

Consider the cross between the two genotypes M/M ; D/D ; w/w and M/M ; d/d ; W/W. The phenotype of the first genotype is dark purple because it has the D modifier and the ability to deposit pigment. The second phenotype is white with purple spots because, although the plant has the ability to synthesize pigment (conferred by the allele M), the W allele prevents deposition except in the throat spots. Let us consider the usual type of pedigree but eliminate the M allele because it will be homozygous in all individuals.

$(M/M)\ D/D$; w/w $\times$ $(M/M)\ d/d$; W/W
    (dark purple)   (white with purple spots)

$\downarrow$

F_1 D/d ; W/w (white with purple spots)

$\qquad D/d$; W/w $\times$ D/d ; W/w
(white with purple spots)   (white with purple spots)

$\downarrow$

F_2 9 $D/-$; $W/-$ (white with purple spots) ⎫
        3 d/d ; $W/-$ (white with purple spots) ⎭ 12
        3 $D/-$; w/w (dark purple)   3
        1 d/d ; w/w (light purple)   1

Overall, a 12 : 3 : 1 phenotypic ratio is produced. This kind of interaction is called **dominant epistasis** because, as can be seen from the F_2 results, the dominant allele W eliminates the two alternatives expressed by D and d, dark and light purple, and replaces them with another phenotype, white with purple spots.

The other two dihybrids, in which the third gene is homozygous, result in 9 : 3 : 4 recessive epistasis ratios because both are affected by the m allele, which wipes out all pigment production. Hence the dihybrid M/m ; D/d ; w/w (for example) results in the following progeny ratio:

$M/-$; $D/-$ (dark purple)   9
$M/-$; d/d (light purple)   3
m/m ; $D/-$ (white) ⎫
m/m ; d/d (white) ⎭ 4

A trihybrid M/m ; W/w ; D/d would produce the following progeny ratio:

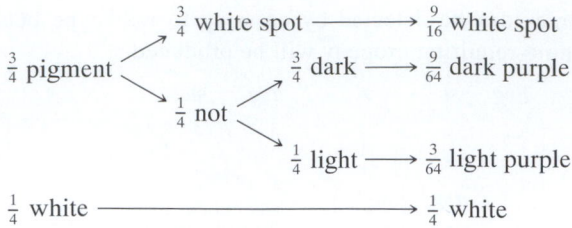

When converted into $\frac{1}{64}$ths, the ratio is:

$\frac{36}{64}$ *white with purple spots*	36
$\frac{9}{64}$ *dark purple*	9
$\frac{3}{64}$ *light purple*	3
$\frac{16}{64}$ *white*	16

Gene interaction in coat color of mammals

Studies of coat color in mammals reveal beautifully how different genes cooperate in the determination of one character. The mouse is a good mammal for genetic studies because it is small and thus easy to maintain in the laboratory and because its reproductive cycle is short. It is the best-studied mammal in regard to the genetic determination of coat color. The genetic determination of coat color in other mammals closely parallels that of mice and, for this reason, the mouse acts as a model system. We shall look at examples from other mammals as we proceed. At least five major genes interact to determine the coat color of mice: the genes are A, B, C, D, and S.

A gene

This gene determines the distribution of pigment in the hair. The wild-type allele A produces a phenotype called agouti. Agouti is an overall grayish color with a brindled, or "salt and pepper," appearance. It is a common color of mammals in nature. The effect is caused by a band of yellow on the otherwise dark hair shaft. In the nonagouti phenotype (determined by the allele a), the yellow band is absent, so there is solid dark pigment throughout (Figure 4-18).

The lethal allele A^Y, discussed in an earlier section, is another allele of this gene; it makes the entire shaft yellow. Still another allele is a^t, which results in a "black and tan" effect, a yellow belly with dark pigmentation elsewhere. For simplicity, we shall not include these two alleles in the following discussion.

B gene

This gene determines the color of pigment. There are two major alleles: B coding for black pigment and b for brown. The allele B gives the normal agouti color in combination with A but gives solid black with a/a. The genotype $A/-$; b/b gives a streaked brown color called cinnamon, and a/a ; b/b gives solid brown.

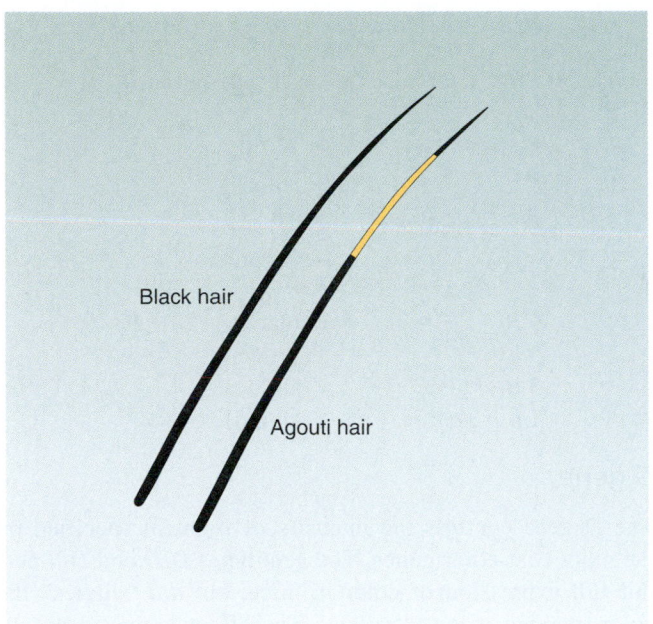

Figure 4-18 Individual hairs from an agouti mouse and a black mouse. The yellow band on each hair gives the agouti pattern its brindled appearance.

The following cross illustrates the inheritance pattern of the A and B genes:

A/A ; b/b (cinnamon) $\times$ a/a ; B/B (black)
or A/A ; B/B (agouti) $\times$ a/a ; b/b (brown)

$\downarrow$

F_1 all A/a ; B/b (agouti)
A/a ; B/b (agouti) $\times$ A/a ; B/b (agouti)

$\downarrow$

F_2 9 $A/-$; $B/-$ (agouti)
3 $A/-$; b/b (cinnamon)
3 a/a ; $B/-$ (black)
1 a/a ; b/b (brown)

The breeding of domestic horses seems to have eliminated the A allele that determines the agouti phenotype, although certain wild relatives of the horse do have this allele. The color that we have called brown in mice is called chestnut in horses, and this phenotype also is recessive to black.

C gene

The wild-type allele C permits color expression, and the allele c prevents color expression. The c/c constitution is epistatic to the other color genes. The c/c animals, lacking coat pigment, are called albinos. Common in many mammalian species, albinos have also been reported among birds, snakes, fish (Figure 4-19), and humans (Chapter 1). Another allele of the C gene is the c^h (Himalayan) allele.

(a)

(b)

Figure 4-19 Albinism in reptiles and birds. In each case, the phenotype is produced by a recessive allele that determines an inability to produce the dark pigment melanin in skin cell. (The normal allele determines the ability to synthesize melanin.) (a) In this rattlesnake species, the normal dark coloration is due entirely to melanin, so the albino allele results in a completely unpigmented appearance. (b) In this penguin species, melanin normally makes dorsal feathers black, but the reddish orange colors in the head feathers and beak are due to another pigment chemically unrelated to melanin. The recessive albino allele results in no melanin, but the reddish parts are unaffected and retain their normal coloration. (Part a from K. H. Switak/NHPA; part b from A.N.T./NHPA.)

This allele can be considered a heat-sensitive version of the c allele. Only at the colder body extremities is c^h functional and able to make pigment. In warm parts of the body, it behaves just like the albino allele c. A Himalayan mouse is shown in Figure 4-20, which also shows the action of the same allele in rabbits and in cats, where it produces the Siamese phenotype.

The c/c constitution produces the standard recessive epistasis modified ratio, as seen in the following cross (in which both parents are a/a):

(a)

(b)

(c)

B/B ; c/c (albino) × b/b ; C/C (brown)

or B/B ; C/C (black) × b/b ; c/c (albino)

$\downarrow$

F₁   all B/b ; C/c (black)
B/b ; C/c (black) × B/b ; C/c (black)

$\downarrow$

F₂   9 $B/-$; $C/-$ (black)   9
      3 b/b ; $C/-$ (brown)   3
      3 $B/-$; c/c (albino) ⎤
      1 b/b ; c/c (albino) ⎦ 4

D gene

The D gene controls the intensity of pigment specified by the other coat-color genes. The genotypes D/D and D/d permit full expression of color in mice, but d/d "dilutes" the color, making it look "milky." The effect is due to an uneven distribution of pigment in the hair shaft. Dilute agouti, dilute cinnamon, dilute brown, and dilute black coats are all possible. This is another example of a modifier gene. In the following cross, we assume that both parents are a/a ; C/C:

B/B ; d/d (dilute black) × b/b ; D/D (brown)

or B/B ; D/D (black) × b/b ; d/d (dilute brown)

$\downarrow$

F₁   all B/b ; D/d (black)
B/b ; D/d (black) × B/b ; D/d (black)

$\downarrow$

F₂ 9 $B/-$; $D/-$ (black)
    3 $B/-$; d/d (dilute black)
    3 b/b ; $D/-$ (brown)
    1 b/b ; d/d (dilute brown)

In horses, the D allele shows incomplete dominance. Figure 4-21 shows how dilution affects the appearance of chestnut and bay horses. The milky effect of D is often seen in domestic cats.

S gene

The S gene controls the distribution of coat pigment throughout the body. In effect, it controls the presence or

Figure 4-20 Temperature-sensitive alleles of the C gene result in similar phenotypes in several different mammals. These alleles result in very much reduced or no synthesis of the dark pigment melanin in the skin covering warmer parts of the body. At lower temperatures, such as those found at the body extremities, melanin is synthesized, producing darker snout, ears, tail, and feet. (a) Himalayan mouse. (b) Siamese cat. (c) Himalayan rabbits, which are often sold as pets. All three are of genotype c^h/c^h. (Part a from Anthony Griffiths; part b from Walter Chandoha; part c from Dan McCoy/Rainbow.)

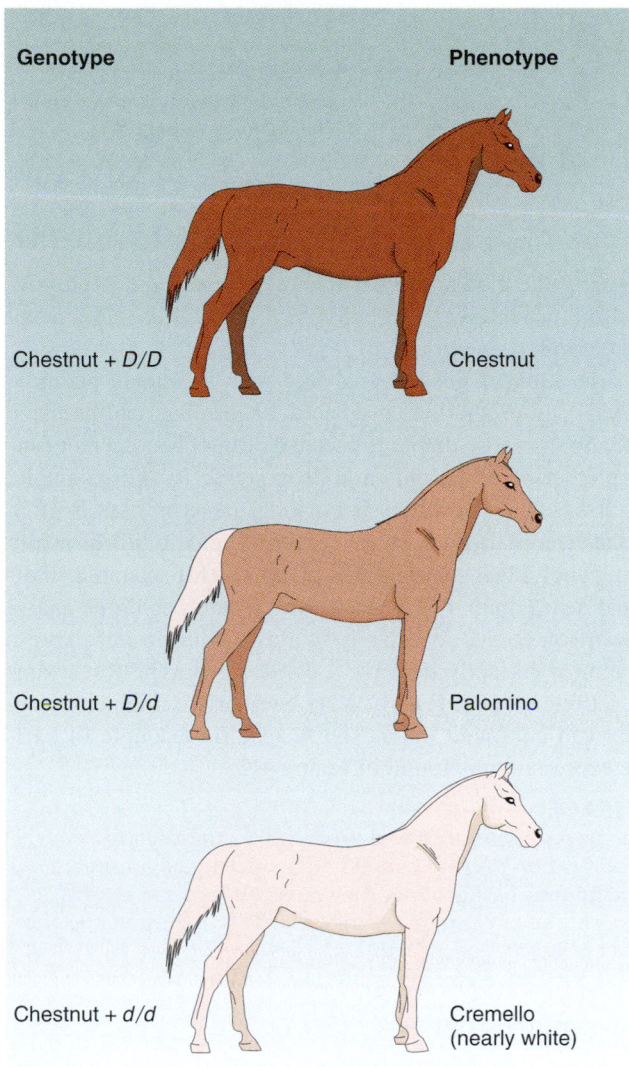

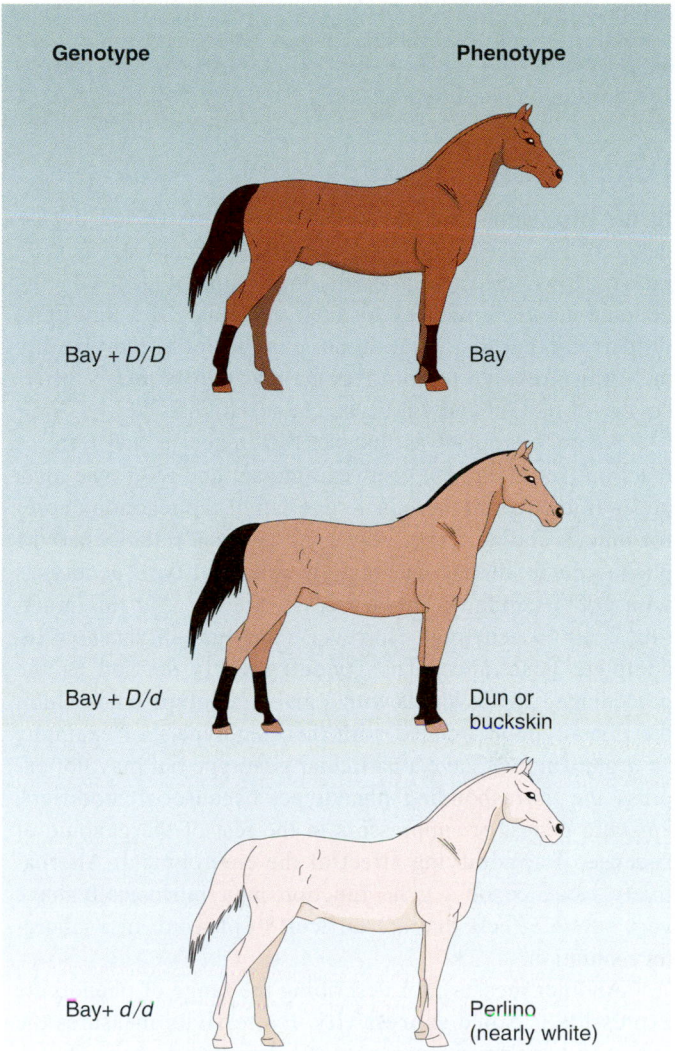

Figure 4-21   The modifying effect of the dilution allele on basic chestnut and bay genotypes in horses. Note the incomplete dominance shown by *D*. (From J. W. Evans et al., *The Horse*. Copyright © 1977 by W. H. Freeman and Company.)

absence of spots. The genotype *S/−* results in no spots, and *s/s* produces a spotting pattern called piebald in both mice and horses. This pattern can be superimposed on any of the coat colors considered so far — with the exception of albino.

Let us summarize the foregoing discussion of coat color in mice. The normal coat appearance in wild mice is produced by a complex set of interacting genes determining pigment type, pigment distribution in the individual hairs, pigment distribution on the animal's body, and the presence or absence of pigment. Such interactions are deduced from crosses in which two or more of the interacting genes are heterozygous for alleles that modify the normal coat color and pattern. Figure 4-22 illustrates some of the pigment patterns in mice. Interacting genes such as those in mice determine most characters in any organism.

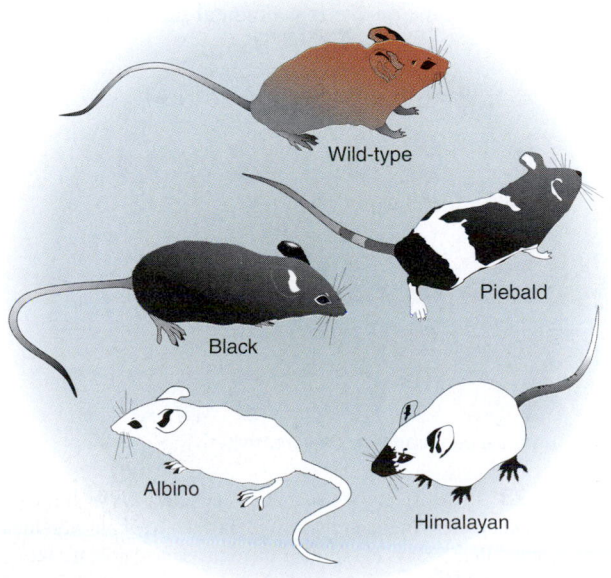

Figure 4-22   Some coat phenotypes in mice.

MESSAGE ·······································
Different kinds of modified dihybrid ratios point to
different ways in which genes can interact with each other
to determine phenotype.

Penetrance and expressivity

In the preceding example, the genetic basis of the dependence of one gene on another is deduced from clear genetic
ratios. However, only a small proportion of genes in the
genome lend themselves to such analysis. One important
property is that the mutation not exhibit decreased viability
or fertility relative to wild type so that the frequency of recovery of mutant and wild-type classes are not skewed.

Another property is that the difference in the norm of
reaction (see Chapter 1) between mutant and wild type must
be so dramatic that there is no overlap of the reaction curves
for mutant and wild type, and hence we can reliably use the
phenotype to distinguish mutant and wild-type genotypes
with 100% certainty. In such cases, we say that this mutation is 100% penetrant. However, many mutations show incomplete penetrance. Thus **penetrance** is defined as the
percentage of individuals with a given genotype who exhibit
the phenotype associated with that genotype. For example,
an organism may have a particular genotype but may not express the corresponding phenotype, because of modifiers,
epistatic genes, or suppressors in the rest of the genome or
because of a modifying effect of the environment. Alternatively, absence of a gene function may intrinsically have
very subtle effects that are difficult to measure in a laboratory situation.

Another measure for describing the range of phenotypic
expression is called **expressivity.** Expressivity measures the
extent to which a given genotype is expressed at the phenotypic level. Different degrees of expression in different individuals may be due to variation in the allelic constitution of
the rest of the genome or to environmental factors. Figure 4-23

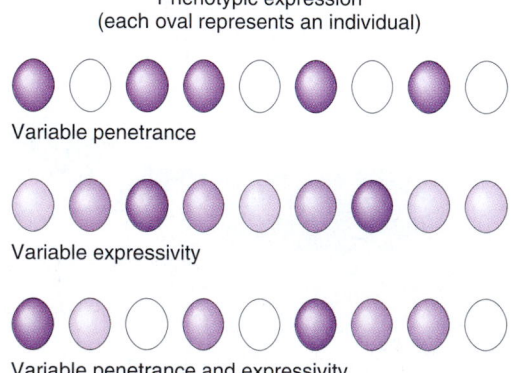

Phenotypic expression
(each oval represents an individual)

Variable penetrance

Variable expressivity

Variable penetrance and expressivity

Figure 4-23 The effects of penetrance and expressivity through a
hypothetical character "pigment intensity." In each row, all individuals
have the same allele — say, *P* — giving them the same "potential to
produce pigment." However, effects deriving from the rest of the
genome and from the environment may suppress or modify pigment
production in an individual.

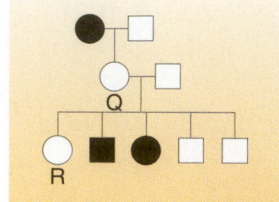

Figure 4-24 Lack of penetrance
illustrated by a pedigree for a
dominant allele. Individual Q must
have the allele (because it was passed
on to her progeny), but it was not
expressed in her phenotype. An
individual such as R cannot be sure
that her genotype lacks the allele.

illustrates the distinction between penetrance and expressivity. Like penetrance, expressivity is integral to the concept
of the norm of reaction.

Any kind of genetic analysis, such as human pedigree
analysis and predictions in genetic counseling, can be made
substantially more difficult because of the phenomena of incomplete penetrance and variable expressivity. For example,
if a disease-causing allele is not fully penetrant (as is often
the case), it is difficult to give a clean genetic bill of health
to any individual in a disease pedigree (for example, individual *R* in Figure 4-24). On the other hand, pedigree analysis can sometimes identify individuals who do not express
but almost certainly do have a disease genotype (for example, individual *Q* in Figure 4-24). Similarly, variable expressivity can confound diagnosis. A specific example of variable expressivity is found in Figure 4-25.

MESSAGE ···
The terms *penetrance* and *expressivity* quantify the
modification of gene expression by varying environment
and genetic background; they measure respectively the
percentage of cases in which the gene is expressed and the
level of expression.

Chi-square test

The topic of gene interaction includes a sometimes bewil-
dering array of different phenotypic ratios. Although these
ratios are easily demonstrated in established systems such as
the ones illustrated in this chapter, in an experimental setting a researcher may observe an array of different progeny
phenotypes and not initially know the meaning of this ratio.
At this stage, a hypothesis is devised to explain the observed
ratio. The next step is to determine whether the observed
data are compatible with the expectations of the hypothesis.

In research generally, it is often necessary to compare
experimentally observed numbers of items in several different categories with numbers that are predicted on the basis
of some hypothesis. For example, you might want to determine whether the sex ratio in some specific population of
insects is 1 : 1 as expected. If there is a close match, then the
hypothesis is upheld, whereas, if there is a poor match, then
the hypothesis is rejected. As part of this process, a judgment has to be made about whether the observed numbers
are a close enough match to those expected. Very close
matches and blatant mismatches generally present no problem in judgment, but inevitably there are gray areas in
which the match is not obvious. Genetic analysis often requires the interpretation of numbers in various phenotypic

Figure 4-25   Variable expressivity shown by 10 grades of piebald spotting in beagles. Each of these dogs has S^P, the allele responsible for piebald spots in dogs. (After Clarence C. Little, *The Inheritance of Coat Color in Dogs.* Cornell University Press, 1957; and Giorgio Schreiber, *Journal of Heredity* 9, 1930, 403.)

classes. In such cases, a statistical procedure called the χ^2 (chi-square) test is used to help in making the decision to hold onto or reject the hypothesis.

The χ^2 test is simply a way of quantifying the various deviations expected by chance if a hypothesis is true. For example, consider a simple hypothesis that a certain plant is a heterozygote (monohybrid) of genotype A/a. To test this hypothesis, we would make a testcross to a/a and predict a 1:1 ratio of A/a and a/a in the progeny. Even if the hypothesis is true, we do not always expect an exact 1:1 ratio. We

can model this experiment with a barrel full of equal numbers of red and blue marbles. If we blindly removed samples of 100 marbles, on the basis of chance we would expect samples to show small deviations such as 52 red:48 blue quite commonly and larger deviations such as 60 red:40 blue less commonly. The χ^2 test allows us to calculate the probability of such chance deviations from expectations if the hypothesis is true. But, if all levels of deviation are expected with different probabilities even if the hypothesis is true, how can we ever reject a hypothesis? It has become a general scientific convention that a probability value of less than 5 percent is to be taken as the criterion for rejecting the hypothesis. The hypothesis might still be true, but we have to make a decision somewhere, and the 5 percent level is the conventional decision line. The logic is that, although results this far from expectations are expected 5 percent of the time even when the hypothesis is true, we will mistakenly reject the hypothesis in only 5% of cases and we are willing to take this chance of error.

Let's consider an example taken from gene interaction. We cross two pure lines of plants, one with yellow petals and one with red. The F_1 are all orange. When the F_1 is selfed to give an F_2, we find the following result:

orange	182
yellow	61
red	<u>77</u>
Total	320

What hypothesis can we invent to explain the results? There are at least two possibilities:

Hypothesis 1.   Incomplete dominance

    (yellow) $G1/G1 \times G2/G2$ (red)

F_1 $G1/G2$ (orange)

			Expected numbers
F_2	$\frac{1}{4}\ G1/G1$	(yellow)	80
	$\frac{1}{2}\ G1/G2$	(orange)	160
	$\frac{1}{4}\ G2/G2$	(red)	80

Hypothesis 2.   Recessive epistasis of r (red) on Y (orange) and y (yellow)

    (yellow) $y/y\ ;\ R/R \times Y/Y\ ;\ r/r$ (red)

F_1 $Y/y\ ;\ R/r$ (orange)

			Expected numbers
F_2	$\frac{9}{16}\ Y/-\ ;\ R/-$	(orange)	180
	$\frac{3}{16}\ y/y\ ;\ R/-$	(yellow)	60
	$\frac{3}{16}\ Y/-\ ;\ r/r$	(red)	80
	$\frac{1}{16}\ y/y\ ;\ r/r$	(red)	

The statistic χ^2 is always calculated from actual numbers, not from percentages, proportions, or fractions. Sample size

4-1 ⟩ TABLE				Critical Values of the χ^2 Distribution					

					P					
df	0.995	0.975	0.9	0.5	0.1	0.05	0.025	0.01	0.005	df
1	.000	.000	0.016	0.455	2.706	3.841	5.024	6.635	7.879	1
2	0.010	0.051	0.211	1.386	4.605	5.991	7.378	9.210	10.597	2
3	0.072	0.216	0.584	2.366	6.251	7.815	9.348	11.345	12.838	3
4	0.207	0.484	1.064	3.357	7.779	9.488	11.143	13.277	14.860	4
5	0.412	0.831	1.610	4.351	9.236	11.070	12.832	15.086	16.750	5
6	0.676	1.237	2.204	5.348	10.645	12.592	14.449	16.812	18.548	6
7	0.989	1.690	2.833	6.346	12.017	14.067	16.013	18.475	20.278	7
8	1.344	2.180	3.490	7.344	13.362	15.507	17.535	20.090	21.955	8
9	1.735	2.700	4.168	8.343	14.684	16.919	19.023	21.666	23.589	9
10	2.156	3.247	4.865	9.342	15.987	18.307	20.483	23.209	25.188	10
11	2.603	3.816	5.578	10.341	17.275	19.675	21.920	24.725	26.757	11
12	3.074	4.404	6.304	11.340	18.549	21.026	23.337	26.217	28.300	12
13	3.565	5.009	7.042	12.340	19.812	22.362	24.736	27.688	29.819	13
14	4.075	5.629	7.790	13.339	21.064	23.685	26.119	29.141	31.319	14
15	4.601	6.262	8.547	14.339	22.307	24.996	27.488	30.578	32.801	15

is therefore very important in the χ^2 test, as it is in most considerations of chance phenomena. Samples to be tested generally consist of several classes. The letter O is used to represent the observed number in a class, and E represents the expected number for the same class based on the predictions of the hypothesis. The general formula for calculating χ^2 is as follows:

Sum of $(O − E)^2/E$ for all classes

For hypothesis 1, the calculation is as follows:

	O	E	$(O − E)^2$	$(O − E)^2/E$
orange	182	160	484	3.0
yellow	61	80	361	4.5
red	77	80	9	0.1
				$\chi^2 = 7.6$

To convert the χ^2 value into a probability, we use Table 4-1, which shows χ^2 values for different degrees of freedom (df). For any total number of progeny, if the number of individuals in two of the three phenotypic classes is known, then the size of the third class is automatically determined. Hence, there are only *2 degrees of freedom* in the distribution of individuals among the three classes. Generally, the number of degrees of freedom (shown as the different rows

of Table 4-1) is the number of classes minus 1. In this case, it is $3 − 1 = 2$. Looking along the 2-df line, we find that the χ^2 value places the probability at less than 0.025, or 2.5 percent. This means that, if the hypothesis is true, then deviations from expectations this large or larger are expected approximately 2.5 percent of the time. As mentioned earlier, by convention the 5 percent level is used as the cutoff line. When values of less than 5 percent are obtained, the hypothesis is rejected as being too unlikely. Hence the incomplete dominance hypothesis must be rejected.

For hypothesis 2, the calculation is set up as follows.

	O	E	$(O − E)^2$	$(O − E)^2/E$
orange	182	180	4	0.02
yellow	61	60	1	0.02
red	77	80	9	0.11
				$\chi^2 = 0.15$

The probability value (for 2 df) this time is greater than 0.9, or 90 percent. Hence a deviation this large or larger is expected approximately 90 percent of the time — in other words, very frequently. Formally, because 90 percent is greater than 5 percent, we conclude that the results uphold the hypothesis of recessive epistasis.

SUMMARY

Although it is possible by genetic analysis to isolate a single gene whose alleles dictate two alternative phenotypes for one character, this gene does not control that character by itself; the gene must interact with many other genes in the genome. There are two types of interaction: (1) between alleles of a single gene and (2) between alleles of different genes. The complementation test decides if two related mutant phenotypes determined by recessive alleles are due to mutations of one gene or of two different genes. The mutant genotypes are brought together in an F_1 individual and, if the phenotype is mutant, then no complementation has occurred and the two alleles must be of the same gene. If complementation is observed, the alleles must be of different genes.

Alleles of a single gene interact in several ways. First, their specific cellular actions result in variations of dominance: incomplete dominance results from intermediate amounts of the protein in the heterozygote, and codominance results from the production of two distinct protein products. Alleles also interact through a kind of dosage effect: one mutant allele in heterozygous condition might produce a mildly deleterious effect, whereas two doses in the homozygous mutant might cause severe effects or even lethality.

The interaction of alleles of different genes can be detected through modified 9:3:3:1 dihybrid ratios. Mutations of genes in a single pathway give a 9:7 ratio if the mutant phenotypes are the same or a 9:3:4 ratio if the mutant phenotypes are different. The 9:3:4 ratio is said to result from recessive epistasis, a situation in which the mutation in the first-acting protein in the sequence obliterates the expression of a later-acting gene. Another two-gene interaction is suppression: suppressor mutations convert a mutant phenotype caused by a mutant allele of another locus into a wild-type phenotype.

Penetrance is the proportion of individuals in which a mutation is expressed, and expressivity is the severity of the phenotype attributed to one allele. Variable penetrance and expressivity are caused by interactions with other genes and with the environment.

An observed phenotypic ratio can be assessed against that expected from a specific hypothesis by using the chi-square test.

CONCEPT MAP

Draw a concept map interrelating as many of the following terms as possible. Note that the terms are listed in no particular order.

dihybrid / 9:3:4 ratio / independent assortment / biochemical pathway / recessive epistasis / gene interaction / meiosis / wild type

CHAPTER INTEGRATION PROBLEM

Most pedigrees show polydactyly (Figure 2-22) to be inherited as a rare autosomal dominant, but the pedigrees of some families do not fully conform to the patterns expected for such inheritance. Such a pedigree is shown below. (The unshaded diamonds stand for the specified number of unaffected persons of unknown sex.)

a. What irregularity does this pedigree show?

b. What genetic phenomenon does this pedigree illustrate?

c. Suggest a specific gene-interaction mechanism that could produce such a pedigree, showing genotypes of pertinent family members.

◆ Solution ◆

a. The normal expectation for an autosomal dominant is for each affected individual to have an affected parent, but this expectation is not seen in this pedigree, which constitutes the irregularity. What are some possible explanations?

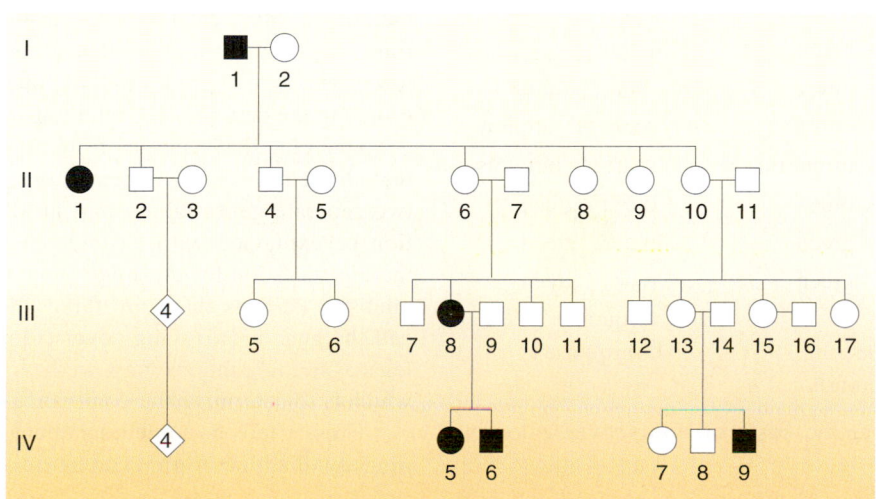

Could some cases of polydactyly be caused by a different gene, one that is an X-linked dominant? This suggestion is not useful, because we still have to explain the absence of the condition in persons II-6 and II-10. Furthermore, postulating recessive inheritance, whether autosomal or sex linked, requires many people in the pedigree to be heterozygotes, which is inappropriate because polydactyly is a rare condition.

b. Thus we are left with the conclusion that polydactyly must sometimes be incompletely penetrant. We have learned in this chapter that some individuals who have the genotype for a particular phenotype do not express it. In this pedigree, II-6 and II-10 seem to belong in this category; they must carry the polydactyly gene inherited from I-1 because they transmit it to their progeny.

c. We have seen in the chapter that environmental suppression of gene expression can cause incomplete penetrance, as can suppression by another gene. To give the requested genetic explanation, we must come up with a genetic hypothesis. What do we need to explain? The key is that I-1 passes the gene on to two types of progeny, represented by II-1, who expresses the gene, and by II-6 and II-10, who do not. (From the pedigree, we cannot tell whether the other children of I-1 have the gene.) Is genetic suppression at work? I-1 does not have a suppressor allele, because he expresses polydactyly. So the only person from whom a suppressor could come is I-2. Furthermore, I-2 must be heterozygous for the suppressor gene because at least one of her children does express polydactyly. We have thus formulated the hypothesis that the mating in generation I must have been

$$\text{(I-1) } P/p \cdot s/s \times \text{(I-2) } p/p \cdot S/s$$

where S is the suppressor and P is the allele responsible for polydactyly. From this hypothesis, we predict that the progeny will comprise the following four types if the genes assort:

Genotype	Phenotype	Example
$P/p \cdot S/s$	normal (suppressed)	II-6, II-10
$P/p \cdot s/s$	polydactylous	II-1
$p/p \cdot S/s$	normal	
$p/p \cdot s/s$	normal	

If S is rare, the matings and progenies of II-6 and II-10 are probably giving

Progeny genotype	Example
$P/p \cdot S/s$	III-13
$P/p \cdot s/s$	III-8
$p/p \cdot S/s$	
$p/p \cdot s/s$	

We cannot rule out the possibilities that II-2 and II-4 have the genotype $P/p \cdot S/s$ and that by chance none of their descendants are affected.

Note that we have just used concepts from Chapter 1 (environmental effects), Chapter 2 (Mendelian inheritance), Chapter 3 (chromosomal locations of genes), and Chapter 4 (gene interactions).

SOLVED PROBLEMS

1. Beetles of a certain species may have green, blue, or turquoise wing covers. Virgin beetles were selected from a polymorphic laboratory population and mated to determine the inheritance of wing-cover color. The crosses and results were as follows:

Cross	Parents	Progeny
1	blue × green	all blue
2	blue × blue	$\frac{3}{4}$ blue : $\frac{1}{4}$ turquoise
3	green × green	$\frac{3}{4}$ green : $\frac{1}{4}$ turquoise
4	blue × turquoise	$\frac{1}{2}$ blue : $\frac{1}{2}$ turquoise
5	blue × blue	$\frac{3}{4}$ blue : $\frac{1}{4}$ green
6	blue × green	$\frac{1}{2}$ blue : $\frac{1}{2}$ green
7	blue × green	$\frac{1}{2}$ blue : $\frac{1}{4}$ green $\frac{1}{4}$ turquoise
8	turquoise × turquoise	all turquoise

a. Deduce the genetic basis of wing-cover color in this species.

b. Write the genotypes of all parents and progeny as completely as possible.

◆ Solution ◆

a. These data seem complex at first, but the inheritance pattern becomes clear if we consider the crosses one at a time. A general principle of solving such problems, as we have seen, is to begin by looking over all the crosses and by grouping the data to bring out the patterns.

One clue that emerges from an overview of the data is that all the ratios are one-gene ratios: there is no evidence of two separate genes taking part at all. How can such variation be explained with a single gene? The answer is that there is variation for the single gene itself — that is, multiple allelism. Perhaps there are three alleles of one gene; let's call the gene w (for wing-cover color) and represent the alleles as w^g, w^b, and w^t. Now we have an additional problem, which is to determine the dominance of these alleles.

Cross 1 tells us something about dominance because the progeny of a blue × green cross are all blue; hence, blue appears to be dominant to green. This conclusion is supported by cross 5, because the green determinant must have been present in the parental stock to appear in the progeny. Cross 3 informs us about the turquoise determinants, which must have been present, although unexpressed, in the parental stock because there are turquoise wing covers in the progeny. So green must be dominant to turquoise. Hence, we have formed a model in which the dominance is $w^b > w^g > w^t$. Indeed, the inferred position of the w^t allele at the bottom of the dominance series is supported by the results of cross 7, where turquoise shows up in the progeny of a blue $\times$ green cross.

b. Now it is just a matter of deducing the specific genotypes. Notice that the question states that the parents were taken from a polymorphic population; this means that they could be either homozygous or heterozygous. A parent with blue wing covers, for example, might be homozygous (w^b/w^b) or heterozygous (w^b/w^g or w^b/w^t). Here, a little trial and error and common sense are called for, but, by this stage, the question has essentially been answered and all that remains is to "cross the t's and dot the i's." The following genotypes explain the results. A dash indicates that the genotype may be *either* homozygous *or* heterozygous in having a second allele farther down the allelic series.

Cross	Parents	Progeny
1	$w^b/w^b \times w^g/-$	w^b/w^g *or* $w^b/-$
2	$w^b/w^t \times w^b/w^t$	$\frac{3}{4} w^b/- : \frac{1}{4} w^t/w^t$
3	$w^g/w^t \times w^g/w^t$	$\frac{3}{4} w^g/- : \frac{1}{4} w^t/w^t$
4	$w^b/w^t \times w^t/w^t$	$\frac{1}{2} w^b/w^t : \frac{1}{2} w^t/w^t$
5	$w^b/w^g \times w^b/w^g$	$\frac{3}{4} w^b/- : \frac{1}{4} w^g/w^g$
6	$w^b/w^g \times w^g/w^g$	$\frac{1}{2} w^b/w^g : \frac{1}{2} w^g/w^g$
7	$w^b/w^t \times w^g/w^t$	$\frac{1}{2} w^b/- : \frac{1}{4} w^g/w^t : \frac{1}{4} w^t/w^t$
8	$w^t/w^t \times w^t/w^t$	*all* w^t/w^t

2. The leaves of pineapples can be classified into three types: spiny (S), spine tip (ST), and piping (nonspiny) (P). In crosses between pure strains followed by intercrosses of the F$_1$, the following results appeared:

Cross	Parental	F$_1$	F$_2$
		PHENOTYPES	
1	ST $\times$ S	ST	99 ST : 34 S
2	P $\times$ ST	P	120 P : 39 ST
3	P $\times$ S	P	95 P : 25 ST : 8 S

a. Assign gene symbols. Explain these results in regard to the genotypes produced and their ratios.

b. Using the model from part a, give the phenotypic ratios that you would expect if you crossed (1) the F$_1$ progeny from piping $\times$ spiny with the spiny parental

stock and (2) the F$_1$ progeny of piping $\times$ spiny with the F$_1$ progeny of spiny $\times$ spiny tip.

◆ Solution ◆

a. First, let's look at the F$_2$ ratios. We have clear 3:1 ratios in crosses 1 and 2, indicating single-gene segregations. Cross 3, however, shows a ratio that is almost certainly a 12:3:1 ratio. How do we know this? Well, there are simply not that many complex ratios in genetics, and trial and error brings us to the 12:3:1 quite quickly. In the 128 progeny total, the numbers of 96:24:8 are expected, but the actual numbers fit these expectations remarkably well.

One of the principles of this chapter is that modified Mendelian ratios reveal gene interactions. Cross 3 gives F$_2$ numbers appropriate for a modified dihybrid Mendelian ratio, so it looks as if we are dealing with a two-gene interaction. This seems the most promising place to start; we can go back to crosses 1 and 2 and try to fit them in later.

Any dihybrid ratio is based on the phenotypic proportions 9:3:3:1. Our observed modification groups them as follows:

$$\left.\begin{array}{l} 9\ A/- \ ; B/- \\ 3\ A/- \ ; b/b \end{array}\right\} \quad 12 \text{ piping}$$
$$3\ a/a \ ; B/- \qquad\qquad 3 \text{ spiny tip}$$
$$1\ a/a \ ; b/b \qquad\qquad 1 \text{ spiny}$$

So without worrying about the name of the type of gene interaction (we are not asked to supply this anyway), we can already define our three pineapple-leaf phenotypes in relation to the proposed allelic pairs A/a and B/b:

$$\begin{array}{l} \text{piping} = A/- \ (B/b \text{ irrelevant}) \\ \text{spiny-tip} = a/a \ ; B/- \\ \text{spiny} = a/a \ ; b/b \end{array}$$

What about the parents of cross 3? The spiny parent must be $a/a \ ; b/b$, and, because the B gene is needed to produce F$_2$ spiny-tip individuals, the piping parent must be $A/A \ ; B/B$. (Note that we are *told* that all parents are pure, or homozygous.) The F$_1$ must therefore be $A/a \ ; B/b$.

Without further thought, we can write out cross 1 as follows:

$$a/a \ ; B/B \times a/a \ ; b/b \longrightarrow a/a \ ; B/b \left\langle\begin{array}{l} \frac{3}{4}\ a/a \ ; B/b \\ \frac{1}{4}\ a/a \ ; b/b \end{array}\right.$$

Cross 2 can be partly written out without further thought by using our arbitrary gene symbols:

$$A/A \ ; -/- \times a/a \ ; B/B \longrightarrow A/a \ ; B/- \left\langle\begin{array}{l} \frac{3}{4}\ A/- \ ; -/- \\ \frac{1}{4}\ a/a \ ; B/- \end{array}\right.$$

We know that the F$_2$ of cross 2 shows single-gene segregation, and it seems certain now that the A/a allelic pair has a role. But the B allele is needed to produce the spiny tip phenotype, so all individuals must be homozygous B/B:

$$A/A ; B/B \times a/a ; B/B \longrightarrow A/a ; B/B \begin{cases} \tfrac{3}{4} A/- ; B/B \\ \tfrac{1}{4} a/a ; B/B \end{cases}$$

(1) $A/a ; B/b \times a/a ; b/b \longrightarrow$ $\left. \begin{array}{l} \tfrac{1}{4} A/a ; B/b \\ \tfrac{1}{4} A/a ; b/b \end{array} \right\}$ piping

(independent $\tfrac{1}{4} a/a ; B/b$ spiny tip

assortment in $\tfrac{1}{4} a/a ; b/b$ spiny

a standard

testcross)

(2) $A/a ; B/b \times a/a ; B/b \longrightarrow$

$$\tfrac{1}{2} A/a \begin{cases} \tfrac{3}{4} B/- \longrightarrow \tfrac{3}{8} \\ \tfrac{1}{4} b/b \longrightarrow \tfrac{1}{8} \end{cases} \left. \right\} \tfrac{1}{2} \text{ piping}$$

$$\tfrac{1}{2} a/a \begin{cases} \tfrac{3}{4} B/- \longrightarrow \tfrac{3}{8} \quad \text{spiny tip} \\ \tfrac{1}{4} b/b \longrightarrow \tfrac{1}{8} \quad \text{spiny} \end{cases}$$

Notice that the two single-gene segregations in crosses 1 and 2 do not show that the genes are *not* interacting. What is shown is that the two-gene interaction is not *revealed* by these crosses — only by cross 3, in which the F_1 is heterozygous for both genes.

b. Now it is simply a matter of using Mendel's laws to predict cross outcomes:

PROBLEMS

1. If a man of blood group AB marries a woman of blood group A whose father was of blood group O, to what different blood groups can this man and woman expect their children to belong?

2. Erminette fowls have mostly light colored feathers with an occasional black one, giving a flecked appearance. A cross of two erminettes produced a total of 48 progeny, consisting of 22 erminettes, 14 blacks, and 12 pure whites. What genetic basis of the erminette pattern is suggested? How would you test your hypotheses?

3. Radishes may be long, round, or oval, and they may be red, white, or purple. You cross a long, white variety with a round, red one and obtain an oval, purple F_1. The F_2 show nine phenotypic classes as follows: 9 long, red; 15 long, purple; 19 oval, red; 32 oval, purple; 8 long, white; 16 round, purple; 8 round, white; 16 oval, white; and 9 round, red.

   a. Provide a genetic explanation of these results. Be sure to define the genotypes and show the constitution of parents, F_1, and F_2.

   b. Predict the genotypic and phenotypic proportions in the progeny of a cross between a long, purple radish and an oval, purple one.

4. In the multiple allele series that determines coat color in rabbits, $C^+ > C^{ch} > C^h$, dominance is from left to right as shown. In a cross of $C^+/C^{ch} \times C^{ch}/C^h$, what proportion of progeny will be Himalayan?

5. Black, sepia, cream, and albino are all coat colors of guinea pigs. Individual animals (not necessarily from pure lines) showing these colors were intercrossed; the results are tabulated as follows, where the abbreviations A (albino), B (black), C (cream), and S (sepia) represent the phenotypes:

Cross	Parental pheno-types	PHENOTYPES OF PROGENY B	S	C	A
1	B × B	22	0	0	7
2	B × A	10	9	0	0
3	C × C	0	0	34	11
4	S × C	0	24	11	12
5	B × A	13	0	12	0
6	B × C	19	20	0	0
7	B × S	18	20	0	0
8	B × S	14	8	6	0
9	S × S	0	26	9	0
10	C × A	0	0	15	17

   a. Deduce the inheritance of these coat colors and use gene symbols of your own choosing. Show all parent and progeny genotypes.

   b. If the black animals in crosses 7 and 8 are crossed, what progeny proportions can you predict by using your model?

6. In a maternity ward, four babies become accidentally mixed up. The ABO types of the four babies are known to be O, A, B, and AB. The ABO types of the four sets of parents are determined. Indicate which baby belongs to each set of parents: **(a)** AB × O, **(b)** A × O, **(c)** A × AB, **(d)** O × O.

7. Consider two blood polymorphisms that humans have in addition to the ABO system. Two alleles L^M and L^N determine the M, N, and MN blood groups. The dominant allele R of a different gene causes a person to have the Rh$^+$ (rhesus positive) phenotype, whereas the homozygote for r is Rh$^-$ (rhesus negative). Two men took a paternity dispute to court, each claiming three children to be his own. The blood groups of the men, the children, and their mother were as follows:

Person		Blood group	
husband	O	M	Rh$^+$
wife's lover	AB	MN	Rh$^-$
wife	A	N	Rh$^+$
child 1	O	MN	Rh$^+$
child 2	A	N	Rh$^+$
child 3	A	MN	Rh$^-$

From this evidence, can the paternity of the children be established?

8. On a fox ranch in Wisconsin, a mutation arose that gave a "platinum" coat color. The platinum color proved very popular with buyers of fox coats, but the breeders could not develop a pure-breeding platinum strain. Every time two platinums were crossed, some normal foxes appeared in the progeny. For example, the repeated matings of the same pair of platinums produced 82 platinum and 38 normal progeny. All other such matings gave similar progeny ratios. State a concise genetic hypothesis that accounts for these results.

9. For a period of several years, Hans Nachtsheim investigated an inherited anomaly of the white blood cells of rabbits. This anomaly, termed the *Pelger anomaly,* is the arrest of the segmentation of the nuclei of certain white cells. This anomaly does not appear to seriously inconvenience the rabbits.

    a. When rabbits showing the typical Pelger anomaly were mated with rabbits from a true-breeding normal stock, Nachtsheim counted 217 offspring showing the Pelger anomaly and 237 normal progeny. What appears to be the genetic basis of the Pelger anomaly?

    b. When rabbits with the Pelger anomaly were mated to each other, Nachtsheim found 223 normal progeny, 439 showing the Pelger anomaly, and 39 extremely abnormal progeny. These very abnormal progeny not only had defective white blood cells, but also showed severe deformities of the skeletal system; almost all of them died soon after birth. In genetic terms, what do you suppose these extremely defective rabbits represented? Why do you suppose there were only 39 of them?

    c. What additional experimental evidence might you collect to support or disprove your answers to part b?

    d. In Berlin, about one human in 1000 shows a Pelger anomaly of white blood cells very similar to that described in rabbits. The anomaly is inherited as a simple dominant, but the homozygous type has not been observed in humans. Can you suggest why if you are permitted an analogy with the condition in rabbits?

    e. Again by analogy with rabbits, what phenotypes and genotypes might be expected among the children of a man and woman who both show the Pelger anomaly?

    (Problem 9 from A. M. Srb, R. D. Owen, and R. S. Edgar, *General Genetics,* 2d ed. W. H. Freeman and Company, 1965.)

10. Two normal-looking fruit flies were crossed and, in the progeny, there were 202 females and 98 males.

    a. What is unusual about this result?

    b. Provide a genetic explanation for this anomaly.

    c. Provide a test of your hypothesis.

11. You have been given a virgin *Drosophila* female. You notice that the bristles on her thorax are much shorter than normal. You mate her with a normal male (with long bristles) and obtain the following F_1 progeny: $\frac{1}{3}$ short-bristled females, $\frac{1}{3}$ long-bristled females, and $\frac{1}{3}$ long-bristled males. A cross of the F_1 long-bristled females with their brothers gives only long-bristled F_2. A cross of short-bristled females with their brothers gives $\frac{1}{3}$ short-bristled females, $\frac{1}{3}$ long-bristled females, and $\frac{1}{3}$ long-bristled males. Provide a genetic hypothesis to account for all these results, showing genotypes in every cross.

12. A dominant allele H reduces the number of body bristles that *Drosophila* flies have, giving rise to a "hairless" phenotype. In the homozygous condition, H is lethal. An independently assorting dominant allele S has no effect on bristle number except in the presence of H, in which case a single dose of S suppresses the hairless phenotype, thus restoring the hairy phenotype. However, S also is lethal in the homozygous (S/S) condition.

    a. What ratio of hairy to hairless flies would you find in the live progeny of a cross between two hairy flies both carrying H in the suppressed condition?

    b. When the hairless progeny are backcrossed with a parental hairy fly, what phenotypic ratio would you expect to find among their live progeny?

13. A pure-breeding strain of squash that produced disk-shaped fruits (see the illustration on page 132) was crossed with a pure-breeding strain having long fruits. The F_1 had disk fruits, but the F_2 showed a new phenotype, sphere, and was composed of the following proportions:

disk	270
sphere	178
long	32

Propose an explanation for these results, and show the genotypes of P, F_1, and F_2 generations. (Illustration from P. J. Russell, *Genetics,* 3d ed. HarperCollins, 1992.)

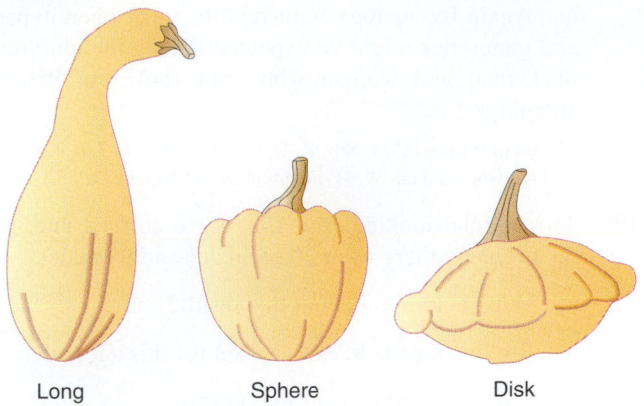

Long   Sphere   Disk

14. Because snapdragons *(Antirrhinum)* possess the pigment anthocyanin, they have reddish purple petals. Two pure anthocyaninless lines of *Antirrhinum* were developed, one in California and one in Holland. They looked identical in having no red pigment at all, manifested as white (albino) flowers. However, when petals from the two lines were ground up together in buffer in the same test tube, the solution, which appeared colorless at first, gradually turned red.

a. What control experiments should an investigator conduct before proceeding with further analysis.

b. What could account for the production of the red color in the test tube?

c. According to your explanation for part b, what would be the genotypes of the two lines?

d. If the two white lines were crossed, what would you predict the phenotypes of the F_1 and F_2 to be?

15. The frizzle fowl is much admired by poultry fanciers. It gets its name from the unusual way that its feathers curl up, giving the impression that it has been (in the memorable words of animal geneticist F. B. Hutt) "pulled backwards through a knothole." Unfortunately, frizzle fowls do not breed true; when two frizzles are intercrossed, they always produce 50 percent frizzles, 25 percent normal, and 25 percent with peculiar woolly feathers that soon fall out, leaving the birds naked.

a. Give a genetic explanation for these results, showing genotypes of all phenotypes, and provide a statement of how your explanation works.

b. If you wanted to mass-produce frizzle fowls for sale, which types would be best to use as a breeding pair?

16. Marfan's syndrome is a disorder of the fibrous connective tissue, characterized by many symptoms including long, thin digits, eye defects, heart disease, and long limbs. (Flo Hyman, the American volleyball star, suffered from Marfan's syndrome. She died soon after a match from a ruptured aorta.)

a. Use the pedigree shown at the bottom of the page to propose a mode of inheritance for Marfan's syndrome.

b. What genetic phenomenon is shown by this pedigree?

c. Speculate on a reason for such a phenomenon.

(Illustration from J. V. Neel and W. J. Schull, *Human Heredity*. University of Chicago Press, 1954.)

17. The petals of the plant *Collinsia parviflora* are normally blue, giving the species its common name, blue-eyed Mary. Two pure-breeding lines were obtained from color variants found in nature; the first line had pink petals and the second line had white petals. The following crosses were made between pure lines, with the results shown:

Parents	F_1	F_2
blue × white	blue	101 blue, 33 white
blue × pink	blue	192 blue, 63 pink
pink × white	blue	272 blue, 121 white, 89 pink

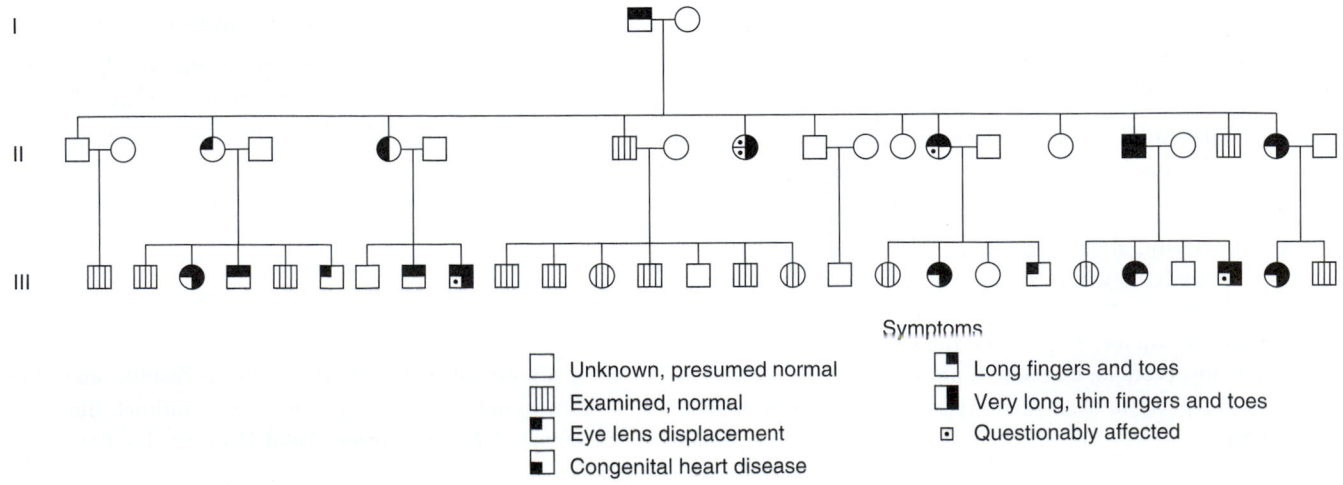

Symptoms

☐ Unknown, presumed normal   ▐ Long fingers and toes
▥ Examined, normal   ▊ Very long, thin fingers and toes
◪ Eye lens displacement   ⊡ Questionably affected
■ Congenital heart disease

a. Explain these results genetically. Define the allele symbols that you use and show the genetic constitution of parents, F_1, and F_2.

b. A cross between a certain blue F_2 plant and a certain white F_2 plant gave progeny of which $\frac{3}{8}$ were blue, $\frac{1}{8}$ were pink, and $\frac{1}{2}$ were white. What must the genotypes of these two F_2 plants have been?

 Unpacking the Problem
....

1. What is the character being studied?

2. What is the wild-type phenotype?

3. What is a variant?

4. What are the variants in this problem?

5. What does "in nature" mean?

6. In what way would the variants have been found in nature? (Describe the scene.)

7. At which stages in the experiments would seeds be used?

8. Would the way of writing a cross "blue × white" (for example) mean the same as "white × blue"? Would you expect similar results? Why or why not?

9. In what way do the first two rows in the table differ from the third row?

10. Which phenotypes are dominant?

11. What is complementation?

12. Where does the blueness come from in the progeny of the pink × white cross?

13. What genetic phenomenon does the production of a blue F_1 from pink and white parents represent?

14. List any ratios that you can see.

16. Are there any monohybrid ratios?

16. Are there any dihybrid ratios?

17. What does observing monohybrid and dihybrid ratios tell you?

18. List four modified Mendelian ratios that you can think of.

19. Are there any modified Mendelian ratios in the problem?

20. What do modified Mendelian ratios indicate generally?

21. What does the specific modified ratio or ratios in this problem indicate?

22. Draw chromosomes representing the meioses in the parents in the cross blue × white and meiosis in the F_1.

23. Repeat for the cross blue × pink.

***18.** In peas *(Pisum sativum)*, the chemical pisatin is associated with defense against parasitic fungi: normal plants are resistant to fungi and contain pisatin. Two pure lines were obtained, both of which lacked pisatin and were highly susceptible to fungal attack. Line 1 was from California and line 2 was from Sweden. The lines were investigated as follows. (**Note:** The normal lines also were pure breeding.)

Cross	F_1 phenotypes	F_2 phenotypes
line 1 × normal	pisatin	$\frac{3}{4}$ pisatin
		$\frac{1}{4}$ no pisatin
line 2 × normal	no pisatin	$\frac{3}{4}$ no pisatin
		$\frac{1}{4}$ pisatin
line 1 × line 2	no pisatin	$\frac{13}{16}$ no pisatin
		$\frac{3}{16}$ pisatin

a. Propose a model that explains the results of these crosses. Make sure that you precisely define any allele symbols that you use.

b. Show the genotypes underlying the parents, F_1, and F_2 in each cross.

c. How do lines 1 and 2 differ in the genetic reason for their lacking pisatin?

19. A woman who owned a purebred albino poodle (an autosomal recessive phenotype) wanted white puppies, so she took the dog to a breeder, who said he would mate the female with an albino stud male, also from a pure stock. When six puppies were born, they were all black, so the woman sued the breeder, claiming that he replaced the stud male with a black dog, giving her six unwanted puppies. You are called in as an expert witness, and the defense asks you if it is possible to produce black offspring from two pure-breeding recessive albino parents. What testimony do you give?

20. A snapdragon plant that bred true for white petals was crossed to a plant that bred true for purple petals, and all the F_1 had white petals. The F_1 was selfed. Among the F_2, three phenotypes were observed in the following numbers:

white	240
solid purple	61
spotted purple	19
Total	320

a. Propose an explanation for these results, showing genotypes of all generations (make up and explain your symbols).

b. A white F_2 plant was crossed to a solid-purple F_2 plant, and the progeny were:

white	50%	
solid purple	25%	
spotted purple	25%	

What were the genotypes of the F_2 plants crossed?

21. Most flour beetles are black, but several color variants are known. Crosses of pure-breeding parents produced the following results in the F_1 generation, and intercrossing the F_1 from each cross gave the ratios shown for the F_2 generation. The phenotypes are abbreviated Bl, black; Br, brown; Y, yellow; and W, white.

Cross	Parents	F_1	F_2
1	Br × Y	Br	3 Br : 1 Y
2	Bl × Br	Bl	3 Bl : 1 Br
3	Bl × Y	Bl	3 Bl : 1 Y
4	W × Y	Bl	9 Bl : 3 Y : 4 W
5	W × Br	Bl	9 Bl : 3 Br : 4 W
6	Bl × W	Bl	9 Bl : 3 Y : 4 W

a. From these results, deduce and explain the inheritance of these colors.

b. Write the genotypes of each of the parents, the F_1, and the F_2 in all crosses.

22. Two albinos marry and have four normal children. How is this possible?

*23. Plant breeders obtained three differently derived pure lines of white-flowered *Petunia* plants. They performed crosses and observed progeny phenotypes as follows:

Cross	Parents	Progeny
1	line 1 × line 2	F_1 all white
2	line 1 × line 3	F_1 all red
3	line 2 × line 3	F_1 all white
4	red F_1 × line 1	$\frac{1}{4}$ red : $\frac{3}{4}$ white
5	red F_1 × line 2	$\frac{1}{8}$ red : $\frac{7}{8}$ white
6	red F_1 × line 3	$\frac{1}{2}$ red : $\frac{1}{2}$ white

a. Explain these results, using gene symbols of your own choosing. (Show parental and progeny genotypes in each cross.)

b. If a red F_1 individual from cross 2 is crossed to a white F_1 individual from cross 3, what proportion of progeny will be red?

24. Consider production of flower color in the Japanese morning glory (*Pharbitis nil*). Dominant alleles of either of two separate genes ($A/-\cdot b/b$ or $a/a\cdot B/-$) produce purple petals. $A/-\cdot B/-$ produces blue petals, and $a/a\cdot b/b$ produces scarlet petals. Deduce the genotypes of parents and progeny in the following crosses:

Cross	Parents	Progeny
1	blue × scarlet	$\frac{1}{4}$ blue : $\frac{1}{2}$ purple : $\frac{1}{4}$ scarlet
2	purple × purple	$\frac{1}{4}$ blue : $\frac{1}{2}$ purple : $\frac{1}{4}$ scarlet
3	blue × blue	$\frac{3}{4}$ blue : $\frac{1}{4}$ purple
4	blue × purple	$\frac{3}{8}$ blue : $\frac{4}{8}$ purple : $\frac{1}{8}$ scarlet
5	purple × scarlet	$\frac{1}{2}$ purple : $\frac{1}{2}$ scarlet

25. Corn breeders obtained pure lines whose kernels turn sun red, pink, scarlet, or orange when exposed to sunlight (normal kernels remain yellow in sunlight). Some crosses between these lines produced the following results. The phenotypes are abbreviated O, orange; P, pink; Sc, scarlet; and SR, sun red.

		PHENOTYPES	
Cross	Parents	F_1	F_2
1	SR × P	all SR	66 SR : 20 P
2	O × SR	all SR	998 SR : 314 O
3	O × P	all O	1300 O : 429 P
4	O × Sc	all Y	182 Y : 80 O : 58 Sc

Analyze the results of each cross, and provide a unifying hypothesis to account for *all* the results. (Explain all symbols that you use.)

26. Many kinds of wild animals have the agouti coloring pattern, in which each hair has a yellow band around it (see Figure 4-18).

a. Black mice and other black animals do not have the yellow band; each of their hairs is all black. This absence of wild agouti pattern is called *nonagouti*. When mice of a true-breeding agouti line are crossed with nonagoutis, the F_1 is all agouti and the F_2 has a 3 : 1 ratio of agoutis to nonagoutis. Diagram this cross, letting A represent the allelel responsible for the agouti phenotype and a, nonagouti. Show the phenotypes and genotypes of the parents, their gametes, the F_1, their gametes, and the F_2.

b. Another inherited color deviation in mice substitutes brown for the black color in the wild-type hair. Such brown-agouti mice are called *cinnamons*. When wild-type mice are crossed with cinnamons, the F_1 is all wild type and the F_2 has a 3 : 1 ratio of wild type to cinnamon. Diagram this cross as in part a, letting B stand for the wild-type black allele and b stand for the cinnamon brown allele.

c. When mice of a true-breeding cinnamon line are crossed with mice of a true-breeding nonagouti (black) line, the F_1 is all wild type. Use a genetic diagram to explain this result.

d. In the F_2 of the cross in part c, a fourth color called *chocolate* appears in addition to the parental cinnamon and nonagouti and the wild type of the F$_1$. Chocolate mice have a solid, rich-brown color. What is the genetic constitution of the chocolates?

e. Assuming that the *A/a* and *B/b* allelic pairs assort independently of each other, what do you expect to be the relative frequencies of the four color types in the F$_2$ described in part d? Diagram the cross of parts c and d, showing phenotypes and genotypes (including gametes).

f. What phenotypes would be observed in what proportions in the progeny of a backcross of F$_1$ mice from part c to the cinnamon parental stock? to the nonagouti (black) parental stock? Diagram these backcrosses.

g. Diagram a testcross for the F$_1$ of part c. What colors would result and in what proportions?

h. Albino (pink-eyed white) mice are homozygous for the recessive member of an allelic pair *C/c* which assorts independently of the *A/a* and *B/b* pairs. Suppose that you have four different highly inbred (and therefore presumably homozygous) albino lines. You cross each of these lines with a true-breeding wild-type line, and you raise a large F$_2$ progeny from each cross. What genotypes for the albino lines can you deduce from the following F$_2$ phenotypes?

PHENOTYPES OF PROGENY

F$_2$ of line	Wild type	Black	Cinna-mon	Choco-late	Albino
1	87	0	32	0	39
2	62	0	0	0	18
3	96	30	0	0	41
4	287	86	92	29	164

(Problem 26 adapted from A. M. Srb, R. D. Owen, and R. S. Edgar, *General Genetics*, 2d ed. W. H. Freeman and Company, 1965.)

27. An allele *A* that is not lethal when homozygous causes rats to have yellow coats. The allele *R* of a separate gene that assorts independently produces a black coat. Together, *A* and *R* produce a grayish coat, whereas *a* and *r* produce a white coat. A gray male is crossed with a yellow female, and the F$_1$ is $\frac{3}{8}$ yellow, $\frac{3}{8}$ gray, $\frac{1}{8}$ black, and $\frac{1}{8}$ white. Determine the genotypes of the parents.

28. The genotype *r/r* ; *p/p* gives fowl a single comb, *R/−* ; *P/−* gives a walnut comb, *r/r* ; *P/−* gives a pea comb, and *R/−* ; *p/p* gives a rose comb (see the illustrations).

a. What comb types will appear in the F$_1$ and in the F$_2$ in what proportions if single-combed birds are crossed with birds of a true-breeding walnut strain?

b. What are the genotypes of the parents in a walnut × rose mating from which the progeny are $\frac{3}{8}$ rose, $\frac{3}{8}$ walnut, $\frac{1}{8}$ pea, and $\frac{1}{8}$ single?

c. What are the genotypes of the parents in a walnut × rose mating from which all the progeny are walnut?

d. How many genotypes produce a walnut phenotype? Write them out.

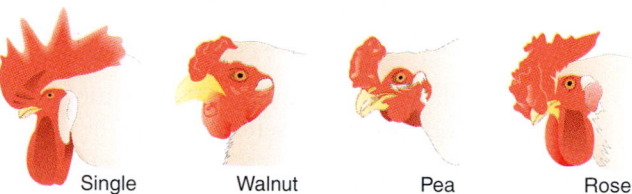

Single   Walnut   Pea   Rose

29. The production of eye-color pigment in *Drosophila* requires the dominant allele *A*. The dominant allele *P* of a second independent gene turns the pigment to purple, but its recessive allele leaves it red. A fly producing no pigment has white eyes. Two pure lines were crossed with the following results:

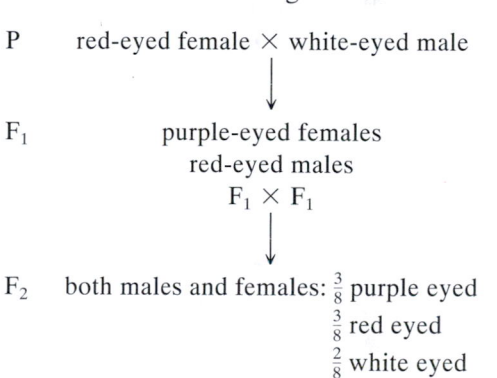

P   red-eyed female × white-eyed male

F$_1$ purple-eyed females
red-eyed males
F$_1$ × F$_1$

F$_2$ both males and females: $\frac{3}{8}$ purple eyed
$\frac{3}{8}$ red eyed
$\frac{2}{8}$ white eyed

Explain this mode of inheritance and show the genotypes of the parents, the F$_1$, and the F$_2$.

30. When true-breeding brown dogs are mated with certain true-breeding white dogs, all the F$_1$ pups are white. The F$_2$ progeny from some F$_1$ × F$_1$ crosses were 118 white, 32 black, and 10 brown pups. What is the genetic basis for these results?

31. In corn, three dominant alleles, called *A*, *C*, and *R*, must be present to produce colored seeds. Genotype *A/−* ; *C/−* ; *R/−* is colored; all others are colorless. A colored plant is crossed with three tester plants of known genotype. With tester *a/a* ; *c/c* ; *R/R*, the colored plant produces 50 percent colored seeds; with *a/a* ; *C/C* ; *r/r*, it produces 25 percent colored; and with *A/A* ; *c/c* ; *r/r*, it produces 50 percent colored. What is the genotype of the colored plant?

32. The production of pigment in the outer layer of seeds of corn requires each of the three independently assorting genes *A*, *C*, and *R* to be represented by at least one dominant allele, as specified in problem 31. The dominant allele *Pr* of a fourth independently assorting gene is required to convert the biochemical precursor into a purple pigment, and its recessive allele *pr* makes the pigment red. Plants that do not produce pigment have yellow seeds. Consider a cross of a strain of genotype A/A ; C/C ; R/R ; pr/pr with a strain of genotype a/a ; c/c ; r/r ; Pr/Pr.

a. What are the phenotypes of the parents?

b. What will be the phenotype of the F_1?

c. What phenotypes, and in what proportions, will appear in the progeny of a selfed F_1?

d. What progeny proportions do you predict from the testcross of an F_1?

33. Wild-type strains of the haploid fungus *Neurospora* can make their own tryptophan. An abnormal allele *td* renders the fungus incapable of making its own tryptophan. An individual of genotype *td* grows only when its medium supplies tryptophan. The allele *su* assorts independently of *td*; its only known effect is to suppress the *td* phenotype. Therefore, strains carrying both *td* and *su* do not require tryptophan for growth.

a. If a *td* ; *su* strain is crossed with a genotypically wild-type strain, what genotypes are expected in the progeny, and in what proportions?

b. What will be the ratio of tryptophan-dependent to tryptophan-independent progeny in the cross of part a?

34. The allele *B* gives mice a black coat, and *b* gives a brown one. The genotype *e/e* of another, independently assorting gene prevents expression of *B* and *b*, making the coat color beige, whereas $E/-$ permits expression of *B* and *b*. Both genes are autosomal. In the following pedigree, black symbols indicate a black coat, pink symbols indicate brown, and white symbols indicate beige.

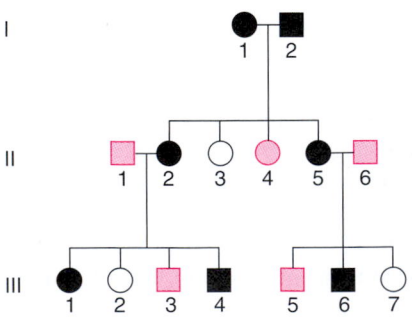

a. What is the name given to the type of gene interaction in this example?

b. What are the genotypes of the individuals in the pedigree? (If there are alternative possibilities, state them.)

35. Mice of the genotypes A/A ; B/B ; C/C ; D/D ; S/S and a/a ; b/b ; c/c ; d/d ; s/s are crossed. (These gene symbols are explained in the text of this chapter.) The progeny are intercrossed. What phenotypes will be produced in the F_2, and in what proportions?

36. Consider the genotypes of two lines of chickens: the pure-line mottled Honduran is i/i ; D/D ; M/M ; W/W, and the pure-line leghorn is I/I ; d/d ; m/m ; w/w, where:

> I = white feathers, i = colored feathers
> D = duplex comb, d = simplex comb
> M = bearded, m = beardless
> W = white skin, w = yellow skin

These four genes assort independently. Starting with these two pure lines, what is the fastest and most convenient way of generating a pure line of birds that has colored feathers, has a simplex comb, has yellow skin, and is beardless? Make sure that you show

a. The breeding pedigree.

b. The genotype of each animal represented.

c. How many eggs to hatch in each cross, and why this number.

d. Why your scheme is the fastest and most convenient.

37. The following pedigree is for a dominant phenotype governed by an autosomal gene. What does this pedigree suggest about the phenotype, and what can you deduce about the genotype of individual A?

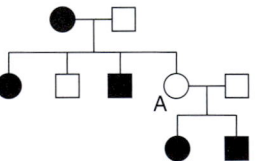

38. The genetic determination of petal coloration in foxgloves is given in the legend of Figure 4-17. Consider the following two crosses:

Cross	Parents	Progeny
1	dark purple × white with yellowish spots	$\frac{1}{2}$ dark purple: $\frac{1}{2}$ light purple
2	white with yellowish spots × light purple	$\frac{1}{2}$ white with purple spots: $\frac{1}{4}$ dark purple: $\frac{1}{4}$ light purple

In each case, give the genotypes of parents and progeny with respect to the three genes.

39. A researcher crosses two white-flowered lines of *Antirrhinum* plants as follows and obtains the following results:

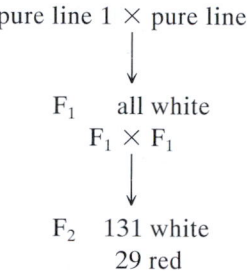

pure line 1 × pure line 2

F_1 all white

$F_1 \times F_1$

F_2 131 white

29 red

a. Deduce the inheritance of these phenotypes, using clearly defined gene symbols. Give the genotypes of the parents, F_1, and F_2.

b. Predict the outcome of crosses of the F_1 to each parental line.

40. Assume that two pigments, red and blue, mix to give the normal purple color of petunia petals. Separate biochemical pathways synthesize the two pigments, as shown in the top two rows of the following diagram. "White" refers to compounds that are not pigments. (Total lack of pigment results in a white petal.) Red pigment forms from a yellow intermediate that normally is at a concentration too low to color petals.

A third pathway whose compounds do not contribute pigment to petals normally does not affect the blue and red pathways; but, if one of its intermediates (white$_3$) should build up in concentration, it can be converted into the yellow intermediate of the red pathway.

In the diagram, A to E represent enzymes; their corresponding genes, all of which are unlinked, may be symbolized by the same letters.

pathway I $\cdots \longrightarrow$ white$_1$ $\xrightarrow{\text{E}}$ blue

pathway II $\cdots \longrightarrow$ white$_2$ $\xrightarrow{\text{A}}$ yellow $\xrightarrow{\text{B}}$ red

$\uparrow$ C

pathway III $\cdots \longrightarrow$ white$_3$ $\xrightarrow{\text{D}}$ white$_4$

Assume that wild-type alleles are dominant and code for enzyme function and that recessive alleles result in lack of enzyme function. Deduce which combinations of true-breeding parental genotypes could be crossed to produce F_2 progenies in the following ratios:

a. 9 purple : 3 green : 4 blue

b. 9 purple : 3 red : 3 blue : 1 white

c. 13 purple : 3 blue

d. 9 purple : 3 red : 3 green : 1 yellow

(**Note:** Blue mixed with yellow makes green; assume that no mutations are lethal.)

41. The flowers of nasturtiums *(Tropaeolum majus)* may be single (S), double (D), or superdouble (Sd). Superdoubles are female sterile; they originated from a double-flowered variety. Crosses between varieties gave the progenies as listed in the following table, where *pure* means "pure breeding."

Cross	Parents	Progeny
1	pure S × pure D	All S
2	cross 1 F_1 × cross 1 F_1	78 S : 27 D
3	pure D × Sd	112 Sd : 108 D
4	pure S × Sd	8 Sd : 7 S
5	pure D × cross 4 Sd progeny	18 Sd : 19 S
6	pure D 3 cross 4 S progeny	14 D : 16 S

Using your own genetic symbols, propose an explanation for these results, showing

a. all the genotypes in each of the six rows.

b. the proposed origin of the superdouble.

*42. In a certain species of fly, the normal eye color is red (R). Four abnormal phenotypes for eye color were found: two were yellow (Y1 and Y2), one was brown (B), and one was orange (O). A pure line was established for each phenotype, and all possible combinations of the pure lines were crossed. Flies of each F_1 were intercrossed to produce an F_2. The F_1's and F_2's are shown within the following square; the pure lines are given in the margins.

		Y1	Y2	B	O
Y1	F_1	all y	all r	all r	all r
	F_2	all y	9 r	9 r	9 r
			7 y	4 y	4 o
				3 b	3 y
Y2	F_1		all y	all r	all r
	F_2		all y	9 r	9 r
				4 y	4 y
				3 b	3 o
B	F_1			all b	all r
	F_2			all b	9 r
					4 o
					3 b
O	F_1				all o
	F_2				all o

a. Define your own symbols and show genotypes of all four pure lines.

b. Show how the F_1 phenotypes and the F_2 ratios are produced.

c. Show a biochemical pathway that explains the genetic results, indicating which gene controls which enzyme.

43. In common wheat, *Triticum aestivum,* kernel color is determined by multiply duplicated genes, each with an *R* and an *r* allele. Any number of *R* alleles will give red, and the complete lack of *R* alleles will give the white phenotype. In one cross between a red pure line and a white pure line, the F_2 was $\frac{63}{64}$ red and $\frac{1}{64}$ white.

a. How many *R* genes are segregating in this system?

b. Show genotypes of the parents, the F_1, and the F_2.

c. Different F_2 plants are backcrossed to the white parent. Give examples of genotypes that would give the following progeny ratios in such backcrosses: (1) 1 red:1 white, (2) 3 red:1 white, (3) 7 red:1 white.

*d. What is the formula that generally relates the number of segregating genes to the proportion of red individuals in the F_2 in such systems?

44. The following pedigree shows the inheritance of deaf-mutism.

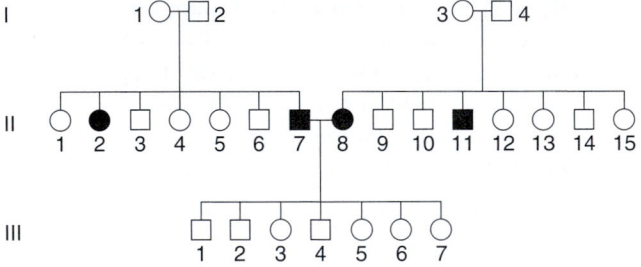

a. Provide an explanation for the inheritance of this rare condition in the two families in generations I and II, showing genotypes of as many individuals as possible by using symbols of your own choosing.

b. Provide an explanation for the production of only normal individuals in generation III, making sure that your explanation is compatible with the answer to part a.

45. The following pedigree is for blue sclera (bluish thin outer wall to the eye) and brittle bones:

a. Are these two abnormalities caused by the same gene or separate genes? State your reasons clearly.

b. Is the gene (or genes) autosomal or sex linked?

c. Does the pedigree show any evidence of incomplete penetrance or expressivity? If so, make the best calculations that you can of these measures.

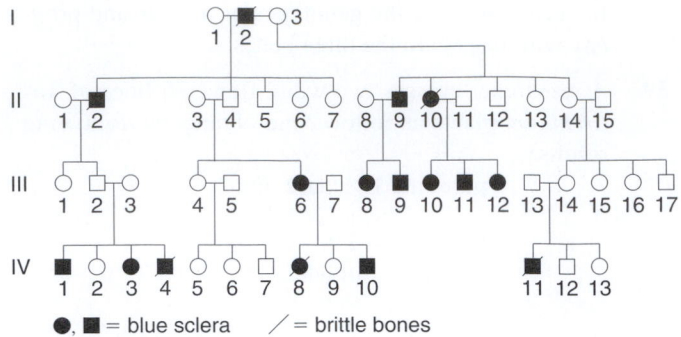

● , ■ = blue sclera   / = brittle bones

46. Workers of the honeybee line known as *Brown* (nothing to do with color) show what is called "hygienic behavior"; that is, they uncap hive compartments containing dead pupae and then remove the dead pupae. This behavior prevents the spread of infectious bacteria through the colony. Workers of the *Van Scoy* line, however, do not perform these actions, and therefore this line is said to be "nonhygienic." When a queen from the *Brown* line was mated with *Van Scoy* drones, the F_1 were all nonhygienic. When drones from this F_1 inseminated a queen from the *Brown* line, the progeny behaviors were as follows:

$\frac{1}{4}$ hygienic

$\frac{1}{4}$ uncapping but no removing of pupae

$\frac{1}{2}$ nonhygienic

However, when the nonhygienic individuals were examined further, it was found that, if the compartment of dead pupae was uncapped by the beekeeper, then about half the individuals removed the dead pupae, but the other half did not.

a. Propose a genetic hypothesis to explain these behavioral patterns.

b. Discuss the data in relation to epistasis, dominance, and environmental interaction.

(**Note:** Workers are sterile, and all bees from one line carry the same alleles.)

47. In one species of *Drosophila,* the wings are normally round in shape, but you have obtained two pure lines, one of which has oval wings and the other sickle-shaped wings. Crosses between pure lines reveal the following results:

PARENTS		F_1	
Female	Male	Female	Male
sickle	round	sickle	sickle
round	sickle	sickle	round
sickle	oval	oval	sickle

a. Provide a genetic explanation of these results, defining all allele symbols.

b. If the F_1 oval females from cross 3 are crossed to the F_1 round males from cross 2, what phenotypic proportions are expected in each sex of progeny?

48. Mice normally have one yellow band on their hairs, but variants with two or three bands are known. A female mouse with one band was crossed to a male who had three bands. (Neither animal was from a pure line.) The progeny were

Females	$\frac{1}{2}$	one band
	$\frac{1}{2}$	three bands
Males	$\frac{1}{2}$	one band
	$\frac{1}{2}$	two bands

a. Provide a clear explanation of the inheritance of these phenotypes.

b. Under your model, what would be the outcome of a cross between a three-banded daughter and a one-banded son?

49. In minks, wild types have an almost black coat. Breeders have developed many pure lines of color variants for the mink coat industry. Two such pure lines are platinum (blue gray) and aleutian (steel gray). These lines were used in crosses, with the following results:

Cross	Parents	F_1	F_2
1	wild × platinum	wild	18 wild, 5 platinum
2	wild × aleutian	wild	27 wild, 10 aleutian
3	platinum × aleutian	wild	133 wild
			41 platinum
			46 aleutian
			17 sapphire (new)

a. Devise a genetic explanation of these three crosses. Show genotypes for parents, F_1, and F_2 in the three crosses, and make sure that you show the alleles of each gene that you hypothesize in every individual.

b. Predict the F_1 and F_2 phenotypic ratios from crossing sapphire with platinum and aleutian pure lines.

50. In *Drosophila,* an autosomal gene determines the shape of the hair, with B giving straight and b bent hairs. On another autosome, there is a gene of which a dominant allele I inhibits hair formation so that the fly is hairless (i has no known phenotypic effect).

a. If a straight-haired fly from a pure line is crossed with a fly from a pure-breeding hairless line known to be an inhibited bent genotype, what will the genotypes and phenotypes of the F_1 and the F_2 be?

b. What cross would give the ratio 4 hairless:3 straight:1 bent?

51. The following pedigree concerns eye phenotypes in *Tribolium* beetles. The solid symbols represent black eyes, the open symbols represent brown eyes, and the cross symbols (X) represent the "eyeless" phenotype, in which eyes are totally absent.

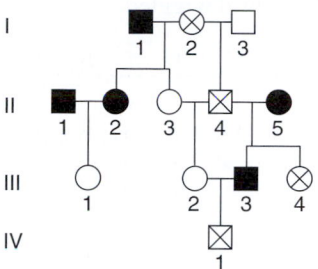

a. From these data, deduce the mode of inheritance of these three phenotypes.

b. Using defined gene symbols, show the genotype of individual II-3.

52. The normal color of snapdragons is red. Some pure lines showing variations of flower color have been found. When these pure lines were crossed, they gave the following results:

Parents	F_1	F_2
1. orange × yellow	orange	3 orange:1 yellow
2. red × orange	red	3 red:1 orange
3. red × yellow	red	3 red:1 yellow
4. red × white	red	3 red:1 white
5. yellow × white	red	9 red:3 yellow:4 white
6. orange × white	red	9 red:3 orange:4 white
7. red × white	red	9 red:3 yellow:4 white

a. Explain the inheritance of these colors.

b. Write the genotypes of the parents, the F_1, and the F_2.

53. Consider the following F_1 individuals in different species and the F_2 ratios produced by selfing:

F_1	Phenotypic ratio in the F_2		
1. cream	$\frac{12}{16}$ cream	$\frac{3}{16}$ black	$\frac{1}{16}$ gray
2. orange	$\frac{9}{16}$ orange	$\frac{7}{16}$ yellow	
3. black	$\frac{13}{16}$ black	$\frac{3}{16}$ white	
4. solid red	$\frac{9}{16}$ solid red	$\frac{3}{16}$ mottled red	$\frac{4}{16}$ small red dots

If each F_1 were testcrossed, what phenotypic ratios would result in the progeny of the testcross?

54. To understand the genetic basis of locomotion in the diploid nematode *Caenorhabditis elegans,* recessive mutations were obtained, all making the worm "wiggle" ineffectually instead of moving with its usual smooth gliding motion. These mutations presumably affect the nervous or muscle systems. Twelve homozygous mutants were intercrossed, and the F_1 hybrids were examined to see if they wiggled. The results were as follows, where a plus sign means that the F_1 hybrid was wild type (gliding) and "w" means that the hybrid wiggled.

	1	2	3	4	5	6	7	8	9	10	11	12
1	w	+	+	+	w	+	+	+	+	+	+	+
2		w	+	+	+	w	+	w	+	w	+	+
3			w	w	+	+	+	+	+	+	+	+
4				w	+	+	+	+	+	+	+	+
5					w	+	+	+	+	+	+	+
6						w	+	w	+	w	+	+
7							w	+	+	+	w	w
8								w	+	w	+	+
9									w	+	+	+
10										w	+	+
11											w	w
12												w

a. Explain what this experiment was designed to test.

b. Use this reasoning to assign genotypes to all 12 mutants.

c. Explain why the F_1 hybrids between mutants 1 and 2 had a different phenotype from that of the hybrids between mutants 1 and 5.

55. In corn, synthesis of purple pigment is controlled by two genes acting sequentially through colorless (white) intermediates:

$$\text{white 1} \xrightarrow{\text{gene } A} \text{white 2} \xrightarrow{\text{gene } B} \text{purple}$$

Recessive nonsense mutations (a^n and b^n) were obtained in genes A and B. Each of these mutations gave a white phenotype, and each could be suppressed by the nonsense-suppressor mutation T^S (wild-type allele T^+)

a. Would you expect T^S to be dominant to T^+? Explain.

b. A trihybrid A/a^n ; B/b^n ; T^S/T^+ is selfed. If all the genes are unlinked, what phenotypic ratio do you expect in the progeny? Explain, preferably with a diagram.

56. A plant believed to be heterozygous for a pair of alleles B/b (where B encodes yellow and b encodes bronze) was selfed and in the progeny there were 280 yellow and 120 bronze individuals. Do these results support the hypothesis that the plant is B/b?

57. A plant thought to be heterozygous for two independently assorting genes (P/p ; Q/q) was selfed and the progeny were:

88	$P-$; $Q-$
32	$P-$; q/q
25	p/p ; $Q-$
14	p/p ; q/q

Do these results support the hypothesis that the original plant was P/p ; Q/q?

58. A plant of phenotype 1 was selfed and in the progeny there were 100 individuals of phenotype 1 and 60 of an alternative phenotype 2. Are these numbers compatible with expected ratios of $9:7$, $13:3$, and $3:1$? Formulate a genetic hypothesis based on your calculations.

5

BASIC
EUKARYOTIC
CHROMOSOME
MAPPING

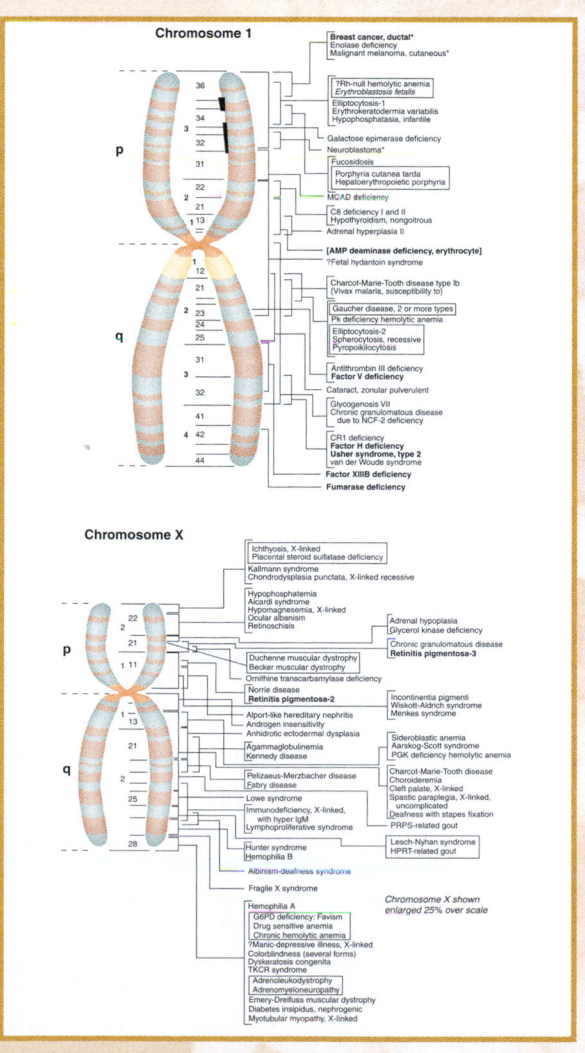

Key Concepts

Two genes close together on the same chromosome pair do not assort independently at meiosis.

Recombination produces genotypes with new combinations of parental alleles.

A pair of homologous chromosomes can exchange segments by crossing-over.

Recombination results from either independent assortment or crossing-over.

Gene loci on a chromosome can be mapped by measuring the frequencies of recombinants produced by crossing-over.

Interlocus map distances based on recombination measurements are roughly additive.

The occurrence of a crossover can influence the occurrence of a second crossover in an adjacent region.

Map of the human chromosome 1 and the X chromosome.
The genes have been positioned by using several techniques, including those covered in Chapters 5 and 6. (*From* Journal of NIH Research, 1992.)

We have already established the basic principles of segregation and assortment, and we have correlated them with chromosome behavior during meiosis. Thus, from the cross $A/a \; ; B/b \times A/a \; ; B/b$, we expect a $9:3:3:1$ ratio of phenotypes. As we learned from Bridges's study of nondisjunction (page 76), exceptions to simple Mendelian expectations can direct the experimenter's attention to new discoveries. Just such an exception observed in the progeny of a dihybrid cross provided the clue to the important concepts considered in this chapter.

MESSAGE

In genetic analysis, exceptions to predicted behavior are often sources of important new insights.

The discovery of linkage

In the early 1900s, William Bateson and R. C. Punnett were studying inheritance in the sweet pea. They studied two genes: one affecting flower color (P, purple, and p, red) and the other affecting the shape of pollen grains (L, long, and l, round). They crossed pure lines $P/P \cdot L/L$ (purple, long) $\times$ $p/p \cdot l/l$ (red, round), and selfed the F_1 $P/p \cdot L/l$ heterozygotes to obtain an F_2. Table 5-1 shows the proportions of each phenotype in the F_2 plants.

The F_2 phenotypes deviated strikingly from the expected $9:3:3:1$ ratio. What is going on? This does not appear to be explainable as a modified Mendelian ratio. Note that two phenotypic classes are larger than expected: the purple, long phenotype and the red, round phenotype. As a possible explanation for this, Bateson and Punnett proposed that the F_1 had actually produced more $P \cdot L$ and $p \cdot l$ gametes than would be produced by Mendelian independent assortment. Because these genotypes were the gametic types in the original pure lines, the researchers thought that physical **coupling** between the dominant alleles P and L and between the recessive alleles p and l might have prevented their independent assortment in the F_1. However, they did not know what the nature of this coupling could be.

The confirmation of Bateson and Punnett's hypothesis had to await the development of *Drosophila* as a genetic

tool. After the idea of coupling was first proposed, Thomas Hunt Morgan found a similar deviation from Mendel's second law while studying two autosomal genes in *Drosophila*. One of these genes affects eye color (pr, purple, and pr^+, red), and the other affects wing length (vg, vestigial, and vg^+, normal). The wild-type alleles of both genes are dominant. Morgan crossed $pr/pr \cdot vg/vg$ flies with $pr^+/pr^+ \cdot vg^+/vg^+$ and then testcrossed the doubly heterozygous F_1 females: $pr^+/pr \cdot vg^+/vg \; \female \times pr/pr \cdot vg/vg \; \male$.

The use of the testcross is extremely important. Because one parent (the tester) contributes gametes carrying only recessive alleles, the phenotypes of the offspring reveal the gametic contribution of the other, doubly heterozygous parent. Hence, the analyst can concentrate on meiosis in one parent and forget about the other. This contrasts with the analysis of progeny from an F_1 self, where there are two sets of meioses to consider: one in the male parent and one in the female. Morgan's results follow; the alleles contributed by the F_1 female specify the F_2 classes:

$pr^+ \cdot vg^+$	1339
$pr \cdot vg$	1195
$pr^+ \cdot vg$	151
$pr \cdot vg^+$	154
	2839

Obviously, these numbers deviate drastically from the Mendelian prediction of a $1:1:1:1$ ratio, and they indicate a coupling of genes. The two largest classes are the combinations $pr^+ \cdot vg^+$ and $pr \cdot vg$, originally introduced by the homozygous parental flies. You can see that the testcross clarifies the situation. It directly reveals the allelic combinations in the gametes from one sex in the F_1, thus clearly showing the coupling that could only be inferred from Bateson and Punnett's F_1 self. The testcross also reveals something new: there is approximately a $1:1$ ratio not only between the two parental types, but also between the two nonparental types.

Now let us consider what may be learned by repeating the crossing experiments but changing the combinations of alleles contributed as gametes by the homozygous parents in the first cross. In this cross, each parent was homozygous for one dominant allele and for one recessive allele. Again F_1 females were testcrossed:

P $pr^+/pr^+ \cdot vg/vg \times pr/pr \cdot vg^+/vg^+$

$\downarrow$

F_1 $pr^+/pr \cdot vg^+/vg$

$pr^+/pr \cdot vg^+/vg \; \female \times pr/pr \cdot vg/vg \; \male$

The following progeny were obtained from the testcross:

$pr^+ \cdot vg^+$	157
$pr \cdot vg$	146
$pr^+ \cdot vg$	965
$pr \cdot vg^+$	1067
	2335

5-1 TABLE	Sweet Pea Phenotypes Observed in the F_2 by Bateson and Punnett	
	NUMBER OF PROGENY	
Phenotype (and genotype)	Observed	Expected from $9:3:3:1$ ratio
purple, long ($P/- \cdot L/-$)	4831	3911
purple, round ($P/- \cdot l/l$)	390	1303
red, long ($p/p \cdot L/-$)	393	1303
red, round ($p/p \cdot l/l$)	1338	435
	6952	6952

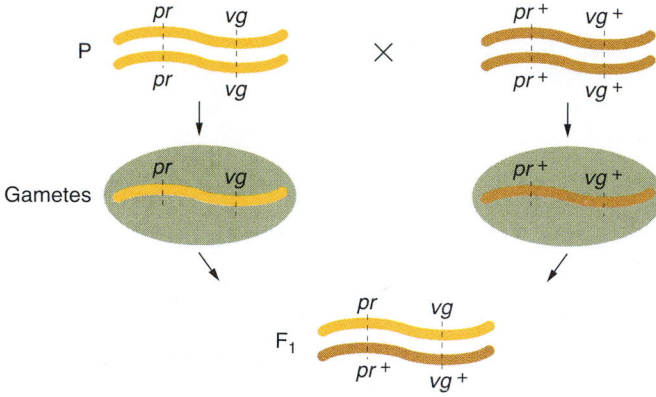

Figure 5-1　Simple inheritance of two pairs of alleles located on the same chromosome pair.

Again, these results are not even close to a 1:1:1:1 Mendelian ratio. Now, however, the largest classes are those that have one dominant allele or the other rather than, as before, two dominant alleles or two recessives. But notice that once again the allelic combinations that were originally contributed to the F₁ by the parental flies provide the most frequent classes in the testcross progeny. In the early work on coupling, Bateson and Punnett coined the term **repulsion** to describe this situation, because it seemed to them that, in this case, the nonallelic dominant alleles "repelled" each other — the opposite of the situation in coupling, where the dominant alleles seemed to "stick together." What is the explanation of these two phenomena: coupling and repulsion?

Morgan suggested that the genes governing both phenotypes are located *on the same pair of homologous chromosomes.* Thus, when *pr* and *vg* are introduced from one parent, they are physically located on the same chromosome, whereas *pr⁺* and *vg⁺* are on the homologous chromosome from the other parent (Figure 5-1). This hypothesis also explains repulsion. In that case, one parental chromosome carries *pr* and *vg⁺* and the other carries *pr⁺* and *vg*. Repulsion, then, is just another case of coupling: in this case, the domi-

nant allele of one gene is coupled with the recessive allele of the other gene. This hypothesis explains why allelic combinations from P remain together, but how do we explain the appearance of nonparental combinations?

Morgan suggested that, when homologous chromosomes pair in meiosis, the chromosomes occasionally exchange parts in a process called **crossing-over.** Figure 5-2 illustrates this physical exchange of chromosome segments. The two new combinations are called **crossover products.**

Morgan's hypothesis that homologs may exchange parts may seem a bit farfetched. Is there any cytologically observable process that could account for crossing-over? We saw in Chapter 3 that in meiosis, when duplicated homologous chromosomes pair with each other, two nonsister chromatids often appear to cross each other, as diagrammed in Figure 5-3. Recall that the resulting cross-shaped structure is called a *chiasma.* To Morgan, the appearance of the chiasmata visually corroborated the concepts of crossing-over. (Note that the chiasmata seem to indicate that it is chromatids, not unduplicated chromosomes, that cross over. We shall return to this point later.) Note that Morgan did not arrive at this interpretation out of nowhere; he was looking for a *physical* explanation for his *genetic* results. His achievement in correlating the results of breeding experiments with cytological phenomena thus emphasizes the importance of the chromosome theory as a powerful basis for research.

MESSAGE ·····
Chiasmata are the visible manifestations of crossovers.

Data like those just presented, showing coupling and repulsion in testcrosses and in F₁ selfs, are commonly encountered in genetics. Clearly, results of this kind are a departure from independent assortment. Such exceptions, in fact, constitute a major addition to Mendel's view of the genetic world.

MESSAGE ·····
When two genes are close together on the same chromosome pair, they do not assort independently.

The residing of genes on the same chromosome pair is termed **linkage.** Two genes on the same chromosome pair are said to be *linked.* It is also proper to refer to the linkage of specific alleles: for example, in one *A/a · B/b* individual, *A* might be linked to *b*; *a* would then of necessity be linked to *B*. These terms graphically allude to the existence of a

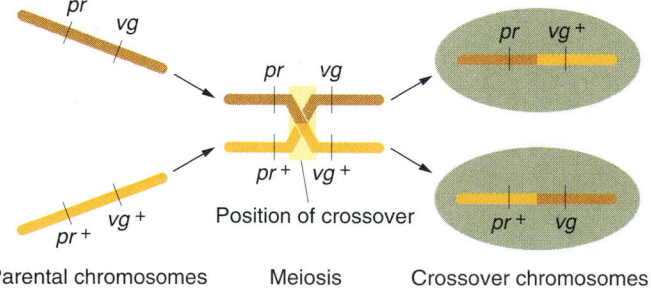

Figure 5-2　Crossing-over in meiosis. An individual offspring receives one homolog from each parent. The exchange of parts by crossing-over may produce gametic chromosomes whose allelic combinations differ from the parental combinations.

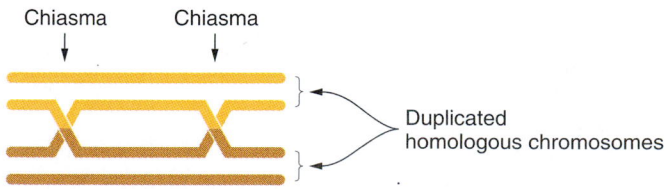

Figure 5-3　Diagrammatic representation of chiasmata at meiosis. Each line represents a chromatid of a pair of synapsed chromosomes.

physical entity linking the genes—that is, the chromosome itself. You may wonder why we refer to such genes as "linked" rather than "coupled"; the answer is that the words coupling and repulsion are now used to indicate two different types of linkage conformation in a double heterozygote, as follows:

Coupling conformation $\dfrac{pr \qquad vg}{pr^+ \qquad vg^+}$

Repulsion conformation $\dfrac{pr \qquad vg^+}{pr^+ \qquad vg}$

In other words, coupling refers to the linkage of two dominant or two recessive alleles, whereas repulsion indicates that dominant alleles are linked with recessive alleles. To ascertain whether a double heterozygote is in coupling or repulsion conformation, an investigator must testcross the double heterozygote or consider the genotypes of its parents.

Recombination

In modern genetic analysis, the main test for determining whether two genes are linked is based on the concept of recombination. Recombination is observed in a variety of situations but, for the present, let's define it in relation to meiosis. **Meiotic recombination** is any meiotic process that generates a haploid product with a genotype that differs from both haploid genotypes that constituted the meiotic diploid cell. The product of meiosis so generated is called a **recombinant.** This definition makes the important point that we detect recombination by comparing the *output* genotypes of meiosis and the parental *input* genotypes (Figure 5-4).

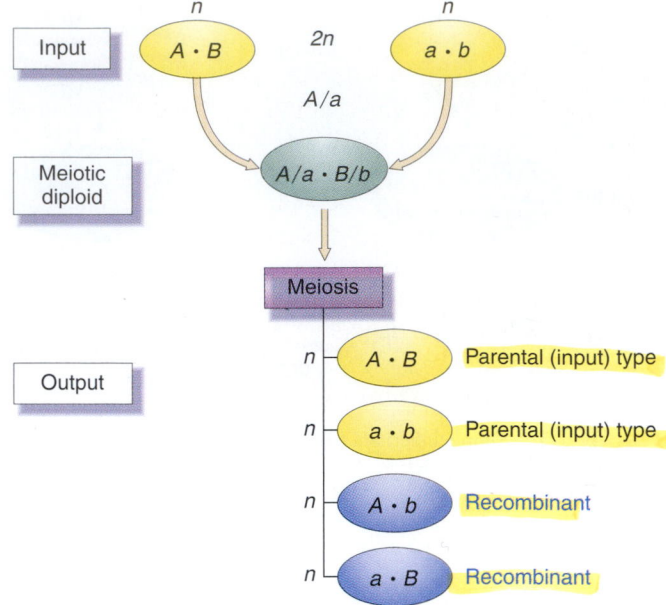

Figure 5-4 Recombinants are those products of meiosis with allelic combinations different from those of the haploid cells that formed the meiotic diploid.

The input genotypes are the two haploid genotypes that combined to make the genetic constitution of the meiocyte, the diploid cell that undergoes meiosis.

MESSAGE ·······································

In meiosis, recombination generates haploid genotypes differing from the haploid parental genotypes.

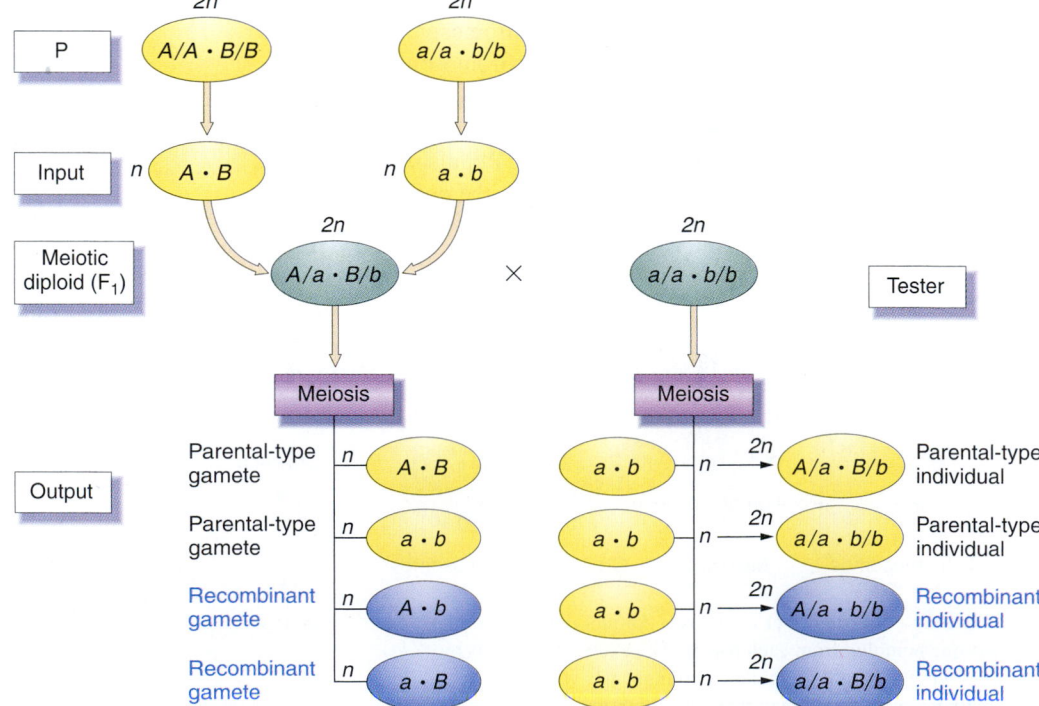

Figure 5-5 The detection of recombination in diploid organisms. Note that Figure 5-4 is a part of this diagram. Recombinant products of a diploid meiosis are most readily detected in a cross of a heterozygote and a recessive tester.

Meiotic recombination is a part of both haploid and diploid life cycles; however, detecting recombinants in haploid cycles is straightforward, whereas detecting them in diploid cycles is more complex. The input and output types in haploid cycles are the genotypes of individuals and may thus be inferred directly from phenotypes. Figure 5-4 can be viewed as summarizing the simple detection of recombinants in haploid life cycles. The input and output types in diploid life cycles are gametes. Because we must know the input gametes to detect recombinants in a diploid cycle, it is preferable to have pure-breeding parents. Furthermore, we cannot detect recombinant output gametes directly: we must testcross the diploid individual and observe its progeny (Figure 5-5). If a testcross offspring is shown to have been constituted from a recombinant product of meiosis, it too is called a *recombinant*. Notice again that the testcross allows us to concentrate on *one* meiosis and prevent ambiguity. From a self of the F_1 in Figure 5-5, for example, a recombinant $A/A \cdot B/b$ offspring cannot be distinguished from $A/A \cdot B/B$ without further crosses. Recombinants are produced by two different cellular processes: independent assortment and crossing-over.

Recombination by independent assortment

Mendelian independent assortment is viewed with regard to *recombination* in Figure 5-6. In a testcross, the two recombinant classes always make up 50 percent of the progeny; that is, there is 25 percent of each recombinant type among the progeny.

If we observe a recombinant frequency of 50 percent in a testcross, we can infer that the two genes under study assort independently. The simplest interpretation of such a result is that the two genes are on separate chromosome pairs. However, genes that are far apart on the *same* chromosome pair can act virtually independently and produce the same result.

Recombination by crossing-over

Crossing-over also can produce recombinants. Any two nonsister chromatids can cross over. (We shall show proof of this in Chapter 6.) There is not a crossover between two specific genes in all meioses, but, when there is, half the products of *that* meiosis are recombinant, as shown in Figure 5-7. Meiosis with no crossover between the genes under study produces only parental genotypes for these genes.

For genes close together on the same chromosome pair, the physical linkage of parental allele combinations makes independent assortment impossible and hence produces recombinant frequencies significantly lower than 50 percent (Figure 5-8). We saw an example of this situation in Morgan's data (page 142), where the recombinant frequency was $(151 + 154) \div 2839 = 10.7$ percent. This is obviously much less than the 50 percent that we would expect with independent assortment. The recombinant frequency arising from linked genes ranges from 0 to 50 percent, depending

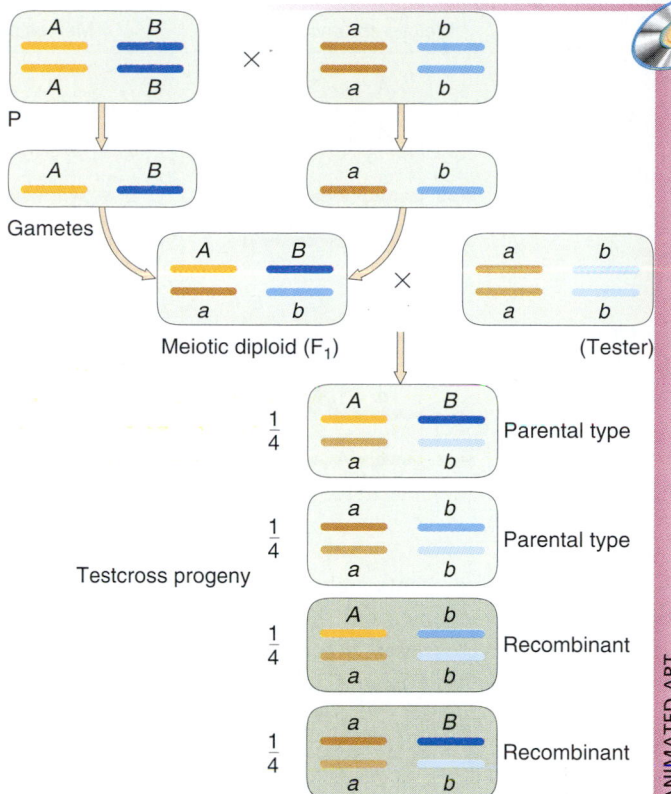

Figure 5-6   Independent assortment always produces a recombinant frequency of 50 percent. This diagram shows two chromosome pairs of a diploid organism with *A* and *a* on one pair and *B* and *b* on the other. Note that we could represent the haploid situation by removing the parental (P) cross and the testcross.

on their closeness. What about recombinant frequencies greater than 50 percent? The answer is that such frequencies are *never* observed, as we shall see in Chapter 6.

Note in Figure 5-7 that crossing-over generates two reciprocal products, which explains why the reciprocal recombinant classes are generally approximately equal in frequency.

MESSAGE ····························
A recombinant frequency significantly less than 50 percent shows that the genes are linked. A recombinant frequency of 50 percent generally means that the genes are unlinked on separate chromosomes.
····························

The remainder of this chapter focuses mainly on linked genes and recombinants arising from crossing-over.

Linkage symbolism

Our symbolism for describing crosses becomes cumbersome with the introduction of linkage. We can depict the genetic constitution of each chromosome in the *Drosophila* cross as in the following example:

$$\frac{pr \qquad vg}{pr^+ \qquad vg^+}$$

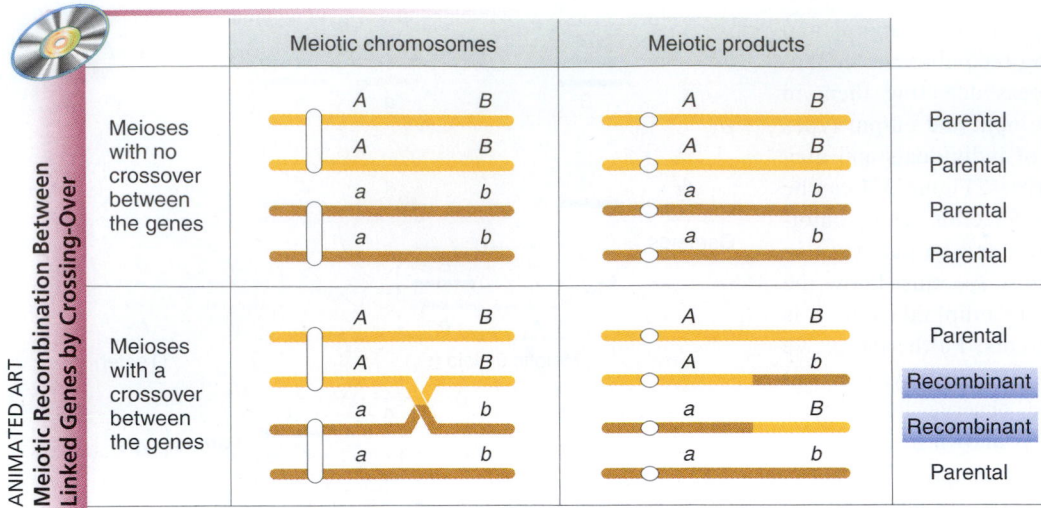

Figure 5-7 Recombinants arise from meioses in which nonsister chromatids cross over between the genes under study.

where each line represents a chromosome; the alleles above are on one chromosome, and those below are on the other chromosome. A crossover is represented by placing an **X** between the two chromosomes, so that

$$\frac{pr \qquad vg}{\quad\; \mathsf{X} \quad\;}$$
$$\overline{pr^+ \qquad vg^+}$$

is the same as

$$\frac{pr \qquad\qquad vg}{pr^+ \qquad\qquad vg^+}$$

We can simplify the genotypic designation of linked genes by drawing a single line, with the genes on each side being on the same chromosome; now our symbol is

$$\frac{pr \qquad vg}{pr^+ \qquad vg^+}$$

But this is still inconvenient for typing and writing, so let's tip the line to give us $pr\ vg/pr^+\ vg^+$, still keeping the genes of one chromosome on one side of the line and those of its homolog on the other. We always designate linked genes on each side in the same order; it is always $a\ b/a\ b$, never $a\ b/b\ a$. The rule that genes are always written in the same order permits geneticists to use a shorter notation in which the wild-type allele is written with a plus sign alone. In this notation the genotype $pr\ vg/pr^+\ vg^+$ becomes $pr\ vg/+\ +$. You may see this notation in other books or in research papers.

As we have seen in earlier chapters, genes known to be on different chromosome pairs are shown separated by a semicolon, for example, $A/a\ ;\ B/b$. In this book, genes of unknown linkage are shown separated by a dot, $A/a \cdot B/b$.

Now, if we reconsider the results obtained by Bateson and Punnett, we can easily explain the coupling phenomenon by using the concept of linkage. Their results are com-

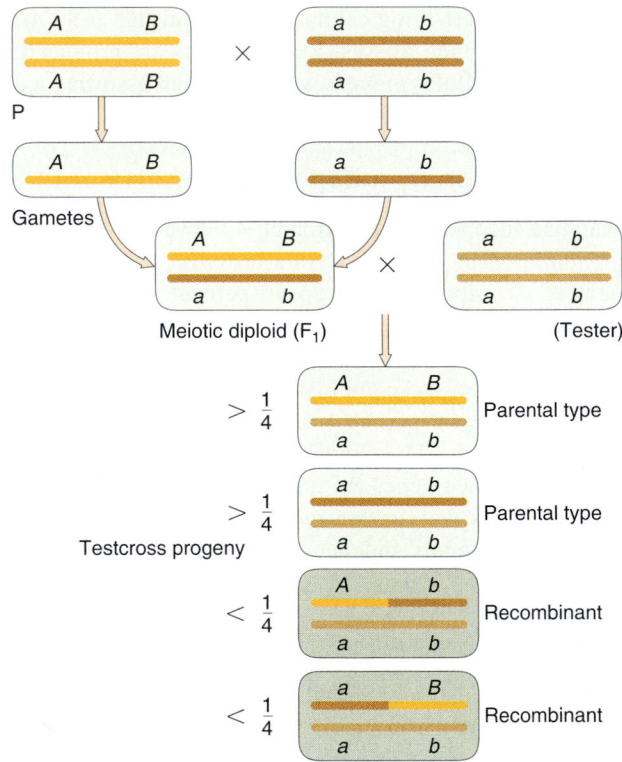

Figure 5-8 Recombination from crossing-over. Notice that the frequencies of the recombinants add up to less than 50 percent.

plex because they did not do a testcross. However, we will see later that in fact it is possible to derive estimated numbers for recombinant and parental types in a dihybrid cross.

Linkage of genes on the X chromosome

Until now, we have been considering recombination of autosomal genes. What are the consequences of nonsister chromatids of the X chromosome crossing over between two genes of interest? Recall that a human or *Drosophila* female produces male progeny hemizygous for the genes of the X chromosome, so the genotype of the gamete that a mother contributes to her son is the sole determinant of the son's phenotype. Let's consider an example in which we first observe the F_1 progeny from the mating of two *Drosophila* flies and then the F_2 progeny from intercrossing the F_1. We use here the following symbols: y and y^+ for the alleles governing yellow body and brown body, respectively; w and w^+ for alleles for white eye and red eye; and Y for the Y chromosome.

P $y\,w^+/y\,w^+$ ♀ × $y^+\,w/$Y ♂

F₁
          $y\,w^+/y^+\,w$ ♀ × $y\,w^+/$Y ♂

In the F_1 the numbers of males in the phenotypic classes are:

$y\,w$	43	recombinant
$y^+\,w$	2146	parental
$y\,w^+$	2302	parental
$y^+\,w^+$	22	recombinant
	4513	

Because the F_2 males obtain only a Y chromosome from the F_1 males, these classes represent perfectly the products of meiosis in the F_1 females. Notice that this fact eliminates the need for a testcross; we can follow meiosis in a single parent, just as we can in a testcross. The total frequency of the recombinants in this example is $(43 + 22) \div 4513 = 1.4$ percent.

Linkage maps

The frequency of recombinants for the *Drosophila* autosomal genes that we studied (*pr* and *vg*) was 10.7 percent of the progeny — a frequency much greater than that for the linked genes on the X chromosome just studied. Apparently, the amount of crossing-over between various linked genes differs. Indeed, there is no reason to expect that chromatids would cross over between different linked genes with the same frequency. As Morgan studied more linked genes, he saw that the proportion of recombinant progeny varied considerably, depending on which linked genes were being studied, and he thought that these variations in crossover frequency might somehow indicate the actual distances separating genes on the chromosomes. Morgan assigned the study of this problem to a student, Alfred Sturtevant, who (like Bridges) became a great geneticist. Morgan asked Sturtevant, still an undergraduate at the time, to make some sense of the data on crossing-over between different linked genes. In one night, Sturtevant developed a method for describing relations between genes that is still used today. In Sturtevant's own words, "In the latter part of 1911, in conversation with Morgan, I suddenly realized that the variations in strength of linkage, already attributed by Morgan to differences in the spatial separation of genes, offered the possibility of determining sequences in the linear dimension of a chromosome. I went home and spent most of the night (to the neglect of my undergraduate homework) in producing the first chromosome map."

As an example of Sturtevant's logic, consider a testcross from which we obtain the following results:

$pr\,vg/pr\,vg$	165	parental
$pr^+\,vg^+/pr\,vg$	191	
$pr\,vg^+/pr\,vg$	23	recombinant
$pr^+\,vg/pr\,vg$	21	
	400	

The progeny in this example represent 400 female gametes, of which 44 (11 percent) are recombinant. Sturtevant suggested that we can use the percentage of recombinants as a quantitative index of the linear distance between two genes on a genetic map, or **linkage map,** as it is sometimes called.

The basic idea here is quite simple. Imagine two specific genes positioned a certain fixed distance apart. Now imagine random crossing-over along the paired homologs. In some meiotic divisions, nonsister chromatids cross over by chance in the chromosomal region between these genes; from these meioses, recombinants are produced. In other meiotic divisions, there are no crossovers between these genes; no recombinants result from these meioses. Sturtevant postulated a rough proportionality: the greater the distance between the linked genes, the greater the chance that nonsister chromatids would cross over in the region between the genes and, hence, the greater the proportion of recombinants that would be produced. Thus, by determining the frequency of recombinants, we can obtain a measure of the map distance between the genes (Figure 5-9). In fact, we

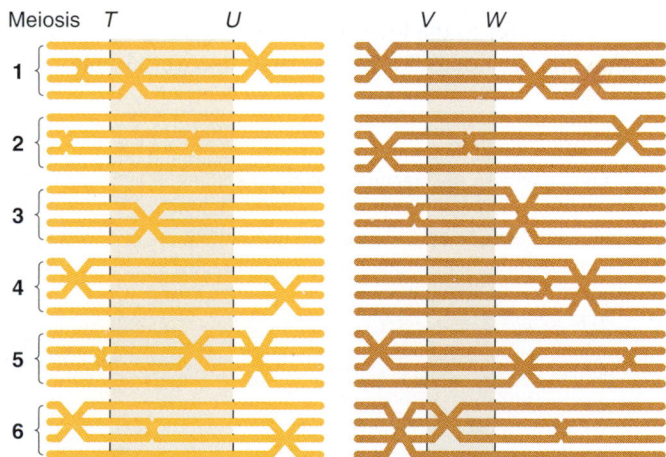

Figure 5-9 Proportionality between chromosome distance and recombinant frequency. In every meiosis, chromatids cross over at random along the chromosome. The two genes *T* and *U* are farther apart on a chromosome than *V* and *W*. Chromatids cross over between *T* and *U* in a larger proportion of meioses than between *V* and *W*, so the recombinant frequency for *T* and *U* is higher than that for *V* and *W*. As we will learn later in the chapter, a crossover can occur between any two nonsister chromatids within a homologous pairing.

can define one **genetic map unit (m.u.)** as that distance between genes for which one product of meiosis in 100 is recombinant. Put another way, a **recombinant frequency (RF)** of 0.01 (1 percent) is defined as 1 m.u. [A map unit is sometimes referred to as a centimorgan (cM) in honor of Thomas Hunt Morgan.]

A direct consequence of the way in which map distance is measured is that, if 5 map units (5 m.u.) separate genes A and B whereas 3 m.u. separate genes A and C, then B and C should be either 8 or 2 m.u. apart (Figure 5-10). Sturtevant found this to be the case. In other words, his analysis strongly suggested that genes are arranged in some linear order.

The place on the map — and on the chromosome — where a gene is located is called the **gene locus** (plural, **loci**). The locus of the eye-color gene and the locus of the wing-length gene, for example, are 11 m.u. apart. The relation is usually diagrammed this way:

$$\underset{pr}{\vdash}\hspace{3em}\overset{11.0}{}\hspace{3em}\underset{vg}{\dashv}$$

although it could be diagrammed equally well like this:

$$\underset{pr^+}{\vdash}\hspace{3em}\overset{11.0}{}\hspace{3em}\underset{vg^+}{\dashv}$$

or like this:

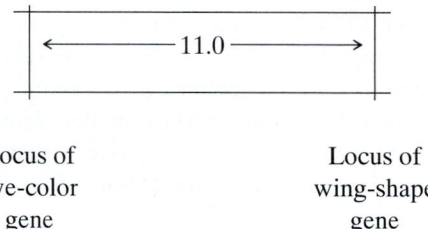

Locus of   Locus of
eye-color   wing-shape
gene   gene

Usually we refer to the locus of this eye-color gene in shorthand as the "*pr* locus," after the first discovered non-wild-

type allele, but we mean the place on the chromosome where any allele of this gene will be found.

Given a genetic distance in map units, we can predict frequencies of progeny in different classes. For example, in the progeny from a testcross of a female $pr\ vg/pr^+\ vg^+$ heterozygote, we know that there will be 11 percent recombinants, of which $5\frac{1}{2}$ percent will be $pr\ vg^+/pr\ vg$ and $5\frac{1}{2}$ percent will be $pr^+\ vg/pr\ vg$; of the progeny from a testcross of a female $pr\ vg^+/pr^+\ vg$ heterozygote, $5\frac{1}{2}$ percent will be $pr\ vg/pr\ vg$ and $5\frac{1}{2}$ percent will be $pr^+\ vg^+/pr\ vg$.

There is a strong implication that the "distance" on a linkage map is a physical distance along a chromosome, and Morgan and Sturtevant certainly intended to imply just that. But we should realize that the linkage map is another example of an entity constructed from a purely genetic analysis. The linkage map could have been derived without even knowing that chromosomes existed. Furthermore, at this point in our discussion, we cannot say whether the "genetic distances" calculated by means of recombinant frequencies in any way represent actual physical distances on chromosomes, although cytogenetic and molecular analysis has shown that genetic distances are, in fact, roughly proportional to chromosome distances. Nevertheless, it must be emphasized that the hypothetical structure (the linkage map) was developed with a very real structure (the chromosome) in mind. In other words, the chromosome theory provided the framework for the development of linkage mapping.

MESSAGE ·······························

Recombination between linked genes can be used to map their distance apart on the chromosome. The unit of mapping (1 m.u.) is defined as a recombinant frequency of 1 percent.

The stage of analysis that we have reached in our discussion is well illustrated by linkage maps of the screwworm (*Cochliomyia hominivorax*). The larval stage of this

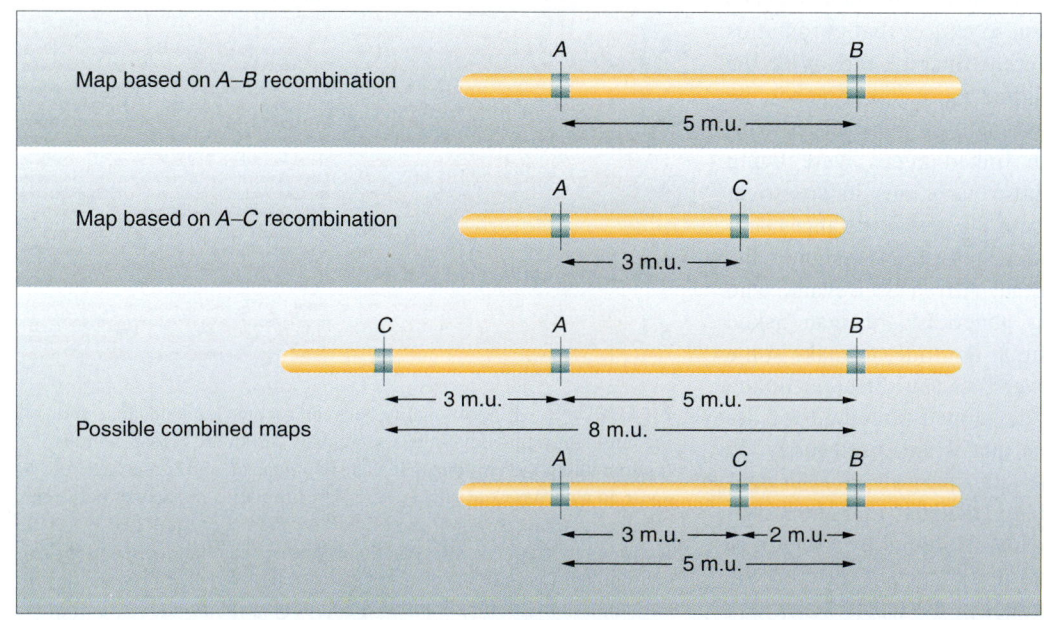

Figure 5-10 Because map distances are additive, calculation of the $A-B$ and $A-C$ distances leaves us with the two possibilities shown for the $B-C$ distance.

insect — the worm — is parasitic on mammalian wounds and is a costly pest of livestock in some parts of the world. A genetic system of population control has been proposed, of a type that has been successful in other insects. To accomplish this goal, an understanding of the basic genetics of the insect is needed, one important part of which is to prepare a map of the chromosomes. This animal has six chromosome pairs, and mapping has begun.

The job of general mapping starts by finding and analyzing as many variant phenotypes as possible. The adult stage of this insect is a fly, and geneticists have found phenotypic variants among screwworm flies. They found flies of six different eye colors, all different from the brown-eyed, wild-type flies, as Figure 5-11a shows. They also found five variant phenotypes for some other characters.

Eleven mutant alleles were shown to determine the 11 variant phenotypes, each at a different autosomal locus. Pure lines of each phenotype were intercrossed to generate dihybrid F$_1$s, and then these were testcrossed. The testcross revealed the set of four linkage groups shown in Figure 5-11b. Notice that the *ye* and *cw* loci are shown tentatively linked, although the recombinant frequency is not significantly different from 50 percent.

A linkage analysis such as the preceding one cannot assign linkage groups to specific chromosomes; this must be done by using the cytogenetic techniques to be considered in Chapter 17. In the present example, such cytogenetic techniques have allowed the linkage groups to be correlated with the chromosomes previously numbered as shown in Figure 5-11b.

Figure 5-11 (a) Wild-type adult of screwworm and six flies whose eye colors are determined by alleles at six different autosomal loci. (b) Linkage maps of the six eye-color loci (highlighted) and five other loci of screwworms. The numbers between the loci give the recombinant frequencies. (D. B. Taylor, USDA.)

Three-point testcross

So far, we have looked at linkage in crosses of double heterozygotes to doubly recessive testers. The next level of complexity is a cross of a triple heterozygote to a triply recessive tester. This kind of cross, called a **three-point testcross,** illustrates the standard approach used in linkage analysis. We shall consider two examples of such crosses here.

First, we focus on three *Drosophila* genes that have the non-wild-type alleles *sc* (short for *scute,* or loss of certain thoracic bristles), *ec* (short for *echinus,* or roughened eye surface), and *vg* (short for *vestigial wing*). We can cross $sc/sc \cdot ec/ec \cdot vg/vg$ triply recessive flies with wild-type flies to generate triple heterozygotes, $sc/sc^+ \cdot ec/ec^+ \cdot vg/vg^+$. We analyze recombination in these heterozygotes by testcrossing heterozygous females with triply recessive tester males. The results of such a testcross follow. The progeny are listed as gametic genotypes derived from the heterozygous females. Eight gametic types are possible, and they were counted in the following numbers in a sample of 1008 progeny flies:

$sc \cdot ec \cdot vg$	235
$sc^+ \cdot ec^+ \cdot vg^+$	241
$sc \cdot ec \cdot vg^+$	243
$sc^+ \cdot ec^+ \cdot vg$	233
$sc \cdot ec^+ \cdot vg$	12
$sc^+ \cdot ec \cdot vg^+$	14
$sc \cdot ec^+ \cdot vg^+$	14
$sc^+ \cdot ec \cdot vg$	16
	1008

The systematic way to analyze such crosses is to calculate all possible recombinant frequencies, but it is always worthwhile to inspect the data for obvious patterns before doing so. At first glance, we can note in the preceding data that there is a considerable deviation from the 1:1:1:1:1:1:1:1 ratio that is expected if the genes are all unlinked. So let's begin to calculate recombinant frequency values, taking the loci a pair at a time. Starting with the *sc* and *ec* loci (ignoring the *vg* locus for the time being), we determine which of the gametic genotypes are recombinant for *sc* and *ec*. Because we know that the heterozygotes were established from $sc \cdot ec$ and $sc^+ \cdot ec^+$ gametes, we know that the recombinant products of meiosis must be $sc \cdot ec^+$ and $sc^+ \cdot ec$. We note from the list that there are $12 + 14 + 14 + 16 = 56$ of these types; therefore, RF = (56/1008) × 100 = 5.5 m.u. This frequency tells us that these loci must be linked on the same chromosome, as follows:

Now let's look at recombination between the *sc* and the *vg* loci. The "input" parental genotypes were *sc vg* and $sc^+ vg^+$, so we must calculate the frequency of $sc\ vg^+$ and $sc^+\ vg$ progeny types (this time, we ignore *ec*). We see that

there are $243 + 233 + 14 + 16 = 506$ recombinants; because 506/1008 is very close to an RF of 50 percent, we conclude that the *sc* and *vg* loci are not linked and are probably not on the same chromosome. We can summarize the linkage relationship as follows:

Now you should see that the *ec* and *vg* loci must also be unlinked. Adding up the recombinants in the list and calculating the RF will confirm this. (Try it.) Having made those deductions about linkage, we can rewrite the parents of the testcross as $sc^+ ec^+/sc\ ec\ ;\ vg^+/vg \times sc\ ec/sc\ ec\ ;\ vg/vg$.

A second example, in which some other loci of *Drosophila* are used, will introduce some more important genetic concepts. Here the non-wild-type alleles are *v* (vermilion eyes), *cv* (crossveinless, or absence of a crossvein on the wing), and *ct* (cut, or snipped, wing edges). This time the parental stocks are homozygous doubly recessive flies of genotype $v^+/v^+ \cdot cv/cv \cdot ct/ct$ and homozygous singly recessive flies of genotype $v/v \cdot cv^+/cv^+ \cdot ct^+/ct^+$. From this cross, triply heterozygous progeny of genotype $v/v^+ \cdot cv/cv^+ \cdot ct/ct^+$ are obtained, and females of this genotype are testcrossed to triple recessives of genotype $v/v \cdot cv/cv \cdot ct/ct$. The female gametic genotypes determining the eight progeny types from this testcross are shown here with their numbers, out of a total sample of 1448 flies:

$v \cdot cv^+ \cdot ct^+$	580
$v^+ \cdot cv \cdot ct$	592
$v \cdot cv \cdot ct^+$	45
$v^+ \cdot cv^+ \cdot ct$	40
$v \cdot cv \cdot ct$	89
$v^+ \cdot cv^+ \cdot ct^+$	94
$v \cdot cv^+ \cdot ct$	3
$v^+ \cdot cv \cdot ct^+$	5
	1448

Once again, the standard recombination approach is called for, but we must be careful in our classification of parental and recombinant types. Note that the parental input genotypes for the triple heterozygotes are $v^+ \cdot cv \cdot ct$ and $v \cdot cv^+ \cdot ct^+$; we must take this into consideration when we decide what constitutes a recombinant.

Starting with the *v* and *cv* loci, we see that the recombinants are of genotype $v \cdot cv$ and $v^+ \cdot cv^+$ and that there are $45 + 40 + 89 + 94 = 268$ of these recombinants. Of a total of 1448 flies, this number gives an RF of 18.5 percent.

For the *v* and *ct* loci, the recombinants are $v \cdot ct$ and $v^+ \cdot ct^+$. There are $89 + 94 + 3 + 5 = 191$ of these recombinants among 1448 flies, so RF = 13.2 percent.

For *ct* and *cv*, the recombinants are $cv \cdot ct^+$ and $cv^+ \cdot ct$. There are $45 + 40 + 3 + 5 = 93$ of these recombinants among the 1448, so RF = 6.4 percent.

All the loci are linked on the same chromosome, because the RF values are all considerably less than 50 percent. Because the *v* and *cv* loci show the largest RF value, they must be farthest apart; therefore, the *ct* locus must be between them. A map can be drawn as follows:

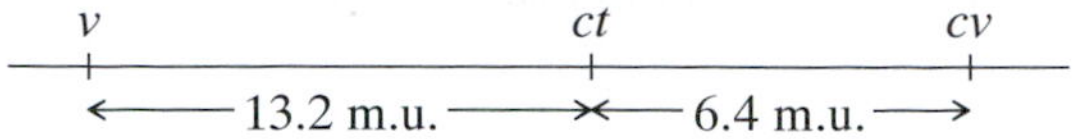

```
    v              ct             cv
    |               |              |
    ←——— 13.2 m.u. ———→←— 6.4 m.u. —→
```

The testcross can be rewritten as $v^+ cv\ ct/v\ cv^+\ ct^+ \times v\ ct\ cv/v\ ct\ cv$.

Note several important points here. First, we have deduced a different gene order from that of our listing of the progeny genotypes. Because the point of the exercise was to determine the linkage relation of these genes, the original listing was of necessity arbitrary; the order simply was not known before the data were analyzed.

Second, we have definitely established that *ct* is between *v* and *cv* and what the distances are between *ct* and these loci in map units. But we have arbitrarily placed *v* to the left and *cv* to the right; the map could equally well be inverted.

A third point to note is that the two smaller map distances, 13.2 m.u. and 6.4 m.u., add up to 19.6 m.u., which is greater than 18.5 m.u., the distance calculated for *v* and *cv*. Why is this so? The answer to this question lies in the way in which we have analyzed the two rarest classes in our classification of recombination for the *v* and *cv* loci. Now that we have the map, we can see that these two rare classes are in fact double recombinants, arising from two crossovers (Figure 5-12). However, we did not count the $v\ ct\ cv^+$ and $v^+\ ct^+\ cv$ genotypes when we calculated the RF value for *v* and *cv*; after all, with regard to *v* and *cv*, they are parental combinations ($v\ cv^+$ and $v^+\ cv$). In the light of our map, however, we see that this led to an underestimate of the distance between the *v* and *cv* loci. Not only should we have counted the two rarest classes, we should have counted each of them twice because each represents a double recombinant class. Hence, we can correct the value by adding the numbers $45 + 40 + 89 + 94 + 3 + 3 + 5 + 5 = 284$. Of the total of 1448, this number is exactly 19.6 percent, which is identical with the sum of the two component values.

Now that we have had some experience with the data of this cross, we can look back at the progeny listing and see that it is usually possible to deduce gene order by inspection, without a recombinant frequency analysis. Only three gene orders are possible, each with a different gene in the middle position. It is generally true that the double recombinant classes are the smallest ones. Only one order should be

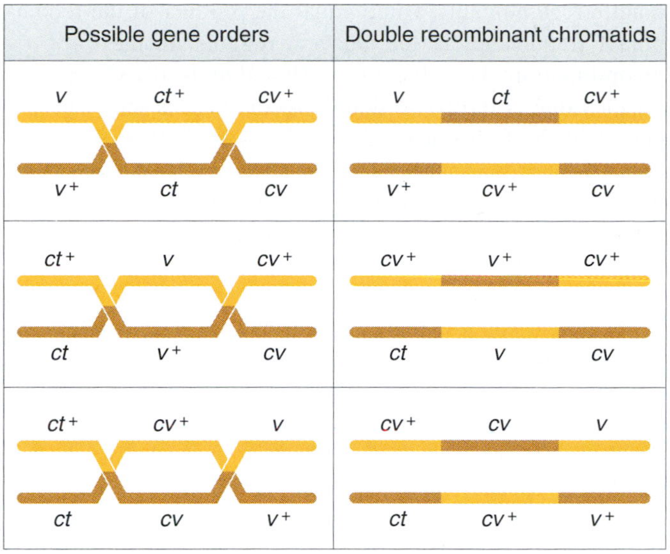

Figure 5-13 With three genes, only three gene orders are possible. Double crossovers create unique double recombinant genotypes for each gene order. Only the first possibility is compatible with the data in the text.

compatible with the smallest classes having been formed by double crossovers, as shown in Figure 5-13. Only one order gives double recombinants of genotype $v\ ct\ cv^+$ and $v^+\ ct^+\ cv$. Notice in passing that the ability to detect a double crossover depends on having a heterozygous gene between the two crossovers; if the mothers of these progeny had not been heterozygous ct/ct^+, we could never have identified the double recombinant classes.

Finally, note that linkage maps merely map the loci in relation to each other, with the use of standard map units. We do not know where the loci are on a chromosome — or even which specific chromosome they are on. The linkage map is essentially an abstract construct that can be correlated with a specific chromosome and with specific chromosome regions only by applying special kinds of cytogenetic analyses, as we shall see in Chapter 17.

MESSAGE ·······
Three (and higher)-point testcrosses enable linkage between three (or more) genes to be evaluated in one cross.

Interference

The detection of the double recombinant classes shows that double crossovers must occur. Knowing this, we can ask, are the crossovers in adjacent chromosome regions independent or does a crossover in one region affect the likelihood of there being a crossover in an adjacent region. It turns out that often they are not independent: the interaction is called **interference.**

The analysis can be approached in the following way. If the crossovers in the two regions are independent, then, according to the product rule (see page 37), the frequency of

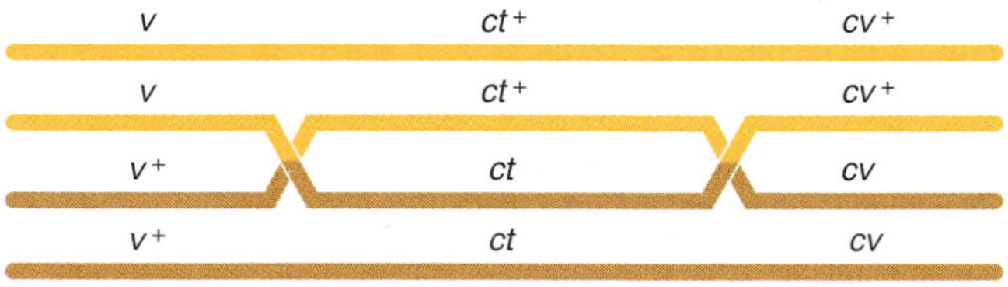

Figure 5-12 An example of a double crossover. Notice that a double crossover produces double recombinant chromatids that have the parental allelic combinations at the outer loci.

double recombinants would equal the product of the recombinant frequencies in the adjacent regions. In the $v-ct-cv$ recombination data, the $v-ct$ RF value is 0.132 and the $ct-cv$ value is 0.064, so double recombinants might be expected at the frequency $0.132 \times 0.064 = 0.0084$ (0.84 percent) if there is independence. In the sample of 1448 flies, $0.0084 \times 1448 = 12$ double recombinants are expected. But the data show that only 8 were actually observed. If this deficiency of double recombinants were consistently observed, it would show us that the two regions are not independent and suggest that the distribution of crossovers favors singles at the expense of doubles. In other words, there is some kind of *interference*: a crossover reduces the probability of a crossover in an adjacent region.

Interference is quantified by first calculating a term called the **coefficient of coincidence (c.o.c.)**, which is the ratio of observed to expected double recombinants subtracted from 1. Hence

Interference (I) = $1 -$ c.o.c. =

$$1 - \left[\frac{\substack{\text{observed frequency, or} \\ \text{number of double recombinants}}}{\substack{\text{expected frequency, or} \\ \text{number of double recombinants}}} \right]$$

In our example

$$I = 1 - \tfrac{8}{12} = \tfrac{4}{12} = \tfrac{1}{3}, \text{ or } 33\%$$

In some regions, there are never any observed double recombinants. In these cases, c.o.c. = 0, so I = 1 and interference is complete. Most of the time, the interference values that are encountered in mapping chromosome loci are between 0 and 1; but, in certain special situations, observed doubles exceed expected, giving negative interference values.

Recombination analysis relies so heavily on three-point testcrosses and extended versions of them that it is worth making a step-by-step summary of the analysis, ending with an interference calculation. We shall use numerical values from the data on the v, ct, and cv loci.

1. Calculate recombinant frequencies for each pair of genes:

$$v-cv = 18.5\%$$
$$cv-ct = 6.4\%$$
$$ct-v = 13.2\%$$

2. Represent linkage relations in a linkage map:

3. Determine the double recombinant classes.

4. Calculate the frequency and number of double recombinants expected if there is no interference:

$$\text{Expected frequency} = 0.132 \times 0.064 = 0.0084$$

$$\text{Expected number} = 0.0084 \times 1448 = 12$$

5. Calculate interference:

$$\text{Observed number of double recombinants} = 8$$

$$\text{Expected number of double recombinants} = 12$$

$$\therefore I = 1 - \tfrac{8}{12} = \tfrac{4}{12} = 0.33, \text{ or } 33\%$$

You may have wondered why we always use heterozygous females for testcrosses in our examples of linkage in *Drosophila*. When $pr \, vg/pr^+ \, vg^+$ males are crossed with $pr \, vg/pr \, vg$ females, only $pr \, vg/pr^+ \, vg^+$ and $pr \, vg/pr \, vg$ progeny are recovered. This result shows that there is no crossing-over in *Drosophila* males. However, this absence of crossing-over in one sex is limited to certain species; it is not the case for males of all species (or for the heterogametic sex). In other organisms, there is crossing-over in XY males and in WZ females. The reason for the absence of crossing-over in *Drosophila* males is that they have an unusual prophase I, with no synaptonemal complexes.

Incidentally, there is a recombination difference between human sexes as well. Women show higher recombinant frequencies for the same loci than do men.

Calculating recombinant frequencies from selfed dihybrids

A testcross is the most convenient way to measure recombinant frequency, but in practice the appropriate tester is not always available. A common situation is that a new phenotype has been identified and shown by Mendelian analysis to be caused by some genotype such as a/a. To map this newly discovered locus, the a/a individual will be crossed to a variety of genotypes such as b/b, where the locus of the B gene is known. In this situation, you can see that no $a/a \cdot b/b$ tester will be available at the outset. In fact, $a/a \cdot b/b$ can be derived only from further breeding experiments.

Nevertheless, it is possible to calculate recombinant frequencies from the self of the dihybrid formed by crossing the two stocks. In this example, the parents would be $a/a \cdot B/B$ and $A/A \cdot b/b$, and the dihybrid will be $A/a \cdot B/b$. This dihybrid will produce parental gametes, $a \cdot B$ and $A \cdot b$, and recombinant gametes $A \cdot B$ and $a \cdot b$. Upon selfing, the gametes will fertilize randomly, and initially it seems that the recombinants cannot be identified in the progeny. However, the $a/a \cdot b/b$ progeny genotype comes to the rescue because this is the only genotype that *must* have been constituted by recombinants — in fact, by fertilization of $a \cdot b$ by another $a \cdot b$. Therefore, if the frequency of the $a \cdot b$ products of meiosis is p, then the frequency of $a/a \cdot b/b$ offspring will be p^2. Therefore, to find p we simply have to take the square root of the frequency of $a/a \cdot b/b$ progeny. Because we know that the frequency of $a \cdot b$ will equal the frequency of $A \cdot B$, we can double p to find the total recombinant frequency.

If the two genes are unlinked, we know that the $a/a \cdot b/b$ class will be formed at a frequency of $\tfrac{1}{16}$ (the "1" of the 9:3:3:1 ratio). In this case, the square root of $\tfrac{1}{16}$ is $\tfrac{1}{4}$; and,

TABLE 5-2	RF Values (Percent) Corresponding to Selected Values of z in Selfs of Repulsion Dihybrids	
	z	RF
	0.001	2.2
	0.005	4.9
	0.020	9.9
	0.040	13.8
	0.100	21.1
	0.200	28.5
	0.300	33.5
	0.500	40.3
	0.700	45.0

on doubling this p value, we would obtain an RF of $\frac{1}{2}$, or 50 percent, as expected. What about linkage? Cases of linkage will produce $a\ b/a\ b$ frequencies significantly less than $\frac{1}{16}$. Let's assume that we observe that the frequency of $a\ b/a\ b$ is 0.01 (1 percent); the frequency of $a\ b$ meiotic products must have been 0.1, or 10 percent, and the RF must be 20 percent, a clear case of linkage. Note also that, if we wanted to calculate the parental frequencies, they must be 40 percent each for $a\ B$ and $A\ b$, calculated as 0.5 $(100 - 20)$ percent.

The above method is theoretically correct, but in practice it is inaccurate because it extrapolates from just one of the F_2 phenotypes and furthermore includes taking a square root. A more accurate formulation has been devised that incorporates all the F_2 phenotypes. A statistic called the *product ratio (z)* is calculated, and a recombinant frequency is derived from a table of values of z. In the aforementioned repulsion dihybrid $(A\ b/a\ B)$, the product ratio is calculated as follows, in which the four components of the calculation are the four F_2 phenotypes:

$$z = \frac{(A/-\ B/-) \times (a/a\ b/b)}{(A/-\ b/b) \times (a/a\ B/-)}$$

(**Note:** For brevity, this formula is not written with the use of linkage symbolism.) The RF values corresponding to selected values of z are shown in Table 5-2. Computer programs that calculate RF values from F_2 data are available.

MESSAGE ·······························

Recombinant frequency can be calculated indirectly from the progeny of selfed dihybrids.

Examples of linkage maps

Linkage maps are an essential aspect of the experimental genetic study of any organism. They are the prelude to any serious piece of genetic manipulation. Why is mapping so important? The types of genes that an organism has and their positions in the chromosome set are fundamental aspects of genetic analysis. The main reasons for mapping

concern gene *function*, gene *evolution*, and gene *isolation*. A gene's location is known to affect its expression in many cases, a phenomenon generally called a "neighborhood effect." Genes of related function are often clustered next to each other in bacterial chromosomes, generally because they are transcribed as one unit. The position of a eukaryotic gene in or near heterochromatin can affect its expression. A knowledge of gene position is useful in evolutionary studies because, from the relative positions of the same genes in related organisms, the rearrangement of chromosomes in the course of evolution can be deduced. Finally, if a gene is to be isolated for molecular analysis, its chromosomal location is often the beginning of the isolation procedure (positional cloning — see Chapter 12). Many organisms have had their chromosomes intensively mapped. The resultant maps represent a vast amount of genetic analysis generally achieved by collaborative efforts of research groups throughout the world. Figures 5-14 and 5-15 show two examples of linkage maps: one from *Drosophila* and one from the tomato. The *Drosophila* genome is one of the most intensively mapped of all model genetic organisms. The map in Figure 5-14 shows only a fraction of the known loci. Notice that genes with related functions (for example, those for eye color) are scattered all over the map. Tomatoes, too, have been interesting from the perspectives of both basic and applied genetic research, and the tomato genome is one of the best mapped of plants.

The different panels of Figure 5-15 illustrate some of the stages of understanding through which research arrives at a comprehensive map. First, although chromosomes are visible under the microscope, there is initially no way to locate genes on them. However, the chromosomes can be individually identified and numbered, on the basis of their inherent landmarks such as staining patterns and centromere positions, as has been done in Figure 5-15a and b. Next, analysis of recombinant frequencies generates a set of linkage groups that must correspond to chromosomes, but specific correlations cannot necessarily be made with the numbered chromosomes. At some stage, as in the screwworm example, cytogenetic analyses allow the linkage groups to be assigned to specific chromosomes. Figure 5-15c shows a tomato map made in 1952, with the linkages of the genes known at that time. Each locus is represented by the two alleles used in the original mapping experiments. As more and more loci became known, they were mapped in relation to the loci shown in Figure 5-15c, so today the map contains hundreds of loci. Some of the chromosome numbers shown in Figure 5-15c are tentative and do not correspond to the modern chromosome numbering system. Notice again that genes with related functions (for example, fruit shape) are scattered.

Chi-square test for linkage

When RF values are close to 50 percent, the χ^2 test can be used as a critical test for linkage. Assume that we have

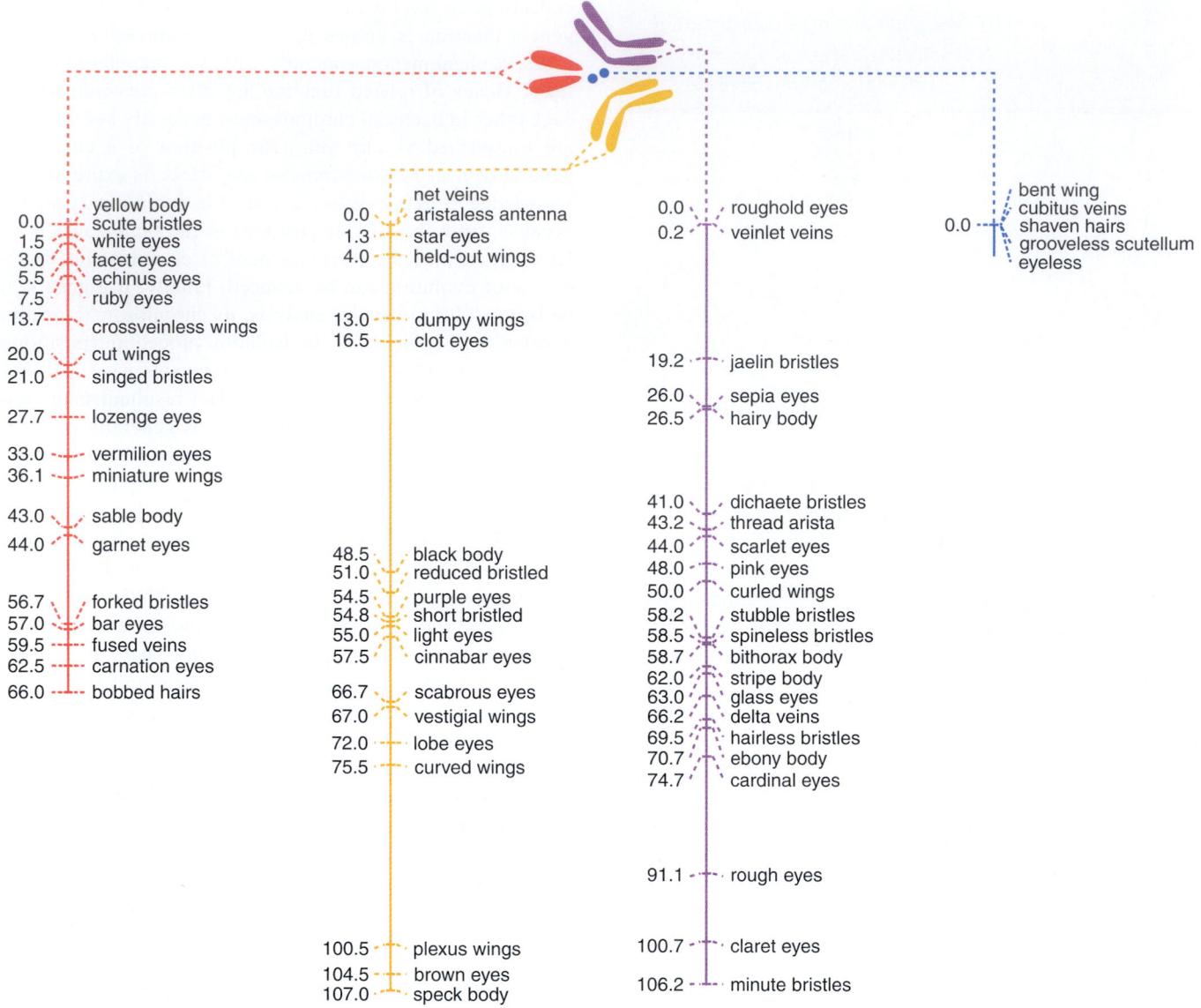

Figure 5-14   Genetic map of the *Drosophila* genome, showing some of the loci and how each linkage group corresponds to one chromosome pair. Values are given in map units measured from the gene closest to one end. Larger values are calculated as sums of shorter intervals because the recombinant frequency for any two loci cannot exceed 50 percent. (From E. W. Sinnott, L. C. Dunn, and T. Dobzhansky, *Principles of Genetics,* 5th ed. Copyright © 1962 by McGraw-Hill.)

Figure 5-15   Mapping the chromosomes of tomatoes.   ▶
(a) Photomicrograph of a meiotic prophase I (pachytene) from anthers, showing the 12 pairs of chromosomes as they appear under the microscope. (b) The currently used chromosome numbering system. The centromeres are orange, and the flanking, densely staining regions (heterochromatin) are shown in green. (c) A linkage map made in 1952 showing the linkage groupings known at the time. Each locus is flanked by drawings of the variant phenotype that first identified that genetic locus (to the right) and the appropriate normal phenotype (to the left). Interlocus map distances are shown in map units. (Parts a and b from C. M. Rick, "The Tomato." Copyright © 1978 by Scientific American, Inc. All rights reserved. Part c from L. A. Butler.)

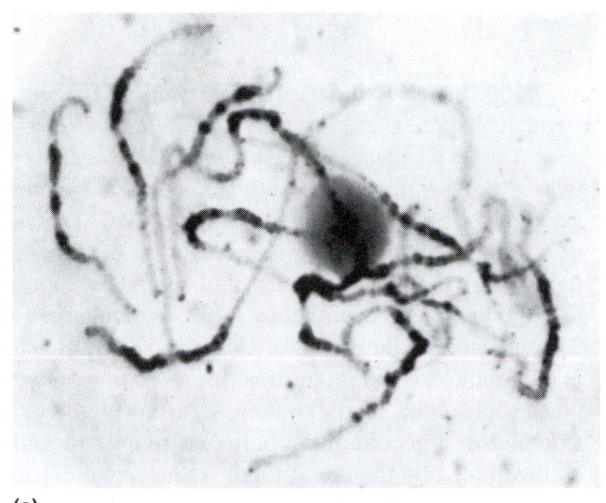

(a)

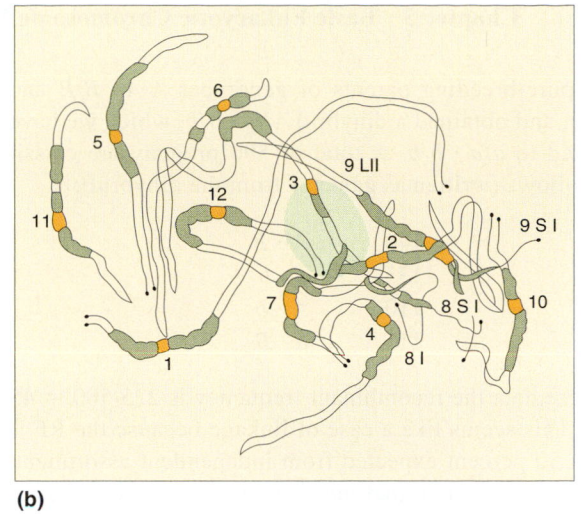

(b)

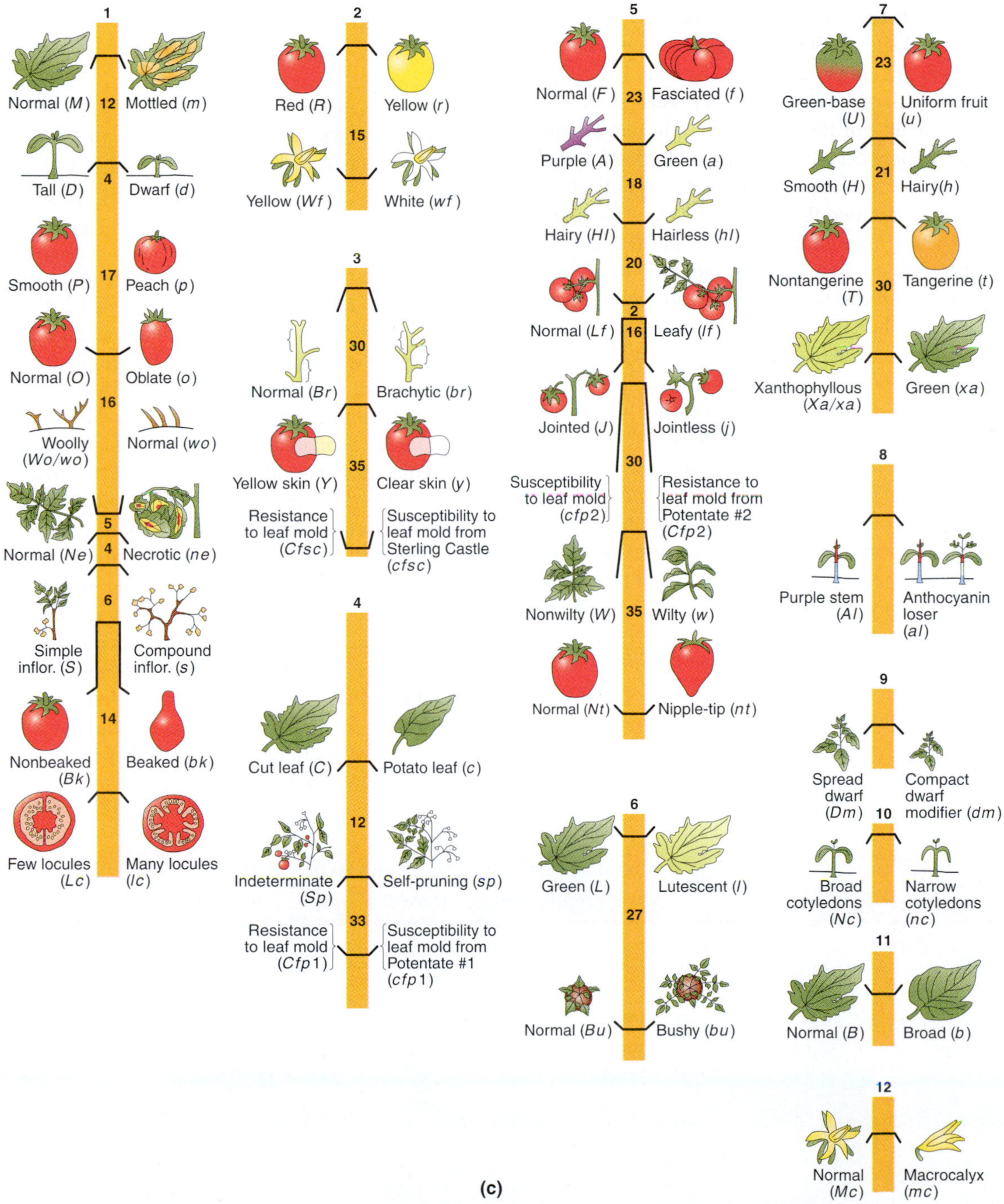

1

Normal (*M*) — **12** — Mottled (*m*)

Tall (*D*) — **4** — Dwarf (*d*)

Smooth (*P*) — **17** — Peach (*p*)

Normal (*O*) — Oblate (*o*)

Woolly (*Wo/wo*) — **16** — Normal (*wo*)

Normal (*Ne*) — **5** — Necrotic (*ne*)
— **4** —

Simple inflor. (*S*) — **6** — Compound inflor. (*s*)

Nonbeaked (*Bk*) — **14** — Beaked (*bk*)

Few locules (*Lc*) — Many locules (*lc*)

2

Red (*R*) — Yellow (*r*)
— **15** —
Yellow (*Wf*) — White (*wf*)

3

Normal (*Br*) — **30** — Brachytic (*br*)

Yellow skin (*Y*) — **35** — Clear skin (*y*)

Resistance to leaf mold (*Cfsc*) — Susceptibility to leaf mold from Sterling Castle (*cfsc*)

4

Cut leaf (*C*) — Potato leaf (*c*)

Indeterminate (*Sp*) — **12** — Self-pruning (*sp*)

Resistance to leaf mold (*Cfp1*) — **33** — Susceptibility to leaf mold from Potentate #1 (*cfp1*)

5

Normal (*F*) — **23** — Fasciated (*f*)

Purple (*A*) — **18** — Green (*a*)

Hairy (*Hl*) — **20** — Hairless (*hl*)

Normal (*Lf*) — **2** — Leafy (*lf*)
— **16** —
Jointed (*J*) — Jointless (*j*)

Susceptibility to leaf mold (*cfp2*) — **30** — Resistance to leaf mold from Potentate #2 (*Cfp2*)

Nonwilty (*W*) — **35** — Wilty (*w*)

Normal (*Nt*) — Nipple-tip (*nt*)

6

Green (*L*) — **27** — Lutescent (*l*)

Normal (*Bu*) — Bushy (*bu*)

7

Green-base (*U*) — **23** — Uniform fruit (*u*)

Smooth (*H*) — **21** — Hairy (*h*)

Nontangerine (*T*) — **30** — Tangerine (*t*)

Xanthophyllous (*Xa/xa*) — Green (*xa*)

8

Purple stem (*Al*) — Anthocyanin loser (*al*)

9

Spread dwarf (*Dm*) — Compact dwarf modifier (*dm*)
— **10** —
Broad cotyledons (*Nc*) — Narrow cotyledons (*nc*)

11

Normal (*B*) — Broad (*b*)

12

Normal (*Mc*) — Macrocalyx (*mc*)

(c)

crossed pure-breeding parents of genotypes $A/A \cdot B/B$ and $a/a \cdot b/b$, and obtained a dihybrid $A/a \cdot B/b$, which we have testcrossed to $a/a \cdot b/b$. A total of 500 progeny are classified as follows (written as gametes from the dihybrid):

140	$A \cdot B$
135	$a \cdot b$
110	$A \cdot b$
115	$a \cdot B$

From these data the recombinant frequency is $225/500 = 45$ percent. This seems like a case of linkage because the RF is less than 50 percent expected from independent assortment. However, it is possible that the two recombinant classes are in the minority merely on the basis of chance; therefore, we need to perform a χ^2 test.

The problem then is to find the expectations, E, for each class. For linkage testing, it might be supposed that these expectations are simply given by the $1:1:1:1$ ratio of the four backcross classes that we expect when there is independent assortment. For the 500 progeny in our example, then, we would expect $500/4 = 125$ in each class. But the $1:1:1:1$ ratio is not the appropriate test for linkage, because to get such a ratio two things must be true. There must be independent assortment between the A and B locus, but, in addition, there must be an equal chance for the different genotypes formed at fertilization to reach the age at which they are scored for the test, which usually means that the four genotypes must have equal chance of survival from egg to adult. However, it is often the case that mutations used in linkage tests have some deleterious effect in the homozygous condition; so a/a or b/b genotypes have a lower probability of survival than do the wild-type heterozygotes A/a and B/b. We might then be led to reject the hypothesis of independent assortment even when it is correct because the differential survivorship of the genotypes causes deviations from the $1:1:1:1$ expected ratio. What we need is a method of calculating the expectations, E, that is insensitive to differences in survivorship.

No matter what the frequencies of a/a or b/b genotypes are among the backcross adults, if there is independent assortment, we expect the frequency of the $a \cdot b$ genotypes to be the product of the frequencies of the a and the b alleles. In our example, the total proportion of a alleles is $(135 +$ $115)/500$, which is indeed the expected 50 percent, but the frequency of b alleles is only $(135 + 110)/500 = 49$ percent. Thus we expect the *proportion* of $a \cdot b$ genotypes to be $0.50 \times 0.49 = 0.245$ and the *number* of $a \cdot b$ genotypes in a sample of 500 to be $500 \times 0.245 = 122.5$. The same kind of calculation can be performed for each of the other genotypes to give all the expected numbers. The comparison is usually done in a **contingency table,** as shown in Table 5-3. The expected number for an entry in the contingency table is the product of the proportion observed in its row, the proportion observed in its column, and the total sample size. But the row and column proportions are the row and column totals divided by the grand total, so the actual calculation of the expected number in each entry is simply to multiply the appropriate row total by the appropriate column total and then divide by the grand total. The value of χ^2 is then calculated as follows:

Class	O	E	$O - E$	$(O - E)^2/E$
$A \cdot B$	140	127.5	12.5	1.23
$A \cdot b$	110	122.5	-12.5	1.28
$a \cdot B$	115	127.5	-12.5	1.23
$a \cdot b$	135	122.5	12.5	1.28
			Total $\chi^2 =$	5.02

The obtained value of χ^2 is converted into a probability by using a χ^2 table (see Table 4-1, page 126). To do so, we need to decide on the degrees of freedom (df) in the test, which, as the name suggests, is the number of independent deviations of observed from expected that have been calculated. We notice that, because of the way that the expectations were calculated in the contingency table from the row and column totals, all deviations are identical in absolute magnitude, 12.5, and that they alternate in sign and so they cancel out when summed in any row or column. Thus, there is really only one independent deviation, so there is only one degree of freedom. Therefore, looking along the one degree of freedom line in Table 4-1, we see that the probability of obtaining a deviation from expectations this large (or larger) by chance alone is 0.025 (2.5 percent). Because this probability is less than 5 percent, the hypothesis of independent assortment must be rejected. Thus, having rejected the hypothesis of no linkage, we are left with the inference that the loci must be linked.

5-3 ▷ TABLE	Contingency Table Comparing Observed and Expected Results of a Testcross to Examine Linkage Between Loci A/a and B/b				
LOCUS 2:		**B**		**b**	
	O	E	O	E	Total
LOCUS 1:					
A	140	$255 \times 250/500 = 127.5$	110	$245 \times 250/500 = 122.5$	250
a	115	$255 \times 250/500 = 127.5$	135	$245 \times 250/500 = 122.5$	250
Total	255		245		500

Mapping with molecular markers

In the first 70 years of building genetic maps, the markers on the maps were genes with variant alleles producing detectably different phenotypes. As organisms became more and more researched, large numbers of such genes could be used as markers on the maps. However, even in those organisms in which the maps appeared to be "full" of loci of known phenotypic effect, measurements showed that the chromosomal intervals between genes had to contain vast amounts of DNA. These gaps could not be mapped by linkage analysis, because there were no markers in those regions. What was needed were large numbers of additional genetic markers that could be used to fill in the gaps to provide a higher-resolution map. This need was met by the discovery of various kinds of molecular markers. A **molecular marker** is a site of heterozygosity for some type of silent DNA variation not associated with any measurable phenotypic variation. Such a "DNA locus," when heterozygous, can be used in mapping analysis just as a conventional heterozygous allele pair can be used. Because molecular markers can be easily detected and are so numerous in a genome, when they are mapped by linkage analysis, they fill the voids between genes of known phenotype. Note that, in mapping, the biological significance of the DNA marker is not important in itself; the heterozygous site is merely a convenient reference point that will be useful in finding one's way around the chromosomes. In this way, markers are being used just as milestones were used by travelers in previous centuries. Travelers were not interested in the milestones (markers) themselves, but they would have been disoriented without them.

The two basic types of molecular markers are those based on restriction-site variation and on repetitive DNA.

Use of restriction fragment length polymorphisms in mapping

Bacterial restriction enzymes cut DNA at specific target sequences that exist by chance in the DNA of other organisms. Generally, the target sites are found in the same position in the DNA of different individuals in a population; that is, in the DNA of homologous chromosomes. However, quite commonly, a specific site might be missing as a result of some silent mutation. The mutation might be within a gene or in a noncoding intergenic area. If an individual is heterozygous for presence and absence ($+/-$), that locus can be used in mapping. The $+/-$ sites are found by Southern analyses using a probe derived from DNA of that region. A typical example follows:

On a Southern hybridization of such an individual, the probe would highlight three fragments, of size 3, 2, and 1 kb. Another individual might be homozygous for the long fragment and show only a 3-kb band in the Southern hybridization.

These multiple forms of this region constitute a **restriction fragment length polymorphism (RFLP).**

In a cross of the aforedescribed two individuals, half the progeny would show three fragments when probed and the other half only one fragment, following Mendel's law of equal segregation just as a gene would. Hence an RFLP can be mapped and treated just like any other chromosomal site. The following situation shows linkage of the heterozygous RFLP in our example to a heterozygous gene, with D in coupling conformation with the 1 plus 2 morph:

Crossovers between these sites would produce recombinant products that are detectable as $D-3$ and $d-2-1$. In this way, the RFLP locus can be mapped in relation to genes or to other molecular markers.

Use of polymorphism of VNTRs in mapping

The number of repeated units in a tandem array is variable. The mechanisms for producing this variation need not concern us at present. The important fact is that individuals that are heterozygous for different numbers of tandem repeats can be detected, and the heterozygous site can be used as a marker in mapping. A probe that binds to the repetitive DNA is needed. The following example uses restriction enzyme target sites that are outside the repetitive array. The basic unit of the array is shown as an arrow.

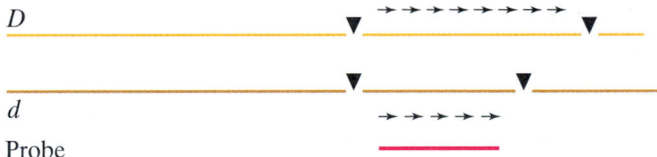

This VNTR locus will form two bands, one long and one short, on a Southern hybridization autoradiogram. Once again, this heterozygous site can be used in mapping just as the RFLP locus was.

Linkage mapping by recombination in humans

Humans have thousands of autosomally inherited phenotypes, and it might seem that it should be relatively straight-

forward to map the loci of the genes causing these phenotypes by using the techniques developed in this chapter.
However, progress in mapping these loci was initially slow
for several reasons. First, it is not possible to make controlled crosses in humans, and geneticists had to try to calculate recombinant frequencies from the occasional dihybrids that were produced by chance in human matings.
Crosses that were the equivalent of testcrosses were extremely rare. Second, human progenies are generally small,
making it difficult to obtain enough data to calculate reliable map distances. Third, the human genome is immense,
which means that on average the distances between the
known genes are large.

DNA markers have been particularly helpful in mapping human chromosomes; an example is shown in Figure
5-16.

Mapping the X chromosome

The human X chromosome has always been more amenable
to mapping by recombination analysis than the autosomes,
and the first human chromosome map was for the X chromosome. The reason for this success is that males are hemizygous for X-linked genes, and, just as we did for
Drosophila, if we look only at male progeny of a dihybrid
female, we are effectively sampling her gametic output. In
other words, we have a close approximation to a testcross.
Consider the following situation concerning the rare X-
linked recessive alleles for defective sugar processing (*g*)
and, at another locus, for color blindness (*c*). A doubly affected male (*c g/Y*) marries a normal woman (who is almost certainly *C G/C G*). The daughters of this mating are
coupling-conformation heterozygotes. The male children of
women of this type will provide an opportunity for geneticists to measure the frequency of recombinants issuing from

Figure 5-16 Linkage map of human chromosome 1, correlated
with chromosome banding pattern. The histogram shows the
distribution of all markers available for chromosome 1. Some markers
are genes of known phenotype, but the majority are DNA markers
based on neutral sequence variation.

A linkage map, based on recombinant frequency analyses of the
type considered in this chapter, is in the center of the figure. It shows
only some of the markers available. Map distances are shown in
centimorgans (= m.u.). The total length of the chromosome 1 map is
356 cM; it is the longest human chromosome.

The positions of some markers are cross-referenced to a diagram
of subregions of chromosome 1 based on standard banding pattern
(such a diagram is called an idiogram). These kinds of correlations can
be made only in situ hybridization (Chapter 17) and by cytogenetic
analysis (Chapter 14). Most of the markers shown on the map are
molecular, but several genes (highlighted in blue) also are included:
APOA2, apolipoprotein; ACTN2, actin protein; CRP, C-reactive
protein; SPTA1, spectrin protein. (From B. R. Jasney et al., *Science,*
September 30, 1994.)

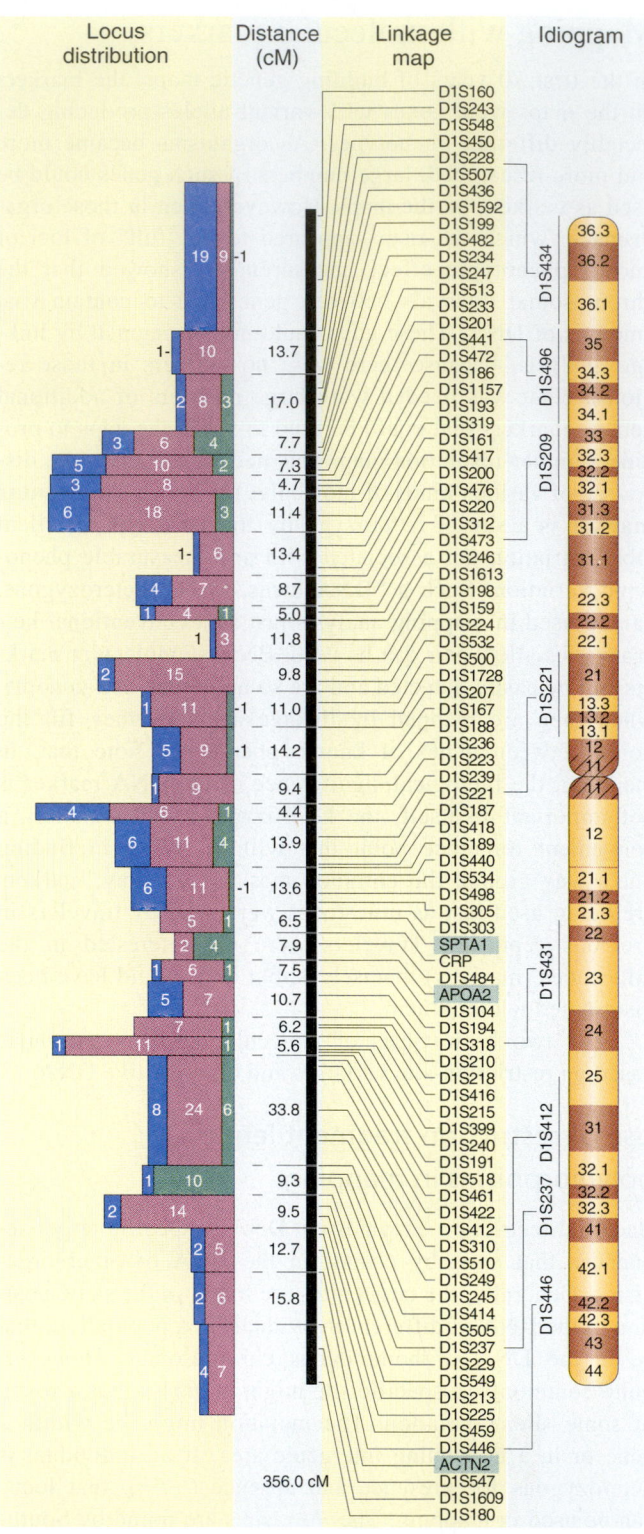

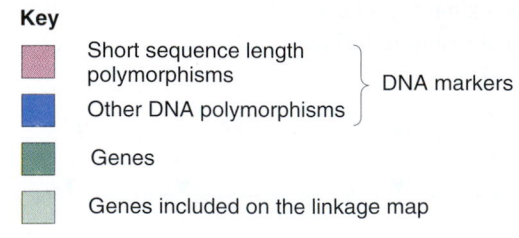

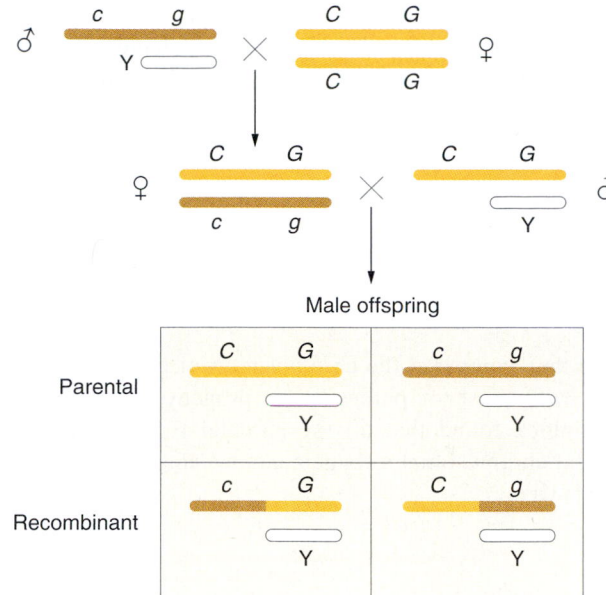

Figure 5-17   The phenotypic proportions in male children of women heterozygous for two X-linked genes can be used to calculate recombinant frequency. Thus, the X chromosome can be mapped by combining such pedigrees.

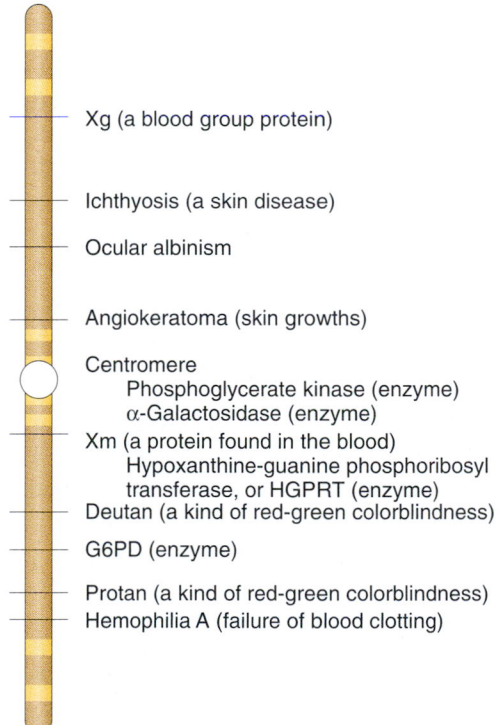

Figure 5-18   A linkage map of the human X chromosome, one of the first human chromosome maps. It is based solely on RF analyses of the loci shown and does not rely on the use of intervening molecular markers or somatic cell hybridization techniques (Chapter 6). The intended loci show tentative map assignments. (From W. F. Bodmer and L. L. Cavalli-Sforza, *Genetics, Evolution, and Man.* Copyright © 1976 by W. H. Freeman and Company.)

the maternal meioses (Figure 5-17). A human X chromosome map of some genes causing X-linked phenotypes is shown in Figure 5-18. Note, however, that silent DNA markers also can be used in this type of X-chromosome mapping.

Lod score for linkage testing by pedigrees

The small numbers of progeny in human families mean that it is impossible to determine linkage on the basis of single matings. To obtain reliable RF values, large sample sizes are necessary. However, if the results of many identical matings can be combined, then a more reliable estimate can be made. The standard way of doing so is to calculate **Lod scores.** Lod stands for "log of odds." The method simply calculates the probability of obtaining a set of results in a family on the basis of independent assortment and a specific degree of linkage. Then the ratio (odds) of the two probabilities is calculated, and the logarithm of this number is calculated, which is the Lod. Because logarithms are exponents, the Lod score has the useful feature that scores from different matings for which the same markers are used can be added, hence providing a cumulative set of data either supporting or not supporting some particular linkage value. Let's look at a simple example of how the calculation works.

Assume that we have a family that amounts to a dihybrid testcross. Also assume that, for the dihybrid individual, we can deduce the input gametes and hence can assess that individual's gametes for recombination. The dihybrid is heterozygous for a dominant disease allele (D/d) and for a molecular marker (M1/M2). Assume that it is a man and that the gametes that united to form him were $D \cdot M1$ and $d \cdot M2$. His wife is d/d · M2/M2. The pedigree in Figure 5-19 shows their six children, categorized with respect to the father's contributing gamete. Of the six, there are two recombinants, which would give an RF of 33 percent. However, it is possible that the genes are assorting independently and the children constitute a nonrandom sample. Let's

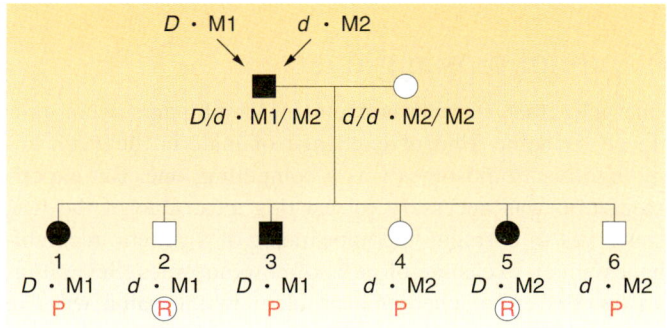

Figure 5-19   Pedigree amounting to a dihybrid testcross. *D/d* are alleles of a disease gene; M1 and M2 are molecular "alleles," such as two forms of an RFLP. P, parental (nonrecombinant); R, recombinant.

calculate the probability of this outcome under several hypotheses. The following display shows the expected proportions of parental (P) and recombinant (R) genotypes under three RF values and of independent assortment:

RF:	0.5	0.4	0.3	0.2
P	0.25	0.3	0.35	0.4
P	0.25	0.3	0.35	0.4
R	0.25	0.2	0.15	0.1
R	0.25	0.2	0.15	0.1

The probability of obtaining the results under independent assortment (RF of 50 percent) will be equal to:

$$0.25 \times 0.25 \times 0.25 \times 0.25 \times 0.25 \times 0.25 \times B$$
$$= 0.00024 \times B$$

where B = the number of possible birth orders for four parental and two recombinant individuals.

For an RF of 0.2, the probability is:

$$0.4 \times 0.1 \times 0.4 \times 0.4 \times 0.1 \times 0.4 \times B = 0.00026 \times B$$

Hence the ratio of the two is $0.00026/0.00024 = 1.08$ (note that the B's cancel out). Hence, on the basis of these data, the hypothesis of an RF of 0.2 is 1.08 times as likely as the hypothesis of independent assortment. Some other ratios and their Lod values are shown in the following display:

RF:	0.5	0.4	0.3	0.2
Probability	0.00024	0.00032	0.00034	0.00026
Ratio	1.0	1.33	1.41	1.08
Lod	0	0.12	0.15	0.03

These numbers confirm our original suspicions that the RF is about 30 to 40 percent. However, the data do not provide convincing support for any model of linkage. Conventionally, cumulative Lod scores of 3 are considered convincing support for a specific RF value. Note that a Lod score of 3 represents an RF value that is 1000 times (that is, 10^3 times) as likely as the hypothesis of no linkage.

Nature of crossing-over

The idea that intrachromosomal recombinants were produced by some kind of exchange of material between homologous chromosomes was a compelling one. But experimentation was necessary to test this idea. One of the first steps was to correlate the appearance of a genetic recombinant with an exchange of parts of chromosomes. Several investigators approached this problem in the same way. In 1931, Harriet Creighton and Barbara McClintock were studying two loci of chromosome 9 of corn: one affecting seed color (*C*, colored; *c*, colorless) and the other affecting endosperm composition (*Wx*, waxy; *wx*, starchy). Further-

more, the chromosome carrying *C* and *Wx* was unusual in that it carried a large, densely staining element (called a *knob*) on the *C* end and a longer piece of chromosome on the *Wx* end; thus, the heterozygote was

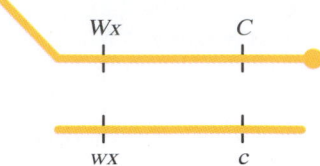

When they compared the chromosomes of genetic recombinants with those of parental-type progeny, Creighton and McClintock found that all the parental types retained the parental chromosomal arrangements, whereas all the recombinants were

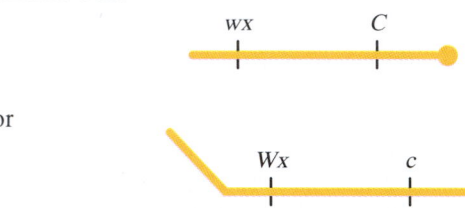

or

Thus, they correlated the genetic and cytological events of intrachromosomal recombination. The chiasmata appeared to be the sites of the exchange, but the final proof of this did not come until 1978.

But what is the mechanism of chromosome exchange in a crossover event? The short answer is that a crossover results from chromosome breakage and reunion. Two parental chromosomes break at the same position, and then join up again in two nonparental combinations. In Chapter 19 we will study models of the molecular processes that allow DNA to break and rejoin in such a precise manner.

MESSAGE ··
Chromosomes cross over by breaking at the same position and rejoining in two reciprocal nonparental combinations.
··

You will notice that, in our diagrammatic representations of crossing-over in this chapter, we have shown crossovers taking place at the four-chromatid stage of meiosis. However, just from studying random recombinant products of meiosis, as in a testcross, it is not possible to distinguish this possibility from crossing-over at the two-chromosome stage. This matter was settled through the genetic analysis of organisms whose four products of meiosis remain together in groups of four called **tetrads.** These organisms are mainly fungi and unicellular algae. The meiotic products in a single tetrad can be isolated, which is equivalent to isolating all four chromatids arising from a single meiosis. Tetrad analyses of crosses in which genes are linked clearly show that in many cases tetrads contain four different genotypes with regard to these loci; for example, from the cross

$$A\ B \times a\ b$$

some tetrads contain four genotypes

$$A\ B$$
$$A\ b$$
$$a\ B$$
$$a\ b$$

This result can be explained only by the occurrence of a crossover at the four-chromatid stage because, if crossing over occurred at the two-chromosome stage, then there could be only two different genotypes in an individual meiosis, as shown in Figure 5-20.

Tetrad analysis allows the exploration of many other aspects of intrachromosomal recombination, which will be considered in detail in Chapter 6, but for the present let us use tetrads to answer two more fundamental questions about crossing-over. First, can multiple crossovers be between more than two chromatids? To answer this question, we need to look at double crossovers; and, to study double crossovers, we need three linked genes. For example, in a cross such as

$$A\ B\ C \times a\ b\ c$$

there are many different tetrads possible, but some of them can be explained only by double crossovers. Consider the following tetrad as an example:

$$A\ B\ c$$
$$A\ b\ C$$
$$a\ B\ C$$
$$a\ b\ c$$

This tetrad must be explained by two crossovers involving three chromatids, as shown in Figure 5-21. Other types of tetrads show that all four of the chromatids can participate in crossing-over in the same meiosis. Therefore, two, three, or four chromatids can take part in crossing-over events in a single meiosis.

If all chromatids can take part, we can ask if there is any **chromatid interference;** in other words, does the occurrence of a crossover between any two nonsister chromatids affect the likelihood of those two chromatids taking part in another crossover in the same meiosis? Tetrad analysis can answer this question and shows that generally the distribution of crossovers between chromatids is random; in other words, there is no chromatid interference.

Before we leave the topic of the involvement of chromatids in crossovers, it is worth raising another question: Is

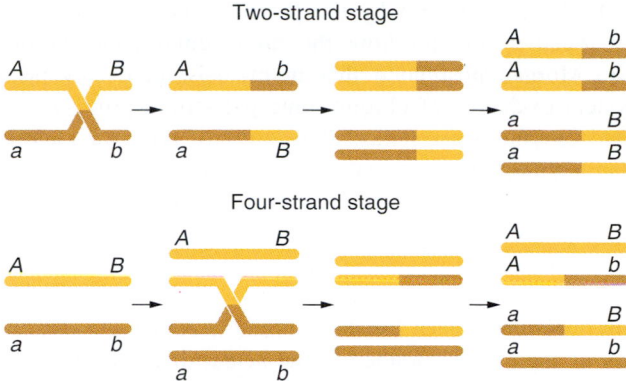

Figure 5-20 Tetrad analysis provides evidence that enabled geneticists to decide whether crossing-over occurs at the two-strand (two-chromosome) or at the four-strand (four-chromatid) stage of meiosis. Because more than two different products of a single meiosis can be seen in some tetrads, crossing-over cannot occur at the two-strand stage.

it possible for crossing-over to occur between sister chromatids? It has been shown in some organisms that indeed there is sister-chromatid crossing-over; but, because it produces no recombinants and, furthermore, because it is not clear whether it occurs in all organisms, it is conventional not to represent this type of exchange in crossover diagrams.

Crossing-over is a remarkably precise process. The synapsis and exchange of chromosomes is such that no segments are lost or gained, and four complete chromosomes emerge in a tetrad. A great deal has been learned about the nature of the molecular events in and around the sites of crossing-over, and these will be explored in Chapter 19.

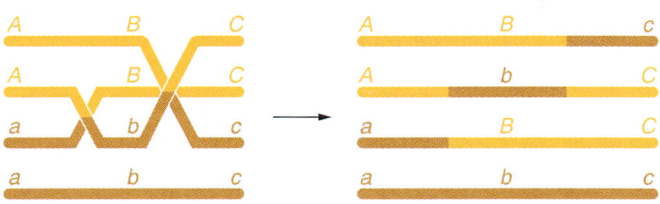

Figure 5-21 One of the several possible types of double-crossover tetrads that are regularly observed. Note that more than two chromatids exchanged parts.

SUMMARY

After making dihybrid crosses of sweet pea plants, William Bateson and R. C. Punnett discovered deviations from the 9:3:3:1 ratio of phenotypes expected in the F_2 generation. The parental gametic types outnumbered the other two classes. Later, in his studies of two different autosomal genes in *Drosophila*, Thomas Hunt Morgan found a similar deviation from Mendel's law of independent assortment. Morgan postulated that the two genes were located on the same pair of homologous chromosomes. This relation is called linkage.

Linkage explains why the parental gene combinations stay together but not how the nonparental combinations arise. Morgan postulated that in meiosis there may be a physical exchange of chromosome parts by a process now called crossing-over. Thus, there are two types of meiotic recombination. Recombination by Mendelian independent assortment results in a recombinant frequency of 50 percent. Crossing-over results in a recombinant frequency of less than 50 percent.

As Morgan studied more linked genes, he discovered many different values for recombinant frequency and wondered if these corresponded to the actual distances between genes on a chromosome. Alfred Sturtevant, a student of Morgan's, developed a method of determining the distance between genes on a linkage map, based on the percentage of recombinants. A linkage map is another example of a hypothetical entity based on genetic analysis.

The easiest way to measure recombinant frequencies is with a testcross of a dihybrid or trihybrid. However, recombinant frequencies can also be deduced from selfs.

Although the basic test for linkage is deviation from the $1:1:1:1$ ratio of progeny types in a testcross, such a deviation may not be all that obvious. The χ^2 test, which tells how often observations deviate from expectations purely by chance, can help us determine whether loci are linked.

There are several theories about how recombinant chromosomes are generated. We now know that crossing-over is the result of physical breakage and reunion of chromosome parts and occurs at the four-strand stage of meiosis.

Silent DNA variation is now used as a source of markers for chromosome mapping.

Sample sizes in pedigree analysis are two small to permit mapping, but cumulative data, expressed as Lod scores, can demonstrate linkage.

CONCEPT MAP

Draw a concept map interrelating as many of the following terms as possible. Note that the terms are listed in no particular order.

crossing-over / chromatids / mapping / chromosomes / testcross / linkage / map units / recombination / additivity / independent assortment

CHAPTER INTEGRATION PROBLEM

In *Drosophila*, the alleles *P* and *p* determine red and purple eyes, respectively. The alleles *B* and *b* determine brown and black body, respectively. A geneticist crossed a purple-eyed, brown-bodied female from a pure line with a red-eyed, black-bodied male, also from a pure line. All the F_1 flies had red eyes and brown bodies. The geneticist produced an F_2 by interbreeding the F_1 flies. The F_2 had the following composition:

brown, red	684
brown, purple	343
black, red	341
	1368

What do these data tell us about the chromosomal location of the genes?

◆ Solution ◆

When we think about chromosomal location, which of the phenomena that we have studied so far come to mind? First, there is the distinction between autosomal and sex-linked inheritance. Then there are the alternative possibilities of linkage or independent assortment. There isn't much else that we should worry about at this stage. How can we sort out these possibilities? Let's start with sex linkage. The characteristic feature of sex linkage is that there is an inheritance pattern that is correlated with sex in some way. There is no evidence of this in our data; because we were given no information to the contrary, we must assume that equal

numbers of males and females were obtained in each class. In the genetics of *Drosophila*, there is, however, a fact touching on the sex of the flies that must always be kept in mind; as we learned in this chapter, during meiosis in *Drosophila* males, there is no crossing-over. We will put that notion on the back burner for the moment; it might become useful later.

So we have concluded that both genes must be autosomal. Now we can reconstruct the crosses by using gene symbols to restate what we know in a slightly different format.

$$P \qquad B/B \cdot p/p \times b/b \cdot P/P$$
$$\downarrow$$
$$F_1 \qquad B/b \cdot P/p$$
$$B/b \cdot P/p \times B/b \cdot P/p$$
$$\downarrow$$
$$F_2 \qquad B/- \cdot P/-$$
$$B/- \cdot p/p$$
$$b/b \cdot P/-$$

Now we can see better what is so unusual about these data. The mating between the F_1 flies is a dihybrid cross and, from Mendel's second law, we expect independently assorting genes, to give four F_2 phenotypic classes in a $9:3:3:1$ ratio. We have only three phenotypic classes, and their ratio is $2:1:1$. Clearly, something else is going on. The missing class is the $b/b \cdot p/p$ genotype, so, for one thing, something must be happening to prevent the production of this class.

Remembering the discussion of lethality in Chapter 4, we might speculate that the $b/b \cdot p/p$ genotype is lethal. This possibility is unlikely, however, because it would leave us with the 9:3:3 part of the ratio, which would reduce to 3:1:1, not 2:1:1. But we have just considered another phenomenon that can keep genes from assorting freely to produce specific genotypes — that is, linkage; so let us explore that possibility. For this hypothesis, we could write the first cross as

$$\text{P} \qquad \frac{B\ p}{B\ p} \times \frac{b\ P}{b\ P}$$

$$\downarrow$$

$$\text{F}_1 \qquad \frac{B\ p}{b\ P}$$

If this is true, what ratios do we expect in the F_2? We might think that to answer this question we must know the number of map units that separate the two loci. That is reasonable, but we must start somewhere, so let's assume that 20 map units separate the loci, and then see what happens. (Do not forget about trial and error in analysis; scientists use it a lot as part of their calculations.) The expectations for the F_2 can be set up as follows, bearing in mind that there is no crossing-over in the males:

EGGS		SPERM			
		$B\ p$	50%	$b\ P$	50%
$B\ P$	10%	$B\ P/B\ p$	5%	$B\ P/b\ P$	5%
$b\ p$	10%	$b\ p/B\ p$	5%	$b\ p/b\ P$	5%
$B\ p$	40%	$B\ p/B\ p$	20%	$B\ p/b\ P$	20%
$b\ P$	40%	$b\ P/B\ p$	20%	$b\ P/b\ P$	20%

Overall, the phenotypic proportions in the progeny are

$B\ P$	$5 + 5 + 20 + 20 = 50\%$
$B\ p$	$5 + 20$	$= 25\%$
$b\ P$	$5 + 20$	$= 25\%$
$b\ p$		$= 0\%$

This gives us precisely the ratio that we need, 2:1:1. How did this happen? Were we just lucky in choosing the 20 percent recombinant frequency? The answer is no. It does not matter what frequency we might have chosen; the same result would have prevailed (try it).

In conclusion then, we can say that the ratio was produced by the inheritance of linked genes in repulsion phase in the parental strains, but we cannot determine how far apart they are. Precise mapping would require a testcross.

Although this was a tricky problem, the path that we took through it has shown how a variety of different concepts can be used to rule out various possibilities and arrive at a solution that works.

SOLVED PROBLEMS

1. The allele b gives *Drosophila* flies a black body and b^+ gives brown, the wild-type phenotype. The allele wx of a separate gene gives waxy wings and wx^+ gives nonwaxy, the wild-type phenotype. The allele cn of a third gene gives cinnabar eyes and cn^+ gives red, the wild-type phenotype. A female heterozygous for these three genes is testcrossed, and 1000 progeny are classified as follows: 5 wild type; 6 black, waxy, cinnabar; 69 waxy, cinnabar; 67 black; 382 cinnabar; 379 black, waxy; 48 waxy; and 44 black, cinnabar.

   a. Explain these numbers.

   b. Draw the alleles in their proper positions on the chromosomes of the triple heterozygote.

   c. If it is appropriate according to your explanation, calculate interference.

   d. If two triple heterozygotes of the type in this problem were crossed, what proportion of progeny would be black, waxy? (Remember, there is no crossing-over in *Drosophila* males.)

◆ Solution ◆

a. One of the general pieces of advice given earlier is to be methodical. Here it is a good idea to write out the genotypes

that may be inferred from the phenotypes. The cross is a testcross of type

$$b^+/b \cdot wx^+/wx \cdot cn^+/cn \times b/b \cdot wx/wx \cdot cn/cn$$

Notice that there are distinct pairs of progeny classes in regard to frequency. Already, we can guess that the two largest classes represent parental chromosomes, that the two classes of about 68 represent single crossovers in one region, that the two classes of about 45 represent single crossovers in the other region, and that the two classes of about 5 represent double crossovers. Note that a progeny group may be specified by listing only the mutant phenotypes. We can write out the progeny as classes derived from the female's gametes, grouped as follows:

$b^+ \cdot wx^+ \cdot cn$	382
$b \cdot wx \cdot cn^+$	379
$b^+ \cdot wx \cdot cn$	69
$b \cdot wx^+ \cdot cn^+$	67
$b^+ \cdot wx \cdot cn^+$	48
$b \cdot wx^+ \cdot cn$	44
$b \cdot wx \cdot cn$	6
$b^+ \cdot wx^+ \cdot cn^+$	5
	1000

Writing the classes out this way confirms that the pairs of classes are in fact reciprocal genotypes arising from zero, one, or two crossovers.

At first, because we do not know the parents of the triple heterozygous female, it looks as if we cannot apply the definition of recombination in which gametic genotypes are compared with the two input genotypes that form an individual. But on reflection, the only parental types that make sense in regard to the data presented are $b^+/b^+ \cdot wx^+/wx^+ \cdot cn/cn$ and $b/b \cdot wx/wx \cdot cn^+/cn^+$ because these are still the most common classes.

Now we can calculate the recombinant frequencies. For

$$b-wx, \text{ the RF} = \frac{69 + 67 + 48 + 44}{1000} = 22.8\%$$

$$b-cn, \text{ the RF} = \frac{48 + 44 + 6 + 5}{1000} = 10.3\%$$

$$wx-cn, \text{ the RF} = \frac{69 + 67 + 6 + 5}{1000} = 14.7\%$$

The map is therefore

$$b \qquad\qquad cn \qquad\qquad\qquad wx$$
$$\overleftarrow{\quad\text{10.3 m.u.}\quad}\times\overleftarrow{\quad\text{14.7 m.u.}\quad\rightarrow}$$

b. The parental chromosomes in the triple heterozygote were

$$b^+ \qquad\qquad cn \qquad\qquad\qquad wx^+$$
$$b \qquad\qquad cn^+ \qquad\qquad\qquad wx$$

c. The expected number of double recombinants is $0.103 \times 0.147 \times 1000 = 15.141$. The observed number is $6 + 5 = 11$, so interference can be calculated as $I = 1 - 11/15.141 = 1 - 0.726 = 0.274 = 27.4$ percent.

d. Here we are asked to use the newly gained knowledge of the linkage of these genes to predict the outcome of a cross. Note that we are not dealing with a testcross here. We are asked for the expected proportion of black, waxy flies among the progeny of two triple heterozygotes. Because there is no crossing-over in the male, one half of his gametes are $b\ cn^+\ wx$ and the other half are $b^+\ cn\ wx^+$ but the latter are incapable of contributing to the black, waxy phenotype. In the female, two gametic types contribute to black, waxy offspring: $b\ cn^+\ wx$ and $b\ cn\ wx$. The contribu-

tions of the parents to the black, waxy phenotype are shown here:

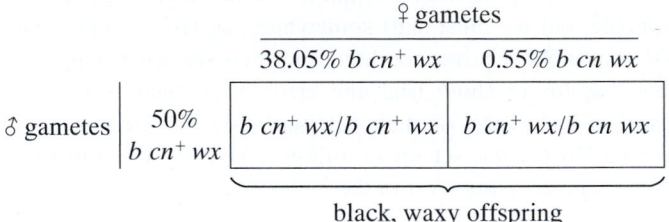

The chromosome carrying $b\ cn^+\ wx$ is a nonrecombinant parental type. There is a total of $(382 + 379)/1000 = 76.1$ percent of nonrecombinant parental chromosomes, half of which, or 38.05 percent, carry $b\ cn^+\ wx$. The frequency of $b\ cn\ wx$ is $(11 \div 2)/1000 = 0.55$ percent. In total, then, $38.05 + 0.55 = 38.6$ percent of the female's gametes are capable of forming a black, waxy offspring; half of these gametes fuse with a $b\ cn^+\ wx$ gamete from the male, so the frequency of black, waxy offspring is $38.6/2 = 19.3$ percent.

In summary, the appropriate gametic fusion can be shown this way:

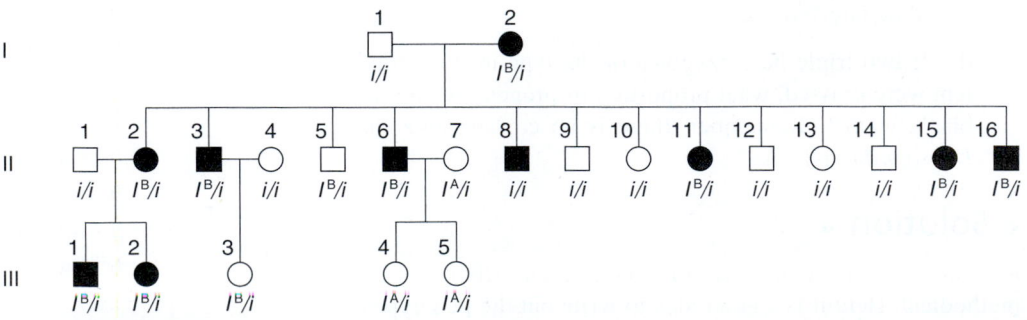

black, waxy offspring

2. A human pedigree shows people affected with the rare nail-patella syndrome (misshapen nails and kneecaps) and gives the ABO blood group genotype of each individual. Both loci concerned are autosomal. Study the pedigree below.

a. Is the nail-patella syndrome a dominant or recessive phenotype? Give reasons to support your answer.

b. Is there evidence of linkage between the nail-patella gene and the gene for ABO blood type, as judged from this pedigree? Why or why not?

c. If there is evidence of linkage, then draw the alleles on the relevant homologs of the grandparents. If there is no evidence of linkage, draw the alleles on two homologous pairs.

d. According to your model, which descendants are recombinants?

e. What is the best estimate of RF?

f. If man III-1 marries a normal woman of blood type O, what is the probability that their first child will be blood type B with nail-patella syndrome?

◆ Solution ◆

a. Nail-patella syndrome is most likely dominant. We are told that it is a rare abnormality, so it is unlikely that the unaffected people marrying into the family carry a presumptive recessive allele for nail-patella syndrome. Let N be the causative allele. Then all people with the syndrome are heterozygotes N/n because all (probably including the grandmother, too) result from a mating to an n/n normal person. Notice that the syndrome appears in all three successive generations — another indication of dominant inheritance.

b. There is evidence of linkage. Notice that most of the affected people — those that carry the N allele — also carry the I^B allele; most likely, these alleles are linked on the same chromosome.

c.
$$\frac{n \quad i}{n \quad i} \times \frac{N \quad I^B}{n \quad i}$$

(The grandmother must carry both recessive alleles to produce offspring of genotype i/i and n/n.)

d. Notice that the grandparental mating is equivalent to a testcross, so the recombinants in generation II are:

II-5: $n\,I^B/n\,i$ and   II-8: $N\,i/n\,i$

whereas all others are nonrecombinants, being either $N\,I^B/n\,i$ or $n\,i/n\,i$.

e. Notice that the grandparental cross and the first two crosses in generation II are identical and are all testcrosses. Three of the total 16 progeny are recombinant (II-5, II-8, and III-3). This gives a recombinant frequency of RF $= \frac{3}{16} = 18.8$ percent. (We cannot include the cross of II-6 × II-7, because the progeny cannot be designated as recombinant or not.)

f. (III-1 ♂) $\dfrac{N \quad I^B}{n \quad i} \times \dfrac{n \quad i}{n \quad i}$ (normal type O ♀)

↓

Gametes

$81.2\% \begin{cases} N\,I^B & 40.6\% \quad \longleftarrow \text{ nail-patella,} \\ n\,i & 40.6\% \qquad\qquad \text{blood type B} \end{cases}$

$18.8\% \begin{cases} N\,i & 9.4\% \\ n\,I^B & 9.4\% \end{cases}$

The two parental classes are always equal, and so are the two recombinant classes. Hence, the probability that the first child will have nail-patella syndrome and blood type B is 40.6 percent.

PROBLEMS

1. A plant of genotype
$$\frac{A \quad B}{a \quad b}$$
is testcrossed to
$$\frac{a \quad b}{a \quad b}$$
If the two loci are 10 m.u. apart, what proportion of progeny will be $A\,B/a\,b$?

2. The A locus and the D locus are so tightly linked that no recombination is ever observed between them. If $A\,d/A\,d$ is crossed to $a\,D/a\,D$, and the F_1 is intercrossed, what phenotypes will be seen in the F_2 and in what proportions?

3. The R and S loci are 35 m.u. apart. If a plant of genotype
$$\frac{R \quad S}{r \quad s}$$
is selfed, what progeny phenotypes will be seen and in what proportions?

4. The cross $E/E \cdot F/F \times e/e \cdot f/f$ is made, and the F_1 is then backcrossed to the recessive parent. The progeny genotypes are inferred from the phenotypes. The progeny genotypes, written as the gametic contributions of the heterozygous parent, are in the following proportions:

$$\begin{array}{ll} E \cdot F & \frac{2}{6} \\ E \cdot f & \frac{1}{6} \\ e \cdot F & \frac{1}{6} \\ e \cdot f & \frac{2}{6} \end{array}$$

Explain these results.

5. A strain of *Neurospora* with the genotype $H \cdot I$ is crossed with a strain with the genotype $h \cdot i$. Half the progeny are $H \cdot I$, and half are $h \cdot i$. Explain how this is possible.

6. A female animal with genotype $A/a \cdot B/b$ is crossed with a double-recessive male ($a/a \cdot b/b$). Their progeny include 442 $A/a \cdot B/b$, 458 $a/a \cdot b/b$, 46 $A/a \cdot b/b$, and 54 $a/a \cdot B/b$. Explain these results.

7. If $A/A \cdot B/B$ is crossed to $a/a \cdot b/b$, and the F_1 is test-crossed, what percent of the testcross progeny will be $a/a \cdot b/b$ if the two genes are (a) unlinked; (b) completely linked (no crossing-over at all); (c) 10 map units apart; (d) 24 map units apart?

8. In a haploid organism, the C and D loci are 8 m.u. apart. From a cross $C\ d \times c\ D$, give the proportion of each of the following progeny classes: (a) $C\ D$, (b) $c\ d$, (c) $C\ d$, (d) all recombinants.

9. A fruit fly of genotype $B\ R/b\ r$ is testcrossed to $b\ r/b\ r$. In 84 percent of the meioses, there are no chiasmata between the linked genes; in 16 percent of the meioses, there is one chiasma between the genes. What proportion of the progeny will be $B\ r/b\ r$?

10. A three-point testcross was made in corn. The results and a partial recombination analysis are shown in the following display, which is typical of three-point testcrosses (p = purple leaves, $+$ = green; v = virus-resistant seedlings, $+$ = sensitive; b = brown midriff to seed, $+$ = plain). Study the display and answer parts a–c.

P	$+/+ \cdot +/+ \cdot +/+ \times p/p \cdot v/v \cdot b/b$	
Gametes	$+ \cdot + \cdot +$	$p \cdot v \cdot b$
F_1	$+/p \cdot +/v \cdot +/b \times p/p \cdot v/v \cdot b/b$ (tester)	

Class	Progeny phenotypes	F_1 gametes	Numbers	RECOMBINANT FOR p–b	p–v	v–b
1	gre sen pla	$+ \cdot + \cdot +$	3,210			
2	pur res bro	$p \cdot v \cdot b$	3,222			
3	gre res pla	$+ \cdot v \cdot +$	1,024		R	R
4	pur sen bro	$p \cdot + \cdot b$	1,044		R	R
5	pur res pla	$p \cdot v \cdot +$	690	R		R
6	gre sen bro	$+ \cdot + \cdot b$	678	R		R
7	gre res bro	$+ \cdot v \cdot b$	72	R	R	
8	pur sen pla	$p \cdot + \cdot +$	60	R	R	
		Total	10,000	1,500	2,200	3,436

a. Determine which genes are linked.
b. Draw a map that shows distances in map units.
c. Calculate interference if appropriate.

Unpacking the Problem

1. Sketch cartoon drawings of the parent (P), F_1, and tester corn plants, and use arrows to show exactly how you would perform this experiment. Show where seeds are collected.

2. Why do all the $+$'s look the same, even for different genes? Why does this not cause confusion?

3. How can a phenotype be purple and brown (for example) at the same time?

4. Is it significant that the genes are written in the order p-v-b in the problem?

5. What is a tester and why is it used in this analysis?

6. What does the column marked "progeny phenotypes" represent? In class 1, for example, state exactly what "gre sen pla" means.

7. What does the line marked "gametes" represent, and how is this different from the column marked "F_1 gametes"? In what way is comparison of these two types of gametes relevant to recombination?

8. Which meiosis is the main focus of study? Label it on your drawing.

9. Why are the gametes from the tester not shown?

10. Why are there only eight phenotypic classes? Are there any classes missing?

11. What classes (and in what proportions) would be expected if all the genes are on separate chromosomes?

12. What do the four class sizes (two very big, two intermediate, two intermediate, two very small) correspond to?

13. What can you tell about gene order from inspecting the phenotypic classes and their frequencies?

14. What will be the expected phenotypic class distribution if only two genes are linked?

15. What does the word "point" refer to in a three-point testcross? Does this word useage imply linkage? What would a four-point testcross be like?

16. What is the definition of *recombinant*, and how is it applied here?

17. What do the "recombinant for" columns mean?

18. Why are there only three "recombinant for" columns?

19. What do the R's mean, and how are they determined?

20. What do the column totals signify? How are they used?

21. What is the diagnostic test for linkage?

22. What is a map unit? Is it the same as a centimorgan?

23. In a three-point testcross such as this one, why aren't the F_1 and the tester considered to be parental in calculating recombination? (They *are* parents in one sense.)

24. What is the formula for interference? How are the "expected" frequencies calculated in the coefficient of coincidence formula?

25. Why does part c of the problem say "if appropriate"?

26. How much work is it to obtain such a large progeny size in corn? Approximately how many progeny are represented by one corn cob?

11. An individual heterozygous for four genes, $A/a \cdot B/b \cdot C/c \cdot D/d$, is testcrossed to $a/a \cdot b/b \cdot c/c \cdot d/d$, and 1000 progeny are classified by the gametic contribution of the heterozygous parent as follows:

$a \cdot B \cdot C \cdot D$	42
$A \cdot b \cdot c \cdot d$	43
$A \cdot B \cdot C \cdot d$	140
$a \cdot b \cdot c \cdot D$	145
$a \cdot B \cdot c \cdot D$	6
$A \cdot b \cdot C \cdot d$	9
$A \cdot B \cdot c \cdot d$	305
$a \cdot b \cdot C \cdot D$	310

   a. Which genes are linked?

   b. If two pure-breeding lines had been crossed to produce the heterozygous individual, what would their genotypes have been?

   c. Draw a linkage map of the linked genes, showing the order and the distances in map units.

   d. Calculate an interference value, if appropriate.

12. The squirting cucumber, *Echballium elaterium,* has two separate sexes (it is dioecious). The sexes are determined, not by heteromorphic sex chromosomes, but by the alleles of two genes. The alleles at the two loci govern sexual phenotypes as follows: *M* determines male fertility; *m* determines male sterility; *F* determines female sterility; *f* determines female fertility. In populations of this plant, individuals can be male (approximately 50 percent) or female (approximately 50 percent). In addition, a hermaphroditic type is found, but only at a very low frequency. The hermaphrodite has male and female sex organs on the same plant.

   a. What must be the full genotype of a male plant? (Indicate linkage relations of the genes.)

   b. What must be the full genotype of a female plant? (Indicate linkage relations of the genes.)

   c. How does the population maintain an approximately equal proportion of males and females?

   d. What is the origin of the rare hermaphrodite?

   e. Why are hermaphrodites rare?

*13. There is an autosomal gene *N* in humans that causes abnormalities in nails and patellae (kneecaps) called the *nail-patella syndrome.* Consider marriages in which one partner has the nail-patella syndrome and blood type A and the other partner has normal nail-patella and blood type O. These marriages produce some children who have both the nail-patella syndrome and blood type A. Assume that unrelated children from this phenotypic group mature, intermarry, and have children. Four phenotypes are observed in the following percentages in this second generation:

nail-patella syndrome, blood type A	66%
normal nail-patella, blood type O	16%
normal nail-patella, blood type A	9%
nail-patella syndrome, blood type O	9%

   Fully analyze these data, explaining the relative frequencies of the four phenotypes.

*14. Using the data obtained by Bateson and Punnett (Table 5-1), calculate the map distance (in m.u.) separating the color and shape genes.

15. You have a *Drosophila* line that is homozygous for autosomal recessive alleles *a*, *b*, and *c*, linked in that order. You cross females of this line with males homozygous for the corresponding wild-type alleles. You then cross the F_1 heterozygous males with their heterozygous sisters. You obtain the following F_2 phenotypes (where letters denote recessive phenotypes and pluses denote wild-type phenotypes): 1364 + + +, 365 *a b c*, 87 *a b* +, 84 + + *c*, 47 *a* + +, 44 + *b c*, 5 *a* + *c*, and 4 + *b* +.

   *a. What is the recombinant frequency between *a* and *b*? Between *b* and *c*?

   b. What is the coefficient of coincidence?

16. R. A. Emerson crossed two different pure-breeding lines of corn and obtained a phenotypically wild-type F_1 that was heterozygous for three alleles that determine recessive phenotypes: *an* determines anther; *br*, brachytic; and *f*, fine. He testcrossed the F_1 to a tester that was homozygous recessive for the three genes and obtained these progeny phenotypes: 355 anther; 339 brachytic, fine; 88 completely wild type; 55 anther, brachytic, fine; 21 fine; 17 anther, brachytic; 2 brachytic; 2 anther, fine.

   a. What were the genotypes of the parental lines?

   b. Draw a linkage map for the three genes (include map distances).

   c. Calculate the interference value.

17. Chromosome 3 of corn carries three loci having alleles *b* and b^+, *v* and v^+, and *lg* and lg^+. The corresponding recessive phenotypes are abbreviated *b* (for plant-color booster), *v* (for virescent), and *lg* (for liguleless); pluses denote wild-type phenotypes. A testcross of triple recessives with F_1 plants heterozygous for the three genes yields progeny having the following genotypes: 305 + *v lg*, 275 *b* + +, 128 *b* + *lg*, 112 + *v* +, 74 + + *lg*, 66 *b v* +, 22 + + +, and 18 *b v lg*. Give the gene sequence on the chromosome, the map distances between genes, and the coefficient of coincidence.

18. Groodies are useful (but fictional) haploid organisms that are pure genetic tools. A wild-type groody has a fat body, a long tail, and flagella. Mutant lines are known that have thin bodies, or are tailless, or do not have flagella. Groodies can mate with each other (although they are so shy that we do not know how) and produce recombinants. A wild-type groody mates with a thin-bodied groody lacking both tail and flagella. The 1000 baby groodies produced are classified as shown in the following illustration. Assign genotypes, and map the three genes. (Problem 18 from Burton S. Guttman.)

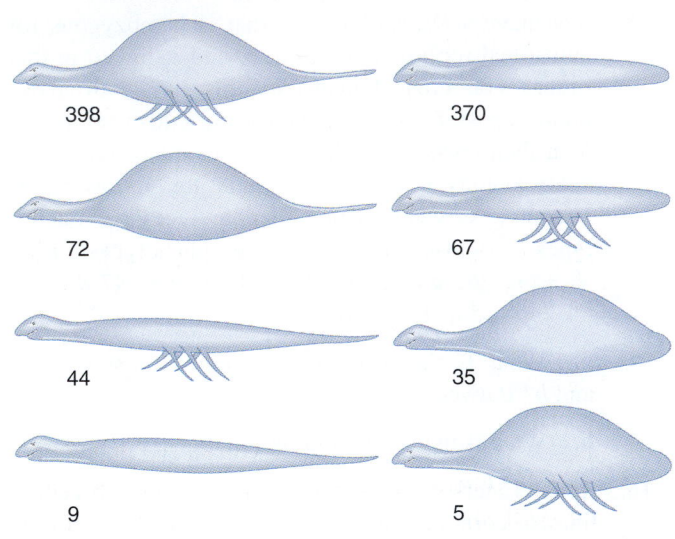

398	370
72	67
44	35
9	5

19. Assume that three pairs of alleles are found in *Drosophila*: x^+ and x, y^+ and y, and z^+ and z. As shown by the symbols, each non-wild-type allele is recessive to its wild-type allele. A cross between females heterozygous at these three loci and wild-type males yields progeny having the following genotypes: 1010 $x^+ \cdot y^+ \cdot z^+$ females, 430 $x \cdot y^+ \cdot z$ males, 441 $x^+ \cdot y \cdot z^+$ males, 39 $x \cdot y \cdot z$ males, 32 $x^+ \cdot y^+ \cdot z$ males, 30 $x^+ \cdot y^+ \cdot z^+$ males, 27 $x \cdot y \cdot z^+$ males, 1 $x^+ \cdot y \cdot z$ male, and 0 $x \cdot y \cdot z^+$ males.

a. In what chromosome of *Drosophila* are the genes carried?

b. Draw the relevant chromosomes in the heterozygous female parent, showing the arrangement of the alleles.

c. Calculate the map distances between the genes and the coefficient of coincidence.

20. From the five sets of data given in the following table determine the order of genes by inspection — that is, without calculating recombination values. Recessive phenotypes are symbolized by lowercase letters and dominant phenotypes by pluses.

Phenotypes observed in 3-point testcross	DATA SET				
	1	2	3	4	5
+ + +	317	1	30	40	305
+ + c	58	4	6	232	0
+ b +	10	31	339	84	28
+ b c	2	77	137	201	107
a + +	0	77	142	194	124
a + c	21	31	291	77	30
a b +	72	4	3	235	1
a b c	203	1	34	46	265

21. From the phenotype data given in the following table for two 3-point testcrosses for (1) *a*, *b*, and *c* and (2) *b*, *c*, and *d*, determine the sequence of the four genes *a*, *b*, *c*, and *d*, and the three map distances between them. Recessive phenotypes are symbolized by lowercase letters and dominant phenotypes by pluses.

1		2	
+ + +	669	b c d	8
a b +	139	b + +	441
a + +	3	b + d	90
+ + c	121	+ c d	376
+ b c	2	+ + +	14
a + c	2280	+ + d	153
a b c	653	+ c +	65
+ b +	2215	b c +	141

22. In *Drosophila*, the allele dp^+ determines long wings and *dp* determines short ("dumpy") wings. At a separate locus, e^+ determines gray body and *e* determines ebony body. Both loci are autosomal. The following crosses were made, starting with pure-breeding parents:

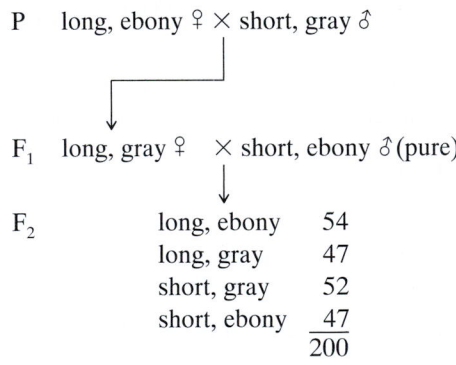

P   long, ebony ♀ × short, gray ♂

F₁   long, gray ♀   × short, ebony ♂ (pure)

F₂
long, ebony	54
long, gray	47
short, gray	52
short, ebony	47
	200

Use the χ^2 test to determine if these loci are linked. In doing so, state (a) hypothesis, (b) calculation of χ^2, (c) *p* value, (d) what the *p* value means, (e) conclusion, (f) inferred chromosomal constitutions of parents, F₁, tester, and progeny.

23. Two strains of haploid yeast each carrying two mutant alleles were crossed to determine gene linkage. Strain A had *asp* and *gal*, which caused it to require aspartate and be unable to utilize galactose; strain B had *rad* and

aro, which caused it to be sensitive to radiation and to require aromatic amino acids. Haploid progeny were obtained and tested for genotype with the following results (pluses denote wild-type alleles):

Genotype	Frequency
$asp \cdot gal \cdot + \cdot +$	0.136
$asp \cdot + \cdot + \cdot rad$	0.136
$asp \cdot gal \cdot + \cdot rad$	0.064
$asp \cdot + \cdot + \cdot +$	0.064
$+ \cdot gal \cdot aro \cdot +$	0.136
$+ \cdot + \cdot aro \cdot rad$	0.136
$+ \cdot gal \cdot aro \cdot rad$	0.064
$+ \cdot + \cdot aro \cdot +$	0.064
$asp \cdot gal \cdot aro \cdot +$	0.034
$asp \cdot + \cdot aro \cdot rad$	0.034
$asp \cdot gal \cdot aro \cdot rad$	0.016
$asp \cdot + \cdot aro \cdot +$	0.016
$+ \cdot gal \cdot + \cdot +$	0.034
$+ \cdot + \cdot + \cdot rad$	0.034
$+ \cdot gal \cdot + \cdot rad$	0.016
$+ \cdot + \cdot + \cdot +$	0.016
	1.000

a. Calculate the six recombinant frequencies.

b. Draw a linkage map to illustrate positions of the four genetic loci and label in map units.

24. A geneticist puts a female mouse from a line that breeds true for wild-type eye and body color with a male mouse from a pure line having apricot eyes (determined by allele *a*) and gray coat (determined by allele *g*). The mice mate and produce an F_1 that is all wild type. The F_1 mice intermate and produce an F_2 of the following composition:

Females	all wild type	
Males	wild type	45%
	apricot, gray	45%
	gray	5%
	apricot	5%

a. Explain these frequencies.

b. Give the genotypes of the parents and of both sexes of the F_1 and F_2.

25. The mother of a family with 10 children has blood type Rh$^+$. She also has a very rare condition (elliptocytosis, phenotype E) that causes red blood cells to be oval rather than round in shape but that produces no adverse clinical effects. The father is Rh$^-$ (lacks the Rh$^+$ antigen) and has normal red cells (phenotype e). The children are 1 Rh$^+$ e, 4 Rh$^+$ E, and 5 Rh$^-$ e. Information is available on the mother's parents, who

are Rh$^+$ E and Rh$^-$ e. One of the 10 children (who is Rh$^+$ E) marries someone who is Rh$^+$ e, and they have an Rh$^+$ E child.

a. Draw the pedigree of this whole family.

b. Is the pedigree in agreement with the hypothesis that the *Rh$^+$* allele is dominant and *Rh$^-$* is recessive?

c. What is the mechanism of transmission of elliptocytosis?

*__d.__ Could the genes governing the E and Rh phenotypes be on the same chromosome? If so, estimate the map distance between them, and comment on your result.

26. The father of Mr. Spock, first officer of the starship *Enterprise,* came from planet Vulcan; Spock's mother came from Earth. A Vulcan has pointed ears (determined by allele *P*), adrenals absent (determined by *A*), and a right-sided heart (determined by *R*). All these alleles are dominant to normal Earth alleles. The three loci are autosomal, and they are linked as shown in this linkage map:

```
P                      A              R
├──────────────────────┼──────────────┤
 ←──── 15 m.u. ────→ ←── 20 m.u. ──→
```

If Mr. Spock marries an Earth woman and there is no (genetic) interference, what proportion of their children will have

a. Vulcan phenotypes for all three characters?

b. Earth phenotypes for all three characters?

c. Vulcan ears and heart but Earth adrenals?

d. Vulcan ears but Earth heart and adrenals?

(Problem 26 from D. Harrison, *Problems in Genetics.* Addison-Wesley, 1970.)

27. In a certain diploid plant, the three loci *A*, *B*, and *C* are linked as follows:

```
A                      B              C
├──────────────────────┼──────────────┤
 ←──── 20 m.u. ────→ ←── 30 m.u. ──→
```

One plant is available to you (call it parental plant). It has the constitution *A b c/a B C.*

a. Assuming no interference, if the plant is selfed, what proportion of the progeny will be of the genotype *a b c/a b c?*

b. Again assuming no interference, if the parental plant is crossed with the *a b c/a b c* plant, what genotypic classes will be found in the progeny? What will be their frequencies if there are 1000 progeny?

*c.   Repeat part b, this time assuming 20 percent interference between the regions.

28.   From several crosses of the general type $A/A \cdot B/B \times a/a \cdot b/b$ the F_1 individuals of type $A/a \cdot B/b$ were testcrossed to $a/a \cdot b/b$. The results are as follows:

Testcross of	TESTCROSS PROGENY			
F_1 from cross	$A/a \cdot B/b$	$a/a \cdot b/b$	$A/a \cdot b/b$	$a/a \cdot B/b$
1	310	315	287	288
2	36	38	23	23
3	360	380	230	230
4	74	72	50	44

For each set of progeny, use the χ^2 test to decide if there is evidence of linkage.

29.   Certain varieties of flax show different resistances to specific races of the fungus called *flax rust*. For example, the flax variety 77OB is resistant to rust race 24 but susceptible to rust race 22, whereas flax variety Bombay is resistant to rust race 22 and susceptible to rust race 24. When 77OB and Bombay were crossed, the F_1 hybrid was resistant to both rust races. When selfed, the F_1 produced an F_2 containing the phenotypic proportions shown here, where R stands for resistant and S stands for susceptible.

		RUST RACE 22	
		R	S
RUST	R	184	63
RACE			
24	S	58	15

a.   Propose a hypothesis to account for the genetic basis of resistance in flax to these particular rust races. Make a concise statement of the hypothesis, and define any gene symbols that you use. Show your proposed genotypes of the 77OB, Bombay, F_1, and F_2 flax plants.

b.   Test your hypothesis by using the χ^2 test. Give the expected values, the value of χ^2 (to two decimal places), and the appropriate probability value. Explain exactly what this value means. Do you accept or reject your hypothesis on the basis of the χ^2 test?

(Problem 29 adapted from M. Strickberger, *Genetics,* Macmillan, 1968.)

30.   In the two pedigrees diagrammed here, a vertical bar in a symbol stands for steroid sulfatase deficiency and a horizontal bar stands for ornithine transcarbamylase deficiency.
a.   Is there any evidence in these pedigrees that the genes determining the deficiencies are linked?

b.   If the genes are linked, is there any evidence in the pedigree of crossing-over between them?

c.   Draw genotypes of these individuals as far as possible.

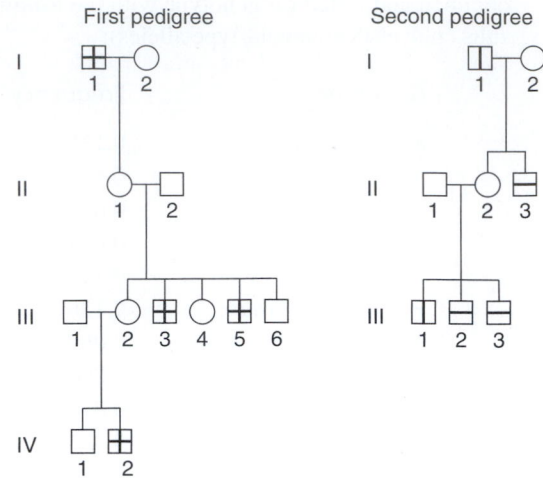

31.   The following pedigree shows a family with two rare abnormal phenotypes: blue sclerotic (a brittle bone defect), represented by a black-bordered symbol, and hemophilia, represented by a black center in a symbol. Individuals represented by completely black symbols have both disorders.

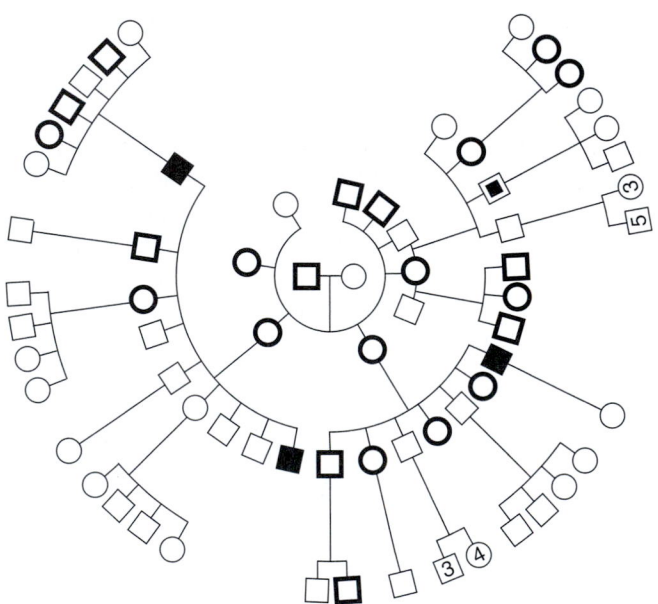

a.   What pattern of inheritance is shown by each condition in this pedigree?

b.   Provide the genotypes of as many family members as possible.

c.   Is there evidence of linkage?

d.   Is there evidence of independent assortment?

e.   Can any of the members be judged as recombinants (that is, formed from at least one recombinant gamete)?

32. In the following pedigree, the vertical lines stand for protan colorblindness, and the horizontal lines stand for deutan colorblindness. These are separate conditions causing different misperceptions of colors; each is determined by a separate gene.

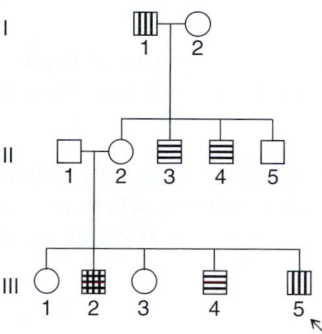

a. Does the pedigree show any evidence that the genes are linked?

b. If there is linkage, does the pedigree show any evidence of crossing-over? Explain both of your answers with the aid of the diagram.

c. Can you calculate a value for the recombination between these genes? Is this recombination by independent assortment or crossing-over?

*33. The human genes for colorblindness and for hemophilia are both on the X chromosome, and they show a recombinant frequency of about 10 percent. Linkage of a pathological gene to a relatively harmless one can be used for genetic prognosis. Here is part of a more extensive pedigree. Blackened symbols indicate that the subjects had hemophilia, and crosses indicate colorblindness. What information could be given to women III-4 and III-5 about the likelihood of their having sons with hemophilia? (Problem 33 adapted from J. F. Crow, *Genetics Notes: An Introduction to Genetics*, Burgess, 1983.)

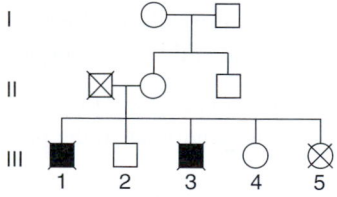

34. A geneticist mapping the genes A, B, C, D, and E makes two 3-point testcrosses. The first cross of pure lines is

$$A/A \cdot B/B \cdot C/C \cdot D/D \cdot E/E$$
$$\times a/a \cdot b/b \cdot C/C \cdot d/d \cdot E/E$$

The geneticist crosses the F_1 with a recessive tester and classifies the progeny by the gametic contribution of the F_1:

$A \cdot B \cdot C \cdot D \cdot E$	316
$a \cdot b \cdot C \cdot d \cdot E$	314
$A \cdot B \cdot C \cdot d \cdot E$	31
$a \cdot b \cdot C \cdot D \cdot E$	39
$A \cdot b \cdot C \cdot d \cdot E$	130
$a \cdot B \cdot C \cdot D \cdot E$	140
$A \cdot b \cdot C \cdot D \cdot E$	17
$a \cdot B \cdot C \cdot d \cdot E$	13
	1000

The second cross of pure lines is

$$A/A \cdot B/B \cdot C/C \cdot D/D \cdot E/E$$
$$\times a/a \cdot B/B \cdot c/c \cdot D/D \cdot e/e$$

The geneticist crosses the F_1 from this cross with a recessive tester and obtains:

$A \cdot B \cdot C \cdot D \cdot E$	243
$a \cdot B \cdot c \cdot D \cdot e$	237
$A \cdot B \cdot c \cdot D \cdot e$	62
$a \cdot B \cdot C \cdot D \cdot E$	58
$A \cdot B \cdot C \cdot D \cdot e$	155
$a \cdot B \cdot c \cdot D \cdot E$	165
$a \cdot B \cdot C \cdot D \cdot e$	46
$A \cdot B \cdot c \cdot D \cdot E$	34
	1000

The geneticist also knows that genes D and E assort independently.

a. Draw a map of these genes, showing distances in map units wherever possible.

b. Is there any evidence of interference?

35. In the tiny crucifer *Arabidopsis thaliana* a new phenotype "glabrous" (lacking hairs) was discovered, inherited as a recessive (g/g). It was suspected that the G locus was on chromosome 4, so a g/g plant was crossed to a plant of genotype $y\ y$ (yellow, caused by reduced amounts of chlorophyll), whose locus is known to be on chromosome 4. The F_1 was wild type in appearance; it was selfed and the following F_2 was obtained:

wild type	514
yellow	237
glabrous	234
yellow, glabrous	15
Total	1000

a. Is the G locus on chromosome 4?

b. If so, estimate how many map units it is from the Y locus.

c. Explain the frequencies of the yellow and the glabrous F_2 classes.

36. In the plant *Arabidopsis,* the loci for pod length (*L*, long; *l*, short) and fruit hairs (*H*, hairy; *h*, smooth) are linked 16 map units apart on the same chromosome. The following crosses were made:

    (i) $L\,H/L\,H \times l\,h/l\,h \longrightarrow F_1$
    (ii) $L\,h/L\,h \times l\,H/l\,H \longrightarrow F_1$

    If the F_1's from (i) and (ii) are crossed,

    a. What proportion of the progeny is expected to be *l h/l h*?

    b. What proportion of the progeny is expected to be *L h/l h*?

37. In corn, a triple heterozygote was obtained carrying the mutant alleles *s* (shrunken), *w* (white aleurone), and *y* (waxy endosperm), all paired with their normal wild-type alleles. This triple heterozygote was test-crossed, and the progeny contained 116 shrunken, white; 4 fully wild type; 2538 shrunken; 601 shrunken, waxy; 626 white; 2708 white, waxy; 2 shrunken, white, waxy; and 113 waxy.

    a. Determine if any of these three loci are linked, and, if so, show map distances.

    b. Show the allele arrangement on the chromosomes of the triple heterozygote used in the testcross.

    c. Calculate interference, if appropriate.

38. The *A* and *B* loci are linked 10 map units apart. Prove that, when a repulsion dihybrid (*A b/a B*) is selfed, the *z* value will be 0.002, as shown in Table 5-2.

39. **a.** A mouse cross is made $A/a \cdot B/b \times a/a \cdot b/b$ and in the progeny there are

    25% $A/a \cdot B/b$, 25% $a/a \cdot b/b$,

    25% $A/a \cdot b/b$, 25% $a/a \cdot B/b$

    Explain these proportions with the aid of simplified meiosis diagrams.

    b. A mouse cross is made $C/c \cdot D/d \times c/c \cdot d/d$ and in the progeny there are

    45% $C/c \cdot d/d$, 45% $c/c \cdot D/d$,

    5% $c/c \cdot d/d$, 5% $C/c \cdot D/d$

    Explain these proportions with the aid of simplified meiosis diagrams.

40. In the tiny plant *Arabidopsis,* the recessive allele *hyg* confers seed resistance to the drug hygromycin, and *her,* a recessive allele of a different gene, confers seed resistance to herbicide. A plant that was homozygous *hyg/hyg · her/her* was crossed to wild type, and the F_1 was selfed. Seeds resulting from the F_1 self were placed on petri dishes containing hygromycin and herbicide.

a. If the two genes are unlinked, what percentage of seeds is expected to grow?

b. In fact, 13 percent of the seeds grew. Does this percentage support the hypothesis of no linkage? Explain. If not, calculate the number of map units between the loci.

c. Under your hypothesis, if the F_1 is testcrossed, what proportion of seeds will grow on the medium containing hygromycin and herbicide?

41. In corn *(Zea mays),* the genetic map of part of chromosome 4 is as follows, where *w*, *s*, and *e* represent recessive mutant alleles affecting the color and shape of the pollen:

w	8 m.u.	*s*	14 m.u.	*e*

    If the following cross is made

    $$+ + +/+ + + \times w\,s\,e/w\,s\,e$$

    and the F_1 is testcrossed to *w s e/w s e,* and if it is assumed that there is no interference on this region of the chromosome, what proportion of progeny will be of genotype

    a. $+ + +$?

    b. *w s e*?

    c. $+ s e$?

    d. $w + +$?

    e. $+ + e$?

    f. $w s +$?

    g. $w + e$?

    h. $+ s +$?

42. In the mosquito, *Anopheles,* the loci for eye color (alleles *B*, black; *b*, white) and body length (*L*, long; *l*, short) are linked 20 map units apart. A cross is made of a dihybrid in coupling conformation to a dihybrid in repulsion conformation. What progeny phenotypes will be produced and in what proportions?

43. A plant of genotype $Q/q \cdot R/r \cdot T/t$ produces the following gamete genotypes in the proportions shown:

$Q \cdot R \cdot T$	$\frac{1}{8}$
$Q \cdot R \cdot t$	$\frac{1}{8}$
$q \cdot R \cdot T$	$\frac{1}{8}$
$Q \cdot r \cdot T$	$\frac{1}{8}$
$q \cdot r \cdot t$	$\frac{1}{8}$
$Q \cdot r \cdot t$	$\frac{1}{8}$
$q \cdot R \cdot t$	$\frac{1}{8}$
$q \cdot r \cdot T$	$\frac{1}{8}$

Draw labeled meiosis diagrams to clearly explain how these gametic proportions were produced. As part of your answer, show the chromosomes (1) before premeiotic S phase (replication phase); (2) after chromatid formation; (3) when paired; (4) at anaphase I; and (5) at anaphase II.

44. In beans, tall (*T*) is dominant to short (*t*), red flowers (*R*) are dominant to white (*r*), and wide leaves (*W*) are dominant to narrow (*w*). The following cross is made and progeny are obtained as shown:

Cross   tall, red, wide × short, white, narrow

Progeny   478 tall, white, wide
                  21 tall, red, wide
                  19 short, white, wide
                  482 short, red, wide

a. Explain why these progeny phenotypes were obtained and in the proportions observed (list all genotypes and show chromosomal positions).

b. Under your hypothesis, if the tall, red, wide parent is selfed, what will be the proportion of short, white, wide progeny?

45. In the pedigree in Figure 5-19, calculate the Lod score for a recombinant frequency of 34 percent.

46. In a diploid organism of genotype *A/a* ; *B/b* ; *D/d*, the allele pairs are all on different chromosome pairs. The following diagrams purport to show anaphases ("pulling apart" stages) in individual cells. A line represents a chromosome or a chromatid, and the dot indicates the position of the centromere. State whether each drawing represents mitosis, meiosis I, and meiosis II or is impossible for this particular genotype.

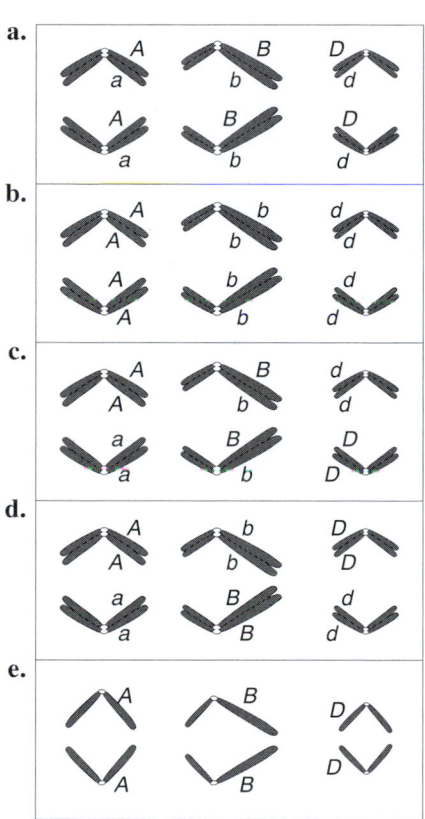

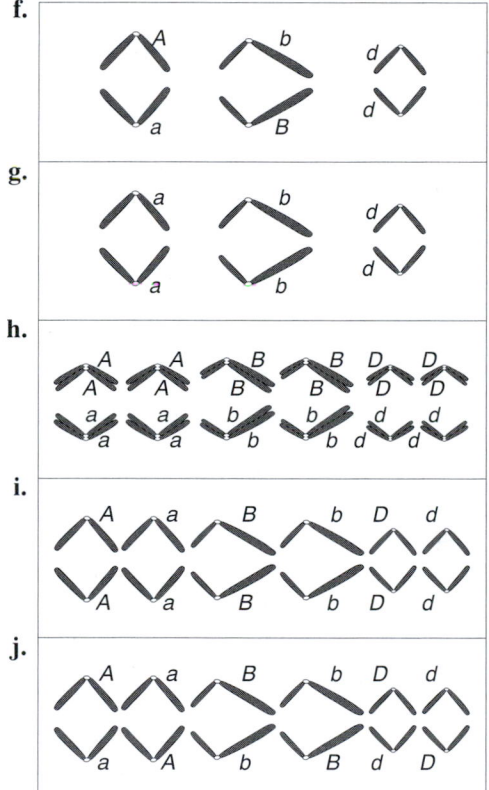

6

SPECIALIZED EUKARYOTIC CHROMOSOME MAPPING TECHNIQUES

Some of the color variants of the ascomycete fungus *Aspergillus*, normally dark green.
The alleles producing such color variants are useful markers in the study of mitotic recombination, one of the topics of this chapter. The variant colors are usually pigments that are intermediates in the biochemical reaction sequences terminating in the dark-green pigment. *(J. Peberdy, Department of Life Sciences, University of Nottingham, England.)*

Key Concepts

Double and higher multiple crossovers cause underestimates of map distances calculated from recombinant frequencies.

Special mapping formulas provide accurate map distances corrected for the effect of multiple crossovers.

In fungi, analysis of individual meioses can position centromeres on genetic maps.

Analysis of individual meioses is a source of insight into the mechanisms of gene segregation and recombination in meiosis.

Occasionally, chromosomes cross over in diploid cells undergoing mitosis.

The DNA of cloned genes binds to specific chromosomal locations, revealing their map position.

Human genes can be mapped by using human–rodent cell hybrids.

In Chapter 5, we considered the basic method for mapping eukaryotic chromosomes. According to that method, the investigator measures the *recombinant frequency* in a *random sample* of products of *meiosis* and, from these measurements, determines the degree of linkage between the genes in question. In this chapter, we shall extend the analysis beyond these three key concepts: recombinant frequency, random products, and meiosis.

First, *recombinant frequency*. You will see that measuring recombinant frequency does not always give an accurate measurement of map distance, and sometimes corrections need to be made. Second, *random products*. Some organisms, because of the way in which they complete their life cycles, allow the investigator to study the products of individual meioses, allowing a different view of recombination that is not possible by using random meiotic products. Third, *meiosis*. It might come as a surprise to learn that genes can recombine at mitosis. This opens up some important mapping methodologies in those situations.

Mapping chromosomes is one of the central activities of geneticists. The goal of genetics, after all, is to understand the structure, function, and evolution of the genome, and, clearly, knowing the locations of genes is central to this task. The concepts in this chapter are all part of the everyday vocabulary of genetics. In the same way that earlier we extended Mendel's basic rules of segregation, now we have to extend the basic rules of mapping.

Accurate calculation of large map distances

In Chapter 5, we learned that the basic genetic method of measuring map distance is based on recombinant frequency (RF). A genetic map unit (m.u.) was defined as a recombinant frequency of 1 percent. This is a useful fundamental unit that has stood the test of time and is still used in genetics. However, the larger the recombinant frequency, the less accurate it is as a measure of map distance. In fact, map units calculated from larger recombinant frequencies are smaller than map units calculated from smaller recombinant frequencies. We encountered this effect in examples in Chapter 5. Typically, when measuring recombination between three linked loci, the sum of the two internal recombinant frequencies is greater than the recombinant frequency between the outside loci. With the use of such data, what is the most accurate estimate that can be made of map distance between the two outside loci *A* and *C* in the following diagram?

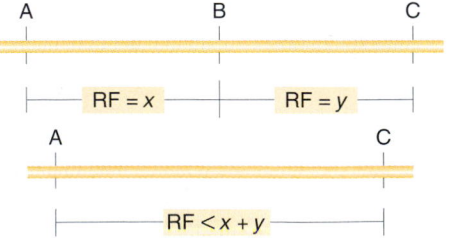

The answer is that $x + y$ is the best estimate, and more accurate than the smaller overall $A - C$ value. This gives us the following useful mapping principle:

MESSAGE ···
The best estimates of map distance are obtained from the sum of the distances calculated for shorter subintervals.
···

However, what if we have no intervening marker loci available to measure recombination in shorter intervals? Such a situation would be commonly encountered when beginning to map a new experimental organism or in cases in which the genome is huge, as it is in human beings. For example, in the preceding diagram, what if there were no known *B* locus? Would we have to make do with the map distance value obtained directly from the $A - C$ recombinant frequency? Furthermore, what about the shorter intervals themselves? If there were other loci between *A* and *B* and between *B* and *C*, then we might obtain even better estimates of the *A* to *C* distance. Luckily, there is a way of taking any recombinant frequency and performing a calculation to make it a more accurate measure of map distance, without studying shorter and shorter intervals.

Before we consider the calculation, let's think about the reason why larger RF values are less accurate measures of map distance. We have already encountered the culprit: multiple crossovers. In Chapter 5, we learned that double crossovers often lead to a parental arrangement of alleles and therefore the resulting meiotic products are not counted when measuring recombinant frequency. The same is true for other types of multiple crossovers: triples, quadruples, and so forth. So it is easy to see that multiple crossovers automatically lead to an underestimate of map distance, and, because the multiples are expected to be relatively more common over longer regions, we can see why the problem is worse for larger recombinant frequencies.

How can we take these multiple crossovers into account when calculating the map distances? What we need is a mathematical function that accurately relates recombination to map distance. In other words, what we need is a **mapping function.**

MESSAGE ···
A mapping function is a formula for using recombinant frequencies to calculate map distances corrected for multiple crossover products.
···

Poisson distribution

To derive a mapping function, we need a mathematical tool widely used in genetic analysis because it is useful in describing many different types of genetic processes. This mathematical tool is the **Poisson distribution.** A distribution is merely a description of the frequencies of the different types of classes that arise from sampling. The Poisson distribution describes the frequency of classes containing 0, 1, 2, 3, 4, . . . , i items when the average number of items per sample is known. The Poisson distribution is particularly useful when the average is small in relation to the total number of items possible. For example, the possible number of tadpoles obtainable in a single dip of a net in a pond is quite large, but most dips yield only one or two or none. The number of dead birds on the side of a highway is potentially very large, but in a sample mile the number is usually small. Such samplings are described well by the Poisson distribution.

Let's consider a numerical example. Suppose that we randomly distribute 100 one-dollar bills to 100 students in a lecture room, perhaps by scattering them over the class from some point near the ceiling. The average (or mean) number of bills per student is 1.0, but common sense tells us that it is very unlikely that each of the 100 students will capture one bill. We would expect a few lucky students to grab three or four bills each and quite a few students to come up with two bills each. However, we would expect most students to get either one bill or none. The Poisson distribution provides a quantitative prediction of the results.

In this example, the item being considered is the capture of a bill by a student. We want to divide the students into classes according to the number of bills each captures and then find the frequency of each class. Let m represent the mean number of items (here, $m = 1.0$ bill per student). Let i represent the number for a particular class (say, $i = 3$ for those students who get three bills each). Let $f(i)$ represent the frequency of the i class — that is, the proportion of the 100 students who each capture i bills. The general expression for the Poisson distribution states that

$$f(i) = \frac{e^{-m} m^i}{i!}$$

where e is the base of natural logarithms (e is approximately 2.7) and ! is the factorial symbol. As examples, $3! = 3 \times 2 \times 1 = 6$ and $4! = 4 \times 3 \times 2 \times 1 = 24$. By definition, $0! = 1$. When computing $f(0)$, recall that any number raised to the power of 0 is defined as 1. Table 6-1 gives values of e^{-m} for m values from 0.000 to 1.000. Values for m greater than 1 can be obtained by calculation.

In our example, $m = 1.0$. Using Table 6-1, we compute the frequencies of the classes of students capturing 0, 1, 2, 3, and 4 bills as follows:

$$f(0) = \frac{e^{-1} 1^0}{0!} = \frac{e^{-1}}{1} = 0.368$$

$$f(1) = \frac{e^{-1} 1^1}{1!} = \frac{e^{-1}}{1} = 0.368$$

$$f(2) = \frac{e^{-1} 1^2}{2!} = \frac{e^{-1}}{2 \times 1} = \frac{e^{-1}}{2} = 0.184$$

$$f(3) = \frac{e^{-1} 1^3}{3!} = \frac{e^{-1}}{3 \times 2 \times 1} = \frac{e^{-1}}{6} = 0.061$$

$$f(4) = \frac{e^{-1} 1^4}{4!} = \frac{e^{-1}}{4 \times 3 \times 2 \times 1} = \frac{e^{-1}}{24} = 0.015$$

Figure 6-1 is a histogram of this distribution. We predict that about 37 students will capture no bills, about 37 will capture one bill, about 18 will capture two bills, about 6 will capture three bills, and about 2 will capture four bills. This accounts for all 100 students; in fact, you can verify that the Poisson distribution yields $f(5) = 0.003$, which makes it likely that no student in this sample of 100 will capture five bills.

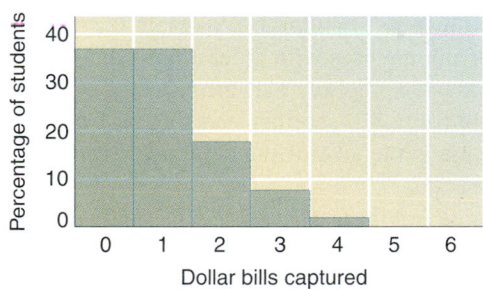

Figure 6-1   Poisson distribution for a mean of 1.0, illustrated by a random distribution of dollar bills to students.

6-1 TABLE	Values of e^{-m} for m Values of 0 to 1*		
m	e^{-m}	m	e^{-m}
0.000	1.00000	0.550	0.57695
0.050	0.95123	0.600	0.54881
0.100	0.90484	0.650	0.52205
0.150	0.86071	0.700	0.49659
0.200	0.81873	0.750	0.47237
0.250	0.77880	0.800	0.44933
0.300	0.74082	0.850	0.42741
0.350	0.70469	0.900	0.40657
0.400	0.67032	0.950	0.38674
0.450	0.63763	1.000	0.36788
0.500	0.60653		

*Values for m greater than 1 can be obtained from an electronic calculator or by using logarithms.

Source: F. James Rohlf and Robert R. Sokal, *Statistical Tables*, 3d ed. W. H. Freeman and Company, 1995.

Similar distributions may be developed for other m values. Some are shown in Figure 6-2 as curves instead of bar histograms.

Derivation of a mapping function

The Poisson distribution can also describe the distribution of crossovers along a chromosome in meiosis. In any chromosomal region, the actual number of crossovers is probably small in relation to the total number of possible

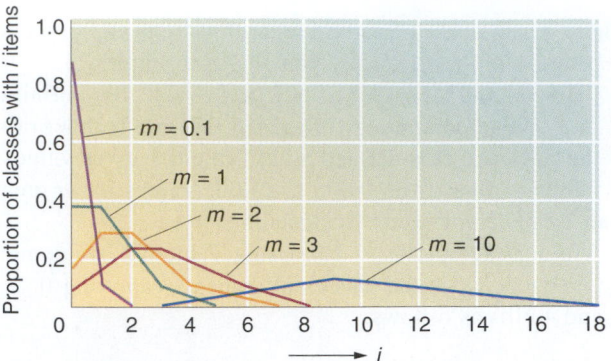

Figure 6-2   Poisson distributions for five different mean values: m is the mean number of items per sample, and i is the actual number of items per sample. (After R. R. Sokal and F. J. Rohlf, *Introduction to Biostatistics*. W. H. Freeman and Company, 1973.)

crossovers in that region. If crossovers are distributed randomly (that is, there is no interference), then, if we knew the mean number of crossovers in the region per meiosis, we could calculate the distribution of meioses with zero, one, two, three, four, and more multiple crossovers. This calculation is unnecessary in the present context because, as we shall see, the only class that is really crucial is the zero class. We want to correlate map distances with observable RF values. Meioses in which there are one, two, three, four, or any finite number of crossovers per meiosis all behave similarly in that they produce an RF of 50 percent among the products of those meioses, whereas the meioses with no crossovers produce an RF of 0 percent. To see how this can be so, consider a series of meioses in which nonsister chromatids do not cross over, cross over once, and cross over twice, as shown in Figure 6-3. We obtain recombinant products only from meioses with at least one crossover in the region, and always precisely half the products of such meioses are recombinant. We see then that the real determinant of the RF value is the size of the zero crossover class in relation to the rest.

As noted in Figure 6-3, we consider only crossovers between nonsister chromatids; sister-chromatid exchange is thought to be rare at meiosis. If it occurs, it can be shown to have no net effect in most meiotic analyses.

At last, we can derive the mapping function. Recombinants make up half the products of those meioses having at least one crossover in the region. The proportion of meioses with at least one crossover is 1 minus the fraction with zero crossovers. The zero-class frequency will be:

$$\frac{e^{-m}m^0}{0!}$$

which equals

$$e^{-m}$$

So the mapping function can be stated as

$$\text{RF} = \tfrac{1}{2}(1 - e^{-m})$$

This formula relates recombinant frequency to m, the mean number of crossovers. Because the whole concept of genetic mapping is based on the occurrence of crossovers, as well as proportionality between crossover frequency and the physical size of a chromosomal region, you can see that m is probably the most fundamental variable in the whole process. In fact, m could be considered to be the ultimate genetic mapping unit.

If we know an RF value, we can calculate m by solving the equation. After obtaining many values of m, we can plot the function as a graph, as in Figure 6-4. Viewing the function plotted as a graph should help us see how it works. First, notice that the function is linear for a certain range

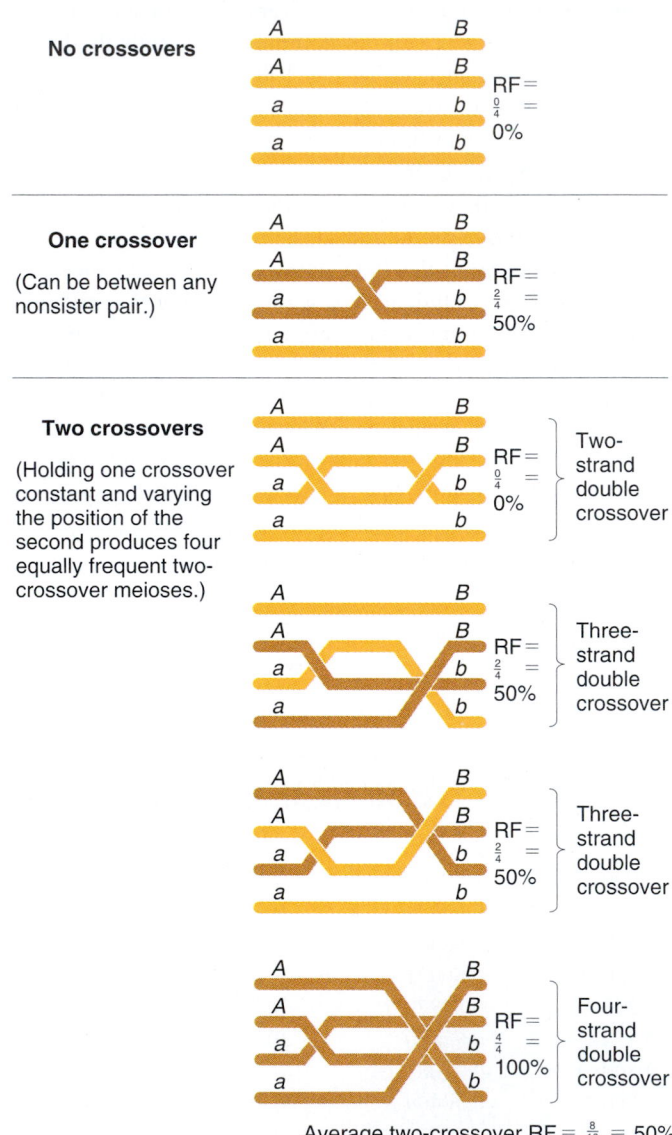

Figure 6-3   Demonstration that the average RF is 50 percent for meioses in which the number of crossovers is not zero. Recombinant chromatids are brown. Two-strand double crossovers produce all parental types, so all the chromatids are orange. Note that all crossovers are between nonsister chromatids. Try the triple crossover class yourself.

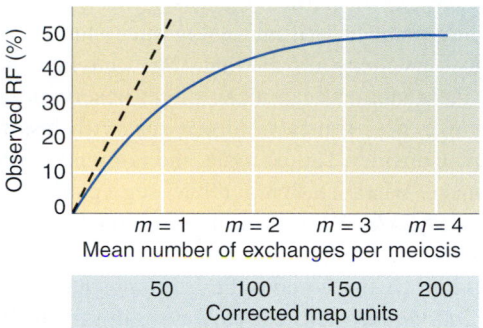

Figure 6-4   The blue line gives the mapping function in graphic form. Where the blue curve and the dashed line coincide, the function is linear and RF's estimate correct map distances well.

corresponding to very small m values. (Remember that m is our best measure of genetic distance.) Therefore, RF is a good measure of distance where the dashed line coincides with the function in Figure 6-4. In this region, the map unit defined as 1 percent RF has real meaning. Therefore, let's use this region of the curve to define corrected map units by considering some small values of m:

For $m = 0.05$, $e^{-m} = 0.95$, and
$$\text{RF} = \tfrac{1}{2}(1 - 0.95) = \tfrac{1}{2}(0.05) = \tfrac{1}{2} \times m$$

For $m = 0.10$, $e^{-m} = 0.90$, and
$$\text{RF} = \tfrac{1}{2}(1 - 0.90) = \tfrac{1}{2}(0.10) = \tfrac{1}{2} \times m$$

We see that $\text{RF} = \frac{m}{2}$, and this relation defines the dashed line in Figure 6-4. It allows us to translate m values into corrected map units. Expressing m as a percentage, we see that an m of 100 percent ($= 1$) is the equivalent of 50 corrected map units. Because an m value of 1 is the equivalent of 50 corrected map units, we can express the horizontal axis of Figure 6-4 in our new map units. Now we can see from the graph that two loci separated by 150 corrected map units show an RF of only 50 percent. We can use the graph of the function to convert any RF into map distance simply by drawing a horizontal line from the RF value to the curve and dropping a perpendicular to the map unit axis — a process equivalent to using the equation $\text{RF} = \tfrac{1}{2}(1 - e^{-m})$ to solve for m.

Let's consider a numerical example of the use of the mapping function. Suppose that we get an RF of 27.5 percent. How many corrected map units does this represent? From the function,

$$0.275 = \tfrac{1}{2}(1 - e^{-m})$$
$$0.55 = 1 - e^{-m}$$

Therefore

$$e^{-m} = 1 - 0.55 = 0.45$$

From e^{-m} tables (or by using a calculator), we find that $m = 0.8$, which is the equivalent of 40 corrected map units. If we had been happy to accept 27.5 percent RF as repre-

senting 27.5 map units, we would have considerably under-estimated the distance between the loci.

MESSAGE ···
To estimate map distances most accurately, put RF values through the mapping function. Alternatively, add distances that are each short enough to be in the region where the mapping function is linear.
···

A corollary of the second statement of this message is that for organisms for which the chromosomes are already well mapped, such as *Drosophila*, a geneticist seldom needs to calculate from the map function to place newly discovered genes on the map. This is because the map is already divided into small, marked regions by known loci. However, when the process of mapping has just begun in a new organism or when the available genetic markers are sparsely distributed, the corrections provided by the function are needed.

Notice that no matter how far apart two loci are on a chromosome, we never observe an RF value of greater than 50 percent. Consequently, an RF value of 50 percent would leave us in doubt about whether two loci are linked or are on separate chromosomes. Stated another way, as m gets larger, e^{-m} gets smaller and RF approaches $\tfrac{1}{2}(1 - 0) = \tfrac{1}{2} \times 1 = 0.5$, or 50 percent. This is an important point: RF values of 100 percent are not observed, no matter how far apart the loci are.

Analysis of single meioses

In Chapter 5, we encountered a type of organism in which the products of single meioses remain together as groups of four cells called a **tetrad**, a term based on the Greek word for "four." Tetrad analysis is possible only in some fungi and single-celled algae. In the mushroomlike fungi (Basidiomycetes), a single meiosis produces a group of four basidiospores — the spores that fall to the earth from the mushroom cap. In ascomycete fungi such as yeasts and molds, the products of a single meiosis are enclosed in a sac called an **ascus.** The spores in the ascus are called *ascospores*. The unicellular alga *Chlamydomonas* also has an ascuslike structure. Basidiospores, ascospores, and the equivalent algal spores are types of **sexual spores.**

Advantages of haploids for genetic analysis

First, let us take note of the fact that fungi and unicellular algae are microbial eukaryotes. This means that, just like the prokaryotic microbes (bacteria), they can be propagated easily and cheaply, take up little space, and can be studied in large numbers. These properties are useful in genetic research. Moreover, recall that fungi and algae are haploid. Haploidy has two basic advantages for genetic studies. First, because there is only one chromosome set, dominance and recessiveness normally do not obscure gene expression, and the phenotype is a direct manifestation of the genotype. Second, in addition to the usefulness of tetrad analysis, even random meiotic product analysis is easier than in diploids. Recall that in the haploid life cycle the two haploid parental cells unite to form the diploid meiocyte, so for the geneticist this is the only meiosis to worry about. In contrast, in diploids the analyst must consider meioses in both parents. As an example, let's look at how easy it is to study linkage in haploids by using random meiotic products. We can make a cross such as $a^+ \cdot b^+ \times a \cdot b$, and test a random sample of meiotic products. This approach is straightforward because each product develops directly into a haploid progeny individual. We might find the following genotypic frequencies in the progeny:

$$
\begin{array}{ll}
a^+ \cdot b^+ & 45\% \\
a \cdot b & 45\% \\
a^+ \cdot b & 5\% \\
a \cdot b^+ & 5\%
\end{array}
$$

From such data, determining which progeny are parental (the first two classes) and which are recombinant (the last two classes) is straightforward. Then it is simple to compute the RF value as 10 percent, indicating linkage of the two genes 10 map units apart.

It is interesting to note at this point that molecular technology has opened up the ability to analyze haploid products of meiosis in humans. As noted in Chapter 5, neutral variant sites in the DNA are detectable by using molecular probes, and "dihybrids" for two such sites can be used to study recombination between them. For example, sperm samples can be diluted so that there is one sperm per test tube. The DNA in one sperm is enough to enable routine screens to be made for neutral variant "alleles." To screen a specific site, the DNA at that site must first be amplified by using a special process called the *polymerase chain reaction* (see Chapter 12). Let's assume a man is heterozygous for sites that we will call *DNA-P* and *DNA-Q*. His genotype can be designated:

$$DNA\text{-}P' \cdot DNA\text{-}P'' \cdot DNA\text{-}Q' \cdot DNA\text{-}Q''$$

His sperm will be:

$$
\begin{array}{l}
DNA\text{-}P' \cdot DNA\text{-}Q' \\
DNA\text{-}P' \cdot DNA\text{-}Q'' \\
DNA\text{-}P'' \cdot DNA\text{-}Q' \\
DNA\text{-}P'' \cdot DNA\text{-}Q''
\end{array}
$$

and their relative frequencies will determine if the *P* and *Q* sites are linked. Hence, for some loci, humans can be analyzed with the same ease as haploids.

Benefits of analyzing individual meioses in genetics

As mentioned earlier, an important feature of some haploid species is that they produce tetrads, which permit the study of individual meioses. You might be wondering why the study of individual meioses is important; after all, we have just developed some of the basic analytical rules of genetics in preceding chapters without using such studies. The first answer to this question is that, by studying individual meioses, the geneticist can make direct observations on the behavior of genes in meiotic processes, with less need for inferences. Consider, for example, the inference that we implicitly make when studying allele segregation in a heterozygote, say A/a. When we observe equal numbers of A and a alleles in a random sample of gametes, we attribute it to segregation of the homologous chromosomes carrying A and a in *individual* meioses. However, this is an inference that cannot be confirmed directly in most eukaryotic organisms. Tetrad analysis shows directly that such segregation does take place. This kind of directness is useful in genetic research. In addition, tetrads permit several kinds of studies that are not possible by using conventional analysis. In Chapter 5, we encountered some of these studies, for example that assessing the distribution of crossovers between four chromatids during the crossover process. Another analysis not possible with random meiotic products is the mapping of centromeres in relation to other loci. Finally, observations of exceptional allele segregations in tetrad analysis gave rise to one of the central molecular models of crossing-over, the heteroduplex (or hybrid) DNA model; we shall consider this model in detail in Chapter 19.

But do fungi and algae have typical eukaryotic chromosomes with typical eukaryotic behavior? The answer is generally yes, so discoveries made in these model systems can be applied with reasonable confidence to plants and animals.

In summary, tetrads provide a direct method for the genetic analysis of individual meioses, and processes can be studied that are not accessible by analyzing random meiotic products. Meiosis is one of the central processes of eukaryotic biology. Although a great deal is known about meiosis, there is still much to be learned. In fact, it is curious that, despite the wealth of molecular knowledge that is available on how cells carry out their day-to-day affairs, the processes that were some of the first ever to be identified by geneticists (pairing, crossing-over, and segregation) remain among the most mysterious of cellular processes.

MESSAGE ··

Tetrad analysis makes it possible to study individual meioses, a useful ability for the investigation of recombination and assortment.

In some fungi and algae, the cells that represent the four products of meiosis develop directly into a tetrad of four sexual spores (Figure 6-5). In certain fungi, each of the four products of meiosis undergoes an additional mitotic division, yielding a group of eight cells called an **octad.** However, an octad is simply a double tetrad composed of four spore pairs. The members of a spore pair are identical, being mitotic daughter cells of one of the four products of meiosis.

The sexual spores, whether four or eight in number, can be found in a variety of arrangements. In some species, the

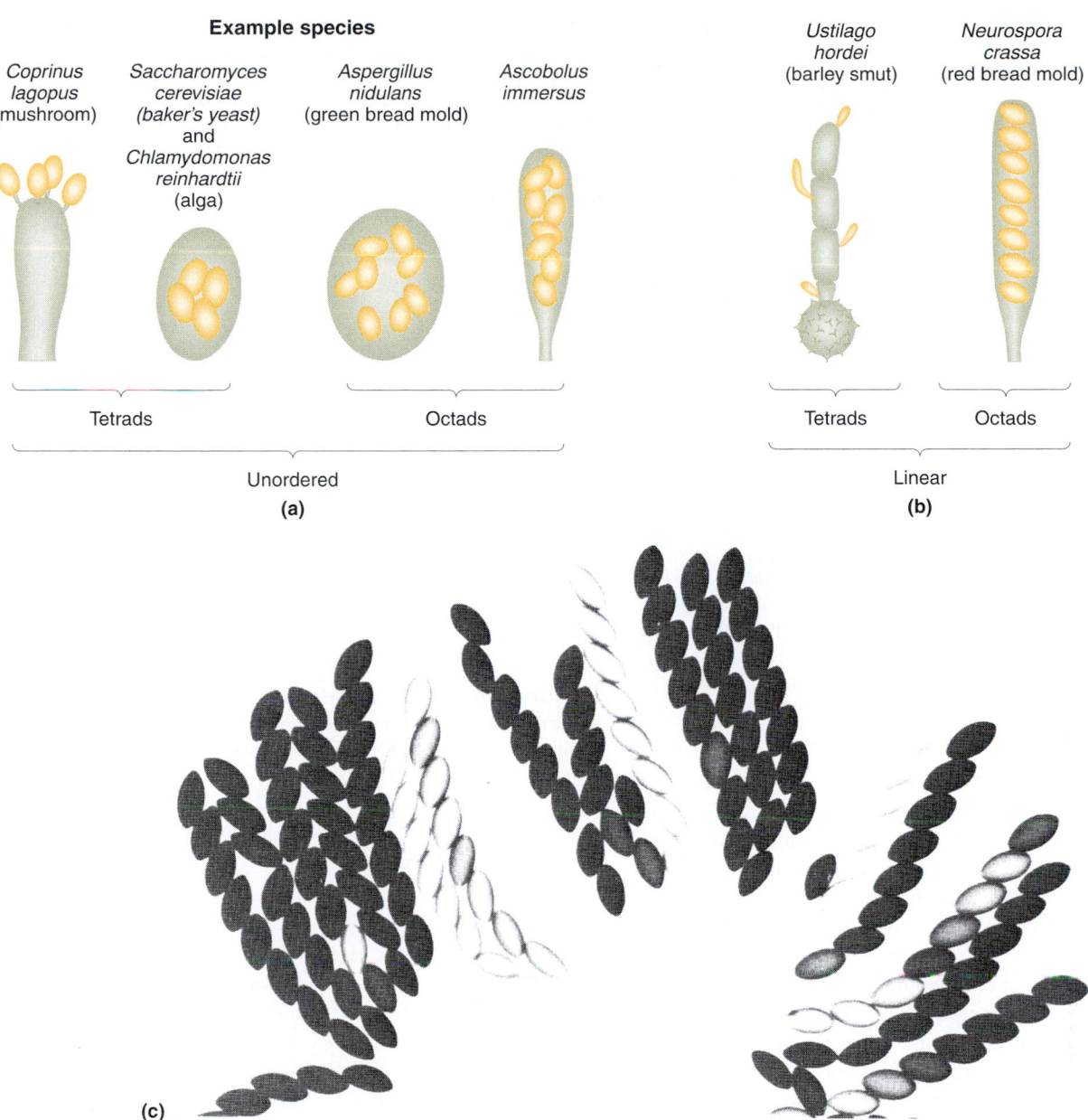

Example species

Coprinus lagopus (mushroom)

Saccharomyces cerevisiae (baker's yeast) and *Chlamydomonas reinhardtii* (alga)

Aspergillus nidulans (green bread mold)

Ascobolus immersus

Tetrads   Octads

Unordered

(a)

Ustilago hordei (barley smut)

Neurospora crassa (red bread mold)

Tetrads   Octads

Linear

(b)

(c)

Figure 6-5   Various forms of tetrads and octads found in different organisms: (a) unordered; (b) linear; (c) normally maturing asci of *Neurospora crassa*. (Part c after Namboori B. Raju, *European Journal of Cell Biology*, 23, 1980, 208–223.)

spores are found in a jumbled arrangement called an **unordered tetrad,** shown in Figure 6-5a. In other species, the spores are arranged in a striking linear arrangement called a **linear tetrad,** shown in Figure 6-5b. We shall deal with the linear types first.

Using linear tetrads to map centromeres

How are linear tetrads produced? The key fact is that the spindles of the first and second meiotic divisions and of the postmeiotic mitosis are positioned end to end in the long ascus sac and do not overlap. The reason is probably related to the fact that these divisions take place in a tubelike struc-

ture, which physically prevents the spindles from overlapping. In any case, the absence of spindle overlap means that the nuclei are laid out in a straight array, and the lineage of each of the eight nuclei of the final octad can therefore be traced back through meiosis, as shown in Figure 6-6.

As you might expect, tetrads directly illustrate the segregation and independent assortment of genes at meiosis. Linear tetrads are also ideally suited to a special kind of analysis that is not possible in most organisms: **centromere mapping,** the locating of centromeres in relation to gene loci on the chromosomes. The centromere is a fascinating region of the chromosome that interacts with the spindle fibers and ensures proper chromosome movement during

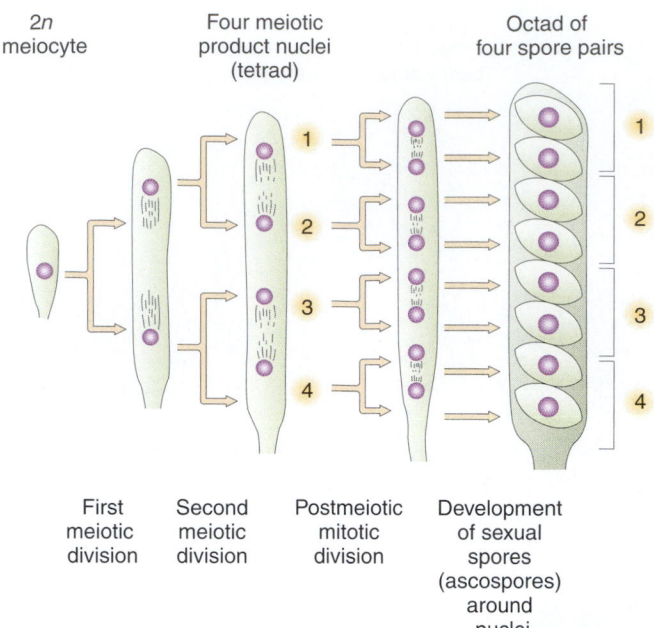

Figure 6-6   Meiosis and postmeiotic mitosis in a linear tetrad. The nuclear spindles do not overlap at any division, so nuclei never pass each other in the sac. The resulting eight nuclei are laid down in a linear array that can be traced back through the divisions.

nuclear division. When this process fails, the daughter cells have abnormal chromosome numbers, which can lead to death or phenotypic abnormality. For example, abnormal chromosome numbers in humans have produced a major class of genetic diseases (considered in Chapter 18). Therefore, the centromere is a subject of considerable attention in genetics.

How does centromere mapping work? We shall use the ascomycete fungus *Neurospora crassa* as an example. *Neurospora* produces linear octads. In its simplest form, centromere mapping considers a gene locus and asks how far this locus is from its centromere. We could pick any *Neurospora* locus to illustrate this technique, but we will use the **mating-type locus** because it was one of the first to be used for centromere mapping. This locus has two alleles, which are represented as *A* and *a*, even though neither is dominant or recessive. These alleles determine the *A* and *a* mating types. Although the mating-type phenomenon is interesting in itself, here we are merely using the locus as a genetic marker to illustrate the analysis, and we will not be concerned with its function.

Centromere mapping is based on the fact that a meiosis in which nonsister chromatids cross over between the centromere and a heterozygous locus produces a different allele pattern in the tetrad or octad from that of a meiosis in which nonsister chromatids do not cross over in that region. Figures 6-7 and 6-8 show examples of these two possibilities. The simpler pattern, shown in Figure 6-7, arises when there is no crossover. This pattern is typified by all the products at one end of the tetrad or octad carrying one allele and the

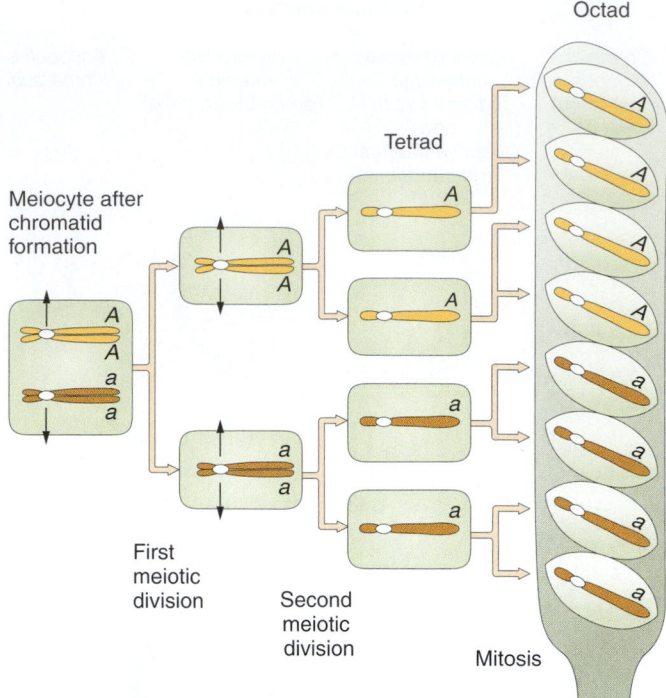

Figure 6-7   *A* and *a* segregate into separate nuclei at the first meiotic division when there is no crossover between the centromere and the locus. The resultant allele pattern in the octad is called a *first-division segregation pattern.*

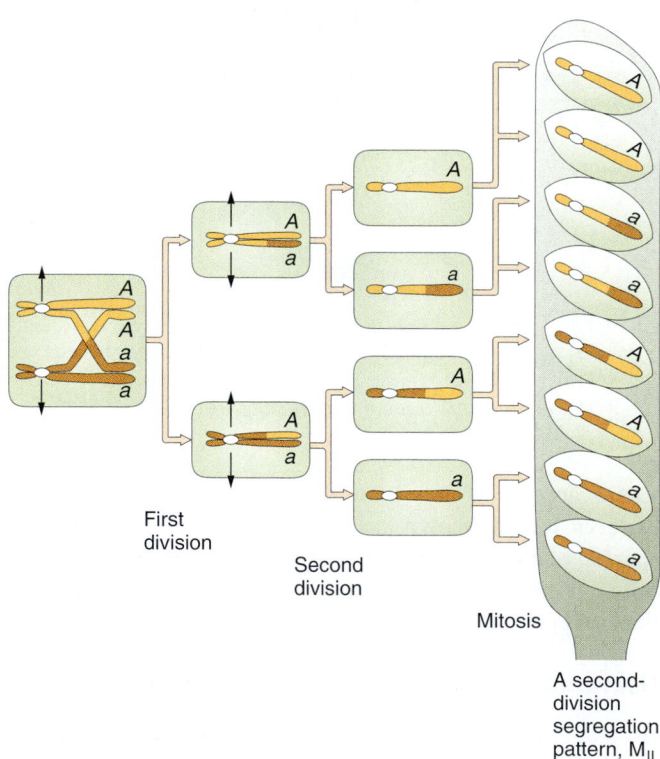

Figure 6-8   *A* and *a* segregate into separate nuclei at the second meiotic division when there is a crossover between the centromere and the locus.

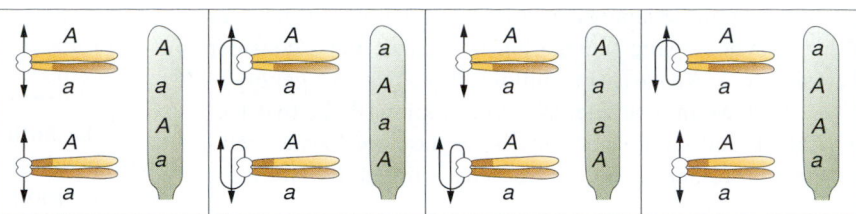

Figure 6-9 Four second-division segregation ascus patterns are equally frequent in linear asci because centromeres attach to spindles at the second meiotic division at random in regard to whether they will be pulled "up" or "down."

products at the other end carrying the other allele. You can see from the diagram how this pattern is produced: because there is no spindle overlap, the *A*-bearing nuclei and the *a*-bearing nuclei never pass each other in the ascus. Also notice that, although the *A* and *a* alleles are together in the diploid meiocyte nucleus, the first meiotic division cleanly segregates the *A* and *a* alleles, and these alleles remain separate throughout the second division of meiosis. This separation gives rise to the term **first-division segregation,** and the allele pattern in the spores is called a **first-division segregation pattern** or **M_I pattern.**

When nonsister chromatids do cross over between the centromere and the locus (see Figure 6-8), the *A* and *a* alleles are still together in the nuclei at the end of the first division of meiosis. There has been no first-division segregation. However, the second meiotic division does move the *A* and *a* alleles into separate nuclei, giving rise to the expression **second-division segregation.** The resulting allele pattern in the tetrad or octad is called a **second-division segregation pattern** or **M_{II} pattern.** This pattern is typified by any arrangement in which a half tetrad or a half octad contains both alleles.

Now let's look at the following experimental results from a *Neurospora* octad analysis and interpret them in light of these ideas. Remember, the cross was *A* × *a*. The octad patterns obtained were as follows:

			Octads		
A	*a*	*A*	*a*	*A*	*a*
A	*a*	*A*	*a*	*A*	*a*
A	*a*	*a*	*A*	*a*	*A*
A	*a*	*a*	*A*	*a*	*A*
a	*A*	*A*	*a*	*a*	*A*
a	*A*	*A*	*a*	*a*	*A*
a	*A*	*a*	*A*	*A*	*a*
a	*A*	*a*	*A*	*A*	*a*
126	132	9	11	10	12

Total = 300

The first two octads have the M_I segregation pattern, with the first octad being simply an upsidedown version of the second. Notice that these two are more or less equal in frequency (126 versus 132). This equality is simply due to the fact that centromeres attach to spindles at the first meiotic division at random in regard to whether the arrangement will pull *A* "up" and *a* "down" or *a* "up" and *A* "down." We can deduce that 126 + 132 = 258 meioses of

300, or 86 percent of meioses, had no crossover in the region between the mating-type gene and the centromere.

The remaining four octads all have both *A* and *a* present in half octads; hence, by the preceding definition *A* and *a* segregated not at the first but at the second division of meiosis. What accounts for these four different variations on the same basic M_{II} theme? Once again, it is the randomness of spindle attachment not only at the first, but also at the second division of meiosis, as illustrated in Figure 6-9. The M_{II} patterns in our example total 9 + 11 + 10 + 12 = 42, or 14 percent, and show that nonsister chromatids crossed over between the mating-type locus and the centromere in 14 percent of the meioses.

We have measured the M_{II} pattern frequency at 14 percent in this example. Does this percentage mean that the mating-type locus is 14 map units from the centromere? The answer is no, but this value can be used to calculate the number of map units. The 14 percent value is a percentage of meioses, which is not the way that map units are defined. Map units are defined in terms of the percentage of recombinant chromosomes issuing from meiosis. Figure 6-10 shows that, when a crossover takes place in the region between the centromere and the locus, only half the chromosomes issuing from that meiosis will be recombinant. So, to specify the length of the region in map units, it is necessary to divide the M_{II} pattern frequency by 2. In our example, the distance of the mating-type locus from the centromere is therefore 14 ÷ 2 = 7 map units.

MESSAGE ···

To calculate the distance of a locus from its centromere in map units, measure the percentage of tetrads showing second-division segregation patterns for that locus and divide by 2.

···

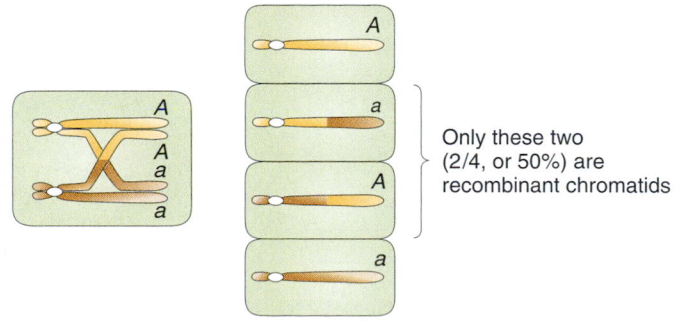

Only these two (2/4, or 50%) are recombinant chromatids

Figure 6-10 Only half the chromatids from a meiosis with a single crossover are recombinant.

The foregoing analysis can be extended to any number of heterozygous pairs segregating in a cross. Let's consider meiosis in a *Neurospora* diploid meiocyte of genotype $a^+/a \cdot b^+/b$ as an example. We do not know if the two loci are linked, and we do not know how they are located relative to centromeres. There are only limited possibilities, however:

1. The loci are on separate chromosomes.

2. The loci are on opposite sides of the centromere on the same chromosome.

3. The loci are on the same side of the centromere on the same chromosome.

The first two possibilities would show independence of the M_{II} patterns for both loci because a crossover in one arm cannot produce an M_{II} pattern in the other arm. The third possibility is more interesting in that a crossover in the region between the centromere and the locus closest to it will produce coinciding M_{II} patterns for both of the two heterozygous loci (barring rare double crossovers). Hence, if a were closer to the centromere than b, for example, we would expect that most asci with M_{II} patterns for a should also show an M_{II} pattern for b (Figure 6-11). However, some M_{II} patterns for gene b will be the result of crossovers in the a-to-b region and will not show a coincident M_{II} pattern for gene a. Such coincidence of M_{II} patterns provides clues about linkage between loci.

MESSAGE ·······································
Coincident second-division segregations can indicate the location and order of genes on the same chromosome arm.
·······································

The farther a locus is from the centromere, the greater will be the M_{II} frequency. But the M_{II} frequency never reaches 100 percent; in fact, the theoretical maximum is 67 percent, or $\frac{2}{3}$. The reason for this limit is that multiple crossovers (especially doubles) become more and more prevalent as the interval becomes larger, and double crossovers can generate M_I patterns as well as M_{II} patterns. For example:

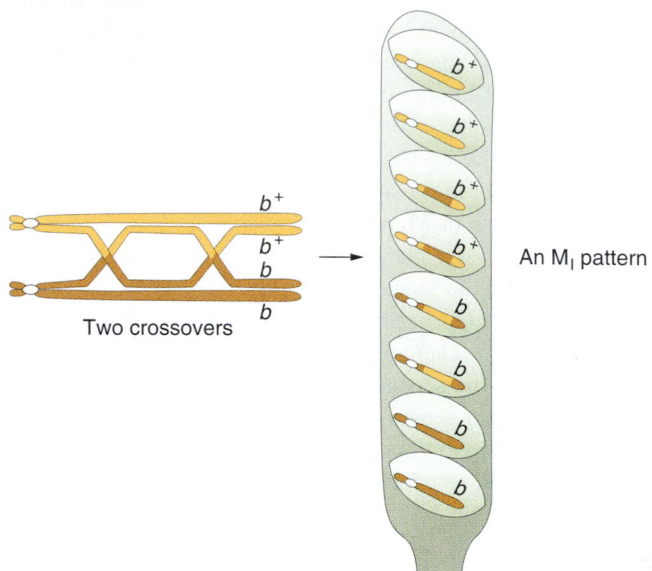

An intuitive way to calculate the maximum M_{II} frequency is by the following "thought experiment." Consider a heterozygous locus, b^+/b, that is so far from the centromere that there are numerous crossovers in the intervening region (Figure 6-12). These many crossovers effectively uncouple the locus from the centromere, and the two b^+ alleles and the two b alleles can end up in the tetrad in any arrangement. We can simulate this uncoupling by considering how many different ways there are of dropping four marbles (two b^+ and two b) into a test tube (follow this simulation by referring to Figure 6-13). The first marble can be b^+ or b; it makes no difference. Let's assume that it is b^+. We then have two b's and one b^+ left. The next marble determines if the pattern will be M_I or M_{II}. If the marbles are dropped at random into the tube over and over again, one-

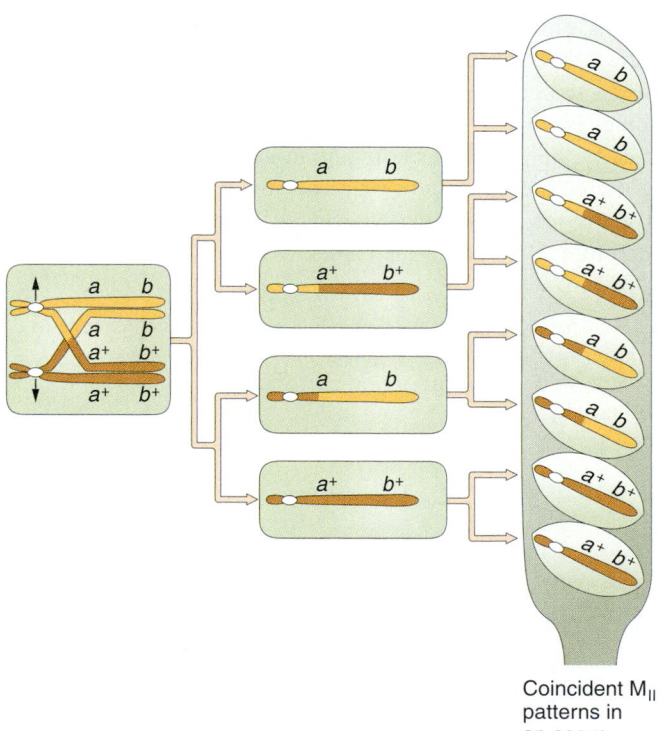

Coincident M_{II} patterns in an ascus

Figure 6-11 When two loci are on the same chromosome arm, a crossover between the centromere and the locus closest to it produces a coincident M_{II} pattern for *both* loci.

Many crossovers

Figure 6-12 When a locus is far from the centromere, a large number of crossovers effectively unlink the locus from the centromere.

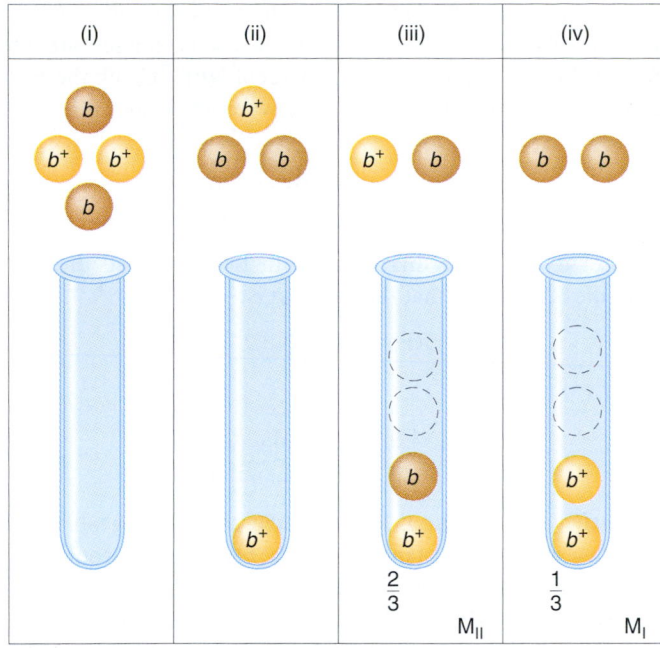

Figure 6-13 Demonstration of the limiting M_I and M_{II} segregation pattern frequencies of $\frac{1}{3}$ and $\frac{2}{3}$ in linear tetrad analysis. (See text for details.)

third of the time the second marble will be b^+; thus, the third and fourth marbles must be b, and an M_I pattern is generated. The other two-thirds of the time, the second marble will be b, generating an M_{II} pattern. Therefore, we see that, even with this very large number of crossovers, the $\frac{2}{3}$ frequency of M_{II} patterns can never be exceeded.

> **MESSAGE** ··
> The maximum frequency possible for second-division segregation patterns is 66.7 percent.

This maximum of 66.7% raises an enigma. We feel intuitively that this M_{II} maximum should be directly equivalent to the 50 percent RF maximum observed for very large map distances, but $66.7 \div 2 = 33.3$ percent, which would be 33.3 m.u. The villain, once again, is the existence of multiple crossovers. We could derive a map function for M_{II} patterns, but, in practice, it is simpler to stick to the analysis of smaller intervals and to recognize that the larger M_{II} frequencies provide increasingly inaccurate estimates of distances between centromeres and loci.

Using tetrad analysis to correct map distance for double crossovers

Both linear and unordered tetrads and octads can be used to map genes in relation to each other. To do so, the tetrads must be classified as **parental ditype (PD), nonparental ditype (NPD),** and **tetratype (T).** In a cross of $a^+ \cdot b^+ \times a \cdot b$, these types are:

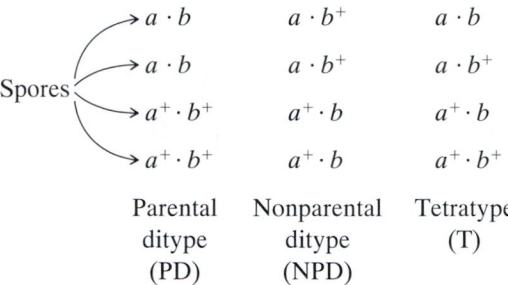

| Parental ditype (PD) | Nonparental ditype (NPD) | Tetratype (T) |

Note that these genotypes are written in no particular order; so, even though the first type (PD) might look as if there has been M_I segregation for both loci, that is not the case: the spores could have been written equally well in any order. These asci have merely been classified according to whether they contain two genotypes (*di*types) or four genotypes (*tetra*types, represented by T). Within the ditype class, both genotypes can be either parental or nonparental. You might try writing out a few asci to convince yourself that no types other than PD, NPD, and T are possible.

Now, what do these types tell us about linkage? You will notice from examining the three types that only the NPD and T types contain recombinants, so they are key classes in determining recombinant frequency. The NPD class contains only recombinants, whereas half the spores in the T class are recombinant. Therefore, we can write a formula for determining RF by using tetrads, where T and NPD represent the percentages of those classes:

$$RF = \frac{1}{2}T + NPD$$

If this formula gives a frequency of 50 percent, then we know that the loci must be unlinked, and correspondingly, if the RF is less than 50 percent, then the genes must be linked and we could use that value to represent the number of map units between them. However, just as with other linkage analyses studied earlier in the chapter, this value is an underestimate, because it does not consider double recombinants and other higher-level crossovers. Nevertheless, the frequencies of PD, NPD, and T can be used to make a correction for doubles. First, we need to understand how the PD, NPD, and T classes are produced in crosses in which there are linked markers. Let us assume that genes a and b are linked. If we assume that individual meioses can have no crossovers (NCO), a single crossover (SCO), or a double crossover (DCO) in the a-to-b region, then we can represent the classes of unordered asci that emerge from such meioses as shown in Figure 6-14. Triples and higher numbers of crossovers might occur, but we may assume that such crossovers are rare and therefore negligible. Notice that Figure 6-14 is merely an extension of Figure 6-3.

The key to the analysis is the NPD class, which arises only from a double crossover between all four chromatids. Because we are assuming that double crossovers occur randomly between the chromatids, we can also assume that the frequencies of the four DCO classes are equal. This assumption

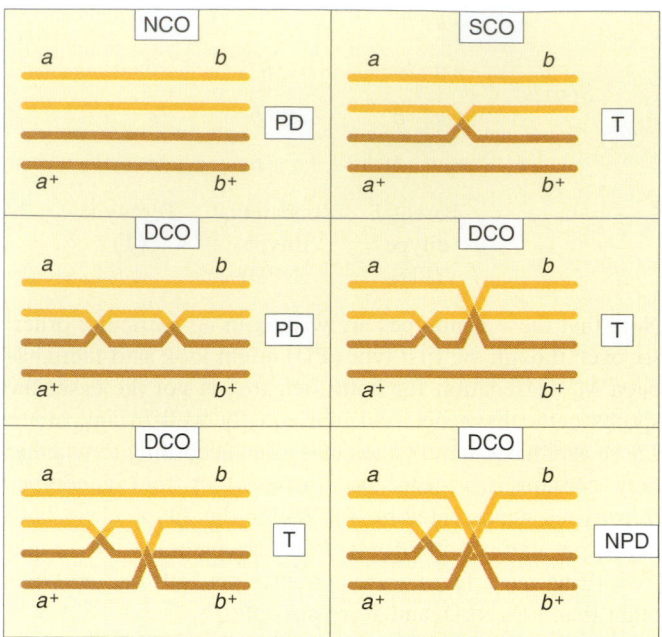

Figure 6-14   The ascus classes produced by crossovers between linked loci. NCO, noncrossover meioses; SCO, single-crossover meioses; DCO, double-crossover meioses.

means that the NPD class should contain $\frac{1}{4}$ of the DCOs, and therefore we can estimate that

$$DCO = 4NPD$$

The single-crossover class also can be calculated by a similar kind of reasoning. Notice that tetratype (T) asci can result from either single-crossover or double-crossover meioses. But we can estimate the component of the T class that comes from DCO meioses to be 2NPD. Hence, the size of the SCO class can be stated as

$$SCO = T - 2NPD$$

Now that we have estimated the sizes of the SCO and DCO classes, the noncrossover class can be estimated as

$$NCO = 1 - (SCO + DCO)$$

Thus, we have estimates of the sizes of the NCO, SCO, and DCO classes in this marked region. We can use these values to derive a value for m, the mean number of crossovers per meiosis in this region. We can calculate the value of m simply by taking the sum of the SCO class plus *twice* the DCO class (because this class contains two crossovers). Hence:

$$m = (T - 2NPD) + 2(4NPD)$$
$$= T + 6NPD$$

In the mapping-function section, we learned that, to convert an m value into map units, m must be multiplied by 50 because each crossover on average produces 50 percent recombinants. So:

$$Map\ distance = 50(T + 6NPD)\ m.u.$$

Let's assume that, in our hypothetical cross of $a^+\ b^+ \times a\ b$, the observed frequencies of the ascus classes are 56 percent PD, 41 percent T, and 3 percent NPD. Using the formula, we find the map distance between the a and b loci to be:

$$50[0.41 + (6 \times 0.03)] = 50(0.59)$$
$$= 29.5\ m.u.$$

Let us compare this value with that obtained directly from the RF. Recall that the formula is:

$$RF = \tfrac{1}{2}T + NPD$$

In our example,

$$RF = 0.205 + 0.03$$
$$= 0.235,\ or\ 23.5\ m.u.$$

This RF value is 6 m.u. less than the estimate we obtained by using the map-distance formula because we could not correct for double crossovers in the RF analysis.

What PD, NPD, and T values are expected when we deal with unlinked genes? The sizes of the PD and NPD classes will be equal as a result of independent assortment. The T class can be produced only from a crossover between either of the two loci and their respective centromeres, and therefore the size of the T class will depend on the total size of these two regions.

MESSAGE ··
Linear and unordered tetrads can be used to calculate the frequencies of single and double crossovers, which can be used to calculate accurate map distances.

Mapping genes by mitotic segregation and recombination

We normally think of segregation and recombination in connection with meiosis, but these processes do take place (although less frequently) in mitosis. Mitotic segregation and recombination can be easily demonstrated if the genetic system is appropriately chosen.

Mitotic segregation

In genetics, *segregation* refers to the separation of two alleles constituting a heterozygous genotype into two different cells or individuals. We have seen segregation repeatedly in meiotic analyses based on Mendel's first law. However, the alleles of a heterozygote can be occasionally seen to segregate when the heterozygous cell undergoes mitotic division. The following example clarifies the nature of mitotic segregation.

In the 1930s, Calvin Bridges was observing *Drosophila* females that were genotypically M^+/M (M is a dominant allele that produces a phenotype of slender bristles). Surprisingly, some females had a patch, or **sector,** of wild-type bristles on a body of predominantly M phenotype. Thus, the

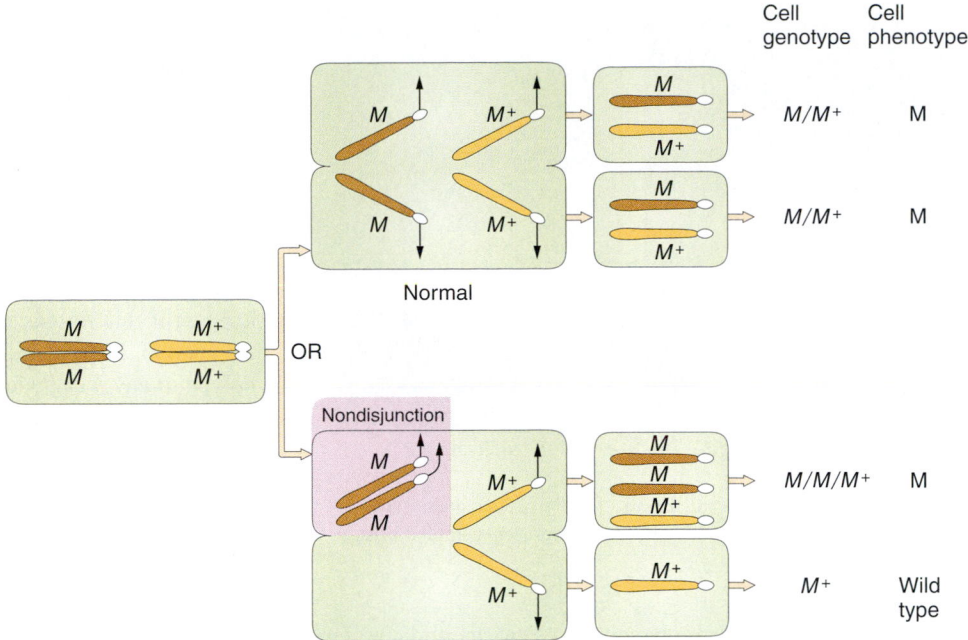

Figure 6-15   Mitotic nondisjunction can lead to phenotypic segregation.

alleles of the heterozygote were showing segregation at the phenotypic level. Bridges concluded that this segregation was the result of **mitotic nondisjunction,** a type of chromosome separation failure that is diagrammed in Figure 6-15. In a similar effect, heterozygotes of autosomal recessive alleles paired with wild-type alleles (a^+/a) produce patches of recessive phenotype on backgrounds of wild-type phenotype. These patches also can be explained by mitotic nondisjunction.

Other cases of segregation in the somatic tissue of a heterozygote have been found to be due to **mitotic chromosome loss.** Here one chromosome somehow gets left behind when the daughter nuclei reconstitute after mitotic division (Figure 6-16).

Geneticists find two other terms useful in relation to such phenomena. First, **variegation** is the coexistence of different-looking sectors of somatic tissue, whatever the cause. Second, a **mosaic** is an individual composed of tissues of two or more different genotypes, often recognizable by their different phenotypes. The following sentence gives an example of their usage: the M^+/M *Drosophila* females studied by Bridges were variegated, because mitotic nondisjunction had produced mosaics containing sectors of two different genotypes.

Mitotic crossing-over

In 1936, Curt Stern found the first case of mitotic segregation caused by **mitotic crossing-over.** Working with the *Drosophila* X-linked genes *y* (which codes for yellow body) and *sn* (singed, because it codes for short, curly bristles), Stern made a cross:

$$y^+ \; sn/y^+ \; sn \times y \; sn/\text{Y}$$

The female progeny were wild type in appearance, as expected from their $y^+ \; sn/y \; sn^+$ genotypes. However, some females had sectors of yellow tissue or of singed tissue; these sectors could be explained by nondisjunction or chromosome loss. Other females showed **twin spots.** A twin spot, in this example, is two adjacent sectors — one of yellow tissue and one of singed tissue — in a background of wild-type tissue (Figure 6-17). Stern noticed that the twin spots were too common to be chance juxtapositions of single spots. For this reason and because the twin sectors of a twin spot were always adjacent, he reasoned that the twin sectors must be reciprocal products of the same event. That event, he realized, must have begun when, by chance, the homologous parental chromosomes had come to lie in a pairing conformation, and chromatids of the different homologs

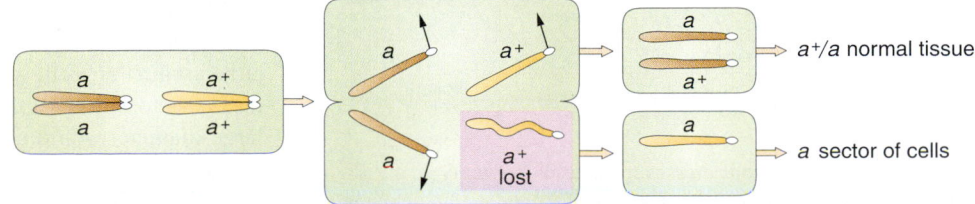

Figure 6-16   Chromosome loss at mitotic division can lead to phenotypic segregation.

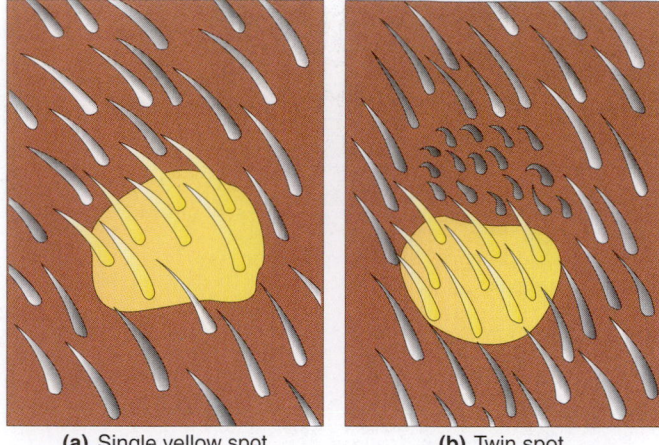

(a) Single yellow spot   (b) Twin spot

Figure 6-17   Unexpected sectors of body surface phenotypes of *Drosophila* genotype *yt sn/y snt*, where *sn* represents singled bristles and *y* represents yellow body.

must have crossed over between the *sn* locus and the centromere. Figure 6-18 diagrams this crossover. The diagram also shows that a mitotic crossover between the *y* and *sn* loci will give rise to a single yellow spot. Hence both twin and single yellow spots can be explained by a common hypothesis of mitotic crossing-over. Note that all heterozygous genes distal to the crossover are rendered homozygous, which is a general characteristic of mitotic crossing-over.

Twin spots have been observed in the cells of other diploid organisms, including plants. Mitotic crossing-over seems to take place regularly (although rarely) in all eukaryotes.

The definition of mitotic recombination is similar to that of meiotic recombination: **mitotic recombination** is any mitotic process that generates a diploid daughter cell with a combination of alleles different from that in the diploid parental cell. Compare this definition with that of meiotic recombination on page 144.

Phenotype

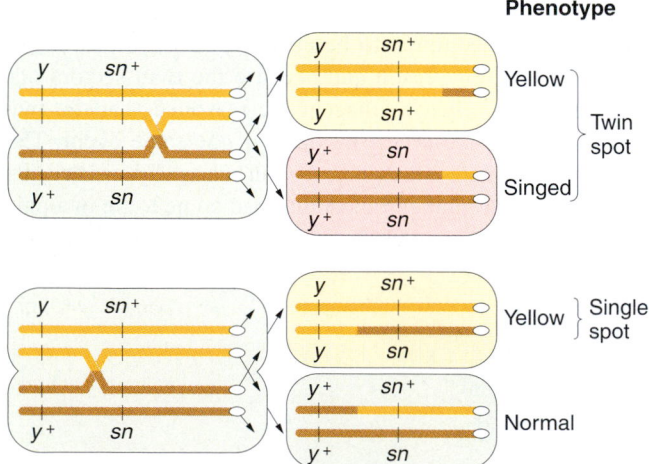

Figure 6-18   A mitotic crossover can lead to phenotypic segregation of the type shown in Figure 6-17.

Mitotic recombination in fungi

Some fungi provide convenient systems for the study of mitotic crossing-over and chromosome assortment. However, to observe these mitotic phenomena in fungi, the geneticist must generate diploid fungal cells because a haploid cell does not provide the opportunity for two genomes to recombine. Diploids form spontaneously in many fungi. The fungus that we shall examine is *Aspergillus nidulans,* a greenish mold. *Aspergillus* is highly suitable for mitotic analysis. The aerial hyphae of this fungus produce long chains of **conidia** (asexual spores). Each conidium has a single nucleus, and the phenotype of any individual spore is dependent only on the genotype of its own nucleus, which makes certain kinds of selective techniques possible. If two haploid strains are mixed, the hyphae fuse and then both types of nuclei are present in a common cytoplasm. Such a strain is called a **heterokaryon.** Genetically, heterokaryons are not diploid; their constitutions can best be thought of as *n* + *n*. Heterokaryons, like other strains, produce uninucleate conidia.

Consider a heterokaryon composed of the following nuclear genotypes:

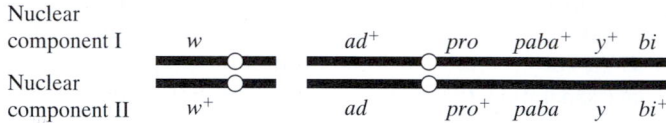

Nuclear component I   *w*   *ad⁺*   *pro*   *paba⁺*   *y⁺*   *bi*

Nuclear component II   *w⁺*   *ad*   *pro⁺*   *paba*   *y*   *bi⁺*

The allele symbols mean the following: *ad* = adenino-requiring; *pro* = proline-requiring; *paba* = *para*-amino benzoic acid–requiring; *bi* = biotin-requiring; *w* = white conidia; *y* = yellow conidia.

The alleles *ad, pro, paba,* and *bi* are all recessive to their wild-type counterparts. Because the heterokaryon has a wild-type allele of each heterozygous gene, it does not require any supplements for growth. Because each conidium produced by a heterokaryon is uninucleate, it has either the yellow or the white phenotype. Therefore, the heterokaryotic colony looks yellowish white and has a kind of "pepper and salt" appearance.

Green sectors appear in some heterokaryons. Green is the normal wild-type color of the fungus, and green coloration in the present example reveals that a diploid nucleus has formed spontaneously and has multiplied to form the green sector. The *complementation* (page 107) of the dominant *y⁺* and *w⁺* alleles in the same nucleus results in the wild-type coloration. Diploid conidia can be removed from a green sector, and a diploid culture can be grown from it. Like the heterokaryon, the diploid cells require no growth supplements, because they have the wild-type alleles. When the cultured diploid is fully grown, rare sectors producing either white or yellow conidia can be observed. Some of these sectors are diploid (recognizable because of their large-diameter conidia) and some are haploid (with small-diameter conidia). Two types of sectors are suitable for illustrating mitotic crossing-over and chromosome assort-

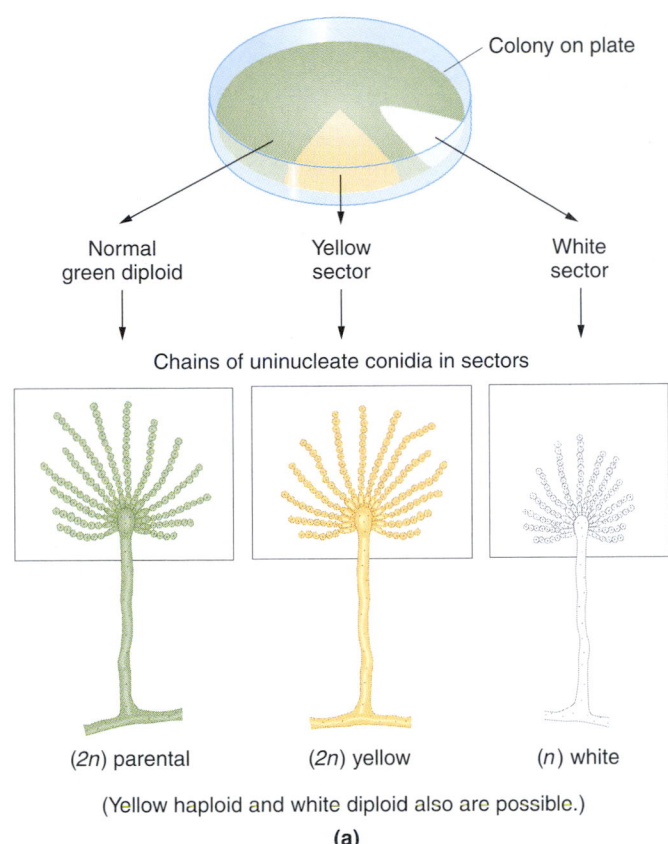

Colony on plate

Normal green diploid   Yellow sector   White sector

Chains of uninucleate conidia in sectors

(2n) parental   (2n) yellow   (n) white

(Yellow haploid and white diploid also are possible.)

(a)

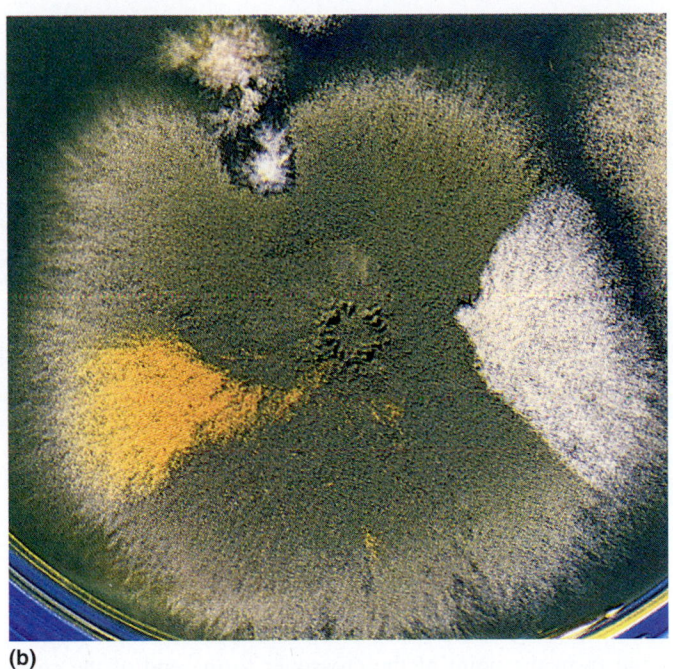

(b)

Figure 6-19 (a) Sectors showing segregation in an *Aspergillus* diploid of genotype $w^+/w \cdot y^+/y$, where *w* and *y* cause conidia to be white and yellow, respectively. Haploids are smaller. (b) Photograph of a white and a yellow sector in a diploid colony. (Part b from Etta Käfer.)

ment: haploid white sectors and diploid yellow sectors (Figure 6-19).

Haploid white sectors. If haploid white sectors are isolated and tested, almost exactly half of them prove to have the genotype *w* ; *ad⁺ pro paba⁺ y⁺ bi* and half have the genotype *w* ; *ad pro⁺ paba y bi⁺*. Thus, the original diploid nucleus has somehow become haploid, a process known as **haploidization,** presumably through the progressive loss of one member of each chromosome pair. By looking at only white haploid conidia, we automatically select for the *w*-bearing chromosome. In one half of the *w*-expressing sectors, the *ad⁺ pro paba⁺ y⁺ bi* chromosome is retained; in the other half, the *ad pro⁺ paba y bi⁺* chromosome is retained (Figure 6-20). In this way, the recessive-color alleles can be used to derive linkage information because haploidization results in an outcome similar to independent assortment. The general procedure is first to select the chromosome bearing the color marker and then find out which genes are retained with it and which are independent of it, and in what groupings. In the present example the groupings show that the *ad, pro, paba, y,* and *bi* loci are linked in a chromosome other than that carrying *w*.

MESSAGE ···

Haploidization produces an effect like that of independent assortment and can be used to deduce which genes are linked and which assort independently.

Diploid yellow sectors. When diploid yellow sectors are tested, they usually prove to contain recombinant chromosomes. For example, one sector type was yellow and required *para*-aminobenzoic acid (PABA) for growth. Remember that mitotic crossing-over can make heterozygous loci homozygous, so mitotic crossing-over explains this type (Figure 6-21). Notice that we must follow two spindle fibers to each pole in mitotic analysis. This yellow, PABA-requiring diploid arose from a mitotic exchange that took place in the centromere-to-*paba* region; other yellow diploids would arise from mitotic exchanges in the *paba*-to-*y* region, but these diploids would not require PABA for growth. The relative

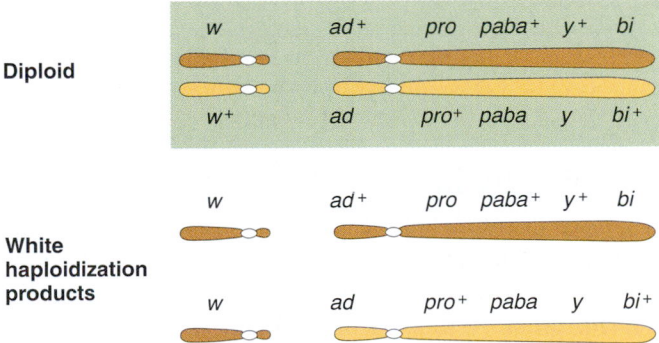

Diploid	*w*	*ad⁺*	*pro*	*paba⁺*	*y⁺*	*bi*
	w⁺	*ad*	*pro⁺*	*paba*	*y*	*bi⁺*

White haploidization products	*w*	*ad⁺*	*pro*	*paba⁺*	*y⁺*	*bi*
	w	*ad*	*pro⁺*	*paba*	*y*	*bi⁺*

Figure 6-20 Genotypes of an *Aspergillus* diploid and two white haploidization products derived from it.

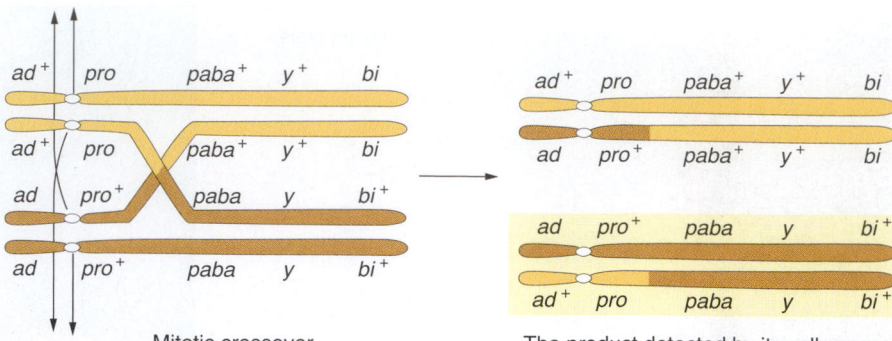

Figure 6-21 A mitotic crossover can produce a diploid yellow sector in the *Aspergillus* diploid shown. Note that the crossover produces homozygosity at all heterozygous loci beyond the crossover.

frequencies of these two types provide a kind of mitotic linkage map for that region. Such mapping can be carried out in several fungi. As expected, gene order in mitotic maps corresponds to gene order in meiotic maps, but, unexpectedly, the relative sizes of many of the intervals are found to differ when meiotic and mitotic maps are compared. The reason for this difference is not known. Note that every heterozygous gene from the point of the crossover to the end of the chromosome arm is made homozygous by a mitotic crossover, which in itself can provide useful mapping information. Because mitotic recombination experiments are done in fast-growing vegetative fungal cultures, the results can be obtained much more quickly than in a meiotic analysis, which requires the slower sexual phase.

MESSAGE ··

A mitotic crossover makes all heterozygous genes distal to the crossover homozygous, and this principle may be used to deduce gene order. The frequencies of different homozygous classes are a measure of relative map distances.

In this section, we have encountered several different processes that can cause a heterozygous allele pair to segregate in mitosis and produce a mosaic of different genotypes. In later chapters, we shall encounter other processes that can produce such variegation, but the following message summarizes the sources of variegation that we have seen in this chapter.

MESSAGE ··

At mitosis, nondisjunction, chromosome loss, crossing-over, and haploidization all cause a heterozygous pair of alleles to segregate in somatic tissue, resulting in a mosaic expressing the phenotypes of both alleles.

Mapping by in situ hybridization

If a gene has been cloned, the clone can be used in a specialized mapping technique to find the gene's chromosomal location. The clone is labeled and used as a probe for **in situ hybridization.** A solution of the probe is used to bathe a chromosomal preparation in which the DNA has been partly denatured, leaving the two strands slightly apart. After random diffusion of the probe, complementary hydrogen bonding takes place between the probe and the gene in its origi-

nal chromosomal position. The label reveals the locus of hybridization and, hence, the chromosomal position.

Commonly used probe labels are radioactivity and fluorescence. In the process of **fluorescence in situ hybridization (FISH),** the clone is labeled with a fluorescent dye, and a partly denatured chromosome preparation is bathed in the probe. The probe binds to the chromosome in situ, and the location of the cloned fragment is revealed by a bright fluorescent spot (Figure 6-22).

Mapping human genes by using human–rodent somatic cell hybrids

We saw in Chapter 5 that it is difficult to map human genes by recombination analysis. Human–rodent cell hybridization circumvents this problem. It can be used to assign genes to specific chromosomes and to determine map positions.

Assigning genes to chromosomes

The technique of somatic cell hybridization is extensively used in human genome mapping, but it can in principle be used in many different animal systems. The procedure uses cells growing in culture. A virus called the *Sendai virus* has a useful property that makes the mapping technique possible. Each Sendai virus has several points of attachment, so it can simultaneously attach to two different cells if they happen to be close together. However, a virus is very small in comparison with a cell, so the two cells to which the virus is attached are held very close together indeed. In fact, the membranes of the two cells may fuse together and the two cells become one — a binucleate heterokaryon.

If suspensions of human and mouse cells are mixed together in the presence of Sendai virus that has been inactivated by ultraviolet light, the virus can mediate fusion of the cells from the different species (Figure 6-23). When the cells have fused, the nuclei subsequently fuse to form a uninucleate cell line composed of both human and mouse chromosome sets. Because the mouse and human chromosomes are recognizably different in number and shape, the two sets in the hybrid cells can be readily distinguished. However, in the course of subsequent cell divisions, for unknown reasons the human chromosomes are gradually eliminated from

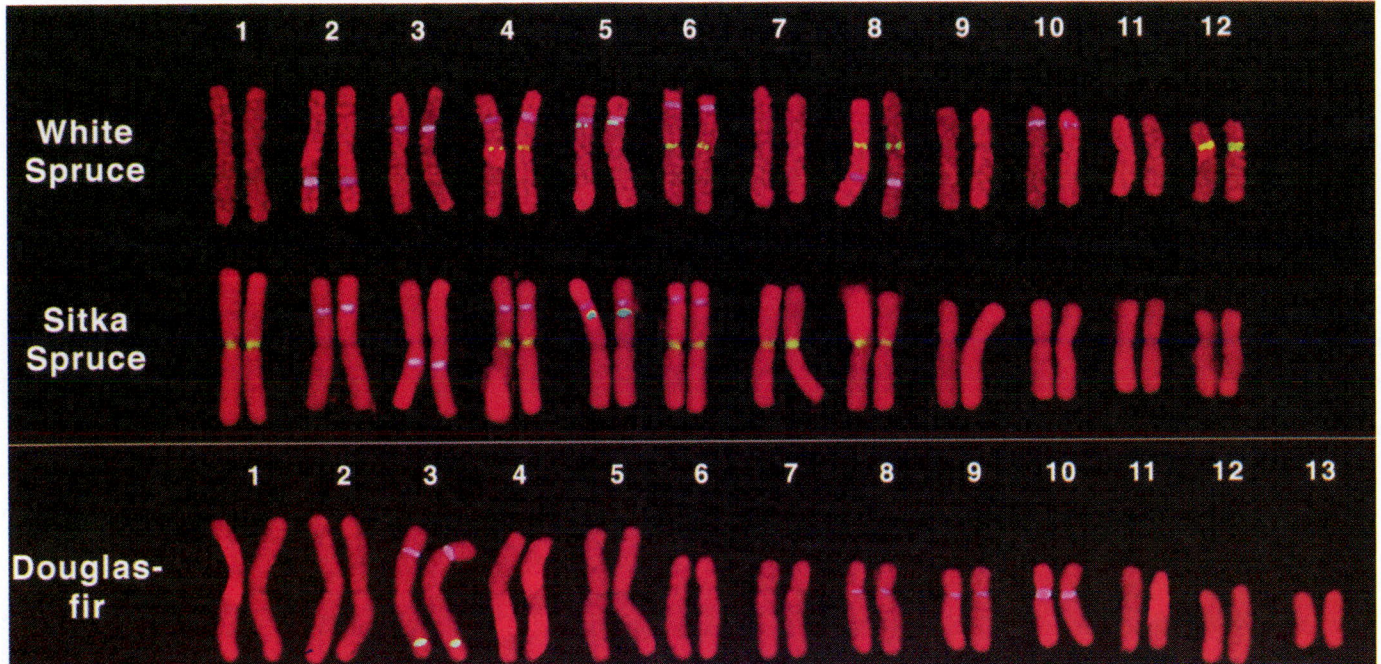

Figure 6-22 Fluorescence in situ hybridization analysis of the chromosomes of three coniferous tree species. The probes used were clones of 5S rDNA (blue), the region coding for 18S, 5.8S, and 26S rDNA (pink), and, in the white and Sitka spruce, a type of satellite DNA code-named SGR-31 (green). In this case, the FISH analysis was undertaken to locate the relevant loci, use the loci as markers to identify specific chromosomes, and make evolutionary comparisons between the species. (Photograph from Garth Brown, Vindhya Amarasinghe, and John Carlson.)

the hybrid at random. Perhaps this process is analogous to haploidization in the fungus *Aspergillus*.

The loss of human chromosomes can be arrested in the following way to encourage the formation of a stable partial hybrid. The cells used are mutant for some biochemical function; so, if the cells are to grow, the missing function must be supplied by the other genome. This selective technique results in the maintenance of hybrid cells that have a complete set of mouse chromosomes and a small number of human chromosomes, which vary in number and type from hybrid to hybrid but which always include the human chromosome carrying the wild-type allele defective in the mouse genome.

Let's look at the specific genes that make the selective system work. In cells, DNA can be made either de novo ("from scratch") or through a salvage pathway that uses molecular skeletons already available. The selective technique

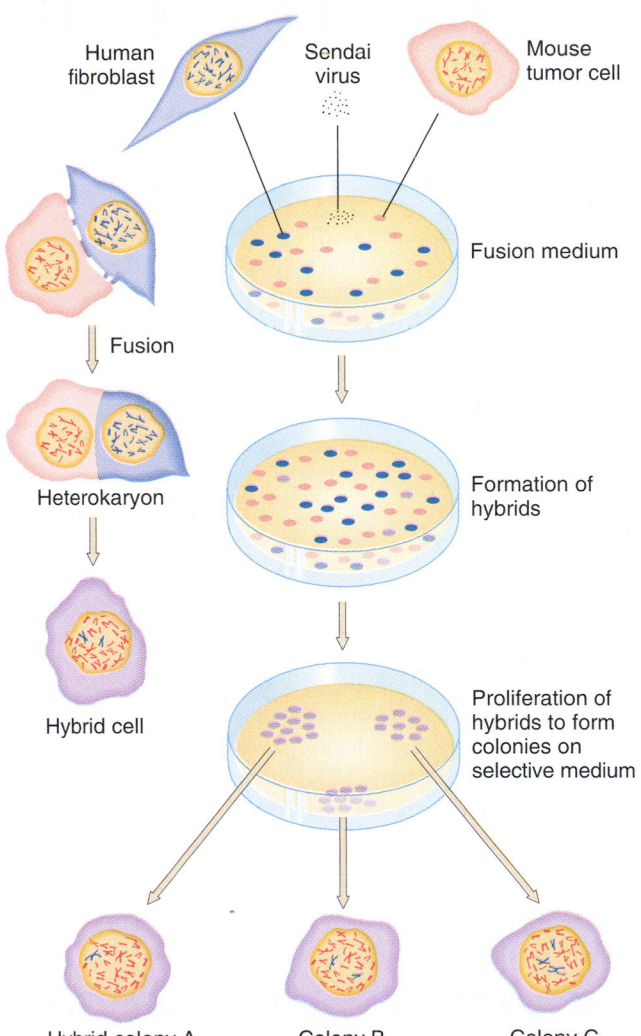

Figure 6-23 Cell-fusion techniques applied to human and mouse cells produce colonies, each of which contains a full mouse genome plus a few human chromosomes (blue). A fibroblast is a cell of fibrous connective tissue. (From F. H. Ruddle and R. S. Kucherlapati, "Hybrid Cells and Human Genes." Copyright © 1974 by Scientific American, Inc. All rights reserved.)

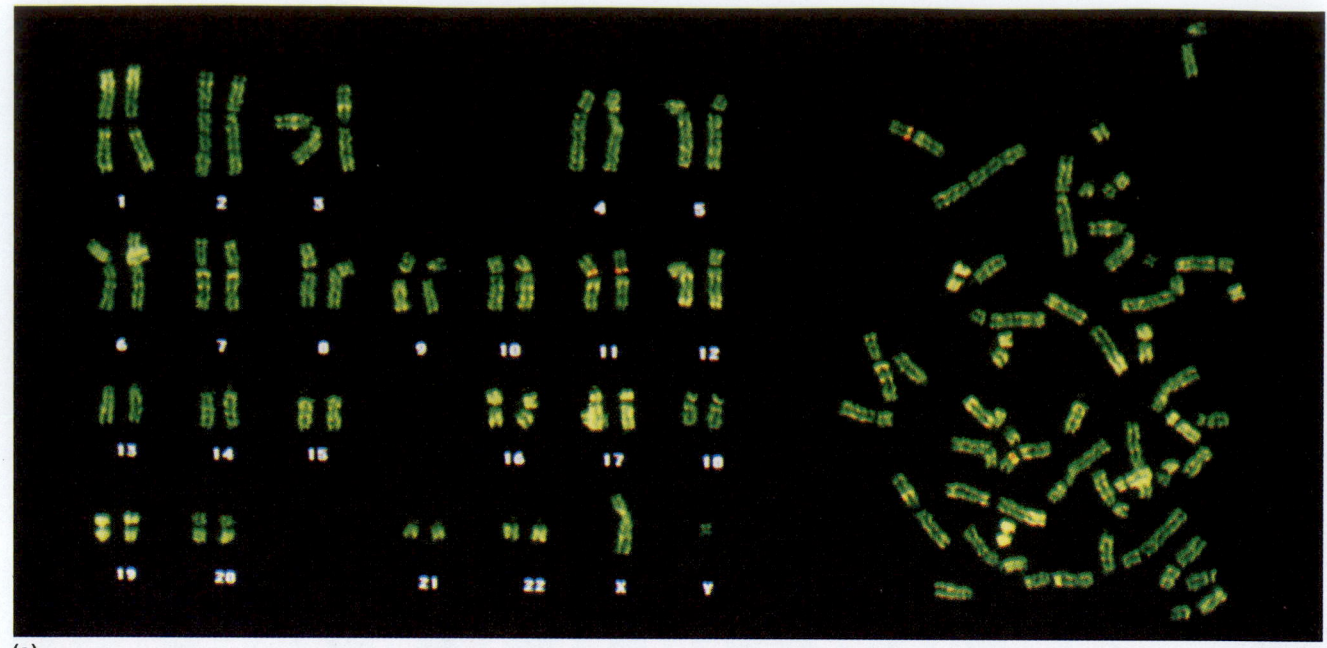

(a)

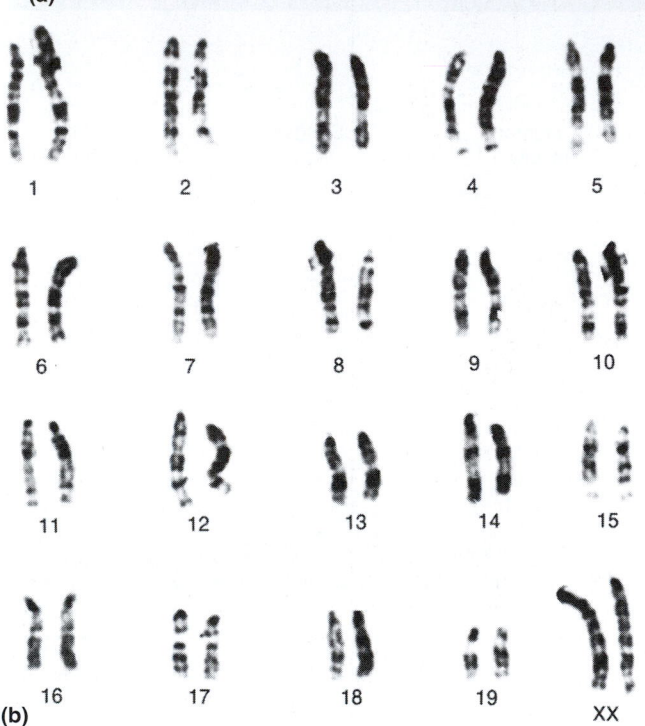

(b)

Figure 6-24   (a) Stained human chromosomes. Under the microscope, the chromosomes appear as a jumbled cluster, as shown at the right. This array is photographed, and the individual chromosomes are cut out of the photograph and then grouped by size and banding pattern, as shown at the left. The chromosome set of a male is shown. (b) The chromosomes of a female mouse, shown for comparison. To the experienced eye, the mouse chromosomes can be easily distinguished from human chromosomes, as required in the human–rodent cell hybrid technique of gene localization. (Part a from David Ward, Yale University School of Medicine; part b from Jackson Laboratory, Bar Harbor, ME.)

involves the application of a chemical, aminopterin, that blocks the de novo synthetic pathway, confining DNA synthesis to the salvage pathway. Two essential salvage enzymes, thymidine kinase (TK) and hypoxanthine-guanine phosphoribosyl transferase (HGPRT), are relevant to the system, as shown in the following two reactions:

$$\text{thymine} \xrightarrow{\text{TK}} \text{thymidilic acid (a DNA precursor)}$$

$$\text{hypoxanthine} \xrightarrow{\text{HGPRT}} \text{inosinic acid (a DNA precursor)}$$

The mouse cell line to be fused is genetically unable to make TK because it is homozygous for the allele tk^-,

whereas the human cell line is genetically unable to make HGPRT because it is homozygous at another locus for the allele $hgprt^-$. So the genotypes of the two fusing cell lines are:

Mouse: tk^-/tk^- ; $hgprt^+/hgprt^+$
Human: tk^+/tk^+ ; $hgprt^-/hgprt^-$

Because each is deficient for one enzyme, neither the mouse nor the human cells are able to make DNA individually. In the hybrid cells, however, the tk^+ allele complements the $hgprt^+$ allele, so the cells can make both enzymes. Therefore, DNA is synthesized and the cells can proliferate. Most human chromosomes are eliminated from the hybrid cell cultures because their loss has no effect on the cultures' ability to grow. But, to continue to grow in medium containing hypoxanthine, aminopterin, and thymidine (*HAT medium),* a hybrid culture must retain at least one of the human chromosomes that carries the tk^+ allele.

Luckily, the progressive elimination of the human chromosomes from the fused cell lines can be followed under the microscope because mouse chromosomes can easily be distinguished from human chromosomes. Moreover, chromosome stains such as quinacrine and Giemsa reveal a pattern of banding within the chromosomes. The size and the

6-2 TABLE	Comparison of Five Hybrid Lines					
		HYBRID CELL LINES				
		A	B	C	D	E
Human genes	1	+	−	−	+	−
	2	−	+	−	+	−
	3	+	−	−	+	−
	4	+	+	+	−	−
Human chromosomes	1	−	+	−	+	−
	2	+	−	−	+	−
	3	−	−	−	+	+

position of these bands vary from chromosome to chromosome, but the banding patterns are highly specific and invariant for each chromosome. Thus, it is easy to identify the human chromosomes that are present in any hybrid cell (Figure 6-24). Different hybrid cells are grown separately into lines; eventually a bank of lines is produced that contains, in total, all the human chromosomes.

With a complete bank of chromosomes, we can begin to assign genes or markers to chromosomes. If the human chromosome set is homozygous for a human molecular marker — such as an allele that controls a cell-surface antigen, drug resistance, a nutritional requirement, a specific protein, or a DNA marker — then the presence or absence of this genetic marker in each line of hybrid cells can be correlated with the presence or absence of certain human chromosomes in each line. Data of this sort are presented in Table 6-2 in which "+" means presence and "−" means absence of the genetic marker. We can see that, in the different hybrid cell lines, genetic markers 1 and 3 are always present or absent together. We can conclude, then, that they are linked. Furthermore, the presence or absence of genes 1 and 3 is correlated with the presence or absence of chromosome 2, so we can assume that these genes are located on chromosome 2. By the same reasoning, gene 2 must be on chromosome 1, but the location of gene 4 cannot be assigned. Large numbers of human genes have now been localized to specific chromosomes in this way.

Chromosome mapping

In the preceding subsection, we discussed the use of human-rodent cell hybrids to assign genes to chromosomes. This technique can be extended to obtain mapping data. One extension of the hybrid cell technique is called **chromosome-mediated gene transfer.** First, samples of individual human chromosomes are isolated by **fluorescence-activated chromosome sorting (FACS)** (Figure 6-25). In this procedure metaphase chromosomes are stained with two dyes, one of which binds to AT-rich regions, and the other to GC-rich regions. Cells are disrupted to liberate whole chromosomes into liquid suspension. This suspension is converted into a spray in which the concentration of chromosomes is such

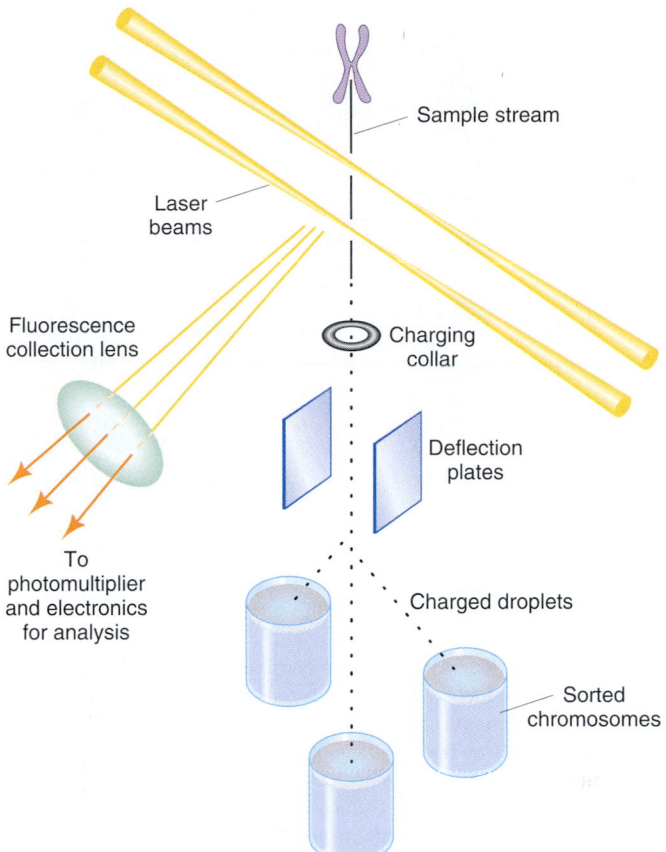

Figure 6-25   Chromosome purification by using flow sorting. Chromosomes stained with a fluorescent dye are passed through a laser beam. Each time, the amount of fluorescence is measured and the chromosome is deflected accordingly. The chromosomes are then collected as droplets.

that each spray droplet contains one chromosome. The spray passes through laser beams tuned to excite the fluorescence. Each chromosome produces its own characteristic fluorescence signal, which is recognized electronically, and two deflector plates direct the droplets containing the specific chromosome needed into a collection tube.

Then a sample of one specific chromosome under study is added to rodent cells. The human chromosomes are engulfed by the rodent cells and whole chromosomes or fragments become incorporated into the rodent nucleus. Correlations are then made between the human fragments present and human markers. The closer two human markers are on a chromosome, the more often they are transferred together. Some results of these kinds of mapping are shown in Figure 6-26.

A second method, called **irradiation and fusion gene transfer (IFGT),** an extension of the chromosome-mediated gene transfer, is designed to generate a higher-resolution map of molecular markers along a chromosome. The procedure is to irradiate human cells with 3000 rads of X rays to fragment the chromosomes, then fuse the irradiated cells with the rodent cells to form a panel of different

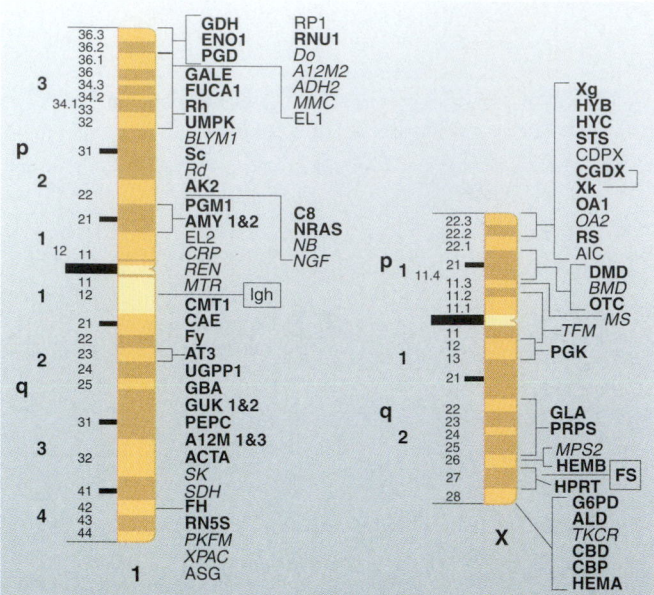

Figure 6-26   Maps of the human chromosome 1 and the X chromosome. The loci are labeled with code names. Many code names are based on the protein specified by the gene; others represent marker loci identified only as DNA variants. Such loci with some molecular tag can be mapped well by somatic cell hybridization methods. Some of these loci are interesting in themselves, and some are useful as chromosomal landmarks to which other loci may be mapped. For autosomes such as chromosome 1, approximately $\frac{2}{3}$ of the loci have been assigned by somatic hybridization, and the remainder from pedigree analysis. Compare these maps with those on the first page of Chapter 5, which list genes that determine various disorders and other phenotypes that map to these chromosomes. Protein variants are the basis of most of these conditions. The letters p and q designate the short and long arms, respectively. (From V. A. McKusick, *Genetic Maps.* Vol. 2. Edited by S. J. O'Brien. Cold Spring Harbor Laboratory Press, 1982.)

hybrids. In this case the hybrids have an assortment of fragments of human chromosomes, as diagrammed in Figure 6-27. Most of the fragments are seen to be embedded into the rodent chromosomes, but truncated human chromosomes also can be found. First, the retention of various molecular markers in the hybrids is calculated. The next step is to calculate the frequency of coretention of pairs of human molecular markers. The assumption is made that closely linked markers will be coretained at high frequencies because there is a low probability that a radiation-induced

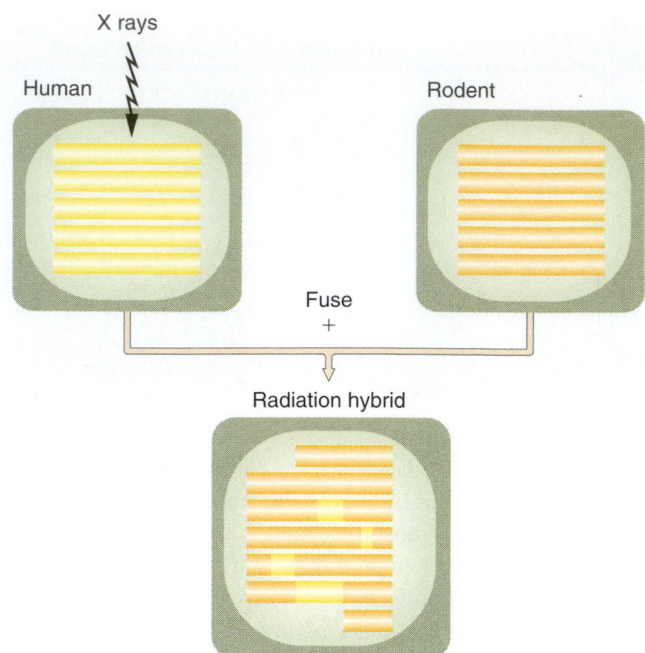

Figure 6-27   Making radiation hybrids by using irradiation by X rays. Fragments of human chromosomes integrate into rodent chromosomes. A panel of different radiation hybrids is analyzed for cotransfer of human markers, which can indicate linkage.

break will occur between the loci. Distant markers and markers on different chromosomes should be retained at frequencies close to the product of individual retention frequencies. A mapping unit cR_{3000} is calculated, which can be calibrated to approximately 0.1 cM (m.u.).

One of the advantages of this method is in sample size; a large number of radiation hybrids can be amassed relatively easily. A standard panel of only 100 to 200 hybrids is enough to generate a high-resolution "cR_{3000} map" of the human genome, with a tenfold better resolution than a map based on RF.

MESSAGE ·······························
Correlation of retention of human markers and chromosomes in hybrid rodent–human cell lines allows chromosomal assignment of the markers. Coretention of different human markers in X-irradiated hybrids allows high-resolution mapping of the chromosomal loci of the markers.

SUMMARY

Because of multiple crossovers, map distance is not linearly related to recombinant frequency. A mathematical relation between map distance and recombinant frequency is called a mapping function. The Poisson distribution of crossovers at meiosis can be used to devise a mapping function.

Tetrad analysis is used to study individual meioses. Tetrad analysis is possible in those fungi and single-celled

algae in which the four products of each meiosis remain together prior to spore dispersal.

Tetrads may be linear or unordered. Analysis of linear tetrads enables us to map loci in relation to their centromeres and to each other. From crosses in which the parents differ at two loci, the asci in a linear or unordered tetrad may be classified as parental ditype (PD), nonparental ditype (NPD), and tetratype (T). From these classes, recombinant frequency can be calculated by using the simple formula RF = NPD + $\frac{1}{2}$T. A different expression 50(T + 6NPD) provides a map distance corrected for the occurrence of double crossovers. In haploid organisms, recombination analysis by random meiotic products is simpler than that in diploids.

Although segregation and recombination are normally thought of as meiotic phenomena, alleles do occasionally segregate and recombine in mitosis. Mitotic segregation was first identified in the 1930s, when Calvin Bridges observed patches of M^+ bristles on the body of an M^+/M female *Drosophila* that was predominantly M in phenotype. He concluded that the patches were the result of abnormal chromosome segregation in mitosis. About the same time, Curt Stern observed twin spots in *Drosophila* and assumed that they must be the reciprocal products of mitotic crossing-over. Mitotic crossing-over makes all heterozygous genes distal to the crossover homozygous, and this principle can be used to deduce map position and distance. Some fungi provide convenient systems for the study of mitotic recombination.

If a gene has been cloned, the clone can be labeled and used for in situ chromosomal hybridization. The position where the probe binds reveals the gene's locus. Human–rodent hybrid cells lose their human chromosomes spontaneously. Correlating retained chromosomes and retained genes allows chromosomal loci to be assigned.

CONCEPT MAP

Draw a concept map interrelating as many of the following terms as possible. Note that the terms are listed in no particular order.

map distance / map function / linkage map / recombinant frequency / tetrad analysis / centromere mapping / second-division segregation / correction for double crossovers / individual meioses

CHAPTER INTEGRATION PROBLEM

Strains of the haploid fungus *Neurospora* bearing the *am* allele do not grow unless alanine is added to the medium. Normal wild-type strains grow without alanine supplementation. Linked to the *am* locus is a suppressor gene *ssu*. The *ssu* allele suppresses the alanine requirement, and *am ssu* strains do not need alanine supplementation. The *ssu* allele has no effect on the wild-type *am*$^+$ allele and, furthermore, the wild-type allele *ssu*$^+$ has no known effect on the *am* locus alleles.

It has been calculated that triple and higher crossovers are negligible in the region between *am* and *ssu*. Meioses thus have no crossover (41 percent), one crossover (39 percent), or two crossovers (20 percent). From this information answer the following questions:

a. In what way is the genetic system described dependent on the environment?

b. What is the recombinant frequency (RF value)?

c. If ascospores from the cross *am ssu* × *am*$^+$ *ssu*$^+$ are sampled at random, what proportion will require alanine to grow?

d. What is the distance between the loci in corrected map units?

e. If asci from the cross *am ssu* × *am*$^+$ *ssu*$^+$ are analyzed as unordered tetrads, what proportion will show the following five phenotypic patterns:

Pattern	Number of ascospores not requiring alanine	Number of ascospores requiring alanine
1	8	0
2	6	2
3	4	4
4	2	6
5	0	8

◆ Solution ◆

a. The only aspect of the environment that is mentioned is alanine as part of the growth medium. If there is alanine in the medium, we cannot identify the two basic phenotypes — the growth phenotype and the no-growth phenotype. Therefore, the ability to study and analyze this interesting case of gene interaction is environmentally dependent, because the starting point for any piece of genetic analysis is variation, and without it there would be no study. Another point that is relevant is that the *am* allele would probably be lethal in a natural environment; only because scientists propagate such phenotypes (in this case on special medium — that is, in a special environment) do we have them for study.

b. The chapter has provided us with the formula for calculating the RF from tetrad analysis. We have learned that RF = T/2 + NPD. However, we do not know what the values of T and NPD are, so they have to be calculated. We have been given the frequencies of meioses with 0, 1, and 2 crossovers, and we have learned how PD, T, and NPD are produced from such meioses. We thus have what we need to calculate the required frequencies. All asci with no crossovers are PD, so this contributes 41 percent to the PD class. Single crossovers also are easy; they produce only T asci, so this contributes 39 percent to the T class. Double crossovers are a bit more complicated, but we have seen that they produce $\frac{1}{4}$PD, $\frac{1}{2}$T, and $\frac{1}{4}$NPD, which gives us 5 percent PD, 10 percent T, and 5 percent NPD. Adding these percentages, we get 46 percent PD, 49 percent T, and 5 percent NPD. Now we can calculate the RF value, which is $\frac{49}{2} + 5 = 29.5$ percent.

c. We now expect 29.5 percent recombinants in the ascospores from the cross, comprising the two recombinant classes, *am ssu*$^+$ and *am*$^+$ *ssu*, each being 14.75 percent. The two parental classes are *am ssu* and *am*$^+$ *ssu*$^+$, and we expect them to make up the difference $100 - 29.5 = 70.5$ percent, each being 35.25 percent. Which of these four genotypes will require alanine? Obviously, *am*$^+$ *ssu*$^+$ will not; neither will *am ssu*. The genotype *am*$^+$ *ssu* also will not require alanine, because it has the wild-type allele at the *am* locus, and we have been told that *ssu* does not affect that allele. However, *am ssu*$^+$ does not have the suppressor allele and will require alanine, so the answer is 14.75 percent, the frequency of this class.

d. We have learned in this chapter that, if we know the frequency of PD, T, and NPD in an unordered tetrad analysis, we can use these values to correct for the effect of double crossovers, which always tend to cause an underestimation of map distance. The formula is $50(T + 6NPD)$. Applying our calculated values, we get $50(0.49 + 0.30)$, which equals 39.5 corrected map units.

e. We have done most of the reasoning to answer this part because we have deduced that the only genotype that will confer a requirement for alanine is the genotype *am ssu*$^+$. Because we know there are only three ascus types (PD, T, and NPD) in an unordered tetrad analysis of two heterozygous loci, we simply need to deduce how many *am ssu*$^+$ ascospores will be seen in each. In PD asci, there will be no *am ssu*$^+$ genotypes, so these asci will give us an 8:0 ratio. Tetratype (T) asci will contain one spore pair of genotype *am ssu*$^+$, so these asci will give us a 6:2 ratio. The NPD class will be composed of four *am ssu*$^+$ ascospores and four *am*$^+$ *ssu* ascospores, giving a 4:4 ratio. And that covers all the possibilities, so we can answer that the proportions for the table are (1) 46 percent, (2) 49 percent, (3) 5 percent, (4) 0 percent, (5) 0 percent.

In summary, notice the concepts that we used from preceding chapters: Mendelian segregation, chromosomal inheritance, gene interaction, lethal effects, environmental effects, crossing-over, and mapping.

SOLVED PROBLEMS

1. A cross is made between a haploid strain of *Neurospora* of genotype *nic*$^+$ · *ad* and another haploid strain of genotype *nic* · *ad*$^+$. From this cross, a total of 1000 linear asci are isolated and categorized as follows:

1	2	3	4	5	6	7
nic$^+$ · *ad*	*nic*$^+$ · *ad*$^+$	*nic*$^+$ · *ad*$^+$	*nic*$^+$ · *ad*	*nic*$^+$ · *ad*	*nic*$^+$ · *ad*$^+$	*nic*$^+$ · *ad*$^+$
nic$^+$ · *ad*	*nic*$^+$ · *ad*$^+$	*nic*$^+$ · *ad*$^+$	*nic*$^+$ · *ad*	*nic*$^+$ · *ad*	*nic*$^+$ · *ad*$^+$	*nic*$^+$ · *ad*$^+$
nic$^+$ · *ad*	*nic*$^+$ · *ad*$^+$	*nic*$^+$ · *ad*	*nic* · *ad*	*nic* · *ad*$^+$	*nic* · *ad*	*nic* · *ad*
nic$^+$ · *ad*	*nic*$^+$ · *ad*$^+$	*nic*$^+$ · *ad*	*nic* · *ad*	*nic* · *ad*$^+$	*nic* · *ad*	*nic* · *ad*
nic · *ad*$^+$	*nic* · *ad*	*nic* · *ad*$^+$	*nic*$^+$ · *ad*$^+$	*nic*$^+$ · *ad*	*nic*$^+$ · *ad*$^+$	*nic*$^+$ · *ad*
nic · *ad*$^+$	*nic* · *ad*	*nic* · *ad*$^+$	*nic*$^+$ · *ad*$^+$	*nic*$^+$ · *ad*	*nic*$^+$ · *ad*$^+$	*nic*$^+$ · *ad*
nic · *ad*$^+$	*nic* · *ad*	*nic* · *ad*	*nic* · *ad*$^+$	*nic* · *ad*$^+$	*nic* · *ad*	*nic* · *ad*$^+$
nic · *ad*$^+$	*nic* · *ad*	*nic* · *ad*	*nic* · *ad*$^+$	*nic* · *ad*$^+$	*nic* · *ad*	*nic* · *ad*$^+$
808	1	90	5	90	1	5

Map the *ad* and *nic* loci in relation to centromeres and to each other.

◆ Solution ◆

What principles can we draw on to solve this problem? It is a good idea to begin by doing something straightforward, which is to calculate the two locus-to-centromere distances. We do not know if the *ad* and the *nic* loci are linked, but we

do not need to know. The frequencies of the M$_{II}$ patterns for each locus give the distance from locus to centromere. (We can worry about whether it is the same centromere later.)

Remember that an M$_{II}$ pattern is any pattern that is not two blocks of four. Let's start with the distance between the *nic* locus and the centromere. All we have to do is add the ascus types 4, 5, 6, and 7, because they are all M$_{II}$ patterns for the *nic* locus. The total is $5 + 90 + 1 + 5 = 101$, or 10.1 percent. In this chapter, we have seen that, to convert this percentage into map units, we must divide by 2, which gives 5.05 m.u.

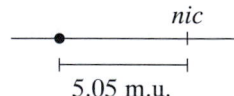

We do the same thing for the *ad* locus. Here the total of the M$_{II}$ patterns is given by types 3, 5, 6, and 7 and is $90 + 90 + 1 + 5 = 186$, or 18.6 percent, which is 9.3 m.u.

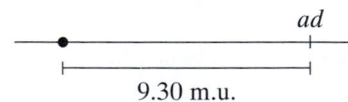

Now we have to put the two together and decide between the following alternatives, all of which are compatible with the preceding locus-to-centromere distances:

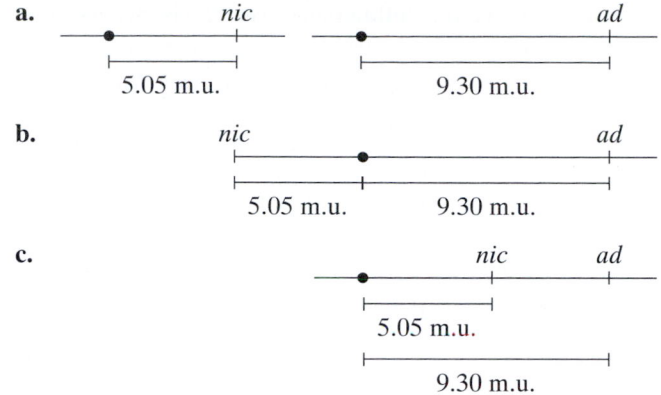

a.

nic   ad

5.05 m.u.   9.30 m.u.

b.

nic   ad

5.05 m.u.   9.30 m.u.

c.

nic   ad

5.05 m.u.

9.30 m.u.

Here a combination of common sense and simple analysis tells us which alternative is correct. First, an inspection of the asci reveals that the most common single type is the one labeled 1, which contains more than 80 percent of all the asci. This type contains only nic^+ ad and nic ad^+ genotypes, and they are *parental* genotypes. So we know that recombination is quite low and the loci are certainly linked. This rules out alternative a.

Now consider alternative c; if this alternative were correct, a crossover between the centromere and the *nic* locus would generate not only an M_{II} pattern for that locus, but also an M_{II} pattern for the *ad* locus, because it is farther from the centromere than *nic*. The ascus pattern produced by alternative c should be:

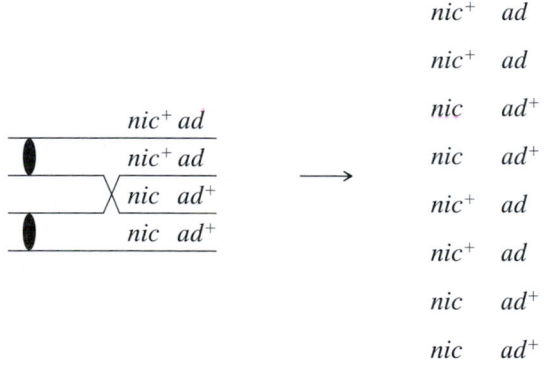

nic$^+$ ad

nic$^+$ ad

nic ad$^+$

nic ad$^+$

nic$^+$ ad

nic$^+$ ad

nic ad$^+$

nic ad$^+$

Remember that the *nic* locus shows M_{II} patterns in asci types 4, 5, 6, and 7 (a total of 101 asci); of them, type 5 is the very one that we are talking about and contains 90 asci. Therefore, alternative c appears to be correct because ascus type 5 comprises about 90 percent of the M_{II} asci for the *nic* locus. This relation would not hold if alternative b were correct, because crossovers on either side of the centromere would generate the M_{II} patterns for the *nic* and the *ad* loci independently.

Is the map distance from *nic* to *ad* simply $9.30 - 5.05 = 4.25$ m.u.? Close, but not quite. The best way of calculating map distances between loci is always by measuring the recombinant frequency (RF). We could go through the asci and count all the recombinant ascospores, but it is simpler to use the formula $RF = \frac{1}{2}T + NPD$. The T asci are classes 3, 4, and 7, and the NPD asci are classes 2 and 6.

Hence, $RF + [\frac{1}{2}(100) + 2]/1000 = 5.2$ percent of 5.2 m.u., and a better map is:

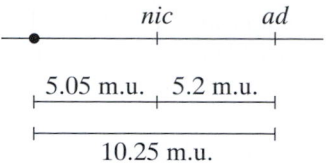

nic   ad

5.05 m.u.   5.2 m.u.

10.25 m.u.

The reason for the underestimate of the *ad*-to-centromere distance calculated from the M_{II} frequency is the occurrence of double crossovers, which can produce an M_I pattern for *ad*, as in ascus type 4:

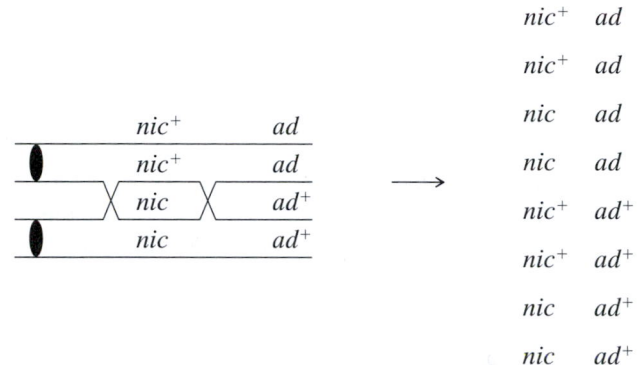

nic$^+$ ad

nic$^+$ ad

nic   ad

nic   ad

nic$^+$ ad$^+$

nic$^+$ ad$^+$

nic   ad$^+$

nic   ad$^+$

2. In *Aspergillus*, the recessive chromosome VI alleles *leu1*, *met5*, *thi3*, *pro2*, and *ad2* confer requirements for leucine, methionine, thiamine, proline, and adenine, respectively, whereas their wild-type alleles confer no such requirements. At the tip of chromosome VI there is a locus with a recessive allele *su* that suppresses *ad2*, so strains expressing *ad2* and *su* require no adenine. A diploid strain is made by combining the following haploid genotypes, where the loci, with the exception of *su*, are written in no particular order:

   su   *leu1*   *met5*   *thi3*   *pro2*   *ad2*

and

   su$^+$ *leu1$^+$*   *met5$^+$*   *thi3$^+$*   *pro2$^+$*   *ad2*

Diploid asexual spores were spread on a medium containing all supplements except adenine. Most spores did not grow, owing to the recessiveness of *su*, but a few did grow into colonies; 100 of these colonies were removed and tested for the phenotypes of the other markers. The following four classes were found (note that they are diploid phenotypes, not haploid genotypes):

(1)	su	leu$^+$	met$^-$	thi$^+$	pro$^+$	ad$^-$	60
(2)	su	leu$^-$	met$^-$	thi$^+$	pro$^+$	ad$^-$	25
(3)	su	leu$^+$	met$^+$	thi$^+$	pro$^+$	ad$^-$	10
(4)	su	leu$^-$	met$^-$	thi$^-$	pro$^+$	ad$^-$	5

a. Explain the production of these four classes and their relative amounts.

b. Why do you think that no colonies expressing *pro2* were recovered?

◆ Solution ◆

a. The experiment concerns the behavior of diploid cells at mitosis, so the four classes must be explained by a mitotic mechanism. It seems likely that the *su* allele in all classes has been made homozygous *su/su*, because only in that condition can it suppress an *ad2/ad2* homozygote to permit growth without adenine. This then was the basis of the original selection of the 100 colonies. But evidently other alleles have become homozygous too — and in different combinations in different classes.

This chapter demonstrates that mitotic crossing-over can produce homozygosity in any recessive allele that is distal to the crossover. Can crossovers in different regions of chromosome VI explain the different classes? Inspection of the classes shows that homozygosity can be produced for either *su* alone (class 3), or *su* and *met* (class 1), or *su, met,* and *leu* (class 2), or *su, met, leu,* and *thi* (class 4). This virtually dictates to us that the order must be *su–met–leu–thi*–centromere. The crossovers responsible for homozygosity must then be in the following regions:

su —— *met* —— *leu* —— *thi* —— centromere
　　|　　　　|　　　　|　　　　|
　Class 3　Class 1　Class 2　Class 4

For example, the following crossover is necessary to produce class 2:

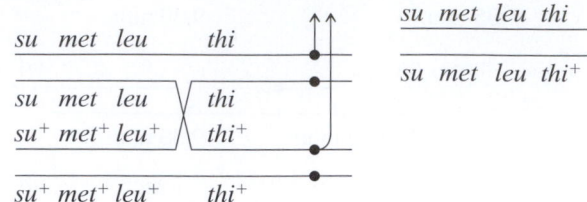

The relative sizes of the classes must be indicative of the relative distances separating the loci as follows:

<pre>
     su met   leu   thi
    ┝━┿━┿━━━━━━━━━━━━━━━┿━━━┿━●
     10   60   25   5
</pre>

Note that these units are relative proportions and are not the same as meiotic map units.

b. We are told that *pro2* is on chromosome VI, yet it is never made homozygous. Note the importance of the fact that, to select for mitotic crossovers, we are making use of an allele *su*, which is at the tip of one arm. Recessive alleles in the other arm would not be simultaneously made homozygous by these crossovers, so *pro2* is probably in the other arm. We have no way of knowing where *ad2* is, because it is homozygous from the outset.

PROBLEMS

1. The *Neurospora* cross *al-2*⁺ × *al-2* is made. A linear tetrad analysis reveals that the second-division segregation frequency is 8 percent.

   a. Draw two examples of second-division segregation patterns in this cross.

   b. What can the 8 percent value be used to calculate?

2. From the fungal cross *arg-6 · al-2* × *arg-6*⁺ · *al-2*⁺, what will the spore genotypes be in unordered tetrads that are **(a)** parental ditypes? **(b)** tetratypes? **(c)** nonparental ditypes?

3. For a certain chromosomal region, the mean number of crossovers at meiosis is calculated to be two per meiosis. In that region, what proportion of meioses are predicted to have **(a)** no crossovers? **(b)** one crossover? **(c)** two crossovers?

4. In a *Drosophila* of genotype

<pre>
      y   sn
    ─────────────────────●
          2   1
    ─────────────────────●
      y⁺   sn⁺
</pre>

   y determines yellow body and *y*⁺ determines brown body; *sn* determines singed hairs and *sn*⁺ unsinged.

What is the detectable outcome if there is a mitotic crossover in the course of development in **(a)** region 1? **(b)** region 2?

5. Every Friday night, genetics student Jean Allele, exhausted by her studies, goes to the students' union bowling lane to relax. But even there, she is haunted by her genetic studies. The rather modest bowling lane has only four bowling balls: two red and two blue. They are bowled at the pins and are then collected and returned down the chute in random order, coming to rest at the end stop. Over the evening, Jean notices familiar patterns of the four balls as they come to rest at the stop. Compulsively, she counts the different patterns. What patterns did she see, what were their frequencies, and what is the relevance of this matter to genetics?

6. **a.** Use the mapping function to calculate the corrected map distance between loci having a recombinant frequency of 20 percent. Remember that an *m* value of 1 is equal to 50 corrected map units.

   b. If you obtain an RF value of 45 percent in one experiment, what can you say about linkage? (The actual numbers that you observed were 58 and 52 parental types and 47 and 43 recombinant types of 200 progeny.)

7. In a haploid yeast, a cross between $arg^- \cdot ad^- \cdot nic^+ \cdot leu^+$ and $arg^+ \cdot ad^+ \cdot nic^- \cdot leu^-$ produces haploid sexual spores, 20 of which are isolated at random. When the resulting cultures are tested on various media, they give the following results, where Arg means arginine; Ad, adenine; Nic, nicotinamide; Leu, leucine; +, growth; and −, no growth.

	MINIMAL MEDIUM, PLUS			
Culture	Arg, Ad, Nic	Arg, Ad, Leu	Arg, Nic, Leu	Ad, Nic, Leu
1	+	+	−	−
2	−	−	+	+
3	−	+	−	+
4	+	−	+	−
5	−	−	+	+
6	+	+	−	−
7	+	+	−	−
8	−	−	+	+
9	+	−	+	−
10	−	+	−	+
11	−	+	−	+
12	+	−	+	−
13	+	+	−	−
14	+	−	+	−
15	−	+	−	+
16	+	−	−	−
17	+	+	−	−
18	−	−	+	+
19	+	+	−	−
20	−	+	−	+

a. What can you say about linkage among these genes?

b. What is the origin of culture 16?

8. You measure the frequency of recombinants between the linked loci waxy (*wx*) and shrunken (*sh*) on chromosome 9 of corn. The RF is 36 percent. In what proportion of meiocytes would there be

a. no crossovers between the *wx* and *sh* loci?

b. one crossover between the *wx* and *sh* loci?

c. two crossovers between the *wx* and *sh* loci?

d. at least one crossover between the *wx* and *sh* loci?

* 9. In a tetrad analysis, the linkage arrangement of the *p* and *q* loci is as follows:

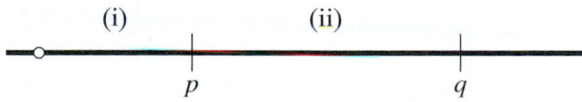

Assume that,

- in region i, there is no crossover in 88 percent of meioses and a single crossover in 12 percent of meioses;

- in region ii, there is no crossover in 80 percent of meioses and a single crossover in 20 percent of meioses;

- there is no interference (in other words, the situation in one region does not affect what is going on in the other region).

What proportions of tetrads will be of the following types? **(a)** $M_I M_I$, PD; **(b)** $M_I M_I$, NPD; **(c)** $M_I M_{II}$, T; **(d)** $M_{II} M_I$, T; **(e)** $M_{II} M_{II}$, PD; **(f)** $M_{II} M_{II}$, NPD; **(g)** $M_{II} M_{II}$, T. (**Note:** Here the M pattern written first is the one that pertains to the *p* locus.) **Hint:** The easiest way to do this problem is to start by calculating the frequencies of asci with crossovers in both regions, region 1, region 2, and neither region. Then determine what M_I and M_{II} patterns result.

10. The following cross is made in *Neurospora*: $a^+ b^+ c^+ d^+ \times a\, b\, c\, d$ (loci *a, b, c,* and *d* are linked in the order written). Construct crossover diagrams to illustrate how the following unordered ascus patterns could arise:

a. $a^+ b^+ c\ d^+$
$a\ b\ c\ d^+$
$a^+ b\ c^+ d$
$a\ b^+ c^+ d$

b. $a^+ b\ c\ d$
$a^+ b^+ c^+ d$
$a\ b\ c\ d^+$
$a\ b^+ c^+ d^+$

c. $a^+ b^+ c\ d^+$
$a^+ b\ c^+ d$
$a\ b^+ c\ d^+$
$a\ b\ c^+ d$

d. $a^+ b\ c^+ d$
$a^+ b\ c^+ d$
$a\ b^+ c\ d^+$
$a\ b^+ c\ d^+$

e. $a^+ b\ c\ d$
$a\ b\ c\ d^+$
$a^+ b^+ c^+ d$
$a\ b^+ c^+ d^+$

f. $a^+ b\ c^+ d$
$a^+ b^+ c^+ d$
$a\ b\ c\ d^+$
$a\ b^+ c\ d^+$

g. $a^+ b\ c^+ d^+$
$a^+ b\ c^+ d^+$
$a\ b^+ c\ d$
$a\ b^+ c\ d$

h. $a^+ b\ c^+ d$
$a\ b^+ c\ d^+$
$a^+ b\ c^+ d$
$a\ b^+ c\ d^+$

11. A *Neurospora* cross was made between one strain that carried the mating-type allele *A* and the mutant allele *arg-1* and another strain that carried the mating-type allele *a* and the wild-type allele for *arg-1* (+). Four hundred linear tetrads were isolated, and they fell into the following seven classes. (Each class contained several different spore orders within the tetrad.)

1	2	3	4	5	6	7
$A \cdot arg$	$A \cdot +$	$A \cdot arg$	$A \cdot arg$	$A \cdot arg$	$A \cdot +$	$A \cdot +$
$A \cdot arg$	$A \cdot +$	$A \cdot +$	$a \cdot arg$	$a \cdot +$	$a \cdot arg$	$a \cdot arg$
$a \cdot +$	$a \cdot arg$	$a \cdot arg$	$A \cdot +$	$A \cdot arg$	$A \cdot +$	$A \cdot arg$
$a \cdot +$	$a \cdot arg$	$a \cdot +$	$a \cdot +$	$a \cdot +$	$a \cdot arg$	$a \cdot +$
127	125	100	36	2	4	6

a. Deduce the linkage arrangement of the mating-type locus and the *arg-1* locus. Include the centromere or centromeres on any map that you draw. Label *all* intervals in map units.

b. Diagram the meiotic divisions that led to tetrad class 6. Label clearly.

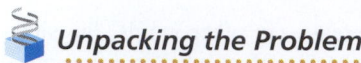

Unpacking the Problem

1. Are fungi generally haploid or diploid?

2. How many ascospores are in the ascus of *Neurospora?* Does your answer match the number presented in this problem? Explain any discrepancy.

3. What is mating type in fungi? How do you think it is determined experimentally?

4. Do the symbols *A* and *a* have anything to do with dominance and recessiveness?

5. What does the symbol *arg-1* mean? How would you test for this genotype?

6. How does the *arg-1* symbol relate to the symbol +?

7. What does the expression *wild type* mean?

8. What does the word *mutant* mean?

9. Does the biological function of the alleles shown have anything to do with the solution of this problem?

10. What does the expression *linear tetrad analysis* mean?

11. What more can be learned from linear tetrad analysis that cannot be learned from unordered tetrad analysis?

12. How is a cross made in a fungus such as *Neurospora?* Explain how to isolate asci and individual ascospores. How does the term *tetrad* relate to the terms *ascus* and *octad?*

13. Where does meiosis take place in the *Neurospora* life cycle? (Show it on a diagram of the life cycle.)

14. What does Problem 11 have to do with meiosis?

15. Can you write out the genotypes of the two parental strains?

16. Why are only four genotypes shown in each class?

17. Why are there only seven classes? How many ways have you learned for classifying tetrads generally? Which of these classifications can be applied to both linear and unordered tetrads? Can you apply these classifications to the tetrads in this problem? (Classify each class in as many ways as possible.) Can you think of more possibilities in this cross? If so, why are they not shown?

18. What does "Each class contained several different spore orders within the tetrad" mean? Why would these different spore orders not change the class?

19. Why is the following class

$$
\begin{array}{l}
a \cdot + \\
a \cdot + \\
A \cdot arg \\
A \cdot arg
\end{array}
$$

not listed?

20. What does the expression *linkage arrangement* mean?

21. What is a genetic *interval?*

22. Why does the problem state "centromere or centromeres" and not just "centromere"? What is the general method for mapping centromeres in tetrad analysis?

23. What is the total frequency of $A \cdot +$ ascospores? (Did you calculate this frequency by using a formula or by inspection? Is this a recombinant genotype? If so, is it the only recombinant genotype?)

24. The first two classes are the most common and are approximately equal in frequency. What does this information tell you? What is their content of parental and recombinant genotypes?

12. In the fungus *Neurospora,* a strain that was auxotrophic for thiamine (mutant allele *t*) was crossed to a strain that was auxotrophic for methionine (mutant allele *m*). Linear asci were isolated and classified into the following groups:

Spore pair	Ascus types					
1 and 2	$t \cdot +$	$t \cdot +$	$t \cdot +$	$t \cdot +$	$t \cdot m$	$t \cdot m$
3 and 4	$t \cdot +$	$t \cdot m$	$+ \cdot m$	$+ \cdot +$	$t \cdot m$	$+ \cdot +$
5 and 6	$+ \cdot m$	$+ \cdot +$	$t \cdot +$	$t \cdot m$	$+ \cdot +$	$t \cdot +$
7 and 8	$+ \cdot m$	$+ \cdot m$	$+ \cdot m$	$+ \cdot m$	$+ \cdot +$	$+ \cdot m$
Number	260	76	4	54	1	5

a. Determine the linkage relation of these two genes to their centromere or centromeres and to each other. Specify distances in map units.

b. Draw diagrams to show the origin of each ascus type.

c. If 1000 randomly selected ascospores are plated on minimal medium, how many are expected to grow into colonies?

13. A geneticist studies 11 different pairs of *Neurospora* loci by making crosses of the type $a \cdot b \times a^+ \cdot b^+$ and then analyzing 100 linear asci from each cross. For the convenience of making a table, the geneticist organizes the data as if all 11 pairs of loci had the same designation — *a* and *b* — as shown here:

NUMBER OF ASCI OF TYPE

Cross	$a \cdot b$ $a \cdot b$ $a^+ \cdot b^+$ $a^+ \cdot b^+$	$a \cdot b^+$ $a \cdot b^+$ $a^+ \cdot b$ $a^+ \cdot b$	$a \cdot b$ $a \cdot b^+$ $a^+ \cdot b^+$ $a^+ \cdot b$	$a \cdot b$ $a^+ \cdot b$ $a^+ \cdot b^+$ $a \cdot b^+$	$a \cdot b$ $a^+ \cdot b^+$ $a^+ \cdot b^+$ $a \cdot b$	$a \cdot b^+$ $a^+ \cdot b$ $a^+ \cdot b$ $a \cdot b^+$	$a \cdot b^+$ $a^+ \cdot b$ $a^+ \cdot b^+$ $a \cdot b$
1	34	34	32	0	0	0	0
2	84	1	15	0	0	0	0
3	55	3	40	0	2	0	0
4	71	1	18	1	8	0	1
5	9	6	24	22	8	10	20
6	31	0	1	3	61	0	4
7	95	0	3	2	0	0	0
8	6	7	20	22	12	11	22
9	69	0	10	18	0	1	2
10	16	14	2	60	1	2	5
11	51	49	0	0	0	0	0

For each cross, map the loci in relation to each other and to centromeres.

14. In *Neurospora,* the *a* locus is 5 m.u. from the centromere on chromosome 1. The *b* locus is 10 m.u. from the centromere on chromosome 7. From the cross of $a; b^+ \times a^+; b$, determine the frequencies of the following types: **(a)** parental ditype asci, **(b)** nonparental ditype asci, **(c)** tetratype asci, **(d)** recombinant ascospores, **(e)** wild-type ascospores. (**Note:** Do not bother with mapping-function complications here.)

15. Three different crosses in *Neurospora* are analyzed on the basis of unordered tetrads. Each cross combines a different pair of linked genes. The results are shown in the following table:

Cross	Parents	Parental ditypes (%)	Tetra- types (%)	Non- parental ditypes (%)
1	$a \cdot b^+ \times a^+ \cdot b$	51	45	4
2	$c \cdot d^+ \times c^+ \cdot d$	64	34	2
3	$e \cdot f^+ \times e^+ \cdot f$	45	50	5

For each cross, calculate:

a. The frequency of recombinants (RF).

b. The uncorrected map distance, based on RF.

c. The corrected map distance, based on tetrad frequencies.

16. A geneticist crosses two yeast strains differing at the linked loci *ura3* (which governs uracil requirement) and *lys4* (which governs lysine requirement):

$$ura3 \; lys4^+ \times ura3^+ \; lys4$$

The geneticist isolates and classifies 300 *unordered* tetrads as follows:

ura3 lys4$^+$ ura3$^+$ lys4 ura3 lys4 ura3$^+$ lys4$^+$	ura3 lys4 ura3 lys4 ura3$^+$ lys4$^+$ ura3$^+$ lys4$^+$	ura3 lys4$^+$ ura3 lys4$^+$ ura3$^+$ lys4 ura3$^+$ lys4
138	12	150

a. What is the recombinant frequency?

b. If it is assumed that there may be zero, one, or two (never more) crossovers between these loci at meiosis, what are the percentages of zero, one, and two crossover meioses?

c. What is the distance between the loci in map units *corrected* for double crossovers?

*17. For an experiment with haploid yeast, you have two different cultures. Each will grow on minimal medium to which arginine has been added, but neither will grow on minimal medium alone. (Minimal medium is inorganic salts plus sugar.) Using appropriate methods, you induce the two cultures to mate. The diploid cells then divide meiotically and form unordered tetrads. Some of the ascospores will grow on minimal medium. You classify a large number of these tetrads for the phenotypes ARG⁻ (arginine-requiring) and ARG⁺ (arginine-independent) and record the following data:

Segregation of ARG⁻:ARG⁺	Frequency (%)
4:0	40
3:1	20
2:2	40

a. Using symbols of your own choosing, assign genotypes to the two parental cultures. For each of the three kinds of segregation, assign genotypes to the segregants.

b. If there is more than one locus governing arginine requirement, are these loci linked?

*18. Four histidine loci are known in *Neurospora.* As shown here, each of the four loci is located on a different chromosome.

In your experiment, you begin with an *ad-3* line from which you recover a cell that also requires histidine. Now you wish to determine which of the four histidine loci determines this requirement. You cross the *ad-3 · his-?* strain with a wild type (*ad-3⁺ his-1⁺* ; *his-2⁺* ; *his-3⁺* ; *his-4⁺*) and analyze ten unordered tetrads: two are PD, six are T, and two are NPD. From this result, which of the four *his* loci is most probably the one that changed from *his⁺* to *his?* (Problem 18 from Luke deLange.)

19. The haploid ascomycete fungus *Sordaria* has ascospores that are normally black. Two ascospore color variants (mutants) are isolated. When mutant 1 is crossed to a wild type, the asci produced contain four black spores and four white spores; when mutant 2 is crossed to a wild type, the asci produced contain four black spores and four tan spores. When mutants 1 and 2 are intercrossed, some asci contain four black and four white spores, some asci contain four tan and four white spores, and some asci contain four white and two black and two tan spores. Explain these results by giving

a. genotypes underlying the three phenotypes.

b. an explanation of the types of asci produced in the crosses to wild type and the intercross.

*20. *Drosophila melanogaster* can have two entire X chromosomes attached to the same centromere:

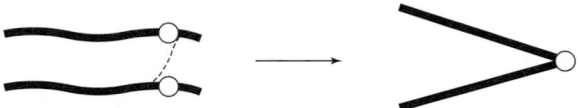

The joined X chromosomes behave as a single chromosome, called an *attached X*. Work out the inheritance of sex chromosomes in crosses of attached-X-bearing females with normal males. In meiosis, an attached-X chromosome duplicates and segregates as follows:

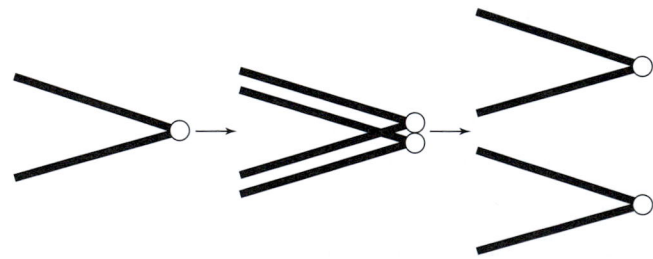

Nonsister chromatids of attached-X chromosomes can cross over. Suppose that you have an attached-X chromosome of the following genotype:

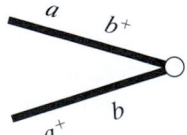

Diagram all the possible genotypic products and their phenotypes **(a)** from single crossovers between *a* and *b* and between *b* and the centromere, and **(b)** from double exchanges between *a* and *b* and between *b* and the centromere. Make sure that you consider all possibilities. Could such a system be considered a kind of tetrad analysis?

21. In *Neurospora*, the mating-type locus, which has alleles *a* and *A*, is 5 m.u. from the centromere on chromosome 1, and the cycloheximide-resistance locus, which has alleles *c* and *C*, is 8 m.u. to the other side of the centromere on the same chromosome. What proportion of *unordered* tetrads from the cross *A C* × *a c* will be of each of the following types?

a. *A c*	**b.** *A C*	**c.** *a C*
A C	*A C*	*a C*
a c	*a c*	*A c*
a C	*a c*	*A c*

22. A cross was made in the fungus *Sordaria*,

$$un \cdot cyh \times un^+ \cdot cyh^+$$

and 100 linear tetrads were isolated. There proved to be six classes in the following proportions:

1	2	3
un · cyh	*un · cyh*	*un · cyh*
un⁺ · cyh	*un · cyh⁺*	*un · cyh*
un · cyh⁺	*un⁺ · cyh*	*un⁺ · cyh⁺*
un⁺ · cyh⁺	*un⁺ · cyh⁺*	*un⁺ · cyh⁺*
15	29	47

4	5	6
un · cyh	*un · cyh⁺*	*un · cyh*
un⁺ · cyh⁺	*un⁺ · cyh*	*un⁺ · cyh⁺*
un · cyh	*un · cyh⁺*	*un · cyh⁺*
un⁺ · cyh⁺	*un⁺ · cyh*	*un⁺ · cyh*
2	2	5

a. Map the genes in relation to each other and to centromeres. Show distances in uncorrected map units.

b. Draw six crossover diagrams, one to explain the origin of each class.

c. Which type of tetrad is missing, and why do you think this is so?

d. Why is class 6 more common than class 4 or 5?

23. *Neurospora's* chromosome 4 carries the *leu3* locus, located near the centromere; its alleles always segregate at the first division. On the other arm of chromosome 4, the *cys2* locus is 8 m.u. away from the centromere. If we cross a *leu3*$^+$ *cys2* strain and a *leu3 cys2*$^+$ strain and if we ignore double and other multiple crossovers, what will we expect as the frequencies of the following seven classes of linear tetrad (where *l* = *leu3* and *c* = *cys2*)?

a. *l c*
    l c
    l$^+$ *c*$^+$
    l$^+$ *c*$^+$

b. *l c*$^+$
    l c$^+$
    l$^+$ *c*
    l$^+$ *c*

c. *l c*
    l c$^+$
    l$^+$ *c*$^+$
    l$^+$ *c*

d. *l c*
    l$^+$ *c*
    l$^+$ *c*$^+$
    l c$^+$

e. *l c*
    l$^+$ *c*$^+$
    l$^+$ *c*$^+$
    l c

f. *l c*$^+$
    l$^+$ *c*
    l$^+$ *c*
    l c$^+$

g. *l c*$^+$
    l$^+$ *c*
    l$^+$ *c*$^+$
    l c

24. Consider the following *Drosophila* alleles:

> *g* codes for gray body (wild type is black)
>
> *c* codes for curly bristles (wild type is straight)
>
> *s* codes for stippled body (wild type is smooth)

These autosomal loci are arranged in the order *g-c-s*-centromere. A homozygous gray, stippled strain was crossed to a homozygous curly strain, and the F$_1$ were generally wild type in their phenotypic appearance, but microscopic examination of the flies revealed that a few rare individuals had one of three basic patterns. The patterns are:

a. A gray, stippled sector next to a curly sector.

b. A gray sector next to a curly sector.

c. Solitary gray sectors.

Explain the origin of these three patterns with a single unifying hypothesis.

25. Previous experiments on the fungus *Aspergillus* had shown that the loci *ad, col, phe, pu, sm,* and *w* are all on the same chromosome, but their order was not known. A diploid was constructed with the following genotype: *ad col phe pu sm w/ad*$^+$ *col*$^+$ *phe*$^+$ *pu*$^+$ *sm*$^+$ *w*$^+$ (the gene order is written alphabetically). When this diploid was cultured, white diploid sectors (*w/w*) were observed and isolated. They were found to be of the following *phenotypes* (none of the white diploid sectors expressed *col*):

ad phe pu sm	41%	
ad phe$^+$ *pu sm*$^+$	30%	
ad phe$^+$ *pu sm*	24%	
ad$^+$ *phe*$^+$ *pu sm*$^+$	5%	

a. What is the likely origin of these sectors?

b. What is the relative order of the six genes and the centromere?

c. Give relative map distances where possible.

d. Why was *col* not expressed by the white diploid cells?

26. An incompletely dominant allele of soybeans, *Y*, causes yellowish leaves in *Y*$^+$/*Y* heterozygotes. However, heterozygotes regularly show rare patches of green adjacent to a patch of very pale yellow, all in the yellowish background. Propose an explanation for these rare patches.

27. You isolate white haploidized sectors from an *Aspergillus* diploid that is *w*$^+$/*w* · *a*$^+$/*a* · *b*$^+$/*b* · *c*$^+$/*c* and score them for *a, b,* and *c*. You find that 25 percent are *w · a · b · c*, 25 percent are *w · a*$^+$ · *b*$^+$ · *c*$^+$, 25 percent are *w · a · b*$^+$ · *c*, and 25 percent are *w · a*$^+$ · *b · c*$^+$. What linkage relations can you deduce from these frequencies? Sketch your conclusions.

28. You know that two loci, *y* and *ribo,* are linked to *Aspergillus,* but you do not know their locations relative to the centromere. You have a diploid culture of genotype *y*$^+$ *ribo*$^+$/*y ribo* that is green and that grows without riboflavin supplementation. You notice some yellow sectors in the culture and study them. They are diploid, and you discover that 80 percent of them can grow without riboflavin, whereas 20 percent require media supplemented with riboflavin. What is the most likely order of the two loci and the centromere?

29. An *Aspergillus* diploid is *pro*$^+$/*pro* · *fpa*$^+$/*fpa* · *paba*$^+$/*paba*; *pro* is a recessive allele for proline requirement, *fpa* is a recessive allele for fluorophenylalanine resistance, and *paba* is a recessive allele for *para*-aminobenzoic acid (PABA) requirement. When conidia are plated on fluorophenylalanine, only resistant colonies develop. Of 154 *diploid* resistant colonies, 35 do not require proline or PABA, 110 require PABA, and 9 require both.

a. What do these figures indicate?

b. Sketch your conclusions in the form of a map.

c. Some resistant colonies (not the ones described) are haploid. What would you predict their genotype to be?

30. An *Aspergillus* diploid was made by fusing haploid strains of genotype *ad*$^+$ · *leu*$^+$ · *ribo*$^+$ · *fpa*$^+$ · *w*$^+$ and *ad* · *leu* · *ribo* · *fpa* · *w*. The mutant alleles are all recessive and determine requirements for adenine, leucine, and riboflavin; resistance to fluorophenylalanine; and white asexual spores, respectively. When thousands of asexual spores of the diploid strain were spread on medium containing fluorophenylalanine, adenine, leucine, and riboflavin, only 93 colonies grew and were then isolated.

a.   It was determined that 38 colonies were haploid: 22 with the genotype $fpa \cdot leu \cdot ribo \cdot ad^+ \cdot w^+$ and 16 with the genotype $fpa \cdot leu \cdot ribo \cdot ad \cdot w$. What does this result tell us about the linkage of the genes? Summarize with a diagram.

b.   The remaining 55 colonies were diploid; of them, 24 required neither leucine nor riboflavin, 17 required riboflavin but not leucine, and 14 required both. What further linkage information does this result provide? Summarize with a diagram.

31.   Seven human-rodent radiation hybrids were obtained and tested for six different human genome molecular markers A through F. The results are shown here, where a "+" shows the presence of a marker.

RADIATION HYBRIDS

Markers	1	2	3	4	5	6	7
A	−	+	−	−	+	+	−
B	+	−	+	−	−	−	−
C	+	−	+	+	−	+	−
D	−	+	−	+	+	+	−
E	+	−	−	+	+	−	+
F	+	−	−	+	+	−	+

a.   What marker linkages are suggested by these results?

b.   Is there any evidence of markers being on separate chromosomes?

32.   A geneticist succeeds in maintaining three colonies of human−mouse hybrid cells. The only human chromosomes retained by the hybrid cells are those indicated by plus signs in the following table:

Hybrid colony	HUMAN CHROMOSOME							
	1	2	3	4	5	6	7	8
A	+	+	+	+	−	−	−	−
B	+	+	−	−	+	+	−	−
C	+	−	+	−	+	−	+	−

The geneticist tests each of the colonies for the presence of five enzymes (α, β, γ, δ, and ϵ) with the following results: α is active only in colony C, β is active in all three colonies, γ is active only in colonies B and C, δ is active only in colony B, and ϵ shows no activity in any colony. What can the geneticist conclude about the locations of the genes responsible for these enzyme activities?

33.   Consider the following set of eight hybridized human−mouse cell lines:

Cell line	HUMAN CHROMOSOME								
	1	2	6	9	12	13	17	21	X
A	+	+	−	q	−	p	+	+	+
B	+	−	p	+	−	+	+	−	−
C	−	+	+	+	p	−	+	−	+
D	+	+	−	+	+	−	q	−	+
E	p	−	+	−	q	−	+	+	q
F	−	p	−	−	q	−	+	+	p
G	q	+	−	+	+	+	+	−	−
H	+	q	+	−	−	q	+	−	+

Each cell line may carry an intact (numbered) chromosome (+), only its long arm (q), or only its short arm (p), or it may lack the chromosome (−).

The following human enzymes were tested for their presence (+) or absence (−) in cell lines A–H:

	CELL LINE							
Enzyme	A	B	C	D	E	F	G	H
Steroid sulfatase	+	−	+	+	−	+	−	+
Phosphoglucomutase-3	−	−	+	−	+	−	−	+
Esterase D	−	+	−	−	−	−	+	+
Phosphofructokinase	+	−	−	−	+	+	−	−
Amylase	+	+	−	+	+	−	−	+
Galactokinase	+	+	+	+	+	+	+	+

Identify the chromosome carrying each enzyme locus. Where possible, identify the chromosome arm. (Problem 33 from L. A. Snyder, D. Freifelder, and D. L. Hartl, *General Genetics,* Jones and Bartlett, 1985.)

34.   Let's say that you undertake a study of hybridized human and mouse cells because you want to map part of human chromosome 17. Three loci — a, b, and c — on this chromosome are concerned with making the compounds a, b, and c — all of which are essential for growth and are present in mice as well as humans. You fuse mouse cells that are phenotypically $a^- b^- c^-$ with $a^+ b^+ c^+$ human cells. Assume that you find a hybrid in which the only human component is the right arm of chromosome 17 (17R), translocated by some unknown mechanism to a mouse chromosome. The hybrid can make the compounds a, b, and c. Then you treat cells with adenovirus, which causes chromosome breaks. Assume that you can isolate 200 lines in which bits of the translocated 17R have been clipped off. You test these lines for the ability to make a, b, and c and obtain the following results:

Number	Can make
0	a only
0	b only
12	c only
0	a and b only

80	b and c only
0	a and c only
60	a, b, and c
48	nothing

a. How would these different types arise?

b. Are *a*, *b*, and *c* all located on the right arm of chromosome 17? If so, draw a map indicating their relative positions.

c. How would banding patterns help you in your analysis?

35. In *Neurospora,* the mutant alleles *thr* and *met* stand for threonine and methionine requirement, respectively. The two genes are located very close together on the right arm of chromosome 4. The *thr* locus is closest to the centromere. However crossovers are almost never observed between the *thr* and *met* loci. They are located 8 map units from the centromere. Assuming that no crossovers occur between *thr* and *met,* in a cross of wild type × *thr met* what proportion of linear tetrads will be (the M pattern for the *thr* locus is written first):

a. $M_I M_I$ PD?

b. $M_I M_I$ NPD?

c. $M_I M_{II}$ T?

d. $M_{II} M_I$ T?

e. $M_{II} M_{II}$ PD?

f. $M_{II} M_{II}$ NPD?

g. $M_{II} M_{II}$ T?

What proportion of random ascospores will grow and form colonies on minimal medium?

36. *Gelasinospora* is a fungal species that produces unordered tetrads. A cross was made between two *Gelasinospora* mutants, one of which was *al* (albino cells instead of the usual orange) and the other was *cp* (compact colony instead of the usual spreading type). Two unordered tetrads were chosen at random and isolated. They contained the following genotypes:

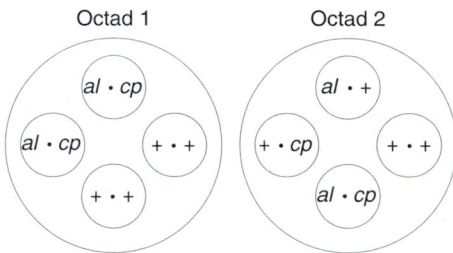

Octad 1   Octad 2

Draw meiosis diagrams that illustrate clearly how these two types originated under the following alternative hypotheses:

a. The two loci are linked.

b. The two loci are on different chromosomes.

Because these two octads were chosen at random, should their genotypes be used to judge which of the two hypotheses is more likely? Explain.

37. In *Drosophila,* the genes for ebony body (*e*) and stubby bristles (*s*) are linked on the same arm of chromosome 2. Flies of genotype $+ s/e +$ develop predominantly as wild type but occasionally show two different types of unexpected abnormalities on their bodies. The first abnormality is pairs of adjacent patches, one with stubby bristles and the other with ebony body. The second abnormality is solitary patches of ebony color.

a. Draw diagrams to show the likely origins of these two types of unexpected abnormalities.

b. Explain why there are no single patches that are stubby.

7

GENE TRANSFER IN BACTERIA AND THEIR VIRUSES

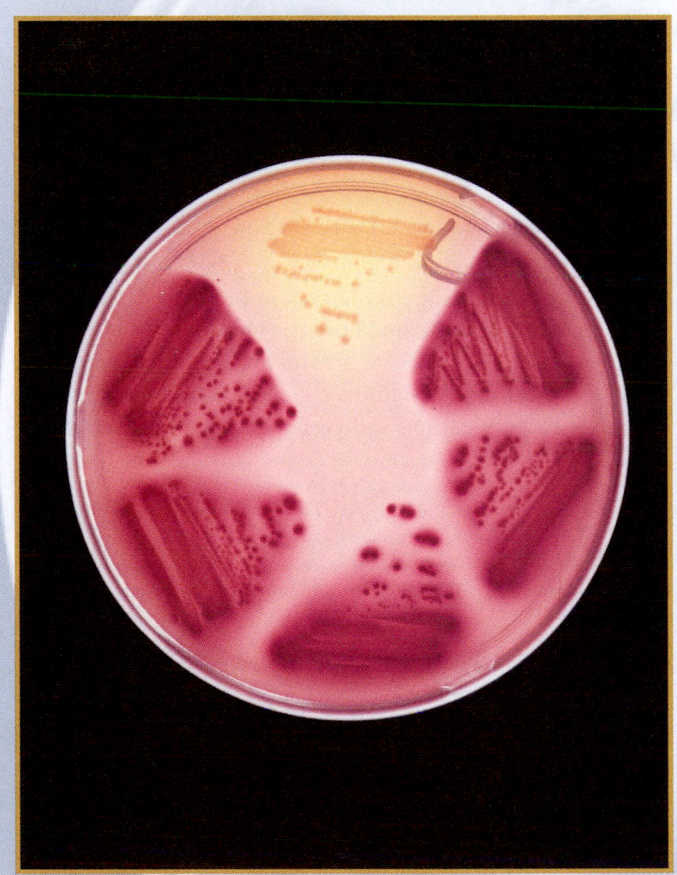

Bacterial colonies on staining medium.
Lac⁻ cells are white. Lac⁺ colonies are stained red.
(Jeffrey H. Miller.)

Key Concepts

The fertility factor (F) permits bacterial cells to transfer DNA to other cells through the process of conjugation.

F can exist in the cytoplasm or can be integrated into the bacterial chromosome.

When F is integrated in the chromosome, chromosomal markers can be transferred during conjugation.

Bacteriophages can also transfer DNA from one bacterial cell to another.

In generalized transduction, random chromosome fragments are incorporated into the heads of certain bacterial phages and transferred to other cells by infection.

In specialized transduction, specific genes near the phage integration sites on the bacterial chromosome are mistakenly incorporated into the phage genome and transferred to other cells by infection.

The different methods of gene transfer in bacteria allow geneticists to make detailed maps of bacterial genes.

Thus far, we have dealt almost exclusively with genes that are packed into chromosomes and enclosed within the nuclei of eukaryotic organisms. However, a very large part of the history of genetics and current genetic analysis (particularly molecular genetics) is concerned with prokaryotic organisms, which have no distinct nuclei, and with viruses. Although **viruses** share some of the definitive properties of organisms, many biologists regard viruses as distinct entities that in some sense are not fully alive. They are not cells; they cannot grow or multiply alone. To reproduce, they must parasitize living cells and use the metabolic machinery of these cells. Nevertheless, viruses do have hereditary properties that can be subjected to genetic analysis. Genetic analysis of bacteria and their viruses has been a source of key insights into the nature and structure of the genetic material, the genetic code, and mutation.

The **prokaryotes** are the blue-green algae, now classified as *cyanobacteria*, and the bacteria. The viruses that parasitize bacteria are called **bacteriophages** or, simply, **phages.** Pioneering work with bacteriophages has led to a great deal of recent research on tumor-causing viruses and other kinds of animal and plant viruses.

Compared with eukaryotes, prokaryotic organisms and viruses have simple chromosomes that are not contained within a nuclear membrane. Because they are monoploid, these chromosomes do not undergo meiosis, but they do go through stages analogous to meiosis. The approach to the genetic analysis of recombination in these organisms is surprisingly similar to that for eukaryotes.

The opportunity for genetic recombination in bacteria can arise in several different ways, as this chapter will detail. In the first process to be examined here, **conjugation,** one bacterial cell transfers DNA segments to another cell by direct cell-to-cell contact. A bacterial cell can also acquire a piece of DNA from the environment and incorporate this DNA into its own chromosome; this procedure is called **transformation.** In addition, certain bacterial viruses can pick up a piece of DNA from one bacterial cell and inject it into another, where it can be incorporated into the chromosome, in a process known as **transduction.**

Working with microorganisms

Bacteria can be grown in a liquid medium or on a solid surface, such as an agar gel, as long as basic nutritive ingredients are supplied. In a liquid medium, the bacteria divide by binary fission: they multiply geometrically until the nutrients are exhausted or until toxic factors (waste products) accumulate to levels that halt the population growth. A small amount of such a liquid culture can be pipetted onto a petri plate containing an agar medium and spread evenly on the surface with a sterile spreader, in a process called **plating** (Figure 7-1). Each cell then reproduces by fission. Because the cells are immobilized in the gel, all the daughter cells remain together in a clump. When this mass reaches more than 10^7 cells, it becomes visible to the naked eye as a

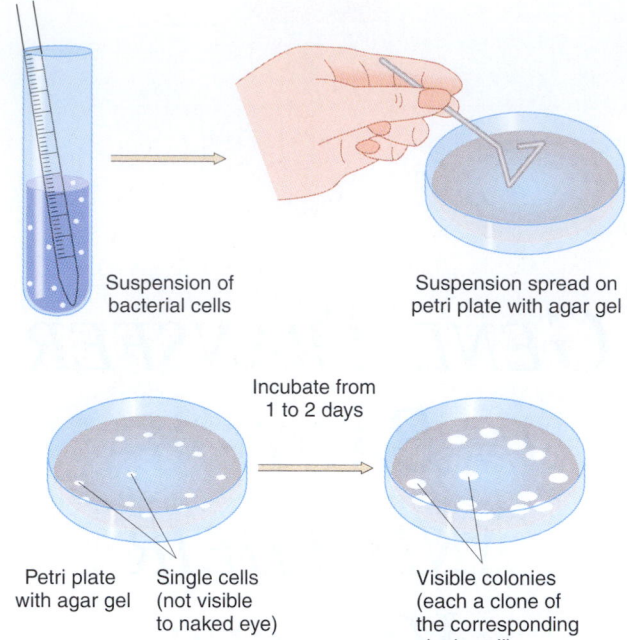

Suspension of bacterial cells

Suspension spread on petri plate with agar gel

Incubate from 1 to 2 days

Petri plate with agar gel

Single cells (not visible to naked eye)

Visible colonies (each a clone of the corresponding single cell)

Figure 7-1   Method of growing bacteria in the laboratory. A few bacterial cells that have been grown in a liquid medium containing nutrients can be spread on an agar medium also containing the appropriate nutrients. Each of these original cells will divide many times by binary fission and eventually give rise to a colony. All cells in a colony, having been derived from a single cell, will have the same genotype and phenotype.

colony. If the initially plated sample contains very few cells, then each distinct colony on the plate will be derived from a single original cell. Members of a colony that share a single genetic ancestor are known as **clones.**

Wild-type bacteria are **prototrophic:** they can grow colonies on **minimal medium** — a substrate containing only inorganic salts, a carbon source for energy, and water. Mu-

7-1 **TABLE**		Some Genotypic Symbols Used in Bacterial Genetics

Symbol	Character or phenotype associated with symbol
bio⁻	Requires biotin added as a supplement to minimal medium
arg⁻	Requires arginine added as a supplement to minimal medium
met⁻	Requires methionine added as a supplement to minimal medium
lac⁻	Cannot utilize lactose as a carbon source
gal⁻	Cannot utilize galactose as a carbon source
*str*ʳ	Resistant to the antibiotic streptomycin
*str*ˢ	Sensitive to the antibiotic streptomycin

Note: Minimal medium is the basic synthetic medium for bacterial growth without nutrient supplements.

tant clones can be identified because they are **auxotrophic:** they will not grow unless the medium contains one or more specific nutrients — say, adenine or threonine and biotin. Furthermore, wild types are susceptible to certain inhibitors, such as streptomycin, whereas **resistant mutants** can form colonies despite the presence of the inhibitor. These properties allow the geneticist to distinguish different phenotypes among plated colonies.

For many characters, the phenotype of a clone can be readily determined through visual inspection or simple chemical tests. This phenotype can then be assigned to the original cell of the clone, and the frequencies of various phenotypes in the pipetted sample can be determined. Table 7-1 lists some bacterial phenotypes and their genetic symbols.

Bacterial conjugation

This section and subsequent sections describe the discovery of gene transfer in bacteria and explain several types of gene transfer and their use in bacterial genetics. First, we shall consider conjugation, which requires cell-to-cell contact. Conjugation was the first extensively studied method of gene transfer.

Discovery of conjugation

Do bacteria possess any processes similar to sexual reproduction and recombination? The question was answered in 1946 by the elegantly simple experimental work of Joshua Lederberg and Edward Tatum, who studied two strains of *Escherichia coli* with different nutritional requirements. Strain A would grow on a minimal medium only if the medium were supplemented with methionine and biotin; strain B would grow on a minimal medium only if it were supplemented with threonine, leucine, and thiamine. Thus, we can designate strain A as $met^-\ bio^-\ thr^+\ leu^+\ thi^+$ and strain B as $met^+\ bio^+\ thr^-\ leu^-\ thi^-$. Figure 7-2a displays in simplified form the concept of their experiment. Here,

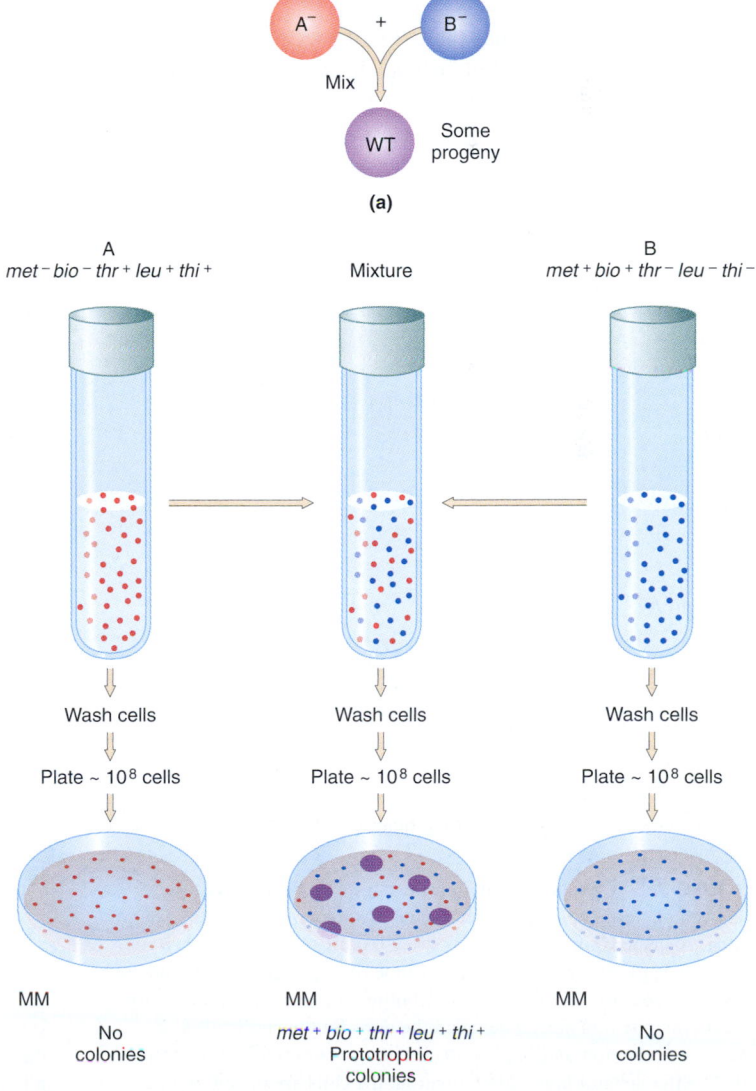

Figure 7-2   Demonstration by Lederberg and Tatum of genetic recombination between bacterial cells. Cells of type A or type B cannot grow on an unsupplemented (minimal) medium (MM), because A and B each carry mutations that cause the inability to synthesize constituents needed for cell growth. When A and B are mixed for a few hours and then plated, however, a few colonies appear on the agar plate. These colonies derive from single cells in which an exchange of genetic material has occurred; they are therefore capable of synthesizing all the required constituents of metabolism.

strains A and B are mixed together, and some of the progeny are now wild type, having regained the ability to grow without added nutrients. Figure 7-2b illustrates their experiment in more detail.

Lederberg and Tatum plated bacteria into dishes containing only unsupplemented minimal medium. Some of the dishes were plated only with strain A bacteria, some only with strain B bacteria, and some with a mixture of strain A and strain B bacteria that had been incubated together for several hours in a liquid medium containing all the supplements. No colonies arose on plates containing either strain A or strain B alone, showing that back mutations cannot restore prototrophy, the ability to grow on unsupplemented minimal medium. However, the plates that received the mixture of the two strains produced growing colonies at a frequency of 1 in every 10,000,000 cells plated (in scientific notation, 1×10^{-7}). This observation suggested that some form of recombination of genes had taken place between the genomes of the two strains to produce prototrophs.

Requirement for physical contact

It could be suggested that the cells of the two strains do not really exchange genes but instead leak substances that the other cells can absorb and use for growing. This possibility of "cross feeding" was ruled out by Bernard Davis. He constructed a U-tube in which the two arms were separated by a fine filter. The pores of the filter were too small to allow bacteria to pass through but large enough to allow easy passage of the fluid medium and any dissolved substances (Figure 7-3). Strain A was put in one arm; strain B in the other. After the strains had been incubated for a while, Davis tested the content of each arm to see if cells had become able to grow on a minimal medium, and none were found. In other words, *physical contact* between the two strains was needed for wild-type cells to form. It looked as though some kind of gene transfer had taken place, and genetic recombinants were indeed produced.

Discovery of the fertility factor (F)

In 1953, William Hayes determined that genetic transfer occurred in one direction in the above types of crosses. Therefore, the transfer of genetic material in *E. coli* is not reciprocal. One cell acts as donor, and the other cell acts as the recipient. This kind of unidirectional transfer of genes was originally compared to a sexual difference, with the donor being termed "male" and the recipient "female." However, this type of gene transfer is not true sexual reproduction. In **bacterial gene transfer,** one organism receives genetic information from a **donor;** the **recipient** is changed by that information. In **sexual reproduction,** two organisms donate equally (or nearly so) to the formation of a new organism, but only in exceptional cases is either of the donors changed.

MESSAGE ··

The transfer of genetic material in *E. coli* is not reciprocal. One cell acts as the donor, and the other cell acts as the recipient.

Loss and regain of ability to transfer. By accident, Hayes discovered a variant of his original donor strain that would not produce recombinants on crossing with the recipient strain. Apparently, the donor-type strains had lost the ability to transfer genetic material and had changed into recipient-type strains. In his analysis of this "sterile" donor variant, Hayes realized that the fertility (ability to donate) of *E. coli* could be lost and regained rather easily. Hayes suggested that donor ability is itself a hereditary state imposed

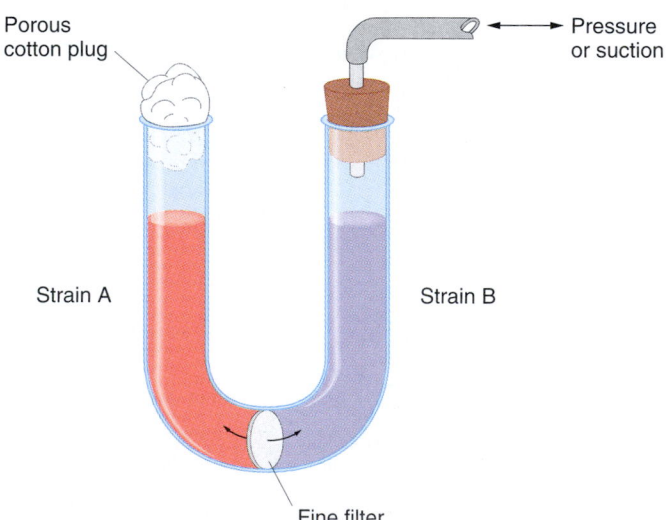

Figure 7-3 Experiment demonstrating that physical contact between bacterial cells is needed for genetic recombination to take place. A suspension of a bacterial strain unable to synthesize certain nutrients is placed in one arm of a U-tube. A strain genetically unable to synthesize different required metabolites is placed in the other arm. Liquid may be transferred between the arms by the application of pressure or suction, but bacterial cells cannot pass through the center filter. After several hours of incubation, the cells are plated, but no colonies grow on the minimal medium.

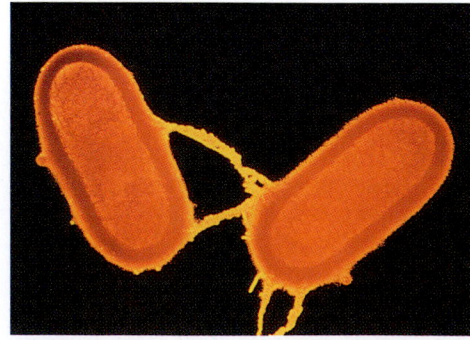

Figure 7-4 Bacteria can transfer plasmids (circles of DNA), through conjugation. A donor cell extends one or more projections — pili — that attach to a recipient cell and pull the two bacteria together. (Oliver Meckes/MPI-Tübingen, Photo Researchers.)

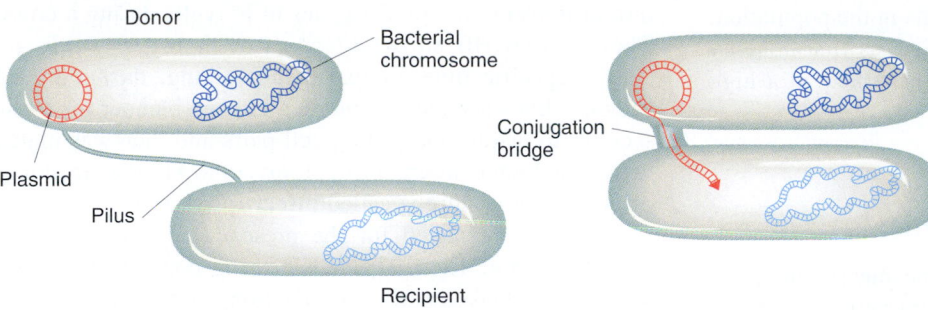

Figure 7-5   (a) During conjugation, the pilus pulls two bacteria together. (b) Next, a bridge (essentially a pore) forms between the two cells. Then one strand of plasmid DNA passes into the recipient bacterium, and each single strand becomes double stranded again.

by a **fertility factor (F).** Strains that carry F can donate, and are designated **F⁺**. Strains that lack F cannot donate and are recipients. These strains are designated **F⁻**.

Transfer of F during conjugation.   Recombinant genotypes for marker genes are relatively rare in bacterial crosses, Hayes noted, but the F factor apparently was transmitted effectively during physical contact, or **conjugation.** A kind of "infectious transfer" of the F factor seemed to be taking place. We now know much more about the process of conjugation and about F, which is an example of a plasmid that can replicate in the cytoplasm independently of the host chromosome. Figures 7-4 and 7-5 show how bacteria can transfer plasmids such as F. The F plasmid directs the synthesis of pili, projections that initiate contact with a recipient (Figure 7-4) and draw it closer, allowing the F DNA to pass through a pore into the recipient cell. One strand of the double-stranded F DNA is transferred and then DNA replication restores the complementary strand in both the donor and the recipient. This replication results in a copy of F remaining in the donor and another appearing in the recipient, as shown in Figure 7-5.

Hfr strains

An important breakthrough came when Luca Cavalli-Sforza discovered a derivative of an F⁺ strain. On crossing with F⁻ strains this new strain produced 1000 times as many recombinants for genetic markers as did a normal F⁺ strain. Cavalli-Sforza designated this derivative an Hfr strain to indicate a high frequency of recombination. In Hfr × F⁻ crosses, virtually none of the F⁻ parents were converted into F⁺ or into Hfr. This result is in contrast with F⁺ × F⁻ crosses, where infectious transfer of F results in a large proportion of the F⁻ parents being converted into F⁺. Figure 7-6 portrays this concept. It became apparent that an Hfr strain results from the integration of the F factor into the chromosome, as pictured in Figure 7-6a.

Now, during conjugation between an Hfr cell and a F⁻ cell a part of the chromosome is transferred with F. Random breakage interrupts the transfer before the entire chromosome is transferred. The chromosomal fragment can then recombine with the recipient chromosome. Clearly, the low level of chromosomal marker transfer observed by Lederberg and Tatum (see Figure 7-2) in an F⁺ × F⁻ cross can be

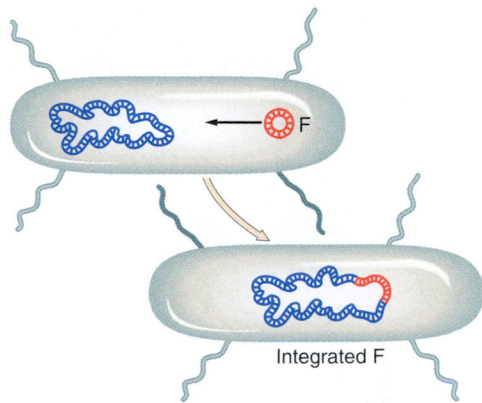

F is integrated into the host chromosome

(a)

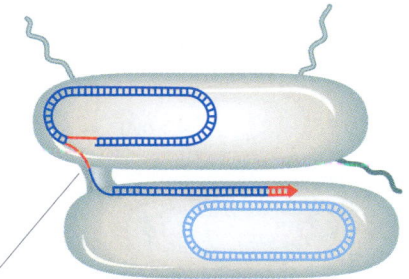

A single strand of F is transferred, along with a copy of part of the host chromosome, to a recipient cell, where a second strand is synthesized.

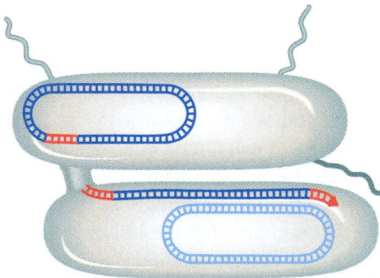

A copy of the host chromosome with F integrated (both generated by replication) remains in the donor cell after replication of the remaining single strand.

(b)

Figure 7-6   The transfer of *E. coli* chromosomal markers mediated by F. (a) Occasionally, the independent F factor combines with the *E. coli* chromosome. (b) When the integrated F transfers to another *E. coli* cell during conjugation, it carries along any *E. coli* DNA that is attached, thus transferring host chromosomal markers to a new cell.

explained by the presence of rare Hfr cells in the population. When these cells are isolated and purified, as first done by Cavalli, they now transfer chromosomal markers at a high frequency, because every cell is an Hfr.

Determining linkage from interrupted-mating experiments

The exact nature of Hfr strains became clearer in 1957, when Elie Wollman and François Jacob investigated the pattern of transmission of Hfr genes to F$^-$ cells during a cross. They crossed Hfr str^s a^+ b^+ c^+ d^+ with F$^-$ str^r a^- b^- c^- d^-. At specific time intervals after mixing, they removed samples. Each sample was put in a kitchen blender for a few seconds to disrupt the mating cell pairs and then was plated onto a medium containing streptomycin to kill the Hfr donor cells. This procedure is called **interrupted mating.** The str^r cells then were tested for the presence of marker alleles from the donor. Those str^r cells bearing donor marker alleles must have taken part in conjugation; such cells are called **exconjugants.** Figure 7-7a shows a plot of the results; azi^r, ton^r, lac^+, and gal^+ correspond to the a^+, b^+, c^+, and d^+ mentioned in our generalized description of the experiment. Figure 7-7b portrays the transfer of markers.

The most striking thing about these results is that each donor allele first appeared in the F$^-$ recipients at a specific time after mating began. Furthermore, the donor alleles appeared in a specific sequence. Finally, the maximal yield of cells containing a specific donor allele was smaller for the donor markers that entered later. Putting all these observations together, Wollman and Jacob concluded that gene transfer occurs from a fixed point on the donor chromosome, termed the **origin (O),** and continues in a linear fashion.

MESSAGE

The Hfr chromosome, originally circular, unwinds and is transferred to the F$^-$ cell in a linear fashion. The unwinding and transfer begin from a specific point at one end of the integrated F, called the origin or O. The farther a gene is from O, the later it is transferred to the F$^-$; the transfer process most likely will stop before the farthermost genes are transferred.

Wollman and Jacob realized that it would be easy to construct linkage maps from the interrupted-mating results, using as a measure of "distance" the times at which the donor alleles first appear after mating. The units of distance in this case are minutes. Thus, if b^+ begins to enter the F$^-$ cell 10 minutes after a^+ begins to enter, then a^+ and b^+ are 10 units apart (Figure 7-8). Like the maps based on crossover frequencies, these linkage maps are purely genetic constructions; at the time, they had no known physical basis.

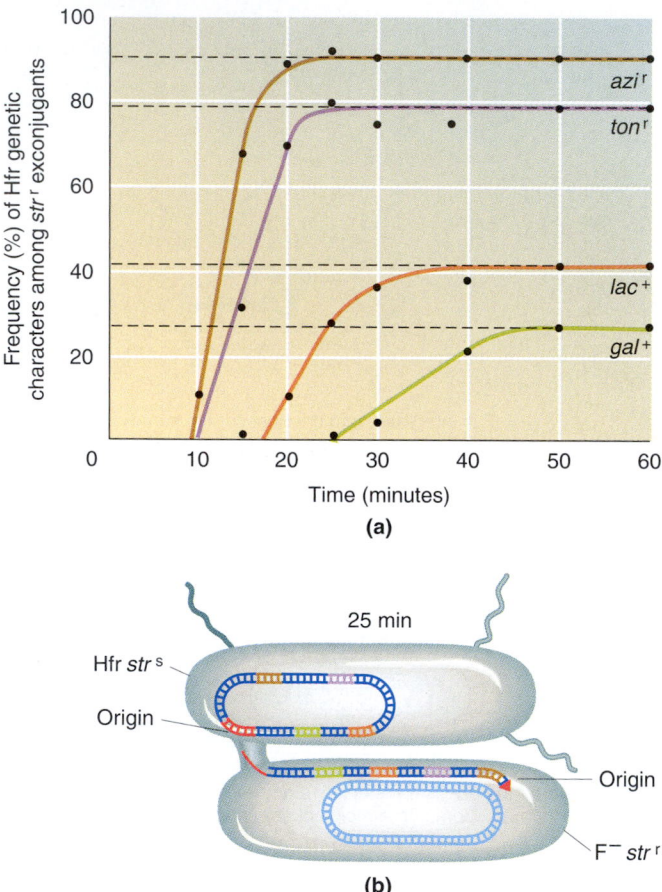

(a)

(b)

Figure 7-7   Interrupted-mating conjugation experiments with *E. coli.* F$^-$ cells that are str^r are crossed with Hfr cells that are str^s. The F$^-$ cells have a number of mutations (indicated by the genetic markers *azi, ton, lac,* and *gal*) that prevent them from carrying out specific metabolic steps. However, the Hfr cells are capable of carrying out all these steps. At different times after the cells are mixed, samples are withdrawn, disrupted in a blender to break conjugation between cells, and plated on media containing streptomycin. The antibiotic kills the Hfr cells but allows the F$^-$ cells to grow and to be tested for their ability to carry out the four metabolic steps. (a) A plot of the frequency of recombinants for each metabolic marker as a function of time after mating. Transfer of the donor allele for each metabolic step depends on how long conjugation is allowed to continue. (b) A schematic view of the transfer of markers over time. (Part a after E. L. Wollman, F. Jacob, and W. Hayes, *Cold Spring Harbor Symposia on Quantitative Biology* 21, 1956, 141.)

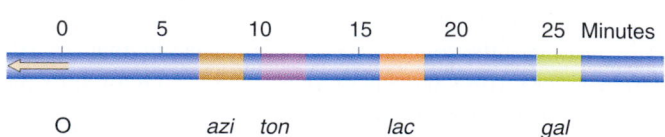

Figure 7-8   Chromosome map of Figure 7-7. A linkage map can be constructed for the *E. coli* chromosome from interrupted-mating studies by using the time at which the donor alleles first appear after mating. The units of distance are given in minutes; arrowhead at left indicates the direction of transfer of the donor alleles.

Chromosome circularity and integration of F

When Wollman and Jacob allowed Hfr × F⁻ crosses to continue for as long as 2 hours before blending, they found that a few of the exconjugants were converted into Hfr. In other words, an important part of F (the terminal part now known to confer "maleness," or donor ability), was eventually being transmitted but at a very low efficiency, and it apparently was transmitted as the last element of the linear chromosome. We now have the following map, in which the arrow indicates the process of transfer, beginning with O:

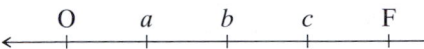

However, when several different Hfr linkage maps were derived by interrupted-mating and time-of-entry studies using different, separately derived Hfr strains, the maps differed from strain to strain.

Hfr strain	
H	O *thr pro lac pur gal his gly thi* F
1	O *thr thi gly his gal pur lac pro* F
2	O *pro thr thi gly his gal pur lac* F
3	O *pur lac pro thr thi gly his gal* F
AB 312	O *thi thr pro lac pur gal his gly* F

At first glance, there seems to be a random reshuffling of genes. However, a pattern does exist; the genes are not thrown together at random in each strain. For example, note

that in every case the *his* gene has *gal* on one side and *gly* on the other. Similar statements can be made about each gene, except when it appears at one end or the other of the linkage map. The order in which the genes are transferred is not constant. In two Hfr strains, for example, the *his* gene is transferred before the *gly* gene (*his* is closer to O), but, in three strains, the *gly* gene is transferred before the *his* gene.

How can we account for these unusual results? Allan Campbell proposed a startling hypothesis: suppose that, in an F⁺ male, F is a small cytoplasmic element (and therefore easily transferred to an F⁻ cell on conjugation). If the *chromosome* of the F⁺ male were a *ring*, any of the linear Hfr chromosomes could be generated simply by inserting F into the ring at the appropriate place and in the appropriate orientation (Figure 7-9).

Several conclusions — later confirmed — follow from this hypothesis.

1. The orientation in which F is inserted would determine the polarity of the Hfr chromosome, as indicated in Figure 7-9a.

2. At one end of the integrated F factor would be the **origin,** where transfer of the Hfr chromosome begins; the **terminus** at the other end of F would not be transferred unless all the chromosome had been transferred. Because the chromosome often breaks before all of it is transferred and because the F terminus is what confers maleness, then only a small fraction of the recipient cells would be converted into male cells.

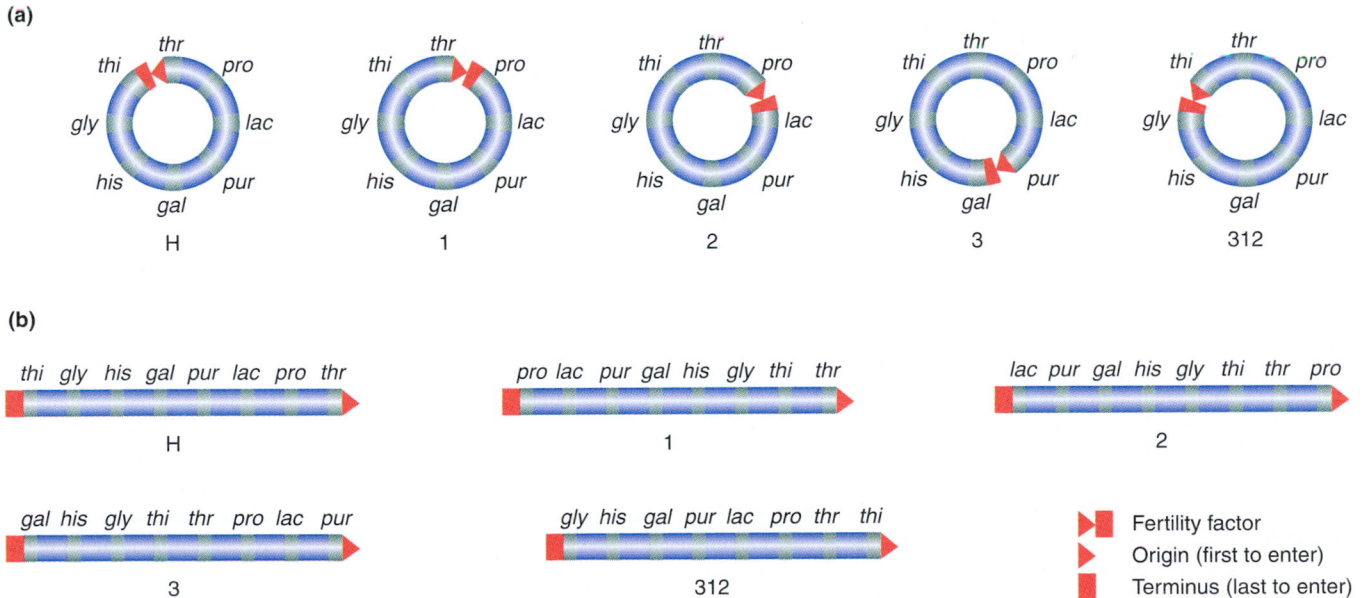

Figure 7-9 Circularity of the *E. coli* chromosome. (a) Through the use of different Hfr strains (H, 1, 2, 3, 312) that have the fertility factor inserted into the chromosome at different points and in different directions, interrupted-mating experiments indicate that the chromosome is circular. The mobilization point (origin) is shown for each strain. (b) The linear order of transfer of markers for each Hfr strain; arrowheads indicate the origin and direction of transfer.

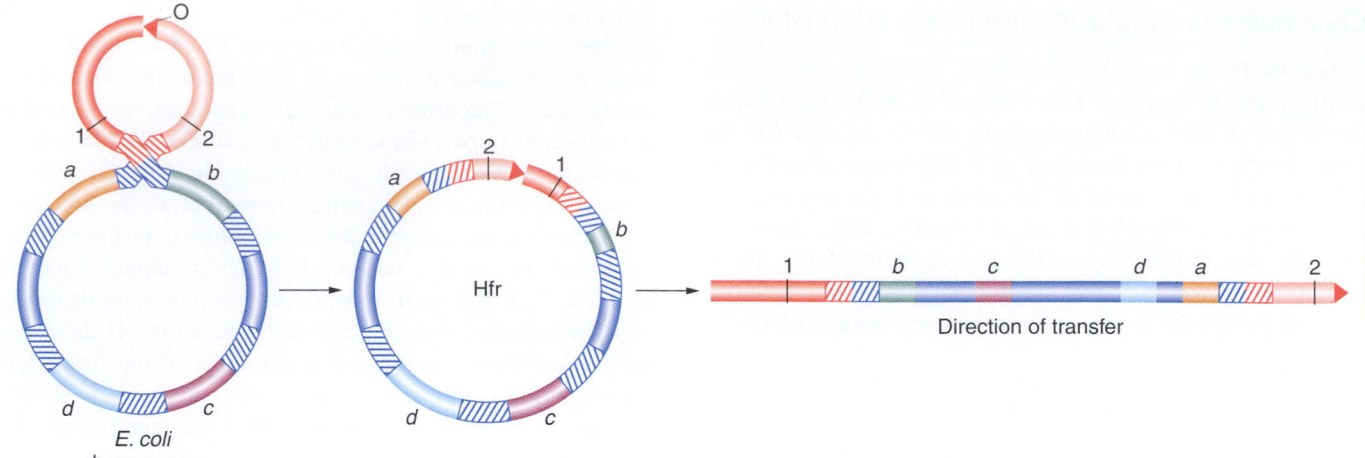

Figure 7-10   Insertion of the F factor into the *E. coli* chromosome by crossing-over. Hypothetical markers 1 and 2 are shown on F to depict the direction of insertion. The origin (O) is the mobilization point where insertion begins; the pairing region is homologous with a region on the *E. coli* chromosome; *a*–*d* are representative genes in the *E. coli* chromosome. Fertility genes on F are responsible for the F⁺ phenotype. Pairing regions (hatched) are identical in plasmid and chromosome. They are derived from mobile elements called insertion sequences (see Chapter 20). In this example, the Hfr cell created by the insertion of F would transfer its genes in the order *a, d, c, b*. What would be the order of transfer if F inserted at the other homologous sites?

How, then, might F integration be explained? Wollman and Jacob suggested that some kind of crossover event between F and the F⁺ chromosome might generate the Hfr chromosome. Campbell then came up with a brilliant extension of that idea. He proposed that, if F, like the chromosome, were circular, then a crossover between the two rings would produce a single larger ring with F inserted (Figure 7-10).

Now suppose that F consists of three different regions, as shown in Figure 7-10. If the bacterial chromosome had several homologous regions that could match up with the pairing region of F, then different Hfr chromosomes could be easily generated by crossovers at these different sites.

Chromosomal and F circularity were wildly implausible concepts initially, inferred solely from the genetic data; confirmation of their physical reality came only a number of years later. The direct-crossover model of integration also was subsequently confirmed.

The fertility factor thus exists in two states: (1) the plasmid state, as a free cytoplasmic element F that is easily transferred to F⁻ recipients, and (2) the integrated state, as a contiguous part of a circular chromosome that is transmitted only very late in conjugation. The word **episome** (literally, "additional body") was coined for a genetic particle having such a pair of states. A cell containing F in the first state is called an F⁺ cell, a cell containing F in the second state is an Hfr cell, and a cell lacking F is an F⁻ cell. Today the term **plasmid** is used to refer to any self-replicating circular element in the cytoplasm and "episome" is rarely used.

R factors

A frightening ability of pathogenic bacteria was discovered in Japanese hospitals in the 1950s. Bacterial dysentery is caused by bacteria of the genus *Shigella*. This bacterium initially proved sensitive to a wide array of antibiotics that were used to control the disease. In the Japanese hospitals, however, *Shigella* isolated from patients with dysentery proved to be simultaneously resistant to many of these drugs, including penicillin, tetracycline, sulfanilamide, streptomycin, and chloramphenicol. This multiple-drug-resistance phenotype was inherited as a single genetic package, and it could be transmitted in an infectious manner — not only to other sensitive *Shigella* strains, but also to other related species of bacteria. This talent is an extraordinarily useful one for the pathogenic bacterium, and its implications for medical science were terrifying. From the point of view of the geneticist, however, the situation is very interesting. The vector carrying these resistances from one cell to another proved to be a self-replicating element similar to the F factor. These **R factors** (for "resistance") are transferred rapidly on cell conjugation, much like the F particle in *E. coli*.

In fact, these R factors proved to be just the first of many similar F-like factors to be discovered. These elements, which exist in the plasmid state in the cytoplasm, have been found to carry many different kinds of genes in bacteria. Table 7-2 shows some of the characteristics that can be borne by plasmids.

7-2 TABLE	Genetic Determinants Borne by Plasmids
Characteristic	Plasmid examples
Fertility	F, R1, Col
Bacteriocin production	Col E1
Heavy-metal resistance	R6
Enterotoxin production	Ent
Metabolism of camphor	Cam
Tumorigenicity in plants	T1 (in *Agrobacterium tumefaciens*)

Mechanics of transfer

Does an Hfr cell die after donating its chromosome to an F⁻ cell? The answer is no (unless the culture is treated with streptomycin). The Hfr chromosome replicates while it is transferring a single strand to the F⁻ cell; this replication ensures a complete chromosome for the donor cell after mating. The transferred strand is replicated in the recipient cell, and donor genes may become incorporated in the recipient's chromosome through crossovers, creating a recombinant cell. Otherwise, transferred fragments of DNA in the recipient are lost in the course of cell division.

We assume that the F⁻ chromosome is also circular, because the recipient F⁻ cell, if it receives the F factor from an F⁺ cell, is readily converted into an F⁺ cell from which an Hfr cell can be derived.

The picture emerges of a circular Hfr chromosome unwinding a copy of itself, which is then transferred in a linear fashion into the F⁻ cell. How is the transfer achieved? Electron microscope studies show that Hfr and F⁺ cells have fibrous structures, **F pili,** protruding from their cell walls, as shown in Figure 7-4. The F pili facilitate cell-to-cell contact, during which DNA is transferred through pores in the F⁻.

E. coli conjugation cycle

We can now summarize the various aspects of the conjugation cycle in *E. coli* (Figure 7-11). We shall review the conjugation cycle in regard to the differences between F⁻, F⁺, and Hfr cells, because these differences epitomize the cycle.

F⁻ strains do not contain the F factor and cannot transfer DNA by conjugation. They are, however, recipients of DNA transferred from F⁺ or Hfr cells by conjugation.

F⁺ cells contain the F factor in the cytoplasm and can therefore transfer F in a highly efficient manner to F⁻ cells during conjugation.

Hfr cells have F integrated into the bacterial chromosome, not in the cytoplasm.

Chromosomal markers are transferred in a strain of F⁺ cells because, in any population of F⁺ cells, a small fraction of cells (about 1 in 1000) have been converted into Hfr cells by the integration of F into the bacterial chromosome. Because conjugation experiments are usually carried out by mixing from 10^7 to 10^8 cells consisting of prospective donors and recipients, the population will contain various different Hfr cells derived from independent integrations of F into the chromosome at various different sites. Therefore, when chromosomal markers are transferred by different cells in the population, transfer will start at different points on the chromosome. This results in an approximately equal transfer of markers all around the chromosome, although at a low frequency. This type of F⁺-mediated transfer is what Lederberg and Tatum observed when they discovered gene transfer in bacteria.

Each of the Hfr cells in an F⁺ population with an integrated F factor can be the source of a new Hfr strain if it is isolated and used to start a clone.

Hfr strains are derived from a clone of Hfr cells in which a specific integration of F into the bacterial chromosome has

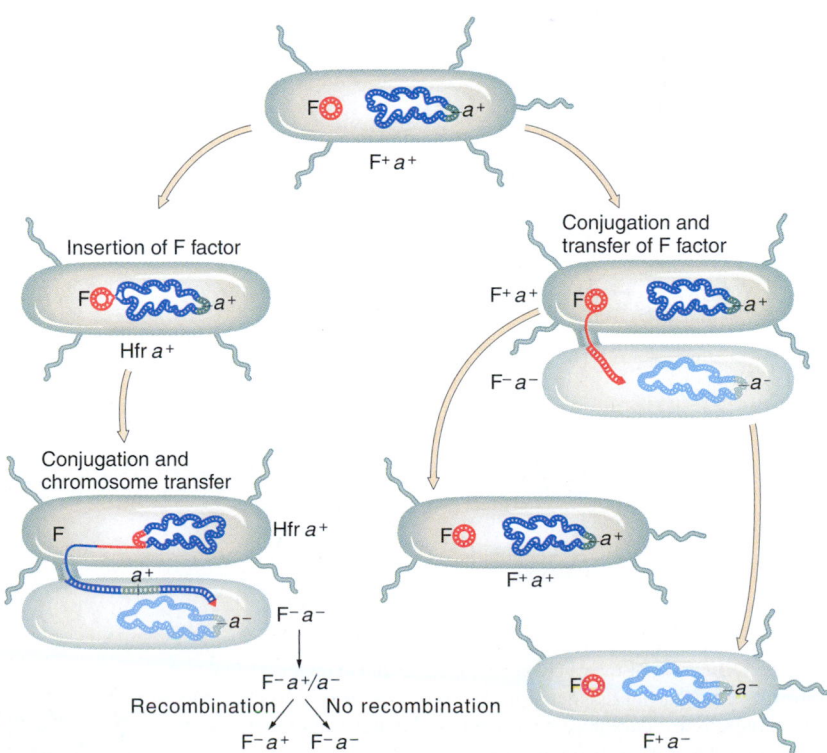

Figure 7-11 Summary of the various events that take place in the conjugational cycle of *E. coli*.

taken place. Therefore, all the cells in any given Hfr strain have F integrated into the chromosome at exactly the same point.

Hfr populations transfer chromosomal markers to F$^-$ cells at a high frequency compared with F$^+$ populations, because only a fraction of cells in an F$^+$ population have F integrated into the chromosome. Further, in any given Hfr strain, the markers are transferred from a fixed point in a specific order. This also contrasts with F$^+$ populations, where the Hfr cells transfer chromosomal markers in no particular fixed order, given that the F factor integrates into the chromosome at different points in different F$^+$ cells.

In an Hfr × F$^-$ cross, the F$^-$ is not converted into Hfr or into F$^+$, except in very rare cases, because the Hfr chromosome nearly always breaks before the F terminus is transferred to the F$^-$ cell.

Recombination between marker genes after transfer

Thus far, we have studied only the process of the transfer of genetic information between individuals in a cross. This transfer is inferred from the existence of recombinants produced from the cross. However, before a stable recombinant can be produced, the transferred genes must be integrated or incorporated into the recipient's genome by an exchange mechanism. We now consider some of the special properties of this exchange event.

Genetic exchange in prokaryotes does not take place between two whole genomes (as it does in eukaryotes);

rather, it takes place between one *complete* genome, derived from F$^-$, called the **endogenote,** and an *incomplete* one, derived from the donor, called the **exogenote.** What we have in fact is a partial diploid, or **merozygote.** Bacterial genetics is merozygous genetics. Figure 7-12a is a diagram of a merozygote.

A single crossover would not be very useful in generating viable recombinants, because the ring is broken to produce a strange, partly diploid linear chromosome (Figure 7-12b). To keep the ring intact, there must be an even number of crossovers (Figure 7-12c). The fragment produced in such a crossover is only a partial genome, which is generally lost in subsequent cell growth. Hence, both reciprocal products of recombination do not survive — only one does. A further unique property of bacterial exchange, then, is that we must forget about reciprocal exchange products in most cases.

> **MESSAGE** ···
> In the genetics of bacteria, we generally are concerned with double crossovers and we do not expect reciprocal recombinants.

Gradient of transfer

Only partial diploids exist in the merozygote. Some genes don't even get into the act. To better understand this fact, let us look again at the consequences of gene transfer. Usually, only a fragment of the donor chromosome appears in the recipient, owing to spontaneous breakage of the mating pairs; so the entire chromosome is rarely transferred. The spontaneous breakage can occur at any time after transfer begins, which creates a natural **gradient of transfer** and makes it less and less likely that a recipient cell will receive later and later genetic markers. ("Later" here refers to markers that are increasingly farther from the origin and hence are donated later in the order of markers transferred.) For example, in a cross of Hfr-donating markers in the order *met, arg, leu*, we would expect a distribution of fragments such as the one represented here:

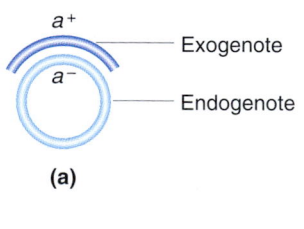

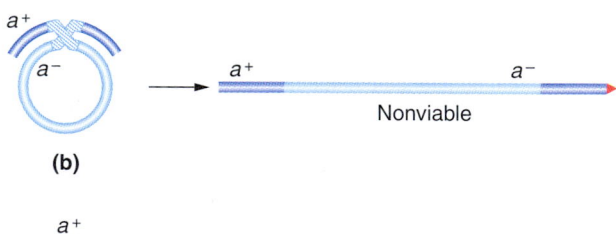

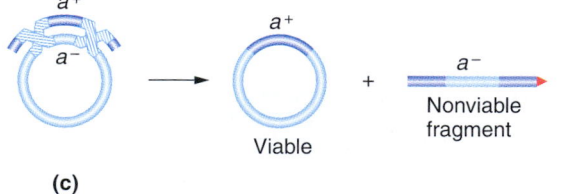

(a)

(b)

(c)

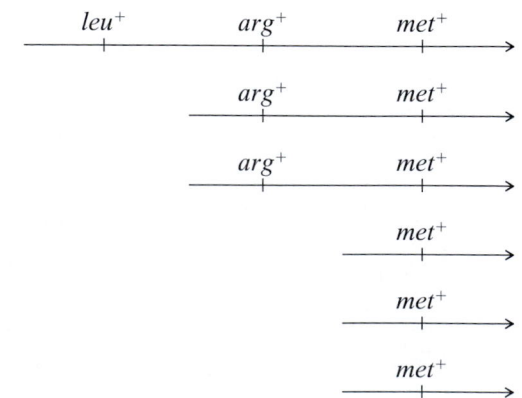

Figure 7-12   Crossover between exogenote and endogenote in a merozygote. (a) The merozygote. (b) A single crossover leads to a partly diploid linear chromosome. (c) An even number of crossovers leads to a ring plus a linear fragment.

Note that many more fragments contain the *met* locus than the *arg* locus and that the *leu* locus is present on only one fragment. It is easy to see that the closer the marker is to the origin, the greater the chance it will be transferred in conjugation.

The concept of the gradient of transfer is the same as the one described earlier for interrupted matings, except that here we are allowing the natural disruption of mating pairs to occur instead of interrupting the pairs mechanically.

Determining gene order from gradient of transfer

We can use the natural gradient of transfer to establish the order of genetic markers, provided we select for an early marker that enters before the markers that we are ordering. Let's see how this works. Suppose that we use an Hfr strain that donates markers in the order *met, arg, aro, his.* In a cross of an Hfr that is *met⁺ arg⁺ aro⁺ his⁺ str*ˢ with an F⁻ that is *met⁻ arg⁻ aro⁻ his⁻ str*ʳ, recombinants are selected that can grow on a minimal medium without methionine but with arginine, aromatic amino acids, and histidine and in the presence of streptomycin. Here we are selecting for recombinants in the F⁻ strain that are *met⁺* in a cross in which the *met* locus is transferred as the earliest marker. We can then score for inheritance of the other markers present in the Hfr by testing on supplemental minimal medium lacking, in turn, one of the required nutrients.

A typical result would be:

$$met^+ = 100\%$$
$$arg^+ = 60\%$$
$$aro^+ = 20\%$$
$$his^+ = 4\%$$

Note how the frequency of inheritance corresponds to the order of transfer. This correspondence is due to the fact that the frequency of inheritance is indicative of the frequency of transfer. For this method to work, it is crucial that it be applied only to genetic markers that enter after the selected marker — in this case, after *met.*

Higher-resolution mapping by recombinant frequency in bacterial crosses

Although interrupted-mating experiments and the natural gradient of transfer can give us a rough set of gene locations over the entire map, other methods are needed to obtain a higher resolution between marker loci that are close together. Here we consider one approach to the problem: using the frequency of recombinants to measure linkage.

Previous attempts to measure linkage in conjugational crosses were hindered by the failure to understand that only fragments of the chromosome are transferred and that the gradient of transfer produces a bias toward the inheritance of early markers. To measure linkage and to attach any meaning to a calculated map distance, it is necessary to produce a situation in which every marker has an equal chance at being transferred so that the recombinant frequencies are dependent only on the distance between the relevant genes.

Suppose that we consider three markers: *met, arg,* and *leu.* If the order is *met, arg, leu* and if *met* is transferred first and *leu* last, then we want to set up the situation diagrammed here to calculate map distances separating these markers:

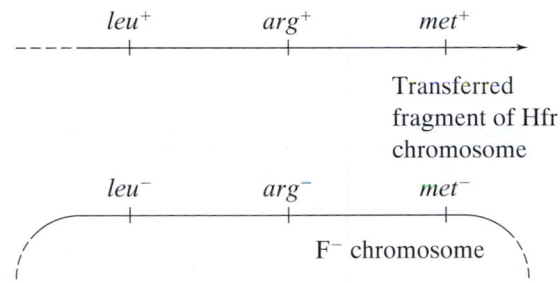

Here, we have to arrange to select the *last* marker to enter, which in this case is *leu.* Why? Because, if we select for the last marker, then we know that every cell that received fragments containing the last marker also received the earlier markers — namely, *arg* and *met* — on the same fragments. We can then proceed to calculate map distance in the classic manner. Rather than using map units, we simply refer to the percentage of crossovers in the respective interval on the map. In practice, this is done by calculating, among the total recovered recombinants, the percentage of recombinants produced by crossovers between two markers. Let's look at an example.

Sample cross

In the cross of the Hfr strain just described (*met⁺ arg⁺ leu⁺ str*ˢ) with an F⁻ that is *met⁻ arg⁻ leu⁻ str*ʳ, we would select *leu⁺* recombinants and then examine them for the *arg* and *met* markers. In this case, the *arg* and *met* markers are called the **unselected markers.** Figure 7-13 depicts the types of crossover events expected. Note how two crossover events are required to incorporate part of the incoming fragment into the F⁻ chromosome. One crossover must be on each side of the selected (*leu*) marker. Thus, in Figure 7-13, one crossover must be on the left side of the *leu* marker and the second must be on the right side. Suppose that the map distance between each marker is 5 percent recombination. In 5 percent of the total *leu⁺* recombinants, the second crossover occurs between *leu* and *arg* (Figure 7-13a); in another 5 percent of the cases, the second crossover occurs between *leu* and *met* (Figure 7-13b). We would then expect 90 percent of the selected *leu⁺* recombinants to be *arg⁺ met⁺*, because the second crossover occurs outside the *leu–arg–met* interval (Figure 7-13c) in 90 percent of the cases. We would also expect 5 percent of the *leu⁺* recombinants to be *arg⁻ met⁻*, resulting from a crossover between *leu* and *arg*, and 5 percent of the *leu⁺* recombinants to be *arg⁺ met⁻*, resulting from a crossover between *arg* and *met.* In reality, then, we are simply determining the percentage of the time that the second crossover occurs in each of the three possible intervals.

(a) Insertion of late marker only

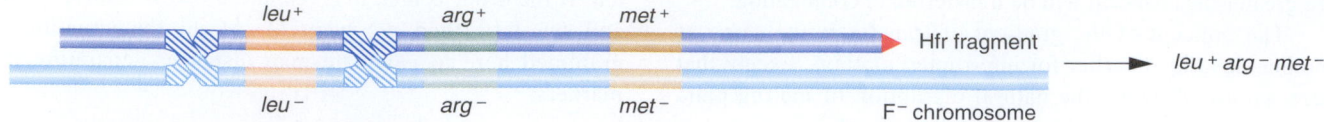

(b) Insertion of late marker and one early marker

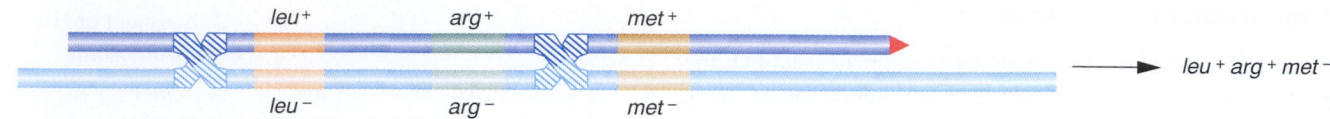

(c) Insertion of late and early markers

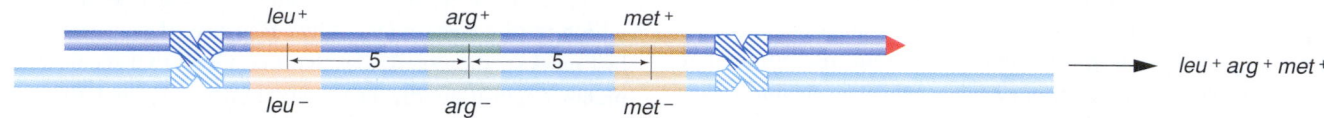

(d) Insertion of late and early markers, but not of marker in between

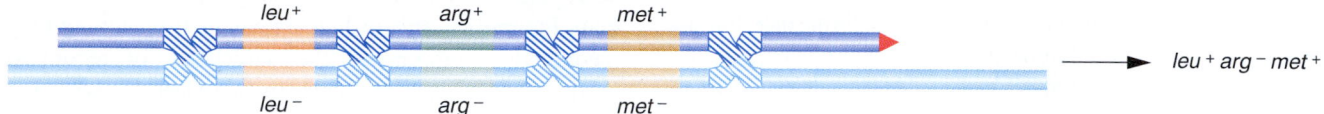

Figure 7-13 Incorporation of a late marker into the F⁻ *E. coli* chromosome. After an Hfr cross, selection is made for the *leu⁺* marker, which is donated late. The early markers (*arg⁺* and *met⁺*) may or may not be inserted, depending on the site where recombination between the Hfr fragment and the F⁻ chromosome takes place. As indicated in part c, crossovers will occur in each of these intervals 5 percent of the time, resulting in the *leu⁺ arg⁻ met⁻* (a) and *leu⁺ arg⁺ met⁻* (b) recombinants. Crossovers occurring outside of these intervals can result in the *leu⁺ arg⁺ met⁺* recombinant (c). The *leu⁺ arg⁻ met⁺* recombinant requires an additional two crossovers (d).

In a cross such as the one just described, one class of potential recombinants requires an additional two crossover events (Figure 7-13d). In this case, the *leu⁺ arg⁻ met⁺* recombinants would require four crossovers instead of two. These recombinants are rarely recovered, because their frequency is sharply reduced compared with the other classes of recombinants.

Infectious marker-gene transfer by episomes

Edward Adelberg's work led to the discovery of gene transfer at high frequency by episomes. When he began his recombination experiments in 1959, the particular Hfr strain that he used kept producing F⁺ cells, so the recombination frequencies were not very large. Adelberg called this particular fertility factor F′ to distinguish it from the normal F, for the following reasons:

1. The F′-bearing F⁺ strain reverted back to an Hfr strain much more frequently than do typical F⁺ strains.

2. F′ always integrated at the *same place* to give back the original Hfr chromosome. (Remember that randomly selected Hfr derivatives from F⁺ males have origins at many different positions.)

How could these properties of F′ be explained? The answer came from the recovery of an F′ from an Hfr strain in which the *lac⁺* locus was near the end of the Hfr chromosome (it was transferred very late). Using this Hfr *lac⁺* strain, François Jacob and Adelberg found an F⁺ derivative that transferred *lac⁺* to F⁻ *lac⁻* recipients at a very high frequency. Furthermore, the recipients that behaved like F⁺ *lac⁺* occasionally produced F⁻ *lac⁻* daughter cells, at a frequency of 1×10^{-3}. Thus, the genotype of these recipients appeared to be F *lac⁺/lac⁻*.

Now we have the clue: F′ is a cytoplasmic element that carries a part of the bacterial chromosome. In fact, it is nothing more than F with a piece of the host chromosome incorporated. Its origin and reintegration can be visualized as shown in Figure 7-14. This F′ is known as F′ *lac*, because the piece of host chromosome that it picked up has the *lac* gene on it. F′ factors have been found carrying many different chromosomal genes and have been named accordingly. For example, F′ factors carrying *gal* or *trp* are called F′ *gal* and F′ *trp*, respectively. Because F *lac⁺/lac⁻* cells

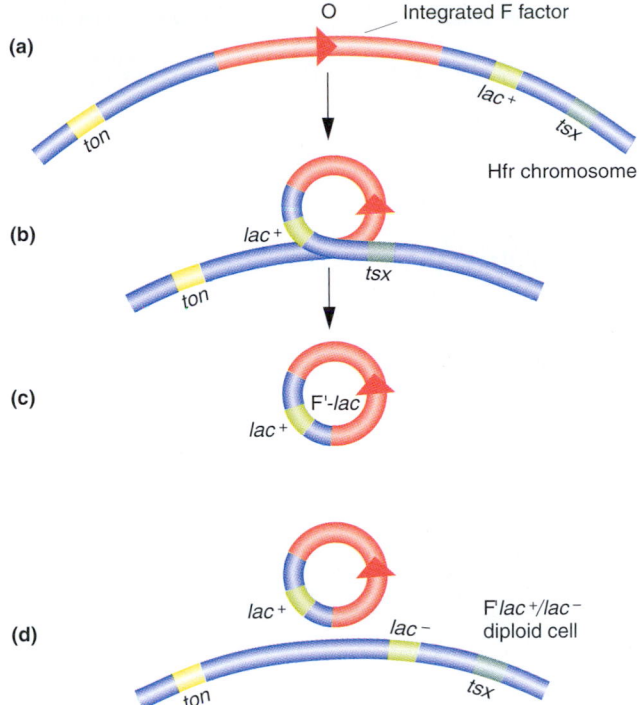

Figure 7-14 Origin and reintegration of the F′ factor, in this case, F′ *lac*. (a) F is inserted in an Hfr strain between the *ton* and *lac*⁺ alleles. (b and c) Abnormal "outlooping" and separation of F occurs to include the *lac* locus, producing the F′ *lac*⁺ particle. (d) An F *lac*⁺/*lac*⁻ partial diploid is produced by the transfer of the F′ *lac* particle to an F⁻ *lac*⁻ recipient. (From G. S. Stent and R. Calendar, *Molecular Genetics*, 2d ed. Copyright © 1978 by W. H. Freeman and Company.)

Bacterial transformation

Some bacteria have another method of transferring DNA and producing recombinants that does not require conjugation. The conversion of one genotype into another by the introduction of exogenous DNA (that is, bits of DNA from an external source) is termed **transformation.** Transformation was discovered in *Streptococcus pneumoniae* in 1928 by Frederick Griffith; in 1944, Oswald T. Avery, Colin M. MacLeod, and Maclyn McCarty demonstrated that the "transforming principle" was DNA. Both results are milestones in the elucidation of the molecular nature of genes. We consider this work in more detail in Chapter 8.

After DNA was shown to be the agent that determines the polysaccharide character of *S. pneumoniae*, transformation was demonstrated for other genes, such as those for drug resistance (Figure 7-15). The transforming principle, exogenous DNA, is incorporated into the bacterial chromosome by a breakage-and-insertion process analogous to that depicted for Hfr × F⁻ crosses in Figure 7-12. Note, however, that, in *conjunction*, DNA is transferred from one living cell to another through close contact, whereas, in *transformation,* isolated pieces of external DNA are taken up by a cell. Figure 7-16 depicts this process.

are Lac⁺ in phenotype, we know that *lac*⁺ is dominant over *lac*⁻. As we shall see in Chapter 11, the dominant–recessive relation between alleles can be a very useful bit of information in interpreting gene function.

Partial diploidy for specific segments of the genome can be made with an array of F′ derivatives from Hfr strains. The F′ cells can be selected by looking for the infectious transfer of normally late genes in a specific Hfr strain. Some F′ strains can carry very large parts (up to one-quarter) of the bacterial chromosome; if appropriate markers are used, the merozygotes generated can be used for recombination studies.

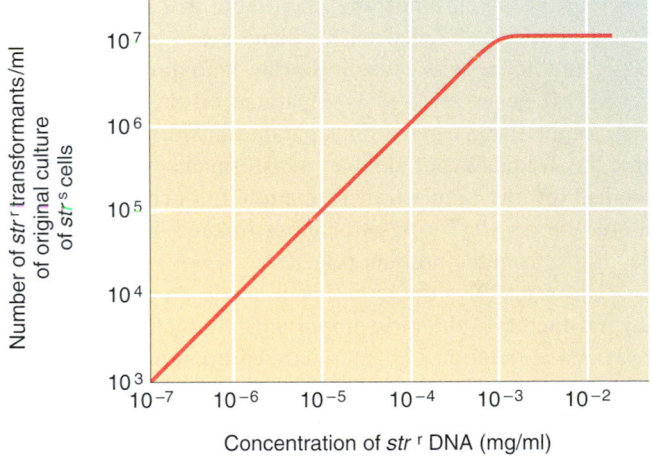

Figure 7-15 The genetic transfer of streptomycin resistance (*str*ʳ) to streptomycin-sensitive (*str*ˢ) cells of *E. coli*. The recovery of *str*ʳ transformants among *str*ˢ cells depends on the concentration of *str*ʳ DNA. (From G. S. Stent and R. Calendar, *Molecular Genetics*, 2d ed. Copyright © 1978 by W. H. Freeman and Company.)

Linkage information from transformation

Transformation has been a very handy tool in several areas of bacterial research. We learn later how it is used in some of the modern techniques of genetic engineering. Here we examine its usefulness in providing linkage information.

When DNA (the bacterial chromosome) is extracted for transformation experiments, some breakage into smaller pieces is inevitable. If two donor genes are located close

> **MESSAGE** ···
> During conjugation between an Hfr donor and an F⁻ recipient, the genes of the donor are transmitted *linearly* to the F⁻ cell, through the bacterial chromosome, with the inserted fertility factor transferring last.
>
> In the course of conjugation between an F⁺ donor carrying an F′ plasmid and an F⁻ recipient, a specific part of the donor genome may be transmitted *infectiously* to the F⁻ cell, through the plasmid. The transmitted part was originally adjacent to the F locus in an Hfr strain from which the F⁺ was derived.

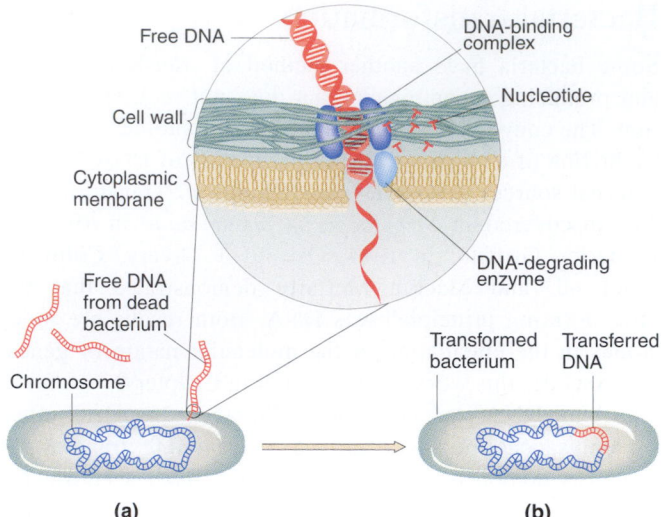

(a)   (b)

Figure 7-16   Bacterium undergoing transformation (a) picks up free DNA released from a dead bacterial cell. As DNA-binding complexes on the bacterial surface take up the DNA (inset), enzymes break down one strand into nucleotides; meanwhile the other strand may integrate into the bacterium's chromosome (b). (After R. V. Miller, "Bacterial Gene Swapping in Nature." Copyright © 1998 by Scientific American, Inc. All rights reserved.)

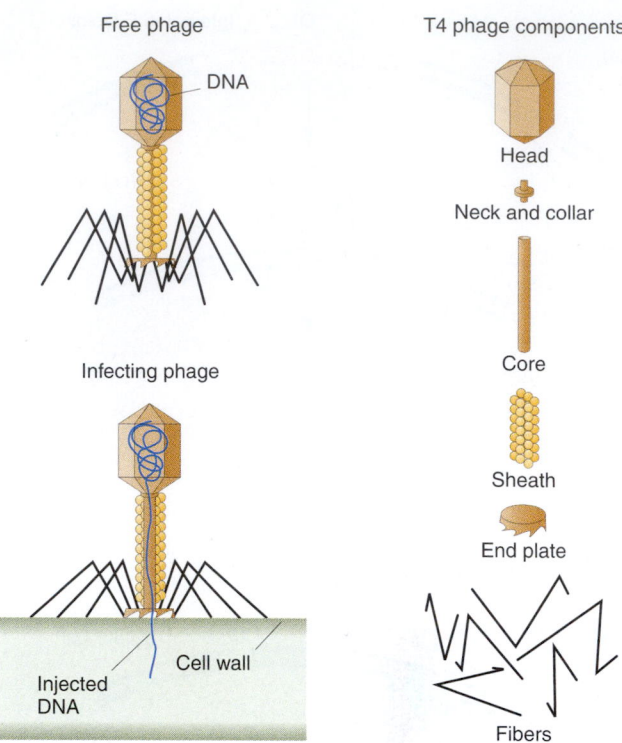

Figure 7-17   Phage T4, shown in its free state and in the process of infecting an *E. coli* cell. The infecting phage injects DNA through its core structure into the cell. On the right, a phage has been diagrammatically exploded to show its highly ordered three-dimensional structure. (After R. S. Edgar and R. H. Epstein, "The Genetics of a Bacterial Virus." Copyright © 1965 by Scientific American, Inc. All rights reserved.)

together on the chromosome, then there is a greater chance that they will be carried on the same piece of transforming DNA and hence will cause a **double transformation.** Conversely, if genes are widely separated on the chromosome, then they will be carried on separate transforming segments and the frequency of double transformants will equal the product of the single-transformation frequencies. Thus, it should be possible to test for close linkage by testing for a departure from the product rule.

Unfortunately, the situation is made more complex by several factors—the most important of which is that not all cells in a population of bacteria are **competent,** or able to be transformed. Because single transformations are expressed as proportions, the success of the product rule depends on the absolute size of these proportions. There are ways of calculating the proportion of competent cells, but we need not detour into that subject now. You can sharpen your skills in transformation analysis in one of the problems at the end of the chapter, which assumes 100 percent competence of the recipient cells.

Bacteriophage genetics

In this section, we shall describe how crosses can actually be done with viruses (phages) that infect bacteria and the experiments that dissect the fine structure of the gene.

Infection of bacteria by phages

Most bacteria are susceptible to attack by bacteriophages, which literally means "eaters of bacteria." A phage consists of a nucleic acid "chromosome" (DNA or RNA) surrounded

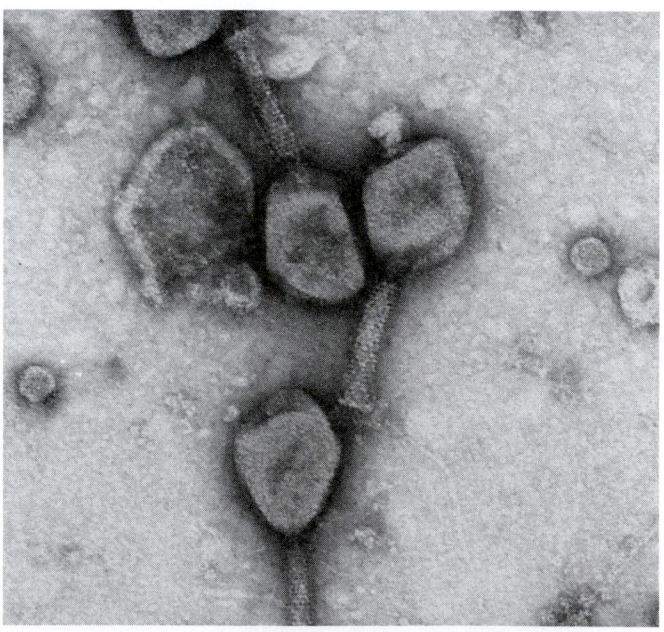

Figure 7-18   Enlargement of the *E. coli* phage T4 showing details of structure: note head, tail, and tail fibers. The T4 phage was used by Benzer in his experiments on the nature of the *rII* (rapid lysis) gene. (Photograph from Jack D. Griffith.)

by a coat of protein molecules. One well-studied set of phage strains are identified as T1, T2, and so forth. Figures 7-17 and 7-18 show the structure of a **T-even phage** (T2, T4, and so forth).

During infection, a phage attaches to a bacterium and injects its genetic material into the bacterial cytoplasm (Figure 7-19). The phage genetic information then takes over the machinery of the bacterial cell by turning off the synthesis of bacterial components and redirecting the bacterial synthetic material to make more phage components (Figure 7-20). (The use of the word *information* is interesting in this connection; it literally means "to give form," which is precisely the role of the genetic material: to provide blueprints for the construction of form. In the present discussion, the form is the elegantly symmetrical structure of the new phages.) Ultimately, many phage descendants are released when the bacterial cell wall breaks open. This breaking-open process is called **lysis.**

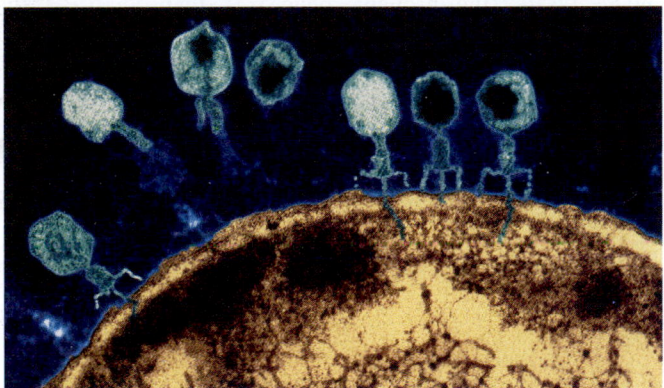

Figure 7-19   Micrograph of a bacteriophage attaching to a bacterium and injecting its DNA. (Dr. L. Caro/Science Photo Library, Photo Researchers.)

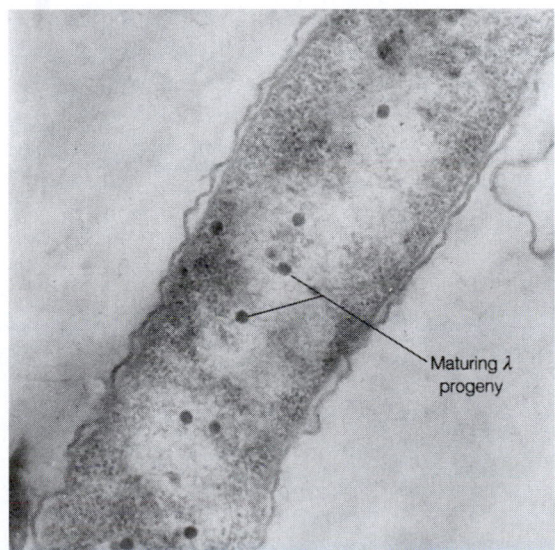

Figure 7-20   Progeny particles of phage λ maturing inside an *E. coli* cell. (Jack D. Griffith.)

How can we study inheritance in phages when they are so small that they are visible only under the electron microscope? In this case, we cannot produce a visible colony by plating, but we can produce a visible manifestation of an infected bacterium by taking advantage of several phage characters. Let's look at the consequences of a phage's infecting a single bacterial cell. Figure 7-21 shows the sequence of

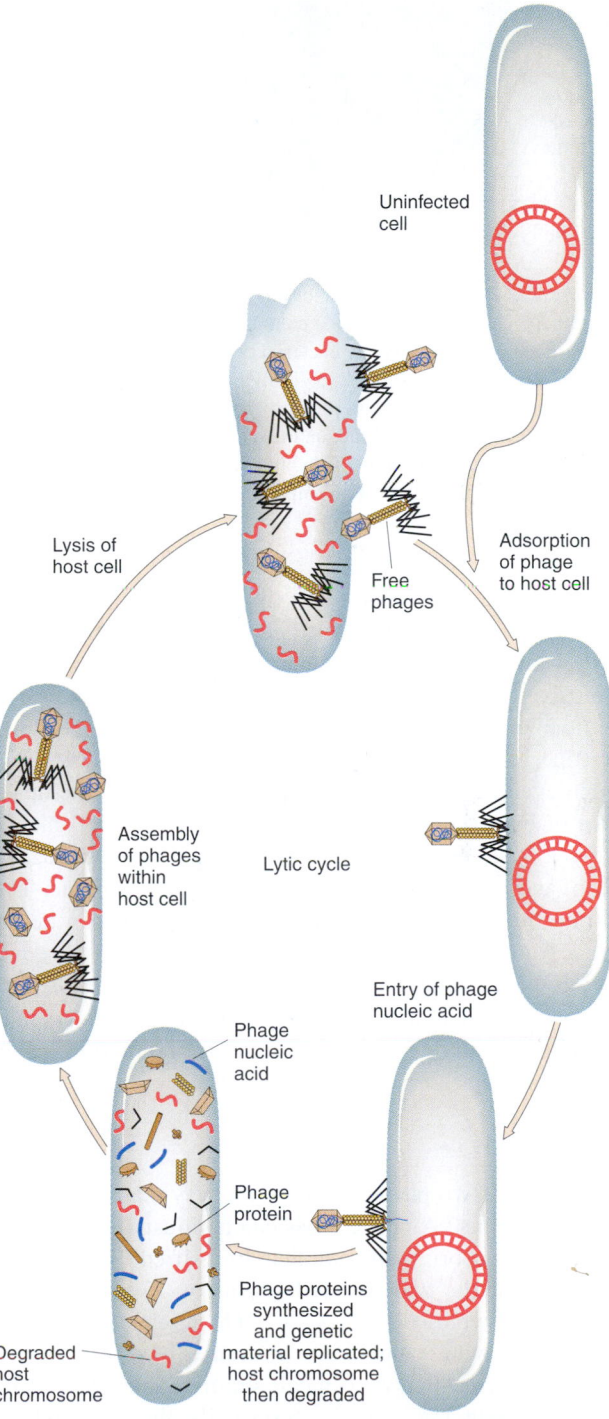

Figure 7-21   A generalized bacteriophage lytic cycle. (After J. Darnell, H. Lodish, and D. Baltimore, *Molecular Cell Biology.* Copyright © 1986 by W. H. Freeman and Company.)

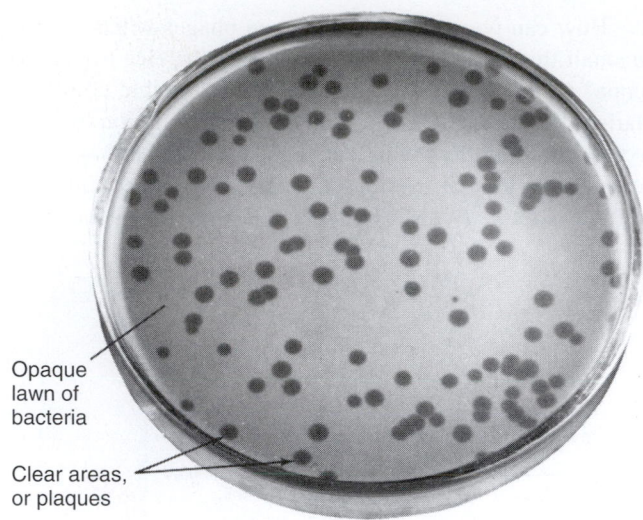

Figure 7-22   The appearance of phage plaques. Individual phages are spread on an agar medium that contains a fully grown "lawn" of *E. coli*. Each phage infects one bacterial cell, producing 100 or more progeny phages that burst the *E. coli* cell and infect neighboring cells. They in turn are exploded with progeny, and the continuing process produces a clear area, or plaque, on the opaque lawn of bacterial cells. (From G. S. Stent, *Molecular Biology of Bacterial Viruses*. Copyright © 1963 by W. H. Freeman and Company.)

events in the infectious cycle that leads to the release of progeny phages from the lysed cell. After lysis, the progeny phages infect neighboring bacteria. This is an exponentially explosive phenomenon (it causes an exponential increase in the number of lysed cells). Within 15 hours after the start of an experiment of this type, the effects are visible to the naked eye: a clear area, or **plaque,** is present on the opaque lawn of bacteria on the surface of a plate of solid medium (Figure 7-22). Such plaques can be large or small, fuzzy or sharp, and so forth, depending on the phage genome. Thus, **plaque morphology** is a phage character that can be analyzed. Another phage phenotype that we can analyze genetically is **host range,** because phages may differ in the spectra of bacterial strains that they can infect and lyse. For example, certain strains of bacteria are immune to adsorption (attachment) or injection by phages.

Phage cross

Can we cross two phages in the same way that we cross two bacterial strains? A phage cross can be illustrated by a cross of T2 phages originally studied by Alfred Hershey. The genotypes of the two parental strains of T2 phage in Hershey's cross were $h^- r^+ \times h^+ r^-$. The alleles are identified by the following characters: h^- can infect two different *E. coli* strains (which we can call strains 1 and 2); h^+ can infect only strain 1; r^- rapidly lyses cells, thereby producing large plaques; and r^+ slowly lyses cells, thus producing small plaques.

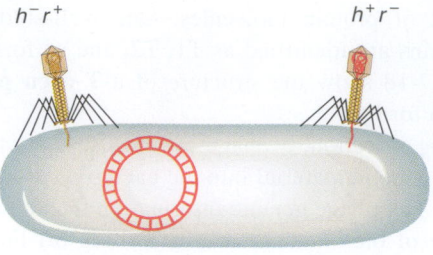

$h^- r^+$ $h^+ r^-$

E. coli strain 1

Figure 7-23   A double infection of *E. coli* by two phages.

In the cross, *E. coli* strain 1 is infected with both parental T2 phage genotypes at a phage : bacteria concentration (called **multiplicity of infection**) that is high enough to ensure that a large percentage of cells are simultaneously infected by both phage types. This kind of infection (Figure 7-23 is called a **mixed infection** or a **double infection.** The phage lysate (the progeny phage) is then analyzed by spreading it onto a bacterial lawn composed of a mixture of *E. coli* strains 1 and 2. Four plaque types are then distinguishable (Figure 7-24 and Table 7-3). These four genotypes can be scored easily as parental ($h^- r^+$ and $h^+ r^-$) and recombinant ($h^+ r^+$ and $h^- r^-$), and a recombinant frequency can be calculated as follows:

$$ RF = \frac{(h^+ r^+) + (h^- r^-)}{\text{total plaques}} $$

If we assume that entire phage genomes recombine, then single exchanges can occur and produce viable reciprocal products, unlike bacterial crosses where two crossover events are required. Nevertheless, phage crosses are subject to complications. First, several rounds of exchange can potentially occur within the host: a recombinant produced shortly after infection may undergo further exchange at later times. Second, recombination can occur between genetically similar phages as well as between different types. Thus, $P_1 \times P_1$ and $P_2 \times P_2$ occur in addition to $P_1 \times P_2$ (P_1 and P_2 refer to phage 1 and phage 2, respectively). For both of these reasons, recombinants from phage crosses are a conse-

7-3 TABLE	Progeny Phage Plaque Types from Cross $h^- r^+ \times h^+ r^-$
Phenotype	**Inferred genotype**
clear and small	$h^- r^+$
cloudy and large	$h^+ r^-$
cloudy and small	$h^+ r^+$
clear and large	$h^- r^-$

Note: Clearness is produced by the h^- allele, which allows infection of *both* bacterial strains in the lawn; cloudiness is produced by the h^+ allele, which limits infection to the cells of strain 1.

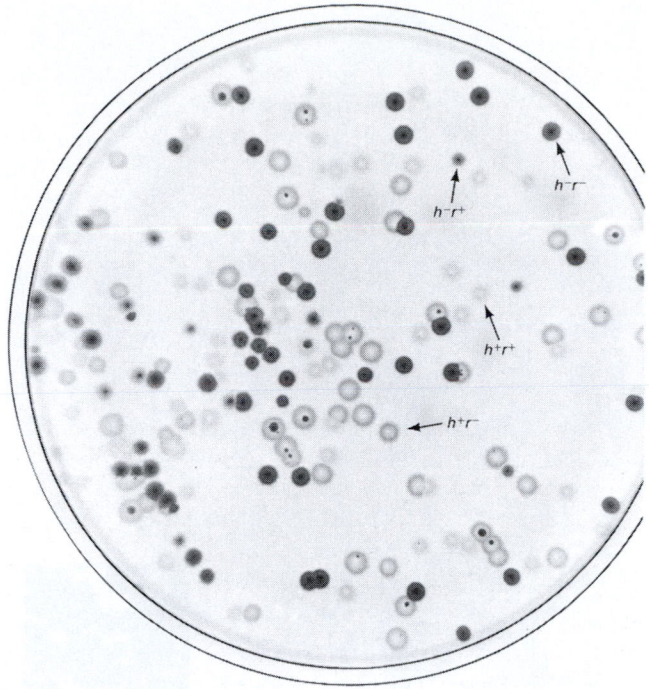

Figure 7-24 Plaque phenotypes produced by progeny of the cross $h^- r^+ \times h^+ r^-$. Enough phages of each genotype are added to ensure that most bacterial cells are infected with at least one phage of each genotype. After lysis, the progeny phages are collected and added to an appropriate *E. coli* lawn. Four plaque phenotypes can be differentiated, representing two parental types and two recombinants. (From G. S. Stent, *Molecular Biology of Bacterial Viruses.* Copyright © 1963 by W. H. Freeman and Company.)

	7-4 ⟩ TABLE	Plaque Phenotypes Produced by Different Combinations of *E. coli* and Phage Strains	
		E. coli STRAIN	
T4 phage strain		B	K
rII		large, round	no plaques
rII+		small, ragged	small, ragged

quence of a *population* of events rather than defined, single-step exchange events. Nevertheless, *all other things being equal,* the RF calculation does represent a valid index of map distance in phages.

MESSAGE ·············

Recombination between phage chromosomes can be studied by bringing the parental chromosomes together in one host cell through mixed infection. Progeny phages can be examined for parental versus recombinant genotypes.

rII system

Seymour Benzer's work in the 1950s refined the phage cross to the point where extremely small levels of recombination could be detected. This work led to a greater understanding of the nature of the fine structure of the gene, which we consider in detail in Chapter 9. The key to this work was the development of a system that allowed the selection of rare recombinants. This system used the *rII* genes of phage T4.

One type of mutant T4 phage produced larger, ragged plaques: these were *r* (rapid lysis) **mutants.** Benzer mapped

the mutations responsible for the r phenotype to two loci: *rI* and *rII*. He then studied the *rII* mutants intensively.

One extraordinary property of *rII* mutants made all of Benzer's work possible: *rII* mutants have a different host range from that of wild-type phages. Two related but different strains of *E. coli,* termed B and K(λ), can be used as different hosts for phage T4. Both bacterial strains can distinguish *rII* mutants from wild-type phages. *E. coli* B allows both to grow, but plaques of different sizes result: wild-type phages produce small plaques, and *rII* mutants produce large plaques. *E. coli* K, an abbreviation for *E. coli* K(λ), does not permit the growth of *rII* mutants, but it does allow wild-type phages to grow. The *rII* mutants are then **conditional mutants** — namely, mutants that can grow under one set of conditions but not another. *E. coli* B is said to be **permissive** for *rII* mutants, because it allows phage growth, whereas *E. coli* K is said to be **nonpermissive** for *rII* mutants, because it does not allow phage growth. Table 7-4 shows the growth characteristics and plaque morphology of these phages on each host strain.

Selection in genetic crosses of bacteriophages

Benzer crossed various *rII* mutants of the T4 phage and obtained recombination frequencies, which he then used to map mutations *within* the *rII* gene region.

Let's see how this works. Suppose that we wish to cross two *rII* mutants and recover wild-type recombinants. Because wild-type and *rII* mutants make plaques that can be distinguished from each other, we could cross two different *rII* mutants in *E. coli* B and examine the progeny on *E. coli* B (Figure 7-25, top photograph at lower right), hoping to find small wild-type plaques among the large parental *rII* plaques. If the recombination frequency is high enough to yield from 2 to 3 percent or more wild-type plaques, then this method would suffice. However, for recombination that is less frequent than 1 percent, a lot of work would be required to generate a map of numerous *rII* mutations.

Instead of plating the progeny phages from the cross on *E. coli* B, however, we could plate the progeny on *E. coli* K (Figure 7-25, bottom photograph at lower right), so that only the wild-type recombinant phages could grow. Even if the recombination frequency is very low (say, 0.01 percent),

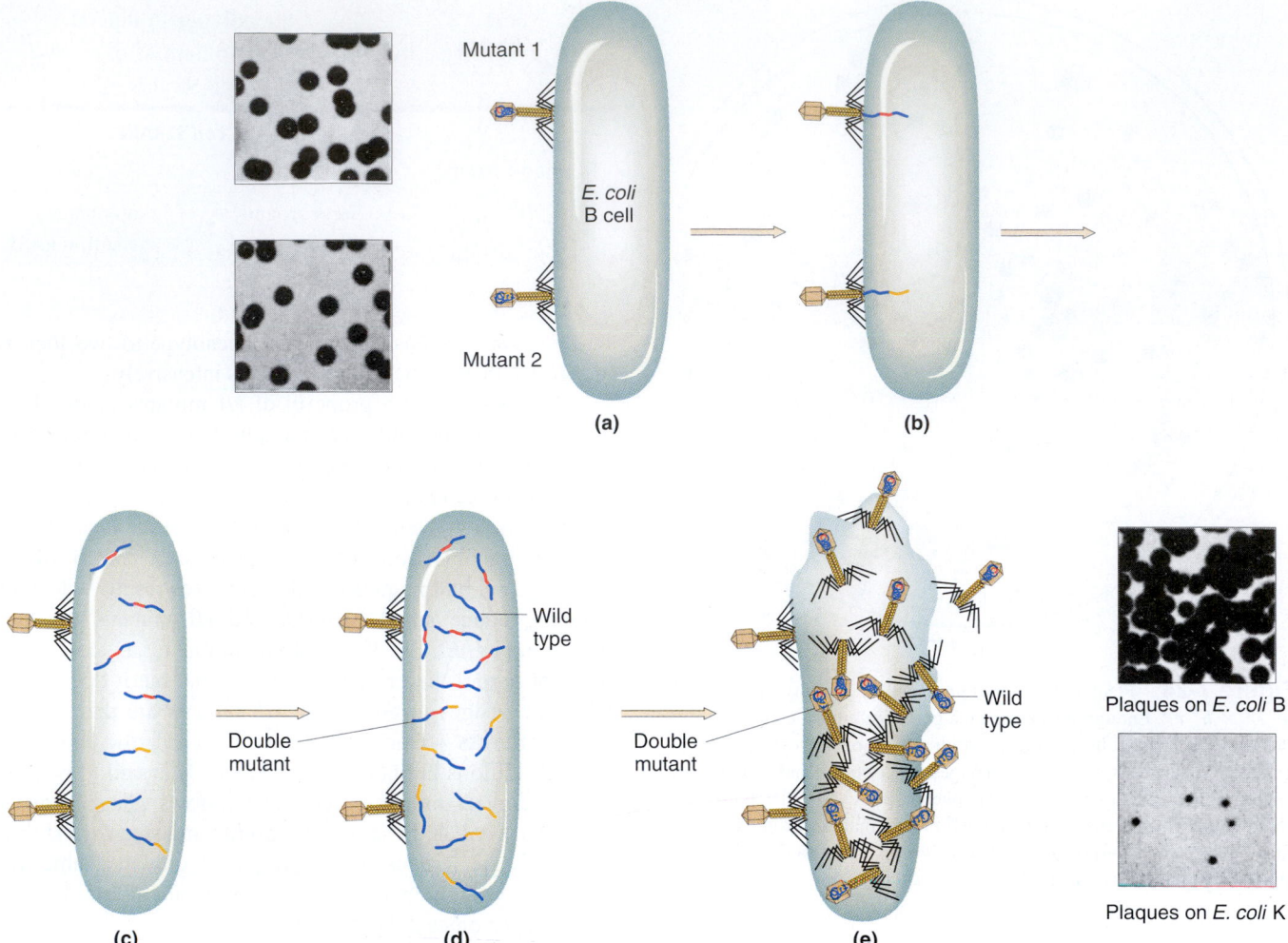

Figure 7-25   The process of recombination permits parts of the DNA of two different phage mutants to be reassembled in a new DNA molecule that may contain both mutations or neither of them. Mutants obtained from two different cultures are introduced into a broth of *E. coli* strain B. Crossing occurs when DNA from each mutant type infects a single bacillus. Most of the DNA replicas are of one type or the other, but occasionally recombination will produce either a double mutant or a wild-type recombinant containing neither mutation. When the progeny of the cross are plated on strain B, all grow successfully, producing many plaques. Plated on strain K, only the wild-type recombinants are able to grow. A single wild-type recombinant can be detected among as many as 100 million progeny. (After S. Benzer, "The Fine Structure of the Gene." Copyright © 1962 by Scientific American, Inc. All rights reserved.)

we could easily detect the recombinant wild-type phages. Why? Because a typical phage lysate (the phage mixture released after lysis of the bacteria) from such an infection (whether it includes a cross or not) contains in excess of 10^9 phages per milliliter. If we mix 0.1 ml of such a phage lysate with 0.1 ml of *E. coli* K bacteria, then we will have more than 10^5 (100,000) wild-type recombinant phage infecting the bacteria when the recombination frequency is 0.01 percent. (In practice, increasing dilutions of the phage lysate are used until one yields a countable number of plaques.) Now we can see the power of Benzer's *rII*−*E. coli*

B/K system. In a single milliliter, it can find one recombinant or revertant per 10^9 organisms. Contrast this with trying to find one recombinant in 10^9 *Drosophila* or 10^9 mice!

After we have made our cross, we need to determine the recombinant frequency. First, we count the number of active virus particles, or plaque-forming units (**pfu**), that grew on *E. coli* K (these plaque-forming units, remember, are only wild-type recombinant phages) and the number that grew on *E. coli* B (which represent the total progeny phages, because all of the virus particles can grow on strain B). The recombinant frequency can then be calculated as twice the number of pfu on *E. coli* K divided by the number of pfu on *E. coli* B. Why do we use twice the pfu frequency for *E. coli* K? To account for the recombinants that are double mutants and that we cannot detect; such mutants should be present at the same frequency as the wild-type recombinants.

Finally, in any cross of this type, we need to plate each parental lysate on *E. coli* K to see how many revertants to wild type there were in the population. **Back (reverse) mutations** occur at some very low but real frequency. It is important to monitor this frequency and to compare it with our calculated frequency of recombination to be sure that recombination — not back reversion of the parental types — has occurred.

In summary, Benzer's use of the *rII* system and two different bacterial hosts provided him with a method for selecting for rare crossover events within the gene without having to screen large numbers of plaques.

MESSAGE ···

Benzer capitalized on the fantastic resolving power made possible by a system that selects for rare events in rapidly multiplying phages; this system allowed him to map a gene in molecular detail.

Transduction

Some phages are able to "mobilize" bacterial genes and carry them from one bacterial cell to another through the process of *transduction*. Thus, transduction joins the battery of modes of genetic transfer in bacteria — along with conjugation, infectious transfer of episomes, and transformation.

Discovery of transduction

In 1951, Joshua Lederberg and Norton Zinder were testing for recombination in the bacterium *Salmonella typhimurium* by using the techniques that had been successful with *E. coli.* The researchers used two different strains: one was *phe⁻ trp⁻ tyr⁻*, and the other was *met⁻ his⁻*. (We won't worry about the nature of these markers except to note that the mutant alleles confer nutritional requirements.) When either strain was plated on a minimal medium, no wild-type cells were observed. However, after the two strains were mixed, wild-type cells appeared at a frequency of about 1 in 10^5. Thus far, the situation seems similar to that for recombination in *E. coli.*

However, in this case, the researchers also recovered recombinants from a U-tube experiment, in which cell contact (conjugation) was prevented by a filter separating the two arms. By varying the size of the pores in the filter, they found that the agent responsible for recombination was about the size of the virus P22, a known temperate phage of *Salmonella.* Further studies supported the suggestion that the vector of recombination is indeed P22. The filterable agent and P22 are identical in properties of size, sensitivity to antiserum, and immunity to hydrolytic enzymes. Thus,

Lederberg and Zinder, instead of confirming conjugation in *Salmonella,* had discovered a new type of gene transfer mediated by a virus. They called this process **transduction.** In the lytic cycle, some virus particles somehow pick up bacterial genes that are then transferred to another host, where the virus inserts its contents. Transduction has subsequently been shown to be quite common among both temperate and virulent phages.

There are two kinds of transduction: generalized and specialized. *Generalized* transducing phages can carry any part of the chromosome, whereas *specialized* transducing phages carry only restricted parts of the bacterial chromosome.

Transducing phages and generalized transduction

How are transducing phages produced? In 1965, K. Ikeda and J. Tomizawa threw light on this question in some experiments on the temperate *E. coli* phage P1. They found that, when a donor cell is lysed by P1, the bacterial chromosome is broken up into small pieces. Occasionally, the forming phage particles mistakenly incorporate a piece of the bacterial DNA into a phage head in place of phage DNA. This event is the origin of the transducing phage.

Because the phage coat proteins determine a phage's ability to attack a cell, transducing phages can bind to a bacterial cell and inject their contents, which now happen to be donor bacterial genes. When a transducing phage injects its contents into a recipient cell, a merodiploid situation is created in which the transduced bacterial genes can be incorporated by recombination (Figure 7-26). Because any of the host markers can be transduced, this type of transduction is termed **generalized transduction.**

Phages P1 and P22 both belong to a phage group that shows generalized transduction (that is, they transfer virtually any gene of the host chromosome). During their cycles, P22 probably inserts into the host chromosome, whereas P1 remains free, like a large plasmid. But both transduce by faulty head stuffing in lysis.

Linkage data from transduction

Generalized transduction allows us to derive linkage information about bacterial genes when markers are close enough that the phage can pick them up and transduce them in a single piece of DNA. For example, suppose that we wanted to find the linkage between *met* and *arg* in *E. coli.* We might set up a cross of a *met⁺ arg⁺* strain with a *met⁻ arg⁻* strain. We could grow phage P1 on the donor *met⁺ arg⁺* strain, allow P1 to infect the *met⁻ arg⁻* strain, and select for *met⁺* colonies. Then, we could note the percentage of *met⁺* colonies that became *arg⁺*. Strains transduced to both *met⁺* and *arg⁺* are called **cotransductants.**

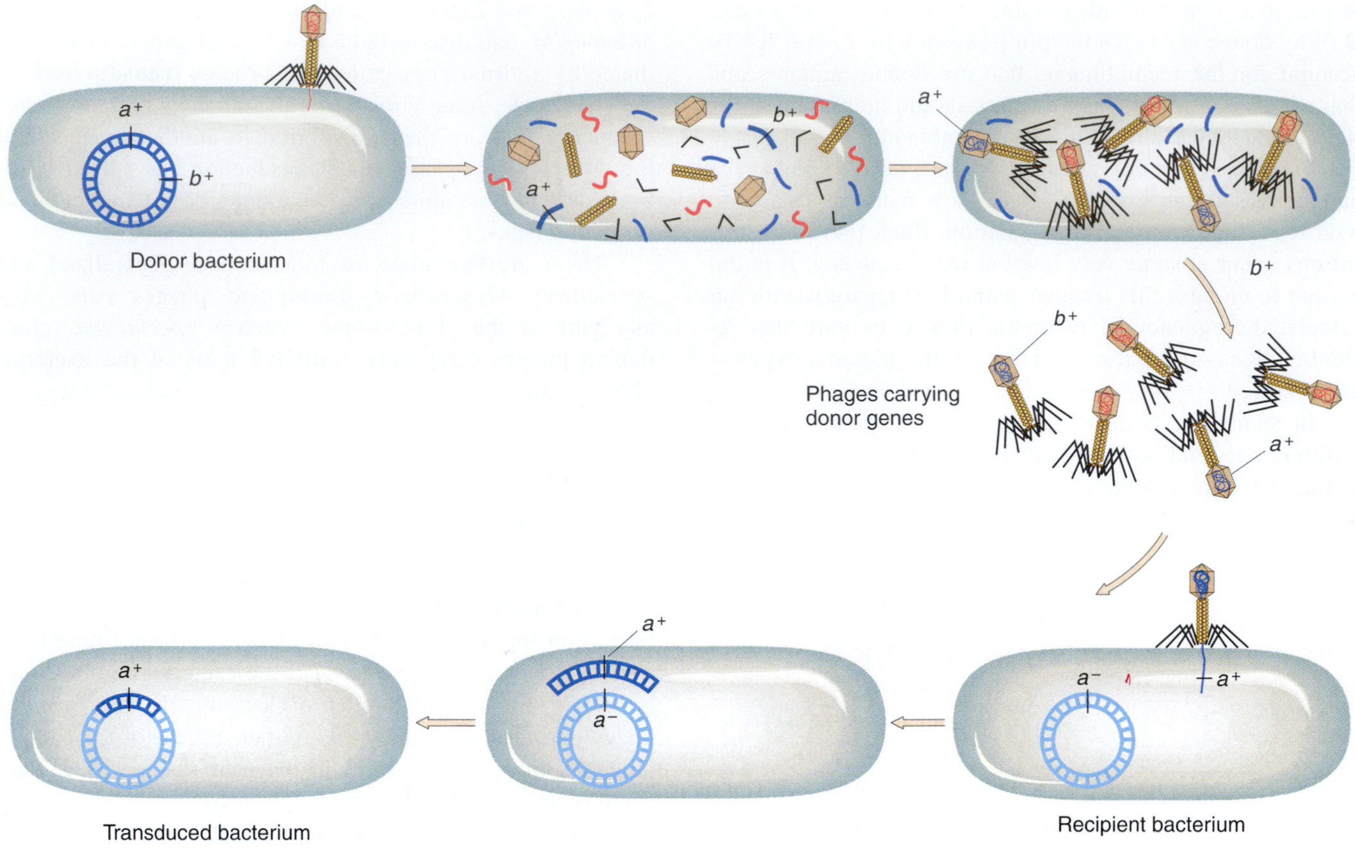

a+
b+
Donor bacterium
Phages carrying donor genes
Transduced bacterium
a−
Recipient bacterium

Figure 7-26 The mechanism of generalized transduction. In reality, only a very small minority of phage progeny (1 in 10,000) carries donor genes.

Linkage values are usually expressed as cotransduction frequencies (Figure 7-27). The greater the cotransduction frequency, the closer two genetic markers are.

Using an extension of this approach, we can estimate the size of the piece of host chromosome that a phage can pick up. The following type of experiment uses P1 phage:

$$\text{donor } leu^{+}\ thr^{+}\ azi^{\mathrm{r}} \longrightarrow \text{recipient } leu^{-}\ thr^{-}\ azi^{\mathrm{s}}$$

We can select for one or more donor markers in the recipient and then (in true merozygous genetics style) look for the presence of the other unselected markers, as outlined in Table 7-5. Experiment 1 in Table 7-5 tells us that *leu* is relatively close to *azi* and distant from *thr,* leaving us with two possibilities:

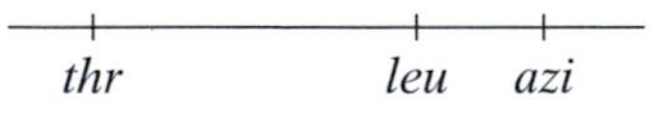

thr
leu
azi

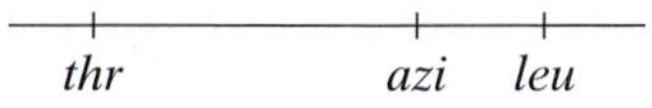

thr
azi
leu

Experiment 2 tells us that *leu* is closer to *thr* than *azi* is, so the map must be:

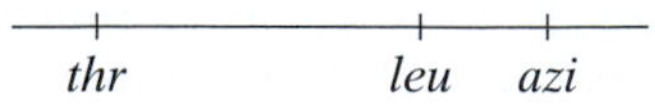

thr
leu
azi

TABLE 7-5 Accompanying Markers in Specific P1 Transductions

Experiment	Selected marker	Unselected markers
1	leu^{+}	50% are azi^{r}; 2% are thr^{+}
2	thr^{+}	3% are leu^{+}; 0% are azi^{r}
3	leu^{+} and thr^{+}	0% are azi^{r}

By selecting for thr^{+} and leu^{+} in the transducing phages in experiment 3, we see that the transduced piece of genetic material never includes the *azi* locus.

If enough markers were studied to produce a more complete linkage map, we could estimate the size of a transduced segment. Such experiments indicate that P1 cotransduction occurs within approximately 1.5 minutes of the *E. coli* chromosome map (1 minute equals the length of chromosome transferred by an Hfr in 1 minute's time at 37°C).

Lysogeny

In the 1920s, long before *E. coli* became the favorite organism of microbial geneticists, some interesting results were

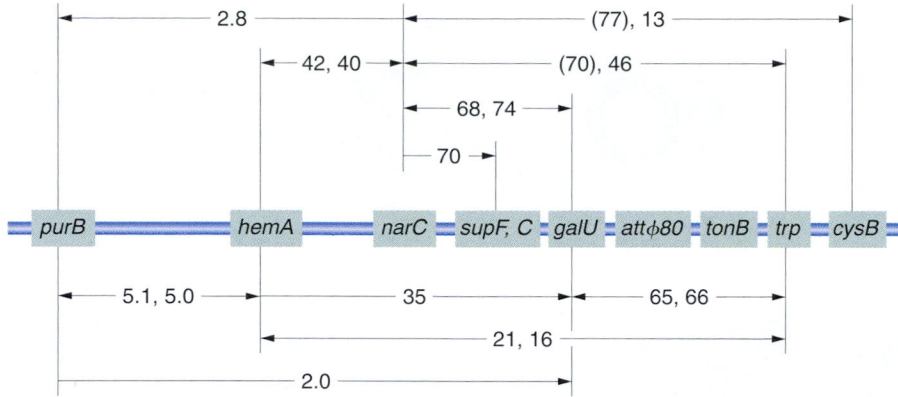

Figure 7-27 Genetic map of the *purB*-to-*cysB* region of *E. coli* determined by P1 cotransduction. The numbers given are the averages in percent for cotransduction frequencies obtained in several experiments. Where transduction crosses were performed in both directions, the head of each arrow points to the selective marker with the corresponding linkage nearest to each arrow. The values in parentheses are considered unreliable owing to interference from the nonselective marker. (After J. R. Guest, *Molecular and General Genetics* 105, 1969, 285.)

obtained in the study of phage infections of E. coli. Some bacterial strains were found to be resistant to infection by certain phages, but these resistant bacteria caused lysis of nonresistant bacteria when the two bacterial strains were mixed together. The resistant bacteria that induced lysis in other cells were said to be lysogenic bacteria or lysogens. When non-lysogenic bacteria were infected with phages derived from a lysogenic strain, a small fraction of the infected cells did not lyse but instead became lysogenic themselves.

Apparently, the lysogenic bacteria could somehow "carry" the phages while remaining immune to their lysing action. Initially, little attention was paid to this phenomenon after some studies seemed to show that the lysogenic bacteria were simply contaminated with external phages that could be removed by careful purification. However, in the mid-1940s, André Lwoff examined lysogenic strains of *Bacillus megaterium* and followed the behavior of a lysogenic strain through many cell divisions. Carefully observing his culture, he separated each pair of daughter cells immediately after division. One cell was put into a culture; the other was observed until it divided. In this way, Lwoff obtained 19 cultures representing 19 generations (19 consecutive cell divisions). All 19 cultures were lysogenic, but tests of the medium showed no free phages at any time during these divisions, thereby confirming that lysogenic behavior is a character that persists through reproduction in the absence of free phages.

On rare occasions, Lwoff observed spontaneous lysis in his cultures. When the medium was spread on a lawn of nonlysogenic cells after one of these spontaneous lyses, plaques appeared, showing that free phages had been released in the lysis. Lwoff was able to propose a hypothesis to explain all his observations: each bacterium of the lysogenic strain contains a noninfective factor that is passed from bacterial generation to generation, but this factor occasionally gives rise to the production of infective phages (without the presence of free phages in the medium). Lwoff called this factor the **prophage** because it somehow seemed to be able to *induce* the formation of a "litter" of infective

phages. Later studies showed that a variety of agents, such as ultraviolet light or certain chemicals, could activate the prophage, inducing lysis and infective phage release in a large fraction of a population of lysogenic bacteria.

We now know exactly how Lwoff's observations occur. A lysogenic bacterium contains a prophage, which somehow protects the cell against additional infection, or **superinfection**, from free phages and which is duplicated and passed on to daughter cells in division. In a small fraction of the lysogenic cells, the prophage is **induced**, or **activated**, producing infective phages. This process robs the cell of its protection against the phage; it lyses and releases infective phages into the medium, thus infecting any nonlysogenic cells present in the culture.

Phages can be categorized into two types. **Virulent phages** have an infectious cycle that is always **lytic** — for these phages, there are no lysogenic bacteria. (Resistant bacterial mutants may exist for virulent phages, but their resistance is not due to lysogeny.) **Temperate phages** follow a lytic cycle under some circumstances, but they usually initiate a **lysogenic cycle,** in which the phage exists as a prophage within the bacterial cell. In this case, the lysogenic bacterium becomes resistant to superinfection, an "immunity" conferred by the presence of the prophage, which is transmitted genetically through many bacterial generations. Temperate phages also cause lysis when the prophage is induced, or activated. Figure 7-28 diagrams the lytic and lysogenic infectious cycles of a typical temperate phage.

MESSAGE ···
Virulent phages cannot become prophages; they are always lytic. Temperate phages can exist within the bacterial cell as prophages, allowing their hosts to survive as lysogenic bacteria; they are also capable of direct bacterial lysis.

Genetic basis of lysogeny

What is the nature of the prophage? On induction, the prophage is capable of directing the production of a complete mature phage, so all of the phage genome must be

Figure 7-28   Alternative cell cycles of a temperate phage and its host. (After A. Lwoff, *Bacteriological Reviews* 17, 1953, 269.)

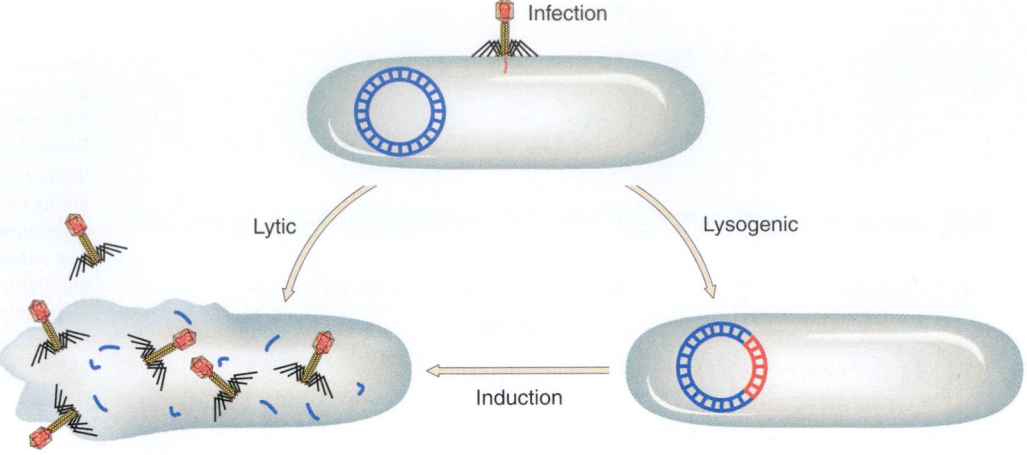

present in the prophage. But is the prophage a small particle free in the bacterial cytoplasm — a plasmid — or is it somehow associated with the bacterial genome? Fortuitously, the original strain of *E. coli* used by Lederberg and Tatum (page 209) proved to be lysogenic for a temperate phage called **lambda (λ)**. Phage λ has become the most intensively studied and best-characterized phage. Crosses between F$^+$ and F$^-$ cells have yielded interesting results. It turns out that F$^+$ × F$^-$(λ) crosses yield recombinant lysogenic recipients, whereas the reciprocal cross F$^+$(λ) × F$^-$ almost never gives lysogenic recombinants.

These results became more understandable when Hfr strains were discovered. In the cross Hfr × F$^-$(λ), lysogenic F$^-$ exconjugants with Hfr genes are readily recovered. However, in the reciprocal cross Hfr(λ) × F$^-$, the early genes from the Hfr chromosome are recovered among the exconjugants, but recombinants for late markers (those expected to transfer after a certain time in mating) are not recovered. Furthermore, lysogenic exconjugants are almost never recovered from this reciprocal cross. What is the explanation? The observations make sense if the λ prophage is behaving like a bacterial gene locus (that is, like part of the bacterial chromosome). In interrupted-mating experiments, the λ prophage always enters the F$^-$ cell at a specific time, closely linked to the *gal* locus. Thus, we can assign the λ prophage to a specific locus next to the *gal* region.

In the cross of a lysogenic Hfr with a nonlysogenic (nonimmune) F$^-$ recipient, the entry of the λ prophage into the nonimmune cell immediately triggers the prophage into a lytic cycle; this process is called **zygotic induction.** But, in the cross Hfr(λ) × F$^-$(λ), any recombinants are readily recovered; that is, no induction of the prophage, and consequently lysis, occurs (Figure 7-29). It would seem that the cytoplasm of the F$^-$ cell must exist in two different states (depending on whether the cell contains a λ prophage), so contact between an entering prophage and the cytoplasm of a nonimmune cell immediately induces the lytic cycle. We now know that a cytoplasmic factor specified by the prophage represses the multiplication of the virus. Entry of the prophage into a nonlysogenic environment immediately dilutes this repressing factor, and therefore the virus reproduces. But, if the virus specifies the repressing factor, then why doesn't the virus shut itself off again? Clearly it does, because a fraction of infected cells do become lysogenic. There is a race between the λ gene signals for reproduction and those specifying a shutdown. The model of a phage-directed cytoplasmic repressor nicely explains the immunity of the lysogenic bacteria, because any superinfecting phage would immediately encounter a repressor and be inactivated. We present this model in more detail in Chapter 11.

Prophage attachment

How is the prophage attached to the bacterial genome? Allan Campbell proposed in 1962 that λ attaches to the bacterial chromosome by a reciprocal crossover between the cir-

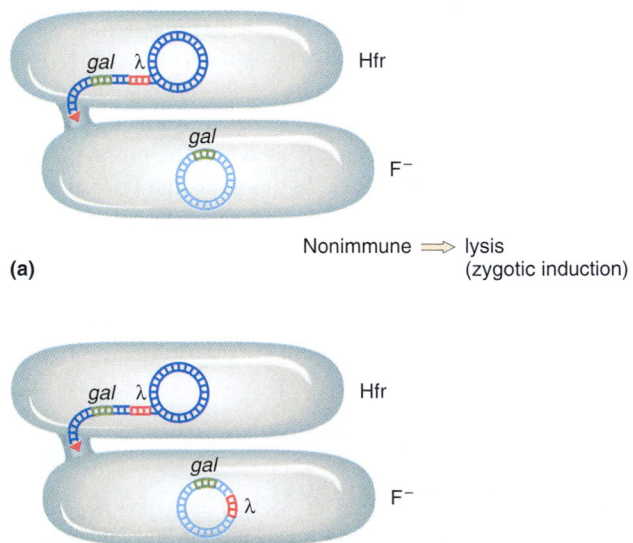

Nonimmune ⟹ lysis
(zygotic induction)

(a)

Immune ⟹ no lysis

(b)

Figure 7-29   Zygotic induction.

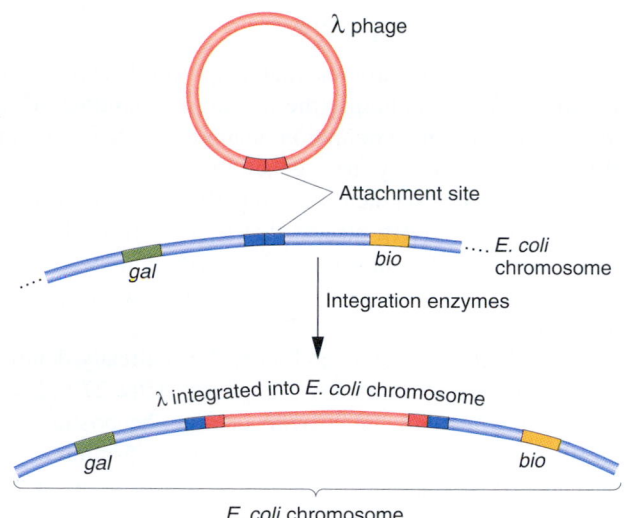

Figure 7-30 Campbell's model for the integration of phage λ into the *E. coli* chromosome. Reciprocal recombination takes place between a specific attachment site on the circular λ DNA and a specific region on the bacterial chromosome between the *gal* and *bio* genes.

cular λ chromosome and the circular *E. coli* chromosome, as shown in Figure 7-30. The crossover point would be between a specific site in λ, the **λ attachment site,** and a site in the bacterial chromosome located between the genes *gal* and *bio,* because λ integrates at that position in the *E. coli* chromosome.

One attraction of Campbell's proposal is that it allows predictions that geneticists can test by using phage λ:

1. Integration of the prophage into the *E. coli* chromosome should increase the genetic distance between flanking bacterial markers, as can be seen in Figure 7-30 for *gal*

and *bio.* In fact, studies show that time-of-entry or recombination distances between the bacterial genes *are* increased by lysogeny.

2. Deleting bacterial segments adjacent to the prophage site should delete phage genes at least some of the time. Experimental studies also confirm this prediction.

Specialized transduction

We can now understand the process of **specialized transduction,** in which only certain host markers can be transduced.

Lambda is a good example of a specialized transducing phage. As a prophage, λ always inserts between the *gal* region and the *bio* region of the host chromosome (see Figure 7-30). In transduction experiments, λ can transduce only the *gal* and *bio* genes. Let's visualize the mechanism of λ transduction.

The recombination between regions of λ and the bacterial chromosome is catalyzed by a specific enzyme system. This system normally ensures that λ integrates at the same point in the chromosome and, when the lytic cycle is induced (for instance, by ultraviolet light), it ensures that the λ prophage excises at precisely the correct point to produce a normal circular λ chromosome. Very rarely, excision is abnormal and can result in phage particles that now carry a nearby gene and leave behind some phage genes (Figure 7-31a). In λ, the nearby genes are *gal* on one side and *bio* on the other. The resulting particles are defective due to the genes left behind and are referred to as **λdgal** (λ-defective *gal*), or **λdbio.** These defective particles carrying nearby genes can be packaged into phage heads and can infect other bacteria. In the presence of a second, normal phage particle in a double infection, the λdgal can integrate into the

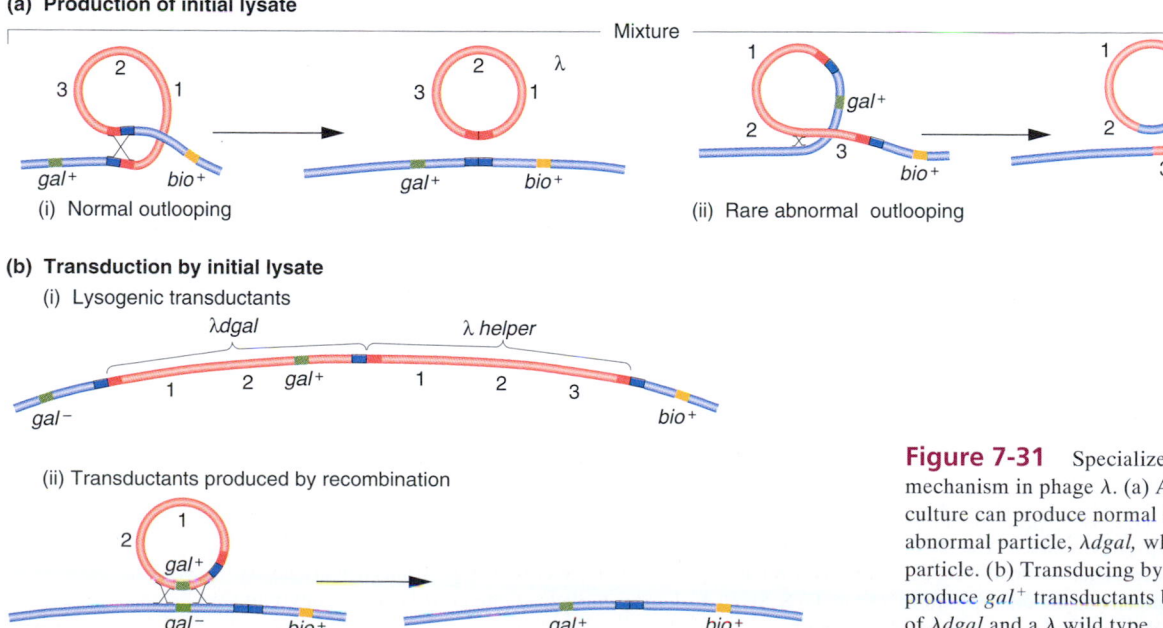

(a) **Production of initial lysate**

(i) Normal outlooping

(ii) Rare abnormal outlooping

(b) **Transduction by initial lysate**

(i) Lysogenic transductants

(ii) Transductants produced by recombination

Figure 7-31 Specialized transduction mechanism in phage λ. (a) A lysogenic bacterial culture can produce normal λ or, rarely, an abnormal particle, λdgal, which is the transducing particle. (b) Transducing by the mixed lysate can produce *gal*⁺ transductants by the coincorporation of λdgal and a λ wild type.

chromosome at the λ-attachment site (Figure 7-31b). In this manner, the *gal* genes in this case are transduced into the second host. Because this transduction mechanism is limited to genes very near the original integrated prophage, it is called specialized transduction.

> **MESSAGE** ··
>
> Transduction occurs when newly forming phages acquire host genes and transfer them to other bacterial cells. *Generalized transduction* can transfer any host gene. It occurs when phage packaging accidentally incorporates bacterial DNA instead of phage DNA. *Specialized transduction* is due to faulty separation of the prophage from the bacterial chromosome, so the new phage includes both phage and bacterial genes. The transducing phage can transfer only specific host genes.

Chromosome mapping

Some very detailed chromosomal maps for bacteria have been obtained by combining the mapping techniques of interrupted mating, recombination mapping, transformation, and transduction. Today, new genetic markers are typically mapped first into a segment of about 10 to 15 map minutes by using a series of Hfr strains that transfer from different points around the chromosome. This method allows the selection of markers within the interval to be used for P1 cotransduction.

By 1963, the *E. coli* map (Figure 7-32) already detailed the positions of approximately 100 genes. After 27 years of further refinement, the 1990 map depicts the positions of more than 1400 genes. Figure 7-33 shows a 5-minute section of the 1990 map (which is adjusted to a scale of 100

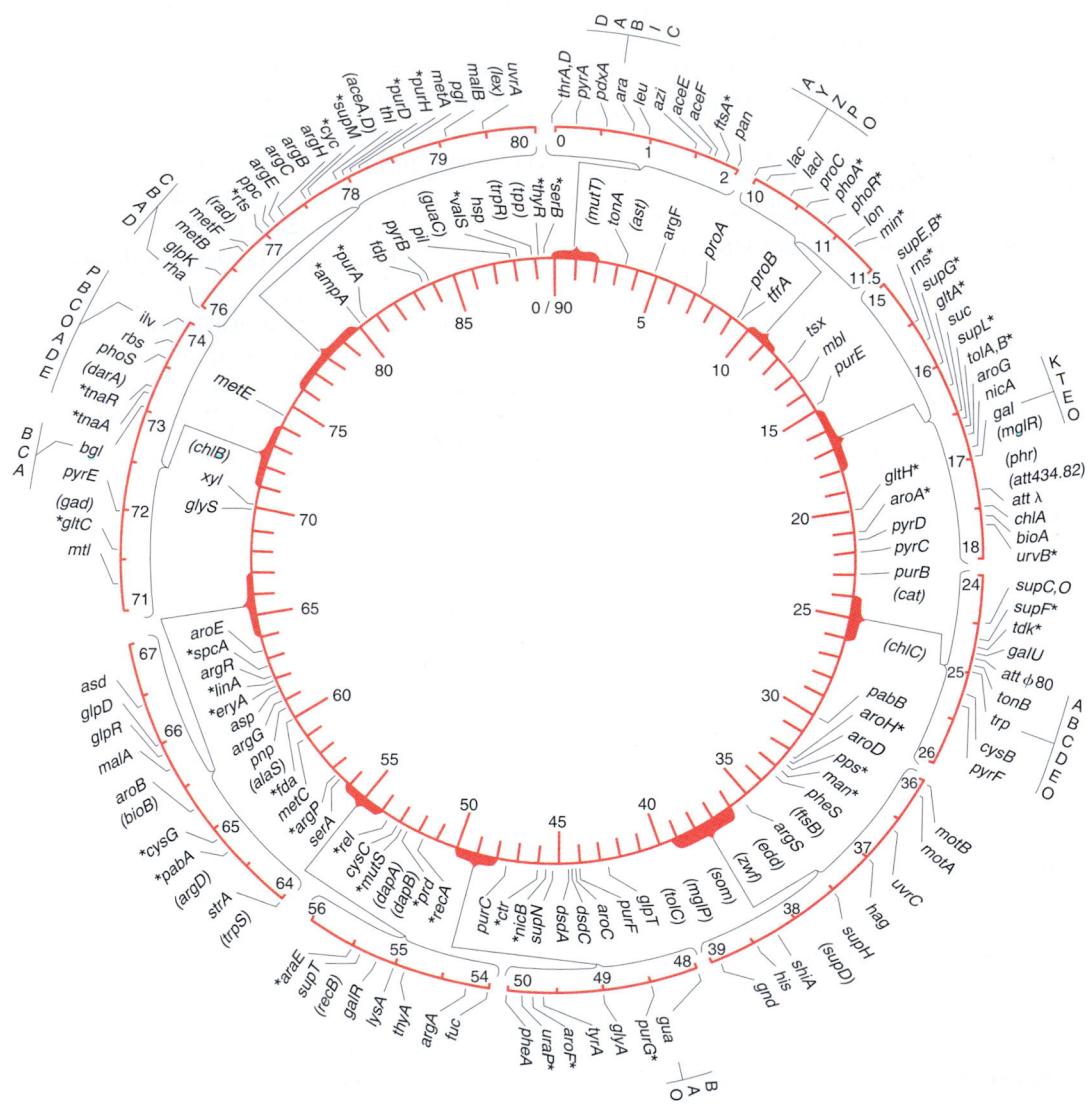

Figure 7-32   The 1963 genetic map of *E. coli*. Units are minutes, based on interrupted-mating experiments, timed from an arbitrarily located origin. Asterisks refer to map positions that are not as precise as the other positions. (From G. S. Stent, *Molecular Biology of Bacterial Viruses*. Copyright © 1963 by W. H. Freeman and Company.)

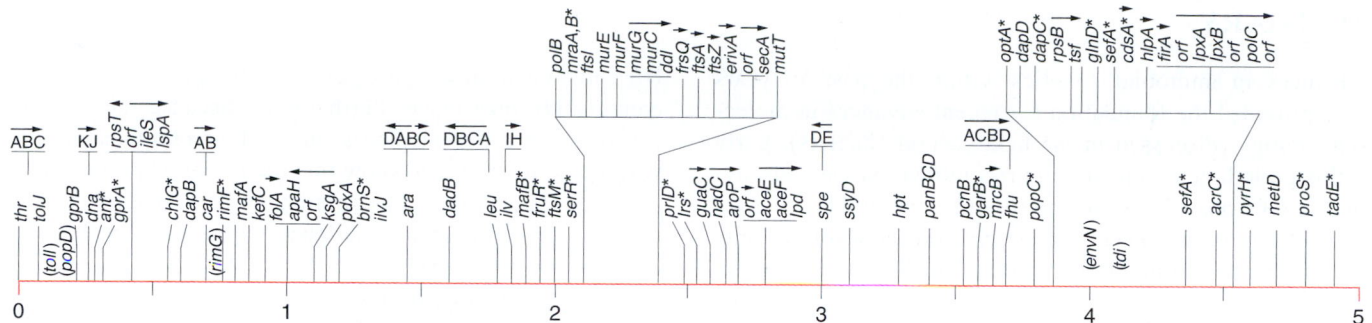

Figure 7-33   Linear scale drawing of a 5-minute section of the 100-minute 1990 *E. coli* linkage map. Markers in parentheses are not precisely mapped; those marked by asterisks are more precisely mapped than those within parentheses but are still not exactly known. Arrows above genes and groups of genes indicate the direction of transcription of these loci. (From B. J. Bachmann, "Linkage Map of *Escherichia coli* K-12, Edition 8." *Microbiological Reviews,* 54, 1990, 130–197.)

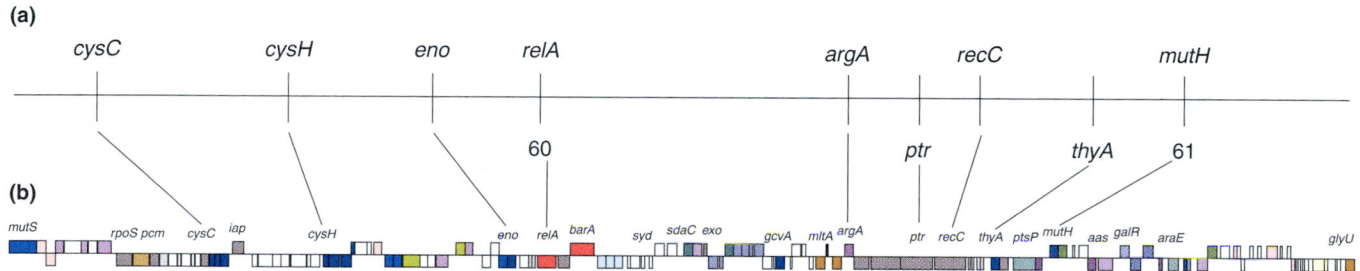

Figure 7-34   Correlation of the genetic and physical maps. (a) Markers on the 1990 genetic map in the region near 60 and 61 minutes. (b) The exact positions of every gene based on the complete sequence of the *E. coli* genome. (Not every gene is named in this figure, for simplicity.) The elongated boxes are genes and putative genes. Each color represents a different type of function. For example, red denotes regulatory functions, and dark blue denotes functions in DNA replication, recombination, and repair. The correspondence of the order of genes on both maps is indicated.

minutes). The complexity of these maps illustrates the power and sophistication of genetic analysis. How well do these maps correspond to physical reality? In 1997, the DNA sequence of the entire *E. coli* genome was completed, allowing us to compare the exact position of genes on the genetic map with the corresponding position of the respective coding sequence on the linear DNA sequence. Figure 7-34 makes this comparison for a segment of both maps. Clearly, the genetic map accurately corresponds to the relative positions on the physical map.

Bacterial gene transfer in review

1. Gene transfer in bacteria can be achieved through conjugation, transformation, and viral transduction.

2. The inheritance of genetic markers through the conjugative transfer of DNA by Hfr strains, the transformation of parts of the donor chromosome, and generalized transduction all share one important property. Each process introduces a DNA fragment into the recipient cell; then a double-crossover event must take place if the fragment is to be incorporated into the recipient

genome and subsequently inherited. Unincorporated fragments cannot replicate and are diluted out and lost from the population of daughter cells.

3. The conjugative transfer of F′ factors that carry bacterial genes and the specialized transduction of certain genetic markers are similar processes in that a specific and limited set of bacterial genes in each case is efficiently introduced into the recipient cell. Inheritance does not require normal recombination, as in the inheritance of DNA fragments. After the F′ transfer, the F′ factor replicates in the bacterial cytoplasm as a separate entity. The specialized transducing phage DNA is recombined into the bacterial chromosome by a recombination system specific for that phage. In both cases, a partial diploid (merodiploid) results, because each process allows the inheritance of the transferred gene and of the recipient's counterpart.

4. Gene transfer can be used to map the chromosome. Hfr crosses are first used to localize a mutation to a region of the chromosome. Then, generalized transduction provides a more exact localization.

SUMMARY

Advances in microbial genetics within the past 50 years have provided the foundation for recent advances in molecular biology (discussed in the next several chapters). Early in this period, gene transfer and recombination were discovered to take place between certain different strains of bacteria. In bacteria, however, genetic material is passed in only one direction — from a donor cell (F⁺ or Hfr) to a recipient cell (F⁻). Donor ability is determined by a presence in the cell of a fertility (F) factor acting as an episome.

On occasion, the F factor present in a free state in F⁺ cells can integrate into the *E. coli* chromosome and form an Hfr cell. When this occurs, gene transfer and subsequent recombination take place. Furthermore, because the F factor can insert at different places on the host chromosome, investigators were able to show that the *E. coli* chromosome is a single circle, or ring. Interruptions of the transfer at different times has provided geneticists with a new method for constructing a linkage map of the single chromosome of *E. coli* and other similar bacteria.

Genetic traits can also be transferred from one bacterial cell to another in the form of purified DNA. This process of transformation in bacterial cells was the first demonstration

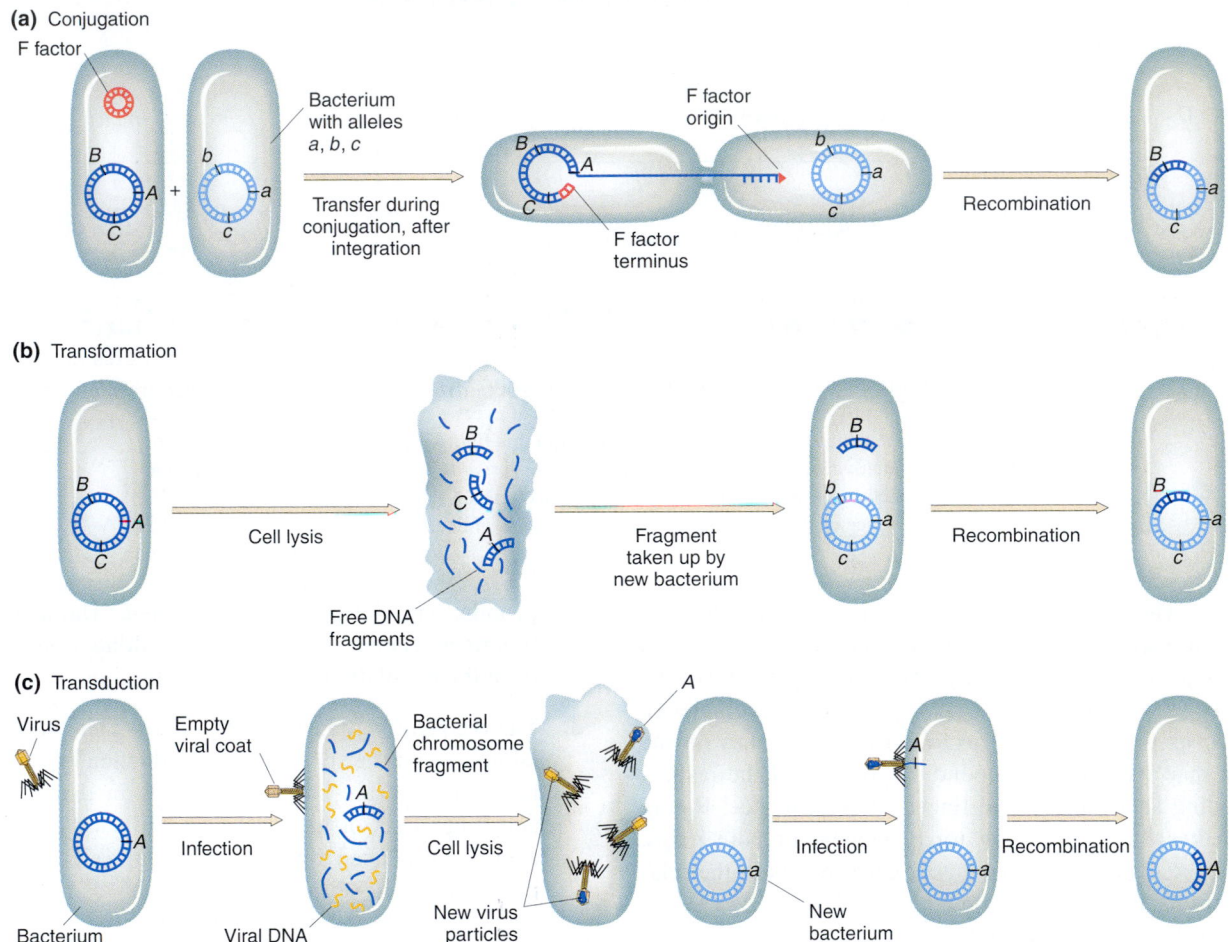

Figure 7-35   Recombination processes in bacteria. Bacterial recombination requires that a bacterial cell receive an allele obtained from another cell. (a) In conjugation, a cytoplasmic element such as the fertility factor (F) integrates into the chromosome of a bacterial cell. During cell-to-cell contact, the integrated factor can transfer part or all of that chromosome to another cell whose chromosome carries alleles of genes on the transferred chromosome. The transferred segment recombines with a homologous segment in the recipient cell's chromosome; in the example shown here, allele *B* thereby replaces allele *b*. (b) In transformation, a DNA segment bearing a particular allele is taken up from the environment by a cell whose chromosome carries a matching allele; the alleles (in our example, *B* and *b*) are then exchanged by homologous recombination. (c) In transduction, after a phage has infected a bacterial cell, one of the newly forming phage particles picks up a bacterial DNA segment instead of viral DNA. When this phage particle infects another cell, it injects its bacterial DNA, which recombines with a homologous segment in the second cell, thereby exchanging any corresponding alleles (in our example, *A* and *a*).

that DNA is the genetic material. For transformation to occur, DNA must be taken into a recipient cell, and recombination between a recipient chromosome and the incorporated DNA must then take place.

Bacteria can also be infected by bacteriophages. In one method of infection, the phage chromosome may enter the bacterial cell and, using the bacterial metabolic machinery, produce progeny phage that burst the host bacterium. The new phages can then infect other cells. If two phages of different genotypes infect the same host, recombination between their chromosomes can take place in this lytic process. Mapping the genetic loci through these recombinational events has led to the discovery that some phage chromosomes also are circular.

In another infection method, lysogeny, the injected phage lies dormant in the bacterial cell. In many cases, this dormant phage (the prophage) incorporates into the host chromosome and replicates with it. Either spontaneously or under appropriate stimulation, the prophage can arise from its latency and can lyse the bacterial host cell.

Phages can carry bacterial genes from a donor to a recipient. In generalized transduction, random host DNA is incorporated alone into the phage head during lysis. In specialized transduction, faulty excision of the prophage from a unique chromosomal locus results in the inclusion of specific host genes as well as phage DNA in the phage head.

Figure 7-35 summarizes the processes of conjugation, transformation, and transduction.

CONCEPT MAP

Draw a concept map interrelating as many of the following terms as possible. Note that the terms are listed in no particular order.

bacteria / conjugation / recombination / F plasmid / Hfr / F⁻ / donor / recipient / interrupted mating / chromosome map / pilus / merozygote / gene

CHAPTER INTEGRATION PROBLEM

Suppose that a cell were unable to carry out generalized recombination (*rec⁻*). How would this cell behave as a recipient in generalized and in specialized transduction? First compare each type of transduction and then determine the effect of the *rec⁻* mutation on the inheritance of genes by each process.

◆ Solution ◆

Generalized transduction entails the incorporation of chromosomal fragments into phage heads, which then infect recipient strains. Fragments of the chromosome are incorporated randomly into phage heads, so any marker on the bacterial host chromosome can be transduced to another strain by generalized transduction. In contrast, specialized transduction entails the integration of the phage at a specific point on the chromosome and the rare incorporation of chromosomal markers near the integration site into the phage genome. Therefore, only those markers that are near the

specific integration site of the phage on the host chromosome can be transduced.

Markers are inherited by different routes in generalized and specialized transduction. A generalized transducing phage injects a fragment of the donor chromosome into the recipient. This fragment must be incorporated into the recipient's chromosome by recombination, with the use of the recipient recombination system. Therefore, a *rec⁻* recipient will not be able to incorporate fragments of DNA and cannot inherit markers by generalized transduction. On the other hand, the major route for the inheritance of markers by specialized transduction is by integration of the specialized transducing particle into the host chromosome at the specific phage integration site. This integration, which sometimes requires an additional wild-type (helper) phage, is mediated by a phage-specific enzyme system that is independent of the normal recombination enzymes. Therefore, a *rec⁻* recipient can still inherit genetic markers by specialized transduction.

SOLVED PROBLEMS

1. In *E. coli,* four Hfr strains donate the following genetic markers shown in the order donated:

Strain 1:	Q	W	D	M	T
Strain 2:	A	X	P	T	M
Strain 3:	B	N	C	A	X
Strain 4:	B	Q	W	D	M

All of these Hfr strains are derived from the same F⁺ strain. What is the order of these markers on the circular chromosome of the original F⁺?

◆ Solution ◆

Recall the two-step approach that works well: (1) determine the underlying principle, and (2) draw a diagram. Here the principle is clearly that each Hfr strain donates genetic markers from a fixed point on the circular chromosome and that the earliest markers are donated with the highest frequency. Because not all markers are donated by each Hfr, only the early markers must be donated for each Hfr. Each strain allows us to draw the following circles:

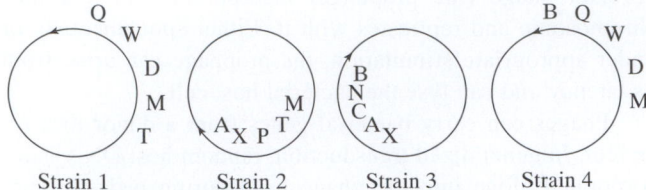

Strain 1   Strain 2   Strain 3   Strain 4

From this information, we can consolidate each circle into one circular linkage map of the order Q, W, D, M, T, P, X, A, C, N, B, Q.

2. In an Hfr × F⁻ cross, *leu⁺* enters as the first marker, but the order of the other markers is unknown. If the Hfr is wild type and the F⁻ is auxotrophic for each marker in question, what is the order of the markers in a cross where *leu⁺* recombinants are selected if 27 percent are *ile⁺*, 13 percent are *mal⁺*, 82 percent are *thr⁺*, and 1 percent are *trp⁺*?

◆ Solution ◆

Recall that spontaneous breakage creates a natural gradient of transfer, which makes it less and less likely for a recipient to receive later and later markers. Because we have selected for the earliest marker in this cross, the frequency of recombinants is a function of the order of entry for each marker. Therefore, we can immediately determine the order of the genetic markers simply by looking at the percentage of recombinants for any marker among the *leu⁺* recombinants. Because the inheritance of *thr⁺* is the highest, this must be the first marker to enter after *leu*. The complete order is *leu, thr, ile, mal, trp*.

3. A cross is made between an Hfr that is *met⁺ thi⁺ pur⁺* and an F⁻ that is *met⁻ thi⁻ pur⁻*. Interrupted-mating studies show that *met⁺* enters the recipient last, so *met⁺* recombinants are selected on a medium containing supplements that satisfy only the *pur* and *thi* requirements. These recombinants are tested for the presence of the *thi⁺* and *pur⁺* alleles. The following numbers of individuals are found for each genotype:

met⁺ thi⁺ pur⁺	280
met⁺ thi⁺ pur⁻	0
met⁺ thi⁻ pur⁺	6
met⁺ thi⁻ pur⁻	52

a. Why was methionine (Met) left out of the selection medium?

b. What is the gene order?

c. What are the map distances in recombination units?

◆ Solution ◆

a. Methionine was left out of the medium to allow selection for *met⁺* recombinants, because *met⁺* is the last marker

to enter the recipient. The selection for *met⁺* ensures that all the loci that we are considering in the cross will have already entered each recombinant that we analyze.

b. Here it is helpful to diagram the possible gene orders. Because we know that *met* enters the recipient last, there are only two possible gene orders if the first marker enters on the right: *met, thi, pur* or *met, pur, thi*. How can we distinguish between these two orders? Fortunately, one of the four possible classes of recombinants requires two additional crossovers. Each possible order predicts a different class that arises by four crossovers rather than two. For instance, if the order were *met, thi, pur*, then *met⁺ thi⁻ pur⁺* recombinants would be very rare. On the other hand, if the order were *met, pur, thi*, then the four-crossover class would be *met⁺ pur⁻ thi⁺*. From the information given in the table, it is clear that the *met⁺ pur⁻ thi⁺* class is the four-crossover class and therefore that the gene order *met, pur, thi* is correct.

c. Refer to the following diagram:

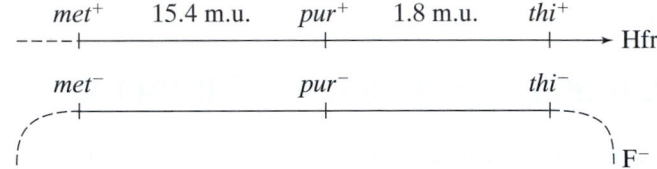

To compute the distance between *met* and *pur*, we compute the percentage of *met⁺ pur⁻ thi⁻*, which is 52/388 = 15.4 m.u. The distance between *pur* and *thi* is, similarly, 6/388 = 1.8 m.u.

4. Compare the mechanism of transfer and inheritance of the *lac⁺* genes in crosses with Hfr, F⁺, and F′-*lac⁺* strains. How would an F⁻ cell that cannot undergo normal homologous recombination (*rec⁻*) behave in crosses with each of these three strains? Would the cell be able to inherit the *lac⁺* genes?

◆ Solution ◆

Each of these three strains donates genes by conjugation. In the Hfr and F⁺ strains, the *lac⁺* genes on the host chromosome are donated. In the Hfr strain, the F factor is integrated into the chromosome in every cell, so efficient donation of chromosomal markers can occur, particularly if the marker is near the integration site of F and is donated early. The F⁺ cell population contains a small percentage of Hfr cells, in which F is integrated into the chromosome. These cells are responsible for the gene transfer displayed by cultures of F⁺ cells. In the Hfr- and F⁺-mediated gene transfer, inheritance requires the incorporation of a transferred fragment by recombination (recall that two crossovers are needed) into the F⁻ chromosome. Therefore, an F⁻ strain that cannot undergo recombination cannot inherit donor chromosomal markers even though they are transferred by Hfr strains or Hfr cells in F⁺ strains. The fragment cannot be incorporated into the chromosome by recombination. Because these fragments do not possess the ability to replicate within the F⁻ cell, they are rapidly diluted out during cell division.

Unlike Hfr cells, F′ cells transfer genes carried on the F′ factor, a process that does not require chromosome transfer. In this case, the lac⁺ genes are linked to the F′ and are transferred with the F′ at a high efficiency. In the F⁻ cell, no recombination is required, because the F′ lac⁺ strain can replicate and be maintained in the dividing F⁻ cell population. Therefore, the lac⁺ genes are inherited even in a rec⁻ strain.

PROBLEMS

1. Describe the state of the F factor in an Hfr, F⁺, and F⁻ strain.

2. How does a culture of F⁺ cells transfer markers from the host chromosome to a recipient?

3. Draw an analogy between gene transfer and integration of the transferred gene into the recipient genome in

   a. Hfr crosses by conjugation and generalized transduction.

   b. F′ derivatives such as F′*lac* and specialized transduction.

4. Why can generalized transduction transfer any gene, but specialized transduction is restricted to only a small set?

5. A microbial geneticist isolates a new mutation in *E. coli* and wishes to map its chromosomal location. She uses interrupted-mating experiments with Hfr strains and generalized-transduction experiments with phage P1. Explain why each technique, by itself, is insufficient for accurate mapping.

6. In *E. coli,* four Hfr strains donate the following markers, shown in the order donated:

Strain 1:	M	Z	X	W	C
Strain 2:	L	A	N	C	W
Strain 3:	A	L	B	R	U
Strain 4:	Z	M	U	R	B

   All these Hfr strains are derived from the same F⁺ strain. What is the order of these markers on the circular chromosome of the original F⁺?

7. Four *E. coli* strains of genotype a⁺ b⁻ are labeled 1, 2, 3, and 4. Four strains of genotype a⁻ b⁺ are labeled 5, 6, 7, and 8. The two genotypes are mixed in all possible combinations and (after incubation) are plated to determine the frequency of a⁺ b⁺ recombinants. The following results are obtained, where M = many recombinants, L = low numbers of recombinants, and 0 = no recombinants.

	1	2	3	4
5	0	M	M	0
6	0	M	M	0
7	L	0	0	M
8	0	L	L	0

   On the basis of these results, assign a sex type (either Hfr, F⁺, or F⁻) to each strain.

8. An Hfr strain of genotype a⁺ b⁺ c⁺ d⁻ str^s is mated with a female strain of genotype a⁻ b⁻ c⁻ d⁺ str^r. At various times, the culture is shaken vigorously to separate mating pairs. The cells are then plated on agar of the following three types, where nutrient A allows the growth of a⁻ cells; nutrient B, of b⁻ cells; nutrient C, of c⁻ cells; and nutrient D, of d⁻ cells (a plus indicates the presence of streptomycin and a nutrient, and a minus indicates its absence):

Agar type	Str	A	B	C	D
1	+	+	+	−	+
2	+	−	+	+	+
3	+	+	−	+	+

   a. What donor genes are being selected on each type of agar?

   b. The table below shows the number of colonies on each type of agar for samples taken at various times after the strains are mixed. Use this information to determine the order of the genes a, b, and c.

Time of sampling (minutes)	NUMBER OF COLONIES ON AGAR OF TYPE		
	1	2	3
0	0	0	0
5	0	0	0
7.5	100	0	0
10	200	0	0
12.5	300	0	75
15	400	0	150
17.5	400	50	225
20	400	100	250
25	400	100	250

   c. From each of the 25-minute plates, 100 colonies are picked and transferred to a dish containing agar with all of the nutrients except D. The numbers of colonies that grow on this medium are 89 for the sample from agar type 1, 51 for the sample from agar type 2, and 8 for the sample from agar type 3. Using these data, fit gene d into the sequence of a, b, and c.

   d. At what sampling time would you expect colonies to first appear on agar containing C and streptomycin but no A or B?

   (Problem 8 is from D. Freifelder, *Molecular Biology and Biochemistry.* Copyright © 1978 by W. H. Freeman and Company.)

9. You are given two strains of *E. coli*. The Hfr strain is *arg⁺ ala⁺ glu⁺ pro⁺ leu⁺ Tˢ*; the F⁻ strain is *arg⁻ ala⁻ glu⁻ pro⁻ leu⁻ Tʳ*. The markers are all nutritional except *T*, which determines sensitivity or resistance to phage T1. The order of entry is as given, with *arg⁺* entering the recipient first and *Tˢ* last. You find that the F⁻ strain dies when exposed to penicillin (*penˢ*) but the Hfr strain does not (*penʳ*). How would you locate the locus for *pen* on the bacterial chromosome with respect to *arg, ala, glu, pro,* and *leu*? Formulate your answer in logical, well-explained steps and draw explicit diagrams where possible.

10. A cross is made between two *E. coli* strains: Hfr *arg⁺ bio⁺ leu⁺* × F⁻ *arg⁻ bio⁻ leu⁻*. Interrupted-mating studies show that *arg⁺* enters the recipient last, so *arg⁺* recombinants are selected on a medium containing *bio* and *leu* only. These recombinants are tested for the presence of *bio⁺* and *leu⁺*. The following numbers of individuals are found for each genotype:

arg⁺ bio⁺ leu⁺	320
arg⁺ bio⁺ leu⁻	8
arg⁺ bio⁻ leu⁺	0
arg⁺ bio⁻ leu⁻	48

a. What is the gene order?

b. What are the map distances in recombination percentages?

11. Linkage maps in an Hfr bacterial strain are calculated in units of minutes (the number of minutes between genes indicates the length of time it takes for the second gene to follow the first in conjugation). In making such maps, microbial geneticists assume that the bacterial chromosome is transferred from Hfr to F⁻ at a constant rate. Thus, two genes separated by 10 minutes near the origin end are assumed to be the same physical distance apart as two genes separated by 10 minutes near the F-attachment end. Suggest a critical experiment to test the validity of this assumption.

12. In the cross Hfr *aro⁺ arg⁺ eryʳ strˢ* × F⁻ *aro⁻ arg⁻ eryˢ strʳ*, the markers are transferred in the order given (with *aro⁺* entering first), but the first three genes are very close together. Exconjugants are plated on a medium containing Str (streptomycin, to counterselect Hfr cells), Ery (erythromycin), Arg (arginine), and Aro (aromatic amino acids). The following results are obtained for 300 colonies from these plates isolated and tested for growth on various media: on Ery only, 263 strains grow; on Ery + Arg, 264 strains grow; on Ery + Aro, 290 strains grow; on Ery + Arg + Aro, 300 strains grow.

a. Draw up a list of genotypes, and indicate the number of individuals in each genotype.

b. Calculate the recombination frequencies.

c. Calculate the ratio of the size of the *arg*-to-*aro* region to the size of the *ery*-to-*arg* region.

13. A particular Hfr strain normally transmits the *pro⁺* marker as the last one in conjugation. In a cross of this strain with an F⁻ strain, some *pro⁺* recombinants are recovered early in the mating process. When these *pro⁺* cells are mixed with F⁻ cells, the majority of the F⁻ cells are converted into *pro⁺* cells that also carry the F factor. Explain these results.

14. F′ strains in *E. coli* are derived from Hfr strains. In some cases, these F′ strains show a high rate of integration back into the bacterial chromosome of a second strain. Furthermore, the site of integration is often the same site that the sex factor occupied in the original Hfr strain (before production of the F′ strains). Explain these results.

15. You have two *E. coli* strains, F⁻ *strʳ ala⁻* and Hfr *strˢ ala⁺*, in which the F factor is inserted close to *ala⁺*. Devise a screening test to detect strains carrying F′ *ala⁺*.

16. Five Hfr strains A through E are derived from a single F⁺ strain of *E. coli*. The following chart shows the entry times of the first five markers into an F⁻ strain when each is used in an interrupted-conjugation experiment:

A		B		C		D		E	
mal⁺	(1)	*ade⁺*	(13)	*pro⁺*	(3)	*pro⁺*	(10)	*his⁺*	(7)
strˢ	(11)	*his⁺*	(28)	*met⁺*	(29)	*gal⁺*	(16)	*gal⁺*	(17)
ser⁺	(16)	*gal⁺*	(38)	*xyl⁺*	(32)	*his⁺*	(26)	*pro⁺*	(23)
ade⁺	(36)	*pro⁺*	(44)	*mal⁺*	(37)	*ade⁺*	(41)	*met⁺*	(49)
his⁺	(51)	*met⁺*	(70)	*strˢ*	(47)	*ser⁺*	(61)	*xyl⁺*	(52)

a. Draw a map of the F⁺ strain, indicating the positions of all genes and their distances apart in minutes.

b. Show the insertion point and orientation of the F plasmid in each Hfr strain.

c. In the use of each of these Hfr strains, state which gene you would select to obtain the highest proportion of Hfr exconjugants.

17. *Streptococcus pneumoniae* cells of genotype *strˢ mtl⁻* are transformed by donor DNA of genotype *strʳ mtl⁺* and (in a separate experiment) by a mixture of two DNA's with genotypes *strʳ mtl⁻* and *strˢ mtl⁺*. The adjoining table shows the results.

Transforming DNA	PERCENTAGE OF CELLS TRANSFORMED INTO		
	strʳ mtl⁻	*strˢ mtl⁺*	*strʳ mtl⁺*
strʳ mtl⁺	4.3	0.40	0.17
strʳ mtl⁻ + *strˢ mtl⁺*	2.8	0.85	0.0066

a. What does the first line of the table tell you? Why?

b. What does the second line of the table tell you? Why?

18. A transformation experiment is performed with a donor strain that is resistant to four drugs: A, B, C, and D. The recipient is sensitive to all four drugs. The treated recipient cell population is divided up and plated on media containing various combinations of the drugs. The table below shows the results.

Drug(s) added	Number of colonies	Drugs(s) added	Number of colonies
None	10,000	BC	51
A	1156	BC	49
B	1148	CD	786
C	1161	ABC	30
D	1139	ABD	42
AB	46	ACD	630
AC	640	BCD	36
AD	942	ABCD	30

a. One of the genes is obviously quite distant from the other three, which appear to be tightly (closely) linked. Which is the distant gene?

b. What is the probable order of the three tightly linked genes?

(Problem 18 is from Franklin Stahl, *The Mechanics of Inheritance*, 2d ed. Copyright © 1969, Prentice Hall, Englewood Cliffs, New Jersey. Reprinted by permission.)

19. Recall that in Chapter 5 we considered the possibility that a crossover event may affect the likelihood of another crossover. In the bacteriophage T4, gene *a* is 1.0 m.u. from gene *b*, which is 0.2 m.u. from gene *c*. The gene order is *a*, *b*, *c*. In a recombination experiment, you recover five double crossovers between *a* and *c* from 100,000 progeny viruses. Is it correct to conclude that interference is negative? Explain your answer.

20. You have infected *E. coli* cells with two strains of T4 virus. One strain is minute (*m*), rapid-lysis (*r*), and turbid (*tu*); the other is wild type for all three markers. The lytic products of this infection are plated and classified. Of 10,342 plaques, the following numbers are classified as each genotype:

m r tu	3467	*m + +*	520
+ + +	3729	*+ r tu*	474
m r +	853	*+ r +*	172
m + tu	162	*+ + tu*	965

a. Determine the linkage distances between *m* and *r*, between *r* and *tu*, and between *m* and *tu*.

b. What linkage order would you suggest for the three genes?

c. What is the coefficient of coincidence (see Chapter 6) in this cross, and what does it signify?

(Problem 20 is reprinted with the permission of Macmillan Publishing Co., Inc., from Monroe W. Strickberger, *Genetics*. Copyright © 1968 by Monroe W. Strickberger.)

21. With the use of P22 as a generalized transducing phage grown on a *pur⁺ pro⁺ his⁺* bacterial donor, a recipient strain of genotype *pur⁻ pro⁻ his⁻* is infected and incubated. Afterward, transductants for *pur⁺*, *pro⁺*, and *his⁺* are selected individually in experiments I, II, and III, respectively.

a. What media are used for these selection experiments?

b. The transductants are examined for the presence of unselected donor markers, with the following results:

I		II		III	
pro⁻ his⁻	87%	*pur⁻ his⁻*	43%	*pur⁻ pro⁻*	21%
pro⁺ his⁻	0%	*pur⁺ his⁻*	0%	*pur⁺ pro⁻*	15%
pro⁻ his⁺	10%	*pur⁻ his⁺*	55%	*pur⁻ pro⁺*	60%
pro⁺ his⁺	3%	*pur⁺ his⁺*	2%	*pur⁺ pro⁺*	4%

What is the order of the bacterial genes?

c. Which two genes are closest together?

d. On the basis of the order that you proposed in part c, explain the relative proportions of genotypes observed in experiment II.

(Problem 21 is from D. Freifelder, *Molecular Biology and Biochemistry*. Copyright © 1978 by W. H. Freeman and Company, New York.)

22. Although most λ-mediated *gal⁺* transductants are inducible lysogens, a small percentage of these transductants in fact are not lysogens (that is, they contain no integrated λ). Control experiments show that these transductants are not produced by mutation. What is the likely origin of these types?

23. An *ade⁺ arg⁺ cys⁺ his⁺ leu⁺ pro⁺* bacterial strain is known to be lysogenic for a newly discovered phage, but the site of the prophage is not known. The bacterial map is:

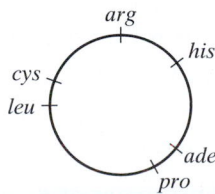

The lysogenic strain is used as a source of the phage, and the phages are added to a bacterial strain of genotype *ade⁻ arg⁻ cys⁻ his⁻ leu⁻ pro⁻*. After a short incubation, samples of these bacteria are plated on six different media, with the supplementations indicated in the table below. The table also shows whether colonies were observed on the various media.

	NUTRIENT SUPPLEMENTATION IN MEDIUM						Presence of
Medium	Ade	Arg	Cys	His	Leu	Pro	colonies
1	−	+	+	+	+	+	N
2	+	−	+	+	+	+	N
3	+	+	−	+	+	+	C
4	+	+	+	−	+	+	N
5	+	+	+	+	−	+	C
6	+	+	+	+	+	−	N

(In this table, a plus sign indicates the presence of a nutrient supplement, a minus sign indicates supplement not present, N indicates no colonies, and C indicates colonies present.)

a. What genetic process is at work here?

b. What is the approximate locus of the prophage?

24. You have two strains of λ that can lysogenize *E. coli;* the following figure shows their linkage maps:

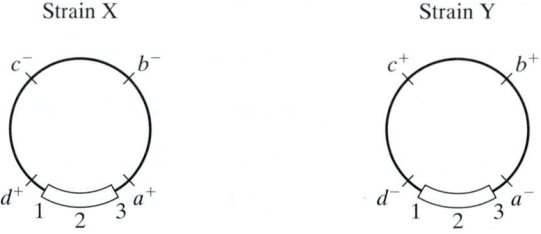

The segment shown at the bottom of the chromosome, designated 1−2−3, is the region responsible for pairing and crossing over with the *E. coli* chromosome. (Keep the markers on all your drawings.)

a. Diagram the way in which λ strain X is inserted into the *E. coli* chromosome (so that the *E. coli* is lysogenized).

b. It is possible to superinfect the bacteria that are lysogenic for strain X by using strain Y. A certain percentage of these superinfected bacteria become "doubly" lysogenic (that is, lysogenic for both strains). Diagram how this will occur. (Don't worry about how double lysogens are detected.)

c. Diagram how the two λ prophages can pair.

d. It is possible to recover crossover products between the two prophages. Diagram a crossover event and the consequences.

25. You have three strains of *E. coli.* Strain A is F′ *cys⁺ trp1/cys⁺ trp1* (that is, both the F′ and the chromosome carry *cys⁺* and *trp1,* an allele for tryptophan re-

quirement). Strain B is F⁻ *cys⁻ trp2 Z* (this strain requires cysteine for growth and carries *trp2,* another allele causing a tryptophan requirement; strain B is lysogenic for the generalized transducing phage *Z*). Strain C is F⁻ *cys⁺ trp1* (it is an F⁻ derivative of strain A that has lost the F′). How would you determine whether *trp1* and *trp2* are alleles of the same locus? (Describe the crosses and the results expected.)

26. A generalized transducing phage is used to transduce an *a⁻ b⁻ c⁻ d⁻ e⁻* recipient strain of *E. coli* with an *a⁺ b⁺ c⁺ d⁺ e⁺* donor. The recipient culture is plated on various media with the results shown in the table below. (Note that *a⁻* determines a requirement for A as a nutrient, and so forth.) What can you conclude about the linkage and order of the genes?

Compounds added to minimal medium	Presence (+) or absence (−) of colonies
C D E	−
B D E	−
B C E	+
B C D	+
A D E	−
A C E	−
A C D	−
A B E	−
A B D	+
A B C	−

27. In a generalized transduction system using P1 phage, the donor is *pur⁺ nad⁺ pdx⁻* and the recipient is *pur⁻ nad⁻ pdx⁺*. The donor allele *pur⁺* is initially selected after transduction, and 50 *pur⁺* transductants are then scored for the other alleles present. Here are the results:

Genotype	Number of colonies
nad⁺ pdx⁺	3
nad⁺ pdx⁻	10
nad⁻ pdx⁺	24
nad⁻ pdx⁻	13
	50

a. What is the cotransduction frequency for *pur* and *nad?*

b. What is the cotransduction frequency for *pur* and *pdx?*

c. Which of the unselected loci is closest to *pur?*

d. Are *nad* and *pdx* on the same side or on opposite sides of *par?* Explain. (Draw the exchanges needed to produce the various transformant classes under either order to see which requires the minimum number to produce the results obtained.)

28. In a generalized transduction experiment, phages are collected from an *E. coli* donor strain of genotype *cys*⁺ *leu*⁺ *thr*⁺ and used to transduce a recipient of genotype *cys*⁻ *leu*⁻ *thr*⁻. Initially, the treated recipient population is plated on a minimal medium supplemented with leucine and threonine. Many colonies are obtained.

   a. What are the possible genotypes of these colonies?

   b. These colonies are then replica plated onto three different media: (1) minimal plus threonine only, (2) minimal plus leucine only, and (3) minimal. What geno- types could, in theory, grow on these three media?

   c. It is observed that 56 percent of the original colonies grow on medium 1, 5 percent grow on medium 2, and no colonies grow on medium 3. What are the actual genotypes of the colonies on media 1, 2, and 3?

   d. Draw a map showing the order of the three genes and which of the two outer genes is closer to the middle gene.

*29. In 1965, Jon Beckwith and Ethan Signer devised a method of obtaining specialized transducing phages carrying the *lac* region. In a two-step approach, the researchers first "transposed" the *lac* genes to a new region of the chromosome and then isolated the specialized transducing particles. They noted that the integration site, designated *att80,* for the temperate phage φ80 (a relative of phage λ) was located near one of the genes, termed *tonB,* that confer resistance to the virulent phage T1:

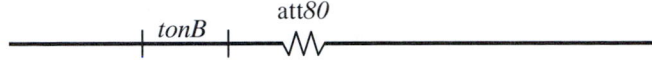

Beckwith and Signer used an F′ *lac* episome that could not replicate at high temperatures in a strain carrying a deletion of the *lac* genes. By forcing the cell to remain *lac*⁺ at high temperatures, the researchers could select strains in which the episome had integrated into the chromosome, thereby allowing the F′ *lac* to be maintained at high temperatures. By combining this selection with a simultaneous selection for resistance to T1 phage infection, they found that the only survivors were cells in which the F′ *lac* had integrated into the *tonB* locus, as shown in the accompanying figure.

This placed the *lac* region near the integration site for phage φ80. Describe the subsequent steps that the researchers must have followed to isolate the specialized transducing particles of phage φ80 that carried the *lac* region.

8

THE STRUCTURE AND REPLICATION OF DNA

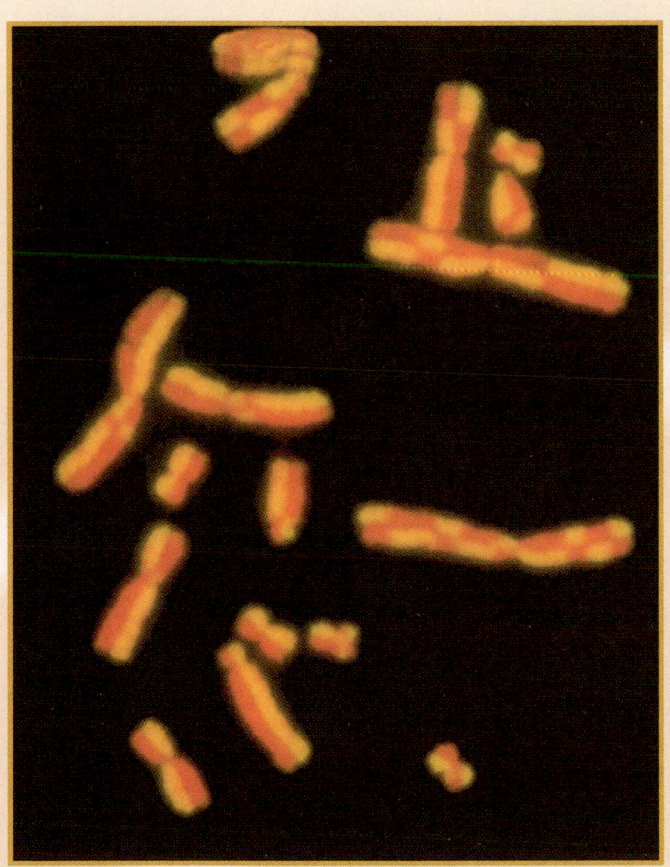

Harlequin chromosomes. *(Sheldon Wolff/University of California, San Francisco.)*

Key Concepts

Bacterial cells that express one phenotype can be transformed into cells that express a different phenotype; the transforming agent is DNA.

Experiments with labeled T2 phage have established that DNA is the hereditary material.

James Watson and Francis Crick showed that the structure of DNA is a double helix. Each helix is a chain of nucleotides linked by phosphodiester bonds. The helices are held together by specific hydrogen bonds between pairs of bases.

DNA structure ensures the fidelity of replication because the complementary base of each base is specified by hydrogen bonding.

The replication of DNA is semiconservative in that each daughter duplex contains one parental and one newly synthesized strand.

Many of the enzymes taking part in DNA synthesis have been characterized.

The elucidation of the structure of DNA in 1953 by James Watson and Francis Crick was one of the most exciting discoveries in the history of genetics. It paved the way for an understanding of gene action and heredity in molecular terms. Before we see how the solution of DNA structure was achieved, let's review what was known about genes and DNA at the time that Watson and Crick began their historic collaboration:

1. Genes — the hereditary "factors" described by Mendel — were known to be associated with specific character traits, but their physical nature was not understood.

2. The one-gene–one-enzyme theory (described more fully in Chapter 9) postulated that genes control the structure of proteins.

3. Genes were known to be carried on chromosomes.

4. The chromosomes were found to consist of DNA and protein.

5. Research by Frederick Griffith and subsequently by Oswald Avery and his co-workers pointed to DNA as

the genetic material. These experiments, described here, showed that bacterial cells that express one phenotype can be transformed into cells that express a different phenotype and that the transforming agent is DNA.

DNA: The genetic material

The physical nature of the gene fascinated scientists for many years. A series of experiments beginning in the 1920s finally revealed that DNA was the genetic material.

Discovery of transformation

A puzzling observation was made by Frederick Griffith in the course of experiments on the bacterium *Streptococcus pneumoniae* in 1928. This bacterium, which causes pneumonia in humans, is normally lethal in mice. However, different strains of this bacterial species have evolved that differ in virulence (in the ability to cause disease or death). In his experiments, Griffith used two strains that are distinguishable by the appearance of their colonies when grown in laboratory cultures. In one strain, a normal virulent type, the cells are enclosed in a polysaccharide capsule, giving

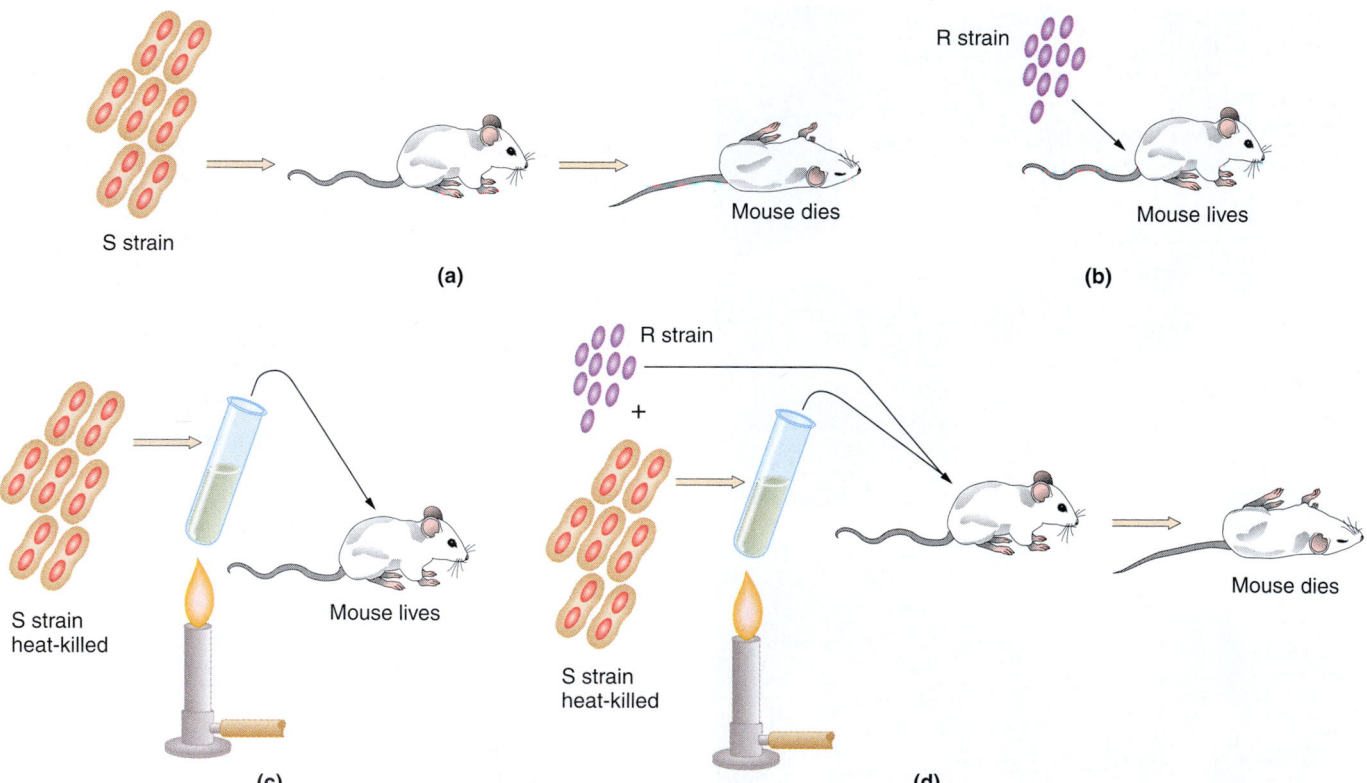

Figure 8-1   The first demonstration of bacterial transformation. (a) Mouse dies after injection with the virulent S strain. (b) Mouse survives after injection with the R strain. (c) Mouse survives after injection with heat-killed S strain. (d) Mouse dies after injection with a mixture of heat-killed S strain and live R strain. The heat-killed S strain somehow transforms the R strain into virulence. Parts a, b, and c act as control experiments for this demonstration. (From G. S. Stent and R. Calendar, *Molecular Genetics*, 2d ed. Copyright © 1978 by W. H. Freeman and Company. After R. Sager and F. J. Ryan, *Cell Heredity*. Wiley, 1961.)

colonies a smooth appearance; hence, this strain is labeled *S*. In Griffith's other strain, a mutant nonvirulent type that grows in mice but is not lethal, the polysaccharide coat is absent, giving colonies a rough appearance; this strain is called *R*.

Griffith killed some virulent cells by boiling them and injected the heat-killed cells into mice. The mice survived, showing that the carcasses of the cells do not cause death. However, mice injected with a mixture of heat-killed virulent cells and live nonvirulent cells did die. Furthermore, live cells could be recovered from the dead mice; these cells gave smooth colonies and were virulent on subsequent injection. Somehow, the cell debris of the boiled S cells had converted the live R cells into live S cells. The process is called **transformation.** Griffith's experiment is summarized in Figure 8-1.

This same basic technique was then used to determine the nature of the *transforming principle* — the agent in the cell debris that is specifically responsible for transformation. In 1944, Oswald Avery, C. M. MacLeod, and M. McCarty separated the classes of molecules found in the debris of the dead S cells and tested them for transforming ability, one at a time. These tests showed that the polysaccharides themselves do not transform the rough cells. Therefore, the polysaccharide coat, although undoubtedly concerned with the pathogenic reaction, is only the phenotypic expression of virulence. In screening the different groups, Avery and his colleagues found that only one class of molecules, DNA, induced the transformation of R cells (Figure 8-2). They deduced that DNA is the agent that determines the polysaccharide character and hence the pathogenic character (see pages 219–220 for a description of the mechanism of transformation). Furthermore, it seemed that providing R cells with S DNA was tantamount to providing these cells with S genes.

MESSAGE ·····························

The demonstration that DNA is the transforming principle was the first demonstration that genes are composed of DNA.

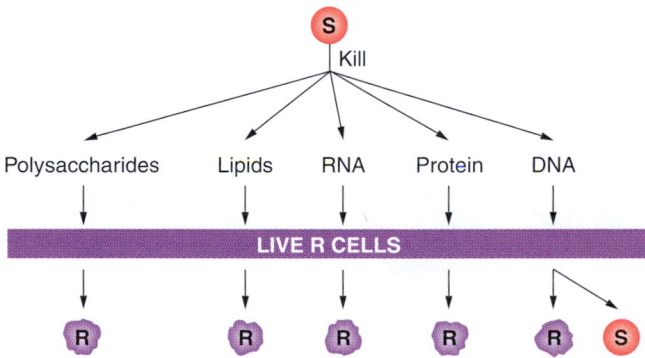

Figure 8-2   Demonstration that DNA is the transforming agent. DNA is the only agent that produces smooth (S) colonies when added to live rough (R) cells.

Hershey-Chase experiment

The experiments conducted by Avery and his colleagues were definitive, but many scientists were very reluctant to accept DNA (rather than proteins) as the genetic material. The clincher was provided in 1952 by Alfred Hershey and Martha Chase with the use of the phage (virus) T2. They reasoned that phage infection must entail the introduction (injection) into the bacterium of the specific information that dictates viral reproduction. The phage is relatively simple in molecular constitution. Most of its structure is protein, with DNA contained inside the protein sheath of its "head."

Phosphorus is not found in proteins but is an integral part of DNA; conversely, sulfur is present in proteins but never in DNA. Hershey and Chase incorporated the radioisotope of phosphorus (^{32}P) into phage DNA and that of sulfur (^{35}S) into the proteins of a separate phage culture. They then used each phage culture independently to infect *E. coli* with many virus particles per cell. After sufficient time for injection to take place, they sheared the empty phage carcasses (called *ghosts*) off the bacterial cells by agitation in a kitchen blender. They used centrifugation to separate the bacterial cells from the phage ghosts and then measured the radioactivity in the two fractions. When the ^{32}P-labeled phages were used, most of the radioactivity ended up inside the bacterial cells, indicating that the phage DNA entered the cells. ^{32}P can also be recovered from phage progeny. When the ^{35}S-labeled phages were used, most of the radioactive material ended up in the phage ghosts, indicating that the phage protein never entered the bacterial cell (Figure 8-3). The conclusion is inescapable: DNA is the hereditary material; the phage proteins are mere structural packaging that is discarded after delivering the viral DNA to the bacterial cell.

Why such reluctance to accept this conclusion? DNA was thought to be a rather simple chemical. How could all the information about an organism's features be stored in such a simple molecule? How could such information be passed on from one generation to the next? Clearly, the genetic material must have both the ability to encode specific information and the capacity to duplicate that information precisely. What kind of structure could allow such complex functions in so simple a molecule?

Structure of DNA

Although the DNA structure was not known, the basic building blocks of DNA had been known for many years. The basic elements of DNA had been isolated and determined by partly breaking up purified DNA. These studies showed that DNA is composed of only four basic molecules called **nucleotides,** which are identical except that each contains a different nitrogen base. Each nucleotide contains phosphate, sugar (of the deoxyribose type), and one of the

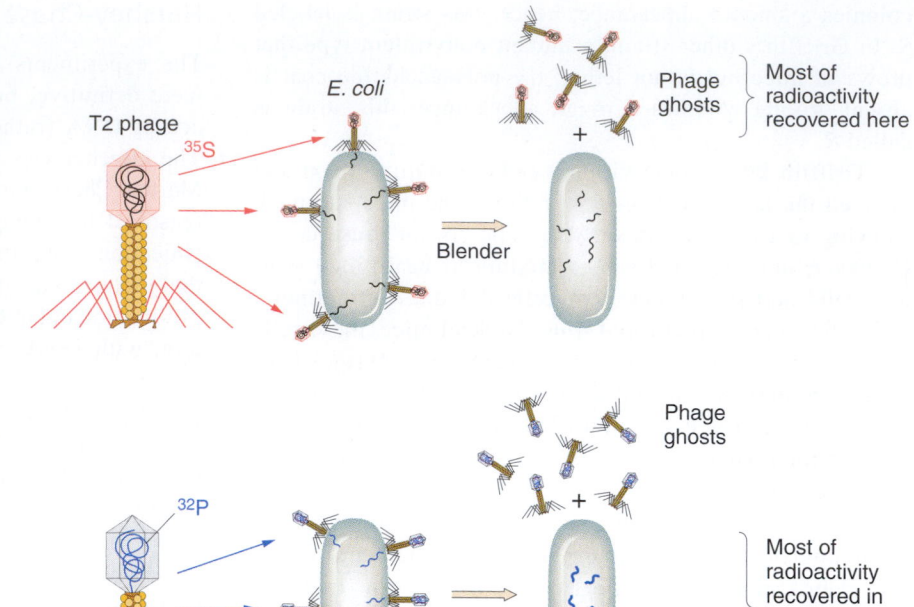

Figure 8-3   The Hershey-Chase experiment, which demonstrated that the genetic material of phage is DNA, not protein. The experiment uses two sets of T2 bacteriophages. In one set, the protein coat is labeled with radioactive sulfur (^{35}S), not found in DNA. In the other set, the DNA is labeled with radioactive phosphorus (^{32}P), not found in protein. Only the ^{32}P is injected into the *E. coli,* indicating that DNA is the agent necessary for the production of new phages.

Purine nucleotides

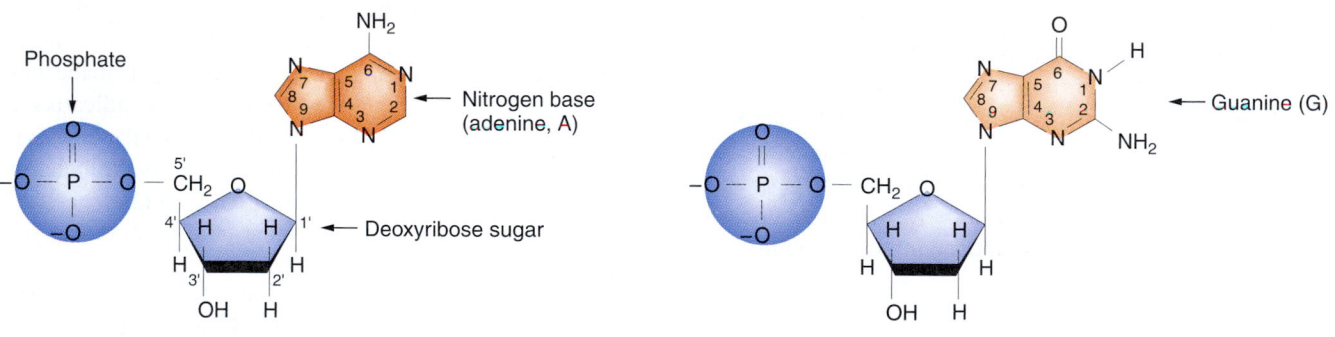

Pyrimidine nucleotides

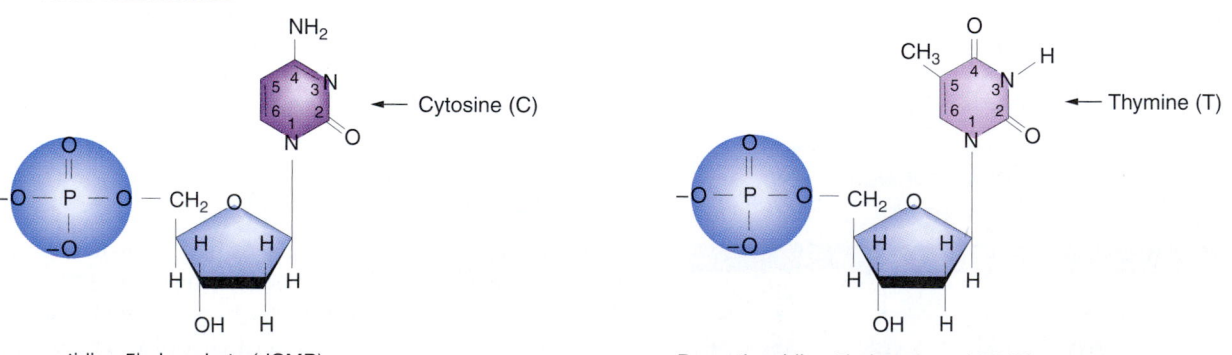

Figure 8-4   Chemical structure of the four nucleotides (two with purine bases and two with pyrimidine bases) that are the fundamental building blocks of DNA. The sugar is called *deoxyribose* because it is a variation of a common sugar, ribose, that has one more oxygen atom.

four bases (Figure 8-4). When the phosphate group is not present, the base and the deoxyribose form a **nucleoside** rather than a nucleotide. The four bases are **adenine, guanine, cytosine,** and **thymine.** The full chemical names of the nucleotides are deoxyadenosine 5′-monophosphate (deoxyadenylate, or dAMP), deoxyguanosine 5′-monophosphate (deoxyguanylate, or dGMP), deoxycytidine 5′-monophosphate (deoxycytidylate, or dCMP), and deoxythymidine 5′-monophosphate (deoxythymidylate, or dTMP). However, it is more convenient just to refer to each nucleotide by the abbreviation of its base (A, G, C, and T, respectively). Two of the bases, adenine and guanine, are similar in structure and are called **purines.** The other two bases, cytosine and thymine, also are similar and are called **pyrimidines.**

After the central role of DNA in heredity became clear, many scientists set out to determine the exact structure of DNA. How can a molecule with such a limited range of different components possibly store the vast range of information about all the protein primary structures of the living organism? The first to succeed in putting the building blocks together and finding a reasonable DNA structure — Watson and Crick in 1953 — worked from two kinds of clues. First, Rosalind Franklin and Maurice Wilkins had amassed X-ray diffraction data on DNA structure. In such experiments, X rays are fired at DNA fibers, and the scatter of the rays from the fiber is observed by catching them on photographic film, where the X rays produce spots. The angle of scatter represented by each spot on the film gives information about the position of an atom or certain groups of atoms in the DNA molecule. This procedure is not simple to carry out (or to explain), and the interpretation of the spot patterns is very difficult. The available data suggested that DNA is long and skinny and that it has two similar parts that are parallel to each other and run along the length of the molecule. The X-ray data showed the molecule to be helical (spiral-like). Other regularities were present in the spot patterns, but no one had yet thought of a three-dimensional structure that could account for just those spot patterns.

The second set of clues available to Watson and Crick came from work done several years earlier by Erwin Chargaff. Studying a large selection of DNAs from different organisms (Table 8-1), Chargaff established certain empirical rules about the amounts of each component of DNA:

1. The total amount of pyrimidine nucleotides (T + C) always equals the total amount of purine nucleotides (A + G).

2. The amount of T always equals the amount of A, and the amount of C always equals the amount of G. But the amount of A + T is not necessarily equal to the amount of G + C, as can be seen in the last column of Table 8-1. This ratio varies among different organisms.

Double helix

The structure that Watson and Crick derived from these clues is a **double helix,** which looks rather like two interlocked bedsprings. Each bedspring (helix) is a chain of nucleotides held together by **phosphodiester bonds,** in which

8-1 TABLE				Molar Properties of Bases* in DNAs from Various Sources		
Organism	Tissue	Adenine	Thymine	Guanine	Cytosine	$\dfrac{A + T}{G + C}$
Escherichia coli (K12)	—	26.0	23.9	24.9	25.2	1.00
Diplococcus pneumoniae	—	29.8	31.6	20.5	18.0	1.59
Mycobacterium tuberculosis	—	15.1	14.6	34.9	35.4	0.42
Yeast	—	31.3	32.9	18.7	17.1	1.79
Paracentrotus lividus (sea urchin)	Sperm	32.8	32.1	17.7	18.4	1.85
Herring	Sperm	27.8	27.5	22.2	22.6	1.23
Rat	Bone marrow	28.6	28.4	21.4	21.5	1.33
Human	Thymus	30.9	29.4	19.9	19.8	1.52
Human	Liver	30.3	30.3	19.5	19.9	1.53
Human	Sperm	30.7	31.2	19.3	18.8	1.62

*Defined as moles of nitrogenous constituents per 100 g-atoms phosphate in hydrolysate.

Source: E. Chargaff and J. Davidson, eds., *The Nucleic Acids.* Academic Press, 1955.

a phosphate group forms a bridge between —OH groups on two adjacent sugar residues. The two "bedsprings" (helices) are held together by **hydrogen bonds,** in which two electronegative atoms "share" a proton, between the bases. Hydrogen bonds form between hydrogen atoms with a small positive charge and acceptor atoms with a small negative charge. For example,

Each hydrogen atom in the NH_2 group is slightly positive (δ^+) because the nitrogen atom tends to attract the electrons of the N–H bond, thereby leaving the hydrogen atom

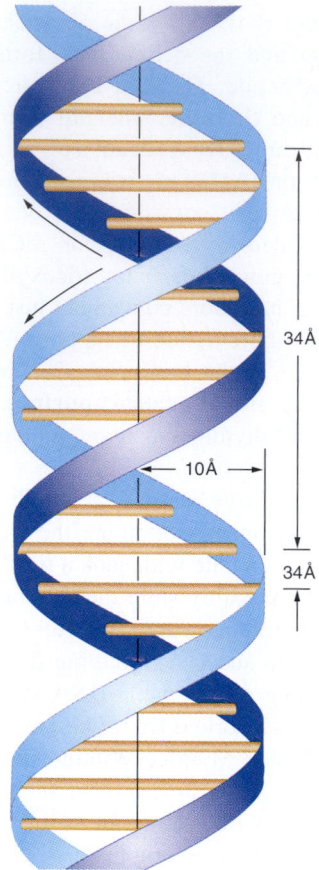

Figure 8-6   A simplified model showing the helical structure of DNA. The sticks represent base pairs, and the ribbons represent the sugar-phosphate backbones of the two antiparallel chains. The various measurements are given in angstroms (1 Å = 0.1 nm).

slightly short of electrons. The oxygen atom has six unbonded electrons in its outer shell, making it slightly negative (δ^-). A hydrogen bond forms between one H and the O. Hydrogen bonds are quite weak (only about 3 percent of the strength of a covalent chemical bond), but this weakness (as we shall see) plays an important role in the function of the DNA molecule in heredity. One further important chemical fact: the hydrogen bond is much stronger if the participating atoms are "pointing at each other" in the ideal orientations.

The hydrogen bonds are formed by pairs of bases and are indicated by dotted lines in Figure 8-5, which shows a part of this paired structure with the helices uncoiled. Each base pair consists of one purine base and one pyrimidine based, paired according to the following rule: G pairs with C, and A pairs with T. In Figure 8-6, a simplified picture of the coiling, each of the base pairs is represented by a "stick" between the "ribbons," or so-called sugar-phosphate backbones of the chains. In Figure 8-5, note that the two backbones run in opposite directions; they are thus said to be **antiparallel,** and (for reasons apparent in the figure) one is called the 5′ → 3′ strand and the other the 3′ → 5′ strand.

The double helix accounted nicely for the X-ray data and tied in very nicely with Chargaff's data. Studying mod-

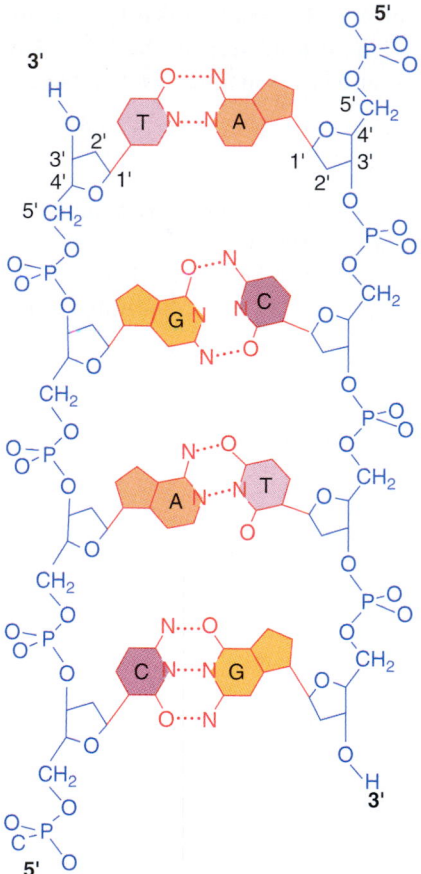

Figure 8-5   The DNA double helix, unrolled to show the sugar-phosphate backbones (blue) and base-pair rungs (red). The backbones run in opposite directions; the 5′ and 3′ ends are named for the orientation of the 5′ and 3′ carbon atoms of the sugar rings. Each base pair has one purine base, adenine (A) or guanine (G), and one pyrimidine base, thymine (T) or cytosine (C), connected by hydrogen bonds *(dotted lines).* (From R. E. Dickerson, "The DNA Helix and How It Is Read." Copyright © 1983 by Scientific American, Inc. All rights reserved.)

els that they made of the structure, Watson and Crick realized that the observed radius of the double helix (known from the X-ray data) would be explained if a purine base always pairs (by hydrogen bonding) with a pyrimidine base (Figure 8-7). Such pairing would account for the (A + G) = (T + C) regularity observed by Chargaff, but it would predict four possible pairings: T···A, T···G,

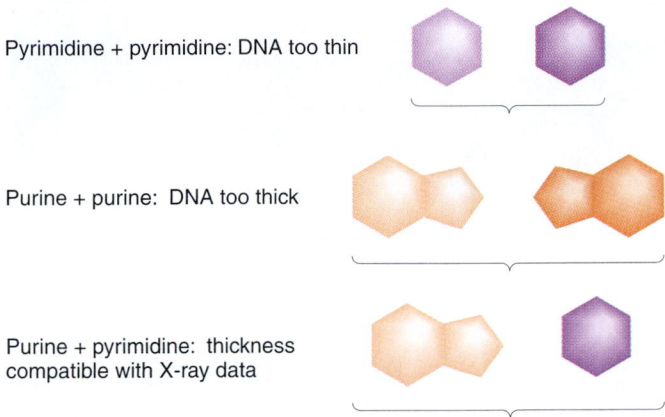

Pyrimidine + pyrimidine: DNA too thin

Purine + purine: DNA too thick

Purine + pyrimidine: thickness compatible with X-ray data

Figure 8-7 The pairing of purines with pyrimidines accounts exactly for the diameter of the DNA double helix determined from X-ray data. (From R. E. Dickerson, "The DNA Helix and How It Is Read." Copyright © 1983 by Scientific American, Inc. All rights reserved.)

C···A, and C···G. Chargaff's data, however, indicate that T pairs only with A and C pairs only with G. Watson and Crick showed that only these two pairings have the necessary complementary "lock and key" shapes to permit efficient hydrogen bonding (Figure 8-8).

Note that the G–C pair has three hydrogen bonds, whereas the A–T pair has only two. We would predict that DNA containing many G–C pairs would be more stable than DNA containing many A–T pairs. In fact, this prediction is confirmed. DNA structure neatly explains Chargaff's data (Figure 8-9), and that structure is consistent with the X-ray data.

Three-dimensional view of the double helix

In three dimensions, the bases form rather flat structures, and these flat bases partly stack on top of one another in the twisted structure of the double helix. This stacking of bases adds tremendously to the stability of the molecule by excluding water molecules from the spaces between the base pairs. (This phenomenon is very much like the stabilizing force that you can feel when you squeeze two plates of glass together underwater and then try to separate them.) Subsequently, it was realized that there were two forms of DNA in the fiber analyzed by diffraction. The **A form** is less hydrated than the **B form** and is more compact. It is believed

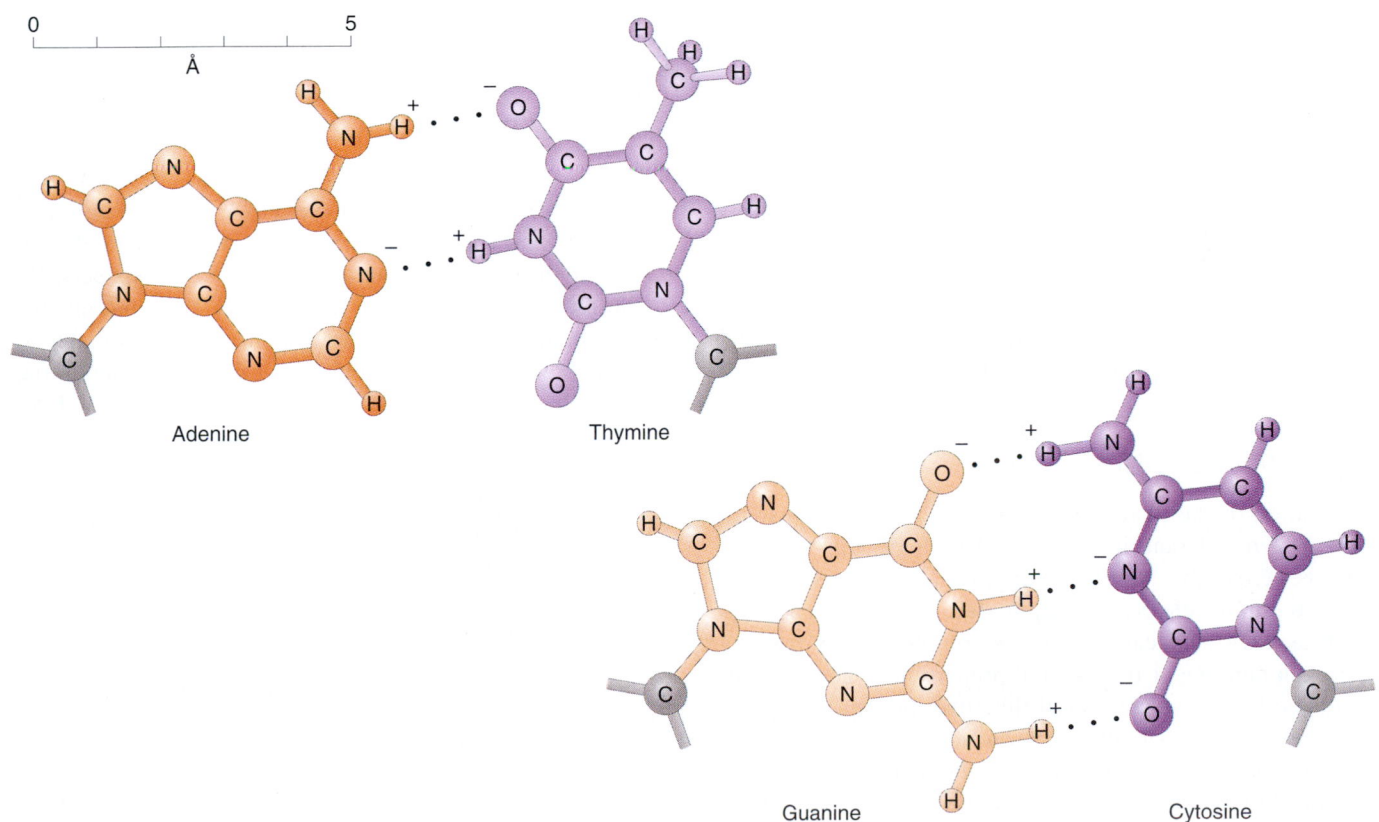

Figure 8-8 The lock-and-key hydrogen bonding between A and T and between G and C. (From G. S. Stent, *Molecular Biology of Bacterial Viruses.* Copyright © 1963 by W. H. Freeman and Company.)

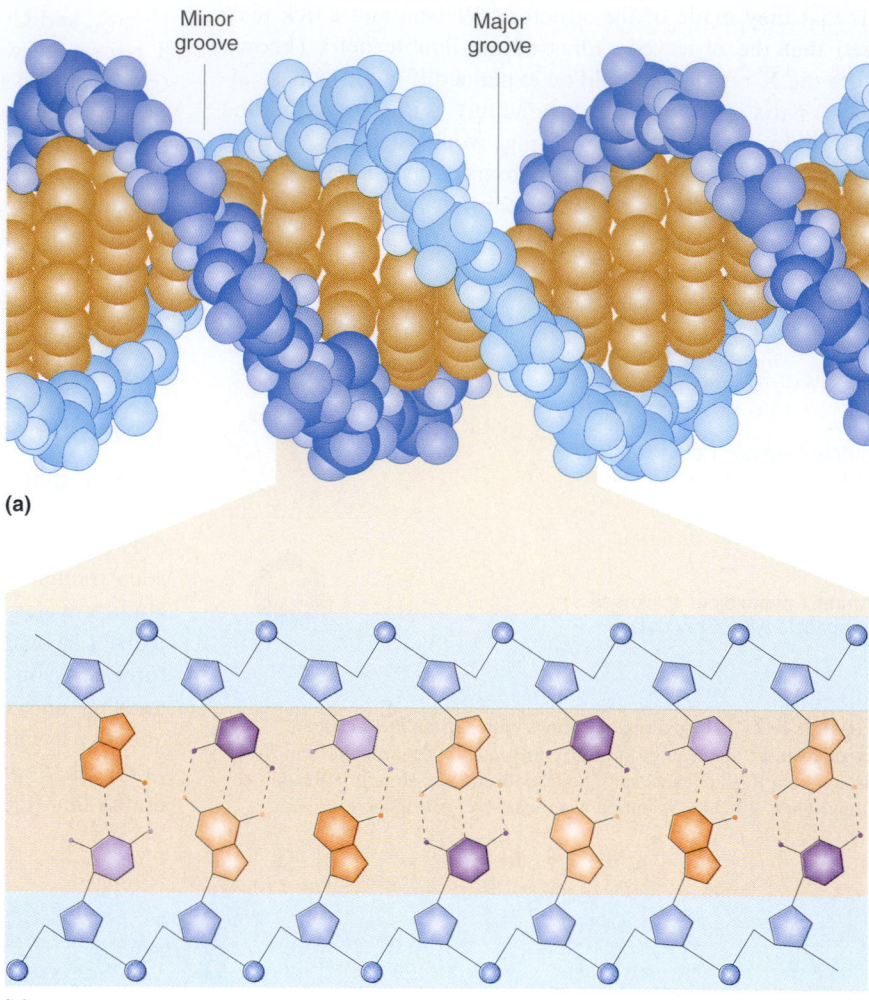

Minor groove

Major groove

(a)

(b)

Figure 8-9 (a) A space-filling model of the DNA double helix. (b) An unwound representation of a short stretch of nucleotide pairs, showing how A−T and G−C pairing produces the Chargaff ratios. This model is of one of several forms of DNA, termed the *B form.* (Part a from C. Yanofsky, "Gene Structure and Protein Structure." Copyright © 1967 by Scientific American, Inc. All rights reserved. Part b after A. Kornberg, "The Synthesis of DNA." Copyright © 1968 by Scientific American, Inc. All rights reserved.)

that the B form of DNA is the form found most frequently in living cells.

The stacking of the base pairs in the double helix results in two grooves in the sugar-phosphate backbones. These grooves are termed the **major** and **minor grooves** and can be readily seen in the space-filling (three-dimensional) model in Figure 8-9a.

Implications of DNA structure

Elucidation of the structure of DNA caused a lot of excitement in genetics (and in all areas of biology) for two basic reasons. First, the structure suggests an obvious way in which the molecule can be **duplicated,** or **replicated,** inasmuch as each base can specify its complementary base by hydrogen bonding. This essential property of a genetic molecule had been a mystery until this time. Second, the structure suggests that perhaps the *sequence* of nucleotide pairs in DNA dictates the sequence of amino acids in the protein organized by that gene. In other words, some sort of **genetic code** may write information in DNA as a sequence of nucleotide pairs and then translate it into a different language of amino acid sequences in protein.

This basic information about DNA is now familiar to almost anyone who has read a biology textbook in elementary or high school, or even magazines and newspapers. But try to put yourself back into the scene in 1953 and imagine the excitement. Until then, the evidence that the uninteresting DNA was the genetic molecule had been disappointing and discouraging. But the Watson-Crick structure of DNA suddenly opened up the possibility of explaining two of the biggest "secrets" of life. James Watson told the story of this discovery (from his own point of view, strongly questioned by other participants) in a fascinating book called *The Double Helix,* which reveals the intricate interplay of personality clashes, clever insights, hard work, and simple luck in such important scientific advances.

Alternative structures

In addition to the A and B forms of DNA, a new form was found in crystals of synthetically prepared DNA that contain alternating G's and C's on the same strand. This **Z DNA** form has a zigzaglike backbone and generates a left-handed helix, whereas both A and B DNA form right-handed helices.

Replication of DNA

The faithful transmission of hereditary information depends on accurate replication of the genetic material. This section examines the mechanism of DNA replication.

Semiconservative replication

Figure 8-10 diagrams the possible basic mechanism for DNA replication proposed by Watson and Crick. The sugar-phosphate backbones are represented by lines, and the sequence of base pairs is random. Let's imagine that the double helix is like a zipper that unzips, starting at one end (at the bottom in Figure 8-10). We can see that, if this zipper analogy is valid, the unwinding of the two strands will expose single bases on each strand. Because the pairing requirements imposed by the DNA structure are strict, each exposed base will pair only with its complementary base. Because of this base complementarity, each of the two single strands will act as a **template,** or mold, and will begin to reform a double helix identical with the one from which it was unzipped. The newly added nucleotides are assumed to come from a pool of free nucleotides that must be present in the cell.

If this model is correct, then each daughter molecule should contain one parental nucleotide chain and one newly synthesized nucleotide chain. This prediction has been tested in both prokaryotes and eukaryotes. A little thought shows that there are at least three different ways in which a parental DNA molecule might be related to the daughter molecules. These hypothetical modes are called semiconservative (the Watson-Crick model), conservative, and dispersive (Figure 8-11). In **semiconservative replication,** each daughter duplex contains one parental and one newly synthesized strand. However, in **conservative replication,** one daughter duplex consists of two newly synthesized strands, and the parent duplex is conserved. **Dispersive replication** results in daughter duplexes that consist of strands containing only *segments* of parental DNA and newly synthesized DNA.

Meselson-Stahl experiment

In 1958, Matthew Meselson and Franklin Stahl set out to distinguish among these possibilities. They grew *E. coli* cells in a medium containing the heavy isotope of nitrogen (^{15}N) rather than the normal light (^{14}N) form. This isotope was inserted into the nitrogen bases, which then were incorporated into newly synthesized DNA strands. After many cell divisions in ^{15}N, the DNA of the cells were well labeled with the heavy isotope. The cells were then removed from the ^{15}N medium and put into a ^{14}N medium; after one and two cell divisions, samples were taken. DNA was extracted from the cells in each of these samples and put into a solution of cesium chloride (CsCl) in an ultracentrifuge.

If cesium chloride is spun in a centrifuge at tremendously high speeds (50,000 rpm) for many hours, the cesium

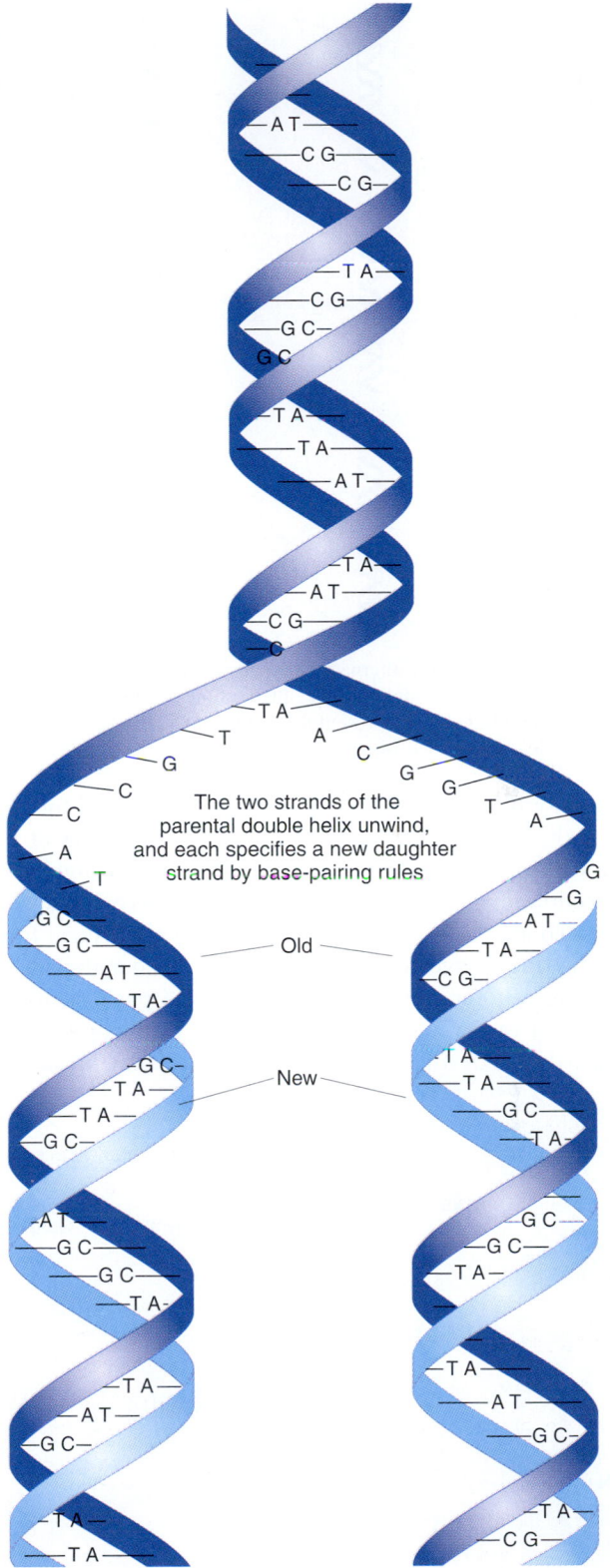

The two strands of the parental double helix unwind, and each specifies a new daughter strand by base-pairing rules

Old

New

Figure 8-10   The model of DNA replication proposed by Watson and Crick is based on the hydrogen-bonded specificity of the base pairs. Complementary strands are shown in different colors. This drawing is a simplified version of the current picture of replication but represents the basic concept suggested by the Watson-Crick structure. The fact that new strands can grow only in the 5′-to-3′ direction adds complexities to the detailed mechanism of replication.

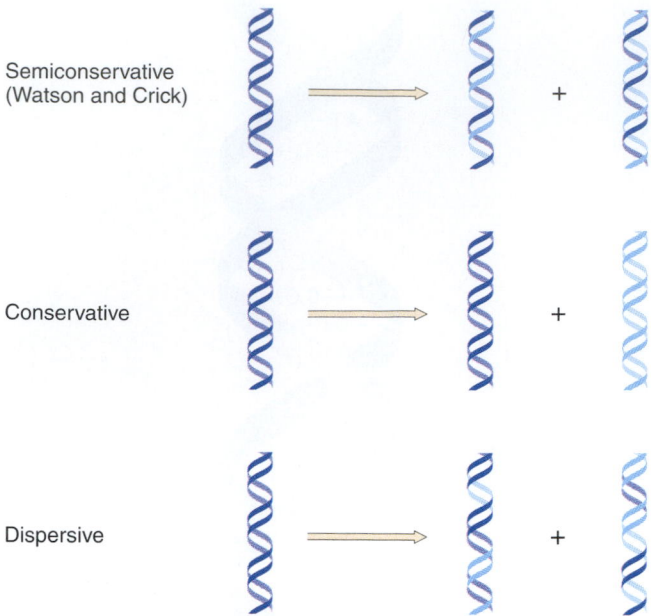

Figure 8-11   Three alternative patterns for DNA replication. The Watson-Crick model would produce the first (semiconservative) pattern. Light blue lines represent the newly synthesized strands.

and chloride ions tend to be pushed by centrifugal force toward the bottom of the tube. Ultimately, a gradient of Cs^+ and Cl^- ions is established in the tube, with the highest ion concentration at the bottom. Molecules of DNA in the solution also are pushed toward the bottom by centrifugal force. But, as they travel down the tube, they encounter the increasing salt concentration, which tends to push them back up owing to the buoyancy of DNA (its tendency to float). Thus, the DNA finally "settles" at some point in the tube

where the centrifugal forces just balance the buoyancy of the molecules in the cesium chloride gradient. The buoyancy of DNA depends on its density (which in turn depends on the ratio of G–C to A–T base pairs). The presence of the heavier isotope of nitrogen changes the buoyant density of DNA. The DNA extracted from cells grown for several generations on ^{15}N medium can be readily distinguished from the DNA of cells grown on ^{14}N medium by the equilibrium position reached in a cesium chloride gradient. Such samples are commonly called *heavy* and *light* DNA, respectively.

Meselson and Stahl found that, one generation after the heavy cells were moved to ^{14}N medium, the DNA formed a single band of an intermediate density between the densities of the heavy and light controls. After two generations in ^{14}N medium, the DNA formed two bands: one at the intermediate position, the other at the light position (Figure 8-12). This result would be expected from the semiconservative mode of replication; in fact, the result is compatible with *only* this mode *if* the experiment begins with chromosomes composed of individual double helices (Figure 8-13).

Incubation of heavy cells in ^{14}N

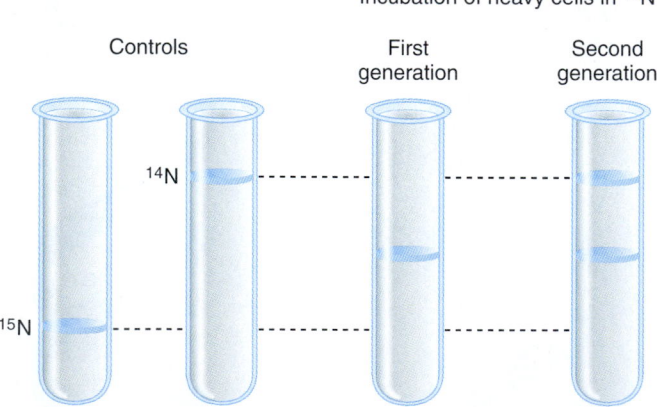

Figure 8-12   Centrifugation of DNA in a cesium chloride (CsCl) gradient. Cultures grown for many generations in ^{15}N and ^{14}N media provide control positions for heavy and light DNA bands, respectively. When the cells grown in ^{15}N are transferred to a ^{14}N medium, the first generation produces an intermediate DNA band and the second generation produces two bands: one intermediate and one light.

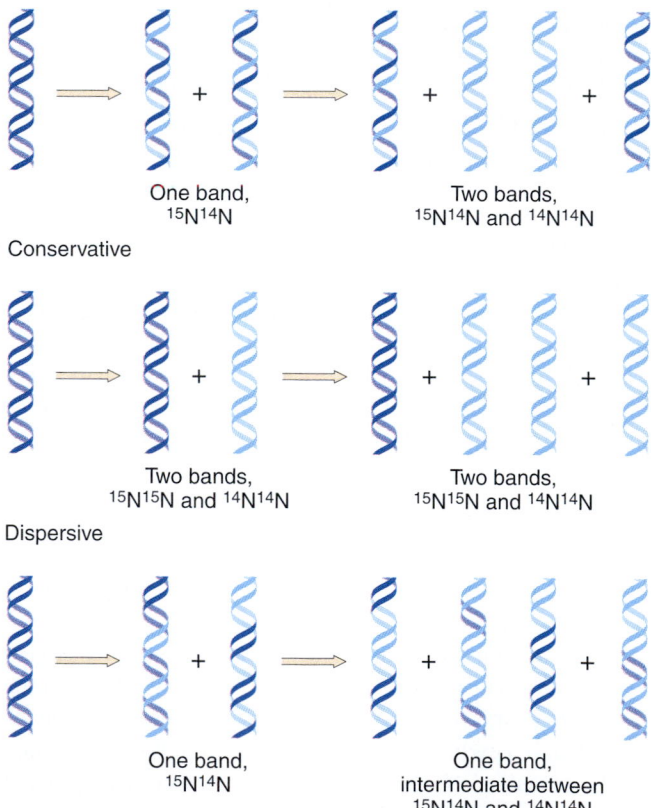

Figure 8-13   Only the semiconservative model of DNA replication predicts results like those shown in Figure 8-12: a single intermediate band in the first generation and one intermediate and one light band in the second generation. (See Figure 8-11 for explanation of colors.) Recall that, after growth in ^{15}N, the cells are switched to ^{14}N.

Autoradiography

The Meselson-Stahl experiment on *E. coli* was essentially duplicated in 1958 by Herbert Taylor on the chromosomes of bean root-tip cells, by using a cytological technique. Taylor put root cells into a solution containing tritiated thymidine ([³H]thymidine) — the thymine nucleotide labeled with a radioactive hydrogen isotope called *tritium*. He allowed the cells to undergo mitosis in this solution so that the [³H]thymidine could be incorporated into DNA. He then washed the tips and transferred them to a solution containing nonradioactive thymidine. Addition of colchicine to such a preparation inhibits the spindle apparatus so that chromosomes in metaphase fail to separate and sister chromatids remain "tied together" by the centromere.

The cellular location of ³H can be determined by **autoradiography**. As ³H decays, it emits a beta particle (an energetic electron). If a layer of photographic emulsion is spread over a cell that contains ³H, a chemical reaction takes place wherever a beta particle strikes the emulsion. The emulsion can then be developed like a photographic print so that the emission track of the beta particle appears as a black spot or grain. The cell can also be stained, making the structure of the cell visible, to identify the location of the radioactivity. In effect, autoradiography is a process in which radioactive cell structures "take their own pictures."

Figure 8-14 shows the results observed when colchicine is added during the division in [³H]thymidine or during the subsequent mitotic division. It is possible to interpret these results by representing each chromatid as a single DNA molecule that replicates semiconservatively (Figure 8-15).

Harlequin chromosomes

With the use of a more modern staining technique, it is now possible to visualize the semiconservative replication of chromosomes at mitosis without the aid of autoradiography. In this procedure, the chromosomes are allowed to go

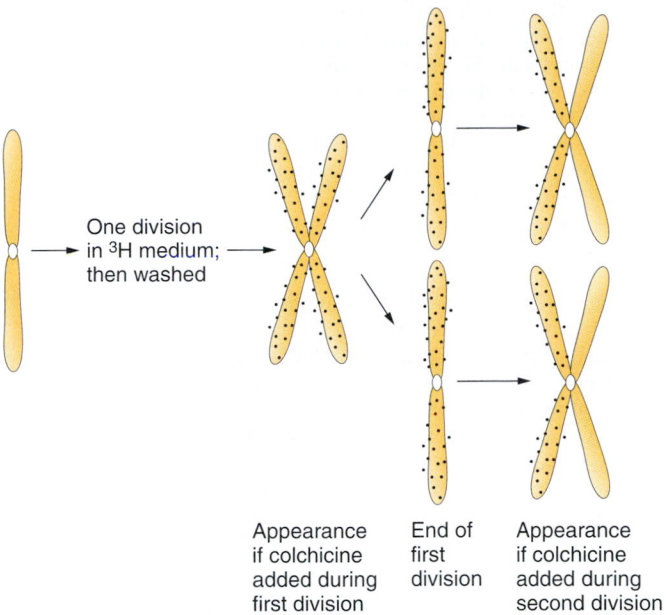

One division in ³H medium; then washed

| Appearance if colchicine added during first division | End of first division | Appearance if colchicine added during second division |

Figure 8-14 Diagrammatic representation of the autoradiography of chromosomes from cells grown for one cell division in the presence of the radioactive hydrogen isotope ³H (tritium) and then grown in a nonradioactive medium for a second mitotic division. Each dot represents the track of a particle of radioactivity.

through two rounds of replication in bromodeoxyuridine (BUdR). The BUdR labeling pattern, shown in Figure 8-16a, is the reciprocal of that of Figure 8-15, because the BUdR is used in both replications, rather than being replaced by normal thymidine for the second replication as in the autoradiographic process. The chromosomes are then stained with fluorescent dye and Giemsa stain; this process distinguishes hybrid chromatids with one BUdR-containing strand and one original strand (dark stain) from those in which both strands contain BUdR (light stain), and generates so-called

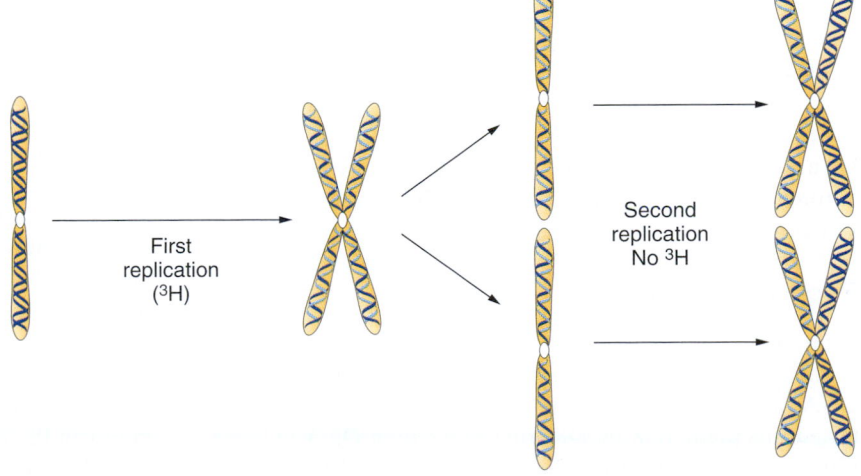

First replication (³H)

Second replication No ³H

Figure 8-15 An explanation of Figure 8-14 at the DNA level. Light blue lines represent radioactive strands. In the second replication (which takes place in nontritiated solution), both the ³H strand and the nontritiated strand incorporate nonradioactive nucleotides, yielding one hybrid and one nontritiated chromatid.

Figure 8-16 (a) A diagrammatic representation of the production of harlequin chromosomes. In this procedure, the chromosomes go through two rounds of replication in the presence of bromodeoxyuridine (BUdR), which replaces thymidine in the newly synthesized DNA. The chromosomes are then stained with a fluorescent dye and Giemsa stain, producing the appearance shown. (The light blue lines represent the BUdR-substituted strands.) (b) Photograph of harlequin chromosomes in a Chinese hamster ovary (CHO) cell. The chromatids with two strands containing BUdR are light in this photograph, whereas those with one BUdR strand and one original strand are dark. A chromosome at the top (see arrows) has two sister-chromatid exchanges. (Photograph courtesy of Sheldon Wolff and Judy Bodycote.)

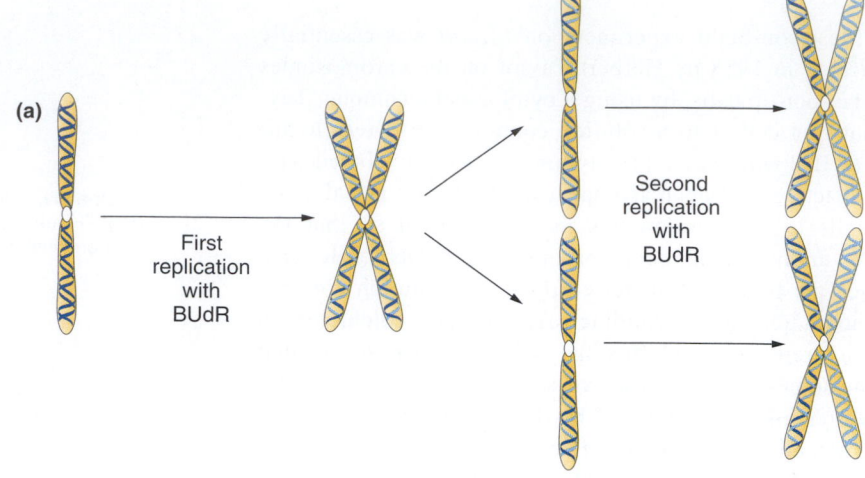

(a)

First replication with BUdR

Second replication with BUdR

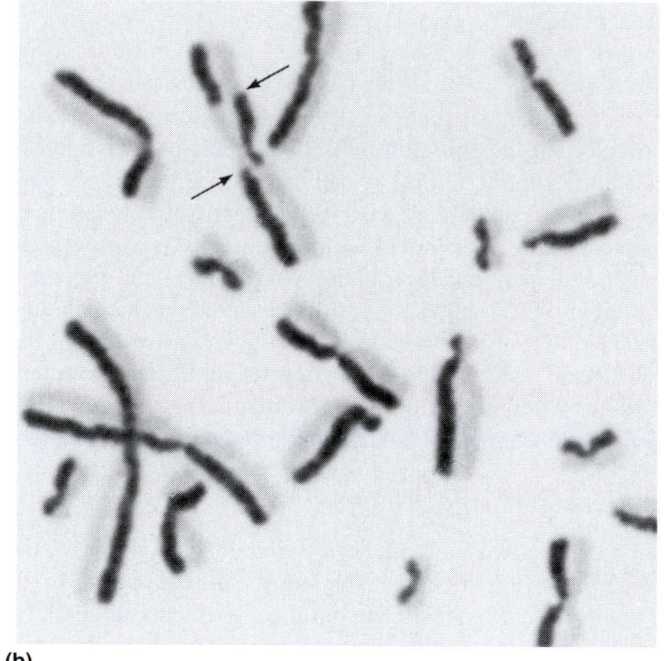

(b)

harlequin chromosomes (Figure 8-16b). (Note, in passing, that harlequin chromosomes are particularly favorable for the detection of sister-chromatid exchange at mitosis; two examples are seen in Figure 8-16b). Using similar techniques. Taylor showed that chromosome replication at meiosis also is semiconservative.

Chromosome structure

Figures 8-14 and 8-15 bring up one of the remaining great unsolved questions of genetics: Is a eukaryotic chromosome basically a single DNA molecule surrounded by a protein matrix? Two things strongly suggest that this is, in fact, the case. First, if there were many DNA molecules in the chromosome (whether they were side by side, end to end, or randomly oriented), it would be almost impossible for the chromosome to replicate semiconservatively (with all the label going into one chromatid, as in Taylor's results). Studies on isolated chromosomes and long DNA molecules are consistent with the suggestion that each chromatid is a single molecule of DNA. The second fact supporting a single-molecule hypothesis is that DNA and genes behave as though they are attached end to end in a single string, or thread, that we call a *linkage group*. All genetic linkage data (Chapter 5) tell us that we need nothing more than a single linear array of genes per chromosome to explain the genetic facts. It has been convincingly demonstrated that a chromosome or a chromatid in fact contains just one DNA molecule, as we saw in Chapter 3.

Replication fork

A prediction of the Watson-Crick model of DNA replication is that a replication zipper, or **fork,** will be found in the DNA molecule during replication. In 1963, John Cairns tested this prediction by allowing replicating DNA in bacterial cells to incorporate tritiated thymidine. Theoretically,

each newly synthesized daughter molecule should then contain one radioactive ("hot") strand and another nonradioactive ("cold") strand. After varying intervals and varying numbers of replication cycles in a "hot" medium, Cairns extracted the DNA from the cells, put it on a slide, and autoradiographed it for examination under the light microscope. After one replication cycle in [³H]thymidine, rings of dots appeared in the autoradiograph. Cairns interpreted these rings as shown in Figure 8-17. It is also apparent from Figure 8-17 that the bacterial chromosome is circular — a fact that also emerged from genetic analysis described earlier (Chapter 7).

In the second replication cycle, the forks predicted by the model were indeed seen. Furthermore, the density of grains in the three segments was such that the interpretation shown in Figure 8-18 could be made. Cairns saw all sizes of

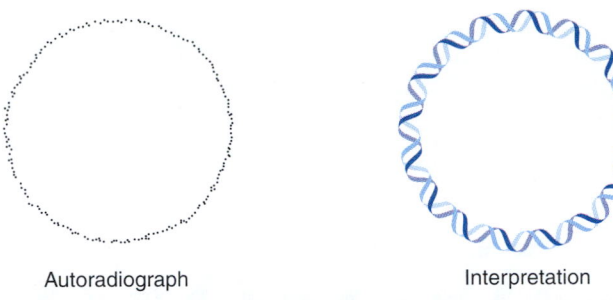

Autoradiograph   Interpretation

Figure 8-17 *Left:* Autoradiograph of a bacterial chromosome after one replication in tritiated thymidine. According to the semiconservative model of replication, one of the two strands should be radioactive. *Right:* Interpretation of the autoradiograph. The light blue line represents the tritiated strand.

these moon-shaped, autoradiographic patterns, corresponding to the progressive movement of the replication zipper, or fork, around the ring. Structures of the sort shown in Figure 8-18 are called **theta (θ) structures.**

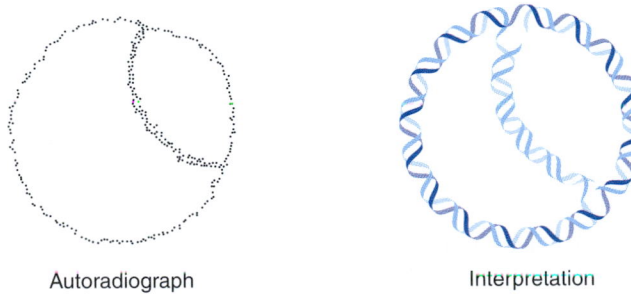

Autoradiograph   Interpretation

Figure 8-18 *Left:* Autoradiograph of a bacterial chromosome in the second round of replication in tritiated thymidine. In this theta (θ) structure, the newly replicated double helix that crosses the circle could consist of two radioactive strands (if the parental strand was the radioactive one). *Right:* The double thickness of the radioactive tracing on the autoradiogram appears to confirm the interpretation shown here. The light blue helices represent the "hot" strands.

Rolling-circle replication

The replication of some circular molecules, such as plasmids and certain viruses, proceeds by the mechanism depicted in Figure 8-19. Here, a nuclease cut provides a free 3′-OH end onto which nucleotides are added. As can be seen from Figure 8-19, as the synthesis proceeds, the other end of the template strand is displaced from the double-stranded circle and then copied. We can envision this displacement as the strand rolling off the circle. Because there is no termination point, synthesis often continues beyond a single circle unit, producing concatamers (a series of linked chains) of several circle lengths, which are then processed by recombination to yield normal-length circles.

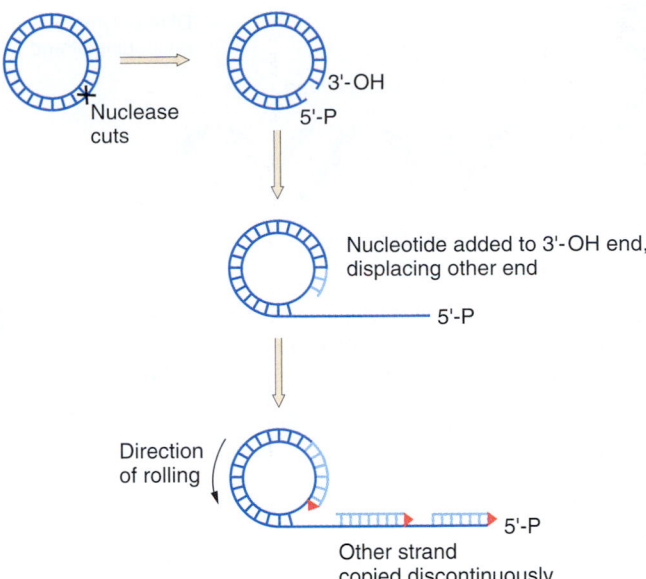

Figure 8-19 Rolling-circle replication. Newly synthesized DNA is light blue. The displaced strand is replicated discontinuously, as described in the text. (After D. L. Hartl and E. W. Jones, *Genetics: Principles and Analysis,* 4th ed. Jones and Bartlett, 1998.)

Mechanism of DNA replication

Watson and Crick first reasoned that complementary base pairing provides the basis of fidelity in DNA replication; that is, that each base in the template strand dictates the complementary base in the new strand. However, we now know that the process of DNA replication is very complex and requires the participation of many different components. Let's examine each of these components and see how they fit together to produce our current picture of DNA synthesis in *E. coli,* the best-studied cellular replication system. In the preceding section, we introduced the concept of the replication fork. Figure 8-20 gives a detailed schematic view of fork movement during DNA replication; we can refer to this illustration as we consider each component of the process.

DNA polymerases

In the late 1950s, Arthur Kornberg successfully identified and purified the first DNA polymerase, an enzyme that catalyzes the replication reaction.

$$\text{Primer (parental) DNA} + \left\{ \begin{array}{c} \text{dATP} \\ + \\ \text{dGT} \\ + \\ \text{dCTP} \\ + \\ \text{dTTP} \end{array} \right\} \xrightarrow[\text{polymerase}]{\text{DNA}} \text{progeny DNA}$$

This reaction works only with the triphosphate forms of the nucleotides (such as deoxyadenosine triphosphate, or dATP). The total amount of DNA at the end of the reaction can be as much as 20 times the amount of original input DNA, so most of the DNA present at the end must be progeny

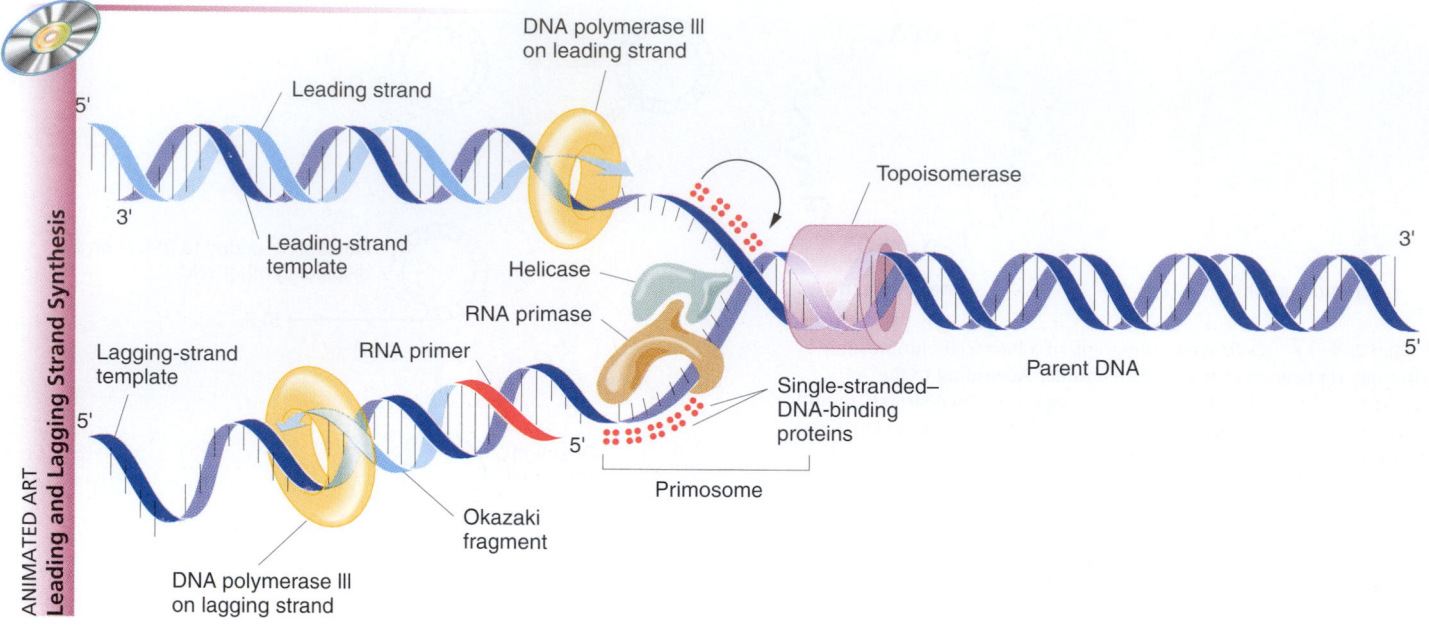

Figure 8-20   DNA replication fork.

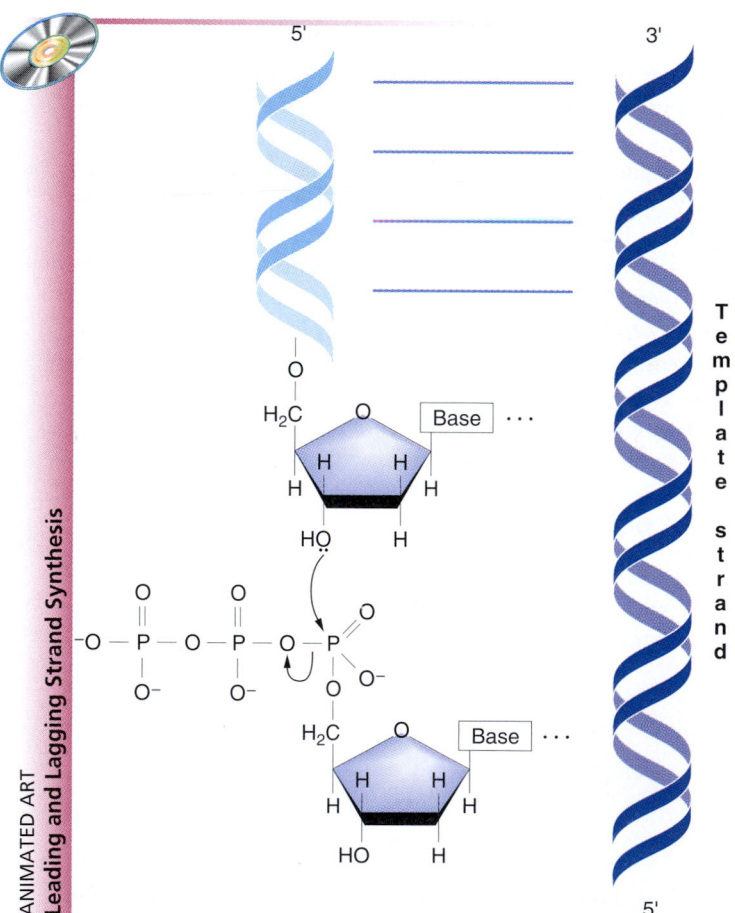

Figure 8-21   Chain-elongation reaction catalyzed by DNA polymerase.

DNA. Figure 8-21 depicts the chain-elongation reaction, or **polymerization** reaction, catalyzed by DNA polymerases. We now know that there are three DNA polymerases in *E. coli.* The first enzyme that Kornberg purified is called **DNA polymerase I** or **pol I.** This enzyme has three activities, which appear to be located in different parts of the molecule:

1. a polymerase activity, which catalyzes chain growth in the $5' \rightarrow 3'$ direction;

2. a $3' \rightarrow 5'$ exonuclease activity, which removes mismatched bases; and

3. A $5' \rightarrow 3'$ exonuclease activity, which degrades double-stranded DNA.

Subsequently, two additional polymerases, pol II and pol III, were identified in *E. coli.* Pol II may repair damaged DNA, although no particular role has been assigned to this enzyme. Pol III, together with pol I, has a role in the replication of *E. coli* DNA (Figure 8-20). The complete complex, or **holoenzyme,** of pol III contains at least 20 different polypeptide subunits, although the catalytic "core" consists of only three subunits, alpha (α), epsilon (ϵ), and theta (θ). The pol III complex will complete the replication of single-stranded DNA if there is at least a short segment of duplex already present. The short oligonucleotide that creates the duplex is termed a **primer.**

Prokaryotic origins of replication

E. coli replication begins from a fixed **origin** but then proceeds **bidirectionally** (with moving forks at both ends

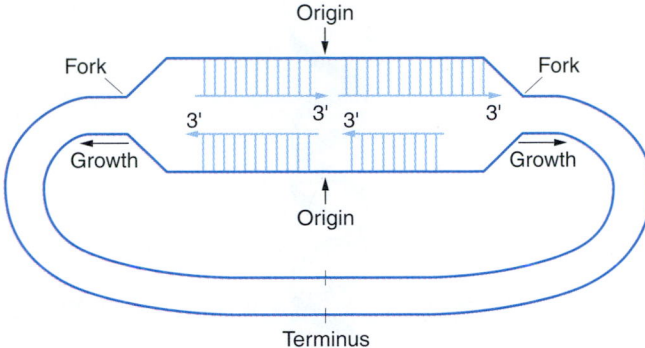

Figure 8-22 Chain-elongation reaction catalyzed by DNA polymerase. (From L. Stryer, *Biochemistry*, 4th ed. Copyright © 1995 by Lubert Stryer.)

of the replicating piece), as shown in Figure 8-22, ending at a site called the **terminus.** The unique origin is termed *oriC* and is located at 83 minutes on the genetic map. It is 245 bp long and has several components, as illustrated in Figure 8-23. First, there is a side-by-side, or tandem, set of 13-bp sequences, which are nearly identical. There is also a set of binding sites for a protein, the DnaA protein. An initial step in DNA synthesis is the unwinding of the DNA at the origin in response to binding of the DnaA protein. The consequences of bidirectional replication can be seen in Figure 8-24, which gives a larger view of DNA replication.

Eukaryotic origins of replication

Bacteria such as *E. coli* usually require a 40-minute replication-division cycle, but, in eukaryotes, the cycle can vary from 1.4 hours in yeast to 24 hours in cultured animal

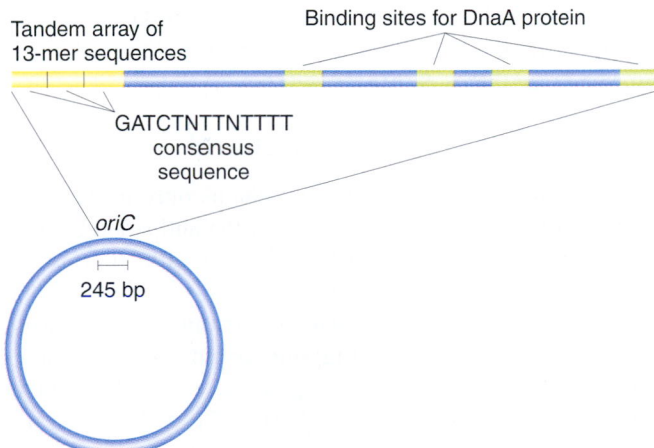

Figure 8-23 *OriC,* the origin of replication in *E. coli,* has a length of 245 bp. It contains a tandem array of three nearly identical 13-nucleotide sequences and four binding sites for DNA protein.

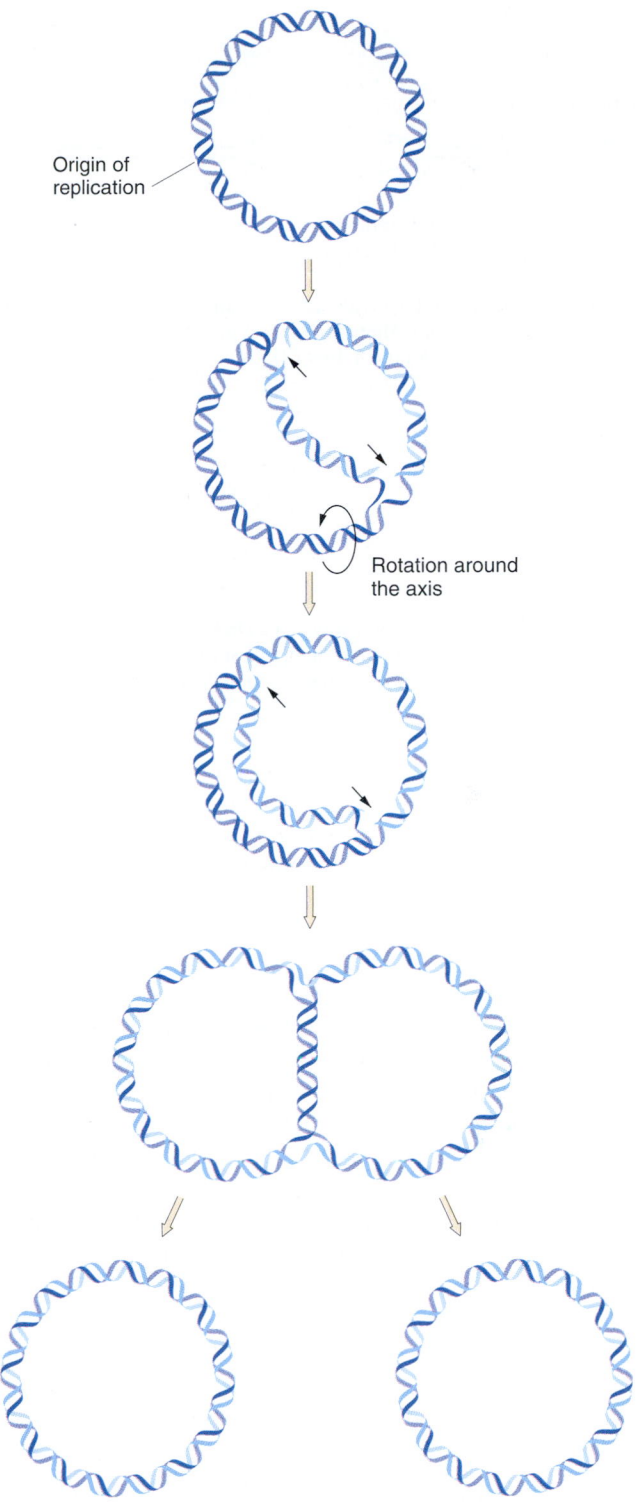

Figure 8-24 Bidirectional replication of a circular DNA molecule.

cells and may last from 100 to 200 hours in some cells. Eukaryotes have to solve the problem of coordinating the replication of more than one chromosome, as well as replicating the complex structure of the chromosome itself (see Chapter 3 for a description of chromosome structure).

Autoradiogram

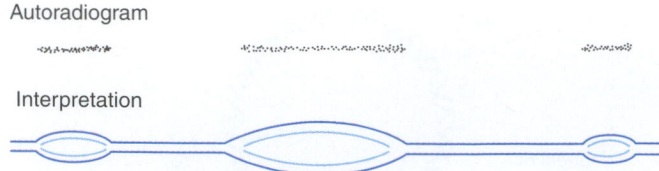

Interpretation

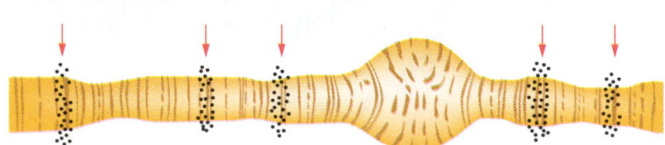

Figure 8-25   A replication pattern in DNA revealed by autoradiography. A cell is briefly exposed to [³H]thymidine (pulse) and then provided with an excess of nonradioactive ("cold") thymidine (chase). DNA is spread on a slide and autoradiographed. In the interpretation shown here, there are several initiation points for replication within one double helix of DNA.

Figure 8-26   Replication pattern in a *Drosophila* polytene chromosome revealed by autoradiography. Several points of replication are seen within a single chromosome, as indicated by the arrows.

In eukaryotes, replication proceeds from multiple points of origin. This process can be demonstrated by a procedure in which a eukaryotic cell is briefly exposed to [³H]thymidine, in a step called a **pulse exposure,** and is then provided an excess of "cold" (unlabeled) thymidine, in a step called the **chase;** the DNA is then extracted, and autoradiographs are made. Figure 8-25 shows the results of such a procedure, with what appear to be distinct, simultaneously replicating regions along the DNA molecule. Replication appears to begin at several different sites on these eukaryotic chromosomes. Similarly, a pulse-and-chase study of DNA replication in polytene (giant) chromosomes of *Drosophila* by autoradiography reveals many replication regions within single chromosome arms (Figure 8-26). As yet there is no firm proof that these regions are indeed different starting points on a single DNA molecule. However, experiments in yeast indicate the existence of approximately 400 replication origins distributed among the 17 yeast chromosomes, and in humans there are estimated to be more than 10,000 growing forks.

Priming DNA synthesis

DNA polymerases can extend a chain but cannot start a chain. Therefore, as already mentioned, DNA synthesis must first be initiated with a primer, or short oligonucleotide, that generates a segment of duplex DNA. The primer in DNA replication can be seen in Figure 8-27 (see also Figure 8-20). RNA primers are synthesized either by RNA polymerase or by an enzyme termed **primase.** Primase synthesizes a short (approximately 30 bp long) stretch of RNA complementary to a specific region of the chromosome. The RNA chain is then extended with DNA by DNA

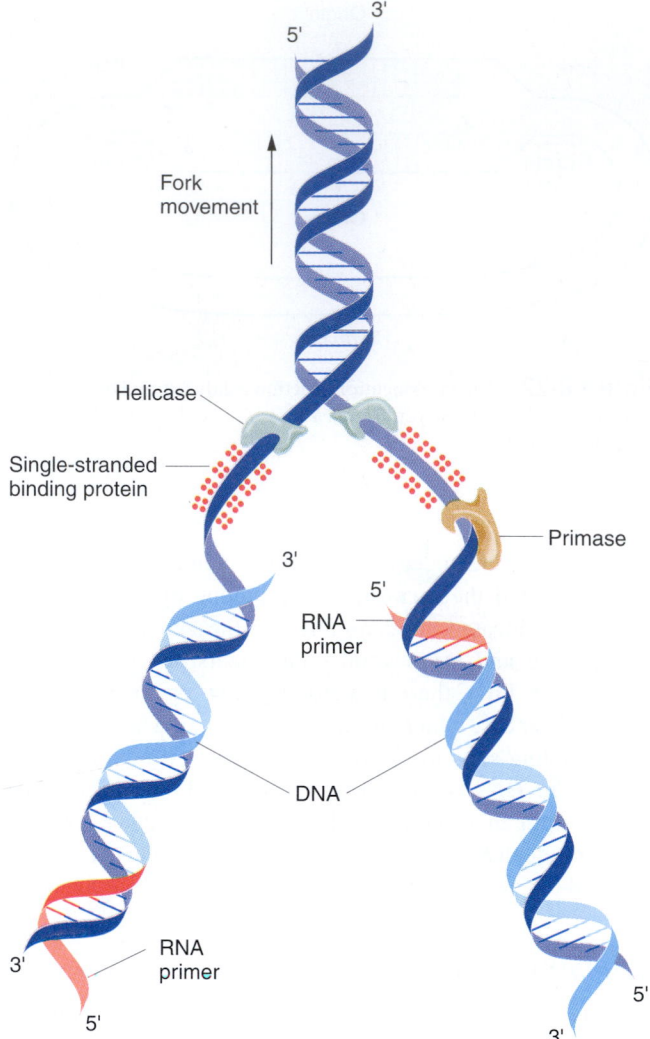

Figure 8-27   Initiation of DNA synthesis by an RNA primer.

polymerase. *E. coli* primase forms a complex with the template DNA, and additional proteins, such as DnaB, DnaT, Pri A, Pri B, and Pri C. The entire complex is termed a **primosome** (see Figure 8-20).

Leading strand and lagging strand

DNA polymerases synthesize new chains only in the 5′ → 3′ direction and therefore, because of the antiparallel nature of the DNA molecule, move in a 3′ → 5′ direction *on the template strand.* The consequence of this polarity is that while one new strand, the **leading strand,** is synthesized continuously, the other, the **lagging strand,** must be synthesized in short, discontinuous segments, as can be seen in Figure 8-28 (see also Figure 8-20). The addition of nucleotides along the template for the lagging strand must proceed toward the template's 5′ end (because replication *always* moves along the template in a 3′ → 5′ direction so that the new strand can grow 5′ → 3′). Thus, the new strand must grow in a direction opposite that of the movement of

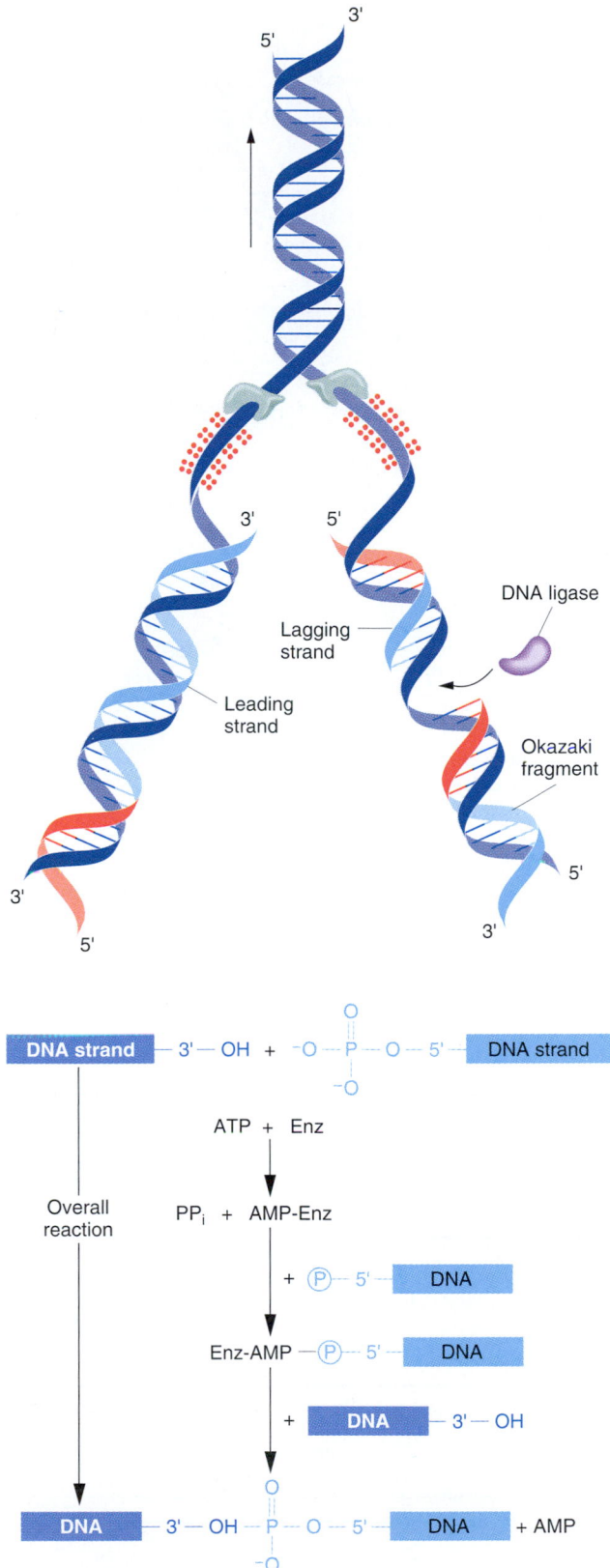

the replication fork. As fork movement exposes a new section of lagging-strand template, a new lagging-strand fragment is begun and proceeds away from the fork until it is stopped by the preceding fragment. In *E. coli,* pol III carries out most of the DNA synthesis on both strands, and pol I fills in the gaps left in the lagging strand, which are then sealed by the enzyme **DNA ligase.** DNA ligases join broken pieces of DNA by catalyzing the formation of a phosphodiester bond between the 5′ phosphate end of a hydrogen-bonded nucleotide and an adjacent 3′ OH group, as shown in Figure 8-29. It is the only enzyme that can seal DNA chains. Figure 8-30 shows the lagging-strand synthesis and gap repair in detail. The primers for the discontinuous

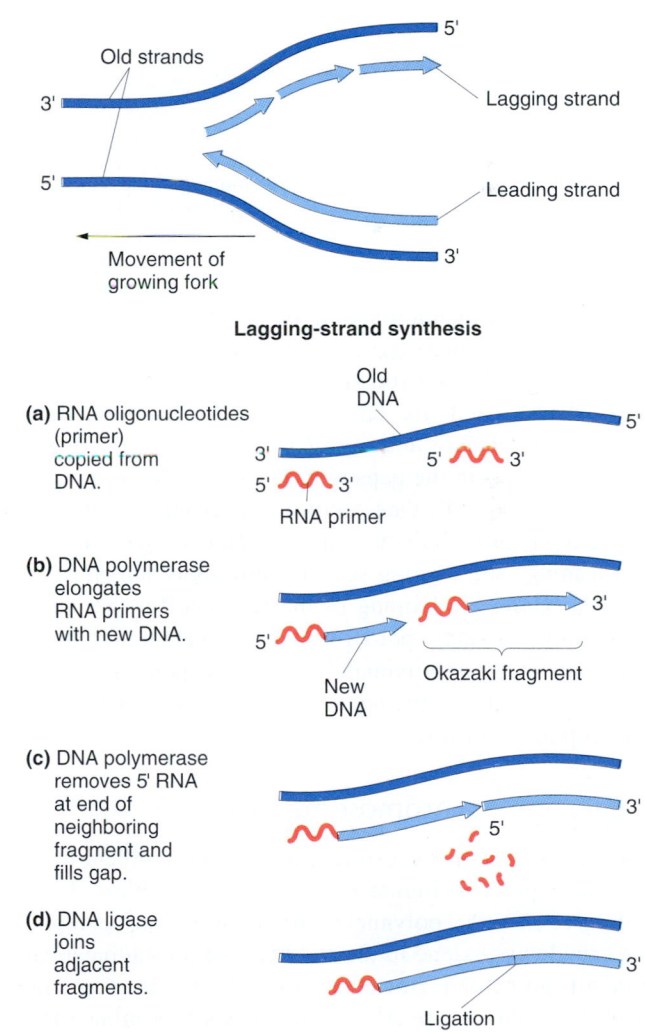

Lagging-strand synthesis

(a) RNA oligonucleotides (primer) copied from DNA.

(b) DNA polymerase elongates RNA primers with new DNA.

(c) DNA polymerase removes 5′ RNA at end of neighboring fragment and fills gap.

(d) DNA ligase joins adjacent fragments.

Figure 8-29 The reaction catalyzed by DNA ligase (Enz) joins the 3′-OH end of one fragment to the 5′ phosphate of the adjacent fragment (From H. Lodish, D. Baltimore, A. Berk, S. L. Zipursky, P. Matsudaira, and J. Darnell, *Molecular Cell Biology,* 3d ed. Copyright © 1995 by Scientific American Books, Inc.)

Figure 8-30 The overall structure of a growing fork (*top*) and steps in the synthesis of the lagging strand. (From H. Lodish, D. Baltimore, A. Berk, S. L. Zipursky, P. Matsudaira, and J. Darnell, *Molecular Cell Biology,* 3d ed. Copyright © 1995 by Scientific American Books, Inc.)

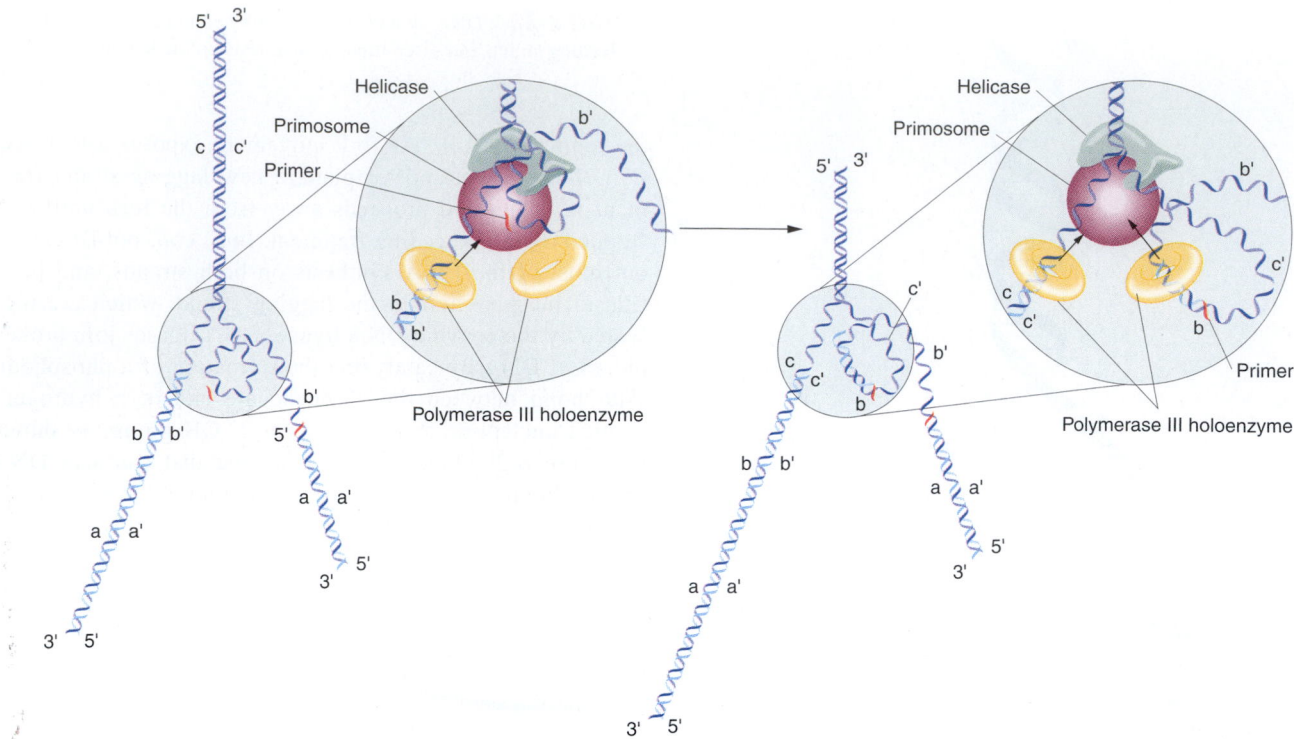

Figure 8-31   The looping of the template for the lagging strand enables a dimeric DNA polymerase III holoenzyme at the replication fork to synthesize both daughter strands. (Courtesy of A. Kornberg. From L. Stryer, *Biochemistry*, 4th ed. Copyright © 1995 by Lubert Stryer.)

synthesis on the lagging strand are synthesized by primase (step a). The primers are extended by DNA polymerase (step b) to yield DNA fragments that were first detected by Reiji Okazaki and are termed **Okazaki fragments.** The $5' \rightarrow 3'$ exonuclease activity of pol I removes the primers (step c) and fills in the gaps with DNA, which are sealed by DNA ligase (step d). One proposed mechanism that allows the same dimeric holoenzyme molecule to participate in both leading- and lagging-strand synthesis is shown in Figure 8-31. Here, the looping of the template for the lagging strand allows a single pol III dimer to generate both daughter strands. After approximately 1000 base pairs, pol III will release the segment of lagging-strand duplex and allow a new loop to be formed.

Replication at chromosome tips

The ends of chromosomes present a special problem for the replication process. Figure 8-32 shows the problem: for the leading strand, the polynucleotide addition during replication can always extend to the end because it is automatically primed from behind. However, at the tip, the lagging strand reaches a point where its system of RNA priming cannot work, and an unpolymerized section remains and a shortened chromosome would be the result. To solve this problem, the tips of chromosomes, called **telomeres,** have adjacent repeats of simple DNA sequences. For example, in the ciliate *Tetrahymena,* there is repetition of the sequence

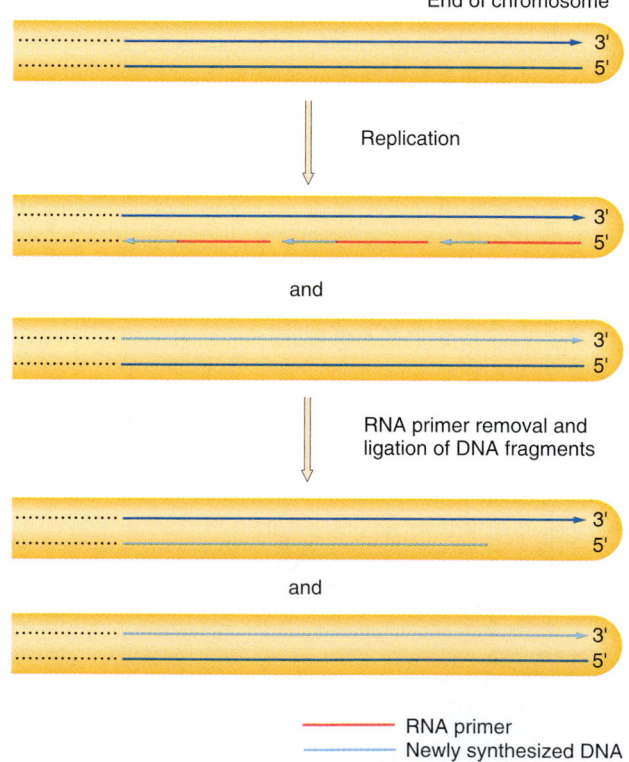

Figure 8-32   The replication problem at chromosome ends. There is no way of priming the last section of the lagging strand, and a shortened chromosome would result.

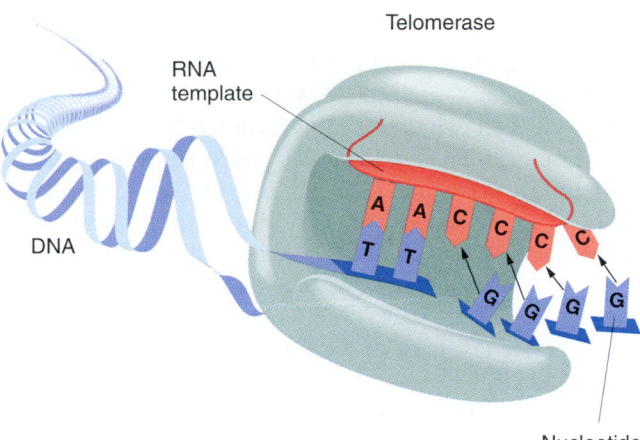

Telomerase

RNA template

DNA

Nucleotide

Figure 8-33 Telomerase carries a short RNA molecule that acts as a template for the addition of the complementary DNA sequence at the 3′ end of the double helix. In the ciliate *Tetrahymena,* the DNA sequence added is TTGGGG.

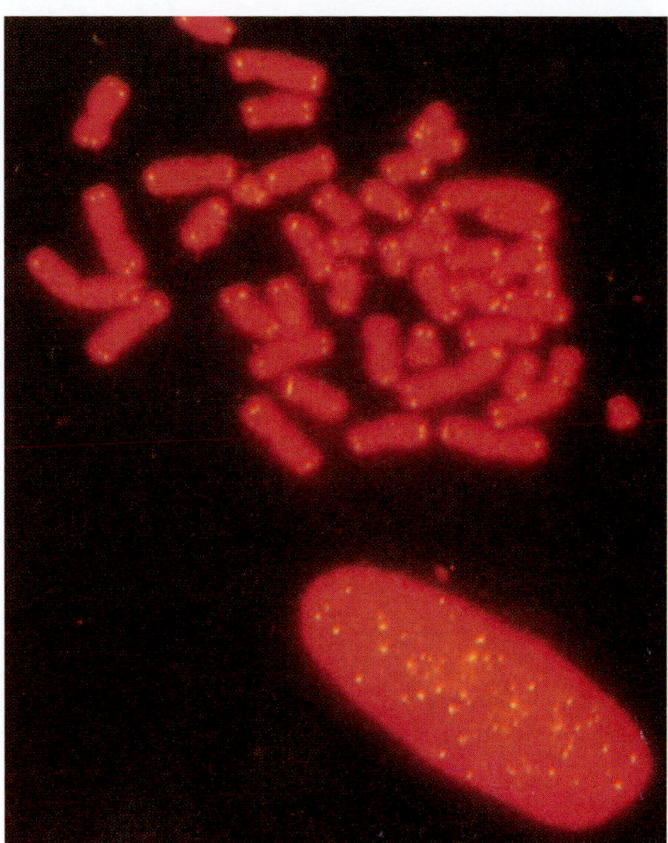

Figure 8-34 Chromosomes probed in situ with a telomere-specific DNA probe that has been coupled to a substance that can fluoresce yellow under the microscope. Each sister chromatid binds the probe at both ends. An unbroken nucleus is shown at the bottom of the photograph. (Robert Moyzis.)

TTGGGG; in humans, it is TTAGGG. These repeats do not code for an RNA or a protein product but nevertheless serve a definite function in replication. An enzyme called **telomerase** adds these simple repeat units to the chromosome ends. The telomerase protein is a member of a class of enzymes called reverse transcriptases, which are used in specialized situations to synthesize DNA from RNA. The telomerase carries a small RNA molecule, part of which acts as a template for the polymerization of the telomeric repeat unit that is added to the 3′ end. For example, in *Tetrahymena,* the RNA is 3′-AACCCC-5′, which acts as the template for the repeat unit, which is 5′-TTGGGG-3′ (Figure 8-33). The additional DNA is then able to act as template for synthesis on the lagging strand. This process counteracts the tendency to shorten at replication. Figure 8-34 demonstrates the positions of the telomeric DNA through in situ hybridization. An age-dependent decline in telomere length has been found in several somatic tissues in humans. In addition, human fibroblasts in culture show progressive telomere shortening up to their eventual death. Such observations have led to the telomere theory of aging, and the validity of this theory is now being tested.

Helicases and topoisomerases

Helicases are enzymes that disrupt the hydrogen bonds that hold the two DNA strands together in a double helix. Hydrolysis of ATP drives the reaction. Among *E. coli* helicases are the DnaB protein and the Rep protein. The Rep protein may help to unwind the double helix ahead of the polymerase (refer to Figure 8-20). The unwound DNA is stabilized by the single-stranded binding (SSB) protein, which binds to the single-stranded DNA and retards reformation of the duplex.

The action of helicases during DNA replication generates twists in the circular DNA that need to be removed to allow replication to continue. Circular DNA can be twisted and coiled, much like the extra coils that can be introduced into a rubber band. This **supercoiling** can be created or relaxed by enzymes termed **topoisomerases,** an example of which is DNA gyrase (Figure 8-35). Topoisomerases can also induce (*catenate*) or remove (*decatenate*) knots, or links in a chain. There are two basic types of isomerases. Type I enzymes induce a single-stranded break into the DNA duplex. Type II enzymes cause a break in both strands. In *E. coli,* topo I and topo III are examples of type I enzymes, whereas gyrase is an example of a type II enzyme.

Untwisting of the DNA strands to open the replication fork causes extra twisting at other regions, and the supercoiling releases the strain of the extra twisting (Figure 8-36). During replication, gyrase is needed to remove positive supercoils ahead of the replication fork.

Exonuclease editing

Both DNA polymerase I and DNA polymerase III also possess 3′ → 5′ exonuclease activity, which serves a "proofreading" and "editing" function by searching for mismatched bases that were inserted erroneously during

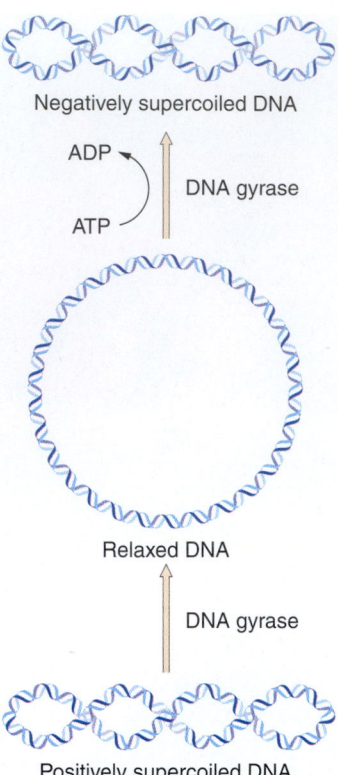

Figure 8-35 DNA-gyrase-catalyzed supercoiling. Replicating DNA generates "positive" supercoils, depicted at the bottom of the diagram, as a result of rapid rotation of the DNA at the replication fork. DNA gyrase can nick and close phosphodiester bonds, relieving the supercoiling, as shown here (relaxed DNA). Gyrase can also generate supercoils twisted in the opposite direction, termed *negative* supercoils; this arrangement facilitates the unwinding of the helix. (After L. Stryer, *Biochemistry*, 4th ed. Copyright © 1995 by Lubert Stryer.)

polymerization and excising them. The proofreading activity of pol III is the ϵ subunit, which must be bound to α for full proofreading activity (Figure 8-37). Strains lacking a functional ϵ have a higher mutation rate (see Chapter 16). Figure 8-38 shows the excision of a cytosine residue that has erroneously been paired with an adenine. As can be seen, hydrolysis takes place at the 5′ end of the mismatched base; removal of the incorrect base leaves a 3′-OH group on the preceding base, which is then free to continue the growing strand by accepting the correct nucleotide triphosphate (thymidine, in this case).

Note that this exonuclease activity takes place at the 3′ end of the growing strand (and is therefore 3′ → 5′). The coordination of exonuclease activity with strand growth helps to explain why replication is in the 5′ → 3′ direction. As we saw earlier, new bases are added when the 3′ OH on the terminal deoxyribose of the growing strand attacks the high-energy phosphate of the nucleotide triphosphate that is being added (see Figure 8-21). Chain growth is thus 5′ → 3′. It is conceivable that replication could be in the 3′ → 5′ direction (in Figure 8-21, the 5′ triphosphate at the bottom would be the last base on the chain, and the 3′ OH that attacks it would be on the free nucleotide triphosphate about to be added to the strand). However, if replication were in this direction, there would be exonuclease excisions at the 5′ end of the strand. When a mismatched base was removed, a 5′ OH would be left at the end of the growing strand. The 3′ OH of an incoming nucleotide triphosphate would thus be facing this 5′ OH instead of the high-energy 5′ triphosphate necessary for bond formation. No bond would form and strand growth would stop. Therefore, replication is not in the 3′ → 5′ direction.

Eukaryotic DNA polymerases

There are at least five DNA polymerases, α, β, γ, δ, and ϵ, in higher eukaryotes. Polymerases α and δ in the nucleus have roles similar to pol I in *E. coli*. Polymerase β has a role in DNA repair and gap filling. The γ polymerase is found in mitochondria and appears to take part in replication of mitochondrial DNA.

Experimental applications of base-sequence complementarity

In 1960, Paul Doty and Julius Marmur observed that, when DNA is heated to 100°C, all the hydrogen bonds between the complementary strands are destroyed, and the DNA becomes single stranded (Figure 8-39). If the solution is cooled slowly, some double-stranded DNA is formed. This **reannealing** process occurs when two single strands happen to collide in such a way that the complementing base sequences can align and reconstitute the original double helix. The annealing of complementary strands is very specific, as shown in Figure 8-39, and forms the basis of many important techniques in molecular biology, such as the identifica-

Figure 8-36 Swivel function of topoisomerase during replication. Extra-twisted (positively supercoiled) regions accumulate ahead of the fork as the parental strands separate for replication. A topoisomerase is required to remove these regions, acting as a swivel to allow extensive replication. (From A. Kornberg and T. A. Baker, *DNA Replication*, 2d ed. Copyright © 1992 by W. H. Freeman and Company.)

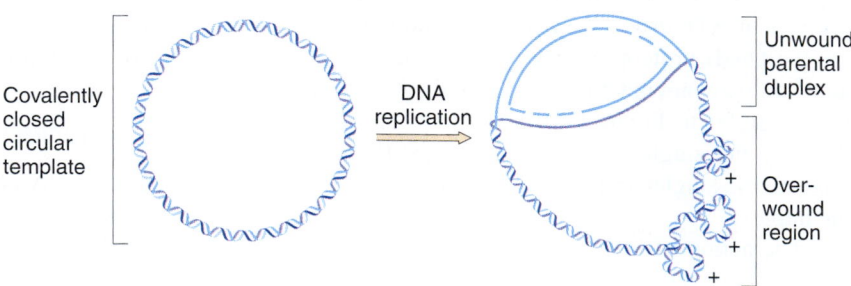

Covalently closed circular template

DNA replication

Unwound parental duplex

Overwound region

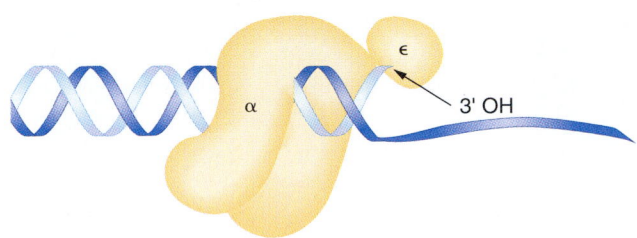

Figure 8-37 Proofreading by the pol III $\alpha - \epsilon$ complex. (From A. Kornberg and T. A Baker, *DNA Replication,* 2d ed. Copyright © 1992 by W. H. Freeman and Company.)

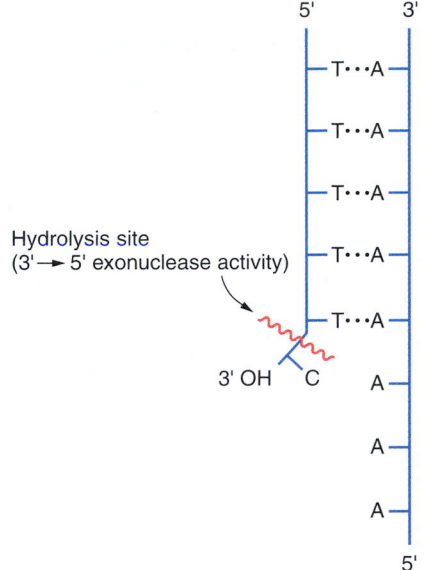

Figure 8-38 The $3' \rightarrow 5'$ exonuclease action of DNA polymerase III.

tion of specific DNA segments by hybridization and the isolation of specific DNA fragments that are used in cloning, as explained in Chapter 12. Figure 8-40a shows the basic profile of a typical DNA strand as the temperature increases. At a temperature characteristic for each DNA segment, the DNA starts to denature. The melting temperature, T_m, is defined as the temperature at which half the molecules are denatured into single strands. The melting temperature depends on the proportion of G:C base pairs, because they are held together by three hydrogen bonds, whereas A:T base pairs are held together by two hydrogen bonds. The higher the G:C content, the higher the melting temperature, as shown in Figure 8-40b.

Doty and Marmur's finding that in solutions the separated single strands of a double helix will find each other because of complementary base pairing led to several experimental applications that have had an enormous effect on research in molecular genetics.

1. *Analysis of genome structure.* If total genomic DNA is melted and allowed to reanneal, several distinct stages are observed in the annealing process. First, there is a

stage of very rapid annealing. This stage represents highly repetitive DNA because, for this type of DNA, there are many copies per genome. These copies can find each other faster than can unique genes that are present in only one copy per haploid genome. Later annealing fractions contain progressively less repetitive DNA, and unique sequences anneal last. Hence this property allowed scientists to isolate and characterize the various repetitive categories. This characterization in turn allowed the overall characterization of genomes from organisms of most taxonomic groups, providing another approach to evolutionary comparison at the genetic level.

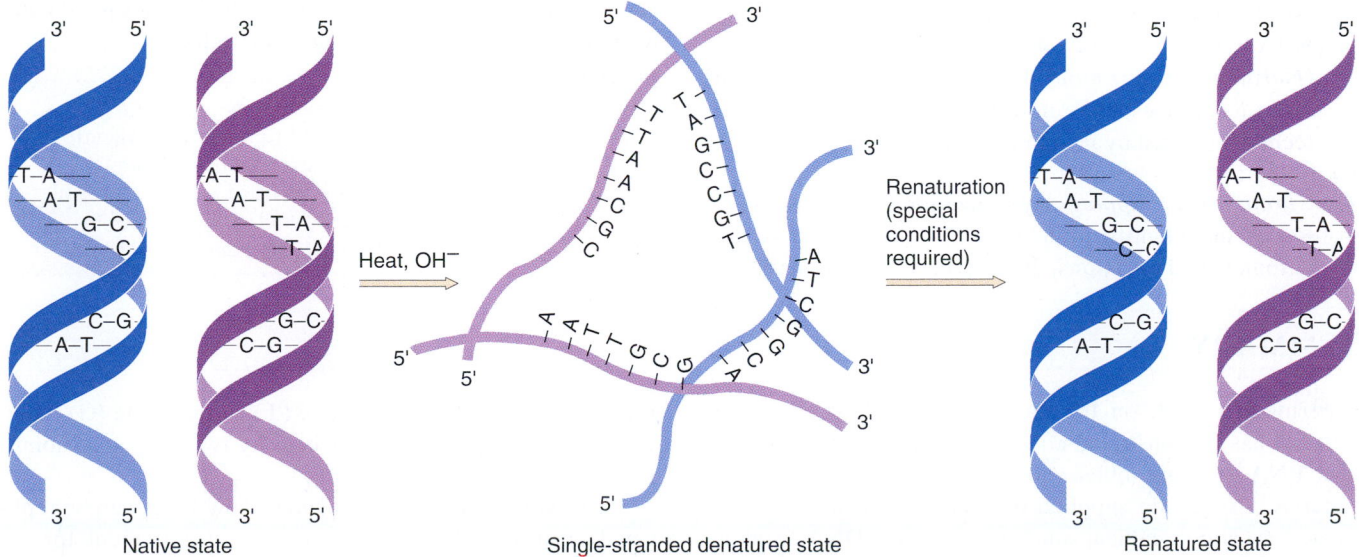

Figure 8-39 The denaturation and renaturation of double-stranded DNA molecules.

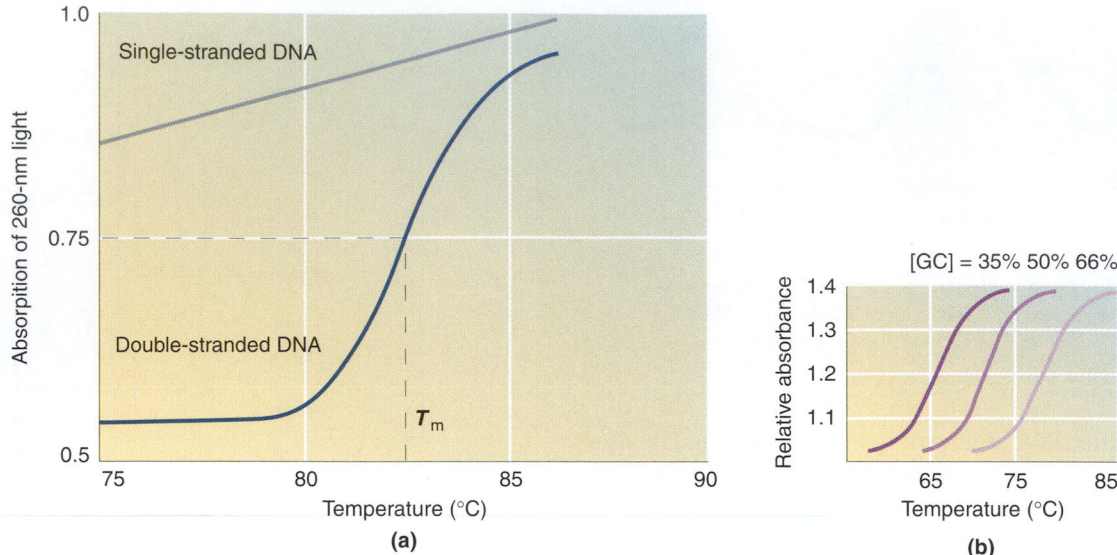

Figure 8-40   (a) The absorption of ultraviolet light of 260-nm wavelength by solutions of single-stranded and double-stranded DNA. As regions of double-stranded DNA unpair, the absorption of light by those regions increases almost twofold. The temperature at which half the bases in a double-stranded DNA sample have denatured is denoted T_m. (b) DNA melting curves. The absorbance relative to that at 25°C is plotted against temperature. (The wavelength of the incident light was 260 nm.) The T_m is 69°C for *E. coli* DNA (50 percent GC pairs) and 76°C for *Pseudomonas aeruginosa* DNA (68 percent GC pairs). (Part b from L. Stryer, *Biochemistry,* 4th ed. Copyright © 1995 by Lubert Stryer.)

2. *Gene isolation.* Many of the techniques for isolating genes (gene cloning; see Chapter 12) are based on DNA hybridization by base complementarity. The most common type uses a radioactive denatured DNA fragment as a probe to find a clone of some specific gene of interest in a mixture of clones representing the whole genome.

3. *Southern and Northern hybridization.* We learned in Chapter 1 that a denatured labeled probe can be used to identify specific genomic fragments in a mixture separated on an electrophoretic gel (the technique of Southern hybridization); in a parallel technique, specific RNA transcripts can be detected on electrophoretic gels (Northern hybridization). Because of the incisiveness of these techniques, they now form part of the everyday technology used by geneticists everywhere.

4. *Chromosome mapping.* Hybridization by probes has allowed the identification of *DNA markers* at specific chromosomal locations. [One type of DNA marker is the restriction fragment length polymorphism (RFLP); see Chapter 13.] DNA markers have provided many thousands of new loci to saturate the chromosomal map. Furthermore, such markers have provided linked markers for the diagnosis of disease alleles in humans. In a related technique, labeled probes can be added to partly denatured DNA still in chromosomes, revealing the chromosomal position of the DNA homologous to the probe (in situ hybridization; see Chapter 14).

Hence we see that the structure of DNA provides not only two key properties for biological function (replication and information storage), but also key techniques for the genetic dissection of organisms and their cells.

MESSAGE ·······
The specificity of base complementarity forms the foundation for the continuity of life through replication and the foundation for information transfer from DNA into protein—the main determinant of biological form. This same specificity is used by geneticists as a tool to investigate gene and genome structure and function.

SUMMARY

Experimental work on the molecular nature of hereditary material has demonstrated conclusively that DNA (not protein, RNA, or some other substance) is indeed the genetic material. Using data supplied by others, Watson and Crick created a double-helical model with two DNA strands, wound around each other, running in antiparallel fashion. Specificity of binding the two strands together is based on the fit of adenine (A) to thymine (T) and guanine (G) to cytosine (C). The former pair is held by two hydrogen bonds; the latter, by three.

The Watson-Crick model shows how DNA can be replicated in an orderly fashion—a prime requirement for genetic material. Replication is accomplished semiconservatively in both prokaryotes and eukaryotes. One double helix is replicated into two identical helices, each with identical linear orders of nucleotides; each of the two new double helices is composed of one old and one newly polymerized strand of DNA.

Replication is achieved with the aid of several enzymes, including DNA polymerase, gyrase, and helicase. Replication starts at special regions of the DNA called origins of replication and proceeds down the DNA in both directions. Because DNA polymerase acts only in a 5′ → 3′ direction, one of the newly synthesized strands at each replication fork must be synthesized in short segments and then joined by the enzyme ligase. DNA polymerization cannot begin without a short primer, which is also synthesized with special enzymes.

CONCEPT MAP

Draw a concept map interrelating as many of the following terms as possible. Note that the terms are listed in no particular order.

DNA double helix / nucleotides / hydrogen bonding / semiconservative / 5′ / replication / chromatid / mitosis / DNA polymerase / meiosis / S phase / gene / 3′

CHAPTER INTEGRATION PROBLEM

Mitosis and meiosis were presented in Chapter 3. Considering what we have covered in this chapter concerning DNA replication, draw a graph showing DNA content against time in a cell that undergoes mitosis and then meiosis. Assume a diploid cell.

◆ Solution ◆

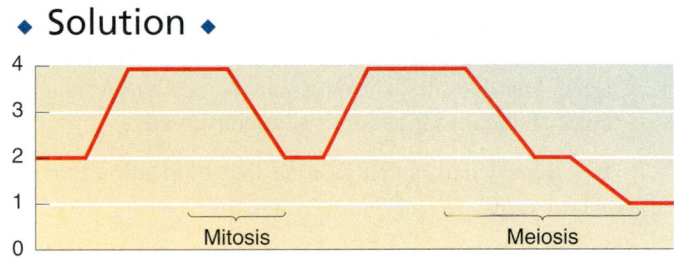

SOLVED PROBLEMS

1. If the GC content of a DNA molecule is 56 percent, what are the percentages of the four bases (A, T, G, and C) in this molecule?

◆ Solution ◆

If the GC content is 56 percent, then, because G = C, the content of G is 28 percent and the content of C is 28 percent. The content of AT is 100 − 56 = 44 percent. Because A = T, the content of A is 22 percent and the content of T is 22 percent.

2. Describe the expected pattern of bands in a CsCl gradient for *conservative* replication in the Meselson-Stahl experiment. Draw a diagram.

◆ Solution ◆

Refer to Figure 8-13 for an additional explanation. In conservative replication, if bacteria are grown in the presence of ^{15}N and then shifted to ^{14}N, one DNA molecule will be all ^{15}N after the first generation and the other molecule will be all ^{14}N, resulting in one heavy band and one light band in the gradient. After the second generation, the ^{15}N DNA will yield one molecule with all ^{15}N and one molecule with all ^{14}N, whereas the ^{14}N DNA will yield only ^{14}N DNA. Thus, only all ^{14}N or all ^{15}N DNA is generated, again, yielding a light band and a heavy band:

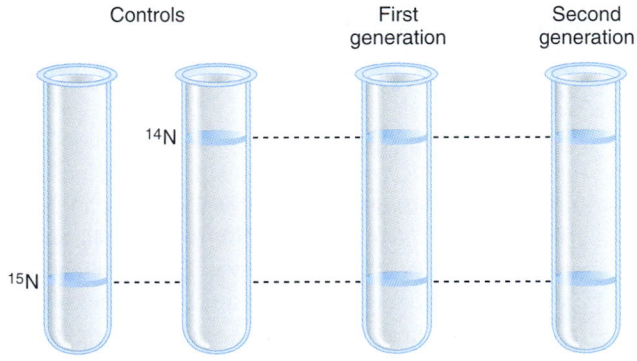

PROBLEMS

1. Describe the types of chemical bonds and forces of the DNA double helix.

2. Explain what is meant by the terms *conservative* and *semiconservative replication.*

3. What is meant by a *primer,* and why are primers necessary for DNA replication?

4. What are helicases and topoisomerases?

5. Why is DNA synthesis continuous on one strand and discontinuous on the opposite strand?

6. If thymine makes up 15 percent of the bases in a specific DNA molecule, what percentage of the bases is cytosine?

7. If the GC content of a DNA molecule is 48 percent, what are the percentages of the four bases (A, T, G, and C) in this molecule?

8. *E. coli* chromosomes in which every nitrogen atom is labeled (that is, every nitrogen atom is the heavy isotope ^{15}N instead of the normal isotope ^{14}N) are allowed to replicate in an environment in which all the nitrogen is ^{14}N. Using a solid line to represent a heavy polynucleotide chain and a dashed line for a light chain, sketch the following:

    a. The heavy parental chromosome and the products of the first replication after transfer to a ^{14}N medium, assuming that the chromosome is one DNA double helix and that replication is semiconservative.

    b. Repeat part a, but assume that replication is conservative.

    c. Repeat part a, but assume that the chromosome is in fact two side-by-side double helices, each of which replicates semiconservatively.

    d. Repeat part c, but assume that each side-by-side double helix replicates conservatively and that the overall *chromosome* replication is semiconservative.

    e. Repeat part d, but assume that the overall chromosome replication is conservative.

    f. If the daughter chromosomes from the first division in ^{14}N are spun in a cesium chloride (CsCl) density gradient and a single band is obtained, which of possibilities in parts a through e can be ruled out? Reconsider the Meselson-Stahl experiment: what does it *prove?*

9. R. Okazaki found that the immediate products of DNA replication in *E. coli* include single-stranded DNA fragments approximately 1000 nucleotides in length after the newly synthesized DNA is extracted and denatured (melted). When he allowed DNA replication to proceed for a longer period of time, he found a lower frequency of these short fragments and long single-stranded DNA chains after extraction and denaturation. Explain how this result might be related to the fact that all known DNA polymerases synthesize DNA only in a $5' \rightarrow 3'$ direction.

10. When plant and animal cells are given pulses of $[^{3}H]$thymidine at different times during the cell cycle, heterochromatic regions on the chromosomes are in-

variably shown to be "late replicating." Can you suggest what, if any, biological significance this observation might have?

11. On the planet of Rama, the DNA is of six nucleotide types: A, B, C, D, E, and F. A and B are called *marzines,* C and D are *orsines,* and E and F are *pirines*. The following rules are valid in all Raman DNAs:

$$\text{Total marzines} = \text{total orsines} = \text{total pirines}$$
$$A = C = E$$
$$B = D = F$$

    a. Prepare a model for the structure of Raman DNA.

    b. On Rama, mitosis produces three daughter cells. Bearing this fact in mind, propose a replication pattern for your DNA model.

    c. Consider the process of meiosis on Rama. What comments or conclusions can you suggest?

12. If you extract the DNA of the coliphage ϕX174, you will find that its composition is 25 percent A, 33 percent T, 24 percent G, and 18 percent C. Does this composition make sense in regard to Chargaff's rules? How would you interpret this result? How might such a phage replicate its DNA?

13. The temperature at which a DNA sample denatures can be used to estimate the proportion of its nucleotide pairs that are G–C. What would the basis for this determination be, and what would a high denaturation temperature for a DNA sample indicate?

14. Suppose that you extract DNA from a small virus, denature it, and allow it to reanneal with DNA taken from other strains that carry either a deletion, an inversion, or a duplication. What would you expect to see on inspection with an electron microscope?

15. DNA extracted from a mammal is heat denatured and then slowly cooled to allow reannealing. The following graph shows the results obtained. These are two "shoulders" in the curve. The first shoulder indicates the presence of a very rapidly annealing part of the DNA — so rapid, in fact, that the annealing occurs before strand interactions take place.

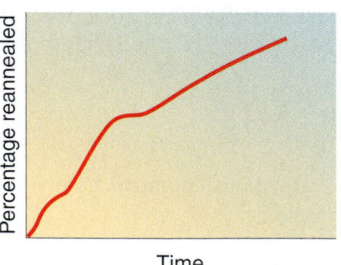

a. What could this part of the DNA be?

b. The second shoulder is a rapidly reannealing part as well. What does this evidence suggest?

16. Design tests to determine the physical relation between highly repetitive and unique DNA sequences in chromosomes. (**Hint:** It is possible to vary the size of DNA molecules by the amount of shearing to which they are subjected.)

17. Viruses are known to cause cancer in mice. You have a pure preparation of virus DNA, a pure preparation of DNA from the chromosomes of mouse cancer cells, and pure DNA from the chromosomes of normal mouse cells. Viral DNA will specifically anneal with cancer-cell DNA, but not with normal-cell DNA. Explore the possible genetic significance of this observation, its significance at the molecular level, and its medical significance.

18. Ruth Kavenaugh and Bruno Zimm devised a technique to measure the maximal length of the longest DNA molecules in solution. They studied DNA samples from the three *Drosophila* karyotypes shown at the right. They found the longest molecules in karyotypes a and b to be of similar length and about twice the length of the longest molecule in karyotype c. Interpret these results.

19. In the harlequin chromosome technique, you allow *three* rounds of replication in bromodeoxyuridine and then stain the chromosomes. What result do you expect?

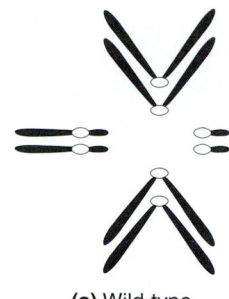

(a) Wild-type

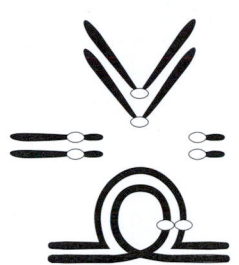

(b) Pericentric inversion

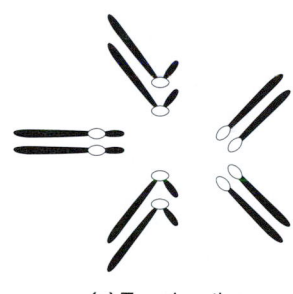

(c) Translocation

9

GENETICS OF DNA FUNCTION

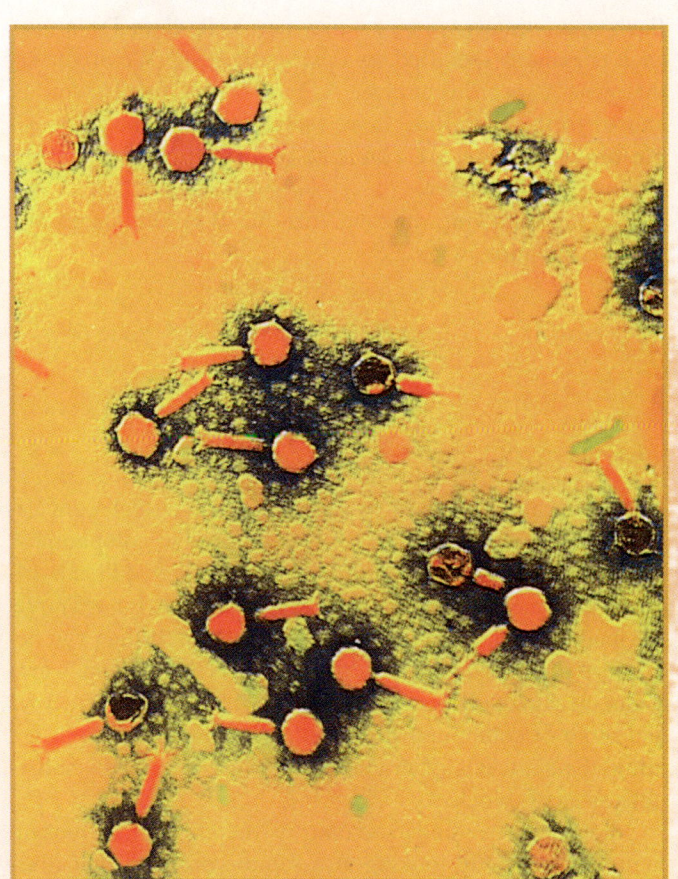

False-color transmission electron micrograph of unidentified phage.
(Magnification, 58,000×.) *(CNRI/Science Photo Library/Photo Researchers.)*

Key Concepts

The one-gene–one-enzyme hypothesis implies that genes control the structure of proteins.

Studies on hemoglobin demonstrated that a gene mutation causes an altered amino acid sequence in a protein.

The linear sequence of nucleotides in a gene determines the linear sequence of amino acids in a protein.

Fine structural analysis of the *rII* genes in phage T4 showed that the gene consists of a linear array of subelements that can mutate and recombine with one another; these subelements were later correlated with nucleotide pairs.

A gene can be defined as a unit of function by a complementation test.

What is the nature of the gene, and how do genes control phenotypes? For example, how can one allele of a gene produce a wrinkled pea and another produce a round, smooth pea? Today we realize that all biochemical reactions of a cell are catalyzed by enzymes, which have a specific three-dimensional configuration that is crucial to their function. We now know that genes specify the structures of proteins, some of which are enzymes, and we can even relate the structure of the genetic material to the structure of proteins. Table 9-1 summarizes our current model of the relation between genotype and phenotype.

How did we arrive at this point? By following two lines of inquiry:

1. What is the physical structure of the genetic material?

2. How does the genetic material exert its effect? (Or, in detail, how does the structure function?)

Chapter 8 recounted the demonstration that DNA is the genetic material and detailed the unraveling of the structure of DNA. In Chapters 9 and 10, we examine how genes function.

TABLE 9-1   Model of the Relation Between Genotype and Phenotype

1. The characteristic features of an organism are determined by the phenotype of its parts, which are in turn determined by the phenotype of the component cells.
2. The phenotype of a cell is determined by its internal chemistry, which is controlled by enzymes that catalyze its metabolic reactions.
3. The function of an enzyme depends on its specific three-dimensional structure, which in turn depends on the specific linear sequence of amino acids in the enzyme.
4. The enzymes present in a cell, and structural proteins as well, are determined by the genotype of the cell.
5. Genes specify the linear sequence of amino acids in polypeptides and hence in proteins; thus, genes determine phenotypes.

How genes work

The first clues about the nature of primary gene function came from studies of humans. Early in the twentieth century, Archibald Garrod, a physician, noted that several hereditary human defects are produced by recessive mutations. Some of these defects can be traced directly to metabolic defects that affect the basic body chemistry — an observation that led to the notion of "inborn errors" in metabolism. For example, phenylketonuria, which is caused by an autosomal recessive allele, results from an inability to convert phenylalanine into tyrosine. Consequently, phenylalanine accumulates and is spontaneously converted into a toxic compound, phenylpyruvic acid. In a different example, the inability to convert tyrosine into the pigment melanin produces an albino. Garrod's observations focused attention on metabolic control by genes.

One-gene–one-enzyme hypothesis

Clarification of the actual function of genes came from research in the 1940s on *Neurospora* by George Beadle and Edward Tatum, who later received a Nobel Prize for their work. Before we describe their actual experiments, let's jump ahead and examine some aspects of biosynthetic pathways, on the basis of our current understanding. We now know that molecules are synthesized as a series of steps, each one catalyzed by an enzyme. For instance, a biosynthetic pathway might have four steps, where 1 is the starting material and 5 is the final product:

$$1 \xrightarrow{A} 2 \xrightarrow{B} 3 \xrightarrow{C} 4 \xrightarrow{D} 5$$

Each step is catalyzed by an enzyme: A, B, C, or D. In turn, each enzyme is specified by a particular gene. We might say that gene *A* specifies enzyme A, gene *B* specifies enzyme B, and so forth. Therefore, if we inactivate the gene responsible for an enzyme, we eliminate one required step and the pathway is interrupted.

In the following diagram, enzyme B is eliminated owing to a mutation in gene *B*:

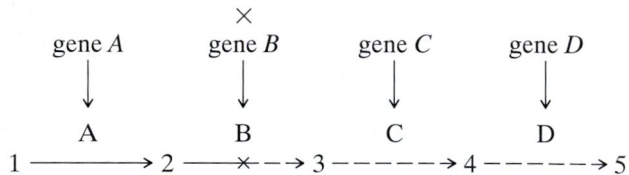

Now the cell cannot carry out the reaction that converts compound 2 into compound 3. It is blocked at compound 2, and cannot go further. But what happens if we add back different intermediate compounds? Suppose that, for example, we feed the cell compound 3 or 4. Can the cell then synthesize the final product, 5? Yes, with either compound 3 or 4 it can, because subsequent steps are not blocked. What if we add more of compound 1? No, adding compound 1 will not allow the synthesis of product 5, because one of the subsequent steps is blocked.

Let's test our understanding of the concept of biosynthetic pathways by looking at a sample problem that illustrates this principle:

Let's assume that we have isolated five mutants, 1 to 5, that cannot synthesize compound G for growth. We know the five compounds, A to E, that are required in the biosynthetic pathway, but we do not know the order in which they are synthesized by the wild-type cell. We have tested each compound for its ability to support the growth of each mutant, with the following results (a plus sign means growth and a minus sign means no growth):

| | Compound tested | | | | | |
Mutant	A	B	C	D	E	G
1	−	−	−	+	−	+
2	−	+	−	+	−	+
3	−	−	−	−	−	+
4	−	+	+	+	−	+
5	+	+	+	+	−	+

1. **What is the order of compounds A to E in the pathway?** How do we approach this type of problem? First, let's work out the underlying principle and then draw a diagram. The main points are: a mutation blocks a biosynthetic pathway by canceling out one enzyme needed in the pathway. The mutant therefore lacks one compound needed in the pathway and thus cannot make any of the compounds that come after the blocked step. If we add a compound that is normally synthesized before the block, it will not help matters: the mutant will still be unable to synthesize the rest of the compounds in the pathway. However, the mutant does have the enzymes to make these later compounds; thus, if we add either the blocked compound or any compound that comes after the block, the mutant will grow. So, in our testing of mutants with blocks at different steps in the pathway, the compounds that are used latest in the pathway will support the growth of the most mutants, and compounds that are used earliest in the pathway will allow the growth of the fewest mutants. A look at our table shows us that the compound supporting the fewest mutants is compound E, with no mutants supported, followed by A (one mutant), then by C, B, D, and G, the final product. Now we can construct our diagram.

$$E \longrightarrow A \longrightarrow C \longrightarrow B \longrightarrow D \longrightarrow G$$

2. **At which point in the pathway is each mutant blocked?** Clearly, a mutant that is blocked between E and A cannot be supported by E but can be supported by all the other compounds. Thus, we see that mutant 5 must be blocked in the E−A conversion. We also see that mutant 4 cannot be supported by E or A, so it must be blocked in the A−C conversion. By similar logic, we obtain the order 5−4−2−1−3, which we can insert into the diagram as follows:

$$\overset{5}{E} \longrightarrow \overset{4}{A} \longrightarrow \overset{2}{C} \longrightarrow \overset{1}{B} \longrightarrow \overset{3}{D} \longrightarrow G$$

Now we can comprehend how Beadle and Tatum first worked out this experiment, using one particular biosynthetic pathway of *Neurospora*.

Experiments of Beadle and Tatum

Beadle and Tatum analyzed mutants of *Neurospora*, a fungus with a haploid genome. They first irradiated *Neurospora* to produce mutations and then tested cultures from ascospores for interesting mutant phenotypes. They detected numerous **auxotrophs** — strains that cannot grow on a minimal medium unless the medium is supplemented with one or more specific nutrients. In each case, the mutation that generated the auxotrophic requirement was inherited as a single-gene mutation: each gave a 1:1 ratio when crossed with a wild type. Figure 9-1 depicts the procedure that Beadle and Tatum used.

One set of mutant strains required arginine to grow on a minimal medium. These strains provided the focus for much of Beadle and Tatum's further work. First, they found that the mutations mapped into three different locations on separate chromosomes, even though the same supplement (arginine) satisfied the growth requirement for each mutant. Let's call the three loci the *arg-1*, *arg-2*, and *arg-3* genes. Beadle and Tatum discovered that the auxotrophs for each of the three loci differed in their response to the chemical compounds ornithine and citrulline, which are related to arginine (Figure 9-2). The *arg-1* mutants grew when supplied with ornithine, citrulline, or arginine in addition to the minimal medium. The *arg-2* mutants grew on either arginine or citrulline but not on ornithine. The *arg-3* mutants grew only when arginine was supplied. We can see this more easily by looking at Table 9-2.

It was already known that cellular enzymes often interconvert related compounds such as these. On the basis of the properties of the *arg* mutants, Beadle and Tatum and their colleagues proposed a biochemical model for such conversions in *Neurospora*:

$$\text{precursor} \xrightarrow{\text{enzyme X}} \text{ornithine} \xrightarrow{\text{enzyme Y}}$$
$$\text{citrulline} \xrightarrow{\text{enzyme Z}} \text{arginine}$$

Note how this relation easily explains the three classes of mutants shown in Table 9-2. The *arg-1* mutants have a defective enzyme X, so they are unable to convert the precursor into ornithine as the first step in producing arginine. However, they have normal enzymes Y and Z, and so the *arg-1* mutants are able to produce arginine if supplied with either ornithine or citrulline. The *arg-2* mutants lack enzyme Y, and the *arg-3* mutants lack enzyme Z. Thus, a mutation at a particular gene is assumed to interfere with the production of a single enzyme. The defective enzyme, then, creates a block in some biosynthetic pathway. The block can be circumvented by supplying to the cells any compound that normally comes after the block in the pathway.

9-2 TABLE   Growth of *arg* Mutants in Response to Supplements

| | SUPPLEMENT | | |
Mutant	Ornithine	Citrulline	Arginine
arg-1	+	+	+
arg-2	−	+	+
arg-3	−	−	+

Note: A plus sign means growth; a minus sign means no growth.

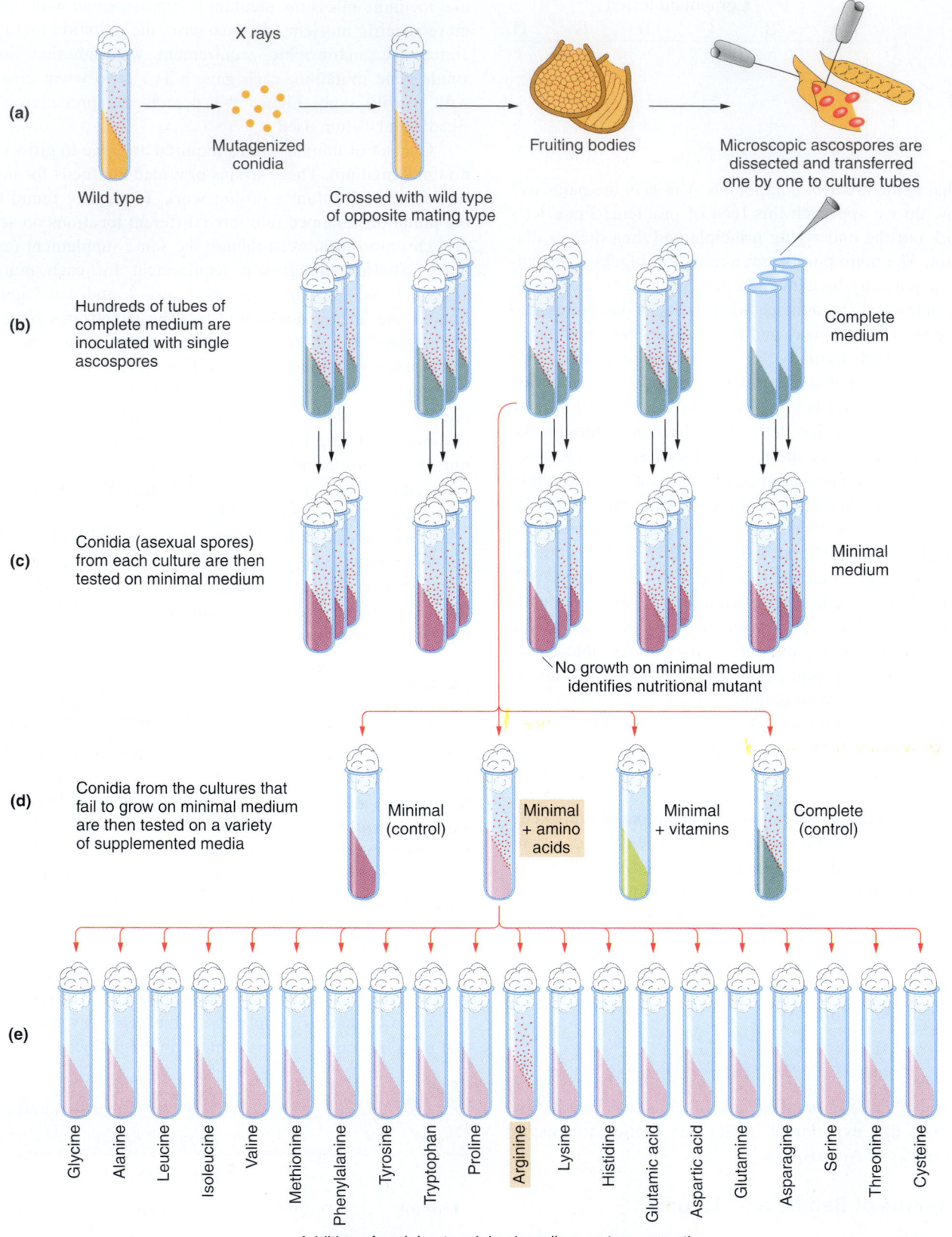

Figure 9-1 The procedure used by Beadle and Tatum. See Chapter 6 for a review of the *Neurospora* life cycle and genetics. (After Peter J. Russell, *Genetics,* 2d ed. Scott, Foresman.)

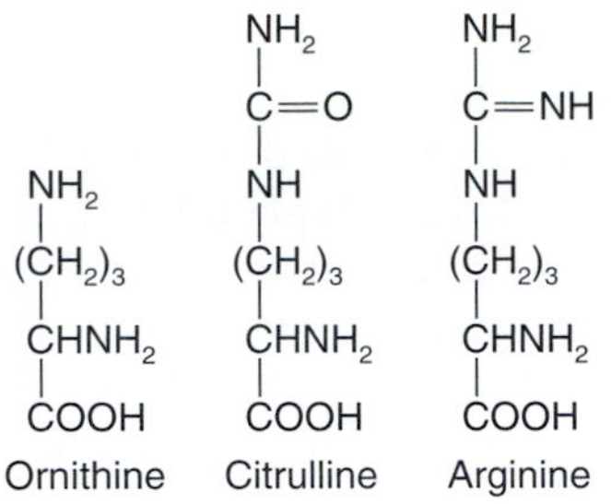

Figure 9-2 Chemical structures of arginine and the related compounds citrulline and ornithine. In the work of Beadle and Tatum, different *arg* auxotrophic mutants of *Neurospora* were found to grow when the medium was supplemented with citrulline or with ornithine as an alternative to arginine.

We can now diagram a more complete biochemical model:

$$\text{precursor} \xrightarrow[\text{enzyme X}]{arg\text{-}1^+} \text{ornithine} \xrightarrow[\text{enzyme Y}]{arg\text{-}2^+}$$

$$\text{citrulline} \xrightarrow[\text{enzyme Z}]{arg\text{-}3^+} \text{arginine}$$

Note that this entire model was inferred from the properties of the mutant classes detected through genetic analysis. Only later were the existence of the biosynthetic pathway and the presence of defective enzymes demonstrated through independent biochemical evidence.

This model, which has become known as the one-gene–one-enzyme hypothesis, was the source of the first exciting insight into the functions of genes: genes somehow were responsible for the function of enzymes, and each gene apparently controlled one specific enzyme. Other researchers obtained similar results for other biosynthetic pathways, and the hypothesis soon achieved general acceptance. It was subsequently refined, as we shall see later in this chapter. Nevertheless, the one-gene–one-enzyme hypothesis became one of the great unifying concepts in biology, because it provided a bridge that brought together the concepts and research techniques of genetics and biochemistry.

MESSAGE ⋯⋯⋯⋯⋯⋯⋯⋯⋯⋯⋯⋯⋯⋯⋯⋯⋯⋯
Genes control biochemical reactions by producing enzymes.

We should pause to ponder the significance of this discovery. Let's summarize what it established:

1. Biochemical reactions in vivo (in the living cell) consist of a series of discrete, step-by-step reactions.

2. Each reaction is specifically catalyzed by a single enzyme.

3. Each enzyme is specified by a single gene.

Gene-protein relations

The one-gene–one-enzyme hypothesis was an impressive step forward in our understanding of gene function, but just *how* do genes control the functioning of enzymes? Virtually all enzymes are proteins, and thus we must review the basic facts of protein structure to follow the next step in the study of gene function.

Protein structure

In simple terms, a **protein** is a macromolecule composed of **amino acids** attached end to end in a linear string. The general formula for an amino acid is $H_2N-CHR-COOH$, in which the side chain, or R (reactive) group, can be anything from a hydrogen atom (as in the amino acid glycine) to a complex ring (as in the amino acid tryptophan). There are 20 common amino acids in living organisms (Table 9-3), each having a different R group. Amino acids are linked together in proteins by covalent (chemical) bonds called peptide bonds. A **peptide bond** is formed through a

9-3 TABLE	The 20 Amino Acids Common in Living Organisms					
	ABBREVIATION				**ABBREVIATION**	
Amino acid	3-letter	1-letter		Amino acid	3-letter	1-letter
Alanine	Ala	A		Leucine	Leu	L
Arginine	Arg	R		Lysine	Lys	K
Asparagine	Asn	N		Methionine	Met	M
Aspartic acid	Asp	D		Phenylalanine	Phe	F
Cysteine	Cys	C		Proline	Pro	P
Glutamic acid	Glu	E		Serine	Ser	S
Glutamine	Gln	Q		Threonine	Thr	T
Glycine	Gly	G		Tryptophan	Trp	W
Histidine	His	H		Tyrosine	Tyr	Y
Isoleucine	Ile	I		Valine	Val	V

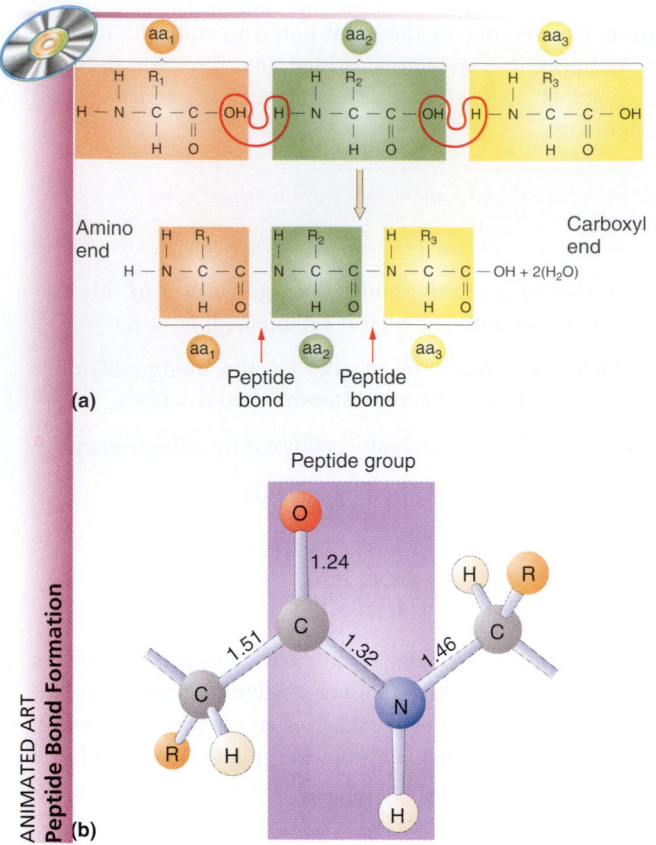

condensation reaction that includes the removal of a water molecule (Figure 9-3).

Several amino acids linked together by peptide bonds form a molecule called a **polypeptide;** the proteins found in living organisms are large polypeptides. For instance, the α chain of human hemoglobin contains 141 amino acids, and some proteins consist of more than 1000 amino acids.

The properties of the amino acid side chains are responsible for the structure and function of each protein. These side chains vary in many chemical properties. A key property is the hydrophobic, or water repelling, character of each amino acid. Hydrophobic amino acids tend to avoid contact with water and are often turned toward the inside of the protein, whereas the amino acids that are charged or that can form hydrogen bonds with water are excluded from the interior of the protein and are turned toward the exterior or surface of the protein.

Figure 9-3   The peptide bond. (a) A polypeptide is formed by the removal of water between amino acids to form peptide bonds. Each aa indicates an amino acid. R_1, R_2, and R_3 represent R groups (side chains) that differentiate the amino acids. R can be anything from a hydrogen atom (as in glycine) to a complex ring (as in tryptophan). (b) The peptide group is a rigid planar unit with the R groups projecting out from the C—N backbone. Standard bond distances (in angstroms) are shown. (Part b from L. Stryer, *Biochemistry*, 4th ed. Copyright © 1995 by Lubert Stryer.)

(a)

								10										20										30	
Met-Glu-Arg-Tyr-Glu-Ser-Leu-Phe-Ala-Gln-Leu-Lys-Glu-Arg-Lys-Glu-Gly-Ala-Phe-Val-Pro-Phe-Val-Thr-Leu-Gly-Asp-Pro-Gly-Ile -

40 ... 50 ... 60
Glu-Gln-Ser-Leu-Lys-Ile-Ile-Asp-Thr-Leu-Ile-Glu-Ala-Gly-Ala-Asp-Ala-Leu-Glu-Leu-Gly-Ile-Pro-Phe-Ser-Asp-Pro-Leu-Ala-Asp-

70 ... 80 ... 90
Gly-Pro-Thr-Ile-Gln-Asn-Ala-Thr-Leu-Arg-Ala-Phe-Ala-Ala-Gly-Val-Thr-Pro-Ala-Gln-Cys-Phe-Glu-Met-Leu-Ala-Leu-Ile-Arg-Gln-

100 ... 110 ... 120
Lys-His-Pro-Thr-Ile-Pro-Ile-Gly-Leu-Leu-Met-Tyr-Ala-Asn-Leu-Val-Phe-Asn-Lys-Gly-Ile-Asp-Glu-Phe-Tyr-Ala-Gln-Cys-Glu-Lys-

130 ... 140 ... 150
Val-Gly-Val-Asp-Ser-Val-Leu-Val-Ala-Asp-Val-Pro-Val-Gln-Glu-Ser-Ala-Pro-Phe-Arg-Gln-Ala-Ala-Leu-Arg-His-Asn-Val-Ala-Pro-

160 ... 170 ... 180
Ile-Phe-Ile-Cys-Pro-Pro-Asn-Ala-Asp-Asp-Asp-Leu-Leu-Arg-Gln-Ile-Ala-Ser-Tyr-Gly-Arg-Gly-Tyr-Thr-Tyr-Leu-Leu-Ser-Arg-Ala-

190 ... 200 ... 210
Gly-Val-Thr-Gly-Ala-Glu-Asn-Arg-Ala-Ala-Leu-Pro-Leu-Asn-His-Leu-Val-Ala-Lys-Leu-Lys-Glu-Tyr-Asn-Ala-Ala-Pro-Pro-Leu-Gln-

220 ... 230 ... 240
Gly-Phe-Gly-Ile-Ser-Ala-Pro-Asp-Gln-Val-Lys-Ala-Ala-Ile-Asp-Ala-Gly-Ala-Ala-Gly-Ala-Ile-Ser-Gly-Ser-Ala-Ile-Val-Lys-Ile-

250 ... 260 ... 268
Ile-Glu-Gln-His-Asn-Ile-Glu-Pro-Glu-Lys-Met-Leu-Ala-Ala-Leu-Lys-Val-Phe-Val-Gln-Pro-Met-Lys-Ala-Ala-Thr-Arg-Ser

(b)

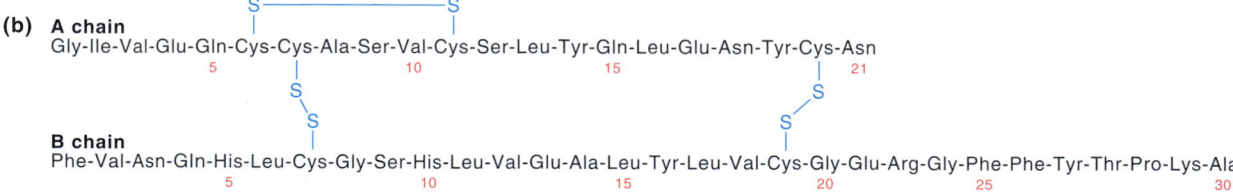

A chain
```
      S————————S
      |         |
Gly-Ile-Val-Glu-Gln-Cys-Cys-Ala-Ser-Val-Cys-Ser-Leu-Tyr-Gln-Leu-Glu-Asn-Tyr-Cys-Asn
          5              10              15                      21
              S                                     S
              |                                     |
              S                                     S
**B chain**
Phe-Val-Asn-Gln-His-Leu-Cys-Gly-Ser-His-Leu-Val-Glu-Ala-Leu-Tyr-Leu-Val-Cys-Gly-Glu-Arg-Gly-Phe-Phe-Tyr-Thr-Pro-Lys-Ala
          5              10              15              20              25              30
```

Figure 9-4   Linear sequences of two proteins. (a) The *E. coli* tryptophan synthetase A protein, 268 amino acids long. (b) Bovine insulin protein. Note that the amino acid cysteine can form unique "sulfur bridges," because it contains sulfur.

Proteins have a complex structure that is traditionally thought of as having four levels. The linear sequence of the amino acids in a polypeptide chain is called the **primary structure** of the protein. Figure 9-4 shows the linear sequence of tryptophan synthetase (an enzyme) and beef insulin (a hormonal protein).

The **secondary structure** of a protein refers to the interrelations of amino acids that are close together in the linear sequence. This spatial arrangement often results from the fact that polypeptides can bend into regularly repeating (periodic) structures, created by hydrogen bonds between the CO and NH groups of different residues. Two of the basic periodic structures are the α helix (Figure 9-5) and the β pleated sheet (Figure 9-6).

A protein also has a three-dimensional architecture, termed the **tertiary structure,** which is created by electrostatic, hydrogen, and Van der Waals bonds that form between the various amino acid R groups, causing the protein chain to fold back on itself. In many cases, amino acids that are far apart in the linear sequence are brought close together in the tertiary structure. Often, two or more folded structures will bind together to form a **quaternary structure;** this structure is **multimeric** because it is composed of several separate polypeptide chains, or monomers.

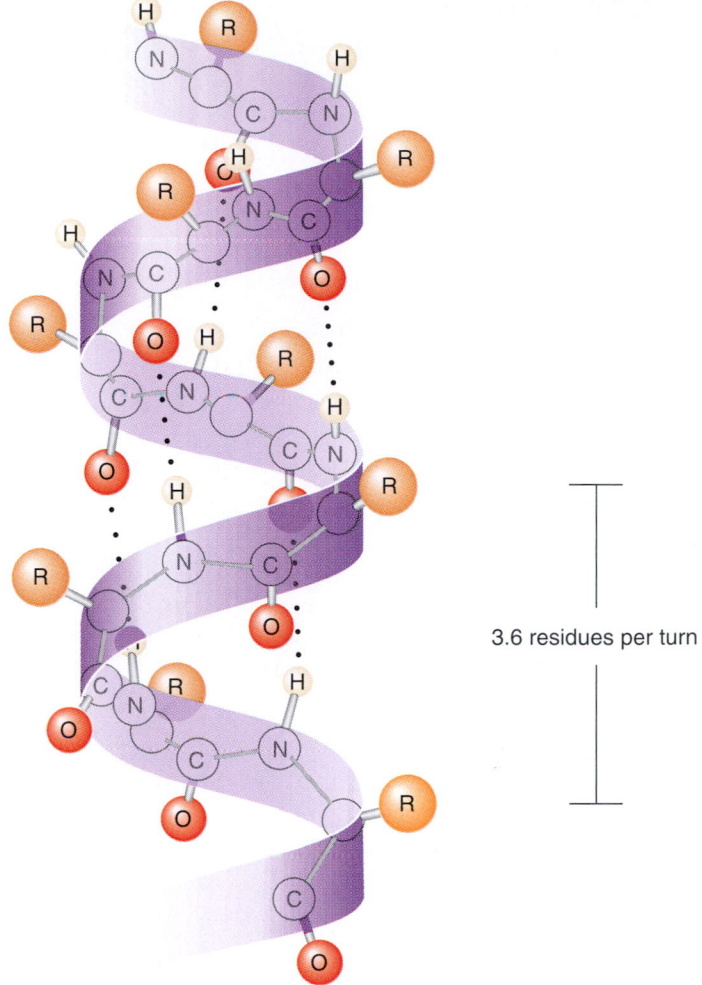

3.6 residues per turn

Figure 9-5   The α helix, a common basis of secondary protein structure. Each R is a specific side chain on one amino acid. The black dots represent weak hydrogen bonds that bond the CO group of residue n to the NH group of residue $n + 4$, thereby stabilizing the helical shape. (From H. Lodish, D. Baltimore, A. Berk, S. L. Zipursky, P. Matsudaira, and J. Darnell, *Molecular Cell Biology,* 3d ed. Copyright © 1995 by Scientific American Books, Inc.)

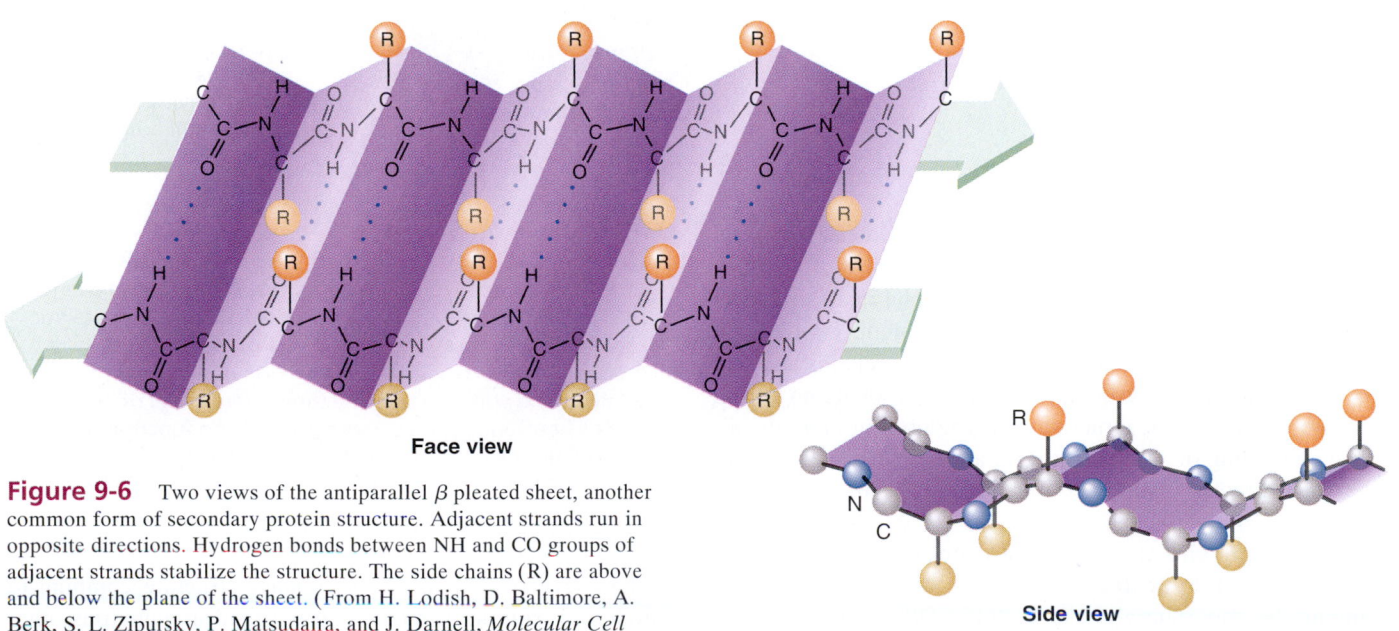

Face view

Side view

Figure 9-6   Two views of the antiparallel β pleated sheet, another common form of secondary protein structure. Adjacent strands run in opposite directions. Hydrogen bonds between NH and CO groups of adjacent strands stabilize the structure. The side chains (R) are above and below the plane of the sheet. (From H. Lodish, D. Baltimore, A. Berk, S. L. Zipursky, P. Matsudaira, and J. Darnell, *Molecular Cell Biology,* 3d ed. Copyright ©1995 by Scientific American Books, Inc.)

(a) Primary structure

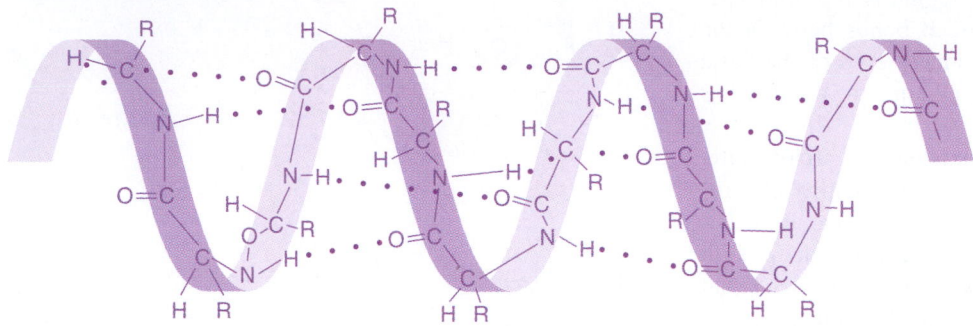

(b) Secondary structure

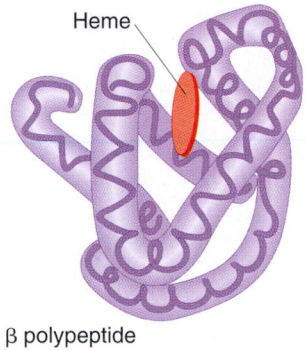

Hydrogen bonds between amino acids
at different locations in polypeptide chain

(c) Tertiary structure

Heme

β polypeptide

(d) Quaternary structure

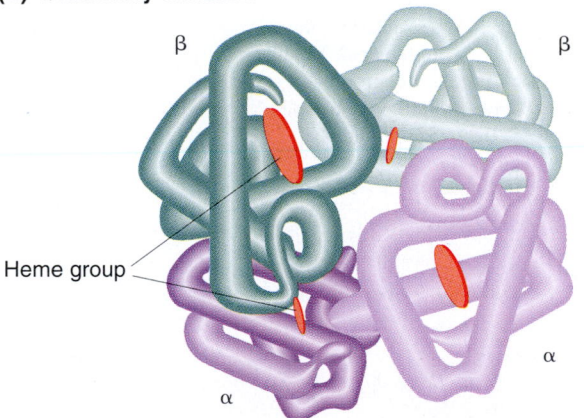

Heme group

Figure 9-7   Different
levels of protein structure.
(a) Primary structure.
(b) Secondary structure. The
polypeptide shown in part a
is drawn into an α helix by
hydrogen bonds. (c) Tertiary
structure: the three-
dimensional structure of
myoglobin. (d) Quaternary
structure: the arrangement of
two α subunits and two β
subunits to form the
complete quaternary
structure of hemoglobin.

Figure 9-7 depicts the four levels of protein structure. In Figure 9-7c, we can see the tertiary structure of myoglobin. Note how the α helix is folded back on itself to generate the three-dimensional shape of the protein. Figure 9-7d shows the combining of four subunits (two α chains and two β chains) to form the quaternary structure of hemoglobin. Figure 9-8 shows the structure of myoglobin in more detail. The combining of subunits to form a multimeric enzyme can be seen directly in the electron microscope in some cases (Figure 9-9).

Many proteins are compact structures; such proteins are called **globular proteins.** Enzymes and antibodies are among the important globular proteins. Other, unfolded proteins, called **fibrous proteins,** are important components of such structures as hair and muscle.

MESSAGE ·····································

The linear sequence of a protein folds up to yield a unique three-dimensional configuration. This configuration creates specific sites to which substrates bind and at which catalytic reactions take place. The three-dimensional structure of a protein, which is crucial for its function, is determined solely by the primary structure (linear sequence) of amino acids. Therefore, genes can control enzyme function by controlling the primary structure of proteins.

Protein motifs

Often, several elements of secondary structure combine to produce a pattern, or motif, that is found in numerous other proteins. We can recognize motifs sometimes by their amino acid sequence pattern and other times by observing the

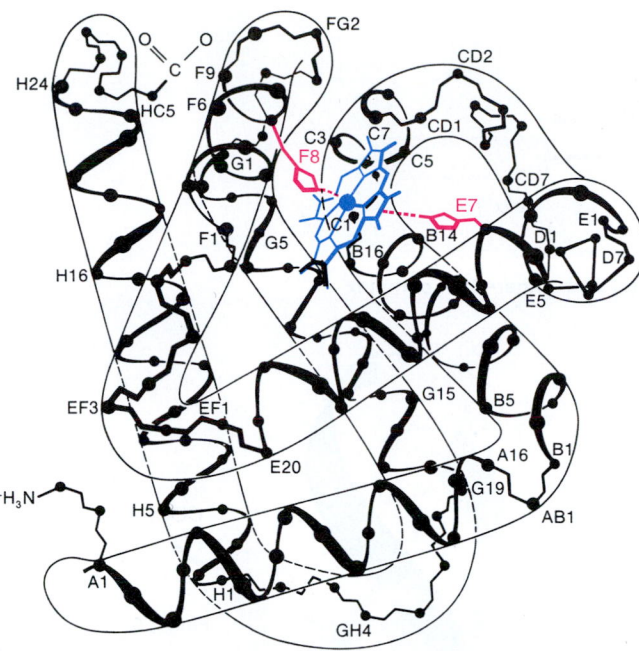

Figure 9-8 Folded tertiary structure of myoglobin, an oxygen-storage protein. Each dot represents an amino acid. The heme group, a cofactor that facilitates the binding of oxygen, is shown in blue. (From L. Stryer, *Biochemistry,* 4th ed. Copyright © 1995 by Lubert Stryer. Based on R. E. Dickerson, *The Proteins,* 2d ed., vol. 2. Edited by H. Neurath. Copyright ©1964 by Academic Press.)

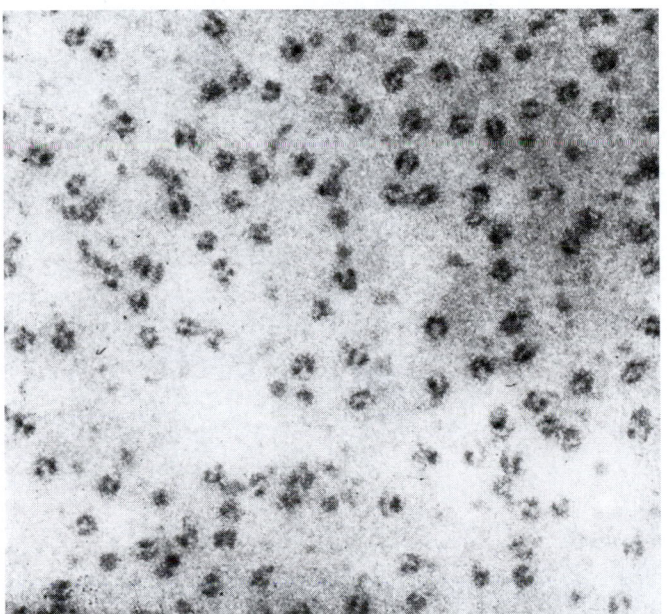

Figure 9-9 Electron micrograph of the enzyme aspartate transcarbamylase. Each small "glob" is an enzyme molecule. Note the quaternary structure: the enzyme is composed of subunits. (Photograph from Jack D. Griffith.)

three-dimensional structure. Figure 9-10 shows two examples. The helix-loop-helix motif is found in calcium binding proteins, and a variant of it is found in regulatory proteins

that bind DNA. The zinc-binding motif, also found in DNA binding proteins, is termed the zinc finger, because of the way that the residues protrude outward, like a finger.

Determining protein sequence

If we purify a particular protein, we find that we can specify a particular ratio of the various amino acids that make up that specific protein. But the protein is not formed by a random hookup of fixed amounts of the various amino acids; each protein has a unique, characteristic sequence. For a small polypeptide, the amino acid sequence can be determined by clipping off one amino acid at a time and identifying it. However, large polypeptides cannot be readily "sequenced" in this way.

Frederick Sanger worked out a brilliant method for deducing the sequence of large polypeptides. There are several different **proteolytic enzymes**—enzymes that can break peptide bonds only between specific amino acids in proteins. Proteolytic enzymes can break a large protein into a number of smaller fragments, which can then be separated according to their migration speeds in a solvent on chromatographic paper. Because different fragments will move at different speeds in various solvents, **two-dimensional chromatography** can be used to enhance the separation of the fragments (Figure 9-11). In this technique, a mixture of fragments is separated in one solvent; then the paper is turned 90° and another solvent is used. When the paper is stained, the polypeptides appear as spots in a characteristic chromatographic pattern called the **fingerprint** of the protein. Each of the spots can be cut out, and the polypeptide fragments can be washed from the paper. Because each spot contains only small polypeptides, their amino acid sequences can be easily determined.

Using different proteolytic enzymes to cleave the protein at different points, we can repeat the experiment to obtain other sets of fragments. The fragments from the different treatments overlap, because the breaks are made in different places with each treatment. The problem of solving the overall sequence then becomes one of fitting together the small-fragment sequences—almost like solving a tricky jigsaw or crossword puzzle (Figure 9-12).

Using this elegant technique, Sanger confirmed that the sequence of amino acids (as well as the amounts of the various amino acids) is specific to a particular protein. In other words, the amino acid sequence is what makes insulin insulin.

Relation between gene mutations and altered proteins

We now know that the change of just one amino acid is sometimes enough to alter protein function. This was first shown in 1957 by Vernon Ingram, who studied the globular protein hemoglobin—the molecule that transports oxygen in red blood cells. As shown in Figure 9-7d, hemoglobin is

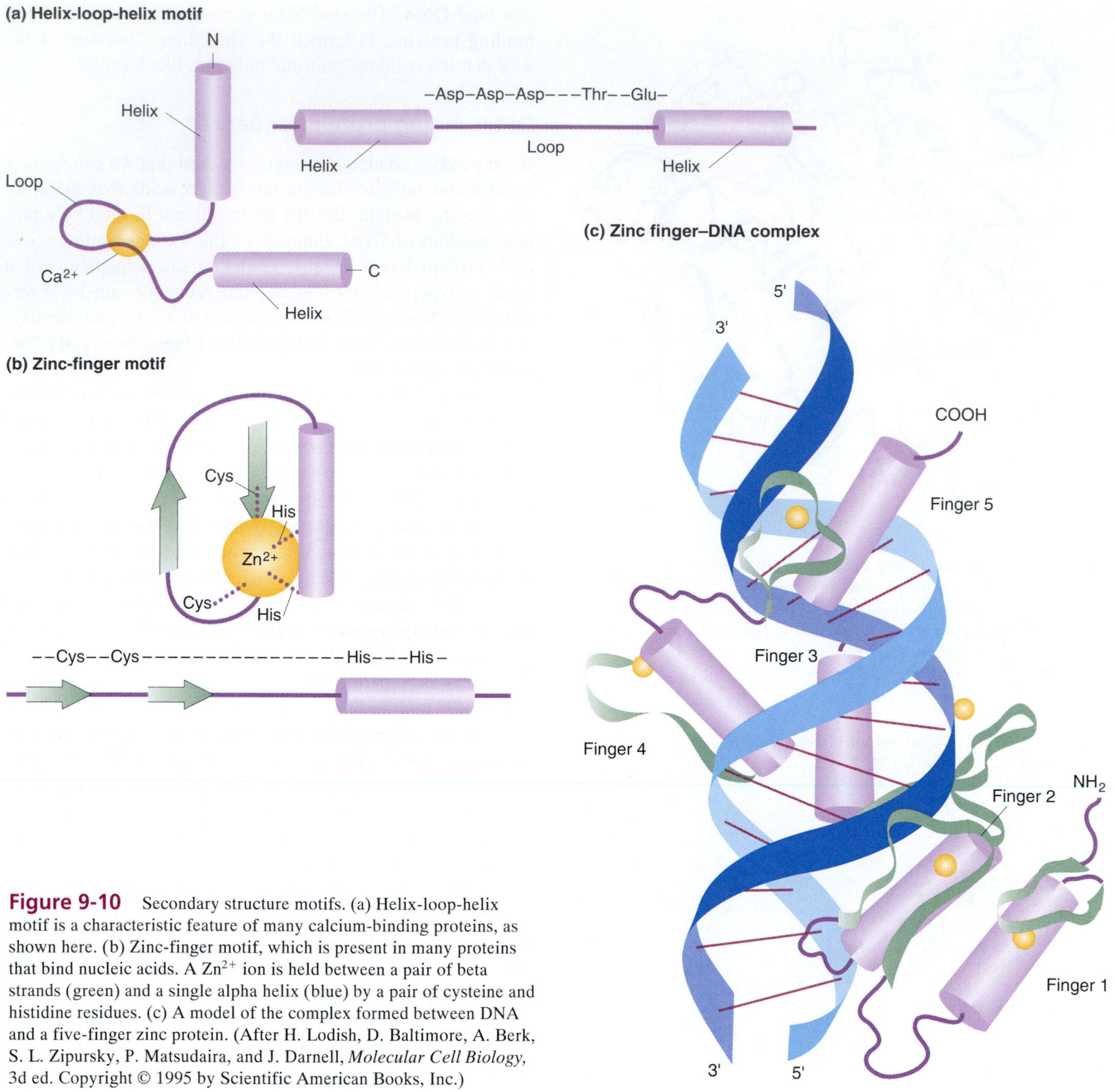

(a) Helix-loop-helix motif

N

Helix

Helix

Helix

—Asp–Asp–Asp———Thr——Glu—

Loop

Helix

Loop

Ca²⁺

Helix

C

(b) Zinc-finger motif

Cys

His

Zn²⁺

Cys

His

—–Cys—–Cys———————————His–––His–

(c) Zinc finger–DNA complex

5'

3'

COOH

Finger 5

Finger 3

Finger 4

Finger 2

NH₂

Finger 1

3'   5'

Figure 9-10   Secondary structure motifs. (a) Helix-loop-helix motif is a characteristic feature of many calcium-binding proteins, as shown here. (b) Zinc-finger motif, which is present in many proteins that bind nucleic acids. A Zn²⁺ ion is held between a pair of beta strands (green) and a single alpha helix (blue) by a pair of cysteine and histidine residues. (c) A model of the complex formed between DNA and a five-finger zinc protein. (After H. Lodish, D. Baltimore, A. Berk, S. L. Zipursky, P. Matsudaira, and J. Darnell, *Molecular Cell Biology,* 3d ed. Copyright © 1995 by Scientific American Books, Inc.)

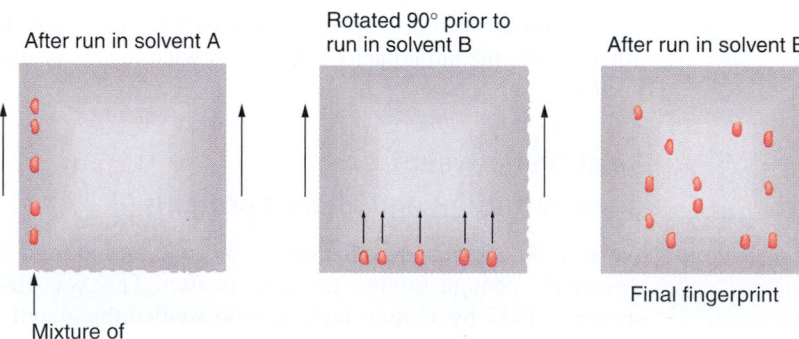

After run in solvent A

Rotated 90° prior to run in solvent B

After run in solvent B

Final fingerprint

Mixture of fragments spotted here

Figure 9-11   Two-dimensional chromatographic fingerprinting of a polypeptide fragment mixture. A protein is digested by a proteolytic enzyme into fragments that are only a few amino acids long. A piece of chromatographic filter paper is then spotted with this mixture and dipped into solvent A. As solvent A ascends the paper, some of the fragments become separated. The paper is then turned 90° and further resolution of the fragments is obtained as solvent B ascends.

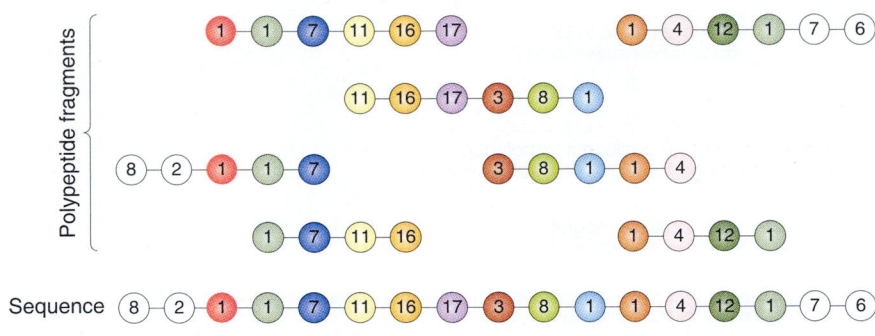

Figure 9-12 Alignment of polypeptide fragments to reconstruct an entire amino acid sequence. Different proteolytic enzymes can be used on the same protein to form different fingerprints, as shown here. The amino acid sequence of each fragment can be determined rather easily and, owing to the overlap of amino acid sequences from different fingerprints, the entire amino acid sequence of the original protein can be determined. Using this procedure, Sanger took about six years to determine the sequence of the insulin molecule, a relatively small protein.

made up of four polypeptide chains: two identical α chains, each containing 141 amino acids, and two identical β chains, each containing 146 amino acids.

Ingram compared hemoglobin A (HbA), the hemoglobin from normal adults, with hemoglobin S (HbS), the protein from people homozygous for the mutant gene that causes sickle-cell anemia, the disease in which red blood cells take on a sickle-cell shape (see Figure 4-2). Using Sanger's technique, Ingram found that the fingerprint of HbS differs from that of HbA in only one spot. Sequencing that spot from the two kinds of hemoglobin, Ingram found that only one amino acid in the fragment differs in the two kinds. Apparently, of all the amino acids known to make up a hemoglobin molecule, a substitution of valine for glutamic acid at just one point, position 6 in the β chain, is all that is needed to produce the defective hemoglobin (Figure 9-13). Unless patients with HbS receive medical attention, this single error in one amino acid in one protein will hasten their

death. Figure 9-14 shows how this gene mutation ultimately leads to the pattern of sickle-cell disease.

Notice what Ingram accomplished. A gene mutation that had been well established through genetic studies was connected with an altered amino acid sequence in a protein. Subsequent studies identified numerous changes in hemoglobin, and each one is the consequence of a single amino acid difference. (Figure 9-15 shows a few examples.) We can conclude that one mutation in a gene corresponds to a change of one amino acid in the sequence of a protein.

MESSAGE ·····························
Genes determine the specific primary sequences of amino acids in specific proteins.

Colinearity of gene and protein

Once the structure of DNA had been determined by Watson and Crick, it became apparent that the structure of proteins must be encoded in the linear sequence of nucleotides in the

Normal hemoglobin

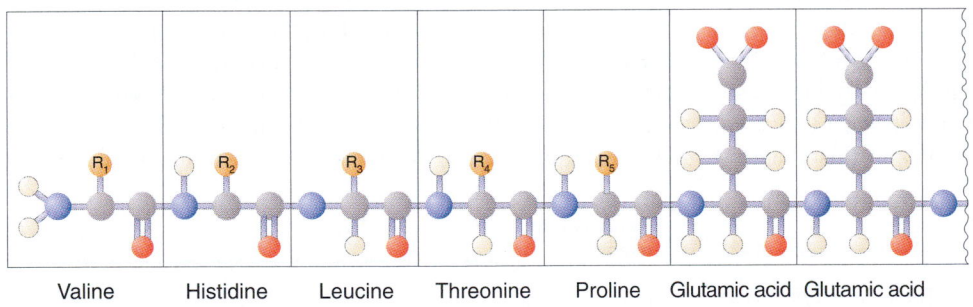

Hemoglobin S

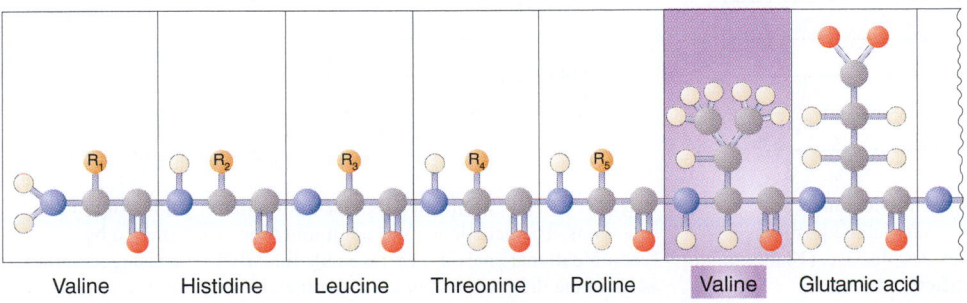

Figure 9-13 The difference at the molecular level between normalcy and sickle-cell disease. Shown are only the first seven amino acids; all the rest not shown are identical. (From Anthony Cerami and Charles M. Peterson, "Cyanate and Sickle-Cell Disease." Copyright ©1975 by Scientific American, Inc. All rights reserved.)

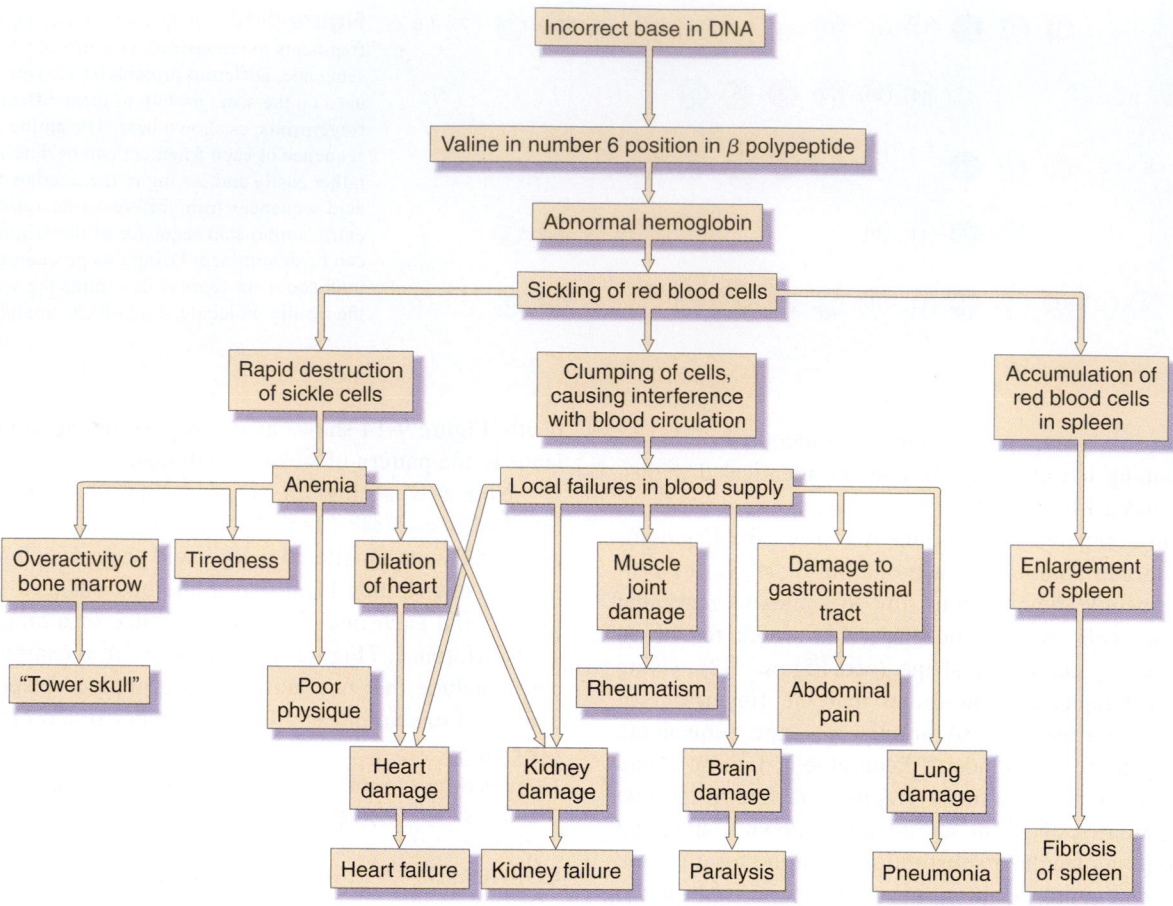

Figure 9-14   The compounded consequences of one amino acid substitution in hemoglobin to produce sickle-cell anemia.

DNA. (We shall see in Chapter 10 how this genetic code was deciphered.) After Ingram's demonstration that one mutation alters one amino acid in a protein, a relation was sought between the linear sequence of mutant sites in a gene and the linear sequence of amino acids in a protein. (It is possible to map mutational sites within a gene by high-resolution recombination analysis, as we saw in the *rII* system, described in Chapter 7 and expanded later in this chapter.)

Charles Yanofsky probed the relation between altered genes and altered proteins by studying the enzyme tryptophan synthetase in *E. coli*. This enzyme catalyzes the conversion of indole glycerol phosphate into tryptophan. Two genes, *trpA* and *trpB*, control the enzyme. Each gene controls a separate polypeptide; after the A and B polypeptides are produced, they combine to form the active enzyme (a multimeric protein). Yanofsky analyzed mutations in the

Amino acid position	- 2-----16---------43--------67------87------------132----143
Normal	His----Gly------Glu-------Val------Thr----------Lys----His
Tokuchi	Tyr———————————————————————
Baltimore	——Asp———————————————————
Galveston	————————Ala—————————————
Milwaukee	————————————Glu—————————
Woolwich	———————————————————Gln———
Kenwood	——————————————————————————Asp

Figure 9-15   Some single amino acid substitutions found in human hemoglobin. Amino acids are normal at all residue positions except those indicated. Each type of change causes disease. (Names indicate areas in which cases were first identified.)

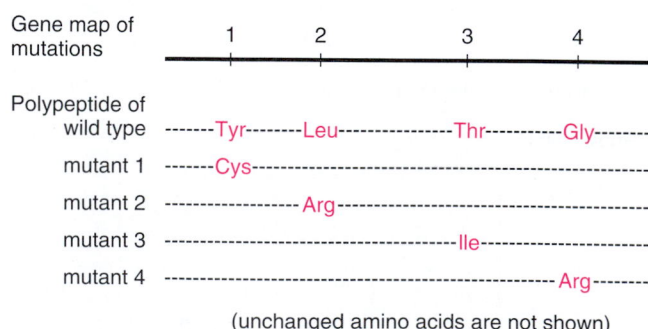

Gene map of mutations	1	2	3	4
Polypeptide of wild type	------Tyr--------Leu----------------Thr----------Gly--------
mutant 1	------Cys--
mutant 2	------------------Arg--
mutant 3	--Ile-----------------
mutant 4	---Arg-------

(unchanged amino acids are not shown)

Figure 9-16   Simplified representation of the colinearity of gene mutations. The genetic map of point mutations (determined by recombinational analysis) corresponds linearly to the changed amino acids in the different mutants (determined by fingerprint analysis).

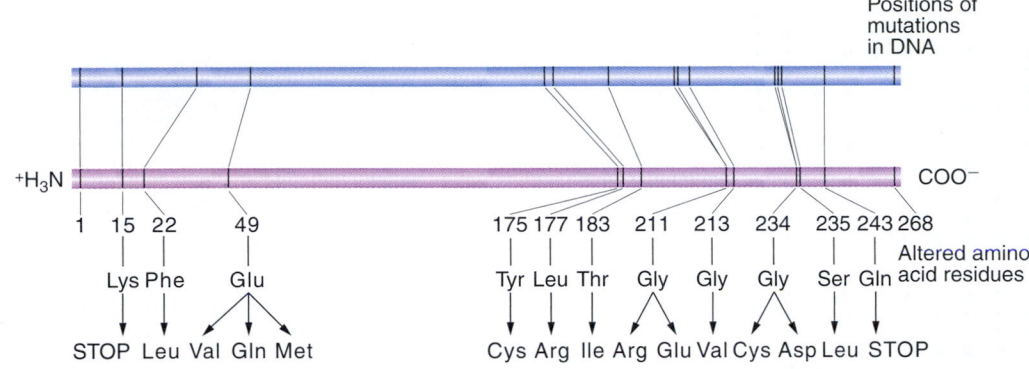

trpA gene that resulted in alterations of the tryptophan synthetase A subunit. He produced a detailed map of the mutations, and he then determined the amino acid sequence of each respective altered tryptophan synthetase. His results were similar to Ingram's for hemoglobin: each mutant had a defective polypeptide associated with a specific amino acid substitution at a specific point. However, Yanofsky was able to show an exciting correlation that Ingram was not able to observe, owing to the limitations of his system. Yanofsky found an exact match between the sequence of the mutational sites in the gene map of the *trpA* gene and the location of the corresponding altered amino acids in the A polypeptide chain. The farther apart two mutational sites were in map units, the more amino acids there were between the corresponding substitutions in the polypeptide (Figure 9-16). Thus, Yanofsky demonstrated **colinearity** — the correspondence between the linear sequence of the gene and that of the polypeptide. Figure 9-17 shows the complete set of data.

> **MESSAGE**
> The linear sequence of nucleotides in a gene determines the linear sequence of amino acids in a protein.

X-ray determination of three-dimensional structure of proteins

X-ray crystallography is a powerful method for determining the three-dimensional structure of a protein in atomic detail. John Kendrew first applied this method to a molecule as complex as a protein to elucidate the structure of myoglobin in 1957, and Max Perutz succeeded in unraveling the complexities of hemoglobin several years later. Now, the structures of hundreds of proteins are known. In this technique, crystals of the protein are obtained in a concentrated salt solution of the pure protein. Then, a narrow beam of X rays is passed through the crystal. The repeating pattern of atoms in the protein complex scatters (or diffracts) the X-ray beams, giving a pattern of spots on X-ray film, as depicted in Figure 9-18a (left); see also Figure 9-18b. Information about the electron density in different parts of the protein is contained in the position and intensity of each spot.

Sophisticated mathematical analysis is used to generate electron-density maps (Figure 9-17a; right), which are in turn used to derive contour maps of the protein. Ultimately, detailed models of a protein can be built, as shown in the space-filling model of chymotrypsin (Figure 9-17c). Amazingly, this model stems from the pattern of spots seen in Figure 9-17b.

Enzyme function

How can a single amino acid substitution, such as that in sickle-cell hemoglobin (Figure 9-13), have such an enormous effect on protein function and the phenotype of an organism? Take enzymes, for example. Enzymes are known to do their job of catalysis by physically grappling with their substrate molecules, twisting or bending the molecules to make or break chemical bonds. Figure 9-19 shows the gastric digestion enzyme carboxypeptidase in its relaxed position and after grappling with its substrate molecule, glycyltyrosine. The substrate molecule fits into a notch in the enzyme structure; this notch is called the **active site.**

Figure 9-20 diagrams the general concept. Note that there are two basic types of reactions performed by enzymes: (1) the breakdown of a substrate into simpler products and (2) the synthesis of a complex product from one or more simpler substrates.

Much of the globular structure of an enzyme is nonreactive material that simply supports the active site. So we might expect that amino acid substitutions throughout most of the structure would have little effect, whereas very specific amino acids would be required for the part of the enzyme molecule that gives the precise shape to the active site. Hence, the possibility arises that a functional enzyme does not always require a unique amino acid sequence for the *entire* polypeptide. This possibility has proved to be the case: in a number of systems, numerous positions in a polypeptide can be filled by several alternative amino acids, and enzyme function is retained. But, at certain other positions in the polypeptide, only the wild-type amino acid will preserve activity; in all likelihood, these amino acids form critical parts of the active sites. Some of these critical amino acids in carboxypeptidase are indicated in red in Figure 9-19.

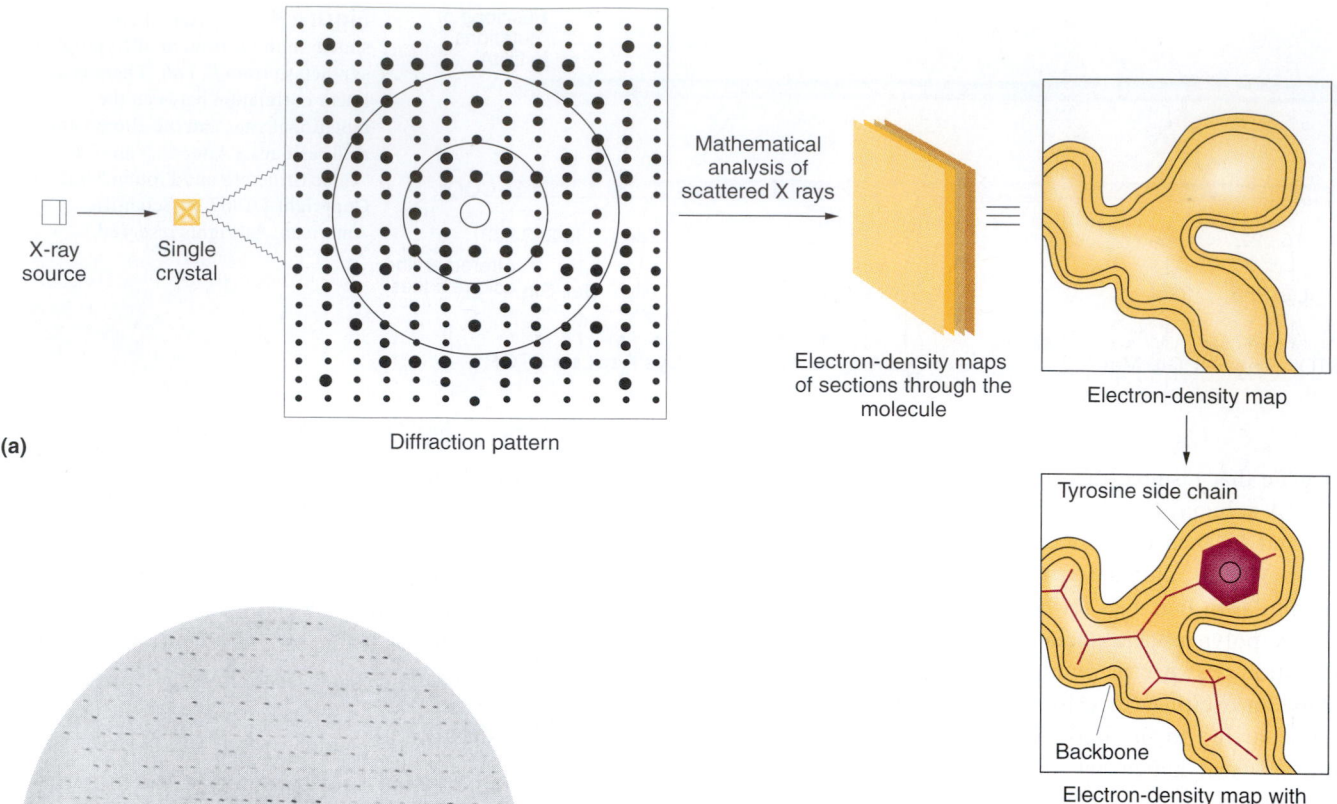

(a)   Diffraction pattern

Mathematical analysis of scattered X rays

Electron-density maps of sections through the molecule

Electron-density map

Tyrosine side chain

Backbone

Electron-density map with inferred polypeptide backbone and tyrosine side chain fitted in

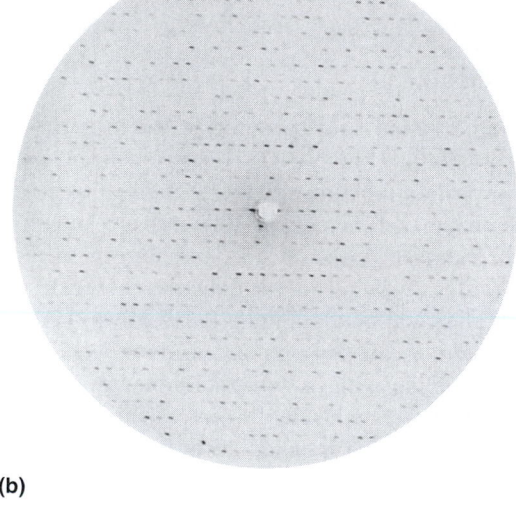

(b)

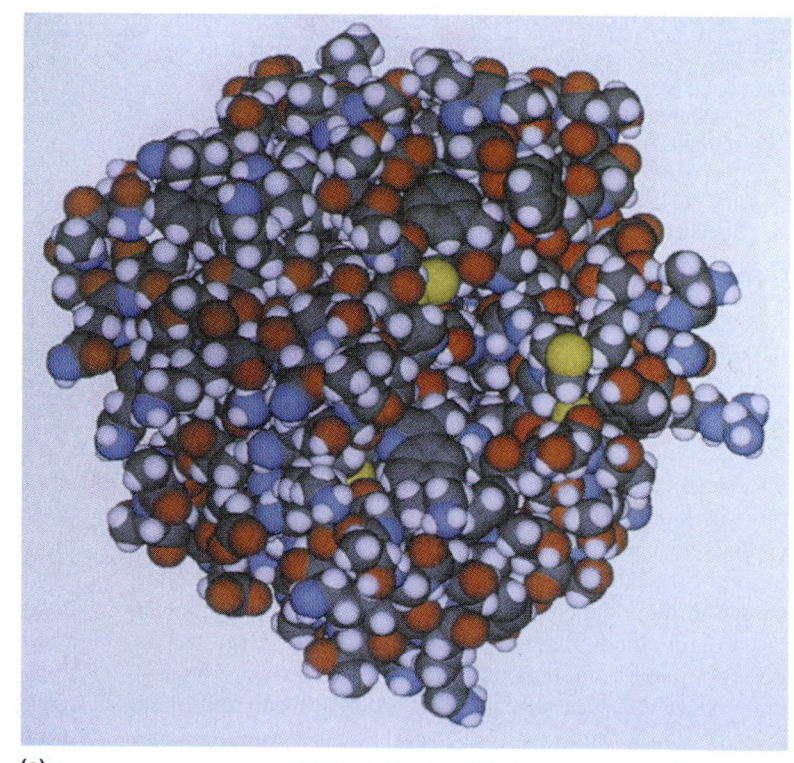

(c)

Figure 9-18   The use of X rays to determine the structure of enzymes. (a) A beam of X rays is passed through a crystal, and the pattern of spots on an X-ray film is used to generate electron-density maps and contour maps. (b) The X-ray diffraction pattern given by the enzyme chymotrypsin. (c) The three-dimensional structure of chymotrypsin. Carbon atoms are shown in black, nitrogen in blue, oxygen in red, hydrogen in white, and sulfur in yellow. The diameter of the enzyme is about 45 Å (somewhat less than a millionth of an inch). The hydrophilic side chain of arginine-145 is clearly visible projecting outward from the right side of the molecule. The ridges and grooves on the surface of the chymotrypsin molecule are as unique as the mountains and craters of the moon, and herein lies the fulfillment of the lock-and-key hypothesis. (From D. Dressler and H. Potter, *Discovering Enzymes.* Copyright © 1991 by Scientific American Library.)

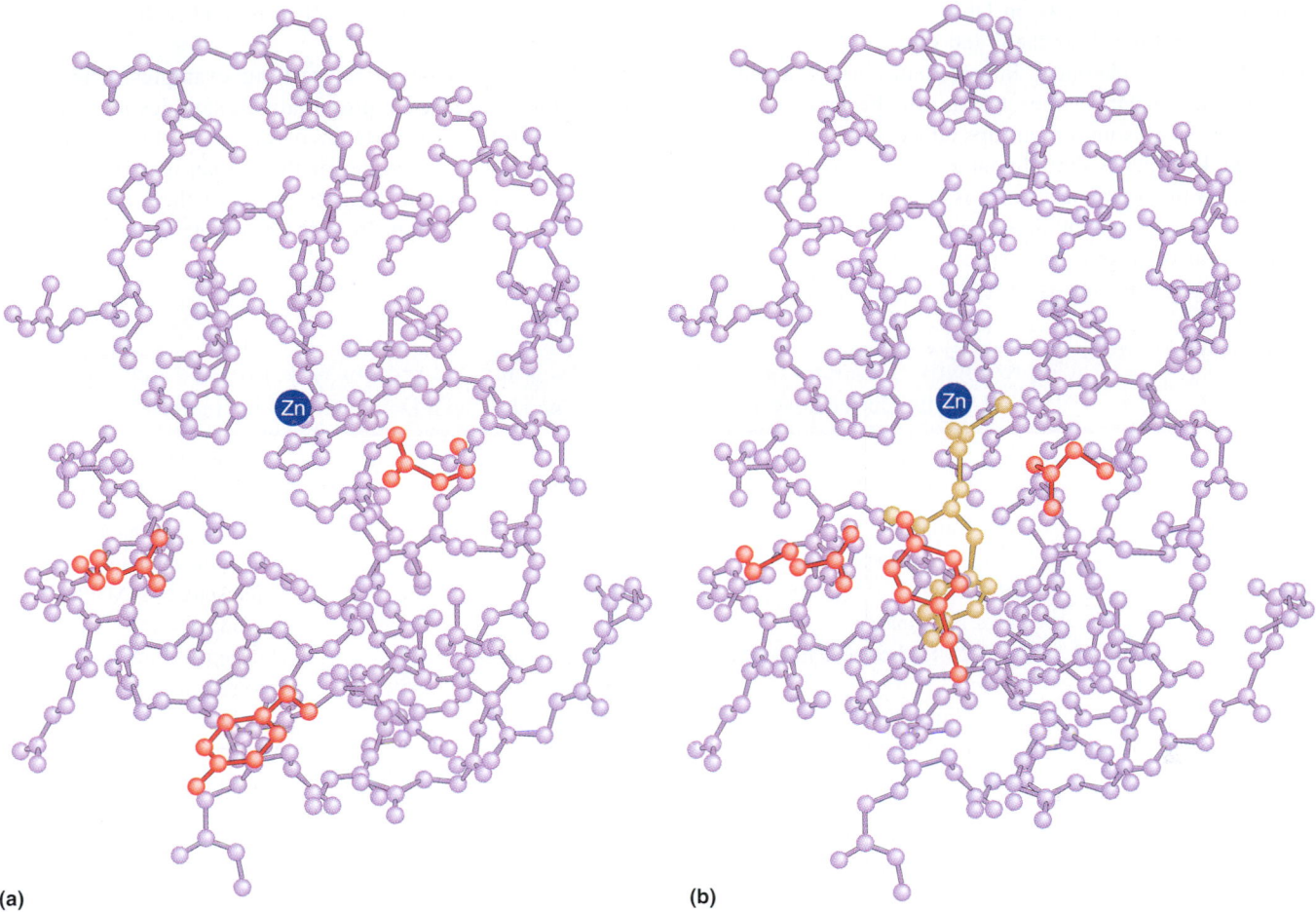

(a)

(b)

Figure 9-19 The active site of the digestive enzyme carboxypeptidase. (a) The enzyme without substrate. (b) The enzyme with its substrate (gold) in position. Three crucial amino acids (red) have changed positions to move closer to the substrate. Carboxypeptidase carves up proteins in the diet. (From W. N. Lipscomb, *Proceedings of the Robert A. Welch Foundation Conferences on Chemical Research* 15, 1971, 140–141.)

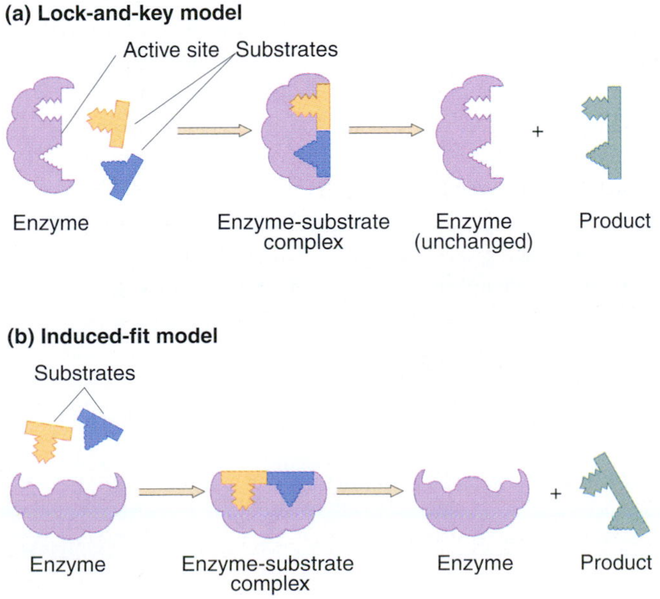

(a) Lock-and-key model

Active site   Substrates

Enzyme   Enzyme-substrate complex   Enzyme (unchanged)   Product

(b) Induced-fit model

Substrates

Enzyme   Enzyme-substrate complex   Enzyme   Product

MESSAGE ···

Protein architecture is the key to gene function. A gene mutation typically results in the substitution of a different amino acid into the polypeptide sequence of a protein. The new amino acid may have chemical properties that are incompatible with the proper protein architecture at that particular position; in such a case, the mutation will lead to a nonfunctional protein.

Genes and cellular metabolism: genetic diseases

When we think of enzyme activity in relation to cellular metabolism, we realize that the inactivation of one or more enzymes can have staggering consequences. Most of us have

Figure 9-20 Schematic representation of the action of a hypothetical enzyme in putting two substrate molecules together. (a) In the lock-and-key mechanism the substrates have a complementary fit to the enzyme's active site. (b) In the induced-fit model, binding of substrates induces a conformational change in the enzyme.

been amazed by the charts on laboratory walls showing the myriad interlocking, branched, and circular pathways along which the cell's chemical intermediates are shunted like parts on an assembly line. Bonds are broken, molecules cleaved, molecules united, groups added or removed, and so forth. The key fact is that almost every step, represented by an arrow on the metabolic chart, is controlled (mediated) by an enzyme, and each of these enzymes is produced under the direction of a gene that specifies its function. Change

one critical gene, and the entire assembly line can break down.

Humans provide some startling examples. The list in Table 9-4 gives some representative examples and suggests the magnitude of genetic involvement in human disease. Figure 9-21 shows a corner of the human metabolic map to illustrate how a set of diseases, some of them common and familiar to us, can stem from the blockage of adjacent steps in biosynthetic pathways.

9-4 ▸ TABLE   Representative Examples of Enzymopathies: Inherited Disorders in Which Altered Activity (Usually Deficiency) of a Specific Enzyme Has Been Demonstrated in Humans

Condition	Enzyme with deficient activity*	Condition	Enzyme with deficient activity*
Acatalasia	Catalase	Granulomatous disease	Reduced nicotinamide adenine dinucleotide phosphate (NADPH) oxidase
Acid phosphatase deficiency	Acid phosphatase		
Albinism	Tyrosinase		
Aldosterone deficiency	18-Hydroxydehydrogenase	Hydroxyprolinemia	Hydroxyproline oxidase
		Hyperlysinemia	Lysine-ketoglutarate reductase
Alkaptonuria	Homogentisic acid oxidase		
Angiokeratoma, diffuse (Fabry disease)	Ceramide trihexosidase	Hypophosphatasia	Alkaline phosphatase
		Immunodeficiency disease	Adenosine deaminase
Apnea, drug-induced	Pseudocholinesterase		Uridine monophosphate kinase
Argininemia	Arginase		
Argininosuccinic aciduria	Argininosuccinase	Krabbe disease	Galactosylceramide β-galactosidase
Ataxia, intermittent	Pyruvate decarboxylase		
Citrullinemia	Argininosuccinic acid synthetase	Leigh necrotizing encephalomyelopathy	Pyruvate carboxylase
Crigler-Najjar syndrome	Glucuronyl transferase	Maple-sugar urine disease	Keto acid decarboxylase
Cystathioninuria	Cystathionase	Niemann-Pick disease	Sphingomyelinase
Ehlers-Danlos syndrome, type V	Lysyl oxidase	Ornithinemia	Ornithine ketoacid aminotransferase
Farber lipogranulomatosis	Ceramidase	Pentosuria	Xylitol dehydrogenase (L-xylulose reductase)
Galactosemia	Galactose 1-phosphate uridyl transferase		
		Phenylketonuria	Phenylalanine hydroxylase
Gangliosidosis, GM$_1$; generalized, type I, or infantile form	β-Galactosidase A, B, C	Refsum disease	Phytanic acid oxidase
		Richner-Hanhart syndrome	Tyrosine aminotransferase
Gangliosidosis, GM$_1$; type II, or juvenile form	β-Galactosidase B, C	Sandhoff disease (GM$_2$ gangliosidosis, type II)	Hexosaminidase A, B
Gaucher disease	Glucocerebrosidase	Tay-Sachs disease	Hexosaminidase A
Gout	Hypoxanthine-guanine phosphoribosyl-transferase	Wolman disease	Acid lipase
		Xeroderma pigmentosum	Ultraviolet-specific endonuclease
	Phosphoribosyl pyrophosphate (PRPP) synthetase (increased activity)		

* The form of gout due to increased activity of PRPP is the only disorder listed that is due to *increased* enzymatic activity.

Source: Victor A. McKusick, *Mendelian Inheritance in Man,* 4th ed. Copyright © 1975 by Johns Hopkins University Press.

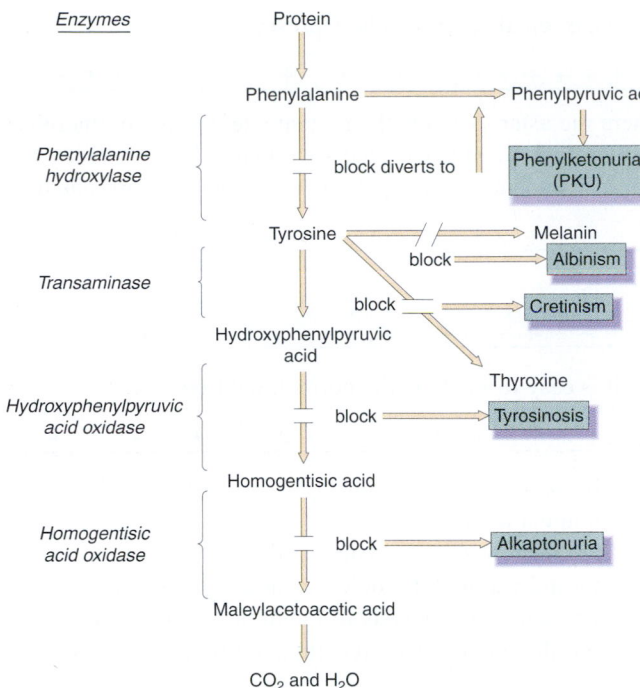

Figure 9-21 One small part of the human metabolic map, showing the consequences of various specific enzyme failures. (Disease phenotypes are shown in colored boxes.) (After I. M. Lerner and W. J. Libby, *Heredity, Evolution, and Society,* 2d ed. Copyright © 1976 by W. H. Freeman and Company.)

Genetic observations
explained by enzyme structure

Armed with an understanding of the gene-protein relation and how enzymes function, we can now reexamine some of the genetic findings presented in earlier chapters and look at them in regard to the biochemistry.

A good example can be found in **temperature-sensitive alleles.** Recall that some mutants appear to be wild type at normal temperatures but can be detected as mutants at high or low temperatures. We now know that such mutations result from the substitution of an amino acid that produces a protein that is functional at normal temperatures, called **permissive** temperatures, but is distorted and nonfunctional at high or low temperatures, called **restrictive** temperatures (Figure 9-22).

As we have seen, conditional mutations such as temperature-sensitive mutations can be very useful to geneticists. Stocks of the mutant culture can be easily maintained under permissive conditions, and the mutant phenotype can be studied intensively under restrictive conditions. Such mutants can be very handy in the genetic dissection of biological systems. For example, with a temperature-sensitive allele, we can shift to a restrictive temperature at various times in the course of development to determine the time at which a gene is active.

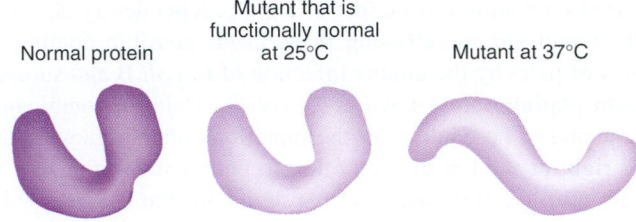

Figure 9-22 Schematic representation of protein conformational distortion, which is probably the basis for temperature sensitivity in certain mutants. An amino acid substitution that has no significant effect at normal (permissive) temperatures may cause significant distortion at abnormal (restrictive) temperatures.

Genetic fine structure

Until the 1950s, our genetic and cytological analysis led to the concept of the chromosome as a linear (one-dimensional) array of genes, strung rather like beads on an unfastened necklace. Indeed, this model is sometimes called the **bead theory.** According to the bead theory, the existence of a gene as a unit of inheritance is recognized through its mutant alleles. A mutant allele affects a single phenotypic character, maps to one chromosome locus, gives a mutant phenotype when paired, and shows a Mendelian ratio when intercrossed. Several tenets of the bead theory are worth emphasizing:

1. The gene is viewed as a fundamental unit of *structure,* indivisible by crossing-over. Crossing-over takes place between genes (the beads in this model) but never within them.

2. The gene is viewed as the fundamental unit of *change,* or mutation. It changes in toto from one allelic form into another; there are no smaller components within it that can change.

3. The gene is viewed as the fundamental unit of *function* (although the precise function of the gene is not specified in this model). Parts of a gene, if they exist, cannot function.

Yet, how can we reconcile the fact that the gene consists of a series of nucleotides with the view that the gene is the smallest unit of mutation and recombination? Seymour Benzer's work in the 1950s showed that the bead theory was not correct. Benzer demonstrated that, whereas a gene can be defined as a unit of function, a gene can be subdivided into a linear array of sites that are mutable and that can be recombined. The smallest units of mutation and recombination are now known to be correlated with single nucleotide pairs.

Intragenic recombination

Benzer used the *rII* system, described in Chapter 7, to detect very low levels of recombination in phage T4. Benzer started with an initial sample of eight independently derived *rII* mutant strains, crossing them in all possible combinations of pairs by the double infection of *E. coli* B and subsequent plating onto a lawn of *E. coli* K. Using recombinant frequencies, he could map the mutations unambiguously to the right or the left of each other to get what we now call a **gene map** (in this case, the map units are the frequency of *rII*⁺ plaques):

rII¹	*rII⁷*	*rII⁴*	*rII⁵*	*rII²*	*rII⁸*	*rII³*	*rII⁶*

Recombination within a gene, called **intragenic recombination,** seems to be the rule rather than the exception. It can virtually always be found at any locus if a suitable selection system is available to detect recombinants. In other words, a mutant allele can be pictured as a length of genetic material (the gene) that contains a damaged or non-wild-type part, a **mutational site.** This partial damage is what causes the non-wild-type phenotype. Different alleles produce different phenotypic effects because damage is at different sites within the wild-type allele.

Thus, an allele a^1 can be represented as

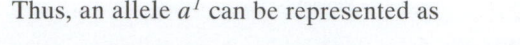

where the asterisk (*) is the mutant site within an otherwise normal gene (denoted by the sites marked +). A cross between a^1 and another mutant allele, a^2, can be represented as

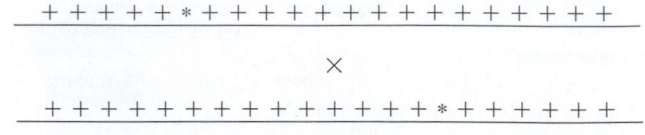

and it is easy to see how the normal, wild-type gene

$$+ +$$

could be generated by a simple crossover anywhere between the two mutational sites.

Benzer showed that, contrary to the classical view, genes were not indivisible but could be subdivided by recombination. Extending his analysis to hundreds of *rII* alleles, Benzer found that the minimal recombinant frequency in a cross

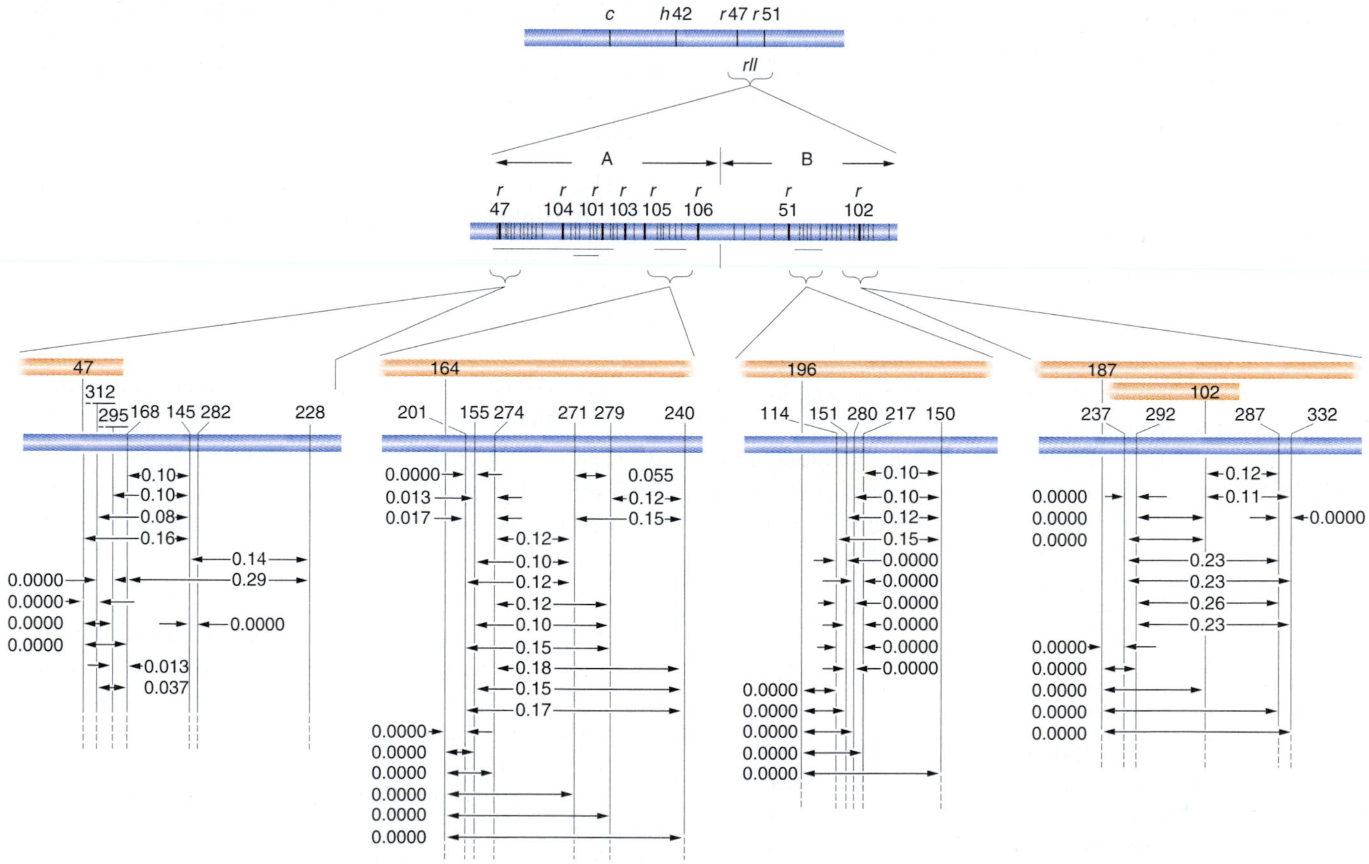

Figure 9-23   Detailed recombination map of the *rII* region of the phage T4 chromosome. The map unit is the percentage of *rII*⁺ recombinants in crosses between the *rII* mutants. Typical regions are progressively enlarged. Numbers on the map represent mutational sites; numbers 47, 64, 196, 187, and 102 represent deletions. Note that the two parts of the *rII* region, A and B, are two different functional units of the region, as described later in the text. (After S. Benzer, *Proceedings of the National Academy of Sciences USA* 41, 1955, 344. From G. S. Stent and R. Calendar, *Molecular Genetics,* 2d ed. Copyright ©1978 by W. H. Freeman and Company.)

between a pair of different mutant alleles was 0.01 percent, even though his analytical system was capable of detecting recombinant frequencies as low as 0.0001 percent if they occurred. This led to the idea that genes were composed of small units and that recombination could take place between but not within these units. We now know that the single nucleotide pair is the smallest unit of recombination.

MESSAGE ·························
A gene is composed of subelements that can recombine. The smallest subunit not divisible by recombination is the single nucleotide pair.

Recombination within genes permits us to construct detailed gene maps, such as the map for the *rII* region shown in Figure 9-23. Making crosses between various mutations provides us with the relative frequencies of intragenic recombinants, and these recombinants reveal the order and relative positions of the mutational sites within a gene. It should be noted that recombination *within* genes is the same as recombination *between* genes, except that the scale is different.

Mutational sites

Benzer exploited the properties of partial deletions to rapidly map new point mutations to specific segments within the *rII* gene, a procedure called **deletion mapping.**

Using deletions in mapping mutational sites

Deletions are mutations that result from the elimination of segments of DNA. Deletions can be intercrossed and mapped just like point mutations. The deleted region is represented diagrammatically by a bar. If no wild-type recombinants are produced in a cross between different deletions, then the bars are shown as overlapping. A typical deletion map might be as follows:

Such deletion maps are useful in delineating regions of the gene to which new point mutations can be assigned. The logic is that a mutation cannot give wild-type recombinants in a cross to deletion in which the DNA corresponding to the wild-type region for that particular mutation is no longer present. Therefore, mutations can be ordered against a set of deletions by rapid tests that do not depend on extensive quantitative measurements.

Deletion mapping of the *rII* region

Benzer carried out the analysis in the following way. For example, consider the following gene map, showing 12 identifiable mutational sites:

Let us suppose that one mutant, D_1, fails to give rII^+ recombinants when crossed with mutants carrying altered sites *1, 2, 3, 4, 5, 6, 7,* or *8*; therefore, D_1 behaves as if it has a deletion of sites 1 to 8:

Another mutant, D_2, fails to give rII^+ recombinants when crossed with *5, 6, 7, 8, 9, 10, 11,* or *12*; therefore, D_2 behaves as if it has a deletion of sites 5 to 12:

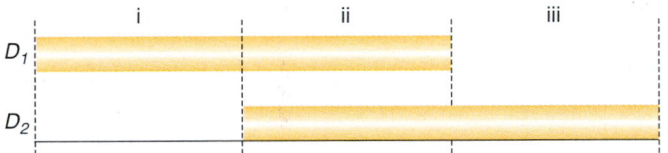

These overlapping deletions now define three areas of the gene. Let's call them i, ii, and iii:

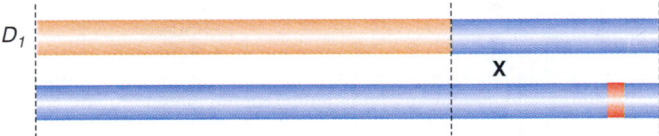

A new mutant that gives rII^+ recombinants when crossed with D_1 but not when crossed with D_2 must have its mutational site in area iii. One that gives rII^+ recombinants when crossed with D_2 but not with D_1 must have its mutational site in area i. A new mutant that does not give rII^+ recombinants with either D_1 or D_2 must have its mutational site in area ii. For example, assume that a mutant in area iii is crossed with D_1. We would envision the cross schematically as follows, where the mutant site is shown with a red bar:

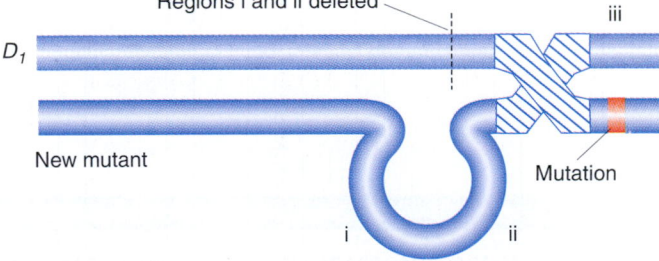

and draw the actual pairing as follows:

Regions i and ii deleted

New mutant

Mutation

The more deletions there are in the tester set, the more areas can be uniquely designated and the more rapidly new mutational sites can be located (Figure 9-24). When assigned to a region, a mutation can be mapped against other alleles in the same region to obtain an accurate position. Figure 9-25 shows the complexity of Benzer's actual map.

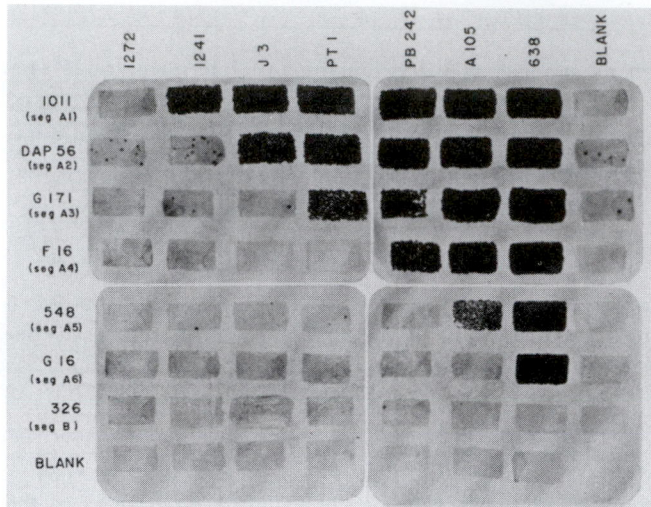

Analysis of mutational sites

The use of deletions enabled Benzer to define the **topology** of the gene — the manner in which the parts are interconnected. His genetic experiments showed the gene to consist of a linear array of mutable subelements. Benzer's next step was to examine the **topography** of the gene — differences in the properties of the subelements. Operationally, he determined these differences by asking whether all the subelements or sites were equally mutable or whether mutations were prevalent at some sites and rare at others. For this

Figure 9-24   Crosses for mapping *rII* mutations. The photograph is a composite of four plates. Each row shows a given mutant tested against the reference deletions of Figure 9-25. The results show each of these mutations to be located in a different segment. Plaques appearing in the blanks are due to revertants present in the mutant stock.

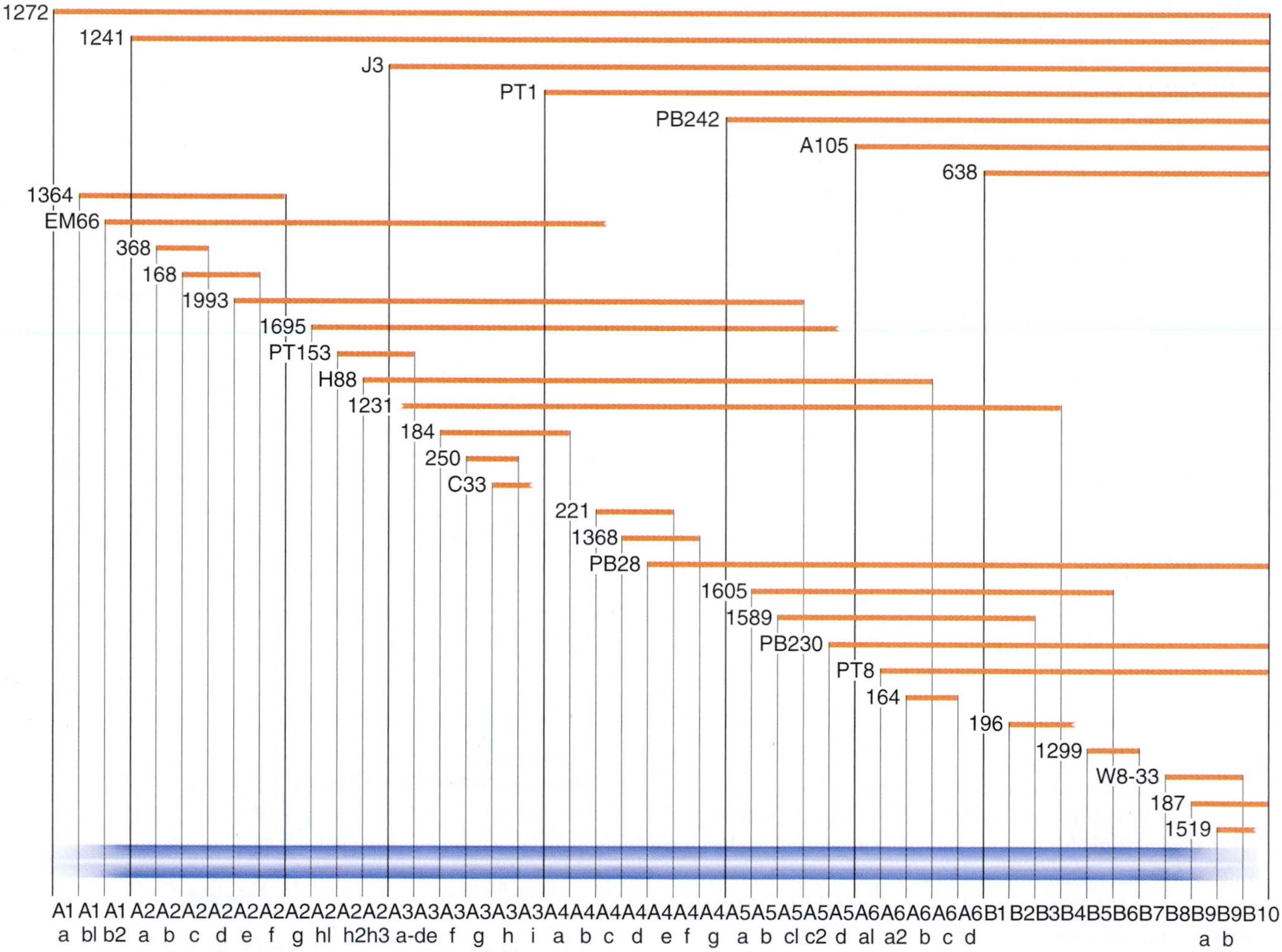

Figure 9-25   Detailed deletion map of the *rII* gene. Each deletion (*horizontal orange bar*) has an identification number. Along the bottom are the arbitrary identification numbers of the regions defined by the deletions. Note that some deletions extend out of the *rII* gene. (From G. S. Stent and R. Calendar, *Molecular Genetics,* 2d ed. Copyright ©1978 by W. H. Freeman and Company. After S. Benzer, *Proceedings of the National Academy of Sciences USA* 47, 1961, 403–416.)

study, it was essential to work with the smallest mutable subelements possible. Instead of multisite mutations (deletions) that exhibited no reversion, Benzer used revertible mutations, because they probably represented small alterations, or point mutations. Additionally, he discarded mutants with high reversion rates, because high reversion interferes with recombination tests.

Benzer used his deletion mutants to rapidly map the set of point mutations. He first localized each mutation into short deletion segments and then crossed all the point mutations within a segment against one another; any two revertible mutations that failed to recombine with each other represented mutations at the same site.

Figure 9-26 shows the distribution of 1612 spontaneous mutations in the *rII* locus. In Benzer's own words, "That the distribution is not random leaps to the eye." This extraordinary nonrandom distribution demonstrates that all sites are not equally mutable. Benzer termed sites that are more mutable than other sites **hot spots.** The most prominent hot spot was represented by more than 500 repeated occurrences in the collection of 1612 mutations. By examining a Poisson distribution calculated to fit the number of sites having only one or two occurrences (Figure 9-27), Benzer could show that at least 60 sites were truly more mutable than those with only one or two occurrences. Mutations were not observed at all (by chance) in at least 129 sites in this collection, even though these sites were as mutable as those represented by one or two occurrences. When Benzer extended the analysis to include mutagen-induced mutations, results were similar to the spontaneous mutation analysis: there were also hot spots, although often different ones from the spontaneous hot spots, among the mutations generated by mutagenic agents (see Chapter 16).

The size of a mutational site also was of interest. The physical size of the *rII* region can be used to calculate the approximate number of base pairs in the region. From this figure, the number of mutational sites was determined to be approximately one-fifth of the number of nucleotide pairs. In other words, the smallest mutable site was five nucleotide pairs or less. The deciphering of the genetic code (Chapter 10), together with the demonstrations by Ingram, Yanofsky, and others (described earlier in this chapter) that single amino acid substitutions resulted from single mutations, allowed Benzer to conclude that a mutation could result from the alteration of a single nucleotide pair. (The direct sequencing of DNA, described in Chapter 12, has since confirmed these conclusions in many examples.)

MESSAGE ···

The gene can be divided into a linear array of mutable subelements that correspond to individual nucleotide pairs.

Destruction of the bead theory

Let's look again at the bead theory in light of Benzer's work. With the aid of deletion mapping, Benzer was able to map an extraordinary number of mutations in the *rII* locus against one another. His experiments showed that mutations in the same gene can indeed recombine with one another. This result contradicts the bead theory of classical genetics, which held that recombination could take place between genes but not within genes.

Benzer's analysis of the fine structure of the gene demonstrated that each gene consists of a linear array of subelements and that these sites within a gene can be altered by mutation and can undergo recombination. This finding also contradicts the bead theory, one tenet of which implies that only the gene as a whole is mutable, not parts of the gene.

Subsequent work by several investigators identified each genetic site as a base pair in double-stranded DNA. Therefore, Benzer's contribution bridged the gulf between classical genetics and the knowledge of the chemical structure of DNA revealed by Watson and Crick. According to the bead theory, the Watson-Crick structure made no sense. However, Benzer's demonstration that genes do indeed have fine structures that can be revealed solely by genetic analysis allowed a fusion of the two disciplines and helped to launch the modern era of molecular genetics. Figure 9-28 illustrates the fine structure analysis of the *rII* locus and its correspondence with the DNA structure.

Complementation

In another part of his studies, Benzer carried out a series of experiments designed to define the gene in regard to function. Benzer studied the concept of **complementation;** he wanted to find out whether the entire *rII* region of phage T4 acts as a single functional unit or whether it is made up of subunits that function independently. Therefore, he tested the mutations that he had mapped in the *rII* region to see whether various combinations of pairs of the mutations would restore the wild-type phenotype. In other words, Benzer looked for complementation in *E. coli* host cells that were temporarily "diploid" for the T4 chromosome. To do so, he carried out a mixed infection with different *rII* mutants (Figure 9-29). His criterion for the wild-type phenotype was the ability to lyse *E. coli* K hosts. (Recall that *rII* mutants cannot do so but that wild-type phages can.)

Complementation tests such as the one conducted by Benzer are carried out in one cycle of infection; they do not require the multiple cycles of reinfection needed for plaque formation. Samples of the two phages to be tested are spread over a strip of host bacteria on a section of a petri plate at a high ratio of phages to bacteria to ensure that essentially every bacterium is infected with both phages. After a period of incubation, the growth or the absence of growth of the bacteria in the strip indicates whether the bacteria have lysed as a result of the phage infection.

It is important to understand the difference between recombination and complementation. Recombination represents the creation of new combinations of genes through the

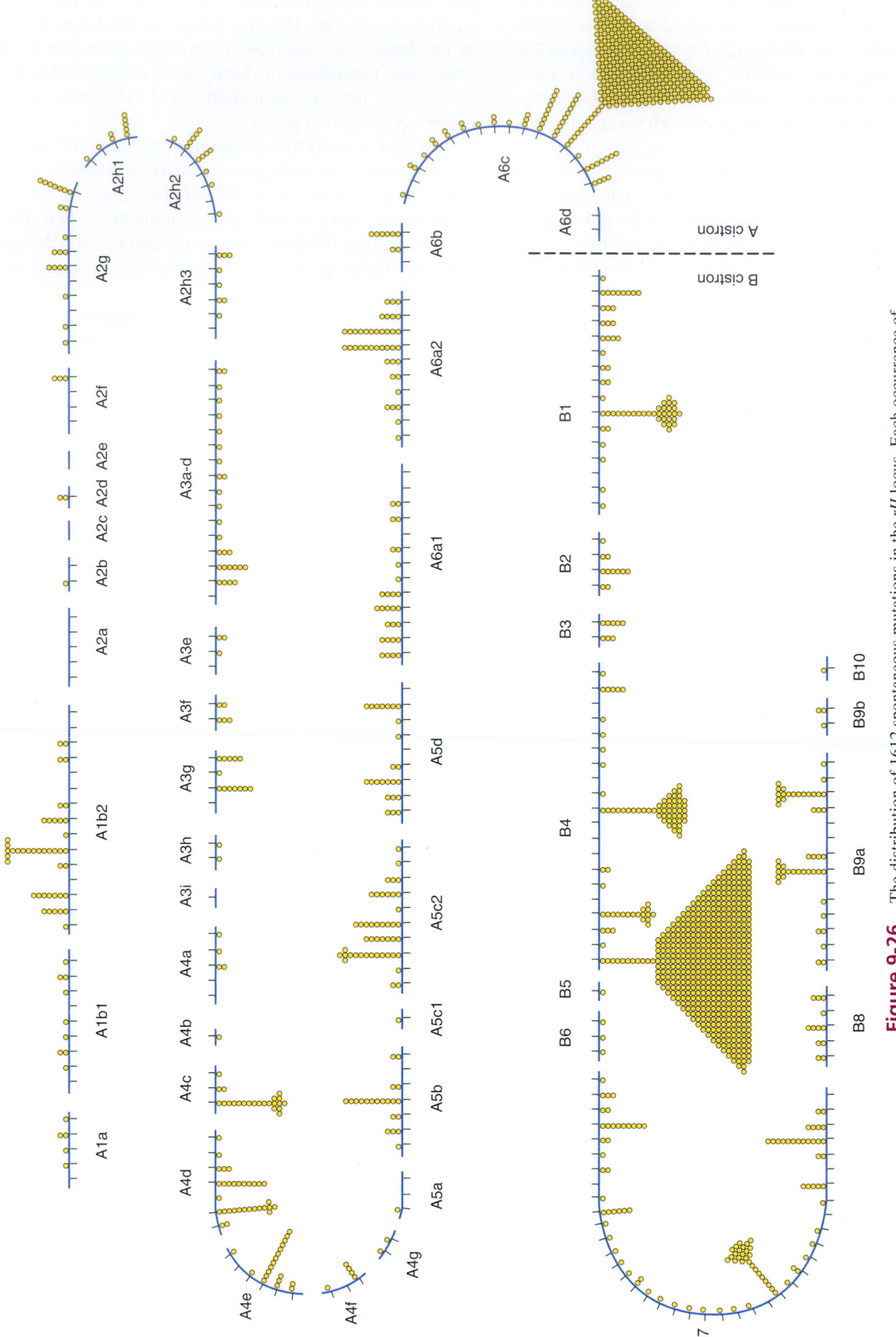

288

Figure 9-26 The distribution of 1612 spontaneous mutations in the *rII* locus. Each occurrence of an independent mutation is depicted by a circle. When mutations are identical, the circles are drawn on top of one another. Although many positions are represented by only a single circle, others have a large number of circles; these sites are called *hot spots*. The two genes (formerly called cistrons), *A* and *B*, of the *rII* region are indicated in this diagram.

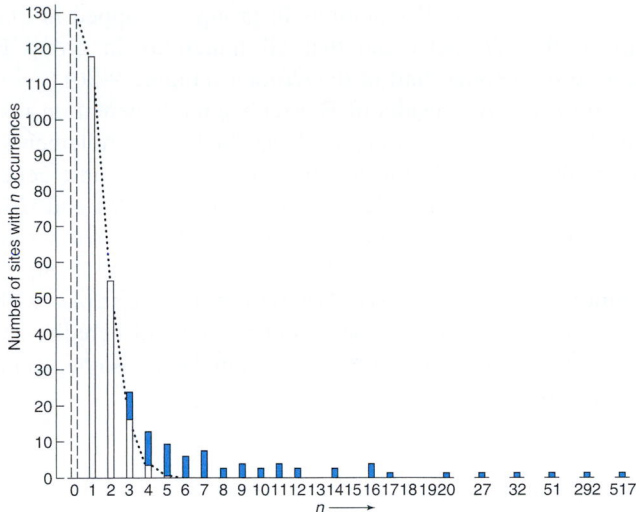

Figure 9-27 Distribution of occurrences of spontaneous mutations at various sites. The dotted curve indicates a Poisson distribution fitted to the number of sites having one and two occurrences. It predicts a minimum estimate for the number of sites of comparable mutability that have zero occurrences owing to chance (dashed column at $n = 0$). Blue bars indicate the minimum numbers of sites that have mutation rates significantly higher than the one- and two-occurrence class. (From S. Benzer, *Proceedings of the National Academy of Sciences USA* 47, 1961, 403–416.)

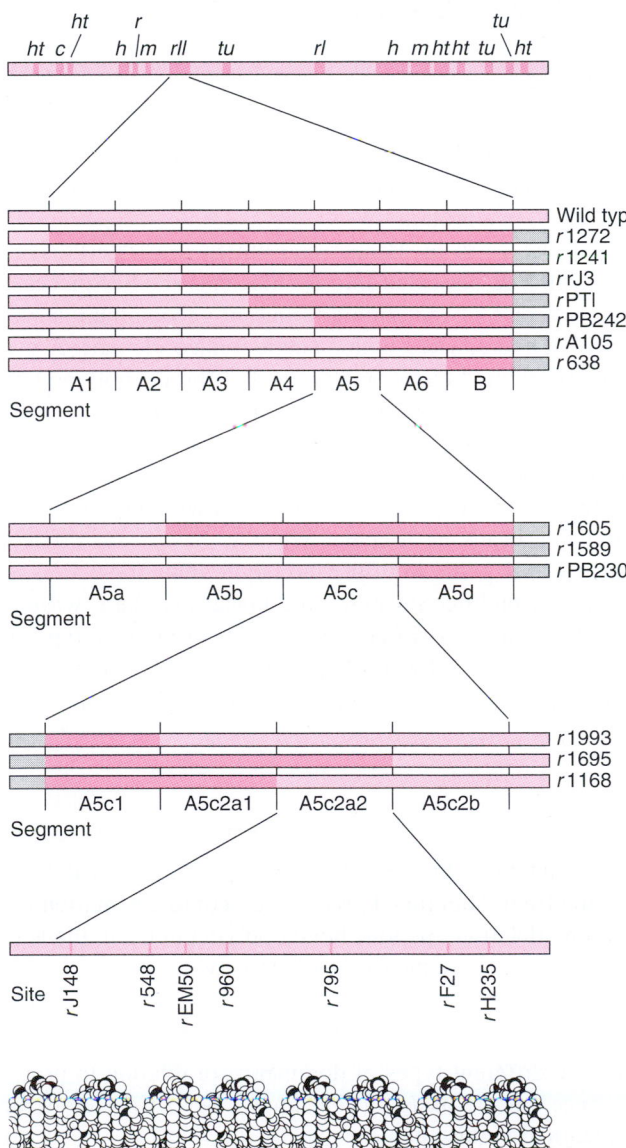

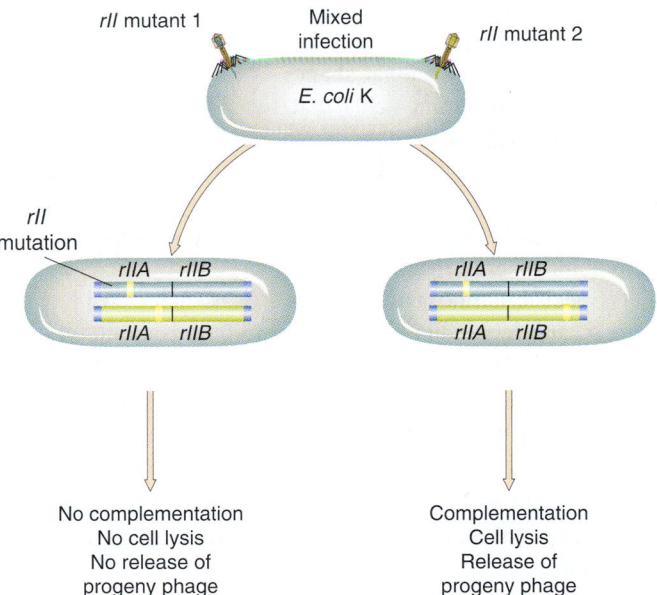

Figure 9-28 Fine-structure analysis of the *rII* locus. This mapping technique localizes the position of a given mutation in progressively smaller segments of the DNA molecule contained in phage T4. The *rII* region represents only a few percent of the entire molecule *(top)*. The mapping is done by crossing an unknown mutant with reference mutants having deletions (darker red) of known extent in the *rII* region. Each site represents the smallest mutable unit in the DNA molecule, a single base pair. The molecular segment *(extreme bottom)*, estimated to be roughly in proper scale, contains a total of about 40 base pairs. (From S. Benzer, "The Fine Structure of the Gene." Copyright © 1962 by Scientific American, Inc. All rights reserved.)

Figure 9-29 Complementation test: a schematic view of *rII* complementation. Two different mutants of *rII* are used to simultaneously infect *E. coli* K (mixed infection). Normally, an *rII* mutant cannot lyse *E. coli* K or generate progeny phages. However, if the two different mutants can complement each other, then lysis and phage growth will result. If the two *rII* mutants cannot complement each other, then no lysis or phage growth will result.

physical breakage and rejoining of chromosomes. The progeny from a cross in which recombination has occurred have *new genotypes* that are different from the parental genotypes. Complementation, on the other hand, does not involve any change in the genotypes of individual chromosomes; rather it represents the *mixing of gene products.* Complementation occurs during the time that two chromosomes are in the same cell and can each supply a function. Afterward, each respective chromosome remains unaltered. In the case of *rII* mutants, complementation occurs when two different phage chromosomes, with mutations in different *rII* genes, are in the same host cell. However, progeny that result from this complementation carry only the parental genotypes.

Tests of many different paired mutants allowed Benzer to separate the mutations into two groups, labeled A and B. All mutations in the A group complemented those in the B group, whereas no mutations in the A group complemented any other mutations in the A group and no mutations in the B group complemented any other mutations in the B group.

Benzer found that all mutations in group A mapped in one half of the *rII* locus and that all mutations in group B mapped in the other half of the *rII* locus (Figure 9-30).

What did the results of Benzer's complementation test mean? We learned in Chapter 4 that lack of complementation is the diagnostic for mutations being in the same gene. Hence it must be concluded that the groupings *rIIA* and *rIIB* must represent different genes, located next to each other on the chromosome. Benzer called the *rIIA* and *rIIB* complementation groups *cistrons.* We now equate cistrons with genes and recognize that the functional units of heredity are genes. The term cistron, which was popular for some years, is now rarely used.

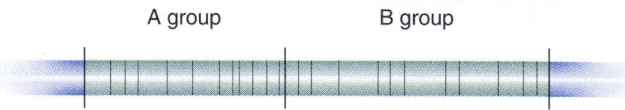

A group   B group

Figure 9-30   *Gene map of rII.*

SUMMARY

The work of Beadle and Tatum in the 1940s showed that one gene codes for one protein, often an enzyme. When an enzyme fails to function normally owing to a mutation, a variant phenotype results. These variant phenotypes are often the basis of genetic diseases in any organism, including humans. To understand how abnormal enzymes can cause phenotypic change, we need to understand the structure of proteins. Composed of a specific linear sequence of amino acids connected through peptide bonds, the proteins assume specific three-dimensional shapes as a result of the interaction of the 20 amino acids that in different combinations constitute the polypeptide chain. Different areas of this folded chain are sites for the attachment and interaction of substrates. Furthermore, many functional enzymes and other proteins are built by combining several polypeptide chains into a multimeric form.

Specific amino acid changes can be detected in a protein by the technique of fingerprinting and amino acid sequencing. Work of this type by Sanger, Ingram, Yanofsky,

and others has demonstrated colinearity between a mutational site on the genetic map and the position of an altered amino acid in a protein.

Further work by Benzer and others illustrated that the gene could be dissected into smaller and smaller pieces. A structural unit of mutation and one of recombination were identified and equated with a single nucleotide pair. A gene (formerly termed cistron) is defined at the phenotypic level as a genetic region within which there is no complementation between mutations. This unit codes for the structure of a single functional polypeptide.

In the early days of genetics, genes were represented as indivisible beads on a chain. We now have a far different picture of the gene. Multiple intragenic mutational sites exist, and recombination may take place anywhere within a gene. In addition, a closer connection between genotype and phenotype was realized when it was established that one gene is responsible for the synthesis of one polypeptide.

CONCEPT MAP

Draw a concept map interrelating as many of the following terms as possible. Note that the terms are listed in no particular order.

allele / mutation / biosynthetic pathway / enzyme / catalysis / auxotroph / dominant / recessive / complementation / cistron / deletion / peptide bond / polypeptide / amino acid sequence / recombination

CHAPTER INTEGRATION PROBLEM

We learned in Chapter 4 that the wild-type hemoglobin allele, *Hb*A, displayed dominance to *Hb*S (the allele of sickle-cell anemia) with regard to anemia, but codominance with regard to the electrophoretic properties of hemoglobin. Ex-

plain the different types of dominance in relation to protein sequence, on the basis of your knowledge of protein structure learned in this chapter.

◆ Solution ◆

The protein hemoglobin has two α and two β polypeptide chains, closely fitted together in an elaborate quaternary structure that holds the heme groups that carry oxygen to the tissues. The primary structure of a protein—the linear sequence of its component amino acids—determines the protein's ultimate tertiary and quaternary structures and hence its activity. Because of the change from glutamic acid to valine at position 6 in the β chain, the hemoglobin of sickle-cell anemia (HbS) is changed structurally, so it car-

ries less oxygen that does normal adult hemoglobin (HbA). Heterozygotes (Hb^A/Hb^S) still maintain sufficient oxygen-carrying capacity to prevent severe symptoms of anemia; hence, Hb^A shows dominance to Hb^S with regard to anemia. Substituting valine for glutamic acid also changes the electric charge of the molecule, so HbS moves more slowly than HbA on gel electrophoresis (see Figure 4-4). Thus, the two hemoglobin molecules (HbA and HbS) show codominance with respect to electrophoretic properties, because both can be resolved and detected by this technique.

SOLVED PROBLEMS

1. Various pairs of *rII* mutants of phage T4 are tested in *E. coli* in both the cis (= coupling) and trans (= repulsion) conformations, and the "burst size" (the average number of phage particles produced per bacterium) for each test pair is compared. Results for six different *r* mutants — *rM, rN, rO, rP, rR,* and *rS* — are as follows (+ indicates the wild-type allele):

Cis genotype	Burst size	Trans genotype	Burst size
rM rN/+ +	245	*rM* +/+ *rN*	250
rO rP/+ +	256	*rO* +/+ *rP*	268
rR rS/+ +	248	*rR* +/+ *rS*	242
rM rO/+ +	270	*rM* +/+ *rO*	0
rM rP/+ +	255	*rM* +/+ *rP*	255
rM rR/+ +	264	*rM* +/+ *rR*	0
rM rS/+ +	240	*rM* +/+ *rS*	240
rN rO/+ +	257	*rN* +/+ *rO*	268
rN rP/+ +	250	*rN* +/+ *rP*	0
rN rR/+ +	245	*rN* +/+ *rR*	255
rN rS/+ +	259	*rN* +/+ *rS*	0
rP rR/+ +	260	*rP* +/+ *rR*	245
rP rS/+ +	253	*rP* +/+ *rS*	0

If we assign the mutation *rO* to the A gene, what are the locations of the other five mutations with respect to the A and B genes? (NOTE: This problem is similar to Problem 21 in the problem set for this chapter.)

◆ Solution ◆

In solving problems such as this one, we first look for the underlying principle and then attempt to draw a diagram to help us work out the solution. The key principle is that, in the trans position, mutations in the same gene will not complement one another and thus will yield no progeny phage (no burst), whereas mutations in different genes will complement and thus will yield progeny phages of normal burst size. Because *rO* is the A gene, we start with the following initial diagram:

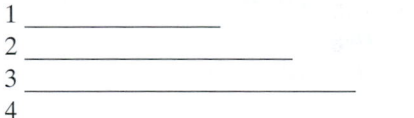

We can now look at the test results in the problem and assign each mutation to a gene on the basis of whether complementation occurs in a mixed infection with *rO* mutants. From the table, it is evident that *N* and *P* complement *O* and must be in a different gene (the *B* gene) whereas *M* does not complement *O* and must be in the same gene (the *A* gene). This leaves us with *R* and *S*, which can be assigned to the *A* and *B* genes, respectively, on the basis of their complementation tests with *M, N,* and *P*. Therefore, we have the arrangement

We can check this assignment by examining the other crosses to find out if the results are consistent with our answer.

2. The following deletion map shows four deletions (1 to 4) in the *rIIA* gene of phage T4:

```
1 _____
2 _____
3 _____
4 _____
```

Five point mutations (*a* to *e*) are tested against these four deletion mutants for their ability (+) or inability (−) to give wild-type (r^+) recombinants; the results are

		a	*b*	*c*	*d*	*e*
Mutant	1	+	+	+	+	+
	2	+	+	+	−	−
	3	+	−	+	−	−
	4	−	−	+	−	−

What is the order of the point mutations?

◆ Solution ◆

The key principle here is that point mutations can recombine with deletions that *do not* extend past the mutation, but they cannot recombine to yield wild-type phages with deletions that *do* extend past the mutation. For the test results given in the problem, any mutation that recombines with deletion 1 must be to the right of the deletion, any mutation that recombines with deletion 2 must be to the right of deletion 2, and so forth. Let's look at point mutation *a*. It recombines with deletions 1, 2, and 3 but not with deletion 4. Therefore, it is to the right of deletions 1, 2, and 3 but not to the right of deletion 4. We can therefore place point mutation *a* in the interval between deletions 3 and 4. Point mutation *b* recombines with deletions 1 and 2 and must be to the right of them. It does not recombine with deletions 3 and 4, so it is in the interval between deletion 2 and 3. Point mutation *c* recombines with all the deletions and is to the right of

all deletions, even deletion 4. Finally, point mutations *d* and *e* recombine only with deletion 1 and must therefore be in the interval between deletions 1 and 2. The solution that we have just derived can be summarized as follows:

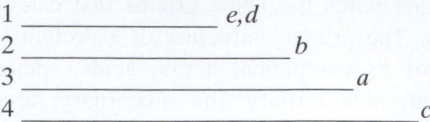

Problem 25 in the following problem set is similar to this sample problem. See if you can apply the reasoning set forth here when you solve Problem 25.

PROBLEMS

1. Describe the primary, secondary, tertiary, and quaternary structures of a protein. Give examples of each type of structure, where possible.

2. What is the one-gene–one-enzyme hypothesis?

3. What is an auxotroph?

4. What is the exact change in a protein that results in sickle-cell anemia?

5. What experiment did Yanofsky do to demonstrate colinearity between the gene and the protein?

6. What is the difference between complementation and recombination?

7. Complete the following table by filling in the plaque types:

T4 phage	E. coli STRAIN TYPE	
	B	K
rII	?	?
rII⁺ (wild type)	?	?

8. A common weed, St.-John's-wort, is toxic to albino animals. It also causes blisters on animals that have white areas of fur. Suggest a possible genetic basis for this reaction.

9. In humans, the disease galactosemia causes mental retardation at an early age because lactose in milk cannot be broken down, and this failure affects brain function. How would you provide a secondary cure for galactosemia? Would you expect this phenotype to be dominant or recessive?

10. Aminocentesis is a technique in which a hypodermic needle is inserted through the abdominal wall of a pregnant woman and into the amnion, the sac that surrounds the developing embryo, to withdraw a small amount of amniotic fluid. This fluid contains cells that come from the embryo (not from the woman). The cells can be cultured; they will divide

and grow to form a population of cells on which enzyme analyses and karyotypic analyses can be performed. Of what use would this technique be to a genetic counselor? Name at least three specific conditions under which amniocentesis might be useful. (**Note:** This technique entails a small but real risk to the health of both the woman and the embryo; take this fact into account in your answer.)

11. The table below shows the ranges of enzymatic activity (in units that we need not worry about) observed for enzymes having roles in two recessive human metabolic diseases. Similar information is available for many metabolic genetic diseases.

   a. Of what use is such information to a genetic counselor?

   b. Indicate any possible sources of ambiguity in interpreting studies of an individual patient.

   c. Reevaluate the concept of dominance in the light of such data.

Disease	Enzyme	Patients	RANGE OF ENZYME ACTIVITY Parents of patients	Normal persons
Acatalasia	Catalase	0	1.2–2.7	4.3–6.2
Galactosemia	Gal-1-P uridyl transferase	0–6	9–30	25–40

12. Two albinos marry and have a normal child. How is this possible? Suggest at least two ways.

13. In humans, PKU (phenylketonuria) is a disease caused by an enzyme inefficiency at step A in the following simplified reaction sequence, and AKU (alkaptonuria) is due to an enzyme inefficiency in one of the steps summarized as step B here:

$$\text{phenylalanine} \xrightarrow{\text{A}} \text{tyrosine} \xrightarrow{\text{B}} CO_2 + H_2O$$

A person with PKU marries a person with AKU. What phenotypes do you expect for their children? All normal, all having PKU only, all having AKU only, all having both PKU and AKU, or some having AKU and some having PKU.

14. Three independently isolated tryptophan-requiring strains of yeast are called TrpB, TrpD, and TrpE. Cell suspensions of each strain are streaked on a plate supplemented with just enough tryptophan to permit weak growth for a Trp⁻ strain. The streaks are arranged in a triangular pattern so that they do not touch one another. Luxuriant growth is noted at both ends of the TrpE streak and at one end of the TrpD streak.

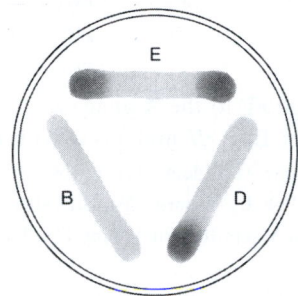

*a. Do you think that complementation is involved?

b. Briefly explain the patterns of luxuriant growth.

c. In what order in the tryptophan-synthesizing pathway are the enzymatic steps defective in TrpB, TrpC, and TrpE?

d. Why was it necessary to add a small amount of tryptophan to the medium to demonstrate such a growth pattern?

15. In *Drosophila* pupae, certain structures called *imaginal disks* can be detected as thickenings of the skin; after metamorphosis, these imaginal disks develop into specific organs of the adult fly. George Beadle and Boris Ephrussi devised a means of transplanting eye imaginal disks from one larva into another larval host. When the host metamorphoses into an adult, the transplant can be found as a colored eye located in its abdomen. The researchers took two strains of flies that were phenotypically identical in their bright scarlet eyes: one owing to the sex-linked mutant vermilion (*v*); the other owing to cinnabar (*cn*) on chromosome 2. If *v* disks are transplanted into *v* hosts or *cn* disks into *cn* hosts, then the transplants develop as mutant scarlet eyes. Transplanted *cn* or *v* disks in wild-type hosts develop wild-type eye colors. A *cn* disk in a *v* host develops a mutant eye color, but a *v* disk in a *cn* host develops wild-type eye color. Explain these results, and outline the experiments that you would propose to test your explanation.

16. In *Drosophila,* the autosomal recessive *bw* causes a dark-brown eye, and the unlinked autosomal recessive *st* causes a bright scarlet eye. A homozygote for both genes has a white eye. Thus, we have the following correspondences between genotypes and phenotypes:

$$st^+/st^+; bw^+/bw^+ = \text{red eye (wild type)}$$
$$st^+/st^+; bw/bw = \text{brown eye}$$
$$st/st; bw^+/bw^+ = \text{scarlet eye}$$
$$st/st; bw/bw = \text{white eye}$$

Construct a hypothetical biosynthetic pathway showing how the gene products interact and why the different mutant combinations have different phenotypes.

17. Several mutants are isolated, all of which require compound G for growth. The compounds (A to E) in the biosynthetic pathway to G are known but their order in the pathway is not known. Each compound is tested for its ability to support the growth of each mutant (1 to 5). In the following table, a plus sign indicates growth and a minus sign indicates no growth:

| | Compound tested | | | | | |
	A	B	C	D	E	G
Mutant 1	−	−	−	+	−	+
2	−	+	−	+	−	+
3	−	−	−	−	−	+
4	−	+	+	+	−	+
5	+	+	+	+	−	+

a. What is the order of compounds A to E in the pathway?

b. At which point in the pathway is each mutant blocked?

c. Would a heterokaryon composed of double mutants 1,3 and 2,4 grow on a minimal medium? 1,3 and 3,4? 1,2 and 2,4 and 1,4?

18. In *Neurospora* (a haploid), assume that two genes participate in the synthesis of valine. Their mutant alleles are *val-1* and *val-2,* and their wild-type alleles are *val-1⁺* and *val-2⁺*. The two genes are linked on the same chromosome, and a crossover takes place between them, on average, in one of every two meioses.

a. In what proportion of meioses are there no crossovers between the genes?

b. Use the map function to determine the recombinant frequency between these two genes.

c. Progeny from the cross *val-1 val-2⁺* × *val-1⁺ val-2* are plated on a medium containing no valine. What proportion of the progeny will grow?

d. The *val-1 val-2*⁺ strains accumulate intermediate compound B, and the *val-1*⁺ *val-2* strains accumulate intermediate compound A. The *val-1 val-2*⁺ strains grow on valine or A, but the *val-1*⁺ *val-2* strains grow only on valine and not on B. Show the pathway order of A and B in relation to valine, and indicate which gene controls each conversion.

19. In a certain plant, the flower petals are normally purple. Two recessive mutations arise in separate plants and are found to be on different chromosomes. Mutation 1 (m_1) gives blue petals when homozygous ($m_1 m_1$). Mutation 2 (m_2) gives red petals when homozygous ($m_2 m_2$). Biochemists working on the synthesis of flower pigments in this species have already described the following pathway:

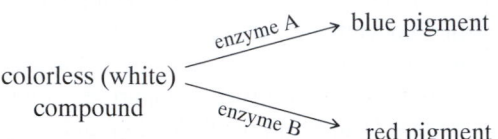

a. Which mutant would you expect to be deficient in enzyme A activity?

b. A plant has the genotype M_1/m_1 ; M_2/m_2. What would you expect its phenotype to be?

c. If the plant in part b is selfed, what colors of progeny would you expect, and in what proportions?

d. Why are these mutants recessive?

20. In sweet peas, the synthesis of purple anthocyanin pigment in the petals is controlled by two genes, B and D. The pathway is

| white intermediate | gene B enzyme → | blue intermediate | gene D enzyme → | anthocyanin (purple) |

a. What color petals would you expect in a pure-breeding plant unable to catalyze the first reaction?

b. What color petals would you expect in a pure-breeding plant unable to catalyze the second reaction?

c. If the plants in parts a and b are crossed, what color petals will the F_1 plants have?

d. What ratio of purple:blue:white plants would you expect in the F_2?

21. Various pairs of *rII* mutants of phage T4 are tested in *E. coli* in both the cis (coupling) and trans (repulsion) conformations. Comparisons are made of the "burst size" (the average number of phage particles produced per bacterium). Results for six different *r* mutants — *rU, rV, rW, rX, rY,* and *rZ* — are as follows:

Cis genotype	Burst size	Trans genotype	Burst size
rU rV/+ +	250	rU +/+ rV	258
rW rX/+ +	255	rW +/+ rX	252
rY rZ/+ +	245	rY +/+ rZ	0
rU rW/+ +	260	rU +/+ rW	250
rU rX/+ +	270	rU +/+ rX	0
rU rY/+ +	253	rU +/+ rY	0
rU rZ/+ +	250	rU +/+ rZ	0
rV rW/+ +	270	rV +/+ rW	0
rV rX/+ +	263	rV +/+ rX	270
rV rY/+ +	240	rV +/+ rY	250
rV rZ/+ +	274	rV +/+ rZ	260
rW rY/+ +	260	rW +/+ rY	240
rW rZ/+ +	250	rW +/+ rZ	255

If we assign *rV* to the *A* gene, what are the locations of the other five *rII* mutations with respect to the *A* and *B* genes? (Problem 21 from M. Strickberger, *Genetics.* Copyright 1968 by Monroe W. Strickberger. Reprinted with permission of Macmillan Publishing Co., Inc.)

22. There is evidence that occasionally in meiosis either one or both homologous centromeres will divide and segregate precociously at the first division rather than at the second division (as is the normal situation). In *Neurospora, pan2* alleles produce a pale ascospore, aborted ascospores are completely colorless, and normal ascospores are black. In a cross between two complementing mutations *pan2x × pan2y,* what ratios of black:pale:colorless would you expect in asci resulting from the precocious division of **(a)** one centromere? **(b)** both centromeres? (Assume that *pan2* is near the centromere.)

23. *Protozoon mirabilis* is a hypothetical single-celled haploid green alga that orients to light by means of a red eyespot. By the selection of cells that do not move toward the light, 14 white-eyespot mutants (*eye⁻*) are isolated after mutation. It is possible to fuse haploid cells to make diploid individuals. The 14 *eye⁻* mutants are paired in all combinations, and the color of the eyespot is scored in each. The table (see top of next page) shows the results, where a plus sign indicates a red eyespot and a minus sign indicates a white eyespot.

a. Mutant 14 is obviously different from the rest. Why might this be?

b. Excluding mutant 14, how many complementation groups are there, and which mutants are in which group?

c. Three crosses are made, with the following results:

	1	2	3	4	5	6	7	8	9	10	11	12	13	14
1	−	+	+	+	−	+	+	−	−	+	+	+	+	−
2	+	−	−	−	+	+	+	+	+	+	+	−	+	−
3	+	−	−	−	+	+	+	+	+	+	+	−	+	−
4	+	−	−	−	+	+	+	+	+	+	+	−	+	−
5	−	+	+	+	−	+	+	−	−	+	+	+	+	−
6	+	+	+	+	+	−	−	+	+	−	−	+	−	−
7	+	+	+	+	+	−	−	+	+	−	−	+	−	−
8	−	+	+	+	+	+	+	−	−	+	+	+	+	−
9	−	+	+	+	−	+	+	−	−	+	+	+	+	−
10	+	+	+	+	+	−	−	+	+	−	−	+	−	−
11	+	+	+	+	+	−	−	+	+	−	−	+	−	−
12	+	−	−	−	+	+	+	+	+	+	+	−	+	−
13	+	+	+	+	+	−	−	+	+	−	−	+	−	−
14	−	−	−	−	−	−	−	−	−	−	−	−	−	−

Mutants crossed	NUMBER OF PROGENY		
	eye$^+$	eye$^-$	Total
1 × 2	31	89	120
2 × 6	5	113	118
1 × 14	0	97	97

	a	b	c	d	e
1	+	+	−	+	+
2	+	+	−	−	−
3	−	−	+	−	+
4	+	−	+	+	+

Explain these genetic ratios with symbols.

d. How many genetic loci take part altogether, and which of the 14 mutants are at each locus?

e. What is the linkage arrangement of the loci? (Draw a map.)

24. You have the following map of the *rII* locus:

A gene   B gene

rX   rY   rZ rD   rE   rF

You detect a new mutation, *rW*, and you find that it does not complement any of the mutants in the *A* or *B* gene. You find that wild-type recombinants are obtained in crosses with *rX*, *rY*, *rE*, and *rF* but not with *rZ* or *rD*. Suggest possible explanations for these results. Describe tests that you would use to choose between the explanations.

25. The following map shows four deletions (1 to 4) in the *rIIA* gene of phage T4:

1 ———
2 ————————
3   ————————
4   ————

Five point mutations (*a* to *e*) in *rIIA* are tested against these four deletion mutants for their ability (+) or inability (−) to give r^+ recombinants, with the following results:

a. What is the order of the point mutants?

b. Another strain of T4 has a point mutation in the *rIIB* gene. This strain is mixed in turn with each of the *rIIA* deletion mutants, and the mixtures are used to infect *E. coli* K at a multiplicity of infection great enough that each host cell will be infected by at least one *rIIA* and one *rIIB* mutant. A normal plaque is formed with deletions 1, 2, and 3, but no plaque forms with deletion 4. Given that the *B* gene is to the right of A, explain the behavior of deletion 4. Does your explanation affect your answer to part a?

26. In a phage, a set of deletions is intercrossed in paired combinations. The following results are obtained (a + indicates that wild-type recombinants are obtained from that cross):

	1	2	3	4	5
1	−	+	−	+	−
2	+	−	+	+	−
3	−	+	−	−	−
4	+	+	−	−	+
5	−	−	−	+	−

a. Construct a deletion map from this table.

b. The first geneticists to do a deletion-mapping analysis in the mythical schmoo phage SH4 (which lyses schmoos) came up with this unique set of data:

	1	2	3	4
1	−	−	+	−
2	−	−	−	+
3	+	−	−	−
4	−	+	−	−

Show why this is a unique result by drawing the only deletion map that is compatible with this table. (Do not let your mind be shackled by conventional expectations.)

27. In a haploid eukaryote, four alleles of the *cys-2* gene are obtained. Each allele requires cysteine, and all alleles map to the same locus. The four strains bearing these mutant alleles are crossed to wild types to obtain a set of eight cultures representing the four mutant alleles in association with each mating type. Then the mutant alleles are intercrossed in all possible combinations of pairs. The haploid meiotic products from each cross are plated on a medium containing no cysteine. In some crosses, cys^+ prototrophs are observed at low frequencies. The results follow, where the numbers represent the frequencies of cys^+ colonies per 10^4 meiotic products plated:

		Mating type A′		
	1	2	3	4
Mating type A″ 1	0	14	2	20
2	14	0	12	6
3	2	12	0	18
4	20	6	18	0

a. Draw a map of the four mutant sites within the *cys-2* gene. Provide a measurement of the relative intersite distances.

b. Do you see any evidence that mutation might have a role in the production of the prototrophs?

*28. In *Neurospora,* there is a gene controlling the production of adenine, and mutants in this gene are called *ad-3* mutants. The *his-2* locus is 2 m.u. to the left, and the *nic-2* locus is 3 m.u. to the right of the *ad-3* locus (*his-2* controls histidine, and *nic-2* controls nicotinamide). Thus, the genetic map is:

```
  his-2          ad-3                    nic-2
  ──┼─────────────┼──────────────────────┼──
      2 m.u.            3 m.u.
```

Three different *ad-3* auxotrophs are detected: *ad-3*ᵃ, *ad-3*ᵇ, and *ad-3*ᶜ. (Use *a, b,* and *c* as their labels.) The following crosses are made:

Cross 1   *his-2⁺ a nic-2⁺ × his-2 b nic-2*
Cross 2   *his-2⁺ a nic-2 × his-2 c nic-2⁺*
Cross 3   *his-2 b nic-2 × his-2⁺ c nic-2⁺*

The ascospores are then plated on a minimal medium containing histidine and nicotinamide, and *ad-3⁺* prototrophs are picked up. The table shows the results obtained. What is the map order of the *ad-3* mutants and the genetic distance between them?

Genotype of *ad-3⁺*	NUMBER OF *ad-3⁺* SPORES PICKED UP		
recombinants	Cross 1	Cross 2	Cross 3
his-2 nic-2	0	6	0
his-2⁺ nic-2⁺	0	0	0
his-2 nic-2⁺	15	0	5
his-2⁺ nic-2	0	0	0
Total ascospores scored	41,236	38,421	43,600

29. In a hypothetical diploid organism, squareness of cells is due to a threshold effect: more than 50 units of "square factor" per cell will produce a square phenotype and less than 50 units will produce a round phenotype. Allele s^f is a functional gene that causes the synthesis of the square factor. Each s^f allele contributes 40 units of square factor; thus, s^f/s^f homozygotes have 80 units and are phenotypically square. A mutant allele (s^n) arises; it is nonfunctional, contributing no square factor at all.

a. Which allele will show dominance, s^f or s^n?

b. Are functional alleles necessarily always dominant? Explain your answer.

c. In a system such as this one, how might a specific allele become changed in evolution so that its phenotype shows recessive inheritance at generation 0 and dominant inheritance at a later generation?

*30. Explain how Benzer was able to calculate the number of sites with zero occurrences, as depicted in Figure 9-27. (Assume, as Benzer did, that the sites with 0, 1, and 2 occurrences are a set of sites with equal mutability.)

31. In *Collinsia parviflora,* the petal color is normally purple. Four recessive mutations are induced, each of which produces white petals. The four pure-breeding lines are then intercrossed in the following combinations, with the results indicated:

Mutant crosses	F₁	F₂
1 × 2	all purple	$\frac{1}{2}$ purple, $\frac{1}{2}$ white
1 × 3	all purple	$\frac{9}{16}$ purple, $\frac{7}{16}$ white
1 × 4	all white	all white

a. Explain all these results clearly, using diagrams wherever possible.

b. What F₁ and F₂ do you predict from crosses of 2 × 3 and 2 × 4?

*32. A biologist is interested in the genetic control of leucine synthesis in the haploid filamentous fungus *Aspergillus.* He treats spores with mutagen and obtains five point mutations (*a* to *e*), all of which are leucine auxotrophs. He first makes heterokaryons between them to check on their functional relations. He determines the following results, where a plus sign indicates that the heterokaryon grew and a minus sign indicates that the heterokaryon did not grow, on a medium lacking leucine:

	a	*b*	*c*	*d*	*e*
a	−	+	+	+	−
b		−	+	+	+
c			−	+	+
d				−	+
e					−

The biologist then intercrosses the mutations in all possible combinations. From each cross, he tests 500 ascospore progeny by inoculating them onto a medium lacking leucine. The results follow (the numbers indicate the number of leucine prototrophs in the 500 progeny):

	a	*b*	*c*	*d*	*e*
a	0	125	128	126	0
b		0	124	2	125
c			0	124	127
d				0	123
e					0

Explain both sets of data genetically. (Note that the two leucine prototrophs from the *b* × *d* cross were found *not* to be due to reversion.)

33. In *Drosophila*, the eye phenotype "star" is caused by recessive mutations (*s*) mapping to one location on the second chromosome. This region is flanked at the left by the *A* locus (allele *A* or *a*) and at the right by the *B* locus (allele *B* or *b*):

$$A/a \qquad star \qquad B/b$$

Six independently induced *star* mutations are each made homozygous, with both *A B/A B* and *a b/a b* constitutions, and the six are intercrossed to study complementation at the *star* locus. The results follow, where "+" indicates wild eye and "s" indicates star eye, both being phenotypes of the F_1.

a s b/a s b	*A s B/A s B*					
	1	2	3	4	5	6
1	s	+	s	s	+	[+]
2		s	+	[+]	s	+
3			s	s	+	+
4				s	+	+
5					s	+
6						s

a. How many genes are at the *star* location, and which mutational sites are in each gene?

b. The heterozygotes in brackets are allowed to produce gametes. In both cases, s^+ recombinant gametes are identified. The gametes are tested for the flanking marker conformation, which is *a B* in both the 1 × 6 heterozygote and the 2 × 4 heterozygote. Order the genes in relation to the *A/a* and *B/b* loci.

34. In Norway in 1934, a mother with two mentally retarded children consulted the physician Asbjørn Følling. In the course of the interview, Følling learned that the urine of the children had a curious odor. He later tested their urine with ferric chloride and found out that, whereas normal urine gives a brownish color, the children's urine stained green. He deduced that the chemical responsible must be phenylpyruvic acid. Because of chemical similarity to phenylalanine, it seemed likely that this substance had been formed from phenylalanine in the blood, but there was no assay for phenylalanine. However, a certain bacterium could convert phenylalanine into phenylpyruvic acid so that the level of phenylalanine could be measured by using the ferric chloride test. It was found that indeed the children had high levels of phenylalanine in their blood, and it was probably the source of the phenylpyruvic acid. Følling noted that, in general in families with retarded children whose urine stained green with ferric chloride, there was a proportion of $\frac{1}{4}$ affected children and $\frac{3}{4}$ unaffected. This disease, which came to be known as phenylketonuria, seemed to have a genetic basis and was inherited as a simple Mendelian recessive.

It became clear that phenylalanine was the culprit and that this chemical built up in the PKU patients and was converted into high levels of phenylpyruvic acid, which then interfered with the normal development of nervous tissue. This finding led to the formulation of a special diet low in phenylalanine, which could be fed to newborn babies diagnosed with PKU and which allowed normal development to continue without retardation. Indeed, it was found that, after the child's nervous system had developed, the patient could be taken off the special diet. However, tragically, PKU women who had developed normally with the special diet were found to have babies who were born mentally retarded, and the special diet had no effect on these children.

a. Why do you think the babies of the PKU mothers were born retarded?

b. Explain the reason for the difference in the results of the PKU babies and those of the babies of PKU mothers.

c.   Propose a treatment that might allow PKU mothers to have unaffected children.

d.   Write a short essay on PKU integrating concepts at the genetic, diagnostic, enzymatic, physiological, pedigree, and population levels.

 Unpacking the Problem

1. Draw the pedigree of the Norwegian family as far as possible. Do you think that some siblings might have been unaffected?

2. Could the original children brought to Følling (assume that they were 10 years old) have benefited from the special diet if started at that age?

3. Why were the bacteria needed by Følling?

4. Why was a blood test considered important?

5. Why was there phenylpyruvic acid in the urine?

6. What is the relation expected between blood concentrations and urine concentrations?

7. Was the green substance in the ferric chloride test itself important?

8. What phenylalanine and phenylpyruvic acid concentrations are expected in the blood and urine of unaffected children?

9. What was the source of the odor in the children's urine?

10. What were the genotypes of the parents?

11. Would families with parents having these genotypes be common?

12. What genotypes of offspring are expected and in what proportions?

13. What would be the genotypes of parents in most families in the population?

14. Why was it inferred that the disease was inherited?

15. Why was it inferred to be inherited as a Mendelian recessive?

16. At which developmental stage(s) is most of the nervous system developing?

17. Why was it believed that PKU adults no longer needed to be on the special diet?

18. What are the relative volumes of the maternal and fetal blood circulation systems?

19. Which types of entities are able to pass the placental barrier (cells, macromolecules, small molecules)?

20. Which PKU-associated substances might pass into the mother from the child? into the child from the mother?

21. What is an essential amino acid?

22. Is phenylalanine an essential amino acid?

23. How does PKU illustrate the one-gene–one-enzyme concept?

24. Why is PKU recessive?

25. What is the relevance of biochemical pathways to PKU?

26. In the chapter, find a chart that shows the position of the PKU-associated chemicals in the chemistry of the cell.

27. Draw a diagram relating chromosome, gene, mutant site, mRNA, enzyme, substrate, and product in normal individuals and in PKU sufferers.

10

MOLECULAR BIOLOGY OF GENE FUNCTION

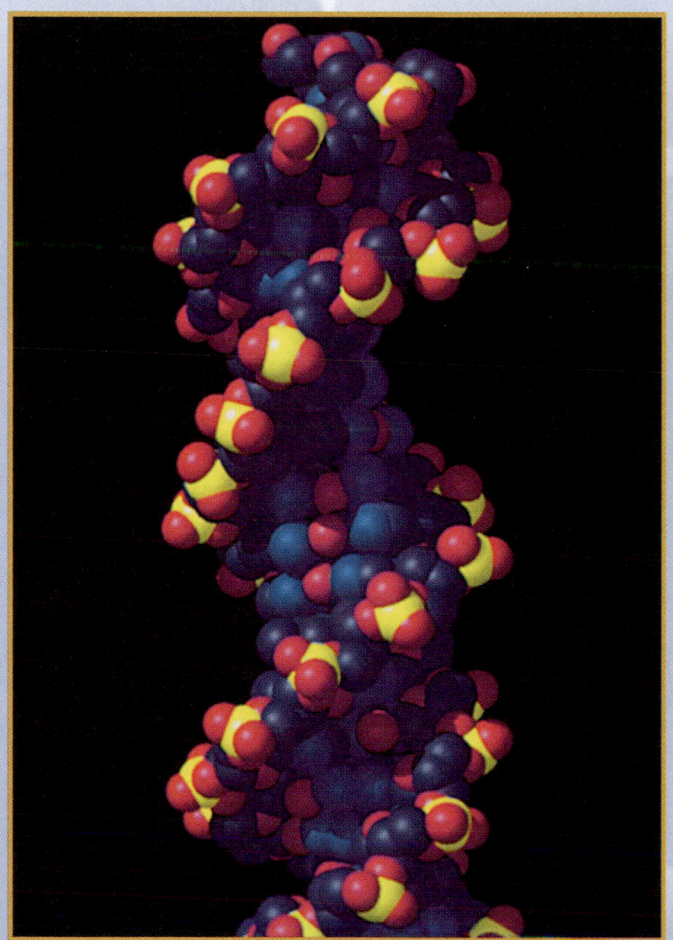

Computer model of DNA.

(J. Newdol, Computer Graphics Laboratory, University of California, San Francisco. Copyright © Regents, University of California.)

Key Concepts

DNA is transcribed into an mRNA molecule, which is then translated during protein synthesis.

Translation requires transfer RNAs and ribosomes.

The genetic code is a nonoverlapping triplet code.

Special sequences signal the initiation and termination of both transcription and translation.

In eukaryotes, the initial RNA transcript is processed in several ways to generate the final mRNA.

Many eukaryotic genes contain segments of DNA, termed introns, that interrupt the normal gene coding sequence.

The primary eukaryotic transcript is spliced in one of a variety of ways to remove the RNA encoded by the intron and to yield the final mRNA.

In this chapter, we shall see how genetic information is turned into functional macromolecules. The initial products of all genes are ribonucleic acids (RNAs). RNA is similar to DNA, except that ribose is the sugar used in RNA, and uracil replaces thymine. RNA is produced by a process that copies the nucleotide sequence in DNA. Because this process is reminiscent of transcribing (copying) written words, the synthesis of RNA is called **transcription,** and the RNA product is termed a **transcript.** We shall see in this chapter that one of the first clues to how DNA directs the synthesis of proteins came from bacteriophages, when it was shown that gene expression resulted in the transcription of RNA molecules from a DNA template. Transcription is catalyzed by an enzyme, **RNA polymerase,** and follows rules similar to those of replication.

There are several classes of RNA that we shall deal with in this chapter. *Informational RNAs* are intermediates in the process of decoding genes into polypeptide chains. The informational RNA from which proteins are directly synthesized is **messenger RNA (mRNA).** In prokaryotes, the transcript, as it is synthesized directly from the DNA (the primary transcript), is the mRNA. In eukaryotes, however, the primary transcript is processed through modification of the 5′ and 3′ ends and removal of pieces (introns) of the primary transcript. At the end of this pre-mRNA processing, an mRNA is produced. These steps in producing mRNA are considered later in the chapter. The sequence of nucleotides in mRNA is converted into the sequence of amino acids in a polypeptide chain by a process called **translation.**

Functional RNAs are never translated into polypeptides. Their action is purely at the level of the RNA, and they play many diverse roles. Two classes are found in all organisms. **Transfer RNA (tRNA)** molecules transport amino acids to the mRNA during protein synthesis. The tRNAs are general components of the translation machinery. **Ribosomal RNAs (rRNAs)** combine with an array of different proteins to form ribosomes, the "machines" used for protein synthesis. Two other classes of functional RNAs in information processing are specific to eukaryotes. **Small nuclear RNAs (snRNAs)** take part in the splicing of primary transcripts into messenger RNAs in the nucleus. Specific proteins combine with snRNAs to form small ribonucleoprotein particles **(snRNPs),** which serve as a platform for splicing reactions. **Small cytoplasmic RNAs (scRNAs)** direct protein traffic within the eukaryotic cell. Specifically, they ensure that polypeptides destined, for example, to be secreted from the cell are inserted into one of the membrane compartments of the cell (the rough endoplasmic reticulum). This begins the process of protein secretion.

All DNA and RNA function is based on two key elements:

1. Complementary bases in single-stranded nucleotide chains can hydrogen bond to form double-stranded structures.

2. Particular base sequences in single-stranded or double-stranded nucleic acids can be recognized by specific nucleic acid-binding proteins. Look for the application of this principle in the following sections.

Properties of RNA

Although both RNA and DNA are nucleic acids, RNA differs in several important ways:

1. RNA is usually single stranded, not a double helix. One consequence of its being single stranded is that RNA can form a much greater variety of complex three-dimensional molecular shapes than can double-stranded DNA. We shall consider this ability in more detail later in the chapter.

2. RNA has the sugar **ribose** in its nucleotides, rather than deoxyribose. The two sugars differ in the presence or absence of just one oxygen atom. Analogous to the individual strands of DNA, RNA has a phosphate-ribose backbone, with a base covalently linked to the 1′ position on each ribose.

Ribose   Deoxyribose

3. RNA nucleotides carry the bases adenine, guanine, and cytosine, but the pyrimidine base **uracil (U)** is found in place of thymine. However, uracil does form hydrogen bonds with adenine, just as thymine does.

Uracil

Transcription

Early investigators had good reason for thinking that information is not transferred directly from DNA to protein. In a eukaryotic cell, DNA is found in the nucleus, whereas protein is known to be synthesized in the cytoplasm. An intermediate is needed.

Early experiments suggesting RNA intermediate

If cells are fed radioactive RNA precursors, then the labeled RNA shows up first of all in the nucleus, indicating that the RNA is synthesized there. In a **pulse-chase** experiment, a

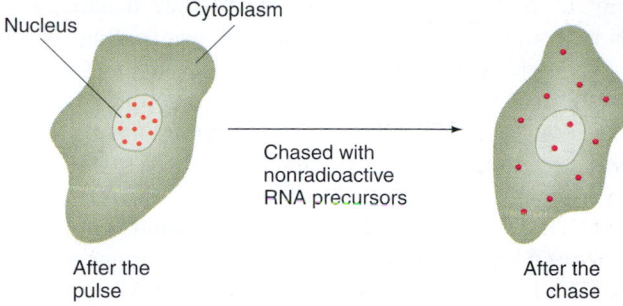

Nucleus — Cytoplasm

Chased with
nonradioactive
RNA precursors

After the pulse | After the chase

Figure 10-1 RNA synthesized during one short time period is labeled by feeding the cell a brief "pulse" of radioactive RNA precursors, followed by a "chase" of nonradioactive precursors. In an autoradiograph, the labeled RNA appears as dark grains. Apparently, the RNA is synthesized in the nucleus and then moves out into the cytoplasm.

brief pulse of labeled RNA precursors is given. These precursors are incorporated into RNA molecules. The cells are then transferred to medium with unlabeled RNA precursors. This "chases" the label out of the RNA because, as the RNA breaks down, only the unlabeled precursors are used to synthesize new RNA molecules. The pulse-chase protocol enables one to track a population of RNA molecules, synthesized almost simultaneously, over time. In samples taken after the chase, the labeled RNA is found in the cytoplasm (Figure 10-1). Apparently, the RNA is synthesized in the nucleus and then moves into the cytoplasm, where proteins are synthesized. Thus, RNA is a good candidate as an information-transfer intermediary between DNA and protein.

In 1957, Elliot Volkin and Lawrence Astrachan made a significant observation. They found that one of the most striking molecular changes when *E. coli* is infected with the phage T2 is a rapid burst of RNA synthesis. Furthermore, this phage-induced RNA "turns over" rapidly, as shown in the following experiment. The infected bacteria are first pulsed with radioactive uracil (a specific precursor of RNA); the bacteria are then chased with cold uracil. The

RNA recovered shortly after the pulse is labeled, but that recovered somewhat longer after the chase is unlabeled, indicating that the RNA has a very short lifetime. Finally, when the nucleotide contents of *E. coli* and T2 DNA are compared with the nucleotide content of the phage-induced RNA, the RNA is found to be very similar to the phage DNA.

The tentative conclusion from the two aforedescribed experiments is that RNA is synthesized from DNA and that it is somehow used to synthesize protein. We can now outline three stages of information transfer (Figure 10-2): *replication* (the synthesis of DNA), *transcription* (the synthesis of an RNA copy of a part of the DNA), and *translation* (the synthesis of a polypeptide directed by the RNA sequence).

Complementarity and asymmetry in RNA synthesis

The similarity of RNA to DNA suggested that transcription may be based on the complementarity of bases, which is also the key to DNA replication. A transcription enzyme, RNA polymerase, could carry out transcription from a DNA template strand in a fashion quite similar to replication.

In fact, this model of transcription is confirmed cytologically (Figure 10-3). The fact that RNA can be synthesized with DNA acting as a template is also demonstrated

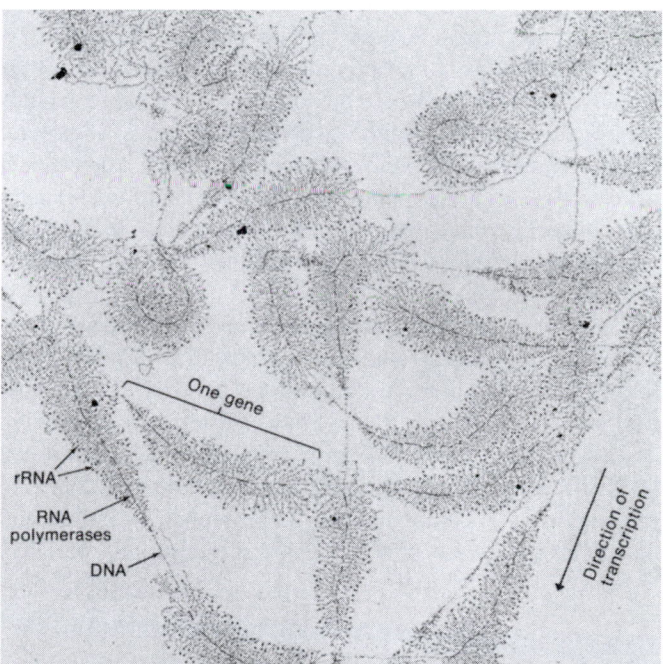

One gene

rRNA
RNA polymerases
DNA

Direction of transcription

Figure 10-3 Tandem repeats of ribosomal RNA (rRNA) genes being transcribed in the nucleolus of *Triturus viridiscens* (an amphibian). (rRNA is a component of the ribosome, a cellular organelle.) Along each gene, many RNA polymerase molecules are attached and transcribing in one direction. The growing RNA molecules appear as threads extending out from the DNA backbone. The shorter RNA molecules are nearer the beginning of transcription; the longer ones have almost been completed. Hence, the "Christmas tree" appearance. (Photograph from O. L. Miller, Jr., and Barbara A. Hamkalo.)

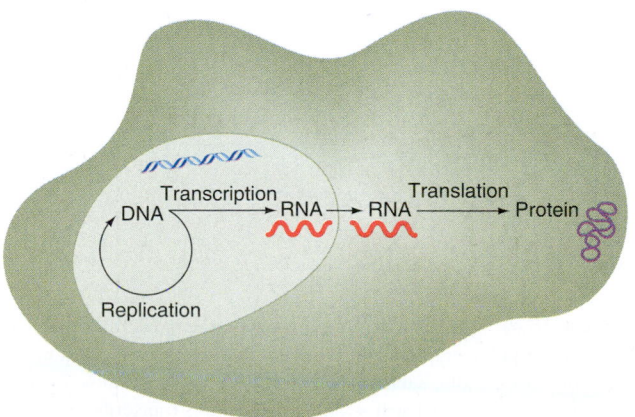

Transcription — RNA | Translation
DNA — RNA — RNA — Protein

Replication

Figure 10-2 The three processes of information transfer: replication, transcription, and translation.

by synthesis in vitro of RNA from nucleotides in the presence of DNA by using an extractable RNA polymerase. Whatever source of DNA is used, the RNA synthesized has an (A + U)/(G + C) ratio similar to the (A + T)/(G + C) ratio of the DNA (Table 10-1). This experiment does not indicate whether the RNA is synthesized from both DNA strands or from just one, but it does indicate that the linear frequency of the A−T pairs (in comparison with the G−C pairs) in the DNA precisely corresponds to the relative abundance of (A + U) in the RNA. (These points are difficult to grasp without drawing some diagrams; Problem 2 at the end of this chapter provides some opportunities to clarify these notions.)

To test the complementarity of DNA with RNA, investigators can apply the specificity and precision of nucleic acid hybridization. DNA can be denatured and mixed with the RNA formed from it. On slow cooling, some of the RNA strands anneal with complementary DNA to form a DNA-RNA hybrid. The DNA-RNA hybrid differs in density from the DNA-DNA duplex, so its presence can be detected by ultracentrifugation in cesium chloride (CsCl). Nucleic acids will anneal in this way only if there are stretches of base-sequence complementarity, so the experiment does prove that the RNA transcript is complementary in base sequence to the parent DNA.

Can we determine whether RNA is synthesized from both or only one of the DNA strands? It seems reasonable that only one strand is used, because transcription of RNA from both strands would produce two complementary RNA strands from the same stretch of DNA, and these strands presumably would produce two different kinds of protein (with different amino acid sequences). In fact, a great deal of chemical evidence confirms that transcription usually takes place on only one of the DNA strands (although not necessarily the same strand throughout the entire chromosome).

The hybridization experiment can be extended to explore this problem. If the two strands of DNA have distinctly different purine:pyrimidine ratios, they can be purified separately because they have different densities in cesium chloride. The RNA made from a stretch of DNA can be purified and annealed separately to each of the strands to see whether it is complementary to only one. J. Marmur and his colleagues were able to separate the strands of DNA

from the *Bacillus subtilis* phage SP8. They denatured the DNA, cooled it rapidly to prevent reannealing of the strands, and then separated the strands in CsCl. They showed that the SP8 RNA hybridizes to only one of the two strands, proving that transcription is **asymmetrical** — that it takes place only on one DNA strand.

Although, for each gene, RNA is transcribed from only one of the DNA strands, the same DNA strand is not necessarily transcribed throughout the entire chromosome or through all stages of the life cycle. The RNA produced at different stages in the cycle of a phage hybridizes to different segments of the chromosome, showing the different genes that are activated at each stage (Figure 10-4). In λ phage, each of the two DNA strands is partly transcribed at a different stage. In phage T7, however, the same strand is transcribed for both early-acting and late-acting genes. Figure 10-5 shows a sequence of RNA made from the DNA template strand. The DNA strand that is transcribed for a

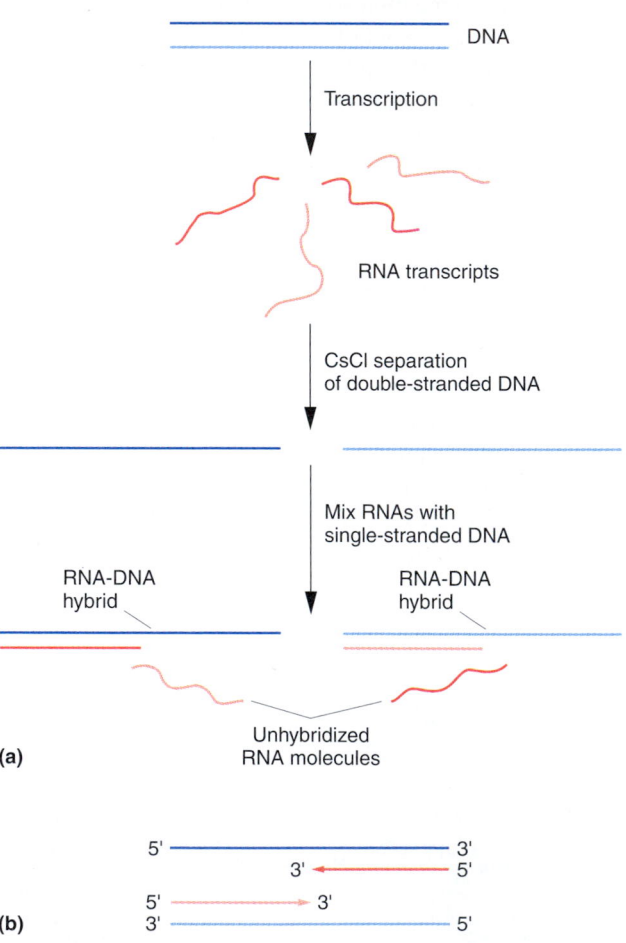

(a)

(b)

Figure 10-4 (a) DNA-RNA hybridization demonstrates that each RNA transcript is complementary to only one strand of the parent DNA. In this example, each of the two DNA strands is transcribed, but transcription is asymmetrical — only one strand is transcribed at any particular location. (b) A map of this hypothetical genome, showing the direction of transcription for the two transcripts from opposite DNA strands.

10-1 TABLE	Nucleotide Ratios in Various DNAs and in Their Transcripts (in Vitro)	
DNA source	$\frac{(A + T)}{(G + C)}$ of DNA	$\frac{(A + U)}{(G + C)}$ of RNA
T2 phage	1.84	1.86
Cow	1.35	1.40
Micrococcus (bacterium)	0.39	0.49

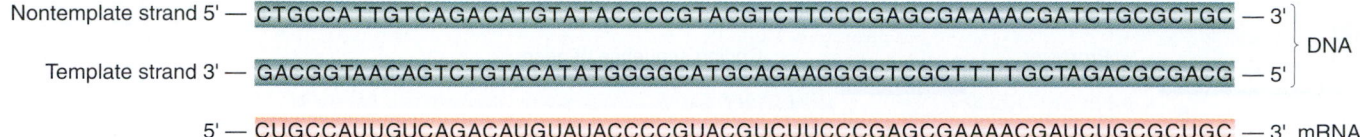

Nontemplate strand 5′ — CTGCCATTGTCAGACATGTATACCCCGTACGTCTTCCCGAGCGAAAACGATCTGCGCTGC — 3′ ⎱ DNA

Template strand 3′ — GACGGTAACAGTCTGTACATATGGGGCATGCAGAAGGGCTCGCTTTTGCTAGACGCGACG — 5′ ⎰

5′ — CUGCCAUUGUCAGACAUGUAUACCCCGUACGUCUUCCCGAGCGAAAACGAUCUGCGCUGC — 3′ mRNA

Figure 10-5 The mRNA sequence is complementary to the DNA template strand from which it is synthesized. The sequence shown here is from the gene for the enzyme β-galactosidase, which takes part in lactose metabolism.

given mRNA is termed the **template strand.** The complementary DNA strand is called the **nontemplate strand.** Note that the mRNA has the same sequence (with U substituted for T) as that of the nontemplate strand.

Transcription and RNA polymerase

As described earlier, transcription relies on the complementary pairing of bases. The two strands of the double helix separate locally, and one of the separated strands acts as a template. Next, free nucleotides are aligned on the DNA template by their complementary bases in the template. The free ribonucleotide A aligns with T in the DNA, G with C, C with G, and U with A. The process is catalyzed by the enzyme **RNA polymerase,** which attaches and moves along the DNA adding ribonucleotides in the growing RNA as shown in Figure 10-6a. Hence, already we see the two principles of base complementarity and binding proteins (in this case, the RNA polymerase) in action.

RNA growth is always in the 5′ → 3′ direction: in other words, nucleotides are always added at a 3′ growing tip, as shown in Figure 10-6b. Because of the antiparallel nature of the nucleotide pairing, the fact that RNA is synthe-

sized 5′ → 3′ means that the template strand must be oriented 3′ → 5′.

RNA polymerase

In most prokaryotes, a single RNA polymerase species transcribes all types of RNA. Figure 10-7 shows the structure of RNA polymerase from *E. coli.* We can see that the enzyme consists of four different subunit types. The beta (β) subunit

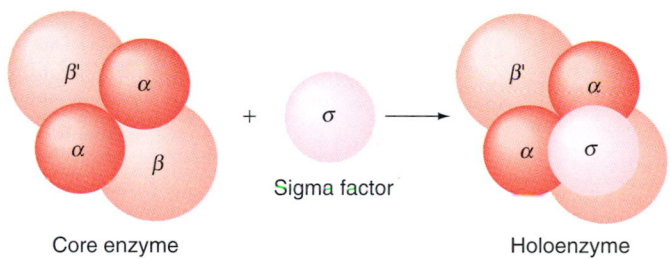

Core enzyme + Sigma factor → Holoenzyme

Figure 10-7 The structure of RNA polymerase. The core enzyme contains two α polypeptides, one β polypeptide, and one β′ polypeptide. The addition of the σ subunit allows initiation at promoter sites. (Promoters are discussed in the next subsection.)

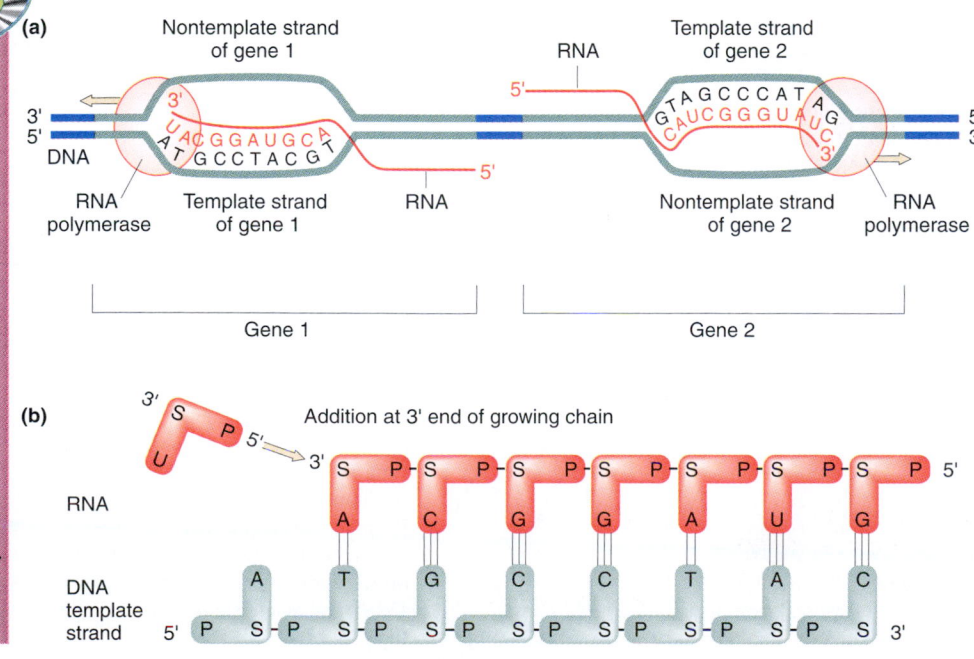

(a)

(b)

Figure 10-6 Transcription of two genes. (a) RNA polymerase moves from the 3′ end of the template strand, creating an RNA strand that grows in a 5′ → 3′ direction (because it must be antiparallel to the template strand). Note that some genes are transcribed from one strand of the DNA double helix; other genes use the other strand as the template. (b) A uracil is being added to the 3′ end of the transcript for gene 1. Growth is thus 5′ → 3′.

ANIMATED ART
Transcription

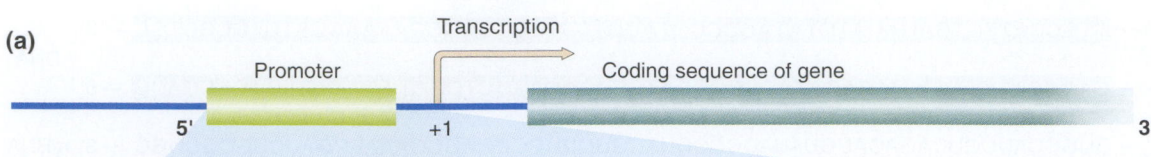

(a)

Transcription

Promoter   Coding sequence of gene

5'   +1   3'

(b) Strong *E. coli* promoters

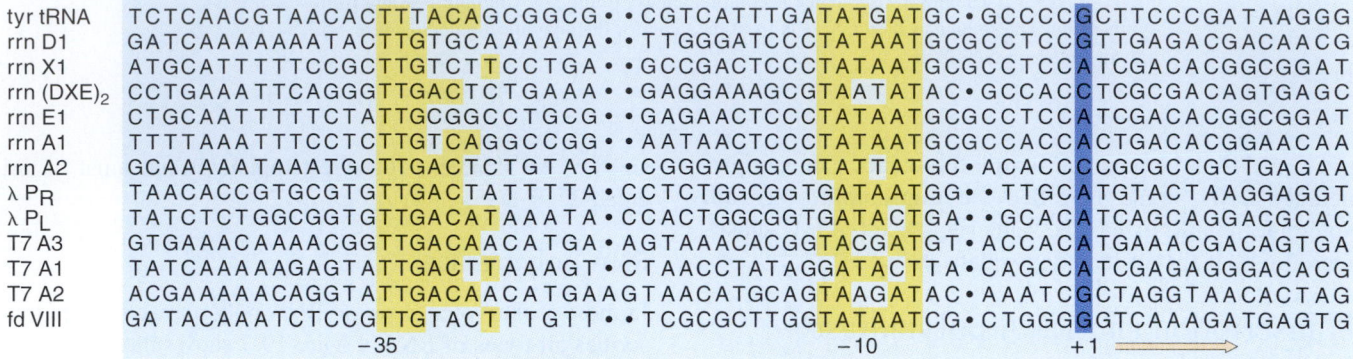

tyr tRNA   TCTCAACGTAACACTTTACAGCGGCG • • CGTCATTTGATATGATGC • GCCCCGCTTCCCGATAAGGG
rrn D1   GATCAAAAAAATACTTGTGCAAAAAA • • TTGGGGATCCCTATAATGCGCCTCCGTTGAGACGACAACG
rrn X1   ATGCATTTTTCCGCTTGTCTTCCTGA • • GCCGACTCCCTATAATGCGCCTCCATCGACACGGCGGAT
rrn (DXE)₂   CCTGAAATTCAGGGTTGACTCTGAAA • • GAGGAAAGCGTAATATAC • GCCACCTCGCGACAGTGAGC
rrn E1   CTGCAATTTTTCTATTGCGGCCTGCG • • GAGAACTCCCTATAATGCGCCTCCATCGACACGGCGGAT
rrn A1   TTTTAAATTTCCTCTTGTCAGGCCGG • • AATAACTCCCTATAATGCGCCACCACTGACACGGAACAA
rrn A2   GCAAAAATAAATGCTTGACTCTGTAG • • CGGGAAGGCGTATTATGC • ACACCCCGCGCCGCTGAGAA
λ P_R   TAACACCGTGCGTGTTGACTATTTTA • CCTCTGGCGGTGATAATGG • • TTGCATGTACTAAGGAGGT
λ P_L   TATCTCTGGCGGTGTTGACATAAATA • CCACTGGCGGTGATACTGA • • GCACATCAGCAGGACGCAC
T7 A3   GTGAAACAAAACGGTTGACAACATGA • AGTAAACACGGTACGATGT • ACCACATGAAACGACAGTGA
T7 A1   TATCAAAAAGAGTATTGACTTAAAGT • CTAACCTATAGGATACTTA • CAGCCATCGAGAGGGACACG
T7 A2   ACGAAAAACAGGTATTGACAACATGAAGTAACATGCAGTAAGATAC • AAATCGCTAGGTAACACTAG
fd VIII   GATACAAATCTCCGTTGTACTTTGTT • • TCGCGCTTGGTATAATCG • CTGGGGGGTCAAAGATGAGTG

                        −35   −10   +1 →

(c) Consensus sequences for all *E. coli* promoters

−35 region   15–17 bp   −10 region
TTGACAT   **TA**TAAT

Figure 10-8   Promoter sequence. (a) The promoter lies "upstream" (toward 5′ end) of the initiation point and coding sequences. (Remember, transcription actually takes place on the complementary strand.) (b) Promoter sites have regions of similar sequences, as indicated by the yellow region in the 13 different promoter sequences in *E. coli*. Spaces (dots) are included to maximize homology at consensus sequences. The gene governed by each promoter sequence is indicated on the left. Numbering is according to the number of bases before (−) or after (+) the RNA synthesis initiation point. (c) Color coding in the consensus sequences for *all E. coli* promoters is as follows: blue letters, > 75%; boldface black letters, 50–75%; black letters, 40–50%. [From H. Lodish, D. Baltimore, A. Berk, S. L. Zipursky, P. Matsudaira, and J. Darnell, *Molecular Cell Biology*, 3d ed. Copyright © 1995 by Scientific American Books, Inc. See W. R. McClure, *Annual Review of Biochemistry*, 54, 1985, 171 (Consensus Sequences).]

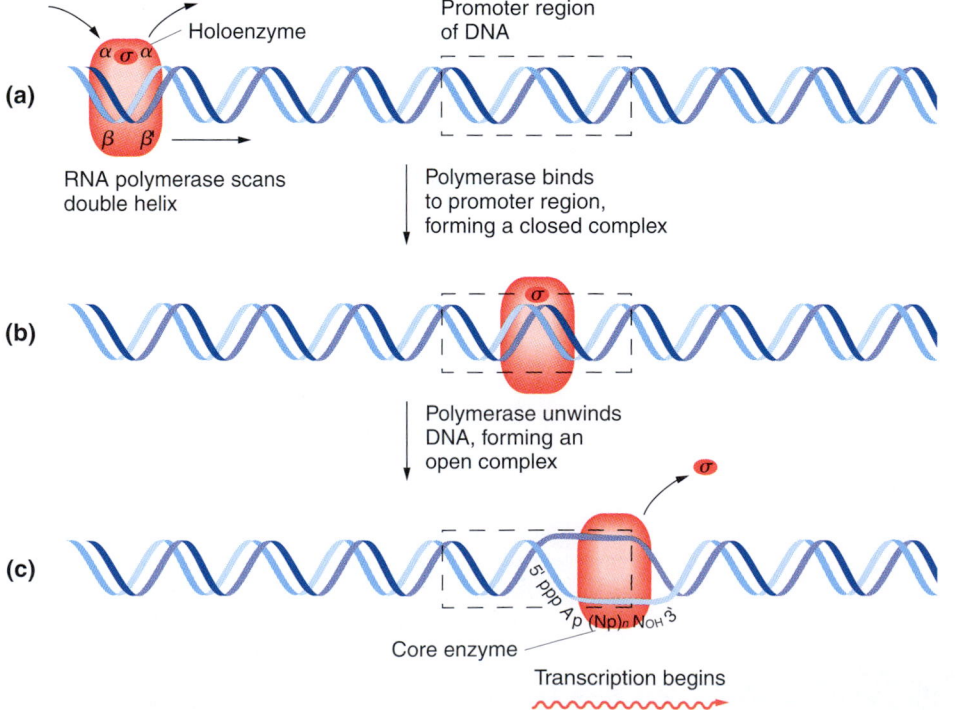

(a)

Holoenzyme   Promoter region of DNA

α σ α
β β'

RNA polymerase scans double helix

Polymerase binds to promoter region, forming a closed complex

(b)

σ

Polymerase unwinds DNA, forming an open complex

(c)

σ

5′ pppAp(Np)ₙN_OH 3′

Core enzyme

Transcription begins

Figure 10-9   Initiation of transcription. (a) RNA polymerase searches for a promoter site. (b) It recognizes a promoter site and binds tightly, forming a closed complex. (c) The holoenzyme unwinds a short stretch of DNA, forming an open complex. Transcription begins, and the σ factor is released. The RNA transcript is shown, beginning with adenosine triphosphate (pppA), carrying on through an indeterminate number of nucleotides [(Np)ₙ], and ending with N_OH. (After J. Darnell, H. Lodish, and D. Baltimore, *Molecular Cell Biology*, 2d ed. Copyright © 1990 by Scientific American Books, Inc.)

has a molecular weight of 150,000, beta prime (β') 160,000, alpha (α) 40,000, and sigma (σ) 70,000. The σ subunit can dissociate from the rest of the complex, leaving the **core enzyme.** The complete enzyme with σ is termed the **RNA polymerase holoenzyme** and is necessary for correct initiation of transcription, whereas the core enzyme can continue transcription after initiation.

Next, let's look at the three distinct stages of transcription: **initiation, elongation,** and **termination.**

Initiation

The regions of the DNA that signal initiation of transcription in prokaryotes are termed **promoters.** (We consider their role in gene regulation in Chapter 11.) Figure 10-8 shows the promoter sequences from 13 different transcription initiation points on the *E. coli* genome. The bases are aligned according to homologies, or similar base sequences, that appear just before the first base transcribed (designated the "initiation site" in Figure 10-8).

Note in Figure 10-8 that two regions of partial homology appear in virtually each case. These regions have been termed the -35 and -10 regions because of their locations relative to the transcription initiation point. At the bottom of Figure 10-8, an ideal, or consensus, sequence of a promoter is given. Physical experiments have confirmed that RNA polymerase makes contact with these two regions when binding to the DNA. The enzyme then unwinds DNA and begins the synthesis of an RNA molecule.

The dissociative subunit of RNA polymerase, the σ factor, allows RNA polymerase to recognize and bind specifically to promoter regions. First, the holoenzyme searches for a promoter (Figure 10-9a) and initially binds loosely to it, recognizing the -35 and -10 regions. The resulting structure is termed a *closed promoter complex* (Figure 10-9b). Then, the enzyme binds more tightly, unwinding bases near the -10 region. When the bound polymerase causes this local denaturation of the DNA duplex, it is said to form an *open promoter complex* (Figure 10-9c). This initiation step, the formation of an open complex, requires the sigma factor.

Elongation

Shortly after initiating transcription, the sigma factor dissociates from the RNA polymerase. The RNA is always synthesized in the $5' \rightarrow 3'$ direction (Figures 10-10 and 10-11),

Figure 10-10 The sequential addition of nucleotides takes place one at a time in the 5'-to-3' direction. The chain grows by the formation of a bond between the 3' hydroxyl end of the growing strand and a nucleoside triphosphate, releasing one pyrophosphate ion (PP_i). This results in the net addition of one phosphate, which is incorporated into the backbone of the new strand. DNA grows by reaction with deoxyribonucleoside triphosphates, and RNA grows by reaction with ribonucleoside triphosphates. (After H. Lodish, D. Baltimore, A. Berk, S. L. Zipursky, P. Matsudaira, and J. Darnell, *Molecular Cell Biology,* 3d ed. Copyright © 1995 by Scientific American Books, Inc.)

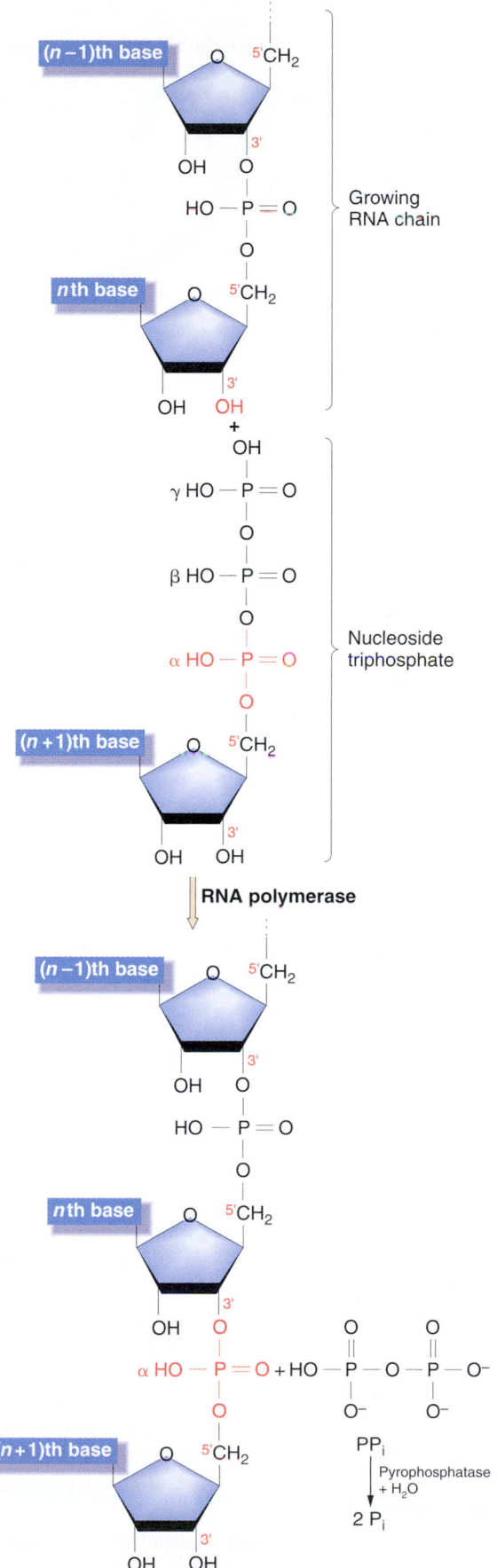

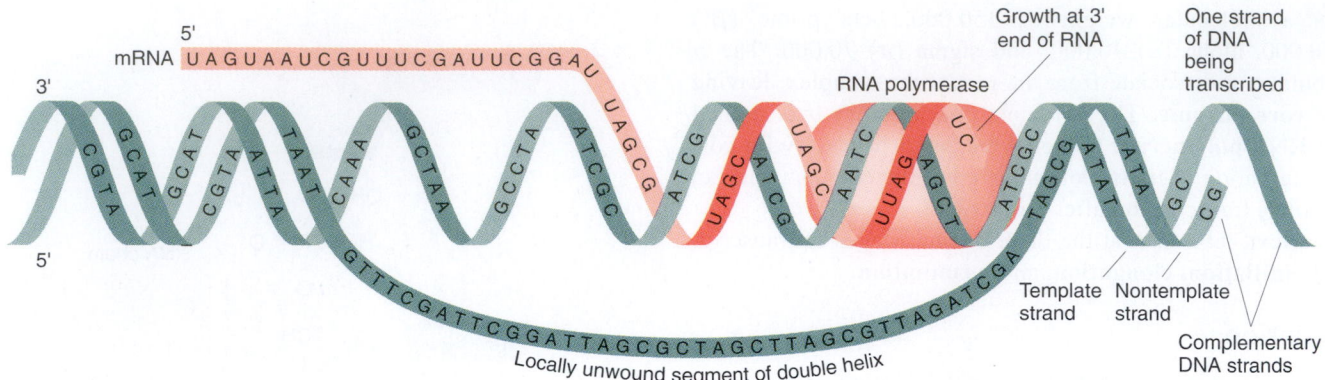

Figure 10-11   Transcription by RNA polymerase. An RNA strand is synthesized in the 5′ → 3′ direction from a locally single stranded region of DNA. (After E. J. Gardner, M. J. Simmons, and D. P. Snustad, *Principles of Genetics,* 8th ed. Copyright © 1991 by John Wiley and Sons, Inc.)

with nucleoside triphosphates (NTPs) acting as substrates for the enzyme. The following equation represents the addition of each ribonucleotide.

$$\mathrm{NTP} + (\mathrm{NMP})_n \xrightarrow[\substack{\mathrm{Mg^{2+}}\\ \mathrm{RNA}\\ \mathrm{polymerase}}]{\mathrm{DNA}} (\mathrm{NMP})_{n+1} + \mathrm{PP_i}$$

The energy for the reaction is derived from splitting the high-energy triphosphate into the monophosphate and releasing the inorganic diphosphates (PP$_i$), as shown in Figure 10-10. Figure 10-11 gives a physical picture of elongation. Note how a "transcription bubble" must be maintained, because the transcription takes place on a double-stranded template. The bubble must move along the DNA duplex during elongation. Certain sequences may cause stalling or pausing, which becomes critical for termination of transcription.

Termination

RNA polymerase also recognizes signals for chain termination, which includes the release of the nascent RNA and the enzyme from the template. There are two major mechanisms for termination in *E. coli*.

In the first mechanism, the termination is direct. The terminator sequences contain about 40 bp, ending in a GC-rich stretch that is followed by a run of six or more A's on the template strand. The corresponding GC sequences on the RNA are so arranged that the transcript in this region is able to form complementary bonds *with itself,* as can be seen in Figure 10-12. The resulting double-stranded RNA section is called a **hairpin loop.** It is followed by the terminal run of U's that correspond to the A residues on the DNA template. The hairpin loop and section of U residues appear to serve as a signal for the release of RNA polymerase and termination of transcription.

In the second type, the help of an additional protein factor, termed **rho,** is required for RNA polymerase to recog-

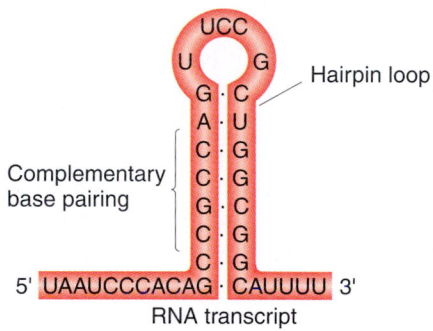

Figure 10-12   The structure of a termination site for RNA polymerase in bacteria. The hairpin structure forms by complementary base pairing *within* the RNA strand.

nize the termination signals. mRNAs with rho-dependent termination signals do not have the string of U residues at the end of the RNA and usually do not have hairpin loops. A model for rho-dependent termination is shown in Figure 10-13. Rho is a hexamer consisting of six identical subunits; the hydrolysis of ATP to ADP and P$_i$ drives the termination reaction. The first step in termination is the binding of rho to a specific site on the RNA termed *rut* (Figure 10-13a and b). After binding, rho pulls the RNA off the RNA polymerase, probably by translocating along the mRNA, as depicted in Figure 10-13b and c. The *rut* sites are located just upstream from (that is, 5′ from) sequences at which the RNA polymerase tends to pause.

The efficiency of both mechanisms of termination is influenced by surrounding sequences and other protein factors, as well.

Eukaryotic RNA

Several aspects of RNA synthesis and processing in eukaryotes are distinctly different from their counterparts in prokaryotes.

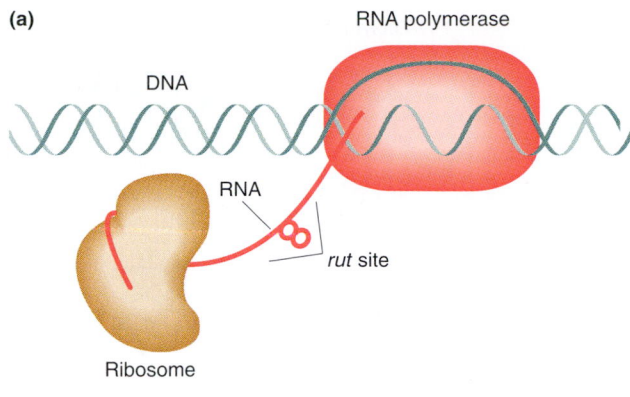

(a)

DNA

RNA polymerase

RNA

rut site

Ribosome

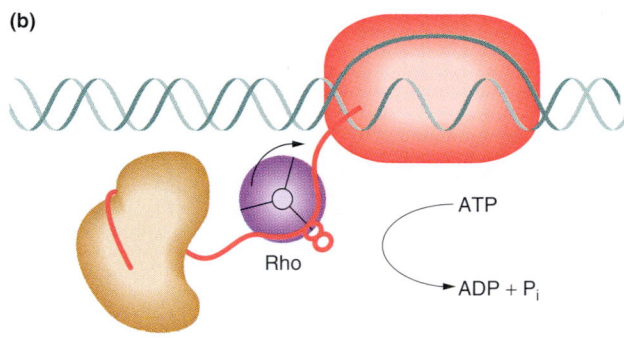

(b)

ATP

ADP + P_i

Rho

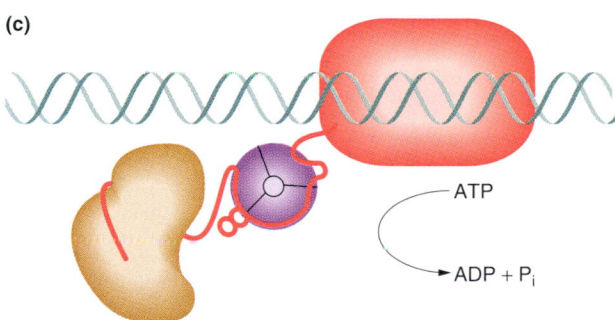

(c)

ATP

ADP + P_i

Figure 10-13   A model for rho action on a nascent cotranslated mRNA. (After J. P. Richardson, *Cell* 64, 1991, 1047–1049.)

RNA synthesis

Whereas a single RNA polymerase species synthesizes all RNAs in prokaryotes, there are three different RNA polymerases in eukaryotic systems:

1. RNA polymerase I synthesizes rRNA.

2. RNA polymerase II synthesizes mRNA. In eukaryotes, the mRNA molecules always code for one protein, whereas in prokaryotes, many mRNAs code for several proteins.

3. RNA polymerase III synthesizes tRNAs as well as small nuclear and cellular RNA molecules.

The eukaryotic polymerases have a more complex subunit structure than that of prokaryotic polymerases. Some of the subunits are similar to the corresponding *E. coli* proteins, but others are not.

RNA processing

The primary RNA transcript produced in the nucleus is usually processed in several ways before its transport to the cytoplasm, where it is used to program the translation machinery (Figure 10-14). Figure 10-15 depicts these processing events in detail. First a **cap** consisting of a 7-methylguanosine residue linked to the 5′ end of the transcript by a triphosphate bond is added during transcription. Then stretches of adenosine residues are added at the 3′ ends. These **poly(A) tails** are 150 to 200 residues long. After these modifications, a crucial **splicing** step removes internal parts of the RNA transcript. The uncovering of this process, and the corresponding realization that genes are "split," with coding regions interrupted by "intervening sequences," constitutes one of the most important discoveries in molecular genetics in the past 25 years.

Split genes

Studies of mammalian viral DNA transcripts first suggested a lack of correspondence between the viral DNA and specific mRNA molecules. As recombinant DNA techniques (see Chapter 12) facilitated the physical analysis of eukaryotic genes, it became apparent that *primary* RNA transcripts were being shortened by the elimination of internal segments

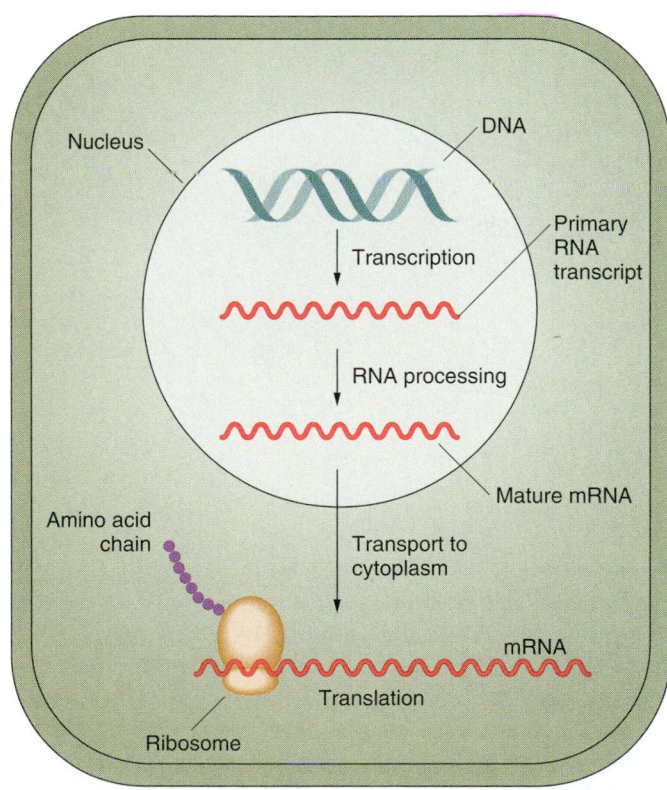

Nucleus

DNA

Transcription

Primary RNA transcript

RNA processing

Mature mRNA

Amino acid chain

Transport to cytoplasm

mRNA

Translation

Ribosome

Figure 10-14   Gene expression in eukaryotes. The mRNA is processed in the nucleus before transport to the cytoplasm. (From J. E. Darnell, Jr., "The Processing of RNA." Copyright © 1983 by Scientific American, Inc. All rights reserved.)

Figure 10-15 Processing of primary transcript. (a) Transcription is mediated by RNA polymerase. (b) Early in transcription, an enzyme, guanyltransferase, adds 7-methylguanosine (m⁷Gppp) to the 5′ end of the mRNA. (c) The sequence AAUAAA, near the 3′ end, helps signal a cleavage event (d) by an endonuclease approximately 20 bp farther downstream. (e) An enzyme, poly(A) polymerase, then adds a poly(A) tail, made up of 150 to 200 adenosine residues, to the site of this cleavage at the 3′ end, yielding (f) the complete primary mRNA. (From J. E. Darnell, Jr., "The Processing of RNA." Copyright © 1983 by Scientific American, Inc. All rights reserved.)

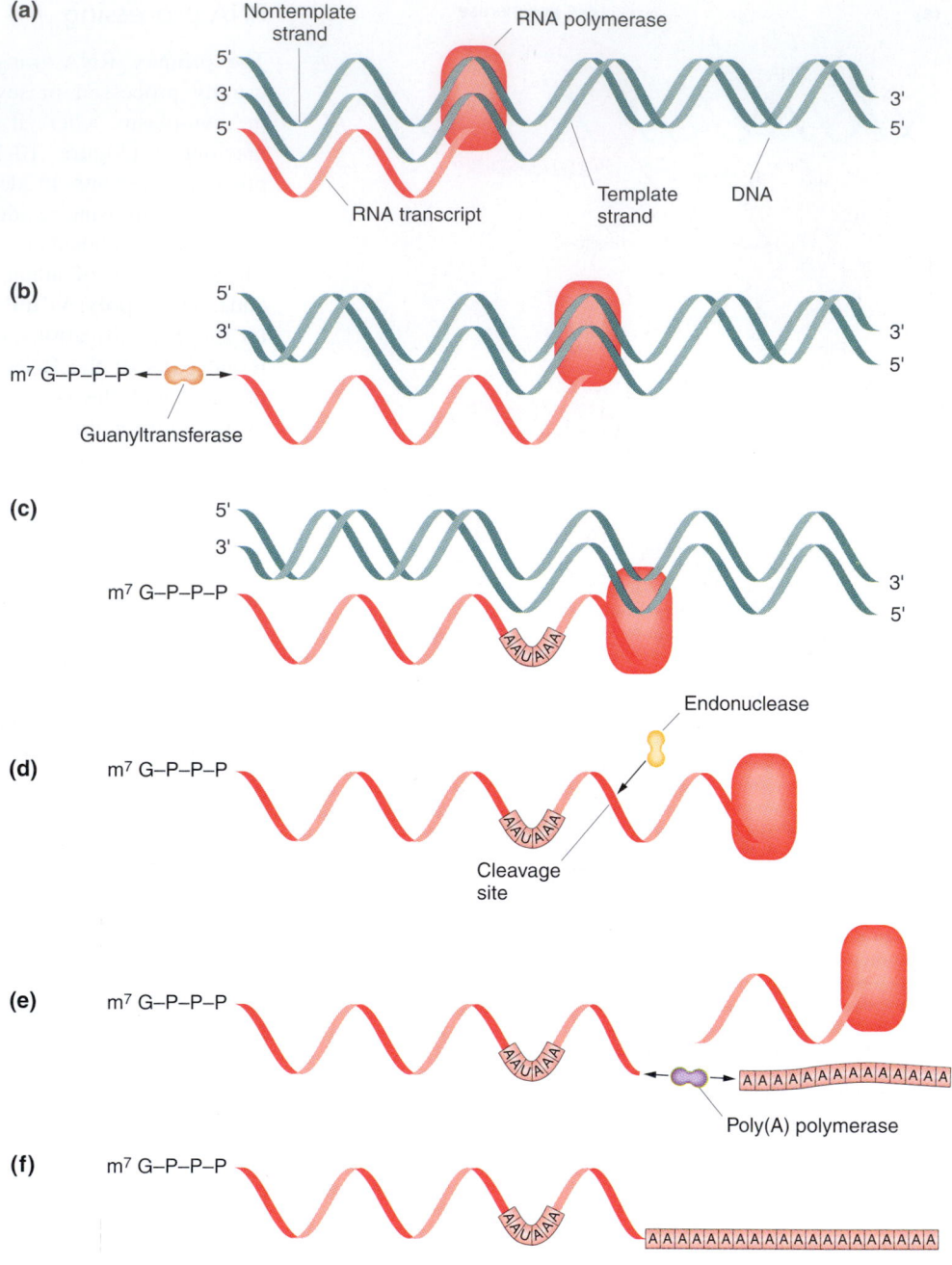

before transport into the cytoplasm. In most higher eukaryotes studied, this was found to be true not only for mRNA, but also for rRNA — and even for tRNA in some cases.

Figure 10-16 shows the organization of the gene for chicken ovalbumin, a polypeptide consisting of 386 amino acids. The DNA segments that code for the structure of the protein are interrupted by intervening sequences, termed **introns.** In Figure 10-16, these segments are designated with the letters A to G. The primary transcript is processed by a series of splicing reactions, much in the same way that a taperecorded message can be cut and pasted back together. Splicing removes the introns and brings together the coding

regions, termed **exons,** to form an mRNA, which now consists of a sequence that is completely colinear with the ovalbumin protein. The exons are indicated by the letter *L* and numbers 1 to 7 in Figure 10-16. In different genes, introns have been detected that are as large as 2000 base pairs in length. Some genes have as many as 16 introns.

It is clear that splicing occurs after transcription and in several steps, because RNA transcripts (formerly termed heterogeneous nuclear RNA, or HnRNA) that correspond to the entire genetic region (introns + exons), as well as transcripts intermediate in length, can be isolated. In these intermediate-length RNA molecules, certain introns have already been removed, but

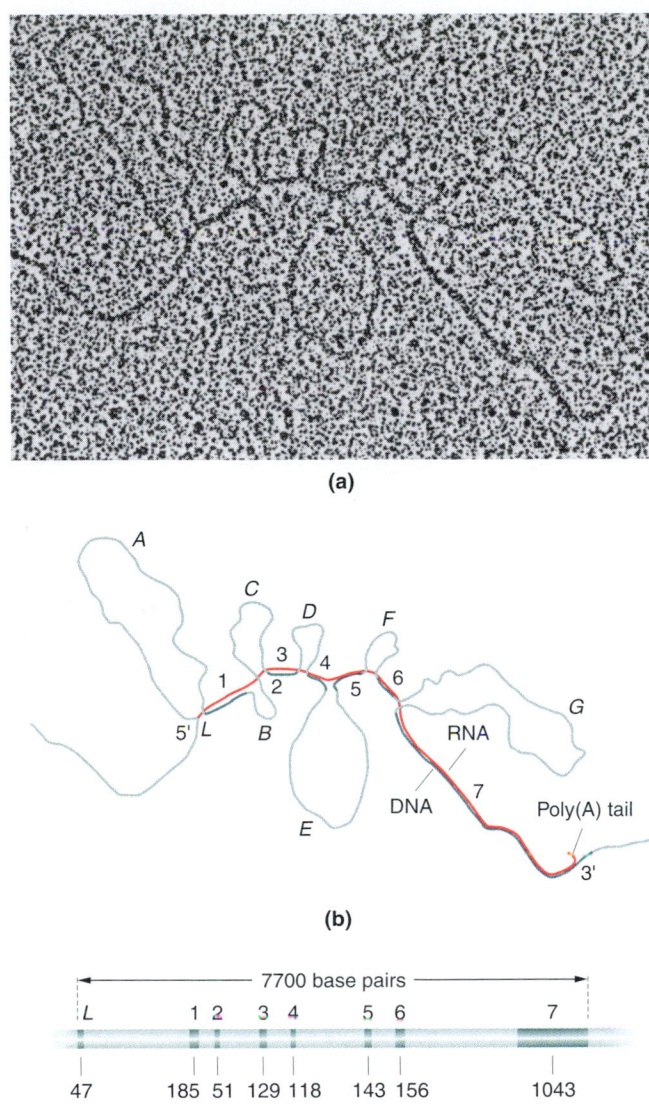

(a)

(b)

(c)

Figure 10-16 Split-gene organization of the gene for the protein ovalbumin. (a) The electron micrograph and (b) its map show the result of an experiment in which a single strand of the DNA incorporating the gene for the egg white protein ovalbumin was allowed to hybridize with ovalbumin mRNA, the molecule from which the protein is translated. The looped-out single-stranded segments of DNA represent the introns (c). The schematic representation of the gene shows the seven introns (light green), the eight exons (dark green), and the number of base pairs in each of the exons; the size of the introns ranges from 251 base pairs for intron *B* to about 1600 (for *G*). (From P. Chambon, "Split Genes." Copyright © 1981 by Scientific American, Inc. All rights reserved.)

others are retained. The entire sequence of events for RNA processing and splicing is summarized in Figure 10-17.

Alternative splicing

Alternative pathways of splicing can produce different mRNAs and subsequently different proteins from the same

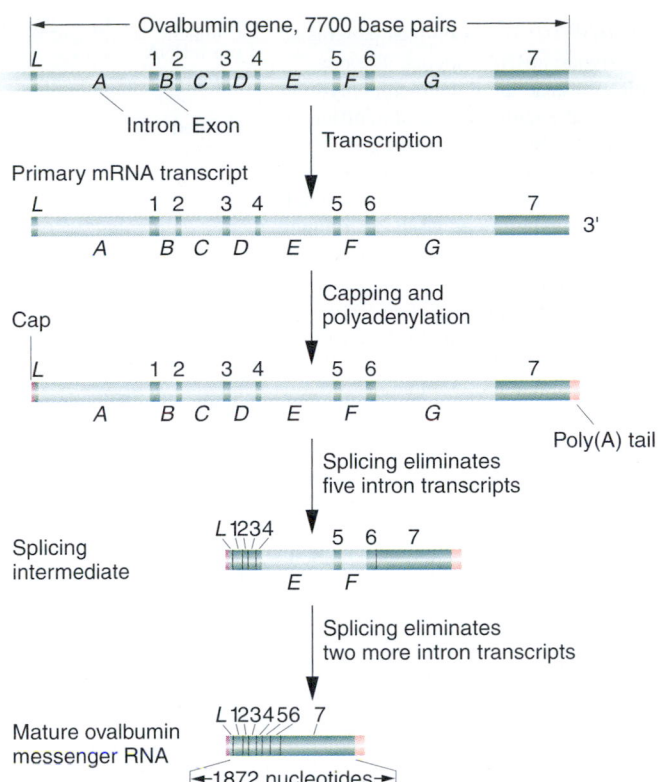

Figure 10-17 Mature mRNA is produced in a number of steps. (From P. Chambon, "Split Genes." Copyright © 1981 by Scientific American, Inc. All rights reserved.)

primary transcript. The altered forms of the same protein that are generated by alternative splicing are usually used in different cell types or at different stages of development. Figure 10-18 shows the myriad combinations produced by the differential splicing of the primary RNA transcript of the α-tropomyosin gene to ultimately generate a set of related proteins that function optimally in each cell type.

Mechanism of gene splicing

Sequencing of many exon–intron junctions has revealed sequence homologies at these points. As Figure 10-19 shows, a GU is at the 5′ splice site and AG is at the 3′ splice site. It has now been demonstrated that the interaction of small nuclear RNA molecules (snRNAs) interact with the splice site in reactions in which complementary base pairs are coordinated with the splicing enzymes. The splicing reaction itself is diagrammed in Figure 10-20, which shows an intron being cut out as a branched "lariat" structure, resulting from two successive transesterification reactions. The reactions result in the exchange of one phosphoester bond for another — fusing, or ligating, two exons.

The sRNAs associate with proteins to form small ribonuclear particles (snRNPs). In higher cells, complexes form between the snRNPs, the primary transcript, and

Figure 10-18   Complex patterns of eukaryotic mRNA splicing. The pre-mRNA transcript of the α-tropomyosin gene is alternatively spliced in different cell types. The light green boxes represent introns; the other colors represent exons. Polyadenylation signals are indicated by an A. Dashed lines in the mature mRNAs indicate regions that have been removed by splicing. TM, tropomyosin. (After J. P. Lees et al., *Molecular and Cellular Biology* 10, 1990, 1729–1742.)

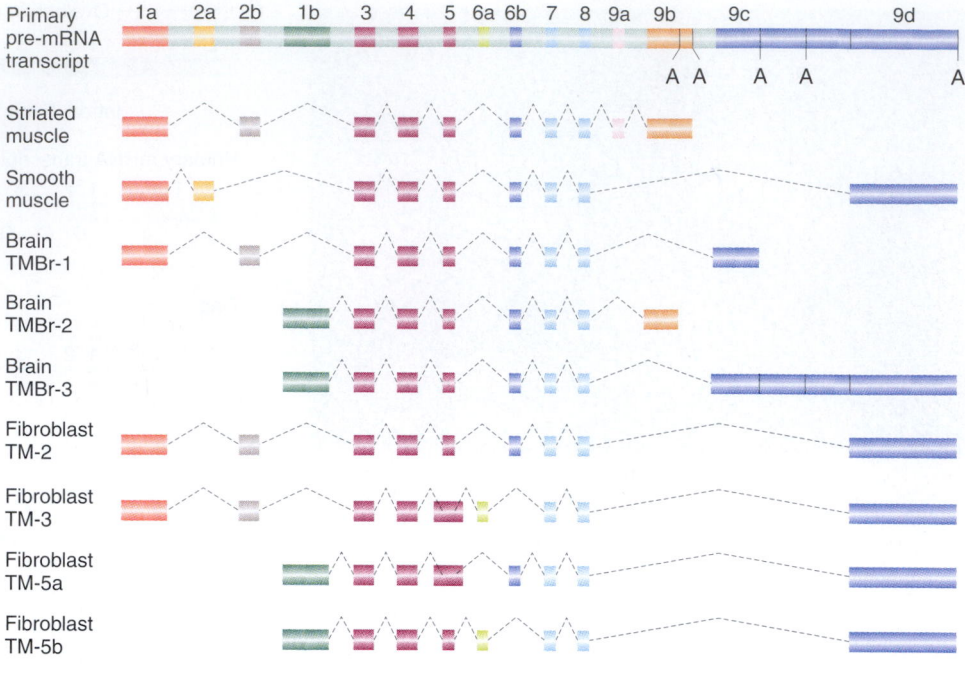

Figure 10-19   Consensus sequences of 5′ and 3′ splice junctions in eukaryotic mRNAs. Almost all introns begin with GU and end with AG. From the analysis of many exon–intron boundaries, extended consensus sequences of preferred nucleotides at the 5′ and 3′ ends have been established. In addition to AG, other nucleotides just upstream of the 3′ splice junction also are important for precise splicing. (From J. D. Watson, M. Gilman, J. Witkowski, and M. Zoller, *Recombinant DNA,* 2d ed. Copyright © 1992 by James D. Watson, Michael Gilman, Jan Witkowski, and Mark Zoller.)

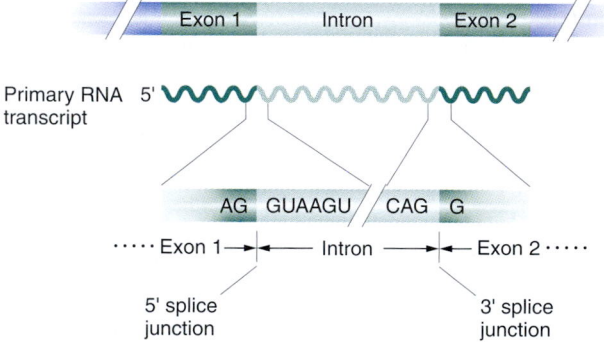

Figure 10-20   Splicing of exons in the primary transcript pre-mRNA takes place in two transesterification reactions. In the first reaction, the ester bond between the 5′ phosphorus of the intron and the 3′ oxygen (red) of exon 1 is exchanged for an ester bond with the 2′ oxygen (dark blue) of the branch-site **A** residue. In the second reaction, the ester bond between the 5′ phosphorus of exon 2 and the 3′ oxygen (light blue) of the intron is exchanged for an ester bond with the 3′ oxygen of exon 1, releasing the intron as a lariat structure and joining the two exons. Arrows show where the activated hydroxyl oxygens react with phosphorus atoms. (From H. Lodish, D. Baltimore, A. Berk, S. L. Zipursky, P. Matsudaira, and J. Darnell, *Molecular Biology of the Cell,* 3d ed. Copyright © 1995 by Scientific American Books, Inc.)

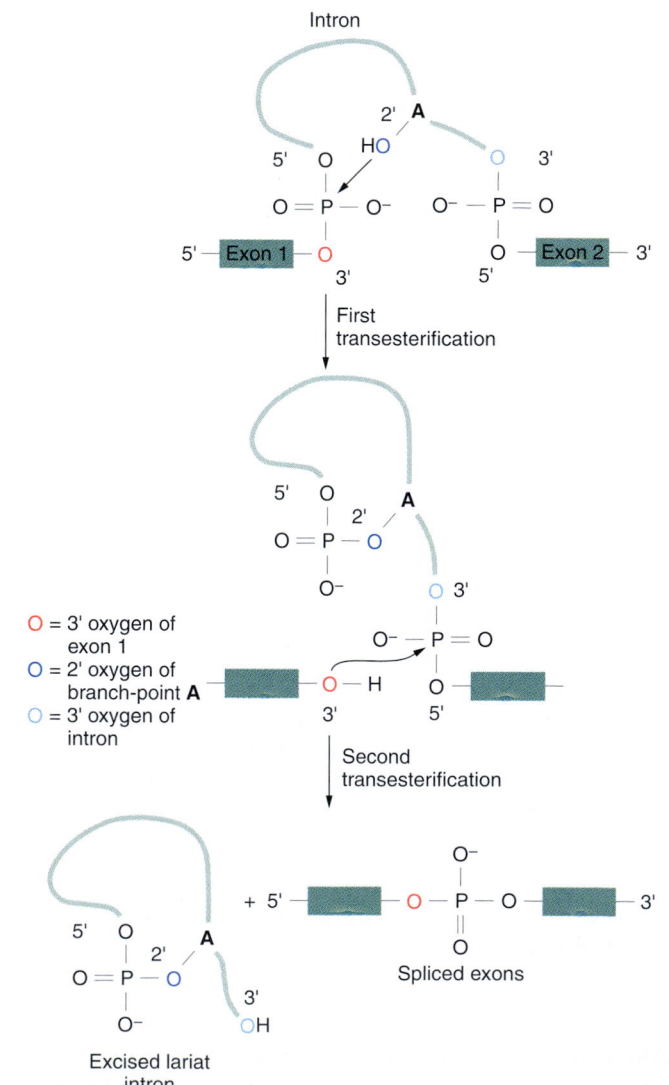

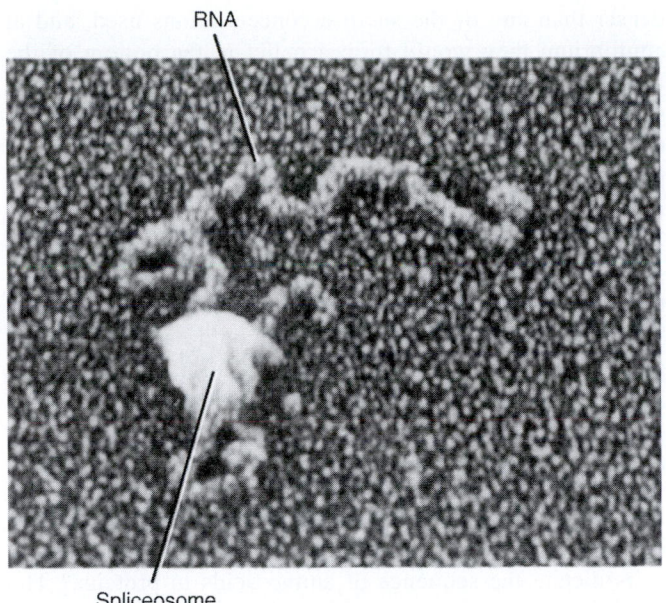

RNA

Spliceosome

Figure 10-21 Electron micrograph of a spliceosome. (From H. Lodish, D. Baltimore, A. Berk, S. L. Zipursky, P. Matsudaira, and J. Darnell, *Molecular Cell Biology,* 3d ed. Copyright © 1995 by Scientific American Books, Inc.)

Self-splicing RNA

There are now numerous examples of RNA molecules that can catalyze the splicing of their introns without the aid of any proteins. This self-splicing was first shown by Thomas Cech and his co-workers in *Tetrahymena* and was the first demonstration that an RNA molecule can catalyze a specific biological reaction. These RNAs with enzymatic activity have been termed *ribozymes.* On the basis of the detailed mechanism of splicing, the introns that are self-spliced are classified as group I or group II introns. The group I introns are found in primary transcripts from some *E. coli* viruses, *Tetrahymena,* and certain other single-cell organisms, mitochondria, and chloroplasts, as well as in some tRNAs from bacteria. Group II introns are found in some tRNA primary transcripts and in some chloroplast and mitochondrial primary transcripts. A schematic view of the differences in the splicing mechanism of group I, group II, and spliceosome-dependent introns is shown in Figure 10-22. The product of

Figure 10-22 Splicing mechanisms in group I and group II self-splicing introns and spliceosome-catalyzed splicing of pre-mRNA. The intron is shown in blue; the exons to be joined in red. In group I introns *(left),* a guanosine cofactor (**G**) associates with the active site. The 3′-hydroxyl group of this guanosine participates in a transesterification reaction with the phosphate at the 5′ end of the intron; this reaction is analogous to that of the 2′-hydroxyl groups of the branch-site **A** in group II introns and pre-mRNA introns spliced in spliceosomes. The subsequent transesterification that links the 5′ and 3′ exons is similar in all three splicing mechanisms. Note that spliced-out group I introns are linear structures, unlike the branched intron products in the other two cases. (After P. A. Sharp, *Science* 235, 1987, 769.)

associated factors to form a high-molecular-weight (60S) ribonucleoprotein complex, called a **spliceosome** (Figure 10-21), which catalyzes the splicing transesterification reactions.

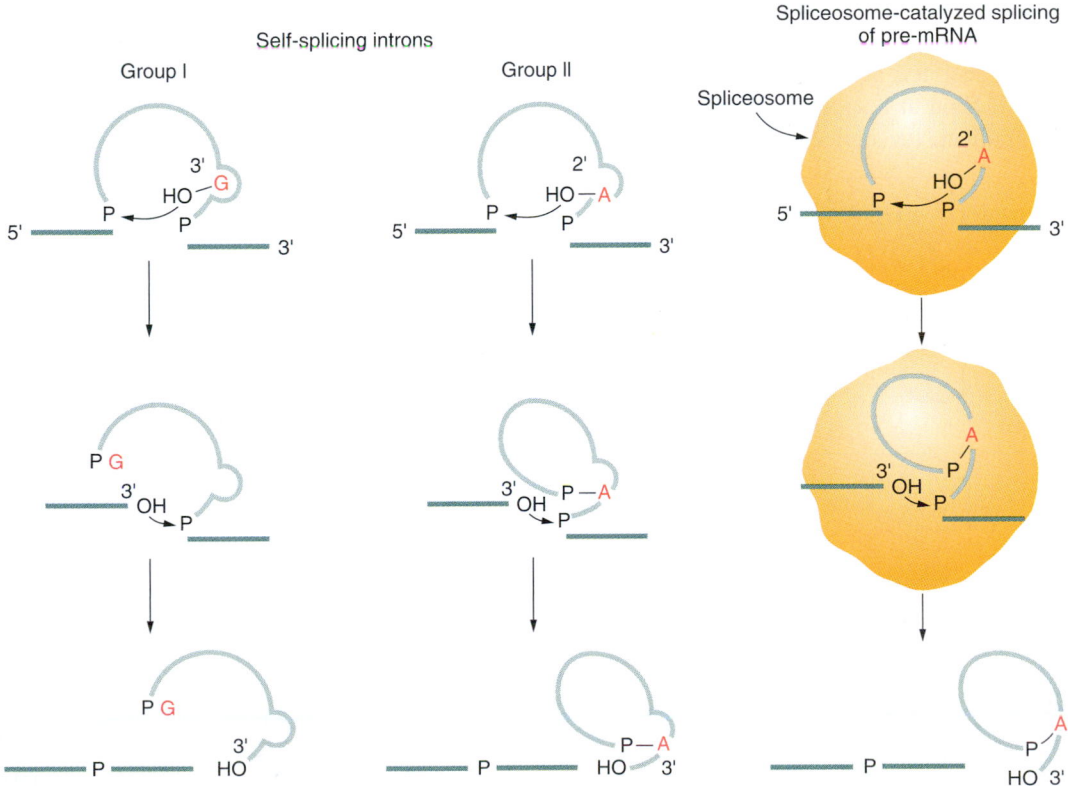

the spliced-out group I intron is not a lariat; rather it is a linear molecule.

Translation

Although mRNA directs protein synthesis, if you mix mRNA and all 20 amino acids in a test tube and hope to make protein, you will be disappointed. Other components are needed; the discovery of the nature of these components provided the key to understanding the mechanism of translation. A simple but elegant centrifugation technique helped to reveal these additional components.

Application of sucrose gradients

The **sucrose density gradient** is created in a test tube by layering successively lower concentrations of sucrose solution, one on top of the other. The material to be studied is carefully placed on top. When the solution is centrifuged in a machine that allows the test tube to swivel freely, the sedimenting material travels through the gradient at different rates that are related to the sizes and shapes of the molecules. Large molecules migrate farther in a given period of time than smaller molecules do. The separated molecules can be collected individually by capturing sequential drops from a small opening in the bottom of the tube (Figure 10-23). The velocity with which a fraction moves the fixed distance to the tube bottom indicates its sedimentation (S) value, which is a measure of the size of the molecules in the fraction.

It is important to note the difference between a sucrose gradient and the CsCl gradient that we considered in Chapter 8. In a CsCl gradient, the molecules being studied have a density somewhere in between the lowest and highest concentrations of CsCl generated in the gradient. Therefore, at equilibrium they will band at a specific point on the gradient. In a sucrose gradient, the molecules being studied are denser than any of the sucrose concentrations used, and at equilibrium they would form a pellet in the bottom of the tube. However, they migrate toward the bottom at varying speeds, depending on their size and shape. By comparing the different positions of each molecule in the gradient at a particular time, we can determine the relative sizes of the molecules.

With the use of the separatory powers of the sucrose-gradient technique (Figure 10-23), the main components in a typical protein-synthesizing system were easily separated by size. Transfer RNA molecules (4S) could easily be distinguished from ribosomal RNA, which forms three classes: 23S, 16S, and 5S. These sizes are summarized in Table 10-2.

Genetic code

If genes are segments of DNA and if DNA is just a string of nucleotide pairs, then how does the sequence of nucleotide pairs dictate the sequence of amino acids in proteins? The analogy to a code springs to mind at once. The cracking of the genetic code is the story told in this section. The experimentation was sophisticated and swift, and it did not take long for the code to be deciphered once its existence was strongly indicated.

Figure 10-23   The sucrose-gradient technique. (a) A sucrose density gradient is created in a centrifuge tube by layering solutions of differing densities. (b) The sample to be tested is placed on top of the gradient. (c) Centrifugation causes the various components (fractions) of the sample to sediment differentially. (d) The different fractions appear as bands in the centrifuged gradient. (e) The different bands can be collected separately by collecting samples from the bottom of the tube at fixed time intervals. The S value for the fraction is based on its position in the gradient, which is related to the time at which it drips from the bottom of the tube. (From A. Rich, "Polyribosomes." Copyright © 1963 by Scientific American, Inc. All rights reserved.)

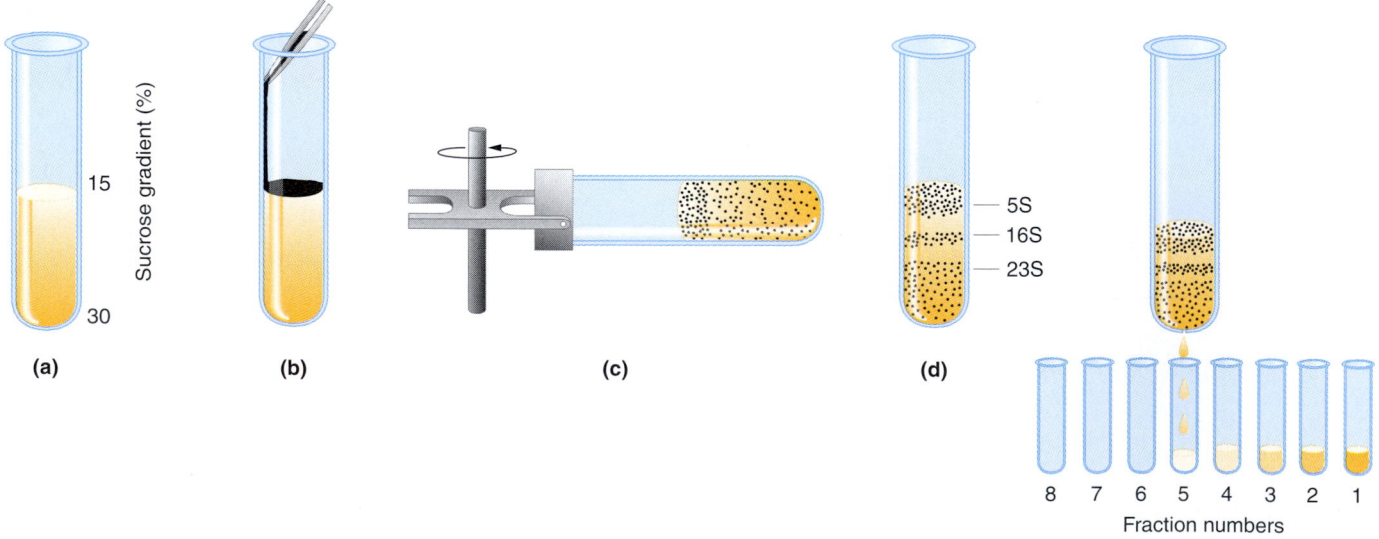

10-2 TABLE		RNA Molecules in *E. coli*		
Type	Percentage of cell RNA	Sedimentation coefficient, S	Molecular weight	Number of nucleotides
Ribosomal RNA (rRNA)	80	23	1.2×10^6	3700
		16	0.55×10^6	1700
		5	3.6×10^4	1700
Transfer RNA (tRNA)	15	4	2.5×10^4	75
Messenger RNA (mRNA)	5		Heterogeneous	Varies

Source: After L. Stryer, *Biochemistry,* 4th ed. Copyright © 1995 by Lubert Stryer.

Simple logic tells us that, if nucleotide pairs are the "letters" in a code, then a combination of letters can form "words" representing different amino acids. We must ask how the code is read. Is it overlapping or nonoverlapping? Then we must ask how many letters in the mRNA make up a word, or **codon,** and which specific codon or codons represent each specific amino acid.

Overlapping versus nonoverlapping codes

Figure 10-24 shows the difference between an overlapping and a nonoverlapping code. In the example, a three-letter, or **triplet,** code is shown. For the nonoverlapping code, consecutive amino acids are specified by consecutive code words (codons), as shown at the bottom of Figure 10-24. For an overlapping code, consecutive amino acids are en-

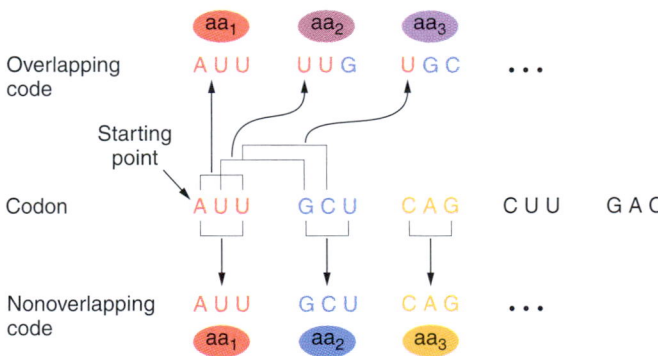

Figure 10-24 The difference between an overlapping and a nonoverlapping code. The case illustrated is for a code with three letters (a triplet code). An overlapping code uses codons that employ some of the same nucleotides as those of other codons for the translation of a single protein, as shown at the top of the diagram (for the mRNA sequence shown at the bottom of the diagram). In a nonoverlapping code, a protein is translated by reading codons that do not share any of the same nucleotides. Note that the designation of an amino acid as aa$_2$, aa$_3$, and so forth, in the nonoverlapping model does not mean it is necessarily the same amino acid as its numerical counterpart in the overlapping model. The reason is that the triplets making up the respective codons for the two aa$_3$'s, for example, are different and would more than likely encode different amino acids. Identical amino acid numbers between the two models merely indicate the same amino acid position on the protein chain.

coded in the mRNA by codons that share some consecutive bases; for example, the last two bases of one codon may also be the first two bases of the next codon. Overlapping codons are shown in the upper part of Figure 10-24. Thus, for the sequence AUUGCUCAG in a nonoverlapping code, the first three amino acids are encoded by the three triplets AUU, GCU, and CAG, respectively. However, in an overlapping code, the first three amino acids are encoded by the triplets AUU, UUG, and UGC if the overlap is two bases, as shown in Figure 10-24.

By 1961, it was already clear that the genetic code was nonoverlapping. The analysis of mutationally altered proteins, in particular, the nitrous acid–generated mutants of tobacco mosaic virus, showed that only a single amino acid changes at one time in one region of the protein. This result is predicted by a nonoverlapping code. As you can see from Figure 10-24, an overlapping code predicts that a single base change will alter as many as three amino acids at adjacent positions in the protein.

It should be noted that, although the use of an overlapping *code* was ruled out by the analysis of single proteins, nothing precluded the use of alternative reading frames to encode amino acids in two different proteins. In the example here, one protein might be encoded by the series of codons that reads AUU, GCU, CAG, CUU, and so forth. A second protein might be encoded by codons that are shifted over by one base and therefore read UUG, CUC, AGC, UUG, and so forth. This is an example of storing the information encoding two different proteins in two different reading frames, while still using a genetic code that is read in a nonoverlapping manner during the translation of a *specific* protein. Some examples of such shifts in reading frame have been found.

Number of letters in the code

In reading an mRNA molecule from one particular end, only one of four different bases, A, U, G, or C, can be found at each position. Thus, if the words were one letter long, only four words would be possible. This vocabulary cannot be the genetic code, because we must have a word for each of the 20 amino acids commonly found in cellular proteins. If the words were two letters long, then $4^2 = 16$ words would be possible; for example, AU, CU, or CC. This vocabulary is still not large enough.

If the words are three letters long, then $4^3 = 64$ words are possible; for example, AUU, GCG, or UGC. This vocabulary provides more than enough words to describe the amino acids. We can conclude that the code word must consist of at least three nucleotide pairs. However, if all words are "triplets," then we have a considerable excess of possible words over the 20 needed to name the common amino acids.

Use of suppressors to demonstrate a triplet code

Convincing proof that a codon is, in fact, three letters long (and no more than three) came from beautiful genetic experiments first reported in 1961 by Francis Crick, Sidney Brenner, and their co-workers, who used mutants in the *rII* locus of T4 phage. Mutations causing the rII phenotype (see Chapter 9) were induced by using a chemical called *proflavin,* which was thought to act by the addition or deletion of single nucleotide pairs in DNA. (This assumption is based on experimental evidence not presented here.) The following examples illustrate the action of proflavin on double-stranded DNA.

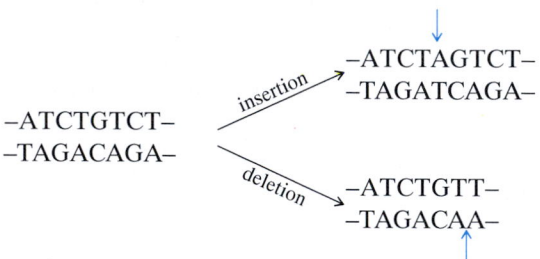

Then, starting with one particular proflavin-induced mutation called FCO, Crick and his colleagues found "reversions" (reversals of the mutation) that were detected by their wild-type plaques on *E. coli* strain K(λ). Genetic analysis of these plaques revealed that the "revertants" were not identical true wild types, thereby suggesting that the back mutation was not an exact reversal of the original forward mutation. In fact, the reversion was found to be caused by the presence of a *second mutation* at a different site from — but in the same gene as — that of FCO; this second mutation "suppressed" mutant expression of the original FCO. Recall from Chapter 4 that a **suppressor mutation** counteracts or suppresses the effects of another mutation.

The suppressor mutation could be separated from the original forward mutation by recombination, and, as we have seen, when this was done, the suppressor was shown to be an *rII* mutation itself (Figure 10-25).

How can we explain these results? If we assume that reading is polarized — that is, if the gene is read from one end only — then the original proflavin-induced addition or deletion could be mutant because it interrupts a normal

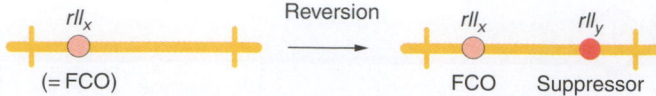

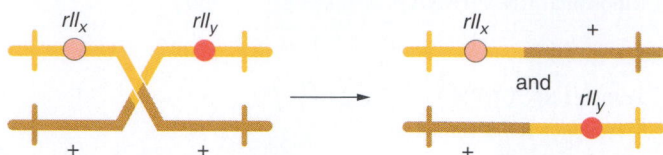

Figure 10-25   The suppressor of an initial *rII* mutation is shown to be an *rII* mutation itself after separation by crossing over. The original mutant, FCO, was induced by proflavin. Later, when the FCO strain was treated with proflavin again, a revertant was found, which on first appearance seemed to be wild type. However, a second mutation within the *rII* region was found to have been induced, and the double mutant *rII*$_x$*rII*$_y$ was shown not to be quite identical with the original wild type.

reading mechanism that establishes the group of bases to be read as words. For example, if each three bases on the resulting mRNA make a word, then the "reading frame" might be established by taking the first three bases from the end as the first word, the next three as the second word, and so forth. In that case, a proflavin-induced addition or deletion of a single pair on the DNA would shift the reading frame on the mRNA from that corresponding point on, causing all following words to be misread. Such a **frameshift mutation** could reduce most of the genetic message to gibberish. However, the proper reading frame could be restored by a compensatory insertion or deletion somewhere else, leaving only a short stretch of gibberish between the two. Consider the following example in which three-letter English words are used to represent the codons:

		THE FAT CAT ATE THE BIG RAT
Delete C:	THE FAT ATA TET HEB IGR AT
Insert A:	THE FAT ATA ATE THE BIG RAT

The insertion suppresses the effect of the deletion by restoring most of the sense of the sentence. By itself, however, the insertion also disrupts the sentence:

THE FAT CAT AAT ETH EBI GRA T

If we assume that the FCO mutant is caused by an addition, then the second (suppressor) mutant would have to be a deletion because, as we have seen, this would restore the reading frame of the resulting message (a second insertion would not correct the frame). In the following diagrams, we use a hypothetical nucleotide chain to represent RNA for simplicity. We also assume that the code words are three letters long and are read in one direction (left to right in our diagrams).

1. Wild-type message

CAU   CAU   CAU   CAU   CAU

2. *rII$_a$* message: distal words changed (**x**) by frameshift mutation (words marked ✔ are unaffected)

Addition ─┐
          ↓
CAU   ACA   UCA   UCA   UCA   U___
 ✔   x   x   x   x

3. *rII$_a$rII$_b$* message: few words wrong, but reading frame restored for later words

Deletion ──────┐
               ↓
CAU   ACA   UCU   CAU   CAU
 ✔   x   x   ✔   ✔

The few wrong words in the suppressed genotype could account for the fact that the "revertants" (suppressed phenotypes) that Crick and his associates recovered did not look exactly like the true wild types phenotypically.

We have assumed here that the original frameshift mutation was an addition, but the explanation works just as well if we assume that the original FCO mutation is a deletion and the suppressor is an addition. If the FCO is defined as plus, then suppressor mutations are automatically minus. Experiments have confirmed that a plus cannot suppress a plus and a minus cannot suppress a minus. In other words, two mutations of the same sign never act as suppressors of each other. However, very interestingly, combinations of *three* pluses or *three* minuses have been shown to act together to restore a wild-type phenotype.

This observation provided the first experimental confirmation that a word in the genetic code consists of three successive nucleotide pairs, or a triplet. The reason is that three additions or three deletions within a gene automatically restore the reading frame in the mRNA if the words are triplets. For example,

Deletions
    ↓   ↓   ↓
CAU CAU CAU CAU CAU CAU CAU

CAU ACA UAU CAU CAU CAU
 ✔   x   x   ✔   ✔   ✔

Proof that the genetic deductions about proflavin were correct came from an analysis of proflavin-induced mutations in a gene with a protein product that could be analyzed. George Streisinger worked with the gene that controls the enzyme lysozyme, which has a known amino acid sequence. He induced a mutation in the gene with proflavin and selected for proflavin-induced revertants, which were shown genetically to be double mutants (with mutations of opposite sign). When the protein of the double mutant was analyzed, a stretch of different amino acids lay between two wild-type ends, just as predicted:

Wild type:

–Thr–Lys–Ser–Pro–Ser–Leu–Asn–Ala–
                └─────────────────┘

Suppressed mutant:

–Thr–Lys–Val–His–His–Leu–Met–Ala–
                └─────────────────┘

Degeneracy of the genetic code

Crick's work also suggested that the genetic code is **degenerate.** That expression is not a moral indictment. It simply means that each of the 64 triplets must have some meaning within the code; so at least some amino acids must be specified by two or more different triplets. If only 20 triplets are used (with the other 44 being nonsense, in that they do not code for any amino acid), then most frameshift mutations can be expected to produce nonsense words, which presumably stops the protein-building process. If this were the case, then the suppression of frameshift mutations would rarely, if ever, work. However, if all triplets specified some amino acid, then the changed words would simply result in the insertion of incorrect amino acids into the protein. Thus, Crick reasoned that many or all amino acids must have several different names in the base-pair code; this hypothesis was later confirmed biochemically.

MESSAGE ···
The discussion up to this point demonstrates that
1. The genetic code is nonoverlapping.
2. Three bases encode an amino acid. These triplets are termed codons.
3. The code is read from a fixed starting point and continues to the end of the coding sequence. We know this because a single frameshift mutation anywhere in the coding sequence alters the codon alignment for the rest of the sequence.
4. The code is degenerate in that some amino acids are specified by more than one codon.

Cracking the code

The deciphering of the genetic code — determining the amino acid specified by each triplet — was one of the most exciting genetic breakthroughs of the past 50 years. Once the necessary experimental techniques became available, the genetic code was broken in a rush.

The first breakthrough was the discovery of how to make synthetic mRNA. If the nucleotides of RNA are mixed with a special enzyme (polynucleotide phosphorylase), a single-stranded RNA is formed in the reaction. No DNA is needed for this synthesis, and so the nucleotides are incorporated at random. The ability to synthesize mRNA offered the exciting prospect of creating specific mRNA sequences and then seeing which amino acids they would specify. The first synthetic messenger obtained, poly(U), was made by reacting only uracil nucleotides with the RNA-synthesizing enzyme, producing –UUUU–. In 1961, Marshall Nirenberg and Heinrich Matthaei mixed poly(U) with the protein-synthesizing machinery of *E. coli* in vitro and *observed the formation of a protein!* The main excitement centered on the question of the amino acid sequence of this protein. It proved to be polyphenylalanine — a string of phenylalanine molecules attached to form a polypeptide. Thus, the triplet UUU must code for phenylalanine:

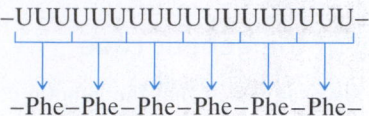

This type of analysis was extended by mixing nucleotides in a known fixed proportion when making synthetic mRNA. In one experiment, the nucleotides uracil and guanine were mixed in a ratio of 3:1. When nucleotides are incorporated at random into synthetic mRNA, the relative frequency at which each triplet will appear in the sequence can be calculated on the basis of the relative proportion of the various nucleotides present (Table 10-3). Note that, in Table 10-3, UUU is used as the baseline frequency against which the other frequencies are measured in determining their respective ratios. For example, UUG, with a probability of $p(UUG) = 9/64$, would be expected only one-third as often as UUU, with its probability of $p(UUU) = 27/64$. Stated alternatively, $p(UUG)/p(UUU) = 9/27 = 1/3 = 0.33$, which is the ratio for UUG given in Table 10-3.

If these codons each encode a different amino acid (that is, are not redundant), we expect the amino acids generated by this particular mix of guanine and uracil to be in ratios similar to those of the various codons. Although there is some redundancy among these codons, the ratios of the amino acids actually obtained from this mix of bases (Table 10-4) are indeed quite similar to the ratios seen for the codon frequencies in Table 10-3. (In Table 10-4, phenylalanine is used as the baseline in determining ratios.)

From this evidence, we can deduce that codons consisting of one guanine and two uracils (G + 2 U) code for va-

10-4 TABLE	Observed Frequencies of Various Amino Acids in Protein Translated from mRNA Composed of $\frac{3}{4}$ Uracil and $\frac{1}{4}$ Guanine	
Amino acid	**Ratio***
Phenylalanine	1.00
Leucine	0.37
Valine	0.36
Cysteine	0.35
Tryptophan	0.14
Glycine	0.12

* Phenylalanine is used as the baseline concentration against which the concentrations of other amino acids are measured in deriving their respective ratios. Note the correlations with the ratios in Table 10-3.

line, leucine, and cysteine, although we cannot distinguish the specific sequence for each of these amino acids. Similarly, one uracil and two guanines (U + 2 G) must code for tryptophan, glycine, and perhaps one other. It looks as though the Watson-Crick model is correct in predicting the importance of the precise sequence (not just the ratios of bases). Many provisional assignments (such as those just outlined for G and U) were soon obtained, primarily by groups working with Nirenberg or with Severo Ochoa.

Before we consider other code words, we will examine tRNA molecules, which further explain the link between the mRNA codon and amino acid recognition.

tRNA recognition of the codon

Is it the tRNA or the amino acid itself that recognizes the mRNA that encodes a specific amino acid? A very convincing experiment answered this question. In the experiment, an aminoacyl-tRNA (aa-tRNA), cysteinyl-tRNA (tRNA^{Cys}, the tRNA specific for cysteine) "charged" with cysteine was treated with nickel hydride, which converted the cysteine (while still bound to tRNA^{Cys}) into another amino acid, alanine, without affecting the tRNA:

$$\text{cysteine}-\text{tRNA}^{Cys} \xrightarrow{\text{nickel hydride}} \text{alanine}-\text{tRNA}^{Cys}$$

Protein synthesized with this hybrid species had alanine wherever we would expect cysteine. Thus, the experiment demonstrated that the amino acids are "illiterate"; they are inserted at the proper position because the tRNA "adapters" recognize the mRNA codons and insert their attached amino acids appropriately. We would expect, then, to find some site on the tRNA that recognizes the mRNA codon by complementary base pairing.

Figure 10-26a shows several functional sites of the tRNA molecule. The site that recognizes an mRNA codon is called the **anticodon;** its bases are complementary and antiparallel to the bases of the codon. Another operationally identifiable site is the amino acid attachment site. The other

10-3 TABLE	Expected Frequencies of Various Codons in Synthetic mRNA Composed of $\frac{3}{4}$ Uracil and $\frac{1}{4}$ Guanine	
Codon	**Probability**	**Ratio***
UUU	$p(UUU) = \frac{3}{4} \times \frac{3}{4} \times \frac{3}{4} = \frac{27}{64}$	1.00
UUG	$p(UUG) = \frac{3}{4} \times \frac{3}{4} \times \frac{1}{4} = \frac{9}{64}$	0.33
UGU	$p(UGU) = \frac{3}{4} \times \frac{1}{4} \times \frac{3}{4} = \frac{9}{64}$	0.33
GUU	$p(GUU) = \frac{1}{4} \times \frac{3}{4} \times \frac{3}{4} = \frac{9}{64}$	0.33
UGG	$p(UGG) = \frac{3}{4} \times \frac{1}{4} \times \frac{1}{4} = \frac{3}{64}$	0.11
GGU	$p(GGU) = \frac{1}{4} \times \frac{1}{4} \times \frac{3}{4} = \frac{3}{64}$	0.11
GUG	$p(GUG) = \frac{1}{4} \times \frac{3}{4} \times \frac{1}{4} = \frac{3}{64}$	0.11
GGG	$p(GGG) = \frac{1}{4} \times \frac{1}{4} \times \frac{1}{4} = \frac{1}{64}$	0.03

* UUU is used as the baseline frequency against which the frequencies of the other codons are measured in establishing the respective ratios. For example, the ratio for UUG is derived from $p(UUG)/p(UUU) = 0.33$.

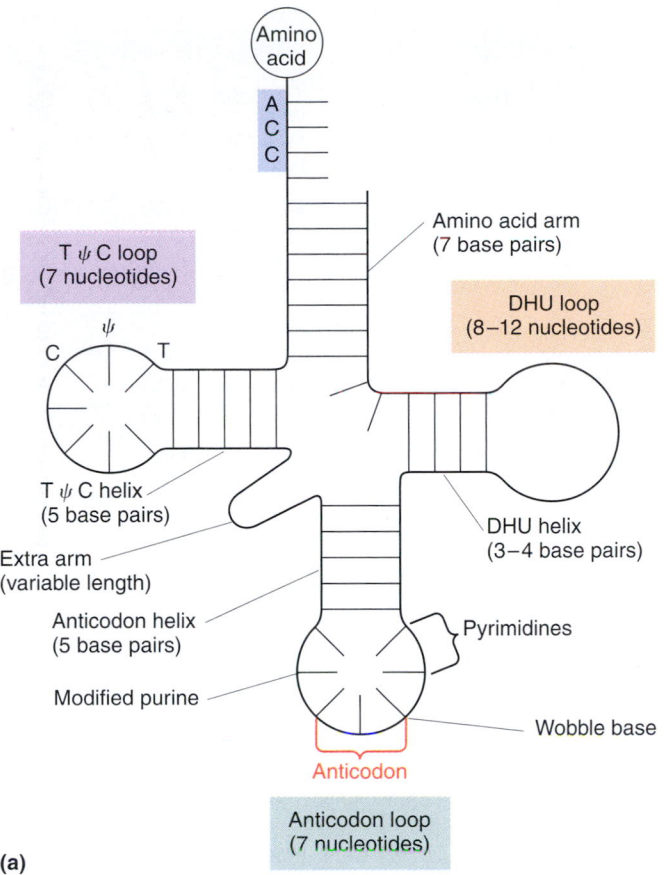

(a)

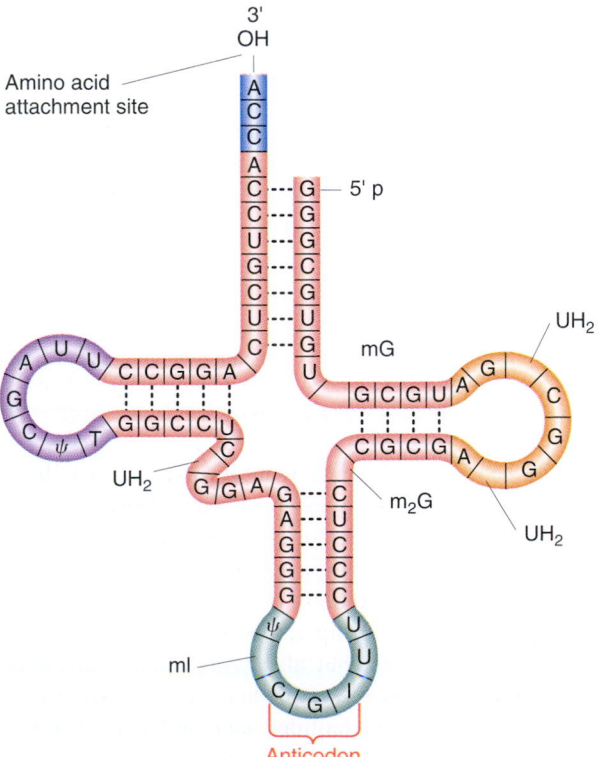

(b)

arms probably assist in binding the tRNA to the ribosome. Figure 10-26b shows a specific tRNA (yeast alanine tRNA). The "flattened" cloverleafs shown in these diagrams are not the normal conformation of tRNA molecules; tRNA normally exists as an L-shaped folded cloverleaf, as shown in Figure 10-26c. These diagrams are supported by very sophisticated chemical analysis of tRNA nucleotide sequences and by X-ray crystallographic data on the overall shape of the molecule. Although tRNA molecules have many structural similarities, each has a unique three-dimensional shape that allows recognition by the correct synthetase, which catalyzes the joining of a tRNA with its specific amino acid to form an aminoacyl-tRNA. (Synthetases will be considered in this chapter under "Protein Synthesis.") The specificity of charging the tRNAs is crucial to the integrity of protein synthesis.

Where does tRNA come from? If radioactive tRNA is put into a cell nucleus in which the DNA has been partly denatured by heating, the radioactivity appears (by autoradiography) in localized regions of the chromosomes. These regions probably indicate the location of genes that specify

Figure 10-26 The structure of transfer RNA. (a) The functional areas of a generalized tRNA molecule. (b) The specific sequence of yeast alanine tRNA. Arrows indicate several kinds of rare modified bases. (c) Diagram of the actual three-dimensional structure of yeast phenylalanine tRNA. The abbreviations ψ, mG, m_2G, mI, and DHU (or UH₂) are abbreviations for modified bases pseudouridine, methylguanosine, dimethylguanosine, methylinosine, and dihydrouridine, respectively. (Part a from S. Arnott, "The Structure of Transfer RNA," *Progress in Biophysics and Molecular Biology* 22, 1971, 186; parts b and c from L. Stryer, *Biochemistry*, 4th ed. Copyright © 1995 by Lubert Stryer; part c based on a drawing by Sung-Hou Kim.)

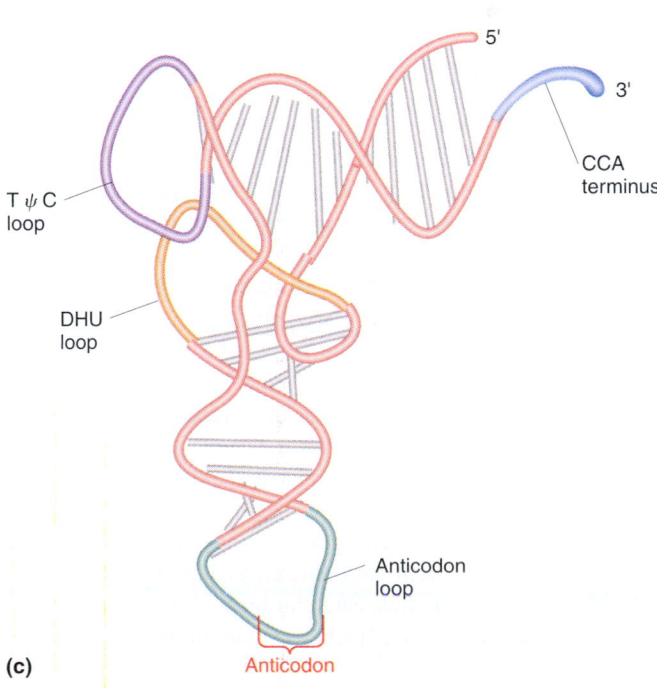

(c)

tRNA; they are regions of DNA that produce tRNA rather than mRNA, which produces a protein. The labeled tRNA hybridizes to these sites because of the complementarity of base sequences between the tRNA and its parent gene. A similar situation holds for rRNA. Thus, we see that even the one-gene–one-polypeptide idea is not completely valid. Some genes do not code for protein; rather, they specify RNA components of the translational apparatus.

MESSAGE Some genes encode proteins; other genes specify RNA (for example, tRNA or rRNA) as their final product.

How does tRNA get its fancy shape? It probably folds up spontaneously into a conformation that produces maximal stability. Transfer RNA contains many "odd" or modified bases (such as pseudouracil, ψ) in its nucleotides; these bases play a direct role in folding and have been implicated in other tRNA functions. You may have noticed some unusual base pairing within the loops of the tRNA in Figure 10-26b; G is hydrogen bonded to U (instead of C). This apparent mismatching is considered next.

The complete code

Specific code words were finally deciphered through two kinds of experiments. The first required making "mini mRNAs," each only three nucleotides in length. These mini mRNAs are too short to promote translation into protein, but they do stimulate the binding of aminoacyl-tRNAs to ribosomes in a kind of abortive attempt at translation. It is possible to make a specific mini mRNA and determine *which* aminoacyl-tRNA that it will bind to ribosomes. For example, the G + 2 U problem described earlier can be resolved by using the following mini mRNAs:

GUU stimulates binding of valyl-tRNA
UUG stimulates binding of leucyl-tRNA
UGU stimulates binding of cysteinyl-tRNA

Analogous mini RNAs provided 64 possible codons.

The second kind of experiment that was useful in cracking the genetic code required the use of *repeating copolymers.* For instance, the copolymer designated $(AGA)_n$, which is a long sequence of AGAAGAAGAAGAAGA, was used to stimulate polypeptide synthesis in vitro. From the sequence of the resulting polypeptides and the possible triplets that could reside in the respective RNA copolymer, many code words could be verified. (This kind of experiment is detailed in Problem 10 at the end of this chapter. In solving it, you can put yourself in the place of H. Gobind Khorana, who received a Nobel Prize for directing the experiments.)

Figure 10-27 gives the genetic code dictionary of 64 words. Inspect this dictionary carefully, and ponder the miracle of molecular genetics. Such an inspection should reveal several points that require further explanation.

Second letter

First letter	U	C	A	G	Third letter
U	UUU, UUC: Phe UUA, UUG: Leu	UCU, UCC, UCA, UCG: Ser	UAU, UAC: Tyr UAA Stop UAG Stop	UGU, UGC: Cys UGA Stop UGG Trp	U C A G
C	CUU, CUC, CUA, CUG: Leu	CCU, CCC, CCA, CCG: Pro	CAU, CAC: His CAA, CAG: Gln	CGU, CGC, CGA, CGG: Arg	U C A G
A	AUU, AUC, AUA: Ile AUG Met	ACU, ACC, ACA, ACG: Thr	AAU, AAC: Asn AAA, AAG: Lys	AGU, AGC: Ser AGA, AGG: Arg	U C A G
G	GUU, GUC, GUA, GUG: Val	GCU, GCC, GCA, GCG: Ala	GAU, GAC: Asp GAA, GAG: Glu	GGU, GGC, GGA, GGG: Gly	U C A G

Figure 10-27 The genetic code.

Multiple codons for a single amino acid

From the discussion of degeneracy, we know that the number of codons for a single amino acid varies, ranging from one (tryptophan = UGG) to as many as six (serine = UCU or UCC or UCA or UCG or AGU or AGC). Why? The answer is complex but not difficult; it can be divided into two parts:

1. Certain amino acids can be brought to the ribosome by several *alternative* tRNA types (species) having different anticodons, whereas certain other amino acids are brought to the ribosome by only one tRNA.
2. Certain tRNA species can bring their specific amino acids in response to several codons, not just one, through a loose kind of base pairing at one end of the codon and anticodon. This sloppy pairing is called **wobble.**

MESSAGE The degree of degeneracy for a given amino acid is determined by the number of codons for that amino acid that have only one tRNA each plus the number of codons for amino acids that share a tRNA through wobble.

We had better consider wobble first, and it will lead us into a discussion of the various species of tRNA. Wobble is caused by the third nucleotide of an anticodon (at the 5′ end) that is not quite aligned (Figure 10-28). This out-of-line nucleotide can sometimes form hydrogen bonds not only with its normal complementary nucleotide in the third position of the codon, but also with a different nucleotide in that position. Crick established certain "wobble rules" that dictate which nucleotides can and cannot form new

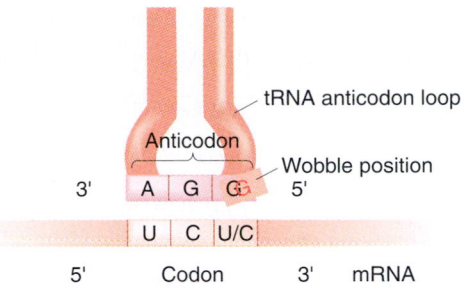

Figure 10-28 In the third site (5′ end) of the anticodon, G can take either of two wobble positions, thus being able to pair with either U or C. This ability means that a single tRNA species carrying an amino acid (in this case, serine) can recognize two codons — UCU and UCC — in the mRNA.

hydrogen-bonded associations through wobble (Table 10-5). In Table 10-5, I (inosine) is one of the rare bases found in tRNA, often in the anticodon.

Figure 10-28 shows the possible codons that one tRNA serine species can recognize. As the wobble rules indicate, G can pair with U or with C. Table 10-6 lists all the codons for serine and shows how different tRNAs can service these codons. Serine affords a good example of the effects of wobble on the genetic code.

Sometimes there can be an additional tRNA species that we represent as tRNA^{Ser_4}; it has an anticodon identical with any of the three anticodons shown in Table 10-6, but it differs in its nucleotide sequence elsewhere in the tRNA molecule. These four tRNAs are called **isoaccepting tRNAs** because they accept the same amino acid, but they are probably all transcribed from different tRNA genes.

Stop codons

The second point that you may have noticed in Figure 10-27 is that some codons do not specify an amino acid at all. These codons are labeled as **stop** or **termination codons.** They can be regarded as being similar to periods or commas punctuating the message encoded in the DNA.

One of the first indications of the existence of stop codons came in 1965 from Brenner's work with the T4 phage. Brenner analyzed certain mutations (m_1–m_6) in a single gene that controls the head protein of the phage. These mutants had two things in common. First, the head protein of each mutant was a shorter polypeptide chain than that of the wild type. Second, the presence of a suppressor mutation (*su*) in the host chromosome would cause the phage to develop a head protein of normal (wild-type) chain length despite the presence of the *m* mutation (Figure 10-29).

Brenner examined the ends of the shortened proteins and compared them with wild-type protein, recording for each mutant the next amino acid that *would* have been inserted to continue the wild-type chain. These amino acids for the six mutations were glutamine, lysine, glutamic acid, tyrosine, tryptophan, and serine. There is no immediately obvious pattern to these results, but Brenner brilliantly deduced that certain codons for each of these amino acids are similar in that each of them can mutate to the codon UAG by a single change in a DNA nucleotide pair. He therefore postulated that UAG is a stop (termination) codon — a signal to the translation mechanism that the protein is now complete.

UAG was the first stop codon deciphered; it is called the **amber codon.** Mutants that are defective owing to the presence of an abnormal amber codon are called *amber mutants,* and their suppressors are *amber suppressors.* UGA, the **opal codon,** and UAA, the **ochre codon,** also are stop codons and also have suppressors. Stop codons are often called **nonsense codons** because they designate no amino acid. Not surprisingly, stop codons do not act as mini

10-5 TABLE	Codon-Anticodon Pairings Allowed by the Wobble Rules	
5′ end of anticodon	**3′ end of codon**	
G	U or C	
C	G only	
A	U only	
U	A or G	
I	U, C, or A	

10-6 TABLE	Different tRNAs That Can Service Codons for Serine	
Codon	**tRNA**	**Anticodon**
UCU	tRNA^{Ser_1}	AGG + wobble
UCC		
UCA	tRNA^{Ser_2}	AGU + wobble
UCG		
AGU	tRNA^{Ser_3}	UCG + wobble
AGC		

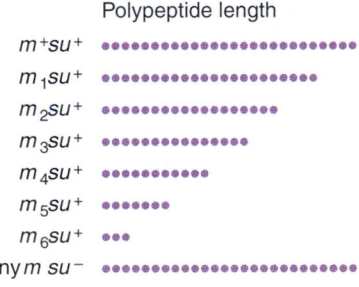

Figure 10-29 Polypeptide chain lengths of phage T4 head protein in wild type *(top)* and various amber mutants *(m).* An amber suppressor *(su)* leads to phenotypic development of the wild-type chain.

mRNAs in binding aa-tRNA to ribosomes in vitro. We shall consider stop codons and their suppressors further after we have dealt with the process of protein synthesis.

Protein synthesis

We can regard **protein synthesis** as a chemical reaction, and we shall take this approach at first. Then we shall take a three-dimensional look at the physical interactions of the major components.

In protein synthesis as a chemical reaction:

1. Each amino acid is attached to a tRNA molecule specific to that amino acid by a high-energy bond derived from ATP. The process is catalyzed by a specific enzyme called a **synthetase** (the tRNA is said to be "charged" when the amino acid is attached):

$$aa_1 + tRNA_1 = ATP \xrightarrow{synthetase_1}$$
$$aa_1 - tRNA_1 + AMP + PP_i$$

There is a separate synthetase for each amino acid.

2. The energy of the charged tRNA is converted into a peptide bond linking the amino acid to another one on the ribosome:

$$aa_1 - tRNA_1 + aa_2 - tRNA_2 \xrightarrow{\substack{peptidyl\ transferase \\ on\ a\ ribosome}}$$
$$\underbrace{aa_1 - aa_2}_{\substack{small \\ polypeptide}} - tRNA_2 + tRNA_1 \text{ (released)}$$

3. New amino acids are linked by means of a peptide bond to the growing chain:

$$aa_3 - tRNA_3 + aa_1 - aa_2 - tRNA_2 \longrightarrow$$
$$\underbrace{aa_1 - aa_2 - aa_3}_{\substack{larger \\ polypeptide}} - tRNA_3 + tRNA_2 \text{ (released)}$$

4. This process continues until aa_n (the final amino acid) is added. The whole thing works only in the presence of mRNA, ribosomes, several additional protein factors, enzymes, and inorganic ions.

Ribosomes

Ribosomes consist of two subunits that, in prokaryotes, sediment as 50S and 30S particles and associate to form a 70S particle, as seen in Figure 10-30a. The eukaryotic counterparts are 60S and 40S for the large and small subunits, and

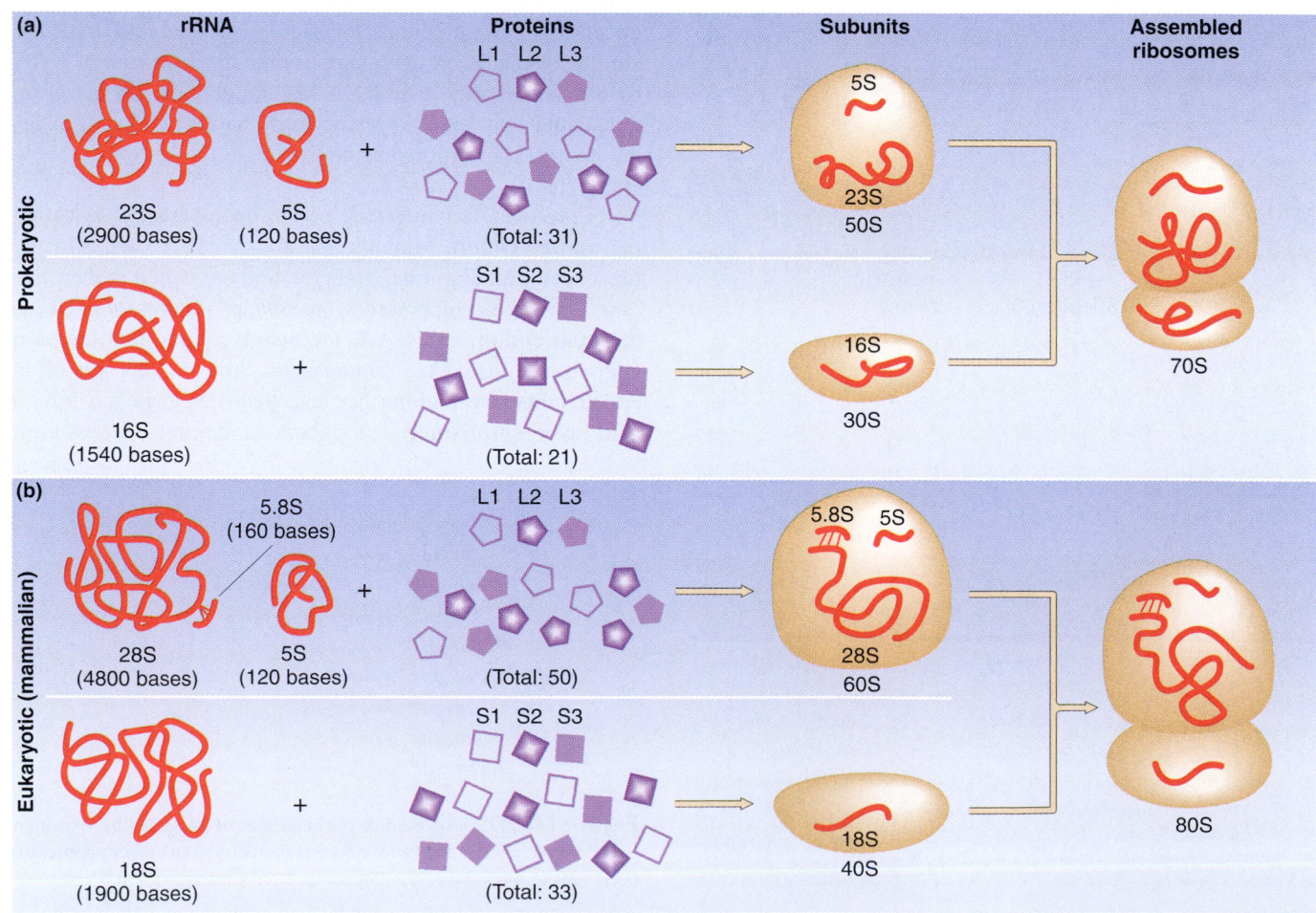

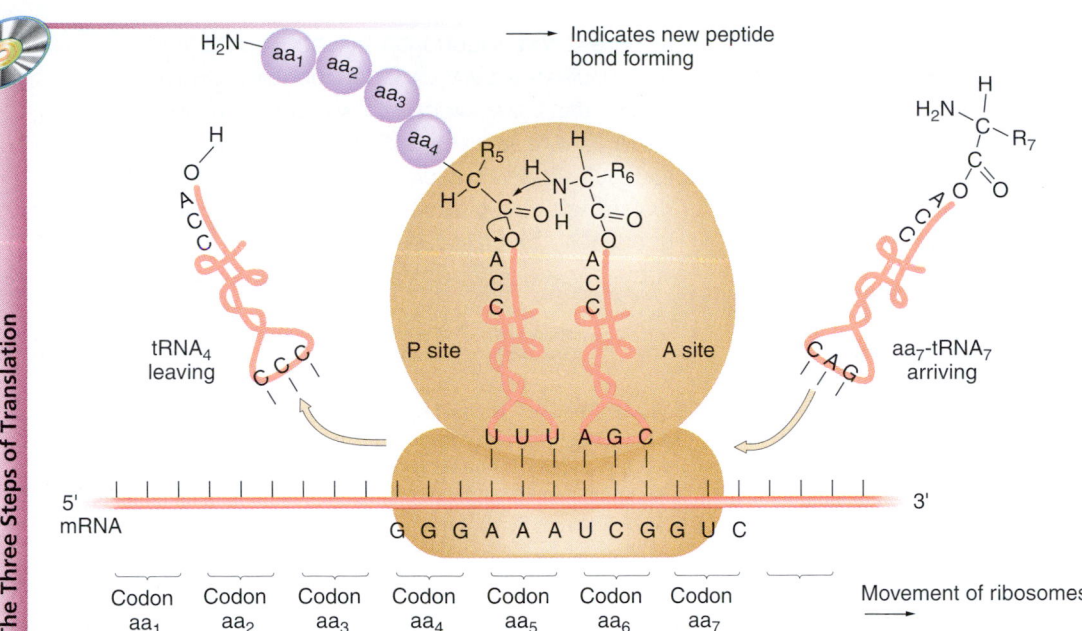

Indicates new peptide bond forming

Figure 10-31   The addition of a single amino acid to the growing polypeptide chain in the course of translation of mRNA.

80S for the complete ribosome (Figure 10-30b). Ribosomes contain specific sites that enable them to bind to the mRNA, the tRNAs, and specific protein factors required for protein synthesis. Let's look at a general picture of protein synthesis on the ribosome and then examine each of the steps in the process in more detail.

Figure 10-31 shows a polypeptide being synthesized on the ribosome. The mRNA binds to the 30S subunit. The tRNAs bind to two sites on the ribosome. These sites overlap the subunits. The **A site** is the entry site for an aminoacyl-tRNA (a tRNA carrying a single amino acid). The peptidyl-tRNA carrying the growing polypeptide chain binds at the **P site.** Each new amino acid is added by the transfer of the growing chain to the new aminoacyl-tRNA, forming a new peptide bond. The deacylated tRNA is then released from the P site, and the ribosome moves one codon farther along the message, transferring the new peptidyl-tRNA to the P site and leaving the A site vacant for the next incoming aminoacyl-tRNA.

We can separate the process of protein synthesis into three distinct steps. **Initiation, elongation,** and **termination.** Let's examine each of these steps in detail, by using prokaryotes as an example.

Figure 10-30   A ribosome contains a large and a small subunit. Each subunit contains both rRNA of varying lengths and a set of proteins (designated by different shapes and shading). There are two principal rRNA molecules in all ribosomes. (a) Ribosomes from prokaryotes also contain one 120-base-long rRNA that sediments at 5S. (b) Eukaryotic ribosomes have two small rRNAs: a 5S RNA molecule similar to the prokaryotic 5S, and a 5.8S molecule 160 bases long. The large subunit proteins are named L1, L2, and so forth, and the small subunit proteins are named S1, S2, and so forth. (From H. Lodish, D. Baltimore, A. Berk, S. L. Zipursky, P. Matsudaira, and J. Darnell, *Molecular Cell Biology,* 3d ed. Copyright © 1995 by Scientific American Books, Inc.)

Initiation

Three steps of initiation.   In addition to mRNA, ribosomes, and specific tRNA molecules, initiation requires the participation of several factors, termed **initiation factors IF1, IF2,** and **IF3.** In *E. coli* and in most other prokaryotic organisms, the first amino acid in any newly synthesized polypeptide is *N*-formylmethionine. It is inserted not by $tRNA^{Met}$, however, but by an **initiator tRNA** called $tRNA^{fMET}$. This initiator tRNA has the normal methionine anticodon but inserts *N*-formylmethionine rather than methionine (Figure 10-32). In *E. coli,* AUG and GUG, and on rare occasions UUG, serve as initiation codons. When one of these triplets is present in the initiation position, it is recognized by *N*-formylMet-tRNA, and methionine appears as the first amino acid in the chain. Let's examine the steps in initiation in detail.

1. The first step in initiation is the binding of the mRNA to the 30S subunit (Figure 10-33). The binding is stimulated by the initiation factor IF3. When not engaged in protein synthesis, the ribosomal subunits exist in the free form; they assemble into complete ribosomes as a result of the initiation process.

2. The initiation factor IF2 binds to GTP and to the initiator fMet-tRNA and stimulates the binding of fMet-tRNA to the initiation complex, leading the fMet-tRNA into the P site, as shown in the middle of Figure 10-33.

3. A ribosomal protein splits the GTP bound to IF2, helping to drive the assembly of the two ribosomal subunits (Figure 10-33, bottom). At this stage, the factors IF2 and IF3 are released. (The exact role of IF1 is not completely clear, although it seems to take part in the recycling of the ribosome.)

NH2
H—C—COOH
(CH2)2
S
CH3
Methionine
O
H—C
H
N
H—C—COOH
(CH2)2
S
CH3
N-Formylmethionine

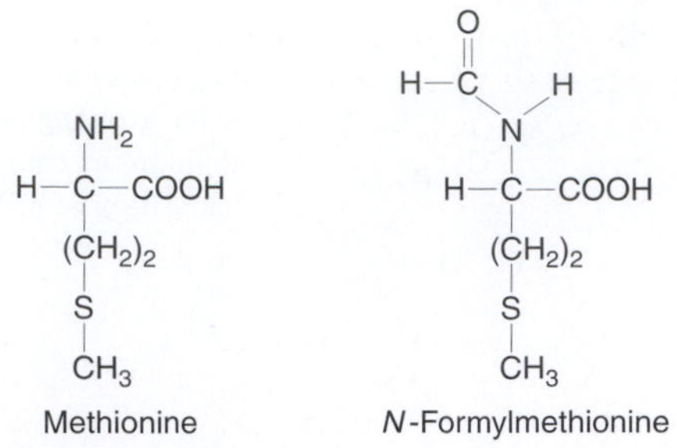

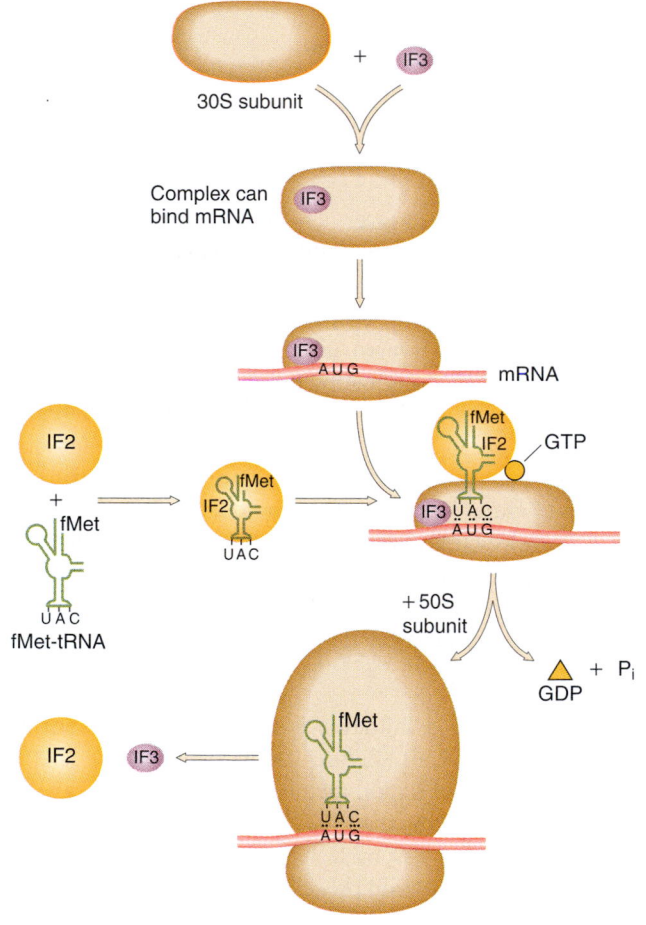

Figure 10-32 The structures of methionine (Met) and *N*-formylmethionine (fMet). A tRNA bearing fMet can initiate a polypeptide chain in prokaryotes but cannot be inserted in a growing chain; a tRNA bearing Met can be inserted in a growing chain but will not initiate a new chain. Both these tRNAs bear the same anticodon complementing the codon AUG.

AGCACGAGGGGAAAUCUGAUGGAACGCUAC E. coli trpA
UUUGGAUGGAGUGAAACGAUGGCGAUUGCA E. coli araB
GGUAACCAGGUAACAACCAUGCGAGUGUUG E. coli thrA
CAAUUCAGGGUGGUGAAUGUGAAACCAGUA E. coli lacI
AAUCUUGGAGGCUUUUUUAUGGUUCGUUCU φX174 phage A protein
UAACUAAGGAUGAAAUGCAUGUCUAAGACA Qβ phage replicase
UCCUAGGAGGUUUGACCUAUGCGAGCUUUU R17 phage A protein
AUGUACUAAGGAGGUUGUAUGGAACAACGC λ phage cro
Pairs with 16S rRNA
Pairs with initiator tRNA

Figure 10-34 Ribosomal binding-site sequences in *E. coli* and its bacteriophages have certain features in common, which are shown in the colored regions. The initiation codon (color) is separated by several bases from a short sequence (color) that is complementary to the 3′ end of 16S rRNA. (After L. Stryer, *Biochemistry,* 4th ed. Copyright © 1995 by Lubert Stryer.)

+
IF3
30S subunit
Complex can bind mRNA
IF3
IF3
AUG
mRNA
IF2
+
fMet
UAC
fMet-tRNA
IF2
fMet
UAC
fMet
IF2
GTP
IF3
UAC
AUG
+50S subunit
+ Pi
GDP
IF2
IF3
fMet
UAC
AUG

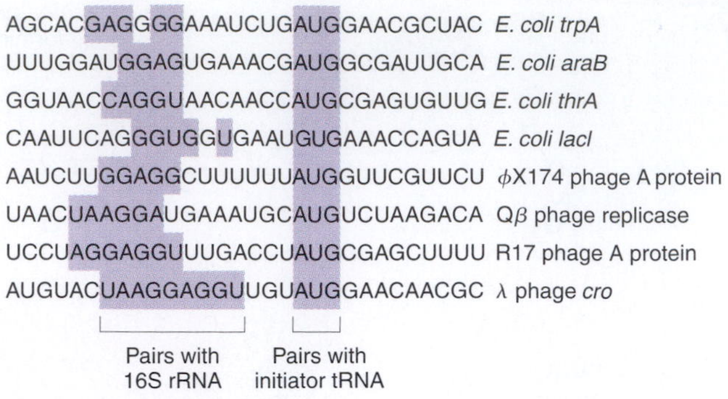

Figure 10-33 Steps in the initiation of translation (see text).

Ribosome-binding sites. How are the correct initiation codons selected from the many AUG and GUG codons in an mRNA molecule? John Shine and Lynn Dalgarno first noticed that true initiation codons were preceded by sequences that paired well with the 3′ end of 16S rRNA. Figure 10-34 shows some of these sequences. There is a short but variable separation between the Shine-Dalgarno sequence and the initiation codon. Figure 10-35 depicts the base pairing between idealized mRNA and the 16S rRNA that results in ribosome-mRNA complexes leading to protein initiation in the presence of fMet-tRNA.

Elongation

Figure 10-36 details the steps in elongation, which are aided by three protein factors, **EF-Tu, EF-Ts,** and **EF-G.** The steps are as follows:

1. Elongation factor EF-Tu mediates the entry of amino-acyl-tRNAs into the A site. To do so, EF-Tu first binds to GTP. This activated EF-Tu–GTP complex binds to the tRNA. Next, hydrolysis of the GTP of the complex to GDP helps drive the binding of the aminoacyl-tRNA to the A site, at which point the EF-Tu is released (Figure 10-36a), leaving the new tRNA in the A site (Figure 10-36b).
2. Elongation factor EF-Ts mediates the release of EF-Tu–GDP from the ribosome and the regeneration of EF-Tu–GTP.

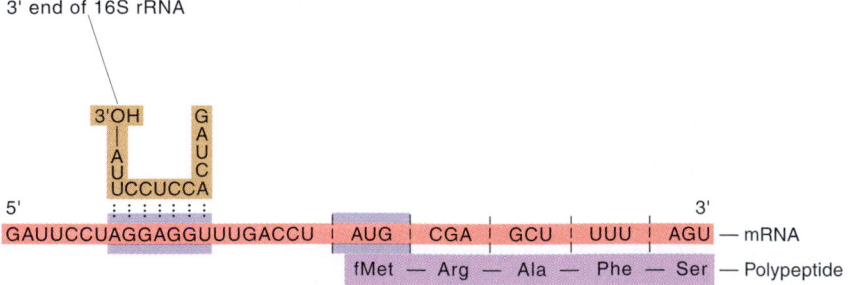

3' end of 16S rRNA
3'OH
A
U
UCCUCCA
G
A
U
C
5'
3'
GAUUCCUAGGAGGUUUGACCU AUG CGA GCU UUU AGU — mRNA
fMet — Arg — Ala — Phe — Ser — Polypeptide

Figure 10-35 Binding of the Shine-Dalgarno sequence on an mRNA to the 3′ end of 16S rRNA. (After L. Stryer, *Biochemistry,* 4th ed. Copyright © 1995 by Lubert Stryer.)

3. In the translocation step, the polypeptide chain on the peptidyl-tRNA is transferred to the aminoacyl-tRNA on the A site in a reaction catalyzed by the enzyme peptidyltransferase (Figure 10-36c). The ribosome then translocates by moving one codon farther along the mRNA, going in the $5' \rightarrow 3'$ direction. This step is mediated by the elongation factor EF-G (Figure 10-36d) and is driven by splitting a GTP to GDP. This action releases the uncharged tRNA from the P site and transfers the newly formed peptidyl-tRNA from the A site to the P site (Figure 10-36e).

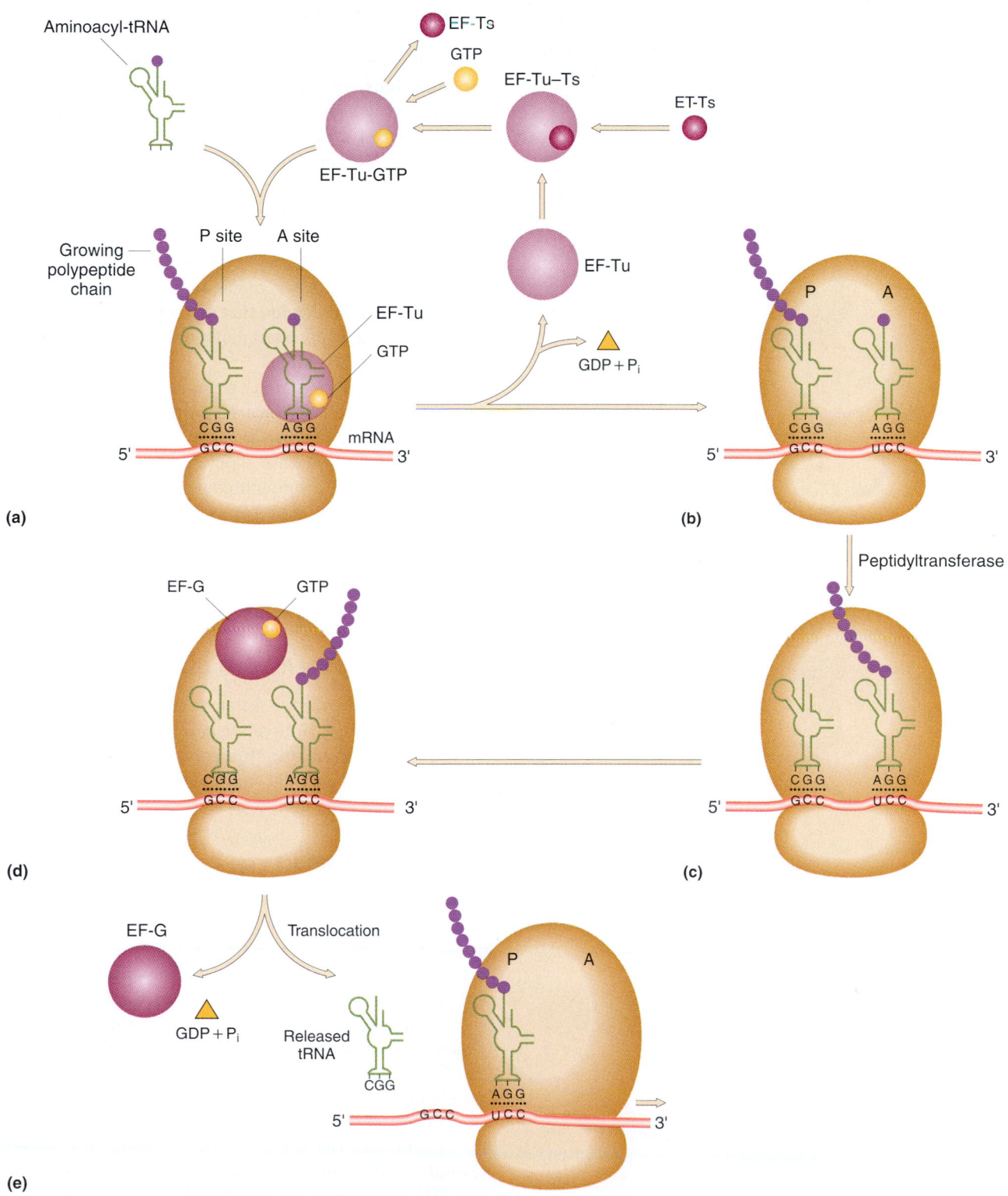

Figure 10-36 Steps in elongation (see text).

Termination

Release factors.   In the earlier discussion of the genetic code, we described the three chain-termination codons UAG, UAA, and UGA. Interestingly, these three triplets are not recognized by a tRNA, but instead by protein factors, termed **release factors,** which are abbreviated **RF1** and **RF2**. RF1 recognizes the triplets UAA and UAG, and RF2 recognizes UAA and UGA. A third factor, **RF3**, also helps to catalyze chain termination. When the peptidyl-tRNA is in the P site, the release factors, in response to the chain-terminating codons, bind to the A site. The polypeptide is then released from the P site, and the ribosomes dissociate into two subunits in a reaction driven by the hydrolysis of a GTP molecule. Figure 10-37 provides a schematic view of this process.

Nonsense suppressor mutations.   It is interesting to consider the suppressors of the nonsense mutations that Brenner and co-workers defined. Many of these **nonsense suppressor mutations** are known to alter the anticodon loop of specific tRNAs in such a way as to allow recognition of a nonsense codon in mRNA. Thus, an amino acid is inserted in response to the nonsense codon, and translation continues past that triplet. In Figure 10-38, the amber mutation replaces a wild-type codon with the chain-terminating nonsense codon UAG. By itself, the UAG would result in prematurely cutting off the protein at the corresponding position. The suppressor mutation in this case produces a tRNA^{Tyr} with an anticodon that recognizes the mutant UAG stop codon. The suppressed mutant thus contains tyrosine at that position in the protein.

What happens to normal termination signals at the ends of proteins in the presesnce of a suppressor? Many of the natural termination signals consist of two chain-termination signals in a row. Nonsense suppressors are sufficiently inefficient in translating through chain-terminating triplets, because of competition with release factors, that the probability of suppression at two codons in a row is small. Consequently, very few protein copies that carry many extraneous amino acids resulting from translation beyond the natural stop codon are produced.

Overview of protein synthesis

Figure 10-39 summarizes the steps in protein synthesis covered in this section. A direct visualization of protein synthesis can be seen in the electron micrograph shown in Figure 10-40, which shows the simultaneous transcription and translation of a gene in *E. coli.*

Protein processing

Even after mRNA has been successfully translated into its protein product, processing may continue. For example, membrane proteins or proteins that are secreted from the cell are synthesized with a short leader peptide, called a **signal sequence,** at the amino-terminal (N-terminal) end. This signal sequence is a stretch of 15 to 25 amino acids, most of which are hydrophobic. It allows for recognition by factors and protein receptors that mediate transport through the cell membrane; in this process, the signal sequence is cleaved by a peptidase (Figure 10-41). (A similar phenomenon exists for certain bacterial proteins that are secreted.) Moreover, several small peptide hormones, such as corticotropin (ACTH), result from the specific cleavage of a single, large polypeptide precursor.

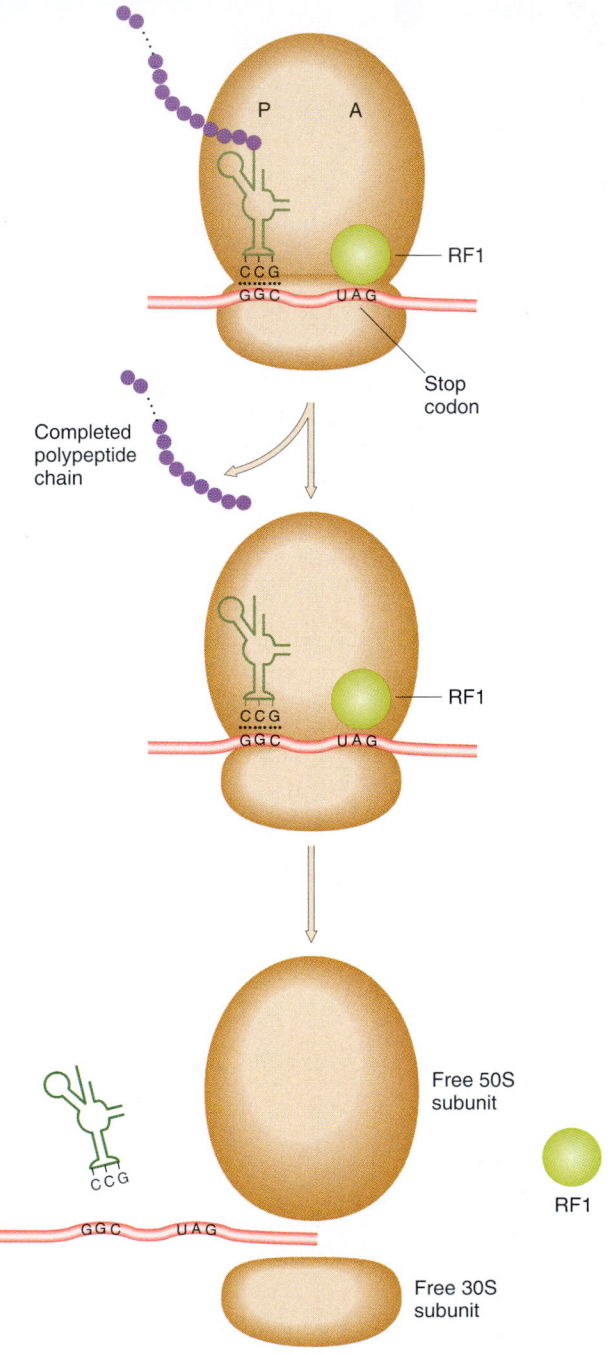

Figure 10-37   Steps leading to termination of protein synthesis (see text).

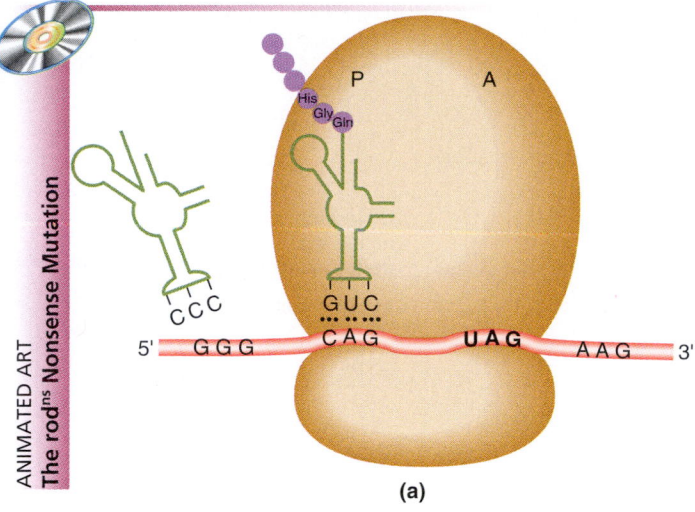

(a)

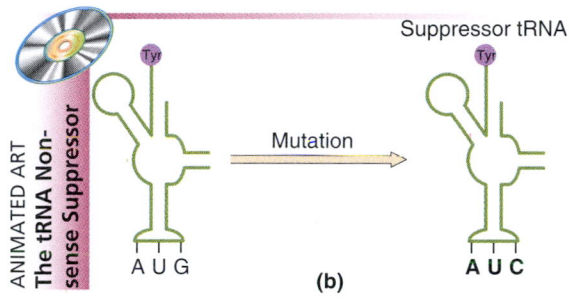

Suppressor tRNA

(b)

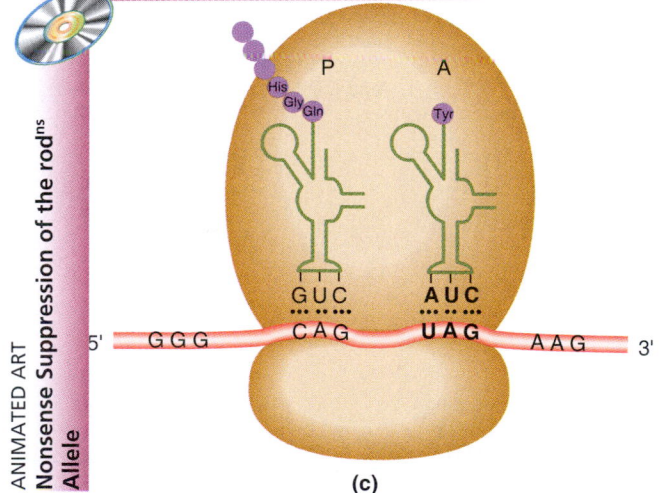

(c)

Figure 10-38 (a) Termination of translation. Here the translation apparatus cannot go past a nonsense codon (UAG in this case), because there is no tRNA that can recognize the UAG triplet. This leads to the termination of protein synthesis and to the subsequent release of the polypeptide fragment. The release factors are not shown here. (b) The molecular consequences of a mutation that alters the anticodon of a tyrosine tRNA. This tRNA can now read the UAG codon. (c) The suppression of the UAG codon by the altered tRNA, which now permits chain elongation. (After D. Watson, J. Tooze, and D. T. Kurtz, *Recombinant DNA: A Short Course.* Copyright © 1983 by W. H. Freeman and Company.)

Protein splicing

An extraordinary process that splices out an internal segment of certain proteins has been described in a variety of organisms, including prokaryotes and eukaryotes. This internal segment is termed an **intervening protein sequence,** or **IVPS.** The essential facet of this process is the formation of a new peptide bond between the two sequences flanking the IVPS. This reaction is autocatalytic and can take place in vitro. Figure 10-42 depicts protein splicing in schematic form. Interestingly, all IVPS segments studied so far contain an endonuclease activity, although this activity is unrelated to the protein-splicing reaction.

Universality of genetic information transfer

Thus far, our discussion has focused on bacteria, but the amazing fact is that the information-transfer and coding processes are virtually identical in all organisms that have been studied. For example, all the different single amino acid substitutions known to occur in human hemoglobin result from single nucleotide-pair substitutions based on the genetic code derived from *E. coli* (Table 10-7). Such observations suggest that the genetic code is common to all organisms. Furthermore, an information-bearing molecule, such as rabbit red blood cell mRNA, which is predominantly hemoglobin gene transcript, will be translated in an alien environment (such as a frog egg) into rabbit hemoglobin (Figure 10-43). Apparently, the translation apparatus is functionally the same in a wide range of different organisms.

Sequencing techniques at the protein, RNA, and DNA levels (see Chapter 12) have verified that the genetic code is universal in all organisms studied to date, ranging from viruses and bacteria to humans. One exception is mitochondrial DNA (see also Chapter 21). Two codons are translated

10-7 TABLE	Mutations and Their Inferred Codon Changes	
Protein*	Amino acid substitution	Inferred codon change
Hemoglobin	Glu → Val	GAA → GUA
Hemoglobin	Glu → Lys	GAA → AAA
Hemoglobin	Glu → Gly	GAA → GGA
Tryptophan synthetase	Gly → Arg	GGA → AGA
Tryptophan synthetase	Gly → Glu	GGA → GAA
Tryptophan synthetase	Glu → Ala	GAA → GCA
TMV coat protein	Leu → Phe	CUU → UUU
TMV coat protein	Glu → Gly	GAA → GGA
TMV coat protein	Pro → Ser	CCC → UCC

* Mutations are for human hemoglobins, *E. coli* tryptophan synthetase, and tobacco mosaic virus (TMV) coat protein.
Source: L. Stryer, *Biochemistry,* 2d ed. Copyright © 1981 by Lubert Stryer.

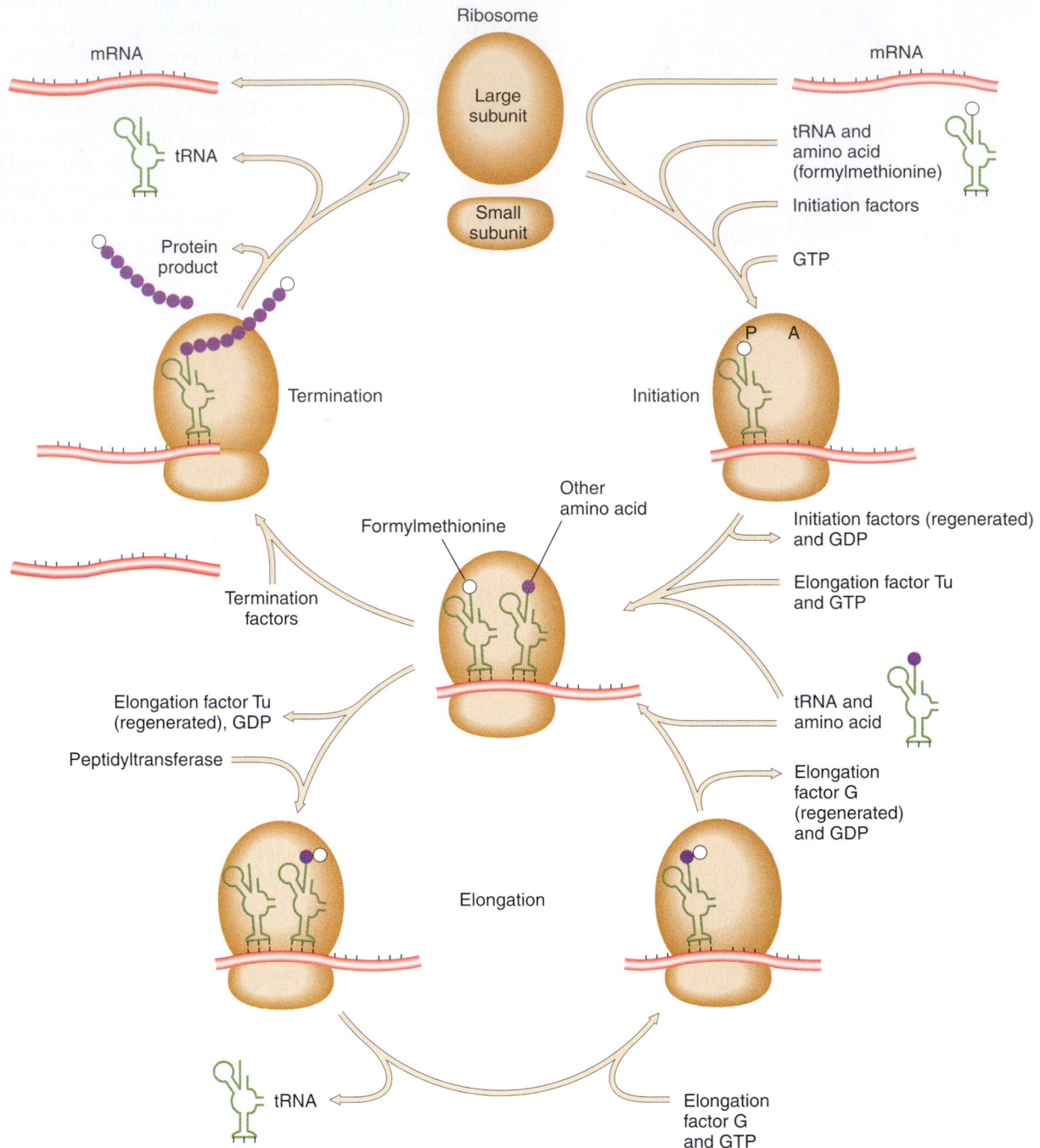

Figure 10-39   The transactions of the ribosome. At initiation, the ribosome recognizes the starting point in a segment of mRNA and binds a molecule of tRNA bearing a single amino acid. In all bacterial proteins, this first amino acid is *N*-formylmethionine. In elongation, a second amino acid is linked to the first one. The ribosome then shifts its position on the mRNA molecule, and the elongation cycle is repeated. When the stop codon is reached, the chain of amino acids folds spontaneously to form a protein. Subsequently, the ribosome splits into its two subunits, which rejoin before a new segment of mRNA is translated. Protein synthesis is facilitated by a number of catalytic proteins (initiation, elongation, and termination factors) and by guanosine triphosphate (GTP), a small molecule that releases energy when it is converted into guanosine diphosphate (GDP). (After D. M. Engleman and P. B. Moore, "Neutron-Scattering Studies of the Ribosome." Copyright © 1976 by Scientific American, Inc. All rights reserved.)

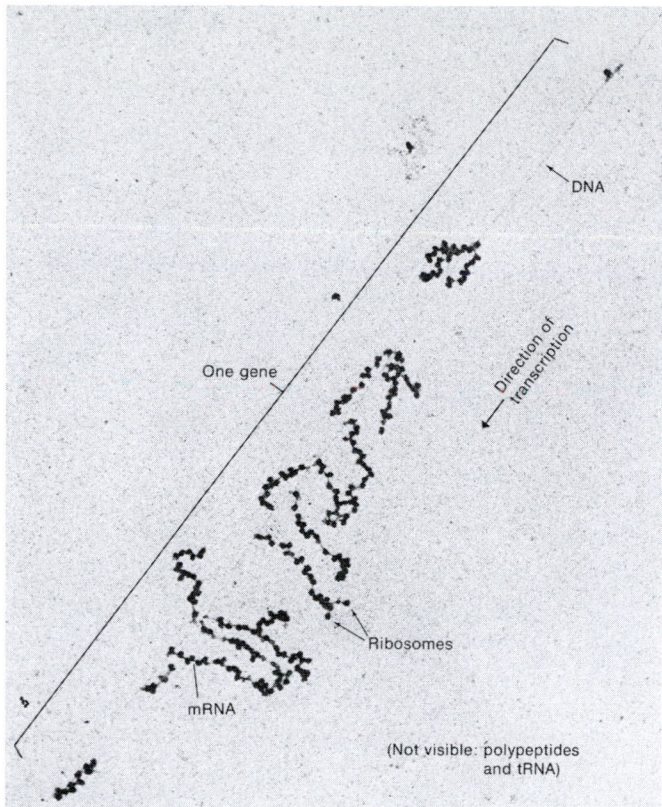

Figure 10-40   A gene of *E. coli* being simultaneously transcribed and translated. (Electron micrograph by O. L. Miller, Jr., and Barbara A. Hamkalo.)

Figure 10-41   Signal sequences. Proteins destined to be secreted from the cell have an amino-terminal sequence that is rich in hydrophobic residues. This signal sequence binds to the membrane and draws the remainder of the protein through the lipid bilayer. The signal sequence is cleaved from the protein in this process by an enzyme called *signal peptidase.* (After J. D. Watson, J. Tooze, and D. T. Kurtz, *Recombinant DNA: A Short Course.* Copyright © 1983 by W. H. Freeman and Company.)

differently here, owing to the properties of tRNAs that are confined to the mitochondrial system. Thus, whereas AUA is normally translated as isoleucine, it is read as methionine in mitochondria. In addition, mammalian mitochondria translate UGA as tryptophan, although UGA normally specifies a chain-terminating codon. In yeast, mitochondria translate UGA as tryptophan, as in mammalian mitochondria, but they translate AUA as isoleucine, as in bacterial (nonmitochondrial) systems. Other exceptions to the universal code are found in the nuclear genome of some protozoans. As we shall see from genetic engineering experiments in Chapter 13, DNA is DNA no matter what its origin. The nature and message of DNA represent a universal language of life on earth.

Does this interspecific equivalence of parts in the genetic apparatus indicate a common evolutionary ancestry for all life forms on earth? Or is it simply due to the fact that this is the only workable biochemical option in the earth environment (biochemical predestination)? Whatever the answer, the wonderful uniformity of the molecular basis of life is firmly established. Minor variations do exist, but they do not detract from the central uniformity of the mechanism that we have described.

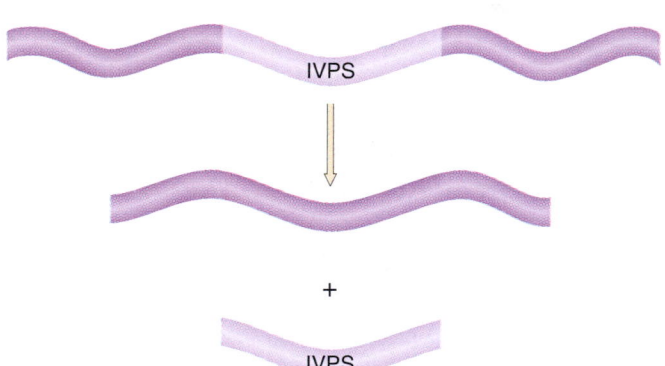

Figure 10-42   Protein splicing results in the removal of an internal segment (IVPS) and the formation of a new peptide bond that links the two regions that originally flanked the IVPS.

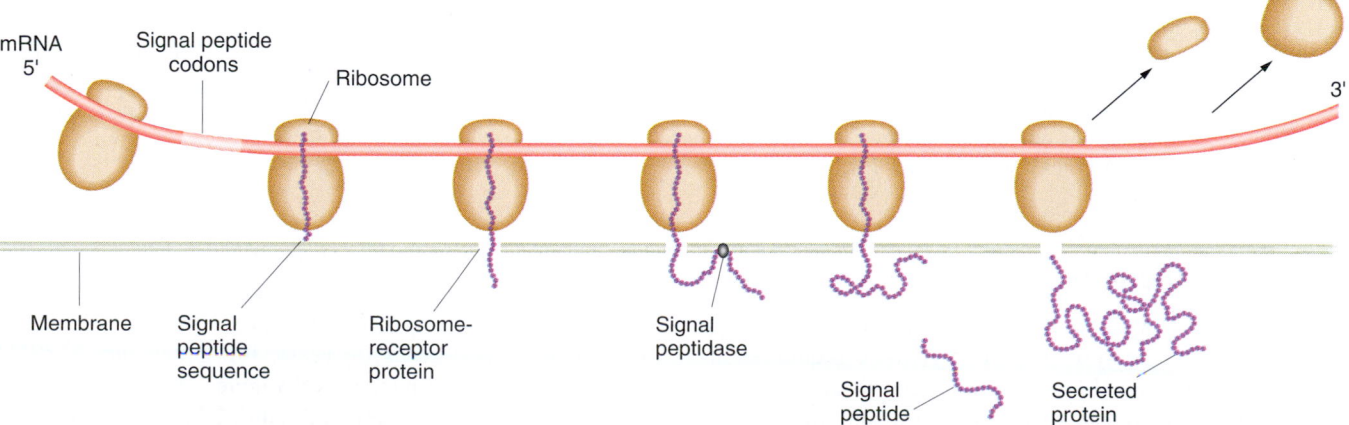

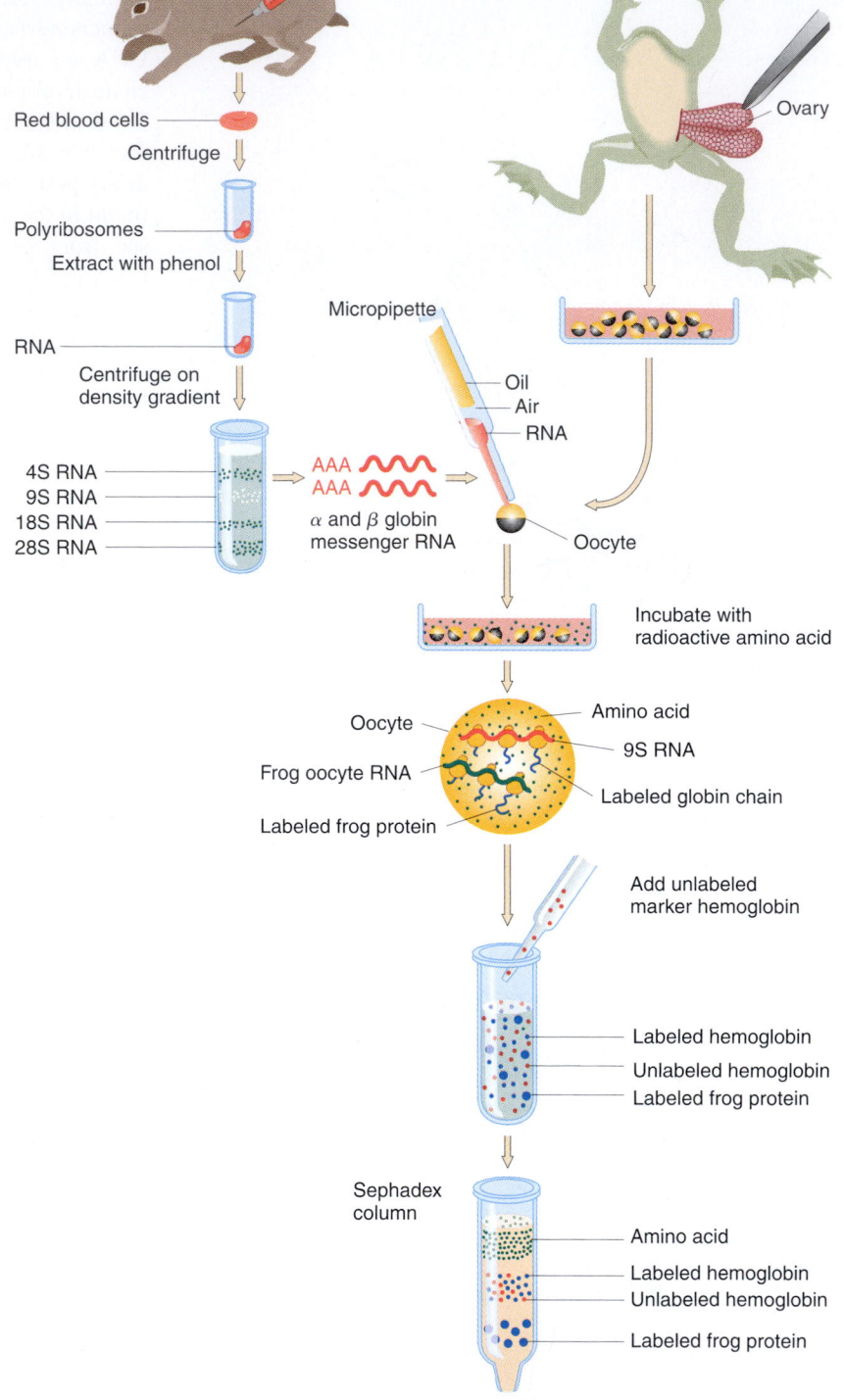

Figure 10-43 Rabbit hemoglobin mRNA is translated into rabbit hemoglobin in a frog *(Xenopus)* egg. This translation is accomplished with the use of the *Xenopus* translation apparatus. The mRNA for rabbit hemoglobin is injected into a *Xenopus* oocyte, which is then incubated in radioactively labeled amino acids. The products of translation are next separated in a chromatography column. Rabbit hemoglobin is identified by its position relative to an unlabeled marker (rabbit) hemoglobin. This experiment is one of many that point toward the uniformity of the genetic molecular mechanism in all forms of life. (Sephadex is a separatory material used in chromatography.) (From C. Lane, "Rabbit Hemoglobin from Frog Eggs." Copyright © 1976 by Scientific American, Inc. All rights reserved.)

Functional division of labor in the gene set

What different types of genes are needed to construct an organism? Recent analyses of whole genome sequences allow us to begin to tackle this question and to give us some ideas of the general categories of genes and the relative sizes of these categories. The complete nucleotide sequence of two eukaryotes has been deduced; baker's yeast, *Saccharomyces cerevisiae,* which has more than 6300 genes, and the nematode worm *Caenorhabditis elegans,* which has more than 19,000 genes. In yeast, 140 genes encode rRNA and 40 encode small nuclear RNA genes. There are also 275 tRNA genes in yeast, compared with 877 tRNA genes in the nematode.

Exactly 6217 genes in yeast and 19,099 genes in nematodes encode proteins. The set of protein-coding genes of an organism is called its **proteome.** Division of labor within the proteome is important for understanding what types of genes are needed for an organism to function. A comparison of the proteins identified by analyzing the sequences of both yeast and the nematode shows that the proteins participating in core biological functions are very similar in both organisms, but the other proteins found in each organism are not similar. Figure 10-44 shows that a similar number of proteins take part in the same core biological function in both organisms. Therefore, although many core biological processes are carried out by closely related proteins in these organisms, the large differences in the total number of proteins are due to proteins that provide functions specific to each organism.

From a large sequenced part of the genome of the model plant *Arabidopsis thaliana,* a tentative view of a plant proteome has been deduced and is represented diagramatically in Figure 10-45.

We are now at the threshold of an era in which outlines are appearing of the complete genetic blueprints for life on this planet, including the blueprint for the human species.

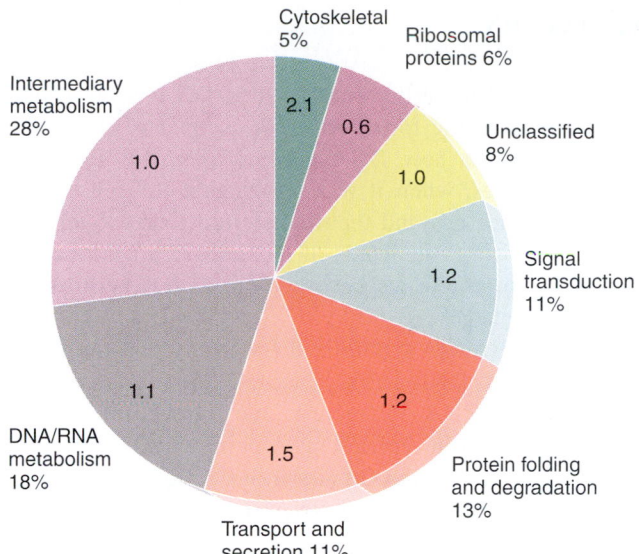

Figure 10-44   Distribution of core biological functions conserved in both yeast and worm. Yeast and worm protein sequences are clustered into closely related groups. Each sequence group (including groups with two or more sequences) is assigned to a single functional category. The number within each category is the ratio of worm to yeast proteins for that category. (After S. A. Chervitz et al., "Comparison of the complete protein sets of worm and yeast: orthology and divergence." *Science* 282, 1998, 2022–2028.)

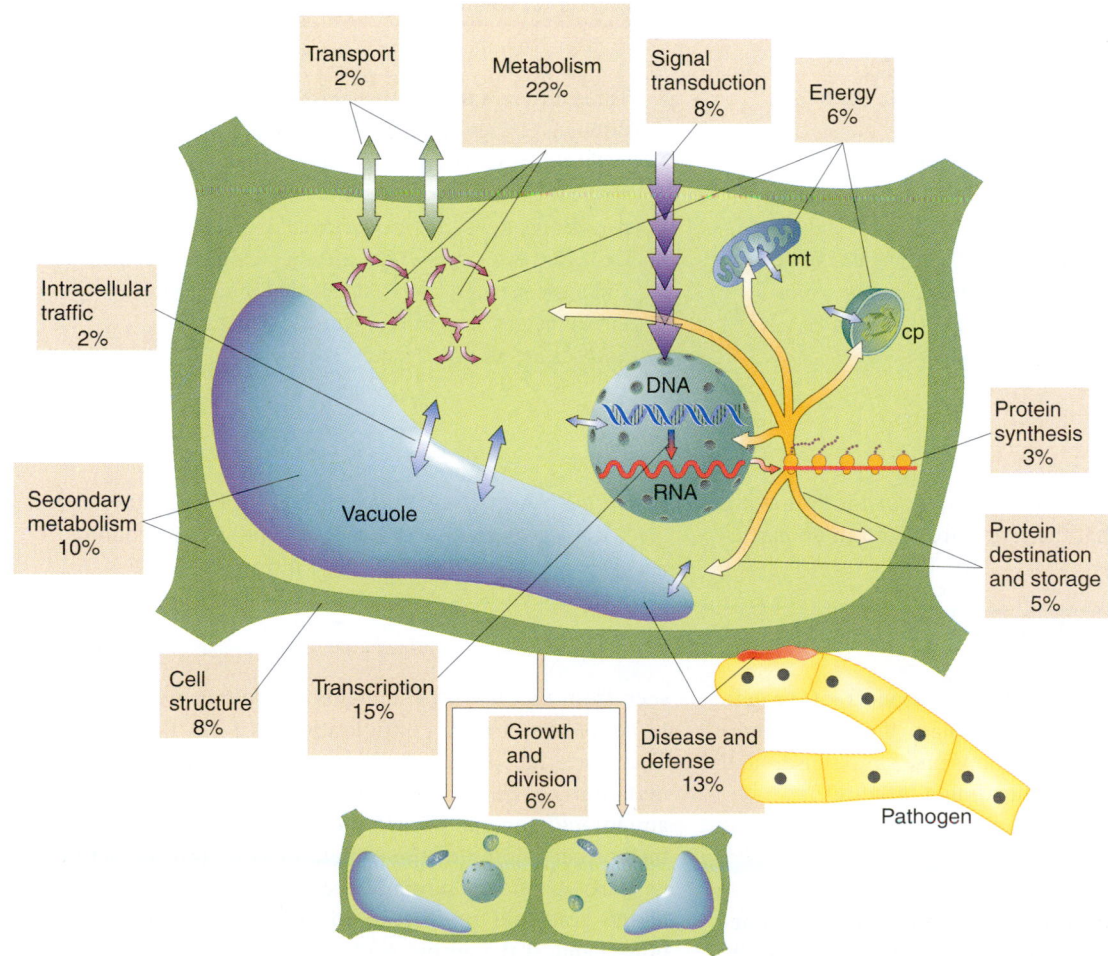

Figure 10-45   Sizes of various categories of protein-coding genes, estimated from currently known genes in the plant *Arabidopsis thaliana.* mt, mitochondrion; cp, chloroplast.

SUMMARY

We discovered in earlier chapters that DNA is the genetic material responsible for directing the synthesis of proteins. The first clue to how DNA accomplishes this feat came from eukaryotes, when it was shown that RNA is synthesized in the nucleus and then transferred to the cytoplasm. However, most of the details of this transfer of information from DNA to protein were worked out with experiments in bacteria and phages.

A given RNA is synthesized from only one strand of a double-stranded DNA helix. This transcription is catalyzed by an enzyme, RNA polymerase, and follows rules similar to those followed in replication: A pairs with U (instead of with T, as in DNA), and G pairs with C. Ribose is the sugar used in RNA. Extraction of RNA from a cell yields three basic varieties: ribosomal, transfer, and messenger RNA. The three sizes of ribosomal RNA (rRNA) combine with an array of specific proteins to form ribosomes, which are the machines used for protein synthesis (translation). Transfer RNAs (tRNAs) are a group of rather small RNA molecules, each with specificity for a particular amino acid; they carry the amino acids to the ribosome for attachment to the growing polypeptide.

Messenger RNA (mRNA) molecules are of many sizes and base sequences. These molecules contain information for the structure of proteins. The sequence of codons in mRNA determines the sequence of amino acids that will constitute a polypeptide. Each codon is specific for one amino acid, but several different codons may encode the same amino acid; that is, there is redundancy in the genetic code. In addition, there are three codons for which there are no tRNAs; these stop codons terminate the process of translation.

RNA is processed in eukaryotes before transport to the cytoplasm. Caps and tails are added, and internal parts of the primary transcript are removed. Many genes are therefore "split" in eukaryotes, and the coding segments of a gene are not colinear with the processed mRNA. Table 10-8 summarizes some of the differences between RNA synthesis in prokaryotes and eukaryotes.

The processes of information storage, replication, transcription, and translation are fundamentally similar in all living organisms. In demonstrating this similarity, molecular genetics has provided a powerful unifying force in biology.

10-8 ▸ TABLE	Differences in Gene Expression Between Prokaryotes and Eukaryotes
Prokaryotes	**Eukaryotes**
1. All RNA species are synthesized by a single RNA polymerase.	1. Three different RNA polymerases are responsible for the different classes of RNA molecules.
2. mRNA is translated during transcription.	2. mRNA is processed before transport to the cytoplasm, where it is translated. Caps and tails are added, and internal parts of the transcript are removed.
3. Genes are contiguous segments of DNA that are colinear with the mRNA that is translated into a protein.	3. Genes are often split. They are not contiguous segments of coding sequences; rather, the coding sequences are interrupted by intervening sequences (introns).
4. mRNAs are often polycistronic.	4. mRNAs are monocistronic.

CONCEPT MAP

Draw a concept map interrelating as many of the following terms as possible. Note that the terms are listed in no particular order.

gene / mRNA / tRNA / transcription / splicing / translation / RNA polymerase / ribosome / codon / anticodon / amino acid / intron

CHAPTER INTEGRATION PROBLEM

In Chapter 9, we considered different aspects of protein structure. Refer to Figure 9-8 and explain the effects individually on protein activity of nonsense, missense, and frameshift mutations in a gene encoding a protein.

◆ Solution ◆

Nonsense mutations result in termination of protein synthesis. Only a fragment of the protein is completed. As can be seen from Figure 9-8, fragments of proteins will not be able to adopt the correct configuration to form the active site. Therefore, the protein fragment will be inactive. Missense mutations, on the other hand, result in the exchange of one amino acid for another. If the amino acid is important for the correct folding of the protein or is part of the active site or participates in subunit interaction, then changing that amino acid will often lead to an inactive protein. On the other hand, if the amino acid is on the outside of the protein and does not take part in any of these activities or functions, then many such substitutions are acceptable and will not affect activity. Frameshift mutations change the reading frame and result in incorporation of amino acids that are different from those encoded in the wild-type reading frame. Unless the frameshift mutation is at the end of the protein, the resulting altered protein will not be able to fold into the correct configuration and will not have activity.

SOLVED PROBLEMS

1. Using Figure 10-26, show the consequences on subsequent translation of the addition of an adenine base to the beginning of the following coding sequence:

$$\overset{\text{(A)}}{\underset{\downarrow}{}}$$

$$-\overset{}{\text{CGA}}-\text{UCG}-\text{GAA}-\text{CCA}-\text{CGU}-\text{GAU}-\text{AAG}-\text{CAU}-$$
$$-\ \text{Arg}\ -\ \text{Ser}\ -\ \text{Glu}\ -\ \text{Pro}\ -\ \text{Arg}\ -\ \text{Asp}\ -\ \text{Lys}\ -\ \text{His}\ -$$

◆ Solution ◆

With the addition of A at the beginning of the coding sequence, the reading frame shifts, and a different set of amino acids is specified by the sequence, as shown here (note that a set of nonsense codons is encountered, which results in chain termination):

$$-\text{ACG}-\text{AUC}-\text{GGA}-\text{ACC}-\text{ACG}-\text{UGA}-\text{UAA}-\text{GCA}-$$
$$-\ \text{Thr}\ -\ \text{Ile}\ -\ \text{Gly}\ -\ \text{Thr}\ -\ \text{Thr}\ -\ \text{stop}\ -\ \text{stop}\ -$$

2. A single nucleotide addition followed by a single nucleotide deletion approximately 20 bp apart in DNA causes a change in the protein sequence from

$$-\text{His}-\text{Thr}-\text{Glu}-\text{Asp}-\text{Trp}-\text{Leu}-\text{His}-\text{Gln}-\text{Asp}-$$

to

$$-\text{His}-\text{Asp}-\text{Arg}-\text{Gly}-\text{Leu}-\text{Ala}-\text{Thr}-\text{Ser}-\text{Asp}-$$

Which nucleotide has been added and which nucleotide has been deleted? What are the original and the new mRNA sequences? (**Hint:** Consult Figure 10-27.)

◆ Solution ◆

We can draw the mRNA sequence for the original protein sequence (with the inherent ambiguities at this stage):

$$-\text{His}-\text{Thr}-\text{Glu}-\text{Asp}-\text{Trp}-\text{Leu}-\text{His}-\text{Gln}-\text{Asp}$$

$$-\text{CA}^{\text{U}}_{\text{C}}-\text{ACC}-\text{GA}^{\text{A}}_{\text{G}}-\text{GA}^{\text{U}}_{\text{C}}-\text{UGG}-\text{CUC}-\text{CA}^{\text{U}}_{\text{C}}-\text{CA}^{\text{A}}_{\text{G}}-\text{GA}^{\text{U}}_{\text{C}}$$

(with the additional alternative codons A, G below ACC; A, G, UUA, G below CUC)

Because the protein-sequence change given to us at the beginning of the problem begins after the first amino acid (His) owing to a single nucleotide addition, we can deduce that a Thr codon must change to an Asp codon. This change must result from the addition of a G directly before the Thr codon (indicated by a box), which shifts the reading frame, as shown here:

$$-\text{CA}^{\text{U}}_{\text{C}}-\boxed{\text{G}}\text{AC}-\text{UGA}-\text{GA}-\text{UUG}-\text{GCU}-\text{UCA}-\text{UCA}\uparrow-\text{GA}^{\text{U}}_{\text{C}}-$$

$$-\ \text{His}\ -\ \text{Asp}\ -\ \text{Arg}\ -\ \text{Gly}\ -\ \text{Leu}\ -\ \text{Ala}\ -\ \text{Thr}\ -\ \text{Ser}\ -\ \text{Asp}\ -$$

Additionally, because a deletion of a nucleotide must restore the final Asp codon to the correct reading frame, an A or G must have been deleted from the end of the original next-to-last codon, as shown by the arrow. The original protein sequence permits us to draw the mRNA with a number of ambiguities. However, the protein sequence resulting from the frameshift allows us to determine which nucleotide was in the original mRNA at most of these points of ambiguity. The nucleotide that must have appeared in the original sequence is circled. In only a few cases does the ambiguity remain.

PROBLEMS

1. The two strands of λ phage differ from each other in their GC content. Owing to this property, they can be separated in an alkaline cesium chloride gradient (the alkalinity denatures the double helix). When RNA synthesized by λ phage is isolated from infected cells, it is found to form DNA-RNA hybrids with both strands of λ DNA. What does this tell you? Formulate some testable predictions.

2. The data in the following table represent the base compositions of two double-stranded DNA sources and their RNA products in experiments conducted in vitro.

Species	$\dfrac{A + T}{G + C}$	$\dfrac{A + U}{G + C}$	$\dfrac{A + G}{U + C}$
Bacillus subtilis	1.36	1.30	1.02
E. coli	1.00	0.98	0.80

a. From these data, can you determine whether the RNA of these species is copied from a single strand or from both strands of the DNA? How? Drawing a diagram will make it easier to solve this problem.

b. Explain how you can tell whether the RNA itself is single stranded or double stranded.

(Problem 2 is reprinted with the permission of Macmillan Publishing Co., Inc., from M. Strickberger, *Genetics.* Copyright ©1968, Monroe W. Strickberger.)

3. Before the true nature of the genetic coding process was fully understood, it was proposed that the message might be read in overlapping triplets. For example, the sequence GCAUC might be read as GCA CAU AUC:

$$G \quad C \quad A \quad U \quad C$$

Devise an experimental test of this idea.

4. In protein-synthesizing systems in vitro, the addition of a specific human mRNA to the *E. coli* translational apparatus (ribosomes, tRNA, and so forth) stimulates the synthesis of a protein very much like that specified by the mRNA. What does this result show?

5. Which anticodon would you predict for a tRNA species carrying isoleucine? Is there more than one possible answer? If so, state any alternative answers.

6. **a.** In how many cases in the genetic code would you *fail* to know the amino acid specified by a codon if you knew only the first two nucleotides of the codon?

b. In how many cases would you fail to know the first two nucleotides of the codon if you knew which amino acid is specified by it?

7. Deduce what the six wild-type codons may have been in the mutants that led Brenner to infer the nature of the amber codon UAG.

8. If a polyribonucleotide contains equal amounts of randomly positioned adenine and uracil bases, what proportion of its triplets will code for **(a)** phenylalanine, **(b)** isoleucine, **(c)** leucine, **(d)** tyrosine?

9. You have synthesized three different messenger RNAs with bases incorporated in random sequence in the following ratios: **(a)** 1U:5C, **(b)** 1A:1C:4U, **(c)** 1A:1C:1G:1U. In a protein-synthesizing system in vitro, indicate the identities and proportions of amino acids that will be incorporated into proteins when each of these mRNAs is tested. (Refer to Figure 10-27.)

*10. One of the techniques used to decipher the genetic code was to synthesize polypeptides in vitro, with the use of synthetic mRNA with various repeating base sequences — for example, $(AGA)_n$, which can be

written out as AGAAGAAGAAGAAGA.... Sometimes the synthesized polypeptide contained just one amino acid (a homopolymer), and sometimes it contained more than one (a heteropolymer), depending on the repeating sequence used. Furthermore, sometimes different polypeptides were made from the same synthetic mRNA, suggesting that the initiation of protein synthesis in the system in vitro does not always start on the end nucleotide of the messenger. For example, from $(AGA)_n$, three polypeptides may have been made: aa_1 homopolymer (abbreviated aa_1-aa_1), aa_2 homopolymer (aa_2-aa_2), and aa_3 homopolymer (aa_3-aa_3). These probably correspond to the following readings derived by starting at different places in the sequence:

AGA AGA AGA AGA . . .
GAA GAA GAA GAA . . .
AAG AAG AAG AAG . . .

The table below shows the actual results obtained from the experiment done by Khorana.

Synthetic mRNA	Polypeptide(s) synthesized
$(UG)_n$	(Ser-Leu)
$(UG)_n$	(Val-Cys)
$(AC)_n$	(Thr-His)
$(AG)_n$	(Arg-Glu)
$(UUC)_n$	(Ser-Ser) and (Leu-Leu) and (Phe-Phe)
$(UUG)_n$	(Leu-Leu) and (Val-Val) and (Cys-Cys)
$(AAG)_n$	(Arg-Arg) and (Lys-Lys) and (Glu-Glu)
$(CAA)_n$	(Thr-Thr) and (Asn-Asn) and (Gln-Gln)
$(UAC)_n$	(Thr-Thr) and (Leu-Leu) and (Tyr-Tyr)
$(AUC)_n$	(Ile-Ile) and (Ser-Ser) and (His-His)
$(GUA)_n$	(Ser-Ser) and (Val-Val)
$(GAU)_n$	(Asp-Asp) and (Met-Met)
$(UAUC)_n$	(Tyr-Leu-Ser-Ile)
$(UUAC)_n$	(Leu-Leu-Thr-Tyr)
$(GAUA)_n$	None
$(GUAA)_n$	None

[**Note:** The order in which the polypeptides or amino acids are listed in the table is not significant except for $(UAUC)_n$ and $(UUAC)_n$.]

a. Why do $(GUA)_n$ and $(GAU)_n$ each encode only two homopolypeptides?

b. Why do $(GAUA)_n$ and $(GUAA)_n$ fail to stimulate synthesis?

c. Assign an amino acid to each triplet in the following list. Bear in mind that there often are several codons for a single amino acid and that the first two letters in a codon usually are the important ones (but that the third letter is occasionally significant). Also remember that some very different looking codons sometimes encode the same amino acid. Try to carry out this task without consulting Figure 10-27.

AUG   GAU   UUG   AAC
GUG   UUC   UUA   CAA
GUU   CUC   AUC   AGA
GUA   CUU   UAU   GAG
UGU   CUA   UAC   GAA
CAC   UCU   ACU   UAG
ACA   AGU   AAG   UGA

To solve this problem requires both logic and trial and error. Don't be disheartened: Khorana received a Nobel Prize for doing it. Good luck!

(Problem 10 is from J. Kuspira and G. W. Walker, *Genetics: Questions and Problems.* McGraw-Hill, 1973.)

11. You are studying a gene in *E. coli* that specifies a protein. A part of its sequence is

−Ala−Pro−Trp−Ser−Glu−Lys−Cys−His−

You recover a series of mutants for this gene that show no enzymatic activity. By isolating the mutant enzyme products, you find the following sequences:

Mutant 1:
−Ala−Pro−Trp−Arg−Glu−Lys−Cys−His−
Mutant 2:
−Ala−Pro−
Mutant 3:
−Ala−Pro−Gly−Val−Lys−Asn−Cys−His−
Mutant 4:
−Ala−Pro−Trp−Phe−Phe−Thr−Cys−His−

What is the molecular basis for each mutation? What is the DNA sequence that specifies this part of the protein?

12. A single nucleotide addition and a single nucleotide deletion approximately 15 sites apart in the DNA cause a protein change in sequence from

−Lys−Ser−Pro−Ser−Leu−Asn−Ala−Ala−Lys−

to

−Lys−Val−His−His−Leu−Met−Ala−Ala−Lys−

a. What are the old and the new mRNA nucleotide sequences? (Use Figure 10-26).

b. Which nucleotide has been added and which has been deleted?

(Problem 12 is from W. D. Stansfield, *Theory and Problems of Genetics.* McGraw-Hill, 1969.)

13. Suppressors of frameshift mutations are now known. Propose a mechanism for their action.

14. Use Figure 10-26 to complete the following table. Assume that reading is from left to right and that the columns represent transcriptional and translational alignments.

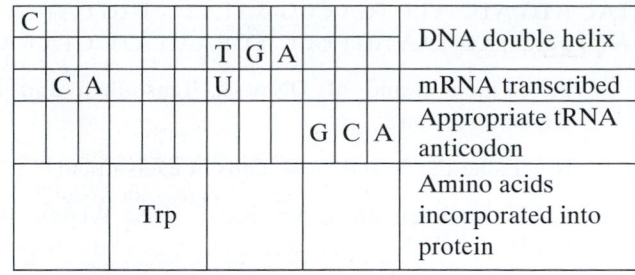

15. A mutational event inserts an extra nucleotide pair into DNA. Which of the following outcomes do you expect? (1) No protein at all; (2) a protein in which one amino acid is changed; (3) a protein in which three amino acids are changed; (4) a protein in which two amino acids are changed; (5) a protein in which most amino acids after the site of the insertion are changed.

16. Consider the gene that specifies the structure of hemoglobin. Arrange the following events in the most likely sequence in which they would occur.

a. Anemia is observed.

b. The shape of the oxygen-binding site is altered.

c. An incorrect codon is transcribed into hemoglobin mRNA.

d. The ovum (female gamete) receives a high radiation dose.

e. An incorrect codon is generated in the DNA of the hemoglobin gene.

f. A mother (an X-ray technician) accidentally steps in front of an operating X-ray generator.

g. A child dies.

h. The oxygen-transport capacity of the body is severely impaired.

i. The tRNA anticodon that lines up is one of a type that brings an unsuitable amino acid.

j. Nucleotide-pair substitution occurs in the DNA of the gene for hemoglobin.

17. An induced cell mutant is isolated from a hamster tissue culture because of its resistance to α-amanitin (a poison derived from a fungus). Electrophoresis shows that the mutant has an altered RNA polymerase; *just one* electrophoretic band is in a position different from that of the wild-type polymerase. The cells are presumed to be diploid. What does this experiment tell you about ways in which to detect recessive mutants in such cells?

18. A double-stranded DNA molecule with the sequence shown here produces, in vivo, a polypeptide that is five amino acids long.

TAC ATG ATC ATT TCA CGG AAT TTC TAG CAT GTA
ATG TAC TAG TAA AGT GCC TTA AAG ATC GTA CAT

    a. Which strand of DNA is transcribed and in which direction?

    b. Label the 5′ and the 3′ ends of each strand.

    c. If an inversion occurs between the second and third triplets from the left and right ends, respec-

tively, and the same strand of DNA is transcribed, how long will the resultant polypeptide be?

    d. Assume that the original molecule is intact and that transcription is of the bottom strand from left to right. Give the base sequence, and label the 5′ and 3′ ends of the anticodon that inserts the *fourth* amino acid into the nascent polypeptide. What is this amino acid?

11

REGULATION OF GENE TRANSCRIPTION

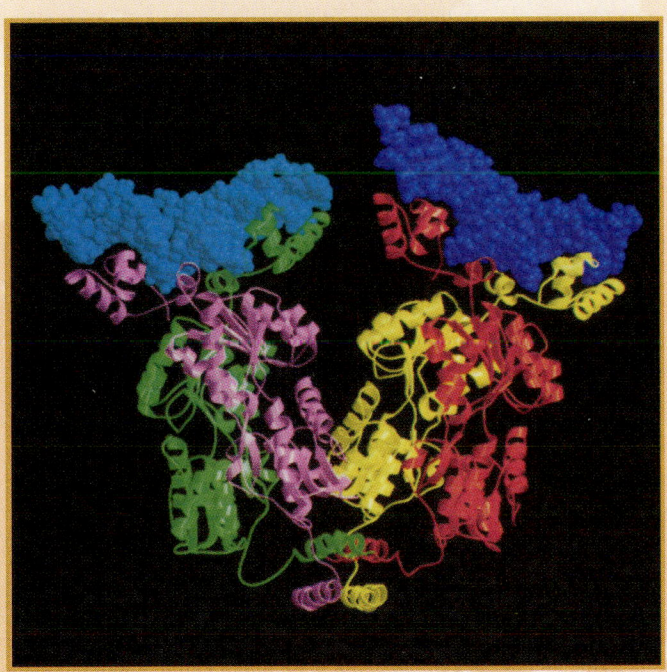

View of the *lac* repressor-DNA complex, as determined by X-ray crystallography.
Here the repressor tetramer is shown binding to two operators. Each dimer binds to one operator. The operators are 21 base pairs long here and are shown in dark and light blue. The monomers are in green, violet, red, and yellow. Note how the amino-terminal headpiece on one monomer (violet) crosses the other monomer (green) to bind the first part of the operator, whereas the second monomer headpiece (green) crosses over to bind the second part of the operator. Thus, the recognition helix from each of two subunits binds to the consecutive parts of the same operator. *(M. Lewis, Geoffrey Chang, N. C. Horton, M. A. Kercher, H. C. Pace, M. A. Schumacher, R. G. Brennan, and P. Lu, University of Pennsylvania.)*

Key Concepts

Gene regulation is most often mediated by proteins that react to environmental signals by raising or lowering the transcription rates of specific genes.

In prokaryotes, coordinate gene control is achieved by clustering coordinated structural genes on the chromosome so that they are transcribed into multigenic mRNAs.

Negative regulatory control is exemplified by the *lac* system, in which a repressor protein blocks transcription by binding to DNA at a site termed the operator.

Positive regulatory control requires protein factors to activate transcription.

Many regulatory proteins have common structural features.

In eukaryotes, additional regulatory DNA sites, such as enhancers, can act at considerable distance from the transcription start site to modulate gene expression by interacting with specific regulatory proteins.

In Chapters 9 and 10, we considered what genes are and how the cell machinery transcribes them into RNA molecules, many of which are translated into proteins. But how does the cell control the synthesis of its proteins? We can see that controlling the level of different proteins is crucial to an organism. In higher cells, specific cell types have differentiated to the point that they are highly specialized. A human eye cell synthesizes the proteins important for eye color but does not produce the detoxification enzymes that are synthesized in liver cells. Each cell type has arranged to express only some of its proteins.

Bacteria also have a need to regulate the expression of their proteins. Enzymes in sugar metabolism provide just one example. Metabolic enzymes are required to break down different carbon sources to yield energy. However, there are many different types of compounds that bacteria could use as carbon sources. Sugars such as lactose, glucose, maltose, rhamnose, galactose, and xylose, among others, are just a fraction of the possible energy sources. Several enzymes allow each of these compounds to enter the cell and to catalyze different steps in sugar breakdown. If a cell were to simultaneously synthesize all the enzymes that it might possibly need, it would cost the cell too much energy. Therefore, the synthesis of these enzymes needs to be strictly regulated.

Protein levels can be regulated by mechanisms that operate at the transcription or translation steps. This chapter deals with the control of transcription, which is the major control point for most regulation. Although the regulation of transcription is usually more complex in higher cells, some of the aspects are similar to those in prokaryotes. Here we shall first consider prokaryotic transcription control and then describe transcription regulation in eukaryotes.

Basic control circuits

As just defined, the basic problem facing the cell is to devise mechanisms to repress or shut down the transcription of all the genes encoding enzymes that are not needed and to activate the transcription of those genes when the enzymes are needed. To do this, two requirements must be met:

1. Cells need to be able to turn on or off the transcription of each specific gene or group of genes.

2. Cells must be able to recognize environmental conditions in which they should activate or repress transcription of the relevant genes.

Let's review the current model for one example of transcription control in prokaryotes and then consider its development. The first system that we shall focus on concerns lactose metabolism in *E. coli*. The detailed genetic analysis of this system by François Jacob and Jacques Monod in the 1950s provided the first major breakthrough in understanding transcription control. We have now learned a lot about how this system works. Figure 11-1 shows a physical model for the control of the lactose enzymes.

The metabolism of lactose requires two enzymes: a permease to transport lactose into the cell and β-galactosidase to cleave the lactose molecule to yield glucose and galactose. Permease and β-galactosidase are encoded by two contiguous genes, *Z* and *Y*, respectively. A third gene, the *A* gene, encodes an additional enzyme, termed *transacetylase*, but this enzyme is not required for lactose metabolism, and we will not concentrate on it for now. All three genes are transcribed into a single multigenic, messenger RNA (mRNA) molecule. Thus, it can be seen that, by regulation of the production of this mRNA, the regulation of the synthesis of all three enzymes can be coordinated. A fourth gene, the *I* gene, which maps near but not directly adjacent to the *Z*, *Y*, and *A* genes, encodes a **repressor** protein, so named because it can block the expression of the *Z*, *Y*, and *A* genes. The repressor binds to a region of DNA near the beginning of the *Z* gene and near the point at which transcription of the mRNA begins. The site on the DNA to which the repressor binds is termed the **operator**. One necessary

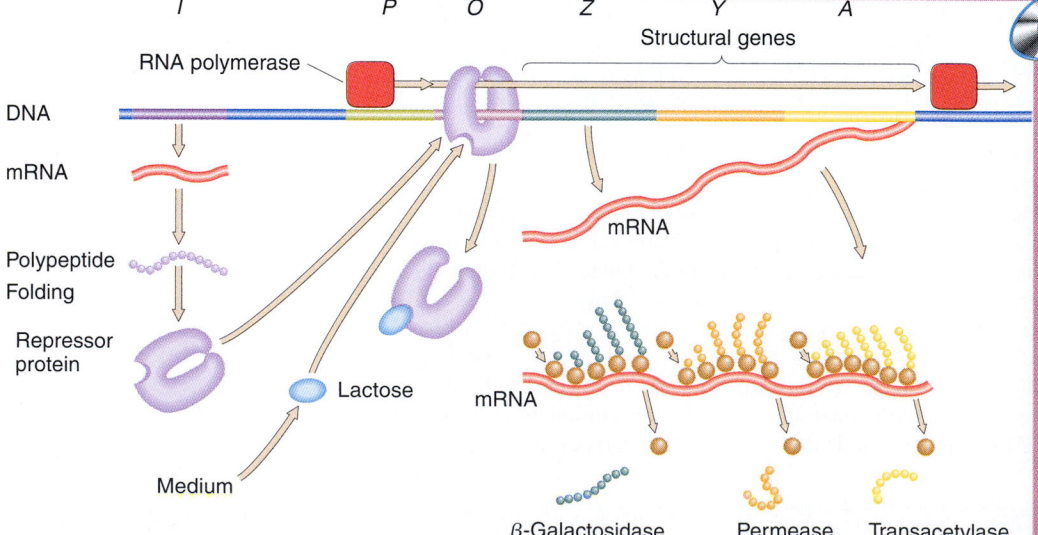

Figure 11-1 Regulation of the *lac* operon. The *I* gene continually makes repressor. The repressor binds to the *O* (operator) region, blocking the RNA polymerase bound to *P* (the promoter region) from transcribing the adjacent structural genes. When lactose is present, it binds to the repressor and changes its shape so that the repressor no longer binds to *O*. The RNA polymerase is then able to transcribe the *Z*, *Y*, and *A* structural genes, and the three enzymes are produced.

property of the repressor is that it be able to recognize a specific short sequence of DNA—namely, a specific operator. This property ensures that the repressor will bind only to the site on the DNA near the genes that it is controlling and not to other random sites all over the chromosome. By binding to the operator, the repressor prevents the initiation of transcription by RNA polymerase. Normally, RNA polymerase binds to specific regions of the DNA at the beginning of genes or groups of genes, termed **promoters** (see Chapter 10) so that it can initiate transcription at the proper starting points. The *POZYA* segments shown in Figure 11-1 constitute an **operon**, which is a genetic unit of coordinate expression.

The *lac* repressor is a molecule with two recognition sites—one that can recognize the specific operator sequence for the *lac* operon and another that can recognize lactose and certain analogs of lactose. When the repressor binds to lactose derivatives, it undergoes a conformational change; this slight alteration in shape changes the operator binding site so that the repressor loses affinity for the operator. Thus, in response to binding lactose derivatives, the repressor falls off the DNA. This satisfies the second requirement for such a control system—the ability to recognize conditions under which it is worthwhile to activate expression of the *lac* genes. The relief of repression for systems such as *lac* is termed **induction;** derivatives of lactose that inactivate the repressor and lead to expression of the *lac* genes are termed **inducers.**

Other bacterial systems operate by using protein *activator* molecules, which must bind to DNA as a prerequisite of transcription. Still additional mechanisms of control require proteins that allow the continuation of transcription in response to intracellular signals. Before we examine some of these control circuits in detail, let's review the classic work that initially described bacterial control systems, because these studies are landmarks in the use of genetic analysis.

Discovery of the *lac* system: negative control

Jacob and Monod used the lactose metabolism system of *E. coli* (Figure 11-2) to attack the problem of enzyme induction (originally termed *adaptation*); that is, the appearance of a specific enzyme only in the presence of its substrates. This phenomenon had been observed in bacteria for many years. How could a cell possibly "know" precisely which enzymes to synthesize? How could a particular substrate induce the appearance of a specific enzyme?

For the *lac* system, such an induction phenomenon could be illustrated when, in the presence of certain galactosides termed *inducers,* cells produced more than 1000 times as much of the enzyme β-galactosidase, which cleaves β-galactosides, as they produced when grown in the absence of such sugars. What role did the inducer play in the induction phenomenon? One idea was that the inducer was simply activating a pre-β-galactosidase intermediate that had accu-

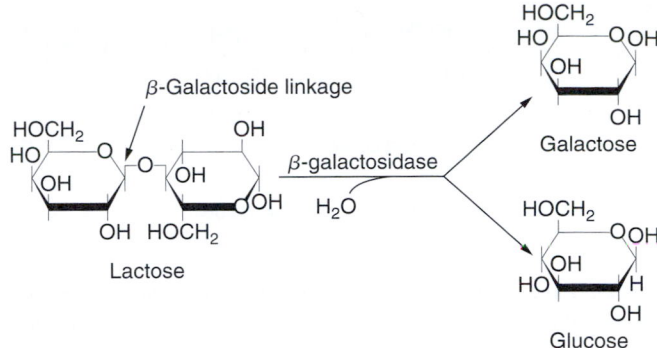

Figure 11-2 The metabolism of lactose. The enzyme β-galactosidase catalyzes a reaction in which water is added to the β-galactosidase linkage to break lactose into separate molecules of galactose and glucose. The enzyme lactose permease is required to transport lactose into the cell.

mulated in the cell. However, when Jacob and Monod followed the fate of radioactively labeled amino acids added to growing cells either before or after the addition of an inducer, they could show that induction represented the synthesis of new enzyme molecules. Kinetic studies established that these molecules could be detected as early as 3 minutes after the addition of an inducer. Additionally, withdrawal of the inducer brought about an abrupt halt in the synthesis of the new enzyme. Therefore, it became clear that the cell has a mechanism for turning gene expression on and off in response to environmental signals.

Genes controlled together

When Jacob and Monod induced β-galactosidase, they also induced the enzyme permease, which is required to transport lactose into the cell. The analysis of mutants indicated that each enzyme was encoded by a different gene. The enzyme transacetylase (with a dispensable and as yet unknown function) also was characterized and later shown to be encoded by a separate gene, although it was induced together with β-galactosidase and permease. Therefore, Jacob and Monod could identify three **coordinately controlled genes:** the *Z* gene encoding β-galactosidase, the *Y* gene encoding permease, and the *A* gene encoding transacetylase. Mapping defined the *Z*, *Y*, and *A* genes as being closely linked on the chromosome. Later studies of these and other coordinately controlled genes showed that in many cases a single mRNA molecule is produced by a contiguous set of genes. A frequently used inducer is the synthetic β-galactoside isopropyl-β-D-thiogalactoside, IPTG (Figure 11-3), which is not cleaved by β-galactosidase, allowing the control of the concentration of inducer inside the cell.

The *I* gene

Further genetic analysis sheds more light on the control circuit. Jacob and Monod characterized a new class of mutant, which synthesized all three enzymes at full levels,

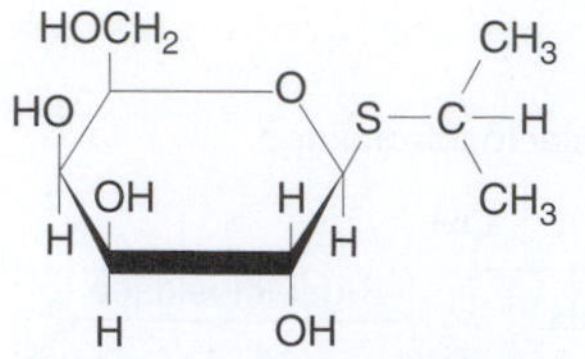

HOCH₂ ... CH₃ ... Isopropyl-β-D-thiogalactoside (IPTG)

Figure 11-3 Structure of IPTG, the inducer of the *lac* operon. The β-D-thiogalactoside linkage is not cleaved by β-galactosidase, allowing manipulation of the intracellular concentration of this inducer.

even in the absence of an inducer. For the first time, a mutant had been found with a defect not in the *activity* of an enzyme but in the control of enzyme *production*. These **constitutive** (always expressed in an unregulated fashion) **mutants** were found to have mutations mapping close to but distinct from the *Z*, *Y*, and *A* genes, permitting the definition of the *I* **locus** as the region controlling the inducibility of the *lac* enzymes. I^+ cells synthesize full levels of the *lac* enzymes only in the presence of an inducer, whereas I^- cells synthesize full levels in the presence or absence of an inducer. Figure 11-4 depicts the *lac* region defined by these experiments.

The repressor

The discovery of F′ factors (see Chapter 7) carrying the *lac* region allowed the construction of stable partial diploids and thus direct complementation tests. Tests with I^+ and I^- alleles showed that I^+ is dominant in trans over I^- alleles, demonstrating that the *I* gene product acts through the cyto-

plasm as a repressor. Table 11-1 shows the effect of various combinations of mutations, in the induced and noninduced state, on the production of β-galactosidase and permease.

A piece of evidence in support of the repressor model was the characterization of I^s mutations. Although mapping within the *I* gene, these mutations prevented induction of the *lac* enzymes by lactose or by the synthetic inducer IPTG (see Figure 11-3). Moreover, they were dominant to both an I^+ and an I^- allele (Table 11-2). The I^s mutation eliminates response to an inducer, presumably by altering the stereospecific binding site and destroying inducer binding. Therefore, even in the presence of IPTG, these molecules can still block *lac* enzyme synthesis. This ability would also explain their dominance, because the I^s repressor would be active, even in the presence of the wild-type repressor that was inactivated by the inducer. The I^s mutations clearly pointed to a direct interaction between the *I* gene product and the inducer.

MESSAGE ·····

The I^- mutation affects the *DNA-binding region* of the repressor, thus preventing binding and allowing transcription, even in the *absence* of inducer; the I^s mutation affects the *inducer-binding region* of the repressor, thus repressing transcription, even in the *presence* of inducer.

The operator and the operon

The specificity of the repressor, which results in turning off *lac* enzyme synthesis, suggests a stereospecific complex with an element that Jacob and Monod termed the *operator*. The operator was postulated to be a region of DNA near the beginning of the set of genes that it controlled. The researchers sought mutations in the operator that would allow

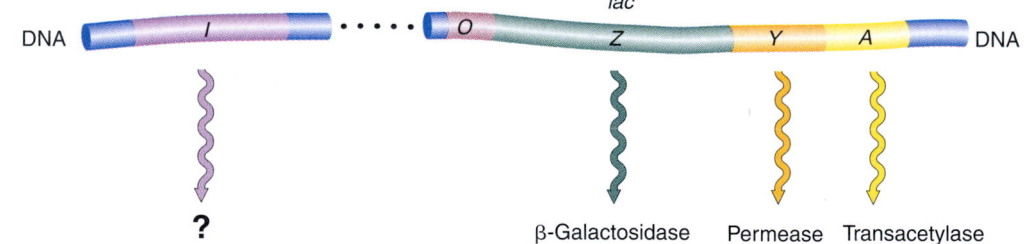

Figure 11-4 The *I* locus is the region controlling the inducibility of the *lac* enzymes.

		β-GALACTO-SIDASE (Z)		**PERMEASE (Y)**		
Strain	Genotype	Noninduced	Induced	Noninduced	Induced	Conclusions
1	$O^+ Z^+ Y^+$	−	+	−	+	Wild type is inducible
2	$O^+ Z^+ Y^+/F' O^+ Z^- Y^+$	−	+	−	+	Z^+ is dominant to Z^-
3	$O^c Z^+ Y^+$	+	+	+	+	O^c is constitutive
4	$O^+ Z^- Y^+/F' O^c Z^+ Y^-$	+	+	−	+	Operator is cis acting

11-1 TABLE   Synthesis of β-Galactosidase and Permease in Haploid and Heterozygous Diploid Operator Mutants

Note: Bacteria were grown in glycerol (no glucose present) with and without the inducer IPTG. The presence or absence of enzyme is indicated by a plus sign or a minus sign, respectively. All strains are I^+.

11-2 TABLE	Synthesis of β-Galactosidase and Permease in Haploid and Heterozygous Diploid Strains Carrying I^+ and I^-				

Strain	Genotype	β-GALACTO-SIDASE (Z)		PERMEASE (Y)		Conclusions
		Noninduced	Induced	Noninduced	Induced	
1	$I^+\ Z^+\ Y^+$	–	+	–	+	I^+ is inducible
2	$I^-\ Z^+\ Y^+$	+	+	+	+	I^- is constitutive
3	$I^+\ Z^-\ Y^+/F'\ I^-\ Z^+\ Y^+$	–	+	–	+	I^+ is dominant to I^-
4	$I^-\ Z^-\ Y^+/F'\ I^+\ Z^+\ Y^-$	–	+	–	+	I^+ is trans acting

Note: Bacteria were grown in glycerol (no glucose present) and induced with IPTG. The presence of the maximal level of the enzyme is indicated by a plus sign; the absence or very low level of an enzyme is indicated by a minus sign. (All strains are O^+.)

synthesis of the *lac* enzymes even in the presence of an active repressor. These mutations should exert their effect on other genes adjacent to them in the operon, an effect known as **cis dominance.** (The word *cis* means "adjacent"; *trans*, used below, means "across.") Regular (**trans**) dominance implies the action of a diffusible product; cis dominance implies the physical interaction of an element with the genes directly in contact with it. No diffusible product is altered by the mutation. By selecting for constitutivity (unrepressed synthesis) in cells with two copies of the *lac* region, to eliminate the effects of single I^- mutations, Jacob and Monod detected such mutations and labeled them O^c, for **operator constitutive.** As Table 11-3 indicates, strains carrying these

mutations are capable of synthesizing maximal amounts of enzyme in the presence of IPTG, but they can also synthesize from 10 to 20 percent of these levels in the absence of an inducer. The O^c mutations are indeed dominant in the cis position, as shown in Table 11-3. Mapping experiments have located the operator locus between I and Z.

As we have seen, the *OZYA* segment constitutes a genetic unit of coordinate expression that Jacob and Monod termed the *operon*. Figure 11-5 depicts a simplified operon model for the *lac* system. The *lac* operon is said to be under the **negative control** of the *lac* repressor, because the repressor normally blocks expression of the *lac* enzymes in the absence of an inducer.

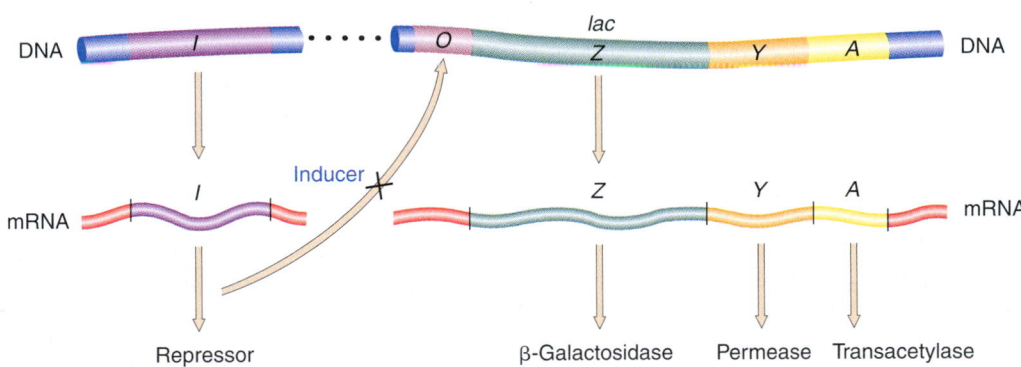

Figure 11-5 A simplified *lac* operon model. The three genes Z, Y, and A are coordinately expressed. The product of the I gene, the repressor, blocks the expression of the Z, Y, and A genes by interacting with the operator (O). The inducer can inactivate the repressor, thereby preventing interaction with the operator. When this happens, the operon is fully expressed.

11-3 TABLE	Synthesis of β-Galactosidase and Permease by the Wild Type and by Strains Carrying Different Alleles of the I Gene				

Strain	Genotype	β-GALACTO-SIDASE (Z)		PERMEASE (Y)		Conclusions
		Uninduced	Induced	Uninduced	Induced	
1	$I^+\ Z^+\ Y^+$	–	+	–	+	I^+ is inducible
2	$I^s\ Z^+\ Y^+$	–	–	–	–	I^s is always repressed
3	$I^s\ Z^+\ Y^+/F'\ I^+$	–	–	–	+	I^s is dominant to I^+

Note: Bacteria were grown in glycerol (no glucose present) with and without the inducer IPTG. Presence of the indicated enzyme is represented by a plus sign; absence or low levels, a minus sign.

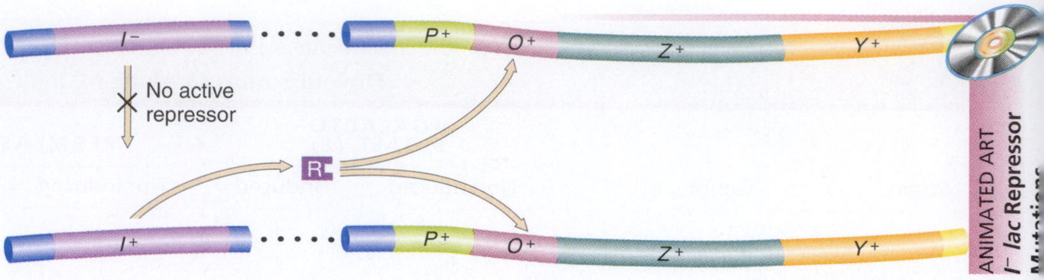

Figure 11-6 The recessive nature of I^- mutations demonstrates that the repressor is trans acting. Although no active repressor is synthesized from the I^- gene, the wild-type (I^+) gene provides a functional repressor that binds to both operators in a diploid cell and blocks *lac* operon expression (in the absence of an inducer).

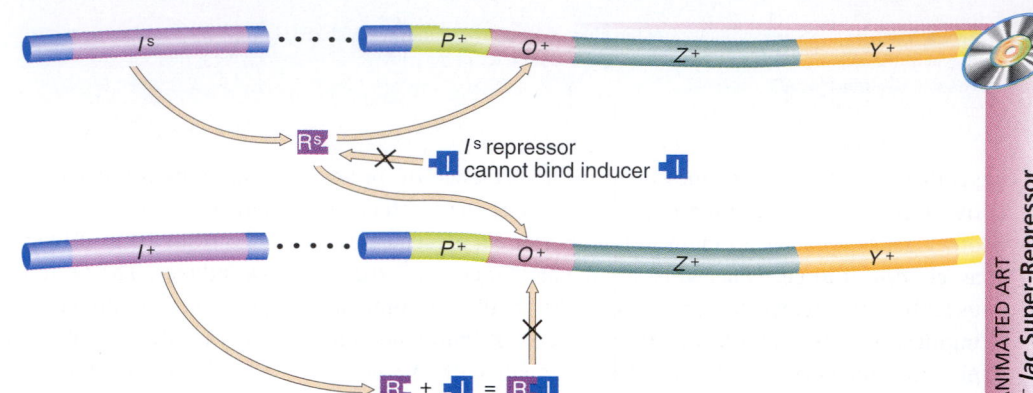

Figure 11-7 The dominance of I^s mutations is due to the inactivation of the allosteric site on the Lac repressor. In an I^s/I^+ diploid cell, none of the *lac* structural genes are transcribed, even in the presence of an inducer. In contrast with the wild-type repressor, the I^s repressor lacks a functional lactose-binding site (the allosteric site) and thus is not inactivated by an inducer. Thus, even in the presence of an inducer, the I^s repressor binds irreversibly to all operators in a cell, thereby blocking transcription of the *lac* operon.

Let's review the model in Figure 11-5, as postulated by Jacob and Monod. The *Z* and *Y* genes encode the structure of two enzymes required for the metabolism of the sugar lactose, β-galactosidase and permease, respectively. The *A* gene encodes transacetylase. All three genes are linked together on the chromosome. Their transcription into a polycistronic (single) mRNA provides the basis for coordinate control at the level of mRNA synthesis. The synthesis of the polycistronic *lac* mRNA can be blocked by the action of a repressor protein molecule, which binds to an operator region near the start point for transcription. The repressor is the product of the *I* gene. Therefore, mutations in the *I* gene that prevent the synthesis of a functional repressor result in unrepressed, or constitutive, synthesis of the *lac* enzymes. Repression can also be overcome by certain galactosides, termed *inducers,* which inactivate the repressor by binding to it and altering its affinity for the operator. In this manner, the inducer can pull the repressor off the DNA.

We can now understand the properties of some of the diploids used for complementation tests, in light of the operon model. Figure 11-6 shows how I^- mutations are recessive to wild type, because one functional *I* gene is all that is needed to produce a repressor that can bind to both operators in a diploid. On the other hand, Figure 11-7 shows a diploid cell carrying one copy of a wild-type *I* gene and one copy of an I^s gene. The I^s mutation alters the inducer binding site so that repressor no longer binds to inducer, although it still recognizes the operator. These diploid cells are Lac$^-$, because the altered repressor always binds to the operator, even in the presence of an inducer, and blocks synthesis of the *lac* enzymes.

Operator mutations (O^c) are cis dominant, because they are dominant only for genes directly linked to them on the same chromosome, as diagrammed in Figure 11-8. Thus, if an altered operator is in the same cell with a second chromosome that contains a wild-type operator, the repressor

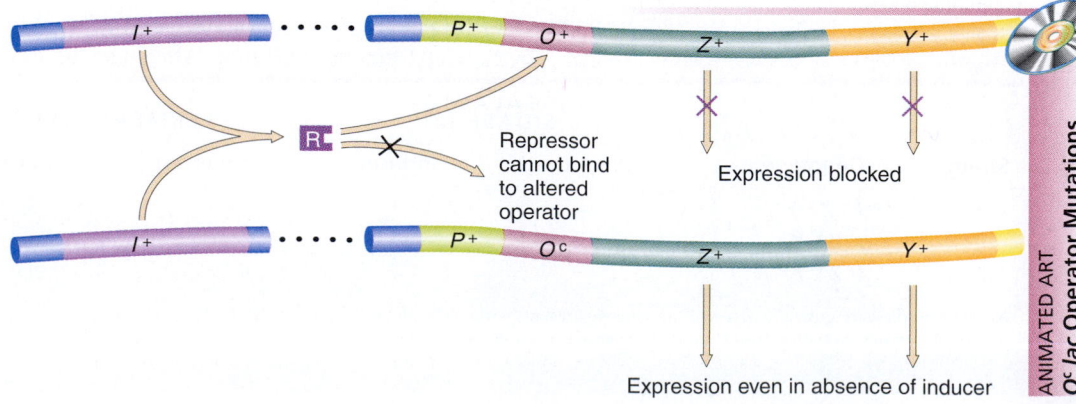

Figure 11-8 O^+/O^c heterozygotes demonstrate that operators are cis acting. Because a repressor cannot bind to O^c operators, the *lac* structural genes linked to an O^c operator are expressed even in the absence of an inducer. However, the *lac* genes adjacent to an O^+ operator are still subject to repression.

will recognize the wild-type operator and repress the genes linked to it but will not recognize the altered operator. Therefore, the genes linked to the altered operator are expressed even in the absence of inducer.

The *lac* promoter

Genetic experiments demonstrated that an element essential for *lac* transcription is located between *I* and *O* in the operon model for the *lac* system. This element, termed the *promoter* (*P*), serves as an initiation site for transcription. Promoter mutations affect the transcription of all genes in the operon in a similar manner. Promoter mutations are cis dominant, as would be expected for a site on the DNA that serves as a recognition element for transcription initiation, because each promoter governs transcription only for those genes in the operon adjacent to it on the *same* DNA molecule. As outlined in Chapter 10, in vitro experiments demonstrated that RNA polymerase binds to the promoter region and that repressor binding to the operator can block RNA polymerase from binding to the promoter. Mutant analysis, physical experiments, and comparisons with other promoters identified two binding regions for RNA polymerase in a typical prokaryotic promoter. Figure 11-9 summarizes this body of information.

> **MESSAGE** ··
> The *lac* operon is a cluster of structural genes that specify enzymes having roles in lactose metabolism. These genes are controlled by the coordinated actions of cis-dominant promoter and operator regions. The activity of these regions is, in turn, determined by a repressor molecule specified by a separate regulator gene. Figure 11-1 integrates all this information into a single picture.

Characterization of the *lac* repressor and the *lac* operator

The decisive experiment was provided by Walter Gilbert and Benno Müller-Hill, who in 1966 isolated and purified the repressor by monitoring the binding of the radioactively labeled inducer IPTG. They demonstrated that the repressor is a protein consisting of four identical subunits, each with a molecular weight of approximately 38,000. Each molecule contains four IPTG-binding sites. (A more detailed description of the repressor is given later in the chapter.) In vitro,

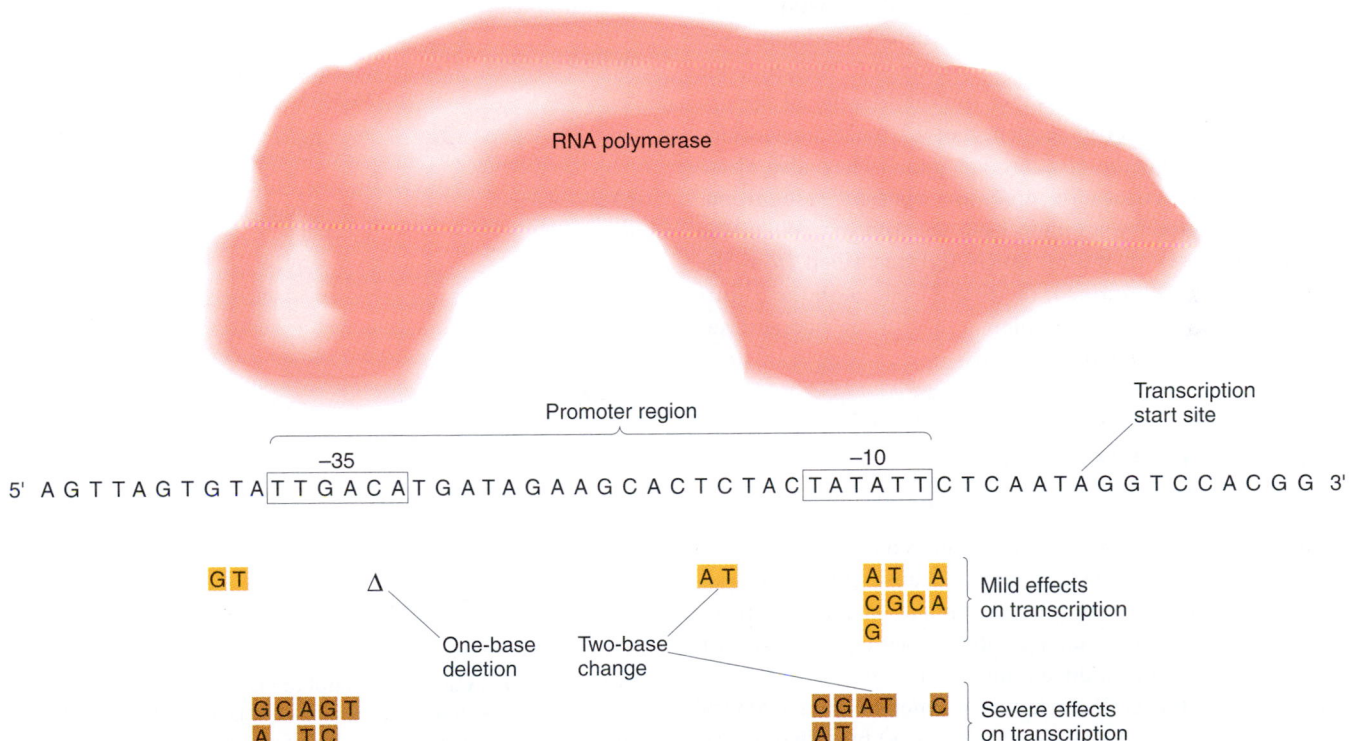

Figure 11-9 Specific DNA sequences are important for the efficient transcription of *E. coli* genes by RNA polymerase. The boxed sequences at approximately 35 and 10 nucleotides before the transcription start site are highly conserved in all *E. coli* promoters, an indication of their role as contact sites on the DNA for RNA polymerase binding. Mutations in these regions have mild (gold) and severe (brown) effects on transcription. The mutations may be changes of single nucleotides or pairs of nucleotides, or a deletion (Δ) may occur. (From J. D. Watson, M. Gilman, J. Witkowski, and M. Zoller, *Recombinant DNA*, 2d ed. Copyright © 1992 by James D. Watson, Michael Gilman, Jan Witkowski, and Mark Zoller.)

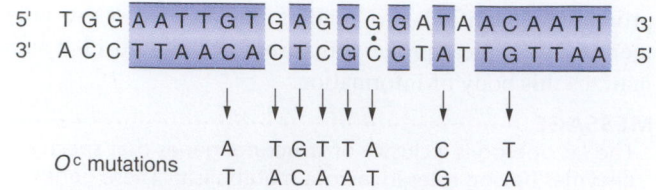

O^c mutations

Figure 11-10   The DNA base sequence of the lactose operator and the base changes associated with eight O^c mutations. Regions of twofold rotational symmetry are indicated by color and by a dot at their axis of symmetry. (From W. Gilbert, A. Maxam, and A. Mirzabekov, in N. O. Kjeldgaard and O. Malløe, eds., *Control of Ribosome Synthesis.* Academic Press, 1976. Used by permission of Munksgaard International Publishers, Ltd., Copenhagen.)

repressor binds to DNA containing the operator and comes off the DNA in the presence of IPTG. Gilbert and his co-workers have shown that the repressor can protect specific bases in the operator from chemical reagents. These experiments provide crucial proofs of the mechanism of repressor action formulated by Jacob and Monod.

Gilbert used the enzyme DNase to break apart the DNA bound to the repressor. He was able to recover short DNA strands that had been shielded from the enzyme activity by the repressor molecule and that presumably constituted the operator sequence. This sequence was determined, and each operator mutation was shown to be a change in the sequence (Figure 11-10). These results confirm the identity of the operator locus as a specific sequence of 17 to 25 nucleotides situated just before the structural *Z* gene. They also show the incredible specificity of repressor–operator recognition, which is disrupted by a single base substitution. When the sequence of bases in the *lac* mRNA (transcribed from the *lac* operon) was determined, the first 21 bases on the 5′ initiation end proved to be complementary to the operator sequence that Gilbert had determined.

Catabolite repression of the *lac* operon: positive control

An additional control system is superimposed on the repressor–operator system. This system exists because cells have specific enzymes that favor glucose uptake and metabolism. If both lactose *and* glucose are present, synthesis of β-galactosidase is not induced until all the glucose has been utilized. Thus, the cell conserves its metabolic machinery (that, for example, induces the *lac* enzymes) by utilizing any existing glucose before going through the steps of creating new machinery to metabolize the lactose. The operon model outlined earlier in this chapter will not account for the suppression of induction by glucose, so we must modify it.

Studies indicate that in fact some catabolic breakdown product of glucose (no exact identity is yet known) prevents activation of the *lac* operon by lactose, so this effect was

originally called **catabolite repression.** The effect of the glucose catabolite is exerted on an important cellular constituent called *cyclic adenosine monophosphate (cAMP).*

When glucose is present in high concentrations, the cAMP concentration is low; as the glucose concentration decreases, the concentration of cAMP increases correspondingly. The high concentration of cAMP is necessary for activation of the *lac* operon. Mutants that cannot convert ATP into cAMP cannot be induced to produce β-galactosidase, because the concentration of cAMP is not great enough to activate the *lac* operon. In addition, there are other mutants that do make cAMP but cannot activate the *lac* enzymes, because these mutants lack yet another protein, called catabolite activator protein (CAP), made by the *crp* gene. CAP forms a complex with cAMP, and it is this complex that is able to bind to the CAP site of the operon. The DNA-bound CAP is then able to interact physically with RNA polymerase and essentially increase the affinity of RNA polymerase for the *lac* promoter. In this way, the catabolite repression system contributes to the selective activation of the *lac* operon (Figure 11-11).

Glucose control is accomplished because a glucose breakdown product inhibits formation of the CAP-cAMP complex required for RNA polymerase to attach at the *lac*

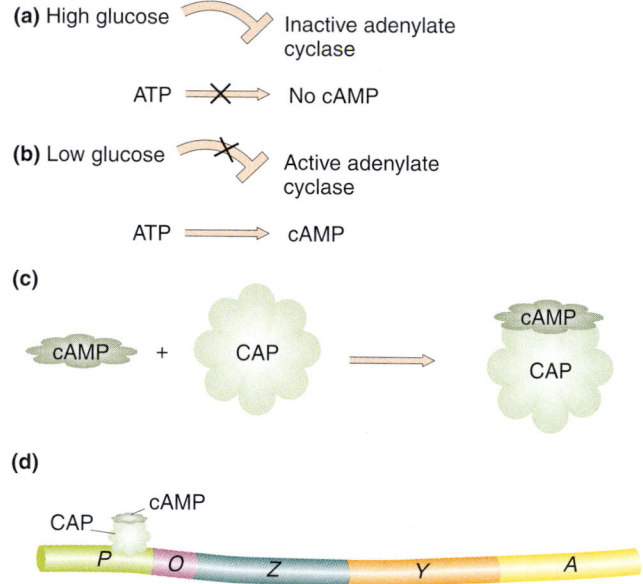

Figure 11-11   Catabolite control of the *lac* operon. The operon is inducible by lactose to the maximal levels when cAMP and CAP form a complex. (a) Under conditions of high glucose, a glucose breakdown product inhibits the enzyme adenylate cyclase, preventing the conversion of ATP into cAMP. (b) Under conditions of low glucose, there is no breakdown product, and therefore adenylate cyclase is active and cAMP is formed. (c) When cAMP is present, it acts as an allosteric effector, complexing with CAP. (d) The cAMP-CAP complex acts as an activator of *lac* operon transcription by binding to a region within the *lac* promoter. (CAP, catabolite activator protein; cAMP, cyclic adenosine monophosphate.)

promoter site. Even when there is a shortage of glucose catabolites and CAP-cAMP forms, the enzymes taking part in lactose transport and metabolism are produced only if lactose is present. This level of control is accomplished because *lac* operon inducers must bind to the repressor protein to remove it from the operator site and permit transcription of the *lac* operon. Thus, the cell conserves its energy and resources by producing the lactose-metabolizing enzymes only when they are both needed and useful. These concepts are summarized in Figure 11-12, which also depicts the

bending of the DNA resulting from CAP-cAMP binding to the CAP site, presumably increasing the affinity of RNA polymerase to the promoter.

MESSAGE ··························

The *lac* operon has an added level of control so that the operon remains inactive in the presence of glucose even if lactose also is present. High concentrations of glucose catabolites produce low concentrations of cAMP, which must form a complex with CAP to permit the induction of the *lac* operon.

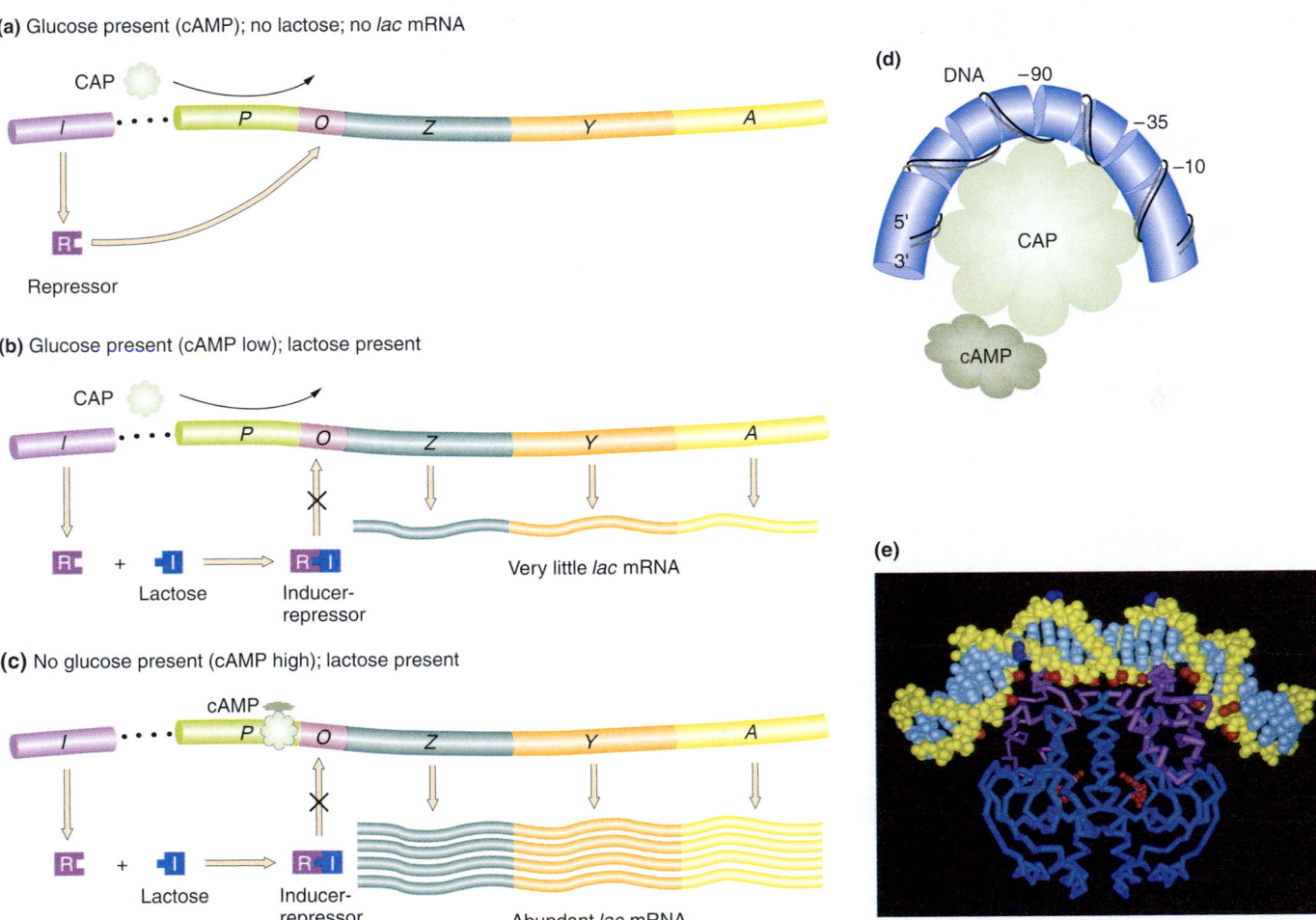

(a) Glucose present (cAMP); no lactose; no *lac* mRNA

(b) Glucose present (cAMP low); lactose present

Very little *lac* mRNA

(c) No glucose present (cAMP high); lactose present

Abundant *lac* mRNA

(d)

(e)

Figure 11-12   Negative and positive control of the *lac* operon by the Lac repressor and the catabolite activator protein (CAP), respectively. (a) In the absence of lactose to serve as an inducer, the Lac repressor is able to bind the operator; regardless of the levels of cAMP and the presence of CAP, mRNA production is repressed. (b) With lactose present to bind the repressor, the repressor is unable to bind the operator; however, only small amounts of mRNA are produced because the presence of glucose keeps the levels of cAMP low, and thus the cAMP-CAP complex does not form and bind the promoter. (c) With the repressor inactivated by lactose and with high levels of cAMP present (owing to the absence of glucose), cAMP binds CAP. The cAMP-CAP complex is then able to bind the promoter; the *lac* operon is thus activated, and large amounts of mRNA are produced. (d) When CAP binds the promoter, it creates a bend greater than 90° in the DNA. Apparently, RNA polymerase binds more effectively when the promoter is in this bent configuration. (e) CAP bound to its DNA recognition site. This part is derived from the structural analysis of the CAP-DNA complex. [Parts a–d after B. Gartenberg and D. M. Crothers, *Nature* 333, 1988, 824. (See H. N. Lie-Johnson et al., *Cell* 47, 1986, 995.) Adapted from H. Lodish, D. Baltimore, A. Berk, S. L. Zipursky, P. Matsudaira, and J. Darnell, *Molecular Cell Biology,* 3d ed. Copyright © 1995 by Scientific American Books. Part e from L. Schultz and T. A. Steitz.]

Positive and negative control

The inducer–repressor control of the *lac* operon is an example of **negative control,** in which expression is normally blocked. In contrast, the CAP-cAMP system is an example of **positive control,** because expression of the *lac* operon requires the *presence* of an activating signal — in this case, the interaction of the CAP-cAMP complex with the CAP region. Figure 11-13 distinguishes between these two basic types of control systems.

For activators or repressor proteins to do their job, each must be able to exist in two states: one that can bind its DNA targets and one that cannot. The binding state must be appropriate for a given set of environmental conditions. For

(a) Negative control

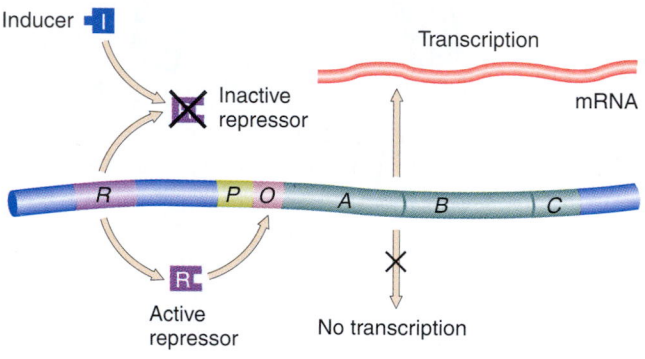

(b) Positive control

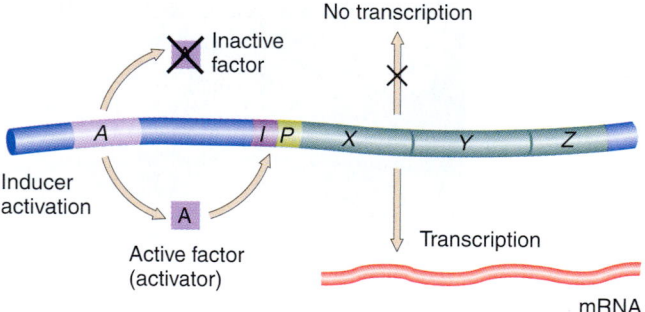

Figure 11-13   Comparison of positive and negative control. The basic aspects of negative and positive control are depicted. (a) In negative control, an active repressor (encoded by the *R* gene in the example shown here) blocks gene expression of the *A B C* operon by binding to an operator site (*O*). An inactive repressor allows gene expression. The repressor can be inactivated either by an inducer or by mutation. (b) In positive control, an active factor is required for gene expression, as shown for the *X Y Z* operon here. Small molecules can convert an inactive factor into an active one, as in the case of cyclic AMP and the CAP protein. An inactive positive control factor results in no gene expression. The activator binds to the control region of the operon, termed *I* in this case. (The positions of both *O* and *I* with respect to the promoter, *P*, in the two examples are arbitrarily drawn.)

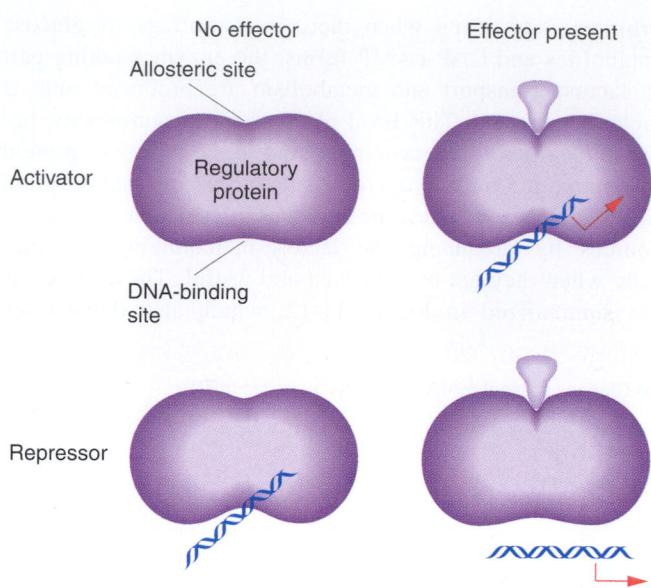

Figure 11-14   The effects of allosteric effectors on the DNA-binding activities of activators and repressors.

many activator or repressor proteins, the way that DNA binding is regulated is through the interaction of two different sites in the three-dimensional structure of the protein. One site is the DNA-binding domain. The other site, the **allosteric site,** acts as a switch that sets the DNA-binding domain in one of two modes: functional or nonfunctional. The allosteric site interacts with small molecules called allosteric effectors. In regard to the *lac* operon, the *lac* inducers are allosteric effectors. When allosteric effectors bind to the allosteric site, they cause a conformational change in the regulatory protein so that it alters the structure of the DNA-binding domain. Some activator or repressor proteins must bind to their allosteric effectors to bind DNA. Others can bind DNA only in the absence of their allosteric effectors. Figure 11-14 depicts these concepts.

By using different combinations of controlling elements, bacteria have evolved numerous strategies for regulating gene expression. Some examples follow.

Dual positive and negative control: the arabinose operon

The metabolism of the sugar arabinose is catalyzed by three enzymes encoded by the *araB, araA,* and *araD* genes. Figure 11-15 depicts this operon. Expression is activated at *araI*, the **initiator** region. Within this region, the product of the *araC* gene, when bound to arabinose, can activate transcription, perhaps by directly affecting RNA polymerase binding in the *araI* region, which contains the promoter for the *araB, araA,* and *araD* genes (Figure 11-16a). This activation exemplifies positive control, because the product of the regulatory gene (*araC*) must be active for the operon to be expressed.

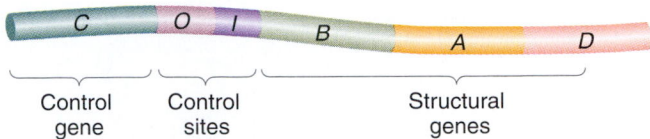

Control gene   Control sites   Structural genes

Figure 11-15 Map of the *ara* region. The *B*, *A*, and *D* genes together with the *I* and *O* sites constitute the *ara* operon.

An additional positive control is mediated by the same CAP-cAMP system that regulates *lac* expression. In the presence of arabinose, both CAP and the binding of the *araC* product to the initiator region are required to allow RNA polymerase to bind to the promoter for the *araB*, *araA*, and *araD* genes. In the absence of arabinose, the *araC* product assumes a different conformation and represses the *ara* operon by binding both to *araI* and to an operator region, *araO*, thereby forming a loop (Figure 11-16b) that prevents transcription. Thus, the AraC protein has two conformations that promote two opposing functions at two alternative binding sites. The conformation depends on whether the inducer, arabinose, is bound to the protein.

Metabolic pathways

Coordinate control of genes in bacteria is widespread. In the early 1960s, when Milislav Demerec studied the distribution of loci affecting a common biosynthetic pathway, he found that the genes controlling steps in the synthesis of the amino acid tryptophan in *Salmonella typhimurium* are clustered together in a restricted part of the genome. Demerec then looked at the distribution of genes having roles in a number of different metabolic pathways. Analyzing auxotrophic mutations representing 87 different genes, he found that 63 genes could be located in 17 functionally similar clusters. A **cluster** is defined as two or more loci that control related functions, where the loci are not separated by an unrelated gene.

Furthermore, in cases in which the sequence of catalytic activity is known, there is a remarkable congruence between the sequence of genes on the chromosome and the sequence in which their products act in the metabolic pathway. This congruence is illustrated by the tryptophan cluster in *E. coli* (Figure 11-17), characterized by Charles Yanofsky.

MESSAGE ···
Genes having roles in the same metabolic pathway are frequently tightly clustered on prokaryotic chromosomes, often in the same sequence as the reactions that they control. Furthermore, the genes within a cluster are expressed at the same time.
··

Additional examples of control: attenuation

The *lac* operon is an example of an inducible system in the sense that the synthesis of an enzyme is induced by the presence of its substrate. Repressible systems also exist, in which an excess of product leads to a shutdown of the production of the enzymes that synthesize that product. Such a control system has been identified for a cluster of genes controlling enzymes in the pathway for tryptophan production. Synthesis of tryptophan is shut off when there is an excess of tryptophan in the medium. Jacob and Monod suggested that the cluster of five *trp* genes in *E. coli* forms another operon, differing from the *lac* operon in that the tryptophan repressor will bind to the *trp* operator only when it *is* bound to tryptophan (see Figure 11-17). (Recall that the *lac* repressor binds to the operator *except* when it is bound to lactose.) A second control pathway also modulates tryptophan biosynthesis at the level of enzyme activity. This is termed **feedback inhibition.** Here, the first enzyme in the pathway, encoded by the *trpE* and *trpD* genes, is inhibited by tryptophan itself.

As with the *lac* operon, further analysis of the *trp* operon revealed yet another level of control superimposed

(a) Positive control

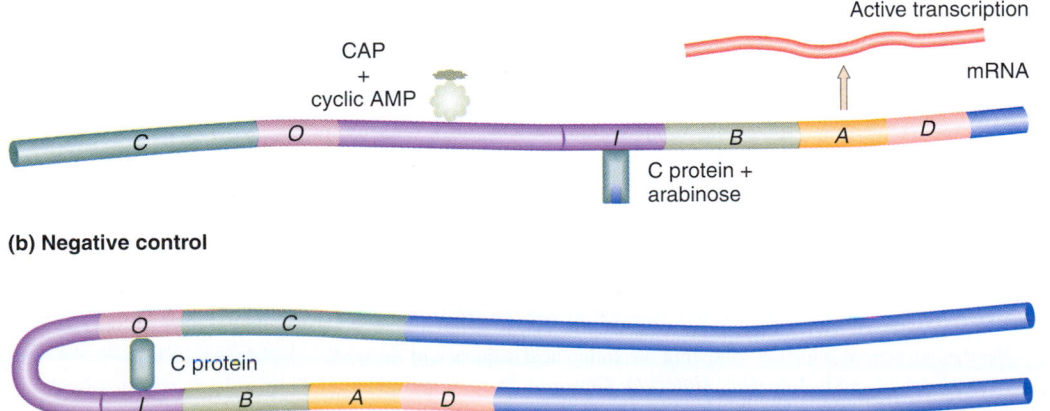

(b) Negative control

Figure 11-16 Dual control of the *ara* operon. (a) In the presence of arabinose, the AraC protein binds to the *araI* region and, when bound to cAMP, the CAP protein binds to a site adjacent to *araI*. This binding stimulates the transcription of the *araB*, *araA*, and *araD* genes. (b) In the absence of arabinose, the AraC protein binds to both the *araI* and *araO* regions, forming a DNA loop. This binding prevents transcription of the *ara* operon.

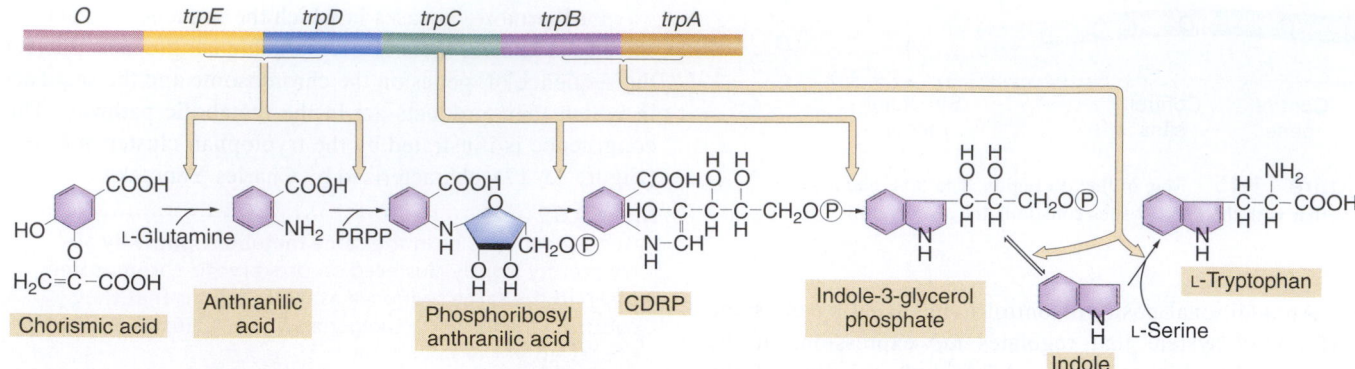

Figure 11-17 The chromosomal order of genes in the *trp* operon of *E. coli* and the sequence of reactions catalyzed by the enzyme products of the *trp* structural genes. The products of genes *trpD* and *trpE* form a complex that catalyzes specific steps, as do the products of genes *trpB* and *trpA*. Tryptophan synthetase is a tetrameric enzyme formed by the products of *trpB* and *trpA*. It catalyzes a two-step process leading to the formation of tryptophan. (PRPP, phosphoribosyl pyrophosphate; CDRP, 1-(*o*-carboxyphenylamino)-1-deoxyribulose 5-phosphate.) (After S. Tanemura and R. H. Bauerle, *Genetics* 95, 1980, 545.)

on the basic repressor−operator mechanism. In studying mutant strains (carrying a mutation in *trpR,* the repressor locus) that continue to produce *trp* mRNA in the presence of tryptophan, Yanofsky found that removal of tryptophan from the medium leads to almost a tenfold increase in *trp* mRNA production in these strains, even though the Trp repressor was inactive, and thus could not account for the increase through normal derepression of the operator because of low tryptophan levels. Furthermore, Yanofsky identified the region responsible for this increase by isolating a totally constitutive mutant strain that produces *trp* mRNA at this tenfold maximal level, even in the presence of tryptophan, and he showed that this mutation has a deletion located between the operator and the *trpE* gene (see the map in Figure 11-17).

Yanofsky was able to isolate the multigene *trp* operon mRNA. On sequencing it, he found a long sequence, termed the **leader sequence,** of approximately 160 bases at the 5′ end before the first triplet in the *trpE* gene. The deletion

mutant that always produces *trp* mRNA at maximal levels has a deletion extending from base 130 to base 160 (Figure 11-18). Yanofsky called the element inactivated by the deletion the **attenuator,** because its presence apparently leads to a reduction in the rate of mRNA transcription when tryptophan is present. Figure 11-19 shows the position of these elements in the *trp* operon. But what is the role of the leader sequence in bases 1 to 130? A surprising observation provides the key to solving this problem.

While studying mRNAs transcribed by the *trp* operon (using *trpR⁻* mutants), Yanofsky discovered that, even in the presence of high levels of tryptophan (which should cause the attenuator region to reduce the rate of transcription tenfold), the first 141 bases of the leader sequence were always transcribed at the maximal rate, though the full-length mRNA was found, as expected, at levels only one-tenth as great. Another way of stating this is that, even in the presence of high concentrations of tryptophan, the first 141 bases are transcribed in maximal numbers but, because

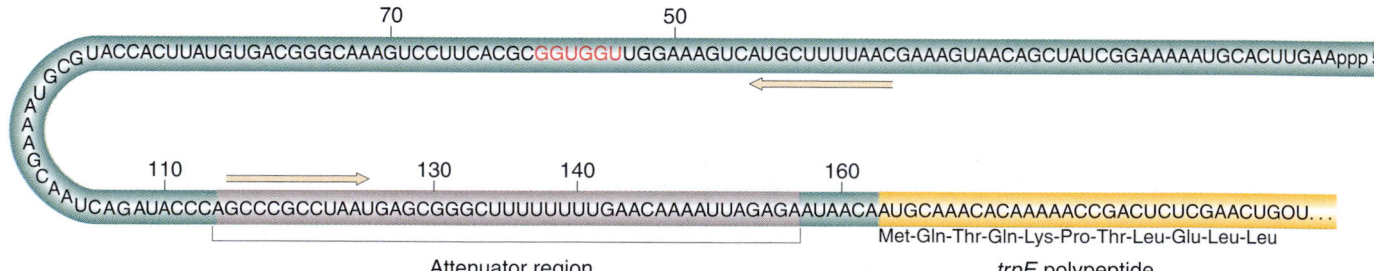

Figure 11-18 The leader sequence, showing the attenuator segment of the *trp* operon, along with the beginning of the *trpE* structural sequence (showing the amino acid sequence of the *trpE* polypeptide). (After G. S. Stent and R. Calendar, *Molecular Genetics,* 2d ed. Copyright © 1978 by W. H. Freeman and Company. Based on unpublished data provided by Charles Yanofsky.)

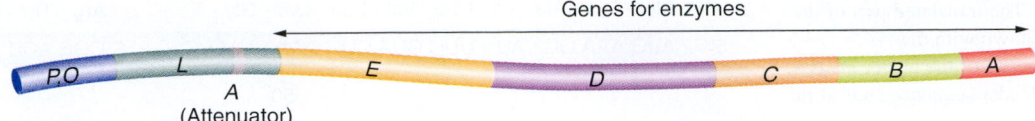

Figure 11-19 Diagram of the *trp* operon showing the promoter (*P*), operator (*O*), and attenuator (*A*) control sites and the genes for the leader sequence (*L*) and the enzymes of the tryptophan pathway (*E, D, C, B,* and *A*). (After L. Stryer, *Biochemistry,* 4th ed. Copyright © 1995 by Lubert Stryer.)

of some attenuating mechanism in this region, only 1 in 10 of the mRNAs can be transcribed farther (to completion). This suggests that the end of the attenuator acts as an mRNA chain terminator that, *in the presence of tryptophan,* halts transcription of 9 of 10 mRNAs. In the absence of tryptophan, this attenuator is somehow deactivated, and every mRNA goes to completion — hence, the tenfold increase. In those *trpR⁻* mutants in which the attenuator also is deleted, there is no block to extension of the mRNA, so transcription is carried through in every case (that is, pro-

duction is at its maximal rate) regardless of whether tryptophan is present.

What causes the interference with termination at the attenuator in the absence of tryptophan? Figure 11-20 presents a model based on alternative secondary structures formed by the mRNA in the leader region. The model proposes that one of the two conformations favors transcription termination and that the other favors elongation. Translation of part of the leader sequence would promote the conformation that favors termination.

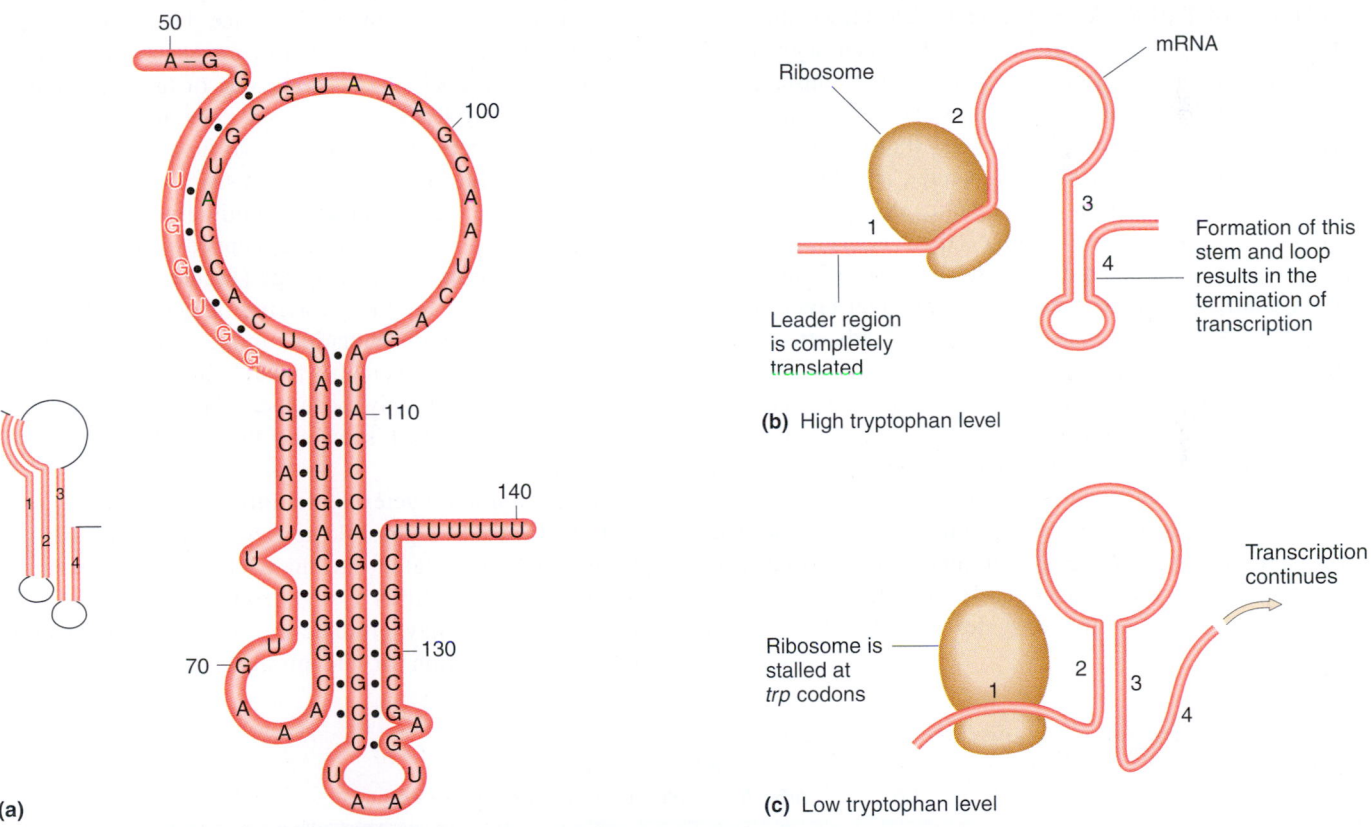

Figure 11-20 Model for attenuation in the *trp* operon. (a) Proposed secondary structures in *E. coli* terminated *trp* leader RNA. Four regions can base pair to form three stem-and-loop structures. (b) When tryptophan is abundant, segment 1 of the *trp* mRNA is fully translated. Segment 2 enters the ribosome (although it is not translated), which enables segments 3 and 4 to base pair. This base-paired region somehow signals RNA polymerase to terminate transcription. In contrast, when tryptophan is scarce (c), the ribosome is stalled at the codons of segment 1. Segment 2 interacts with segment 3 instead of being drawn into the ribosome, and so segments 3 and 4 cannot pair. Consequently, transcription continues. (After D. L. Oxender, G. Zurawski, and C. Yanofsky, *Proceedings of the National Academy of Sciences,* 76, 1979, 5524.)

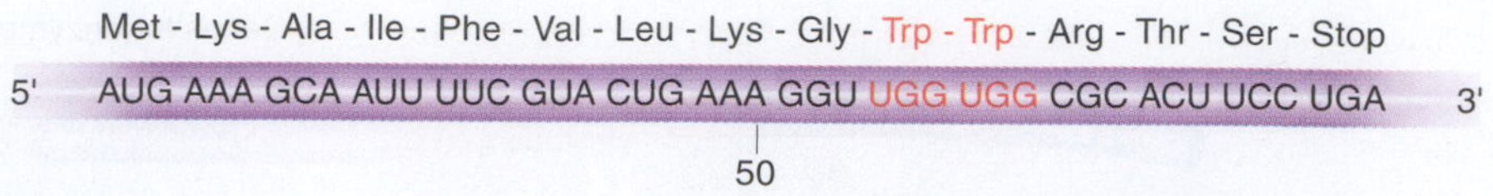

Figure 11-21   The translated part of the *trp* leader region, shown with the corresponding sequence of the leader mRNA. Translation of the leader sequence ends at the stop codon.

Met - Lys - Ala - Ile - Phe - Val - Leu - Lys - Gly - Trp - Trp - Arg - Thr - Ser - Stop

5'   AUG AAA GCA AUU UUC GUA CUG AAA GGU UGG UGG CGC ACU UCC UGA   3'

50

It is known that a part of the leader sequence near the beginning is in fact translated and yields a short peptide of 14 amino acids. There are two tryptophan codons in the translated stretch of the leader mRNA (Figure 11-21). When excess tryptophan is present, there is a sufficient supply of Trp-tRNA to allow efficient translation through the relevant part of the leader mRNA. The mRNA passes through the ribosome at a sufficiently fast rate that segment 2 of the leader section is drawn into the ribosome before it can form a stem-loop structure with segment 3, as shown in Figure 11-20b. Segment 3 is thus able to form a transcription-termination stem-loop structure with segment 4. In conditions of low tryptophan levels, however, translation is slowed in segment 1 at the Trp codons by the relative unavailability of Trp-tRNA. As Figure 11-20c shows, this condition allows the stem-loop structure between segments 2 and 3 to form, preventing the formation of the transcription-terminating loop between segments 3 and 4. Consequently, under conditions of low tryptophan, transcription is not stopped by the attenuator. In this manner, an additional tenfold range of tryptophan biosynthetic enzymes is superimposed on the normal range that is achieved by repressor–operator interaction. The analysis of numerous point mutations in the *trp* leader sequence that favor or disfavor the respective secondary structures lends strong support to Yanofsky's attenuation model.

Several operons for enzymes in biosynthetic pathways have attenuation controls similar to the one described for tryptophan. For instance, the leader region of the *his* operon, which encodes the enzymes of the histidine biosynthetic pathway, contains a translated region with seven consecutive histidine codons. Mutations at outside loci that result in lowering levels of normal charged His-tRNA produce partly constitutive levels of the enzymes encoded by the *his* operon.

> **MESSAGE**
>
> The *trp* operon is regulated by a negative repressor–operator control system that represses the synthesis of tryptophan enzymes when tryptophan is present in the medium. A second level of control is an attenuator region where termination of transcription is induced by the presence of tryptophan.

Lambda phage: a complex of operons

When they proposed the operon model, Jacob and Monod suggested that the genetic activity of temperate phages might be controlled by a system analogous to the *lac* operon. In the lysogenic state, the prophage genome is inactive (repressed). In the lytic phase, the phage genes for reproduction are active (induced). Since Jacob and Monod proposed the idea of operon control for phages, the genetic system of the λ phage has become one of the better understood control systems. This phage does indeed have an operon-type system controlling its two functional states. By now, you should not be surprised to learn that this system has proved to be more complex than initially suggested.

Allan Campbell induced and mapped many conditionally lethal mutations in the λ phage (Figure 11-22), providing clear evidence for the clustering of genes with related functions. Furthermore, mutations in the *N*, *O*, and *P* genes prevent most of the genome from being expressed after phage infection, with only those loci lying between *N* and *O* being active. We shall soon see the significance of this observation.

When normal bacteria are infected by wild-type λ phage particles, two possible sequences may follow: (1) a phage may be integrated into the bacterial chromosome as an inert prophage (thus lysogenizing the bacterial cell) or (2) the phage may produce the necessary enzymes to guide the production of the products needed for phage

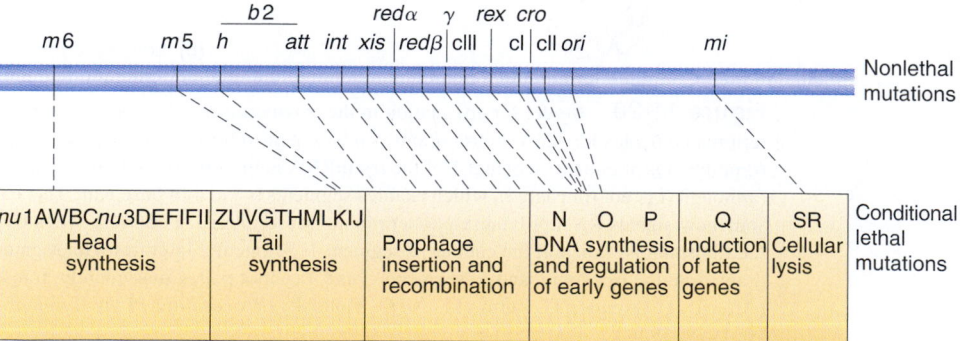

Figure 11-22   The genetic map of the λ phage. The positions of nonlethal and conditionally lethal mutations are indicated, and the characteristic clusters of genes with related functions are shown. (From A. Campbell, *The Episomes.* Harper & Row, 1969.)

maturation and cell lysis. When wild-type phage particles are placed on a lawn of sensitive bacteria, clearings (plaques) appear where bacterial cells are infected and lysed, but these plaques are turbid because lysogenized bacteria (which are resistant to phage superinfection) grow within the plaques.

Mutant phages that form clear plaques can be selected as a source of phages that are unable to lysogenize cells. Such *clear* (*c*) mutants prove to be analogous to *I* and *O* mutants in *E. coli*. For example, conditional mutants for a site called *cI* are unable to establish a lysogenic state under restrictive conditions, suggesting either a defective repressor or a defective repressor-binding site (operator). However, these mutants also fail to induce lysis after superinfecting a cell that has been lysogenized by a wild-type prophage, so the operator is functional, because repressor from the wild-type phage is clearly able to prevent the mutant from entering the lytic phase. Apparently, the *cI* mutation produces a defective repressor in the phage control system.

Virulent mutant λ phages have also been isolated that do not lysogenize cells but that do grow and enter the lytic cycle after superinfecting a lysogenized cell. Thus, these mutant phages are insensitive to the λ repressor. Genetic mapping has revealed *two* operators, designated O_L and O_R, located on the left and right of *cI*, respectively. When the λ phage infects a nonimmune host, RNA polymerase begins transcription from two promoters, P_L and P_R. A race begins to determine whether the phage will express its virulent functions and enter the lytic cycle or whether it will repress

these functions and establish lysogeny. Figure 11-23 depicts part of the map of λ and shows the extent of some of the different transcripts. P_L governs the leftward transcript that encodes the antiterminator protein N, plus a number of lytic functions. P_R governs the rightward transcript, including the *cro, cII,* and *Q* genes, among others. These transcripts normally terminate at specific critical points, before key genes necessary for lytic development can be transcribed. However, if the N protein is synthesized, it prevents termination of the transcript (see Figure 11-23), and the lytic functions are expressed. Transcription from P_L and P_R can be blocked by repressor binding to the operators O_L and O_R. In the former case, this blocking prevents synthesis of the N protein and thus also blocks the extension of transcripts from P_L and P_R.

The critical point in establishing lysogeny is the synthesis of the λ repressor, encoded by the *cI* gene. The *cI* gene has two promoters: one, P_E, for establishing lysogeny, and the second, P_M, for maintaining lysogeny. Transcription initiated at each promoter requires an activator protein. For P_M, the activator is the *cI*-encoded repressor itself. Therefore, to synthesize repressor in a cell without preexisting repressor molecules, P_E must be used. P_E is activated by the *cII* product. Thus, *cII* must be transcribed to establish lysogeny. One additional protein is required to establish lysogeny, the *integrase*, encoded by *int*. The *int* gene is transcribed from its own promoter, P_I (Figure 11-23), which also requires activation by the *cII* gene product.

In the lysogenic state, the λ repressor controls its own synthesis from the promoter P_M by interacting with the three

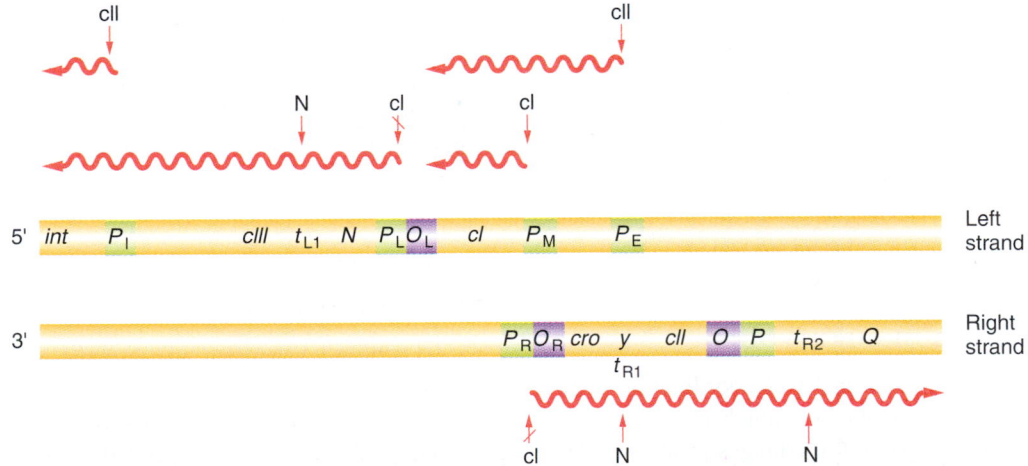

Figure 11-23 Partial genetic map of λ phage, showing the primary genes and proteins of regulation. The wavy lines represent mRNA and show the origin, direction, and extent of transcription. A vertical arrow at the *beginning* of an mRNA marks the site where the indicated gene product activates a promoter. If the arrow has a slash through it, the gene product represses transcription. Where an arrow marks a site *within* an mRNA, the indicated gene product prevents termination of transcription at the site. t_{L1} is a potential terminator for P_L, and t_{R1} is a potential terminator for P_R, when N does not act on them, as the diagram indicates. (After D. Wulff and M. Rosenberg, in R. W. Hendrix et al., eds., *Lambda II*. Cold Spring Harbor Laboratory Press, 1983.)

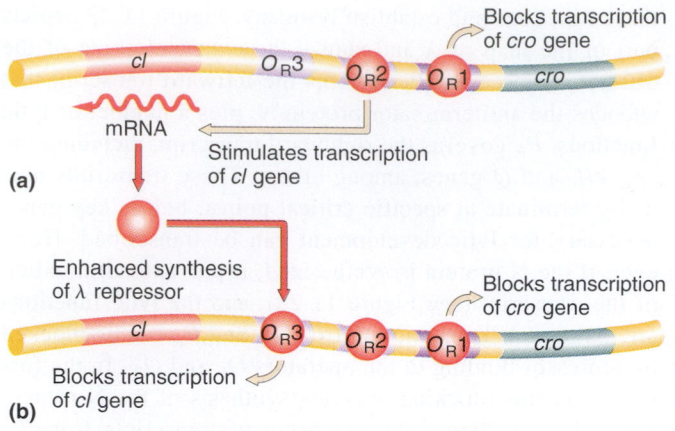

(a)

Blocks transcription of cro gene

mRNA

Stimulates transcription of cI gene

Enhanced synthesis of λ repressor

Blocks transcription of cro gene

Blocks transcription of cI gene

(b)

Figure 11-24 Self-regulation of cI repressor levels. (a) Binding of cI repressor to O_R1 both represses transcription of *cro* and potentiates the binding of a cI molecule at O_R2, which in turn stimulates transcription of cI repressor. (b) When cI repressor levels are high, the third site, O_R3, is bound by cI repressor. Because this binding overlaps the *cI* promoter, further production of cI repressor is blocked until its levels drop and the O_R3 site is again free. (After L. Stryer, *Biochemistry,* 4th ed. Copyright © 1995 by Lubert Stryer.)

operators O_R1, O_R2, and O_R3, as diagrammed in Figure 11-24. These operators have different affinities for the λ repressor, with $O_R1 > O_R2 > O_R3$. When there is too little λ repressor, only the O_R1 site, which blocks *cro* expression, is filled. (The Cro protein prevents synthesis of λ repressor by binding to O_R3.) The binding of repressor to O_R1 facilitates the binding of a second molecule of repressor to O_R2. When λ repressor binds to O_R2, it serves to activate transcription of the *cI* gene. However, when too much λ repressor is present, then binding to the low-affinity O_R3 blocks further transcription of the *cI* gene.

MESSAGE ·······································

The regulation of the lytic and lysogenic states of the λ phage provides a model for interacting control systems that may be useful for interpreting gene regulation in eukaryotes.

Transcription: an overview of gene regulation in eukaryotes

Eukaryotes face the same basic tasks of coordinating gene expression as do prokaryotes but in a much more intricate way. Some genes have to respond to changes in physiological conditions. Many others are parts of developmentally triggered genetic circuits that organize cells into tissues and tissues into an entire organism (except for unicellular eukaryotes). In these cases, the signals controlling gene expression are the products of developmental regulatory genes, rather than signals from the external environment.

Most eukaryotic genes are controlled at the level of transcription, and the mechanisms are similar in concept to those found for bacteria. Trans-acting regulatory proteins

work through sequence-specific DNA binding to their cis-acting regulatory target sequences. Because of the much more complex regulation that is required to coordinate proper gene activity throughout the lifetime of a multicellular organism, there are some considerable novelties as well. We shall see examples of these novelties in this and subsequent chapters.

Typically, eukaryotes have many more genes than do prokaryotes, sometimes by several orders of magnitude. The genes of higher organisms also tend to be larger, owing to the facts that cis-acting sequences on the DNA can be located tens of thousands of base pairs away from the transcription start site and that a battery of regulatory factors is sometimes needed to bring about proper regulation of certain genes.

Cis-acting sequences in transcriptional regulation

As mentioned in Chapter 10, eukaryotes have three different classes of RNA polymerases (distinguished by roman numerals I, II, and III). All mRNA molecules are synthesized by RNA polymerase II, and the rest of this chapter will focus on the transcription of mRNAs. To achieve maximal rates of transcription by RNA polymerase II, the cooperation of multiple cis-acting regulatory elements is required. We can distinguish three classes of elements on the basis of their relative locations. Near the transcription initiation site are the core promoter (the RNA polymerase II–binding region) and promoter-proximal cis-acting sequences that bind to proteins that in turn assist in the binding of RNA polymerase II to its promoter. Additional cis-acting sequence elements can act at considerable distance — these elements are termed enhancers and silencers. Often, an enhancer or silencer element will act only in one or a few cell types in a multicellular eukaryote. The promoters, promoter-proximal elements, and distance-independent elements are all targets for binding by different trans-acting DNA-binding proteins.

Core promoter and promoter-proximal elements

Figure 11-25 is a schematic view of the core promoter and promoter-proximal sequence elements. The **core promoter** usually refers to the region from the transcription start site including the TATA box, which resides approximately 30 bp upstream of the transcription initiation site. This core promoter is unable to mediate efficient transcription by itself. Some important elements near the promoter, the **promoter-proximal elements,** are found within 100 to 200 bp of the transcription initiation site. The CCAAT box functions as one of these promoter-proximal cis-acting sequences, and a GC-rich segment often functions as another. An example of the consequences of mutating these sequence elements on transcription rates is shown in Figure 11-26.

mRNA
GGGCGG
CCAAT
TATA
~200 bp
~100 bp
−30 bp

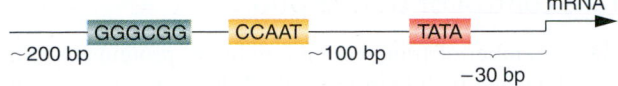

Figure 11-25 The promoter region in higher eukaryotes. The TATA box is located approximately 30 base pairs from the mRNA start site. Usually, two or more promoter-proximal elements are found 100 and 200 bp upstream of the mRNA start site. The CCAAT box and the GC-rich box are shown here. Other upstream elements include the sequences GCCACACCC and ATGCAAAT.

Distance-independent cis-acting elements

In eukaryotes, we distinguish between two classes of cis-acting elements that can exert their effects at considerable distance from the promoter. **Enhancers** are cis-acting sequences that can greatly increase transcription rates from promoters on the same DNA molecule; thus, they act to activate, or positively regulate, transcription. Silencers have the opposite effect. **Silencers** are cis-acting sequences that are bound by repressors, thereby inhibiting activators and reducing transcription. Enhancers and silencers are similar to promoter-proximal regions in that they are organized as a series of cis-acting sequences that are bound by trans-acting regulatory proteins. However, they are distinguished from promoter-proximal elements by being able to act at a distance, sometimes 50 kb or more, and by being able to operate either upstream or downstream from the promoter that they control. Enhancer and silencer elements are intricately structured. Figure 11-27 shows the DNA sequence of the SV40 (simian virus 40) enhancer which is required for high-level expression of SV40 transcripts. Within the enhancer, there are five sequence elements required for maximal enhancement of transcription. Enhancers that are

Relative transcription level
3.5
3.0
1.0
0
GCCACACCC
GGCCAATC
ATATAA

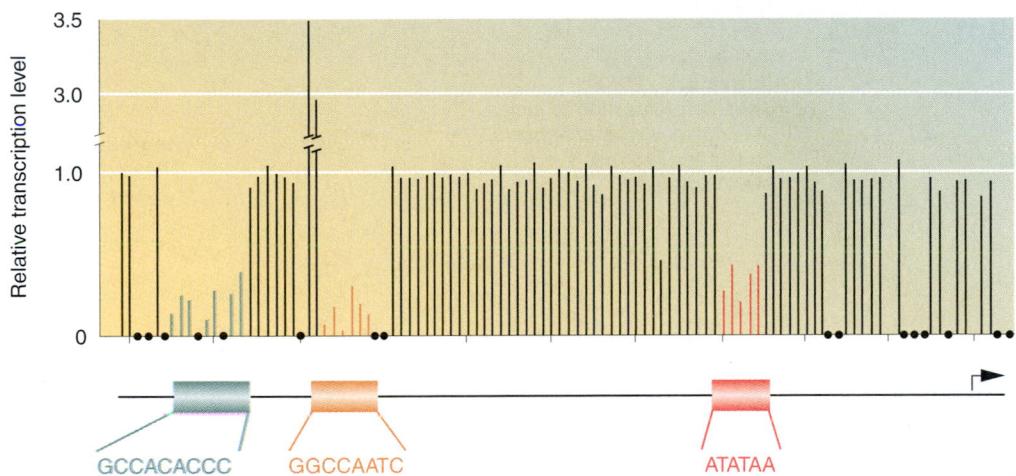

Figure 11-26 Consequences of point mutations in the promoter for the *β*-globin gene. Point mutations through the promoter region were analyzed for their effects on transcription rates. The height of each line represents the transcription level, relative to a wild-type promoter, that results from promoters with base changes at that point. A level of 1.0 means that the rates are equal to the wild-type rate; reductions in transcription rates yield levels less than 1.0. Almost every nucleotide throughout the promoter was tested, except for the points shown with black dots. The diagram below the bar graph shows the position of the TATA box and two upstream elements of the promoter. Only the base substitutions that lie within the three promoter elements change the level of transcription. (From T. Maniatis, S. Goodbourn, and J. A. Fischer, *Science* 236, 1987, 1237.)

1
2
3' TTGGTCGACACCTTACACACAGTCAATCCCACACCTTTCAGGGGTCCGAGGGGTCGT
5' AACCAGCTGTGGAATGTGTGTCAGTTAGGGTGTGGAAAGTCCCCAGGCTCCCCAGCA
3
4
5
CCGTCTTCATACGTTTCGTACGTAGAGTTAATCAGTCGTTGGTC 5' Early mRNA
GGCAGAAGTATGCAAAGCATGCATCTCAATTAGTCAGCAACCAG 3' Late mRNA

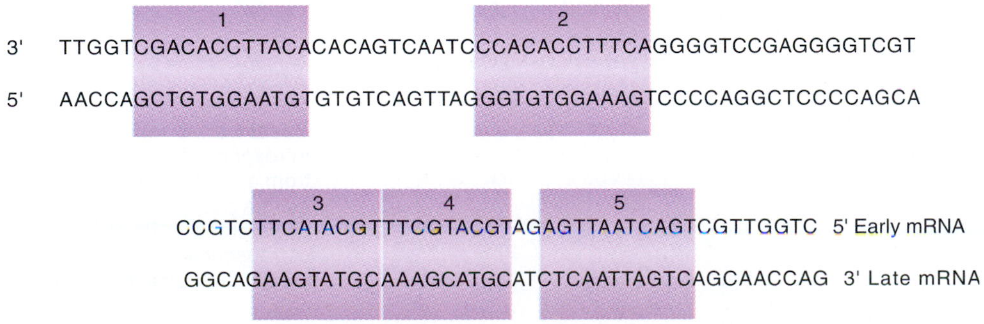

Figure 11-27 Organization of the SV40 enhancer. SV40 is a virus that infects primates, and its regulatory sequences interact with the eukaryotic cell's transcriptional regulatory machinery. Boxed sequences 1 to 5 indicate the sequences that are required for maximum levels of enhancer activity. (From T. Maniatis, S. Goodbourn, and J. A. Fischer, *Science* 236, 1987, 1237.)

themselves composed of multiple copies of a DNA-binding element are common. Different DNA sequences serve as target-recognition sites for specific trans-acting regulatory proteins.

Mechanisms for action at a distance

How do enhancer and silencer elements many thousands of base pairs away regulate transcription? Most models for such action at a distance include some type of DNA looping. Figure 11-28 details a DNA-looping model for activation of the initiation complex (see Figure 11-29). In this model, a DNA loop brings activator proteins bound to distant enhancer elements into proximity to protein complexes associated with promoter-proximal cis-acting sequences.

MESSAGE ···
Eukaryotic enhancers and silencers can act at great distance.
··

Trans control of transcription

A large number of trans-acting regulatory proteins have now been identified in eukaryotic cells. Like their counterparts in prokaryotes, these regulatory proteins act by binding to specific target DNA sequences.

Regulatory proteins that bind the core promoter and promoter-proximal elements help RNA polymerase II to initiate transcription and, together with the polymerase, they form an initiation complex, as pictured in Figure 11-29a. Several different transcription factor complexes (TFII complexes) interact with RNA polymerase II. For example, the TFIID complex consists of a TATA-box-binding protein (TBP) and more than eight additional subunits (TAFs). The TFII complexes are often referred to as *basal* or *general* transcription factors, because they are the minimal requirement for RNA polymerase II to initiate transcription (usually very weakly) at a promoter. Figure 11-29b shows the structure of the TATA-box-binding protein binding to DNA.

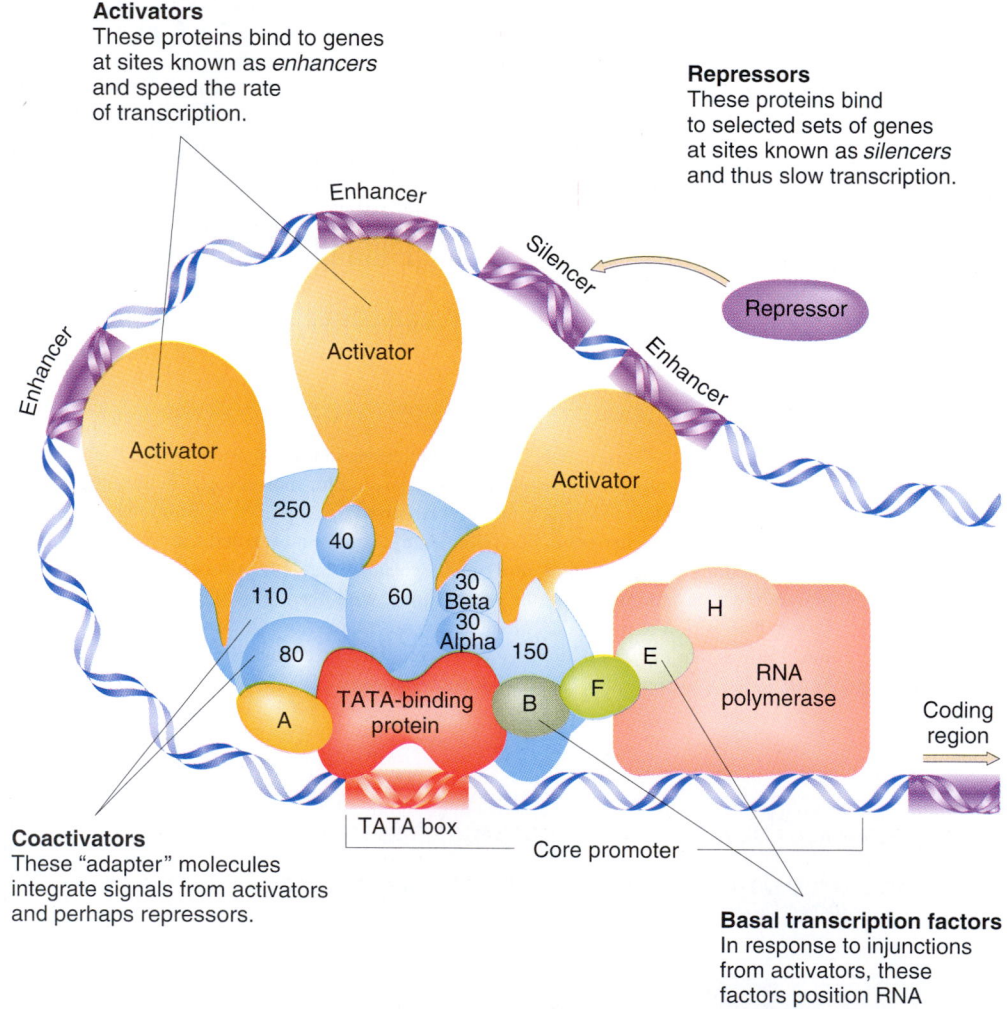

Figure 11-28 The molecular apparatus controlling transcription in human cells consists of four kinds of components. (The numbered proteins are the names of the subunits of RNA polymerase II. Each subunit is named according to its molecular mass in kilodaltons.) Basal transcription factors (labeled A, B, F, E, H) are essential for transcription but cannot by themselves increase or decrease its rate. That task falls to regulatory molecules known as activators and repressors. Activators, and possibly repressors, communicate with the basal factors through coactivators — proteins that are linked in a tight complex to the TATA-binding protein, the first of the basal transcription factors to land on the core promoter. (From R. Tjian, "Molecular Machines That Control Genes." Copyright © 1995 by Scientific American, Inc. All rights reserved.)

Activators
These proteins bind to genes at sites known as *enhancers* and speed the rate of transcription.

Repressors
These proteins bind to selected sets of genes at sites known as *silencers* and thus slow transcription.

Coactivators
These "adapter" molecules integrate signals from activators and perhaps repressors.

Basal transcription factors
In response to injunctions from activators, these factors position RNA polymerase at the start of transcription and initiate the transcription process.

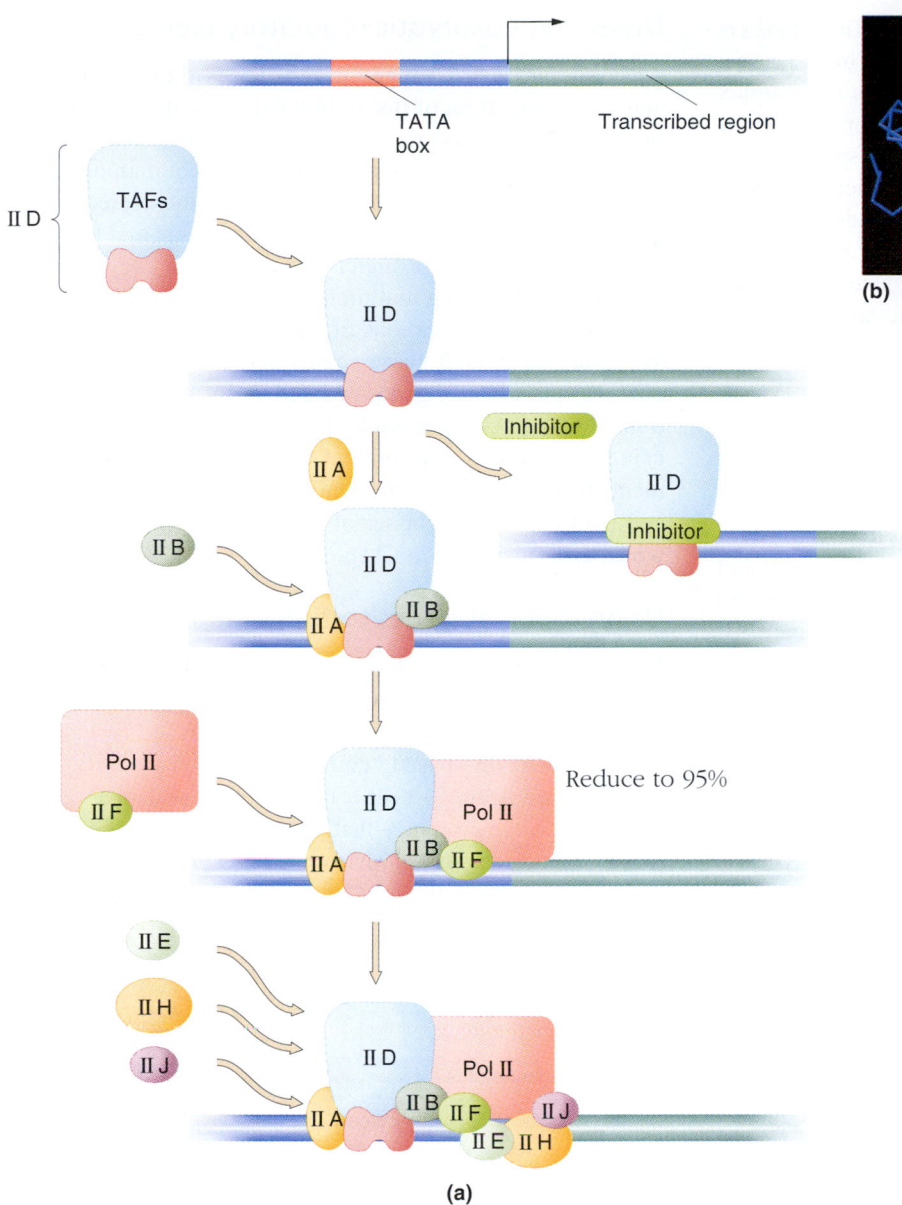

(a)

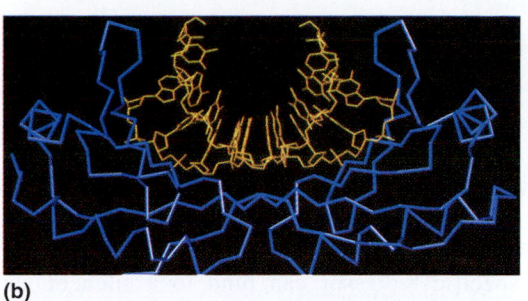

(b)

Figure 11-29 (a) Assembly of the RNA polymerase II initiation complex begins with the binding of transcription factor TFIID to the TATA box. TFIID is composed of one TATA box-binding subunit called TBP (dark blue) and more than eight other subunits (TAFs), represented by one large symbol (light blue). Inhibitors can bind to the TFIID-promoter complex, blocking the binding of other general transcription factors. Binding of TFIIA to the TFIID-promoter complex (to form the D-A complex) prevents inhibitor binding. TFIIB then binds to the D-A complex, followed by binding of a preformed complex between TFIIF and RNA polymerase II. Finally, TFIIE, TFIIH, and TFIIJ must add to the complex, in that order, for transcription to be initiated. (b) TATA-binding protein (blue) is a remarkably symmetrical, saddle-shaped molecule. Its underside rides on DNA and seems to bend it. This bending may somehow facilitate assembly of the complex that initiates transcription. (Part a from H. Lodish, D. Baltimore, A. Berk, S. L. Zipursky, P. Matsudaira, and J. Darnell, *Molecular Cell Biology,* 3d ed. Copyright © 1995 by Scientific American Books; part b courtesy of J. L. Kim and S. K. Berley, from *Nature* 365, 1993, 520.)

CCAAT and GC boxes are recognized by additional DNA-binding proteins.

Some of the proteins that bind distance-independent elements also have been identified. The protein encoded by the yeast *GCN4* gene is an example of a trans-acting enhancer-binding protein. It binds enhancers called upstream activating sequences (UASs). GCN4 activates the transcription of many yeast genes that encode enzymes of amino acid biosynthetic pathways. In response to amino acid starvation, the level of GCN4 protein rises and, in turn, increases the expression levels of the amino acid biosynthetic genes. The UASs recognized by GCN4 contain the principal recognition sequence element ATGACTCAT.

> **MESSAGE**
> The core promoter, promoter-proximal elements, and distance-independent elements are all DNA sites that are recognized by sequence-specific DNA-binding proteins. The proper constellation of such trans-acting proteins is required for RNA polymerase II to initiate transcription and to achieve maximal rates of transcription.

Tissue-specific regulation of transcription

Many enhancer elements in higher eukaryotes activate transcription in a tissue-specific manner — that is, they induce expression of a gene in one or a few cell types. For example, antibody genes are flanked by powerful enhancers that operate only in the B lymphocytes of the immune system. Many enhancers are integral components of complex tissue-specific genetic circuits that underlie complex events in development in higher eukaryotes. Tissue specificity is conferred in one of two ways. An enhancer can act in a tissue-specific manner if the activator that binds to it is present in only some types of cells. Alternatively, a tissue-specific repressor can bind to a silencer element located very near the enhancer element, making the enhancer inaccessible to its transcription factor.

Properties of tissue-specific enhancers

In some genes, regulation can be controlled by simple sets of enhancers. For example, in *Drosophila,* vitellogenins are large egg yolk proteins made in the female adult's ovary and fat body (an organ that is essentially the fly's liver) and transported into the developing oocyte. Two distinct enhancers located within a few hundred base pairs of the promoter regulate the vitellogenin gene, one driving expression in the ovaries and the other in the fat body.

The array of enhancers for a gene can be quite complex, controlling similarly complex patterns of gene expression. The *dpp* (decapentaplegic) gene in *Drosophila,* for example, encodes a protein that mediates signals between cells (see Chapter 23). It contains numerous enhancers, perhaps numbering in the tens or hundreds, dispersed along a 50-kb interval of DNA. Some of these enhancers are located 5′ (upstream) of the transcription initiation site of *dpp,* others are downstream of the promoter, some are in introns, and still others are 3′ of the polyadenylation site of the gene. Each of these enhancers regulates the expression of *dpp* in a different site in the developing animal. Some of the better characterized *dpp* enhancers are shown in Figure 11-30.

The requirement for multiple enhancer elements to regulate tissue-specific expression helps to explain the large size of genes in higher eukaryotes. The tissue-specific regulation of a gene may be quite complex, requiring the action of numerous, distantly located enhancer elements.

Dissecting eukaryotic regulatory elements

An important part of modern genetics is the identification and characterization of distantly located regulatory elements by means of transgenic constructs, in which recombinant DNA molecules are inserted into the genome of an organism (see Chapter 13). In these constructs, isolated pieces of a gene are incorporated to determine what tissue and temporal regulatory patterns are under the control of the gene. By means of such slicing and dicing experiments, it is possible to home in on the locations of specific regulatory elements. We can also exploit these regulatory elements to develop new ways to identify genes of interest. In the following sections, we shall see how such experiments are carried out, by using studies of transcriptional enhancers as examples. Remember, however, that the same techniques and logic can be applied to any other classes of regulatory elements, some of which are considered in Chapter 23.

Using reporter genes to find enhancers

Enhancers of a cloned gene are typically identified by means of transgenic **reporter genes.** In reporter-gene constructs, pieces of cis-regulatory DNA are fused (usually by restriction-enzyme-based "cutting and pasting" of recombinant DNA molecules) near a transcription unit that can express a reporter protein — that is, a protein whose presence can be monitored. Our old friend, the *E. coli* β-galactosidase enzyme encoded by the *lacZ* gene, is a very popular reporter

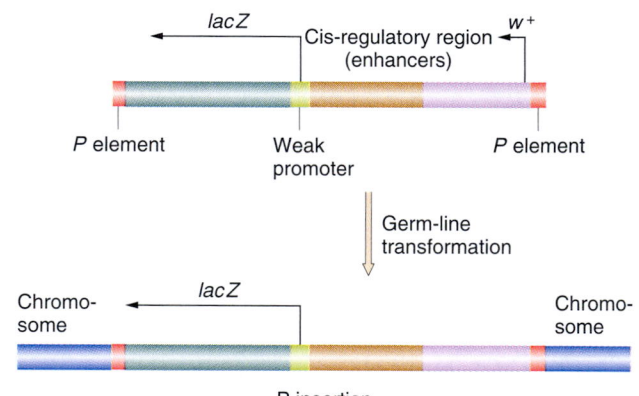

Figure 11-31   Use of a reporter-gene construct in *Drosophila* to identify enhancers. The top line represents a part of a plasmid, bracketed by *P*-element ends so that the material in between can be inserted into the genome by *P*-element transformation (see Chapter 20). A region of DNA thought to contain one or more enhancers is inserted immediately adjacent to a "weak" promoter — that is, a promoter that by itself cannot initiate transcription. The promoter is joined to the *E. coli lacZ* structural gene (the reporter gene), which encodes β-galactosidase. If there are any enhancers in the construct, they will induce tissue-specific expression of *lacZ*. *w⁺* is the wild-type allele of the *Drosophila* white gene, used to detect the presence of the reporter-gene construct in the fly. The embryos and imaginal disks in Figures 11-32 and 11-33 are examples of expression from reporter-gene constructs in *Drosophila* and in mice.

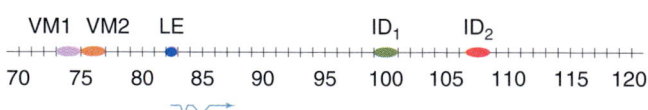

Figure 11-30   A molecular map of a complex gene — the *dpp* gene of *Drosophila.* Units on the map are in kilobases. The basic transcription unit of the gene is shown below the map coordinate line. The abbreviations above the line mark the sites of a few of the many tissue-specific enhancers that regulate this transcript in different stages of development. Tissue-specific expression patterns conferred by these enhancers are shown in Figure 11-32, and the abbreviations are explained there.

protein, because it is very easy to detect the presence of β-galactosidase histochemically by adding a synthetic substrate, X-gal, to the medium and observing which tissues turn blue. The reporter-gene construct transcription unit contains a "weak" promoter — one that cannot initiate transcription without the assistance of an enhancer (Figure 11-31). The construct is then introduced by DNA transformation into the germ line of a host organism, and appropriate cells are histochemically assayed for the presence of the reporter protein. Two examples of reporter-gene expression are shown, one for enhancers of the *dpp* gene in *Drosophila* (Figure 11-32) and the other for a mouse enhancer expressed in muscle precursor cells (Figure 11-33).

Reporter protein expression in a tissue indicates the presence of one or more enhancers within the tested piece of DNA. When a piece of DNA has been found to act as an enhancer, the enhancer can be further localized by testing smaller and smaller subfragments of the original DNA segment, by using the same reporter-gene assay.

Ultimately, the DNA sequence of the enhancer can be identified by whittling down the piece of cis-regulatory DNA. With this sequence known, the next question of importance is the identity of the transcription-factor proteins that bind to the enhancer. Methods now exist to identify enhancer-binding proteins and to clone the genes that encode these proteins. When these genes have been cloned, they can be characterized by genetic and molecular techniques. With the use of these approaches, it is possible to build detailed circuit diagrams of the genetic pathways that regulate gene expression in eukaryotes.

MESSAGE ···
Reporter-gene techniques can be used to isolate individual regulatory elements of genes.
···

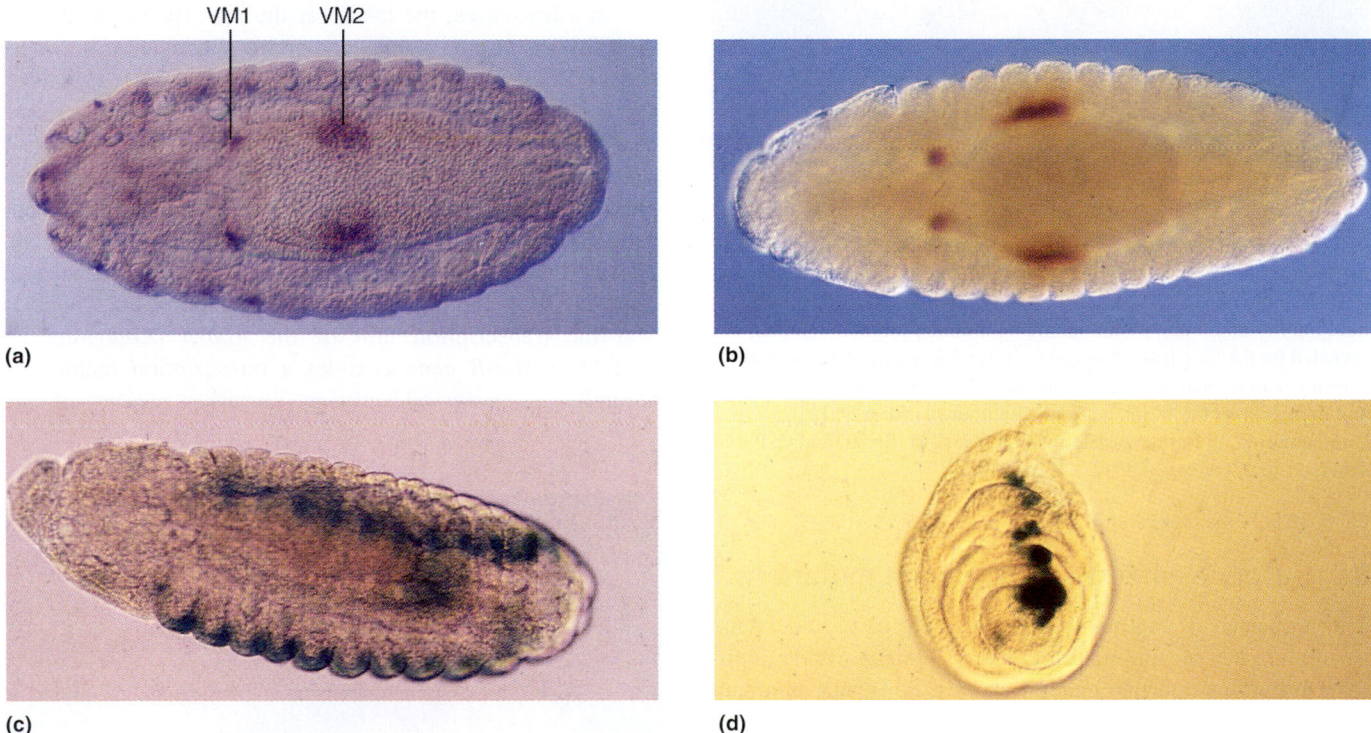

(a)   **(b)**   **(c)**   **(d)**

Figure 11-32 Examples of the complex tissue-specific regulation of the *dpp* gene. In parts a, c, and d, the blue staining is due to a histochemical assay for *E. coli* β-galactosidase activity (the protein encoded by the *lacZ* reporter gene). The map positions of the *dpp* enhancers responsible for the staining patterns seen here are shown in Figure 11-30. (a) Reporter-gene assay for expression of the *Drosophila dpp* gene in two parts of the embryonic visceral mesoderm, the precursor of the gut musculature. The left-hand block of blue staining indicates the enhancer VM1, which drives anterior visceral mesoderm expression; the right-hand block of staining indicates the enhancer VM2, which drives posterior visceral mesoderm expression. (b) RNA in situ hybridization assay of *dpp* expression in the embryonic visceral mesoderm. Note that the blue reporter-gene expression pattern in part a is the same as the brown *dpp* RNA expression pattern shown here, confirming the reliability of the reporter-gene assay. (c) Reporter-gene expression driven by a different enhancer (LE) of *dpp* in the lateral ectoderm of an embryo. (d) Reporter-gene expression driven by ID, one of many enhancer elements driving imaginal disk expression of *dpp*. (An imaginal disk is a flat circle of cells in the larva that gives rise to one of the adult appendages.) A blue sector of *dpp* reporter-gene expression in a leg imaginal disk is shown. (Parts a and b courtesy of D. Hursh; part c courtesy of R. W. Padgett; part d courtesy of R. Blackman and M. Sanicola.)

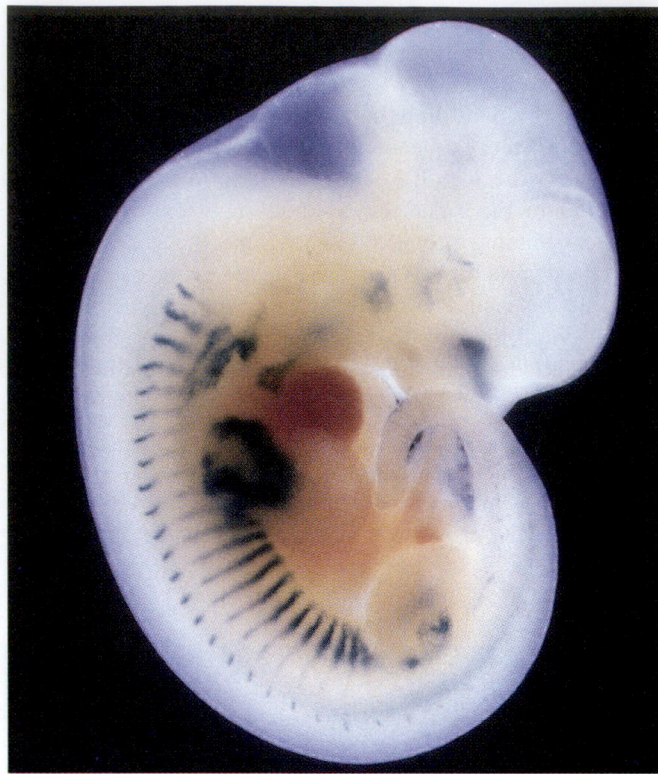

Figure 11-33 An 11.5 transgenic mouse embryo contains recombinant DNA composed of a 258-bp mouse DNA sequence fused to the *E. coli lacZ* gene, which encodes the enzyme β-galactosidase. The 258-bp mouse DNA contains all of the cis-regulatory sequences necessary to direct expression of *lacZ* in muscle precursor cells, as revealed by the blue histochemical stain for β-galactosidase enzyme activity. The stained muscle precursors include the somites, limb buds, and branchial arches of the embryo. (Reproduced from D. J. Goldhamer, B. P. Brunk, A. Faerman, A. King, M. Shani, and C. P. Emerson, Jr., *Development* 121, 1995, 644.)

Regulatory elements and dominant mutations

The properties of regulatory elements help us to understand certain classes of dominant mutations. We can divide dominant mutations into two general classes. For some dominant mutations, inactivation of one of the two copies of a gene reduces the gene product below some critical threshold for producing a normal phenotype; we can think of these mutations as **loss-of-function dominant mutations** (referred to as haploinsufficient mutations in earlier chapters). In other cases, the dominant phenotype is due to some new property of the mutant gene, not to a reduction in its normal activity; this class comprises **gain-of-function dominant mutations.**

Commonly, gain-of-function dominant mutations arise through the fusion of parts of two genes to each other. (Be aware that there are also other mechanisms for producing a gain-of-function dominant mutation.) Such fusions can occur at the breakpoints of chromosomal rearrangements such as inversions, translocations, duplications, or deletions (see Chapter 17). Because enhancers can act at long distance and

can activate many different promoters, misregulation of a gene can occur if a chromosomal rearrangement juxtaposes enhancers of one gene and a transcription unit of another gene. In such cases, the enhancers of the gene at one breakpoint can now regulate the transcription of a gene near the other breakpoint. Often, this misregulation leads to the misexpression of the mRNA encoded by the transcription unit in question. Such fusions can lead to gain-of-function dominant mutant phenotypes, depending on the nature of the protein product of the misexpressed mRNA and the tissues in which it is misexpressed.

The classic *Bar* dominant mutation in *Drosophila* is an example of such misregulation through gene fusion. In the *Bar* mutation, cis-regulatory elements that promote expression in the developing eye are fused to a gene that is ordinarily not expressed in the eye. This latter gene encodes a transcription factor, and misexpression of that transcription factor in the developing eye leads to the death of many cells of the developing eye and thus to the small eye Bar phenotype.

In a few cases, the basis for the misexpression in such gene fusions is quite well understood. One such case is the *Tab (Transabdominal)* mutation in *Drosophila*. *Tab* causes part of the thorax of the adult fly to develop instead as tissue normally characteristic of the sixth abdominal segment (A6) (Figure 11-34). *Tab* is associated with a chromosomal inversion. One breakpoint of the inversion is within an enhancer region of a different gene, the *sr* (striped) gene. These enhancers of the *sr* gene induce gene expression in certain parts of the thorax of the fly. The other breakpoint is near the transcription unit of the *Abd-B (Abdominal-B)* gene. The *Abd-B* gene encodes a transcription factor that normally is expressed only in posterior regions of the

Figure 11-34 The *Tab* mutation. The fly on the left is a wild-type male. The fly on the right is a *Tab/+* heterozygous mutant male. In the mutant fly, part of the thorax (the black tissue) is changed into tissue normally found in the dorsal part of one of the posterior abdominal segments. The rest of the thorax is normal. (From S. Celniker and E. B. Lewis, *Genes and Development* 1, 1987, 111.)

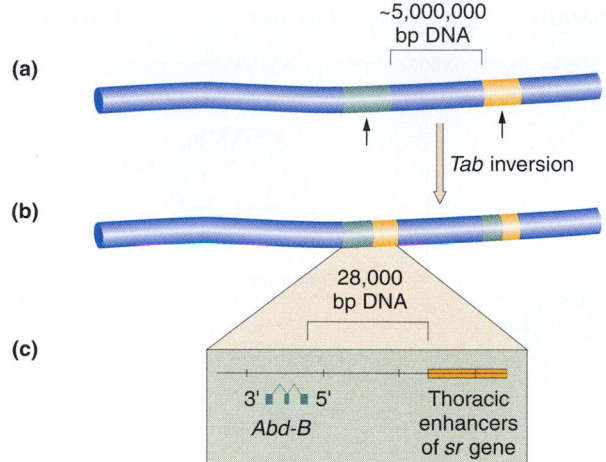

(a)

~5,000,000
bp DNA

Tab inversion

(b)

28,000
bp DNA

(c)

3' 5'
Abd-B

Thoracic
enhancers
of *sr* gene

Figure 11-35 *Tab* is due to a gene fusion. (a) The locations of the *Abd-B* and *sr* genes on a map of a normal chromosome 3 are depicted. The two genes are (very) approximately 5 million base pairs apart, and there are normally hundreds of genes in between them. (b) The *Tab* inversion has one chromosomal breakpoint within the *Abd-B* gene and the other within *sr* (arrows in part a). As a result of misfusion of these breakpoints, an inversion arises in which pieces of the *Abd-B* and *sr* genes are fused at either end of the inversion. (c) A magnified view of the centromere-proximal breakpoint of the inversion. The inversion breakpoints have produced a DNA molecule in which the promoter region of the *Abd-B* gene is now only 28,000 base pairs away from enhancer elements of the *sr* gene. This proximity causes the *Abd-B* transcript to be ectopically expressed in some parts of the fly's thorax, where the wild-type *Abd-B* gene is not transcribed.

animal, and this *Abd-B* transcription factor is responsible for conferring an abdominal phenotype on any tissues that express it. (We shall have more to say about genes such as *Abd-B* in the treatment of homeotic genes in Chapter 23.) In the *Tab* inversion, the *sr* enhancer elements controlling thoracic expression are juxtaposed to the *Abd-B* transcription unit, causing the *Abd-B* gene to be activated in exactly those parts of the thorax where *sr* would ordinarily be expressed (Figure 11-35). Because of the function of the *Abd-B* transcription factor, its activation in these thoracic cells changes their fate to that of posterior abdomen. In this way, we can understand the molecular basis of a dominant mutation.

Gene fusions are an extremely important source of genetic variation. Through chromosomal rearrangements, novel patterns of gene expression can be generated. In fact, we can imagine that such fusions might play an important role in the shifts of gene expression pattern in the divergence and evolution of species. In addition to affecting development (Chapter 23), such mutations can play a pivotal role in the formation and progression of many cancers (Chapter 22).

MESSAGE ••

The fusion of tissue-specific enhancers to genes not normally under their control can produce dominant gain-of-function mutant phenotypes.

Regulation of transcription factors

Some transcription factors are synthesized only in specific tissues. The activities of the factors themselves are also regulated in different cell types.

Steroid hormones: linking enhancers to the physiology of the organism

Just as it was crucial in prokaryotes to wed the activity of transcriptional regulatory proteins to the physiological state of the bacterium, the tie between transcriptional regulation and physiology is crucial in eukaryotes as well. In Chapter 23, we shall see examples of how pathways of differentiation and pattern development provide the physiological context for activation of specific transcription factors. Sometimes, the regulatory signal that activates eukaryotic transcription factors comes from a very distant source in the body. For instance, hormones released into the circulatory system by an organ that is part of the endocrine system (Figure 11-36) can travel through the circulation to essentially all parts of the body. The endocrine system can thus serve as a master regulator to coordinate changes in transcription in cells of many different tissues. This mechanism underlies sex determination in mammals, which will be discussed in Chapter 23.

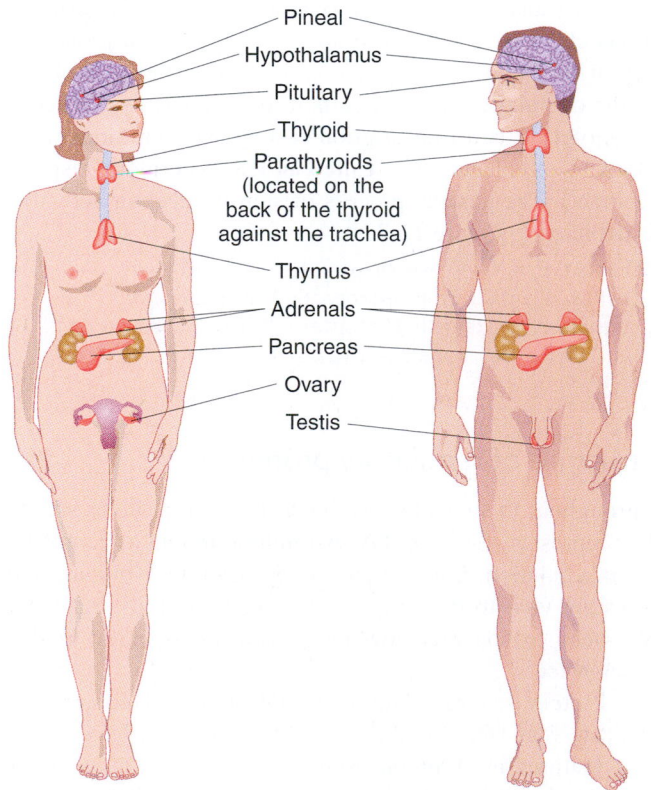

Pineal
Hypothalamus
Pituitary
Thyroid
Parathyroids
(located on the
back of the thyroid
against the trachea)
Thymus
Adrenals
Pancreas
Ovary
Testis

Figure 11-36 The human endocrine organs. (Reprinted from W. K. Purves, G. H. Orians, and H. C. Heller, *Life: The Science of Biology,* 4th ed. Sinauer Associates, Inc., and W. H. Freeman and Company, 1995.)

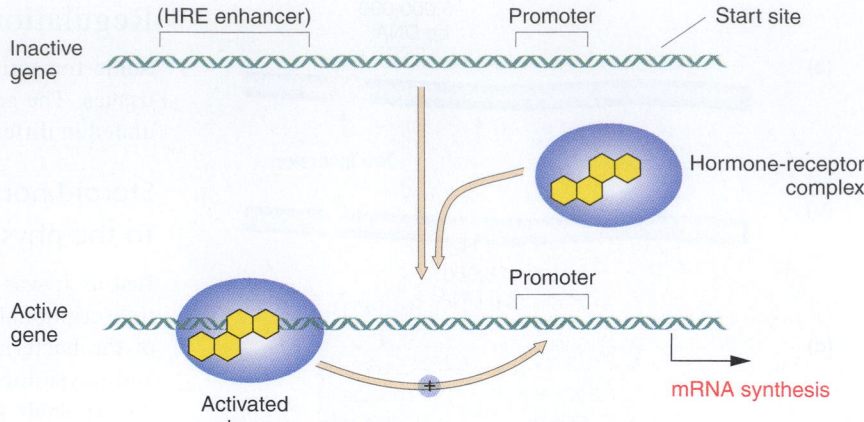

Figure 11-37 The action of steroid hormones at enhancer sequences. A steroid hormone (blue) binds to a soluble receptor protein. This complex, in turn, binds to enhancer sequences and enables them to stimulate the transcription of hormone-responsive genes. (From L. Stryer, *Biochemistry,* 4th ed. Copyright © 1995 by Lubert Stryer.)

Some hormones are small molecules that, because of their lipid-solubility properties, can directly pass through the plasma membrane of the cell—examples being various steroid hormones, such as glucocorticoid, testosterone, and estrogen. In the cell, steroid hormones bind to and regulate specific transcription factors in the nucleus. In this respect, we can think of steroids as being analogous to allosteric effectors regulating some bacterial operons.

One example is the female sex hormone estrogen. In chicken oviducts, the egg white protein ovalbumin is specifically synthesized in response to estrogen, which causes increased transcription of the ovalbumin gene. The estrogen molecule activates transcription by binding to a protein receptor molecule that first recognizes the estrogen molecule in the cytoplasm and transports it to the nucleus. The receptor molecule is a transcription factor that then binds to the DNA at an enhancer site called an HRE (hormone-response element). The general process of steroid-receptor activation is depicted in Figure 11-37.

MESSAGE ···
Just as in prokaryotes, eukaryotic transcriptional regulatory proteins must contain domains that interact with molecular signals of the physiological state of the cell.

Structure of regulatory proteins

Throughout this chapter, we have seen that proteins such as Lac repressor, CAP, or TATA-binding protein are crucial to gene regulation. Such sequence-specific DNA-binding proteins are vital to transcriptional regulation in all organisms. We need to consider how DNA sequence-specific binding takes place.

Protein sequence analyses and structural comparisons indicate that DNA-binding regulatory proteins have important features in common. Many consist of a DNA-binding domain, located at one end of the protein, that protrudes from the main "core" of the protein. In certain cases, the core protein contains the allosteric effector site. This arrangement holds for the Lac repressor (Figure 11-38) and CAP, as well as for many other regulatory proteins such as the steroid re-

ceptors. For such proteins, certain protruding α helices fit into the major groove of the DNA. This fit has been visualized by solving the three-dimensional structures of various protein–DNA complexes, such as the Lac repressor–operator complex (Figure 11-39). Here two α helices from the repressor protein interact with the two consecutive major grooves of the DNA of the operator site. The helices are connected by a turn in the protein secondary structure. This helix-turn-helix motif (Figure 11-40) is common to many regulatory proteins. However, many other DNA-binding motifs abound as well. As just one example, Figure 11-41 shows the structure of a part of a *zinc-finger* protein, in which a zinc atom is conjugated to four amino acids of a small part of a polypeptide chain [two cysteines (C) and two histidines (H)]. Zinc-finger proteins generally have several such zinc fingers. Each zinc finger appears to be able to interact with a specific DNA sequence.

MESSAGE ···
The structures of DNA-binding proteins help us understand how they contact specific DNA sequences through polypeptide domains that fit into the major groove of the DNA double helix.

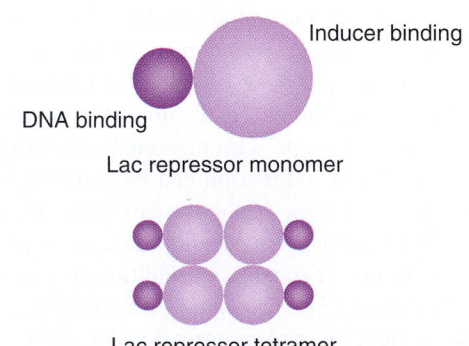

Figure 11-38 Schematic representation of the arrangement of domains in the Lac repressor. All mutations affecting DNA and operator binding result in alterations in the amino-terminal end of the protein, whereas mutants defective in inducer binding of aggregation result in alterations in the remainder of the protein.

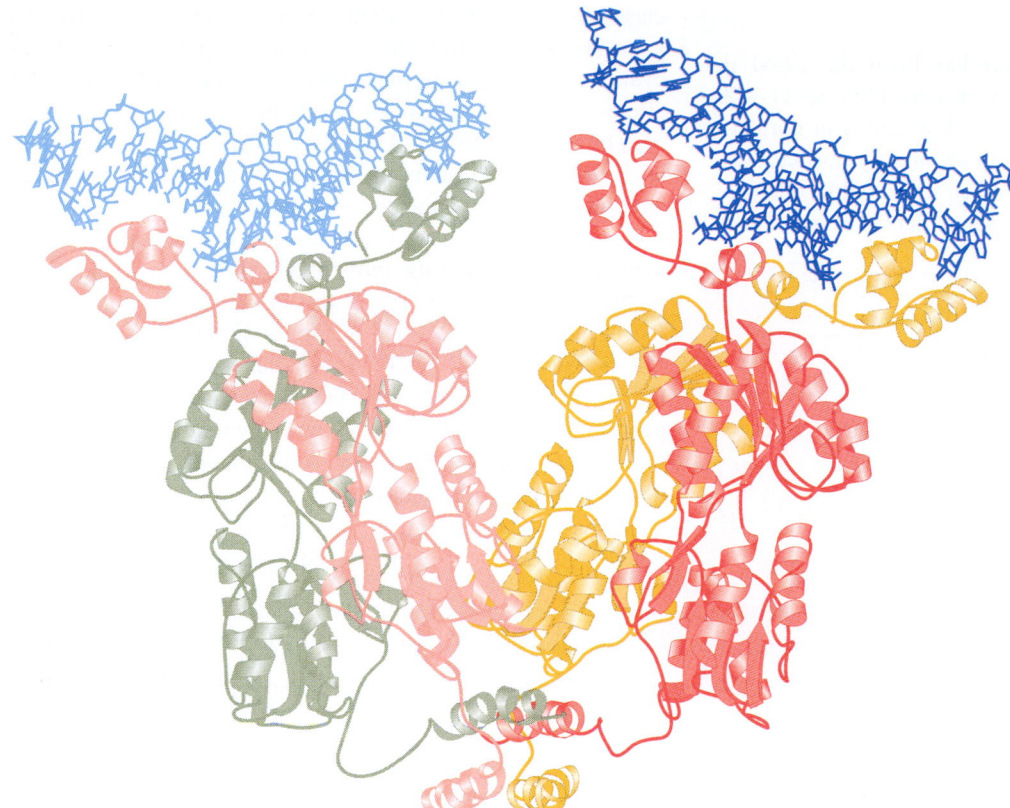

Figure 11-39 View of the Lac repressor–DNA complex, as determined by X-ray crystallography. Here the repressor tetramer is shown binding to two operators. Each dimer binds to one operator. The operators are 21 base pairs long here and are shown in dark and light blue. The monomers are colored green, pink, red, and yellow. Note how the amino-terminal headpiece on one monomer (pink) crosses the other monomer (green) to bind the first part of the operator, whereas the second monomer headpiece (green) crosses over to bind the second part of the operator. Thus, the recognition helix from each of two subunits binds to the consecutive parts of the same operator. (Illustration supplied by M. Lewis, Geoffrey Chang, N. C. Horton, M. A. Kercher, H. C. Pace, M. A. Schumacher, R. G. Brennan, and P. Lu, University of Pennsylvania.)

Epigenetic inheritance

We now have a general view of transcriptional regulation that can account for most observations that geneticists have made in the past century. However, there are still some phenomena that beg for explanation. An important set of phenomena, termed *epigenetic inheritance,* seems to be due to heritable alterations in which the DNA sequence itself is unchanged. Indeed, it is likely that these phenomena constitute another, poorly understood level of gene control. Examples of epigenetic inheritance in which the activity state of a gene depends on its genealogical history are paramutation and parental imprinting.

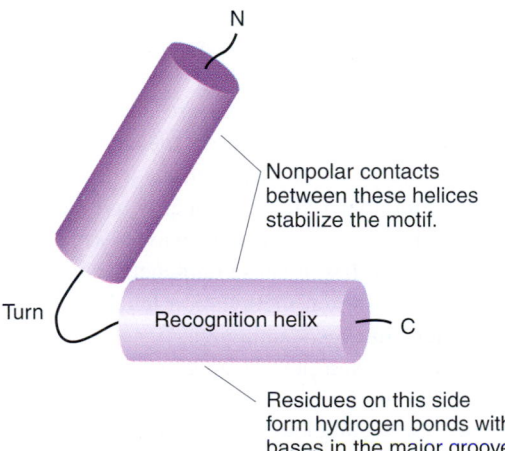

Figure 11-40 Helix-turn-helix motif of DNA-binding proteins. Each monomer of these dimeric proteins contains a helix-turn-helix; the two units are separated by 34 A — the pitch of the DNA helix. (From L. Stryer, *Biochemistry,* 4th ed. Copyright © 1995 by Lubert Stryer.)

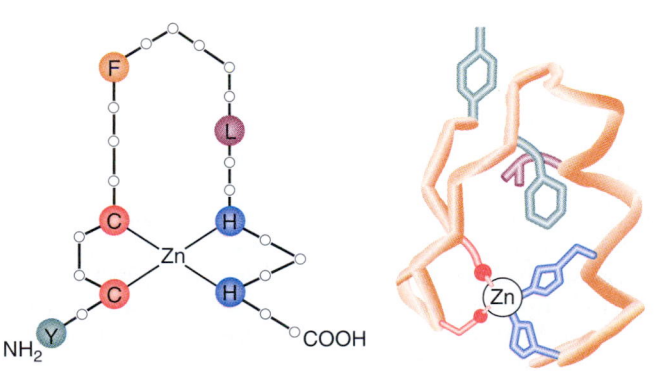

C_2H_2 zinc finger

Figure 11-41 Structural model for the zinc-finger DNA-binding domain. (From J. D. Watson, M. Gilman, J. Witkowski, and M. Zoller, *Recombinant DNA,* 2d ed. Copyright © 1992 by James D. Watson, Michael Gilman, Jan Witkowski, and Mark Zoller.)

Paramutation

The phenomenon of *paramutation* has been described in several plant species, most notably in corn (Figure 11-42). Paramutation was observed at only a few genes in corn. In this phenomenon, certain special but seemingly normal alleles, called paramutable alleles, suffer irreversible changes after having been present in the same genome as another

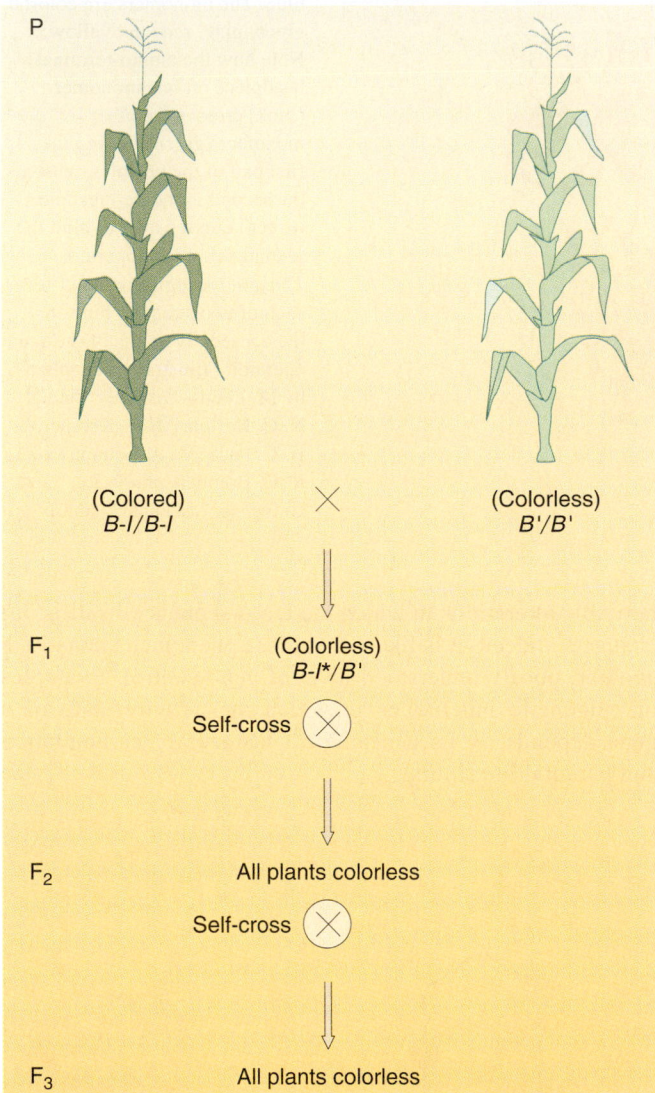

P

(Colored)
B-I/B-I   ×   (Colorless)
                              B'/B'

F₁   (Colorless)
                         B-I*/B'

                    Self-cross ⊗

F₂   All plants colorless

                    Self-cross ⊗

F₃   All plants colorless

Figure 11-42   A series of crosses depicting paramutation. The *B-I* mutation produces pigmented plants, whereas the *B'* mutation produces nearly unpigmented plants. Normally, when *B-I* is crossed with recessive colorless alleles of the *b* gene, the resulting plants are pigmented. However, when *B-I* and *B'* plants are intercrossed, the F₁ plants are essentially unpigmented, like the *B'* homozygotes. Thus, *B-I* is altered by being in the same genome as *B'*, indicated by the *B-I** designation. If this outcome were due simply to the dominance of *B'* to *B-I*, then a self-cross of the F₁ plants should generate *B-I*—colored homozygotes as approximately 1/4 of the F₂ progeny. Instead, no F₂ are pigmented. Intercrosses of the F₂ and of further generations do not restore the pigmented phenotype. Thus, *B-I* is said to have been paramutated by virtue of being in the same nucleus with the *B'* allele.

class of special alleles, called paramutagenic alleles. The *B-I* gene in corn encodes an enzyme in the pathway of anthocyanin pigments in various tissues in corn. Ordinary null *b* alleles lack these pigments, and these *b* alleles are completely recessive to *B-I*. There is a special paramutagenic allele, called *B'*, that confers the ability to make only a small amount of anthocyanin pigment. In crosses of *B-I* with *B'* homozygotes, the resulting heterozygotes are weakly pigmented, thus appearing indistinguishable from the *B'* homozygous plants. This result would seemingly suggest that *B-I* is recessive to *B'*. If this simple explanation were true, self-crosses of these heterozygous plants would generate homozygous *B-I* plants. However, instead, only *B'* alleles appear in the next (and subsequent) generations, indicating that the *B-I* allele has been paramutated. Somehow, by virtue of having been exposed to the paramutagenic *B'* allele by being in the same genotype for but a single generation, the *B-I* allele has been permanently crippled in its activity.

Parental imprinting

Another example of epigenetic inheritance, discovered about 15 years ago in mammals, is *parental imprinting*. In parental imprinting, certain autosomal genes have seemingly unusual inheritance patterns. For example, the mouse *Igf2* gene is expressed in a mouse only if it was inherited from the mouse's father. It is said to be maternally imprinted, inasmuch as a copy of the gene derived from the mother is inactive. Conversely, the mouse *H19* gene is expressed only if it was inherited from the mother; *H19* is paternally imprinted. The consequence of parental imprinting is that imprinted genes are expressed as if they were hemizygous, even though there are two copies of each of these autosomal genes in each cell. Furthermore, when these genes are examined at the molecular level, no changes in their DNA sequences are observed. Rather, the only changes that are seen are extra methyl ($-CH_3$) groups present on certain bases of the DNA of the imprinted genes. Occasional bases of the DNA of most higher organisms are methylated (an exception being *Drosophila*). These methyl groups are enzymatically added and removed, through the action of special methylases and demethylases. The level of methylation generally correlates with the transcriptional state of a gene: active genes are less methylated than inactive genes. However, whether altered levels of DNA methylation cause epigenetic changes in gene activity or whether altered methylation levels arise as a consequence of such changes is unknown.

What do these examples of epigenetic inheritance have in common? The main thread is that, somehow, a piece of a chromosome can be labeled as different on the basis of its ancestry or on the basis of which other genes were in the same genome. For many of these examples, differences in DNA methylation have been associated with differences in gene activity. Nonetheless, the underlying mechanisms and rationales for why such systems evolved still seem rather mysterious.

SUMMARY

The operon model explains how prokaryotic genes are controlled through a mechanism that coordinates the activity of a number of related genes. In negative control, the initiation of transcription is controlled at the operator by a repressor with binding affinities to the operator that may be altered by inducer molecules. Inactivation of the repressor — the negative control element — is required for active transcription. In positive control, transcription initiation requires the activation of a factor. Sometimes one control system is superimposed on another. For instance, superimposed on the repressor–operator system for the *lac* operon is the cAMP-CAP positive control system.

A major problem in transcriptional control in eukaryotes is understanding how tens or hundreds of thousands of promoters can be regulated to yield desired levels of mRNA. It is now clear that promoters are governed both by the number and type of promoter-proximal and enhancer elements and by the action of regulatory proteins that recognize these elements. In some aspects, transcriptional control in eukaryotes is similar to that found in prokaryotes; namely, trans-acting factors recognize cis-acting sites in both cases. Layered on top of these similarities are differences imposed by the nature of unicellular versus multicellular life, such as the coordination of appropriate gene function in different tissues.

CONCEPT MAP

Draw a concept map interrelating as many of the following terms as possible. Note that the terms are listed in no particular order.

environment / promoter / operator / gene / operon / RNA polymerase / mRNA / enhancer / trans-acting factors / regulation

CHAPTER INTEGRATION PROBLEM

In Chapter 7 we learned how an Hfr strain transfers to a recipient strain a segment of the chromosome, which then breaks off and remains for a period of time as a fragment in the cytoplasm. Before F′ plasmids were available, Jacob, Monod, and their co-worker Arthur Pardee exploited the properties of Hfr transfer to create temporary diploids of the *lac* region, inasmuch as the Hfr with which they worked donated the *lac* region early in the course of conjugational transfer. For several hours, the *lac* region on the chromosomal fragment became the second copy of the *lac* region in the recipient strain. They used an Hfr strain that transferred an $I^+ Z^+$ *lac* region into a strain that was $I^- Z^-$, and they studied the expression of β-galactosidase as a function of time. The graph here shows the results of their experiment. They observed an initial burst of β-galactosidase synthesis followed by the onset of repression. Can you offer an explanation for their results?

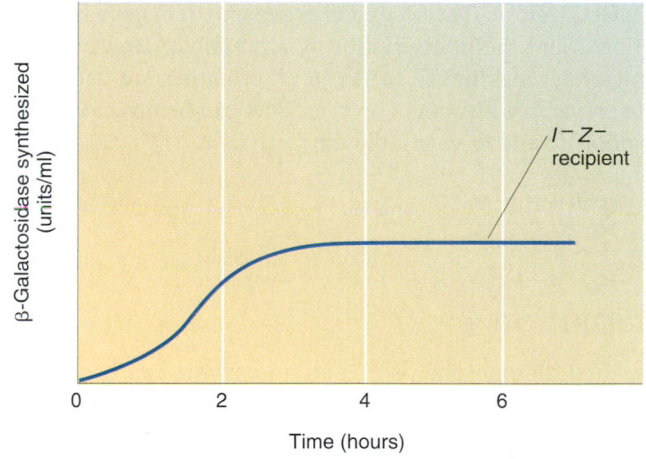

◆ Solution ◆

At the point at which the $I^+ Z^+$ genes are introduced into the cytoplasm of the recipient, there is no repressor in the cytoplasm, so initially β-galactosidase synthesis takes place at its maximal rate. However, the newly introduced I^+ gene directs the synthesis of repressor; after a period of time, repressor builds up in the cytoplasm in sufficient concentration to block further β-galactosidase synthesis, which explains why the curve flattens after 2 hours.

SOLVED PROBLEMS

This set of four solved problems, which are similar to Problem 4 at the end of this chapter, is designed to test understanding of the operon model. Here we are given several diploids and are asked to determine whether Z and Y gene products are made in the presence or absence of an inducer.

Use a table similar to the one in Problem 4 as a basis for your answers, except that the column headings in order will be: Z gene — inducer absent; Z gene — inducer present; Y gene — inducer absent; Y gene — inducer present.

1. $$\frac{I^-P^-O^cZ^+Y^+}{I^+P^+O^+Z^-Y^-}$$

◆ Solution ◆

One way to approach these problems is first to consider each chromosome separately and then to construct a diagram. The following illustration diagrams this diploid:

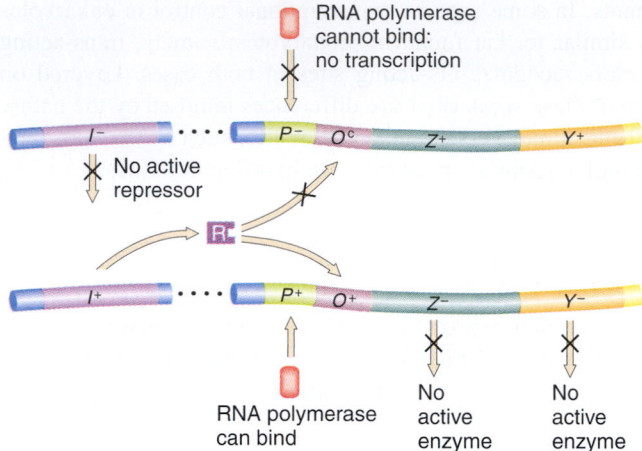

The first chromosome is P^-, so no Lac enzyme can be synthesized from it. The second chromosome (O^+) can be transcribed, and this transcription is repressible. However, the structural genes linked to the good promoter are defective; thus, no active Z product or Y product can be produced. The symbols to add to your table are "$-, -, -, -$."

2. $$\frac{I^+P^-O^+Z^+Y^+}{I^-P^+O^+Z^+Y^-}$$

◆ Solution ◆

The first chromosome is P^-, so no enzyme can be synthesized from it:

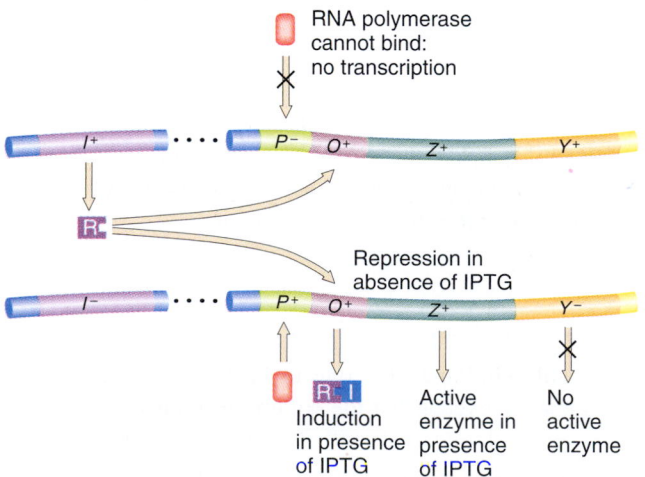

The second chromosome is O^+, so transcription will be repressed by the repressor supplied from the first chromosome, which can act in trans through the cytoplasm. However, only the Z gene from this chromosome is intact. Therefore, in the absence of an inducer, no enzyme will be made; in the presence of an inducer, only the Z gene product, β-galactosidase, will be produced. The symbols to add to the table are "$-, +, -, -$."

3. $$\frac{I^+P^+O^cZ^-Y^+}{I^+P^-O^+Z^+Y^-}$$

◆ Solution ◆

Because the second chromosome is P^-, we need consider only the first chromosome:

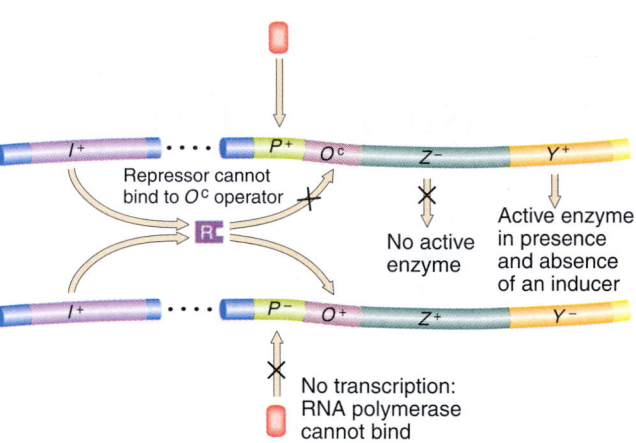

This chromosome is O^c, so enzyme is made in the absence of an inducer, although only the Y gene is active. The entries in the table should be "$-, -, +, +$."

4. $$\frac{I^sP^+O^+Z^+Y^-}{I^-P^+O^cZ^-Y^+}$$

◆ Solution ◆

In the presence of an I^s repressor, all wild-type operators are shut off, both with and without an inducer. Therefore, the first chromosome will be unable to produce any enzyme. However, the second chromosome has an altered (O^c) operator and can produce enzyme in both the absence and the presence of an inducer. Only the Y gene is active on this chromosome, so the entries in the table should be "$-, -, +, +$."

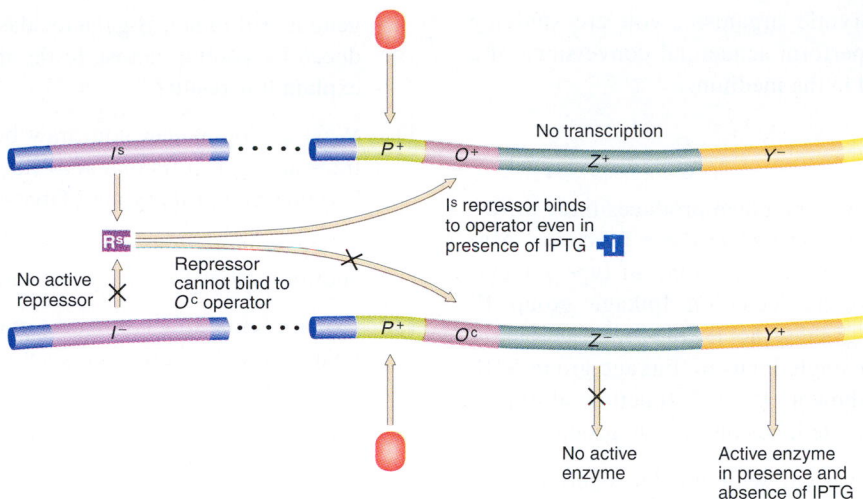

PROBLEMS

1. Explain why I^- mutations in the *lac* system are normally recessive to I^+ mutations and why I^+ mutations are recessive to I^s mutations.

2. What do we mean when we say that O^c mutations in the *lac* system are cis dominant?

3. The genes shown in the table below are from the *lac* operon system of *E. coli*. The symbols *a*, *b*, and *c* represent the repressor (*I*) gene, the operator (*O*) region, and the structural gene (*Z*) for β-galactosidase, although not necessarily in that order. Furthermore, the order in which the symbols are written in the genotypes is not necessarily the actual sequence in the *lac* operon.

ACTIVITY (+) OR INACTIVITY (−) OF *Z* GENE

Genotype	Inducer absent	Inducer present
$a^- \ b^+ \ c^+$	+	+
$a^+ \ b^+ \ c^-$	+	+
$a^+ \ b^- \ c^-$	−	−
$a^+ \ b^- \ c^+/a^- \ b^+ \ c^-$	+	+
$a^+ \ b^+ \ c^+/a^- \ b^- \ c^-$	−	+
$a^+ \ b^+ \ c^-/a^- \ b^- \ c^+$	−	+
$a^- \ b^+ \ c^+/a^+b^- \ c^-$	+	+

a. State which symbol (*a*, *b*, or *c*) represents each of the *lac* genes *I*, *O*, and *Z*.

b. In the table, a superscript minus sign on a gene symbol merely indicates a mutant, but you know that some mutant behaviors in this system are given special mutant designations. Use the conventional gene symbols for the *lac* operon to designate each genotype in the table.

(Problem 3 is from J. Kuspira and G. W. Walker, *Genetics: Questions and Problems*. Copyright © 1973 by McGraw-Hill.)

4. The map of the *lac* operon is

$$I \qquad P\,O\,Z\,Y$$

The promoter (*P*) region is the start site of transcription through the binding of the RNA polymerase molecule before actual mRNA production. Mutationally altered promoters (P^-) apparently cannot bind the RNA polymerase molecule. Certain predictions can be made about the effect of P^- mutations. Use your predictions and your knowledge of the lactose system to complete the table below. Insert a "+" where an enzyme is produced and a "−" where no enzyme is produced.

Part	Genotype	β-GALACTOSIDASE No lactose	Lactose	PERMEASE No lactose	Lactose
Example	$I^+ \ P^+ \ O^+ \ Z^+ \ Y^+/I^+ \ P^+ \ O^+ \ Z^+ \ Y^+$	−	+	−	+
a.	$I^- \ P^+ \ O^c \ Z^+ \ Y^-/I^+ \ P^+ \ O^+ \ Z^- \ Y^+$				
b.	$I^+ \ P^- \ O^c \ Z^- \ Y^+/I^- \ P^+ \ O^c \ Z^+ \ Y^-$				
c.	$I^s \ P^+ \ O^+ \ Z^+ \ Y^-/I^+ \ P^+ \ O^+ \ Z^- \ Y^+$				
d.	$I^s \ P^+ \ O^+ \ Z^+ \ Y^+/I^- \ P^+ \ O^+ \ Z^+ \ Y^+$				
e.	$I^- \ P^+ \ O^c \ Z^+ \ Y^-/I^- \ P^+ \ O^+ \ Z^- \ Y^+$				
f.	$I^- \ P^- \ O^+ \ Z^+ \ Y^+/I^- \ P^+ \ O^c \ Z^+ \ Y^-$				
g.	$I^+ \ P^+ \ O^+ \ Z^- \ Y^+/I^- \ P^+ \ O^+ \ Z^+ \ Y^-$				

5. In a haploid eukaryotic organism, you are studying two enzymes that perform sequential conversions of a nutrient A supplied in the medium:

$$A \xrightarrow{E_1} B \xrightarrow{E_2} C$$

Treatment of cells with mutagen produces three different mutant types with respect to these functions. Mutants of type 1 show no E_1 function; all type 1 mutations map to a single locus on linkage group II. Mutations of type 2 show no E_2 function; all type 2 mutations map to a single locus on linkage group VIII. Mutants of type 3 show no E_1 or E_2 function; all type 3 mutants map to a single locus on linkage group I.

a. Compare this system with the *lac* operon of *E. coli,* pointing out the similarities and the differences. (Be sure to account for each mutant type at the molecular level.)

b. If you were to intensify the mutant hunt, would you expect to find any other mutant types on the basis of your model? Explain.

6. In *Neurospora,* all mutants affecting the enzymes carbamyl phosphate synthetase and aspartate transcarbamylase map at the *pyr-3* locus. If you induce *pyr-3* mutations by ICR-170 (a chemical mutagen), you find that either both enzyme functions are lacking or only the transcarbamylase function is lacking; in no case is the synthetase activity lacking when the transcarbamylase activity is present. (ICR-170 is assumed to induce frameshifts.) Interpret these results in regard to a possible operon.

7. In 1972, Suzanne Bourgeois and Alan Jobe showed that a derivative of lactose, allolactose, is the true natural inducer of the *lac* operon, rather than lactose itself. Lactose is converted into allolactose by the enzyme β-galactosidase. How does this result explain the early finding that many Z^- mutations that are not polar still do not allow the induction of *lac* permease and transacetylase by lactose?

8. Certain *lacI* mutations eliminate operator binding by the *lac* repressor but do not affect the aggregation of subunits to make a tetramer, the active form of the repressor. These mutations are partly dominant to wild type. Can you explain the partly I^- phenotype of the I^-/I^+ heterodiploids?

9. Explain the fundamental differences between negative control and positive control.

10. Mutants that are *lacY*$^-$ retain the capacity to synthesize β-galactosidase. However, even though the *lacI* gene is still intact, β-galactosidase can no longer be induced by adding lactose to the medium. How can you explain this result?

11. What analogies can you draw between transcriptional trans-acting factors that activate gene expression in eukaryotes and prokaryotes? Give an example.

12. Compare the arrangement of cis-acting sites in the control regions of eukaryotes and prokaryotes.

13. Explain how models for bacterial operons such as *ara* relate to eukaryotic trans-acting proteins and their mechanism of action.

14. It is now known that the Lac repressor of *E. coli* has a leucine zipper at the carboxyl-terminal region of the protein, which is required to allow dimers to associate into tetramers. It is also known that a weak operator sequence, O_2, exists early in the Z gene, to which the repressor can also bind, in addition to the normal O_1. The elements are as shown here:

Using Figures 11-16 and 11-38 for reference, devise a model that explains why mutations in *lacI* that eliminate the leucine zipper reduce the ability of the repressor to block *lac* operon transcription completely. Draw a diagram.

15. One interesting mutation in *lacI* results in repressors with 100-fold increased binding to both operator and nonoperator DNA. These repressors display a "reverse" induction curve, allowing β-galactosidase synthesis in the *absence* of inducer (IPTG), but partly repressing β-galactosidase expression in the *presence* of IPTG. How can you explain this result? Note that, when IPTG binds repressor, it does not completely destroy operator affinity, but rather it reduces affinity 1000-fold. In addition, as cells divide and new operators are generated by the synthesis of daughter strands, repressor must find the new operators by searching along the DNA, rapidly binding to and dissociating from nonoperator sequences.

16. You are studying a mouse gene that is expressed in the kidneys of male mice. You have already cloned this gene. Now you wish to identify the segments of DNA that control the tissue-specific and sex-specific expression of this gene. Describe an experimental approach that would allow you to do so.

12

RECOMBINANT DNA TECHNOLOGY

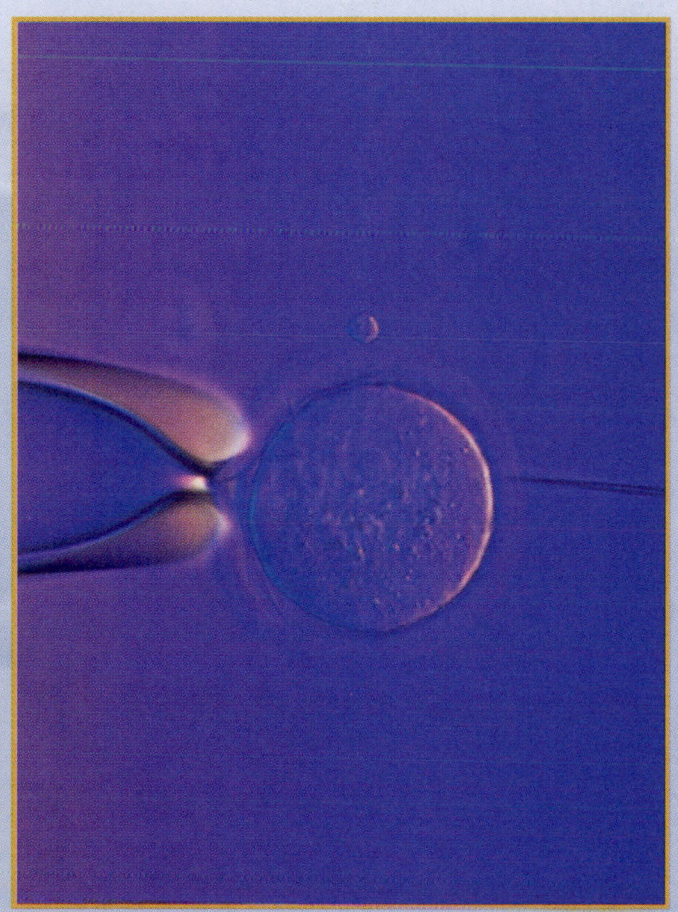

Injection of foreign DNA into an animal cell.
The microneedle used for injection is shown at right and a cell-holding pipette at left. *(© M. Baret/Rapho/Photo Researchers, Inc.)*

Key Concepts

Recombinant DNA is made by splicing a foreign DNA fragment into a small replicating molecule (such as a bacterial plasmid), which will then amplify that fragment along with itself and result in a molecular clone of the inserted DNA.

Restriction enzymes cut DNA at specific target sites, resulting in defined fragments with sticky ends suitable for insertion into a vector that has been cut open by the same enzyme.

A collection of DNA clones that encompasses the entire genome of an organism is called a genomic library.

An individual DNA clone can be selected from a library by using a specific probe for the DNA or its protein product or by its ability to transform a null mutant.

DNA fragments of different sizes produced by restriction-enzyme digestion can be fractionated because they migrate to different positions on an electrophoretic gel.

RNA or restriction-enzyme-cut DNA molecules that have been fractionated by size in an electrophoretic gel can be probed to detect a specific molecule.

Restriction-enzyme target sites can be mapped, providing useful markers for DNA manipulation.

A gene can be found by testing overlapping clones radiating outward from a linked marker.

After a gene has been cloned, its nucleotide sequence can be determined, and the sequence can be used to study gene function and evolution.

A pair of replication primers spanning a DNA sequence can be used to amplify that sequence for isolation.

The goal of genetics is to study the structure and function of genes and genomes. Since Mendel's time, genes have been identified by observing standard phenotypic ratios in controlled crosses. Clues about gene function first came from correlating specific mutations with enzyme and other protein deficiencies. The correlation of mutant sites within a gene with amino acid substitutions in the appropriate protein led to a better understanding of gene structure and function. To these ideas were added discoveries about the nature of DNA and the genetic code, leading to a fairly comprehensive understanding of the basic nature of the gene. However, all were *indirect* inferences about genes; no gene had ever been isolated and its DNA sequence examined directly. Indeed, it seemed impossible to isolate an individual gene from the genome.

Although it is relatively easy to isolate DNA from living tissue, DNA in a test tube looks like a glob of mucus. How could it be possible to isolate a single gene from this tangled mass of DNA threads? Recombinant DNA technology provides the techniques for doing just that, and today individual genes and other parts of genomes are isolated routinely.

Why is gene isolation so important? First, isolation of a gene enables the determination of its nucleotide sequence. From this information, the internal landmarks of the gene can be determined — for example, intron number and position. A comparison of DNA sequences between genes also can lead to insights in gene evolution. Converting the DNA sequence of a gene into amino acid sequence by using the genetic code leads to comparisons with the protein products of known genes; and, from this knowledge, the function of the gene can often be inferred. Function can also be studied by direct modification of part of the gene's DNA sequence followed by the reintroduction of the gene into the genome. Furthermore, a gene can be moved from one organism to another. An organism containing a foreign gene is called **transgenic.** Transgenic organisms can be used either for basic research or for specialized commercial applications. One application has been to make valuable human gene products such as insulin in transgenic bacteria carrying the appropriate human gene. From this brief overview, we see that gene isolation has become an indispensable tool of modern genetic analysis.

What are some examples of interesting genes that could be isolated? The answer depends very much on which biological process is being studied. Let's look at a few cases. A fungal geneticist studying the cellular pathway for synthesizing tryptophan would be interested in the genes that, when mutated, confer an auxotrophic requirement for tryptophan, because each gene would represent a step in the synthetic pathway (see Chapter 10). These genes can be identified through mutation, segregation, and mapping analysis. They would be named *trp1, trp2, trp3,* and so forth. This geneticist would be very interested in isolating and characterizing one or more of these genes. Likewise, human genes that have mutant alleles conferring

some type of functional disorder are interesting for medical and biological reasons. We have seen that these genes are identified by pedigree analysis. Two examples covered in Chapters 2 and 9 are the recessive autosomal conditions albinism and alkaptonuria. In these cases, the general nature of the defect has been understood for some time (both are enzyme defects), but it would be very useful to isolate the genes themselves. Other human genes are known from pedigree analysis, but no biochemical function is known for them. Isolating such genes would be particularly useful because the characterization of gene structure might lead to a determination of gene function and the nature of the disease. A good example is cystic fibrosis, a disease known from pedigree analysis to be caused by an autosomal recessive allele of a gene for which no function was known until the gene was isolated and sequenced. Cases such as these would be raised in all the organisms used in genetic research.

In our consideration of gene isolation, we shall first examine the nature of recombinant DNA and the principle whereby recombinant DNA technology can be used to isolate a gene. Next, we shall examine the methods for isolating specific genes such as those just discussed.

Making recombinant DNA

How does recombinant DNA technology work? The organism under study, which will be used to donate DNA for the analysis, is called the **donor organism.** The basic procedure is to extract and cut up DNA from a donor genome into fragments containing from one to several genes and allow these fragments to insert themselves individually into opened-up small autonomously replicating DNA molecules such as bacterial plasmids. These small circular molecules act as carriers, or **vectors,** for the DNA fragments. The vector molecules with their inserts are called **recombinant DNA** because they consist of novel combinations of DNA from the donor genome (which can be from any organism) with vector DNA from a completely different source (generally a bacterial plasmid or a virus). The recombinant DNA mixture is then used to transform bacterial cells, and it is common for single recombinant vector molecules to find their way into individual bacterial cells. Bacterial cells are plated and allowed to grow into colonies. An individual transformed cell with a single recombinant vector will divide into a colony with millions of cells, all carrying the same recombinant vector. Therefore an individual colony contains a very large population of identical DNA inserts, and this population is called a **DNA clone.** A great deal of the analysis of the cloned DNA fragment can be performed at the stage when it is in the bacterial host. Later, however, it is often desirable to reintroduce the cloned DNA back into cells of the original donor organism to carry out specific manipulations of genome structure and function. Hence the protocol is often as follows:

Identify donor gene of interest by using crosses

↓

Clone donor gene of interest in bacterium

↓

Characterize donor gene in bacterial system

↓

Modify donor gene

↓

Reintroduce donor gene into donor cells

MESSAGE ···

Cloning allows the amplification and recovery of a specific DNA segment from a large, complex DNA sample such as a genome.

Inasmuch as the donor DNA was cut into many different fragments, most colonies will carry a different recombinant DNA (that is, a different cloned insert). Therefore, the next step is to find a way to select the clone with the insert containing the specific gene in which we are interested. When this clone has been obtained, the DNA is isolated in bulk and the cloned gene of interest can be subjected to a variety of analyses, which we shall consider later in the chapter. Notice that the cloning method works because individual recombinant DNA molecules enter individual bacterial host cells, and then these cells do the job of amplifying the single molecules into large populations of molecules that can be treated as chemical reagents. Figure 12-1 gives a general outline of the approach.

The term *recombinant DNA* must be distinguished from the natural DNA recombinants that result from crossing-over between homologous chromosomes in both eukaryotes

Figure 12-1 Recombinant DNA technology enables individual fragments of DNA from any genome to be inserted into vector DNA molecules, such as plasmids, and individually amplified in bacteria. Each amplified fragment is called a *DNA clone.*

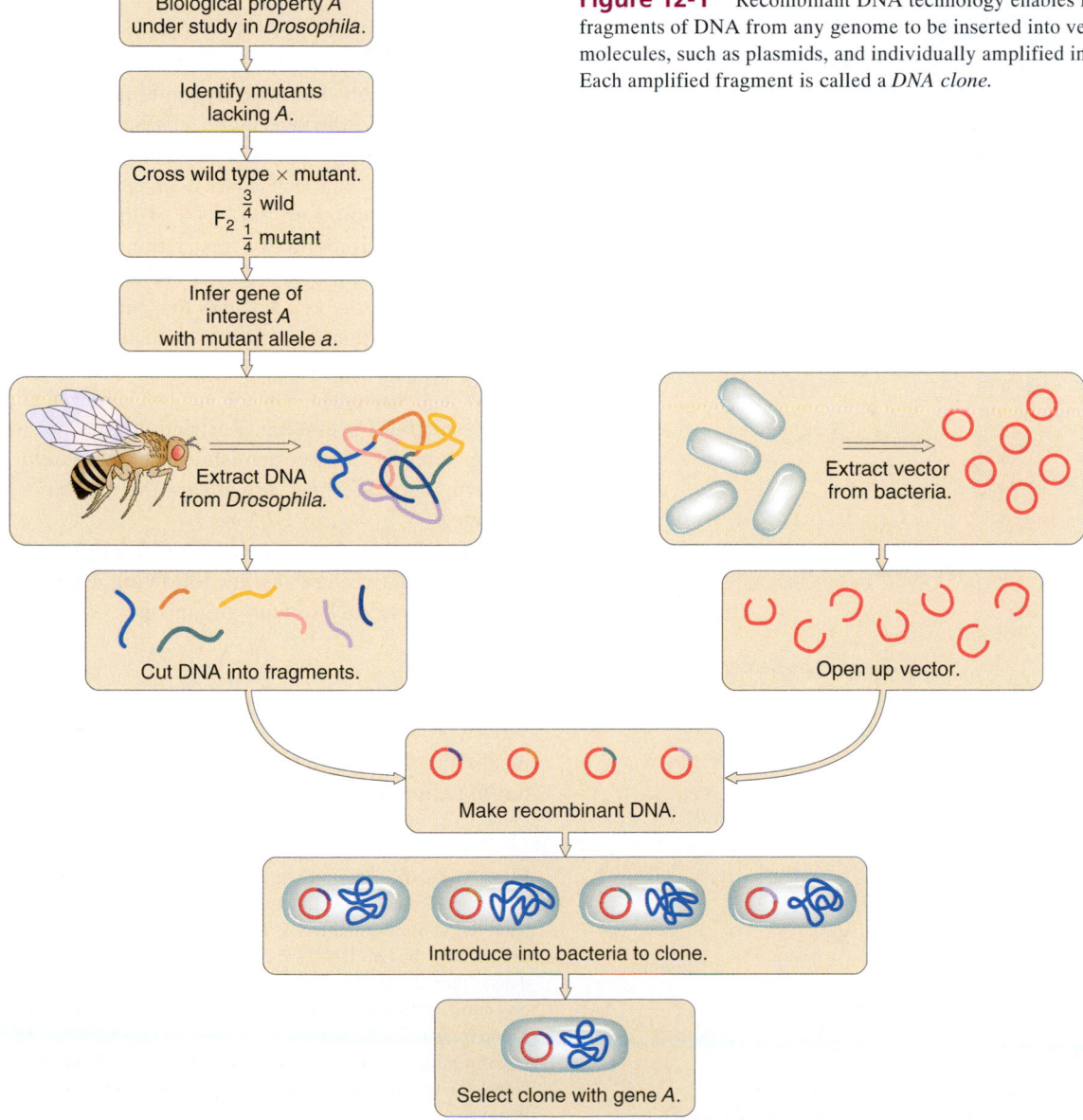

Biological property *A* under study in *Drosophila*.

Identify mutants lacking *A*.

Cross wild type × mutant.
F_2 $\frac{3}{4}$ wild
$\frac{1}{4}$ mutant

Infer gene of interest *A* with mutant allele *a*.

Extract DNA from *Drosophila*.

Extract vector from bacteria.

Cut DNA into fragments.

Open up vector.

Make recombinant DNA.

Introduce into bacteria to clone.

Select clone with gene *A*.

and prokaryotes. Recombinant DNA in the sense being used in this chapter is an unnatural union of DNAs from nonhomologous sources, usually from different organisms. Some geneticists prefer the alternative name **chimeric DNA,** after the mythological Greek monster Chimera. Through the ages, the Chimera has stood as the symbol of an impossible biological union, a combination of parts of different animals. Likewise, recombinant DNA is a DNA chimera and would be impossible without the experimental manipulation that we call *recombinant DNA technology.*

Isolating DNA

The first step in making recombinant DNA is to isolate donor and vector DNA. General protocols for DNA isolation were available many decades before the advent of recombinant DNA technology. With the use of such methods,

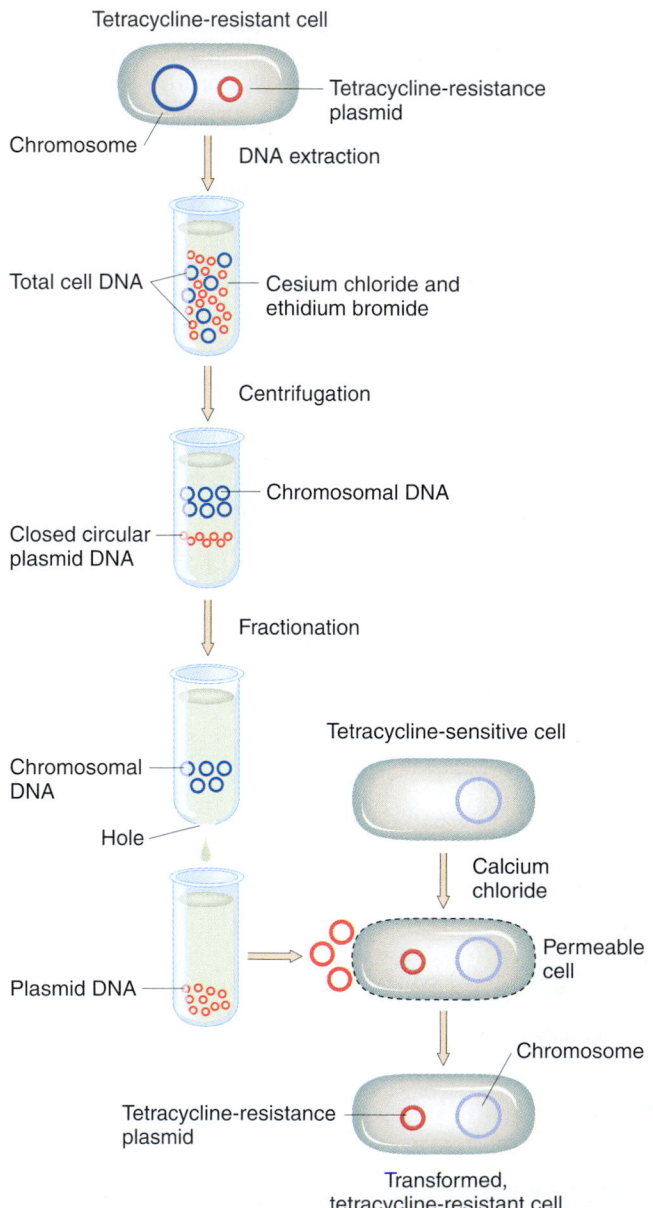

Tetracycline-resistant cell

Chromosome — Tetracycline-resistance plasmid

DNA extraction

Total cell DNA — Cesium chloride and ethidium bromide

Centrifugation

Chromosomal DNA

Closed circular plasmid DNA

Fractionation

Chromosomal DNA

Hole

Tetracycline-sensitive cell

Calcium chloride

Plasmid DNA

Permeable cell

Chromosome

Tetracycline-resistance plasmid

Transformed, tetracycline-resistant cell

the bulk of DNA extracted from the donor will be nuclear genomic DNA in eukaryotes or the main genomic DNA in prokaryotes; these types are generally the ones required for analysis. The procedure used for obtaining vector DNA depends on the nature of the vector. Bacterial plasmids are commonly used vectors, and these plasmids must be purified away from the bacterial genomic DNA. A protocol for extracting plasmid DNA by ultracentrifugation is summarized in Figure 12-2. Plasmid DNA forms a distinct band after ultracentrifugation in a cesium chloride density gradient containing ethidium bromide. The plasmid band is collected by punching a hole in the plastic centrifuge tube. Another protocol relies on the observation that, at a specific alkaline pH, bacterial genomic DNA denatures but plasmids do not. Subsequent neutralization precipitates the genomic DNA, but plasmids stay in solution. Phages such as λ also can be used as vectors for cloning DNA in bacterial systems. Phage DNA is isolated from a pure suspension of phages recovered from a phage lysate.

Cutting DNA

The breakthrough that made recombinant DNA technology possible was the discovery and characterization of **restriction enzymes.** Restriction enzymes are produced by bacteria as a defense mechanism against phages. The enzymes act like scissors, cutting up the DNA of the phage and thereby inactivating it. Importantly, restriction enzymes do not cut randomly; rather, they cut at specific DNA target sequences, which is one of the key features that make them suitable for DNA manipulation. Any DNA molecule, from viral to human, contains restriction-enzyme target sites purely by chance and therefore may be cut into defined fragments of a size suitable for cloning. Restriction sites are not relevant to the function of the organism, and they would not be cut in vivo, because most organisms do not have restriction enzymes.

Let's look at an example: the restriction enzyme *Eco*RI (from *E. coli*) recognizes the following six-nucleotide-pair sequence in the DNA of any organism:

$$5'\text{-GAATTC-}3'$$
$$3'\text{-CTTAAG-}5'$$

This type of segment is called a DNA **palindrome,** which means that both strands have the same nucleotide sequence but in antiparallel orientation. Many different restriction en-

Figure 12-2 Plasmids such as those carrying genes for resistance to the antibiotic tetracycline (*top left*) can be separated from the bacterial chromosomal DNA. Because differential binding of ethidium bromide by the two DNA species makes the circular plasmid DNA denser than the chromosomal DNA, the plasmids form a distinct band on centrifugation in a cesium chloride gradient and can be separated (*bottom left*). They can then be introduced into bacterial cells by transformation (*right*). (After S. N. Cohen, "The Manipulation of Genes." Copyright © 1975 by Scientific American, Inc. All rights reserved.)

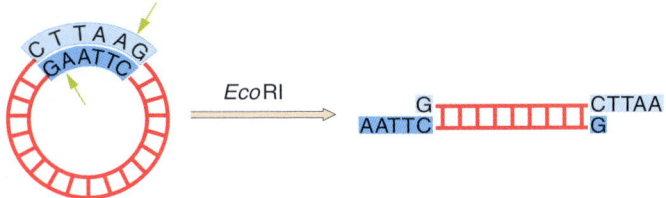

Figure 12-3 The restriction enzyme *Eco*RI cuts a circular DNA molecule bearing one target sequence, resulting in a linear molecule with single-stranded sticky ends.

zymes recognize and cut specific palindromes. The enzyme *Eco*RI cuts within this sequence but in a pair of staggered cuts between the G and the A nucleotides.

$$5'\text{-G}\mid\text{AATTC-}3' \longrightarrow 5'\text{-G} \qquad \text{AATTC-}5'$$
$$3'\text{-CTTAA}\mid\text{G-}5' \qquad 3'\text{-CTTAA} \qquad \text{G-}5'$$

This staggered cut leaves a pair of identical single-stranded "sticky ends." The ends are called *sticky* because they can hydrogen bond (stick) to a complementary sequence. Figure 12-3 shows *Eco*RI making a single cut in a circular DNA molecule such as a plasmid: the cut opens up the circle, and the linear molecule formed has two sticky ends. Production of these sticky ends is another feature of restriction enzymes that makes them suitable for recombinant DNA technology. The principle is simply that, if two different DNA molecules are cut with the same restriction enzyme, both will produce fragments with the same complementary sticky ends, making it possible for DNA chimeras to form. Hence, if both vector DNA and donor DNA are cut with *Eco*RI, the sticky ends of the vector can bond to the sticky ends of a donor fragment when the two are mixed.

MESSAGE ···

Restriction enzymes have two properties useful in recombinant DNA technology. First, they cut DNA into fragments of a size suitable for cloning. Second, many restriction enzymes make staggered cuts that create single-stranded sticky ends conducive to the formation of recombinant DNA.

Dozens of restriction enzymes with different sequence specificities have now been identified, some of which are shown in Table 12-1. You will notice that all the target sequences are palindromes, but, like *Eco*RI, some enzymes make staggered cuts, whereas others make flush cuts. Even flush cuts, which lack sticky ends, can be used for making recombinant DNA.

DNA can also be cut by mechanical shearing. For example, agitating DNA in a blender will break up the long chromosome-sized molecules into flush-ended clonable segments.

Joining DNA

Most commonly, both donor DNA and vector DNA are digested with the use of a restriction enzyme that produces

sticky ends and then mixed in a test tube to allow the sticky ends of vector and donor DNA to bind to each other and form recombinant molecules. Figure 12-4a shows a plasmid vector that carries a single *Eco*RI restriction site; so digestion with the restriction enzyme *Eco*RI converts the circular DNA into a linear molecule with sticky ends. Donor DNA

12-1 TABLE		Recognition, Cleavage, and Modification Sites of Various Restriction Enzymes
Enzyme	**Source organism**	**Restriction site in double-stranded DNA**

*Eco*RI	*Echerichia coli*	↓ m
		5′ -G-A-A-T-T-C-
		   -C-T-T-A-A-G- 5′
		     m ↑

*Eco*RII	*E. coli*	↓ m
		5′ -G-C-C-T-G-G-C-
		   -C-G-G-A-C-C-G - 5′
		       m ↑

*Hind*II	*Haemophilus influenzae*	↓ m
		5′ -G-T-Py-Pu-A-C-
		   -C-A-Pu-Py-T-G - 5′
		     m ↑

*Hind*III	*H. influenzae*	m↓
		5′ -A-A-G-C-T-T-
		   -T-T-C-G-A-A - 5′
		           ↑m

*Hae*III	*H. aegyptius*	↓
		5′ -G-G-C-C-
		   -C-C-G-G - 5′
		     ↑

*Hpa*II	*H. parainfluenzae*	↓
		5′ -C-C-G-G-
		   -G-G-C-C - 5′
		       ↑

*Pst*I	*Providencia stuartii*	↓
		5′ -C-T-G-C-A-G-
		   -G-A-C-G-T-C - 5′
		         ↑

*Sma*I	*Serratia marcescens*	↓
		5′ -C-C-C-G-G-G-
		   -G-G-G-C-C-C - 5′
		       ↑

*Bam*I	*Bacillus amyloliquefaciens*	↓
		5′ -G-G-A-T-C-C-
		   -C-C-T-A-G-G - 5′
		           ↑

*Bgl*II	*B. globiggi*	↓
		5′ -A-G-A-T-C-T-
		   -T-C-T-A-G-A - 5′
		           ↑

Note: An asterisk (*) is commonly used to indicate methylation sites, but an "m" is used here to prevent confusion with radioactive labeling.

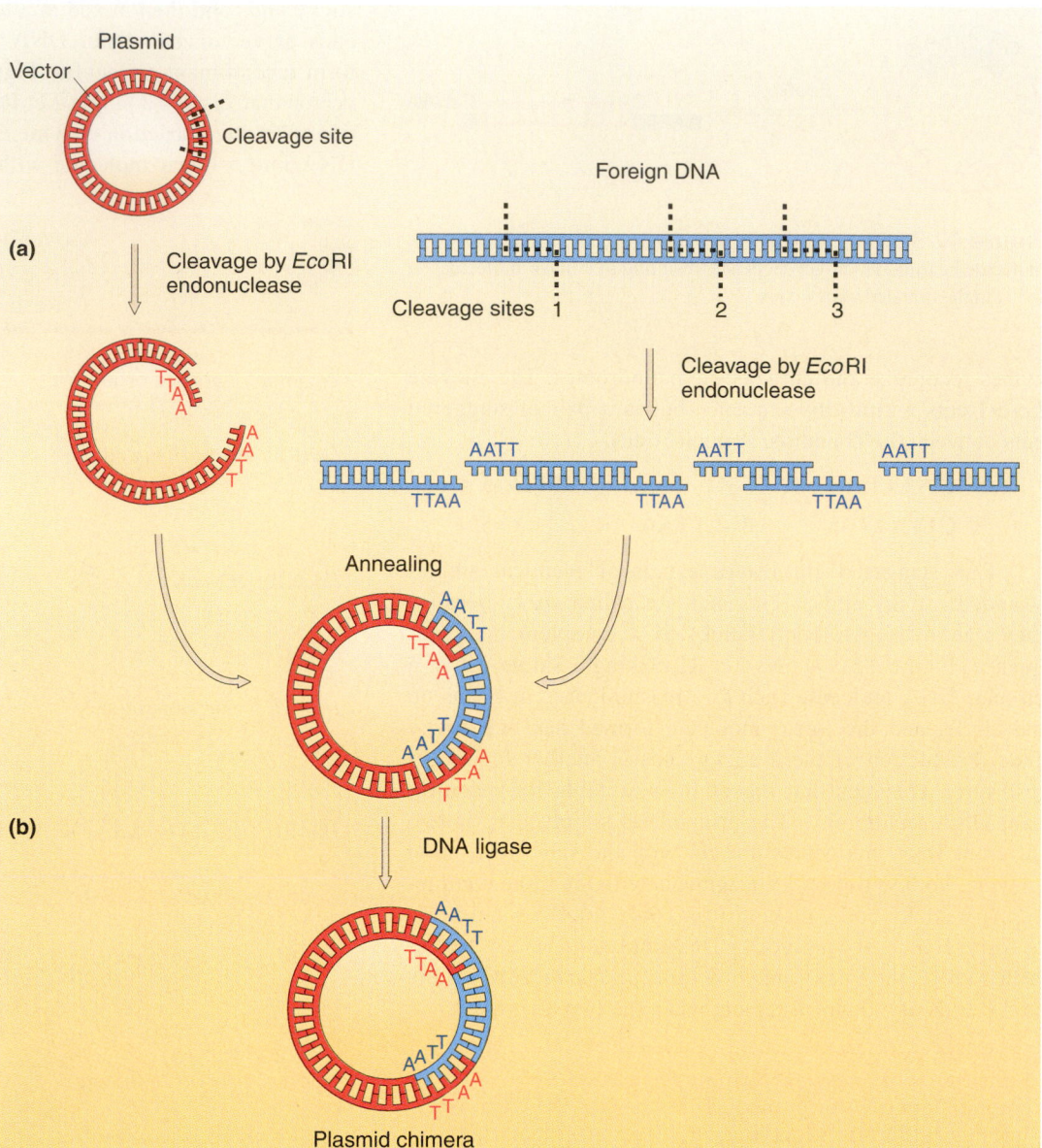

Figure 12-4   Method for generating a chimeric DNA plasmid containing genes derived from foreign DNA. (From S. N. Cohen, "The Manipulation of Genes." Copyright © 1975 by Scientific American, Inc. All rights reserved.)

from any other source (say, *Drosophila*) also is treated with the *Eco*RI enzyme to produce a population of fragments carrying the same sticky ends. When the two populations are mixed, DNA fragments from the two sources can unite, because double helices form between their sticky ends. There are many opened-up vector molecules in the solution, and many different *Eco*RI fragments of donor DNA. Therefore a diverse array of vectors carrying different donor inserts will be produced. At this stage, although sticky ends have united to generate a population of chimeric molecules, the sugar-phosphate backbones are still not complete at two positions at each junction. However, the backbones can be sealed by the addition of the enzyme **DNA ligase,** which create phosphodiester bonds at the junctions (Figure 12-4b). Certain ligases are even capable of joining DNA fragments with blunt-cut ends.

Amplifying recombinant DNA

The ligated recombinant DNA enters a bacterial cell by transformation. After it is in the host cell, the plasmid vector is able to replicate because plasmids normally have a replication origin. However, now that the donor DNA insert is part of the vector's length, the donor DNA is automatically replicated along with the vector. Each recombinant plasmid that enters a cell will form multiple copies of itself in that cell. Subsequently, many cycles of cell division will take place, and the recombinant vectors will undergo more rounds of replication. The resulting colony of bacteria will contain billions of copies of the single donor DNA insert. This set of amplified copies of the single donor DNA fragment is the DNA clone (Figure 12-5).

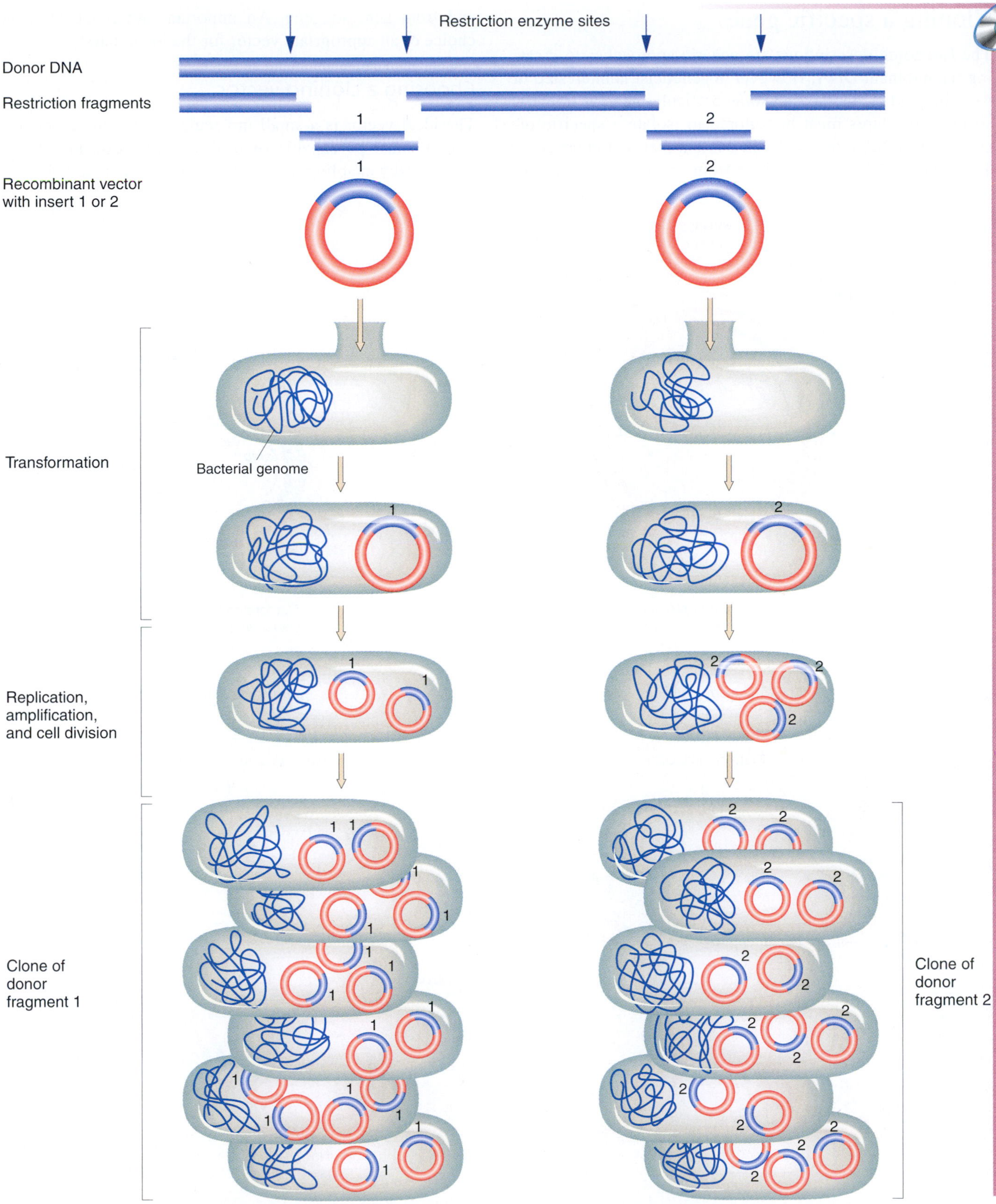

Figure 12-5 How amplification works. Restriction-enzyme treatment of donor DNA and vector allows insertion of single fragments into vectors. A single vector enters a bacterial host, where replication and cell division result in a large number of copies of the donor fragment.

ANIMATED ART
Making a Library of Wild-Type Yeast DNA

Cloning a specific gene

The foregoing descriptions are generic approaches to creating recombinant DNA. However, a geneticist is interested in isolating and characterizing some particular gene of interest, so the procedures must be tailored to isolate a specific recombinant DNA clone that will contain that particular gene. The details of the process differ from organism to organism and from gene to gene. An important initial factor is the choice of an appropriate vector for the job at hand.

Choosing a cloning vector

The ideal vector is a small molecule, facilitating manipulation. It must be capable of prolific replication in a living cell, thereby enabling the amplification of the inserted donor

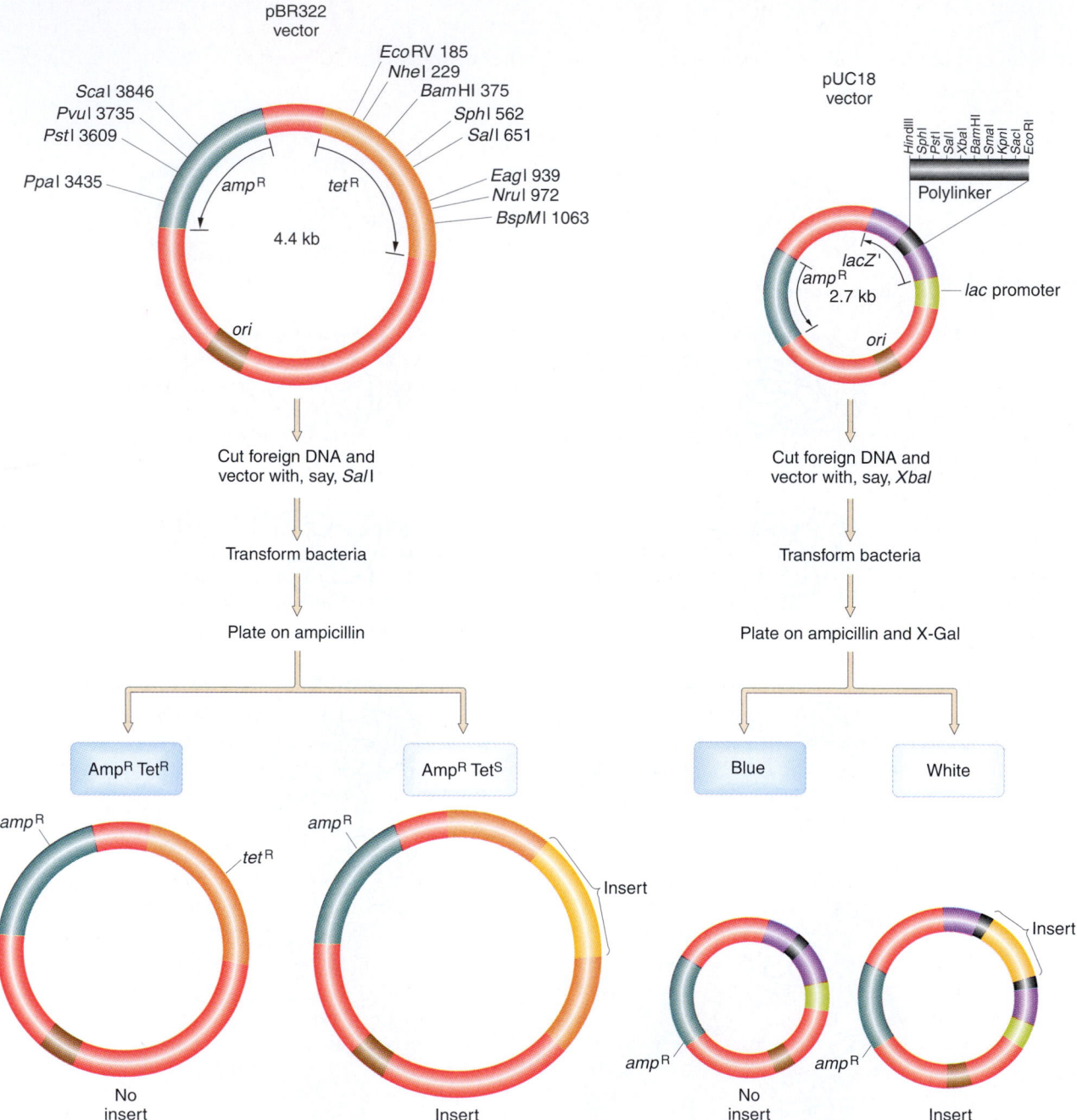

Figure 12-6   Two plasmids designed as vectors for DNA cloning, showing general structure and restriction sites. Insertion into pBR322 is detected by inactivation of one drug-resistance gene (tet^R), indicated by the Tet^S (sensitive) phenotype. Insertion into pUC18 is detected by inactivation of the β-galactosidase function of Z', resulting in an inability to convert the artificial substrate X-Gal into a blue dye.

fragment. Another important requirement is to have convenient restriction sites that can be used for insertion of the DNA to be cloned. Unique sites are most useful because then the insert can be targeted to one site in the vector. It is also important to have a method for easily identifying and recovering the recombinant molecule. Numerous cloning vectors are in current use, and the choice between them often depends on the size of the DNA segment that needs to be cloned and on the intended application for the cloned gene. We shall consider several commonly used types.

Plasmids. As described earlier, bacterial plasmids are small circular DNA molecules that are not only distinct from the main bacterial chromosome, but also additional to it. They replicate their DNA independently of the bacterial chromosome. Many different types of plasmids have been found in bacteria. The distribution of any one plasmid within a species is generally sporadic; some cells have the plasmid, whereas others do not. In Chapter 7, we encountered the F plasmid, which confers certain types of conjugative behavior to cells of *E. coli.* The F plasmid can be used as a vector for carrying large donor DNA inserts, as we shall see in Chapter 14. However, the plasmids that are routinely used as vectors are those that carry genes for drug resistance. The drug-resistance genes are useful because the drug-resistant phenotype can be used to select not only for cells transformed by plasmids, but also for vectors containing recombinant DNA. Plasmids are also an efficient means of amplifying cloned DNA because there are many copies per cell, as many as several hundred for some plasmids.

Two plasmid vectors that have been extensively used in genetics are shown in Figure 12-6. These vectors are derived from natural plasmids, but both have been genetically modified for convenient use as recombinant DNA vectors. Plasmid pBR322 is simpler in structure; it has two drug-resistance genes, tet^R and amp^R. Both genes contain unique restriction target sites that are useful in cloning. For example, donor DNA could be inserted into the tet^R gene. A successful insertion will split and inactivate the tet^R gene, which will then no longer confer tetracycline resistance, and the cell will be sensitive to that drug. Therefore, the cloning procedure is to mix the samples of cut plasmid and donor DNA, transform bacteria, and select first for ampicillin-resistant colonies, which must have been successfully transformed by a plasmid molecule. Of the Amp^R colonies, only those that prove to be tetracycline sensitive have inserts; in other words, the Amp^R Tet^S colonies are the ones that contain recombinant DNA. Further experiments are needed to find the clones with the specific insert required.

The pUC plasmid is a more advanced vector, whose structure allows direct visual selection of colonies containing vectors with donor DNA inserts. The key element is a small part of the *E. coli* β-galactosidase gene. Into this region has been inserted a piece of DNA called a **polylinker** or **multiple cloning site,** which contains many unique restriction target sites useful for insert-

ing donor fragments. The polylinker is in frame translationally with the β-galactosidase fragment and does not interfere with its translation. The transformation protocol uses recipient cells that contain a β-galactosidase gene lacking the fragment present on the plasmid. An unusual type of complementation takes place in which the partial proteins encoded by the two fragments unite to form a functional β-galactosidase. A colorless substrate for β-galactosidase called X-Gal is added to the medium, and the functional enzyme converts this substrate into a blue dye, which colors the colony blue. If donor DNA is inserted into the polylinker, the enzyme fragment borne on the vector is disrupted, no complete β-galactosidase protein is formed, and the colony is white. Hence, selection for white Amp^R colonies selects directly for vectors bearing inserts, and such colonies are isolated for further study.

Small plasmids that contain large inserts of foreign DNA tend to spontaneously lose the insert; therefore, these plasmids are not useful for cloning DNA fragments larger than 20 kb.

Viral vectors. Viral vectors, in which the gene or genes of interest are incorporated into the genome of a virus, offer many advantages for cloning and the subsequent applications of cloned genes. Because viruses infect cells with high efficiency, the cloned gene can be introduced into cells at a significantly higher frequency than by simple transformation. Some viral vectors are specialized for producing high levels of proteins encoded by the cloned genes, as exemplified by the use of insect baculovirus to express foreign proteins in a eukaryotic cell system, which is detailed in Chapter 13. Other viral vectors, such as the bacterial M13-based vectors, are designed to facilitate sequencing and the generation of mutations in cloned genes. Vectors derived from retroviruses can effect the stable integration into mammalian chromosomes of cloned DNA, allowing continued expression of the gene. Viral vectors are also the vehicles of choice for gene-therapy strategies. Some examples of viral vectors used in bacteria are described next.

Phage lambda. Phage λ is a convenient cloning vector for several reasons. First, λ phage heads will selectively package a chromosome about 50 kb in length, and, as will be seen, this property can be used to select for λ molecules with inserts of donor DNA. The central part of the phage genome is not required for replication or packaging of λ DNA molecules in *E. coli,* so the central part can be cut out by using restriction enzymes and discarded. The two "arms" are ligated to restriction-enzyme-cut donor DNA. The chimeric molecules can be either introduced into *E. coli* directly by transformation or packaged into phage heads in vitro. In the in vitro system, DNA and phage-head components are mixed together, and infective λ phages form spontaneously. In either method, recombinant molecules with 10- to 15-kb inserts are the ones that will be most effectively packaged into phage heads, because this size of insert substitutes for the deleted central part of the phage genome and brings the total molecule size to 50 kb. Therefore the presence of a phage plaque on the bacterial lawn automatically signals the presence of recombinant phage bearing an insert (Figure 12-7). A second useful property of a phage vector is that recombinant molecules are automatically packaged into infective phage particles, which can be conveniently stored and handled experimentally.

Single-stranded phages. Some phages contain only single-stranded DNA molecules. On infection of bacteria, the single infecting strand is converted into a double-stranded replicative form, which can be isolated and used

for cloning. The advantage of using these phages as cloning vectors is that single-stranded DNA is the very substrate required for the Sanger DNA-sequencing technique currently in widespread use (page 387). Phage M13 is the one most widely used for this purpose.

Cosmids. Cosmids are vectors that are hybrids of λ phages and plasmids, and their DNA can replicate in the cell like that of a plasmid or be packaged like that of a phage. However, cosmids can carry DNA inserts about three times as large as those carried by λ itself (as large as about 45 kb). The key is that most of the λ phage structure has been deleted, but the signal sequences that promote phage-head

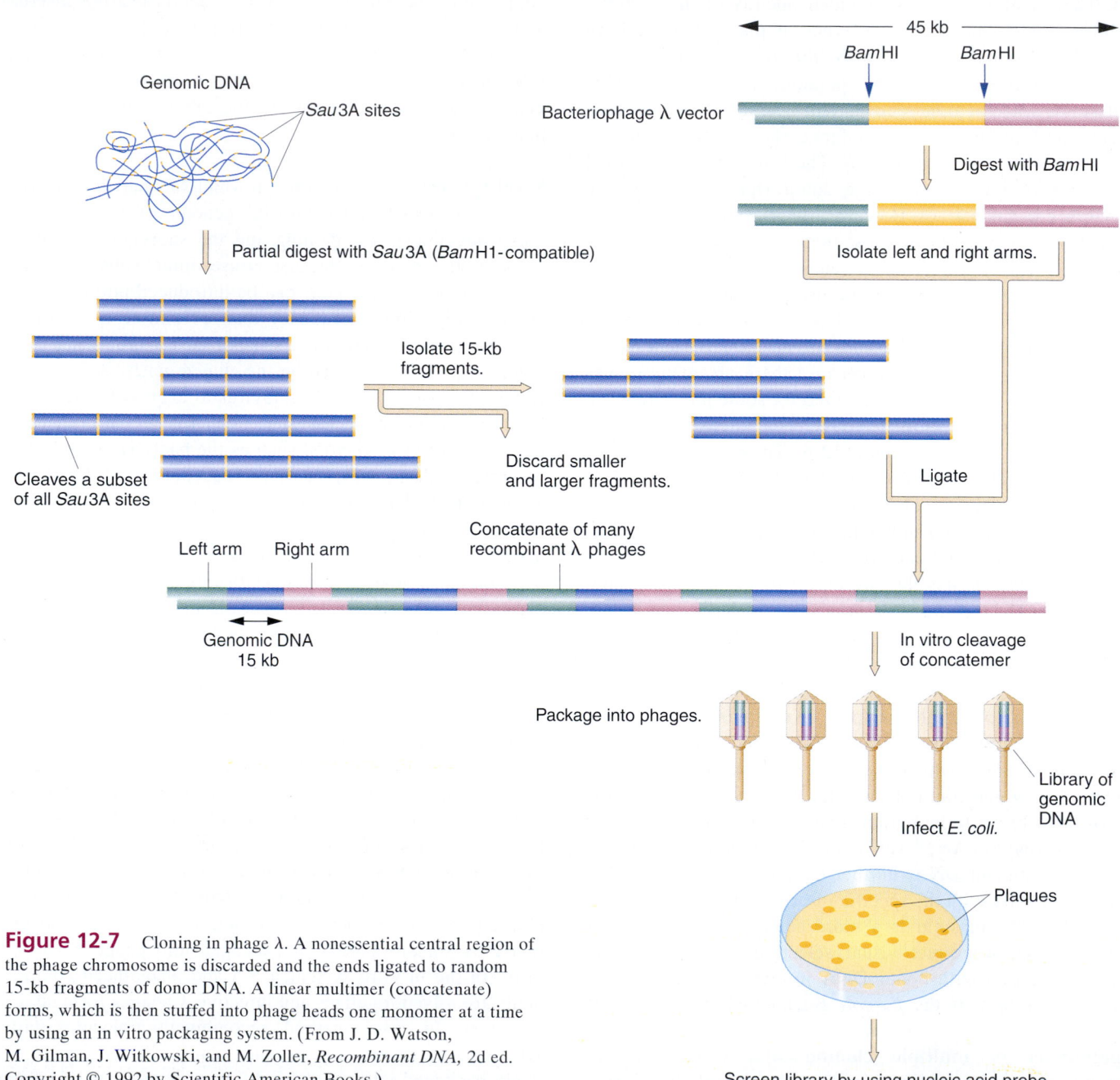

Figure 12-7 Cloning in phage λ. A nonessential central region of the phage chromosome is discarded and the ends ligated to random 15-kb fragments of donor DNA. A linear multimer (concatenate) forms, which is then stuffed into phage heads one monomer at a time by using an in vitro packaging system. (From J. D. Watson, M. Gilman, J. Witkowski, and M. Zoller, *Recombinant DNA,* 2d ed. Copyright © 1992 by Scientific American Books.)

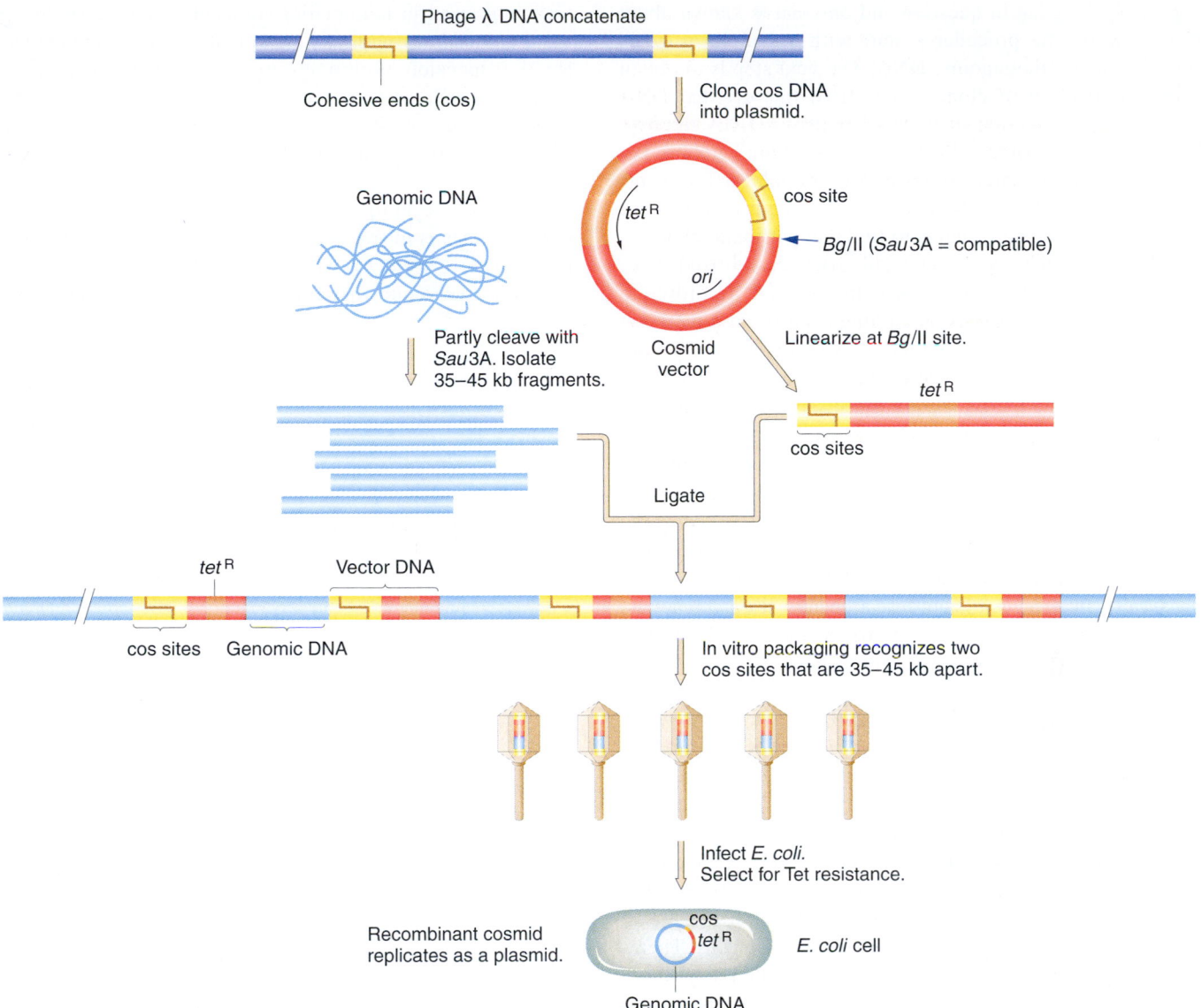

Figure 12-8 Cloning by cosmids. The cosmid is cut at a *Bgl*II site next to the cos site. Donor genomic DNA is cut by using *Sau*3A, which gives sticky ends compatible with *Bgl*II. A tandem array of donor and vector DNA results from mixing. Phage is packaged in vitro by cutting at the cos site. The cosmid with insert recircularizes after it is in the bacterial cell. (From J. D. Watson, M. Gilman, J. Witkowski, and M. Zoller, *Recombinant DNA*, 2d ed. Copyright © 1992 by Scientific American Books.)

stuffing (**cos sites**) remain. This modified structure enables phage heads to be stuffed with almost all donor DNA. Cosmid DNA can be packaged into phage particles by using the in vitro system. Cloning by cosmids is illustrated in Figure 12-8.

Expression vectors. One way of detecting a specific cloned gene is by detecting its protein product expressed in the bacterial cell. Therefore, in these cases, it is necessary to be able to express the gene in bacteria; that is, to transcribe it and translate the mRNA into protein. Most cloning vectors do not permit expression of cloned genes, but such expression is possible if special vectors are used. However, because bacteria cannot process introns, the cloned se-

quences must be stripped of introns. The cloned gene is inserted next to appropriate bacterial transcription and translation start signals. Some expression vectors have been designed with restriction sites located just next to a *lac* regulatory region. These restriction sites permit foreign DNA to be spliced into the vector for expression under the control of the *lac* regulatory system.

Making a DNA library

We have learned that the most important goal of recombinant DNA technology is to clone a particular gene or other genomic fragment of interest to the researcher. The approach used to clone a specific gene depends to a large degree on the gene in question and on what is known about it. Generally, the procedures start with a sample of DNA such as eukaryotic genomic DNA. The next step is to obtain a large collection of clones made from this original DNA sample. The collection of clones is called a **DNA library.** This step is sometimes referred to as "shotgun" cloning because the experimenter clones a large sample of fragments and hopes that one of the clones will contain a "hit"—the desired gene. The task then is to find that particular clone.

There are different types of libraries, categorized, first, according to which vector is used and, second, according to the source of DNA. Different cloning vectors carry different amounts of DNA, so the choice of vector for library construction depends on the size of the genome (or other DNA sample) being made into the library. Plasmid and phage vectors carry small amounts of DNA, so these vectors are suitable for cloning genes from organisms with small genomes. Cosmids carry larger amounts of DNA, and other vectors such as YACs and BACs (see Chapters 13 and 14) carry the largest amounts of all. Ease of manipulation is another important factor in choosing a vector. A phage library is a suspension of phages. A plasmid or a cosmid library is a suspension of bacteria or a set of defined bacterial cultures stored in culture tubes or microtiter dishes.

The second important decision is whether to make a **genomic library** or a **cDNA library. cDNA,** or **complementary DNA,** is synthetic DNA made from mRNA with the use of a special enzyme called *reverse transcriptase* originally isolated from retroviruses. With the use of an mRNA as a template, reverse transcriptase synthesizes a single-stranded DNA molecule that can then be used as a template for double-stranded DNA synthesis (Figure 12-9). Because it is made from mRNA, cDNA is devoid of both upstream and downstream regulatory sequences and of introns. Therefore cDNA from eukaryotes can be translated into functional protein in bacteria—an important feature when expressing eukaryotic genes in bacterial hosts.

The choice between genomic DNA and cDNA depends on the situation. If a specific gene that is active in a specific type of tissue in a plant or animal is being sought, then it makes sense to use that tissue to prepare mRNA to be converted into cDNA and then make a cDNA library from that sample. This library should be enriched for the gene in question. A cDNA library is based on the regions of the genome transcribed, so it will inevitably be smaller than a complete genomic library, which should contain all of the genome. Although genomic libraries are bigger, they do have the benefit of containing genes in their native form, including introns and regulatory sequences. If the purpose of constructing the library is a prelude to cloning an entire genome, then a genomic library is necessary at some stage.

In some cases, it is possible to narrow the genomic fraction used in library construction to more easily detect the desired gene. This approach is possible if the experimenter already knows which chromosome contains the gene. One

technique used in mammalian molecular genetics is to sort the chromosomes with an instrument called a *flow cytometer.* A suspension of chromosomes is passed through the apparatus, which sorts the chromosomes according to size (this procedure is discussed in more detail in Chapter 14). The appropriate chromosomal fraction is then used to make the library.

Another technique possible in organisms with small chromosomes is to fractionate whole chromosomes by using *pulsed field gel electrophoresis (PFGE).* Electrophoresis is a general technique that fractionates nucleic acids or pro-

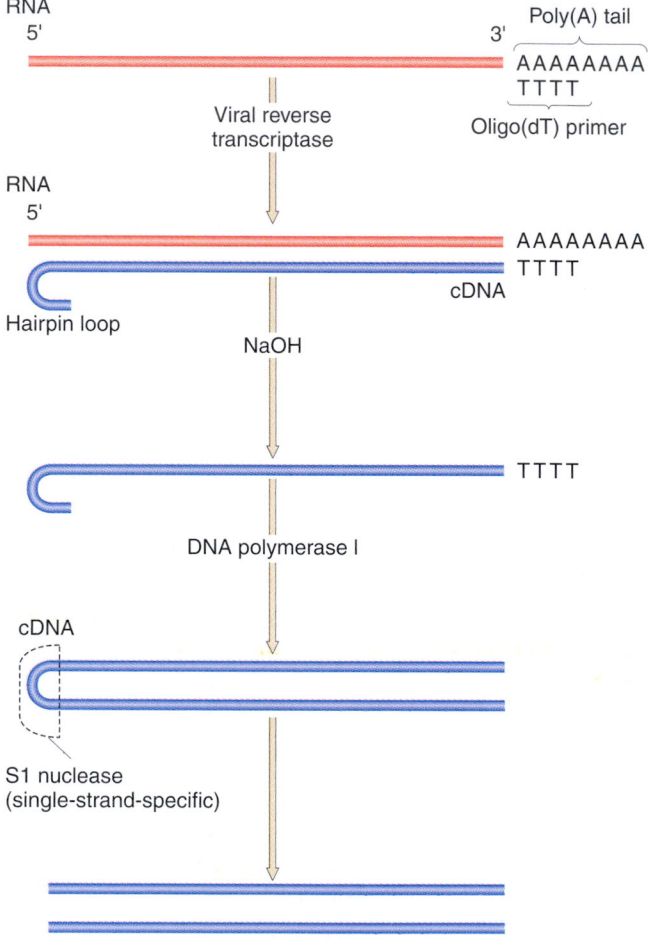

Figure 12-9 The synthesis of double-stranded cDNA from mRNA. A short oligo(dT) chain is hybridized to the poly(A) tail of an mRNA strand. The oligo(dT) segment serves as a primer for the action of reverse transcriptase, which uses the mRNA as a template for the synthesis of a complementary DNA strand. The resulting cDNA ends in a hairpin loop. When the mRNA strand has been degraded by treatment with NaOH, the hairpin loop becomes a primer for DNA polymerase I, which completes the paired DNA strand. The loop is then cleaved by S1 nuclease (which acts only on the single-stranded loop) to produce a double-stranded cDNA molecule. (From J. D. Watson, J. Tooze, and D. T. Kurtz, *Recombinant DNA: A Short Course.* Copyright © 1983 by W. H. Freeman and Company.)

teins according to size on gels under the influence of a strong electric field. This type of procedure separates shorter DNA fragments. PFGE is a specialized type of electrophoresis useful for very long DNA molecules. It uses several oscillating electric fields oriented in several different directions. These electric fields enable large DNA molecules such as whole chromosomes to snake through the gel to different positions according to their size. The appropriate chromosome can be identified on the gel by probing with a chromosome-specific probe (see the next subsection). Then the desired chromosome can be cut out, eluted from the gel, and used to make a chromosome-specific library.

How can an experimenter determine whether a library is large enough to contain any one unique sequence of interest with a reasonable degree of certainty? There are formulas for calculating the minimum number of clones needed, but a rough idea of the general order of magnitude of the library can be obtained simply by taking the total genome size and dividing by the average size of the inserts carried by the vector being used. Generally, this number will be at least doubled, but it does provide a rough estimate of the magnitude of the job of library construction.

MESSAGE ·······························
The task of isolating a clone of a specific gene begins with making a library of genomic DNA or cDNA—if possible, enriched for sequences containing the gene in question.

Finding specific clones by using probes

The library, which might contain as many as hundreds of thousands of cloned fragments, must be screened to find the recombinant DNA molecule containing the gene of interest. Such screening is accomplished by using a specific **probe** that will find and mark the clone for the researcher to identify. Broadly speaking, there are two types of probes: those that recognize DNA and those that recognize protein.

Probes for finding DNA. These probes depend on the natural tendency of a single strand of nucleic acid to find and hybridize to another single strand with a complementary base sequence. A probe that is itself DNA, when denatured (made single stranded by unwinding the two halves of the double helix), will therefore find and bind to other similar denatured DNAs in the library.

Probe

3'-AAGCCTATTTATGGGCAAT-5'

Clone

5'- . . . AGCTAGGGATCTTCGGATAAATACCCGTTACGTACTATTGGAAGGA . . . -3'

Identification of a specific clone in a library is a two-step procedure (Figure 12-10). First, colonies or plaques of the library on a petri plate are transferred to an absorbent membrane (often nitrocellulose) by simply laying the membrane on the surface of the medium. The membrane is peeled off, and colonies or plaques clinging to the surface are lysed in situ and the DNA denatured. The next step is to bathe the membrane with a solution of a probe that is specific for the DNA being sought. The probe must be labeled either with radioactivity or with a fluorescent dye. Generally, the probe is itself a cloned piece of DNA that has a sequence homologous to the desired gene. The probe DNA must be denatured; it will then bind only to the DNA of the clone being sought. The position of a positive clone will become clear from the position of the concentrated label, often as a spot on an autoradiogram.

Where does the DNA to make a probe come from? The DNA can be from one of several sources. One source is cDNA from tissue that expresses the gene of interest. The idea is that, because the mRNA of a gene is abundant, many of the cDNAs made from this tissue and inserted individually into vectors will very likely be for the desired gene. For example, in mammalian reticulocytes, 90 percent of the mRNA is known to be transcribed from the β-globin gene, so reticulocytes would be a good source of mRNA for making a cDNA probe to find a genomic globin gene. In this case, a genomic library would be probed. The need for this kind of analysis depends on which questions are to be asked about the gene. If only the transcribed sequence is of interest, then the cDNA clone itself could provide that information just as well. However, if introns and control regions are needed, the genomic clone must be obtained.

Another source of DNA for a probe might be a homologous gene from a related organism. For example, if a certain gene has been cloned in the ascomycete fungus *Neurospora*, then it is very likely that this gene can be used as a probe to find the homologous gene in the related fungus *Podospora*. This method depends on the evolutionary conservation of DNA sequences through time. Even though the probe DNA and the DNA of the desired clone might not be identical, they are often similar enough to promote hybridization. The method is jokingly called "clone by phone" because, if you can telephone a colleague who has a clone of your gene of interest but from a related organism, then your job of cloning is made relatively easy.

Probe DNA can be synthesized if the protein product of the gene of interest is known and an amino acid sequence has been obtained. Synthetic DNA probes are designed on the basis of knowledge of the genetic code, so an amino acid sequence merely has to be translated backward to obtain the DNA sequence that encoded it. However, because of the redundancy of the code—in other words, the fact that most amino acids are coded by more than one codon—several possible DNA sequences could have encoded the protein in question. To get around this problem, a short stretch of amino acids with minimal redundancy is selected. The nucleotide sequence is calculated by using the codon dictionary. The chemical DNA synthesizing reaction is a step-by-step process; so, wherever in the sequence there are alternative nucleotides, a mixture of those alternative nucleotides is fed into the reaction and all possible DNA

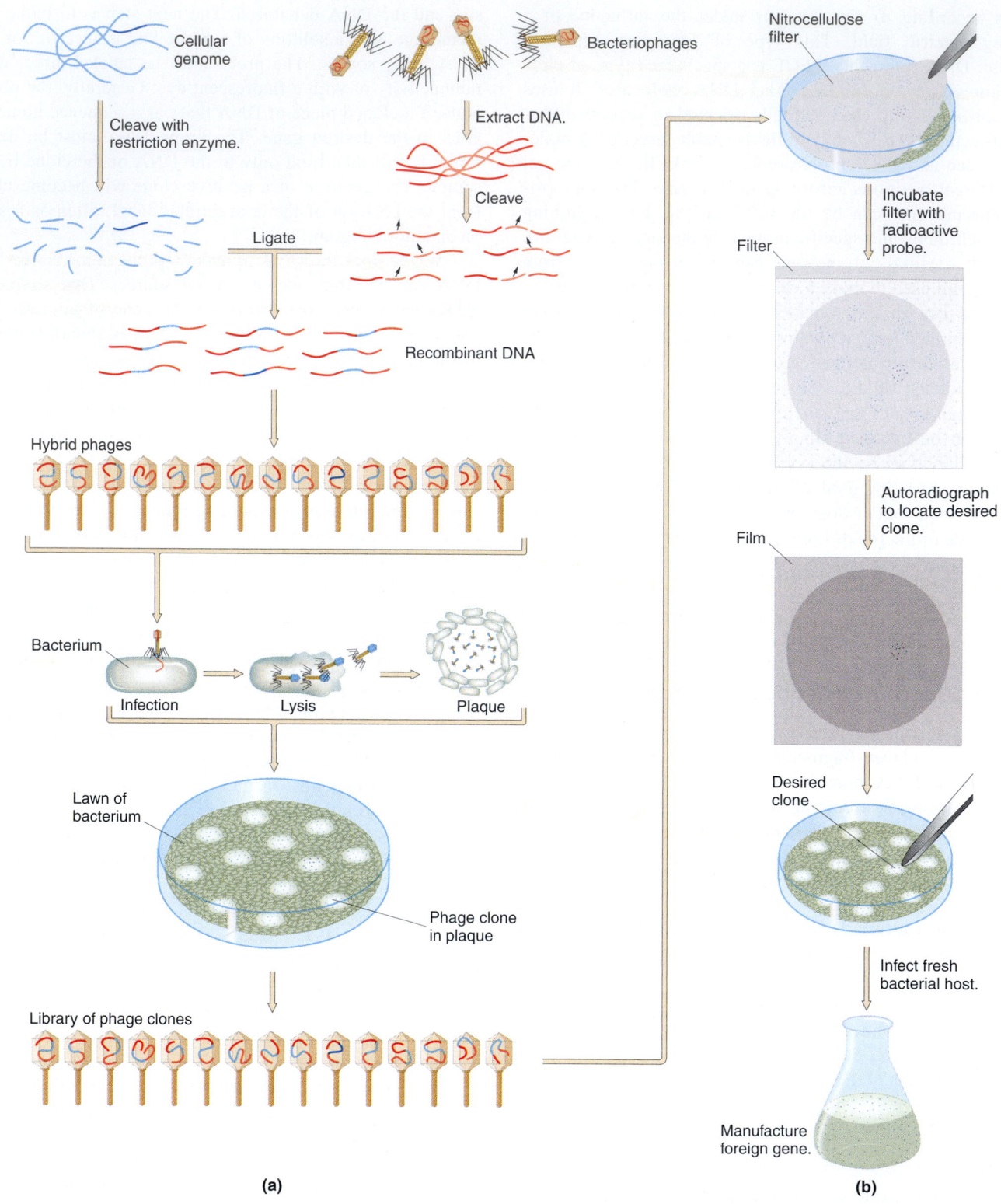

(a)

(b)

strands are synthesized. Figure 12-11 shows an example in which there are five positions of redundancy, showing 2, 3, 2, 2, and 2 alternatives, respectively. The reaction would make $2 \times 3 \times 2 \times 2 \times 2 = 48$ **oligonucleotide** strands at the same time. This "cocktail" of oligonucleotides would be used as a probe. The correct strand within this cocktail would find the gene of interest. Twenty nucleotides embody

enough specificity to find one unique DNA sequence in the library.

Additionally, free RNA can be radioactively labeled and used as a probe. This use of labeled free RNA as a probe is possible only when a relatively pure population of identical molecules of RNA, such as rRNA or fractionated tRNAs, can be isolated.

Figure 12-10 (a) A genomic library can be made by cloning genes in λ bacteriophages. When a lawn of bacteria on a petri plate is infected by a large number of different hybrid phages, each plaque in the lawn is inhabited by a single clone of phages descended from the original infecting phage. Each clone carries a different fragment of cellular DNA. The problem now is to identify the clone carrying a particular gene of interest (dark blue) by probing the clones with DNA or RNA known to be related to the desired gene. (b) The plaque pattern is transferred to a nitrocellulose filter, and the phage protein is dissolved, leaving the recombinant DNA, which is then denatured so that it will stick to the filter. The filter is incubated with a radioactively labeled probe; that is, a DNA copy of the messenger RNA representing the desired gene. The probe hybridizes with any recombinant DNA incorporating a matching DNA sequence, and the position of the clone having the DNA is revealed by autoradiography. Now the desired clone can be selected from the culture medium and transferred to a fresh bacterial host, so that a pure gene can be manufactured. (After R. A. Weinberg, "A Molecular Basis of Cancer," and P. Leder, "The Genetics of Antibody Diversity." Copyright © 1983, 1982 by Scientific American, Inc. All rights reserved.)

Probes for finding proteins. If the protein product of a gene is known and isolated in pure form, then this protein can be used to detect the clone of the corresponding gene in a library. The process is described in Figure 12-12. An antibody to the protein is prepared, and this antibody is used to screen an expression library. These libraries are made by using expression vectors designed to express high levels of a specific bacterial protein. To make the library, cDNA is inserted into the vector in frame with the bacterial protein, and the cells will make a fusion protein. A membrane is laid over the surface of the medium and removed with an imprint of colonies. It is dried and bathed in a solution of the antibody. Positive clones are revealed by making an antibody to the first antibody; the second antibody is labeled by a radioactive isotope or a chemical that will fluoresce or become a colored dye. By detecting the correct protein, the an-

tibody effectively identifies the clone containing the gene that must have synthesized that protein.

At the beginning of the chapter, we asked how it might be possible to find the gene for human albinism. It was in fact cloned by using an antibody to the enzyme that is known to be defective in this condition, the enzyme tyrosinase. This enzyme, like any protein, can be purified by standard biochemical procedures, and subsequently an antibody to the enzyme was prepared in rabbits. From tyrosinase-producing cells, mRNA was isolated and used to make cDNA. This cDNA was used to make an expression-vector library. The library was probed with the antibody to tyrosinase, and several positive clones were detected. The cDNA in the positive clone was sequenced and found to contain a gene whose exons total 1590 nucleotide pairs. The cDNA was used to probe a library of human genomic DNA, and, in this process, the intact tyrosinase gene was found. It proved to have five exons and four introns.

MESSAGE ·······································

A cloned gene can be selected from a library by using probes for the gene's DNA sequence or for the gene's protein product.

··

Finding specific clones by functional complementation

Specific clones in a bacterial or phage library can be detected through their ability to confer a missing function on a mutant line of the donor organism, which acts as the transformation recipient. This procedure is called **functional complementation.** Here the protocol is:

Make a bacterial or phage library by using wild-type donor DNA

↓

Use the library to transform cells of mutant donor-line a^-

↓

Select the a^+ phenotype among cells of donor organism

↓

Recover the a^+ gene from the successful bacterial or phage clone or from transformed donor cells

This method depends on the ability to transform the donor organism, often a eukaryote. We have already considered transformation in prokaryotes (Chapter 7), but eukaryotes can be transformed, too. The procedure differs among eukaryotes, but generally some special treatment of recipient cells is required. For example, to transform fungi, generally the cell walls must be removed enzymatically. Let's assume that we have isolated a mutant that is relevant to some biological process that interests us. For the present purpose, we will assume that it is an auxotrophic mutation in a fungus.

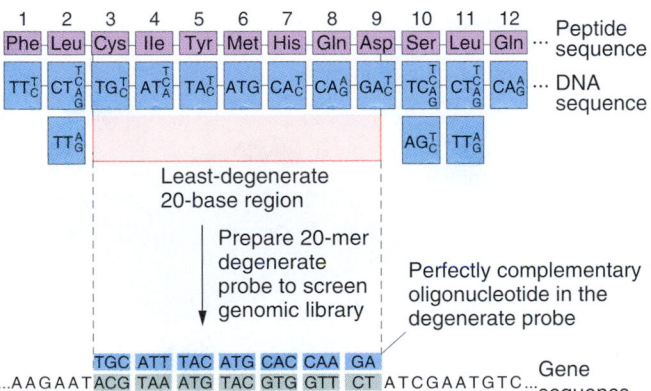

Figure 12-11 A short sequence of a protein is used to design a set of redundant oligonucleotides for use as a probe to recover the gene that encoded the protein. One of the set of probes will be a perfect match for the gene. (From H. Lodish, D. Baltimore, A. Berk, S. L. Zipursky, P. Matsudaira, and J. Darnell, *Molecular Cell Biology,* 3d ed. Copyright © 1995 by Scientific American Books, Inc.)

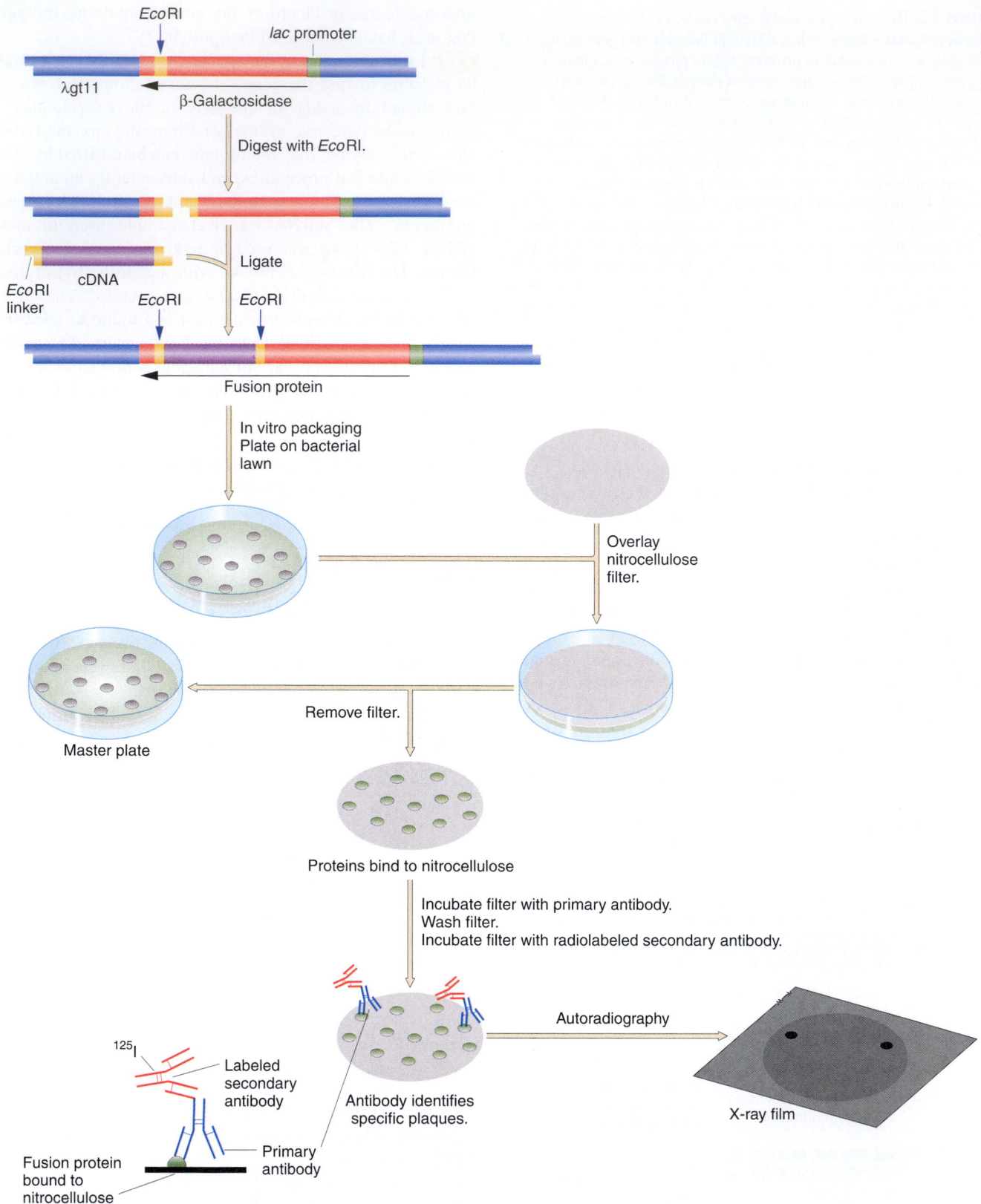

Figure 12-12   Finding the clone of interest by using antibody. An expression library made with phage derivative λgt11 is screened with a protein-specific antibody. After the unbound antibodies have been washed off the filter, the bound antibodies are visualized through binding of a radioactive secondary antibody. (From J. D. Watson, M. Gilman, J. Witkowski, and M. Zoller, *Recombinant DNA,* 2d ed. Copyright © 1992 by Scientific American Books.)

We shall use DNA from the library to transform the auxotrophic mutant strain and then plate these recipient fungal cells on minimal medium. Fungal cells that contain the wild-type allele (from the wild-type culture used to make the library) will transform the auxotroph to prototrophy and allow growth on minimal medium.

The reason that this transformation method works is that the transforming fragment functionally complements the deficiency caused by the mutant allele in the recipient. It might seem at first that this view of complementation is not the same as the one developed in Chapter 4; that is, the production of a wild-type phenotype from the union of two mutant genomes. However, the transforming vector contributes something that the recipient genome lacks (the wild-type allele being sought), and the recipient genome contributes something that the vector lacks (the entire remainder of the genome); so a type of complementation is accomplished.

If the transformation recipient is an organism in which plasmid vectors replicate autonomously (mainly bacteria and yeasts), then the transforming insert can be recovered simply by isolating the plasmid. However, as we shall see, in most eukaryotic organisms, the bacterial or phage vector cannot replicate and must insert into the genome to achieve stable transformation. In these cases, the transforming fragment is relatively inaccessible and must be retrieved from the successful clone in the library. This method uses a library in which the clones are laid out as a collection of numbered bacterial cultures in tubes or microtiter dishes. DNA is isolated in bulk from all the strains in specific subsets of the library, and transformation is attempted. By a process of narrowing down the library subsets that successfully transform, the clone with the wild-type allele can be identified. The process is illustrated in Figure 12-13, using as an example the *Neurospora trp3* gene mentioned at the beginning of the chapter. In this case, a cosmid library was used. The cosmid must also carry a marker gene that can be used to select for successful transformants of the fungus. A gene for hygromycin resistance is commonly used in fungi,

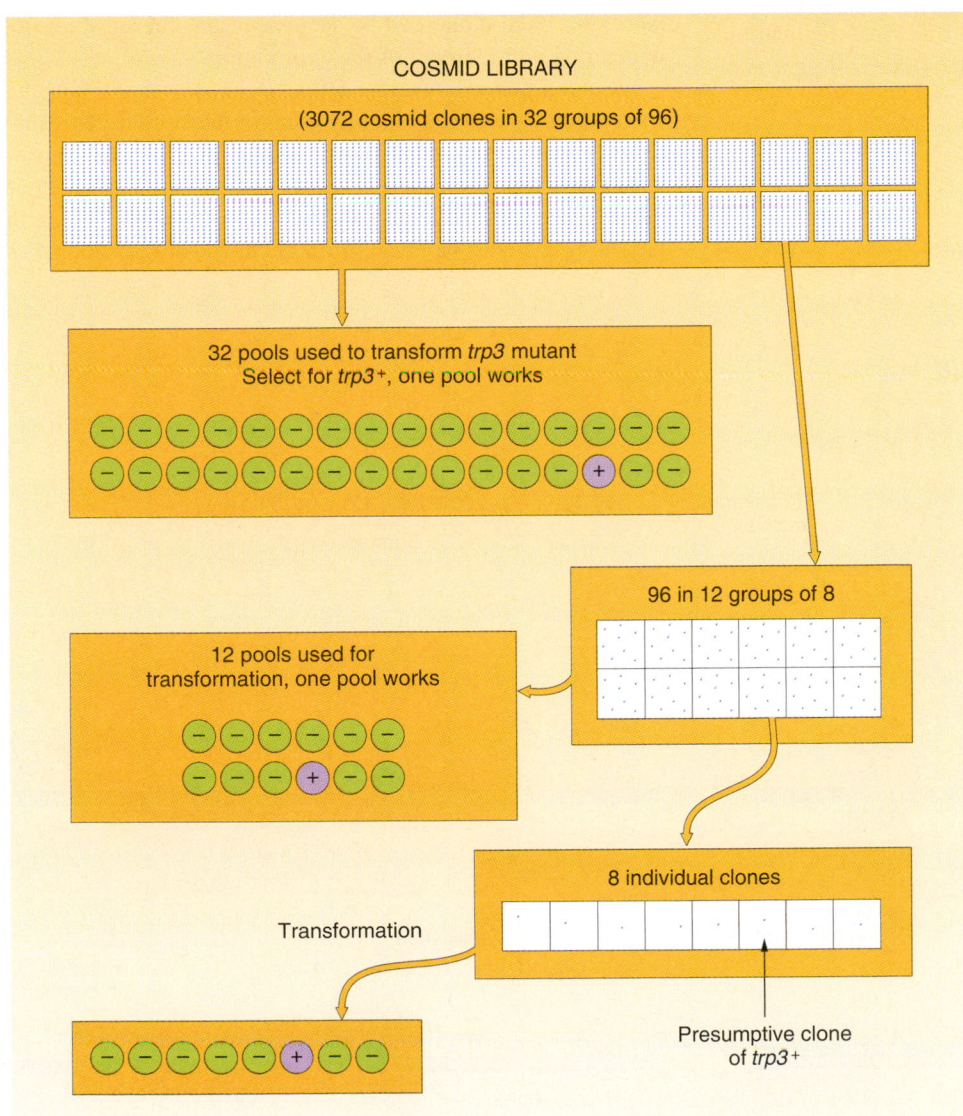

Figure 12-13 Finding a cloned gene by using progressively smaller pooled DNA samples in transformation. In this example, the quest is for the *trp3* gene of *Neurospora*. (After J. R. S. Fincham, *Genetic Analysis*. Copyright © 1995 by Blackwell.)

which are normally sensitive to this drug. Subsets of the cosmid library made from wild-type *Neurospora* DNA were used to transform *trp3* mutant cells, and *trp3*⁺ clones were selected by plating transformed cells on medium containing hygromycin but lacking tryptophan. Colonies that grow are likely to contain the *trp3*⁺ allele and are isolated from the plate.

In most cases, transformants are found to contain the vector carrying the wild-type allele inserted into one of the recipient's chromosomes at a location that is different from the mutant locus in the recipient. This is called **ectopic insertion** (Figure 12-14). Less commonly, the transforming wild-type allele replaces the resident auxotrophic mutation by a double-crossover-like process.

If a eukaryotic gene is cloned on a prokaryotic vector but a specific eukaryotic sequence is known that can act as an origin of replication, this sequence can be added to the vector. Then the vector will be able to replicate in both bacterial and eukaryotic cells, and insertion into the chromosome is not essential. These types of vectors are called *shuttle vectors* because they can be moved back and forth between different hosts. Without an origin of replication, the donor DNA *must* integrate into the eukaryotic chromosome to effect stable transformation.

MESSAGE ··
Specific cloned donor genes can be selected by using their DNA to transform and complement null alleles in recipient cells of the donor organism.
··

Positional cloning

Information about a gene's position in the genome can be used to circumvent the hard work of assaying an entire library to find the clone of interest. **Positional cloning** is a term that can be applied to any method that makes use of such information. Often both probing and complementation are part of positional cloning. A common starting point is the availability of another cloned gene or other marker known to be closely linked to the gene being sought. The linked marker acts as the departure point in a process, called **chromosome walking,** that will terminate at the target gene. Figure 12-15 summarizes the procedure of chromosome walking. End fragments of a clone of the linked marker are used as probes to select other clones from the library. These probes will detect clones of DNA regions that overlap with the initial clone. Restriction maps (see discussion on pages 389–392) are made of the DNA of this second set of clones, and, again, outward fragments are used for a new round of selection of overlapping clones from the library. Hence the walking process moves outward in two directions from the start site. Each clone can be sequenced or otherwise tested, depending on the intent of the exploration.

Sometimes a large insert that is known to contain the linked marker will also luckily contain the sought gene, and subcloning and transformation will narrow down the appropriate region of the cosmid. The availability of a large number of neutral DNA markers (restriction fragment length polymorphisms) dispersed throughout most genomes has

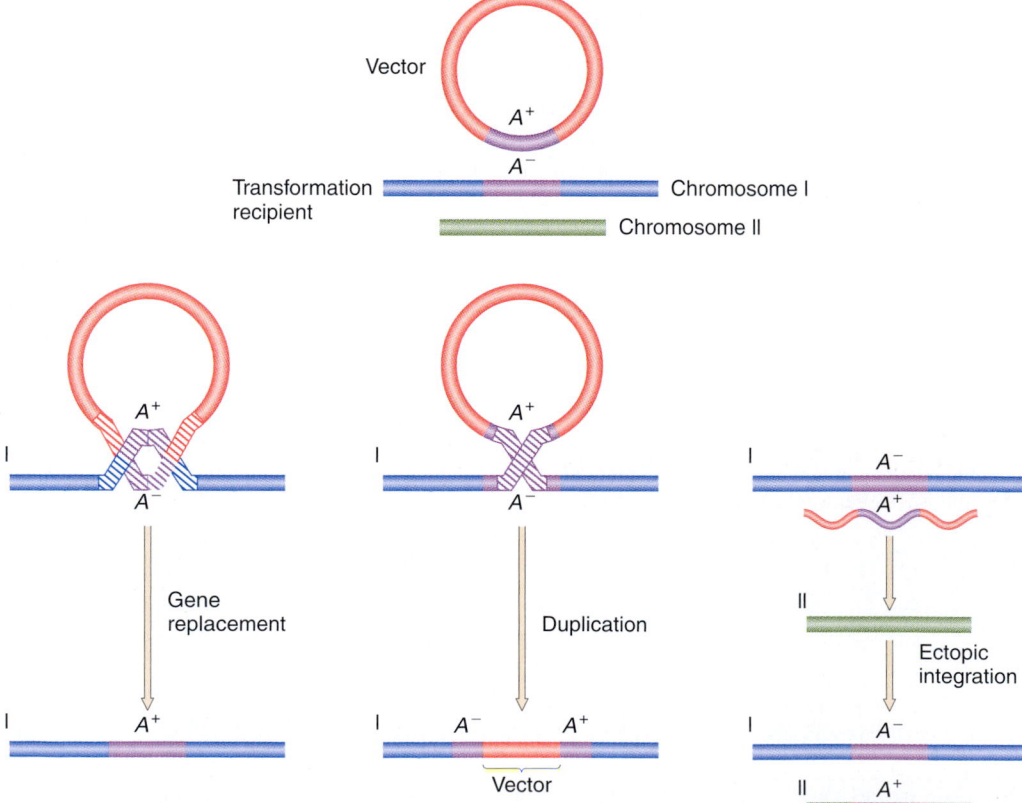

Figure 12-14 The possible fates of transforming DNA. A donor wild-type allele *A*⁺ (cloned in a bacterial vector) transforms an *A*⁻ recipient by one of three different types of insertion. **Note:** The recipient is generally of the same species as the donor DNA, either prokaryotic or eukaryotic. Two recipient chromosomes are shown, I and II.

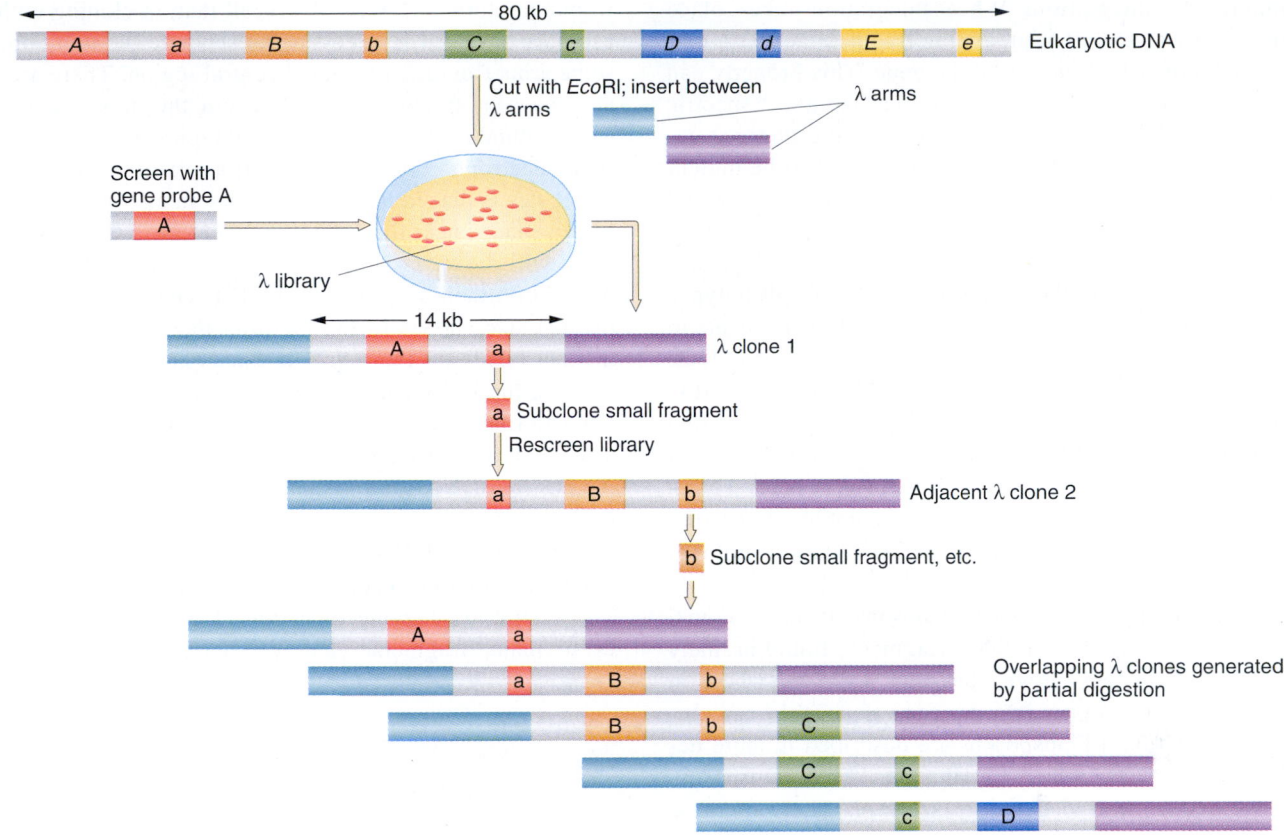

Figure 12-15   Chromosome walking. One recombinant phage obtained from a phage library made by the partial *Eco*RI digest of a eukaryotic genome can be used to isolate another recombinant phage containing a neighboring segment of eukaryotic DNA, as described in the text. (From J. D. Watson, J. Tooze, and D. T. Kurtz, *Recombinant DNA: A Short Course*. Copyright © 1983 by W. H. Freeman and Company.)

provided many useful start points. Positional cloning has been particularly useful for cloning human genes, many of which have no known biochemical function and cannot be easily selected by functional complementation. The human gene for cystic fibrosis, mentioned at the beginning of the chapter, was cloned by chromosome walking, and we shall examine its cloning in more detail in Chapter 14). For any case of chromosome walking, there must be some type of criterion to assess each step of the walk for the gene of interest, and these criteria depend on the individual gene concerned.

Cloning a gene by tagging

Tagging is a cloning method that zeros in on the desired gene directly by inducing a mutation in that gene with the use of a specific piece of DNA as an insertional mutagen. The specific sequence is then used as a tag to recover the gene. The approach is summarized in Figure 12-16. One type of tag is transforming DNA. When exogenous DNA is added by transformation or by other methods such as injection, it can integrate into the genome and become part of the chromosome. Ectopic integration is random throughout the genome, and apparently no segment of chromosomal DNA

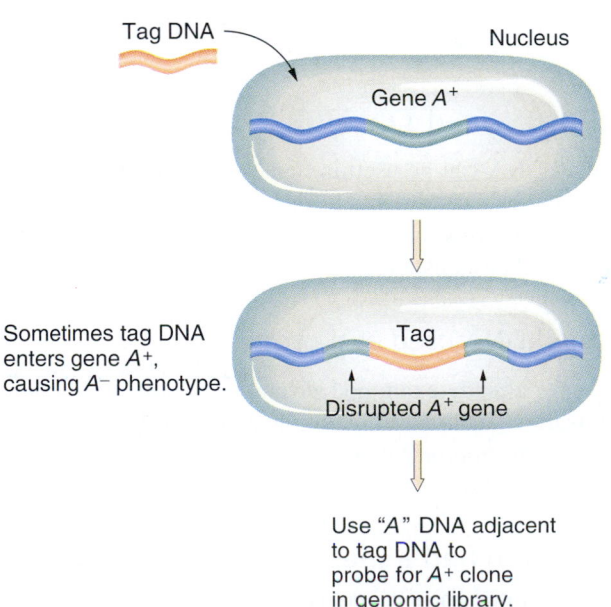

Figure 12-16   Using DNA insertion as a tag for marking and recovering a gene from the genome. The tag DNA can be transforming DNA or an endogenous transposon (movable element).

is immune to integration. When integration takes place within or near a gene, the integrating fragment acts as a mutagen, disrupting the function of the gene. This property can be used to good advantage. Suppose that we use a specific cloned gene x^+ and transform x^- cells of the donor organism into x^+. Many of the x^+ transformants will be mutant for the genes into which the transforming DNA has inserted ectopically. A subset of such x^+ cells will be mutant for the target gene a^+, the gene of interest, and will be of phenotype a^-. Hence among the x^+ transformants, a^- phenotypes are identified. The next step is to cross the transformants to determine if the a^- phenotype segregates with x^+. If it does, the mutation is likely to have been caused by the integration of the fragment containing x^+. The DNA of this mutant line is used to construct a library, and gene x^+ can be used as a probe to recover the clone of the disrupted a gene. To recover the intact wild-type a gene, a fragment of the disrupted a gene sequence is used in another round of probing, this time with a wild-type library.

A similar approach uses transposons as tags. Transposons are naturally mobile DNA fragments found in many organisms. When they move, they can insert anywhere in the genome. If they insert into or near a gene, they can create a null mutation. (Transposons are described in more detail in Chapter 20.) In a line containing an active transposon, mutants for the desired gene are selected. Many of these mutants will have been caused by the insertion of the transposon. This mutant line is used to make a library. A cloned part of the transposon DNA can then be used as a tag to recover the gene, in a manner similar to that shown in Figure 12-16.

MESSAGE ·······································
Mutating a gene by the insertion of transforming DNA or a transposon allows the gene to be tagged as a prelude to its isolation.
·······································

Using cloned DNA

Cloned DNA can be used in numerous ways dictated by the needs of the experiment. In this section, we consider some of the basic uses that are applicable in a wide range of experimental circumstances.

Cloned DNA used as a probe

We have already examined several examples of this type of application, such as using a cloned gene from one organism to select the appropriate clone from another organism. More examples follow.

Probing to find a specific nucleic acid in a mixture

In the course of gene and genome manipulation, it is often necessary to detect and isolate specific DNA molecules

from a mixture. For example, recall that, in cloning with the use of λ phage, it is necessary to separate the two chromosome arms from the unwanted central region. There are several ways of fractionating DNA, but the most extensively used method is electrophoresis (see Figure 4-5 for a drawing of an electrophoretic apparatus). If a mixture of linear DNA molecules is placed in a well cut into an agarose gel and the well is placed near the cathode of an electric field, the molecules will move through the gel to the anode at speeds dependent on their size (Figure 12-17). Therefore, if there are distinct size classes in the mixture, these classes will form distinct bands on the gel. The bands can be visualized by staining the DNA with ethidium bromide, which causes the DNA to fluoresce in ultraviolet (UV) light. If bands are well separated, an individual band can be cut from the gel and the purified DNA sample can be removed from the gel matrix. Therefore DNA electrophoresis can be either diagnostic (showing which DNA fragments are present) or preparative (useful in isolating specific DNA fragments).

Restriction-enzyme digestion of genomic DNA results in so many fragments that a stained electrophoretic gel

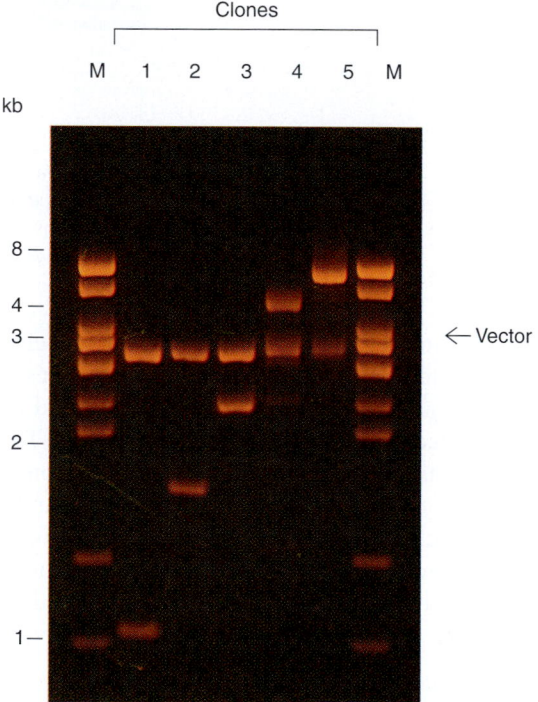

Figure 12-17   Mixtures of different-size DNA fragments separated electrophoretically on an agarose gel. In this case, the samples are five recombinant vectors treated with *Eco*RI. The mixtures are applied to wells at the top of the gel, and fragments move under the influence of an electric field to different positions dependent on size (and, therefore, number of charges). The DNA bands have been visualized by staining with ethidium bromide and photographing under UV light. (M represents lanes containing standard fragments acting as markers for estimating DNA length.) (From H. Lodish, D. Baltimore, A. Berk, S. L. Zipursky, P. Matsudaira, and J. Darnell, *Molecular Cell Biology,* 3d ed. Copyright © 1995 by Scientific American Books, Inc.)

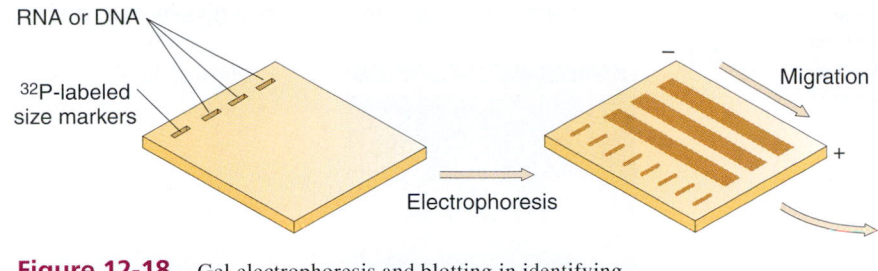

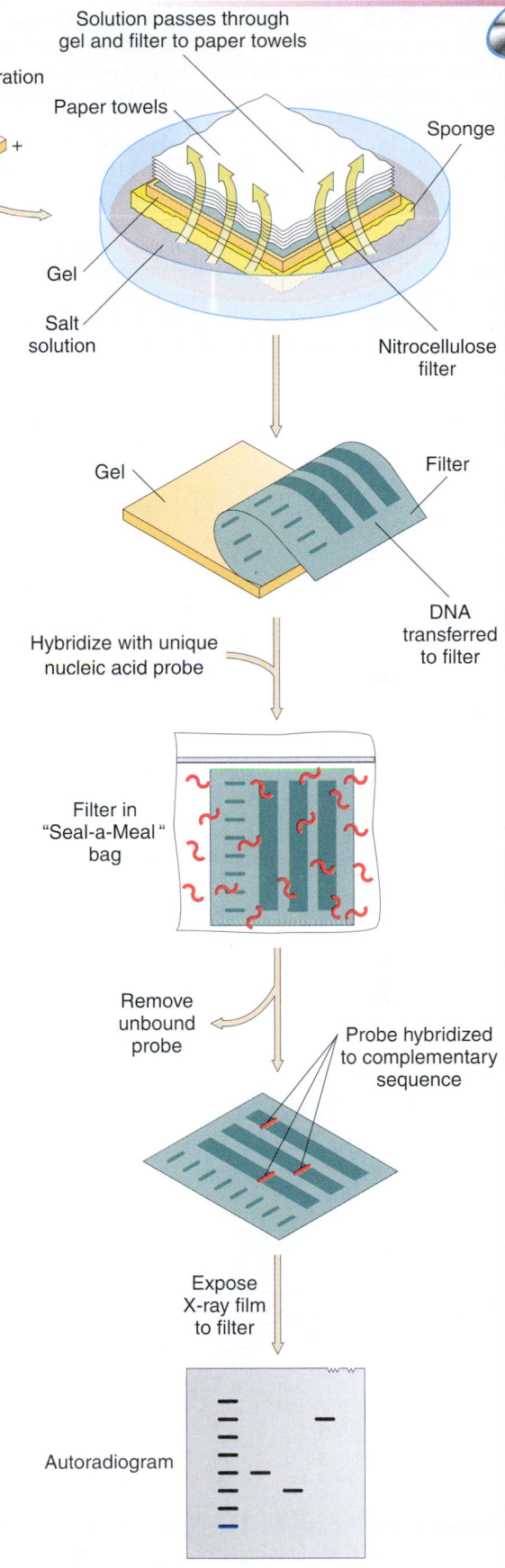

Figure 12-18   Gel electrophoresis and blotting in identifying specific cloned genes. RNA or DNA restriction fragments are applied to an agarose gel and electrophoresed. The various fragments migrate at differing rates according to their respective sizes. The gel is placed in buffer and covered by a nitrocellulose filter and a stack of paper towels. The fragments are denatured to single strands so that they can stick to the filter. They are carried to the filter by the buffer, which is wicked up by the towels. The filter is then removed and incubated with a radioactively labeled single-stranded probe that is complementary to the targeted sequence. Unbound probe is washed away, and X-ray film is exposed to the filter. Because the radioactive probe has hybridized only with its complementary restriction fragments, the film will be exposed only in bands corresponding to those fragments. Comparison of these bands with labeled markers reveals the number and size of the fragments in which the targeted sequences are found. This information can be used to position the sequences along restriction maps. This procedure is termed *Southern blotting* when DNA is transferred to nitrocellulose and *Northern blotting* when RNA is transferred. (From J. D. Watson, M. Gilman, J. Witkowski, and M. Zoller, *Recombinant DNA,* 2d ed. Copyright © 1992 by Scientific American Books.)

shows a smear of DNA. A probe can identify one fragment in this mixture, with the use of a technique developed by E. M. Southern called **Southern blotting.** (Recall that this technique, and the parallel technique to be detailed below, were introduced briefly in Chapter 1.) After DNA fragments are fractionated on the gel, an absorbent membrane is laid over the gel and the DNA bands are transferred ("blotted") onto the membrane by capillary action. When transferred to the membrane, the DNA bands stay in the same relative positions as on the gel. The membrane is bathed in a labeled probe, and an autoradiogram is used to reveal the presence of any bands on the gel that are homologous to the probe. If appropriate, those bands can be cut out of the gel and further processed. The gel can be calibrated for DNA fragment size by running a standard "ladder" of fragments of known size on the same gel. Hence the sizes of any interesting fragments in the experimental sample can be inferred. Figure 12-18 shows the general procedure of Southern blotting. Figure 12-19 shows an application in which a cloned fragment of a fungal plasmid was used to detect homologous plasmids in a variety of strains.

The Southern blotting technique can be extended to detect a specific *RNA* molecule from a mixture of RNAs fractionated on a gel. This technique is called **Northern blotting** to contrast it with the Southern technique for *DNA* analysis. The fractionated RNA is blotted onto a membrane and probed in the same way as in Southern blotting. One

Figure 12-19   An example of a Southern analysis. The DNA from the mitochondria of 11 strains of *Neurospora* from different geographical locations was electrophoresed, and the gel was stained with ethidium bromide. The bright bands are linear plasmids, and the positions of some of these plasmids are marked. The lane marked "L" is a calibration ladder composed of DNA fragments of known size. The sizes of the rungs visible in this figure are (from the bottom up) 1.5 kb, 2 kb, and then one additional kilobase for each rung, up to 12 kb. The plasmid in strain 11 was cloned and used as a probe in a Southern hybridization. The autoradiogram shows that several other plasmids have sequences homologous to the strain 11 plasmid. Indeed, strain 7 has a family of three plasmids, all homologous to the strain 11 plasmid. (From Simon Yang.)

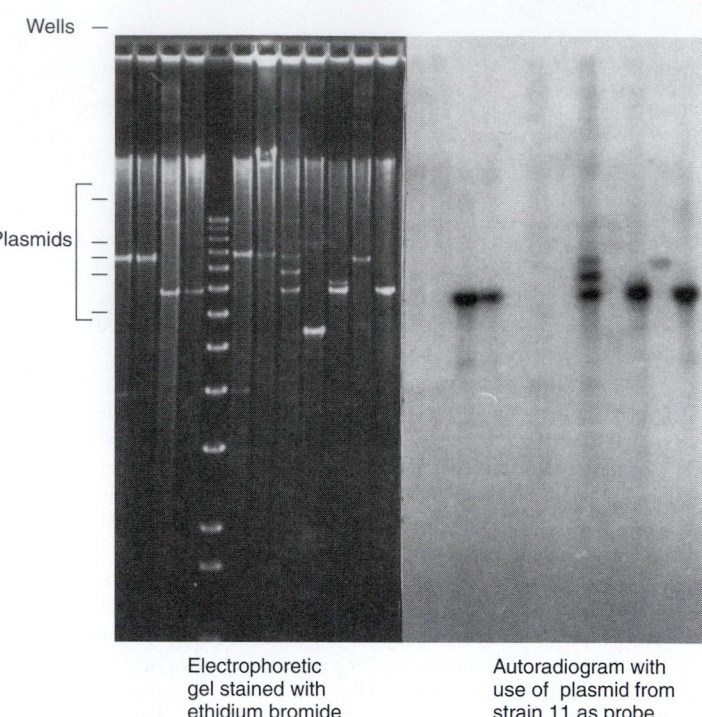

Electrophoretic gel stained with ethidium bromide

Autoradiogram with use of plasmid from strain 11 as probe

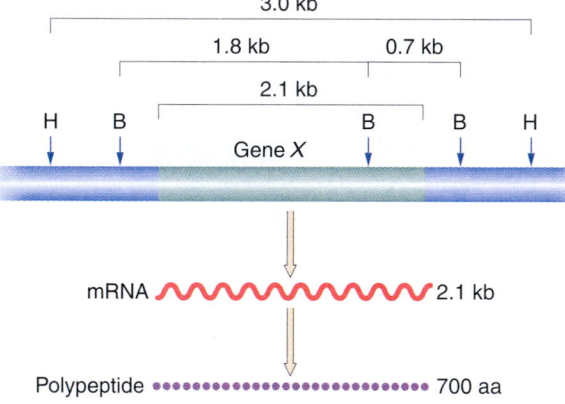

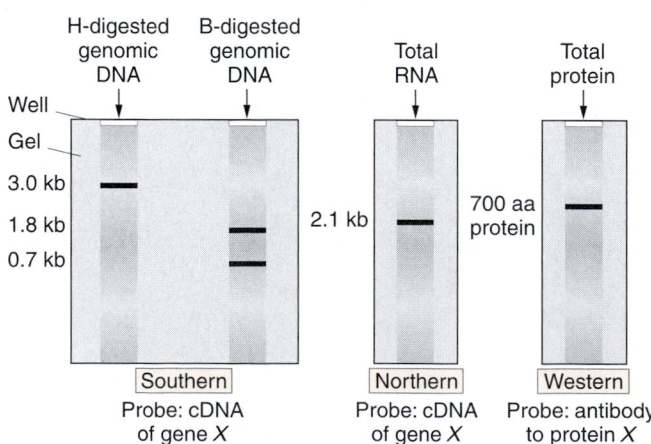

Figure 12-20   Comparison of Southern, Northern, and Western analyses of gene *X*. B and H represent target sites for restriction enzymes B and H. (**Note:** Positions of bands on Southern blot cannot be compared with positions of bands on Northern and Western gels.)

application of Northern analysis is in determining whether a specific gene is transcribed in a certain tissue or under certain environmental conditions. RNA is extracted from the appropriate cell sample and then electrophoresed, blotted, and probed with the cloned gene in question. A positive signal shows the presence of the transcript.

Hence we see that cloned DNA finds widespread application as a probe for detecting either a specific clone or a specific DNA fragment or a specific RNA. In all these cases, the ability of nucleic acids with complementary nucleotide sequences to find and bind to each other in solution is being exploited.

A parallel technique called **Western blotting** has been developed to transfer electrophoretically fractionated *protein* from a gel and then to visualize specific proteins with antibodies. However, note that the probe here is not DNA but a labeled antibody. These three techniques are very powerful experimental tools in detecting and sizing specific macromolecules in molecular genetics. They are compared in Figure 12-20.

MESSAGE ·····

DNA or RNA fractionated by size in an electrophoretic gel can be blotted onto an absorbent membrane, which in turn can be probed to detect the position of specific fragments.

DNA sequence determination

Success in cloning a desired gene is merely the beginning of a second round of analysis in which the aim is to characterize the structure and function of that gene. The gene's nucleotide sequence is the ultimate characterization of its genetic structure. One of the requirements for DNA sequencing is the ability to obtain defined fragments of DNA. Thus, there is a strong interdependence of DNA cloning and DNA sequencing technology, inasmuch as DNA cloning provides amplified samples of defined DNA fragments.

Any method of DNA sequencing starts with a population of a defined fragment of DNA labeled at one end. From this population, sets of molecules are generated that differ in size by one base at the other end. These molecules are then fractionated. Fractionation is by acrylamide or agarose gel electrophoresis of single-stranded DNA molecules. The mobility of a strand is inversely proportional to the logarithm of its length. This technique is so sensitive that fragments differing in length by only a single nucleotide can be separated. The base at the truncated end of each of the fractionated molecules is determined and hence the nucleotide sequence is established.

The sequencing method now most commonly used was developed by Fred Sanger. His method is based on DNA synthesis in the presence of dideoxy nucleotides, which differ from normal deoxynucleotides in that they lack a 3'-hydroxyl group (Figure 12-21). The respective dideoxynucleotide triphosphates (ddNTPs) can be incorporated into a growing chain, but, when incorporated, they terminate synthesis because they lack the 3'-hydroxyl group necessary to bond with the next nucleotide triphosphate. Each of four reaction tubes is prepared with a single-stranded DNA template for the sequence of interest, plus DNA polymerase and a section of labeled primer. Each tube receives a small amount of a different ddNTP (ddATP, ddTTP, ddCTP, or ddGTP), together with the four normal deoxynucleotide triphosphates (dNTPs). A dideoxynucleotide will be incorporated randomly, at different sites in different syntheses in the reaction tube. Therefore, in any given tube, various truncated chain lengths will be produced, each corresponding to the point at which the respective ddNTP for that tube was incorporated and terminated chain growth. These lengths, in turn, are a clear indication of where the bases complementary to the ddNTPs are on the template strand. Because incorporation is random, all possible truncated fragments will be produced, corresponding to all the various positions of that particular base. The fragments can be visualized by electrophoresis of the four samples in

four lanes on an acrylamide gel, where the fragments form bands. The base sequence can be determined by scanning up the gel, encompassing all four lanes, and recording whichever base occupies the terminus in the next band. The procedure is illustrated in Figures 12-22 and 12-23.

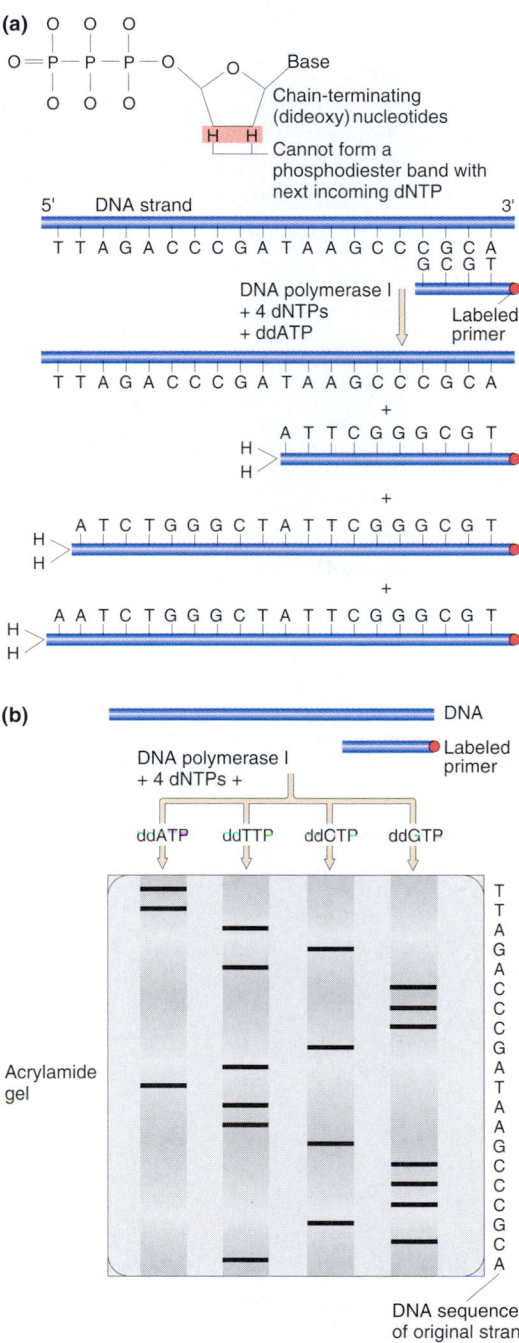

Figure 12-22 The dideoxy sequencing method. (a) A labeled primer (designed from the flanking vector sequence) is used to initiate DNA synthesis. The addition of four different dideoxynucleotides (ddATP is shown here) randomly arrests synthesis. (b) The resulting fragments are separated electrophoretically and subjected to autoradiography. (From J. D. Watson, M. Gilman, J. Witkowski, and M. Zoller, *Recombinant DNA*, 2d ed. Copyright © 1992 by Scientific American Books.)

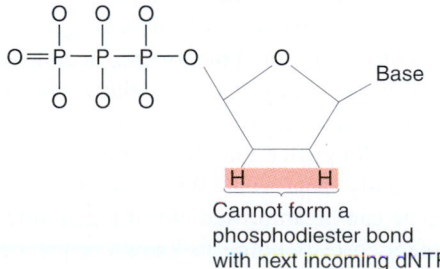

Figure 12-21 The structure of 2',3'-dideoxynucleotides, which are employed in the Sanger DNA-sequencing method.

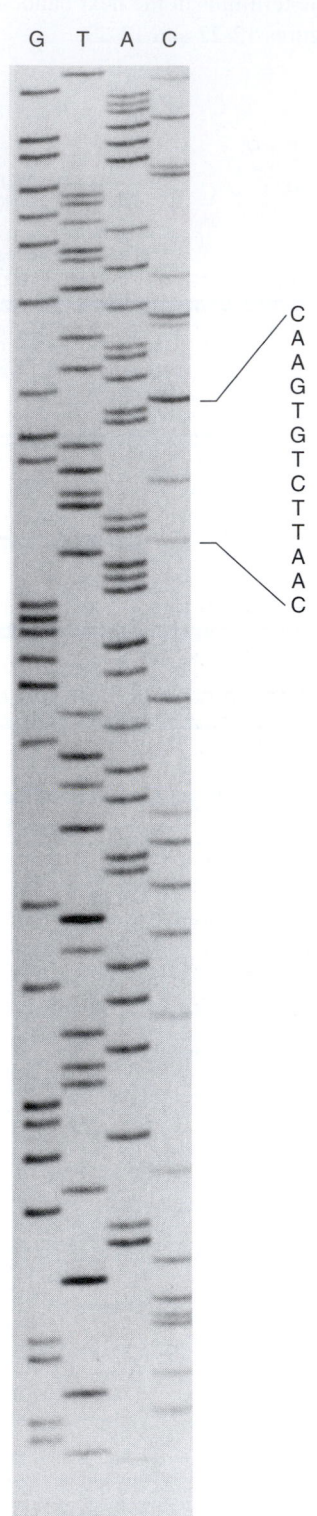

G T A C

C
A
A
G
T
G
T
C
T
T
A
A
C

Figure 12-23   Sanger sequencing gel. The inferred sequence is shown at the right. (Photomicrograph from Loida Escote-Carlson.)

Instead of radioactive labels, the tags attached to oligonucleotide primers can be fluorescent dyes. A different fluorescent color emitter is used for each of the four reactions, and the four mixtures are electrophoresed together. Fluorescence detection is used in automated DNA sequencing machines, which can read as many as 1000 bases in one separation. Figure 12-24 illustrates a readout of automated sequencing.

MESSAGE ··

A cloned DNA fragment can be sequenced by generating a set of labeled single-stranded DNAs differing in length by one nucleotide; when these DNAs are electrophoresed, the nucleotide sequence can be read directly from an autoradiogram of the gel.

···

The nucleotide sequence of a cloned DNA fragment can be used to find the cloned gene or genes that it contains. The nucleotide sequence is fed into a computer, which then scans all six reading frames (three in each direction) in the search for a gene-sized stretch of DNA beginning with an ATG initiation codon and ending with a stop codon. These stretches are called **open reading frames (ORFs).** They represent sequences that are candidate genes. Figure 12-25 shows such an analysis in which two candidate genes have been identified as ORFs. It is interesting to note that ORF detection and Mendelian progeny ratio analysis, although poles apart in their approaches, are both ways of identifying genes.

Detecting and amplifying sequences by the polymerase chain reaction

If a region of DNA has already been cloned and sequenced, the sequence can be used to retrieve parts of the equivalent region from a specific individual organism or from other species, all without cloning. The technique uses a procedure called the **polymerase chain reaction (PCR).** Figure 12-26 illustrates the principle of the technique. A temperature-resistant DNA polymerase, *Taq* polymerase, from the bacterium *Thermus aquaticus* is used to catalyze growth from DNA primers. Pairs of primers on opposite strands are extended toward each other, as shown in Figure 12-26. After completion of the replication of the segment between the two primers (one cycle), the two new duplexes are heat denatured to generate single-stranded templates, and a second cycle of replication is carried out by lowering the temperature in the presence of all the components necessary for the polymerization. Repeated cycles of synthesis and denaturation result in an exponential increase in the number of segments replicated. Amplifications by as much as a million-fold can be readily achieved. With the use of PCR, a single-copy gene can be amplified out of a genomic sample, provided primers corresponding to known sequences of the gene can be synthesized. Because of the exponential amplification, PCR is very sensitive and can detect target se-

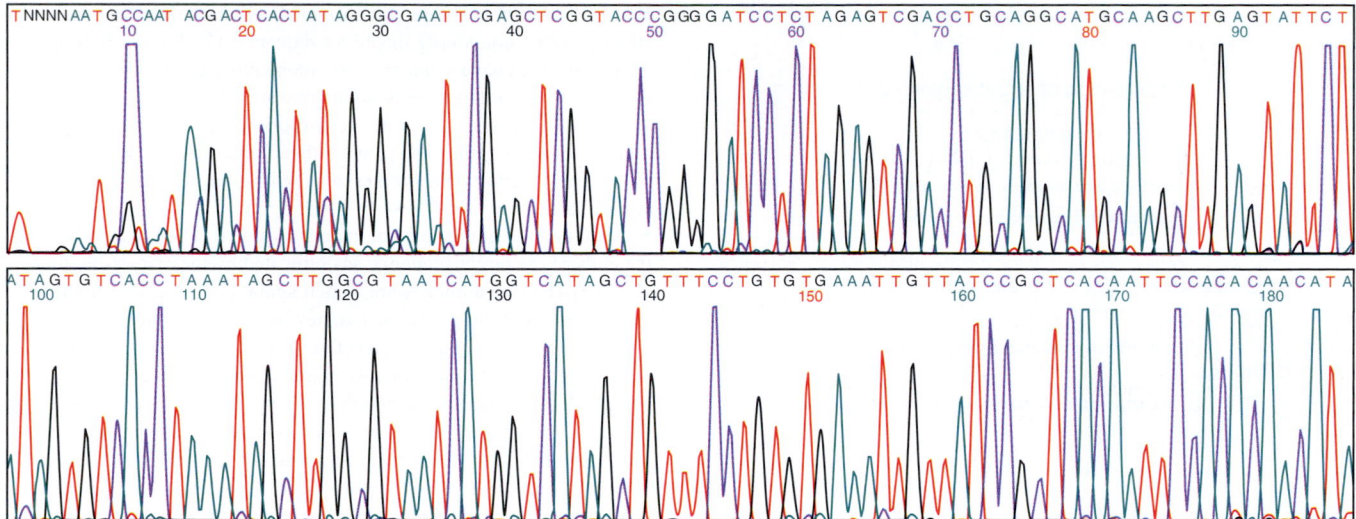

Figure 12-24 Printout from an automatic sequencer that uses fluorescent dyes. N represents a base that cannot be assigned.

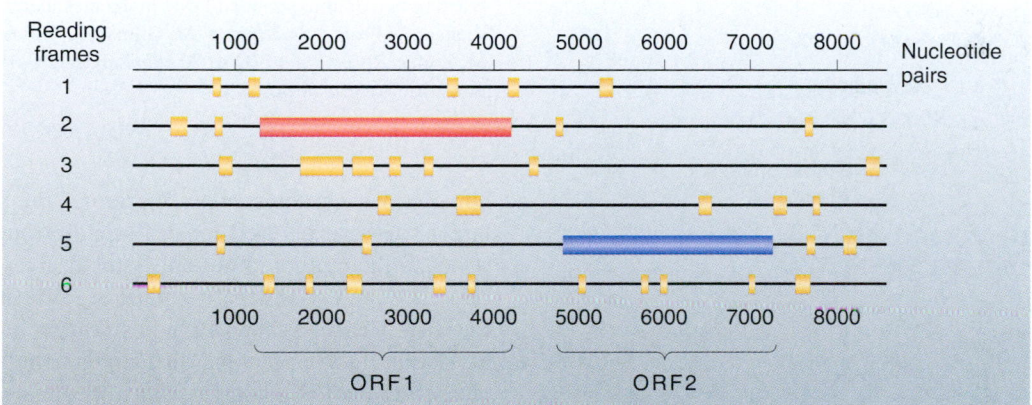

Figure 12-25 Any piece of DNA has six possible reading frames, three in each direction. Here the computer has scanned a 9-kb fungal plasmid sequence in looking for ORFs (potential genes). Two large ORFs, 1 and 2, are the most likely candidates as potential genes.

quences that are in extremely low copy number in a sample. For example, segments of human DNA can be amplified by using just the few follicle cells surrounding a single pulled-out hair. PCR is generally useful in DNA diagnostics — in other words, in checking for the presence of a gene or for the mutational state of a specific gene or, preparatively, in amplifying a defined segment.

Note that, in the PCR technique, no restriction digestion of the substance DNA is needed, because the primers will home in on the appropriate sequence of native DNA. Furthermore, no lengthy cloning procedures are necessary, because enough DNA is amplified that a clear band on a gel is produced. In addition, only very small amounts of substrate DNA are needed.

> **MESSAGE** ·
> The polymerase chain reaction uses specially designed primers to amplify specific short regions of DNA without cloning.

Locating genes
on restriction maps

We have already seen the importance of restriction enzymes in cutting DNA for use in cloning. Another useful feature of restriction enzymes is that the positions of their target sites along a DNA molecule (that is, along a chromosome) can be used as DNA markers. Even though the sites exist by chance, they are generally in the same positions on

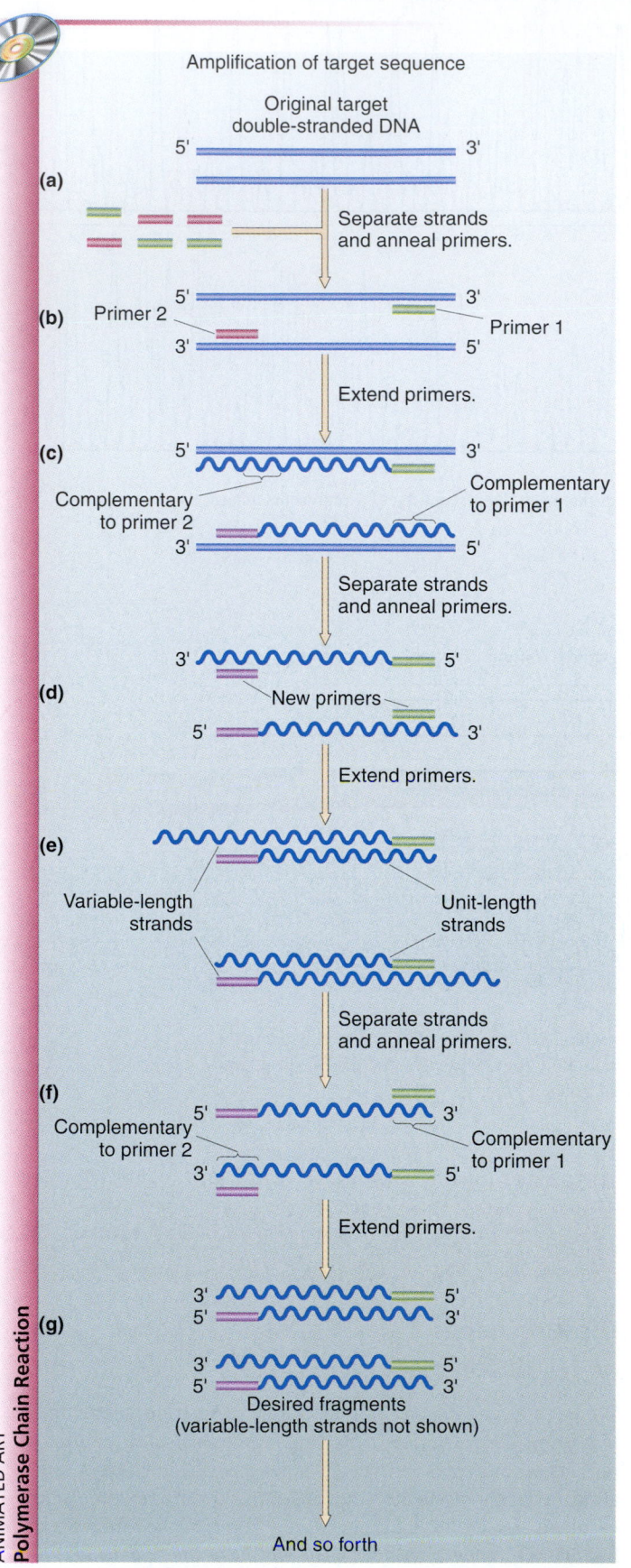

Amplification of target sequence

Original target
double-stranded DNA

(a)

Separate strands
and anneal primers.

(b) Primer 2 / Primer 1

Extend primers.

(c) Complementary to primer 2 / Complementary to primer 1

Separate strands
and anneal primers.

(d) New primers

Extend primers.

(e) Variable-length strands / Unit-length strands

Separate strands
and anneal primers.

(f) Complementary to primer 2 / Complementary to primer 1

Extend primers.

(g) Desired fragments
(variable-length strands not shown)

And so forth

Figure 12-26 The polymerase chain reaction. (a) Double-stranded DNA containing the target sequence. (b) Two chosen or created primers have sequences complementing primer-binding sites at the 3′ ends of the target gene on the two strands. The strands are separated by heating, allowing the two primers to anneal to the primer-binding sites. Together, the primers thus flank the targeted sequence. (c) *Taq* polymerase then synthesizes the first set of complementary strands in the reaction. These first two strands are of varying length, because they do not have a common stop signal. They extend beyond the ends of the target sequence as delineated by the primer-binding sites. (d) The two duplexes are heated again, exposing four binding sites. (For simplicity, only the two new strands are shown.) The two primers again bind to their respective strands at the 3′ ends of the target region. (e) *Taq* polymerase again synthesizes two complementary strands. Although the template strands at this stage are variable in length, the two strands just synthesized from them are precisely the length of the target sequence desired. This is because each new strand begins at the primer-binding site, *at one end of the target sequence,* and proceeds until it runs out of template, *at the other end of the sequence.* (f) Each new strand now begins with one primer sequence and ends with the primer-binding sequence for the other primer. Subsequent to strand separation, the primers again anneal and the strands are extended to the length of the target sequence. (The variable-length strands from part c are also producing target-length strands.) (g) The process can be repeated indefinitely, each time creating two double-stranded DNA molecules identical with the target sequence. (From J. D. Watson, M. Gilman, J. Witkowski, and M. Zoller, *Recombinant DNA,* 2d ed. Copyright © 1992 by Scientific American Books.)

homologous chromosomes. Therefore, the positions of the target sites can be used much like milestones along a road, which, although not important in themselves, can be used as reference points for locating more significant features along the road. Hence, a map of the positions of restriction sites is a valuable tool in genome analysis generally. Most often, a restriction map is made for a localized chromosomal region that is of particular interest or for a relatively small chromosome such as that of a plasmid or an organelle.

One way of making a restriction map is to compare single-enzyme digests with double digests. Two restriction enzymes are applied in separate digestions, and then the two enzymes are used together. After cutting, the fragment sizes are determined by electrophoresis. The double digest shows whether a fragment produced by one enzyme contains sites for the other; if so, the fragment disappears and is replaced by two or more smaller fragments. Comparison of the sizes of fragments produced in the different digestions allows an approximate localization of the restriction-enzyme target sites. The procedure is illustrated in Figure 12-27.

Another method for making restriction maps is shown in Figure 12-28. In this method, a piece of DNA is labeled at each 5′ end with ^{32}P and cut to generate fragments with the label at one end. The longest fragment is isolated and digested with a second restriction enzyme, but the reaction is not allowed to go to completion; therefore, a population of variously sized labeled fragments is produced, each with one

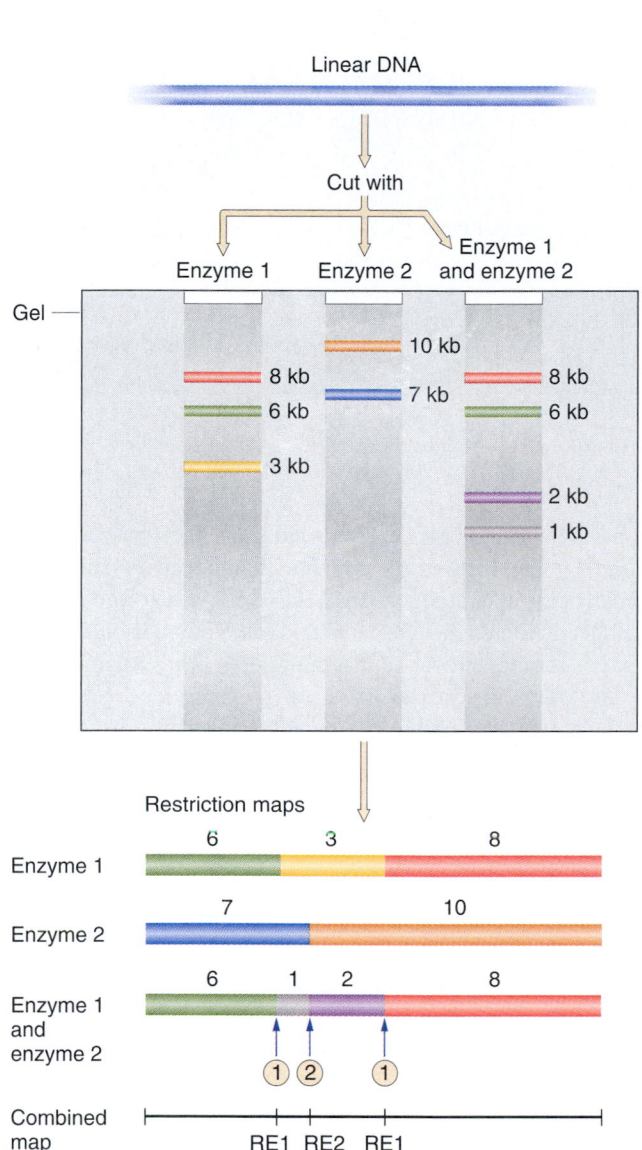

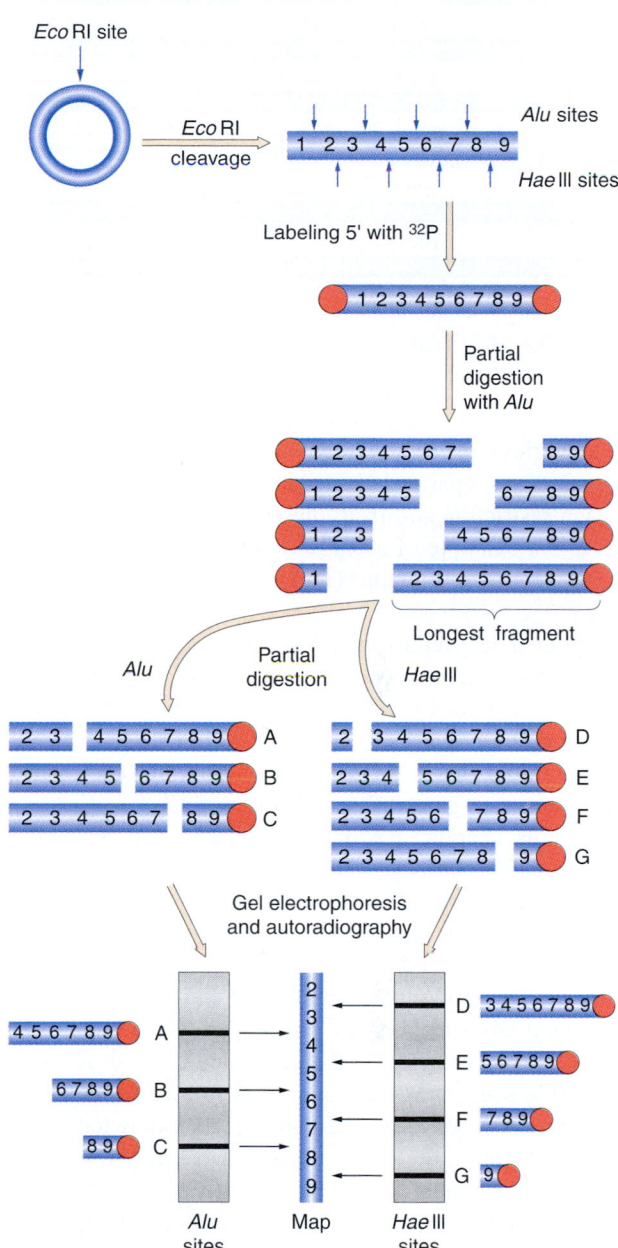

Figure 12-27   Restriction mapping by comparing electrophoretic separations of single and multiple digests. In this simplified example, digestion with enzyme 1 shows that there are two restriction sites for this enzyme but does not reveal whether the 3-kb segment is in the middle or on the end of the digested sequence, which is 17 kb long. Combined digestion by both enzyme 1 and enzyme 2 leaves the 6- and 8-kb segments intact but cleaves the 3-kb fragment, showing that enzyme 2 cuts at a site within this enzyme 1 fragment. If the 3-kb section were on the outside of the sequence being studied, digestion by enzyme 2 alone would yield a 1- or 2-kb fragment. Because this is not the case, of the three restriction fragments produced by enzyme 1, the 3-kb fragment must lie in the middle. That the restriction enzyme 2 (RE2) site lies closer to the 6-kb section than the 8-kb section can be inferred from the 7- and 10-kb lengths of the enzyme 2 digestion.

Figure 12-28   Identification of the sequence of restriction-enzyme sites in a circular DNA molecule. In this example, the ring of DNA with a single *Eco*RI site is opened into a linear molecule by cleavage with *Eco*RI. The molecule is labeled on its 5′ end with ^{32}P and then partly digested with *Alu* (restriction enzyme from *Arthrobacter lutens*, which could just as easily be *Hae*III, the restriction enzyme from *Haemophilus aegyptius*) to produce fragments of varying lengths. Each fragment now carries the radioactive label at only one end. The longest fragment is selected (after testing on a gel to compare lengths) and is divided into two samples. One sample is partly digested with *Alu*, and the other is partly digested with *Hae*III. The new sets of fragments are separated on gels. The location of the labeled fragment can be determined by autoradiography of the gels. The sequence of restriction sites can now be read in order as the two gels are compared.

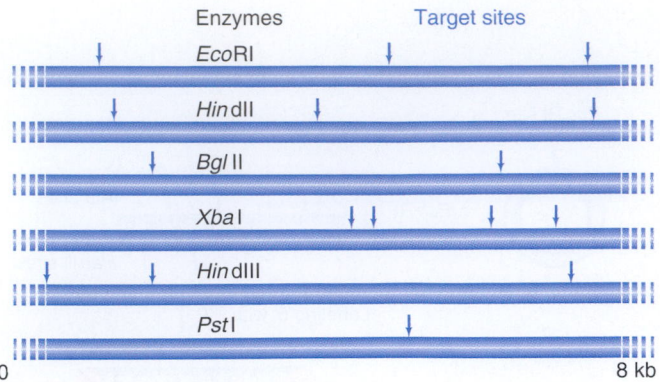

Figure 12-29 Restriction map of a linear fungal plasmid, showing sites for six different enzymes.

Figure 12-30 Archibald Garrod.

of the various restriction sites at the nonlabeled end. These fragments are separated electrophoretically, and the distance of each restriction site from the labeled end can thus be deduced. An example of a restriction map is shown in Figure 12-29. This map is of the DNA of a linear plasmid from a fungus.

After a restriction map has been made, genes can be placed on the map by using Southern analysis. For example, if a cloned gene hybridizes with a fragment A produced from restriction enzyme 1 and with fragment B produced by restriction enzyme 2, then the gene must be located in the region where fragments A and B overlap.

In this chapter, we have introduced the fundamental techniques that have revolutionized genetics. In the next section, we describe the genetic analysis of the disease alkaptonuria, which incorporates and integrates many of the techniques introduced in this chapter.

Genes are no longer hypothetical entities. Thanks to recombinant DNA technology and related technologies, genes can be isolated in a test tube and characterized as specific nucleotide sequences. But even this is not the end of the story. We shall see in the next chapter that knowledge of the sequence is often the beginning of a fresh round of genetic manipulation of that sequence, giving rise to sophisticated new approaches for altering an organism's phenotype.

A century of genetic research on alkaptonuria

The story of the cloning of the human gene defective in alkaptonuria brings together many of the techniques introduced in this chapter. Alkaptonuria is a disease with several symptoms, of which the most conspicuous is that the urine turns black when exposed to air. In 1898, an English doctor named Archibald Garrod (Figure 12-30) showed that the substance responsible is homogentisic acid, which is excreted in abnormally large amounts into the urine of alkaptonuria patients. In 1902, early in the post-Mendelian era, Garrod suggested, on the basis of pedigree patterns, that alkaptonuria is inherited as a Mendelian recessive (Figure 12-31).

Soon after, in 1908, he proposed that the disorder was caused by the lack of an enzyme that normally splits the aromatic ring of homogentisic acid to convert it into maleylacetoacetic acid. Because of this enzyme deficiency, he reasoned, homogentisic acid accumulates. Thus alkaptonuria is among the earliest proposed cases of an "inborn error of metabolism," an enzyme deficiency caused by a defective gene. There was a 50-year delay before it was shown by others that, in the liver of patients with alkaptonuria, activity for the enzyme that normally splits homogentisic acid, an enzyme called homogentisate 1,2-dioxygenase (HGO), is indeed totally absent. It seemed likely that the enzyme HGO was normally encoded by the alkaptonuria gene. In 1992, the alkaptonuria gene was mapped genetically to band 2 of the long arm of chromosome 3 (band 3q2).

In 1995, Jose Fernandez-Canon and colleagues characterized a gene encoding HGO activity in the fungus *Aspergillus nidulans,* and, in 1996, they used the deduced amino acid sequence of this gene to do a computer search through a large number of sequenced fragments of a human cDNA library. They identified a positive clone that contained a human gene that encodes 445 amino acids, which showed 52 percent similarity to the *Aspergillus* gene. When the human gene was expressed in an *E. coli* expression vector, it showed HGO activity. Furthermore, when the gene was used as a probe in a Northern analysis of liver RNA, a single RNA of the expected size was hybridized.

When the cloned gene was used as a probe on chromosomes in which the DNA had been partly denatured (in situ hybridization — see Chapter 14), the probe bound to band 3q2, showing that it was indeed the gene for alkaptonuria.

The cDNA clone was used to recover the full-length gene from a genomic library. The gene was found to have 14 exons and spanned a total of 60 kb. A family of seven in which three children suffered from alkaptonuria was tested for mutations. PCR analysis was used to amplify all the exons individually. The amplified products were sequenced. One parent was heterozygous for a proline → serine substitution at position 230 in exon 10 (mutation P230S). The other parent was heterozygous for a valine → glycine substitution at position 300 in exon 12 (mutation V300G). All three children with alkaptonuria were of the constitution P230S V300G, as expected if these were the mutant sites inactivating the HGO enzyme.

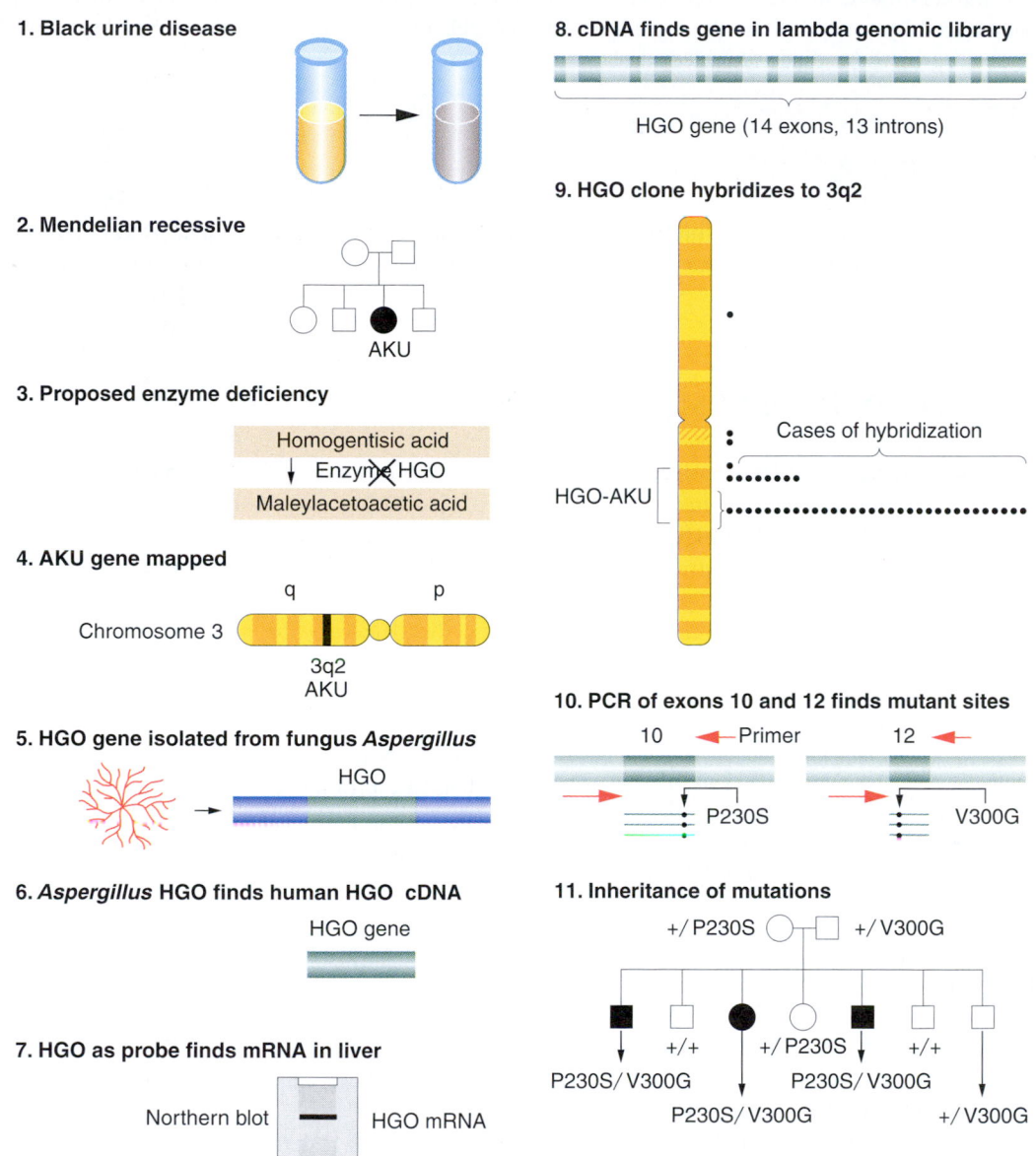

Figure 12-31 The analysis of alkaptonuria (black urine disease).

SUMMARY

Genetics focuses on the nature of genes, and a major goal is to characterize their structure and function. Recombinant DNA technology has allowed individual genes to be isolated in a test tube and then characterized at the molecular level. The technology is based on restriction enzymes, which cut DNA into defined fragments. Restriction target sites can be mapped and act as DNA landmarks. Restriction fragments often have sticky ends, enabling them to be inserted into a vector capable of replicating in a bacterial cell. Such molecular hybrids are known as recombinant DNA. Bacteria amplify a single recombinant DNA molecule to form a DNA clone. Common vectors are plasmids, phages, and cosmids. An entire genome can be cloned in a set of clones known as a library. A specific clone can be found in a library by using a probe that specifically binds to the DNA or to the protein of the desired clone. Specific clones can also be isolated by their ability to transform null mutants. Tagging also is useful for cloning a gene: transforming DNA or a transposon is used to cause a mutation by insertion, and the DNA adjacent to the tag is isolated. Chromosome walking provides a way of isolating a gene by sequential isolation of overlapping clones, starting from a marker linked to the desired gene. Cloned DNA can be sequenced by several methods, including the arrest of DNA chain growth by dideoxynucleotides.

The polymerase chain reaction uses primers to amplify DNA sequences. It is a way of rapidly isolating DNA whose structure is already partly sequenced and of detecting small amounts of one specific type of DNA. Gel electrophoresis separates variously sized DNA or RNA molecules from a mixture. Probes can detect specific DNA or RNA molecules on the gel, in procedures known as Southern and Northern analyses, respectively.

CONCEPT MAP

Draw a concept map interrelating as many of the following terms as possible. Note that the terms are listed in no particular order.

restriction enzyme / recombinant DNA / DNA clone / probe / transformation / vector / complementation / sequencing / library

CHAPTER INTEGRATION PROBLEM

In Chapter 10, we studied the structure of tRNA molecules. Suppose that you want to clone a fungal gene that encodes a certain tRNA. You have a sample of the purified tRNA and an *E. coli* plasmid that contains a single *Eco*RI cutting site in a *tet*^{R} (tetracycline-resistance) gene, as well as a gene for resistance to ampicillin (*amp*^{R}). How can you clone the gene of interest?

◆ Solution ◆

You can use the tRNA itself or a cloned cDNA copy of it to probe for the DNA containing the gene. One method is to digest the genomic DNA with *Eco*RI and then mix it with the plasmid, which you also have cut with *Eco*RI. After

transformation of an *amp*^{R} *tet*^{S} recipient, Amp^R colonies are selected, indicating successful transformation. Of these Amp^R colonies, select the colonies that are Tet^S. These Tet^S colonies will contain vectors with inserts, and a great number of them are needed to make the library. Test the library by using the tRNA as the probe. You can examine those colonies hybridizing with the probe further to find out how much of the tRNA sequence they contain.

Alternatively, you can subject *Eco*RI-digested DNA from organism X to gel electrophoresis and then identify the correct band by probing with the tRNA. This region of the gel can be cut out and used as a source of enriched DNA to clone into the plasmid cut with *Eco*RI.

SOLVED PROBLEM

1. The restriction enzyme *Hin*dIII cuts DNA at the sequence AAGCTT, and the restriction enzyme *Hpa*II cuts DNA at the sequence CCGG. On average, how frequently will each enzyme cut double-stranded DNA? (In other words, what is the average spacing between restriction sites?)

◆ Solution ◆

We need to consider only one strand of DNA, because both sequences will be present on the opposite strand at the same site owing to the symmetry of the sequences. The frequency of the six-base-long *Hin*dIII sequence is $(1/4)^6 = 1/4096$, because there are four possibilities at each of the six positions. Therefore, the average spacing between *Hin*dIII sites is approximately 4 kb. For *Hpa*II, the frequency of the four-

base-long sequence is $(1/4)^4$, or 1/256. The average spacing between *Hpa*II sites is approximately 0.25 kb.

2. From Table 12-1, determine whether the *Eco*RI enzyme or the *Sma*I enzyme would be more useful for cloning. Explain your answer.

◆ Solution ◆

Both restriction enzymes recognize a six-base-pair sequence, so both would be expected to have approximately the same number of recognition sites per genome. The major difference between the two is that *Eco*RI leaves staggered ends, whereas *Sma*I leaves blunt ends. Staggered ends are much easier to manipulate during cloning because of the base-pairing capacity inherent in them. (Solution from Diane K. Lavett.)

PROBLEMS

1. A circular bacterial plasmid (pBP1) has a single *Hin*dIII restriction-enzyme site in the middle of a tetracycline-resistance gene (*tet*^{R}). Fruit fly genomic DNA is digested with *Hin*dIII, and a library is made in pBP1. Probing reveals that clone 15 contains a specific *Drosophila* gene of interest. Clone 15 is studied by restriction analysis with *Hin*dIII and another restriction enzyme, *Eco*RV. The ethidium bromide-stained electrophoretic gel shows bands as illustrated in the diagram below (the control was an uninserted plasmid pBP1). The sizes of the bands (in kilobases) are shown alongside. (**Note:** Circular molecules do not give intense bands on this type of gel, so you can assume that all bands represent linear molecules.)

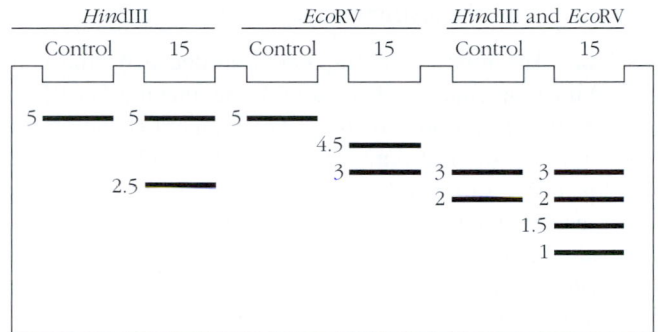

a. Draw restriction maps for plasmid pBP1 with and without the insert, showing the sites of the target sequences and the approximate position of the *tet*^{R} gene.

b. If the same *tet*^{R} gene cloned in a completely nonhomologous vector is made radioactive and used as a probe in a Southern blot of this gel, which bands do you expect to appear radioactive on an autoradiogram?

c. If the same gene of interest from a fly closely related to *Drosophila* has been cloned in a nonhomologous vector and this clone is used as a probe for the same gel, what bands do you expect to see on the autoradiogram?

 ## Unpacking the Problem

1. Which plasmid described in this chapter seems closest in type to pBP1?

2. What is the importance of the single *Hin*dIII restriction site?

3. Why is it important that the single site is in a resistance gene? Would it be useful if not?

4. What is the effect of insertion of donor DNA into the resistance gene? Is this effect important for this problem?

5. What is a library? What type was used in this experiment and does it matter for this problem?

6. What kind of probing would have shown that clone 15 contains the gene of interest? And is this relevant to the present problem?

7. What is an electrophoretic gel?

8. What function does ethidium bromide serve in this experiment?

9. Does the gel shown represent a Southern or a Northern blot or neither?

10. Genetically, what types of molecules are visible on the gel?

11. How many fragments are produced if a circular molecule is cut once?

12. How many fragments are produced if a circular molecule is cut twice?

13. Can you write a simple formula relating the number of restriction-enzyme sites in a circular molecule to the number of fragments produced?

14. If one enzyme produces *n* fragments and another produces *m* fragments, how many fragments are produced if both enzymes are used?

15. In the diagram, at what positions were the DNA samples loaded into the gel?

16. Why are the smaller-molecular-weight fragments at the bottom of the gel?

17. What is the total molecular weight of the fragments in all the lanes? What patterns do you see?

18. Is it a coincidence that the 3- and 2-kb fragments together equal the 5-kb fragment in size?

19. Is it a coincidence that the 1.5- and 1-kb fragments together equal the 2.5-kb fragment in size?

20. If a fragment produced by one enzyme disappears when the DNA is treated with that same enzyme plus another enzyme, what does this signify?

21. What determines whether a probe will hybridize to a DNA blot (denatured)?

22. In part c, why is it stressed that a nonhomologous vector is used?

Now attempt to solve the problem.

2. The restriction enzyme *Eco*RI cuts DNA at the sequence GTTAAC, and the enzyme *Hae*III cuts DNA at

the sequence GGCC. On average, how frequently will each enzyme cut double-stranded DNA? (In other words, what is the average spacing between restriction sites?)

3. You have a purified DNA molecule, and you wish to map restriction-enzyme sites along its length. After digestion with *Eco*RI, you obtain four fragments: 1, 2, 3, and 4. After digestion of each of these fragments with *Hind*II, you find that fragment 3 yields two subfragments (3_1 and 3_2) and that fragment 2 yields three (2_1, 2_2, and 2_3). After digestion of the entire DNA molecule with *Hind*II, you recover four pieces: A, B, C, and D. When these pieces are treated with *Eco*RI, piece D yields fragments 1 and 3_1, A yields 3_2 and 2_1, and B yields 2_3 and 4. Piece C is identical with 2_2. Draw a restriction map of this DNA.

4. After *Drosophila* DNA has been treated with a restriction enzyme, the fragments are attached to plasmids and selected as clones in *E. coli*. By using this "shotgun" technique, David Hogness recovered every DNA sequence of *Drosophila* in a library.

   a. How would you identify the clone that contains DNA encoding the protein actin, whose amino acid sequence is known?

   b. How would you identify a clone encoding a specific tRNA?

5. You have isolated and cloned a segment of DNA that is known to be a unique sequence in the genome. It maps near the tip of the X chromosome and is about 10 kb in length. You label the 5′end with ^{32}P and cleave the molecule with *Eco*RI. You obtain two fragments: one is 8.5 kb long; the other is 1.5 kb. You separate the 8.5-kb fragment into two fractions, partly digesting one with *Hae*III and the other with *Hind*II. You then separate each sample on an agarose gel. By autoradiography, you obtain the following results:

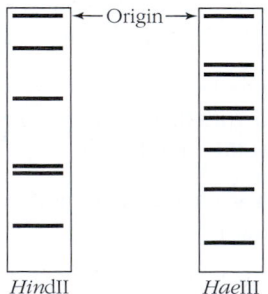

Draw a restriction-enzyme map of the complete 10-kb molecule.

6. Calculate the average distances (in nucleotide pairs) between the restriction sites in organism X for the following restriction enzymes:

*Alu*I	5′	AGCT	3′
	3′	TCGA	5′
*Eco*RI	5′	GAATTC	3′
	3′	CTTAAG	5′
*Acy*I	5′	G Pu CG Py C	3′
	3′	C Py GC Pu G	5′

(**Note:** Py = any pyrimidine; Pu = any purine.)

7. A linear fragment of DNA is cleaved with the individual restriction enzymes *Hind*III and *Sma*I and then with a combination of the two enzymes. The fragments obtained are:

*Hind*III	2.5 kb, 5.0 kb
*Sma*I	2.0 kb, 5.5 kb
*Hind*III and *Sma*I	2.5 kb, 3.0 kb, 2.0 kb

a. Draw the restriction map.

b. The mixture of fragments produced by the combined enzymes is cleaved with the enzyme *Eco*RI, resulting in the loss of the 3-kb fragment (band stained with ethidium bromide on an agarose gel) and the appearance of a band stained with ethidium bromide representing a 1.5-kb fragment. Mark the *Eco*RI cleavage site on the restriction map.

(Problem 7 courtesy of Joan McPherson. From A. J. F. Griffiths and J. McPherson, *100 + Principles of Genetics*. W. H. Freeman and Company, 1989.)

8. A viral DNA fragment carrying a specific gene *V* is introduced into a muscle cell culture by transformation. After incubation with ^{32}P-labeled ribonucleotides, the virus-encoded RNA product is isolated at two timed intervals. The radiolabeled viral RNA is treated as follows. First, it is hybridized to a specific cDNA previously constructed from viral-gene-*V* mature mRNA. Second, the hybrid is treated with RNase. Finally, the hybrid is denatured and electrophoresed on a gel, which is then subjected to autoradiography. The following results suggest that the pathologic nature of the virus is time related (the number of nucleotides is indicated on the bands observed):

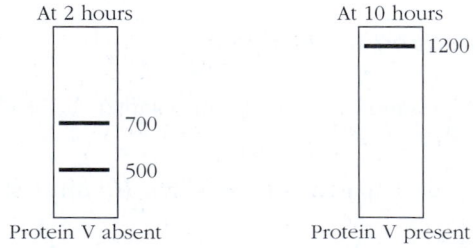

a. What is the size of the mature mRNA for gene *V*?

b. Draw a diagram of each hybrid and indicate what the illustrated bands represent.

c. Why is protein V not produced until after 2 hours?

(Problem 8 courtesy of Joan McPherson. From A. J. F. Griffiths and J. McPherson, *100 + Principles of Genetics*. W. H. Freeman and Company, 1989.)

9. The gene for β-tubulin has been cloned from *Neurospora* and is available. List a step-by-step procedure for cloning the same gene from the related fungus *Podospora*, using as the cloning vector the pBR *E. coli* plasmid shown here, where *kan* encodes kanamycin and *tet* encodes tetracycline:

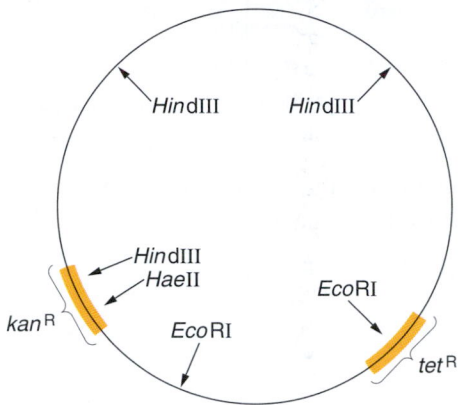

10. A circular bacterial plasmid containing a gene for tetracycline resistance was cut with restriction enzyme *Bgl*II. Electrophoresis showed one band of 14 kb.

    a. What can be deduced from this result?

    The plasmid was cut with *Eco*RV and electrophoresis produced two bands, of 2.5 and 11.5 kb.

    b. What can be deduced from this result?

    Digestion with both enzymes together resulted in three bands of 2.5, 5.5, and 6 kb.

    c. What can be deduced from this result?

    Plasmid DNA cut with *Bgl*II was mixed and ligated with donor DNA fragments also cut with *Bgl*II to make recombinant DNA molecules. All recombinant clones proved to be tetracycline sensitive.

    d. What can be deduced from this result?

    One recombinant clone was cut with *Bgl*II, and fragments of 4 and 14 kb were observed.

    e. Explain this result.

    The same clone was treated with *Eco*RV and fragments of 2.5, 7, and 8.5 were observed.

    f. Explain these results by showing a restriction map of the recombinant DNA.

11. **a.** A fragment of mouse DNA with *Eco*RI sticky ends carries the gene *M*. This DNA fragment, which is 8 kb long, is inserted into the bacterial plasmid pBR322 at the *Eco*RI site. The recombinant plasmid is cut with three different restriction-enzyme treatments.

The patterns of ethidium bromide fragments, after electrophoresis on agarose gels, are shown in this diagram:

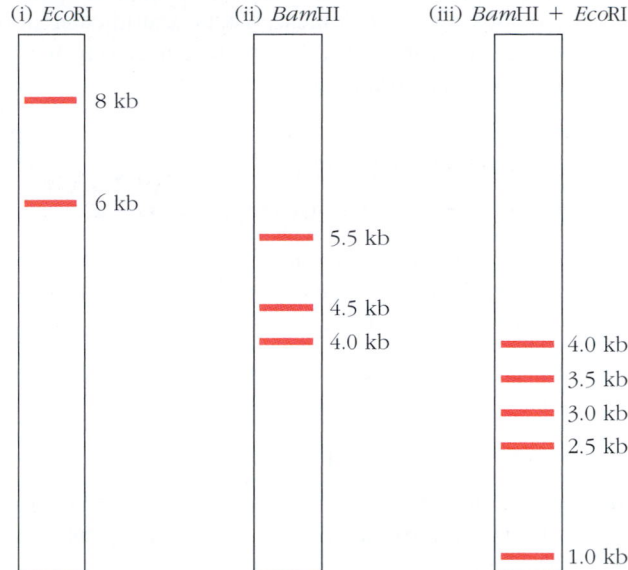

A Southern blot is prepared from gel iii. Which fragments will hybridize to a probe (^{32}P) of pBR plasmid DNA?

b. Gene *X* is carried on a plasmid consisting of 5300 nucleotide pairs (5300 bp). Cleavage of the plasmid with the restriction enzyme *Bam*HI gives fragments 1, 2, and 3, as indicated in the following diagram. (B = *Bam*HI restriction site.) Tandem copies of gene *X* are contained within a single *Bam*HI fragment. If gene *X* encodes a protein X of 400 amino acids, indicate the approximate positions and orientations of the gene *X* copies.

(Problem 11 courtesy of Joan McPherson. From A. J. F. Griffiths and J. McPherson, *100 + Principles of Genetics*. W. H. Freeman and Company, 1989.)

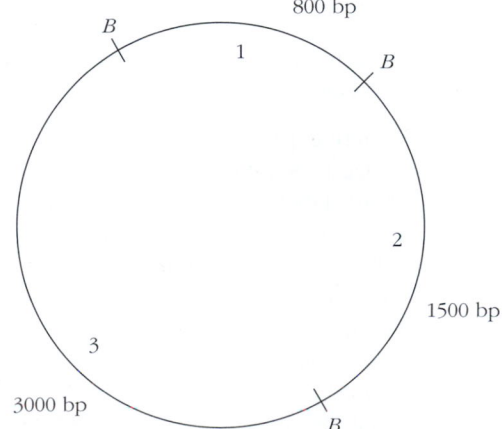

12. In functional complementation in both prokaryotes and eukaryotes, prototrophy is often the phenotype selected to detect transformants. Prototrophic cells are

used for donor DNA extraction; then this DNA is cloned and the clones are added to an auxotrophic recipient culture. Successful transformants are identified by plating the recipient culture on minimal medium and looking for colonies. What experimental design would you use to make sure that a colony that you hope is a transformant is not, in fact,

a. a prototrophic cell that has entered the recipient culture as a contaminant?

b. a revertant of the auxotrophic mutation?

13. In any particular transformed eukaryotic cell (say, of *Neurospora*), how could you tell if the transforming DNA (carried on a circular bacterial vector)

a. replaced the resident gene of the recipient by double crossing-over or single crossing-over?

b. was inserted ectopically?

14. In an electrophoretic gel across which is applied a powerful electric alternating pulsed field, the DNA of the haploid fungus *Neurospora crassa* ($n = 7$) moves slowly but eventually forms seven bands, which represent DNA fractions that are of different sizes and hence have moved at different speeds. These bands are presumed to be the seven chromosomes. How would you show which band corresponds to which chromosome?

15. A linear phage chromosome is labeled at both ends with ^{32}P and digested with restriction enzymes. *Eco*RI produces fragments of sizes 2.9, 4.5, 6.2, 7.4, and 8.0 kb. An autoradiogram developed from a Southern blot of this digest shows radioactivity associated with the 6.2 and 8.0 fragments. *Bam*HI cleaves the same molecule into fragments of sizes 6.0, 10.1, and 12.9, and the label is found associated with the 6.0 and 10.1 fragments. When *Eco*RI and *Bam*HI are used together, fragments of sizes 1.0, 2.0, 2.9, 3.5, 6.0, 6.2, and 7.4 kb are produced.

a. Draw a restriction-enzyme–target-site map of this molecule, showing relative positions and distances apart.

b. A radioactive probe made from a cloned phage gene *X* is added to Southern blots of single-enzyme digests of phage DNA. The autoradiograms show hybridization associated with the 4.5-, 10.1-, and 12.9-kb fragments. Draw the approximate location of gene *X* on the restriction map.

(Problem 15 courtesy of Joan McPherson. From A. J. F. Griffiths and J. McPherson, *100+ Principles of Genetics.* W. H. Freeman and Company, 1989.)

16. The cDNA clone and genomic clone for a phosphatase enzyme have been isolated. From the following data, the structural characteristics of the gene and its transcript can be determined:

cDNA Map

The fragment of cDNA was excised from a plasmid and end labeled with ^{32}P.

Combined digest with HaeIII and TaqI enzymes

Electrophoretic fragments stained with ethidium bromide:

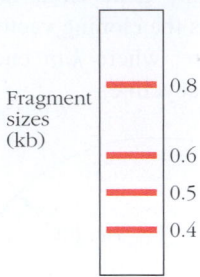

Gel from the combined digestion exposed to X-ray film; that is, the autoradiogram pattern:

Digestion with TaqI enzyme alone

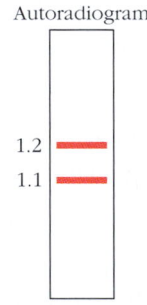

a. Determine the cDNA map.

Genomic DNA Map

The fragment of genomic DNA was cleaved from a λ phage clone with *Eco*RI enzyme, and its restriction characteristics were determined. The fragment was end labeled and digested.

Complete BamII digest

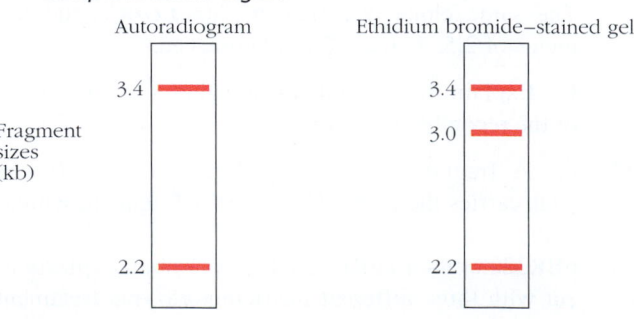

b. Draw the genomic map, marking the restriction sites.

c. A labeled cDNA probe hybridized to the 3.4- and 2.2-kb genomic fragments. The 1.2-kb *Taq*I fragment hybridized to the 3.4-kb genomic fragment. If the phosphatase gene is a single-copy gene, to which genomic fragment might the 1.1-kb *Taq*I fragment hybridize?

d. Which part of the gene does the 3.0-kb genomic fragment represent?

(Problem 16 courtesy of Joan McPherson. From A. J. F. Griffiths and J. McPherson, *100 + Principles of Genetics.* W. H. Freeman and Company, 1989.)

17. α-Interferon is encoded by a gene that does not contain introns. The *Bam*HI restriction fragment containing the complete gene can be identified on a Southern blot by hybridization to a specific interferon cDNA probe (^{32}P), which, under the conditions of hybridization, will detect only interferon sequences.

To determine the cause of unknown immune deficiencies, blood samples from patients and unaffected people are screened for the α-interferon gene and its expression. The following drawing represents autoradiograms for Southern and Northern blots and a Western blot probed by antibody that recognizes α-interferon sequences. Persons 1 and 2 are normal for immune capacity; persons 3, 4, and 5 have immune deficiencies.

Autoradiograms

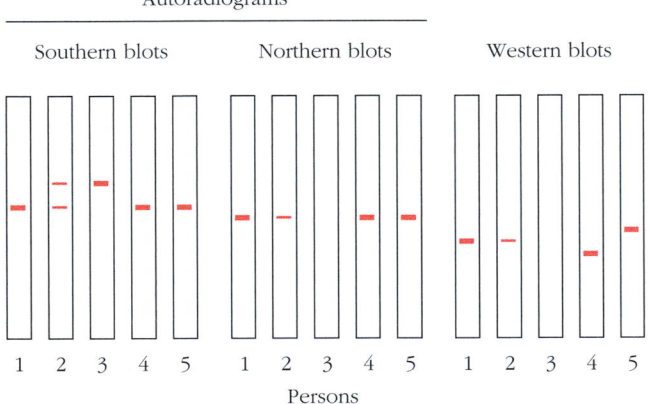

Southern blots   Northern blots   Western blots

1 2 3 4 5   1 2 3 4 5   1 2 3 4 5

Persons

a. Which persons are homozygous for the α-interferon gene?

b. What do you think are the causes of the immune deficiencies of persons 3, 4, and 5? Describe the type of gene mutation for each person.

(Problem 17 courtesy of Joan McPherson. From A. J. F. Griffiths and J. McPherson, *100 + Principles of Genetics.* W. H. Freeman and Company, 1989.)

18. A fragment of human genomic DNA for gene *P* is excised from a λ vector by using *Hin*dIII, and the ends of

the gene fragment (GP) are labeled with ^{32}P. The fragment is initially digested with *Bam,* giving two fragments, of 1.0 kb and 10 kb. The 10-kb fragment is *partly* digested with *Hpa*II and the products are electrophoresed, giving the following results:

Autoradiogram of *Hpa*II digest of GP fragment

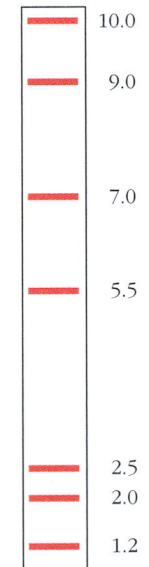

10.0
9.0
7.0
5.5
2.5
2.0
1.2

a. Map the genomic fragment, indicating the *Bam* site and the *Hpa*II sites.

The cDNA for gene *P* can be cut by *Hpa*II into only two fragments. Both are labeled, and each is used as a ^{32}P probe to investigate fragments from a complete digest of GP by using *Hpa*II and *Bam* enzymes, on Southern blots.

Southern blots of complete digest

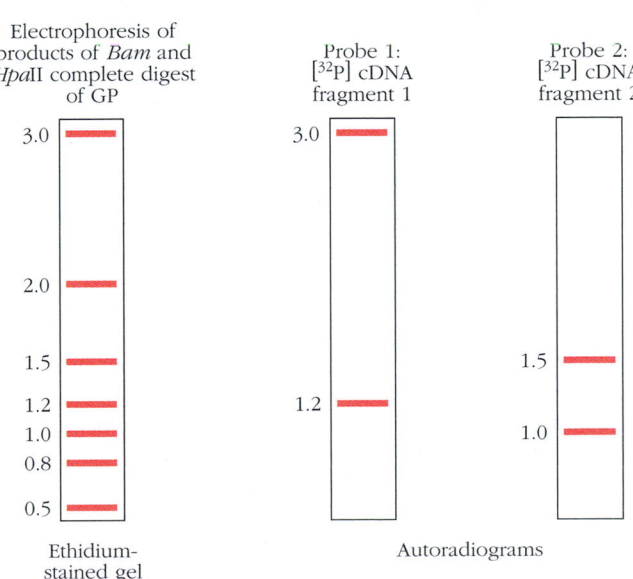

Electrophoresis of products of *Bam* and *Hpa*II complete digest of GP

3.0
2.0
1.5
1.2
1.0
0.8
0.5

Ethidium-stained gel

Probe 1: [^{32}P] cDNA fragment 1

3.0
1.2

Probe 2: [^{32}P] cDNA fragment 2

1.5
1.0

Autoradiograms

b. Draw a diagram comparing genomic DNA with cDNA. Indicate the restriction sites. Explain the differences.

c. Mark the *Hpa*II site on the cDNA and the orientation of the two *Hpa* fragments.

(Problem 18 courtesy of Joan McPherson. From A. J. F. Griffiths and J. McPherson, *100+ Principles of Genetics.* W. H. Freeman and Company, 1989.)

19. Two children are investigated for the expression of a gene *(D)* that encodes an important enzyme for muscle development. The results of the studies of the gene and its product follow.

Child 1

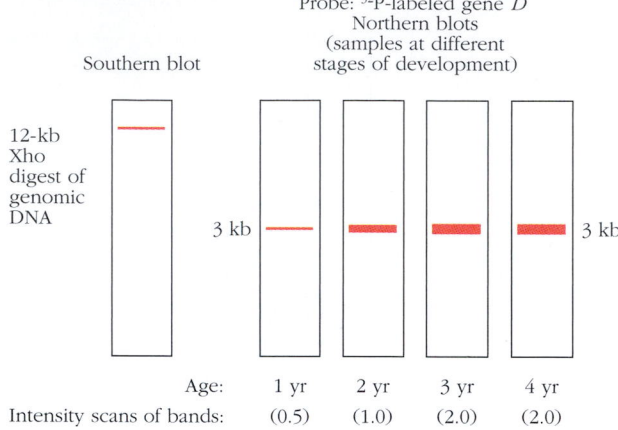

Southern blot

Probe: ³²P-labeled gene *D*
Northern blots
(samples at different stages of development)

12-kb
Xho
digest of
genomic
DNA

3 kb   3 kb

Age:	1 yr	2 yr	3 yr	4 yr
Intensity scans of bands:	(0.5)	(1.0)	(2.0)	(2.0)

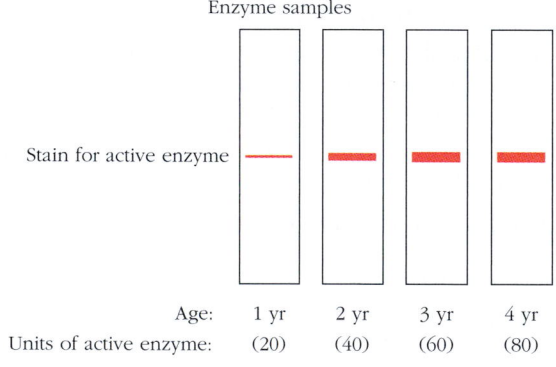

Enzyme samples

Stain for active enzyme

Age:	1 yr	2 yr	3 yr	4 yr
Units of active enzyme:	(20)	(40)	(60)	(80)

For child 2, the enzyme activity of each stage was very low and could be estimated only at approximately 0.1 unit at ages 1, 2, 3, and 4.

Child 2

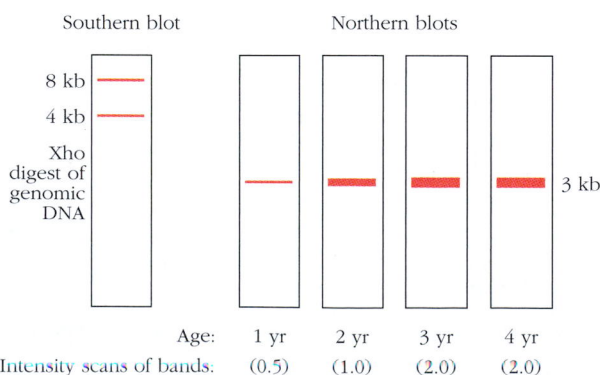

Southern blot   Northern blots

8 kb
4 kb
Xho
digest of
genomic
DNA

3 kb

Age:	1 yr	2 yr	3 yr	4 yr
Intensity scans of bands:	(0.5)	(1.0)	(2.0)	(2.0)

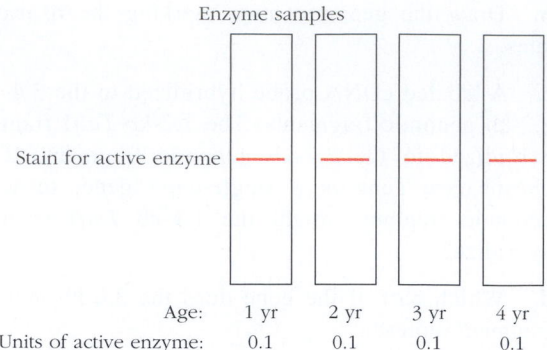

Enzyme samples

Stain for active enzyme

Age:	1 yr	2 yr	3 yr	4 yr
Units of active enzyme:	0.1	0.1	0.1	0.1

a. For both children, draw graphs representing the developmental expression of the gene. (Fully label both axes.)

b. How can you explain the very low levels of active enzyme for child 2? (Protein degradation is only one possibility.)

c. How might you explain the change in the Southern blot for child 2 compared with that for child 1?

d. If only one mutant gene has been detected in family studies of the two children, define the individual children as either homozygous or heterozygous for gene *D*.

(Problem 19 courtesy of Joan McPherson. From A. J. F. Griffiths and J. McPherson, *100+ Principles of Genetics.* W. H. Freeman and Company, 1989.)

20. A cloned fragment of DNA was sequenced by using the dideoxy method. A part of the autoradiogram of the sequencing gel is represented here.

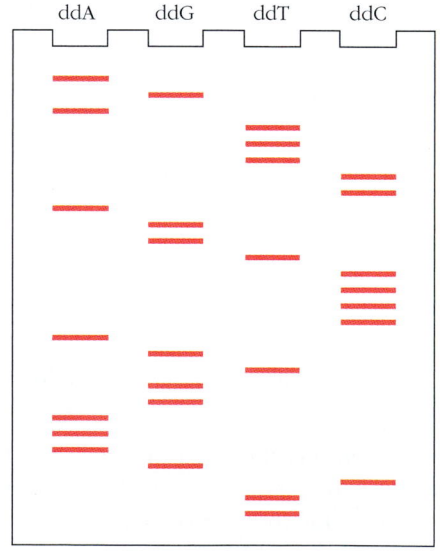

ddA   ddG   ddT   ddC

a. Deduce the nucleotide sequence of the DNA nucleotide chain synthesized from the primer. Label the 5′ and 3′ ends.

b. Deduce the nucleotide sequence of the DNA nucleotide chain used as the template strand. Label the 5′ and 3′ ends.

c. Write the nucleotide sequence of the DNA double helix (label the 5′ and 3′ ends).

d. How many of the six reading frames are *open* as far as you can tell?

21. The cDNA clone for the human gene encoding tyrosinase was used in a Southern analysis of *Eco*RI-digested genomic DNA of wild-type mice. Three mouse fragments were found to be radioactive (were bound by the probe). Albino mice were used in a Southern analysis, but, in this case, no genomic fragments bound to the probe. Explain these results in relation to the nature of the wild-type and mutant mouse alleles.

22. The protein encoded by the alkaptonuria gene is 445 amino acids long, yet the gene spans 60 kb (see the section "A Century of Genetic Research on Alkaptonuria"). How is this possible?

13

APPLICATIONS OF RECOMBINANT DNA TECHNOLOGY

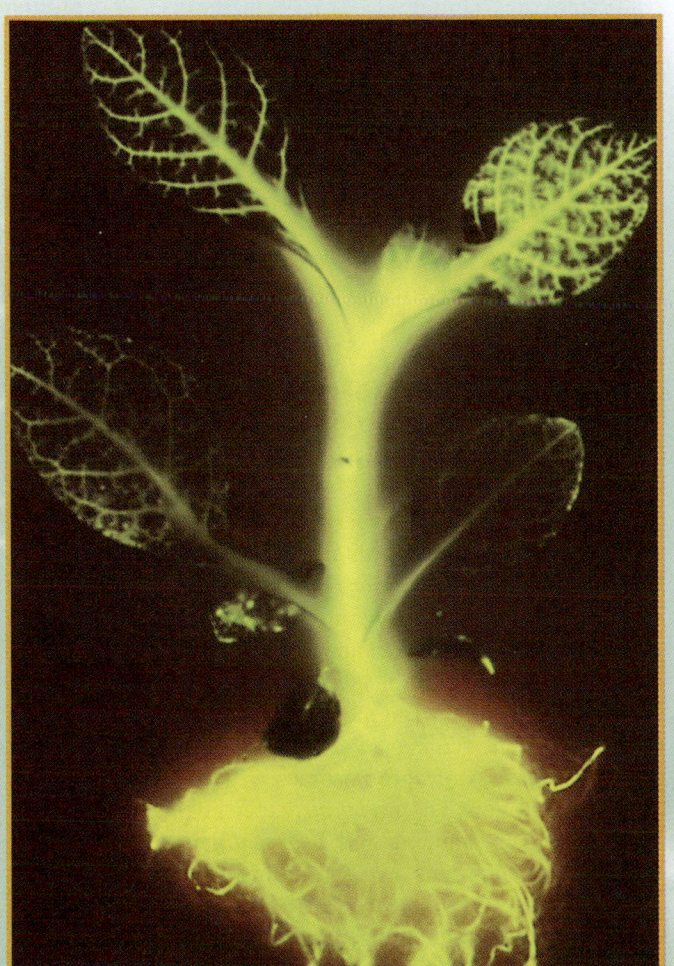

Transgenic tobacco plant expressing the luciferase gene from a firefly.

(Keith Wood, Promega, Madison, WI)

Key Concepts

In vitro mutagenesis allows highly specific changes to be made at specific positions within a gene.

In the chromosomes of an individual organism, specific restriction sites can be either present or absent, resulting in restriction fragment length polymorphisms (RFLPs).

RFLPs can be used as loci for genome mapping, as well as in the diagnosis of linked disease genes.

Reverse genetics works in a direction opposite that of traditional genetics; it enables the function of an unknown DNA or protein sequence to be identified at the phenotypic level.

A gene with an easily detectable product can be spliced to the regulatory regions of another gene and thereby act as a reporter for that gene's function.

Recombinant DNA technology is used to make microbes, plants, and animals that carry genes from other species.

Recombinant DNA technology can be used in the prenatal diagnosis of human genetic disease.

In the preceding chapter, we examined the basic techniques for using recombinant DNA technology to isolate and characterize genes. In this chapter, we expand this discussion to see how having an isolated gene in hand as a DNA clone opens up diverse experimental possibilities. For example, a cloned gene can be mutated in a highly specific way in the test tube and then reintroduced into its original host to study the gene's normal functional properties. This procedure mirrors the standard genetic procedure of genetic dissection. However, in spontaneous and induced mutations, the nature of the DNA change is largely unpredictable, and the sites of the mutations can be in any number of locations within the gene. In contrast, mutagenesis in vitro can be directed toward a very specific location and type of change. Hence in vitro mutagenesis is a powerful approach to studying gene function.

The techniques of DNA cloning and restriction-enzyme analysis have been combined to create a powerful new method used in genetic mapping. In this method, the variation of restriction-enzyme target sites within a species provides large numbers of molecular "alleles," called **molecular markers,** for chromosome mapping. Such markers are detected by probing with cloned DNA fragments. The ability to detect such molecular markers in large numbers has revolutionized mapping in most organisms, including humans.

Cloned genes (wild type or mutated) can be moved into a new host organism to create transgenic microbes, plants, and animals that could never have been made by using standard breeding procedures. This technique not only finds extensive use in basic research, but also opens up a plethora of applications in plant and animal breeding, industrial microbiology, and medicine.

The availability of cloned genes to be used as probes also opens up new frontiers in the diagnosis of human genetic diseases in prospective parents and in fetuses in utero.

First, we examine some ways in which specific mutations can be made in cloned genes in the test tube (in vitro); in other words, how to fabricate "designer genes."

In vitro mutagenesis

One of the most established techniques is **site-directed mutagenesis.** By using this method, we can create mutations at any specific site in a gene whose wild-type sequence is already known. This known sequence is used to chemically synthesize short DNA segments called **oligonucleotides.** Through single-strand hybridization, these oligonucleotides can be directed to any chosen site in the gene. In one approach, the gene of interest is inserted into a single-stranded phage vector, such as the phage M13. A synthetic oligonucleotide containing the desired mutation is designed. This oligonucleotide is allowed to hybridize to the mutant site by complementary base pairing. The oligonucleotide serves as a primer for the in vitro synthesis of the complementary strand of the M13 vector (Figure 13-1a). Any desired specific base change can be programmed into the sequence of the synthetic primer. Although there will be a mispaired base when the synthetic oligonucleotide hybridizes with the complementary sequence on the M13 vector, a few mismatched bases can be tolerated when hybridization takes place at a low temperature and a high salt concentration. After DNA synthesis has been mediated by DNA polymerase in vitro, the M13 DNA is allowed to replicate in *E. coli,* in which case many of the resulting phages will be the desired mutant. The synthetic oligonucleotide can be used as a labeled probe to distinguish wild-type from mutant phages. Although, at low temperature, the mismatched base will not prevent the primer from hybridizing with both types of phage, at high temperature, the primer will hybridize only with the mutant phage. Oligonucleotides with deletions or insertions will direct comparable mutations in the resident gene. The site-directed method can also be used on genes cloned in double-stranded vectors if the DNA is first denatured.

A knowledge of restriction sites is also useful in modifying a cloned gene. For example, a small deletion can be made by removing the fragment liberated by cutting at two restriction sites (Figure 13-1b). With the use of a similar double cut, a fragment, or "cassette," can be inserted at a single restriction cut to create a duplication or other modification (Figure 13-1c). Another approach is to enzymatically erode a cut end created by a restriction enzyme, to create deletions of various lengths (Figure 13-1d).

The polymerase chain reaction (PCR) also can be used (Figure 13-1e). A primer containing a mutation is used in a first round of PCR. The product of the first round is used as a primer for a second round of PCR, whose product is the mutant gene.

MESSAGE ⋯⋯⋯⋯⋯⋯⋯⋯⋯⋯⋯⋯⋯⋯⋯⋯⋯⋯⋯⋯⋯⋯⋯⋯
Site-directed mutagenesis allows various types of directed changes to be made at specific sites in a gene.

RFLP mapping

In Chapter 12, we learned that, if a cloned DNA fragment is used as a probe of genomic DNA that has been cut with a restriction enzyme, then the probe will bind to one or more genomic fragments. For example, if the restriction enzyme used does not cut within the chromosomal region encompassed by the cloned fragment, then the probe should bind to one fragment flanked by restriction sites on each side. Because the DNA of chromosomes within a species is generally homologous, it might be expected that a constant-sized genomic fragment will be bound in all individuals. However, when probes are used in this way, the bound fragments are often found to be of different sizes in different individuals. The explanation is that a given restriction site is not always found in all individuals. The absence of a site is usually caused by a single nucleotide difference that is most likely biologically neutral. Hence, for example, if a probe binds a 2-kb fragment in individual A of a haploid species

(a) Oligonucleotide-directed mutagenesis

(i) Base-pair substitution

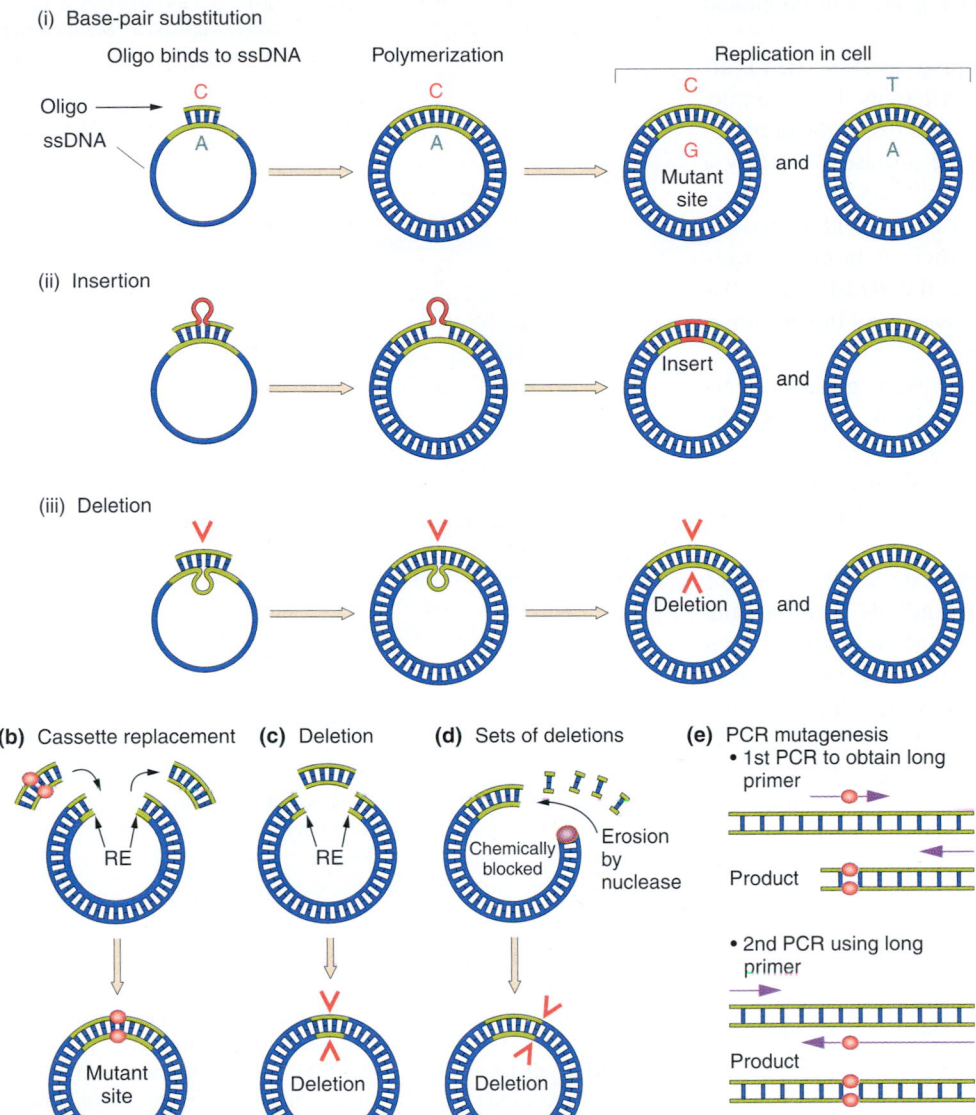

(b) Cassette replacement   **(c)** Deletion   **(d)** Sets of deletions   **(e)** PCR mutagenesis
- 1st PCR to obtain long primer

- 2nd PCR using long primer

Figure 13-1 In vitro mutagenesis. (Oligo, oligonucleotide; PCR, polymerase chain reaction; RE, restriction enzyme; ssDNA, single-stranded DNA.)

and it binds a fragment of 2.3 kb in individual B, the reason is usually that one of the sites that flanked the 2-kb fragment is missing in B, and the next site is 0.3 kb away, making the hybridized fragment 2.3 kb in size.

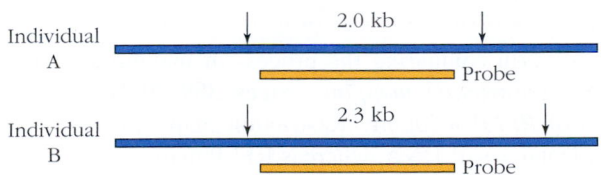

The presence and absence of the restriction site can be treated as two alleles that can be thought of as + and − alleles. The presence of the + in some individuals in the population and the absence (−) in others generates a **restriction fragment length polymorphism,** or **RFLP.** (In the case just considered, there was a dimorphism — two "morphs,"

one short and one long.) Geneticists were surprised to discover that RFLPs are quite common in populations and that a large proportion of probes will detect one. RFLPs are identified by a rather hit-or-miss method of hybridizing panels of randomly cloned genomic fragments to genomic restriction digests of several different individuals in a family or a population. Because RFLPs are a relatively common type of variation in nature, this method succeeds in finding RFLPs in most cases.

The significance of RFLPs is threefold. First, if an individual is heterozygous for two morphs of an RFLP, this heterozygous "locus" can be used as a marker in chromosomal mapping. Although at first the locus of the RFLP is not necessarily known, as more and more RFLPs are found, they can be mapped in relation to gene loci and in relation to other RFLP loci, and their positions gradually saturate the genetic map. The RFLPs are not biologically significant in most cases, but they can be used to map interesting genes and act as positions from which these genes can be cloned by positional cloning.

Second, in an extension of mapping analysis, RFLP alleles (morphs) can be used as diagnostic tools. For example, in a family with a record of a certain disease, if it can be established that the people who have the disease also carry a specific allele of an RFLP, then this fact suggests not only that the RFLP locus is linked to the disease gene locus, but furthermore that the specific RFLP allele is in cis arrangement with the disease allele. Hence the RFLP allele becomes a diagnostic marker for the disease, and this information can be used in genetic counseling.

Third, RFLPs can be used to measure genetic divergence between different populations or related species. The restriction-site difference is effectively a DNA difference, so a measure of the total number of RFLP differences represents a measure of genetic difference. Hence RFLPs are important in studies of evolution.

RFLP mapping is often performed on a defined set of strains or individuals that become "standards" for mapping that species. For example, in the fungus *Neurospora*, two wild-type strains, Oak Ridge and Mauriceville, are known to show many RFLP differences, so these strains have become standards used in a RFLP mapping. The RFLPs can be mapped relative to one another or to genes of known phenotypic expression. For example, let *ad* stand for an allele for adenine requirement, and 1 and 2 stand for RFLP loci with either the Oak Ridge (OR) or Mauriceville (M) "alleles." A cross can be made of the type $ad \cdot 1^{OR} \cdot 2^{OR} \times ad^{+} \cdot 1^{M} \cdot 2^{M}$. Progeny are tested for all three loci. Adenine requirement is tested by inoculating strains on medium lacking adenine, and the RFLP alleles are tested by probing with the relevant probes. Recombinant frequencies are calculated in the usual way. Most mutants in *Neurospora* have been induced in Oak Ridge wild-type strains, so it is a simple matter to map the mutant alleles to RFLPs simply by crossing the mutant Oak Ridge strain to the wild-type Mauriceville strain. An example of mapping a phenotypic mutant by using RFLP markers is shown in Figure 13-2.

Similar standard strains have been established in other organisms. An analogous approach has been used in human genome mapping by collecting DNA from a defined set of individuals in 61 families with an average of eight children per family and making this DNA available throughout the world to provide a standard for RFLP mapping.

Figure 13-3 shows an example of linkage of a human disease allele to an RFLP locus and the potential for using this information in diagnostics. Because of the close linkage, future generations of persons showing the RFLP morph 1 can be predicted to have a high chance of inheriting the disease allele *D*. This sort of predictive power can be used in prenatal diagnoses of the genotypes of fetuses, with the use of amniocentesis or chorionic villus sampling (considered later in this chapter).

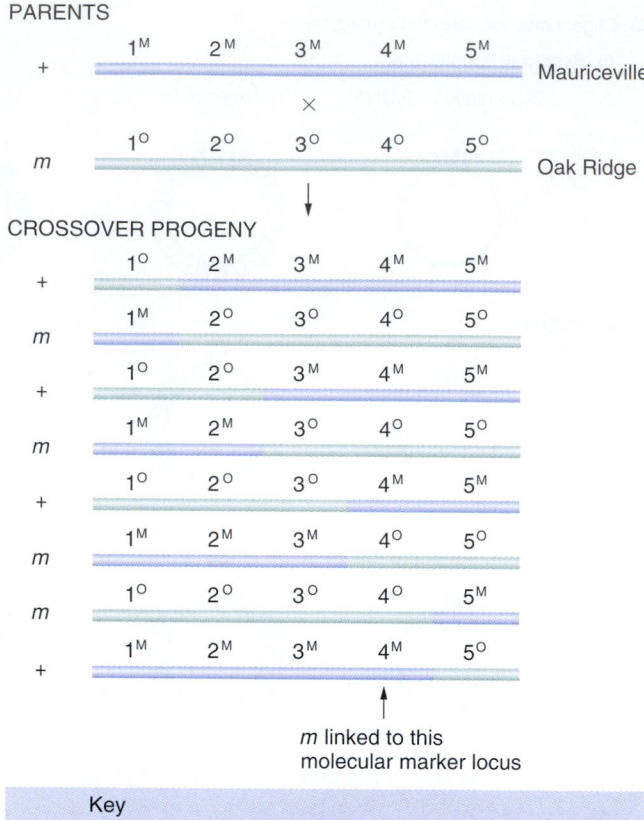

Figure 13-2   Mapping a gene (*m*) by RFLP analysis in *Neurospora*. The two parental strains show many RFLPs ranged along the chromosomes; their loci are labeled 1 through 5. The two parental strains are from Oak Ridge (Tennessee) and Mauriceville (Texas), and their RFLP alleles are labeled O and M. Many different progeny types are recovered, and some of the more common types are shown. The results show that the + allele always segregates with 4^{M} and the *m* allele always segregates with 4°, suggesting linkage of *m* to RFLP locus 4.

MESSAGE ·······································
RFLPs provide useful molecular marker loci for chromosome mapping and for diagnosis of human disease alleles.

It is worth comparing the process of making a restriction map (*restriction mapping*, pages 389–392) with the process of *RFLP mapping*. Restriction maps are based on *physical* analysis of DNA, whereas RFLP maps are based on *recombination* analysis of matings. Note also that restriction mapping is based on restriction sites with no variation, whereas RFLP mapping is based on restriction-site variation between homologous chromosomes. Most restriction maps are short-range (fine-scale) maps, although long-range maps can be constructed with rare-cutting restriction enzymes. In

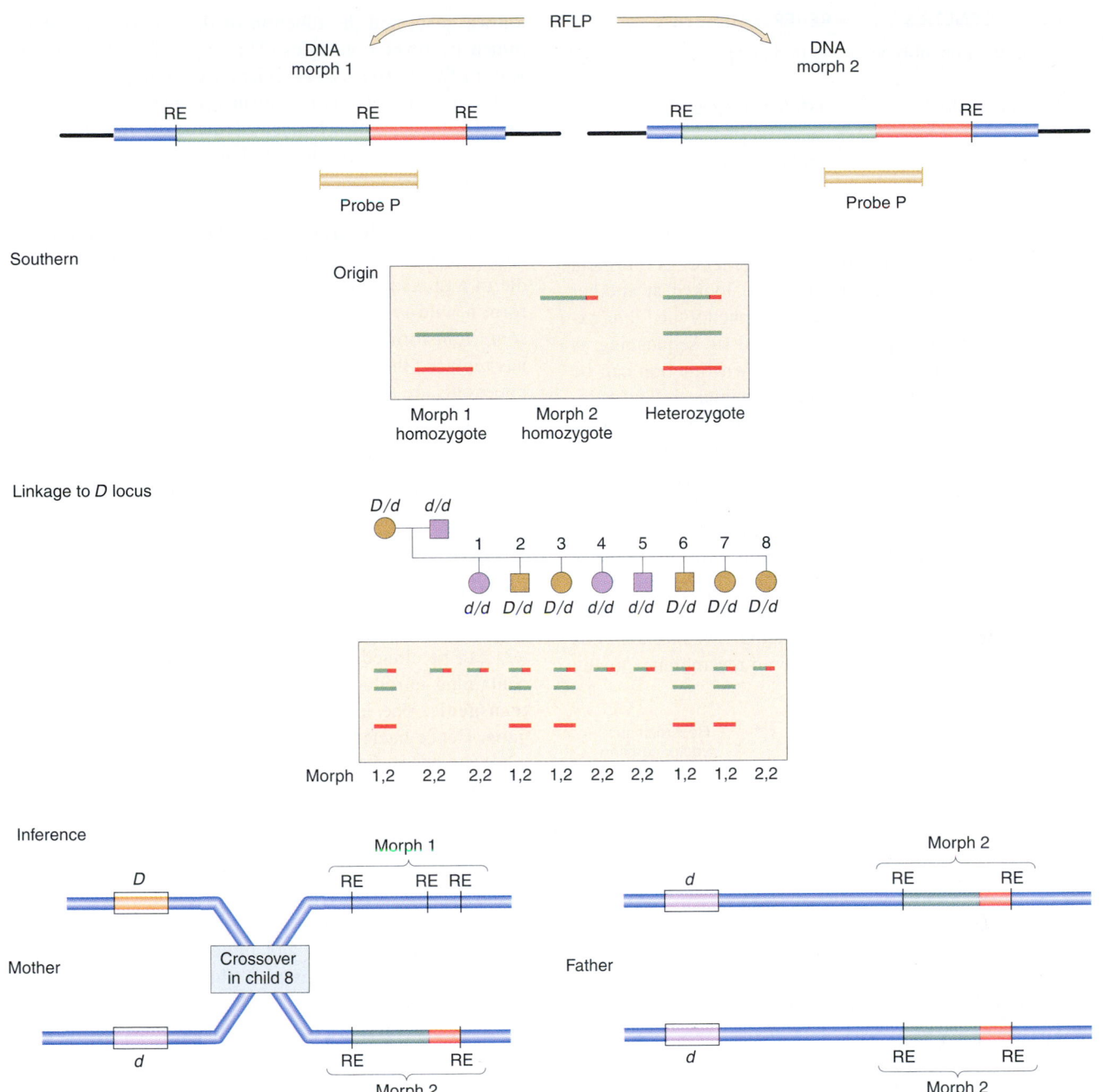

Figure 13-3 The detection and inheritance of a restriction fragment length polymorphism (RFLP). A probe P detects two DNA morphs when the DNA is cut by a certain restriction enzyme (RE). The pedigree of the dominant disease phenotype D shows linkage of the *D* locus to the RFLP locus; only child 8 is recombinant.

contrast, RFLP mapping generally produces long-range (coarse-scale) maps. RFLP mapping of whole genomes will be covered in detail in Chapter 14.

Reverse genetics

"Regular" genetics begins, as we have seen, with a mutant phenotype, proceeds to show the existence of the relevant gene by progeny ratio analysis, and finally clones and sequences the gene to determine its DNA and protein sequence. However, **reverse genetics,** a new approach made possible by recombinant DNA technology, works in the opposite direction. Reverse genetics starts from a protein or DNA for which there is no genetic information and then works backward to make a mutant gene, ending up with a mutant phenotype.

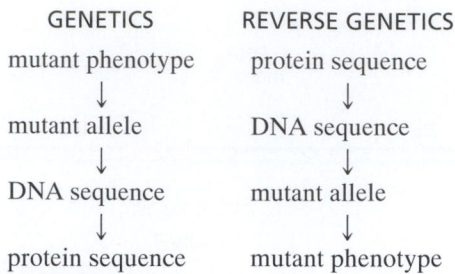

GENETICS	REVERSE GENETICS
mutant phenotype	protein sequence
↓	↓
mutant allele	DNA sequence
↓	↓
DNA sequence	mutant allele
↓	↓
protein sequence	mutant phenotype

Often the starting point of reverse genetics is a protein or a gene "in search of a function," or, looked at another way, a protein or a gene "in search of a phenotype." For example, a cloned wild-type gene detected by sequencing as an open reading frame (ORF) of unknown function can be subjected to some form of in vitro mutagenesis and reinserted back into the organism to determine the phenotypic outcome. In regard to a protein of unknown function, the amino acid sequence can be translated backward into a DNA sequence, which can then be synthesized in vitro. This sequence is then used as a probe to find the relevant gene, which is cloned and itself subjected to in vitro mutagenesis to determine the phenotypic effect. This approach will become even more significant as data accumulate in whole-genome-sequencing projects. Sequencing reveals numerous unknown ORFs, many of which have no resemblance to any

known gene, and the function of these genes can be determined by reverse genetics. This type of analysis is underway in *Saccharomyces cerevisiae;* the complete sequencing of the genome of this organism revealed many unassigned ORFs, which are now being systematically mutated into null alleles to detect a phenotype that might provide some clue about function.

Important tools for reverse genetics are in vitro mutagenesis and **gene disruption,** also known as **gene knockout.** One approach is to insert a selectable marker into the middle of a cloned gene and then to use this construct to transform a wild-type recipient. Selecting for the marker yields some transformants in which the disrupted (mutated) gene has replaced the in situ wild-type allele (Figure 13-4). Gene knockouts are also considered later in the chapter.

MESSAGE ··
Reverse genetics discovers the normal role of cryptic DNA or protein sequences by mutating the sequence in vitro and looking for changes at the phenotypic level.
···

Expressing eukaryotic genes in bacteria

We learned in Chapter 12 that genes for any eukaryotic protein can be cloned in *E. coli* expression vectors. Organisms containing introduced foreign DNA are referred to as being **transgenic.** The introduced foreign gene is called a **transgene.** Hence bacteria containing eukaryotic genes are transgenic bacteria. Such transgenic bacterial cultures can be used as "factories" for the synthesis of valuable eukaryotic proteins that are otherwise difficult to obtain. Broadly speaking, the term **biotechnology** is the name given to this area of research — the application of recombinant DNA techniques in commercial enterprises. Patented "designer organisms," both prokaryotic and eukaryotic, are already finding widespread use, and biotechnology is expected to become a dominant sector of the economy early in the twenty-first century.

When bacteria are used to produce a eukaryotic protein, it is desirable to design the system so as to produce as large an amount of the protein as possible. There are several such systems for overproducing foreign proteins in *E. coli.* One system uses phage T7 RNA polymerase operating on a T7 promoter. Late in infection, phage T7 synthesizes enormous amounts of gene products from several sites termed *late promoters.* Because host chromosomal genes are not synthesized at this point, these products are the only proteins synthesized. Figure 13-5 depicts the two components of this system. In the first component, the T7 RNA polymerase gene is arranged so that it is transcribed from the *lac* promoter, from which transcription is normally shut off by the *lac* repressor, a negative regulatory protein (see Chapter 11). However, on the addition of the lactose analog IPTG (isopropyl β-D-thiogalactoside), the repressor is inactivated and large amounts of T7 RNA polymerase are synthesized.

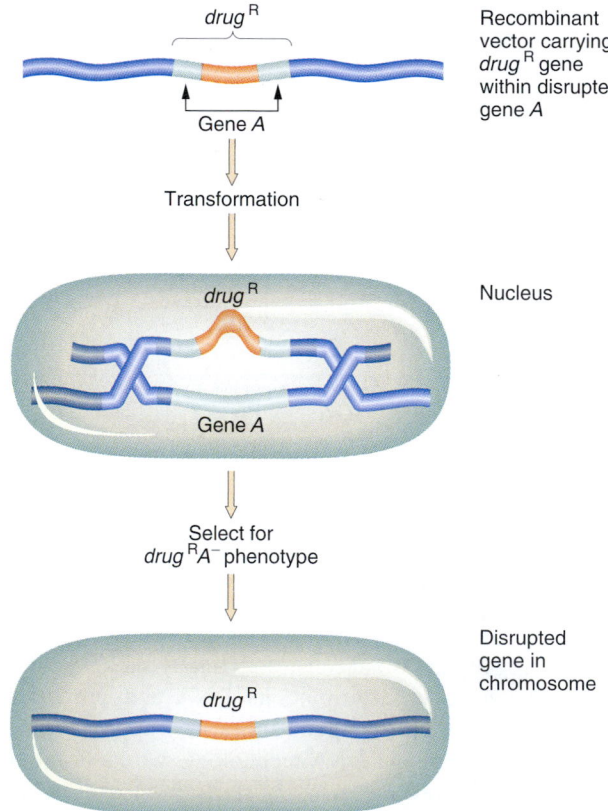

Figure 13-4   Gene disruption by homologous integration of a sequence containing a selectable marker.

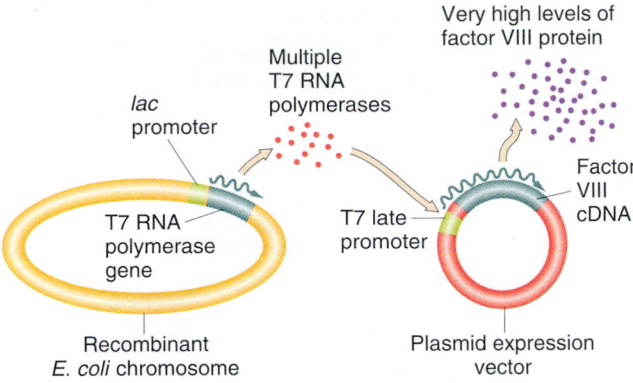

Figure 13-5   Two-step expression vector system based on bacteriophage T7 RNA polymerase and T7 late promoter. The chromosome of a specially engineered *E. coli* cell contains a copy of the T7 RNA polymerase gene under the transcriptional control of the *lac* promoter. When transcription from the *lac* promoter is induced by addition of IPTG, the T7 RNA polymerase gene is transcribed, and the mRNA is translated into the enzyme. The T7 RNA polymerase molecules produced then initiate transcription at a very high rate from the T7 late promoter on the expression vector. Multiple copies of the expression vector are present in such cells, although only one copy is diagrammed here. The large quantity of mRNA transcribed from the cDNA cloned next to the T7 late promoter is translated into abundant protein product. (After H. Lodish, D. Baltimore, A. Berk, S. L. Zipursky, P. Matsudaira, and J. Darnell, *Molecular Cell Biology,* 3d ed. Copyright © 1995 by Scientific American Books, Inc.)

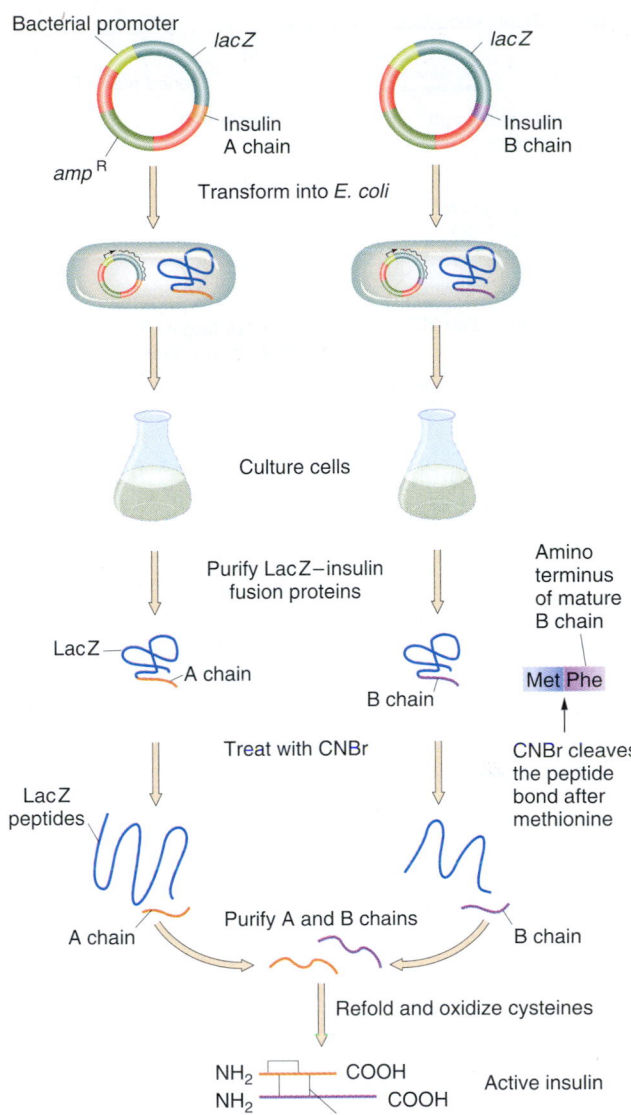

Figure 13-6   Expression of human insulin in *E. coli.* The two chains of insulin are made separately as fusion proteins with β-galactosidase. They are processed chemically and then mixed, and active insulin forms. (Copyright © 1992 by J. D. Watson, M. Gilman, J. Witkowski, and M. Zoller, *Recombinant DNA,* 2d ed. Copyright © Scientific American Books.)

In the second component of the system, the gene of choice [in this example, a human blood-clotting factor, factor VIII (defective in hemophiliacs)] is inserted (as a cDNA) adjacent to a late T7 promoter. The T7 RNA polymerase then transcribes this gene at high levels, resulting in high production of the factor VIII protein in the bacteria.

Now, many proteins such as human insulin (Figure 13-6), human growth hormone (Figure 13-7), and a wide range of pharmaceuticals are mass-produced from genetically engineered bacteria and fungi. Figure 13-8 shows an example in which the genetic engineering resulted in a drug that happened to give a strikingly different color to the bacterial colony when streaked onto a petri plate.

Recombinant DNA technology in eukaryotes

The techniques for gene manipulation, cloning, and expression were first developed in bacteria but are now applied routinely in a variety of model eukaryotes. The genomes of eukaryotes are larger and more complex than those of bacteria, so modifications of the techniques are needed to handle the larger amounts of DNA and the array of different cells and life cycles of eukaryotes. For instance, some eukaryotic proteins cannot be easily expressed in large amounts in bacteria, and eukaryotic expression systems need to be employed. A widely used vector–expression system for eu-

karyotic proteins is insect baculovirus, into which genes are inserted and expressed at high rates in cultured insect cells, as depicted in Figure 13-9). Although eukaryotic genes are cloned and sequenced in bacterial hosts, it is often desirable to introduce such genes back into the original eukaryotic host or into another eukaryote — in other words, to make a transgenic eukaryote.

Transgenic eukaryotes

DNA is introduced into a eukaryotic cell by a variety of techniques, such as transformation, injection, viral infection,

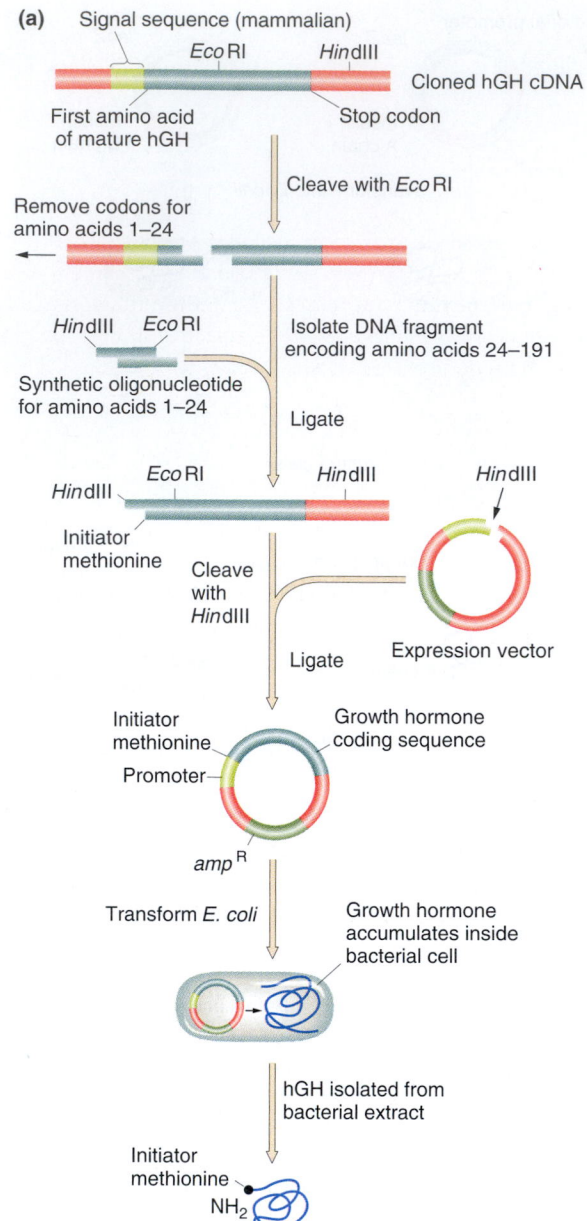

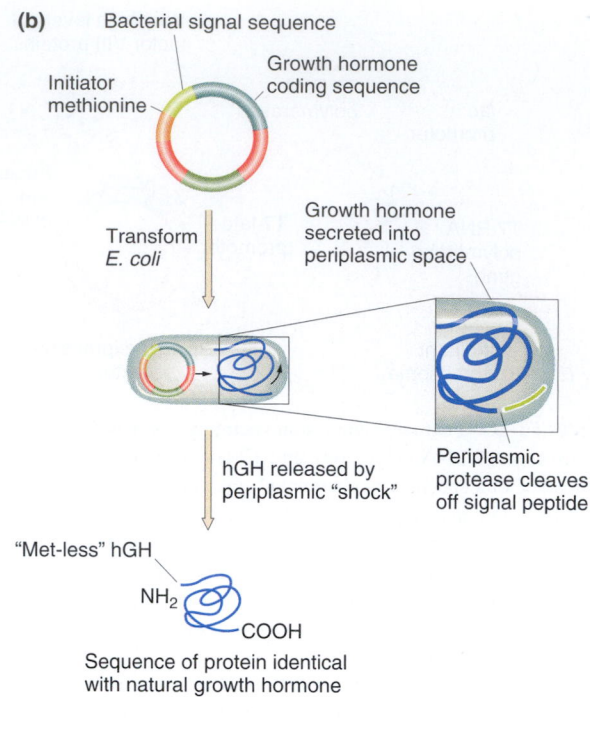

Figure 13-7   *Expression of human growth hormone (hGH) in* E. coli. (a) The human signal sequence is removed, enabling the protein to be produced in bacterial cells. The product contains an extra bacterial methionine. (b) A bacterial signal sequence that targets the protein for secretion to the outside can be added. In this method, the product has no extra methionine. (Copyright © 1992 by J. D. Watson, M. Gilman, J. Witkowski, and M. Zoller, *Recombinant DNA,* 2d ed. Copyright © Scientific American Books.)

or bombardment with DNA-coated tungsten particles (Figure 13-10). As we learned in Chapter 12, when exogenously added DNA that is originally from that organism inserts into the genome, it can either replace the resident gene or insert ectopically. If the DNA is a transgene from another species, it inserts ectopically. (Vectors that replicate autonomously in eukaryotic cells are rare; so, in most cases, chromosomal integration is the route followed.)

The possibility of transgenic modification of eukaryotes such as plants and animals (including humans) opens up many new approaches to research because genotypes can be genetically engineered to make them suitable for some specific experiment. (An example in basic research is in the use

of reporter genes. Sometimes it is difficult to detect the activity of a particular gene in the tissue where it normally functions. This problem can be circumvented by splicing the promoter of the gene in question to the coding region of a gene, known as a **reporter gene,** whose product *is* easily detectable. Wherever and whenever the gene in question is active, the reporter gene will announce that activity in the appropriate tissue. Examples will be given later in the chapter.)

Furthermore, because plants, animals, and fungi form the basis for a large part of the economy, transgenic "designer" genotypes are finding extensive use in applied research. A particularly exciting application of transgenesis is

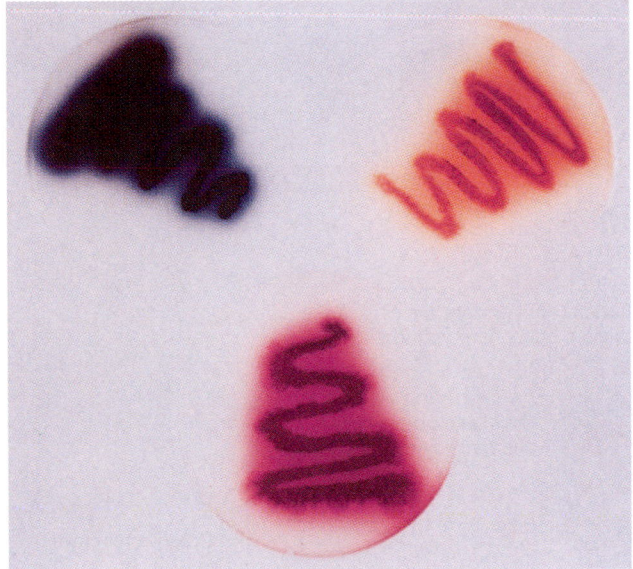

Actinorhodin

Medermycin

Mederrhodin A

Figure 13-8 Genetically engineered bacteria. New antibiotics can be synthesized by genetically engineered bacteria. Different species of *Streptomyces* produce related but different antibiotics. One species produces actinorhodin, which is blue at alkaline pH. A second species produces medermycin, which is brown. When the genes from the second species were cloned and transformed on a plasmid into the first species, some of the transformed bacteria had acquired the capacity to synthesize a new antibiotic, mederrhodin A, which gives a reddish-purple color. The structures of the three compounds are shown here, together with the streaks of the bacteria synthesizing each of the three compounds. The new antibiotic, mederrhodin A, is similar to medermycin but carries a hydroxyl group at the 6 position. The cloned segment probably contains a gene that encodes a hydrolyase that is expressed when transformed into a different species. (Mervyn J. Bibb, John Innes Institute, United Kingdom.)

in human gene therapy — the introduction of a normally functional transgene that can replace or compensate for a resident malfunctioning allele.

MESSAGE ··

Transgenesis, the design of a specific genotype by the addition of exogenous DNA to the genome, has increased the scope of breeding in basic genetic research and in commercial applications.

We now turn to some of the specialized recombinant DNA techniques used in the eukaryotes baker's yeast, plants, and animals and in human gene therapy.

Genetic engineering in baker's yeast

The yeast *Saccharomyces cerevisiae* has become the most sophisticated eukaryotic model for recombinant DNA technology. One of the main reasons is that the transmission genetics of yeast is extremely well understood, and the stockpile of thousands of mutants affecting hundreds of different phenotypes is a valuable resource when using yeast as a molecular system. In yeast, another important advantage is the availability of a circular 6.3-kb natural yeast plasmid. This plasmid, which has a circumference of 2 μm, has become known as the "2-micron" plasmid. It forms the basis for several sophisticated cloning vectors. This plasmid is transmitted to the cellular products of meiosis and mitosis.

The simplest yeast vectors are derivatives of bacterial plasmids into which the yeast locus of interest has been inserted (Figure 13-11a). When transformed into yeast cells, these plasmids insert into yeast chromosomes generally by homologous recombination with the resident gene and by either a single or a double crossover (Figure 13-12). As a result, either the entire plasmid is inserted or the targeted allele is replaced by the allele on the plasmid. Such integrations can be detected by plating cells on a medium that selects for the allele on the plasmid. Because bacterial plasmids do not replicate in yeast, integration is the only way to generate a stable modified genotype with the use of these vectors.

If the 2-μm plasmid is used as the basic vector and other bacterial and yeast segments are spliced into it (Figure 13-11b), then a construct having several useful properties is obtained. First, the 2-μm segment confers the ability to replicate autonomously in the yeast cell, and insertion is not necessary for a stable transformation. Second, genes can be introduced into yeast, and their effects can be studied in that organism; then the plasmid can be recovered and put back into *E. coli*, provided that a bacterial replication origin and a selectable bacterial marker are on the plasmid. Such **shuttle vectors** are very useful in the routine cloning and manipulation of yeast genes.

With any autonomously replicating plasmid, there is the possibility that a daughter cell will not inherit a copy, because the partitioning of plasmid copies to daughter cells is essentially a random process depending on where the plasmids are in the cell when the new cell wall is formed. However, if the section of yeast DNA containing a centromere is added to the plasmid (Figure 13-11c), then the nuclear spindle that ensures the proper segregation of chromosomes will treat the plasmid in somewhat the same way and partition it into daughter cells at cell division. The addition of a centromere is one step toward the creation of an artificial chromosome. A further step has been made by linearizing a plasmid containing a centromere and adding the DNA from yeast telomeres to the ends (Figure 13-11d). If this construct contains yeast replication origins (autonomous replication sequences, ARSs), then it constitutes a **yeast artificial chromosome (YAC),** which behaves in many ways like a small yeast chromosome at mitosis and meiosis. For example, when two

haploid cells — one bearing a *ura*$^+$ YAC and another bearing a *ura*$^-$ YAC — are brought together to form a diploid, many tetrads will show the clean 2:2 segregations expected if these two elements are behaving as regular chromosomes.

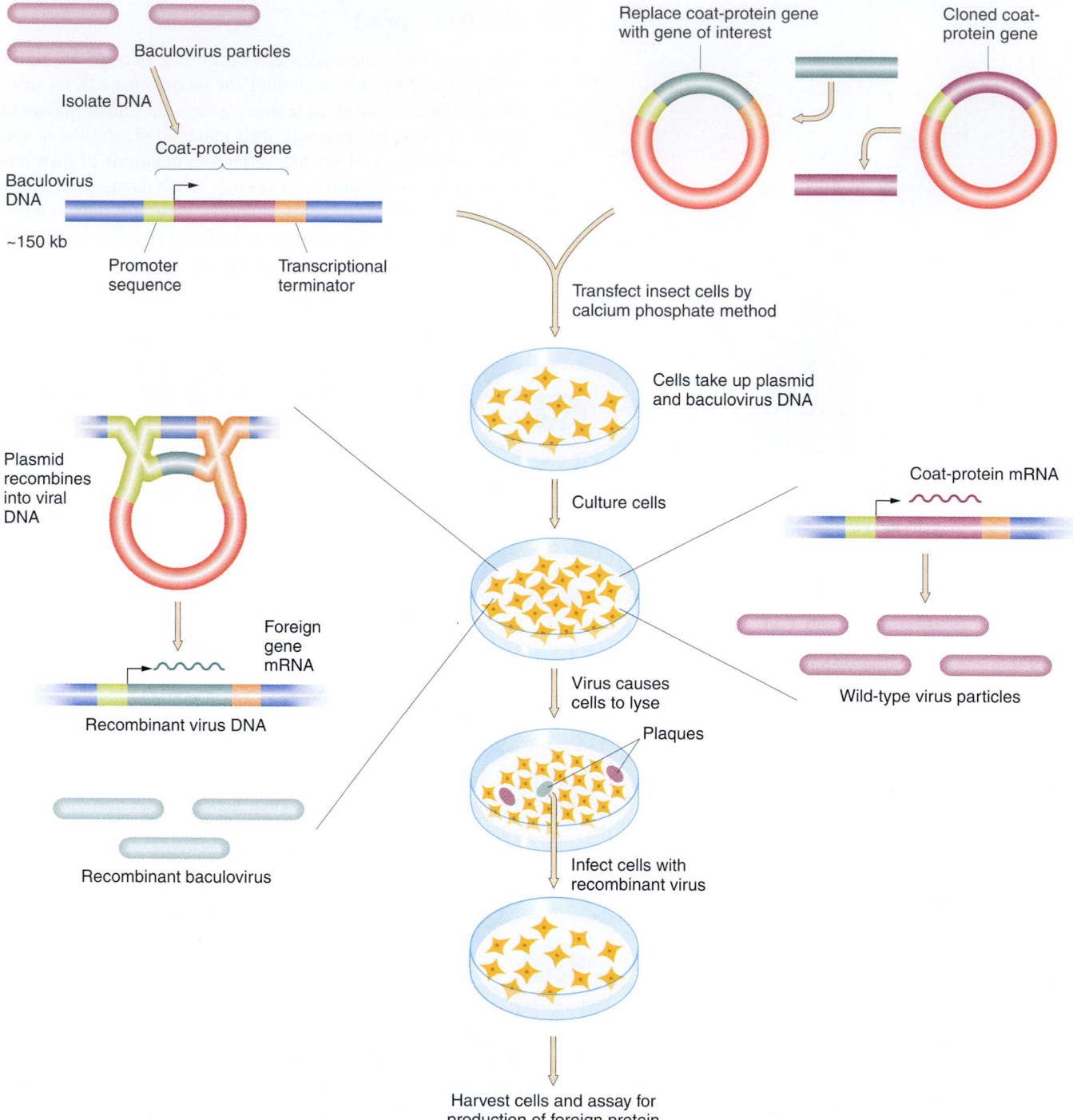

Figure 13-9   Baculovirus is a very large DNA virus (genome of about 150 kb) that infects insect cells. To express a foreign gene in baculovirus, the gene of interest is cloned in place of the viral coat-protein gene in a plasmid carrying a small part of the viral genome. The recombinant plasmid is cotransfected into insect cells with wild-type baculovirus DNA. At a low frequency, the plasmid and viral DNAs recombine through homologous sequences, resulting in the insertion of the foreign gene into the viral genome. Virus plaques develop, and the plaques containing recombinant virus look different because they lack the coat protein. The plaques with recombinant virus are picked and expanded. This virus stock is then used to infect a fresh culture of insect cells, resulting in high expression of the foreign protein. (Copyright © 1992 by J. D. Watson, M. Gilman, J. Witkowski, and M. Zoller, *Recombinant DNA,* 2d ed. Copyright © Scientific American Books.)

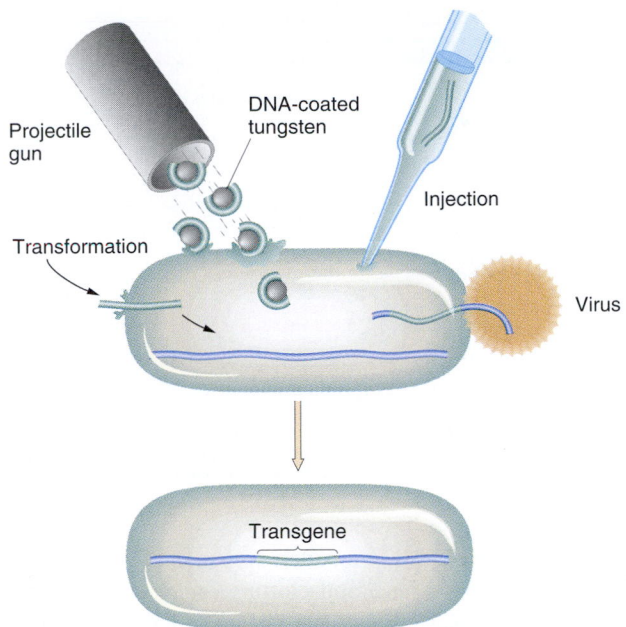

Figure 13-10 Some of the different ways of introducing foreign DNA into a cell.

Centromeric plasmids can be used to study the regulatory elements upstream of a gene (Figure 13-13). The relevant coding region and its upstream region can be spliced into a plasmid, which can be selected by a separate yeast marker such as $ura3^+$. The upstream region can be manipulated by inducing a series of deletions, which are achieved by cutting the DNA, using a special exonuclease to chew away the DNA in one direction to different extents, and then rejoining it. The experimental objective is then to determine which of these deletions still permits normal functioning of the gene. Proper function is assayed by transforming the plasmid into a recipient in which the chromosome locus carries a defective mutant allele and then monitoring for the return of gene function in the recipient. The results generally define a specific region that is necessary for normal function and regulation of the gene.

In such regulatory studies, it is often more convenient to use a reporter gene instead of the gene of interest. Therefore, if the regulation of gene X is of interest, the promoter region of gene X is spliced to the reporter gene. A gene that has been extensively used as a reporter in yeast is the bacterial *lacZ* gene, which encodes the enzyme β-galactosidase.

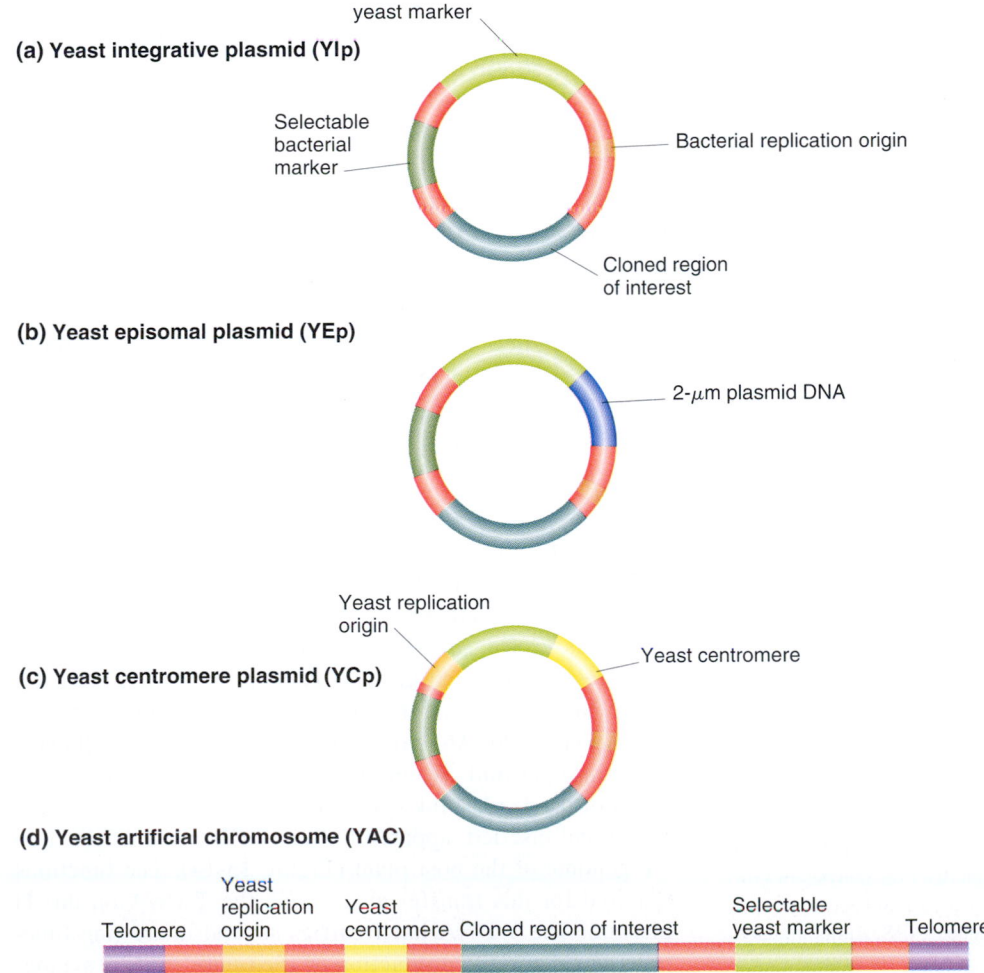

Figure 13-11 Simplified representations of four different kinds of plasmids used in yeast. Each is shown acting as a vector for some genetic region of interest, which has been inserted into the vector. The function of such segments can be studied by transforming a yeast strain of suitable genotype. Selectable markers are needed for the routine detection of the plasmid in bacteria or yeast. Origins of replication are sites needed for the bacterial or yeast replication enzymes to initiate the replication process. (DNA derived from the 2-μm natural yeast plasmid has its own origins of replication.)

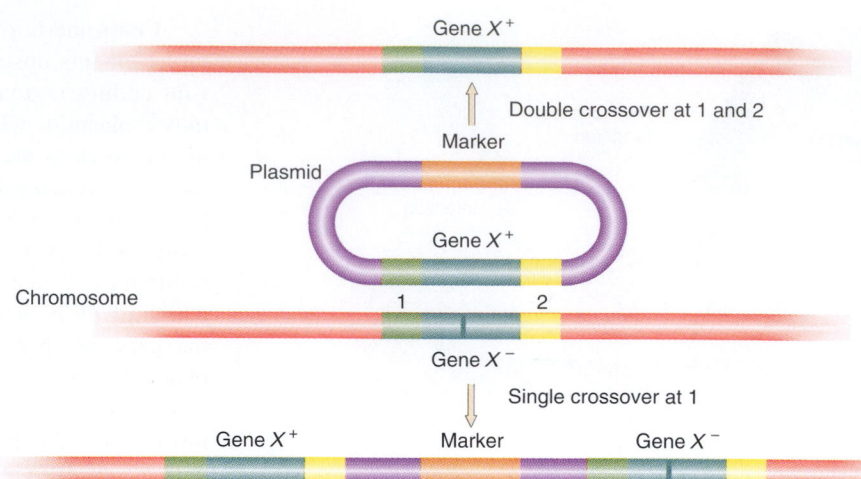

Figure 13-12 Two ways in which a recipient yeast strain bearing a defective X^- can be transformed by a plasmid bearing an active allele (gene X^+). The mutant site of gene X^- is represented as a vertical dark green bar. Single crossovers at position 2 also are possible but are not shown.

This enzyme normally breaks down lactose, but it can also break down an analog of lactose, X-Gal (5-bromo-4-chloro-indolyl-β, D-galactoside), to yield 5-bromo-4-chloroindigo, which is bright blue. The blue color is expressed as a blue yeast colony whenever it is active. (In Chapter 12, we learned that this same enzyme is used to select for inte-grants into the pUC18 plasmid.) Generally, the fusions are constructed in such a way that the promoter of gene X plus a few codons from the structural gene X are fused in the cor-rect reading frame to the region encoding the enzymatically active part of β-galactosidase. These constructs can be transformed on a nonintegrative vector.

Yeast artificial chromosomes have been extensively used as cloning vectors for large sections of eukaryotic (es-pecially human) DNA. Consider, for example, that the size of the region encoding blood-clotting factor VIII in humans is known to span about 190 kb and that the gene for Duchenne muscular dystrophy spans more than 1000 kb. Furthermore, the large size of mammalian genomes in gen-eral means that libraries made in bacterial vectors would be huge. Yeast artificial chromosomes carry much longer in-serts, as many as 1000 kb, and the library size is corre-spondingly smaller. We return to this topic in Chapter 14.

MESSAGE ···

Yeast vectors can be integrative, can autonomously replicate, or can resemble artificial chromosomes, allowing genes to be isolated, manipulated, and reinserted in molecular genetic analysis.

··

Genetic engineering in plants

Because of their economic significance, plants have long been the subject of genetic analysis aimed at developing im-proved varieties. Recombinant DNA technology has intro-duced a new dimension to this effort because the genome modifications made possible by this technology are almost limitless. No longer is breeding confined to selecting vari-ants within a given species. DNA can now be introduced from other species of plants, animals, or even bacteria.

Ti plasmid. The only vectors routinely used to produce transgenic plants are derived from a soil bacterium called *Agrobacterium tumefaciens*. This bacterium causes what is known as *crown gall disease*, in which the infected plant produces uncontrolled growths (tumors, or galls), normally at the base (crown) of the plant. The key to tumor produc-tion is a large (200-kb) circular DNA plasmid — the Ti (tumor-inducting) plasmid. When the bacterium infects a plant cell, a part of the Ti plasmid — a region called T-DNA — is trans-ferred and inserted, apparently more or less at random, into the genome of the host plant (Figure 13-14). The functions required for this transfer are outside the T-DNA on the Ti plasmid. The T-DNA itself carries several interesting func-tions, including the production of the tumor and the synthe-

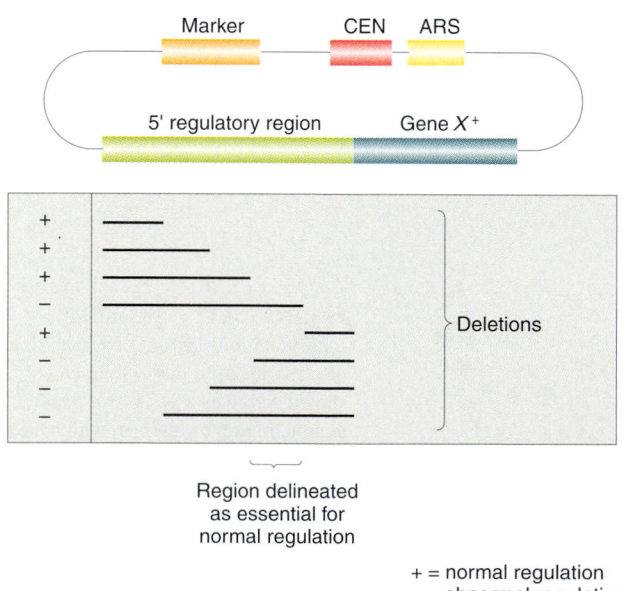

Figure 13-13 The regulation of the yeast gene X^+ can be studied by manipulating its regulatory region through deletion analysis in vitro and then transforming the constructs into a yeast strain bearing a defective allele X^-. (CEN, centromere sequence; ARS, autonomously replicating sequence.)

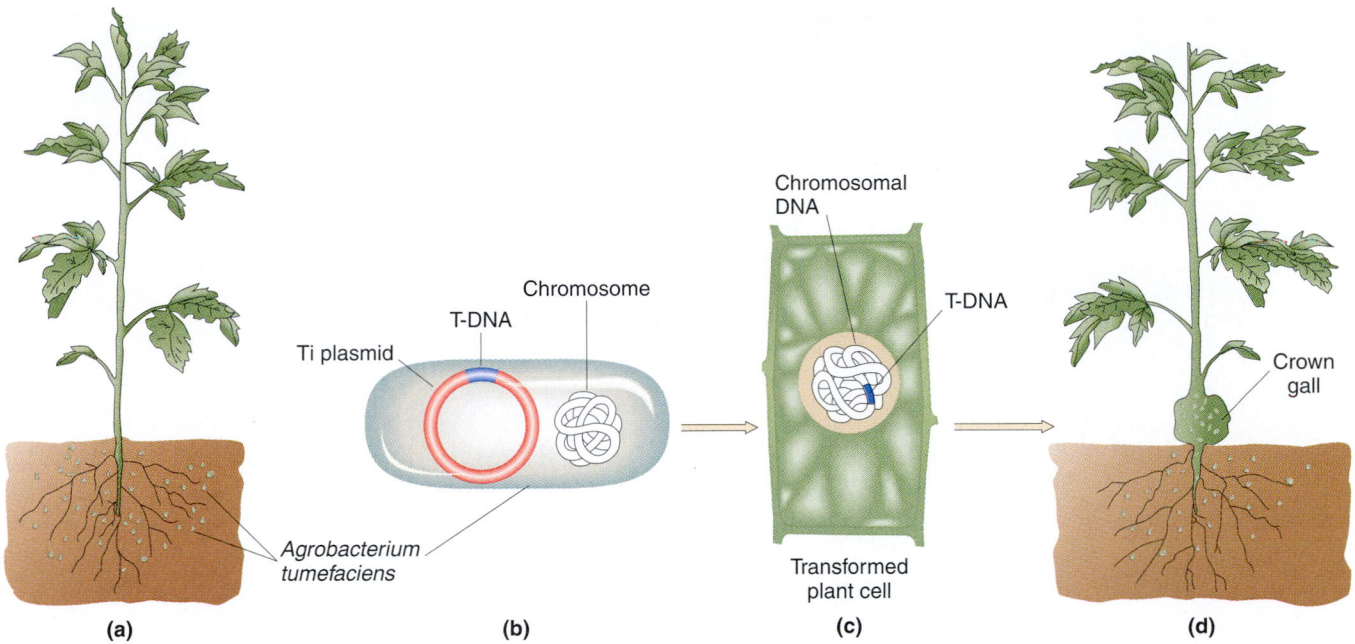

Figure 13-14 In the process of causing crown gall disease, the bacterium *A. tumefaciens* inserts a part of its Ti plasmid — a region called T-DNA — into a chromosome of the host plant.

sis of compounds called *opines*. Opines are actually synthesized to the host plant under the direction of the T-DNA. The bacterium then uses the opines for its own purposes, calling on opine-utilizing genes on the Ti plasmid. Two important opines are nopaline and octopine; two separate Ti plasmids produce them. The structure of Ti is shown in Figure 13-15.

The natural behavior of the Ti plasmid makes it well suited to the role of a plant vector. If the DNA of interest could be spliced into the T-DNA, then the whole package would be inserted in a stable state into a plant chromosome. This system has indeed been made to work essentially in this way, but with some necessary modifications. Let's examine one protocol.

Ti plasmids are too large to be easily manipulated and cannot be readily made smaller, because they contain few unique restriction sites. Consequently, a smaller, intermediate vector initially receives the insert of interest and the various other genes and segments necessary for recombination, replication, and antibiotic resistance. When engineered with the desired gene elements, this intermediate vector can then be inserted into the Ti plasmid, forming a cointegrate plasmid that can be introduced into a plant cell by transformation. Figure 13-16a shows one method of creating the cointegrate. The Ti plasmid that will receive the intermediate vector is first attenuated; that is, it has the entire right-hand region of its T-DNA, including tumor genes and nopaline-synthesis genes, deleted, rendering it incapable of tumor formation — a "nuisance" aspect of the T-DNA function. It retains the left-hand border of its T-DNA, which will be used as the crossover site for incorporation of the intermediate vector. The intermediate vector has had a convenient

cloning segment spliced in, containing a variety of unique restriction sites. The gene of interest has been inserted at this site in Figure 13-16. Also spliced into the vector are a selectable bacterial gene (*spc*^R) for spectinomycin resistance; a bacterial kanamycin-resistance gene (*kan*^R), engineered for expression in plants; and two segments of T-DNA. One segment carries the nopaline-synthesis gene (*nos*) plus the right-hand T-DNA border sequence. The

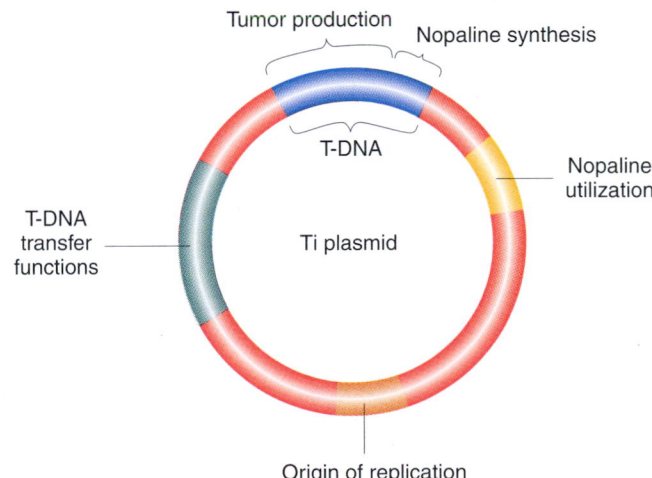

Figure 13-15 Simplified representation of the major regions of the Ti plasmid of *A. tumefaciens*. The T-DNA, when inserted into the chromosomal DNA of the host plant, directs the synthesis of nopaline, which is then utilized by the bacterium for its own purposes. T-DNA also directs the plant cell to divide in an uncontrolled manner, producing a tumor.

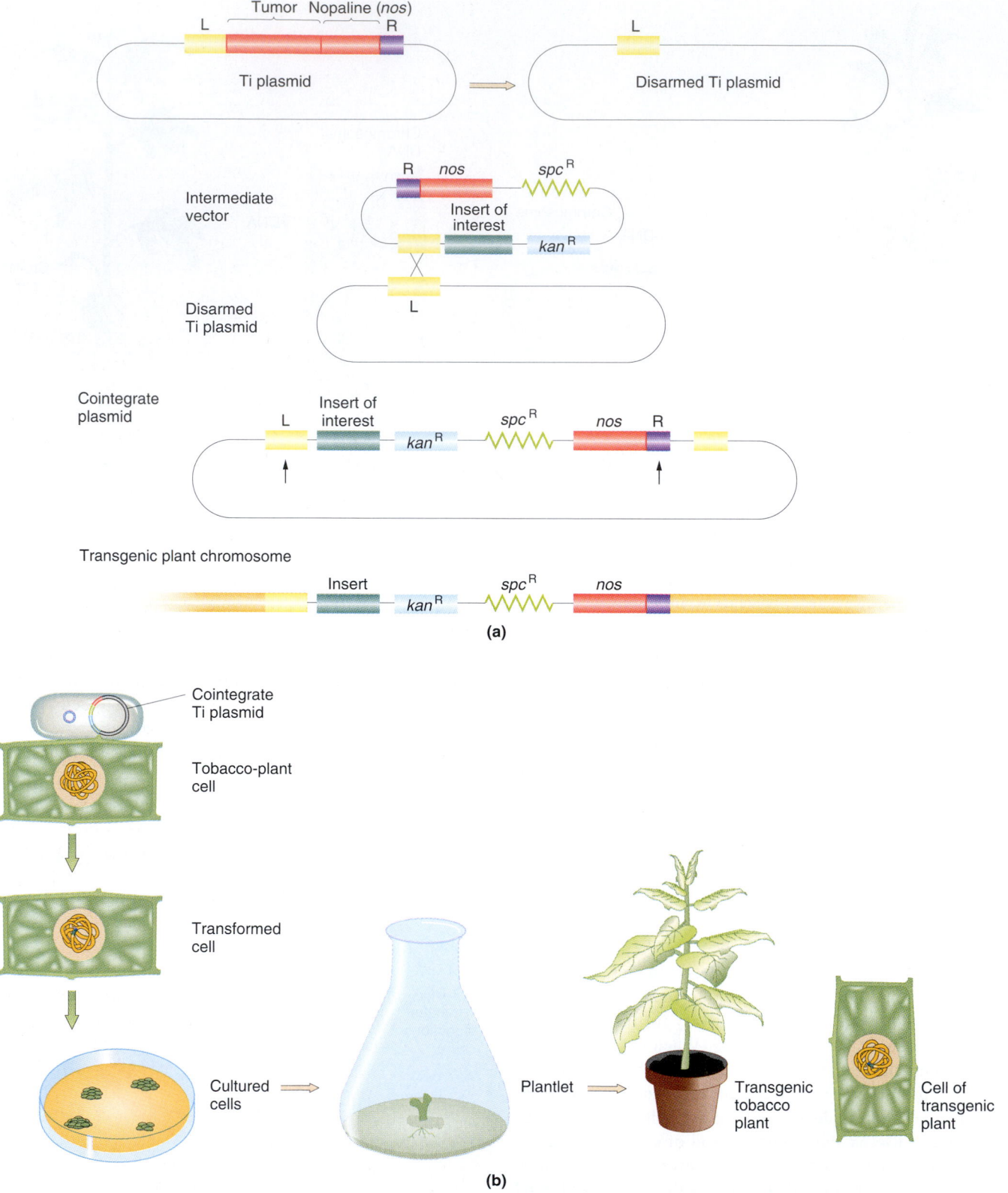

(a)

(b)

Figure 13-16 (a) To produce transgenic plants, an intermediate vector of manageable size is used to clone the segment of interest. In the method shown here, the intermediate vector is then recombined with an attenuated ("disarmed") Ti plasmid to generate a cointegrate structure bearing the insert of interest and a selectable plant kanamycin-resistance marker between the T-DNA borders, which is all the T-DNA necessary to promote insertion. (L, left-hand region; R, right-hand region.) (b) The generation of a transgenic plant through the growth of a cell transformed by T-DNA.

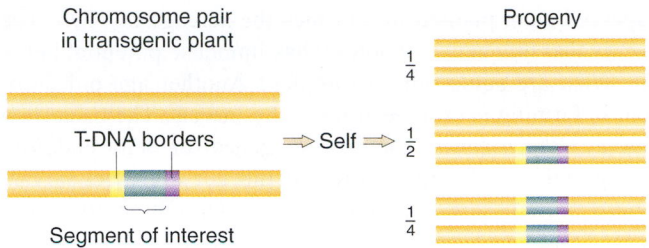

Chromosome pair
in transgenic plant

T-DNA borders

Segment of interest

Progeny

$\frac{1}{4}$

$\Rightarrow$ Self $\Rightarrow$ $\frac{1}{2}$

$\frac{1}{4}$

Figure 13-17 T-DNA and any DNA contained within it are inserted into a plant chromosome in the transgenic plant and then transmitted in a Mendelian pattern of inheritance.

second T-DNA segment comes from near the left-hand border and provides a section for recombination with a homologous part of the left-hand region, which was retained in the disarmed Ti plasmid. After the intermediate vectors have been introduced into *Agrobacterium* cells containing the disarmed Ti plasmids (by conjugation with *E. coli*), plasmid recombinants (cointegrates) can be selected by plating on spectinomycin. The selected bacterial colonies will contain only the Ti plasmid, because the intermediate vector is incapable of replication in *Agrobacterium.*

As Figure 13-16b shows, after spectinomycin selection for the cointegrates, bacteria containing the recombinant double, or cointegrant, plasmid are then used to infect cut segments of plant tissue, such as punched-out leaf disks. If bacterial infection of plant cells takes place, any genetic material between the left and right T-DNA border sequences can be inserted into the plant chromosomes. If the leaf disks are placed on a medium containing kanamycin, the only plant cells that will undergo cell division are those that have acquired the *kan*^{R} gene from T-DNA transfer. The growth of such cells results in a clump, or callus, which is an indication that transformation has taken place. These calli can be induced to form shoots and roots, at which time they are transferred to soil where they develop into transgenic plants (Figure 13-16b). Often only one T-DNA insert is detectable in such plants, where it segregates at meiosis like a regular Mendelian allele (Figure 13-17). The insert can be detected by a T-DNA probe in a Southern hybridization or can be verified by the detection of the chemical nopaline in the transgenic tissue.

Expression of cloned DNA. The DNA cloned into the T-DNA can be any DNA that the investigator wants to insert into the subject plant. A particularly striking foreign DNA that has been inserted with the use of T-DNA is the gene for the enzyme luciferase, which is isolated from fireflies. The enzyme catalyzes the reaction of a chemical called *luciferin* with ATP; in this process, light is emitted, which explains why fireflies glow in the dark. A transgenic tobacco plant expressing the luciferase gene also will glow in the dark when watered with a solution of luciferin (see photograph at the beginning of this chapter). Such manipulation might seem like an attempt to develop technology for

making Christmas trees that do not need lights, but in fact the luciferase gene is useful as a reporter to monitor the function of any gene during development. In other words, the upstream promoter sequences of any gene of interest can be fused to the luciferase gene and put into a plant by T-DNA. Then the luciferase gene will follow the same developmental pattern as that of the normally regulated gene, but the luciferase gene will announce its activity prominently by glowing at various times or in various tissues, depending on the regulatory sequence.

Other genes used as reporters in plants are the bacterial *GUS* (β-glucuronidase) gene, which turns the compound X-Gluc blue, and the bacterial *lac* (β-galactosidase) gene, which turns X-Gal blue. Cells in which these reporters are expressed turn blue, and this blueness can be easily seen either by the naked eye or under the microscope.

Transgenic plants carrying any one of a variety of foreign genes are in current use, and many more are in development. Not only are the qualities of plants themselves being manipulated, but, like microorganisms, plants are also being used as convenient "factories" to produce proteins encoded by foreign genes.

Transgenic crop plants

In November 1996, several British newspapers ran front-page photographs of inflatable rafts belonging to the Greenpeace organization blocking the entry of a freighter into Liverpool harbor. Nearby, Greenpeace had erected a huge banner on a barge, showing the words "Floodgate: Genetic Pollution." The protest was against the first shipment of genetically engineered soybeans from the United States to Britain. In the next month, Greenpeace was in Hamburg harbor, Germany, using a powerful slide projector to beam the words "Genetic experiment — don't buy it" onto the side of a similar freighter. The soybeans contained a bacterial transgene that conferred resistance to a weedkiller called glyphosate, inserted so that fields of crops could be sprayed and easily kept weed free. The advantages to the farmer are obvious, inasmuch as weeds are a considerable problem in agriculture. The objection of the protesters to such transgenic plants was that the transgenes might be deleterious to human health. One concern is possible allergic reactions to the transgene's protein product. The company that developed the soybeans claimed that there was no danger, and this view was supported by regulatory agencies. Another fear was that, as a result of the crop being resistant, consumers might be exposed to higher levels of herbicides. An additional concern was that the engineered plants might escape and introduce the transgenes into related species.

Whether these objections are valid or not, the reference to a "floodgate" in the Greenpeace banner was correct because transgenic crops are flooding into farms and hence into the marketplace. The popular term for transgenic crop plants is "GM foods" (standing for "genetically modified foods"). In 1998, transgenic plants of corn, soybeans, and

Figure 13-18   *(Top)* European corn borer. (Mycogen.) *(Middle)* Spraying the rows at left with the glyphosate herbicide Roundup has killed the weeds, but the genetically engineered resistant corn survives. (Monsanto Corporation.) *(Bottom)* Bt cotton, genetically engineered to be insect resistant *(left)*, compared with unprotected cotton *(right)*. (Monsanto Corporation.)

rapeseed were planted in 20 times the acreage of 1996. The fact is that this new technology has immense potential benefit. What are some other examples? Another age-old problem in farming is crop destruction by insects. The bacterium *Bacillus thuringensis* contains 96 genes for delta endotoxins, proteins that punch holes in the gut of insect larvae. These proteins are called Bt toxins. The Bt toxin genes, when introduced into plants next to a powerful plant promoter, confer resistance to specific insect pests. For example, a certain Bt gene makes corn resistant to one of its most devastating insect pests, the European corn borer (Figure 13-18). This type of resistance is particularly desirable because it is specific; it has no adverse effects on people, on other mammals, or on harmless insects. Bt toxin genes, like the genes for glyphosate resistance, are inserted into the plant genome by using the T-DNA method. Millions of acres of corn, cotton, and potatoes bearing Bt toxin genes are currently under cultivation in North America.

Another area of great promise is "molecular farming," making transgenic plants that contain health-care products such as pharmaceuticals and vaccines. Plants containing vaccines are particularly important for developing countries because the vaccines are inexpensive, orally administered without specialized techniques (possibly eating one seed is all that is necessary), and easy to deliver to remote areas (vaccines currently require refrigeration and often decay en route).

Genetic engineering in animals

Some of the animals most extensively used as model systems for DNA manipulation are *Caenorhabditis elegans* (a nematode), *Drosophila,* and mice. Versions of many of the techniques considered so far can also be applied in these animal systems.

Transgenic animals.   There are several ways of producing transgenic animals. An example is the production of transgenic *Drosophila* by the injection of plasmid vectors containing P elements into the fly egg (described in detail in Chapter 20). Transgenic *Drosophila* provide us with another illustration of the use of the bacterial *lacZ* gene as a reporter in the study of gene regulation during development. The *lacZ* gene is fused to the promoter region of a *Drosophila* heat-shock gene, which is normally activated by high temperatures. This construct is then used to generate transgenic flies. Subsequent to heat shock, the flies are killed and bathed in X-Gal. The resulting pattern of blue tissues provides information on the major sites of action of the heat-shock gene (Figure 13-19). Another approach to the production of transgenic mammals is by injecting special plasmid vectors into a fertilized egg (considered later in this chapter).

In both these cases, because it is the egg that is initially made transgenic, the extra DNA can find its way into germ-line cells, is then passed on to the progeny derived from these cells, and behaves from then on rather like a regular

Figures 13-21 and 13-22 illustrate, step-by-step, the generation of a knockout mouse. First, a cloned, disrupted gene is used to produce embryonic stem (ES) cells containing a gene knockout (Figure 13-21a). Although recombination of

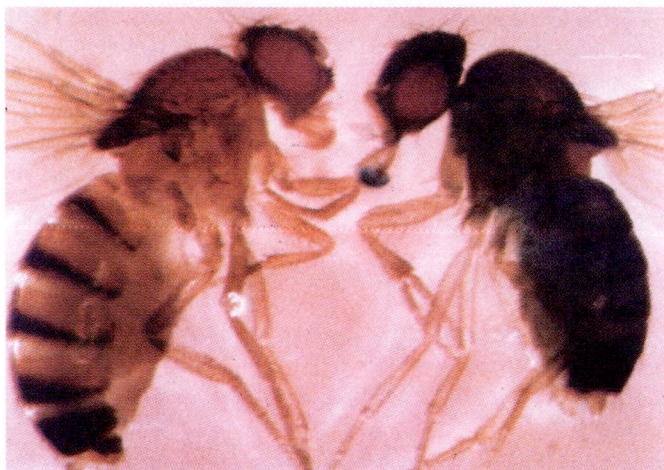

Figure 13-19 Transgenic *Drosophila* expressing a bacterial β-galactosidase gene. *Drosophila* was transformed with a construct consisting of the *E. coli lacZ* gene driven by a *Drosophila* heat-shock promoter. The resulting flies were heat shocked, killed immediately, and stained for β-galactosidase activity, detected by production of a blue pigment. Transformed fly is at right, normal fly at left. (John Lis.)

nuclear gene. Like plants, animals are being manipulated not only to improve the qualities of the animal itself, but also to act as convenient producers of foreign proteins. For example, mammalian milk is easily obtained, so it is a convenient medium for the collection of proteins that are otherwise more difficult to obtain without sacrificing the animal (Figure 13-20).

Gene disruptions and gene replacement in mice. Mice are the most important models for mammals generally. Furthermore, much of the general technology developed in mice can be applied to humans. Two key techniques are the abilities to disrupt a gene (perhaps for a reverse genetics study) and to replace one allele with another. We shall consider examples of these techniques.

As mentioned earlier, gene disruptions are sometimes called *knockouts*. An organism carrying the gene knockout can then be examined for altered phenotypes. Knockout mice are invaluable models for the study of mutants similar to those found in humans. For instance, knockout mice that lack vital DNA repair enzymes (see Chapter 7) have been created to study whether these enzymes control cancer rates.

Figure 13-20 Production of a pharmaceutically important protein in the milk of transgenic sheep. The gene of interest encodes a protein that is of therapeutic importance, such as tissue plasminogen activator used to dissolve blood clots in humans. The gene is placed under control of the β-lactoglobulin promoter, active only in mammary tissue, and is introduced into sheep ova by microinjection of the expression vector into the nucleus. The injected ova are implanted into foster mothers, and progeny expressing the transgene are identified by PCR amplification of chromosomal DNA by using primers from the sequence of the gene of interest. Transgenic sheep express this gene only in mammary tissue and secrete high levels of the gene's protein into the milk, from which it can be purified.

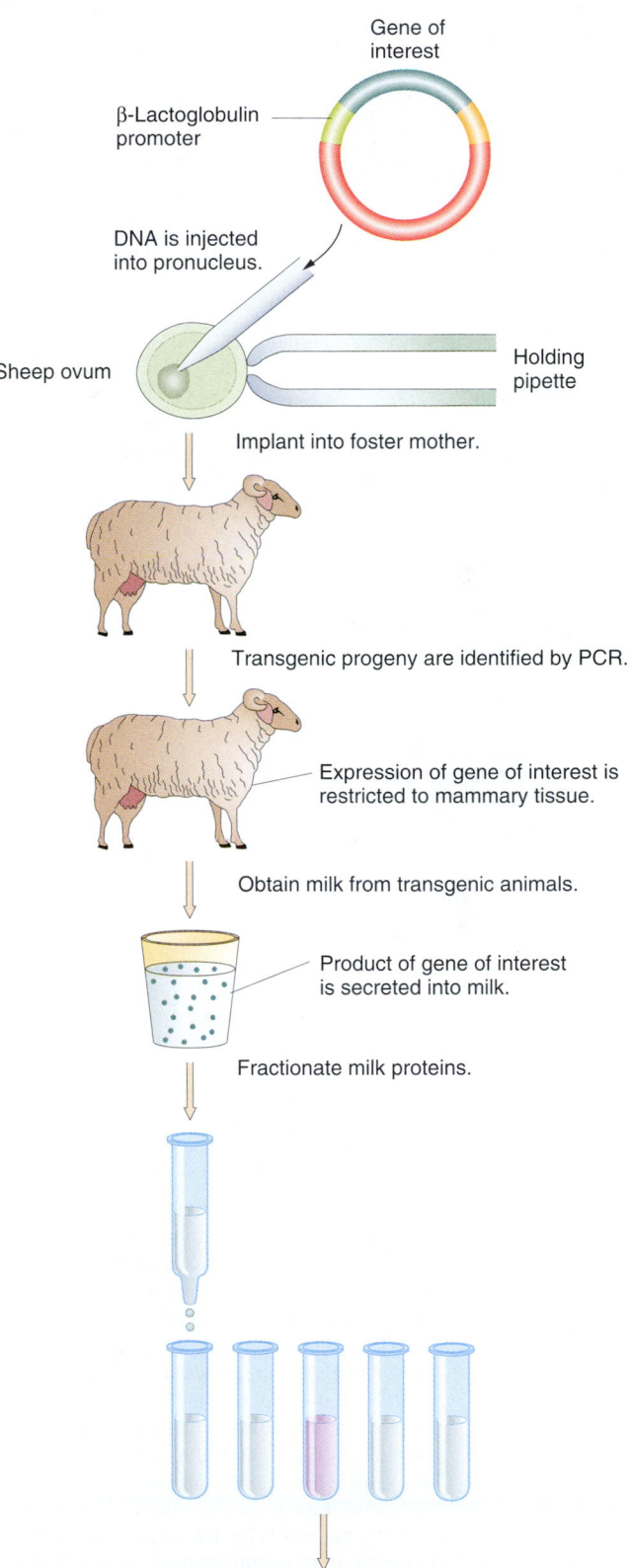

Gene of interest

β-Lactoglobulin promoter

DNA is injected into pronucleus.

Sheep ovum

Holding pipette

Implant into foster mother.

Transgenic progeny are identified by PCR.

Expression of gene of interest is restricted to mammary tissue.

Obtain milk from transgenic animals.

Product of gene of interest is secreted into milk.

Fractionate milk proteins.

Pure product of gene of interest

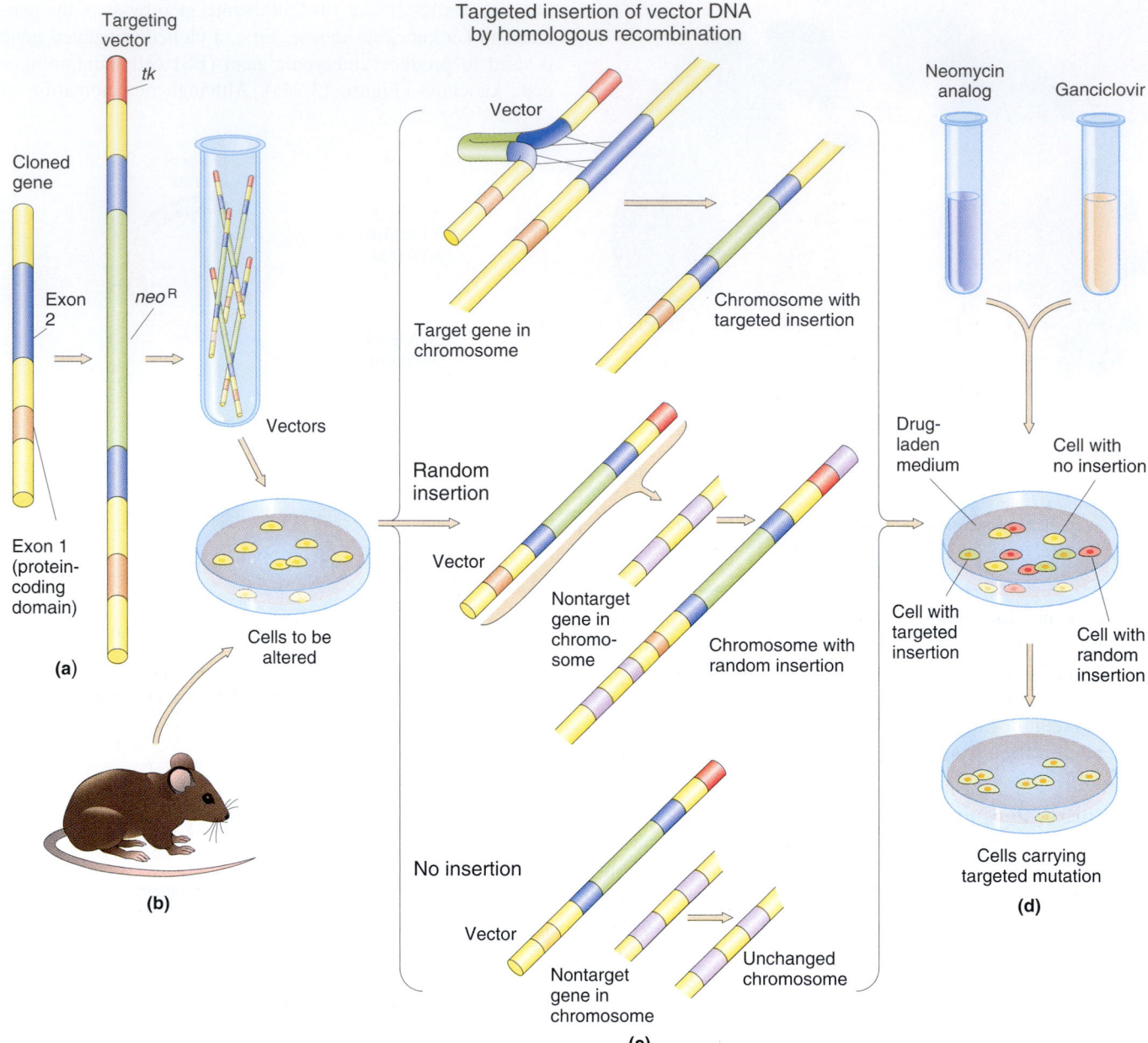

Figure 13-21   The production of cells that contain a mutation in one specific gene, known as a targeted mutation or a gene knockout. (a) Copies of a cloned gene are altered in vitro to produce the targeting vector. The gene shown here has been inactivated by insertion of the neo^R gene (green) into a protein-coding region (blue). The neo^R gene will serve later as a marker to indicate that the vector DNA took up residence in a chromosome. The vector has also been engineered to carry a second marker at one end: the herpes *tk* gene (red). These markers are standard, but others could be used instead. (b) When a vector, with its dual markers, is complete, it is introduced into cells isolated from a mouse embryo. (c) When homologous recombination occurs *(top)*, those regions on the vector (together with any DNA in between) take the place of the original gene, excluding the marker at the tip (red). In many cells, though, the full vector (complete with the extra marker) fits itself randomly into a chromosome *(middle)* or does not become integrated at all *(bottom)*. (d) To isolate cells carrying a targeted mutation, all the cells are put into a medium containing selected drugs, here a neomycin analog (G418) and ganciclovir. G418 is lethal to cells unless they carry a functional neo^R gene, and so it eliminates cells in which there has been no integration of vector DNA (yellow). Meanwhile, ganciclovir kills any cells that harbor the *tk* gene, thereby eliminating cells bearing a randomly integrated vector (red). Consequently, virtually the only cells that survive and proliferate are those harboring the targeted insertion (green). (After M. R. Capecchi, "Targeted Gene Replacement." Copyright © 1994 by Scientific American, Inc. All rights reserved.)

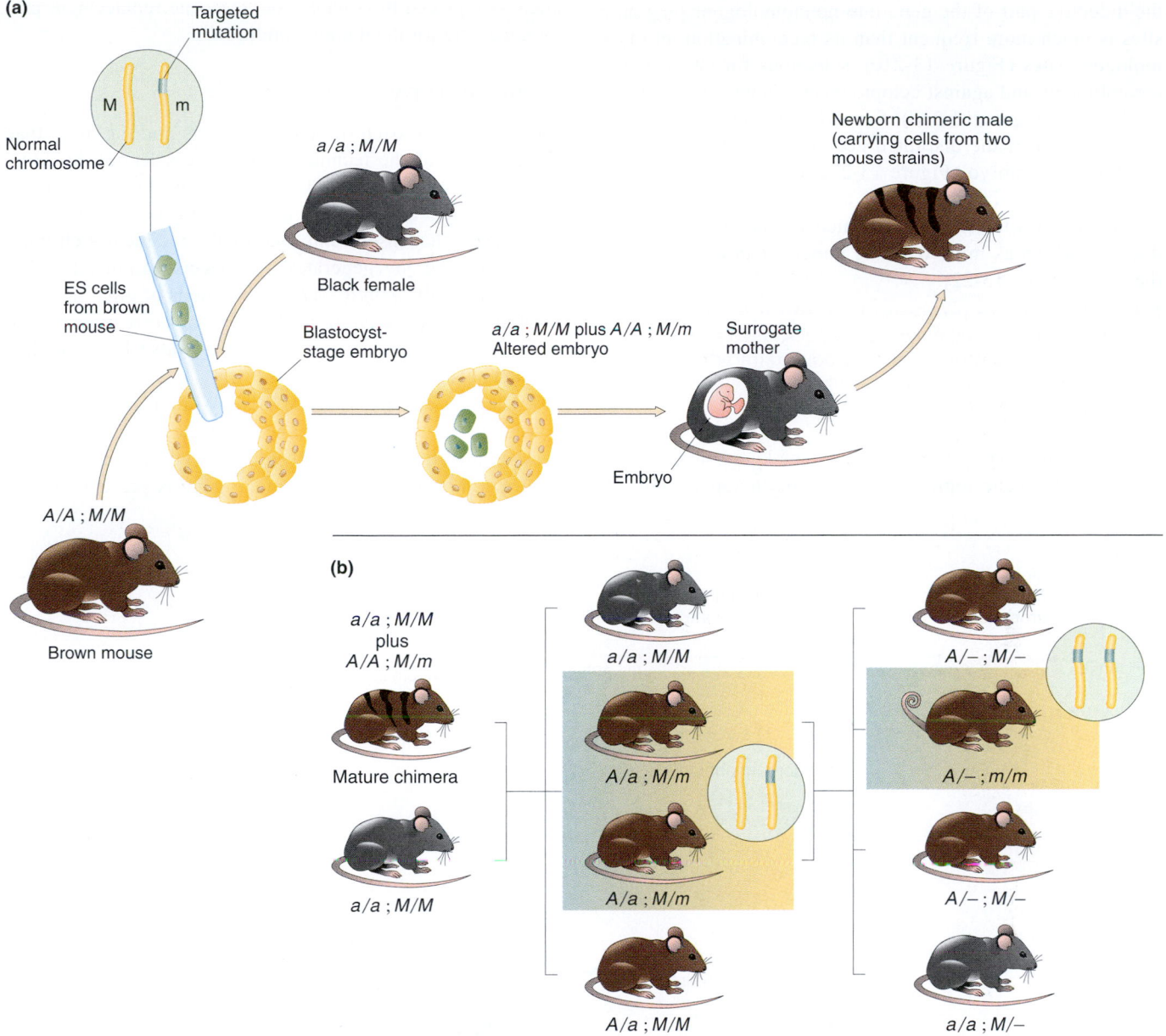

Figure 13-22 Producing a knockout mouse carrying the targeted mutation. (a) Embryonic stem (ES) cells *(green at far left)* are isolated from an agouti mouse strain and altered to carry a targeted mutation in one chromosome. The ES cells are then inserted into young embryos, one of which is shown. Coat color of the future newborns is a guide to whether the ES cells have survived in the embryo. Hence, ES cells are typically put into embryos that, in the absence of the ES cells, would acquire a totally black coat. Such embryos are obtained from a black strain that lacks the dominant agouti allele (see Chapter 4). The embryos containing the ES cells grow to term in surrogate mothers. Newborns with agouti shading intermixed with black indicate that the ES cells have survived and proliferated in an animal. (Such mice are called *chimeras* because they contain cells derived from two different strains of mice.) Solid black coloring, in contrast, would indicate that the ES cells had perished, and these mice are excluded. *A* represents agouti, *a* black; *m* is the targeted mutation and *M* is its wild-type allele. (b) Chimeric males are mated to black (nonagouti) females. Progeny are screened for evidence of the targeted mutation (green in *inset*) in the gene of interest. Direct examination of the genes in the agouti mice reveals which of those animals *(boxed)* inherited the targeted mutation. Males and females carrying the mutation are mated to one another to produce mice whose cells carry the chosen mutation in both copies of the target gene *(insert)* and thus lack a functional gene. Such animals *(boxed)* are identified definitively by direct analyses of their DNA. The knockout results in a curly-tail phenotype. (After M. R. Capecchi, "Targeted Gene Replacement." Copyright © 1994 by Scientific American, Inc. All rights reserved.)

the defective part of the gene into nonhomologous (ectopic) sites is much more frequent than its recombination into homologous sites (Figure 13-21c), selections for site-specific recombinants and against ectopic recombinants can be used, as shown in Figure 13-21d). Second, the ES cells that contain one copy of the disrupted gene of interest are injected into an early embryo (Figure 13-22). The resulting progeny are chimeric, having tissue derived from either recipient or transplanted ES lines. Chimeric mice are then mated to produce homozygous mice with the knockout in each copy of the gene (Figure 13-22).

> **MESSAGE** ·······························
> Fungal, plant, and animal genes can be cloned and manipulated in bacteria and reintroduced into the eukaryote cell, where they generally integrate into chromosomal DNA.

The technology for mammalian gene knockouts is similar to that for gene replacement. In gene therapy, a mutant allele is replaced by a wild type, the gene replacement providing a cure for the mutant condition.

Gene therapy

The general approach of gene therapy is nothing more than an extension of the technique for clone selection by functional complementation (Chapter 12). The functions absent in the recipient as a result of a defective gene are introduced on a vector that inserts into one of the recipient's chromosomes and thereby generates a transgenic animal that has been genetically "cured." The technique is of great potential in humans because it offers the hope of correcting hereditary diseases. However, gene therapy is also being applied to mammals other than humans.

The first example of gene therapy in a mammal was the correction of a growth-hormone deficiency in mice. The recessive mutation *little (lit)* results in dwarf mice. Even though a mouse's growth-hormone gene is present and ap-

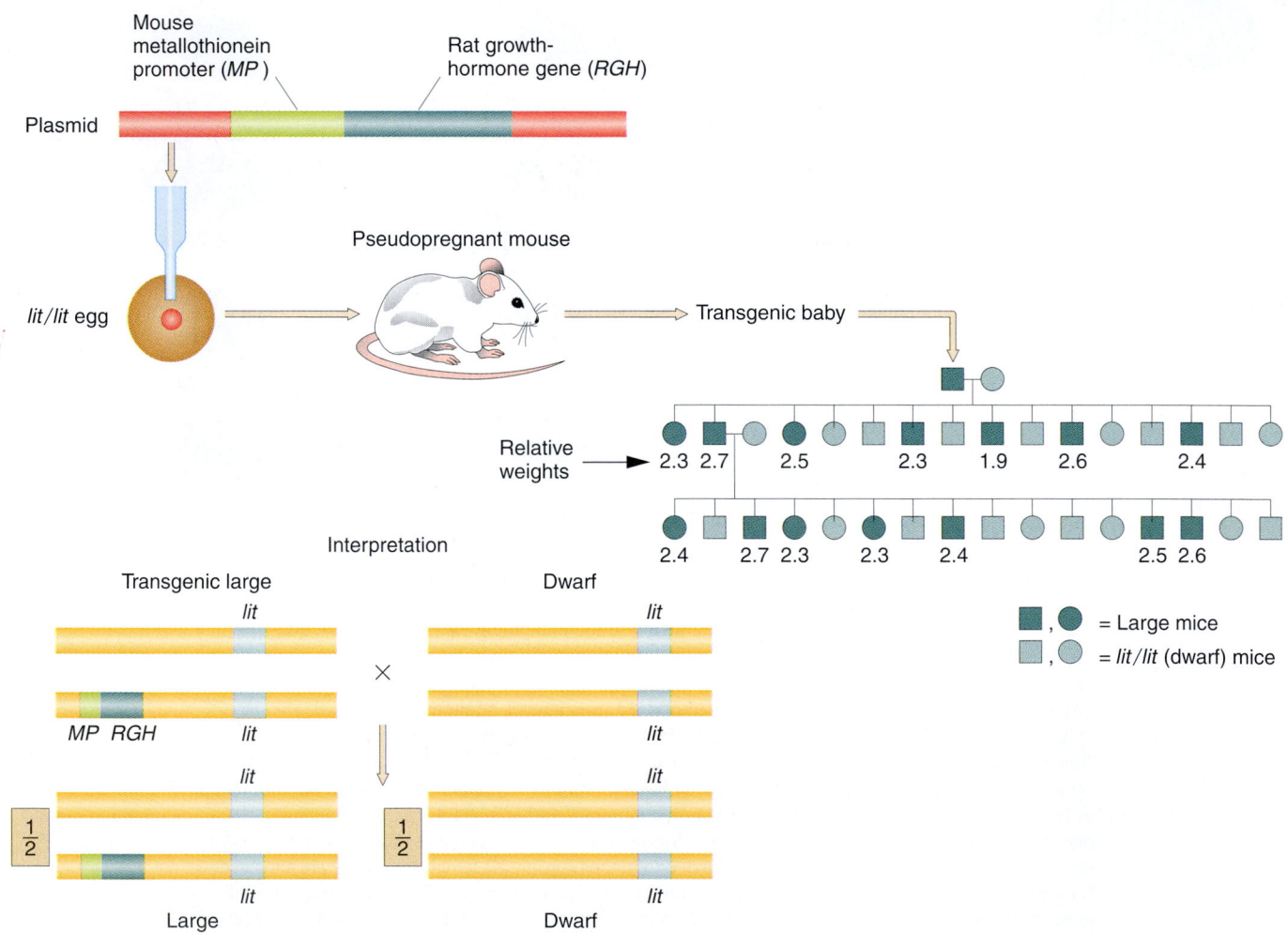

Figure 13-23   The rat growth-hormone gene *(RGH)*, under the control of a mouse promoter region that is responsive to heavy metals, is inserted into a plasmid and used to produce a transgenic mouse. *RGH* compensates for the inherent dwarfism *(lit/lit)* in the mouse. *RGH* is inherited in a Mendelian dominant pattern in the ensuing mouse pedigree.

parently normal, no mRNA for this gene is produced. The initial step in correcting this deficiency was to inject homozygous *lit/lit* eggs with about 5000 copies of a 5-kb linear DNA fragment that contained the rat growth-hormone structural gene *(RGH)* fused to a regulator–promoter sequence from a mouse metallothionein gene *(MP)*. The normal job of metallothionein is to detoxify heavy metals, so the regulatory sequence is responsive to the presence of heavy metals in the animal. The eggs were then implanted into pseudopregnant mice, and the baby mice were raised. About 1 percent of these babies turned out to be transgenic, showing increased size when heavy metals were administered in the course of development. A representative transgenic mouse was then crossed with a homozygous *lit/lit* female. The ensuing pedigree is shown in Figure 13-23. We can see in Figure 13-23 that mice two to three times the weight of their *lit/lit* relatives are produced in subsequent generations, with the rat growth-hormone transgene acting as a dominant allele, always heterozygous in this pedigree. The rat growth-hormone transgene also makes *lit*$^+$ mice bigger (Figure 13-24).

The site of insertion of the introduced DNA in mammals is highly variable, and the DNA is generally not found at the homologous locus. Hence, gene therapy most often provides not a genuine correction of the original problem but a masking of it.

Similar technology has been used to develop transgenic fast-growing strains of Pacific salmon, with spectacular results. A plasmid containing a growth-hormone gene placed next to a metallothionein promoter (all derived from salmon) was microinjected into salmon eggs. A small pro-

Figure 13-25 Effect of introducing a hormone transgene complex with a strong promoter into Pacific salmon. All salmon shown are the same age. (R. H. Devlin, T. Y. Yesaki, C. A. Biagi, E. M. Donaldson, P. Swanson, and W.-K. Chan, "Extraordinary Growth," *Nature* 371, 1994, 209–210.)

portion of the resulting fish proved to be transgenic, testing positive when their DNA was probed with the plasmid construct. These fish were on average 11-fold heavier than the nontransgenic controls (Figure 13-25). Progeny inherited the transgene in the same manner as the mice in the earlier example.

Human gene therapy

Perhaps the most exciting and controversial application of transgenic technology is in human gene therapy, the treatment and alleviation of human genetic disease by adding exogenous wild-type genes to correct the defective function of mutations. We have seen that the first case of gene therapy in mammals was to "cure" a genotypically dwarf fertilized mouse egg by injecting the appropriate wild-type allele for normal growth. This technique (Figure 13-26a) has little application in humans, because it is currently impossible to diagnose whether a fertilized egg cell carries a defective genotype without destroying the cell. (However, in an early embryo containing only a few cells, one cell can be removed and analyzed with no ill effects on the remainder.)

Two basic types of gene therapy can be applied to humans, germ line and somatic. The goal of **germ-line gene therapy** (Figure 13-26b) is the more ambitious: to introduce transgenic cells into the germ line as well as into the somatic cell population. Not only should this therapy achieve a cure of the person treated, but some gametes could also carry the corrected genotype. We have seen that such germinal therapy has been achieved by injecting mice eggs. However, the protocol that is relevant for application to humans is the removal of an early embryo (blastocyst) with a defective genotype from a pregnant mouse and injection with transgenic cells containing the wild-type allele. These cells become part of many tissues of the body, often including the

Figure 13-24 Transgenic mouse. The mice are siblings, but the mouse on the left was derived from an egg transformed by injection with a new gene composed of the mouse metallothionein promoter fused to the rat growth-hormone structural gene. (This mouse weighs 44 g, and its untreated sibling weighs 29 g.) The new gene is passed on to progeny in a Mendelian manner and so is proved to be chromosomally integrated. (R. L. Brinster.)

(a) Injection of fertilized egg

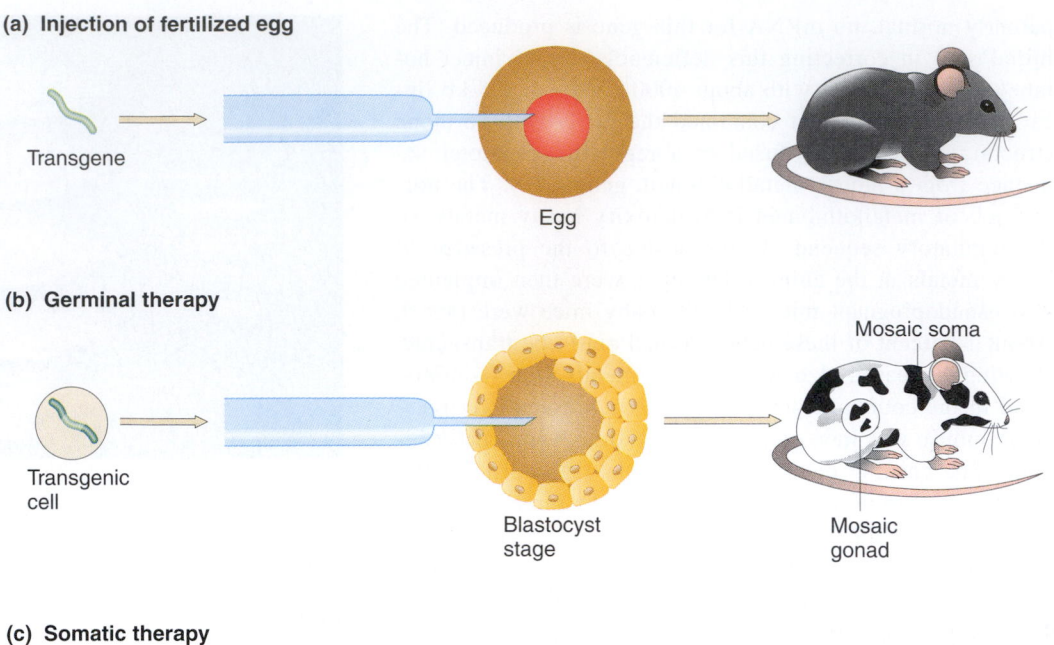

Transgene

Egg

(b) Germinal therapy

Transgenic
cell

Blastocyst
stage

Mosaic soma

Mosaic
gonad

(c) Somatic therapy

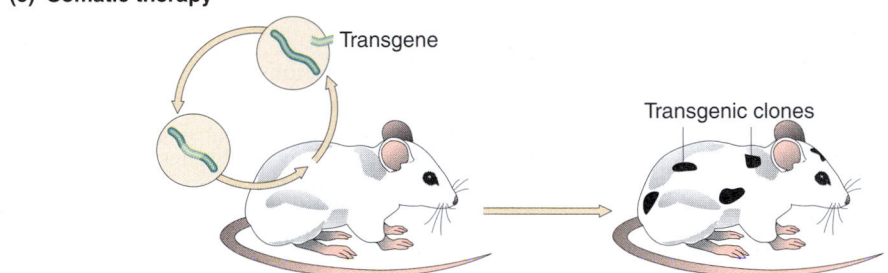

Transgene

Transgenic clones

Figure 13-26   Types of
gene therapy in mammals.

germ line, which will give rise to the gonads. Then the gene can be passed on to some or all progeny, depending on the size of the clone of transgenic cells that lodges in the germinal area. However, no human germ-line gene therapy has been performed to date.

We have seen that most transforming fragments will insert ectopically throughout the genome. This is a disadvantage in human gene therapy not only because of the possibility of the ectopic insert causing gene disruption, but also because, even if the disease phenotype is reversed, the defective allele is still present and can segregate away from the transgene in future generations. Therefore, for effective germinal gene therapy, an efficient targeted gene replacement will be necessary, in which case the wild-type transgene replaces the resident defective copy by a double crossover.

Somatic gene therapy (Figure 13-26c) focuses only on the body (soma). The approach is to attempt to correct a disease phenotype by treating *some* somatic cells in the affected person. At present, it is not possible to render an entire body transgenic, so the method addresses diseases whose phenotype is caused by genes that are expressed predominantly in one tissue. In such cases, it is likely that not all the cells of that tissue need to become transgenic; a percentage of cells being transgenic can ameliorate the overall disease symptoms. The method proceeds by removing some

cells from a patient with the defective genotype and making these cells transgenic through the introduction of copies of the cloned wild-type gene. The transgenic cells are then reintroduced into the patient's body, where they provide normal gene function.

Currently, there are two ways of getting the transgene into the defective somatic cells. Both methods use viruses. The older method uses a disarmed retrovirus with the transgene spliced into its genome, replacing most of the viral genes. The natural cycle of retroviruses includes the integration of the viral genome at some location in one of the host cell's chromosomes. The recombinant retrovirus will carry the transgene along with it into the chromosome. This type of vector poses a potential problem, because the integrating virus can act as an insertional mutagen and inactivate some unknown resident gene, causing a mutation. Another problem with this type of vector is that a retrovirus attacks only proliferating cells such as blood cells. This procedure has been used for somatic gene therapy of severe combined immunodeficiency disease (SCID), otherwise known as *bubble-boy disease*. This disease is caused by a mutation in the gene encoding the blood enzyme adenosine deaminase (ADA). In an attempt at gene therapy, blood stem cells are removed from the bone marrow, the transgene is added, and the transgenic cells are reintroduced into the blood system. Prognosis for such patients is currently good.

Even solid tissues seem to be accessible to somatic gene therapy. In a dramatic case, gene therapy was administered to a patient homozygous for a recessive mutant allele of the *LDLR* gene for low-density-lipoprotein receptor (genotype *LDLR*⁻/*LDLR*⁻). This mutant allele increases the risk of atherosclerosis and coronary disease. The receptor protein is made in liver cells, so 15 percent of the patient's liver was removed, and the liver cells were dissociated and treated with retrovirus carrying the *LDLR*⁺ allele. Transgenic cells were reintroduced back into the body by injection into the portal venous system, which takes blood from the intestine to the liver. The transgenic cells took up residence in the liver. The latest reports are that the procedure seems to be working and the patient's lipid profile has improved.

The other vector used in human gene therapy is the adenovirus. This virus normally attacks respiratory epithelia, injecting its genome into the epithelial cells. The viral genome does not integrate into a chromosome but persists extrachromosomally in the cells, which eliminates the problem of insertional mutagenesis by the vector. Another advantage of the adenovirus as a vector is that it attacks nondividing cells, making most tissues susceptible in principle. Inasmuch as cystic fibrosis is a disease of the respiratory epithelium, adenovirus is an appropriate choice of vector for treating this disease, and gene therapy for cystic fibrosis is currently being attempted with the use of this vector. Viruses bearing the wild-type cystic fibrosis allele are introduced through the nose as a spray. It is also possible to use the adenovirus to attack cells of the nervous system, muscle, and liver.

A promising type of construct that should find use in gene therapy is the human artificial chromosome (HAC). HACs contain essentially the same components as YACs. They were made by mixing human telomeric DNA, genomic DNA, and arrays of repetitive α-satellite DNA (thought to have centrometric activity). To this unjoined mixture was added lipofectin, a substance needed for passage through the membrane, and the complete mixture was added to cultured cells. Some cells were observed to contain small new chromosomes that seemed to have assembled de novo inside the cell from the added components (Figure 13-27). When the technology has been perfected, these HACs should be potent vectors capable of transferring large amounts of human DNA into cells in a stable replicating form.

MESSAGE ···
Gene therapy introduces transgenic cells either into somatic tissue to correct defective function (somatic therapy) or into the germ line for transmission to descendants (germ-line therapy).
···

Using recombinant DNA to detect disease alleles directly

Recessive mutant phenotypes that follow single-gene inheritance are responsible for more than 500 human genetic dis-

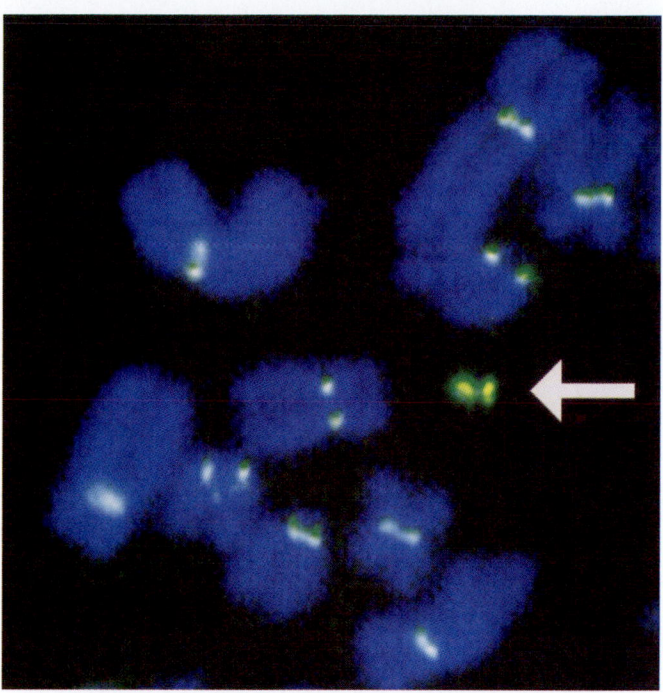

Figure 13-27 An artificial human chromosome (at arrow). (John J. Harrington, G. Van Bokkelen, R. W. Mays, K. Gustashaw, and H. F. Willard, "Formation of de Novo Centromeres and Construction of First-Generation Human Artificial Microchromosomes," *Nature Genetics* 15, 1997, 345–355. Image courtesy of Athersys Inc. and Case Western Reserve.)

eases. A homozygous person resulting from a marriage between two heterozygotes for the same recessive allele will be affected by the disease. Cells derived from the fetus can be screened at an early enough stage to allow the option of abortion of afflicted fetuses. Many of the enzymes or other proteins that are altered or missing in genetic diseases are now known (refer to the list of inborn errors of metabolism, Table 9-4).

To detect such genetic defects, fetal cells can be taken from the amniotic fluid, separated from other components, and cultured to allow the analysis of chromosomes, proteins, enzymatic reactions, and other biochemical properties. This process, **amniocentesis** (Figure 13-28), can identify a number of known disorders; Table 13-1 lists some examples. **Chorionic villus sampling (CVS)** is a related technique in which a small sample of cells from the placenta is aspirated out with a long syringe. CVS can be performed earlier in the pregnancy than can amniocentesis, which must await the development of a large enough volume of amniotic fluid.

Testing physiological properties or enzymatic activity in cultured fetal cells limits the screening procedure to those disorders that affect characters or proteins expressed in the cultured cells. However, recombinant DNA technology improves the screening for genetic diseases in utero, because the DNA can be analyzed directly. In principle, the gene being

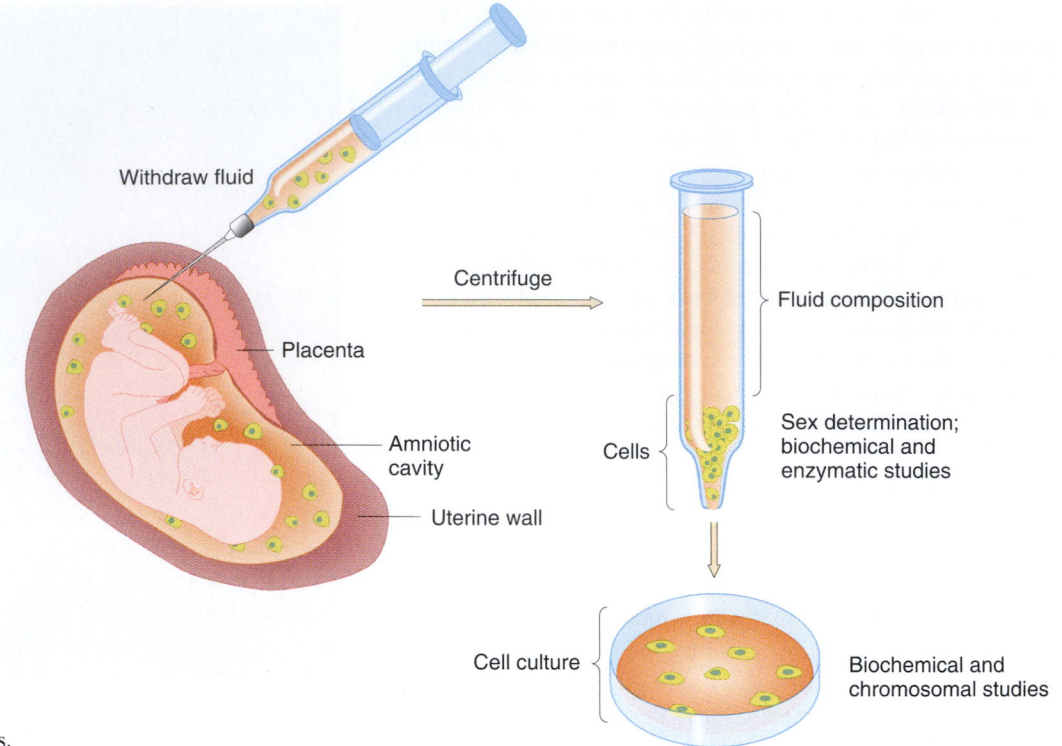

Figure 13-28   Amniocentesis.

tested could be cloned and its sequence compared with that of a cloned normal gene. However, this procedure would be lengthy and impractical, so shortcuts have to be devised to allow more rapid screening. Several of the useful techniques that have been developed for this purpose are explained in this section.

Alteration of restriction site by mutation

Sickle-cell anemia is a genetic disease that is caused by a well-characterized mutational alteration. Affecting approximately 0.25 percent of African Americans, the disease results from an altered hemoglobin, in which the amino acid

13-1 TABLE	Some Common Genetic Diseases
Inborn errors of metabolism	**Approximate incidence among live births**
Cystic fibrosis	1/1600 Caucasians
Duchenne muscular dystrophy	1/3000 boys (X linked)
Gaucher disease (defective glucocerebrosidase)	1/2500 Ashkenazi Jews; 1/75,000 others
Tay-Sachs disease (defective hexosaminidase A)	1/3500 Ashkenazi Jews; 1/35,000 others
Essential pentosuria (a benign condition)	1/2000 Ashkenazi Jews; 1/50,000 others
Classic hemophilia (defective clotting factor VIII)	1/10,000 boys (X linked)
Phenylketonuria (defective phenylalanine hydroxylase)	1/5000 Celtic Irish; 1/15,000 others
Cystinuria (mutated gene unknown)	1/15,000
Metachromatic leukodystrophy (defective arylsulfatase A)	1/40,000
Galactosemia (defective galactose 1-phosphate uridyl transferase)	1/40,000
Hemoglobinopathies	**Approximate incidence among live births**
Sickle-cell anemia (defective β-globin chain)	1/400 U.S. blacks. In some West African populations, the frequency of heterozygotes is 40%.
β-Thalassemia (defective β-globin chain)	1/400 among some Mediterranean populations

Note: Although a vast majority of more than 500 recognized recessive genetic diseases are extremely rare, in combination they constitute an enormous burden of human suffering. As is consistent with Mendelian mutations, the incidence of some of these diseases is much higher in certain racial groups than in others.

Source: J. D. Watson, M. Gilman, J. Witkowski, and D. T. Kurtz, *Recombinant DNA,* 2d ed. Scientific American Books. Copyright © 1992 by J. D. Watson, M. Gilman, J. Witkowski, and D. T. Kurtz.

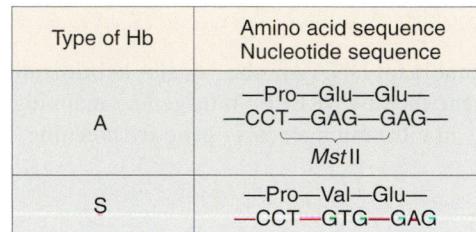

Type of Hb	Amino acid sequence Nucleotide sequence
A	—Pro—Glu—Glu— —CCT—GAG—GAG— *Mst*II
S	—Pro—Val—Glu— —CCT—GTG—GAG

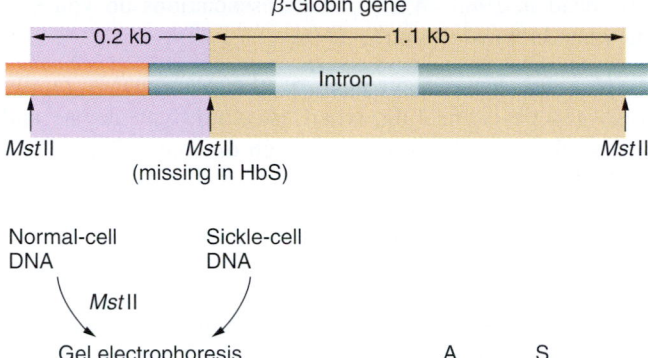

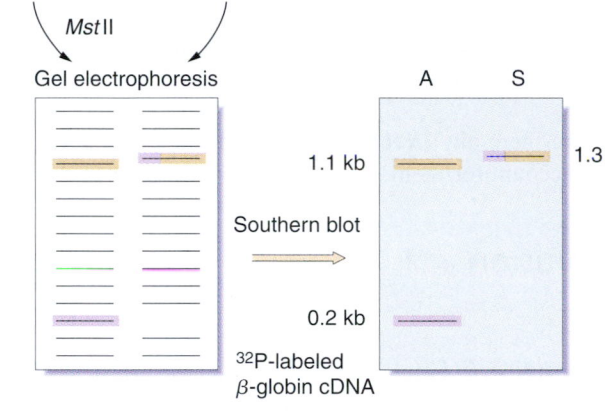

Figure 13-29 Detection of the sickle-cell globin gene by Southern blotting. The base change (A → T) that causes sickle-cell anemia destroys a *Mst*II target site that is present in the normal β-globin gene. This difference can be detected by Southern blotting. (After *Recombinant DNA*, 2d ed. Scientific American Books. Copyright © 1992 by J. D. Watson, M. Gilman, J. Witkowski, and M. Zoller.)

valine substitutes for glutamic acid at position 6 in the β-globin chain. The GAG-to-GTG change eliminates a cleavage site for the restriction enzyme *Mst*II, which cuts the sequence CCTNAGG (in which N represents any of the four bases). The change from CCTGAGG to CCTGTGG can thus be recognized by Southern analysis with the use of labeled β-globin cDNA as a probe, because the DNA derived from persons with sickle-cell disease lacks one fragment contained in the DNA of normal persons and contains a large (uncleaved) fragment not seen in normal DNA (Figure 13-29).

Probing for altered sequences

When a genetic disorder can be attributed to a change in a specific nucleotide, synthetic oligonucleotide probes can identify that change. The best example is α_1-antitrypsin deficiency, which leads to a greatly increased probability of developing pulmonary emphysema. The condition results from a single base change at a known nucleotide position. A synthetic oligonucleotide probe that contains the wild-type sequence in the relevant region of the gene can be used in a Southern blot analysis to determine whether the DNA contains the wild-type or the mutant sequence. At higher temperatures, a complementary sequence will hybridize, whereas a sequence containing even a single mismatched base will not.

PCR tests

Because PCR allows an investigator to zero in on a potentially defective DNA sequence, it can be used effectively in the diagnosis of diseases in which a specific mutational site is in question. Primers spanning the site are used, and the amplified DNA can be isolated and sequenced or otherwise compared with the wild type.

We have already covered the detection of linked RFLPs as a more indirect diagnosis of the presence of a disease allele.

MESSAGE ···
Recombinant DNA technology provides sensitive techniques for testing for mutant alleles in people or in embryos in utero.
···

SUMMARY

Cloned genes can be modified with specific mutational changes at specific sites by using the technique of site-directed mutagenesis. RFLPs are useful as markers in chromosome mapping generally. Linkage of an RFLP to a human disease gene is a useful tool in counseling prospective parents and in diagnosing disease in utero. Reverse genetics starts with a gene or protein of unknown function, induces a specific change such as a gene knockout, and then monitors the phenotype to try to understand the gene's function. Biotechnology is the application of recombinant DNA techniques to modify animals, plants, and microbes important in commerce. Transgenic organisms, modified by insertion of specific exogenous DNA, have found a central place in biotechnology because they allow highly specific genome modifications. Human gene therapy is a special application of transgenic technology: germ-line therapy aims at incorporating some transgenic cells into the germ line so that they can be passed on to descendants, and somatic therapy introduces genetically modified cells into the body. Both amniocentesis and chorionic villus sampling use recombinant DNA techniques for direct diagnosis of a disease allele.

CONCEPT MAP

Draw a concept map interrelating as many of the following terms as possible. Note that the terms are listed in no particular order.

recombinant DNA / probe / in situ hybridization / gene therapy / RFLP / transgenic / mapping / in vitro mutagenesis / genetic screening

CHAPTER INTEGRATION PROBLEM

In Chapter 2 we learned how pedigrees can be used to trace a family's genetic history, and in Chapter 12 we examined Southern blots. We can incorporate these ideas into the concepts discussed in this chapter to help find information about human genetic diseases. Huntington's disease (HD) is a lethal neurodegenerative disorder that exhibits autosomal dominant inheritance. Because the onset of symptoms is usually not until the third, fourth, or fifth decade of life, patients with HD usually have already had their children, and some of them inherit the disease. There has been little hope of a reliable pre-onset diagnosis until recently, when a team of scientists searched for and found a cloned probe (called G8) that revealed a DNA polymorphism (actually a tetramorphism) relevant to HD. The probe and its four hybridizing DNA types are shown here; the vertical lines represent *Hin*dIII cutting sites:

```
                                          Extent of
        _____  homology of
                                          G8 probe

     17.5        3.7  1.2  2.3      8.4
    |_____|___|___|_____|____|  DNA A

     17.5          4.9       2.3    8.4
    |_____|_____|_____|____|  DNA B

     15.0        3.7  1.2  2.3      8.4
    |_____|___|___|_____|____|  DNA C

     15.0          4.9       2.3    8.4
    |_____|_____|_____|____|  DNA D
```

a. Draw the Southern blots expected from the cells of people who are homozygous (*A/A*, *B/B*, *C/C*, and *D/D*) and all who are heterozygous (*A/B*, *A/C*, and so on). Are they all different?

b. What do the DNA differences result from in terms of restriction sites? Do you think they are probably trivial or potentially adaptive? Explain.

c. When human-mouse cell lines were studied, the G8 probe bound only to DNA containing human chromosome 4. What does this tell you?

d. Two families showing HD — one from Venezuela, and one from the United States — are checked to determine their G8-hybridizing DNA type. The results are shown in the pedigree at the top of the next page, where solid black symbols indicate HD and slashes indicate family members who

were dead in 1983. What linkage associations do you see, and what do they tell you?

e. If a 20-year-old member of the Venezuelan family needs genetic counseling, what test would you devise and what advice would you give for each outcome? Repeat this for the U.S. family.

f. How might these data be helpful in finding the primary defect of HD?

g. Could these results be useful in counseling other HD families? Explain.

h. Are there any exceptional individuals in the pedigrees? If so, account for them.

◆ Solution ◆

a.

	A/A	B/B	C/C	D/D	A/B	A/C	A/D	B/C	B/D	C/D
17.5	—	—			—	—	—	—		
15.0			—	—	—	—		—	—	—
8.4	—	—	—	—	—	—	—	—	—	—
4.9		—		—	—		—	—	—	—
3.7	—		—		—	—	—	—		—
2.3	—	—	—	—	—	—	—	—	—	—
1.2	—		—		—	—	—	—		—

A/D and *B/C* are identical. The rest are different.

b. The differences in restriction sites come from differences in DNA sequence. There is no evidence on which to base a judgment of either trivial or potentially adaptive differences.

c. The sequence that gave rise to the G8 probe is located on chromosome 4.

d. In the Venezuelan family there is a strong association of the disease with RFLP morph C, suggesting tight linkage. In the U.S. family there is a weak association of this disease with another morph, A. However, this again suggests linkage. The difference between the families might be explained by statistical variation.

e. In each case, test for the relevant polymorphism by digesting with *Hin*dIII and probing with G8.

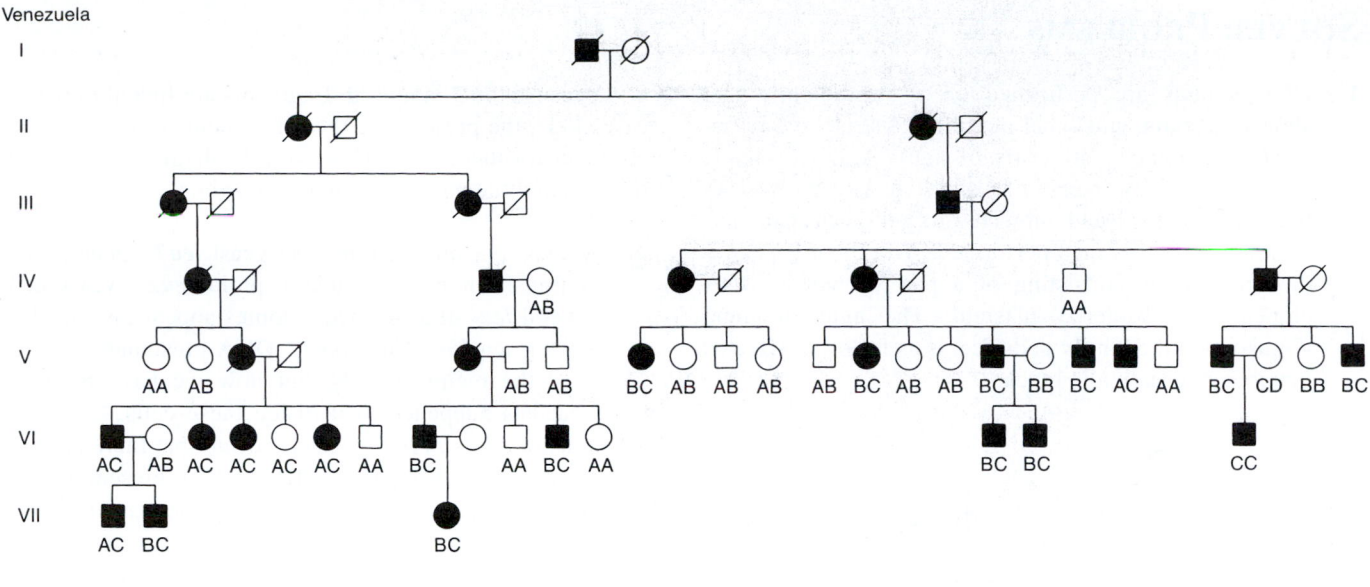

Venezuela

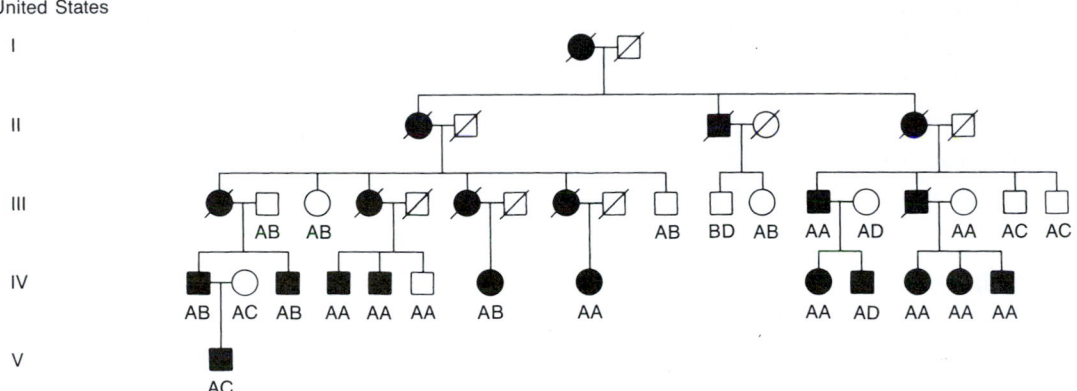

United States

In the family from Venezuela, there is one crossover individual (VI-5) among the 20 that carry the *C* polymorphism. Therefore, the G8 probe is 100% $\left(\frac{1}{20}\right) = 5$ m.u. from the Huntington's disease gene. If the person tests positive for the *C* polymorphism, there is a 95 percent chance of having the gene for Huntington's disease and a 5 percent chance of not having it. If the person tests negative for the *C* polymorphism, there is a 5 percent chance that he has the Huntington's disease gene and a 95 percent chance that he does not have the gene.

The situation in the family from the United States is unclear in comparison with the situation in the family from Venezuela. In the American family, the *A* polymorphism is present in four people with no family history of Huntington's disease who married into the family in question. Their presence makes it impossible to identify crossover individuals unambiguously. It can be assumed that the G8 probe is identifying a polymorphism that is approximately 5 m.u. from the Huntington locus, just as it did in the family from Venezuela. However, the conclusions from testing of the individual in question would vary with the polymorphism genotype of his affected parent.

For instance, if the affected parent were *AA* and the individual in question were *A⁻*, the chance of the person's having inherited the Huntington's disease gene would be 50 percent. If, however, the affected parent were *AD* and the individual in question were *A⁻*, the chance of having inherited the Huntington's disease gene would be 95 percent, unless the unaffected parent also carried *A*. In that case, the risk would be 50 percent.

f. The G8 probe can be used to identify the region in which the Huntington's disease gene is located. The locus can be isolated by means of chromosome walking. The gene can be transcribed and translated, and the protein product can be identified.

g. The family must show the RFLP. Then linkage of HD to one of the RFLP morphs would have to be established.

h. One exceptional person was identified already: Venezuela VI-5, who is a crossover between the polymorphism and the Huntington's disease gene. In the U.S. pedigree there is no individual who is an obligate crossover.

(Solution from Diane K. Lavett.)

SOLVED PROBLEMS

1. DNA studies are performed on a large family that shows a certain autosomal dominant disease of late onset (approximately 40 years of age). A DNA sample from each family member is digested with the restriction enzyme *Taq*I and subjected to gel electrophoresis. A Southern blot is then performed, with the use of a radioactive probe consisting of a part of human DNA cloned in a bacterial plasmid. The autoradiogram, aligned with the family pedigree, is as follows. Affected members are shown in black.

Pedigree

Autoradiogram

Migration

| |
| 5 kb |
| |
| 3 kb |
| 2 kb |

a. Analyze fully the relation between the DNA variation, the probe DNA, and the gene for the disease. Draw the relevant chromosome regions.

b. How do you explain the last son?

c. Of what use would these results be in counseling people in this family who subsequently married?

◆ **Solution** ◆

a. The probe detects three bands of 5, 3, and 2 kb. All family members have the 5-kb band. Five of the six affected persons have the 3- and 2-kb bands, but no unaffected persons have them; therefore, these fragments are likely to be linked in cis arrangement to the disease allele. Because 3 plus 2 adds up to 5, a likely arrangement in the father is as follows, where D is the disease allele and an arrowhead represents the restriction-enzyme target site:

| D | 3 kb | 2 kb |

| d | 5 kb | |

The mother is as follows:

| d | 5 kb |

| d | 5 kb |

b. The last son most likely represents a crossover between the disease locus and the RFLP locus, resulting in a chromosome with D and the 5-kb fragment.

c. Because the 3- and 2-kb fragments are linked to the disease allele, the presence of these fragments can be used as a predictor for the disease. However, the diagnosis has to take into account the possibility of a crossover.

2. A yeast plasmid carrying the yeast $leu2^+$ gene is used to transform nonrevertible haploid $leu2^-$ yeast cells. Several leu^+-transformed colonies appear on a medium lacking leucine. Thus, $leu2^+$ DNA presumably has entered the recipient cells, but now we have to decide what has happened to it inside these cells. Crosses of transformants to $leu2^-$ testers reveal that there are three types of transformants, A, B, and C, representing three different fates of the $leu2^+$ gene in the transformation. The results are:

$$\text{Type A} \times leu2^- \longrightarrow \tfrac{1}{2}\ leu^-$$
$$\tfrac{1}{2}\ leu^+,\ \times \text{ standard } leu2^+$$
$$\longrightarrow \tfrac{3}{4}\ leu^+$$
$$\tfrac{1}{4}\ leu^-$$
$$\text{Type B} \times leu2^- \longrightarrow \tfrac{1}{2}\ leu^-$$
$$\tfrac{1}{2}\ leu^+,\ \times \text{ standard } leu2^+$$
$$\longrightarrow 100\%\ leu^+$$
$$0\%\ leu^-$$
$$\text{Type C} \times leu2^- \longrightarrow 100\%\ leu^+$$

What three different fates of the $leu2^+$ DNA do these results suggest? Be sure to explain *all* the results according to your hypotheses. Use diagrams if possible.

◆ **Solution** ◆

If the yeast plasmid does not integrate, then it replicates independently of the chromosomes. In meiosis, the daughter plasmids would be distributed to the daughter cells, resulting in 100 percent transmission. This percentage was observed in transformant type C.

If one copy of the plasmid is inserted, in a cross with a $leu2^-$ line, the resulting offspring would have a ratio of 1 leu^+ : 1 leu^-. This ratio is seen in type A and type B.

When the resulting leu^+ cells are crossed with standard $leu2^-$ lines, the data from type A cells suggest that the inserted gene is segregating independently of the standard $leu2^+$ gene; so the $leu2^+$ transgene has inserted ectopically into another chromosome.

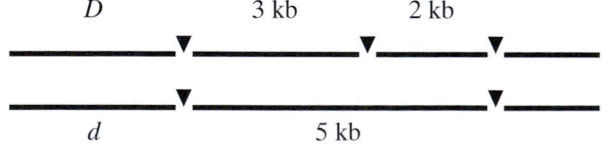

When this new strain is crossed with a standard wild-type strain,

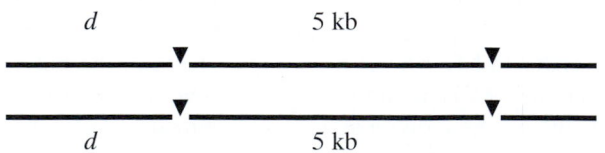

then the following segregation results:

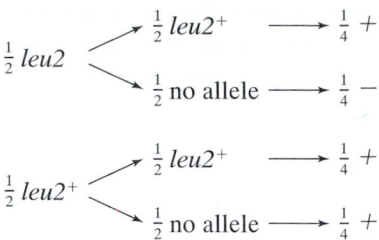

The data from type B cells suggest that the inserted gene has replaced the standard *leu2*$^+$ allele at its normal locus.

Type B   *leu2*$^+$

When type B is crossed with a standard wild type, all the progeny will be *leu*$^+$.

PROBLEMS

1. Transgenic tobacco plants were obtained in which the vector Ti plasmid was designed to insert the gene of interest plus an adjacent kanamycin-resistance gene. The inheritance of chromosomal insertion was followed by testing progeny for kanamycin resistance. Two plants typified the results obtained generally. When plant 1 was backcrossed with wild-type tobacco, 50 percent of the progeny were kanamycin resistant and 50 percent were sensitive. When plant 2 was backcrossed with the wild type, 75 percent of the progeny were kanamycin resistant, and 25 percent were sensitive. What must have been the difference between the two transgenic plants? What would you predict about the situation regarding the gene of interest?

2. In *Neurospora,* which has seven chromosomes, the following chromosomal rearrangements were obtained in different strains:

   a. A paracentric inversion of chromosome 1 (the largest chromosome)

   b. A pericentric inversion of chromosome 1

   c. A reciprocal translocation in which about half of chromosome 1 was exchanged with about half of chromosome 7 (the smallest chromosome)

   d. A unidirectional insertional translocation, in which a part of one chromosome was inserted into another

   e. A disomic ($n + 1$)

   f. A monosomic ($2n - 1$)

   g. A tandem duplication of a large part of chromosome 1

   From all these strains and a normal wild type, DNA was isolated carefully to prevent mechanical breakage, and the samples were subjected to pulsed field gel electrophoresis. Predict the bands that you would expect to see in each case.

3. In a bacterial vector, you have cloned a plant gene that encodes a photosynthesis protein. Now you wish to find out if this gene is active in roots and other non-photosynthetic tissue. How would you go about doing this? Describe the experimental details as well as you can.

4. In *Neurospora,* you have two chromosome 5 probes that detect RFLP loci approximately 20 m.u. apart. In Southern blots of *Pst*I digest of strain 1, probe A picks up two fragments of 1 and 2 kb, and probe B picks up two fragments of 4 and 1.5 kb. In Southern blots of strain 2, probe A detects one band of 3 kb, and probe B detects one band of 5.5 kb. Draw the Southern banding patterns that you would see in *Pst*I digests of individual ascospore cultures from asci, using both probes simultaneously. Be sure to state how many different ascus types that you expect, and be sure to account for the occurrence of crossovers.

5. A cystic fibrosis mutation in a certain pedigree is due to a single nucleotide-pair change. This change destroys an *Eco*RI restriction site normally found in this position. How would you use this information in counseling members of this family about their likelihood of being carriers? State the precise experiments needed. Assume that you detect that a woman in this family is a carrier, and it transpires that she is married to an unrelated man who is also a heterozygote for cystic fibrosis, but, in his case, it is a different mutation in the same gene. How would you counsel this couple about the risks of a child's having cystic fibrosis?

6. In yeast, you have sequenced a piece of wild-type DNA and it clearly contains a gene, but you do not know what gene it is. Therefore, to investigate further, you would like to find out its mutant phenotype. How would you use the cloned wild-type gene to do this? Show your experimental steps clearly.

7. How would you use pulsed field gel electrophoresis to find out what chromosome a cloned gene is on?

8. Bacterial glucuronidase converts a colorless substance called X-Gluc into a bright-blue indigo pigment. The gene for glucuronidase also works in plants if given a plant promoter region. How would you use this gene as a reporter gene to find out in which tissues a plant gene that you have just cloned is normally active? (Assume X-Gluc is easily taken up by the plant tissues.)

9. In mouse *Hin*dIII restriction digests, a certain probe picks a simple RFLP consisting of two alternative alleles of 1.7 kb and 3.8 kb. A mouse heterozygous for a

dominant allele for bent tail and heterozygous for the just described RFLP is mated with a wild-type mouse that shows only the 3.8-kb fragment. Forty percent of the bent-tail progeny are homozygous for the 3.8-kb allele, and 60 percent are heterozygous for the 3.8- and 1.7-kb forms.

a. Is the bent-tail locus linked to the RFLP locus? Draw the parental and progeny chromosomes to illustrate your answer.

b. What RFLP types do you predict among the wild-type offspring, and in what proportions?

10. Probes A and B are used to detect RFLPs in a study of two different haploid populations of yeast. In population 1, gene *A* gives morph *A1* and gene *B* gives morph *B1*; in population 2, gene *A* gives *A2* and gene *B* gives *B2*. These morphs are distinguished by the *Hin*dIII fragments bound by the probe (size is in kilobases).

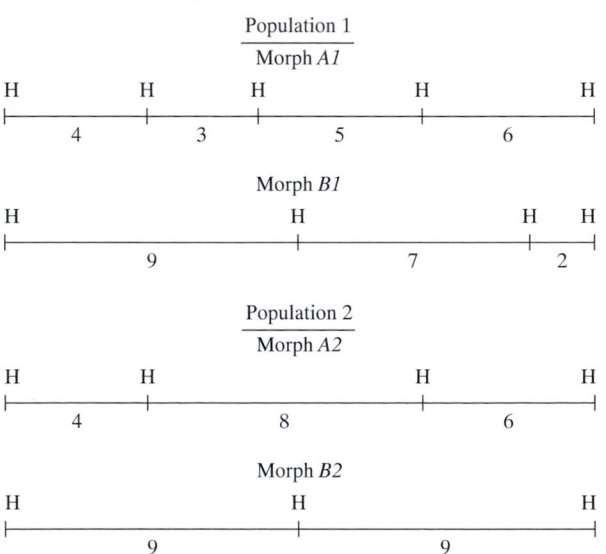

H = *Hin*dIII restriction site

Meiotic products are examined, and the DNA fragments hybridizing to probes A and B are given for each type:

Spore type	Frequency	DNA (*HindIII* FRAGMENTS) Probe A	Probe B
1	15%	4, 3, 5, 6	9, 9
2	15%	4, 8, 6	9, 7, 2
3	35%	4, 3, 5, 6	9, 7, 2
4	35%	4, 8, 6	9, 9

a. How were the different spore DNAs produced?

b. Draw the appropriate chromosomal region(s).

(Problem 10 courtesy of Joan McPherson.)

11. The plant *Arabidopsis thaliana* was transformed by using the Ti plasmid into which a kanamycin-resistance gene had been inserted in the T-DNA region. Two

kanamycin-resistant colonies (A and B) were selected, and plants were regenerated from them. The plants were allowed to self, and the results were as follows:

Plant A selfed ⟶ $\frac{3}{4}$ progeny resistant to kanamycin

$\frac{1}{4}$ progeny sensitive to kanamycin

Plant B selfed ⟶ $\frac{15}{16}$ progeny resistant to kanamycin

$\frac{1}{16}$ progeny sensitive to kanamycin

a. Draw the relevant plant chromosomes in both plants.

b. Explain the two different ratios.

12. Two different circular yeast plasmid vectors (YP1 and YP2) were used to transform *leu$^-$* cells into *leu$^+$*. The resulting *leu$^+$* cultures from both experiments were crossed with the same *leu$^-$* cell of opposite mating type. Typical results were as follows:

YP1 *leu$^+$* × *leu$^-$* ⟶ all progeny *leu$^+$* and the DNA of all these progeny showed positive hybridization to a probe specific to the vector YP1

YP2 *leu$^+$* × *leu$^-$* ⟶ $\frac{1}{2}$ progeny *leu$^+$* and hybridize to vector probe specific to YP2

$\frac{1}{2}$ progeny *leu$^-$* and do not hybridize to YP2 probe

a. Explain the different action of these two plasmids during transformation.

b. If total DNA is extracted from transformants from YP1 and YP2 and digested with an enzyme that cuts once within the vector (and not the insert), predict the results of electrophoresis and Southern analyses of the DNA; use the specific plasmid as a probe in each case.

13. A linear 9-kb *Neurospora* plasmid mar1 seems to be a neutral passenger, causing no apparent harm to its host. It has a restriction map as follows:

*Bgl*II   *Xba*I   *Bgl*II

However, it was suspected that this plasmid sometimes integrates into genomic DNA. To test this idea, the central large *Bgl*II fragment was cloned into a pUC vector, and the resulting vector was used as a probe in a Southern analysis of *Xba*I-digested genomic DNA from a mar1-containing strain. Predict what the autoradiogram would look like if

a. the plasmid never integrates.

b. the plasmid occasionally integrates into genomic DNA.

14. The following pedigree is that of a man affected with dominantly inherited Huntington disease.

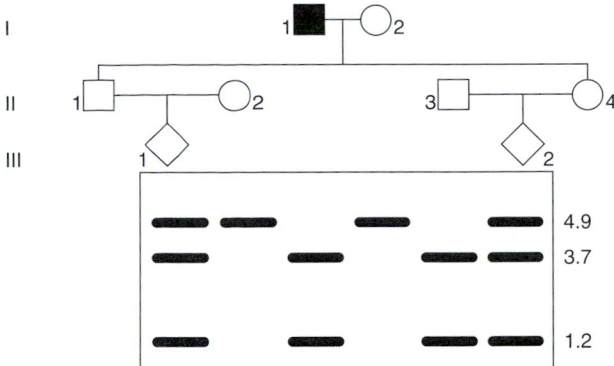

Both of his at-risk children have elected prenatal diagnosis, although they do not wish their status to be evaluated. The Southern blot has been typed for a two-allele system (with allelic fragments at 4.9 kb or 3.7 and 1.2 kb) that shows 4 percent recombination with Huntington disease. For each of the two fetal samples tested, calculate the chance that the fetus will inherit the allele for Huntington. (Problem 14 from Steve Wood.)

15. A *Neurospora* geneticist is interested in the genes that control hyphal extension. He decides to clone a sample of these genes. It is known from previous mutational analysis of *Neurospora* that a common type of mutant has a small-colony ("colonial") phenotype on plates, caused by abnormal hyphal extension. Therefore, he decides to do a tagging experiment by using transforming DNA to produce colonial mutants by insertional mutagenesis. He transforms *Neurospora* cells by using a bacterial plasmid carrying a gene for benomyl resistance *(ben-R)* and recovers resistant colonies on benomyl-containing medium. Some colonies show the colonial phenotype being sought, so he isolates and tests a sample of these colonies. The colonial isolates prove to be of two types as follows:

> Type 1   *col · ben-R* × wild type *(+ · ben-S)*
>
> Progeny $\frac{1}{2}$ *col · ben-R*
>
> $\frac{1}{2}$ *+ · ben-S*
>
> Type 2   *col · ben-R* × wild type *(+ · ben-S)*
>
> Progeny $\frac{1}{4}$ *col · ben-R*
>
> $\frac{1}{4}$ *col · ben-S*
>
> $\frac{1}{4}$ *+ · ben-R*
>
> $\frac{1}{4}$ *+ · ben-S*

a. Explain the difference between these two types of results.

b. Which type should the geneticist use to try to clone the genes affecting hyphal extension?

c. How should he proceed with the tagging protocol?

d. If a probe specific for the bacterial plasmid is available, which progeny should be hybridized by this probe?

 Unpacking the Problem

1. What is hyphal extension, and why do you think anyone would find it interesting?

2. How does the general approach of this experiment fit in with the general genetic approach of mutational dissection?

3. Is *Neurospora* haploid or diploid, and is it relevant to the problem?

4. Is it appropriate to transform a fungus (a eukaryote) with a bacterial plasmid? Does it matter?

5. What is transformation, and in what way is it useful in molecular genetics?

6. How are cells prepared for transformation?

7. What is the fate of transforming DNA if successful transformation takes place?

8. Draw a successful plasmid entry into the host cell, and draw a representation of a successful stable transformation.

9. Does it make any difference to the protocol to know what benomyl is? What role is the benomyl-resistance gene playing in this experiment? Would the experiment work with another resistance marker?

10. What does the word *colonial* mean in the present context? Why did the experimenter think that finding and characterizing colonial mutations would help in understanding hyphal extension?

11. What kind of "previous mutational studies" do you think are being referred to?

12. Draw the appearance of a typical petri plate after transformation and selection. Pay attention to the appearance of the colonies.

13. What is tagging? How does it relate to insertional mutagenesis? How does insertion cause mutation?

14. How are crosses made and progeny isolated in *Neurospora?*

15. Is recombination relevant to this problem? Can an RF value be calculated? What does it mean?

16. Why are there only two colonial types? Is it possible to predict which type will be more common?

17. What is a probe? How are probes useful in molecular genetics? How would the probing experiment be done?

18. How would you obtain a probe specific to the bacterial plasmid?

14

GENOMICS

Key Concepts

Genomics is the molecular characterization of whole genomes.

Genomic analysis begins by using several different techniques to assign genes to specific chromosomes.

The second level of analysis is low-resolution chromosome mapping, based mainly on meiotic recombination analysis. In analysis of humans, special cell hybrids can also be used.

Repetitive DNA variation has provided numerous heterozygous molecular marker loci for use in high-resolution recombination mapping.

The highest level of resolution in genomic mapping analysis is physical mapping of cloned DNA fragments.

The arrangement of cloned DNA fragments into overlapping sets is facilitated by special molecular procedures for tagging the clones.

Ultimately, the sequence of the genome can be obtained by sequencing a representative set of overlapping clones.

Genomics provides an information base for isolating specific genes of interest, including human disease genes.

Functional genomics attempts to understand the broad sweep of genome function at different developmental stages and under different environmental conditions.

Pigs used in the international pig genome mapping project PiGMaP.
Two pure parental strains, Large White and Meisham, differ by many molecular and morphological markers. These lines are crossed and the F_1 hybrids are interbred to produce an F_2. In the F_2, the various patterns of marker assortment are used to determine linkage and to map the genome. In this picture, the reclining sow is an F_1 hybrid and the variously colored nursing piglets are F_2 progeny. *(Reproduced by kind permission of the Roslin Institute, Edinburgh.)*

Genomics aims to understand the molecular organization and information content of the entire genome and of the gene products that the genome encodes. This subdiscipline of genetics takes many of the modern analytical techniques that the geneticist applies to individual genes or to small chromosomal regions and extends them globally to the entire genome. Thus questions of large-scale gene and chromosomal organization and of global gene regulation can be directly addressed. Even though considerable technical hurdles remain, we can nonetheless be sure that, in a very few years, we will have a complete catalog at the nucleotide and amino acid sequence levels of all of the genes and gene products encoded by the genomes of many complex organisms, including humans. Having such catalogs will provide the raw material that can serve as sources of insight into everything from practical matters of human disease and agricultural genetics to basic biological phenomena such as those underlying cell physiology, development, behavior, ecology, and evolution. The availability of such catalogs is an exciting scientific event worthy of the millenial transition and one that promises to have dramatic effects on the process of scientific investigation in biology.

Genomics: an overview

Genomics is divided into two basic areas: **structural genomics,** characterizing the physical nature of whole genomes; and **functional genomics,** characterizing the **transcriptome** (the entire range of transcripts produced by a given organism) and the **proteome** (the entire array of encoded proteins).

The prime directive of structural genomic analysis is the complete and accurate elucidation of the DNA sequence of a representative haploid genome of a given species. When this sequence is known, it opens the door to numerous possibilities. By computational analysis of the sequence, using principles developed by genetic and molecular biological analysis of transcripts and proteins, we can make predictions of all of the encoded proteins. We can analyze other haploid genomes from the same species and develop a statistical picture of the genetic variation within populations of that species. We can compare the genomic sequence of different species and thereby gain an understanding of how the genome has been remodeled in the course of evolution. Studies of **comparative genomics** have already proceeded far enough to reveal that, in related species (for example, within all mammals), there is considerable **synteny** (conserved gene location within large blocks of the genome). Studies of comparative genomics also offer a powerful opportunity to identify highly conserved and therefore functionally important sequence motifs in coding and noncoding genomic DNA. This identification helps researchers confirm predictions of protein-coding regions of the genome and identify important regulatory elements within DNA.

Even though structural genomics is only a little more than a decade old and is already fulfilling the promise of providing complete sequences of many genomes, the leap from classical genetic maps to complete DNA sequence maps did not happen in a single bound. Rather, quite analogous to the way in which one proceeds through several increases in magnification on a light microscope, there was a step-by-step progression in genome-wide map resolution in the development of genomic technologies. In this chapter, we will focus considerable attention on the development of high-resolution genetic and physical mapping technologies that ultimately permitted sequencing of complex genomes. Not only were these technologies invaluable steps on the way to the establishment of sequence-level maps, but they also proved to be extremely important tools in themselves for disease-gene identification and positional cloning.

It quickly became apparent that the availability of completely sequenced genomes merely whetted the scientific appetite for additional global information. In particular, turning the "Rosetta stone" of genomic sequence into rigorous predictions of transcript and protein sequence proved to be a challenge in itself, and so projects to directly characterize the structures and sequences of all RNAs and all polypeptides have evolved. These projects have formed the foundation of **functional genomics.** Typically, transcript structures have been characterized by sequencing full-length cDNAs (see Chapters 12 and 13) and comparing these sequences with those of the corresponding genomic DNA. As we will see toward the end of this chapter, the availability of these cDNA sequences has permitted the development of very dense microspot arrays in which each microspot represents a different mRNA. These microspot arrays, constituting an entire *transcriptome,* can be kept on a single microscope slide and can then be probed by hybridization for the concentrations of transcripts in a given cell type under a given set of environmental conditions. These hybridization experiments permit the assay of literally hundreds of thousands of data points in a single afternoon and provide global information on how a given condition perturbates gene activities in a systematic way.

Similar to approaches used for the transcriptome, ways to systematically and globally identify the *proteome* (that is, all proteins that a species can produce) are under development. Because, as we shall see later in the book (Chapter 23), many biological decision-making processes require protein modifications and changes in protein–protein interactions, understanding the proteome (and the transcriptome for that matter) is just as important as understanding the genome.

Genome projects: practical considerations

Genome projects are in progress in a range of different organisms, including humans and several model organisms. The model systems are the same ones that have been intensively exploited for standard genetic analysis. They include *Mus musculus* (the mouse), *Drosophila melanogaster* (the fruit fly), *Saccharomyces cerevisiae* (baker's yeast), *Caenorhabditis elegans* (a nematode), *Arabidopsis thaliana* (a plant), and several bacteria. The first genomes to have been completely sequenced were the smaller ones. The first were complete viral genome sequences, followed by those of mitochondrial and chloroplast genomes. Then the first of a series of bacterial genomes was sequenced. Here, some of the genomes were chosen for their genetic interests, others for analyzing evolutionary diversity within prokaryotes, and still others because the organisms are important human pathogens. In 1996, the first complete eukaryotic genome sequence, that of the budding yeast, *Saccharomyces cerevesiae,* was published. Because of the scope of these tasks, many of the projects are international ventures, with hundreds of researchers collaborating and sharing data about different regions of the genome. Often groups or even whole nations specialize in analyzing certain specific chromosomes. Because these efforts entail experimentation on a much larger scale than an individual laboratory can mount, genome projects have succeeded by bringing together geneticists, molecular biologists, chemists, physicists, engineers, and computer scientists to develop the necessary technologies including automation of many steps of the process. This interdisciplinary effort in regard to genome analysis is a continuation of the scientific history of genetics, which has benefitted in many ways from intellectual cross-fertilization from other disciplines (Mendel himself was a physicist by training).

MESSAGE ···

Characterizing whole genomes is important to a fundamental understanding of the design principles of living organisms and for the discovery of new genes such as those responsible for human genetic disease.

Prior to genomic analysis, the genetic knowledge of an organism is usually based on relatively low resolution chromosomal maps of genes producing known mutant phenotypes and some molecular markers. From this point, genomic analysis generally proceeds through several steps of increasing resolution:

1. Position genes and molecular markers on high-resolution genetic maps of each chromosome.

2. Physically characterize and position individual cloned DNA fragments relative to one another to create physical maps of each chromosome. The genetic map of the genome can then be anchored to the physical map.

3. Conduct large-scale genomic DNA sequence analysis to produce a complete sequence map of each chromosome. The genetic and physical maps can then be anchored to the sequence map.

This progressively increasing resolution of analysis is paralleled by the increasing resolution of analysis needed to find a specific gene. These general approaches are illustrated in Figure 14-1.

Structural genomics

As its name suggests, the aim of structural genomics is to characterize the structure of the genome. Knowledge of the structure of an individual genome can be useful in manipulating genes and DNA segments in that particular species. For example, genes can be cloned on the basis of knowing where they are in the genome. When a number of genomes have been characterized at the structural level, the hope is that, through comparative genomics, it will become possible to deduce the general rules that govern the overall structural organization of all genomes.

Structural genomics proceeds through increasing levels of analytic resolution, starting with the assignment of genes and markers to individual chromosomes, then the mapping of these genes and markers within a chromosome, and finally the preparation of a physical map culminating in sequencing.

Assigning loci to specific chromosomes

Several different methods are useful in assigning genes or markers to individual chromosomes. Some of these methods have been covered elsewhere in this book but are included here for completeness.

Linkage to known loci.   In well-studied organisms, it is a simple matter to cross a strain carrying the "new," unmapped allele with a set of strains carrying markers spread throughout the genome, each one of known chromosomal location. Meiotic recombinant frequencies of less than 50 percent indicate that the unmapped allele and a specific marker are linked and therefore reside on the same chromosome. Often such linkage data give a rough idea of chromosomal position, perhaps in a specific chromosomal arm or even a specific band.

Pulsed field gel electrophoresis.   If chromosomes are small enough to be separated by PFGE (Figure 14-2), the DNA bands on the gel can be used to locate new genes by hybridization. First, correlations must be made to establish which DNA band corresponds to which chromosome. Chromosome size, translocations between known chromosomes, and hybridization to probes of known location are useful for this purpose. Then a new cloned gene can be used as a probe in a Southern blot of the PFGE gel, and hence its chromosomal locus can be determined.

Human–rodent somatic cell hybrids.   The technique of somatic cell hybridization is used extensively in human genome mapping, but it can in principle be used in many different animal systems. The methodology was outlined in Chapter 6.

High-resolution chromosome maps

The next level of increasing resolution is to determine the position of a gene or molecular marker on the chromosome. This step is important because the genetic maps that are

(a)

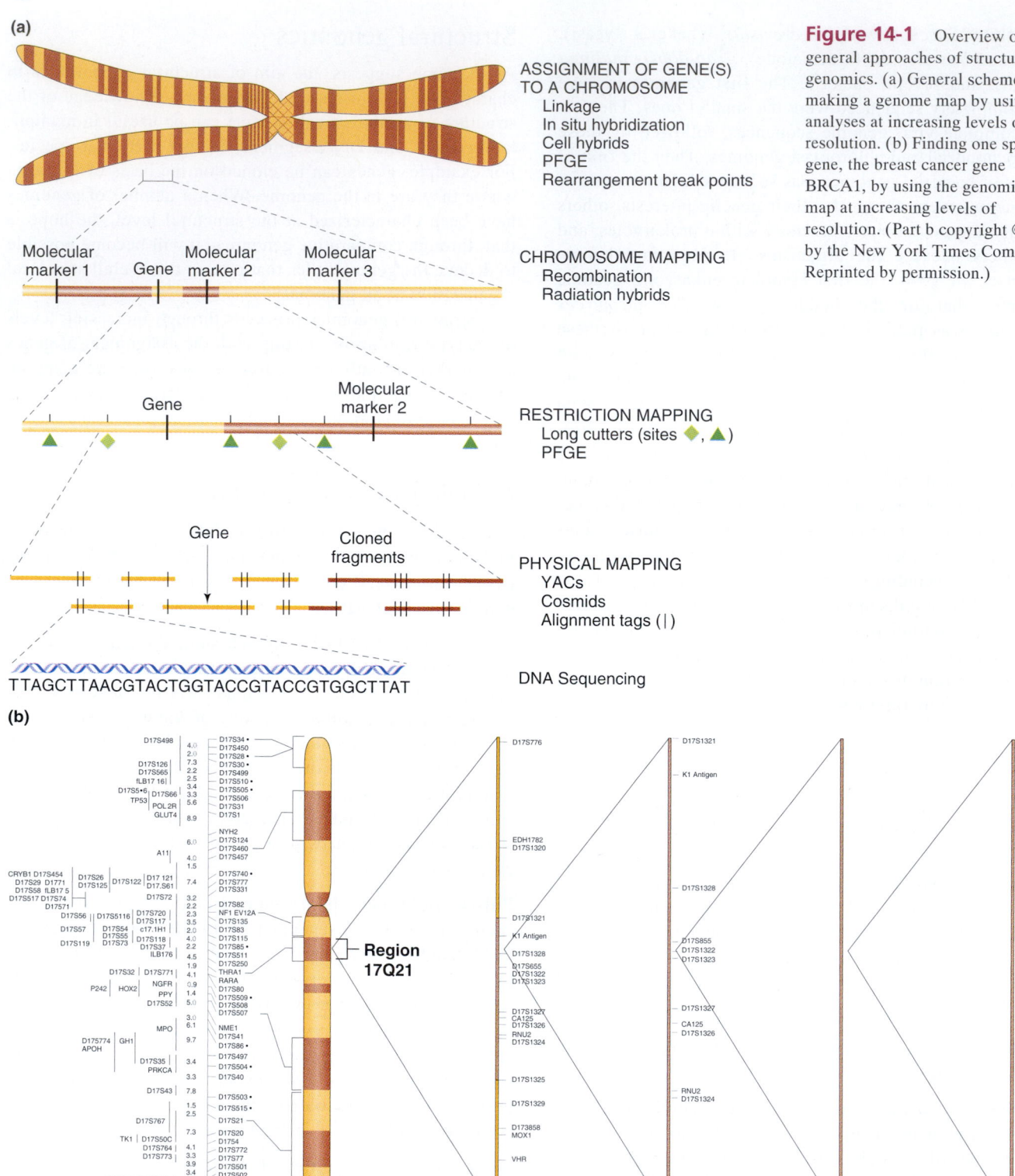

ASSIGNMENT OF GENE(S)
TO A CHROMOSOME
  Linkage
  In situ hybridization
  Cell hybrids
  PFGE
  Rearrangement break points

CHROMOSOME MAPPING
  Recombination
  Radiation hybrids

RESTRICTION MAPPING
  Long cutters (sites ◆, ▲)
  PFGE

PHYSICAL MAPPING
  YACs
  Cosmids
  Alignment tags (|)

DNA Sequencing

TTAGCTTAACGTACTGGTACCGTACCGTGGCTTAT

Figure 14-1 Overview of the general approaches of structural genomics. (a) General scheme for making a genome map by using analyses at increasing levels of resolution. (b) Finding one specific gene, the breast cancer gene BRCA1, by using the genomic map at increasing levels of resolution. (Part b copyright © 1994 by the New York Times Company. Reprinted by permission.)

(b)

produced can be aligned with the physical maps considered in the next section and used to validate the physical maps. In addition, the clones generated as parts of the physical map can be used to help identify the genomic DNA corresponding to the genes on the genetic map. Several different methods are used in localizing genes or markers.

Meiotic mapping by recombination. Meiotic linkage mapping used in genomics is based on the principles of mapping covered in Chapter 5—in other words, on the analysis of recombinant frequency in dihybrid and multihybrid crosses. In experimental organisms such as yeast, *Neurospora, Drosophila,* and *Arabidopsis,* the genes that deter-

(a)

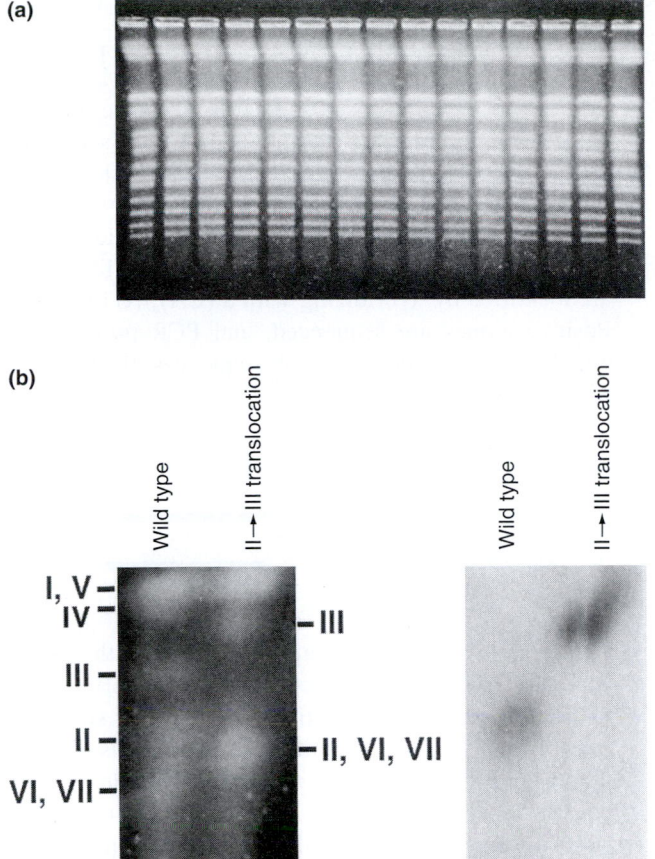

(b)

Stained pulsed-field gel

Autoradiogram of
Southern blot using
a cloned gene probe

Figure 14-2 (a) Pulsed field gel electrophoresis of uncut chromosomal DNA from different strains of yeast. Sixteen bands are resolved. There are 22 chromosomes in yeast, so some have obviously comigrated in the gel. (All lanes were loaded identically.) (b) Pulsed field gel electrophoresis separates *Neurospora* chromosomes and locates a cloned gene to one chromosomal segment: *(left)* the seven chromosomes of wild type and a translocation in which a 100-kb section of chromosome II has been inserted into chromosome III; *(right)* autoradiogram with the use of a probe consisting of a cloned gene of formerly unknown location. The results show that the gene is located on the region of chromosome II that is moved in the translocation. (Part a from Bio-Rad Laboratories; part b from Myron Smith.)

mine qualitative phenotypic differences can be mapped in a straightforward way because of the ease with which controlled experimental crosses (such as testcrosses) can be made. Therefore, in these organisms, the chromosome maps built over the years appear to be full of genes with known phenotypic effect, all mapped to their respective loci.

This is not the case for humans. First, informative crosses are lacking. Second, progeny sample sizes are too small for accurate statistical determination of linkage. Third, the human genome is enormous. In fact, even the as-

signment of a human disease gene to an individual autosome by linkage analysis was a difficult task. (Most genes with known phenotypes were assigned not by RF analysis but by human–rodent cell hybrid mapping.)

Even in those organisms in which the maps appeared to be "full" of loci of known phenotypic effect, measurements showed that the recombinational intervals between known genes had to contain vast amounts of DNA. These intervals, or gaps, could not be mapped by linkage analysis, because there were no markers in those regions. Large numbers of additional genetic markers were needed, which could be used to fill in the gaps to provide a higher-resolution map. This need was met by the discovery of various kinds of **molecular markers.** A molecular marker is a site of heterozygosity for some type of **neutral DNA variation.** Neutral variation is that which is not associated with any measurable phenotypic variation. Such a "DNA locus," when heterozygous, can be used in mapping analysis just as a conventional heterozygous allele pair can be used. Because molecular markers can be easily detected and are so numerous in a genome, when mapped by linkage analysis, they fill the voids between genes of known phenotype. Note that, in mapping, the biological significance of the DNA marker is not important in itself; the heterozygous site is merely a convenient reference point that will be useful in finding one's way around the genome. In this way, markers are being used just as milestones were used by travelers in earlier centuries. Travelers were not interested in the milestones (markers) themselves, but they would have been disoriented without them.

Restriction fragment length polymorphisms. RFLPs (described in Chapters 1, 5, and 13) were the first neutral DNA markers to be applied to genome mapping by recombinant frequency.

DNA markers based on variable numbers of short-sequence repeats. Although RFLPs were the first DNA markers to have been generally used in genomic characterization, in the analysis of animal and plant genomes, they have now been largely replaced by markers based on variation in the number of short tandem repeats. These markers are collectively called **simple-sequence length polymorphisms (SSLPs).** SSLPs have two basic advantages over RFLPs. First, in regard to RFLPs, usually only one or two "alleles," or morphs, are found in a pedigree or population under study. This limits their usefulness; it would be better to have a larger number of alleles that could act as specific tags for a larger variety of homologous chromosomal regions. The SSLPs fill this need because multiple allelism is much more common, and as many as 15 alleles have been found for one locus. Second, the heterozygosity for RFLPs can be low; in other words, if one allele of a locus is relatively uncommon in relation to the other allele, the proportion of heterozygotes (the crucial individuals useful in mapping) will be low. However, SSLPs, in addition to having

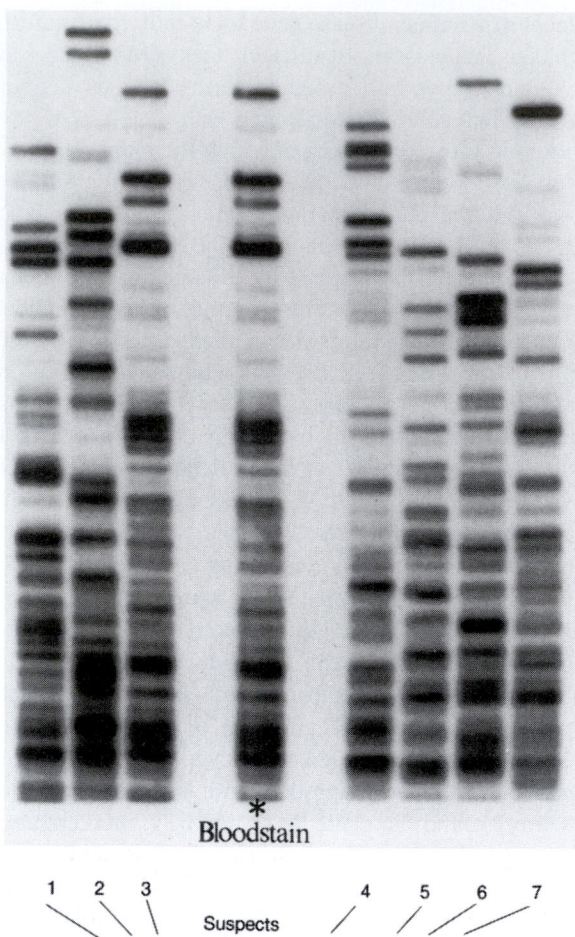

Figure 14-3   DNA fingerprints from a bloodstain at the scene of a crime and from the blood of seven suspects. (Cellmark Diagnostics, Germantown, MD.)

5′ . . . . . . C-A-C-A-C-A-C-A-C-A-C-A-C-A-C-A . . . 3′

3′ . . . . . . G-T-G-T-G-T-G-T-G-T-G-T-G-T-G-T . . . 5′

Probes for detecting these segments are made with the help of the polymerase chain reaction (PCR; see Chapter 12). First, digestion of human DNA with the restriction enzyme *Alu*I results in fragments with an average length of 400 bp, and these fragments are cloned into an M13 phage vector. Phages with $(CA)_n/(GT)_n$ inserts are identified by hybridizing with a $(CA)_n/(GT)_n$ probe. Positive clones are sequenced, and PCR primer pairs are designed on the basis of sequences flanking the repetitive tract:

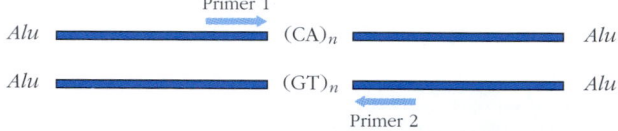

The primers are used to amplify DNA with the use of genomic DNA as a substrate. An individual primer pair will amplify its own repetitive tract and any size variants of it in DNAs from different individuals. A high proportion of PCR primer pairs reveals at least three marker "alleles" of different-sized amplification products. An example of the microsatellite mapping technique is shown in Figure 14-6. Thousands of primer pairs can be made that likewise detect thousands of marker loci. The latest microsatellite marker map of human chromosome 1 is shown in Figure 14-7.

Note some differences in the convenience of RFLP and SSLP analyses. RFLP analysis requires a specific cloned probe to be on hand in the laboratory for the detection of each individual marker locus. Microsatellite analysis requires a primer pair for each marker locus, but these primer sequences can be easily shared throughout the world — distributed by electronic mail and rapidly constructed by using a DNA synthesizer. Minisatellite analysis requires just one probe that detects the core sequence of the repetitive element at loci anywhere in the genome.

Together, the discovery of RFLP and SSLP markers has enabled the construction of a human genetic map with centimorgan [cM, or map unit (m.u.)] density. Although such resolution is a remarkable achievement, a centimorgan is still a huge segment of DNA, estimated in humans to be 1 megabase (1 Mb = 1 million base pairs, or 1000 kb). Currently, even higher resolution genetic maps are being developed on the basis of single-nucleotide polymorphisms (SNPs). An SNP is a single base-pair site within the genome at which more than one of the four possible base pairs is commonly found in natural populations. Several hundred thousand SNP sites are being identified and mapped on the sequence of the genome, providing the densest possible map of genetic differences.

more alleles, show much higher levels of heterozygosity, which makes them more useful than RFLPs in mapping because heterozygotes are the basis for recombination analysis. Two types of SSLPs are now routinely used in genomics.

1. *Minisatellite markers.* Minisatellite markers are based on variation in the number of tandem repeats (VNTRs). A DNA fingerprint is an array of bands in a Southern hybridization of a restriction digest (Figures 14-3 and 14-4). The individual bands of the DNA fingerprint represent different-sized DNA sequences at many different chromosomal positions. If parents differ for a particular band, then this difference becomes a heterozygous ("plus/minus") locus that can be used in mapping. A simple example is shown in Figure 14-5. This same technique can be applied in most organisms with repetitive DNA.

2. *Microsatellite markers.* Recall that microsatellite DNA is a class of repetitive DNA based on dinucleotide repeats. The most common type consists of repeats of CA and its complement GT, as in the following example:

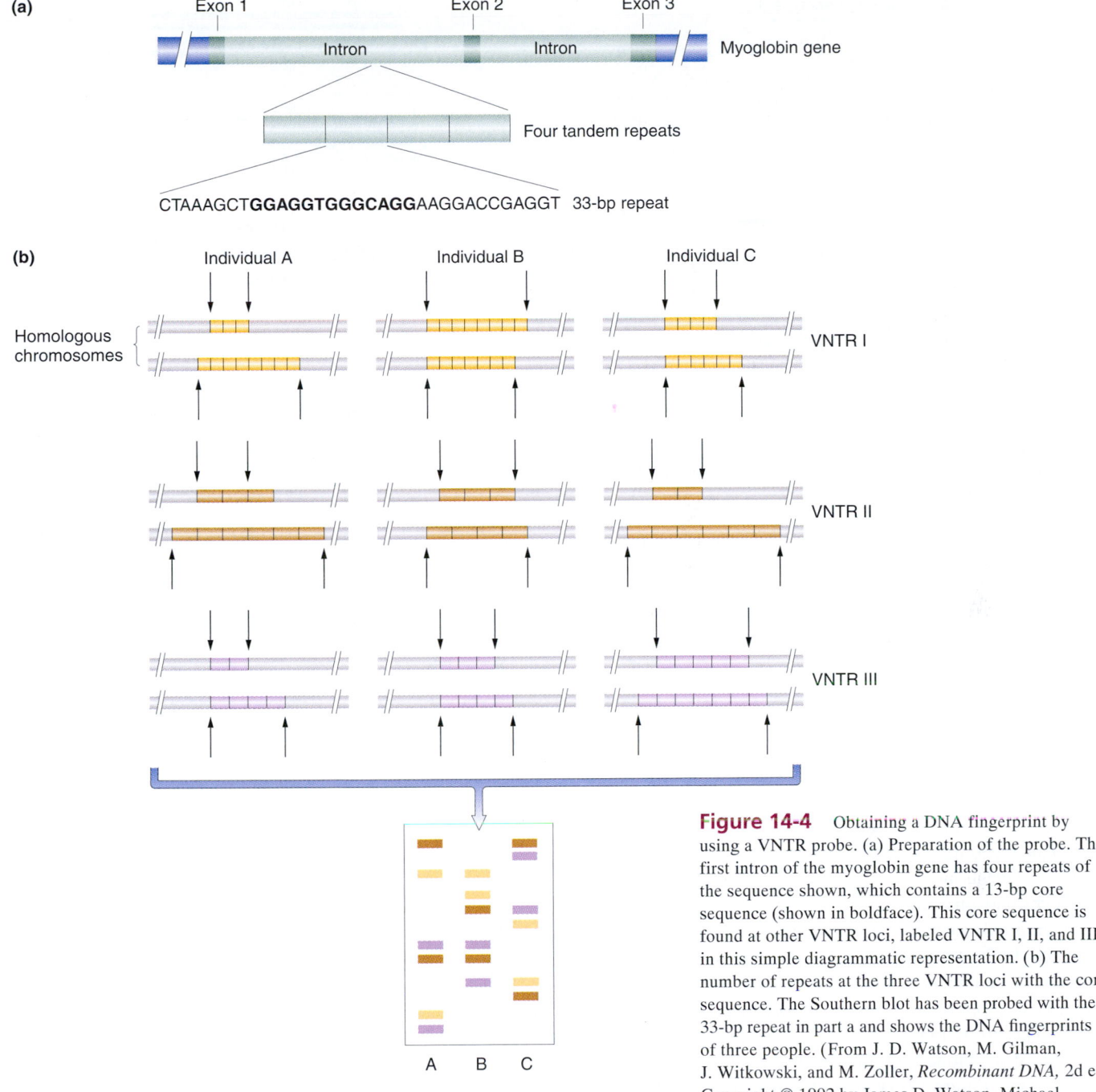

(a)

Exon 1   Exon 2   Exon 3

Intron   Intron   Myoglobin gene

Four tandem repeats

CTAAAGCT**GGAGGTGGGCAGG**AAGGACCGAGGT 33-bp repeat

(b)

Individual A   Individual B   Individual C

Homologous chromosomes

VNTR I

VNTR II

VNTR III

A   B   C

Figure 14-4 Obtaining a DNA fingerprint by using a VNTR probe. (a) Preparation of the probe. The first intron of the myoglobin gene has four repeats of the sequence shown, which contains a 13-bp core sequence (shown in boldface). This core sequence is found at other VNTR loci, labeled VNTR I, II, and III in this simple diagrammatic representation. (b) The number of repeats at the three VNTR loci with the core sequence. The Southern blot has been probed with the 33-bp repeat in part a and shows the DNA fingerprints of three people. (From J. D. Watson, M. Gilman, J. Witkowski, and M. Zoller, *Recombinant DNA,* 2d ed. Copyright © 1992 by James D. Watson, Michael Gilman, Jan Witkowski, and Mark Zoller.)

MESSAGE ···································
Meiotic recombination analysis of loci of genes with known phenotypic effect, RFLP markers, and SSLP markers has resulted in a map of the human genome that is saturated down to the 1 centimorgan (1 map unit) level. SNP analysis promises even greater resolution.

Randomly amplified polymorphic DNAs (RAPDs). A single PCR primer designed at random will often by chance amplify several different regions of the genome. The single sequence "finds" DNA bracketed by two inverted copies of the primer sequence. The result is a set of different-sized amplified bands of DNA (Figure 14-8). In a cross, some of the amplified bands may be unique to one parent, in which case they can be treated as heterozygous (+/−) loci and used as molecular markers in mapping analysis. Notice, too, that the set of amplified DNA fragments (called a **RAPD,** pronounced "rapid") is yet another type of DNA fingerprint that can be used to characterize an individual organism. Such identity tags can be very useful in routine genetic analysis or in population studies.

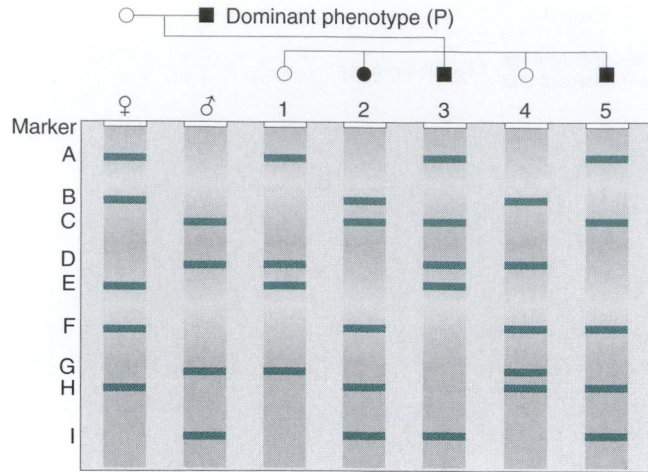

Dominant phenotype (P)
♀ ♂ 1 2 3 4 5
Marker
A
B
C
D
E
F
G
H
I
ANALYSIS EXAMPLES
F and H Always inherited together — linked?
A and B In progeny, always *either* A *or* B — "allelic"?
A and D Four combinations; A and D, A, D, or neither — unlinked?
F, H, and E Always *either* F and H *or* E — closely linked in trans?
Allele *P* Possibly linked to I and C.

Figure 14-5 Using DNA fingerprint bands as molecular markers in mapping. Simplified fingerprints are shown for parents and five progeny. Examples illustrate methods of linkage analysis. Molecular markers can be mapped to one another or to a locus with known phenotypic expression.

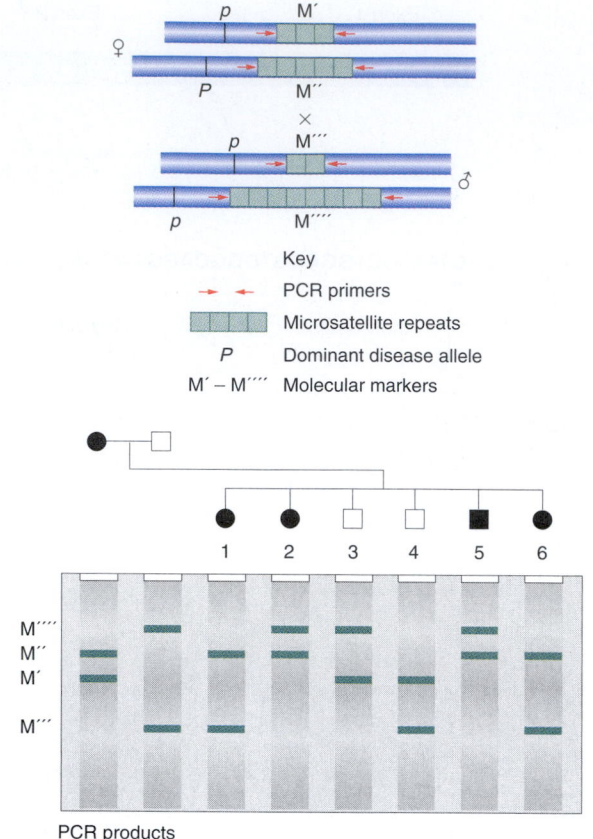

p M′
♀
P M″
×
p M‴
♂
p M⁗
Key
PCR primers
Microsatellite repeats
P Dominant disease allele
M′ – M⁗ Molecular markers
1 2 3 4 5 6
M⁗
M″
M′
M‴
PCR products

Figure 14-6 Using microsatellite repeats as molecular markers for mapping. A hybridization pattern is shown for a family with six children, and this pattern is interpreted at the top of the illustration with the use of four different-sized microsatellite "alleles," M′ through M⁗, one of which (M″) is probably linked in cis configuration to the disease allele *P*.

In situ hybridization. If a cloned gene is available, it can be used to make a labeled probe for hybridization to chromosomes in situ. If the individual chromosomes of the genomic set are recognizable through their banding patterns, size, arm ratio, or other cytological feature, then the new gene can be assigned to the chromosome to which it hybridizes. Furthermore, the locus of hybridization reveals a rough chromosomal position.

Commonly used probe labels are radioactivity and fluorescence. In the process of fluorescence in situ hybridization (FISH), the clone is labeled with a fluorescent dye, and a partially denatured chromosome preparation is bathed in the probe. The probe binds to the chromosome in situ, and the location of the cloned fragment is revealed by a bright fluorescent spot (Figure 14-9). An extension of FISH is chromosome painting. Sets of cloned DNA known to be from specific chromosomes or specific chromosome regions are labeled with different fluorescent dyes. These dyes then "paint" specific regions and identify them under the microscope (Figure 14-10). If a clone of a gene of unknown location is labeled with yet another dye, its position can be established in the painted array.

Rearrangement breakpoints. We shall see in Chapter 17 that mutant alleles giving a new observable phenotype are sometimes caused by a chromosomal rearrangement. Usually such mutations trace to a chromosome break that is part of the rearrangement and that splits the gene in two, disrupt-

Figure 14-7 Linkage map of human chromosome 1, correlated with chromosome banding pattern. The histogram shows the distribution of all markers available for chromosome 1. Some markers are genes of known phenotype, but most are DNA markers based on neutral sequence variation. ▶

A linkage map, based on recombinant frequency analyses of the type described in this chapter, is in the center of the illustration. It shows only some of the markers available. Map distances are shown in centimorgans (cM, or m.u.). The total length of the chromosome 1 map is 356 cM; it is the longest human chromosome.

The positions of some markers are cross-referenced to a diagram of subregions of chromosome 1 based on a standard banding pattern (such a diagram is called an idiogram). These kinds of correlations can be made only by using cytogenetic analysis (Chapter 17) and in situ hybridization. Most of the markers shown on the map are molecular, but several genes (highlighted in light green) also are included:

APOA2	apolipoprotein
ACTN2	actin protein
CRP	C-reactive protein
SPTA1	spectrin protein

(From B. R. Jasney et al., "The Genome Maps 1994," *Science,* 265, 1994, 2055–2070.)

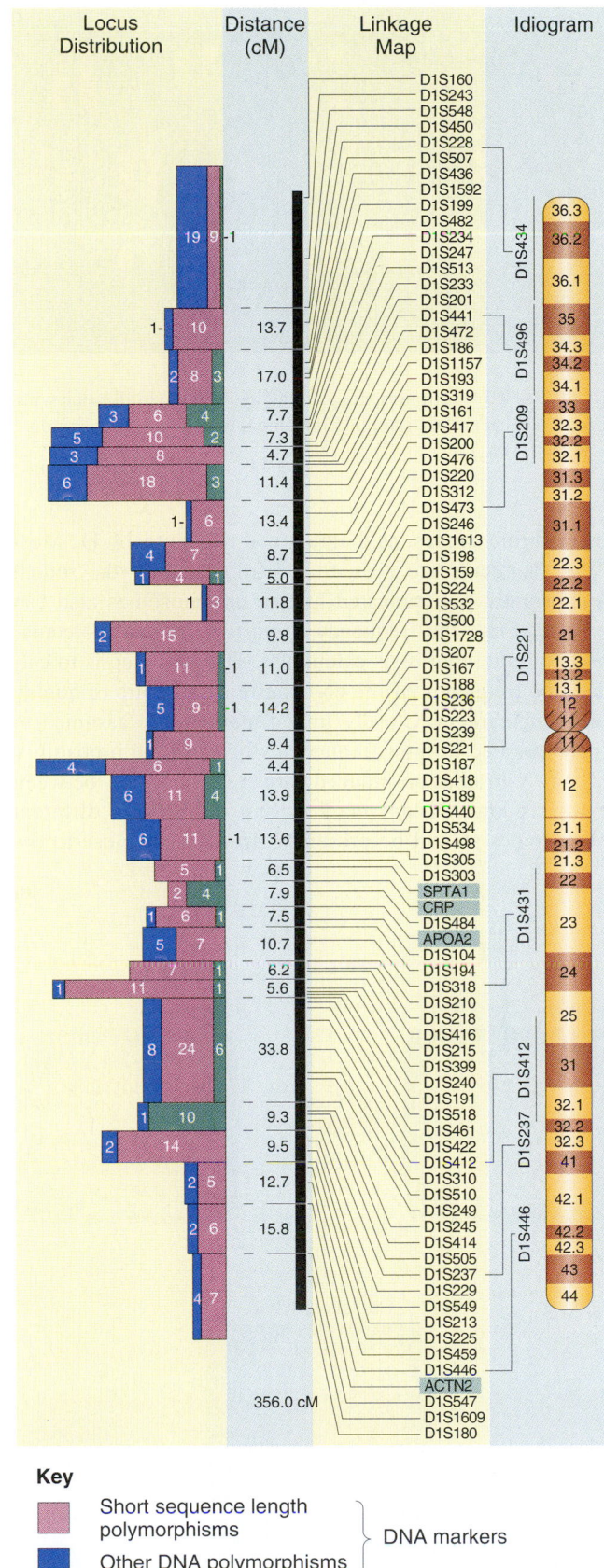

Key

▨ (pink)	Short sequence length polymorphisms
▨ (blue)	Other DNA polymorphisms
▨ (green)	Genes
▨ (light blue)	Genes included on the linkage map

DNA markers } (brackets Short sequence length polymorphisms and Other DNA polymorphisms)

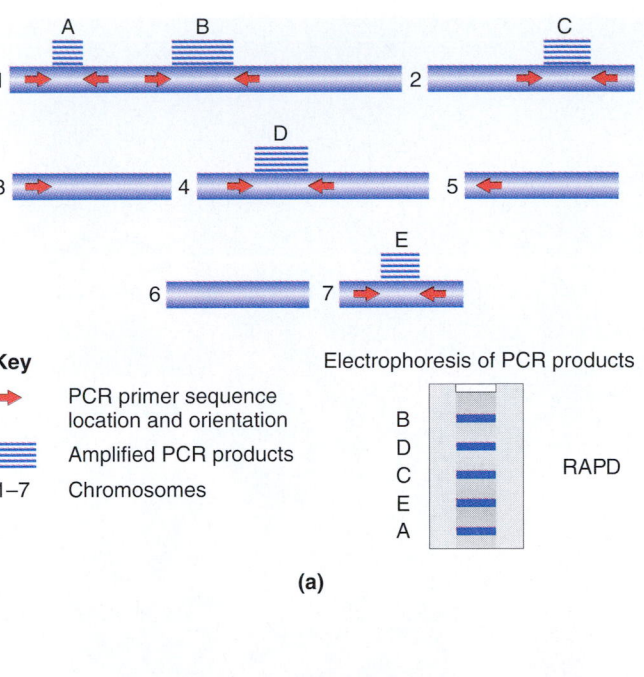

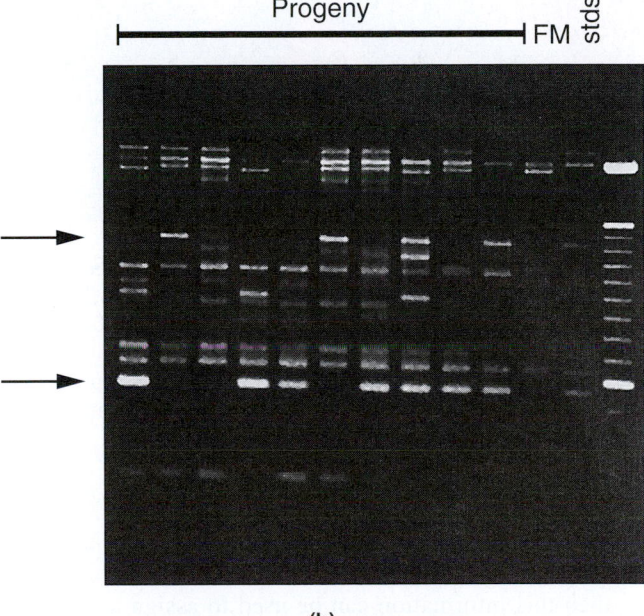

Figure 14-8 (a) Randomly amplified polymorphic DNA analysis (RAPD analysis) provides molecular chromosome markers. If another strain lacks one of the bands, this band can be considered a heterozygous marker locus and used in mapping, as shown in the example in part b. (b) RAPD analysis of a cross in a species of tree. A 10-nucleotide primer was used to amplify regions of genomic DNA in the parental trees (M, male; F, female) and 10 progeny. DNA standards for size calibration are shown in the right lane, marked "stds." The arrows point to two bands that represent two loci in a "dihybrid" testcross. The two alleles of both these loci are expressed as the presence and the absence of a RAPD band. Of the two parents, only the male showed these bands, but the male must have been heterozygous for the presence ($+$ allele) and absence ($-$ allele) of bands at both loci. The male could be designated $1^+/1^- \cdot 2^+/2^-$, and the female $1^-/1^- \cdot 2^-/2^-$. The progeny show various parental and nonparental combinations of these alleles. (John E. Carlson.)

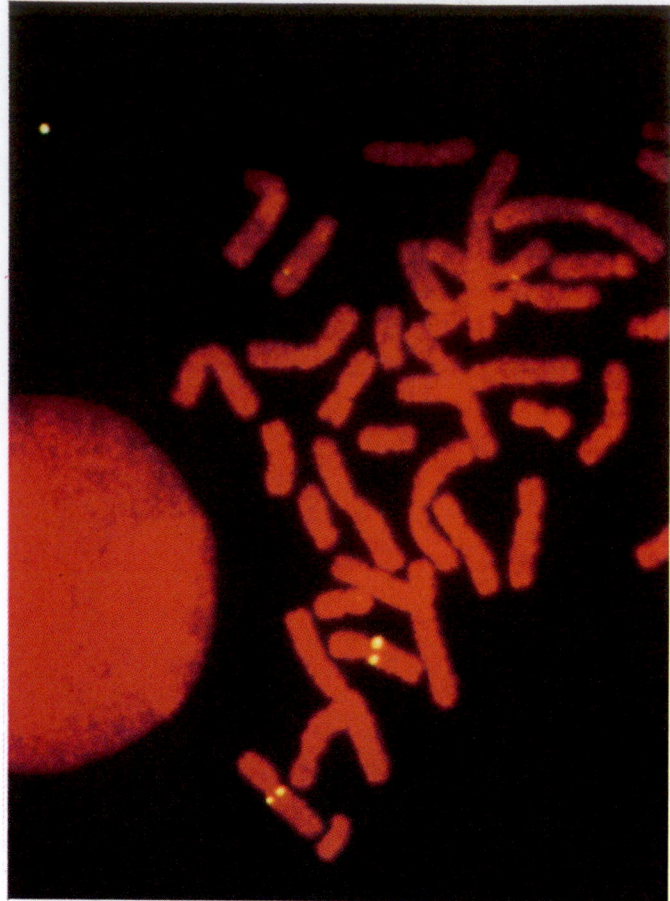

Figure 14-9   FISH analysis. Chromosomes probed in situ with a fluorescent probe specific for a gene present in a single copy in each chromosome set — in this case, a muscle protein. Only one locus shows a fluorescent spot corresponding to the probe bound to the muscle protein gene. (From P. Lichter et al., "High-Resolution Mapping of Human Chromosome 11 by in Situ Hybridization with Cosmid Clones," *Science* 247, 1990, 64.)

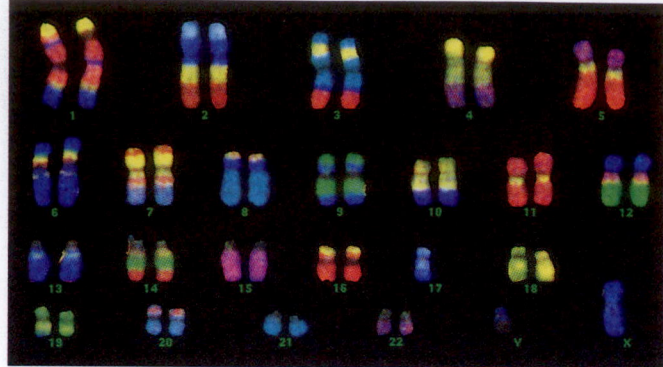

Figure 14-10   Chromosome painting by in situ hybridization with different-labeled probes. (Applied Imaging, Hylton Park, Wessington, Sunderland, U.K.)

human chromosomes, as diagrammed in Figure 14-11. Most of the fragments are seen to be embedded in the rodent chromosomes, but truncated human chromosomes also can be found. First, the frequency of various human molecular markers in the hybrids is calculated. The next step is to calculate the frequency of the co-occurrence of pairs of human molecular markers. Closely linked markers are assumed to be incorporated at high frequencies because the probability that an X-irradiation-induced break will occur between the loci is low. Distant markers and markers on different chromosomes should be present at frequencies close to the

ing vital coding or regulatory sequences. If the break can be seen or mapped to known markers by recombination analysis, then this information can be used to assign a gene to a position on a cytogenetic map of a chromosome. One helpful feature of rearrangement breaks is that they also serve as molecular landmarks. When cloned DNA spanning a break has been identified, the break is easily detected on Southern blots as the loss of an expected band and the appearance of two novel bands.

Radiation hybrid mapping.   The technique that is used to localize genes to individual chromosomes can be extended to obtain map loci. One important extension is **radiation hybrid mapping.** This technique was designed to produce a higher-resolution map of molecular markers along a chromosome. The procedure is to X-ray treat human cells to fragment the chromosomes and then fuse the irradiated cells with the rodent cells to form a panel of different hybrids. In this case, the hybrids have an assortment of *fragments* of

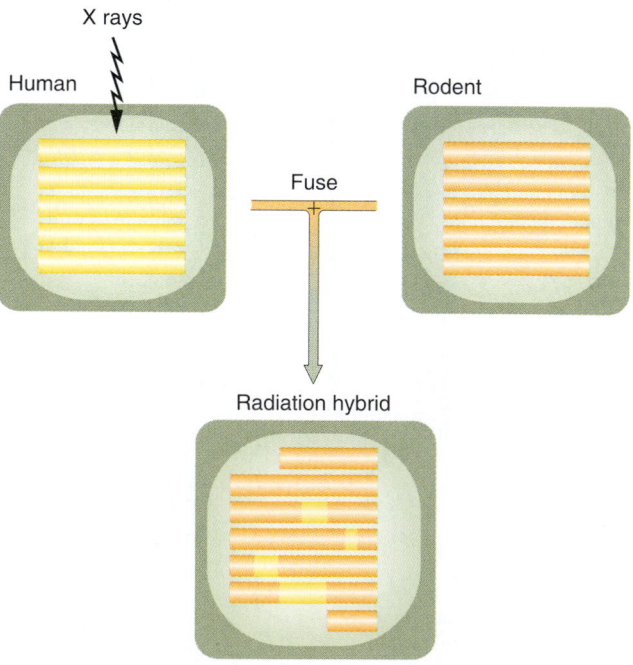

Figure 14-11   Making radiation hybrids by using X rays. Fragments of human chromosomes integrate into rodent chromosomes. A panel of different radiation hybrids is analyzed for cotransfer of human markers, which can indicate linkage.

product of individual frequencies. A mapping unit cR_{3000} is calculated, which has been calibrated to approximately 0.1 cM (m.u.).

A standard panel in the range of 100 to 200 radiation hybrids is quite straightforward to obtain. Such a panel is sufficient to obtain a high-resolution cR_{3000} map of the human genome, which would have 10-fold greater resolution than the current centimorgan genetic map. One downside of the technique is that it is limited to those markers for which human–rodent differences are available.

> **MESSAGE** ···
> Correlation of human markers and chromosomes in hybrid rodent–human cell lines allows chromosomal assignment of the markers. The co-occurrence of different human markers in X-irradiated hybrids allows high-resolution mapping of the chromosomal loci of the markers.

In this section on marker mapping, we have encountered techniques based on widely differing premises — for example, meiotic crossover frequency and radiation-induced breakage. Hence, even though these maps give the same order of markers, distances between markers on one map may not be proportional to distances between markers on another map.

Physical mapping of genomes

A further increase in mapping resolution is accomplished by manipulating cloned DNA fragments directly. Because DNA is the physical material of the genome, the procedures are generally called **physical mapping.** One goal of physical mapping is to identify a set of *overlapping* cloned fragments that together encompass an entire chromosome or an entire genome. The resulting physical map is useful in three ways. First, the genetic markers carried on the clones can be ordered and hence contribute to the overall genome mapping process. Second, when the contiguous clones have been obtained, they represent an ordered library of DNA sequences that can be exploited for future genetic analysis — for example, to correlate mutant phenotypes with disruptions of specific molecular regions. Third, these clones form the raw material that will be sequenced in large-scale genome projects.

In the preparation of physical maps of genomes, vectors that can carry very large inserts are naturally the most useful. Cosmids, YACs (yeast artificial chromosomes), BACs (bacterial artificial chromosomes), and PACs (phage P1-based artificial chromosomes) have been the main types. Cosmids and YACs were introduced in Chapters 12 and 13. **BACs** (Figure 14-12) are based on the 7-kb F plasmid of *E. coli.* Recall that F can carry large fragments of *E. coli* DNA as F′ derivatives (Chapter 7). In a similar manner, as cloning vectors, they can also carry inserts of fragments of foreign DNA as large as 300 kb, although the average is about 100 kb. **PACs** are produced by a type of engineering similar to that of phage P1; they carry inserts comparable to those of BACs.

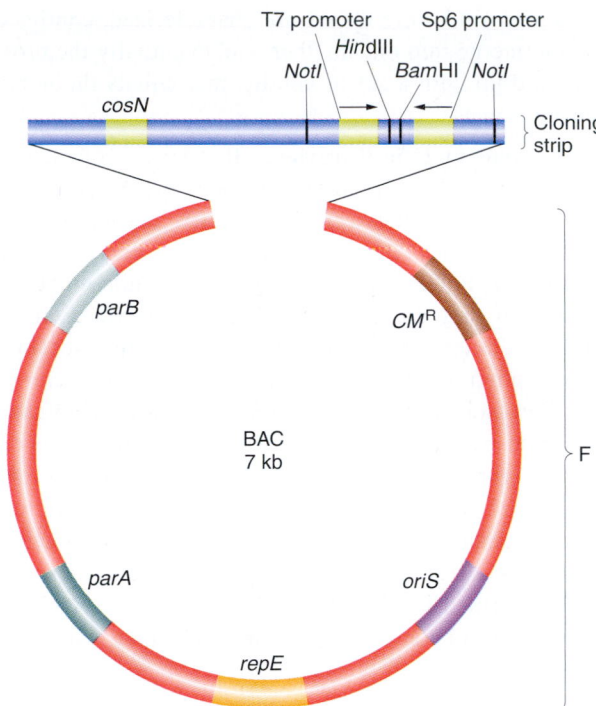

Figure 14-12 Structure of a bacterial artificial chromosome (BAC), used for cloning large fragments of donor DNA. CM^R is a selectable marker for chloramphenicol resistance. *oriS, repE, parA,* and *parB* are F genes for replication and regulation of copy number. *cosN* is the *cos* site from λ phage. *Hin*dIII and *Bam*III are cloning sites at which foreign DNA is inserted. The two promoters are for transcribing the inserted fragment. The *Not*I sites are used for cutting out the inserted fragment.

Although the maximum insert sizes of BACs and PACs are not as large as those of YACs, the former types have several advantages over YACs. First, they can be amplified in bacteria and isolated and manipulated simply with basic bacterial plasmid technology. Second, BACs and PACs form fewer hybrid inserts than YACs do. Hybrid inserts are composed of several different fragments; their presence can thwart attempts to order the clones.

However, despite these useful vectors, the task of genomic cloning is a daunting one. Even so-called small genomes still contain huge amounts of DNA. Consider, for example, the 100-Mb genome of the tiny nematode *Caenorhabditis elegans;* because an average cosmid insert is about 40 kb, at least 2500 cosmids would be required to embrace this genome, and many more would be required to narrow the number to such a complete set. YACs can contain on the order of 1 Mb, so here the task is somewhat simpler.

Cloning a whole genome begins by amassing a large number of randomly cloned inserts. The contents of these clones must be characterized in some way, and overlaps must be determined. A set of overlapping clones is called a **contig.** In the early phases of a genome project, contigs are numerous and represent cloned "islands" of the genome.

But, as more and more clones are characterized, contigs enlarge and merge into one another, and eventually the project should end up with a set of contigs that equals the number of chromosomes.

Chromosome-specific libraries.   If a library of clones is prepared from total genomic DNA, then contig development is relatively slow. However, if a specific chromosome can be used to develop the library of clones, contigs emerge more rapidly. PFGE can be used to isolate individual chromosomes (if they are small) or chromosome fragments cut with "long-cutter" enzymes such as NotI. **Flow sorting** is another option for preparing DNA of a specific chromosome. Chromosomes (such as human chromosomes) can be flow-sorted by fluorescence-activated chromosome sorting (FACS; Figure 14-13). In this procedure, metaphase chromosomes are stained with two dyes, one of which binds to AT-rich regions and the other to GC-rich regions. Cells are disrupted to liberate whole chromosomes into liquid suspension. This suspension is converted into a spray in which the concentration of chromosomes is such that each spray droplet contains one chromosome. The spray passes through laser beams tuned to excite the fluorescence. Each chromosome produces its own characteristic fluorescence signal, which is recognized electronically, and two deflector plates direct the droplets containing the specific chromosome needed into a collection tube.

MESSAGE ....
Genomic cloning proceeds by assembling clones into overlapping groups called contigs. As more data accumulate, the contigs become equivalent to whole chromosomes.

Several different techniques are used to order genomic clones into contigs. We shall consider some of the main ones.

Ordering by FISH.   If good chromosomal landmarks are known, FISH analysis can be used to locate the approximate positions of the large inserts. Figure 14-14 shows results of a FISH analysis that generates a rough ordering of BACs and PACs in human chromosomes.

Ordering by clone fingerprints.   The genomic insert carried by a vector has its own unique sequence, which can be used to generate a DNA fingerprint. For example, a multiple restriction-enzyme digestion can generate a set of bands whose number and positions are a unique "fingerprint" of that clone. The different bands generated by separate clones can be aligned either visually or by using a computer program to determine if there is any overlap between the inserted DNAs. In this way, the contig can be built up.

Ordering by sequence-tagged sites.   Unique short sequences of large cloned inserts can be used as tags to align the various clones into contigs. For example, if clone A has tags 1 and 2 and clone B has tags 2 and 3, clones A and B must overlap in the region of tag 2. The practical procedure is to amass a large set of random clones with small genomic inserts (say, in λ phage) and sequence short regions of each. From these sequences, pairs of PCR primers are designed that will amplify the short specific sequence of DNA flanked by the primers. These short DNA sequences are known as **sequence-tagged sites (STSs)**. Even though initially the location of these STSs in the genome is not known, a panel of many STSs can be used to characterize clones with large genomic inserts (such as YAC clones). The clones that are shown to have specific STSs in common must have overlapping inserts and therefore can be aligned into contigs. An example of this process is shown in Figure 14-15.

Short stretches of sequence are sometimes obtained from cDNA clones. These stretches are known as **expressed sequence tags (ESTs)**. ESTs are obtained by sequencing into the cDNA insert by using a primer based on the vector sequence. They can be used to align the cDNAs on the contig, thus anchoring the gene map to the physical map. Further, if part of the open reading frame (ORF) of the transcript

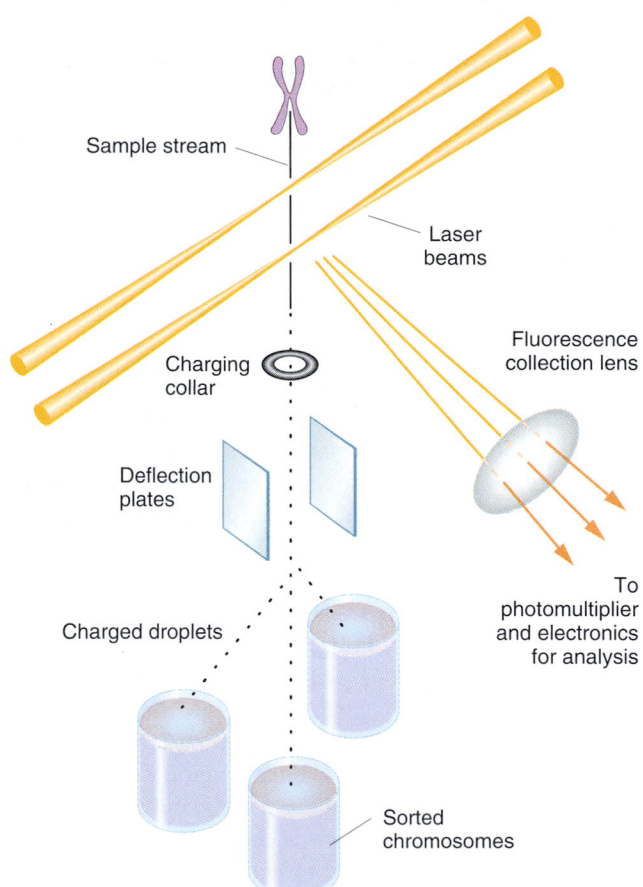

Figure 14-13   Chromosome purification by using flow sorting. Chromosomes stained with a fluorescent dye are passed through a laser beam. Each time, the amount of fluorescence is measured, and the chromosome is deflected accordingly. The chromosomes are then collected as droplets.

Human Genome Integrated BAC/PAC Map
Korenberg et al., 1998

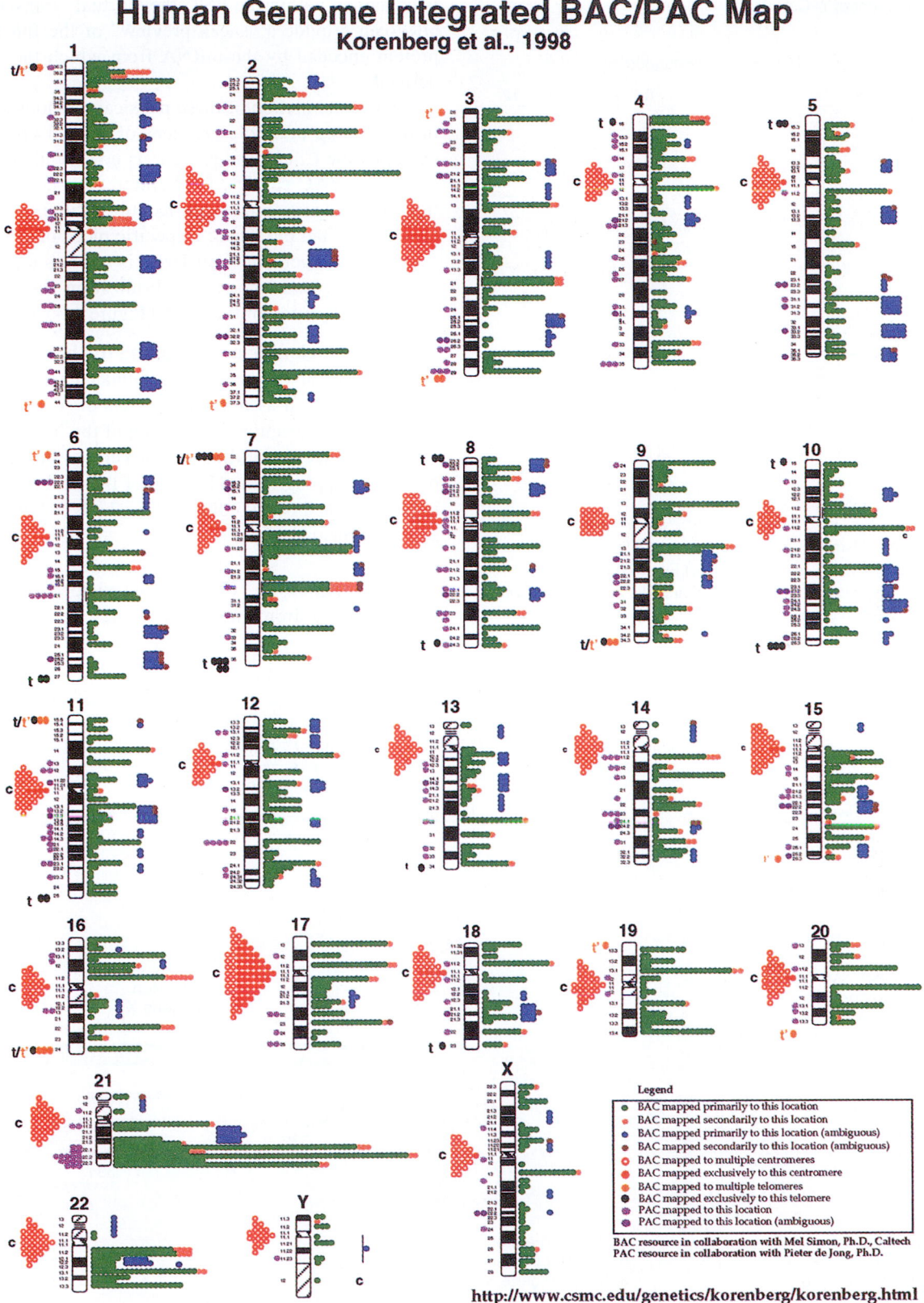

Figure 14-14 Ordering BAC and PAC clones by FISH. Each color represents a type of vector; each circle represents hybridization by a specific clone; red arrows represent centromere-specific binding. (Julie Korenberg, Ph.D., M.D., and Xiao-Ning Chen, M.D., Cedars Sinai Medical Center, Los Angeles, CA.)

Legend
- BAC mapped primarily to this location
- BAC mapped secondarily to this location
- BAC mapped primarily to this location (ambiguous)
- BAC mapped secondarily to this location (ambiguous)
- BAC mapped to multiple centromeres
- BAC mapped exclusively to this centromere
- BAC mapped to multiple telomeres
- BAC mapped exclusively to this telomere
- PAC mapped to this location
- PAC mapped to this location (ambiguous)

BAC resource in collaboration with Mel Simon, Ph.D., Caltech
PAC resource in collaboration with Pieter de Jong, Ph.D.

http://www.csmc.edu/genetics/korenberg/korenberg.html

DATA (STS content of YACs)

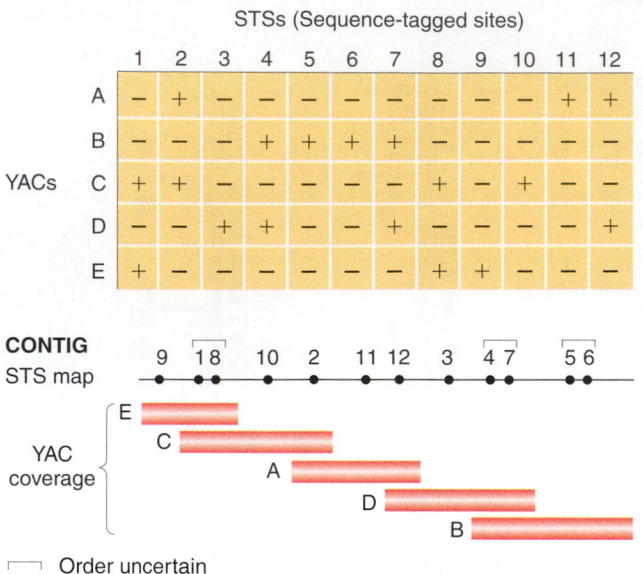

STSs (Sequence-tagged sites)

	1	2	3	4	5	6	7	8	9	10	11	12
A	−	+	−	−	−	−	−	−	−	−	+	+
B	−	−	−	+	+	+	+	−	−	−	−	−
C	+	+	−	−	−	−	−	+	−	+	−	−
D	−	−	+	+	−	−	+	−	−	−	−	+
E	+	−	−	−	−	−	−	+	+	−	−	−

YACs

CONTIG
STS map

9 1 8 10 2 11 12 3 4 7 5 6

YAC coverage
E
C
A
D
B

☐ Order uncertain

Figure 14-15 Using sequence-tagged sites (STSs) to order overlapping clones (YACs, in this example) into a contig. Five different YACs are tested to determine which STSs they contain (top), and these data are used to assemble a physical map (bottom).

is contained within the EST, the "virtual" translation of the ORF can provide a "sneak preview" of the function of the protein encoded by the mRNA from which the cDNA was derived.

The combination of these physical methods has resulted in the cloning of whole genomes of several organisms. For example, the *C. elegans* genome is now available as sets of cosmid or YAC contigs. Furthermore, the DNA of the contigs has been arranged on nitrocellulose filters in ordered arrays; so, to find out where a specific piece of DNA of interest lies in the genome, that DNA is used as a probe on the contig filters, and a positive hybridization signal announces the precise location of the DNA (Figure 14-16).

An example: cloning and mapping the human Y chromosome. Several of the smaller human chromosomes have been fully cloned as overlapping sets of YAC clones (contigs). We shall examine the cloning of the Y chromosome as an example because it illustrates several of the techniques of physical mapping. The STS map of the Y chromosome was in fact obtained by two different methods — YAC alignment and deletion analysis.

YAC alignment. Flow sorting yielded a sample of Y chromosomes, from which λ clones were made. From

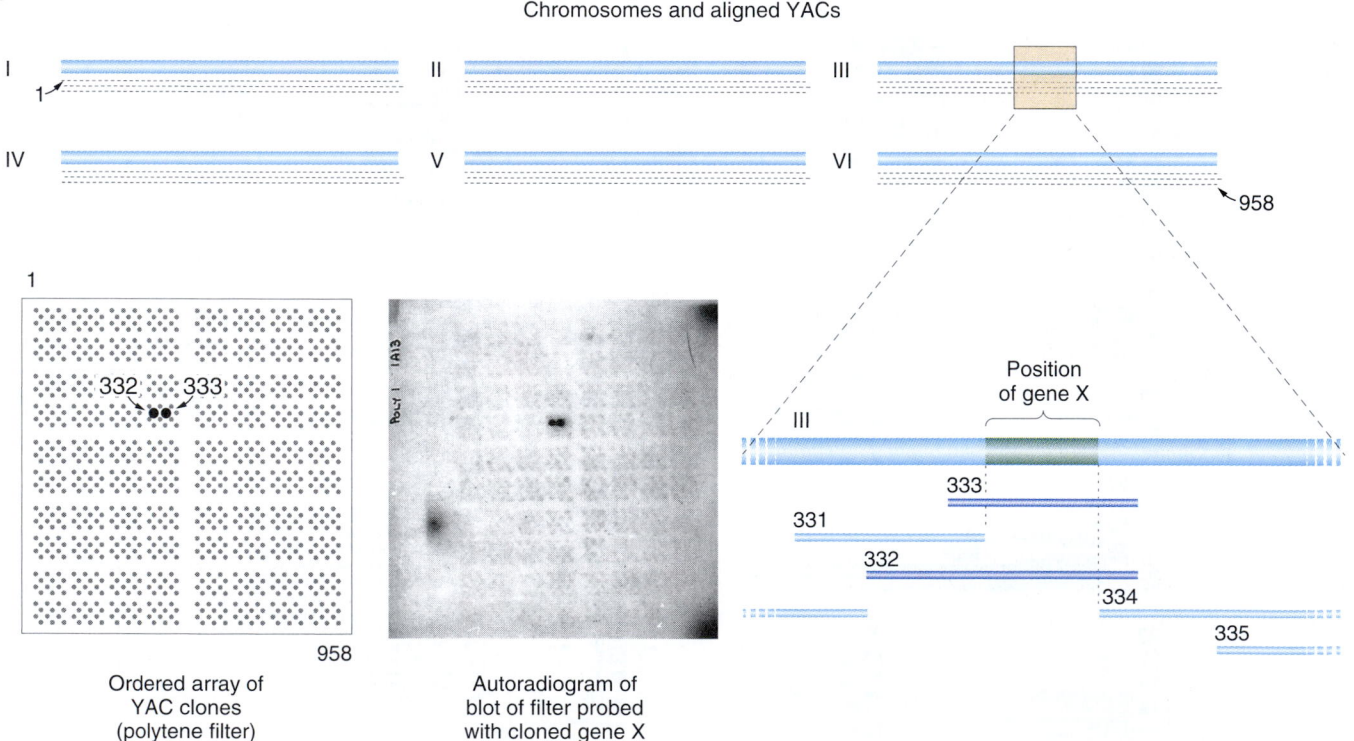

Chromosomes and aligned YACs

Ordered array of YAC clones (polytene filter)

Autoradiogram of blot of filter probed with cloned gene X

Position of gene X

Figure 14-16 Using an ordered array of YACs to locate the map position of a newly cloned gene in *C. elegans*. The YAC clones are placed in order on a polytene filter. DNA from this filter is blotted and probed with the cloned gene, giving the autoradiogram with two positive spots corresponding to two adjacent YACs numbered 332 and 333. Because of the hybridization pattern obtained, the location of the cloned gene can be narrowed down to a small region on chromosome III. (Autoradiogram from Alan Coulson.)

clones that did not contain repetitive DNA, STS primers were designed. In all, 160 primer pairs were made. A Y chromosome YAC library of 10,368 clones was obtained in which the average insert size was 650 kb. From these numbers, each point on the Y chromosome was estimated to have been sampled an average of four times. The YAC clones were divided into 18 pools of 576 YACs, and the pools were screened with the STS primers. Subdivision of positive pools led rapidly to the assignment of a particular STS to specific YACs. The total STS content of each YAC was assessed, and overlaps between the YACs were determined in the same way as that shown in the generalized example in Figure 14-15.

Deletion analysis. Various types of Y chromosome deletions occur naturally. For example, some XX males contain truncated fragments of the Y, whereas some XY females have deletions of the region containing the maleness (testis-determining) gene (see Chapters 2 and 23). These Y deletions were maintained in cell culture and formed the basis for aligning the Y chromosome STSs. Each deletion was tested for STS content. Because by nature the deletions were nested sets, the STS content could be used not only to develop an STS map, but also to map the coverage of the deletions. The principle is illustrated in Figure 14-17. The STS maps produced by YAC alignment and by deletion analysis were identical.

MESSAGE ·····························

Clones can be arranged into contigs by matching DNA fingerprints, by matching short sequences within cloned segments, and by analyzing deletions.

Genome sequencing

Several different strategies have been successfully applied to genome projects. Their advantages and disadvantages depend on the size and complexity of the genome. Of particular importance is the frequency of repetitive DNA in the genome.

Random clone sequencing. The first genome to be cloned was that of the bacterium *Haemophilus influenzae*. Genomic DNA was mechanically sheared and used to obtain a large number of random clones that were presumed to overlap each other in numerous ways. Primers based on adjacent vector DNA were used to sequence short regions at the ends of the cloned *Haemophilus* inserts. Then these short sequences were used (much like sequence-tagged sites) to align the genomic clones. Because so many random short sequences were obtained, together they encompassed most of the *Haemophilus* genome. Gaps were filled in by **"primer walking"**; that is, by using the end of a cloned sequence as a primer to sequence into adjacent uncloned fragments.

Sequencing ordered clones. Most genomic sequencing programs start with a set of ordered clones. We have seen that an ordered set of YAC clones was developed for the

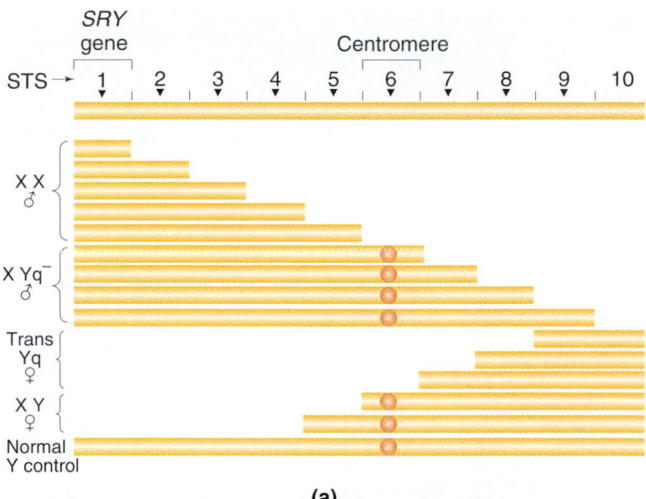

(a)

(b)

Figure 14-17 (a) The STS content of naturally occurring Y chromosome fragments was used to order the STSs on a Y chromosome map. **Note**: "q⁻" means "lacking the q arm"; "trans Yq" means "translocation of Yq to an autosome." (b) Deep freeze containing plates of YAC clones used in the human genome project. (Part b from Roger Bessmeyer — © 1995 Corbis. All rights reserved Library of Human Genes.)

human Y chromosome and other human chromosomes. However, YAC clones are not suitable for sequencing directly. YACs are subcloned into overlapping BACs or PACs. The BACs or PACs are again aligned into contigs by using STSs or the alignment of clone fingerprints. The BAC

Figure 14-18   Automatic DNA isolation. (Courtesy of Fred Rick, Los Alamos National Laboratory, Los Alamos, New Mexico.)

or PAC clones are again subcloned into smaller inserts for sequencing. At this level, multiple overlapping clones are sequenced randomly (without establishing clone alignment) so that any BAC or PAC clone is sequenced as many as five times in all.

Sequencing unordered clones.   One current strategy is to sequence the two ends of cloned genomic fragments from sequencing primers at the ends of the vector. If the length of the sequenced stretches and the lengths of the cloned fragments are sufficiently long, these sequences can be compiled to create long contiguous stretches of sequence that can extend over repetitive DNAs contained within the genome (see Chapters 3 and 20 for a discussion of transposable elements and other repetitive DNAs). The advantage of such a strategy is that the time- and labor-intensive process of clone mapping is avoided. This strategy is currently being tested for the *Drosophila* and human genomes.

Automation.   All stages of genomic analysis can be speeded up by automation. The preparation of clones, DNA isolation, electrophoresis, and sequencing protocols have all been adapted to machines. An example of this "high throughput" machinery is shown in Figure 14-18.

Using genome maps for genetic analysis

Genetic and physical maps are an important starting point for several types of genetic analysis, including gene isolation (including human disease genes) and functional genomics.

Isolating human disease genes by positional cloning. We shall follow the methods used to identify the genomic sequence of the cystic fibrosis (CF) gene as an example. No primary biochemical defect was known at the time that the gene was isolated, so it was very much a gene in search of a function. Linkage to molecular markers had located the gene to the long arm of chromosome 7, between bands 7q22

and 7q31.1. The *CF* gene was thought to be inside this region, flanked by the gene *met* (a proto-oncogene; see Chapter 22) at one end and a molecular marker, D788, at the other end. But between these markers lay 1.5 centimorgans (map units) of DNA, a vast uncharted terrain of 1.5 million bases. Additional markers within the region were obtained by using new probes derived from a chromosome 7 library made by flow sorting.

However, the two key techniques that were used to traverse the huge genetic distances were chromosome walking (Chapter 13) and a related technique called **chromosome jumping.** The latter technique provides a way of jumping across potentially unclonable areas of DNA and generates widely spaced landmarks along the sequence that can be used as initiation points for multiple bidirectional chromosomal walks.

Chromosome jumping is illustrated in Figure 14-19. In this procedure, large fragments are created by partial restriction cleavage of the DNA in the region believed to contain the gene of interest. Each DNA fragment is then circularized, thus bringing the beginning and end of the fragment together. This junction is cut out and cloned into a phage vector, which together with the other junction segments make up a *jumping library*. A probe from the beginning of the stretch of DNA under investigation can be used to screen the jumping library to find the clone that contains the beginning sequence. When this clone is found, the other end of the junction sequence is excised and used to screen the library again to make a second jump. From each jump position, chromosome walks can be made in both directions to search for genelike sequences.

A restriction map of the overall region was obtained with rare-cutting restriction enzymes, and the restriction sites were used to position and orient the sequences obtained from jumping and walking. When enough sequencing had been done to cover representative parts of the overall region, the hunt for any genes along this stretch began. Genes were sought by several techniques. First, human genes were known to be generally preceded at the 5′ end by clusters of cytosines and guanines, called *CpG islands,* and several of these clusters were found. Second, it was reasoned that a gene would show homology to the DNA of other animals, because of evolutionary conservation, so candidate sequences were used to probe what were called *zoo* blots of genomic DNA from a range of animals. Third, genes should have appropriate start and stop signals. Fourth, genes should be transcribed, and transcripts should be found.

Ultimately, a strong candidate gene was found spanning 250 kb of the region. Some CF symptoms are expressed in sweat glands; so, from cultured sweat gland cells, cDNA was prepared, and a 6500-nucleotide cDNA homologous to the candidate gene was detected. On sequencing this cDNA in normal and CF patients, the cDNA of the patients showed the deletion of three base pairs, eliminating a phenylalanine from the protein. Therefore it was very likely that this was

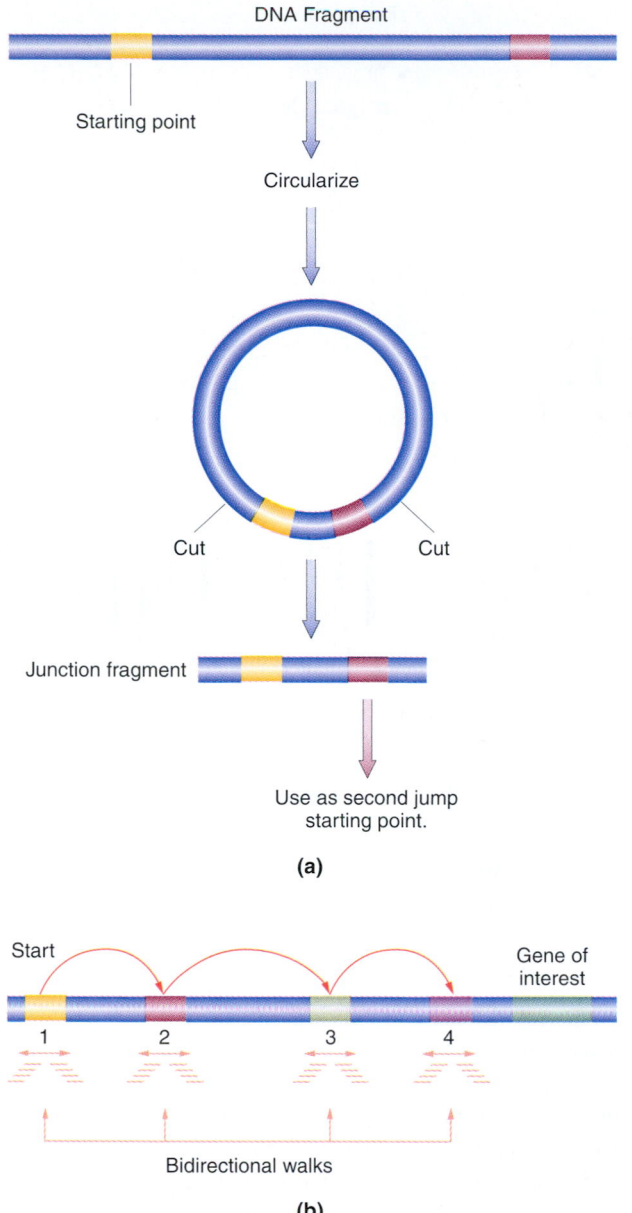

Figure 14-19 Manipulating cloned genomic fragments for chromosome jumping, a modified type of chromosome walking that can bypass regions difficult to clone, such as those containing repetitive DNA (see text).

the CF coding sequence. Thus the *CF* gene had been found. From its cDNA nucleotide sequence, an amino acid sequence was inferred. In turn, from this inferred sequence, the three-dimensional structure of the protein was predicted. This protein is structurally similar to ion-transport proteins in other systems, suggesting that a transport defect is the primary cause of CF. When used to transform mutant cell lines from CF patients, the wild-type gene restored normal function; this phenotypic "rescue" was the final confirmation that the isolated sequence was in fact the *CF* gene.

The candidate-gene approach. Inevitably, intensive cloning and sequence-level characterization of a chromoso-

mal region reveal the presence of genes of unknown function. If a gene of interest such as a disease gene is mapped to that chromosomal region, then these "orphan" gene sequences become **candidate genes** for the disease gene. This procedure is termed the **candidate-gene approach** to gene isolation. Knowledge about the gene's phenotype such as biochemical defect and pattern of tissue expression can be matched to the sequence domains and tissue expression of the candidate gene. The method works in the opposite direction, too; the domains and tissue expression of randomly sequenced genes often suggest a possible disease-gene phenotype.

> **MESSAGE** ·······························
> Cloning is made easier by the availability of a set of overlapping genomic clones.

Genes underlying complex inheritance patterns. Most of the contrasting phenotypes analyzed in this book are determined *simply* by alleles of a single gene. However, many phenotypes are determined in a *complex* manner. Here two situations can be distinguished.

First the phenotypic variation may be quantitative (Chapters 1 and 25), and the characters (traits) are called **quantitative traits.** Examples are metric characters such as height and weight. This type of variation is thought to be based on the cumulative interaction between + and − alleles of several genes and the environment. The availability of thousands of molecular markers such as SSLPs arranged along all the chromosomes of a genome has made it possible to map some of the genes that contribute to quantitative variation, whose loci are called **quantitative trait loci,** abbreviated **QTLs.** The approach is to take two lines that show widely contrasting phenotypes for a quantitative trait and to interbreed these lines to generate homozygous descendants that contain only one segment or a small number of segments from one line, as shown in Figure 14-20. (These segments can be identified by the SSLP alleles that they carry.) Such hybrid individuals are then assessed for their quantitative phenotype, and estimates are made of the contributions (or lack of contribution) of specific segments to the observed variation. The average phenotype of lines with, say, region A is compared with the average of lines lacking region A; if there is a difference, region A becomes a candidate for containing a QTL. Ideally, a derived pure line would carry only one QTL, and then in backcrosses to the appropriate parent this QTL would segregate in a monohybrid manner. The QTL can then be mapped precisely by recombination with SSLP markers.

The second situation is a type of discontinuous variation that is not inherited as a simple Mendelian allele. Examples are all-or-none phenotypes such as epilepsy, heart disease, diabetes, and Alzheimer disease. Here the model for inheritance is again alleles of one to several contributing genes plus a large environmental component. However, to produce discontinuous phenotypes, these factors seem to contribute to a type of cellular or organismal "threshold" beyond which the disorder is expressed. These conditions also

Figure 14-20 Producing lines for QTL identification and mapping. Two pure lines that differ significantly in some quantitative trait (character) are crossed, and, after many generations of inbreeding, pure recombinant lines are produced. The comparison of phenotypes of these lines leads to the identification of specific chromosomal segments that consistently contribute to the parental difference. These regions contain presumptive QTLs. (After W. N. Frankel, "Taking Stock of Complex Trait Genetics in Mice," *Trends in Genetics* 12, 1995, Figure 1.)

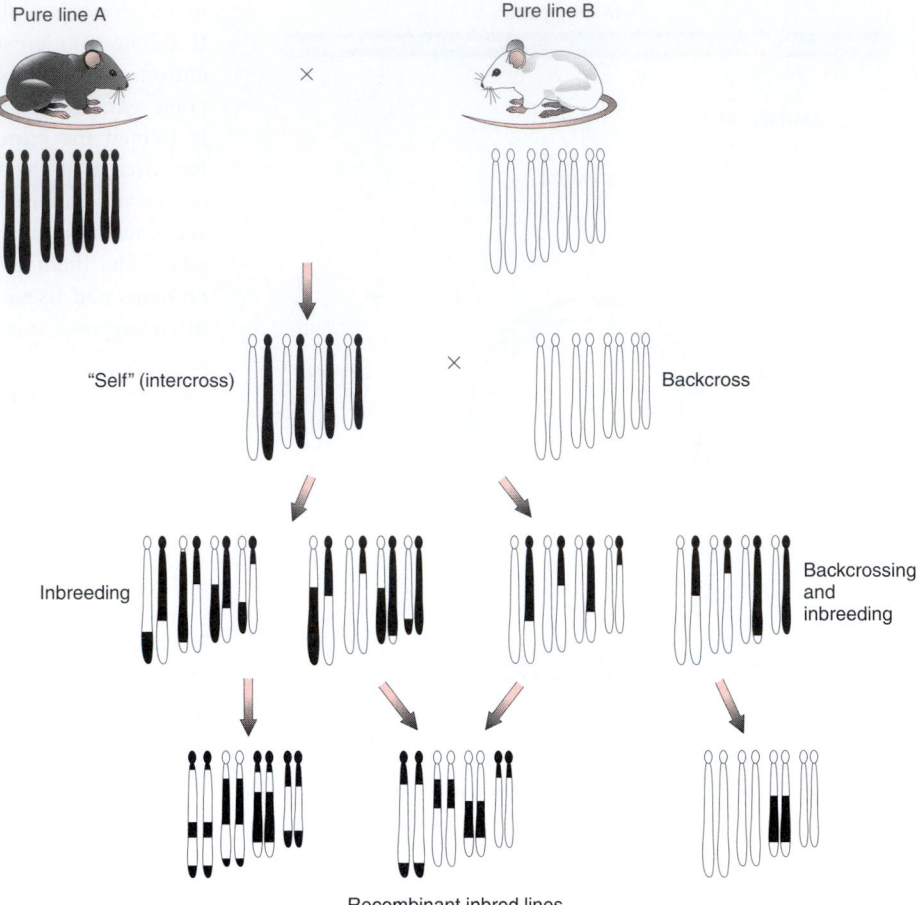

are amenable to gene identification by using the approach shown in Figure 14-20, and several complex trait loci have been identified in experimental organisms and humans. In humans, studies on isolated populations with little genetic variation are particularly useful in identifying the contributing loci. In the future, SNP analysis promises to accelerate the mapping of complex traits.

Functional genomics

Large-scale sequence data are the beginning of functional genomics. The following sections show some of the analyses that can be performed to investigate function.

Characterize the proteome by ORF analysis

The genomic DNA sequence is analyzed by computer gene-prediction software that, among other things, examines each of the six reading frames of all sequences and searches for segments beginning with the translational start codon AUG and ending with a stop codon. Any open reading frames of at least 100 codons are candidates for genes. Most ORFs are completely novel, not corresponding to any familiar gene with alleles producing identifiable phenotypes. The ORFs can be analyzed for function initially by using the computer

to search data bases to look for full or partial homology to known genes characterized in other organisms. The location, orientation, and clustering of ORFs also are important genomic information. Examples from *Haemophilus* and *Saccharomyces* are shown in Figures 14-21 and 14-22. A provisional proteome gene distribution can be deduced from such analysis, as shown in Figure 14-23. In higher eukaryotes, in which introns are common features of transcripts, predicting from ORF genomic DNA is more difficult.

Gene disruption knockouts

ORF function can be investigated by systematically knocking out the gene by in vitro mutagenesis and then looking for any possible mutant phenotype that might provide clues about function. This process is underway in the fully sequenced genomes. Interestingly, many knocked-out ORFs show no phenotypic effects. More than half of the predicted ORFs may fall into this category.

The study of gene interactions by the yeast two-hybrid system

This method investigates interaction with the use of a two-plasmid system in yeast. The basis for the test is the yeast

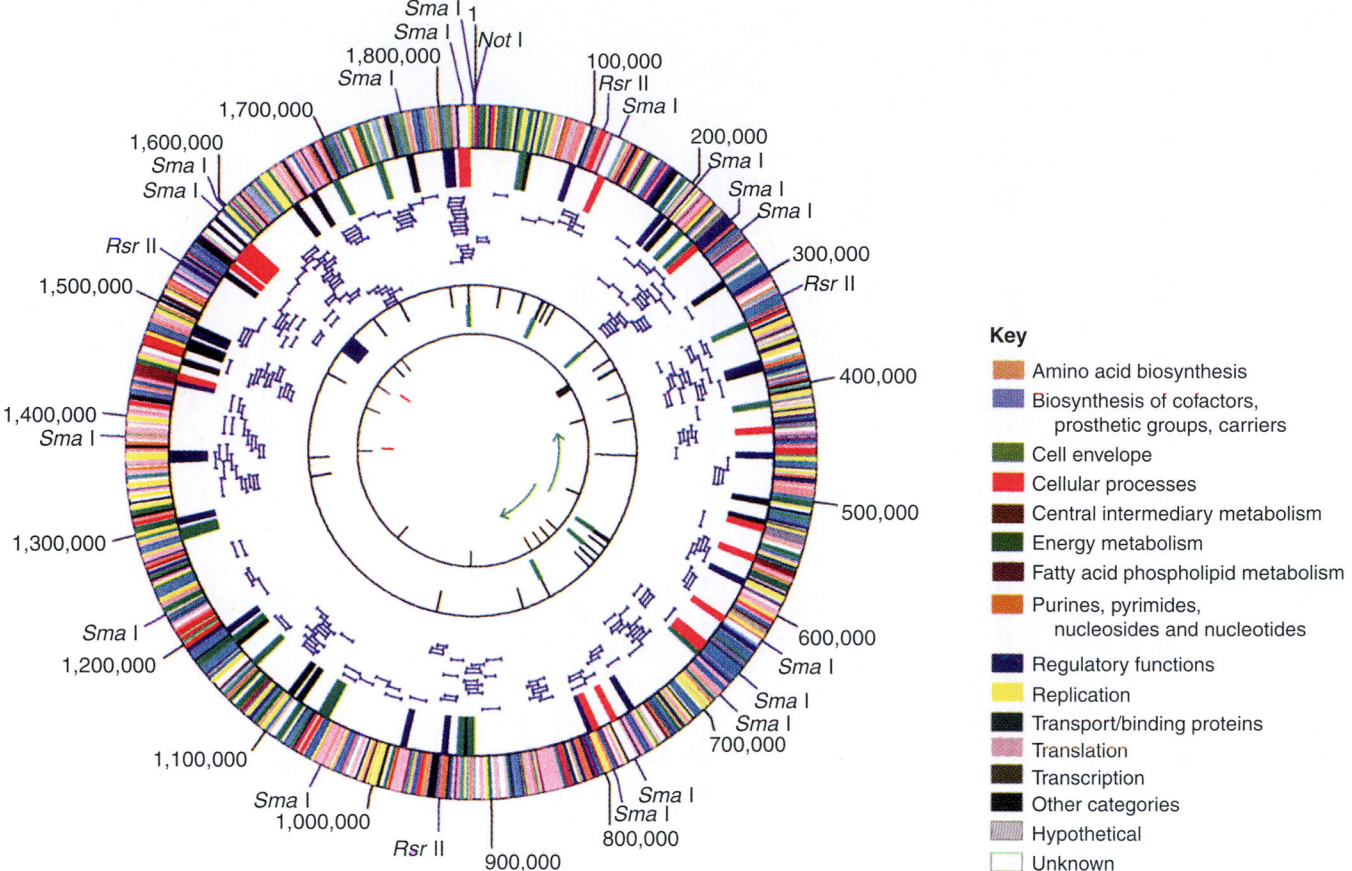

Key

- Amino acid biosynthesis
- Biosynthesis of cofactors, prosthetic groups, carriers
- Cell envelope
- Cellular processes
- Central intermediary metabolism
- Energy metabolism
- Fatty acid phospholipid metabolism
- Purines, pyrimides, nucleosides and nucleotides
- Regulatory functions
- Replication
- Transport/binding proteins
- Translation
- Transcription
- Other categories
- Hypothetical
- Unknown

Figure 14-21 The architecture of the genome of the bacterium *Haemophilus influenzae,* based on the complete genomic sequence reported in 1995. The circles have the following meanings, starting with the outer circle:

- Restriction sites and genomic positions, starting from the *Not*I site
- Genes color coded by their functional groups (see key)
- Regions of high GC content: red > 42%; blue > 40%. Regions of high AT content: black > 66%; green > 64%.
- Coverage of some clones used in sequencing
- Ribosomal operons (green), tRNAs (black), μ-like prophage (blue)
- Positions of simple tandem repeats
- Replication origins (green arrows) and termination sequences (red)

(Illustration by Dr. Anthony R. Kerlavage; Institute for Genomic Research. From R. D. Fleischmann et al., "Whole-Genome Random Sequencing and Assembly of *Haemophilus influenzae* Rd," *Science* 1995, 269, 496–512.)

GAL4 transcriptional activator. This protein has two domains, a DNA-binding domain and an activation domain, both of which must be in close juxtaposition in order for the protein to initiate transcription. A gene for one protein under investigation is spliced next to the GAL4 DNA-binding domain on one plasmid and acts as "bait." On another plasmid a gene for another protein being tested is spliced to the activation domain; this protein is said to be the "target" (Figure 14-24). The two plasmids are then introduced into the same cell. One way of doing so is to mate haploid cells containing bait and target. The only way in which the GAL4 binding and activation domains can come together is if the bait and target proteins bind to each other, demonstrating a physical interaction. The two-hybrid system can be automated to facilitate large-scale hunting for protein interactions throughout the proteome.

The study of developmental regulation by using DNA chips

DNA chips are about to revolutionize genetics in the same way that silicone chips revolutionized the computer industry.

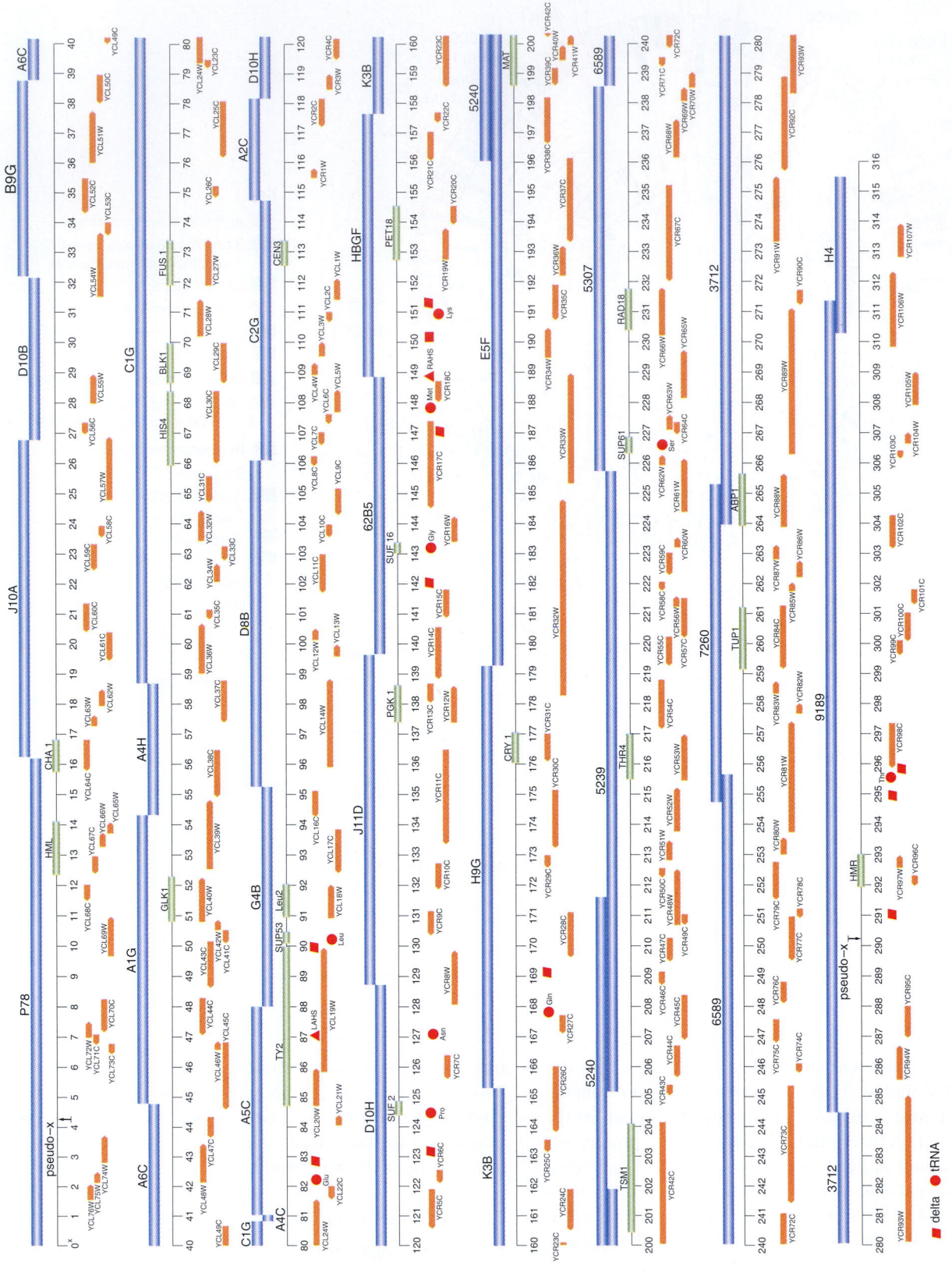

Figure 14-22 The genetic landscape of chromosome 3 in yeast, determined by sequencing the entire chromosome. Genes previously detected from mutant phenotypes are shown in green. Open reading frames, which likely are protein-encoding genes, were detected by sequence analysis and are shown below the line in orange. The clones used in sequencing are shown above the line in blue. Delta sequences are derived from the transposon Ty. (S. G. Oliver et al., "The Complete DNA Sequence of Yeast Chromosome III," *Nature* 357, 1992, 38–46.)

■ delta ● tRNA

454

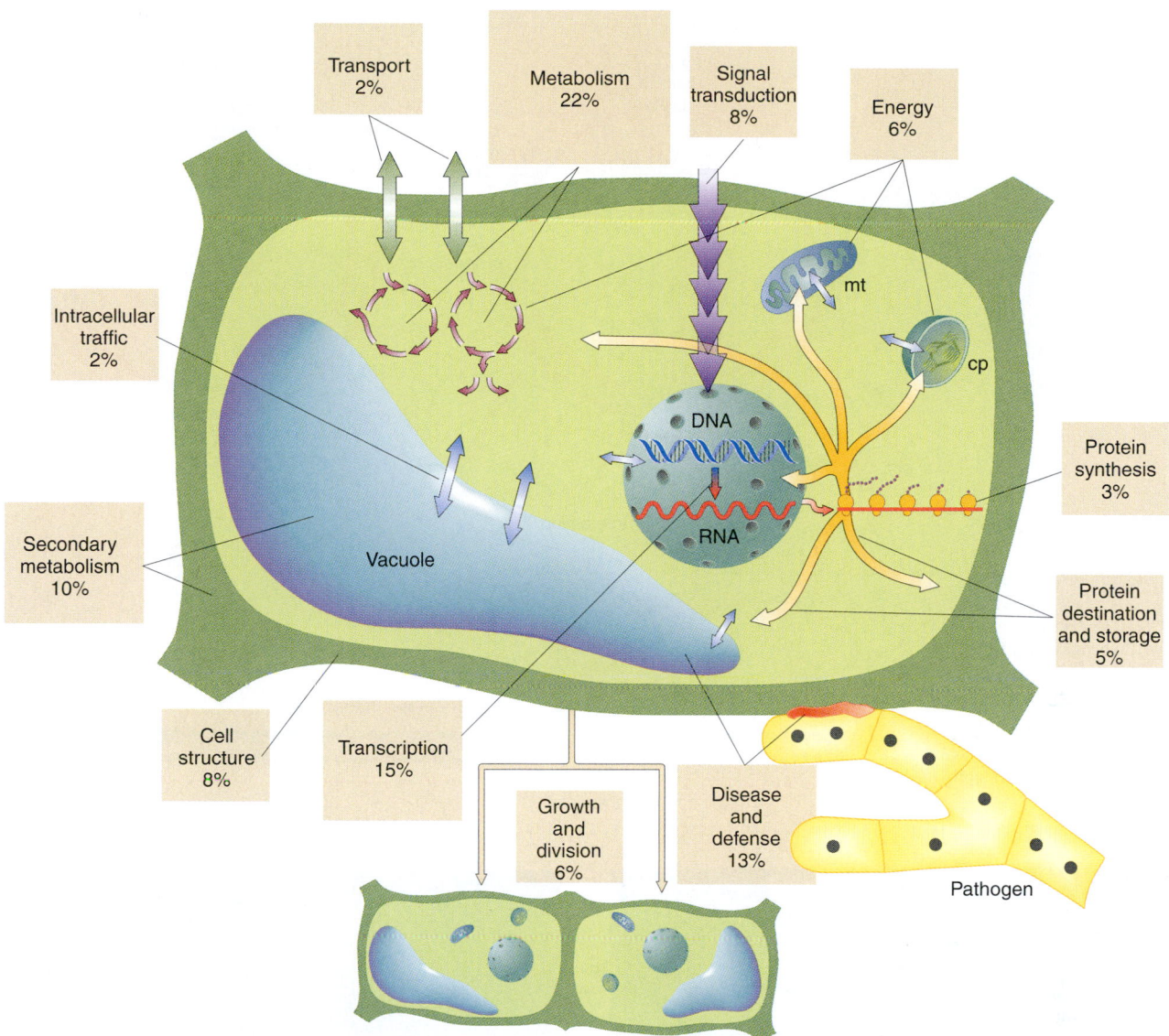

Figure 14-23   Distributions of various categories of protein-encoding genes, estimated from currently known genes in the plant *Arabidopsis thaliana*.

DNA chips are samples of DNA laid out in regimented arrays bound to a glass "chip" the size of a microscope cover slip.

One protocol is as follows. Robotic machines with multiple printing tips resembling miniature fountain pen nibs deliver microscopic droplets of DNA solution to specific positions (addresses) on the chip. The DNA is dried and treated so that it will bind to the glass. Thousands of samples can be applied to one chip. Commonly, the array of DNAs are known cDNAs from different genes. In principle, all the cDNAs of the entire genome could be arrayed on chips. The chips are exposed to a heterogeneous *labeled* cDNA sample made from total mRNA isolated at some specific stage of development. Fluorescent label is used, and the binding of the probe molecules to the glass chip is moni-

tored automatically by laser beams. A typical result is shown in Figure 14-25a. In this way, the genes that are active at any stage of development or under any environmental condition can be assayed. Once again the idea is to identify protein networks that are active in the cell at any particular stage of interest. Figure 14-26 shows an example of a developmental expression sequence.

Another protocol loads the chip with an array of oligonucleotides synthesized nucleotide by nucleotide on the chip itself (Figure 14-27). The glass is first covered with protecting groups that prevent DNA deposition. A mask is placed on the glass with holes corresponding to the sites of deposition. Then laser beams are shone onto the holes where synthesis is to begin. The light knocks off the protect-

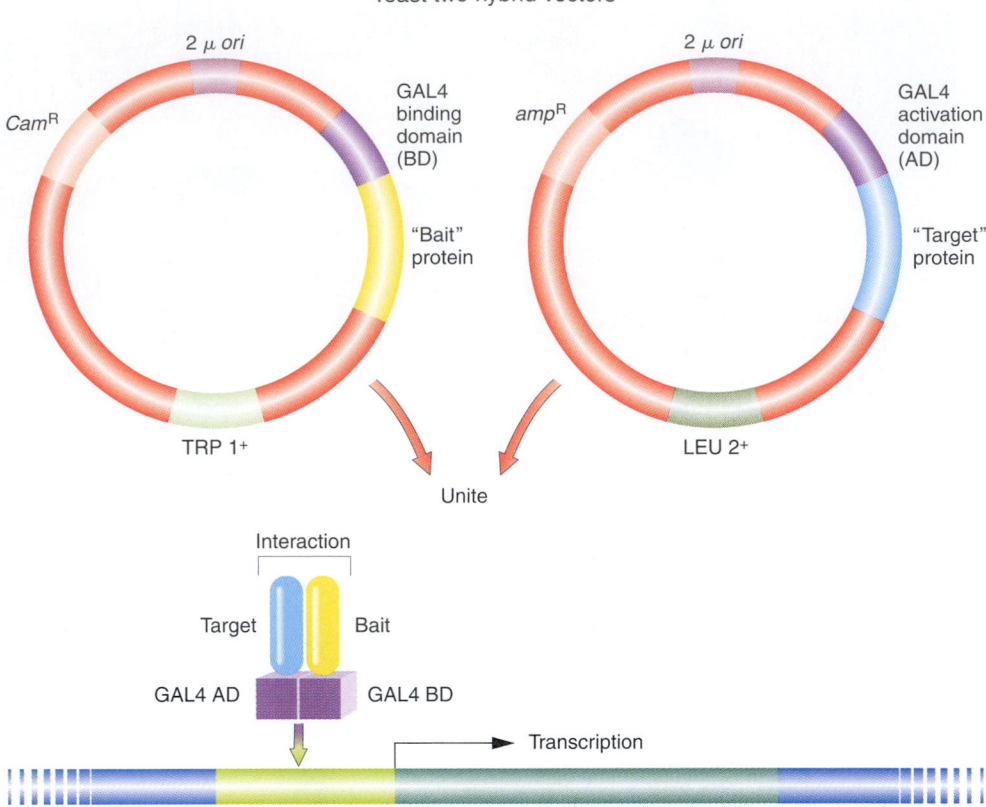

Figure 14-24 The yeast two-hybrid system for detecting gene interaction. The system uses the binding of two proteins under test to restore the function of the GAL4 protein, which activates a reporter gene.

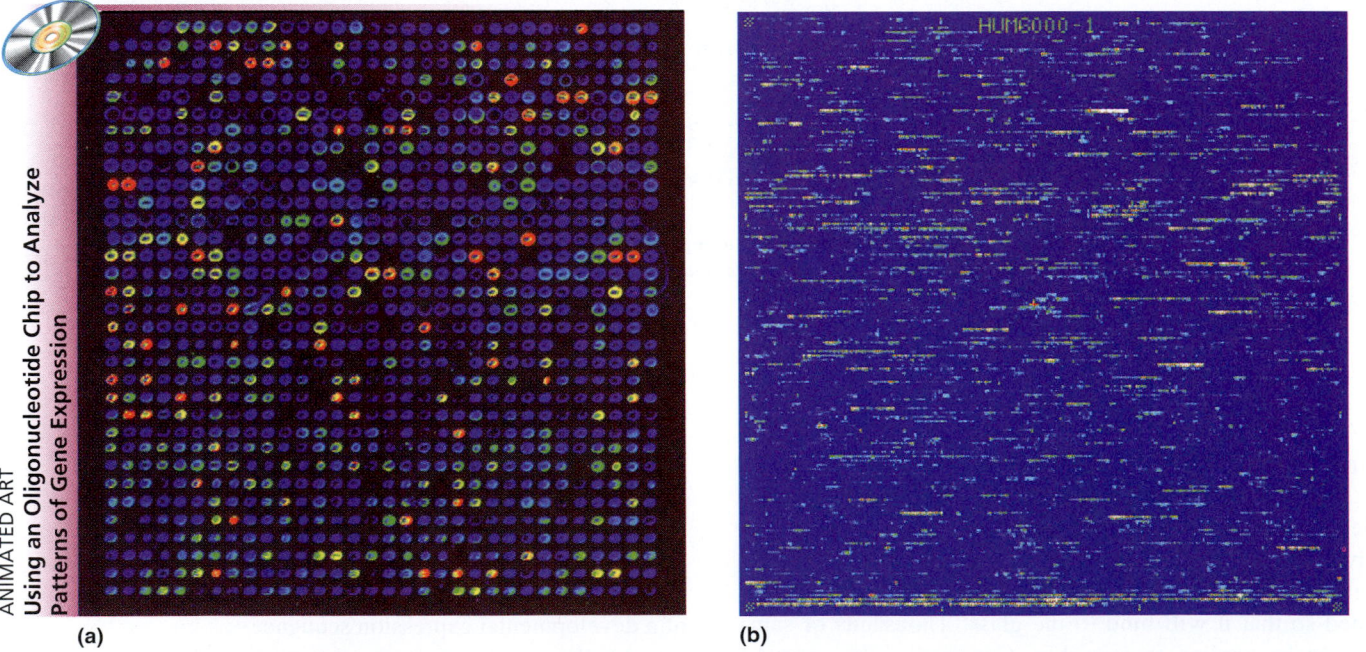

ANIMATED ART
Using an Oligonucleotide Chip to Analyze Patterns of Gene Expression

(a)

(b)

Figure 14-25 Fluorescence detection of binding to DNA microarrays: (a) Array of 1046 cDNAs probed with bone-marrow-based cDNA probe. Intensity of signal follows the colors of the spectrum with red highest and blue lowest. (b) Affymetrix GeneChip, a 65,000-oligonucleotide array representing 1641 genes, probed with tissue-specific cDNAs. (Part a courtesy of Mark Scheria, Stanford; image appeared in *Nature Genetics,* 16, 1997, 127, Figure 1a; part b courtesy of Affymetrix Inc., Santa Clara, CA. Image generated by David Lockhart. Affymetrix and GeneChip are U.S. registered trademarks used by Affymetrix. This image appeared in *Nature Genetics,* 16, 1997, 127, Figure 1b.)

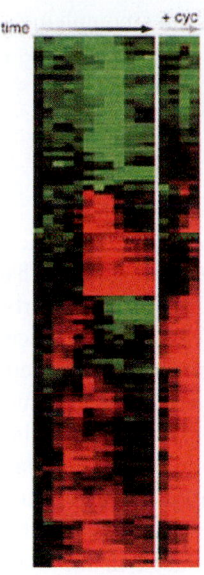

Figure 14-26 Display of gene-expression patterns detected by DNA microarrays. Each row is a different gene and each column a different time point. Red means active; green inactive. The four columns labeled +cyc are from cells grown on cycloheximide. (Mike Eisen and Vishy Iyer, Stanford; image appeared in *Nature Genetics* 18, 1998, 196, Figure 1.)

ing groups. Then the glass is bathed in the first nucleotide to be deposited. Each nucleotide carries its own protection group, which can be knocked off for the second round of deposition. Hence, by the sequential application of the appropriate masks and bathing sequences, arrays of different nucleotides can be built up. For studying genomic function, these oligonucleotides could be identifying sequences of genes, such as ESTs. As before, the completed array is bathed in fluorescent probe. Binding to an oligonucleotide array is shown in Figure 14-23b.

Note that these DNA array methods basically take an approach to genetic dissection that is an alternative to mutational analysis. Under either method, the goal is to define the set of genes or proteins that are important to any specific process under study. Traditional mutational analysis does this by amassing mutations that affect a specific process under study; chip technology does it by detecting the specific mRNAs that are transcribed during that process.

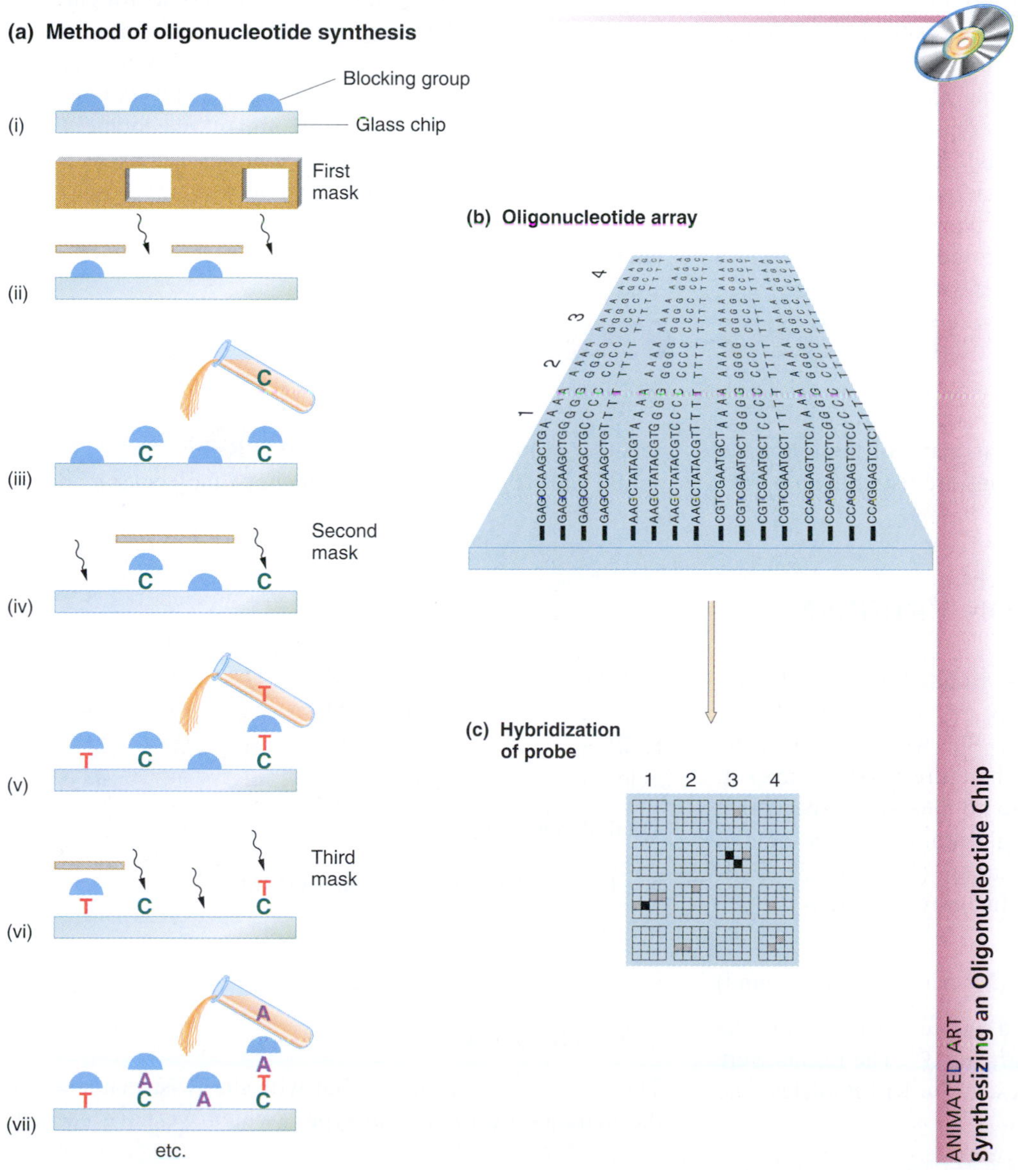

(a) Method of oligonucleotide synthesis

(i)

Blocking group

Glass chip

(ii)

First mask

(iii)

(iv)

Second mask

(v)

(vi)

Third mask

(vii)

etc.

(b) Oligonucleotide array

(c) Hybridization of probe

1 2 3 4

ANIMATED ART
Synthesizing an Oligonucleotide Chip

Figure 14-27 One method for the synthesis of a large array of oligonucleotides on a glass chip in situ. Nucleotides are deposited one at a time at addresses activated by light shining through a pattern of holes in a mask. Each nucleotide carries a blocking group that prevents further polymerization unless activated by light.

DNA chips can also be used to detect mutations. Oligonucleotides can be prepared that are complementary to all possible simple mutational changes in a genetic region under analysis. Alternatively, oligonucleotides complementary to all the known mutations in a human gene (such as a breast cancer gene) can be arrayed on the chip.

SUMMARY

Genomics is the branch of genetics that deals with the systematic molecular characterization of genomes. Some of the methods used are traditional genetic-mapping procedures; in addition, specialized techniques have been developed for manipulating the large amounts of DNA in a genome. Genomic analysis is important for two reasons: (1) it represents a way of obtaining an overview of the genetic architecture of an organism and (2) it forms a set of basic information that can be used to find new genes such as those responsible for human disease. Genomic analysis generally proceeds from low-resolution analysis to techniques with higher resolution. Initially, genes must be assigned to chromosomes, which can be achieved with a variety of techniques including linkage to standard markers, in situ hybridization, pulsed field gel electrophoresis, and human–rodent cell hybridization. The arrangement of loci along a chromosome can be determined by using various types of meiotic recombination mapping and (in humans) X-irradiated hybrids. Particularly useful are molecular markers that can fill in the gaps between genes of known phenotypic association. RFLPs, SSLPs, and RAPDs all provide heterozygous loci that can be used as molecular marker loci in mapping.

The highest level of genomic resolution is physical mapping of DNA fragments. Most useful are fragments that have been cloned in vectors that carry large DNA inserts, such as cosmids, YACs, and BACs. The goal of physical mapping is to produce a set of overlapping clones that encompass an entire chromosome or an entire genome. Sequence-tagged sites are particularly useful in aligning overlapping cosmids into contigs. As more clones are characterized, contigs grow to the size of entire chromosomes. Expressed sequence tags can slot functional genes into the genomic map. Genomic maps have been used in the positional cloning of human disease genes of unknown function. The maps have provided suitable starting points for chromosome walking and jumping. Genomic sequencing often reveals genes that have never been associated with a phenotype; such genes must be investigated by doing gene disruptions to check for a possible mutant phenotype. Functional genomics uses a variety of approaches such as defining all ORFs, the use of gene knockouts to probe gene function, the yeast two-hybrid system to look for gene interaction, and DNA microarrays to determine which genes are transcribed.

CONCEPT MAP

Draw a concept map interrelating as many of the following terms as possible. Note that the terms are listed in no particular order.

contig / physical map / YAC / BAC / RFLP / SSLP / STS / EST / FISH / recombinant frequency / molecular marker

CHAPTER INTEGRATION PROBLEM

A *Neurospora* geneticist has just isolated a new mutation that causes aluminum insensitivity (*al*) in a strain of Oak Ridge background (see page 406). She wishes to clone the gene by positional cloning and therefore needs to map it. For reasons that we do not need to go into, she suspects that it is located near the tip of the right arm of chromosome 4. Luckily, there are three RFLP markers (1, 2, and 3) available in that vicinity, so the following cross is made:

al (Oak Ridge background) ×

al^+ (Mauriceville background)

One hundred progeny are isolated and tested for *al* and the six RFLP alleles 1°, 2°, 3°, 1^M, 2^M, and 3^M. The results were as follows, where O and M represent the RFLP alleles, and *al* and + represent *al* and al^+:

RFLP 1	O	M	O	M	O	M
RFLP 2	O	M	M	O	O	M
RFLP 3	O	M	M	O	M	O
al locus	*al*	+	*al*	+	*al*	+
Total of genotype	34	36	6	4	12	8

a. Is the *al* locus in fact in this vicinity?

b. If so, to which RFLP is it closest?

c. How many map units separate the three RFLP loci?

◆ Solution ◆

This is a mapping problem, but with the twist that some of the markers are traditional types (which we have encountered in preceding chapters) and others are molecular markers (in this case, RFLPs). Nevertheless, the principle of mapping is the same as we used before; in other words, it is based on recombinant frequency. In any recombination analysis, we must be clear about the genotype of the parents before we can classify progeny into recombinant classes. In this case, we know that the Oak Ridge parent must contain all O alleles, and Mauriceville must contain all M alleles; therefore, the parents are:

$$al\ 1^O\ 2^O\ 3^O \times +\ 1^M\ 2^M\ 3^M$$

and knowing this makes determining recombinant classes easy. We see from the data that the parental classes are the two most common (34 and 36). We first of all notice that the *al* alleles are tightly linked to RFLP 1 (all progeny are *al* $1°$ or $+ 1^M$). Therefore, the *al* locus is definitely on this part of chromosome 4. There are $6 + 4 = 10$ recombinants

between RFLP 1 and 2, so these loci must be 10 map units apart. There are $12 + 8 = 20$ recombinants between RFLP 2 and 3; that is, they are 20 map units apart. There are $6 + 4 + 12 + 8 = 30$ recombinants between RFLP 1 and 3, showing that these loci must flank RFLP 2. Therefore, the map is:

| *al* RFLP 1 | 10 m.u. | RFLP 2 | 20 m.u. | RFLP 3 |

There are evidently no double recombinants, which would have been of the type M O M and O M O.

Notice that there is really no new principle at work in the solution of this problem; the real challenge is to understand the nature of RFLPs and to translate this understanding into genotypes from which to study recombination. If you still don't understand RFLPs, you might ask yourself how exactly the RFLP alleles are tested experimentally.

SOLVED PROBLEMS

1. Duchenne muscular dystrophy (DMD) is an X-linked recessive human disease affecting muscles. Six small boys had DMD, together with various other disorders, and they were found to have small deletions of the X chromosome, as shown here:

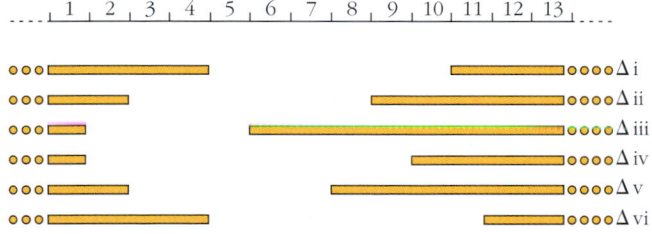

a. On the basis of this information, which chromosomal region most likely contains the gene for DMD?

b. Why did the boys show other symptoms in addition to DMD?

c. How would you use DNA samples from these six boys and DNA from unaffected boys to obtain an en-

riched sample of DNA containing the gene for DMD, as a prelude to cloning the gene?

◆ Solution ◆

a. The only region that all the deletions are lacking is the chromosomal region labeled 5, so this region presumably contains the gene for DMD.

b. The other symptoms probably result from the deletion of the other regions surrounding the DMD region.

c. If the DNA from all the DMD deletions is denatured (that is, its strands separated) and bound to some kind of filter, the normal DNA can be cut by shearing or by restriction-enzyme treatment, denatured, and passed through the filter containing the deleted DNA. Most DNA will bind to the filter, but the region-5 DNA will pass through. This process can be repeated several times. The filtrate DNA can be cloned and then used in a FISH analysis to see if it binds to the DMD X chromosomes. If not, it becomes a candidate for the DMD-containing sequence.

PROBLEMS

1. From in situ hybridizations, five different YACs containing genomic fragments were known to hybridize to one specific chromosome band of the human genome. Genomic DNA was digested with a long-cutter restriction enzyme, and radioactively labeled YACs were each hybridized to blots of the digest. The autoradiogram was as follows:

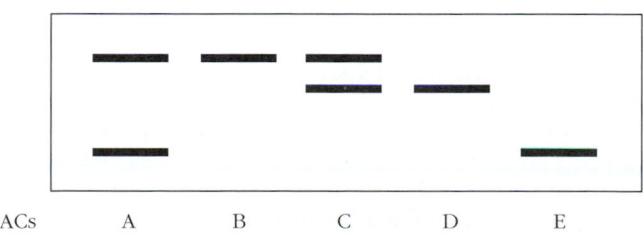

a. Use these results to order the three hybridized restriction fragments.

b. Show the locations of the YACs in relation to the three genomic restriction fragments in part a.

 Unpacking the Problem

1. State two types of hybridization used in genetics. What type of hybridizations are used in this problem, and what is the molecular basis for such hybridizations? (Draw a rough sketch of what happens at the molecular level during hybridization.)

2. How are in situ hybridizations done in general? How would the specific in situ hybridizations in this problem be done (as in the first sentence)?

3. What is a YAC?

4. What are chromosome bands, and what procedure is used to produce them? Sketch a chromosome with some bands and show how the in situ hybridizations would look.

5. How would five different YACs have been shown to hybridize to one band?

6. What is a genomic fragment? Would you expect the five YACs to contain the same genomic fragment or different ones? How do you think these genomic fragments were produced (what are some general ways of fragmenting DNA)? Does it matter how the DNA was fragmented?

7. What is a restriction enzyme?

8. What is a long cutter? If you do not know what a long cutter is, what do you think it might be, and does your guess make sense of this part of the problem? If not, refer to discussions of long cutters in the chapter.

9. Why were the YACs radioactively labeled? (What does it mean to radioactively label something?)

10. What is an autoradiogram?

11. Write a sentence that uses the words *DNA, digestion, restriction enzyme, blot, autoradiogram.*

12. Explain exactly how the pattern of dark bands shown in the problem was obtained.

13. Approximately how many kilobases of DNA are in a human genome?

14. If human genomic DNA were digested with a restriction enzyme, roughly how many fragments would be produced? Tens? Hundreds? Thousands? Tens of thousands? Hundreds of thousands?

15. Would all these DNA fragments be different? Would most of them be different?

16. If these fragments were separated on an electrophoretic gel, what would you see if you added a DNA stain to the gel?

17. How does your answer to question 16 compare with the number of autoradiogram bands in the diagram?

18. Part a of the problem mentions "three hybridized restriction fragments." Point to them in the diagram.

19. Would there actually be any restriction fragments on an autoradiogram?

20. Which YACs hybridize to one restriction fragment and which YACs hybridize to two DNA fragments?

21. How is it possible for a YAC to hybridize to two DNA fragments? Suggest two explanations, and decide which makes more sense in this problem. Does the fact that all the YACs in this problem bind to one chromosome band (and apparently nothing else) help you in deciding? Could a YAC hybridize to more than two fragments?

22. Distinguish the use of the word *band* by cytogeneticists (chromosome microscopists) from the use of the word *band* by molecular geneticists. In what way do these uses come together in this problem?

2. Three genes, *leu2, ade3,* and *mata,* were cloned in yeast. A *Neurospora* geneticist wanted to find out if *Neurospora* had these three genes and, if so, wanted to clone the *Neurospora* equivalents. As a first step to this analysis, he hybridized the clones as radioactive probes to PFGE preparations of *Neurospora* chromosomes, with the following results. Which genes are present in *Neurospora* and what chromosomes are they on?

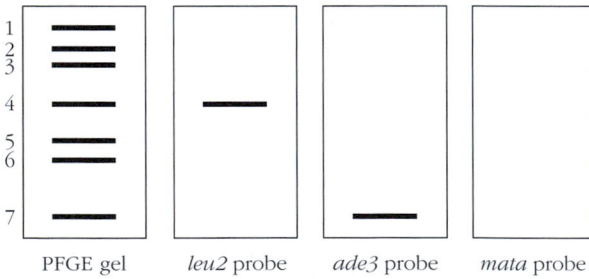

3. A *Neurospora* geneticist wanted to clone the gene *cys-1,* which was believed to be near the centromere on chromosome 5. Two RFLP markers (RFLP 1 and RFLP 2) were available in that vicinity, so he made the following cross:

Oak Ridge *cys-1* × Mauriceville *cys-1*[+]

Then 100 ascospores were tested for RFLP and *cys-1* genotypes, and the results were:

RFLP 1	O	M	O	M	O	M
RFLP 2	O	M	M	O	M	O
cys locus	cys	+	+	cys	cys	+
Total of genotype	40	43	2	3	7	5

a. Is *cys-1* in this region of the chromosome?

b. If so, draw a map of the loci in this region, labeled with map units.

c. What would be a suitable next step in cloning the *cys-1* gene?

4. In a certain haploid fungus, there had been extensive genetic analysis, including genetic mapping, and four linkage groups had been developed, suggesting four chromosomes. However, the chromosomes were very small and difficult to see under the microscope, so it was not known if there really were four chromosomes. The advent of PFGE technology showed that there are four chromosomes. However, the linkage groups still needed to be assigned to these chromosomes. To begin this process, a cloned gene *P* was used in a Southern analysis with a PFGE preparation from wild type and from two translocations that were known from genetic studies to be between linkage groups 1 and 4 in one case and between 2 and 4 in the other. The results follow:

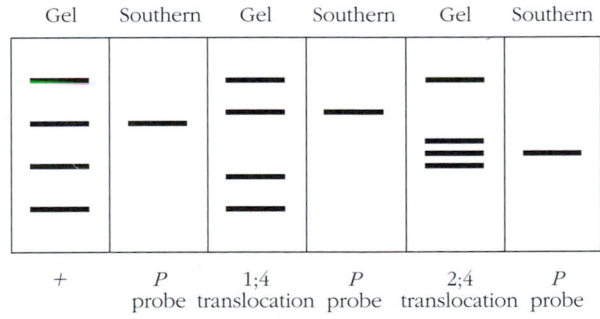

Gel	Southern	Gel	Southern	Gel	Southern
+	*P* probe	1;4 translocation	*P* probe	2;4 translocation	*P* probe

a. From these data, determine which of the four real chromosomes (bands) the gene is on.

b. Determine which chromosome (band) corresponds to each of the four linkage groups.

5. A cloned gene from *Arabidopsis* is used as a radioactive probe against DNA samples from cabbage (in the same plant family) digested by three different restriction enzymes. For enzyme 1, there were three radioactive bands on the autoradiogram; for enzyme 2, there was one band; and, for enzyme 3, there were two bands. How can these results be explained?

6. Five YAC clones of human DNA (YAC-A through YAC-E) were tested for sequence-tagged sites STS 1 through STS 7. The results are shown in the following table, in which a plus sign shows that the YAC contains that STS.

	STS						
YAC	1	2	3	4	5	6	7
A	+	−	+	+	−	−	−
B	+	−	−	−	+	−	−
C	−	−	+	+	−	−	+
D	−	+	−	−	+	+	−
E	−	−	+	−	−	−	+

a. Draw a physical map showing the STS order.

b. Align the YACs into a contig.

7. Seven human–rodent radiation hybrids were obtained and tested for six different human genome molecular markers A through F. The results are shown here, where a plus sign indicates the presence of a marker.

	RADIATION HYBRIDS						
Markers	1	2	3	4	5	6	7
A	−	+	−	−	+	+	−
B	+	−	+	−	−	−	−
C	+	−	+	+	−	+	−
D	−	+	−	+	+	+	−
E	+	−	−	+	+	−	+
F	+	−	−	+	+	−	+

a. What marker linkages are suggested by these results?

b. Is there any evidence of markers being on separate chromosomes?

8. A RAPD primer amplified two bands in *Aspergillus nidulans* haploid strain 1 and no bands in *A. nidulans* strain 2 (which was from a different country). These strains were crossed, and seven progeny were analyzed. The results were as follows:

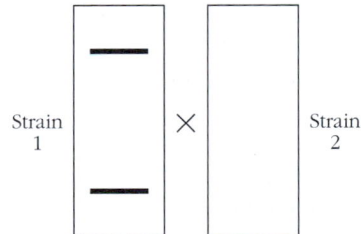

Progeny

a. Draw diagrams that explain the difference between the parents.

b. Explain the origin of the progeny and their relative frequencies.

c. Draw an example of a single tetrad from this cross, showing RAPD bands.

9. A *Caenorhabditis* contig for one region of chromosome 2 is as follows, where A through H are cosmids:

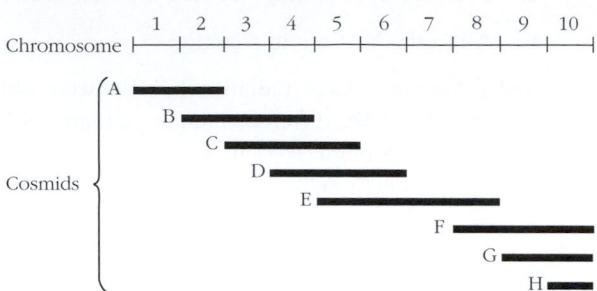

a. A cloned gene *pBR322-x* hybridized to cosmids C, D, and E. What is the approximate location of this gene *x* on the chromosome?

b. A cloned gene *pUC18-y* hybridized only to cosmids E and F. What is its location?

c. Explain exactly how it is possible for both probes to hybridize to cosmid E?

10. A certain disease is inherited as an autosomal dominant *N*. It was noted that some patients carry reciprocal translocations in which one of the chromosomes is always chromosome 3 and the break is always in band 3q3.1. Four molecular probes (a through d) are known to hybridize in situ to this band, but their order is not known. In the translocations, only probe c hybridizes to chromosome 3 carrying a part of another chromosome, and probes a, b, and d always hybridize to the translocated fragment of chromosome 3.

a. Draw diagrams that illustrate the meaning of these findings.

b. How would you use this information for positional cloning of the normal allele *n*?

c. Once *n* is cloned, how would you use this clone to investigate the nature of the mutations in disease patients who do not have translocations?

11. The gene for the autosomal dominant disease shown in this pedigree is thought to be on chromosome 4, so

five RFLPs (1–5) mapped on chromosome 4 were tested in all family members. The results are shown in the diagram; the superscripts represent different alleles of the RFLP loci.

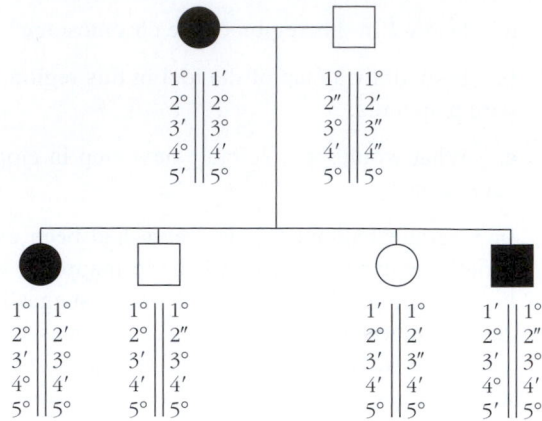

a. Explain how this experiment was carried out.

b. Decide which RFLP locus is closest to the disease gene (explain your logic).

c. How would you use this information to clone the disease gene?

12. A couple has three children with cystic fibrosis (CF). Their oldest son has recently married his second cousin. He has molecular testing done to determine if there is a chance that he may have children with CF. Three probes detecting RFLPs known to be very closely linked to the *CF* gene were used to assess the genotypes in this family. Answer the following questions, explaining your reasoning.

a. Is this man homozygous normal or a carrier?

b. Are his three normal siblings homozygous normal or carriers?

c. From which parent did each carrier inherit the disease allele?

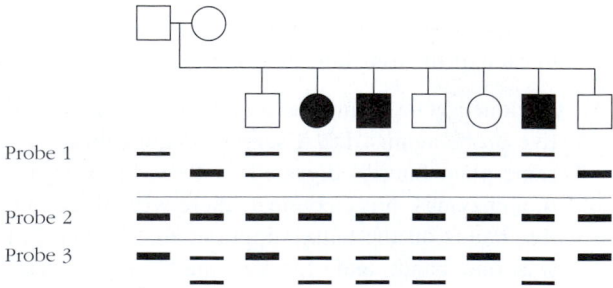

(Problem 12 is from Tamera Western.)

15

GENE MUTATION

The phenotype produced by an unstable mutation in *Zinnia*.
The mutation eliminates red pigment, resulting in white tissue.
However, in the course of development, the mutation frequently
reverts back to the normal allele, which permits synthesis of red
pigment. Stripes are produced because cell division in the petals
takes place mainly on the long axis, so the daughters of the
revertant cells tend to be arranged longitudinally.

(From M. A. L. Smith, Department of Horticulture, University of Illinois; see
Journal of Heredity *80, 1989.)*

Key Concepts

Mutation is the process whereby genes change from one
allelic form to another.

Forward mutations are changes away from the wild-type
allele, and reverse mutations are changes to the wild-type
allele.

Mutations can lead to loss of function of a gene or to new
function.

Mutations in germ-line cells can be transmitted to progeny,
but somatic mutations cannot.

Selective systems make it easier to obtain mutations.

Mutagens are agents that increase normally low rates of
mutation.

Genes mutate randomly, at any time and in any cell of an
organism.

A biological process can be dissected genetically if
mutations that affect that process can be obtained. Each
gene identified by a mutation identifies a separate
component of the process.

enetic analysis would not be possible without *variants*—organisms that differ in some particular character. We have considered many examples in which we could analyze differing phenotypes for particular characters. Now we consider the origin of the variants. How, in fact, do genetic variants arise?

The simple answer to this question is that organisms have an inherent tendency to undergo change from one hereditary state to another. Such hereditary change is called **mutation.** Geneticists recognize two different levels at which mutation takes place. In **gene mutation,** an allele of a gene changes, becoming a different allele. Because such a change takes place within a single gene and maps to one chromosomal locus ("point"), a gene mutation is sometimes called a **point mutation.** At the other level of hereditary change—**chromosome mutation**—segments of chromosomes, whole chromosomes, or even entire sets of chromosomes change. Gene mutation is not necessarily a part of such a process; the effects of chromosome mutation are due more to the new arrangements of chromosomes and of the genes that they contain. Nevertheless, some chromosome mutations, in particular those proceeding from chromosome breaks, are accompanied by gene mutations caused by the disruption at the breakpoint. In this chapter, we explore gene mutation; in Chapters 17 and 18, we shall consider chromosome mutation.

To consider change, we must have a fixed reference point, or standard. In genetics, the **wild type** provides the standard. Remember that the wild-type allele may be either the form found in nature or the form found in a standard laboratory stock. Any change away from the wild-type allele is called **forward mutation;** any change back to the

15-1 ▷ TABLE	Gene Mutations at the Molecular Level
Type of mutation	Result and example(s)

Forward mutations
Single-nucleotide-pair (base-pair) substitutions

At DNA level
   Transition — Purine replaced by a different purine; pyrimidine replaced by a different pyrimidine:

$$AT \longrightarrow GC \qquad GC \longrightarrow AT \qquad CG \longrightarrow TA \qquad TA \longrightarrow CG$$

   Transversion — Purine replaced by a pyrimidine; pyrimidine replaced by a purine:

$$AT \longrightarrow CG \qquad AT \longrightarrow TA \qquad GC \longrightarrow TA \qquad GC \longrightarrow CG$$
$$TA \longrightarrow GC \qquad TA \longrightarrow AT \qquad CG \longrightarrow AT \qquad CG \longrightarrow GC$$

At protein level
   Silent mutation — Triplet encodes same amino acid:

$$AGG \longrightarrow CGG$$
Both encode Arg

   Synonymous mutation — Codon specifies different but functionally equivalent amino acid:

$$AAA \longrightarrow AGA$$
Changing basic Lys to basic Arg
(will not alter protein function in many cases)

   Missense mutation — Codon specifies a different and nonfunctional amino acid.
   Nonsense mutation — Codon signals chain termination:

$$CAG \longrightarrow UAG$$

changing from a codon for Gln to a termination codon

Single-nucleotide-pair addition or deletion:
   frameshift mutation
Addition or deletion of several to
   many nucleotide pairs
— Any addition or deletion of base pairs that is not a multiple of 3 changes the reading frame in DNA segments that encode proteins, resulting in different amino acids from that point on and frequently chain termination.

Reverse mutations
Exact reversion

$$AAA\ (Lys) \xrightarrow{\text{forward}} GAA\ (Glu) \xrightarrow{\text{reverse}} AAA\ (Lys)$$
Wild type   Mutant   Wild type

wild-type allele is called **reverse mutation** (or **reversion** or **back mutation**). For example,

$$a^+ \longrightarrow a$$
$$D^+ \longrightarrow D \quad \Big\} \text{ forward mutation}$$
$$a \longrightarrow a^+$$
$$D \longrightarrow D^+ \quad \Big\} \text{ reverse mutation}$$

The non-wild-type allele of a gene is sometimes called a *mutation*. To use the same word for the process and the product may seem confusing at first, but in practice little confusion arises. Thus, we can speak of a dominant mutation (such as D in the example) or a recessive mutation (such as a). Bear in mind how arbitrary these gene states are; the wild type of today may have been a mutation in the evolutionary past, and vice versa.

Another useful term is **mutant.** This term is, strictly speaking, an adjective and should properly precede a noun. A mutant organism or cell is one whose changed phenotype is attributable to the possession of a mutation. Sometimes the noun is left unstated; in this case, a mutant always means an individual or cell with a phenotype that shows that it bears a mutation.

Two other useful terms are **mutation event,** which is the actual occurrence of a mutation, and **mutation frequency,** the proportion of mutations in a population of cells or individual organisms.

How DNA changes affect phenotype

The term **point mutation** originated before the advent of DNA sequencing and therefore before it was routinely possible to discover the molecular basis for a mutational event. Nowadays, point mutations typically refer to alterations of single base pairs of DNA or to a small number of adjacent base pairs. In this section, we shall consider the effect of such changes at the phenotypic level. Point mutations are classified in molecular terms in Table 15-1, which shows the main types of DNA changes and their functional effects at the protein level.

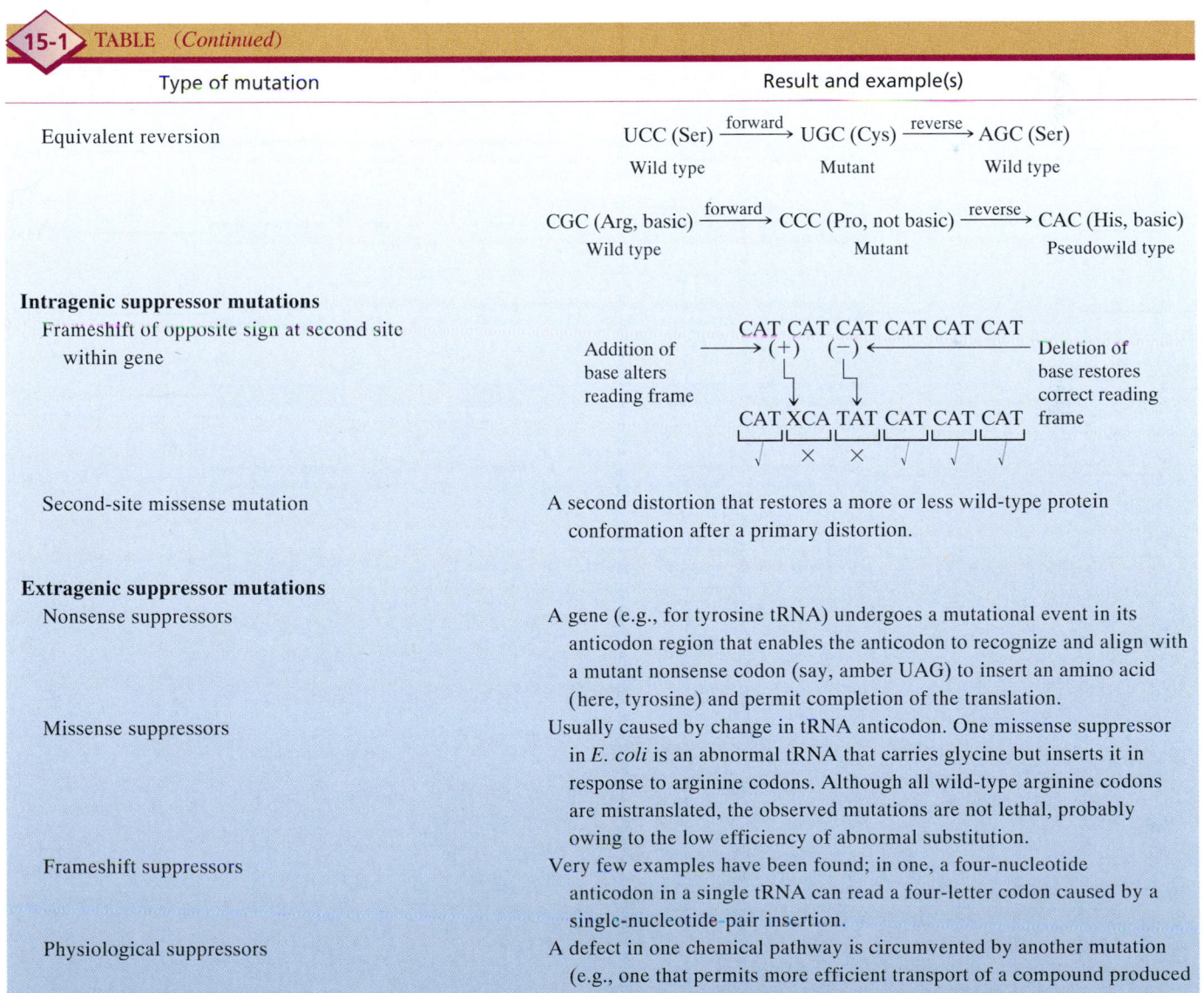

15-1 TABLE *(Continued)*	
Type of mutation	**Result and example(s)**
Equivalent reversion	UCC (Ser) $\xrightarrow{\text{forward}}$ UGC (Cys) $\xrightarrow{\text{reverse}}$ AGC (Ser)
	Wild type · · · · Mutant · · · · Wild type
	CGC (Arg, basic) $\xrightarrow{\text{forward}}$ CCC (Pro, not basic) $\xrightarrow{\text{reverse}}$ CAC (His, basic)
	Wild type · · · · Mutant · · · · Pseudowild type
Intragenic suppressor mutations	
Frameshift of opposite sign at second site within gene	CAT CAT CAT CAT CAT CAT
	Addition of base alters reading frame → (+) (−) ← Deletion of base restores correct reading frame
	CAT XCA TAT CAT CAT CAT
Second-site missense mutation	A second distortion that restores a more or less wild-type protein conformation after a primary distortion.
Extragenic suppressor mutations	
Nonsense suppressors	A gene (e.g., for tyrosine tRNA) undergoes a mutational event in its anticodon region that enables the anticodon to recognize and align with a mutant nonsense codon (say, amber UAG) to insert an amino acid (here, tyrosine) and permit completion of the translation.
Missense suppressors	Usually caused by change in tRNA anticodon. One missense suppressor in *E. coli* is an abnormal tRNA that carries glycine but inserts it in response to arginine codons. Although all wild-type arginine codons are mistranslated, the observed mutations are not lethal, probably owing to the low efficiency of abnormal substitution.
Frameshift suppressors	Very few examples have been found; in one, a four-nucleotide anticodon in a single tRNA can read a four-letter codon caused by a single-nucleotide-pair insertion.
Physiological suppressors	A defect in one chemical pathway is circumvented by another mutation (e.g., one that permits more efficient transport of a compound produced in smaller quantities owing to the original mutation).

At the DNA level, there are two main types of point mutational changes: **base substitutions** and **base additions or deletions.** Base substitutions are those mutations in which one base pair substitutes for another. They, again, can be divided into two subcategories: transitions and transversions. To describe these subcategories, we consider how a mutation alters the sequence on one DNA strand (the complementary change will take place on the other strand). A **transition** is the replacement of a base by another base of the same chemical category (purine replaced by purine: A → G or G → A; pyrimidine replaced by pyrimidine: C → T or T → C). A **transversion** is the opposite — the replacement of a base of one chemical category by a base of the other (pyrimidine replaced by purine: C → A, C → G, T → A, T → G; purine replaced by pyrimidine: A → C,

A → T, G → C, G → T). In a description of the same changes at the double-stranded level of DNA, both members of a base pair must be stated: an example of a transition is GC → AT; that of a transversion is GC → TA.

Addition or deletion mutations are actually of *nucleotide* pairs, but nevertheless the convention is to call them *base*-pair additions or deletions. The simplest are single-base-pair additions or single-base-pair deletions. There are examples in which mutations arise through simultaneous addition or deletion of multiple base pairs.

What are the functional consequences of these different categories of point mutations? First, consider what happens when a mutation arises in a polypeptide-encoding part of a gene. For single-base substitutions, there are several possible outcomes, which are direct consequences of two aspects

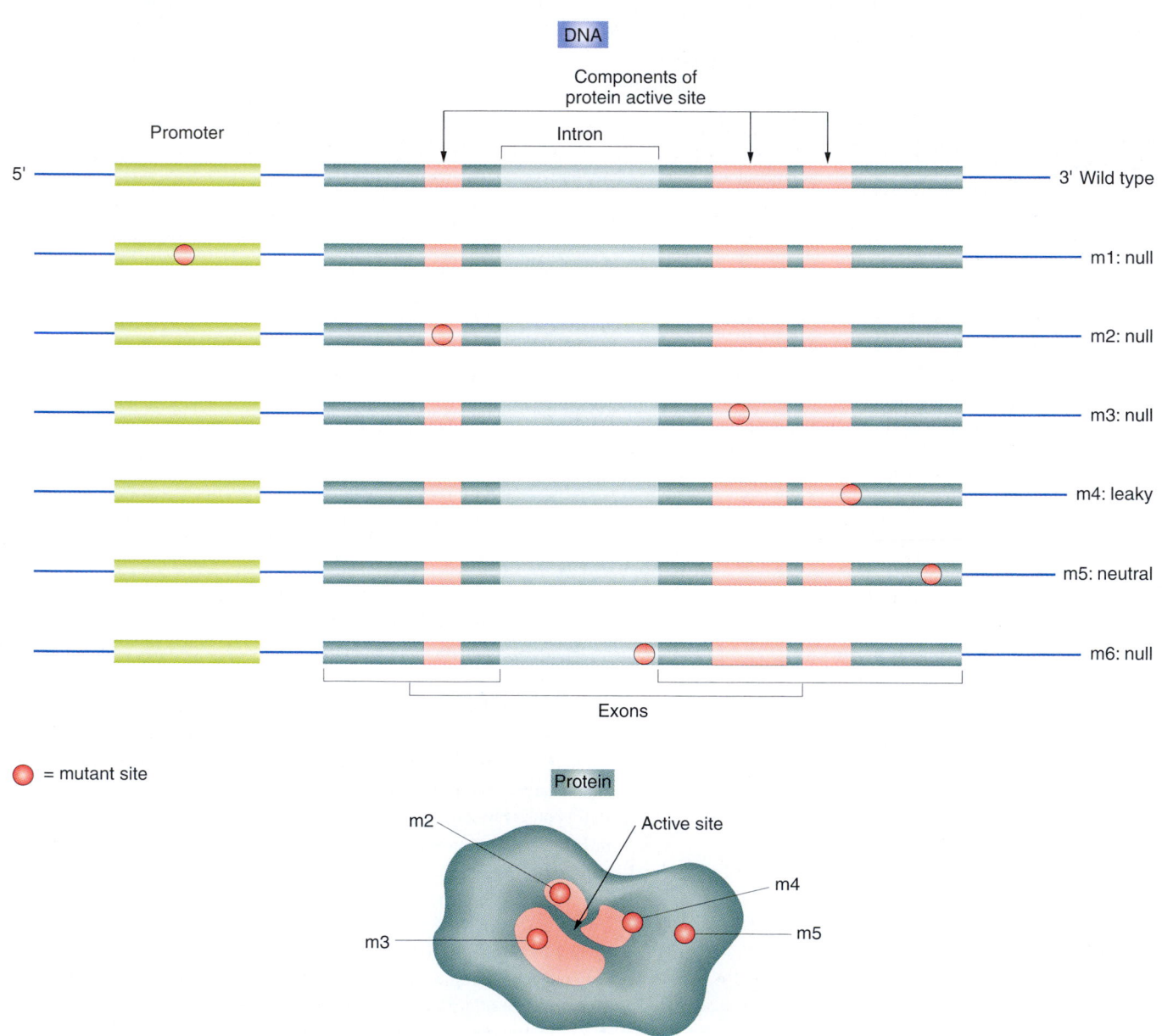

Figure 15-1 Positions of mutant sites and their functional consequences.

of the genetic code: (1) degeneracy of the code and (2) the existence of translation-termination codons. The outcomes are:

1. **Silent substitution.** The mutation changes one codon for an amino acid into another codon for that same amino acid.

2. **Missense mutation.** The codon for one amino acid is replaced by a codon for another amino acid.

3. **Nonsense mutation.** The codon for one amino acid is replaced by a translation-termination (stop) codon.

Silent substitutions never alter the amino acid sequence of the polypeptide chain. The severity of the effect of missense and nonsense mutations on the polypeptide will differ on a case-by-case basis. For example, if a missense mutation causes the substitution of a chemically similar amino acid (**synonymous substitution**), then it is likely that the alteration will have a less severe effect on the protein's structure and function. Alternatively, chemically different amino acid substitutions (**nonsynonymous substitutions**) are more likely to produce severe changes in protein structure and function. The effects of missense mutations in various parts of a gene are shown in Figure 15-1. Mutations that are in or close to the active site of a protein will most likely lead to lack of function; such mutations are called **null** (nothing) **mutations.** Mutations in less crucial areas of a protein will most likely have less deleterious effect, often resulting in "leaky," or partly inactivated, mutants.

Nonsense mutations will lead to the premature termination of translation. Thus, they have a considerable effect on protein function. Typically, unless it is very close to the 3' end of the open reading frame so that only a partly functional truncated polypeptide is produced, a nonsense mutation will produce a completely inactive protein product.

Like nonsense mutations, single-base additions or deletions have consequences on polypeptide sequence that extend far beyond the site of the mutation itself. Because the sequence of mRNA is "read" by the translational apparatus in groups of three base pairs (codons), the addition or deletion of a single base pair of DNA will cause a change of reading frame starting from the location of the addition or deletion and extending through to the carboxyl terminus of the protein. Hence, these lesions are called **frameshift** mutations. These mutations cause the entire amino acid sequence translationally downstream of the mutant site to bear no relation to the original amino acid sequence. Thus, frameshift mutations typically exhibit complete loss of normal protein structure and function.

Changes from mutant to wild-type phenotype may be due to changes within the mutant allele itself or to changes in some other gene that will result in a suppressor allele. Several specific examples of suppressors are given in Table 15-1.

Now let's turn to those mutations that take place in regulatory and other noncoding sequences. In essence, those parts of a gene that are not protein coding consist of a series of DNA and RNA target sites for binding by a large array of proteins and RNAs, interspersed with nonessential sequences. Thus, the ramifications of mutations in parts of a gene other than the polypeptide-coding segments are much harder to predict.

In general, the functional consequences of any point mutation (substitution or addition or deletion) in such a region depends on its location and on whether it disrupts a particular target site for protein or RNA binding. Some examples of the kinds of target sites for protein or RNA binding are splice donor and acceptor sites for introns in eukaryotic pre-mRNAs, RNA polymerase-binding sites in promoters, and 16S rRNA-binding sites (Shine-Delgarno sequences) upstream of the translation-initiation sites of prokaryotic mRNAs. Mutations that disrupt such target sites have the potential to change the expression pattern of a gene in regard to the amount of product expressed at a certain time, the response to certain environmental cues, the expression in certain tissues, or all three. We will see numerous additional examples of such target sites as we explore mechanisms of development later in the book. It is most important to realize that regulatory mutations affect the amount of the protein product of a gene, but they do not alter the structure of the protein. Figure 15-2 shows some examples of the effects of different mutational types on mRNA and protein.

Somatic versus germinal mutation

Genes and chromosomes can mutate in either somatic or germinal tissue, and these changes are called **somatic mutations** and **germinal mutations,** respectively. These two different types are shown diagrammatically in Figure 15-3.

Somatic mutation

If a somatic mutation occurs in a single cell in developing somatic tissue, that cell is the progenitor of a population of identical mutant cells, all of which have descended from the cell that mutated. A population of identical cells derived asexually from one progenitor cell is called a **clone.** Because the members of a clone tend to stay close to one another during development, an observable outcome of a somatic mutation is often a patch of phenotypically mutant cells called a **mutant sector.** The earlier in development the mutation event, the larger the mutant sector will be (Figure 15-4). Mutant sectors can be identified by eye only if their phenotype contrasts visually with the phenotype of the surrounding wild-type cells (Figure 15-5).

In diploids, a dominant mutation is expected to show up in the phenotype of the cell or clone of cells containing it. On the other hand, a recessive mutation will not be expressed, because it is masked by a wild-type allele that is by definition dominant to the recessive mutation. A second mutation could create a homozygous recessive mutation, but this event would be rare.

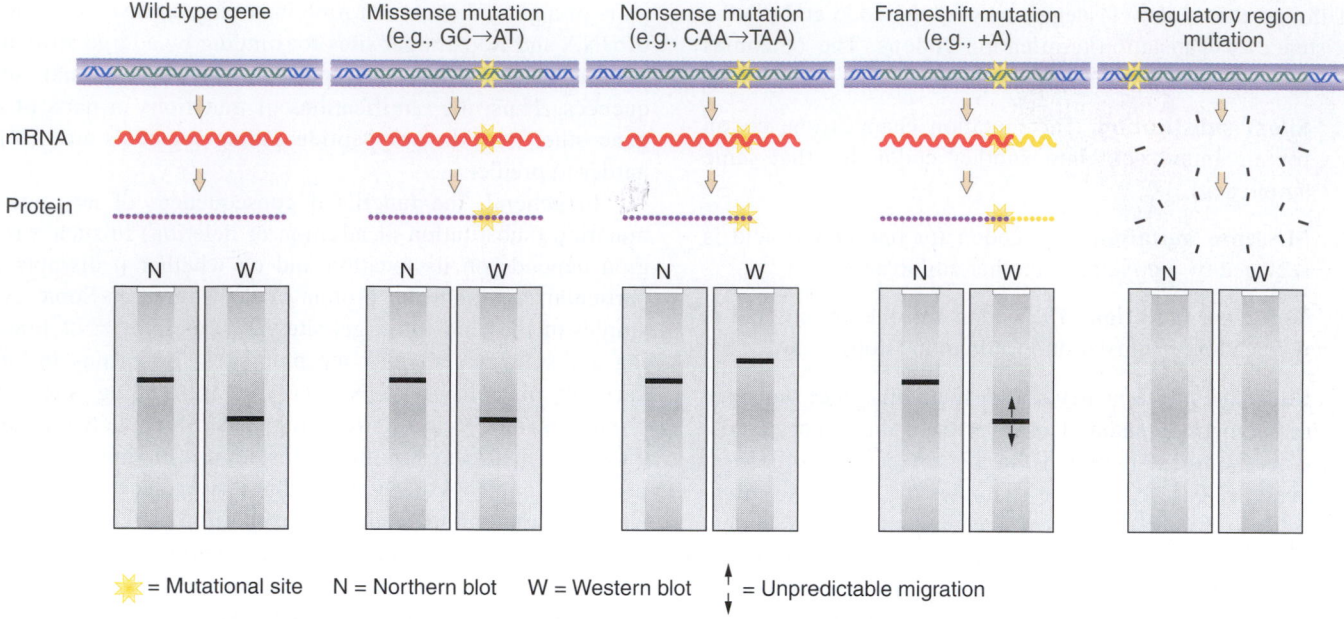

Figure 15-2   The effect of some common types of mutations at the RNA and protein levels. The lower diagrams represent Northern (N) and Western (W) analyses.

What would be the consequences of a somatic mutation in a cell of a fully developed organism? If the mutation is in tissue in which the cells are still dividing, then there is the possibility of a mutant clone's arising. If the mutation is in a postmitotic cell — that is, one that is no longer dividing — then the effect on phenotype is likely to be negligible. Even when dominant mutations result in a cell that is either dead or defective, this loss of function will be compensated by other normal cells in that tissue. However, mutations that give rise to cancer are a special case. Cancer mutations arise in a special category of genes called **proto-oncogenes,** many of which regulate cell division. When mutated, such cells enter a state of uncontrolled division, resulting in a cluster of cells called a **tumor.** We shall look at some examples later in this chapter.

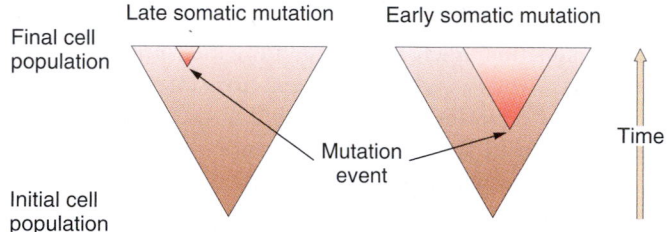

Figure 15-4   Early mutation produces a larger proportion of mutant cells in the growing population than does later mutation.

Figure 15-5   Somatic mutation in the red Delicious apple. The mutant allele determining the golden color arose in a flower's ovary wall, which eventually developed into the fleshy part of the apple. The seeds are not mutant and will give rise to red-appled trees. (Note that, in fact, the golden Delicious apple originally arose as a mutant branch on a red Delicious tree.) (Anthony Griffiths.)

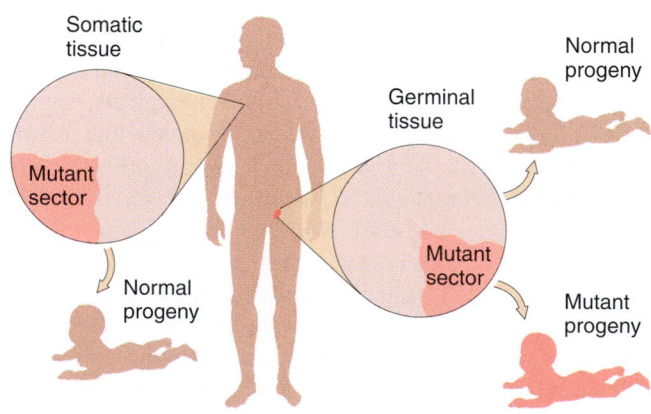

Figure 15-3   Somatic mutations are not transmitted to progeny, but germinal mutations may be transmitted to some or all progeny.

Figure 15-6   A mutation producing an allele for white petals that arose originally in somatic tissue but eventually became part of germinal tissue and could be transmitted through seeds. The mutation arose in the primordium of a side branch of the rose. The branch grew long and eventually produced flowers. (From Harper Horticultural Slide Library.)

Are somatic mutations ever passed on to progeny? No. It is impossible, because somatic cells by definition are those that are never transmitted to progeny. However, note that, if we take a plant cutting from a stem or leaf that includes a mutant somatic sector, the plant that grows from the cutting may develop germinal tissue out of the mutant sector. Put another way, a branch bearing flowers can grow out of the mutant somatic sector. Hence, what arose as a somatic mutation can be transmitted sexually. An example is shown in Figure 15-6.

Any method for the detection of somatic mutation must be able to rule out the possibility that the sector is due to mitotic segregation or recombination (Chapter 6). If the individual is a homozygous diploid, somatic sectoring is almost certainly due to mutation.

Germinal mutation

A *germinal mutation* occurs in the germ line, special tissue that is set aside in the course of development to form sex cells. If a mutant sex cell participates in fertilization, then the mutation will be passed on to the next generation. An individual of perfectly normal phenotype and of normal ancestry can harbor undetected mutant sex cells. These mutations can be detected only if they are included in a zygote (Figures 15-7 and 15-8). Remember from Chapter 2 that the X-linked hemophilia mutation in European royal families is thought to have arisen in the germ cells of Queen Victoria or one of her parents. The mutation was expressed only in her male descendants.

The experimental detection of germinal mutation depends on the ability to rule out meiotic segregation and recombination as possible causes of phenotypic differences between parents and offspring.

Figure 15-7   Germinal mutation determining white petals in viper's bugloss *(Echium vulgare)*. A recessive germinal mutation, *a*, arose in an *A/A* blue plant of the preceding generation, making its germinal tissue *A/a*. On selfing, the mutation was transmitted to progeny, some of which were *a/a* and expressed the mutant phenotype. (Anthony Griffiths.)

Figure 15-8   A mutation to an allele determining curled ears arose in the germ line of a normal straight-eared cat and was expressed in progeny such as the cat shown here. This mutation arose in a population in Lakewood, California, in 1981. It is an autosomal dominant. (From R. Robinson, *Journal of Heredity* 80, 1989, 474.)

MESSAGE ··
Before a new heritable phenotype can be attributed to mutation, both segregation and recombination must be ruled out as possible causes. This requirement is true for both somatic and germinal mutations.

Mutant types

The phenotypic consequences of mutation may be so subtle as to require refined biochemical techniques to detect a difference from the phenotype conferred by the wild-type allele. Alternatively, the mutation may be so severe as to produce gross morphological defects or death. A rough classification follows, based only on the ways in which the mutations are recognized. This classification is not meant to be complete. Furthermore, the various ways of categorizing mutations often overlap.

Morphological mutations

Morph means "form." Morphological mutations affect the outwardly visible properties of an organism, such as shape, color, or size. Albino ascospores in *Neurospora,* curly wings in *Drosophila,* and dwarf peas are all morphological mutations. Some additional examples of morphological mutants are shown in Figure 15-9.

Lethal mutations

A new lethal mutant allele is recognized by its effects on the survival of the organism. Sometimes a primary cause of death from a lethal mutation is easy to identify (for example, in certain blood abnormalities). But often the cause of death is hidden, and the mutant allele is recognizable *only* by its effects on viability. An example of a lethal mutation is shown in Figure 15-10.

Conditional mutations

In the class of conditional mutations, a mutant allele causes a mutant phenotype in only a certain environment, called the **restrictive condition,** but causes a wild-type phenotype in some different environment, called the **permissive condition.** Geneticists have studied many temperature-conditional mutations. For example, certain *Drosophila* mutations are known as *dominant heat-sensitive lethals.* Heterozygotes (say, H^+/H) are wild type at 20°C (the permissive condition) but die if the temperature is raised to 30°C (the restrictive condition).

Many mutant organisms are less vigorous than their normal counterparts and thus more troublesome as experimental subjects. For this reason, conditional mutants are useful because they can be grown under permissive conditions and then shifted to restrictive conditions for study. Another advantage of conditional mutations is that they allow the determination of a developmental **sensitive period** at which specific time the gene acts. In these studies, orga-

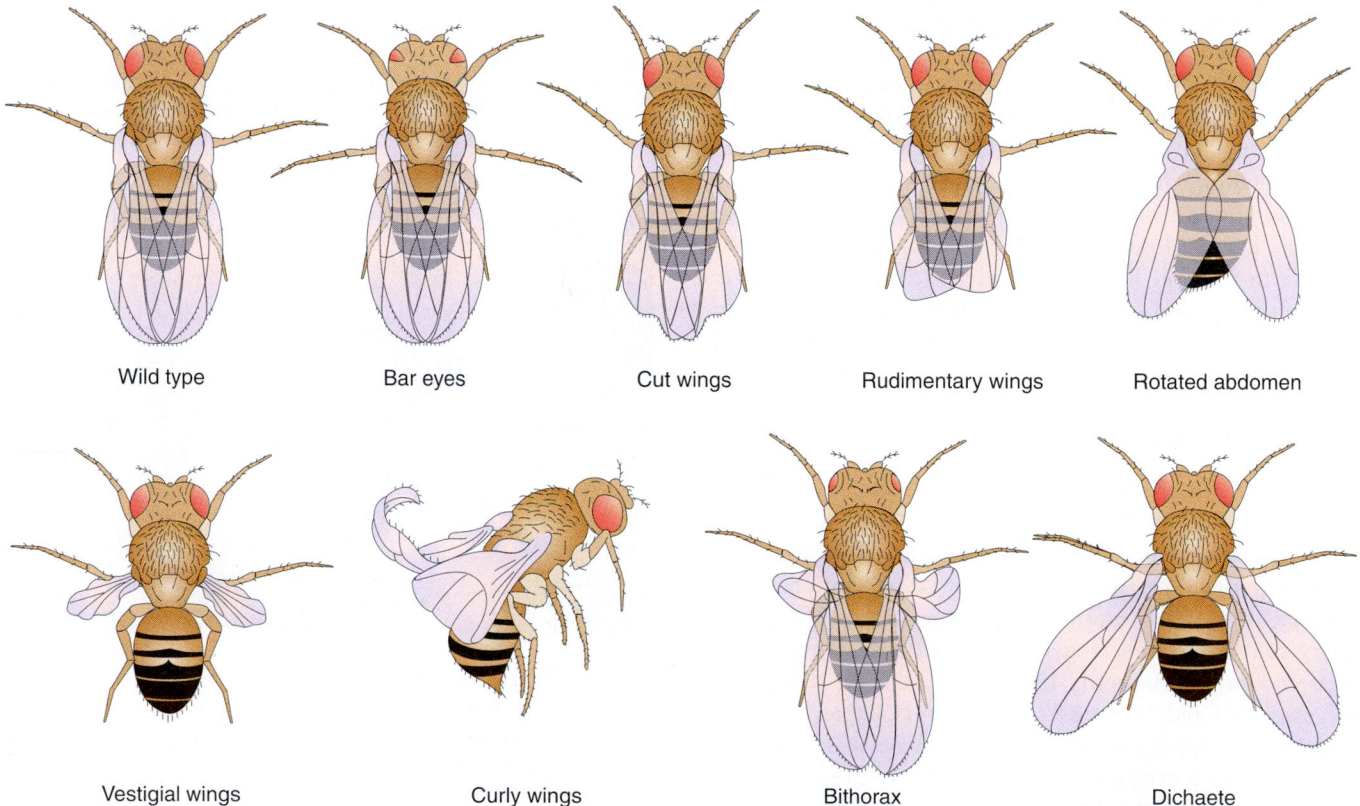

Wild type   Bar eyes   Cut wings   Rudimentary wings   Rotated abdomen

Vestigial wings   Curly wings   Bithorax   Dichaete

Figure 15-9   Eight morphological mutations of *Drosophila,* and the wild type for comparison. Most of the mutant phenotypes are self-explanatory; bithorax is an abnormality of the thorax featuring small wings instead of balancers; the most prominent feature of dichaete is that the wings are held at 45 degrees to the body.

Figure 15-10 Phenotypes of the wild type (left) and a mutation affecting plumage of Japanese quail (right). This mutation arose in a laboratory colony of quail and could be maintained as an interesting subject for genetic analysis. However, if such a mutation had arisen in nature, it would almost certainly have been lethal. (Janet Fulton.)

nisms carrying some specific conditional mutation are shifted from permissive to restrictive conditions at different times in the course of development. Some shifts will lead to mutants, others to wild types, and from these results the sensitive period of gene action is assessed.

Biochemical mutations

Microbial cultures are convenient material for the study of biochemical mutations, which are identified by the loss or change of some biochemical function of the cells. This change typically results in an inability to grow and proliferate. In many cases, however, growth of a mutant cell can be restored by supplementing the growth medium with a specific nutrient. Microorganisms, by and large, are **prototrophic:** they can exist on a substrate of simple inorganic salts and an energy source; such a growth medium is called a **minimal medium.** Biochemical mutants, however, often are **auxotrophic:** they must be supplied certain additional nutrients if they are to grow. For example, one class of biochemically mutant fungi will not grow unless supplied with the nitrogenous base adenine. They are called *ad* mutations, whereas the wild-type allele is ad^+. Mutant *ad* alleles determine the auxotrophic, adenine-requiring phenotype. The practical method of testing for the auxotrophic or prototrophic phenotype is shown in Figure 15-11.

Although microbial cultures are used for experimental induction of biochemical mutations, we should note that many human hereditary diseases are biochemical mutations

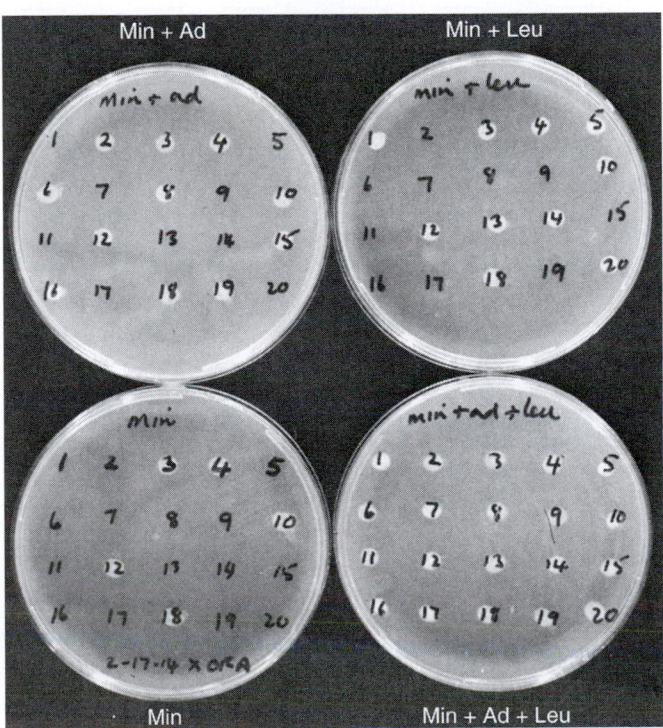

Figure 15-11 Testing strains of *Neurospora crassa* for auxotrophy and prototrophy. In this experiment, the test utilizes 20 progeny from a cross of an adenine-requiring auxotroph and a leucine-requiring auxotroph. Genotypically, the cross is $ad \cdot leu^+ \times ad^+ \cdot leu$, and the progeny can carry any of the four possible combinations of these alleles. To test the progeny, the geneticist attempts to grow cells on various kinds of gelled media in petri dishes. The media are minimal medium (Min) with either adenine *(ad, top left)*, leucine *(leu, top right)*, neither *(bottom left)*, or both *(bottom right)*. Growth appears as a small circular colony (white in the photograph). Any culture growing on minimal must be $ad^+ \cdot leu^+$, one growing on adenine and no leucine must be leu^+, and one on leucine and no adenine must be ad^+. All should grow on adenine plus leucine; it is a kind of control to check viability. As examples, culture 8 must be $ad \cdot leu^+$, 9 must be $ad \cdot leu$, 10 must be $ad^+ \cdot leu^+$, and 13 must be $ad^+ \cdot leu$. (Anthony Griffiths.)

defective in some step of cellular chemistry. The expression **inborn errors of metabolism** has been used to describe such biochemical disorders. Phenylketonuria and galactosemia are two examples.

Loss-of-function mutations

Generally, loss-of-function (null) mutations are found to be recessive. In a wild-type diploid cell, there are two wild-type alleles of a gene, both making normal gene product. In heterozygotes (the crucial genotypes for testing dominance or recessiveness), the single wild-type allele may be able to provide enough normal gene product to produce a wild-type phenotype. In such cases, loss-of-function mutations are recessive. In some cases, the cell is able to "upregulate" the level of activity of the single wild-type allele so that in the heterozygote the total amount of wild-type gene product is more than half that found in the homozygous wild type. However, some loss-of-function mutations are dominant. In such cases, the single wild-type allele in the heterozygote cannot provide the amount of gene product needed for the cells and the organism to be wild type. The action of loss-of-function mutations is represented diagrammatically in Figure 15-12a and b.

Gain-of-function mutations

Because mutation events introduce random genetic changes, most of the time they result in loss of function. The mutation events are like bullets being fired at a complex machine; most of the time they will inactivate it. However, it is conceivable that in rare cases a bullet will strike the machine in such a way that it produces some new function. So it is with mutation events; sometimes the random change by pure chance confers some new function on the gene. In a heterozygote, the new function will be expressed, and therefore the gain-of-function mutation most likely will act like a dominant allele and produce some kind of new phenotype. Gain of function is represented in Figure 15-12c.

Occurrence of mutations

Mutation is a biological process that has characterized life from its beginning. This process is fascinating and worthy of study in its own right. Mutant alleles such as the ones mentioned in the preceding sections are invaluable in the study of the process of mutation itself. For instance, they allow us to measure the frequency of mutation. In this connection, they are used as genetic markers, or representative genes: their precise function is not particularly important, except as a way to follow the process.

In modern genetics, however, mutant genes have another role in which their precise function *is* important. Mutant genes can be used as probes to disassemble the constituent parts of a biological function and to examine their workings and interrelations. Thus, it is of considerable inter-

(a) Null loss-of-function mutation (_m_)

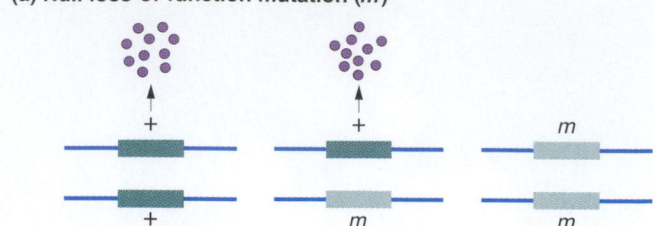

(b) Leaky loss-of-function mutation (_m'_)

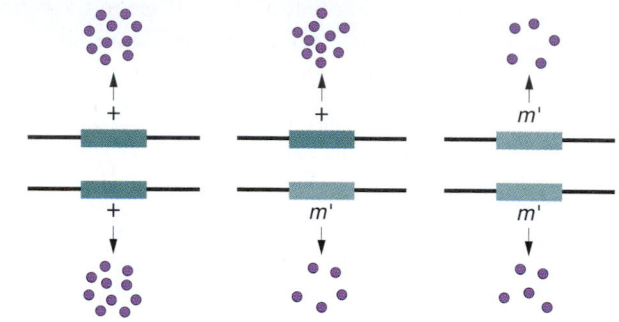

(c) Gain-of-function mutation (_M_)

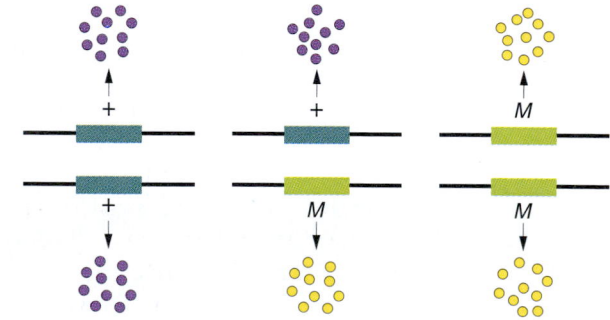

Figure 15-12   (a) Mutation _m_ has completely lost its function (it is a null mutation). In the heterozygote, wild-type gene product is still being made, and often the amount is enough to result in a wild-type phenotype, in which case, _m_ will act as a recessive. If the wild-type gene product is insufficient, the mutation will be seen as dominant. (b) Mutation _m'_ still retains some function, but in the homozygote there is not enough to produce a wild-type phenotype. (c) Mutation _M_ has acquired a new cellular function represented by the gene product colored green. _M_ will be expressed in the heterozygote and most likely will act as a dominant. The homozygous mutant may or may not be viable, depending on the role of the + allele.

est to a biologist studying a particular function to have as many mutant forms that affect the function as possible. This need has led to "mutant hunts" as an important prelude to any genetic dissection in biology. To identify a genetic variant is to identify a component of the biological process. We explore this idea further later in the chapter.

Mutation detection systems

The tremendous stability and constancy of form of species from generation to generation suggest that mutation must be a rare process. This supposition has been confirmed, creating a problem for the geneticist who is trying to demonstrate mutation.

The prime need is for a detection system — a set of circumstances in which a mutant allele will make its presence known at the phenotypic level. Such a system allows rare mutations to be found in large-scale searches called **screens.**

One of the main considerations here is that of dominance. The system must be set up so that mutations that create new recessive alleles, which are the most common type, will not be masked by a dominant wild-type allele. (Dominant mutations are less of a problem.) As an example, we can use one of the first detection systems ever set up — the one used by Lewis Stadler in the 1920s to study mutation in corn from C, expressed phenotypically as a colored kernel, to c, expressed as a white kernel. Here we are dealing with the phenotype of the endosperm of the seed. If you refer to Figure 3-22, you will see that this tissue forms when two identical haploid female nuclei fuse with one haploid nucleus from the male pollen cell. Hence, the tissue has three chromosome sets (it is $3n$). One dominant C allele in combination with two c alleles causes the kernel to be colored.

Stadler crossed $c/c \; ♀ \times C/C \; ♂$ and examined thousands of individual kernels on the corn ears that resulted from this cross. Each kernel is a progeny individual. In the absence of mutation, every kernel is $C/c/c$ and shows the colored phenotype. Therefore, a white kernel reveals that C mutated to c in a male reproductive cell of the C/C parent. The system has thus detected a germinal mutation. Although laborious, this is a very straightforward and reliable method of mutant detection (Figure 15-13).

This basic system can be extended to as many loci as can be conveniently made heterozygous in the same cross. For example,

$$a^+/a^+ \cdot b^+/b^+ \cdot c^+/c^+ \cdot d^+/d^+ \times a/a \cdot b/b \cdot c/c \cdot d/d$$

Mutant 1

$$a^+/a \cdot \boxed{b/b} \cdot c^+/c \cdot d^+/d$$

Normal   Mutant 2

$$a^+/a \cdot b^+/b \cdot c^+/c \cdot d^+/d \qquad a^+/a \cdot b^+/b \cdot c^+/c \cdot \boxed{d/d}$$

By increasing the number of loci under study, the investigator increases the likelihood of detecting a mutation in the experiment.

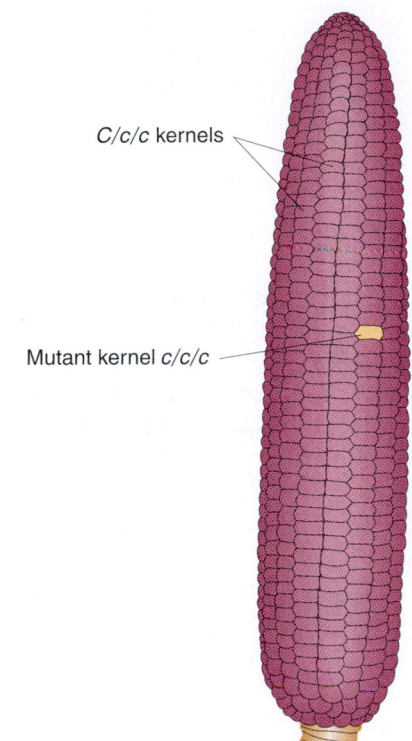

C/c/c kernels

Mutant kernel *c/c/c*

Figure 15-13   The detection system for mutations at a specific locus of corn. The C allele determines the presence of a purple pigment in kernels, whereas c results in none. The geneticist makes the cross $c/c \; ♀ \times C/C \; ♂$, and $C \to c$ mutations in the male germ line show up as unpigmented kernels on the cobs. (**Note:** The colored layer of the corn seed is triploid, formed by two identical female nuclei plus a male nucleus.)

Stadler's method for detecting germinal mutations is now called the **specific-locus test.** The same principle can be applied to somatic mutations too. Once again, to increase the likelihood of finding and quantifying mutations, the number of loci under study can be increased to any level that is feasible. Let's look at an example in which detectable mutations affect the coat colors of mice. The phenotypes and genes responsible for them are leaden *(ln)*, pallid *(pa)*, brown *(b)*, chinchilla *(ch)*, pink-eyed dilution *(p)*, dilution *(d)*, and pearl *(pe)*. Mice from pure lines of the following genotypes are crossed:

$$ln/ln \cdot pa/pa \cdot b^+/b^+ \cdot ch^+/ch^+ \cdot p^+/p^+ \cdot d^+/d^+ \cdot pe/pe$$
$$\times \; ln^+/ln^+ \cdot pa^+/pa^+ \cdot b/b \cdot ch/ch \cdot p/p \cdot d/d \cdot pe^+/pe^+$$

and the females bear embryos of the multiply heterozygous genotype

$$ln/ln^+ \cdot pa/pa^+ \cdot b/b^+ \cdot ch/ch^+ \cdot p/p^+ \cdot d/d^+ \cdot pe/pe^+$$

Because all the mutant alleles are recessive, the coats of the F_1 are expected to be wild type. However, somatic mutations from wild type to mutant at any of the heterozygous

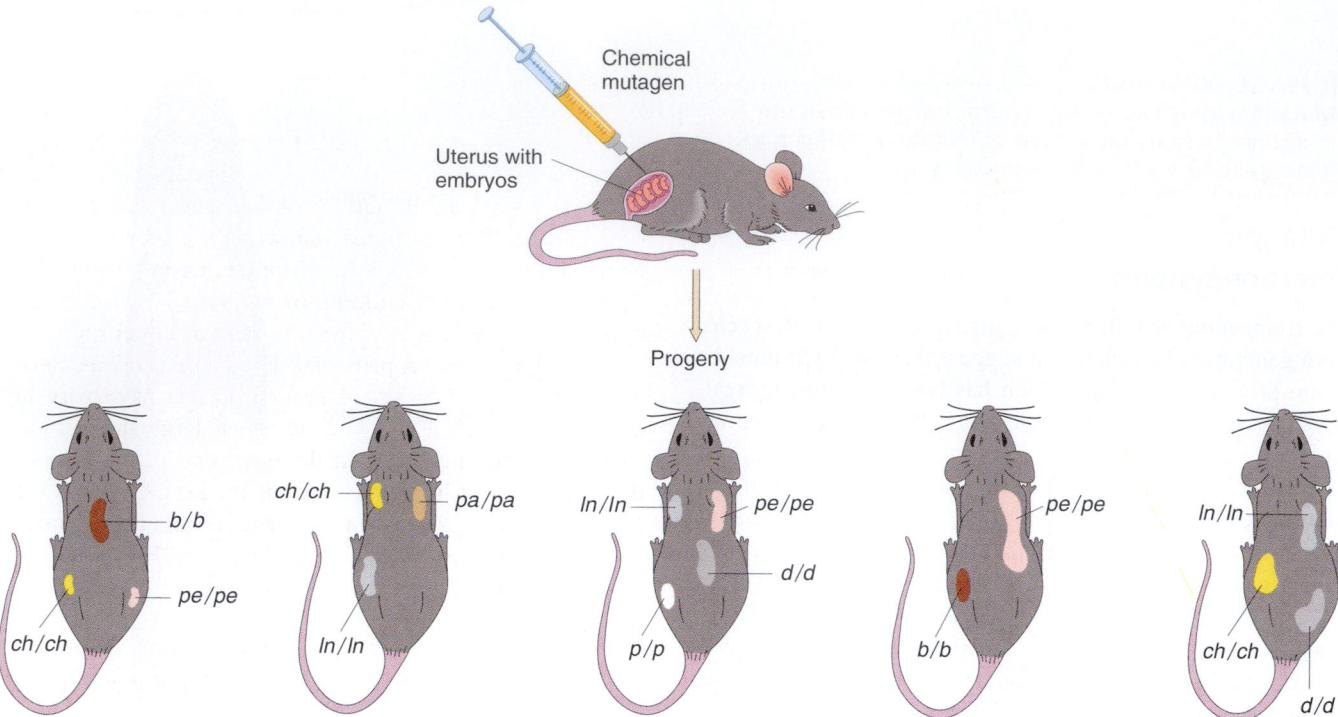

Genotype of progeny is $ln/ln^+ \cdot pa/pa^+ \cdot b/b^+ \cdot ch/ch^+ \cdot p/p^+ \cdot d/d^+ \cdot pe/pe^+$
Phenotype of progeny is wild type with mutant sectors

Figure 15-14 A detection system for recessive somatic mutations at seven coat color loci in mice. The cross $ln/ln \cdot pa/pa \cdot b^+/b^+ \cdot ch^+/ch^+ \cdot p^+/p^+ \cdot d^+/d^+ \cdot pe/pe \times ln^+/ln^+ \cdot pa^+/pa^+ \cdot b/b \cdot ch/ch \cdot p/p \cdot d/d \cdot pe^+/pe^+$ results in progeny heterozygous for all seven genes and predominantly wild type in appearance. Any somatic mutation from wild type to mutant at any of the loci causes a mutant sector in the coat of the offspring. The mutant colors are leaden *(ln)*, pallid *(pa)*, brown *(b)*, pink-eyed dilution *(p)*, dilution *(d)*, and pearl *(pe)*. The frequency of mutations can be increased by administering to developing embryos a chemical that is known to produce mutations (a chemical mutagen).

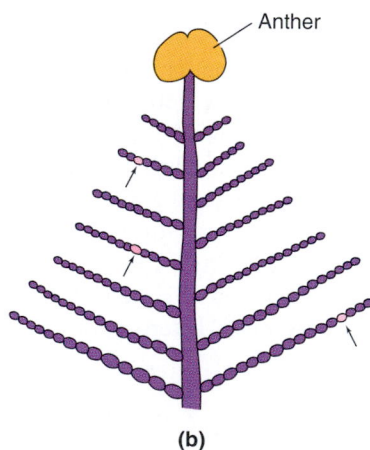

Figure 15-15 (a) Stamen of a *Tradescantia* plant heterozygous for *P* and *p* alleles. (b) In the chains that constitute the lateral hairs, some cells (marked by arrows) are mutant. The darker color (purple pigmentation) is determined by *P* and the paler color (pink pigmentation) by *p/p*. (Part a from Runk/Schoenberger/Grant Heilman.)

loci cause mutant sectors on the coat (Figure 15-14). The frequency of these mutant sectors increases dramatically if the investigator injects a mutation-inducing chemical into the uterus of the pregnant mother at the eighth day of pregnancy. This injection exposes the developing embryos to the chemical. The progeny show many more mutant sectors than do mice that were not exposed to the mutagen as embryos. The potency of the chemical can be quantified simply by counting the number of mutant sectors and comparing this number with that observed in untreated animals.

In the plant *Tradescantia,* an effective somatic specific-locus test was developed by using a single heterozygous gene. A vegetatively propagated line of *Tradescantia* called 02 is heterozygous *P/p*. These alleles determine the dominant blue and recessive pink pigmentation phenotypes. The pigmentation is expressed in the petals and the stamens. The stamens of this plant have hairs that are chains of single cells. Therefore millions of single somatic cells can easily be scanned under the microscope to look for pink cells, which must have arisen from a $P \rightarrow p$ mutation, converting *P/p* into *p/p* (Figure 15-15).

Human geneticists detect germinal mutations by the sudden appearance of a novel phenotype in a pedigree in which there is no previous record of such a phenotype. Dominant mutations are easily detected in this way because they are expected to be expressed in the phenotype. An example is given in Figure 15-16, which shows a pedigree for neurofibromatosis, an autosomal dominant disorder characterized by abnormal skin pigmentation ("café au lait" spots) and by numerous tumors, called *neurofibromas,* that associate with the peripheral or central nervous system. The tumors are visible on the surface of the body as shown in Figure 15-16. The pedigree shows an ancestry free of neurofibromatosis, with the disorder arising suddenly in generation IV in the children of one couple. The mutation must have arisen in the germinal tissue of either the mother or the father. It so happens that the gene mutations causing both neurofibromatosis and achondroplasia have very high mutation frequencies in humans, so a large proportion of cases are from mutation. At the other end of the spectrum, the gene for Huntington disease has the lowest mutation frequency, so most cases of this disease are inherited from preceding generations. In fact, most cases of Huntington disease in North America can be traced to two immigrant families.

A human recessive mutation is more difficult to detect. Because of its recessiveness, the mutation is not expressed in the heterozygous state and can go unnoticed for many generations. It will become expressed only through inbreeding or the chance mating of two separate heterozygotes. X-linked recessive mutations can be identified more easily than autosomal recessive mutations. We have already considered the X-linked hemophilia allele present in the royal families of Europe. The disease was not recorded in the

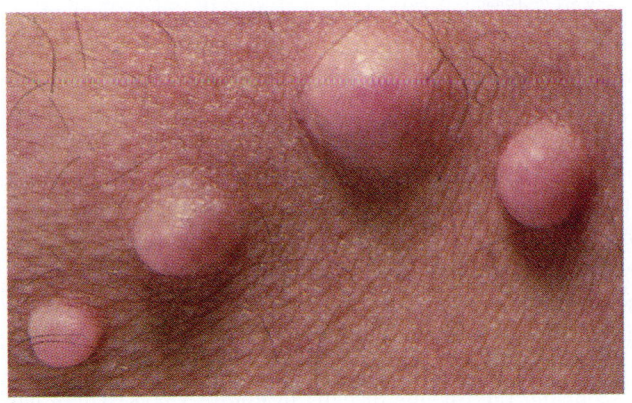

(a)

Figure 15-16 Mutation to neurofibromatosis. (a) Neurofibromas. (b) Pedigree showing that the neurofibromatosis mutation must have been in the germ line of III-8 or III-9. (Part a from Michael English/Custom Medical Stock.)

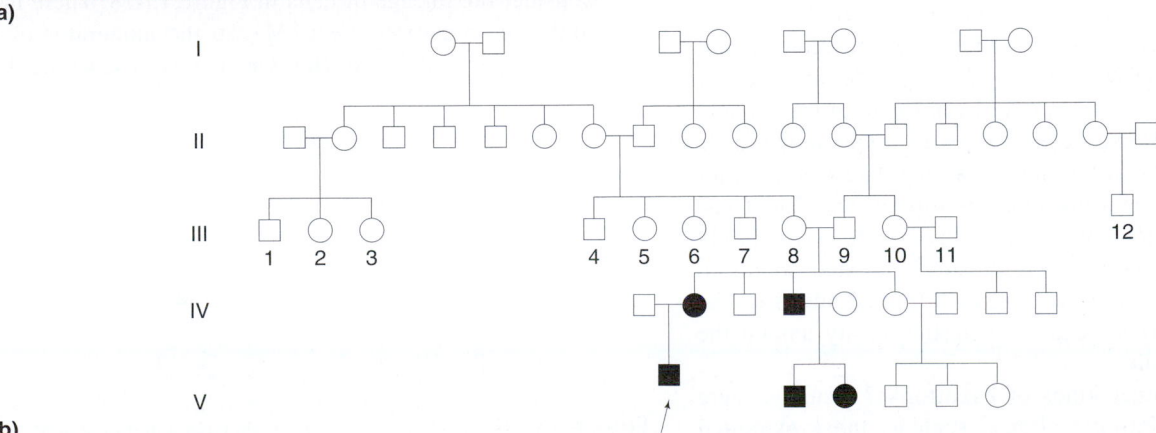

(b)

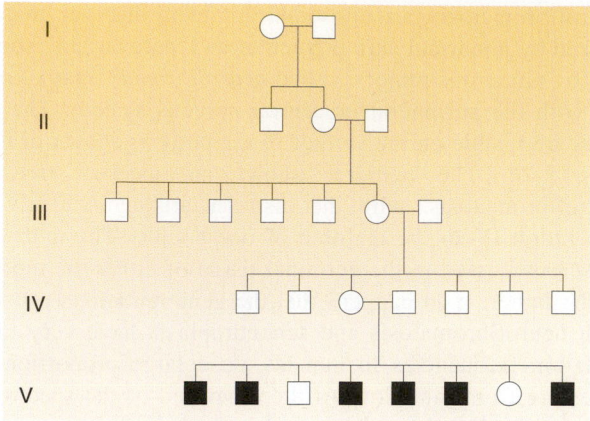

Figure 15-17   Mutation to X-linked recessive hemophilia.

			Average number of mutations
Gene	Number of gametes tested	Number of mutations	per million gametes
$R \rightarrow r$	554,786	273	492.0
$I \rightarrow i$	265,391	28	106.0
$Pr \rightarrow pr$	647,102	7	11.0
$Su \rightarrow su$	1,678,736	4	2.4
$Y \rightarrow y$	1,745,280	4	2.2
$Sh \rightarrow sh$	2,469,285	3	1.2
$Wx \rightarrow wx$	1,503,744	0	0.0

15-2 TABLE   Forward Mutation Frequencies at Some Specific Corn Loci

pedigree until expressed by one of the sons of Queen Victoria, and therefore the mutation event must have taken place either in the germinal tissue of Queen Victoria herself or in the germinal tissue of one of her parents. A similar case is shown in the hemophilia pedigree in Figure 15-17. It is possible that the allele was present in the female ancestors, but the absence of hemophilia in a total of 11 male predecessors makes this possibility unlikely.

How can the frequency of autosomal recessive mutation events be measured? The answer begins with the idea that, for any particular autosomal mutant phenotype, mutations are constantly occurring, entering the population like water drops from a dripping faucet. On the other hand, the mutations are removed from the population by reverse mutation and by selection, which are like leaks in the water vessel. The losses and gains eventually become equal, in a type of equilibrium. Measuring the frequency of the mutant phenotype at equilibrium provides a way of calculating the frequency of mutation, as we shall see in Chapter 24 on population genetics.

Haploids present a great advantage over diploids in mutation studies. The system of detecting mutations in haploids is quite straightforward: any newly arisen recessive allele announces its presence unobscured by a dominant partner allele. In fact, the question of dominance or recessiveness need never arise: there is what amounts to a built-in detection facility. In some cases, a direct identification of mutants is possible. In *Neurospora*, for example, auxotrophic adenine-requiring mutations have been found to map at several loci. Of these loci, the *ad-3* locus is unique in that cells carrying auxotrophic *ad-3* mutants accumulate a purple pigment when grown on a low concentration of adenine. Thus, auxotrophic mutations of this gene may be detected simply by allowing large populations of single asexual spores to grow into colonies on a medium with limited adenine. The purple mutant colonies can be identified easily among the normal white colonies.

What about other kinds of auxotrophs? Usually, there are no visual pleiotropic effects, such as those exhibited

with *ad-3* mutations. We shall examine the most commonly used detection technique, called *replica plating,* later in this chapter.

How common are mutations?

A detection system must be designed before an investigator can find mutations. If a detection system is available, the investigator can set out on a mutant hunt. One thing will become readily apparent: mutations are, in general, very rare. This is shown in Table 15-2, which presents some data collected by Lewis Stadler while working with several corn loci. Mutation studies of this sort are a lot of work. Counting a million of *anything* is no small task. Another feature shown by these data is that different genes seem to generate different frequencies of mutations; a 500-fold range is seen in the corn results. One of the prime requisites of mutation analysis is to be able to measure the tendency of different genes to mutate. Two terms are commonly used to quantify mutation: mutation rate and mutation frequency. The **mutation rate** is expressed as the number of mutations occurring in some unit of time. The units commonly used are a cell generation span, an organismal generation span, or a cell division. You can see that these spans are all biological units of time.

Consider the lineage of cells in Figure 15-18. There has been only one mutation event (M), so the numerator of a mutation rate is established. But what can we use as the de-

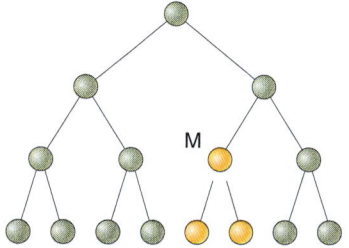

Figure 15-18   A simple cell lineage, showing a mutation at M.

15-3 TABLE		Mutation Rates or Frequencies in Various Organisms	
Organism	Mutation	Value	Units
Bacteriophage T2 (bacterial virus)	Lysis inhibition $r \rightarrow r^+$	1×10^{-8}	*Rate:* mutant genes per gene replication
	Host range $h^+ \rightarrow h$	3×10^{-9}	
Escherichia coli (bacterium)	Lactose fermentation $lac \rightarrow lac^+$	2×10^{-7}	*Rate:* mutant cells per cell division
	Histidine requirement $his^- \rightarrow his^+$	4×10^{-8}	
	$his^+ \rightarrow his^-$	2×10^{-6}	
Chlamydomonas reinhardtii (alga)	Streptomycin sensitivity $str^s \rightarrow str^r$	1×10^{-6}	
Neurospora crassa (fungus)	Inositol requirement $inos^- \rightarrow inos^+$	8×10^{-8}	*Frequency* per asexual spore
	adenine requirement $ad^- \rightarrow ad^+$	4×10^{-8}	
Corn	See Table 15-2		
Drosophila melanogaster (fruit fly)	Eye color W $\rightarrow$ w	4×10^{-5}	
Mouse	Dilution $D \rightarrow d$	3×10^{-5}	
Human			
To autosomal dominants	Huntington disease	0.1×10^{-5}	
	Nail-patella syndrome	0.2×10^{-5}	
	Epiloia (predisposition to type of brain tumor)	$0.4 - 0.8 \times 10^{-5}$	*Frequency* per gamete
	Multiple polyposis of large intestine	$1 - 3 \times 10^{-5}$	
	Achondroplasia (dwarfism)	$4 - 12 \times 10^{-5}$	
	Neurofibromatosis (predisposition to tumors of nervous system)	$3 - 25 \times 10^{-5}$	
To X-linked recessives	Hemophilia A	$2 - 4 \times 10^{-5}$	
	Duchenne muscular dystrophy	$4 - 10 \times 10^{-5}$	
Bone-marrow tissue-culture cells	Normal $\rightarrow$ azaguanine resistance	7×10^{-4}	*Rate:* mutant cells per cell division

Source: After R. Sager and F. J. Ryan, *Heredity*. John Wiley, 1961.

nominator? The total time for mutation may be represented either by the total number of straight lines in Figure 15-18 (14 total generation spans) or, alternatively, by the total number of cell divisions (7). We might state the mutation rate, for example, as one mutation per seven cell divisions.

The **mutation frequency** is the frequency at which a specific kind of mutation (or mutant) is found in a population of cells or individuals. The cell population can be gametes, asexual spores, or almost any other cell type. In our example in Figure 15-18, the mutation frequency in the final population of eight cells would be $2/8 = 0.25$.

Some mutation rates and frequencies are shown in Table 15-3.

Selective systems

The rarity of mutations is a problem if an investigator is trying to amass a collection of a specific type of mutation for genetic study. Geneticists respond to this problem in two ways. One approach is to use a **selective system,** an experimental protocol designed to separate the desired mutant types from wild-type individuals. The other approach is to increase the mutation rate using **mutagens,** agents that have

the biological effect of inducing mutations above the background, or spontaneous, rate. Mutation induction will be considered later in the chapter.

MESSAGE ···
Selective systems and mutagens offer two ways to improve the recovery of rare mutations.
···

Most of the examples of selective systems presented here are in microorganisms, which does not mean that selective systems are impossible for studies of more complex organisms. It merely means that selection can be used to much better advantage in microbes. A million spores or bacterial cells are easy to produce; a million mice or even a million fruit flies require a large-scale commitment of money, time, and laboratory space.

Reversion of auxotrophs

For the reversion of auxotrophy to prototrophy, the selective system is direct. Take an adenine-requiring auxotroph, for example. Grow a culture of this strain on an adenine-containing medium. Then plate a large number of cells on a solid nutritive medium containing no adenine. The only

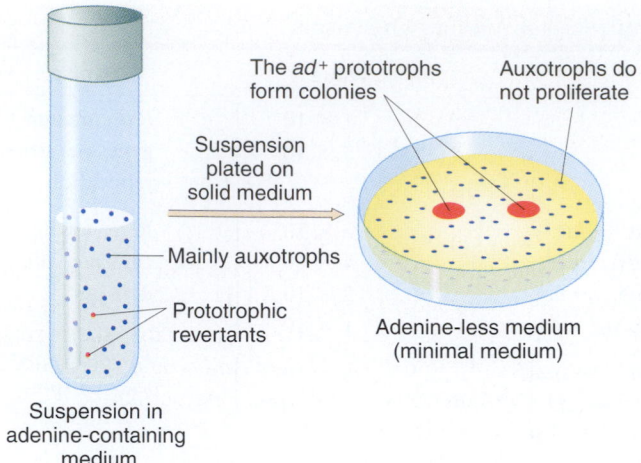

Figure 15-19   Selection for prototropic revertants of auxotrophic adenine-requiring mutants.

cells that can grow and divide on this medium are adenine prototrophs, which must have arisen by reverse mutation in the original culture (Figure 15-19). For most genes (not just those concerned with nutrition), the rate of reverse mutation is generally lower than the rate of forward mutation. We shall explore the reason for this later.

Early studies on the reversion of auxotrophic mutations revealed a complication that must be dealt with in any reversion experiment. Some prototrophic colonies turn out not to be revertants at all but, rather, cases in which a gain-of-function mutation has occurred at a different locus. The second mutation acts as a suppressor of auxotrophy. If we represent the original mutation as *m* and the suppressor as *s*, then the apparent revertant was in fact of genotype *m s*. Suppressor mutations are generally not as common as revertants. They can be distinguished from revertants by crossing to a wild-type strain and observing the reappearance of the original mutant phenotype in the progeny:

$$m \cdot s \times + \cdot +$$
$$\downarrow$$

$m \cdot s$ (prototrophic)
$+ \cdot +$ (prototrophic)
$+ \cdot s$ (prototrophic)
$m \cdot +$ (auxotrophic)

A true revertant, which would be m^+, would produce only prototrophic progeny when crossed to a wild type.

MESSAGE ··

In any reversion experiment, true revertants must be distinguished from suppressor mutations at a separate locus.

···

Filtration enrichment

Filamentous fungi such as *Neurospora* and *Aspergillus* grow as a mass of branching threadlike hyphae. Fungal geneticists

can make use of this property when selecting auxotrophic mutants, by using a procedure called **filtration enrichment.** Let us assume that the investigator is interested in obtaining auxotrophs that specifically require the amino acid leucine. These auxotrophs must result from mutations of the type *leu*$^+$ to *leu* at one of several genetic loci. The investigator begins by growing a wild-type culture and, from it, obtains large numbers of conidia. Most of these conidia will be prototrophic in regard to leucine, as expected of a wild-type culture. However, spontaneous forward mutations occur constantly, and some of the cells will be leucine-requiring auxotrophs bearing a mutant *leu* allele. These mutants are selected as follows. The conidia are placed in a flask of liquid minimal medium and allowed to grow overnight with gentle shaking. Each prototrophic conidium acts as a growth center and produces a spiderlike colony of hyphal threads. However, auxotrophs of all kinds, including the leucine auxotrophs, will not grow but, rather, remain in suspension as live conidia. The contents of the flask are next poured through several layers of fine gauze, which act as a filter. The prototrophic colonies become caught on the fibers of the filter, but the single-celled auxotrophic mutants pass through (Figure 15-20). The cells that pass through the filter are then plated onto minimal medium supplemented with

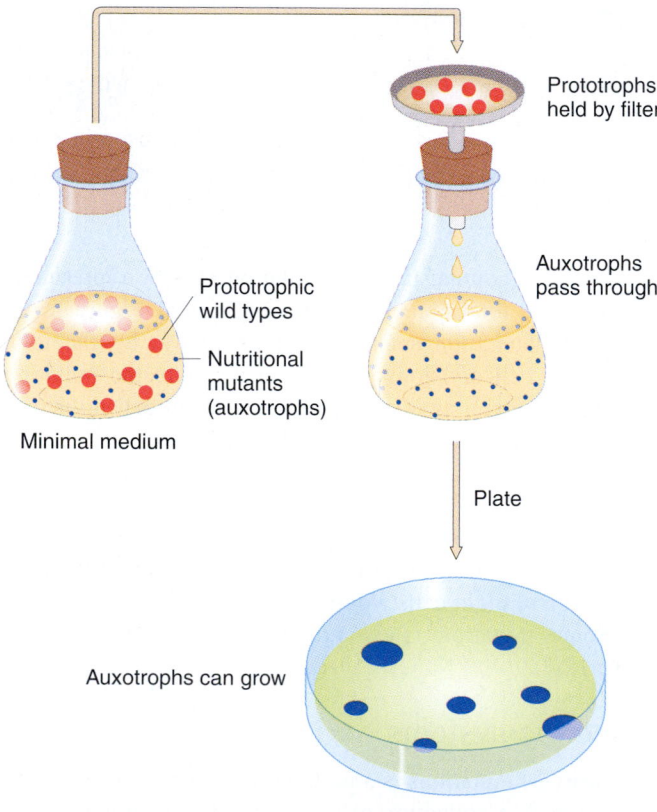

Figure 15-20   The filtration-enrichment method for selecting forward mutations to auxotrophy in filamentous fungi. (This example illustrates mutation to a leucine requirement.)

leucine. Each leucine auxotroph will grow and form a visible colony, but all other kinds of auxotrophs will remain invisible as single cells. Next, the colonies are isolated with a needle and transferred to permanent culture tubes containing supplemented medium. In this way, it is relatively easy to amass a large number of auxotrophic mutants.

Penicillin enrichment

A parallel technique is available for auxotroph selection in bacteria. Many species of bacteria are highly sensitive to the antibiotic penicillin, but only during their proliferating stage. If penicillin is added to a suspension of cells in liquid minimal medium, all the prototrophic cells are killed because they divide, but the auxotrophs survive because they cannot divide without supplementation. After treatment, the penicillin can be removed by washing the cells on a filter. Plating the washed cells on a minimum medium supplemented with some specific chemical such as the amino acid leucine will selectively allow leucine auxotrophs to divide and produce visible colonies.

Resistance

Mutations that confer resistance to specific environmental agents not normally tolerated by wild types are easily demonstrated in microorganisms. We shall use an example that was important historically in determining the nature of mutation.

The intestinal bacterium *E. coli* is parasitized by many specific phages. One of them, called T1, was used in early bacterial mutation studies. Phage T1 attacks and kills most *E. coli* cells, liberating a large brood of newly synthesized progeny viruses from each dead bacterial cell. If a plate is spread with large numbers of bacteria (about 10^9) mixed with phages, most of the bacteria will be killed. However, some rare bacterial cells survive and produce colonies that can be isolated.

When T1-resistant bacteria were first observed, their origin was not known, but there were two main hypotheses. First, the resistance could be due to random mutation from a wild-type allele conferring sensitivity (Ton^s) to a mutant allele conferring resistance (Ton^r). Second, the bacteria could somehow sense the presence of phages and adjust their cellular physiology in such a way that some cells might succeed in becoming resistant. This adjustment could be similar to the manner in which bacteria can shift their physiology to utilize a new nutrient in the medium. In 1943, Salvadore Luria and Max Delbrück designed a classic experiment to distinguish between these two hypotheses. Their experimental design was called a **fluctuation test.** Not only was the fluctuation test historically important in determining the origin of phage-resistant bacteria, but it also provided a method for calculating mutation rates that is still used today.

Luria and Delbrück reasoned that the two hypotheses of mutation and physiological change gave different predictions about the numbers of resistant bacteria found in a sample of cultures. Under the mutation hypothesis, in the cultures where a rare mutation event occurs, the mutation event could occur relatively early or relatively late in the growth of the culture because, presumably, mutation is random over time. An early event would produce a larger clone of descendant resistant cells than would a late event. Hence, considerable variation in the number of resistant cells from culture to culture is predicted (Luria and Delbrück used the term *fluctuation* instead of *variation*). Under the physiological change hypothesis, there is no reason to expect such variation; the physiological gear shifting would presumably be quite constant. These different predictions are diagrammed in Figure 15-21.

The practical details of the experiment were as follows. Into each of 20 culture tubes containing 0.2 ml of medium and into one containing 10 ml, they introduced 10^3 *E. coli* cells per milliliter and incubated them until about 10^8 cells per milliliter were obtained. Each of the 0.2-ml cultures was then spread on a plate that had a dense layer of T1 phages. From the 10-ml "bulk" culture, 10 separate 0.2-ml volumes were withdrawn and plated. Many T1-resistant colonies were found, as shown in Table 15-4.

If resistance were due to random mutation during the incubation period, each culture tube would produce a

15-4 TABLE	Results of Luria and Delbrück's Test		
INDIVIDUAL CULTURES		BULK CULTURE	
Culture number	Number of T1-resistant colonies	Culture number	Number of T1-resistant colonies
1	1	1	14
2	0		
3	3	2	15
4	0		
5	0	3	13
6	5		
7	0	4	21
8	5		
9	0	5	15
10	6		
11	107	6	14
12	0		
13	0	7	26
14	0		
15	1	8	16
16	0		
17	0	9	20
18	64		
19	0	10	13
20	35		
Mean 11.3		Mean 16.7	

(a) Physiological change

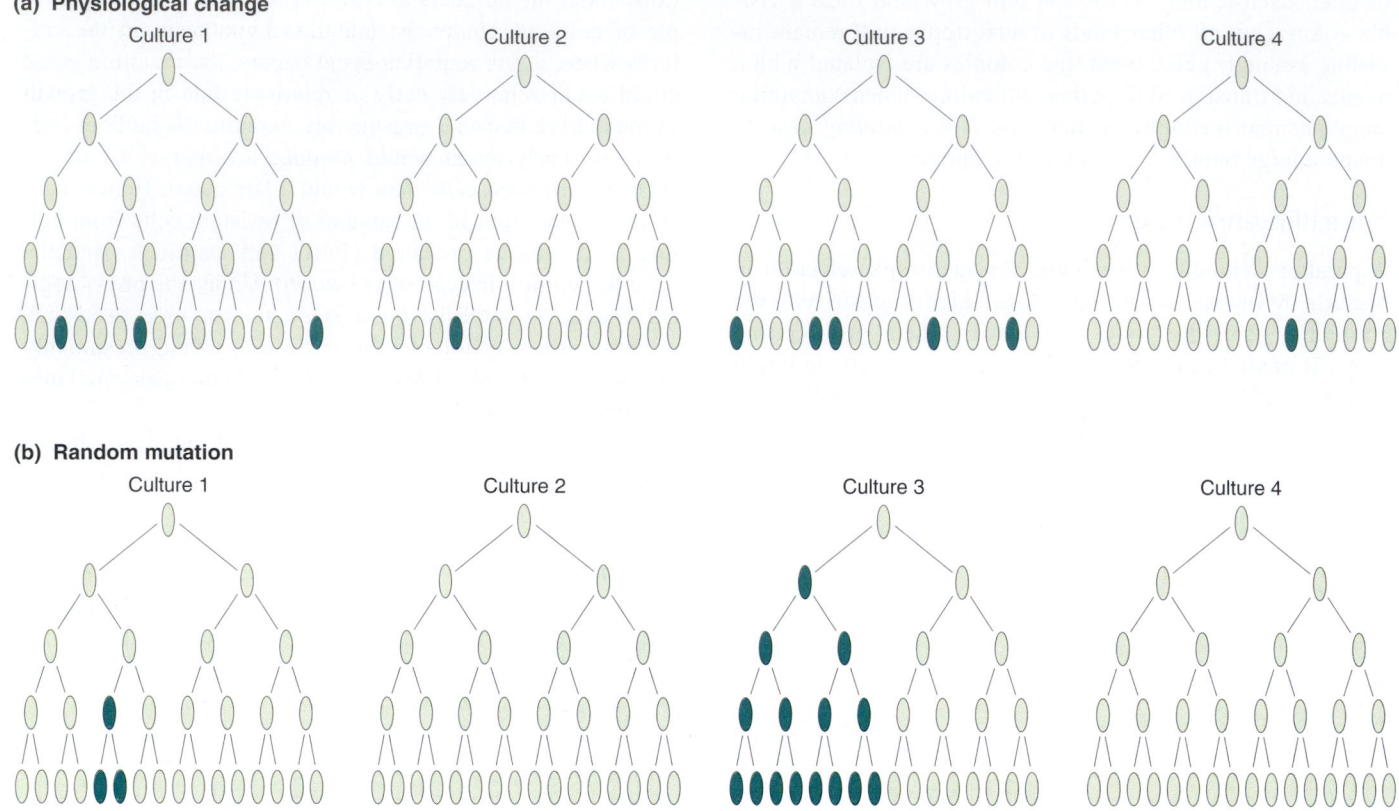

(b) Random mutation

Figure 15-21   Cell pedigrees illustrating the expectations from two contrasting hypotheses about the origin of resistant cells. (From G. S. Stent and R. Calendar, *Molecular Genetics,* 2d ed. W. H. Freeman and Company, 1978.)

different number of resistant cells; the number would vary depending on how early in the cascade of growing cells the mutation occurred. These resistant cells would each produce a separate colony when plated with the T1 phage. If resistance were due to a physiological adaptation occurring after exposure of the cells on the T1 phage, all cultures would be expected to generate resistant cells at roughly the same rate and little variation from culture to culture would be expected.

Variation from plate to plate was indeed observed in the individual 0.2-ml cultures but not in the samples from the bulk culture (which represent a kind of control experiment). This situation cannot be explained by physiological change, because all the samples spread had the same approximate number of cells. The simplest explanation is random mutation, occurring early in the incubation of the 0.2-ml cultures (producing a large number of resistant cells and, therefore, a large number of colonies) or late (producing few resistant cells and colonies) or not at all (producing no resistant cells).

This elegant analysis suggests that the resistant cells are selected by the environmental agent (here, phages) rather than produced by it. Can the existence of mutants in a popu-

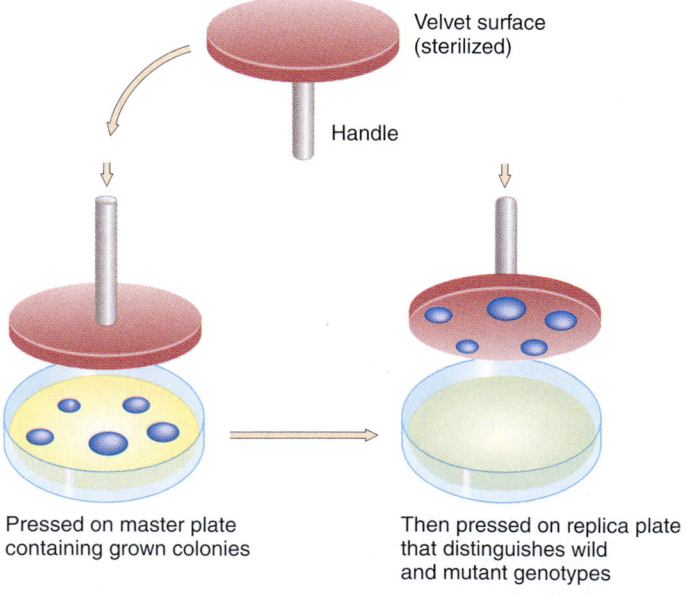

Figure 15-22   Replica plating reveals mutant colonies on a master plate through their behavior on selective replica plates. (From G. S. Stent and R. Calendar, *Molecular Genetics,* 2d ed. W. H. Freeman and Company, 1978.)

lation before selection be demonstrated directly? This demonstration was made possible by the use of a technique called **replica plating,** developed by Joshua and Esther Lederberg in 1952. A population of bacteria was plated on nonselective medium — that is, containing no phages — and from each cell a colony grew. This plate was called the *master plate.* A sterile piece of velvet was pressed down lightly on the surface of the master plate, and the velvet picked up cells wherever there was a colony (Figure 15-22). In this way, the velvet picked up a colony "imprint" from the whole plate. On touching the velvet to replica plates containing selective medium (that is, containing T1 phages), cells clinging to the velvet are inoculated onto the replica plates in the same relative positions as those of the colonies on the original master plate. As expected, rare Ton^r mutant colonies were found on the replica plates, but the multiple replica plates showed identical patterns of resistant colonies (Figure 15-23). If the mutations had occurred *after* exposure to the selective agents, the patterns for each plate would have been as random as the mutations themselves. The mutation events must have occurred before exposure to the selective agent.

MESSAGE

Mutation is a random process. Any allele in any cell may mutate at any time.

Replica plating has become an important technique of microbial genetics. It is useful in screening for mutants that *fail* to grow under the selective regime. The position of an absent colony on the replica plate is used to retrieve the mu-

Master plate containing
10^7 colonies of Ton^s *E. coli* (T1-sensitive)

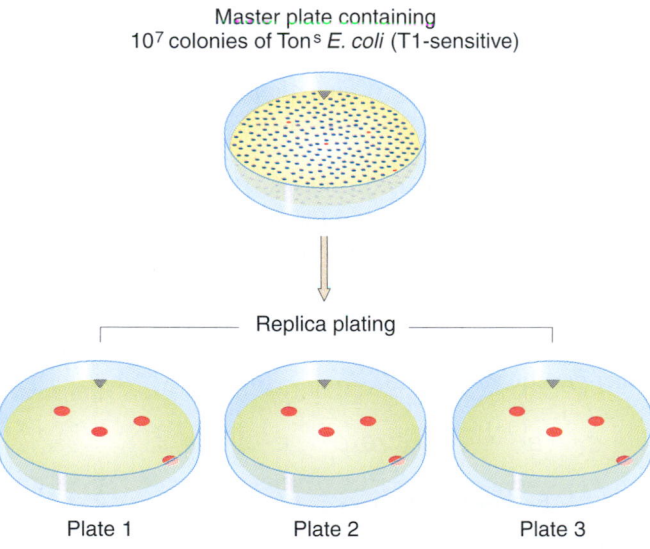

Replica plating

Plate 1   Plate 2   Plate 3

Series of replica plates containing high concentrations
of T1 phage and four Ton^r colonies

Figure 15-23 Replica plating demonstrates the presence of mutants before selection. The identical patterns on the replicas show that the resistant colonies are from the master. (From G. S. Stent and R. Calendar, *Molecular Genetics,* 2d ed. W. H. Freeman and Company, 1978.)

tant from the master. For example, replica plating can be used to screen auxotrophic mutants in precisely this way.

In general, replica plating is a way of retaining an original set of strains on a master plate while simultaneously subjecting replicas to various kinds of tests on different media or under different environmental conditions.

Measuring mutation rate

Let us see how Luria and Delbrück's fluctuation test provides a way of measuring the mutation rate. Consider the 20 cultures that were tested for phage T1 resistance. Here we have a situation that is well described by the Poisson distribution — a great opportunity for mutations to be found, but a large class of samples having no mutations at all in the culture (11 of the 20). Look back at Figure 15-18 and convince yourself that the number of cell divisions necessary to produce *n* cells is *n* minus the original number in the culture (it would be $8 - 1 = 7$ in the example in Figure 15-18). If *n* is very large, which it is in the fluctuation test cultures, and if the original number of cells was relatively very small, then a sufficiently accurate estimate of the number of cell divisions is given by *n* itself. If the mutation rate per cell division is μ, then each culture tube may be expected to have contained an average of μn mutation events. The Poisson distribution (see page 176) then tells us that the zero class with no mutants will equal $e^{-\mu n}$. Because we know that the zero class $= 11/20 = 0.55$ and that $n = 0.2 \times 10^8$, we can solve the following equation for μ:

$$0.55 = e^{-\mu(0.2 \times 10^8)}$$

We find that $\mu = 3 \times 10^{-8}$ mutation events per cell division. [Note that, in the same data, the mutation frequency is $(1 + 3 + 5 + 5 + 6 + 107 + 1 + 64 + 35)/(20 \times 0.2 \times 10^8)$, which equals $227/(4 \times 10^8)$, or 5.7×10^{-7}.]

Isolating mutations from cultures of plant and animal cells

Like microbial cells, cells taken from certain animal and plant tissues, including cancerous tissue, divide mitotically in culture. Furthermore, many of the techniques used to induce and select mutations in microbial cell cultures can also be used with these plant and animal cells, so **somatic cell genetics** of plants and animals is made possible. The combination of these mutation-selection and screening techniques with sophisticated techniques of cell fusion and hybridization now makes possible a wide variety of in vitro manipulations on plant and animal cells, including human cells.

One extensively used cell-culture system uses Chinese hamster ovary (CHO) cells. Many mutant cell types have been isolated from this system and from others. Let's look at some examples. In the first place, dominant mutations can be selected. In this context, dominance is defined by the phenotype observed in hybrid cells formed by fusing a mutant cell with a normal cell. Some mutations that confer resistance to a particular drug, such as ouabain, α-amanitin, methotrexate, or colchicine, are dominant. Finding dominant mutations is not surprising: we might expect to find *only* dominant mutations because the cells are diploid and recessives would be hidden.

Nevertheless, recessive mutants are commonly seen, too. Alleles that confer resistance to lectins, 8-azaguanine, or methotrexate or confer the auxotrophic requirement for glycine, proline, or lysine, as well as various temperature-sensitive lethals, are all examples of recessive mutants that have been found in CHO cells. (Note that certain mutant phenotypes are dominant as determined by one mutant allele and recessive as determined by another mutant allele.) How is it possible for recessive mutant phenotypes to show up in diploid mammalian cells in culture? No one has the complete answer to this question. However, there is some evidence that, although cultured CHO cells are predominantly diploid, they are hemizygous for very small sections of the genome. Thus, recessive mutations in such regions would be expressed. But whatever accounts for the appearance of such recessive phenotypes, the mutant alleles themselves provide useful markers for somatic cell genetics.

To grow single cells of plants in culture, a geneticist plates cells on a medium containing appropriate combinations of plant hormones. The resulting cell colonies, called **calluses** or **calli,** can be subdivided for further growth in culture or they can be transferred to a different medium that promotes differentiation of roots and shoots. Such *plantlets* can then be potted in soil and grown into mature plants. In one application of this technique, geneticists selected cells resistant to a toxin produced by a fungus that is parasitic on certain plants. From calli growing on the toxin-containing plates, the geneticists produced plants that also showed resistance to the fungal toxin and hence to fungal disease.

Mutation induction

The task of finding rare mutations in multicellular organisms is difficult compared with that in microorganisms. So techniques have been designed to open the selective net as wide as possible. In 1928, Hermann J. Muller devised a method of searching for any lethal mutation on the X chromosome in *Drosophila.* He first "constructed" an X chromosome called *ClB.* The *C* stands for a chromosomal rearrangement called an *inversion,* a type that we shall learn about in Chapter 17. For the moment, just think of it as standing for crossover suppressor because, in a female fly carrying this special *ClB* chromosome and a normal X, the X chromosome chromatids do not cross over. The *ClB* chromosome also bears *l*, a recessive lethal allele, and the allele *B*, which determines the dominant bar-eye phenotype. $C\,l\,B/Y$ males die because of hemizygosity for the lethal allele, but the chromosome can be maintained in heterozygous $C\,l\,B/C^+\,l^+\,B^+$ females. This special *ClB* system allowed Muller to screen for lethal mutations anywhere on the X chromosomes in samples of male gametes. The experimental protocol is diagrammed in Figure 15-24. Bar-eyed daughters from the cross of females heterozygous for the *ClB* chromosome and wild-type males are crossed individually with wild-type males. Each bar-eyed daughter lays her eggs in a separate culture vial. When the progeny hatch, the vials are examined for the presence of males. We can see that, if there was a new lethal recessive mutation on an X chromosome in one of the original male gametes, then the F_1 female carrying that chromosome will not produce any viable male progeny. Note that only rarely will a new lethal mutation *m* on the X chromosome be an allele of the *l* locus. (When it is, the $C\,l\,B/l$ female will die.) The absence of males in a vial is easy to determine under low-power magnification and readily allows scoring for the presence of lethal recessive mutations on the X chromosome. Muller found the recessive frequency of such mutations occurring spontaneously to be about 1.5 per 1000 chromosomes, still a relatively low value for an entire chromosome.

Muller then asked whether there were any agents that would increase the rate of mutation. Using the *ClB* test, he measured X-linked lethal frequencies after irradiating males with X rays and observed frequencies that were much higher than those in unirradiated controls. His results supplied the first experimental evidence of a mutagen — in this case, the X rays. Recall that a **mutagen** is an agent that causes mutation at higher than spontaneous levels. Mutagens have been invaluable tools not only for studying the mechanism of mutation itself, but also for increasing the yield of mutants for other genetic studies.

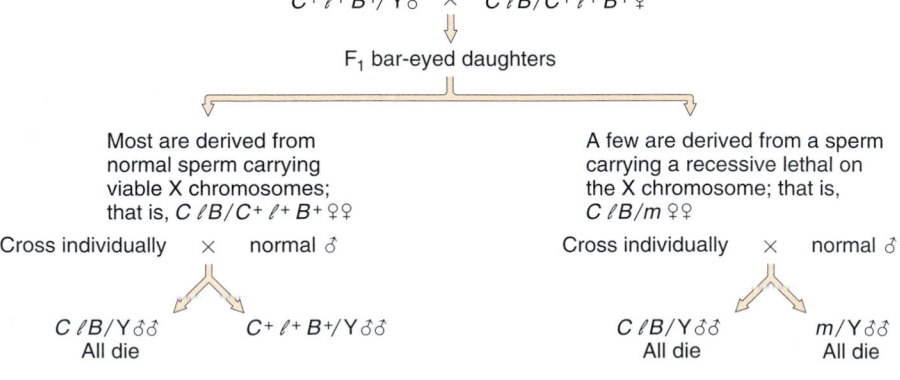

Figure 15-24   The *ClB* test for new X chromosome mutations in *Drosophila.* The symbol *m* represents a recessive lethal mutation anywhere on the X chromosome. Observing presence versus absence of males in each individual progeny amounts to scanning — by genetic analysis — a sample of gametes from the male parent of the bar-eyed daughter.

15-5 TABLE	Relative Efficiencies of Various Types of Radiation in Producing Mutations in *Drosophila*
Type of radiation	Percentage of male X chromosomes bearing recessive lethal mutations after a dose of 1000 roentgens*
Visible light (spontaneous)	0.15
X rays (25 Mev)	1.70
β rays, γ rays, hard X rays	2.90
Soft X rays	2.50
Neutrons	1.90
α rays	0.84

* The roentgen (r) is a unit of radiation energy.

It is now known that many kinds of radiation increase mutations. Table 15-5 shows the effects of several types of radiation on increasing mutation frequencies in *Drosophila*. To the list in Table 15-5 must be added ultraviolet (UV) radiation which is also mutagenic. Radiation is often categorized as *ionizing* or *nonionizing,* depending on whether ions are produced in the tissue through which it passes. X rays and γ (gamma) rays, for example, do produce ions, and UV radiation does not.

The harnessing of nuclear energy has become a social issue because of the powerful mutagenic effect of nuclear radiation. Even in the absence of nuclear war, many peacetime uses of nuclear energy — especially as fuel — carry their own hazards. No containment system is infallible, and recent decades have seen many examples of accidental release of nuclear isotopes, for example, through explosions and leaks at power plants. Furthermore, disposal of nuclear wastes is not as easy as had been supposed. Nuclear energy illustrates well the ecological principle that there is no such thing as a free lunch, and the pros and cons of the use of nuclear energy must be weighed by each of us individually.

It is known that, within a certain range of radiation dosage, induction of point mutations is linear; that is, if we double or halve the radiation level, the number of mutants produced will vary accordingly. From the kind of graph shown in Figure 15-25, it is possible to extrapolate to very low radiation levels and to infer very low frequencies of mutation induction. Because exposure to cosmic radiation, radiation from X-ray machines, radioactive fallout from bomb testing, and contamination from nuclear plants is very low, it is possible to conclude that the effects are negligible. Yet, even though increases in mutation rates might be low, population numbers are large: every year 200 million new gametes unite to form 100 million babies in the world. In this very large annual "mutation experiment," even low mutation frequencies are potentially translatable into large numbers of mutations.

Radiation doses generally are cumulative. If a population of organisms is repeatedly exposed to radiation, the frequency of mutations induced will be in direct proportion to the total amount of radiation absorbed over time. However, there are exceptions to the cumulative effect. For example, if mice given *x* rads (a biological measure of radiation dose) in one short burst, called an *acute dose,* are compared with those given the same dose gradually over a protracted period of weeks or months, a *chronic dose,* significantly fewer mutations are found in the chronically exposed group. This finding has been interpreted to mean that there is some form of repair of radiation-induced genetic damage over time. Repair mechanisms for mutations are considered in the next chapter.

In 1947, Charlotte Auerbach and J. M. Robson conducted experiments on mustard gases, which had been used as toxins in gas warfare. They found that, in addition to being toxic, the mustard gases were potently mutagenic. This pioneer study provided the impetus for research into the mutagenic effects of other chemicals, and it is now known that many of the chemicals that we are exposed to in our daily environments are mutagens. Cigarette smoke is an obvious example, but there are many others, including certain preservatives, pesticides, herbicides, industrial chemicals, and pollutants. There are even natural chemical mutagens in our foods.

Indeed, some chemicals are so potent that they are now routinely used by geneticists to induce mutations in

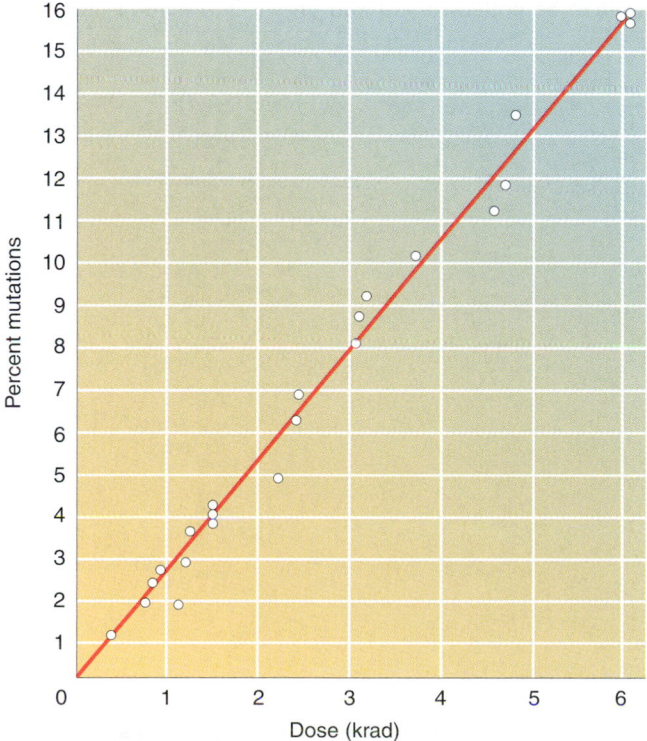

Figure 15-25 Linear relation between X-ray dose to which *D. melanogaster* were exposed and the percentage of mutations (mainly sex-linked recessive lethals).

experimental organisms. The mutagenicity of chemicals is important experimentally because, in many cases, the chemical reactions responsible for mutagenic action are understood, and this knowledge can be used to produce specific types of mutations. Furthermore, many chemicals are much less toxic to an organism than radiation is and yet result in much higher frequencies of mutation. So, as a research tool, chemical mutagens have been useful in producing the wide array of mutations now available for genetic studies. The mechanisms of action of chemical mutagens are considered in Chapter 16.

Mutation and cancer

It is now clear that cancer is a genetic disease. Cancer has many causes, but ultimately all these causes exert their effects on a special class of genes called *cancer genes* or *proto-oncogenes*. Many of these proto-oncogenes have now been identified and mapped, as shown in Figure 15-26. Oncogenes normally carry out basic cellular functions, generally related to the regulation of cell division. However, several types of events can change a proto-oncogene into an oncogene — that is, into a state in which it promotes the two main characteristics of cancer: (1) uncontrolled cell division leading to an overgrown group of cells called a *tumor* and (2) the spread of tumor cells throughout the body to form

new tumors, a process called **metastasis.** One of the main ways in which proto-oncogenes can be changed into their cancer-causing (oncogenic) state is by mutation. Spontaneous or environmentally induced mutation occurs in a proto-oncogene of a single cell, which then undergoes multiple cell divisions to form a tumor. Because all the cells of the tumor carry the mutated oncogene, you can see that a tumor is a mutant clone. This, then, is the sense in which cancer is a genetic disease caused by somatic mutation.

Many of the mutations that can cause cancer are repaired by cellular repair systems (Chapter 19). Those that escape repair can cause a tumor. Some genetic aspects of cancer as a somatic genetic disease are summarized in Figure 15-27. The genetic aspects of cancer are considered in Chapter 22.

Mutagens in genetic dissection

Mutations are the basis of genetic analysis. In the same way that we can learn how the engine of a car works by tinkering with its parts one at a time to see what effect each part has, we can see how a cell or an organism works by altering its parts one at a time by inducing mutations. The mutation analysis is a further aspect of the process that we have called *genetic dissection* of living systems. As we have seen, the first stage of any mutational dissection is the

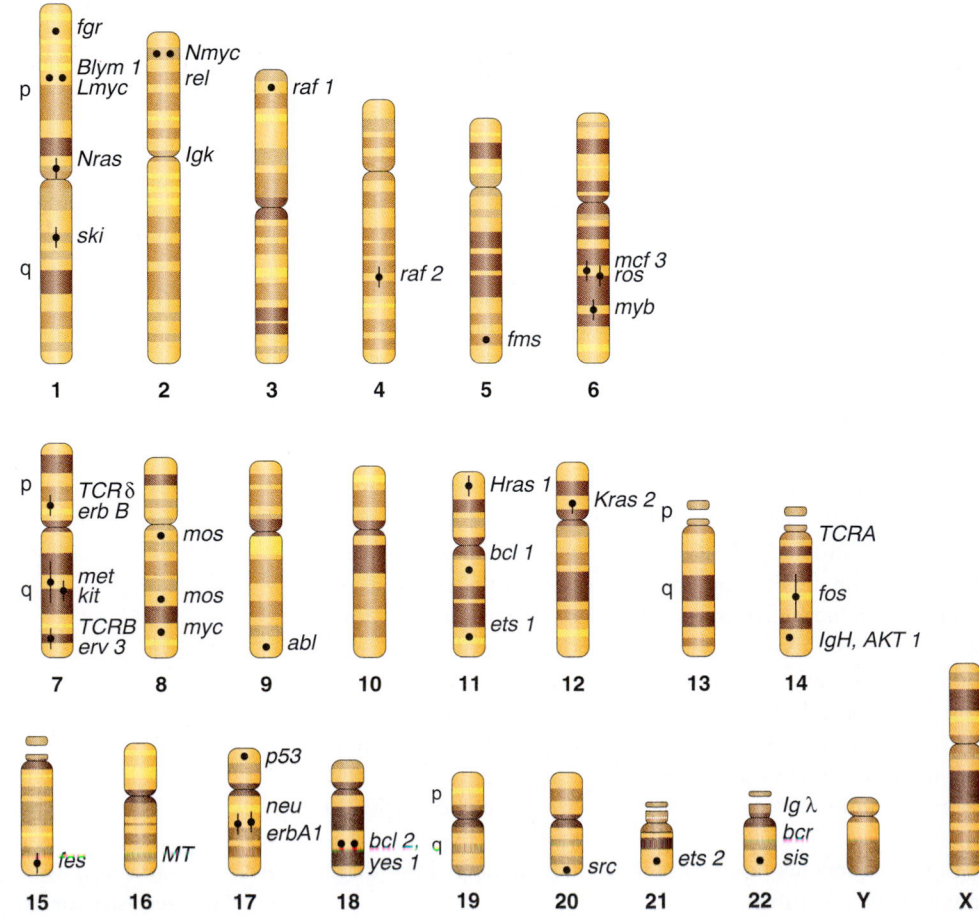

Figure 15-26 Human chromosomes showing bands from Giemsa staining and the positions (shown by black dots) of known proto-oncogenes; mutations in proto-oncogenes lead to cancer.

				Number of
15-6 TABLE		Forward Mutation Frequencies Obtained with Various Mutagens in *Neurospora*		
Mutagenic treatment	Exposure time (minutes)		Survival (%)	ad-3 mutants per 10^6 survivors
No treatment (spontaneous rate)	—		100	~0.4
Amino purine (1–5 mg/ml)	During growth		100	3
Ethylmethanesulfonate (1%)	90		56	25
Nitrous acid (0.05 M)	160		23	128
X rays (2000 r/min)	18		16	259
Methylmethanesulfonate (20 mM)	300		26	350
UV rays (600 erg/mm²/min)	6		18	375
Nitrosoguanidine (25 µM)	240		65	1500
ICR-170 acridine mustard (5 µg/ml)	480		28	2287

Note: The assay measures the frequency of purple *ad-3* colonies among the white colonies produced by the wild-type *ad-3⁺*.

mutant hunt. The effectiveness of the hunt can be improved dramatically by using mutagens. Let's look at some examples.

First, Table 15-6 shows the relative frequencies of *ad-3* forward mutants in *Neurospora* after various mutagenic treatments. Note that an investigator would have to test about 2.5 million untreated cells to find a single mutant cell but could hope to find a mutant cell among only 450 treated with ICR-170. Table 15-7 provides another example of the improved recovery of mutant cells after treatment with an appropriate mutagen. The term **supermutagen** has been introduced to describe some of these highly potent agents.

For a third example, we turn to *Drosophila*. Until the mid-1960s, most researchers working with *Drosophila* used radiation to induce mutations. Doses of 4000 roentgens produce lethal mutations in perhaps 10 to 11 percent of all X chromosomes of irradiated males. Higher doses, however, make the flies increasingly infertile. Another problem is that many of the mutations induced by X rays include chromosomal rearrangements, and these rearrangements can complicate the genetic analysis.

In contrast, the chemical ethylmethane sulfonate (EMS) induces a vastly higher proportion of point mutations. This

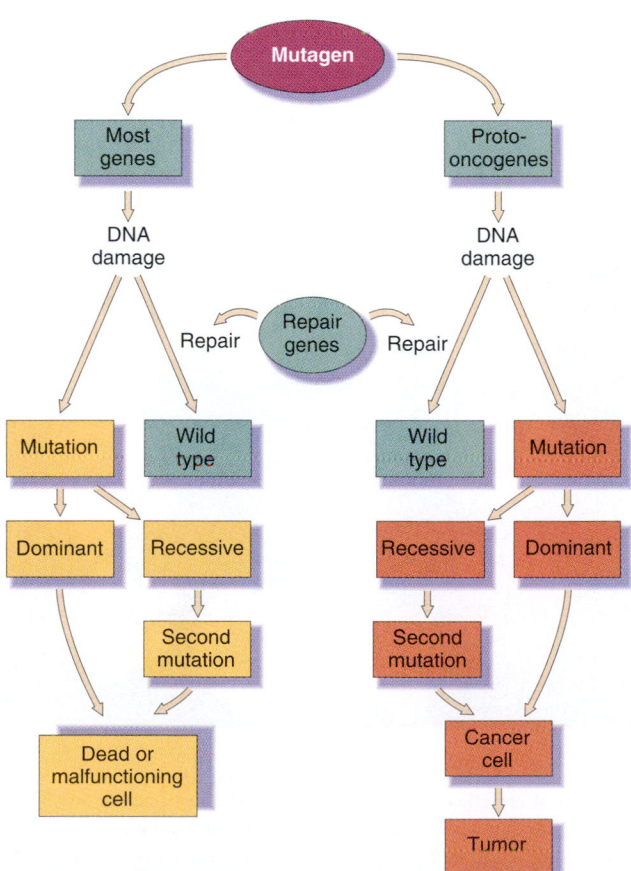

Figure 15-27   Comparison of the various outcomes of somatic mutations in proto-oncogenes and other genes.

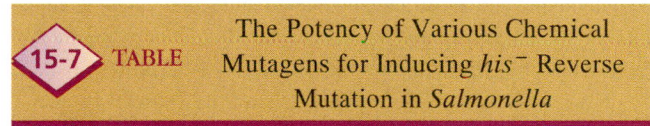

15-7 TABLE	The Potency of Various Chemical Mutagens for Inducing *his⁻* Reverse Mutation in *Salmonella*	
Mutagen	Revertants per nanomole	Ratio of effectiveness*
1,2-Epoxybutane	0.006	1
Benzylchloride	0.02	3
Methylmethanesulfonate	0.63	105
2-Naphthylamine	8.5	1400
2-Acetylaminofluorene	108	18,000
Aflatoxin B₁	7057	1,200,000
Furylfuramide (AF-2)	20,800	3,500,000

* Relative effectiveness of each mutagen, compared with 1,2-epoxybutane.
Source: After J. McCann and B. N. Ames, in W. G. Flamm and M. A. Mehlman, eds., *Advances in Modern Toxicology*, vol. 5. Hemisphere Publishing Corp., 1976.

mutagen is very easily administered simply by placing adult flies on a filter pad saturated with a mixture of sugar and EMS. Simple ingestion of the EMS produces large numbers of mutations. For example, males fed 0.025 M EMS produce sperm carrying lethal alleles on more than 70 percent of all X chromosomes and on virtually every chromosome 2 and chromosome 3. At these levels of mutation induction, it becomes quite feasible to screen for many mutations at specific loci or for defects with unusual phenotypes.

In the *ClB* test, to screen for X-linked recessive lethals, it was necessary to test for lethality in the second generation of descendants. However, it is possible to screen an X chromosome for viable recessive mutations immediately in the F_1 generation, by making use of a special X-chromosome rearrangement called an **attached X** (represented as $\hat{X}X$). An attached X is a compound chromosome formed by the fusing of two separate X chromosomes. It is inherited as a single unit. EMS-treated males are crossed with females bearing an attached X and a Y ($\hat{X}X/Y$), as diagrammed in Figure 15-28. (If you need to review sex determination in *Drosophila,* turn back to Chapter 2.) The gametes of the female are $\hat{X}X$ and Y. All F_1 zygotes carrying sex-linked lethals die, but newly induced, sex-linked mutations governing observable phenotypes are expressed in the males. Because of the effectiveness of the method, it is possible to select a wide range of mutations in a particular region of the genome, an approach known as **saturation mutagenesis.** All F_1 males carry a mutagenized paternal X chromosome, each from a different treated sperm.

Geneticists can then take the study of the mutations further: if individual F_1 males are crossed with $\hat{X}X/Y$ females, each resulting progeny represents a single "cloned" X. The F_1 flies can be raised at 22°C, and then progeny clones of each individual X can be incubated at 29°C and 17°C to detect heat-sensitive and cold-sensitive lethals. Such temperature-sensitive (ts) lethals make up about 10 to 12 percent of all EMS-induced lethals. We have already learned how useful such conditional mutants are for genetic analysis. Many laboratories now routinely test for ts mutants in *Drosophila* screening tests. The approach is possible because large numbers of mutants are easily obtained.

The availability of ts lethals has a side benefit in that it simplifies laboratory procedures. A time-consuming task in any large-scale *Drosophila* experiment is separating females from males within 12 hours after emergence from the pupae to prevent undesired mating. (The flies do not mate during the first 12 to 14 hours after emergence.) The task can be eliminated by using ts lethal alleles (represented as l^{ts}) to produce cultures of a single sex at will. For example, the cross $\hat{X}X/Y$ ♀ × l^{ts}/Y ♂ produces progeny of both sexes at permissive temperatures. If such a culture is shifted to restrictive temperatures before hatching, the l^{ts}/Y males die and only the females hatch. Similarly, homozygosity of an l^{ts} allele in an $\hat{X}X$ chromosome determines that only wild-type males hatch from cultures kept at restrictive temperatures.

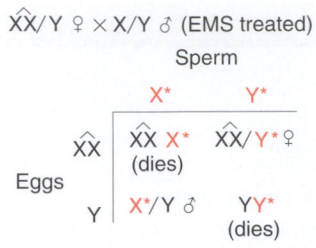

$\hat{X}X/Y$ ♀ × X/Y ♂ (EMS treated)

Sperm

		X*	Y*
Eggs	$\hat{X}X$	$\hat{X}X$ X* (dies)	$\hat{X}X/Y*$ ♀
	Y	X*/Y ♂	YY* (dies)

X*, Y* = chromosomes exposed to mutagen

Figure 15-28   The use of attached-X chromosomes ($\hat{X}X$) in *Drosophila* to facilitate the search for X-linked mutations. Sperm previously exposed to EMS fertilize eggs containing either the attached-X chromosome or a Y chromosome. The treated X chromosomes from the males show up as the hemizygous X's of the sons, revealing phenotypically recessive mutations.

MESSAGE ···
The mutagen EMS has revolutionized the genetic versatility of *Drosophila* by providing a potent method for the isolation of a wide range of point mutants.

Mutation breeding

In addition to the mutational dissection of biological systems, geneticists use mutation in other ways. We learned in Chapter 2 that one way of breeding a better crop plant is to make a hybrid and then to select the desired recombinants from the progeny generations. That approach makes use of the variation naturally found between available stocks or isolates from nature. Figure 15-29 shows examples of the

Figure 15-29   *Potentilla fruticosa* wild type *(center)* and horticultural varieties arranged in a circle. All the attractive novel shapes, colors, patterns, and sizes arose through mutations, which were then selected and bred by the horticulturalists. Most flowering plants in our gardens and parks and most food plants in use today have been produced by such a procedure, beginning with either spontaneous or induced mutations. (Anthony Griffiths.)

outcome of such breeding. Another way to generate variability for selection is to treat with a mutagen. In this way, the variability is increased through human intervention.

A variety of procedures may be used. Pollen may be mutagenized and then used in pollination. Dominant mutations will be expressed in the next generation, and further generations of selfing reveal recessives. Alternatively, seeds may be mutagenized. A cell in the enclosed embryo of a seed may become mutant, and then it may become part of germinal tissue or somatic tissue. If the mutation is in somatic tissue, any dominant mutations will show up in the plant derived from that seed, but this generation will be the end of the road for such mutations. Germinal mutations will show up in later generations, where they can be selected as appropriate. Figure 15-30 summarizes mutation breeding.

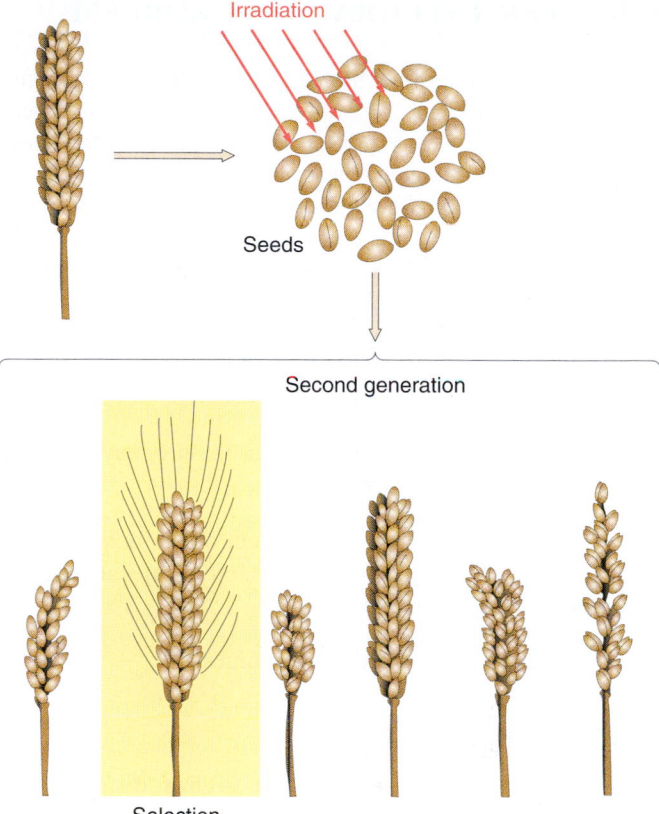

Figure 15-30 Mutation breeding in crops. (From Bjorn Sigurbjornsson, "Induced Mutations in Plants." Copyright © 1970 by Scientific American, Inc. All rights reserved.)

SUMMARY

Genes may mutate in somatic cells or germinal cells. Within these two broad categories, there are several kinds of mutations, including morphological, lethal, conditional, biochemical, and resistance mutations.

Mutations can be used to study the process of mutation itself or to permit genetic dissection of biological functions. To carry out such studies, however, it is necessary to have a system for detecting mutant alleles at the phenotypic level. In diploids, dominant mutations are easily detected. On the other hand, the detection of recessive mutations requires the use of special genotypes; otherwise, the mutations may never be manifested in the phenotype. For this reason, detection systems are much more straightforward in haploids, where the question of dominance does not arise.

Mutation events occur randomly at any time and in any cell. Therefore the size of a mutant clone is proportional to the developmental stage at which the mutation event occurred.

Mutations often result in loss of the normal function of the affected allele, either fully (null mutations) or partly (leaky mutations). Loss-of-function mutations are generally recessive, which can be demonstrated in heterozygotes, in which the wild-type allele provides enough function to result in normal phenotype. Gain-of-function mutations are often dominant.

Somatic mutation in oncogenes is the major cause of cancer. Some environmental chemicals and radiation can influence mutation rates in cancer genes.

Because mutations are rare, geneticists use selective systems or mutagens or both to obtain mutations. Selective systems automatically distinguish between mutant and nonmutant states. Mutagens are valuable not only in studying the mechanisms of mutation, but also in inducing mutations to be used in other genetic studies. In addition, mutagens are frequently used in crop breeding.

CONCEPT MAP

Draw a concept map interrelating as many of the following terms as possible. Note that the terms are listed in no particular order.

mutation / allele / reversion / wild type / recessive / somatic / mutagen / germinal / genetic dissection / progeny

CHAPTER INTEGRATION PROBLEM

Some recessive mutations in maize affect the color, texture, or shape of the kernel. To consider these mutations in general, we shall use m to represent the recessive allele. These mutations can be detected in a straightforward manner by crossing an m^+/m^+ male with an m/m female. Any new mutations are found by visually scanning many seeds — in other words, millions of kernels on thousands of corncobs. Finding a seed with the recessive phenotype shows that the gene mutated in the male. When maize geneticists sow seeds with the mutant phenotype and self the resulting plants, they find one or the other of two kinds of results:

1. If the mutation was spontaneous, the progeny from the selfing are generally all mutant.

2. If the mutation followed treatment of the pollen with a mutagen, the progeny from the selfing are generally $\frac{3}{4}$ wild type and $\frac{1}{4}$ mutant.

   Explain these two different outcomes.

◆ Solution ◆

As usual, restating the results in a slightly different way gives us a clue to what is going on. The difference between outcomes seems to indicate some difference between spontaneous and induced mutation. From a spontaneous mutation, we obtain an individual that seems to be of genotype m/m because it breeds true for the mutant phenotype. From an induced mutation, we obtain an individual that seems, from the 3:1 phenotypic ratio on selfing, to have an m^+/m genotype. Clearly, the latter deduction does not mesh with the observation of a mutant phenotype. This is the paradoxical case. How is it possible for a seed to have a recessive mutant phenotype but to mature into a plant that produces the selfed progeny expected from a heterozygote?

The experimenter should always keep in mind the peculiarities of the organism that is being used. What do we know about corn that might be relevant to this situation? The explanation hinges on mutational events in and around the pollen grain in the anthers of the plant. Recall from Chapter 3 that the mature pollen grain of maize is not just a cell with a single haploid nucleus but that there are three important components: a haploid tube nucleus and two haploid nuclei that act as sperm. One of these sperm fuses with one female nucleus to create the zygote, and the other sperm fuses with two other female nuclei to create the triploid endosperm. When mature pollen cells are treated with a mutagen, it is likely that only one of the sperm nuclei mutates in any one cell; the only mutations that we see are those in sperm that produce the endosperm, because the endosperm forms the bulk of the mature kernel. This is the solution to the problem, but let's write out the genotypes: The endosperm must be $m/m/m$, and the embryo must be m^+/m. This is why selfing produces a 3:1 ratio.

What about the spontaneous mutations? A gene can mutate at any time. There is only a relatively short time between the mitosis that produces the three nuclei of a mature pollen cell and fertilization. A gene is thus most likely to mutate in the relatively long time that precedes the pollen cell's maturation, possibly even a long time before, in tissue that is still somatic. If a gene does mutate in this time to become allele m, the pollen cell's mitoses will produce three nuclei of genotype m. When such a cell unites with a gamete from the m/m female, the endosperm is m m m and the embryo is m/m.

In summary, we have used concepts from the discussion of germinal and somatic mutation and of detection systems (all from this chapter); from preceding chapters, we have used our knowledge of dominance relations, Mendelian genetics, mitosis and meiosis, and organismal life cycles.

SOLVED PROBLEMS

1. A certain plant species normally has the purple pigment anthocyanin dispersed throughout the plant. This makes the green parts of the plant look brown (green plus purple) and the parts of the flower that lack chlorophyll (petals, ovary, anthers) look bright purple. The allele A is essential for anthocyanin production, and, in a homozygote, the recessive allele a results in a plant that lacks anthocyanin. An interesting new allele called a^u arises, which is unstable. The a^u allele reverts to A at a frequency thousands of times greater than regular (stable) a alleles do.

   a. What phenotype would you expect in plants of genotype (1) a^u/a^u; (2) a^u/a; (3) A/a^u?

   b. How can you confirm that the reversions are true mutations?

◆ Solution ◆

a. (1) We have seen that mutations tend to be very rare; in the normal course of events, they are seldom observed unless they are seriously sought out. However, an allele that is reverting as frequently as a^u is likely to announce its reversion behavior prominently if a proper detection system is available. In a plant containing billions of somatic cells, many cells will undoubtedly revert at some stage of development. As development proceeds, each initial revertant cell will give rise to a clone of revertant cells that should be visible as a purple or brown sector. Therefore, a plant of a^u/a^u genotype should have white flowers with purple sectors of varying size and photosynthetic tissues that are basically green with brown sectors.

(2) Each sector is a clone derived from a single cell with a^u reverting to A. The allele a^u is expected to be dominant over a, because a is essentially inactive and will not prevent a^u from reverting. Hence, a^u/a plants will look the same as a^u/a^u but with fewer purple or brown sectors because there are half the chances for reversion.

(3) Because A produces pigment in all cells, A/a^u will be indistinguishable from A/A. Even though a^u is reverting, the reversions will not show up—a detection system is lacking. Such plants will be purple/brown throughout.

b.   How can we prove that a pigmented sector is due to a mutation? The key is that genes are hereditary units and mutant alleles should be transmitted to subsequent generations of cells or individuals. How can we detect the transmission of revertant A alleles to progeny cells or individuals? If we were able to take or "scrape" a few cells out of a sector and grow them into a plant and that plant turned out to be purple, then the reversion hypothesis would be proved. The technology for doing such an experiment is available. Is there any other way? If revertant sectors appear in the flower, then presumably the germinal tissue (anthers and ovaries) should sometimes be part of a sector. Therefore, we could collect pollen from such a flower and pollinate a plant of genotype a/a. If there is some pollen carrying a revertant A, then the progeny from this pollen should be A/a and purple/brown throughout. Finding some progeny of this phenotype would prove the reversion is a true mutation because the A gene could have come from nowhere else.

2.   A mutation experiment is performed on the *tryp4* locus in yeast. A *tryp4* mutation confers a requirement for the amino acid tryptophan. A *tryp4* mutant allele named *tryp4-1* is known to be revertible; the experiments want to measure reversion frequency in a population of haploid cells. A culture of mating type α and of genotype *tryp4-1* is grown and 10 million cells are plated on a medium lacking tryptophan; 120 colonies are obtained. The genotypes of these prototrophic colonies are checked by crossing each one to a wild-type culture of

mating type a. From the results of these crosses, the prototrophic colonies were found to be of two types: two-thirds of type 1 and one-third of type 2:

Type $1\alpha \times$ wild-type $a \longrightarrow$
    progeny all tryptophan independent

Type $2\alpha \times$ wild-type $a \longrightarrow$
    $\frac{3}{4}$ progeny tryptophan independent
    $\frac{1}{4}$ progeny tryptophan requiring

a.   Propose a genetic explanation for the two types.

b.   Calculate the frequency of revertants.

◆ Solution ◆

a.   The technique of plating on a medium lacking tryptophan should be a selection system for $tryp4^+$ revertants, because they do not require tryptophan for growth. When backcrossed to a wild type ($tryp4^+$), all progeny of revertants should be tryptophan independent. This behavior is exhibited by type 1 colonies, so these prototrophs are revertants.

Now what about the type 2 colonies? The fact that some progeny are tryptophan requiring shows that the *tryp4-1* mutation could not have genuinely reverted. Rather, the requirement for tryptophan appears to have been merely masked or suppressed. We have already studied several examples of the suppression of a mutant allele by a new mutation at a separate locus. But, even if we had not remembered these examples, the $3:1$ ratio in a haploid organism should provide the clue that two independent loci are involved. Let's designate the suppressor mutation as *su* and its inactive wild-type allele as su^+. Type 2 colonies must be *tryp4-1*; *su*, and the wild types are $tryp4^+$; su^+. A cross of these strains produces

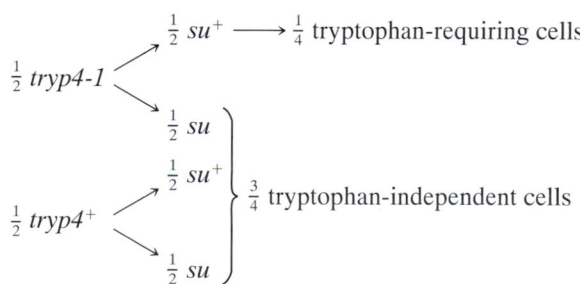

(Notice that *su* has no effect on the $tryp4^+$ allele.)

b.   We are told that two-thirds of the colonies are type 1, or true revertants. Thus, there are $120 \times \frac{2}{3} = 80$ revertants of a total of 10^7, or a revertant frequency of 8×10^{-6} cells.

PROBLEMS

1.   A certain species of plant produces flowers with petals that are normally blue. Plants with the mutation w produce white petals. In a plant of genotype w/w, one w allele reverts in the course of the development of a petal. What detectable outcome would this reversion produce in the resulting petal?

2. *Penicillium* (a commercially important filamentous fungus) can normally synthesize its own leucine (an amino acid). How would you go about selecting mutants that are leucine requiring (that cannot synthesize their own leucine)? (**Note:** Like many filamentous fungi, *Penicillium* produces profuse numbers of asexual spores.)

3. How would you select revertants of the yeast allele *pro-1*? This allele confers an inability to synthesize the amino acid proline, which can be synthesized by wild-type yeast and which is necessary for growth.

4. How would you use the replica-plating technique to select arginine-requiring mutants of haploid yeast?

5. Using the filtration-enrichment technique, you do all your filtering with a minimal medium and do your final plating on a complete medium that contains every known nutritional compound. How would you find out what *specific* nutrient is required? After replica plating onto every kind of medium supplement known to science, you still cannot identify the nutritional requirement of your new yeast mutant. What could be the reason or reasons?

6. An experiment is initiated to measure the reversion rate of an *ad-3* mutant allele in haploid yeast cells. Each of 100 tubes of a liquid, adenine-containing medium are inoculated with a very small number of mutant cells and incubated until there are 10^6 cells per tube. Then the contents of each tube are spread over a separate plate of solid medium containing no adenine. The plates are observed after 1 day, and colonies are seen on 63 plates. Calculate the reversion rate of this allele per cell division.

7. Suppose that you want to determine whether caffeine induces mutations in higher organisms. Describe how you might do so (include control tests).

8. In corn, alleles at a single locus determine the presence (*Wx*) or absence (*wx*) of amylose in the cell's starch. Cells that have *Wx* stain blue with iodine; those that have only *wx* stain red. Design a system for studying the frequency of rare mutations from $Wx \rightarrow wx$ without using acres of plants. (**Hint:** You might start by thinking of an easily studied cell type.)

9. Suppose that you cross a single male mouse from a homozygous wild-type stock with several virgin females homozygous for the allele that determines black coat color. The F_1 consists of 38 wild-type mice and five black mice. Both sexes are represented equally in both phenotypic classes. How can you explain this result?

10. A man employed for several years in a nuclear power plant becomes the father of a hemophiliac boy. There is no hemophilia in the extensive pedigrees of the man's ancestors and his wife's ancestors. Another man, also employed for several years in the same plant, has an achondroplastic dwarf child — a condition nowhere recorded in his ancestry or in that of his wife. Both men sue their employer for damages. As a geneticist, you are asked to testify in court. What do you say about each situation? (**Note:** Hemophilia is an X-linked recessive; achondroplasia is an autosomal dominant.)

11. In a large maternity hospital in Copenhagen, there were 94,075 births. Ten of the infants were achondroplastic dwarfs. (Achondroplasia is an autosomal dominant, showing full penetrance.) Only two of the dwarfs had a dwarf parent. What is the mutation frequency to achondroplasia in gametes? Do you have to worry about reversion rates in this problem? Explain.

12. One of the jobs of the Hiroshima-Nagasaki Atomic Bomb Casualty Commission was to assess the genetic consequences of the blast. One of the first things that they studied was the sex ratio in the offspring of the survivors. Why do you suppose they did so?

13. Many organisms present examples of unstable recessive alleles that revert to wild types at very high frequencies. Different unstable alleles often revert in different ways. One gene in corn (*C*), which produces a reddish pigment in the kernels, has several unstable null alleles. The unstable allele c^{m1} reverts late in the development of the kernel and at a very high rate. Another unstable allele, c^{m2}, reverts earlier in the development of the kernel and at a lower rate. Assuming that lack of pigment leaves the cell looking yellowish, what phenotypes are expected in plants of the following genotypes?

   a. C/c^{m1} **e.** c^{m1}/c (*c* is a stable mutant allele)

   b. C/c^{m2} **f.** c^{m2}/c

   c. c^{m1}/c^{m1} **g.** c^{m1}/c^{m2}

   d. c^{m2}/c^{m2}

14. A haploid strain of *Aspergillus nidulans* carried an auxotrophic *met-8* mutation conferring a requirement for methionine. Several million asexual spores were plated on minimal medium, and two prototrophic colonies grew and were isolated. These prototrophs were crossed sexually with two different strains, with the progeny shown in the body of the following table, where *met*$^+$ means methionine is not required for growth, and *met*$^-$ means methionine is required for growth.

	Crossed to a wild-type strain	Crossed to a strain carrying the original *met-8* allele
Prototroph 1	All *met*⁺	$\frac{1}{2}$ *met*⁺ $\frac{1}{2}$ *met*⁻
Prototroph 2	$\frac{3}{4}$ *met*⁺ $\frac{1}{4}$ *met*⁻	$\frac{1}{2}$ *met*⁺ $\frac{1}{2}$ *met*⁻

a. Explain the origin of both of the original prototrophic colonies.

b. Use clearly defined gene symbols to explain the results of all four crosses.

Unpacking the Problem

Before trying to solve this problem, try these questions that are relevant to the experimental system.

1. Draw a labeled diagram that shows how this experiment was done. Show test tubes, plates, and so forth.

2. Define all the genetic terms in this problem.

3. Many problems show a number next to the auxotrophic mutation's symbol — here the number *8* next to *met*. What does the number mean? Is it necessary to know its meaning to solve the problem?

4. How many crosses were actually made? What were they?

5. Use genetic symbols to represent the crosses.

6. Is the problem about somatic mutation or germinal mutation?

7. Is the problem about forward mutation or reversion?

8. Why was such a small number of prototrophic colonies (two) found on the plate?

9. Why didn't the several million asexual spores grow?

10. Do you think that any of the millions of spores that did not grow were mutant? dead?

11. Do you think that the wild type used in the crosses was prototrophic or auxotrophic? Explain.

12. If you had the two prototrophs from the plates and the wild-type strain in three different culture tubes, could you tell them apart just by looking at them?

13. How do you think the *met-8* mutation was obtained in the first place? (Show with a simple diagram. **Note:** *Aspergillus* is a filamentous fungus.)

14. In Chapter 6, we learned that *Aspergillus* is a convenient organism in which to study mitotic crossing-over and haploidization. Do you think that those processes apply to this problem? Can they be ruled out as explanations of the results?

15. What progeny do you predict from crossing *met-8* with a wild type? What about *met-8* × *met-8*? What about wild type × wild type?

16. Do you think that this is a random meiotic progeny analysis or a tetrad analysis?

17. Draw a simple life-cycle diagram of a haploid organism showing where meiosis takes place in the cycle.

18. Consider the $\frac{3}{4}:\frac{1}{4}$ ratio. In haploids, crosses heterozygous for one gene generally give progeny ratios based on halves. How can this idea be extended to give ratios based on quarters?

15. A diploid strain of yeast was heterozygous for several genes as shown below:

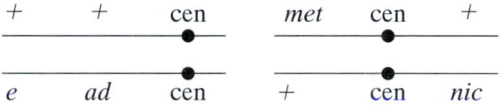

The recessive *e* allele gives an elongated, egg-shaped cell instead of the usual round one, and the recessive *ad* (adenine-requiring) allele causes accumulation of a red pigment in the medium under conditions of limiting adenine (the wild-type allele gives a white color). The *met* and *nic* alleles are recessive auxotrophic "forcing markers," whose job is to force the cells to stay diploid so that they can grow on minimal medium. Cen represents centromere. The original culture was white and grew well on minimal medium and, on limiting adenine, produced no red pigment. Cells were irradiated and plated onto medium with limiting adenine. There were several phenotypes on these plates:

a. Most colonies were white.

b. Several diploid red colonies appeared. The red colonies were examined under the microscope.

   (1) Most colonies were composed of round cells.

   (2) Some had elongated cells.

c. There were many cells on the plates that did not grow into colonies. Some were transferred to complete medium containing yeast extract (all possible nutrients included).

   (1) Some then grew into colonies.

   (2) Some did not.

What are the most likely explanations of the origins of these five types?

16. Cells of a haploid wild-type *Neurospora* strain were mutagenized with EMS. Large numbers of these cells were plated and grown into colonies at 25°C on complete medium (containing all possible nutrients). These plates were used as masters; then replicas were imprinted onto plates of minimal medium and complete medium at both 25°C and at 37°C. There were several mutant phenotypes, as shown in the accompanying diagram. The large circles represent luxuriant growth, the spidery symbols represent weak growth, and a blank means no growth at all. How would you categorize the types of mutants represented by isolates 1 through 5?

Strain	Minimal		Complete	
	25°C	37°C	25°C	37°C
Mutant 1			◯	◯
Mutant 2	◯		◯	◯
Mutant 3	✳	✳	◯	◯
Mutant 4	◯	✳	◯	✳
Mutant 5	◯		◯	◯
Wild type (control)	◯	◯	◯	◯

17. **a.** You have isolated three tryptophan-requiring mutants of a haploid fungus. You cross them (in pairs) and score 1000 spores from each cross for the ability to grow without added tryptophan.

Cross	Parents	Trp$^+$/1000
1	A × B	135
2	A × C	179
3	B × C	50

What can you conclude from these results?

b. You grow cultures from five of the Trp-requiring spores from cross 2 (you label them "2-1," "2-2," and so forth), and you cross each of these cultures with the original mutant B. Again, you score 1000 spores from each cross for the ability to grow without added tryptophan.

Cross	Parents	Trp$^+$/1000
4	2-1 × B	132
5	2-2 × B	13
6	2-3 × B	53
7	2-4 × B	137
8	2-5 × B	47

What can you conclude about each of strains 2-1, 2-2, etc.? Your explanation should include each strain's genotype and how it arose.

(Problem 17 by Rosie Redfield.)

18. To understand the genetic basis of locomotion in the diploid nematode *Caenorhabditis elegans*, recessive mutations were obtained, all making the worm "twitch" ineffectually instead of moving with its usual smooth gliding motion. These mutations presumably affect the nervous or muscle systems. Twelve homozygous mutants were intercrossed, and the hybrids were examined to see if they twitched or not. The results were as follows, where "+" means that the hybrid was wild type (gliding) and "t" means that the hybrid twitched:

	1	2	3	4	5	6	7	8	9	10	11	12
1	t	+	+	+	t	+	+	+	+	+	+	+
2		t	+	+	+	t	+	t	+	t	+	+
3			t	t	+	+	+	+	+	+	+	+
4				t	+	+	+	+	+	+	+	+
5					t	+	+	+	+	+	+	+
6						t	+	t	+	t	+	+
7							t	+	+	+	t	t
8								t	+	t	+	+
9									t	+	+	+
10										t	+	+
11											t	t
12												t

a. Explain what the intercrossing experiment was designed to test.

b. How many genes affecting movement have been identified, and which mutations are in which gene? Give the genes symbols of your choice.

c. As illustrative examples, show the relevant genotypes of mutants 1, 2, and 5, and of the 1/5 and 1/2 hybrids, using the gene symbols you defined in part b, and explain why different results are obtained in these two hybrids.

19. **Note:** This problem requires knowledge of the genetic code (see Figure 10-27). In corn, the synthesis of purple pigment is controlled by two genes acting sequentially through colorless ("white") intermediates:

$$\text{white 1} \xrightarrow{\text{gene } A} \text{white 2} \xrightarrow{\text{gene } B} \text{purple}$$

UAG nonsense mutations were obtained in genes *A* and *B*, and we can call them a^n and b^n. Both lacked enzyme activity and were recessive to their wild-type alleles *A* and *B*.

A nonsense suppressor is also available; it is a mutation in a duplicate copy of a tRNA gene. The mutational site is in the anticodon, and this mutation causes the tRNA to insert the amino acid tryptophan at the nonsense codon UAG, thereby allowing the protein synthesis machinery to complete translation. The result is a wild-type phenotype. We shall represent the nonsense suppressor allele as T^s and its wild-type allele T^+. Neither T^s nor T^+ has any detectable effect on the wild-type alleles A and B or on any other aspect of the phenotype.

a. Would you expect T^s to be dominant to T^+ or not? Explain.

b. A trihybrid A/a^n ; B/b^n ; T^s/T^+ is selfed. If all the genes are unlinked, what phenotypic ratio do you expect in the progeny. Explain carefully. (Assume that absence of purple results in a light color, which we can call *white*.)

20. Outline the precise steps that you would take to find the chromosomal locus of a new recessive mutation that you have just found in tomatoes. Show all steps, including what crosses you would make, which lines you would use, how you would obtain the lines, how much field or greenhouse space is needed, how long it would take, and so forth. (You may refer to the tomato map, which is shown in Figure 5-15.)

21. A *Neurospora* strain unable to synthesize arginine produces a revertant, arginine-independent colony. A cross is made between the revertant and a wild-type strain. What proportion of the progeny from this cross would be arginine-independent if

a. the reversion precisely reversed the original change that produced the arg^- mutant allele?

b. the revertant phenotype was produced by a mutation in a second gene located on a different chromosome (the new mutation suppresses arg^-)?

c. a suppressor mutation occurred in a second gene located 10 map units from the arg^- locus on the same chromosome?

22. A certain haploid fungus is normally red owing to a carotenoid pigment. Mutants were obtained that were different colors owing to the presence of different pigments: orange (o^-), pink (p^-), white (w^-), yellow (y^-), and beige (b^-). Each phenotype was inherited as if a single gene mutation governed it. To determine

what these mutations signified, double mutants were constructed with all possible combinations, and the results were as follows:

	p^-	w^-	y^-	b^-
o^-	pink	white	yellow	beige
p^-	—	white	pink	pink
w^-		—	white	white
y^-			—	yellow

Interpret this table as follows, using the first entry as an example: the double mutant $o^- \cdot p^-$ has a pink phenotype.

a. What do the results in the table mean? Explain in detail.

b. What phenotypes would result if the mutants were paired as heterokaryons in all the combinations possible?

c. What phenotypic proportions would be found in the progeny of a cross between an $o^- \cdot p^-$ double mutant and a wild type if the loci are linked 16 map units apart?

23. A strain of *Aspergillus* was subjected to mutagenesis by X rays and two tryptophan-requiring mutants (A and B) were isolated by the replica-plating technique. These tryptophan-requiring strains were plated in large numbers to obtain revertants to wild type. You failed to recover any revertants from mutant A and recovered one revertant from mutant B. This revertant was crossed with a normal wild-type strain. What proportion of the progeny from this cross would be tryptophan independent if

a. the reversion precisely reversed the original change that produced the trp^- mutant allele?

b. the revertant phenotype was produced by a mutation in a second gene located on a different chromosome (the new mutation suppresses trp^-)?

c. a suppressor mutation occurred in a second gene located 24 map units from the trp^- locus on the same side of the centromere?

d. Comment on the nature of mutant A.

16

MECHANISMS OF GENE MUTATION

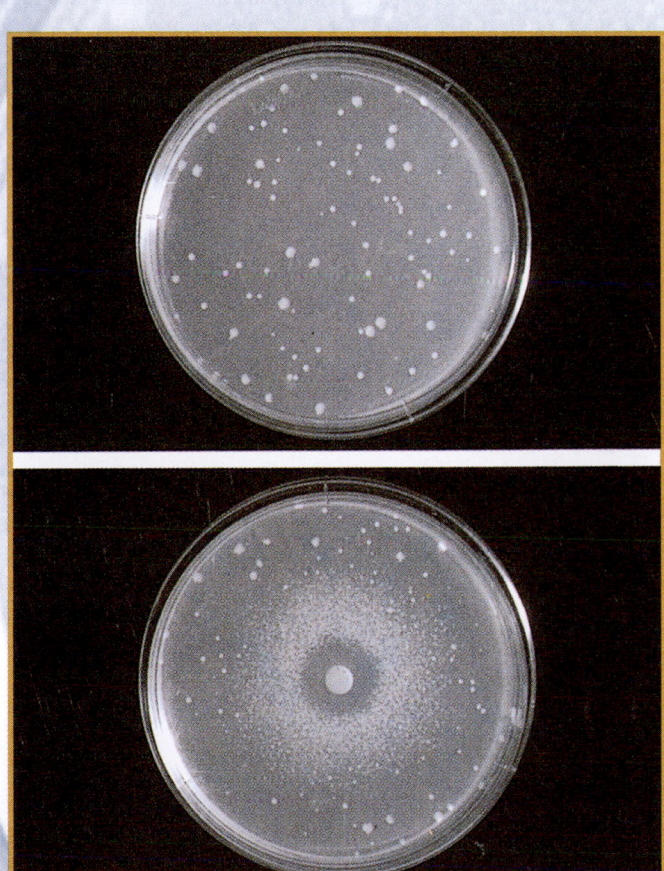

Salmonella test (Ames test) for mutagens.
Top: Petri plate containing about 10^9 bacteria that cannot synthesize histidine (the small number of visible colonies are spontaneous revertants). *Bottom:* Plate containing a disk with a mutagen, which produces a large number of revertants that can synthesize histidine; revertants appear as a ring of colonies around the disk. *(Kristien Mortelmans, SRI International, Menlo Park, CA.)*

Key Concepts

Mutations can occur spontaneously owing to several different mechanisms, including errors of DNA replication and spontaneous damage to the DNA.

Mutagens are agents that increase the frequency of mutagenesis, usually by altering the DNA.

Potentially mutagenic and carcinogenic compounds can be detected easily by mutagenesis tests with bacterial systems.

Biological repair systems eliminate many potentially mutagenic alterations in the DNA.

Cells lacking certain repair systems have higher than normal mutation rates.

This chapter describes the main types of molecular processes that give rise to the mutant alleles discussed in the previous chapter. Such processes are relevant not only to experimental genetics but also have direct bearing on human health.

Molecular basis of gene mutations

Gene mutations can arise spontaneously or they can be induced. **Spontaneous mutations** are naturally occurring mutations and arise in all cells. **Induced mutations** are produced when an organism is exposed to a mutagenic agent, or mutagen; such mutations typically occur at much higher frequencies than spontaneous mutations do.

To understand the mechanisms of gene mutation requires analysis at the level of DNA and protein molecules. Preceding chapters have described models for DNA and protein structure and have considered the nature of mutations that alter these structures (see, for instance, Table 15-1).

Molecular genetics techniques can be used to determine the sequence of large segments of DNA and the sequence changes resulting from mutations. Sequencing has greatly increased our understanding of the pathways that lead to mutagenesis and has even helped to unravel the mysteries of mutational hot spots — genetic sites with a penchant for mutating.

Much work on the molecular basis of mutation has been carried out in single-celled bacteria and their viruses. However, many mutations leading to inherited diseases in humans also have been analyzed. We shall review some of the findings of these studies. We shall also consider biological repair mechanisms, because repair systems play a key role in mutagenesis, operating to lower the final observed mutation rates. For example, in *Escherichia coli*, with all repair systems functioning, base substitutions occur at rates of 10^{-10} to 10^{-9} per base pair per cell per generation. As a general principle, base substitutions arise by a perturbation of the normal pairing of complementary bases.

Spontaneous mutations

Spontaneous mutations arise from a variety of sources, including errors in DNA replication, spontaneous lesions, and transposable genetic elements. The first two are considered in this section; the third is examined in Chapter 20.

Errors in DNA replication

An error in DNA replication can occur when an illegitimate nucleotide pair (say, A–C) forms in DNA synthesis, leading to a base substitution.

Each of the bases in DNA can appear in one of several forms, called **tautomers,** which are isomers that differ in the positions of their atoms and in the bonds between the atoms. The forms are in equilibrium. The **keto** form of each base is normally present in DNA (Figure 16-1), whereas the **imino** and **enol** forms of the bases are rare. The ability of the wrong tautomer of one of the standard bases to mispair and cause a mutation in the course of DNA replication was first noted by Watson and Crick when they formulated their model for the structure of DNA (Chapter 8). Figure 16-2

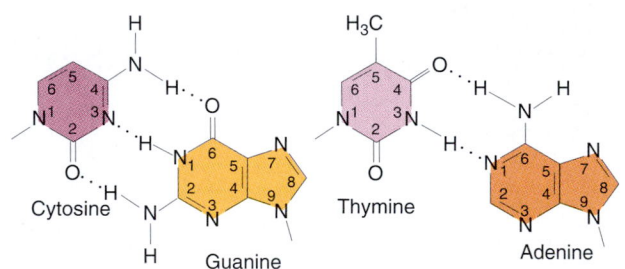

Figure 16-1   Pairing between the normal (keto) forms of the bases.

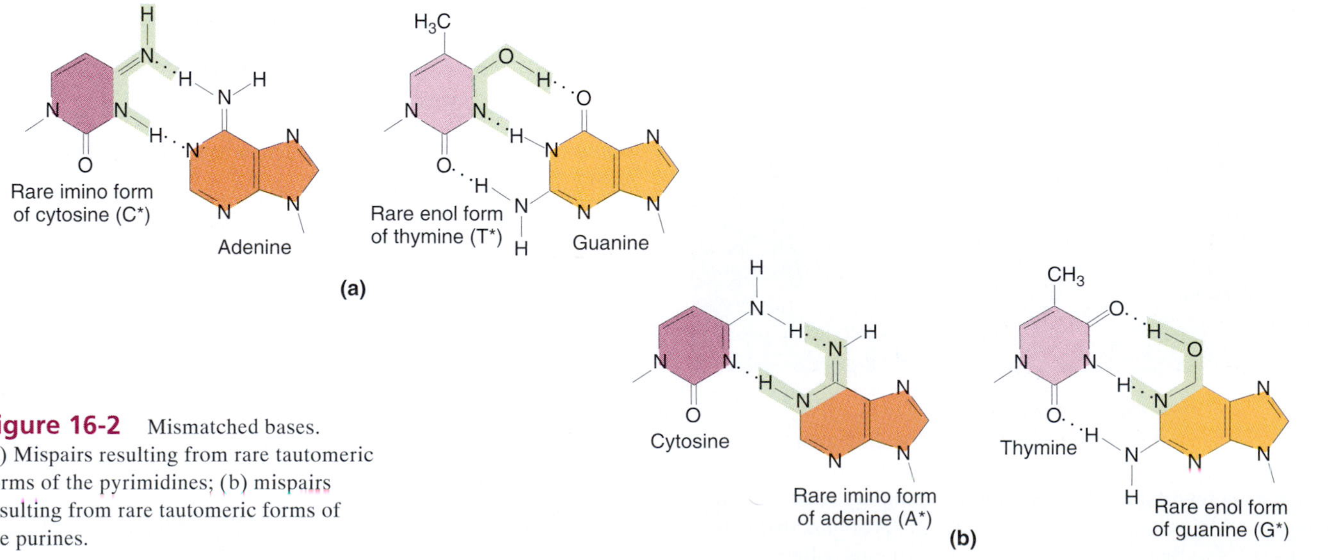

Figure 16-2   Mismatched bases. (a) Mispairs resulting from rare tautomeric forms of the pyrimidines; (b) mispairs resulting from rare tautomeric forms of the purines.

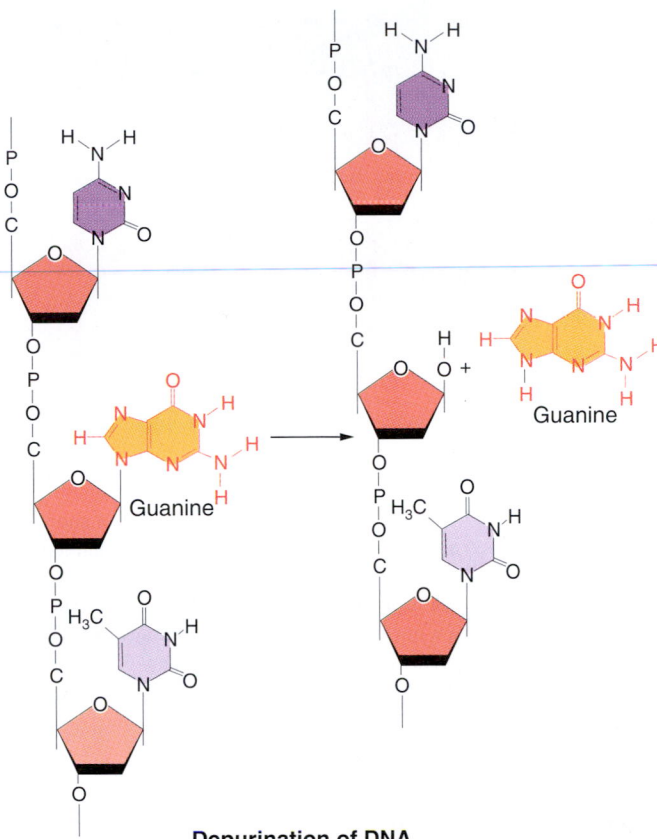

Site (no. of bp)	Sequence repeat	No. of bases deleted	Occurrences
20 to 95	GTGGTGAA	75	2 S74, S112
146 to 269	GCGGCGAT	123	1 S23
331 to 351	AAGCGGCG	20	2 S10, S136
316 to 338	GTCGA	22	2 S32, S65
694 to 707	CA	13	1 S24
694 to 719	CA	25	1 S56
943 to 956	G	13	1 S42
322 to 393	None	71	1 S120
658 to 685	None	27	1 S86

Figure 16-6 Deletions in *lacI*. Deletions occurring in *S74* and *S112* are shown at the top of the figure. As indicated by the gold bars, one of the sequence repeats (aqua) and all the intervening DNA are deleted, leaving one copy of the repeated sequence. All mutations were analyzed by direct DNA sequence determination. (From P. J. Farabaugh, U. Schmeissner, M. Hofer, and J. H. Miller, *Journal of Molecular Biology* 126, 1978, 847.)

Deletions and duplications. Large **deletions** (more than a few base pairs) constitute a sizable fraction of spontaneous mutations, as shown in Figure 16-5. The majority, although not all, of the deletions occur at repeated sequences. Figure 16-6 shows the results for the first 12 deletions analyzed at the DNA sequence level, presented by Miller and his co-workers in 1978. Further studies showed that hot spots for deletions are in the longest repeated sequences. **Duplications** of segments of DNA have been observed in many organisms. Like deletions, they often occur at sequence repeats.

How do deletions and duplications form? Several mechanisms could account for their formation. Deletions may be generated as replication errors. For example, an extension of the Streisinger model of slipped mispairing (Figure 16-4) could explain why deletions predominate at short repeated sequences. Alternatively, deletions and duplications could be generated by recombinational mechanisms (to be described in Chapter 19).

Spontaneous lesions

In addition to replication errors, **spontaneous lesions,** naturally occurring damage to the DNA, can generate mutations. Two of the most frequent spontaneous lesions result from depurination and deamination.

Depurination, the more common of the two, consists of the interruption of the glycosidic bond between the base and deoxyribose and the subsequent loss of a guanine or an adenine residue from the DNA (Figure 16-7). A mammalian cell spontaneously loses about 10,000 purines from its DNA in a 20-hour cell-generation period at 37°C. If these lesions were to persist, they would result in significant genetic damage because, in replication, the resulting **apurinic sites** cannot specify a base complementary to the original purine. However, as

Depurination of DNA

Figure 16-7 The loss of a purine residue (guanine) from a single strand of DNA. The sugar-phosphate backbone is left intact.

we shall see later in the chapter, efficient repair systems remove apurinic sites. Under certain conditions (to be described later), a base can be inserted across from an apurinic site; this insertion will frequently result in a mutation.

The **deamination** of cytosine yields uracil (Figure 16-8a). Unrepaired uracil residues will pair with adenine

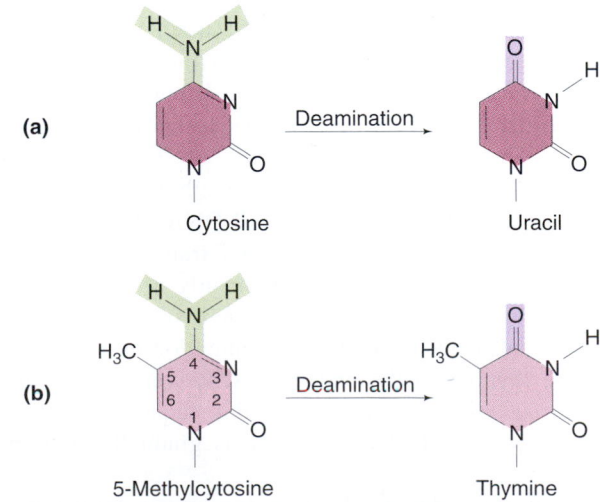

Figure 16-8 Deamination of (a) cytosine and (b) 5-methylcytosine.

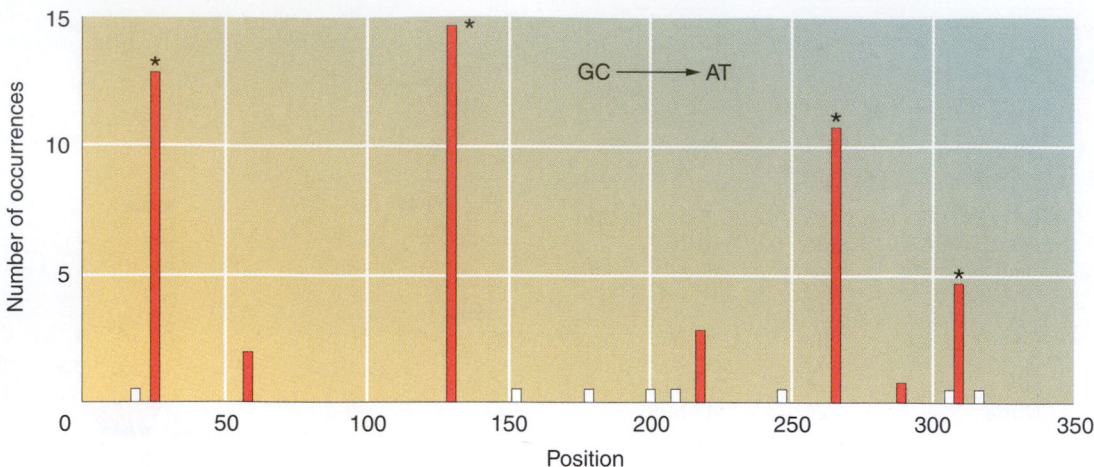

Figure 16-9   5-Methylcytosine hot spots in *E. coli.* Nonsense mutations occurring at 15 different sites in *lacI* were scored. All result from the GC → AT transition. The asterisks (*) mark the positions of 5-methylcytosines. Open bars depict sites at which the GC → AT change could be detected but at which no mutations occurred in this particular collection. It can be seen that 5-methylcytosine residues are hot spots for the GC → AT transition. Of 50 independently occurring mutations, 44 were at the 4 methylated cytosine sites and only 6 were at the 11 unmethylated cytosines. (From C. Coulondre et al., *Nature* 274, 1978, 775.).

in replication, resulting in the conversion of a G–C pair into an A–T pair (a **GC → AT transition**). In 1978, deaminations at certain cytosine residues were found to be the cause of one type of mutational hot spot. DNA sequence analysis of GC → AT transition hot spots in the *lacI* gene showed that 5-methylcytosine residues are present at each hot spot. (Certain bases in prokaryotes and eukaryotes are methylated.) Some of the data from this *lacI* study are shown in Figure 16-9. The height of each bar on the graph represents the frequency of mutations at each of a number of sites. It can be seen that the positions of 5-methylcytosine residues correlate nicely with the most mutable sites.

Why are 5-methylcytosines hot spots for mutations? One of the repair enzymes in the cell, uracil-DNA glycosylase, recognizes the uracil residues in the DNA that arise from deaminations and excises them, leaving a gap that is subsequently filled (a process to be described later in the chapter). However, the deamination of 5-methylcytosine (Figure 16-8b) generates thymine (5-methyluracil), which is not recognized by the enzyme uracil-DNA glycosylase and thus is not repaired. Therefore, C → T transitions generated by deamination are seen more frequently at 5-methylcytosine sites, because they escape this repair system.

A consequence of the frequent mutation of 5-methylcytosine to thymine is the underrepresentation of CpG dinucleotides in higher cells, because this sequence is methylated to give 5-methyl-CpG, which is gradually converted into TpG.

Oxidatively damaged bases represent a third type of spontaneous lesion implicated in mutagenesis. Active oxy-

gen species, such as superoxide radicals ($O_2\cdot$), hydrogen peroxide (H_2O_2), and hydroxyl radicals (OH$\cdot$), are produced as by-products of normal aerobic metabolism. They can cause oxidative damage to DNA, as well as to precursors of DNA (such as GTP), which results in mutation and which has been implicated in a number of human diseases. Figure 16-10 shows two products of oxidative damage. The 8-oxo-7-hydrodeoxyguanosine (8-oxodG, or GO) product frequently mispairs with A, resulting in a high level of G → T transversions. Thymidine glycol blocks DNA replication if unrepaired but has not yet been implicated in mutagenesis.

MESSAGE ·······································
Spontaneous mutations can be generated by different processes. Replication errors and spontaneous lesions generate most of the base-substitution and frameshift mutations. Replication errors may also cause some deletions that occur in the absence of mutagenic treatment.

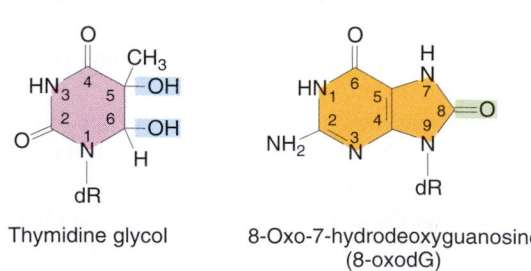

Thymidine glycol   8-Oxo-7-hydrodeoxyguanosine (8-oxodG)

Figure 16-10   DNA damage products formed after attack by oxygen radicals. dR = deoxyribose.

Spontaneous mutations and human diseases

DNA sequence analysis has revealed the mutations responsible for a number of human hereditary diseases. The previously discussed studies of bacterial mutations allow us to suggest mechanisms that cause these human disorders.

A number of these disorders are due to **deletions** or **duplications** involving repeated sequences. For example, mitochondrial encephalomyopathies are a group of disorders affecting the central nervous system or the muscles (Kearns-Sayre syndrome). They are characterized by dysfunction of oxidation phosphorylation (a function of the mitochondria) and by changes in mitochondrial structure. These disorders have been shown to result from deletions that occur between repeated sequences. Figure 16-11 depicts one of these deletions. Note how similar it is in form to the spontaneous *E. coli* deletions shown in Figure 16-6.

A common mechanism that is responsible for a number of genetic diseases is the **expansion of a three-base-pair repeat,** as in fragile X syndrome (Figure 16-12). This syndrome is the most common form of inherited mental retardation, occurring in close to 1 of 1500 males and 1 of 2500 females. It is evidenced cytologically by a fragile site in the X chromosome that results in a break in vitro.

The inheritance of fragile X syndrome is unusual in that 20 percent of the males with a fragile X chromosome are

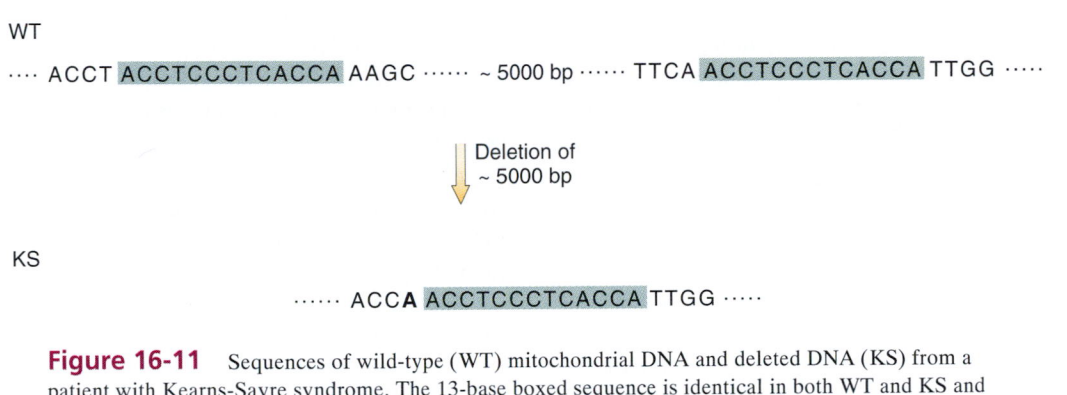

Figure 16-11 Sequences of wild-type (WT) mitochondrial DNA and deleted DNA (KS) from a patient with Kearns-Sayre syndrome. The 13-base boxed sequence is identical in both WT and KS and serves as a breakpoint for the DNA deletion. A single base (boldface type) is altered in KS, aside from the deleted segment.

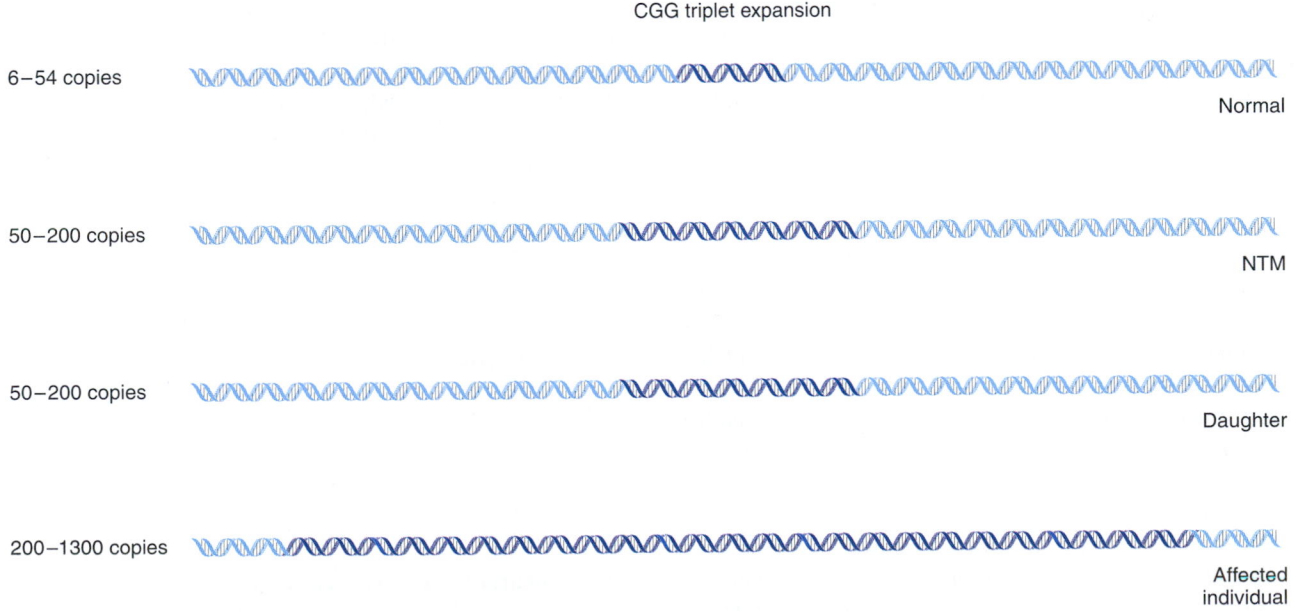

Figure 16-12 Expansion of the CGG triplet in the *FMR-1* gene seen in the fragile X syndrome. Normal persons have from 6 to 54 copies of the CGG repeat, whereas members of susceptible families display an increase (premutation) in the number of repeats: normally transmitting males (NTMs) and their daughters are phenotypically normal but display from 50 to 200 copies of the CGG triplet; the number of repeats expands to some 200 to 1300 in those showing full symptoms of the disease.

phenotypically normal but transmit the affected chromosome to their daughters, who also appear normal. These males are said to be normally transmitting males (NTMs). However, the sons of the daughters of the NTMs frequently display symptoms. The fragile X syndrome results from mutations in a (CGG)$_n$ repeat in the coding sequence of the *FMR-1* gene. Patients with the disease show specific methylation, induced by the mutation, at a nearby CpG cluster, resulting in reduced *FMR-1* expression.

Why do symptoms develop in some persons with a fragile X chromosome and not in others? The answer seems to lie in the number of CGG repeats in the *FMR-1* gene. Humans normally show a considerable variation in the number of CGG repeats in the *FMR-1* gene, ranging from 6 to 54, with 29 repeats in the most frequent allele. [The variation in CGG repeats produces a corresponding variation in the number of arginine residues (CGG is an arginine codon) in the *FMR-1*-encoded protein.] Both NTMs and their daughters have a much larger number of repeats, ranging from 50 to 200. These increased repeats have been termed **premutations.** All premutation alleles are unstable. The males and females with symptoms of the disease, as well as many carrier females, have additional insertions of DNA, suggesting repeat numbers of 200 to 1300. The frequency of expansion has been shown to increase with the size of the DNA insertion (and thus, presumably, with the number of repeats). Apparently, the number of repeats in the premutation alleles found in NTMs and their daughters is above a certain threshold and thus is much more likely to expand to a full mutation than is the case for normal persons.

The proposed mechanism for these repeats is a slipped mispairing in DNA synthesis (as shown in Figure 16-6) involving a one-step expansion of the four-base-pair sequence CTGG. However, the extraordinarily high frequency of mutation at the three-base-pair repeats in the fragile X syndrome suggests that in human cells, after a threshold level of about 50 repeats, the replication machinery cannot faithfully replicate the correct sequence, and large variations in repeat numbers result.

A second inherited disease, X-linked spinal and bulbar muscular atrophy (known as Kennedy disease), also results from the amplification of a three-base-pair repeat, in this case a repeat of the CAG triplet. Kennedy disease, which is characterized by progressive muscle weakness and atrophy, results from mutations in the gene that encodes the androgen receptor. Normal persons have an average of 21 CAG repeats in this gene, whereas affected patients have repeats ranging from 40 to 52.

Myotonic dystrophy, the most common form of adult muscular dystrophy, is yet another example of sequence expansion causing a human disease. Susceptible families display an increase in severity of the disease in successive generations; this increase is caused by the progressive amplification of a CTG triplet at the 3′ end of a transcript. Normal people possess, on average, five copies of the CTG repeat; mildly affected people have approximately 50 copies,

and severely affected people have more than 1000 repeats of the CTG triplet. Additional examples of triplet expansion are still appearing — for instance, Huntington disease.

Induced mutations
Mutational specificity

When we observe the distribution of mutations induced by different mutagens, we see a distinct specificity that is characteristic of each mutagen. Such **mutational specificity** was first noted in the phage T4 *rII* system by Benzer in 1961. Specificity arises from a given mutagen's "preference" both for a certain *type* of mutation (for example, GC → AT transitions) and for certain mutational *sites* (hot spots). Figure 16-13 shows the mutational specificity in *lacI* of three mutagens described later: ethylmethanesulfonate (EMS), ultraviolet (UV) light, and aflatoxin B$_1$ (AFB$_1$). The graphs show the distribution of base-substitution mutations that create chain-terminating UAG codons. Figure 16-13 is similar to Figure 9-26, which shows the distribution of mutations in *rII*, except that the specific sequence changes are known for each *lacI* site, allowing the graphs to be broken down into each category of substitution.

Figure 16-13 reveals the two components of mutational specificity. First, each mutagen shown favors a specific category of substitution. For example, EMS and UV favor GC → AT transitions, whereas AFB$_1$ favors GC → TA transversions. These preferences are related to the different mechanisms of mutagenesis. Second, even within the same category, there are large differences in mutation rate. These differences can be seen best with UV light for the GC → AT changes. Some aspect of the surrounding DNA sequence must cause these differences. In some cases, the cause of mutational hot spots can be determined by DNA sequence studies, as previously described for 5-methylcytosine residues and for certain frameshift sites (Figures 16-5 and 16-9). In many examples of mutagen-induced hot spots, the precise reason for the high mutability of specific sites is still unknown. However, high lesion frequency at some sites and reduced repair at certain sites are sometimes causes of hot spots.

Mechanisms of mutagenesis

Mutagens induce mutations by at least three different mechanisms. They can replace a base in the DNA, alter a base so that it specifically mispairs with another base, or damage a base so that it can no longer pair with any base under normal conditions.

Incorporation of base analogs. Some chemical compounds are sufficiently similar to the normal nitrogen bases of DNA that they occasionally are incorporated into DNA in place of normal bases; such compounds are called **base analogs.** Once in place, these analogs have pairing properties unlike those of the normal bases; thus, they can produce mutations by causing incorrect nucleotides to be inserted

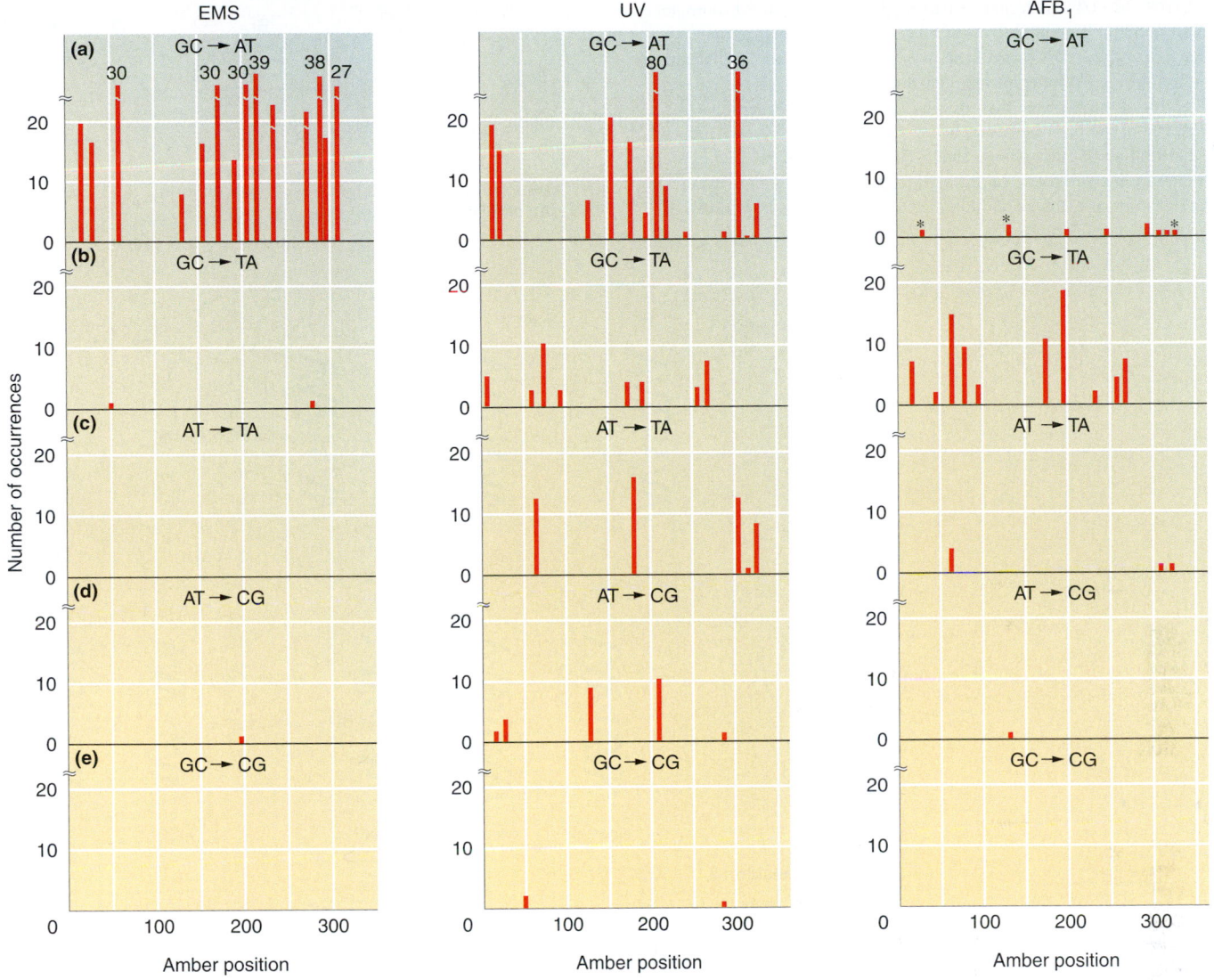

Figure 16-13 Specificity of mutagens. The distribution of mutations among 36 sites in the *lacI* gene is shown for three mutagens: EMS, UV light, and aflatoxin B_1. The height of each bar represents the number of occurrences of mutations at the respective site. Some hot spots are shown off-scale, with the number of occurrences indicated directly above the respective peak. For instance, in the UV-generated collection, one site resulting from a GC → AT transition is represented by 80 occurrences. Each mutational site represented in the figure generates an amber (UAG) codon in the corresponding mRNA. The mutations are arranged according to the type of base substitution. Asterisks mark the positions of 5-methylcytosines. (After C. Coulondre and J. H. Miller, *Journal of Molecular Biology* 117, 1977, 577; and P. L. Foster et al., *Proceedings of the National Academy of Sciences USA* 80, 1983, 2695.)

opposite them in replication. The original base analog exists in only a single strand, but it can cause a nucleotide-pair substitution that is replicated in all DNA copies descended from the original strand.

For example, **5-bromouracil (5-BU)** is an analog of thymine that has bromine at the C-5 position in place of the CH$_3$ group found in thymine. This change does not affect the atoms that take part in hydrogen bonding in base pair-

ing, but the presence of the bromine significantly alters the distribution of electrons in the base. The normal structure (the keto form) of 5-BU pairs with adenine, as shown in Figure 16-14a. 5-BU can frequently change to either the enol form or an ionized form; the latter pairs in vivo with guanine (Figure 16-14b). Thus, the nature of the pair formed in replication will depend on the form of 5-BU at the moment of pairing (Figure 16-15). 5-BU causes

Figure 16-14   Alternative pairing possibilities for 5-bromouracil (5-BU). 5-BU is an analog of thymine that can be mistakenly incorporated into DNA as a base. It has a bromine atom in place of the methyl group. (a) In its normal keto state, 5-BU mimics the pairing behavior of the thymine that it replaces, pairing with adenine. (b) The presence of the bromine atom, however, causes a relatively frequent redistribution of electrons, so that 5-BU can spend part of its existence in the rare ionized form. In this state, it pairs with guanine, mimicking the behavior of cytosine and thus inducing mutations in replication.

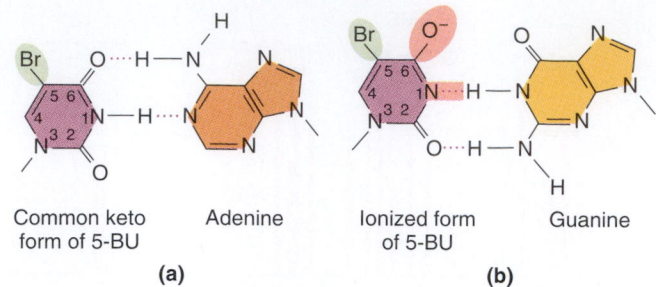

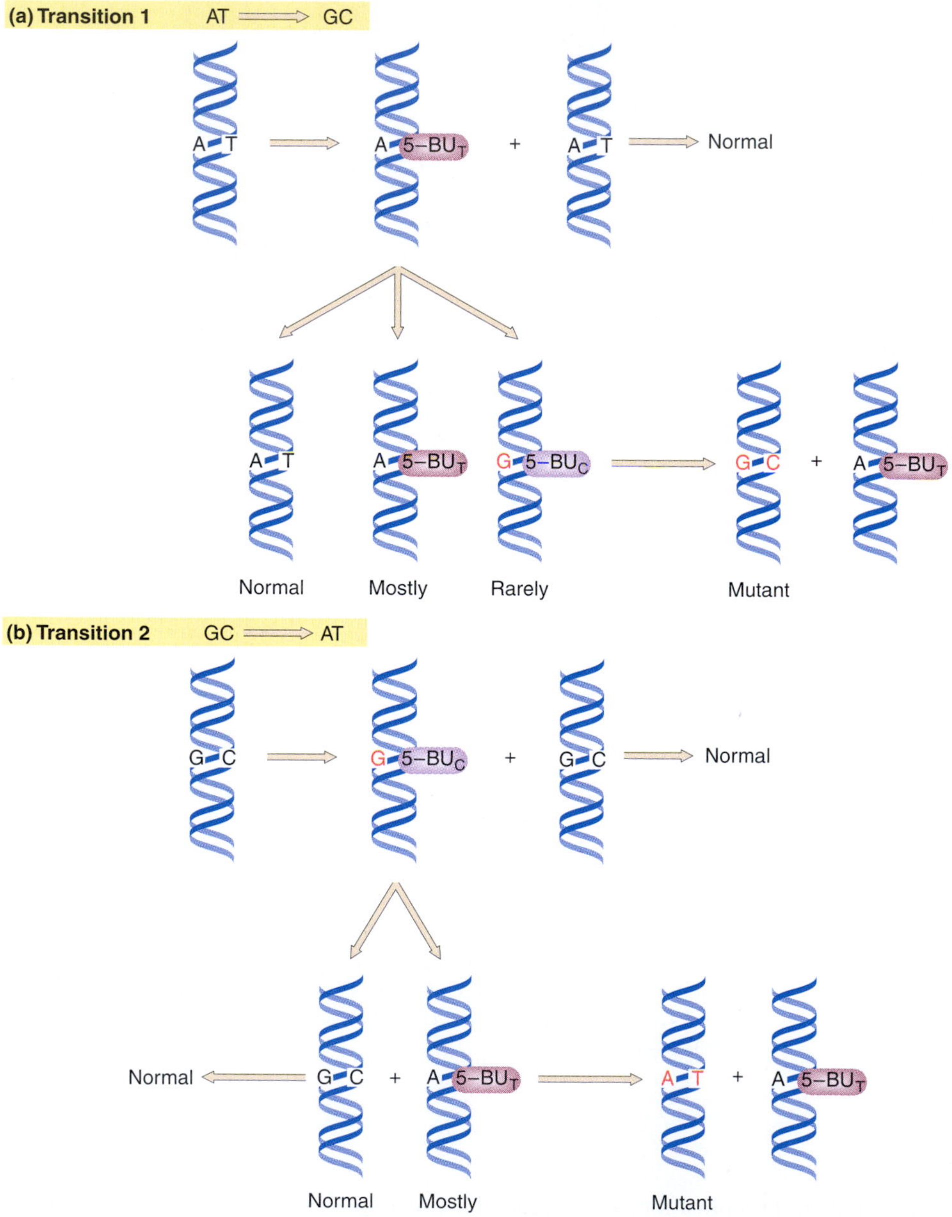

Common keto form of 5-BU   Adenine   Ionized form of 5-BU   Guanine

(a)   (b)

(a) Transition 1   AT ⟶ GC

A–T ⟶ A–5-BU_T + A–T ⟶ Normal

A–T   A–5-BU_T   G–5-BU_C ⟶ G–C + A–5-BU_T

Normal   Mostly   Rarely   Mutant

(b) Transition 2   GC ⟶ AT

G–C ⟶ G–5-BU_C + G–C ⟶ Normal

Normal ⟵ G–C + A–5-BU_T ⟶ A–T + A–5-BU_T

Normal   Mostly   Mutant

Figure 16-15   The mechanism of 5-BU mutagenesis. 5-BU causes mutations when it is incorporated in one form and then shifts to another form. (a) In its normal keto state, 5-BU pairs like thymine (5-BU_T). Thus, 5-BU is incorporated across from adenine and subsequently mispairs with guanine, resulting in AT → GC transitions. (b) In its ionized form, 5-BU pairs like cytosine (5-BU_C). Thus, 5-BU is misincorporated across from guanine and subsequently pairs with adenine, resulting in GC → AT transitions.

transitions almost exclusively, as predicted in Figures 16-14 and 16-15.

Another analog widely used in research is **2-amino-purine (2-AP),** which is an analog of adenine that can pair with thymine but can also mispair with cytosine when protonated, as shown in Figure 16-16. Therefore, when 2-AP is incorporated into DNA by pairing with thymine, it can generate AT → GC transitions by mispairing with cytosine in subsequent replications. Or, if 2-AP is incorporated by mispairing with cytosine, then GC → AT transitions will result when it pairs with thymine. Genetic studies have shown that 2-AP, like 5-BU, is very specific for transitions.

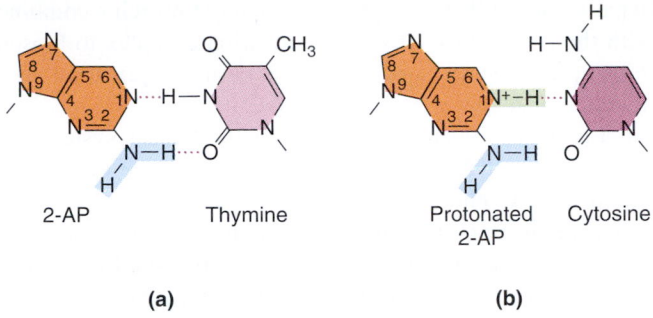

Figure 16-16 Alternative pairing possibilities for 2-aminopurine (2-AP), an analog of adenine. Normally, 2-AP pairs with thymine (a), but in its protonated state it can pair with cytosine (b).

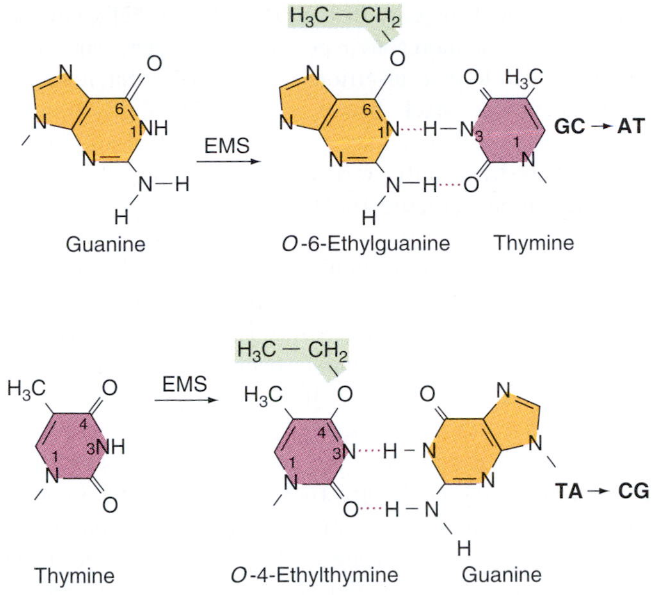

Specific mispairing. Some mutagens are not incorporated into the DNA but instead alter a base, causing specific mispairing. Certain **alkylating agents,** such as **ethylmethanesulfonate (EMS)** and the widely used **nitrosoguanidine (NG),** operate by this pathway:

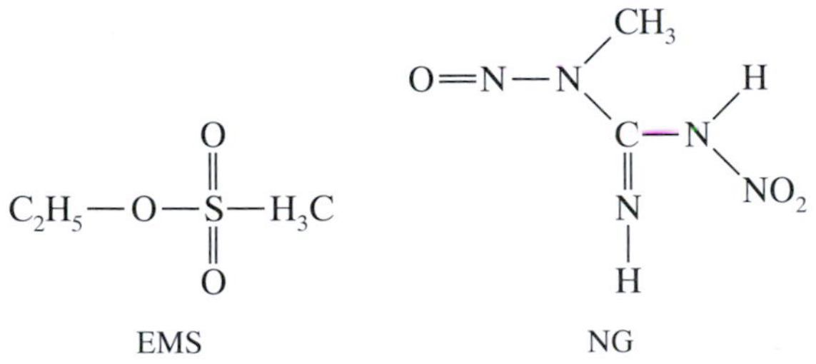

EMS   NG

Although such agents add alkyl groups (an ethyl group in EMS and a methyl group in NG) to many positions on all four bases, mutagenicity is best correlated with an addition to the oxygen at the 6 position of guanine to create an *O*-6-alkylguanine. This addition leads to direct mispairing with thymine, as shown in Figure 16-17, and would result in GC → AT transitions at the next round of replication. As expected, determinations of mutagenic specificity for EMS

Figure 16-17 Alkylation-induced specific mispairing. The alkylation (in this case, EMS-generated ethylation) of the O-6 position of guanine and the O-4 position of thymine can lead to direct mispairing with thymine and guanine, respectively, as shown here. In bacteria, where mutations have been analyzed in great detail, the principal mutations detected are GC → AT transitions, indicating that the O-6 alkylation of guanine is most relevant to mutagenesis.

and NG show a strong preference for GC → AT transitions (see the data for EMS shown in Figure 16-13). Alkylating agents can also modify the bases in dNTPs (where N is any base), which are precursors in DNA synthesis.

The **intercalating agents** form another important class of DNA modifiers. This group of compounds includes **proflavin, acridine orange,** and a class of chemicals termed **ICR compounds** (Figure 16-18a). These agents are planar molecules, which mimic base pairs and are able to slip themselves in (**intercalate**) between the stacked

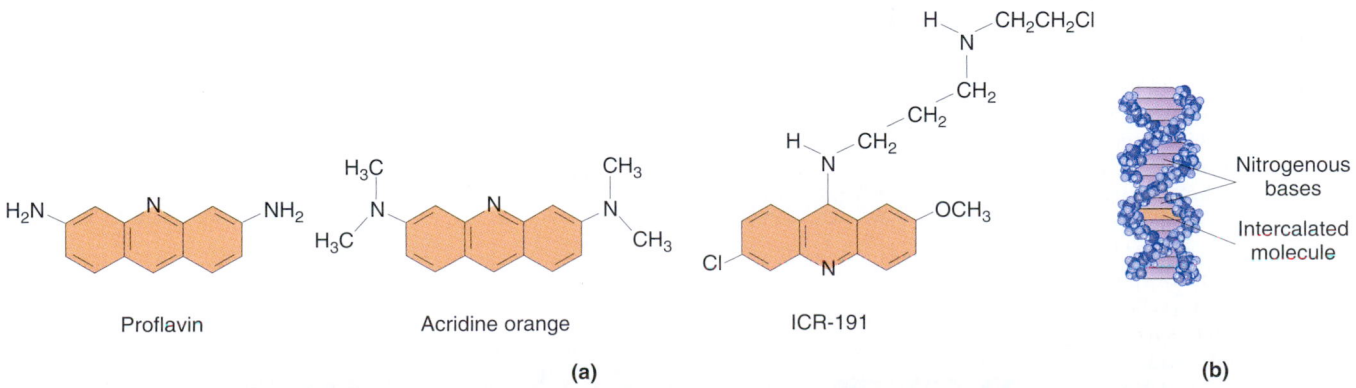

Figure 16-18 Intercalating agents. (a) Structures of the common agents proflavin, acridine orange, and ICR-191. (b) An intercalating agent slips between the nitrogenous bases stacked at the center of the DNA molecule. This occurrence can lead to single-nucleotide-pair insertions and deletions. (From L. S. Lerman, *Proceedings of the National Academy of Sciences USA* 49, 1963, 94.)

nitrogen bases at the core of the DNA double helix (Figure 16-18b). In this intercalated position, the agent can cause single-nucleotide-pair insertions or deletions. Intercalating agents may also stack between bases in single-stranded DNA; in so doing, they may stabilize bases that are looped out during frameshift formation, as depicted in the Streisinger model (Figure 16-4).

Base damage.   A large number of mutagens damage one or more bases, so no specific base pairing is possible. The result is a replication block, because DNA synthesis will not proceed past a base that cannot specify its complementary partner by hydrogen bonding. In bacterial cells, such replication blocks can be *bypassed* by inserting nonspecific bases. The process requires the activation of a special system, the **SOS system** (Figure 16-19). The name SOS comes from the idea that this system is induced as an emergency response to prevent cell death in the presence of significant DNA damage. SOS induction is a last resort, allowing the cell to trade death for a certain level of mutagenesis.

Exactly how the SOS bypass system functions is not clear, although in *E. coli* it is known to be dependent on at least three genes, *recA* (which also has a role in general recombination), *umuC,* and *umuD.* Current models for SOS

bypass suggest that the UmuC and UmuD proteins combine with the polymerase III DNA replication complex to loosen its otherwise strict specificity and permit replication past noncoding lesions.

Figure 16-19 shows a model for the bypass system operating after DNA polymerase III stalls at a type of damage called a T–C photodimer. Because replication can restart downstream from the dimer, a single-stranded region of DNA is generated. This region attracts the stabilizing protein, called single-strand-binding protein (Ssb), as well as the RecA protein, which forms filaments and signals the cell to synthesize the UmuC and UmuD proteins. The UmuD protein binds to the filaments and is cleaved by the RecA protein to yield a shortened version termed UmuD′, which then recruits the UmuC protein to form a complex that allows DNA polymerization to continue past the dimer, adding bases across from the dimer with a high error frequency (see Figure 16-19).

Therefore mutagens that damage specific base-pairing sites are dependent on the SOS system for their action. The category of SOS-dependent mutagens is important, because it includes most cancer-causing agents (carcinogens), such as ultraviolet light, aflatoxin B_1, and benzo(a)pyrene (discussed later).

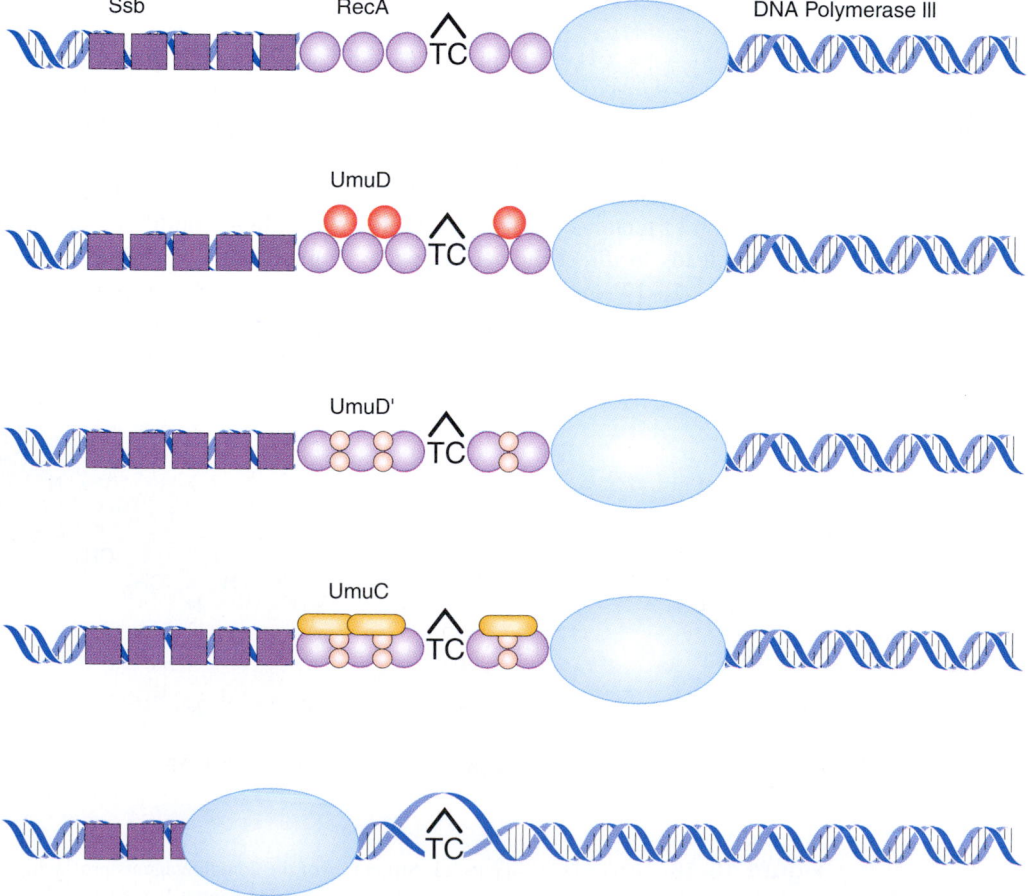

Figure 16-19   The SOS system. DNA polymerase III, shown in blue, stops at a noncoding lesion, such as the T–C photodimer shown here, generating single-stranded regions that attract the Ssb protein (dark purple) and RecA (light purple), which forms filaments. The presence of RecA filaments helps to signal the cell to synthesize UmuD (red circles), which is cleaved by RecA to yield UmuD′ (pink circles) and UmuC (yellow ovals). The UmuC is recruited to form a complex with UmuD′ that permits DNA polymerization to proceed past the blocking lesion.

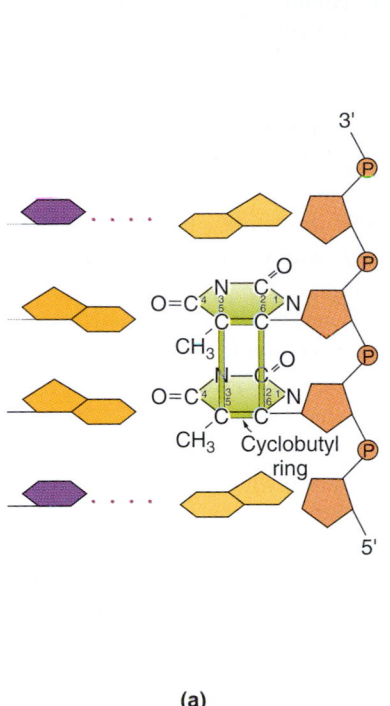

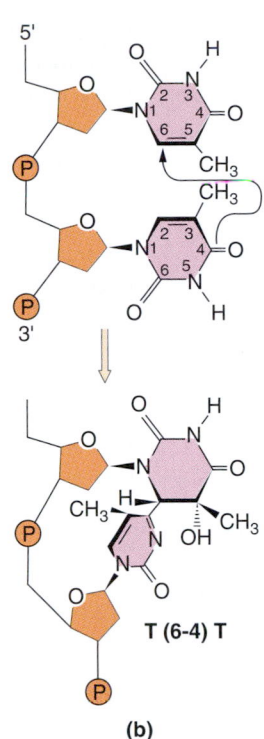

T (6-4) T

(a)   (b)

Figure 16-20   (a) Structure of a cyclobutane pyrimidine dimer. Ultraviolet light stimulates the formation of a four-membered cyclobutane ring (green) between two adjacent pyrimidines on the same DNA strand by acting on the 5,6 double bonds. (b) Structure of the 6-4 photoproduct. The structure forms most prevalently with 5′-CC-3′ and 5′-TC-3′, between the C-6 and the C-4 positions of two adjacent pyrimidines, causing a significant perturbation in local structure of the double helix. (Part a after E. C. Friedberg, *DNA Repair.* Copyright © 1985 by W. H. Freeman and Company; part b after E. C. Friedberg, *DNA Repair.* Copyright © 1985 by W. H. Freeman and Company. As adapted from A. M. R. Taylor, C. M. Rosney, and J. B. Campbell, *Cancer Research* 39, 1979, 1046–1050.

How does the SOS system take part in the recovery of mutations after mutagenesis? Does the SOS system lower the fidelity of DNA replications so much (to permit the bypass of noncoding lesions) that many replication errors occur, even for undamaged DNA? If this hypothesis were correct, most mutations generated by different SOS-dependent mutagens would be similar, rather than specific to each mutagen. Most mutations would result from the action of the SOS system itself on undamaged DNA. The mutagen, then, would play the indirect role of inducing the SOS system. Studies of mutational specificity, however, have shown that this is not the case. Instead, a series of different SOS-dependent mutagens have markedly different specificities, as seen for UV light and aflatoxin B_1 in Figure 16-13. Each mutagen induces a unique distribution of mutations. Therefore, the mutations must be generated in response to specific damaged base pairs. The type of lesion differs in many cases. Some of the most widely studied lesions include UV photoproducts and apurinic sites.

Ultraviolet light generates a number of photoproducts in DNA. Two different lesions that occur at adjacent pyrimidine residues — the cyclobutane pyrimidine photodimer and the 6-4 photoproduct (Figure 16-20) — have been most strongly correlated with mutagenesis. These lesions interfere with normal base pairing; hence, induction of the SOS system is required for mutagenesis. The insertion of incorrect bases across from UV photoproducts is at the 3′ position of the dimer, and more frequently for 5′-CC-3′ and 5′-TC-3′ dimers. The C → T transition is the most frequent mutation, but other base substitutions (transversions) and frameshifts also are stimulated by UV light, as are duplica-

tions and deletions. The mutagenic specificity of UV light is illustrated in Figure 16-13.

Ionizing radiation

Ionizing radiation results in the formation of ionized and excited molecules that can cause damage to cellular components and to DNA. Because of the aqueous nature of biological systems, the molecules generated by the effects of ionizing radiation on water produce the most damage. Many different types of reactive oxygen specials are produced, including superoxide radicals, such as ·OH. The most biologically relevant reaction products are ·OH, O_2^-, and H_2O_2. These species can damage bases and cause different adducts and degradation products. Among the most prevalent, which result in mutations, are thymine glycol and 8-oxodG, pictured in Figure 16-10. Ionizing radiation can cause breakage of the N-glycosydic bond, leading to the formation of AP sites, and can cause strand breaks that are responsible for most of the lethal effects of such radiation.

Aflatoxin B_1 is a powerful carcinogen. It generates **apurinic sites** following the formation of an addition product at the N-7 position of guanine (Figure 16-21). Studies with apurinic sites generated in vitro demonstrated a requirement for the SOS system and showed that the SOS bypass of these sites leads to the preferential insertion of an adenine across from an apurinic site. Thus agents that cause depurination at guanine residues should preferentially induce GC → TA transversions. Can you see why the insertion of an adenine across from an apurinic site derived from a guanine would generate this substitution at the next round

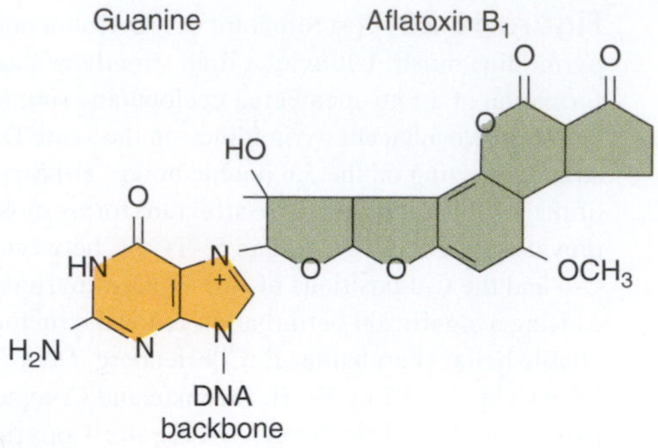

Figure 16-21   The binding of metabolically activated aflatoxin B_1 to DNA.

of replication? Figure 16-13 shows the genetic analysis of many base substitutions induced by AFB_1. You can verify that most of the substitutions are indeed $GC \rightarrow TA$ transversions.

AFB_1 is a member of a class of chemical carcinogens known as **bulky addition products** when they bind covalently to DNA. Other examples include the diol epoxides of **benzo(a)pyrene,** a compound produced by internal combustion engines. For many different compounds, it is not yet clear which DNA addition products play the principal role in mutagenesis. In some cases, the mutagenic specificity suggests that depurination may be an intermediate step in mutagenesis; in others, the question of which mechanism is operating is completely open.

MESSAGE ···
Mutagens induce mutations by a variety of mechanisms. Some mutagens mimic normal bases and are incorporated into DNA, where they can mispair. Others damage bases and either cause specific mispairing or destroy pairing by causing nonrecognition of bases. In the latter case, a bypass system, the SOS system, must be induced to allow replication past the lesion.

Reversion analysis

Testing for the reversion of a mutation can tell us something about the nature of the mutation or the action of a mutagen.

For example, if a mutation cannot be reverted by action of the mutagen that induced it, then the mutagen must have some relatively specific unidirectional action. In a mutation induced by hydroxylamine (HA), for instance, it would be reasonable to expect that the original mutation is $GC \rightarrow AT$, which cannot be reverted by another specific $GC \rightarrow AT$ event. Similarly, mutations that can be reverted by proflavin are in all likelihood frameshift mutations; thus mutations induced by nitrous acid (NA), which are transitions, should not be revertible by proflavin.

Transversions cannot be induced by the aforementioned agents, but they are definitely known to be common among spontaneous mutations, as shown by studies of DNA and protein sequencing. Thus, in the reversion test, if a mutation reverts spontaneously but does not revert in response to a transition mutagen or a frameshift mutagen, then, by elimination, it is probably a transversion.

Table 16-1 summarizes some reversion expectations based on simple assumptions from reversion analysis. Recall that mutagen specificities depend on the organism, the genotype, the gene studied, and the action of biological repair systems. Note that the kinds of logic employed in the reversion test rely heavily on the assumption that the reversion events are not due to suppressors or transposable elements; either of them would make inference from reversion more difficult.

Relation between mutagens and carcinogens

Mutagenicity and carcinogenicity are clearly correlated. One study showed that 157 of 175 known carcinogens (approximately 90 percent) are also mutagens. The **somatic mutation theory** of cancer holds that these agents cause cancer by inducing the mutation of somatic cells. Thus, understanding mutagenesis is of great relevance to our society.

Induced mutations and human cancer

Understanding the specificity of mutagens in bacteria has led to the direct implication of certain environmental mutagens in the causation of human cancers. Ultraviolet light

16-1 TABLE	Reversion Tests			
	REVERSION MUTAGENS			
Mutation	NA	HA or EMS	Proflavin	Spontaneous reversion
Transition ($GC \rightarrow AT$)	+	−	−	+
Transition ($AT \rightarrow GC$)	+	+	−	+
Transversion	−	−	−	+
Frameshift	−	−	+	+

Note: A plus sign indicates a measurable rate of reversion due to a given mutagen.
NA, nitrous acid; HA, hydroxylamine; EMS, ethyl methanesulfonate.

and aflatoxin B_1 have long been suspected of causing skin cancer and liver cancer, respectively. Now, DNA sequence analysis of mutations in a human cancer gene has provided direct evidence of their involvement. The gene in question is termed *p53* and is one of a number of **tumor-suppressor genes** — genes that encode proteins that suppress tumor formation. (We will learn more about these genes in Chapter 23.) A sizable proportion of human cancer patients have mutated tumor-suppressor genes.

Liver cancer is prevalent in southern Africa and East Asia, and a high exposure to AFB_1 in these regions has been correlated with the high incidence of liver cancer. When *p53* mutations in cancer patients were analyzed, $G \rightarrow T$ transversions, the signature of AFB_1-induced mutations, were found in liver cancer patients from South Africa and East Asia but not in patients from these regions with lung, colon, or breast cancer. On the other hand, *p53* mutations in liver cancer patients from areas of low AFB_1 exposure did not result from $G \rightarrow T$ transversions. These findings, together with the results from the mutagenic specificity studies of AFB_1 (see Figure 16-13), allow us to conclude that AFB_1-induced mutations are a prime cause of liver cancer in South Africa and East Asia.

Sequencing *p53* mutations has also strengthened the link between UV and human skin cancers. The majority of invasive human squamous cell carcinomas analyzed so far have *p53* mutations, all of them mutations at dipyrimidine sites, most of which are $C \rightarrow T$ substitutions when the C is the $3'$ pyrimidine of a TC dimer. This is the profile of UV-induced mutations. In addition, several tumors have *p53* mutations resulting from a $CC \rightarrow TT$ double base change, which is found most frequently among UV-induced mutations.

The modern environment exposes everyone to a wide variety of chemicals in drugs, cosmetics, food preservatives, pesticides, compounds used in industry, pollutants, and so forth. Many of these compounds have been shown to be carcinogenic and mutagenic. Examples include the food preservative AF-2, the food fumigant ethylene dibromide, the antischistosome drug hycanthone, several hair-dye additives, and the industrial compound vinyl chloride; all are potent, and some have subsequently been subjected to government control. However, hundreds of new chemicals and products appear on the market each week. How can such vast numbers of new agents be tested for carcinogenicity before much of the population has been exposed to them?

Ames test

Many test systems have been devised to screen for carcinogenicity. These tests are time consuming, typically requiring laborious research with small mammals. More rapid tests do exist that make use of microbes (such as fungi or bacteria) and test for mutagenicity rather than carcinogenicity. The most widely used test was developed in the 1970s by Bruce Ames, who worked with *Salmonella typhimurium*. This

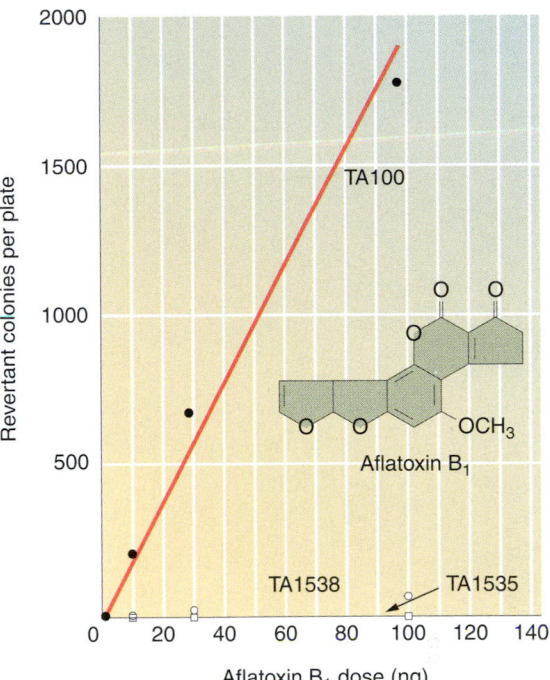

Figure 16-22   Ames test results showing the mutagenicity of aflatoxin B_1, which is also a potent carcinogen. TA100, TA1538, and TA1535 are strains of *Salmonella* bearing different *his* auxotrophic mutations. The TA100 strain is highly sensitive to reversion through base-pair substitution. The TA1535 and TA1538 strains are sensitive to reversion through frameshift mutation. The test results show that aflatoxin B_1 is a potent mutagen that causes base-pair substitutions but not frameshifts. (From J. McCann and B. N. Ames, in W. G. Flamm and M. A. Mehlman, eds., *Advances in Modern Technology*, Vol. 5. Copyright © by Hemisphere Publishing Corp., Washington, DC.)

Ames test uses two auxotrophic histidine mutations, which revert by different molecular mechanisms (Figure 16-22). Further properties were genetically engineered into these strains to make them suitable for mutagen detection. First, they carry a mutation that inactivates the excision-repair system (described later). Second, they carry a mutation that eliminates the protective lipopolysaccharide coating of wild-type *Salmonella* to facilitate the entry of many different chemicals into the cell.

Bacteria are evolutionarily a long way removed from humans. Can the results of a test on bacteria have any real significance in detecting chemicals that are dangerous for humans? First, we have seen that the genetic and chemical nature of DNA is identical in all organisms, so a compound acting as a mutagen in one organism is likely to have some mutagenic effects in other organisms. Second, Ames devised a way to simulate the human metabolism in the bacterial system. In mammals, much of the important processing of ingested chemicals takes place in the liver, where externally derived compounds normally are detoxified or broken down. In some cases, the action of liver enzymes can create a toxic or mutagenic compound from a substance that was

(a)
Benzo(a)pyrene

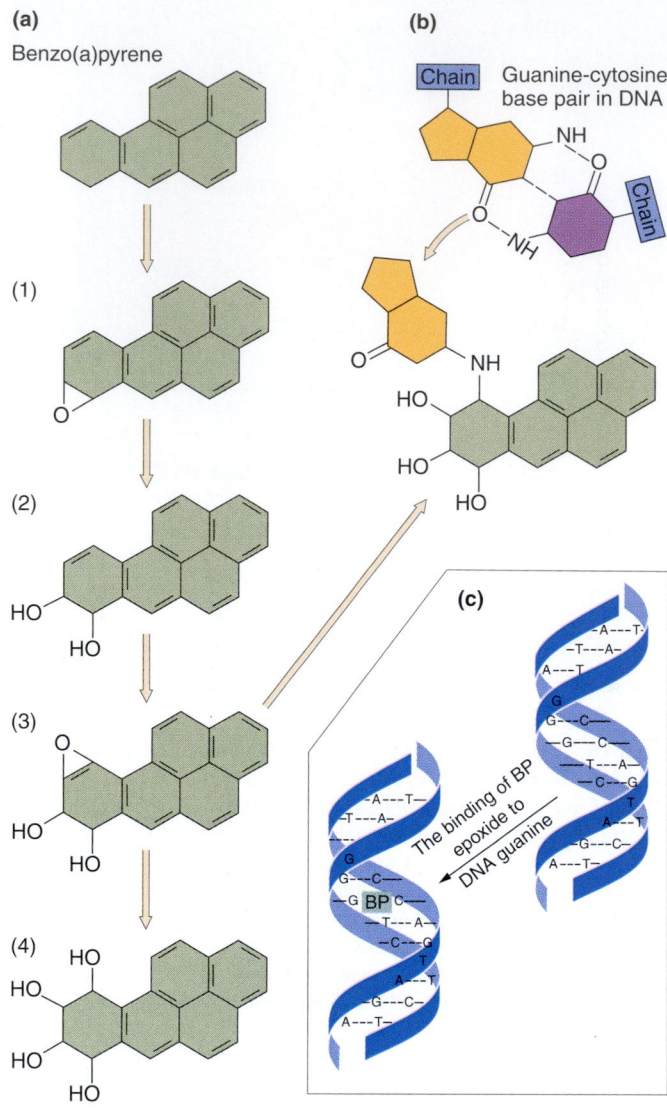

(b)

Figure 16-23   The metabolic conversion of benzo(a)pyrene (BP) into a mutagen (and a carcinogen). Benzo(a)pyrene goes through several steps (a) as it is made more water soluble prior to excretion. One of the intermediates in this process, a diol epoxide (3), is capable of reacting with guanine in DNA (b). This reaction leads to a distortion of the DNA molecule (c) and mutations. Benzo(a)pyrene is therefore a mutagen for any cell that has the enzymes that produce this intermediate. (After I. B. Weinstein et al., *Science* 193, 1976, 592; from J. Cairns, *Cancer: Science and Society.* Copyright © 1978 by W. H. Freeman and Company.)

not originally dangerous (Figure 16-23). Ames incorporated mammalian liver enzymes in his bacterial test system, using rat livers for this purpose. Figure 16-24 outlines the procedure used in the Ames test.

Chemicals detected by this test can be regarded not only as potential carcinogens (sources of somatic mutations), but also as possible causes of mutations in germinal cells. Because the test system is so simple and inexpensive, many laboratories throughout the world now routinely test large numbers of potentially hazardous compounds for mutagenicity and potential carcinogenicity.

Biological repair mechanisms

As we have seen in the preceding discussions, there are many potential threats to the fidelity of DNA replication.

Figure 16-24   Summary of the procedure used for the Ames test. First, rat liver enzymes are mobilized by injecting the animals with Arochlor. (Enzymes from the liver are used because they carry out the metabolic processes of detoxifying and toxifying body chemicals.) The rat liver is then homogenized, and the supernatant of solubilized liver enzymes (S9) is added to a suspension of auxotrophic bacteria in a solution of the potential carcinogen (X). This mixture is plated on a medium containing no histidine, and revertants of mutant strains 1 and 2 are looked for. A control experiment containing no potential carcinogen is always run simultaneously. The presence of revertants indicates that the chemical is a mutagen and possibly a carcinogen as well.

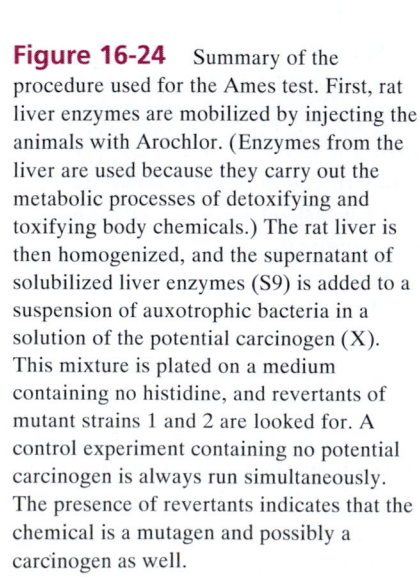

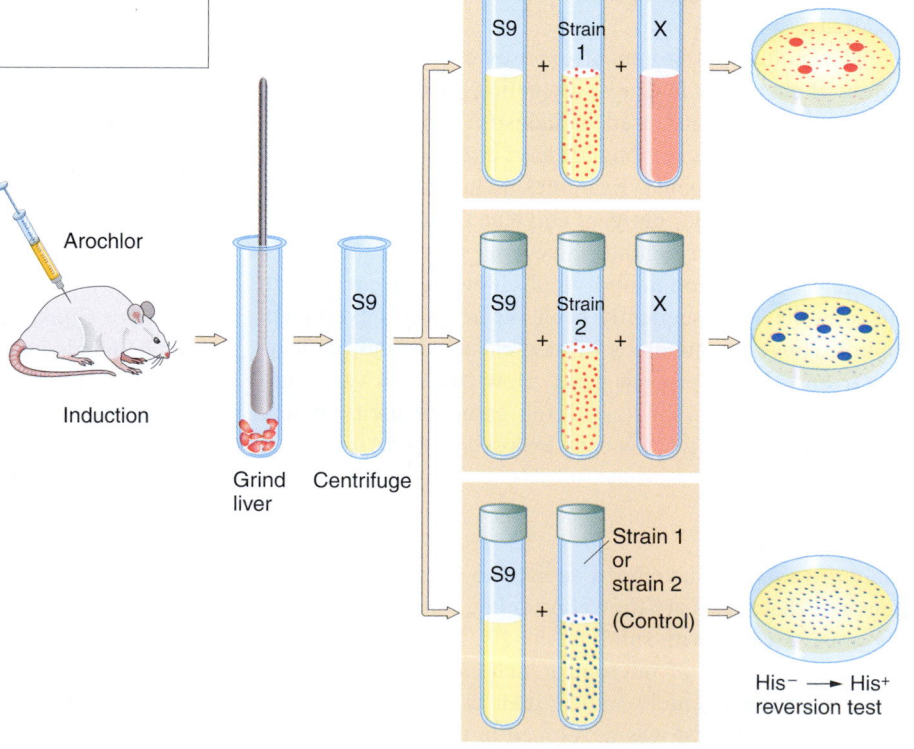

Not only is there an inherent error rate for the replication of DNA, but there are also spontaneous lesions that can provoke additional errors. Moreover, mutagens in the environment can damage DNA and greatly increase the mutation rate.

Living cells have evolved a series of enzymatic systems that repair DNA damage in a variety of ways. Failure of these systems can lead to a higher mutation rate. A number of human diseases including certain types of cancer can be attributed to defects in DNA repair, as we shall see later. Let's first examine some of the characterized repair pathways and then consider how the cell integrates these systems into an overall strategy for repair.

We can divide repair pathways into several categories.

Prevention of errors before they happen

Some enzymatic systems neutralize potentially damaging compounds before they even react with DNA. One example of such a system is the detoxification of superoxide radicals produced during oxidative damage to DNA: the enzyme **superoxide dismutase** catalyzes the conversion of the superoxide radicals into hydrogen peroxide, and the enzyme **catalase,** in turn, converts the hydrogen peroxide into water. Another error-prevention pathway depends on the protein product of the *mutT* gene: this enzyme prevents the incorporation of 8-oxodG (see Figure 16-10), which arises by oxidation of dGTP, into DNA by hydrolyzing the triphosphate of 8-oxodG back to the monophosphate.

Direct reversal of damage

The most straightforward way to repair a lesion, once it occurs, is to reverse it directly, thereby regenerating the normal base. Reversal is not always possible, because some types of damage are essentially irreversible. In a few cases, however, lesions can be repaired in this way. One case is a mutagenic photodimer caused by UV light (see Figure 16-20). The cyclobutane pyrimidine photodimer can be repaired by a **photolyase** that has been found in bacteria and lower eukaryotes but not in humans. The enzyme binds to the photodimer and splits it, in the presence of certain wavelengths of visible light, to generate the original bases (Figure 16-25). This enzyme cannot operate in the dark, so other repair pathways are required to remove UV damage. A photolyase that reverses the 6-4 photoproducts has been detected in plants and *Drosophila*.

Alkyltransferases also are enzymes taking part in the direct reversal of lesions. They remove certain alkyl groups that have been added to the O-6 positions of guanine (Figure 16-17) by such agents as NG and EMS. The methyltransferase from *E. coli* has been well studied. This enzyme transfers the methyl group from *O*-6-methylguanine to a cysteine residue on the protein. When this happens, the enzyme is inactivated, so this repair system can be saturated if the level of alkylation is high enough.

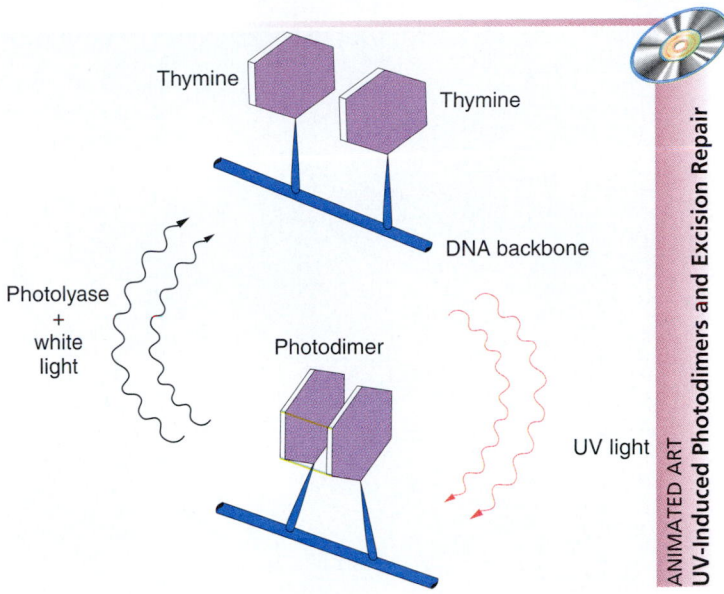

Figure 16-25 Repair of a UV-induced pyrimidine photodimer by a photoreactivating enzyme, or photolyase. The enzyme recognizes the photodimer (here, a thymine dimer) and binds to it. When light is present, the photolyase uses its energy to split the dimer into the original monomers. (After J. D. Watson, *Molecular Biology of the Gene*, 3d ed. Copyright © 1976 by W. A. Benjamin.)

Excision-repair pathways

General excision repair. Also termed *nucleotide excision repair*, this system includes the breaking of a phosphodiester bond on either side of the lesion, on the same strand, resulting in the excision of an oligonucleotide. This excision leaves a gap that is filled by repair synthesis, and a ligase seals the breaks. In prokaryotes, 12 or 13 nucleotides are removed; whereas, in eukaryotes, from 27 to 29 nucleotides are eliminated. Figure 16-26 depicts the incision pattern in each case.

In *E. coli*, the products of the *uvrA, B,* and *C* genes constitute the excinuclease. The UvrA protein, which recognizes the damaged DNA, forms a complex with UvrB and leads the UvrB subunit to the damage site before dissociating. The UvrC protein then binds to UvrB. Each of these subunits makes an incision. The short DNA 12-mer is unwound and released by another protein, helicase II. Figure 16-27 shows a detailed view of these excision events.

The human excinuclease is considerably more complex than its bacterial counterpart and includes at least 17 proteins. However, the basic steps are the same as those in *E. coli*.

Coupling of transcription and repair. The involvement of TFIIH, a transcription factor, in excision repair underscores the fact that transcription and repair are coupled. In both eukaryotes and prokaryotes, there is a preferential repair of the transcribed strand of DNA for actively expressed genes. Figure 16-28 portrays a mechanism for this coupling.

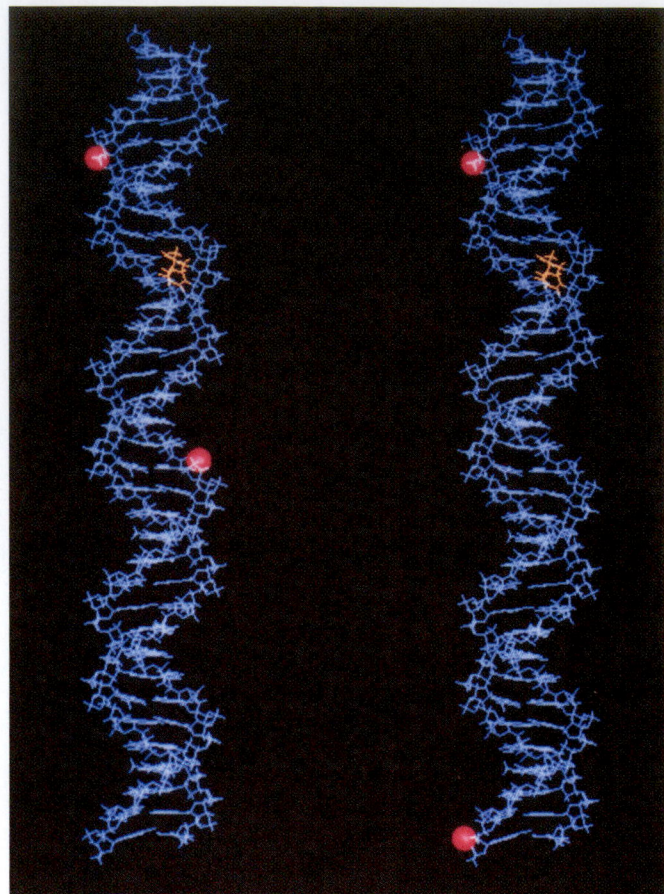

Figure 16-26 Excinuclease incision patterns by *E. coli (left)* and human enzymes. The red points indicate the incision patterns of a lesion — in this case, a thymine dimer — which is shown in orange. (Courtesy of J. E. Hearst, in A. Sancar, *Science* 266, 1974, 1954.)

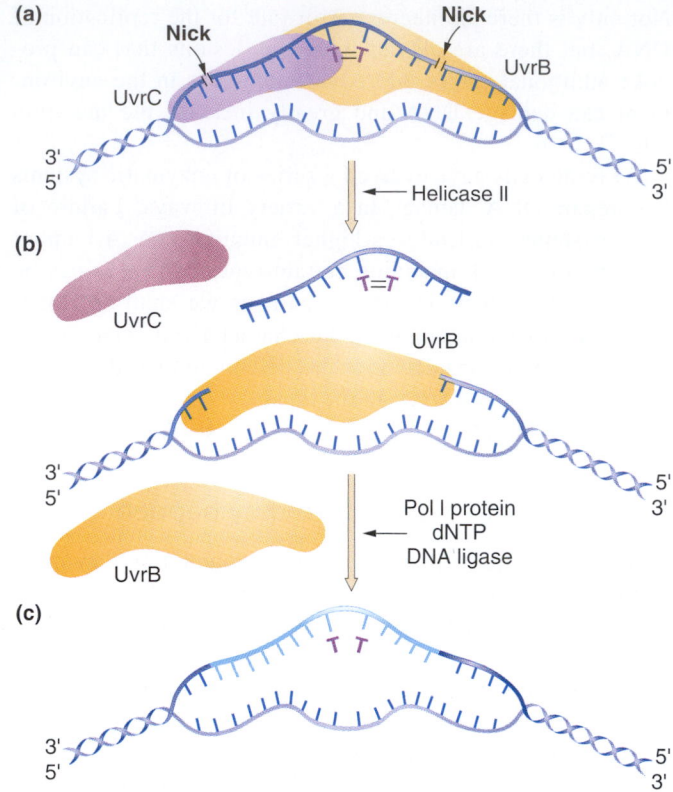

Figure 16-27 Schematic representation of events following incision by UvrABC exinuclease in *E. coli.* First, the UvrA subunit recruits the UvrB subunit to the damage site before dissociation. Then, as shown here, DNA helicase II mediates the release of a segment of the DNA bounded by two nicks in the same strand of DNA. The UvrC protein is also displaced at this point. The subsequent repair synthesis displaces UvrB. (From E. C. Friedberg, G. C. Walker, and W. Seide, *DNA Repair and Mutagenesis.* Copyright © 1995 by the American Society for Microbiology.)

Specific excision pathways. Certain lesions are too subtle to cause a distortion large enough to be recognized by the *uvrABC*-encoded general excision-repair system and its counterparts in higher cells. Thus, additional excision pathways are necessary.

DNA glycosylase repair pathway (base-excision repair). **DNA glycosylases** do not cleave phosphodiester bonds, but instead cleave N-glycosidic (base–sugar) bonds, liberating the altered base and generating an apurinic or an apyrimidinic site, both called **AP sites,** because they are biochemically equivalent (see Figure 16-7). This initial step is shown in Figure 16-29. The resulting Ap site is then repaired by an AP endonuclease repair pathway (described in the next subsection).

Figure 16-28 Nucleotide excision repair is coupled to transcription. This model for coupled repair in mammalian cells shows RNA polymerase (pink) pausing when encountering a lesion. It undergoes a conformational change, allowing the DNA strands at the lesion site to reanneal. Protein factors aid in coupling by bringing TFIIH and other factors to the site to carry out the incision, excision, and repair reactions. Then transcription can continue normally. (From P. C. Hanawalt, *Science* 266, 2994, 2957.)

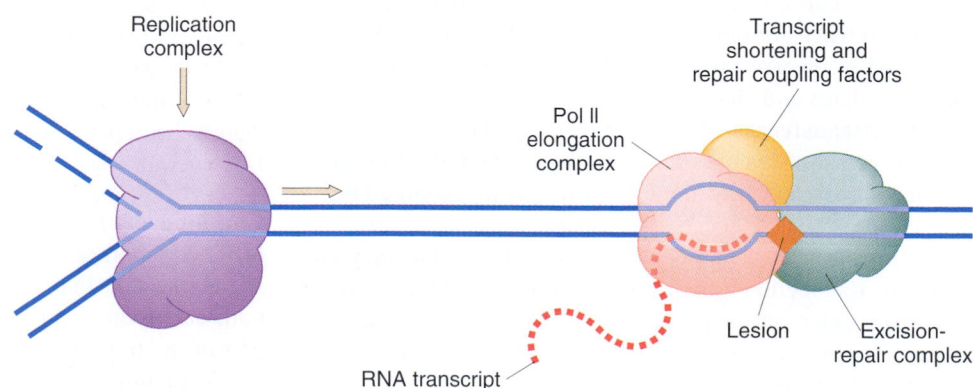

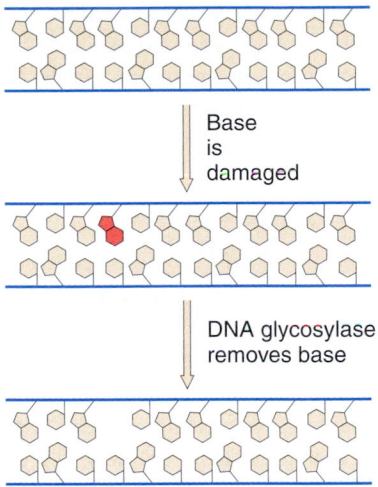

Figure 16-29 Action of DNA glycosylases. Glycosylases remove altered bases and leave an AP site. The AP site is subsequently excised by the AP endonucleases diagrammed in Figure 16-30. (After B. Lewin, *Genes.* Copyright © 1983 by John Wiley.)

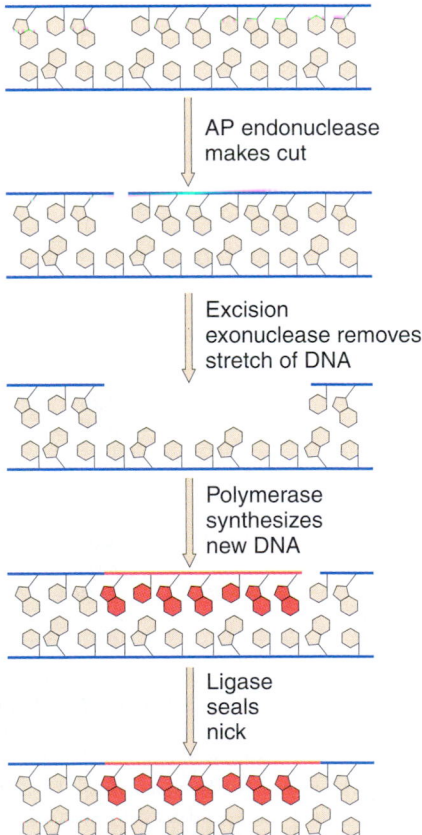

Figure 16-30 Repair of AP (apurinic or apyrimidinic) sites. AP endonucleases recognize AP sites and cut the phosphodiester bond. A stretch of DNA is removed by an exonuclease, and the resulting gap is filled by DNA polymerase I and DNA ligase. (After B. Lewin, *Genes.* Copyright © 1983 by John Wiley.)

Numerous DNA glycosylases exist. One, uracil-DNA glycosylase, removes uracil from DNA. Uracil residues, which result from the spontaneous deamination of cytosine (Figure 16-8), can lead to a C → T transition if unrepaired. It is possible that the natural pairing partner of adenine in DNA is thymine (5-methyluracil), rather than uracil, to allow the recognition and excision of these uracil residues. If uracil were a normal constituent of DNA, such repair would not be possible.

There is also a glycosylase that recognizes and excises hypoxanthine, the deamination product of adenine. Other glycosylases remove alkylated bases (such as 3-methyladenine, 3-methylguanine, and 7-methylguanine), ring-opened purines, oxidatively damaged bases, and, in some organisms, UV photodimers. New glycosylases are still being discovered.

AP endonuclease repair pathway. All cells have endonucleases that attack the sites left after the spontaneous loss of single purine or pyrimidine residues. The **AP endonucleases** are vital to the cell, because, as noted earlier, spontaneous depurination is a relatively frequent event. These enzymes introduce chain breaks by cleaving the phosphodiester bonds at AP sites. This bond cleavage initiates an excision-repair process mediated by three further enzymes — an exonuclease, DNA polymerase I, and DNA ligase (Figure 16-30).

Owing to the efficiency of the AP endonuclease repair pathway, it can be the final step of other repair pathways. Thus, if damaged base pairs can be excised, leaving an AP site, the AP endonucleases can complete the restoration to the wild type. This is what happens in the DNA glycosylase repair pathway.

GO system. Two glycosylases, the products of the *mutM* and *mutY* genes, work in concert to prevent mutations arising from the 8-oxodG, or GO, lesion in DNA (see Figure 16-10). Together with the product of the *mutT* gene mentioned earlier, these glycosylases form the GO system. When GO lesions are generated in DNA by spontaneous oxidative damage, a glycosylase encoded by *mutM* removes the lesion (Figure 16-31). Still, some GO lesions persist and mispair with adenine. A second glycosylase, the product of the *mutY* gene, removes the adenine from this specific mispair, leading to restoration of the correct cytosine by repair synthesis (mediated by DNA polymerase I) and allowing subsequent removal of the GO lesion by the *mutM* product.

The *mutT* product prevents incorporation of GO across from A. The human counterparts of the *mutT*, *mutY*, and *mutM* gene products have been detected.

Postreplication repair

Mismatch repair. Some repair pathways are capable of recognizing errors even after DNA replication has already occurred. One such system, termed the **mismatch repair system,** can detect mismatches that occur in DNA replication. Suppose you were to design an enzyme system that

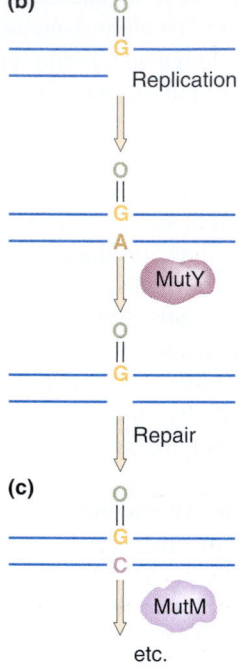

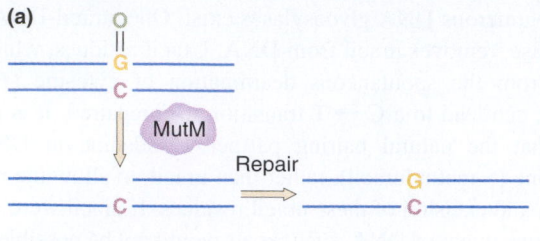

on the newly synthesized strand, so it is the base on this strand that must be recognized and excised.

To distinguish the old, template strand from the newly synthesized strand, the mismatch repair system in bacteria takes advantage of the normal delay in the postreplication methylation of the sequence

$$5'—G—A—T—C—3'$$
$$3'—C—T—A—G—5'$$

The methylating enzyme is **adenine methylase,** which creates 6-methyladenine on each strand. However, it takes the adenine methylase several minutes to recognize and modify the newly synthesized GATC stretches. During that interval, the mismatch repair system can operate because it can now distinguish the old strand from the new one by the methylation pattern. Methylating the 6-position of adenine does not affect base pairing, and it provides a convenient tag that can be detected by other enzyme systems. Figure 16-32 shows the replication fork during mismatch correction. Note that

Figure 16-31 The GO system. (a) 8-OxodG lesions are removed by the MutM protein, leaving an AP site that is repaired by endonucleases and repair synthesis. (b) However, when replicating polymerases are allowed to operate across from the lesion, they usually add an A residue. This mispair would result in the GC → TA change, but the MutY protein removes the A, allowing repair of the resulting AP site. (c) When repair polymerases operate across from the 8-oxodG lesion, they preferentially restore a C across from the lesion, allowing the MutM protein another opportunity to remove the lesion.

could repair replication errors. What would this system have to be able to do? At least three things:

1. Recognize mismatched base pairs.

2. Determine which base in the mismatch is the incorrect one.

3. Excise the incorrect base and carry out repair synthesis.

The second point is the crucial property of such a system. Unless it is capable of discriminating between the correct and the incorrect bases, the mismatch repair system could not determine which base to excise. If, for example, a G–T mismatch occurs as a replication error, how can the system determine whether G or T is incorrect? Both are normal bases in DNA. But replication errors produce mismatches

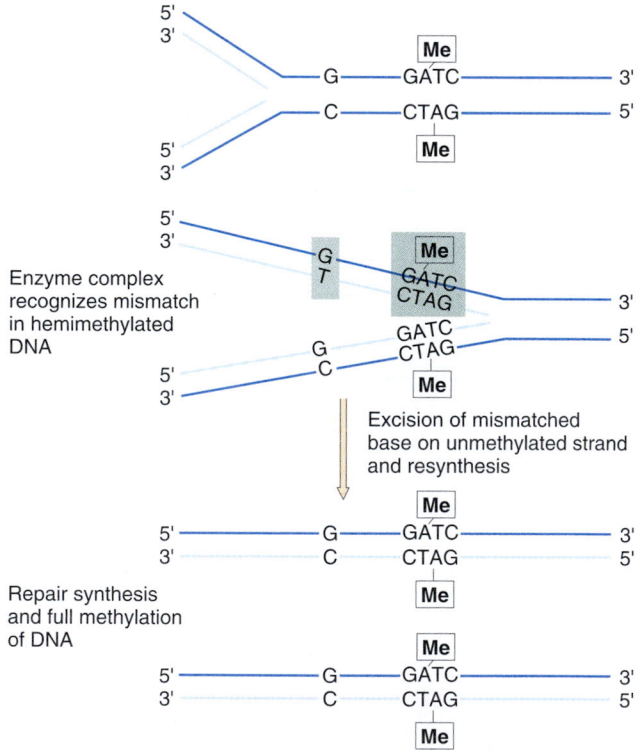

Figure 16-32 Model for mismatch repair in *E. coli.* Because DNA is methylated by enzymatic reactions that recognize the A in a GATC sequence, the newly synthesized strand will not be methylated directly after DNA replication. The *hemimethylated* DNA duplex serves as a recognition point for the mismatch repair system in discerning the old from the new strand. Here a G–T mismatch is shown. The mismatch repair system can recognize and bind to this mismatch, determine the correct (old) strand because it is the methylated strand of a hemimethylated duplex, and then excise the mismatched base from the new strand. Repair synthesis restores the normal base pair. (After E. C. Friedberg, *DNA Repair.* Copyright © 1985 by W. H. Freeman and Company.)

only the old strand is methylated at GATC sequences right after replication.

When the mismatched site has been identified, the mismatch repair system corrects the error. Figure 16-33 depicts a model of how the mismatch repair system carries out the correction in *E. coli*.

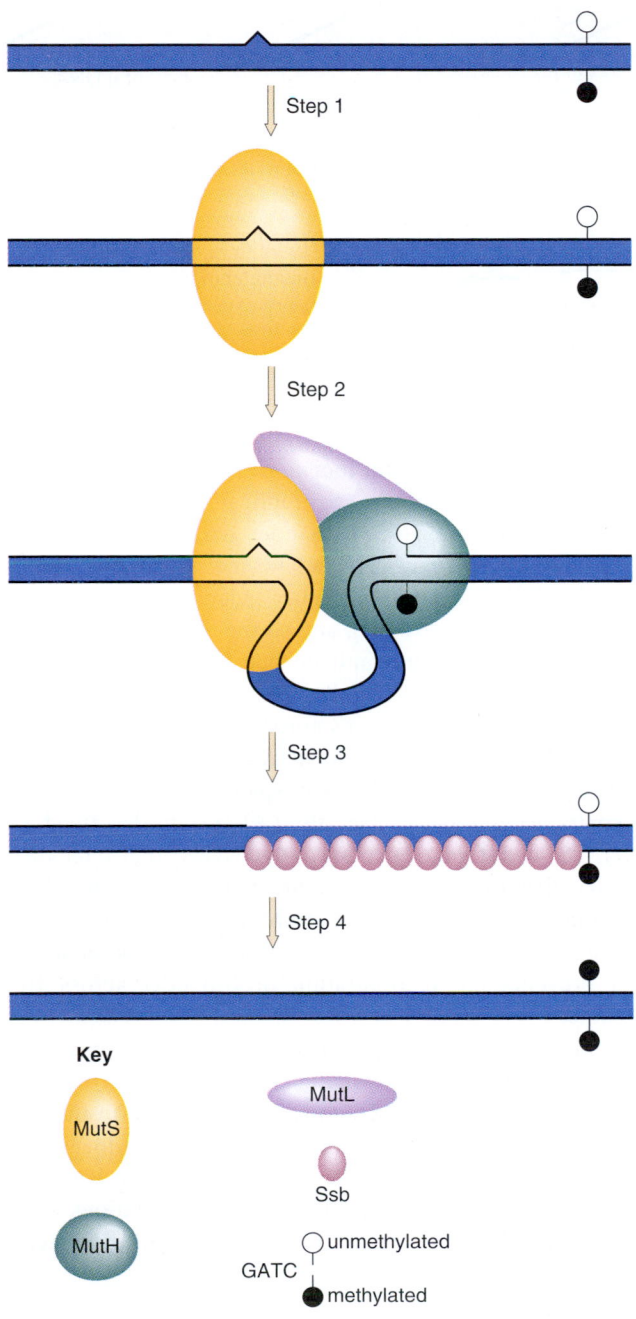

Key

MutS

MutL

MutH

Ssb

GATC ⊙ unmethylated
⬤ methylated

Figure 16-33   Steps in *E. coli* mismatch repair. (1) MutS binds to mispair. (2) MutH and MutL are recruited to form a complex. MutH cuts the newly synthesized (unmethylated) strand, and exonuclease degradation goes past the point of the mismatch, leaving a patch. (3) Single-strand-binding protein (Ssb) protects the single-stranded region across from the missing patch. (4) Repair synthesis and ligation fill in the gap. (From J. Jiricny, *Trends in Genetics,* Vol. 10. Elsevier Trends Journals, Cambridge, UK, 1995.)

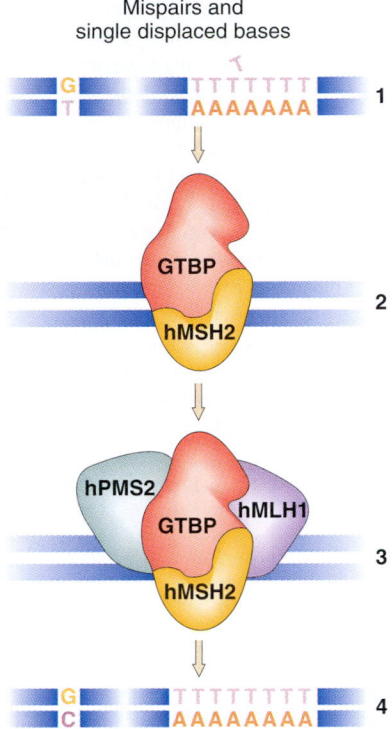

Mispairs and single displaced bases

Figure 16-34   Mismatch repair in humans. (1) Mispairs and misaligned bases arise in the course of replication. (2) The G–T-binding protein (GTBP) and the human MutS homolog (hMSH2) recognize the incorrect matches. (3) Two additional proteins, hPMS2 and hMLH1, are recruited and form a larger repair complex. (4) The mismatch is repaired after removal, DNA synthesis, and ligation. (From P. Karran, *Science* 268, 1995, 1857.)

The mismatch repair system has also been characterized in humans. Two of the proteins, hMSH2 and hMLH1, are very similar to their bacterial counterparts, MutS and MutL, respectively. Figure 16-34 depicts how the hMSH2 protein, together with the G–T-binding protein (GTBP), binds to the mismatches and then recruits the other components of the system, hPMS2 and hMLH1, to effect repair of the mismatch.

Recombinational repair.   The *recA* gene, which has a role in SOS bypass (Figure 16-19), also takes part in postreplication repair. Here the DNA replication system stalls at a UV photodimer or other blocking lesion and then restarts past the block, leaving a single-stranded gap. In recombinational repair, this gap is patched by DNA cut from the sister molecule (Figure 16-35a). This process seems to lead to few errors. SOS bypass, in contrast, is highly mutagenic, as described earlier. Here the replication system continues past the lesion (Figures 16-19 and 16-35b), accepting noncomplementary nucleotides for new strand synthesis.

Strategy for repair

We can now assess the overall repair system strategy used by the cell. The many different repair systems available to

Figure 16-35 Schemes for postreplication repair. (a) In recombinational repair, replication jumps across a blocking lesion, leaving a gap in the new strand. A *recA*-directed protein then fills the gap, using a piece from the opposite parental strand (because of DNA complementarity, this filler will supply the correct bases for the gap). Finally, the RecA protein repairs the gap in the parental strand. (b) In SOS bypass, when replication reaches a blocking lesion, the SOS system inserts the necessary number of bases (often incorrect ones) directly across from the lesion and replication continues without a gap. Note that with either pathway the original blocking lesion is still there and must be repaired by some other repair pathway. (After A. Kornberg and T. Baker, *DNA Replication*, 2d ed. Copyright © 1992 by W. H. Freeman and Company.)

(a)   Postreplication recombination repair

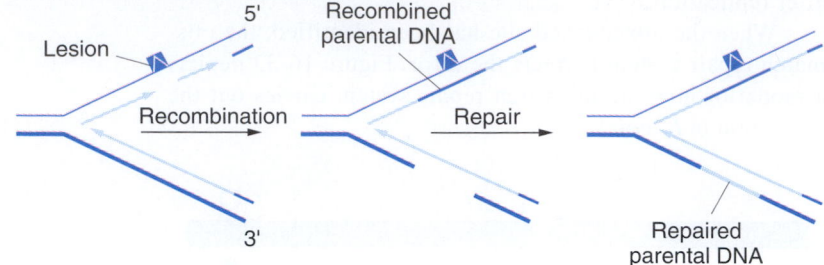

(b)   Error-prone (SOS) replication (lesion bypass)

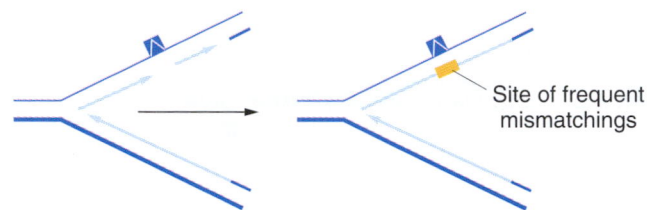

the cell are summarized in Table 16-2. It would be convenient if enzymes could be used to directly reverse each specific lesion. However, sometimes that is not chemically possible, and not every possible type of DNA damage can be anticipated. Therefore, a general excision repair system is used to remove any type of damaged base that causes a recognizable distortion in the double helix. When lesions are too subtle to cause such a distortion, specific excision systems, glycosylases, or removal systems are designed. To eliminate replication errors, a postreplication mismatch repair system operates; finally, postreplication recombinational systems eliminate gaps across from blocking lesions that have escaped the other repair systems.

A number of repair pathways are induced in response to damage, such as the SOS system (see Figure 16-19), and many of the proteins participating in the repair of alkylation damage were discussed previously.

> **MESSAGE** ···
> Repair enzymes play a crucial role in reducing genetic damage in living cells. The cell has many different repair pathways at its command to eliminate potentially mutagenic errors.

Mutators

As the preceding description of repair processes indicates, normal cells are programmed for error prevention. The repair processes are so efficient that the observed base-substitution rate is as low as 10^{-10} to 10^{-9} per base pair per cell per generation in *E. coli*. However, mutant strains with increased spontaneous mutation rates have been detected. Such strains are termed **mutators.** In many cases, the mutator phenotype is due to a defective repair system. In humans, these repair defects often lead to serious diseases.

In *E. coli,* the mutator loci *mutH, mutL, mutU,* and *mutS* affect components of the postreplication mismatch repair system (see Figure 16-33), as does the *dam* locus, which specifies the enzyme deoxyadenosine methylase. Strains that are Dam⁻ cannot methylate adenines at GATC sequences (see Figure 16-32), and so the mismatch repair system can no longer discriminate between the template and the newly synthesized strands. This failure to discriminate leads to a higher spontaneous mutation rate.

Mutations in the *mutY* locus result in GC → TA transversions, because many G–A mispairs and all 8-oxodG–A mispairs are unrepaired (see the GO system described earlier). The *mutM* gene encodes a glycosylase that removes 8-oxodG. Strains lacking *mutM* are mutators for the GC → TA transversion. Strains that are MutT⁻ have elevated rates of the AT → CG transversion, because they lack an activity that prevents the incorporation of 8-oxodG across from adenine.

Strains that are Ung⁻ are missing the enzyme uracil DNA glycosylase. These mutants cannot excise the uracil resulting from cytosine deaminations and, as a result, have elevated levels of C → T transitions. The *mutD* locus is responsible for a very high rate of mutagenesis (at least three orders of magnitude higher than normal). Mutations at this locus affect the proofreading functions of DNA polymerase III (Chapter 8).

Repair defects and human diseases

Several human genetic diseases are known or suspected to be due to repair defects. These defects often lead to an increased incidence of cancer. Table 16-3 summarizes information about these diseases, which are usually autosomal recessive disorders. We consider two examples.

16-2 TABLE		Repair Systems in *E. coli*	
General mode of operation	Example	Type of lesion repaired	Mechanism
Detoxification	Superoxide dismutase	Prevents formation of oxidative lesion	Converts peroxides into hydrogen peroxide, which is neutralized by catalase
Direct removal of lesions	Alkyltransferases	*O*-6-alkylguanine	Transfers alkyl group from *O*-6-alkylguanine to cysteine residue on transferase
	Photolyase	6-4 photoproduct	Breaks 6−4 bond and restores bases to normal
	Photolyase	UV photodimers	Splits dimers in the presence of white light
General excision	*uvrABC*-encoded exonuclease system	Lesions causing distortions in double helix, such as UV photoproducts and bulky chemical additions	Makes endonucleolytic cut on either side of lesion; resulting gap is repaired by DNA polymerase I and DNA ligase
Specific excision	AP endonucleases	AP sites	Makes endonucleolytic cut; exonuclease creates gap, which is repaired by DNA polymerase I and DNA ligase
	DNA glycosylases	Deaminated bases (uracil, hypoxanthine), certain methylated bases, ring-opened purines, oxydatively damaged bases; and certain other modified bases	Removes base, creating AP site, which is repaired by AP endonucleases
	GO system	8-oxodG	A glycosylase removes 8-oxodG from DNA; another glycosylase removes the A from 8-oxodG−A mispairs, leading to re-creation of an 8-oxodG−C pair, and the first glycosylase then removes the 8-oxodG
Postreplication	Mismatch repair system	Replication errors resulting in base-pair mismatches	Recognizes newly synthesized strand by detecting nonmethylated adenine residues in 5′-GATC-3′ sequences; then excises bases from the new strand when a mismatch is detected
	Recombinational repair	Lesions that block replication and result in single-stranded gaps	Recombinational exchange
	SOS system	Lesions that block replication	Allows replication bypass of blocking lesion, resulting in frequent mutations across from lesion

Xeroderma pigmentosum (XP) results from a defect in any of the genes (complementation groups) effecting nucleotide excision repair (see Figures 16-27 and 16-28). People suffering from this disorder are extremely prone to UV-induced skin cancers (Figure 16-36) as a result of exposure to sunlight and have frequent neurological abnormalities. The difference in UV photosensitivity between normal and diseased cells is evident from the survival curves in Figure 16-37.

Hereditary nonpolyposis colorectal cancer (HNPCC) is one of the most common inherited predispositions to

16-3 TABLE			Human Diseases with DNA-Repair Defects	
Disease	Sensitivity	Cancer susceptibility	Complementation groups	Symptoms
Ataxia telangiectasia	γ irradiation	Lymphomas	5	Ataxia, dilation of blood vessels (telangiectases) in skin and eyes, chromosome aberrations, immune dysfunction
Bloom syndrome	Mild alkylating agents	Carcinomas, leukemias, lymphomas	1	Photosensitivity, facial telangiectases, chromosome alterations
Cockayne syndrome	UV light		2	Dwarfism, retinal atrophy, photosensitivity, progeria, deafness, trisomy 10
Fanconi anemia	Cross-linking agents	Leukemias	3	Hypoplastic pancytopenia, congenital anomalies
Xeroderma pigmentosum	UV light, chemical mutagens	Skin carcinomas and melanomas	8	Skin and eye photosensitivity, keratoses
HNPCC		Colon, ovary	4	Early development of tumors

Note: Other human hereditary disorders that may be related to DNA-repair defects include dyskeratosis congenita (Zinsser-Cole-Engman syndrome), progeria (Hutchinson-Gilford syndrome), and trichothiodystrophy. HNPCC, hereditary nonpolyposis colorectal cancer.
Source: After A. Kornberg and T. Baker, *DNA Replication,* 2d ed. Copyright © 1992 by W. H. Freeman and Company.

cancer, affecting as many as 1 in 200 people in the Western world. Studies have shown that this syndrome results from a loss of the mismatch repair system. Most HNPCC results from a defect in genes that encode the human counterparts (and homologs) of the bacterial MutS and MutL proteins (see Figure 16-34). The inheritance of HNPCC is autosomal dominant. Cells with one functional copy of the mismatch repair genes have normal mismatch repair activity, but tumor cell lines arise from cells that have lost the one functional copy and are thus mismatch repair deficient. These cells display high mutation rates that eventually result in tumor growth and proliferation.

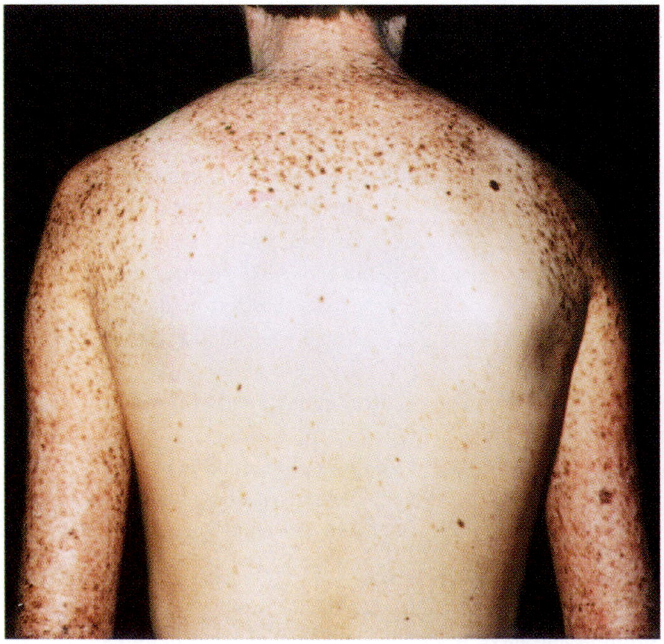

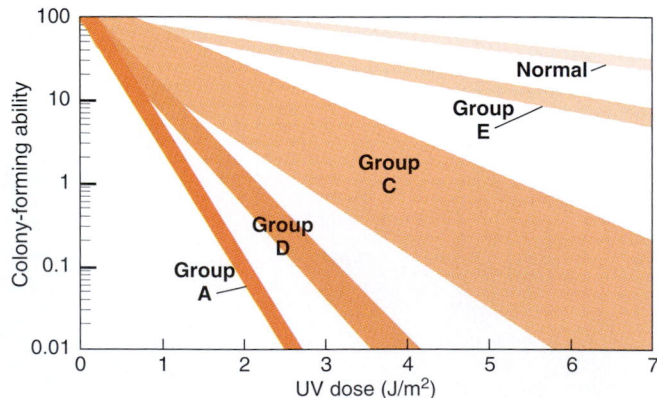

Figure 16-36 Skin cancer in xeroderma pigmentosum. This recessive hereditary disease is caused by a deficiency in one of the excision-repair enzymes, which leads to the formation of skin cancers on exposure of the skin to the UV rays in sunlight. (Photograph courtesy of Dirk Bootsma, Erasmus University, Rotterdam.)

Figure 16-37 Hypersensitivity to UV radiation of XP cells in culture. Here the cells from a number of complementation groups are shown. There is a variation between complementation groups, but all are more sensitive to UV radiation than are normal cells. (After J. E. Cleaver and K. H. Kraemer, Xeroderma pigmentosum, in C. R. Scriver et al., eds., *The Metabolic Basis of Inherited Disease.* Copyright © 1989 by McGraw-Hill Book Co., New York.)

SUMMARY

Gene mutations can arise through many different processes. Spontaneous mutations can result from replication errors or from spontaneous lesions, such as those generated by deamination or depurination. (Recombination and transposable elements also can result in altered genes, as described in Chapters 19 and 20.) Mutagens can increase the frequency of mutations. Some of these agents act by mimicking a base and then mispairing in the course of DNA replication. Others alter bases in the DNA and convert them into derivatives that mispair. A third class of mutagens, which includes most carcinogens, damages DNA in such a way that replication is blocked. The activation of an enzymatic pathway, termed the SOS system, is required to replicate past the blocking lesions, resulting in mutations across from the blocking lesion.

Molecular studies of mutagenesis in bacteria can identify agents responsible for human cancers. DNA sequencing carried out on mutations in the tumor-suppressor genes of patients with liver or skin cancer has identified changes that closely resemble the characteristic mutational changes caused by specific bacterial mutagens.

Our knowledge of the molecular basis of mutation can be exploited for useful purposes. One example is the Ames test, which utilizes mutant bacterial strains to test compounds in the environment for mutagenic activity. Owing to the correlation between mutagenicity and carcinogenicity, the identification of potential carcinogens in the environment can be achieved by this rapid assay.

Repair enzymes present in living cells greatly minimize genetic damage, thus preventing many mutations. Mutant organisms lacking certain repair enzymes have higher than normal mutation rates. In humans, repair deficiency leads to a variety of diseases and cancer susceptibilities.

CONCEPT MAP

Draw a concept map interrelating as many of the following terms as possible. Note that the terms are listed in no particular order.

mutation / auxotroph / transition / frameshift / transversion / mutagen / photoreactivation / mutator / SOS / genetic disease / amino acid

CHAPTER INTEGRATION PROBLEM

In Chapter 10, we learned that amber and ochre codons were two of the chain-terminating nonsense triplets. Based on the specificity given in the test for aflatoxin B_1 and ethyl methanesulfonate (EMS), describe whether each mutagen would be able to revert amber (UAG) and ochre (UAA) codons back to wild type.

◆ Solution ◆

EMS induces primarily GC → AT transitions. UAG codons could not be reverted back to wild type, because only the UAG → UAA change would be stimulated by EMS and that generates a nonsense (ochre) codon. UAA codons would not be acted on by EMS. Aflatoxin B_1 induces primarily GC → TA transversions. Only the third position of the UAG codons would be acted on, resulting in a UAG → UAU change (at the mRNA level), which produces tyrosine. Therefore, if tyrosine were an acceptable amino acid at the corresponding site in the protein, aflatoxin B_1 could revert UAG codons. Aflatoxin B_1 would not revert UAA codons, because no G–C base pairs appear at the corresponding position in the DNA.

SOLVED PROBLEM

1. A mutant of *E. coli* is highly resistant to mutagenesis by a variety of agents, including ultraviolet light, aflatoxin B_1, and benzo(a)pyrene. Explain one possible cause of this mutant phenotype.

◆ Solution ◆

The mutant might lack the SOS system and perhaps carry a defect in the *umuC* gene. Such strains would not be able to bypass replication-blocking lesions of the type caused by the three mutagens listed. Without the processing of premutational lesions, mutations would not be recovered in viable cells.

PROBLEMS

1. The continuous sequence presented here (see the following page), which is not necessarily translated, is from one of the two DNA strands in a genetic region. From the sequence, find the mutations that would be expected to occur most frequently at the DNA level spontaneously (without the aid of mutagens). Explain what each mutation is and why it would be frequent.

```
A A G G C T A G C T T T A G G A G A T C C C G
A T C T C A A A G C T A T C T A G C T T T A G G
T A T A T A G A T C T A T G C T C T C T G A T C
T A G C A T C C C T A G C A T C A T A T C G G G
A T C C T A C G A A T C T T T G A T C G G T A T
C G G G A T A C G T A T G A A G G C T A G C C
T C A T C C A T C C A T C C A A G C T T A A T A
T C G A T C G G A T C C T G G A A T T G G A T T
C C C A G A G A T C T T T T T A G C T A G C T C
C C G C C T A G C T T T C G G A G C T T A A T C
C T A A T G A G C A A C C A C C G G T A T A T A
G C C A A T A C A A G C C G G A T T C G G G A
T C C T T A G G A T C C A A G A A T C T T C G A
G A A A T C G G A T C G G G A T C T T A T C G T
C T C C A A G A A T C T C C T A G T T C C A A T
C T T T A T C G G A T C G G A A G G C T A T
```

(Problem 1 from *Discovering Molecular Genetics* by Jeffrey H. Miller. Cold Spring Harbor Laboratory Press, Cold Spring Harbor, NY, 1996.)

2. Describe the Streisinger model for frameshift formation. Show how this model can explain mutational hot spots in the *lacI* gene of *E. coli*.

3. Diagram two different mechanisms for deletion formation. How do DNA-sequencing experiments suggest these possibilities?

4. The bases in the DNA of *Escherichia classicus* are named after the musical keys A, B, C, and E. Here, A pairs with E, and B pairs with C. Thus, there are A–E base pairs and B–C base pairs in the double-stranded DNA.

   You have the following eight sequences read 5′ to 3′. Here, just a single strand of DNA is given; the second line is a continuation of the first line for each entry.

   (1) A B B E C E A A B B C B B A C C B C B A A
   C A B A C C C A B A E A C B E A A C A E B E B B
   C E B A C A E E C A A E C E C

   (2) B C E E C A A C A B C E C B A E C B E C E B
   C E E A B C A A C C B E B A B C B B E A B C A E
   A C E B A E E A B B E A C B

   (3) B B E C A A A C B A C B E C A B C E A A B
   C B E C B C A C C E E C A B B E A A A B A E A C
   A E A E C B E C B A C B A E A

   (4) A C C A E A C B B B C E E A B A E C B A C C
   C B A C C C B A C C C B A C C C B A C C C E B C
   A A C B A E A B C E B A E C

   (5) C C B E C E E B A A E C E E A B B C E A B C
   E B C A B A E C A E B B C B A A E B A C C B E B
   E A C C B A E B C E A E C A

   (6) B C E A A E E C A B B E B C E C C A E C A A
   B C B B C E E C B A C A E C C E C B C E A A E E
   C A B B E B C C A E

5. (7) A A E C B B A B C C E B A C B C E A E B E E
   A E C C A C E C B E A E C A C B B E A A C B A A
   C E B C E A C A E B B C C E

   (8) B B E A C C A B A C A E B E A C A A C B E E
   C E A B C E B A E C C E A B B E A B C E C C A B
   C C B E A A E A C E B C E C

   a. Which of the eight sequences would have the biggest hot spot for frameshift sequences?

   b. Which of the eight sequences would have the biggest hot spot for deletions of more than a few base pairs?

5. Suppose that an organism from outer space was analyzed and found to have DNA with six bases, J, K, L, M, P, and Q. These bases have specific pairing partners; so, in the DNA double helix, J pairs with K (J–K), L pairs with M (L–M), and P pairs with Q (P–Q).

   a. Write out a sequence of a single strand of this DNA with more than 30 bases, and build into it a hot spot for small mutations (of only a few base pairs or less), and explain why this sequence would be a hotspot relative to other sequences.

   b. If you had to design a 1000-base-pair gene with these six bases, what principles would you use to prevent deletions of different segments from occurring with high frequency throughout the gene. This includes not only short deletions but also long deletions (of greater than 50 base pairs).

6. Describe two spontaneous lesions that can lead to mutations.

7. Compare the mechanism of action of 5-bromouracil (5-BU) with ethyl methanesulfonate (EMS) in causing mutations. Explain the specificity of mutagenesis for each agent in light of the proposed mechanism.

8. Fill in the following table, using a plus sign (+) to indicate that the mutagenic lesion (base damage) induces the indicated base change and a minus sign (−) if it does not.

| | LESION | | |
Base Change	O^{-6}- Methy G	8-OxodG	C–C photodimer
AT → GC			
GC → TA			
GC → AT			

9. Compare the two different systems used to repair AP sites and to remove bulky chemical adducts.

10. Describe the repair systems that operate after replication.

11. Normal *(tight)* auxotrophic mutants will not grow at all in the absence of the appropriate supplement to the

medium. However, in mutant hunts for auxotrophic mutants, it is common to find some mutants (called *leaky*) that grow very slowly in the absence of the appropriate supplement but normally in the presence of the supplement. Propose an explanation for the molecular nature of the leaky mutants.

12. Strain A of *Neurospora* contains an *ad-3* mutation that reverts spontaneously at a rate of 10^{-6}. Strain A is crossed with a newly acquired wild-type isolate, and Ad-3 strains are recovered from the progeny. When 28 different Ad-3 progeny strains are examined, 13 lines are found to revert at the rate of 10^{-6}, but the remaining 15 lines revert at the rate of 10^{-3}. Formulate a hypothesis to account for these findings, and outline an experimental program to test your hypothesis.

13. **a.** Why is it impossible to induce nonsense mutations (represented at the mRNA level by the triplets UAG, UAA, and UGA) by treating wild-type strains with mutagens that cause only AT $\rightarrow$ GC transitions in DNA?

b. Hydroxylamine (HA) causes only GC $\rightarrow$ AT transitions in DNA. Will HA produce nonsense mutations in wild-type strains?

c. Will HA treatment revert nonsense mutations?

14. Several auxotrophic point mutants in *Neurospora* are treated with various agents to see if reversion will occur. The following results were obtained (a plus sign indicates reversion):

Mutant	5-BU	HA	Proflavin	Spontaneous reversion
1	−	−	−	−
2	−	−	+	+
3	+	−	−	+
4	−	−	−	+
5	+	+	−	+

a. For each of the five mutants, describe the nature of the original mutation event (not the reversion) at the molecular level. Be as specific as possible.

b. For each of the five mutants, name a possible mutagen that could have caused the original mutation event. (Spontaneous mutation is not an acceptable answer.)

c. In the reversion experiment for mutant 5, a particularly interesting prototrophic derivative is obtained. When this type is crossed with a standard wild-type strain, the progeny consists of 90 percent prototrophs and 10 percent auxotrophs. Provide a full explanation for these results, including a precise reason for the frequencies observed.

15. You are using nitrous acid to "revert" mutant *nic-2* alleles in *Neurospora*. You treat cells, plate them on a medium without nicotinamide, and look for prototrophic colonies. You obtain the following results for two mutant alleles. Explain these results at the molecular level, and indicate how you would test your hypotheses.

a. With *nic-2* allele 1, you obtain no prototrophs at all.

b. With *nic-2* allele 2, you obtain three prototrophic colonies, and you cross each separately with a wild-type strain. From the cross prototroph A × wild type, you obtain 100 progeny, all of which are prototrophic. From the cross prototroph B × wild type, you obtain 100 progeny, of which 78 are prototrophic and 22 are nicotinamide requiring. From the cross prototroph C × wild type, you obtain 1000 progeny, of which 996 are prototrophic and 4 are nicotinamide requiring.

16. Devise imaginative screening procedures for detecting the following mutants:

a. Nerve mutants in *Drosophila*.

b. Mutants lacking flagella in a haploid unicellular alga.

c. Supercolossal-sized mutants in bacteria.

d. Mutants that overproduce the black compound melanine in normally white haploid fungus cultures.

e. Individual humans (in large populations) whose eyes polarize incoming light.

f. Negatively phototrophic *Drosophila* or unicellular algae.

g. UV-sensitive mutants in haploid yeast.

17. For each lesion in parts a–g, indicate which of the following repair systems repairs that lesion.
   (1) alkyl transferase
   (2) endonuclease
   (3) photolyase
   (4) MutY glycosylase
   (5) MutM glycosylase
   (6) uracil DNA glycosylase
   (7) general nucleotide excision repair
   (8) methyl-directed mismatch repair

a. Deamination of cytosine

b. 8-oxodG

c. Aflatoxin B_1 adduct

d. G–T mispair as replication error

e. 5'-CC-3' dimer

f. AP site

g. O^6-methyl guanine

17

CHROMOSOME MUTATION I: CHANGES IN CHROMOSOME STRUCTURE

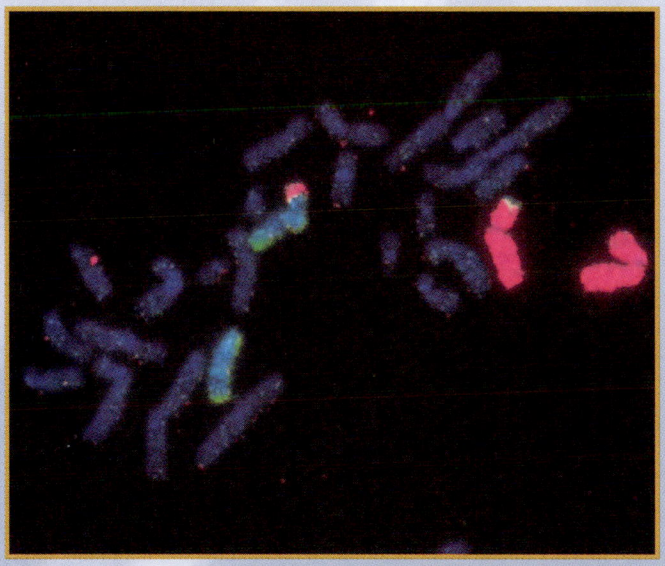

A reciprocal translocation demonstrated by a technique called chromosome painting.
A suspension of chromosomes from many cells is passed through an electronic device that sorts them by size. DNA is extracted from individual chromosomes, joined to one of several fluorescent dyes, then added to partly denatured chromosomes on a slide. The fluorescent DNA "finds" its own chromosome and binds along its length to paint it. In this preparation, a light-blue and a pink dye have been used to paint different chromosomes. The preparation shows one normal pink chromosome, one normal blue, and two that have exchanged their tips. *(Lawrence Berkeley Laboratory.)*

Key Concepts

Owing to the strong meiotic pairing affinity of homologous regions, diploids with one standard and one rearranged chromosome set produce pairing structures that have shapes and properties unique to that rearrangement.

A deletion in one chromosome set is generally deleterious as a result of gene imbalance and the unmasking of deleterious alleles in the other chromosome set.

Duplications can lead to gene imbalance but also provide extra material for evolutionary divergence.

Heterozygous inversions show reduced fertility and reduced recombination in the region spanned by the inversion.

A heterozygous translocation shows 50 percent sterility and linkage of genes on the chromosome involved in the translocation.

Chromosome mutation is the process of change that results in rearranged chromosome parts, abnormal numbers of individual chromosomes, or abnormal numbers of chromosome sets. As with gene mutation, the term *chromosome mutation* is applied both to the process and to the product, so the novel genomic arrangements may be called *chromosome mutations.* Sometimes chromosome mutation can be detected by microscopic examination, sometimes by genetic analysis, and sometimes by both. In contrast, gene mutations are never detectable microscopically on the chromosome; a chromosome bearing a gene mutation looks the same under the microscope as one carrying the wild-type allele.

Many chromosome mutations lead to abnormalities in cell and organismal function. There are two basic reasons for this outcome. First, the chromosome mutations can result in abnormal gene number or position. Second, if chromosome mutation includes chromosome breakage, which is often the case, then the break may occur in the middle of a gene, thereby disrupting its function.

Chromosome mutations are important at several different levels of biology. First, in research, they provide ways of designing special arrangements of genes, uniquely suited to answer certain biological questions. Second, chromosome mutations are important at the applied level, especially in medicine and in plant and animal breeding. Finally, chromosome mutations have been instrumental in shaping genomes as part of the evolutionary process.

Let's consider some important properties of chromosomes that are useful in understanding structural chromosome mutations.

1. In prophase I of meiosis, homologous regions of chromosomes have a strong pairing affinity and, if necessary, will contort to pair. Consequently, many curious structures may be seen in a cell that has one standard set of chromosomes and one aberrant set. Remember that in polytene chromosomes the homologs also pair (even though they are not in meiotic cells), so comparable shapes result.

2. Changes in structure are usually due to chromosome breakage, and the broken chromosome ends are highly "reactive," tending strongly to join with other broken ends. The telomeres (the regular chromosome ends), however, do not tend to join.

3. In a diploid, if parts of chromosomes are lost or gained, the result is almost always lethal. The chromosome set is exquisitely sensitive to changes in gene content, even if one complete set is present. Examples of such disturbed gene balance will appear throughout this chapter and the next.

Origin of changes in chromosome structure

A chromosome rearrangement can occur de novo in a cell of any tissue, and the result will be a heterozygous karyotype

consisting of one normal set plus one rearranged set. Rearrangements occurring in somatic tissue may have phenotypic effects in one cell or a somatic sector of cells. Rearrangements occurring in germinal tissue may generate heterozygous meiocytes. In this case, the original rearrangement may enter a proportion of the gametes; but, in addition, various kinds of abnormal meiotic chromosome segregations can occur, resulting in novel types of chromosome rearrangements, and they also will enter the gametes. If a rearrangement takes place de novo within the gamete population, then it too may end up in one of the offspring.

Gametes with rearrangements would most likely unite with normal gametes from the opposite sex, so the zygote and all the cells of the F_1 individual developing from this zygote would be heterozygous for the rearrangement. The potential phenotypic effect of the rearrangement therefore is much more severe. Furthermore, *all* the meiocytes in such an individual will be heterozygous for the rearrangement, and a large proportion of gametes would contain the parental or a derived rearrangement.

Most of the following discussions will concern individuals whose entire body is heterozygous for a rearrangement: in these individuals, the rearrangement may have occurred de novo in parental gonads or gametes or may have been derived from a parent whose body also was heterozygous in every cell.

Types of changes

In discussions of chromosome rearrangements, it is convenient to use letters to represent different chromosome regions. These letters therefore represent large segments of DNA, each containing many genes.

The simple loss of a chromosomal segment is called a **deletion** or **deficiency.** In the following diagram, region B has been deleted:

A B C D E F → A C D E F

The presence of two copies of a chromosomal region is called a **duplication:**

A B B C D E F

A segment of a chromosome can rotate 180 degrees and rejoin the chromosome, resulting in a chromosomal mutation called an **inversion:**

A E D C B F

Finally, two nonhomologous chromosomes can exchange parts to produce a chromosomal mutation called a **translocation:**

A B C D J K   G H I E F

Mechanisms of change

Chromosome rearrangements can arise through physical breakage and rejoining of the DNA molecule that constitutes the chromosome. Such processes can either occur spontaneously or be induced by treatment with high-energy radiation such as X or γ radiation. Another mechanism is by "illegitimate" crossing-over between repetitive elements in the genome. This type of crossover takes place in somatic cells after illegitimate pairing between the repeated segments. The two mechanisms are compared in Figure 17-1. We shall consider the properties of each of the chromosomal rearrangements in turn.

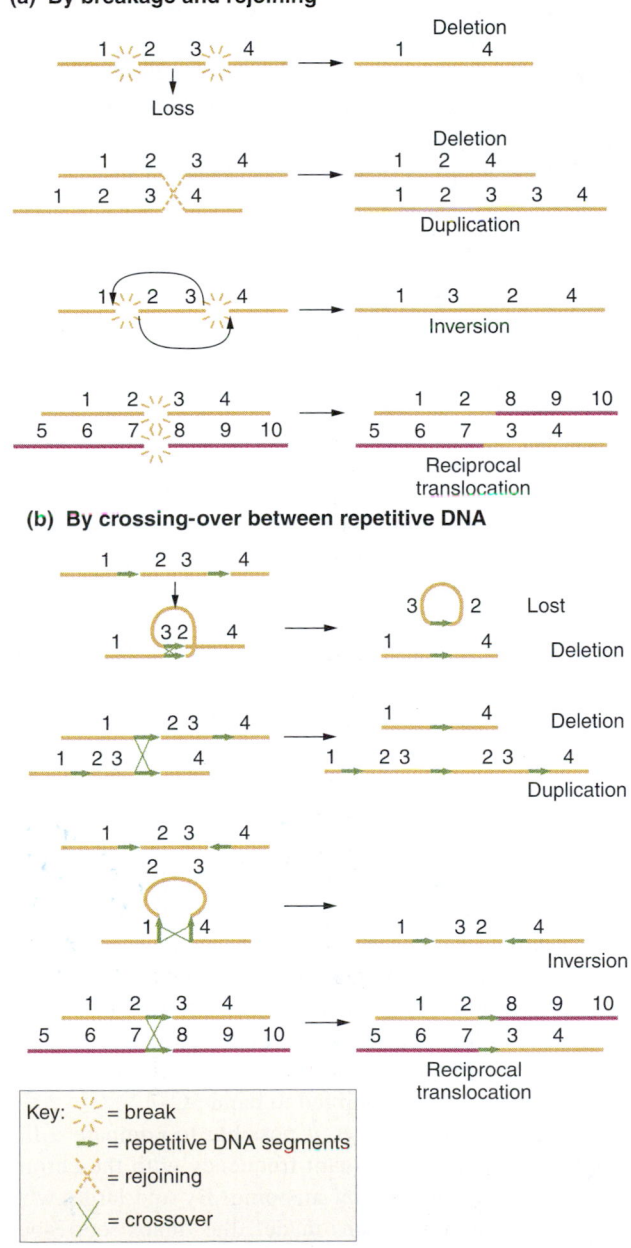

(a) By breakage and rejoining

(b) By crossing-over between repetitive DNA

Key: = break
= repetitive DNA segments
= rejoining
= crossover

Figure 17-1   Origins of chromosomal rearrangements.

Deletions

The process of spontaneously occurring deletion must include two chromosome breaks to cut out the intervening segment. If the two ends join and one of them bears the centromere, a shortened chromosome results, which is said to carry a deletion. The deleted fragment is acentric; consequently it is immobile and will be lost. An effective mutagen for inducing chromosomal rearrangements of all kinds is ionizing radiation. This kind of radiation, of which X rays and γ rays are examples, is highly energetic and causes chromosome breaks. The way in which the breaks rejoin determines the kind of rearrangement produced. Two types of deletion are possible. Two breaks can produce an **interstitial deletion,** as shown in Figure 17-2. In principle, a single break can cause a **terminal deletion;** but, because of the need for the special chromosome tips (telomeres), it is likely that apparently terminal deletions include two breaks, one close to the telomere.

The effects of deletions depend on their size. A small deletion within a gene, called an **intragenic deletion,** inactivates the gene and has the same effect as other null mutations of that gene. If the homozygous null phenotype is viable (as, for example, in human albinism), then the homozygous deletion also will be viable. Intragenic deletions can be distinguished from single nucleotide changes because they are nonrevertible.

For most of this section, we shall be dealing with **multigenic deletions,** those that remove from two to several thousand genes. Multigenic deletions have severe consequences. If by inbreeding such a deletion is made homozygous (that is, if both homologs have the same deletion), then the combination is almost always lethal. This outcome suggests that most regions of the chromosomes are essential for normal viability and that complete elimination of any segment from the genome is deleterious. Even individuals heterozygous for a multigenic deletion — those with one normal homolog and one that carries the deletion — may not survive. There are several possible reasons for this failure to

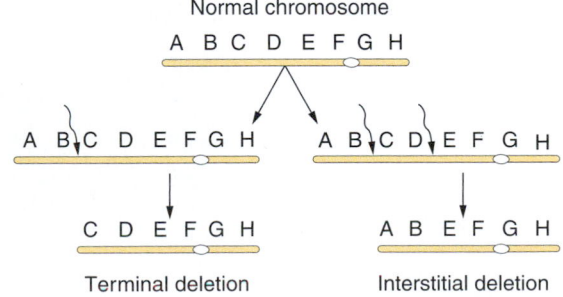

Normal chromosome

Terminal deletion   Interstitial deletion

Figure 17-2   Terminal and interstitial deletions. Chromosome can be broken when struck by ionizing radiation (wavy arrows). A terminal deletion is the loss of the end of a chromosome. An interstitial deletion results after two breaks are induced if the terminal part (AB) rejoins the main body of the chromosome, with the acentric fragment (CD) being lost.

survive. First, a genome has been "fine tuned" during evolution to require a specific balance of genes, and the deletion upsets this balance. We shall encounter this balance notion several times in this chapter and the next, because several different types of chromosome mutations upset the ratio, or balance, of genes in a genome. Second, in many organisms there are recessive lethal and other deleterious mutations throughout the genome. If "covered" by wild-type alleles on the other homolog, these recessives are not expressed. However, a deletion can "uncover" recessives, allowing their expression at the phenotypic level.

MESSAGE ···

The lethality of heterozygous deletions can be explained by genome imbalance and by unmasking of recessive lethal alleles.

Nevertheless, some small deletions are viable in combination with a normal homolog. In these cases, the deletion can sometimes be identified by cytogenetic analysis. If meiotic chromosomes are examined in an individual carrying a heterozygous deletion, the region of the deletion can be determined by the failure of the corresponding segment on the normal homolog to pair, resulting in a **deletion loop** (Figure 17-3a). In insects, deletion loops are detected in the polytene chromosomes, in which the homologs are fused (Figure 17-3b). A deletion can be assigned to a specific chromosome location by determining which chromosome shows the deletion loop and the position of the loop along the chromosome.

Deletions of some chromosomal regions produce their own unique phenotypes. A good example is a deletion of one specific small chromosome region of *Drosophila*. When one homolog carries the deletion, the fly shows a unique notch-wing phenotype, so the deletion acts as a dominant mutation in this regard. But the deletion is lethal when homozygous and therefore acts as a recessive in regard to its lethal effect. The specific dominant phenotypic effect of certain deletions might be caused by one of the chromosome

breaks being inside a gene, which, when disrupted, will act as a dominant mutation.

What are the genetic properties of deletions? In addition to cytogenetic criteria, there are several purely genetic criteria for inferring the presence of a deletion. These criteria are particularly useful in species whose chromosomes are not easily analyzed cytogenetically.

Two genetic criteria we have encountered already. The first is the failure of the chromosome to survive as a homozygote; however, this effect could also be produced by any lethal mutation. Second, chromosomes with deletions can never revert to a normal condition. This criterion is useful only if there is some specific phenotype associated with the deletion.

A third criterion is that, in heterozygous deletions, recombinant frequencies between genes flanking the deletion are lower than in control crosses. This makes intuitive sense because part of the region contains an unpaired chromosomal region, which cannot participate in crossing-over. We will see that inversions have a similar effect on recombinant frequencies but can be distinguished in other ways.

A fourth criterion for inferring the presence of a deletion is that deletion of a segment on one homolog sometimes unmasks recessive alleles present on the other homolog, leading to their unexpected expression. Consider, for example, the deletion shown in the following diagram:

$$\left.\begin{array}{l} \underline{a\ b\ c\quad d\ e\ f} \\[6pt] \underline{a^+ \quad\quad d^+\ e^+\ f^+} \end{array}\right\} \text{Phenotype is } a^+\ b\ c\ d^+\ e^+\ f^+$$

In this case, none of the six recessive alleles is expected to be expressed, but, if b and c are expressed, then a deletion is suggested to have occurred on the other homolog spanning the b^+ and c^+ loci. Because in such cases it seems that recessive alleles are showing dominance, the effect is called **pseudodominance.**

The pseudodominance effect can also be used in the opposite direction. A known set of overlapping deletions is used to locate the map positions of new mutant alleles. This procedure is called **deletion mapping.** An example from the fruit fly *Drosophila* is shown in Figure 17-4. In this diagram, the recombination map is shown at the top, marked with distances in map units from the left end. The horizontal bars below the chromosome show the extent of the deletions identified at the left. The mutation prune (*pn*), for example, shows pseudodominance only with deletion 264-38, which determines its location in the 2D-4 to 3A-2 region. However, *fa* shows pseudodominance with all but two deletions, so its position can be pinpointed to band 3C-7.

Deletion analysis makes it possible to compare a linkage map based on recombinant frequency with the chromosome map based on deletion mapping. By and large, where this comparison has been made, the maps correspond well—a satisfying cytological endorsement of a purely genetic creation.

(a) Meiotic chromosomes   **(b) Polytene chromosomes**

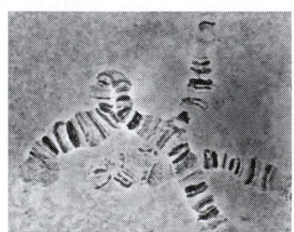

Figure 17-3   Looped configurations in a *Drosophila* deletion heterozygote. In the meiotic pairing, the normal homolog forms a loop. The genes in this loop have no alleles with which to synapse. Because polytene chromosomes in *Drosophila* have specific banding patterns, we can infer which bands are missing from the homolog with the deletion by observing which bands appear in the loop of the normal homolog. (Part b from William M. Gelbart.)

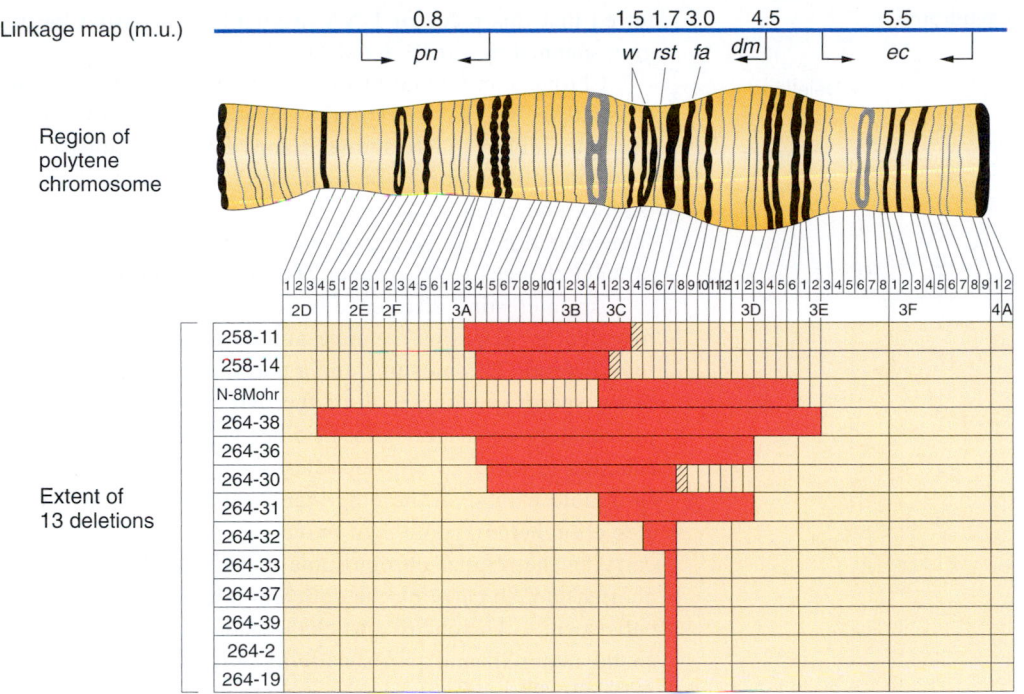

Figure 17-4 Locating genes to chromosomal regions by observing pseudodominance in *Drosophila* heterozygous for deletion and normal chromosomes. The red bars show the extent of the deleted segments in 13 deletions. All recessive alleles spanned by a deletion will be expressed.

MESSAGE ·································

Chromosome maps made by analyzing deletion coverage are congruent with linkage maps made by analyzing recombinant frequency.

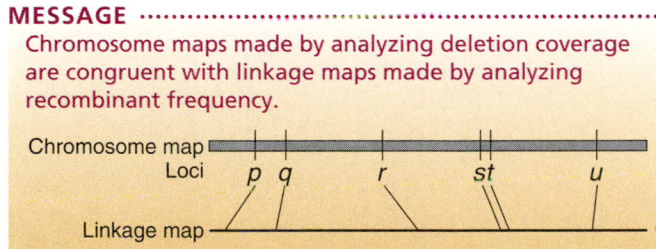

Moreover, pseudodominance can be used to map a small deletion that cannot be visualized microscopically. Let's consider an X chromosome in *Drosophila* that carries a recessive lethal suspected of being a deletion; we call this chromosome "X*." We can cross X*-bearing females with males carrying recessive alleles of loci on that chromosome. For example, a map of loci in the tip region is

y	dor	br	gt	swa	w	rst	vt
0.3	0.3	0.3	0.4	0.2	0.2	0.6	

Suppose we obtain all wild-type flies in crosses between X*/X females and males carrying *y, dor, br, gt, rst,* and *vt* but obtain pseudodominance of *swa* and *w* with X* (that is, X*/*swa* shows the recessive swa phenotype and X*/*w* shows the recessive w phenotype). Then we have good genetic evidence for a deletion of the chromosome that includes at least the *swa* and *w* loci but not *gt* or *rst*.

MESSAGE ·································

Deletions are recognized genetically by (1) reduced RF, (2) pseudodominance, (3) recessive lethality, and (4) lack of reverse mutation and cytologically by (5) deletion loops.

Clinicians regularly find deletions in human chromosomes. In most cases, the deletions are relatively small, but they nevertheless have an adverse phenotypic effect, even though heterozygous. Deletions of specific human chromosome regions cause unique syndromes of phenotypic abnormalities. An example is the *cri du chat* syndrome, caused by a heterozygous deletion of the tip of the short arm of chromosome 5 (Figure 17-5). It is the convention to call the short arm of a chromosome p and to call the long arm q. The specific bands deleted in cri du chat syndrome are 5p15.2 and 5p15.3, the two most distal bands identifiable on 5p. The most characteristic phenotype in the syndrome is the one that gives it its name, the distinctive catlike mewing cries made by infants with this deletion. Other phenotypic manifestations of the syndrome are microencephaly (abnormally small head) and a moonlike face. Like syndromes caused by other deletions, the cri du chat syndrome also includes mental retardation.

Most human deletions, such as those that we have just considered, arise spontaneously in the germ line of a normal parent of an affected person; thus no signs of the deletions are found in the somatic chromosomes of the parents. However, as we shall see in a later section, some human deletions are produced by meiotic irregularities in a parent heterozygous for another type of rearrangement. Cri du chat syndrome, for example, can result from a parent heterozygous for a translocation.

Geneticists have mapped human genes from deletions by using a molecular technique called **in situ hybridization.** This technique was introduced in Chapters 3 and 6, but for now we can review the basics to show the usefulness of deletions. If an interesting gene or other DNA fragment has been isolated with the use of modern molecular technology,

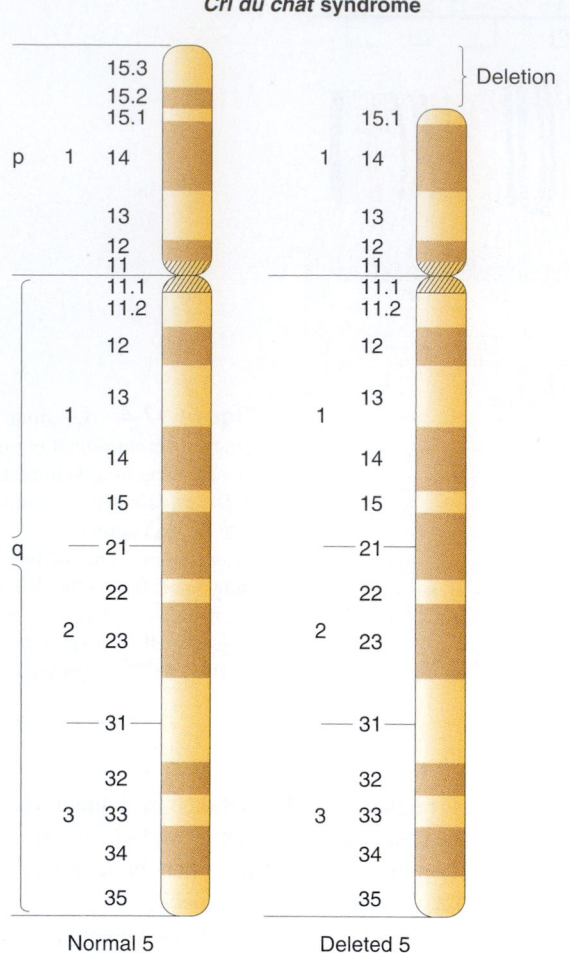

***Cri du chat* syndrome**

Normal 5   Deleted 5

Figure 17-5   The cause of the *cri du chat* syndrome of abnormalities in humans is loss of the tip of the short arm of one of the homologs of chromosome 5.

ferred that this piece of DNA originally came from the region spanned by 11q13.5 and 11q21.

Chromosome mutations often arise in cancer cells, and we shall see several cases in this chapter and the next. As an example, Figure 17-8 shows some deletions consistently found in solid tumors. Not all the cells in a tumor show the deletion indicated, and often a mixture of different chromosome mutations can be found in one tumor. The contribution of such changes to the cancer phenotype is not understood.

An interesting difference between animals and plants is revealed by deletions. A male animal that is heterozygous for a deletion chromosome and a normal one produces functional sperm carrying each of the two chromosomes in approximately equal numbers. In other words, sperm seem to function to some extent regardless of their genetic content. In diploid plants, on the other hand, the pollen produced by a deletion heterozygote is of two types: (1) functional pollen carrying the normal chromosome and (2) nonfunctional (or aborted) pollen carrying the deficient homolog. Thus, pollen cells seem to be sensitive to changes in amount of chromosomal material, and this sensitivity might act to weed out deletions. The situation is somewhat different for polyploid plants, which are far more tolerant of pollen deletions. This tolerance is due to the fact that even the pollen carries sev-

it can be tagged with a radioactive or chemical label and then added to a chromosomal preparation under the microscope. In such a situation, the DNA recognizes and physically binds to its normal chromosomal counterpart by nucleotide pairing and is recognized as a spot of radioactivity or dye. The precise location of such spots is difficult to correlate with specific bands, but the deletion technique comes to the rescue. If a deletion happens to span the locus in question, no spot will appear when the test is run with the chromosome carrying the deletion, because the region for binding simply is not present (Figure 17-6). By saving cell lines from patients with deletions, geneticists develop test panels of overlapping deletions spanning specific chromosomal regions, and these test panels can be used to pinpoint a gene's position. An example from chromosome 11 is shown in Figure 17-7. The extent of the deletions in the test panel are shown as vertical bars, and the coded DNA fragments under test are shown at the right. If fragment 270, for example, failed to bind to deletions 35, 8, 10, 7, 9, 23, 24, A2, 27A, and 4D but did bind to the other deletions, it can be in-

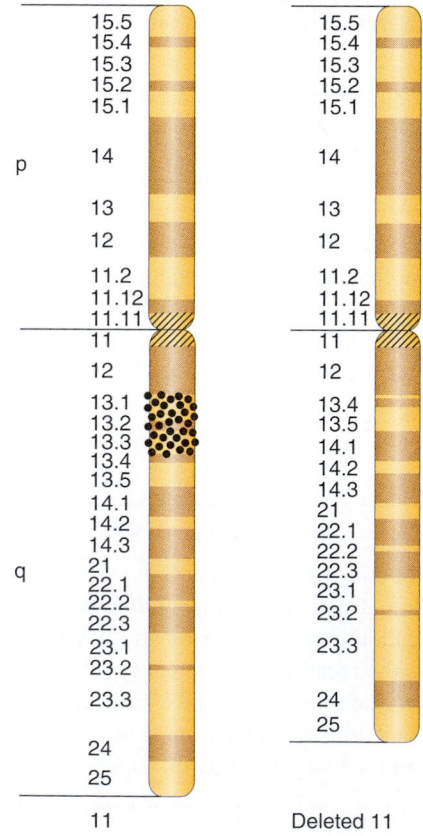

11   Deleted 11

Figure 17-6   Radioactive spots show up on only one chromosome 11, because the other one has a deletion in the region where the radioactive DNA binds.

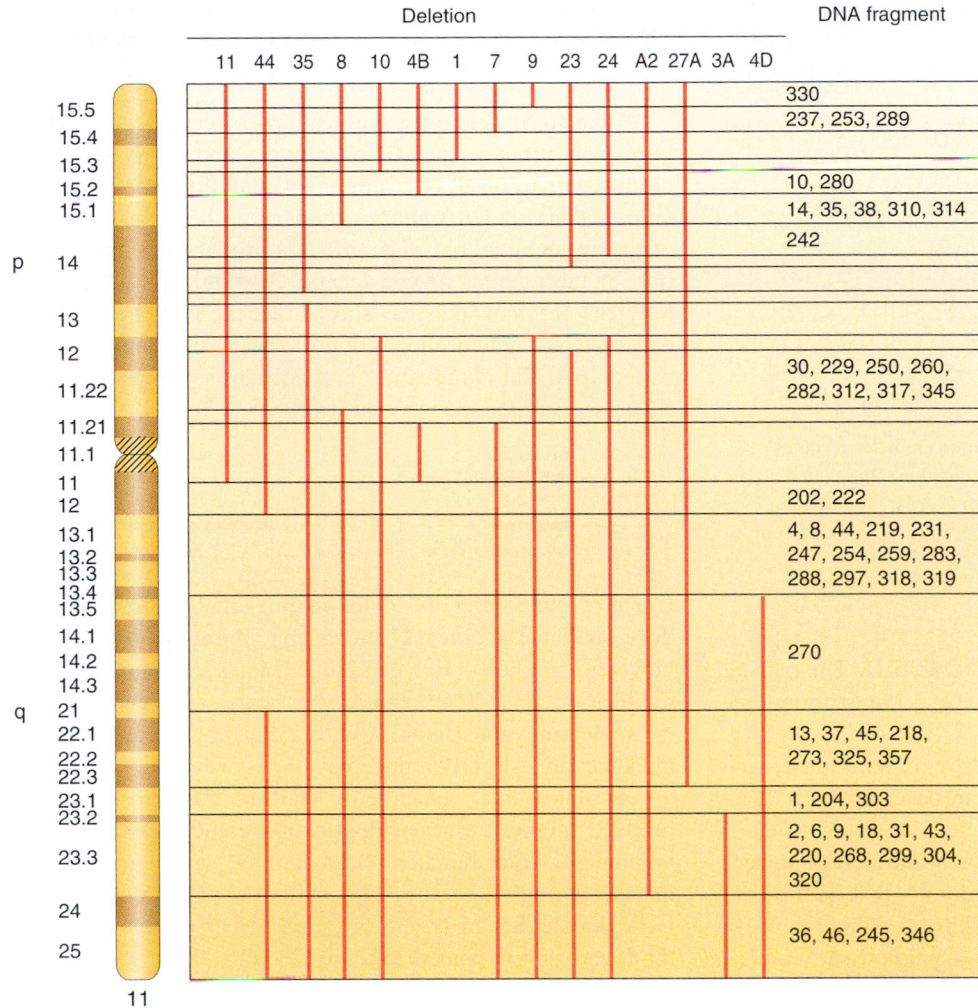

Figure 17-7 Human DNA fragments mapped to regions of chromosome 11 by their failure to bind to particular deletions. The red bars show the extent of the deletions, and the DNA fragments that were mapped are identified at the right. Notice that fragment 270, for example, failed to bind to deletions 35, 8, 10, 7, 9, 23, 24, A2, 27A, and 4D but did bind to the other deletions. The region spanned by 11q13.5 and 11q21 is missing from all the deletions that did not bind 270; 270 was thus inferred to lie within this region. (From Y. Nakamura.)

eral chromosome sets, and the loss of a segment in one of these sets is less crucial than it would be in a haploid pollen cell. Ovules in either diploid or polyploid plants also are quite tolerant of deletions, presumably because of the nurturing effect of the surrounding maternal tissues.

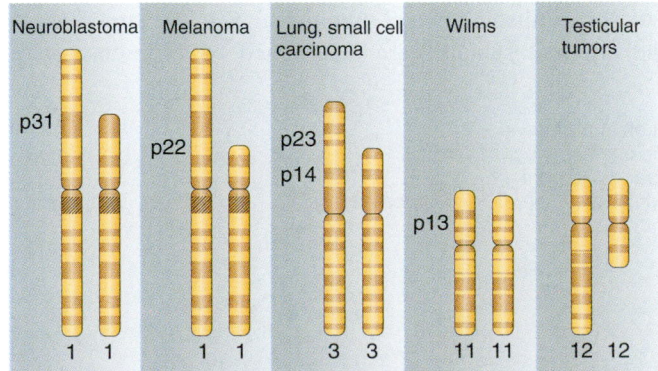

Figure 17-8 Deletions found consistently in several different types of solid tumors in humans. Band numbers indicate recurrent breakpoints. (After Jorge Yunis.)

Duplications

The processes of chromosome mutation sometimes produce an extra copy of some chromosome region. In considering a haploid organism, which has one chromosome set, we can easily see why such a product is called a *duplication* because the region is now present in duplicate. The duplicate regions can be located adjacent to each other or one of the duplicate regions can be in its normal location and the other in a novel location on a different part of the same chromosome or even on another chromosome. In a diploid organism, the chromosome set containing the duplication is generally present together with a standard chromosome set. The cells of such an organism will thus have three copies of the chromosome region in question, but nevertheless such duplication heterozygotes are generally referred to as duplications because they carry the product of one duplication event.

Duplication heterozygotes also show interesting pairing structures at meiosis or in salivary gland chromosomes. The precise structure that forms depends on the type of duplication. For the present we shall consider only adjacent duplications, which can be **tandem,** as in the example

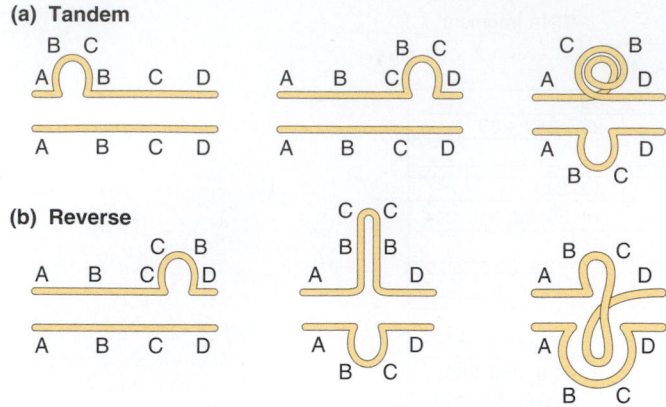

(a) Tandem

(b) Reverse

Figure 17-9   Possible pairing configurations in heterozygotes of a standard chromosome and a side-by-side duplication. Duplicated segments may be (a) in tandem or (b) in reverse order. A particular duplicated segment may assume different configurations in different meioses.

A B C B C D, or **reverse,** as in A B C C B D. The pairing structures in heterozygotes for adjacent duplications are shown in Figure 17-9.

When a tandem duplication arises in an individual, then by inbreeding it is possible that some descendants will be duplication homozygotes, carrying a total of four copies of the duplicated chromosome region. Such individuals present another interesting feature of duplications — that is, the possibility of asymmetrical pairing, as shown in Figure 17-10. The crossing-over at meiosis of such asymmetrically paired regions may create a tandem triplication of the chromosome region.

The extra region of a duplication is free to undergo gene mutation because the necessary basic functions of the region will be provided by the other copy. Mutation in the extra region provides an opportunity for divergence in the function of the duplicated genes, which could be advantageous in genome evolution. Indeed, from situations in which different gene products with related functions can be compared, such as the globins (to be considered later), there is good evidence that these products arose as duplicates of one another.

MESSAGE ·······································

Duplications supply additional genetic material capable of evolving new functions.

Like some deletions, duplications of certain genetic regions may produce specific phenotypes and act as gene mutations. For example, the dominant mutation *Bar* on the X chromosome of *Drosophila* produces a slitlike eye instead of the normal oval one. The effect is produced by the reduction of the number of eye facets. Cytological study of the polytene chromosomes has shown that the Bar phenotype is caused by a tandem duplication of the chromosomal region called 16A. The duplication probably arose from asymmetric crossover at meiosis as shown here:

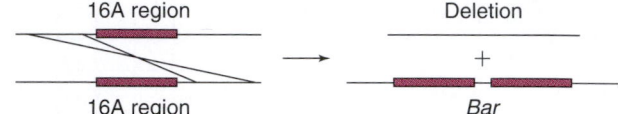

Gametes containing the deletion presumably died or produced inviable zygotes. Gametes containing the duplication, however, produced Bar offspring. Males carrying the *Bar* duplication in a hemizygous state have severely reduced eyes. Heterozygous females with a *Bar* and a normal chromosome have slightly reduced eyes.

Evidence that asymmetric pairing and crossing-over produces higher orders of duplication comes from studying homozygous *Bar* females. Occasionally, such females produce offspring with extremely small eyes called *double Bar*. Each double Bar offspring is found to carry three doses of the *Bar* region in tandem (Figure 17-11).

Some of the best evidence that tandem duplications (and their reciprocal deletions) derive from unequal crossovers comes from studies of the genes that determine the structure of human hemoglobin, the oxygen-transport molecule. The hemoglobin molecule is composed of two different kinds of subunits. Furthermore, there are always two of each kind of subunit, giving a total of four subunits. A person has different hemoglobin at different ages — that is, the kinds of subunits differ. For example, a fetus has two α subunits and two γ subunits ($\alpha_2\gamma_2$), whereas an adult has two α subunits and two β subunits ($\alpha_2\beta_2$). The structures of the different subunits are determined by different genes,

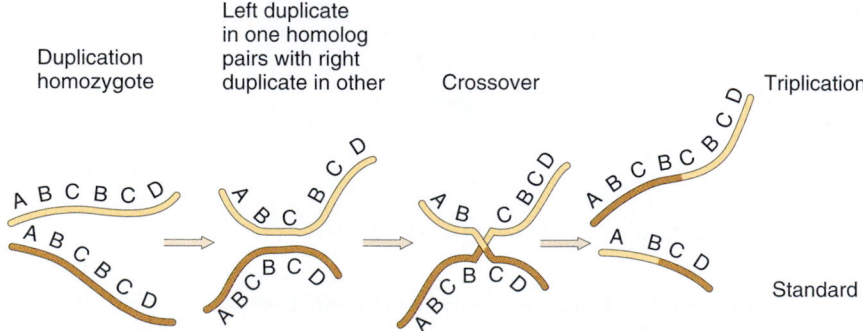

Figure 17-10   Generation of higher orders of duplications by asymmetric pairing followed by crossing-over in a duplication homozygote.

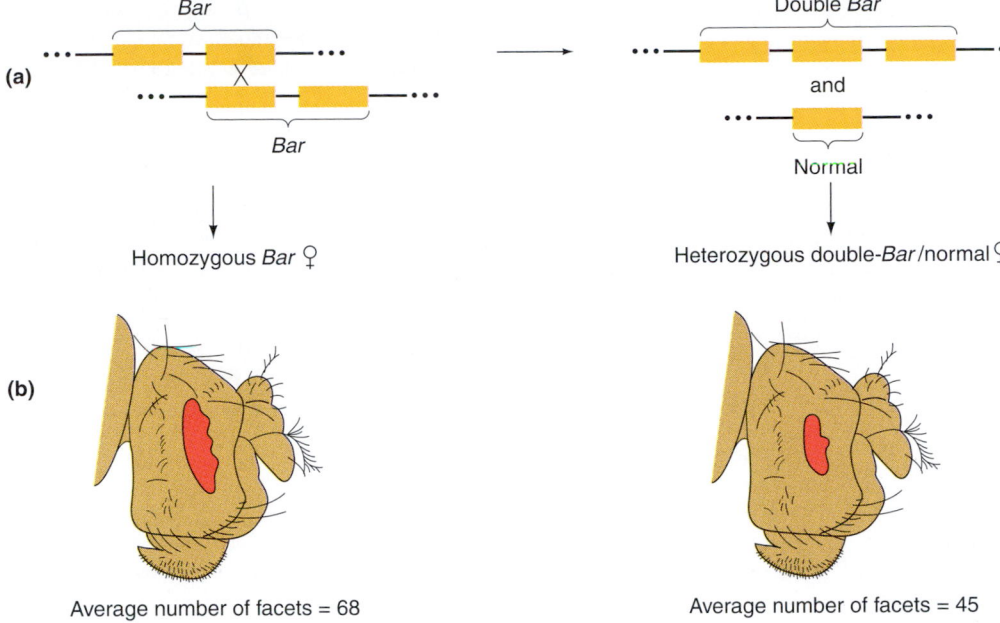

(a)

Bar — Double *Bar* — and — Normal

Homozygous *Bar* ♀ — Heterozygous double-*Bar*/normal ♀

(b)

Average number of facets = 68

Average number of facets = 45

Figure 17-11 Production of double-*Bar* (triplication) and *Bar*-revertant (normal) chromosomes by asymmetric pairing followed by crossing-over in a duplication homozygote. (a) Diagrammatic representation. (b) Comparison of Bar and double-Bar phenotypes. (Part b from C. Bridges, *Science* 83, 1936, 210.)

some of which are linked and some of which are not. The situation is summarized in Figure 17-12.

The linked γ-δ-β group provides the data that we need on unequal crossover. Some people afflicted with certain thalassemias (a kind of inherited blood disease) have a hemoglobin subunit that is part δ and part β (Lepore hemoglobin) or one that is part γ and part β (Kenya hemoglobin). The origin of these rare hemoglobin subunits can be explained by the unequal crossover models shown in Figure 17-13. Deletion chromosomes determine both Lepore and Kenya hemoglobins, causing their bearers to have the blood diseases. Also diagrammed in Figure 7-13 are the origins of the reciprocal crossover products called *anti-Lepore* and *anti-Kenya*.

Tandem duplications are rare in humans. Most duplications consist of an extra chromosomal arm or part of an arm, generally attached to a nonhomologous chromosome. In discussions that follow about other chromosomal rearrangements — inversions and translocations — we shall learn about the typical origins of such duplications. Like some deletions, duplications of some human chromosomes cause syndromes of phenotypic abnormalities. A person afflicted with a duplication syndrome has three copies of the duplicated region, whereas other chromosome regions are present in two copies as usual. Humans homozygous for duplications are unknown in medical genetics. Unlike deletions, duplications do not unmask deleterious recessives, so the abnormalities associated with duplications may be attributed

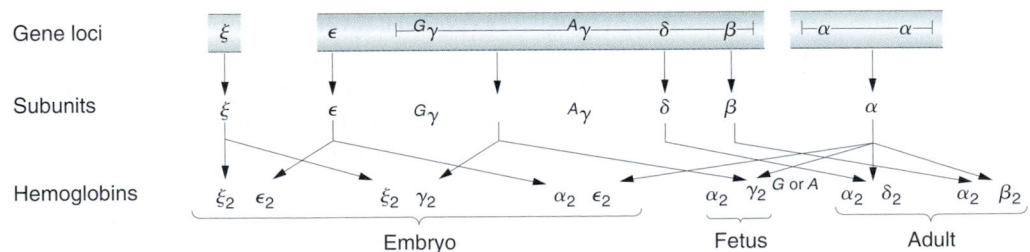

Figure 17-12 Human hemoglobin genes. Each Greek letter represents a different gene locus and the hemoglobin subunit determined by that gene. *A* and *G* are duplicates of the δ hemoglobin gene. These subunits combine in different ways at different stages in development, as indicated, to yield functioning hemoglobin molecules. Notice that adults have a minor hemoglobin, $\alpha_2\delta_2$ (<2 percent) in addition to $\alpha_2\beta_2$. (From D. J. Weatherall and J. B. Clegg, "Recent Developments in Molecular Genetics of Human Hemoglobin," *Cell* 16, 1979.)

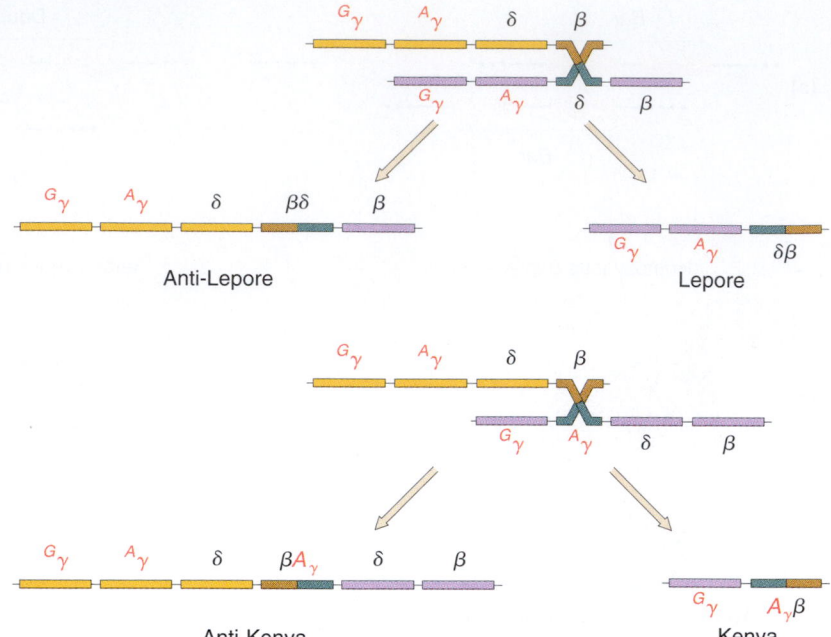

Figure 17-13   Proposed generation of variant human hemoglobin subunits by unequal crossing-over in the γ-δ-β genetic region. Notice that both Lepore and Kenya are deletions. (From D. J. Weatherall and J. B. Clegg, "Recent Developments in Molecular Genetics of Human Hemoglobin," *Cell* 16, 1979.)

solely to the imbalance generated by the extra copy of the chromosome region.

In general, duplications are hard to detect and are rare. However, they are useful tools in research and can be generated from other chromosomal aberrations by manipulations that we shall learn later.

Duplications can also be created by using modern DNA technology. Cells can be induced to take up specific fragments of DNA from their own species, and this DNA can insert into the linear structure of a chromosome. It can insert either at its regular locus, replacing the resident sequence, or at a completely different site, called an **ectopic site.** Insertion of the DNA at an ectopic site creates a duplication of the fragment that was taken up. Insertion at an ectopic site is a direct and convenient way of making small duplications.

Inversions

If two breaks occur in one chromosome, sometimes the region between the breaks rotates 180 degrees before rejoining with the two end fragments. Such an event creates a chromosomal mutation called an **inversion.** Unlike deletions and duplications, inversions do not change the overall amount of the genetic material, so inversions are generally viable and show no particular abnormalities at the phenotypic level. In some cases, one of the chromosome breaks is within a gene of essential function, and then that breakpoint acts as a lethal gene mutation linked to the inversion. In such a case, the inversion could not be bred to homozygosity. However, many inversions can be made homozygous; furthermore, inversions can be detected in haploid organisms. In these cases, the breakpoint is clearly not in an essential region. Some of the possible outcomes of inversion at the DNA level are shown in Figure 17-14.

Most analyses of inversions use heterozygous inversions—diploids in which one chromosome has the standard sequence and one carries the inversion. Microscopic observation of meioses in inversion heterozygotes reveals the location of the inverted segment because one chromosome twists once at the ends of the inversion to pair with the other, untwisted chromosome; in this way the paired homologs form an **inversion loop** (Figure 17-15).

The location of the centromere relative to the inverted segment determines the genetic behavior of the chromosome. If the centromere is outside the inversion, then the inversion is said to be **paracentric,** whereas inversions spanning the centromere are **pericentric:**

Paracentric   A B C D E F → A B E D C F

Pericentric   A B C D E F → A D C B E F

How do inversions behave genetically? Crossing-over within the inversion loop of a paracentric inversion connects homologous centromeres in a **dicentric bridge** while also producing an **acentric fragment**—a fragment without a centromere. Then, as the chromosomes separate in anaphase I, the centromeres remain linked by the bridge, which orients the centromeres so that the noncrossover chromatids lie farthest apart. The acentric fragment cannot align itself or move and is, consequently, lost. Tension eventually breaks the bridge, forming two chromosomes with terminal deletions (Figure 17-16). The gametes containing such deleted chromosomes may be inviable but, even if viable, the zygotes that they eventually form are inviable. Hence, a crossover event, which normally generates the recombinant class of meiotic products, instead produces lethal products. The overall result is a lower recombinant frequency. In fact,

Breakpoints *between* genes

Normal sequence

Breaks in DNA

Inverted alignment

Joining of breaks to complete inversion

Inversion

One breakpoint *between* genes
One *within* gene C (C disrupted)

Inversion

Breakpoints *in* genes *A* and *D*
Creating gene fusions

Inversion

Figure 17-14 Effects of inversions at the DNA level. Genes are represented by *A*, *B*, *C*, and *D*. Template strand is dark green; nontemplate strand is light green; jagged lines indicate break in DNA. The letter P stands for promoter; thick arrow indicates the position of the breakpoint.

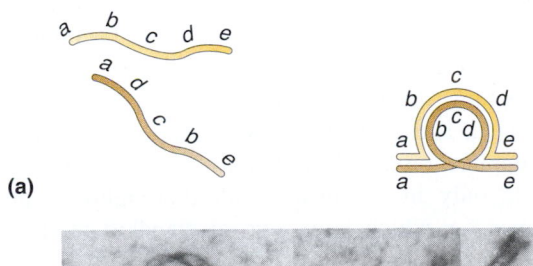

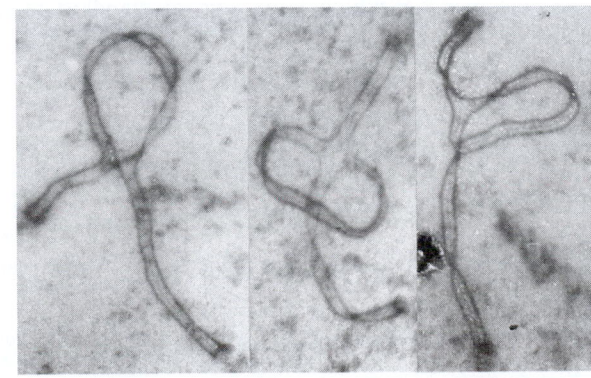

(a)

(b)

for genes within the inversion, the RF is zero. For genes flanking the inversion, the RF is reduced in proportion to the relative size of the inversion.

Inversions affect recombination in another way too. Inversion heterozygotes often have mechanical pairing problems in the region of the inversion; these pairing problems reduce the frequency of crossing-over and hence the recombinant frequency in the region.

The net genetic effect of a pericentric inversion is the same as that of a paracentric one — crossover products are not recovered — but for different reasons. In a pericentric inversion,

Figure 17-15 The chromosomes of inversion heterozygotes pair in a loop at meiosis. (a) Diagrammatic representation; each chromosome is actually a pair of sister chromatids. (b) Electron micrographs of synaptonemal complexes at prophase I of meiosis in a mouse heterozygous for a paracentric inversion. Three different meiocytes are shown. (Part b from M. J. Moses, Department of Anatomy, Duke Medical Center.)

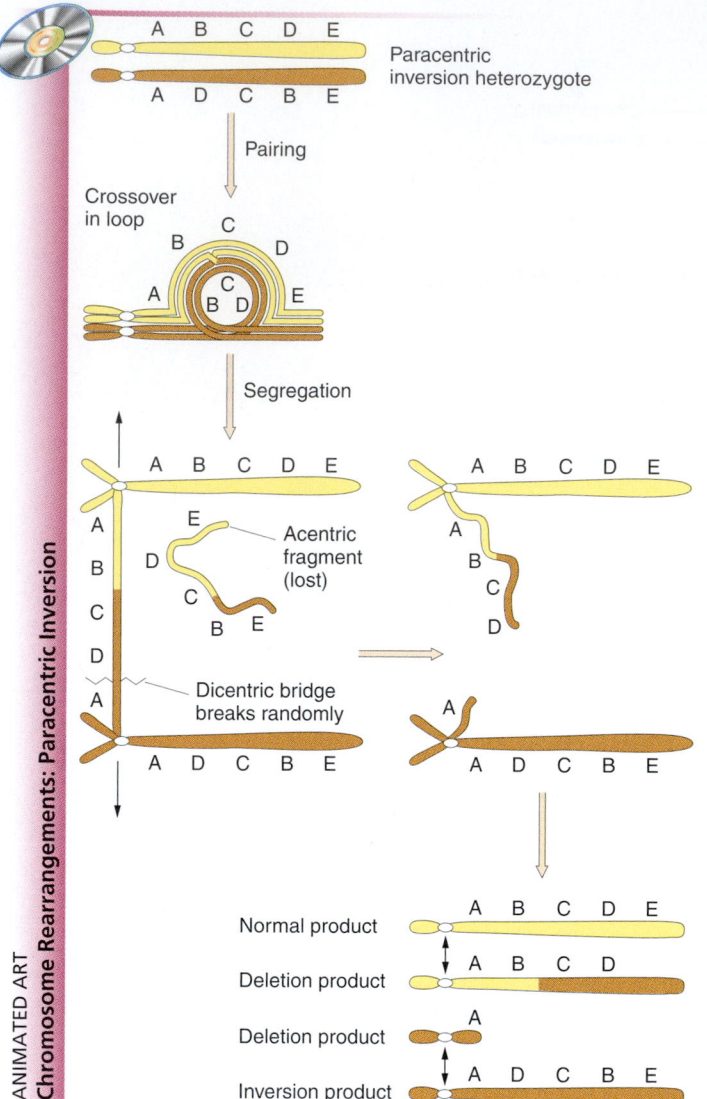

Figure 17-16   Meiotic products resulting from a single crossover within a paracentric inversion loop. Two nonsister chromatids cross over within the loop.

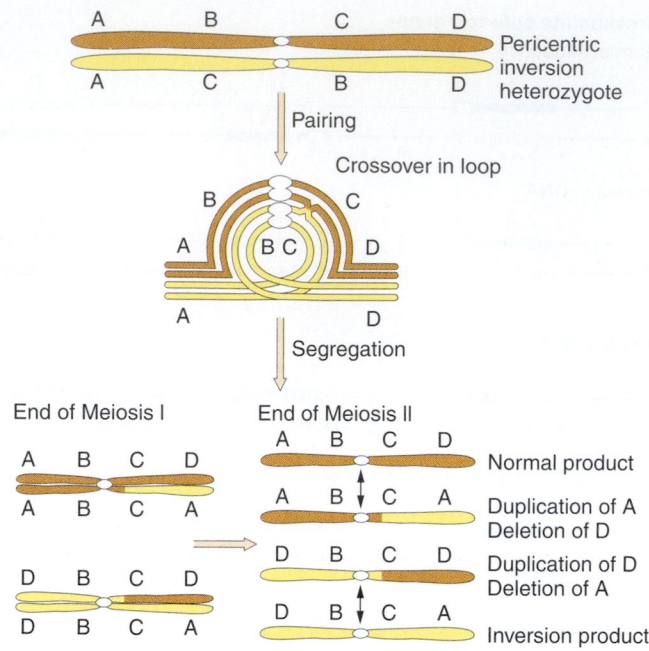

Figure 17-17   Meiotic products resulting from a meiosis with a single crossover within a pericentric inversion loop.

because the centromeres are contained within the inverted region, the chromosomes that have crossed over disjoin in the normal fashion, without the creation of a bridge. However, the crossover produces chromatids that contain a duplication and a deficiency for different parts of the chromosome (Figure 17-17). In this case, if a nucleus carrying a crossover chromosome is fertilized, the zygote dies because of its genetic imbalance. Again, the result is the selective recovery of noncrossover chromosomes in viable progeny.

MESSAGE ··
Two mechanisms reduce the number of recombinant products among the progeny of inversion heterozygotes: elimination of the products of crossovers in the inversion loop and inhibition of pairing in the region of the inversion.

It is worth adding a note about homozygous inversions. In such cases the homologous inverted chromosomes pair and cross over normally, there are no bridges, and the meiotic products are viable. However, an interesting effect is that the linkage map will show the inverted gene order.

Geneticists use inversions to create duplications of specific chromosome regions for various experimental purposes. For example, consider a heterozygous pericentric inversion with one breakpoint at the tip (T) of the chromosome, as shown in Figure 17-18. A crossover in the loop produces a chromatid type in which the entire left arm is duplicated; if the tip is nonessential, a duplication stock is generated for investigation. Another way to make a duplication (and a deficiency) is to use two paracentric inversions with overlapping breakpoints (Figure 17-19). A complex loop is formed, and a crossover within the inversion produces the duplication and the deletion. These manipulations are possible only in organisms with thoroughly mapped chromosomes for which large sets of standard rearrangements are available.

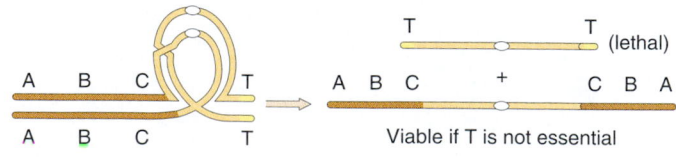

Figure 17-18   Generation of a viable nontandem duplication from a pericentric inversion close to a dispensable chromosome tip.

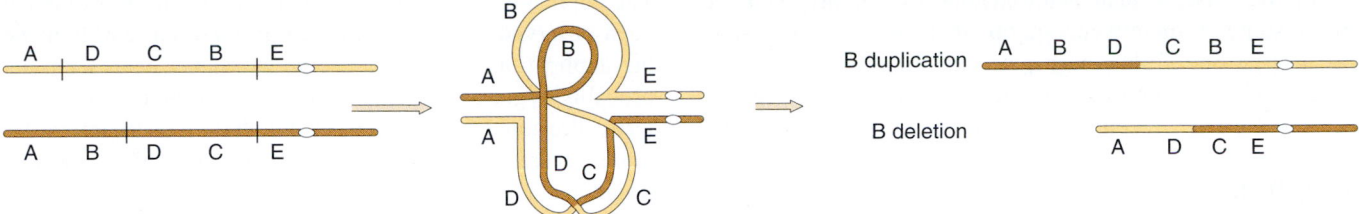

Figure 17-19 Generation of a nontandem duplication by crossing-over between two overlapping inversions.

We have seen that genetic analysis and meiotic chromosome cytology are both good ways of detecting inversions. As with most rearrangements, there is also the possibility of detection through mitotic chromosome analysis. A key operational feature is to look for new arm ratios. Consider a chromosome that has mutated as follows:

Note that the ratio of the long to the short arm has been changed from about 4 to about 1 by the pericentric inversion. Paracentric inversions do not alter the arm ratio, but they may be detected microscopically if banding or other chromosome landmarks are available.

MESSAGE ·······························

The main diagnostic features of inversions are inversion loops, reduction of recombinant frequency, and reduced fertility from unbalanced or deleted meiotic products, all observed in individuals heterozygous for inversions. Some inversions may be directly observed as an inverted arrangement of chromosomal landmarks.

···

Inversions are found in about 2 percent of humans. The heterozygous inversion carriers generally show no adverse phenotype but produce the expected array of abnormal meiotic products from crossing-over in the inversion loop. Let us consider pericentric inversions as an example. Persons heterozygous for pericentric inversions produce offspring with the duplication–deletion chromosomes predicted; these offspring show varying degrees of abnormalities depending on the lengths of the chromosome regions affected. Some phenotypes caused by duplication–deletion chromosomes are so abnormal as to be incapable of survival to birth and are lost as spontaneous abortions. However, there is a way to study the abnormal meiotic products that does not depend on survival to term. Human sperm placed in contact with unfertilized eggs of the golden hamster penetrate the eggs but fail to fertilize them. The sperm nucleus does not fuse with the egg nucleus, and, if the cell is prepared for cytogenetic examination, the human chromosomes are easily visible as a distinct group (Figure 17-20). This technique makes it possible to study the chromosomal products of a male meiosis directly and is particularly useful in the study of meiotic products of men who have chromosome mutations.

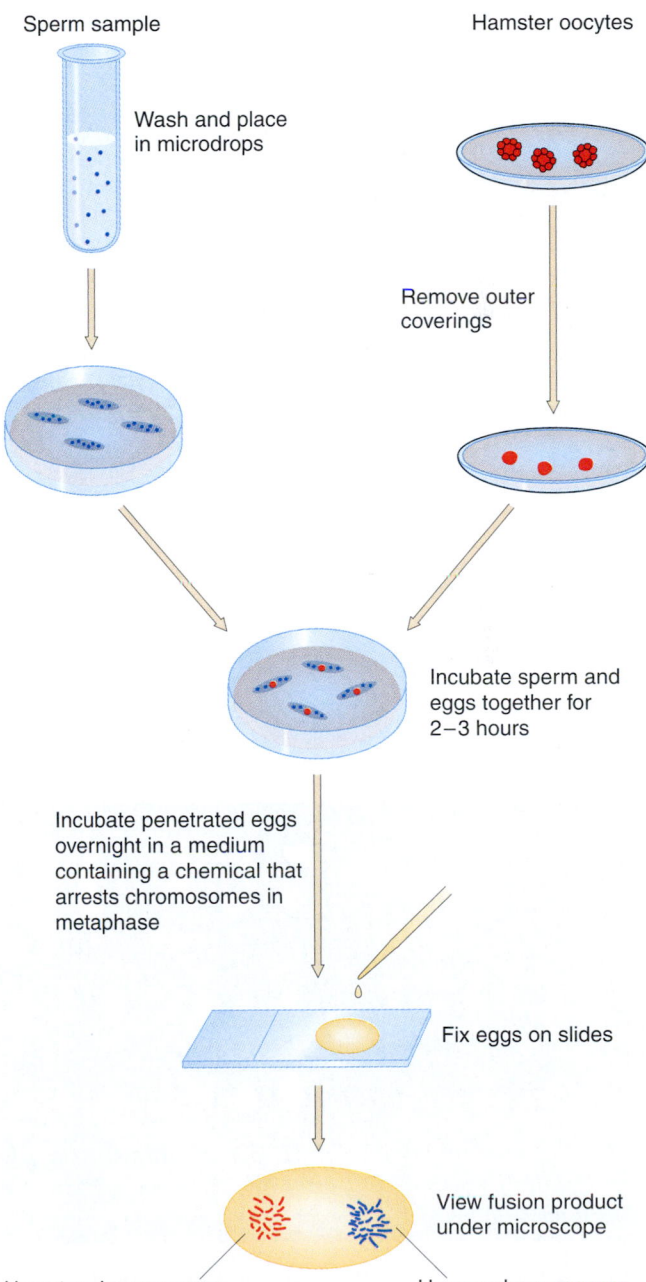

Figure 17-20 Human sperm and hamster oocytes are fused to permit study of the chromosomes in the meiotic products of human males. (After original art by Renée Martin.)

In one case, a man heterozygous for an inversion of chromosome 3 underwent sperm analysis. The inversion was a large one with a high potential for crossing-over in the loop. Four chromosome 3 types were found in the man's sperm — normal, inversion, and two recombinant types (Figure 17-21). The sperm contained the four types in the following frequencies:

Normal	38%
Inversion	32%
Duplication q, deletion p	17%
Duplication p, deletion q	13%

(a)

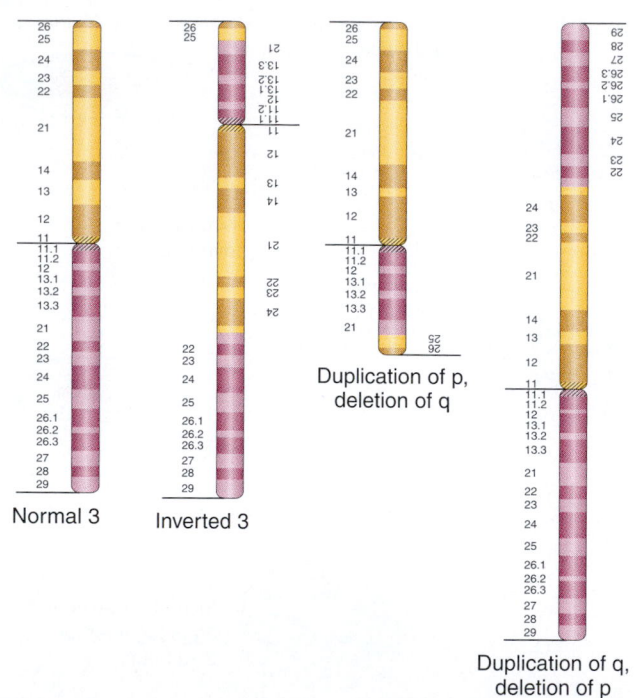

Normal 3   Inverted 3   Duplication of p, deletion of q   Duplication of q, deletion of p

(b)

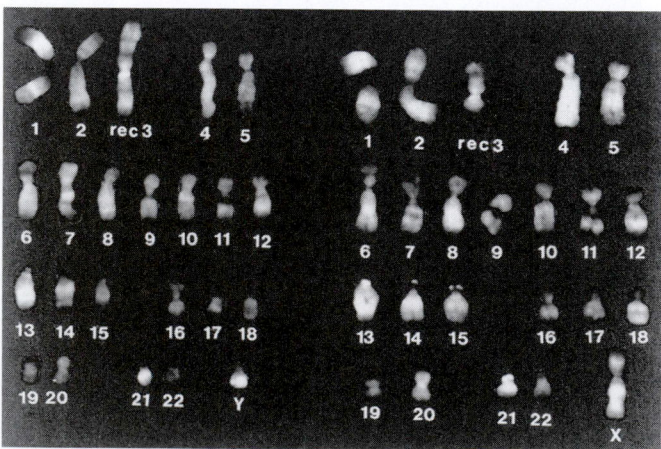

Figure 17-21   (a) Four different chromosomes 3 found in sperm of a man heterozygous for a large pericentric inversion. The duplication-deletion types result from a crossover in the inversion loop. (b) Two complete sperm chromosome sets containing the two duplication–deletion types (denoted rec 3). (Renée Martin.)

The duplication-q–deletion-p recombinant chromosome had been observed previously in several abnormal children, but the duplication-p–deletion-q type had never been seen, and probably zygotes receiving it are too abnormal to survive to term. Presumably, deletion of the larger q fragment has more severe consequences than deletion of the smaller p fragment.

Translocations

When two nonhomologous chromosomes mutate by exchanging parts, the resulting chromosomal rearrangements are translocations. Here we consider **reciprocal translocations,** the most common type. A segment from one chromosome is exchanged with a segment from another nonhomologous one, so two translocation chromosomes are generated simultaneously.

The exchange of chromosome parts between nonhomologs establishes new linkage relations. These new linkages are revealed if the translocated chromosomes are homozygous and, as we shall see, even when they are heterozygous. Furthermore, translocations may drastically alter the size of a chromosome as well as the position of its centromere. For example,

Here a large metacentric chromosome is shortened by half its length to an acrocentric one, and the small chromosome becomes a large one. Examples from natural populations are known in which chromosome numbers have been changed by translocation between acrocentric chromosomes and the subsequent loss of the resulting small chromosome elements (Figure 17-22).

In heterozygotes having two translocated chromosomes and their normal counterparts, there are important genetic and cytological effects. Again, the pairing affinities of homologous regions dictate a characteristic configuration for chromosomes synapsed in meiosis. Figure 17-23, which illustrates meiosis in a reciprocally translocated heterozygote, shows that the configuration is that of a cross.

Note that the configuration presented in Figure 17-23 lies on the equatorial plate of the cell at metaphase, with the spindle fibers perpendicular to the page. Thus, the centromeres would migrate upward above the page or downward under it. Homologous paired centromeres disjoin, whether or not a translocation is present. Because Mendel's second law still applies to different paired centromeres, there are two common patterns of disjunction. The segregation of each of the structurally normal chromosomes with one of the translocated ones ($T_1 + N_2$ and $T_2 + N_1$) is called **adjacent-1 segregation.** Both meiotic products are duplicated and deficient for different regions. These products are inviable. On the other hand, the two normal chromosomes may segregate together, as do the reciprocal parts of the translocated ones, to produce $N_1 + N_2$ and $T_1 + T_2$

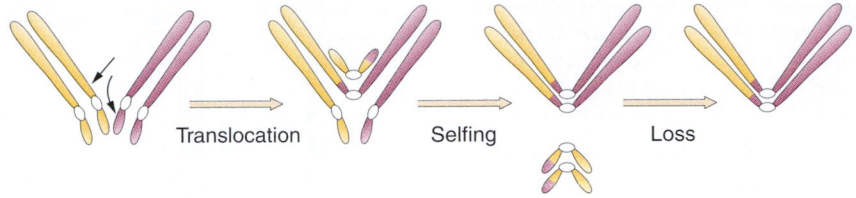

Figure 17-22   Genome restructuring by translocations. Short arrows indicate breakpoints in one homolog of each of two pairs of acrocentric chromosomes. The resulting fusion of the breaks yields one short and one long metacentric chromosome. If, as in plants, self-fertilization (selfing) takes place, an offspring could be formed with only one pair of long and only one pair of short metacentric chromosomes. Under appropriate conditions, the short metacentric chromosome may be lost. Thus, we see a conversion from two acrocentric pairs of chromosomes into one pair of metacentrics.

products. This type is called **alternate segregation.** These products are viable. Because the adjacent-1 and alternate segregation patterns are equally frequent. There is another event, called **adjacent-2 segregation,** in which homologous centromeres migrate to the same pole, but in general this event is rare.

As a result of the equality of adjacent and alternate segregations, half the gametes will be incapable of contributing to the next generation, a condition known as **semisterility.** The condition of semisterility, or "half sterility," is an important diagnostic for translocation heterozygotes. However, semisterility is defined differently for plants and animals. In plants, the 50 percent unbalanced meiotic products from the adjacent-1 segregation generally abort at the gametic stage (Figure 17-24). In animals, however, the duplication–deletion products are viable as gametes but lethal to the zygote.

Remember that heterozygotes for the other rearrangements such as deletions and inversions may show some reduction of fertility, but the precise 50 percent reduction in viable gametes or zygotes is usually a reliable diagnostic clue for a translocation.

Translocations are economically important. In agriculture, translocations in certain crop strains can reduce yields considerably owing to the number of unbalanced zygotes that form. On the other hand, translocations are potentially useful: it has been proposed that insect pests could be controlled by introducing translocations into their wild populations. According to the proposal, 50 percent of the offspring of crosses between insects carrying the translocation and wild types would die, and 10/16 of the progeny of crosses between translocation-bearing insects would die.

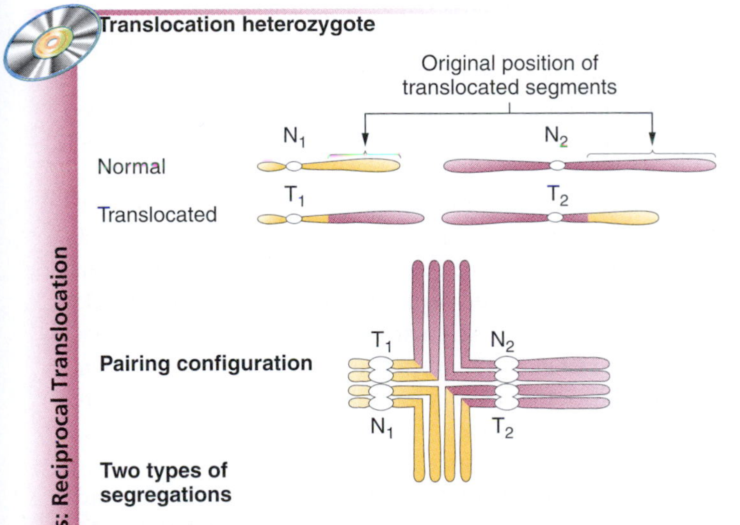

Translocation heterozygote

Original position of translocated segments

Normal
N_1 N_2

Translocated
T_1 T_2

Pairing configuration

T_1 N_2

N_1 T_2

Two types of segregations

Adjacent-1		Products	
Up	$T_1 + N_2$	Duplication of purple, deletion of orange translocated segment	Often inviable
Down	$T_2 + N_1$	Duplication of orange, deletion of purple translocated segment	
Alternate			
Up	$T_1 + T_2$	Translocation genotype	Both complete and viable
Down	$N_1 + N_2$	Normal	

Figure 17-23   The meiotic products resulting from the two most commonly encountered chromosome-segregation patterns in a reciprocal translocation heterozygote.

Figure 17-24   Photomicrograph of normal and aborted pollen of a semisterile corn plant. The clear pollen grains contain chromosomally unbalanced meiotic products of a reciprocal translocation heterozygote. The opaque pollen grains, which contain either the complete translocation genotype or normal chromosomes, are functional in fertilization and development. (William Sheridan.)

Genetically, markers on nonhomologous chromosomes appear to be linked if these chromosomes take part in a translocation and the loci are close to the translocation breakpoint. Figure 17-25 shows a situation in which a translocation heterozygote has been established by crossing an a/a ; b/b individual with a translocation homozygote bearing the wild-type alleles. We shall assume that a and b are close to the translocation breakpoint. On testcrossing the heterozygote, the only viable progeny are those bearing the parental genotypes, so linkage is seen between loci that were originally on different chromosomes. In fact, if all four arms of the meiotic pairing structure are genetically marked, recombination studies should result in a cross-shaped linkage map. Apparent linkage of genes known to be on separate nonhomologous chromosomes is a genetic giveaway for the presence of a translocation.

Translocations in humans are always carried in the heterozygous state. An example between chromosomes 5 and 11 is shown in Figure 17-26. The offspring of the person with this particular translocation had a duplication of 11q and a deletion of 5p. These children showed symptoms of both cri du chat syndrome, which is caused by the deletion of 5p, and the syndrome associated with the duplication of 11q. The reciprocal duplication-deficiency chromosome was not observed.

Down syndrome is a constellation of human disorders usually caused by the presence of an extra chromosome 21 that failed to segregate from its homolog at meiosis (see Chapter 18). This common type of Down syndrome (ap-

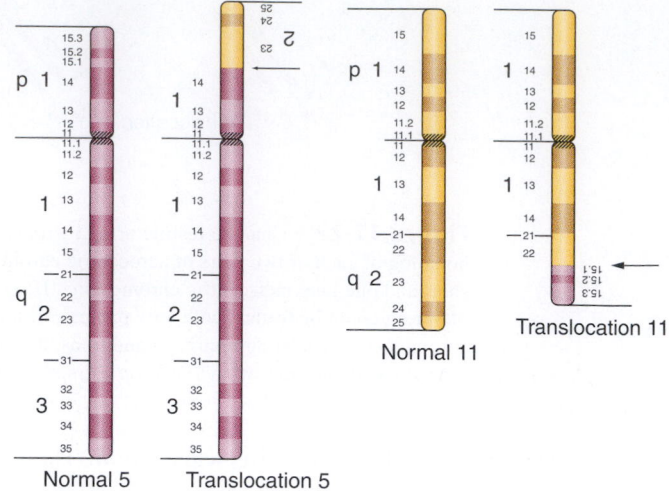

Figure 17-26   A human translocation heterozygote with reciprocal exchange of 5p and 11q (5p15; 11q23).

proximately 95% of all cases) is sporadic and shows no recurrence in the family. However, there is a less common type of Down syndrome caused by a special type of translocation called a **Robertsonian translocation,** and this form can recur in a family. A Robertsonian translocation is one that combines the long arms of two acrocentric chromosomes, as shown in Figure 17-27. Initially, a small chromosome composed of the two short arms also forms; however, this small chromosome is generally not present. The material in the short arms must be nonessential because their loss has no effect on phenotype. Translocation Down syndrome is caused by a Robertsonian fusion between chromosomes 21 and 14. The translocated chromosome passes down through the generations in unaffected carriers. However, meiotic segregation in the translocation carriers can result in offspring that carry three copies of most of chromosome 21, as shown in Figure 17-27, and these offspring have Down syndrome. The factors responsible for numerous other hereditary disorders have been traced to translocation heterozygosity in the parents.

Translocations also appear in cancer cells, and some examples are shown for solid tumors in Figure 17-28. In solid tumors, translocations are not as common as deletions. As with other rearrangements in cancer cells, the involvement with the cancer phenotype is generally not clear. However, in a later section we shall study an example in which the relocation of a specific proto-oncogene seems to be causally connected to cancer.

As is true for other rearrangements, the translocation breakpoints can sometimes disrupt an essential gene, and the gene is thereby inactivated and behaves as a point mutation. Molecular geneticists can use this effect to find the exact location of a human gene and can then proceed to isolate the gene. For example, information from translocations helped in the isolation of the gene for the X-linked recessive disease Duchenne muscular dystrophy. Some rare female

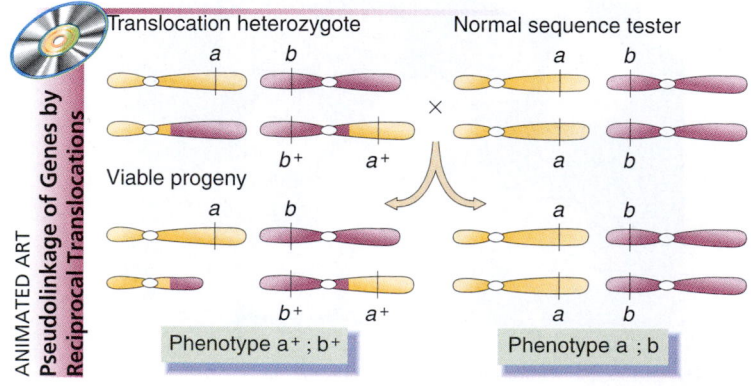

Figure 17-25   When a translocated fragment carries a marker gene, this marker can show linkage to genes on the other chromosome because the recombinant genotypes (in this case, a^+ ; b and a ; b^+) tend to be in duplication–deletion gametes and do not survive.

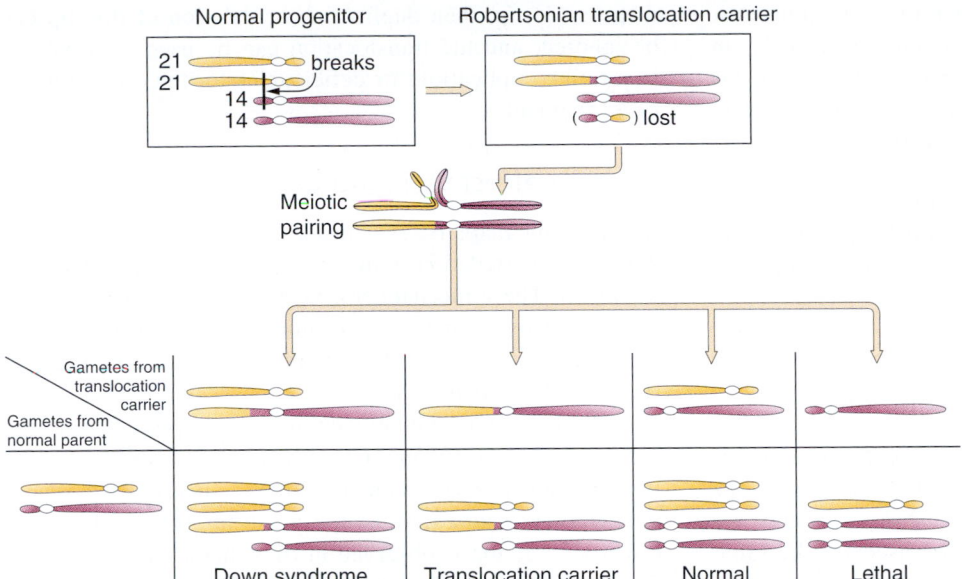

Figure 17-27   How Down syndrome arises in the children of an unaffected carrier of a special type of translocation, called a Robertsonian translocation, in which the long arms of two acrocentric chromosomes have fused. The specific Robertsonian translocation in translocation Down syndrome is between the Down syndrome chromosome 21 and chromosome 14. Translocation Down syndrome accounts for less than 5% of cases.

cases of Duchenne muscular dystrophy were also heterozygous for translocations between the X chromosome and a variety of different autosomes. The X chromosome breakpoint was always in the band Xp21, so the gene for muscular dystrophy, already known to be X-linked, was evidently in this band and had been disrupted by the break. The hunt for the gene could begin by focusing on that area. In passing, note that the expression of the mutant phenotype in a female must have been because the normal X chromosome was inactivated (Figure 17-29).

One specific autosomal breakpoint proved to be useful in providing a molecular "tag" for the Duchenne gene. The particular breakpoint that advanced the research was in the ribosomal RNA locus on chromosome 21. (Do not confuse X chromosome band 21 with chromosome 21.) A DNA probe was already available for the ribosomal locus. It was

reasoned that, in the X–21 translocation, the Duchenne gene must be disrupted and attached next to the ribosomal RNA gene. The geneticists therefore used the ribosomal probe to isolate a DNA segment that had part of the Duchenne gene on it.

Translocations were also used to isolate the human gene for neurofibromatosis. Once again, the critical chromosomal material came from people who not only had the disease,

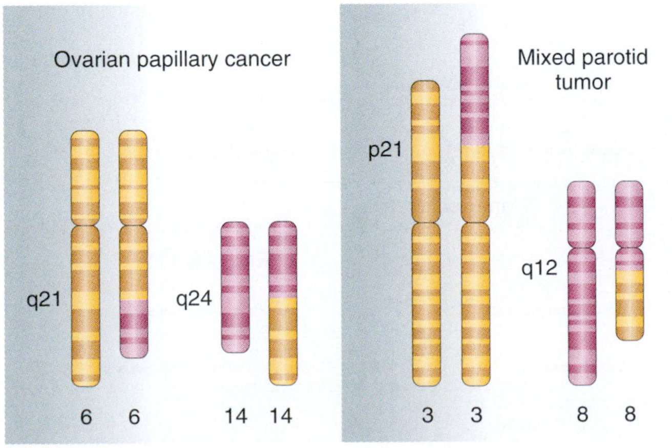

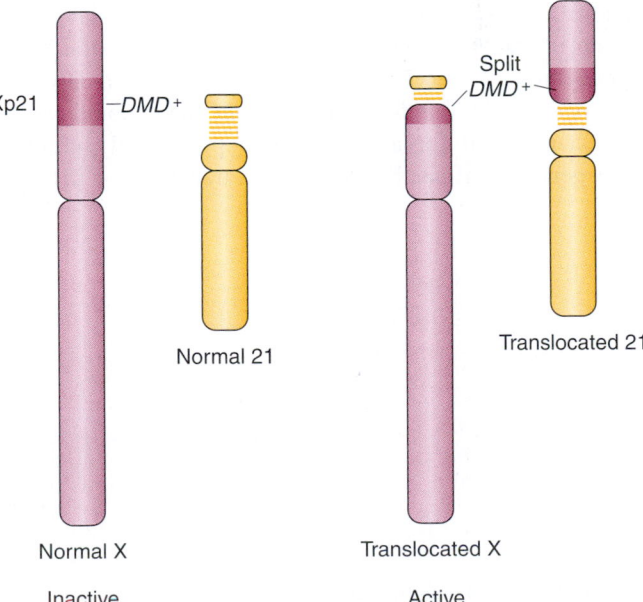

Figure 17-29   Diagram of the chromosomes of a woman with Duchenne muscular dystrophy and heterozygous for a reciprocal translocation between the X chromosome and chromosome 21. The translocation breakpoint disrupted one *DMD*⁺ allele, rendering it nonfunctional, and X chromosome inactivation made its undisrupted partner *DMD*⁺ allele also nonfunctional.

Figure 17-28   Translocations found consistently in several different types of solid tumors in humans. Band numbers indicate breakpoints. (After Jorge Yunis.)

but also carried chromosomal translocations. The translocations all had one of their breakpoints in chromosome 17, in a band close to the centromere. Hence it appeared that this band must be the locus of the neurofibromatosis gene, which had been disrupted by the translocation breakage. Subsequent analysis showed that the chromosome 17 breakpoints were not identical, and their positions helped to map the region occupied by the neurofibromatosis gene. Isolation of DNA fragments from this region eventually led to the recovery of the gene itself.

Use of translocations
in producing duplications and deletions

Geneticists regularly need to make specific duplications or deletions to answer specific experimental questions. We have seen that they use inversions to do so, and now we shall see how translocations also can be used for the same purpose. Let's take *Drosophila* as an example. For reasons that are still unclear, the densely staining chromosomal regions near the centromere — regions called *heterochromatin* — are physically extensive but contain few genes. In fact, for a long time, heterochromatin was considered useless and inert material. In any case, for our purposes, *Drosophila* can tolerate a loss or an excess of heterochromatin with little effect on viability or fertility.

Now let's select two different reciprocal translocations of the same two chromosomes. Each translocation has a breakpoint somewhere in heterochromatin, and each has another breakpoint in *euchromatin* (nonheterochromatin) on opposite sides of the region we want to duplicate or delete (Figure 17–30). It can be seen that, if we have a large collection of translocations having one heterochromatic break and euchromatic breaks at many different sites, then duplications and deletions for many parts of the genome can be produced at will for a variety of experimental purposes. More generally, if one breakpoint of a translocation is near a

dispensable tip, then duplication or deletion of this tip can be ignored, and the translocation can be used as a way of generating duplications or deficiencies for the other translocated segment.

Position-effect variegation

In preceding chapters, we considered several mechanisms of generating variegation in the somatic cells of a multicellular organism. These mechanisms were somatic segregation, somatic crossover, and somatic mutation. Another cause of variegation is associated with translocations and is called *position-effect variegation.*

The locus for white eye color in *Drosophila* is near the tip of the X chromosome. Consider a translocation in which the tip of an X chromosome carrying w^+ is relocated next to the heterochromatic region of, say, chromosome 4. Position-effect variegation is observed in flies that are heterozygotes for such a translocation and that have the normal X chromosome carrying the recessive allele w. The expected eye phenotype is red because the wild-type allele is dominant to w. However, in such cases, the observed phenotype is a variegated mixture of red and white eye facets. How can we explain the white areas? We could suppose that, when the chromosomes broke and rejoined in the translocation, the w^+ allele was somehow changed to a state that made it more mutable in somatic cells; so the white eye tissue is due to cells in which w^+ has mutated to w.

In 1972, Burke Judd tested this hypothesis by recombining the w^+ allele out of the translocation and onto a normal X chromosome and by recombining a w allele from the normal X chromosome onto the translocation (Figure 17–31). Judd found that, when the w^+ allele on the translocation was crossed onto a normal X chromosome and w was then inserted into the translocation, the eye color was red; so obviously the w^+ allele was not defective. When he crossed the w^+ allele back onto the translocation, the pheno-

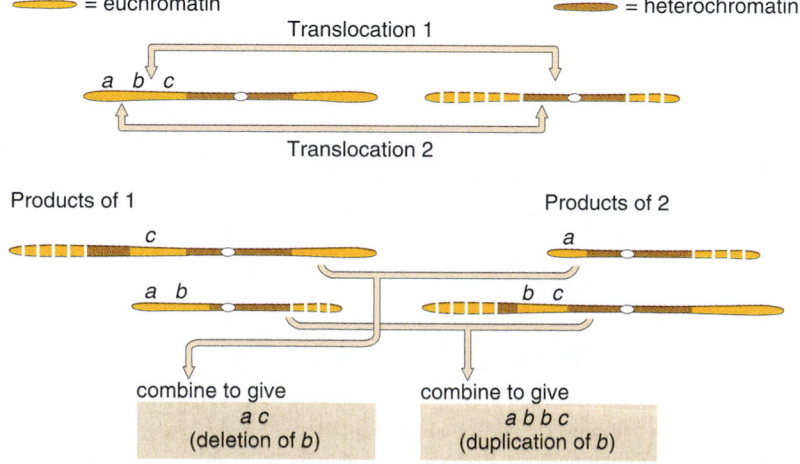

Figure 17-30 Using translocations with one breakpoint in heterochromatin to produce a duplication and a deletion. If the upper product of translocation 1 is combined with the upper product of translocation 2 by means of an appropriate mating, a deletion of *b* results. If the lower products of the two translocations are combined, the genotype *a b b c* is produced.

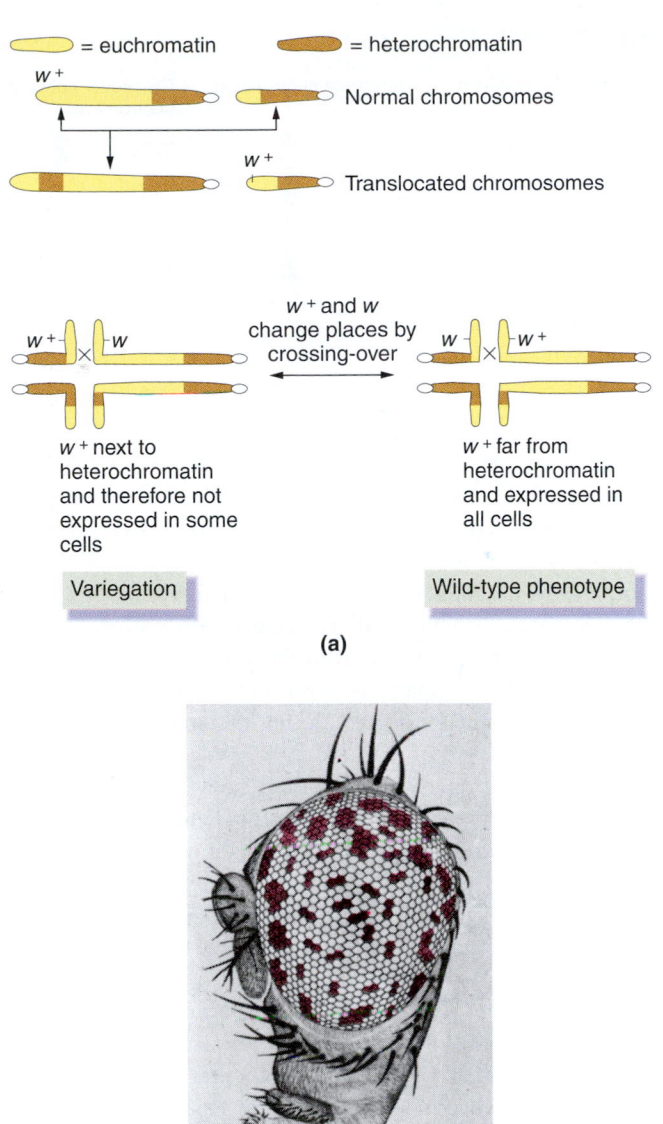

(a)

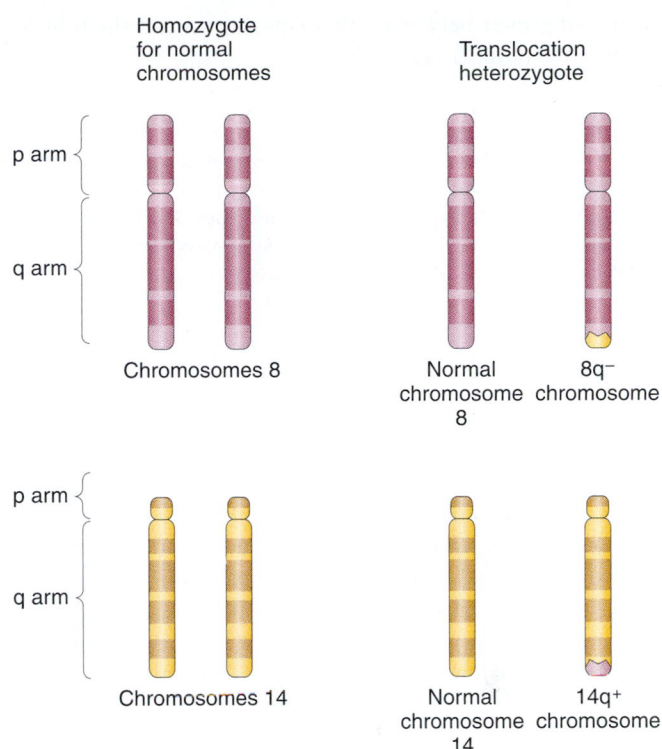

Figure 17-31 Position-effect variegation. (a) The translocation of w^+ to a position next to heterochromatin causes the w^+ function to fail in some cells, producing position-effect variegation. (b) A *Drosophila* eye showing the position-effect variegation. (Part b from Randy Mottus.)

Figure 17-32 Reciprocal translocations between chromosomes 8 and 14 cause most cases of Burkitt's lymphoma. An oncogene on the tip of chromosome 8 becomes relocated next to an antibody gene enhancer region on chromosome 14.

tion of antibodies (Figure 17-32). The oncogene is then activated, resulting in cancer.

MESSAGE ....
The expression of a gene can be affected by its position in the genome.

type was again variegated. Thus, we can conclude that, for some reason, the w^+ allele in the translocation is not expressed in some cells, thereby allowing the expression of w. This kind of variegation is called **position-effect variegation** because the unstable expression of a gene is due to its position in a rearrangement.

Such position effects can affect the genes that cause cancer. For example, most cases of Burkitt's lymphoma, a cancer of certain human antibody-producing cells called *B cells,* are caused by the relocation of a proto-oncogene to a position next to a region that normally enhances the produc-

Diagnosis of rearrangements by tetrad analysis

In fungi, tetrad analysis can be useful in detecting chromosomal aberrations. Any *Neurospora* genome containing a deletion produces an ascospore that does not ripen to the normal black color and is incapable of growth. The parents of crosses with a high proportion of such "aborted" white ascospores usually contain chromosome rearrangements. Duplications are generally recovered as black, viable ascospores.

Specific spore abortion patterns sometimes identify specific rearrangements. For example, a translocation causes three types of asci to be produced in certain proportions. Most asci have either eight black spores (produced from alternate-segregation meioses) or eight white spores (produced from adjacent-1-segregation meioses). A few asci have four black and four white spores, which are produced by crossing-over between either centromere and the translocation breakpoint (Figure 17-33).

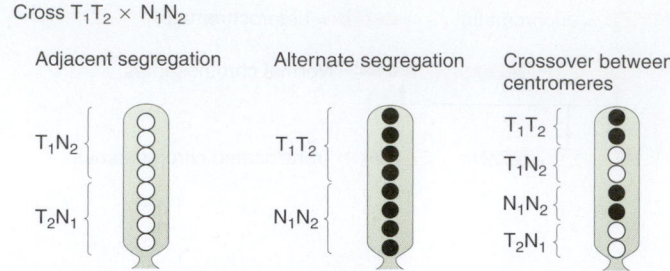

Figure 17-33 Abortion patterns in asci from a *Neurospora* cross between a normal strain and one with a reciprocal translocation. White sexual spores abort; black are viable. T_1 and T_2 are translocated chromosomes; N_1 and N_2 are normal chromosomes. The four black plus four white spores are produced by crossing-over between either centromere and the translocation breakpoint.

SUMMARY

The chromosome set mutates spontaneously to produce chromosome rearrangements or changes in chromosome number. All are called chromosome mutations.

Four rearrangements of chromosome structure are deletions, duplications, inversions, and translocations. Deletions are missing sections of chromosomes. If the region removed is essential to life, a homozygous deletion is lethal. Heterozygous deletions can be nonlethal or lethal and can uncover recessive alleles, causing them to be expressed phenotypically. Heterozygous deletions lower recombination between flanking loci. Deletions in one homolog often allow expression of recessive alleles in the other.

Tandem duplications can result from unequal crossover, and nontandem duplications arise from other rearrangements. Duplications can unbalance the genetic material, thereby producing a deleterious phenotypic effect in the organism or death. However, there is good evidence from a number of species, including humans, that duplications can lead to an increased variety of gene functions. In other words, duplications can be a source of new material for evolution.

An inversion is a 180-degree turn of a part of a chromosome. In the homozygous state, inversions may cause little

problem for an organism unless heterochromatin brings about a position effect or one of the breaks disrupts a gene. On the other hand, inversion heterozygotes have pairing difficulties at meiosis, and an inversion loop may result. Crossing-over within the loop results in inviable products. The crossover products of pericentric inversions, which span the centromere, differ from those of paracentric inversions, which do not span the centromere, but both show reduced recombinant frequency in the affected region and reduced fertility.

A translocation moves a chromosome segment to another position in the genome. In the heterozygous state, translocations produce duplication–deletion meiotic products, which can be lethal or cause lethality in the zygotes. New gene linkages can be produced by translocations. Translocation heterozygotes show 50 percent sterility (semisterility).

Chromosomal rearrangements are an important cause of ill health in human populations and are useful in engineering special strains of organisms in pure and applied biology. Extensive chromosomal rearrangement has occurred during evolution.

CONCEPT MAP

Draw a concept map interrelating as many of the following terms as possible. Note that the terms are listed in no particular order.

genome / deletion / loops / duplication / pairing / recombinant frequency / heterozygotes / inversions / lethality / phenotypic ratios

CHAPTER INTEGRATION PROBLEM

Geneticists conducted a *Drosophila* mutant hunt to find recessive lethal mutations on chromosome 2. Ten mutations of this type were obtained. The geneticists then crossed stocks carrying these mutations to produce hybrid zygotes that each bore a lethal chromosome from one stock plus a lethal chromosome from another. All paired combinations were crossed, and in every case the viability of the hybrid zygotes was recorded. The results were as shown in the table on the following page, where a plus means that the hybrids were viable and a minus means that the combination of the two independently isolated, recessive mutations was lethal.

Interpret these results as far as possible showing

♦ how a viable hybrid and an inviable hybrid can be produced genetically.

♦ how many genes are implicated in generating these results.

♦ what you can say about their location.

♦ the meaning of any departures from a pattern in the data.

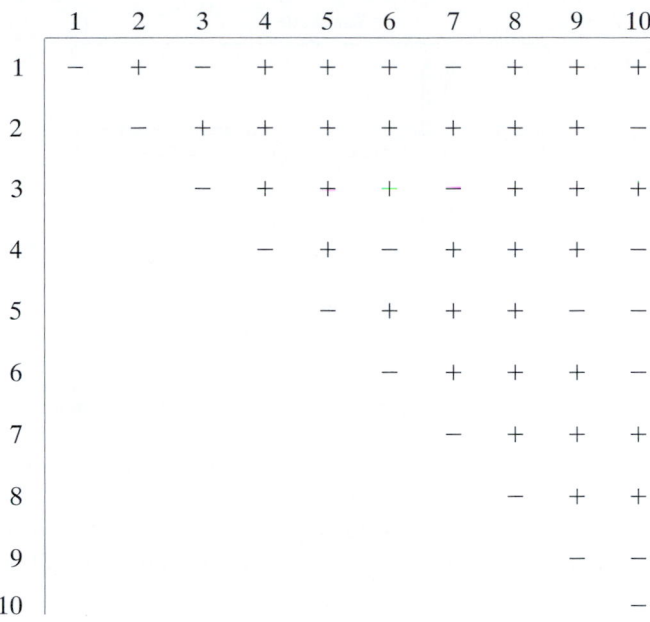

Because we know that the mutations are recessive, which implies that their wild-type alleles are dominant, we would predict that this type would be viable. In fact, in Chapter 4 we called this type of effect *genetic complementation,* based on the idea that the two wild-type alleles are "helping each other out." This effect then could explain some of the viability in the grid.

On the basis of these musings, one of the next things that can be done is to find out which mutations are in the same gene, and this information should also tell us something about how many genes are implicated.

Starting with mutation 1 and looking across the row, we find that it fails to complement with mutations 3 and 7, so we can tentatively infer that mutations 1, 3, and 7 are all in the same gene. Doing the same for all mutations, we arrive at the following groupings, which we write in columns:

1	2	4	5	8
3	10	6	9	
7		10	10	

We certainly see a pattern emerging, one that suggests that five genes had mutated in the stocks. But a striking inconsistency to this pattern is the behavior of mutation 10. This mutation fails to complement the mutations in three different groupings. What type of mutation could show this behavior? If we are to believe our hypothesis that five genes had mutated, mutation 10 must have damage that spans three separate genes. One possibility is that it is a deletion. Furthermore, we have learned in this chapter that deletions essentially behave as recessive lethals, which is consistent with the geneticists' finding mutation 10 in their search for recessive lethals.

So we could diagram what we have learned about chromosome 2 as follows:

		GENES		
A	*B*	*C*	*D*	*E*
1	2	4	5	8
3		6	9	
7				

Deletion

But note that this diagram is only one of many possible arrangements; we have no data that allow the ordering of *B*, *C*, and *D* within the deleted region, and furthermore we do not know whether *A* and *E* are on the same side or on opposite sides of the deletion.

Note the concepts used in this solution: dominance and recessiveness from Chapter 2, complementation from Chapter 4, gene mutation from Chapter 15, and deletion from the present chapter.

◆ Solution ◆

To begin, let's restate the system under investigation; in each cross the geneticists put together on a pair of homologous chromosomes two recessive lethal mutations that had been independently isolated. We can think of several possible underlying situations.

First, the mutations could be point mutations in the same gene:

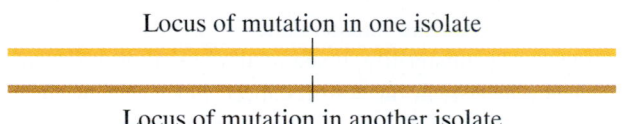

Locus of mutation in one isolate

Locus of mutation in another isolate

In this case we would predict that the hybrid genotype would be lethal because we have learned from Chapter 4 that this is how recessive lethals work. So perhaps some of the inviability recorded in the grid can be explained by the pairing of point mutations in the same gene.

Another possibility is that the two mutations are in different genes on chromosome 2:

Locus of mutation in one isolate

Wild-type allele

Wild-type allele

Locus of mutation in another isolate

SOLVED PROBLEMS

1. A corn plant is heterozygous for a reciprocal translocation and therefore is semisterile. This plant is backcrossed to a chromosomally normal strain that is homozygous for the recessive allele brachytic (*b*), located on chromosome 2. A semisterile F₁ plant is then backcrossed to the homozygous brachytic strain. The progeny obtained show the following phenotypes (see page 544):

NONBRACHYTIC		BRACHYTIC	
Semisterile	Fertile	Semisterile	Fertile
334	27	42	279

a. What ratio would you expect if the chromosome carrying the brachytic allele does not take part in the translocation?

b. Do you think that chromosome 2 took part in the translocation? Explain your answer, showing the conformation of the relevant chromosomes of the semisterile F_1 and the reason for the specific numbers obtained.

◆ Solution ◆

a. We should start with the methodical approach and simply restate the data in the form of a diagram, where

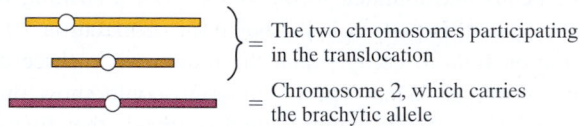

} = The two chromosomes participating in the translocation

= Chromosome 2, which carries the brachytic allele

To simplify the diagram, we do not show the chromosomes divided into chromatids (although they would be at this stage of meiosis):

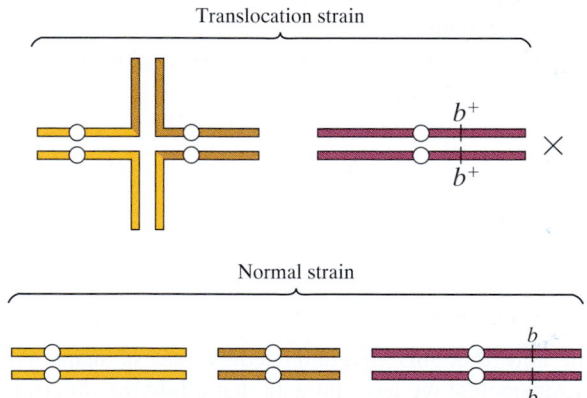

Translocation strain

Normal strain

All the progeny from this cross will be heterozygous for the chromosome carrying the brachytic allele, but what about the chromosomes participating in the translocation? In this chapter, we have seen that only alternate-segregation products survive and that half these survivors will be chromosomally normal and half will carry the two rearranged chromosomes. The rearranged combination will regenerate a translocation heterozygote when it combines with the chromosomally normal complement from the normal parent. These types—the semisterile F_1s—are diagrammed as part of the backcross to the parental brachytic strain:

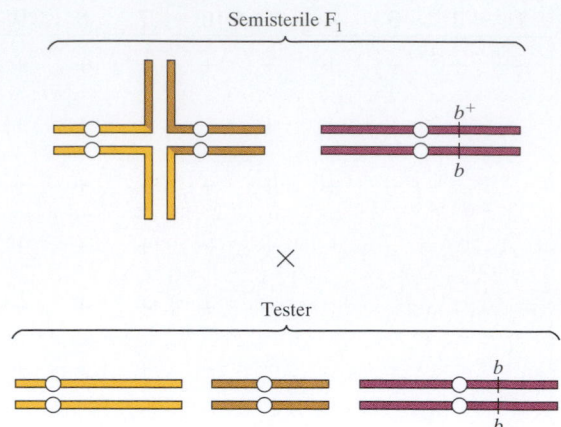

Semisterile F_1

×

Tester

In calculating the expected ratio of phenotypes from this cross, we can treat the behavior of the translocated chromosomes independently of the behavior of chromosome 2. Hence, we can predict that the progeny will be:

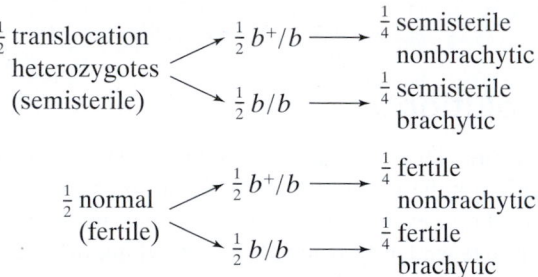

$\frac{1}{2}$ translocation heterozygotes (semisterile) → $\frac{1}{2} b^+/b$ → $\frac{1}{4}$ semisterile nonbrachytic
$\frac{1}{2} b/b$ → $\frac{1}{4}$ semisterile brachytic

$\frac{1}{2}$ normal (fertile) → $\frac{1}{2} b^+/b$ → $\frac{1}{4}$ fertile nonbrachytic
$\frac{1}{2} b/b$ → $\frac{1}{4}$ fertile brachytic

This predicted $1:1:1:1$ ratio is quite different from that obtained in the actual cross.

b. Because we observe a departure from the expected ratio based on the independence of the brachytic phenotype and semisterility, it seems likely that chromosome 2 *does* take part in the translocation. Let's assume that the brachytic locus (b) is on the orange chromosome. But where? For the purpose of the diagram, it does not matter where we put it, but it does matter genetically because the position of the b locus affects the ratios in the progeny. If we assume that the b locus is near the tip of the piece that is translocated, we can redraw the pedigree:

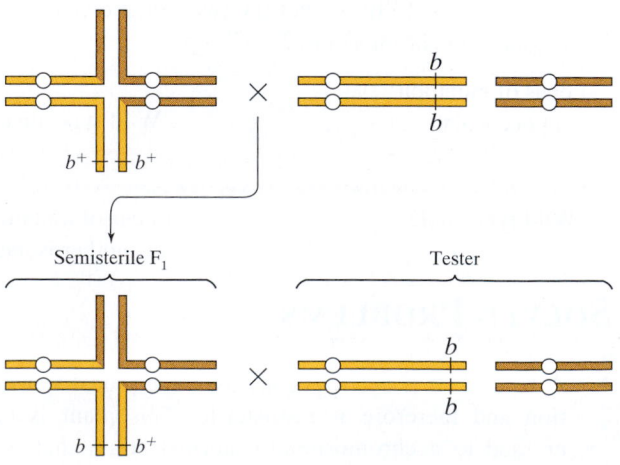

Semisterile F_1

Tester

If the chromosomes of the semisterile F_1 segregate as diagrammed here, we could then predict

$\frac{1}{2}$ fertile brachytic

$\frac{1}{2}$ semisterile nonbrachytic

Most progeny are certainly of this type, so we must be on the right track. How are the two less frequent types produced? Somehow we have to get the b^+ allele onto the normal orange chromosome and the b allele onto the translocated chromosome. This must be achieved by crossing-over between the translocation breakpoint (the center of the cross-shaped structure) and the brachytic locus. To represent this crossover, we must show chromatids, because crossing-over occurs at the chromatid stage:

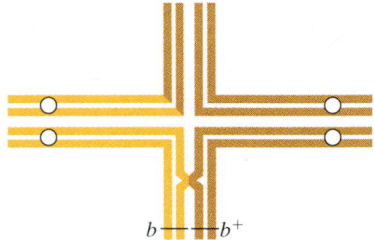

The recombinant chromosomes produce some progeny that are fertile and nonbrachytic and some that are semisterile and brachytic (these two classes together constitute 69 progeny of a total of 682, or a frequency of about 10 percent). We can see that this frequency is really a measure of the map distance (10 m.u.) of the brachytic locus from the breakpoint. (The same basic result would have been obtained if we had drawn the brachytic locus in the part of the chromosome on the other side of the breakpoint.)

2. A maize geneticist is studying recombination between two genes b and l, which are 18 m.u. apart on the same arm of chromosome 12. She is particularly interested in one pure-breeding line (M) isolated from nature. Crosses within line M give the expected RF of 18 percent between b and l, exactly the same RF obtained when she works within the conventional line (P). However, when an appropriately marked stock from M is crossed to an appropriately marked stock from P ($b^+ l^+/b^+ l^+ \times b\,l/b\,l$) and the F_1 plants are testcrossed to a $b\,l/b\,l$ tester from M, the RF value drops to 2 percent. The same value is obtained when the F_1 is crossed to a $b\,l/b\,l$ tester from P. Formulate a model and use it to explain

a.   the RF value within line M.

b.   the RF value within line P.

c.   the RF value in the $F_1 \times$ M testcross.

d.   the RF value in the $F_1 \times$ P testcross.

e.   the *origin* of recombinants in the testcrosses.

f.   how you would test your model.

◆ Solution ◆

In this problem, as in many genetic analyses, the data contain an important clue; when the significance of the clue has been realized, the details of the experimental results fall rapidly into place. Here the clue is the drastically reduced recombinant frequency. Not very many genetic mechanisms can cause such a reduction. In this chapter, we have studied two major ones: inversions and deletions. In this problem, however, the deletion hypothesis is highly unlikely for two reasons. First, it is clear that the unusual pure line M is viable, and we have seen that large deletions generally are not viable as homozygotes. Second, the recombinant frequency is normal within line M, which would not be true for a deletion. So we are left with the basic hypothesis that line M has a large inversion spanning all or most of the $b-l$ region. Initially, we might consider the possibility that one locus is located inside and one locus is located outside the inversion, but this possibility does not explain the values from crosses within line M (see answer to part a). We may draw diagrams consistent with our model, using R and L to indicate the normal right and left ends of the segment that may be inverted in M and b and l to indicate positions of the loci.

a.

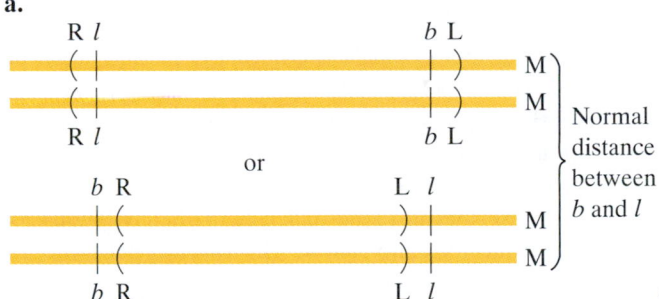

Notice that, if only one locus were inside the inversion and the other locus were outside, then a normal map distance would not prevail. For example,

b.

c.

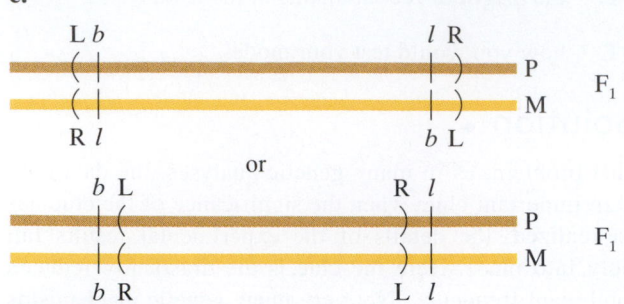

d.   Same F_1 as in part c. The tester makes no difference.

e.   We have been carrying two basic alternatives. In the first, the recombinants would have to come from double crossovers: one between the loci and one outside the loci but within the inversion. (Note that double crossovers such as two-strand doubles do not create a dicentric bridge.) In the second alternative, the recombinants could come from single crossovers in the small part of the $b-l$ region that is not spanned by the inversion.

f.   A simple cytological test is to look for inversion loops at meiosis or, if there are chromosome markers such as constrictions or staining bands, to look for inversion in mitotic chromosomes. For a genetic test, you could map the inverted loci in relation to chromosome markers outside the inversion, as shown here for r and s:

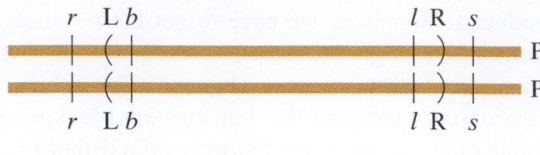

In the normal line P, b would map close to r; in line M, b would map close to s.

PROBLEMS

1. List the features (genetic or cytological) that identify and distinguish between the following: **(a)** deletions, **(b)** duplications, **(c)** inversions, **(d)** reciprocal translocations.

2. The two loci P and Bz are normally 36 m.u. apart on the same arm of a certain plant chromosome. A paracentric inversion spans about one-quarter of this region but does not include either of the loci. What approximate recombinant frequency would you predict for these loci in plants that are **(a)** heterozygous for the paracentric inversion? **(b)** Homozygous for the paracentric inversion?

3. Assume that the following loci in *Drosophila* are linked in the order $a-b-c-d-e-f$. A fly of genotype $a\,b\,c\,d\,e\,f/a\,b\,c\,d\,e\,f$ is crossed to a wild-type fly. About half the progeny are fully wild type in phenotype, but the other half show the recessive phenotype corresponding to d and e. Propose an explanation for these results.

4. The normal sequence of markers on a certain *Drosophila* chromosome is $123 \cdot 456789$, where the dot represents the centromere. Some flies were isolated with chromosome aberrations that have the following structures: **(a)** $123 \cdot 476589$, **(b)** $123 \cdot 46789$, **(c)** $1654 \cdot 32789$, **(d)** $123 \cdot 4566789$. Name each type of chromosome mutation and draw diagrams to show how each would synapse with the normal chromosome.

5. A *Neurospora* heterokaryon is established between cells of the genotypes shown in the diagram below. Here *leu, his, ad, nic,* and *met* are all recessive alleles causing specific nutritional requirements for growth. *A* and *a* are the mating-type alleles (one parent must be *A* and the other *a* for a cross to occur). Usually a nucleus carrying *A* will not pair with a nucleus carrying *a* to form a heterokaryon, but the recessive mutant *tol* suppresses this incompatibility and permits heterokaryons to grow on vegetative medium. The recessive allele *un* prevents the fungus from growing at 37°C (*un* is a temperature-sensitive allele) and cannot be corrected nutritionally. The heterokaryon grows well on a minimal medium, as do most of the cells derived mitotically from it. However, some rare cells show the following traits:

◆   They will not grow on a minimal medium unless it is supplemented with leucine.

◆   When the cells are transferred to a special medium that induces the reproductive cycle, a cross does not occur (they will not self).

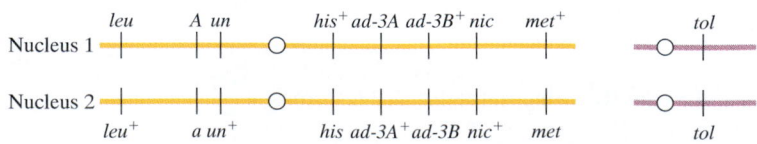

◆ They will not grow when moved into a 37°C temperature *even* when supplied with leucine.

◆ When haploid wild-type cells carrying the mating-type allele *a* are added to these aberrant cells, a cross occurs, but the addition of cells carrying the mating-type allele *A* does not cause a cross.

◆ From the cross with wild-type *a*, progeny with the genotype of nucleus 1 are recovered, but no alleles from nucleus 2 ever emerge from the cross.

Formulate an explanation for the origin of these strange cells in the original heterokaryon, and account for the observations concerning them.

6. Certain mice called *waltzers* execute bizarre steps in contrast with the normal gait for mice. The difference between normal mice and the waltzers is genetic, with waltzing being a recessive characteristic. W. H. Gates crossed waltzers with homozygous normals and found among several hundred normal progeny a single waltzing mouse (♀). When mated to a waltzing ♂, she produced all waltzing offspring. Mated to a homozygous normal ♂, she produced all normal progeny. Some ♂♂ and ♀♀ of this normal progeny were intercrossed, and there were no waltzing offspring among their progeny. T. S. Painter examined the chromosomes of waltzing mice that were derived from some of Gates's crosses and that showed a breeding behavior similar to that of the original, unusual waltzing ♀. He found that these individuals had 40 chromosomes, just as in normal mice or the usual waltzing mice. In the unusual waltzers, however, one member of a chromosome pair was abnormally short. Interpret these observations as completely as possible, both genetically and cytologically. (Problem 6 from A. M. Srb, R. D. Owen, and R. S. Edgar, *General Genetics,* 2d ed. W. H. Freeman and Company, 1965.)

7. *Neurospora crassa* mutations of the *ad-3B* gene are relatively easy to amass because they have a purple coloration in addition to a requirement for adenine. From haploid cultures, 100 spontaneous *ad-3B* mutants were obtained. Cells from each mutant strain were then plated on a minimal medium (containing no adenine) to test for reversion. Even after extensive platings with a wide array of mutagens, 13 cultures produced no colonies. What kind of mutations are they? Account for both the lack of revertibility and the viability of these strains.

8. Six bands in a salivary gland chromosome of *Drosophila* are shown here, along with the extent of five deletions (Del 1 to Del 5):

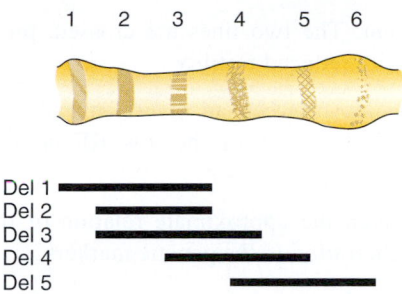

Recessive alleles *a, b, c, d, e,* and *f* are known to be in the region, but their order is unknown. When the deletions are combined with each allele, the following results are obtained:

	a	*b*	*c*	*d*	*e*	*f*
Del 1	−	−	−	+	+	+
Del 2	−	+	−	+	+	+
Del 3	−	+	−	+	−	+
Del 4	+	+	−	−	−	+
Del 5	+	+	+	−	−	−

In this table, a minus sign means that the deletion is missing the corresponding wild-type allele (the deletion uncovers the recessive) and a plus sign means that the corresponding wild-type allele is still present. Use these data to infer which salivary band corresponds to each gene. (Problem 8 from D. L. Hartl, D. Friefelder, and L. A. Snyder, *Basic Genetics.* Jones and Bartlett, 1988.)

9. Five recessive lethal mutations in *Drosophila* are all shown to map onto chromosome 2. Flies bearing each of the lethal mutations are then paired with flies from a stock having six recessive mutations (*h, i, j, k, l,* and *m*) that are distributed along chromosome 2 in that order. The appearance of the resulting flies is listed in the following diagram, where M stands for mutant for any particular phenotype and W stands for wild type.

Lethal mutation	Chromosome 2 marker					
	h	*i*	*j*	*k*	*l*	*m*
1	M	M	W	W	W	W
2	W	W	W	M	M	W
3	W	W	W	W	W	M
4	W	W	W	M	M	M
5	W	W	W	W	W	W

a. What is the probable basis of lethal mutations 1 through 4?

b. What can you say about lethal mutation 5?

10. Two pure lines of corn show different recombinant frequencies in the region from *P1* to *sm* on chromosome 6. The normal strain (A) shows an RF of 26 percent, and the abnormal strain (B) shows an RF of

8 percent. The two lines are crossed, producing hybrids with reduced fertility.

a. Decide between an inversion and a deletion as the possible cause of the low RF in the abnormal strain. Give your reason.

b. Sketch the approximate relation of the chromosome aberration to the genetic markers.

c. Why do the hybrids show reduced fertility?

11. In a fruit fly of genotype *P Bar Q/p Bar q*, the *P* and *Q* loci represent markers that closely flank the homozygous bar-eye mutation. In a testcross designed to reveal the meiotic products of such a fly, some normal-eye and some double-bar types are recovered at low frequencies. These show the phenotypes associated with the flanking marker combinations *P q* or *p Q*. Explain the following with diagrams:

a. The origin of the rare normal and double-bar types

b. The association of normal and bar-eye types with the flanking marker genotypes

12. A pure line of *Drosophila* is developed that carries a duplication of the X chromosome segment that contains the vermilion-eye gene. The stock is

$$\frac{v^+ \qquad v^-}{v^+ \qquad v^-}$$

and has wild-type color. Females of this stock are mated to vermilion-eyed males that lack the duplication:

$$\begin{array}{l} X \dfrac{v^+ \qquad v^-}{} \\ X \dfrac{}{v^+ \qquad v^-} \end{array}$$

$$\times$$

$$\begin{array}{l} X \dfrac{v^-}{} \\ Y \rule{1cm}{0.4pt} \end{array}$$

The male offspring all have wild-type eye color, and the female offspring all have vermilion eyes. **(a)** Explain why these results are surprising in regard to the theory of dominance. Explain the phenotypes of the following flies: **(b)** female and male parents; **(c)** female and male progeny.

13. A fruit fly was found to be heterozygous for a paracentric inversion. However, it was impossible to obtain flies that were homozygous for the inversion even after extensive crosses. What is the most likely explanation of this result?

14. The *Neurospora un3* locus is near the centromere on chromosome 1 and always segregates at the first mei-

otic division. The *ad3* locus is 10 m.u. to the other side of the same centromere.

a. What linear asci do you predict and in what frequencies in a normal cross of *un3 ad3* × wild type? Assume that only single crossovers or no crossovers occur in the *un3–ad3* region.

b. Most of the time such crosses behave predictably, but in one case a standard *un3 ad3* strain is crossed to a wild type isolated from a field of sugar cane in Hawaii. The results follow:

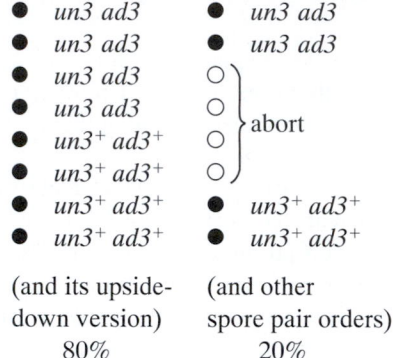

● *un3 ad3*	● *un3 ad3*
● *un3 ad3*	● *un3 ad3*
● *un3 ad3*	○ ⎫
● *un3 ad3*	○ ⎪
● *un3⁺ ad3⁺*	○ ⎬ abort
● *un3⁺ ad3⁺*	○ ⎭
● *un3⁺ ad3⁺*	● *un3⁺ ad3⁺*
● *un3⁺ ad3⁺*	● *un3⁺ ad3⁺*

(and its upside-
down version)
80%

(and other
spore pair orders)
20%

Explain these results, and state how you could test your idea.

15. A petunia is heterozygous for the following autosomal homologs:

A	B	C	D	E	F	G	H	I
a	b	c	d	h	g	f	e	i

a. Draw the pairing configuration that you would see at metaphase I and label all parts of your diagram. Number the chromatids sequentially from top to bottom of the page.

b. A three-strand double crossover occurs with one crossover between the *C* and *D* loci on chromatids 1 and 3, and the second crossover between *G* and *H* loci on chromatids 2 and 3. Diagram the results of these recombination events as you would see them at anaphase I and label your diagram.

c. Draw the chromosome pattern that you would see at anaphase II after the crossovers described in part b.

d. Give the genotypes of the gametes from this meiosis that will lead to the formation of viable progeny. Assume that all gametes are fertilized by pollen that has the gene order *A B C D E F G H I*.

16. An inversion heterozygote has one chromosome with the following linkage arrangement:

a	b	c	d	e	f	g

The other chromosome possesses an inversion that spans genes *c, d, e,* and *f.*

a. What type of inversion is it?

b. Draw the configuration of these chromosomes at synapsis.

c. If a four-strand double crossover occurs in the areas *c–d* and *e–f,* what are the meiotic products?

d. Repeat for a three-strand double crossover in these regions.

e. Will any of the meiotic products have a normal karyotype? Explain.

17. Orangutans are an endangered species in their natural environment (the islands of Borneo and Sumatra), so a captive-breeding program has been established by using orangutans currently held in zoos around the world. One component of this program is research into orangutan cytogenetics. This research has shown that all orangutans from Borneo carry one form of chromosome 2, as shown in the following diagram, and all orangutans from Sumatra carry the other form. Before this cytogenetic difference became known, some matings were carried out between animals from different islands, and 14 hybrid progeny are now being raised in captivity.

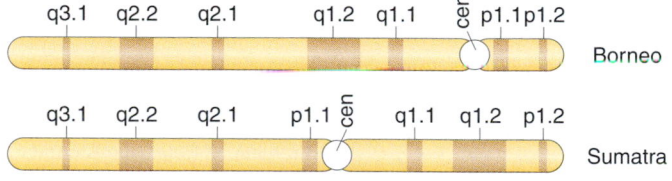

a. What term or terms describe the differences between these chromosomes?

b. Draw the chromosomes 2 of such a hybrid individual, paired during the first meiotic prophase. Be sure to show all the landmarks indicated in the figure, and label all parts of your drawing.

c. In 30 percent of meioses, there will be a crossover somewhere in the region between bands p1.1 and q1.2. Draw the gamete chromosomes 2 that would result from a meiosis in which a single crossover occurred within band q1.1.

d. What fraction of the gametes produced by a hybrid orangutan will give rise to viable progeny, if only these chromosomes differ between the parents?

(Problem 17 from Rosemary Redfield.)

18. Two groups of geneticists, in California and in Chile, begin work to develop a linkage map of the medfly. They both independently find that the loci for body color (*B* = black, *b* = gray) and eye shape (*R* = round, *r* = star) are linked 28 m.u. apart. They send strains to each other; a summary of all their findings is shown here:

Cross	F₁	Progeny of F₁ × any *b r*/*b r*	
B R/*B R* (Calif.)	*B R*/*b r*	*B R*/*b r*	36%
		b r/*b r*	36
× *b r*/*b r* (Calif.)		*B r*/*b r*	14
		b R/*b r*	14
B R/*B R* (Chile)	*B R*/*b r*	*B R*/*b r*	36
		b r/*b r*	36
× *b r*/*b r* (Chile)		*B r*/*b r*	14
		b R/*b r*	14
B R/*B R* (Calif.)	*B R*/*b r*	*B R*/*b r*	48
		b r/*b r*	48
× *b r*/*b r* (Chile)		*B r*/*b r*	2
or		*b R*/*b r*	2
b r/*b r* (Calif.)			
× *B R*/*B R* (Chile)			

a. Provide a genetic hypothesis that explains the three sets of testcross results.

b. Draw the key chromosomal features of meiosis in the F₁ from a cross of the Californian and Chilean lines.

19. *Drosophila* can be found throughout the world. A geneticist studying linkage on chromosome 2 collects pure-breeding strains in the Okanagan Valley (Canada) and in Spain. He crosses flies within each strain and scores the two sets of progeny for traits associated with six loci *a–f* on chromosome 2. He records the following recombinant frequencies (in percent), where I represents independent assortment, or RF = 50:

Okanagan:

	a	*b*	*c*	*d*	*e*	*f*
a	0	12	14	23	3	29
b		0	2	35	15	17
c			0	37	17	15
d				0	20	I
e					0	32
f						0

Spain:

	a	*b*	*c*	*d*	*e*	*f*
a	0	12	7	30	3	22
b		0	19	18	15	34
c			0	37	4	15
d				0	33	I
e					0	19
f						0

a. Analyze these data with regard to the arrangement of these loci in flies from each geographic location, and explain any difference or differences. Draw a map for each set of data, including map distances.

b. Show what recombinant frequencies might be expected in the five chromosomal regions delineated by these loci in Okanagan-Spanish hybrids.

20. An aberrant corn plant gives the following RF values when testcrossed:

	INTERVAL				
	d–f	f–b	b–x	x–y	y–p
Control	5	18	23	12	6
Aberrant plant	5	2	2	0	6

(The locus order is centromere–d–f–b–x–y–p.) The aberrant plant is a healthy plant, but it produces far fewer normal ovules and pollen than the control plant.

a. Propose a hypothesis to account for the abnormal recombination values and the reduced fertility in the aberrant plant.

b. Use diagrams to explain the origin of the recombinants according to your hypothesis.

21. The following corn loci are on one arm of chromosome 9, in the order indicated (the distances between them are shown in map units):

$$c \text{——} bz \text{——} wx \text{——} sh \text{——} d \text{——} centromere$$
$$\quad 12 \quad\quad 8 \quad\quad 10 \quad\quad 20 \quad\quad 10$$

C gives colored aleurone; c, white aleurone.
Bz gives green leaves; bz, bronze leaves.
Wx gives starchy seeds; wx, waxy seeds.
Sh gives smooth seeds; sh, shrunken seeds.
D gives tall plants; d, dwarf.

A plant from a standard stock that is homozygous for all five recessive alleles is crossed to a wild-type plant from Mexico that is homozygous for all five dominant alleles. The F_1 plants express all the dominant alleles and, when backcrossed to the recessive parent, give the following progeny phenotypes:

colored, green, starchy, smooth, tall	360
white, bronze, waxy, shrunk, dwarf	355
colored, bronze, waxy, shrunk, dwarf	40
white, green, starchy, smooth, tall	46
colored, green, starchy, smooth, dwarf	85
white, bronze, waxy, shrunk, tall	84
colored, bronze, waxy, shrunk, tall	8
white, green, starchy, smooth, dwarf	9
colored, green, waxy, smooth, tall	7
white, bronze, starchy, shrunk, dwarf	6

Propose a hypothesis to explain these results. Include:

a. a general statement of your hypothesis, with diagrams if necessary.

b. why there are 10 classes.

c. an account of the origin of each class, including its frequency.

d. at least one test of your hypothesis.

22. A *Drosophila* geneticist has a strain of fruit flies that is true breeding and wild type. She crosses this strain with a multiply marked X chromosome strain carrying the recessive alleles y (yellow), cv (crossveinless), v (vermilion), f (forked), and car (carnation), which are evenly spaced along the X chromosome from one end to the other. She collects the heterozygous F_1 female offspring and mates them with $y\ cv\ v\ f\ B\ car$ males (B gives bar, or slitlike, eye). The geneticist obtains the following classes among the male offspring:

1. y	cv	v	f	car	
2. y^+	cv^+	v^+	f^+	car^+	
3. y	cv^+	v^+	f^+	car	
4. y^+	cv	v	f	car^+	
5. y	cv	v^+	f	car	
6. y^+	cv^+	v	f^+	car^+	
7. y	cv	v^+	f^+	car	
8. y^+	cv^+	v	f	car^+	
9. y	cv^+	v^+	f	car	
10. y^+	cv	v	f^+	car^+	
11. y	cv	v	f	$B\ car$	

a. Account for the results in classes 1 to 10.

b. How can you account for class 11?

c. How would you test your hypothesis?

(Problem 22 from Tom Kaufman.)

*23. Suppose that you are given a *Drosophila* line from which you can get males or virgin females at any time. The line is homozygous for chromosome 2, which has an inversion to prevent crossing-over, a dominant allele (Cu) for curled wings, and a recessive allele (pr, purple) for dark eyes. The chromosome can be drawn as

You have irradiated sperm in a wild-type male and wish to determine whether recessive lethal mutations have been induced in chromosome 2. How would you go about doing this? (**Hint:** Remember that each sperm carries a *different* irradiated chromosome 2.) Indicate the kinds and numbers of flies used in each cross.

24. Predict the chromosome shapes that will be produced at anaphase 1 of meiosis in a reciprocal translocation heterozygote undergoing (**a**) alternate segregations and (**b**) adjacent-1 segregations.

25. The *Neurospora* loci *a* and *b* are on separate chromosomes. In a cross of a standard *a* ; *b* strain with a wild type obtained from nature, the progeny are as follows: *a* ; *b*, 45 percent; *a*⁺ ; *b*⁺, 45 percent; *a* ; *b*⁺, 5 percent; *a*⁺ ; *b*, 5 percent. Interpret these results, and explain the origin of all the progeny types according to your hypothesis.

26. Curly wings (*Cy*) is a dominant mutation on chromosome 2 of *Drosophila*. A *Cy/Cy*⁺ male is irradiated with X rays and crossed with *Cy*⁺/*Cy*⁺ females. The *Cy/Cy*⁺ sons are then mated individually with *Cy*⁺/*Cy*⁺ females. From one cross, the progeny are

curly males	146
wild-type males	0
curly females	0
wild-type females	163

What abnormality in chromosome structure is the most likely explanation for these results? Use chromosome diagrams of all strains referred to in the question in your explanation. (**Hint:** Remember that crossing-over does not occur in male *Drosophila*.)

27. You discover a *Drosophila* male that is heterozygous for a reciprocal translocation between chromosomes 2 and 3. Each chromosomal breakpoint is near the centromere, which is near the center of each of these metacentric chromosomes.

a. Draw a diagram showing how these chromosomes would synapse at meiosis.

b. You find that this fly has the recessive alleles *bw* (brown eye) and *e* (ebony body) on the nontranslocated chromosomes 2 and 3, respectively, and wild-type alleles on the translocated chromosomes. The fly is mated with a female that has normal chromosomes and is homozygous for *bw* and *e*. What type of offspring would you expect and in what ratio? (**Hint:** Remember that zygotes that have an extra chromosome arm or are deficient for one chromosome do not survive. There is no crossing-over in *Drosophila* males.)

28. An *insertional* translocation consists of a piece from the center of one chromosome inserted into the middle of another (nonhomologous) chromosome. Thus

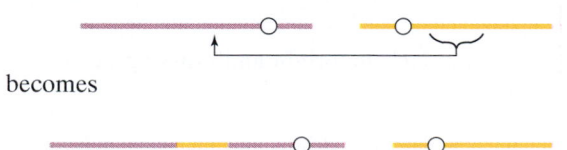

becomes

How will genomes that are heterozygous for such translocations pair at meiosis? What spore abortion patterns will be produced and in what relative proportions in a *Neurospora* cross between a wild type and an insertional translocation? (**Hint:** Remember that spores with chromosomal duplications survive and become dark, whereas those with deficiencies are light-spored and inviable.)

29. In corn, the genes for tassel length (alleles *T* and *t*) and rust resistance (alleles *R* and *r*) are known to be on separate chromosomes. In the course of making routine crosses, a breeder noticed that one *T/t* ; *R/r* plant gave unusual results in a testcross to the double-recessive pollen parent *t/t* ; *r/r*. The results were

Progeny:	*T/t* ; *R/r*	98
	t/t ; *r/r*	104
	T/t ; *r/r*	3
	t/t ; *R/r*	5

Corn cobs: only about half as many seeds as usual

a. What are the key features of the data that are different from expected results?

b. State a concise hypothesis that explains the results.

c. Show genotypes of parents and progeny.

d. Draw a diagram showing the arrangement of alleles on the chromosomes.

e. Explain the origin of the two classes of progeny having 3 and 5 members.

 Unpacking the Problem

1. What do "a gene for tassel length" and "a gene for rust resistance" mean?

2. Does it matter that the precise meanings of the allelic symbols *T*, *t*, *R*, and *r* are not given? Why or why not?

3. How do the terms *gene* and *allele*, as used here, relate to the concepts of locus and gene pair? (A concept map would be one way of answering this question.)

4. What prior experimental evidence would give the corn geneticists the idea that the two genes are on separate chromosomes?

5. What do you imagine "routine crosses" are to a corn breeder?

6. What term is used to describe genotypes of the type *T/t* ; *R/r*?

7. What is a "pollen parent"?

8. What are testcrosses and why do geneticists find them so useful?

9. What progeny types and frequencies might the breeder have been expecting from the testcross?

10. Describe how the observed progeny differ from expectations?

11. What does the approximate equality of the first two progeny classes tell you?

12. What does the approximate equality of the second two progeny classes tell you?

13. What were the gametes from the unusual plant and what were their proportions?

14. Which gametes were in the majority?

15. Which gametes were in the minority?

16. Which of the progeny types seem to be recombinant?

17. Which allele combinations appear to be linked in some way?

18. How can there be linkage of genes supposedly on separate chromosomes?

19. What do these majority and minority classes tell us about the genotypes of the parents of the unusual plant?

20. What is a corn cob?

21. What does a normal corn cob look like? (Sketch one and label it.)

22. What do the corn cobs from this cross look like? (Sketch one.)

23. What exactly is a kernel?

24. What effect could lead to the absence of half the kernels?

25. Did half the kernels die? If so, was the female or the male parent the reason for the death?

Now try to solve the problem.

30. The ascomycete fungus *Schizosaccharomyces pombe* produces unordered tetrads. A cross was made between a standard laboratory strain of genotype *nic-2; leu-3* and a wild-type strain isolated from grapes. Previous genetic studies on laboratory strains showed that *nic-2* is on chromosome 1 and *leu-3* is on chromosome 4. Therefore, it was surprising that, when approximately 100 tetrads were analyzed, there were only two tetrad types, as shown here, in equal frequency.

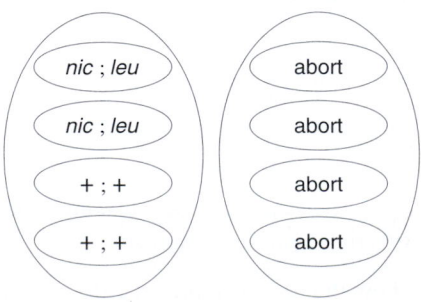

a. What was the expected result of this cross?

b. Provide a genetic explanation for the observed results, clearly showing the genotypes of the parents and ascospores under your model and explaining why there are only two ascus types of equal frequency.

(**Note:** In this fungus, genomes with deletions or duplications and deletions all abort at meiosis.)

31. In corn, the *g* locus (+ = purple, *g* = green) is on chromosome 2, and the *w* locus (+ = starchy, *w* = waxy) is on chromosome 9. A homozygous wild type was giving unusual results in routine crosses involving these markers. In one analysis, it was crossed to green, waxy, and the F_1 looked wild type, but about half its eggs and pollen aborted. The F_1 was testcrossed, and in 50 progeny there were 28 green, waxy showing normal gamete production, and 22 wild-type showing about 50 percent abortion of gametes.

a. Propose an explanation for these results by using diagrams of parents, F_1, and progeny.

b. If the F_1 were selfed, what phenotypes would result and in what proportions?

32. Yellow body in *Drosophila* is caused by a mutant allele *e* of a gene located at the tip of the X chromosome (the wild-type allele causes gray body). In a radiation experiment, a wild-type male was irradiated with X rays and then crossed with a yellow-bodied female. Most of the male progeny were yellow, as expected, but scanning of thousands of flies revealed two gray-bodied (phenotypically wild-type) males. These two males were crossed with yellow females, with the following results:

	Progeny
gray male 1 × yellow female	females all yellow males all gray
gray male 2 × yellow female	$\frac{1}{2}$ females yellow $\frac{1}{2}$ females gray $\frac{1}{2}$ males yellow $\frac{1}{2}$ males gray

a. Explain the origin and crossing behavior of gray male 1.

b. Explain the origin and crossing behavior of gray male 2.

33. Two auxotrophic mutations in *Neurospora*, *ad-3* and *pan-2*, are located on chromosomes 1 and 6, respectively. An unusual *ad-3* line arises in the laboratory, giving the following results:

	Ascospore appearance	RF between *ad-3* and *pan-2*
1. normal *ad-3* × normal *pan-2*	all black	50%
2. abnormal *ad-3* × normal *pan-2*	about $\frac{1}{2}$ black and $\frac{1}{2}$ white (inviable)	1%

3. Of the black spores from cross 2, about half were completely normal and half repeated the behavior of the original abnormal *ad-3* strain.

Explain all three results with the aid of clearly labeled diagrams. (**Note:** In *Neurospora*, ascospores with extra chromosomal material survive and are the normal black color, whereas ascospores lacking any chromosome region are white and inviable.)

34. Corn plants that carry the allele *P* produce purple leaves, and plants homozygous for *p* produce green leaves. The *P* locus is 20 m.u. from the centromere on the long arm of chromosome 9. A homozygous *p/p* plant is crossed with a *P/P* plant homozygous for a reciprocal translocation that has one breakpoint 30 m.u. from the centromere on the long arm of chromosome 9. The F$_1$ is obtained and backcrossed to the *p/p* plant. Predict what proportions of progeny from the backcross will be

a. green, semisterile.

b. green, fully fertile.

c. purple, semisterile.

d. purple, fully fertile.

e. If the F$_1$ is selfed, what proportion of progeny will be green, fully fertile?

35. Chromosomally normal corn plants have a *p* locus on chromosome 1 and an *s* locus on chromosome 5.

> *P* gives dark-green leaves; *p*, pale-green.
> *S* gives large ears; *s*, shrunken ears.

An original plant of genotype *P/p* ; *S/s* has the expected phenotype (large ears, dark-green) but gives unexpected results in crosses as follows:

♦ On selfing, fertility is normal, but the frequency of *p/p* ; *s/s* types is $\frac{1}{4}$ (not $\frac{1}{16}$, as expected).

♦ When crossed with a normal tester of genotype *p/p* ; *s/s*, the F$_1$ progeny are $\frac{1}{2}$ *P/p* ; *S/s* and $\frac{1}{2}$ *p/p* ; *s/s*; fertility is normal.

♦ When an F$_1$ *P/p* ; *S/s* plant is crossed with a normal *p/p* ; *s/s* tester, it proves to be semisterile, but again the progeny are $\frac{1}{2}$ *P/p* ; *S/s* and $\frac{1}{2}$ *p/p* ; *s/s*.

Explain these results, showing the full genotypes of the original plant, the tester, and the F$_1$ plants. How would you test your hypothesis?

36. A corn plant of genotype *pr/pr* that has standard chromosomes is crossed with a *Pr/Pr* plant that is homozygous for a reciprocal translocation between chromosomes 2 and 5. The F$_1$ is semisterile and phenotypically Pr (a seed color). A backcross to the parent with standard chromosomes gives 764 semisterile Pr; 145 semisterile pr; 186 normal Pr; and 727 normal pr. What is the map distance between the *Pr* locus and the translocation point?

37. A reciprocal translocation of the following types is obtained in *Neurospora*:

The following cross is then made:

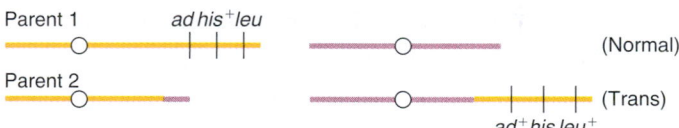

Assume that the small, purple piece of the chromosome in the translocation does not carry any essential genes. How would you select products of meiosis that are duplicated for the translocated part of the solid chromosome?

38. A male rat that is phenotypically normal shows reproductive anomalies when compared with normal male rats, as shown in the table below. Propose a genetic explanation of these unusual results, and say how your idea could be tested.

	EMBRYOS (MEAN NUMBER)			
Mating	Implanted in uterine wall	Degenerating after implantation	Normal	Degeneration (%)
exceptional ♂ × normal ♀	8.7	5.0	3.7	57.5
normal ♂ × normal ♀	9.5	0.6	8.9	6.5

39. In ascus formation in *Neurospora* any ascospore with a chromosomal deletion aborts and appears white and any ascospore with a duplication survives as a dark spore. What patterns of ascospore abortion would you observe, and in what approximate proportions, in octads from *Neurospora* crosses between a strain with normal chromosomes and one with

a. a reciprocal translocation?

b. a pericentric inversion?

c. a paracentric inversion?

Be sure to consider the effect of crossing-over in each one.

40. Suppose that you are studying the cytogenetics of five closely related species of *Drosophila*. The figure below gives the arrangement of loci (the letters indicate loci that are identical in all five species) and the chromosomal patterns that you find in each species. Show how these species probably evolved from each other, describing the changes occurring at each step. (**Note:** Be sure to compare gene order carefully.)

41. Show how the $\frac{10}{16}$ ratio on page 537 was derived.

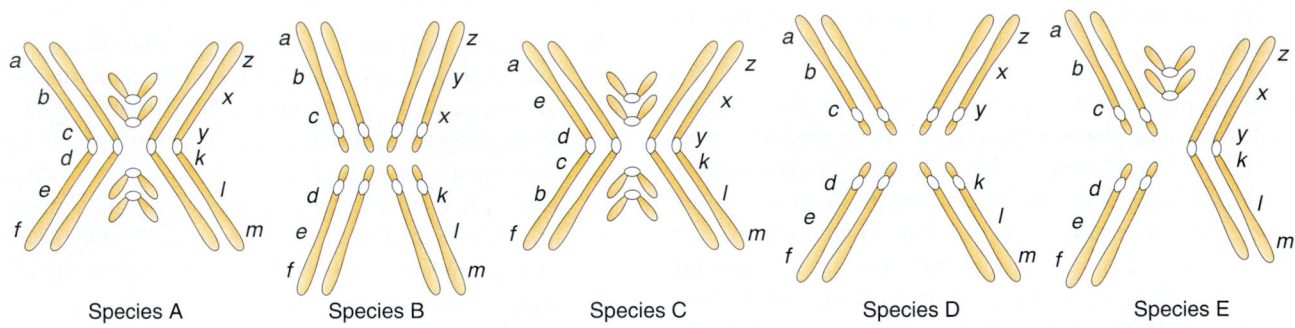

Species A   Species B   Species C   Species D   Species E

18

CHROMOSOME MUTATION II: CHANGES IN CHROMOSOME NUMBER

Key Concepts

Organisms with multiple chromosome sets (polyploids) are generally larger than diploid organisms, but meiotic pairing anomalies make some polyploid organisms sterile.

An even number of polyploid sets is generally more likely to result in fertility. Then the single-locus segregation ratios are different from those of diploids.

Crosses between two different species followed by the doubling of the chromosome number in the hybrid produces a special kind of fertile interspecific polyploid.

Variants in which a single chromosome has been gained or lost generally arise by nondisjunction (abnormal chromosome segregation at meiosis or mitosis).

Such variants tend to be sterile and show the abnormalities attributable to gene imbalance.

When fertile, such variants show abnormal gene segregation ratios for the misrepresented chromosome only.

A geneticist examining an orchid with multiple chromosome sets (a polyploid).
Compare the dimensions of the polyploid with those of its diploid parents shown in the foreground. *(Runk/Schoenberger/Grant Heilman.)*

The second major type of chromosome mutation is change in chromosome number. Few aspects of genetics impinge on human affairs quite so directly as this one. Chromosome numbers change spontaneously as accidents within cells, and this process has been going on for as long as life has existed on the planet. In fact, changes in chromosome number have been instrumental in molding genomes during evolution. For examples of such changes, we have to look no farther than the food on our dining tables because many of the plants (and some of the animals) that we eat arose through spontaneous changes in chromosome number in the course of the evolution of those species. Today, breeders emulate this process by manipulating chromosome number to improve productivity or some other useful feature of the organism. But perhaps the main relevance of chromosome numbers is for members of our own species in that a large proportion of genetically determined ill health in humans is caused by abnormal chromosome numbers.

In this chapter, we shall investigate the processes that produce new chromosome numbers, the diagnostic tests for detecting such changes, and the properties of cells and individual organisms carrying the different kinds of variant chromosomal complements. As with any area of cytogenetics, the techniques are a combination of genetics and microscopy. Changes in chromosome number are usually classified into two types — changes in whole chromosome sets and changes in parts of chromosome sets — and these two types are dealt with in the next two sections.

Aberrant euploidy

The number of chromosomes in a basic set is called the **monoploid number** (x). Organisms with multiples of the monoploid number of chromosomes are called **euploid.** We learned in earlier chapters that eukaryotes normally carry either one chromosome set (haploids) or two sets (diploids). Haploids and diploids, then, are both cases of normal euploidy. Euploid types that have more than two sets of chromosomes are called **polyploid.** The polyploid types are named **triploid** ($3x$), **tetraploid** ($4x$), **pentaploid** ($5x$), **hexaploid** ($6x$), and so forth. Polyploids arise naturally as spontaneous chromosomal mutations and, as such, they must be considered aberrations because they differ from the previous norm. However, many species of plants and animals have clearly arisen through polyploidy, so evidently evolution can take advantage of polyploidy when it arises. It is worth noting that organisms with one chromosome set sometimes arise as variants of diploids; such variants are called **monoploid** ($1x$). In some species, monoploid stages are part of the regular life cycle, but other monoploids are spontaneous aberrations.

The haploid number (n), which we have already used extensively, refers strictly to the number of chromosomes in gametes. In most animals and many plants with which we are familiar, the haploid number and monoploid number are the same. Hence, n or x (or $2n$ or $2x$) can be used inter-

changeably. However, in certain plants, such as modern wheat, n and x are different. Wheat has 42 chromosomes, but careful study reveals that it is hexaploid, with six rather similar but not identical sets of seven chromosomes. Hence, $6x = 42$ and $x = 7$. However, the gametes of wheat contain 21 chromosomes, so $n = 21$ and $2n = 42$.

Monoploids

Male bees, wasps, and ants are monoploid. In the normal life cycles of these insects, males develop parthenogenetically — that is, they develop from unfertilized eggs. However, in most species, monoploid individuals are abnormal, arising in natural populations as rare aberrations. The germ cells of a monoploid cannot proceed through meiosis normally, because the chromosomes have no pairing partners. Thus, monoploids are characteristically sterile. (Male bees, wasps, and ants bypass meiosis in forming gametes; here, mitosis produces the gametes.) If a monoploid cell does undergo meiosis, the single chromosomes segregate randomly, and the probability of all chromosomes going to one pole is $\left(\frac{1}{2}\right)^{x-1}$, where x is the number of chromosomes. This formula estimates the frequency of viable (whole-set) gametes, which is a small number if x is large.

Monoploids play an important role in modern approaches to plant breeding. Diploidy is an inherent nuisance when breeders want to induce and select new gene mutations that are favorable and to find new combinations of favorable alleles at different loci. New recessive mutations must be made homozygous before they can be expressed, and favorable allelic combinations in heterozygotes are broken up by meiosis. Monoploids provide a way around some of these problems. In some plant species, monoploids can be artificially derived from the products of meiosis in a plant's anthers. A cell destined to become a pollen grain can instead be induced by cold treatment to grow into an embryoid, a small dividing mass of cells. The embryoid can be grown on agar to form a monoploid plantlet, which can then be potted in soil and allowed to mature (Figure 18-1).

Plant monoploids can be exploited in several ways. In one, they are first examined for favorable traits or allelic combinations, which may arise from heterozygosity already present in the parent or induced in the parent by mutagens. The monoploid can then be subjected to chromosome doubling to achieve a completely homozygous diploid with a normal meiosis, capable of providing seed. How is this achieved? Quite simply, by the application of a compound called colchicine to meristematic tissue. **Colchicine** — an alkaloid drug extracted from the autumn crocus — inhibits the formation of the mitotic spindle, so cells with two chromosome sets are produced (Figure 18-2). These cells may proliferate to form a sector of diploid tissue that can be identified cytologically.

Another way in which the monoploid may be used is to treat its cells basically like a population of haploid organisms in a mutagenesis-and-selection procedure. A popula-

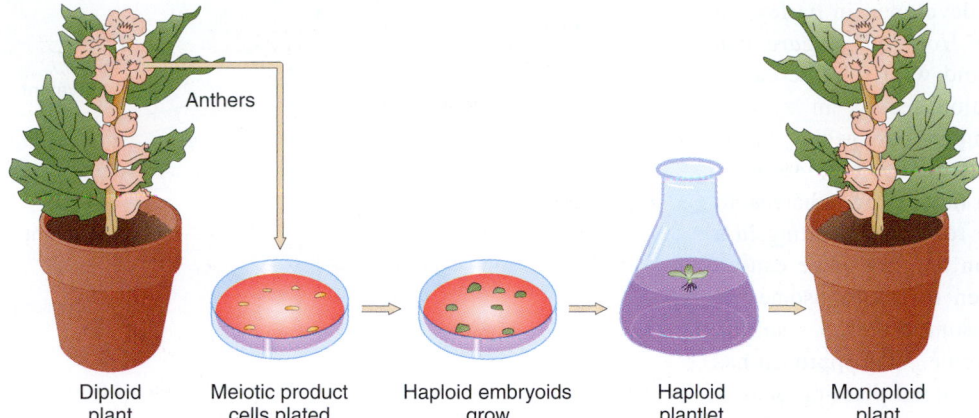

Figure 18-1   Generating monoploid plants by tissue culture. Pollen grains (haploid) are treated so that they will grow and are placed on agar plates containing certain plant hormones. Under these conditions, haploid embryoids will grow into monoploid plantlets. After having been moved to a medium containing different plant hormones, these plantlets will grow into mature monoploid plants with roots, stems, leaves, and flowers.

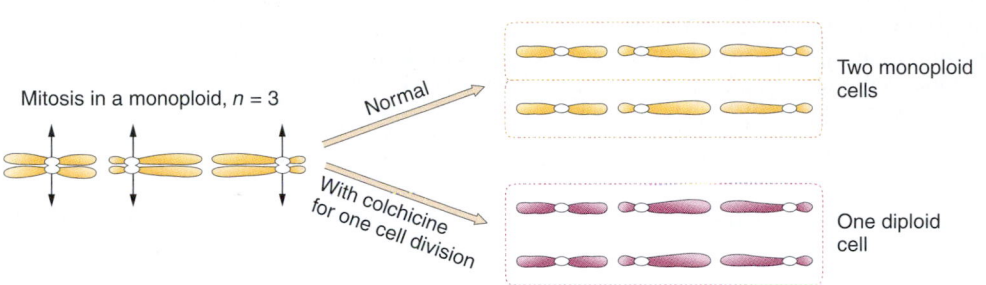

Figure 18-2   The use of colchicine to generate a diploid from a monoploid. Colchicine added to mitotic cells during metaphase and anaphase disrupts spindle-fiber formation, preventing the migration of chromatids after the centromere is split. A single cell is created that contains pairs of identical chromosomes that are homozygous at all loci.

tion of cells is isolated, their walls are removed by enzymatic treatment, and they are treated with mutagen. They are then plated on a medium that selects for some desirable phenotype. This approach has been used to select for resistance to toxic compounds produced by one of the plant's parasites and to select for resistance to herbicides being used by farmers to kill weeds. Resistant plantlets eventually grow into haploid plants, which can then be doubled (with

the use of colchicine) into a pure-breeding, diploid, resistant type (Figure 18-3).

These powerful techniques can circumvent the normally slow process of meiosis-based plant breeding. The techniques have been successfully applied to several important crop plants, such as soybeans and tobacco.

The anther technique for producing monoploids does not work in all organisms or in all genotypes of an organism.

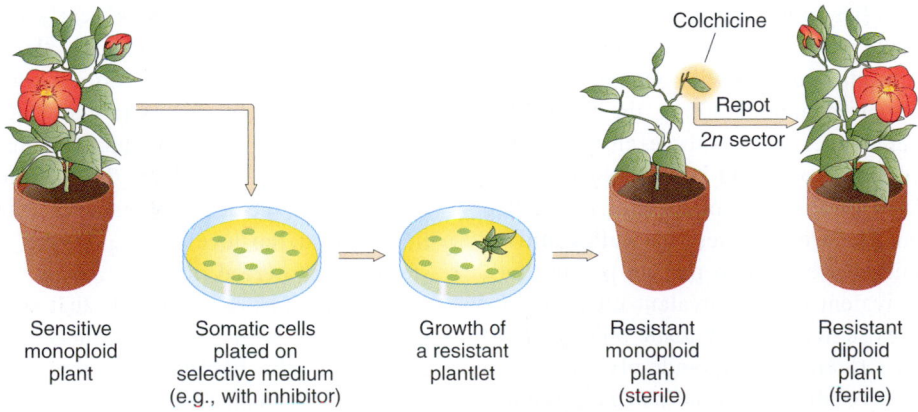

Figure 18-3   Using microbial techniques in plant engineering. The cell walls of haploid cells are removed enzymatically. The cells are then exposed to a mutagen and plated on an agar medium containing a selective agent, such as a toxic compound produced by a plant parasite. Only those cells containing a resistance mutation that allows them to live within the presence of this toxin will grow. After treatment with the appropriate plant hormones, these cells will grow into mature monoploid plants and, with proper colchicine treatment, can be converted into homozygous diploid plants.

Another useful technique has been developed in barley, an important crop plant. Diploid barley, *Hordeum vulgare,* can be fertilized by pollen from a diploid wild relative called *Hordeum bulbosum.* This fertilization results in zygotes with one chromosome set from each parental species. In the ensuing somatic cell divisions, however, the chromosomes of *H. bulbosum* are eliminated from the zygote, whereas all the chromosomes of *H. vulgare* are retained, resulting in a haploid embryo. (The haploidization appears to be caused by a genetic incompatibility between the chromosomes of the different species.) The chromosomes of the resulting haploids can be doubled with colchicine. This approach has led to the rapid production and widespread planting of several new barley varieties, and it is being used successfully in other species too.

MESSAGE ··
To create new plant lines, geneticists produce monoploids with favorable genotypes and then double the chromosomes to form fertile, homozygous diploids.

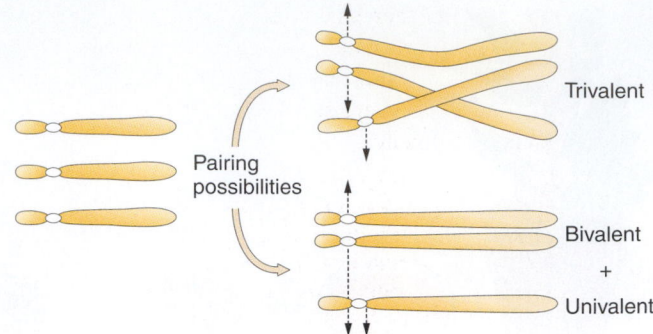

Figure 18-4 Two possibilities for the pairing of three homologous chromosomes before the first meiotic division in a triploid. Notice that the outcome will be the same in both cases: one resulting cell will receive two chromosomes and the other will receive just one. The probability that the latter cell can become a functional haploid gamete is very small, however, because, to do so, it would also have to receive only one of the three homologous chromosomes of every other set in the organism. Note that each chromosome is really a pair of chromatids.

Polyploids

In the realm of polyploids, we must distinguish between **autopolyploids,** which are composed of multiple sets from within one species, and **allopolyploids,** which are composed of sets from different species. Allopolyploids form only between closely related species; however, the different chromosome sets are **homeologous** (only partly homologous) — not fully homologous, as they are in autopolyploids.

Triploids

Triploids are usually autopolyploids. They arise spontaneously in nature or are constructed by geneticists from the cross of a $4x$ (tetraploid) and a $2x$ (diploid). The $2x$ and the x gametes unite to form a $3x$ triploid.

Triploids are characteristically sterile. The problem, like that of monoploids, lies in pairing at meiosis. Synapsis, or true pairing, can take place only between two chromosomes, but one chromosome can pair with one partner along part of its length and with another along the remainder, which gives rise to an association of three chromosomes. Paired chromosomes of the type found in diploids are called **bivalents.** Associations of three chromosomes are called **trivalents,** and unpaired chromosomes are called **univalents.** Hence in triploids there are two pairing possibilities, resulting either in a trivalent or in a bivalent plus a univalent. Paired centromeres segregate to opposite poles, but unpaired centromeres pass to either pole randomly. We see in Figure 18-4 that the net result of both the pairing possibilities is an uneven segregation, with two chromosomes going in one direction and one in the other. This happens for every chromosome threesome.

If all the single chromosomes pass to the *same* pole and simultaneously the other two chromosomes pass to the op-

posite pole, then the gametes formed will be haploid and diploid. The probability of this type of meiosis will be $(\frac{1}{2})^{x-1}$, and this proportion is likely to be low. All other possibilities will give gametes with chromosome numbers intermediate between the haploid and diploid number; such genomes are **aneuploid** — "not euploid." It is likely that these aneuploid gametes will not lead to viable progeny; in fact, this category is responsible for the almost complete lack of fertility of triploids. The problem is one of **genome imbalance,** a phenomenon that we shall encounter repeatedly in this chapter. For most organisms, the euploid chromosome set is a finely tuned set of genes in relative proportions that seem to be functionally significant. Multiples of this set are tolerated because there is no change in the relative proportions of genes. However, the addition of one or more extra chromosomes is nearly always deleterious because the proportions of genes in those extra chromosomes are altered. Although the action of some genes can be regulated to compensate for extra gene "dosage," the overall effect of the extra genetic material seems too great to be overcome by gene regulation. The deleterious effect can be expressed at the level of gametes, making them nonfunctional, or at the level of the zygote, resulting in lethality, sterility, or lowered fitness.

In triploids, it is possible that some haploid or diploid gametes will form, and some may unite to form a euploid zygote, but the likelihood of this possibility is inherently low. Consider bananas. The bananas that are widely available commercially are triploids with 11 chromosomes in each set ($3x = 33$). The probability of a meiosis in which all univalents pass to the same pole is $(\frac{1}{2})^{x-1}$, or $(\frac{1}{2})^{10} = \frac{1}{1024}$, so bananas are effectively sterile. The most obvious expression of the sterility of bananas is that there are no seeds in the fruit that we eat. Another example of the commercial exploitation of triploidy in plants is the production of triploid watermelons. For the same reasons that bananas are seedless, triploid watermelons are seedless, a phenotype favored by some for its convenience.

> **MESSAGE** ···
> Some types of chromosome mutations are themselves aneuploid; other types produce aneuploid gametes or zygotes. Aneuploidy is nearly always deleterious because of genetic imbalance—the ratio of genes is different from that in euploids and interferes with the normal operation of the genome.

Autotetraploids

Autotetraploids arise naturally by the spontaneous accidental doubling of a $2x$ genome to a $4x$ genome, and autotetraploidy can be induced artificially through the use of colchicine. Autotetraploid plants are advantageous as commercial crops because, in plants, the larger number of chromosome sets often leads to increased size. Cell size, fruit

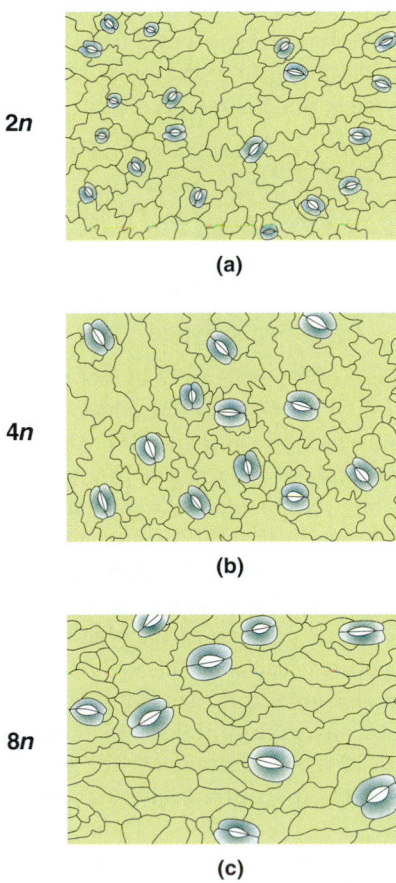

Figure 18-5 Epidermal leaf cells of tobacco plants, showing an increase in cell size, particularly evident in stoma size, with an increase in autopolyploidy. (a) Diploid, (b) tetraploid, (c) octoploid. (From W. Williams, *Genetic Principles and Plant Breeding.* Blackwell Scientific Publications, Ltd.)

size, flower size, stomata size, and so forth, can be larger in the polyploid (Figure 18-5). Here we see another effect that must be explained by gene numbers. Presumably the amount of gene product (protein or RNA) is proportional to the number of genes in the cell, and this number is higher in the cells of polyploids compared with diploids.

> **MESSAGE** ···
> Polyploid plants are often larger and have larger organs than their diploid relatives.

Because 4 is an even number, autotetraploids can have a regular meiosis, although this is by no means always the case. The crucial factor is how the four homologous chromosomes, one from each of the four sets, pair and segregate. There are several possibilities, as shown in Figure 18-6. Pairings between four chromosomes are called **quadrivalents.** In tetraploids, the two-bivalent and the quadrivalent pairing modes tend to be most regular in segregation, but even here there is no guarantee of a $2:2$ segregation. If all chromosome sets segregate $2:2$ as they do in some species, then the gametes will be functional and a formal genetic analysis can be developed for such autotetraploids.

Let's consider the genetics of a fertile tetraploid. We can consider an experiment in which colchicine is used to double the chromosomes of an A/a plant to form an $A/A/a/a$ autotetraploid, which we will assume shows $2:2$ segregation. We now have a further concern because polyploids such as tetraploids give different phenotypic ratios in their progeny, depending on whether the locus in question is tightly linked to the centromere. First, we consider a centromere-linked gene. The three possible pairing and segregation patterns are

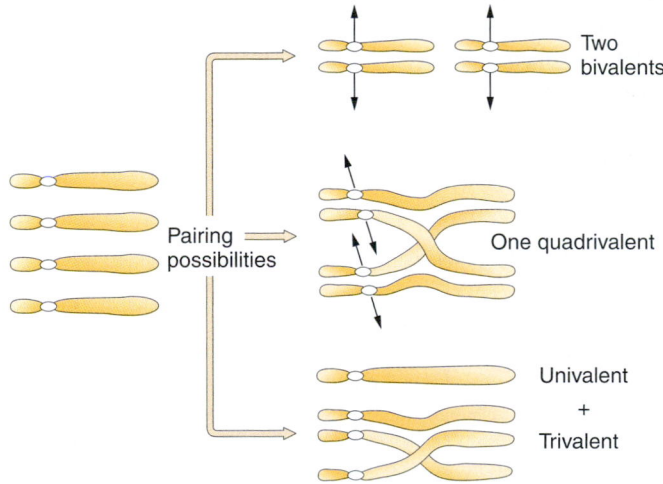

Figure 18-6 Meiotic pairing possibilities in tetraploids. (Each chromosome is really two chromatids.) The four homologous chromosomes may pair as two bivalents or as a quadrivalent. Both possibilities can yield functional gametes. However, the four chromosomes may also pair in a univalent-trivalent combination, yielding nonfunctional gametes. A specific tetraploid can show one or more of these pairings.

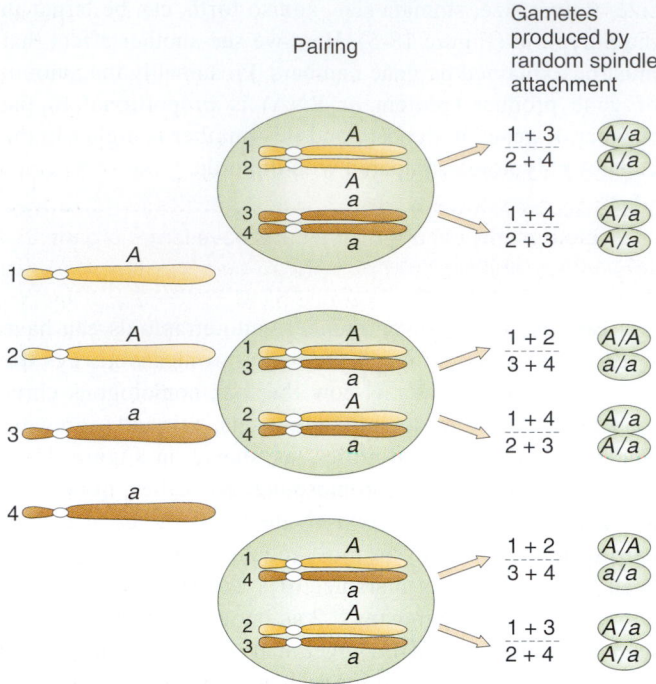

Pairing

Gametes produced by random spindle attachment

Figure 18-7 Gene segregation in a tetraploid showing orderly pairing by bivalents. (Each chromosome is really two chromatids.) The locus is assumed to be close to the centromere. Self-fertilization could yield a variety of genotypes, including $a/a/a/a$.

presented in Figure 18-7; these patterns occur by chance and with equal frequency. As Figure 18-7 shows, the $2x$ gametes produced are A/a, A/A, or a/a, in a ratio of 8:2:2, or 4:1:1. If such a plant is selfed, the probability of an $a/a/a/a$ phenotype in the offspring is $\frac{1}{6} \times \frac{1}{6} = \frac{1}{36}$. In other words, a 35:1 phenotypic ratio of $A/-/-/- : a/a/a/a$ will be observed if A is fully dominant over three a alleles.

If, in the same kind of plant, a genetic locus having the alleles B and b is very far removed from the centromere, crossing-over must be considered. This example forces us to

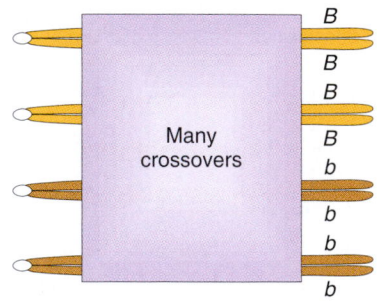

Figure 18-8 Highly diagrammatic representation of a tetraploid meiosis in which a heterozygous locus is distant from the centromere. The net effect of multiple crossovers in such a long region will be that the genes become effectively unlinked from their original centromeres. Genes are packaged two at a time into gametes, much as two balls may be grabbed at random from a bag containing eight balls: four of one kind and four of another.

think in terms of chromatids instead of chromosomes; there are four B chromatids and four b chromatids (Figure 18-8). Because the number of crossovers in such a long region will be large, the genes will become effectively unlinked from their original centromeres. The packaging of genes two at a time into gametes is very much like grabbing two balls at random from a bag of eight balls: four of one kind and four of another. The probability of picking two b genes is then

$$\frac{4}{8} \text{ (the first one)} \times \frac{3}{7} \text{ (the second one)} = \frac{12}{56} = \frac{3}{14}$$

So, in a selfing, the probability of a $b/b/b/b$ phenotype is $\frac{3}{14} \times \frac{3}{14} = \frac{9}{196}$, which is approximately $\frac{1}{22}$. Hence, there will be a 21:1 phenotypic ratio of $B/-/-/- : b/b/b/b$. For genetic loci of intermediate position, intermediate ratios will result.

Allopolyploids

The "classic" **allopolyploid** was synthesized by G. Karpechenko in 1928. He wanted to make a fertile hybrid that would have the leaves of the cabbage *(Brassica)* and the roots of the radish *(Raphanus)*. Each of these species has 18 chromosomes, and they are related closely enough to allow intercrossing. A viable hybrid progeny individual was produced from seed. However, this hybrid was functionally sterile because the nine chromosomes from the cabbage parent were different enough from the radish chromosomes that pairs did not synapse and disjoin normally:

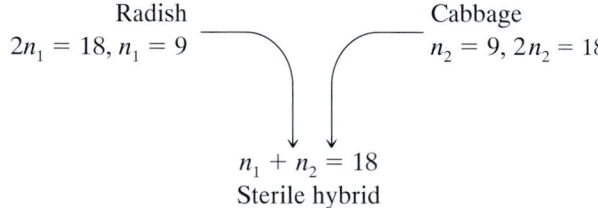

Radish   Cabbage

$2n_1 = 18, n_1 = 9$ $n_2 = 9, 2n_2 = 18$

$n_1 + n_2 = 18$
Sterile hybrid

However, one day a few seeds were in fact produced by this (almost) sterile hybrid. On planting, these seeds produced fertile individuals with 36 chromosomes. All these individuals were allopolyploids. They had apparently been derived from spontaneous, accidental chromosome doubling to $2n_1 + 2n_2$ in the sterile hybrid, presumably in tissue that eventually became germinal and underwent meiosis. Thus, in $2n_1 + 2n_2$ tissue, there is a pairing partner for each chromosome and balanced gametes of the type $n_1 + n_2$ are produced. These gametes fuse to give $2n_1 + 2n_2$ allopolyploid progeny, which also are fertile. This kind of allopolyploid is sometimes called an **amphidiploid,** which means "doubled diploid" (Figure 18-9). (Unfortunately for Karpechenko, his amphidiploid had the roots of a cabbage and the leaves of a radish.)

When the allopolyploid was crossed with either parental species, sterile offspring resulted. The offspring of the cross with radish were $2n_1 + n_2$, constituted from an $n_1 + n_2$ gamete from the allopolyploid and an n_1 gamete from the radish. The n_2 chromosomes had no pairing partners, so sterility resulted. Consequently, Karpechenko had effectively created a new species, with no possibility of

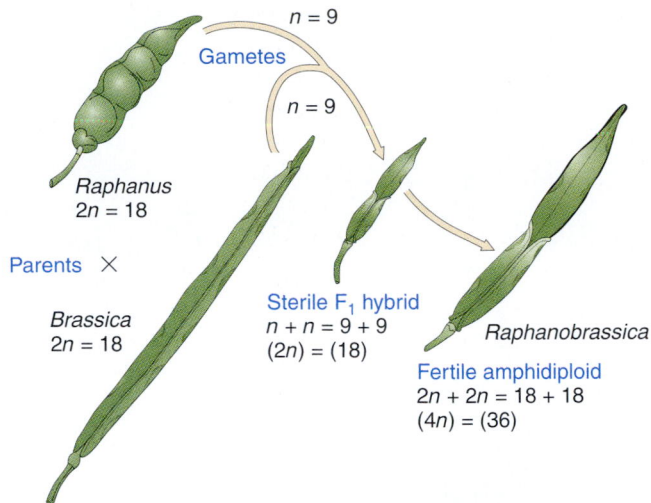

Figure 18-9 The origin of the amphidiploid *(Raphanobrassica)* formed from cabbage *(Brassica)* and radish *(Raphanus)*. The fertile amphidiploid arose in this case from spontaneous doubling in the $2n = 18$ sterile hybrid. Colchicine can be used to promote doubling. (From A. M. Srb, R. D. Owen, and R. S. Edgar, *General Genetics,* 2d ed. Copyright © 1965 by W. H. Freeman and Company. After G. Karpechenko, *Z. Indukt. Abst. Vererb.* 48, 1928, 27.)

gene exchange with its parents. He called his new species *Raphanobrassica.*

Today, allopolyploids are routinely synthesized in plant breeding. Instead of waiting for spontaneous doubling to occur in the sterile hybrid, the plant breeder adds colchicine to induce doubling. The goal of the breeder is to combine some of the useful features of both parental species into one type. This kind of endeavor is very unpredictable, as Karpechenko learned. In fact, only one synthetic amphidiploid has ever been widely used. This amphidiploid is *Triticale,* an amphidiploid between wheat *(Triticum,* $2n = 6x = 42$) and rye *(Secale,* $2n = 2x = 14$). *Triticale* combines the high yields of wheat with the ruggedness of rye. Figure 18-10 shows the procedure for synthesizing *Triticale.*

In nature, allopolyploidy seems to have been a major force in speciation of plants. There are many different examples. One particularly satisfying one is shown by the genus *Brassica,* as illustrated in Figure 18-11. Here three different parent species have hybridized in all possible pair combinations to form new amphidiploid species.

A particularly interesting natural allopolyploid is bread wheat, *Triticum aestivum* ($2n = 6x = 42$). By studying various wild relatives, geneticists have reconstructed a probable

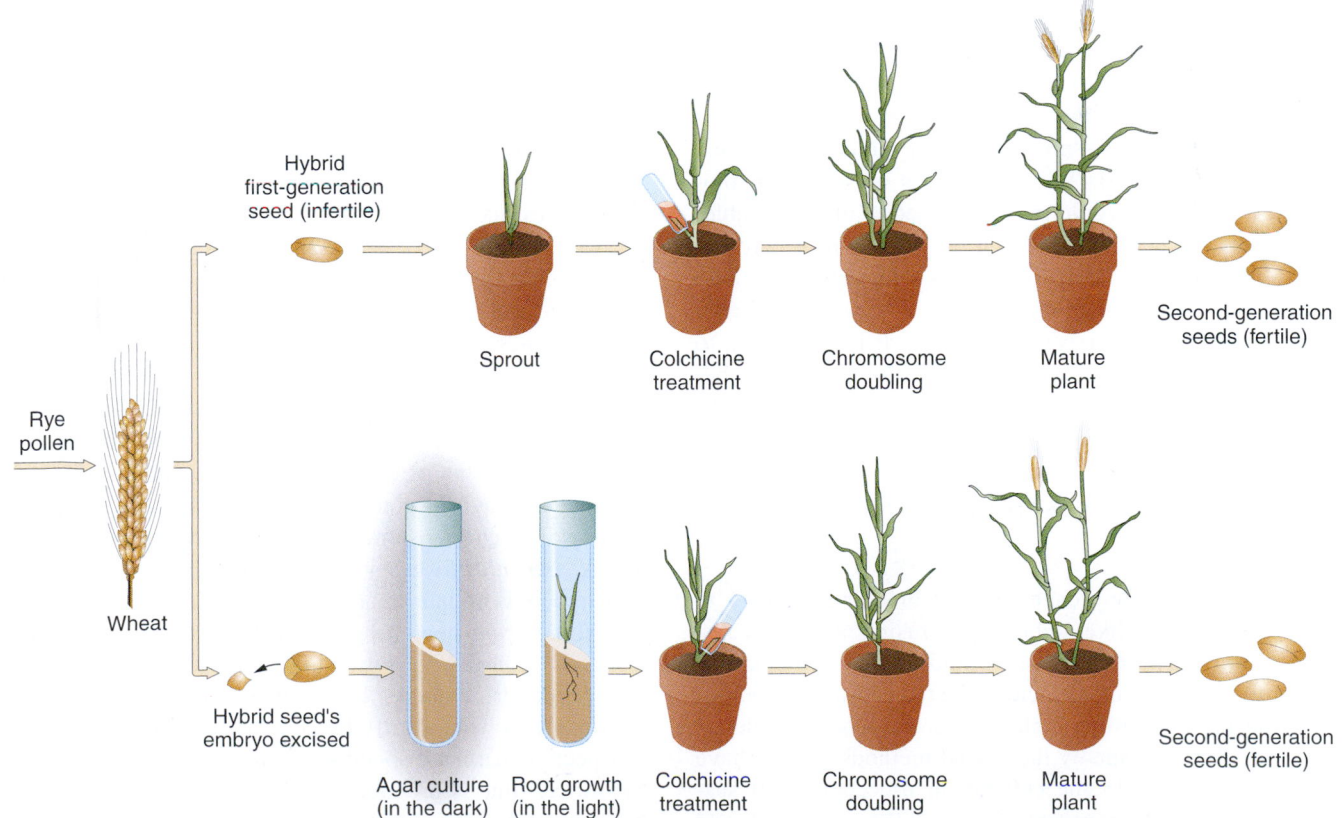

Figure 18-10 Techniques for the production of the amphidiploid *Triticale.* If the hybrid seed does not germinate, then tissue culture *(lower path)* may be used to obtain a hybrid plant. (From Joseph H. Hulse and David Spurgeon, "Triticale." Copyright © 1974 by Scientific American, Inc. All rights reserved.)

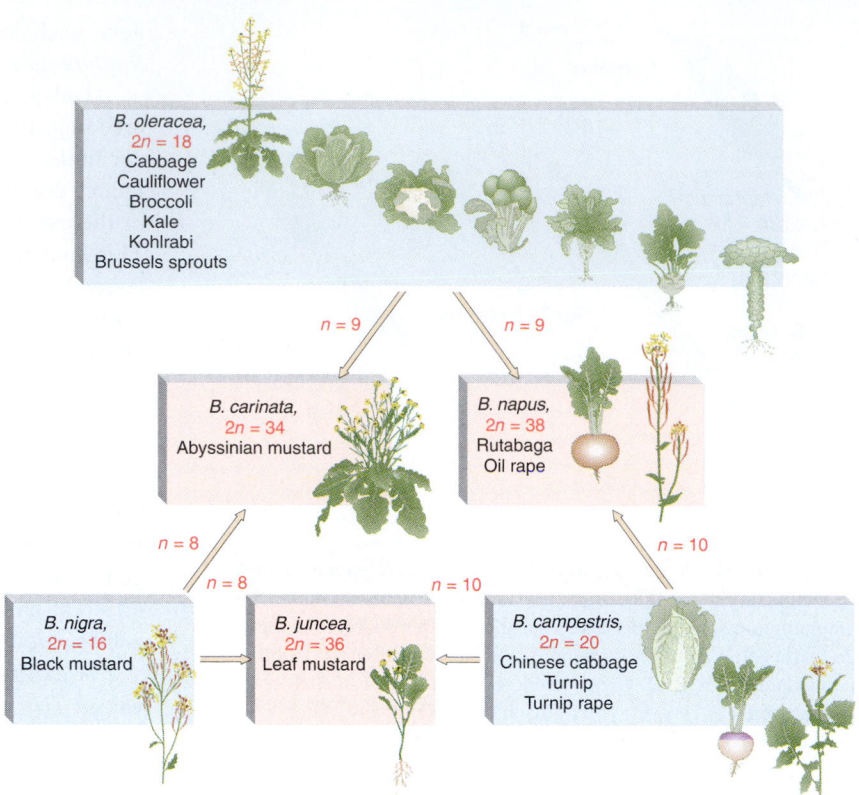

Figure 18-11 A species triangle, showing how amphidiploidy has been important in the production of new species of *Brassica*.

evolutionary history of bread wheat (Figure 18-12). In a bread wheat meiosis, there are always 21 pairs of chromosomes. Furthermore, it has been possible to establish that any given chromosome has only one specific pairing partner (homologous pairing) — not five other potential partners (homeologous pairing). The suppression of such homeologous pairing (which would make the species more unstable) is maintained by an allele, *Ph*, on the long arm of chromosome 5 of the B set. Thus, *Ph* ensures a diploidlike meiotic behavior for this hexaploid species. Without *Ph*, bread wheat could probably never have arisen. It is interesting to speculate about whether Western civilization could have begun or progressed without this species — in other words, without the *Ph* mutation.

Somatic allopolyploids from cell hybridization

Another innovative approach to plant breeding is to try to make allopolyploid-like hybrids by fusing asexual cells. Theoretically, such a technique would allow us to combine widely differing parental species. The technique does indeed work, but the only allopolyploids that have been produced so far can also be made by the sexual methods that we have considered already. In the cell-fusion procedure, cell suspensions of the two parental species are prepared and stripped of their cell walls by special enzyme treatments. The stripped cells are called **protoplasts.** The two protoplast suspensions are then combined with polyethylene glycol, which enhances protoplast fusion. The parental cells

and the fused cells proliferate on an agar medium to form colonies (in much the same way as microbes). If these colonies, or *calluses* as they are called, are examined, a fair percentage of them are found to be allopolyploid-like hybrids with chromosome numbers equal to the sum of the parental numbers. Thus, not only do the protoplast cell membranes fuse to form a kind of heterokaryon, but the nuclei fuse, too, to give rise to a $2n_1 + 2n_2$ amphidiploid.

Another good example of an allopolyploid-like hybrid is commercial tobacco, *Nicotiana tabacum,* which has 48 chromosomes. This species of tobacco was originally found in nature as a spontaneous amphidiploid. The two probable parents are *N. sylvestris* and *N. tomentosiformis,* each of which has 24 chromosomes. A sexual cross between *N. tabacum* and either of these two probable parents give a 36-chromosome hybrid containing 12 chromosome pairs plus 12 unpaired chromosomes. A cross between *N. sylvestris* and *N. tomentosiformis* yields a 24-chromosome hybrid in which there is no pairing at all. Hence, it appears that part of the *N. tabacum* genome is from *N. sylvestris* and part is from *N. tomentosiformis.* This amphidiploid can be recreated either sexually, by using colchicine as described previously, or somatically by cell fusion. When cells of the prospective parental species are fused, a 48-chromosome hybrid cell line is produced, from which plants identical in behavior to *N. tabacum* may be grown. (Note that, in the latter method, colchicine is not required, because the fusion product is already amphidiploid.)

The recovery of somatic hybrids may be enhanced if a selective system is available. In one example in *N. tabacum,*

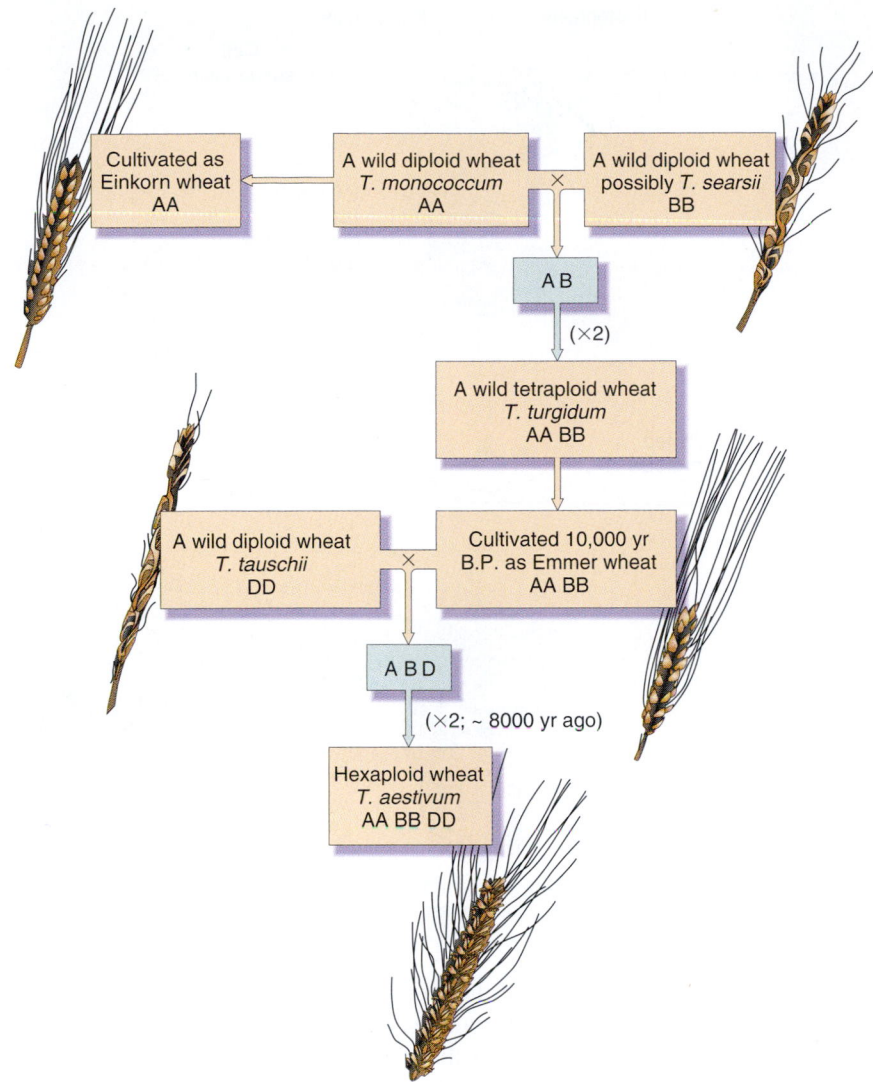

Figure 18-12 Diagram of the proposed evolution of modern hexaploid wheat, in which amphidiploids are produced at two points. A, B, and D are different chromosome sets.

two monoploid lines were fused to form a diploid hybrid culture by using complementation as the selection system. The first line had whitish, light-sensitive leaves due to a recessive mutation *w*. The other had yellowish, light-sensitive leaves due to a mutation *y* at a separate locus. When the cells were combined in a petri dish, the diploid $w^+/w \cdot y^+/y$ calluses could be selected by their resistance to light and their normal green color. The calluses can be grown into plantlets, which then are either grafted onto a mature plant to develop or potted themselves. The protocol for this experiment is illustrated in Figure 18-13.

Two allotetraploids of *Petunia,* one produced by sexual hybridization and the other by somatic hybridization, are compared in Figure 18-14. Note that the two are identical in appearance and produce the same range of progeny types.

MESSAGE ....
Allopolyploids can be synthesized either by crossing related species and doubling the chromosomes of the hybrid or by asexually fusing the cells of different species.

Polyploidy in animals

You may have noticed that most of the discussion of polyploidy so far has concerned plants. Indeed, polyploidy is more common in plants than in animals; nevertheless there are many cases of polyploid animals. Examples are found in flatworms, leeches, and brine shrimp. In these cases, reproduction is by parthenogenesis, the development of a special type of unfertilized egg into an embryo, without the need for fertilization. However, examples are not confined to these so-called lower forms. Polyploid amphibians and reptiles are surprisingly common. They show several modes of reproduction. The males and females of polyploid frogs and toads participate in their sexual cycles, whereas polyploid salamanders and lizards are parthenogenetic.

Some fish also are polyploid; in two cases, it appears that a single polyploid event gave rise to an entire taxonomic family in evolution. This situation contrasts with that of amphibians and reptiles because in those cases the polyploids all have closely related diploid species, and, hence, the polyploid events do not seem to have been important in

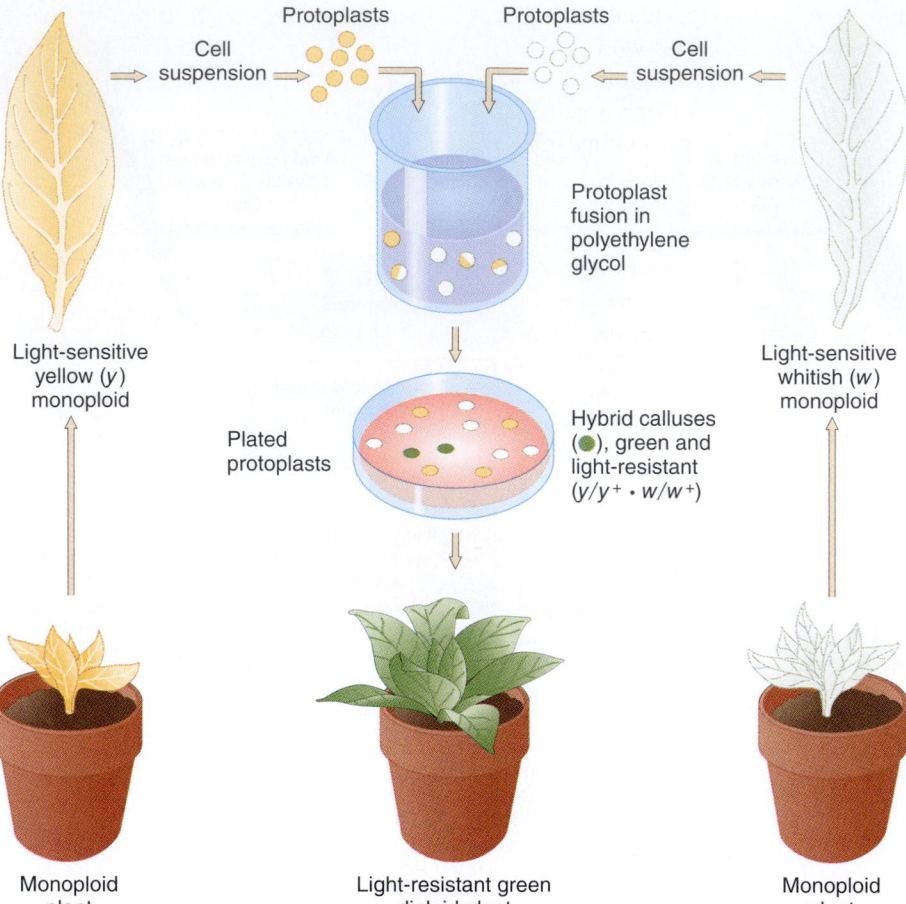

Figure 18-13 Creating a hybrid of two monoploid lines of *Nicotiana tabacum* by cell fusion. One line has light-sensitive, yellowish leaves, and the other has light-sensitive, whitish leaves. Protoplasts are produced by enzymatically stripping the cell walls from the leaf cells of each strain. The protoplasts can fuse, as indicated; those that fuse as hybrids can be grown into calluses that are light-resistant as a result of recessiveness of the parental genotypes. The calluses, under the appropriate hormone regime, can be grown into green diploid plants.

Figure 18-14 Sexual hybridization and somatic hybridization produce identical *Petunia* allotetraploids. The two parental lines are illustrated at the top, the red *P. hybrida* and the white *P. parodii*. The F$_1$ allotetraploid in the second row resulted from crossing the two diploid parents and doubling the chromosome number of the hybrid; the allotetraploid next to it resulted from fusion of somatic cells of the two parental lines. The two allotetraploids produce the same range of progeny, illustrated in the third and fourth rows. The numbers identify plants used in further experiments. (J. B. Power.)

the evolution of the group as a whole. The Salmonidae family of fishes (containing salmon and trout) is a familiar example of a group that appears to have originated through polyploidy. Salmonids have twice as much DNA as have related fish. Different salmonid species have different chromosome numbers, but the group has an almost invariant number of chromosome arms (some fused in some species) that is twice the number of arms in related groups. Hence, the evidence points to the salmonids' having evolved from a single event that gave rise to a tetraploid.

You might also be interested to know that the sterility of triploids has been commercially exploited in animals as well as plants. Triploid oysters have been developed, and they have a commercial advantage over their diploid relatives. The diploids go through a spawning season, during which they are unpalatable. Triploids, however, because of their sterility, do not spawn and are palatable the whole year round.

Human polyploid zygotes do arise through various kinds of mistakes in cell division. Most die in utero. Occasionally, triploid babies are born, but none survive.

Aneuploidy

Aneuploidy is the second major category of chromosome mutations in which chromosome number is abnormal. An

aneuploid is an individual organism whose chromosome number differs from the wild type by part of a chromosome set. Generally, the aneuploid chromosome set differs from wild type by only one or a small number of chromosomes. Aneuploids can have a chromosome number either greater or smaller than that of the wild type. Aneuploid nomenclature is based on the number of copies of the specific chromosome in the aneuploid state. For example, the aneuploid condition $2n - 1$ is called **monosomic** (meaning "one chromosome") because only one copy of some specific chromosome is present instead of the usual two found in its diploid progenitor. The aneuploid $2n + 1$ is called **trisomic,** $2n - 2$ is **nullisomic,** and $n + 1$ is **disomic.**

Nullisomics ($2n - 2$)

Although nullisomy is a lethal condition in diploids, an organism such as bread wheat, which behaves meiotically like a diploid although it is a hexaploid, can tolerate nullisomy. The four homoeologous chromosomes apparently compensate for a missing pair of homologs. In fact, all the possible 21 bread wheat nullisomics have been produced; they are illustrated in Figure 18-15. Their appearances differ from the normal hexaploids; furthermore, most of the nullisomics grow less vigorously.

Monosomics ($2n - 1$)

Monosomic chromosome complements are generally deleterious for two main reasons. First, the missing chromosome perturbs the overall gene balance in the chromosome set. (We encountered this effect earlier). Second, having a chromosome missing allows any deleterious recessive allele on the single chromosome to be hemizygous and thus to be directly expressed phenotypically. Notice that these are the same effects produced by deletions.

Nondisjunction in mitosis or meiosis is the cause of most aneuploids. Disjunction is the normal separation of homologous chromosomes or chromatids to opposite poles at nuclear division. Nondisjunction is a failure of this disjoining process, and two chromosomes (or chromatids) go to one pole and none to the other. Nondisjunction occurs spontaneously; it is another example of a chance failure of a basic cellular process.

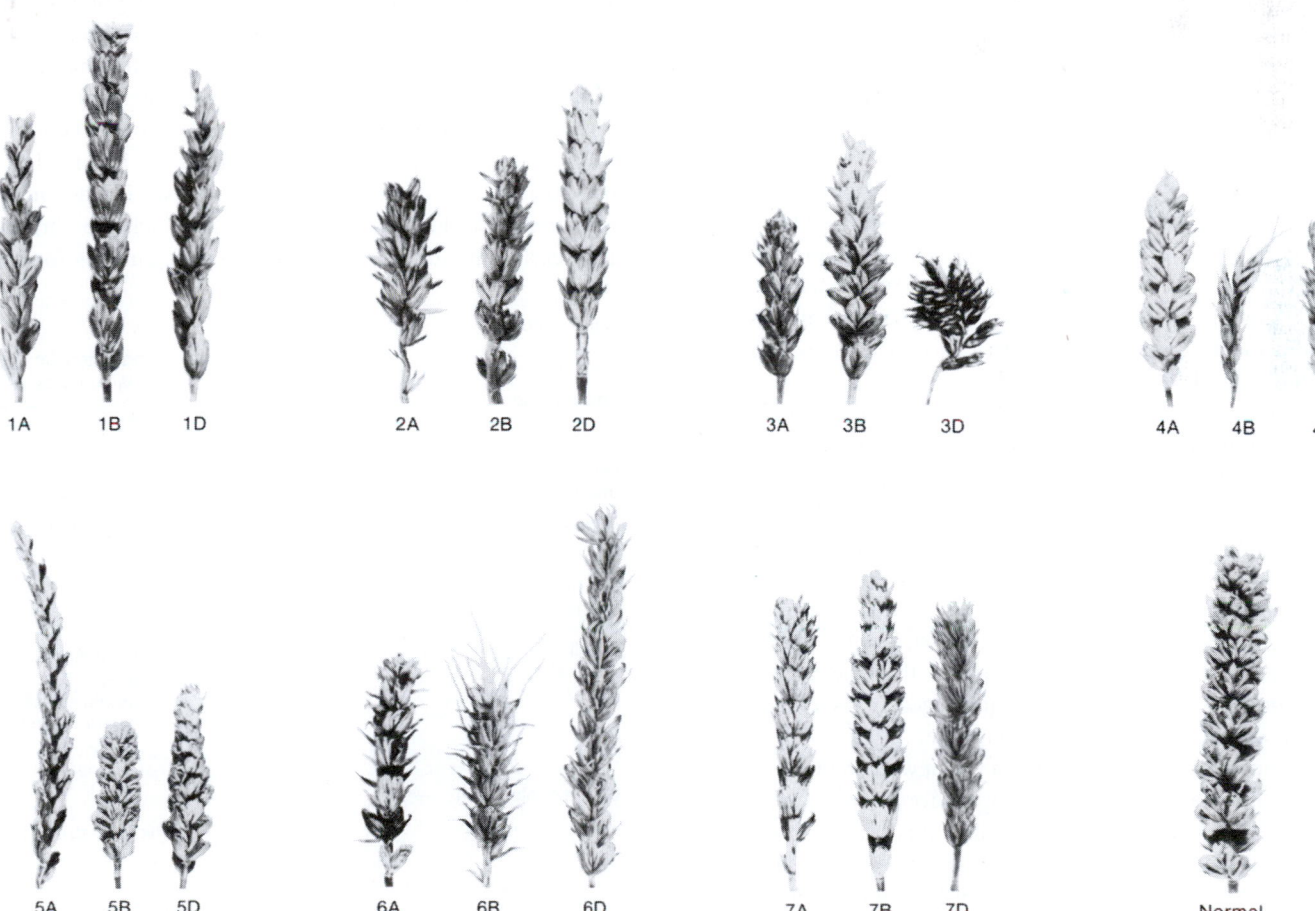

1A 1B 1D   2A 2B 2D   3A 3B 3D   4A 4B 4D

5A 5B 5D   6A 6B 6D   7A 7B 7D   Normal

Figure 18-15   Ears of the nullisomics of wheat. The number and letter under each ear designates the absent chromosome. Although nullisomics are usually lethal in regular diploids, organisms such as wheat, which "pretends" to be diploid but is hexaploid, can tolerate nullisomy. Nullisomics, however, are less vigorous growers. (E. R. Sears.)

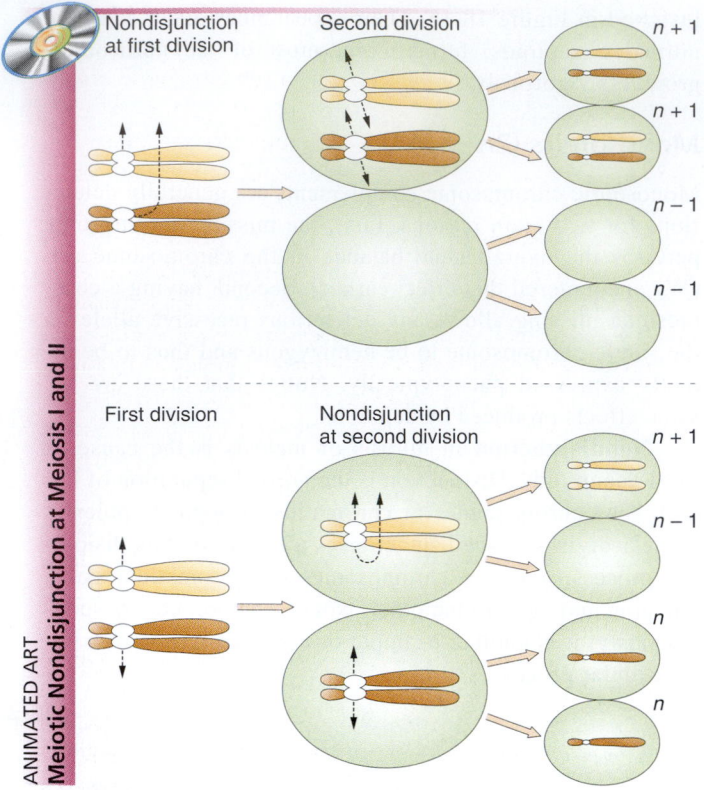

Figure 18-16   The origin of aneuploid gametes by nondisjunction at the first or second meiotic division.

In meiotic nondisjunction, the chromosomes may fail to disjoin at either the first or second division (Figure 18-16). Either way, $n + 1$ and $n - 1$ gametes are produced. If an $n - 1$ gamete is fertilized by an n gamete, a monosomic $(2n - 1)$ zygote is produced. The fusion of an $n + 1$ and an n gamete yields a trisomic $2n + 1$.

The precise molecular processes that fail in nondisjunction are not known, but, in experimental systems, the frequency of nondisjunction can be increased by interference with microtubule action. It appears that disjunction is more likely to go awry in meiosis I. This likelihood may not be surprising, because normal anaphase I disjunction requires that proper homologous associations be maintained during prophase I and metaphase I. In contrast, proper disjunction at anaphase II or at mitosis requires that the centromere splits properly but does not require nearly as elaborate a process during prophase and metaphase.

Meiosis I nondisjunction can be viewed as the failure to form or maintain a tetrad (a group of four chromatids) until anaphase I. Crossovers are implicit in this process normally. In most organisms, the amount of crossing-over is sufficient to ensure that all tetrads will have at least one exchange per meiosis. In *Drosophila,* many of the nondisjunctional chromosomes in newly arising disomic gametes are nonrecombinant, with one nondisjunctional homolog carrying the markers of one input chromosome and the other homolog carrying the markers of the other chromosome. Similar observa-

tions have been made in human trisomies. In addition, in several different experimental organisms, mutations that interfere with recombination have the effect of massively increasing the frequency of meiosis I nondisjunction. This effect points to an important role of crossing-over in maintaining chromosomal associations in the tetrad; in the absence of these associations, chromosomes are vulnerable to anaphase I nondisjunction.

In humans, a sex-chromosome monosomic complement of 44 autosomes + 1 X produces a phenotype known as *Turner syndrome.* Affected people have a characteristic, easily recognizable phenotype: they are sterile females, are short in stature, and often have a web of skin extending between the neck and shoulders (Figure 18-17). Although their intelligence is near normal, some of their specific cognitive functions are defective. About 1 in 5000 female births have this monosomic chromosomal complement. Monosomics for all human autosomes die in utero.

Geneticists have used viable plant nullisomics and monosomics to identify the chromosomes that carry the loci of newly found recessive mutant alleles. For example, a geneticist may obtain different monosomic lines, each of which lacks a different chromosome. Homozygotes for the new mutant allele are crossed with each monosomic line, and the progeny of each cross are inspected for expression of the recessive phenotype. The phenotype appears in some of the progeny of the parent that is monosomic for the locus-bearing chromosome and thus identifies it. Figure 18-18 shows that this test works because meiosis in the monosomic parent produces some gametes that lack the chromosome bearing the locus. When one of these gametes forms a

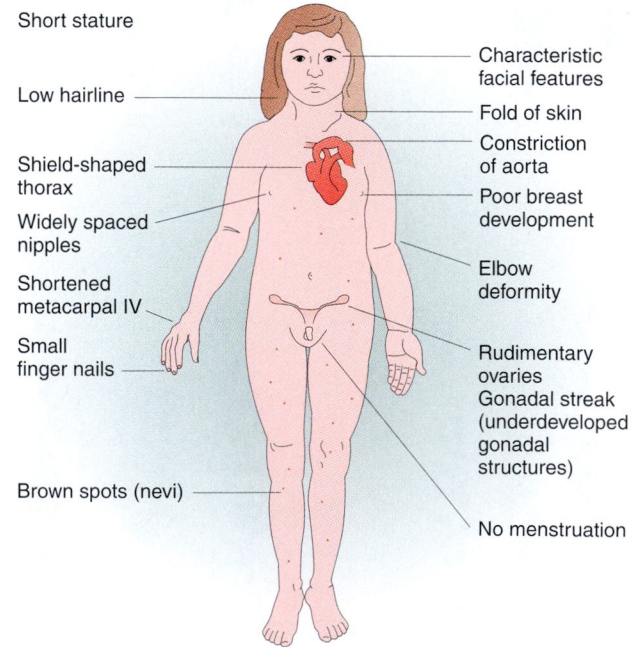

Figure 18-17   Characteristics of Turner syndrome, which results from having a single X chromosome (XO). (After F. Vogel and A. G. Motulsky, *Human Genetics.* Springer-Verlag, 1982.)

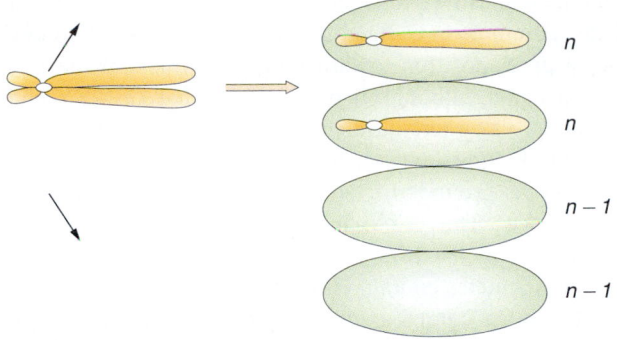

Figure 18-18 Meiosis in which the chromosome of interest is monosomic. Two of the resulting gametes are haploid (*n*); the other two gametes contain a set lacking a chromosome (*n* − 1).

zygote, the single chromosome contributed by the other homozygous recessive parent determines the phenotype.

Genetic analysis of humans occasionally reveals a similar unmasking of a recessive phenotype by an *n* − 1 gamete. For example, two people whose vision is normal may produce a daughter who has *Turner syndrome* and who is also red-green colorblind. This coincidence is interpreted as follows. First, because the father is not colorblind, the mother must be heterozygous for the recessive allele and must have passed this allele on to the Turner daughter. Nondisjunction must have occurred in the father, resulting in an *n* − 1 sperm, which combined with an egg bearing the colorblind allele from the mother.

MESSAGE ·
Monosomics show the deleterious effects of genome inbalance, as well as unexpected expression of recessive alleles carried on the monosomic chromosome.

Trisomics (2*n* + 1)

The trisomic condition also is one of chromosomal imbalance and can result in abnormality or death. However, there are many examples of viable trisomics. You might remem-

ber that we studied the trisomics of the Jimson weed *Datura stramonium* in Chapter 3 (see Figure 3-7). Furthermore, trisomics can be fertile. When cells from some trisomic organisms are observed under the microscope at the time of meiotic chromosome pairing, the trisomic chromosomes are seen to form a trivalent, an associated group of three, whereas the other chromosomes form regular bivalents. What genetic ratios might we expect for genes on the trisomic chromosome? Let us consider a gene, *A*, that is close to the centromere on that chromosome, and let us assume that the genotype is *A/a/a*. Furthermore, if we postulate that two of the centromeres disjoin to opposite poles as in a normal bivalent and that the other chromosome passes randomly to either pole, then we can predict the three equally frequent segregations shown in Figure 18-19. These segregations result in an overall gametic ratio of 1*A* : 2*A/a* : 2*a* : 1*a/a*. This ratio and the one corresponding to a trisomic of genotype *A/A/a* are observed in practice. If a trisomic tester set is available (much like the nullisomic tester set described earlier), then a new mutation can be located to a chromosome by determining which of the testers gives the special ratio.

There are several examples of viable trisomics in humans. The combination XXY (1 in 1000 male births) results in *Klinefelter syndrome,* males with lanky builds who are mentally retarded and sterile (Figure 18-20). Another

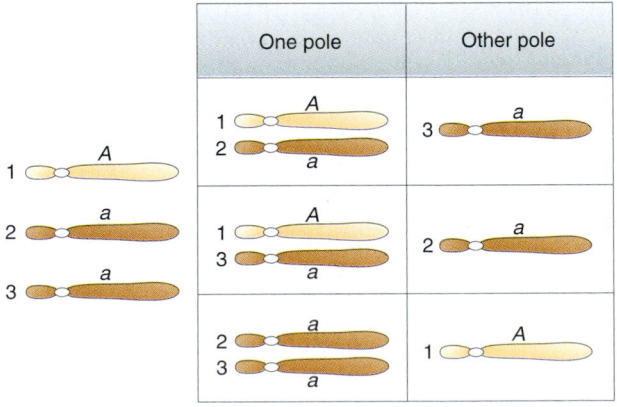

Figure 18-19 Genotypes of the meiotic products of an *A/a/a* trisomic. Three segregations are equally likely.

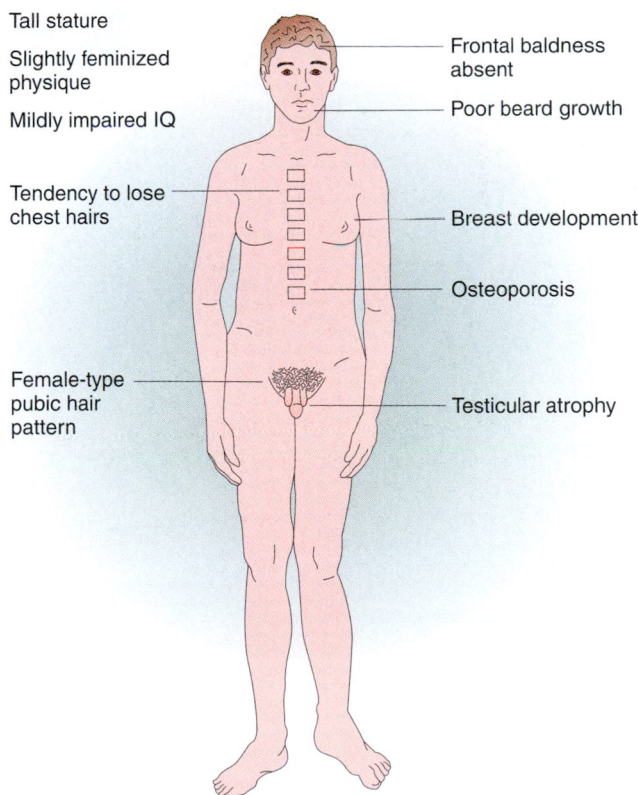

Figure 18-20 Characteristics of Klinefelter syndrome (XXY). (After F. Vogel and A. G. Motulsky, *Human Genetics.* Springer-Verlag, 1982.)

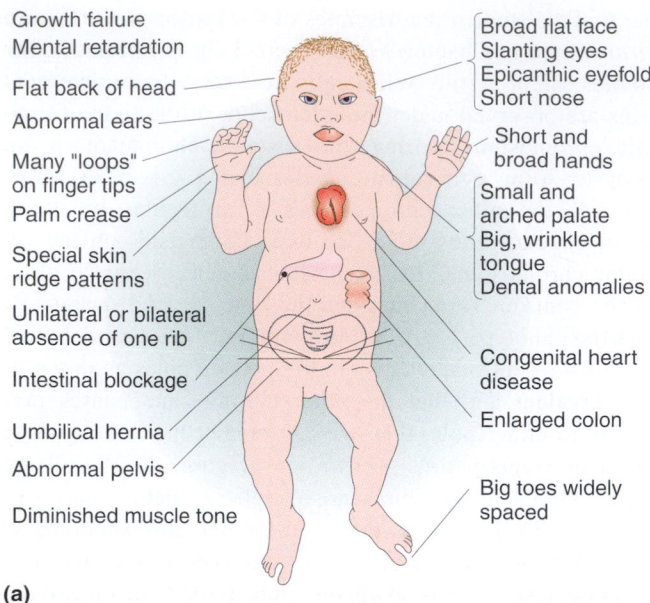

(a)

(b)

Figure 18-21   Characteristics of Down syndrome (trisomy 21). (a) Diagrammatic representation of the syndrome in an infant. (b) Athletes with Down syndrome. (Part a after F. Vogel and A. G. Motulsky, *Human Genetics*. Springer-Verlag, 1982; part b from Bob Daemmrich/The Image Works.)

combination, XYY, also occurs in about 1 in 1000 male births. Attempts have been made to link the XYY condition with a predisposition toward violence. This linkage is still hotly debated, although it is now clear that an XYY condition in no way guarantees such behavior. Nevertheless, several enterprising lawyers have attempted to use the XYY genotype as grounds for acquittal or compassion in crimes of violence. The XYY males are usually fertile. Their meioses are of the XY type; the extra Y is not transmitted, and their gametes contain either X or Y, never YY or XY.

The most common type of viable human aneuploid is *Down syndrome* (Figure 18-21), occurring at a frequency of about 0.15 percent of all live births. We have already encountered the translocation form of Down syndrome in Chapter 17. However, by far the most common type of Down syndrome is trisomy 21, caused by nondisjunction of chromosome 21 in a parent who is chromosomally normal. Like any mechanism, chromosome disjunction is error prone and sometimes produces aneuploid gametes. In this type of Down syndrome, there is no family history of aneuploidy, unlike the translocation type described earlier.

Down syndrome is related to maternal age; older mothers run a greatly elevated risk of having Down-syndrome children (Figure 18-22). For this reason, fetal chromosome analysis (by amniocentesis or by chorionic villus sampling) is now recommended for older mothers. A less pronounced paternal-age effect also has been demonstrated.

Even though the maternal-age effect has been known for many years, its cause is still not known. Nonetheless, there are some interesting biological correlations. It is possible that one aspect of the strong maternal-age effect on nondisjunction is an age-dependent decrease in the probability

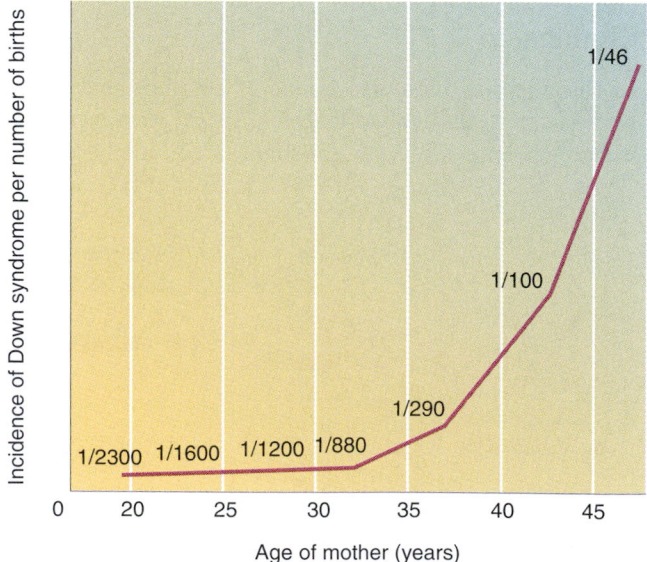

Figure 18-22   Maternal age and the production of Down-syndrome offspring. (From L. S. Penrose and G. F. Smith, *Down's Anomaly*. Little, Brown and Company, 1966.)

of keeping the chromosomal tetrad together during prophase I of meiosis. Meiotic arrest of oocytes (female meiocytes) in late prophase I is a common phenomenon in many animals. In female humans, all oocytes are arrested at diplotene before birth. Meiosis continues only at menstruation, which means that proper chromosome associations in the tetrad must be maintained for decades. If we speculate that, by accident through time, these associations have an increasing probability of breaking down, we can envision a mechanism contributing to increased maternal nondisjunction with age. Consistent with this speculation, most nondisjunction related to the maternal-age effect is due to nondisjunction at anaphase I, not anaphase II.

The multiple phenotypes that make up Down syndrome include mental retardation, with an IQ in the 20 to 50 range; broad, flat face; eyes with an epicanthic fold; short stature; short hands with a crease across the middle; and a large, wrinkled tongue. Females may be fertile and may produce normal or trisomic progeny, but males have never reproduced. Mean life expectancy is about 17 years, and only 8 percent survive past age 40.

The only other human autosomal trisomics to survive to birth are afflicted with either trisomy 13 *(Patau syndrome)* or trisomy 18 *(Edwards syndrome)*. Both show severe physical and mental abnormalities. The generalized phenotype of trisomy 13 includes a harelip; a small, malformed head; "rockerbottom" feet; and a mean life expectancy of 130 days. That of trisomy 18 includes "faunlike" ears, a small jaw, a narrow pelvis, and rockerbottom feet; almost all babies with trisomy 18 die within the first few weeks after birth.

Chromosome mutation in general plays a prominent role in determining genetic ill health in humans. Figure 18-23 summarizes the surprisingly high levels of various chromosome abnormalities at different human developmental stages. In fact, the incidence of chromosome mutations ranks close to that of gene mutations in human live births (Table 18-1). This fact is particularly surprising when we realize that virtually all the chromosome mutations listed in Table 18-1 arise anew with each generation. In contrast, gene mutations (as we shall see in Chapter 24) owe their level of incidence to a complex interplay of mutation rates and environmental selection that spans many human generations.

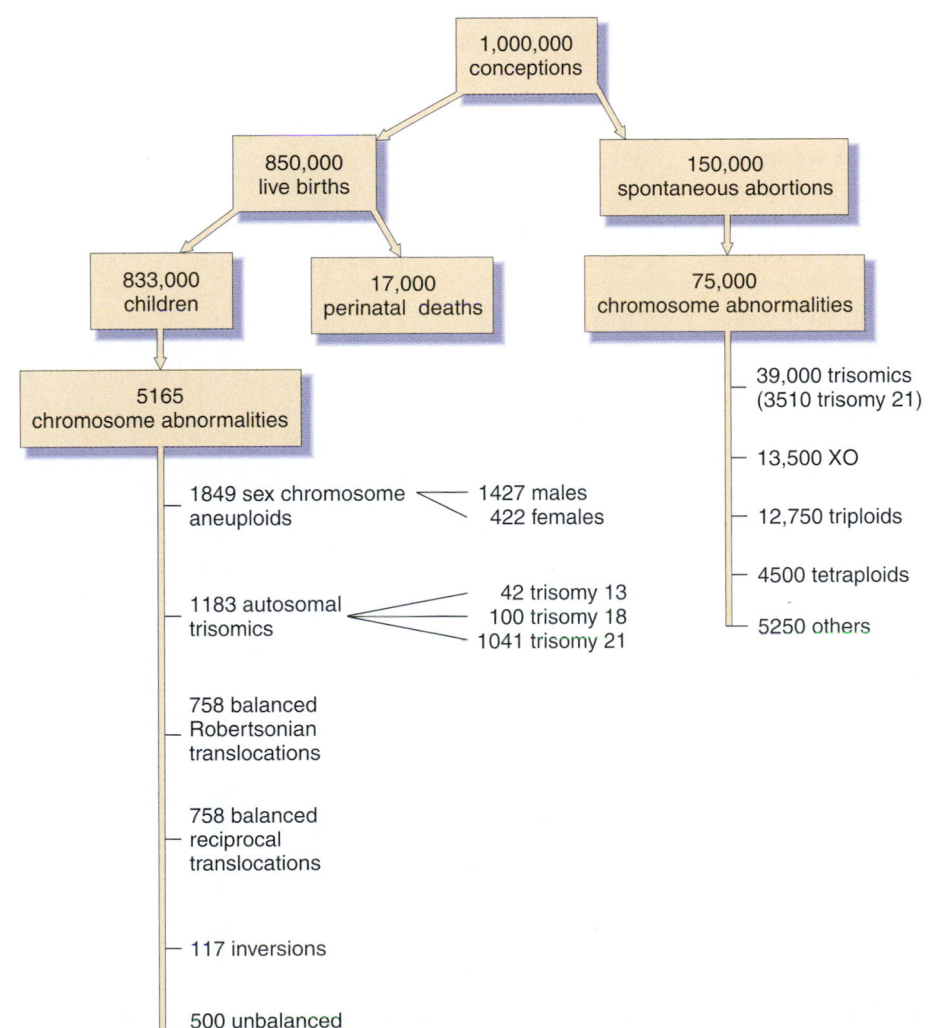

Figure 18-23 The fate of a million implanted human zygotes. (Robertsonian translocations are due to fusion or dissociation of centromeres.) (From K. Sankaranarayanan, *Mutation Research* 61, 1979.)

18-1 TABLE — Relative Incidence of Human Ill Health Due to Gene Mutation and to Chromosome Mutation

Type of mutation	Percentage of live births
Gene mutation	
Autosomal dominant	0.90
Autosomal recessive	0.25
X linked	0.05
Total gene mutation:	1.20
Chromosome mutation	
Autosomal trisomies (mainly Down syndrome)	0.14
Other unbalanced autosomal aberrations	0.06
Balanced autosomal aberrations	0.19
Sex chromosomes	
XYY, XXY, and other ♂♂	0.17
XO, XXX, and other ♀♀	0.05
Total chromosome mutation:	0.61

18-2 TABLE — Number and Type of Chromosomal Abnormalities Among Spontaneous Abortions and Live Births in 100,000 Pregnancies

	100,000 PREGNANCIES	
	15,000 spontaneous abortions 7,500 chromosomally abnormal	85,000 live births 550 chromosomally abnormal
Trisomy		
1	0	0
2	159	0
3	53	0
4	95	0
5	0	0
6–12	561	0
13	128	17
14	275	0
15	318	0
16	1229	0
17	10	0
18	223	13
19–20	52	0
21	350	113
22	424	0
Sex chromosomes		
XYY	4	46
XXY	4	44
XO	1350	8
XXX	21	44
Translocations		
Balanced	14	164
Unbalanced	225	52
Polyploid		
Triploid	1275	0
Tetraploid	450	0
Other (mosaics, etc.)	280	49
Total:	7500	550

When the frequencies of various chromosome mutations in live births are compared with the corresponding frequencies found in spontaneous abortions (Table 18-2), it becomes clear that the chromosome mutations that we know about as clinical abnormalities are just the tip of an iceberg of chromosome mutations. First, we see that many more types of abnormalities are produced than survive to birth; for example, trisomies of chromosome 2, 16, and 22 are relatively common in abortuses but never survive to birth. Second, the specific aberrations that survive are part of a much larger number that do not survive; for example, Down syndrome (trisomy 21) is produced at almost 20 times the frequency in live births. The comparison is even more striking for Turner syndrome (XO). An estimated minimum of 10 percent of conceptions have a major chromosome abnormality; our reproductive success depends on the natural weeding-out process that eliminates most of these abnormalities before birth. Incidentally, no evidence suggests that these aberrations are produced by environmental insult to our reproductive systems or that the frequency of the aberrations is increasing.

MESSAGE ·······························

Trisomics show the deleterious effects of genome inbalance and produce chromosome-specific modified phenotypic ratios.

Disomics (n + 1)

A disomic is an aberration of a haploid organism. In fungi, they can result from meiotic nondisjunction. In the fungus *Neurospora* (a haploid), an $n - 1$ meiotic product aborts and does not darken like a normal ascospore; so we may detect MI and MII nondisjunctions by observing asci with 4:4 and 6:2 ratios of normal to aborted spores, respectively, as shown here.

These diagrams correspond exactly to the outcomes of the chromosomal events shown in Figure 18-16. In these organisms, the disomic ($n + 1$) meiotic product becomes a disomic strain directly. The abortion patterns themselves are diagnostic for the presence of disomics in the asci. Another way of detecting disomics in fungi is to cross two strains with homologous chromosomes bearing multiple auxotrophic mutations; for example:

$$ade\ his^+\ nic\ ala^+\ leu\ arg^+$$

$$\times$$

$$ade^+\ his\ nic^+\ ala\ leu^+\ arg$$

From such a cross, large populations of ascospores are plated onto minimal medium. Only ascospores of genotype $+ + +$ $+ + +$ can grow and form colonies. Most of these colonies are found to be disomics and not multiple crossover types.

MESSAGE ··
Disomics in fungi can be selected from asci showing special spore abortion patterns or as meiotic progeny that must contain homologous chromosomes from both parents.
··

Somatic aneuploids

Aneuploid cells can arise spontaneously in somatic tissue or in cell culture. In such cases, the initial result is a genetic mosaic of cell types.

Human sexual mosaics — individuals whose bodies are a mixture of male and female tissue — are good examples. One type of sexual mosaic, (XO)(XYY), can be explained by postulating an XY zygote in which the Y chromatids fail to disjoin at an early mitotic division, so both go to one pole:

The phenotypic sex of such individuals depends on where the male and female sectors end up in the body. In the type of nondisjunction being considered, nondisjunction at a later mitotic division would produce a three-way mosaic (XY)(XO)(XYY), which contains a clone of normal male cells. Other sexual mosaics have different explanations; as examples, XO/XY is probably due to early X-chromosome loss in a male zygote (Figure 18-24), and (XX)(XY) is probably the result of a double fertilization (fused twins). In general, sexual mosaics are called *gynandromorphs*.

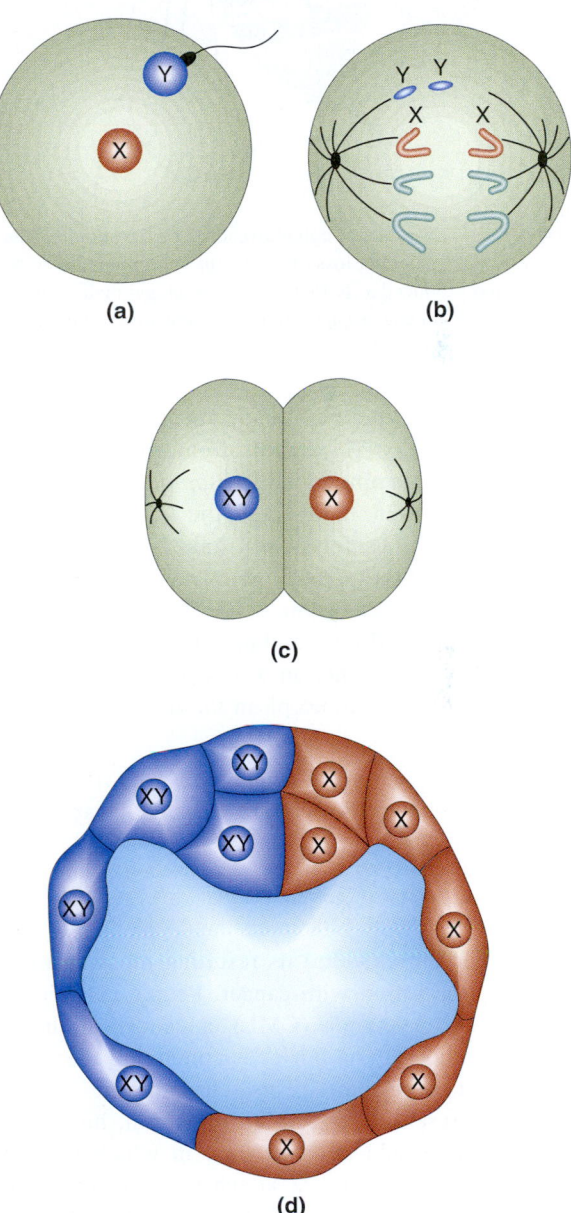

(a)   **(b)**

(c)

(d)

Figure 18-24   Origin of a human sexual mosaic (XY)(XO) by Y chromosome loss at the first mitotic division of the zygote. (a) Fertilization. (b) Chromosome loss. (c) Resulting male and female cells. (d) Mosaic blastocyst. (After C. Stern, *Principles of Human Genetics*, 3d ed. Copyright © 1973 by W. H. Freeman and Company.)

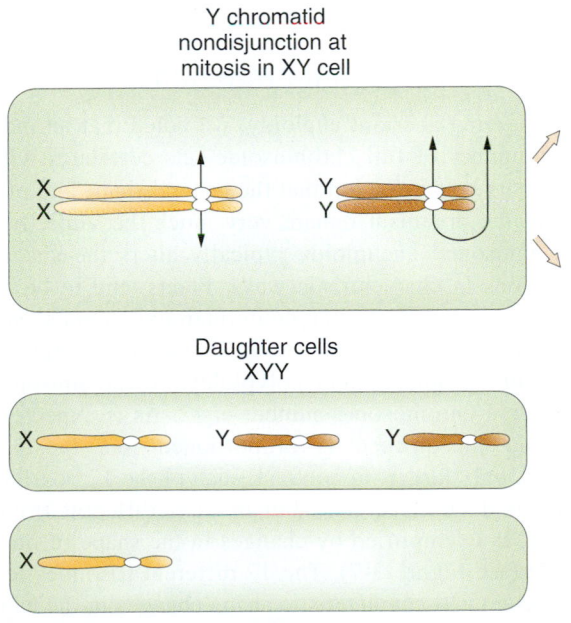

Y chromatid nondisjunction at mitosis in XY cell

Daughter cells
XYY

XO

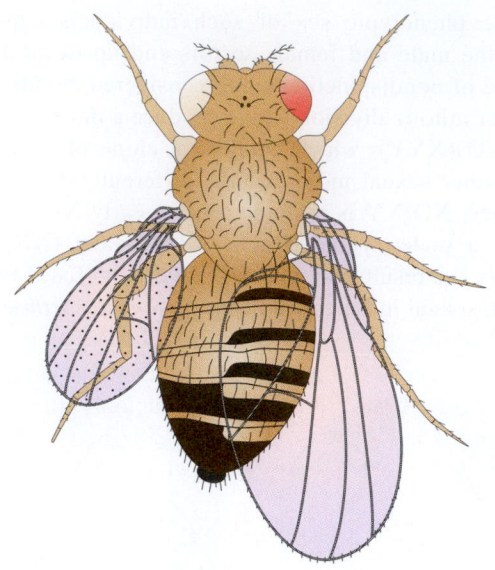

Figure 18-25   A bilateral gynandromorph of *Drosophila*. The zygote was $w^+ m^+/w\ m$, but loss of the $w^+ m^+$ chromosome in the first mitotic division produced a fly that was $\frac{1}{2}$ O/$w\ m$ and male *(left)* and $\frac{1}{2}$ $w^+ m^+/w\ m$ and female *(right)*. Mutant allele *w* gives white eye, and *m* gives miniature wing.

Figure 18-26   A bilateral gynandromorph in the Io moth, *Automeris io io.* One half of the body is female and happens to carry the sex chromosome mutation "broken eye"; the other half of the body is male and carries the normal allele of the broken-eye gene. This moth resulted from an event similar to that which produced the *Drosophila* gynandromorph in Figure 18-25. The color and size differences are both sex differences. (T. R. Manley.)

Geneticists working with many species of experimental animals occasionally find gynandromorphs among their stocks. A classic example is the *Drosophila* gynandromorph shown in Figure 18-25. In this case, the zygote started out as a female heterozygous for two X-linked genes, white eye and miniature wing ($w^+ m^+/w\ m$). Loss of the wild-type allele–bearing X chromosome at the first mitotic division resulted in the two cell lines and ultimately in a fly differing from one side to the other in sex, eye color, and size of wing. A similar gynandromorph in the Io moth is shown in Figure 18-26.

MESSAGE ⸱⸱⸱

Mitotic nondisjunction and other types of aberrant mitotic chromosome behavior can give rise to mosaics consisting of two or more chromosomally distinct cell types, including aneuploids.

Somatic aneuploidy and its resulting mosaics are often observed in association with cancer. People suffering from chronic myeloid leukemia (CML), a cancer of the white blood cells, frequently harbor cells containing the so-called Philadelphia chromosome. This chromosome was once thought to represent an aneuploid condition, but it is now known to be a translocation product in which part of the long arm of chromosome 22 attaches to the long arm of chromosome 9. However, CML patients often show aneuploidy in addition to the Philadelphia chromosome. In one study of 67 people with CML, 33 proved to have an extra Philadelphia chromosome and the remainder had various aneuploidies; the most common aneuploidy was trisomy for the long arm of chromosome 17, which was detected in 28 people. Of 58 people with acute myeloid leukemia, 21 were

shown to have aneuploidy for chromosome 8; 16 for chromosome 9; and 10 for chromosome 21. In another study of 15 patients with intestinal tumors, 12 had cells with abnormal chromosomes, at least some with trisomy for chromosome 8, 13, 15, 17, or 21. Such studies merely established correlations, and it is not clear whether the abnormalities are best thought of as a cause or as an effect of cancer.

MESSAGE ⸱⸱⸱

Aneuploids are produced by nondisjunction or some other type of chromosome misdivision at either meiosis or mitosis.

Mechanisms of gene imbalance

In considering aberrant euploidy, we noted that an increase in the number of full chromosome sets correlates with increased organism size but that the general shape and proportions of the organism remain very much the same. In contrast, autosomal aneuploidy typically alters the shape and proportions in characteristic ways. Plants tend to be somewhat more tolerant of aneuploidy than are animals. Studies of the jimsonweed, *Datura,* provide a classic example of the effects of aneuploidy and polyploidy. In the jimsonweed, the haploid chromosome number is 12. As is expected, the polyploid jimsonweed is proportioned like the normal diploids, only larger. In contrast, each of the 12 possible trisomics is disproportionate, but in ways different from one another, as exemplified by changes in the shape of the seed capsule (see Figure 3-7). The 12 different trisomies lead to 12 different and characteristic shape changes in the capsule. Indeed, these and other characteristics of the individual trisomies are so reliable that the phenotypic syndrome can be used to identify plants carrying a particular trisomy. Similarly, the 12 monosomies are themselves different from one another and from each of the trisomies. In general, a monosomic for a particular chromosome is more severely abnormal than is the corresponding trisomic.

We see similar trends in aneuploids of animals as well. In the fruit fly, *Drosophila,* the only autosomal aneuploids that survive to adulthood are trisomics and monosomics for chromosome 4, which is the smallest *Drosophila* chromosome, representing from only about 1 to 2 percent of the genome. Trisomics for chromosome 4 are only very mildly affected and are much less severely abnormal than are monosomic-4 flys. In humans, no autosomal monosomic survives to birth, whereas three autosomal trisomies survive, as mentioned earlier. As is true with aneuploid jimsonweed, the three surviving trisomies produce unique phenotypic syndromes, owing to the special effects of altered dosages of each of these chromosomes.

Why are aneuploids so much more abnormal than polyploids? Why do aneuploids for different chromosomes each have their own characteristic phenotypic effects? And why are monosomics typically more severely affected than the corresponding trisomics? The answers seem certain to be a matter of **gene balance.** In a euploid, the ratio of genes on any one chromosome to genes on other chromosomes is 1 : 1 (that is, 100 percent), regardless of whether we are considering a monoploid, diploid, triploid, or tetraploid. In contrast, in an aneuploid, the ratio of genes on the aneuploid chromosome to genes on the other chromosomes differs from wild type by 50 percent (50 percent for monosomics; 150 percent for trisomics). Thus, we can see that the aneuploid genes are out of balance. How does this help us to answer the questions raised?

A key fact is that, in general, the amount of transcript produced by a gene is directly proportional to the number of copies of that gene in a cell. That is, for a given gene, the rate of transcription is directly related to the number of DNA templates. Thus, the more copies of that gene, the more transcripts are produced. Because of this gene-dosage relation, *segmental aneuploids,* in which *pieces* of individual chromosomes are trisomic or monosomic (effectively, duplications and deletions), have proved to be very useful in locating the positions of genes encoding various cellular enzymes. The approach is to look for segments of the genome that change the amount of an enzyme proportionally to the dosage of that genomic segment. This approach has been exploited extensively in *Drosophila,* where there has been about a 90 percent success rate in identifying enzyme-coding gene loci by this method.

We can infer that normal physiology in a cell depends on the proper ratio of gene products in the euploid cell. This ratio is the normal gene balance. If the relative dosage of certain genes changes — for example, owing to the removal of one of the two copies of a chromosome or a segment thereof — physiological imbalances in cellular pathways can arise.

In some cases, the imbalances of aneuploidy are due to a few "major" genes. Such genes can be viewed as haploabnormal or triploabnormal or both and contribute significantly to the aneuploid phenotypic syndrome. For example, the study of persons trisomic for only part of human chromosome 21 has made it possible to localize determinants specific to Down syndrome to various regions of chromosome 21, hinting that some aspects of the phenotype might be due to trisomy for single major genes in these chromosomal regions. In addition to these major-gene effects, other aspects of aneuploid syndromes are likely to be due to cumulative effects of aneuploidy for numerous genes whose products are all out of balance. Undoubtedly, the entire aneuploid phenotype is a synthesis of the imbalance effects of a few major genes, together with a cumulative imbalance of many minor genes.

However, the gene-balance idea does not tell us why having too few gene products (monosomy) is much worse for an organism than having too many gene products (trisomy). Along the same lines, in well-studied organisms, there are many more haploabnormal genes than triploabnormal ones. An important factor in explaining the abnormality of monosomics is that any deleterious recessives present on the autosome will be automatically expressed. This same effect is relevant to deletion mutations.

How do we apply the idea of gene balance to cases of sex-chromosome aneuploidy? Gene balance holds for sex chromosomes as well, but we also have to take into account the special properties of the sex chromosomes. In organisms with X-Y sex determination, the Y chromosome seems to be a degenerate X chromosome in which there are very few functional genes other than some involved in sex determination itself or in sperm production or both. On the other hand, the X chromosome contains many genes involved in basic cellular processes ("housekeeping genes") that just happen to reside on the chromosome that eventually evolved into the X chromosome. X-Y sex-determination mechanisms have probably evolved independently from 10 to 20 times in different taxonomic groups. Thus, there appears to be one sex-determination mechanism for all mammals, but it is completely different from the mechanism governing X-Y sex determination in fruit flies.

In a sense, X chromosomes are naturally aneuploid. Females have two of them, whereas males have only one. Nonetheless, it has been found that the X chromosome's housekeeping genes are expressed to equal extents per cell in both females and males. How is this accomplished? The answer depends on the organism. In fruit flies, the male's X chromosome appears to be hyperactivated, allowing it to be transcribed at twice the rate of either X chromosome in the female. In mammals, in contrast, the rule is that no matter how many X chromosomes are present, there is only one transcriptionally active X chromosome in each somatic cell. This dosage compensation is achieved by random X-chromosome inactivation. (A person with two or more X chromosomes is a mosaic of two cell types in which one or the other X is active.) Thus, XY and XX mammals produce the same amount of X-chromosome housekeeping gene products. X-chromosome inactivation also explains why triplo-X humans are phenotypically normal, inasmuch as only one of the three X chromosomes is transcriptionally active in a given cell. Similarly, an XXY male is only moderately affected because only one of his two X chromosomes is active in each cell.

Why are XXY persons abnormal at all, given that triplo-X persons are phenotypically normal? It turns out that a small part of the X chromosome near one telomere — the pseudoautosomal region — is not inactivated by the mechanism of dosage compensation. In XXY males, this region is active at twice the level of the pseudoautosomal region in XY males. This level appears to have the consequence of slightly feminizing the phenotype of XXY males, although exactly which pseudoautosomal genes contribute to this effect is currently unknown. In XXX females, on the other hand, the pseudoautosomal region is active at only 1.5 times the level that it is in XX females. This lower level of functional aneuploidy in XXX than that in XXY plus the fact that the pseudoautosomal genes appear to lead to feminization may explain the feminized phenotype of XXY males. The severity of XO, Turner syndrome, can be interpreted as being due to the considerable deleterious effects of monosomy for the pseudoautosomal region of the X. As is usually observed for aneuploids, monosomy for this segment of the X chromosome produces a more abnormal phenotype than does having an extra copy of the same region (triplo-X females or XXY males).

As noted earlier, occasionally human polyploids are born, but none survive. This fact seems to violate the principle discussed earlier in this section — namely, that poly-

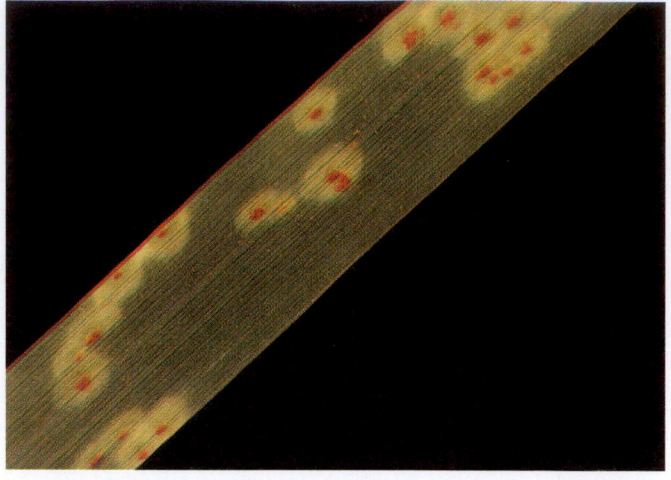

Figure 18-27   A wheat leaf infected by rust fungus, which derives its name from the rust-colored patches of fungal spores produced at the infection centers. Various kinds of rust fungi are pathogens on many important crop species including cereals, pines, and coffee. Rust infection results in billions of dollars of damage yearly, and a large part of plant breeding is directed at producing genetically resistant plant lines. (V. A. Wilmot/Visuals Unlimited.)

ploids are more normal than aneuploids. The explanation for this violation seems to lie with X-chromosome dosage compensation. Part of the rule for the single active X seems to be that there is one active X for every two copies of the autosomal chromosome complement. Thus, some cells in triploid mammals are found to have one active X, whereas others have two. Neither situation is in balance with autosomal genes. Presumably this functional underactivation of housekeeping genes (in $3n$ cells with one active X) or func-

(a)

(b)

Figure 18-28   (a) Wheat. (b) *Aegilops umbellulata.* Both whole plants and seed heads are shown. (E. R. Sears.)

tional overactivation (in 3*n* cells with two active X's) leads to substantial functional aneuploidy and the inviability of triploid humans.

Chromosome mechanics in plant breeding

What we know about chromosome mutation can be used in the type of genetic engineering that produces and maintains new crop types in our hungry world. The classic example, performed by E. R. Sears in the 1950s, concerns the transfer of a gene for leaf-rust resistance from a wild grass, *Aegilops umbellulata,* to bread wheat, which is highly susceptible to this disease, to offset a major problem in the wheat industry (Figure 18-27).

The first problem that Sears encountered was that these two species (Figure 18-28) are not interfertile, so the feat of gene transfer seemed impossible. Sears sidestepped this problem with a bridging cross, in which he crossed *A. umbellulata* to a wild relative of bread wheat called *emmer, Triticum dicoccoides.* (Follow the process in Figure 18-29.)

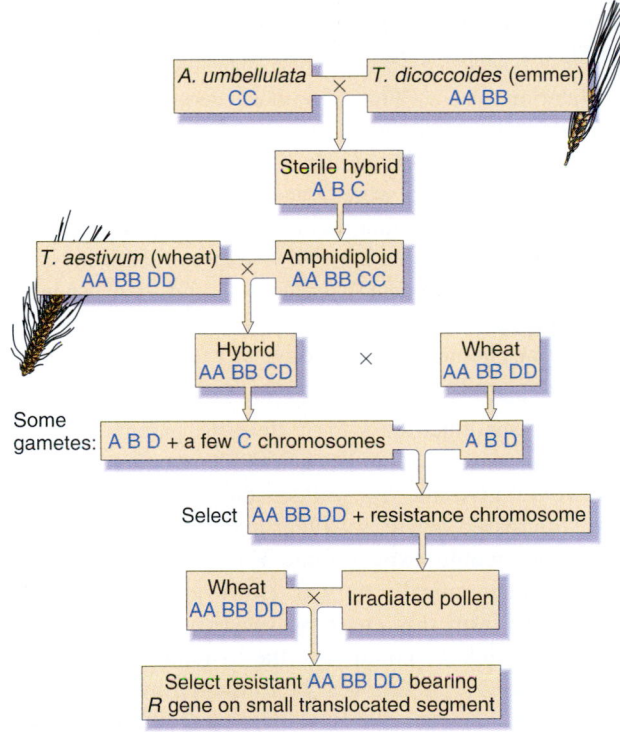

Figure 18-29 Summary of Sears's program for transferring rust resistance from *Aegilops* to wheat. A, B, C, and D represent chromosome sets of diverse origin. *R* represents the genetic determinant (single gene?) of resistance.

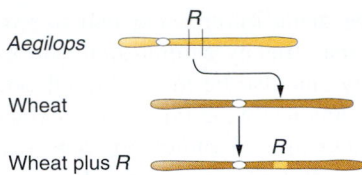

Figure 18-30   Translocation of *Aegilops R* segment to wheat by using radiation as a means of breaking the chromosomes.

A. umbellulata is a diploid, $2n = 2x = 14$. We will call its chromosome sets CC. *T. dicoccoides* is a tetraploid, $2n = 4x = 28$, with sets AA BB. From this cross, the resulting sterile hybrid A B C was doubled into a fertile amphidiploid AA BB CC with 42 chromosomes. This amphidiploid was fertile in crosses with bread wheat *(T. aestivum),* which is represented as $2n = 6x = 42$, AA BB DD. The offspring, AA BB CD, were almost completely sterile owing to pairing irregularities between the C and D sets. However, backcrosses to wheat did produce a few rare seeds, some of which grew into plants that were resistant to leaf rust, like the wild parent in the first cross. Some of these plants were almost the desired type, having 43 chromosomes (42 of which were wheat, plus one *Aegilops* chromosome bearing the resistance gene). Thus, in the AA BB CD hybrid, some aberrant form of chromosome assortment had produced a gamete with 22 chromosomes: A B D plus one from the C group.

Unfortunately, the extra chromosome carried just too many undesirable *Aegilops* genes along with the good one, and the plants were weedy and low producers. So the *Aegilops* gene linkage had to be broken. Sears did so by using irradiated pollen from these plants to pollinate wheat. He was looking for translocations of part of the *Aegilops* chromosome tacked onto the wheat chromosomes. These translocations were quite common, but only one turned out to be ideal—a very small, insertional, unidirectional translocation (Figure 18-30). When bred to homozygosity, the resistant plants were indistinguishable from wheat.

The study of chromosome mutation is of great importance—not only in pure biology, where there are ramifications in many areas (especially in evolution), but also in applied biology. The application of such studies is especially important in agriculture and medicine. Rather curiously, very little is known about the mechanisms of chromosome mutation, especially at the molecular level. More details are needed about the normal chemical architecture of chromosomes.

SUMMARY

When chromosome mutation changes the number of whole chromosome sets present, aberrant euploids result; when chromosome mutation changes parts of sets, aneuploids result.

Polyploids such as triploids (3*x*) and tetraploids (4*x*) are common in the plant kingdom and are even represented in the animal kingdom. An odd number of chromosome sets makes an organism sterile because there is not a partner for each chromosome at meiosis, whereas even numbers of sets can produce standard segregation ratios. Allopolyploids (polyploids formed by combining sets from different species) can be made by crossing two related species and then doubling the progeny chromosomes through the use of colchicine; they can also be made through somatic cell fusion. These techniques have important applications in crop breeding, because allopolyploids combine features of both parental species. Polyploidy can result in an organism of larger dimensions; this discovery has permitted important advances to be made in horticulture and in crop breeding.

Aneuploids have also been important in the engineering of specific crop genotypes, although aneuploidy itself usually results in an unbalanced genome with an abnormal phenotype. Examples of aneuploids include monosomics $(2n - 1)$, trisomics $(2n + 1)$, and disomics $(n + 1)$. Aneuploid conditions are well studied in humans. Down syndrome (trisomy 21), Klinefelter syndrome (XXY), and Turner syndrome (XO) are well-documented examples. The spontaneous level of aneuploidy in humans is quite high and produces a major percentage of genetically based ill health in human populations. Most instances of aneuploidy result from chromosome nondisjunction at meiosis or chromatid nondisjunction at mitosis. Mitotic nondisjunction can lead to somatic mosaics.

CONCEPT MAP

Draw a concept map interrelating as many of the following terms as possible. Note that the terms are listed in no particular order.

euploid / imbalance / nondisjunction / meiosis / gametes / sterility / aneuploid / phenotypic ratios / abortion / mitosis / genes

CHAPTER INTEGRATION PROBLEM

We have lines of mice that breed true for two alternative behavioral phenotypes that we know are determined by two alleles at a single locus: v causes a mouse to move with a "waltzing" gait, whereas V determines a normal gait. After crossing the true-breeding waltzers and normals, we observe that most of the F_1 is normal, but, unexpectedly, there is one waltzing female. We mate the F_1 waltzer with two different waltzer males and note that she produces only waltzer progeny. When we mate her with normal males, she produces normal progeny and no waltzers. We mate three of her normal female progeny with two of their brothers and these mice produce 60 progeny, all normal. When, however, we mate one of these same three females with a third brother, we get six normals and two waltzers in a litter of eight. By thinking about the parents of the F_1 waltzer, we can consider some possible explanations of these results:

a. A dominant allele may have mutated to a recessive allele in her normal parent.

b. There may have been a mutation to a dominant suppressor of V in one parent.

c. Meiotic nondisjunction of the chromosome carrying V in her normal parent may have given a viable aneuploid.

d. There may have been a viable deletion spanning V in the meiocyte from her normal parent.

Which of these explanations are possible, and which are eliminated by the genetic analysis? Explain in detail.

◆ Solution ◆

This question will force us to use concepts from Chapter 4 (suppression), Chapter 15 (mutation), Chapter 17 (deletion), and the present chapter (nondisjunction). The best way to answer it is to take the explanations one at a time and see if each fits the results given.

a. Mutation V to v. This hypothesis requires that the exceptional waltzer female be homozygous v/v. It is compatible with the results of mating her both to waltzer males, which would — if she were v/v — produce all waltzer offspring (v/v), and to normal males, which would produce all normal offspring of genotype V/v. However, brother-sister matings within this normal progeny should then produce a $3:1$ normal to waltzer ratio. Because some of the brother-sister matings actually produced no waltzers, this hypothesis does not explain the data.

b. Suppressor mutation s to S. Here the parents would be V/V ; s/s and v/v ; s/s, and a germinal mutation in one of them would give the F_1 waltzer the genotype V/v ; S/s. When we crossed her to a waltzer male, whose genotype is v/v ; s/s, we would expect some V/v ; s/s progeny, which would be phenotypically normal. However, we saw no normal progeny from this cross, so the hypothesis is already overthrown. Linkage could save the hypothesis temporarily if we assumed the mutation to be in the normal parent, giving a gamete $V\,S$. Then the F_1 waltzer would be $V\,S/v\,s$, and if linkage were tight enough, few or no $V\,s$ gametes would be produced, the type that are necessary to combine with the $v\,s$ gamete from the male to give $V\,s/v\,s$ normals. However, if this were true, the cross to the normal males would be $V\,S/v\,s \times V\,s/V\,s$ and this would give a high percentage of $V\,S/V\,s$ progeny, which would be waltzers, and of course none were seen.

c. Nondisjunction in the normal parent. This explanation would give a nullisomic gamete that would combine with v to give the F_1 waltzer the hemizygous genotype v. The subsequent matings would be as follows:

◆ $v \times v/v$ gives v/v and v progeny, all waltzers. This fits.

◆ $v \times V/V$ gives V/v and V progeny, all normals. This also fits.

◆ First intercrosses of normal progeny: $V \times V$. This gives V and V/V, which are all normal. This fits.

◆ Second intercrosses of normal progeny: $V \times V/v$. This gives 25 percent each of V/V, V/v, V (all normals), and v (waltzers). This also fits.

This hypothesis is therefore consistent with the data.

d. Deletion of V in normal parent. Let's call the deletion D. The F_1 waltzer would be D/v, and the subsequent matings would be as follows:

◆ $D/v \times v/v$. This gives v/v and D/v, which are all waltzers. This fits.

◆ $D/v \times V/V$. This gives V/v and D/V, which are all normal. This fits.

◆ First intercrosses of normal progeny: $D/V \times D/V$. This gives D/V and V/V, which are all normal. This fits.

◆ Second intercrosses of normal progeny: $D/V \times V/v$. This gives 25 percent of each of V/V, V/v, D/V (all normals), and D/v (waltzers). This also fits.

Once again, the hypothesis fits the data provided, so we are left with two hypotheses that are compatible with the results, and further experiments would be necessary to distinguish them. One obvious way would be to examine the chromosomes of the exceptional female under the microscope; aneuploidy should be relatively easy to distinguish from deletion.

SOLVED PROBLEMS

1. There is a controversy about the type of chromosome pairing seen in autotetraploids formed between two geographical races of a certain plant. It is known that chromosomes associate by pairs, but three hypotheses are put forward concerning how pairs form:

a. Pair formation is random.

b. Pairs form only between chromosomes of the same race.

c. Pairs form only between chromosomes of different races.

Consider a locus, A, that is closely linked to the centromere. The following cross is made:

Race 1: $A/A/A/A$ × Race 2: $a/a/a/a$

↓

Autotetraploid: $A/A/a/a$

The autotetraploid is then selfed. What ratio of phenotypes can be expected under each hypothesis of chromosome pairing? Explain your answer.

◆ Solution ◆

a. Under random pairing, all possible pairing combinations are equally likely. If we label the four homologous chromosomes 1, 2, 3, and 4, then the equally probable combinations are 1-2/3-4, 1-3/2-4, and 1-4/2-3. This kind of situation in an $A/A/a/a$ tetraploid was examined in the chapter (see page 560). We saw that a/a gametes are produced at a frequency of $\frac{1}{6}$ and that the frequency of $a/a/a/a$ progeny from selfing must therefore be $\frac{1}{36}$. All other progeny types contain at least one A allele. Hence, the expected ratio is 35 $A/-/-/-$: 1 $a/a/a/a$.

b. In this alternative, the chromosome pairs look like this:

because both the chromosomes carrying A come from race 1 and both the chromosomes carrying a come from race 2. This diagram shows how being methodical can help in problem solving, because it makes it very clear that the only possible segregation is that of one A and a to each pole. Hence, all gametes are A/a and, on self-fertilization, all genotypes would be $A/A/a/a$; only the A phenotype would be seen.

c. Here the pairing looks like this:

Because segregation of the pairs is independent, the gametic population can be represented as follows:

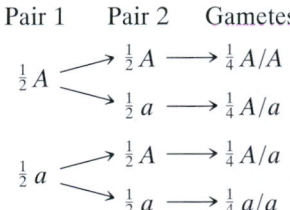

Self-fertilization would produce $a/a/a/a$ progeny at a frequency of $\frac{1}{4} \times \frac{1}{4} = \frac{1}{16}$. Therefore, $\frac{15}{16}$ of the progeny are $A/-/-/-$ and a 15 : 1 phenotypic ratio is predicted.

In conclusion, the ratio of phenotype A to phenotype a in each case is
  a. 35 : 1
  b. 1 : 0
  c. 15 : 1

2. Owing to the small size of the *Drosophila* chromosome 4, monosomics and trisomics for this chromosome are viable, but tetrasomics and nullisomics are not. A fly that is trisomic for chromosome 4 and carries the recessive gene for bent bristles (*b*) on all copies of chromosome 4 is crossed to a phenotypically normal fly that is monosomic for chromosome 4.

a. What genotypes and phenotypes can be expected in the progeny and in what proportions?

b. If trisomics from these progeny are interbred, what phenotypic ratio can be expected in the next generation? Assume that only one copy of b^+ is needed to produce normal (nonbent) bristles, that unpaired chromosomes pass to either pole at random, and that aneuploid gametes of any kind survive.

♦ **Solution** ♦

a. The cross is

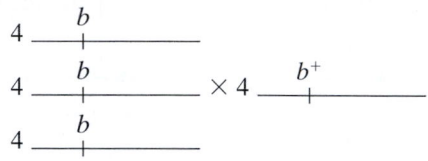

From the bent-bristled parent, the gametes will be $\frac{1}{2}$ b/b and $\frac{1}{2}b$. From the monosomic parent, the gametes will be $\frac{1}{2}b^+$

and $\frac{1}{2}0$ (the latter contain no chromosome 4). Hence, the progeny will be $\frac{1}{4}$ $b^+/b/b$, $\frac{1}{4}$ b/b, $\frac{1}{4}$ b^+/b, and $\frac{1}{4}$ b, which provides a phenotypic ratio of $1:1$.

b. The trisomics referred to are of genotype $b^+/b/b$. If we label these chromosomes 1, 2, and 3, then

$$1 = b$$
$$2 = b$$
$$3 = b^+$$

and segregation produces

$$1, 2 \ (b/b)/3 \ (b^+)$$
$$1, 3 \ (b/b^+)/2 \ (b)$$
$$2, 3 \ (b/b^+)/1 \ (b)$$

Fertilization can be represented as follows:

	$b/b\left(\frac{1}{6}\right)$	$b^+/b\left(\frac{2}{6}\right)$	$b\left(\frac{2}{6}\right)$	$b^+\left(\frac{1}{6}\right)$
$b/b\left(\frac{1}{6}\right)$	×	×	$b/b/b$ $\frac{2}{36}$	$\frac{1}{36}$
$b^+/b\left(\frac{2}{6}\right)$	×	×	$\frac{4}{36}$	$\frac{2}{36}$
$b\left(\frac{2}{6}\right)$	$b/b/b$ $\frac{2}{36}$	$\frac{4}{36}$	b/b $\frac{4}{36}$	$\frac{2}{36}$
$b^+\left(\frac{1}{6}\right)$	$\frac{1}{36}$	$\frac{2}{36}$	$\frac{2}{36}$	$\frac{1}{36}$

The ×'s represent tetrasomics, which, we are told, are not viable. Of the remaining $\frac{27}{36}$, $\frac{8}{36}$ are of *b* phenotype; therefore, a ratio of $19:8$ is predicted.

PROBLEMS

1. Distinguish between Klinefelter, Down, and Turner syndromes in humans.

2. List two ways in which you could make an allotetraploid between two related diploid plant species $2n = 28$.

3. **a.** Suppose that nondisjunction of *Neurospora* chromosome 3 occurs at the second division of meiosis. Show the content of each of the eight resultant ascospores with regard to chromosome 3.

   b. *Neurospora* normally has seven chromosomes. How many chromosomes are present in each of the ascospores in part a?

4. **a.** How would you synthesize a pentaploid (5*x*)?

   b. How would you synthesize a triploid (3*x*) of genotype *A/a/a*?

   c. You have just obtained a rare recessive mutation *a** in a diploid plant, which Mendelian analysis tells you is *A/a**. From this plant, how would you synthesize a tetraploid (4*x*) of genotype *A/A/a*/a**?

   d. How would you synthesize a tetraploid of genotype *A/a/a/a*?

   e. How would you synthesize a plant that is resistant to a chemical herbicide? (Assume that mutation to this trait is very infrequent, and you are lucky to get one mutant.)

5. Which of the following phrases correctly completes the sentence? Allopolyploids are **(a)** not fertile at all; **(b)** fertile only among themselves; **(c)** fertile with one parent only; **(d)** fertile with both parents only; **(e)** fertile with both parents and themselves.

6. Tetraploid yeast can be created by fusing two diploid cells. These tetraploids undergo meiosis like any other tetraploid and produce four diploid products of meiosis. Assume that homologous chromosomes synapse randomly in pairs and that there is no crossing-over in the region between the *B* locus and the centromere. What nonlinear tetrads are produced by a tetraploid of genotype *B/B/b/b*? What are the frequencies of the ascus types? (**Note:** This question requires tetrad analysis of tetraploid cells instead of the usual diploid cells.)

7. Consider a tetraploid of genotype *A/A/a/a* in which there is no pairing between chromosomes from the same parent. In another tetraploid of genotype *B/B/b/b*, pairs form only between chromosomes from

the same parent. What phenotypic ratios result from selfing in each of these cases? (Assume that one parent carries the dominant allele and that the other carries the recessive allele in each case.)

8. In the plant genus *Triticum,* there are many different polyploid species, as well as diploid species. Crosses were made between some different species, and hybrids were obtained. The meiotic pairing was observed in each hybrid and is recorded in the following table. (A bivalent is two homologous chromosomes paired at meiosis, and a univalent is an unpaired chromosome at meiosis.)

Species crossed to make hybrid	Pairing in hybrid
T. turgidum × *T. monococcum*	7 bivalents + 7 univalents
T. aestivum × *T. monococcum*	7 bivalents + 14 univalents
T. aestivum × *T. turgidum*	14 bivalents + 7 univalents

Explain these results and, in doing so,

a. deduce the somatic chromosome number of each species used.

b. state which species are polyploid, and whether they are auto- or allopolyploids.

c. account for the chromosome pairing pattern in the three hybrids.

9. The New World cotton species *Gossypium hirsutum* has a 2*n* chromosome number of 52. The Old World species *G. thurberi* and *G. herbaceum* each have a 2*n* number of 26. Hybrids between these species show the following chromosome pairing arrangements at meiosis:

Hybrid	Pairing arrangement
G. hirsutum × *G. thurberi*	13 small bivalents + 13 large univalents
G. hirsutum × *G. herbaceum*	13 large bivalents + 13 small univalents
G. thurberi × *G. herbaceum*	13 large univalents + 13 small univalents

Draw diagrams to interpret these observations phylogenetically, clearly indicating the relations between the species. How would you go about proving that your interpretation is correct? (Problem 9 after A. M. Srb, R. D. Owen, and R. S. Edgar, *General Genetics,* 2d ed. W. H. Freeman and Company, 1965.)

10. The genotype of an autotetraploid that is heterozygous at two loci is *F/F/f/f* ; *G/G/g/g*. Each locus affects a different character, and the two loci are located very close to the centromere on different (nonhomologous) chromosomes.

a. What gametic genotypes are produced by this individual and in what proportions?

b. If the individual is self-fertilized, what proportion of the progeny will have the genotype *F/F/F/f* ; *G/G/g/g*? the genotype *f/f/f/f* ; *g/g/g/g*?

11. Which of the following abnormalities are *not* caused by meiotic nondisjunction? (a) Turner syndrome; (b) cri du chat syndrome; (c) Down syndrome; (d) Klinefelter syndrome; (e) XYY syndrome; (f) achondroplastic dwarfism; (g) Edwards syndrome; (h) Marfan syndrome.

12. A woman with Turner syndrome is found to be colorblind. Both her mother and father have normal vision. How can her colorblindness be explained? Does this outcome tell us whether nondisjunction occurred in the father or in the mother? If the colorblindness gene were close to the centromere (it is not, in fact), would the available facts tell us whether the nondisjunction occurred at the first or at the second meiotic division? Repeat the question for a colorblind man with Klinefelter syndrome.

13. Individuals have been found who are colorblind in one eye but not in the other. What would this finding suggest if these individuals were (a) only or mostly females? (b) only or mostly males? (Assume that this trait is X-linked recessive.)

14. When geneticists treat human sperm with quinacrine dihydrochloride, about half the sperm show a fluorescent spot thought to be the Y chromosome. About 1.2 percent of normal human sperm show two fluorescent spots. Geneticists examined the sperm of some industrial workmen who were exposed for about 1 year to the chemical dibromochloropropane. They found the frequency of sperm with two spots to be on average 3.8 percent. Propose an explanation for this result, and explain how you would test your explanation.

15. A person with Turner syndrome is also found to have the phenotype nystagmus (an eye disorder).

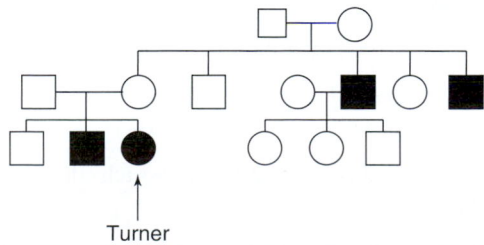

Turner

a. What event or events can explain the coincidence of the two hereditary conditions?

b. In which parent and at which stage did the event or events occur?

16. In British Columbia, the mean age of women giving birth to Down syndrome babies fell from 34 years to 28 years between 1952 and 1972. What are two possible causes of this population trend, and how could the two hypotheses be tested?

17. Several kinds of sexual mosaicism are well documented in humans. Suggest how each of the following examples may have arisen:

    a. (XX)(XO) (that is, there are two cell types in the body, (XX and XO)

    b. (XX)(XXYY)   **c.** (XO)(XXX)

    d. (XX)(XY)   **e.** (XO)(XX)(XXX)

18. The discovery of chromosome banding in eukaryotes has greatly improved the ability to distinguish various cytogenetic events. Particularly useful are banding polymorphisms, because the "morphs" can be used as chromosome markers. These morphs make chromosomes cytologically distinguishable by virtue of minor variation in band size, position, and so forth. Consider chromosome 21 in humans. Assume that one set of parents is $21^a 21^b$ ♀ × $21^c 21^d$ ♂, where a, b, c, and d represent morphs of a polymorphism for this chromosome. Also assume that fetuses of the following types are produced (where 42A stands for the rest of the autosomes):

    1. $42A + 21^b 21^b 21^c + XY$

    2. $42A + 21^a 21^b 21^d + XX$

    3. $42A + 21^b 21^d + XY$

    4. $42A + 21^a 21^c 21^c + XX$

    5. $42A + 21^a 21^b + XY$

    6. $42A + 21^a 21^c + XYY$

    7. $(42A + 21^a + XY)(42A + 21^a 21^c 21^c + XY)$ (mosaic)

    8. $(42A + 21^a 21^c + XY)(42A + 21^b 21^d + XX)$ (mosaic)

    In each case:

    a. State the genetic term for the condition.

    b. Diagram the events that give rise to the condition.

    c. State the individual in which the events took place.

19. Suppose that you have a line of mice that has cytologically distinct forms of chromosome 4. The tip of the chromosome can have a knob (called 4^K), or a satellite (4^S), or neither (4). Here are sketches of the three types:

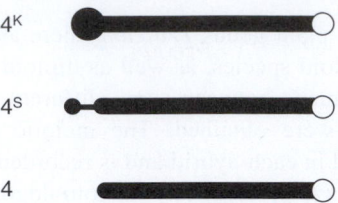

You cross a $4^K 4^S$ female with a 4 4 male and find that most of the progeny are $4^K 4$ or $4^S 4$, as expected. However, you occasionally find some rare types as follows (all other chromosomes are normal):

a. $4^K 4^K 4$ **b.** $4^K 4^S 4$ **c.** 4^K

d. $4/4^K 4^K 4$ mosaic   **e.** $4^K 4/4^S 4$ mosaic

Explain the rare types that you have found. Give as precisely as possible the stages at which they originate, and state if they are present in the male parent, the female parent, or the zygote. (Give reasons briefly.)

20. In *Drosophila,* a cross (cross 1) is made between two mutant flies, one homozygous for the recessive mutation bent wing (*b*) and the other homozygous for the recessive mutation eyeless (*e*). The mutations *e* and *b* are alleles of two different genes that are known to be very closely linked on the tiny autosomal chromosome 4. All the progeny were wild-type phenotype. One of the female progeny was crossed to a male of genotype *b e/b e*; call this cross 2. The progeny of cross 2 were mostly of the expected types, but there was also one rare female of wild-type phenotype.

    a. Explain what the common progeny are expected to be from cross 2.

    b. Could the rare wild-type female have arisen by

    (1) crossing-over?

    (2) nondisjunction?

    Explain.

    c. The rare wild-type female was testcrossed to a male of genotype *b e/b e* (cross 3). The progeny were

    $\frac{1}{6}$ wild type

    $\frac{1}{6}$ bent, eyeless

    $\frac{1}{3}$ bent

    $\frac{1}{3}$ eyeless

Which of the explanations in part b are compatible with this result? Explain the genotypes and phenotypes of progeny of cross 3 and their proportions.

 ### Unpacking the Problem

1. Define *homozygous, mutation, allele, closely linked, recessive, wild type, crossing-over, nondisjunction, testcross, phenotype,* and *genotype.*

2. Does this problem concern sex linkage? Explain.

3. How many chromosomes does *Drosophila* have?

4. Draw a clear pedigree summarizing the results of crosses 1, 2, and 3.

5. Draw the gametes produced by both parents in cross 1.

6. Draw the chromosome 4 constitution of the progeny of cross 1.

7. Is it surprising that the progeny of cross 1 are wild-type phenotype? What does this outcome tell you?

8. Draw the chromosome 4 constitution of the male tester used in cross 2, and the gametes that he can produce.

9. With respect to chromosome 4, what gametes can the female parent in cross 2 produce in the absence of nondisjunction? Which would be common and which rare?

10. Draw first- and second-division meiotic nondisjunction in the female parent of cross 2, as well as the resulting gametes.

11. Are any of the gametes from part 10 aneuploid?

12. Would you expect the aneuploid gametes to give rise to viable progeny? Would these progeny be nullisomic, monosomic, disomic, or trisomic?

13. What progeny phenotypes would be produced by the various gametes considered in parts 9 and 10?

14. Consider the phenotypic ratio in the progeny of cross 3. Many genetic ratios are based on halves and quarters, but this ratio is based on thirds and sixths. What might this point to?

15. Could there be any significance to the fact that the crosses concern genes on a very small chromosome? When is chromosome size relevant in genetics?

16. Draw the progeny expected from cross 3 under the two hypotheses, and give some idea of relative proportions.

21. A cross is made in tomatoes between a female plant that is trisomic for chromosome 6 and a normal diploid male plant that is homozygous for the recessive allele for potato leaf (*p/p*).

a. A trisomic F₁ plant is backcrossed to the potato-leaved male. What is the ratio of normal-leaved plants to potato-leaved plants when you assume that *p* is on chromosome 6?

b. What is the ratio of normal- to potato-leaved plants when you assume that *p* is not located on chromosome 6?

22. A tomato geneticist attempts to assign five recessive genes to specific chromosomes using trisomics. She crosses each homozygous mutant (2*n*) to each of three trisomics—chromosomes 1, 7, and 10. From these crosses, the geneticist selects trisomic progeny (which are less vigorous) and backcrosses them to the appropriate homozygous recessive. The *diploid* progeny from these crosses are examined. Her results follow, in which the ratios are wild type : mutant.

Trisomic chromosome	GENE				
	d	*y*	*c*	*h*	*cot*
1	48:55	72:29	56:50	53:54	32:28
7	52:56	52:48	52:51	58:56	81:40
10	45:42	36:33	28:32	96:50	20:17

Which of the genes can the geneticist assign to which chromosomes? (Explain your answer fully.)

23. *Petunia* plants have four loci, *A, B, C,* and *D,* that are known to be very closely linked. A plant of genotype *a B c D/A b C d* is irradiated with γ rays and then crossed to an *a b c d/a b c d* plant. In the progeny, plants of phenotype A B C D are occasionally found. What are two possible modes of origin, and which is the more likely?

24. A cross in *Neurospora* between strains carrying the multiply marked chromosomes *a b⁺ c d⁺ e* and *a⁺ b c⁺ d e⁺* produces one product of meiosis that grows on minimal medium. (Assume that *a, b, c, d,* and *e* determine nutritional requirements.) When the rare colony grows up, some *asexual* spores are *a b⁺ c d⁺ e* in genotype, some are *a⁺ b c⁺ d e⁺*, and the remainder grow on minimal medium. Explain the origin of the rare product of meiosis and the origin of the three types of asexual spores.

25. Two *Neurospora* auxotrophs are crossed: *pan1* × *leu2*. These loci are linked on the same chromosome arm, with *leu2* between the centromere and *pan1*. Most octads from this cross were of the type expected, but the following unexpected types also were obtained:

a. fully prototrophic **b.** pan-requiring
   fully prototrophic   pan-requiring
   fully prototrophic   abort
   fully prototrophic   abort
   abort   Leu-requiring
   abort   Leu-requiring
   abort   Leu-requiring
   abort   Leu-requiring

c. fully prototrophic
   fully prototrophic
   abort
   abort
   Leu-requiring
   Leu-requiring
   pan-requiring
   pan-requiring

Give explanations for these three types.

*26. The ascomycete *Sordaria brevicollis* has two closely linked complementing markers, b_1 and b_2, that result in buff-colored (light-brown) ascospores. In the cross $b_1 \times b_2$, if one or both centromeres divide and separate precociously at the first division of meiosis, what patterns of spore colors are produced in those asci? How can these asci be distinguished from normal asci and from asci in which nondisjunction has occurred? (**Note:** Ascospores in this fungus are normally black, and nullisomic ascospores are white; assume that there is no crossing-over between b_1 and b_2.)

27. Design a test system for detecting agents in the human environment that are potentially capable of causing aneuploidy in eukaryotes.

28. Two alleles determine flower color in the plant *Datura:* P determines purple; and p, white. The P locus is on the smallest chromosome, number 3. In plants trisomic for this chromosome, $n + 1$ pollen is never functional, and only half the $n + 1$ eggs are functional. Assume that P is always completely dominant to any number of p's, and that the locus is close to the centromere. What is the expected ratio of purple:white in the progeny of the following crosses?

a. $P/P/p \; ♀ \times P/P/p \; ♂$

b. $P/p/p \; ♀ \times P/p/p \; ♂$

c. $P/p/p \; ♀ \times p/p \; ♂$

29. There are six main species in the *Brassica* genus: *B. carinata, B. campestris, B. nigra, B. oleracea, B. juncea,* and *B. napus.* You can deduce the interrelationships between these six species from the following table:

Species or F$_1$ hybrid	Chromosome number	Number of bivalents	Number of univalents
B. juncea	36	18	0
B. carinata	34	17	0
B. napus	38	19	0
B. juncea × *B. nigra*	26	8	10
B. napus × *B. campestris*	29	10	9
B. carinata × *B. oleracea*	26	9	8
B. juncea × *B. oleracea*	27	0	27
B. carinata × *B. campestris*	27	0	27
B. napus × *B. nigra*	27	0	27

a. Deduce the chromosome number of *B. campestris, B. nigra,* and *B. oleracea.*

b. Show clearly any evolutionary relationships between the six species that you can deduce at the chromosomal level.

30. In the fungus *Ascobolus* (similar to *Neurospora*), ascospores are normally black. The mutation *f*, producing fawn ascospores, is in a gene just to the right of the centromere on chromosome 6, whereas mutation *b*, producing beige ascospores, is in a gene just to the left of the same centromere. In a cross of fawn with beige parents $(+ f \times b +)$ most octads show four fawn and four beige ascospores, but three rare exceptional octads were found as shown here. In the sketch, black represents the wild-type phenotype, a vertical line represents fawn, a horizontal line represents beige, and an empty circle represents an aborted (dead) ascospore.

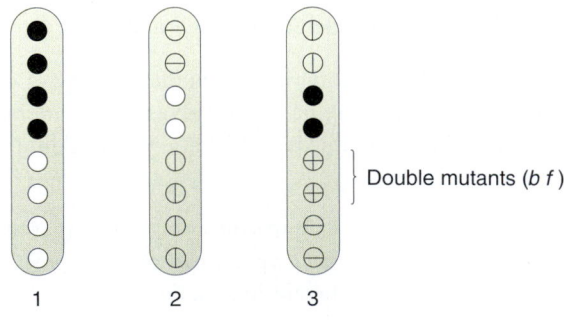

a. Provide reasonable explanations for these three exceptional octads.

b. Diagram the meiosis that gave rise to octad 2.

31. The life cycle of the haploid fungus *Ascobolus* is similar to that of *Neurospora*. A mutational treatment produced two mutant strains, 1 and 2, both of which when crossed to wild type gave unordered tetrads all of the following type (fawn is a light-brown color; normally, crosses produce all black ascospores):

spore pair 1	black
spore pair 2	black
spore pair 3	fawn
spore pair 4	fawn

a. What does this result show? Explain.

The two mutant strains were crossed. Most of the unordered tetrads were of the following type:

spore pair 1	fawn
spore pair 2	fawn
spore pair 3	fawn
spore pair 4	fawn

b. What does this result suggest? Explain.

When large numbers of unordered tetrads were screened under the microscope, some rare ones that contained black spores were found. Four cases are shown here:

	Case A	Case B	Case C	Case D
spore pair 1	black	black	black	black
spore pair 2	black	fawn	black	abort
spore pair 3	fawn	fawn	abort	fawn
spore pair 4	fawn	fawn	abort	fawn

(**Note:** Ascospores with extra genetic material survive, but those with less than a haploid genome abort.)

c. Propose reasonable genetic explanations for each of these four rare cases.

d. Do you think the mutations in the two original mutant strains were in one single gene? Explain.

32. In a certain tropical plant there are two genes, each of which produces color pigment. The R (red color; $R > r$) and B (blue color; $B > b$) loci are centromere

linked and on separate chromosomes. Tetraploid plants are made in which the chromosomes pair in twos at meiosis. The following cross is made:

$$R/R/R/R \; ; \; b/b/b/b \qquad \times \qquad r/r/r/r \; ; \; B/B/B/B$$

and an F_1 is obtained and then selfed to give an F_2. If only one dominant allele is required to give the dominant phenotype at each locus, what phenotypes are expected in the F_2 and in what proportions?

33. A tomato geneticist working on Fr, a dominant mutant allele that causes rapid fruit ripening, decides to find out which chromosome contains this gene by using a set of lines, each of which is trisomic for one chromosome. To do so, she crosses a homozygous diploid mutant to each of the wild-type trisomic lines.

a. A trisomic F_1 plant is crossed to a diploid wild-type plant. What is the ratio of fast- to slow-ripening plants in the *diploid* progeny of this second cross if Fr is on the trisomic chromosome? Use diagrams to explain.

b. What is the ratio of fast- to slow-ripening plants in the *diploid* progeny of this second cross if Fr is not located on the trisomic chromosome? Use diagrams to explain.

c. Here are the results of the crosses. On which chromosome is Fr and why?

Trisomic chromosome	Fast ripening : slow ripening in diploid progeny
1	45 : 47
2	33 : 34
3	55 : 52
4	26 : 30
5	31 : 32
6	37 : 41
7	44 : 79
8	49 : 53
9	34 : 34
10	37 : 39

(Problem 33 from Tamara Western.)

19

MECHANISMS OF RECOMBINATION

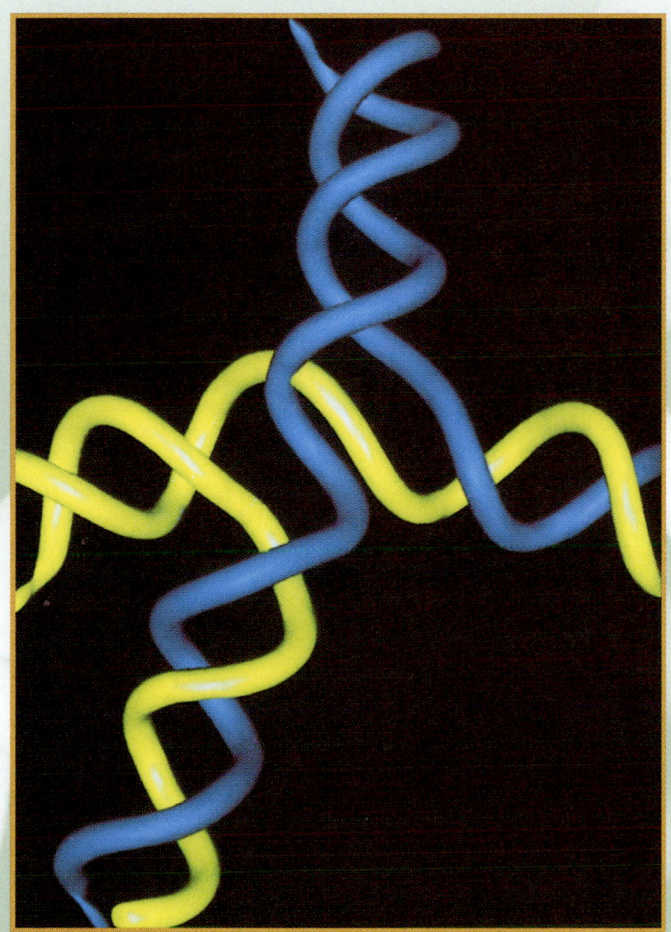

Computer model of exchange site.
(Julie Newdol, Computer Graphics Laboratory, University of California, San Francisco. Copyright by Regents, University of California.)

Key Concepts

Recombination occurs at regions of homology between chromosomes through the breakage and reunion of DNA molecules.

Models for recombination, such as the Holliday model, involve the creation of a heteroduplex branch, or cross bridge, that can migrate and the subsequent splicing of the intermediate structure to yield different types of recombinant DNA molecules.

Recombination models can be applied to explain genetic crosses.

Many of the enzymes participating in recombination in bacteria have been identified.

Throughout our analysis of linkage, we studied the recombination of genes by crossing-over. In this chapter, we consider molecular mechanisms for generating recombination by crossing-over. Figure 19-1 depicts a basic crossover event, in which two homologous molecules are aligned and subsequently undergo recombination. When Benzer's work and that of others revealed that recombination occurs within genes, it became evident that recombination had to be very precise, because even single-base-pair errors could disrupt the integrity of the gene. How can recognition of homologous chromosomes and recombination events be so precise? The answer lies in the power of base-pair complementarity. We shall see how base-pair complementarity and the formation of heteroduplex regions between complementary regions of homologous chromosomes lead to the recombination events that we have been studying.

Breakage and reunion of DNA molecules

The experiments discussed in Chapter 5 provide good *indirect* evidence in favor of breakage and reunion. One of the first *direct* proofs that chromosomes (in this case, viral chromosomes) can break and rejoin came from experiments on λ phage done in 1961 by Matthew Meselson and Jean Weigle. They simultaneously infected *E. coli* with two strains of λ. One strain, which had the genetic markers *c* and *mi* at one end of the chromosome, was "heavy" because the phages were produced from cells grown in heavy isotopes of carbon (^{13}C) and nitrogen (^{15}N). The other strain was $c^+ mi^+$ for the markers and had "light" DNA because it was harvested from cells grown on the normal light isotopes ^{12}C and ^{14}N. The two DNAs (chromosomes) can be represented as shown in Figure 19-2a. The multiply infected cells were then incubated in a light medium until they lysed.

The progeny phages released from the cells were spun in a cesium chloride density gradient. A wide band was obtained, indicating that the viral DNAs ranged in density from the heavy parental value to the light parental value, with a great many intermediate densities (Figure 19-2b). Interestingly, some recombinant phages were recovered with density values very close to the heavy parental value. They were of genotype $c mi^+$, and they must have arisen through an exchange event between the two markers (Figure 19-2c). The heavy density of the chromosome would be expected because only the small tip of the chromosome carrying the mi^+ allele would come from the light parental chromosome. In the reciprocal cross of heavy $c^+ mi^+$ phages to light *c mi*,

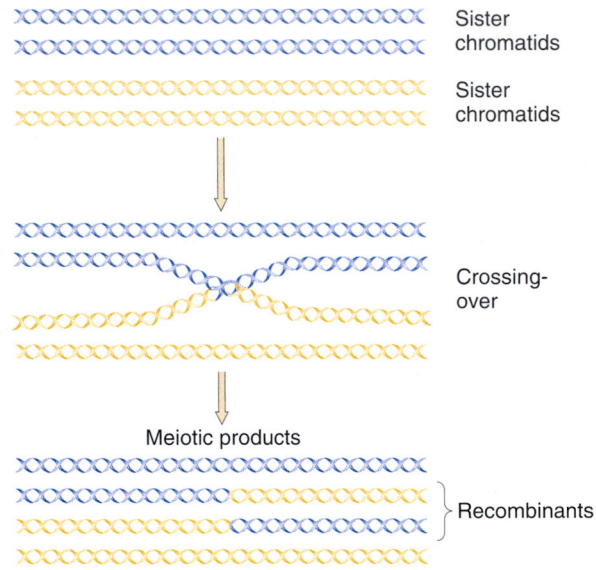

Sister chromatids

Sister chromatids

Crossing-over

Meiotic products

Recombinants

Figure 19-1 The molecular event of recombination may be schematically represented by two double-stranded molecules breaking and rejoining.

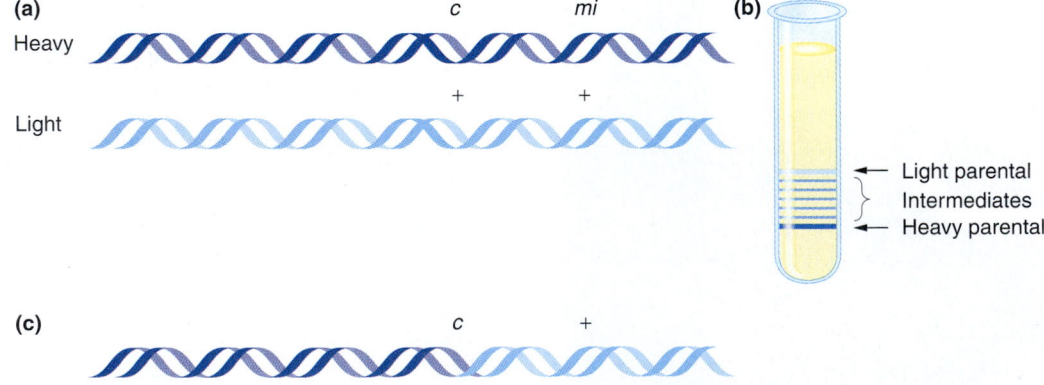

(a) Heavy   *c*   *mi*

Light   +   +

(b) Light parental / Intermediates / Heavy parental

(c)   *c*   +

Figure 19-2 Evidence for chromosome breakage and reunion in λ phages. (a) The chromosomes of the two λ strains used to multiply infect *E. coli*. (b) Bands produced when progeny phages are spun in a cesium chloride density gradient. The fact that intermediate densities are obtained indicates a range of chromosome compositions with partly light and partly heavy components. (c) The chromosome of the heavy $c mi^+$ progeny resulting from crossover between the two markers. The density of this crossover product confirms that the crossover entailed a physical breakage and reunion of the DNA.

the heavy recombinants were found to be c^+ mi, as expected. These results can be explained in only one way: the recombination event must have occurred through the physical breakage and reunion of DNA. Although we have to be careful about extrapolating from viral to eukaryotic chromosomes, this evidence shows that the breakage and reunion of DNA strands does occur.

Chiasmata: the crossover points

In Chapter 5, we made the simple assumption that chiasmata are the actual sites of crossovers. Mapping analysis indirectly supports this idea: because an average of one crossover per meiosis produces 50 genetic map units, there should be correlation between the size of the genetic map of a chromosome and the observed mean number of chiasmata per meiosis. The correlation has been made in well-mapped organisms.

However, the harlequin chromosome-staining technique (see Chapter 8) has made it possible to test the idea directly. In 1978, C. Tease and G. H. Jones prepared harlequin chromosomes in meioses of the locust. Remember that the harlequin technique produces sister chromatids: one dark and the other light. When a crossover occurs, it can be between two dark, two light, or nonsister dark and light chromatids, as shown in Figure 8-16. This last situation is crucial because mixed (part dark and part light) crossover chromatids are produced. Tease and Jones found that the dark–light transition is right at the chiasma — proving that the chiasmata are the crossover sites and settling a question that had been unresolved since the early 1900s (Figure 19-3).

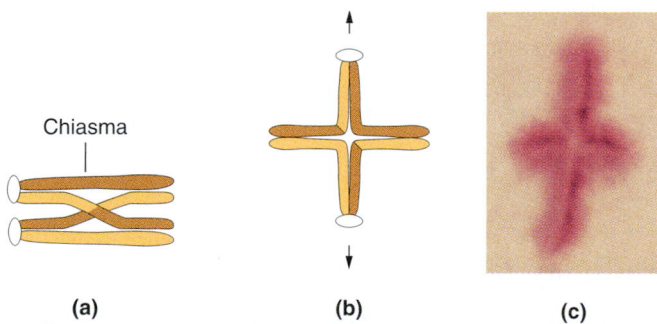

(a)   **(b)**   **(c)**

Figure 19-3   Crossing-over between dark- and light-stained nonsister chromatids in a meiosis in the locust. (a) Representation of the chiasma. (b) The best stage for observing is when the centromeres (white ovals) have pulled apart slightly, forming a cross-shaped structure with the chiasma at the center. (c) Photograph of the stage shown in part b. (Photograph courtesy of C. Tease and G. H. Jones, *Chromosoma* 69, 1978, 163–178.)

Genetic results leading to recombination models

Tetrad analysis in filamentous fungi, such as *Neurospora crassa,* where all four products of a single meiosis can be recovered and examined (see Chapter 6), provided the impe-

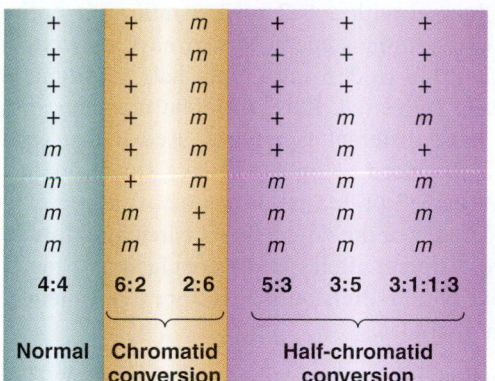

+	+	m	+	+	+
+	+	m	+	+	+
+	+	m	+	+	+
+	+	m	+	m	m
m	+	m	+	m	+
m	+	m	m	m	m
m	m	+	m	m	m
m	m	+	m	m	m
4:4	**6:2**	**2:6**	**5:3**	**3:5**	**3:1:1:3**
Normal	**Chromatid conversion**		**Half-chromatid conversion**		

(*Note*: 3:1:1:3 = aberrant 4:4.)

Figure 19-4   Rare aberrant allele ratios observed in a cross of type $+ \times m$ in fungi. (Ascus genotypes are represented here.) When the Mendelian ratio of 4:4 is not obtained, some of the alleles in the cross have been converted into the opposite alleles. In some asci, it appears that the entire chromatid has been converted (6:2 or 2:6 ratios). In others, it appears that only half-chromatids have been converted (5:3, 3:5, or 3:1:1:3 ratios).

tus for the first models of intragenic recombination. These crucial findings, reviewed in the following list, were gene conversion, evidence for postmeiotic segregation of gene-conversion events, polarity, and the association of gene conversion with crossing-over.

1. *Gene conversion.* Departures from the predicted Mendelian 4:4 segregation ratios are detectable in some asci (0.1–1.0 percent in filamentous fungi, but as high as 4 percent in yeast). Figure 19-4 gives the most common aberrant ratios obtained. It appears as though some alleles in the cross have been "converted" into the opposite alleles (Figure 19-5). The process therefore has become known as **gene conversion;** it can occur only where there is heterozygosity for two different alleles

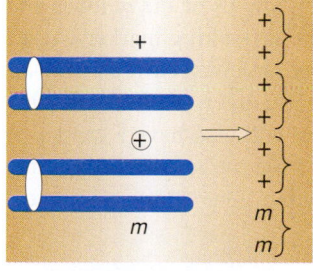 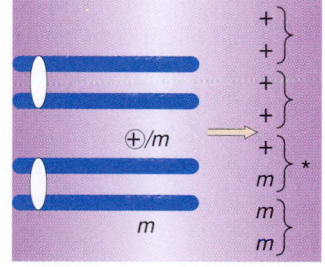

(a) Chromatid conversion   **(b) Half-chromatid conversion**

Figure 19-5   Gene conversions are inferred from the patterns of alleles observed in asci. (a) In a chromatid conversion, the allele on one chromatid seems somehow to have been converted into an allele like those on the other chromatid pair. The converted allele is shown by the symbol ⊕. One spore pair is of the opposite genotype from that expected in Mendelian segregation. (b) In a half-chromatid conversion, one spore pair (*) has nonidentical alleles. Somehow, one chromatid seems to be "half converted," giving rise to one spore of the original genotype and one spore converted into the other allele.

of a gene. In asci with a 6:2 or 2:6 ratio, one entire chromatid of a chromosome seems to have converted. In asci with a 5:3 or 3:5 ratio, only half a chromatid seems to have converted. Here, different members of a spore pair have different genotypes. Recall that each spore pair is produced by mitosis from a single product of meiosis. The 5:3 or 3:5 ratios can be explained only by the two strands of the double helix carrying information for two different alleles at the conclusion of meiosis. The next mitotic division is therefore a postmeiotic segregation of alleles.

Conversion cannot be explained by mutation, because the allele that is converted always changes into the other specific allele *taking part in the cross,* not to some other allele known for the locus but not a part of the cross.

2. *Polarity.* In genes for which accurate allele maps are available, we can compare the conversion frequencies of alleles at various positions within the gene. In almost every case, the sites closer to one end show higher frequencies than do the sites farther away from that end. In other words, there is a gradient, or polarity, of conversion frequencies along the gene (Figure 19-6).

3. *Conversion and crossing-over.* In heteroallelic crosses where the locus under study is closely flanked by other genetically marked loci, the conversion event is very often (about 50 percent of the time) accompanied by an exchange in one of the flanking regions. This exchange nearly always takes place on the side nearer the allele that has converted and almost always includes the chromatid in which conversion has occurred.

For example, consider the chromatids diagrammed in Figure 19-6. Suppose that the polarity is such that the alleles toward the left end of the chromatid convert more often than those toward the right end. The cross diagrammed here is between $a^+ m_2 b^+$ and $a m_1 b$, where m_1 and m_2 are different alleles of the m locus and a and b represent closely linked flanking markers. If we look at asci in which conversion has occurred at the m_1 site (the most frequent kind of conversion in this locus), we find that half of these asci will also have a crossover in region I and half will have no crossover. In the smaller number of asci showing gene conversion at the

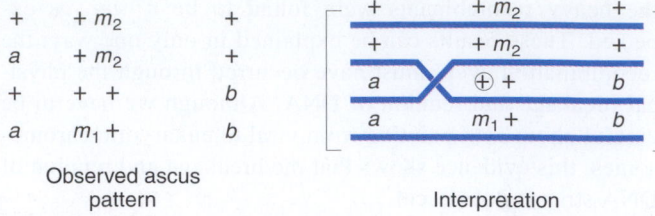

Figure 19-7 A specific ascus pattern can be explained by both crossover and a chromatid conversion. In this case, a conversion of $m_1 \rightarrow +$ is accomplished by a crossover in the region between a and m_1.

m_2 site, half will also have a crossover in region II and half will have no crossover. Such events are detected in ascus genotypes like the one shown in Figure 19-7, which can be interpreted as a conversion of $m_1 \rightarrow +$, accompanied by a crossover in region I.

In some asci, a single conversion event seems to include several sites at once. In a heteroallelic cross, this event is called a **co-conversion** (Figure 19-8). The frequency of co-conversion increases as the distance between alleles decreases.

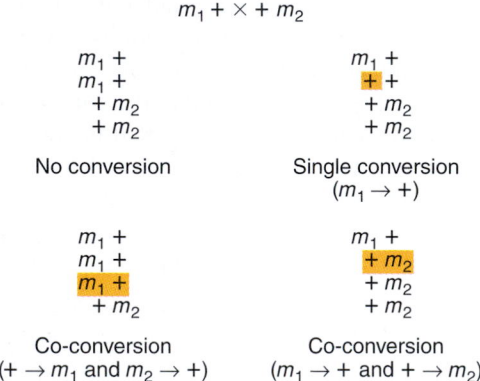

Figure 19-8 Sample ascus patterns obtained from a single conversion and co-conversions.

Holliday model

One of the first plausible models to account for the preceding observations was formulated by Robin Holliday. The key features of the **Holliday model** are the formation of **heteroduplex DNA;** the creation of a **cross bridge;** its migration along the two heteroduplex strands, termed **branch migration;** the occurrence of **mismatch repair;** and the subsequent **resolution,** or splicing, of the intermediate structure to yield different types of recombinant molecules. The model is depicted in Figure 19-9.

Enzymatic cleavage and the creation of heteroduplex DNA

Looking at Figure 19-9a, we can see that two homologous double helices are aligned, although note that they have

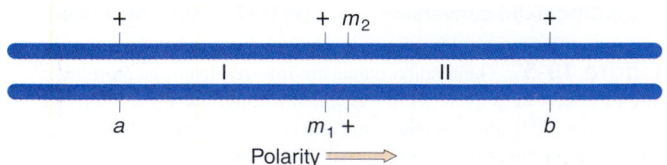

Figure 19-6 Diagram of chromatids taking part in a cross in regions I and II of a gene. The arrow indicates the polarity of gene conversion in the m locus, pointing toward the end with lower conversion frequency.

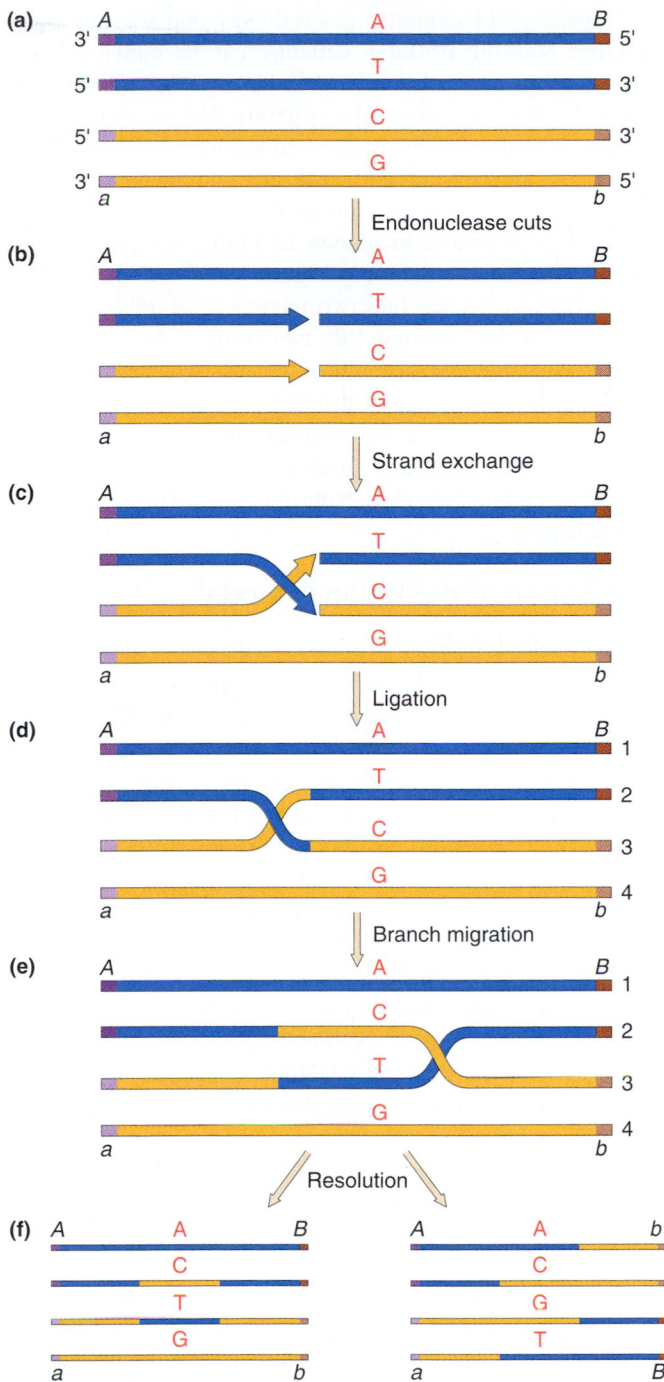

Figure 19-9 A prototype mechanism for genetic recombination. (a) Two homologous double helices are shown. Each pair represents a chromatid and the two pairs represent two nonsister chromatids. The helices are aligned so that the bottom strand of the first helix has the same polarity as the top strand of the second helix. (b) Two parallel or two antiparallel strands are cut. (c) The free ends become associated with the complementary strands in the homologous double helix. (d) Ligation creates partly heteroduplex double helices, the Holliday structure. (e) Migration of the branch point is by continued strand transfer by the two polynucleotide chains taking part in the crossover. (f) Resolution can occur in one of two ways, which will be described in detail later. (After H. Potter and D. Dressler, *Cold Spring Harbor Symposium on Quantitative Biology* 43, 1979, 970. Cold Spring Harbor Laboratory, Cold Spring Harbor, NY.)

been rotated so that the bottom strand of the first helix has the same polarity as the top strand of the second helix ($5' \rightarrow 3'$ in this case). Then a nuclease cleaves the two strands that have the same polarity (Figure 19-9b). The free ends leave their original complementary strands and undergo hydrogen bonding with the complementary strands in the homologous double helix (Figure 19-9c). Ligation produces the structure shown in Figure 19-9d. This partially heteroduplex double helix is a crucial intermediate in recombination, and has been termed the **Holliday structure.**

Branch migration

The Holliday structure creates a cross bridge, or branch, that can move, or migrate, along the heteroduplex (Figure 19-9d and e). This phenomenon of branch migration is a distinctive property of the Holliday structure. Figure 19-10 portrays a more realistic view of this structure as it might appear during branch migration.

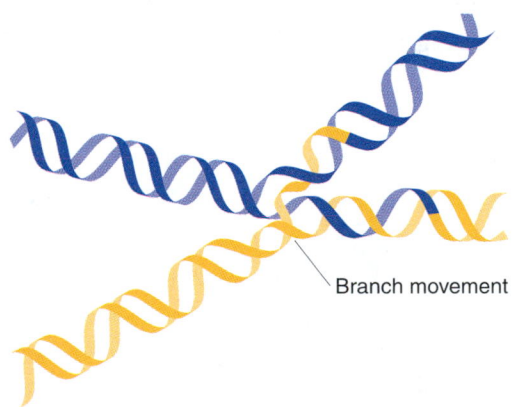

Figure 19-10 Branch migration, the movement of the crossover point between DNA complexes. (After T. Broker, *Journal of Molecular Biology* 81, 1973, 1; from J. D. Watson et al., *Molecular Biology of the Gene,* 4th ed. Copyright © 1987 by Benjamin Cummings.)

Resolution of the Holliday structure

The Holliday structure can be resolved by cutting and ligating either the two originally exchanged strands (Figure 19-9f, left) or the originally unexchanged strands (Figure 19-9f, right). The former generates a pair of duplexes that are parental, except for a stretch in the middle containing one strand from each parent. If the two parents had different alleles in this stretch, as indicated here, then the DNA will be heteroduplex. The latter resolution step generates two duplexes that are recombinant, with a stretch of heteroduplex DNA. The Holliday model also postulated that the heteroduplex DNA mismatches can be repaired by an enzymatic correction system that recognizes mismatches and excises the mismatched base from one of the two strands, filling in the excised base with the correct complementary base. The resulting molecules will carry either the wild-type or the mutant allele, depending on which allele is excised.

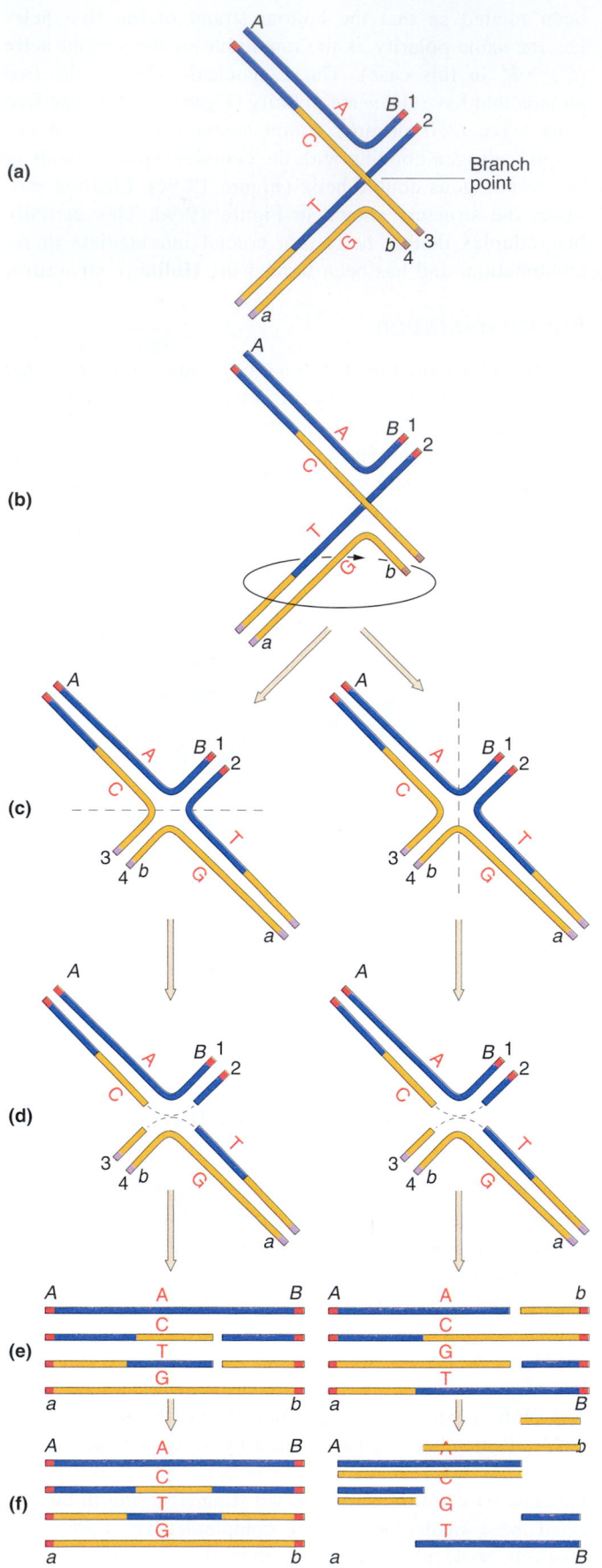

Figure 19-11 demonstrates one way that we can easily visualize how the Holliday structure can be converted into the recombinant structures with which we are familiar. In Figure 19-11a, we can see the structure that we arrived at in Figure 19-4e drawn out in an extended form. Compare Figures 19-9e and 19-11a until you are convinced that these two structures are indeed equivalent. If we rotate the bottom part of this structure, as shown in Figure 19-11b, we can generate the form depicted in Figure 19-11c. This last form can be converted back into two unconnected double helices by enzymatically cleaving only two strands. As indicated in 19-11c, cleavage can occur in either of two ways, each of which generates a different product (Figure 19-11d). These cleaved structures can be viewed more simply (Figure 19-11e). Repair synthesis produces the final recombinant molecules (Figure 19-11f). Note the two different types of recombinants.

Application of the Holliday model to genetic crosses

The Holliday model nicely explained the phenomena that we described previously. Gene conversion and the aberrant ratios depicted in Figure 19-4 can result as a consequence of mismatch repair, as shown in Figure 19-12 and Table 19-1. In Table 19-1, the symbols + for wild type and *m* for mutant are used for simplicity. When both mismatches are corrected to yield the same parental type, then 6:2 or 2:6 ratios result; when only one heteroduplex is corrected, then a 5:3 ratio results; and, when there is no correction, an aberrant 4:4 ratio is the product.

The Holliday model also accounted for polarity of gene-conversion events, because conversion takes place only within the heteroduplex DNA between the break point and the branch point at which the Holliday structure is resolved. The farther a gene locus is from the breakage point, the more likely it is to be beyond the branch point and thus not part of the heteroduplex.

It should be noted that the phenomenon of gene conversion and its association with about half the cases of crossing-over was a driving force for the formulation of the Holliday model, which entails a strand exchange that results in reciprocal crossovers almost half the time. This 50 percent reciprocal crossover result is because of the resolution of the ex-

Figure 19-11   (a) The Holliday structure shown in an extended form. (b) The rotation of the structure shown in part a can yield the form depicted in part c. Resolution of the structure shown in part c can proceed in two ways, depending on the points of enzymatic cleavage, yielding the structures shown in part d. The dotted lines show which segments will rejoin to form recombinant strands for each particular cleavage scheme. The strands are shown linearly in part e and can be repaired to the forms shown in part f. (From H. Potter and D. Dressler, *Cold Spring Harbor Symposium on Quantitative Biology* 43, 1970, 970. Cold Spring Harbor Laboratory, Cold Spring Harbor, NY.)

19-1 TABLE		Correction of One or Both Heteroduplex DNA Molecules		
DNA strands at start of meiosis	Strands at heteroduplex DNA stage	No correction	One heteroduplex DNA corrected to +	Both heteroduplex DNAs corrected to +
Two homologous chromosomes { _m_ _m_ _m_ _m_ _+_ _+_ _+_ _+_ }	Four chromatids, two strands each { _m_ _m_ _m_ _+_ _m_ _+_ _+_ _+_ }	} 3 m } 1+ } 1 m } 3+	} 3 m } 5+	} 2 m } 6+
		Aberrant 4:4	5:3	6:2

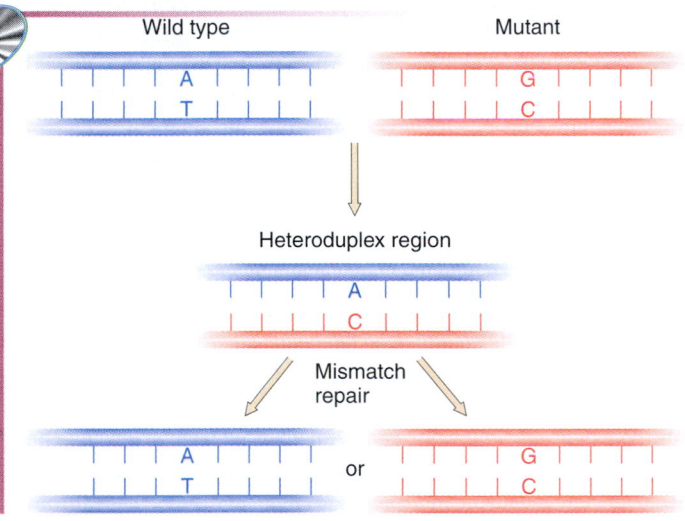

Figure 19-12 When a mismatch is generated within a heteroduplex region, mismatch repair converts the mismatch into either the wild-type or the mutant sequence.

change point in two equally likely ways, as seen in Figure 19-9f, one of which produces crossing-over of markers outside the region of heteroduplex DNA. Coconversion is explained by the location of both sites in the region of heteroduplex DNA and by the excision of both sites in the same excision-repair act. This double excision converts both sites into the same parental type.

Meselson-Radding model

As the data from tetrad analyses accumulated, it became clear that the Holliday model could not explain everything. For instance, the two mismatches resulting from the two heteroduplexes (see Figures 19-9e and 19-12) should be manifested in the progeny from a cross, yielding aberrant

4:4 tetrads. Yet, tetrad analyses in yeast and other organisms showed that, whereas 6:2 tetrads were frequent among gene-conversion events, aberrant 4:4 tetrads were very rare. It seemed as if gene conversion and the formation of heteroduplex DNA occurred primarily in only one chromatid. The model proposed by Meselson and Radding (shown in Figure 19-13) generates the Holliday structure with one single-strand cut in only one chromosome (Figure 19-13a), in contrast with the Holliday model, in which a nick is made in one strand in each of the two homologous chromatids. This single-strand cut is followed by DNA synthesis (Figure 19-13b). After the nick, the displaced single strand invades the second duplex (Figure 19-13c), generating a loop, which is excised (Figure 19-13d). After ligation to produce a Holliday structure, followed by branch migration (Figure 19-13e), a heteroduplex is generated in each chromosome. Resolution of this intermediate (Figure 19-13f) occurs exactly as depicted in Figure 19-9 (or after rotation, as in Figure 19-11). Note the lack of symmetry in the heteroduplex DNA at resolution in Figure 19-13f (left), compared with Figures 19-9f and 19-11f. Thus, in the Meselson-Radding model (left side), the bottom chromatid duplex (in Figure 19-13) has a heteroduplex region, instead of the two chromatids having heteroduplex regions, as in the Holliday model. However, branch migration and isomerization can generate a structure that has heteroduplex regions on both duplexes (Figure 19-13, right side), which is required to explain aberrant 4:4 ratios (Table 19-1).

Double-strand break-repair model for recombination

In the Holliday and Meselson-Radding models for genetic recombination, the initiation events for recombination are single-strand nicks that result in the generation of

Genetic Consequences of the Meselson-Radding Model

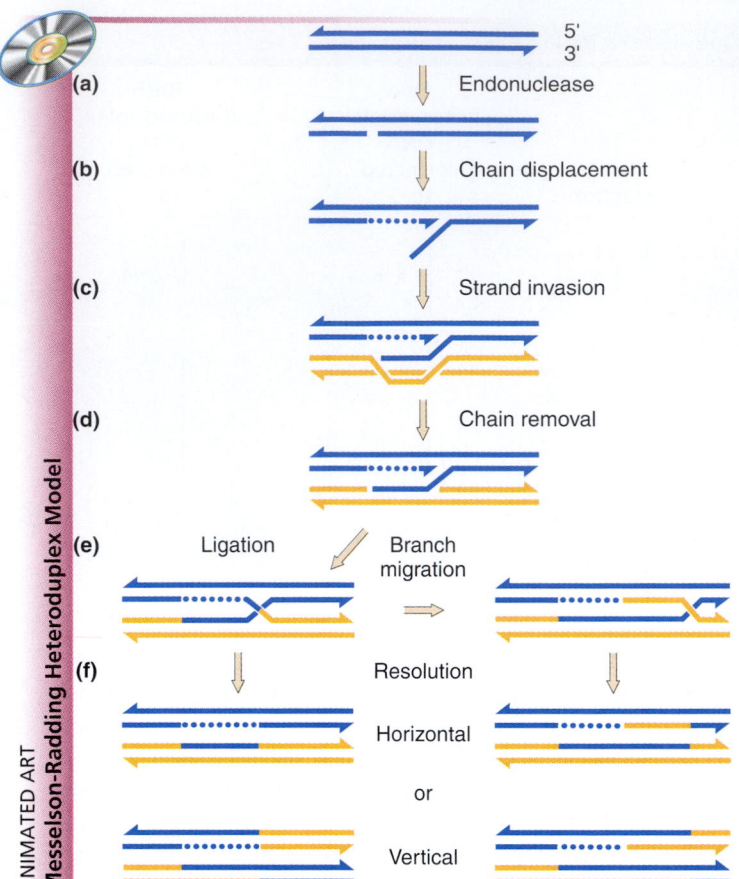

Figure 19-13 The Meselson-Radding heteroduplex model. (a) A duplex is cut on one chain. (b) DNA polymerase displaces one chain. (c) The resulting single chain displaces its counterpart in the homolog. (d) This displaced chain is enzymatically digested. (e) Ligation completes the formation of a Holliday junction, which is genetically asymmetric in that only one of the two duplexes has a region of potentially heteroduplex DNA. If the junction migrates, heteroduplex DNA can arise on both duplexes. (f) Resolution of the junction occurs as in the Holliday model. (From F. W. Stahl, "The Holliday Junction on Its Thirtieth Anniversary," *Genetics* 138, 1994, 241–246.)

heteroduplex DNA. However, the finding that yeast transformation is stimulated 1000-fold when a double-strand break is introduced into a circular donor plasmid provided the impetus for an additional model, the double-strand-break model shown in Figure 19-14. Originally formulated by Jack Szostak, Terry Orr-Weaver, and Rodney Rothstein, this model invokes double-strand breaks to initiate recombination. The breaks are enlarged to gaps, and the repair of the double-stranded gap results in gene conversion. The key features of this model are diagrammed in the steps in Figure 19-14: (1) a double-strand break, followed by digestion of the 5′ end of both cut sites; (2) the invasion by a remaining 3′ tail of the uncut other duplex; (3) the repair synthesis of one strand; (4) the repair synthesis of the other strand, and ligation to form two Holliday junctions; (5) resolution in one of two ways, one of which generates a reciprocal

crossover; and (6) mismatch repair correction to yield gene conversion.

MESSAGE ························

The phenomenon of gene conversion led to the development of heteroduplex models to explain the mechanism of crossing-over. Mendelian (1:1) allele ratios are normally observed in crosses because it is only rarely that a heterozygous locus is the precise point of chromosome exchange. Asci showing gene conversion at a typical heterozygous locus are relatively rare (on the order of 1 percent).

Visualization of recombination intermediates

Several of the individual steps that constitute the Holliday model have been demonstrated to occur in vivo or in vitro, such as nicking, strand displacement, branch migration, repair synthesis, and ligation. H. Potter and D. Dressler showed that DNA intermediates of the type predicted by the Holliday model can be found in recombining phages or plasmids. Figure 19-15 shows an electron micrograph of a recombinant molecule. It is formally equivalent to the central pair of DNA double helices shown in Figure 19-11, with two arms rotated to produce the central single-stranded "diamond," as in Figure 19-11c, which runs between the sections of double-stranded DNA.

Enzymatic mechanism of recombination

By the isolation of mutants defective in some stage of recombination, much light has been shed on the enzymology of recombination. In *E. coli,* the products of several genes involved in general recombination — the *recA, recB, recC,* and *recD* genes — have been well characterized, as has the single-strand-DNA-binding (Ssb) protein. Mutants deficient in any of these proteins have reduced levels of recombination. In fact, three distinct recombination pathways have been identified. In addition to the major RecBCD pathway, two minor pathways, RecF and RecE, are activated in certain situations. All three pathways use the RecA protein.

Figure 19-14 Double-strand-break model of meiotic ▶ recombination in the yeast *S. cerevisiae.* The two strands of each of two chromatids (in blue and yellow) are shown. The darker and lighter colors indicate complementary strands, as do the matching primes (for example, *C* and *C′* or *c* and *c′* for the allele designations). A key point of this model is shown in steps 3 and 4, which picture the generation of a Holliday structure with two crossover points. Two different ways of resolving this structure lead to the recombinant or nonrecombinant duplexes seen in step 5a and b. The red patch shows the heteroduplex mismatched regions, for the alleles *d* and *D′.* Mismatch repair results in the "conversion" of *d* into *D.* (See Table 19-1 for the possible segregation patterns resulting from mismatch repair.) (From H. Lodish, D. Baltimore, A. Berk, S. L. Zipursky, P. Matsudaira, and J. Darnell, *Molecular Cell Biology,* 3d ed. Copyright © 1995 by Scientific American Books.)

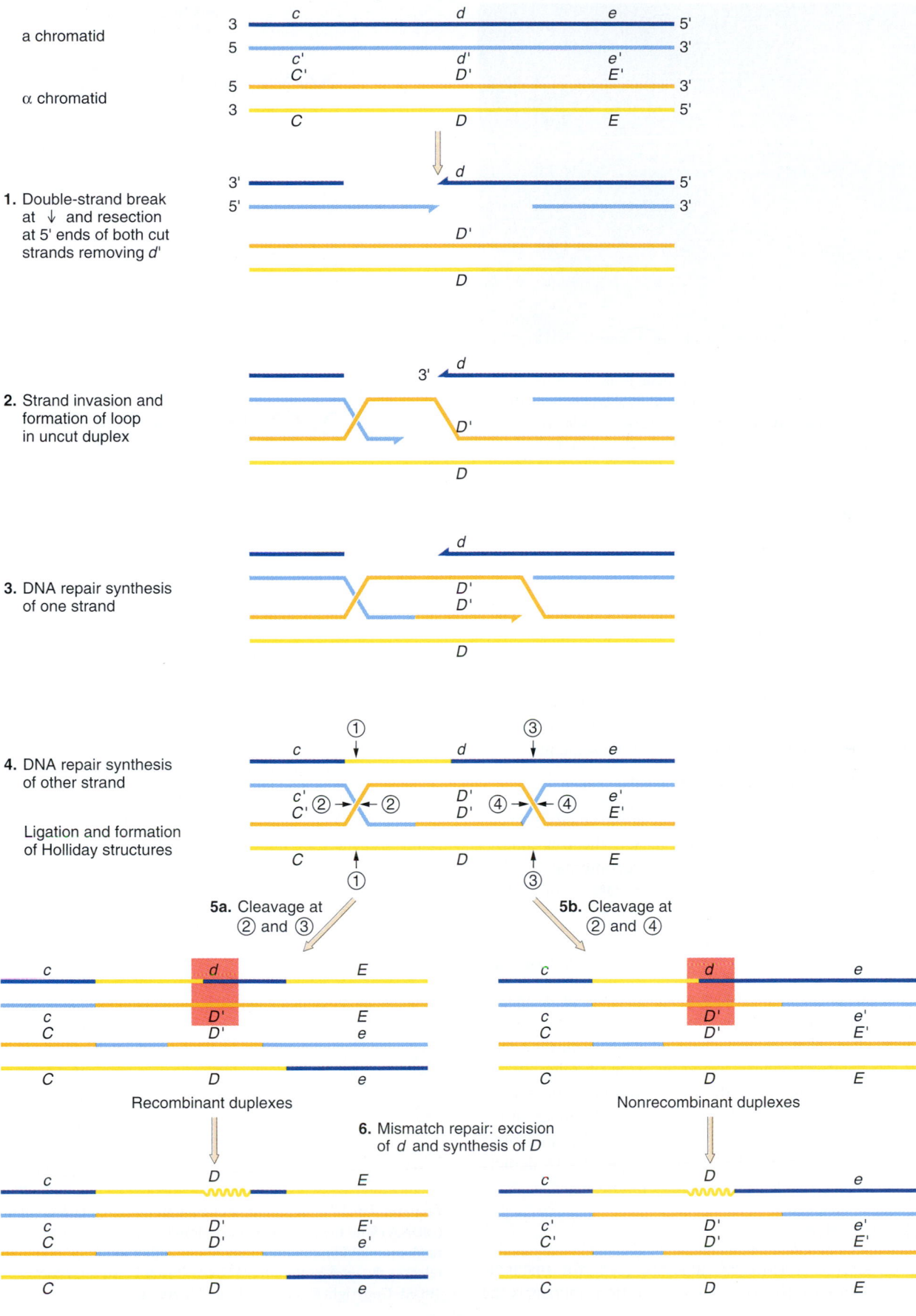

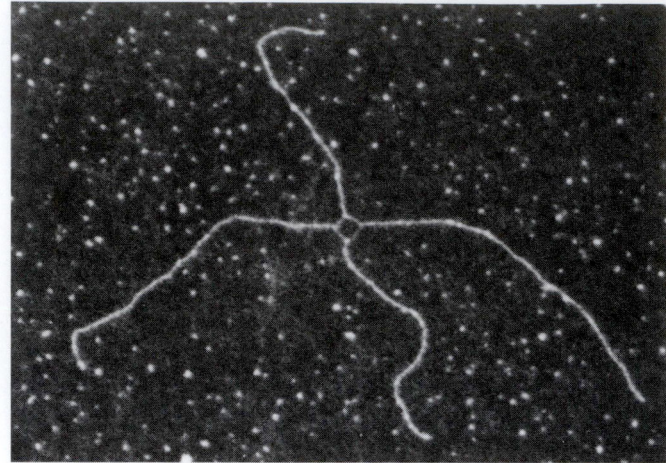

Figure 19-15   Recombination intermediate from recombination between plasmids. Shown are four double-helical arms and a single-stranded diamond uniting them. This is the same kind of intermediate as that postulated in the heteroduplex DNA model for recombination in eukaryotes and is formally equivalent to the structure shown in Figure 19-11c. (Photograph courtesy of H. Potter and D. Dressler.)

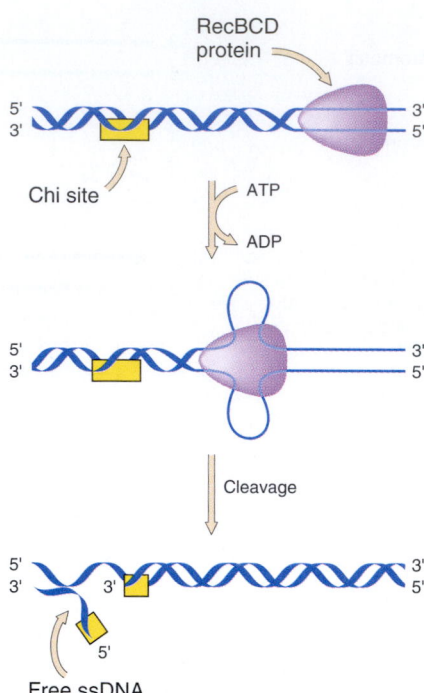

Figure 19-16   Model for the generation of single-stranded DNA by the RecBCD complex (purple), an ATP-driven helicase and nuclease. Two single-stranded loops are formed as the complex moves forward. The enzyme cleaves one of the strands when it encounters a chi sequence (yellow) to form single-stranded DNA with a free end. Recombination can then take place. (After A. Taylor and G. R. Smith, *Cell* 22, 1980, 447; from L. Stryer, *Biochemistry*, 4th ed. Copyright © 1995 by Lubert Stryer.)

Production of single-stranded DNA

An initial step in recombination is probably the nicking and unwinding of a DNA duplex by a protein complex consisting of the RecB, RecC, and RecD proteins. The complex has both helicase and nuclease activity. Figure 19-16 shows how this complex unwinds the DNA, driven by the hydrolysis of ATP as it moves and generates single strands from a duplex molecule. The nuclease activity recognizes an 8-bp sequence:

$$5' \text{ G C T G G T G G } 3'$$

called a **chi site;** these chi sites appear approximately every 64 kb. As the complex unwinds the DNA, the free single-stranded DNA can be used to initiate recombination. The Ssb protein, which also takes part in DNA replication (Chapter 8), can bind to and stabilize the single strands.

RecA-protein-mediated single-strand exchange

The RecA protein, which also plays a role in the induction of the SOS repair system (see Chapter 16), can bind to single strands along their length, forming a nucleoprotein filament. RecA catalyzes single-strand invasion of a duplex and subsequent displacement of the corresponding strand from the duplex. The invasion and displacement take place in the presence of ATP, as shown in Figure 19-17. The displaced strand forms what is termed a **D loop.** Figure 19-18 depicts how this sequence can lead to Holliday junction.

Branch migration

The movement of a Holliday junction (see, for instance, Figures 19-9 and 19-10), or branch migration, increases the length of heteroduplex DNA. The RuvA and RuvB proteins catalyze branch migration, driving the reaction by the hydrolysis of ATP. The RuvA protein binds to the crossover point and then is flanked by two RuvB ATPase hexameric

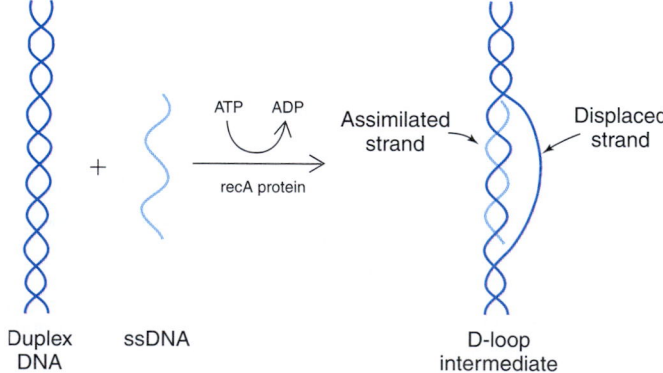

Figure 19-17   The pairing of a single-stranded DNA molecule (ssDNA) with the complementary strand of a duplex is catalyzed by the recA protein. The resulting structure is called a *D loop*. ATP hydrolysis releases the recA protein from DNA. (From L. Stryer, *Biochemistry*, 4th ed. Copyright © 1995 by Lubert Stryer.)

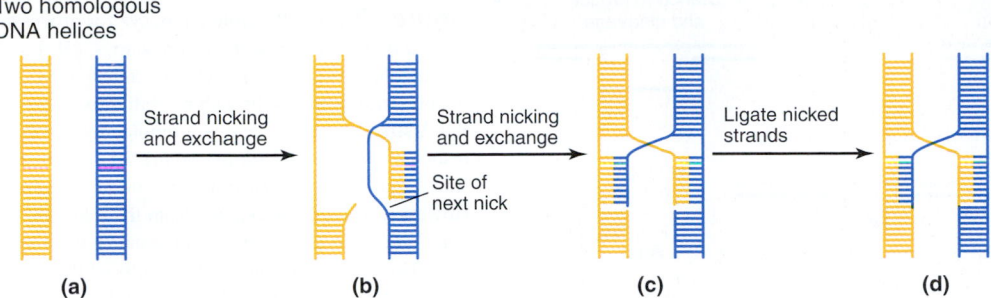

Figure 19-18   A schematic representation of some of the steps in recombination. (a) The pairing of two homologous duplexes. (b) A nick is made by the RecB,C nuclease. The helix is partly unwound, and the single-stranded region is extended and stabilized by the Ssb protein. The RecA protein catalyzes the invasion by the single strand of the duplex. The Ssb protein aids in keeping the single strand free. (c) After nicking by the RecB,C nuclease, the free single strand from the second duplex can anneal with the first duplex. (d) RNA ligase can seal this structure. (After B. Alberts et al., *Molecular Biology of the Cell.* Copyright © 1983 by Garland Publishing.)

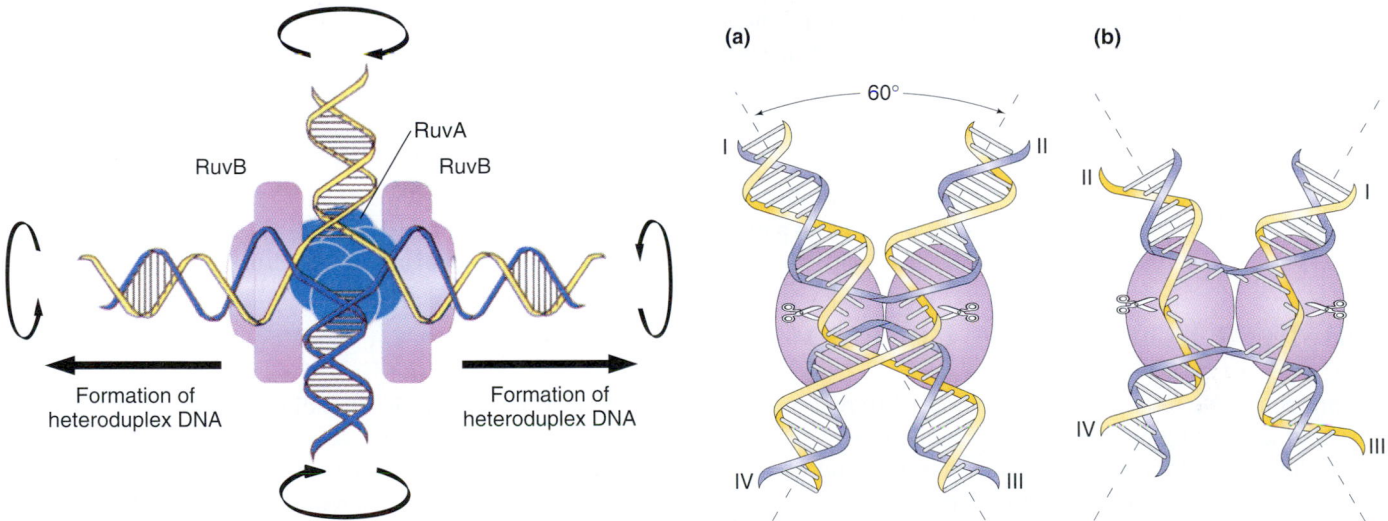

Figure 19-19   Model for RuvAB-mediated branch migration. The blue spheres indicate the RuvA protein, which binds to the crossover in the Holliday junction–RuvAB complex. Two hexameric rings of RuvB flank RuvA. The two RuvB ring motors lie in opposite orientation and affect branch migration by promoting the passage of DNA. (From Carol A. Parsons, Andrzej Stasiak, Richard J. Bennett, and Stephen C. West, "Structures of a Multisubunit Complex That Promotes DNA Branch Migration," *Nature* 374, 1995, 377.)

Figure 19-20   Model showing RuvC cleavage of two of the four strands of an antiparallel Holliday junction. The scissors depict sites where the nicking is symmetric. (a) Here nicking resolves a stacked X-structure. (b) Nicking resolves an unfolded junction. A twofold-symmetric unfolded structure shown here is formed by a 180° rotation of arms I and II of the structure shown in part a. (From Richard J. Bennett and Stephen C. West, "RuvC Protein Resolves Holliday Junctions via Cleavage of the Continuous (Noncrossover) Strands," *Proceedings of the National Academy of Sciences USA* 92, 1995, 5639.)

rings, as seen by electron microscopy. Figure 19-19 depicts a model for the action of these proteins in branch migration.

Resolution of Holliday junctions

Several enzymatic pathways have been identified that cleave across the point of strand exchange in a Holliday

structure to yield two duplexes. RuvC is an endonuclease that resolves Holliday junctions by symmetric cleavage of the continuous (noncrossing) pair of DNA strands, as seen in Figure 19-20. In addition, the RecG and Rus proteins may provide alternative routes to cleavage. These reactions are summarized in Figure 19-21.

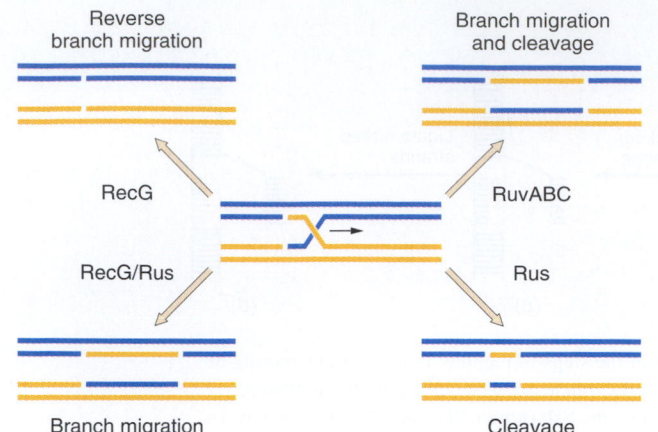

Reverse branch migration   Branch migration and cleavage

RecG   RuvABC

RecG/Rus   Rus

Branch migration   Cleavage

Figure 19-21   Possible pathways for processing Holliday junction in *E. coli*. The center of the figure shows a Holliday junction formed by homologous pairing and strand exchange by the RecA protein. In the presence of RecA, branch migration is in the same direction as the RecA strand exchange and is promoted by RuvAB. RuvC cleavage resolves this structure (*upper right*). Also in the presence of RecA, RecG could drive the Holliday junction backward (*upper right*). However, if RecA dissociates from the junction, RecG could drive the reaction forward, with resolution coming from the Rus protein (*lower left*). The lower right shows the results of Rus cleavage alone, after strand exchange and branch migration by RecA. (From Gary Sharples, Sau Chan, Akeel Mahdi, Matthew Whitby, and Robert Lloyd, "Processing of Intermediates in Recombination and DNA Repair: Identification of a New Endonuclease That Specifically Cleaves Holliday Junctions," *EMBO Journal* 13, 1994, 6140.)

SUMMARY

We have gained increasing knowledge of the molecular processes behind recombination, which produces new gene combinations by exchanging homologous chromosomes. Both genetic and physical evidence has led to several models of recombination that rely on common features: hetero- duplex DNA, mismatch repair, and resolution, or splicing, of the intermediate structure to yield recombinant molecules. The process of recombination itself is under genetic control, and numerous genes that affect the process have been identified.

CONCEPT MAP

Draw a concept map interrelating as many of the following terms as possible. Note that the terms are listed in no particular order.

crossing-over / heteroduplex / chiasmata / gene conversion / correction / non-Mendelian ratio / recombination / 5 : 3 / base mismatch / heterozygote

CHAPTER INTEGRATION PROBLEM

In previous chapters, we learned about different mechanisms of generating mutations, both spontaneously and with mutagens. Describe gene conversion and how you can distinguish gene conversion from mutation.

into its own form. Instead of ending with equal numbers of both alleles in meiosis, gene conversion results in an excess of one allele and a deficiency of the other. Models of recombination employing heteroduplex DNA offer plausible mechanisms for gene conversion. In contrast, mutation is undirected change that can occur in both meiosis and mitosis.

◆ Solution ◆

Gene conversion is a meiotic process of directed change in which one allele directs the conversion of a partner allele

SOLVED PROBLEMS

1. In *Neurospora*, an *ad3* double mutant consisted of two mutant sites within the *ad3* gene — site *1* on the left and site *2* on the right. This mutant was crossed to wild type, with the use of parental stocks heterozygous for two closely linked flanking loci, *A* and *B*, as follows, where + represents wild-type sequence at the mutant positions:

$$A \; 1 \; 2 \; B \times a + + b$$

Most asci were of the expected type showing regular Mendelian segregations, but there were also some

unexpected types, of which several examples are represented here as I through III.

I	II	III
A 1 2 B	*A 1 2 B*	*A 1 2 B*
A 1 2 B	*A 1 2 B*	*A 1 2 B*
A 1 2 b	*A 1 2 B*	*A 1 2 B*
A + + b	*A 1 2 B*	*A + + B*
a 1 2 B	*a + + b*	*a + 2 b*
a + + B	*a 1 2 b*	*a + + b*
a + + b	*a + + b*	*a + + b*
a + + b	*a + + b*	*a + + b*

Explain the likely origin of the rare types I through III according to molecular recombination models.

◆ Solution ◆

Ascus type I has two nonidentical sister spore pairs. Because the members of a spore pair are derived from a post-meiotic mitosis (see Chapter 6), this proves that the meiotic products in these cases must have contained both *1 2* and *+ +* information; in other words, they must have contained heteroduplex DNA. Notice also that these spore pairs are recombinant for *A* and *B*. Therefore it is likely that a crossover occurred between the *A* and *B* loci, that the heteroduplex DNA that constituted the crossover spanned both the *ad3* mutant sites *1* and *2*, and that there was no correction of the heteroduplex at those sites.

Type II shows a 5:3 ratio of *1 2* doubles to *+ +*. Here again the heteroduplexes must have spanned both sites (in a noncrossover configuration), but this time there was correction of a *+ +/1 2* heteroduplex to *1 2/1 2*, presumably by excision and repair of the *+ +* information (co- or double conversion).

Type III reveals another noncrossover heteroduplex configuration, and this time correction occurred only at site 1; this is revealed as a 5:3 ratio for site *1*, with *1 → +* conversion in ascospore 5.

2. In fungal crosses of the following general type,

$$\frac{M \qquad \lceil + 2 \rceil \qquad N}{\times}$$
$$\frac{}{m \qquad \lfloor 1 + \rfloor \qquad n}$$

where *1* and *2* are mutant sites of a nutritional gene and *M* and *N* are flanking loci, it is possible to select rare *+ +* prototrophic recombinants by plating on minimal medium. These prototrophs are then examined for the alleles of the flanking loci. As might be expected, the combination

$$M + + n$$

is commonly encountered, but so, somewhat surprisingly, are

$$M + + N$$

and

$$m + + n$$

a. Explain the origin of these two genotypes in relation to molecular recombination models.

b. If *M + + N* were more common than *m + + n*, what would that mean?

◆ Solution ◆

a. It is likely that *M + + N* arose from a noncrossover heteroduplex that spanned site 2, followed by a correction of *2 → +*.

$$\text{DNA} \left[\begin{array}{cccc} \underline{+} & \underline{2} & \underline{+} & \underline{+} \\ & & \longrightarrow & \\ \underline{+} & \underline{+} & \underline{+} & \underline{+} \end{array} \right.$$

Similarly, *m + + n* would be explained by a heteroduplex spanning site *1*, and corrected *1 → +*.

b. Because *M + + N* arises from gene conversion at the right-hand site, it is likely that heteroduplex DNA is formed more commonly from the right than from the left, possibly because of a closer fixed break point. (**Note:** Prototrophs may also be formed by single-site correction in a heteroduplex spanning both sites, but then the inequality would require another explanation.)

PROBLEMS

1. Which of the following linear asci shows gene conversion at the *arg2* locus?

1	2	3	4	5	6
+	+	+	+	+	+
+	+	+	+	+	+
+	arg	+	+	arg	arg
+	arg	arg	arg	arg	arg
arg	arg	arg	+	+	arg
arg	arg	arg	arg	+	arg
arg	+	arg	arg	arg	arg
arg	+	arg	arg	arg	arg

2. At the light-spore locus of *Ascobolus*, the *1′* mutant site is in the left part of the gene and the *1″* mutant site is more to the right. When crosses are made between *1′*- and *1″*-bearing strains,

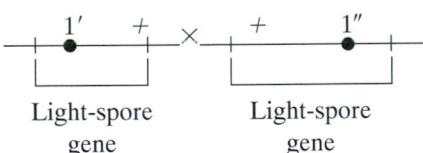

asci with six light and two black spores can be selected visually. They are shown to be caused by gene conversion mostly of the following type:

+ 1″		but very rarely + 1″
+ 1″		+ 1″
+ 1″		+ + ⎫
+ 1″		+ + ⎭ black
+ + ⎫		1′ +
+ + ⎭ black		1′ +
1′ +		1′ +
1′ +		1′ +

What can account for this irregularity?

3. It has been proposed that the "fixed break point" of the heteroduplex recombination model might correspond to a promoter sequence. The following data relate to this idea. In the fungus *Podospora*, mutants were available in adjacent spore color genes 1 through 4, and a cross was made as follows:

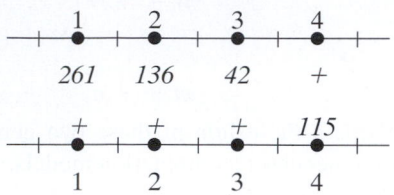

(The numbers *261*, *136*, *42*, and *115* are merely names for mutant sites.) Many asci were obtained showing gene conversion, but one that was relevant to the preceding suggestion was as follows:

Spore 1	*261*	*136*	*42*	+
2	+	+	+	+
3	+	+	+	*115*
4	+	+	+	*115*

Interpret this ascus in relation to the promoter idea.

4. Many mutagens increase the frequency of sister-chromatid exchange. Give possible explanations for this observation.

5. Mutations in locus *46* of the Ascomycete fungus *Ascobolus* produce light-colored ascospores (let's call them *a* mutants). In the following crosses between different *a* mutants, asci are observed for the appearance of dark wild-type spores. In each cross, all such asci were of the genotypes indicated:

$$
\begin{array}{ccc}
a_1 \times a_2 & a_1 \times a_3 & a_2 \times a_3 \\
\downarrow & \downarrow & \downarrow \\
a_1\ + & a_1\ + & a_2\ + \\
+\ + & a_1\ + & a_2\ + \\
+\ a_2 & +\ + & +\ + \\
+\ a_2 & +\ a_3 & +\ a_3
\end{array}
$$

Interpret these results in light of the models discussed in this chapter.

6. In the cross $A\ m_1\ m_2\ B \times a\ m_3\ b$, the order of the mutant sites m_1, m_2, and m_3 is unknown in relation to one another and to A/a and B/b. One nonlinear conversion ascus is obtained:

$$
\begin{array}{l}
A\ m_1\ m_2\ B \\
A\ m_1\ b \\
a\ m_1\ m_2\ B \\
a\ m_3\ b
\end{array}
$$

Interpret this result in light of the heteroduplex DNA theory, and derive as much information as possible about the order of the sites.

7. In the cross $a_1 \times a_2$ (alleles of one locus) the following ascus is obtained:

$$
\begin{array}{l}
a_1\ + \\
a_1\ a_2 \\
a_1\ a_2 \\
+\ a_2
\end{array}
$$

Deduce what events may have produced this ascus (at the molecular level).

8. G. Leblon and J.-L. Rossignol made the following observations in *Ascobolus*. Single-nucleotide-pair insertion or deletion mutations show gene conversions of the 6:2 or 2:6 type and only rarely of the 5:3, 3:5, or 3:1:1:3 type. Base-pair transition mutations show gene conversion of the 3:5, 5:3, or 3:1:1:3 type and only rarely of the 6:2 or 2:6 type.

    a. In relation to the hybrid DNA model, propose an explanation for these observations.

    b. Leblon and Rossignol also showed that there are far fewer 6:2 than 2:6 conversions for insertions and far more 6:2 than 2:6 conversions for deletions (where the ratios are $+:m$). Explain these results in relation to heteroduplex DNA. (You might also think about the excision of thymine photodimers.)

    c. Finally, the researchers showed that, when a frameshift mutation is combined in a meiosis with a transition mutation at the same locus in a cis configuration, the asci showing *joint* conversion are all 6:2 or 2:6 for *both* sites (that is, the frameshift conversion pattern seems to have "imposed its will" on the transition site). Propose an explanation for this result.

9. At the *gray* locus in the Ascomycete fungus *Sordaria*, the cross $+ \times g_1$ is made. In this cross, heteroduplex DNA sometimes extends across the site of heterozygosity, and two heteroduplex DNA molecules are formed (as discussed in this chapter). However, correction of heteroduplex DNA is not 100 percent efficient. In fact, 30 percent of all heteroduplex DNA is not corrected at all, whereas 50 percent is corrected to + and 20 percent is corrected to g_1. What proportion of aberrant-ratio asci will be **(a)** 6:2? **(b)** 2:6? **(c)** 3:1:1:3? **(d)** 5:3? **(e)** 3:5?

10. Noreen Murray crossed α and β, two alleles of the *me-2* locus in *Neurospora*. Included in the cross were two markers, *trp* and *pan*, which each flank *me-2* at a distance of 5 m.u. The ascospores were plated onto a medium containing tryptophan and pantothenate but no methionine. The methionine prototrophs that grew

were isolated and scored for the flanking markers, yielding the results shown in the table below.

	GENOTYPE OF *me-2⁺* PROTOTROPHS			
Cross	*trp +*	*+ pan*	*trp pan*	*+ +*
trp α + × + β pan	26	59	16	56
trp β + × + α pan	84	23	87	15

Interpret these results in light of the models presented in this chapter. Be sure to account for the asymmetries in the classes.

11. In *Neurospora,* the cross *A x × a y* is made, in which *x* and *y* are alleles of the *his-1* locus and *A* and *a* are mating-type alleles. The recombinant frequency between the *his-1* alleles is measured by the prototroph frequency when ascospores are plated on a medium lacking histidine; the recombinant frequency is measured as 10^{-5}. Progeny of parental genotype are backcrossed to the parents, with the following results. All *a y* progeny backcrossed to the *A x* parent show prototroph frequencies of 10^{-5}. When *A x* progeny are backcrossed to the *a y* parent, two prototroph frequencies are obtained: half of the crosses show 10^{-5}, but the other half show the much higher frequency of 10^{-2}. Propose an explanation for these results, and describe a research program to test your hypothesis. (**Note:** Intragenic recombination is a *meiotic* function that occurs in a diploid cell. Thus, even though this organism is haploid, dominance and recessiveness could have roles in this problem.)

20

TRANSPOSABLE GENETIC ELEMENTS

Phenotypic manifestation of a "jumping gene" in snapdragon (*Antirrhinum majus*).
If the jumping gene inserts into a pigment gene, white color results; when it exits, pigment is restored. *(Heinz Saedler, Max-Planck-Institut.)*

Key Concepts

A series of genetic elements can occasionally move, or transpose, from one position on a chromosome to another position on the same chromosome or on a different chromosome.

In bacteria, insertion sequences, transposons, and phage mu are examples of transposable genetic elements.

Transposable elements can mediate chromosomal rearrangements.

In higher cells, transposable elements have been extensively characterized in yeast, *Drosophila,* and maize and in mammalian systems.

In eukaryotes, some transposable elements utilize an RNA intermediate during transposition, whereas, in prokaryotes, transposition is exclusively at the DNA level.

tarting in the 1930s, genetic studies of maize, undertaken independently by Barbara McClintock and Marcus Rhoades, yielded results that greatly upset the classical genetic picture of genes residing only at fixed loci on the main chromosome. The research literature began to carry reports suggesting the existence of genetic elements of the main chromosomes that can somehow mobilize themselves and move from one location to another. These findings were viewed with skepticism for many years, but it is now clear that such mobile elements are widespread in nature.

A variety of colorful names (some of which help to describe their respective properties) have been applied to these genetic elements: controlling elements, cassettes, jumping genes, roving genes, mobile genes, mobile genetic elements, and transposons. We choose the term **transposable genetic element,** which is formally most correct and embraces the entire family of types. The term *transposition* has long been used in genetics to describe transfer of chromosomal segments from one position to another in major structural rearrangements. In the present context, what is being transposed seems to be a gene or a small number of linked genes or a gene-sized fragment. Any genetic entity of this size can be called a *genetic element.*

Transposable genetic elements can move to new positions within the same chromosome or even to a different chromosome. The normal genetic role of these elements is not known with certainty. They have been detected genetically through the abnormalities that they produce in the activities and structures of the genes near the sites to which they move. A variety of physical techniques have been used to detect them as well, including DNA sequencing. Transposable genetic elements have been found in most organisms in which they have been sought.

Today, transposable elements provide valuable tools both in prokaryotes and in eukaryotes for genetic mapping, creating mutants, cloning genes, and even producing transgenic organisms. Let us reconstruct some of the steps in the evolution of our present understanding of transposable elements. In doing so, we will uncover the principles relevant to these fascinating genetic units.

Controlling elements in maize

In 1938, Marcus Rhoades analyzed an ear of Mexican black corn. The ear came from a selfing of a pure-breeding pigmented genotype, but it showed a surprising modified Mendelian dihybrid segregation ratio of $12:3:1$ among pigmented, dotted, and colorless kernels. Analysis showed that two events had occurred at unlinked loci. At one locus, a pigment gene A_1 had mutated to a_1, an allele for the colorless phenotype; at another locus, a dominant allele Dt (*Dotted*) had appeared. The effect of Dt was to produce pigmented dots in the otherwise colorless phenotype of a_1/a_1 (Figure 20-1). Thus, the original line was very probably

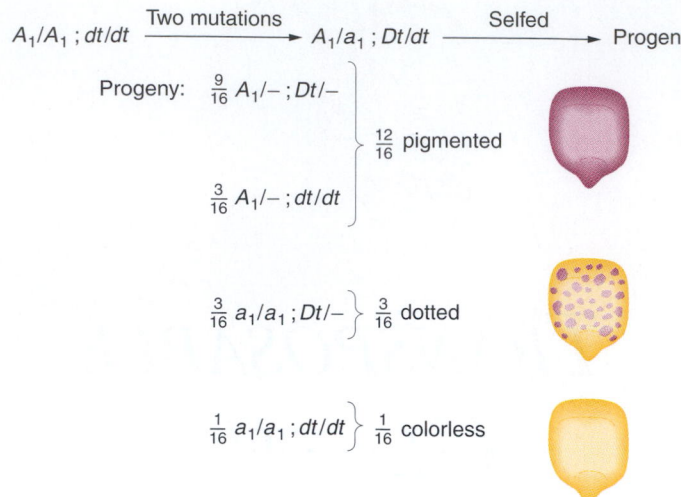

Figure 20-1 A formal genetic explanation of the appearance of the dotted phenotype in corn. The A_1/a_1 ; Dt/dt genotype is created by simultaneous mutations of $A_1 \rightarrow a_1$ and $dt \rightarrow Dt$. On selfing, this genotype yields the observed $12:3:1$ ratio of kernel phenotypes.

A_1/A_1 ; dt/dt, and the mutations generated an A_1/a_1 ; Dt/dt plant, which on selfing gave the observed ration of progeny.

But what was causing the dotted phenotype? A reverse mutation of $a_1 \rightarrow A_1$ in somatic cells would be an obvious possibility, but the large numbers of dots in the *Dotted* kernels would require extremely high reversion rates. Using special stocks, Rhoades was able to find anthers in the flowers of a_1/a_1 ; $Dt/-$ plants that showed patches of pigment (Figure 20-2). He reasoned that these anthers might contain pollen grains bearing the reverted pigment genotype, and he used the pollen from these anthers to fertilize a_1/a_1 tested

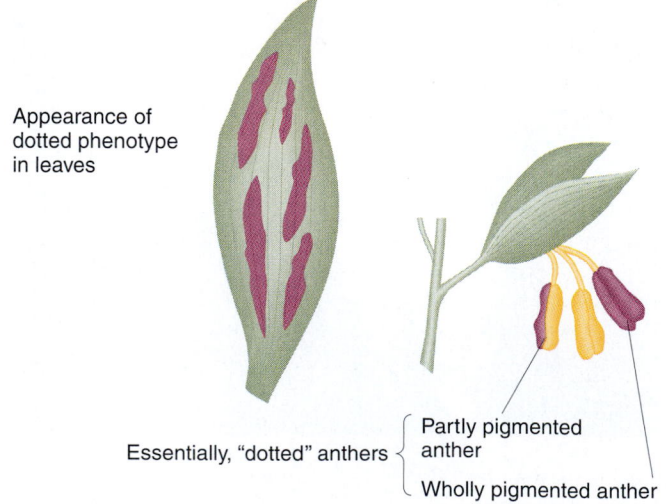

Figure 20-2 Rhoades used special stocks of a_1/a_1 ; $Dt/-$ corn plants carrying certain genes that allow pigmented sectors to be detected in tissue other than the kernels. (After M. M. Rhoades, *Genetics* 23, 1938, 382.)

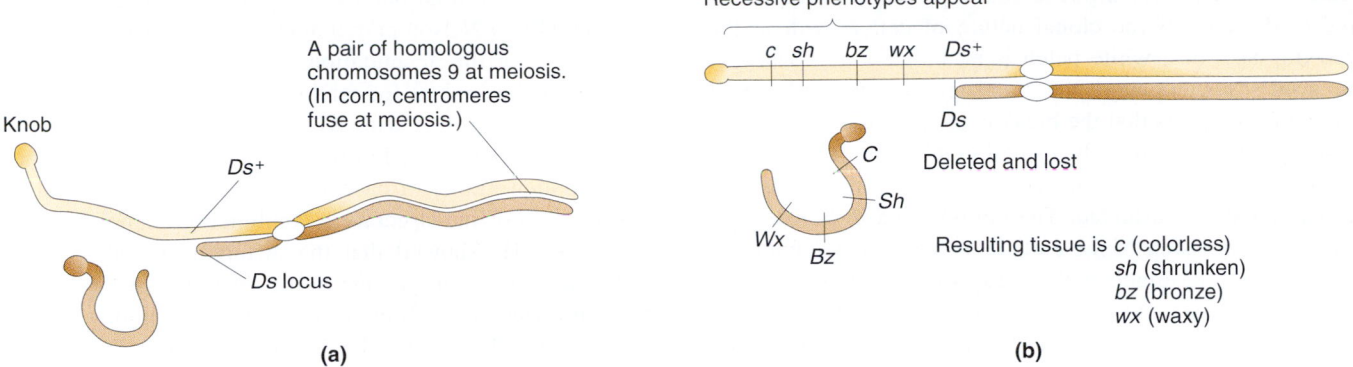

Knob

Ds^+

Ds locus

A pair of homologous
chromosomes 9 at meiosis.
(In corn, centromeres
fuse at meiosis.)

(a)

Recessive phenotypes appear

c sh bz wx Ds^+

Ds

Deleted and lost

C

Sh

Wx Bz

Resulting tissue is *c* (colorless)
sh (shrunken)
bz (bronze)
wx (waxy)

(b)

Figure 20-3 Detection of chromosomal breakage (instability) due to action of the *Ds* element in corn. (a) Cytological detection of the breakage. (Ds^+ indicates a lack of *Ds*.) (b) Genetic detection of the breakage.

females. Sure enough, some of the progeny were completely pigmented, showing that each dot in the parental plants was in fact the phenotypic manifestation of a genetic reversion event. Thus, a_1 is one of the first known examples of an **unstable mutant allele** — an allele for which reverse mutation occurs at a very high rate. However, the allelic instability is dependent on the presence of the unlinked *Dt* gene. When the reverse mutations occur, they are stable; the *Dt* gene can be crossed out of the line with no loss of the A_1 character. This outcome would not be surprising if the a_1 phenotype arose from the insertion of a defective transposable element that by itself is unable to move. In the presence of a transfactor produced by the *Dt* locus, however, the element can move, yielding reversion to A_1. The A_1 allele remains stable in the absence of *Dt*.

McClintock's experiments:
the *Ds* element

In the 1950s, Barbara McClintock demonstrated an analogous situation in another study of corn. She found a genetic factor *Ds (Dissociation)* that causes a high tendency toward chromosome breakage at the location at which it appears. These breaks can be located either cytologically (Figure 20-3a) or genetically by the uncovering of recessive genes (Figure 20-3b). This action of *Ds* is another kind of instability. Once again, this instability proves to be dependent on the presence of an unlinked gene, *Ac (Activator)*, in the same way that the instability of a_1 is dependent on *Dt*.

McClintock found it impossible to map *Ac*. In some plants, it mapped to one position; in other plants of the same line, it mapped to different positions. As if this were not enough of a curiosity, the *Ds* locus itself (Figure 20-3) was constantly changing position on the chromosome arm, as indicated by the differing phenotypes of the variegated sections of the seeds (as different recessive gene combinations were uncovered in a system such as the one illustrated in Figure 20-3b).

The wanderings of the *Ds* element take on new meaning for us in the context of this chapter when we consider the results of the following cross:

$$\male \; C\,Ds/C\,Ds \; ; \; Ac/Ac^+ \times c\,Ds^+/c\,Ds^+ \; ; \; Ac^+/Ac^+ \; \female$$

Here *C* allows color expression and Ds^+ and Ac^+ indicate the lack of the element. Most of the kernels from this cross were of the expected types (Figure 20-4), but one exceptional kernel was very interesting. In Figure 20-4, the first seed shows the normal solid pigment pattern owing to the presence of the dominant *C* allele. The second seed shows the same basic background pigmentation but with the expected white mottling caused by the loss of the *C* allele through chromosome breakage in some of the cell lines

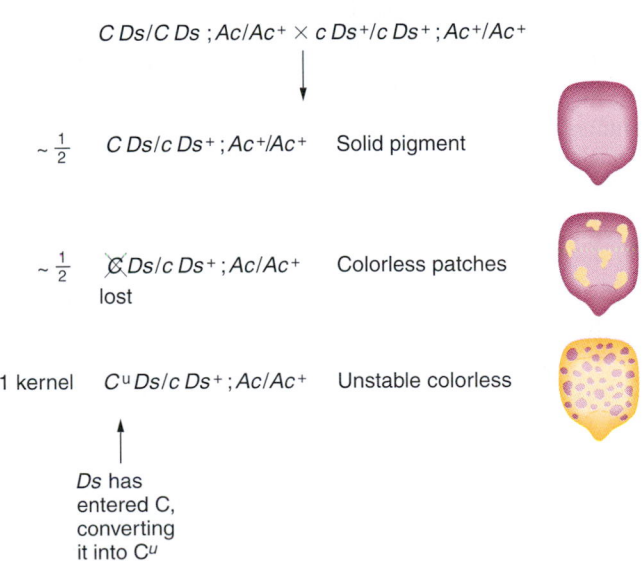

$C\,Ds/C\,Ds \; ; Ac/Ac^+ \times c\,Ds^+/c\,Ds^+ \; ;Ac^+/Ac^+$

~$\frac{1}{2}$ $C\,Ds/c\,Ds^+ \; ;Ac^+/Ac^+$ Solid pigment

~$\frac{1}{2}$ $\cancel{C}\,Ds/c\,Ds^+ \; ;Ac/Ac^+$ Colorless patches
lost

1 kernel $C^u\,Ds/c\,Ds^+ \; ;Ac/Ac^+$ Unstable colorless

Ds has
entered C,
converting
it into C^u

Figure 20-4 Results that indicate the transposition of *Ds* into the *C* gene in corn. (*C* allows color expression; *c* does not. *Ac* = activator.) The action of *Ds* is dependent on the presence of the unlinked gene *Ac*.

within the seed, with the resultant expression of the recessive *c*. Because of the clonal nature of cell growth in the seed, the size of a white patch is an indication of when in the seed's development the breakage occurred. A small white area suggests that the break came late in development, because it gave rise to only a small number of affected cells. A large patch suggests an early break, because many descendant cells are affected. The bottom seed in Figure 20-4 indicates expression of the *c* allele at the very beginning of development, inasmuch as the background is white, not pigmented. However, the presence of pigmented blotches on a white background suggests a *reversible* process at work that allows the *C* allele to be reexpressed. Chromosomal breakage could not be the explanation in this case, because, on breakage, the *C* alleles are left on acentric fragments that are lost in mitosis, and therefore the white phenotype is not reversible in such cases. The white coloration in the bottom seed appears to be the result of a second type of action in which the *Ds* gene transposes into the *C* allele, disrupting its function but not inducing the chromosome to break. If the *Ds* allele is then excised and transposes elsewhere, the *C* allele regains its function, yielding the pigmented cells that form the patches. The *Ac* allele is still required, but in this case it has the effect of mediating a reversible instability at the *C* locus. If *Ac* is crossed out of the line, the *c*^u allele becomes a stable mutant.

The analogy of this system with the *a₁ Dt* system is obvious. Perhaps the earlier situation also is due to the insertion of a *Ds*-like element into the *A₁* gene. It is natural to ask whether *a₁* will respond to *Ac* or *c*^u will respond to *Dt*. The answer is no; some kind of specificity prevents this crossactivation of mutational instability.

The *wx (waxy)* locus

The *Ds* element can wander not only into the middle of the *C* gene, but also into other genes, rendering them unstable mutants dependent on *Ac*. One such locus, *wx (waxy)*, has

been the subject of an intense study on the effects of the *Ds* element. Oliver Nelson paired many unstable *waxy* alleles in the absence of the *Ac* mutation. In such wx^{m-1}/wx^{m-2} heterozygotes, he looked for rare wild-type *Wx* recombinants by staining the pollen with KI-I₂ reagent, which stains *Wx* pollen black and *wx* pollen red. By counting the frequency of *Wx* pollen grains in each kind of heterozygote, Nelson was able to do fine-structure recombination mapping of the *waxy* gene. He showed that the different "mutable *waxy*" mutant alleles are in fact due to the insertion of the *Ds* element in different positions in the gene. Continuing the experiment, he allowed the *Wx*-bearing pollen to fertilize *wx* plants and produce rare *Wx* kernels, which could be detected by staining sliced-off slivers. The *Wx* kernels were then raised into plants, and the exchange of flanking markers expressed in the adults showed that the *Wx* pollen grains had arisen from chromosome exchange.

General characteristics of controlling elements

Several systems like *a₁-Dt* and *Ds-Ac* have now been found in corn. Each shows similar action, having a **target gene** that is inactivated, presumably by the insertion of some **receptor element** into it, and a distant **regulator gene** that maintains the mutational instability of the locus, presumably through its ability to "unhook" the receptor element from the target locus and return the locus to normal function. The receptor and the regulator are termed **controlling elements.**

In the examples considered so far, the unstable allele is said to be **nonautonomous:** it can revert only in the presence of the regulator. Sometimes, however, a system such as the *Ac-Ds* system can produce an unstable allele that is **autonomous.** Such mutants are recognizable because they show Mendelian ratios (such as 3:1 for pigmented to dotted) that apparently are independent of any other element. In fact, such alleles appear to be caused by the insertion of *Ac* itself into the target gene. An allele of this type can subsequently be transformed into a nonautonomous allele. In such

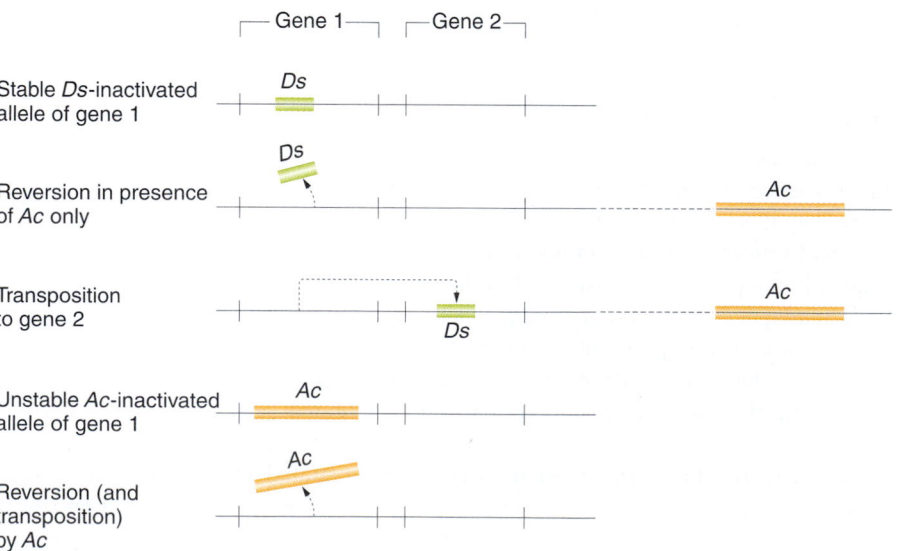

Figure 20-5 Summary of the main effects of controlling elements in corn. *Ac* and *Ds* are used as examples, acting on two hypothetical genes 1 and 2.

cases, the nonautonomy seems to result from the spontaneous generation of a *Ds* element from the inserted *Ac* element. In other words, *Ds* is in all likelihood an incomplete version of *Ac* itself.

Figure 20-5 summarizes the overall behavior of the controlling elements in maize as inferred from genetic data. Note that the mutation events that gave rise to Rhoades's original ratio are nicely explained by this model: a nonautonomous *Ds*-like element is generated from an *Ac*-like progenitor; transposition of the *Ds*-like element into the A_1 gene produces an inactive but mutationally unstable allele a_1. Molecular studies in the past few years on *Ac*, *Ds,* and other controlling elements in maize have confirmed McClintock's genetic model.

MESSAGE ··

Controlling elements in maize can inactivate a gene in which they reside, cause chromosome breaks, and transpose to new locations within the genome. Complete elements can perform these functions unaided; other forms with partial deletions can transpose only with the help of a complete element elsewhere in the genome. Figures 20-6 and 20-7 show examples of the effects of transposons in maize and similar effects in the snapdragon.

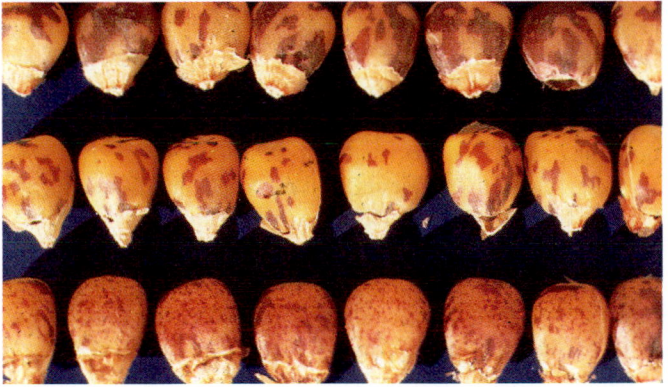

Figure 20-6 Mosaicism through transposon mutagenesis in corn. These seeds represent genotypes in which a transposon has inserted into a gene that produces anthocyanin. Therefore, the cells of these seeds are predominantly lacking anthocyanin and are yellow; let us call the genotype A^T/A^T. However, in the course of development, the transposon can occasionally exit from the gene, forming a revertant cell of genotype A/A^T. Cell division will result in a clone of revertant cells and hence a patch of pigmented cells. The three different rows represent corn lines in which the transposon exits early (large spots), late (small spots), or in between (intermediate spots). (Anthony Griffiths.)

Bacterial insertion sequences

Insertion sequences, or **insertion-sequence (IS) elements,** are now known to be segments of bacterial DNA that can move from one position on a chromosome to a different position on the same chromosome or on a different chromosome. When IS elements appear in the middle of genes, they

Figure 20-7 Mosaicism through transposon mutagenesis in snapdragon *(Antirrhinum)*. As in the corn example (Figure 20-6), the plant has both copies of an anthocyanin gene inactivated by a transposon, and albino tissue results. Revertant sectors (red spots or stripes) appear when the transposon spontaneously exits from that gene. (Photograph from Rosemary Carpenter and Enrico Coen.)

interrupt the coding sequence and inactivate the expression of that gene. Owing to their size and in some cases the presence of transcription and translation termination signals, IS elements can also block the expression of other genes in the same operon if those genes are downstream from the promoter of the operon. IS elements were first found in *E. coli* in the *gal* operon—a set of three genes taking part in the metabolism of the sugar galactose.

Physical demonstration of DNA insertion

Recall that phage λ inserts next to the *gal* operon and that it is a simple matter to obtain λ*dgal* phage particles that have picked up the *gal* region (page 229). When the IS mutations in *gal* are incorporated into λ*dgal* phages and the buoyant density in a cesium chloride (CsCl) gradient of the phages is compared with that of the normal λ*dgal* phages, it is evident that the DNA carrying the IS mutation is longer than the wild-type DNA. This experiment clearly demonstrates that the mutations are caused by the insertion of a significant amount of DNA into the *gal* operon. Figure 20-8 depicts this experiment in more detail.

Direct visualization of inserted DNA

When denatured λ*dgal* DNA containing the insertion mutation is hybridized to denatured wild-type λ*dgal* DNA, the extra piece of DNA can be located under the electron microscope. In such experiments, some of the DNA molecules that form in the mixture are not the parent duplexes but are heteroduplexes between one mutant and one wild-type strand. When point mutations are analyzed, the heteroduplexes are indistinguishable from the parental DNA molecules.

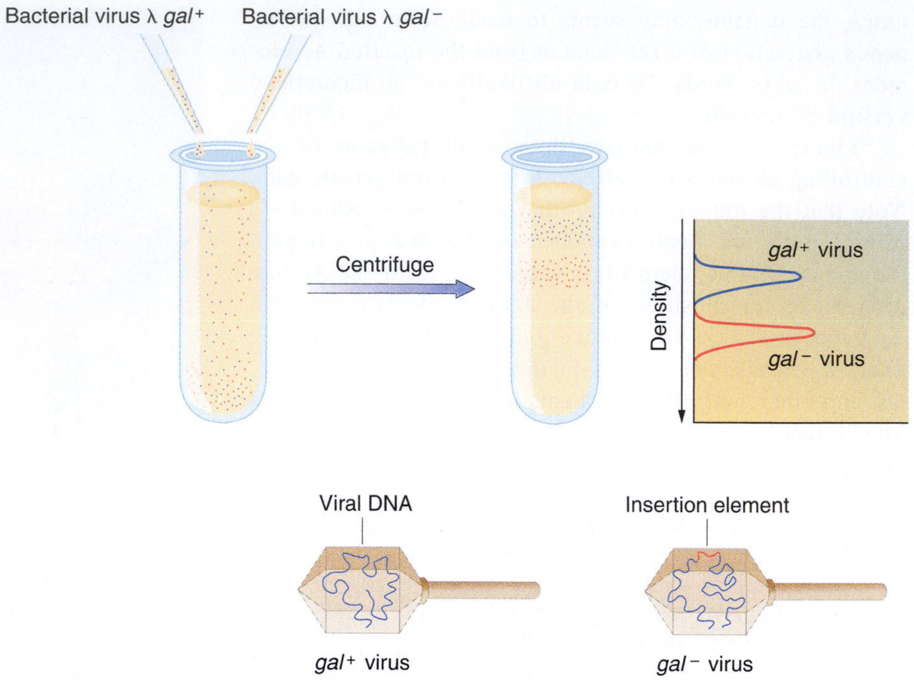

Figure 20-8   Mutation by insertion is
demonstrated with phage λ particles carrying
the bacterial gene for galactose utilization
(*gal*⁺) or the mutant gene *gal*⁻. The viruses
are centrifuged in a cesium chloride solution.
The *gal*⁻ particles are found to be denser.
Because the virus particles all have the same
volume and their outer shells have the same
mass, the increased density of *gal*⁻ particles
shows that they must contain a larger DNA
molecule; the *gal*⁻ mutation was caused by the
insertion of DNA. (From S. M. Cohen and
J. A. Shapiro, "Transposable Genetic
Elements." Copyright © 1980 by Scientific
American, Inc. All rights reserved.)

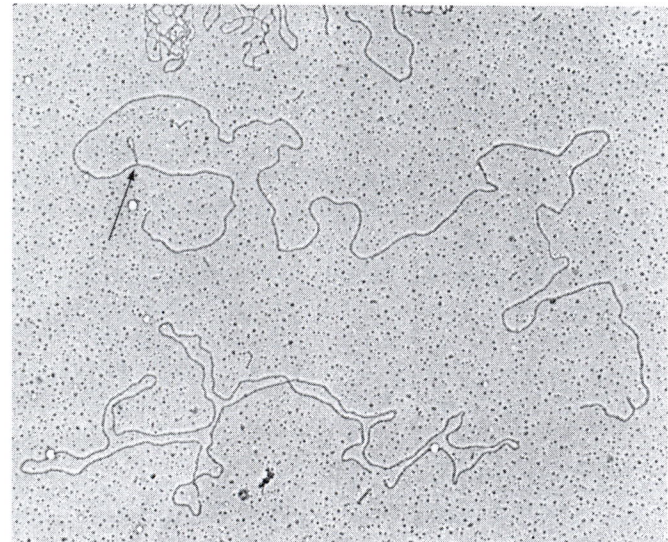

However, in DNA containing the IS mutations, each heteroduplex shows a single-stranded buckle, or loop (Figure
20-9). This single-stranded buckle confirms the presence of
an inserted sequence in the mutated DNA that has no complementary sequence in the wild-type DNA. The length of
this single-stranded loop can be calibrated by including standardized marker DNA in the preparation. It proves to be approximately 800 nucleotides in length.

Figure 20-9   Electron micrograph of a λ*dgal*⁺/λ*dgal*ᵐ DNA
heteroduplex. The single-stranded loop (arrow) is caused by the
presence of an insertion sequence in λ*dgal*ᵐ. (From A. Ahmed and
D. Scraba, "The Nature of the *gal3* Mutation of *Escherichia coli*,"
Molecular and General Genetics 136, 1975, 233.)

Identification of discrete IS elements

Are the segments of DNA that insert into genes merely random DNA fragments or are they distinct genetic entities?
Hybridization experiments show that many different insertion mutations are caused by a small set of insertion sequences. In these experiments, the λ*dgal* phages, which
contain the *gal*⁻ gene, are isolated from the IS mutant bacteria, and their DNA is used to synthesize radioactive RNA
in vitro. Certain fragments of this RNA are found to hybridize with the mutant DNA but not with wild-type DNA,
indicative of the fact that the mutant contains an extra piece
of DNA. These particular RNA fragments also hybridize to
DNA from other IS mutants, showing that the same bit of
DNA is inserted in different places in the different IS mutants.

On the basis of their patterns of cross-hybridization, the
insertion mutants are placed into categories. The first sequence, the 800-bp segment identified in *gal,* is termed *IS1*.
A second sequence, termed *IS2,* is 1350 bp long. Table 20-1
lists some of the insertion sequences and their sizes. The inverted repeats listed in Table 20-1 will be dealt with shortly
under "Transposons."

We now know that the genome of the standard wild-
type *E. coli* is rich in IS elements: it contains eight copies of
IS1, five copies of IS2, and copies of other less well studied
IS types. It should be emphasized that the sudden appearance of an insertion sequence at any given locus under study
means that these elements are truly mobile, with a capability
for transposition throughout the genome. Presumably, they
produce a mutation or some other detectable alteration of
normal cell function only when they happen to end up in an
"abnormal" position, such as the middle of a structural gene.

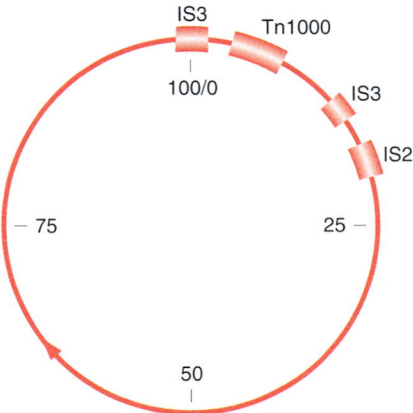

20-1 TABLE	Prokaryotic Insertion Elements		
Insertion sequence	Normal occurrence In *E. coli*	Length (bp)	Inverted repeat* (bp)
IS1	5–8 copies on chromosome	768	18/23
IS2	5 copies on chromosome; 1 on F	1327	32/41
IS3	5 copies on chromosome; 2 on F	1400	32/38
IS4	1 or 2 copies on chromosome	1400	16/18
IS5	Unknown	1250	Short
γ-δ (TN1000)	1 or more copies on chromosome; 1 on F	5700	35
pSC101 segment	On plasmid pSC101	200	30/36

* Fraction of base pairs; for example, 18 of 23 bp, and so forth.
Source: M. P. Calos and J. H. Miller, *Cell* 20, 1980, 579–595.

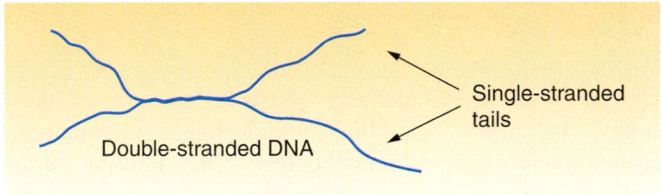

Figure 20-11   Diagrammatic representation of the appearance of a heteroduplex DNA molecule formed by annealing corresponding strands (that is, strands that have the same base sequence, excepting at the site of the IS) of λ*dgal*ᵐ DNA from two specific mutants caused by insertion sequences. This unexpected hybridization between corresponding strands (normally having the same rather than complementary base sequences) can be explained by the model shown in Figure 20-12.

DNA denaturation, they can be recovered separately in an ultracentrifuge. In some cases, the *same* strands (that is, parallel strands) from two different IS1 mutants form an unexpected hybrid with each other. Under the electron microscope, these hybrids have a peculiar appearance: each is a double-stranded region with four single-stranded tails (Figure 20-11). This observation is explained by assuming that the IS1 elements are inserted in opposite directions in the two mutants (Figure 20-12).

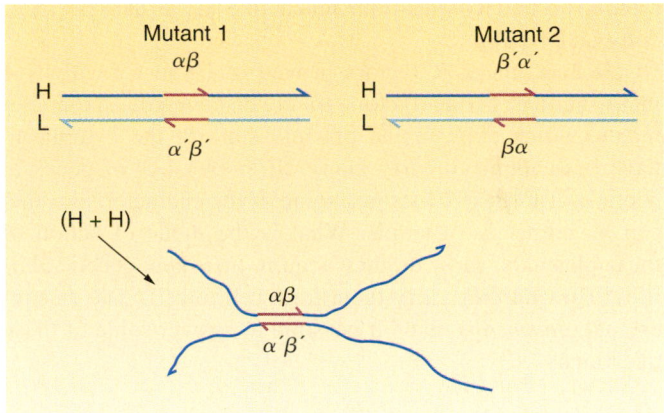

Figure 20-12   A model to explain the heteroduplex DNA structure shown in Figure 20-11. IS1 is represented by the maroon lines; α and β represent the ends of the IS1 molecule; H and L represent the strands of λ*dgal* DNA with high and low buoyant densities, respectively. The hybrid can be explained by assuming that the IS1 sequence is inserted in opposite directions in the two mutants.

Figure 20-10   Genetic and physical map of the F factor. The positions of the resident insertion sequences, IS2, IS3, and Tn1000 (boxes), are shown relative to the map coordinates 0–100. The arrow indicates the origin and direction of F DNA transfer during conjugation.

Insertion sequences also are commonly observed in the F factor. Figure 20-10 shows an example of an F-*lac* episome.

MESSAGE ···
The bacterial genome contains segments of DNA, termed IS elements, that can move from one position on the chromosome to a different position on the same chromosome or on a different chromosome.

Orientation of IS elements

Because of the base sequence, the two strands of λ*dgal*⁺ DNA happen to have different buoyant densities. After

Prokaryotic transposons

A frightening ability of pathogenic bacteria was discovered in Japanese hospitals in the 1950s. Bacterial dysentery is caused by bacteria of the genus *Shigella*. This bacterium initially proved to be sensitive to a wide array of antibiotics that were used to control the disease. In the Japanese hospitals, however, *Shigella* isolated from patients with dysentery

20-2 TABLE Genetic Determinants Borne by Plasmids	
Characteristic	Plasmid examples
Fertility	F, R1, Col
Bacteriocin production	Col E1
Heavy-metal resistance	R6
Enterotoxin production	Ent
Metabolism of camphor	Cam
Tumorigenicity in plants	T1 (in *Agrobacterium tumefaciens*)

Figure 20-13   Peculiar structure formed when denatured plasmid DNA is reannealed. The double-stranded IR region separates a large circular loop from the small "lollipop" loop. (Photograph courtesy of S. N. Cohen.)

proved to be simultaneously resistant to many of these drugs, including penicillin, tetracycline, sulfanilamide, streptomycin, and chloramphenicol. This multiple drug-resistance phenotype was inherited as a single genetic package, and it could be transmitted in an infectious manner — not only to other sensitive *Shigella* strains, but also to other related species of bacteria. This talent is an extraordinarily useful one for the pathogenic bacterium, and its implications for medical science were terrifying. From the point of view of the geneticist, however, the situation is very interesting. The vector carrying these resistances from one cell to another proved to be a self-replicating element similar to the F factor. These **R factors** (for resistance) are transferred rapidly on cell conjugation, much like the F particle in *E. coli* (Chapter 7).

In fact, these R factors proved to be just the first of many similar F-like factors to be discovered. These elements, which exist in the plasmid state in the cytoplasm, have been found to carry many different kinds of genes in bacteria. Table 20-2 lists just some of the characteristics that can be borne by plasmids. What is the mode of action of these plasmids? How do they acquire their new genetic abilities? How do they carry them from cell to cell? The discovery of *transposons* made it possible to answer some of these questions.

Physical structure of transposons

If the DNA of a plasmid conferring drug resistance (carrying the genes for kanamycin resistance, for example) is denatured to single-stranded forms and then allowed to renature slowly, some of the strands form an unusual shape under the electron microscope: a large, circular DNA ring is attached to a "lollipop"-shaped structure (Figure 20-13). The "stick" of the lollipop is double-stranded DNA, which has formed through the annealing of two **inverted repeat (IR) sequences** in the plasmid (Figure 20-14). Subsequent studies have shown that the IR sequences are a pair of IS elements in many cases. For instance, IS10 is present at the ends of the region carrying the genes for tetracycline resistance (Figure 20-15). In some cases, however, the IR sequences are much smaller.

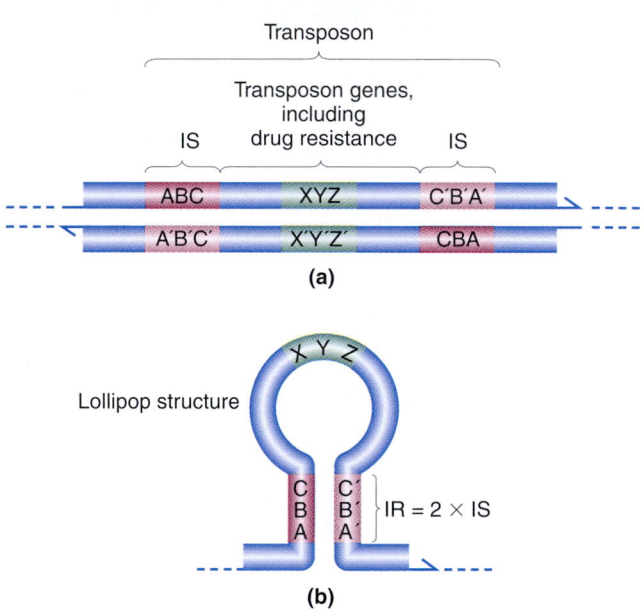

Figure 20-14   An explanation at the nucleotide level for the lollipop structure seen in Figure 20-13. The structure is called a *transposon*. (a) The transposon in its double-stranded form before denaturing. Note the presence of oppositely oriented copies of an insertion sequence (IS). (b) The lollipop structure formed by intrastrand annealing after denaturing. The two IS regions anneal to form the inverted repeat (IR) of the transposon; the transposon genes are carried in the lollipop loop.

The genes for drug resistance or other genetic abilities carried by the plasmid are located between the IR sequences in the lollipop head. The IR sequences together with their contained genes have been collectively called a **transposon (Tn).** (Transposons are therefore longer than IS elements, because they contain extra protein-encoding genes.) The remainder of the plasmid, bearing the genes encoding resistance-transfer functions (RTF), is called the **RTF**

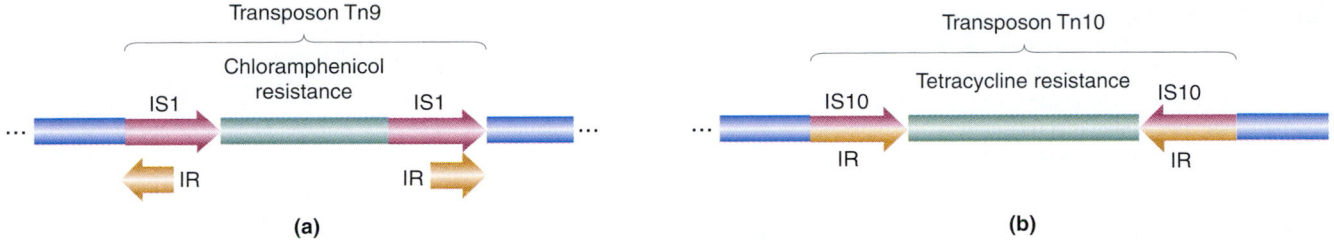

Figure 20-15 Two different transposons having different inverted repeat (IR) regions and carrying different drug-resistance genes. (a) Tn9 has a short IR region, because the two IS1 elements are in the same orientation and each element has a short inverted repeat. (b) Tn10 has a large IR region because the two IS10 components have opposite orientations, and the entire IS10 sequence constitutes the inverted repeat.

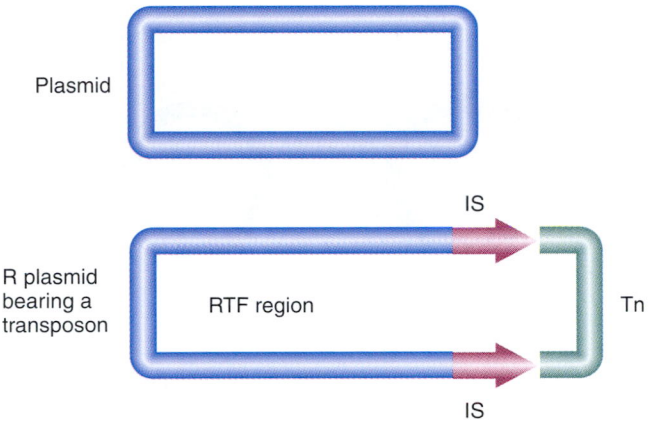

Figure 20-16 The insertion of a transposon (Tn) into a plasmid. RTF represents the resistance-transfer functional genes of the plasmid. Tn includes both the IS elements and the drug-resistance genes.

20-3 TABLE	Some Prokaryotic Transposons		
Transposon	Marker	Length (bp)	Inverted repeat
Tn1	Ampicillin	4,957	38
Tn2	Ampicillin		
Tn3	Ampicillin		
Tn4	Ampicillin, streptomycin, sulfanilamide	20,500	Short
Tn5	Kanamycin	5,400	1500
Tn6	Kanamycin	4,200	Not detectable with electron microscopy
Tn7	Trimethoprim, streptomycin	14,000	Not detectable with electron microscopy
Tn9	Chloramphenicol	2,638	18/23*
Tn10	Tetracycline	9,300	1400

* 18 of 23 base pairs.
Source: M. P. Calos and J. H. Miller, *Cell* 20, 1980, 579–595.

region (Figure 20-16). Table 20-3 lists some of the known transposons.

Movement of transposons

A transposon can jump from a plasmid to a bacterial chromosome or from one plasmid to another plasmid. In this manner, multiple drug-resistant plasmids are generated. Figure 20-17 is a composite diagram of an R plasmid, indicating the various places at which transposons can be located.

MESSAGE ·······················
Transposons were originally detected as mobile genetic elements that confer drug resistance. Many of these elements consist of recognizable IS elements flanking a gene that encodes drug resistance. IS elements and transposons are now grouped together under the single term *transposable elements*.
·······················

Phage mu

Phage mu is a normal-appearing phage. We consider it here because, although it is a true virus, it has many features in common with IS elements. The DNA double helix of this phage is 36,000 nucleotides long—much larger than an IS element. However, it does appear to be able to insert itself anywhere in a bacterial or plasmid genome in either orientation. Once inserted, it causes mutation at the locus of insertion—again like an IS element. (The phage was named for this ability: *mu* stands for "mutator.") Normally, these mutations cannot be reverted, but reversion can be produced by certain kinds of genetic manipulation. When this reversion is produced, the phages that can be recovered show no deletion, proving that excision is exact and that the insertion of the phage therefore does not involve any loss of phage material either.

Each mature phage particle has on each end a piece of flanking DNA from its previous host (Figure 20-18). However, this DNA is not inserted anew into the next host. Its function is unclear. Phage mu also has an IR sequence, but neither of the repeated elements is at a terminus.

Mu can also act like a genetic snap fastener, mobilizing any kind of DNA and transposing it anywhere in a genome.

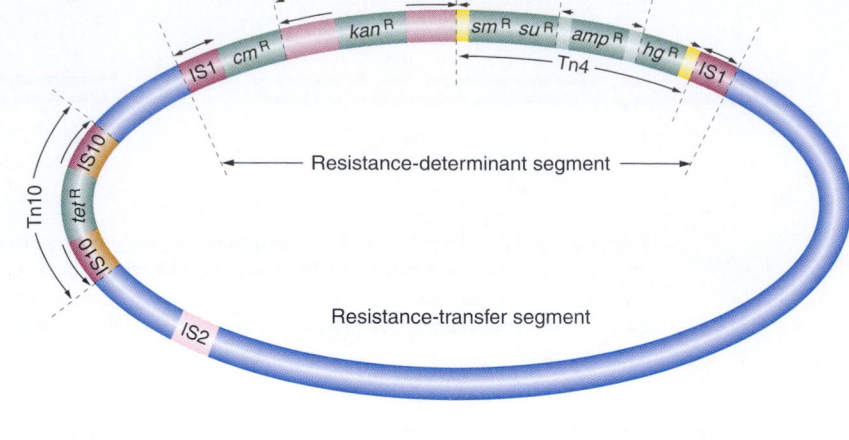

Figure 20-17   The role of transposable elements in the evolution of antibiotic-resistance plasmids is illustrated by a schematic map of a plasmid carrying many resistance genes. The plasmid appears to have been formed by the joining of a resistance-determinant segment and a resistance-transfer segment; there are insertion elements (IS1) at the junctions, where the two segments sometimes dissociate reversibly. Genes encoding resistance to the antibiotics chloramphenicol (cm^R), kanamycin (kan^R), streptomycin (sm^R), sulfonamide (su^R), and ampicillin (amp^R) and to mercury (Hg^R) are clustered on the resistance-determinant segment, which consists of multiple transposable elements; inverted repeat termini are designated by arrows pointing outward from the element. A transposon encoding resistance to tetracycline (tet^R) is on the resistance-transfer segment. Transposon Tn3 is within Tn4. Each transposon can be transferred independently. (From S. N. Cohen and J. A. Shapiro, "Transposable Genetic Elements." Copyright © 1980 by Scientific American, Inc. All rights reserved.)

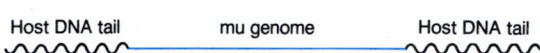

Figure 20-18   The DNA of a free mu phage has tails derived from its previous host. (From S. N. Cohen and J. A. Shapiro, "Transposable Genetic Elements." Copyright © 1980 by Scientific American, Inc. All rights reserved.)

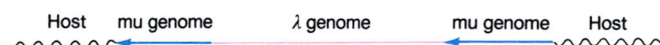

Figure 20-19   Phage mu can mediate the insertion of phage λ into a bacterial chromosome, resulting in a structure like the one shown here.

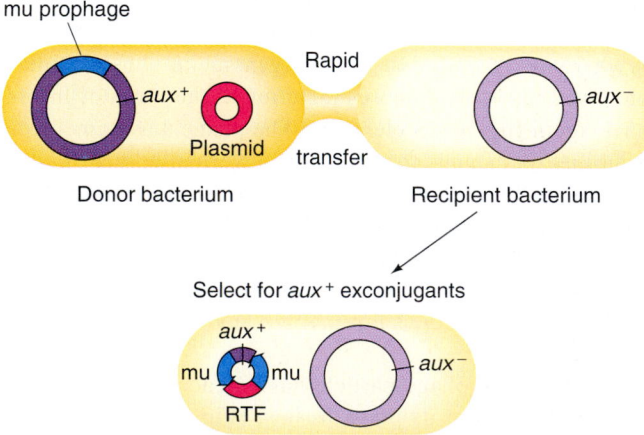

Figure 20-20   Phage mu can mediate the transposition of a bacterial gene into a plasmid. The selection procedure for detecting the transposition is indicated here. An auxotrophic mutant gene is indicated by aux^-.

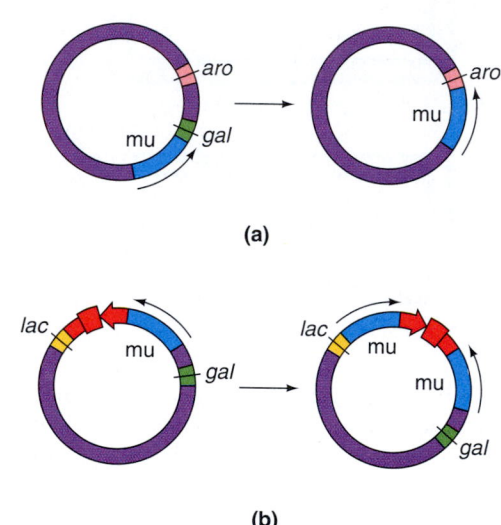

(a)

(b)

Figure 20-21   Phage mu can cause the deletion or inversion of adjacent bacterial segments. (a) The *gal* region is deleted by transposition of phage mu. (b) The F-factor region of an Hfr strain is inverted by transposition of phage mu.

For example, it can mobilize another phage (such as λ) or the F factor. In such situations, the inserted DNA is flanked by two mu genomes (Figure 20-19). It can also transfer bacterial markers onto a plasmid; here again, the transferred region is flanked by a pair of mu genomes (Figure 20-20). Finally, the phage mu can mediate various kinds of structural chromosome rearrangements (Figure 20-21).

Mechanism of transposition in prokaryotes

Several different mechanisms of transposition are employed by prokaryotic transposable elements. And, as we shall see later, eukaryotic elements exhibit still additional mechanisms of transposition.

In *E. coli*, we can identify **replicative** and **conservative** (nonreplicative) modes of transposition. In the replicative pathway, a new copy of the transposable element is generated in the transposition event. The results of the transposition are that one copy appears at the new site and one copy remains at the old site. In the conservative pathway, there is no replication. Instead, the element is excised from the chromosome or plasmid and is integrated into the new site.

Replicative transposition

When transposition is from one locus to a second locus for certain transposons, a copy of the transposable element is left behind at the first locus. An analysis of transposon mutants revealed an interesting fact about the mechanism of transposition. Using the transposon Tn3 (Figure 20-22), researchers grouped the mutations that prevent transposition into two categories. A trans-recessive class maps in the gene that encodes the transposase enzyme, a catalyst of transposition. A second class of cis-dominant mutations results in the buildup of an intermediate in the transposition process. Figure 20-23 diagrams the transposition pathway in the Tn3 transposition from one plasmid to another. The intermediate is a double plasmid, with both donor and recipient plasmid being fused together. The combined circle resulting from the fusion of two circular elements is termed a **cointegrate**. Apparently, the mutations in this second class delete a region on the transposon at which a recombination event takes place that resolves cointegrates into two smaller circles. This region, called the **internal resolution site (IRS)**, appears in Figure 20-22.

The finding of a cointegrate structure as an intermediate in transposition helped establish a replicative mode of transposition for certain elements. In Figure 20-23, note how the transposable element is duplicated in the fusion event and how the recombination event that resolves the cointegrate into two smaller circles leaves one copy of the transposable element in each plasmid.

Conservative transposition

Some transposons, such as Tn10, excise from the chromosome and integrate into the target DNA. In these cases, DNA replication of the element does not occur, and the element is lost from the site of the original chromosome. Re-

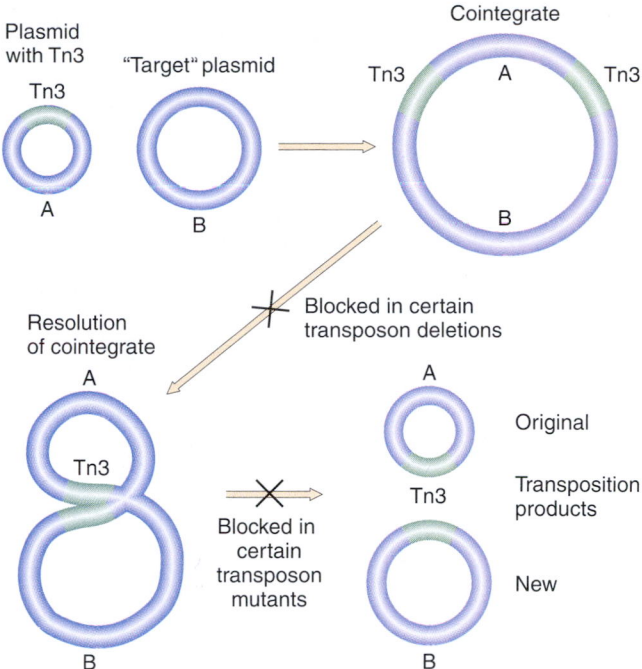

Figure 20-23   Transposition of Tn3 takes place through a cointegrate intermediate. Cointegrates in Tn3 transposition are observed for some internal deletions in the transposon. The correct explanation for this observation is that the cointegrates are intermediates in Tn3 transposition and that their resolution is blocked because the internal deletion has removed an internal resolution site, where recombination takes place. (From F. Heffron, in J. A. Shapiro, ed., *Mobile Genetic Elements*, pp. 223–260. Copyright © 1983 by Academic Press.)

searchers demonstrated this lack of replication by constructing heteroduplexes of λTn10 derivatives containing the *lac* region of *E. coli*. The researchers used DNA from Tn10-*lacZ*$^+$ and Tn10-*lacZ*$^-$ derivatives. The heteroduplexes, therefore, contain one strand with the wild-type *lac* region and a second strand with the mutated (Z^-) *lac* region. Figure 20-24 diagrams this part of the experiment. The heteroduplex DNA is used to infect cells that have no *lac* genes, and transpositions of the Tet^R Tn10 are selected. Different types of colonies arise from the transposition of a heteroduplex Z^-/Z^+ carrying transposon (Figure 20-25). If replication takes place (the replicative mode of transposition), all colonies are either completely Lac$^+$ or completely Lac$^-$, because the replication will convert the heteroduplex DNA into two homoduplex daughter molecules. The mechanism by which this conversion takes place will be examined in detail in the next section. However, if the transposition is conservative and does not include replication, each colony arises from a *lacZ$^+$/lacZ$^-$* heteroduplex. Such colonies are partly Lac$^+$ and partly Lac$^-$. By using media that stain Lac$^+$ and Lac$^-$ cells different colors, researchers can observe the Lac$^+$ and Lac$^-$ sectors in colonies.

Therefore, the determination of whether Tn10 undergoes replicative or conservative transposition can be made by

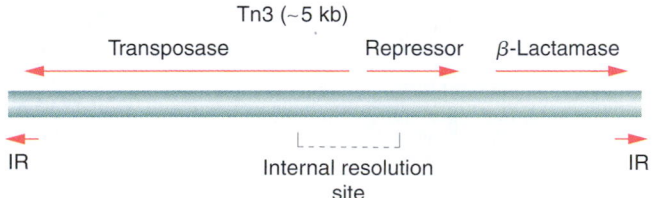

Figure 20-22   The structure of Tn3. Tn3 contains 4957 base pairs and encodes three polypeptides: the transposase is required for transposition, the repressor is a protein that regulates the transposase gene (see Chapter 11), and β-lactamase confers ampicillin resistance. Tn3 is flanked by inverted repeats (IR) of 38 base pairs and contains a site designated the *internal resolution site* necessary for the resolution of Tn3 cointegrates.

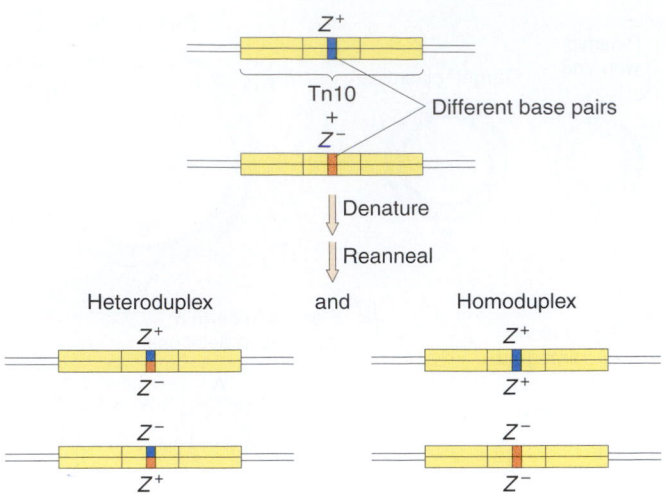

Figure 20-24 Generation of heteroduplex and homoduplex Tn10 elements. The denaturation and reannealing of a mixture of two parental λ phages carrying Tn10 elements that differ only at three single bases in the transposon yields a mixture of heteroduplex and homoduplex products. The base differences in the Z gene allow the ultimate determination of the heteroduplex or homoduplex nature of a cell that has received the Z gene through transposition. (After J. Bender and N. Kleckner, *Cell* 45, 1986, 801–815.)

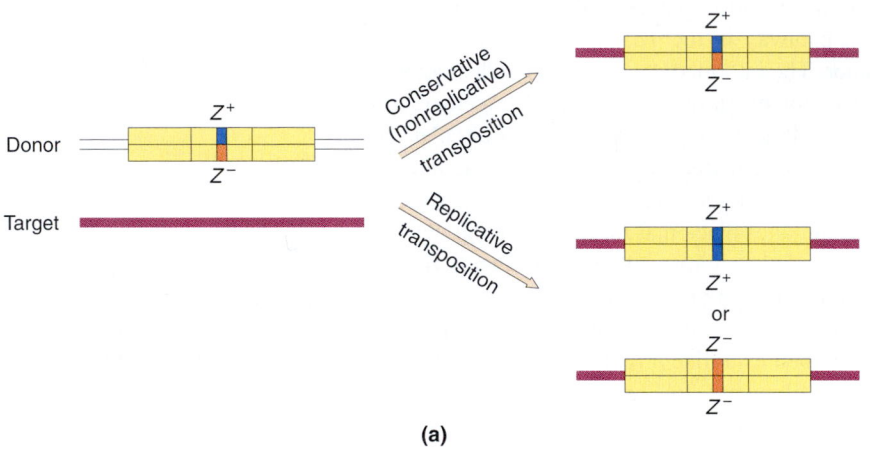

(a)

Figure 20-25 Consequences of conservative and replicative transposition. (a) The heteroduplex or homoduplex nature of DNA (see Figure 20-24) is transposed into a target gene. If the starting DNA is heteroduplex, then the resulting DNA will still be heteroduplex only in a conservative, or nonreplicative, pathway. (b) Because the heteroduplex results in a transposed cell that maintains the heteroduplex nature of the DNA during conservative transposition, colonies will arise that are partly Z^+ and partly Z^-. However, in a replicative pathway, transposition results in individual cells that are either all Z^+ or all Z^-, and all the colonies will be either Z^+ or Z^-.

(b)

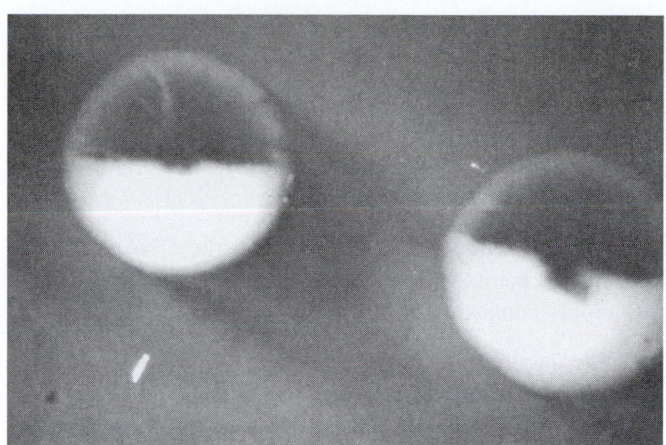

Figure 20-26 Colonies resulting from the experiment illustrated in Figures 20-24 and 20-25. A dye is used that stains Z^+ cells blue. One-half of this colony is Z^+ (dark area), and the other is Z^- (white). (Photograph courtesy of N. Kleckner.)

observing whether differently colored sectors exist within the same colony resulting from the transposition. Sectored colonies are observed in a majority of cases (Figure 20-26). Thus, Tn10 — and perhaps other transposable elements in *E. coli* — transpose by excising themselves from the donor DNA and integrating directly into the recipient DNA.

Molecular consequences of transposition

The molecular consequences of transposition reveal an additional piece of evidence concerning the mechanism of trans-

position: on integration into a new target site, transposable elements generate a repeated sequence of the target DNA in both replicative and conservative transposition. Figure 20-27 depicts the integration of IS1 into a gene. In the example shown, the integration event results in the repetition of a 9-bp target sequence. Analysis of many integration events reveals that the repeated sequence does not result from reciprocal site-specific recombination (as is the case in phage λ integration; see page 229); rather, it is generated in the process of integration itself. The number of base pairs is a characteristic of each element. In bacteria, 9-bp and 5-bp repeats are most common.

The preceding observations have been incorporated into somewhat complicated models of transposition. Most models postulate that staggered cleavages are made at the target site and at the ends of the transposable element by a transposase enzyme that is encoded by the element. One end of the transposable element is then attached by a single strand to each protruding end of the staggered cut. Subsequent steps depend on the mode of transposition (replicative or conservative).

Rearrangements mediated by transposable elements

Transposable elements generate a high incidence of deletions in their vicinity. These deletions emanate from one end of the element into the surrounding DNA (Figure 20-28). Such events, as well as element-induced inversions, can be viewed as aberrant transposition events. Transposons

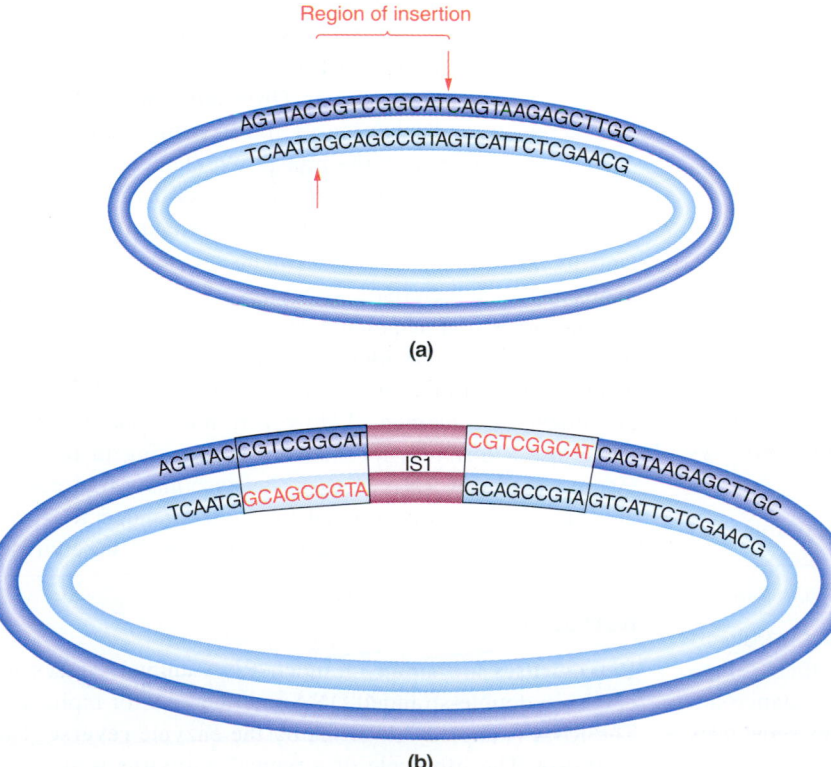

(a)

(b)

Figure 20-27 Duplication of a short sequence of nucleotides in the recipient DNA is associated with the insertion of a transposable element; the two copies bracket the inserted element. Here the duplication that attends the insertion of IS1 is illustrated in a way that indicates how the duplication may come about. IS1 insertion causes a nine-nucleotide duplication. If the two strands of the recipient DNA are cleaved (arrow) at staggered sites that are nine nucleotides apart, as shown in part a, followed by insertion of IS1 between the resulting single-stranded ends, then the subsequent filling in of single strands on each side of the newly inserted element, indicated by red letters in part b, with the right complementary nucleotides could account for the duplicated sequences (boxes). (From S. N. Cohen and J. A. Shapiro, "Transposable Genetic Elements." Copyright © 1980 by Scientific American, Inc. All rights reserved.)

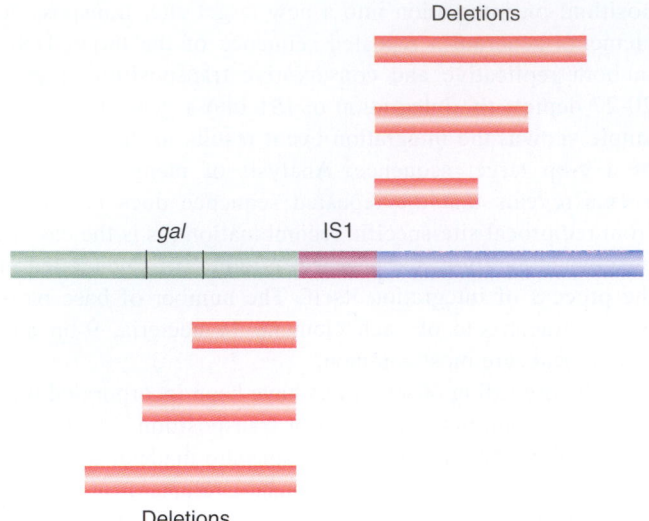

Figure 20-28   Deletion formation mediated by a transposable element. In this example, the transposable element IS1 is shown at a point in the *E. coli* chromosome near the *gal* genes. Deletions can be generated from each end of the IS1 element, extending into the neighboring DNA sequences. In cases where the deletions extend into the *gal* regions, they can be detected as a result of the Gal⁻ phenotype.

also give rise to readily detectable deletions in which part of the element is deleted together with varying lengths of the surrounding DNA. This process of **imprecise excision** is now recognized as deletions or inversions emanating from the internal ends of the IR segments of the transposon. The process of **precise excision** — the loss of the transposable element and the restoration of the gene that was disrupted by the insertion — also occurs, although at very low rates compared with the frequencies of the events just described.

MESSAGE ⋯⋯⋯⋯⋯⋯⋯⋯⋯⋯⋯⋯⋯⋯⋯⋯⋯⋯⋯⋯⋯
Some DNA sequences in bacteria and phages act as mobile genetic elements. They are capable of joining different pieces of DNA and are thus capable of splicing DNA fragments into or out of the middle of a DNA molecule. Some naturally occurring mobile or transposable elements carry antibiotic-resistance genes.

Review of transposable elements in prokaryotes

Let's examine what we have learned so far about prokaryotic transposable elements:

1.   There are several different types of transposable elements, including insertion sequences (IS1, IS2, . . .) and transposons (Tn1, Tn2, . . .).

2.   Two copies of a transposable element can act in concert to transpose the DNA segments in between them. Some of the transposons that confer antibiotic resistance are formed in this manner, with two insertion sequences flanking the genes for antibiotic resistance.

3.   Most of the transposable elements have recognizable inverted repeat (IR) structures, some of which can be observed under the electron microscope after denaturation and renaturation.

4.   Transposable elements are found in bacterial chromosomes, as well as in plasmids.

5.   After insertion into a new site on the DNA, transposable elements generate a short repeated sequence, commonly consisting of 9 or 5 bp.

6.   The detailed mechanism of transposition is not known, but two different pathways for transposition have been identified. In some cases, transposition takes place by replication of a new copy of the element into the target site, with one copy being left behind at the original site. In other cases, transposition consists of the excision of the element from the original site and its reintegration into a new site. These two modes of transposition are called *replicative* and *conservative*, respectively.

Transposable elements have been found in eukaryotes, and some have close similarities to those observed in bacteria, transposing through DNA intermediates. Interestingly, other mobile elements transpose through RNA intermediates and, in certain cases, resemble mammalian retroviruses.

Molecular nature of transposable elements in eukaryotes

Transposable genetic elements are even more prevalent in eukaryotic chromosomes than in bacterial chromosomes. For instance, from 25 percent to 40 percent of mammalian chromosomes consist of transposable elements that have accumulated in the course of evolution. In addition, half of the spontaneous mutations seen in *Drosophila* are attributed to the movement and insertion of transposable elements. The phenotype conferred by the transposed element was used to detect the first elements in maize, as well as in bacteria — for instance, the patched kernels in corn and the Gal⁻ mutants in bacteria. Now, most elements are detected by recognizing their characteristic structures after sequencing large chromosomal regions. Some eukaryotic elements, such as the maize *Ac* and *Ds* elements, and the *Drosophila P* elements discussed in a later section, transpose as DNA, as do all prokaryotic elements. However, most eukaryotic elements use a different mechanism, transposing in the same manner as that of RNA viruses. Because many transposable elements appear to be related to single-stranded RNA animal viruses, we shall examine some aspects of these viruses.

Retroviruses

Retroviruses are single-stranded RNA animal viruses that employ a double-stranded DNA intermediate for replication. The RNA is copied into DNA by the enzyme **reverse transcriptase.** The life cycle of a typical retrovirus is shown in

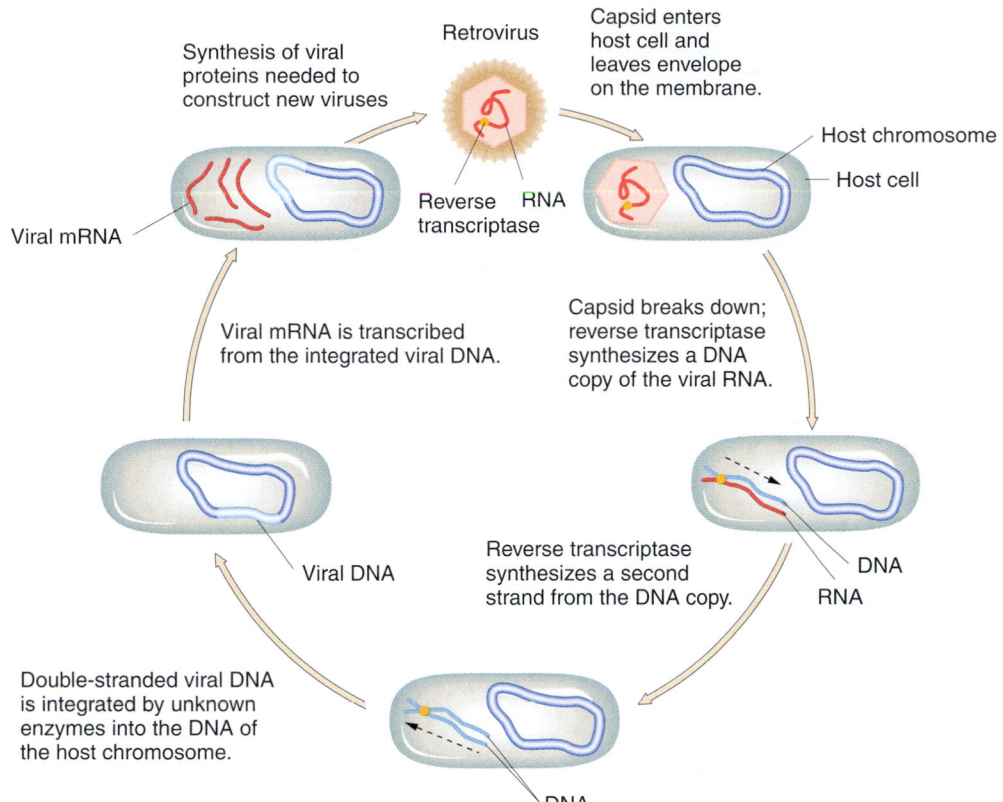

Figure 20-29 The life cycle of a retrovirus. Viral RNA is shown in red; DNA, in blue.

Figure 20-29. Some retroviruses, such as mouse mammary tumor virus (MMTV) and Rous sarcoma virus (RSV), are responsible for the induction of cancerous tumors. When integrated into host chromosomes as double-stranded DNA, these retroviruses are termed **proviruses.** Proviruses, like the mu phage in bacteria, can be considered transposable elements, because they can in effect transpose from one location to another. Figure 20-30 depicts transposition by a retroviral mechanism.

Retroviruses are structurally similar to transposable elements in other organisms, as we shall see later. Additionally, integration results in the duplication of a short target sequence in the host chromosome.

Retrotransposons

Transposable elements that utilize reverse transcriptase to transpose through an RNA intermediate are termed **retrotransposons.** They are widespread among eukaryotes and are generally divided into two classes. *Viral retrotransposons* have properties similar to those of retroviruses. For instance, they display long terminal repeats, or LTRs, as shown in Figure 20-31. Two examples of viral retrotransposons are the yeast *Ty* elements, and the *Drosophila* **copia** elements.

Figure 20-32 shows the structure of one of the *Ty* elements in yeast: the *Ty1* sequence, which is present in approximately 35 copies in the yeast genome. The 330-bp-long termini (terminal sequences), called δ *(delta) sequences,* are present in about 100 copies in the genome. Yeast δ sequences, as well as *Ty* elements as a whole, show significant sequence divergence. *Ty* elements generate a repeated sequence of target DNA (in this case, 5 bp) during

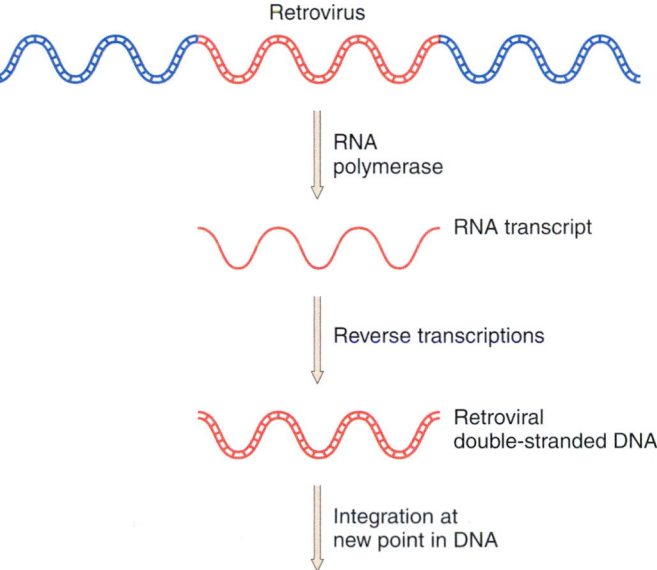

Figure 20-30 Transposition by a retrovirus. By making an RNA transcript that is reverse transcribed into double-stranded DNA, retroviruses effectively transpose on insertion of the new DNA into a different point in the chromosome.

Viral retrotransposon

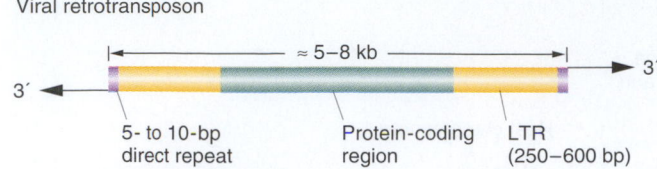

Figure 20-31 A schematic representation of a viral retrotransposon. Such elements have two LTRs (long terminal repeats that flank a central region that encodes specific protein functions). (From H. Lodish, D. Baltimore, A. Berk, S. L. Zipursky, P. Matsudaira, and J. Darnell, *Molecular Cell Biology,* 3d ed., p. 329. Copyright © 1995 by Scientific American Books.)

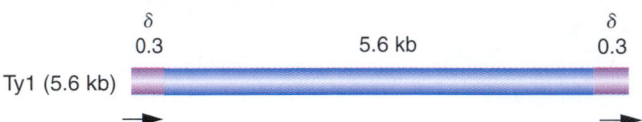

Figure 20-32 The structure of a yeast transposable element. The *Ty1* sequence appears approximately 35 times in the yeast genome. It contains two copies of the delta (δ) sequence in direct orientation at each end. Delta appears approximately 100 times in the yeast genome.

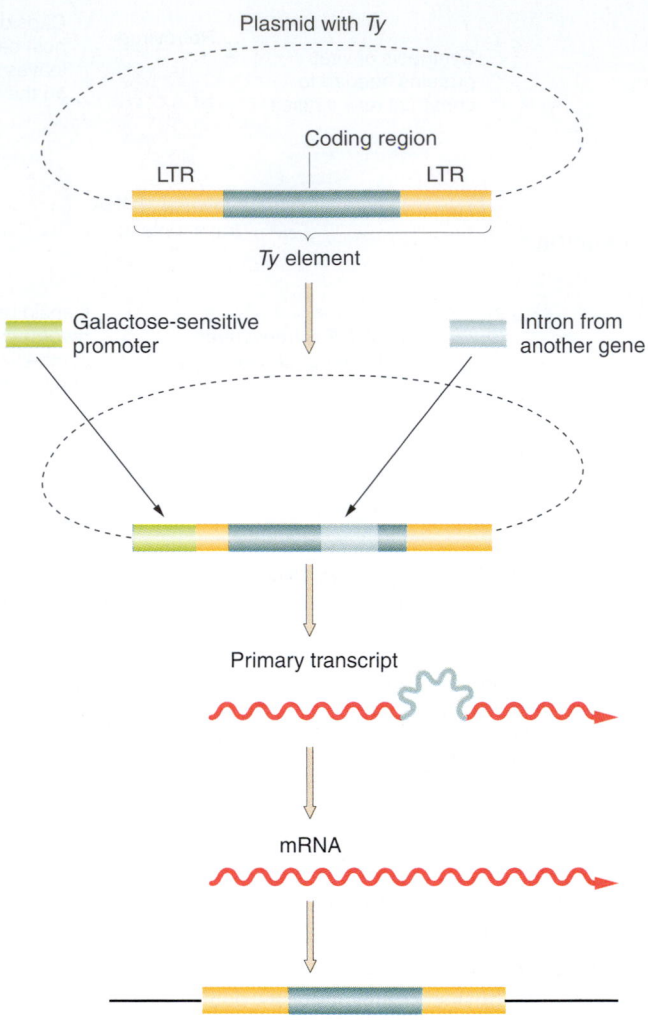

Figure 20-33 Demonstration of transposition through an RNA intermediate. A *Ty* element is altered by adding a promoter that can be activated by the addition of galactose. Activation of the promoter will increase transcription through the *Ty* element. Then an intron from another gene is inserted into the *Ty* element. Because the final product of transposition contains no intron, the intron must have been spliced out of an RNA transcript (see Chapter 10). This splicing must have taken place as shown here, where the primary transcript contains the intron but the final processed mRNA does not. This RNA is then copied by reverse transcriptase and integrated into the chromosomal DNA. (After H. Lodish, D. Baltimore, A. Berk, S. L. Zipursky, P. Matsudaira, and J. Darnell, *Molecular Cell Biology,* 3d ed., p. 332. Copyright © 1995 by Scientific American Books.)

transposition, like prokaryotic transposons. *Ty* elements cause mutations by insertion into different genes in the yeast chromosome. In 1985, Jef Boeke and Gerald Fink use *Ty1* to show that the transposition of *Ty* elements is through an RNA intermediate.

Figure 20-33 diagrams the experimental design used by Boeke and Fink and their colleagues to alter a yeast *Ty* element, cloned on a plasmid. First, a promoter was inserted near the end of an element that could be activated by the addition of galactose to the medium. The use of a galactose-sensitive promoter allows the manipulation of the expression of *Ty* RNA. Galactose enhances the transcription of *Ty* RNA. Second, an intron from another yeast gene was introduced into the coding region of the *Ty* transposon.

The addition of galactose greatly increases the frequency of transposition of the altered *Ty* element. This increased frequency suggests the involvement of RNA, because galactose-stimulated transcription begins at the galactose-sensitive promoter (Figure 20-33). The key experimental result, however, is the fate of the transposed *Ty* DNA. When the researchers examined the *Ty* DNA resulting from transpositions, they found that the intron had been removed. Because introns are excised only in the course of RNA processing (see Chapter 10), the transposed *Ty* DNA must have been copied from an RNA intermediate transcribed from the original *Ty* element and then processed by RNA–RNA splicing. The DNA copy of the spliced mRNA is then integrated into the yeast chromosome.

The ***copia*-like elements** of *Drosophila* constitute at least seven families, ranging in size from 5 kb to 8.5 kb. Members of each family appear at 10–100 positions in the *Drosophila* genome. Each member carries a long, direct ter-

minal repeat and a short, imperfect inverted repeat (Figure 20-34) and is structurally similar to a yeast *Ty* element. *Copia*-like elements also cause a duplication of a characteristic number of base pairs of *Drosophila* DNA on insertion. Certain classic *Drosophila* mutations result from the insertion of *copia*-like and other elements. For example, the white-apricot (w^a) mutation for eye color is caused by the insertion of an element from the *copia* family into the white locus.

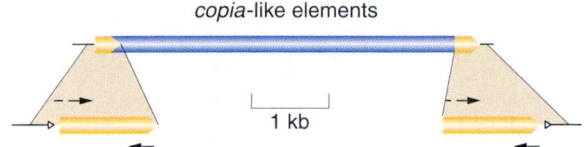

Figure 20-34 *Copia*-like elements carry long direct terminal repeats. Each repeat makes up about 5 percent of the length of the element. These repeats are shown on an expanded scale below the element to illustrate the presence of short, imperfect inverted repeats (→) at the ends of each long direct repeat and the presence of a few base pairs of duplicate target sequence (▷) flanking the element after insertion. The different genomic copies of the family elements are very similar in structure to one another. (From G. Robin, in J. A. Shapiro, ed., *Mobile Genetic Elements*, pp. 329–361. Copyright 1983 by Academic Press.)

Figure 20-35 shows the similarities in the arrangements of genes between integrated retrovirus DNA and the *Ty* and *copia* elements. Note how each of the proteins encoded by the retrovirus have counterparts in *Ty* and *copia* transposons.

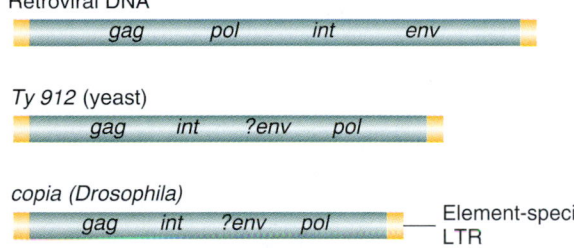

Figure 20-35 A comparison of the genes of integrated retrovirus DNA and the yeast *Ty* elements and *Drosophila copia* elements. The four functions encoded by the retroviral DNA have counterparts in the yeast and *Drosophila* elements shown here. The LTRs are represented by the colored ends of the elements. (From H. Lodish, D. Baltimore, A. Berk, S. L. Zipursky, P. Matsudaira, and J. Darnell, *Molecular Cell Biology*, 3d ed., p. 329. Copyright © 1995 by Scientific American Books.)

Nonviral retrotransposons: LINES and SINES

Nonviral retrotransposons are the most frequently encountered transposons in mammals. The LINEs (long interspersed elements) and SINEs (short interspersed elements; including the *Alu* sequences) discussed in Chapter 3 are the two most abundant. All of these elements are repeated many times in mammalian genomes. The human genome has from 20,000 to 40,0000 LINE elements and about 500,000 SINE elements, most of which display some sequence divergence. Some idea of the profusion of repetitive elements in the human genome can be obtained from Figure 20-36, which shows the distribution of repetitive elements in the gene for homogentisate 1,2-dioxygenase [the alkaptonuria gene (*AKU* or *HGO*)].

Function of transposable elements

Transposable elements play a role in the biology of organisms. They cause mutations by insertion into genes and affect the regulation of genes by inserting near promoters. They also provide substrates for genetic rearrangements and thus act as agents of genome evolution. For instance, in *E. coli* growing in natural environments, a sizable fraction of spontaneous mutations are caused by the IS series of transposable elements. As mentioned earlier, more than half of the spontaneous mutations in *Drosophila* result from transposable-element insertion. Insertions of IS elements near the promoter region of the *bgl* operon, which encodes proteins participating in β-glucosidase metabolism, activates transcription in this normally "cryptic" operon. There are numerous other examples of insertions that turn on gene expression.

Insertion elements provide portable regions of homology that can serve as substrates for recombination enzymes, creating deletions, duplications, and inversions. There is evidence that transposable element-mediated rearrangements may have played an important role in generating the genomes of different organisms, as well as contributing to new functions by stimulating gene duplication. Transposable elements have accumulated during evolution to the point where they constitute much of the genome of higher cells. For instance, in humans, approximately 30 percent of the genomic DNA consists of transposable elements. The evolution of different antibiotic-resistance-carrying microorganisms is influenced by transposable elements. Recall the transposons that can generate different combinations of resistance-encoding genes on R plasmids that are then transferred to different bacteria.

Uses of transposable elements

Transposable elements have many uses. In prokaryotes, the antibiotic-resistance marker carried by different transposons serves as a convenient marker. For instance, derivatives of Tn10 are often used to create insertions in the bacterial chromosome. After selection for tetracycline-resistant (Tet^r) cells, each of which contains an insertion somewhere in the gene, the position of the insert can be mapped genetically, by following the Tet^r phenotype, or physically, by using primers matching the ends of Tn10 and then matching the sequence of the adjacent bacterial DNA with that from the known sequence of the entire genome (in *S. typhimurium* and *E. coli* for example). With the use of linked transposons, genes can be transferred from one strain to another easily and can be cloned by selecting for the antibiotic-resistance marker that is either inserted into or very near the gene of interest. In eukaryotes, transposable elements are also used for generating insertion mutations, mapping them, and facilitating both the cloning of genes and the generation of transgenic organisms.

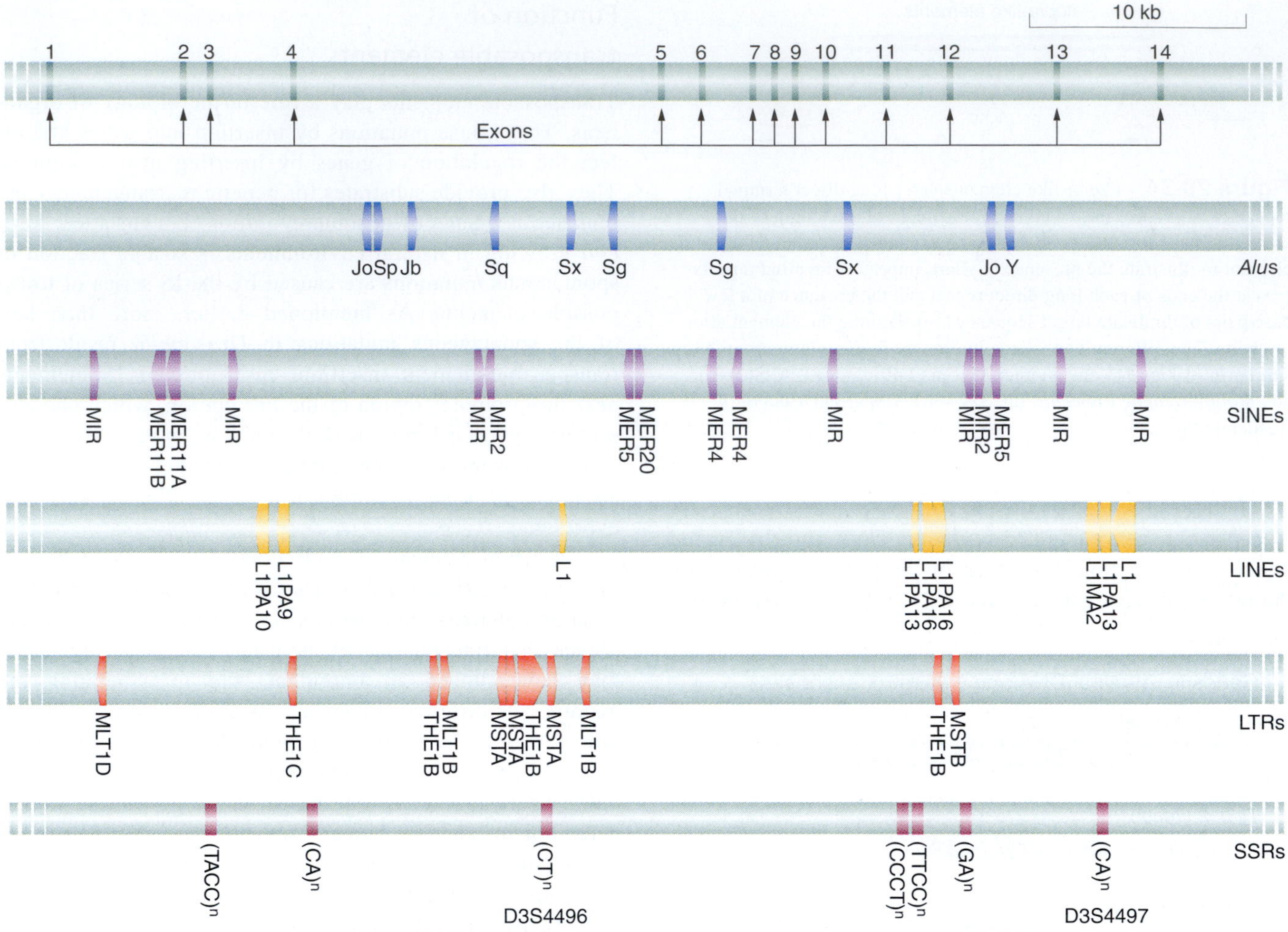

10 kb

1   2 3   4   5   6   7 8 9 10   11   12   13   14

Exons

JoSp Jb   Sq   Sx Sg   Sg   Sx   Jo Y   *Alus*

MIR   MER11A MER11B   MIR   MIR2 MIR   MER20 MER5   MER4 MER4   MIR   MER5 MIR2 MIR   MIR   MIR   SINEs

L1PA9 L1PA10   L1   L1PA16 L1PA13   L1 L1PA13 L1MA2   LINEs

MLT1D   THE1C   MLT1B THE1B   THE1B MSTA MSTA   MSTA   MLT1B   MSTB THE1B   LTRs

(TACC)n   (CA)n   (CT)n   (TTCC)n (CCCT)n   (GA)n   (CA)n   SSRs

D3S4496   D3S4497

Figure 20-36   Repetitive elements found in the human gene *(HGO)* encoding homogentisate 1,2-dioxygenase, the enzyme whose deficiency causes alkaptonuria. The first line diagrams the position of the *HGO* exons. The location and direction of *Alus* (blue), SINEs (purple), LINEs (orange), retrotransposon-derived sequences (LTRs, red), and short-sequence repeats (SSRs, maroon) in the *HGO* sequence are indicated by color. (After B. Granadino, D. Beltrán-Valero de Bernabé, J. M. Fernández-Cañón, M. A. Peñalva, and S. Rodríguez de Córdoba, "The Human Homogentisate 1,2-Dioxygenase *(HGO)* Gene," *Genomics* 43, 1997, 115.)

P elements

The *P* element in *Drosophila* is one of the best examples of exploiting the properties of transposable elements in eukaryotes. This element, shown in Figure 20-37, is 2907 bp long and features a 31-bp inverted repeat at each end. The element encodes a transposase. Although the transposase is required for transposition, it can be supplied by a second element. Therefore, *P* elements with internal deletions can be mobilized and then remain fixed in the new position in the absence of the second element; thus the *P* element can serve as a convenient marker. *P* elements do not utilize an RNA intermediate during transposition and can insert at many different positions in the *Drosophila* chromosome. The transposition of a *P* element is controlled by repressors encoded by the element.

P elements have been developed as tools for *Drosophila* much in the same way that transposons such as

Tn10 have for bacteria. Namely, *P* elements can be used to create mutations by insertion, to mark the position of genes, and to facilitate the cloning of genes. *P* elements can be inserted into genes in vivo, and different phenotypes can be selected. Then, the interrupted gene can be cloned, with the

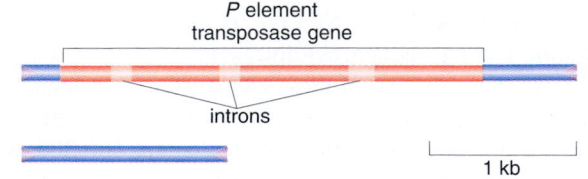

P element transposase gene

introns

1 kb

Figure 20-37   *P* element structure. DNA sequence analysis of the 2.9-kb element reveals a gene, composed of four exons and three introns, that encodes transposase. There is a perfect 31-bp inverted repeat at each terminus. (From G. Robin, in J. A. Shapiro, ed., *Mobile Genetic Elements*, pp. 329–361. Copyright 1983 by Academic Press.)

use of *P* element segments as a probe, a method termed *transposon tagging.* Primers matching the 31-bp sequence at each end can be used to sequence chromosomal regions adjacent to *P* element insertion sites.

Using *P* elements to insert genes

Gerald Rubin and Allan Spradling showed that *P* element DNA can be used as an effective vehicle for transferring donor genes into the germ line of a recipient fly. Rubin and Spradling devised the following experimental procedure (Figure 20-38). The recipient genotype is homozygous for the *rosy* (*ry⁻*) mutation, which confers a characteristic eye color. From this strain, embryos are collected at the completion of about nine nuclear divisions. At this stage, the embryo is one multinucleate cell, and the nuclei destined to form the germ cells are clustered at one end. (*P* elements mobilize only in germ-line cells.) Two types of DNA are injected into embryos of this type. The first is a bacterial plasmid carrying a deleted *P* element into which the *ry⁺* gene has been spliced. This deleted element is not able to transpose, owing to the deletion; so, as mentioned earlier, a helper plasmid bearing a complete element also is injected. Flies developing from these embryos are phenotypically still *rosy* mutants, but their offspring include a large proportion of *ry⁺* flies. These *ry⁺* descendants show Mendelian inheritance of the newly acquired *ry⁺* gene, suggesting that it is located on a chromosome. This location was confirmed by in situ hybridization, which shows that the *ry⁺* gene, together with the deleted *P* element, has been inserted into one of several distinct chromosome locations. None appears exactly at the normal locus of the *rosy* gene. These new *ry⁺*

genes are found to be inherited in a stable fashion. A variation of this method, described in Chapter 13, uses *P* elements to make transgenic *Drosophila* by transferring foreign genes into the germ line and then monitoring their expression pattern.

Review of transposable elements in eukaryotes

Let's examine some of the essential points about eukaryotic transposable elements:

1. Transposable elements exist in all cells. Elements in yeast, *Drosophila,* and maize have been well studied, as have retroviruses in mammalian cells.

2. Some transposable elements can be used as tools for cloning and gene manipulation. For instance, the *P* elements of *Drosophila* can be employed to transfer genes into the germ line of a recipient fly. As another example, the T-DNA segment of Ti plasmids (described in Chapter 13) can be used to introduce cloned genes into certain plants.

3. A similarity between eukaryotic transposable elements and their counterparts in prokaryotes is that transposition into a new site generates a short repeated sequence at the target site.

4. A difference between certain eukaryotic and prokaryotic transposable elements lies in the mechanism of transposition. Some eukaryotic transposable elements transpose through an RNA intermediate; prokaryotic elements do not use an RNA intermediate.

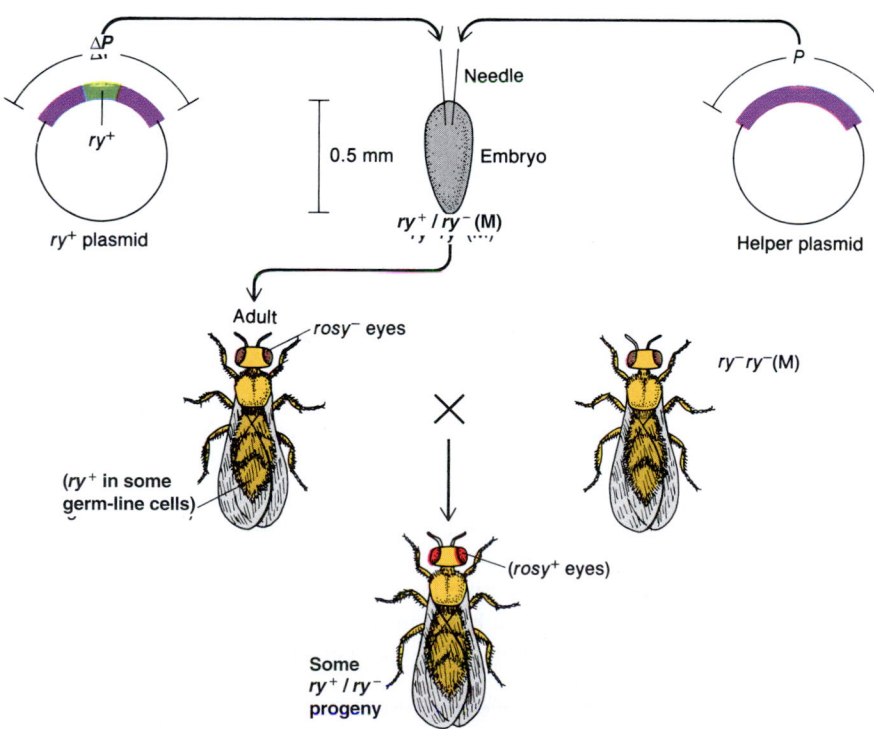

Figure 20-38 *P* element-mediated gene transfer in *Drosophila.* The *rosy⁺* (*ry⁺*) eye-color gene is inserted into a deleted *P* element (Δ*P*) carried on a bacterial plasmid. At the same time, a helper plasmid bearing an intact *P* element is used. Both are injected into a *ry⁻* embryo, where *ry⁺* transposes with the Δ*P* element into the chromosomes of the germ-line cells.

SUMMARY

Nature has devised many different ways of changing the genetic architecture of organisms. We are now beginning to understand the molecular processes behind some of these phenomena. Gene mutation, recombination between chromosomes, and transposition can all be reasonably explained at the DNA level. Far from merely producing genetic waste, all these processes undoubtedly have important roles in evolution. This idea is strengthened through the knowledge that the processes themselves are to a large extent under genetic control: there are genes that affect the efficiency of mutation, recombination, and transposition.

Although different mechanisms of transposition are sometimes used, the analogies between the transposable elements of phages, bacteria, and eukaryotes are striking. At present, it is not known if transposons are elements that normally play a role in the day-to-day transactions of the genome, as originally proposed by Barbara McClintock in the 1950s, or if they are pieces of "selfish DNA" that exist for no purpose other than their own survival. Whatever the truth of this matter is, transposons certainly constitute a completely unexpected element of chaos in the genome, which geneticists have already harnessed into their team of analytical procedures. At the evolutionary level, transposons may be important in the sudden leaps that characterize the fossil record.

CONCEPT MAP

Draw a concept map interrelating as many of the following terms as possible. Note that the terms are listed in no particular order.

IS element / plasmids / transposons / inverted repeats / rearrangements / phenotype / tagging / unstable mutations / sectors

CHAPTER INTEGRATION PROBLEM

In Chapter 11 we studied the operon model. Note that for the *gal* operon the order of transcription of the genes in the operon is *E-T-K*. Suppose we have five different mutations in *galT:gal-1, gal-2, gal-3, gal-4,* and *gal-5*. The following table shows the expression of *galE* and *galK* in mutants carrying each of these mutations:

galT mutation	Expression of *galE*	Expression of *galK*
gal-1	+	−
gal-2	+	−
gal-3	+	−
gal-4	+	+
gal-5	+	+

In addition, the reversion patterns of these mutations with several mutagens we studied in Chapter 19 are shown in the following table. Here, a "+" indicates a high rate of reversion in the presence of each mutagen, a "−" depicts no reversion, and a "low" indicates a low rate of reversion.

	REVERSION				
Mutation	Spontaneous	2-Amino- purine	ICR191	UV	EMS
gal-1	−	−	+	+	−
gal-2	−	−	+	−	−
gal-3	Low	Low	Low	Low	Low
gal-4	−	−	−	−	−
gal-5	Low	+	Low	+	+

Which mutation is most likely to result from the insertion of a transposable element such as IS1 and why? Can you assign the other mutations to other categories?

◆ Solution ◆

Transposable elements will cause polarity, preventing expression of downstream genes from the point of insertion, but not of upstream genes. Therefore, we would expect the insertion mutation to prevent expression of the *galK* gene. Three mutations are in this category, *gal-1, gal-2,* and *gal-3*. These could be frameshifts, nonsense mutations, or insertions, since each of these mutations can lead to polarity. If we examine the reversion data, however, we can distinguish among these possibilities. Transposable elements revert at low rates spontaneously, and this rate is not stimulated by base analogs, frameshift mutagens, alkylating agents, or UV. On the basis of these criteria, the *gal-3* mutation is most likely to result from an insertion, since it reverts at a low rate that is not stimulated by any of the mutagens, *gal-1* might be a frameshift, since it does not revert with 2-AP and EMS, but does with ICR191, a frameshift mutagen, and UV. (Refer to Chapter 16 for details of each mutagen.) Likewise, *gal-2* is probably a frameshift, since it reverts only with ICR191. The *gal-4* mutation is probably a deletion, since it is not stimulated to revert at all. The *gal-5* mutation appears to be a base substitution, since it reverts with 2-AP, but not above the spontaneous background rate with ICR191.

SOLVED PROBLEM

Transposable elements have been referred to as "jumping genes" because they appear to jump from one position to another, leaving the old locus and appearing at a new locus. In light of what we now know concerning the mechanism of transposition, how appropriate is the term "jumping genes" for bacterial transposable elements?

◆ Solution ◆

In bacteria, transposition takes place by two different modes. The conservative mode results in true jumping genes, because, in this case, the transposable element excises from its original position and inserts at a new position. A second mode is termed the *replicative mode.* In this pathway, transposable elements move to a new location by replicating into the target DNA, leaving behind a copy of the transposable element at the original site. When operating by the replicative mode, transposable elements are not really jumping genes, because a copy does remain at the original site.

PROBLEMS

1. Suppose that you want to determine whether a new mutation in the *gal* region of *E. coli* is the result of an insertion of DNA. Describe two physical experiments that would allow you to demonstrate the presence of an insertion.

2. Explain the difference between the replicative and conservative modes of transposition. Briefly describe an experiment demonstrating each of these modes in prokaryotes.

3. Describe the generation of multiple drug-resistance plasmids.

4. Briefly describe the experiment that demonstrates that the transposition of the *Ty1* element in yeast takes place through an RNA intermediate.

5. Explain how the properties of *P* elements in *Drosophila* make possible gene-transfer experiments in this organism.

6. When Rhoades took pollen from wholly pigmented anthers on plants of genotype a_1/a_1 ; *Dt/Dt* and used this pollen to pollinate a_1/a_1 ; *dt/dt* tested females, he found wholly pigmented kernels and, in addition, some dotted kernels. Explain the origin of *both* phenotypes.

7. In *Drosophila,* M. Green found a *singed* allele *(sn)* with some unusual characteristics. Females homozygous for this X-linked allele have singed bristles, but they have numerous patches of sn^+ (wild-type) bristles on their heads, thoraxes, and abdomens. When these flies are mated with *sn* males, some females give only singed progeny, but others give both singed and wild-type progeny in variable proportions. Explain these results.

8. Crown gall tumors are found in many dicotyledonous plants infected by the bacterium *Agrobacterium tumefaciens.* The tumors are caused by the insertion of DNA from a large plasmid carried by the bacterium into the plant DNA. Suppose that a tobacco plant of type A (there are many types of tobacco plants) is infected, and it produces tumors. You remove tumor tissue and grow it on a synthetic medium. Some of these tumor cultures produce aerial shoots. You graft one of these shoots onto a normal tobacco plant of type B, and the graft grows to an apparently normal A-type shoot and flowers.

   a. You remove cells from the graft and place them in synthetic medium, where they grow like tumor cells. Explain why the graft appears to be normal.

   b. When seeds are produced by the graft, the resulting progeny are normal A-type plants. No trace of the inserted plasmid DNA remains. Propose a possible explanation for this "reversal."

9. Consider two maize plants:

   a. Genotype C/c^m ; Ac/Ac^+ where c^m is an unstable allele caused by *Ds* insertion

   b. Genotype C/c^m, where c^m is an unstable allele caused by *Ac* insertion

   What phenotypes would be produced and in what proportions when (1) each plant is crossed with a base-pair-substitution mutant *c/c* and when (2) the plant in part a is crossed with the plant in part b? Assume that *Ac* and *c* are unlinked, that the chromosome breakage frequency is negligible, and that mutant *c/C* is Ac^+.

21

EXTRANUCLEAR GENES

A variegated mosaic of *Euonymus fortunei.*
This mosaic is caused by segregation in cell division of a mixture of two chloroplast DNA types, one normal and another that leads to albino tissue. The segregation results in variegated leaves or, less commonly, in whole branches that are green or albino.
(Anthony Griffiths.)

Key Concepts

Chloroplasts and mitochondria each contain multiple copies of their own unique "chromosome" of genes.

Generally, organelle DNA — and any variant phenotype encoded therein — is inherited through the maternal parent in a cross.

In mixtures of two genetically different mitochondrial DNAs or chloroplast DNAs, it is commonly observed that a sorting-out process results in descendant cells of one type or the other.

In "dihybrid" organelle mixtures, recombination can be detected.

Organelle genes encode mainly organelle translation components and components of energy-producing systems.

Most organelle-encoded polypeptides unite with nucleus-encoded polypeptides to produce active proteins, which function in the organelle.

By far the larger proportion of the DNA of eukaryotic organisms is found in the nuclear chromosomes. However, two types of organelles, mitochondria and chloroplasts (Figure 21-1), each contain a unique type of "chromosome" of genes that encode specific functions of that organelle. The mitochondrial chromosome is called **mtDNA,** and the chloroplast chromosome is **cpDNA.** The functions of mitochondrial genes are directed at making ATP ("chemical energy") by oxidative phosphorylation, which takes place in the mitochondrion itself. Chloroplast genes are ultimately concerned with making ATP by photosynthesis.

The number of genes in organellar chromosomes is small relative to the number in the nucleus. For example, the human nuclear genome consists of 3,000,000 kb of DNA containing about 100,000 genes, whereas human mtDNA is only 17 kb and has only 37 genes. In any one organism, a gene found in an organelle chromosome is generally not found in the nuclear chromosomes, although a few may be present in the nucleus as inactive pseudogenes. In structure and function, organelle genes show many generic similarities with nuclear genes, but there are enough differences in their action and inheritance to make them worthy of specialized treatment in a chapter of their own.

MESSAGE ··

Mitochondria and chloroplasts each contain a relatively small unique type of chromosome containing some of the genes needed for the function of that organelle.

Origin of extranuclear genes

The question of how the mitochondria and chloroplasts came to have these specific sets of genes is still a matter of experimentation and debate in biology. Part of the answer is to be found in the origin of the chloroplasts and mitochondria themselves. It is generally believed that these two organelles arose in the course of evolution as endosymbionts. Specifically, cells of ancestors of eukaryotes were "invaded" at different times by prokaryotic cells, one of which was photosynthetic, giving rise to chloroplasts, and one of which was nonphotosynthetic, giving rise to mitochondria. These invasions set up mutually beneficial symbioses. This symbiosis was a key event in the origin of the lines that eventually became modern eukaryotes.

However, the ancestral invading prokaryotes must have contained many more genes than are found in modern mitochondria and chloroplasts. The evidence suggests that some of these genes were lost, whereas others found their way into the nucleus. The precise set of genes that remains in the organelles of modern eukaryotes is somewhat variable, although a core set tends to be found in most organisms. It is likely that there is adaptive advantage in having some organelle genes located in the organelle itself. The differences between organisms presumably are due to different migration patterns of the organelle genes in the evolution of dif-

ferent eukaryotes. The precise reasons for the differences are not known.

Most modern eukaryotic cells are fully dependent on the organelle genes for their normal function; hence what arose originally as an optional symbiosis is now obligatory. Nevertheless, it is known that some organisms can survive without their mitochondria or their chloroplasts. For example, the yeast *Saccharomyces cerevisiae* can obtain energy from fermentation, a type of chemistry that does not need the mitochondrial genes. Hence mutants that lack mitochondrial genes survive. In another example, some plants can survive saprophytically without their chloroplast genes.

Structure of organelle chromosomes

Mitochondria and chloroplasts can be isolated by various cell fractionation methods, and from these fractions organelle DNA is isolated in the usual way. With the use of standard recombinant DNA technology (Chapters 12 and 13), several organelle chromosomes have now been fully sequenced. The functions of the organelle genes have been determined by a combination of mutation analysis (see later section) and homology with DNA databank sequences of known function.

Overall organization

Here we see the first big difference between organelle and nuclear chromosomes. Most organelle chromosomes appear to be fundamentally circular. The evidence for their circularity is that restrictions maps of organelle DNA are circular, and furthermore DNA circles can be seen in organelle preparations under the electron microscope. There is evidence that some organelle chromosomes can take on linear forms, but by and large most geneticists treat them as though they are circles.

Another important difference in overall organization is that organelle chromosomes are not in the highly condensed form found in eukaryotic chromosomes; that is, they are not in a euchromatic state.

How many copies?

Here we see another difference: whereas nuclear chromosomes are present as either one copy per cell (haploid) or two copies (diploid), organelle chromosomes are present in many copies per cell, often in the hundreds or thousands. The regulation of this cellular copy number is relatively loose; so in different cells of the same organism there is some variation around a mean value.

The leaf cells of the garden beet have about 40 chloroplasts per cell. The chloroplasts themselves contain specific areas that stain heavily with DNA stains; these areas are called **nucleoids,** and they are a feature commonly found in many organelles. Each beet chloroplast contains from 4 to 8

Figure 21-1 Structures of (a) a typical animal cell and (b) a typical plant cell.

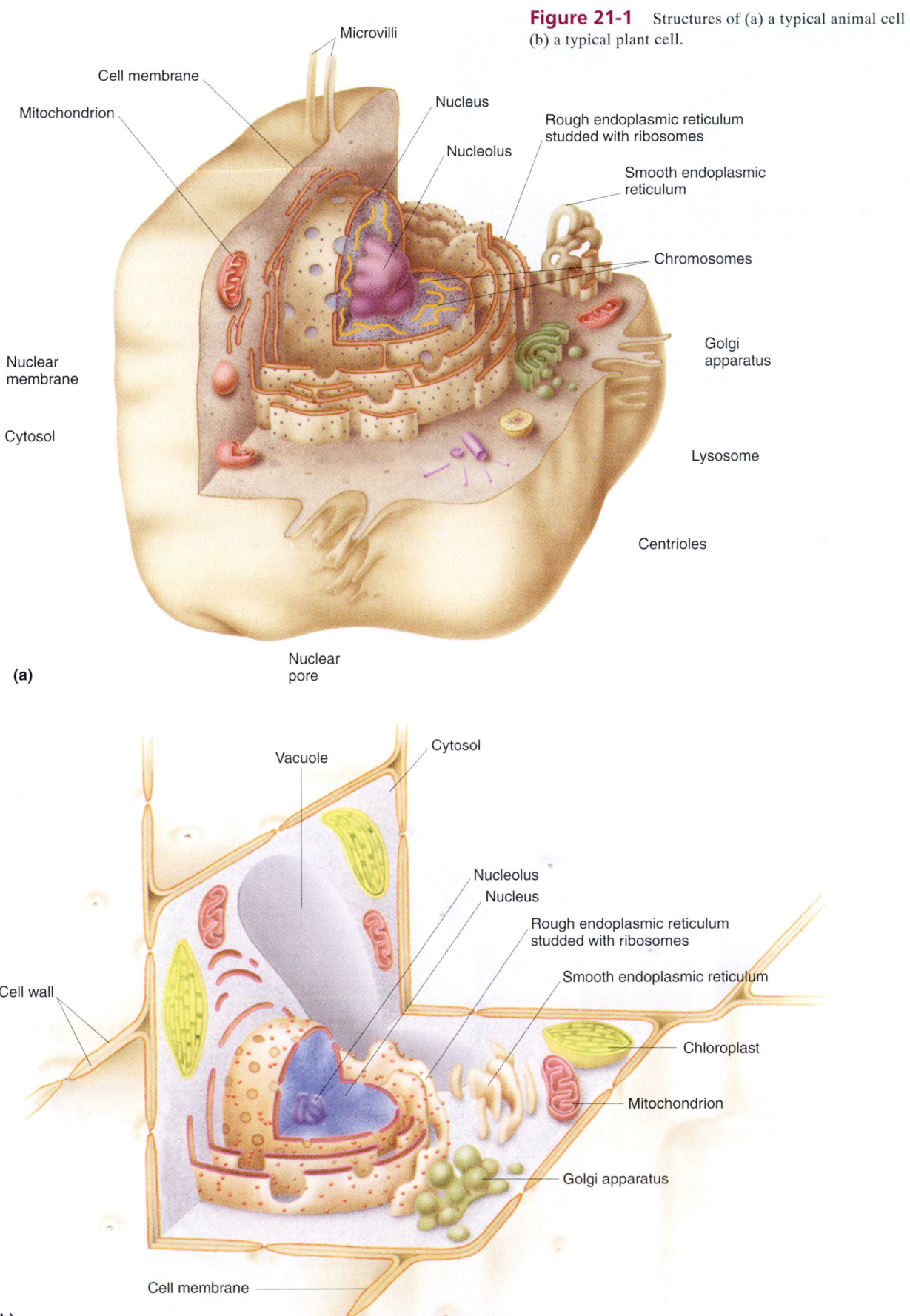

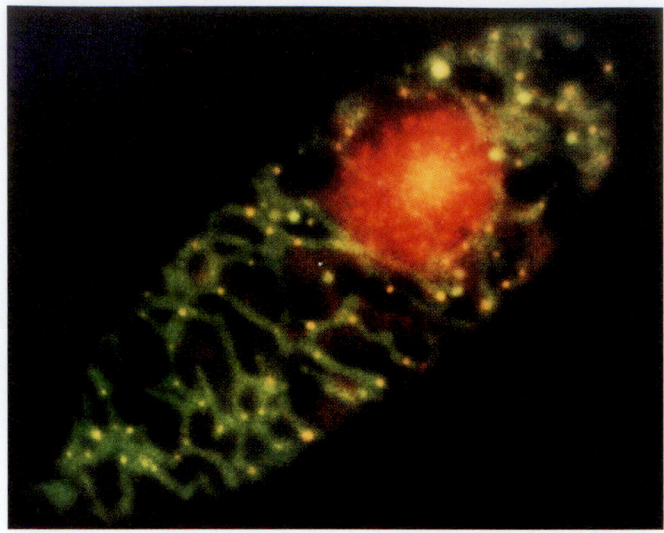

Figure 21-2 Fluorescent staining of a cell of *Euglena gracilis*. With the dyes used, the nucleus appears red because of the fluorescence of large amounts of nuclear DNA. The mitochondria fluoresce green, and within mitochondria the concentrations of mtDNA (nucleoids) fluoresce yellow. (From Y. Huyashi and K. Veda, *Journal of Cell Science* 93, 1989, 565.)

nucleoids, and each nucleoid contains from 4 to 18 cpDNA molecules. Thus, single cells of a beet leaf can contain as many as $40 \times 8 \times 18 = 5760$ copies of the chloroplast genome. Although the photosynthetic protist *Chlamydomonas* has only one chloroplast per cell, the chloroplast contains from 500 to 1500 cpDNA molecules, commonly observed to be packed in nucleoids.

What about mitochondria? A "typical" haploid yeast cell can contain from 1 to 45 mitochondria, each having from 10 to 30 nucleoids, with 4 or 5 molecules in each nucleoid. The mitochondrial nucleoids of the unicellular *Euglena gracilis* are shown in Figure 21-2. In human cells, there can be from 2 to 10 mtDNA molecules per mitochondrion. The number of mitochondria per cell is different in different cell types. Hence both number per mitochondrion and number of mitochondria vary. There are several hundred mtDNA molecules in human fibroblast cells and approximately 100,000 in human oocytes.

Mitochondrial genomes

The yeast and human mitochondrial genomes are shown in Figure 21-3. This view of the mitochondrial genome shows

Figure 21-3 Maps of yeast and human mtDNAs. Each map is shown as two concentric circles corresponding to the two strands of the DNA helix. Note that the mutants used in yeast mtDNA analysis are shown opposite their corresponding structural genes. Green = exons and uninterrupted genes, red = tRNA genes, and yellow = URFs (unassigned reading frames). tRNA genes are shown by their amino acid abbreviations; ND genes encode subunits of NADH dehydrogenase. (Note that the human map is not drawn to the same scale as the yeast map.)

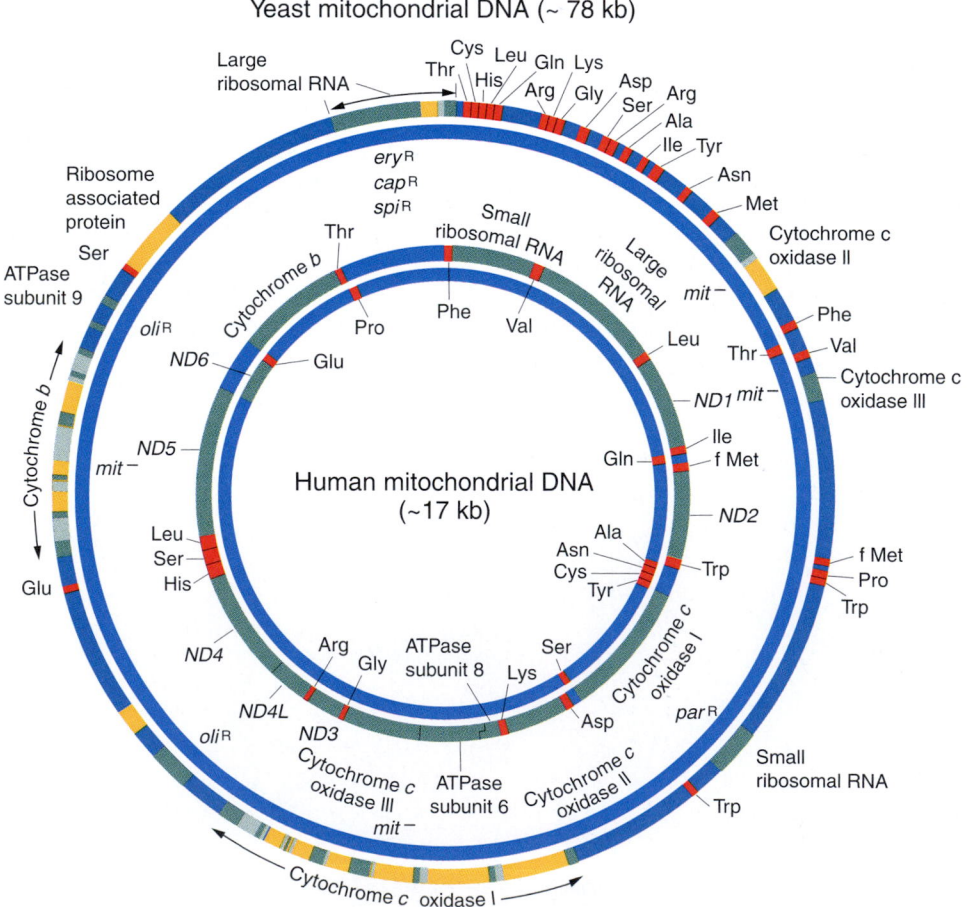

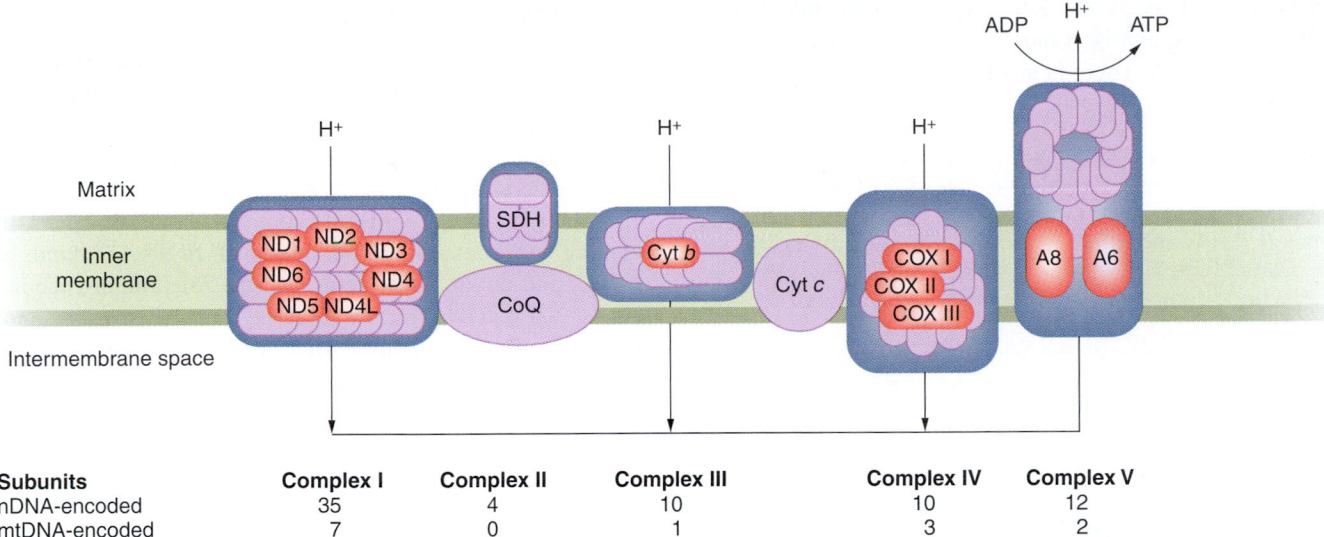

Subunits	Complex I	Complex II	Complex III	Complex IV	Complex V
nDNA-encoded	35	4	10	10	12
mtDNA-encoded	7	0	1	3	2

Figure 21-4 The mitochondrial respiratory chain. Nuclear DNA-encoded subunits are in red, mtDNA-encoded subunits are in purple. Protons (H^+) are pumped from the matrix to the intermembranous space through complexes I, III, and IV. They flow back into the matrix through complex V (ATP synthetase) with the concomitant production of ATP. Coenzyme Q10 (CoQ) and cytochrome c (Cyt c) are nucleus-encoded electron-transfer proteins. ND, NADH dehydrogenase; SDH, succinate dehydrogenase; COX, cytochrome oxidase; A6 and A8, subunits of the ATP synthetase complex V. (After S. DiMauro et al., "Mitochondria in Neuromuscular Disorders," *Biochimica et Biophysica Acta* 1366, 1998, 199–210.)

it to have two main functions: (1) it encodes some of the proteins that constitute the oxidative phosphorylation system and (2) it encodes tRNAs, rRNAs, and some proteins used in mitochondrial protein synthesis.

Yet it is striking that not all the components of the oxidative phosphorylation system are encoded in the mtDNA. The remaining proteins are encoded by nuclear genes, and the mRNA is translated outside the mitochondrion on cytosolic ribosomes. Proteins made on these cytosolic ribosomes are transported into the mitochondrion, and the complete system is assembled in the mitochondrial inner membrane (Figure 21-4).

Genes for 25 yeast and 22 human mitochondrial tRNAs are shown on the maps in Figure 21-3. These tRNAs carry out all the translation that takes place in mitochondria. They are far fewer than the minimum of 32 required to translate nucleus-derived mRNA. The economy is achieved by a "more wobbly" wobble pairing of the tRNA anticodons (see Chapter 10). The tRNA specificities in human mtDNA are shown in Figure 21-5. Notice that the codon assignments are in some cases different from those of the nuclear code. The code also varies in different species. Hence, the genetic code is not universal, as had been supposed for many years.

The map contains some other surprises. Most prominent are the introns in several mitochondrial genes of yeast. Subunit I of cytochrome oxidase contains nine introns. The discovery of introns in the mitochondrial genes of yeast is particularly surprising because they are relatively rare in yeast nu-

Second letter

First letter		U	C	A	G	Third letter
U		Phe	Ser	Tyr	Cys	U
		Phe	Ser	Tyr	Cys	C
		Leu	Ser	*Stop*	*(Stop)* Trp	A
		Leu	Ser	*Stop*	Trp	G
C		Leu	Pro	His	Arg	U
		Leu	Pro	His	Arg	C
		Leu	Pro	Gln	Arg	A
		Leu	Pro	Gln	Arg	G
A		Ile (Met)	Thr	Asn	Ser	U
		Ile	Thr	Asn	Ser	C
		(Ile) Met	Thr	Lys	(Arg) *Stop*	A
		Ile	Thr	Lys	(Arg) *Stop*	G
G		Val	Ala	Asp	Gly	U
		Val	Ala	Asp	Gly	C
		Val	Ala	Glu	Gly	A
		Val	Ala	Glu	Gly	G

Figure 21-5 The genetic code of the human mitochondrion. The functions of the 22 tRNA types are shown by the 22 boxes that do not contain stop codons. Cases in which the nuclear code differs are given in parentheses.

clear genes. Another surprise is the existence of unassigned reading frames (URFs) within the yeast introns. URFs are sequences that have correct initiation codons and are uninterrupted by stop codons. Some URFs within introns appear to specify proteins important in the splicing out of the introns themselves at the RNA level. Notice that human mtDNA is by comparison much smaller and more compact than yeast mtDNA. There is much less spacer DNA between the genes.

> **MESSAGE** ·····································
> Mitochondria contain multiple copies of mtDNA, a circular molecule with genes for mitochondrial protein synthesis (mainly rRNAs and tRNAs) and for subunits of the proteins associated with mitochondrial ATP production. Less understood regions include the introns, unassigned reading frames, and spacer DNA.

Chloroplast genomes

Figure 21-6 shows the organization and functions of most of the genes of the cpDNA from the liverwort *Marchantia polymorpha*. Typically, cpDNA molecules range from 120 to 200 kb in different plant species. In *Marchantia*, the molecular size is 121 kb.

The *Marchantia* molecule contains about 136 genes, including those encoding four kinds of rRNA, 31 kinds of tRNA, and about 90 proteins. Of the 90 protein-encoding genes, 20 encode photosynthesis and electron-transport functions. Genes encoding translational functions take up

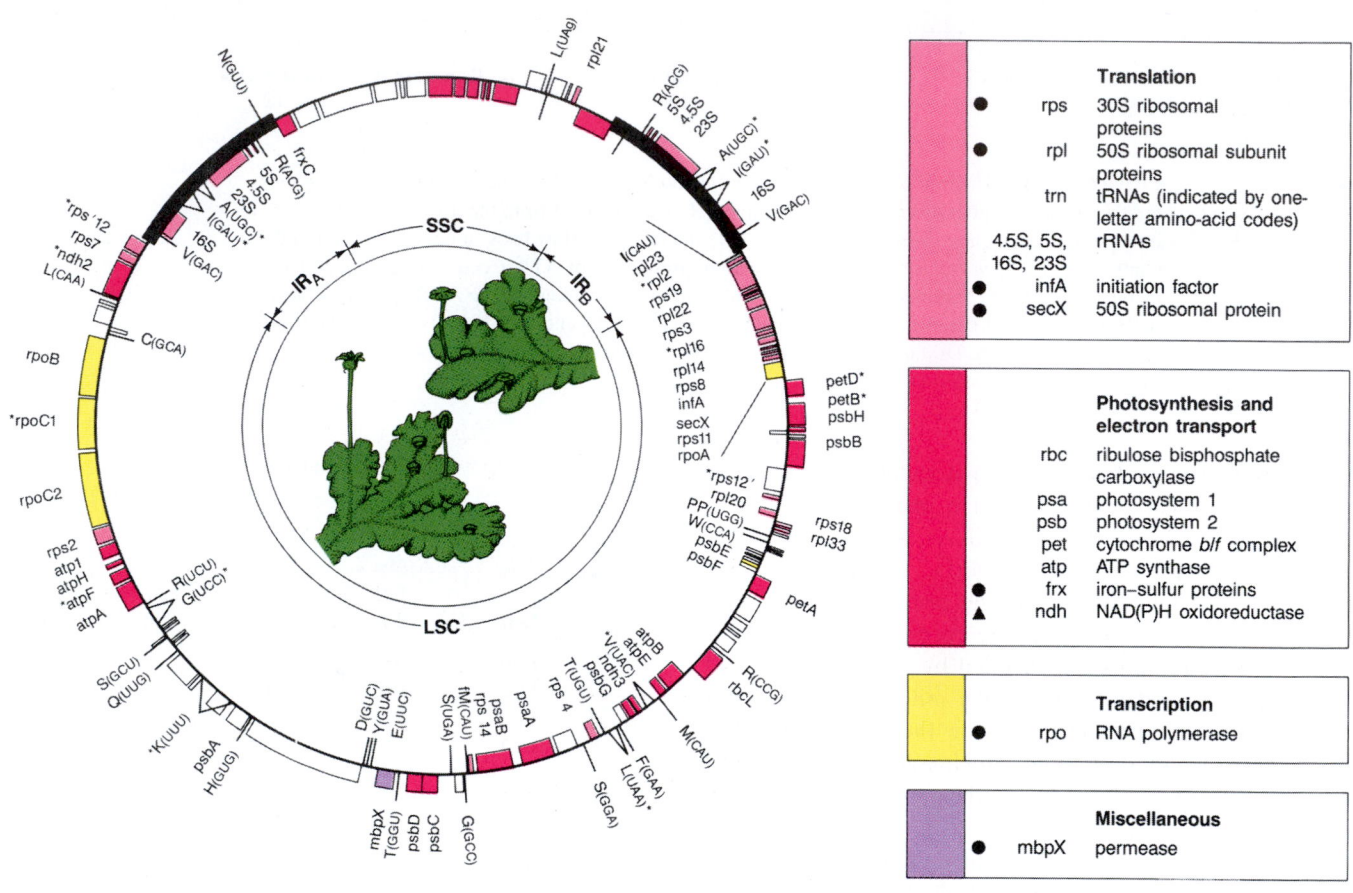

Figure 21-6   The chloroplast genome of the liverwort *Marchantia polymorpha*. IR_A and IR_B, LSC, and SSC on the inner circle indicate the inverted repeats, large single-copy regions, and small single-copy regions, respectively. Genes shown inside the map are transcribed clockwise, and those outside are transcribed counterclockwise. Genes for rRNAs in the IR regions are represented by 16S, 23S, 4.5S, and 5S. Genes for tRNAs are indicated by the one-letter amino acid code with the unmodified anticodon. Protein genes identified are indicated by gene symbols, and the remaining open boxes represent unidentified ORFs, approximately to scale. Genes containing introns are marked with asterisks. The boxes to the right of the gene map summarize the functions of the genes identified to date; groups with related functions are shown in different shades. ORFs are shown in color. Genes identified on the basis of homology with genes from bacterial or mitochondrial genomes are marked ● and ▲, respectively. The drawing in the center of the map depicts a male *(upper)* and a female *(lower) Marchantia* plant. The antheridia and archegonia are elevated on specialized stalks above the thallus, which contains the chloroplasts. *Machantia* can also reproduce asexually: disks of green tissue (gemmae) grow from the bottom of cup-shaped structures on the thallus surface. When mature, the gemmae separate from the thallus and grow to produce new gametophyte plants. (From K. Umesono and H. Ozeki, *Trends in Genetics* 3, 1987.)

about half the chloroplast genome and include those encoding the proteins and RNA types necessary for translation in the organelle.

Notice the presence of a large inverted repeat in Figure 21-6. Such inverted repeats are found in the cpDNA of virtually all species of plants. However, there is some variation in regard to which genes are included in the inverted repeat region and therefore in the relative size of that region. One of the mysteries of the inverted repeat is that the duplicates have exactly the same sequence, yet to date no mechanism is known that ensures this complete identity.

Like mtDNA, cpDNA cooperates with nuclear DNA to provide subunits for functional proteins used inside the organelle. The nuclear components are translated outside in the cytosol and then transported into the chloroplast, where they are assembled together with the components synthesized in the organelle.

MESSAGE ···

Chloroplasts contain multiple copies of cpDNA, a circular molecule that contains genes for photosynthesis, electron transport, and chloroplast protein synthesis. Most cpDNAs contain an inverted repeat.

Organelle mutations

Like nuclear genes, the organelle genes are mutable. Indeed, in mammalian mtDNA, the base-pair-substitution rate is approximately 10 times as high as that of nuclear genes. (Plant organelle DNA does not show these high rates.) Deletions and other rearrangements also are found.

Many of these DNA changes are expressed as abnormal phenotypes at the cellular and organismal level. Because organelles produce energy, a typical mutant phenotype is energy deficiency and hence slow growth or sickly appearance. Mutations in the genes for electron-transport components are often of this type. Mutations in rRNA- or ribosomal protein-encoding genes often lead to resistance to specific drugs such as streptomycin and erythromycin, antibiotics whose effect is exerted by binding to ribosomes. Plant mutations in chloroplast DNA sometimes lead to a white phenotype, indicating a lack of the green photosynthetic pigment chlorophyll. Some examples of mitochondrial and chloroplast mutants follow.

One of the first mitochondrial mutants discovered was a slow-growing mutant of the fungus *Neurospora*. Because of its slow growth, this mutant became known as poky. Although at the time of its discovery the mitochondrial basis of poky was inferred from its inheritance pattern (see next section), it is now known to be a deletion of four base pairs in the gene for the small subunit of mitochondrial rRNA. Because *Neurospora* is an obligate aerobe, it cannot survive without functional mitochondria; so none of the *Neurospora* mutants are nulls, and they retain some function.

In budding yeast, point mutations in some electron-transport proteins cause a slow rate of cell division resulting in small colonies (*mit* mutations — see Figure 21-3 for their location). Deletions of part or even all of the mtDNA also produce small colonies (called petites). Yeast cells can obtain ATP by fermentation, which does not rely on the mitochondrial oxidative phosphorylation system, so yeast with these drastically deleted genotypes can survive, albeit at a reduced activity level. Petites in which part of the mtDNA has been deleted regenerate full-sized mtDNA molecules, as shown in Figure 21-7.

The photosynthetic protist *Chlamydomonas reinhardtii* has been the subject of extensive analysis of mutations in the cpDNA, starting with the work of Ruth Sager in 1954. Sager isolated a large number of antibiotic-resistant and other abnormal phenotypes and, by correlating their unusual inheritance patterns with those of cpDNA, showed that they were almost certainly mutations in cpDNA itself.

In humans, several diseases have been shown to be caused by mutations in mtDNA. In general these diseases are called **mitochondrial cytopathies.** The organs most affected by these mutations are those in which there is a high energy demand, notably muscles and nerves. The mutations are either point mutations in individual mitochondrial genes or large deletions. The positions of some of these mutations are shown in Figure 21-8. The common deletions almost certainly arise from a crossover between direct repeats (Figure 21-9). (This same mechanism has also been detected in fungal mtDNA and plant cpDNA, where both the large and small circular products have been found.)

Myoclonic epilepsy and ragged red fiber (MERRF) disease is an example of a human disease resulting from a mitochondrial point mutation — in this case, nearly always a substitution of G for A at position 8344 in the tRNA gene for lysine. It is a muscle disease (myopathy), but symptoms also include eye and hearing disorders. The ragged red muscle fibers show absence of oxidative phosphorylation.

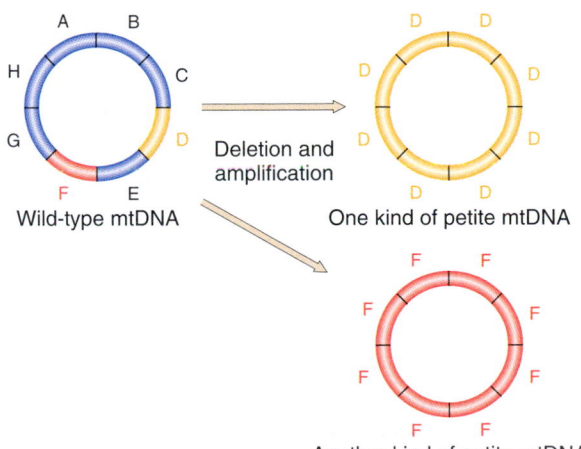

Figure 21-7   When a petite is produced from a wild-type cell, a large region of mtDNA may be deleted. The DNA region retained by the petite (D or F in these examples) is amplified through tandem duplication to provide a chromosome of approximately the normal size.

Figure 21-8 Map of human mitochondrial DNA (mtDNA) showing loci of mutations leading to cytopathies. Single letters are one-letter abbreviations for amino acids. ND, NADH dehydrogenase; COX, cytochrome oxidase; 12S and 16S, ribosomal RNAs. (After S. DiMauro et al., "Mitochondria in Neuromuscular Disorders," *Biochimica et Biophysica Acta* 1366, 1998, 199–210.)

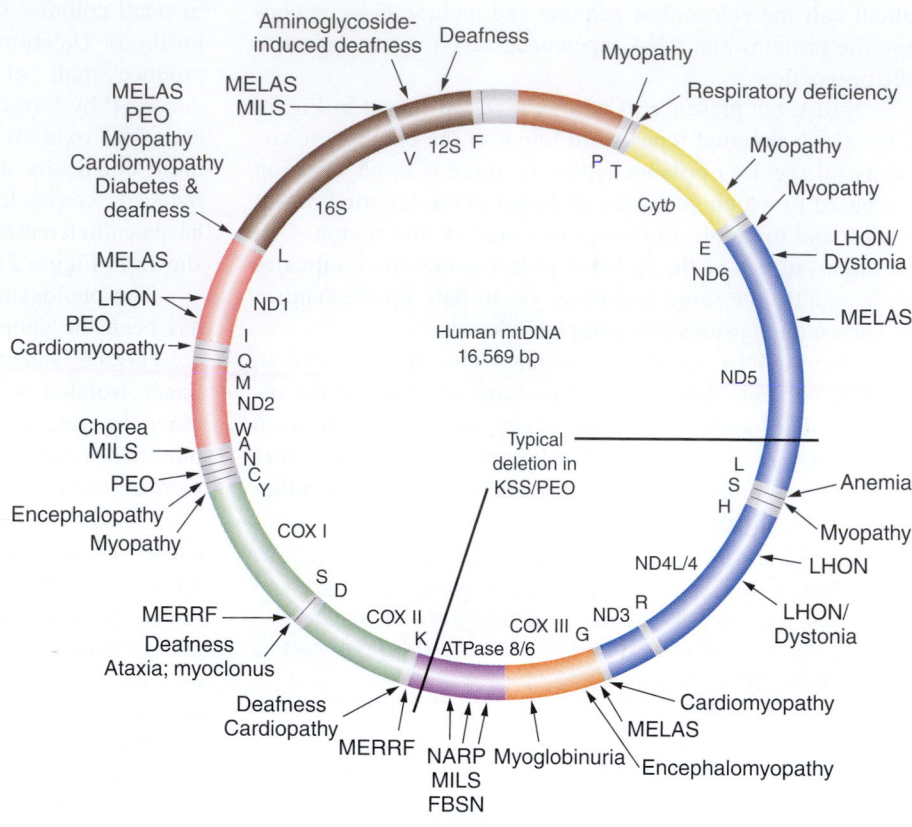

Diseases:

MERRF	Myoclonic epilepsy and ragged red fiber disease
LHON	Leber hereditary optic neuropathy
NARP	Neurogenic muscle weakness, ataxia, and retinitis pigmentosum
MELAS	Mitochondrial encephalomyopathy, lactic acidosis, and strokelike symptoms
MMC	Maternally inherited myopathy and cardiomyopathy
PEO	Progressive external opthalmoplegia
KSS	Kearns-Sayre syndrome
MILS	Maternally inherited Leigh syndrome

Kearns-Sayre (KS) syndrome is a constellation of symptoms affecting eyes, muscles, heart, and brain. It is nearly always associated with a deletion in mtDNA.

Mutations in the nuclear components of the oxidative phosphorylation system of mitochondria and the photosynthetic complexes of chloroplasts result in many of the same types of phenotypic expression as those of their organelle-encoded counterparts. For example, chlorophyll-less mutants of plants are often caused by defective nuclear genes that encode some aspect of chlorophyll structure or function (Figure 21-10). These mutations, however, are inherited in a strict Mendelian manner, as expected of nuclear genes, not in the non-Mendelian manner to be described in the next section.

MESSAGE ·······························

Organelle mutations can result in abnormal growth, abnormal amounts of organelle proteins, defective electron transport, antibiotic resistance, and (in cpDNA) abnormal photosynthesis.

Inheritance of organelle genes and mutations

We shall consider inheritance at three levels — expression, cytoplasmic segregation, and maternal inheritance (Figure 21-11). First, because a cell contains many copies of organelle DNA, it is intuitively difficult to see how a mutation affecting expression can rise to a position where it will influence the phenotype of the cell and the organism. Yet we must remember that organelle DNA replicates even in cells that are no longer dividing. We shall consider some possible fates of mutant organelle DNAs within a cell.

Second, a cell in which both wild-type and mutant organelle DNA coexist is called a **heteroplasmon** or, sometimes, a **cytohet.** When heteroplasmons divide asexually, daughter cells are commonly observed to contain only one or the other organelle DNA type. This type of inheritance is called **cytoplasmic segregation.** We shall consider how this segregation might take place.

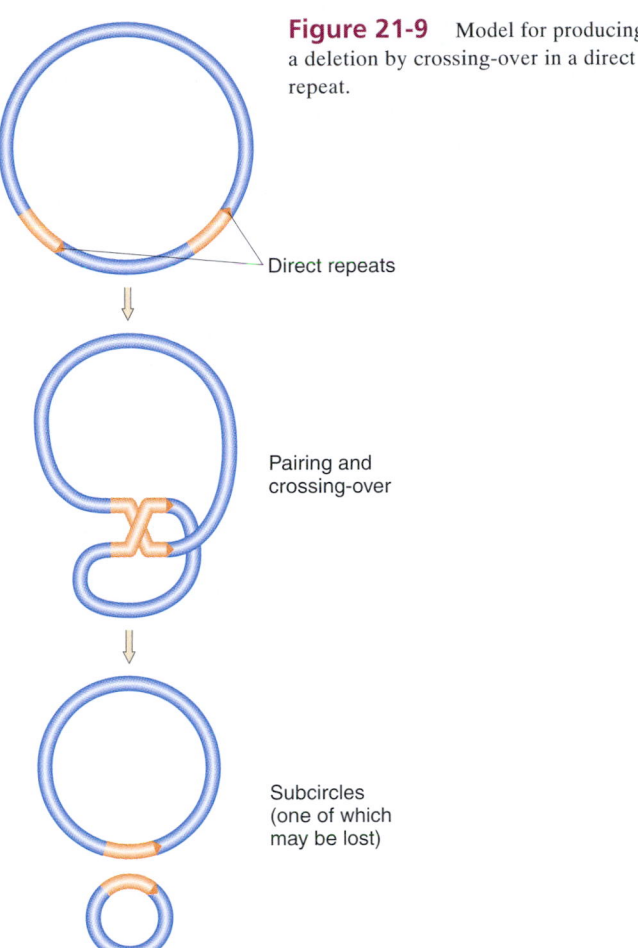

Figure 21-9 Model for producing a deletion by crossing-over in a direct repeat.

Direct repeats

Pairing and crossing-over

Subcircles (one of which may be lost)

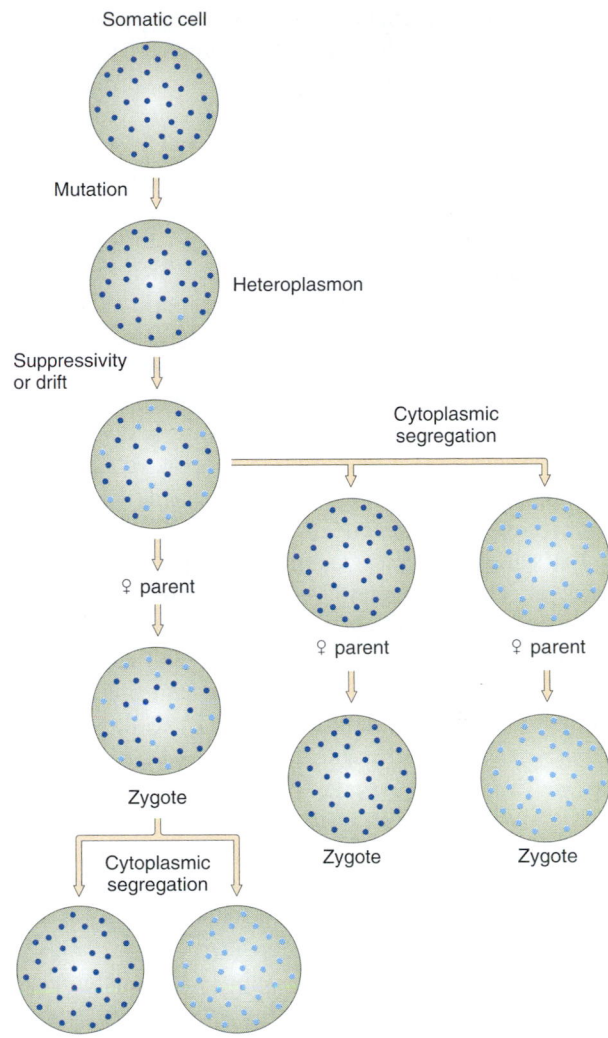

Figure 21-11 The genetic fate of an organelle DNA mutation.

Figure 21-10 Mutations in nuclear genes can result in white leaves. Here a green C/c corn plant, heterozygous for a recessive allele causing albino leaves, was selfed and the progeny were $\frac{1}{4}$ c/c and fully albino. Because they cannot photosynthesize, these albinos die as soon as the food reserve in the seed (laid down by the maternal plant) is exhausted. Such mutations can be in the synthesis of chlorophyll itself or in one of the other nuclear proteins that interact with the chloroplast-encoded proteins to produce a functional photosynthetic light reaction. (Anthony Griffiths.)

The third level of inheritance concerns transmission during the sexual cycle. Organelles are located in the cytoplasm, so it is expected that they will show an inheritance pattern characteristic of that location. In the zygote of an organism that is heterogametic (has different-sized male and female gametes), virtually all the cytoplasm is derived from the egg of the mother. Hence we expect the organelles, organelle DNA, and organelle mutations to follow this cytoplasmic line of descent. This type of inheritance is called **maternal inheritance.**

These three processes are now considered in more detail.

Expression of organelle mutations

How can a mutation rise from a frequency of 1 in several hundred or several thousand to a state in which it can express itself in the phenotype? There are at least three hypotheses that account for this rise.

First, some organelle mutations are **suppressive.** This word does not mean that they act as suppressors (see Chapter 4). Suppressivity means that they can outreplicate the wild-type organelle genomes within a cell. This renegade activity of mitochondrial mutations is unexpected; after all, our experience with nuclear genes is that mutants either die or stay at a low frequency. However, it seems to be a characteristic of certain types of mutant organelle genomes that they can gain ascendancy in the cell in which they arise. A second possibility, for which there is experimental support, is that the frequency of mtDNA types can rise and fall entirely on the basis of chance (called **random drift**). Sometimes the frequency drifts so far that one mtDNA type is completely eliminated. A third possibility is that mitochondria containing certain types of mutations have a mechanism for recognizing that there is a potential energy deficiency and start replicating faster.

Let's look at some examples of suppressive organelle mutations. The first example ever found was a class of *petite* mutations in budding yeast called *suppressive petites.* When petite cells of this type are fused with wild-type cells to make a heteroplasmon, the petite-causing mitochondria gain ascendancy in the mixture and most of the subsequent cells are generally petite in phenotype. These *petite* mutations are the deleted type in which the remaining fragment is tandemly repeated. It is likely that the cause of the suppressivity is that replication origins are duplicated.

In *Neurospora,* the mitochondrial mutation *abn* (abnormal) has the same effect. If cells of this type are fused with wild-type cells even at very lopsided fusion ratios (say 10,000 : 1 in favor of wild type), the heteroplasmic mycelium (which is really one big cell) quite rapidly expresses the mutant abnormal phenotype, and the underlying mtDNA becomes predominant. The same effect can be observed by injecting a small number of *abn* mitochondria into a wild-type cell.

In humans, a parallel situation is found. The frequency of mutant mtDNA forms associated with the mitochondrial cytopathies often changes throughout life, and different parts of the body contain different proportions. This situation is especially pronounced in cells that are postmitotic (that is, will never again undergo mitosis). For example, in one patient an mtDNA deletion was found in the following proportions:

smooth muscle	4%
liver	14%
heart	40%
kidney	40%
skeletal muscle	50%
frontal lobe of brain	44%
cerebellum of brain	14%

There is evidence of a **threshold effect** for some cytopathies; when the frequency of the mutant type rises above this threshold level, the disease symptoms are expressed. Localized high levels of mutant mtDNA can severely disrupt function of organs such as muscle.

Cytoplasmic segregation

The term *cytoplasmic segregation* is used to describe the production of mutant and wild-type descendant cells of a heteroplasmon. Some of the mechanisms proposed for mutant expression in the preceding section also pertain to cytoplasmic segregation. For example, if a mutant type can drift to ascendency in one part of a heteroplasmon, it seems likely that this area can give rise to mutant descendant cells.

Cytoplasmic segregation has been used as the basis for the **heterokaryon test** for mitochondrial mutations in filamentous fungi. If a new mutation has arisen and is suspected of being mitochondrial, a heterokaryon is forced with a wild-type strain. (Forcing means that both components must carry a nuclear auxotrophic marker that prevents growth as a separate strain.) In most fungi, the nuclei never or rarely fuse; so, if cells bearing the nuclear auxotrophic marker of the wild-type partner can be recovered with the new mutant phenotype, that nucleus has likely acquired the mutant phenotype by cytoplasmic contact and subsequent cytoplasmic segregation. Hence the mutation is most likely mitochondrial. Figure 21-12 shows this process.

MESSAGE ···

When a heteroplasmic cell divides, cytoplasmic segregation can occur to produce daughter cells with one organelle DNA or the other.

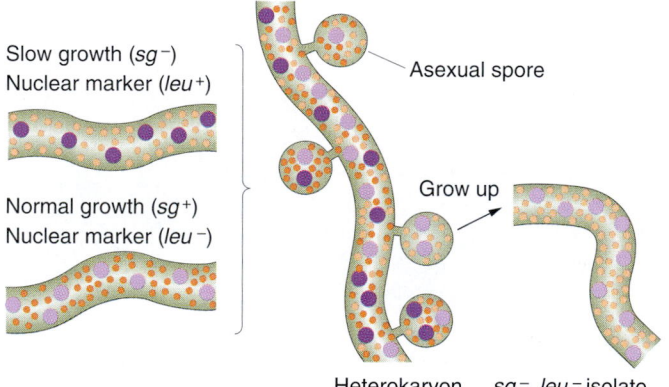

Slow growth (*sg⁻*)
Nuclear marker (*leu⁺*)

Normal growth (*sg⁺*)
Nuclear marker (*leu⁻*)

Asexual spore

Grow up

Heterokaryon   *sg⁻ leu⁻* isolate

Figure 21-12   The heterokaryon test is used to detect extranuclear inheritance in filamentous (threadlike) fungi. A strain with a possible extranuclear mutation (here, *sg⁻*, causing slow growth, indicated by light orange dots) is combined with a strain having a nuclear mutation (*leu⁻*, indicated by light purple circles) to form a heterokaryon. Cultures having the phenotype caused by *leu⁻* can be derived from the heterokaryon. If some of these cultures are also *sg⁻* in phenotype, then *sg* is very likely to be an extranuclear gene, borne in an organelle. Because no nuclear recombination normally occurs in a heterokaryon, the *sg⁻* phenotype must have been acquired by cytoplasmic contact.

Maternal inheritance

This type of inheritance pattern can be illustrated by several landmark crosses in the study of mitochondrial genomes.

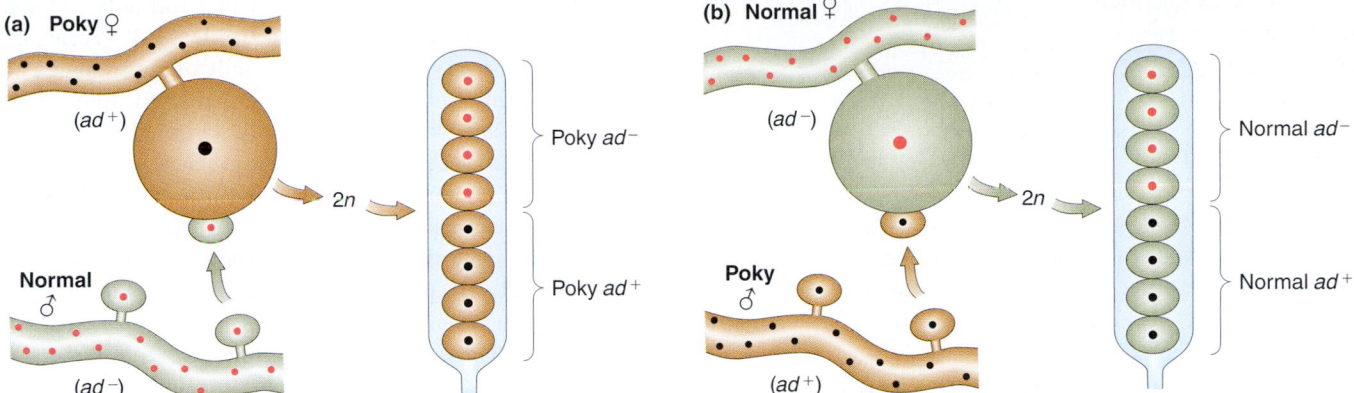

(a) Poky ♀

(ad⁺)

Normal ♂

(ad⁻)

2n →

Poky ad⁻

Poky ad⁺

(b) Normal ♀

(ad⁻)

Poky ♂

(ad⁺)

2n →

Normal ad⁻

Normal ad⁺

Figure 21-13   Explanation of the different results from reciprocal crosses of poky and normal *Neurospora*. The parent contributing most of the cytoplasm of the progeny cells is called *female*. Brown shading represents cytoplasm with the poky determinants. The nuclear locus with the alleles *ad⁺* and *ad⁻* is used to illustrate the segregation of the nuclear genes in the expected 1:1 Mendelian ratio.

Maternal inheritance of poky *Neurospora*. In 1952, Mary Mitchell isolated the poky mutant strain of *Neurospora*. This mutant differs from the wild-type fungus in a number of ways: it is slow growing (as mentioned earlier), and it has abnormal amounts of mitochondrial cytochrome proteins. Like most organisms, wild-type *Neurospora* has three main types of cytochrome: *a*, *b*, and *c*. Poky, however, lacks cytochromes *a* and *b* and has an excess of cytochrome *c*.

Mitchell established the cytoplasmic basis for the *poky* mutation by showing that it was inherited maternally. It is possible to cross some fungi in such a way that one parent contributes the bulk of the cytoplasm to the progeny; this cytoplasm-contributing parent is called the *maternal parent,* even though no genuine male or female parents exist. Mitchell demonstrated maternal inheritance for the poky phenotype in the following reciprocal crosses:

poky female × wild-type male ⟶ all progeny poky
wild-type female × poky male ⟶ all progeny wild type

In such crosses, any nuclear genes that differed between the parental strains were observed to segregate in the normal Mendelian manner and to produce 1:1 ratios in the progeny (Figure 21-13). All poky progeny behaved like the original poky strain, transmitting the poky phenotype down through many generations when crossed as females.

Maternal inheritance of chloroplast pigments in plants. In 1909, Carl Correns reported some surprising results from his studies on variegated four-o'clock plants *(Mirabilis jalapa)*. He noted that most of the leaves of these variegated plants show patches of green and white tissue, but some branches carry only green leaves and others carry only white leaves (Figure 21-14).

Flowers appeared on all types of branches, so Correns intercrossed a variety of different combinations by transferring pollen from one flower to another. Table 21-1 shows the results of such crosses. Two features of these results are

relevant. First, there is a difference between reciprocal crosses: for example, white female × green male gives a different result from green female × white male. Overall, the phenotype of the maternal parent is solely responsible for determining the phenotype of all progeny. The phenotype of the male parent appears to be irrelevant, and its contribution to the progeny appears to be zero. This is a case of strict maternal inheritance. Although the white progeny plants do not live long, because they lack chlorophyll, the other progeny types do survive and can be used in further generations of crosses. In these subsequent generations, maternal inheritance always appears in the same patterns as those observed in the original crosses.

All-white branch

All-green branch

Main shoot is variegated

Figure 21-14   Leaf variegation in *Mirabilis jalapa,* the four-o'clock plant. Flowers can form on any branch (variegated, green, or white), and these flowers can be used in crosses.

	TABLE	Results of Crossing Flowers on Variegated Four-o'clock Plants	
Phenotype of branch bearing egg parent (♀)	Phenotype of branch bearing pollen parent (♂)	Phenotype of progeny
White	White	White
White	Green	White
White	Variegated	White
Green	White	Green
Green	Green	Green
Green	Variegated	Green
Variegated	Green	Variegated, green, or white
Variegated	Green	Variegated, green, or white
Variegated	Variegated	Variegated, green, or white

Figure 21-15 diagrams a model that formally accounts for all the inheritance patterns in Table 21-1. Variegated branches apparently produce three kinds of eggs: some contain only white chloroplasts, some contain only green chloroplasts, and some contain both kinds. The egg type containing both green and white chloroplasts produces a zygote that also contains both kinds of chloroplasts. In subsequent mitotic divisions, the white and green chloroplasts segregate in some cell lines, thus producing the variegated phenotype. Here, again, we see the phenomenon of cytoplasmic segregation.

In most plants, organelles are inherited from the maternal parent. However there are some glaring exceptions in which inheritance is strictly paternal.

Uniparental inheritance in *Chlamydomonas reinhardtii*. The life cycle of *Chlamydomonas* is shown in Figure 21-16. In 1954, Ruth Sager isolated a streptomycin-sensitive mutant with an inheritance pattern that at the time was highly unexpected. In the following crosses, *sm-r* and *sm-s* indicate streptomycin resistance and sensitivity, respectively. The

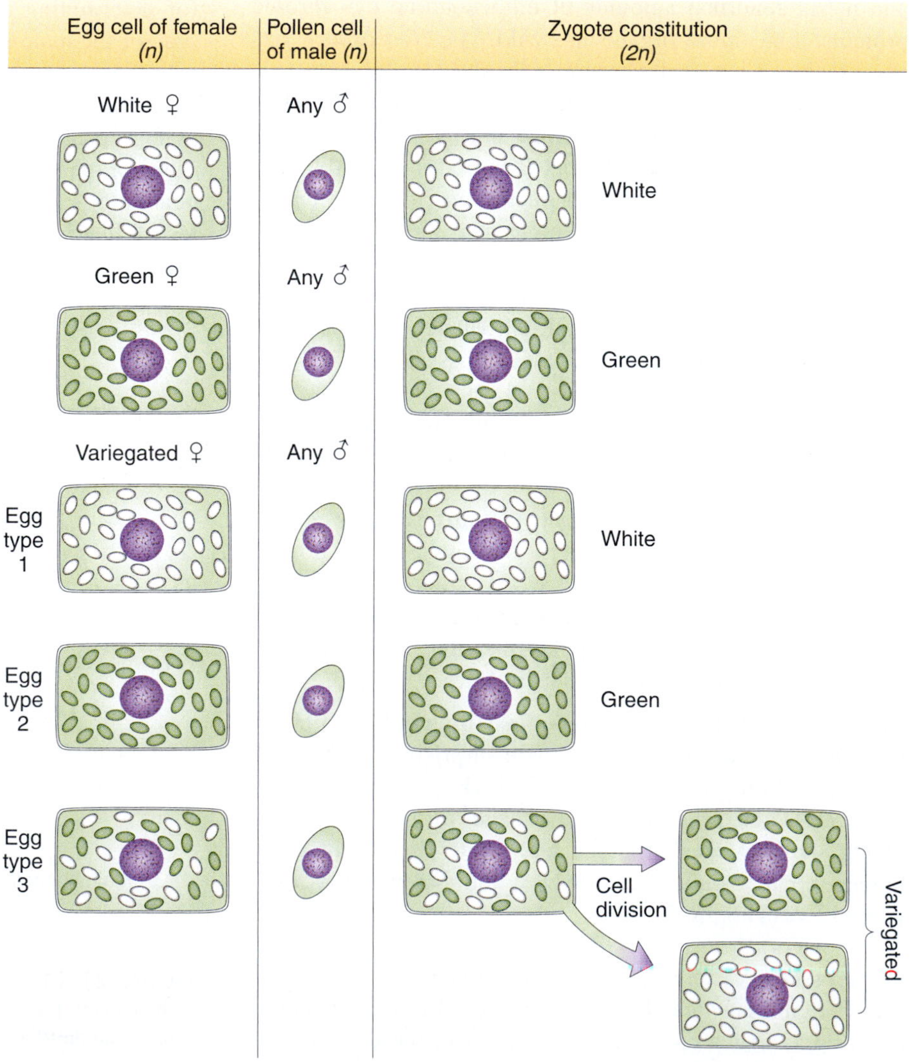

Figure 21-15 A model explaining the results of the *Mirabilis jalapa* crosses in regard to autonomous chloroplast inheritance. The large, dark spheres are nuclei. The smaller bodies are chloroplasts, either green or white. Each egg cell is assumed to contain many chloroplasts, and each pollen cell is assumed to contain no chloroplasts. The first two crosses exhibit strict maternal inheritance. If the maternal branch is variegated, three types of zygotes can result, depending on whether the egg cell contains only white, only green, or both green and white chloroplasts. In the last case, the resulting zygote can produce both green and white tissue, so a variegated plant results.

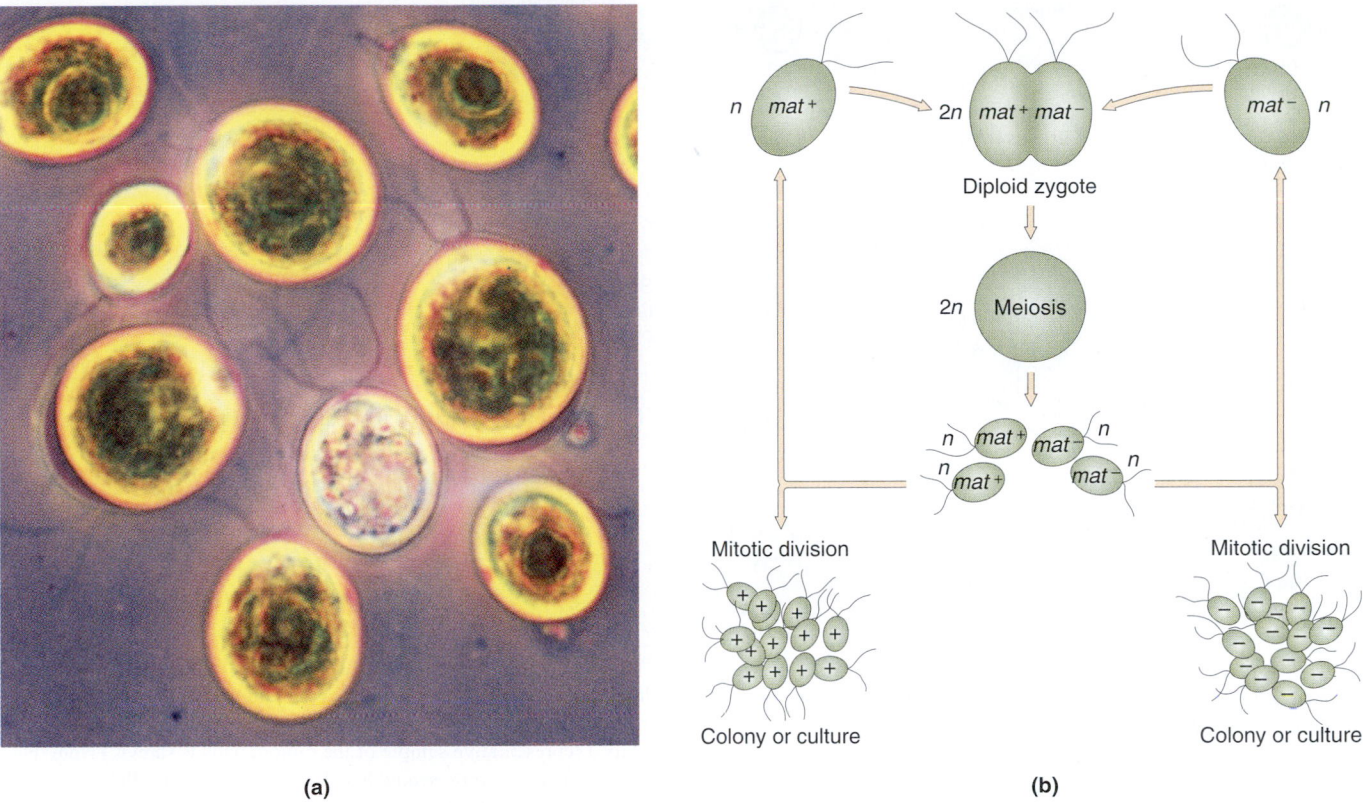

Figure 21-16 (a) Living cells of the photosynthetic protist *Chlamydomonas reinhardtii*. Note the pair of flagella and the large single chloroplast. (b) The life cycle of *Chlamydomonas*. All diploid zygotes are heterozygous for the mating-type alleles *mat*⁺ and *mat*⁻ because only algae differing in these alleles can mate. (Part a from M. I. Walker/Science Source/Photo Researchers.)

mating type gene is *mat,* with alleles + and −. (Crosses can take place only between + and − cultures.)

$$sm\text{-}r\ mat^+ \times sm\text{-}s\ mat^- \longrightarrow \text{all progeny } sm\text{-}r$$
$$sm\text{-}s\ mat^+ \times sm\text{-}r\ mat^- \longrightarrow \text{all progeny } sm\text{-}s$$

Here we see again a difference in reciprocal crosses; all progeny cells show the streptomycin phenotype of the *mat*⁺ parent. Like the maternal-inheritance phenomenon, this is a case of **uniparental inheritance.** In fact, Sager referred to the *mat*⁺ mating type as the female, even though there is no observable physical distinction between the mating types; nor is there a difference in the contribution of cytoplasm as seen in *Neurospora.* In these crosses, the conventional nuclear marker genes (such as *mat* itself) all behave in a Mendelian manner and give 1 : 1 progeny ratios.

Several other mutants (referred to earlier) showed uniparental inheritance. These experiments revealed to Sager the existence of a mysterious "uniparental genome" in *Chlamydomonas* — that is, a group of genes that all show uniparental transmission only through the *mat*⁺ parents in crosses. This uniparental genome is the chloroplast DNA (cpDNA). In a zygote the cpDNA of the *mat*⁻ parent was shown to be destroyed in some way. This destruction can be demonstrated easily by showing that the restriction pattern of cpDNA of the progeny is always the same as the *mat*⁺

parent. This loss of cpDNA from the *mat*⁻ parent parallels the loss of cpDNA genes (such as the *sm* genes) borne by the *mat*⁻ parent.

In *Chlamydomonas,* the mtDNA and its mutations also are inherited uniparentally, but, perversely, the mtDNA is inherited from the *mat*⁻ parent. In other words, in a cross, all the progeny carry the mtDNA genotype of the *mat*⁻ parent.

Uniparental inheritance of mitochondrial mutations in budding yeast. In a yeast cross, the two parental cells fuse and apparently contribute equally to the cytoplasm of the resulting diploid cell (Figure 21-17). Hence a *Neurospora*-style "maternal" type of inheritance based on unequal contribution of cytoplasm is not expected or observed. Furthermore, the inheritance patterns of the mtDNA are independent of mating type. In this sense, then, organelle inheritance in yeast is quite different from that of cpDNA mutations in *Chlamydomonas.* However, a type of uniparental inheritance is shown, as the following examples illustrate.

If a petite of the type that lacks mtDNA is crossed with wild type, none of the progeny are petite. For this reason, these petites are called **neutral petites.** This is a type of uniparental inheritance. However, suppressive petites do produce petite progeny in proportions that correlate with the degree of suppressiveness.

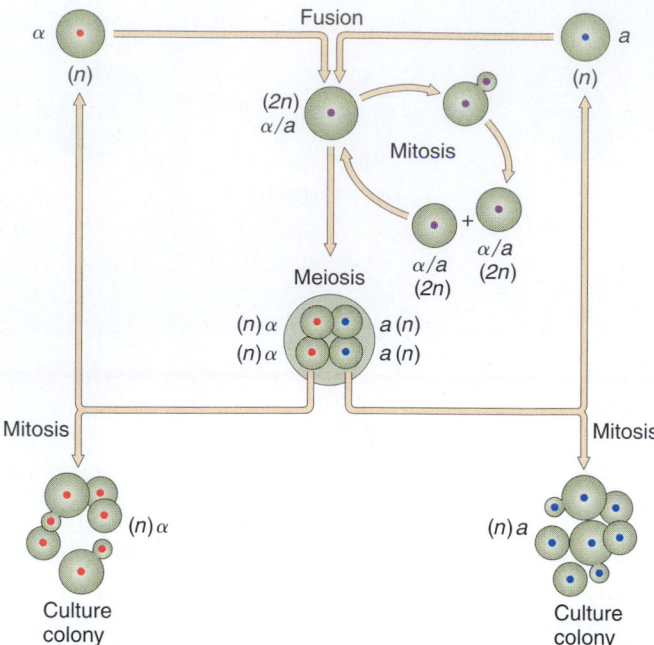

Figure 21-17   The life cycle of baker's yeast (*Saccharomyces cerevisiae*). The nuclear alleles *a* and α determine mating type. Cell fusion between haploid *a* and α cells produces a diploid cell. Normally, the cell then goes through a diploid mitotic cycle (budding). However, the cell can be induced (by plating on a special medium) to undergo sporulation, producing haploid products. Meiosis takes place during sporulation. Note that budding consists of the formation of a small growth on the side of the parent cell; this bud eventually enlarges and separates to become one of the daughter cells.

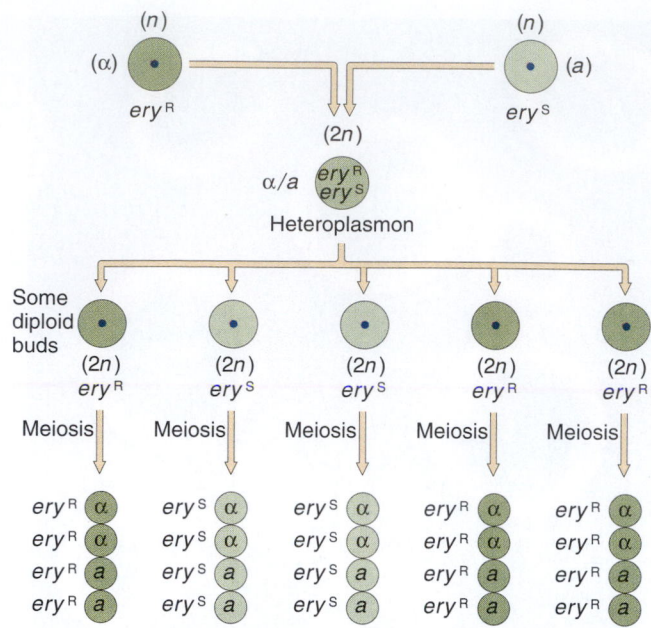

Figure 21-18   The special inheritance pattern shown by certain drug-resistant phenotypes in yeast. When diploid buds undergo meiosis, the products of each meiosis show uniparental inheritance. (Only a representative sample of the diploid buds is shown.) Note that the nuclear genes, represented here by the mating-type alleles *a* and α, segregate in a strictly Mendelian pattern. The alleles ery^R and ery^S determine erythromycin resistance and sensitivity, respectively.

MESSAGE ··
Organelle DNA and associated phenotypes are inherited uniparentally, most often through the maternal parent.
··

Recombination of extranuclear DNA

Recombination of mtDNA can take place in mitochondrially "dihybrid" heteroplasmons. It must be through mitochondrial fusion and a crossing-over-like process, although few molecular details of the process are known. In budding yeast, drug-resistance markers can be used to demonstrate this process, as illustrated in Figure 21-19. The diagram shows that recombination takes place in the zygote (heteroplasmon). When this cell divides mitotically, cytoplasmic segregation results in tetrads that contain either one of the parental mtDNA genotypes or one of the reciprocal recombinant types. The frequency of "recombinant tetrads" was used historically to map mtDNA by using recombination units, but the technique had only limited success and was largely supplanted by physical mapping techniques.

Recombination mapping has also been attempted in *Chlamydomonas*. Sager discovered that, if the *mat⁻* parent is irradiated with UV light, its cpDNA is not inactivated and the zygote is a heteroplasmon. Starting with such heteroplasmons, she was able to obtain recombinant descendant cells, and their frequency in standardized procedures was used to partly map the cpDNA.

In a sense, petites are quite atypical of mitochondrial mutations generally. The drug-resistance and *mit* point mutations more clearly show the inheritance pattern of mitochondrial genomes in this organism.

The inheritance of an erythromycin-resistant mutation is shown in Figure 21-18. The original zygote is effectively a heteroplasmon, consisting of a mixture of the parental cytoplasms. In a yeast cross, zygotes often divide mitotically as diploids before meiosis takes place. In the mitotic divisions, the two mtDNAs undergo cytoplasmic segregation, so the meiocytes are "pure" in regard to their mtDNA type, and all spores in one tetrad will be either ery^R or ery^S. Hence uniparental inheritance is shown at the level of individual meiocytes.

Maternal inheritance of human cytopathies.   Human mtDNA deletions tend to be de novo in origin and are not inherited maternally. However, the various point mutations are inherited maternally. For example, the MERRF mutation can be detected throughout a maternal lineage over several generations. However, because of heteroplasmy, cytoplasmic segregation, and the threshold effect, the members of the family may be severely affected, show only weak symptoms, or show no clinical symptoms at all.

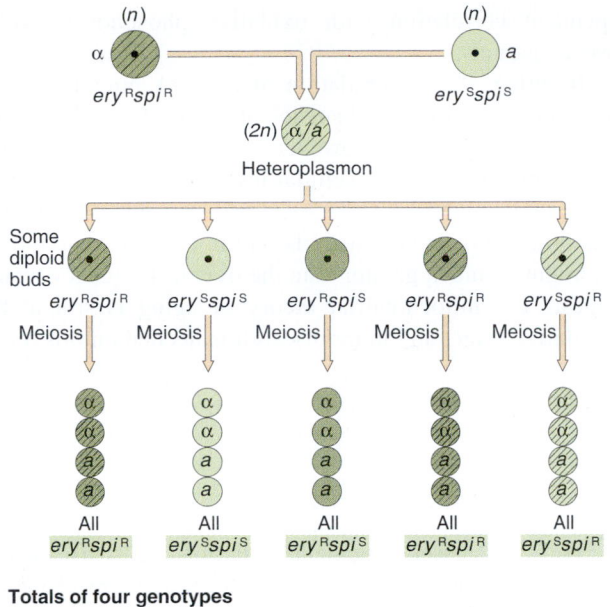

Totals of four genotypes

$ery^R spi^R$	63 tetrads	$ery^S spi^R$	7 tetrads
$ery^S spi^S$	48 tetrads	$ery^R spi^S$	1 tetrad

Figure 21-19 The study of inheritance in a cross between yeast cells differing with respect to two different drug-resistance alleles (ery^R = erythromycin resistance; spi^R = spiramycin resistance). Each diploid bud can be classified as parental or recombinant on the basis of these results. Note that the identity of the extranuclear genotype for all four products of meiosis confirms that cytoplasmic segregation and recombination must occur in the production of the diploid buds. The mitochondrial RF is equal to $(7 + 1)$/total.

Cytoplasmic male sterility

Male sterility in plants is often cytoplasmically based and maternally inherited. Male sterile plants produce no functional pollen, but do produce viable eggs. **Cytoplasmic male sterility** is used in agriculture to facilitate the production of hybrid seed. Hybrid seed is produced from a cross between two genetically different lines; such seeds usually result in larger, more vigorous plants. The main practical problem in producing hybrid seed is to prevent self-pollination, which would produce seeds that are not hybrid. One breeding scheme is illustrated in Figure 21-20.

Mitochondria and aging

Among the theories of the mechanism of aging is the *wear-and-tear theory*. Cells are likened to machines that over time accumulate damage that cannot be fully repaired and, in the end, the machine cannot function and "dies."

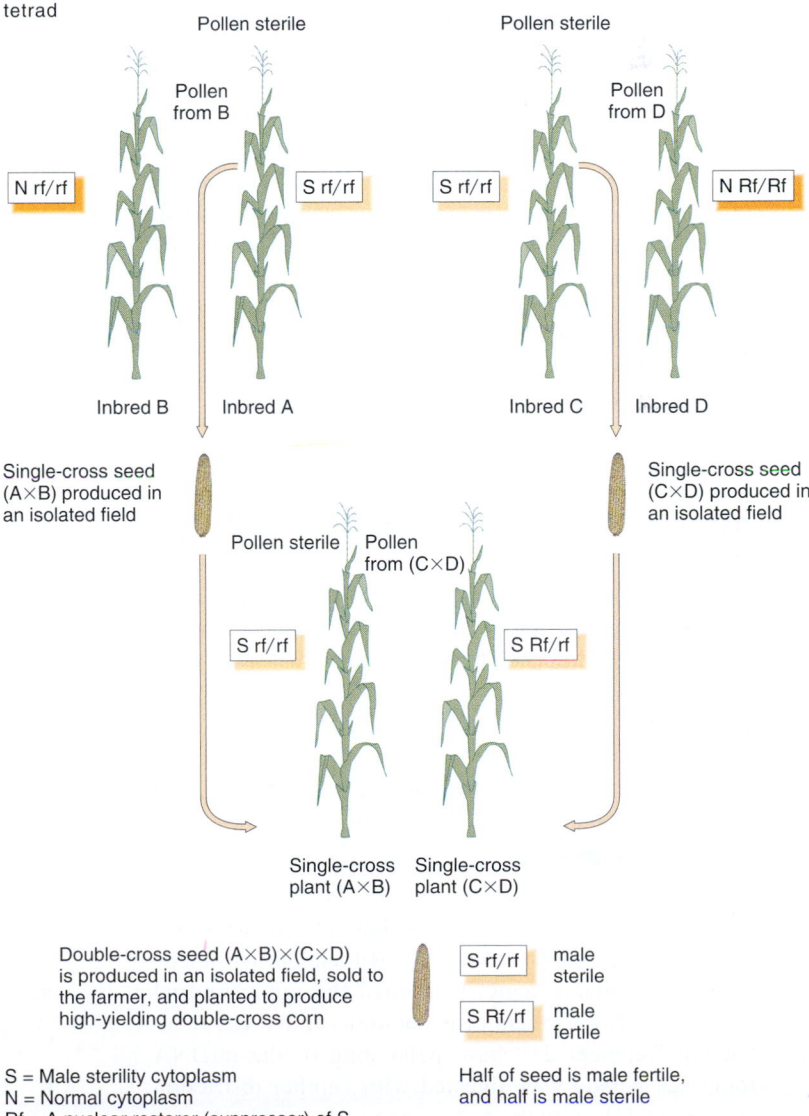

Figure 21-20 The use of cytoplasmic male sterility to facilitate the production of hybrid corn. In this scheme, the hybrid corn is generated from four pure parental lines: A, B, C, and D. Such hybrids are called *double-cross hybrids*. At each step, appropriate combinations of cytoplasmic genes and nuclear restorer genes ensure that the female parents will not self and that male parents will have fertile pollen. (After J. Janick et al., *Plant Science*. Copyright © 1974 by W. H. Freeman and Company.)

Throughout the aging process of animals, there is a reduction in oxidative phosphorylation, the function housed in mitochondria. Furthermore, there is an accumulation of a certain deletion (the "common" 5-kb deletion) and certain point mutations throughout aging. These kinds of observations have suggested the possibility of a connection between mitochondrial mutation and mitochondrial wear-and-tear in aging. Indeed aging might be the ultimate mitochondrial disease. There is some support for this theory. For example, if the mtDNA of human cells is removed and replaced with mtDNA from people of different ages, there is an age-dependent correlation with oxidative phosphorylation in these constructs.

In aging, the accumulation of mtDNA mutations does not seem to rise to a level sufficient to interfere with oxidative phosphorylation. However, the effect of defective mtDNA molecules at the cellular level is poorly understood. For example, if there were localized accumulation within parts of a cell or in certain cells within a tissue, then the effect might be much greater than the overall frequency would suggest. The mitochondrial theory of aging is still at the speculative stage, and more research is needed on this topic.

SUMMARY

Mitochondria contain multiple copies of a small circular "chromosome" of genes whose functions relate to mitochondrial oxidative phosphorylation and to mitochondrial protein synthesis. The set of mitochondrial genes does not provide an adequate set of proteins to carry out these functions. The other necessary proteins are encoded in the nucleus, translated outside the mitochondrion, and imported into the inner mitochondrial membrane. Chloroplasts also contain many copies of a unique circular "chromosome" of DNA that contains genes mainly related to photosynthesis and chloroplast protein synthesis. These genes also interact with nuclear genes to become fully functional. Organelle translation uses a modified genetic code.

Mutations in some organelle genes lead to defects in the energy-producing systems and hence slow or abnormal growth. Mutations in rRNAs and ribosomal proteins are often resistant to specific drugs that bind to the ribosome.

To be expressed, organelle mutations must rise in frequency above an expression threshold. Several mechanisms have been proposed that allow this to happen. In cells that are mixtures of mutant and wild-type organelle DNA, cytoplasmic segregation leads to daughter cells of one type or the other. When individuals carrying organelle mutations are used as parents in a sexual cross, the mutations are generally transmitted exclusively through the maternal parent. In some organisms, there exists a specialized type of uniparental inheritance that is not maternal in nature.

When two different organelle genotypes are combined in the same cell, a crossing-over type of process can take place, leading to recombinant molecules. The frequency of such recombinants has some use in mapping organelle chromosomes.

CHAPTER INTEGRATION PROBLEM

1. The human pedigree below concerns a rare visual abnormality in which the person affected loses central vision while retaining peripheral vision.

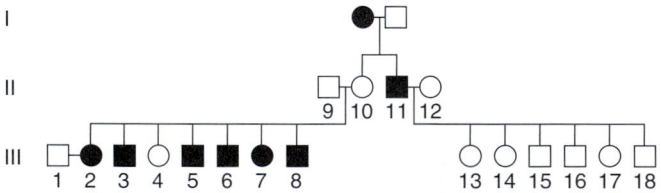

a. What inheritance pattern is shown? Can it be explained by nuclear inheritance? Mitochondrial inheritance? Molecular geneticists studied the mitochondrial DNA of the 18 members of generations II and III. A restriction fragment 212 base pairs long of the mtDNA from each person was digested with another restriction enzyme, *Sfa*N1, with the following results:

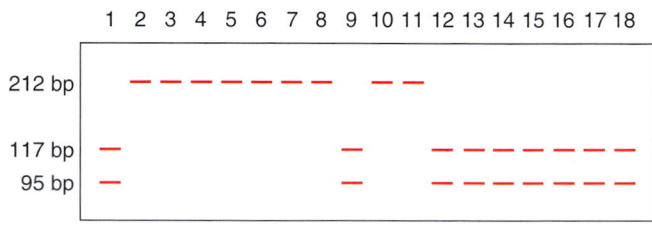

b. What inheritance pattern is shown by these restriction fragments?

c. How does the restriction-pattern inheritance relate to the inheritance of the disease?

d. How can you explain individuals 4 and 10?

e. What is the likely nature of the mutation?

f. How would this analysis be useful in counseling this family?

◆ Solution ◆

a. On the basis of the pedigree alone, it is possible, but unlikely, that the disease is caused by a dominant nuclear allele. But we would have to invoke lack of penetrance in individual 10, who would have to carry the allele because it is passed on to her children. In addition, we have to explain the ratios in generation III. The matings 9 × 10 and 11 × 12 would have to be *A/a* × *a/a*, and the phenotypic ratio of affected to normal then expected among the children in each family is 1:1. So overall this model is not an attractive one for explaining the results.

The results can also be explained by maternal inheritance of the disease. Individuals 4 and 10, however, require special explanation. Once again, we can invoke incomplete penetrance. But, alternatively, we could invoke cytoplasmic segregation; we have learned that cells can be mixtures of normal and abnormal cytoplasmic determinants (here, mitochondria), and cytoplasmic segregation can skew the ratio from cell to cell. The mother in the first generation would have to be a heteroplasmon and by chance pass along predominantly normal mitochondria to her daughter (10) who does not express the disease. Then, by a skewing in the other direction, 10 would pass along mainly abnormal mitochondria to six of her seven children.

b. The restriction patterns clearly show maternal inheritance, which is expected because we are dealing with mtDNA.

c. There is obviously a close correlation between the presence of the large 212-bp fragment and the disease. If this same correlation were to be found in other similar pedigrees, one could formulate a model in which the mutation

that causes the disease simultaneously causes the loss of an *Sfa*N1 restriction site.

d. The possibility that 4 and 10 are heteroplasmons is now less attractive because, if there were mixtures, we would expect to see that the restriction-enzyme patterns of some persons in the family have all three bands — 95, 117, and 212 — but none were seen. Therefore the most likely explanation is the incomplete penetrance of a mitochondrial disease.

e. According to the model, the most likely kind of mutation would be a nucleotide-pair substitution because, if the mutation is at the *Sfa*N1 site, no nucleotides are lost or gained, because 117 + 95 = 212.

f. If the model is upheld by other studies, appearance of the 212-bp fragment after *Sfa*N1 digestion would be a diagnostic marker for the mutation. All women with this marker could pass the disease on to their children, whereas men with the marker could not transmit the disease. Note that, in solving this problem, we have combined concepts of Mendelian inheritance, cytoplasmic inheritance, mutation, and DNA restriction analysis. The problem is based on patterns shown in a true pedigree for the disease Leber hereditary optic neuropathy (LHON), which is mitochondrially based.

SOLVED PROBLEMS

1. In a strain of *Chlamydomonas* that carries the *mt*⁺ allele, a temperature-sensitive mutation arises that renders cells unable to grow at higher temperatures. This mutant strain is crossed to a wild-type stock, and all the progeny of both mating types are temperature sensitive. What can you conclude about the mutation?

◆ Solution ◆

We are told that the mutation arose in a *mat*⁻ stock. Therefore, the cross must have been

$$mat^+ \; ts \times mat^- \; ts^+$$

and the progeny must have been *mat*⁺ *ts* and *mat*⁻ *ts*. This is a clear-cut case of uniparental inheritance from the *mat*⁺ parent to all the progeny. In *Chlamydomonas,* this type of inheritance pattern is diagnostic of genes in chloroplast DNA, so the mutation must have occurred in the chloroplast DNA.

2. Owing to evolutionary conservation, organelle DNA shows homology across a wide range of organisms.

Consequently, DNA probes derived from one organism often hybridize with the DNA of other species. Two probes derived from the cpDNA and mtDNA of a fir tree are hybridized to a Southern blot of the restriction digests of the cpDNA and mtDNA of two pine trees, R and S, that had been used as parents in a cross. The autoradiograms follow (numbers are in kb):

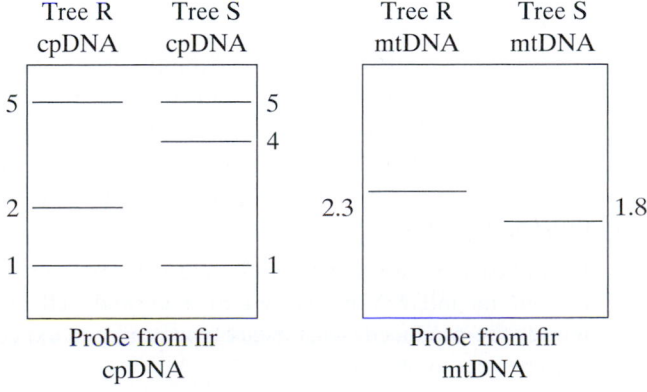

The cross R♀ × S♂ is made, and 20 progeny are isolated. All are identical in regard to their hybridization to the two probes. The autoradiogram for each progeny is:

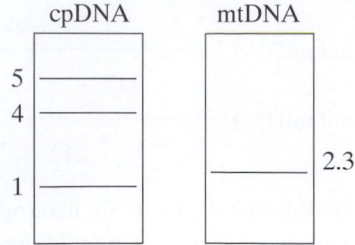

a. Explain the probe hybridization of parents and progeny.

b. Explain the progeny results. Compare and contrast them with the results in this chapter.

c. What do you predict from the cross S♀ × R♂?

Note: This problem is based on results shown in several conifer species.

◆ **Solution** ◆

a. For both cpDNA and mtDNA, the total amount of DNA hybridized by the probe is different in plants R and S. Hence, we can represent the DNA something like this (other fragment arrangements are possible):

Thus, the probes reveal a (presumably neutral) restriction fragment length polymorphism of both the cpDNA and the mtDNA. These RFLPs are useful organelle markers in the cross.

b. We can see that all the progeny have inherited their mtDNA from the maternal parent R, because all show the same R mtDNA fragment hybridized by the probe. This outcome is what we might have predicted, on the basis of the predominantly maternal inheritance encountered in this chapter. However, cpDNA is apparently inherited exclusively paternally, because all progeny show the 5/4/1 pattern of the paternal plant S. This paternal inheritance is surprising, but it is the only explanation of the data. In fact, all the gymnosperms studied so far show paternal inheritance of cpDNA. The cause is unknown, but the phenomenon contrasts with that in angiosperms.

c. From this cross, we can predict that all progeny will show the paternal 5/2/1 pattern for cpDNA and the maternal 1.8-kb band for mtDNA.

PROBLEMS

1. How do the nuclear and organelle genomes cooperate at the protein level?

2. Name and describe two tests for organelle inheritance.

3. What is the basis for the green-white color variegation in the leaves of *Mirabilis?* If the following cross is made,

variegated ♀ × green ♂

what progeny types can be predicted? What about the reciprocal cross?

4. In *Neurospora*, the mutant *stp* exhibits erratic stop-start growth. The mutant site is known to be in the mitochondrial DNA. If an *stp* strain is used as the female parent in a cross to a normal strain acting as the male, what type of progeny can be expected? What about the progeny from the reciprocal cross?

5. If a yeast cell carrying an antibiotic-resistance mutation in its mtDNA is crossed to a normal cell and tetrads are produced, what ascus types can you expect with respect to resistance?

6. A new antibiotic mutation (ant^R) is discovered in a certain yeast. Cells of genotype ant^R are mutagenized, and petite colonies are obtained. Some of these petites prove to have lost the ant^R determinant.

a. What can you conclude about the location of the ant^R gene?

b. Why didn't all the petites lose the ant^R gene?

7. Two corn plants are studied. One is resistant (R) and the other is susceptible (S) to a certain pathogenic fungus. The following crosses are made, with the results shown:

$$S♀ × R♂ \longrightarrow \text{all progeny S}$$
$$R♀ × S♂ \longrightarrow \text{all progeny R}$$

What can you conclude about the location of the genetic determinants of R and S?

8. In *Chlamydomonas,* a certain probe picks up a restriction fragment length polymorphism in cpDNA. There are two morphs, as follows:

Morph 1: two bands, sizes 2 and 3 kb
Morph 2: two bands, sizes 3 and 5 kb

If the following crosses are made,

mat^+ morph 1 × mat^- morph 2
mat^+ morph 2 × mat^- morph 1

what progeny types can be predicted from these crosses? Be sure to draw the DNA morphs with their restriction sites. Also draw a sketch of the autoradiogram.

9. In yeast, the following cross is dihybrid for two mitochondrial antibiotic genes:

$MATa$ oli^R cap^R × $MAT\alpha$ oli^S cap^S

($MATa$ and $MAT\alpha$ are the mating-type alleles in yeast.) What types of tetrads can be predicted from this cross?

10. In the genus *Antirrhinum,* a yellowish leaf phenotype called prazinizans (pr) is inherited as follows:

normal♀ × pr♂ ⟶ 41,203 normal + 13 variegated
pr♀ × normal♂ ⟶ 42,235 pr + 8 variegated

Explain these results on the basis of a hypothesis of cytoplasmic inheritance. (Explain both the majority *and* the minority classes of progeny.)

11. You are studying a plant with tissue comprising both green and white sectors. You wish to decide whether this phenomenon is due (1) to a chloroplast mutation of the type considered in this chapter or (2) to a dominant nuclear mutation that inhibits chlorophyll production and is present only in certain tissue layers of the plant as a mosaic. Outline the experimental approach that you would use to resolve this problem.

12. A dwarf variant of tomato appears in a research line. The dwarf is crossed as female to normal plants, and all the F_1 progeny are dwarfs. These F_1 individuals are selfed, and all the F_2 progeny are normal. Each of the F_2 individuals is selfed, and the resulting F_3 generation is $\frac{3}{4}$ normal and $\frac{1}{4}$ dwarf. How can these results be explained?

13. Assume that diploid plant A has a cytoplasm genetically different from that of plant B. To study nuclear–cytoplasmic relations, you wish to obtain a plant with the cytoplasm of plant A and the nuclear genome predominantly of plant B. How would you go about producing such a plant?

14. Two species of *Epilobium* (fireweed) are intercrossed reciprocally as follows:

♀*E. luteum* × ♂*E. hirsutum* ⟶ all very tall
♀*E. hirsutum* × ♂*E. luteum* ⟶ all very short

The progeny from the first cross are backcrossed as females to *E. hirsutum* for 24 successive generations. At the end of this crossing program, all the progeny still are tall, like the initial hybrids.

a. Interpret the reciprocal crosses.

b. Explain why the program of backcrosses was performed.

15. One form of male sterility in corn is maternally transmitted. Plants of a male-sterile line crossed with normal pollen give male-sterile plants. In addition, some lines of corn are known to carry a dominant nuclear restorer gene (*Rf*) that restores pollen fertility in male-sterile lines.

a. Research shows that the introduction of restorer genes into male-sterile lines does not alter or affect the maintenance of the cytoplasmic factors for male sterility. What kind of research results would lead to such a conclusion?

b. A male-sterile plant is crossed with pollen from a plant homozygous for gene *Rf*. What is the genotype of the F_1? The phenotype?

c. The F_1 plants from part b are used as females in a testcross with pollen from a normal plant (*rf*/*rf*). What would be the result of this testcross? Give genotypes and phenotypes, and designate the kind of cytoplasm.

d. The restorer gene already described can be called *Rf-1*. Another dominant restorer, *Rf-2*, has been found. *Rf-1* and *Rf-2* are located on different chromosomes. Either or both of the restorer alleles will give pollen fertility. With the use of a male-sterile plant as a tester, what would be the result of a cross in which the male parent is:

(i) heterozygous at both restorer loci?

(ii) homozygous dominant at one restorer locus and homozygous recessive at the other?

(iii) heterozygous at one restorer locus and homozygous recessive at the other?

(iv) heterozygous at one restorer locus and homozygous dominant at the other?

16. Treatment with streptomycin induces the formation of streptomycin-resistant mutant cells in *Chlamydomonas*. In the course of subsequent mitotic divisions, some of the daughter cells produced from some of these mutant cells show streptomycin sensitivity. Suggest a possible explanation for this phenomenon.

17. In *Aspergillus,* a "red" mycelium arises in a haploid strain. You make a heterokaryon with a nonred haploid that requires *para*-aminobenzoic acid (PABA). From this heterokaryon, you obtain some PABA-requiring progeny cultures that are red, along with several other phenotypes. What does this information tell you about the gene determining the red phenotype?

18. Adrian Srb crossed two closely related species, *Neurospora crassa* and *N. sitophila*. In the progeny of some of these crosses, a phenotype called aconidial (ac) appeared that lacks conidia (asexual spores). The observed inheritance was:

$$♀N.\ sitophila \times ♂N.\ crassa \longrightarrow \tfrac{1}{2}\ ac, \tfrac{1}{2}\ normal$$
$$♀N.\ crassa \times ♂N.\ sitophila \longrightarrow all\ normal$$

a. What is the explanation for this result? Use symbols to explain all components of your model.

b. From which parent(s) did the genetic determinants for the ac phenotype originate?

c. Why were neither of the parental types ac?

19. Several crosses between poky or nonpoky strains A, B, C, D, and E were made in *Neurospora*. Explain the results of the following crosses, and assign genetic symbols for each of the strains. (Note that poky strain D behaves just like poky strain A in all crosses.)

Cross	Progeny
a. nonpoky B♀ × poky A♂	all nonpoky
b. nonpoky C♀ × poky A♂	all nonpoky
c. poky A♀ × nonpoky B♂	all poky
d. poky A♀ × nonpoky C♂	$\tfrac{1}{2}$ poky, all identical (e.g., D); $\tfrac{1}{2}$ nonpoky, all identical (e.g., E)
e. nonpoky E♀ × nonpoky C♂	all nonpoky
f. nonpoky E♀ × nonpoky B♂	$\tfrac{1}{2}$ poky $\tfrac{1}{2}$ nonpoky

20. The mtDNA of seven cytoplasmic petites was characterized by restriction-enzyme analysis. The results showed that the mtDNA retained in each of these petites was as indicated by the arcs in the diagram below.

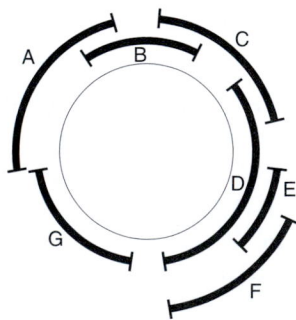

Ten *mit⁻* mutants were fused with each of the petites to make heteroplasmons, and cells from these heteroplasmon cultures were plated onto standard growth medium. Of the 70 combinations, some showed only petite phenotype colonies on the plate (represented by a minus sign in the following grid), but the remainder showed some wild-type colonies as well as petites (represented by a plus sign in the grid). These wild types must have arisen by crossing-over between the petite and the *mit* point mutations. Use these results to locate the approximate positions in the mtDNA of the genes that mutated to give the original *mit⁻* cultures.

	1	2	3	4	5	6	7	8	9	10
A	+	−	−	−	−	−	−	−	+	+
B	−	+	+	−	−	−	−	−	+	−
C	−	−	+	+	+	−	−	−	−	−
D	−	−	−	−	+	−	+	+	−	−
E	−	−	−	−	−	−	+	−	−	−
F	−	−	−	−	−	−	+	−	−	−
G	−	−	−	−	−	−	−	−	−	−

Unpacking the Problem

1. What is mtDNA?

2. Draw an mtDNA molecule, showing at least five specific genes found in mtDNA.

3. What is a cytoplasmic petite?

4. What is the nature of a cytoplasmic petite at the mtDNA level?

5. Why is it appropriate to represent petite DNA by an arc of a circle?

6. Briefly describe a restriction-enzyme analysis that might have been used to determine the extent of the DNA retained by a petite.

7. What is a *mit⁻* mutant? How does a *mit⁻* mutant compare with a petite mutant? Sketch a *mit⁻*, a petite, and a normal colony.

8. Are *mit⁻* mutants drug resistant?

9. Give another word for a heteroplasmon or make up a description term for the concept yourself.

10. How would you make the heteroplasmons in this experiment? Would auxotrophic markers be useful?

11. In what sense is the word *fused* used in this problem?

12. What happens to yeast cells on growth medium?

13. Draw a typical plate representing a + result in the grid.

14. Draw a typical plate representing a − result in the grid.

15. What exactly do the + and − results mean? Do they refer to complementation or recombination?

16. Are all the *mit⁻* mutants different in their behavior?

17. Do all the petites show different behavior in combination with the *mit*⁻ mutants?

18. If some *mit*⁻ mutants show the same behavior, how is this possible if the petites are different according to the restriction analysis?

21. In yeast, an antibiotic-resistant haploid strain *ant*^R arises spontaneously. It is combined with a normal *ant*^S strain of opposite mating type to form a diploid culture that is then allowed to go through meiosis. Three tetrads are isolated:

Tetrad 1	Tetrad 2	Tetrad 3
α *ant*^R	α *ant*^R	*a ant*^S
α *ant*^R	*a ant*^R	*a ant*^S
a ant^R	*a ant*^R	α *ant*^S
a ant^R	α *ant*^R	α *ant*^S

a. Interpret these results.

b. Explain the origin of each ascus.

c. If an *ant*^R grande strain were used to generate petites, would you expect some of the petites to be *ant*^S? Explain your answer.

22. In yeast, two haploid strains are obtained that are both defective in their cytochromes; the mutants are designated *cyt1* and *cyt2*. The following crosses are made:

$$cyt1^- \times cyt1^+$$
$$cyt2^- \times cyt2^+$$

One tetrad is isolated from each cross:

cyt1⁻	*cyt2*⁻
cyt1⁻	*cyt2*⁻
cyt1⁺	*cyt2*⁻
cyt1⁺	*cyt2*⁻

a. From these tetrad patterns, explain the differences in the two underlying mutations.

b. What other ascus types could be expected from each cross?

c. How might the two genes interact at the functional level?

23. The mtDNAs from two haploid strains of baker's yeast are compared. Strain 1 (mating-type α) is from North America, and strain 2 (mating-type *a*) is from Europe. A single restriction enzyme is used to fragment the DNAs, and the fragments are separated on an electrophoretic gel. The sample from strain 1 produces two bands, corresponding to one very large and one very small fragment. Strain 2 also produces two bands, but they are of more intermediate sizes. If a standard diploid budding analysis is performed, what results do you expect to observe in the resulting cells and in the

tetrads derived from them? In other words, what kinds of restriction fragment patterns do you expect?

24. In yeast, some strains are found to have in their cytoplasm circular DNA plasmids that are 2 micrometers (μm) in circumference. In some of these strains, this 2-μm DNA has a single *Eco*RI restriction site; in other strains, there are two such sites. A strain with one site is mated to a strain with two sites. All the resulting diploid buds are found to contain both kinds of 2-μm DNA.

a. Is the 2-μm DNA inherited in the same fashion as mtDNA?

b. If you used radioactive 2-μm DNA as a probe, predict the results of hybridizing it to a Southern blot of *Eco*RI-treated DNA of ascospores from these diploid cells.

25. Circular mitochondrial DNA is cut with two restriction enzymes, A and B, with the following results:

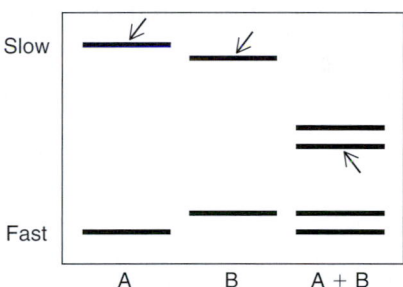

(The arrows indicate the bands that bound a radioactive mt rRNA-derived cDNA probe in a Southern blot.) Draw a rough map of the positions of the restriction site(s) of A and B, and show approximately where the mt rRNA gene is located.

26. You are interested in the mitochondrial genome of a fungal species for which genetic analysis is very difficult but from which mtDNA can be extracted easily. How would you go about finding the positions of the major mitochondrially encoded genes in this species? (Assume some evolutionary conservation for such genes.)

27. a. Early in the development of a plant, a mutation in cpDNA removes a specific *Bg*III restriction site (*B*) as follows:

Normal cpDNA
Mutant cpDNA

In this species, cpDNA is inherited maternally. Seeds from the plant are grown, and the resulting progeny plants are sampled for cpDNA. The cpDNAs are cut with *Bg*III, and Southern blots are hybridized with the probe P shown. The autoradiograms show three patterns of hybridization:

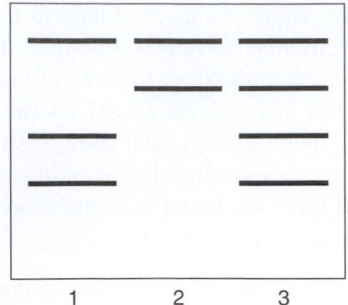

Explain the production of these three seed types.

b. In *Gryllus* (a cricket) and *Drosophila,* rare females have been found that show mixtures of two mtDNA types, differing by the presence or absence of one specific restriction site. In the progeny of these females, each and every individual also proves to be a mixture of the parental mtDNA types. Contrast these results with the results from part a and with other results in this chapter.

28. The mitochondrial genome of the turnip is a large, circular molecule 218 kb in size with a pair of 2-kb direct repeats located about 83 kb apart. However, when turnip mtDNA is examined carefully, three molecular types are seen: the 218-kb circle just described, a 135-kb circle bearing a single 2-kb repeat, and an 83-kb circle also bearing only one copy of the 2-kb repeat. Propose a model to explain the presence of the two smaller molecular types.

29. Reciprocal crosses and selfs were performed between the two moss species *Funaria mediterranea* and *F. hygrometrica.* The appearance of the sporophytes and the leaves of the gametophytes are shown in the accompanying diagram. The crosses are written with the female parent first.

Gametophytes   Sporophytes
(leaves)

F. mediterranea

F. hygrometrica

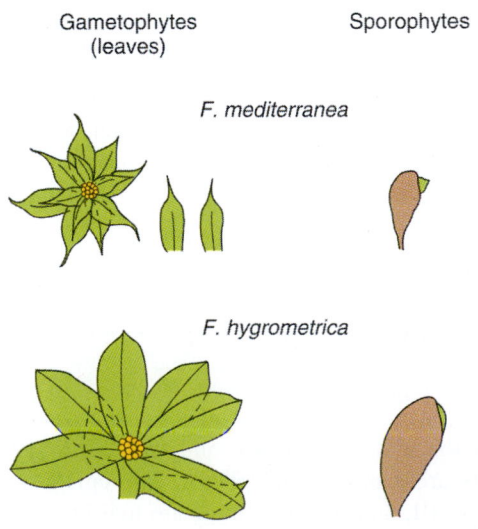

a. Describe the results presented, summarizing the main findings.

b. Propose an explanation for the results.

c. Show how you would test your explanation; be sure to show how it could be distinguished from other explanations.

Gametophytes   Sporophytes
(leaves)

F. mediterranea X *F. hygrometrica*

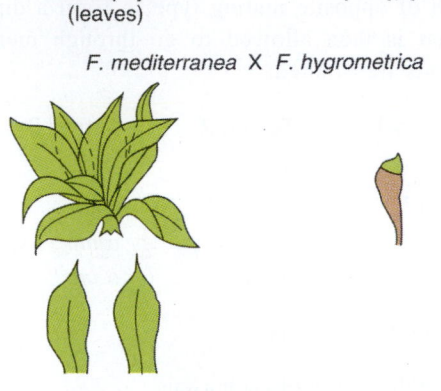

F. hygrometrica X *F. mediterranea*

(Diagrams after C. H. Waddington, *An Introduction to Modern Genetics,* Macmillan, 1939.)

30. The pedigree below shows a very unusual inheritance pattern that actually did exist. All progeny are shown, but the fathers in each mating have been omitted to draw attention to the remarkable pattern.

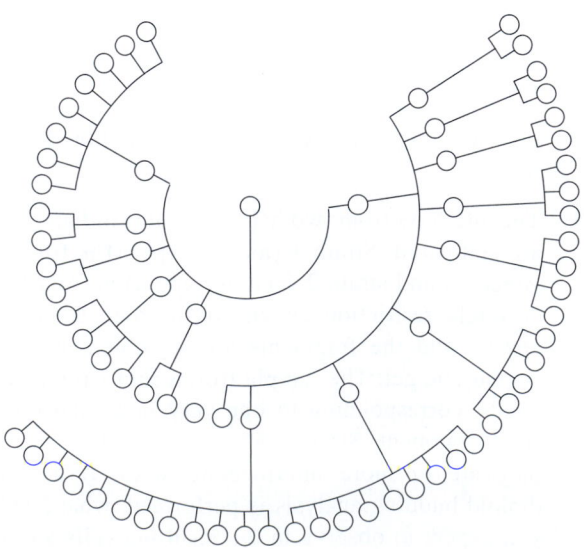

a. State concisely exactly what is unusual about this pedigree.

b. Can the pattern be explained by

    (i)   chance (if so, what is the probability)?

    (ii)   cytoplasmic factors?

    (iii)   Mendelian inheritance?

Explain.

31. Consider the following pedigree for a rare human muscle disease:

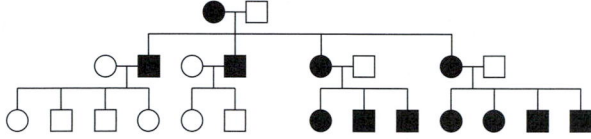

a. What is the unusual feature that distinguishes this pedigree from those studied earlier in this book?

b. Where in the cell do you think the mutant DNA resides that is responsible for this phenotype?

32. The following pedigree shows the recurrence of a rare neurological disease (large black symbols) and spontaneous fetal abortion (small black symbols) in one family. (Slashes mean that the individual is deceased.) Provide an explanation for this pedigree in regard to cytoplasmic segregation of defective mitochondria.

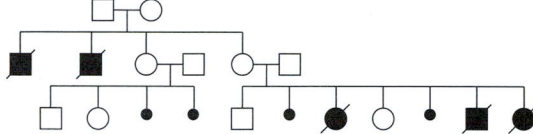

22

CANCER AS A GENETIC DISEASE

Key Concepts

Normal cell proliferation is modulated by cell cycle regulation.

Apoptosis is a normal self-destruction mechanism that eliminates damaged and potentially harmful cells.

Signaling systems permit proliferation and apoptosis to be coordinated within a population of cells.

In cancer, cells proliferate out of control and avoid fail-safe destruction mechanisms through the accumulation of a series of special mutations in the same somatic cell.

Many of the classes of genes that are mutated to cause cancers are important components of the cell that directly or indirectly contribute to growth control and differentiation.

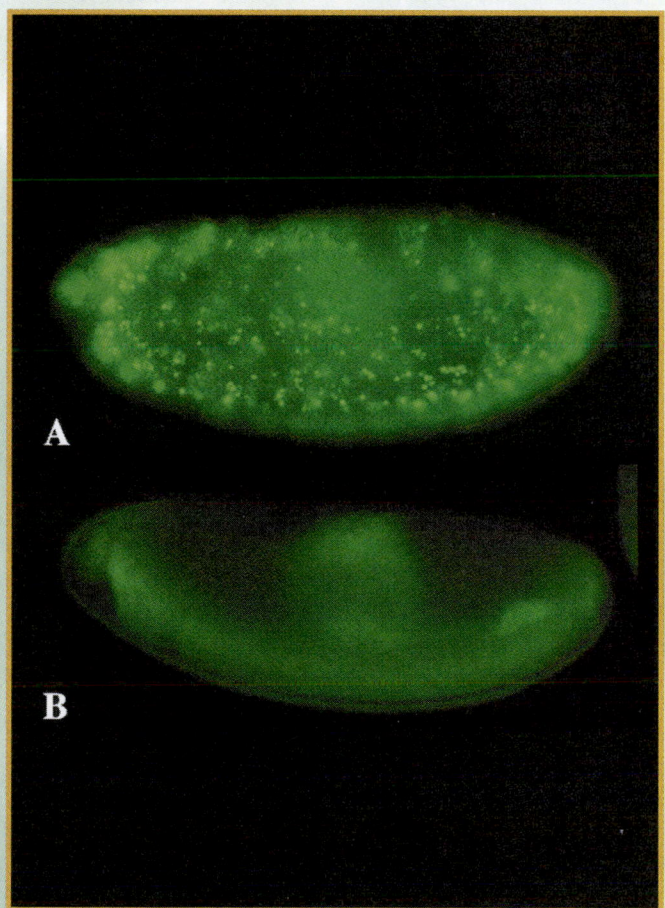

Cell death in *Drosophila* embryo.
(A) Wild-type embryo in which the bright spots are cells carrying out a genetic program to die (apoptosis). (B) Mutant embryo in which this genetic program does not occur. *(Kristin White, Massachusetts General Hospital and Harvard Medical School.)*

In Chapter 11, we learned about some ways in which a cell monitors its status relative to its environment and responds accordingly. For example, by utilizing certain metabolites as allosteric effectors of transcriptional regulatory proteins, an *E. coli* cell can make decisions about which sugar metabolic pathways to implement at any given time. Metazoa (multitissued animals) use steroids and other low-molecular-weight hormones as allosteric effectors of transcriptional regulatory molecules to coordinate appropriate responses of different organs to a particular physiological event.

A major point to remember is that cells have evolved mechanisms that modulate the activity of key target proteins by relatively minor modifications — in the two preceding examples, by forming complexes with allosteric effectors. Much of genetics, indeed much of the biology of a cell, depends on such modulations, in which key proteins are toggled between active and inactive states.

In this chapter and the next one, we shall see this theme exploited in a variety of situations: control of cell numbers, control of developmental pathways, and formation of complex biological patterns. In this chapter, we focus on how such modulations achieve proper control of cell number and how the systems can be overcome by certain classes of mutations to produce uncontrolled proliferation — the diseases that we call cancers.

Cancer and the control of cell number: an overview

Cancer is now clearly understood as a genetic disease of somatic cells. In cancer, the fail-safe mechanisms that are in place to ensure that cell number remains balanced to the needs of the whole organism are subverted, and cancerous cells proliferate out of control. To understand how cells can mutate to a cancerous state, we must first understand the basic mechanisms governing the control of normal cell numbers.

Machinery of cell proliferation

Certain aspects of proliferation control are general to all organisms. Universally, the cell division process has numerous events that must take place sequentially to produce viable progeny cells. Moreover, the cell division cycle has evolved so that there are checks and balances to prevent a subsequent event from taking place before the prerequisite events have been achieved. For example, it would be a lethal event if mitosis occurred before DNA replication was completed. Mechanisms have evolved that prevent such cellular disasters. We shall explore the regulation of the eukaryotic cell cycle. Protein kinases, enzymes that specifically phosphorylate certain amino acid residues on target proteins, and protein phosphatases, enzymes that specifically remove phosphate groups from such amino acid residues, modulate the activities of key proteins in the cell

division cycle. These phosphorylation–dephosphorylation pathways ultimately converge to determine which key proteins are active for a fraction of the entire cell division cycle. Put another way, it is the cyclical variations in these key proteins that determine which parts of the cell cycle are currently being executed.

Machinery of cell death

Some aspects of cell control appear to have evolved only in multicellular organisms. To develop and maintain themselves normally, multicellular organisms must properly balance the numbers of the cell types in their various tissues. Almost all of these cell types are somatic — that is, they do not contribute to the germ line. Loss of such somatic cells is not a problem for the organism from the point of view of propagation of the species, as long as proliferation of the remaining cells of that type in a particular tissue compensates for the cells that are eliminated. Furthermore, abnormal cells have the potential to do considerable harm. Thus, mechanisms have evolved to eliminate certain cells — through a process called programmed cell death or apoptosis. A cascade of enzymes called caspases kill by disrupting numerous structural and functional systems within the cell. Subsequently, the carcasses of the dead cells are removed by scavenger cells.

Linking cell proliferation and death to the environment

The cell proliferation and cell death machinery must be interconnected so that each is activated only under the appropriate environmental circumstances. For example, in adult organs, maintenance of proper cell number requires proper balance between the birth of new cells and the loss of existing ones. Eukaryotic cells have evolved elaborate intercellular signaling pathways to serve as status indicators of the environment. Some signals stimulate proliferation, whereas others inhibit it. Furthermore, other signals can activate apoptosis, whereas still others block activation. Intercellular signaling pathways typically consist of several components: the signals themselves, the receptors that receive the signals, and the signal transduction systems responsible for relaying the signal to various regions of the cell. Just as allosteric effectors regulate the activity of many DNA-binding proteins in bacteria, modifications to the various components of the intercellular signaling systems — protein phosphorylation, allosteric interactions between proteins and small molecules, or interaction between protein subunits — control the activity of these pathways.

Cell proliferation machinery

Cell cycle

There are four main parts to the cell cycle: M phase — mitosis — and the three parts that are components of interphase; G1, the gap period between the end of mitosis and the start of DNA replication; S, the period during which DNA synthesis occurs; and G2, the gap period following DNA replication and preceding the initiation of the mitotic prophase. In mammals, where the cell cycle is particularly well studied, differences in the rate of cell division are largely due to differences in the length of time between entering and exiting G1. This variation is due to an optional G0 resting phase into which G1-phase cells can shunt and remain for variable lengths of time, depending on the cell type and on environmental conditions. Conversely, S, G2, and M phases are normally quite fixed in duration. In this section, we consider the molecules that drive the cell cycle. In a later section, we shall consider how these molecules are integrated into the overall biology of the cell.

Cyclins and cyclin-dependent protein kinases

The engines that drive progression from one step of the cell cycle to the next are a series of protein complexes composed of two subunits: a **cyclin** and a **cyclin-dependent protein kinase** (abbreviated **CDK**). In every eukaryote, there is a family of structurally and functionally related cyclin proteins. Cyclins are so named because each is found only during one or another segment of the cell cycle. The onset of the appearance of a specific cyclin is due to cell-cycle-controlled transcription, in which the previously active cyclin–CDK com-

plex leads to the activation of a transcription factor that activates the transcription of this new cyclin. The disappearance of a cyclin depends on three events: rapid inactivation of the activator of transcription of this cyclin's gene (so that no new mRNA is produced), a high degree of instability of the cyclin mRNA (so that the existing pool of mRNA is eliminated), and a high level of instability of the cyclin itself (so that the pool of cyclin protein is destroyed).

Cyclin-dependent protein kinases also constitute a family of structurally and functionally related proteins. **Kinases are enzymes that add phosphate groups to target substrates; for protein kinases such as CDKs, the substrates are proteins.** CDKs are so named because their activities are regulated by cyclins and because they catalyze the phosphorylation of specific serine and threonine residues of specific target proteins.

The target proteins for CDK phosphorylation are determined by the associated cyclin. In other words, the cyclin tethers the target protein so that the CDK can phosphorylate it (Figure 22-1), thereby changing the activity of each target protein. Because different cyclins are present at different phases of the cell cycle (Figure 22-2), different phases of the cell cycle are characterized by the phosphorylation of different target proteins. The phosphorylation events are transient and reversible. When the cyclin–CDK complex disappears, the phosphorylated substrate proteins are rapidly dephosphorylated by protein phosphatases.

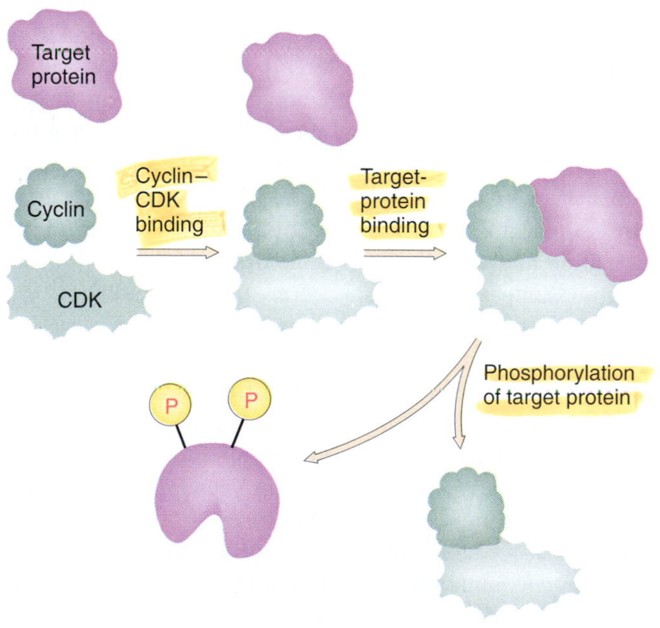

Figure 22-1   The steps in phosphorylation of target proteins by the cyclin–CDK complex. First, a cyclin and a CDK subunit bind to form an active cyclin–CDK complex. Then, the target protein binds to the cyclin part of the complex, placing the target phosphorylation sites in proximity to the active site of CDK. The target protein is then phosphorylated; it is then no longer able to bind to cyclin and is released from the complex.

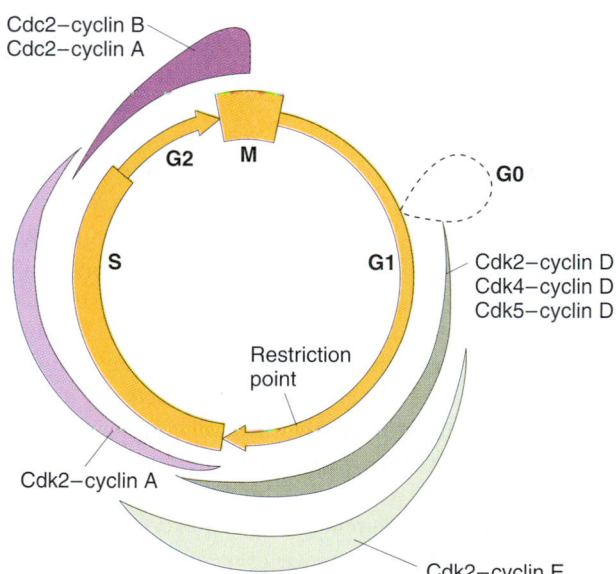

Figure 22-2   A current view of the variations in cyclin–CDK activities throughout the cell cycle of a mammalian cell. The widths of the bands indicate the relative kinase activities of the various cyclin–CDK complexes. Note that several different cyclins (A, B, D, and E) and several different CDKs (Cdc2, Cdk2, Cdk4, and Cdk5) can bind to one another to form different complexes, increasing the array of combinations of cyclin–CDK complexes that can form in the course of the cell cycle. (From H. Lodish, D. Baltimore, A. Berk, S. L. Zipursky, P. Matsudaira, and J. Darnell, *Molecular Cell Biology,* 3d ed. Copyright © 1995 by Scientific American Books.)

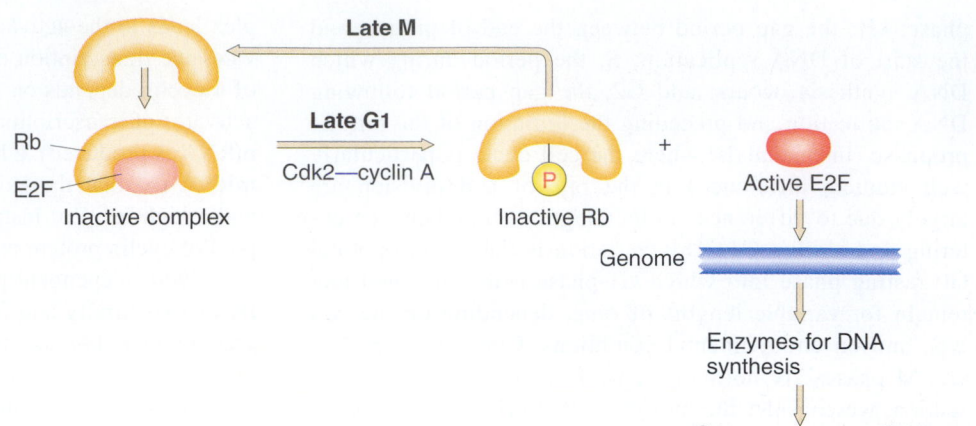

Figure 22-3 The contributions of the Rb and E2F proteins in the regulation of the transition from G1 to S phase in a mammalian cell. (From H. Lodish, D. Baltimore, A. Berk, S. L. Zipursky, P. Matsudaira, and J. Darnell, *Molecular Cell Biology,* 3d ed. Copyright © 1995 by Scientific American Books.)

CDK targets

How does the phosphorylation of some target proteins control the cell cycle? Phosphorylation initiates a chain of events that culminates in the activation of certain transcription factors. These transcription factors promote the transcription of certain genes whose products are required for the next stage of the cell cycle. Much of our knowledge of the cell cycle comes from both genetic studies in yeast (see next section) and from biochemical studies of cultured mammalian cells. A well-understood example is the Rb–E2F pathway in mammalian cells. Rb is the target protein of a CDK–cyclin complex called Cdk2–cyclin A, and E2F is the transcription factor that Rb regulates (Figure 22-3). From late M phase through the middle of G1, the Rb and E2F proteins are combined in a protein complex that is inactive in promoting transcription. In late G1, the Cdk2–cyclin A complex is produced and phosphorylates the Rb protein. This phosphorylation produces a change in the shape of Rb such that it can no longer bind to the E2F protein. The free E2F protein is then able to promote transcription of certain genes that encode enzymes vital for DNA synthesis. This allows the next phase of the cell cycle — S phase — to proceed.

Rb and E2F are in fact representatives of two families of related proteins. In mammals, different cyclin–CDK complexes (Figure 22-2) are thought to selectively phosphorylate different proteins of the Rb family, each of which in turn releases the specific E2F family member to which it is bound. The different E2F transcription factors then promote the transcription of different genes that execute different aspects of the cell cycle.

MESSAGE ••
Sequential activation of different CDK–cyclin complexes ultimately controls progression of the cell cycle.

Yeasts: genetic models for the cell cycle

Genetic contributions to our understanding of the cell cycle have largely come from studies of two fungi: the budding yeast *Saccharomyces cerevesiae* (Figure 22-4) and the fis-

sion yeast *Schizosaccharomyces pombe* (Figure 22-5). In each of these species, cell cycle genetics has revealed a large array of genetic functions needed to maintain the proper cell cycle. These functions are identified as a special class of *ts* (temperature-sensitive) mutations called *cdc* (cell division cycle) mutations

When cultured at low temperature, yeasts with these *cdc* mutations grow normally. When shifted to higher, restrictive temperatures, these *cdc* mutant yeasts no longer grow. What makes these *cdc* mutations novel among the more general class of *ts* mutations is that a particular *cdc* mutant stops growing at a specific time in the cell cycle, and all the yeast cells look alike. Consider some examples in *S. cerevesiae,* a yeast that normally divides through "budding" (Figure 22-4), a process in which a mother cell develops a small outpocketing, a "bud." The bud grows and mitosis occurs such that one spindle pole is in the mother cell and the other is in the bud. The bud continues to grow until it is as big as the mother cell. The mother cell and the bud then separate into two daughter cells. Any run-of-the-mill *ts* mutation in *S. cerevesiae,* when shifted to restrictive temperature, stops growth at variable times in the cycle of bud formation and cell division. In contrast, after a shift to restrictive temperature, one *S. cerevesiae cdc* mutation produces yeast cells that arrest with only tiny buds, whereas another produces yeast cells that arrest with larger buds, half the size of the mother cell. Such different Cdc phenotypes are indicative of different defects in the machinery required to execute specific events in the progression of the cell cycle. In a similar fashion, the fission yeast *S. pombe,* which divides in the more usual symmetrical (fission) fashion to produce two equivalent daughter cells, has been used to generate *cdc* mutation and characterize the cell cycle. Interestingly, the *cdc* genes identified in genetic screens in these two very different yeasts encode the same sets of proteins. In other words, the cell cycle machinery in these two species is essentially identical.

With the completion of the sequencing of the *S. cerevesiae* genome (Chapter 14), we are in the unprecedented position of being able to identify the entire array of proteins of the cyclin and CDK families (22 and 5 members,

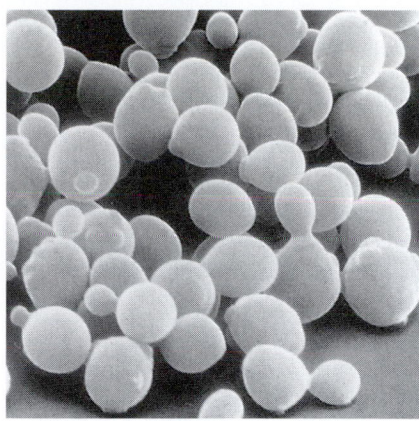

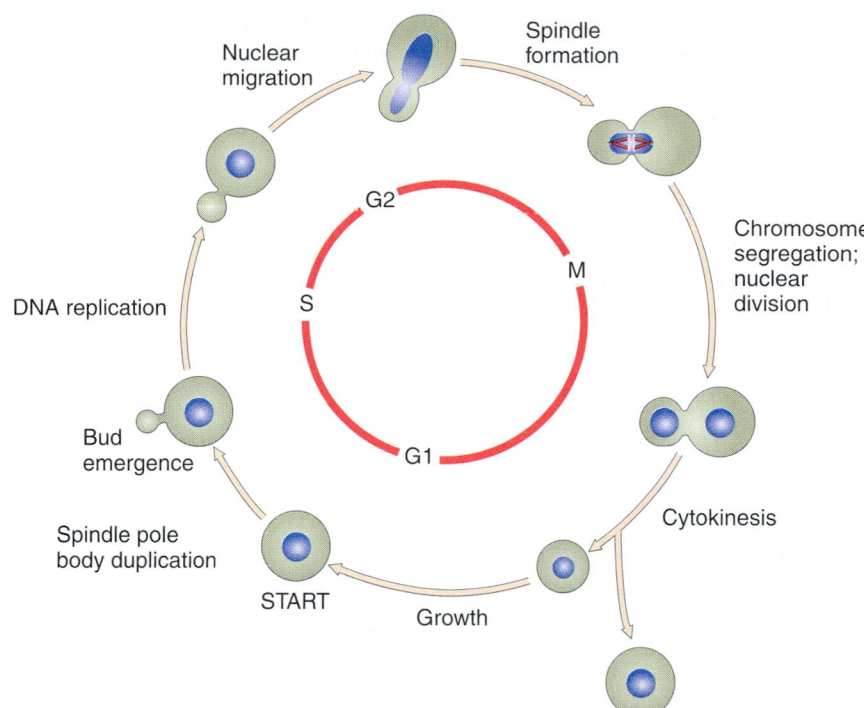

Figure 22-4 The cell cycle of the budding yeast *Saccharomyces cerevisiae.* The scanning electron micrograph shows cells at different points in the cell cycle, as indicated by different bud sizes. The principal events in the cell cycle are shown. (Courtesy of E. Schachtbach and I. Herskowitz. From H. Lodish, D. Baltimore, A. Berk, S. L. Zipursky, P. Matsudaira, and J. Darnell, *Molecular Cell Biology,* 3d ed. Copyright © 1995 by Scientific American Books.)

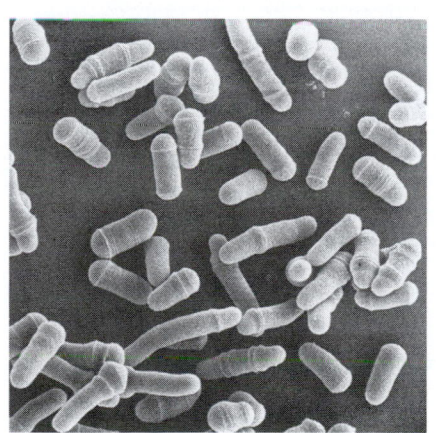

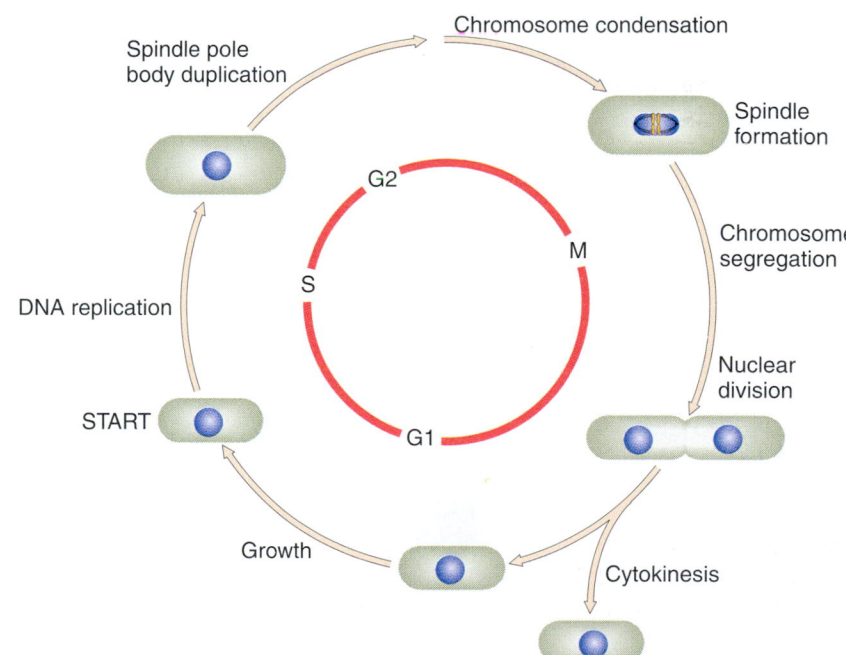

Figure 22-5 The cell cycle of the fission yeast *Schizosaccharomyces pombe.* The scanning electron micrograph shows cells dividing symmetrically and at different points in the cell cycle. The diagram indicates the timing of different major events in the *S. pombe* cell cycle. (Courtesy of E. Schachtbach and I. Herskowitz. From H. Lodish, D. Baltimore, A. Berk, S. L. Zipursky, P. Matsudaira, and J. Darnell, *Molecular Cell Biology,* 3d ed. Copyright © 1995 by Scientific American Books.)

respectively). These genes are now being systematically mutated and genetically characterized to understand how each contributes to the cell cycle.

Machinery for programmed cell death

Apoptosis pathway

In multicellular organisms, systems have evolved to eliminate damaged (and, hence, potentially harmful) cells through a self-destruct and disposal mechanism: **programmed cell death,** or **apoptosis.** This self-destruct mechanism can be activated under many different circumstances. Regardless, the

events in apoptosis seem to be the same (Figure 22-6). First, there is fragmentation of the DNA of the chromosomes, disruption of organelle structure, and loss of normal cell shape (apoptotic cells become spherical). Then, the cells break up into small cell fragments called apoptotic bodies that are phagocytosed (literally, eaten up) by motile scavenger cells.

In this section, we deal with the molecules responsible for carrying out apoptosis. In a later section, we shall consider how these responses are regulated within the cell.

Caspases

The engines of self-destruction are a series of enzymes called **caspases** (cysteine-containing aspartate-specific proteases). Proteases are enzymes that cleave other proteins. Each caspase is a protein rich in cysteines that, when activated, cleaves certain target proteins at specific aspartate residues in the target polypeptide chains. In a given organism, there is a family of caspase proteins, once again related to each other by polypeptide sequence; for example, in humans, 10 caspases have so far been identified. In normal cells, each caspase is present in an enzymatically inactive state, called the **zymogen** form. In general, a zymogen is an

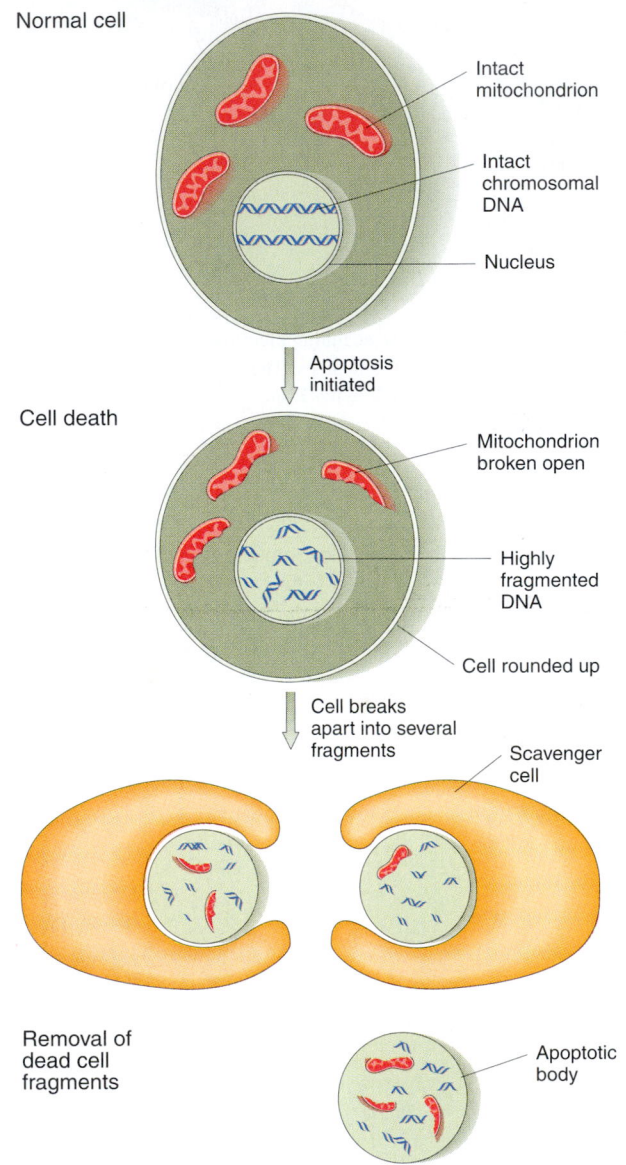

Figure 22-6   The sequence of phenotypic events in apoptosis. First, the membranes of organelles such as mitochondria are disrupted and their contents leak into the cytoplasm, chromosomal DNA becomes fragmented into small pieces, and the cell loses its normal shape. Then, the cell breaks apart into small cell fragments and is disposed of by phagocytotic scavenger cells.

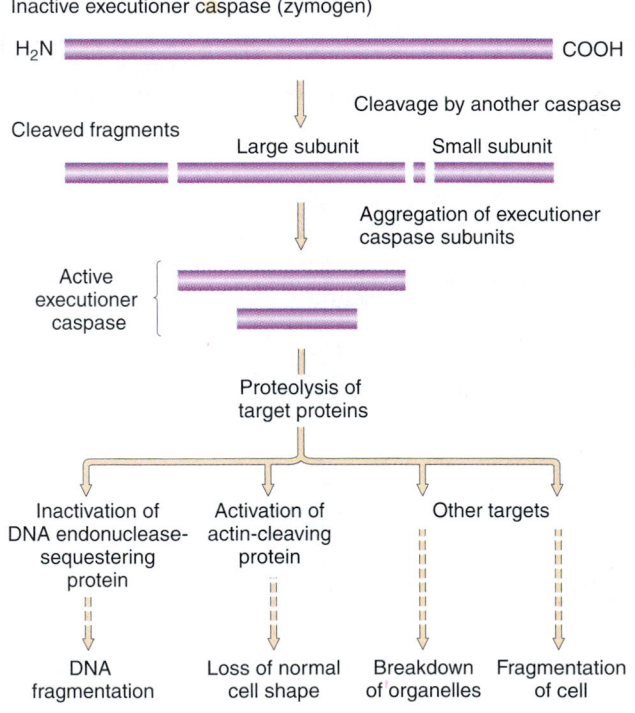

Figure 22-7   The role of executioner caspases in apoptosis. A cascade of caspase activation leads to the activation of the executioner caspases. Cleavage of the zymogen (inactive precursor) form of the executioner caspase by another caspase takes place at several aspartate residues, producing several protein fragments. Two of these fragments, the large and small subunits, bind and form the enzymatically active caspase. Through cleavage of a series of target proteins (also by cleavage at aspartate residues in the target proteins), the various cellular breakdown events take place, leading to cell death and removal.

inactive precursor form of an enzyme; the zymogen contains a longer polypeptide chain than does the final active enzyme. To turn the zymogen form into the active caspase, a part of the polypeptide is removed by enzyme cleavage (also known as proteolysis).

The current view is that there are two classes of caspases: *initiators* and *executioners*. Exactly how they are organized into a cascade of cleavage events is currently unclear. One scenario is that the initiator caspases are cleaved in response to activation signals coming from other classes of proteins. They in turn cleave one of the executioner caspases, which in turn cleaves another, and so forth.

> **MESSAGE** ·····························
> Programmed cell death is mediated by a sequential cascade of proteolysis events that activate enzymes targeted to destroy several key cellular components.

How do the executioner caspases carry out the cellular sentence of death? In addition to activating other caspases, executioner caspases enzymatically cut the target proteins (Figure 22-7). One target is a "sequestering" protein that forms a complex with a DNA endonuclease, thereby holding (seques-

tering) the endonuclease in the cytoplasm. On cleavage of the sequestering protein, the endonuclease is then free to enter the nucleus and chop up the cell's DNA. Another target is a protein that, when cut by the caspases, cleaves actin, a major component of the cytoskeleton, causing disruption of actin filaments and thus leading to a loss of normal cell shape. In similar fashion, all other aspects of the apoptosis phenotype are thought to be mediated by caspase-activated proteases.

The nematode *Caenorhabditis elegans*: a genetic model for programmed cell death

Programmed cell death has been described in a variety of organisms. However, genetic studies in the past 10 years in the nematode (roundworm) *Caenorhabditis elegans* have propelled the field forward. Researchers have mapped the entire series of cell divisions that produce the 1000 or so somatic cells of the adult worm. Interestingly, for some of the embryonic and larval cell divisions, particularly those that will contribute to the worm's nervous system, a progenitor cell gives rise to two progeny cells, one of which then undergoes programmed cell death (Figure 22-8). These

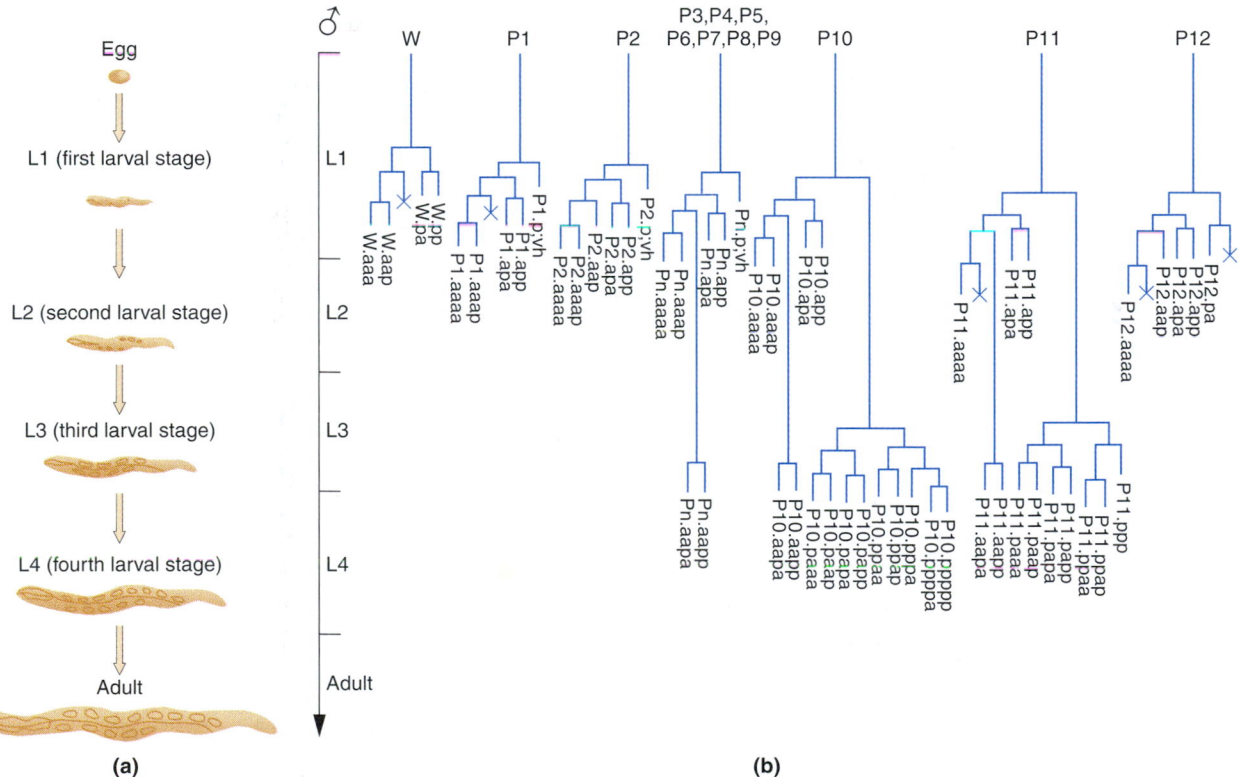

Figure 22-8   Examples of programmed cell death in the development of *C. elegans*. (a) The nematode's life cycle is shown. (b) A symbolic representation of the cell lineages of 13 cells (the W cell, the P1 cell, and so forth) produced during embryogenesis. The vertical axis is developmental time, beginning with the hatching of the egg into the first larval stage (L1). In each lineage, a vertical blue bar connects the various cell division events (horizontal blue bars). The names of the final cells are shown, such as W.aaa or P1.apa. In several cases, a cell division produces one viable cell and one cell that undergoes apoptosis. A cell that undergoes programmed cell death is indicated with a blue X at the end of a branch of a lineage. In homozygotes for mutations such as *ced-3*, these cell deaths do not occur.

divisions, in which one progenitor cell gives rise to only one viable progeny cell, are necessary for the progeny cell to fulfill its normal developmental role.

A set of mutations identified in the worm block this cell death phenotype. Some of these mutations knock out the functions of genes that encode caspases. An example is *ced-3* (cell death gene number 3), clearly implicating these caspases in the apoptosis process. The analysis of other genes with mutant cell death phenotypes is being carried out in worms and other experimental systems and is uncovering other key players in this process. Still other key players are being identified among the tumor suppressor genes that have been uncovered in studies of cancer formation and progress (discussed in the second half of this chapter).

Controlling the cell-proliferation and death machinery

We have used the term *engine* to describe the role of the cyclin–CDK complex or the caspase cascade in cell proliferation or programmed cell death, respectively. To continue the analogy, ignition switches and accelerators (positive controls) start up the engines and get these processes moving, and brakes (negative controls) slow down or halt the processes when necessary. Like the cell cycle and apoptosis, the positive and negative controls comprise a series of modulations of protein activities through protein–protein interactions and protein modifications.

Intracellular signals

Some of the elements of the positive and negative control loops consist of signals that originate within the cell.

The cell cycle: negative intracellular controls. Through activation of proteins that can inhibit the protein kinase activity of CDK–cyclin complexes, the cell cycle can be held in check until various monitoring mechanisms give a "green light," indicating that the cell is properly prepared to proceed to the next phase of the cycle.

One example of how this "checkpoint" system operates begins with damaged DNA (Figure 22-9). When DNA is damaged during G1 (for example, by X-irradiation), the CDK activity of CDK–cyclin complexes is inhibited. The inhibition seems to be mediated by a protein called p53. Part of the p53 protein recognizes certain kinds of DNA mismatches. In the presence of such mismatches, p53 is able to activate another protein, p21. When its levels are high, p21 binds to the CDK–cyclin complex and inhibits its protein kinase enzymatic activity. In the absence of its protein kinase activity, CDK's target proteins are not phosphorylated, and the cell cycle is unable to progress. When the DNA mismatches have been repaired, the inhibiting processes are reversed. This reversal is accomplished by a post-DNA-repair drop in p53 levels and a cessation of inhibition of the CDK–cyclin protein kinase activity and leads to removal of the G1-to-S checkpoint block.

In this manner, checkpoints monitor the status of DNA replication, the spindle apparatus, and other key components of the cell cycle and can operate as braking systems when necessary. The key is the existence of regulatory proteins that can modulate the protein kinase activity of the cyclin–CDK complex.

MESSAGE ······································
Fail-safe systems (checkpoints) ensure that the cell cycle does not progress until the cell is competent.

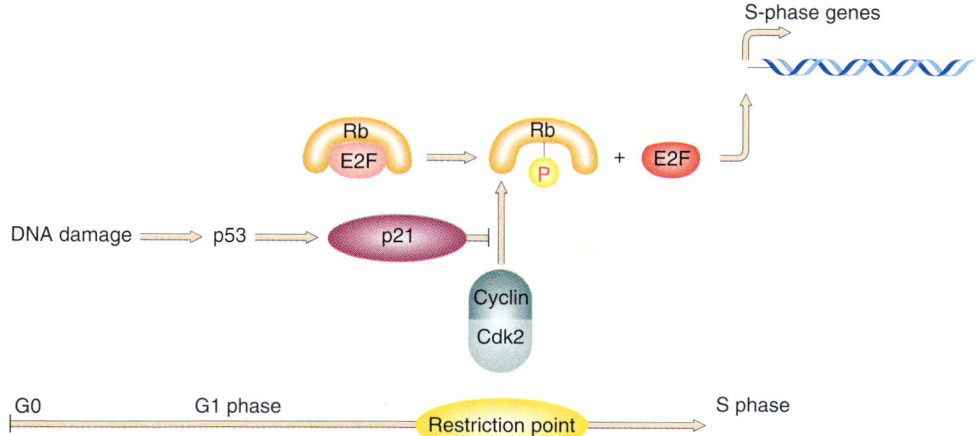

Figure 22-9   An example of inhibitory control of cell cycle progression. In mammals, the transition from the G1 to S phase requires the phosphorylation of Rb protein by the CDK2–cyclin complex. In the presence of damaged DNA, p53 protein is induced, which in turn induces p21 protein. The elevated levels of p21 protein inhibit the protein kinase activity of the CDK2–cyclin complex. When the damaged DNA has been repaired, p53 levels drop. In turn, p21 levels decrease, and the inhibition of the CDK2–cyclin protein kinase activity is relieved, which allows Rb to be phosphorylated and E2F to become an active transcription factor, permitting the cell to enter S phase. (Adapted from C. J. Sherr and J. M. Roberts, *Genes and Development* 9, 1995, 1150.)

The cell cycle: positive intracellular controls. It is necessary not only to release the cell cycle "brake," but also to engage the "transmission" and the "engine" to advance the cell cycle. When the brake is released, independent signals from within or outside the cell induce a cascade of protein kinases that phosphorylate the appropriate cyclin–CDK complex, thereby activating the complex. This activation in turn allows the complex to phosphorylate its target proteins.

Apoptosis: positive intracellular controls. It has been known for several years that, in some manner, many forms of cellular damage trigger leakage of mitochondria and that this leakage somehow induces the apoptotic response. Indeed, it now appears that one of the ignition switches is cytochrome *c*, one of the mitochondrial proteins normally taking part in cell respiration. Leakage of cytochrome *c* into the cytoplasm is detected and triggers the activation of initiator caspases. This detection is thought to happen through the binding of cytochrome *c* to another protein called Apaf (apoptotic protease-activating factor). The cytochrome *c*–Apaf complex then binds to and activates the initiator caspase.

Apoptosis: negative intracellular controls. The irreversibility of cell death has probably been the compelling factor in the evolution of backup systems to make sure that the apoptosis pathway remains "off" under normal conditions. Proteins such as Bcl-2 and Bcl-x in mammals accomplish this. Among the possible actions for these Bcl proteins is that they block the release of cytochrome *c* from mitochondria (possibly by making it more difficult for mitochondria to burst) and by binding to Apaf and preventing its interaction with the initiator caspase.

Extracellular signals

A cell in a multicellular organism continually assesses its own internal status regarding proliferation and survival. Nonetheless, the proliferative and survival abilities of a cell must be subservient to the needs of the population of cells of which it is a member (populations such as the entire early embryo, a tissue, or a body part such as a limb or an organ). For example, in many adult organs, stem cells divide to produce replacement cells only when there is a depletion of cell numbers. Without such homeostatic mechanisms, organs would not be proportioned appropriately for the size of a given individual organism.

Mechanisms for cell-to-cell communication. Many kinds of signals need to be transmitted between cells to coordinate virtually all aspects of the development and physiology of complex multicellular organisms. The major routes of cell-to-cell communication are briefly outlined here.

All systems for intercellular communication have several components. A molecule called a **ligand** is produced by secretion from signaling cells (Figure 22-10). Some ligands, called hormones, are long-range **endocrine signals** that are

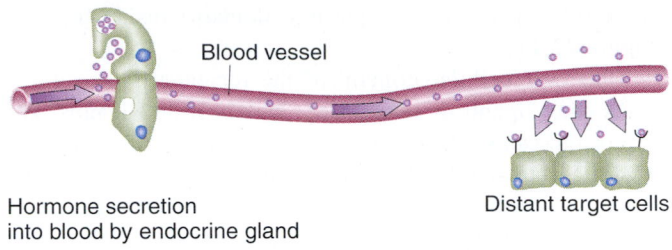

Blood vessel

Hormone secretion into blood by endocrine gland

Distant target cells

(a) Endocrine signaling

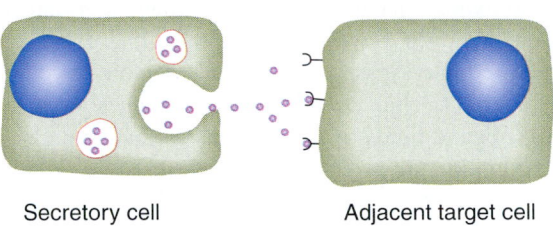

Secretory cell

Adjacent target cell

(b) Paracrine signaling

Figure 22-10   Modes of intercellular signaling. (a) Endocrine signals enter the circulatory system and can be received by distant target cells. (b) Paracrine signals act locally and are received by nearby target cells. (From H. Lodish, D. Baltimore, A. Berk, S. L. Zipursky, P. Matsudaira, and J. Darnell, *Molecular Cell Biology,* 3d ed. Copyright © 1995 by Scientific American Books.)

transmitted throughout the body by being released from endocrine organs into the circulatory system. (Recall the discussion of steroid hormones and their receptors in Chapter 11.) Hormones can act as master control switches for many different tissues, which can then respond in a coordinated fashion. Other secreted ligands act as **paracrine signals;** that is, they do not enter the circulatory system but act only locally, in some cases only on immediately adjacent cells. We shall have more to say about paracrine and endocrine signals in Chapter 23. Some ligands are proteins, whereas others are small molecules such as steroids or vitamin D. Most (but not all) endocrine signals are small molecules, such as the mammalian steroid hormones that are responsible for male (androgen) or female (estrogen) sex-specific phenotypes. In contrast, most paracrine signals are proteins. Here we focus on paracrine signaling through protein ligands.

Protein ligands act as signals by binding to and thereby activating transmembrane receptor proteins that are embedded in the plasma membrane at the surface of the cell. These ligand–receptor complexes initiate chemical signals in the cytoplasm just inside the plasma membrane of the cell. Such signals are passed through a series of intermediary molecules until they finally alter the structure of transcription factors in the nucleus, leading to the activation of transcription of some genes and repression of others.

Transmembrane receptors have one part (the extracellular domain) outside of the cell, a middle part that passes once or several times through the plasma membrane, and another part (the cytoplasmic domain) inside the cell (Figure 22-11).

The extracellular domain of the receptor is the site to which the ligand binds. Many polypeptide ligands are dimers and can simultaneously bind two receptor monomers. This simultaneous binding brings the cytoplasmic domains of the two receptor subunits into close proximity and activates the signaling activity of these cytoplasmic domains. Some receptors for polypeptide ligands are receptor tyrosine kinases (RTKs, Figure 22-11b). Their cytoplasmic domains, when activated, have the ability to phosphorylate certain tyrosine residues on target proteins. Others are receptor serine/threonine kinases. Still other receptors have no enzymatic activity, but conformational changes in a receptor (when a ligand binds to it) cause conformational changes in (and activation of) receptor-bound cytoplasmic proteins.

Perhaps the best understood of the receptors for polypeptide ligands are the receptor tyrosine kinases (Figure 22-12). RTK is a monomer essentially "floating" within the plane of the plasma membrane. When ligand and RTK bind to form a ligand–RTK complex, two RTK monomers bind to form a dimer. RTK dimerization activates the protein ki-

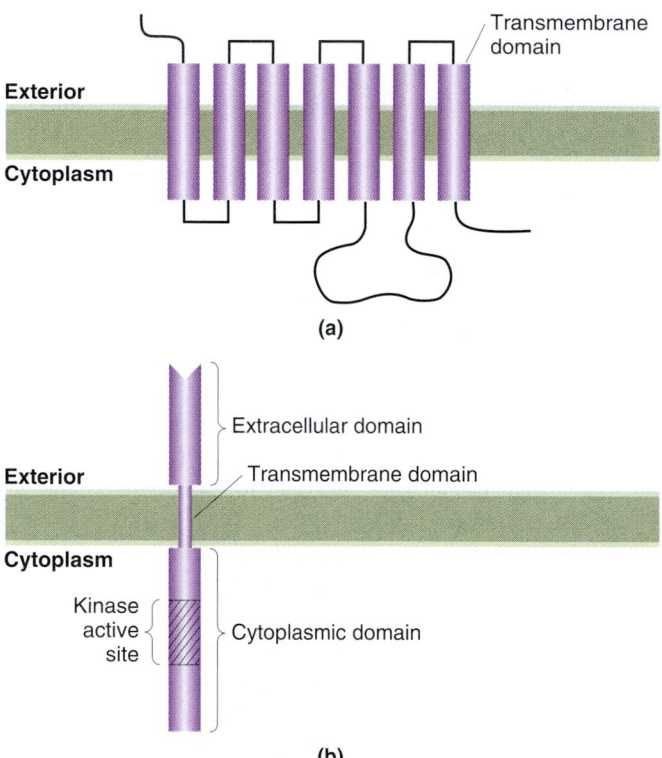

(a)

(b)

Figure 22-11   Examples of transmembrane receptors. (a) A receptor that passes through the cell membrane seven times. (b) Receptor tyrosine kinase (RTK), which has a single transmembrane domain. The extracellular domain binds to ligand. The active site of the tyrosine kinase is in the cytoplasmic domain. (After H. Lodish, D. Baltimore, A. Berk, S. L. Zipursky, P. Matsudaira, and J. Darnell, *Molecular Cell Biology*, 3d ed. Copyright © 1995 by Scientific American Books.)

nase enzymatic activity of the cytoplasmic domain of the RTK. The first phosphorylation targets of the kinase are several tyrosines in the cytoplasmic domain of the RTK itself; this process is called *autophosphorylation* because the kinase acts on itself. Autophosphorylation initiates a signal transduction cascade, in which, sequentially, modifications in the conformations of one protein lead to modifications in the conformations of others. Eventually, the signal transduction cascade leads to the modification of transcriptional activators and repressors and hence to changes in the activities of many genes in the target cell.

RTK autophosphorylation activates signal transduction cascades in two ways. In one process, phosphorylated sites on the RTK are targets for binding by various *adaptor* proteins (Figure 22-12a). Multiple adaptor proteins "dock" on phosphorylated sites on the RTK in the vicinity of one another. These adaptor proteins in turn have affinity for other proteins — elements of signal transduction cascades. By bringing these other signal transduction elements into proximity with one another, protein–protein interactions lead to activation of these cascades. In the other process, the phosphorylated RTK is conformationally changed so that its tyrosine kinase active site phosphorylates other target proteins (Figure 22-12b). These phosphorylated target proteins are then changed in conformation, allowing them to participate in a signal transduction cascade. By these two processes, activation of one RTK can lead to the simultaneous activation of multiple signal transduction pathways.

Quite often, the next step in propagating the signal is to activate a G-protein. G-proteins cycle between being bound by GDP (the inactive state) and being bound by GTP (the activated state). The propagation of the signal from the RTK leads to the activation of a protein that binds to the inactive GDP-bound G-protein, changing its conformation so that it then binds to a molecule of GTP (Figure 22-13). The specific G-protein called Ras is of special importance in carcinogenesis, as discussed later.

The activated GTP-bound G-protein then binds to a cytoplasmic protein kinase, in turn changing its conformation and activating its protein kinase activity. This protein kinase then phosphorylates other proteins, including other protein kinases. (In the example in Figure 22-14, the protein kinases farther down the cascade are called Raf, MEK, and MAP kinase.) The targets of some of these protein kinases are transcriptional activators and repressors. The phosphorylation of the transcription factors changes their conformations, leading to the activation of transcription of some genes and the repression of others (Figure 22-14).

Cell-to-cell signaling depends on conformational changes.   We have seen that the steps in ligand–receptor binding and in signaling within the cell depend on conformational changes. For example, the conformational changes caused by the binding of ligands to receptors activate the signaling pathways. Likewise, conformational changes in protein kinases enable them to phosphorylate specific amino acids on specific proteins, and other proteins undergo con-

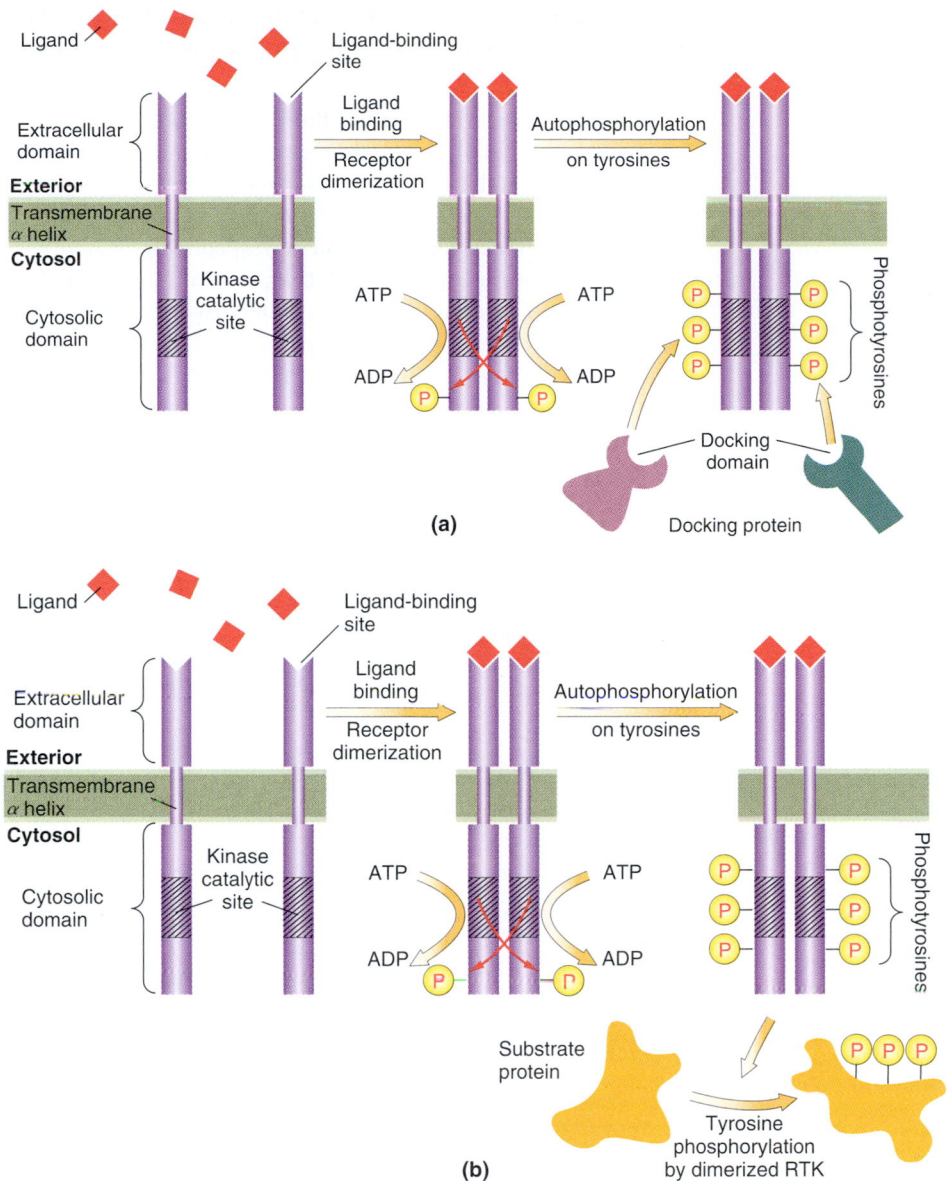

Ligand

Ligand-binding site

Extracellular domain

Exterior

Transmembrane α helix

Cytosol

Cytosolic domain

Kinase catalytic site

Ligand binding

Receptor dimerization

Autophosphorylation on tyrosines

ATP

ADP

ATP

ADP

Docking domain

Docking protein

Phosphotyrosines

(a)

Ligand

Ligand-binding site

Extracellular domain

Exterior

Transmembrane α helix

Cytosol

Cytosolic domain

Kinase catalytic site

Ligand binding

Receptor dimerization

Autophosphorylation on tyrosines

ATP

ADP

ATP

ADP

Substrate protein

Tyrosine phosphorylation by dimerized RTK

Phosphotyrosines

(b)

Figure 22-12   The consequences of ligand binding on RTK activity and the initiation of the signal transduction cascade. On dimerization of the RTK, autophosphorylation occurs. The ability of activated RTK to initiate signal transduction is through two different routes: (a) by the activated RTK serving as an anchor site for some signal transduction proteins and (b) by the activated RTK phosphorylating other proteins. (a) Autophosphorylation creates binding sites for *adaptor*, or "docking," proteins that bind to sites that include specific phosphorylated tyrosine residues. Through the docking of different adaptor proteins in close proximity to one another, protein–protein conformational changes lead to induction of the signal transduction activities of these proteins. (b) The other consequence of dimerization and autophosphorylation is that the RTK directly phosphorylates *substrate* proteins and, through this phosphorylation, induces their signal transduction activities. (After H. Lodish, D. Baltimore, A. Berk, S. L. Zipursky, P. Matsudaira, and J. Darnell, *Molecular Cell Biology,* 3d ed. Copyright © 1995 by Scientific American Books.)

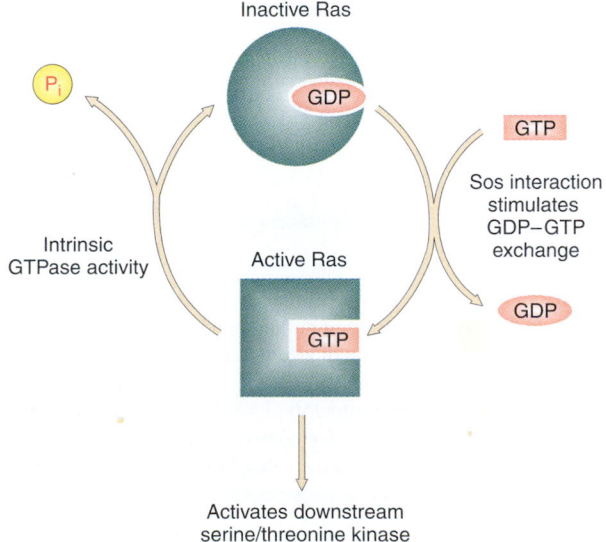

Inactive Ras

GDP

Pi

Intrinsic GTPase activity

GTP

Sos interaction stimulates GDP–GTP exchange

GDP

Active Ras

GTP

Activates downstream serine/threonine kinase

Figure 22-13   An example of the G-protein activity cycle. Ras is a member of the G-protein family. When Ras binds GDP, it does not signal. Through direct interactions with Ras, another protein called Sos causes conformational changes in Ras so that it preferentially binds GTP. The Ras–GTP complex is able to interact in turn with a cytoplasmic serine/threonine kinase, activating its kinase activity, and thus transmits the signal to the next step in the signal transduction pathway. When Ras–GTP is released from Sos, it hydrolyzes GTP to GDP and reassumes the inactive Ras–GDP state. (After J. D. Watson, M. Gilman, J. Witkowski, and M. Zoller, *Recombinant DNA,* 2d ed. Copyright © 1992 by James D. Watson, Michael Gilman, Jan Witkowski, and Mark Zoller.)

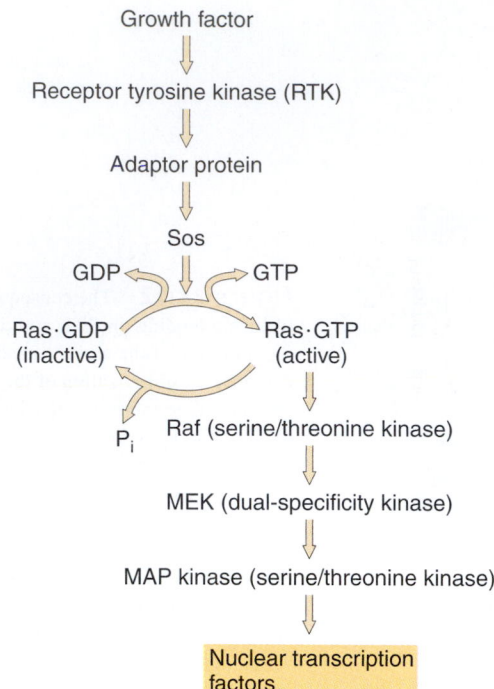

Figure 22-14   One pathway for RTK signaling. Raf, MEK, and MAP kinase are three cytoplasmic protein kinases that are sequentially activated in the signal transduction cascade. (After H. Lodish, D. Baltimore, A. Berk, S. L. Zipursky, P. Matsudaira, and J. Darnell, *Molecular Cell Biology,* 3d ed. Copyright © 1995 by Scientific American Books.)

formational changes when they bind to GTP. Not only do these conformational changes permit rapid response to an initial signal, but they also are readily reversible, enabling signals to be shut down rapidly and permitting recycling of the components of the signaling system so that they are ready to receive further signals.

The cell cycle: positive extracellular controls.   Cell division is promoted by the action of **mitogens,** polypeptide ligands released usually from a paracrine (nearby) source. Many mitogens, also called growth factors, such as EGF (epidermal growth factor), activate RTKs and initiate exactly the sort of signal transduction pathway heretofore described.

The cell cycle: negative extracellular controls.   Certain secreted proteins are known to inhibit cells from dividing. One example is TGF-β, a ligand that is thought to be secreted in a variety of tissues under growth inhibitory conditions. The TGF-β ligand binds to and activates the TGF-β receptor's serine/threonine kinase activity. This activation in turn leads to phosphorylation of proteins called SMADs, which cause changes in transcriptional activities, and perhaps to phosphorylation of other substrates as well. As a result of this signal transduction cascade, the phosphorylation and inactivation of the Rb protein are eventually blocked. Recall, from earlier in the chapter, Rb's cell cycle role of

preventing activation of the E2F transcription factor. This block to Rb inactivation thus keeps E2F off and blocks progression of the cell cycle.

Apoptosis: positive extracellular controls.   Often, the command for self-destruction comes from a neighboring cell. For example, within the immune system, only a small percentage of B cells and T cells mature to make functional antibody or T-cell receptor protein, respectively. If nonfunctional, immature B cells and T cells were not eliminated by induced self-destruction, the vast majority of them would clog up the immune system. The self-destruction signal is activated through the Fas system (Figure 22-15). A cell-surface membrane-bound protein called FasL (Fas ligand) binds to Fas cell-surface receptors on an adjacent cell. This binding induces trimerization of the ligand–receptor complex and trimerization of a cytoplasmic domain of the Fas transmembrane receptor. This trimerization in turn, directly or indirectly, activates a molecule such as Apaf (discussed earlier in this chapter), which activates an initiator caspase and thus the caspase cascade.

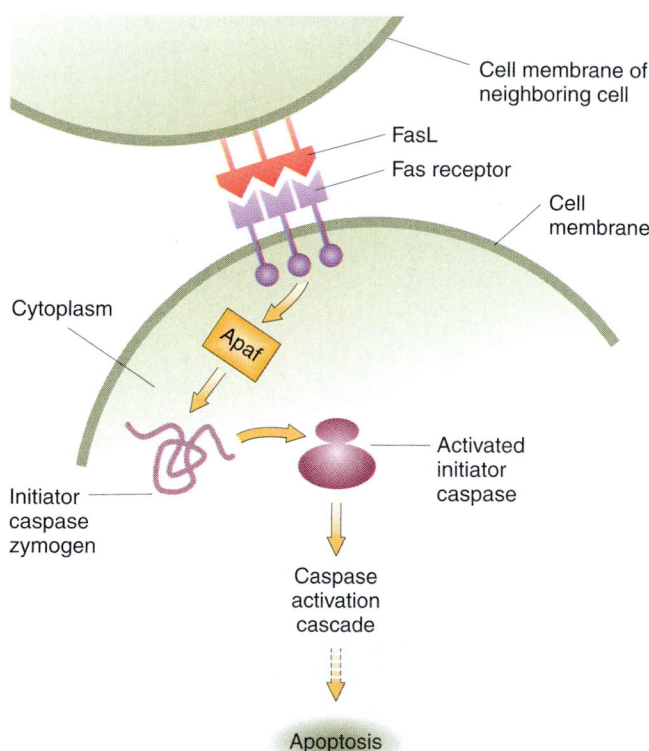

Figure 22-15   Positive extracellular control of apoptosis. A molecule such as Apaf is activated through binding of FasL (Fas ligand), anchored on the outside of an adjacent cell. FasL binds to the Fas transmembrane receptor on the cell that will undergo apoptosis. Apaf activation in turn causes proteolysis and activation of the initiator caspase. A series of caspases are then proteolysed and activated in turn, ultimately leading to apoptosis of the cell. (After S. Nagata, *Cell* 88, 1997, 357.)

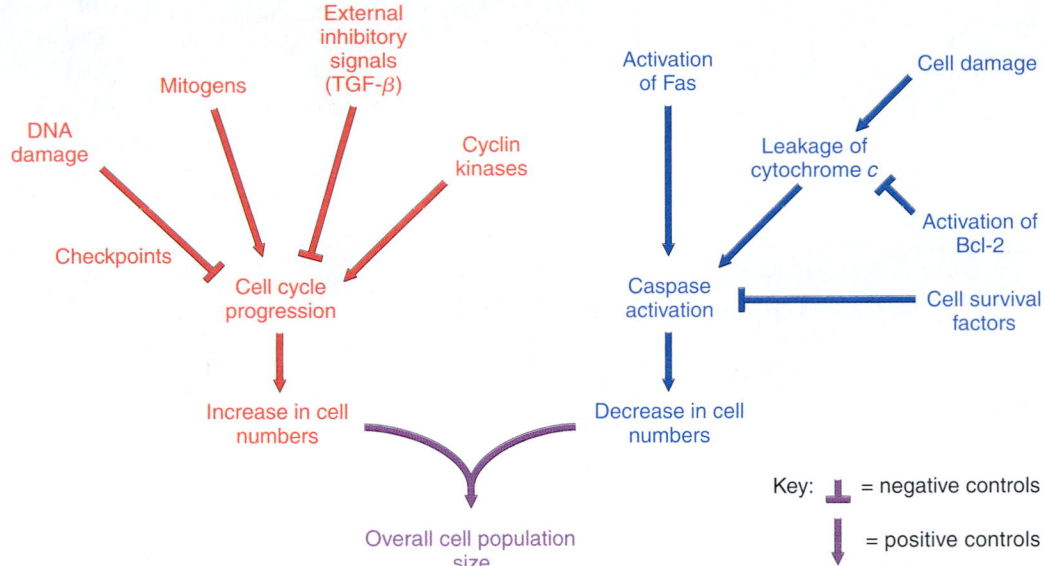

Figure 22-16 The inputs into control of cell number.

Apoptosis: negative extracellular controls. Negative secreted factors that are necessary to block activation of the apoptosis pathway also exist, and they are sometimes referred to as **survival factors.** How they impinge on the apoptosis pathway is not clear at present.

> **MESSAGE** ···
> Intercellular signaling systems communicate instructions between cells to proliferate or to arrest the cell cycle and to initiate or postpone self-destruction.

An integrated view of the control of cell numbers

We have seen in the preceding sections that there are numerous ways to modulate cell number. The general theme is that pathways exist for controlling cell proliferation and self-destruction and that activation of these pathways requires the correct array of positive inputs and the absence of negative, or inhibitory, inputs. Not only do cells have mechanisms for self-assessment of their status regarding proliferation ability or viability, but neighboring cells can play instructive roles through cell-to-cell signaling (Figure 22-16).

Cancer: the genetics of aberrant cell control

A basic article of faith in genetic analysis is that we learn a great deal about normal biology and about the disease state by studying the properties of mutations that disrupt normal processes. This has certainly been true in regard to cancer. It has become clear that virtually all cancers of somatic cells are due to a series of special mutations that accumulate in a cell. We are seeing that these mutations fall into a few major categories: increasing the ability of a cell to proliferate,

decreasing the susceptibility of a cell to apoptosis, or increasing the general mutation rate of the cell so that proliferation or apoptotic mutation is more likely to occur. We can hope that these insights into the basic events in cancer biology will translate into improved diagnosis, treatment, and control of this major group of diseases.

How cancer cells differ from normal cells

Malignant tumors, or cancers, are clonal. **Cancers** are aggregates of cells, all derived from an initial aberrant founder cell that, although surrounded by normal tissue, is no longer integrated into that environment. Cancer cells often differ from their normal neighbors by a host of specific phenotypic changes, such as rapid division rate, invasion of new cellular territories, high metabolic rate, and abnormal shape. For example, when cells from normal epithelial cell sheets are placed in cell culture, they can grow only when anchored to the culture dish itself. In addition, normal epithelial cells in culture divide until they form a continuous monolayer. Then, they somehow recognize that they have formed a single epithelial sheet, and stop dividing. In contrast, malignant cells derived from epithelial tissue continue to proliferate, piling up on one another (Figure 22-17). Clearly, the factors regulating normal cell differentiation have been altered. What, then, is the underlying cause of cancer? Many different cell types can be converted into a malignant state. Is there a common theme to the ontogeny of these different types of cancer or do they each arise in quite different ways? Indeed, we can think about cancer in a general way: as occurring by the production of multiple mutations in a single cell that cause it to proliferate out of control. Some of those mutations may be transmitted from the parents through the germ line. Others arise de novo in the somatic cell lineage of a particular cell.

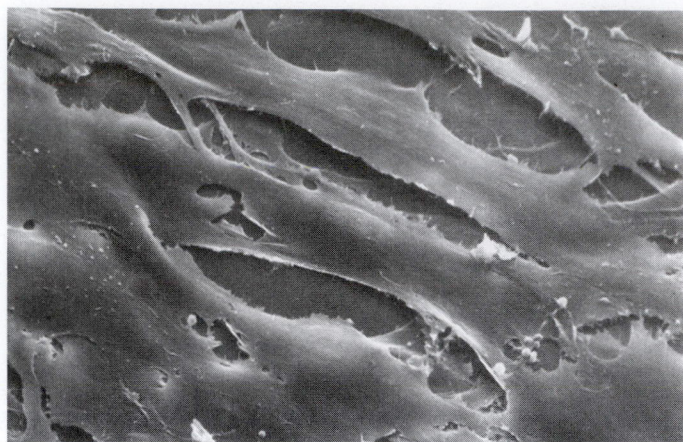

(a)

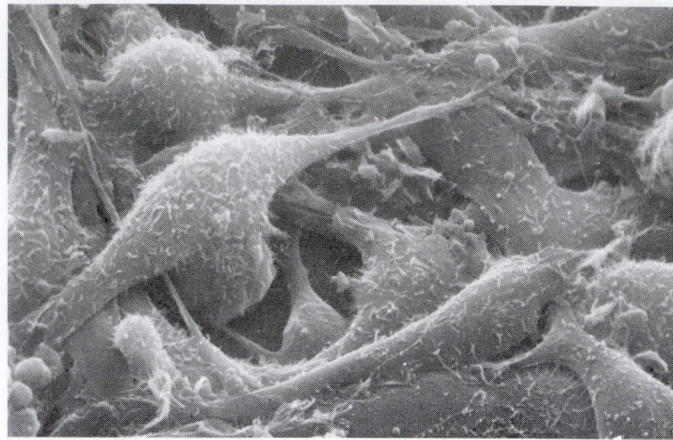

(b)

Figure 22-17 Scanning electron micrographs of (a) normal cells and (b) cells transformed with Rous sarcoma virus, a virus that infects cells with the *src* oncogene. (a) A normal cell line called 3T3. Note the organized monolayer structure of the cells. (b) A transformed derivative of 3T3. Note how the cells are rounder and piled up on one another. (From H. Lodish, D. Baltimore, A. Berk, S. L. Zipursky, P. Matsudaira, and J. Darnell, *Molecular Cell Biology,* 3d ed. Copyright © 1995 by Scientific American Books.)

Evidence for the genetic origin of cancers

Several lines of evidence have pointed to a genetic origin for the transformation of cells from the benign into the cancerous state. Most carcinogenic agents (chemicals and radiation) are also mutagenic. There are occasional instances in which certain cancers are inherited as highly penetrant single Mendelian factors; an example is familial retinoblastoma. Perhaps representing the more general case are less penetrant susceptibility alleles that increase the probability of developing a particular type of cancer. In the past few years, several susceptibility genes have been recombinationally mapped and molecularly cloned and localized with the use of RFLP mapping or related techniques. **Oncogenes,** dominant mutant genes that contribute to cancer in animals, have been isolated from tumor viruses — viruses that can transform normal cells in certain animals into tumor-forming cells. Such dominant oncogenes can also be isolated from tumor cells by using cell-culture assays that can distinguish between some types of benign and malignant cells. Tumors do not arise as a result of single genetic events but rather are the result of multiple-hit processes, in which several mutations must arise within a single cell for it to become cancerous. In some of the best-studied cases, the progression of colon cancer and astrocytoma (a brain cancer) has been shown to entail the sequential accumulation of several different mutations in the malignant cells (Figure 22-18). In the next sections, we shall further consider the genetic origin of cancers and the nature of the proteins that are altered by cancer-producing mutations. We shall see that many of these proteins take part in intercellular communication and the regulation of the cell cycle.

MESSAGE ···
Tumors arise through a series of sequential mutational events that lead to a state of uncontrolled proliferation.
···

Mutations in cancer cells

Two general kinds of mutations are associated with tumors: oncogene mutations and mutations in tumor suppressor genes. Oncogenes are mutated in such a way that the proteins that they encode are activated in tumor cells carrying the dominant mutant allele. A tumor cell will typically be heterozygous for an oncogene mutation and its normal allelic counterpart. Tumor-promoting mutant alleles of **tumor-suppressor genes** mutationally inactivate the proteins that they encode. For such mutations, the tumor cell will lack any copy of the corresponding wild-type allele; in essence, tumor-suppressor mutations that are found in a tumor cell are recessive.

How have tumor-promoting mutations been identified? Several approaches have been used. It is well known that certain types of cancer can "run in families." With modern pedigree analysis techniques, familial tendencies toward certain kinds of cancer can be mapped relative to molecular markers such as microsatellites, and, in several cases, this mapping has led to the successful identification of the mutated genes. Cytogenetic analysis of tumor cells themselves also has proved invaluable. Many types of tumors are typified by characteristic chromosomal translocations or by deletions of particular chromosomal regions. In some cases, these chromosomal rearrangements are so reliably a part of a particular cancer that they can be used for diagnostic purposes. For example, 95 percent of patients with chronic myelogenous leukemia (CML) have a characteristic translocation between chromosomes 9 and 22. This translocation, called the Philadelphia chromosome after the city where this translocation was first described, is a critical part of the CML diagnosis. The Philadelphia chromosome will be considered in more detail later in this chapter. Other translocations characterize other sorts of tumors; diagnostic translocations are most often found associated with cancers of the white blood cells—

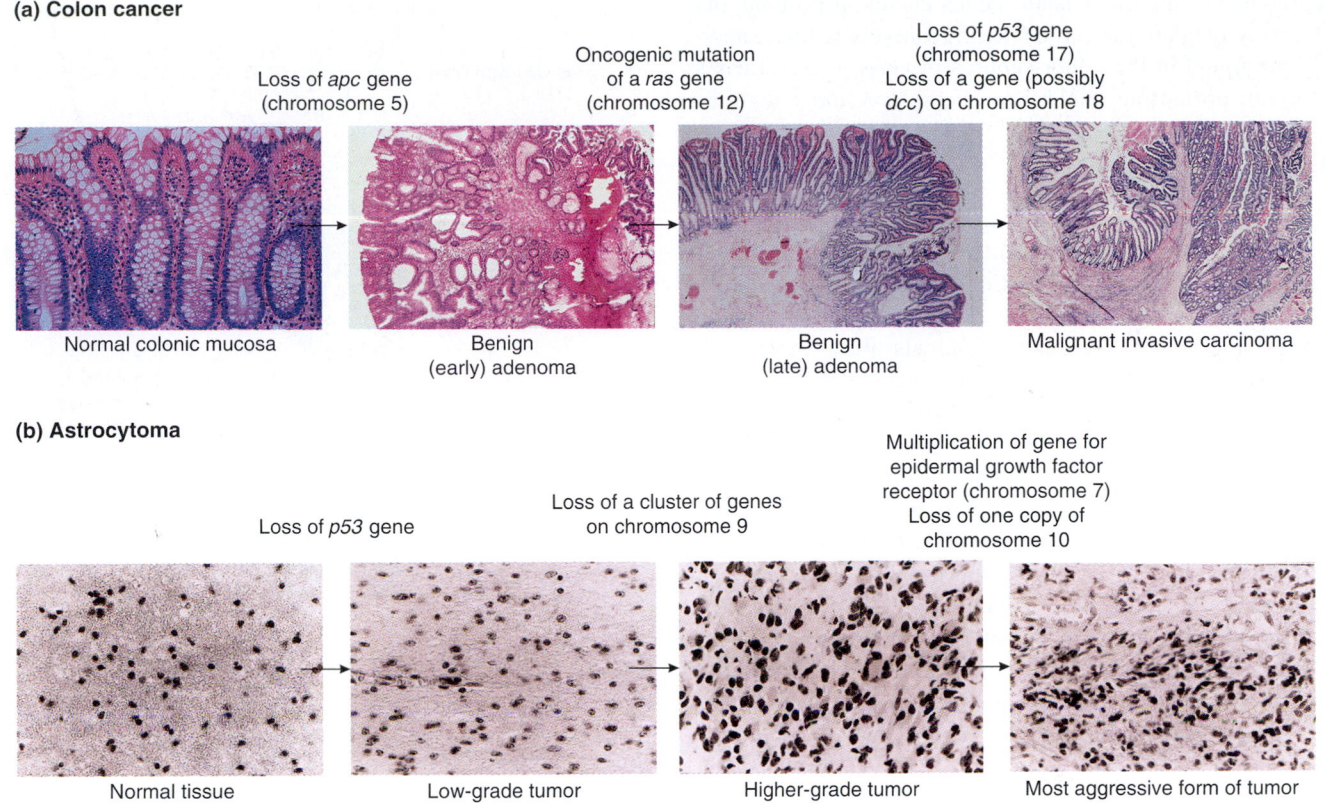

Figure 22-18 The multistep progression to malignancy in cancers of the colon and brain. Several histologically distinct stages can be distinguished in the progression of these tissues from the normal state to benign tumors to a malignant cancer. (a) A common sequence of mutational events in the progression to colon cancer. Note that the tissue becomes more disorganized as the tumor progresses toward malignancy. (b) A different characteristic series of mutations marks the progression toward a malignant astrocytoma, a form of brain cancer. (Micrographs by E. R. Fearon and K. Cho. From W. K. Cavanee and R. L. White, "The Genetic Basis of Cancer." Copyright © 1995 by Scientific American, Inc. All rights reserved.)

leukemias and lymphomas. Not all tumor-promoting mutations are specific to a given type of cancer, however. Rather, the same mutations seem to be tumor promoting for a variety of cell types and thus are seen in many different cancers.

MESSAGE ··

Tumor-promoting mutations can be identified in a variety of ways. When located, they can be cloned and studied to learn how they contribute to the malignant state.

··

It is obvious why mutations that increase the rate of cell proliferation cause tumors. It is not so immediately obvious why mutations that decrease the chances that a cell will undergo apoptosis cause them. The reason seems to be twofold: (1) a cell that cannot undergo apoptosis has a much longer lifetime within which to accumulate proliferation-promoting mutations and (2) the sorts of damage and unusual physiological changes that occur inside a tumor cell will ordinarily induce the self-destruction pathway to engage.

Whether an element of the cell cycle or the apoptosis pathway is due to a dominant oncogene mutation or to a recessive tumor-suppressor gene mutation is a function of how that normal protein contributes to the regulation of cell proliferation or programmed cell death (Table 22-1). Genes

encoding proteins that positively control the cell cycle or block apoptosis can typically be mutated to become oncogenes; these tumor-promoting alleles are *gain-of-function*

22-1 TABLE Relations Between Function of the Wild-Type Protein Product and the Types of Tumor-Promoting Mutations That Can Arise in the Genes Encoding Those Products

Type of wild-type protein function	Properties of tumor-promoting mutations
Promotes cell cycle progression	Oncogene (gain of function)
Inhibits cell cycle progression	Tumor-suppressor gene (loss of function)
Promotes apoptosis	Tumor-suppressor gene (loss of function)
Inhibits apoptosis	Oncogene (gain of function)
Promotes DNA repair	Tumor-suppressor gene (loss of function)

mutations. On the other hand, genes encoding proteins that negatively regulate the cell cycle or positively regulate apoptosis are found in the tumor-suppressor class; in these cases, the tumor-promoting alleles are *loss-of-function* mutations.

Classes of oncogenes

Roughly 100 different oncogenes have been identified (examples are given in Table 22-2). How do their normal counterparts, proto-oncogenes, function? Proto-oncogenes generally encode a class of proteins that are selectively active only when the proper regulatory signals allow them to be activated. As mentioned, many proto-oncogene products are elements of cell cycle positive control pathways, including growth-factor receptors, signal transduction proteins, and transcriptional regulators. Other proto-oncogene products function to negatively regulate the apoptotic pathway. However, in an oncogene mutation, the activity of the mutant oncoprotein has been uncoupled from the regulatory pathway that ought to be controlling its activation, leading to continuous unregulated expression of the oncoprotein (Figure 22-19). Several categories of oncogenes depict different ways in which the regulatory functions have been uncoupled. We will look at examples of some of them.

MESSAGE ·
Oncogenes encode oncoprotein-deregulated forms of proteins whose wild-type function is to participate in the positive control of the cell cycle or in the negative control of apoptosis.
· ·

Types of oncogene mutations

Point mutations.   The change from normal protein to oncoprotein often includes structural modifications to the pro-

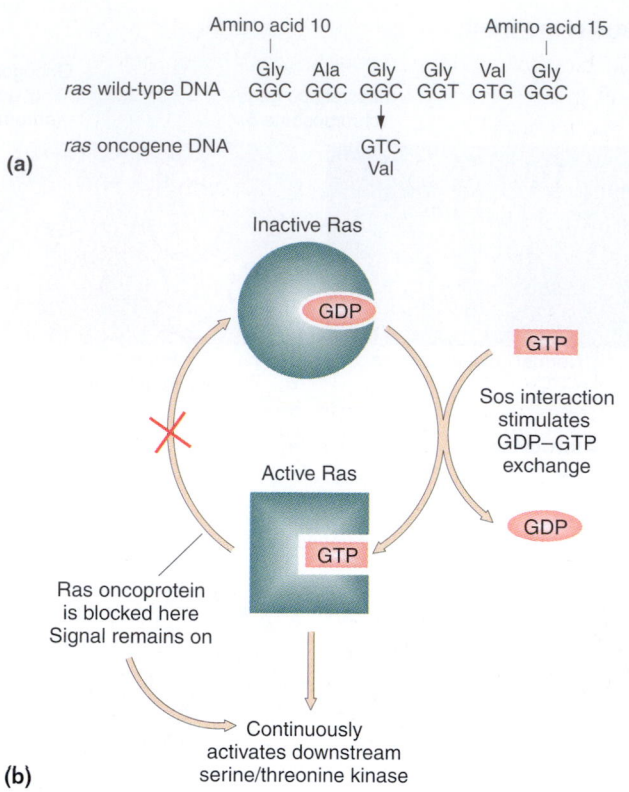

Figure 22-19   The Ras oncoprotein. (a) The *ras* oncogene differs from the wild type by a single base pair, producing a Ras oncoprotein that differs from the wild type in one amino acid, at position 12 in the *ras* open reading frame. (b) The effect of this missense mutation is to create a Ras oncoprotein that cannot hydrolyze GTP to GDP (compare with the normal Ras cycle depicted in Figure 22-13). Because of this defect, the Ras oncoprotein remains in the active Ras–GTP complex and continuously activates the downstream serine/threonine kinase (see Figure 22-14).

22-2 TABLE	Some Well-Characterized Oncogenes and the Proteins That They Encode	
	PROPERTIES OF PROTEIN	
Oncogene	Location	Function
Nuclear transcription regulators		
jun	Nucleus	Transcription factor
fos	Nucleus	Transcription factor
erbA	Nucleus	Member of steroid receptor family
Intracellular signal transducers		
abl	Cytoplasm	Protein tyrosine kinase
raf	Cytoplasm	Protein serine kinase
gsp	Cytoplasm	G-protein α subunit
ras	Cytoplasm	GTP/GDP-binding protein
Mitogen		
sis	Extracellular	Secreted growth factor
Mitogen receptors		
erbB	Transmembrane	Receptor tyrosine kinase
fms	Transmembrane	Receptor tyrosine kinase
Apoptosis inhibitor		
bcl2	Cytoplasm	Upstream inhibitor of caspase cascade

tein itself, such as those caused by simple point mutation. A single base-pair substitution that converts glycine into valine at amino acid number 12 of the Ras protein, for example, creates the oncoprotein found in human bladder cancer (Figure 22-19a). Recall that the normal Ras protein is a G-protein subunit that takes part in signal transduction and, as described earlier in this chapter, normally functions by cycling between the active GTP-bound state and the inactive GDP-bound state (see Figure 22-13). The amino acid change caused by the *ras* oncogene missense mutation produces an oncoprotein that always binds GTP (Figure 22-19b), even in the absence of the normal signals such as phosphorylation of Ras, required for such binding by a wild-type Ras protein. In this way, the Ras oncoprotein continually propagates a signal that promotes cell proliferation.

Loss of protein domains. Structural alterations can also be due to the deletion of parts of a protein. The v-*erbB* oncogene encodes a mutated form of an RTK known as the EGFR, a receptor for the epidermal growth factor (EGF) ligand (Figure 22-20). The mutant form of the EGFR lacks the extracellular, ligand-binding domain as well as some regulatory components of the cytoplasmic domain. The result of these deletions is that the truncated v-*erbB*-encoded EGFR oncoprotein is able to dimerize even in the absence

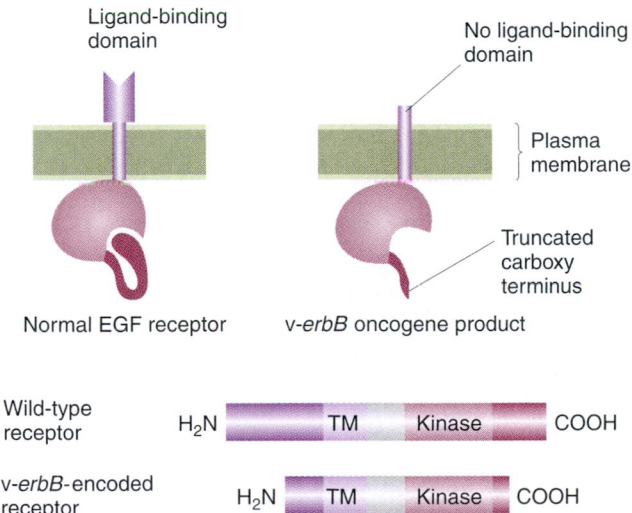

Figure 22-20 An oncogenic mutation affecting signaling between cells, EGFR, the normal receptor for epidermal growth factor (EGF), has a ligand-binding domain outside the cell, a transmembrane (TM) domain that allows the protein to span the plasma membrane, and an intracellular domain that has tyrosine-specific protein kinase activity. Normally, the kinase activity is activated only when EGF binds to the ligand-binding domain, and this activity is only transitory. The erythroblastosis tumor virus carries the v-*erbB* oncogene, which encodes a mutant form of EGFR that lacks pieces at both ends. These deletions allow the mutant protein to constitutively dimerize, leading to continuous autophosphorylation, which results in continuous transduction of a signal from the receptor. (After J. D. Watson, M. Gilman, J. Witkowski, and M. Zoller, *Recombinant DNA,* 2d ed. Copyright © 1992 by James D. Watson, Michael Gilman, Jan Witkowski, and Mark Zoller.)

of the EGF ligand. The constitutive EGFR oncoprotein dimer is always autophosphorylated through its tyrosine kinase activity and thus continuously initiates a signal transduction cascade.

Gene fusions. Perhaps the most remarkable type of structurally altered oncoprotein is one caused by a gene fusion. The classic example of fused genes emerged from studies of the Philadelphia chromosome, which, as already mentioned, is a translocation between chromosomes 9 and 22 that is a diagnostic feature of chronic myelogenous leukemia (CML). Recombinant DNA methods have shown that the breakpoints of the Philadelphia chromosome translocation in different CML patients are quite similar and cause the fusion of two genes, *bcr1* and *abl* (Figure 22-21). The *abl* proto-oncogene encodes a cytoplasmic tyrosine-specific protein kinase. The Brc1-Abl fusion oncoprotein has an activated protein kinase activity that is responsible for its oncogenic state.

Some oncogenes produce an oncoprotein that is identical in structure with the normal protein. In these cases, the oncogene mutation induces misexpression of the protein — that is, it is expressed in cell types from which it is ordinarily absent. Several oncogenes that cause misexpression are also associated with chromosomal translocations diagnostic of various B-lymphocyte tumors. B lymphocytes and their descendants, plasma cells, are the cells that synthesize antibodies, or immunoglobulins. In these B-cell oncogene translocations, no protein fusion is produced; rather, the chromosomal rearrangement causes a gene near one breakpoint to be turned on in the wrong tissue. In follicular lymphoma, 85 percent of patients have a translocation between chromosomes 14 and 18 (Figure 22-22). Near the chromosome 14 breakpoint is located a transcriptional enhancer from one of the immunoglobulin genes. This translocated enhancer element is fused to the *bcl2* gene, which is a negative regulator of apoptosis. This enhancer–*bcl2* fusion causes large amounts of Bcl2 to be expressed in B lymphocytes. These large amounts of Bcl2 essentially block apoptosis in these mutant B lymphocytes and provide them with an unusually long lifetime in which to accumulate cell proliferation-promoting mutations. There are strong parallels between this sort of dominant oncogene mutation and the dominant gain-of-function phenotypes caused by the fusion of the enhancer of one gene to the transcription unit of another in producing the *Tab* allele of the *Abd-B* gene (see Chapter 23). In each case, the introduction of an enhancer causes a dominant gain-of-function phenotype by misregulation of the transcription unit. Mutations such as *Tab* arise in the germ line and are transmitted from one generation to the next, whereas most oncogene mutations arise in somatic cells and are not inherited by offspring.

MESSAGE ··
Dominant oncogenes contribute to the oncogenic state by causing a protein to be expressed in an activated form or in the wrong cells.

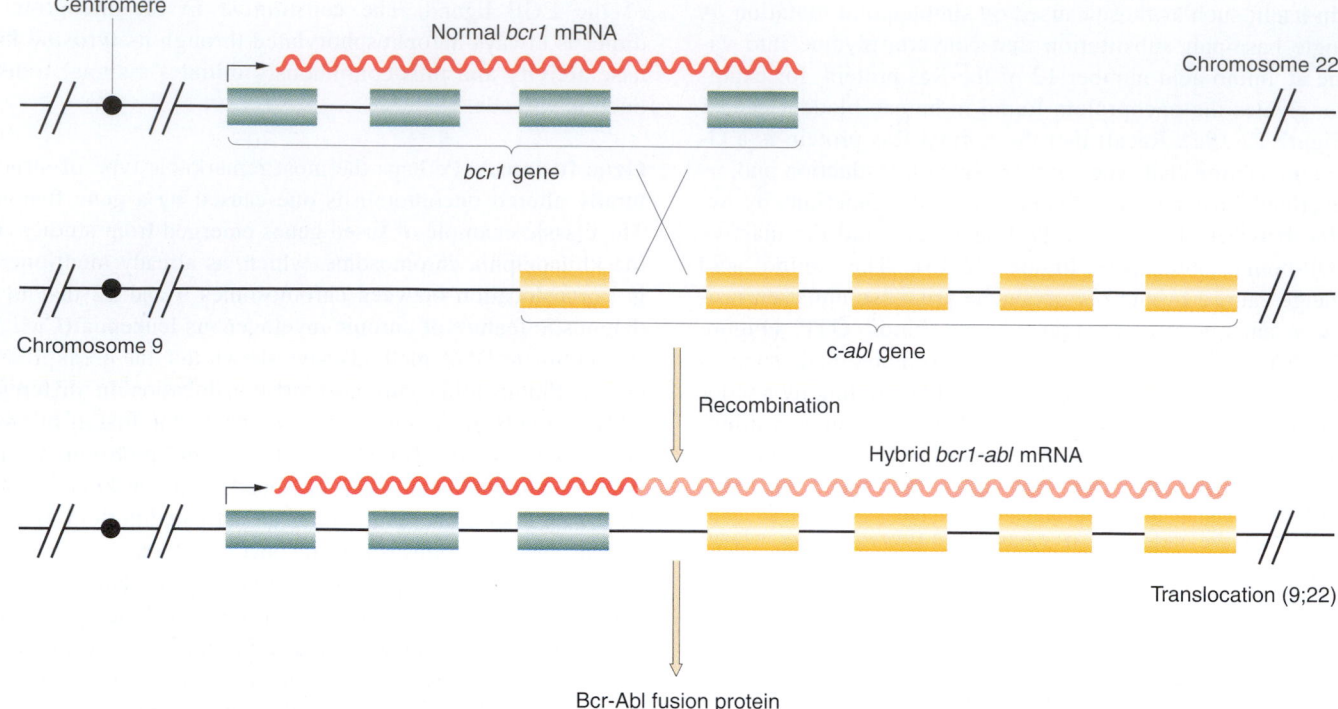

Figure 22-21   The chromosome rearrangement in chronic myelogenous leukemia. The Philadelphia chromosome, which is diagnostic of CML, is a translocation between chromosomes 9 and 22. The translocation breakpoints are in the middle of the c-*abl* gene, which encodes a cytoplasmic protein tyrosine kinase, and the *bcr1* gene, which also is thought to be a protein kinase. The translocation produces a hybrid Bcr1-Abl protein that lacks the normal controls repressing the c-*abl*-encoded protein tyrosine kinase activity. Only one of the two rearranged chromosomes of the reciprocal translocation is shown. (After J. D. Watson, M. Gilman, J. Witkowski, and M. Zoller, *Recombinant DNA,* 2d ed. Copyright © 1992 by James D. Watson, Michael Gilman, Jan Witkowski, and Mark Zoller.)

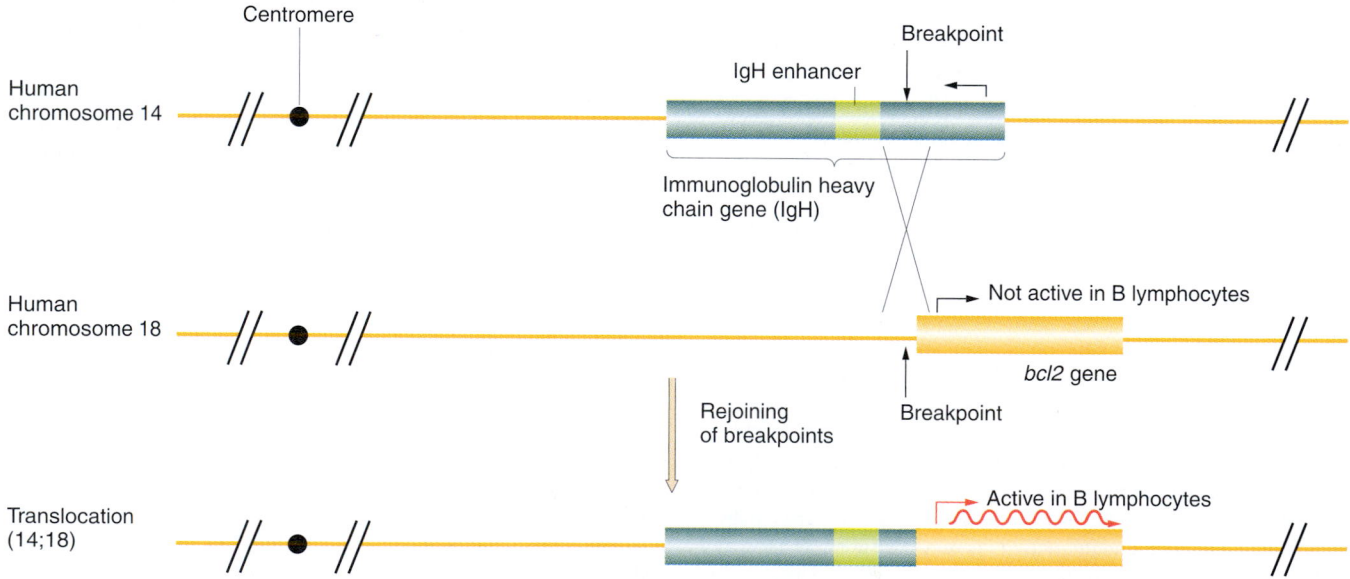

Figure 22-22   The chromosomal rearrangement in follicular lymphoma. The translocation fuses the transcriptional enhancer element of a gene, on chromosome 14, that makes one protein subunit of an antibody (the IgH subunit, also called the immunoglobulin heavy chain) to the transcription unit of a gene, on chromosome 18, that encodes Bcr2, a negative regulator of apoptosis. In this way, the Bcr2 protein is produced in antibody-producing cells, blocking any self-destruction signals from inducing apoptosis in those cells.

Classes of tumor-suppressor genes

The normal functions of tumor-suppressor genes fall into categories complementary to those of proto-oncogenes (see Table 22-1). Some tumor-suppressor genes encode negative regulators of the cell cycle, such as the Rb protein or elements of the TGF-β signaling pathway. Others encode positive regulators of apoptosis (at least part of the function of p53 falls into this category). Still others act indirectly, through a general elevation in the mutation rate. We shall consider two examples here.

Inheritance of the tumor phenotype

In retinoblastoma, the gene encoding the Rb protein, considered in the regulation of the cell cycle, is mutated. In retinoblastoma, a cancer typically affecting young children, retinal cells lacking a functional *RB* gene proliferate out of control. These *rb* null cells are either homozygous for a sin-

gle mutant *rb* allele or are heterozygous for two different *rb* mutations. Most patients have one or a few tumors localized to one site in one eye, and the condition is sporadic — in other words, there is no history of retinoblastoma in the family and the affected person does not transmit it to his or her offspring. Retinoblastoma is not transmitted in this case, because the *rb* mutation or mutations that inactivate both alleles of this autosomal gene arise in a somatic cell whose descendants populate the retina (Figure 22-23). Presumably, the mutations arise by chance at different times in development in the same cell lineage.

A few patients, however, have an inherited form of the disease, called hereditary binocular retinoblastoma (HBR). Such patients have many tumors, and the retinas of both eyes are affected. Paradoxically, even though *rb* is a recessive trait at the cellular level, the transmission of HBR is as an autosomal dominant (Figure 22-23). How do we resolve this paradox? In the presence of a germ-line mutation that

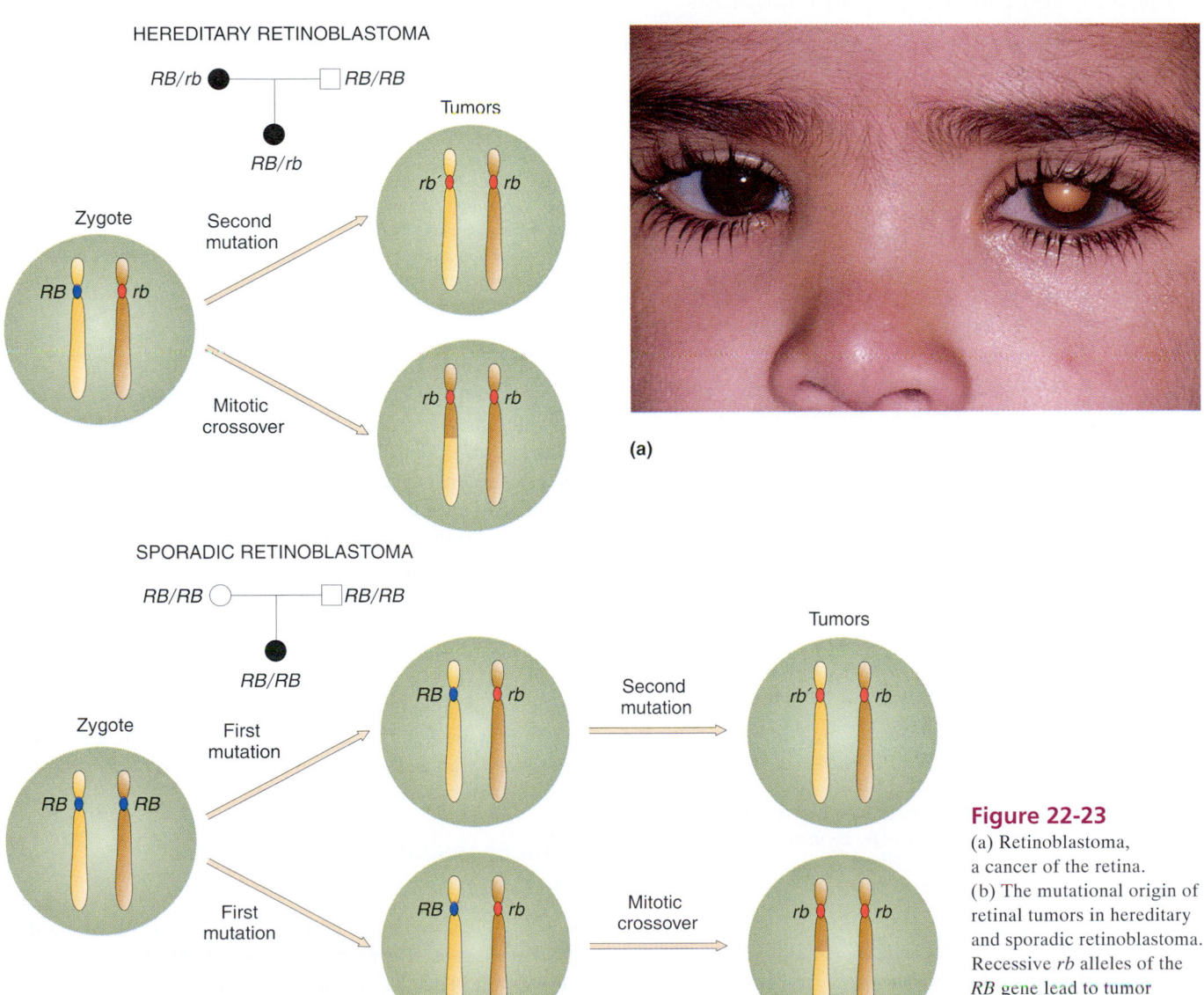

Figure 22-23
(a) Retinoblastoma, a cancer of the retina. (b) The mutational origin of retinal tumors in hereditary and sporadic retinoblastoma. Recessive *rb* alleles of the *RB* gene lead to tumor development. (Part a from Custom Medical Stock.)

knocks out one of the two copies of the *RB* gene, the mutation rate for *RB* makes it virtually certain that at least some of the retinal cells of patients with HBR will have acquired an *rb* mutation in the single remaining normal *RB* gene, thereby producing cells with no functional Rb protein.

Why does the absence of *RB* promote tumor growth? Recall from our consideration of the cell cycle that Rb protein functions by binding the E2F transcription factor. Bound Rb prevents E2F from promoting the transcription of genes whose products are needed for S-phase functions such as DNA replication. An inactive Rb is unable to bind E2F, and so E2F can promote the transcription of S-phase genes. In homozygous null *rb* cells, Rb protein is permanently inactive. Thus, E2F is always able to promote S phase, and the arrest of normal cells in late G1 does not occur in retinoblastoma cells.

p53 tumor-suppressor gene:
a link between the cell cycle and apoptosis

Another very important recessive tumor-promoting mutation has identified the *p53* gene as a tumor-suppressor gene. Mutations in *p53* are associated with many types of tumors, and estimates are that 50 percent of human tumors lack a functional *p53* gene. The active p53 protein is a transcriptional regulator that is activated in response to DNA damage. Activated wild-type p53 serves double duty, preventing progression of the cell cycle until the DNA damage is repaired and, under some circumstances, inducing apoptosis. In the absence of a functional *p53* gene, the p53 apoptosis pathway does not become activated, and the cell cycle progresses even in the absence of DNA repair. This progression elevates the overall frequency of mutations, chromosomal rearrangements, and aneuploidy and thus increases the chances that other mutations promoting cell proliferation or blocking apoptosis will arise. Other recessive tumor-promoting genes that have been identified also are implicated in the repair of DNA damage. Research suggests that genes that, when inactivated, produce the phenotype of elevated mutation rates are very important contributors to the progression of tumors in humans. Such recessive tumor-suppressor mutations that interfere with DNA repair promote tumor growth indirectly, because their elevated mutation rates make it much more likely that a series of oncogene and tumor-suppressor gene mutations will arise, corrupting the normal regulation of the cell cycle and programmed cell death.

MESSAGE ···
Mutations in tumor-suppressor genes, like mutations in oncogenes, act directly or indirectly to promote the cell cycle or block apoptosis.

Complexities of cancer

As discussed in this chapter, numerous mutations that promote tumor growth can arise. These mutations are themati-

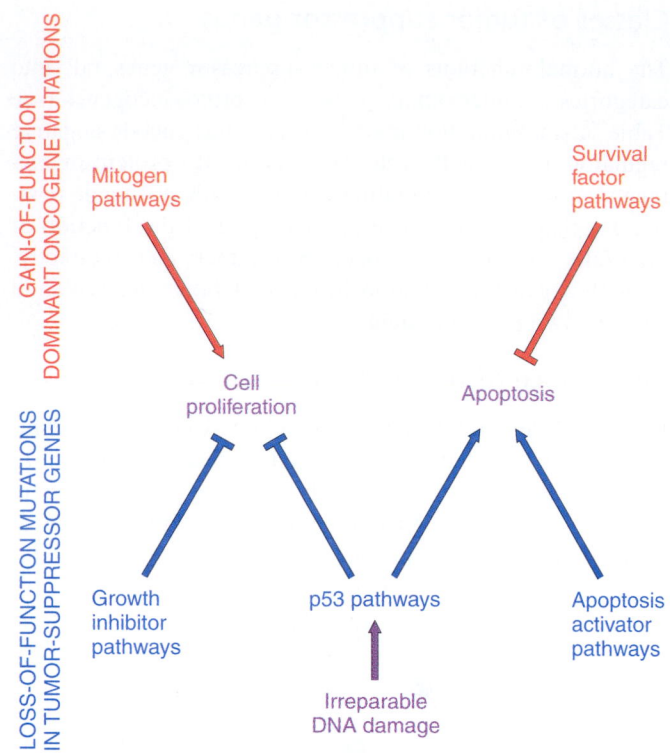

Figure 22-24   The major pathways that are mutated to contribute to cancer formation and progression. The main events that contribute to tumor formation are increased cell proliferation and cell survival (decreased apoptosis). The pathways in red are susceptible to gain-of-function oncogene mutation. The pathways in blue are susceptible to loss-of-function tumor-suppressor gene mutation.

cally related and can be understood in relation to the ways in which they alter the normal processes that govern proliferation and apoptosis (Figure 22-24). In some instances, such as colon cancer (Figure 22-18), we are even able to identify a series of independent mutations that contribute to the progression of a cell from a normal state through various stages of a benign tumor to a truly malignant state. The story does not stop there, however. Even among malignant tumors, their rates of proliferation and their abilities to invade other tissues, or metastasize, are quite different. Undoubtedly, even after a malignant state is achieved, more mutations accumulate in the tumor cell that further promote its proliferation and invasiveness. Thus, there is a considerable way to go before we have a truly comprehensive view of how tumors arise and progress.

Cancer research in the genomic analysis era

It is likely that by the year 2002, we shall have the complete DNA sequence of a human genome. With this information in hand, in principle we shall be able to deduce the coding information for all gene products (RNAs and proteins) encoded by the genome. It will then be possible to survey the expression levels of all gene products during the formation and progression of a particular type of tumor. The hope is that this systematic information will be a source of much greater insight into the panoply of gene-expression disturbances that characterize the malignant state. Indeed, such surveys are already underway, albeit on incomplete samplings of the total array of transcripts encoded by the human genome. These surveys already point to a complex set of modulations in gene expression during tumor progression. From these studies, new and unexpected oncogenes and tumor-suppressor genes will be identified, and factors with subtle contributions to tumor susceptibility will be uncovered. In addition, these surveys will form the basis of assay systems to determine the efficacy of possible anticancer drugs. We can expect that whole-genome approaches to cancer biology will be an important part of cancer research in the twenty-first century.

SUMMARY

Higher eukaryotic cells have evolved mechanisms that control their structure and their ability to proliferate. These controls are all highly integrated and depend on the continual evaluation of the state of the cell and the continual communication of information among neighboring cells and between different tissues. In particular, elaborate mechanisms maintain coordination of the various stages of the cell cycle and permit cell cycle progression only under the proper environmental circumstances. Other elaborate mechanisms essentially continually assess the status of surviving cells, and, if abnormal cells are detected, these mechanisms induce a program of self-destruction called apoptosis. Cancer is a genetic disease of somatic cells. In cancer cells, multiple mutations arise that disrupt both of these processes — in some cases, the cancer-promoting mutations directly affect these processes, whereas, in other cases, cancer-promoting mutations elevate the overall mutation rate of the cell. Both gain-of-function oncogene mutations and loss-of-function tumor-suppressor gene mutations can contribute to the tumor progression process through the uncoupling of the normal controls that hold the cell cycle in check or promote apoptosis.

CONCEPT MAP

Draw a concept map interrelating as many of the following terms as possible. Note that the terms are listed in no particular order.

DNA repair / enhancer / oncogene /
gain-of-function mutations /
cyclin-dependent protein kinase / apoptosis / checkpoints /
signal transduction pathway / cell proliferation / RTK /
tumor-suppressor gene

CHAPTER INTEGRATION PROBLEM

In the inherited form of retinoblastoma, an affected child is heterozygous for an *rb* mutation, which either is passed on by a parent or has newly arisen in the sperm or the oocyte nucleus that gave rise to the child. The heterozygous *RB/rb* cells are nonmalignant, however. The *RB* allele of the heterozygote must also be knocked out in the developing retinal tissue to create a tumorous cell. Such a knockout can occur through an independent mutation of the *RB* allele or by mitotic crossing-over such that the original *rb* mutation would now be homozygous.

a. If retinoblastoma were passed on to other siblings as well, could we determine if the original mutation was derived from the mother or the father? How?

b. Could we determine if the *rb* mutation was maternally or paternally derived if it arose de novo in a germ cell of one parent?

◆ Solution ◆

a. If the trait is inherited, we can determine from which parent it came. The most straightforward approach is to identify DNA polymorphisms, such as restriction fragment length polymorphisms, that map within or near the *RB* gene. *RB* has the curious property of being inherited as an autosomal dominant, even though it is recessive on a cellular basis. Given the dominant pattern of inheritance, by finding DNA differences in the parental genomes that map near each of the four parental alleles, we should be able to determine which allele is passed on to all affected offspring. That allele is the mutant one.

b. If the trait arose de novo in the sperm or the oocyte, we can still possibly determine, but with considerable difficulty, from which parent it came. One way to do so is to use recombinant DNA cloning techniques to isolate each of the two copies of the *RB* gene from normal cells. Only one of these alleles should be mutant. When the gene has been cloned, by DNA sequencing of the two alleles we may be able to identify the mutation that inactivates *RB*. If it arose de novo in the sperm or the oocyte, this mutation would not be present in the somatic cells of the parents. If sequencing also reveals some polymorphisms (for example, in restriction-enzyme recognition sites) that distinguish the alleles, we should then be able to go back to the parents' DNA and find out if the mother or the father carries the polymorphisms in the cloned mutant allele. (Whether this approach will work depends on the exact nature of the parental alleles and the mutation.)

If the mutation arose by mitotic crossing-over, we have additional tools available to us. In this case, the entire region around the *rb* gene will be homozygous for the mutant chromosome. By examining DNA polymorphisms known to map in this region, we should be able to determine if this chromosome derived from the mother or the father. The solution to this problem then becomes a standard exercise in DNA fingerprinting similar to that described in part a.

SOLVED PROBLEMS

MMTV, mouse mammary tumor virus, is an oncogenic retrovirus. It specifically causes tumors in the mammary glands of female mice and nowhere else. Unlike some other oncogenic retroviruses, it does not appear to produce its own oncogenic protein product (in contrast to the *v-erbB* oncogene carried by the erythroblastosis virus). Rather, MMTV encodes just the proteins necessary for its own reverse transcription into a DNA copy that integrates into the host genome, and for packaging into virion particles. It seems puzzling then that MMTV is an oncogenic virus.

Studies have provided two clues to how MMTV produces tumors. First, it turns out that MMTV carries a hormone response element (HRE) that causes strong increases in transcription in response to the presence of certain steroid hormones. Second, it turns out that there is usually just one MMTV insertion in the genome in mouse mammary gland tumors. The DNA surrounding the insertion sites of MMTV in many independently induced tumors was cloned out by standard recombinant DNA techniques, allowing the chromosomal DNA adjacent to the insertion site to be studied. This analysis revealed that in a mouse mammary gland tumor, MMTV is found integrated next to one of only a small number of sites (named *Int sites*) in the genome. (In contrast, in nontumorous infections, MMTV can integrate in many different locations in the genome.) *Int1*, the first of the sites to be studied, is immediately adjacent to the promoter region of a gene that encodes a secreted protein very similar to the *wg* protein of *Drosophila,* which is involved in cell-to-cell signaling during segmentation in the embryo.

a. Bearing in mind that female mammary gland development and lactation are dependent on certain female-specific steroid hormone signals, and based on the clues we have discussed, propose an explanation for how MMTV produces its oncogenic effects.

b. Propose an experimental approach for testing this hypothesis.

◆ Solution ◆

a. Because of the HRE that MMTV carries, it acts as a portable enhancer element. The steroid hormone receptor to which the HRE responds is probably expressed only in mammary glands, so that the HRE is a tissue-specific enhancer for female mammary glands. Thus, if the MMTV integrates near a gene that, when activated in female mammary gland cells, will deregulate cell proliferation, it has the potential to cause a tumor. In the case of *Int1*, for example, the protein is not normally expressed in mammary glands, but becomes expressed at high levels under the influence of the MMTV's HRE. In principle then, this is no different from dominant oncogenes arising from chromosomal rearrangements, as in Burkitt's lymphoma, in which an *Ig* enhancer is fused to the *c-myc* gene, activating *c-myc* in B lymphocytes. In the case of MMTV, however, a viral insertion rather than a chromosomal translocation causes the gene fusion.

b. This hypothesis postulates that the HRE is the only essential portion of MMTV with regard to oncogenesis, and that it acts via misregulation of the *Int1* gene. To test this, we could isolate a small DNA segment that included only the HRE of MMTV, and fuse this segment in vitro to a wild-type *Int1* gene. We could then directly inject this DNA into mouse blastocysts and integrate it randomly in the genome by germ-line transformation. Our prediction would be that each of these insertions should cause mammary gland tumors in females. We should also include two constructs as controls: first, germ-line transformation of a similar construct except that we place a nonsense mutation in the *Int1* coding sequences so that the protein product is inactive, and second, in another construct, we mutate the HRE so that it can no longer bind its steroid receptor. Neither of these control constructs should turn out to be oncogenic.

PROBLEMS

1. Cancer is thought to be caused by the accumulation of two or more "hits" — that is, two or more mutations affecting cell proliferation and survival within the same cell. Many of these oncogenic mutations are dominant: one mutant copy of the pertinent gene is sufficient to change the proliferative properties of a cell. Which of the following general types of mutations have the potential to be dominant oncogenes? Justify each answer.

    a. A mutation increasing the number of copies of a transcriptional activator of cyclin A.

    b. A nonsense mutation occurring shortly after the beginning of translation in a gene encoding a growth-factor receptor.

    c. A mutation increasing the level of FasL.

    d. A mutation that disrupts the active site of a cytoplasmic tyrosine-specific protein kinase.

    e. A translocation joining a gene encoding an inhibitor of apoptosis to an enhancer element for gene expression in the liver.

2. Many of the proteins that participate in the cell cycle progression pathway are reversibly modified, whereas, in the apoptosis pathway, the modification events are irreversible. Rationalize these observations in relation to the nature and end result of the two pathways.

3. Normally, FasL is present on cells only when an instruction needs to be sent to neighboring cells for them to undergo apoptosis. Suppose that you have a mutation that produces FasL on the cell surfaces of all liver cells.

    a. If the mutation were present in the germ line, would you predict such a mutation to be dominant or recessive?

    b. If such a mutant arose in somatic tissues, would you expect it to be tumor promoting? Why or why not?

4. Provide three types of evidence that cancer is a genetic disease.

5. Some genes can be mutated to become oncogenes by increasing the copy number of the gene. This, for example, is true of the gene encoding the Myc transcription factor. On the other hand, oncogenic mutations of *ras* are always point mutations that alter the protein structure of Ras. Rationalize these observations in relation to the roles of normal and oncogenic versions of Ras and Myc.

6. We now understand that mutations that cause the inhibition of apoptosis are found in tumors. Because proliferation itself is not induced by the inhibition of apoptosis, explain how this inhibition might contribute to tumor formation.

7. Suppose that you had the ability to introduce normal copies of a gene into a tumor cell that had mutations in the gene that caused it to promote tumor growth.

    a. If the mutations were in a tumor suppressor gene, would you expect that these normal transgenes would block the tumor-producing activity of the mutations? Why or why not?

    b. If the mutations were of the oncogene type, would you expect that the normal transgenes would block their tumor-promoting activity? Why or why not?

8. Insulin is a protein that is secreted by the pancreas (an endocrine organ) when blood sugar levels are high. Insulin acts on many distant tissues by binding and activating a receptor tyrosine kinase (RTK), leading to a reduction in blood sugar by appropriately storing the products of sugar metabolism.

    Diabetes is a disease in which blood sugar levels remain high because some aspect of the insulin pathway is defective. One kind of diabetes (let's call this type A) can be treated by giving the patient insulin. Another kind of diabetes (call it type B) is not ameliorated by insulin treatment.

    a. Which type of diabetes is likely to be due to a defect in the pancreas, and which type is likely to be due to a defect in the target cells? Justify your answer.

    b. Type B diabetes can be due to mutations in any of several different genes. Explain this observation.

9. Irreparable DNA damage can have consequences for both the cell cycle and apoptosis. Explain what the consequences are, as well as the pathways by which the cell implements them.

23

DEVELOPMENTAL GENETICS

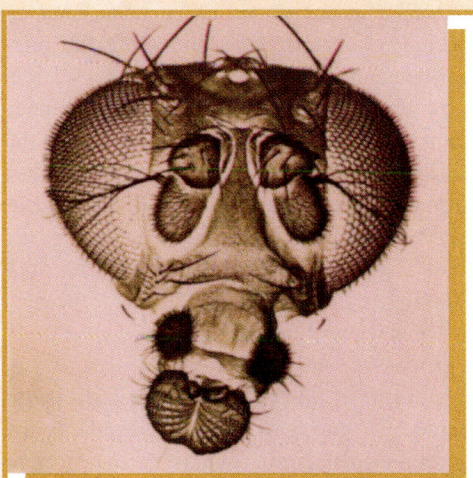

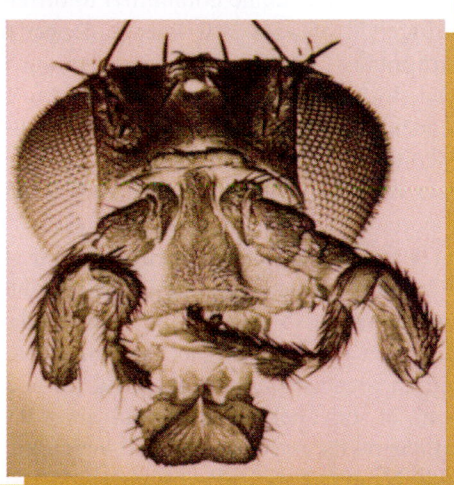

A homeotic mutation that alters the basic *Drosophila* body plan.
Homeosis is the replacement of one body part by another. In place of the normal antennae (shown at the top), an *Antennapedia* mutation causes the antennal precursor cells to develop into a leg.
(F. R. Turner/BPS.)

Key Concepts

A programmed set of instructions in the genome of a higher organism establishes the developmental fates of cells with respect to the major features of the basic body plan.

Developmental pathways are formed by the sequential implementation of various regulatory steps.

The zygote is totipotent, giving rise to every adult cell type; as development proceeds, successive decisions restrict each cell lineage to its particular fate.

Gradients of maternally derived regulatory proteins establish polarity along the major body axes of the egg; these proteins control the local transcriptional activation of genes encoding master regulatory proteins in the zygote.

Many proteins that act as master regulators of early development are transcription factors; others are components of pathways that mediate signaling between cells.

Some fate decisions are made autonomously by individual cells; many fate decisions require communication and collaboration between cells.

The same basic set of genes identified in *Drosophila* and the regulatory proteins that they encode are conserved in mammals and appear to govern major developmental events in many — perhaps all — higher animals.

In all higher organisms, life begins with a single cell, the newly fertilized egg. It reaches maturity with thousands, millions, or even trillions of cells combined into a complex organism with many integrated organ systems. The goal of developmental biology is to unravel the fascinating and mysterious processes that achieve the transfiguration of egg into adult. Because we know more about development in these organisms, we will restrict most of our discussion to animal systems.

The different cell types of the body are distinguished by the variety and amounts of the proteins that they express — the protein profile of each cell (that is, the quantitative and qualitative array of proteins that it contains). The protein profile of a cell in a multicellular organism is the end result of a series of genetic regulatory decisions that determine the "when, where, and how much" of gene expression. Thus, for a particular gene, we are interested in which tissues and at what developmental times the gene is transcribed and how much of the gene product is synthesized. From a geneticist's point of view, all developmental programming that controls an organism's protein profiles is determined by the regulatory information encoded in the DNA. We can look at the genome as a parts list of all the gene products (RNAs and polypeptides) that can be potentially produced and as an instruction manual of when, where, and how much of these products are to be expressed. Thus, one aspect of developmental genetics is to understand how this instruction manual operates to send cells down different developmental pathways, ultimately producing a large constellation of characteristic cell types.

This aspect is not all that we want to understand about developmental genetics and production of cellular diversity, however. We also want to understand how these different cell types are deployed in a coherent and constructive distribution — in other words, how they become organized into organs and tissues and how those organ systems and tissues are organized into an integrated, coherently functioning individual organism.

Central themes of developmental genetics

The general body plan is common to all members of a species and, indeed, is common to many very different species. All mammalian species have four limbs, whereas all insects have six. But all mammals and insects must, in the course of their development, differentiate the anterior from the posterior end and the dorsal from the ventral side. Eyes and legs always appear in the appropriate places. Except for severe disturbances that interfere with development, the basic body plan of a species appears to be quite immune from environmental modification. The study of the basic development of body plan can then be carried out by studying the internal genetic program of the organism without reference to the environment. We should not forget, however, that the study of the genetic determination of these basic developmental processes does not provide us with an expla-

nation of the phenotypic differences between individual members of a species. This chapter focuses on the processes underlying pattern formation, the construction of complex form, and how these processes operate reliably to execute the developmental program for the basic body plan.

Logic of building the body plan

During the elaboration of the body plan, cells commit to specific **cell fates;** that is, the capacity to differentiate into particular kinds of cells. The cell fate commitments have to make sense in regard to the location of the cell, because all organs and tissues are made up of many cells and the entire structure of an organ or tissue requires a cooperative division of labor among the participating cells. Thus, somehow, cell position must be identified, and fate assignments must be parceled out among a cooperating group of cells — called a **developmental field.**

Positional information is generally established through protein signals that emanate from a localized source within a cell (the initial one-cell zygote) or within a developmental field. It is the molecular equivalent to establishing the rules for geographic longitudes and latitudes. Just as we need longitudes and latitudes to navigate on the earth, cells need positional information to determine their location within a developmental field and to respond by executing the appropriate developmental program. When that positional information has been captured, generally a few intermediate cell types are created within a field. Through further processes of cell division and decision making, a population of cells with the necessary final diversity of fates will be established.

These further processes — **fate refinement** — can be of two types. In some situations, through asymmetric divisions of one of the intermediate types of cells, descendants are created that have received different regulatory instructions and therefore become committed to different fates. This can be thought of as a cell lineage-dependent mechanism for partitioning fates. In other situations, such fate decisions are made by committee — that is, the fate of a cell becomes dependent on input from neighboring cells and feedback to them through paracrine signaling mechanisms (see Chapter 22). Such neighborhood-dependent decisions are extremely important, because the chemical dialog between cells ensures that all fates have been allocated and that the pattern of allocation is coherent. The cell neighborhood-dependent mechanisms also provide for a certain developmental flexibility. Developmental mechanisms need to be flexible so that an organism can compensate for accidents such as the death of some cell. If some cells are lost through accidental cell death, the normal paracrine intercellular communication is then aborted and the surviving neighbors can become reprogrammed to divide and instruct a subset of their descendants to adopt the fates of the deceased cells. Indeed, the regeneration of severed limbs, as occurs in some animals, is a manifestation of the power of specification versus hardwired determination in building pattern.

The consequence of the preceding scenario for development is that the process of commitment to a particular fate is a gradual one. A cell does not go in one step from being totally uncommitted, or **totipotent,** to becoming earmarked for a single fate. Each major patterning decision is, in actuality, a series of events in which multiple cells that are at the same level of fate commitment are, in step-by-step fashion, assigned different fates. As these events unfold, cell proliferation is generally occurring. Thus, if we examine a **cell lineage** — that is, a family tree for a somatic cell and its descendants — we see that parental cells in the tree are less committed than their descendants.

Major decisions in building the embryo

A variety of developmental decisions are undertaken in the early embryo to give cells their proper identities and to build the body plan. Some of them are simple binary decisions:

- Separation of the germ line (the gamete-forming cells) from the soma (everything else).

- Establishment of the sex of the organism. (Ordinarily, all cells of the body make the same choice.)

These binary decisions tend to be made at one developmental stage and, as we shall see later, are examples of irreversible fate determination.

The other major decisions concern multiple fate options and far more intricate decision-making pathways; they lead to the complex pattern elements of the body plan, composed almost entirely of the somatic cells. Most of them are specification decisions taken by local populations of cells:

- Establishment of the positional information necessary to orient and organize the two major body axes of the embryo: anterior–posterior (from head to tail) and dorsal–ventral (from back to front).

- Subdivision of the embryonic anterior–posterior axis into a series of distinct units called **segments** or metameres and assignment of distinct roles to each segment according to its location in the developing animal.

- Subdivision of the embryonic dorsal–ventral axis into the outer, middle, and inner sheets of cells, called the **germ layers,** and assignment of distinct roles to each of these layers.

- Production of the various organs, tissues, systems, and appendages of the body through the coordinated

and cooperative action of localized groups of cells of characteristic segmental and germ-layer origin.

Applying regulatory mechanisms to developmental decisions

In Chapter 11, we learned the ways in which transcriptional regulation of biochemical pathways controls the production of specific proteins at the correct time, in the correct place, and in the correct amounts. In a biochemical pathway, the regulatory "switch" that activates or blocks the synthesis of the enzymes of that pathway is usually some nutrient being supplied to the organism externally. In contrast, in the developmental pathway, the regulatory switch depends on some key molecule that is produced internally by the organism itself: either a molecule synthesized by the very cell that makes the decision or a molecule produced by other cells. In a simple developmental pathway, the concentration of such a key molecule will determine if the "on" or the "off" binary choice is made (Figure 23-1). Above some threshold level,

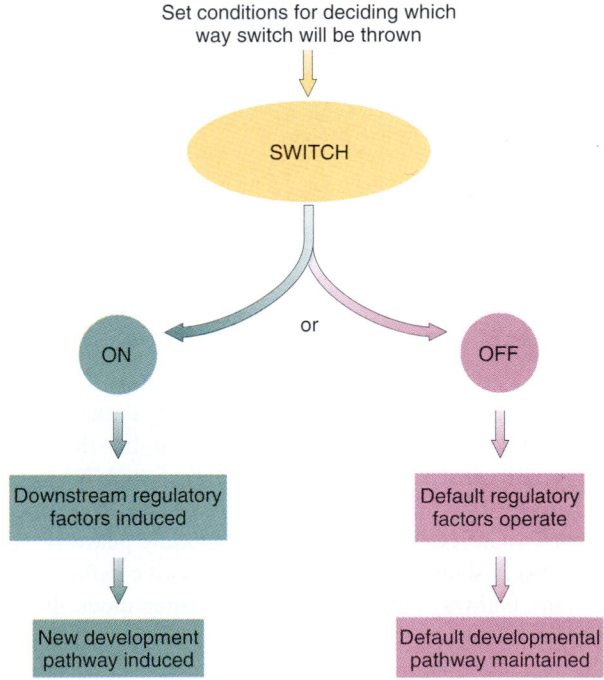

Figure 23-1 Decision making in developmental pathways. In developmental pathways, cells have to achieve different roles through a series of binary (on–off) decisions. Conditions within the cell allow a master switch to be regulated. When the master switch is activated, it sets in motion a cascade of "downstream" regulatory events that eventually lead the cell down a new developmental pathway. In the absence of the activation of the master switch, a set of default signals remains in place and maintains the cell in the default pathway.

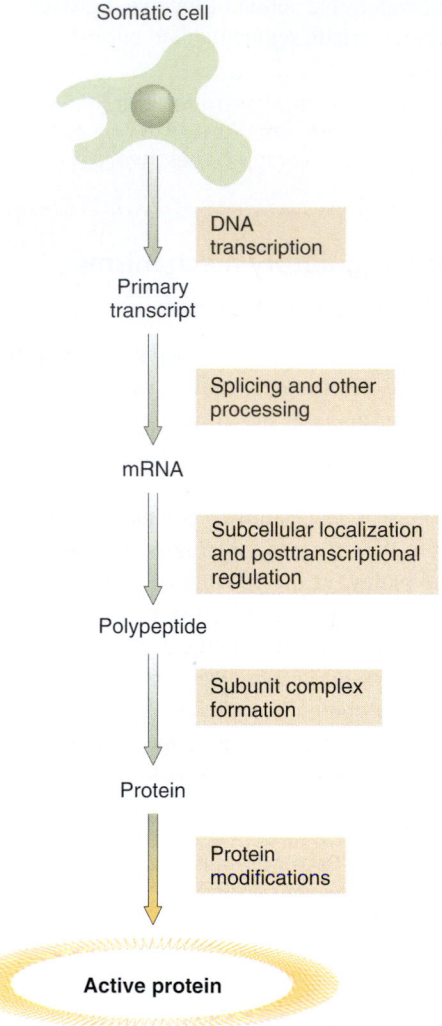

Somatic cell

DNA transcription

Primary transcript

Splicing and other processing

mRNA

Subcellular localization and posttranscriptional regulation

Polypeptide

Subunit complex formation

Protein

Protein modifications

Active protein

Figure 23-2   Regulation of protein activity. In a cell, there are many steps in which information is transferred between the gene and the active protein that it encodes. During development, regulatory mechanisms at any of these levels can control the production of active protein products.

one decision will be taken; below this threshold, the opposite decision will be made. The "off" decision implies that development will proceed along the path that had been programmed by previous decisions in the history of that cell; this "off" state is usually called the *default* pathway. The "on" decision shunts the cell into an alternative pathway.

Many pathways also have a maintenance mode that ensures that the "on" or "off" decision is permanently locked in place. Making a pathway decision and subsequently remembering that decision are both key to cell fate commitment.

When we dissect a developmental pathway, we find that its regulatory decisions might be at any level in the process of transferring information from the gene to the active protein (Figure 23-2). In this chapter, we shall encounter this process at levels ranging from transcriptional regulation to protein modification and subunit interactions to control the active protein profiles within cells.

Gene regulation at levels other than transcription initiation: examples

Tissue-specific regulation at the level of DNA structure

For many years, the structure of somatic cell and germ-line DNA was believed to be the same. It is indeed true for the bulk of DNA in a cell. However, some important exceptions have been observed in somatic cells of higher organisms. Because these somatic cells never contribute genetic material to offspring, regulatory mechanisms that alter the structure of the DNA in these dead-end cells have been able to evolve without altering germ-line gene organization.

Here, we shall examine one instance of gene rearrangement that has important regulatory consequences. In this case, tissue-specific amplification of the number of copies of a gene leads to high levels of gene expression in that tissue. Later in this chapter, we shall see how somatic DNA rearrangement plays a crucial role in a highly specialized and important developmental pathway — the production of antibodies by cells of the immune system.

Egg production (oogenesis) takes place in the ovary of the *Drosophila* adult female. Both a mature oocyte and a proteinaceous eggshell must be constructed during this process. Follicle cells, somatic cells that surround each oocyte, face the task of building the eggshell very rapidly and, hence, must synthesize large amounts of the eggshell proteins in a brief period of time. *Drosophila* has evolved an

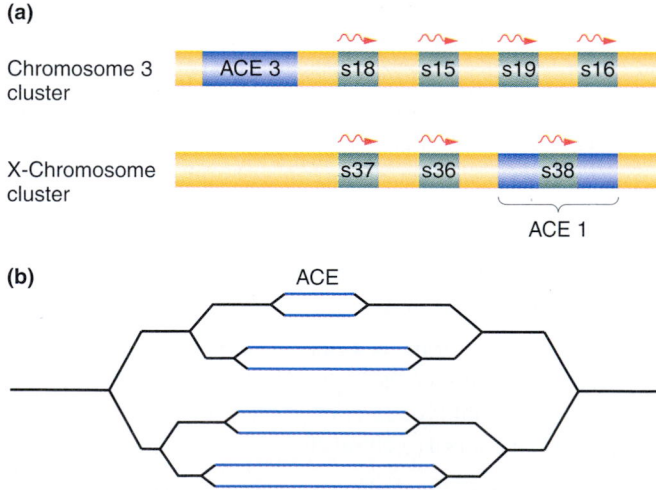

(a)

Chromosome 3 cluster — ACE 3 | s18 | s15 | s19 | s16

X-Chromosome cluster — s37 | s36 | s38 — ACE 1

(b)   ACE

Figure 23-3   The structure of the two eggshell-gene clusters in *Drosophila melanogaster*. (a) The cluster on chromosome 3 contains four eggshell genes, and the cluster on the X chromosome contains three. Each cluster is only about 10 kb long. Each ACE region contains an amplification control element. This region is thought to act as a special site for the initiation of tissue-specific DNA replication in follicle cells of the ovary. (b) A model of how amplification might occur through multiple DNA replication initiations at an ACE. Each line represents a double helix of DNA. (After T. Orr-Weaver, "*Drosophila* Chorion Genes: Cracking the Eggshell's Secrets," *Bioessays* 13, 1991, 97.)

elegant solution to this problem: the copy number of the eggshell genes is increased through somatic DNA rearrangements that occur only in the follicle cells that make the eggshell. In *Drosophila,* there are only seven major eggshell genes in most cells. Specifically during follicle cell development, by a special replication mechanism (Figure 23-3), the eggshell genes in *Drosophila* are amplified between 20-fold and 80-fold in the follicle cells just before they need to be transcribed. Thus, in *Drosophila,* extra DNA templates for the eggshell genes are present only when and where they are needed — in the follicle cells. We can imagine that this tissue-specific amplification is much more efficient than having to carry around the multiple copies of the eggshell genes in every cell in the body.

MESSAGE ··

Somatic changes in gene structure or copy number can be used to regulate tissue-specific gene activity.

Transcript processing and tissue-specific regulation

The transcripts of most higher eukaryotic genes are extensively processed during their maturation to mRNAs. Regulated mRNA splicing can be an important developmental control point. One striking example is the regulation of the **P element,** a transposable element (Figure 23-4). Recall from Chapter 20 that the P element is a family of related DNA sequences found in many strains of *Drosophila melanogaster.*

The intact P element is a 3-kb-long piece of DNA that encodes its own transposase, an enzyme that catalyzes the transposition of the element from one site in the genome to another. In strains of flies carrying the P element, it transposes only in the germ line, not in somatic cells, because P transposase is found only in the germ line. P transcription, however, occurs in every cell. How can we reconcile ubiquitous P transcription with germ-line-specific transposase activity? The splicing of P transposase is different in germ-line cells and soma. The germ-line P transposase mRNA contains four exons numbered 0, 1, 2, and 3 (see Figure 23-4). Frank Laski, Don Rio, and Gerry Rubin hypothesized that, in somatic cells, the P transposase mRNA retains the intron between exons 2 and 3; in other words, during splicing of the nascent P transposase transcript in somatic cells, the 2,3 intron is not spliced out. A stop codon present in this intron creates a defective somatic transposase essentially encoded only by exons 0, 1, and 2. As a consequence, somatic P transposase is missing its normal carboxyl terminus and is nonfunctional. To test this hypothesis, Laski, Rio, and Rubin constructed a P element in which the intron between exons 2 and 3 was removed in vitro by the cutting and pasting of the P-element DNA, and the resulting P{Δ2, 3} transgenic construct was introduced into flies by DNA transformation. The presence of clones of mutant tissue in flies carrying the P{Δ2, 3} transgene together with another P element containing a marker expressed in somatic tissue

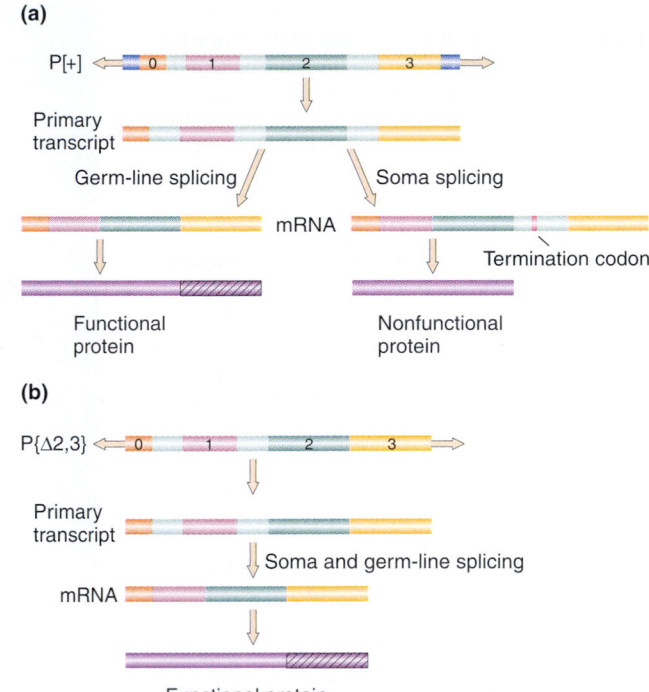

Figure 23-4 Germ-line-specific expression of *Drosophila* P elements: an example of tissue-specific regulation by RNA splicing. (a) The somatic and germ-line mRNA structures for a wild-type P element are shown. The germ-line mRNA is formed by splicing of four exons (numbered 0 to 3) and encodes P transposase. The somatic mRNA lacks the splice between exons 2 and 3; that is, the intron between these two exons is retained in the mRNA. Because the intron between exons 2 and 3 contains a termination codon, the resulting shorter somatic protein is defective, lacking the transposase activity. (b) A modified P element transgene, P {Δ2, 3}, that lacks the intron between exons 2 and 3 is depicted. In all tissues, P{Δ2, 3} produces an mRNA indistinguishable from that of the wild-type P element, resulting in P transposase activity in both germ line and soma.

(the eye) did indeed prove able to induce somatic transposition of other P elements (Figure 23-5), confirming the idea that the germ-line specificity of P transposase activity is at the level of regulated splicing.

MESSAGE ··

The production of an active protein can be regulated by controlling the pattern of splicing of an initial transcript into an mRNA.

Posttranscriptional regulation

A variety of mechanisms modulate the ability to translate mRNAs. Many of these mechanisms operate through interactions of regulatory molecules with sequences in the 3′ ends of transcripts. An mRNA can be divided into three parts: a 5′ untranslated region (5′ UTR), the polypeptide coding region, sometimes called the open reading frame (ORF), and the 3′ translated region (3′ UTR). If certain sequences are present within the 3′ UTR of an mRNA, that mRNA will be very rapidly degraded. In other cases,

Figure 23-5 Somatic expression of P transposase in *Drosophila*. In both panels, the flies contain a source of P transposase and a second P element lacking transposase activity but capable of being transposed by a source of P transposase. This second P element is marked with the w^+ gene, which confers red eye color. The standard white genes in the fly contain w null mutations, so any pigmentation depends on the presence of the P[w^+] transposon. (a) P transposase is derived from a wild-type P element (see Figure 23-4a). The resulting eye is a homogeneous red color. The absence of white eye tissue indicates that the P element is inactive in soma. Thus, the wild-type transposase mRNA was spliced to yield the inactive form of the enzyme. (b) Here, P transposase derives from the P{Δ2, 3} transgene. The resulting eyes are a mixture of red and white sectors. The white sectors represent tissue in which the P[w^+] element has "jumped out of" the chromosome, indicating that P transposase was active in those somatic cells. (Courtesy of S. Misra and G. Rubin, University of California, Berkeley.)

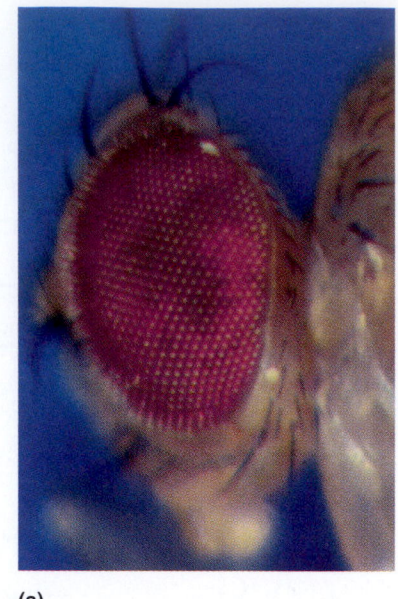

(a)

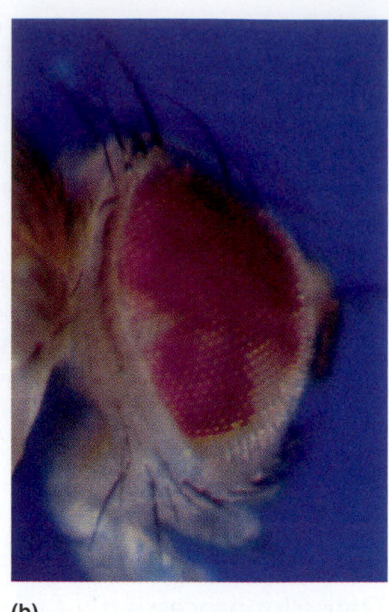

(b)

sequences in the 3′ UTR do not cause mRNA degradation but nonetheless can lead to lower levels of translation. Such sequences have been identified in the 3′ UTRs of mRNAs encoded by genes having roles in sex determination of the nematode *Caenorhabditis elegans*. Mutations that alter these 3′ UTR sequences lead to higher than normal levels of synthesis of the proteins encoded by these genes. Presumably, these 3′ UTR sequences are the target sites for proteins that digest mRNA molecules or that block their translation.

A novel phenomenon, in which 3′ UTR sequences interact with a regulatory RNA molecule, has recently been uncovered in *C. elegans*. This phenomenon was discovered in studies of heterochrony; that is, in studies of mutations that alter the timing of development. In this nematode, the wild-type adult worm develops after four larval stages. Some heterochronic mutations cause the adult worm to develop prematurely, after only three larval stages. Others cause the adult worm to be delayed, with the adult arising only after an additional fifth larval stage. The basis for the altered timing of development is that the normal chronology of cell divisions in the various cell lineages is altered such that certain divisions in a lineage are either skipped, causing premature adult development, or reiterated, producing delayed adulthood. Two of the heterochronic genes are called *lin-4* and *lin-14* (*lin* is an abbreviation for *lineage* defective). The product of the *lin-4* gene is known to repress translation of *lin-14* mRNA. It turns out that the repressor molecule encoded by the *lin-4* gene is not a protein but an RNA that has sequences complementary to sequences within the 3′ UTR of *lin-14*. It is thought that a double-stranded RNA structure is formed between the 3′ UTR of *lin-14* mRNA and the complementary *lin-4* RNA and that this double-stranded RNA somehow leads to repression of translation. A 3′ UTR may also contain sequences that act

as sites for anchoring an mRNA to particular structures within a cell. The ability to localize an mRNA within a cell can lead to differences in the concentration of the protein product of that transcript in that cell and in its progeny. Later in this chapter, we will see how this localization mechanism contributes to the formation of one of the body axes of the *Drosophila* embryo.

MESSAGE ···

Regulatory instructions are also contained within noncoding regions of mRNAs.

Posttranslational regulation

After polypeptides have been synthesized, there are still many opportunities for regulatory events to take place, and posttranslational regulation is of major importance to the cell. Enzymatic modifications to proteins might change their biological activities. We shall see examples later in this chapter, where we consider how phosphorylation of certain amino acids can change the activity of a protein. Interactions between different polypeptides to form multiprotein complexes might also produce changes in the activities of the constituent polypeptides. For example, in the section on the *Drosophila* sex-determination pathway, we shall see that transcription-factor activity is controlled through competitive interactions between different possible subunits to form dimeric proteins.

Binary fate decisions: pathways of sex determination

In many species, sex determination is associated with the inheritance of a heteromorphic chromosome pair in one sex. However, not all species have evolved from a common ancestor that possessed such a heteromorphic sex-chromosome

23-1 TABLE	Comparison of the Features of Sex Determination in *Drosophila* and Mammals	
Aspect of sex determination	*Drosophila*	Mammals
Chromosomal basis of XX versus XY	X:A ratio	Presence or absence of Y
Molecular basis of XX versus XY	Numerator:denominator transcription factor subunit interactions	Transcription factor for testis determination located on Y chromosome
Decision-making process	Cell by cell	Testis acts as source of male-determining organ instructing other tissues
Switch	Transcription of *Sxl(Sex lethal)* in early embryo	Presence or absence of *SRY*(human) ~ *Sry* (mouse) gene
Level of downstream regulation	Alternative mRNA splicing	Transcription
Positive feedback mechanism development	Positive feedback loop of SXL protein on *Sxl* mRNA splicing	Continued presence of testis throughout
Elaboration of sex-specific gene expression	Presence of male- or female-specific form of *dsx (doublesex)*-encoded protein. Generally, DSX-F is a repressor of expression of male-specific genes, and DSX-M is a repressor of female-specific genes. For a few genes, DSX-F or DSX-M may act as activators of female-specific or male-specific gene expression.	Testosterone-mediated activation of androgen receptor in target tissues. Testosterone–androgen receptor regulates transcription by binding to testosterone-responsive elements (enhancers and silencers).

set. Rather, XX-XY sex-determination mechanisms appear to have arisen independently many times in evolution. The XX-XY sex chromosomes of flies and mammals arose independently and, as we shall see, the underlying mechanisms for sex determination are quite different. The main features of sex-determination mechanisms in flies and mammals are contrasted in Table 23-1.

In both flies and mammals, many areas of the body display sexually dimorphic characteristics; that is, these areas differ in males and in females. For example, in *Drosophila*, the two sexes differ in the structure of the sex organs themselves and in pigmentation of the abdomen. The sex-determination mechanisms and how they ensure expression of such sexually dimorphic characteristics are the main focus of the discussions that follow.

Drosophila sex determination: every cell for itself

Every cell lineage in *Drosophila* makes its own sexual decision. One of the best ways to demonstrate this is by analyzing XX-XY mosaic flies; that is, individual flies containing a mixture of XX and XY cells. Such mosaics show a mixture of male and female phenotypes, according to the genotype of each individual cell. The interpretation of this difference is that every cell in *Drosophila* independently determines its sex.

Phenotypic consequences of different X-chromosome-to-autosome ratios

In *Drosophila,* the chromosomal basis of sex determination is due to the ratio of X chromosomes to sets of auto-

23-2 TABLE	Effect of Sex-Chromosome Dosage on Somatic Sexual Phenotype in Diploid *Drosophila* and Humans	
Sex chromosome constitution	SOMATIC SEXUAL PHENOTYPE	
	Drosophila	Humans
Euploidy		
XX	♀	♀
XY	♂	♂
Aneuploidy		
XXY	♀	♂
XO	♂	♀

somes. Recall that, in *Drosophila*, $n = 4$: one sex chromosome and three different autosomes. Hence, one autosomal set, which we shall represent as A, comprises the three different autosomes and, in a diploid fly, A = 2. The effect of this X:A ratio can best be seen by examining sex-chromosome aneuploids (Table 23-2). A normal 2X *Drosophila* diploid (XX AA) has an X:A ratio of 1.0 and is phenotypically female. An XY diploid (XY AA) has an X:A ratio of 0.5 and is male; an XO diploid also is male (although sterile). Triploids with three X chromosomes (XXX AAA) are females, those with one X (XYY AAA) are male, and those with two X's (XXY AAA) are "in between" (intersexes).

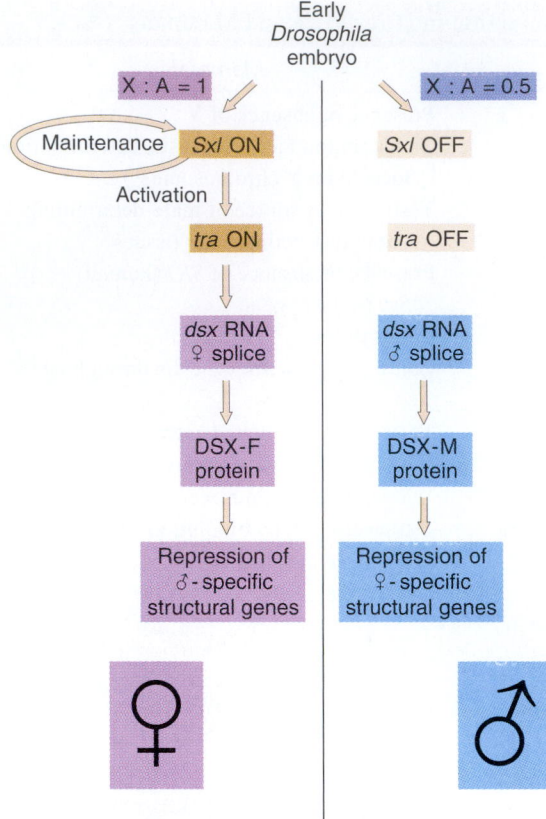

Figure 23-6   The pathway of sex determination and differentiation in *Drosophila*. The X : A ratio is evaluated through the interaction of numerator and denominator monomeric protein subunits that interact to produce an active complex referred to as the NUM–NUM transcription factor. The level of active numerator transcription factor determines whether *Sxl (Sex lethal)* is to be permanently turned on or is to remain off. If *Sxl* is on, then the female sex-differentiation pathway is turned on, ultimately causing splicing of a form of the *dsx (doublesex)* mRNA that produces a transcription factor that represses male-specific genes. If *Sxl* is off, then the sex-differentiation pathway is not activated and the default *dsx* splicing pattern creates an mRNA encoding a female-repressing transcription factor.

Basics of the regulatory pathway

Let's look first at an overview of the pathway (Figure 23-6). The X : A ratio in the early embryo establishes whether a fly will become male or female. This directive for establishing the sexual phenotype is carried out by a master regulatory switch and several downstream sex-specific genes. The "off" position of the switch produces the male pathway of determination, whereas the "on" position shunts cells into the female mode of sexual determination. The pathway is initiated by differential transcription, the direction of the switch is maintained, and the decision is propagated by differential RNA splicing. The default mode of the pathway culminates in the production of male-specific transcription factors, whereas the alternative shunt culminates in the production of female-specific transcription factors.

Regulatory switch

Through genetic analysis, we know that the regulatory switch is the activity of a gene called *Sxl (Sex lethal)*. In a fly with an X : A ratio of 1.0, SXL protein is synthesized and the fly develops as a female. In a fly with an X : A ratio of 0.5, no functional SXL protein is produced and, consequently, the fly develops as a male.

Setting the switch in the "on" or "off" position.   The X : A ratio sets in motion the sex-determination pathway by an interaction of the protein products of a series of X-chromosomal, zygotically expressed *numerator* genes and autosomal, maternally and zygotically expressed *denominator* genes. At least some of the numerator and denominator genes encode transcription factors of a type called **basic helix-loop-helix (bHLH) proteins.** bHLH proteins are known to function as transcription factors only when two bHLH monomers complex to form a dimeric protein. For our purposes here, we shall use NUM to designate the X-chromosome-encoded bHLH numerator proteins and DEM to indicate the autosomally encoded bHLH denominator proteins.

These transcription factors have only one role: in a narrow time window in the early *Drosophila* embryo — roughly from 2 to 3 hours after fertilization — they determine if the *Sxl* regulatory switch gets flipped on. The *Sxl* gene is essentially a "toggle switch" that is permanently locked into an "on" position in females or an "off" position in males (Figure 23-7a). To set the *Sxl* switch in the "on" position, the

Figure 23-7   The initiation and maintenance of the *Sxl* switch.   ▶
(a) The *Sxl* switch. High levels of the NUM–NUM transcription-factor dimers activate transcription from the early promoter (P_E) of *Sxl*. (Activated promoters are represented by blue rectangles, inactive promoters by brown rectangles.) By midembryogenesis, P_E is turned off, and *Sxl* is transcribed in all cells of the animal from the constitutively active late promoter (P_L). Through binding to the primary P_L transcript and regulating its splicing pattern (preventing the male exon from being included in the final mRNA), preexisting active SXL protein ensures the further production of active SXL protein, which continues the splicing pattern leading to its own formation, thus creating an autoregulatory loop. On the other hand, when the X : A ratio is 0.5, P_E is not activated. Thus, no SXL protein is present in the early embryo, and the RNA splicing pattern of the P_L transcript generates an mRNA that includes the male exon. This exon contains a stop codon (UGA), terminating the SXL polypeptide prematurely. The short SXL protein made in males is completely nonfunctional. AUG denotes the location of the translation initiation codon for the SXL polypeptide. (b) A plausible mechanism for the molecular basis of the X : A ratio. During early embryogenesis, the numerator and denominator genes are expressed. NUM subunits (red circles) encoded by X-chromosome genes and DEM polypeptide subunits (green circles) encoded by autosomal genes form dimers at random. Only NUM–NUM dimers form active transcription factor. If the X : A ratio is 1.0, high levels of these NUM–NUM dimers form, bind to the P_E enhancer, and activate transcription factor P_E. If, by chance, the X : A ratio is 0.5, most NUM subunits are part of NUM–DEM heterodimers and do not function as active transcription factors.

level of the active X : A NUM transcription factors must be high (owing to an X : A ratio of 1.0). With high (female) levels, the X : A transcription factors present in the early embryo bind to enhancers of the *Sxl* gene, activating its transcription from the *Sxl* early promoter. The transcript made from the early promoter then produces active SXL protein.

In contrast, if the levels of the NUM factors are too low, as is the case when the X : A ratio is 0.5, then there is insufficient transcription factor to activate *Sxl* transcription and no SXL protein is made.

The NUM proteins very likely measure the X : A ratio by competing for dimer formation with the DEM proteins to

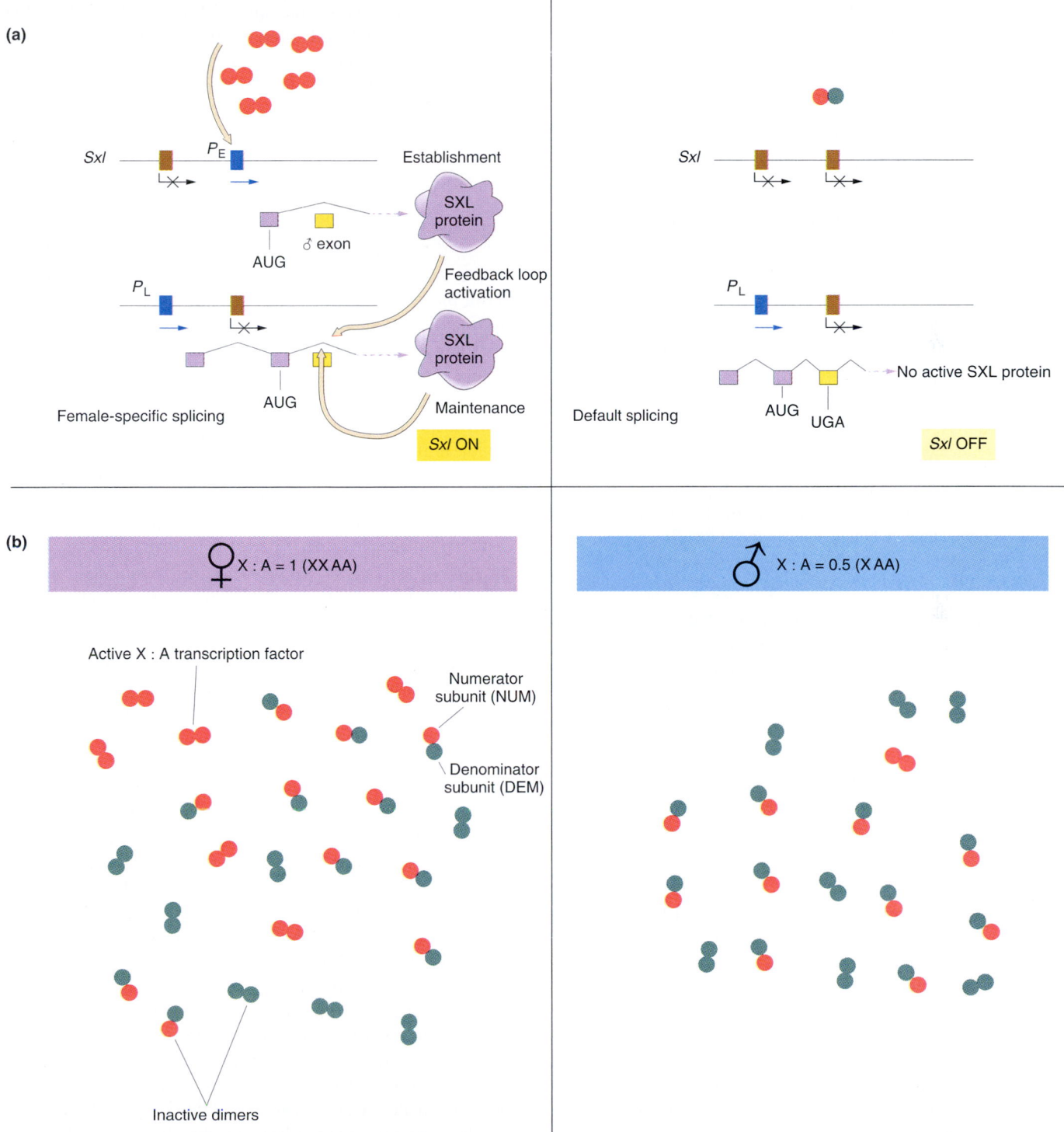

form active NUM transcription-factor dimeric protein complexes (Figure 23-7b). Although we do not know for sure how this works, here is a plausible mechanism.

◆ The NUM monomers have a sequence-specific DNA-binding site, whereas the DEM proteins lack a DNA-binding site.

◆ The DNA-binding site of NUM recognizes an enhancer sequence that regulates transcription from the promoter of the *Sxl* regulatory-switch gene. As we shall see in the next section, transcription from this promoter is required for establishment of *Sxl* gene expression in the early embryo.

◆ Both NUM and DEM polypeptides are synthesized at levels proportionate to the number of copies of each bHLH-encoding numerator or denominator gene in the cell. In this way, embryos with an X : A ratio of 1.0 have twice as much NUM polypeptide per cell as do embryos with an X : A ratio of 0.5. In contrast, regardless of the X : A ratio, these cells have the same level of DEM.

◆ All possible combinations of dimers can form, in proportion to the relative concentrations of NUM and DEM monomers in a cell: NUM–NUM homodimers, NUM–DEM heterodimers, and DEM–DEM homodimers.

◆ To be an active transcription factor, both subunits of a bHLH dimer must possess sequence-specific DNA-binding sites. This is true only for NUM homodimers. In a sense, then, when present in the same dimer, the DEM monomers are inhibitory to the transcription-factor activity of the NUM subunits.

The outcome of this scenario is that the higher the NUM : DEM ratio, the more active NUM–NUM transcription factor will be present in a cell. Thus, in early embryos with an X : A ratio of 1.0, we can expect that much more active numerator transcription factor will accumulate than in embryos with an X : A ratio of 0.5.

MESSAGE ••

Protein–protein interactions, such as competition between normal and inhibitory subunits for dimer formation, can be triggers for controlling developmental switches.

Maintaining the switch in a stable position. The *Sxl* gene has two promoters. The early promoter is the only one that is activated by the NUM–NUM transcription factors. The early promoter (P_E) is active only early in embryogenesis. Later in embryogenesis and for the remainder of the life cycle, the *Sxl* gene is transcribed from the late promoter (P_L) regardless of the X : A ratio or any other condition. This late promoter is active in every cell in the animal, beginning with midembryogenesis and persisting for the lifetime of the organism. The primary transcript produced by *Sxl* transcription from the late promoter is much larger than the primary transcript from the early promoter and is subject to alternative mRNA splicing, depending on the presence or absence of preexisting active SXL protein in the cell. The SXL protein is an RNA-binding protein that alters the splicing of the nascent *Sxl* transcript coming from this late promoter. When mRNA splicing occurs in the presence of bound SXL protein, splicing of *Sxl* produces an mRNA that encodes more active SXL RNA-binding protein. This SXL protein in turn binds to more *Sxl* primary transcript from the late promoter, creating the spliced form of the mRNA that encodes functional SXL protein, and so forth. Thus a feedback, or autoregulatory, loop, controlled at the level of RNA splicing, maintains SXL activity throughout development in flies with an X : A ratio of 1.0.

MESSAGE ••

The autoregulatory loop exemplifies how an early developmental decision can be "remembered" for the rest of development, even after the initial signals that established the decision have long disappeared.

In contrast, when the X : A ratio is 0.5, the *Sxl* switch is set in the "off" position. The early promoter is not activated early in embryogenesis and hence the early X : A = 0.5 embryo has no SXL protein. As a consequence, in the absence of any active SXL protein, the primary *Sxl* transcript of the late, constitutive *Sxl* promoter is processed in the default mRNA splicing pattern. This default *Sxl* mRNA is nonfunctional, in the sense that it encodes a stop codon shortly after the translation-initiation codon of its protein-coding region. The small protein produced from this male-specific spliced mRNA has no biological activity. Thus, in *Drosophila* with a low level of active NUM–NUM transcription factor, the absence of active SXL protein early in development predestines that there will be no SXL activity throughout the remainder of development.

Propagating the decision. Not only does SXL have to have an autoregulatory maintenance function, but it must be capable of activating the shunt pathway that will lead to female-specific gene expression. It accomplishes this activation through the same RNA-binding activity. Only in the presence of SXL protein is the primary *tra (transformer)* transcript spliced to produce an mRNA-encoding active TRA protein (Figure 23-8a). In turn, TRA protein is an RNA-binding protein that produces female-specific splicing of the *dsx (doublesex)* nascent RNA. The mRNA produced by this splicing pattern encodes a DSX-F protein, a transcription factor that globally represses male-specific gene expression (Figure 23-8b).

In the absence of active SXL protein, the splicing pattern of *tra* primary transcript produces an mRNA that does not encode functional TRA protein. In the absence of active TRA protein, splicing of the *dsx* primary transcript leads to the production of a DSX-M transcription factor that represses female-specific gene expression (Figure 23-8b).

How genetic analysis has contributed to this understanding of sex determination is described in the next section.

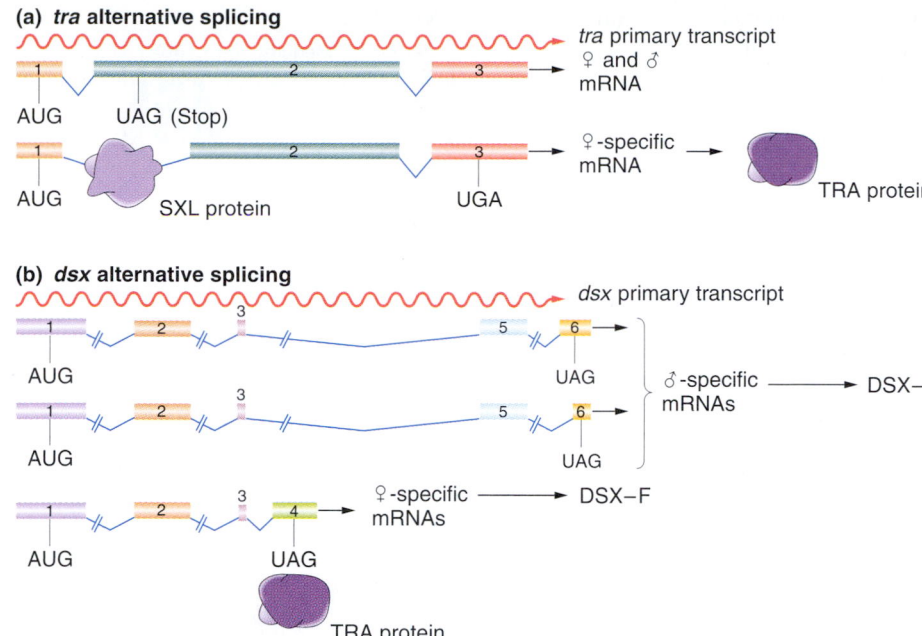

(a) *tra* alternative splicing

(b) *dsx* alternative splicing

Figure 23-8   Alternative splicing of *tra* and *dsx* transcripts. (a) Two forms of *tra* mRNA are produced. One form is present in both sexes and, because of a stop codon (UAG) in exon 2, does not encode a functional protein. The other form is female specific and encodes the active TRA polypeptide. (b) Different *dsx* mRNAs are produced in both sexes. In males, exon 4 is not included in two related male-specific mRNAs. Both male-specific mRNAs make related polypeptides, DSX-M, that act to repress transcription of female-specific genes. In females, exons 5 and 6 are not included in the *dsx* mRNA. The DSX-F polypeptide that is produced acts to repress transcription of male-specific genes. (After M. McKeown, "Sex Differentiation: The Role of Alternative Splicing," *Current Opinion in Genetics and Development 2,* 1992, 301.)

Mutational analysis of *Drosophila* sex determination

Thomas Hunt Morgan, the founding father of *Drosophila* genetics, was quoted as saying: "Treasure your exceptions." This statement has been the guiding principle of genetic analysis of any biological process during the twentieth century. This approach, studying the properties of rare mutant individual organisms and using these observations to make inferences about what the wild-type process is doing, has greatly enhanced our understanding of sex determination in several species from very different taxonomic groups.

Insights into *Drosophila* sex determination have emerged through molecular and genetic analysis of mutations altering the phenotypic sex of the fly, especially by Thomas Cline, Bruce Baker, and their colleagues. What kinds of mutations have been encountered? In regard to sexually dimorphic phenotypes, the effects of null mutations in several of the genes in the pathway are to transform females into phenotypic males. Males homozygous for these mutations are completely normal. These mutated genes include *sis-b (sisterless-b)*, *Sxl (Sex-lethal)*, and *tra (transformer)*. These genes are dispensable in males because the male developmental pathway seems to be the default state of the developmental switch. In other words, the sex-determination pathway in *Drosophila* is constructed so that the activities of several gene products are needed to shunt the animal from the default state into the female developmental pathway. The *sis-b* gene, a numerator gene encoding a bHLH protein, must be active to achieve an X:A ratio of 1.0. The mRNA-splicing regulators — the RNA-binding proteins encoded by the *Sxl* and *tra* genes — must be active for female development to ensue. They are ordinarily "off" in males

anyway, so it is of no consequence to male development to have mutations knocking out the functions of these genes.

The exceptional gene is *dsx (doublesex)*. The knockout of the *dsx* gene leads to the production of flies that simultaneously have male and female attributes. The reason for this phenotype is that each of the two alternative DSX proteins, DSX-F and DSX-M, represses the gene products that produce the phenotypic structures characteristic of the other sex. In the absence of repression, the gene products that build the structures characteristic of each of the two sexes operate simultaneously, and a fly that is simultaneously male and female develops.

(An aside: You may wonder why *Sxl* is called *Sex-lethal,* because phenotypic sex is a dispensable trait. The answer is that the phenomenon of dosage compensation — equalizing the expression of X-linked genes in 2X females and 1X males — also operates through the numerator/denominator balance and through *Sxl* (but not through *tra* or *dsx*). When proper dosage compensation is impaired, lethality ensues. Special genetic tricks that circumvent this lethality problem are used to be able to study the sex-determination-specific aspects of *Sxl*.)

Sex determination in mammals: coordinated control by the endocrine system

An analysis of sex-chromosome aneuploids demonstrates that mammalian sex determination and differentiation are quite different from those of *Drosophila* (see Table 23-2). An XXY human is phenotypically male, with a syndrome of moderate abnormalities (Klinefelter syndrome; see Figure 18-20). XO humans have a number of abnormalities (Turner syndrome; see Figure 18-17), including short stature, mental retardation, and mere traces of gonads, but they are clearly female in morphology. These data are consistent with a mammalian sex-determination mechanism based on the presence or absence of a Y chromosome. Without a Y, the person develops as a female; with it, as a male.

Mammalian reproductive development and endocrine organ control

In contrast with flies, each individual human cell does not make an independent determination of its sex. Humans mosaic for XX and XY tissues typically have a generalized appearance characteristic of one or the other sex. The observation of nonautonomy in mammalian sex determination can be understood in view of the biology of the reproductive system: sex-specific phenotypes are driven by the presence or absence of the testes.

The gonad forms within the first 2 months of human gestation. Primordial germ cells migrate into the genital ridge, which sits atop the rudimentary kidney. The chromosomal sex of the germ cells determines whether they will migrate superficially or deeply into the gonadal ridge and whether they will organize into a testis or an ovary (Figure 23-9). If they form a testis, the Leydig cells of the testis secrete testosterone, an androgenic (male-determining) steroid hormone. (Recall the discussion of steroid hormone receptor transcription factors in Chapter 11.) This hormone binds to androgen receptors. These receptors function as transcription factors; their transcription-factor activity, however, depends on binding to their cognate hormone. Thus, the androgen–receptor complex binds to androgen-responsive enhancer elements, leading to the activation of male-specific gene expression. In chromosomally female embryos, no Leydig cells form in the gonad, no testosterone is produced, androgen receptor is not activated, and the embryos continue along the default female pathway of development. Hence, it is the presence or absence of a testis that determines the sexual phenotype, through the endocrine release of testosterone. Indeed, in XY embryos lacking the androgen receptor, development proceeds along a completely female pathway even though the embryos have testes.

Setting the switch in the "on" or "off" position

What initiates the sex-determination pathway? Molecular genetic analysis has focused on identifying the locus on the Y chromosome that drives testis formation. This hypothetical gene has been called the testis-determining factor on the Y chromosome (*TDF* in humans, *Tdy* in mice) and is now known to be the same gene as the *SRY* (humans) ~ *Sry* (mice) gene, first identified through its gain-of-function dominant sex-reversal effect in which heterozygous mutant XX individuals develop as phenotypic males (see the next section). Furthermore, because the wild-type *SRY* ~ *Sry* gene is on the Y chromosome, we can easily understand

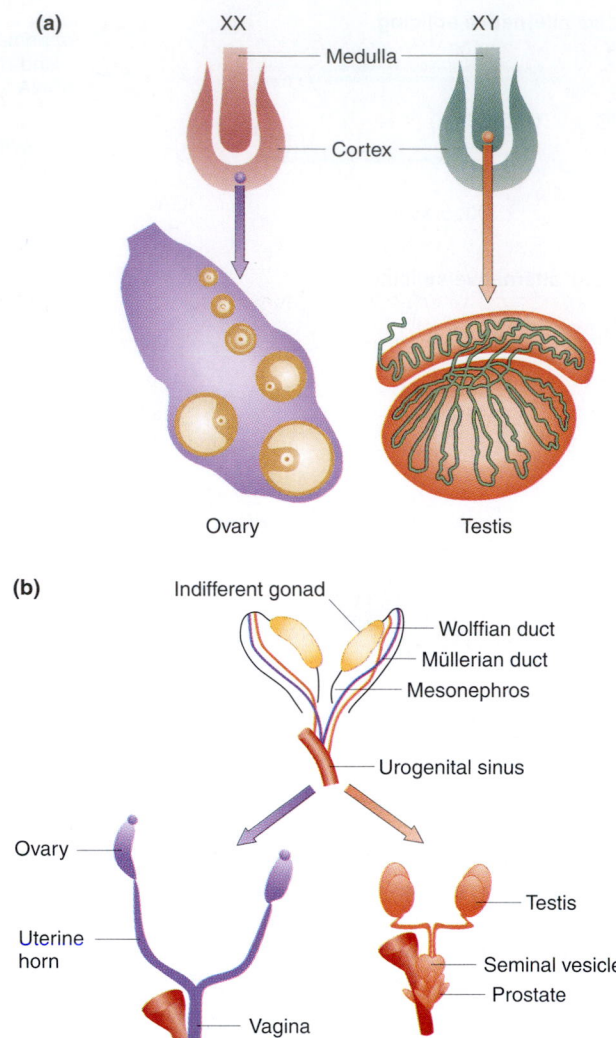

Figure 23-9 Development of the mammalian urogenital system. (a) The embryonic genital ridge consists of a medulla surrounded by a cortex. Female germ cells migrate into the cortex and become organized into an ovary. Male germ cells migrate into the medulla and become organized into a testis. (b) In the initial urogenital organization at the indifferent gonad stage, precursors of both male (Wolffian) and female (Müllerian) ducts are present. If a testis is present, it secretes two hormones, testosterone and a polypeptide hormone called Müllerian-inhibiting substance (MIS, or anti-Müllerian hormone). These hormones cause the Müllerian ducts to regress and the Wolffian ducts to develop into the male reproductive ducts. If an ovary is present, testosterone and MIS are absent and the opposite happens: the Wolffian ducts regress and the Müllerian ducts develop into the female reproductive ducts. (From U. Mittwoch, *Genetics of Sex Differentiation.* Copyright © 1973 by Academic Press.)

how the "on–off" switch is set. The wild-type XY individual has an *SRY* ~ *Sry* gene, which causes the male shunt pathway to be activated, whereas the normal XX individual lacking *SRY* ~ *Sry* remains in the female default pathway.

How does *SRY* ~ *Sry* contribute to sex determination? The SRY ~ Sry protein is a transcription factor and is expressed in the primitive male gonad. Exactly how the SRY ~ Sry protein initiates testis formation is not understood. However, with the SRY ~ Sry protein sequence in hand, many avenues for answering this and other age-old questions about the biological basis of sexual phenotype can be pursued.

Mutational analysis of mammalian sex determination

The Y chromosome testis-determining gene was identified through mapping and characterization by Robin Lovell-Badge and Peter Goodfellow of a genetic syndrome common to mice and humans that almost certainly affects this factor (see the molecular map in Figure 23-10). This syndrome is called **sex reversal.** Sex-reversed XX individuals are phenotypic males and have been shown to carry a fragment of the Y chromosome in their genomes. In general, these Y-chromosome duplications arise by an illegitimate recombination between the X and Y chromosomes that fuses a piece of the Y chromosome to a tip of one of the X chromosomes. The part of the Y chromosome that includes these duplications was cloned; by subsequent molecular analysis, Lovell-Badge and Goodfellow identified from this region a transcript that is expressed in the appropriate location of the developing kidney capsule.

The gene encoding this transcript was named the *sex reversal on Y* gene (*SRY* in humans, *Sry* in mice), because it

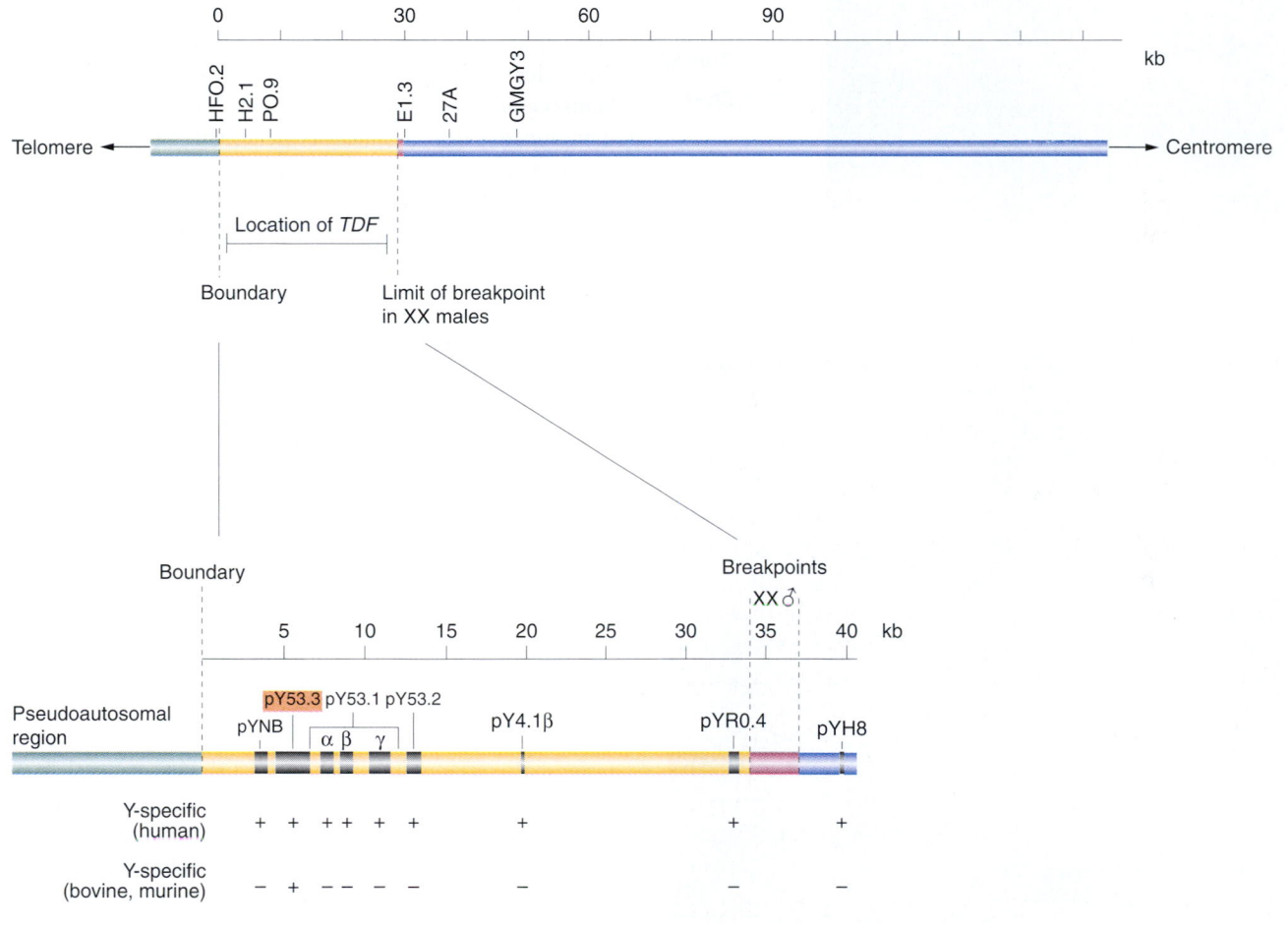

Figure 23-10   A molecular map of the distal part of the short arm of the human Y chromosome. The gene for testis-determining factor *(TDF)* was localized to a 34-kb region on the basis of the following logic: to the left of this region, the DNA sequences are the same on the X and Y chromosomes. (These sequences to the left are called the *pseudoautosomal region.)* Pieces of the Y chromosome starting at the telomere and ending at position 34 (kb) of the map are sufficient to cause the sex-reversal syndrome. Within this region, much of the DNA is repeated many times throughout the genome, but eight blocks of DNA unique to the Y chromosome were identified (the black regions). Cross-hybridization tests showed that only one of these (pY53.3) is present in bovine and mouse Y chromosomes. This fragment encodes SRY. In some humans with sex reversal, point mutations in *SRY* have been found. (From A. H. Sinclair et al., "A Gene from the Human Sex-Determining Region Encodes a Protein with Homology to a Conserved DNA-Binding Motif," *Nature* 346, 1990, 240.)

was identified on the basis of the sex-reversal syndrome, but it is certainly the same gene as *TDY ~ Tdy*. Lovell-Badge and Goodfellow used a transgene to provide spectacular evidence in support of this identity (Figure 23-11). A cloned

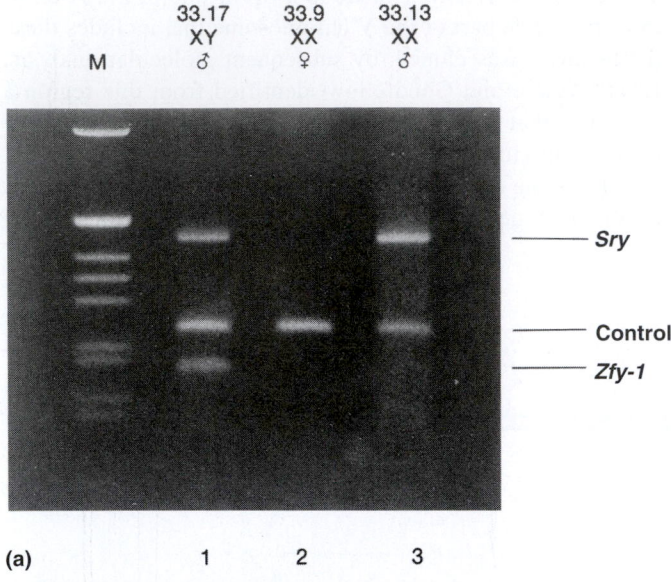

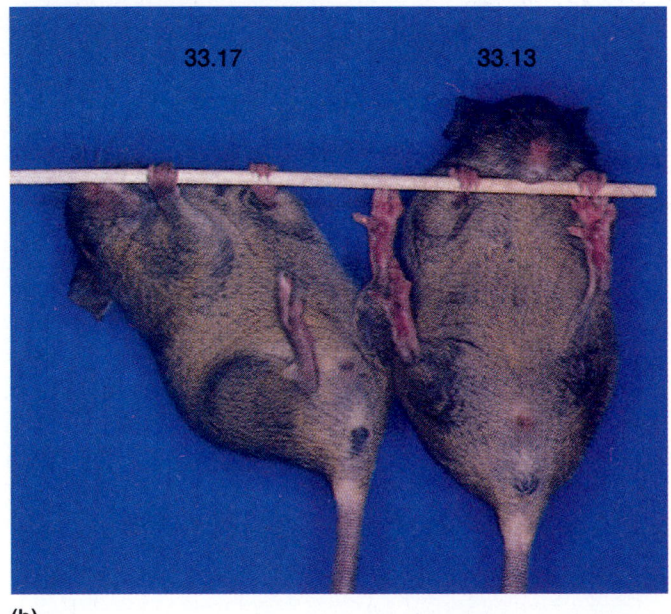

Figure 23-11 A transgenic mouse that proves that *Sry* can cause the sex-reversal syndrome. (a) Amplification of genomic DNA by the polymerase chain reaction shows that mouse 33.13 lacks a DNA marker *(Zfy-1)* for the presence of a Y chromosome but that it does contain the *Sry* transgene. M = marker bands. (b) The external genitalia of sex-reversed XX transgenic mouse 33.13 are indistinguishable from those of a normal XY male sib (33.17), demonstrating that the *Sry* gene is sufficient to cause the sex-reversal phenotype. (From P. Koopman, J. Gubbay, N. Vivian, P. Goodfellow, and R. Lovell-Badge, "Male Development of Chromosomally Female Mice Transgenic for *Sry*," *Nature* 351, 1991, 117.)

14-kb genomic fragment of the mouse Y chromosome, including the *Sry* gene, was inserted into the mouse genome by germ-line transformation. An XX offspring containing this inserted *Sry* DNA (the transgene) was completely male in external and internal phenotype and, as predicted, possessed the somatic tissues of the testis (Figure 23-11b), including the Leydig cells that make testosterone. (It should be noted, however, that this mouse was sterile. The sterility is probably a consequence of having two X chromosomes in a male germ cell, because XXY male mice are similarly sterile.) Thus, a single genetic unit was directly shown to greatly alter the mammalian sexual phenotype, completely consistent with the role of *SRY ~ Sry* as the gene that determines testis development.

The role of the androgen receptor in receiving the testosterone signal and establishing the male secondary sexual characteristics was elucidated through the study of rare *Tfm* mice lacking this receptor. Chromosomally XY mice hemizygous for the X-linked *Tfm (Testicular feminization)* mutation develop as phenotypic females (see Figure 2-26) except that they are infertile and are typically diagnosed at puberty because of their failure to menstruate. *Tfm* XY mice have testes, but the target cells that must decide between alternative pathways regarding sexually dimorphic characteristics lack androgen receptors and so are completely insensitive to the presence of testosterone. Thus, these mice develop along the default developmental pathway, which leads to phenotypic feminization.

> **MESSAGE** ··
>
> In mammals, a Y-chromosome gene encodes a transcription factor that causes the gonad to become a testis. The testis serves as a command organ that, through testosterone release, directs male phenotypic development throughout the body.

Binary fate decisions: the germ line versus the soma

In animal development, the earliest developmental decision is that of separating the germ line from the soma. After this separation occurs, it is irreversible. Germ cells do not contribute to somatic structures. Somatic cells cannot form gametes, and thus their descendants never contribute genetic material to the next generation. This early separation means that genetic or regulatory modifications of somatic cells that occur in the course of development have no consequences on gamete formation.

In making this decision of germ line versus soma, the embryo exploits its machinery for creating asymmetries — the **cytoskeleton,** the girders that support and shape the cell — to localize a germ-line determinant to a subset of early embryonic cells. Before directly addressing the question of how the germ-line versus soma decision is made, we need to consider the nature of cytoskeletal and cellular asymmetries.

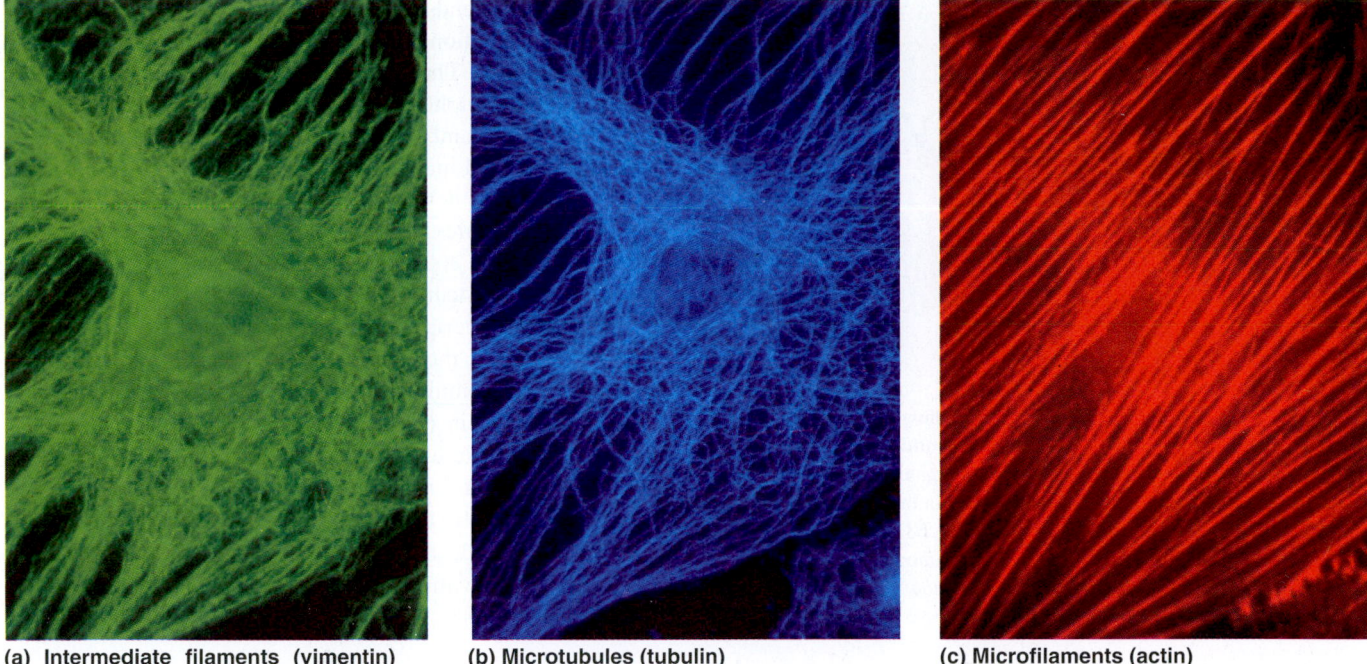

(a) Intermediate filaments (vimentin)   **(b) Microtubules (tubulin)**   **(c) Microfilaments (actin)**

Figure 23-12   Different cytoskeletal systems in the same cell. The distribution of (a) the intermediate filament protein vimentin. (b) the microtubulin protein tubulin, and (c) the microfilament protein actin are shown. (Courtesy of V. Small. Reprinted from H. Lodish, D. Baltimore, A. Berk, S. L. Zipursky, P. Matsudaira, and J. Darnell, *Molecular Cell Biology,* 3d ed. Copyright © 1995 by Scientific American Books.)

Cytoskeleton of the cell

The cytoskeleton consists of several networks of highly organized structural rods that run within each cell: microfilaments, intermediate filaments, and microtubules (Figure 23-12). Each has its own architecture formed of unique sets of protein subunits and proteins that promote production or disassembly of the rods. Furthermore, each type of rod forms higher-order networks through different sets of proteins that reversibly cross-link the individual rods to one another.

Several roles of the cytoskeleton are relevant to our consideration of pattern formation: control of the location of the mitotic cleavage plane within the cell, control of cell shape, and directed transport of molecules and organelles within the cell. All of these roles depend on the fact that the cytoskeletal rods are polar structures. The contributions to

pattern formation of microfilaments and microtubules—polymers of actin and tubulin subunits, respectively—are better documented, and so we will focus on these two classes of cytoskeletal elements.

Intrinsic asymmetry of cytoskeletal filaments

The polarity of microfilaments and microtubules is crucial to their roles as intracellular "highways." The ability of other molecules to move up and down these highways is an important aspect of all of their cellular roles. Microfilaments and microtubules are linear polymers with polarity—conceptually like the 5′-to-3′ polarity of DNA and RNA strands, even though the molecular basis of polarity is quite different (Figure 23-13). Furthermore, the polarity of the cytoskeletal elements can be organized within a cell. Consider microtubules. Near the center of most cell types, all

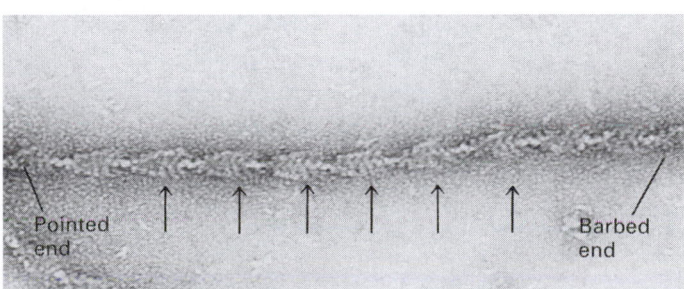

Figure 23-13   The polarity of subunits in an actin microfilament. An actin microfilament does not ordinarily have this appearance, but the microfilament has been coated with a protein that binds in a fashion that reveals the underlying polarity of the actin microfilament itself. (Courtesy of R. Craig. Reprinted from H. Lodish, D. Baltimore, A. Berk, S. I. Zipursky, P. Matsudaira, and J. Darnell, *Molecular Cell Biology,* 3d ed. Copyright © 1995 by Scientific American Books.)

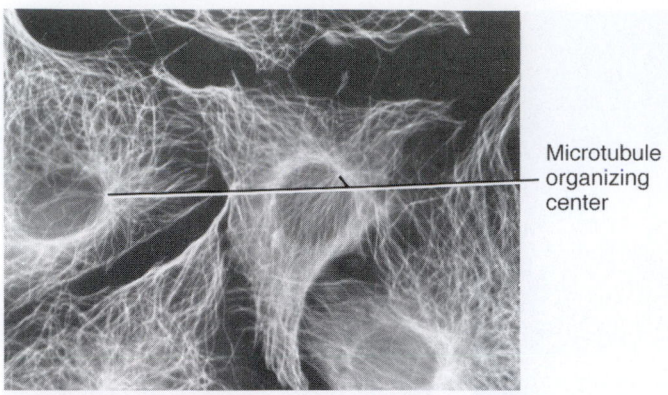

Figure 23-14 Fluorescence micrograph showing the distribution of tubulin in an interphase animal fibroblast. Notice that the microtubules radiate out from a microtubule organizing center. The negative (minus) ends of the microtubules are in the center, and the positive (plus) ends are at the periphery of the cell. (Courtesy of M. Osborn. Reprinted from H. Lodish, D. Baltimore, A. Berk, S. L. Zipursky, P. Matsudaira, and J. Darnell, *Molecular Cell Biology,* 3d ed. Copyright © 1995 by Scientific American Books.)

the "−" (minus) ends of the microtubules are found (Figure 23-14). This location is called the **microtubule organizing center (MTOC).** The "+" (plus) ends of microtubules are located at the periphery of the cell. Very much as an automobile uses the combustion of gasoline to create energy that is then transduced into motion, special "motor" proteins hydrolyze ATP for energy that is utilized to propel movement along a microtubule. For example, a protein called kinesin is able to move in a minus-to-plus direction along microtubules, carrying "cargoes" such as vesicles from the center of the cell to its periphery (Figure 23-15a and b). The "motor"—the part of the kinesin protein that directly interacts with the mocrotubule rod—is contained in the globular head of the protein (Figure 23-15c). The tail of kinesin is thought to be where the cargo is attached. These cargoes might be individual molecules, organelles, or other subcellular particles to be towed from one part of the cell to another. (Comparable motors exist for actin microfilaments.)

What is the value of having multiple independent cytoskeletal systems? A part of the answer is probably divi-

Figure 23-15 Movement of vesicles along microtubules. (a) A scanning electron micrograph of two small vesicles attached to a microtubule. (b) A diagram of how kinesin is thought to attach to cellular cargoes such as vesicles at its tail and to move the cargoes along the microtubule in the − to + direction by using the motor domain in the kinesin head. (c) A diagram of the kinesin protein showing the functions associated with various parts of the molecule. (Part a from B. J. Schnapp et al., *Cell* 40, 1985, 455; courtesy of B. J. Schnapp, R. D. Valle, M. P. Sheetz, and T. S. Reese. All parts: Reprinted from H. Lodish, D. Baltimore, A. Berk, S. L. Zipursky, P. Matsudaira, and J. Darnell, *Molecular Cell Biology,* 3d ed. Copyright © 1995 by Scientific American Books.)

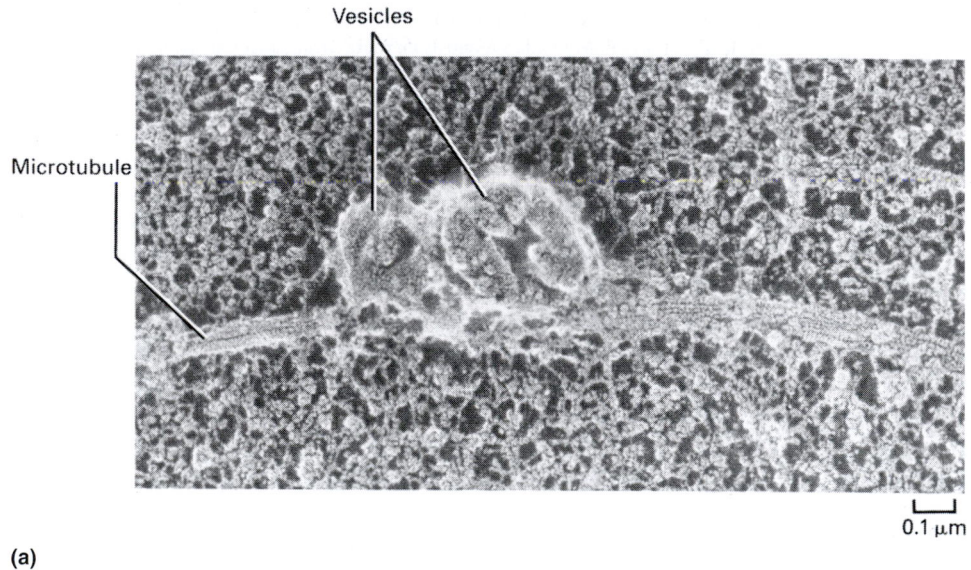

(a)

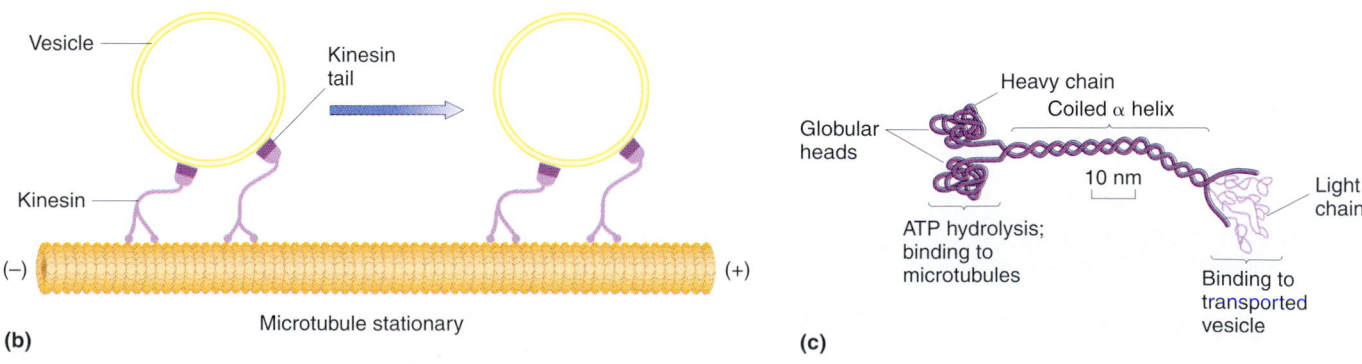

(b)

(c)

sion of labor. Just as cities have complex grids of intersecting streets to permit travel from a starting point to any other location, cells use multiple cytoskeletal systems to move cargo from one part of the cell to any other.

MESSAGE ···
The cytoskeleton serves as a highway system for the directed movement of subcellular particles and organelles.
···

Localizing determinants through cytoskeletal asymmetries: the germ line

In many organisms, a visible particle is asymmetrically distributed to the cells that will form the germ line. These particles — called P granules in *Caenorhabditis elegans,* polar granules in *Drosophila,* and nuage in frogs — are thought to be transport vehicles that ride on specific cytoskeletal highways to deliver the attached germ-cell determinants (regulatory molecules) to the appropriate cell. In *C. elegans* and *Drosophila,* the evidence relating germ-line determination to the cytoskeleton is particularly strong. We shall consider both of these cases.

The early cell divisions of the *C. elegans* zygote provide an example of how cytoskeletal asymmetries help form the germ line. One of the favorable properties of *C. elegans* as an experimental system is that every animal undergoes the same pattern of cell divisions — a pattern that can be readily followed under the microscope. A lineage tree that traces the descent of each of the thousand or so somatic cells of the worm can then be constructed (see Chapter 22).

The one-cell zygote of *C. elegans* that is produced on fertilization is called the P_0 cell. It divides asymmetrically across the long axis of the ellipsoidal P_0 cell to produce a larger, anterior AB cell and a smaller, posterior P_1 cell (Figure 23-16). This division is very important in that it already sets up specialized roles for the descendants of these first two cells. The AB cell descendants will produce most of the skin cells of the worm (the hypoderm) and most of the neurons of the nervous system, whereas most of the muscles and all the digestive system and the germ-line cells will come from the P_1 cell.

The germ-cell fate in the earliest divisions of the P_0 cell and its posterior descendants (P_1, P_2, and so forth) correlate with the distribution of certain fluorescent cytoplasmic particles called P granules. These granules are incorporated exclusively into the P_1 cell at the first division. When the P_1 cell divides, also asymmetrically, the P granules are incorporated into the progeny P_2 cells and, similarly, at the next division into the P_3 cells, and so forth. Only the P_x cell that has these P granules becomes the germ line of the worm — all other cells are somatic. The asymmetric distribution of P granules is microfilament dependent. When applied at the right time to the P_0 cell, drugs such as cytochalasin disrupt actin subunit polymerization into microfilaments. After disruption of microfilament polymerization, P granules are dis-

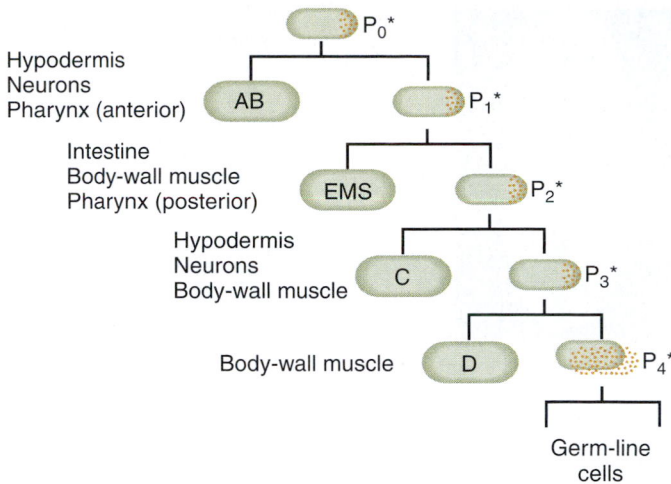

Figure 23-16   The early development of *C. elegans,* showing the early lineage divisions of the zygote. The mature cell types that arise from the various daughter cells of the early divisions are indicated. Note that the entire germ line comes from the P_4 cell. Each of the P-cell divisions indicated by an asterisk is asymmetric, and each of the posterior daughter cells inherits all the P granules, which are thought to be germ-line determinants in the worm. The letters (for example, AB, EMS) are symbols for names of daughter cells.

tributed symmetrically between the two progeny cells. (Presumably because other fate determinants are abnormally distributed owing to the actin disruption, the resulting embryos are quite "confused" and die as masses of cells that look nothing like a normal worm.)

In early *Drosophila* development, the cytoskeleton also is exploited to localize the structure containing the germ-line determinants: the polar granules. In the course of oogenesis in the ovary of the mother, the polar granules are constructed and become tethered to the posterior pole of the oocyte by virtue of their attachment to one end of the microtubules. They remain in this location throughout early embryogenesis until nuclear division 9, when a few nuclei migrate to the posterior pole. (Note that an unusual feature of early *Drosophila* development is that the first 13 mitoses are nuclear divisions without concomitant cytoplasmic division, making the early embryo a syncitium — a multinucleate cell.) After nuclear division 9, the plasma membrane of the oocyte evaginates at the posterior pole to surround each nucleus and pinches off some of the polar-granule-containing cytoplasm. This event creates the pole cells, the first mononucleate cells of the embryo and the cells that will uniquely form the fly's germ line (Figure 23-17).

How do the polar granules get tethered to the posterior pole of the oocyte? Again, the subcellular localization is accomplished by one of the cytoskeletal networks. In contrast with *C. elegans,* in which the actin-based microfilaments seem to form the polar structure to which the P granules attach, here the tubulin-based microtubules provide the essential asymmetry, which is probably just an accident of the

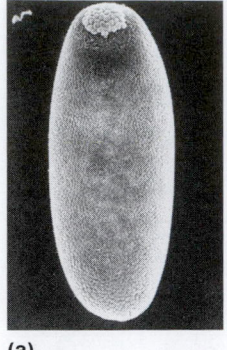

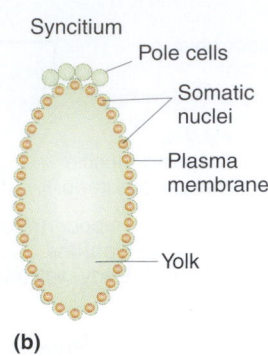

Syncitium

Pole cells

Somatic nuclei

Plasma membrane

Yolk

(a)   **(b)**

Figure 23-17   Pole-cell formation at the syncitial stage of the early *Drosophila* embryo. (a) A scanning electron micrograph of a *Drosophila* embryo with the egg shell removed. Note that the pole cells (the cap of cells on top of the embryo) lie outside the somatic syncitium. (b) A diagram of a longitudinal section through the embryo in part a, showing that the germ-line cells — the pole cells — have formed, whereas the soma still consists of syncitial nuclei. (After F. R. Turner and A. P. Mahowald, "Scanning Electron Microscopy of *Drosophila* Embryogenesis," *Developmental Biology* 50, 1976, 95.)

evolutionary history of these organisms. In each case, the germ-line determinant evolved to co-opt any cytoskeletal system that had the appropriately oriented asymmetry.

Forming complex pattern: establishing positional information

This section and subsequent sections about the early development of *Drosophila* summarize the results of a good deal of mutational analysis. The logic of such analysis is that, by seeing what goes wrong with development in individuals mutant for a particular gene, we can learn how the protein encoded by that gene contributes to normal development. Especially powerful is the combination of these genetic analyses with the cloning and molecular biological analysis of the protein products of these developmental genes. Out of such combined analyses, important insights into the circuitry of development have emerged.

Mutational analysis of early *Drosophila* development

The initial insights into the genetic control of pattern formation came from studies of the fruit fly *Drosophila melanogaster*. The reason that *Drosophila* development has proved to be a gold mine to researchers is that developmental problems can be simultaneously approached by the use of genetic and molecular techniques. Let's consider the basic genetic and molecular techniques that are employed.

The *Drosophila* embryo has been especially important in understanding the formation of the basic animal body plan. One important reason is that the formation of the larval exoskeleton in the *Drosophila* embryo lends itself to easy identification of body plan mutant phenotype. The ex-

oskeleton of the *Drosophila* larva is laid down as a mosaic in the embryo. Each structure of the exoskeleton is built by the epidermal cell or cells underlying that structure. With its intricate pattern of hairs, indentations, and other structures, the exoskeleton provides numerous landmarks to serve as indicators of the fates acquired by the many epidermal cells. In particular, there are many anatomical structures that are distinct along the anterior–posterior (A–P) and dorsal–ventral (D–V) axes. Furthermore, because all the nutrients necessary to develop to the larval stage are prepackaged in the egg, mutant embryos in which the A–P or D–V cell fates are drastically altered can nonetheless develop to the end of embryogenesis and produce a mutant larva. The exoskeleton of such a mutant larva mirrors the mutant fates assigned to subsets of the epidermal cells and can thus identify genes worthy of detailed analysis.

Researchers, most notably Christiane Nüsslein-Volhard, Eric Wieschaus, and their colleagues, have performed extensive mutational screens, essentially saturating the genome for mutations that alter the A–P or D–V patterns of the larval exoskeleton. These mutational screens identified two broad classes of genes affecting the basic body plan: zygotically acting genes and maternal-effect genes (see Figure 23-18). The zygotically acting genes are those in which the gene products contributing to early development are expressed exclusively in the zygote. They are part of the DNA of the zygote itself and are the "standard" sorts of genes that we are used to thinking about. Recessive mutations in zygotically acting genes elicit mutant phenotypes only in homozygous mutant animals. The alternative category — the maternal-effect genes — affects early development through contributions of gene products from the ovary of the mother to the developing oocyte. In maternal-

(a) Zygotically acting genes

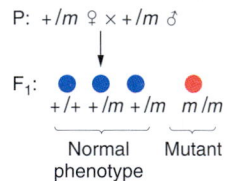

P: +/m ♀ × +/m ♂

F₁: +/+   +/m +/m   m/m
      └─────────────┘   └──┘
       Normal   Mutant
       phenotype

(b) Maternal-effect genes

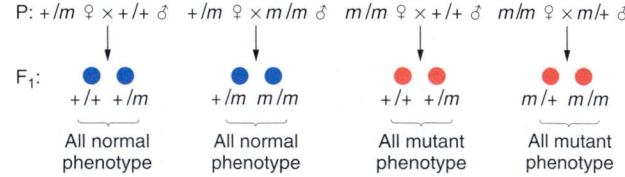

P: +/m ♀ × +/+ ♂   +/m ♀ × m/m ♂   m/m ♀ × +/+ ♂   m/m ♀ × m/+ ♂

F₁: +/+ +/m   +/m m/m   +/+ +/m   m/+ m/m

    All normal   All normal   All mutant   All mutant
    phenotype   phenotype   phenotype   phenotype

Figure 23-18   Genetic assays for recessive zygotic and maternal-effect mutations. The genetic distinction between recessive (a) zygotically acting and (b) maternal-effect mutations is depicted. (a) The phenotypes of offspring are purely a manifestation of their own (that is, the zygote's) genotype. (b) The phenotypes of offspring are purely a manifestation of their mother's genotype. Hence these mutations are said to be material effect, or maternally acting.

effect mutations, the phenotype of the offspring depends on the genotype of the mother, not of the offspring. A recessive maternal-effect mutation will produce mutant animals only when the mother is a mutant homozygote.

Equally important to the mutational identification of genes affecting the body plan is the ease with which these genes can be cloned and characterized at the molecular level. Any *Drosophila* gene can be cloned if its chromosomal map location has been well established, by using recombinant DNA techniques such as those described in Chapters 12 and 13. The analysis of the cloned genes often provides valuable information on the function of the protein product — usually by identifying close relatives in amino acid sequence of the encoded polypeptide through comparisons with all the protein sequences stored in public data bases. In addition, one can investigate the spatial and temporal pattern of expression of (1) an mRNA, by using histochemically tagged single-stranded DNA sequences complementary to the mRNA to perform RNA in situ hybridization, or (2) a protein, by using histochemically tagged antibodies that specifically bind to that protein.

Extensive use is also made of in vitro mutagenesis techniques. P elements are used for germ-line transformation in *Drosophila* (see Chapter 13). A cloned pattern-formation gene is mutated in a test tube and put back into the fly. The mutated gene is then analyzed to see how the mutation alters the gene's function.

Cytoskeletal asymmetries and the *Drosophila* anterior–posterior axis

As we shall see here, not only is the *Drosophila* germ line established through a localized determinant anchored to microtubules, but the same is true for formation of the anterior–posterior axis of the soma. Positional information along the A–P axis of the syncitial *Drosophila* embryo is initially established through the creation of concentration gradients of two transcription factors: the BCD and HB-M proteins. The BCD protein, encoded by the bicoid *(bcd)* gene, is distributed in a steeper gradient in the early embryo, whereas the HB-M protein, encoded by the *hunchback (hb)* gene, is distributed in a shallower but longer gradient (Figure 23-19). Both gradients have their high points at the anterior pole. In somewhat different ways, the gradients of both these proteins depend on the diffusion of protein from a localized origin: localized translation of two mRNA species, one tethered to microtubules at the anterior pole and the other at the posterior pole of the syncitial embryo.

The origin of the BCD gradient is quite straightforward. The *bcd* mRNA, packaged during oogenesis into the developing oocyte, is tethered to the − (minus) ends of microtubules, which are located at the anterior pole (Figure 23-20a). Translation of BCD protein begins midway through the early nuclear divisions of the syncitial embryo. The protein diffuses in the common cytoplasm of the syncitium. Because the protein is a transcription factor, it con-

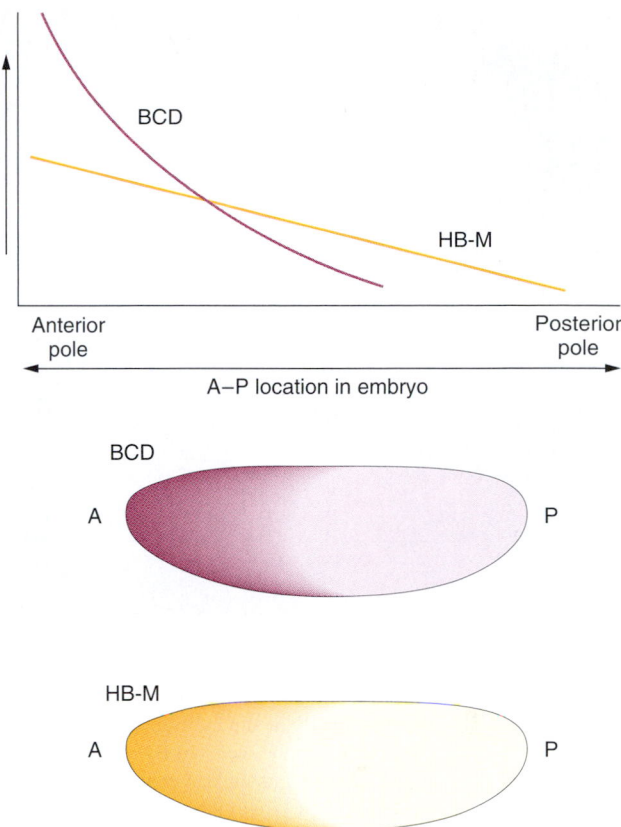

Figure 23-19   The gradients of BCD and HB-M proteins. The BCD gradient is steeper, and the BCD protein is not detectable in the posterior half of the early *Drosophila* embryo. The HB-M gradient is shallower, and HB-M protein can be detected well into the posterior half of the embryo.

tains signals to become localized in nuclei. By diffusion, those nuclei nearer to the source of translation (the anterior pole) incorporate a higher concentration of BCD protein than do those farther away; this difference results in the steep BCD protein gradient (Figure 23-20b).

The origin of the HB-M protein gradient is more complex. The HB-M protein gradient is produced by posttranscriptional regulation. The *hb-m* mRNA is maternal in origin, being packaged during oogenesis into the developing oocyte, and is uniformly distributed throughout the oocyte and the syncitial embryo. However, translation of *hb-m* mRNA is blocked by a translational repressor protein — the NOS protein product, encoded by the *nanos (nos)* gene. Like *bcd* mRNA, *nos* mRNA is maternal in origin. However, in contrast with *bcd* mRNA, *nos* mRNA is localized at the posterior pole, through its association with the + (plus) ends of microtubules (Figure 23-20c). When translation of *nos* mRNA begins midway through the syncitial stage of early embryogenesis, NOS protein becomes distributed by diffusion in a gradient opposite that of BCD. The NOS gradient has a high point at the posterior pole and drops down to background levels in the middle of the A–P axis of the embryo (Figure 23-20d). NOS protein has the ability to

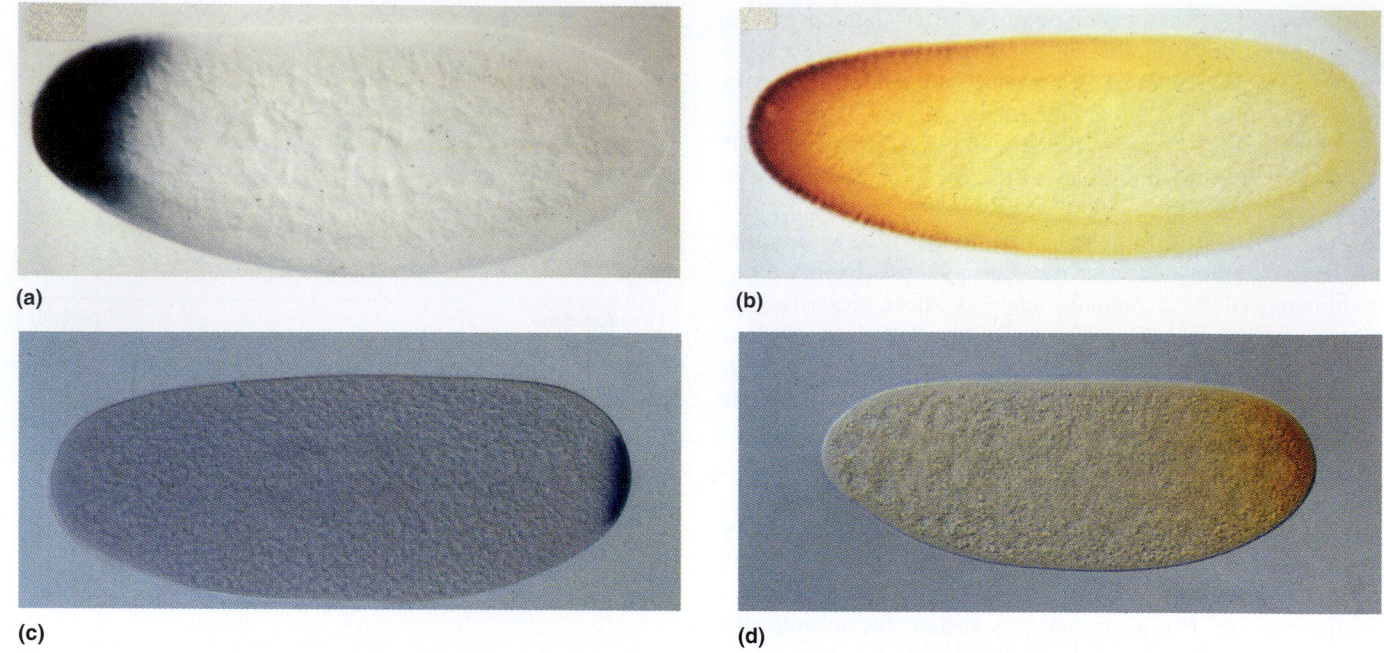

Figure 23-20   Photomicrographs showing the expression of localized A–P determinants in the embryo. (a) By in situ hybridization to RNA, the localization of *bcd* mRNA to the anterior (left) tip of the embryo can be seen. (b) By antibody staining, a gradient of BCD protein (brown stain) can be visualized, with its highest concentration at the anterior tip. (c) Similarly, *nos* mRNA localizes to the posterior (right) tip of the embryo, and (d) NOS protein is in a gradient with a high point at the posterior tip. (Parts a and b from Christiane Nüsslein-Volhard, "Determination of the Embryonic Axes of *Drosophila*," *Development,* Suppl. 1, 1991, 1. Parts c and d provided by E. R. Gavis, L. K. Dickinson, and R. Lehmann, Massachusetts Institute of Technology.)

specifically block translation of *hb-m* mRNA. Through this ability, the NOS translation repressor posterior-to-anterior gradient produces the shallow anterior-to-posterior gradient of HB-M protein.

MESSAGE ·······························

Localization of mRNAs within a cell is accomplished by anchoring the mRNAs to one end of polarized cytoskeleton chains.

·······························

How do the *bcd* and *nos* mRNAs get tethered to opposite ends of the polarized microtubules of the oocyte and early syncitial embryo? The answer is that there are specific microtubule-association sequences located within the 3′ UTRs — untranslated regions of the mRNA 3′ to the translation-termination codon. (Eukaryotic mRNAs always contain some sequence 5′ to the translation-initiation codon, the 5′ UTR, and some sequence 3′ to the translation-termination codon, the 3′ UTR. In some mRNAs, these regions are quite short, but, in others, they can be several kilobases long. We are learning that, in many cases, as here, specific regulatory functions are carried out by sequences within these 3′ UTRs.)

The *bcd* mRNA 3′ UTR localization sequences are bound by a protein that can also bind the − ends of the microtubules. In contrast, the 3′ UTR of *nos* mRNA has localization sequences that can bind other proteins, which also

bind to the + ends of microtubules. (In actuality, there are more intermediary steps in anchoring *nos* mRNA at the posterior end, with the 3′ UTR localization sequences of *nos* mRNA being anchored to the molecules in the polar granules, which are in turn anchored to the + end of the microtubules.)

How can we demonstrate that the 3′ UTRs of the mRNAs are where the localization sequences reside? This determination has been made in part by "swapping" experiments. For example, when a synthetic transgene that produces an mRNA with 5′ UTR and protein-encoding regions of the normal *nos* mRNA glued to the 3′ UTR of the normal *bcd* mRNA is inserted into the fly genome, this fused *nos-bcd* mRNA will be localized at the anterior pole of the oocyte. An otherwise normal embryo containing this transgene has a double gradient of NOS: one from anterior to posterior (due to the transgene's mRNA) and one from posterior to anterior (due to the normal *nos* gene's mRNA). This procedure produces a very weird embryo, with two mirror-image posteriors and no anterior (Figure 23-21). This double-abdomen embryo arises because NOS protein is now present throughout the embryo and translationally represses *hb-m* mRNA (it also represses *bcd* mRNA, although it is not clear that this repression is its normal function in wild-type animals). More detailed information about how A–P positional information is generated is presented in the next section.

(a)

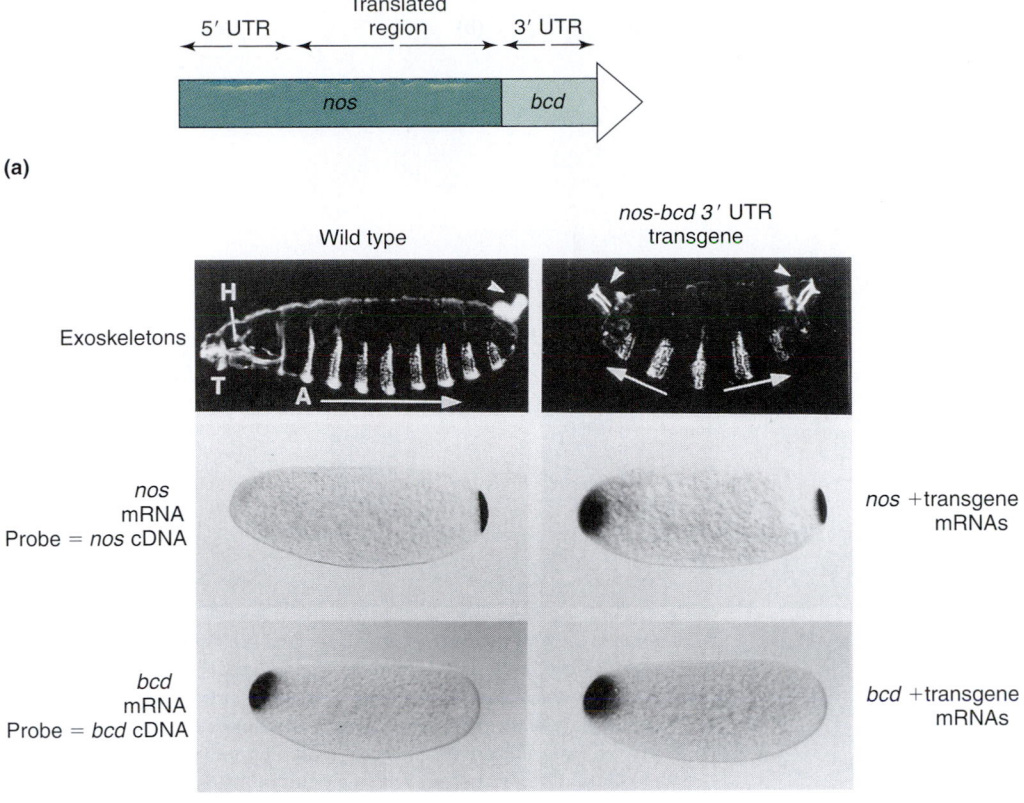

(b)

Figure 23-21   The effect of replacement of the 3′ UTR of the *nanos* mRNA with the 3′ UTR of the *bicoid* mRNA on mRNA localization and embryonic phenotypes. (a) The structure of the *nos-bcd* 3′ UTR transgene. (b) The effects on embryonic development of the transgene. The embryos and larva in the left-hand column are derived from wild-type mothers. Those in the right-hand column are derived from transgenic mothers. All embryos are shown with anterior to the left and posterior to the right. The exoskeletons of the larvae are shown for comparison. The transgene causes a perfect mirror-image double abdomen. In the embryo from a transgenic mother, mRNAs encoding NOS protein are now present at both poles of the embryo. NOS protein will inhibit translation of *hb-m* mRNA (and of *bcd* mRNA as well). (From E. R. Gavis and R. Lehmann, "Localization of *nanos* RNA Controls Embryonic Polarity," *Cell* 71, 1992, 303.)

> **MESSAGE** ··
> The positional information of the *Drosophila* A–P axis is generated by protein gradients. The gradients ultimately depend on diffusion of newly translated protein from localized sources of specific mRNAs anchored through their 3′ UTRs to ends of cytoskeletal filaments.

Studying the BCD gradient

How do we know that molecules such as BCD and HB-M contribute A–P positional information? Let's consider the example of BCD in detail. First, genetic changes in the *bcd* gene alter anterior fates. Embryos derived from *bcd* homozygous null mutant mothers lack anterior segments (Figure 23-22). If, on the other hand, we overexpress *bcd* in the mother, by increasing the number of copies of the *bcd*+ gene from the normal two copies to three, four, or more, we "push" fates that ordinarily appear in anterior positions into increasingly posterior zones of the resulting embryos (Figure 23-23). These observations suggest that BCD protein exerts global control of anterior positional information.

Second, *bcd* mRNA can completely substitute for the anterior determinant activity of anterior cytoplasm (Figure 23-24). If the anterior cytoplasm is removed from a punctured syncitial embryo, anterior segments (head and thorax) are lost. Injection of anterior cytoplasm from another embryo into the anterior region of the anterior cytoplasm-depleted embryo restores normal anterior segment formation, and a normal larva is produced. Similarly, synthetic

bcd mRNA can be made in a test tube and injected into the anterior region of an anterior cytoplasm-depleted embryo. Again, a normal larva is produced. Unlike that of anterior cytoplasm, transplantation of cytoplasm from middle or

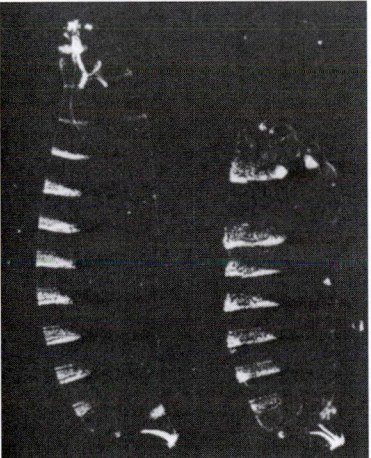

Figure 23-22   Photomicrographs of the exoskeletons of larvae derived from wild-type and *bcd* maternal-effect-lethal mutant mothers. These photomicrographs are in darkfield, so dense structures appear white, as in a photographic negative. Note the bright, segmentally repeated denticle bands present on the ventral side of the embryo. Maternal genotypes and larval phenotypes (and class of mutation) are as follows: (left) wild-type, normal phenotype; (right) *bcd (bicoid)*, anterior head and thoracic structures missing (anterior).

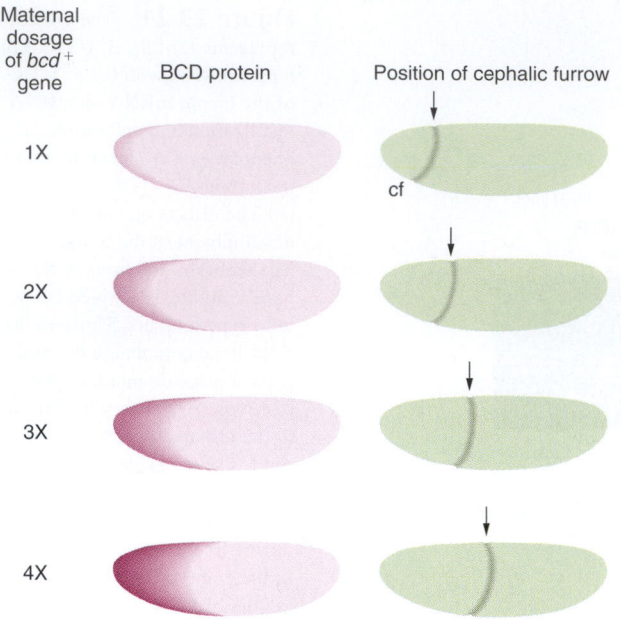

Figure 23-23   The concentration of BCD protein affects A–P cell fates. The amount of BCD protein can be changed by varying the number of copies of the *bcd*⁺ gene in the mother. Embryos derived from mothers carrying from one to four copies of *bcd*⁺ have increasing amounts of BCD protein. (Two copies is the normal diploid gene dose.) Cells that invaginate to form the cephalic furrow (cf) are determined by a specific concentration of BCD protein. In the progression from one maternal copy to four copies of *bcd*⁺, this specific concentration is present more and more posteriorly in the embryo. Thus, the position of the cephalic furrow (marked dorsally by the arrow) arises farther toward the posterior according to *bcd*⁺ gene dosage. (After W. Driever and C. Nüsslein-Volhard, "The Bicoid Protein Determines Position in the *Drosophila* Embryo in a Concentration-Dependent Manner," *Cell* 54, 1988, 100.)

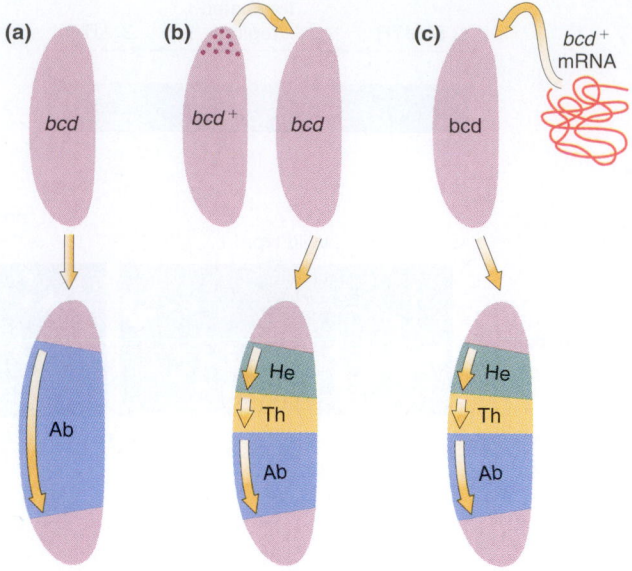

Figure 23-24   The *bcd* "anteriorless" mutant phenotype can be rescued by wild-type cytoplasm or purified *bcd*⁺ mRNA. (a) In embryos derived from *bcd* mothers, the anterior (head and thoracic) segments do not form. (b) If anterior cytoplasm from an early wild-type donor embryo is injected into the anterior of a recipient embryo derived from a *bcd* mutant mother, a fully normal embryo and larva are produced. Cytoplasm from any other part of the donor embryo does not rescue. (c) *bcd*⁺ mRNA injected into the anterior of an embryo derived from a *bcd* mutant mother also rescues the wild-type segmentation pattern. (Ab, abdomen; He, head; Th, thorax.) (After C. Nüsslein-Volhard, H. G. Fronhüfer, and R. Lehmann, "Determination of Anteroposterior Polarity in *Drosophila*," *Science* 238, 1987, 1679.)

posterior regions of a syncitial embryo does not restore normal anterior formation. Thus, the anterior determinant should be located *only* at the anterior end of the egg. As already stated, this location is exactly where *bcd* mRNA is found.

Third, also as already stated, the BCD protein shows the predicted asymmetric and graded distribution to fulfill its role of establishing positional information.

Cell–cell signaling and the *Drosophila* dorsal–ventral axis

In the examples considered thus far, the determinants were intracellular products: mRNAs or larger macromolecular assemblies packaged into the oocyte. In many circumstances, positional information must depend on proteins secreted from a localized subset of cells within a developing field. These secreted proteins diffuse in the extracellular space to form a concentration gradient of ligand. The ligand then activates target cells through a receptor–signal transduction system in a concentration-dependent fashion.

An example of such a mechanism for position information is the establishment of the dorsal–ventral (D–V) axis

in the early *Drosophila* embryo. The proximate effect of the D–V positional information will be to create a gradient of DL protein activity in cells along the D–V axis. The DL protein is a transcription factor encoded by the *dorsal (dl)* gene. It exists in two forms: (1) active transcription factor located in the nucleus and (2) inactive protein located in the cytoplasm, where it is sequestered in a complex bound to the CACT protein encoded by the *cactus (cact)* gene. The concentration of *active* DL protein will determine cell fate along the D–V axis. Both *dl* mRNA and DL protein are distributed uniformly in the oocyte and the very early embryo. However, late in the syncitial embryo stage, there develops a gradient of active DL protein, with its high point at the ventral midline of the embryo (Figure 23-25).

How does positional information generate the gradient of active DL protein? The key events take place in oogenesis, through an interaction between the oocyte itself and the layer of surrounding somatic cells—the follicle cells (Figure 23-26). The follicle cells on the ventral side of the oocyte–follicle-cell complex synthesize and secrete some proteins that lead to a gradient of activation of a secreted precursor of the SPZ ligand, encoded by the *spaetzle (spz)* gene. The follicle cells also make the eggshell (shown in Figure 23-26), the inner boundary of which is the vitelline membrane (shown in Figure 23-27a). The SPZ ligand is

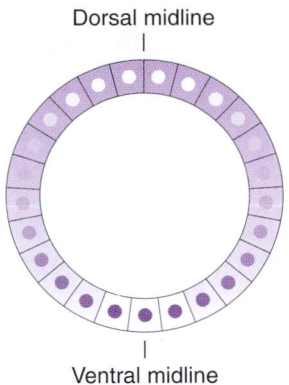

Figure 23-25   The distribution of DL in response to the SPZ signal [encoded by the *spaetzle (spz)* gene]. A cross section of the cellular blastoderm-stage *Drosophila* embryo is depicted, with the dorsal midline at the top and the ventral midline at the bottom. Note that the DL protein is in the nucleus ventrally, throughout the cell laterally, and in the cytoplasm dorsally. (After S. Roth, D. Stein, and C. Nüsslein-Volhard, "A Gradient of Nuclear Localization of the Dorsal Protein Determines Dorsoventral Pattern in the *Drosophila* Embryo," *Cell* 59, 1989, 1196.)

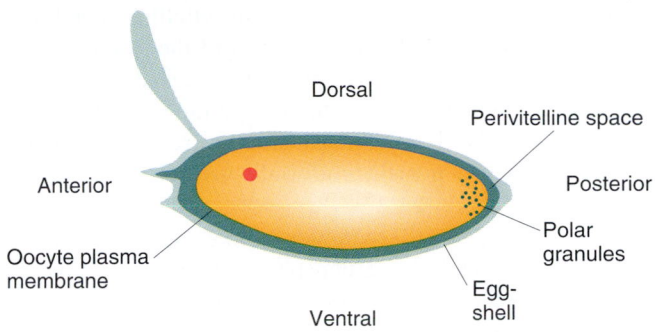

Figure 23-26   A mature oocyte. The follicle cells that built and surrounded the eggshell have been discarded, and the plasma membrane of the oocyte is interior to the eggshell. Note the polar granules at the posterior end of the oocyte. (After Christiane Nüsslein-Volhard, "Determination of the Embryonic Axes of *Drosophila,*" *Development,* Suppl. 1, 1991, 1.)

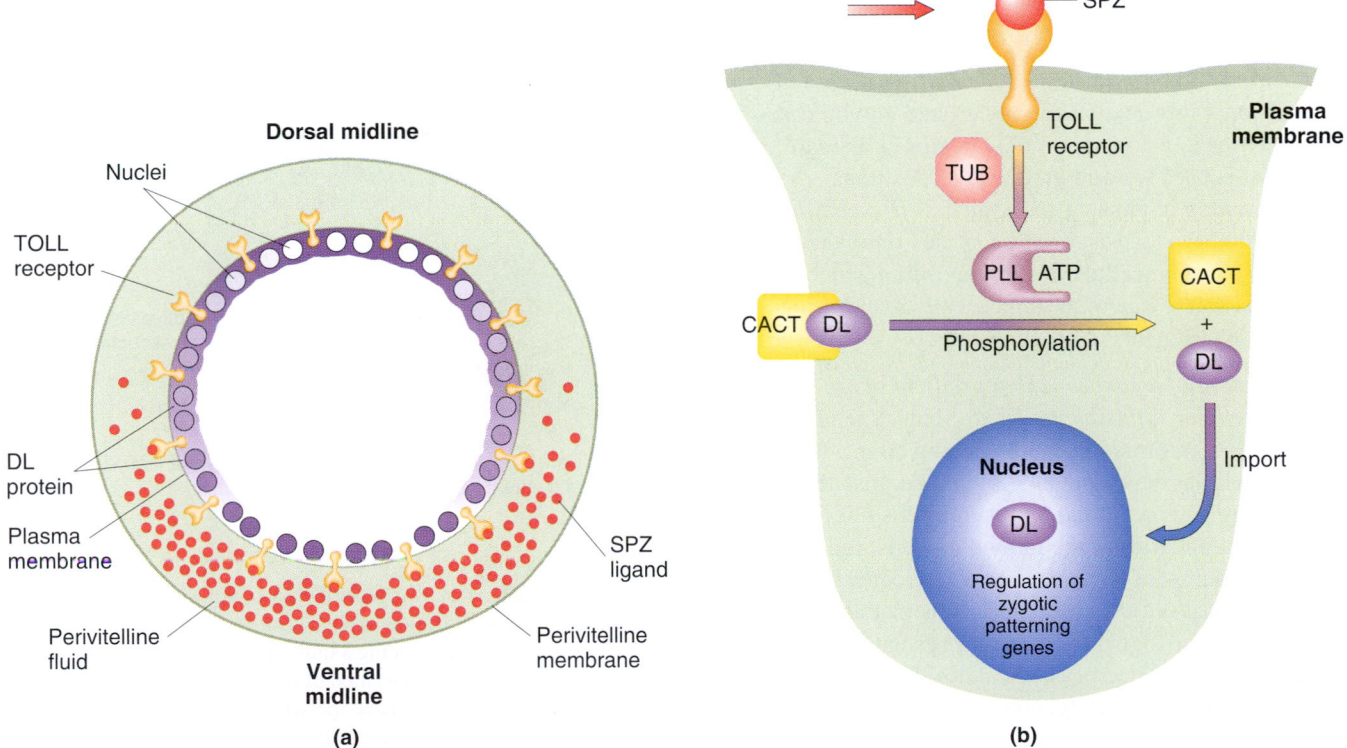

Figure 23-27   The signaling pathway that leads to the gradient of nuclear versus cytoplasmic localization of DL protein. (a) A cross section of a *Drosophila* embryo showing the blastoderm cells inside the plasma membrane and the space (perivitelline space) between the inside boundary of the eggshell (perivitelline membrane) and the plasma membrane, where the SPZ ligand is produced on the ventral side of the embryo. (b) The SPZ ligand binds to the TOLL receptor, activating a signal transduction cascade through two proteins called TUB and PLL, leading to the phosphorylation of DL and its release from CACT. DL then is able to migrate into the nucleus, where it serves as a transcription factor for D–V cardinal genes. (After H. Lodish, D. Baltimore, A. Berk, S. L. Zipursky, P. Matsudaira, and J. Darnell, *Molecular Cell Biology,* 3d ed. Copyright © 1995 by Scientific American Books.)

temporarily bound to structures in the vitelline membrane, sequestering them until almost the end of the syncitial stage of early embryogenesis, when they are released. Active SPZ ligand (with its highest concentration at the ventral midline) then binds to the TOLL transmembrane receptor, encoded by the *Toll* gene, present uniformly in the oocyte plasma membrane (Figure 23-27). In a concentration-dependent manner, the SPZ–TOLL complex triggers a signal transduction pathway that ends up phosphorylating the inactive DL and CACT cytoplasmic proteins of the DL–CACT complex (Figure 23-27b). Phosphorylation of DL and CACT causes conformational changes that break apart the cytoplasmic complex. The free phosphorylated DL protein is then able to migrate into the nucleus, where it serves as a transcription factor activating genes necessary for establishing the ventral fates.

> **MESSAGE** ·······························
> Positional information can be established through cell–cell signaling by means of a concentration gradient of a secreted molecule.

The two classes of positional information

To summarize this section, the most important message is that the specific examples that we considered fall into two general categories of positional information (Figure 23-28).

◆ *Localization of mRNAs within a cell.* This type of positional information can be utilized only in cases where the developmental field begins as a single cell. It is used to form gradients of positional information in unusual cases such as *Drosophila* early embryogenesis, because, at this time, the embryo is a unicellular syncitium. More generally, it is used as a way of asymmetrically distributing local determinants to progeny cells.

◆ *Formation of a concentration gradient of an extracellular diffusible molecule.* This type of positional information can be employed in multicellular developmental fields, because the gradient is extracellular and therefore is not limited by the boundaries of the individual cells. Indeed, we now know of several cases where concentration gradients of secreted protein ligands that activate receptors and signal transduction systems fulfill the properties expected of classical developmental **morphogens** — literally, concentration-dependent determinants of form.

Forming complex pattern: utilizing positional information to establish cell fates

When positional information has been provided, a system that can interpret the positional data must be in place. To use a geographical analogy, it is not sufficient to have a sys-

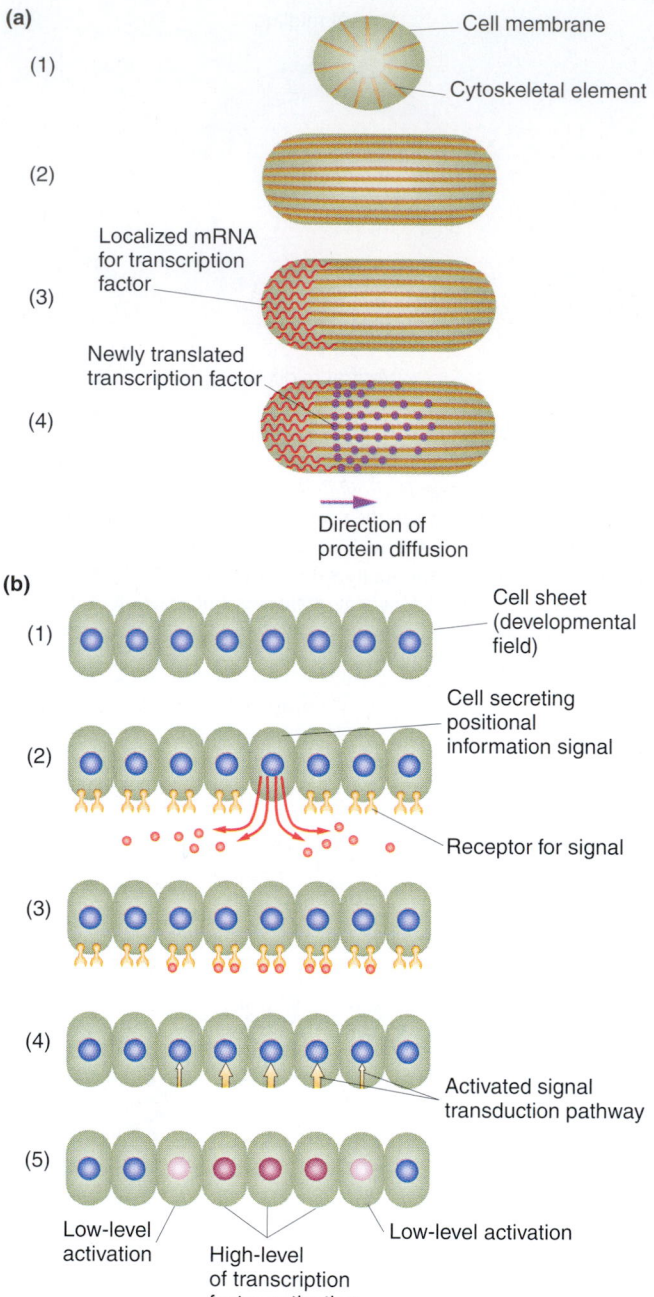

Figure 23-28 The two general classes of positional information. (a) Asymmetric organization of the cytoskeletal system permits localization of mRNA encoding a transcription factor that will provide positional information. Translation of the anchored mRNA will lead to diffusion of the newly translated transcription factor and the formation of a transcription factor gradient with a high point near the site of the mRNA. (b) Secretion of a positional information-signaling molecule from a localized source activates the signal transduction apparatus and the target transcription factors according to the level of signaling molecule that binds to its transmembrane receptor.

tem of longitudes and latitudes; we also need equipment that can receive longitude and latitude information, whether it is through special instruments to read the positions of stars or through receivers that can triangulate signals transmitted from radio beacons. In the same way, the developmental positional information system requires that the signals transmitted be interpretable by elements within the cell.

Initial interpretation of positional information

As described earlier, two very different kinds of positional signal can be produced. However, both lead to the same outcome: a gradient in the amount of one or more specific transcription factors within the cells of the developmental field. In our examples of localized mRNAs in *Drosophila* A–P axis development, this outcome is a direct consequence of the fact that the positional information gradients are of transcription factors themselves (BCD and HB-M). For diffusible extracellular sources of positional information, this outcome requires several intermediary steps. Such cases typically include a gradient of a secreted protein ligand that binds to a transmembrane receptor in a concentration-dependent fashion. In turn, this binding activates a signal transduction pathway proportionately to the level of receptor activation. Eventually, this signal transduction pathway proportionately activates the key transcription factor(s) in the target cells. This is exactly what happens in regard to the *Drosophila* D–V axis, where the SPZ extracellular gradient leads to a graded activation of the DL transcription factor.

Given that positional information leads to a gradient of transcription-factor activities, we would naturally expect that the receivers are regulatory elements (enhancer and silencer elements) of genes whose protein products can begin the gradual process of specifying cell fate. This is exactly what we see. The genes targeted by the A–P and D–V transcription factors are zygotically expressed genes collectively known as the **cardinal genes** (Table 23-3) because they are the first genes to respond to the maternally supplied positional information. As an example, we shall consider the A–P cardinal genes. (The logic by which the D–V axis is divided initially into three domains through the action of the DL transcription-factor activity gradient and then into numerous finer subdivisions is identical with that described for the A–P axis.) Before considering cell fating of the A–P axis, we need to review a bit of *Drosophila* embryology.

After the early embryonic syncitial stage, all somatic nuclei migrate to the surface of the egg and cellularize (Figures 23-29 and 23-30a). A few hours later, the first morphological manifestations of segmentation are apparent. At the end of 10 hours of development, the embryo is already externally divided into 14 segments from anterior to posterior: 3 head, 3 thoracic, and 8 abdominal segments (Figure 23-30b). By this time, each segment has developed a unique set of anatomical structures, corresponding to its special identity and role in the biology of the animal. At the end of 12 hours, organogenesis occurs. At 15 hours, the exoskeleton of the larva begins to form, with its specialized hairs and other external structures. Only 24 hours after development began at fertilization, a fully formed larva hatches out of the eggshell (Figure 23-30c). Of special note in considering the A–P pattern, the segmental arrangement of spikes, hairs, and other sensory structures on the larval exoskeleton makes each segment distinct and recognizable. Now let's return to a consideration of the A–P cardinal genes.

23-3 TABLE		Examples of *Drosophila* A–P Axis Genes Contributing to Pattern Formation	
Gene symbol	Gene name	Protein function	Role(s) in early development
hb-z	*hunchback-zygotic*	Transcription factor–Zn-finger protein	Gap gene
Kr	*Krüppel*	Transcription factor–Zn-finger protein	Gap gene
kni	*knirps*	Transcription factor–steroid receptor type protein	Gap gene
eve	*even-skipped*	Transcription factor–homeodomain protein	Pair-rule gene
ftz	*fushi tarazu*	Transcription factor–homeodomain protein	Pair-rule gene
opa	*odd-paired*	Transcription factor–Zn-finger protein	Pair-rule gene
prd	*paired*	Transcription factor–PHOX protein	Pair-rule gene
en	*engrailed*	Transcription factor–homeodomain protein	Segment-polarity gene
ci	*cubitus-interruptus*	Transcription factor–Zn-finger protein	Segment-polarity gene
wg	*wingless*	Signaling WG protein	Segment-polarity gene
hh	*hedgehog*	Signaling HH protein	Segment-polarity gene
fu	*fused*	Cytoplasmic serine/threonine kinase	Segment-polarity gene
ptc	*patched*	Transmembrane protein	Segment-polarity gene
arm	*armadillo*	Cell–cell junction protein	Segment-polarity gene
lab	*labial*	Transcription factor–homeodomain protein	Segment-identity gene
Dfd	*Deformed*	Transcription factor–homeodomain protein	Segment-identity gene
Antp	*Antennapedia*	Transcription factor–homeodomain protein	Segment-identity gene
Ubx	*Ultrabithorax*	Transcription factor–homeodomain protein	Segment-identity gene

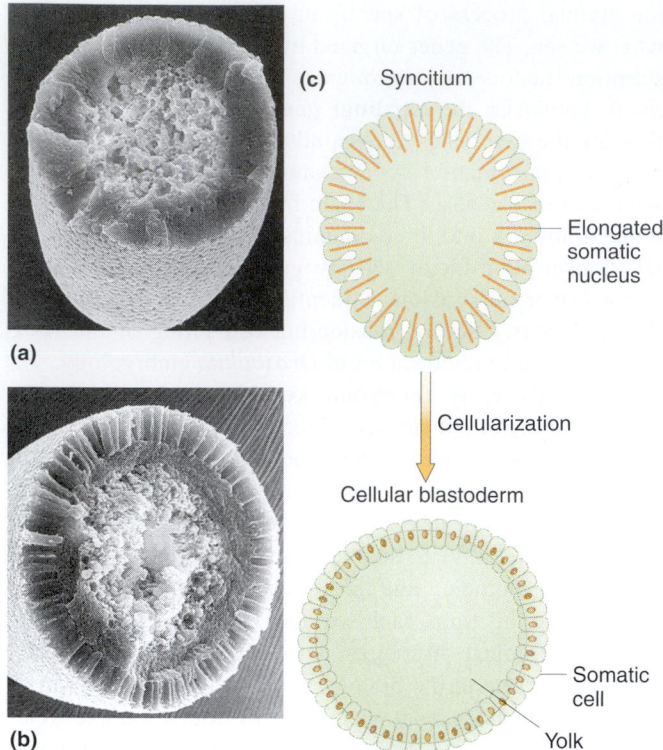

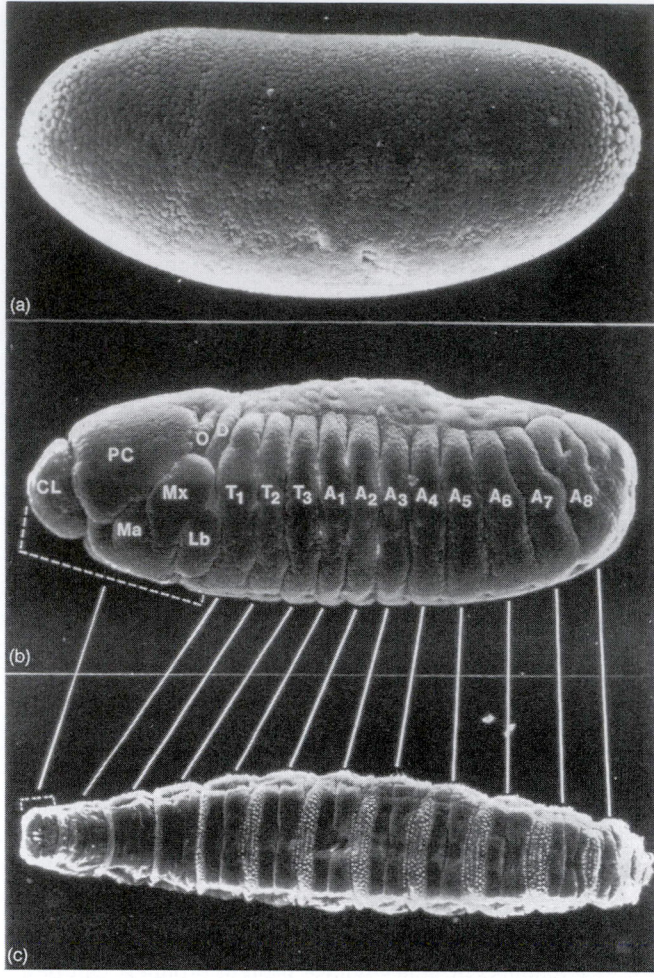

Figure 23-29   Cellularization of the *Drosophila* embryo. The embryo begins as a syncitium; cellularization is not completed until there are about 6000 nuclei. The scanning electron micrographs are of embryos removed from the eggshell. (a) A syncitium-stage embryo, fractured to reveal the common cytoplasm toward the periphery and the central yolk-filled region. The bumps on the outside of the embryo are the beginning of cellularization, in which the plasma membrane of the egg moves from the outside inward. (b) A cellular blastoderm embryo, fractured to reveal the columnar cells that have formed by cell membranes being drawn down between the elongated nuclei to create some 6000 mononucleate somatic cells. (c) Schematic representation of the changes taking place during cellularization. (After F. R. Turner and A. P. Mahowald, "Scanning Electron Microscopy of *Drosophila* Embryogenesis," *Developmental Biology* 50, 1976, 95.)

Figure 23-30   Scanning electron micrographs of (a) a 3-hour *Drosophila* embryo, (b) a 10-hour embryo, and (c) a newly hatched larva. Note that, by 10 hours, the segmentation of the embryo is obvious. Lines are drawn to indicate that the segmental identity of cells along the A–P axis is already fated early in development. The abbreviations refer to different segments of the head (CL, PC, O, D, Mx, Ml, Lb), thorax (T1 to T3), and abdomen (A1 to A8). (T. C. Kaufman and F. R. Turner, Indiana University.)

The A–P cardinal genes are also known as gap genes, because mutant flies lack a sequential series of larval segments, producing a gap in the normal segmentation pattern (look at the phenotypes of mutations in two gap genes, *Krüppel* and *knirps,* in Figure 23-31). BCD or HB-M protein of both bind to enhancer elements of the promoters of the gap genes, thereby regulating their transcription. For example, transcription of one gene, *Krüppel (Kr),* is repressed by high levels of the BCD transcription factor but is activated by low levels of BCD and HB-M. In contrast, the *knirps (kni)* gene is repressed by the presence of any BCD protein but does require low levels of the HB-M transcription factor for its expression. These enhancer and promoter properties thereby ensure that the *kni* gene is expressed more posteriorly than is *Kr* (Figure 23-32a). By having promoters that are differentially sensitive to the concentrations

of the transcription factors of the A–P positional information system, the gap genes can be expressed in a series of distinct domains, and these domains then become different developmental fields. That is, the cells in the different domains become committed to different A–P fates.

These commitments of the domains to different fates are due to the fact that each of the gap genes encodes a different transcription factor and thereby has the capability of regulating a different set of downstream target genes necessary to refine A–P segmental fate.

Refining fate assignments through transcription-factor interactions

The gap gene expression pattern slices up the A–P axis into several domains. However, gap genes are expressed too

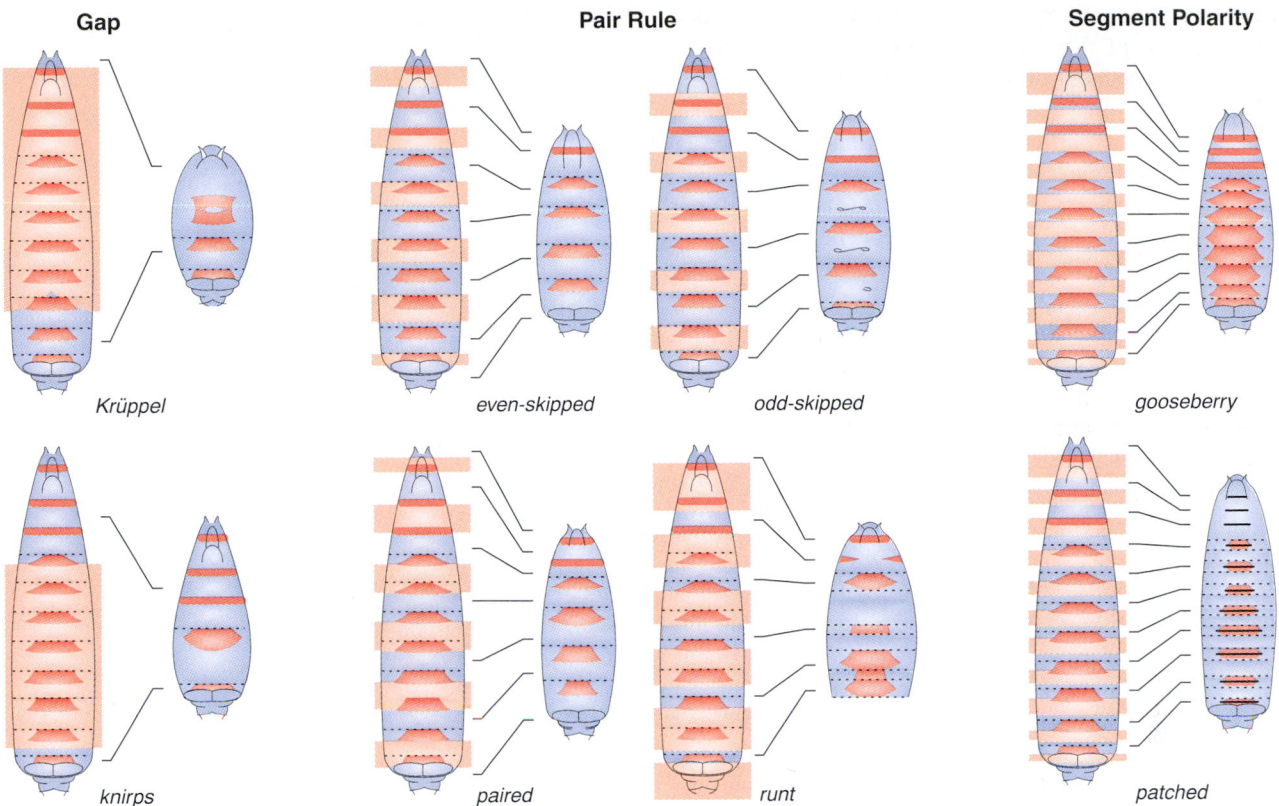

Figure 23-31   A diagram depicting the mutant larval phenotypes due to mutations in the three classes of zygotically acting genes controlling segment number in *Drosophila,* with representative mutants in each class. The denticle belts, shown as red trapezoids, are segmentally repeating swatches of dense projections on the ventral surface of the larval exoskeleton. The boundary of each segment is represented by a dotted line. On the left, the pink regions indicate, on a diagram of a wild-type larva, the A–P domains of the larva that are missing in each mutant. The diagrams on the right are the resulting mutant phenotypes.

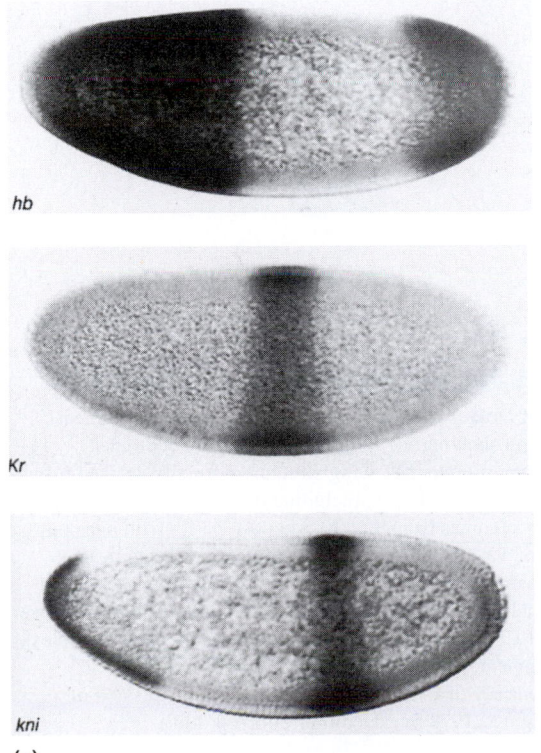

(b)

(a)

Figure 23-32   Photomicrographs showing the early embryonic expression patterns of gap and pair-rule genes. (a) Early blastoderm expression patterns of proteins from three gap genes: *hb-z, Kr,* and *kni.* (b) Late blastoderm expression patterns of proteins from two pair-rule genes: *ftz* (stained gray) and *eve* (stained brown). Note the localized gap gene expression patterns compared with the reiterated pair-rule gene expression pattern. (Part a from M. Hulskamp and D. Tautz, "Gap Genes and Gradients: The Logic Behind the Gaps," *Bioessays* 13, 1991, 261. Part b from Peter Lawrence, *The Making of a Fly.* Copyright © 1992 by Blackwell Scientific Publications.)

coarsely to allocate all the A–P cell fates that are needed. Further, the end of the gap gene expression stage and the beginning of refinement closely coincide when the syncitial embryo becomes fully cellularized. The cytoplasm of each cell then contains a particular concentration of one or perhaps two adjacent gap-gene-encoded proteins. Essentially all further decisions are driven by the particular A–P gap proteins present in the nucleus of a given cell.

The A–P developmental pathway downstream of the gap genes bifurcates. Each of the two branches is instructive in regard to how pattern is refined. One branch establishes the correct number of segments. The other assigns the proper identity to each segment. (These different identities are manifest in the unique patterns of spikes and hairs on each segment of the larva, as described earlier.) The existence of two branches means that there are two different sets of target genes for regulation by the gap-gene-encoded transcription factors.

First, let's briefly consider the formation of segment number. (Refer to Figure 23-31 for a description of the mutant phenotypes that the different classes of segment-number genes produce.) The gap genes activate a set of secondary A–P patterning genes called the **pair-rule genes** in a repeating pattern of seven stripes (Figure 23-32b). There are several different pair-rule genes, and each of them produces a slightly offset pattern of stripes. Additionally, there is a hierarchy within the pair-rule gene class. Some of these genes, called primary pair-rule genes, are regulated directly by the gap genes, whereas others are activated by the primary pair-rule gene-encoded proteins, which are also transcription factors. The pair-rule genes then act combinatorially (several of their proteins are expressed within a given cell) to regulate the transcription of the segment-polarity genes, which are expressed in offset patterns of 14 stripes. Thus, the hierarchy of transcription-factor regulation extends all the way from the positional-information system to the repeating pattern of segment-polarity gene expression. The products of the segment-polarity genes then permit the 14 segments to form and define the individual A–P rows of cells within each segment.

MESSAGE ·······················

Through a hierarchy of transcription-factor-regulation patterns, positional information leads to the formation of the correct number of segments. In the readout of positional information, transcription factors act combinatorially to create the proper segment-number fates.

How do the primary pair-rule genes become activated in a repeating pattern by the asymmetrically expressed gap proteins? The key is that the regulatory elements for the primary pair-rule genes are quite complex. For primary genes such as *eve (even-skipped)*, separate enhancer elements regulate the activation of each *eve* stripe. The *eve* stripe 1 enhancer is activated by high levels of the HB-Z gap transcription factor, the *eve* stripe 2 enhancer by low levels of HB-Z

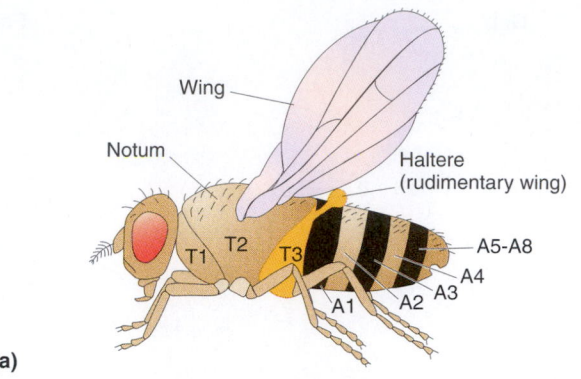

(a)

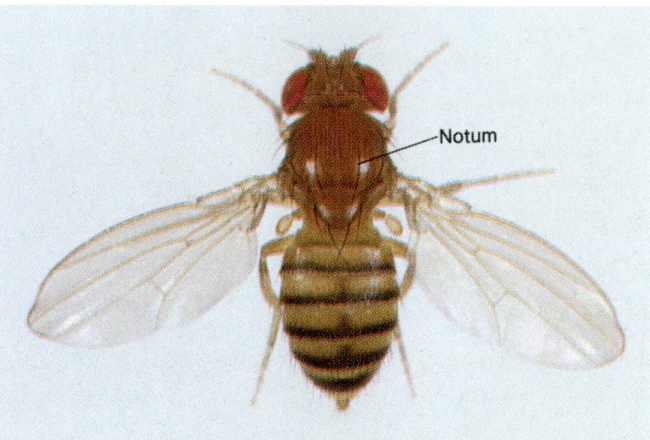

(b)

(c)

Figure 23-33 The homeotic transformation of the third thoracic segment (T3) of *Drosophila* into an extra second thoracic segment (T2). (a) Diagram showing the normal thoracic and abdominal segments; note the rudimentary wing structure normally derived from T3. Most of the thorax of the fly, including the wings and the dorsal part of the thorax, comes from T2. (b) A wild-type fly with one copy of T2 and one of T3. (c) A *bithorax* triple mutant homozygote completely transforms T3 into a second copy of T2. Note the second dorsal thorax and second pair of wings (T2 structures) and the absence of the halteres (T3 structures). (From E. B. Lewis, "A Gene Complex Controlling Segmentation in *Drosophila*," *Nature* 276, 1978, 565. Photographs courtesy of E. B. Lewis. Reprinted by permission of *Nature*. Copyright © 1978 by Macmillan Journals Ltd.)

but high levels of the KR gap transcription factor, and so forth.

Next, let's briefly consider the establishment of segmental identity. The gap genes target a clustered group of genes known as the **homeotic gene complexes.** They are called gene complexes because several of the genes are clustered together on the DNA. *Drosophila* has two homeotic gene clusters. The ANT-C *(Antennapedia* complex) is largely responsible for segmental identity in the head and anterior thorax, whereas the BX-C *(Bithorax* complex) is responsible for segmental identity in the posterior thorax and abdomen.

Homeosis is the conversion of one body part into another. Three examples of body-part-conversion phenotypes due to homeotic gene mutations are (1) the loss-of-function *bithorax* class of mutations that cause the entire third thoracic segment (T3) to be transformed into a second thoracic segment (T2), giving rise to flies with four wings instead of the normal two (Figure 23-33); (2) the gain-of-function dominant *Tab* mutation described in Chapter 11 (Figures 11-34 and 11-35) that transforms part of the adult T2 seg-

ment into the sixth abdominal segment (A6); and (3) the gain-of-function dominant *Antennapedia (Antp)* mutation that transforms antenna into leg (see the photograph on the first page of this chapter). Note that, in each of these cases, the number of segments in the animal remains the same; the *only* change is in the identity of the segments. By studying these homeotic mutations, we have learned much about how segment identity is established.

The domains of expression of the various gap proteins activate the target homeotic genes initially in a series of overlapping domains (Figure 23-34). These homeotic genes encode homeodomain proteins, a class of transcription factors. **Homeodomain proteins** interact with the regulatory elements of the homeotic genes in a specific pattern such that expression patterns of the homeotic genes become mutually exclusive. (We shall consider the structure and function relations within the homeotic gene complexes later in the context of evolution of developmental mechanisms.) These homeodomain proteins also regulate downstream target genes that then are responsible for conferring the specific functions and identities to different regions of each segment.

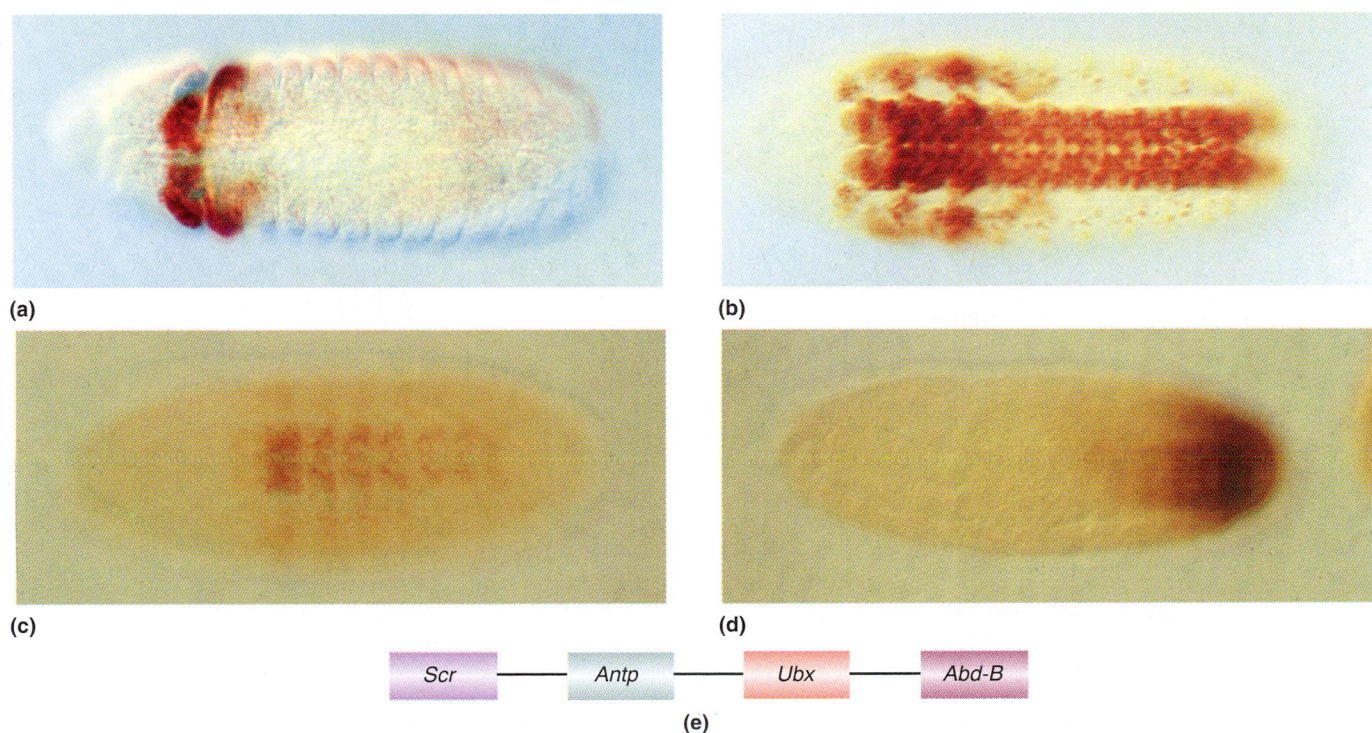

Figure 23-34 Photomicrographs of embryos that exhibit homeotic gene-encoded protein expression patterns in *Drosophila*. (a) SCR; (b) ANTP; (c) UBX; (d) ABD-B. Note that the anterior boundary of homeotic gene expression is ordered from SCR (most anterior) to ANTP, UBX, and ABD-B (most posterior). (e) This order is matched by the linear arrangement of the corresponding genes along chromosome 3. (Parts a and b from T. C. Kaufman, Indiana University. Parts c and d from S. Celniker and E. B. Lewis, California Institute of Technology.)

A cascade of regulatory events

As we have seen, A–P patterning of the *Drosophila* embryo occurs through a sequential triggering of regulatory events. Positional information establishes different concentrations of transcription factors along the A–P axis, and target regulatory genes are then deployed accordingly to execute the increasingly finer subdivisions of the embryo, establishing both segment number and segment identity (Figure 23-35).

Additional aspects of pattern formation

The principles delineated in the preceding sections lay out initial fates, but additional mechanisms must be in place to

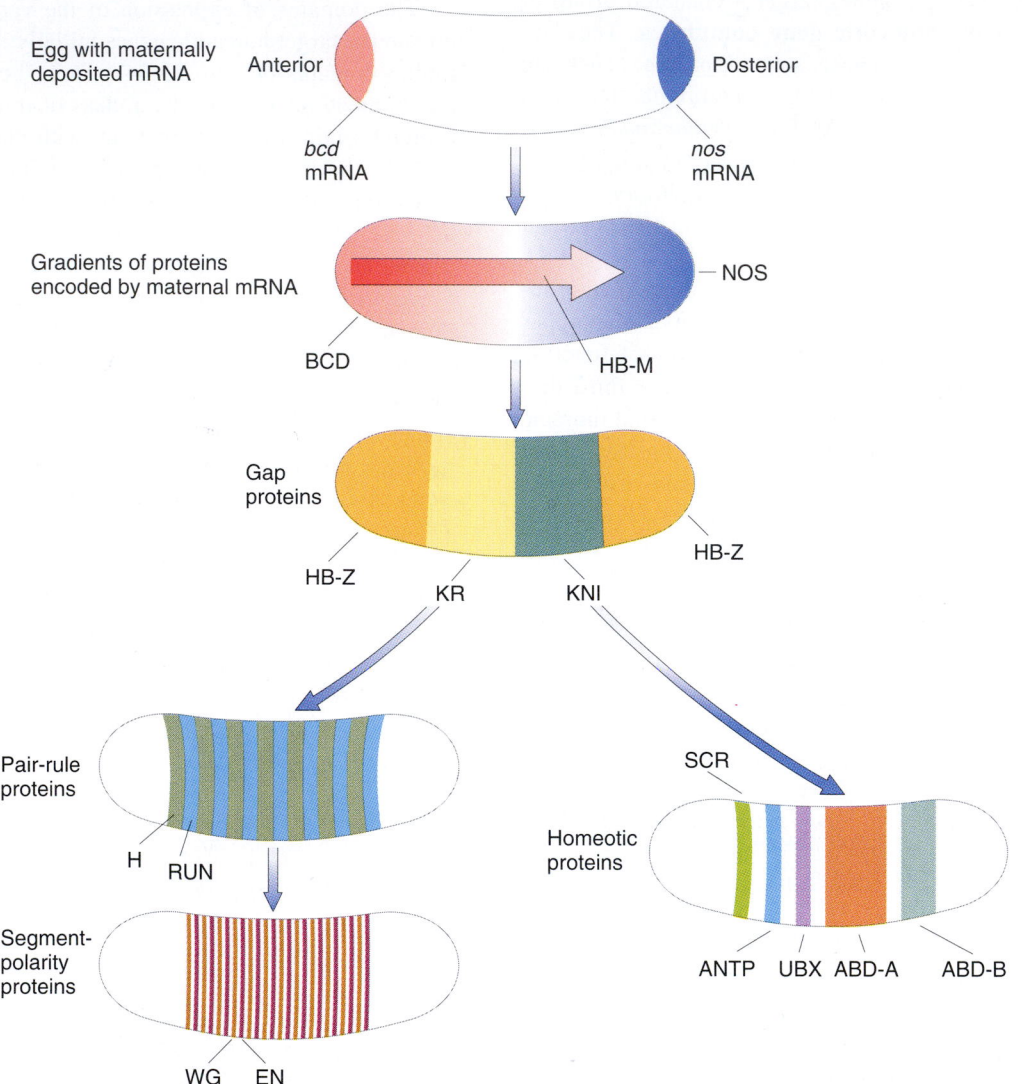

Figure 23-35   A depiction of the hierarchical cascade that activates the elements forming the A–P segmentation pattern in *Drosophila*. The maternally derived *bcd* and *nos* mRNAs are located at the anterior and posterior poles, respectively. Early in embryogenesis, these mRNAs are translated to produce a steep anterior–posterior gradient of BCD transcription factor. The posterior–anterior gradient of NOS inhibits translation of *hb-m* mRNA, thereby creating a shallow anterior–posterior gradient of HB-M transcription factor (shown as an arrow). The gap genes, which are the A–P cardinal genes, are activated in different parts of the embryo in response to the anterior–posterior gradients of the two factors BCD and HB-M. (The posterior band of HB-Z expression gives rise to certain internal organs, not to segments.) The correct number of segments is determined by activation of the pair-rule genes in a zebra-stripe pattern in response to the gap-gene-encoded transcription factors. The segment-polarity genes are then activated in response to the activities of the several pair-rule proteins, leading to further refinement of the organization within each segment. The correct identities of the segments are determined by expression of the homeotic genes due to direct regulation by the transcription factors encoded by the gap genes. (After J. D. Watson, M. Gilman, J. Witkowski, and M. Zoller, *Recombinant DNA*, 2d ed. Copyright ©1992 by James D. Watson, Michael Gilman, Jan Witkowski, and Mark Zoller.)

(a)

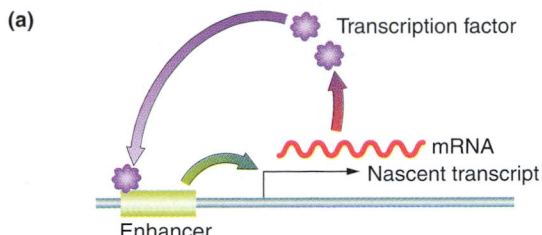

(b)

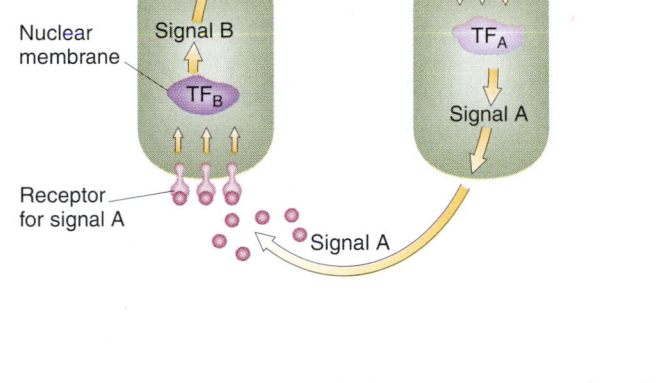

Figure 23-36 Two types of positive feedback loops in maintaining the level of activity of transcription factors determining cell fate. (a) The transcription factor binds to an enhancer of its own gene, maintaining its transcription. (b) Each adjacent cell sends out a signal (different signals from each cell) that activates receptors, signal transduction pathways, and transcription-factor (TF) expression in the other cell. This activation leads to a mutual positive feedback loop between the cells.

ensure that all aspects of patterning are elaborated. Some of these mechanisms are considered in this section.

Memory systems for remembering cell fate

Patterning decisions frequently need to be maintained in a cell lineage for the lifetime of the organism. This requirement is certainly true of the segment-polarity and homeotic gene expression patterns that are set up by the A–P patterning system. Such maintenance is accomplished through intracellular or intercellular positive feedback loops (Figure 23-36).

In several tissues, positive feedback loops are established in which the homeodomain protein that is expressed

binds to enhancer elements in its own gene, ensuring that more of that homeodomain protein will continue to be produced (Figure 23-36a). (This positive feedback loop is reminiscent of that for *Sxl* splicing in the female developmental pathway, discussed earlier in this chapter.)

The other solution requires cell–cell interactions (Figure 23-36b). For example, among the segment-polarity genes, adjacent cells express the WG and EN proteins. The EN protein is a transcription factor that activates HH in the same cells. (See Table 23-4 for more information on these proteins and the genes that encode them.) HH is a secreted protein signaling molecule that induces a receptor-mediated signal transduction cascade in the WG cell, including *wg* *(wingless)* gene expression and more WG protein to be

23-4 TABLE		Examples of Evolutionarily Conserved Genes Contributing to *Drosophila* Pattern Formation		
Drosophila gene symbol	*Drosophila* gene name	Mammalian counterpart(s)	Protein function	Role(s) in early development
dl	*dorsal*	NF$_\kappa$B	DL transcription factor	D–V transcription factor
cact	*cactus*	I$_\kappa$B	Binds to DL protein in cytoplasm	Regulates DL expression
spz	*spaetzle*	IL-1	Signaling SPZ ligand	Creates D–V positional information
Tl	*Toll*	IL-1 receptor	Receptor for SPZ ligand	Regulates DL expression
snk	*snake*	Many	Serine protease	Activates SPZ protein
dpp	*decapentaplegic*	BMP2, BMP4	Signaling DPP ligand	Cardinal gene for dorsal ectoderm
sax	*saxophone*		Receptor serine/	Contribute to dorsal ectodermal
tkv	*thick veins*	BRK	theonine kinases	pattern formation
pnt	*punt*		for DPP ligand	
sog	*short gastrulation*	Chordin	Signaling SOG ligand	Contributes to ectodermal patterning
spi	*spitz*	TGF-α	Signaling SPI ligand	Secondary regulatory gene for ventral ectoderm
en	*engrailed*	En1,En2	Homeodomain protein	Segment-polarity gene
hh	*hedgehog*	Sonic hedgehog	Signaling HH ligand	Segment-polarity gene
wg	*wingless*	Wnt1, Wnt2, etc.	Signaling WG ligand	Segment-polarity gene
arm	*armadillo*	β-Catenin	Transduces WG signal	Segment-polarity gene

Note: Many other conservations are not listed here.

expressed. WG similarly is a secreted protein that activates *en (engrailed)* expression in the adjacent cell, inducing more EN protein in that cell.

> **MESSAGE** ···
> When the fate of a cell lineage has been established, it must be remembered.

Ensuring that all fates are allocated: decisions by committee

Ultimately, for a developmental field to mature into a functional organ or tissue, cells must be committed in appropriate numbers and locations to the full range of fates that are needed. Cell–cell interactions ensure that these proper allocations are made. We should be aware of two types of interactions, both of which operate in the development of the vulva, the opening to the outside of the reproductive tract of the nematode *C. elegans* (Figure 23-37). One type is the ability of one cell to induce a developmental commitment in one neighbor of many, and the other is the ability of one cell to inhibit its neighbors from adopting its fate.

Vulva development has been studied in detail through the analysis of mutants that have either no vulva or too many. Within the hypoderm (the body wall of the worm), several cells have the potential to build certain parts of the vulva. To make an intact vulva, one of the cells must become the primary vulva cell, and two others must become secondary vulva cells (Figure 23-38a); yet others become tertiary cells that contribute to the surrounding hypoderm (Figure 23-38b).

Initially, all these cells can adopt any of these roles and so are called an **equivalence group** — in essence, a developmental field. The key to allocating the different roles to these cells is another single cell, called the *anchor cell,* which lies underneath the cells of the equivalence group

(Figure 23-38c). The anchor cell secretes a polypeptide ligand that binds to a receptor tyrosine kinase (RTK) present on all the cells of the equivalence group. Only the cell that receives the highest level of this signal (the equivalence-group cell nearest the anchor cell) activates the signal transduction pathway at a sufficient level to activate the transcription factors necessary for that cell to become a primary vulva cell (Figure 23-38d). Thus we can say that the anchor cell operates through an **inductive interaction** to commit a cell to the primary vulva fate.

Having acquired its fate, the primary vulva cell sends out a different paracrine signal to its immediate neighbors in the equivalence group, inhibiting those cells from similarly interpreting the anchor cell signal to also adopt the primary role. This process of **lateral inhibition** leads these neighboring cells to adopt the secondary fate. The remaining cells of the equivalence group develop as tertiary cells and contribute to the hypoderm surrounding the vulva. For each of the three cell types into which the equivalence group develops, there is a specific constellation of transcription factors that are activated and that typify the state of the cell: primary, secondary, or tertiary. Thus, through a series of paracrine intercellular signals, a group of equivalent cells can develop into the three necessary cell types.

> **MESSAGE** ···
> Fate allocation can be made through a combination of inductive and lateral inhibitory interactions between cells.

Developmental pathways are composed of plug and play modules

Many developmental pathways are under active investigation in model organisms. From these studies, it is clear that the components of developmental pathways contribute over

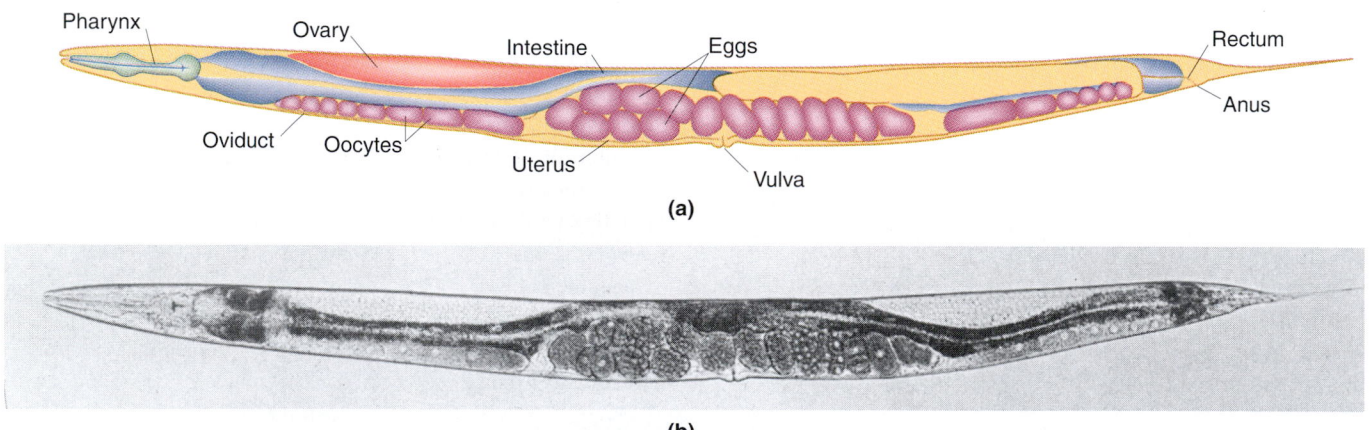

(a)

(b)

Figure 23-37   Adult *Caenorhabditis elegans.* Drawing and photomicrograph of an adult hermaphrodite, showing various organs and nuclei readily identified by their location. Note the position of the vulva midway along the anterior–posterior axis of the worm. (From J. E. Sulston and H. R. Horvitz, "Post-embryonic Cell Lineages of the Nematode, *Caenorhabditis elegans,*" *Developmental Biology* 56, 1977, 111.)

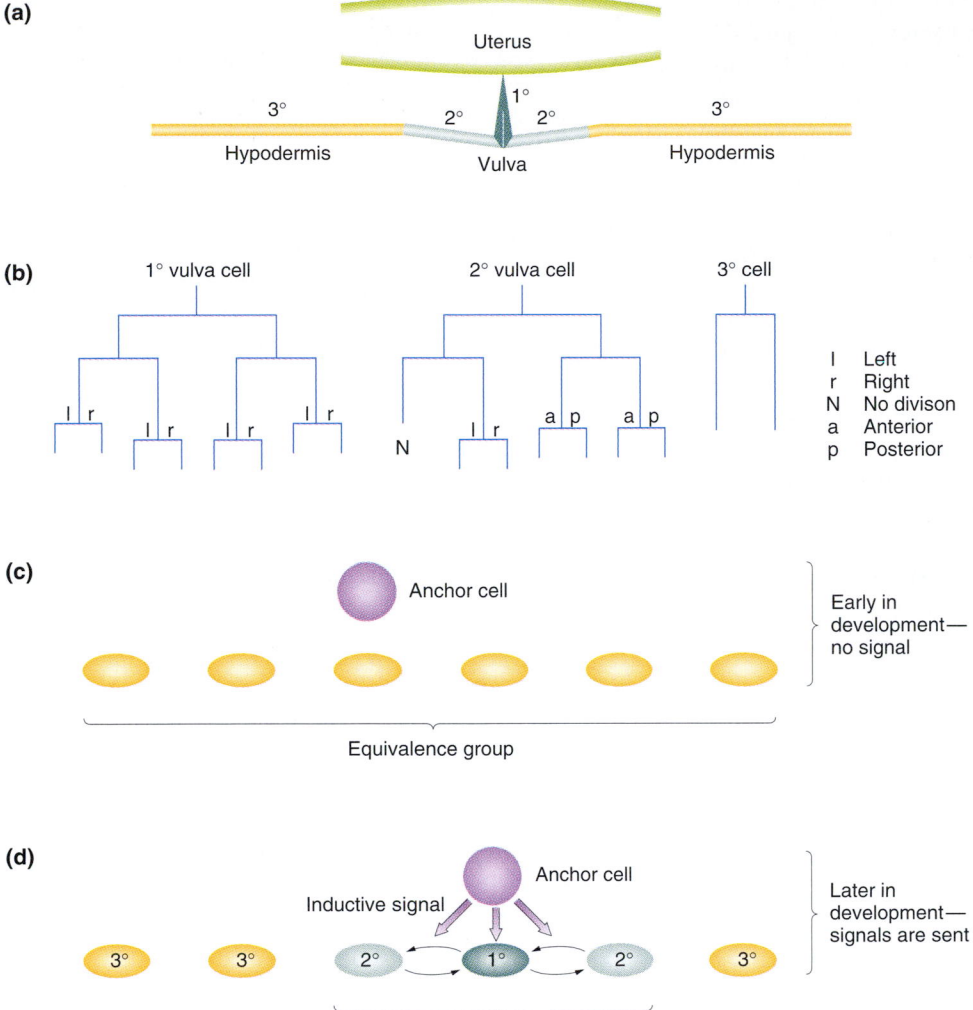

Figure 23-38 The production of the *C. elegans* vulva from the equivalence group by cell–cell interactions. (a) The parts of the vulva anatomy occupied by the descendants of the equivalence-group cells. (b) The primary, secondary, and tertiary cell types are distinguished by the cell-division patterns that they undergo. (c) Early in development, there is no signal from the anchor cell, and all the equivalence-group cells are in the default tertiary cell state. (d) Later in development, the anchor cell sends a signal that activates a receptor tyrosine kinase signal transduction cascade. The equivalence-group cell nearest the anchor cell receives the strongest signal and becomes the primary vulva cell. It then sends out lateral inhibition signals to its neighbors, preventing them from also becoming primary vulva cells and shunting them into the secondary vulva cell pathway. (Part b from H. R. Horvitz and P. W. Sternberg, "Multiple Intercellular Signalling Systems Control the Development of the *Caenorhabditis elegans* Vulva," *Trends in Genetics* 7, 1991, 366.)

and over again to the development of a given species. There are rather few examples of a gene product taking part in pattern formation that contributes to only one developmental decision. Instead, bits and pieces of pathways are combined in different ways to determine different outcomes. It is as if, once an effective solution to a particular developmental problem evolved, it was then applied to solve many other problems. This can be thought of as a molecular correlate of the general evolutionary point of view that new structures arise by modifications and adaptations of existing structures rather than by the invention of something totally new.

MESSAGE ··

Components of one developmental pathway are also found in many others but are often mixed and matched as if they are reusable cartridges.

···

Typically, a part of a pathway has different inputs, in regard to the signals that regulate it, and sometimes the outputs are different as well (Figure 23-39). As an example of

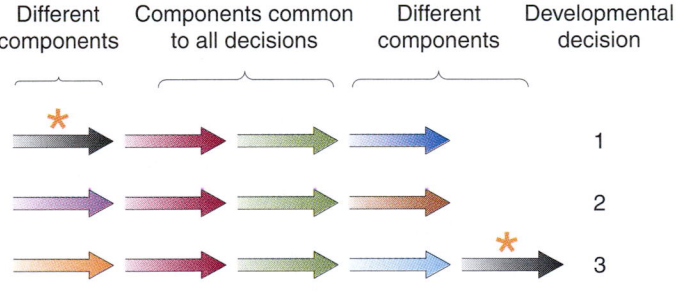

Figure 23-39 Different developmental decisions are made by using mixed and matched combinations of pathway components. Each colored arrow represents a different component. These pathways are joined together in different combinations to achieve different developmental decisions in different cell lineages or in the same cell lineage at different developmental stages. Some components of these decision processes are common; others are unique to each decision process. It is also possible that some components (denoted by an asterisk) are used in multiple decision processes but occupy different positions in the two processes.

different inputs, the *dpp (decapentaplegic)* gene, whose complex structure was described in Chapter 11, is regulated during embryonic D–V pattern formation by the DL transcription factor, whereas, during visceral mesoderm development, it is regulated by the UBX transcription factor. In regard to different outputs, in maintenance of proper segment-polarity gene expression and cell fates, for example, the EN–HH component of the positive feedback loop activates WG in adjacent cells (described earlier). During the larval development of the *Drosophila* adult wing, the EN–HH component activates a different secreted signaling protein—DPP—in cells adjacent to EN–HH expression. As another example of output differences, the DPP signal transduction pathway leads to activation of completely different transcription factors in D–V patterning, wing, and visceral mesoderm development. Indeed, the complicated regulatory element structure common to many higher eukaryotic genes is probably a necessary consequence of the need to respond to many different tissue-specific inputs.

MESSAGE ···
Differences in the developmental context of different cell lineages—that is, the transcription factors active in these cells—permit different inputs to, and outputs from, a given developmental circuit.
···

The many parallels in vertebrate and insect pattern formation

How universal are the developmental principles uncovered in *Drosophila*? Until recently, the type of genetic analysis possible in *Drosophila* was not feasible in most other organisms, at least not without a huge investment to develop comparable genetic tools. However, in the past few years, recombinant DNA technology has provided the tools for addressing the generality of the *Drosophila* findings.

Consider the homeobox genes. With the discovery that there were numerous homeobox genes within the *Drosophila* genome, similarities among the DNA sequences

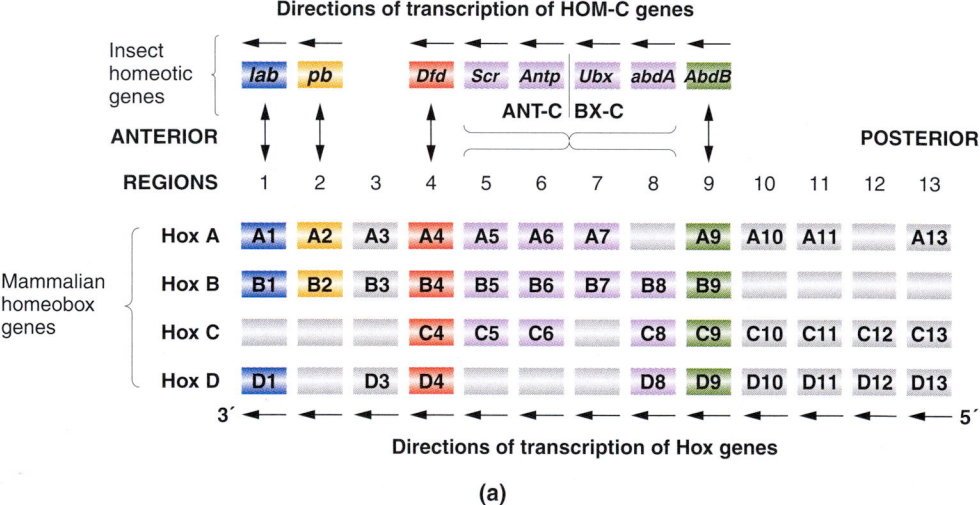

(a)

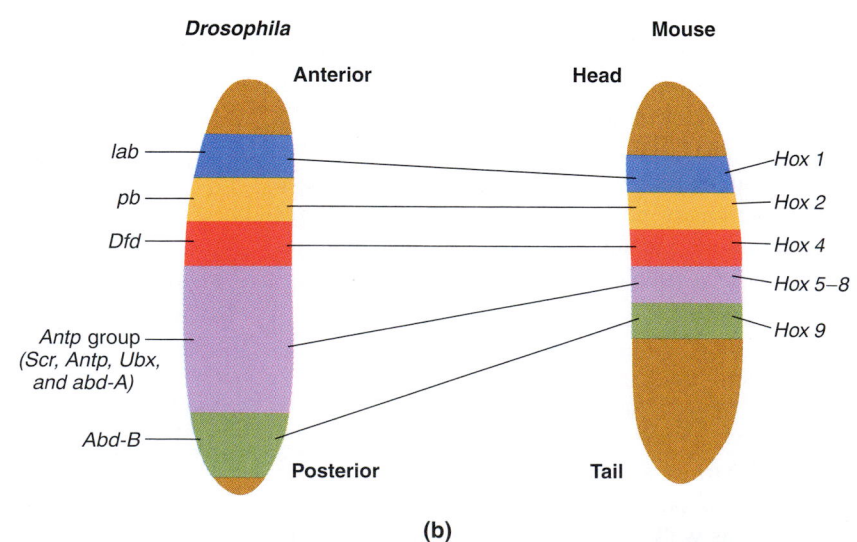

(b)

Figure 23-40   Comparisons of the structures and functions of the insect and mammalian homeotic genes. (a) The comparative anatomy of the HOM-C and Hox gene clusters. The genes of the HOM-C are shown at the top. Each of the four paralogous (see text) Hox clusters maps on a different chromosome. Genes shown in the same color are most closely related to one another in structure and function. (b) The expression domains and regions of the *Drosophila* and mouse embryos that require the various HOM-C and Hox genes. The color scheme parallels that in part a. Note that the order of domains in the two embryos is the same. (After H. Lodish, D. Baltimore, A. Berk, S. L. Zipursky, P. Matsudaira, and J. Darnell, *Molecular Cell Biology,* 3d ed. Copyright © 1995 by Scientific American Books.)

of these genes could be exploited in treasure hunts for other members of the homeotic gene family. These hunts depend on DNA base-pair complementarity. For this purpose, DNA hybridizations were carried out under *moderate stringency conditions,* in which some mismatch of bases between the hybridizing strands could occur without disrupting the proper hydrogen bonding of nearby base pairs. Some of these treasure hunts were carried out in the *Drosophila* genome itself, looking for more family members. Others searched for homeobox genes in other animals, by means of *zoo blots* (Southern blots of restriction-enzyme-digested DNA from different animals), by using radioactive *Drosophila* homeobox DNA as the probe. This approach has led to the discovery of homologous homeobox sequences in many different animals, including humans and mice. (Indeed, it is a very powerful approach to go "fishing" for relatives of almost any gene in your favorite organism.) Some of these mammalian homeobox genes are *very* similar in sequence to the *Drosophila* genes.

Perhaps the most striking case is the similarity between the clusters of mammalian homeobox genes called the Hox complexes and the insect ANT-C and BX-C homeotic gene clusters, now collectively called the HOM-C (homeotic gene complex) (Figure 23-40). The ANT-C and BX-C clusters, which are far apart on chromosome 3 of *Drosophila,* are in one cluster in more primitive insects such as the flour beetle *Tribolium castaneum.* This indicates that the typical case in insects is that there is only one homeotic gene cluster — HOM-C. Moreover, the genes of the HOM-C cluster are arranged on the chromosome in an order that is colin-

ear with their spatial pattern of expression: the genes at the left-hand end of the complex are transcribed near the anterior end of the embryo; rightward along the chromosome, the genes are transcribed progressively more posteriorly (compare Figure 23-40a and b).

We still do not know why the insect genes are clustered or organized in this colinear fashion, but, regardless of the roles of these features, the same structural organization — clustering and colinearity — is seen for the equivalent genes in mammals, which are organized into the Hox clusters (Figure 23-40a). The major difference between flies and mammals is that there is only one HOM-C cluster in the insect genome, whereas there are four Hox clusters, each located on a different chromosome, in mammals. These four Hox clusters are *paralogous,* meaning that the structure (order of genes) in each cluster is very similar, as if the entire cluster had been quadruplicated in the course of vertebrate evolution. The genes near the left end of each Hox cluster are quite similar not only to each other, but also to one of the insect HOM-C genes at the left end of the cluster. Similar relations hold throughout the clusters. Finally, and most notably, the Hox genes are expressed in a segmental fashion in the developing somites and central nervous system of the mouse (and presumably the human) embryo. Each Hox gene is expressed in a continuous block beginning at a specific anterior limit and running posteriorly to the end of the developing vertebral column (Figure 23-41). The anterior limit differs for different Hox genes. Within each Hox cluster, the leftmost genes have the most anterior limits. These limits proceed more and more posteriorly in the rightward direction

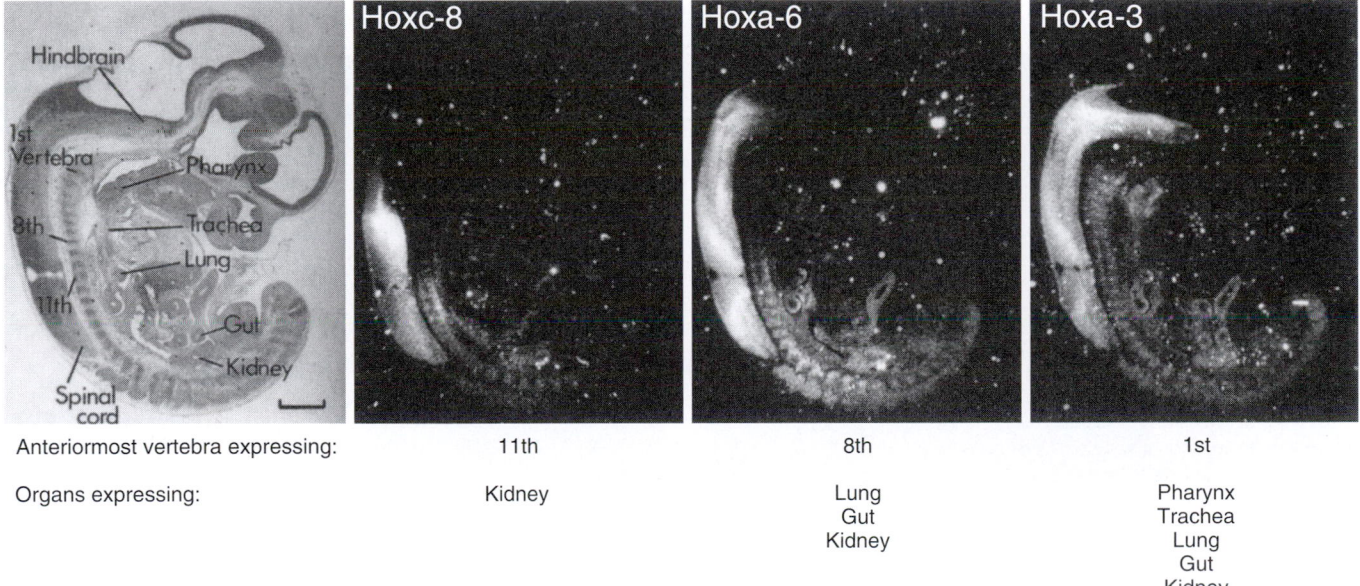

	Hoxc-8	Hoxa-6	Hoxa-3
Anteriormost vertebra expressing:	11th	8th	1st
Organs expressing:	Kidney	Lung Gut Kidney	Pharynx Trachea Lung Gut Kidney

Labels on first panel: Hindbrain, 1st Vertebra, Pharynx, 8th, Trachea, Lung, 11th, Gut, Kidney, Spinal cord

Figure 23-41   Photomicrographs showing the RNA expression patterns of three mouse Hox genes in the vertebral column of a sectioned 12.5-day-old mouse embryo. Note that the anterior limit of each of the expression patterns is different. (From S. J. Gaunt and P. B. Singh, "Homeogene Expression Patterns and Chromosomal Imprinting," *Trends in Genetics* 6, 1990, 208.)

in each Hox cluster. Thus, the Hox gene clusters appear to be arranged and expressed in an order that is strikingly similar to that of the insect HOM-C genes (see Figure 23-40b).

The correlations on structure and expression pattern are further strengthened by considerations of mutant phenotypes. In vitro mutagenesis techniques permit efficient gene knockouts in the mouse. Many of the Hox genes have now been knocked out, and the striking result is that the phenotypes of the homozygous knockout mice are thematically parallel to the phenotypes of homozygous null HOM-C flies. For example, the *Hoxc-8* knockout causes ribs to be produced on the first lumbar vertebra, L1, which ordinarily is the first nonribbed vertebra behind those vertebrae-bearing ribs (Figure 23-42). Thus, the L1 vertebra is homeotically transformed into the segmental identity of a more anterior vertebra. To use geneticists' jargon, *Hoxc-8⁻* has caused a fate shift toward anterior. Clearly, this Hox gene seems to control segmental fate in a manner quite similar to the HOM-C genes, because, for example, the absence of the *Drosophila Ubx* gene also causes a fate shift toward anterior in which T3 and A1 are transformed into T2.

How can such disparate organisms — fly, mouse, and human — have such similar gene sequences? (The same is true for the worm *C. elegans*.) The simplest interpretation is that the Hox and HOM-C genes are the vertebrate and insect descendants of a homeobox gene cluster present in a common ancestor some 600 million years ago. The evolutionary conservation of the HOM-C and Hox genes is not a singular occurrence. Many examples of strong evolutionary and functional conservation of genes and entire pathways have been uncovered. In Figure 23-43, essentially completely

conserved pathways for activating the *Drosophila* DL and mammalian NF$_\kappa$B transcription factors are presented. The *Drosophila* protein at any step in the DL activation pathway is similar in amino acid sequence to its counterpart in the mammalian NF$_\kappa$B activation pathway. (Don't worry about what the particular proteins do; just note the incredible conservation of cellular and developmental pathways as indicated by the similarly shaped objects representing components of the pathways in the diagrams. We do indeed know that DL and NF$_\kappa$B participate in some equivalent developmental decisions.) Indeed, as can be seen from a selection from the known examples (see Table 23-4), such evolutionary and functional conservation seems to be the norm rather than the exception.

MESSAGE ···
Developmental strategies in animals are quite ancient and highly conserved. In essence, a mammal, a worm, and a fly are put together with the same basic building blocks and regulatory devices. *Plus ça change, plus c'est la même chose!*
···

Do the lessons of animal development apply to plants?

The evidence emerging from comparative studies of pattern formation in a variety of animals, such as the insect–mammalian comparisons considered in preceding sections, indicates that many important developmental pathways are ancient inventions conserved and maintained in many, if not all, animal species. The life history, cell biology, and evolutionary origins of plants would, in contrast, argue against

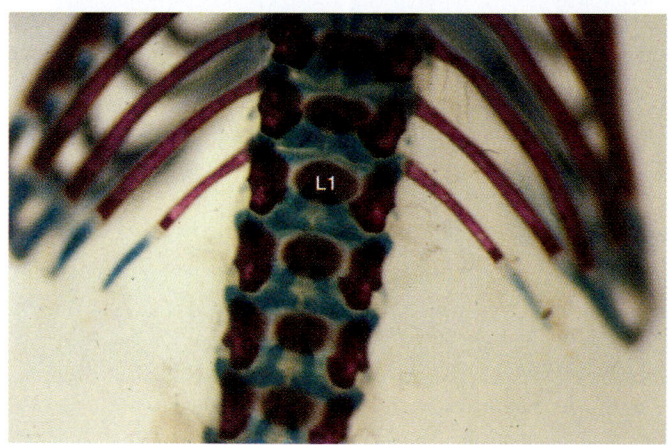

(a)

(b)

Figure 23-42   The phenotype of a homeotic mutant mouse. Mice homozygous for a targeted knockout of the *Hoxc-8* gene were created by using cultured embryonic stem cells. (a) An enlargement of the thoracic and lumbar vertebrae of a homozygous *Hoxc-8⁻* mouse. Note the ribs coming from L1, the first lumbar vertebra. L1 in wild-type mice has no ribs. (b) An unexpected second phenotype of the *Hoxc-8⁻* knockout. Note that the homozygous mutant mouse on the right has clenched fingers, whereas the wild-type mouse on the left has normal fingers. (From H. Le Mouellic, Y. Lallemand, and P. Brulet, "Homeosis in the Mouse Induced by a Null Mutation in the *Hox-3.1* Gene," *Cell* 69, 1992, 251.)

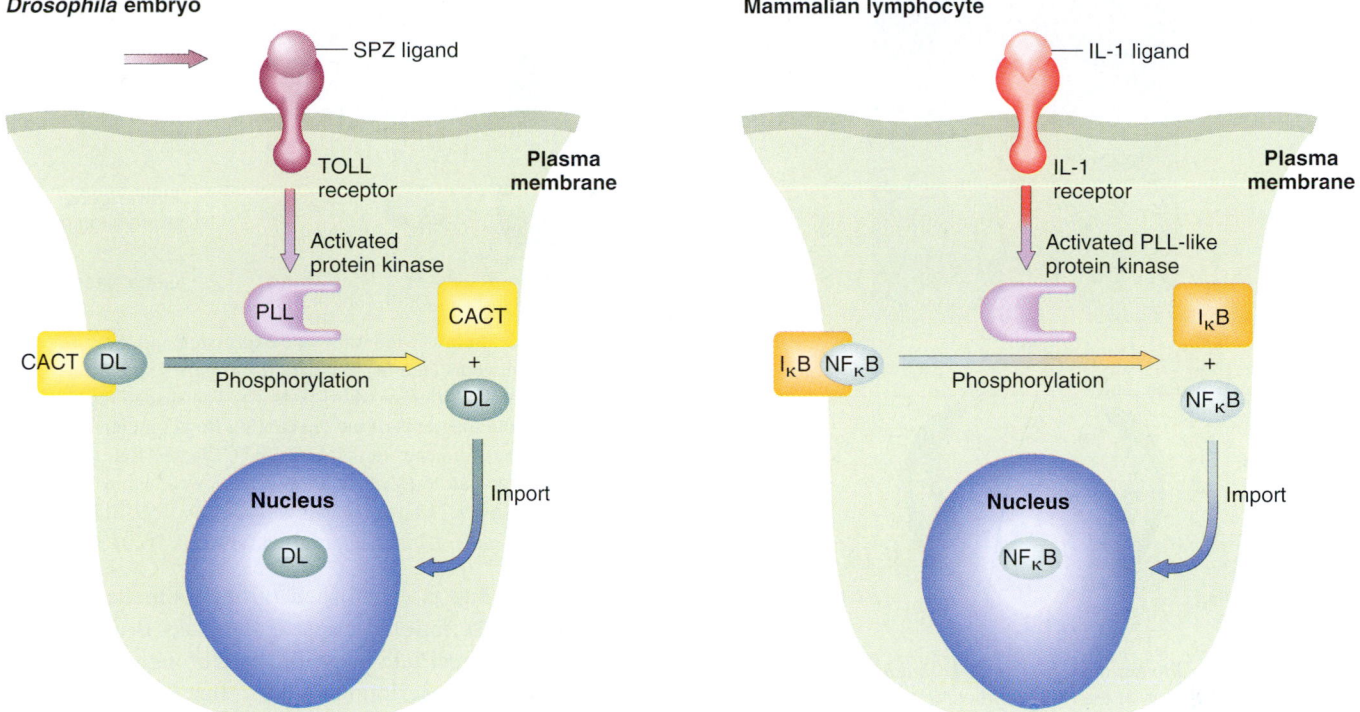

Figure 23-43   The signaling pathway for activation of the *Drosophila* DL morphogen parallels a mammalian signaling pathway for activation of NF$_\kappa$B, the transcription factor that activates the transcription of genes encoding antibody subunits. There are structural protein similarities between SPZ and IL-1, TOLL and IL-1R, CACT and I$_\kappa$B, and DL and NF$_\kappa$B. (After H. Lodish, D. Baltimore, A. Berk, S. L. Zipursky, P. Matsudaira, and J. Darnell, *Molecular Cell Biology*, 3d ed. Copyright © 1995 by Scientific American Books.)

utilization of the same sets of pathways in the regulation of plant development. Plants have very different organ systems from those of animals, depend on rigid cell walls for structural rigidity, separate germ line from soma very late in development, and are very dependent on light intensity and duration to trigger various developmental events. Certainly, plants use hormones to regulate gene activity, to signal locally between cells utilizing as yet unknown signals, and to create cell-fate differences by means of transcription factors. The general themes for establishing cell fates in animals are likely to be seen in plants as well, but the participating molecules in these developmental pathways are likely to be considerably different from those encountered in animal development.

An active area of plant developmental genetic research utilizes a small flowering plant called *Arabidopsis thaliana* as a model system. The genome of *Arabidopsis* is relatively small for a plant (120 megabase pairs of DNA) and is organized into a haploid complement of five chromosomes. Thus, its genomic size and complexity rivals that of the fruit fly, *Drosophila melanogaster*. It is easy to grow in the laboratory in culture tubes or on petri plates, and, because it is a

self-fertilizing plant, F$_2$ mutagenesis surveys can be done in a straightforward manner. Thus, many mutations with interesting phenotypes affecting a variety of developmental events have been obtained.

Perhaps one of the most intensively studied events in *Arabidopsis* development is flower development. Just as the HOM/Hox-encoded transcription factors control segmental identity in animal development, a series of transcriptional regulators determine the fate of the four layers (whorls) of the flower. The outermost whorl of the flower normally develops into the sepal; the next whorl, the petals; the next, the stamen; and the innermost develops into the carpel (Figure 23-44). Several genes have been identified that, when knocked out or ectopically expressed, transform one or more of these whorls into another. For example, the gene *AP1 (Apetala-1)* causes the homeotic transformation of the outer two whorls into the inner two. Analogously to the homeotic mutants in animals, the number of whorls remains the same (four), but the identities of the whorls are transformed. The study of the spatial expression patterns and mutant phenotypes of the various flower-identity genes has produced a model in which whorl fate is established through

(a)

(b)

Figure 23-44   Flower development in *Arabidopsis thaliana*. (a) The mature products of the four whorls of a flower. (b) A cross-sectional diagram of the developing flower, with the normal fates of the four whorls indicated. From outermost to innermost, they are sepal (se), petal (pe), stamen (st), and carpel (ca). (Photograph courtesy of Vivian F. Irish; from V. F. Irish, "Patterning the Flower," *Developmental Biology* 209, 1999, 211–222.)

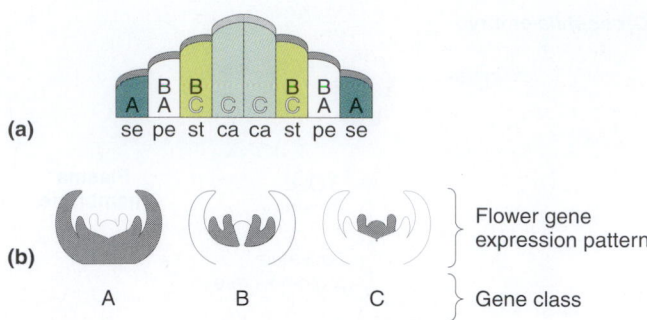

(a)

(b)

Figure 23-45   Flower-identity gene expression and the establishment of whorl fate. (a) The patterns of gene expression corresponding to the different whorl fates. (b) The shaded regions of the cross-sectional diagrams of the developing flower indicate the gene expression patterns for genes of the A, B, and C classes. Refer to Figure 23-44 for the normal anatomy of the developing flower. (From V. F. Irish, "Patterning the Flower," *Developmental Biology* 209, 1999, 211–222.)

combinatorial action of multiple transcription factors (Figure 23-45). Thus, sepal (outermost whorl fate) is established through the expression of transcription factors of the class A type only. Petal fate is established through simultaneous expression of transcription factors of the class A and class B types. Stamen fate is established through simultaneous expression of transcription factors of the class B and class C types. Finally, carpel fate is established through sole expression of class C transcription factors. Just as the homeotic segment-identity genes in animals encode a series of structurally related (homeodomain-containing) transcription factors, the flower-identity genes encode a series of structurally related (MADS-domain) transcription factors. Thus, although different in detail, the overall solution of differentially expressed transcription factors is one of the approaches by which plant cell fate is established. With the combination of sophisticated genetics and a genome that will have been sequenced by the end of the year 2000, studies of *Arabidopsis* development should reveal much about the ways in which plants develop.

SUMMARY

The details of how animal development proceeds undoubtedly differ from species to species. However, the examples that we have looked at here do portray general themes.

In developmental pathways, regulation is at all possible levels of information transfer between the gene and its active protein product.

Early in life, each cell is totipotent: its lineage has the potential to differentiate into a number of cell types. But, in later stages of development, the type of cell that it can become is increasingly restricted. In other words, the lineage's fate is successively restricted to ever-narrowing possibilities until one and only one fate is specified.

The first step in pattern formation is to establish positional information. In the early embryo, positional information defines the two major body axes: A–P and D–V. The cytoskeleton plays a key role in establishing positional information and assigning fates through the subcellular localization of specific gene products. The consequences of the positional information are to create a gradient of activity of one or more transcription factors. These transcription factors then initiate a hierarchical decision-making process that takes different cell lineages down different developmental paths, until all necessary fates are specified. Cell–cell signaling is crucial to ensuring that all fates are apportioned within a developmental field. When fates have been apportioned, feedback loops ensure that these fate decisions are maintained within a lineage.

Components of developmental pathways are reutilized extensively in the course of development and are highly conserved within the animal kingdom. We therefore surmise that these pathways evolved several hundred million years ago in an ancestor common to all modern animals and, thus, that much of development is thematically comparable in all animals.

Keeping in mind the important differences in the cell and developmental biology of plants and animals, we see that common elements of gene regulation during development extend to the plant kingdom as well.

CONCEPT MAP

Draw a concept map interrelating as many of the following terms as possible. Note that the terms are listed in no particular order.

targeted gene knockouts / homeotic genes / totipotency / microtubules / A–P axis / gap genes / segment-polarity genes / morphogen gradients / paralogous genes / lateral inhibition / feedback loops / positional information

CHAPTER INTEGRATION PROBLEM

In developmental pathways, the crucial event seems to be the activation of master switches that set in motion a programmed cascade of regulatory responses. Identify the master switches and explain how they operate in the two examples of sex determination and differentiation discussed in this chapter: **(a)** sex determination in *Drosophila* and **(b)** sex determination in mammals.

◆ Solution ◆

a. In *Drosophila* sex determination, the master switch is the transcriptional regulation of the *Sxl* gene during early embryogenesis. The properties of the SXL protein and the existence of a constitutive late promoter for *Sxl* ensure that, when active SXL protein is produced, an autoregulatory loop will continue to maintain SXL protein activity in the cell. SXL protein will also initiate a downstream set of regulated RNA splicing events, culminating in the production of *dsx-F* mRNA. In the absence of SXL, the default RNA splicing machinery will produce *dsx-M* mRNAs. The key then is whether the *Sxl* switch is flipped on or stays off. This switch is controlled by the level of active numerator-encoded transcription factors. The higher level of these transcription factors conferred by an X : A ratio of 1 is necessary to activate early *Sxl* transcription and set the female developmental pathway in motion. In the absence of higher levels of numerator-encoded transcription factors, development continues along the default pathway, leading to male development.

b. In mammalian sex determination, the master switch is the presence or absence of the *SRY ~ Sry* gene, which is ordinarily located on the Y chromosome. In the presence of the protein product of this gene, which acts as a DNA-binding protein, certain cells of the gonad (Leydig cells) synthesize androgens, male-inducing steroid hormones. These hormones are secreted into the bloodstream and act on target tissues to induce the transcription-factor activity of the androgen receptors. In the absence of androgen-receptor activation, development proceeds along the default pathway, leading to female development. The factors that activate *SRY ~ Sry* expression in the testis are not understood. Because the master switch here is the actual presence or absence of the *SRY ~ Sry* gene itself, it is likely that the regulatory molecules that activate *SRY ~ Sry* are present in the indifferent gonad early in development.

SOLVED PROBLEMS

1. In *Drosophila*, you have identified a new mutation, *mll (malelike)*, that causes flies with an X : A ratio of 1 to develop as phenotypic males. You want to understand how the product of the *mll* gene acts in the developmental pathway for *Drosophila* sex determination and differentiation. You measure the presence of functional SXL, TRA, DSX-F, and DSX-M protein in animals with an X : A ratio of 1. In each of the following cases, propose a role for MLL protein in the sex-determination and -differentiation pathway.

a. You observe that functional early SXL, late SXL, TRA, and DSX-M proteins are produced.

b. You observe that functional early SXL and DSX-M are produced. No functional late SXL or TRA is produced.

c. The only functional protein that you observe is DSX-M.

◆ Solution ◆

It is through experiments like these that the sequential pathway of sex determination and differentiation in *Drosophila* was elucidated. More generally, it is through observations like these, where the effects of a mutant in one gene on the mRNA or protein expression of another gene are studied, that developmental pathways are pieced together.

a. Given that SXL and TRA proteins are still operating normally in the context of an X : A ratio of 1, whereas the alternative splicing of *dsx* mRNA is leading to the male-specific form of the protein, it is likely that *mll* contributes to *dsx* RNA-splicing regulation.

b. Here, the block seems to be between early and late SXL protein production. This finding may indicate that *mll* plays a role in the autoregulation of late *Sxl* alternative splicing.

c. In this example, the block is before *Sxl* expression in the early embryo. One possible explanation is that MLL protein contributes to the numerator function in interpreting the X : A ratio.

2. In the embryogenesis of mammals, the inner cell mass, or ICM (the prospective fetus), quickly separates from the cells that will serve as enclosing membranes and respiratory, nutritive, and excretory channels between the mother and the fetus.

   a. Design experiments using mosaics in mice to determine when the two fates are decided.

   b. How would you trace the formation of different fetal membranes?

◆ Solution ◆

a. We must have markers that enable us to distinguish different cell lineages. Such distinctions can be made by using strains of mice that differ in chromosomal or biochemical markers. (Other ways would be to use differences in sex chromosomes of XX and XY cells or to induce chromosome loss or aberrations by irradiating embryos.)

When you have decided on the marker difference to be used, one way to answer the question is to inject a single cell from one of two strains into embryos of the other at various developmental stages. Another approach is to fuse embryos of defined cell numbers from the two strains. In either case, you would inspect the embryos when the ICM and membranes are distinct and recognizable. When cell insertion or fusion results in membranes and ICM that are exclusively made up of one cell type and never a mosaic of the two, the two developmental fates have been set.

b. Carry out the same injection or fusion experiment on early embryos. Now look for the pattern of mosaicism. Correlate the presence of cells of similar genotype in different membranes. It should be possible to determine the cell lineage of cells in each set of membranes.

PROBLEMS

1. In *Drosophila,* individual flies with two X chromosomes and three sets of autosomes (X : A ratio = 0.67) are intersexes. When examined more closely, their intersexual phenotype turns out to be due to mosaicism. Some cells differentiate with an entirely male phenotype and other cells with an entirely female phenotype. Explain this observation in regard to how the *Sxl* on−off switch is established and maintained.

2. In *Drosophila,* homozygotes for mutations in the *tra* gene transform XX flies into phenotypic males with regard to somatic secondary sexual characteristics. The gonad in *Drosophila* forms from somatic mesoderm tissue and germ-line cells. XX ; *tra* homozygotes are sterile and have rudimentary gonads. You suspect that the reason for this sterility is that the somatic tissue is sexually transformed into male by *tra* but the germ-line cells remain female. Design an experiment that tests this prediction.

3. XYY humans are fertile males. XXX humans are fertile females. What do these observations reveal about the mechanisms of sex determination and dosage compensation?

4. There are, occasionally, humans who are mosaics of XX and XY tissue. They generally exhibit a uniform sexual phenotype. Some of them are phenotypically female, others male. Explain these observations in regard to the mechanism of sex determination in mammals.

5. There are dominant mutations of the *Sxl* gene, called Sxl^M alleles, that transform XY animals into females but do not affect XX animals. Reversions of these Sxl^M alleles can be readily induced with mutagen treatment. These reversions, called Sxl^f alleles, yield normal indi-

viduals in XY animals, but, in homozygotes, they transform XX animals into phenotypic males. Provide a possible explanation for these observations, keeping in mind that *Sxl* is ordinarily dispensable for male development.

6. In *C. elegans* vulva development, one anchor cell in the gonad interacts with six equivalence-group cells (cells with the potential to become parts of the vulva). The six equivalence-group cells have three distinct phenotypic fates: primary, secondary, and tertiary. The equivalence-group cell closest to the anchor cell develops the primary vulva phenotype. If the anchor cell is ablated, all six equivalence-group lineages differentiate into the tertiary state.

   a. Set up a model to explain these results.

   b. The anchor cell and the six equivalence-group cells can be isolated and grown in vitro; design an experiment to test your model.

7. There are two types of muscle cell in *C. elegans:* pharyngeal muscles and body-wall muscles. They can be distinguished from each other, even as single cells. In the following figure, letters designate particular muscle precursor cells, black triangles represent pharyngeal muscles, and white triangles represent body-wall muscles.

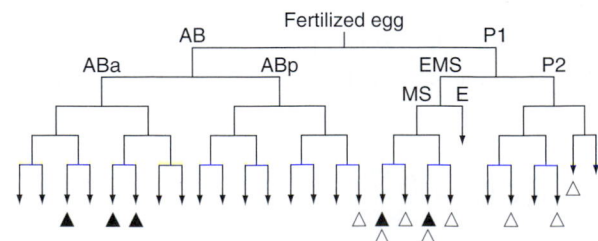

a. It is possible to move these cells around in the course of development. When the positions of ABa and ABp are physically interchanged, the cells develop according to their new position. In other words, the cell that was originally in the ABa position now gives rise to one body-wall muscle cell, whereas the cell that was originally in the ABp position now gives rise to three pharyngeal muscle cells. What does that tell us about the developmental processes controlling the fates of ABa and ABp?

b. If EMS is ablated (by heat inactivation with a laser beam aimed through a microscope lens), no AB descendants make muscles. What does that suggest?

c. If P_2 is ablated, all AB descendants turn into muscle cells. What does this suggest?

(Figure from J. Priess and N. Thomson, *Cell* 48, 1987, 241.)

8. In humans, the male steroid hormone testosterone binds to the testosterone receptor. The hormone–receptor complex then activates transcription of male-specific genes. A mutation in the testosterone receptor prevents binding to testosterone. What would be the phenotype of a person who is XY and is homozygous for such a mutation? Explain your conclusion in relation to how steroid receptors act.

9. For many of the mammalian Hox genes, it has been possible to determine that some of them are more similar to one of the insect HOM-C genes than to the others. Describe an experimental approach using the tools of molecular biology that would allow such a determination.

10. **a.** When you remove the anterior 20 percent of the cytoplasm of a newly formed *Drosophila* embryo, you can cause a bicaudal phenotype, in which there is a mirror-image duplication of the abdominal segments. In other words, from the anterior tip of the embryo to the posterior tip, the order of segments is A8-A7-A6-A5-A4-A4-A5-A6-A7-A8. Explain this phenotype in regard to the action of the anterior and posterior determinants and how they affect gap gene expression.

b. Females homozygous for the maternally acting mutation *nanos (nos)* produce embryos in which the abdominal segments are absent and in which the head and thoracic segments are broader. In regard to the action of the anterior and posterior determinants and gap gene action, explain how *nos* produces this mutant phenotype. In your answer, explain why there is a loss of segments rather than a mirror-image duplication of anterior segments.

11. The three homeodomain proteins ABD-B, ABD-A, and UBX are encoded by genes within the BX-C of *Drosophila*. In wild-type embryos, the *Abd-B* gene is expressed in the posterior abdominal segments, *abd-A*

in the middle abdominal segments, and *Ubx* in the anterior abdominal and posterior thoracic segments. When the *Abd-B* gene is deleted, *abd-A* is expressed in both the middle and the posterior abdominal segments. When *abd-A* is deleted, *Ubx* is expressed in the posterior thorax and in the anterior and middle abdominal segments. When *Ubx* is deleted, the patterns of *abd-A* and *Abd-B* expression are unchanged from wild type. When both *abd-A* and *Abd-B* are deleted, *Ubx* is expressed in all segments from the posterior thorax to the posterior end of the embryo. Explain these observations, taking into consideration the fact that the gap genes control the initial expression patterns of the homeotic genes.

12. When an embryo is homozygous mutant for the gap gene *Kr*, the fourth and fifth stripes of the pair-rule gene *ftz* (counting from the anterior end) do not form normally. When the gap gene *kni* is mutant, the fifth and sixth *ftz* stripes do not form normally. Explain these results in regard to how segment number is established in the embryo.

13. In considering the formation of the A–P and D–V axes in *Drosophila*, we noted that, for mutations such as *bcd*, homozygous mothers uniformly produce mutant offspring with segmentation defects. This outcome is true regardless of whether the offspring themselves are bcd^+/bcd or bcd/bcd. Some other maternal-effect lethal mutations are different, in that the mutant phenotype can be "rescued" by introducing a wild-type allele of the gene from the father. In other words, for such rescuable maternal-effect lethals, mut^+/mut animals are normal, whereas mut/mut animals have the mutant defect. Rationalize the difference between rescuable and nonrescuable maternal-effect lethal mutations.

14. The anterior determinant in the *Drosophila* egg is *bcd*. A mother heterozygous for a *bcd* deletion has only one copy of the bcd^+ gene. With the use of P elements to insert copies of the cloned bcd^+ gene into the genome by transformation, it is possible to produce mothers with extra copies of the gene.

Shortly after the blastoderm has formed, the *Drosophila* embryo develops an indentation called the cephalic furrow that is more or less perpendicular to the longitudinal body axis. In the progeny of bcd^+ monosomics, this furrow is very close to the anterior tip, lying at a position one-sixth of the distance from the anterior to the posterior tip. In the progeny of standard wild-type diploids (disomic for bcd^+), the cephalic furrow arises more posteriorly, at a position one-fifth of the distance from the anterior to the posterior tip of the embryo. In the progeny of bcd^+ trisomics, it is even more posterior. As additional gene doses are added, it moves more and more posteriorly, until, in the progeny of hexasomics, the cephalic furrow is midway along the A–P axis of the embryo.

a. Explain the gene dose effect of bcd^+ on cephalic furrow formation in relation to bcd's contribution to A–P pattern formation.

b. Diagram the relative expression patterns of mRNAs from the gap genes Kr and kni in blastoderm embryos derived from bcd monosomic, trisomic, and hexasomic mothers.

15. In the *Drosophila* embryo, the 3′ untranslated regions (3′ UTRs) of the mRNAs [the regions between the translation-termination codons and the poly(A) tails] are responsible for localizing bcd and nos to the anterior and posterior poles, respectively. Experiments have been done in which the 3′ UTRs of bcd and nos have been swapped. Suppose that we make P-element transformation constructs with both swaps (nos mRNA with bcd 3′ UTR and bcd mRNA with nos 3′ UTR) and transform them into the *Drosophila* genome. We then make a female that is homozygous mutant for bcd and nos and carries both swap constructs. What phenotype would you expect for her embryos in regard to A–P axis development?

24

POPULATION GENETICS

Shell-color polymorphism in *Liguus fascitus.*
(*From David Hillis,* Journal of Heredity, *July–August 1991.*)

Key Concepts

The goal of population genetics is to understand the genetic composition of a population and the forces that determine and change that composition.

In any species, a great deal of genetic variation within and between populations arises from the existence of various alleles at different gene loci.

A fundamental measurement in population genetics is the frequency at which the alleles are found at any gene locus of interest.

The frequency of a given allele in a population can be changed by recurrent mutation, selection, or migration or by random sampling effects.

In an idealized population, in which no forces of change are acting, a randomly interbreeding population would show constant genotypic frequencies for a given locus.

So far in our investigation of genetics, we have been concerned with processes that take place in individual organisms and cells. How does the cell copy DNA and what causes mutations? How do the mechanisms of segregation and recombination affect the kinds and proportions of gametes produced by an individual organism? How is the development of an organism affected by the interactions between its DNA, the cell machinery of protein synthesis, cellular metabolic processes, and the external environment? But organisms do not live only as isolated individuals. They interact with each other in groups, **populations,** and there are questions about the genetic composition of those populations that cannot be answered only from a knowledge of the basic individual-level genetic processes. Why are the alleles of the protein Factor VIII and Factor IX genes that cause hemophilia so rare in all human populations, whereas sickle-cell anemia is very common in some parts of Africa? What changes in the frequency of sickle-cell anemia are to be expected in the descendants of Africans in North America as a consequence of the change in environment and of the interbreeding between Africans and Europeans and Native Americans? What genetic changes occur in a population of insects subject to insecticides generation after generation? What is the consequence of an increase or decrease in the rate of mating between close relatives? All are questions of what determines the genetic composition of populations and how that composition may be expected to change in time. These questions are the domain of **population genetics.**

MESSAGE ·······························

Population genetics relates the processes of individual heredity and development to the genetic composition of populations and to changes in that composition in time.

To relate the basic individual-level processes to population genetic composition, we must investigate the following phenomena:

1. The effect of the *mating pattern* on different genotypes in the population. Individuals may mate at random or they may mate preferentially with close relatives (*inbreeding*) or preferentially on the basis of their genotypic or phenotypic similarity (*assortative mating*).

2. The changes in population composition due to immigration of *individuals* from other populations.

3. The rate of introduction of genetic variation into the population by *mutation* and *recombination.*

4. The effect of the differential rate of reproduction by different genotypes and the differential chance of survival of genetically different offspring of these matings. These differential rates are the result of *natural selection.*

5. The consequences of *random fluctuations* in the actual reproductive rates of different genotypes because any given individual has only a few offspring and the total population size is limited.

Variation and its modulation

Population genetics is both an experimental and a theoretical science. On the experimental side, it provides descriptions of the actual patterns of genetic variation in populations and estimates the parameters of the processes of mating, mutation, natural selection, and random variation in reproductive rates. On the theoretical side, it makes predictions of what the genetic composition of populations should be and how they can be expected to change as a consequence of the various forces operating on them.

MESSAGE ·······························

Population genetics is the experimental and theoretical study of the pattern of inherited variation in populations and its modulation in time and space.

Observations of variation

Population genetics necessarily deals with genotypic variation, but, by definition, only phenotypic variation can be observed. The relation between phenotype and genotype varies in simplicity from character to character. At one extreme, the phenotype may be the observed DNA sequence of a stretch of the genome. In this case, the distinction between genotype and phenotype disappears, and we can say that we are, in fact, directly observing the genotype. At the other extreme lie the bulk of characters of interest to plant and animal breeders and to most evolutionists — the variations in yield, growth rate, body shape, metabolic ratio, and behavior that constitute the obvious differences between varieties and species. These characters have a very complex relation to genotype, and we must use the methods introduced in Chapter 25 to say anything at all about the genotypes. But, as we shall see in Chapter 25, it is not possible to make very precise statements about the genotypic variation underlying quantitative characters. For that reason, most of the study of experimental population genetics has concentrated on characters with simple relations to the genotype, much like the characters studied by Mendel. A favorite object of study for human population geneticists, for example, has been the various human blood groups. The qualitatively distinct phenotypes of a given blood group — say, the MN group — are encoded by alternative alleles at a single locus, and the phenotypes are insensitive to environmental variations.

The study of variation, then, consists of two stages. The first is a description of the phenotypic variation. The second is a translation of these phenotypes into genetic terms and the redescription of the variation genetically. If there is a perfect one-to-one correspondence between genotype and phenotype, then these two steps merge into one, as in the MN blood group. If the relation is more complex — for example, as the result of dominance, heterozygotes resemble homozygotes — it may be necessary to carry out experimental crosses or to observe pedigrees to translate phenotypes into genotypes. This is the case for the human ABO blood group (see page 110).

24-1 TABLE	Frequencies of Genotypes for Alleles at MN Blood Group Locus in Various Human Populations				
	GENOTYPE			ALLELE FREQUEN-CIES	
Population	M/M	M/N	N/N	p(M)	q(N)
Eskimo	0.835	0.156	0.009	0.913	0.087
Australian aborigine	0.024	0.304	0.672	0.176	0.824
Egyptian	0.278	0.489	0.233	0.523	0.477
German	0.297	0.507	0.196	0.550	0.450
Chinese	0.332	0.486	0.182	0.575	0.425
Nigerian	0.301	0.495	0.204	0.548	0.452

Source: W. C. Boyd, *Genetics and the Races of Man.* D. C. Heath, 1950.

The simplest description of Mendelian variation is the frequency distribution of genotypes in a population. Table 24-1 shows the frequency distribution of the three genotypes at the MN blood group locus in several human populations. Note that there is variation between individuals in each population, because there are different genotypes present, and there is variation in the frequencies of these genotypes from population to population. More typically, instead of the frequencies of the diploid genotypes, the frequencies of the alternative alleles are used. The frequency of an allele is simply the proportion of that allelic form of the gene among all copies of the gene in the population. There are twice as many gene copies in the population as there are individuals, because every individual is diploid and homozygotes for an allele have two copies of that allele, whereas heterozygotes have only one copy. So we calculate the frequency of an allele by counting homozygotes and adding half the heterozygotes. Thus, if the frequency of A/A individuals were, say, 0.36 and the frequency of A/a individuals were 0.48, the allele frequency of A would be $0.36 + 0.48/2 = 0.60$. Box 24-1 gives the general form of this cal-

culation. Table 24-1 shows the values of p and q, the **gene frequency** or **allele frequency** of the two alleles in the different populations.

A *measure* of genetic variation (in contrast with its *description* by gene frequencies) is the amount of **heterozygosity** at a locus in a population, which is given by the total frequency of heterozygotes at a locus. If one allele is in very high frequency and all others are near zero, then there will be very little heterozygosity because, by necessity, most individuals will be homozygous for the common allele. We expect heterozygosity to be greatest when there are many alleles at a locus, all at equal frequency. In Table 24-1, the heterozygosity is simply equal to the frequency of the M/N genotype in each population. When more than one locus is considered, there are two possible ways of calculating heterozygosity. Locus S (the secretor factor, determining whether the M and N proteins are also contained in the saliva) is closely linked to the $M N$ locus in humans. Table 24-2 shows the frequencies, commonly symbolized by g's, of the four haplotypes ($M S$, $M s$, $N S$, and $N s$) in various populations. First, we can calculate the frequency of heterozygotes at each locus separately. Alternatively, we can consider each haplotype as a unit, as in Table 24-2, and calculate the proportion of all individuals who carry two different haplotypic or gametic forms. This form of heterozygosity is also referred to as *haplotype diversity* or *gametic diversity*. The results of both calculations are given in Table 24-2. Note that the haplotype diversity is always greater than the average heterozygosity of the separate loci, because an individual is a haplotypic heterozygote if *either* of its loci is heterozygous. (See the discussion of the Hardy-Weinberg equilibrium in Box 24-2 on page 722 for the calculation of heterozygosity.)

Simple Mendelian variation can be observed within and between populations of any species at various levels of phenotype, from external morphology down to the amino acid sequence of enzymes and other proteins. Indeed, with the new methods of DNA sequencing, variations in DNA sequence (such as third-position variants that are not differentially coded in amino acid sequences and even variations in

Box 24-1 Calculation of Allele Frequency

If $f_{A/A}$, $f_{A/a}$, and $f_{a/a}$ are the proportions of the three genotypes at a locus with two alleles, then the frequency $p(A)$ of the A allele and the frequency $q(a)$ of the a allele are obtained by counting alleles. Because each homozygote A/A consists only of A alleles and only half the alleles of each heterozygote A/a are type A, the total frequency (p) of A alleles in the population is:

$$p = f_{A/A} + \tfrac{1}{2}f_{A/a} = \text{frequency of } A$$

Similarly, the frequency q of a alleles is given by:

$$q = f_{a/a} + \tfrac{1}{2}f_{A/a} = \text{frequency of } a$$
$$p + q = f_{A/A} + f_{a/a} + f_{A/a} = 1.00$$

So,

$$q = 1 - p$$

If there are multiple alleles, the frequency for each allele is simply the frequency of its homozygote plus half the sum of the frequencies for all the heterozygotes in which it appears.

24-2 TABLE   Frequencies of Gametic Types for MNS System in Various Human Populations

Population	GAMETIC TYPE				HETEROZYGOSITY (H)	
	M S	M s	N S	N s	From gametes	From alleles
Ainu	0.024	0.381	0.247	0.348	0.672	0.438
Ugandan	0.134	0.357	0.071	0.438	0.658	0.412
Pakistani	0.177	0.405	0.127	0.291	0.704	0.455
English	0.247	0.283	0.080	0.290	0.700	0.469
Navaho	0.185	0.702	0.062	0.051	0.467	0.286

Source: A. E. Mourant, *The Distribution of the Human Blood Groups.* Blackwell Scientific, 1954.

nontranslated intervening sequences) have been observed. Every species of organism ever examined has revealed considerable genetic variation, or **polymorphism,** manifested at one or more levels of phenotype, within populations, between populations, or both. A gene or a phenotypic trait is said to be *polymorphic* if there is more than one form of the gene or trait in a population. Genetic variation that might be the basis for evolutionary change is ubiquitous. The tasks for population geneticists are to describe that ubiquitous variation quantitatively and to build a theory of evolutionary change that can use these observations in prediction.

It is impossible in this text to provide an adequate picture of the immense richness of genetic variation that exists in species. We can consider only a few examples of the different kinds of Mendelian variation to gain a sense of the genetic diversity within species. Each of these examples can be multiplied many times over in other species and with other traits.

Morphologic variation. The shell of the land snail *Cepaea nemoralis* may be pink or yellow, depending on two alleles at a single locus, with pink dominant to yellow. In addition, the shell may be banded or unbanded (Figure 24-1) as a result of segregation at a second linked locus, with unbanded dominant to banded. Table 24-3 shows the variation of these two loci in several European colonies of the snail. The populations also show polymorphism for the number of bands and the height of the shells, but these characters have complex genetic bases.

24-3 TABLE   Frequencies of Snails (*Cepaea nemoralis*) with Different Shell Colors and Banding Patterns in Three Populations in France

Population	YELLOW		PINK	
	Banded	Unbanded	Banded	Unbanded
Guyancourt	0.440	0.040	0.337	0.183
Lonchez	0.196	0.145	0.564	0.095
Peyresourde	0.175	0.662	0.100	0.062

Source: Maxime Lamotte, *Bulletin Biologique de France et Belgique* 35 (suppl.), 1951.

Examples of naturally occurring morphologic variation within plant species are *Plectritis* (see Figure 1-14), *Collinsia* (blue-eyed Mary, page 58), and clover (see Figure 4-5).

Chromosomal polymorphism. Although the karyotype is often regarded as a distinctive characteristic of a species, in fact, numerous species are polymorphic for chromosome number and morphology. Extra chromosomes (supernumeraries), reciprocal translocations, and inversions segregate in many populations of plants, insects, and even mammals.

Table 24-4 gives the frequencies of supernumerary chromosomes and translocation heterozygotes in a population of the plant *Clarkia elegans* from California. The "typical" species karyotype would be hard to identify.

Immunologic polymorphism. A number of loci in vertebrates encode antigenic specificities such as the ABO blood types. More than 40 different specificities on human red cells are known, and several hundred are known in cattle. Another major polymorphism in humans is the HLA system of cellular antigens, which are implicated in tissue graft compatibility. Table 24-5 gives the allelic frequencies for the ABO blood group locus in some very different human populations. The polymorphism for the HLA system is vastly greater. There appear to be two main loci, each with five distinguishable alleles. Thus, there are $5^2 = 25$ different possible gametic types, making 25 different homozygous forms and $(25)(24)/2 = 300$ different heterozygotes. All

Figure 24-1   Shell patterns of the snail *Cepaea nemoralis:* (a) banded yellow; (b) unbanded pink.

24-4 TABLE	Frequencies of Plants with Supernumerary Chromosomes and of Translocation Heterozygotes in a Population of *Clarkia elegans* from California	
No supernumeraries or translocations	Translocations	
0.560	0.133	
Supernumeraries	Both translocations and supernumeraries	
0.265	0.042	

Source: H. Lewis, *Evolution* 5, 1951, 142–157.

genotypes are not phenotypically distinguishable, however; so only 121 phenotypic classes can be seen. L. L. Cavalli-Sforza and W. F. Bodmer report that, in a sample of only 100 Europeans, 53 of the 121 possible phenotypes were actually observed!

Protein polymorphism. Studies of genetic polymorphism have been carried down to the level of the polypeptides encoded by the structural genes themselves. If there is a nonredundant codon change in a structural gene (say, GGU to GAU), the result is an amino acid substitution in the polypeptide produced at translation (in this case, aspartic acid is substituted for glycine). If a specific protein could be purified and sequenced from separate individuals, then it would be possible to detect genetic variation in a population at this level. In practice, such detection is tedious for large organisms and impossible for small ones unless a large mass of protein can be produced from a homozygous line.

There is, however, a practical substitute for sequencing that makes use of the change in the physical properties of a protein when an amino acid is substituted. Five amino acids (glutamic acid, aspartic acid, arginine, lysine, and histidine) have ionizable side chains that give a protein a characteristic net charge, depending on the pH of the surrounding

medium. Amino acid substitutions may directly replace one of these charged amino acids or a noncharged substitution near one of them in the polypeptide chain may affect the degree of ionization of the charged amino acid or a substitution at the joining between two α helices may cause a slight shift in the three-dimensional packing of the folded polypeptide. In all these cases, the net charge on the polypeptide will be altered because the net charge on a protein is not simply the sum of all the individual charges on its amino acids but depends on their exposure to the liquid medium surrounding them.

To detect the change in net charge, protein can be subjected to gel electrophoresis. Figure 24-2 shows the outcome of such an electrophoretic separation of variants of an esterase enzyme in *Drosophila pseudoobscura,* where each track is the protein from a different individual fly. Figure 24-3 shows a similar gel for different variant human hemoglobins. In this case, most individuals are heterozygous for the variant and normal hemoglobin A. Table 24-6 shows the frequencies of different alleles for three enzyme-encoding loci in *D. pseudoobscura* in several populations: a nearly monomorphic locus (malic dehydrogenase), a moderately polymorphic locus (α-amylase), and a highly polymorphic locus (xanthine dehydrogenase).

The technique of gel electrophoresis (or sequencing) differs fundamentally from other methods of genetic analysis in allowing the study of loci that are not segregating, because the presence of a polypeptide is prima facie evidence of a structural gene — that is, a DNA sequence encoding a protein. Thus, it has been possible to ask what proportion of all structural genes in the genome of a species is polymorphic

24-5 TABLE	Frequencies of the Alleles I^A, I^B, and i at the ABO Blood Group Locus in Various Human Populations		
Population	I^A	I^B	i
Eskimo	0.333	0.026	0.641
Sioux	0.035	0.010	0.955
Belgian	0.257	0.058	0.684
Japanese	0.279	0.172	0.549
Pygmy	0.227	0.219	0.554

Source: W. C. Boyd, *Genetics and the Races of Man.* D. C. Heath, 1950.

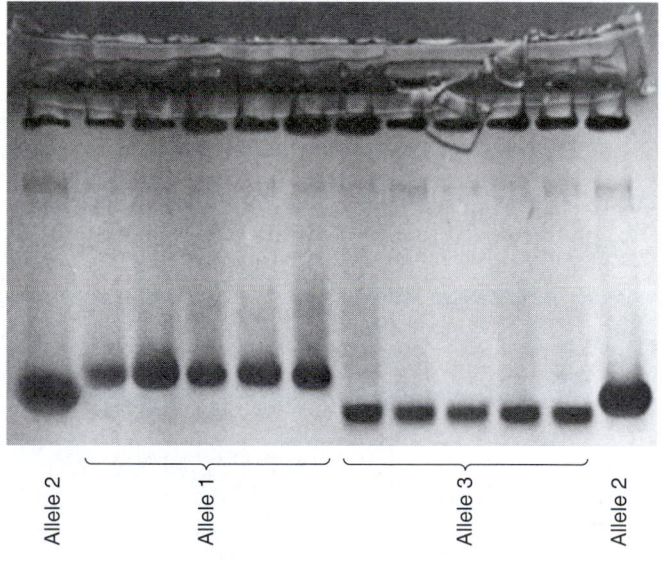

Allele 2   Allele 1   Allele 3   Allele 2

Figure 24-2 Electrophoretic gel showing homozygotes for three different alleles at the *esterase-5* locus in *Drosophila pseudoobscura.* Repeated samples of the same allele are identical, but there are repeatable differences between alleles.

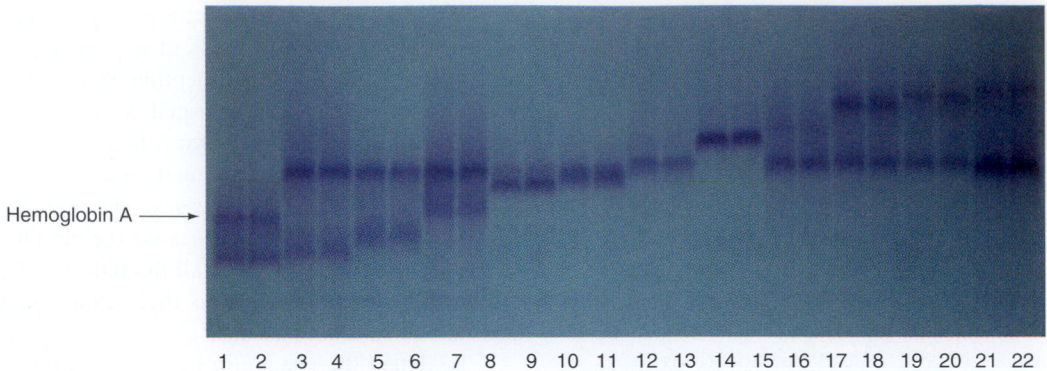

Hemoglobin A →

1   2   3   4   5   6   7   8   9   10 11 12 13 14 15 16 17 18 19 20 21 22

Figure 24-3   Electrophoretic gel showing heterozygotes of normal hemoglobin A and a number of variant hemoglobin alleles. One of the dark-staining bands is marked as hemoglobin A. The second dark-staining band in each track (seen most clearly in tracks 3 and 4) represents the second protein derived from the second allele of the heterozygote. Hemoglobin A is missing from tracks 9 and 10 because the individuals are homozygotes for the variant allele. (The light-staining band near the top of the gel in all tracks represents a protein other than hemoglobin.)

and what the average heterozygosity is in a population. Very large numbers of species have been sampled by this method, including bacteria, fungi, higher plants, vertebrates, and invertebrates. The results are remarkably consistent over species. About one-third of structural-gene loci are polymorphic, and the average heterozygosity in a population over all loci sampled is about 10 percent. This means that scanning the genome in virtually any species would show that about 1 in every 10 loci is in heterozygous condition and that about one-third of all loci have two or more alleles segregating in any population. Thus the potential of variation for evolution is immense. The disadvantage of the electrophoretic technique is that it detects variation only in structural genes. If most of the evolution of shape, physiology, and behavior rests on changes in regulatory genetic elements, then the observed variation in structural genes is beside the point.

DNA sequence polymorphism

DNA analysis makes it possible to examine variation among individuals and between species in their DNA sequences. There are two levels at which such studies can be done. Studying variation in the sites recognized by restriction enzymes provides a coarse view of base-pair variation. At a finer level, methods of DNA sequencing allow variation to be observed base pair by base pair.

Restriction-site variation.   A restriction enzyme that recognizes six-base sequences (a "six cutter") will recognize

		POPULATION			
Locus (enzyme-encoding)	Allele	Berkeley	Mesa Verde	Austin	Bogotá
Malic dehydrogenase	A	0.969	0.948	0.957	1.00
	B	0.031	0.052	0.043	0.00
α-Amylase	A	0.030	0.000	0.000	0.00
	B	0.290	0.211	0.125	1.00
	C	0.680	0.789	0.875	0.00
Xanthine dehydrogenase	A	0.053	0.016	0.018	0.00
	B	0.074	0.073	0.036	0.00
	C	0.263	0.300	0.232	0.00
	D	0.600	0.581	0.661	1.00
	E	0.010	0.030	0.053	0.00

TABLE 24-6   Frequencies of Various Alleles at Three Enzyme-Encoding Loci in Four Populations of *Drosophila pseudoobscura*

Source: R. C. Lewontin, *The Genetic Basis of Evolutionary Change*. Columbia University Press, 1974.

an appropriate sequence approximately once every 4^6 = 4096 base pairs along a DNA molecule [determined from the probability that a specific base (of which there are four) will be found at each of the six positions]. If there is polymorphism in the population for one of the six bases at the recognition site, then there will be a restriction fragment length polymorphism (RFLP) in the population, because in one variant the enzyme will recognize and cut the DNA, whereas in the other variant it will not (see pages 398–399). A panel of, say, eight enzymes will then sample every 4096/8 ≅ 500 base pairs for such polymorphisms. However, when one is found, we do not know which of the six base pairs at the recognition site is polymorphic.

If we use enzymes that recognize four-base sequences ("four cutters"), there is a recognition site every 4^4 = 256 base pairs; so a panel of eight different enzymes can sample about once every 32 base pairs along the enzyme. In addition to single base-pair changes that destroy restriction-

enzyme recognition sites, there are insertions and deletions of stretches of DNA that also cause restriction fragment lengths to vary.

Extensive samples have been made for several regions of the genome in a number of species of *Drosophila* with the use of both four-cutting and six-cutting enzymes. The result of one such study of the xanthine dehydrogenase gene in *Drosophila pseudoobscura* is shown in Figure 24-4. The figure shows, symbolically, the restriction pattern of 53 chromosomes (haplotypes) sampled from nature, polymorphic for 78 restriction sites along a sequence 4.5 kb in length. Among the 53 haplotypes, there are 48 different ones. (Try to find the identical pairs.) Clearly there is an immense amount of nucleotide variation at the xanthine dehydrogenase locus in nature.

Twenty restriction-enzyme studies of different regions of the X chromosome and the two large autosomes of *Drosophila melanogaster* have found between 0.1 and 1.0 percent heterozygosity per nucleotide site, with an average

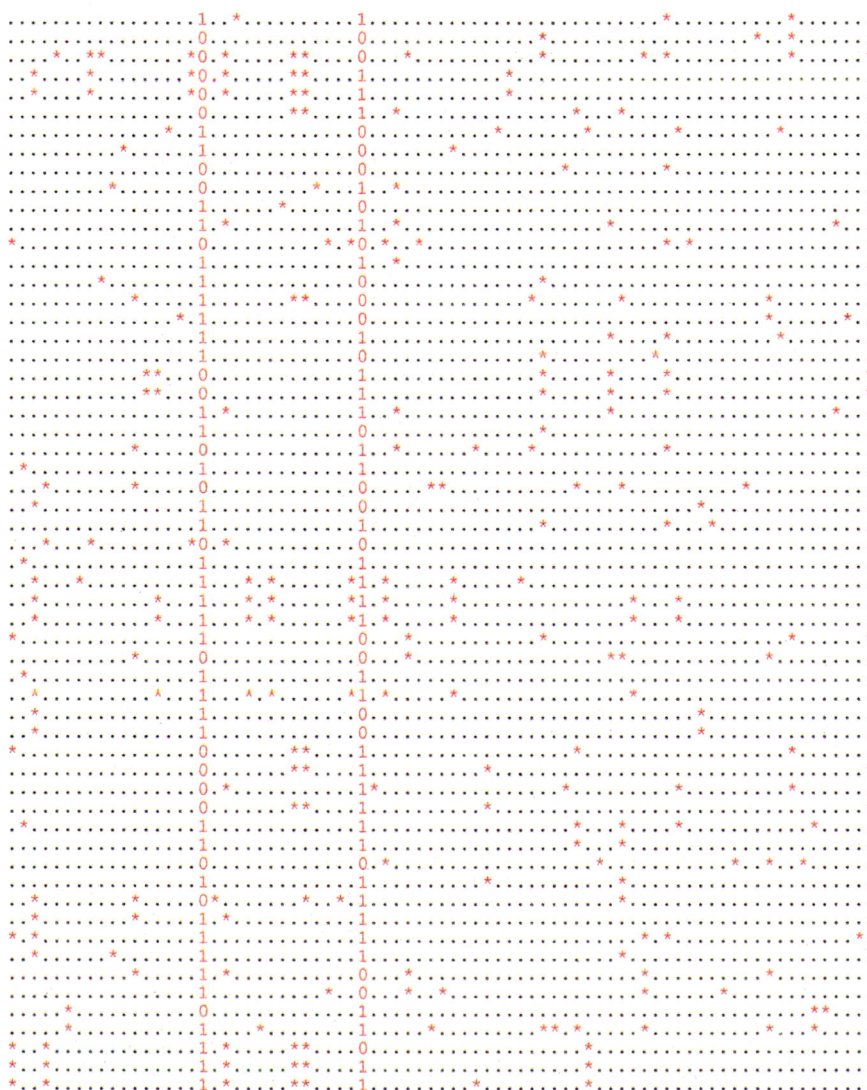

Figure 24-4 The result of a four-cutter survey of 53 chromosomes, probed for the xanthine dehydrogenase gene in *Drosophila pseudoobscura*. Each line is a chromosome (haplotype) sampled from a natural population. Each position along the line is a polymorphic restriction rate along the 4.5-kb sequence studied. Where an asterisk appears, the haplotype differs from the majority, either cutting where most haplotypes are not cut or not cutting where most haplotypes are cut. At two sites, there is no clear majority type, so a 0 or 1 is used to show whether the site is absent or present.

of 0.4 percent. A study of the very small fourth chromosome, however, found no polymorphism at all.

Tandem repeats. Another form of DNA sequence variation that can be revealed by restriction fragment surveys arises from the occurrence of multiply repeated DNA sequences. In the human genome, there are a variety of different short DNA sequences dispersed throughout the genome, each one of which is multiply repeated in a tandem row. The number of repeats may vary from a dozen to more than 100 in different individual genomes. Such sequences are known as **variable number tandem repeats (VNTRs).** If the restriction enzymes cut sequences that flank either side of such a tandem array, a fragment will be produced whose size is proportional to the number of repeated elements. The different-sized fragments will migrate at different rates in an electrophoretic gel. Unfortunately, the individual elements are too short to allow distinguishing between, say, 64 and 68 repeats, but size classes (*bins*) can be established, and a population can be assayed for the frequencies of the different classes. Table 24-7 shows the data for two different VNTRs sampled in two American Indian groups from Brazil. In one case, D14S1, the Karitiana are nearly homozygous, whereas the Surui are very variable; in the other case, D14S13, both populations are variable but with different frequency patterns.

Complete sequence variation. Studies of variation at the level of single base pairs by DNA sequencing can provide information of two kinds. First, translating the sequences of coding regions obtained from different individuals in a population or from different species allows the exact amino acid

sequence differences to be determined. Electrophoretic studies can show that there is variation in amino acid sequences but cannot identify how many or which amino acids differ between individuals. So, when DNA sequences were obtained for the various electrophoretic variants of esterase-5 in *Drosophila pseudoobscura* (see Figure 24-2), electrophoretic classes were found to differ from each other by an average of 8 amino acids, and the 20 different kinds of amino acids were involved in polymorphisms at about the frequency that they were present in the protein. Such studies also show that different regions of the same protein have different amounts of polymorphism. For the esterase-5 protein, consisting of 545 amino acids, 7 percent of amino acid positions are polymorphic, but the last amino acids at the carboxyl terminus of the protein are totally invariant between individuals, probably because of functional constraints on these amino acids.

Second, DNA base-pair variation can also be studied for those base pairs that do not determine or change the protein sequence. Such base-pair variation can be found in DNA in introns, in 5'-flanking sequences that may be regulatory, in nontranscribed DNA 3' to the gene, and in those nucleotide positions within codons (usually third positions) whose variation does not result in amino acid substitutions. Within coding sequences, these so-called silent or synonymous base-pair polymorphisms are much more common than are changes that result in amino acid polymorphism, presumably because many amino acid changes interfere with normal function of the protein and are eliminated by natural selection. An examination of the codon translation table (see Figure 10-27) shows that approximately 25 percent of all random base-pair changes would be synonymous, giving an alternative codon for the same amino acid, whereas 75 percent of random changes would change the amino acid coded. For example, a change from AAT to AAC still encodes asparagine, but a change to ATT, ACT, AAA, AAG, AGT, TAT, CAT, or GAT, all single-base-pair changes from AAT, changes the amino acid encoded. So, if mutations of base pairs are at random and if the substitution of an amino acid made no difference to function, we would expect a 3:1 ratio of amino acid replacement to silent polymorphisms. The actual ratios found in *Drosophila* vary from 2:1 to 1:10. Clearly, there is a great excess of synonymous polymorphism, showing that most amino acid changes are subject to natural selection. It should not be assumed, however, that silent sites in coding sequences are entirely free from constraints. Different alternative triplet codings for the same amino acid may differ in speed and accuracy of transcription, and the mRNA corresponding to different alternative triplets may have different accuracy and speed of translation because of limitations on the pool of tRNAs available. Evidence for the latter effect is that alternative synonymous triplets for an amino acid are not used equally, and the inequality of use is much more pronounced for genes that are transcribed at a very high rate.

| **24-7** TABLE | Size Class Frequencies for Two Different VNTR Sequences, D14S1 and D14S13, in the Karitiana and Surui of Brazil |

	D14S1		D14S13	
Size class	Karitiana	Surui	Karitiana	Surui
3–4	105	4	0	0
4–5	0	3	3	14
5–6	0	11	1	4
6–7	0	2	1	2
7–8	0	1	1	2
8–9	3	3	8	16
9–10	0	11	28	9
10–11	0	2	22	0
11–12	0	4	18	8
12–13	0	0	13	18
13–14	0	0	13	3
>14	0	0	0	2
	108	78	108	78

Source: Data from J. Kidd and K. Kidd, *American Journal of Physical Anthropology* 81, 1992, 249.

There are also constraints on 5′ and 3′ noncoding sequences and on intron sequences. Both 5′ and 3′ noncoding DNA sequences contain signals for transcription, and introns may contain enhancers of transcription (see Chapter 11).

MESSAGE ·······································
Within species, there is great genetic variation. This variation is manifest at the morphologic level of chromosome form and number and at the level of DNA segments that may have no observable developmental effects.
·······································

Quantitative variation

Not all variation in traits can be described in terms of allelic frequencies, because many characteristics, such as height, vary continuously over a range rather than falling into a few qualitatively distinct classes. There is no allele for being 5′8″ or 5′4″ tall. Such characters, if they are varying as a consequence of genetic variation, will be affected by several or many genes and by environmental variation as well. Special techniques are needed for the study of such *quantitative traits,* and these techniques are presented in Chapter 25. For the moment, we confine ourselves to the question of whether genetic differences between individuals affect the trait at all. In experimental organisms, a simple way to answer this question is to choose two groups of parents that differ markedly in the trait and to raise offspring from both groups in the same environment. If the offspring of the two groups are different, then the trait is said to be *heritable* (see Chapter 25 for a more detailed discussion of the concept and estimation of heritability). A simple measure of the degree of heritability of the variation is the ratio of the difference between the offspring groups to the difference between the parental groups. So, if two groups of *Drosophila* parents differed by, say, 0.1 mg in weight, whereas the offspring groups, raised in identical environments, differed by 0.03 mg, the heritability of weight difference would be estimated as 30 percent. When this technique is applied to morphologic variation in *Drosophila,* virtually every variable trait is found to have some heritability. It is important to note that this method cannot be applied to organisms for which no rigorous control over developmental environment is possible. In humans, for example, children of different parental groups differ from one another not only because their genetic makeup is different, but also because the environments of different families, social classes, and nations are different. Japanese are, on the average, shorter than Europeans, but the difference between children of Japanese ancestry and children of European ancestry, both born in North America, is less and becomes even less in the second generation, presumably because of diet. It is not clear whether all the differences in height would disappear or even be reversed if the family environments were identical.

Effect of sexual reproduction on variation

Before Mendel, blending inheritance was the standard model. This concept has powerful consequences for population variation.

Suppose that some trait (say, height) has a distribution in the population and that individuals mate more or less at random. If intermediate individuals mated with each other, they would produce only intermediate offspring, according to a blending model. The mating of a tall with a short individual also would produce only intermediate offspring. Only the mating of tall with tall individuals and short with short individuals would preserve extreme types. The net result of all matings would be an increase in intermediate types and a decrease in extreme types. The variance of the distribution would shrink, simply as a result of sexual reproduction. In fact, it can be shown that the variance is *cut in half* in each generation, so that the population would be essentially uniformly intermediate in height before very many generations had passed.

The particulate nature of inheritance changes this picture completely. Because of the discrete nature of the Mendelian genes and the segregation of alleles at meiosis, a cross of intermediate with intermediate individuals does *not* result in all intermediate offspring. On the contrary, extreme types (homozygotes) segregate out of the cross. To see the consequence of Mendelian inheritance for genetic variation, consider a population in which males and females mate with each other at random with respect to some gene locus *A;* that is, individuals do not choose their mates preferentially with respect to the partial genotype at the locus. Such random mating is equivalent to mixing all the sperm and all the eggs in the population together and then matching randomly drawn sperm with randomly drawn eggs.

The outcome of such a random pairing of sperm and eggs is easy to calculate. If, in some population, the allele frequency of *A* is 0.60 in sperm and eggs, then the chance that a randomly chosen sperm and a randomly chosen egg are both *A* is $0.60 \times 0.60 = 0.36$. Thus, in a random-mating population with this allele frequency, offspring will be 36 percent *A/A.* In the same way, the frequency of *a/a* offspring will be $0.40 \times 0.40 = 0.16$. Heterozygotes will be produced by the fusion either of an *A* sperm with an *a* egg or of an *a* sperm with an *A* egg. If gametes pair at random, then the chance of an *A* sperm and an *a* egg is 0.60×0.40, and the reverse combination has the same probability, so the frequency of heterozygous offspring is $2 \times 0.6 \times 0.4 = 0.48$. Moreover, the process of random mating has done nothing to change *allele* frequencies, as can be easily checked by calculating the frequencies of the alleles *A* and *a* among the offspring by using the method described on page 715. So the proportions of homozygotes and heterozygotes in each successive generation will remain the same. Box 24-2 gives a general form of this equilibrium result.

Box 24-2 Hardy-Weinberg Equilibrium

If the frequency of allele A is p in both the sperm and the eggs and the frequency of allele a is $q = 1 - p$, then the consequences of random unions of sperm and eggs are shown in the adjoining Punnett square. The probability that both the sperm and the egg will carry A is $p \times p = p^2$, so p^2 will be the frequency of A/A homozygotes in the next generation. In like manner, the chance of heterozygotes A/a will be $(p \times q) + (q \times p) = 2pq$, and the chance of homozygotes a/a will be $q \times q = q^2$. The three genotypes, after a generation of random mating, will be in the frequencies $p^2 : 2pq : q^2$. As the Punnett square shows, the allelic frequency of A has not changed and is still p. Therefore, in the second generation, the frequencies of the three genotypes will again be $p^2 : 2pq : q^2$, and so forth, forever.

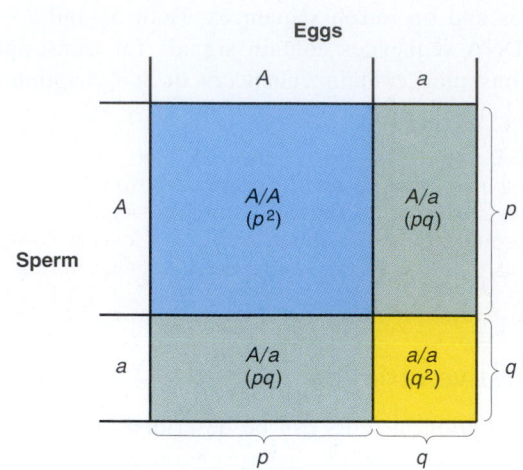

The Hardy-Weinberg equilibrium frequencies that result from random mating. The frequencies of A and a among both eggs and sperm are p and q ($= 1 - p$), respectively. The total frequencies of the zygote genotypes are p^2 for A/A, $2pq$ for A/a, and q^2 for a/a. The frequency of the allele A in the zygotes is the frequency of A/A plus half the frequency of A/a, or $p^2 + pq = p(p + q) = p$.

MESSAGE ···

Mendelian segregation has the property that random mating results in an equilibrium distribution of genotypes after only one generation, so genetic variation is maintained.

The equilibrium distribution

A/A	A/a	a/a
p^2	$2pq$	q^2

is called the **Hardy-Weinberg equilibrium** after those who independently discovered it. (A third independent discovery was made by the Russian geneticist Sergei Chetverikov.)

The Hardy-Weinberg equilibrium means that sexual reproduction does not cause a constant reduction in genetic variation in each generation; on the contrary, the amount of variation remains constant generation after generation, in the absence of other disturbing forces. The equilibrium is the direct consequence of the segregation of alleles at meiosis in heterozygotes.

Numerically, the equilibrium shows that, irrespective of the particular mixture of genotypes in the parental generation, the genotypic distribution after one round of mating is completely specified by the allelic frequency p. For example, consider three hypothetical populations:

	$f(A/A)$	$f(A/a)$	$f(a/a)$
I	0.3	0.0	0.7
II	0.2	0.2	0.6
III	0.1	0.4	0.5

The allele frequency p of A in the three populations is:

I	$p = f(A/A) + \frac{1}{2}f(A/a) = 0.3 + \frac{1}{2}(0)$	$= 0.3$
II	$p =$ $\qquad 0.2 + \frac{1}{2}(0.2) = 0.3$	
III	$p =$ $\qquad 0.1 + \frac{1}{2}(0.4) = 0.3$	

So, despite their very different genotypic compositions, they have the same allele frequency. After one generation of random mating, however, each of the three populations will have the same genotypic frequencies:

A/A	A/a	a/a
$(0.3)^2 = 0.09$	$2(0.3)(0.7) = 0.42$	$(0.7)^2 = 0.49$

and they will remain so indefinitely.

One consequence of the Hardy-Weinberg proportions is that rare alleles are virtually never in homozygous condition. An allele with a frequency of 0.001 is present in homozygotes at a frequency of only 1 in a million; most copies of such rare alleles are found in heterozygotes. In general, because two copies of an allele are in homozygotes but only one copy of that allele is in each heterozygote, the relative frequency of the allele in heterozygotes (in contrast with homozygotes) is, from the Hardy-Weinberg equilibrium frequencies,

$$\frac{2pq}{2q^2} = \frac{p}{q}$$

which for $q = 0.001$ is a ratio of $999 : 1$. The general relation between homozygote and heterozygote frequencies as a function of allele frequencies is shown in Figure 24-5.

In our derivation of the equilibrium, we assumed that the allelic frequency p is the same in sperm and eggs. The Hardy-Weinberg equilibrium theorem does not apply to sex-

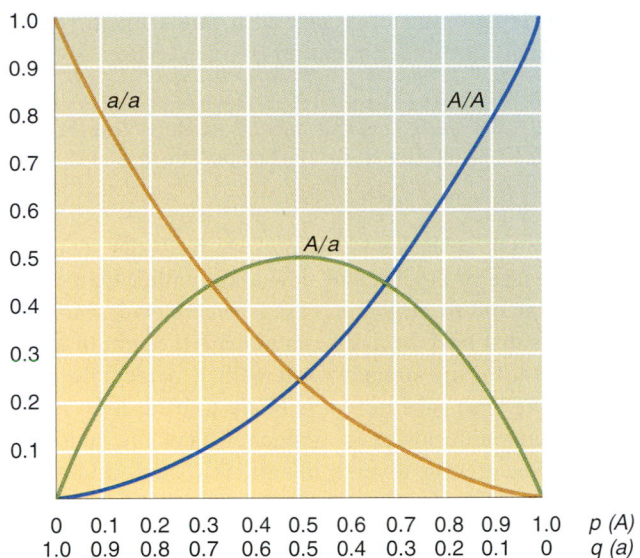

Figure 24-5   Curves showing the proportions of homozygotes A/A (blue line), homozygotes a/a (orange line), and heterozygotes A/a (green line) in populations of different allelic frequencies if the populations are in Hardy-Weinberg equilibrium.

		24-8 TABLE		Comparison Between Observed Frequencies of Genotypes for the MN Blood Group Locus and the Frequencies Expected from Random Mating		
		OBSERVED			EXPECTED	
Population	M/M	M/N	N/N	M/M	M/N	N/N
Eskimo	0.835	0.156	0.009	0.834	0.159	0.008
Egyptian	0.278	0.489	0.233	0.274	0.499	0.228
Chinese	0.332	0.486	0.182	0.331	0.488	0.181
Australian aborigine	0.024	0.304	0.672	0.031	0.290	0.679

Note: The expected frequencies are computed according to the Hardy-Weinberg equilibrium, using the values of p and q computed from the observed frequencies.

linked genes if males and females start with unequal gene frequencies.

The Hardy-Weinberg equilibrium was derived on the assumption of "random mating," but we must carefully distinguish two meanings of that process. First, we may mean that individuals do not choose their mates on the basis of some heritable character. Human beings are random mating with respect to blood groups in this first sense, because they generally do not know the blood type of their prospective mates, and, even if they did, it is unlikely that blood type would be used as a criterion for choice. In the first sense, random mating will occur with respect to genes that have no effect on appearance, behavior, smell, or other characteristics that directly influence mate choice.

The second sense of random mating is relevant when there is any division of a species into subgroups. If there is genetic differentiation between subgroups so that the frequencies of alleles differ from group to group and if individuals tend to mate within their own subgroup (**endogamy**), then, with respect to the species as a whole, mating is not at random and frequencies of genotypes will depart more or less from Hardy-Weinberg frequencies. In this sense, human beings are not random mating, because ethnic and racial groups differ from one another in gene frequencies and people show high rates of endogamy not only within major races, but also within local ethnic groups. Spaniards and Russians differ in their ABO blood group frequencies, Spaniards marry Spaniards and Russians marry Russians, so there is unintentional nonrandom mating with respect to ABO blood groups. Table 24-8 shows random mating in the first sense and nonrandom mating in the second sense for the MN blood group. Within Eskimo, Egyptian, Chinese,

and Australian subpopulations, females do not choose their mates by MN type, and, thus, Hardy-Weinberg equilibrium exists *within* the subpopulations. But Egyptians do not mate with Eskimos or Australian aborigines, so the nonrandom associations in the human species *as a whole* result in large differences in genotype frequencies and departure from Hardy-Weinberg equilibrium.

Sources of variation

The variational theory of evolution has a peculiar self-defeating property. If evolution occurs by the differential reproduction of different variants, we expect the variant with the highest rate of reproduction eventually to take over the population and all other genotypes to disappear. But then there is no longer any variation for further evolution. The possibility of continued evolution therefore is critically dependent on renewed variation.

For a given population, there are three sources of variation: mutation, recombination, and immigration of genes. However, recombination by itself does not produce variation unless alleles are segregating already at different loci; otherwise there is nothing to recombine. Similarly, immigration cannot provide variation if the entire species is homozygous for the same allele. Ultimately, the source of all variation must be mutation.

Variation from mutations

Mutations are the *source* of variation, but the *process* of mutation does not itself drive evolution. The rate of change in gene frequency from the mutation process is very low because spontaneous mutation rates are low (Table 24-9). The mutation rate is defined as the probability that a copy of an allele changes to some other allelic form in one generation. Suppose that a population were completely homozygous A and mutations to a occurred at the rate of $\frac{1}{100,000}$. Then, in the next generation, the frequency of a alleles would be

24-9 TABLE	Point-Mutation Rates in Different Organisms	
Organism	Gene	Mutation rate per generation
Bacteriophage	Host range	2.5×10^{-9}
Escherichia coli	Phage resistance	2×10^{-8}
Zea mays (corn)	R (color factor)	2.9×10^{-4}
	Y (yellow seeds)	2×10^{-6}
Drosophila melanogaster	Average lethal	2.6×10^{-5}

Source: T. Dobzhansky, *Genetics and the Origin of Species*, 3d ed., rev. Columbia University Press, 1951.

only $1.0 \times \frac{1}{100,000} = 0.00001$ and the frequency of A alleles would be 0.99999. After yet another generation of mutation, the frequency of a would be increased by $0.99999 \times \frac{1}{100,000} = 0.000009$ to a new frequency of 0.000019, whereas the original allele would be reduced in frequency to 0.999981. It is obvious that the rate of increase of the new allele is extremely slow and that *it gets slower every generation* because there are fewer copies of the old

allele still left to mutate. A general formula for the change in allele frequency under mutation is given in Box 24-3.

MESSAGE ···
Mutation rates are so low that mutation alone cannot account for the rapid evolution of populations and species.
··

If we look at the mutation process from the standpoint of the increase of a particular new allele rather than the decrease of the old form, the process is even slower. Most mutation rates that have been determined are the sum of all mutations of A to any mutant form with a detectable effect. Any *specific* base substitution is likely to be at least two orders of magnitude lower in frequency than the sum of all changes. So, precise reverse mutations ("back mutations") to the original allele A are unlikely, although many mutations may produce alleles that are *phenotypically* similar to the original.

It is not possible to measure locus-specific mutation rates for continuously varying characters, but the rate of accumulation of genetic variance can be determined. Beginning with a completely homozygous line of *Drosophila* derived from a natural population, $\frac{1}{1000}$ to $\frac{1}{500}$ of the genetic

---

Box 24-3   Effect of Mutation on Allele Frequency

Let μ be the **mutation rate** from allele A to some other allele a (the probability that a gene copy A will become a in the DNA replication preceding meiosis). If p_t is the frequency of the A allele in generation t, if $q_t = 1 - p_t$ is the frequency of the a allele, and if there are no other causes of gene frequency change (no natural selection, for example), then the change in allelic frequency in one generation is:

$$\Delta p = p_t - p_{t-1} = -\mu p_{t-1}$$

where p_{t-1} is the frequency in the preceding generation. This tells us that the frequency of A decreases (and the frequency of a increases) by an amount that is proportional to the mutation rate μ and to the proportion p of all the genes that are still available to mutate. Thus Δp gets smaller as the frequency of p itself decreases, because there are fewer and fewer A alleles to mutate into a alleles. We can make the approximation that, after n generations of mutation,

$$p_n = p_0 e^{-n\mu}$$

where e is the base of the natural logarithms. This relation of gene frequency to number of generations is shown in the graph for $\mu = 10^{-5}$. After 10,000 generations of continued mutation of A to a,

$$p = p_0 e^{-(10^4) \times (10^{-5})} = p_0 e^{-0.1} = 0.904 p_0$$

If the population started with only A alleles ($p_0 = 1.0$), it would still have only 10 percent a alleles after

10,000 generations at this rather high mutation rate and would require 60,000 additional generations to reduce p to 0.5. Even if mutation rates were doubled (say, by environmental mutagens), the rate of evolution would be very slow. For example, radiation levels of sufficient intensity to double the mutation rate over the reproductive lifetime of an individual human are at the limit of occupational safety regulations, and a dose of radiation sufficient to increase mutation rates by an order of magnitude would be lethal; so rapid genetic change in the species would not be one of the effects of increased radiation. Although we have many things to fear from environmental radiation pollution, turning into a species of monsters is not one of them.

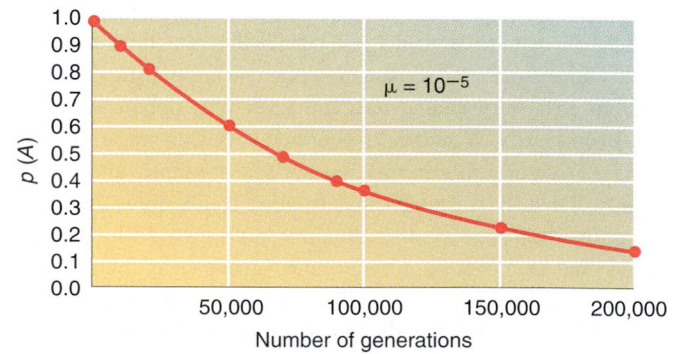

The change over generations in the frequency of a gene A due to mutation from A to a at a constant mutation rate (μ) of 10^{-5}.

variance in bristle number in the original population is restored each generation by spontaneous mutation.

Variation from recombination

The creation of genetic variation by recombination can be a much faster process than its creation by mutation. When just two chromosomes with "normal" survival, taken from a natural population of *Drosophila,* are allowed to recombine for a single generation, they produce an array of chromosomes with 25 to 75 percent as much genetic variation in survival as was present in the entire natural population from which the parent chromosomes were sampled. This outcome is simply a consequence of the very large number of different recombinant chromosomes that can be produced even if we take into account only single crossovers. If a pair of homologous chromosomes is heterozygous at n loci, then a crossover can take place in any one of the $n - 1$ intervals between them, and, because each recombination produces two recombinant products, there are $2(n - 1)$ new unique gametic types from a single generation of crossing-over, even considering only single crossovers. If the heterozygous loci are well spread out on the chromosomes, these new gametic types will be frequent and considerable variation will be generated. Asexual organisms or organisms, such as bacteria, that very seldom undergo sexual recombination do not have this source of variation, so new mutations are the only way in which a change in gene combinations can be achieved. As a result, asexual organisms may evolve more slowly under natural selection than sexual organisms.

Variation from migration

A further source of variation is migration into a population from other populations with different gene frequencies. The resulting mixed population will have an allele frequency that is somewhere intermediate between its original value and the frequency in the donor population. Suppose a population receives a group of migrants whose number is equal to, say, 10 percent of its native population size. Then the newly formed mixed population will have an allele frequency that is a 0.90:0.10 mixture between its original allele frequency and the allele frequency of the donor population. If its original allele frequency of A were, say, 0.70, whereas the donor population had an allele frequency of

only, say, 0.40, the new mixed population would have a frequency of $0.70 \times 0.90 + 0.40 \times 0.10 = 0.67$. Box 24-4 derives the general result. The change in gene frequency is proportional to the difference in frequency between the recipient population and the average of the donor populations. Unlike the mutation rate, the migration rate (m) can be large, so the change in frequency may be substantial.

We must understand *migration* as meaning any form of the introduction of genes from one population into another. So, for example, genes from Europeans have "migrated" into the population of African origin in North America steadily since the Africans were introduced as slaves. We can determine the amount of this migration by looking at the frequency of an allele that is found only in Europeans and not in Africans and comparing its frequency among blacks in North America.

We can use the formula for the change in gene frequency from migration if we modify it slightly to account for the fact that several generations of admixture have taken place. If the rate of admixture has not been too great, then (to a close order of approximation) the sum of the single-generation migration rates over several generations (let's call this M) will be related to the total change in the recipient population after these several generations by the same expression as the one used for changes due to migration. If, as before, P is the allelic frequency in the donor population and p_0 is the original frequency among the recipients, then

$$\Delta p_{\text{total}} = M(P - p_0)$$

so

$$M = \frac{\Delta p_{\text{total}}}{P - p_0}$$

For example, the Duffy blood group allele Fy^a is absent in Africa but has a frequency of 0.42 in whites from the state of Georgia. Among blacks from Georgia, the Fy^a frequency is 0.046. Therefore, the total migration of genes from whites into the black population since the introduction of slaves in the eighteenth century is

$$M = \frac{\Delta p_{\text{total}}}{P - p} = \frac{0.046 - 0}{0.42 - 0} = 0.1095$$

When the same analysis is carried out on American blacks from Oakland (California) and Detroit, M is 0.22 and 0.26,

---

| **Box 24-4** | **Effect of Migration on Allele Frequency** |

If p_t is the frequency of an allele in the recipient population in generation t and P is the allelic frequency in a donor population (or the average over several donor populations) and if m is the proportion of the recipient population that is made up of new migrants from the donor population, then the gene frequency in the recipient population in the next generation, p_{t+1}, is the result of mixing $1 - m$ genes from the recipient with m genes from the donor population. Thus:

$$p_{t+1} = (1 - m)p_t + mP = p_t + m(P - p_t)$$

and

$$\Delta p = p_{t+1} - p_t = m(P - p_t)$$

respectively, showing either greater admixture rates in these cities than in Georgia or differential movement into these cities by American blacks who have more European ancestry. In any case, the genetic variation at the *Fy* locus has been increased by this admixture.

Inbreeding and assortative mating

Random mating with respect to a locus is common, but it is not universal. Two kinds of deviation from random mating must be distinguished. First, individuals may mate with each other nonrandomly because of their degree of common ancestry; that is, their degree of genetic relationship. If mating between relatives occurs more commonly than would occur by pure chance, then the population is **inbreeding.** If mating between relatives is less common than would occur by chance, then the population is said to be undergoing **enforced outbreeding,** or **negative inbreeding.**

Second, individuals may tend to choose each other as mates, not because of their degree of genetic relationship but because of their degree of resemblance to each other at some locus. Bias toward mating of like with like is called **positive assortative mating.** Mating with unlike partners is called **negative assortative mating.** Assortative mating is never complete.

Inbreeding and assortative mating are not the same. Close relatives resemble each other more than unrelated individuals on the average but not necessarily for any particular trait in particular individuals. So inbreeding can result in the mating of quite dissimilar individuals. On the other hand, individuals who resemble each other for some trait may do so because they are relatives, but unrelated individuals also may have specific resemblances. Brothers and sisters do not all have the same eye color, and blue-eyed people are not all related to one another.

Assortative mating for some traits is common. In humans, there is a positive assortative mating bias for skin color and height, for example. An important difference between assortative mating and inbreeding is that the former is

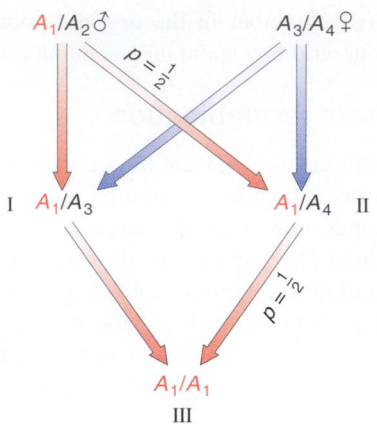

Figure 24-6   Calculation of homozygosity by descent for an offspring (III) of a brother–sister (I–II) mating. The probability that II will receive A_1 from its father is $\frac{1}{2}$; if it does, the probability that II will pass A_1 on to the generation producing III is $\frac{1}{2}$. Thus, the probability that III will receive an A_1 from II is $\frac{1}{2} \times \frac{1}{2} = \frac{1}{4}$.

specific to a trait, whereas the latter applies to the entire genome. Individuals may mate assortatively with respect to height but at random with respect to blood group. Cousins, on the other hand, resemble each other genetically on the average to the same degree at all loci.

For both positive assortative mating and inbreeding, the consequence to population structure is the same: there is an increase in homozygosity above the level predicted by the Hardy-Weinberg equilibrium. If two individuals are related, they have at least one common ancestor. Thus, there is some chance that an allele carried by one of them and an allele carried by the other are both descended from the identical DNA molecule. The result is that there is an extra chance of **homozygosity by descent,** to be added to the chance of homozygosity ($p^2 + q^2$) that arises from the random mating of unrelated individuals. The probability of homozygosity by descent is called the **inbreeding coefficient** (*F*). Figure 24-6 and Box 24-5 illustrate the calculation of the probabil-

---

Box 24-5   **Effect of the Mating of Close Relatives on Homozygosity**

The probability of a homozygous *a/a* offspring from a brother–sister mating is:

probability that one or the other grandparent is *A/a*

   × probability that *a* is passed to male sib

   × probability that *a* is passed to female sib

   × probability of a homozygous *a/a* offspring from *A/a* × *A/a*

$= (2pq + 2pq) \times \frac{1}{2} \times \frac{1}{2} \times \frac{1}{4}$

$= \dfrac{pq}{4}$

We assume that the chance that both grandparents are *A/a* is negligible. If *p* is very small, then *q* is nearly 1.0 and the chance of an affected offspring is close to *p*/4. For $p = \frac{1}{1000}$, there is 1 chance in 4000 of an affected child, compared with the 1-in-a-million chance from a random mating. In general, for full sibs, the ratio of risks will be:

$$\frac{p/4}{p^2} = \frac{1}{4p}$$

ity of homozygosity by descent. Individuals I and II are full sibs because they share both parents. We label each allele in the parents uniquely to keep track of them. Individuals I and II mate to produce individual III. If individual I is A_1/A_3 and the gamete that it contributes to III contains the allele A_1, then we would like to calculate the probability that the gamete produced by II is also A_1. The chance is $\frac{1}{2}$ that II will receive A_1 from its father, and, if it does, the chance is $\frac{1}{2}$ that II will pass A_1 on to the gamete in question. Thus, the probability that III will receive an A_1 from II is $\frac{1}{2} \times \frac{1}{2} = \frac{1}{4}$, and this is the chance that III — the product of a full-sib mating — will be homozygous by descent.

Such close inbreeding can have deleterious consequences. Let's consider a rare deleterious allele a that, when homozygous, causes a metabolic disorder. If the frequency of the allele in the population is p, then the probability that a random couple will produce a homozygous offspring is only p^2 (from the Hardy-Weinberg equilibrium). Thus, if p is, say, $\frac{1}{1000}$, the frequency of homozygotes will be 1 in 1,000,000. Now suppose that the couple are brother and sister. If one of their common parents is a heterozygote for the disease, they may both receive it and may both pass it on to their offspring. As the calculation shows, the rarer the gene, the worse the *relative* risk of a defective offspring from inbreeding. For more-distant relatives, the chance of homozygosity by descent is less but still substantial. For first cousins, for example, the relative risk is $1/16p$ compared with random mating.

Systematic inbreeding between close relatives eventually leads to complete homozygosity of the population but at different rates, depending on the degree of relationship. Which allele is fixed within a line is a matter of chance. If, in the original population from which the inbred lines are taken, allele A has frequency p and allele a has frequency $q = 1 - p$, then a proportion p of the homozygous lines established by inbreeding will be homozygous A/A and a proportion q of the lines will be a/a. Inbreeding takes the genetic variation present *within* the original population and converts it into variation *between* homozygous inbred lines sampled from the population (Figure 24-7).

Suppose that a population is founded by some small number of individuals who mate at random to produce the next generation. Assume that no further immigration into the population ever occurs again. (For example, the rabbits now in Australia probably have descended from a single introduction of a few animals in the nineteenth century.) In later generations, then, everyone is related to everyone else, because their family trees have common ancestors here and there in their pedigrees. Such a population is then inbred, in the sense that there is some probability of a gene's being homozygous by descent. Because the population is, of necessity, finite in size, some of the originally introduced family lines will become extinct in every generation, just as family names disappear in a closed human population because, by chance, no male offspring are left. As original family lines disappear, the population comes to be made up of descen-

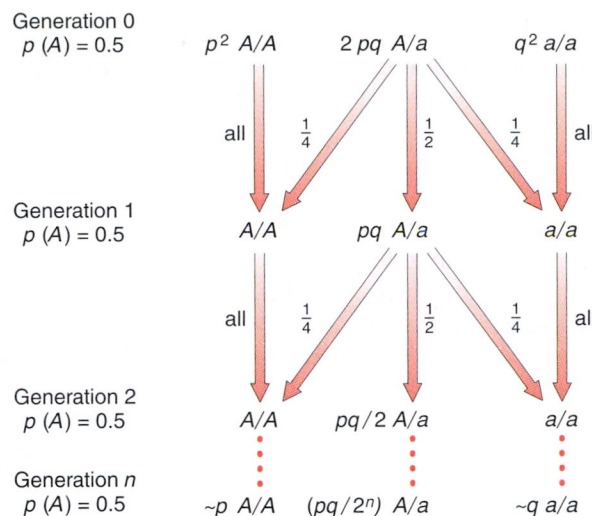

Figure 24-7 Repeated generations of self-fertilization (or inbreeding) will eventually split a heterozygous population into a series of completely homozygous lines. The frequency of A/A lines among the homozygous lines will be equal to the frequency of allele A in the original heterozygous population.

dants of fewer and fewer of the original founder individuals, and all the members of the population become more and more likely to carry the same alleles by descent. In other words, the inbreeding coefficient F increases, and the heterozygosity decreases over time until finally F reaches 1.00 and heterozygosity reaches 0.

The rate of loss of heterozygosity per generation in such a closed, finite, randomly breeding population is inversely proportional to the total number ($2N$) of haploid genomes, where N is the number of diploid individuals in the population. In each generation, $\frac{1}{2N}$ of the remaining heterozygosity is lost, so

$$H_t = H_0 \left(1 - \frac{1}{2N} \right)^t \cong H_0 e^{-t/2N}$$

where H_t and H_0 are the proportions of heterozygotes in the tth and original generations, respectively. As the number t of generations becomes very large, H_t approaches zero.

Balance between inbreeding and new variation

Any population of any species is finite in size, so all populations should eventually become homozygous and differentiated from one another as a result of inbreeding. In nature, however, new variation is always being introduced into populations by mutation and by some migration between localities. Thus, the actual variation available for natural selection is a balance between the introduction of new variation and its loss through local inbreeding. The rate of loss of heterozygosity in a closed population is $\frac{1}{2N}$, so any effective differentiation between populations will be negated if new variation is introduced at this rate or higher.

Selection

So far in this chapter, we have considered changes in a population arising from forces of mutation, migration, recombination, and breeding structure. But these changes are random with respect to the way in which organisms make a living in the environments in which they live. Changes in a species in response to a changing environment occur because the different genotypes produced by mutation and recombination have different abilities to survive and reproduce. The differential rates of survival and reproduction are what is meant by **selection,** and the process of selection results in changes in the frequencies of the various genotypes in the population. Darwin called the process of differential survival and reproduction of different types **natural selection** by analogy with the **artificial selection** carried out by animal and plant breeders when they deliberately select some individuals of a preferred type.

The relative probability of survival and rate of reproduction of a phenotype or genotype is now called its **Darwinian fitness.** Although geneticists sometimes speak loosely of the fitness of an individual, the concept of fitness really applies to the average survival and reproduction of individuals in a phenotypic or genotypic class. Because of chance events in the life histories of individuals, even two organisms with identical genotypes and identical environments will differ in their survival and reproduction rates. It is the fitness of a genotype on average over all its possessors that matters.

Fitness is a consequence of the relation between the phenotype of the organism and the environment in which the organism lives, so the same genotype will have different fitnesses in different environments. In part, this difference is because exposure to different environments during development will result in different phenotypes for the same genotypes. But, even if the phenotype is the same, the success of the organism depends on the environment. Having webbed feet is fine for paddling in water but a positive disadvantage for walking on land, as a few moments spent observing a duck walk will reveal. No genotype is unconditionally superior in fitness to all others in all environments.

Furthermore, the environment is not a fixed situation that is experienced passively by an organism. The environment of an organism is defined by the activities of the organism itself. For example, dry grass is part of the environment of a junco, so juncos that are most efficient at gathering it may waste less energy in nest building and thus have a higher reproductive fitness. But dry grass is part of a junco's environment *because juncos gather it to make nests.* The rocks among which the grass grows are not part of the junco's environment, although the rocks are physically present there. But the rocks are part of the environment of thrushes; these birds use the rocks to break open snails. Moreover, the environment that is defined by the life activities of an organism evolves as a result of those activities. The structure of the soil that is in part determinative of the kinds of plants that will grow is altered by the growth of those very plants. Environment is both the cause and the result of the evolution of organisms. As primitive plants evolved photosynthesis, they changed the earth's atmosphere from one that had had essentially no free oxygen and a high concentration of carbon dioxide to the atmosphere that we know today, which contains 21 percent oxygen and only 0.03 percent carbon dioxide. Plants that evolve today must do so in an environment created by the evolution of their own ancestors.

Darwinian, or reproductive, fitness is not to be confused with "physical fitness" in the everyday sense of the term, although they may be related. No matter how strong, healthy, and mentally alert the possessor of a genotype may be, that genotype has a fitness of zero if for some reason its possessors leave no offspring. Thus such statements as the "unfit are outreproducing the fit, so the species may become extinct" are meaningless. The fitness of a genotype is a consequence of all the phenotypic effects of the genes involved. Thus, an allele that doubles the fecundity of its carriers but at the same time reduces the average lifetime of its possessors by 10 percent will be more fit than its alternatives, despite its life-shortening property. The most common example is parental care. An adult bird that expends a great deal of its energy gathering food for its young will have a lower probability of survival than one that keeps all the food for itself. But a totally selfish bird will leave no offspring, because its young cannot fend for themselves. As a consequence, parental care is favored by natural selection.

Two forms of selection

Because the differences in reproduction and survival between genotypes depend on the environment in which the genotypes live and develop and because organisms may alter their own environments, there are two fundamentally different forms of selection. In the simple case, the fitness of an individual does not depend on the composition of the population; rather it is a fixed property of the individual's phenotype and the external physical environment. For example, the relative ability of two plants that live at the edge of the desert to get sufficient water will depend on how deep their roots grow and how much water they lose through their leaf surfaces. These characteristics are a consequence of their developmental patterns and are not sensitive to the composition of the population in which they live. The fitness of a genotype in such a case does not depend on how rare or how frequent it is in the population. Fitness is then **frequency independent.**

In contrast, consider organisms that are competing to catch prey or to avoid being captured by a predator. Then the relative abundances of two different genotypes will affect their relative fitnesses. An example is *Mullerian mimicry* in butterflies. Some species of brightly colored butterflies (such as monarchs and viceroys) are distasteful to birds, which learn, after a few trials, to avoid attacking butterflies with that pattern. If two species differ in pattern, there will be selection to make them more similar because both will be protected and they share the burden of the birds' initial learning period. The less frequent pattern will be at a disadvantage with respect to the more frequent one, because birds will less often learn to avoid them. Within a species, rarer patterns will be selected against for the same reason. The rarer the pattern, the greater is the selective disadvantage, because birds will be unlikely to have had a prior experience of a low-frequency pattern and therefore will not avoid it. This selection to blend in with the crowd is an example of **frequency-dependent fitness.**

For reasons of mathematical convenience, most models of natural selection are based on frequency-independent fitness. In fact, however, a very large number of selective processes (perhaps most) are frequency dependent. The kinetics of the evolutionary process depend on the exact form of frequency dependence, and, for that reason alone, it is difficult to make any generalizations. The result of *positive* frequency dependence (such as competing predators, where fitness increases with increasing frequency) is quite different from that of *negative* frequency dependence (where fitness of a genotype declines with increasing frequency). For the sake of simplicity and as an illustration of the main qualitative features of selection, we deal only with models of frequency-independent selection in this chapter, but convenience should not be confused with reality.

Measuring fitness differences

For the most part, the differential fitness of different genotypes can be most easily measured when the genotypes differ at many loci. In very few cases (except for laboratory mutants, horticultural varieties, and major metabolic disorders) does the effect of an allelic substitution at a single locus make enough difference to the phenotype to be reflected in measurable fitness differences. Figure 24-8 shows the probability of survival from egg to adult—that is, the **viability**—of a number of second-chromosome homozygotes of *D. pseudoobscura* at three different temperatures. As is generally the case, the fitness (in this case, a component of the total fitness, viability) is different in different

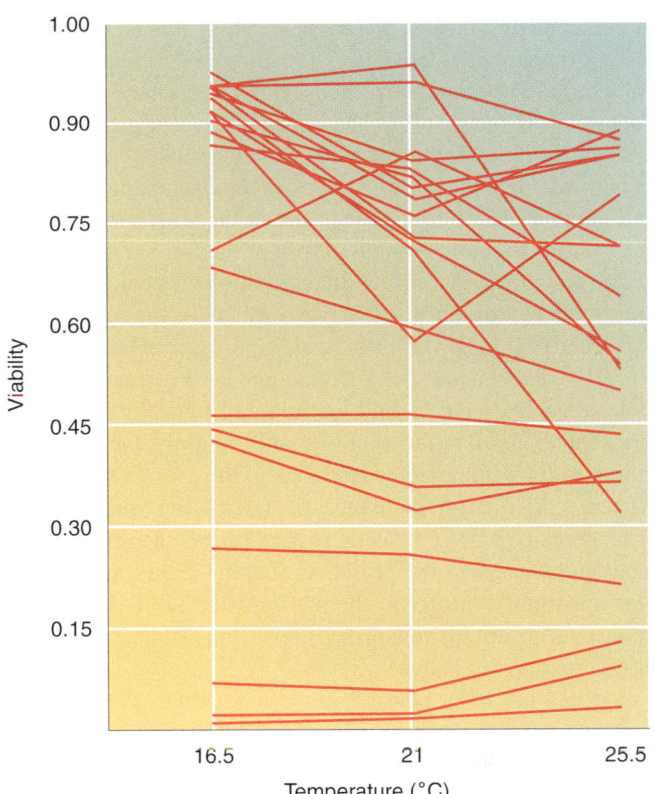

Figure 24-8 Viabilities of various chromosomal homozygotes of *Drosophila pseudoobscura* at three different temperatures.

environments. A few homozygotes are lethal or nearly so at all three temperatures, whereas a few have consistently high viability. Most genotypes, however, are not consistent in viability between temperatures, and no genotype is unconditionally the most fit at all temperatures. The fitness of these chromosomal homozygotes was not measured in competition with each other; all are measured against a common standard, so we do not know whether they are frequency dependent. An example of frequency-dependent fitness is shown in the estimates for inversion homozygotes and heterozygotes of *D. pseudoobscura* in Table 24-10.

Examples of clear-cut fitness differences associated with single-gene substitutions are the many "inborn errors

	HOMOZYGOTES			HETEROZYGOTES		
Experiment	ST/ST	AR/AR	CH/CH	ST/AR	ST/CH	AR/CH
ST and *AR* alone	0.8	0.5	—	1.0	—	—
ST and *CH* alone	0.8	—	0.4	—	1.0	—
AR and *CH* alone	—	0.86	0.48	—	—	1.0
ST, *AR*, and *CH* together	0.83	0.15	0.36	1.0	0.77	0.62

24-10 TABLE Comparison of Fitnesses for Inversion Homozygotes and Heterozygotes in Laboratory Populations of *Drosophila pseudoobscura* When Measured in Different Competitive Combinations

Note: Chromosomal types are: AR, arrowhead; CH, Chiricahua; ST, Standard.

of metabolism," where a recessive allele interferes with a metabolic pathway and causes lethality of the homozygotes. An example in humans is phenylketonuria, where tissue degeneration is the result of the accumulation of a toxic intermediate in the pathway of tyrosine metabolism. A case that illustrates the relation of fitness to environment is sickle-cell anemia. An allelic substitution at the structural-gene locus for the β chain of hemoglobin results in substitution of valine for the normal glutamic acid at chain position 6. The abnormal hemoglobin crystallizes at low oxygen pressure, and the red cells deform and hemolyze. Homozygotes Hb^S/Hb^S have a severe anemia, and survivorship is low. Heterozygotes have a mild anemia and under ordinary circumstances exhibit the same or only slightly lower fitness than normal homozygotes Hb^A/Hb^A. However, in regions of Africa with a high incidence of falciparum malaria, heterozygotes (Hb^A/Hb^S) have a *higher* fitness than normal homozygotes because the presence of some sickling hemoglobin apparently protects them from the malaria. Where malaria is absent, as in North America, the fitness advantage of heterozygosity is lost.

It has not been possible to measure fitness differences for most single-locus polymorphisms. The evidence for differential net fitness for different ABO or MN blood types is shaky at best. The extensive enzyme polymorphism present in all sexually reproducing species has for the most part not been connected with measurable fitness differences, although, in *Drosophila*, clear-cut differences in the fitness of different genotypes have been demonstrated in the laboratory for a few loci such as those encoding α-amylase and alcohol dehydrogenase.

How selection works

The simplest way to see the effect of selection is to consider an allele, *a*, that is completely lethal before reproductive age in homozygous condition, such as the allele that leads to Tay-Sachs disease. Suppose that, in some generation, the allele frequency of this gene is 0.10. Then, in a random-mating population, the proportions of the three genotypes after fertilization are

A/A	A/a	a/a
0.81	0.18	0.01

At reproductive age, however, the homozygotes *a/a* will have already died, leaving the genotypes at this stage as

A/A	A/a	a/a
0.81	0.18	0.00

But these proportions add up to only 0.99 because only 99 percent of the population is still surviving. Among the actual surviving reproducing population, the proportions must be recalculated by dividing by 0.99 so that the total proportions add up to 1.00. After this readjustment, we have

A/A	A/a	a/a
0.818	0.182	0.00

The frequency of the lethal *a* allele among the gametes produced by these survivors is then

$$0.00 + 0.182/2 = 0.091$$

and the change in allelic frequency in one generation, expressed as the new value minus the old one, has been $0.091 - 0.100 = -0.019$. We can repeat this calculation in each successive generation to obtain the predicted frequencies of the lethal and normal alleles in a succession of future generations.

The same kind of calculation can be carried out if genotypes are not simply lethal or normal, but if each genotype has some relative probability of survival. This general calculation is shown in Box 24-6. After one generation of selection, the new value of the frequency of *A* is equal to the old value (*p*) multiplied by the ratio of the average fitness of *A* alleles to the fitness of the whole population. If the fitness of *A* alleles is greater than the average fitness of all alleles, then $\overline{W}_A/\overline{W}$ is greater than unity and p' is larger than *p*. Thus, the allele *A* increases in the population. Conversely, if $\overline{W}_A/\overline{W}$ is less than unity, *A* decreases. But the mean fitness of the population ($\overline{W}$) is the average fitness of the *A* alleles and of the *a* alleles. So if $\overline{W}_A$ is greater than the mean fitness of the population, it must be greater than $\overline{W}_a$, the mean fitness of *a* alleles.

MESSAGE ···
The allele with the higher average fitness increases in the population.

It should be noted that the fitnesses $W_{A/A}$, $W_{A/a}$, and $W_{a/a}$ may be expressed as absolute probabilities of survival and absolute reproduction rates or they may all be rescaled relative to one of the fitnesses, which is given the standard value of 1.0. This rescaling has absolutely no effect on the formula for p', because it cancels out in the numerator and denominator.

MESSAGE ···
The course of selection depends only on relative fitnesses.

An increase in the allele with the higher fitness means that the average fitness of the population as a whole increases, so selection can also be described as a process that *increases mean fitness*. This rule is strictly true only for frequency-independent genotypic fitnesses, but it is close enough to a general rule to be used as a fruitful generalization. This maximization of fitness does not necessarily lead to any optimal property for the species as a whole, because fitnesses are only defined relative to each other within a population. It is relative (not absolute) fitness that is increased by selection. The population does not necessarily become larger or grow faster, nor is it less likely to become extinct.

Rate of change in gene frequency

The general expression for the change in gene frequency, derived in Box 24-6, is particularly illuminating. It says that

Δp will be positive (*A* will increase) if the mean fitness of *A* alleles is greater than the mean fitness of *a* alleles, as we saw before. But it also shows that the speed of the change depends not only on the difference in fitness between the alleles, but also on the factor *pq*, which is proportional to the frequency of heterozygotes (2*pq*). For a given difference in fitness of alleles, gene frequency will change most rapidly when the alleles *A* and *a* are in intermediate frequency, so *pq* is large. If *p* is near 0 or 1 (that is, if *A* or *a* is nearly fixed), then *pq* is nearly 0 and selection will proceed very slowly. Figure 24-9 shows the S-shaped curve that represents the course of selection of a new favorable allele *A* that has recently entered a population of homozygotes *a/a*. At first, the change in frequency is very small because *p* is still close to 0. Then it accelerates as *A* becomes more frequent, but it slows down again as *A* takes over and *a* becomes very

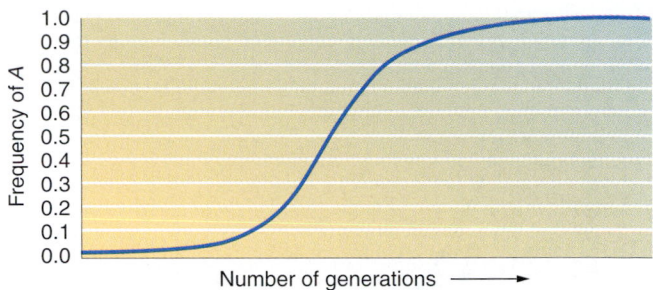

Figure 24-9   The time pattern of increasing frequency of a new favorable allele *A* that has entered a population of *a/a* homozygotes.

rare. This is precisely what is expected from a selection process. When most of the population is of one type, there is nothing to select. For evolution by natural selection to occur,

---

Box 24-6   Effect of Selection on Allele Frequency

Suppose that a population is mating at random with respect to a given locus with two alleles and that the population is so large that (for the moment) we can ignore inbreeding. Just after the eggs have been fertilized, the zygotes will be in Hardy-Weinberg equilibrium:

Genotype	*A/A*	*A/a*	*a/a*
Frequency	p^2	$2pq$	q^2

and $p^2 + 2pq + q^2 = (p + q)^2 = 1.0$, where *p* is the frequency of *A*.

Further suppose that the three genotypes have probabilities of survival to adulthood (viabilities) of $W_{A/A} : W_{A/a} : W_{a/a}$. For simplicity, let us also assume that all selective differences are differences in survivorship between the fertilized egg and the adult stage. Differences in fertility give rise to much more complex mathematical formulations. Then among the progeny having reached adulthood, the frequencies will be:

Genotype	*A/A*	*A/a*	*a/a*
Frequency	$p^2 W_{A/A}$	$2pq W_{A/a}$	$q^2 W_{a/a}$

These adjusted frequencies do not add up to unity, because the *W*'s are all fractions smaller than 1. However, we can readjust them so that they do, without changing their relation to each other, by dividing each frequency by the sum of the frequencies after selection ($\overline{W}$):

$$\overline{W} = p^2 W_{A/A} + 2pq W_{A/a} + q^2 W_{a/a}$$

So defined, $\overline{W}$ is called the **mean fitness** of the population because it is, indeed, the mean of the fitnesses of all individuals in the population. After this adjustment, we have

Genotype	*A/A*	*A/a*	*a/a*
Frequency	$p^2 \dfrac{W_{A/A}}{\overline{W}}$	$2pq \dfrac{W_{A/a}}{\overline{W}}$	$q^2 \dfrac{W_{a/a}}{\overline{W}}$

We can now determine the frequency p' of the allele *A* in the next generation by summing up genes:

$$p' = A/A + (\tfrac{1}{2})A/a = p^2 \frac{W_{A/A}}{\overline{W}} + \frac{pq W_{A/a}}{\overline{W}} =$$

$$p \frac{p W_{A/A} + q W_{A/a}}{\overline{W}}$$

Finally, we note that the expression $p W_{A/A} + q W_{A/a}$ is the mean fitness of *A* alleles because, from the Hardy-Weinberg frequencies, a proportion *p* of all *A* alleles are present in homozygotes with another *A* and then they have a fitness of $W_{A/A}$, whereas a proportion *q* of all the *A* alleles are present in heterozygotes with *a* and have a fitness of $W_{A/a}$. Using $\overline{W}_A$ to denote $p W_{A/A} + q W_{A/a}$ yields the final new gene frequency:

$$p' = p \frac{\overline{W}_A}{\overline{W}}$$

An alternative way to look at the process of selection is to solve for the *change* in allelic frequency in one generation:

$$\Delta p = p' - p = \frac{p \overline{W}_A}{\overline{W}} - p = \frac{p(\overline{W}_A - \overline{W})}{\overline{W}}$$

But $\overline{W}$, the mean fitness of the population, is the average of the allelic fitnesses $\overline{W}_A$ and $\overline{W}_a$, so

$$\overline{W} = p \overline{W}_A + q \overline{W}_a$$

Substituting this expression for $\overline{W}$ in the formula for Δp and remembering that $q = 1 - p$, we obtain (after some algebraic manipulation)

$$\Delta p = \frac{pq(\overline{W}_A - \overline{W}_a)}{\overline{W}}$$

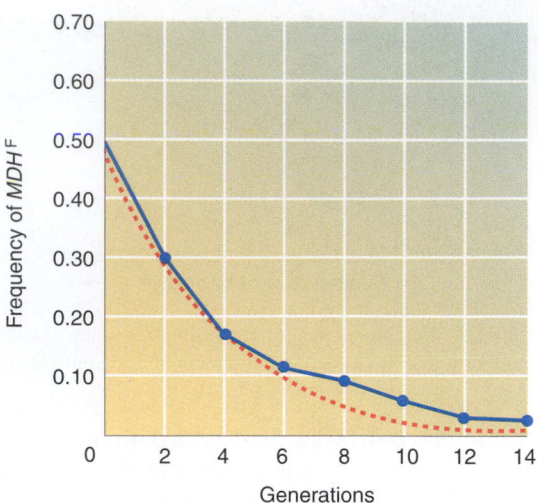

Figure 24-10   The loss of an allele of the malic dehydrogenase locus MDH^F due to selection in a laboratory population of *Drosophila melanogaster*. The red dashed line shows the theoretical curve of change computed for fitnesses $W_{A/A} = 1.0$, $W_{A/a} = 0.75$, $W_{a/a} = 0.4$. (From R. C. Lewontin, *The Genetic Basis of Evolutionary Change*. Copyright © 1974 by Columbia University Press. Data courtesy of E. Berger.)

there must be genetic variance; the more variance, the faster the process.

One consequence of the dynamics shown in Figure 24-9 is that it is extremely difficult to significantly reduce the frequency of an allele that is already rare in a population. Thus, eugenic programs designed to eliminate deleterious recessive genes from human populations by preventing the reproduction of affected persons do not work. Of course, if all heterozygotes could be prevented from reproducing, the gene could be eliminated (except for new mutations) in a single generation. Because every human being is heterozygous for a number of different deleterious genes, however, no one would be allowed to reproduce.

When alternative alleles are not rare, selection can cause quite rapid changes in allelic frequency. Figure 24-10 shows the course of elimination of a malic dehydrogenase allele in a laboratory population of *D. melanogaster*. The fitnesses in this case are:

$$W_{A/A} = 1.0 \qquad W_{A/a} = 0.75 \qquad W_{a/a} = 0.40$$

The frequency of *a* is not reduced to 0, and further reduction in frequency will require longer and longer times, as shown in the negative eugenics case.

MESSAGE ··
Unless alternative alleles are present in intermediate frequencies, selection (especially against recessives) is quite slow. Selection is dependent on genetic variation.
···

Balanced polymorphism

Let's reexamine the general formula for allelic frequency change (see Box 24-6):

$$\Delta p = pq \, \frac{\overline{W}_A - \overline{W}_a}{\overline{W}}$$

Under what conditions will the process stop? When is $\Delta p = 0$? Two answers are: when $p = 0$ or when $q = 0$ (that is, when either allele *A* or allele *a*, respectively, has been eliminated from the population). One of these events will eventually occur if $\overline{W}_A - \overline{W}_a$ is consistently positive or negative, so that Δp is always positive or negative irrespective of the value of *p*. The condition for such unidirectional selection is that the heterozygote fitness be somewhere between the fitnesses of the two homozygotes: If *A/A* homozygotes are most fit, then *A* alleles are more fit than *a* alleles in both the heterozygous and the homozygous condition. Then the mean allelic fitness of *A*, $\overline{W}_A$, is larger than the mean allelic fitness of *a*, $\overline{W}_a$, no matter what the frequencies of the genotypes may be. In this case, $\overline{W}_A - \overline{W}_a$ is positive, and *A* always increases until it reaches $p = 1$. If, on the other hand, *a/a* are most fit, then $\overline{W}_A - \overline{W}_a$ is negative, and *a* always increases until it reaches $q = 1$.

But there is another possibility for $\Delta p = 0$, even when *p* and *q* are not 0:

$$\overline{W}_A = \overline{W}_a$$

which can occur if the heterozygote is not intermediate between the homozygotes but has a fitness that is more extreme than either homozygote. In this case, selection will lead to an intermediate allele frequency, $\hat{p}$ (see Box 24-7).

There are, in fact, two qualitatively different possibilities for $\hat{p}$. One possibility is that $\hat{p}$ is an *unstable* equilibrium. There will be no change in frequency if the population has exactly this value of *p*, but the frequency will move *away* from the equilibrium (toward $p = 0$ or $p = 1$) if the slightest perturbation of frequency occurs. This unstable case will exist when the heterozygote is *lower* in fitness than either homozygote; such a condition is an example of **underdominance.** The alternative possibility is a *stable* equilibrium, or **balanced polymorphism,** in which slight perturbations from the value of $\hat{p}$ will result in a return to $\hat{p}$. The condition for this balance is that the heterozygote be *greater* in fitness than either homozygote — a condition termed **overdominance.**

In nature, the chance that a gene frequency will remain balanced on the knife edge of an unstable equilibrium is negligible, so we should not expect to find naturally occurring polymorphisms in which heterozygotes are less fit than homozygotes. On the contrary, the observation of a long-lasting polymorphism in nature might be taken as evidence of a superior heterozygote.

Unfortunately, life confounds theory. The *Rh* locus (rhesus blood group) in humans has a widespread polymorphism with Rh^+ and Rh^- alleles. In Europeans, the frequency of the Rh^- allele is about 0.4, whereas, in Africans, it is about 0.2. Thus, this human polymorphism must be very old, antedating the origin of modern geographical races. But this polymorphism causes a maternal–fetal incompatibility

| **Box 24-7** | **Natural Selection Leading to Equilibrium of Allele Frequencies** |

The average fitness of an allele is defined as the average of the fitnesses of the genotypes that carry that allele:

$$\overline{W}_A = (pW_{A/A} + qW_{A/a})$$

and

$$\overline{W}_a = (pW_{A/a} + qW_{a/a})$$

Suppose that the fitness of heterozygotes, $W_{A/a}$, is greater than the fitnesses of both homozygotes, $W_{A/A}$ and $W_{a/a}$. Then, for part of the range of values of p, $\overline{W}_A < \overline{W}_a$, whereas $\overline{W}_A < \overline{W}_a$ for the rest of the range. Just between those ranges is a value of p (denoted by $\hat{p}$) for which the mean fitnesses of the two alleles are equal. A little algebraic manipulation of

$$\overline{W}_A - \overline{W}_a = 0 = (\hat{p}W_{A/A} + \hat{q}W_{A/a}) - (\hat{p}W_{A/a} + \hat{q}W_{a/a})$$

gives us the solution for $\hat{p}$:

$$\hat{p} = \frac{W_{a/a} - W_{A/a}}{(W_{a/a} - W_{A/a}) + (W_{A/A} - W_{A/a})}$$

The equilibrium value is a simple ratio of the differences in fitness between the homozygotes and the heterozygote. For example, suppose the fitnesses are:

$W_{A/A}$	$W_{A/a}$	$W_{a/a}$
0.9	1.0	0.8

The equilibrium will be $\hat{p} = \frac{2}{3}$.

---

when an RH$^-$ mother (homozygous Rh^-/Rh^-) produces an RH$^+$ fetus (heterozygous Rh^-/Rh^+). This incompatibility results in hemolytic anemia (from a destruction of red blood cells) and the death of the fetus in a moderate proportion of cases if the mother has been previously sensitized by an earlier pregnancy with an incompatible fetus. Thus, there is selection against heterozygotes, although it is frequency dependent, because it occurs only when the mother is a homozygous recessive. This polymorphism is unstable and should have disappeared from the species, yet it exists in most human populations. Many hypotheses have been proposed to explain its apparent stability, but the mystery remains.

In contrast, no fitness difference at all can be demonstrated for many polymorphisms of blood groups (and for the ubiquitous polymorphism of enzymes revealed by electrophoresis). It has been suggested that such polymorphisms are not under selection at all but that

$$W_{A/A} = W_{A/a} = W_{a/a}$$

This situation of **selective neutrality** would also satisfy the requirement that $\overline{W}_A = \overline{W}_a$, but, instead of a stable equilibrium, it gives rise to a **passive (neutral) equilibrium** such that any allele frequency p is as good as any other. This leaves unanswered the problem of how the populations became highly polymorphic in the first place. The best case of overdominance for fitness at a single locus remains that of sickle-cell anemia, where the two homozygotes are at a disadvantage relative to the heterozygote for quite different reasons.

The best-studied cases of balanced polymorphism in nature and in the laboratory are the inversion polymorphisms in several species of *Drosophila*. Figure 24-11 shows the course of frequency change for the inversion ST (Standard) in competition with the alternative chromosomal type CH (Chiricahua) in a laboratory population of *D. pseudoobscura*. The inversions ST and CH are part of a chromosomal

polymorphism in natural populations of this species. The fitnesses estimated for the three genotypes in the laboratory are

$$W_{ST/ST} = 0.89 \qquad W_{ST/CH} = 1.0 \qquad W_{CH/CH} = 0.41$$

Applying the formula for the equilibrium value $\hat{p}$, we obtain $\hat{p} = 0.85$, which agrees quite well with the observations in Figure 24-11.

Another cause of genetic equilibrium in populations is the balance between the introduction of new alleles by repeated mutation and their removal by natural selection. This balance is probably the cause of many low-level polymorphisms for genetic diseases in human populations. New deleterious mutations are constantly arising spontaneously or as the result of the action of mutagens. These mutations may be completely recessive or partly dominant. Selection removes them from the population, but there will be an equilibrium between their appearance and removal.

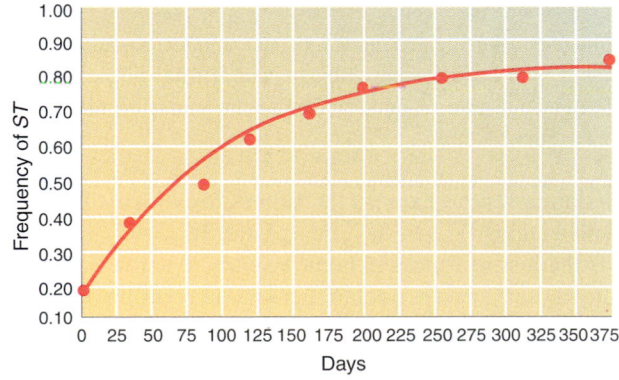

Figure 24-11   Changes in the frequency of the inversion Standard (ST) in competition with Chiricahua (CH) in a laboratory population of *Drosophila pseudoobscura*. The points show the actual frequencies in successive generations. The solid line shows the theoretical course of change if the fitnesses of the genotypes were as given in the text.

The general expression for this equilibrium is that the frequency of the deleterious allele at equilibrium depends on the ratio of the mutation rate, μ, to the intensity of selection, s, against the deleterious genotype. For a completely recessive deleterious allele whose fitness in homozygous state is $1 - s$, the equilibrium frequency is

$$q = \sqrt{(\mu/s)}$$

These results are shown in detail in Box 24-8. So, for example, a recessive lethal ($s = 1$) mutating at the rate of $\mu = 10^{-6}$ will have an equilibrium frequency of 10^{-3}. Indeed, if we knew that a gene was a recessive lethal and had no heterozygous effects, we could estimate its mutation rate as the square of the allelic frequency. But the basis for such calculations must be firm. Sickle-cell anemia was once thought to be a recessive lethal with no heterozygous effects, which led to an estimated mutation rate in Africa of 0.1 for this locus.

A similar result can be obtained for a deleterious gene with some effect in heterozygotes. If we let the fitnesses be $W_{A/A} = 1.0$, $W_{A/a} = 1 - hs$, and $W_{a/a} = 1 - s$ for a partly dominant gene, where h is the degree of dominance of the deleterious allele, then a similar calculation gives us

$$\hat{q} = \frac{\mu}{hs}$$

Thus, if $\mu = 10^{-6}$ and the lethal is not totally recessive but has a 5 percent deleterious effect in heterozygotes ($s = 1.0$, $h = 0.05$), then

$$\hat{q} = \frac{10^{-6}}{5 \times 10^{-2}} = 2 \times 10^{-5}$$

which is smaller by two orders of magnitude than the equilibrium frequency for the purely recessive case. In general, then, we can expect deleterious, completely recessive genes to have frequencies much higher than those of partly dominant genes.

Artificial selection

In contrast with the difficulties of finding simple, well-behaved cases in nature that exemplify the simple formulas of natural selection, there is a vast record of the effectiveness of artificial selection in changing populations phenotypically. These changes have been produced by laboratory selection experiments and by selection of animals and plants in agriculture (as examples, for increased milk production in cows and for rust resistance in wheat). No analysis of these experiments in terms of allelic frequencies is possible, because individual loci have not been identified and followed. Nevertheless, it is clear that genetic changes have occurred in the populations and that some analysis of selected populations has been carried out according to the methods described in Chapter 25. Figure 24-12 shows, as an example,

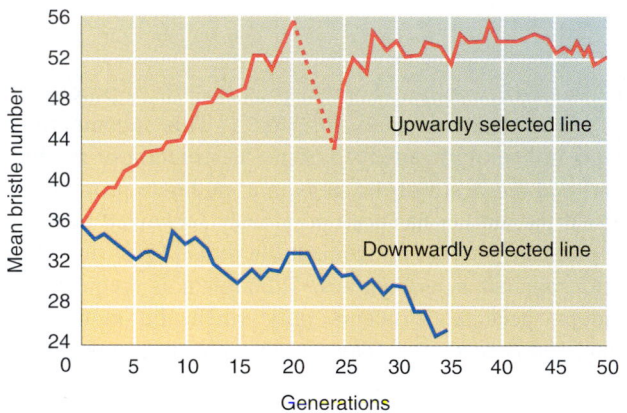

Figure 24-12   Changes in average bristle number obtained in two laboratory populations of *Drosophila melanogaster* through artificial selection for high bristle number in one population and for low bristle number in the other. The dashed segment in the curve for the upwardly selected line indicates a period of five generations in which no selections were performed. (From K. Mather and B. J. Harrison, "The Manifold Effects of Selection," *Heredity* 3, 1949, 1.)

Box 24-8   **Balance Between Selection and Mutation**

If we let q be the frequency of the deleterious allele a and $p = 1 - q$ be the frequency of the normal allele, then the change in allelic frequency due to the mutation rate μ is:

$$\Delta q_{mut} = \mu p$$

A simple way to express the fitness for a recessive deleterious gene is $W_{A/A} = W_{A/a} = 1.0$ and $W_{a/a} = 1 - s$, where s is the loss of fitness in homozygotes. We now can substitute these fitnesses in our general expression for allelic frequency change (see page 731) and obtain

$$\Delta q_{sel} = -\frac{pq(sq)}{1 - sq^2}$$

Equilibrium means that the increase in the allelic frequency due to mutation must exactly balance the decrease

in the allelic frequency due to selection, so

$$\Delta \hat{q}_{mut} + \Delta \hat{q}_{sel} = 0$$

Remembering that $\hat{q}$ at equilibrium will be quite small, so $1 - s\hat{q}^2 \simeq 1$, and substituting the terms for $\Delta \hat{q}_{mut}$ and $\Delta \hat{q}_{sel}$ in the preceding formula, we have

$$\mu\hat{p} - \frac{s\hat{p}\hat{q}^2}{1 - s\hat{q}^2} \simeq \mu\hat{p} - s\hat{p}\hat{q}^2 = 0$$

or

$$q^2 = \frac{\mu}{s} \quad \text{and} \quad \hat{q} = \sqrt{\frac{\mu}{s}}$$

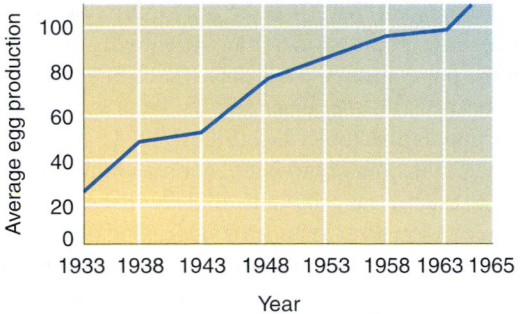

Figure 24-13 Changes in average egg production in a chicken population selected for its increase in egg-laying rate over a period of 30 years. (From I. M. Lerner and W. J. Libby, *Heredity, Evolution, and Society,* 2d ed. Copyright © 1976 by W. H. Freeman and Company. Data courtesy of D. C. Lowry.)

the large changes in average bristle number achieved in a selection experiment with *D. melanogaster.* Figure 24-13 shows the changes in the number of eggs laid per chicken as a consequence of 30 years of selection.

The usual method of selection is **truncation selection.** The individuals in a given generation are pooled (irrespective of their families), a sample is measured, and only those individuals above (or below) a given phenotypic value (the truncation point) are chosen as parents for the next generation. This phenotypic value may be a fixed value over successive generations; then selection is by **constant truncation.** More commonly, a fixed percentage of the population representing the highest (or lowest) value of the selected character is chosen; then selection is by **proportional truncation.** With constant truncation, the intensity of selection decreases with time, as more and more of the population exceeds the fixed truncation point. With proportional truncation, the intensity of selection is constant, but the truncation

point moves upward as the population distribution moves. Figure 24-14 illustrates these two types of truncation.

A common experience in artificial selection programs is that, as the population becomes more and more extreme, its viability and fertility decrease. As a result, eventually no further progress under selection is possible, despite the presence of genetic variance for the character, because the selected individuals do not reproduce. The loss of fitness may be a direct phenotypic effect of the genes for the selected character, in which case nothing much can be done to improve the population further. Often, however, the loss of fitness is tied to linked sterility genes that are carried along with the selected loci. In such cases, a number of generations without selection allow recombinants to be formed, and selection can then be continued, as in the upwardly selected line in Figure 24-12.

We must be very careful in our interpretation of long-term agricultural selection programs. In the real world of agriculture, changes in cultivation methods, machinery, fertilizers, insecticides, herbicides, and so forth are occurring along with the production of genetically improved varieties. Increases in average yields are consequences of all of these changes. For example, the average yield of corn in the United States increased from 40 bushels to 80 bushels per acre between 1940 and 1970. But experiments comparing old and new varieties of corn in common environments show that only about half this increase is a direct result of new corn varieties (the other half being a result of improved farming techniques). Furthermore, the new varieties are superior to the old ones only at the high densities of modern planting for which they were selected.

Random events

If a population is finite in size (as all populations are) and if a given pair of parents has only a small number of offspring,

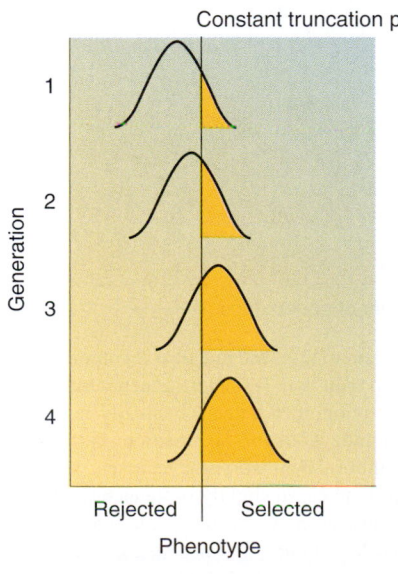

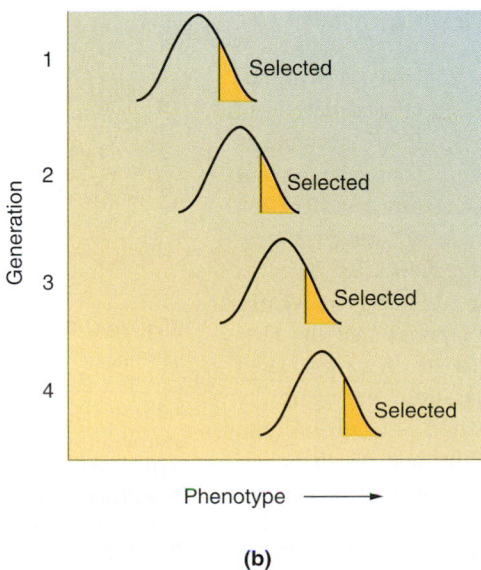

Figure 24-14 Two schemes of truncation selection for a continuously varying trait: (a) constant truncation; (b) proportional truncation.

then, even in the absence of all selective forces, the frequency of a gene will not be exactly reproduced in the next generation, because of sampling error. If, in a population of 1000 individuals, the frequency of *a* is 0.5 in one generation, then it may by chance be 0.493 or 0.505 in the next generation because of the chance production of slightly more or slightly fewer progeny of each genotype. In the second generation, there is another sampling error based on the new gene frequency, so the frequency of *a* may go from 0.505 to 0.511 or back to 0.498. This process of random fluctuation continues generation after generation, with no force pushing the frequency back to its initial state, because the population has no "genetic memory" of its state many generations ago. Each generation is an independent event. The final result of this random change in allelic frequency is that the population eventually drifts to $p = 1$ or $p = 0$. After this point, no further change is possible; the population has become homozygous. A different population, isolated from the first, also undergoes this **random genetic drift,** but it may become homozygous for allele *A,* whereas the first population has become homozygous for allele *a.* As time goes on, isolated populations diverge from each other, each losing heterozygosity. The variation originally present *within* populations now appears as variation *among* populations.

One form of genetic drift occurs when a small group breaks off from a larger population to found a new colony. This "acute drift," called the **founder effect,** results from a single generation of sampling, followed by several generations during which the population remains small. The founder effect is probably responsible for the virtually complete lack of blood group B in Native Americans, whose ancestors arrived in very small numbers across the Bering Strait at the end of the last Ice Age, about 20,000 years ago.

The process of genetic drift should sound familiar. It is, in fact, another way of looking at the inbreeding effect in small populations discussed earlier. Whether regarded as inbreeding or as random sampling of genes, the effect is the same. Populations do not exactly reproduce their genetic constitutions; there is a random component of gene frequency change.

One result of random sampling is that most new mutations, even if they are not selected against, never succeed in entering the population. Suppose that a single individual is heterozygous for a new mutation. There is some chance that the individual in question will have no offspring at all. Even if it has one offspring, there is a chance of $\frac{1}{2}$ that the new mutation will not be transmitted. If the individual has two offspring, the probability that neither offspring will carry the new mutation is $\frac{1}{4}$, and so forth. Suppose that the new mutation is successfully transmitted to an offspring. Then the lottery is repeated in the next generation, and again the allele may be lost. In fact, if a population is of size *N,* the chance that a new mutation is eventually lost by chance is $(2N - 1)/2N$. (For a derivation of this result, which is beyond the scope of this book, see Chapters 2 and 3 of Hartl and Clark, *Principles of Population Genetics.*) But, if the

new mutation is not lost, then the only thing that can happen to it in a finite population is that eventually it will sweep through the population and become fixed. This event has the probability of $\frac{1}{2N}$. In the absence of selection, then, the history of a population looks like Figure 24-15. For some period of time, it is homozygous; then a new mutation appears. In most cases, the new mutant allele will be lost immediately or very soon after it appears. Occasionally, however, a new mutant allele drifts through the population, and the population becomes homozygous for the new allele. The process then begins again.

Even a new mutation that is slightly favorable selectively will usually be lost in the first few generations after it appears in the population, a victim of genetic drift. If a new mutation has a selective advantage of *s* in the heterozygote in which it appears, then the chance is only $2s$ that the mutation will ever succeed in taking over the population. So a mutation that is 1 percent better in fitness than the standard allele in the population will be lost 98 percent of the time by genetic drift.

MESSAGE ···
New mutations can become established in a population even though they are not favored by natural selection simply by a process of random genetic drift. Even new favorable mutations are often lost.
··

Another consequence of the interaction of random and selective forces is that the effectiveness of the selective force in driving population composition depends on population size. The magnitude of the random effect is proportional to the reciprocal of population size, $1/N$, whereas the magnitude of a deterministic force depends on the migration rate, *m,* or mutation rate, μ, or selection coefficient, *s.* Thus

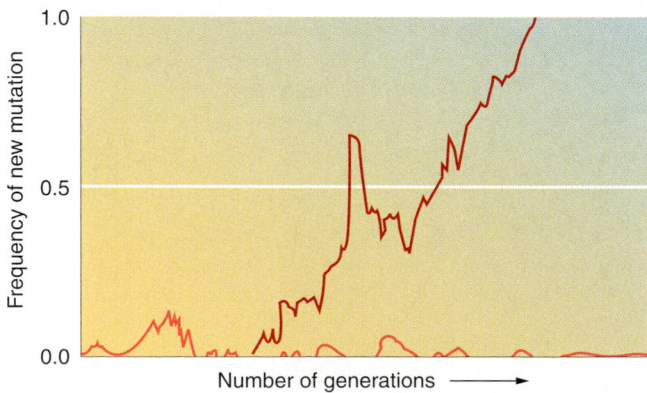

Figure 24-15   The appearance, loss, and eventual incorporation of new mutations in the life of a population. If random genetic drift does not cause the loss of a new mutation, then it must eventually cause the entire population to become homozygous for the mutation (in the absence of selection). In the figure, 10 mutations have arisen, of which 9 (light red at bottom of graph) increased slightly in frequency and then died out. Only the fourth mutation eventually spread into the population. (After J. Crow and M. Kimura, *An Introduction to the Population Genetics Theory.* Copyright © 1970 by Harper & Row.)

we can say, roughly, that migration and mutation are effective if

$$m \geq \frac{1}{N} \quad \text{or} \quad \mu > \frac{1}{N}$$

that is, if

$$Nm \geq 1 \quad \text{or} \quad N\mu > 1$$

The same is true of selection; selection is effective only if $Ns \geq 1$. When Ns is small because selection is weak or pop-

ulation size is small, then mutations are *effectively neutral,* even though there is some selection of them. Small populations will be less affected by selection than large populations even under otherwise identical conditions. For example, human populations were very small for nearly all the history of our species, having grown large only in the past few hundred generations. Thus, we may expect to find that many mutations that are now under selection were effectively neutral for a long time and may have reached high frequency by chance.

SUMMARY

The study of changes within a population, or population genetics, relates the heritable changes in populations or organisms to the underlying individual processes of inheritance and development. Population genetics is the study of inherited variation and its modification in time and space.

Identifiable inherited variation within a population can be studied by observing morphological differences between individuals, examining the differences in specific amino acid sequences of proteins, or even examining, most recently, the differences in nucleotide sequences within the DNA. These kinds of observations have led to the conclusion that there is considerable polymorphism at many loci within a population. A measure of this variation is the amount of heterozygosity in a population. Population studies have shown that, in general, the genetic differences between individuals within human races are much greater than the average differences between races.

The ultimate source of all variation is mutation. However, within a population, the quantitative frequency of specific genotypes can be changed by recombination, immigration of genes, continued mutational events, and chance.

One property of Mendelian segregation is that random mating results in an equilibrium distribution of genotypes after one generation. However, inbreeding is one process that converts genetic variation within a population into differences between populations by making each separate population homozygous for a randomly chosen allele. On the other hand, for most populations, a balance is reached for

any given environment among inbreeding, mutation from one allele to another, and immigration.

"Directed" changes of allelic frequencies within a population occur through the natural selection of a favored genotype. In many cases, such changes lead to homozygosity at a particular locus. On the other hand, the heterozygote may be more suited to a given environment than either of the homozygotes, leading to a balanced polymorphism.

Environmental selection of specific genotypes is rarely this simple, however. More often than not, phenotypes are determined by several interacting genes, and alleles at these different loci will be selected for at different rates. Furthermore, closely linked loci, unrelated to the phenotype in question, may have specific alleles carried along during the selection process. In general, genetic variation is the result of the interaction of evolutionary forces. For instance, a recessive, deleterious mutant may never be totally eliminated from a population, because mutation will continue to resupply it to the population. Immigration also can reintroduce the undesirable allele into the population. And, indeed, a deleterious allele may, under environmental conditions of which we are unaware (including the remaining genetic makeup of the individual), be selected for.

Unless alternative alleles are in intermediate frequencies, selection (especially against recessives) is very slow, requiring many generations. In many populations, especially those of small size, new mutations can become established even though they are not favored by natural selection, simply by a process of random genetic drift.

CONCEPT MAP

Draw a concept map interrelating as many of the following terms as possible. Note that the terms are listed in no particular order.

allelic frequency / heterozygosity / polymorphism / mutation / selection / Hardy-Weinberg equilibrium / immigration / inbreeding / genetic drift / Mendelian ratios

CHAPTER INTEGRATION PROBLEM

The polymorphisms for shell color (yellow or pink) and for the presence or absence of shell banding in the snail *Cepaea nemoralis* are each the result of a pair of segregating alleles

at a separate locus. Design an experimental program that would reveal the forces that determine the frequency and geographical distribution of these polymorphisms.

◆ Solution ◆

a. Describe the frequencies of the different morphs in a large number of populations covering the geographical and ecological range of the species. Each snail must be scored for *both* polymorphisms. At the same time, record a description of the habitat of each population. In addition, estimate the number of snails in a population.

b. Measure migration distances by marking a sample of snails with a spot of paint on the shell, replacing them in the population, and then resampling at a later date.

c. Raise broods from individual snails so that the genotype of male parents can be inferred and nonrandom mating patterns can be observed. The segregation frequencies *within* each family will reveal differences between genotypes in probability of survivorship of early developmental stages.

d. Seek further evidence of selection from (1) geographical clines in the frequencies of the alleles; (2) correlation between allele frequency and ecological variables, including population density; (3) correlation between the frequencies of the two different polymorphisms (are populations with, say, high frequencies of pink shells also characterized by, say, high frequencies of banded shells?); and (4) nonrandom associations *within* populations of the alleles at the two loci, indicating that certain combinations may have a higher fitness.

e. Seek evidence of the importance of random genetic drift by comparing the variation in allele frequencies among small populations with the variation among large populations. If small populations vary more from each other than large ones, then random drift is implicated.

SOLVED PROBLEMS

1. About 70 percent of all white North Americans can taste the chemical phenylthiocarbamide, and the remainder cannot. The ability to taste is determined by the dominant allele T, and the inability to taste is determined by the recessive allele t. If the population is assumed to be in Hardy-Weinberg equilibrium, what are the genotypic and allelic frequencies in this population?

◆ Solution ◆

Because 70 percent are tasters (T/T), 30 percent must be nontasters (t/t). This homozygous recessive frequency is equal to q^2; so, to obtain q, we simply take the square root of 0.30:

$$q = \sqrt{0.30} = 0.55$$

Because $p + q = 1$, we can write

$$p = 1 - q = 1 - 0.55 = 0.45$$

Now we can calculate

$$p^2 = 0.45^2 = 0.20(T/T)$$
$$2pq = 0 \times 0.45 \times 0.55 = 0.50(T/t)$$
$$q^2 = 0.3(t/t)$$

2. In a large natural population of *Mimulus guttatus*, one leaf was sampled from each of a large number of plants. The leaves were crushed and subjected to gel electrophoresis. The gel was then stained for a specific enzyme X. Six different banding patterns were observed, as shown in the diagram.

a. Assuming that these patterns are produced by a single locus, propose a genetic explanation for the six types.

b. How can you test your idea?

c. What are the allelic frequencies in this population?

d. Is the population in Hardy-Weinberg equilibrium?

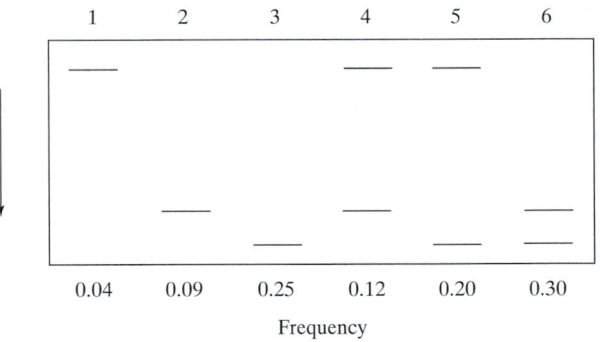

◆ Solution ◆

a. Inspection of the gel reveals that there are only three band positions: we will call them slow, intermediate, and fast. Furthermore, any individual can show either one band or two. The simplest explanation is that there are three alleles of one locus (let's call them A^S, A^I, and A^F) and that the individuals with two bands are heterozygotes. Hence, $1 = S/S$, $2 = I/I$, $3 = F/F$, $4 = S/I$, $5 = S/F$, and $6 = I/F$.

b. The hypothesis can be tested by making controlled crosses. For example, from a self of type 5, we can predict $\frac{1}{4}$ S/S, $\frac{1}{2}$ S/F, and $\frac{1}{4}$ F/F.

c. The frequencies can be calculated by a simple extension of the two-allele formulas. Hence:

$$f(S) = 0.04 + \tfrac{1}{2}(0.12) + \tfrac{1}{2}(0.20) = 0.20$$

$$f(I) = 0.09 + \tfrac{1}{2}(0.12) + \tfrac{1}{2}(0.30) = 0.30$$

$$f(F) = 0.25 + \tfrac{1}{2}(0.20) + \tfrac{1}{2}(0.30) = 0.50$$

d. The Hardy-Weinberg genotypic frequencies are:

$$(p + q + r)^2 = p^2 + q^2 + r^2 \\ + 2pq + 2pr + 2qr$$

$$= 0.04 + 0.09 + 0.25 + 0.12 \\ + 0.20 + 0.30$$

which are precisely the observed frequencies. So it appears that the population is in equilibrium.

3. In a large experimental *Drosophila* population, the fitness of a recessive phenotype is calculated to be 0.90, and the mutation rate to the recessive allele is 5×10^{-5}. If the population is allowed to come to equilibrium, what allelic frequencies can be predicted?

◆ Solution ◆

Here mutation and selection are working in opposite directions, so an equilibrium is predicted. Such an equilibrium is described by the formula

$$\hat{q} = \sqrt{\frac{\mu}{s}}$$

In the present question, $\mu = 5 \times 10^{-5}$ and $s = 1 - W = 1 - 0.9 = 0.1$. Hence

$$\hat{q} = \sqrt{\frac{5 \times 10^{-5}}{10^{-1}}} = 2.2 \times 10^{-2} = 0.022$$

$$\hat{p} = 1 - 0.022 = 0.978$$

PROBLEMS

1. What are the forces that can change the frequency of an allele in a population?

2. In a population of mice, there are two alleles of the A locus ($A1$ and $A2$). Tests showed that in this population there are 384 mice of genotype $A1/A1$, 210 of $A1/A2$, and 260 of $A2/A2$. What are the frequencies of the two alleles in the population?

3. In a randomly mating laboratory population of *Drosophila*, 4 percent of the flies have black bodies (black is the autosomal recessive b) and 96 percent have brown bodies (the normal color B). If this population is assumed to be in Hardy-Weinberg equilibrium, what are the allelic frequencies of B and b and the genotypic frequencies of B/B and B/b?

4. In a population, the $D \rightarrow d$ mutation rate is 4×10^{-6}. If $p = 0.8$ today, what will p be after 50,000 generations?

5. You are studying protein polymorphism in a natural population of a certain species of a sexually reproducing haploid organism. You isolate many strains from various parts of the test area and run extracts from each strain on electrophoretic gels. You stain the gels with a reagent specific for enzyme X and find that in the population there is a total of, say, five electrophoretic variants of enzyme X. You speculate that these variants represent various alleles of the structural gene for enzyme X.

a. How could you demonstrate that the speculation is correct, both genetically and biochemically? (You can make crosses, make diploids, run gels, test enzyme activities, test amino acid sequences, and so forth.) Outline the steps and conclusions precisely.

b. Name at least one other possible way of generating the different electrophoretic variants, and say how you would distinguish this possibility from the one described here.

6. A study made in 1958 in the mining town of Ashibetsu in the Hokkaido province of Japan revealed the frequencies of MN blood type genotypes (for individuals and for married couples) shown in the following table:

Genotype	Number of individuals or couples
Individuals	
L^M/L^M	406
L^M/L^N	744
L^N/L^N	332
Total	1482
Couples	
$L^M/L^M \times L^M/L^M$	58
$L^M/L^M \times L^M/L^N$	202
$L^M/L^N \times L^M/L^N$	190
$L^M/L^M \times L^N/L^N$	88
$L^M/L^N \times L^N/L^N$	162
$L^N/L^N \times L^N/L^N$	41
Total	741

a. Show if the population is in Hardy-Weinberg equilibrium with respect to the MN blood types.

b. Show whether mating is random with respect to MN blood types.

(Problem 6 is from J. Kuspira and G. W. Walker, *Genetics: Questions and Problems.* Copyright © 1973 by McGraw-Hill.)

7. Consider the populations that have the genotypes shown in the following table:

Population	A/A	A/a	a/a
1	1.0	0.0	0.0
2	0.0	1.0	0.0
3	0.0	0.0	1.0
4	0.50	0.25	0.25
5	0.25	0.25	0.50
6	0.25	0.50	0.25
7	0.33	0.33	0.33
8	0.04	0.32	0.64
9	0.64	0.32	0.04
10	0.986049	0.013902	0.000049

a. Which of the populations are in Hardy-Weinberg equilibrium?

b. What are p and q in each population?

c. In population 10, it is discovered that the $A \rightarrow a$ mutation rate is 5×10^{-6} and that reverse mutation is negligible. What must be the fitness of the a/a phenotype?

d. In population 6, the a allele is deleterious; furthermore, the A allele is incompletely dominant, so A/A is perfectly fit, A/a has a fitness of 0.8, and a/a has a fitness of 0.6. If there is no mutation, what will p and q be in the next generation?

8. Colorblindness results from a sex-linked recessive allele. One in every 10 males is colorblind.

a. What proportion of all women are colorblind?

b. By what factor is colorblindness more common in men (or, how many colorblind men are there for each colorblind woman)?

c. In what proportion of marriages would colorblindness affect half the children of each sex?

d. In what proportion of marriages would all children be normal?

e. In a population that is not in equilibrium, the frequency of the allele for colorblindness is 0.2 in women and 0.6 in men. After one generation of random mating, what proportion of the female progeny will be colorblind? What proportion of the male progeny?

f. What will the allelic frequencies be in the male and in the female progeny in part e?

(Problem 8 courtesy of Clayton Person.)

9. In a wild population of beetles of species X, you notice that there is a 3:1 ratio of shiny to dull wing covers. Does this ratio prove that *shiny* is dominant? (Assume that the two states are caused by the alleles of one gene.) If not, what does it prove? How would you elucidate the situation?

10. It seems clear that most new mutations are deleterious. Why?

11. Most mutations are recessive to the wild type. Of those rare mutations that are dominant in *Drosophila*, for example, the majority turn out either to be chromosomal aberrations or to be inseparable from chromosomal aberrations. Explain why the wild type is usually dominant.

12. Ten percent of the males of a large and randomly mating population are colorblind. A representative group of 1000 from this population migrates to a South Pacific island, where there are already 1000 inhabitants and where 30 percent of the males are colorblind. Assuming that Hardy-Weinberg equilibrium applies throughout (in the two original populations before emigration and in the mixed population immediately after immigration), what fraction of males and females can be expected to be colorblind in the generation immediately after the arrival of the immigrants?

13. Using pedigree diagrams, find the probability of homozygosity by descent of the offspring of (a) parent–offspring matings; (b) first-cousin matings; (c) aunt–nephew or uncle–niece matings.

14. In a survey of Native American tribes in Arizona and New Mexico, albinos were completely absent or very rare in most groups (there is one albino per 20,000 North American Caucasians). However, in three Native American populations, albino frequencies are exceptionally high: 1 per 277 Native Americans in Arizona; 1 per 140 Jemez in New Mexico; and 1 per 247 Zuni in New Mexico. All three of these populations are culturally but not linguistically related. What possible factors might explain the high incidence of albinos in these three tribes?

15. In an animal population, 20 percent of the individuals are A/A, 60 percent are A/a, and 20 percent are a/a. What are the allelic frequencies? In this population, mating is always with *like phenotype* but is random within phenotype. What genotypic and allelic frequencies will prevail in the next generation? Such *assortative mating* is common in animal populations. Another type of assortative mating is that which takes place only between *unlike* phenotypes: answer the preceding question with this restriction imposed. What will the end result be after many generations of mating of both types?

16. A *Drosophila* stock isolated from nature has an average of 36 abdominal bristles. By the selective breeding of only those flies with more bristles, the mean is raised to 56 in 20 generations. What is the source of this genetic flexibility? The 56-bristle stock is infertile, so selection is relaxed for several generations and the bristle number drops to about 45. Why doesn't it drop back to 36? When selection is reapplied, 56 bristles are soon attained, but this time the stock is *not* sterile. How can this situation arise?

17. The fitnesses of three genotypes are $W_{A/A} = 0.9$, $W_{A/a} = 1.0$, and $W_{a/a} = 0.7$.

    a. If the population starts at the allelic frequency $p = 0.5$, what is the value of p in the next generation?

    b. What is the predicted equilibrium allelic frequency?

18. A/A and A/a individuals are equally fertile. If 0.1 percent of the population is a/a, what selection pressure exists against a/a if the $A \rightarrow a$ mutation rate is 10^{-5}?

19. Gene B is a deleterious autosomal dominant. The frequency of affected individuals is 4.0×10^{-6}. The reproductive capacity of these individuals is about 30 percent that of normal individuals. Estimate μ, the rate at which b mutates to its deleterious allele B.

20. Of 31 children born of father–daughter matings, 6 died in infancy, 12 were very abnormal and died in childhood, and 13 were normal. From this information, calculate roughly how many recessive lethal genes we have, on average, in our human genomes. For example, if the answer is 1, then a daughter would stand a 50 percent chance of carrying the gene, and the probability of the union's producing a lethal combination would be $\frac{1}{2} \times \frac{1}{4} = \frac{1}{8}$. (So 1 is not the answer.) Consider also the possibility of undetected fatalities in utero in such matings. How would they affect your result?

21. If we define the **total selection cost** to a population of deleterious recessive genes as the loss of fitness per individual affected (s) multiplied by the frequency of affected individuals (q^2), then

$$\text{Genetic cost} = sq^2$$

    a. Suppose that a population is at equilibrium between mutation and selection for a deleterious recessive gene, where $s = 0.5$ and $\mu = 10^{-5}$. What is the equilibrium frequency of the gene? What is the genetic cost?

    b. Suppose that we start irradiating individual members of the population, so the mutation rate doubles. What is the new equilibrium frequency of the gene? What is the new genetic cost?

    c. If we do not change the mutation rate but we lower the selection intensity to $s = 0.3$ instead, what happens to the equilibrium frequency and the genetic cost?

25

QUANTITATIVE GENETICS

The composite flowers of *Gaillardia pulchella*.
Quantitative variation in flower color, flower diameter, and
number of flower parts. (*J. Heywood,* Journal of Heredity, *May/June 1986.*)

Key Concepts

In natural populations, variation in most characters takes
the form of a continuous phenotypic range rather than
discrete phenotypic classes. In other words, the variation is
quantitative, not qualitative.

Mendelian genetic analysis is extremely difficult to apply
to such continuous phenotypic distributions, so statistical
techniques are employed instead.

A major task of quantitative genetics is to determine the
ways in which genes interact with the environment to
contribute to the formation of a given quantitative trait
distribution.

The genetic variation underlying a continuous character
distribution can be the result of segregation at a single
genetic locus or at numerous interacting loci that produce
cumulative effects on the phenotype.

The estimated ratio of genetic to environmental variation is
not a measure of the relative contribution of genes and
environment to phenotype.

Estimates of genetic and environmental variance are
specific to the single population and the particular set of
environments in which the estimates are made.

Ultimately, the goal of genetics is the analysis of the genotype of organisms. But the genotype can be identified — and therefore studied — only through its phenotypic effect. We recognize two genotypes as different from each other because the phenotypes of their carriers are different. Basic genetic experiments, then, depend on the existence of a simple relation between genotype and phenotype. That is why studies of DNA sequences are so important, because we can read off the genotype directly from this most basic of all phenotypes. In general, we hope to find a uniquely distinguishable phenotype for each genotype and only a simple genotype for each phenotype. At worst, when one allele is completely dominant, it may be necessary to perform a simple genetic cross to distinguish the heterozygote from the homozygote. Where possible, geneticists avoid studying genes that have only partial penetrance and incomplete expressivity (see Chapter 4) because of the difficulty of making genetic inferences from such traits. Imagine how difficult (if not impossible) it would have been for Benzer to study the fine structure of the gene in phage, if the only effect of the *rII* mutants was a 5 percent reduction from wild type in their ability to grow on *E. coli* K. For the most part, then, the study of genetics presented in the preceding chapters has been the study of allelic substitutions that cause *qualitative* differences in phenotype.

However, the actual variation between organisms is usually quantitative, not qualitative. Wheat plants in a cultivated field or wild asters at the side of the road are not neatly sorted into categories of "tall" and "short," any more than humans are neatly sorted into categories of "black" and "white." Height, weight, shape, color, metabolic activity, reproductive rate, and behavior are characteristics that vary more or less continuously over a range (Figure 25-1). Even when the character is intrinsically countable (such as eye facet or bristle number in *Drosophila*), the number of distinguishable classes may be so large that the variation is nearly continuous. If we consider extreme individuals — say, a corn plant 8 feet tall and another one 3 feet tall — a cross between them will not produce a Mendelian result. Such a corn cross will produce plants about 6 feet tall, with some clear variation among siblings. The F_2 from selfing the F_1 will not fall into two or three discrete height classes in ratios of $3:1$ or $1:2:1$. Instead, the F_2 will be continuously distributed in height from one parental extreme to the other. This behavior of crosses is not an exception; it is the rule for most characters in most species. Mendel obtained his simple results because he worked with horticultural varieties of the garden pea that differed from one another by single allelic differences that had drastic phenotypic effects. Had Mendel conducted his experiments on the natural variation of the weeds in his garden, instead of abnormal pea varieties, he would never have discovered Mendel's laws. In general, size, shape, color, physiological activity, and behavior do not assort in a simple way in crosses.

The fact that most phenotypic characters vary continuously does not mean that their variation is the result of some genetic mechanisms different from the Mendelian genes with which we have been dealing. The continuity of phenotype is a result of two phenomena. First, each genotype does not have a single phenotypic expression but a norm of reaction (see Chapter 1) that covers a wide phenotypic range. As a result, the phenotypic differences between genotypic classes become blurred, and we are not able to assign a particular phenotype unambiguously to a particular genotype. Second, many segregating loci may have alleles that make a difference in the phenotype under observation. Suppose, for example, that five equally important loci affect the number of flowers that will develop in an annual plant and that each

Figure 25-1   Quantitative inheritance of bract color in Indian paintbrush (*Castilleja hispida*). The left photograph shows the extremes of the color range, and the right one shows examples from throughout the phenotypic range.

locus has two alleles (call them $+$ and $-$). For simplicity, also suppose that there is no dominance and that a $+$ allele adds one flower, whereas a $-$ allele adds nothing. Thus, there are $3^5 = 243$ different possible genotypes [three possible genotypes ($+/+$, $+/-$, and $-/-$) at each of five loci], ranging from

$$
\begin{array}{ccc}
++++ & & ++++ & & ---- \\
++++ & \text{through} & ---- & \text{to} & ---- \\
\end{array}
$$

but there are only 11 phenotypic classes (10, 9, 8, . . . , 0) because many of the genotypes will have the same numbers of $+$ and $-$ alleles. For example, although there is only one genotype with 10 $+$ alleles and therefore an average phenotypic value of 10, there are 51 different genotypes with 5 $+$ alleles and 5 $-$ alleles; for example,

$$
\begin{array}{ccc}
++++- & & ++-+- \\
+---- & \text{and} & ++--- \\
\end{array}
$$

Thus, many different genotypes may have the same average phenotype. At the same time, because of environmental variation, two individuals of the same genotype may not have the same phenotype. This lack of a one-to-one correspondence between genotype and phenotype obscures the underlying Mendelian mechanism. If we cannot study the behavior of the Mendelian factors controlling such traits directly, then what can we learn about their genetics?

Using current experimental techniques, geneticists can answer the following questions about the genetics of a continuously varying character in a population (say, height in a human population). These questions constitute the study of **quantitative genetics** — the study of the genetics of continuously varying characters:

1. Is the observed variation in the character influenced *at all* by genetic variation? Are there alleles segregating in the population that produce some differential effect on the character or is all the variation simply the result of environmental variation and developmental noise (see Chapter 1)?

2. If there is genetic variation, what are the norms of reaction of the various genotypes?

3. How important is genetic variation as a source of total phenotypic variation? Are the norms of reaction and the environments such that nearly all the variation is a consequence of environmental difference and developmental instabilities or does genetic variation predominate?

4. Do many loci (or only a few) vary with respect to the character? How are they distributed over the genome?

5. How do the different loci interact with one another to influence the character? Is there dominance, and is there any epistasis (interaction between genes at different loci)?

6. Is there any nonnuclear inheritance (for example, any maternal effect)?

In the end, the purpose of answering these questions is to be able to predict what kinds of offspring will be produced by crosses of different phenotypes.

The precision with which these questions can be framed and answered varies greatly. In experimental organisms, on the one hand, it is relatively simple to determine whether there is any genetic influence at all, but extremely laborious experiments are required to localize the genes (even approximately). In humans, on the other hand, it is extremely difficult to answer even the question of the presence of genetic influence for most traits, because it is almost impossible to separate environmental from genetic effects in an organism that cannot be manipulated experimentally. As a consequence, we know a relatively large amount about the genetics of bristle number in *Drosophila* but virtually nothing about the genetics of complex human traits; a few (such as skin color) clearly are influenced by genes, whereas others (such as the specific language spoken) clearly are not. The purpose of this chapter is to develop the basic statistical and genetic concepts needed to answer these questions and to provide some examples of the applications of these concepts to particular characters in particular species.

Some basic statistical notions

To consider the answers to these questions about the most common kinds of genetic variation, we must first examine a number of statistical tools that are essential in the study of quantitative genetics.

Distributions

The outcome of a cross for a Mendelian character can be described in terms of the proportions of the offspring that fall into several distinct phenotypic classes or often simply in terms of the presence or absence of a class. For example, a cross between a red-flowered plant and a white-flowered plant might be expected to yield all red-flowered plants or, if it were a backcross, $\frac{1}{2}$ red-flowered plants and $\frac{1}{2}$ white-flowered plants. However, we require a different mode of description for quantitative characters. The basic concept is that of the **statistical distribution.** If the heights of a large number of male undergraduates are measured to the nearest 5 centimeters (cm), they will vary (say, between 145 and 195 cm), but many more male undergraduates will be in the middle categories (say, between 170 and 180 cm) than at the extremes.

Representing each measurement class as a bar, with its height proportional to the number of individuals in each class, we can graph the result as shown in Figure 25-2a. Such a graph of numbers of individuals observed against measurement class is a **frequency histogram.** Now suppose that five times as many individuals are measured, each to the nearest centimeter. The classes in Figure 25-2a are now subdivided to produce a histogram like the one shown in Figure 25-2b. If we continue this process, refining the

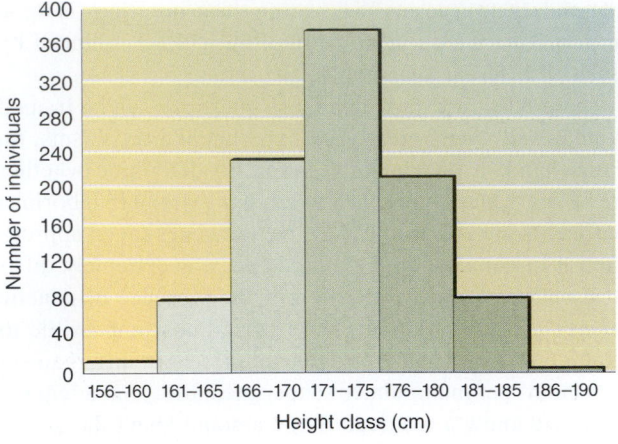

(a)

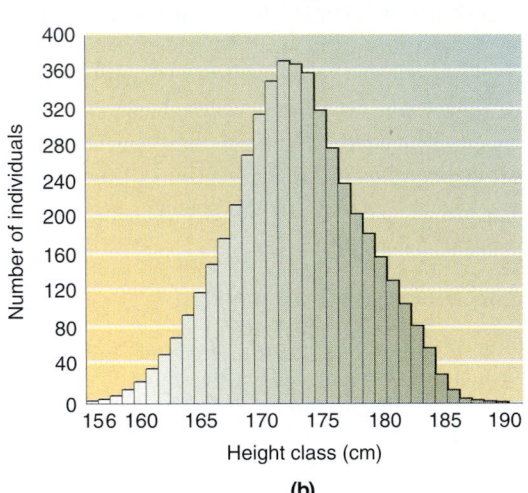

(b)

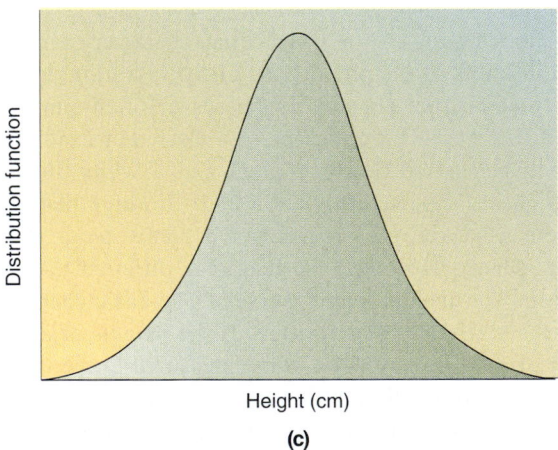

(c)

Figure 25-2   Frequency distributions for height of males: (a) a histogram with 5-cm class intervals; (b) a histogram with 1-cm class intervals; (c) the limiting continuous distribution.

measurement but proportionately increasing the number of individuals measured, then the histogram eventually takes on the continuous appearance of Figure 25-2c, which is the **distribution function** of heights in the population.

This continuous curve is an idealization, because no measurement can be taken with infinite accuracy or on an

unlimited number of individuals. Moreover, the measured variate itself may be intrinsically discontinuous because it is the count of some number of discrete objects such as eye facets or bristles. It is sometimes convenient, however, to develop concepts by using this slightly idealized picture as a shorthand for the more cumbersome observed frequency histogram (Figure 25-2a).

Statistical measures

Although a distribution contains all of the information about a set of measurements, we would like to be able to distill this information into a few characteristic numbers that convey the necessary information about the distribution without giving it in detail. The characteristics of a distribution that we would like to specify are as follows:

1. Where is the distribution located along the range of possible values? Are the observed values near 100 or near 1000? Therefore we need a measure of **central tendency.**

2. How much variation is there among individual measurements? Are they all concentrated around the central measurement or do they vary widely across a large range? That is, we need a measure of **dispersion.**

3. If we are considering more than one characteristic, how are the values of these different characteristics related? Is there any relation between body size and fertility? Is it positive or negative? Do larger parents have larger offspring? If so, we would regard this as evidence that genes influence body size. Thus, we need measures of **relation** between measurements.

Among the most commonly used measures of central tendency are the most frequent observation, the **mode,** and the arithmetic average of the observations, the **mean.** The dispersion of a distribution is almost always measured by the **variance,** which is the average squared distance of the observations from their mean. The relation between different variables is measured by their **correlation,** which is the average product of the deviation of one variable from its own mean times the deviation of the other variable from its own mean.

These most commonly used measures of central tendency, dispersion, and relation are considered in detail in the Statistical Appendix to this chapter. The detailed discussion is in a separate section in order not to interrupt the flow of logic of the consideration of quantitative genetics. It should not be assumed, however, that an understanding of those statistical concepts is somehow secondary. A proper understanding of quantitative genetics requires a grasp of that material.

Genotypes and phenotypic distribution

Using the concepts of distribution, mean, and variance, we can understand the difference between quantitative and

Mendelian genetic traits. Suppose that a population of plants contains three genotypes, each of which has some differential effect on growth rate. Furthermore, assume that there is some environmental variation from plant to plant because of inhomogeneity in the soil in which the population is growing and that there is some developmental noise (see Chapter 1). For each genotype, there will be a separate distribution of phenotypes with a mean and a standard deviation that depend on the genotype and the set of environments. Suppose that these distributions look like the three height distributions in Figure 25-3a. Finally, assume that the population consists of a mixture of the three genotypes but in the unequal proportions $1:2:3$ ($a/a:A/a:A/A$). Then the phenotypic distribution of individuals in the population as a whole will look like the black line in Figure 25-3b, which is the result of summing the three underlying separate genotypic distributions, weighted by their frequencies in the population. This weighting by frequency is indicated in Figure 25-3b by the different heights of the component distributions that add up to the total distribution. The mean of this total distribution is the average of the three genotypic means, again weighted by the frequencies of the genotypes in the population. The variance of the total distribution is produced partly by the environmental variation within each genotype and partly by the slightly different means of the three genotypes.

Two features of the total distribution are noteworthy. First, there is only a single mode. Despite the existence of three separate genotypic distributions underlying it, the population distribution as a whole does not reveal the separate modes. Second, any individual whose height lies between the two arrows could have come from any one of the three genotypes, because they overlap so much. The result is that we cannot carry out any simple Mendelian analysis to determine the genotype of an individual organism. For example, suppose that the three genotypes are the two homozygotes and the heterozygote for a pair of alleles at a locus. Let a/a be the short homozygote and A/A be the tall one, with the heterozygote being of intermediate height. Because the phenotypic distributions overlap so much, we cannot know to which genotype a given individual belongs. Conversely, if we cross a homozygote a/a and a heterozygote A/a, the offspring will not fall into two discrete classes in a $1:1$ ratio

but will cover almost the entire range of phenotypes smoothly. Thus, we cannot know that the cross is in fact $a/a \times A/a$ and not $a/a \times A/A$ or $A/a \times A/a$.

Suppose we grew the hypothetical plants in Figure 25-3 in an environment that exaggerated the differences between genotypes — for example, by doubling the growth rate of all genotypes. At the same time, we were very careful to provide all plants with exactly the same environment. Then, the phenotypic variance of each separate genotype would be reduced because all the plants were grown under identical conditions; at the same time, the differences between genotypes would be exaggerated by the more rapid growth. The result (Figure 25-4b) would be a separation of the population as a whole into three nonoverlapping phenotypic distributions, each characteristic of one genotype. We could now carry out a perfectly conventional Mendelian analysis of plant height. A "quantitative" character has been converted into a "qualitative" one. This conversion has been accomplished by finding a way to make the differences between the means of the genotypes large compared with the variation within genotypes.

MESSAGE ··

A quantitative character is one for which the average phenotypic differences between genotypes are small compared with the variation between individuals within genotypes.

··

It is sometimes assumed that continuous variation in a character is necessarily caused by a large number of segregating genes, so continuous variation is taken as evidence for control of the character by many genes. But, as we have just shown, this is not necessarily true. If the difference between genotypic means is small compared with the environmental variance, then even a simple one-gene–two-allele case can result in continuous phenotypic variation.

If the range of a character is limited and if many segregating loci influence it, then we expect the character to show continuous variation, because each allelic substitution must account for only a small difference in the trait. This **multiple-factor hypothesis** (that large numbers of genes, each with a small effect, are segregating to produce quantitative variation) has long been the basic model of quantitative

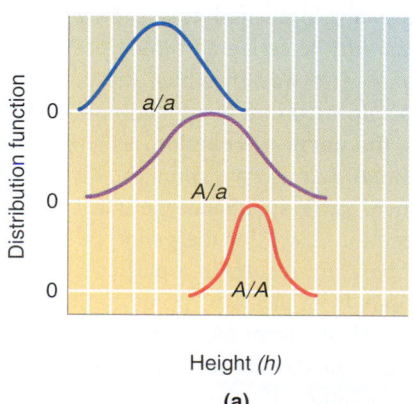

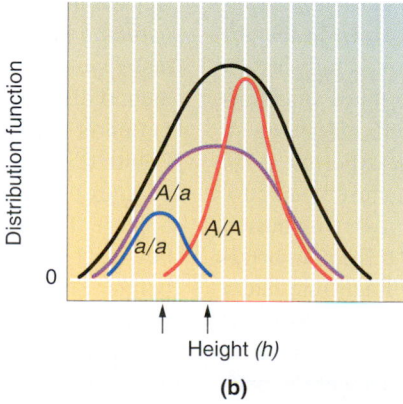

Figure 25-3 (a) Phenotypic distributions of three genotypes. (b) A population phenotypic distribution results from mixing individuals of the three genotypes in a proportion $1:2:3$ ($a/a:A/a:A/A$).

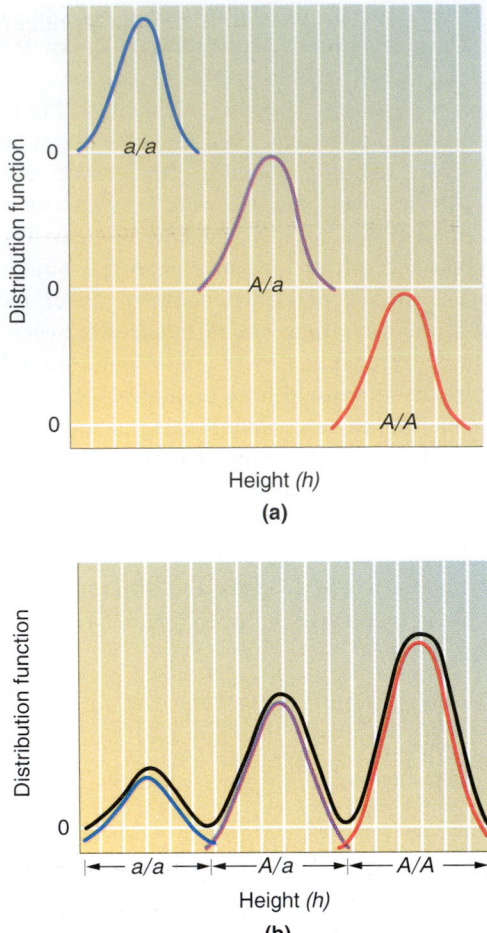

Figure 25-4   When the same genotypes as those in Figure 25-3 are grown in carefully controlled stress environments, the result is a smaller phenotypic variation in each genotype and a greater difference between genotypes. The heights of the individual distributions in part b are proportional to the frequencies of the genotypes in the mixture.

cally different (for example, five of the lines had seed weights of 0.450, 0.453, 0.454, 0.454, and 0.455 g), but let's take the most extreme position — that the lines *were* all different. These observations would be incompatible with a simple one-locus–two-allele model of gene action. In that case, if the original population were segregating for the two alleles *A* and *a*, all inbred lines derived from that population would have to fall into one of two classes: *A/A* or *a/a*. If, in contrast, there were, say, 100 loci, each of small effect, segregating in the original population, then a vast number of different inbred lines could be produced, each with a different combination of homozygotes at different loci.

However, we do not need such a large number of loci to obtain the result observed by Johannsen. If there were only five loci, each with three alleles, then $3^5 = 243$ different kinds of homozygotes could be produced from the inbreeding process. If we make 19 inbred lines at random, there is a good chance (about 50 percent) that each of the 19 lines will belong to a different one of the 243 classes. So Johannsen's experimental results can be easily explained by a relatively small number of genes. Thus, there is no real dividing line between polygenic traits and other traits. It is safe to say that no phenotypic trait above the level of the amino acid sequence in a polypeptide is influenced by only one gene. Moreover, traits influenced by many genes are not equally influenced by all of them. Some genes will have major effects on a trait; others, minor effects.

MESSAGE ·······································
The critical difference between Mendelian and quantitative traits is not the number of segregating loci but the size of phenotypic differences between genotypes compared with the individual variation within genotypic classes.

Norm of reaction and phenotypic distribution

The phenotypic distribution of a trait, as we have seen, is a function of the average differences between genotypes and of the variation between genotypically identical individuals. But both are in turn functions of the sequence of environments in which the organisms develop and live. For a given genotype, each environment will result in a given phenotype (for the moment, ignoring developmental noise). Then a *distribution of environments* will be reflected biologically as a *distribution of phenotypes*. The way in which the environmental distribution is transformed into the phenotypic distribution is determined by the **norm of reaction**, as shown in Figure 25-5. The horizontal axis is environment (say, temperature) and the vertical axis is phenotype (say, plant height). The norm of reaction curve for the genotype shows how each particular temperature results in a particular plant height. So, the dashed lines from the 18°C point on the temperature axis is reflected off the norm of reaction curve to a corresponding plant height on the vertical phenotype axis, and so forth for each temperature. If a large number of indi-

genetics, although there is no convincing evidence that such groups of genes really exist. A special name, **polygenes,** has been coined for these hypothetical factors of small-but-equal effect, in contrast to the genes of simple Mendelian analysis.

It is important to remember, however, that the *number* of segregating loci that influence a trait is not what separates quantitative and qualitative characters. Even in the absence of large environmental variation, it takes only a few genetically varying loci to produce variation that is indistinguishable from the effect of many loci of small effect. As an example, we can consider one of the earliest experiments in quantitative genetics, that of Wilhelm Johannsen on pure lines. By **inbreeding** (mating close relatives), Johannsen produced 19 homozygous lines of bean plants from an originally genetically heterogeneous population. Each line had a characteristic average seed weight ranging from 0.64 g per seed for the heaviest line to 0.35 g per seed for the lightest line. It is by no means clear that all these lines were geneti-

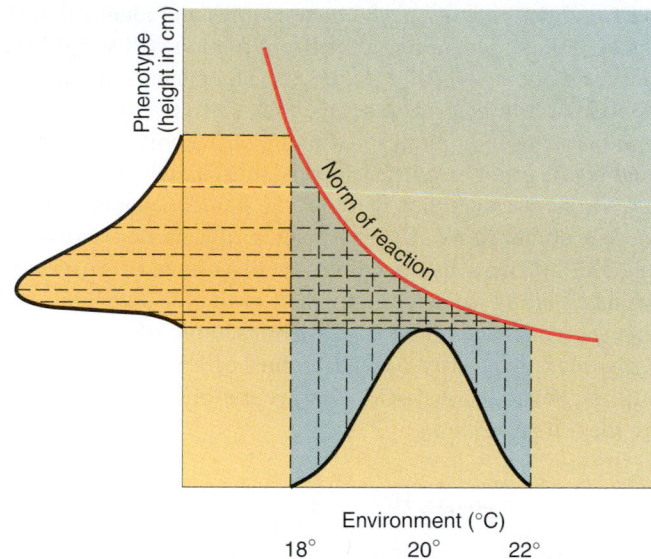

Figure 25-5   The distribution of environments on the horizontal axis is converted into the distribution of phenotypes on the vertical axis by the norm of reaction of a genotype.

viduals develop at, say, 20°C, then a large number of individuals will have the phenotype that corresponds to 20°C, as shown by the dashed line; and, if only small numbers develop at 18°C, few plants will have the corresponding plant height. Then the frequency distribution of developmental environments will be reflected as a frequency distribution of phenotypes as determined by the shape of the norm of reaction curve. It is as if an observer, standing at the vertical phenotype axis, were seeing the environmental distribution,

not directly, but reflected in the curved mirror of the norm of reaction. The shape of the curvature will determine how the environmental distribution is distorted on the phenotype axis. So, the norm of reaction in Figure 25-5 falls very rapidly at lower temperatures (the phenotype changes rapidly with small changes in temperature) but flattens out at higher temperatures, so the plant height is much less sensitive to temperature differences at the higher temperatures. The result is that the symmetric environmental distribution is converted into an asymmetric phenotype distribution with a long tail at the larger plant heights, corresponding to the lower temperatures.

By means of the same analysis, Figure 25-6 shows how a population consisting of two genotypes with different norms of reaction has a phenotypic distribution that depends on the distribution of environments. If the environments are distributed as shown by the black distribution curve, then the resulting population of plants will have a unimodal distribution, because the difference between genotypes is very small in this range of environments compared with the sensitivity of the norms of reaction to small changes in temperature. If the distribution of environments is shifted to the right, however, as shown by the gray distribution curve, a bimodal distribution of phenotypes results, because the norms of reaction are nearly flat in this environmental range but very different from each other.

MESSAGE ··

A distribution of environments is reflected biologically as a distribution of phenotypes. The transformation of environmental distribution into phenotypic distribution is determined by the norm of reaction.

··

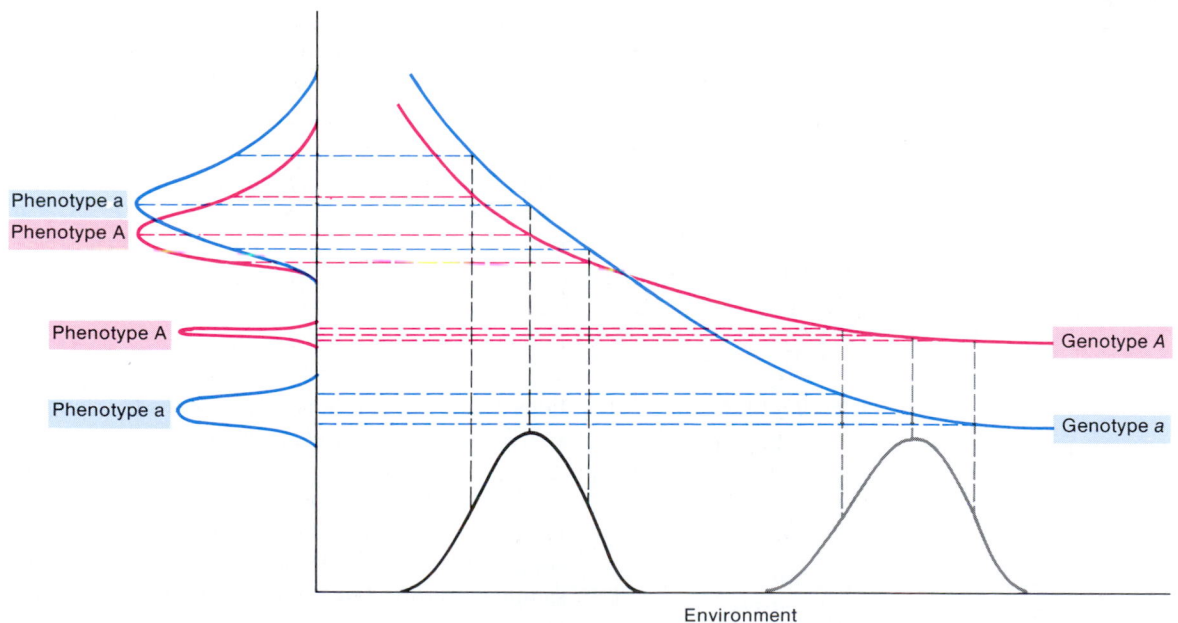

Figure 25-6   Two different environmental distributions are converted into different phenotypic distributions by two different genotypes.

Determining norms of reaction

Remarkably little is known about the norms of reaction for any quantitative traits in any species — partly because it is difficult in most sexually reproducing species to replicate a genotype so that it can be tested in different environments. For this reason, for example, we do not have a norm of reaction for any genotype for any human quantitative trait.

In domesticated plants and animals

A few norm of reaction studies have been carried out with plants that can be clonally propagated. The results of one of these experiments are presented in Chapter 1. It is possible to replicate genotypes in sexually reproducing organisms by the technique of mating close relatives, or inbreeding. By selfing (where possible) or by mating brother and sister repeatedly generation after generation, a **segregating line** (one that contains both homozygotes and heterozygotes at a locus) can be made homozygous.

The purpose of creating homozygous lines is to produce groups of organisms within which all individuals are genetically identical. These genetically identical individuals can then be allowed to develop in different environments to produce a norm of reaction. Alternatively, two different homozygous lines can be crossed and the F_1 offspring, all genetically identical with one another, can be characterized in different environments.

Ideally for a norm of reaction study all the individuals should be absolutely identical genetically, but the process of inbreeding increases the homozygosity of the group slowly, generation after generation, depending on the closeness of the relatives that are mated. In corn, for example, a single individual is chosen and self-pollinated. Then in the next generation, a single one of its offspring is chosen and self-pollinated. In the third generation, a single one of *its* offspring is chosen and self-pollinated, and so forth. Suppose that the original individual in the first generation is already a homozygote at some locus. Then all of its offspring from

self-pollination will also be homozygous and identical at the locus. Future generations of self-pollination will simply preserve the homozygosity. If, on the other hand, the original individual is a heterozygote, then the selfing $A/a \times A/a$ will produce $\frac{1}{4}$ A/A homozygotes and $\frac{1}{4}$ a/a homozygotes. If a single offspring is chosen in this subsequent generation to propagate the line, then there is a 50 percent chance that it is now a homozygote. If, by bad luck, the chosen individual should still be a heterozygote, there is another 50 percent chance that the selected individual in the third generation is homozygous, and so forth. Of the ensemble of all heterozygous loci, then, after one generation of selfing, only $\frac{1}{2}$ will still be heterozygous; after two generations, $\frac{1}{4}$; after three, $\frac{1}{8}$. In the nth generation,

$$\text{Het}_n = \frac{1}{2^n} \text{Het}_0$$

where Het_n is the proportion of heterozygous loci in the nth generation and Het_0 is the proportion in the 0 generation. When selfing is not possible, brother–sister mating will accomplish the same end, although more slowly. Table 25-1 is a comparison of the amount of heterozygosity left after n generations of selfing and brother–sister mating.

In natural populations

To carry out a norm of reaction study of a natural population, a large number of lines are sampled from the population and inbred for a sufficient number of generations to guarantee that each line is virtually homozygous at all its loci. Each line is then homozygous at each locus for a randomly selected allele present in the original population. The inbred lines themselves cannot be used to characterize norms of reaction in the natural population, because such totally homozygous genotypes do not exist in the original population. Each inbred line can be crossed to every other inbred line to produce heterozygotes that reconstitute the original population, and an arbitrary number of individuals from each cross can be produced. If inbred line 1 has the genetic constitution $A/A \cdot B/B \cdot c/c \cdot d/d \cdot E/E$ . . . and inbred line 2 is $a/a \cdot B/B \cdot C/C \cdot d/d \cdot e/e$ . . . , then a cross between them will produce a large number of offspring, all of whom are identically $A/a \cdot B/B \cdot C/c \cdot d/d \cdot E/e$ . . . and can be raised in different environments.

Inbreeding by mating of close relatives for many generations results in total homozygosity for the entire genome. In species such as *Drosophila* in which the necessary dominant markers and crossover suppressors are available, it is possible to produce lines that are homozygous for only a single chromosome, rather than for the whole set, as shown for an autosome in Figure 25-7. A single male from the population to be sampled is crossed to a female carrying a chromosome with a crossover suppressor C (usually a complex inversion), a recessive lethal l, and a dominant visible marker M_1 heterozygous with a second dominant visible M_2. In the F_1, a *single* male carrying the ClM_1 chromosome is chosen. This male,

25-1 TABLE	Heterozygosity Remaining After Various Generations of Inbreeding for Two Systems of Mating

	REMAINING HETEROZYGOSITY	
Generation	Selfing	Brother–sister mating
0	1.000	1.000
1	0.500	0.750
2	0.250	0.625
3	0.125	0.500
4	0.0625	0.406
5	0.03125	0.338
10	0.000977	0.114
20	1.05×10^{-6}	0.014
n	$\text{Het}_n = \frac{1}{2}\text{Het}_{n-1}$	$\text{Het}_n = \frac{1}{2}\text{Het}_{n-1} + \frac{1}{4}\text{Het}_{n-2}$

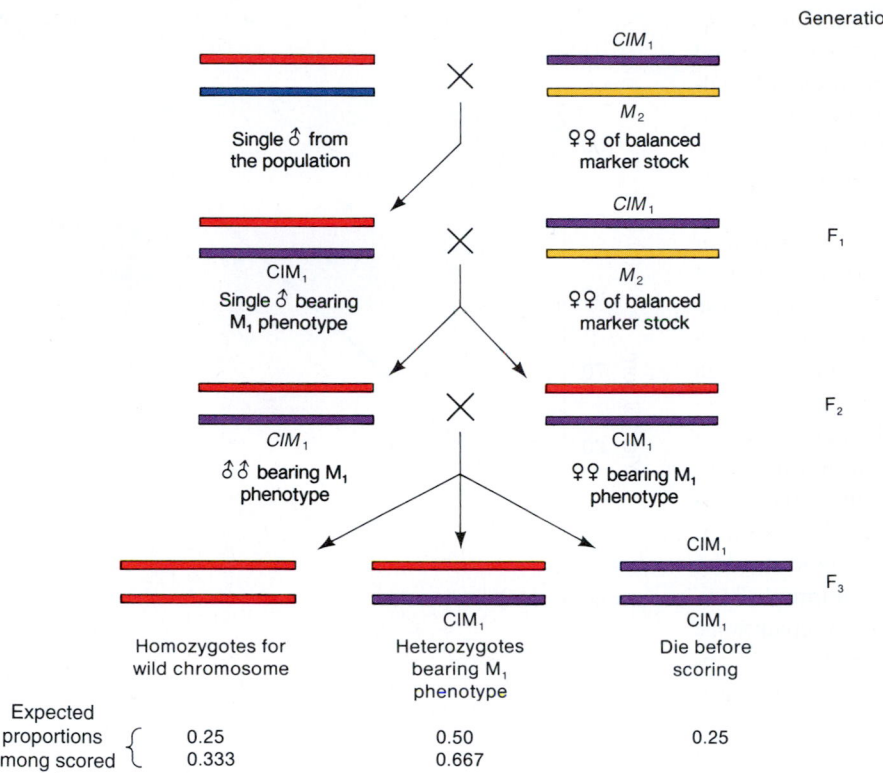

Generation

Single ♂ from the population

♀♀ of balanced marker stock

CIM_1

M_2

F_1

Single ♂ bearing M_1 phenotype

CIM_1

♀♀ of balanced marker stock

CIM_1

M_2

F_2

♂♂ bearing M_1 phenotype

CIM_1

♀♀ bearing M_1 phenotype

CIM_1

F_3

CIM_1

Homozygotes for wild chromosome

Heterozygotes bearing M_1 phenotype

CIM_1

Die before scoring

CIM_1

CIM_1

Expected proportions among scored adults.

| 0.25 | 0.50 | 0.25 |
| 0.333 | 0.667 | |

Figure 25-7   Method for making autosomes homozygous by using dominant marker genes M_1 and M_2, a crossover suppressor C, and a lethal gene l.

which is also carrying a wild-type chromosome from the population, is again crossed to the marker stock. In the F_2, all flies showing the M_1 trait but not M_2 are necessarily all heterozygotes for copies of the original wild-type chromosome because ClM_1/ClM_1 is lethal, and no crossovers have taken place. In the F_3, all wild-type flies are identically homozygous for the wild-type chromosome and are now available to make a stock for norm of reaction studies for crosses. (See Chapter 15 for another use of this technique.)

Results of norm of reaction studies

Very few norm of reaction studies have been carried out for quantitative characters found in natural populations, but many more have been carried out for domesticated species such as corn, which can be self-pollinated, or strawberries, which can be clonally propagated. The outcomes of such studies resemble those given in Figure 25-8, which shows the norms of reaction for abdominal bristle number as a function of temperature for second chromosome homozygotes of *D. pseudoobscura*. No genotype is consistently above or below the others. Instead, there are small differences between genotypes, and the direction of these differences is not consistent over a wide range of environments.

These factors have two important consequences. First, the selection of "superior" genotypes in domesticated animals and cultivated plants will result in very specifically adapted varieties that may not show their superior properties in other environments. To some extent, this problem is

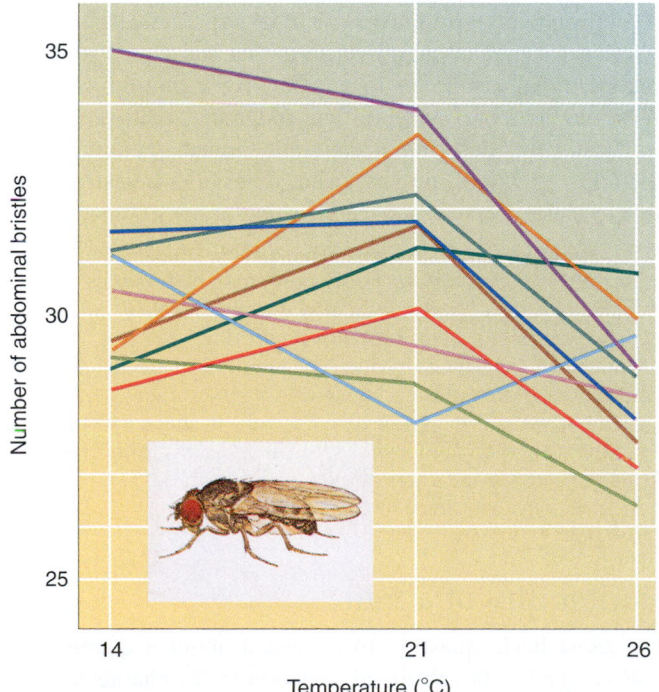

Figure 25-8   The number of abdominal bristles in different homozygous genotypes of *Drosophila pseudoobscura* at three different temperatures. (Data courtesy of A. P. Gupta.) (Image: Plate IV, University of Texas Publication 4313, *Studies in the Genetics of Drosophila III: The Drosophilidae of the Southwest,* by J. T. Patterson. Courtesy of the Life Sciences Library, University of Texas, Austin.)

overcome by deliberately testing genotypes in a range of environments (for example, over several years and in several locations). It would be even better, however, if plant breeders could test their selections in a variety of controlled environments in which different environmental factors could be separately manipulated. The consequences of actual plant-breeding practices can be seen in Figure 25-9, in which the yields of two varieties of corn are shown as a function of different farm environments. Variety 1 is an older variety of hybrid corn; variety 2 is a later "improved" hybrid. These performances are compared at a low planting density, which prevailed when variety 1 was developed, and at a high planting density characteristic of farming practice when hybrid 2 was selected. At the high density, the new variety is clearly superior to the old variety in all environments (Figure 25-9a). At the low density, however, the situation is quite different. First, note that the new variety is less sensitive to environment than is the older hybrid, as evidenced by its flatter norm of reaction. Second, the new "improved" variety is actually poorer under the best farm conditions. Third, the yield improvement of the new variety is not apparent under the low densities characteristic of earlier agricultural practice.

The second consequence of the nature of reaction norms is that, even if it should turn out that there is genetic variation for various mental and emotional traits in the human species, which is by no means clear, this variation is unlikely to favor one genotype over another across a range of environments. We must beware of hypothetical norms of reaction for human cognitive traits that show one genotype unconditionally superior to another. Even putting aside all questions of moral and political judgment, there is simply no basis for describing different human genotypes as "better" or "worse" on any scale, unless the investigator is able to make a very exact specification of environment.

MESSAGE ···

Norm of reaction studies show only small differences between natural genotypes, and these differences are not consistent over a wide range of environments. Thus, "superior" genotypes in domesticated animals and cultivated plants may be superior only in certain environments. If it should turn out that humans exhibit genetic variation for various mental and emotional traits, this variation is unlikely to favor one genotype over another across a range of environments.

Heritability of a trait

The most basic question to be asked about a quantitative trait is whether the observed variation in the character is influenced by genes at all. It is important to note that this is not the same as asking whether genes play any role in the character's development. Gene-mediated developmental processes lie at the base of every character, but *variation* from individual to individual is not necessarily the result of *genetic variation*. Thus, the possibility of speaking any lan-

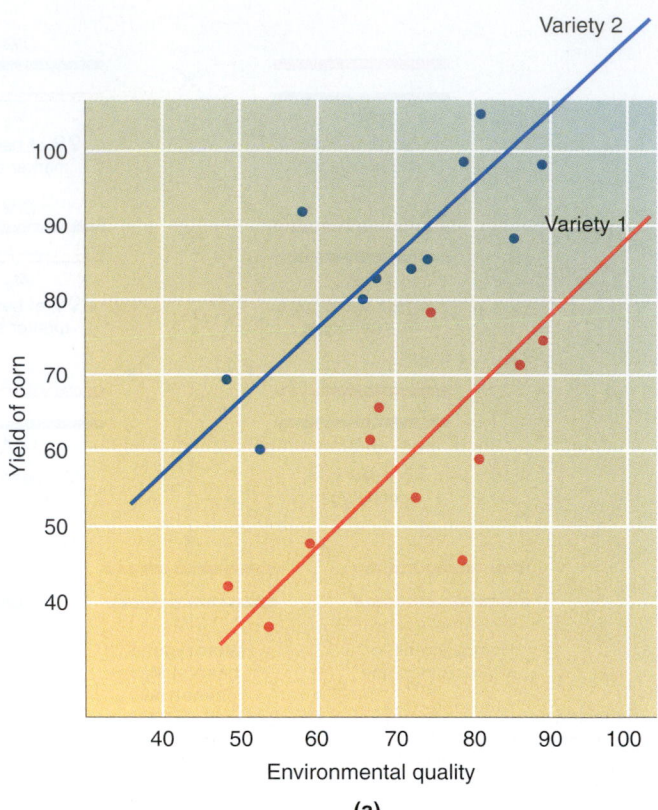

(a)

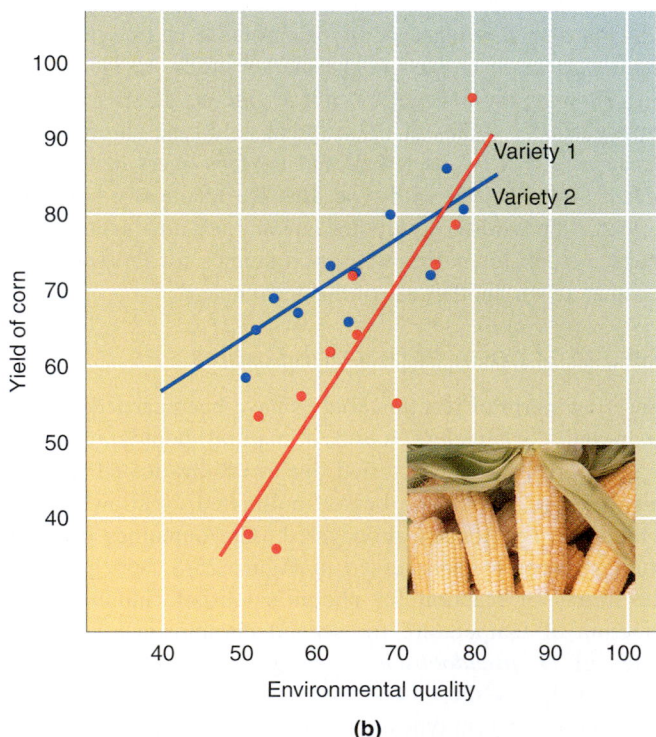

(b)

Figure 25-9   Yields of grain of two varieties of corn in different environments: (a) at a high planting density; (b) at a low planting density. (Data courtesy of W. A. Russell, *Proceedings of the 29th Annual Corn and Sorghum Research Conference,* 1974. Photograph © Bonnie Sue/Photo Researchers.)

guage at all depends critically on the structures of the central nervous system as well as of the vocal cords, tongue, mouth, and ears, which depend in turn on the nature of the human genome. There is no environment in which cows will speak. But, although the particular language that is spoken by humans varies from nation to nation, that variation is totally nongenetic.

MESSAGE ··
> The question of whether a trait is heritable is a question about the role that differences in genes play in the phenotypic differences between individuals or groups.

Familiality and heritability

In principle, it is easy to determine whether any genetic variation influences the phenotypic variation among organisms for a particular trait. If genes are involved, then (on average) biological relatives should resemble one another more than unrelated individuals do. This resemblance would be seen as a positive correlation between parents and offspring or between siblings (offspring of the same parents). Parents who are larger than the average would have offspring who are larger than the average; the more seeds that a plant produces, the more seeds that its siblings would produce. Such correlations between relatives, however, are evidence for genetic variation only if the relatives do not share common environments *more than nonrelatives do.* It is absolutely fundamental to distinguish *familiality* from *heritability.* Traits are **familial** if members of the same family share them, for whatever reason. Traits are **heritable** only if the similarity arises from shared genotypes.

There are two general methods for establishing the heritability of a trait as distinct from its familial occurrence. The first depends on *phenotypic similarity* between relatives. For most of the history of genetics, this method has been the only one available; so nearly all the evidence about heritability for most traits in experimental organisms and in humans has been established by using this approach. The second method, using *marker-gene segregation,* depends on showing that genotypes carrying different alleles of marker genes also differ in their average phenotype for the quantitative character. If the marker genes (which have nothing to do with the character under study) are seen to vary in relation to the character, presumably they are linked to genes that *do* influence the character and its variation. Thus, heritability is demonstrated even if the actual genes causing the variation are not known. This method requires that the genome of the organism being studied have large numbers of detectable genetically variable marker loci spread throughout the genome. Such marker loci can be observed from electrophoretic studies of protein variation or, in vertebrates, from immunological studies of blood group genes. For example, within flocks, chickens of different blood groups show some difference in egg weight.

Since the introduction of molecular methods for the study of DNA sequence variation, very large numbers of variable nucleotide positions have been discovered in a great variety of organisms. This molecular variation includes both single nucleotide replacements and insertions and deletions of longer nucleotide sequences. These variations are usually detected by the gain or loss of sites of cleavage of restriction enzymes or by length variation of DNA sequences between two fixed restriction sites, both of which are a form of restriction fragment length polymorphisms (RFLPs). In tomatoes, for example, strains carrying different RFLP variants differ in fruit characteristics.

However, because so much of what is known or claimed about heritability still depends on phenotypic similarity between relatives, especially in human genetics, we will begin the examination of the problem of heritability by analyzing phenotypic similarity.

Phenotypic similarity between relatives

In experimental organisms, there is no problem in separating environmental from genetic similarities. The offspring of a cow producing milk at a high rate and the offspring of a cow producing milk at a low rate can be raised together in the same environment to see whether, despite the environmental similarity, each resembles its own parent. In natural populations, and especially in humans, this is difficult to do. Because of the nature of human societies, members of the same family not only share genes, but also have similar environments. Thus, the observation of simple familiality of a trait is genetically uninterpretable. In general, people who speak Hungarian have Hungarian-speaking parents and people who speak Japanese have Japanese-speaking parents. Yet the massive experience of immigration to North America has demonstrated that these linguistic differences, although familial, are nongenetic. The highest correlations between parents and offspring for any social traits in the United States are those for political party and religious sect, but they are not heritable. The distinction between familiality and heredity is not always so obvious. The Public Health Commission, which originally studied the vitamin-deficiency disease pellagra in the southern United States in 1910, came to the conclusion that it was genetic because it ran in families.

To determine whether a trait is heritable in human populations, we must use adoption studies to avoid the usual environmental similarity between biological relatives. The ideal experimental subjects are identical twins reared apart, because they are genetically identical but environmentally different. Such adoption studies must be so contrived that there is no correlation between the social environment of the adopting family and that of the biological family. These requirements are exceedingly difficult to meet; so, in practice, we know very little about whether human quantitative traits that are familial are also heritable. Skin color is clearly heritable, as is adult height — but even for these traits we must be very careful. We know that skin color is affected by genes from studies of cross-racial adoptions and observations that the offspring of black African slaves were black even when they were born and reared in Canada. But are the differences in height between Japanese and Europeans affected by genes? The children of Japanese immigrants who are born and reared in North America are taller than their parents but shorter than the North American average, so we might conclude that there is some influence of genetic difference. However, second-generation Japanese Americans are even taller than their American-born parents. It appears that some environmental–cultural influence or perhaps a maternal effect is still felt in the first generation of births in North America. We cannot yet say whether genetic differences in height distinguish North Americans of, say, Japanese and Swedish ancestry.

Personality traits, temperament, and cognitive performance (including IQ scores), as well as a whole variety of behaviors such as alcoholism and of mental disorders such as schizophrenia, have been the subject of heritability studies in human populations. Many show familiality. There is indeed a positive correlation between the IQ scores of parents and the scores of their children (the correlation is about 0.5 in white American families), but the correlation does not distinguish familiality from heritability. To make that distinction requires that the environmental correlation between parents and children be broken, so adoption studies are common. Because it is difficult to randomize the environments, even in cases of adoption, evidence of heritability for hu-

man personality and behavior traits remains equivocal despite the very large number of studies that exist. Prejudices about the causes of human differences are widespread and deep, and, as a result, the canons of evidence adhered to in studies of the heritability of IQ, for example, have been much more lax than in studies of milk yield in cows.

Figure 25-10 summarizes the usual method for testing heritability in experimental organisms. Individuals from both extremes of the distribution are mated with their own kind, and the offspring are raised in a common controlled environment. If there is an average difference between the two offspring groups, the trait is heritable. Most morphological traits in *Drosophila*, for example, turn out to be heritable — but not all of them. If flies with right wings that are slightly longer than their left wings are mated together, their offspring have no greater tendency to be "right winged" than do the offspring of "left winged" flies. As we shall see later, this method can also be used to obtain quantitative information about heritability.

MESSAGE ··

In experimental organisms, environmental similarity can often be readily distinguished from genetic similarity (heritability). In humans, however, it is very difficult to determine whether a particular trait is heritable.

···

Quantifying heritability

If a trait is shown to have some heritability in a population, then it is possible to quantify the degree of heritability. In Figure 25-3, we saw that the variation between phenotypes in a population arises from two sources. First, there are average differences between the genotypes; second, each genotype exhibits phenotypic variance because of environmental variation. The total **phenotypic variance** of the population (s_p^2) can then be broken into two parts: the variance between genotypic means (s_g^2) and the remaining variance (s_e^2). The former is called the **genetic variance,** and the latter is called the **environmental variance;** however, as we shall see, these names are quite misleading. Moreover, the breakdown of the phenotypic variance into the sum of environmental and genetic variance leaves out the possibility of some covariance between genotype and environment. For example, suppose it were true (we do not know) that there are genes that influence musical ability. Parents with such genes might themselves be musicians, who would create a more musical environment for their children, who would then have both the genes and the environment promoting musical performance. The result would be an increase in the phenotypic variances of musical ability and an erroneous estimate of genetic and environmental variances. If the phenotype is the sum of a genetic and an environmental effect, $P = G + E$, then, as explained on page 768 of the Statistical Appendix, the variance of the phenotype is the sum of the genetic variance, the environmental variance, and twice the covariance between the genotypic and environmental effects.

$$s_p^2 = s_g^2 + s_e^2 + 2 \text{ cov } ge$$

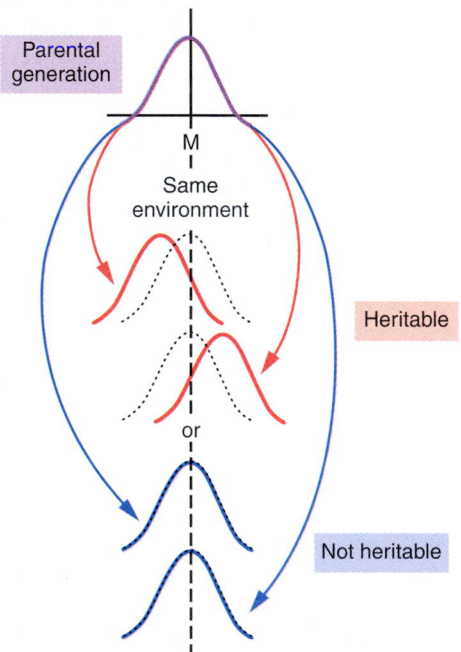

Figure 25-10   Standard method for testing heritability in experimental organisms. Crosses are performed within two populations of individuals selected from the extremes of the phenotypic distribution in the parental generation. If the phenotypic distributions of the two groups of offspring are significantly different from each other, then the trait is heritable. If both offspring distributions resemble the distribution for the parental generation, then the trait is not heritable.

If genotypes are not distributed randomly across environments, there will be some covariance between genotype and environmental values, and the covariance will be hidden in the genetic and environmental variances.

The degree of heritability can be defined as the part of the total variance that is due to genetic variance:

$$H^2 = \frac{s_g^2}{s_p^2} = \frac{s_g^2}{s_g^2 + s_e^2}$$

H^2, so defined, is called the **broad heritability** of the character.

It must be stressed that this measure of "genetic influence" tells us what part of the population's *variation* in phenotype can be assigned to *variation* in genotype. It does not tell us what parts of an *individual's* phenotype can be ascribed to its heredity and to its environment. This latter distinction is not a reasonable one. An individual's phenotype is a consequence of the interaction between its genes and its sequence of environments. It clearly would be silly to say that you owe 60 inches of your height to genes and 10 inches to environment. All measures of the "importance" of genes are framed in terms of the proportion of variance ascribable to their variation. This approach is a special application of the more general technique of the **analysis of variance** for apportioning relative weight to contributing causes. The method was, in fact, invented originally to deal with experiments in which different environmental and genetic factors were influencing the growth of plants. (For a sophisticated but accessible treatment of the analysis of variance written for biologists, see R. Sokal and J. Rohlf, *Biometry*, 3d ed. W. H. Freeman and Company, 1995.)

Methods of estimating H^2

Genetic variance and heritability can be estimated in several ways. Most directly, we can obtain an estimate of s_e^2 by making a number of homozygous lines from the population, crossing them in pairs to reconstitute individual heterozygotes, and measuring the phenotypic variance *within* each heterozygous genotype. Because there is no genetic variance within a genotypic class, these variances will (when averaged) provide an estimate of s_e^2. This value can then be subtracted from the value of s_p^2 in the original population to give s_g^2. With the use of this method, any covariance between genotype and environment in the original population will be hidden in the estimate of genetic variance and will inflate it.

Other estimates of genetic variance can be obtained by considering the genetic similarities between relatives. Using simple Mendelian principles, we can see that half the genes of full siblings will (on average) be identical. For identification purposes, we can label the alleles at a locus carried by the parents differently, so that they are, say, A_1/A_2 and A_3/A_4. Now the older sibling has a probability of $\frac{1}{2}$ of getting A_1 from its father, as does the younger sibling, so the two siblings have a chance of $\frac{1}{2} \times \frac{1}{2} = \frac{1}{4}$ of both carrying A_1. On the other hand, they might both have received an A_2 from their father; so, again, they have a probability of $\frac{1}{4}$ of carrying a gene in common that they inherited from their father. Thus, the chance is $\frac{1}{4} + \frac{1}{4} = \frac{1}{2}$ that both siblings will carry an A_1 or that both siblings will carry an A_2. The other half of the time, one sibling will inherit an A_1 and the other will inherit an A_2. So, as far as paternally inherited genes are concerned, full siblings have a 50 percent chance of carrying the same allele. But the same reasoning applies to their maternally inherited gene. Averaging over their paternally and maternally inherited genes, half the genes of full siblings are identical between them. Their **genetic correlation,** which is equal to the chance that they carry the same allele, is $\frac{1}{2}$.

If we apply this reasoning to half-siblings, say, with a common father but with different mothers, we get a different result. Again, the two siblings have a 50 percent chance of inheriting an identical gene from their father, but this time they have no way of inheriting the same gene from their mothers because they have two different mothers. Averaging the maternally inherited and paternally inherited genes thus gives a probability of $(\frac{1}{2} + 0)/2 = \frac{1}{4}$ that these half-siblings will carry the same gene.

We might be tempted to use the theoretical correlation between, say, siblings to estimate H^2. If the observed phenotypic correlation were, for example, 0.4 and we expect on purely genetic grounds a correlation of .05, then an estimate of heritability would be $0.4/0.5 = 0.8$. But such an estimate fails to take into account the fact that siblings may also be environmentally correlated. Unless we are careful to raise the siblings in independent environments, the estimate of H^2 would be too large and could even exceed 1 if the observed phenotypic correlation were greater than 0.5. To get around this problem, we use the *differences* between phenotypic correlations of different relatives. For example, the difference in genetic correlation between full and half-siblings is $\frac{1}{2} - \frac{1}{4} = \frac{1}{4}$. Let's contrast this with their **phenotypic correlations.** If the environmental similarity is the same for half- and full siblings — a very important condition for estimating heritability — then environmental similarities will cancel out if we take the difference in correlation between the two kinds of siblings. This difference in phenotypic correlation will then be proportional to how much of the variance is genetic. Thus:

$$\left(\begin{array}{c}\text{genetic correlation}\\\text{of full siblings}\end{array}\right) - \left(\begin{array}{c}\text{genetic correlation}\\\text{of half-siblings}\end{array}\right) = \frac{1}{4}$$

but

$$\left(\begin{array}{c}\text{phenotypic}\\\text{correlation}\\\text{of full siblings}\end{array}\right) - \left(\begin{array}{c}\text{phenotypic}\\\text{correlation}\\\text{of half-siblings}\end{array}\right) = H^2 \times \frac{1}{4}$$

so an estimate of H^2 is:

$$H^2 = 4\left[\left(\begin{array}{c}\text{correlation}\\\text{of full siblings}\end{array}\right) - \left(\begin{array}{c}\text{correlation}\\\text{of half-siblings}\end{array}\right)\right]$$

where the correlation here is the *phenotypic* correlation.

We can use similar arguments about genetic similarities between parents and offspring and between twins to obtain two other estimates of H^2:

$$H^2 = 4 \left(\begin{array}{c} \text{correlation} \\ \text{of full siblings} \end{array} \right) - 2 \left(\begin{array}{c} \text{parent–offspring} \\ \text{correlation} \end{array} \right)$$

and

$$H^2 = 2 \left[\left(\begin{array}{c} \text{correlation of} \\ \text{monozygotic twins} \end{array} \right) - \left(\begin{array}{c} \text{correlation of} \\ \text{dizygotic twins} \end{array} \right) \right]$$

These formulas are derived from considering the genetic similarities between relatives. They are only approximate and depend on assumptions about the ways in which genes act. The first two formulas, for example, assume that genes at different loci add together in their effect on the character. The last formula also assumes that the alleles at each locus show no dominance (see the discussion of components of variance on pages 760–762).

All these estimates, as well as others based on correlations between relatives, depend *critically* on the assumption that environmental correlations between individuals are the same for all degrees of relationship. If closer relatives have more similar environments, as they do in humans, the estimates of heritability are biased. It is reasonable to assume that most environmental correlations between relatives are positive, in which case the heritabilities would be overestimated. Negative environmental correlations also can exist. For example, if the members of a litter must compete for food that is in short supply, there could be negative correlations in growth rates among siblings.

The difference in correlation between monozygotic and dizygotic twins is commonly used in human genetics to estimate H^2 for cognitive or personality traits. Here the problem of degree of environmental similarity is very severe. Identical (monozygotic) twins are generally treated more similarly to each other than are fraternal (dizygotic) twins. People often give their identical twins names that are similar, dress them alike, treat them identically, and, in general, accent their similarities. As a result, heritability is overestimated.

Meaning of H^2

Attention to the problems of estimating broad heritability distracts from the deeper questions about the meaning of the ratio when it can be estimated. Despite its widespread use as a measure of how "important" genes are in influencing a trait, H^2 actually has a special and limited meaning.

There are two conclusions that can be drawn from a properly designed heritability study. First, if there is a nonzero heritability, then, in the population measured and in the environments in which the organisms have developed, genetic differences have influenced the variation between individuals, so genetic differences do matter to the trait. This finding is not trivial and is a first step in a more detailed investigation of the role of genes. It is important to notice that the reverse is not true. Finding no heritability for the trait is not a demonstration that genes are irrelevant; rather, it demonstrates that, in the particular population studied, there is no genetic variation at the relevant loci or that the environments in which the population developed were such that different genotypes had the same phenotype. In other populations or other environments, the trait might be heritable.

MESSAGE ··
In general, the heritability of a trait is different in each population and in each set of environments; it cannot be extrapolated from one population and set of environments to another.
··

Moreover, we must distinguish between *genes* being relevant to a trait and *genetic differences* being relevant to *differences* in the trait. The experiment of immigration to North America has proved that the ability to pronounce the sounds of North American English, rather than French, Swedish, or Russian, is not a consequence of genetic differences between our immigrant ancestors. But, without the appropriate genes, we could not speak any language at all.

Second, the value of the H^2 provides a limited prediction of the effect of environmental modification under particular circumstances. If all the relevant environmental variation is eliminated *and the new constant environment is the same as the mean environment in the original population,* then H^2 estimates how much phenotypic variation will still be present. So, if the heritability of performance on an IQ test were, say, 0.4, then, if all children had the same developmental and social environment as the "average child," about 60 percent of the variation in IQ test performance would disappear and 40 percent would remain.

The requirement that the new constant environment be at the mean of the old environmental distribution is absolutely essential to this prediction. If the environment is shifted toward one end or the other of the environmental distribution or a new environment is introduced, nothing at all can be predicted. In the example of IQ performance, the heritability gives us no information at all about how variable performance would be if children's developmental and social environments were generally enriched. To understand why this is so, we must return to the concept of the norm of reaction.

The separation of variance into genetic and environmental components s_g^2 and s_e^2 does not really separate the genetic and environmental causes of variation. Consider Figure 25-9b. When the environment is poor (50), corn variety 2 is much higher yielding than variety 1, so a population made up of a mixture of the two varieties would have a lot of genetic variance for yield. But, in an environment scoring 80, there is no difference in yield between genotypes 1 and 2, so a mixed population would have no genetic variance at all for yield in that environment. Thus, *genetic* variance has been changed by changing the *environment*. On the other hand, variety 2 is less sensitive to environment than variety 1, as shown by the slopes of the lines. So a population made up mostly of genotype 2 would have a lower environmental variance than one made up mostly of genotype 1. So, *environmental* variance in the population is changed by changing the proportion of *genotypes*.

MESSAGE ··
Because genotype and environment interact to produce phenotype, no partition of variation can actually separate causes of variation.

As a consequence of the argument just given, knowledge of the heritability of a trait does not permit us to predict how the distribution of that trait will change if either genotypic frequencies or environmental factors change markedly.

MESSAGE ··
A high heritability does not mean that a trait is unaffected by its environment.

All that high heritability means is that, for the particular population developing in the particular distribution of environments in which the heritability was measured, average differences between genotypes are large compared with environmental variation within genotypes. If the environment is changed, there may be large differences in phenotype.

Perhaps the most well known example of the erroneous use of heritability arguments to make claims about the changeability of a trait is the case of human IQ performance and social success. In 1969, an educational psychologist, A. R. Jensen, published a long paper in the *Harvard Educational Review,* asking the question (in its title) "How much can we boost IQ and scholastic achievement?" Jensen's conclusion was "not much." As an explanation and evidence of this unchangeability, he offered a claim of high heritability for IQ performance. A great deal of criticism has been made of the evidence offered by Jensen for the high heritability of IQ scores. But, irrespective of the correct value of H^2 for IQ performance, the real error of Jensen's argument lies in his equation of high heritability with unchangeability. In fact, the heritability of IQ is *irrelevant* to the question raised in the title of his article.

To see why this is so, let us consider the results of adoption studies in which children are separated from their biological parents in infancy and reared by adoptive parents. Although results may vary quantitatively from study to study, there are three characteristics in common. First, adopting parents generally have higher IQ scores than those of the biological parents. Second, the adopted children have higher IQ scores than those of their biological parents. Third, the adopted children show a higher correlation of IQ scores with their biological parents than with their adoptive families. The following table is a hypothetical data set that shows all these characteristics, in idealized form, to illustrate the concepts. The scores given for parents are meant to be the average of mother and father.

Children	Biological parents	Adoptive parents
110	90	118
112	92	114
114	94	110
116	96	120
118	98	112
120	100	116
Mean 115	95	115

First, we can see that the children have a high correlation with their biological parents but a low correlation with their adoptive parents. In fact, in our hypothetical example, the correlation of children with biological parents is $r = 1.00$, but, with adoptive parents, it is $r = 0$. (The correlation between two sets of numbers does not mean that the two sets are identical but that, for each unit increase in one set, there is a constant proportion increase in the other set. See page 768 of the Statistical Appendix at the end of this chapter.) This perfect correlation with biological parents and zero correlation with adoptive parents means that $H^2 = 1$, given the arguments developed on page 755. All the variation in IQ score between the children is explained by the variation between the biological parents.

Second, however, we notice that each of the IQ scores of the children is 20 points higher than the IQ scores of their respective biological parents and that the mean IQ of the children is equal to the mean IQ of the adoptive parents. Thus, adoption has raised the average IQ of the children 20 points higher than the average IQ of their biological parents; so, as a *group*, the children resemble their adoptive parents. So we have perfect heritability, yet high environmental plasticity.

An investigator who is seriously interested in knowing how genes might constrain or influence the course of development of any trait in any organism must study directly the norms of reaction of the various genotypes in the population over the range of projected environments. No less detailed information will do. Summary measures such as H^2 are not first steps toward a more complete analysis and therefore are not valuable in themselves.

MESSAGE ··
Heritability is not the opposite of phenotypic plasticity. A character may have perfect heritability in a population and still be subject to great changes resulting from environmental variation.

Locating the genes

It is not possible with purely genetic techniques to identify all the genes that influence the development of a given trait. This is true even for simple qualitative traits — for example, the genes that determine the total antigenic configuration of the membrane of the human red blood cell. About 40 loci determining human blood groups are known at present; each has been discovered by finding at least one person with an immunological specificity that differs from the specificities of other people. Many other loci that determine red-cell membrane structure may remain undiscovered because all the individuals studied are genetically identical. *Genetic* analysis detects genes only when there is some allelic variation. In contrast, *molecular* analysis, by dealing directly with DNA and its translated information, can identify genes even when they do not vary — provided the gene products can be identified.

Even though a trait may show continuous phenotypic variation, the genetic basis for the differences may be allelic variation at a single locus. Most of the classical mutations in *Drosophila* are phenotypically variable in their expression, and in many cases the mutant class differs little from wild type, so many individuals that carry the mutation are indistinguishable from normal. Even the genes of the bithorax gene complex, which have dramatic homeotic mutations that turn halteres into wings (see pages 698–699), also have weak alleles that increase the size of the haltere only slightly on the average, so individuals of the mutant genotype may appear to be wild type.

It is sometimes possible to use prior knowledge of the biochemistry and development of an organism to guess that variation at a known locus is responsible for at least some of the variation in phenotype. This locus then is a *candidate gene* for investigation of continuous phenotypic variation. An example is the variation in activity of the enzyme acid phosphatase in human red blood cells. Because we are dealing with variation in enzyme activity, a good hypothesis would be that there is allelic variation at the locus that encodes this enzyme. When H. Harris and D. Hopkinson sampled an English population, they found that there were, indeed, three allelic forms, *A*, *B*, and *C*, with different activities. Table 25-2 shows the mean activity, the variance in activity, and the population frequency of the six geno-

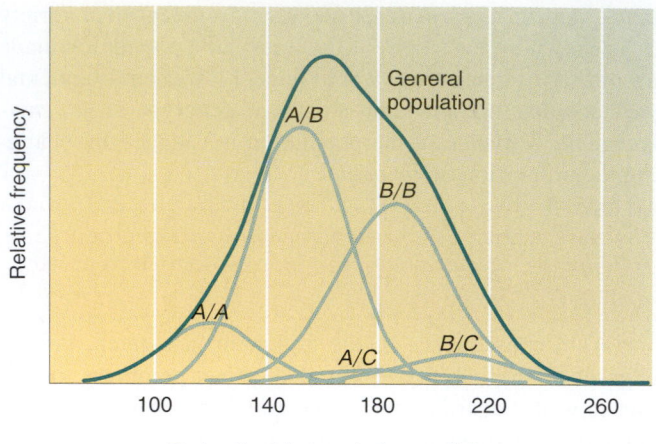

Figure 25-11   Red-cell acid phosphatase activity of different genotypes and the distribution of activity in an English population made up of a mixture of these genotypes. (H. Harris, *The Principles of Human Biochemical Genetics,* 3d ed. Copyright © 1970 by North-Holland.)

types. Figure 25-11 shows the distribution of activity in the entire population and how it is composed of the distributions of the different genotypes. Table 25-2 shows that, of the variance in activity in the total distribution (607.8), about half is explained by the average variance within genotypes (310.7), so half (607.8 − 310.7 = 297.1) is accounted for by the variance between the means of the six genotypes. Although much of the variation in activity is explained by the mean differences between the genotypes, there remains variation within each genotype that may be the result of environmental influences or of the segregation of other, as yet unidentified, genes. This partial explanation of variation by alleles at a single identified locus is typical of what is found by the candidate gene method, and the proportion of variance associated with the single locus is usually less than what was found for acid phosphatase. For example, the three common alleles for the gene *apoE* that encodes the protein apolipoprotein E account for only about 16 percent of the variance in blood levels of low-density lipoproteins that carry cholesterol and are implicated in excess cholesterol levels.

Marker-gene segregation

The genes segregating for a quantitative trait, so-called *quantitative trait loci,* or QTLs, cannot be individually identified in most cases. It is possible, however, to localize those regions of the genome in which the relevant loci lie and to estimate how much of the total variation is accounted for by QTL variation in each region. This analysis is done in experimental organisms by crossing two lines that differ markedly in the quantitative trait and differ in alleles at well-known loci, *marker genes,* where the different genotypes can be distinguished by criteria such as some visible

Genotype	Mean activity	Variance of activity	Frequency in population
A/A	122.4	282.4	0.13
A/B	153.9	229.3	0.43
B/B	188.3	380.3	0.36
A/C	183.8	392.0	0.03
B/C	212.3	533.6	0.05
C/C	~240	—	0.002
Grand average	166.0	310.7	
Total distribution	166.0	607.8	

TABLE 25-2   Red Blood Cell Activity of Different Genotypes of Red-Cell Acid Phosphatase in the English Population

Note: Averages are weighted by frequency in population.
Source: H. Harris, *The Principles of Human Biochemical Genetics,* 3d ed. North-Holland, 1980.

phenotypic effect that is not confused with the quantitative trait (say, eye color in *Drosophila*) or by the electrophoretic mobility of the proteins that they encode or by the DNA sequence of the genes themselves. The F_1 between the two lines is then crossed with itself to make a segregating F_2 or it may be backcrossed to one of the parental lines. If there are QTLs closely linked to a marker gene, then the different marker genotypes in the segregating generation will also carry the QTL alleles that were linked to them in the original parental lines. Thus different marker genotypes in the F_2 or backcross will have different average phenotypes for the quantitative character.

Linkage analysis

The localization of QTLs to small regions within a chromosome requires that there be closely spaced marker loci along the chromosome. Moreover, it must be possible to have parental lines that differ from each other in the alleles carried at these loci. For most of the history of genetics, these requirements could not be met, even in a genetically well known species such as *Drosophila*, because most marker loci were known from severe morphological mutants that had deleterious effects on the viability and fecundity of their carriers. As a result, it was not possible to create a line that carried large numbers of mutant alleles that would distinguish it from an alternative line carrying the wild-type alleles. With the advent of molecular techniques that can detect genetic polymorphism at the DNA level (see pages 718–721), very high densities of variant loci have been discovered along the chromosomes of all species. Especially useful are restriction-site polymorphisms and tandem repeats in DNA (see pages 718–719, and 720–721). Such polymorphisms are so common that any two lines selected for a difference in quantitative traits are also sure to differ from each other at known molecular marker loci spaced a few crossover units from each other along each chromosome.

An experimental protocol for localizing the genes uses groups of individuals that differ markedly in the quantitative trait and differ at marker loci. These groups may be created by a number of generations of divergent selection to create extreme lines or advantage may be taken of already existing varieties of family groups that differ markedly in the trait. These lines must then be surveyed for marker loci that differ between them. A cross is made between the two lines, and the F_1 is crossed with itself to produce a segregating F_2 or is crossed back to one of the parental lines to produce a segregating backcross. A large number of offspring from the segregating generation are then measured for the quantitative trait and characterized for their genotype at the marker loci. A marker locus that is unlinked or very loosely linked to any QTLs will have the same average value of the quantitative trait for all its genotypes, whereas one that is closely linked to some QTLs will differ in its mean phenotype from

one of its genotypes to another. How much difference there is in the mean phenotype between the marker-locus genotypes depends both on the strength of the effect of the QTLs and on the tightness of linkage between the QTLs and the marker locus. Suppose, for example, that there are two selected lines that differ by a total of 100 units in some quantitative character, that the high line is homozygous $+/+$ at a QTL, whereas the low line is homozygous $-/-$, and that each $+$ allele at this QTL accounts for 5 units of the total difference between the lines. Further suppose that the high line is M/M and the low line is m/m at a marker locus 10 crossover units away from the QTL. Then, as shown in Figure 25-12, there are 4 units of difference between the average gamete carrying an M allele and an average gamete carrying an m allele in the segregating F_2, or 8 units of the difference between the two original homozygous lines. Thus we have accounted for 8 percent of the average difference between the lines, although the QTL actually accounts for 10 percent of the difference. The discrepancy comes from the recombination between the marker gene and the QTL.

This technique has been used to locate chromosomal segments associated with traits such as fruit weight in tomatoes, bristle number in *Drosophila*, and vegetative characters in maize. In the maize case, 82 vegetative characters were examined in a cross between lines that differed in 20 DNA markers. On the average, each trait was significantly associated with 14 different markers, but the proportion of the trait difference between the lines that was explained by any given marker was usually very small. Figure 25-13 shows the proportion of the significant marker–trait

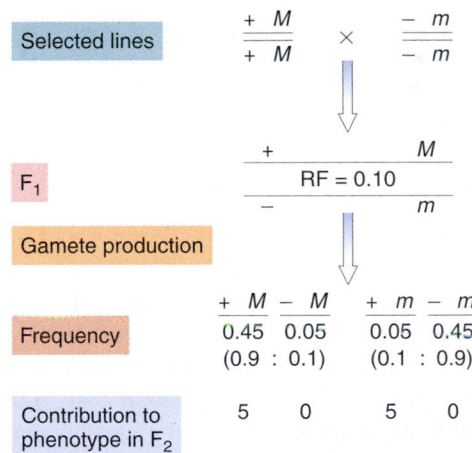

Average phenotypic effect of M class = 5 (0.9) + 0 (0.1) = 4.5
Average phenotypic effect of m class = 5 (0.1) + 0 (0.9) = 0.5
Difference between M-carrying gametes and m-carrying gametes =
4.5 − 0.5 = 4
Difference between M/M homozygotes and m/m homozygotes = 8

Figure 25-12 Result of a cross between two selected lines that differ at a QTL and in a molecular marker 10 crossover units away from the QTL. The QTL $+$ allele adds 5 units to the phenotype.

Figure 25-13 Distribution of associations of the trait differences between two lines of maize with an array of DNA markers. The *x*-axis shows the percentage of the difference explained between the two lines in a given trait that could be associated with any marker gene. The *y*-axis shows the proportion of all the identified QTLs that had the corresponding percentage of its difference explained. Note that 55 percent of all the associations (first two columns) account for less than 1 percent of their trait differences. (After M. Lynch and B. Walsh, *Genetics and Analysis of Quantitative Traits.* Sinauer Associates, 1998. Data from M. D. Edwards, C. W. Stuber, and J. F. Wendel, *Genetics* 116, 1987, 113–125.)

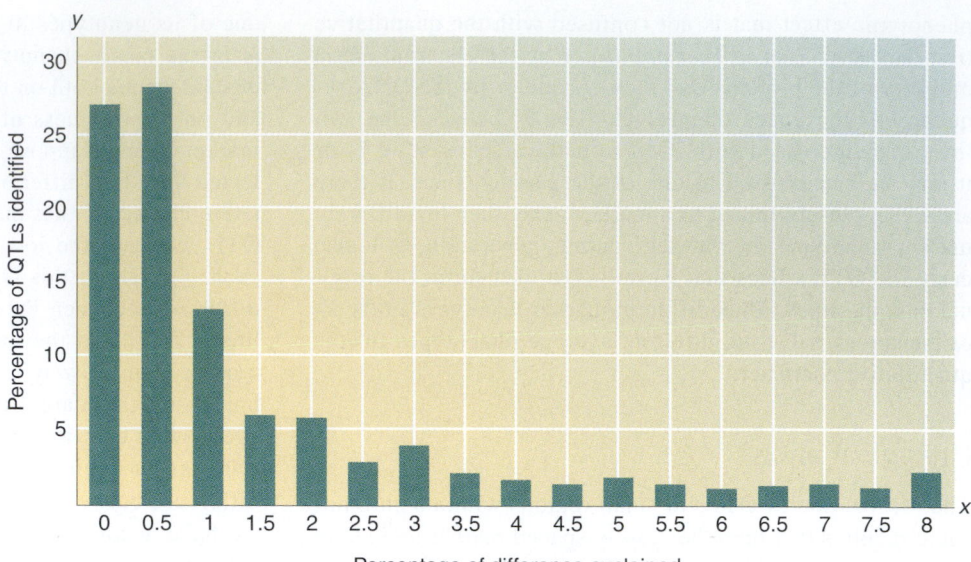

associations (on the *y*-axis) that accounted for different proportions of trait difference between the lines. As Figure 25-13 shows, most associations accounted for less than 1 percent of the trait difference between lines.

For many organisms (for example, humans), it is not possible to make homozygous lines differing in some trait and then cross them to produce a segregating generation. For such organisms, one can use the differences among sibs carrying different marker alleles from heterozygous parents. This method has much less power to find QTLs especially when the number of sibs in any family is small, as it is in human families. As a consequence, the attempts to map QTLs for human traits have not been very successful, although the segregating marker technique has been a success in finding loci whose mutations are responsible for single-gene disorders.

More on analyzing variance

Knowledge of the broad heritability (H^2) of a trait in a population is not very useful in itself, but a finer subdivision of phenotypic variance can provide important information for plant and animal breeders. The genetic variation and the environmental variation can themselves each be further subdivided to provide information about gene action and the possibility of shaping the genetic composition of a population.

Additive and dominance variance

Our previous consideration of gene action suggests that the phenotypes of homozygotes and heterozygotes ought to have a simple relation. If one of the alleles encoded a less active gene product or one with no activity at all and if one unit of gene product were sufficient to allow full physiological activity of the organism, then we would expect complete dominance of one allele over the other, as Mendel observed

for flower color in peas. If, on the other hand, physiological activity were proportional to the amount of active gene product, we would expect the heterozygote phenotype to be exactly intermediate between the homozygotes (show no dominance).

For many quantitative traits, however, neither of these simple cases is the rule. In general, heterozygotes are not exactly intermediate between the two homozygotes but are closer to one or the other (show partial dominance), even though there is an equal mixture of the primary products of the two alleles in the heterozygote. Indeed, in some cases, the heterozygote phenotype may lie outside the phenotypic range of the homozygotes altogether—a feature termed **overdominance.** For example, newborn babies who are intermediate in size have a higher chance of survival than very large or very small newborns. Thus, if survival were the phenotype of interest, heterozygotes for genes influencing growth rate would show overdominance for fitness although not for growth rate.

Suppose that two alleles, *a* and *A*, segregate at a locus influencing height. In the environments encountered by the population, the mean phenotypes (heights) and frequencies of the three genotypes might be:

	a/a	*A/a*	*A/A*
Phenotype	10	18	20
Frequency	0.36	0.48	0.16

There is genetic variance in the population; the phenotypic means of the three genotypic classes are different. Some of the variance arises because there is an average effect on phenotype of substituting an allele *A* for an allele *a*; that is, the average height of all individuals with *A* alleles is greater than that of all individuals with *a* alleles. By defining the average effect of an allele as the average phenotype of all individuals that carry it, we necessarily make the average effect of the allele depend on the frequencies of the genotypes.

The average effect is calculated by simply counting the a and A alleles and multiplying them by the heights of the individuals in which they appear. Thus, 0.36 of all the individuals are homozygous a/a, each a/a individual has two a alleles, and the average height of a/a individuals is 10 cm. Heterozygotes make up 0.48 of the population, each has only one a allele, and the average phenotypic measurement of A/a individuals is 18 cm. The total "number" of a alleles is $2(0.36) + 1(0.48)$. Thus, the average effect of all the a alleles is:

$$\bar{a} = \text{average effect of } a = \frac{2(0.36)(10) + 1(0.48)(18)}{2(0.36) + 1(0.48)}$$

$$= 13.20 \text{ cm}$$

and, by a similar argument,

$$\bar{A} = \text{average effect of } A = \frac{2(0.16)(20) + 1(0.48)(18)}{2(0.16) + 1(0.48)}$$

$$= 18.80 \text{ cm}$$

This average difference in effect between A and a alleles of 5.60 cm accounts for some of the variance in phenotype — but not for all of it. The heterozygote is not exactly intermediate between the homozygotes; there is some dominance.

We would like to separate the so-called **additive effect** caused by substituting a alleles for A alleles from the variation caused by dominance. The reason is that the effect of selective breeding depends on the additive variation and not on the variation caused by dominance. Thus, for purposes of plant and animal breeding or for making predictions about evolution by natural selection, we must determine the additive variation. An extreme example will illustrate the principle. Suppose that there is overdominance and that the phenotypic means and frequencies of three genotypes are:

	A/A	A/a	a/a
Phenotype	10	12	10
Frequency	0.25	0.50	0.25

It is apparent (and a calculation like the preceding one will confirm) that there is no average difference between the a and A alleles, because each has an effect of 11 units. So there is no *additive* variation, although there is obviously variation in phenotype between the genotypes. The largest individuals are heterozygotes. If a breeder attempts to increase height in this population by selective breeding, mating these heterozygotes together will simply reconstitute the original population. Selection will be totally ineffective. This example illustrates the general law that the effect of selection depends on the *additive* genetic variation and not on genetic variation in general.

We partition the total genetic variance in a population into **additive genetic variation** (s_a^2), the variance that arises because there is an average difference between the carriers of a alleles and the carriers of A alleles, and a component called the **dominance variance** (s_d^2), which results from the

fact that heterozygotes are not exactly intermediate between the monozygotes. Thus:

$$s_g^2 = s_a^2 + s_d^2$$

The components of variance in the first example, where $a/a = 10$, $A/a = 18$, and $A/A = 20$, can be calculated by using the definitions of mean and variance developed earlier in this chapter. Remembering that a mean is the sum of the values of a variable, each weighted by the frequency with which that value occurs (see page 765), we can calculate the mean phenotype to be:

$$\bar{x} = \Sigma f_i x_i = (0.36)(10) + (0.48)(18) + (0.16)(20)$$

$$= 15.44 \text{ cm}$$

The total genetic variance that arises from the variation among the mean phenotypes of the three genotypes is:

$$s_g^2 = \Sigma f_i(x_i - \bar{x})^2 = (0.36)(10 - 15.44)^2$$
$$+ (0.48)(18 - 15.44)^2$$
$$+ (0.16)(20 - 15.44)^2$$
$$= 17.13 \text{ cm}^2$$

The frequency of allele a is (by counting alleles):

$$f_a = \frac{2(a/a) + 1(A/a)}{2}$$

$$= \frac{2(0.36) + 1(0.48)}{2} = 0.60$$

and the frequency of the A allele is:

$$f_A = \frac{2(A/A) + 1(A/a)}{2}$$

$$= \frac{2(0.16) + 1(0.48)}{2} = 0.40$$

The variance of allelic means is then

$$s^2 = f_a(\bar{a} - \bar{x})^2 + f_A(\bar{A} - \bar{x})^2$$
$$= (0.60)(13.20 - 15.44)^2 + (0.40)(18.80 - 15.44)^2$$
$$= 7.525 \text{ cm}^2$$

But we want the variance among diploid individuals that results from the allelic effects, and every diploid individual carries two alleles; so:

$$s_a^2 = (2)(7.525) = 15.05 \text{ cm}^2$$

and

$$s_d^2 = s_g^2 - s_a^2 = 17.13 - 15.05 = 2.08 \text{ cm}^2$$

The total phenotypic variance can now be written as

$$s_p^2 = s_g^2 + s_e^2 = s_a^2 + s_d^2 + s_e^2$$

We define a new kind of heritability, the **heritability in the narrow sense (h^2)**, as

$$h^2 = \frac{s_a^2}{s_p^2} = \frac{s_a^2}{s_a^2 + s_d^2 + s_e^2}$$

It is this heritability, not to be confused with H^2, that is useful in determining whether a program of selective breeding will succeed in changing the population. The greater the h^2 is, the greater the difference is between selected parents and the population as a whole that will be preserved in the offspring of the selected parents.

MESSAGE ···
The effect of selection depends on the amount of *additive* genetic variance and not on the genetic variance in general. Therefore, the narrow heritability, h^2, not the broad heritability, H^2, is relevant for a prediction of response to selection.
···

What has been described as the "dominance" variance is really more complicated. It is all the genetic variation that cannot be explained by the average effect of substituting A for a. If more than one locus affects the character, then any epistatic interactions between loci will appear as variance not associated with the average effect of substituting alleles at the A locus. In principle, we can separate this **interaction variance** (s_i^2) from the dominance variance (s_a^2). In practice, however, this separation cannot be done with any semblance of accuracy, so all the nonadditive variance appears as "dominance" variance.

Estimating genetic variance components

Genetic components of variance can be estimated from covariance between relatives, but the derivation of these estimates is beyond the scope of an elementary text.

There is, however, another way to estimate h^2 that reveals its real meaning. If we plot the phenotypes of the offspring against the average phenotypes of their two parents (the midparent value), we may observe a relation like the one illustrated in Figure 25-14. The regression line will pass through the mean of all the parents and the mean of all the offspring, which will be equal to each other because no change has occurred in the population between generations. Moreover, taller parents have taller children and shorter parents have shorter children, so the slope of the line is positive. But the slope is not unity; very short parents have children who are somewhat taller and very tall parents have children who are somewhat shorter than they themselves are. This slope of less than unity for the regression line arises because heritability is less than perfect. If the phenotype were additively inherited with complete fidelity, then the height of the offspring would be identical with the midparent value and the slope of the line would be 1. On the other hand, if the offspring had no heritable similarity to their parents, all parents would have offspring of the same average height and the slope of the line would be 0. This suggests that the slope of the regression line of the offspring value on the midparent value is an estimate of additive heritability. In fact, the relation is precise.

The fact that the slope equals the additive heritability now allows us to use h^2 to predict the effects of artificial se-

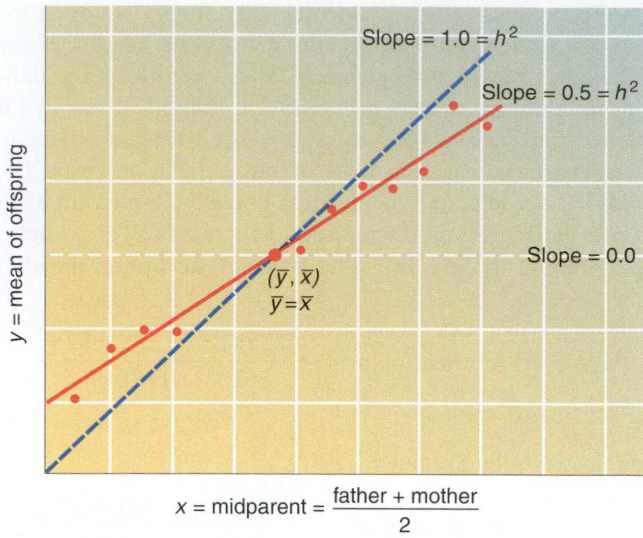

Figure 25-14   The regression (red line) of offspring measurements (y) on midparents (x) for a trait with narrow heritability (h^2) of 0.5. The blue line would be the regression slope if the trait were perfectly heritable.

lection. Suppose that we select parents for the next generation who are on the average 2 units above the general mean of the population from which they were chosen. If $h^2 = 0.5$, then the offspring who form the next, selected generation will lie $0.5(2.0) = 1.0$ unit above the mean of the present population, because the regression coefficient predicts how much increase in y will result from a unit increase in x. We can define the **selection differential** as the difference between the selected parents and the unselected mean and the **selection response** as the difference between their offspring and the preceding generation. Then:

$$\text{selection response} = h^2 \times \text{selection differential}$$

or

$$h^2 = \frac{\text{selection response}}{\text{selection differential}}$$

The second expression provides us with yet another way to estimate h^2: by selecting for one generation and comparing the response with the selection differential. Usually this is carried out for several generations, and the average response is used.

Remember that any estimate of h^2, just as for H^2, depends on the assumption of no greater environmental correlation between closer relatives. Moreover, h^2 in one population in one set of environments will not be the same as h^2 in a different population at a different time. Figure 25-15 shows the range of heritabilities reported in various studies for a number of traits in chickens. The very small ranges are generally close to zero. For most traits for which a substantial heritability has been reported in some population, there are big differences from study to study.

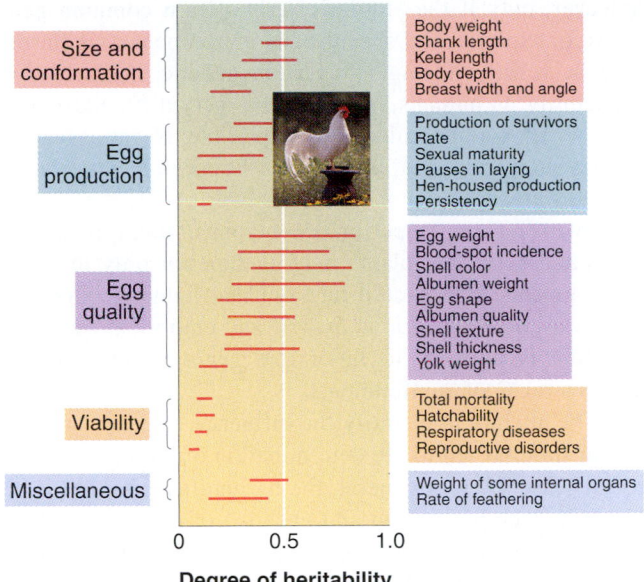

Size and conformation	Body weight Shank length Keel length Body depth Breast width and angle
Egg production	Production of survivors Rate Sexual maturity Pauses in laying Hen-housed production Persistency
Egg quality	Egg weight Blood-spot incidence Shell color Albumen weight Egg shape Albumen quality Shell texture Shell thickness Yolk weight
Viability	Total mortality Hatchability Respiratory diseases Reproductive disorders
Miscellaneous	Weight of some internal organs Rate of feathering

0   0.5   1.0

Degree of heritability

Figure 25-15 Ranges of heritabilities (h^2) reported for a variety of characters in chickens. (From I. M. Lerner and W. J. Libby, *Heredity, Evolution, and Society.* Copyright © 1976 by W. H. Freeman and Company. Photograph © Kenneth Thomas/Photo Researchers.)

Use of h^2 in breeding

Even though h^2 is a number that applies only to a particular population and a given set of environments, it is still of great practical importance to breeders. A poultry geneticist interested in increasing, say, growth rate is not concerned with the genetic variance over all possible flocks and all environmental distributions. Given a particular flock (or a choice between a few particular flocks) under the environmental conditions approximating present husbandry practice, the question becomes: Can a selection scheme be devised to increase growth rate and, if so, how fast? If one flock has a lot of genetic variance and another only a little, the breeder will choose the former to carry out selection. If the heritability in the chosen flock is very high, then the mean of the population will respond quickly to the selection imposed, because most of the superiority of the selected parents will appear in the offspring. The higher h^2 is, the higher the parent–offspring correlation is. If, on the other hand, h^2 is low, then only a small fraction of the increased growth rate of the selected parents will appear in the next generation.

If h^2 is very low, some alternative scheme of selection or husbandry may be needed. In this case, H^2 together with h^2 can be of use to the breeder. Suppose that h^2 and H^2 are both low, which means that there is a lot of environmental variance compared with genetic variance. Some scheme of reducing s_e^2 must be used. One method is to change the husbandry conditions so that environmental variance is lowered. Another is to use **family selection.** Rather than choosing the best individuals, the breeder allows pairs to produce several progeny, and the *mating* is selected on the basis of the average performance of the progeny. Averaging over progeny allows uncontrolled environmental and developmental noise variation to be canceled out, and a better estimate of the genotypic difference between pairs can be made so that the best pairs can be chosen as parents of the next generation.

If, on the other hand, h^2 is low but H^2 is high, then there is not much environmental variance. The low h^2 is the result of a small amount of additive genetic variance compared with dominance and interaction variance. Such a situation calls for special breeding schemes that make use of nonadditive variance. One such scheme is the **hybrid-inbred method,** which is used almost universally for corn. A large number of inbred lines are created by selfing. These inbred lines are then crossed in many different combinations (all possible combinations, if this is economically feasible), and the cross that gives the best hybrid is chosen. Then new inbred lines are developed from this best hybrid, and again crosses are made to find the best second-cycle hybrid. This scheme selects for dominance effects because it takes the best heterozygotes; it has been the basis of major genetic advances in hybrid maize yield in North America since 1930. Yield in corn does not appear to have large amounts of nonadditive genetic variance, so it is debatable whether this technique *ultimately* produces higher-yielding varieties than those that would have resulted from years of simple selection techniques based on additive variance.

The hybrid method has been introduced into the breeding of all kinds of plants and animals. Tomatoes and chickens, as examples, are now almost exclusively hybrids. Attempts also have been made to breed hybrid wheat, but thus far the wheat hybrids obtained do not yield consistently better than the nonhybrid varieties now used.

MESSAGE ······························

The subdivision of genetic variation and environmental variation provides important information about gene action that can be used in plant and animal breeding.

SUMMARY

Many — perhaps most — of the phenotypic traits that we observe in organisms vary continuously. In many cases, the variation of the trait is determined by more than a single segregating locus. Each of these loci may contribute equally to a particular phenotype, but it is more likely that they contribute unequally. The measurement of these phenotypes and the determination of the contributions of specific alleles to the distribution must be made on a statistical basis in these cases. Some of these variations of phenotype (such as height in some plants) may show a normal distribution around a mean value; others (such as seed weight in some plants) will illustrate a skewed distribution around a mean value.

In other characters, the variation in one phenotype may be correlated with the variation in another. A correlation coefficient may be calculated for these two variables.

A quantitative character is one for which the average phenotypic differences between genotypes are small compared with the variation between the individuals within the genotypes. This situation may be true even for characters that are influenced by alleles at one locus. The distribution of environments is reflected biologically as a distribution of phenotypes. The transformation of environmental distribution into phenotypic distribution is determined by the norm of reaction. Norms of reaction can be characterized in organisms in which large numbers of genetically identical individuals can be produced.

With the use of genetically marked chromosomes, it is possible to determine the relative contributions of different chromosomes to variation in a quantitative trait, to observe dominance and epistasis from whole chromosomes, and, in some cases, to map genes that are segregating for a trait.

Traits are familial if they are common to members of the same family, for whatever reason. Traits are heritable, however, only if the similarity arises from common genotypes. In experimental organisms, environmental similarities may be readily distinguished from genetic similarities, or heritability. In humans, however, it is very difficult to determine whether a particular trait is heritable. Norm of reaction studies show only small differences between genotypes, and these differences are not consistent over a wide range of environments. Thus, "superior" genotypes in domesticated animals and cultivated plants may be superior only in certain environments. If it should turn out that humans exhibit genetic variation for various mental and emotional traits, this variation is unlikely to favor one genotype over another across a range of environments.

The attempt to quantify the influence of genes on a particular trait has led to the determination of heritability in the broad sense (H^2). In general, the heritability of a trait is different in each population and each set of environments and cannot be extrapolated from one population and set of environments to another. Because H^2 characterizes present populations in present environments only, it is fundamentally flawed as a predictive device. Heritability in the narrow sense, h^2, measures the proportion of phenotypic variation that results from substituting one allele for another. This quantity, if large, predicts that selection for a trait will succeed rapidly. If h^2 is small, special forms of selection are required.

CONCEPT MAP

Draw a concept map interrelating as many of the following terms as possible. Note that the terms are listed in no particular order.

quantitative variation / polygenes / Mendel's law / norm of reaction / environment / heritability / variance / selection response / additive variance / QTL

STATISTICAL APPENDIX

Measures of central tendency

The mode. Most distributions of phenotypes look roughly like those in Figure 25-2: a single most frequent class, the **mode,** is located near the middle of the distribution, with frequencies decreasing on either side. There are exceptions to this pattern, however. Figure 25-16a shows the very asymmetric distribution of seed weights in the plant *Crinum longifolium*. Figure 25-16b shows a **bimodal** (two-mode) distribution of larval survival probabilities for different second-chromosome homozygotes in *Drosophila willistoni*.

A bimodal distribution may indicate that the population being studied could be better considered a mixture of two populations, each with its own mode. In Figure 25-16b, the left-hand mode probably represents a subpopulation of severe single-locus mutations that are extremely deleterious when homozygous but whose effects are not felt in the heterozygous state in which they usually exist in natural populations. The right-hand mode is part of the distribution of "normal" viability modifiers of small effect.

The mean. Complete information about the distribution of a phenotype in a population can be given only by specifying the frequency of each measured class, but a great deal of information can be summarized in two statistics. First, we need some measure of the location of the distribution along the axis of measurement. (For example, do the individual measurements tend to cluster around 100 cm or 200 cm?) One possibility is to give the measurement of the most common class, the mode. In Figure 25-2b, the mode is 172 cm (for females, the mode would be about 6 cm less). A more common measure of location is the arithmetic average, or the **mean.** The mean of the measurement ($\bar{x}$) is simply the sum of all the measurements (x_i) divided by the number of measurements in the sample (N):

$$\text{mean} = \bar{x} = \frac{x_1 + x_2 + x_3 + \cdots + x_N}{N} = \frac{1}{N} \Sigma x_i$$

where Σ represents summation and x_i is the ith measurement.

In a typical large sample, the same measured value will appear more than once, because several individuals will

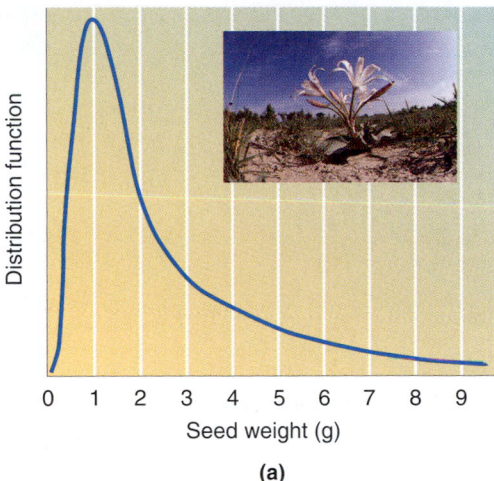

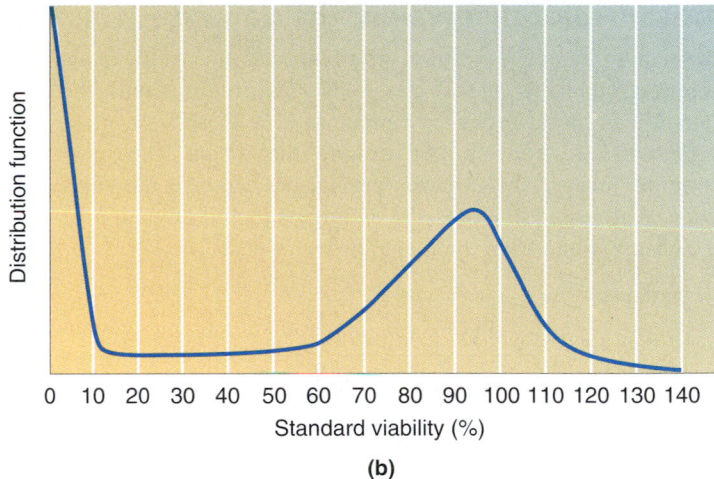

(a)

(b)

Figure 25-16 Asymmetric distribution functions: (a) asymmetric distribution of seed weight in *Crinum longifolium;* (b) bimodal distribution of survival of *Drosophila willistoni* expressed as a percentage of standard survival. (After S. Wright, *Evolution and the Genetics of Populations,* vol. 1. Copyright © 1968 by University of Chicago Press. Photograph: Earth Scenes/© Thompson GOSF.)

have the same value within the accuracy of the measuring instrument.

In such a case, $\bar{x}$ can be rewritten as the sum of all measurement values, each weighted by how frequently it occurs in the population. From a total of N individuals measured, suppose that n_1 fall in the class with value x_1, that n_2 fall in the class with value x_2, and so forth, so that $\Sigma n_i = N$. If we let f_i be the **relative frequency** of the ith measurement class, so that

$$f_i = \frac{n_i}{N}$$

then we can rewrite the mean as

$$\bar{x} = f_1 x_1 + f_2 x_2 + \cdots + f_k x_k = \Sigma f_i x_i$$

where x_i equals the value of the ith measurement class.

Let us apply these calculation methods to the data of Table 25-3, the numbers of toothlike bristles in the sex combs on the right (x) and left (y) front legs and the sum of both legs ($T = x + y$) of 20 *Drosophila*. Looking for the moment only at the sum of the two legs T, we find the mean number of teeth $\bar{T}$ to be:

$$\bar{T} = \frac{11 + 12 + 12 + 12 + 13 + \cdots + 15 + 16 + 16}{20}$$

$$= \frac{274}{20}$$

$$= 13.7$$

Alternatively, using the relative frequencies of the different measurement values, we find that

$$\bar{T} = 0.05(11) + 0.15(12) + 0.20(13) + 0.35(14)$$
$$+ 0.15(15) + 0.10(16)$$
$$= 13.7$$

				TABLE	Number of Teeth in the Sex Comb on the Right (x) and Left (y) Legs and the Sum of the Two (T) for 20 *Drosophila* Males

x	y	T	n_i	$f_i = n_i/N$
6	5	11	1	$\frac{1}{20} = 0.05$
6	6	12	3	$\frac{3}{20} = 0.15$
5	7	12		
6	6	12		
7	6	13	4	$\frac{4}{20} = 0.20$
5	8	13		
6	7	13		
7	6	13		
8	6	14	7	$\frac{7}{20} = 0.35$
6	8	14		
7	7	14		
7	7	14		
7	7	14		
6	8	14		
8	6	14		
8	7	15	3	$\frac{3}{20} = 0.15$
7	8	15		
6	9	15		
8	8	16	2	$\frac{2}{20} = 0.10$
7	9	16		

$N = 20$
$\bar{x} = 6.25$ $s_x^2 = 0.8275$ $s_x = 0.9096$
$\bar{y} = 7.05$ $s_y^2 = 1.1475$ $s_y = 1.0722$
$\bar{T} = 13.70$ $s_T^2 = 1.71$ $s_T = 1.308$
       cov $xy = -0.1325$
       $r_{xy} = -0.1360$

Measures of dispersion: the variance

A second characteristic of a distribution is the width of its spread around the central class. Two distributions with the same mean might differ very much in how closely the measurements are concentrated around the mean. The most common measure of variation around the center is the **variance,** which is defined as the average squared deviation of the observations from the mean, or

$$\text{variance} = s^2$$
$$= \frac{(x_1 - \bar{x})^2 + (x_2 - \bar{x})^2 + \cdots + (x_N - \bar{x})^2}{N}$$
$$= \frac{1}{N} \Sigma\, (x_i - \bar{x})^2$$

When more than one individual has the same measured value, the variance can be written as

$$s^2 = f_1(x_1 - \bar{x})^2 + f_2(x_2 - \bar{x})^2 + \cdots + f_k(x_k - \bar{x})^2$$
$$= \Sigma\, f_i(x_i - \bar{x})^2$$

To avoid subtracting every value of x separately from the mean, we can use an alternative computing formula that is algebraically identical with the preceding equation:

$$s^2 = \left(\frac{1}{N} \Sigma\, x_i^2\right) - \bar{x}^2$$

Because the variance is in squared units (square centimeters, for example), it is common to take the square root of the variance, which then has the same units as the measurement itself. This square-root measure of variation is called the **standard deviation** of the distribution:

$$\text{standard deviation} = s = \sqrt{\text{variance}} = \sqrt{s^2}$$

The data in Table 25-3 can be used to exemplify these calculations:

$$s_T^2 = \frac{(11 - 13.7)^2 + (12 - 13.7)^2 + (12 - 13.7)^2}{20}$$
$$\frac{+ \cdots + (15 - 13.7)^2 + (16 - 13.7)^2}{20}$$
$$= \frac{34.20}{20} = 1.71$$

We can use the computing formula that avoids taking individual deviations:

$$s_T^2 = \frac{1}{N} \Sigma\, T_i^2 - \bar{T}^2 = \frac{3788}{20} - 187.69 = 1.71$$

and

$$s = \sqrt{1.71} = 1.308$$

Figure 25-17 shows two distributions having the same mean but different standard deviations (curves A and B) and two distributions having the same standard deviation but different means (curves A and C).

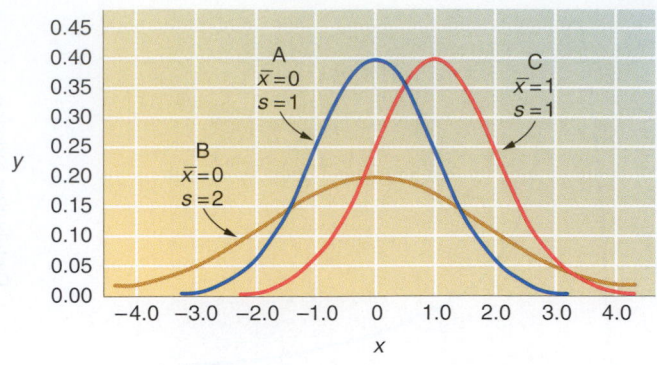

Figure 25-17   Three distribution functions, two of which have the same mean and two of which have the same standard deviation.

The mean and the variance of a distribution do not describe it completely. They do not distinguish a symmetric distribution from an asymmetric one, for example. We can even construct symmetric distributions that have the same mean and variance but still have somewhat different shapes. Nevertheless, for the purposes of dealing with most quantitative genetic problems, the mean and variance suffice to characterize a distribution.

Measures of relation

Covariance and correlation. Another statistical notion that is of use in the study of quantitative genetics is the association, or **correlation,** between variables. As a result of complex paths of causation, many variables in nature vary together but in an imperfect or approximate way. Figure 25-18a provides an example, showing the lengths of two particular teeth in several individual specimens of a fossil mammal, *Phenacodus primaevis*. The longer an individual's first lower molar is, the longer its second molar is, but the relation between the two teeth is imprecise. Figure 25-18b shows that the total length and tail length in individual snakes *(Lampropeltis polyzona)* are quite closely related to each other.

The usual measure of the precision of a relation between two variables x and y is the **correlation coefficient** (r_{xy}). It is calculated in part from the product of the deviation of each observation of x from the mean of the x values and the deviation of each observation of y from the mean of the y values — a quantity called the **covariance** of x and y (cov xy):

$$\text{cov } xy = \frac{(x_1 - \bar{x})(y_1 - \bar{y}) + (x_2 - \bar{x})(y_2 - \bar{y}) + \cdots}{N}$$
$$\frac{+ (x_N - \bar{x})(y_N - \bar{y})}{N}$$
$$= \frac{1}{N} \Sigma\, (x_i - \bar{x})(y_i - \bar{y})$$

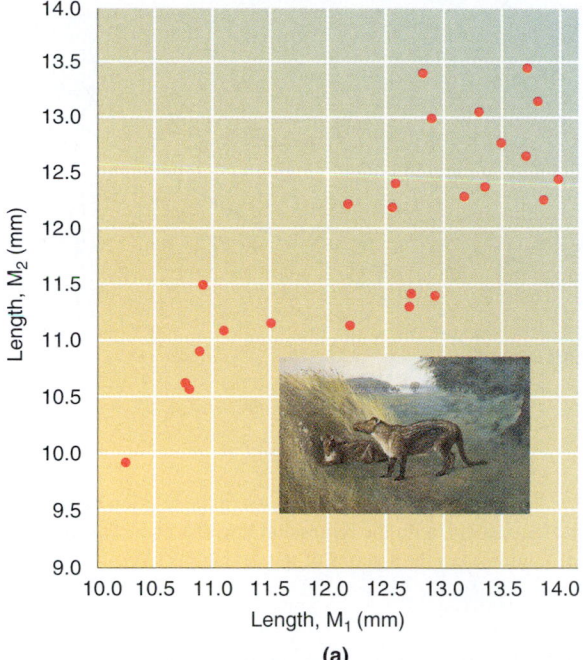

(a)

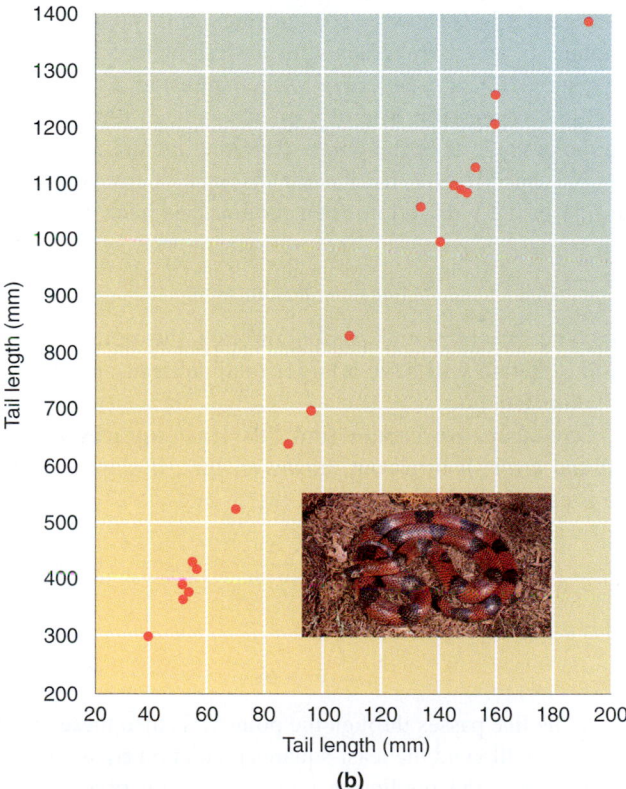

(b)

Figure 25-18 Scatter diagrams of relations between pairs of variables. (a) Relation between the lengths of the first and second lower molars (M_1 and M_2) in the extinct mammal *Phenacodus primaevis*. Each point gives the M_1 and M_2 measurements for one individual. (b) Tail length and body length of 18 individuals of the snake *Lampropeltis polyzona*. (Image: Negative no. 2430, *Phenacodus*, painting by Charles Knight; courtesy Department of Library Services, American Museum of Natural History. Photograph: Animals Animals/© Zig Leszczynski.)

A formula that is exactly algebraically equivalent but that makes computation easier is:

$$\text{cov } xy = \left(\frac{1}{N} \Sigma x_i y_i\right) - \overline{xy}$$

Using this formula, we can calculate the covariance between the right (x) and the left (y) leg counts in Table 25-3.

$$\begin{aligned}\text{cov } xy &= \left(\frac{1}{N} \Sigma xy\right) - \overline{xy}\\ &= \frac{(6)(5) + (6)(6) + \cdots + (8)(8) + (7)(9)}{20}\\ &\quad - (6.65)(7.05)\\ &= -0.1325\end{aligned}$$

The correlation, r_{xy}, is defined as:

$$\text{correlation} = r_{xy} = \frac{\text{cov } xy}{s_x s_y}$$

In the formula for correlation, the products of the deviations are divided by the product of the standard deviations of x and y (s_x and s_y). This normalization by the standard deviations has the effect of making r_{xy} a dimensionless number that is independent of the units in which x and y are measured. So defined, r_{xy} will vary from -1, which signifies a perfectly linear negative relation between x and y, to $+1$, which indicates a perfectly linear positive relation between x and y. If $r_{xy} = 0$, there is no linear relation between the variables. It is important to notice, however, that sometimes when there is no *linear* relation between two variables but there is a regular *nonlinear* relation between them, one variable may be perfectly predicted from the other. Consider, for example, the parabola shown in Figure 25-19. The values of y are perfectly predictable from the values of x; yet $r_{xy} = 0$, because, on average over the whole range of x values, larger x values are not associated with either larger or smaller y values. The data in Figure 25-18a and b have r_{xy} values of 0.82 and 0.99, respectively. In the example of the sex comb teeth of Table 25-3, the correlation between left and right legs is:

$$r_{xy} = \frac{\text{cov } xy}{\sqrt{s_x^2 s_y^2}} = \frac{-0.1325}{\sqrt{(0.8275)(1.1475)}} = -0.1360$$

a very small value.

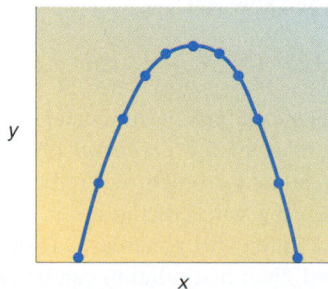

Figure 25-19 A parabola. Each value of y is perfectly predictable from the value of x, but there is no linear correlation.

Correlation and equality. It is important to notice that correlation between two sets of numbers is not the same as numerical identity. For example, two sets of values can be perfectly *correlated*, even though the values in one set are very much larger than the values in the other set. Consider the following pairs of values:

x	y
1	22
2	24
3	26

The variables x and y in the pairs are perfectly correlated ($r = +1.0$), although each value of y is about 20 units greater than the corresponding value of x. Two variables are perfectly correlated if, for a unit increase in one, there is a constant increase in the other (or a constant decrease if r is negative). The importance of the difference between correlation and identity arises when we consider the effect of environment on heritable characters. Parents and offspring can be perfectly correlated in some trait such as height, yet, because of an environmental difference between generations, every child can be taller than the parents. This phenomenon appears in adoption studies, in which children may be correlated with their biological parents but, on the average, may be quite different from the parents as a result of a change in social situation.

Covariance and the variance of a sum. In Table 25-3, the variances of the left and right legs are 0.8275 and 1.1475, which adds up to 1.975, but the variance of the sum of the two legs T is only 1.71. That is, the variance of the whole is less than the sum of the variances of the parts. This discrepancy is a consequence of the negative correlation between left and right sides. Larger left sides are associated with smaller right sides and vice versa, so the sum of the two sides varies less than each side separately. If, on the other hand, there were a positive correlation between sides, then larger left and right sides would go together and the variation of the sum of the two sides would be larger than the sum of the two separate variances. In general, if $x + y = T$, then

$$s_T^2 = s_x^2 + s_y^2 + 2 \text{ cov } xy$$

For the data of Table 25-3,

$$s_T^2 = 1.71 = 0.8275 + 1.1475 - 2(0.1325)$$

Regression. The measurement of correlation provides us with only an estimate of the *precision* of relation between two variables. A related problem is predicting the value of one variable given the value of the other. If x increases by two units, by how much will y increase? If the two variables are linearly related, then that relation can be expressed as

$$y = bx + a$$

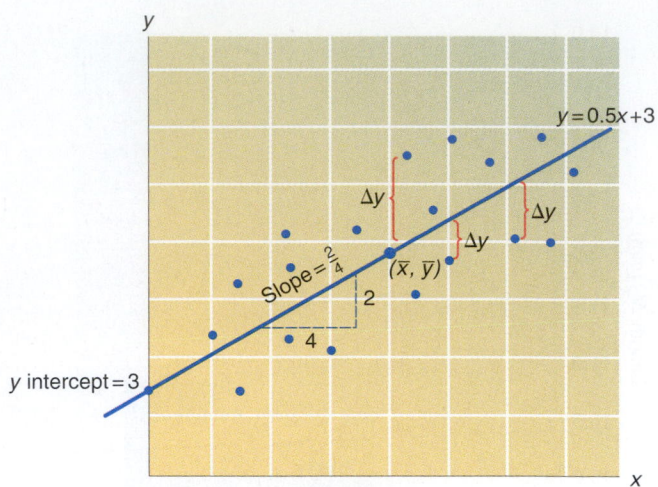

Figure 25-20   A scatter diagram showing the relation between two variables, x and y, with the regression line of y on x. This line, with a slope of $\frac{2}{4}$, minimizes the squares of the deviations (Δy).

where b is the slope of the line relating y to x and a is the y intercept of that line.

Figure 25-20 shows a scatter diagram of points for two variables, y and x, together with a straight line expressing the general linear trend of y with increasing x. This line, called the **regression line of y on x,** has been positioned so that the deviations of the points from the line are as small as possible. Specifically, if Δy is the distance of any point from the line in the y direction, then the line has been chosen so that

$$\Sigma \ (\Delta y)^2 = \text{a minimum}$$

Any other straight line passed through the points on the scatter diagram will have a larger total squared deviation of the points from it.

Obviously, we cannot find this **least-squares line** by trial and error. It turns out, however, that, if slope b of the line is calculated by

$$b = \frac{\text{cov } xy}{s_x^2}$$

and if a is then calculated from

$$a = \bar{y} - b\bar{x}$$

so that the line passes through the point $\bar{x}$, $\bar{y}$, then these values of b and a will yield the least-squares prediction equation.

Note that the prediction equation cannot predict y exactly for a given x, because there is scatter around the line. The equation predicts the *average* y for a given x, if large samples are taken.

Samples and populations. The preceding sections have described the distributions and some statistics of particular assemblages of individuals that have been collected in some experiments or sets of observations. For some purposes, however, we are not really interested in the particular 100 undergraduates or 27 snakes that have been measured. Instead, we are interested in a wider world of phenomena, of which those particular individuals are representative. Thus, we might want to know the average seed weight *in general* of plants of the species *Crinum longifolium.* That is, we are interested in the characteristics of a **universe,** of which our small collection of observations is only a **sample.** The characteristics of any particular sample are not identical with those of the universe but vary from sample to sample.

We can use the sample mean to estimate the true mean of the universe, but the sample variance and covariance are on the average a little smaller than the true value in the universe. This difference is because the deviations from the sample mean are not all independent of one another, inasmuch as the data used to calculate the sample mean are the same as those used to calculate the deviations from that mean. It is simple to correct for this bias. Whenever we are

interested in the variance of a set of measurements — not as a characteristic of the particular collection but as an estimate of a universe that the sample represents — then the appropriate quantity to use, rather than s^2 itself, is $[N/(N-1)]s^2$. Note that this new quantity is equivalent to dividing the sum of squared deviations by $N-1$ instead of N in the first place, so

$$\left(\frac{N}{N-1}\right) s^2 = \left(\frac{N}{N-1}\right) \frac{1}{N} \Sigma (x_i - \bar{x})^2$$

$$= \frac{1}{N-1} \Sigma (x_i - \bar{x})^2$$

All these considerations about bias also apply to the sample covariance. In the formula for the correlation coefficient (page 767), however, the factor $N/(N-1)$ would appear in both the numerator and the denominator and therefore cancel out, so we can ignore it for the purposes of computation.

CHAPTER INTEGRATION PROBLEM

In some species of songbirds, populations living in different geographical regions sing different "local dialects" of the species song. Some people believe that this difference in dialect is the result of genetic differences between populations, whereas others believe that these differences arose from purely individual idiosyncracies in the founders of these populations and have been passed on from generation to generation by learning. Outline an experimental program that would determine the importance of genetic and nongenetic factors and their interaction in this dialect variation. If there is evidence of genetic difference, what experiments could be done to provide a detailed description of the genetic system, including the number of segregating genes, their linkage relations, and their additive and nonadditive phenotypic effects?

◆ Solution ◆

This example has been chosen because it illustrates the very considerable experimental difficulties that arise when we try to examine claims that observed differences in quantitative characters in some species have a genetic basis. To be able to say anything at all about the roles of genes and developmental environment requires, at minimum, that the organisms can be raised from fertilized eggs in a controlled laboratory environment. To be able to make more detailed statements about the genotypes underlying variation in the character requires, further, that the results of crosses between parents of known phenotype and known ancestry be observable and that the offspring of some of those crosses be, in turn, crossed with other individuals of known phenotype and ancestry. Very few animal species can satisfy this requirement, although it is much easier to carry out con-

trolled crosses in plants. We will assume that the songbird species in question can indeed be raised and crossed in captivity, but that is a big assumption.

a. To determine whether there is any genetic difference underlying the observed phenotypic difference in dialect between the populations, we need to raise birds of each population, from the egg, in the absence of auditory input from their own ancestors and in various combinations of auditory environments of other populations. This is done by raising birds from the egg that have been grouped as follows:

1. In isolation

2. Surrounded by hatchlings consisting only of birds derived from the same population

3. Surrounded by hatchlings consisting of birds derived from other populations

4. In the presence of singing adults from other populations

5. In the presence of singing adults from their own population (as a control on the rearing conditions)

If there are no genotypic differences and all dialect differences are learned, then birds from group 5 will sing their population dialect and those from group 4 will sing the foreign dialect. Groups 1, 2, and 3 may not sing at all; they may sing a generalized song not corresponding to any of the dialects; or they may all sing the same song dialect — this dialect would then represent the "intrinsic" developmental program unmodified by learning.

If dialect differences are totally determined by genetic differences, birds from groups 4 and 5 will sing the same dialect, that of their parents. Birds from groups 1, 2, and 3, if they sing at all, will each sing the song dialect of their parent population, irrespective of the other birds in their group. There are then the possibilities of less clear-cut results, indicating that both genetic and learned differences influence the trait. For example, birds in group 4 might sing a song with both population elements. Note that, if the birds in the control group 5 do not sing their normal dialect, the rest of the results are uninterpretable, because the conditions of artificial rearing are interfering with the normal developmental program.

b. If the results of the first experiments show some heritability in the broad sense, then a further analysis is possible. This analysis requires a genetically segregating population, made from a cross between two dialect populations — say, A and B. A cross between males from population A and females from population B and the reciprocal cross will give an estimate of the average degree of dominance of genes influencing the trait and whether there is any sex linkage. (Remember that in birds the female is the heterogametic sex.) The offspring of this cross and all subsequent crosses *must* be raised in conditions that do not confuse the learned and the genetic com-

ponents of the differences, as revealed in the experiments in part a. If learned effects cannot be separated out, this further genetic analysis is impossible.

c. To localize genes influencing dialect differences would require a large number of segregating genetic markers. These markers could be morphological mutants or molecular variants such as restriction-site polymorphisms. Families segregating for the quantitative trait differences would be examined to see if there were cosegregation of any of the marker loci with the quantitative trait. These cosegregated loci would then be candidates for loci linked to the quantitative trait loci. Further crosses between individuals with and without mutant markers and measure of the quantitative trait values in F_2 individuals would establish whether there was actual linkage between the marker and quantitative trait loci. In practice, it is very unlikely that such experiments could be carried out on a songbird species, because of the immense time and effort required to establish lines carrying the large number of different marker genes and molecular polymorphisms.

SOLVED PROBLEMS

1. Two inbred lines of beans are intercrossed. In the F_1, the variance in bean weight is measured at 1.5. The F_1 is selfed; in the F_2, the variance in bean weight is 6.1. Estimate the broad heritability of bean weight in the F_2 population of this experiment.

◆ Solution ◆

The key here is to recognize that all the variance in the F_1 population must be environmental because all individuals must be of identical genotype. Furthermore, the F_2 variance must be a combination of environmental and genetic components, because all the genes that are heterozygous in the F_1 will segregate in the F_2 to give an array of different genotypes that relate to bean weight. Hence, we can estimate

$$s_e^2 = 1.5$$
$$s_e^2 + s_g^2 = 6.1$$

Therefore

$$s_g^2 = 6.1 - 1.5 = 4.6$$

and broad heritability is

$$H^2 = \frac{4.6}{6.1} = 0.75 \ (75\%)$$

2. In an experimental population of *Tribolium* (flour beetles), the body length shows a continuous distribution with a mean of 6 mm. A group of males and females with body lengths of 9 mm are removed and interbred. The body lengths of their offspring average 7.2 mm. From these data, calculate the heritability in the narrow sense for body length in this population.

◆ Solution ◆

The selection differential is $9 - 6 = 3$ mm, and the selection response is $7.2 - 6 = 1.2$ mm. Therefore, the heritability in the narrow sense is:

$$h^2 = \frac{1.2}{3} = 0.4 \ (40\%)$$

PROBLEMS

1. Distinguish between continuous and discontinuous variation in a population, and give some examples of each.

2. In a large herd of cattle, three different characters showing continuous distribution are measured, and the variances in the following table are calculated:

	CHARACTERS		
Variance	Shank length	Neck length	Fat content
Phenotypic	310.2	730.4	106.0
Environmental	248.1	292.2	53.0
Additive genetic	46.5	73.0	42.4
Dominance genetic	15.6	365.2	10.6

a. Calculate the broad- *and* narrow-sense heritabilities for each character.

b. In the population of animals studied, which character would respond best to selection? Why?

c. A project is undertaken to decrease mean fat content in the herd. The mean fat content is currently 10.5 percent. Animals of 6.5 percent fat content are interbred as parents of the next generation. What mean fat content can be expected in the descendants of these animals?

3. Suppose that two triple heterozygotes A/a ; B/b ; C/c are crossed. Assume that the three loci are in different chromosomes.

a. What proportions of the offspring are homozygous at one, two, and three loci, respectively?

b. What proportions of the offspring carry 0, 1, 2, 3, 4, 5, and 6 alleles (represented by capital letters), respectively?

4. In Problem 3, suppose that the average phenotypic effect of the three genotypes at the A locus is $A/A = 4$, $A/a = 3$, and $a/a = 1$ and that similar effects exist for the B and C loci. Moreover, suppose that the effects of loci add to each other. Calculate and graph the distribution of phenotypes in the population (assuming no environmental variance).

5. In Problem 4, suppose that there is a threshold in the phenotypic character so that, when the phenotypic value is above 9, the individual *Drosophila* has three bristles; when it is between 5 and 9, the individual has two bristles; and when the value is 4 or less, the individual has one bristle. Describe the outcome of crosses within and between bristle classes. Given the result, could you infer the underlying genetic situation?

6. Suppose that the general form of a distribution of a trait for a given genotype is:

$$f = 1 - \frac{(x - \bar{x})^2}{s_e^2}$$

over the range of x where f is positive.

a. On the same scale, plot the distributions for three genotypes with the following means and environmental variances:

Genotype	$\bar{x}$	s_e^2	Approximate range of phenotype
1	0.20	0.3	$x = 0.03$ to $x = 0.37$
2	0.22	0.1	$x = 0.12$ to $x = 0.24$
3	0.24	0.2	$x = 0.10$ to $x = 0.38$

b. Plot the phenotypic distribution that would result if the three genotypes were equally frequent in a popu-

lation. Can you see distinct modes? If so, what are they?

7. The following table shows a distribution of bristle number in *Drosophila:*

Bristle number	Number of individuals
1	1
2	4
3	7
4	31
5	56
6	17
7	4

Calculate the mean, variance, and standard deviation of this distribution.

8. The following sets of hypothetical data represent paired observations on two variables (x, y). Plot each set of data pairs as a scatter diagram. Look at the plot of the points, and make an intuitive guess about the correlation between x and y. Then calculate the correlation coefficient for each set of data pairs, and compare this value with your estimate.

a. (1, 1); (2, 2); (3, 3); (4, 4); (5, 5); (6, 6).

b. (1, 2); (2, 1); (3, 4); (4, 3); (5, 6); (6, 5).

c. (1, 3); (2, 1); (3, 2); (4, 6); (5, 4); (6, 5).

d. (1, 5); (2, 3); (3, 1); (4, 6); (5, 4); (6, 2).

9. A book on the problem of heritability of IQ makes the following three statements. Discuss the validity of each statement and its implications about the authors' understanding of h^2 and H^2.

a. "The interesting question then is . . . 'How heritable?' The answer [0.01] has a very different theoretical and practical application from the answer [0.99]." [The authors are talking about H^2.]

b. "As a rule of thumb, when education is at issue, H^2 is usually the more relevant coefficient, and when eugenics and dysgenics (reproduction of selected individuals) are being discussed, h^2 is ordinarily what is called for."

c. "But whether the different ability patterns derive from differences in genes . . . is not relevant to assessing discrimination in hiring. Where it could be relevant is in deciding what, in the long run, might be done to change the situation."

(From J. C. Loehlin, G. Lindzey, and J. N. Spuhler, *Race Differences in Intelligence.* Copyright © 1975 by W. H. Freeman and Company.)

10. Using the concepts of norms of reaction, environmental distribution, genotypic distribution, and phenotypic distribution, try to restate the following statement in more exact terms: "80 percent of the difference in IQ performance between the two groups is genetic." What would it mean to talk about the heritability of a difference between two groups?

11. Describe an experimental protocol for studies of relatives that could estimate the broad heritability of alcoholism. Remember that you must make an adequate observational definition of the trait itself.

12. A line selected for high bristle number in *Drosophila* has a mean of 25 sternopleural bristles, whereas a low-selected line has a mean of only 2. Marker stocks involving the two large autosomes II and III are used to create stocks with various mixtures of chromosomes from the high (h) and low (l) lines. The mean number of bristles for each chromosomal combination is as follows:

$$\frac{h\ h}{h\ h}\ 25.1 \qquad \frac{h\ h}{l\ h}\ 22.2 \qquad \frac{l\ h}{l\ h}\ 19.0$$

$$\frac{h\ h}{h\ l}\ 23.0 \qquad \frac{h\ h}{l\ l}\ 19.9 \qquad \frac{l\ h}{l\ l}\ 14.7$$

$$\frac{h\ l}{h\ l}\ 11.8 \qquad \frac{h\ l}{l\ l}\ 9.1 \qquad \frac{l\ l}{l\ l}\ 2.3$$

What conclusions can you reach about the distribution of genetic factors and their actions from these data?

13. Suppose that number of eye facets is measured in a population of *Drosophila* under various temperature conditions. Further suppose that it is possible to estimate total genetic variance (s_g^2) as well as the phenotypic distribution. Finally, suppose that there are only two genotypes in the population. Draw pairs of norms of reaction that would lead to the following results:

a. An increase in mean temperature decreases the phenotypic variance.

b. An increase in mean temperature increases H^2.

c. An increase in mean temperature increases s_g^2 but decreases H^2.

d. An increase in temperature *variance* changes a unimodal into a bimodal phenotypic distribution (one norm of reaction is sufficient here).

14. Francis Galton compared the heights of male undergraduates with the heights of their fathers, with the results shown in the graph below.

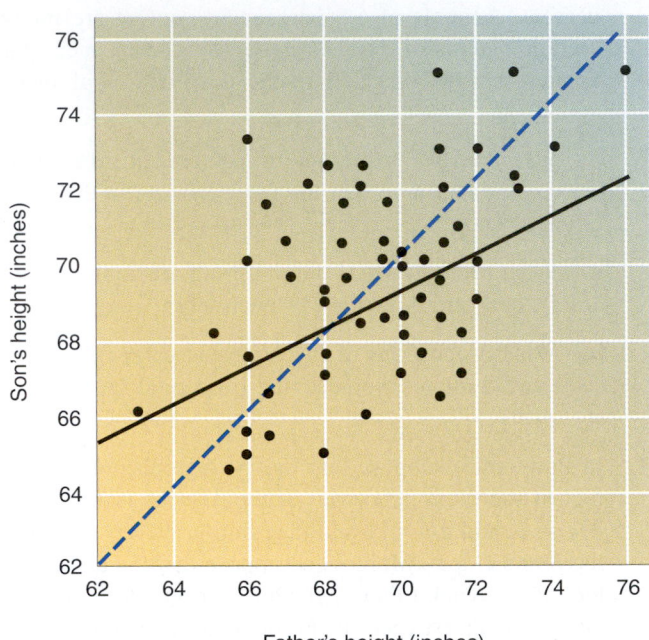

The average height of all fathers is the same as the average height of all sons, but the individual height classes are not equal across generations. The very tallest fathers had somewhat shorter sons, whereas the very short fathers had somewhat taller sons. As a result, the best line that can be drawn through the points on the scatter diagram has a slope of about 0.67 *(solid line)* rather than 1.00 *(dashed line)*. Galton used the term *regression* to describe this tendency for the phenotype of the sons to be closer than the phenotype of their fathers to the population mean.

a. Propose an explanation for this regression.

b. How are regression and heritability related here?

(Graph after W. F. Bodmer and L. L. Cavalli-Sforza, *Genetics, Evolution, and Man.* Copyright © 1976 by W. H. Freeman and Company.)

26

EVOLUTIONARY GENETICS

Charles Darwin. (Corbis/Bettmann.)

Key Concepts

Evolution consists of continuous heritable change of organisms within a single line of descent (phyletic evolution) and the differentiation between different lines of descent to form different species (diversification).

The Darwinian mechanism of evolution rests on three principles: (1) organisms within a species vary from one another, (2) the variation is heritable, and (3) different types leave different numbers of offspring in future generations.

Both phyletic change and diversification are the result of the interaction between the directional force of natural selection and random events.

Natural selection is the differential reproduction of different genotypes that is a consequence of their different physiological, morphological, and behavioral traits.

Random effects include the sampling of gametes each generation in finite populations and the random occurrence of mutations.

A consequence of the random factors in evolution is that the same forces of natural selection do not lead to the same evolutionary result in independent lines of descent.

Species are reproductively isolated populations of organisms that can exchange genes within the group but not with other species, because the groups are physiologically, behaviorally, or developmentally incompatible.

Evolutionary novelties are possible because new DNA is acquired either by duplication and subsequent differentiation of DNA already present in the species or by the introduction of novel DNA from other species.

The modern theory of evolution is so completely identified with the name of Charles Darwin (1809–1882) that many people think that the concept of organic evolution was first proposed by Darwin, but that is certainly not the case. Most scholars had abandoned the notion of fixed species, unchanged since their origin in a grand creation of life, long before publication of Darwin's *The Origin of Species* in 1859. By that time, most biologists agreed that new species arise through some process of evolution from older species; the problem was to explain *how* this evolution could occur.

Darwin's theory of the mechanism of evolution begins with the variation that exists among organisms within a species. Individuals of one generation are qualitatively different from one another. Evolution of the species as a whole results from the differential rates of survival and reproduction of the various types, so the relative frequencies of the types change over time. Evolution, in this view, is a sorting process.

For Darwin, evolution of the group resulted from the differential survival and reproduction of individual variants *already existing* in the group — variants arising in a way unrelated to the environment but whose survival and reproduction do depend on the environment.

MESSAGE ···
Darwin proposed a new explanation to account for the accepted phenomenon of evolution. He argued that the population of a given species at a given time includes individuals of varying characteristics. The population of the next generation will contain a higher frequency of those types that most successfully survive and reproduce under the existing environmental conditions. Thus, the frequencies of various types within the species will change over time.
··

There is an obvious similarity between the process of evolution as Darwin described it and the process by which the plant or animal breeder improves a domestic stock. The plant breeder selects the highest-yielding plants from the current population and (as far as possible) uses them as the parents of the next generation. If the characteristics causing the higher yield are heritable, then the next generation should produce a higher yield. It was no accident that Darwin chose the term **natural selection** to describe his model of evolution through differential rates of reproduction of different variants in the population. As a model for this evolutionary process, he had in mind the selection that breeders exercise on successive generations of domestic plants and animals.

We can summarize Darwin's theory of evolution through natural selection in three principles:

1. **Principle of variation.** Among individuals within any population, there is variation in morphology, physiology, and behavior.

2. **Principle of heredity.** Offspring resemble their parents more than they resemble unrelated individuals.

3. **Principle of selection.** Some forms are more successful at surviving and reproducing than other forms in a given environment.

Clearly, a selective process can produce change in the population composition only if there are some variations among which to select. If all individuals are identical, no amount of differential reproduction of individuals can affect the composition of the population. Furthermore, the variation must be in some part heritable if differential reproduction is to alter the population's genetic composition. If large animals within a population have more offspring than do small ones but their offspring are no larger on average than those of small animals, then no change in population com-

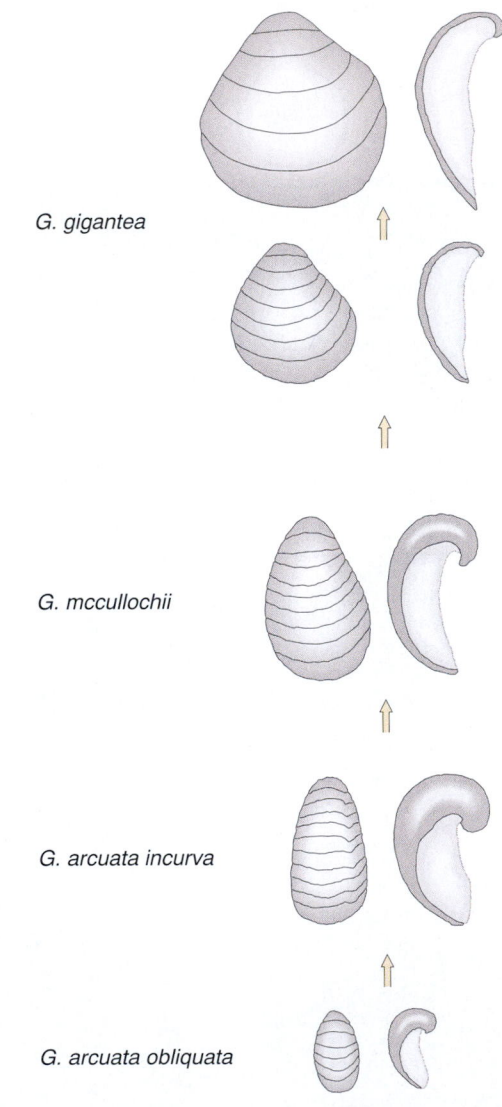

Figure 26-1 Changes in shell size and curvature in the bivalve mollusc *Gryphaea* during its phyletic evolution in the early Jurassic. Only the left shell is shown. In each case, the shell back and a longitudinal section through it are illustrated. (After A. Hallam, "Morphology, Palaeoecology and Evolution of the Genus *Gryphaea* in the British Lias," *Philosophical Transactions of the Royal Society of London Series B,* 254, 1968, 124.)

position can occur from one generation to another. Finally, if all variant types leave, on average, the same number of offspring, then we can expect the population to remain unchanged.

MESSAGE ···
Darwin's principles of variation, heredity, and selection must hold true if there is to be evolution by a variational mechanism.

The Darwinian explanation of evolution must apply to two different aspects of the history of life. One is the successive change of form and function that occurs in a single continuous line of descent time, **phyletic evolution.** Figure 26-1 shows such a continuous change over a period of 40 million years in the size and curvature of the left shell of the oyster, *Gryphea.* The other is the **diversification** that occurs among species: in the history of life on earth, there are many different contemporaneous species having quite different forms and living in different ways. Figure 26-2 shows some of the variety of bivalve mollusc forms that existed at various times in the past 130 million years. Every species eventually becomes extinct and more than 99.9 percent of all the species that have ever existed are already extinct, yet the number of species and the diversity of their forms and functions have increased in the past billion years. Thus species not only must be changing, but must give rise to new and different species in the course of evolution. Both of these processes are the consequences of heritable variation within populations. Heritable variation provides the raw material for successive changes within a species and for the multiplication of new species. The basic mechanisms of those changes (as discussed in Chapter 24) are the origin of new variation by various kinds of mutational mechanisms,

the change in frequency of alleles by selective and random processes, the possibility of divergence of isolated local populations because the selective forces are different or because of random drift, and the reduction of variation between populations by migration. From those basic mechanisms, population genetics, as discussed in Chapter 24, derives a set of principles governing changes in the genetic composition of populations. The application of these principles of population genetics provides an articulated theory of evolution.

MESSAGE ···
Evolution, under the Darwinian scheme, is the conversion of heritable variation between individuals within populations into heritable differences between populations in time and in space, by population genetic mechanisms.

A synthesis of forces: variation and divergence of populations

In evolution, the various forces of breeding structure, mutation, migration, and selection are all acting simultaneously in populations. We need to consider how these forces, operating together, mold the genetic composition of populations to produce both variation within local populations and differences between them.

The genetic variation within and between populations is a result of the interplay of the various evolutionary forces (Figure 26-3). Generally, as Table 26-1 shows, forces that increase or maintain variation within populations prevent the differentiation of populations from each other, whereas the divergence of populations is a result of forces that make each population homozygous. Thus, random drift (or

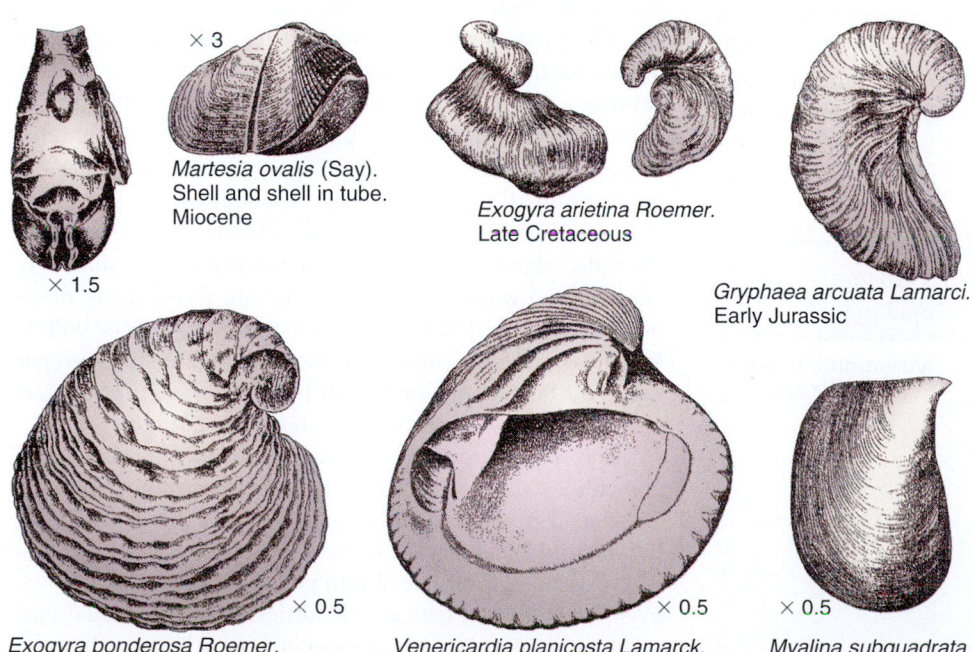

Martesia ovalis (Say).
Shell and shell in tube.
Miocene

×3

×1.5

Exogyra arietina Roemer.
Late Cretaceous

Gryphaea arcuata Lamarci.
Early Jurassic

Exogyra ponderosa Roemer.
Late Cretaceous
×0.5

Venericardia planicosta Lamarck.
Eocene
×0.5

Myalina subquadrata
Shumard
×0.5

Figure 26-2 A variety of bivalve mollusc shell forms that have appeared in the past 300 million years of evolution. (After C. L. Fenton and M. A. Fenton, *The Fossil Book,* Doubleday, 1958.)

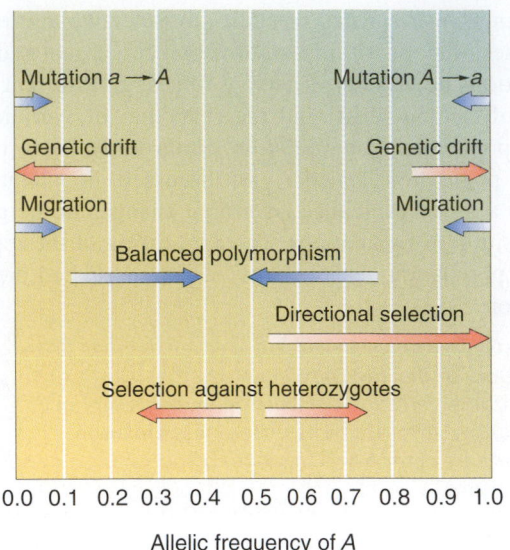

Figure 26-3 The effects on gene frequency of various forces of evolution. The blue arrows show a tendency toward increased variation within the population; the red arrows, decreased variation.

inbreeding) produces homozygosity while causing different populations to diverge. This divergence and homozygosity are counteracted by the constant flux of mutation and the migration between localities, which introduce variation into the populations again and tend to make them more like each other.

When Darwin arrived in the Galapagos Islands in 1835 he found a remarkable group of finchlike birds that provided a very suggestive case for the development of his theory of evolution. The Galapagos archipelago is a cluster of 29 islands and islets of different sizes lying on the equator about 600 miles off the coast of Ecuador. Figure 26-4 shows the 13 Galapagos finch species. Finches are generally ground-feeding seed eaters with stout bills for cracking the tough outer coats of the seeds. The Galapagos species, though

clearly finches, have an immense variation in how they make a living and in their bill shapes and their behaviors, which underly these ecological differences. For example, the vegetarian tree finch eats fruits and leaves, the insectivorous finch has a bill with a biting tip for eating large insects, and, most remarkable of all, the woodpecker finch grasps a twig in its beak and uses it to obtain insect prey by probing holes in trees. This diversity of species arose from an original population of a seed-eating finch that arrived in the Galapagos from the mainland of South America and populated the islands. The descendants of the original colonizers spread to the different islands and to different parts of large islands and formed local populations that diverged from each other and eventually formed different species.

Consider the situation at a genetically variable locus in a group of isolated island populations that were founded by migrants from an initial single population. The original founders of each population are small samples from the donor population and so differ from each other in allele frequencies because of a random sampling effect. This initial variation is called the **founder effect.** In succeeding generations, as a result of random genetic drift within each population, there is a further change in allelic frequency, p_i of each of the i alleles toward either 1 or 0, but average allelic frequency over all such populations remains p_i. Figure 26-5 shows the distribution of allelic frequencies among islands in successive generations, where $p(A_1) = 0.5$. In generation 0, all populations are identical. As time goes on, the gene frequencies among the populations diverge and some become fixed. After about $2N$ generations, every allelic frequency except the fixed classes ($p = 0$ and $p = 1$) is equally likely, and about half the populations are totally homozygous. By the time $4N$ generations have gone by, 80 percent of the populations are fixed, half being homozygous A/A and half being homozygous a/a.

The process of differentiation by inbreeding in island populations is slow, but not on an evolutionary or geological time scale. If an island can support, say, 10,000 individuals of a rodent species, then, after 20,000 generations (about 7000 years, assuming 3 generations per year), the population will be homozygous for about half of all the loci that were initially at the maximum of heterozygosity. Moreover, the island will be differentiated from other similar islands in two ways. For the loci that are fixed, many of the other islands will still be segregating, and others will be fixed at a different allele. For the loci that are still segregating in all the islands, there will be a large variation in gene frequency from island to island, as shown in Figure 26-5.

Any population of any species is finite in size, so all populations should eventually become homozygous and differentiated from one another as a result of inbreeding. Evolution would then cease. In nature, however, new variation is always being introduced into populations by mutation and by some migration between localities. Thus, the actual variation available for natural selection is a balance between the introduction of new variation and its loss through local

	How the Forces of Evolution Increase (+) or Decrease (−) Variation Within and Between Populations		
26-1 TABLE		

Force	Variation within populations	Variation between populations
Inbreeding or genetic drift	−	+
Mutation	+	−
Migration	+	−
Selection		
Directional	−	+/−
Balancing	+	−
Incompatible	−	+

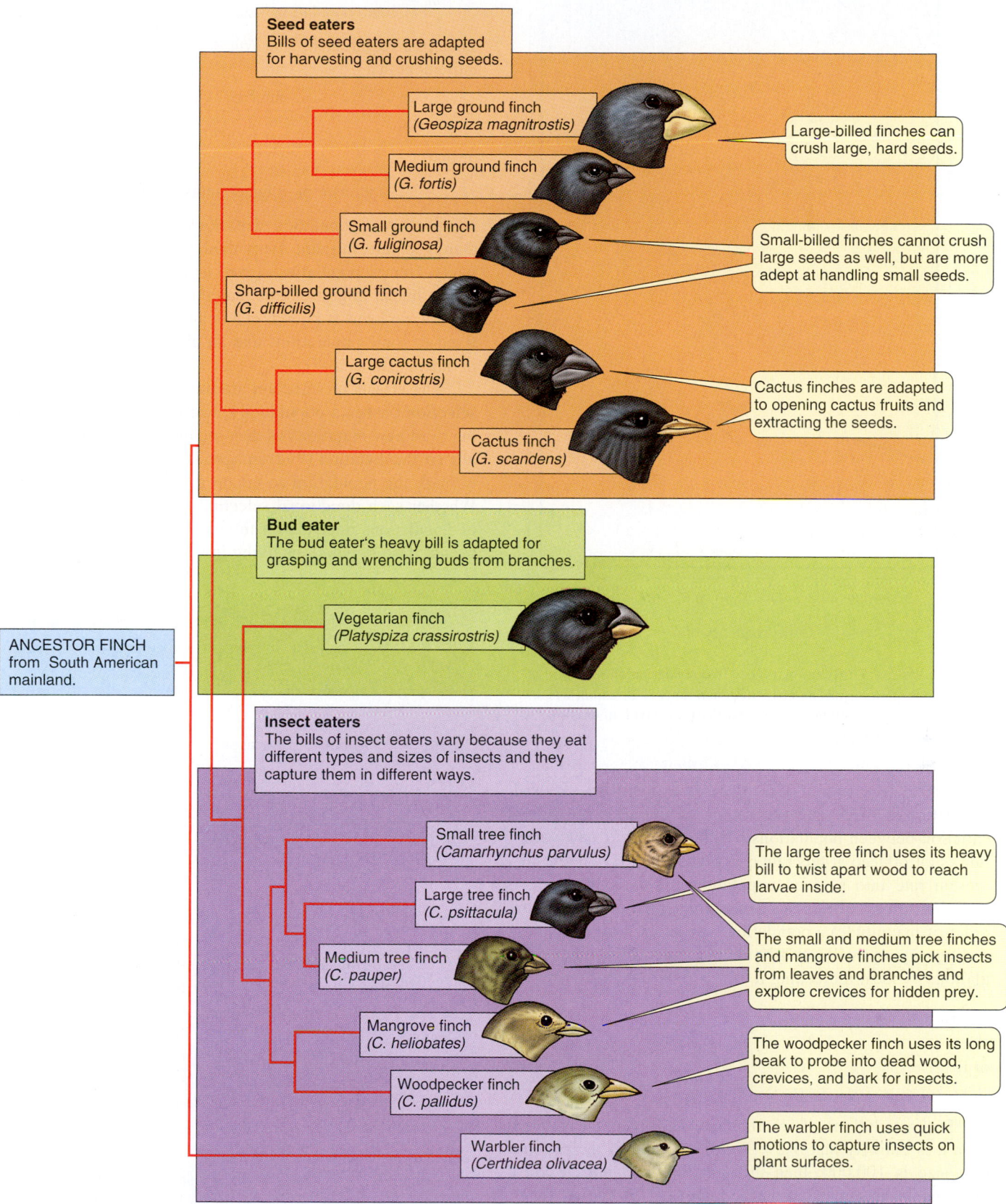

Seed eaters
Bills of seed eaters are adapted for harvesting and crushing seeds.

Large ground finch
(*Geospiza magnitrostis*)

Medium ground finch
(*G. fortis*)

Large-billed finches can crush large, hard seeds.

Small ground finch
(*G. fuliginosa*)

Sharp-billed ground finch
(*G. difficilis*)

Small-billed finches cannot crush large seeds as well, but are more adept at handling small seeds.

Large cactus finch
(*G. conirostris*)

Cactus finches are adapted to opening cactus fruits and extracting the seeds.

Cactus finch
(*G. scandens*)

Bud eater
The bud eater's heavy bill is adapted for grasping and wrenching buds from branches.

Vegetarian finch
(*Platyspiza crassirostris*)

ANCESTOR FINCH from South American mainland.

Insect eaters
The bills of insect eaters vary because they eat different types and sizes of insects and they capture them in different ways.

Small tree finch
(*Camarhynchus parvulus*)

Large tree finch
(*C. psittacula*)

The large tree finch uses its heavy bill to twist apart wood to reach larvae inside.

Medium tree finch
(*C. pauper*)

The small and medium tree finches and mangrove finches pick insects from leaves and branches and explore crevices for hidden prey.

Mangrove finch
(*C. heliobates*)

Woodpecker finch
(*C. pallidus*)

The woodpecker finch uses its long beak to probe into dead wood, crevices, and bark for insects.

Warbler finch
(*Certhidea olivacea*)

The warbler finch uses quick motions to capture insects on plant surfaces.

Figure 26-4 The thirteen species of finches found in the Galapagos Islands. (After W. K. Purves, G. H. Orians, and H. C. Heller, *Life: The Science of Biology,* 4th ed. (New York: Sinauer Associates/W. H. Freeman, 1995), Figure 20.3, p. 450.)

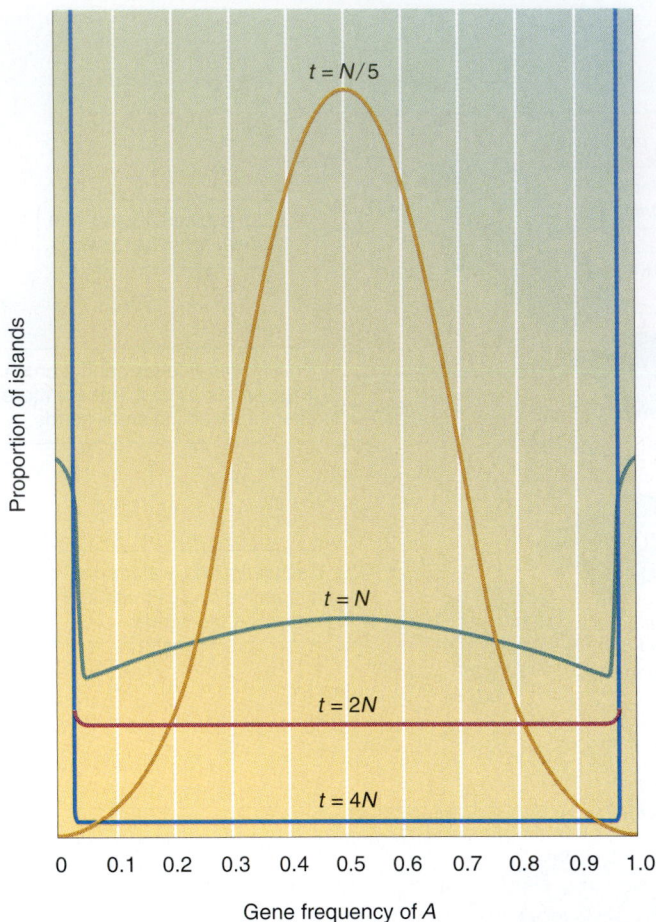

Figure 26-5   Distribution of gene frequencies among island populations after various numbers of generations of isolation, where the number of generations that have passed (t) is given in multiples of the population size (N).

inbreeding. The rate of loss of heterozygosity in a closed population is $1/(2N)$ per generation, so any effective differentiation between populations will be negated if new variation is introduced at this rate or a higher rate. If m is the migration rate into a given population and μ is the rate of mutation to new alleles per generations, then roughly (to an order of magnitude) a population will retain most of its heterozygosity and will not differentiate much from other populations by local inbreeding if

$$m \geq \frac{1}{N} \quad \text{or} \quad \mu \geq \frac{1}{N}$$

or if

$$Nm \geq 1 \quad \text{or} \quad N\mu \geq 1$$

For populations of intermediate and even fairly large size, it is unlikely that $N\mu \geq 1$. For example, if the population size is 100,000, then the mutation rate must exceed 10^{-5}, which is somewhat on the high side for known mutation rates, although it is not an unknown rate. On the other hand, a migration rate of 10^{-5} per generation is not unreasonably large. In fact

$$m = \frac{\text{number of migrants}}{\text{total population size}} = \frac{\text{number of migrants}}{N}$$

Thus, the requirement that $Nm \geq 1$ is equivalent to the requirement that

$$Nm = N \times \frac{\text{number of migrants}}{N} \geq 1$$

or that

$$\text{number of migrant individuals} \geq 1$$

irrespective of population size. For many populations, more than a single migrant individual per generation is quite likely. Human populations (even isolated tribal populations) have a higher migration rate than this minimal value, and, as a result, no locus is known in humans for which one allele is fixed in some populations and an alternative allele is fixed in others (Table 26-2).

The effects of selection are more variable. **Directional selection** pushes a population toward homozygosity, rejecting most new mutations as they are introduced but occasionally (if the mutation is advantageous) spreading a new allele through the population to create a new homozygous state. Whether such directional selection promotes differentiation of populations depends on the environment and on chance events. Two populations living in very similar environments

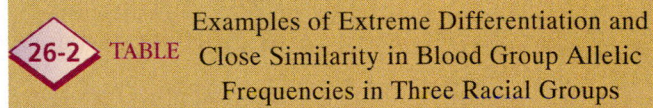

26-2 TABLE   Examples of Extreme Differentiation and Close Similarity in Blood Group Allelic Frequencies in Three Racial Groups

		POPULATION		
Gene	Allele	Caucasoid	Negroid	Mongoloid
Duffy	Fy	0.0300	0.9393	0.0985
	Fy^a	0.4208	0.0000	0.9015
	Fy^b	0.5492	0.0607	0.0000
Rhesus	R_0	0.0186	0.7395	0.0409
	R_1	0.4036	0.0256	0.7591
	R_2	0.1670	0.0427	0.1951
	r	0.3820	0.1184	0.0049
	r'	0.0049	0.0707	0.0000
	Others	0.0239	0.0021	0.0000
P	P_1	0.5161	0.8911	0.1677
	P_2	0.4839	0.1089	0.8323
Auberger	Au^a	0.6213	0.6419	No data
	Au	0.3787	0.3581	No data
Xg	Xg^a	0.67	0.55	0.54
	Xg	0.33	0.45	0.46
Secretor	Se	0.5233	0.5727	No data
	se	0.4767	0.4273	No data

Source: A summary is provided in L. L. Cavalli-Sforza and W. F. Bodmer, *The Genetics of Human Populations* (W. H. Freeman and Company, 1971), pp. 724–731. See L. L. Cavalli-Sforza, P. Menozzi, and A. Piazza, *The History and Geography of Human Genes* (Princeton University Press, 1994), for detailed data.

may be kept genetically similar by directional selection, but, if there are environmental differences, selection may direct the populations toward different compositions.

Selection favoring heterozygotes (balancing selection) will, for the most part, maintain more or less similar polymorphisms in different populations. However, again, if the environments are different enough between them, then the populations will show some divergence. The opposite of balancing selection is selection against heterozygotes, which produces unstable equilibria. Such selection will cause homozygosity and divergence between populations.

Multiple adaptive peaks

We must avoid taking an overly simplified view of the consequences of selection. At the level of the gene — or even at the level of the partial phenotype — the outcome of selection for a trait in a given environment is not unique. Selection to alter a trait (say, to increase size) may be successful in a number of ways. In 1952, F. Robertson and E. Reeve successfully selected to change wing size in *Drosophila* in two different populations. However, in one case, the *number* of cells in the wing changed, whereas, in the other case, the *size* of the wing cells changed. Two different genotypes had been selected, both causing a change in wing size. The initial state of the population at the outset of selection determined which of these selections occurred.

The way in which the same selection can lead to different outcomes can be most easily illustrated by a simple hypothetical case. Suppose that the variation of two loci (there will usually be many more) influences a character and that (in a particular environment) intermediate phenotypes have the highest fitness. (For example, newborn babies have a

higher chance of surviving birth if they are neither too big nor too small.) If the alleles act in a simple way in influencing the phenotype, then the three genetic constitutions *AB/ab*, *Ab/Ab*, and *aB/aB* will produce a high fitness because they will all be intermediate in phenotype. On the other hand, very low fitness will characterize the double homozygotes *AB/AB* and *ab/ab*. What will the result of selection be? We can predict the result by using the mean fitness $\bar{W}$ of a population. As previously discussed, selection acts in most simple cases to increase $\bar{W}$. Therefore, if we calculate $\bar{W}$ for every possible combination of gene frequencies at the two loci, we can determine which combinations yield high values of $\bar{W}$. Then we should be able to predict the course of selection by following a curve of increasing $\bar{W}$.

The surface of mean fitness for all possible combinations of allelic frequency is called an **adaptive surface** or an **adaptive landscape** (Figure 26-6). The figure is like a topographic map. The frequency of allele *A* at one locus is plotted on one axis, and the frequency of allele *B* at the other locus is plotted on the other axis. The height above the plane (represented by topographic lines) is the value of $\bar{W}$ that the population would have for a particular combination of frequencies of *A* and *B*. According to the rule of increasing fitness, selection should carry the population from a low-fitness "valley" to a high-fitness "peak." However, Figure 26-6 shows that there are two adaptive peaks, corresponding to a fixed population of *Ab/Ab* and a fixed population of *aB/aB*, with an adaptive valley between them. Which peak the population will ascend — and therefore what its final genetic composition will be — depends on whether the initial genetic composition of the population is on one side or the other of the dashed "fall line" shown in the figure.

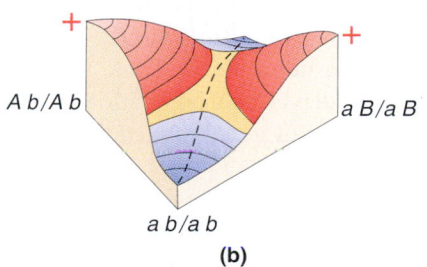

(b)

Figure 26-6 An adaptive landscape with two adaptive peaks (red), two adaptive valleys (blue), and a topographic saddle in the center of the landscape. The topographic lines are lines of equal mean fitness. If the genetic composition of a population always changes in such a way as to move the population "uphill" in the landscape, then the final composition will depend on where the population began with respect to the fall (dashed) line. (a) Topographic map of the adaptive landscape. (b) A perspective sketch of the surface shown in the map.

(a)

(a)

(b)

Figure 26-7   Differences in horn morphology in two geographically separated species of rhinoceros: (a) the African rhinoceros; (b) the Indian rhinoceros. (Part a from Anthony Bannister/NHPA; part b from K. Ghani/NHPA.)

MESSAGE ···
Under identical conditions of natural selection, two populations may arrive at two different genetic compositions as a direct result of natural selection.
··

It is important to note that nothing in the theory of selection requires that the different adaptive peaks be of the same height. The kinetics of selection is such that $\overline{W}$ increases, not that it necessarily reaches the highest possible peak in the field of gene frequencies. Suppose, for example, that a population is near the peak aB/aB in Figure 26-6 and that this peak is lower than the Ab/Ab peak. Selection alone cannot carry the population to Ab/Ab, because that would require a temporary decrease in $\overline{W}$ as the population descended the aB/aB slope, crossed the saddle, and ascended the other slope. Thus, the force of selection is myopic. It drives the population to a *local* maximum of $\overline{W}$ in the field of gene frequencies — not to a *global* one.

The existence of multiple adaptive peaks for a selective process means that some differences between species are the result of history and not of environmental differences. For example, African rhinoceroses have two horns, and Indian rhinoceroses have one (Figure 26-7). We need not invent a special story to explain why it is better to have two horns on the African plains and one in India. It is much more plausible that the trait of having horns was selected but that two long, slender horns and one short, stout horn are simply alternative adaptive features, and historical accident differentiated the species. Explanations of adaptations by natural selection do not require that every difference between species be differentially adaptive.

Exploration of adaptive peaks

Random and selective forces should not be thought of as simple antagonists. Random drift may counteract the force of selection, but it can enhance it as well. The outcome of the evolutionary process is a result of the simultaneous operation of these two forces. Figure 26-8 illustrates these pos-

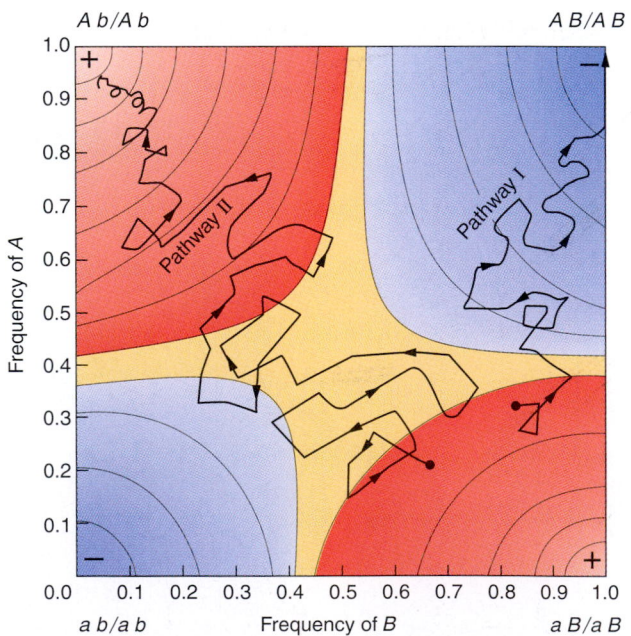

Figure 26-8   Selection and random drift can interact to produce different changes of gene frequency in an adaptive landscape. Without random drift, both populations would have moved toward aB/aB as a result of selection alone.

sibilities. Note that there are multiple adaptive peaks in this landscape. Because of random drift, a population under selection does not ascend an adaptive peak smoothly. Instead, it takes an erratic course in the field of gene frequencies, like an oxygen-starved mountain climber. Pathway I shows a population history where adaptation has failed. The random fluctuations of gene frequency were sufficiently great that the population by chance became fixed at an unfit genotype. In any population, some proportion of loci are fixed at a selectively unfavorable allele because the intensity of selection is insufficient to overcome the random drift to fixation. The existence of multiple adaptive peaks and the random fixation of less fit alleles are integral features of the evolutionary process. Natural selection cannot be relied on to produce the best of all possible worlds.

Pathway II in Figure 26-8, on the other hand, shows how random drift may improve adaptation. The population was originally in the sphere of influence of the lower adaptive peak; however, by random fluctuation in gene frequency, its composition passed over the adaptive saddle, and the population was captured by the higher, steeper adaptive peak. This passage from a lower to a higher adaptive stable state could never have occurred by selection in an infinite population, because, by selection alone, $\overline{W}$ could never decrease temporarily to cross from one slope to another.

An important source of indeterminacy in the outcome of a long selective process is the randomness of the mutational process. After the initial genetic variation is exhausted by the selective and random fixation of alleles, new variation arises from mutation that can be the source of yet

further evolutionary change. The particular direction of this further evolution depends on the particular mutations that occur and the time order in which they take place. A very clear illustration of this historical contingency of the evolutionary process is the selection experiment of H. Wichman and her colleagues to allow the bacteriophage ΦX174 to reproduce at high temperatures and to change its host from *Escherichia coli* to *Salmonella typhimurium*. Two independent selection lines were established and both evolved both the temperature and the host change. In one of the two lines, the ability to reproduce on *E. coli* still existed, but, in the other line, the ability was lost. The bacteriophage has only 11 genes, and the successive changes in the DNA for all these genes and in the proteins encoded by them were recorded during the selection process. The result for the two strains is shown in Table 26-3. There were 15 DNA changes in strain TX, located in 6 different genes; in strain ID, there were 14 changes in 4 of the genes. The strains had identical changes in only 7 cases, including a large deletion, but the temporal order of these identical changes differed between the lines. So, for example, the change at DNA site 1533, causing a substitution of isoleucine for threonine, was the third change in the ID strain, but the fourteenth change in the TX strain.

Heritability of variation

The first rule of any reconstruction or prediction of evolution is that the phenotypic variation must be heritable. It is easy to construct stories of the possible selective advantage of one

| 26-3   TABLE | Molecular Changes in Two Replicated Selection Lines, TX and ID, to Change the Temperature Tolerance and Host Range of the Bacteriophage ΦX174 |

	TX				ID			
Order	Site	Gene	Amino acid	Change	Site	Gene	Amino acid	Change
1	782	*E*	72	T → I	2167	*F*	388	H → Q
2	1727	*F*	242	L → F	1613	*F*	204	T → S
3	2085	*F*	361	A → V	**1533**	*F*	**177**	**T → I**
4	319	*C*	63	V → F	1460	*F*	153	Q → E
5	2973	*H*	15	G → S	1300	*F*	99	Silent
6	323	*C*	64	D → G	**1305**	*F*	**101**	**G → D**
7	**4110**	*A*	**44**	**H → Y**	1308	*F*	102	Y → C
8	1025	*F*	8	E → K	**4110**	*A*	**44**	**H → Y**
9	**3166**	*H*	**79**	**A → V**	4637	*A*	219	Silent
10	5185	*A**	402	T → M	**965–991**		**deletion**	
11	**1305**	*F*	**101**	**G → D**	5365	*A**	462	M → T
12	**965–991**		**deletion**		4168	*A*	63	*Q → R*
13	5365	*A**	462	M → T	3166	*H*	79	A → V
14	**1533**	*F*	**177**	**T → I**	1809	*F*	269	K → R
15	**4168**	*A*	**63**	**Q → R**				

Note: The temporal order of mutations is in the first column. The changes are given as the nucleotide site number, the gene name (*A–H*), the amino acid residue number, and the nature of the substitution. Amino acid abbreviations are as given in Table 9-3. Changes identical in the two strains are shown in boldface.
Sources: Data from H. A. Wichman, M. R. Badgett, L. A. Scott, C. M. Boulianne, and J. J. Bull, *Science* 285, 1999, 422–424.

form of a trait over another, but it is a matter of considerable experimental difficulty to show that the variation in the trait corresponds to genotypic differences (see Chapter 25).

It should not be supposed that all variable traits are heritable. Certain metabolic traits (such as resistance to high salt concentrations in *Drosophila*) show individual variation but no heritability. In general, behavioral traits have lower heritabilities than morphological traits, especially in organisms with more complex nervous systems that exhibit immense individual flexibility in central nervous states. Before any judgment can be made about the evolution of a particular quantitative trait, it is essential to determine if there is genetic variance for it in the population whose evolution is to be predicted. Thus, suggestions that such traits in the human species as performance on IQ tests, temperament, and social organization are in the process of evolving or have evolved at particular epochs in human history depend critically on evidence about whether there is genetic variation for these traits. Reciprocally, traits that appear to be completely invariant in a species may nevertheless evolve.

One of the most important findings in evolutionary genetics has been the discovery of substantial genetic variation underlying characters that show no morphological variation! These are called **canalized characters,** because the final outcome of their development is held within narrow bounds despite disturbing forces. Development is such that all the different genotypes for canalized characters have the same constant phenotype over the range of environments that is usual for the species. The genetic differences are revealed if the organisms are put in a stressful environment or if a severe mutation stresses the developmental system. For example, all wild-type *Drosophila* have exactly four scutellar bristles (Figure 26-9). If the recessive mutant *scute* is present, the number of bristles is reduced, but, in addition, there is variation from fly to fly. This variation is heritable, and lines with zero or one bristle and lines with three or four

bristles can be obtained by selection in the presence of the *scute* mutation. When the mutation is removed, these lines now have two and six bristles, respectively. Similar experiments have been performed by using extremely stressful environments in place of mutants. A consequence of such hidden genetic variation is that a character that is phenotypically uniform in a species may nevertheless undergo rapid evolution if a stressful environment uncovers the genetic variation.

Observed variation within and between populations

In Chapter 24, the existence of genetic variation within populations at the levels of morphology, karyotype, proteins, and DNA were documented. The general conclusion is that about one-third of all protein-encoding loci are polymorphic and that all classes of DNA, including exons, introns, regulatory sequences, and flanking sequences, show nucleotide diversity among individuals within populations. Several of these examples also documented some differences in the genotype frequencies between populations (see Tables 24-1 through 24-3, 24-5, and 24-7). The relative amounts of variation within and between populations vary from species to species, depending on history and environment. In humans, some gene frequencies (for example, those for skin color or hair form) are well differentiated between populations and major geographical groups (so-called geographical races). If, however, we look at single structural genes identified immunologically or by electrophoresis rather than by these outward phenotypic characters, the situation is rather different. Table 26-2 shows the three loci for which Caucasians, Negroids, and Mongoloids are known to be most different from one another (Duffy and Rhesus blood groups and the P antigen) compared with the three polymorphic loci for which the races are most similar (Auberger blood group and Xg and secretor factors). Even for the most divergent loci, no race is homozygous for one allele that is absent in the other two races.

In general, different human populations show rather similar frequencies for polymorphic genes. Figure 26-10 is a **triallelic diagram** for the three main allelic classes, I^A, I^B, and i, of the ABO blood group. Each point represents the allelic composition of a population, where the three allelic frequencies can be read by taking the lengths of the perpendiculars from each side to the point. The diagram shows that all human populations are bunched together in the region of high i, intermediate I^A, and low I^B frequencies. Moreover, neighboring points (enclosed by dashed lines) do not correspond to geographical races, so such races cannot be distinguished from one another by characteristic allelic frequencies for this gene. The study of polymorphic blood groups and enzyme loci in a variety of human populations has shown that about 85 percent of total human genetic diversity is found within local populations, about 7 percent is found among local populations within major geographical races,

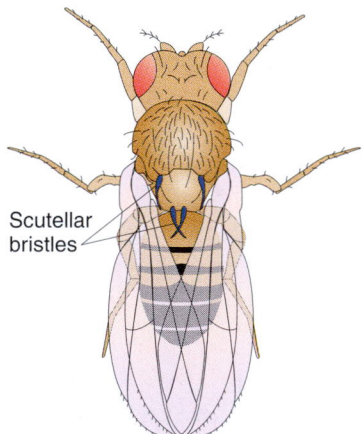

Figure 26-9   The scutellar bristles of the adult *Drosophila*, shown in blue. This is an example of a canalized character; all wild-type *Drosophila* have four scutellar bristles in a very wide range of environments.

Scutellar bristles

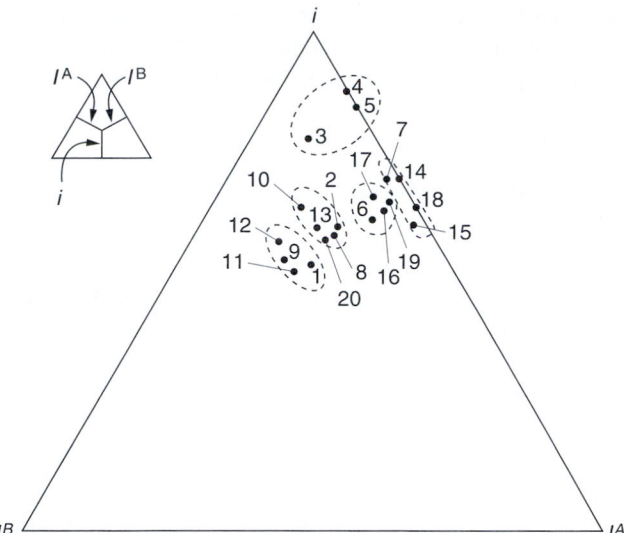

Figure 26-10 Triallelic diagram of the ABO blood group allelic frequencies for human populations. Each point represents a population; the perpendicular distances from the point to the sides represent the allelic frequencies, as indicated in the small triangle. Populations 1 to 3 are African, 4 to 7 are American Indian, 8 to 13 are Asian, 14 to 15 are Australian aborigine, and 16 to 20 are European. Dashed lines enclose arbitrary classes with similar gene frequencies; these groupings do not correspond to "racial" classes. (After A. Jacquard, *Structures génétiques des populations.* Copyright © 1970 by Masson et Cie.)

and the remaining 8 percent is found among major geographical races. Clearly, the genes influencing skin color, hair form, and facial form that are well differentiated among races are not a random sample of structural gene loci.

Process of speciation

When we examine the living world we see that individual organisms are usually clustered into collections that resemble each other more or less closely and are clearly distinct from other clusters. A close examination of a sibship of *Drosophila* will show differences in bristle number, eye size, and details of color pattern from fly to fly, but an entomologist has no difficulty whatsoever in distinguishing *Drosophila melanogaster* from, say, *Drosophila pseudoobscura.* One never sees a fly that is halfway between these two kinds. Clearly, in nature at least, there is no effective interbreeding between these two forms. A group of organisms that exchanges genes within the group but cannot do so with other groups is what is meant by a **species.** Within a species there may exist local populations that are also easily distinguished from one another by some phenotypic characters, but it is also the case that genes can easily be exchanged between them. Thus, no one has any difficulty distinguishing a "typical" Senegalese from a "typical" Swede, but, as a consequence of the migration and mating history of humans in North America in the past 300 years, an immense number of people exist of every degree of intermediacy between these local geographical types. They are not separate

species. A geographically defined population that is genetically distinguishable from other local populations but is capable of exchanging genes with those other local populations is sometimes called a **geographical race.** For example, in regard to the land snail *Cepaea nemoralis,* whose shell color and banding polymorphism was described in Chapter 24, there is a high frequency of albino shells in the Pyrenees at the higher elevations, but nowhere else; so we can distinguish a Pyrenees "race" of *Cepaea.* In general, there is some difference in the frequency of various genes in different geographical populations of a species, so the marking out of a particular population as a distinct race is arbitrary and, as a consequence, the concept of race is no longer much used in biology.

MESSAGE ⋯⋯⋯⋯⋯⋯⋯⋯⋯⋯⋯⋯⋯⋯⋯⋯⋯⋯

A species is a group of organisms that can exchange genes among themselves but are genetically unable to exchange genes in nature with other such groups. A geographical race is a phenotypically distinguishable local population within a species that is capable of exchanging genes with other races within that species. Because nearly all geographical populations are different from others in the frequencies of some genes, race is a concept that makes no clear biological distinction.

All the species now existing are related to each other by common ancestors at various times in the evolutionary past. That means that each of these species has separated out from a previously existing species and has become genetically distinct and genetically isolated from its ancestral line. In extraordinary circumstances, the founding of such a genetically isolated group might occur by a single mutation, but the carrier of that mutation would need to be capable of self-fertilization or vegetative reproduction. Moreover, that mutation would have to cause complete mating incompatibility between its carrier and the original species and to allow the new line to compete successfully with the previously established group. Although not impossible, such events must be rare.

The usual pathway to the formation of new species is through geographical races. As stated earlier in this chapter, populations that are geographically separated will diverge from each other genetically as a consequence of a combination of unique mutations, selection, and genetic drift. Migration between populations will prevent them from diverging too far, however. As shown on page 778, even a single migrant per generation will be sufficient to prevent populations from fixing at alternative alleles by genetic drift alone, and even selection toward different adaptive peaks will not succeed in causing complete divergence unless it is extremely strong. As a consequence, populations that diverge enough to become new, reproductively isolated species must first be virtually totally isolated from each other by some mechanical barrier. This isolation almost always requires some spatial separation, and the separation must be great enough or the natural barriers to the passage of migrants must be strong enough to prevent any effective migration.

Such populations are referred to as **allopatric.** The isolating barrier might be, for example, the extending tongue of a continental glacier during glacial epochs that forces apart a previously continuously distributed population, or the drifting apart of continents that become separated by water, or the infrequent colonization of islands that are far from shore. The critical point is whether the mechanisms of dispersal of the original species will make further migration between the separated populations a very rare event. If so, then the populations are now genetically independent and will continue to diverge by mutation, selection, and genetic drift. Eventually, the genetic differentiation between the populations becomes so great that the formation of hybrids between them would be physiologically, developmentally, or behaviorally impossible even if the geographical separation were abolished. These *biologically* isolated populations are now new species, formed by the process of **allopatric speciation.**

MESSAGE ·

Allopatric speciation occurs through an initial geographical and mechanical isolation of populations that prevents any gene flow between them, followed by genetic divergence of the isolated populations sufficient to make it biologically impossible for them to exchange genes in the future.

· ·

The forms of biological isolating mechanisms that arise between species include:

1. **Prezygotic isolation:** failure to form zygotes

   a. Lack of mating opportunity

      i. Temporal isolation: activity, fertility, or mating at different times or seasons

      ii. Ecological isolation: restriction to different, nonoverlapping habitats or ecological niches

   b. Lack of mating compatability

      i. Sexual, psychological, or behavioral incompatibility

      ii. Mechanical isolation: the failure of genitalia or flower parts to match

      iii. Gametic isolation: physiological incompatibility of sperm with the reproductive tract of the female in animals or of the pollen with the style in plants or a failure of successful fertilization of the egg cell or ovule

Examples of prezygotic isolating mechanisms are well known in plants and animals. The two species of pine growing on the Monterey peninsula, *Pinus radiata* and *P. muricata,* shed their pollen in February and April and so do not exchange genes. The light signals that are emitted by male fireflies and attract females differ in intensity and timing between species. In the tsetse fly, *Glossina,* mechanical in-

compatabilities cause severe injury and even death if males of one species mate with females of another. The pollen of different species of *Nicotiana,* the genus to which tobacco belongs, either fail to germinate or cannot grow down the style of other species.

2. **Postzygotic isolation:** failure of fertilized zygote to contribute gametes to future generations

   a. Hybrid inviability: hybrids either fail to develop or have a lower fitness than individuals of the parental species

   b. Hybrid sterility: partial or complete inability of adult hybrids of either sex to produce gametes in normal numbers

   c. Hybrid breakdown: sterility or inviability of the offspring of matings among hybrids or between hybrids and the parental species

Postzygotic isolation is more common in animals than in plants, apparently because the development of many plants is much more tolerant to genetic incompatabilities and chromosomal variations. When the eggs of the leopard frog, *Rana pipiens,* are fertilized by sperm of the wood frog, *R. sylvatica,* the embryos do not succeed in developing. Horses and asses can easily be crossed to produce mules, but, as is well known, these hybrids are sterile.

Genetics of species isolation

Usually, it is not possible to carry out any genetic analysis of the isolating mechanisms between two species for the simple reason that, by definition, they cannot be crossed with each other. It is possible, however, to make use of very closely related species in which the isolating mechanism is an incomplete hybrid sterility and hybrid breakdown. Then, segregating progeny of hybrid F_2 or backcross generations can be analyzed by using genetic markers. An example is shown in Figure 26-11. *Drosophila pseudoobscura* and *D. persimilis* are closely related species that never exchange genes in nature but can be crossed in the laboratory. The F_1 males are completely sterile, but the F_1 females have normal fertility and can be backcrossed to males of the parental species. A manifestation of hybrid male sterility is that, in the cross between *D. persimilis* females and *D. pseudoobscura* males, the testes of F_1 males are about one-fifth normal size. Genetically marking the chromosomes with visible mutants and backcrossing F_1 females to males of either species permits every combination of X chromosomes and autosomes to be identified and their effects on testis size to be determined. As Figure 26-11 shows, when an X chromosome from one species is present together with a complete diploid set of autosomes from the other species, the testes are at a minimum. As individual autosomes of the species to which the X chromosome belongs are substituted, the testis size increases, up to a complete haploid set of compat-

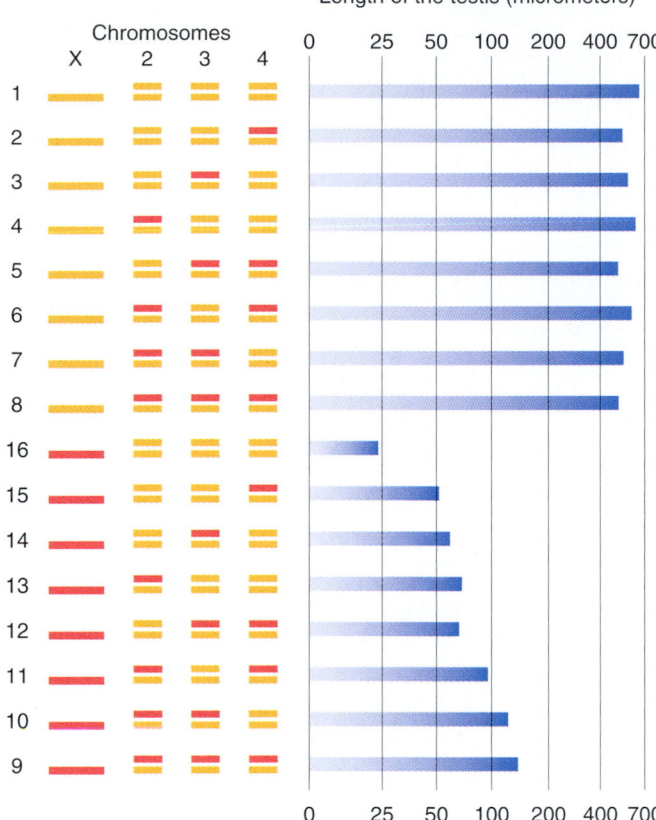

Figure 26-11   Testis size in backcross hybrids between *Drosophila pseudoobscura* and *D. persimilis.* Chromosomes of *D. pseudoobscura* are represented in orange, and those of *D. persimilis* in red.

ible autosomes but not beyond that. There is also evidence (not shown) of an interaction between the source of the cytoplasm and the X chromosome.

When such marker experiments have been performed on other species, mostly in the genus *Drosophila,* the general conclusions are that gene differences responsible for hybrid inviability are on all the chromosomes more or less equally and that, for hybrid sterility, there is some added effect of the X chromosome. For behavioral sexual isolation, the results are variable. In *Drosophila,* all the chromosomes are involved, but, in Lepidoptera, the genes are much more localized, apparently because specific pheromones are involved. The sex chromosome has a very strong effect in butterflies; in the European corn borer, only three loci, one of which is on the sex chromosome, account for the entire isolation between pheromonal races.

Origin of new genes

It is clear that evolution consists of more than the substitution of one allele for another at loci of defined function. New functions have arisen that have resulted in major new radiations of ways of making a living. Many of these new functions—for example, the development of the mammalian inner ear from a transformation of the reptilian jaw bones, result from continuous transformations of shape for which we do not have to invoke totally new genes and proteins. But qualitative novelties arise at the level of genes and proteins, such as the origination of photosynthesis in plants, of cell walls, of contractile proteins, of a variety of cell and tissue types, of oxygenation molecules such as hemoglobin, of the immune system, of chemical detoxification cycles, and of digestive enzymes. Older metabolic functions must have necessarily been maintained while new ones were being developed, which in turn means that old genes had to be preserved while new genes with new functions had to evolve. Where does the DNA for new functions come from?

Polyploidy

One process for the provision of new DNA is the duplication of the entire genome by polyploidization, much more common in plants than in animals (see Chapter 18). The evidence that polyploids have played a major role in the evolution of plant species is shown in Figure 26-12, which shows the frequency distribution of haploid chromosome numbers among dicotyledonous plant species. Above a chromosome number of about 12, even numbers are much more common than odd numbers—a consequence of frequent polyploidy.

Duplications

A second process for the increase in DNA is the duplication of small sections of the genome as a consequence of misreplication of DNA (Chapter 16). At first, after a duplicated segment has arisen, there is the possibility of an increase in the production of the polypeptide, but then functional differentiation between the sequences may occur in one of two directions. In one, no functional change occurs and there is simply a duplication of polypeptide production. The general function of the original sequence is maintained in the new DNA, but there is some differentiation of the sequences by accumulated mutations so that variations on the same protein theme are produced, allowing a somewhat more complex molecular structure. A classic example is the set of gene duplications and divergences that underlie the production of human hemoglobin. Adult hemoglobin is a tetramer consisting of two α polypeptide chains and two β chains, each with its bound heme molecule. The gene encoding the α chain is on chromosome 16 and the gene for the β chain is on chromosome 11, but the two chains are about 49 percent identical in their amino acid sequences, an identity that clearly points to the common origin. However, in fetuses, until birth, about 80 percent of β chains are substituted by a related γ chain. These two polypeptide chains are 75 percent identical, and the gene for the γ chain is close to the β-chain gene on chromosome 11 and has an identical intron-exon structure. This developmental change in globin synthesis is part of a larger set of developmental

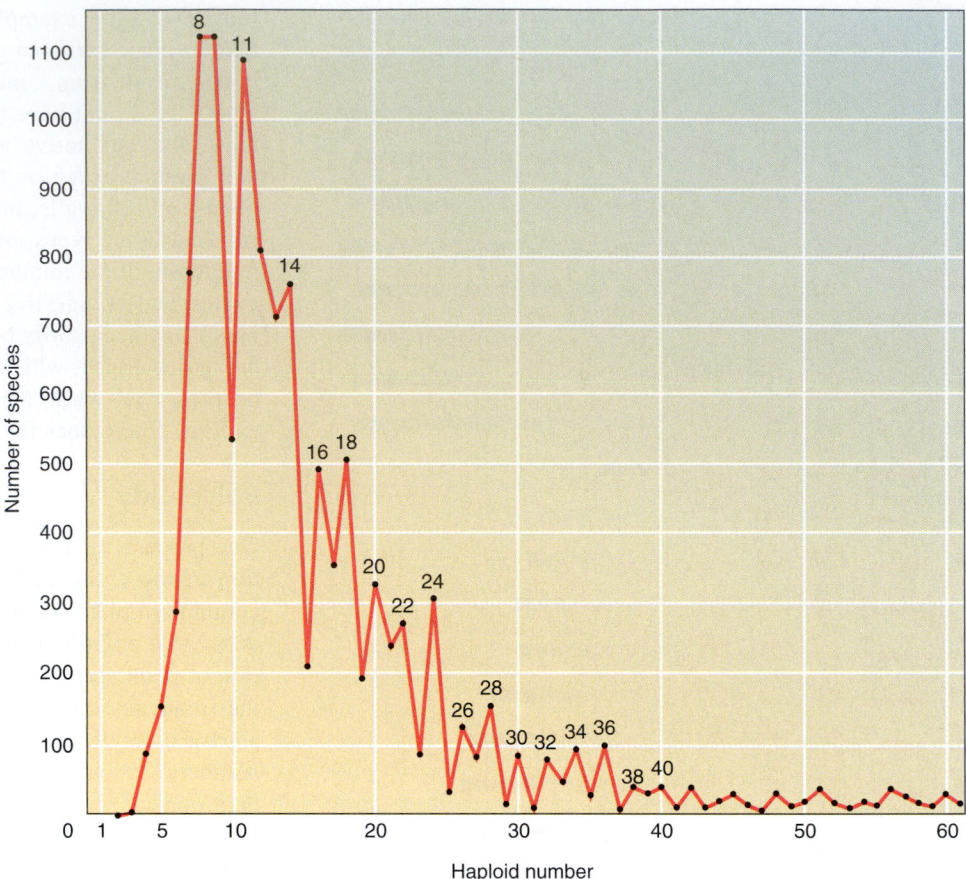

Figure 26-12 Frequency distribution of haploid chromosome numbers in dicotyledonous plants. (After Verne Grant, *The Origin of Adaptations,* Columbia University Press, 1963.)

changes that are shown in Figure 26-13. The early embryo begins with α, γ, ϵ, and ζ chains and, after about 10 weeks, the ϵ and ζ are replaced by α, β, and γ. Near birth, β replaces γ and a small amount of yet a sixth globin, δ, is produced.

Table 26-4 shows the percentage of amino acid identity among these chains, and Figure 26-14 shows the chromosomal locations and intron-exon structures of the genes encoding them. The story is remarkably consistent. The β, δ, γ, and ϵ chains all belong to a "β-like" group; they have very

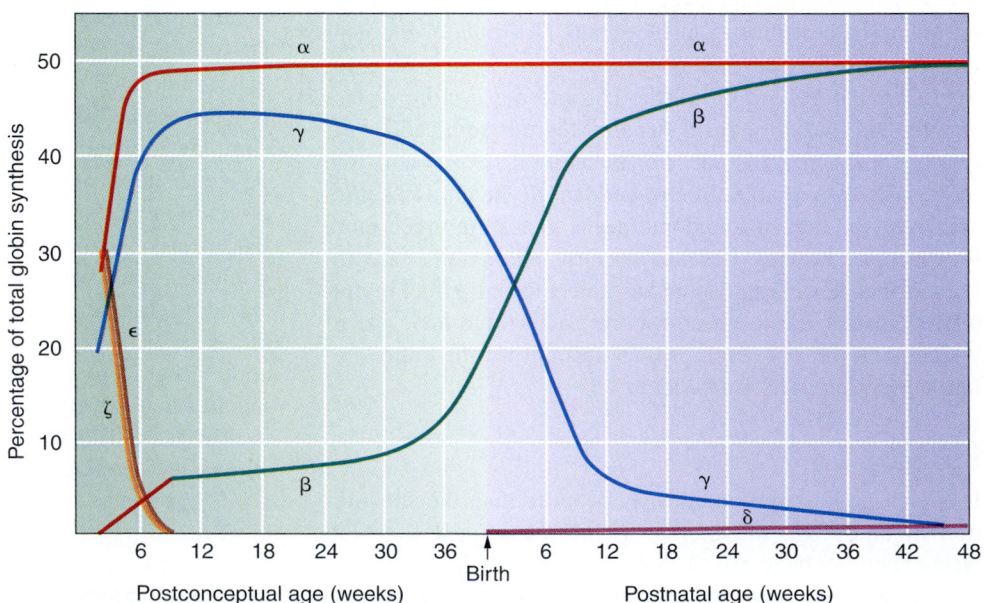

Figure 26-13 Developmental changes in the synthesis of the α-like and β-like globins that make up human hemoglobin.

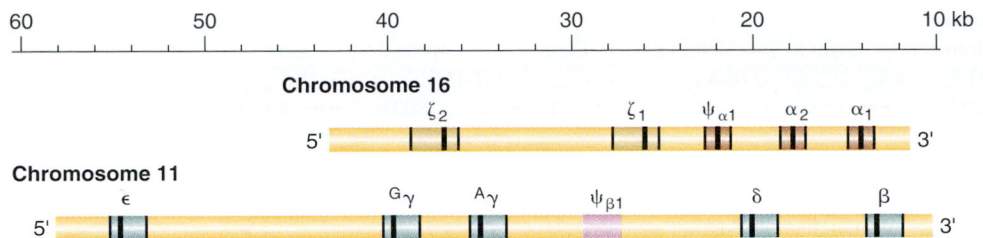

Figure 26-14 Chromosomal distribution of the genes for the α family of globins on chromosome 16 and the β family of globins on chromosome 11 in humans. Gene structure is shown by black bars (exons) and colored bars (introns).

similar amino acid sequences and are encoded by genes of identical intron-exon structure that are all contained in a 60-kb stretch of DNA on chromosome 11. The α and ζ chains belong to an "α-like" group and are encoded by genes contained in a 40-kb region on chromosome 16. Two slightly different forms of the α chain are encoded by neighboring genes with identical intron-exon structure, as are two forms of the ζ chain. In addition, Figure 26-14 shows that on both chromosome 11 and chromosome 16 are pseudogenes, labeled Ψ_α and Ψ_β. These pseudogenes are duplicate copies of the genes that did not acquire new functions but accumulated random mutations that render them nonfunctional. At every moment in development, hemoglobin molecules consist of two chains from the "α-like" group and two from the "β-like" group, but the specific members of the groups change in embryonic, fetal, and newborn life. What is even more remarkable is that the order of genes on each chromosome is the same as the temporal order of appearance of the globin chains in the course of development. This complexity of chain replacement evolved in mammals and is not present in fish, reptiles, birds, and monotremes, which have only the basic α,β system. Figure 26-15 shows the order of origin of the different components of the β-like system.

In the evolution of hemoglobin, the duplicated DNA encodes a function closely related to that served by the original gene from which it arose. The other possibility for evolution of duplicated DNA is a complete qualitative divergence in function. An example of such a divergence is shown in Figure 26-16. Birds and mammals, like other eukaryotic organisms, have a gene encoding lysozyme, a protective enzyme that breaks down the cell wall of bacteria. This gene has been duplicated in mammals to produce a second sequence that encodes a completely different, nonenzy-

matic protein, α-lactalbumin. Figure 26-16 shows that the duplicated gene has the same intron-exon structure as that of the lysozyme gene, whose array of four exons and three introns itself suggests an earlier multiple duplication event in the origin of lysozyme.

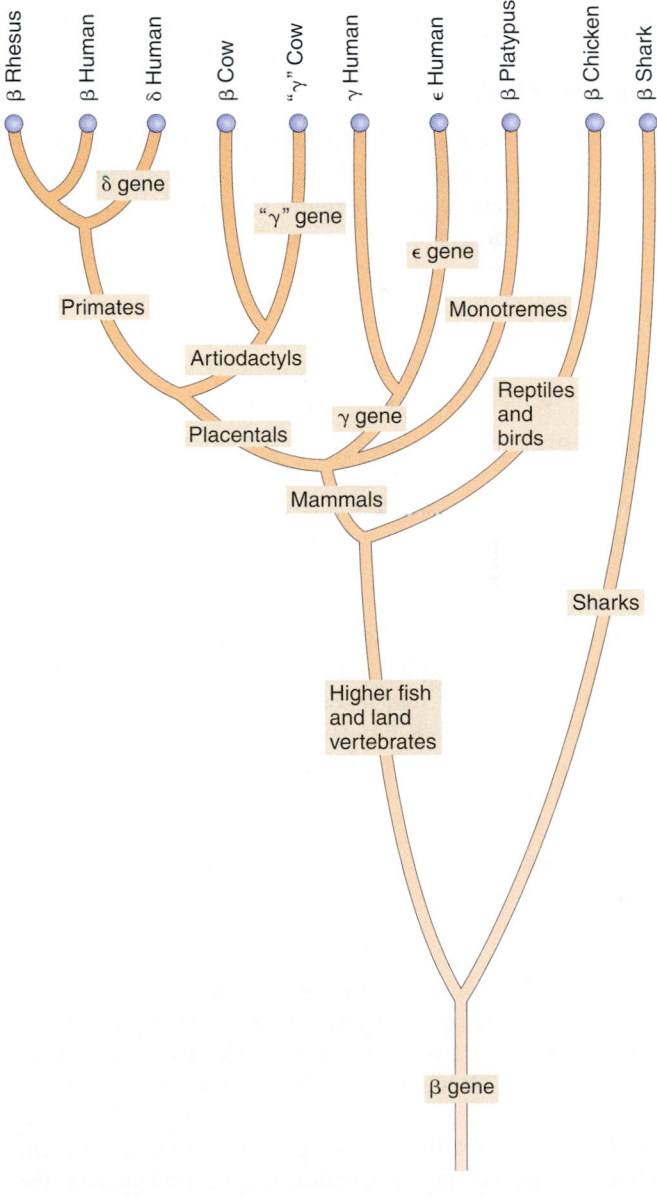

Figure 26-15 Reconstruction of the diversification of the β-globin-gene family in the evolution of the vertebrates.

	α	ζ	β	γ	ϵ
26-4 TABLE	Percentage of Similarity in Amino Acid Sequences Among Globin Chains in Humans				
α		58	42	39	37
ζ			34	38	37
β				73	75
γ					80

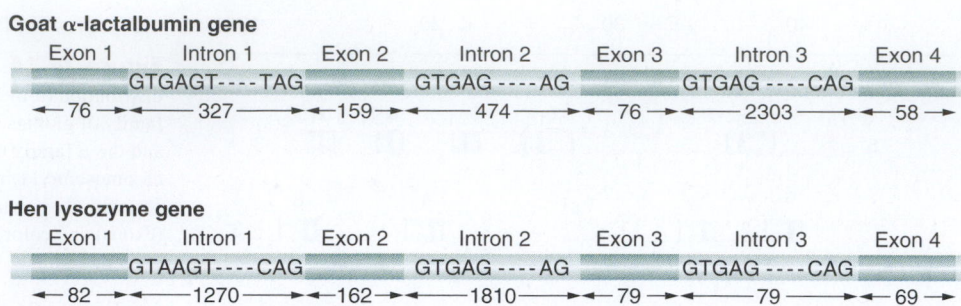

Figure 26-16 Structural homology of the gene for hen lysozyme and mammalian α-lactalbumin. Exons and introns are indicated by dark green bars and light green bars, respectively. Nucleotide sequences at the beginning and end of each intron are indicated, and the numbers refer to the nucleotide lengths of each segment. (After I. Kumagai, S. Takeda, and K.-I. Miura, "Functional Conversion of the Homologous Proteins α-Lactalbumin and Lysozyme by Exon Exchange," *Proceedings of the National Academy of Sciences USA* 89, 1992, 5887–5891.)

Imported DNA

New DNA that is the basis of new functions does not arise only from the duplication of DNA at an adjacent chromosomal location. Repeatedly in evolution, extra DNA has been imported into the genome from outside sources by mechanisms other than normal sexual reproduction. DNA can be inserted into chromosomes from other chromosomal locations and from other species, and genes from totally unrelated organisms can become incorporated into cells to become part of their heredity and function.

Cellular organelles. Eukaryotic cells contain cellular organelles such as mitochondria or the chloroplasts of photosynthetic organisms. Both of these organelles are the descendants of prokaryotes that entered the eukaryotic cells either as infections or from being ingested. These prokaryotes became symbionts, transferring much of their genome to the nuclei of their eukaryotic hosts but retaining genes that are essential to cellular functions. Mitochondria have retained about three dozen genes concerned with cellular respiration as well as some tRNA genes, whereas chloroplast genomes have about 130 genes encoding enzymes of the photosynthetic cycle as well as ribosomal proteins and tRNAs.

Important evidence for the extracellular origin of mitochondria is in their genetic code. The "universal" DNA–RNA code of nuclear genes is not, in fact, universal and differs in some respects from that in mitochondria. Table 26-5 shows that, for 6 of the 64 RNA triplets, mitochondria differ in their coding from the nuclear genome. Moreover, mitochondria in different organisms differ from each other for these coding elements, providing evidence that the invasion of eukaryotic cells by prokaryotes must have occurred at least five times, each time from a prokaryote with a different coding system. For the vertebrates, worms, and insects, the mitochondrial code is more regular than the universal nuclear code. In the nuclear genome, isoleucine is the only threefold

TABLE 26-5 Comparison of the Universal Nuclear DNA Code with Several Mitochondrial Codes for Six Triplets in Which They Differ

	TRIPLET CODE				
	TGA	ATA	AGA	AGG	AAA
Nuclear	Stop	Ile	Arg	Arg	Lys
Mitochondrial					
Mammalia	Trp	Met	Stop	Stop	Lys
Aves	Trp	Met	Stop	Stop	Lys
Amphibia	Trp	Met	Stop	Stop	Lys
Echinoderms	Trp	Ile	Ser	Ser	Asn
Insecta	Trp	Met	Ser	Stop	Lys
Nematodes	Trp	Met	Ser	Ser	Lys
Platyhelminth	Trp	Met	Ser	Ser	Asn
Cnidaria	Trp	Ile	Arg	Arg	Lys

redundantly coded amino acid, with methionine being coded by the fourth member of the codon group, separated by a transition from A to G; whereas, in mitochondria, there are two methionines separated by a transversion from the two isoleucines in this group.

Horizontal transfer. It is now clear that the nuclear genome is open to the insertion of DNA both from other parts of the same genome and from outside (see Chapter 20). The chromosomes of an individual *Drosophila*, for example, contain a large variety of families of transposable elements with multiple copies of each distributed around the genome. As much as 25 percent of the DNA of *Drosophila* may be of transposable origin. It is not clear at present what role this mobile DNA plays in functional evolution. The transposition that occurs when transposable elements are introduced into zygotes at mating, as in the P elements of *Drosophila*, results in an explosive proliferation of the elements in the recipient genome. When a mobile element is inserted into a gene, the resulting mutation usually has a drastic deleterious effect on the organism, but this effect may be an artifact of the methods used to detect the presence of such elements. Laboratory selection experiments on quantitative characters have shown that transposition can act as an added source of selectable variation. Finally, there is the possibility that genes are transferred from the nuclear genome of one species to the nuclear genome of another by the phenomenon of retrotransposition mediated by retroviruses (see Chapter 20). This possibility could be a powerful source of the acquisition of new functions by a species because such retroviruses could be carried between very distantly related species by common disease vectors such as insects or by bacterial infections.

Relation of genetic to functional change

There is no simple relation between the amount of change in DNA and how much functional change occurs in the encoded protein. At one extreme, almost the entire amino acid sequence of a protein can be replaced while maintaining the original function. Eukaryotes, from yeast to humans, produce an enzyme, lysozyme, that breaks down bacterial cell walls. In the evolutionary divergence that has occurred in the yeast and vertebrate lines since their descent from an ancient common ancestor, virtually every amino acid in this protein has been replaced; so an alignment of their two protein or DNA sequences would not reveal any similarity. The evidence that they are descended from an original common ancestral gene comes from comparisons of evolutionarily intermediate forms that show more and more divergence of sequence as species are more divergent. The maintenance of the function despite the replacement of the amino acids has been the result of the maintenance of the three-dimensional structure of the enzyme by the selective substitutions of just the right amino acids to maintain shape.

In contrast, it is possible to change the function of an enzyme by a single amino acid substitution. The sheep blow fly, *Lucilia cuprina*, has developed resistance to organophosphate insecticides used widely to control it. R. Newcombe, P. Campbell, and their colleagues showed that this resistance is the consequence of a single substitution of an aspartic acid for a glycine in the active site of an enzyme that ordinarily is a carboxylesterase. The mutation causes complete loss of the carboxylesterase activity and its replacement by esterase specificity. A three-dimensional modeling of the molecule indicates that the change is the result of the ability of the substituted protein to bind a water molecule close to the site of attachment of the organophosphate, which is then hydrolyzed by the water.

MESSAGE
There is no regular relation between how much DNA change takes place in evolution and how much functional change results.

When more than one mutation is required for the origin of a new function, the order in which these mutations occur in the evolution of the molecule may be critical. B. Hall has experimentally changed a gene to a new function in *E. coli* by a succession of mutations and selection. In addition to the *lacZ* genes specifying the usual lactose-fermenting activity in *E. coli,* another structural gene locus, *ebg,* specifies another β-galactosidase that does not ferment lactose, although it is induced by lactose. The natural function of this second gene is unknown. Hall was able to mutate and select this extra gene to enable *E. coli* to live, without any lactose, on a wholly new substrate, galactobionate. To do so, he first had to mutate the regulatory sequence of *ebg* so that it was constitutive and no longer required lactose to induce its translation. Next, he tried to select mutants that would ferment lactobionate, but he failed. First, it was necessary to select a form that would ferment a related substrate, lactulose, and then the lactulose fermenters could be mutated and selected to operate on lactobionate. Moreover, only some of the independent mutants from lactose fermentation to lactulose utilization could be further mutated and selected to operate on lactobionate. The others were dead ends. Thus, the sequence of evolution had to be (1) from an inducible to a constitutive enzyme, followed by (2) just the right mutation from lactose to lactulose fermentation, followed by (3) a mutation to ferment lactobionate.

MESSAGE
In the evolution of new functions by mutation and selection, there are particular pathways through the array of mutations that must be followed. Other pathways come to dead ends that do not allow further evolution.

Rate of molecular evolution

Although it is possible that only one or a few mutations lead to a change in the specificity of a protein, the more usual situation is that DNA accumulates substitutions over long periods of evolution without making a qualitative change in the functional properties of the proteins that are encoded. There may, however, be smaller effects influencing the kinetic properties, timing of production, or quantities of the encoded proteins that, in turn, will affect the fitness of the organism that carries them. Mutations of DNA can have three effects on fitness. First, they may be deleterious, reducing the probability of survival and reproduction of their carriers. All of the laboratory mutants used by the experimental geneticist have some deleterious effect on fitness. Second, they may actually increase fitness by providing increased efficiency or by expanding the range of environmental conditions in which the species can make a living or by enabling the organism to track changes in the environment. Third, they may have no effect on fitness, leaving the probability of survival and reproduction unchanged; they are the so-called neutral mutations. For the purposes of understanding the rate of molecular evolution, however, we need to make a slightly different distinction — that between *effectively neutral* and *effectively selected* mutations. In Chapter 24, we learned that, in a finite population of N individuals, the process of random genetic drift will not be materially altered if the intensity of selection, s, on an allele is of lower order than $1/N$. That means that the class of evolutionarily neutral mutations includes both those that have absolutely no effect on fitness and those whose fitness effects are less than the reciprocal of population size, so small as to be effectively neutral.

We would like to know how much of molecular evolution is a consequence of new, favorable adaptive mutations sweeping through a species, the picture presented by a simplistic Darwinian view of evolution, and how much is simply the accumulation of the random fixation of effectively neutral mutations. Mutations that are effectively deleterious need not be considered, because they will be kept at low frequencies in populations and will not contribute to evolutionary change. If a newly arisen mutation is effectively neutral then, as pointed out in Chapter 24, there is a probability of $1/(2N)$ that it will replace the previous allele because of random genetic drift. If the rate of appearance of new effectively neutral mutations at a locus per gene copy per generation is μ, then the absolute number of new mutational copies that will appear in a population of N diploid individuals is $2N\mu$. Each one of these new copies has a probability of $1/(2N)$ of eventually taking over the population. Thus, the absolute rate of replacement of old alleles by new ones at a locus per generation is their rate of appearance multiplied by the probability that any one of them will eventually take over by drift

$$\text{Rate of neutral replacement} = 2N\mu \times 1/(2N) = \mu$$

That is, we expect that in every generation there will be μ substitutions of a new allele for an old one at each locus in the population, purely from genetic drift of effectively neutral mutations.

MESSAGE ··
The rate of replacement in evolution resulting from the random genetic drift of effectively neutral mutations is equal to the mutation rate to such alleles, μ.
···

The constant rate of neutral substitution predicts that, if the number of nucleotide differences between two species is plotted against the time since their divergence from a common ancestor, the result should be a straight line with a slope equal to μ. That is, evolution should proceed according to a **molecular clock** that is ticking at the rate μ. Figure 26-17 shows such a plot for the β-globin gene. The results are quite consistent with the claim that nucleotide substitutions have been effectively neutral over the past 500 million years. Two sorts of nucleotide substitutions are plotted: **synonymous substitutions** that are from one alternative codon to another, making no change in the amino acid, and **nonsynonymous substitutions** that result in an amino acid change. Figure 26-17 shows a much lower slope for nonsyn-

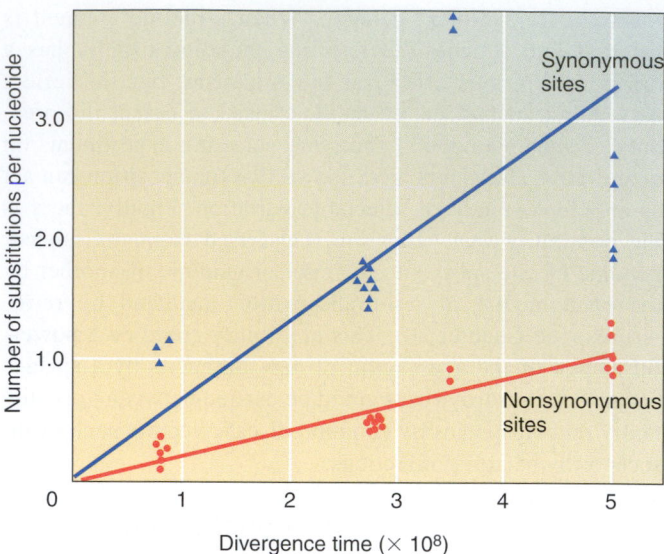

Figure 26-17   The amount of nucleotide divergence at synonymous and nonsynonymous sites of the β-globin gene as a function of time since divergence.

onymous substitutions than that for synonymous changes, which means that the mutation rate to nonsynonymous substitutions is much lower than that to synonymous ones. This is precisely what we expect. The mutation rate to neutral alleles, μ, is the product of the intrinsic nucleotide mutation rate, M, and the proportion of all mutations that are neutral, f. That is,

$$\mu = M \times f$$

and it is reasonable that mutations that cause an amino acid substitution should more often have a deleterious effect, s, above the threshold for neutral evolution and therefore the proportion of neutral changes, f, will be smaller. It is important to note that these observations do not show that synonymous substitutions have *no* selective constraints on them; rather they show that these constraints are, on the average, not as strong as for amino acids. In fact, synonymous changes do have effects on probabilities of correct splicing, on the stability and lifetime of mRNA, on use by the translation apparatus of the available pool of tRNA molecules, and thus on the rate of translation and the folding of the translated polypeptide.

Another prediction of neutral evolution is that different proteins will have different clock rates, because the metabolic function of some proteins will be much more sensitive to changes in their amino acid sequence. Proteins in which every amino acid makes a difference will have smaller values of the effectively neutral mutation rate, Mf, than will proteins that are more tolerant of substitution. Figure 26-18 shows a comparison of the clocks for fibrinopeptides, hemoglobin, and cytochrome c. That fibrinopeptides have a much higher proportion of neutral mutations is reasonable because these peptides are merely a nonmetabolic safety catch, cut

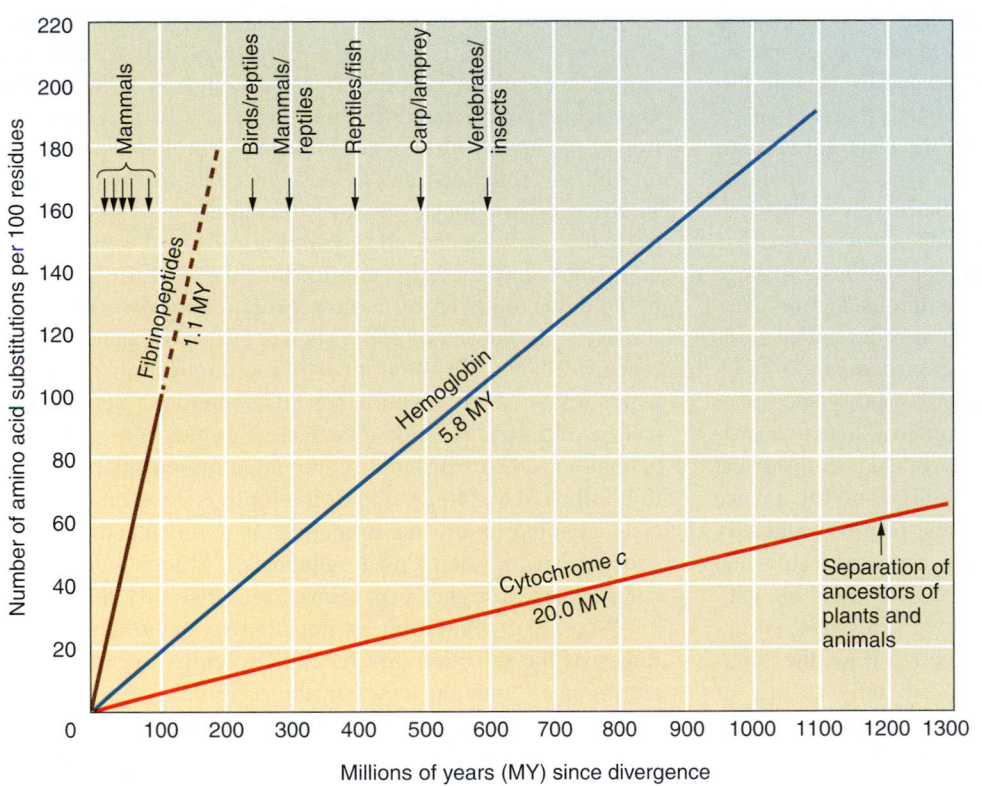

Figure 26-18 Number of amino acid substitutions in the evolution of the vertebrates as a function of time since divergence. The three proteins — fibrinopeptides, hemoglobin, and cytochrome c — differ in rate because different proportions of their amino acid substitutions are selectively neutral.

out of fibrinogen to activate the blood-clotting reaction. From a priori considerations, why hemoglobins are less sensitive to amino acid changes than is cytochrome c is less obvious.

MESSAGE ....

The rate of neutral evolution for the amino acid sequence of a protein depends on the sensitivity of a protein's function to amino acid changes.

....

The demonstration of the molecular clock argues that most nucleotide substitutions are neutral, but it does not tell us how much of molecular evolution is adaptive. One way of detecting adaptive evolution of a protein is by comparing the synonymous and nonsynonymous polymorphisms within species with the synonymous and nonsynonymous changes between species. Under the operation of evolution by random genetic drift, polymorphism within a species is simply a stage in the eventual fixation of a new allele; so the ratio of nonsynonymous to synonymous polymorphisms within a species should be the same as the ratio of nonsynonymous to synonymous substitutions between species. On the other

hand, if the amino acid changes between species have been driven by a positive adaptive selection, there ought to be an excess of nonsynonymous changes between species. Table 26-6 shows an application of this principle by J. MacDonald and M. Kreitman to the alcohol dehydrogenase gene in three closely related species of *Drosophila*. Clearly, there is an excess of amino acid replacements between species over what is expected from the polymorphisms.

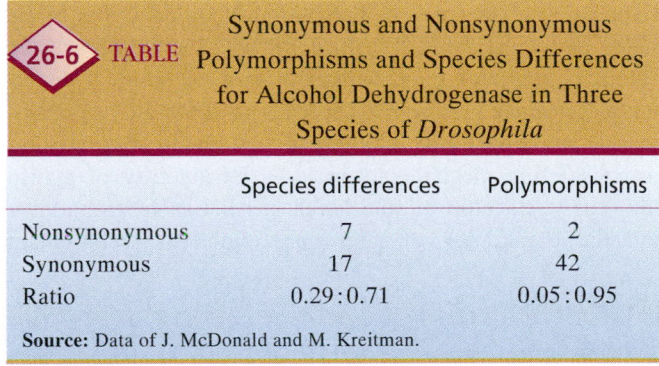

26-6 **TABLE** Synonymous and Nonsynonymous Polymorphisms and Species Differences for Alcohol Dehydrogenase in Three Species of *Drosophila*

	Species differences	Polymorphisms
Nonsynonymous	7	2
Synonymous	17	42
Ratio	0.29:0.71	0.05:0.95

Source: Data of J. McDonald and M. Kreitman.

SUMMARY

The Darwinian theory of evolution is a variational scheme that explains the changes that occur in populations of organisms as being the result of changes in the relative frequencies of different variants in the population. The processes that give rise to the variation within the population are causally independent of the processes that are responsible for

the differential reproduction of the various types. It is this independence that is meant when it is said that mutations are "random." The process of mutation supplies undirected variation, whereas the process of natural selection culls this variation, increasing the frequency of those variants that by chance are better able to survive and reproduce. Many are called, but few are chosen. Thus, if there is no variation within a species for some trait, there can be no evolution. Moreover, that variation must be influenced by genetic differences. If differences are not heritable, they cannot evolve, because the differential reproduction of the different variants will not carry across generational lines. Thus, all hypothetical evolutionary reconstructions depend critically on whether the traits in question are, in fact, heritable.

The evolutionary divergence of populations in space and time is not only a consequence of natural selection. Natural selection is not a globally optimizing process that finds the "best" organisms for a particular environment. Instead, it finds one of a set of alternative "good" solutions to adaptive problems, and the particular outcome of selective evolution in a particular case is subject to chance historical events. Random factors such as genetic drift and the chance occurrence or loss of new mutations may result in radically different outcomes of an evolutionary process even when the force of natural selection is the same. The metaphor usually employed is that there is an "adaptive landscape" of genetic combinations and that natural selection leads the population to a "peak" in that landscape, but only to one of several alternative local peaks.

The vast diversity of different living forms that have existed is a consequence of independent evolutionary histories that have occurred in separate populations. For different populations to diverge from each other, they must not exchange genes; so the independent evolution of large numbers of different species requires that these species be reproductively isolated from each other. Indeed, we define a species as a population of organisms that exchange genes within it but is reproductively isolated from other populations. The mechanisms of reproductive isolation may be prezygotic or postzygotic. Prezygotic isolating mechanisms are those that prevent the union of gametes of two species. These mechanisms may be behavioral incompatability of the males and females of the different species, differences in timing or place of their sexual activity, anatomical differences that make mating mechanically impossible, or physiological incompatability of the gametes themselves. Postzygotic isolating mechanisms include the inability of hybrid embryos to develop to adulthood, sterility of hybrid adults, and the breakdown of later generations of recombinant

genotypes. For the most part, the genetic differences responsible for the isolation between closely related species are spread over all the chromosomes, although in species with chromosomal sex determination there may be a concentration of incompatability genes on the sex chromosome.

If new functions are to arise in evolution without the sacrifice of previously existing functions, new DNA must be made available for the evolution of added genes. This new DNA may arise by duplication of the entire genome, polyploidy, followed by a slow evolutionary divergence of the extra chromosomal set. This has been a frequent occurrence in plants. An alternative is the duplication of single genes followed by selection for differentiation. Yet another source of DNA, recently discovered, is the entry into the genome of DNA from totally unrelated organisms by infection followed by integration of the foreign DNA into the nuclear genome or by the formation of extranuclear cell organelles with their own genomes. Mitochondria and chloroplasts in higher organisms have arisen by this route.

Not all of evolution is impelled by natural selective forces. If the selective difference between two genetic variants is small enough, less than the reciprocal of population size, there may be a replacement of one allele by another purely by genetic drift. A great deal of molecular evolution seems to be the replacement of one protein sequence by another one of equivalent function. The evidence for this neutral evolution is that the number of amino acid differences between two different species in some molecule — for example, hemoglobin — is directly proportional to the number of generations since their divergence from a common ancestor in the evolutionary past. Such a "molecular clock" with a constant rate of change would not be expected if the selection of differences were dependent on particular changes in the environment. Moreover, we expect the clock to run faster for proteins such as fibrinopeptides, in which the amino acid composition is not critical for the function, and this difference in clock rate is, in fact, observed. Thus we cannot assume without evidence that evolutionary changes are the result of adaptive natural selection.

Overall, genetic evolution is a historical process that is subject to historical contingency and chance, but it is constrained by the necessity of organisms to survive and reproduce in a constantly changing world.

CONCEPT MAP

Draw a concept map interrelating as many of the following terms as possible. Note that the terms are listed in no particular order.

mutation / founder effect / species / alternate adaptive peak / migration / reproductive isolation / genetic drift / geographical race / natural selection / hybrid sterility / neutral evolution

CHAPTER INTEGRATION PROBLEM

An entomologist who studies insects that feed on rotting vegetation has discovered an interesting case of diversification of fungus gnats on several islands in an archipelago. Each island has a gnat population that is extremely similar in morphology, although not identical, to those on the other islands, but each lives on a different kind of rotting vegetation that is not present on the other islands. The entomologist postulates that these are closely related species that have diverged by adapting to feeding on slightly different rot conditions.

In order to support this hypothesis, he carries out an electrophoretic study of the alcohol dehydrogenase enzyme in the different populations. He discovers that each population is characterized by a different electrophoretic form of the alcohol dehydrogenase, and then reasons that each of these alcohol dehydrogenase forms is specifically adapted to the particular alcohols that are produced in the fermentation of the vegetation characteristic of a particular island. There is, in addition, some polymorphism of alcohol dehydrogenase within each island, but the frequency of variant alleles is low on each island and can easily be explained as the result of an occasional mutation or rare migrant from another island. These fungus gnats then become a textbook example of how species diversity can come about by natural selection adapting each newly forming species to a different environment.

A skeptical population geneticist reads about the case in a textbook and she immediately has some doubts. It seems to her that, given the evidence, an equally plausible explanation is that these are not species at all but just local geographic races that have become slightly differentiated morphologically by random genetic drift. Moreover the different electrophoretic forms of the alcohol dehydrogenase protein may be physiologically equivalent variants of a gene undergoing neutral molecular evolution in isolated populations.

Outline a program of investigation that could distinguish between these alternative explanations. How could you test whether the different populations are indeed different species? How could you test the hypothesis that the different forms of the alcohol dehydrogenase have diverged selectively?

◆ Solution ◆

To test the species distinctness of the different gnats, it is necessary to be able to manipulate and culture them in captivity. If they cannot be cultured in the laboratory or greenhouse, then their species distinctness cannot be established. The mating behavior compatibility of the different forms can be tested by placing a mixture of males of two different populations with females of one of the forms to see whether there are any female mating preferences. The same experiment can then be repeated with mixed females and one sort

of male and with mixtures of males and females of both forms. From such experiments, patterns of mating preference can be observed. Even if there is some small amount of mating of different forms, this may occur only because of the unnatural conditions in which the test is being carried out. On the other hand, no mating of any kind may occur, even between the same forms, because the necessary cues for mating are missing, in which case nothing can be concluded.

If matings between different forms do occur, the survivorship of the interpopulation hybrids can be compared with that of the intrapopulation matings. If hybrids survive, their fertility can be tested by attempting to backcross them to the two different parental strains. As in the case of the mating tests, under the unnatural conditions of the laboratory or greenhouse, some survivorship or fertility of species hybrids is possible even though the isolation in nature is complete. Any clear reduction in observed survivorship or fertility of the hybrids is strong presumptive evidence that they belong to different species.

To test whether the different amino acid sequences underlying the electrophoretic mobility differences are the result of selective divergence, a program of DNA sequencing of the alcohol dehydrogenase locus is necessary. Replicated samples of *Adh* sequences from each of the island populations must be obtained. The number of such sequences needed from each population depends on the degree of nucleotide polymorphism that is present in the populations, but results from many loci in many species suggest that, as a rule of thumb, at least 10 sequences should be obtained from each population. The polymorphic sites within populations are classified into nonsynonymous (*a*) and synonymous (*b*) sites. The fixed nucleotide differences between populations are also classified into nonsynonymous (*c*) and synonymous (*d*) differences. If the divergence between the populations is purely the result of random genetic drift, then we expect *a*/*b* to be equal to *c*/*d*. If, on the other hand, there has been selective divergence, there should be an excess of fixed nonsynonymous differences, so *a*/*b* should be less than *c*/*d*. The equality of these ratios can be tested by a 2×2 contingency χ^2 test of the form

	POLYMORPHISMS	
	Nonsynonymous	Synonymous
POPULATION	*a*	*b*
DIFFERENCES	*c*	*d*

$$\chi^2 = \frac{(a + b + c + d)(ad - bc)^2}{(a + c)(b + d)(a + b)(c + d)}$$

SOLVED PROBLEMS

1. Two closely related species are found to be fixed for two different electrophoretically detected alleles at a locus encoding an enzyme. How could you demonstrate

that this divergence is a result of natural selection rather than neutral evolution?

◆ Solution ◆

a. Obtain DNA sequences of the gene from a number of separate individuals or strains from each of the two species. Ten or more sequences from each species would be desirable.

b. Tabulate the nucleotide differences among individuals within each species (*polymorphisms*), and classify these differences as either those that result in amino acid changes (replacement polymorphisms) or those that do not change the amino acid (synonymous polymorphisms).

c. Make the same tabulation of replacement and synonymous changes for the differences between the species, counting only those differences that completely differentiate the species. That is, do not count a polymorphism in one species that includes a variant that is seen in the other species.

d. If the ratio of replacement differences between the species to synonymous differences between the species is greater than the ratio of replacement polymorphisms to synonymous polymorphisms, then select for amino acid change.

e. Test the statistical significance of the observed greater ratio by a 2 × 2 chi-square test of the following table:

		POLYMORPHISMS	
		Replacement	Synonymous
SPECIES	Replacement	a	b
DIFFERENCES	Synonymous	c	d

$$\chi^2 = \frac{(a + b + c + d)(ad - bc)^2}{(a + c)(b + d)(a + b)(c + d)}$$

2. How could the molecular evolution of a set of different proteins be used to provide evidence of the relative importance of exact amino acid sequence to the function of each protein?

◆ Solution ◆

Obtain DNA sequences from the genes for each protein from a wide variety of very divergent species whose approximate time to a common ancestor is known from the fossil record. Translate the DNA sequences into amino acid sequences. For each protein, plot the observed amino acid difference for each pair of species against the estimated time of divergence for those species. The line for each protein will have a slope that is proportional to the amount of functional constraint on amino acid substitution in that protein. Highly constrained proteins will have very low rates of substitution, whereas more tolerant proteins will have higher slopes.

PROBLEMS

1. What is the difference between a transformational and a variational scheme of evolution? Give an example of each (not including the Darwinian theory of organic evolution).

2. What are the three principles of Darwin's theory of variational evolution?

3. Why is the Mendelian explanation of inheritance essential to Darwin's variational mechanism for evolution? What would the consequences for evolution be if inheritance were by the mixing of blood? What would the consequence for evolution be if heterozygotes did not segregate exactly 50 percent of each of the two alleles at a locus but were consistently biased toward one or the other allele?

4. If the mutation rate to a new allele is 10^{-5}, how large must isolated populations be to prevent chance differentiation among them to develop in the frequency of this allele?

5. Suppose that a number of local populations of a species are each about 10,000 individuals in size and that there is no migration between them. Suppose, further, that they were originally established from a large population with the frequency of an allele *A* at some locus equal to 0.4. Show by approximate sketches what the distribution of allele frequencies among the local populations would be after 100, 1000, 5000, 10,000, and 100,000 generations of isolation.

6. Show the results for the populations described in Problem 5 if there were an exchange of migrants among the populations at the rate of **(a)** one migrant individual per population every 10 generations; **(b)** one migrant individual per population every generation.

7. Suppose that a population is segregating for two alleles at each of two loci and that the relative probabilities of survival to sexual maturity of zygotes of the nine genotypes are as follows:

	A/A	A/a	a/a
B/B	0.95	0.90	0.80
B/b	0.90	0.85	0.70
b/b	0.90	0.80	0.65

Calculate the mean fitness, *W*, of the population if the allele frequencies are $p(A) = 0.8$ and $p(B) = 0.9$. What direction of change do you expect in allele frequencies in the next generation? Make the same calculation and prediction for the allele frequencies $p(A) = 0.2$ and $p(B) = 0.2$. From inspection of the genotypic fitnesses, how many adaptive peaks are there? What are the allele frequencies at the peak(s)?

8. Suppose the genotypic fitnesses in Problem 7 were:

	A/A	A/a	a/a
B/B	0.9	0.8	0.9
B/b	0.7	0.9	0.7
b/b	0.9	0.8	0.9

Calculate the mean fitness, W, for allelic frequencies $p(A) = 0.5$ and $p(B) = 0.5$. What direction of change do you expect for the allele frequencies in the next generation? Repeat the calculation and prediction for $p(A) = 0.1$ and $p(B) = 0.1$. From inspection of the genotype fitnesses, how many adaptive peaks are there and where are they located?

9. What is the evidence that polyploid formation has been important in plant evolution?

10. What is the evidence that gene duplication has been the source of the α and β gene families in human hemoglobin?

11. The human blood group allele I^B has a frequency of about 0.10 in European and Asian populations but is almost entirely absent in Native American populations. What explanations can account for this difference?

12. What is a geographical race? What is the difference between a geographical race and a separate species? Under what conditions will geographical races of a species become new species?

13. *Drosophila pseudoobscura* and *D. persimilis* are now considered separate species, but originally they were classified as Race A and Race B of a single species. They are morphologically indistinguishable from each other, except for a small difference in the genitalia of the males. When crossed in the laboratory, abundant adult F_1 progeny of both sexes are produced. Outline what program of observations and experiments you would undertake to test the claim that the two forms are different species.

14. Using the data on amino acid similarity of the α-, β-, γ-, ζ-, and ϵ-globin chains given in Table 26-4, draw a branching tree of the evolution of these chains from an original ancestral sequence in which the order of branching in time is as consistent as possible with the observed amino acid similarity on the assumption of a molecular clock.

15. DNA-sequencing studies for a gene in two closely related species produce the following numbers of sites that vary:

Synonymous polymorphisms	50
Nonsynonymous species differences	2
Synonymous species differences	18
Nonsynonymous polymorphisms	20

Does this result support a neutral evolution of the gene? Does it support an adaptive replacement of amino acids? What explanation would you offer for the observations?

Appendix

Genetic nomenclature

There is no universally accepted set of rules for naming genes, alleles, protein products, and associated phenotypes. At first, individual geneticists developed their own symbols for recording their work. Later, groups of people working on any given organism met and decided on a set of conventions that all would use. Because *Drosophila* was one of the first organisms to be used extensively by geneticists, most of the systems now being used are variants of the *Drosophila* system. However, there has been considerable divergence. Some scientists now advocate a standardization of this symbolism, but standardization has not been achieved. Indeed, the situation has been made more complex by the advent of DNA technology. Whereas most genes had previously been named for the phenotypes produced by mutations within them, the new technology has shown the precise nature of the products of many of these genes. Hence it seems more appropriate to refer to them by their cellular function. However, the old names are still in the literature, so many genes have two parallel sets of nomenclature.

The following examples by no means cover all the organisms used in genetics, but most of the nomenclature systems follow one of these types.

Drosophila melanogaster (insect)

ry	A gene that when mutated causes rosy eyes
*ry^502^*	A specific recessive mutant allele producing rosy eyes in homozygotes
*ry^+^*	The wild-type allele of *rosy*
ry	The rosy mutant phenotype
ry^+^	The wild-type phenotype (red eyes)
RY	The protein product of the *rosy* gene
XDH	Xanthine dehydrogenase, an alternative description of the protein product of the *rosy* gene; named for the enzyme that it encodes
D	*Dichaete;* a gene that when mutated causes a loss of certain bristles and wings to be held out laterally in heterozygotes and causes lethality in homozygotes
*D^3^*	A specific mutant allele of the *Dichaete* gene
*D^+^*	The wild-type allele of *Dichaete*
D	The Dichaete mutant phenotype
D^+^	The wild-type phenotype
D	(Depending on context) the protein product of the *Dichaete* gene (a DNA-binding protein)

Neurospora crassa (fungus)

arg	A gene that when mutated causes arginine requirement
arg-1	One specific *arg* gene
arg-1	An unspecified mutant allele of the *arg* gene
arg-1 (1)	A specific mutant allele of the *arg-1* gene
*arg-1^+^*	The wild-type allele
arg-1	The protein product of the *arg-1^+^* gene
Arg^+^	A strain not requiring arginine
Arg^-^	A strain requiring arginine

Saccharomyces cerevisiae (fungus)

ARG	A gene that when mutated causes arginine requirement
ARG1	One specific *ARG* gene
arg1	An unspecified mutant allele of the *ARG* gene
arg1-1	A specific mutant allele of the *ARG* gene
*ARG1^+^*	The wild-type allele
ARG1p	The protein product of the *ARG1^+^* gene
Arg^+^	A strain not requiring arginine
Arg^-^	A strain requiring arginine

Homo sapiens (mammal)

ACH	A gene that when mutated causes achondroplasia
*ACH^1^*	A mutant allele (dominance not specified)
ACH	Protein product of *ACH* gene; nature unknown
FGFR3	Recent name for gene for achondroplasia
*FGFR3^1^* or *FGFR3**1 or *FGFR3<1>*	Mutant allele of *FGR3* (unspecified dominance)

| FGFR3 protein | Fibroblast growth factor receptor 3 |

Mus musculus (mammal)

Tyrc	A gene for tyrosinase
$+^{Tyrc}$	The wild-type allele of this gene
Tyrc^{ch} or *Tyrc-ch*	A mutant allele causing chinchilla color
Tyrc	The protein product of this gene
+TYRC	The wild-type phenotype
TYRCch	The chinchilla phenotype

Escherichia coli (bacterium)

lacZ	A gene for utilizing lactose
lacZ^{+}	The wild-type allele
lacZ1	A mutant allele
LacZ	The protein product of that gene

| Lac^{+} | A strain able to use lactose (phenotype) |
| Lac^{-} | A strain unable to use lactose (phenotype) |

Arabidopsis thaliana (plant)

YGR	A gene that when mutant produces yellow-green leaves
YGR1	A specific *YGR* gene
YGR1	The wild-type allele
ygr1-1	A specific recessive mutant allele of *YGR1*
ygr1-2D	A specific dominant (D) mutant allele of *YGR1*
YGR1	The protein product of *YGR1*
Ygr^{-}	Yellow-green phenotype
Ygr^{+}	Wild-type phenotype

Further Readings

Students interested in pursuing genetics further should start reading original research articles in scientific journals. Some important journals are *Cell, Current Genetics, Evolution, Gene, Genetic Research, Genetics, Heredity, Human Genetics, Journal of Medical Genetics, Journal of Molecular Biology, Molecular and General Genetics, Mutation Research, Nature, Plasmid, Proceedings of the National Academy of Sciences of the United States of America,* and *Science*. Useful review articles may be found in *Annual Review of Genetics, Advances in Genetics,* and *Trends in Genetics*.

Some particularly useful references, mostly general reviews, are listed under the chapters to which they relate.

Chapter 1

Clausen, J., D. D. Keck, and W. W. Hiesey. 1940. *Experimental Studies on the Nature of Species,* vol. 1: *The Effect of Varied Environments on Western North American Plants.* Washington, DC: Carnegie Institute of Washington, Publ. No. 520, 1–452. This publication and the following one by the same authors are the classic studies of norms of reaction of plants from natural populations.

Clausen, J., D. D. Keck, and W. W. Hiesey. 1958. *Experimental Studies on the Nature of Species,* vol. 3: *Environmental Responses of Climatic Races of* Achillea. Washington, DC: Carnegie Institute of Washington, Publ. No. 581, 1–129.

Milunsky, A., and G. J. Annas, eds. 1975. *Genetics and the Law.* New York: Plenum Press. Interesting accounts of the ramifications of genetics in people's lives.

Moore, J. A. 1985. *Science as a Way of Knowing: Genetics. American Zoologist* 25: 1–165. A short book from an excellent series that focuses on the modus operandi of science.

Schmalhausen, I. I. 1949. *Factors of Evolution: The Theory of Stabilizing Selection.* Philadelphia: Blakiston. The most general discussion of the relation of genotype and environment in the formation of phenotypic variation.

Chapter 2

Carlson, E. A. 1966. *The Gene: A Critical History.* Philadelphia: Saunders. A readable history of genetics.

Grant, V. 1975. *Genetics of Flowering Plants.* New York: Columbia University Press. One of the few texts on this subject.

Harpstead, D. 1971. "High-Lysine Corn." *Scientific American* (August). An account of the breeding of lines with increased amounts of normally limiting amino acids.

Hutt, F. B. 1964. *Animal Genetics.* New York: Ronald Press. A standard text on the subject with many interesting examples.

Jennings, P. R. 1976. "The Amplification of Agricultural Production." *Scientific American* (September). A discussion of genetics and the green revolution.

McKusick, V. A., et al. 1990. *Mendelian Inheritance in Man,* 9th ed. Baltimore and London: Johns Hopkins Press. The "bible" of medical genetics; a 2000-page compendium of all the known human inherited disorders, both autosomal and X linked.

Mange, A. P., and E. J. Mange. 1990. *Genetics: Human Aspects,* 2d ed. Sunderland, MA: Sinauer. A useful and up-to-date text for a course in human genetics.

Olby, R. C. 1966. *Origins of Mendelism.* London: Constable. An enjoyable account of Mendel's work and the intellectual climate of his time.

Rousseau, F., et al. 1994. "Mutations in the Gene Encoding Fibroblast Growth Factor Receptor in Achondroplasia." *Nature* 371: 253–254. A short research paper describing the molecular identification of the abnormal allele that causes achondroplasia.

Stern, C., and E. R. Sherwood. 1966. *The Origin of Genetics: A Mendel Source Book.* New York: W. H. Freeman. A short collection of important early papers, including Mendel's papers and correspondence.

Sturtevant, A. H. 1965. *A History of Genetics.* New York: Harper & Row. Another useful historical text.

Thompson, J. S., and M. W. Thompson. 1980. *Genetics in Medicine.* Philadelphia: Saunders. One of the classic textbooks in medical genetics. Short and easy to read.

Todd, N. B. 1977. "Cats and Commerce." *Scientific American* (November). Includes some genetics of domestic cat coat colors and the use of this information to study cat migration throughout history.

Vogel, F., and A. G. Motulsky. 1986. *Human Genetics: Problems and Approaches,* 2d ed. New York: Springer-Verlag. Another classic textbook useful as a reference work.

Chapter 3

Brown, S. W. 1966. "Heterochromatin." *Science* 151: 417–425. A nice review of the classic cytological observations.

Crow, J. F. 1983. *Genetics Notes,* 8th ed. New York: Macmillan. A short and concise review of genetics from Mendel to populations.

Dupraw, E. J. 1970. *DNA and Chromosomes.* New York: Holt, Rinehart & Winston. A useful book on chromosome substructure, containing excellent photographs by the author.

Gasser, S. M., and U. K. Laemmli. 1987. "A Glimpse at Chromosomal Order." *Trends in Genetics* 3: 16–22. A review that focuses on the various levels of chromosome organization.

Grighatti, T. A. 1991. "Position Effect Variegation: An Assay for Nonhistone Chromosomal Proteins and Chromatin Assembly and Modification Factors." *Methods in Cell Biology* 35: 587–627. A readable account of research progress on this topic.

Hilliker, A. J., and C. B. Sharp. "New Perspectives on the Genetics and Molecular Biology of Constitutive Heterochromatin." In *Chromosome Structure and Function,* J. P. Gustaffson and R. Appels, eds. New York: Plenum. A summary of research progress focusing on *Drosophila* heterochromatin. The book generally is useful in regard to the subject matter of this chapter.

Manuelidis, L. 1990. "A View of Interphase Chromosomes." *Science* 250: 1533–1540. An easily readable review that relates metaphase chromosome structure to structures in interphase.

McIntosh, J. R., and K. L. McDonald. 1988. "The Mitotic Spindle." *Scientific American* (October).

McLaren, A. 1988. "Sex Determination in Mammals." *Trends in Genetics* 4: 153–157.

McLeish, J., and B. Snoad. 1958. *Looking at Chromosomes.* New York: Macmillan. A short classic book consisting of many superb photographs of mitosis and meiosis.

Rick, C. M. 1978. "The Tomato." *Scientific American* (August). Includes an account of tomato genes and chromosomes and their role in breeding.

Stern, C. 1973. *Principles of Human Genetics,* 3d ed. New York: W. H. Freeman. A standard text including many examples of the inheritance of human traits.

von Wettstein, D., et al. 1984. "The Synaptonemal Complex in Genetic Segregation." *Annual Review of Genetics* 18: 331–414. An up-to-date review of the structure and function of the complex.

Chapter 4

Bodmer, W. F., and L. L. Cavalli-Sforza. 1976. *Genetics, Evolution, and Man.* New York: W. H. Freeman. A very readable, well-illustrated book, including examples relevant to this chapter.

Griffiths, A. J. F., and F. R. Ganders. 1984. *Wildflower Genetics.* Vancouver: Flight Press. A field guide to plant variation in natural populations and its genetic basis, including examples relevant to this chapter.

Griffiths, A. J. F., and J. McPherson. 1989. *One Hundred + Principles of Genetics.* New York: W. H. Freeman. A short book that distills genetics down to its basic key concepts; useful for rapid review of the whole subject.

Hutt, W. B. 1979. *Genetics for Dog Breeders.* New York: W. H. Freeman. A short book that will make genetics immediately relevant to all dog owners and breeders.

Searle, A. G. 1968. *Comparative Genetics of Coat Color in Mammals.* New York: Academic Press. A classic treatment of the subject with many examples relevant to this and other chapters.

Silvers, W. K. 1979. *The Coat Colors of Mice.* New York: Springer-Verlag. A standard handbook on the subject, including many examples of gene interaction.

Wright, M., and S. Walters, eds. 1981. *The Book of the Cat.* New York: Summit Books. A fascinating book that has an excellent chapter on gene interaction in determining the coat colors of domestic cats.

Chapter 5

O'Brien, S. J., ed. 1984. *Genetic Maps.* Cold Spring Harbor, NY: Cold Spring Harbor Laboratory Press. A compendium of the detailed maps of 80 well-analyzed organisms.

Peters, J. A., ed. 1959. *Classic Papers in Genetics.* Englewood Cliffs, NJ: Prentice Hall. A collection of important papers in the history of genetics.

White, R., et al. 1985. "Construction of Linkage Maps with DNA Markers for Human Chromosomes." *Nature* 313: 101–104. An extension of the techniques of this chapter to DNA markers.

Chapter 6

Finchman, J. R. S., P. R. Day, and A. Radford. 1979. *Fungal Genetics,* 3d ed. London: Blackwell. A large, standard technical work. Good for tetrad analysis.

Kemp, R. 1970. *Cell Division and Heredity.* London: Edward Arnold. A short, clear introduction to genetics. Good for map functions and tetrad analysis.

Murray, A. W., and J. W. Szostak. 1983. "Construction of Artificial Chromosomes in Yeast." *Nature* 305: 189–193. The first creation of new chromosomes by splicing together known telomere, centromere, replicator, and then DNA fragments by recombinant DNA technology. Includes tetrad analysis of markers on the new chromosomes.

Puck, T. T., and F-T. Kao. 1982. "Somatic Cell Genetics and Its Application to Medicine." *Annual Review of Genetics* 16: 225–272. A technical but readable review.

Ruddle, F. H., and R. S. Kucherlapati. 1974. "Hybrid Cells and Human Genes." *Scientific American* (July). A popular account of the use of cell hybridization in mapping human genes.

Stahl, F. W. 1969. *The Mechanics of Inheritance,* 2d ed. Englewood Cliffs, NJ: Prentice Hall. A short introduction to genetics, including some advanced material presented with a novel approach.

Chapter 7

Adelberg, E. A. 1966. *Papers on Bacterial Genetics.* Boston: Little, Brown.

Brock, T. D. 1990. *The Emergence of Bacterial Genetics.* Cold Spring Harbor, NY: Cold Spring Harbor Laboratory Press. The definitive treatise on the beginnings and development of the field of bacterial genetics.

Hayes, W. 1968. *The Genetics of Bacteria and Their Viruses,* 2d ed. New York: Wiley. The standard and classic text, written by a pioneer in the subject.

Levy, S. B. 1998. "The Challenge of Antibiotic Resistance." *Scientific American* (March). This article shows how the different methods of gene transfer can lead to bacteria that are resistant to antibiotics.

Lewin, B. 1977. *Gene Expression,* vol. 1: *Bacterial Genomes.* New York: Wiley. An excellent set of volumes, all of which are relevant to various sections of this text.

Lewin, B. 1977. *Gene Expression,* vol. 3: *Plasmids and Phages.* New York: Wiley.

Miller, J. H. 1992. *A Short Course in Bacterial Genetics.* Cold Spring Harbor, NY: Cold Spring Harbor Laboratory Press. A laboratory manual with extensive historical introductions to each area of bacterial genetics.

Stent, G. S., and R. Calendar. 1978. *Molecular Genetics,* 2d ed. New York: W. H. Freeman. A lucidly written account of the development of our present understanding of the subject, based mainly on experiments in bacteria and phage.

Chapter 8

Dickerson, R. E. 1983. "The DNA Helix and How It Is Read." *Scientific American* (December). An article with some beautiful color models of DNA structures.

Greider, C. W. 1990. "Telomeres, Telomerase, and Senescence." *BioEssays* 12: 363–369. A good review written for nonexperts.

Kornberg, A., and T. Baker. 1992. *DNA Replication,* 2d ed. New York: W. H. Freeman.

Wang, J. C. 1982. "DNA Topoisomerases." *Scientific American* (July). Diagrams different topological forms of DNA.

Watson, J. D. 1981. *The Double Helix.* New York: Atheneum. An enjoyable personal account of Watson and Crick's discovery, including the human dramas.

Chapter 9

Benzer, S. 1962. "The Fine Structure of the Gene." *Scientific American* (January). A popular version of the author's pioneer experiments.

Dressler, D., and H. Potter. 1991. *Discovering Enzymes.* New York: Scientific American Library. A delightful book that describes the discovery of enzymes and shows many of their features in well-planned easy-to-read illustrations.

Felsenfeld, G. 1985. "DNA." *Scientific American* (October).

Prusiner, S. B. 1995. "The Prion Diseases." *Scientific American* (January). A review of an article describing the fascinating phenomenon of infectious protein particles.

Radman, M., and R. Wagner. 1988. "The High Fidelity of DNA Duplication." *Scientific American* (August).

Watson, J. D., et al. 1987. *The Molecular Biology of the Gene,* 4th ed. Menlo Park, CA: Benjamin/Cummings. A superb development of the subject, written in a highly readable style and well illustrated.

Yanofsky, C. 1967. "Gene Structure and Protein Structure." *Scientific American* (May). This article gives the details of colinearity at the molecular level.

Chapter 10

Chambon, P. 1981. "Split Genes." *Scientific American* (May). The discovery of intervening sequences is described.

Crick, F. H. C. 1962. "The Genetic Code." *Scientific American* (October). This article and the following one are popular accounts of code-cracking experiments.

Crick, F. H. C. 1966. "The Genetic Code: III." *Scientific American* (October).

Darnell, J. E., Jr. 1985. "RNA." *Scientific American* (October).

Doolittle, R. F. 1985. "Proteins." *Scientific American* (October).

Lake, J. A. 1981. "The Ribosome." *Scientific American* (August). Three-dimensional model of the ribosome.

Lane, C. 1976. "Rabbit Hemoglobin from Frog Eggs." *Scientific American* (August). This article describes experiments illustrating the universality of the genetic system.

Lawn, R. M., and G. A. Vehar. 1986. "The Molecular Genetics of Hemophilia." *Scientific American* (March).

Miller, O. L. 1973. "The Visualization of Genes in Action." *Scientific American* (March). A discussion of electron microscopy of transcription and translation.

Moore, P. B. 1976. "Neutron-Scattering Studies of the Ribosome." *Scientific American* (October). This article gives the details of ribosome substructure.

Nirenberg, M. W. 1963. "The Genetic Code: II." *Scientific American* (March). Another account of early code-cracking experiments.

Radman, M., and R. Wagner. 1988. "The High Fidelity of DNA Duplication." *Scientific American* (August).

Rich, A., and S. H. Kim. 1978. "The Three-Dimensional Structure of Transfer RNA." *Scientific American* (January). A presentation of the experimental evidence behind the structure described in this chapter.

Weinberg, R. A. 1985. "The Molecules of Life." *Scientific American* (October).

Chapter 11

Hendrix, R. W., J. W. Roberts, F. W. Stahl, and R. A. Weisberg. 1983. *Lambda II.* Cold Spring Harbor, NY: Cold Spring Harbor Laboratory Press. A series of incisive reviews about phage lambda.

Jacob, F., and J. Monod. 1961. "Genetic Regulatory Mechanisms in the Synthesis of Proteins." *Journal of Molecular Biology* 3: 318–356. A classic paper setting forth the elements of an operon and the experimental evidence.

Lewis, M., et al. 1996. "Crystal Structure of the Lactose Operon Repressor and Its Complexes with DNA and Inducer." *Science* 271: 1247–1254. The three-dimensional structure of the *lac* repressor is detailed with color diagrams that show how the repressor functions.

Maniatis, T., S. Goodbourn, and J. A. Fischer. 1987. "Regulation of Inducible and Tissue-Specific Gene Expression." *Science* 236: 1237–1244. A current review of regulatory elements in eukaryotes.

Maniatis, T., and M. Ptashne. 1976. "A DNA Operator-Repressor System." *Scientific American* (January). This article discusses the molecular structures of the components of the *lac* operon.

Miller, J. H., and W. S. Reznikoff, eds. 1980. *The Operon,* 2d ed. Cold Spring Harbor, NY: Cold Spring Harbor Laboratory Press. A valuable set of reviews on gene regulation in bacteria.

Moses, P. B., and N.-H. Chua. 1988. "Light Switches for Plant Genes." *Scientific American* (April).

Ptashne, M., and W. Gilbert. 1970. "Genetic Repressors." *Scientific American* (June). The exciting story of how repressors were identified and purified, thereby confirming the predictions of Jacob and Monod.

Steitz, J. A. 1988. "Snurps." *Scientific American* (June).

Tijan, R. 1995. "Molecular Machines That Control Genes." *Scientific American* (February). An up-to-date review of eukaryotic gene transcription regulation.

Transcription. 1991. *Trends in Biochemical Sciences* (November). A whole issue of this journal consists of reviews about eukaryotic transcription regulation.

Weber, I. T., D. B. McKay, and T. A. Steitz. 1982. "Two Helix DNA Binding Motifs of CAP Found in *lac* Repressor and *gal* Repressor." *Nucleic Acids Research* 10: 5085–5102. Discussion of models for repressor–operator recognition.

Chapter 12

Britten, R. J., and D. Kohne. 1968. "Repeated Sequences in DNA." *Science* 161: 529–540. One of the important summaries of the theoretical basis for distinguishing DNAs by renaturation.

Broda, P. 1979. *Plasmids.* New York: W. H. Freeman. One of the few technical books on the subject.

Brown, D. D. 1973. "The Isolation of Genes." *Scientific American* (August). Illustrates the power of focusing molecular techniques on one specific locus with special properties.

Cohen, S. 1975. "The Manipulation of Genes." *Scientific American* (July). A summary of recombinant DNA techniques by one of the main innovators.

Fiddes, J. C. 1977. "The Nucleotide Sequence of a Viral DNA." *Scientific American* (December). This is a review of a landmark, the DNA sequence of an entire virus genome with an unexpected discovery.

Gilbert, W., and L. Villa-Komaroff. 1980. "Useful Proteins from Recombinant Bacteria." *Scientific American* (April). A description of the method of DNA sequencing used most extensively. Also discusses the potential application of recombinant DNA techniques.

Itakura, K., et al. 1977. "Expression in *E. coli* of a Chemically Synthesized Gene for the Hormone Somatostatin." *Science* 198: 1056–1063. A technical paper well worth reading for its historical significance. It represents the start of bioengineering — using DNA manipulation to modify cells to produce a medically useful human protein.

Khorana, H. G., et al. 1972. "Studies on Polynucleotides CIII: Total Synthesis of the Structural Gene for an Alanine Transfer Ribonucleic Acid from Yeast." *Journal of Molecular Biology* 72: 209–217. A classic technical paper.

Nathans, D., and H. O. Smith. 1975. "Restriction Endonucleases in the Analysis and Restructuring of DNA Molecules." *Annual Review of Biochemistry* 44: 273–293. A technical review of restriction enzymes by two pioneers in the field.

Sinsheimer, R. L. 1977. "Recombinant DNA." *Annual Review of Biochemistry* 46: 415–438. A provocative article by a leading molecular biologist who has expressed concern about potential hazards of DNA manipulation.

Varmus, H. 1987. "Reverse Transcription." *Scientific American* (September).

Watson, J. D., M. Gilman, J. Witkowski, and M. Zoller. 1992. *Recombinant DNA,* 2d ed. New York: W. H. Freeman. A well-written account of recombinant DNA and its applications, with many wonderful illustrations.

Chapter 13

Capecchi, M. R. 1994. "Targeted Gene Replacement." *Scientific American* (March). A popular article describing details of the production of knockout mice by gene replacement.

Mertens, T. R. 1975. *Human Genetics: Readings on the Implications of Genetic Engineering.* New York: Wiley. A collection of popular articles.

Moses, P. B., and N.-H. Chua. 1988. "Light Switches for Plant Genes." *Scientific American* (April).

Murray, A. W., and J. W. Szostak. 1987. "Artificial Chromosomes." *Scientific American* (November).

Ronald, P. C. 1997. "Making Rice Disease-Resistant." *Scientific American* (November). The story of how genetic engineering can improve rice crops.

Velander, W. H., H. Lubon, and W. N. Drohan. 1997. "Transgenic Livestock as Drug Factories." *Scientific American* (January). Description of how genetic engineering makes it possible to breed pigs, cows, sheep, and other animals whose milk contains large amounts of medicinal proteins.

Watson, J. D., M. Gilman, J. Witkowski, and M. Zoller. 1992. *Recombinant DNA,* 2d ed. New York: W. H. Freeman. A well-written account of recombinant DNA and its applications, with many wonderful illustrations.

White R., and J. M. Lalouel. 1988. "Chromosome Mapping with DNA Markers." *Scientific American* (February).

Chapter 14

Britten, R. J., and D. E. Kohne. 1970. "Repeated Segments of DNA." *Scientific American* (April).

Coulson, A., et al. 1991. "YACs and the *C. elegans* Genome." *BioEssays* 13: 413–417. A review describing the use of YACs to build up contigs for this nematode genome.

Darnell, J., H. Lodish, and D. Baltimore. 1990. *Molecular Cell Biology,* 2d ed. New York: Scientific American Books. A useful book that does a good job of translating many of the concepts of genetics into cellular processes, including a particularly strong section on chromosome structure and types of repetitive DNA.

Davidson, E., and R. Britten. 1973. "Organization, Transcription and Regulation in the Animal Genome." *Quarterly Review of Biology* 48: 565–613. The analysis of renaturation kinetics of DNA fragments is a source of insight into chromosome structure.

Fleischmann, M. D., et al. 1995. "Whole-Genome Random Sequencing and Assembly of *Haemophilus influenzae* Rd." *Science* 269: 496–512. The article reports the first full genomic sequence for a free-living organism.

Foote, S., D. Vollrath, A. Hilton, and D. C. Page. 1992. "The Human Y Chromosome: Overlapping DNA Clones Spanning the Euchroatic Region." *Science* 258: 60–66. The use of sequence-tagged sites in producing the physical map of the Y chromosome.

Green, E. D., and M. V. Olson. 1990. "Chromosomal Region of the Cystic Fibrosis Gene in Yeast Artificial Chromosomes: A Model for Human Genome Mapping." *Science* 250: 94–98. An easy-to-understand article on the principles of genome mapping with the use of YACs.

Hayashi, S., et al. 1980. "Hybridization of tRNAs of *Drosophila melanogaster.*" *Chromosoma* 76: 65–84. A technical report showing how specific genes can be located cytologically by hybridization of labeled RNA to chromosomes in situ.

Jeffreys, A. J., V. Wison, and S. L. Thein. 1985. "Hypervariable 'Minisatellite' Regions in Human DNA." *Nature* 314: 67–73. One of the first papers on the subject of DNA fingerprinting.

Jeffreys, A. J., V. Wison, S. L. Thein, D. J. Weatherall, and B. A. J. Ponder. 1986. "DNA Fingerprints and Segregation Analysis of Multiple Markers in Human Pedigrees." *American Journal of Human Genetics* 39: 25–37. A technical article on the inheritance patterns of DNA fingerprints.

Meinke, D. W., et al. 1998. "*Arabidopsis thaliana:* A Model Plant for Genome Analysis." *Science* 282: 662–682.

Murray, J. C., et al. 1994. "A Comprehensive Human Linkage Map with Centimorgan Density." *Science* 265: 2049–2054. Each year *Science* publishes a genome-map issue describing progress in mapping the human genome. This issue describes the use of molecular markers and other methods for obtaining a fine-scale map of the human genome.

Scangos, A. 1997. "Drug Discovery in the Post-genomic Era." *Nature Biotechnology* 15: 1220–1221.

Walter, M., et al. 1994. "A Method for Constructing Radiation Hybrid Maps of Whole Genomes." *Nature Genetics* 7: 22–28. An easy-to-understand research paper describing the technique of radiation hybrid mapping.

Winzeler, E. A., et al. 1998. "Direct Allelic Variation Scanning of the Yeast-Genome." *Science* 281: 1194–1197.

Chapter 15

deSerres, F. J., and A. Hollaender, eds. 1982. *Chemical Mutagens: Principles and Methods for Their Detection,* vol. 7. New York: Plenum Press. One of a set of useful volumes on this important class of mutagens particularly relevant to human mutation.

Induced Mutations: A Tool in Plant Research. 1981. Vienna: International Atomic Energy Agency. A collection of papers by eminent workers in agricultural genetics illustrating the practical uses of mutations in plant breeding.

Lawrence, C. W. 1971. *Cellular Radiobiology.* London: Edward Arnold. A short standard text.

Lindsley, D. L., and E. H. Grell. 1972. *Genetic Variations of* Drosophila melanogaster. Washington, DC: Carnegie Institute of Washington. A reference book for fruitfly researchers, containing all the thousands of known *Drosophila* mutations. Fascinating browsing for anyone interested in genetics generally and in the major contributions made by *Drosophila* research.

Neuffer, M. G., L. Jones, and M. S. Zuber. 1968. *The Mutants of Maize.* Madison, WI: Crop Science Society of America. A color catalog of the many, and often bizarre, mutants used by corn geneticists.

Schull, W. J., et al. 1981. "Genetic Effect of the Atomic Bombs: A Reappraisal." *Science* 213: 1220–1227. A summary of all the indicators of potential genetic effects of the Hiroshima and Nagasaki explosions, concluding that, "In no instance is there a statistically significant effect of parental exposure; but for all indicators the observed effect is in the direction suggested by the hypothesis that genetic damage resulted from the exposure."

Chapter 16

Auerbach, C. 1976. *Mutation Research.* London: Chapman & Hall. Standard text by a pioneer researcher.

Cairns, J. 1978. *Cancer: Science and Society.* New York: W. H. Freeman. A fascinating discussion of the origins of human cancers and of the role played by the environment.

Croce, C. M., and H. Koprowski. 1978. "The Genetics of Human Cancer." *Scientific American* (February). An excellent popular account.

Devoret, R. 1979. "Bacterial Tests for Potential Carcinogens." *Scientific American* (August). This article describes the use of mutation tests to screen for carcinogens and includes some details of DNA reactions.

Drake, J. W. 1970. *The Molecular Basis of Mutation.* San Francisco: Holden-Day. One of the few standard texts on the subject.

Eisenstadt, E. 1987. "Analysis of Mutagenesis." In Escherichia coli *and* Salmonella typhimurium: *Cellular and Molecular Biology,* F. C. Neidhardt, ed. Washington, DC: American Society for Microbiology. A comprehensive review of studies of mutagenesis in bacteria.

Friedberg, E. C., G. C. Walker, and W. Siede. 1995. *DNA Repair and Mutagenesis.* Washington, DC: American Society for Microbiology. A complete review text covering mutagenesis and repair in both bacteria and higher cells. An excellent source for human genetic diseases and cancer in relation to repair genes.

Miller, J. H. 1996. "Spontaneous Mutators in Bacteria: Insights into Pathways of Mutagenesis and Repair." *Annual Review of Microbiology* 50: 625–643. A review of mutagenesis and repair in bacteria.

Walker, G. C. 1984. "Mutagenesis and Inducible Responses to Deoxyribonucleic Acid Damage in *Escherichia coli.*" *Microbiological Reviews* 48: 60–93. A review of mutagenesis and repair, with an excellent discussion of the SOS system.

Chapters 17 and 18

deGrouchy, J., and C. Turleau. 1984. *Clinical Atlas of Human Chromosomes.* New York: Wiley. A systematic examination of all the human chromosomes and the aberrations associated with them.

Dellarco, V. L., P. E. Voytek, and A. Hollaender. 1985. *Aneuploidy: Etiology and Mechanisms.* New York: Plenum Press. A collection of research summaries on aneuploidy in humans and experimental organisms.

Epstein, C. J., et al. 1983. "Recent Developments in Prenatal Diagnosis of Genetic Diseases and Birth Defects." *Annual Review of Genetics* 17: 49–83. Includes amniocentesis.

Feldman, M. G., and E. R. Sears. 1981. "The Wild Gene Resources of Wheat." *Scientific American* (January). A general discussion of the genomes of wheat and its relatives and how new genes can be introduced.

Friedmann, T. 1971. "Prenatal Diagnosis of Genetic Disease." *Scientific American* (November). An early article on amniocentesis and its uses.

Fuchs, F. 1980. "Genetic Amniocentesis." *Scientific American* (August).

German, J., ed. 1974. *Chromosomes and Cancer.* New York: Wiley. A large technical work but readable, describing the relation of chromosome changes and cancer.

Hassold, T. J., and P. A. Jacobs, 1984. "Trisomy in Man." *Annual Review of Genetics* 18: 69–98. A comprehensive summary of trisomy, including a discussion of the maternal-age effect.

Hulse, J. H., and D. Spurgeon. 1984. "Triticale." *Scientific American* (August). An account of the development and possible benefits of this wheat-rye amphidiploid.

Lawrence, W. J. C. 1968. *Studies in Biology,* no. 12: *Plant Breeding.* London: Edward Arnold. A short introduction to the subject.

Mangelsdof, P. C. 1986. "The Origins of Corn." *Scientific American* (August).

Maniatis, T. E., et al. 1980. "The Molecular Genetics of Human Hemoglobins." *Annual Review of Genetics* 14: 145–178. A useful summary, which could be profitably read at this point in the course or after reading the material on molecular genetics.

Patterson, D. 1987. "The Causes of Down Syndrome." *Scientific American* (August).

Shepherd, J. F. 1982. "The Regeneration of Potato Plants from Protoplasts." *Scientific American* (May). A review by one of the leaders in this field.

Swanson, C. P., T. Mertz, and W. J. Young, 1967. *Cytogenetics.* Englewood Cliffs, NJ: Prentice-Hall.

Chapter 19

Alberts, B., D. Bray, J. Lewis, M. Raff, K. Roberts, and J. D. Watson. 1989. *Molecular Biology of the Cell,* 2d ed. New York: Garland. See pages 239 to 248 for an excellent description of recombination at the molecular level, with nice illustrations.

Stahl, F. W. 1979. *Genetic Recombination: Thinking about It in Phage and Fungi.* New York: W. H. Freeman. A rather technical short book on recombination models.

Stahl, F. W. 1994. "The Holliday Junction on Its Thirtieth Anniversary." *Genetics* 138: 241–246. An easy-to-read review of the Holliday model and subsequent variations that are used today to explain recombination.

Whitehouse, H. L. K. 1973. *Towards an Understanding of the Mechanism of Heredity,* 3d ed. London: Edward Arnold. An excellent general introduction to genetics stressing the historical approach and the pivotal experiments. It includes a good section on recombination models.

Chapter 20

Baltimore, D. 1985. "Retroviruses and Retrotransposons: The Role of Reverse Transcription in Shaping the Eukaryotic Genome." *Cell* 40: 481–482. A short review of work on RNA intermediates in transposition.

Berg, D. E., and M. M. Howe, eds. 1989. *Mobile DNA.* Washington, DC: American Society for Microbiology. A collection of incisive reviews on transposable elements from bacteria to humans.

Bukhari, A. I., J. A. Shapiro, and S. L. Adhya, eds. 1977. *DNA Insertion Elements, Plasmids and Episomes.* Cold Spring Harbor, NY: Cold Spring Harbor Laboratory Press. An excellent large collection of short summary papers on lower and higher life forms.

Cohen, S. N., and J. A. Shapiro. 1980. "Transposable Genetic Elements." *Scientific American* (February). A popular account stressing bacteria and phages.

Fedoroff, N. V. 1984. "Transposable Genetic Elements in Maize." *Scientific American* (June). An account of the early experiments in maize, with more recent results on the molecular basis of transposition.

Shapiro, J. A., ed. 1983. *Mobile Genetic Elements.* New York: Academic Press. A comprehensive set of reviews covering the latest developments on transposition in later prokaryotes and eukaryotes.

Chapter 21

Borst, P., and L. A. Grivell. 1978. "The Mitochondrial Genome of Yeast." *Cell* 15: 705–723. A nontechnical review of the molecular biology of mtDNA.

Dujon, B. 1981. "Mitochondrial Genetics and Functions." In *The Molecular Biology of the Yeast* Saccharomyces, J. N. Strathern, E. W. Jones, and J. R. Broach, eds. Cold Spring Harbor, NY: Cold Spring Harbor Laboratory Press. The most comprehensive review available. This book is useful for many other aspects of yeast genetics and molecular biology.

Gillham, N. W. 1978. *Organelle Heredity.* New York: Raven Press. A rather technical work on the subject, especially strong on genetic analysis.

Griffiths, A. J. F. 1992. "Fungal Senescence." *Annual Review of Genetics* 26: 349–370. A review article that shows how most cases of fungal senescence are caused by mitochondrial mutations or mitochondrial plasmids.

Grivell, L. A. 1983. "Mitochondrial DNA." *Scientific American* (March). An excellent and up-to-date review including discussions of the unique mitochondrial translation system and of splicing of mitochondrial introns.

Linnane, A. W., and P. Nagley. 1978. "Mitochondrial Genetics in Perspective: The Derivation of a Genetic and Physical Map of the Yeast Mitochondrial Genome." *Plasmid* 1: 324–345. An excellent review of the genetics and molecular biology of yeast mtDNA.

Sager, R. 1965. "Genes Outside Chromosomes." *Scientific American* (January). A popular description of early *Chlamydomonas* experiments.

Sager, R. 1972. *Cytoplasmic Genes and Organelles.* New York: Academic Press.

Wallace, D. C. 1992. "Mitochondrial Genetics: A Paradigm for Aging and Degenerative Diseases?" *Science* 256: 628–632. A review article on the various types of mtDNA mutations in humans and their disease outcomes.

Chapter 22

Baserga, R. 1994. "Oncogenes and the Strategy of Growth Factors." *Cell* 79: 927–930.

Bowerman, B. 1995. "Determinants of Blastoderm Identity in the Early *C. elegans* Embryo." *BioEssays* 17: 405–414.

Cavenee, W. K., and R. L. White. 1995. "The Genetic Basis of Cancer." *Scientific American* (March).

Cohen, G. B., R. Ren, and D. Baltimore. 1995. "Modular Binding Domains in Signal Transduction Proteins." *Cell* 80: 237–248.

Cox, L. S., and D. P. Lane. 1995. "Tumour Suppressors, Kinases and Clamps: How p53 Regulates the Cell Cycle in Response to DNA Damage." *BioEssays* 17: 501–508.

Draetta, G. F. 1994. "Mammalian G1 Cyclins." *Current Opinion in Cell Biology* 6: 842–846.

Edgar, B. A. 1994. "Cell–Cycle Control in a Developmental Context." *Current Biology* 4: 522–524.

Elledge, S. J., and J. W. Harper. 1994. "Cdk Inhibitors: On the Threshold of Checkpoints and Development." *Current Opinion in Cell Biology* 6: 847–852.

Fishel, R., and R. D. Kolodner. 1995. "Identification of Mismatch Repair Genes and Their Role in the Development of Cancer." *Current Opinion in Genetics and Development* 5: 382–396.

Haffner, R., and M. Oren. 1995. "Biochemical Properties and Biological Effects of p53." *Current Opinion in Genetics and Development* 5: 84–91.

Hartwell, L. H., and M. B. Kastan. 1994. "Cell Cycle Control and Cancer." *Science* 266: 1821–1828.

Jiricny, J. 1994. "Colon Cancer and DNA Repair: Have Mismatches Met Their Match?" *Trends in Genetics* 10: 164–168.

Kamb, A. 1995. "Cell-Cycle Regulators and Cancer." *Trends in Genetics* 11: 136–140.

Karin, M., and T. Hunter. 1995. "Transcriptional Control by Protein Phosphorylation: Signal Transmission from the Cell Surface to the Nucleus." *Current Biology* 5: 747–757.

Kenyon, C. 1995. "A Perfect Vulva Every Time: Gradients and Signaling Cascades in *C. elegans*." *Cell* 82: 171–174.

Maruta, H., and A. W. Burgess. 1994. "Regulation of the Ras Signaling Network." *BioEssays* 16: 489–496.

Murray, A. W. 1995. "The Genetics of Cell Cycle Checkpoints." *Current Opinion in Genetics and Development* 5: 5–11.

Neer, E. J. 1995. "Heterotrimeric G Proteins: Organizers of Transmembrane Signals." *Cell* 80: 249–257.

Nigg, E. A. 1995. "Cyclin-Dependent Protein Kinases: Key Regulators of the Eukaryotic Cell Cycle." *BioEssays* 17: 471–480.

Picksley, S. M., and D. P. Lane. 1994. "p53 and Rb: Their Cellular Roles." *Current Opinion in Cell Biology* 6: 853–858.

Rhyu, M. S., and J. A. Knoblich. 1995. "Spindle Orientation and Asymmetric Cell Fate." *Cell* 82: 523–525.

Sohn, R. H., and P. J. Goldschmidt-Clermont. 1994. "Profilin: At the Crossroads of Signal Transduction and the Actin Cytoskeleton." *BioEssays* 16: 465–472.

Sullivan, W., and W. E. Theurkauf. 1995. "The Cytoskeleton and Morphogenesis of the Early *Drosophila* Embryo." *Current Opinion in Cell Biology* 7: 18–22.

Thomas, B. J., and S. L. Zipursky. 1994. "Early Pattern Formation in the Developing *Drosophila* Eye." *Trends in Cell Biology* 4: 389–394.

Weinberg, R. A. 1995. "The Retinoblastoma Protein and Cell Cycle Control." *Cell* 81: 323–330.

Zipursky, S. L., and G. M. Rubin. 1994. "Determination of Neuronal Cell Fate: Lessons from the R7 Neuron of *Drosophila*." *Annual Review of Neuroscience* 17: 373–397.

Chapter 23

Affara, N. A. 1991. "Sex and the Single Y." *BioEssays* 13: 475–478. Mammalian sex determination.

Ballabio, A., and H. F. Willard. 1992. "Mammalian X-Chromosome Inactivation and the XIST Gene." *Current Opinion in Genetics and Development* 2: 439–448.

Barlow, D. P. 1994. "Imprinting: A Gamete's Point of View." *Trends in Genetics* 10: 194–199.

Capecchi, M. R. 1994. "Targeted Gene Replacement." *Scientific American* (March). A description of approaches to the alteration of gene structure exploiting the cellular machinery for DNA repair and homologous recombination by one of the innovators of the technology.

Curtis, D., R. Lehmann, and P. D. Zamore. 1995. "Translational Regulation in Development." *Cell* 81: 171–178.

DiNardo, S., J. Heemskerk, S. Dougan, and P. H. O'Farrell. 1994. "The Making of a Maggot: Patterning the *Drosophila* Embryonic Epidermis." *Current Opinion in Genetics and Development* 4: 529–534.

Duboule, D., and G. Morata. 1994. "Colinearity and Functional Hierarchy among the Genes of the Homeotic Complexes." *Trends in Genetics* 10: 358–364.

Eden, S., and H. Cedar. 1994. "Role of DNA Methylation in the Regulation of Transcription." *Current Opinion in Genetics and Development* 4: 255–259.

Efstratiadis, A. 1994. "Parental Imprinting of Autosomal Mammalian Genes." *Current Opinion in Genetics and Development* 4: 265–280.

Ehlich, A., and R. Küppers. 1995. "Analysis of Immunoglobulin Rearrangements in Single B Cells." *Current Opinion in Immunology* 4: 281–284.

Englehard, V. H. 1994. "How Cells Present Antigen." *Scientific American* (August).

Gonzalez-Crespo, S., and M. Levine. 1994. "Related Target Enhancers for Dorsal and NF-kappa B-Signaling Pathways." *Science* 264: 255–258.

Hlskamp, M., and D. Tautz. 1991. "Gap Genes and Gradients: The Logic Behind the Gaps." *BioEssays* 13: 261–268.

Ip, Y. T., and M. Levine. 1994. "Molecular Genetics of *Drosophila* Immunity." *Current Opinion in Genetics and Development* 4: 672–677.

Jorgensen, R. A. 1995. "Co-suppression, Flower Color Patterns, and Metastable Gene Expression States." *Science* 268: 686–691.

Kenyon, C. 1994. "If Birds Can Fly, Why Can't We? Homeotic Genes and Evolution." *Cell* 78: 175–180.

Kingston, R. E., and M. R. Green. 1994. "Modeling Eukaryotic Transcriptional Activation." *Current Biology* 4: 325–332.

Klingler, M. 1994. "Segmentation in Insects: How Singular Is *Drosophila*?" *BioEssays* 16: 391–392.

Kornberg, T. B., and T. Tabata. 1993. "Segmentation of the *Drosophila* Embryo." *Current Opinion in Genetics and Development* 3: 585–594.

Krumlauf, R. 1994. "*Hox* Genes in Vertebrate Development." *Cell* 78: 191–201.

Lawrence, P. A., and G. Morata. 1994. "Homeobox Genes: Their Function in *Drosophila* Segmentation and Pattern Formation." *Cell* 78: 181–189.

Lovell-Badge, R. 1992. "Testis Determination: Soft Talk and Kinky Sex." *Current Opinion in Genetics and Development* 2: 596–601.

Marshall-Graves, J. A. 1995. "The Origin and Function of the Mammalian Y Chromosome and Y-Borne Genes: An Evolving Understanding." *BioEssays* 17: 311–320.

McGinnis, W., and M. Kuziora. 1994. "The Molecular Architecture of Body Design." *Scientific American* (February).

McKeown, M. 1992. "Sex Differentiation: The Role of Alternative Splicing." *Current Opinion in Genetics and Development* 2: 299–304.

Meyerowitz, E. M. 1994. "Genetics of Flower Development." *Scientific American* (November). Similar families of transcription factors are found to be responsible for homeotic mutations during plant pattern formation as well.

Moore, T., and D. Haig. 1991. "Genomic Imprinting in Mammalian Development: A Parental Tug-of-War." *Trends in Genetics* 7: 45–49.

Newman, A. J. 1994. "Pre-mRNA Splicing." *Current Opinion in Genetics and Development* 4: 298–304.

Orlando, V., and R. Paro. 1995. "Chromatin Multiprotein Complexes Involved in the Maintenance of Transcription Patterns." *Current Opinion in Genetics and Development* 5: 174–179.

Parkhurst, S. M., and P. M. Meneely. 1994. "Sex Determination and Dosage Compensation: Lessons from Flies and Worms." *Science* 264: 924–932.

Perrimon, N., and C. Desplan. 1994. "Signal Transduction in the Early *Drosophila* Embryo: When Genetics Meets Biochemistry." *Trends in Biochemical Sciences* 19: 509–513.

Pfeifer, K., and S. M. Tilghman. 1994. "Allele-Specific Gene Expression in Mammals: The Curious Case of the Imprinted RNAs." *Genes and Development* 8: 1867–1874.

Quiring, R., U. Walldorf, U. Kloter, and W. Gehring. 1994. "Homology of the Eyeless Gene of *Drosophila* to the Small Eye Gene in Mice and Aniridia in Humans. *Science* 265: 785–789.

Schüpbach, T., and S. Roth. 1994. "Dorsoventral Patterning in *Drosophila* Oogenesis." *Current Opinion in Genetics and Development* 4: 502–507.

St. Johnston, R. D. 1994. "RNA Localization: Getting to the Top." *Current Opinion in Genetics and Development* 4: 54–56.

St. Johnston, R. D. 1995. "The Intracellular Localization of Messenger RNAs." *Cell* 81: 161–170.

Tautz, D., and R. J. Sommer. 1995. "Evolution of Segmentation Genes in Insects." *Trends in Genetics* 11: 23–27.

Tijan, R. 1995. "Molecular Machines That Control Genes." *Scientific American* (February).

Weissman, I. L., and M. D. Cooper. 1993. "How the Immune System Develops." *Scientific American* (September). This entire issue is on the biology of the immune system and contains several other interesting papers.

Wolpert, L. 1994. "Positional Information and Pattern Formation in Development." *Developmental Genetics* 15: 485–490.

Chapter 24

Beadle, G. W. 1980. "The Ancestry of Corn." *Scientific American* (January). A popular account of the various clues that led to the modern version.

Bodmer, W. F., and L. L. Cavalli-Sforza. 1976. *Genetics, Evolution, and Man.* New York: W. H. Freeman.

Cavalli-Sforza, L. L., P. Menozzi, and A. Piazza. 1994. *The History and Geography of Human Genes.* Princeton, NJ: Princeton University Press. A complete discussion of human genetic variation, its history, and the forces molding it, with complete tables of human gene frequencies.

Clarke, B. 1975. "The Causes of Biological Diversity." *Scientific American* (August). A popular account emphasizing genetic polymorphism.

Crow, J. F. 1979. "Genes That Violate Mendel's Rules." *Scientific American* (February). An interesting article on a topic called segregation distortion (not treated in this text) and its effects in populations.

Dobzhansky, T. 1951. *Genetics and the Origin of Species.* New York: Columbia University Press. The classic synthesis of population genetics and the processes of evolution. The most influential book on evolution since Darwin's *Origin of Species.*

Evolution. 1978. *Scientific American* (September). This issue contains several articles relevant to population genetics.

Ford, E. B. 1971. *Ecological Genetics,* 3d ed. London: Chapman & Hall. A nonmathematical treatment of the role of genetic variation in nature, stressing morphological variation.

Futuyma, D. J. 1998. *Evolutionary Biology,* 3d ed. Sunderland, MA: Sinauer. The best modern discussion of population genetics, ecology, and evolution from both a theoretical and an experimental point of view.

Hartl, D. L., and A. G. Clark. 1989. *Principles of Population Genetics,* 2d ed. Sunderland, MA: Sinauer.

Lerner, I. M., and W. J. Libby. 1976. *Hereditary, Evolution, and Society,* 2d ed. New York: W. H. Freeman. A text meant for nonscience students.

Lewontin, R. C. 1974. *The Genetic Basis of Evolutionary Change.* New York: Columbia University Press. A discussion of the prevalence and role of genetic variation in natural populations. Both morphological and protein variations are considered.

Lewontin, R. C. 1982. *Human Diversity.* New York: Scientific American Books. Applies concepts of population genetics to human diversity and evolution.

Li, W. H., and D. Graur. 1991. *Fundamentals of Molecular Evolution.* Sunderland, MA: Sinauer. An excellent review of what is known about the forces and rates of evolution of proteins and DNA sequences.

Wilson, A. C. 1985. "The Molecular Basis of Evolution." *Scientific American* (October).

Chapter 25

Bodmer, W. F., and L. L. Cavalli-Sforza. 1970. "Intelligence and Race." *Scientific American* (October). A popular treatment, including a discussion of heritability.

Falconer, D. S. 1981. *Introduction to Quantitative Genetics,* 2d ed. New York: Ronald Press. A widely read text with a strong mathematical emphasis.

Feldman, M. W., and R. C. Lewontin. 1975. "The Heritability Hangup." *Science* 190: 1163–1168. A discussion of the meaning of heritability and its limitations, especially in relation to human intelligence.

Lewontin, R. C. 1974. "The Analysis of Variance and the Analysis of Causes." *American Journal of Human Genetics* 26: 400–411. A discussion of the meaning of the analysis of variance in genetics as a method for determining the roles of heredity and environment in determining phenotype.

Lewontin, R. C. 1982. *Human Diversity.* New York: Scientific American Books. Includes a discussion of quantitative variation in human populations.

Chapter 26

Dobzhansky, T. 1951. *Genetics and the Origin of Species.* New York: Columbia University Press. The classic synthesis of population genetics and the processes of evolution. The most influential book on evolution since Darwin's *Origin of Species.*

Evolution. 1978. *Scientific American* (September). This issue contains several articles relevant to population genetics.

Futuyma, D. J. 1998. *Evolutionary Biology,* 3d ed. Sunderland, MA: Sinauer. The best modern discussion of population genetics, ecology, and evolution from both a theoretical and an experimental point of view.

Li, W. H., and D. Graur. 1991. *Fundamentals of Molecular Evolution.* Sunderland, MA: Sinauer. An excellent review of what is known about the forces and rates of evolution of proteins and DNA sequences.

Wilson, A. C. 1985. "The Molecular Basis of Evolution." *Scientific American* (October).

Glossary

A Adenine or adenosine

abortive transduction The failure of a transducing DNA segment to be incorporated into the recipient chromosome.

acentric chromosome A chromosome having no centromere.

achondroplasia A type of dwarfism in humans, inherited as an autosomal dominant phenotype.

acrocentric chromosome A chromosome having the centromere located slightly nearer one end than the other.

active site The part of a protein that must be maintained in a specific shape if the protein is to be functional — for example, in an enzyme, the part to which the substrate binds.

adaptation In the evolutionary sense, some heritable feature of an individual's phenotype that improves its chances of survival and reproduction in the existing environment.

adaptive landscape The surface plotted in a three-dimensional graph, with all possible combinations of allele frequencies for different loci plotted in the plane and with mean fitness for each combination plotted in the third dimension.

adaptive peak A high point (perhaps one of several) on an adaptive landscape; selection tends to drive the genotype composition of the population toward a combination corresponding to an adaptive peak.

adaptive surface *See* **adaptive landscape.**

adaptor protein A protein that binds to certain specific phosphorylated amino acid sequences on a second protein, often a transmembrane receptor, and associates with still other proteins, thereby allowing a complex of proteins to "dock" to the receptor. This docking brings proteins of this complex in proximity to one another and, by doing so, permits propagation of an intracellular signal in a signal transduction pathway.

additive genetic variance Genetic variance associated with the average effects of substituting one allele for another.

adenine A purine base that pairs with thymine in the DNA double helix.

adenosine The nucleoside containing adenine as its base.

adenosine triphosphate *See* **ATP.**

adjacent segregation In a reciprocal translocation, the passage of a translocated and a normal chromosome to each of the poles.

ADP Adenosine diphosphate.

Ala Alanine (an amino acid).

albino A pigmentless "white" phenotype, determined by a mutation in a gene coding for a pigment-synthesizing enzyme.

alkylating agent A chemical agent that can add alkyl groups (for example, ethyl or methyl groups) to another molecule; many mutagens act through alkylation.

allele One of the different forms of a gene that can exist at a single locus.

allele frequency A measure of the commonness of an allele in a population; the proportion of all alleles of that gene in the population that are of this specific type.

allopatric speciation The process of the splitting of a species into two or more new species that occurs between populations that are geographically isolated from each other.

allopolyploid *See* **amphidiploid.**

allosteric transition A change from one conformation of a protein into another.

alternate segregation In a reciprocal translocation, the passage of both normal chromosomes to one pole and both translocated chromosomes to the other pole.

alternation of generations The alternation of gametophyte and sporophyte stages in the life cycle of a plant.

alternative splicing The process by which different mRNAs are produced from the same primary transcript, through variations in the splicing pattern of the transcript. Multiple mRNA "isoforms" can be produced in a single cell or the different isoforms can display different tissue-specific patterns of expression. If the alternative exons fall within the open reading frames of the mRNA isoforms, different proteins will be produced by the alternative mRNAs.

amber codon The codon UAG, a nonsense codon.

amber suppressor A mutant allele coding for a tRNA whose anticodon is altered in such a way that the tRNA inserts an amino acid at an amber codon in translation.

Ames test A widely used test to detect possible chemical carcinogens; based on mutagenicity in the bacterium *Salmonella*.

amino acid A peptide; the basic building block of proteins (or polypeptides).

amniocentesis A technique for testing the genotype of an embryo or fetus in utero with minimal risk to the mother or the child.

AMP Adenosine monophosphate.

amphidiploid An allopolyploid; a polyploid formed from the union of two separate chromosome sets and their subsequent doubling.

amplification The production of many DNA copies from one master region of DNA.

anaphase An intermediate stage of nuclear division during which chromosomes are pulled to the poles of the cell.

aneuploid cell A cell having a chromosome number that differs from the normal chromosome number for the species by a small number of chromosomes.

animal breeding   The practical application of genetic analysis for development of lines of domestic animals suited to human purposes.

annealing   Spontaneous alignment of two single DNA strands to form a double helix.

antibody   A protein (immunoglobulin) molecule, produced by the immune system, that recognizes a particular substance (antigen) and binds to it.

anticodon   A nucleotide triplet in a tRNA molecule that aligns with a particular codon in mRNA under the influence of the ribosome so that the amino acid carried by the tRNA is inserted in a growing protein chain.

antigen   A molecule that is recognized by antibody (immunoglobulin) molecules. Generally, multiple antibody molecules can recognize a given antigen.

antiparallel   A term used to describe the opposite orientations of the two strands of a DNA double helix; the 5′ end of one strand aligns with the 3′ end of the other strand.

AP site   Apurinic or apyrimidinic site resulting from the loss of a purine or pyrimidine residue from the DNA.

Arg   Arginine (an amino acid).

ascospore   A sexual spore from certain fungus species in which spores are found in a sac called an ascus.

ascus   In fungi, a sac that encloses a tetrad or an octad of ascospores.

asexual spore   *See* **spore.**

Asn   Asparagine (an amino acid).

Asp   Aspartate (an amino acid).

ATP (adenosine triphosphate)   The "energy molecule" of cells, synthesized mainly in mitochondria and chloroplasts; energy from the breakdown of ATP drives many important cell reactions.

attached X   A pair of *Drosophila* X chromosomes joined at one end and inherited as a single unit.

attenuator   A region adjacent to the structural genes of the *trp* operon; this region acts in the presence of tryptophan to reduce the rate of transcription from the structural genes.

autonomous controlling element   A controlling element that seems to have both regulator and receptor functions combined in a single unit that enters a gene and causes an unstable mutation.

autonomous phenotype   A genetic trait in multicellular organisms in which only genotypically mutant cells exhibit the mutant phenotype. Conversely, a *nonautonomous* trait is one in which genotypically mutant cells cause other cells (regardless of their genotype) to exhibit a mutant phenotype.

autonomous replication sequence (ARS)   A segment of a DNA molecule needed for the initiation of its replication; generally a site recognized and bound by the proteins of the replication system.

autophosphorylation   The process by which a protein kinase phosphorylates specific amino acid residues on itself.

autopolyploid   A polyploid formed from the doubling of a single genome.

autoradiogram   A pattern of dark spots in a developed photographic film or emulsion, in the technique of autoradiography.

autoradiography   A process in which radioactive materials are incorporated into cell structures, which are then placed next to a film or photographic emulsion, thus forming a pattern on the film corresponding to the location of the radioactive compounds within the cell.

autoregulatory loop   The process by which the expression of a gene is controlled by its own gene product.

autosome   Any chromosome that is not a sex chromosome.

auxotroph   A strain of microorganisms that will proliferate only when the medium is supplemented with some specific substance not required by wild-type organisms.

BAC   Bacterial artificial chromosome; an F plasmid engineered to act as a cloning vector that can carry large inserts.

back mutation   *See* **reversion.**

bacteriophage (phage)   A virus that infects bacteria.

balanced polymorphism   Stable genetic polymorphism maintained by natural selection.

balancer   A chromosome with multiple inversions, used to retain favorable allele combinations in the uninverted homolog.

Balbiani ring   A large chromosome puff.

Barr body   A densely staining mass that represents an inactivated X chromosome.

base analog   A chemical whose molecular structure mimics that of a DNA base; because of the mimicry, the analog may act as a mutagen.

bead theory   The disproved hypothesis that genes are arranged on the chromosome like beads on a necklace, indivisible into smaller units of mutation and recombination.

bimodal distribution   A statistical distribution having two modes.

binary fission   The process in which a parent cell splits into two daughter cells of approximately equal size.

biparental zygote   A *Chlamydomonas* zygote that contains cpDNA from both parents; such cells are generally rare.

blast cell   A cell that divides, generally asymmetrically, to give rise to two different progeny cells. One is a blast cell just like the parental cell and the other is a cell that enters a differentiation pathway. In this manner, a continuously propagating cell population can maintain itself and spin off differentiating cells.

blastoderm   In an insect embryo, the layer of cells that completely surrounds an internal mass of yolk.

blastula   An early developmental stage of lower vertebrate embryos, in which the embryo consists of a single layer of cells surrounding the central yolk.

blending inheritance   A discredited model of inheritance suggesting that the characteristics of an individual result from the smooth blending of fluidlike influences from its parents.

brachydactyly   A human phenotype of unusually short digits, generally inherited as an autosomal dominant.

branch migration   The process by which a single "invading" DNA strand extends its partial pairing with its complementary strand as it displaces the resident strand.

bridging cross   A cross made to transfer alleles between two sexually isolated species by first transferring the alleles to an intermediate species that is sexually compatible with both.

broad heritability (H^2)   The proportion of total phenotypic variance at the population level that is contributed by genetic variance.

bud   A daughter cell formed by mitosis in yeast; one daughter cell retains the cell wall of the parent, and the other (the bud) forms a new cell wall.

buoyant density   A measure of the tendency of a substance to float in some other substance; large molecules are distinguished by their differing buoyant densities in some standard fluid. Measured by density-gradient ultracentrifugation.

Burkitt lymphoma   A cancer of the lymphatic system manifested by tumors in the jaw, associated with a chromosomal translocation bringing a specific oncogene next to regulatory elements of one of the immunoglobulin genes.

C   Cytosine or cytidine.

callus   An undifferentiated clone of plant cells.

cAMP (cyclic adenosine monophosphate)   A molecule that plays a key role in the regulation of various processes within the cell.

cancer   The class of disease characterized by rapid and uncontrolled proliferation of cells within a tissue of a multitissued eukaryote. Cancers are generally thought to be genetic diseases of somatic cells, arising through sequential mutations that create oncogenes and inactivate tumor-suppressor genes.

candidate gene   A sequenced gene of previously unknown function that, because of its chromosomal position or some other property, becomes a candidate for a particular function such as disease determination.

CAP (catabolite activator protein)   A protein whose presence is necessary for the activation of the *lac* operon.

cap   A 7-methylguanosine molecule added to the 5′ end of the pre-mRNA.

carbon source   A nutrient (such as sugar) that provides carbon "skeletons" needed in an organism's synthesis of organic molecules.

carcinogen   A substance that causes cancer.

cardinal gene   Those pattern-formation genes in *Drosophila* that are the zygotically acting genes directly responding to the gradients of anterior–posterior and dorsal–ventral positional information created by the maternally expressed pattern-formation genes.

carrier   An individual who possesses a mutant allele but does not express it in the phenotype, because of a dominant allelic partner; thus, an individual of genotype *A a* is a carrier of *a* if there is complete dominance of *A* over *a*.

cassette model   A model to explain mating-type interconversion in yeast. Information for both *a* and *α* mating types is assumed to be present as silent "cassettes"; a copy of either type of cassette may be transposed to the mating-type locus, where it is "played" (transcribed).

catabolite activator protein   *See* **CAP.**

catabolite repression   The inactivation of an operon caused by the presence of large amounts of the metabolic end product of the operon.

cation   A positively charged ion (such as K^+).

CDK   *See* **cyclin-dependent protein kinase.**

cDNA   *See* **complementary DNA.**

cDNA library   A library composed of cDNAs, not necessarily representing all mRNAs.

cell autonomous   Refers to a genetic trait in multicellular organisms in which only genotypically mutant cells exhibit the mutant phenotype. Conversely, a *nonautonomous* trait is one in which genotypically mutant cells cause other cells (regardless of their genotype) to exhibit a mutant phenotype.

cell cycle   The set of events that take place in the divisions of mitotic cells. The cell cycle oscillates between mitosis (M phase) and interphase. Interphase can be subdivided in order into G_1, S phase, and G_2. DNA synthesis takes place during S phase. The length of the cell cycle is regulated through a special option in G_1, in which G_1 cells can enter a resting phase called G_0.

cell division   The process by which two cells are formed from one.

cell fate   The ultimate differentiated state to which a cell has become committed.

cell lineage   A pedigree of cells related through asexual division.

cellular blastoderm   The stage of blastoderm in insects after the nuclei have each been packaged in an individual cellular membrane.

centimorgan (cM)   *See* **map unit.**

central dogma   The hypothesis that information flows only from DNA to RNA to protein; although some exceptions are now known, the rule is generally valid.

centromere   A specialized region of DNA on each eukaryotic chromosome; it acts as a site for the binding of the kinetochore proteins.

character   Some attribute of individuals within a species for which various heritable differences can be defined.

character difference   Alternative forms of the same attribute within a species.

chase   *See* **pulse-chase experiment.**

checkpoint   Stage of the cell cycle at which the completion of a certain event of the cell cycle, such as chromosome replication, must have been successfully completed in order for the cell cycle to progress to the next stage.

chiasma (plural, **chiasmata**)   A cross-shaped structure commonly observed between nonsister chromatids in meiosis; the site of crossing-over.

chimera   *See* **mosaic.**

chimeric DNA   *See* **recombinant DNA.**

chi-square (χ^2) **test**   A statistical test used to determine the probability of obtaining the observed results by chance under a specific hypothesis.

chloroplast   A chlorophyll-containing organelle in plants that is the site of photosynthesis.

chromatid   One of the two side-by-side replicas produced by chromosome division.

chromatid conversion   A type of gene conversion that is inferred from the existence of identical sister-spore pairs in a fungal octad that shows a non-Mendelian allele ratio.

chromatid inference   A situation in which the occurrence of a crossover between any two nonsister chromatids can be shown to affect the probability of those chromatids taking part in other crossovers in the same meiosis.

chromatin   The substance of chromosomes; now known to include DNA, chromosomal proteins, and chromosomal RNA.

chromocenter   The point at which the polytene chromosomes appear to be attached together.

chromomere   A small beadlike structure visible on a chromosome during prophase of meiosis and mitosis.

chromosome   A linear end-to-end arrangement of genes and other DNA, sometimes with associated protein and RNA.

chromosome aberration   Any type of change in the chromosome structure or number.

chromosome loss   Failure of a chromosome to become incorporated into a daughter nucleus at cell division.

chromosome map   *See* **linkage map.**

chromosome mutation   Any type of change in the chromosome structure or number.

chromosome puff   A swelling at a site along the length of a polytene chromosome; the site of active transcription.

chromosome rearrangement   A chromosome mutation in which chromosome parts are newly juxtaposed.

chromosome set   The group of different chromosomes that carries the basic set of genetic information of a particular species.

chromosome theory of inheritance   The unifying theory stating that inheritance patterns may be generally explained by assuming that genes are located in specific sites on chromosomes.

chromosome walking   A method for the dissection of large segments of DNA, in which a cloned segment of DNA, usually eukaryotic, is used to screen recombinant DNA clones from the same genome bank for other clones containing neighboring sequences.

cis conformation   In a heterozygote involving two mutant sites within a gene or within a gene cluster, the arrangement $a_1 \, a_2/+ \; +$.

cis dominance   The ability of a gene to affect genes next to it on the same chromosome.

cis-trans test   A test to determine whether two mutant sites of a gene are in the same functional unit or gene.

cistron   Originally defined as a functional genetic unit within which two mutations cannot complement. Now equated with the term gene as the region of DNA that encodes a single polypeptide (or functional RNA molecule such as tRNA or rRNA).

clone   (1) A group of genetically identical cells or individuals derived by asexual division from a common ancestor. (2) *(colloquial)* An individual formed by some asexual process so that it is genetically identical with its "parent." (3) *See* **DNA clone.**

cM (centimorgan)   *See* **map unit.**

code dictionary   A listing of the 64 possible codons and their translational meanings (the corresponding amino acids).

codominance   The situation in which a heterozygote shows the phenotypic effects of both alleles equally.

codon   A section of DNA (three nucleotide pairs in length) that encodes a single amino acid.

coefficient of coincidence   The ratio of the observed number of double recombinants to the expected number.

cohesive end   End of DNA that is cut in a staggered pattern and then can hydrogen bond with a complementary base sequence from other similarly formed end.

cointegrate   The product of the fusion of two circular elements to form a single, larger circle.

colinearity   The correspondence between the location of a mutant site within a gene and the location of an amino acid substitution within the polypeptide translated from that gene.

colony   A visible clone of cells.

compartmentalization   The existence of boundaries within an organism beyond which a specific clone of cells will never extend during development.

competent   Able to take up exogenous DNA and thereby be transformed.

complementary DNA (cDNA)   Synthetic DNA transcribed from a specific RNA through the action of the enzyme reverse transcriptase.

complementary RNA (cRNA)   Synthetic RNA produced by transcription from a specific DNA single-stranded template.

complementation   The production of a wild-type phenotype when two different mutations are combined in a diploid or a heterokaryon.

complementation test   *See* **cis-trans test.**

complex inheritance   The transmission pattern of discontinuous variants that can be explained only by the interaction of several genes plus the environment.

conditional mutation   A mutation that has the wild-type phenotype under certain (permissive) environmental conditions and a mutant phenotype under other (restrictive) conditions.

conjugation   The union of two bacterial cells, during which chromosomal material is transferred from the donor to the recipient cell.

conjugation tube   *See* **pilus.**

conservative replication   A disproved model of DNA synthesis suggesting that one-half of the daughter DNA molecules should have both strands composed of newly polymerized nucleotides.

constant region   A region of an antibody molecule that is nearly identical with the corresponding regions of antibodies of different specificities.

constitutive   Always expressed in an unregulated fashion (in reference to gene control).

constitutive heterochromatin   Specific regions of heterochromatin always present and in both homologs of a chromosome.

contig   A set of ordered overlapping clones that constitute a chromosomal region or a genome.

continuous variation   Variation showing an unbroken range of phenotypic values.

controlling element   A mobile genetic element capable of producing an unstable mutant target gene; two types exist, the regulator and the receptor elements.

copy-choice model   A model of the mechanism for crossing-over, suggesting that crossing-over occurs in the course of chromosome division and can occur only between two supposedly "new" nonsister chromatids; the experimental evidence does not support this model.

correction   The production (possibly by excision and repair) of a properly paired nucleotide pair from a sequence of hybrid DNA that contains an illegitimate pair.

correlation coefficient   A statistical measure of the extent to which variations in one variable are related to variations in another.

cosegregation   In *Chlamydomonas,* parallel behavior of different chloroplast markers in a cross, due to their close linkage on cpDNA.

cosmid   A cloning vector that, like a plasmid, can replicate autonomously and be packaged into phage.

cotransduction   The simultaneous transduction of two bacterial marker genes.

cotransformation   The simultaneous transformation of two bacterial marker genes.

coupling conformation   Linked heterozygous gene pairs in the arrangement $A \, B/a \, b$.

covariance   A statistical measure used in computing the correlation coefficient between two variables; the covariance is the mean of $(x - \bar{x})(y - \bar{y})$ overall pairs of values for the variables x and y, where $\bar{x}$ is the mean of the x values and $\bar{y}$ is the mean of the y values.

cpDNA   Chloroplast DNA.

cri du chat syndrome   A lethal human condition in infants caused by deletion of part of one homolog of chromosome 5.

crisscross inheritance   Transmission of a gene from male parent to female child to male grandchild — for example, X-linked inheritance.

cRNA   *See* **complementary RNA.**

cross   The deliberate mating of two parental types of organisms in genetic analysis.

crossing-over   The exchange of corresponding chromosome parts between homologs by breakage and reunion.

crossover suppressor   An inversion (usually complex) that makes pairing and crossing-over impossible.

cruciform configuration   A region of DNA with palindromic sequences in both strands; each strand pairs with itself to form a helix extending sideways from the main helix.

culture   Tissue or cells multiplying by asexual division, grown for experimentation.

CVS   Chorionic villus sampling, a placental sampling procedure for obtaining fetal tissue for chromosome and DNA analysis to assist in prenatal diagnosis of genetic disorders.

cyclic adenosine monophosphate   *See* **cAMP.**

cyclin   A family of labile proteins that are synthesized and degraded at specific times within each cell cycle and that regulate cell cycle progression through their interactions with specific cyclin-dependent protein kinases.

cyclin-dependent protein kinase   A family of protein kinases that, on activation by cyclins and an elaborate set of positive and negative regulatory proteins, phosphorylate certain transcription factors whose activity is necessary for a particular stage of the cell cycle.

Cys   Cysteine (an amino acid).

cystic fibrosis   A potentially lethal human disease of secretory glands; the most prominent symptom is excess secretion of lung mucus; inherited as an autosomal recessive.

cytidine   The nucleoside containing cytosine as its base.

cytochromes   A class of proteins, found in mitochondrial membranes, whose main function is oxidative phosphorylation of ADP to form ATP.

cytogenetics   The cytological approach to genetics, mainly entailing microscopic studies of chromosomes.

cytohet   A cell containing two genetically distinct types of a specific organelle.

cytoplasm   The material between the nuclear and cell membranes; includes fluid (cytosol), organelles, and various membranes.

cytoplasmic inheritance   Inheritance through genes found in cytoplasmic organelles.

cytosine   A pyrimidine base that pairs with guanine.

cytoskeleton   The protein cable systems and associated proteins that together form the architecture of a eukaryotic cell.

cytosol   The fluid part of the cytoplasm, outside the organelles.

Darwinian fitness   The relative probability of survival and reproduction for a genotype.

daughter cells   Two identical cells formed by the asexual division of a cell.

daughter chromatids   Two identical chromatids formed by the replication of one chromosome.

default state   The developmental state of a cell (or group of cells) in the absence of the activation of a developmental regulatory switch.

degenerate code   A genetic code in which some amino acids may be encoded by more than one codon each.

deletion   Removal of a chromosomal segment from a chromosome set.

denaturation   The separation of the two strands of a DNA double helix or the severe disruption of the structure of any complex molecule without breaking the major bonds of its chains.

denaturation map   A map of a stretch of DNA showing the locations of local denaturation loops, which correspond to regions of high AT content.

deoxyribonuclease   *See* **DNase.**

deoxyribonucleic acid   *See* **DNA.**

determinant   A spatially localized molecule that causes cells to adopt a particular fate or set of related fates.

determination   The process of commitment of cells to particular fates.

development   The process whereby a single cell becomes a differentiated organism.

developmental field   A set of cells that together interact to form a developing structure (for example, an embryo, a tissue, an organ, a limb, and so forth).

developmental pathway   The chain of molecular events that take a set of equivalent cells and produce the assignment of different fates among those cells.

dicentric chromosome   A chromosome with two centromeres.

differentiation   The changes in cell shape and physiology associated with the production of the final cell types of a particular organ or tissue.

dihybrid cross   A cross between two individuals identically heterozygous at two loci — for example, $A/a \cdot B/b \times A/a \cdot B/b$.

dimorphism   A "polymorphism" including only two forms.

dioecious plant   A plant species in which male and female organs appear on separate individuals.

diploid   A cell having two chromosome sets or an individual having two chromosome sets in each of its cells.

directed mutagenesis   Altering some specific part of a cloned gene and reintroducing the modified gene back into the organism.

directional selection   Selection that changes the frequency of an allele in a constant direction, either toward or away from fixation for that allele.

discontinuous variation   Variation in which there are distinct classes of phenotypes for a particular character.

dispersive replication   Disproved model of DNA synthesis suggesting more or less random interspersion of parental and new segments in daughter DNA molecules.

distribution   *See* **statistical distribution.**

distribution function   A graph of some precise quantitative measure of a character against its frequency of occurrence.

diversification   Differentiation between different lines of descent in the course of evolution.

DNA (deoxyribonucleic acid)   A double chain of linked nucleotides (having deoxyribose as their sugars); the fundamental substance of which genes are composed.

DNA clone   A section of DNA that has been inserted into a vector molecule, such as a plasmid or a phage chromosome, and then replicated to form many copies.

DNA fingerprint The largely individual-specific autoradiographic banding pattern produced when DNA is digested with a restriction enzyme that cuts outside a family of VNTRs and a Southern blot of the electrophoretic gel is probed with a VNTR-specific probe.

DNA polymerase An enzyme that can synthesize new DNA strands from a DNA template; several such enzymes exist.

DNase (deoxyribonuclease) An enzyme that degrades DNA to nucleotides.

dominance variance Genetic variance at a single locus attributable to dominance of one allele over another.

dominant allele An allele that expresses its phenotypic effect even when heterozygous with a recessive allele; thus if A is a dominant over a, then A/A and A/a have the same phenotype.

dominant phenotype The phenotype of a genotype containing the dominant allele; the parental phenotype that is expressed in a heterozygote.

donor DNA Any DNA to be used in cloning.

dosage compensation The process in organisms using a chromosomal sex determination mechanism (such as XX versus XY) that allows standard structural genes on the sex chromosome to be expressed at the same levels in females and males, regardless of the number of sex chromosomes. In mammals, dosage compensation operates by maintaining only a single active X chromosome in each cell; in *Drosophila,* it operates by hyperactivating the male X chromosome.

dose *See* **gene dose.**

double crossover Two crossovers occurring in a chromosomal region under study.

double helix The structure of DNA first proposed by Watson and Crick, with two interlocking helices joined by hydrogen bonds between paired bases.

double infection Infection of a bacterium with two genetically different phages.

Down syndrome An abnormal human phenotype, including mental retardation, due to a trisomy of chromosome 21; more common in babies born to older mothers.

drift *See* **random genetic drift.**

Duchenne muscular dystrophy A lethal muscle disease in humans caused by mutation in a huge gene that encodes the muscle protein dystrophin; inherited as an X-linked recessive phenotype.

duplicate genes Two identical allele pairs in one diploid individual.

duplication More than one copy of a particular chromosomal segment in a chromosome set.

dyad A pair of sister chromatids joined at the centromere, as in the first division of meiosis.

ecdysone A molting hormone in insects.

ectopic expression The occurrence of gene expression in a tissue in which it is normally not expressed. Such ectopic expression can be caused by the juxtaposition of novel enhancer elements to a gene.

ectopic integration In a transgenic organism, the insertion of an introduced gene at a site other than its usual locus.

electrophoresis A technique for separating the components of a mixture of molecules (proteins, DNAs, or RNAs) in an electric field within a gel.

embryonic (or tissue) polarity The production of axes of asymmetry in a developing embryo or tissue primordium.

embryonic stem cells Cultured cell lines that are established from very early embryos and that are essentially totipotent; that is, these cells can be implanted into a host embryo and populate many or all tissues of the developing animal, most importantly including the germ line. Manipulations of these embryonic stem cells (ES cells) are used extensively in mouse genetics to produce targeted gene knockouts.

endocrine system The organs in the body that secrete hormones into the circulatory system.

endogenote *See* **merozygote.**

endonuclease An enzyme that cleaves the phosphodiester bond within a nucleotide chain.

endopolyploidy An increase in the number of chromosome sets caused by replication without cell division.

endosperm Triploid tissue in a seed, formed from the fusion of two haploid female nuclei and one haploid male nucleus.

enforced outbreeding Deliberate avoidance of mating between relatives.

enhancer A cis-regulatory sequence that can elevate levels of transcription from an adjacent promoter. Many *tissue-specific enhancers* can determine spatial patterns of gene expression in higher eukaryotes. Enhancers can act on promoters over many tens of kilobases of DNA and can be 5′ or 3′ to the promoter that they regulate.

enhancer trap A transgenic construction inserted in a chromosome and used to identify tissue-specific enhancers in the genome. In such a construct, a promoter sensitive to enhancer regulation is fused to a reporter gene, such that expression patterns of the reporter gene identify the spatial regulation conferred by nearby enhancers.

enucleate cell A cell having no nucleus.

environment The combination of all the conditions external to the genome that potentially affect its expression and its structure.

environmental variance The variance due to environmental variation.

enzyme A protein that functions as a catalyst.

epigenetic inheritance Processes by which heritable modifications in gene function occur but are not due to changes in the base sequence of the DNA of the organism. Examples of epigenetic inheritance are paramutation, X-chromosome inactivation, and parental imprinting.

episome A genetic element in bacteria that can replicate in the cytoplasm or can be inserted into the main bacterial chromosome and replicate with the chromosome.

epistasis A situation in which the differential phenotypic expression of genotypes at one locus depends on the genotype at another locus.

epitope The part of an antigen molecule that is recognized by a specific immunoglobulin.

equational division A nuclear division that maintains the same ploidy level of the cell.

equivalence group A set of immature cells that all have the same developmental potential. In many cases, cells of an equivalence group end up adopting different fates from one another.

ES cells *See* **embryonic stem cells.**

EST (expressed sequence tag) A sequence-tagged site derived from a cDNA clone; used to position and identify genes in genomic analysis. *See also* **sequence-tagged site.**

ethidium A molecule that can intercalate into a DNA double helix when the helix is under torsional stress.

euchromatin A chromosomal region that stains normally; thought to contain the normally functioning genes.

eugenics Controlled human breeding based on notions of desirable and undesirable genotypes.

eukaryote An organism having eukaryotic cells.

eukaryotic cell A cell containing a nucleus.

euploid A cell having any number of complete chromosome sets or an individual composed of such cells.

excision repair The repair of a DNA lesion by removal of the faulty DNA segment and its replacement with a wild-type segment.

exconjugant A female bacterial cell that has just been in conjugation with a male and that contains a fragment of male DNA.

exogenote *See* **merozygote.**

exon Any nonintron section of the coding sequence of a gene; together, the exons constitute the mRNA and are translated into protein.

exonuclease An enzyme that cleaves nucleotides one at a time from an end of a polynucleotide chain.

expression library A library in which the vector carries transcriptional signals to allow any cloned insert to produce mRNA and ultimately a protein product.

expression vector A vector with the appropriate bacterial regulatory regions located 5′ to the insertion site, allowing transcription and translation of a foreign protein in bacteria.

expressivity The degree to which a particular genotype is expressed in the phenotype.

F⁻ cell In *E. coli,* a cell having no fertility factor; a female cell.

F⁺ cell In *E. coli,* a cell having a free fertility factor; a male cell.

F factor *See* **fertility factor.**

F′ factor A fertility factor into which a part of the bacterial chromosome has been incorporated.

F₁ generation The first filial generation, produced by crossing two parental lines.

F₂ generation The second filial generation, produced by selfing or intercrossing the F_1.

facultative heterochromatin Heterochromatin located in positions that are composed of euchromatin in other individuals of the same species or even in the other homolog of a chromosome pair.

FACS Fluorescence-activated chromosome sorting. The use of specific fluorescence signals of stained chromosomes in droplets to activate deflector plates that sort them into individual tubes of uniform types.

familial trait A trait common to members of a family.

family selection A breeding technique of selecting a pair on the basis of the average performance of their progeny.

fate map A map of an embryo showing areas that are destined to develop into specific adult tissues and organs.

feedback loop *See* **autoregulatory loop.**

fertility factor (F factor) A bacterial episome whose presence confers donor ability (maleness).

filial generations Successive generations of progeny in a controlled series of crosses, starting with two specific parents (the P generation) and selfing or intercrossing the progeny of each new ($F_1, F_2, \ldots$) generation.

filter enrichment A technique for recovering auxotrophic mutants in filamentous fungi.

fingerprint The characteristic spot pattern produced by electrophoresis of the polypeptide fragments obtained through denaturation of a particular protein by a proteolytic enzyme.

first-division segregation pattern A linear pattern of spore phenotypes within the ascus for a particular allele pair, produced when the alleles go into separate nuclei at the first meiotic division, showing that no crossover has occurred between that allele pair and the centromere.

FISH Fluorescent in situ hybridization. In situ hybridization using a probe coupled to a fluorescent molecule.

fitness *See* **Darwinian fitness.**

fixed allele An allele for which all members of the population under study are homozygous, so no other alleles for this locus exist in the population.

fixed breakage point According to the heteroduplex DNA recombination model, the point from which unwinding of the DNA double helix begins, as a prelude to the formation of heteroduplex DNA.

flow sorting *See* **FACS.**

fluctuation test A test used in microbes to establish the random nature of mutation or to measure mutation rates.

fMET *See* **formylmethionine.**

focus map A fate map of areas of the *Drosophila* blastoderm destined to become specific adult structures that is based on the frequencies of specific kinds of mosaics.

foreign DNA DNA from another organism.

formylmethionine (fMet) A specialized amino acid that is the very first one incorporated into the polypeptide chain in the synthesis of proteins.

forward mutation A mutation that converts a wild-type allele into a mutant allele.

founder effect A difference in allele frequencies between a donor population and a small colony derived from it that occurs as a result of sampling in the first generation of colonization.

frame-shift mutation The insertion or deletion of a nucleotide pair or pairs, causing a disruption of the translational reading frame.

frequency-dependent fitness Fitness differences whose intensity changes with changes in the relative frequency of genotypes in the population.

frequency-dependent selection Selection in which the fitnesses of genotypes depend on their relative frequency in the population.

frequency histogram A "step curve" in which the frequencies of various arbitrarily bounded classes are graphed.

frequency-independent fitness Fitness that is not dependent on interactions with other individuals of the same species.

frequency-independent selection Selection in which the fitnesses of genotypes are independent of their relative frequency in the population.

fruiting body In fungi, the organ in which meiosis takes place and sexual spores are produced.

functional complementation The use of a cloned fragment of wild-type DNA to transform a mutant into wild type; used in identifying a clone containing one specific gene.

functional genomics The study of patterns of gene expression and interaction in the genome as a whole.

G   Guanine or guanosine.

G₀ phase   The resting phase that can occur during G_1 of interphase of the cell cycle.

G₁ phase   The part of interphase of the cell cycle that precedes S phase.

G₂ phase   The part of interphase of the cell cycle that follows S phase.

gain-of-function mutation   A mutation that results in a new functional ability for a protein, detectable at the phenotypic level.

gamete   A specialized haploid cell that fuses with a gamete from the opposite sex or mating type to form a diploid zygote; in mammals, an egg or a sperm.

gametophyte   The haploid gamete-producing stage in the life cycle of plants; prominent and independent in some species but reduced or parasitic in others.

gap gene   In *Drosophila*, the class of cardinal genes that are activated in the zygote in response to the anterior–posterior gradients of positional information. Through regulation of the pair-rule and homeotic genes, the patterns of expression of the various gap gene products lead to the specification of the correct number and types of body segments. Gap mutations cause the loss of several adjacent body segments.

gastrulation   The first process of movements and infoldings of the cell sheet in early animal embryos, usually immediately following blastula (or blastoderm).

gene   The fundamental physical and functional unit of heredity, which carries information from one generation to the next; a segment of DNA, composed of a transcribed region and a regulatory sequence that makes possible transcription.

gene amplification   The process by which the number of copies of a chromosomal segment is increased in a somatic cell.

gene conversion   A meiotic process of directed change in which one allele directs the conversion of a partner allele into its own form.

gene disruption   Inactivation of a gene by the integration of a specially engineered introduced DNA fragment.

gene dose   The number of copies of a particular gene present in the genome.

gene family   A set of genes in one genome all descended from the same ancestral gene.

gene frequency   *See* **allele frequency.**

gene fusion   The accidental joining of the DNA of two genes, such as can occur in a translocation. Gene fusions can give rise to hybrid proteins or to the misregulation of the transcription unit of one gene by the cis-regulatory elements (enhancers) of another.

gene interaction   The collaboration of several different genes in the production of one phenotypic character (or related group of characters).

gene locus   The specific place on a chromosome where a gene is located.

gene map   A linear designation of mutant sites within a gene, based on the various frequencies of interallelic (intragenic) recombination.

gene mutation   A point mutation that results from changes within the structure of a gene.

gene pair   The two copies of a particular type of gene present in a diploid cell (one in each chromosome set).

gene rearrangement   The process of programmed changes in the DNA structure of somatic cells, leading to changes in gene number or in the structural and functional properties of the rearranged gene.

gene replacement   The insertion of a genetically engineered transgene in place of a resident gene; often occurs by a double crossover.

gene therapy   The correction of a genetic deficiency in a cell by the addition of new DNA and its insertion into the genome.

generalized transduction   The ability of certain phages to transduce any gene in the bacterial chromosome.

genetic code   The set of correspondences between nucleotide pair triplets in DNA and amino acids in protein.

genetic dissection   The use of recombination and mutation to piece together the various components of a given biological function.

genetic markers   Alleles used as experimental probes to keep track of an individual, a tissue, a cell, a nucleus, a chromosome, or a gene.

genetic variance   Phenotypic variance resulting from the presence of different genotypes in the population.

genetics   (1) The study of genes. (2) The study of inheritance.

genome   The entire complement of genetic material in a chromosome set.

genomic library   A library encompassing an entire genome.

genomics   The cloning and molecular characterization of entire genomes.

genotype   The specific allelic composition of a cell — either of the entire cell or, more commonly, for a certain gene or a set of genes.

geographical race   A population of a species that occupies a different geographical area from other populations of the species and that differs in its allele frequencies from other such populations.

germ line   The cell lineage in a multitissued eukaryote from which the gametes derive.

germinal mutations   Mutations in the cells that are destined to develop into gametes.

Gln   Glutamine (an amino acid).

Glu   Glutamate (an amino acid).

Gly   Glycine (an amino acid).

G-protein   A member of a family of proteins that contribute to signal transduction through protein–protein interactions that occur when the G-protein binds GTP but not when the G-protein binds GDP.

gradient   A gradual change in some quantitative property over a specific distance.

ground state   The developmental state of a cell (or group of cells) in the absence of the activation of a developmental regulatory switch.

growth factors   Paracrine signaling molecules, usually secreted polypeptides, that induce cell division in cells receiving these signals.

guanine   A purine base that pairs with cytosine.

guanosine   The nucleoside having guanine as its base.

gynandromorph   An individual that is a mosaic of male and female structures. The underlying cause is frequently sex-chromosome mosaicism, such that some cells are chromosomal females, whereas others are chromosomal males.

half-chromatid conversion A type of gene conversion that is inferred from the existence of nonidentical sister spores in a fungal octad showing a non-Mendelian allele ratio.

haploid A cell having one chromosome set or an organism composed of such cells.

haploidization Production of a haploid from a diploid by progressive chromosome loss.

haploinsufficient Description of a gene that, in a diploid cell, cannot promote wild-type function in only one copy (dose).

haplosufficient Description of a gene that, in a diploid cell, can promote wild-type function in only one copy (dose).

Hardy-Weinberg equilibrium The stable frequency distribution of genotypes A/A, A/a, and a/a, in the proportions p^2, $2pq$, and q^2, respectively (where p and q are the frequencies of the alleles A and a), that is a consequence of random mating in the absence of mutation, migration, natural selection, or random drift.

harlequin chromosomes Sister chromatids that stain differently, so that one appears dark and the other light.

helicase An enzyme that breaks hydrogen bonds in DNA and unwinds it during movement of the replication fork.

helix-loop-helix (HLH) protein A protein, a part of which forms two α-helices separated by a loop (the HLH domain) that acts as a sequence-specific DNA-binding domain. HLH proteins are thought to act as transcription factors.

hemizygous gene A gene present in only one copy in a diploid organism — for example, X-linked genes in a male mammal.

hemoglobin (Hb) The oxygen-transporting blood protein in most animals.

hemophilia A human disease in which the blood fails to clot, caused by a mutation in a gene coding for a clotting protein; inherited as an X-linked recessive phenotype.

hereditary nonpolyposis colorectal cancer (HNPCC) One of the most common predispositions to cancer.

heredity The biological similarity of offspring and parents.

heritability in the broad sense *See* **broad heritability.**

heritability in the narrow sense (h^2) The proportion of phenotypic variance that can be attributed to additive genetic variance.

hermaphrodite (1) A plant species in which male and female organs are present in the same flower of a single individual (*compare* **monoecious plant**). (2) An animal with both male and female sex organs.

heterochromatin Densely staining condensed chromosomal regions, believed to be for the most part genetically inert.

heteroduplex A DNA double helix formed by annealing single strands from different sources; if there is a structural difference between the strands, the heteroduplex may show such abnormalities as loops or buckles.

heteroduplex DNA model A model that explains both crossing-over and gene conversion by assuming the production of a short stretch of heteroduplex DNA (formed from both parental DNAs) in the vicinity of a chiasma.

heterogametic sex The sex that has heteromorphic sex chromosomes (for example, XY) and hence produces two different kinds of gametes with respect to the sex chromosomes.

heterogeneous nuclear RNA (HnRNA) A diverse assortment of RNA types found in the nucleus, including mRNA precursors and other types of RNA.

heterokaryon A culture of cells composed of two different nuclear types in a common cytoplasm.

heterokaryon test A test for cytoplasmic mutations, based on new associations of phenotypes in cells derived from specially marked heterokaryons.

heteromorphic chromosomes A pair of chromosomes with some homology but differing in size, shape, or staining properties.

heteroplasmon A cell containing a mixture of genetically different cytoplasms, generally different mitochondria, or different chloroplasts.

heterothallic fungus A fungus species in which two different mating types must unite to complete the sexual cycle.

heterozygosity A measure of the genetic variation in a population; with respect to one locus, stated as the frequency of heterozygotes for that locus.

heterozygote An individual having a heterozygous gene pair.

heterozygous gene pair A gene pair having different alleles in the two chromosome sets of the diploid individual — for example, A/a or A^1/A^2.

hexaploid A cell having six chromosome sets or an organism composed of such cells.

high-frequency recombination (Hfr) cell In *E. coli*, a cell having its fertility factor integrated into the bacterial chromosome; a donor (male) cell.

Himalayan A mammalian temperature-dependent coat phenotype, generally albino with pigment only at the cooler tips of the ears, feet, and tail.

His Histidine (an amino acid).

histocompatibility antigens Antigens that determine the acceptance or rejection of a tissue graft.

histocompatibility genes The genes that encode the histocompatibility antigens.

histone A type of basic protein that forms the unit around which DNA is coiled in the nucleosomes of eukaryotic chromosomes.

HLH protein *See* **helix-loop-helix protein.**

homeobox (homeotic box) A family of quite similar 180-base-pair DNA sequences that encode a polypeptide sequence called a homeodomain, a sequence-specific DNA-binding sequence. Although the homeobox was first discovered in all homeotic genes, it is now known to encode a much more widespread DNA-binding motif.

homeodomain A highly conserved family of sequences 60 amino acids in length found within a large number of transcription factors that can form a helix-turn-helix structure and bind DNA in a sequence-specific manner.

homeologous chromosomes Partly homologous chromosomes, usually indicating some original ancestral homology.

homeosis The replacement of one body part by another. Homeosis can be caused by environmental factors leading to developmental anomalies or by mutation.

homeotic gene Genes that control the fate of segments along the anterior–posterior axis of higher animals.

homeotic mutation A mutation that can change the fate of an imaginal disk.

homogametic sex The sex with homologous sex chromosomes (for example, XX).

homolog A member of a pair of homologous chromosomes.

homologous chromosomes Chromosomes that pair with each other at meiosis or chromosomes in different species that have retained most of the same genes during their evolution from a common ancestor.

homothallic fungus   A fungus species in which a single sexual spore can complete the entire sexual cycle (*compare* **heterothallic fungus**).

homozygote   An individual having a homozygous gene pair.

homozygous gene pair   A gene pair having identical alleles in both copies — for example, A/A or A^1/A^1.

hormone   A molecule that is secreted by an endocrine organ into the circulatory system and that acts as a long-range signaling molecule by activating receptors on or within target cells.

hormone response element (HRE)   For hormones that act by binding to receptors that can act as transcription factors, an HRE is a cis-regulatory DNA sequence that is a binding site for a hormone–receptor complex.

host range   The spectrum of strains of a given bacterial species that a given strain of phage can infect.

hot spot   A part of a gene that shows a very high tendency to become a mutant site, either spontaneously or under the action of a particular mutagen.

Huntington disease   A lethal human disease of nerve degeneration, with late-age onset. Inherited as an autosomal dominant phenotype; new mutations rare.

hybrid   (1) A heterozygote. (2) A progeny individual from any cross between parents of differing genotypes.

hybrid dysgenesis   A syndrome of effects including sterility, mutation, chromosome breakage, and male recombination in the hybrid progeny of crosses between certain laboratory and natural isolates of *Drosophila*.

hybridization in situ   Finding the location of a gene by adding specific radioactive probes for the gene and detecting the location of the radioactivity on the chromosome after hybridization.

hybridize   (1) To form a hybrid by performing a cross. (2) To anneal nucleic acid strands from different sources.

hydrogen bond   A weak bond in which an electron is shared with a hydrogen atom; hydrogen bonds are important in the specificity of base pairing in nucleic acids and in the determination of protein shape.

hydroxyapatite   A form of calcium phosphate that binds double-stranded DNA.

hyperploid   Aneupolid containing a small number of extra chromosomes.

hypervariable region   The part of a variable region that determines the specificity of an antibody.

hypha (plural, **hyphae**)   A threadlike structure (composed of cells attached end to end) that forms the main tissue in many fungus species.

hypoploid   Aneuploid with a small number of chromosomes missing.

Ig (immunoglobulin)   *See* **antibody.**

Ile   Isoleucine (an amino acid).

imago   An adult insect.

immune system   The animal cells and tissues that recognize and attack foreign substances within the body.

immunoglobulin   *See* **antibody.**

immunohistochemistry   The use of antibodies or antisera as histological tools for identifying patterns of protein distribution within a tissue or an organism. An antibody (or mixture of antibodies) that binds to a specific protein is tagged with an enzyme that can convert a substrate into a visible dye. The tagged antibody is incubated with the tissue, unbound antibody is washed off, and the enzymatic substrate is then added, revealing the pattern of protein (antigen) localization.

in situ   "In place"; *see* **hybridization in situ.**

in vitro   In an experimental situation outside the organism (literally, "in glass").

in vitro mutagenesis   The production of either random or specific mutations in a piece of cloned DNA. Typically, the DNA will then be repackaged and introduced into a cell or an organism to assess the results of the mutagenesis.

in vivo   In a living cell or organism.

inbreeding   Mating between relatives.

inbreeding coefficient   The probability of homozygosity that results because the zygote obtains copies of the *same* ancestral gene.

incomplete dominance   The situation in which a heterozygote shows a phenotype quantitatively (but not exactly) intermediate between the corresponding homozygote phenotypes. (Exact intermediacy is no dominance.)

independent assortment   *See* **Mendel's second law.**

inducer   An environmental agent that triggers transcription from an operon.

induction   The relief of repression for a gene or set of genes under negative control.

inductive interaction   The interaction between two groups of cells in which a signal passed from one group of cells causes the other group of cells to change their developmental state (or fate).

infectious transfer   The rapid transmission of free episomes (plus any chromosomal genes that they may carry) from donor to recipient cells in a bacterial population.

inosine   A rare base that is important at the wobble position of some tRNA anticodons.

insertion sequence (IS)   A mobile piece of bacterial DNA (several hundred nucleotide pairs in length) that is capable of inactivating a gene into which it inserts.

insertional translocation   The insertion of a segment from one chromosome into another nonhomologous one.

intercalating agent   A chemical that can insert itself between the stacked bases at the center of the DNA double helix, possibly causing a frame-shift mutation.

interchromosomal recombination   Recombination resulting from independent assortment.

interference   A measure of the independence of crossovers from each other, calculated by subtracting the coefficient of coincidence from 1.

intermediate filaments   A heterogeneous class of cytoskeletal elements characterized by an intermediate cable diameter larger than that of microfilaments but smaller than that of microtubules.

interphase   The cell cycle stage between nuclear divisions, when chromosomes are extended and functionally active.

interrupted mating   A technique used to map bacterial genes by determining the sequence in which donor genes enter recipient cells.

interstitial region   The chromosomal region between the centromere and the site of a rearrangement.

intervening sequence   An intron; a segment of largely unknown function within a gene. This segment is initially transcribed, but the transcript is not found in the functional mRNA.

intrachromosomal recombination   Recombination resulting from crossing-over between two gene pairs.

intron   *See* **intervening sequence.**

inversion   A chromosomal mutation in which a chromosome segment is removed, rotated through 180 degrees, and reinserted in the same location.

inverted repeat (IR) sequence   A sequence found in identical (but inverted) form — for example, at the opposite ends of a transposon.

IR   *See* **inverted repeat sequence.**

IS   *See* **insertion sequence.**

isoaccepting tRNAs   The various types of tRNA molecules that carry a specific amino acid.

isotope   One of several forms of an atom having the same atomic number but differing atomic masses.

karyotype   The entire chromosome complement of an individual or cell, as seen in mitotic metaphase.

kb   *See* **kilobase.**

kilobase (kb)   1000 nucleotide pairs.

kinetochore   A complex of proteins to which a nuclear spindle fiber attaches.

Klinefelter syndrome   An abnormal human male phenotype with an extra X chromosome (XXY).

knockout   Inactivation of one specific gene. Same as gene disruption.

lagging strand   In DNA replication, the strand that is synthesized apparently in the 3′-to-5′ direction, by ligating short fragments synthesized individually in the 5′-to-3′ direction.

λ (lambda) phage   One kind ("species") of temperate bacteriophage.

λdgal   A λ phage carrying a *gal* bacterial gene and defective (*d*) for some phage function.

lampbrush chromosome   A large chromosome found in amphibian eggs, with lateral DNA loops that produce a brushlike appearance under the microscope.

lateral inhibition   The signal produced by one cell that prevents adjacent cells from acquiring its fate.

lawn   A continuous layer of bacteria on the surface of an agar medium.

leader   An untranslated segment at the 5′ end of mRNA between the transcriptional and the translational start sites.

leader sequence   At the 5′ end of an mRNA, the sequence that is not translated into protein.

leading strand   In DNA replication, the strand that is made in the 5′-to-3′ direction by continuous polymerization at the 3′ growing tip.

leaky mutant   A mutant (typically, an auxotroph) that results from a partial rather than a complete inactivation of the wild-type function.

leaky mutation   A mutation that confers a mutant phenotype but still retains a low but detectable level of wild-type function.

lesion   A damaged area in a gene (a mutant site), a chromosome, or a protein.

lethal gene   A gene whose expression results in the death of the individual expressing it.

Leu   Leucine (an amino acid).

library   A collection of DNA clones obtained from one DNA donor.

ligand – receptor interaction   The interactions between a molecule (usually of an extracellular origin) and a protein on or within a target cell. One type of ligand – receptor interaction can be between steroid hormones and their cytoplasmic or nuclear receptors. Another can be between secreted polypeptide ligands and transmembrane receptors.

ligase   An enzyme that can rejoin a broken phosphodiester bond in a nucleic acid.

line   A group of identical pure-breeding diploid or polyploid organisms, distinguished from other individuals of the same species by some unique phenotype and genotype.

LINE   Long interspersed element; a type of large repetitive DNA segment found throughout the genome.

linear tetrad   A tetrad that results from the occurrence of the meiotic and postmeiotic nuclear divisions in such a way that sister products remain adjacent to one another (with no passing of nuclei).

linkage   The association of genes on the same chromosome.

linkage group   A group of genes known to be linked; a chromosome.

linkage map   A chromosome map; an abstract map of chromosomal loci, based on recombinant frequencies.

locus (plural, **loci**)   *See* **gene locus.**

Lys   Lysine (an amino acid).

lysis   The rupture and death of a bacterial cell on the release of phage progeny.

lysogen   *See* **lysogenic bacterium.**

lysogenic bacterium   A bacterial cell capable of spontaneous lysis due, for example, to the uncoupling of a prophage from the bacterial chromosome.

M phase   The mitotic phase of the cell cycle.

Mb   *See* **megabase.**

macromolecule   A large polymer such as DNA, a protein, or a polysaccharide.

Manx   Tailless phenotype in cats, caused by an autosomal dominant mutation that is lethal when homozygous.

map unit (m.u.)   The "distance" between two linked gene pairs where 1 percent of the products of meiosis are recombinant; a unit of distance in a linkage map.

mapping function   A formula expressing the relation between distance in a linkage map and recombinant frequency.

Marfan syndrome   A human disorder of the connective tissue expressed as a range of symptoms, including very long limbs and digits as well as heart defects; inherited as an autosomal dominant phenotype.

marker   *See* **genetic markers.**

marker retention   A technique used in yeast to test the degree of linkage between two mitochondrial mutations.

maternal effect   The environmental influence of the mother's tissues on the phenotype of the offspring.

maternal effect lethal   A mutation that is viable in zygotes, but mothers having the mutation produce inviable offspring.

maternal inheritance   A type of uniparental inheritance in which all progeny have the genotype and phenotype of the parent acting as the female.

maternally expressed gene   A gene that contributes to the phenotype of an offspring on the basis of its expression in the mother.

mating types   The equivalent in lower organisms of the sexes in higher organisms; the mating types typically differ only physiologically and not in physical form.

matroclinous inheritance   Inheritance in which all offspring have the nucleus-determined phenotype of the mother.

mean   The arithmetic average.

medium   Any material on (or in) which experimental cultures are grown.

Megabase (Mb)   One million nucleotide pairs.

meiocyte   Cell in which meiosis takes place.

meiosis   Two successive nuclear divisions (with corresponding cell divisions) that produce gametes (in animals) or sexual spores (in plants and fungi) having one-half of the genetic material of the original cell.

meiospore   Cell that is one of the products of meiosis in plants.

melting   Denaturation of DNA.

Mendelian ratio   A ratio of progeny phenotypes in accord with the application of Mendel's laws.

Mendel's first law   The two members of a gene pair segregate from each other in meiosis; each gamete has an equal probability of obtaining either member of the gene pair.

Mendel's second law   The law of independent assortment; unlinked or distantly linked segregating gene pairs assort independently at meiosis.

merozygote   A partly diploid *E. coli* cell formed from a complete chromosome (the endogenote) plus a fragment (the exogenote).

messenger RNA   *See* **mRNA.**

Met   Methionine (an amino acid).

metabolism   The chemical reactions that take place in a living cell.

metacentric chromosome   A chromosome having its centromere in the middle.

metamere   A segmental repeat unit in higher animals.

metaphase   An intermediate stage of nuclear division when chromosomes align along the equatorial plane of the cell.

methylation   Modification of a molecule by the addition of a methyl group.

microfilaments   The smallest-diameter cable system of the cytoskeleton. Microfilament cables are composed of actin polymers.

microsatellite DNA   A type of repetitive DNA based on very short repeats such as dinucleotides.

microtubule organizing center   The part of the microtubule cytoskeleton in which all of the minus ends of the microtubules are clustered. Ordinarily, this cluster is near the center of the cell.

microtubules   The largest-diameter cable system of the cytoskeleton. Microtubules are composed of polymerized tubulin subunits forming a hollow tube.

midparent value   The mean of the values of a quantitative phenotype for two specific parents.

minimal medium   A medium containing only inorganic salts, a carbon source, and water.

minisatellite DNA   A type of repetitive DNA sequence based on short repeat sequences with a unique common core; used for DNA fingerprinting.

misexpression   The occurrence of gene expression in a tissue in which the gene is normally not expressed. Such ectopic expression can be caused by the juxtaposition of novel enhancer elements to a gene.

missense mutation   A mutation that alters a codon so that it encodes a different amino acid.

mitochondrial cytopathies   Human disorders caused by point mutations or deletions in mitochondrial DNA; inherited maternally.

mitochondrion   A eukaryotic organelle that is the site of ATP synthesis and of the citric acid cycle.

mitosis   A type of nuclear division (occurring at cell division) that produces two daughter nuclei identical with the parent nucleus.

mitotic crossover   A crossover resulting from the pairing of homologs in a mitotic diploid.

mobile genetic element   *See* **transposable genetic element.**

mode   The single class in a statistical distribution having the greatest frequency.

modifier gene   A gene that affects the phenotypic expression of another gene.

molecular clock   A constant rate of change in DNA or protein sequences as a consequence of random mutation and genetic drift of unselected molecular variants.

molecular genetics   The study of the molecular processes underlying gene structure and function.

monocistronic mRNA   An mRNA that encodes one protein.

monoecious plant   A plant species in which male and female organs are found on the same plant but in different flowers (for example, corn).

monohybrid cross   A cross between two individuals identically heterozygous at one gene pair — for example, $A/a \times A/a$.

monoploid   A cell having only one chromosome set (usually as an aberration) or an organism composed of such cells.

monosomic   Refers to a cell or individual that is basically diploid but that has only one copy of one particular chromosome type and thus has chromosome number $2n - 1$.

morphogen   A molecule that can induce the acquisition of different cell fates according to the level of morphogen to which a cell is exposed.

mosaic   A chimera; a tissue containing two or more genetically distinct cell types or an individual composed of such tissues.

motor protein   A protein that is able to move unidirectionally along a specific type of cytoskeletal cable. Kinesins and dyneins are microtubule-based and myosins are microfilament-based motor proteins. By attaching to other subcellular components, motor proteins are capable of directed movement of these components within the cell.

mRNA (messenger RNA)   An RNA molecule transcribed from the DNA of a gene and from which a protein is translated by the action of ribosomes.

mtDNA   Mitochondrial DNA.

m.u.   *See* **map unit.**

mu phage   A kind ("species") of phage with properties similar to those of insertion sequences, being able to insert, transpose, inactivate, and cause rearrangements.

multimeric structure   A structure composed of several identical or different subunits held together by weak bonds.

multiple allelism   The existence of several known alleles of a gene.

multiple cloning site   *See* **polylinker.**

multiple-factor hypothesis   A hypothesis to explain quantitative variation by assuming the interaction of a large number of genes (polygenes), each with a small additive effect on the character.

multiple-hit hypothesis   The proposal that a single cell must receive a series of mutational events to become malignant or cancerous.

multiplicity of infection   The average number of phage particles that infect a single bacterial cell in a specific experiment.

mutagen   An agent that is capable of increasing the mutation rate.

mutant   An organism or cell carrying a mutation.

mutant allele   An allele differing from the allele found in the standard, or wild type.

mutant hunt   The process of collecting different mutants showing abnormalities in a certain structure or in a certain function, as a preparation for mutational dissection of that function.

mutant site   The damaged or altered area within a mutated gene.

mutation   (1) The process that produces a gene or a chromosome set differing from the wild type. (2) The gene or chromosome set that results from such a process.

mutation breeding   Use of mutagens to develop variants that can increase agricultural yield.

mutation event   The actual occurrence of a mutation in time and space.

mutation frequency   The frequency of mutants in a population.

mutation rate   The number of mutation events per gene per unit of time (for example, per cell generation).

mutational dissection   The study of the components of a biological function through a study of mutations affecting that function.

muton   The smallest part of a gene that can take part in a mutation event; now known to be a nucleotide pair.

myeloma   A cancer of the bone marrow.

nanometer 10^{-9} meters.

narrow heritability   *See* **heritability in the narrow sense.**

negative assortative mating   Preferential mating between phenotypically unlike partners.

negative control   Regulation mediated by factors that block or turn off transcription.

neurofibromatosis   A human disease with tumors of nerve cells and café au lait spots, both in the skin. The allele generally arises from germinal mutation, but it is inherited as an autosomal dominant.

Neurospora   A pink mold, commonly found growing on old food.

neutral evolution   Nonadaptive evolutionary changes that occur because of random genetic drift.

neutral mutations   Mutations to alleles that are not different enough from each other in their physiological effects to be affected by natural selection.

neutral petite   A petite that produces all wild-type progeny when crossed with wild type.

neutrality   *See* **selective neutrality.**

nicking   Nuclease action to sever the sugar-phosphate backbone in one DNA strand at one specific site.

nitrocellulose filter   A type of filter used to hold DNA for hybridization.

nitrogen bases   Types of molecules that form important parts of nucleic acids, composed of nitrogen-containing ring structures; hydrogen bonds between bases link the two strands of a DNA double helix.

nonautonomous   *See* **autonomous phenotype.**

nondisjunction   The failure of homologs (at meiosis) or sister chromatids (at mitosis) to separate properly to opposite poles.

nonlinear tetrad   A tetrad in which the meiotic products are in no particular order.

non-Mendelian ratio   An unusual ratio of progeny phenotypes that is not in accord with the simple application of Mendel's laws; for example, mutant : wild ratios of 3 : 5, 5 : 3, 6 : 2, or 2 : 6 in tetrads indicate that gene conversion has occurred.

nonparental ditype (NPD)   A tetrad type containing two different genotypes, both of which are recombinant.

nonsense codon   A codon for which no normal tRNA molecule exists; the presence of a nonsense codon causes termination of translation (ending of the polypeptide chain). The three nonsense codons are called amber, ocher, and opal.

nonsense mutation   A mutation that alters a gene so as to produce a nonsense codon.

nonsense suppressor   A mutation that produces an altered tRNA that will insert an amino acid in translation in response to a nonsense codon.

nonsynonymous substitution   Mutational replacement of an amino acid with one of different chemical properties *See also* **replacement substitutions.**

norm of reaction   The pattern of phenotypes produced by a given genotype under different environmental conditions.

Northern blot   Transfer of electrophoretically separated RNA molecules from a gel onto an absorbent sheet, which is then immersed in a labeled probe that will bind to the RNA of interest.

NPD   *See* **nonparental ditype.**

nu body   *See* **nucleosome.**

nuclease   An enzyme that can degrade DNA by breaking its phosphodiester bonds.

nucleoid   A DNA mass within a chloroplast or mitochondrion.

nucleolar organizer   A region (or regions) of the chromosome set physically associated with the nucleolus and containing rRNA genes.

nucleolus   An organelle found in the nucleus, containing rRNA and amplified multiple copies of the genes encoding rRNA.

nucleoside   A nitrogen base bound to a sugar molecule.

nucleosome   A nu body; the basic unit of eukaryotic chromosome structure; a ball of eight histone molecules wrapped about by two coils of DNA.

nucleotide   A molecule composed of a nitrogen base, a sugar, and a phosphate group; the basic building block of nucleic acids.

nucleotide pair   A pair of nucleotides (one in each strand of DNA) that are joined by hydrogen bonds.

nucleotide-pair substitution   The replacement of a specific nucleotide pair by a different pair; often mutagenic.

null allele   An allele whose effect is either an absence of normal gene product at the molecular level or an absence of normal function at the phenotypic level.

null mutation   A mutation that results in complete absence of function for the gene.

nullisomic   Refers to a cell or individual with one chromosomal type missing, with a chromosome number such as $n - 1$ or $2n - 2$.

nurse cells   The sister cells of the oocyte in insects. The nurse cells produce the bulk of the cytoplasmic contents of the mature oocyte.

ocher codon   The codon UAA, a nonsense codon.

octad   An ascus containing eight ascospores, produced in species in which the tetrad normally undergoes a postmeiotic mitotic division.

Okazaki fragment   A small segment of single-stranded DNA synthesized as part of the lagging strand in DNA replication.

oligonucleotide   A short segment of synthetic DNA.

oncogene   A gene that contributes to the production of a cancer. Oncogenes are generally mutated forms of normal cellular genes.

opal codon   The codon UGA, a nonsense codon.

open reading frame   *See* **ORF.**

operator   A DNA region at one end of an operon that acts as the binding site for repressor protein.

operon   A set of adjacent structural genes whose mRNA is synthesized in one piece, plus the adjacent regulatory signals that affect transcription of the structural genes.

ORF (open reading frame)   A section of a sequenced piece of DNA that begins with a start codon and ends with a stop codon; it is presumed to be the coding sequence of a gene.

organelle   A subcellular structure having a specialized function — for example, the mitocondrion, the chloroplast, or the spindle apparatus.

organogenesis   The production of organ systems during animal embryogenesis.

origin of replication   The point of specific sequence at which DNA replication is initiated.

overdominance   A phenotypic relation in which the phenotypic expression of the heterozygote is greater than that of either homozygote.

P element   A *Drosophila* transposable element that has been used as a tool for insertional mutagenesis and for germ-line transformation.

PAC (P1-based artificial chromosome)   A derivative of phage P1 engineered as a cloning vector for carrying large inserts.

pair-rule gene   In *Drosophila,* a member of a class of zygotically expressed genes that act at an intermediary stage in the process of the establishment of the correct numbers of body segments. Pair-rule mutations have half of the normal number of segments, owing to the loss of every other segment.

paracentric inversion   An inversion not involving the centromere.

paracrine signaling   The process by which a secreted molecule binds to receptors on or within nearby cells, thereby inducing a signal transduction pathway within the receiving cell.

paralogous genes   Two genes in the same species that have evolved by gene duplication.

paramutation   An epigenetic phenomenon in plants, in which the genetic activity of a normal allele is heritably reduced by virtue of that allele having been heterozygous with a special "paramutagenic" allele.

parental ditype (PD)   A tetrad type containing two different genotypes, both of which are parental.

parental imprinting   An epigenetic phenomenon in which the activity of a gene is dependent on whether the gene was inherited from the father or the mother. Some genes are maternally imprinted, others paternally.

parthenogenesis   The production of offspring by a female with no genetic contribution from a male.

partial diploid   *See* **merozygote.**

particulate inheritance   The model proposing that genetic information is transmitted from one generation to the next in discrete units ("particles"), so the character of the offspring is not a smooth blend of essences from the parents (*compare* **blending inheritance**).

pathogen   An organism that causes disease in another organism.

patroclinous inheritance   Inheritance in which all offspring have the nucleus-based phenotype of the father.

pattern formation   The processes by which the complex shape and structure of higher organisms develop.

PD   *See* **parental ditype.**

pedigree   A "family tree," drawn with standard genetic symbols, showing inheritance patterns for specific phenotypic characters.

penetrance   The proportion of individuals with a specific genotype who manifest that genotype at the phenotype level.

peptide   *See* **amino acid.**

peptide bond   A bond joining two amino acids.

pericentric inversion   An inversion that involves the centromere.

permissive conditions   Those environmental conditions under which a conditional mutant shows the wild-type phenotype.

petite   A yeast mutation producing small colonies and altered mitochondrial functions. In cytoplasmic petites (neutral and suppressive petites), the mutation is a deletion in mitochondrial DNA; in segregational petites, the mutation is in nuclear DNA.

phage   *See* **bacteriophage.**

Phe   Phenylalanine (an amino acid).

phenocopy   An environmentally induced phenotype that resembles the phenotype produced by a mutation.

phenotype   (1) The form taken by some character (or group of characters) in a specific individual. (2) The detectable outward manifestations of a specific genotype.

phenotypic sex determination   Sex determination by nongenetic means.

phenylketonuria (PKU)   A human metabolic disease caused by a mutation in a gene encoding a phenylalanine-processing enzyme, which leads to mental retardation if not treated; inherited as an autosomal recessive phenotype.

Philadelphia chromosome   A translocation between the long arms of chromosomes 9 and 22, often found in the white blood cells of patients with chronic myeloid leukemia.

phosphodiester bond   A bond between a sugar group and a phosphate group; such bonds form the sugar-phosphate backbone of DNA.

phyletic evolution   Heritable change over time in a continuous line of descent.

physical mapping Mapping the positions of cloned genomic fragments.

piebald A mammalian phenotype in which patches of skin are unpigmented because of lack of melanocytes; generally inherited as an autosomal dominant.

pilus (plural, **pili**) A conjugation tube; a hollow hairlike appendage of a donor *E. coli* cell that acts as a bridge for transmission of donor DNA to the recipient cell during conjugation.

plant breeding The application of genetic analysis to the development of plant lines better suited for human purposes.

plaque A clear area on a bacterial lawn, left by lysis of the bacteria through progressive infections by a phage and its descendants.

plasmid Autonomously replicating extrachromosomal DNA molecule.

plate (1) A flat dish used to culture microbes. (2) To spread cells over the surface of solid medium in a plate.

pleiotropic mutation A mutation that has effects on several different characters.

ploidy The number of chromosome sets.

point mutation A mutation that can be mapped to one specific locus.

Poisson distribution A mathematical expression giving the probability of observing various numbers of a particular event in a sample when the mean probability of an event on any one trial is very small.

poky A slow-growing mitochondrial mutant in *Neurospora.*

polar gene conversion A gradient of conversion frequency along the length of a gene.

polar granules Cytoplasmic granules localized at the posterior end of a *Drosophila* oocyte and early embryo. These granules are associated with the germ-line and posterior determinants.

polar mutation A mutation that affects the transcription or translation of the part of the gene or operon on only one side of the mutant site — for example, nonsense mutations, frame-shift mutations, and IS-induced mutations.

polyacrylamide A material used to make electrophoretic gels for the separation of mixtures of macromolecules.

poly(A)tail A string of adenine nucleotides added to mRNA after transcription.

polycistronic mRNA An mRNA that encodes more than one protein.

polydactyly More than five fingers or toes or both. Inherited as an autosomal dominant phenotype.

polygenes *See* **multiple-factor hypothesis.**

polylinker A vector DNA sequence containing multiple unique restriction-enzyme-cut sites, convenient for inserting foreign DNA.

polymerase chain reaction (PCR) A method for amplifying specific DNA segments that exploits certain features of DNA replication.

polymorphism The occurrence in a population (or among populations) of several phenotypic forms associated with alleles of one gene or homologs of one chromosome.

polypeptide A chain of linked amino acids; a protein.

polyploid A cell having three or more chromosome sets or an organism composed of such cells.

polysaccharide A biological polymer composed of sugar subunits — for example, starch or cellulose.

polytene chromosome A giant chromosome produced by an endomitotic process in which the multiple DNA sets remain bound in a haploid number of chromosomes.

position effect Used to describe a situation in which the phenotypic influence of a gene is altered by changes in the position of the gene within the genome.

position-effect variegation Variegation caused by the inactivation of a gene in some cells through its abnormal juxtaposition with heterochromatin.

positional information The process by which chemical cues that establish cell fate along a geographic axis are established in a developing embryo or tissue primordium.

positive assortative mating A situation in which like phenotypes mate more commonly than expected by chance.

positive control Regulation mediated by a protein that is required for the activation of a transcription unit.

postzygotic isolation Failure to exchange genes between species because zygotes that are formed either cannot develop to adulthood or, if they develop, are sterile in one or both sexes.

prezygotic isolation Failure to exchange genes between species because zygotes cannot be formed. The inability to form zygotes can be a consequence of temporal or ecological differences between the species, behavioral barriers to mating, mechanical lack of fit between the genitalia of the two sexes, or incompatibility of gametes.

primary structure of a protein The sequence of amino acids in the polypeptide chain.

primase An enzyme that makes RNA primers during DNA replication.

primer A short single-stranded RNA or DNA that can act as a start site for 3′ chain growth when bound to a single-stranded template.

primer walking The use of a primer based on a sequenced area of the genome to sequence into a flanking unsequenced area.

prion Proteinaceous infectious particle that causes degenerative disorders of the central nervous system, such as "scrapie" in sheep and Creutzfeldt-Jacob disease in humans.

Pro Proline (an amino acid).

probe Defines a nucleic acid segment that can be used to identify specific DNA molecules bearing the complementary sequence, usually through autoradiography.

product of meiosis One of the (usually four) cells formed by the two meiotic divisions.

product rule The probability of two independent events occurring simultaneously is the product of the individual probabilities.

proflavin A mutagen that tends to produce frame-shift mutations.

prokaryote An organism composed of a prokaryotic cell, such as bacteria and blue-green algae.

prokaryotic cell A cell having no nuclear membrane and hence no separate nucleus.

promoter A regulator region a short distance from the 5′ end of a gene that acts as the binding site for RNA polymerase.

prophage A phage "chromosome" inserted as part of the linear structure of the DNA chromosome of a bacterium.

prophase The early stage of nuclear division during which chromosomes condense and become visible.

propositus In a human pedigree, the person who first came to the attention of the geneticist.

protein kinase   An enzyme that phosphorylates specific amino acid residues on specific target proteins. One major class of protein kinases phosphorylates tyrosines, and the other phosphorylates serines and threonines on target proteins.

proteome   The complete set of protein-encoding genes in a genome.

proto-oncogene   The normal cellular counterpart of a gene that can be mutated to become a dominant oncogene.

protoplast   A plant cell whose wall has been removed.

prototroph   A strain of organisms that will proliferate on minimal medium (*compare* **auxotroph**).

provirus   A virus "chromosome" integrated into the DNA of the host cell.

pseudodominance   The sudden appearance of a recessive phenotype in a pedigree, owing to the deletion of a masking dominant gene.

pseudogene   An inactive gene derived from an ancestral active gene.

puff   *See* **chromosome puff.**

pulse-chase experiment   An experiment in which cells are grown in radioactive medium for a brief period (the pulse) and then transferred to nonradioactive medium for a longer period (the chase).

pulsed-field gel electrophoresis   An electrophoretic technique in which the gel is subjected to electrical fields alternating between different angles, allowing very large DNA fragments to "snake" through the gel and hence permitting efficient separation of mixtures of such large fragments.

Punnett square   A grid used as a graphic representation of the progeny zygotes resulting from different gamete fusions in a specific cross.

pure-breeding line or strain   A group of identical individuals that always produce offspring of the same phenotype when intercrossed.

purine   A type of nitrogen base; the purine bases in DNA are adenine and guanine.

pyrimidine   A type of nitrogen base; the pyrimidine bases in DNA are cytosine and thymine.

QTL (quantitative trait locus)   The position of a gene whose variant alleles contribute to quantitative variation for some trait (character).

quantitative variation   The existence of a range of phenotypes for a specific character, differing by degree rather than by distinct qualitative differences.

quaternary structure of a protein   The multimeric constitution of the protein.

R plasmid   A plasmid containing one or several transposons that bear resistance genes.

radiation hybrid   A type of human–mouse hybrid cell in which human chromosomes have been fragmented by radiation to determine which markers are inherited together and therefore linked.

random genetic drift   Changes in allele frequency that result because the genes appearing in offspring are not a perfectly representative sampling of the parental genes.

random mating   Mating between individuals where the choice of a partner is not influenced by the genotypes (with respect to specific genes under study).

RAPD   Randomly amplified polymorphic DNA. A set of several genomic fragments amplified by a single PCR primer; somewhat variable from individual to individual; +/− heterozygotes for individual fragments can act as markers in genome mapping.

reading frame   The codon sequence that is determined by reading nucleotides in groups of three from some specific start codon.

realized heritability   The ratio of the single-generation progress of selection to the selection differential of the parents.

reannealing   Spontaneous realignment of two single DNA strands to re-form a DNA double helix that had been denatured.

receptor   *See* **ligand–receptor interaction.**

receptor element   A controlling element that can insert into a gene (making it a mutant) and can also excise (thus making the mutation unstable); both of these functions are nonautonomous, being under the influence of the regulator element.

receptor tyrosine kinase   A transmembrane receptor whose cytoplasmic domain includes a tyrosine kinase enzymatic activity. In normal situations, the kinase is activated only on binding of the appropriate ligand to the receptor.

recessive allele   An allele whose phenotypic effect is not expressed in a heterozygote.

recessive phenotype   The phenotype of a homozygote for the recessive allele; the parental phenotype that is not expressed in a heterozygote.

reciprocal crosses   A pair of crosses of the type genotype $A ♀ ×$ genotype $B ♂$ and genotype $B ♀ ×$ genotype $A ♂$.

reciprocal translocation   A translocation in which part of one chromosome is exchanged with a part of a separate nonhomologous chromosome.

recombinant   An individual or cell with a genotype produced by recombination.

recombinant DNA   A novel DNA sequence formed by the combination of two nonhomologous DNA molecules.

recombinant frequency (RF)   The proportion (or percentage) of recombinant cells or individuals.

recombination   (1) In general, any process in a diploid or partly diploid cell that generates new gene or chromosomal combinations not found in that cell or in its progenitors. (2) At meiosis, the process that generates a haploid product of meiosis whose genotype is different from either of the two haploid genotypes that constituted the meiotic diploid.

recombinational repair   The repair of a DNA lesion through a process, similar to recombination, that uses recombination enzymes.

recon   A region of a gene within which there can be no crossing-over; now known to be a nucleotide pair.

reduction division   A nuclear division that produces daughter nuclei each having one-half as many centromeres as the parental nucleus.

redundant DNA   *See* **repetitive DNA.**

regression   A term coined by Galton for the tendency of the quantitative traits of offspring to be closer to the population mean than are their parents' traits. It arises from dominance, gene interaction, and nongenetic influences on traits.

regression coefficient   The slope of the straight line that most closely relates two correlated variables.

regulator element   *See* **receptor element.**

regulatory genes   Genes that have roles in turning on or off the transcription of structural genes.

regulatory region   Upstream (5′) end of the gene to which bind various proteins that cause transcription of the gene at the correct time and place.

repetitive DNA   Redundant DNA; DNA sequences that are present in many copies per chromosome set.

replacement substitutions   Nucleotide changes in the encoding part of a gene that result in a change in the amino acid sequence of the encoded protein.

replication   DNA synthesis.

replication fork   The point at which the two strands of DNA are separated to allow replication of each strand.

replicon   A chromosomal region under the influence of one adjacent replication-initiation locus.

reporter gene   A gene whose phenotypic expression is easy to monitor and is used to study tissue-specific promoter and enhancer activities in transgenes.

repressor protein   A molecule that binds to the operator and prevents transcription of an operon.

repulsion conformation   Two linked heterozygous gene pairs in the arrangement *A b/a B*.

resolving power   The ability of an experimental technique to distinguish between two genetic conditions (typically discussed when one condition is rare and of particular interest).

restriction enzyme   An endonuclease that recognizes specific target nucleotide sequences in DNA and breaks the DNA chain at those points; a variety of these enzymes are known, and they are extensively used in genetic engineering.

restriction map   A map of a chromosomal region showing the positions of target sites of one or more restriction enzymes.

restrictive conditions   Environmental conditions under which a conditional mutant shows the mutant phenotype.

retinoblastoma   A childhood cancer of the human retina.

retrotransposon   A transposable element that utilizes reverse transcriptase to transpose through an RNA intermediate.

retrovirus   An RNA virus that replicates by first being converted into double-stranded DNA.

reverse genetics   The experimental procedure that begins with a cloned segment of DNA or a protein sequence and uses it (through directed mutagenesis) to introduce programmed mutations back into the genome to investigate function.

reverse transcriptase   An enzyme that catalyzes the synthesis of a DNA strand from an RNA template.

reversion   The production of a wild-type gene from a mutant gene.

RF   *See* **recombinant frequency.**

RFLP   Restriction fragment length polymorphism. At some chromosomal locations a probe sometimes detects different sizes or different numbers of restriction fragments (often as a result of presence and absence of restriction sites), and this situation is an RFLP. If an individual is heterozygous for such a chromosomal difference, that region can be used as a marker in chromosome mapping.

RFLP mapping   A technique in which DNA restriction fragment length polymorphisms are used as reference loci for mapping in relation to known genes or other RFLP loci.

rho   A protein factor required to recognize certain transcription termination signals in *E. coli*.

ribonucleic acid   *See* **RNA.**

ribosomal RNA   *See* **rRNA.**

ribosome   A complex organelle that catalyzes the translation of messenger RNA into an amino acid sequence. Composed of proteins plus rRNA.

ribozymes   RNAs with enzymatic activities — for instance, the self-splicing RNA molecules in *Tetrahymena*.

RNA (ribonucleic acid)   A single-stranded nucleic acid similar to DNA but having ribose sugar rather than deoxyribose sugar and uracil rather than thymine as one of the bases.

RNA in situ hybridization   A technique that is used to identify the spatial pattern of expression of a particular transcript (usually an mRNA). In this technique, the DNA probe is labeled, either radioactively or by chemically attaching an enzyme that can convert a substrate into a visible dye. A tissue or organism is soaked in a solution of single-stranded labeled DNA under conditions that allow the DNA to hybridize to complementary RNA sequences in the cells; unhybridized DNA is then removed. Radioactive probe is detected by autoradiography. Enzyme-labeled probe is detected by soaking the tissue in the substrate; the dye develops in sites where the transcript of interest was expressed.

RNA polymerase   An enzyme that catalyzes the synthesis of an RNA strand from a DNA template. In eukaryotes, there are several classes of RNA polymerase. Structural genes for proteins are transcribed by RNA polymerase II.

rRNA (ribosomal RNA)   A class of RNA molecules, encoded in the nucleolar organizer, that have an integral (but poorly understood) role in ribosome structure and function.

S (Svedberg unit)   A unit of sedimentation velocity commonly used to describe molecular units of various sizes (because sedimentation velocity is related to size).

S phase   The part of interphase of the cell cycle in which DNA synthesis takes place.

SARs   Scaffold attachment regions; the positions along DNA where it is anchored to the central scaffold of the chromosome.

satellite   A terminal section of a chromosome separated from the main body of the chromosome by a narrow constriction.

satellite chromosome   A chromosome that seems to be an addition to the normal genome.

satellite DNA   Any type of highly repetitive DNA; formerly defined as DNA forming a satellite band after cesium chloride density gradient centrifugation.

saturation mutagenesis   Induction and recovery of large numbers of mutations in one area of the genome or in one function in the hope of identifying all the genes in that area or affecting that function.

scaffold   The central framework of a chromosome to which the DNA solenoid is attached as loops; composed largely of topoisomerase.

SCE   *See* **sister-chromatid exchange.**

secondary sexual characteristics   The sex-associated phenotypes of somatic tissues in sexually dimorphic animals.

secondary structure of a protein   A spiral or zigzag arrangement of the polypeptide chain.

second-division segregation pattern   A pattern of ascospore genotypes for a gene pair showing that the two alleles separate into different nuclei only at the second meiotic division as a result of a crossover between that gene pair and its centromere; can be detected only in a linear ascus.

second-site mutation   The second mutation of a double mutation within a gene; in many cases, the second-site mutation suppresses the first mutation, so the double mutant has the wild-type phenotype.

sector   An area of tissue whose phenotype is detectably different from the surrounding tissue phenotype.

sedimentation   The sinking of a molecule under the opposing forces of gravitation and buoyancy.

segment-polarity gene   In *Drosophila*, a member of the class of genes that contribute to the final aspects of establishing the correct number of segments. Segment-polarity mutations cause a loss of a comparable part of each of the body segments.

segmentation   The process by which the correct number of segments is established in a developing segmented animal.

segregation   (1) Cytologically, the separation of homologous structures. (2) Genetically, the production of two separate phenotypes, corresponding to two alleles of a gene, either in different individuals (meiotic segregation) or in different tissues (mitotic segregation).

segregational petite   A petite that, in a cross with wild type, produces $\frac{1}{2}$ petite and $\frac{1}{2}$ wild-type progeny; caused by a nuclear mutation.

selection coefficient (*s*)   The proportional excess or deficiency of fitness of one genotype in relation to another genotype.

selection differential   The difference between the mean of a population and the mean of the individuals selected to be parents of the next generation.

selection progress   The difference between the mean of a population and the mean of the offspring in the next generation born to selected parents.

selective neutrality   A situation in which different alleles of a certain gene confer equal fitness.

selective system   An experimental technique that enhances the recovery of specific (usually rare) genotypes.

self   To fertilize eggs with sperm from the same individual.

self-assembly   The ability of certain multimeric biological structures to assemble from their component parts through random movements of the molecules and formation of weak chemical bonds between surfaces with complementary shapes.

semiconservative replication   The established model of DNA replication in which each double-stranded molecule is composed of one parental strand and one newly polymerized strand.

semisterility (half sterility)   The phenotype of individuals heterozygotic for certain types of chromosome aberration; expressed as a reduced number of viable gametes and hence reduced fertility.

sequence-tagged site   A relatively small, sequenced region of a cloned genomic fragment that can be used by a computer to align the cloned fragment into a contig.

Ser   Serine (an amino acid).

sex chromosome   A chromosome whose presence or absence is correlated with the sex of the bearer; a chromosome that plays a role in sex determination.

sex determination   The genetic or environmental process by which the sex of an individual is established.

sex linkage   The location of a gene on a sex chromosome.

sex reversal   A syndrome known in humans and mice in which chromosomally XX individuals develop as males. In some cases, sex reversal is now known to be due to the translocation of the testis-determining region of the Y chromosome to the tip of the X chromosome in such individuals.

sexduction   Sexual transmission of donor *E. coli* chromosomal genes on the fertility factor.

sexual spore   *See* **spore.**

shotgun technique   The cloning of a large number of different DNA fragments as a prelude to selecting one particular clone type for intensive study.

shuttle vector   A vector (for example, a plasmid) constructed in such a way that it can replicate in at least two different host species, allowing a DNA segment to be tested or manipulated in several cellular settings.

sickle-cell anemia   Potentially lethal human disease caused by a mutation in a gene encoding the oxygen-transporting molecule hemoglobin. The altered molecule causes red blood cells to be sickle shaped. Inherited as an autosomal recessive.

signal sequence   The N-terminal sequence of a secreted protein, which is required for transport through the cell membrane.

signal transduction cascade   A series of sequential events, such as protein phosphorylations, that pass a signal received by a transmembrane receptor through a series of intermediate molecules until final regulatory molecules, such as transcription factors, are modified in response to the signal.

silent mutation   Mutation in which the function of the protein product of the gene is unaltered.

SINE   Short interspersed element. A type of small, repetitive DNA sequence found throughout a eukaryotic genome.

sister-chromatid exchange (SCE)   An event similar to crossing-over that can occur between sister chromatids at mitosis or at meiosis; detected in harlequin chromosomes.

site-specific recombination   Recombination between two specific sequences that need not be homologous; mediated by a specific recombination system.

S-9 mix   A liver-derived supernatant used in the Ames test to activate or inactivate mutagens.

solenoid structure   In eukaryotic nuclear chromosomes, the supercoiled arrangement of DNA produced by coiling the continuous string of nucleosomes.

somatic cell   A cell that is not destined to become a gamete; a "body cell," whose genes will not be passed on to future generations.

somatic-cell genetics   Asexual genetics; the study of somatic mutation, assortment, and crossing-over, and of cell fusion.

somatic mutation   A mutation occurring in a somatic cell.

somatostatin   A human growth hormone.

SOS repair   The error-prone process whereby gross structural DNA damage is circumvented by allowing replication to proceed past the damage through imprecise polymerization.

Southern blot   Transfer of electrophoretically separated fragments of DNA from the gel to an absorbent sheet such as paper. This sheet is then immersed in a solution containing a labeled probe that will bind to a fragment of interest.

spacer DNA   DNA found between genes; its function is unknown.

specialized (restricted) transduction   The situation in which a particular phage will transduce only specific regions of the bacterial chromosome.

speciation   The process of forming new species by the splitting of an old species into two or more new species incapable of exchanging genes with each other.

species   A group of organisms that are biologically capable of exchanging genes with each other but are incapable of exchanging genes with other such groups.

specific-locus test   A system for detecting recessive mutations in diploids. Normal individuals treated with mutagen are mated to testers that are homozygous for the recessive alleles at a number of specific loci; the progeny are then screened for recessive phenotypes.

spindle   The set of microtubular fibers that appear to move eukaryotic chromosomes in division.

splicing   The reaction that removes introns and joins together exons in RNA.

spontaneous mutation   A mutation occurring in the absence of mutagens, usually due to errors in the normal functioning of cellular enzymes.

spore   (1) In plants and fungi, sexual spores are the haploid cells produced by meiosis. (2) In fungi, asexual spores are somatic cells that are cast off to act either as gametes or as the initial cells for new haploid individuals.

sporophyte   The diploid sexual-spore-producing generation in the life cycle of plants — that is, the stage in which meiosis occurs.

SSLP   Short sequence length polymorphisms; the presence of different numbers of short repetitive elements (mini- and microsatellite DNA) at one particular locus in different homologous chromosomes; heterozygotes represent useful markers for genome mapping.

stacking   The packing of the flattish nitrogen bases at the center of the DNA double helix.

staggered cuts   The cleavage of two opposite strands of duplex DNA at points near one another.

standard deviation   The square root of the variance.

statistic   A computed quantity, such as the mean, characteristic of a population.

statistical distribution   The array of frequencies of different quantitative or qualitative classes in a population.

stem cell   *See* **blast cell.**

steroid hormones   A class of hormones synthesized by glands of the endocrine system that, by virtue of their nonpolar nature, are able to pass directly through the plasma membrane of a cell. Steroid hormones act by binding to and activating transcription factors called steroid hormone receptors.

steroid hormone receptors   A family of related proteins that act as transcription factors when bound to their cognate hormones. Not all members of this family actually bind to steroids; the name derives from the first family member that was discovered, which was indeed a steroid hormone receptor.

strain   A pure-breeding lineage, usually of haploid organisms, bacteria, or viruses.

structural gene   A gene encoding the amino acid sequence of a protein.

structural genomics   Characterizing and locating the entire set of genes in a genome.

subvital gene   A gene that causes the death of some proportion (but not all) of the individuals that express it.

sum rule   The probability that one or the other of two mutually exclusive events will occur is the sum of their individual probabilities.

supercoil   A closed, double-stranded DNA molecule that is twisted on itself.

superinfection   Phage infection of a cell that already harbors a prophage.

supersuppressor   A mutation that can suppress a variety of other mutations; typically a nonsense suppressor.

suppressive petite   A petite that, in a cross with wild type, produces progeny of which variable non-Mendelian proportions are petite.

suppressor   A secondary mutation that can cancel the effect of a primary mutation and result in wild-type phenotype.

suppressor mutation   A mutation that counteracts the effects of another mutation. A suppressor maps at a different site from the mutation that it counteracts, either within the same gene or at a more distant locus. Different suppressors act in different ways.

synapsis   Close pairing of homologs at meiosis.

synaptonemal complex   A complex structure that unites homologs during the prophase of meiosis.

syncytial blastoderm   In insects, the stage of blastoderm preceding the formation of cell membranes around the individual nuclei of the early embryo.

syncytium   A single cell with many nuclei.

synonymous substitutions   Nucleotide changes in the encoding part of a gene that do not result in a change in the amino acid sequence of the encoded protein.

syntenic   Description of DNA segments in which the gene order is identical in different related species.

T   (1) Thymine, or thymidine. (2) *See* **tetratype.**

tagging   The use of a piece of foreign DNA or a transposon to tag a gene so that a clone of that gene can be identified readily in a library.

tandem duplication   Adjacent identical chromosome segments.

targeted gene knockout   The introduction of a null mutation in a gene by a designed alteration in a cloned DNA sequence that is then introduced into the genome through homologous recombination and replacement of the normal allele.

tautomeric shift   The spontaneous isomerization of a nitrogen base to an alternative hydrogen-bonding condition, possibly resulting in a mutation.

T-DNA   A part of the Ti plasmid that is inserted into the genome of the host plant cell.

telocentric chromosome   A chromosome having the centromere at one end.

telomerase   An enzyme that adds repetitive units to the ends of linear chromosomes to prevent shortening after replication, using a special small RNA as a template.

telomere   The tip (or end) of a chromosome.

telophase   The late stage of nuclear division when daughter nuclei re-form.

temperate phage   A phage that can become a prophage.

temperature-sensitive mutation   A conditional mutation that produces the mutant phenotype in one temperature range and the wild-type phenotype in another temperature range.

template   A molecular "mold" that shapes the structure or sequence of another molecule; for example, the nucleotide sequence of DNA acts as a template to control the nucleotide sequence of RNA during transcription.

teratogen   An agent that interferes with normal development.

terminal redundancy   In a phage, refers to a linear DNA molecule with single-stranded ends that are longer than is necessary to close the DNA circle.

tertiary structure of a protein   The folding or coiling of the secondary structure to form a globular molecule.

testcross   A cross between an individual of unknown genotype or a heterozygote (or a multiple heterozygote) and a tester individual.

tester   An individual homozygous for one or more recessive alleles; used in a testcross.

testicular feminization syndrome   A human condition, caused by a mutation in a gene encoding androgen receptors, in which XY males develop into phenotypic females.

tetrad   (1) Four homologous chromatids in a bundle in the first meiotic prophase and metaphase. (2) The four haploid product cells from a single meiosis.

tetrad analysis   The use of tetrads (definition 2) to study the behavior of chromosomes and genes during meiosis.

tetraparental mouse   A mouse that develops from an embryo created by the experimental fusion of two separate blastulas.

tetraploid   A cell having four chromosome sets; an organism composed of such cells.

tetratype (T)   A tetrad type containing four different genotypes, two parental and two recombinant.

Thr   Threonine (an amino acid).

three-point testcross   A testcross in which one parent has three heterozygous gene pairs.

thymidine   The nucleoside having thymine as its base.

thymine   A pyrimidine base that pairs with adenine.

thymine dimer   A pair of chemically bonded adjacent thymine bases in DNA; the cellular processes that repair this lesion often make errors that create mutations.

Ti plasmid   A circular plasmid of *Agrobacterium tumifaciens* that enables the bacterium to infect plant cells and produce a tumor (crown gall tumor).

tissue-specific gene expression   The expression of a gene in a specific and reproducible subset of tissues and cells during the development of a higher eukaryote.

topoisomerase   An enzyme that can cut and re-form polynucleotide backbones in DNA to allow it to assume a more relaxed configuration.

totipotency   The ability of a cell to proceed through all the stages of development and thus produce a normal adult.

trailer   An untranslated segment at the 3′ end of mRNA.

trans conformation   In a heterozygote involving two mutant sites within a gene or gene cluster, the arrangement $a_1 +/+ a_2$.

transcription   The synthesis of RNA with the use of a DNA template.

transcription factor   A protein that binds to a cis-regulatory element (for example, an enhancer) and thereby, directly or indirectly, affects the initiation of transcription.

transduction   The movement of genes from a bacterial donor to a bacterial recipient with the use of a phage as the vector.

transfection   The process by which exogenous DNA in solution is introduced into cultured cells.

transgene   A gene that has been modified by externally applied recombinant DNA techniques and reintroduced into the genome by germ-line transformation.

transmembrane receptor   A protein that spans the plasma membrane of a cell, with the extracellular part of the protein having the ability to bind to a ligand and the intracellular part having an activity (such as that catalyzed by protein kinase) that can be induced on ligand binding.

transfer RNA   *See* **tRNA.**

transformation   (1) The directed modification of a genome by the external application of DNA from a cell of different genotype. (2) Conversion of normal higher eukaryotic cells in tissue culture into a cancerlike state of uncontrolled division.

transgenic organism   One whose genome has been modified by externally applied new DNA.

transient diploid   The stage of the life cycle of predominantly haploid fungi (and algae) during which meiosis occurs.

transition   A type of nucleotide-pair substitution in which a purine replaces another purine or a pyrimidine replaces another pyrimidine — for example, GC → AT.

translation   The ribosome-mediated production of a polypeptide whose amino acid sequence is derived from the codon sequence of an mRNA molecule.

translocation   The relocation of a chromosomal segment to a different position in the genome.

transmission genetics   The study of the mechanisms of the passage of a gene from one generation to the next.

transposable genetic element   A general term for any genetic unit that can insert into a chromosome, exit, and relocate; includes insertion sequences, transposons, some phages, and controlling elements.

transposition   *See* **translocation.**

transposon   A mobile piece of DNA that is flanked by terminal repeat sequences and typically bears genes encoding transposition functions.

transversion   A type of nucleotide-pair substitution in which a purine replaces a pyrimidine or vice versa — for example, GC → TA.

triplet   The three nucleotide pairs that compose a codon.

triplet expansion   The expansion of a 3-bp repeat from a relatively low number of copies to a high number of copies, which is responsible for a number of genetic diseases, such as fragile X syndrome.

triploid   A cell having three chromosome sets or an organism composed of such cells.

trisomic   Basically a diploid with an extra chromosome of one type, producing a chromosome number of the form $2n + 1$.

tritium   A radioactive isotope of hydrogen.

tRNA (transfer RNA)   A class of small RNA molecules that bear specific amino acids to the ribosome during translation; the amino acid is inserted into the growing polypeptide chain when the anticodon of the tRNA pairs with a codon on the mRNA being translated.

Trp   Tryptophan (an amino acid).

true-breeding line or strain   *See* **pure-breeding line or strain.**

truncation selection   A breeding technique in which individuals in whom quantitative expression of a phenotype is above or below a certain value (the truncation point) are selected as parents for the next generation.

tumor-suppressor gene   A gene encoding a protein that suppresses tumor formation. The wild-type alleles of tumor-suppressor genes are thought to function as negative regulators of cell proliferation.

tumor virus   A virus that is capable of inducing a cancer.

Turner syndrome   An abnormal human female phenotype produced by the presence of only one X chromosome (XO).

twin spot   A pair of mutant sectors within wild-type tissue that are produced by a mitotic crossover in an individual of appropriate heterozygous genotype.

two-hybrid system   A pair of baker's yeast vectors used for detecting protein–protein interaction. Each vector carries the gene for a different foreign protein under test; if the genes unite physically, a reporter gene is transcribed.

2-μm (2-micrometer) plasmid   A naturally occurring extragenomic circular DNA molecule with a circumference of 2 μm, found in some yeast cells. Engineered to form the basis for several types of gene vectors in yeast.

Tyr   Tyrosine (an amino acid).

U   Uracil or uridine.

underdominance   A phenotypic relation in which the phenotypic expression of the heterozygote is less than that of either homozygote.

unequal crossover   A crossover between homologs that are not perfectly aligned.

uniparental inheritance   The transmission of certain phenotypes from one parental type to all the progeny; such inheritance is generally produced by organelle genes.

unstable mutation   A mutation that has a high frequency of reversion; a mutation caused by the insertion of a controlling element whose subsequent exit produces a reversion.

uracil   A pyrimidine base that appears in RNA in place of thymine, found in DNA.

URF   Unassigned reading frame. An open reading frame (ORF) whose function has not yet been determined.

uridine   The nucleoside having uracil as its base.

Val   Valine (an amino acid).

variable   A property that may have different values in various cases.

variable region   In an immunoglobulin molecule, a region that shows many sequence differences between antibodies of different specificities. The variable regions of the light and heavy chains of an immunoglobulin bind antigen.

variance   A measure of the variation around the central class of a distribution; the average squared deviation of the observations from their mean value.

variant   An individual organism that is recognizably different from an arbitrary standard type in that species.

variate   A specific numerical value of a variable.

variation   The differences between parents and their offspring or among individuals in a population.

variegation   The occurrence within a tissue of sectors with differing phenotypes.

vector   In cloning, the plasmid or phage chromosome used to carry the cloned DNA segment.

viability   The probability that a fertilized egg will survive and develop into an adult organism.

viral transforming gene   A gene within a viral genome that can induce abnormal proliferation of cells in culture and, similarly, can induce tumors in infected whole animals.

virulent phage   A phage that cannot become a prophage; infection by such a phage always leads to lysis of the host cell.

VNTR (variable number tandem repeat)   A chromosomal locus at which a particular repetitive sequence is present in different numbers in different individuals or in the two different homologs in one diploid individual.

Western blot   Membrane carrying an imprint of proteins separated by electrophoresis; can be probed with a labeled antibody to detect a specific protein.

wild type   The genotype or phenotype that is found in nature or in the standard laboratory stock for a given organism.

wobble   The ability of certain bases at the third position of an anticodon in tRNA to form hydrogen bonds in various ways, causing alignment with several possible codons.

X:A ratio   The ratio between the X chromosome and the number of sets of autosomes.

X-chromosome inactivation   The process by which the genes of an X chromosome in a mammal can be completely repressed as part of the dosage compensation mechanism. *See also* **dosage compensation; Barr body.**

X hyperactivation   In *Drosophila,* the process by which the structural genes of the male X chromosome are transcribed at the same rate as the two X chromosomes of the female combined.

X linkage   The inheritance pattern of genes found on the X chromosome but not on the Y.

X-and-Y linkage   The inheritance pattern of genes found on both the X and the Y chromosomes (rare).

X-ray crystallography   A technique for deducing molecular structure by aiming a beam of X rays at a crystal of the test compound and measuring the scatter of rays.

Y linkage   The inheritance pattern of genes found on the Y chromosome but not on the X (rare).

zygote   The cell formed by the fusion of an egg and a sperm; the unique diploid cell that will divide mitotically to create a differentiated diploid organism.

zygotic induction   The sudden release of a lysogenic phage from an Hfr chromosome when the prophage enters the F$^-$ cell, and the subsequent lysis of the recipient cell.

zygotically acting gene   A gene whose product is expressed only in the zygote and not included in the maternal contribution to the oocyte.

Answers to Selected Problems

Chapter 1

3. Properties of DNA that are vital to its being the hereditary molecule are: its ability to replicate; its informational content; and its relative stability while retaining the ability to change or mutate. Alien life forms might utilize RNA, just as some viruses do, as a hereditary molecule. However, of the types of molecules that can exist on earth, only the nucleic acids possess the necessary characteristics.

4. 1,048,576

6. If the DNA is double stranded, G = C = 24 percent and A = T = 26 percent.

9. This strand is antiparallel to the strand given, and the sequence must be:

3'-TAACCACGTAATGAAGTCCGAGA-5'

13. 3'-UUGGCACGUCGUAAU-5'

15. Sulfur

19. mRNA size = gene size − (number of introns × average size of introns)

20. 3'-AAU-5'

25. Protein function can be destroyed by a mutation that causes the substitution of a single amino acid even though the protein has the same immunological properties. For example, an enzyme requires very specific amino acids in exact positions within its active site. A substitution of one of these key amino acids might have no effect on the overall size and shape of the protein though it completely destroys the enzymatic activity.

27. a. Recessive. The normal allele provides enough enzyme to be sufficient for normal function (the definition of haplosufficient).

b. There are many ways to mutate a gene to destroy enzyme function. One possible mutation might be a frameshift mutation within an exon of the gene. If a single base pair were deleted, the mutation would completely alter the translational product 3' to the mutation.

c. Hormone replacement could be given to the patient.

d. If the hormone is required before birth, it can be supplied by the mother.

31. Phenotypic variation within a species can be due to genotype, environmental effects, and pure chance (random noise). Showing that variation of a particular trait has a genetic basis, especially for traits that show continuous variation, therefore requires very carefully controlled analyses.

Chapter 2

3. The progeny ratio is approximately 3 : 1, indicating classic heterozygous-by-heterozygous mating. Because black (*B*) is dominant to white (*b*),

Parents: $B/b \times B/b$

Progeny: 3 black : 1 white (1 B/B : 2 B/b : 1 b/b)

5. (a) $\frac{1}{216}$; (b) $\frac{1}{216}$; (c) $\frac{1}{216}$; (d) $\frac{125}{216}$; (e) $\frac{1}{72}$; (f) $\frac{1}{108}$; (g) $\frac{1}{36}$; (h) $\frac{5}{9}$.

7. a. (1) p(nontaster girl) $= \frac{1}{8}$

(2) p(taster girl) $= \frac{3}{8}$

(3) p(taster boy) $= \frac{3}{8}$

b. p(taster for first two children) $= \frac{9}{16}$

10. $\frac{5}{8}$

11. The plants are approximately 3 blotched : 1 unblotched. This outcome suggests that blotched is dominant to unblotched and that the original plant that was selfed was a heterozygote.

a. Let A = blotched and a = unblotched.

P A/a (blotched) $\times$ A/a (blotched)

F₁ 1 A/A : 2 A/a : 1 a/a

3 $A/-$ (blotched) : 1 a/a (unblotched)

b. All unblotched plants should be pure breeding in a testcross with an unblotched plant (*a/a*), and one-third of the blotched plants should be pure breeding.

16. a. The pedigree is:

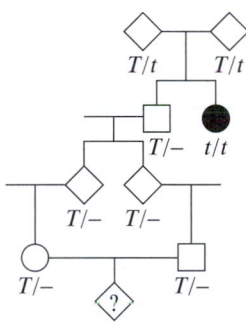

b. $\frac{1}{144}$

25. Let A represent achondroplasia and a represent normal height. Let N represent neurofibromatosis and n represent the normal allele. Because both conditions are extremely rare, the affected persons are assumed to be heterozygous. The genes are assumed to assort independently. The cross is:

P A/a ; $n/n \times a/a$; N/n

F₁ 1 A/a ; n/n : 1 a/a ; n/n : 1 a/a ; N/n : 1 A/a ; N/n

1 anchondoplasia : 1 normal : 1 neurofibromatosis : 1 anchondoplasia, neurofibromatosis

27. a. and b. Cross 2 indicates that purple (*G*) is dominant to green (*g*), and cross 1 indicates that cut (*P*) is dominant to potato (*p*).

Cross	Parental genotypes	Progeny
1	G/g ; $P/p \times g/g$; P/p	$\begin{cases} 3 \text{ cut : 1 potato} \\ 1 \text{ purple : 1 green} \end{cases}$
2	G/g ; $P/p \times G/g$; p/p	$\begin{cases} 3 \text{ purple : 1 green} \\ 1 \text{ cut : 1 potato} \end{cases}$
3	G/G ; $P/p \times g/g$; P/p	$\begin{cases} \text{no green} \\ 3 \text{ cut : 1 potato} \end{cases}$
4	G/g ; $P/P \times g/g$; p/p	$\begin{cases} \text{no potato} \\ 1 \text{ purple : 1 green} \end{cases}$
5	G/g ; $p/p \times g/g$; P/p	$\begin{cases} 1 \text{ cut : 1 potato} \\ 1 \text{ purple : 1 green} \end{cases}$

33. P $s^+/s^+ \times s/Y$

F$_1$ $\frac{1}{2}\, s^+/s$ normal female

          $\frac{1}{2}\, s^+/Y$ normal male

F$_1$ $s^+/s \times s^+/Y$

F$_2$ $\frac{1}{4}\, s^+/s^+$ normal female

          $\frac{1}{4}\, s^+/s$ normal female

          $\frac{1}{4}\, s^+/Y$ normal male

          $\frac{1}{4}\, s/Y$ small male

In the cross: P $s^+/s \times s/Y$

Progeny $\frac{1}{4}\, s^+/s$ normal female

            $\frac{1}{4}\, s/s$ small female

            $\frac{1}{4}\, s^+/Y$ normal male

            $\frac{1}{4}\, s/Y$ small male

36. **a** and **b.** The disease does not appear to be autosomal, because the trait is inherited in a sex-specific way.

c. It cannot be Y linked, because females are affected.

d. Yes. All daughters and none of the sons will be affected if the father has an X-linked dominant trait.

e. Not likely. If the mother is a carrier, it is highly unlikely that she would pass the gene to all her daughters and none of her sons.

39. (a) $\frac{1}{8}$; (b) $\frac{1}{4}$; (c) zero.

43. One-fourth of the sperm will be b ; d

49. On the basis of phenotype, the woman appears to have two different cell lines for G6PD activity in her red blood cells. If G = normal enzyme activity and g = reduced enzyme activity and malaria resistance, then the woman appears to have $G/-$ and g/g cells. This can best be explained by X inactivation in the woman's cells. Assume that she is G/g. In approximately half of her cells, the G allele-containing X chromosomes will be inactivated, leaving only a functional g allele. Those cells will be resistant to the malarial parasite. In the other half of her cells, the g allele-containing X chromosome will be inactivated. Those cells will have a functional G allele and will be susceptible to the parasite.

51. **a.** Note that only males are affected and that they had unaffected parents. For rare traits, this is what is expected for an X-linked recessive disorder.

b. The mothers of all affected sons must be heterozygous for the disorder. In addition, the daughters of all affected men must be heterozygous. Finally, barring mutation, individual I-2 must have been heterozygous or I-1 must have had the trait.

Chapter 3

1. The key function of mitosis is to generate two daughter cells genetically identical with the original parent cell.

2. Two key functions of meiosis are to halve the DNA content and to reshuffle the genetic content of the organism to generate genetic diversity among the progeny.

6. The resulting cells have the identical genotype: A/a ; B/b.

9. The following cells would fit the DNA measurements:

0.7	haploid cells (spores or cells of the gametophyte)
1.4	diploid cells in G$_1$ or haploid cells after S but before cell division
2.1	triploid cells of the endosperm
2.8	diploid cells after S but before cell division
4.2	triploid cells after S but before cell division

10. Assume that both plants are diploid sporophytes, each homozygous for the alleles being studied (r/r for red and b/b for brown). Both can undergo meiosis to generate spores that then divide mitotically into adult haploid gametophytes. If the numbers of spores generated by the original parents are assumed to be equal, an equal number of r (red) and b (brown) plants would result.

14. Pulsed field electrophoresis separates DNA molecules by size. When DNA is carefully isolated from *Neurospora* (which has 7 different chromosomes), 7 bands should be produced by using this technique. Similarly, the pea has 7 different chromosomes and will produce 7 bands (homologous chromosomes will comigrate as a single band).

18. Because the "half" inherited is random, the chances of receiving exactly the same half are vanishingly small. If we ignore recombination and just focus on which chromosomes are inherited from one parent (for example, those inherited from the father or those from the mother), there are 2^{23} = 8,388,608 possible combinations.

21. Remember, the endosperm is formed from two polar nuclei (which are genetically identical) and one sperm nucleus.

Female	Male	Polar nuclei	Sperm	Endosperm
s/s	S/S	s and s	S	$S/s/s$
S/S	s/s	S and S	s	$S/S/s$
S/s	S/s	$\frac{1}{2}\, S$ and S	$\frac{1}{2}\, S$	$\frac{1}{4}\, S/S/S$
				$\frac{1}{4}\, S/S/s$
		$\frac{1}{2}\, s$ and s	$\frac{1}{2}\, s$	$\frac{1}{4}\, S/s/s$
				$\frac{1}{4}\, s/s/s$

28. (5) chromosome pairing

29. $\frac{1}{8}$

32. (1) Impossible; (2) meiosis II; (3) meiosis II; (4) meiosis II; (5) mitosis; (6) impossible; (7) impossible; (8) impossible; (9) impossible; (10) meiosis I; (11) impossible; (12) impossible.

Chapter 4

1. The mating is $A/O \times A/B$. The children are:

Genotype	Phenotype
$\frac{1}{4}\, A/A$	A
$\frac{1}{4}\, A/B$	AB
$\frac{1}{4}\, A/O$	A
$\frac{1}{4}\, B/O$	B

5. a. Order of dominance: $b > s > c > a$
b. 1 b/b black   1 b/c black
    1 b/s black   1 s/c sepia

7. Husband fathered child 1; lover fathered child 2; either could have fathered child 3.

11. The first cross was $A/a \times a/Y$. The expected progeny are:

1 A/a short-bristled females
1 a/a long-bristled females
1 a/Y long-bristled males
1 A/Y nonviable males

13. P $A/A ; B/B$ (disc) $\times a/a ; b/b$ (long)
F$_1$ $A/a ; B/b$ disc
F$_2$ 9 $A/- ; B/-$ disc
      3 $a/a ; B/-$ sphere
      3 $A/- ; b/b$ sphere
      1 $a/a ; b/b$ long

15. a. Frizzle is A/a. Normal and woolly are homozygotes.
b. Cross normal $\times$ woolly.

18. a. In line 1, the wild-type allele makes pisatin, and the recessive deviation does not produce pisatin. In line 2, the wild-type allele allows the expression of the pisatin that is normally made by the first gene, whereas the dominant deviation blocks expression of that wild-type product.

b. Cross 1:   P $a/a ; b/b \times A/A ; b/b$
      F$_1$ $A/a ; b/b$
      F$_2$ 3 $A/- ; b/b$
          1 $a/a ; b/b$
    Cross 2:   P $A/A ; B/B \times A/A ; b/b$
      F$_1$ $A/A ; B/b$
      F$_2$ 3 $A/A ; B/-$
          1 $A/A ; b/b$
    Cross 3:   P $a/a ; b/b \times A/A ; B/B$
      F$_1$ $A/a ; B/b$
      F$_2$ 9 $A/- ; B/-$ no pisatin
          3 $A/- ; b/b$ pisatin
          3 $a/a ; B/-$ no pisatin
          1 $a/a ; b/b$ no pisatin

c. Line 1 does not make pisatin, whereas line 2 blocks the expression of pisatin.

19. It is possible to produce black offspring from two pure-breeding recessive albino parents if albinism results from alleles of two different genes. If the cross is designated

$A/A \cdot b/b \times a/a \cdot B/B$

then all the offspring would be

$A/a \cdot B/b$

and they would have a black phenotype because of complementation.

23. b. $\frac{9}{32}$

26. a. P A/A (agouti) $\times a/a$ (nonagouti)

Gametes A and a

F$_1$ A/a (agouti)
Gametes A and a

F$_2$ 1 A/A (agouti) : 2 A/a (agouti) : 1 a/a (nonagouti)

b. P B/B (wild type) $\times b/b$ (cinnamon)

Gametes B and b

F$_1$ B/b (wild type)
Gametes B and b

F$_2$ 1 B/B (wild type) : 2 B/b (wild type) : 1 b/b (cinnamon)

c. P $A/A ; b/b$ (cinnamon or brown agouti) $\times a/a ; B/B$
                (black nonagouti)
Gametes $A ; b$ and $a ; B$
F$_1$ $A/a ; B/b$ (wild type or black agouti)

d. 9 $A/- ; B/-$ black agouti
    3 $a/a ; B/-$ black nonagouti
    3 $A/- ; b/b$ cinnamon
    1 $a/a ; b/b$ chocolate

e. P $A/A ; b/b$ (cinnamon) $\times a/a ; B/B$ (black nonagouti)
Gametes $A ; b$ and $a ; B$
F$_1$ $A/a ; B/b$ (wild type)
Gametes $A ; B, A ; b, a ; B,$ and $a ; b$
F$_2$ 9 $A/- ; B/-$ wild type
    1 $A/A ; B/B$
    2 $A/a ; B/B$
    2 $A/A ; B/b$
    4 $A/a ; B/b$
    3 $a/a ; B/-$ black nonagouti
    1 $a/a ; B/B$
    2 $a/a ; B/b$
    3 $A/- ; b/b$ cinnamon
    1 $A/A ; b/b$
    2 $A/a ; b/b$
    1 $a/a ; b/b$ chocolate

f. P $A/a ; B/b \times A/A ; b/b$ $A/a ; B/b \times a/a ; B/B$
    (wild type)(cinnamon) (wild type)(black nonagouti)
F$_1$ $A/A ; B/b$ wild type   1 $A/a ; B/B$ wild type
    1 $A/a ; B/b$ wild type   1 $A/a ; B/b$ wild type
    1 $A/A ; b/b$ cinnamon   1 $a/a ; B/B$ black nonagouti
    1 $A/a ; b/b$ cinnamon   1 $a/a ; B/b$ black nonagouti

g. P $A/a ; B/b \times a/a ; b/b$
    (wild type)(chocolate)
F$_1$ 1 $A/a ; B/b$ wild type
    1 $A/a ; b/b$ cinnamon
    1 $a/a ; B/b$ black nonagouti
    1 $a/a ; b/b$ chocolate

h. Cross 1: The F$_1$ parent is A/A, and the original albino must have been $c/c ; A/A ; b/b$.
    Cross 2: The albino parent must be $c/c ; A/A ; B/B$.
    Cross 3: The F$_1$ must be $C/c ; A/a ; B/B$, and the albino parent must be $c/c ; a/a ; B/B$.
    Cross 4: The albino parent must be $c/c ; a/a ; b/b$.

29. The cross is:

P $A/A ; p/p \times a/a ; P/Y$
F$_1$ $A/a ; P/p$ purple-eyed females
      $A/a ; p/Y$ red-eyed males

F$_2$	Females		Males	
	$\frac{3}{8} A/- ; P/p$	purple	$\frac{3}{8} A/- ; P/Y$	purple
	$\frac{3}{8} A/- ; p/p$	red	$\frac{3}{8} A/- ; p/Y$	red
	$\frac{1}{8} a/a ; P/p$	white	$\frac{1}{8} a/a ; P/Y$	white
	$\frac{1}{8} a/a ; p/p$	white	$\frac{1}{8} a/a ; p/Y$	white

31. The seed is A/a ; C/C ; R/r.

33. **a.** The cross is:

P		td ; su (wild type) $\times td^+$; su^+ (wild type)	
F_1	1	td ; su	wild type
	1	td ; su^+	requires tryptophan
	1	td^+ ; su^+	wild type
	1	td^+ ; su	wild type

b. 1 tryptophan-dependent : 3 tryptophan-independent

37. Pedigrees such as the one shown are quite common. They indicate lack of penetrance due to epistasis or environmental effects. Individual A must have the dominant autosomal gene.

39. **a.** This is a dihybrid cross, with only one phenotype in the F_2 being colored. The ratio of white to red indicates that the double recessive is not the colored phenotype. Instead, the general formula for color is represented by $X/-$; y/y.

Let line 1 be A/A ; B/B and line 2 be a/a ; b/b. The F_1 is A/a ; B/b. Assume that A blocks color in line 1 and b/b blocks color in line 2. The F_1 will be white because of the presence of A. The F_2 are:

9 $A/-$; $B/-$	white because of A
3 $A/-$; b/b	white because of A
3 a/a ; $B/-$	red
1 a/a ; b/b	white because of b/b

b. Cross 1: A/A ; $B/B \times A/a$; $B/b \rightarrow$ all $A/-$; $B/-$ white

Cross 2: a/a ; $b/b \times A/a$; $B/b \rightarrow \frac{1}{4}A/a$; B/b white

$\frac{1}{4}A/a$; b/b	white
$\frac{1}{4}a/a$; b/b	white
$\frac{1}{4}a/a$; B/b	red

41. **a.**

Cross	Parents	Progeny
1	$A^S/A^S \times A^D/A^D$	A^S/A^D
CONCLUSION:	A^S is dominant to A^D	
2	$A^S/A^D \times A^S/A^D$	$3 A^S/- : 1 A^D/A^D$
CONCLUSION:	Supports conclusion from cross 1	
3	$A^D/A^D \times A^{Sd}/A^D$	$1 A^{Sd}/A^D : 1 A^D/A^D$
CONCLUSION:	A^{Sd} is dominant to A^D	
4	$A^S/A^S \times A^{Sd}/A^D$	$1 A^{Sd}/A^S : 1 A^S/A^D$
CONCLUSION:	A^{Sd} is dominant to A^S	
5	$A^D/A^D \times A^{Sd}/A^S$	$1 A^{Sd}/A^D : 1 A^D/A^S$
CONCLUSION:	Supports conclusion of heterozygous superdouble	
6	$A^D/A^D \times A^S/A^D$	$1 A^D/A^D : 1 A^D/A^S$
CONCLUSION:	Supports conclusion of heterozygous superdouble	

b. Although this explanation does rationalize all the crosses, it does not take into account either the female sterility or the origin of the superdouble plant from a double-flowered variety.

A number of genetic mechanisms could be proposed to explain the origin of superdouble from the double-flowered variety. Most of the mechanisms will be discussed in later chapters and so will not be mentioned here. However, it can safely be assumed at this point that, whatever the mechanism, it was aberrant enough to block the proper formation of the complex structure of the female flower. Because of female sterility, no homozygote for superdouble can be observed.

43. **a.** A trihybrid cross would give a 63 : 1 ratio. Therefore, there are three R loci segregating in this cross.

b. P R_1/R_1 ; R_2/R_2 ; $R_3/R_3 \times r_1/r_1$; r_2/r_2 ; r_3/r_3

F_1 R_1/r_1 ; R_2/r_2 ; R_3/r_3

F_2		
27 $R_1/-$; $R_2/-$; $R_3/-$	red	
9 $R_1/-$; $R_2/-$; r_3/r_3	red	
9 $R_1/-$; r_2/r_2 ; $R_3/-$	red	
9 r_1/r_1 ; $R_2/-$; $R_3/-$	red	
3 $R_1/-$; r_2/r_2 ; r_3/r_3	red	
3 r_1/r_1 ; $R_2/-$; r_3/r_3	red	
3 r_1/r_1 ; r_2/r_2 ; $R_3/-$	red	
1 r_1/r_1 ; r_2/r_2 ; r_3/r_3	white	

c. (1) To obtain a 1 : 1 ratio, only one of the genes can be heterozygous. A representative cross would be R_1/r_1 ; r_2/r_2 ; $r_3/r_3 \times r_1/r_1$; r_2/r_2 ; r_3/r_3.

(2) To obtain a 3 red : 1 white ratio, two alleles must be segregating, and they cannot be within the same gene. A representative cross would be R_1/r_1 ; R_2/r_2 ; $r_3/r_3 \times r_1/r_1$; r_2/r_2 ; r_3/r_3.

(3) To obtain a 7 red : 1 white ratio, three alleles must be segregating, and they cannot be within the same gene. The cross would be R_1/r_1 ; R_2/r_2 ; $R_3/r_3 \times r_1/r_1$; r_2/r_2 ; r_3/r_3.

d. The formula is $1 - (\frac{1}{2})^N$, where N = number of loci that are segregating in the representative crosses in part c.

45. **a.** The gene causing blue sclera and brittle bones is pleiotropic with variable expressivity.

b. The allele appears to be an autosomal dominant.

c. Both incomplete penetrance and variable expressivity are demonstrated in the pedigree. Individuals II-4, II-14, III-2, and III-14 have descendants with the disorder, although they do not themselves express the disorder. Therefore, 4 of 20 people that can be inferred to carry the gene do not express the trait. That is 80 percent penetrance. (Penetrance could be significantly less than that since many possible carriers have no shown progeny.) The pedigree also exhibits variable expressivity. Of the 16 individuals who have the allele expressed in their phenotypes, 9 do not have brittle bones. Usually, expressivity is put in such terms as none, variable, and highly variable rather than expressed as percentages.

47. **a.** The gene is on the X chromosome. This is a multiple allelic series with oval > sickle > round.

Let W^o = oval, W^S = sickle, and W^r = round. The three crosses are:

Cross 1:	$W^s/W^s \times W^r/Y \rightarrow W^s/W^r : W^s/Y$
Cross 2:	$W^r/W^r \times W^s/Y \rightarrow W^s/W^r : W^r/Y$
Cross 3:	$W^s/W^s \times W^o/Y \rightarrow W^o/W^s : W^s/Y$

b. $W^o/W^s \times W^r/Y \rightarrow 1\ W^o/W^r : 1\ W^s/W^r : 1\ W^o/Y : 1\ W^s/Y$, or 1 female oval : 1 female sickle : 1 male oval : 1 male sickle

49. **a.** Let platinum be a, aleutian be b, and wild type be $A/-$; $B/-$.

Cross 1:	P	A/A ; $B/B \times a/a$; B/B	
	F_1	A/a ; B/B	
	F_2	$3 A/-$; B/B : $1 a/a$; B/B	
Cross 2:	P	A/A ; $B/B \times A/A$; b/b	
	F_1	A/A ; B/b	
	F_2	$3 A/A$; $B/-$: $1 A/A$; b/b	
Cross 3:	P	a/a ; $B/B \times A/A$; b/b	
	F_1	A/a ; B/b	
	F_2	$9 A/-$; $B/-$	wild type
		$3 A/-$; b/b	aleutian
		$3 a/a$; $B/-$	platinum
		$1 a/a$; b/b	sapphire

b. Sapphire × Platinum | Sapphire × Aleutian

P	$a/a \cdot b/b \times a/a \cdot B/B$	$a/a \cdot b/b \times A/A \cdot b/b$
F_1	$a/a \cdot B/b$ platinum	$A/a \cdot b/b$ aleutian
F_2	$3\ a/a \cdot B/-$ platinum	$3\ A/- \cdot b/b$ aleutian
	$1\ a/a \cdot b/b$ sapphire	$1\ a/a \cdot b/b$ sapphire

51. **a.** Let the gene be E and assume the following designations:

$$E^1 = \text{black}$$
$$E^2 = \text{brown}$$
$$E^3 = \text{eyeless}$$

If you next assume, on the basis of the various crosses, that black > brown > eyeless, genotypes in the pedigree become:

I E^1/E^2———E^3/E^3———E^2/E^3

II E^1/E^2—E^1/E^3 E^2/E^3—E^3/E^3 E^1/E^3

III E^2/E^3 E^2/E^3—E^1/E^3 E^3/E^3

IV E^3/E^3

Two genes. If two genes are assumed, then questions arise regarding whether both are autosomal or if one might be X-linked. Eye color appears to be autosomal. The presence or absence of eyes could be X-linked. Let:

$$B = \text{black} \qquad X^E = \text{normal eyes}$$
$$b = \text{brown} \qquad X^e = \text{eyeless}$$

The pedigree then is

I $B/b\ ;\ X^E/Y$———$B/b\ ;\ X^e/X^e$———$b/b\ ;\ X^E/Y$

II $B/b\ ;\ X^E/Y$—$B/b\ ;\ X^E/X^e$ $b/b\ ;\ X^E/X^e$—$-/b\ ;\ X^e/Y$———$B/-\ ;\ X^E/X^e$

III $B/b\ ;\ X^E/X^e$ $b/b\ ;\ X^E/X^e$—$B/-\ ;\ X^E/Y$ $-/-\ ;\ X^e/X^e$

IV $-/b\ ;\ X^e/Y$

With this interpretation of the pedigree, individual II-3 is $b/b\ ;\ X^E/X^e$.

Without further data, it is impossible to choose scientifically between the two possible explanations. A basic "rule" in science is that the more simple answer should be used until such time as data exist that would lead to a rejection of it. In this case, the one-gene explanation is essentially as complex as the two-gene explanation, with one exception. The exception is that eye color is fundamentally different from a lack of eyes. Therefore, the better guess is that two genes are involved.

b. The genotype of individual II-3 is E^2/E^3.

Chapter 5

1. 45 percent of the progeny will be $A\ B/a\ b$

2.

P	$A\ d/A\ d \times a\ D/a\ D$
F_1	$A\ d/a\ D$
F_2	$1\ A\ d/A\ d$
	$2\ A\ d/a\ D$
	$1\ a\ D/a\ D$

4. The two genes are 33.3 map units (m.u.) apart.

5. Because only parental types were recovered, the two genes must be quite close to each other, making recombination quite rare.

8. **(a)** 4 percent; **(b)** 4 percent; **(c)** 46 percent; **(d)** 8 percent.

11. **a.** All four genes are linked.

b. and **c.** The map is:

B ———— A, D ———————— C
10 m.u. ——— 30 m.u.

The parental chromosomes actually were $B\ (A, d)\ c/b\ (c, D)\ C$, where the parentheses indicate that the order of the genes within them is unknown.

d. Interference = 0.5

12. **a.** Males must be heterozygous for both genes, and the two must be closely linked: $M\ F/m\ f$.

b. $m\ f/m\ f$

c. Sex is determined by the male contribution. The two parental gametes are $M\ F$, determining maleness ($M\ F/m\ f$), and $m\ f$, determining femaleness ($m\ f/m\ f$). Occasional recombination would yield $M\ f$, determining a hermaphrodite ($M\ f/m\ f$), and $m\ F$, determining total sterility ($m\ F/m\ f$).

d. Recombination in the male yielding $M\ f$

e. Hermaphrodites are rare because the genes are tightly linked.

15. **a.** $a-b$: 100% (91 + 9)/(1001 = 10 m.u.
$b-c$: 100% (171 + 9)/1001 = 18 m.u.

b. Coefficient of coincidence = 0.5

17. $v-b$: 18.0 m.u.
$b-lg$: 28.0 m.u.
Coefficient of coincidence = 0.79

20. 1. $b\ a\ c$
2. $b\ a\ c$
3. $b\ a\ c$
4. $a\ c\ b$
5. $a\ c\ b$

22. **a.** The hypothesis is that the genes are not linked. Therefore, a 1 : 1 : 1 : 1 ratio is expected.

b. $\chi^2 = 0.76$

c. With 3 degrees of freedom, the p value is between 0.50 and 0.90.

d. Between 50 percent and 90 percent of the time values, this extreme from the prediction would be obtained by chance alone.

e. Accept the initial hypothesis.

f. Because the χ^2 value was insignificant, the two genes are assorting independently. The genotypes of all individuals are:

P	$dp^+/dp^+\ ;\ e/e \times dp/dp\ ;\ e^+/e^+$
F_1	$dp^+/dp\ ;\ e^+/e$
Tester	$dp/dp\ ;\ e/e$
Progeny	long ebony	$dp^+/dp\ ;\ e/e$
	long gray	$dp^+/dp\ ;\ e^+/e$
	short gray	$dp/dp\ ;\ e^+/e$
	short ebony	$dp/dp\ ;\ e/e$

29. Let A = resistance to rust 24, a = susceptibility to rust 24, B = resistance to rust 22, b = susceptibility to rust 22.

a.

P	$A/A\ ;\ b/b$ (770B) $\times a/a\ ;\ B/B$ (Bombay)
F_1	$A/a\ ;\ B/b \times A/a\ ;\ B/b$
F_2	$184\ A/-\ ;\ B/-$
	$63\ A/-\ ;\ b/b$
	$58\ a/a\ ;\ B/-$
	$\underline{15\ a/a\ ;\ b/b}$
	320

b. Expect:

$$180 \ A/- \ ; \ B/-$$
$$60 \ A/- \ ; \ b/b$$
$$60 \ a/a \ ; \ B/-$$
$$20 \ a/a \ ; \ b/b$$

$\chi^2 = 1.555$, nonsignificant; accept the hypothesis.

31. a. Blue sclerotic appears to be an autosomal dominant disorder. Hemophilia appears to be an X-linked recessive disorder.

b. If the individuals in the pedigree are numbered as generations I through IV and the individuals in each generation are numbered clockwise, starting from the top right-hand part of the pedigree, their genotypes are:

I: b/b ; $H/h, B/b$; H/Y

II: B/b ; $H/Y, B/b$; $H/Y, b/b$; $H/Y, B/b$; $H/h, b/b$; $H/Y,$ B/b ; $H/h, B/b$; $H/h, B/b$; $H/-, b/b$; $H/-$

III: b/b ; $H/-, B/b$; $H/-, B/b$; $h/Y, b/b$; $H/Y, B/b$; $H/Y,$ B/b ; $H/-, B/b$; $H/Y, B/b$; $H/-, b/b$; $H/Y,$ B/b ; $H/-, b/b$; $H/Y, B/b$; $H/-, B/b$; $H/Y, B/b$; $h/Y,$ b/b ; $H/Y, b/b$; $H/Y, b/b$; $H/-, b/b$; $H/Y, b/b$; $H/Y,$ B/b ; $H/-, B/b$; $H/Y, B/b$; h/Y

IV: b/b ; $H/-, B/b$; $H/-, B/b$; $H/-, b/b$; $H/h, b/b$; $H/h,$ b/b ; $H/Y, b/b$; $H/H, b/b$; $H/Y, b/b$; $H/h, b/b$; $H/H,$ b/b ; $H/H, b/b$; $H/Y, b/b$; $H/Y, b/b$; $H/H, b/b$; $H/Y,$ b/b ; $H/Y, B/b$; $H/Y, b/b$; $H/Y, b/b$; $H/-, b/b$; $H/Y,$ b/b ; $H/Y, b/b$; $H/-, b/b$; $H/H, b/b$; $H/-, b/b$; $H/-,$ b/b ; $H/Y, b/b$; $H/Y, b/b$; $H/Y, b/b$; $H/h, B/b$; $H/h,$ B/b ; $H/Y, b/b$; $H/Y, B/b$; $H/Y, b/b$; H/h

c. There is no evidence of linkage between these two disorders.

d. The two genes exhibit independent assortment.

e. No individual could be considered intrachromosomally recombinant. However, a number show interchromosomal recombination: all individuals in generation III that have both disorders.

34. a.

	E		C		A		B		D
	40 m.u.		20 m.u.		30 m.u.		10 m.u.		

b. No interference

36. (a) 0.0336; **(b)** 0.182.

44. a. From the data, it can be concluded that the genes for flower color and plant height are linked (on the same chromosome). The position of the gene for leaf width cannot be determined from these data since all progeny express the dominant trait. In the tall, red parent, the alleles for tall and white are in the cis configuration (on the same chromosome) as are the alleles for short and red. Thus, the progeny that are tall, white or short, red represent the parental chromosomes and the tall, red or short, white are recombinants. The two genes are 100% (21 + 19)/total = 4 map units apart.

Tall, Red, Wide × short, white, narrow

$$\frac{T \ r}{t \ R} \cdot \frac{W}{W} \times \frac{t \ r}{t \ r} \cdot \frac{w}{w}$$

b. The chance of obtaining short, white, wide progeny = $p(t \ r)^2 = (4\%/2)^2 = 0.04\%$

Chapter 6

1. a. +, +, al-2, al-2, +, +, al-2, al-2
      al-2, al-2, +, +, al-2, al-2, +, +

b. The 8 percent value can be used to calculate the distance between the gene and its centromere (4 m.u.).

3. a. $f(0) = e^{-2}2^0/0! = e^{-2} = 0.135$

b. $f(1) = e^{-2}2^1/1! = e^{-2}(2) = 0.27$

c. $f(2) = e^{-2}2^2/2! = e^{-2}(2) = 0.27$

6. a. 25.5 m.u.

b. $\chi^2 = 2.52$. With 3 degrees of freedom, the probability is greater than 10 percent that the genes are not linked. Therefore, the hypothesis of no linkage can be accepted.

7. a. The parents were:

$$ad^- \ nic^+ \ ; \ leu^+ \ arg^- \ \times \ ad^+ \ nic^- \ ; \ leu^- \ arg^+$$

b. Culture 16 resulted from a crossover between *ad* and *nic*. The reciprocal did not show up in the small sample.

8. (a) 0.28; **(b)** 0.35; **(c)** 0.22; **(d)** 0.72.

9. (a) 0.704; **(b)** 0; **(c)** 0.176; **(d)** 0.012; **(e)** 0.096; **(f)** 0; **(g)** 0.012.

11. a.

	A		arg-1	
		6 m.u.		14 m.u.

b. Class 6 can be obtained if a single crossover occurred between chromatids 2 and 3 between each gene and its centromere.

13.   Cross 1:

Cross 2:

Cross 3:

The first diagram is the better interpretation of the data.

Cross 4:

Cross 5: PD = NPD (and RF = 49%); the genes are not linked.

a-centromere: $100\% \ (\frac{1}{2})(22 + 8 + 10 + 20)/99 = 30.3$ m.u.

b-centromere: $100\% \ (\frac{1}{2})(24 + 8 + 10 + 20)/99 = 31.3$ m.u.

These values are approaching the 67% theoretical limit of loci exhibiting MU_{II} patterns of segregation and should be considered cautiously.

Cross 6:

Cross 7:

Cross 8: Same as cross 5.
Cross 9:

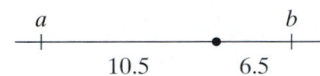

Cross 10:

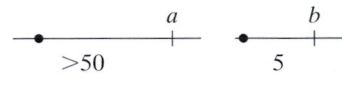

or

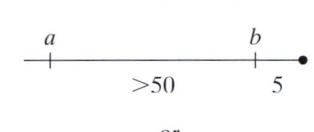

or

Cross 11:

15. Cross 1: recombinant frequency = 26.5%
uncorrected map distance = 26.5 m.u.
corrected map distance = 34.5 m.u.
Cross 2: recombinant frequency = 19%
uncorrected map distance = 19 m.u.
corrected map distance = 29 m.u.
Cross 3: recombinant frequency = 30%
uncorrected map distance = 30 m.u.
corrected map distance = 40 m.u.

22. a. The final map is:

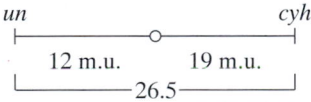

23. (a) 0 percent; **(b)** 84 percent; **(c)** 16 percent; **(d–g)** 0 percent.
27.

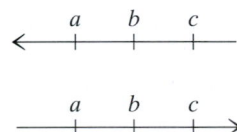

Chapter 7

1. An Hfr strain has the fertility factor, F, integrated into the chromosome. An F$^+$ strain has the fertility factor free in the cytoplasm. An F$^-$ strain lacks the fertility factor.

3. a. Hfr cells undergoing conjugation transfer host genes in a linear fashion. The genes transferred depend on both the Hfr strain and the length of time during which the transfer occurred. Therefore, a population containing several different Hfr strains will appear to have an almost random transfer of host genes. This event is similar to generalized transduction, in which the viral protein coat forms around a specific amount of DNA rather than specific genes. In generalized transduction, any gene can be transferred.

b. F$'$ factors arise from improper excision of an Hfr from the bacterial chromosome. They can have only specific bacterial

genes on them because the integration site is fixed for each strain. Specialized transduction resembles this event in that the viral particle integrates into a specific region of the bacterial chromosome and then, on improper excision, can take with it only specific bacterial genes. In both cases, the transferred gene exists as a second copy.

6. $\overline{\text{MZ}}$XWC

WCNAL

ALBRU

BRU$\underline{\text{MZ}}$

The regions with the bars above or below are identical in sequence (and "close" the circular chromosome). The correct order of markers on this circular map is

$$-M-Z-X-W-C-N-A-L-B-R-U-$$

7. An F$^-$ strain will respond differently to an F$^+$ (L) or an Hfr (M) strain, whereas Hfr × Hfr, Hfr × F$^+$, F$^+$ × F$^+$, and F$^-$ × F$^-$ will give 0. Thus strains 2, 3, and 7 are F$^-$. Strains 1 and 8 are F$^+$, and strains 4, 5, and 6 are Hfr.
10. a. The gene order is *arg bio leu*.
b. *arg–bio:* RF = 12.76 m.u.
bio–leu: RF = 2.12 m.u.
11. The most straightforward way would be to put an Hfr at both ends of the same sequence and measure the time of transfer between two specific genes. For example,

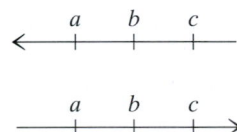

13. The best explanation is that the integrated *pro*$^+$ was incorporated onto an F$'$ factor that was transferred into recipients early in the mating process. These cells now carry the F factor and are able to transmit F$^+$ in the second cross as part of the F$'$ factor, which still carries *pro*$^+$.
14. The high rate of integration and the preference for the same site originally occupied by the sex factor suggest that the F$'$ contains some homology with the original site. The source of homology could be a fragment of the sex factor or it could be the chromosomal copy of the bacterial gene (most likely).
19. The expected number of double recombinants is 2. Interference = −1.5. By definition, the interference is negative.
22. In a small percentage of the cases, *gal*$^+$ transductants can arise by recombination between the *gal*$^+$ DNA of the λdgal transducing phage and the *gal*$^-$ gene on the chromosome. This will generate *gal*$^+$ transductants without phage integration.
23. a. Specialized transduction is at work here. It is characterized by the transduction of one to a few markers.
b. The prophage is located in the *cys–leu* region, which is the only region that gave rise to colonies when tested against the six nutrient markers.
26. Recognize that, if a compound is not added and growth occurs, the *E. coli* has received the genes for it by transduction. Thus, the BCE culture must have received *a*$^+$ and *d*$^+$. The BCD culture received *a*$^+$ and *e*$^+$. The ABD culture received *c*$^+$ and *e*$^+$. The order is thus *d a e c*. Notice that *b* is never cotransduced and is therefore distant from this group of genes.

28. **a.** The colonies are all *cys*⁺ and either + or − for the other two genes.

    b. (1) *cys*⁺ *leu*⁺ *thr*⁺ and *cys*⁺ *leu*⁺ *thr*⁻

       (2) *cys*⁺ *leu*⁺ *thr*⁺ and *cys*⁺ *leu*⁻ *thr*⁺

       (3) *cys*⁺ *leu*⁺ *thr*⁺

    c. Because none grew on minimal medium, no colony was *leu*⁺ *thr*⁺. Therefore, medium 1 had *cys*⁺ *leu*⁺ *thr*⁻, and medium 2 had *cys*⁺ *leu*⁻ *thr*⁺. The remaining cultures were *cys*⁺ *leu*⁻ *thr*⁻, and this genotype was in 39 percent of the colonies.

    d.

Chapter 8

1. The DNA double helix has two types of bonds, covalent and hydrogen. Covalent bonds exist within each linear strand and strongly bond bases, sugars, and phosphate groups (both within each component and between components). Hydrogen bonds exist between the two strands and form between a base, from one strand and a base from the second strand in complementary pairing. These hydrogen bonds are individually weak but collectively quite strong.

3. A primer is a short segment of RNA that is synthesized by primase using the DNA as a template during DNA replication. After the primer has been synthesized, DNA polymerase then adds DNA to the 3′ end of the RNA. Primers are required because the major DNA polymerase catalyzing DNA replication is unable to initiate DNA synthesis and, rather, requires a 3′ end. The RNA is subsequently removed and replaced with DNA so that no gaps exist in the final product.

5. Because the DNA polymerase is capable of adding new nucleotides only at the 3′ end of a DNA strand and because the two strands are antiparallel, at least two molecules of DNA polymerase must take part in the replication of any specific region of DNA. When a region becomes single stranded, the two strands have an opposite orientation. Imagine a single-stranded region that runs from left to right. At the left end, the 3′ end of one strand points to the right, and synthesis can initiate and continue toward the right end of that region. The other strand has a 5′ end pointing toward the right, and synthesis cannot initiate and continue toward the right end of the single-stranded region at the 5′ end. Instead, synthesis must initiate somewhere to the right of the left end of the single-stranded region and move toward the left end of the region. As the first strand continues synthesis (continuous synthesis), the single-stranded region extends toward the right. This now leaves the second strand unreplicated in this new region of single-strandedness, and there must be a second initiation of DNA synthesis moving from the current right end of the single-stranded region toward the first initiation point on that strand. This results in discontinuous synthesis along that strand.

7. The frequency of both A and T is $\frac{1}{2}$ (52%) = 26%.

9. The results suggest that the DNA is replicated in short segments that are subsequently joined by enzymatic action (DNA ligase). Because DNA replication is bidirectional, because there are multiple points along the DNA where replication is initiated, and because DNA polymerases work only in a 5′ → 3′ direction, one strand of the DNA is always in the wrong orientation for the enzyme. This requires synthesis in fragments.

12. Chargaff's rules are that A = T and G = C. Because this composition is not observed, the most likely interpretation is that

the DNA is single stranded. The phage would first have to synthesize a complementary strand before it could begin to make multiple copies of itself.

13. Remember that there are two hydrogen bonds between A and T, whereas there are three hydrogen bonds between G and C. Denaturation necessitates the breaking of these bonds, which requires energy. The more bonds that need to be broken, the more energy that must be supplied. Thus the temperature at which a given DNA molecule denatures is a function of its base composition. The higher the temperature of denaturation, the higher the percentage of G–C pairs.

18. The data suggest that each chromosome is composed of one continuous molecule of DNA and that translocations can alter their size. In part c, it appears that part of the longest chromosome has been translocated to the shortest chromosome.

Chapter 9

1. The primary structure of a protein is the sequence of amino acids along its length. It is held together by covalent bonds. The secondary structure of a protein is caused by hydrogen bonding between CO and NH groups on different amino acids. Frequently, α helices and β pleated sheets result. The tertiary structure of a protein is caused by electrostatic, hydrogen, and Van der Waals bonds between the R groups of amino acids. If the final, functional protein is composed of more than one polypeptide, then the protein has a quaternary structure.

3. An auxotroph is a strain that requires at least one nutrient for growth beyond that normally required for the organism.

5. Yanofsky analyzed mutations in the *trpA* gene and ordered them by using transduction. He also determined the amino acid sequence of the altered gene products. By this means, he demonstrated an exact correlation between the sequence of mutation sites and the sequence of altered amino acids.

7.

	B	K
rII	large plaques	no plaques
rII⁺	small plaques	plaques

9. Lactose is composed of one molecule of galactose and one molecule of glucose. A secondary cure would result if all galactose and lactose were removed from the diet. The disorder would be expected not to be dominant, because one good copy of the gene should allow for at least some, if not all, breakdown of galactose. In fact, the disorder is recessive.

11. **a.** The main use is in detecting carrier parents and in diagnosing a disorder in the fetus.

    b. Because the values for normal individuals and carriers overlap for galactosemia, there is ambiguity if a person has from 25 to 30 units. That person could be either a carrier or normal.

    c. These wild-type genes are phenotypically dominant but are incompletely dominant at the molecular level. A minimal level of enzyme activity apparently is enough to ensure normal function and phenotype.

12. One less likely possibility is a germ-line mutation. More likely is that each parent was blocked at a different point in a metabolic pathway. If one were *A/A · b/b* and the other were *a/a · B/B*, then the child would be *A/a · B/b* and would have sufficient levels of both resulting enzymes to produce pigment.

14. **a.** Complementation refers to genes within a cell and is not what is happening here. Most likely, what is known as cross-feeding is taking place, whereby a product made by one strain diffuses to another strain and allows growth of the second strain. Cross-feeding is equivalent to supplementing the medium. Because cross-feeding seems to be taking place, the suggestion is that the strains are blocked at different points in the metabolic pathway.

b. For cross-feeding to take place, the growing strain must have a block that is earlier in the metabolic pathway than the block in the strain from which it is obtaining the product for growth.

c. $E - D - B$

d. Without some tryptophan, there would be no growth at all, and the cells would not have lived long enough to produce a product that could diffuse.

18. a. Sixty percent of the meioses will not have a crossover.

b. There are 20 m.u. between the two genes.

c. 10 percent

d. $\longrightarrow B \longrightarrow A \longrightarrow$ valine

  val-1   val-2

20. a. White

b. Blue

c. Purple

d. P $b/b \; ; D/D \times B/B \; ; d/d$

  F_1 $B/b \; ; D/d \times B/b \; ; D/d$

  F_2 $9 \; B/- \; ; D/-$ purple

  $3 \; b/b \; ; D/-$ white

  $3 \; B/- \; ; d/d$ blue

  $1 \; b/b \; ; d/d$ white

  or $9 : 3 : 4$

21. The cis and trans burst size should be the same if the mutants are in different cistrons; if they are in the same cistron, the trans burst size should be zero. Therefore, if rV is assumed to be in A, rW also is in A, and rU, rX, rY, and rZ are in B.

28.

$$\overset{\longmapsto 0.072 \text{ m.u.} \longmapsto}{\underset{his \qquad a \qquad c \qquad b \qquad nic}{\vdash\!\!\!\!\!-\!\!\!-\!\!\!+\!\!\!-\!\!\!-\!\!\!+\!\!\!-\!\!\!-\!\!\!+\!\!\!-\!\!\!-\!\!\!+}}$$

  0.031 m.u. 0.023 m.u.

29. a. The allele s^n will show dominance over s^f because there will be only 40 units of square factor in the heterozygote.

b. Here, the functional allele is recessive.

c. The allele s^f may become dominant over time in two ways: (1) it could mutate slightly, so that it produces more than 50 units, or (2) other modifying genes may mutate to increase the production of s^f.

Chapter 10

1. Because RNA can hybridize to both strands, the RNA must be transcribed from both strands. This does not mean, however, that both strands are used as a template *within each gene*. The expectation is that only one strand is used within a gene but that different genes are transcribed in different directions along the DNA. The most direct test would be to purify a specific RNA coding for a specific protein and then hybridize it to the lambda genome. Only one strand should hybridize to the purified RNA.

3. A single nucleotide change should result in three adjacent amino acid changes in a protein. One amino acid change and two

adjacent amino acid changes would be expected to be much rarer than the three changes. This is directly the opposite of what is observed in proteins.

7. The codon for amber is UAG. The following list includes the amino acids that would need to have been inserted to continue the wild-type chain and their codons:

glutamine	CAA, CAG*
lysine	AAA, AAG*
glutamic acid	GAA, GAG*
tyrosine	UAU*, UAC*
tryptophan	UGG*
serine	AGU, AGC, UCU, UCC, UCA, UCG*

In each case, the codon with an asterisk would require a single base change to become UAG.

8. (a) $\frac{1}{8}$; (b) $\frac{1}{4}$; (c) $\frac{1}{8}$; (d) $\frac{1}{8}$.

10. a. (GAU)$_n$ encodes Asp (GAU), Met (AUG), and stop (UGA). (GUA)$_n$ encodes Val (GUA), Ser (AGU), and stop (UAG). One reading frame contains a stop codon.

b. Each of the three reading frames contains a stop codon.

11. Mutant 1: A simple substitution of Arg for Ser exists, suggesting a nucleotide change. Two codons for Arg are AGA and AGG, and one codon for Ser is AGU. The final U for Ser could have been replaced by either an A or a G.

Mutant 2: the Trp codon (UGG) changed to a stop codon (UGA or UAG).

Mutant 3: Two frameshift mutations occurred:

5'-GCN CCN (−U) GGA GUG AAA AA(+U or C)

  UGU/C CAU/C-3'

Mutant 4: An inversion occurred after Trp and before Cys. The DNA original sequence was:

  3'-CGN GGN ACC TCA CTT TTT ACA/G GTA/G-5'

Therefore, the complementary RNA sequence was:

  5'-GCN CCN UGG AGU GAA AAA UGU/C CAU/C-3'

The DNA inverted sequence became:

  3'-CGN GGN ACC∧AAA AAG TGA∧ACA/G GTA/G-5'

Therefore, the complementary RNA sequence was

  5'-GCN CCN UGG∧UUU UUC ACU∧UGU/C CAU/C-3'

12.

a. Old: AA^A_G ⟨A⟩GU CCA UCA CUU AAU GCN GCN AA^A_G

 Base deleted

b. New: AA^A_G GUC CAU CAC UUA AU⟨G⟩ GCN GCN AA^A_G

 Base added

14. 3'-CGT ACC ACT GCA-5'

  5'-GCA TGG TGA CGT-3'

  5'-GCA UGG UGA CGU-3'

  3'-CGU ACC ACU GCA-5'

  NH$_3$-Ala-Trp stop Arg

16. f, d, j, e, c, i, b, h, a, g

17. Cells in long-established culture lines usually are not fully diploid. For reasons that are currently unknown, adaptation to culture frequently results in both karyotypic and gene-dosage changes. Such changes can result in hemizygosity for some genes, which allows for the expression of previously hidden recessive alleles.

Chapter 11

2. O^c mutants are changes in the DNA sequence of the operator that impair the binding of the *lac* repressor. Therefore, the *lac* operon associated with the O^c operator cannot be turned off. Because an operator controls only the genes on the same DNA strand, it is cis (on the same strand) and dominant (cannot be turned off).

7. Nonpolar Z^- mutants cannot convert lactose into allolactose, and, thus, the operon is never induced.

9. An operon is turned off, or inactivated, by the mediator in negative control, and the mediator must be removed for transcription to take place. An operon is turned on by the mediator in positive control, and the mediator must be added or activated for transcription to take place.

10. The *lacY* gene produces a permease that transports lactose into the cell. A *lacY⁻* gene cannot transport lactose into the cell, so β-galactosidase will not be induced.

12. The bacterial operon consists of a promoter region that extends approximately 35 bases upstream of the site where transcription is initiated. Within this region is the promoter. Inducers and repressors, both of which are trans-acting proteins that bind to the promoter region, regulate transcription of associated genes in cis only.

The eukaryotic gene has the same basic organization. However, the promoter region is somewhat larger. In addition, enhancers as distant as several thousand nucleotides upstream or downstream can affect the rate of transcription. A major difference is that eukaryotes have not been demonstrated to have polygene messages.

Chapter 12

2. GTTAAC appears, on average, every 4^6 bases, which is 4.096 kb. GGCC is present, on average, every 4^4 bases, which is 0.256 kb.

3.

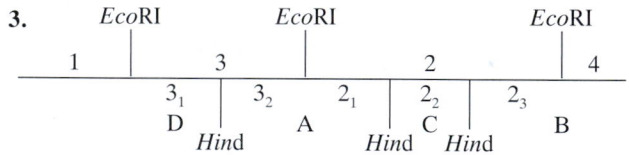

5. Read from the top, the sequence is:

(EcoRI)−HindII−HaeIII−HaeIII−HindII−HaeIII−HaeIII−HaeIII−HindII−HindII−HaeIII−HindII−HaeII−labeled end

6. AluI $(\frac{1}{4})^4$ = every 256 nucleotide pairs

EcoRI $(\frac{1}{4})^6$ = every 4096 nucleotide pairs

AcyI $(\frac{1}{4})^4(\frac{1}{2})^2$ = every 1024 nucleotide pairs

7. a.

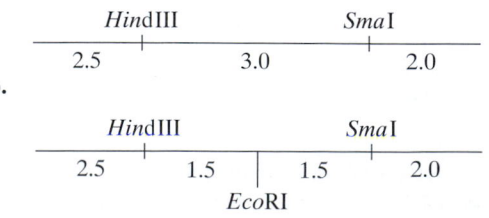

b.

15. a. and b.

| EcoRI | 8.0 | | 7.4 | 4.5 | | 2.9 | 6.2 |

| BamHI | 6.0 | | 12.9 | | 10.1 |

6.0 2.0 7.4 3.5 1.0 2.9 6.2

X

Chapter 13

2. Pulsed field gel electrophoresis consists of the movement through a gel of chromosome-sized pieces of DNA. Unless the overall size of a chromosome is changed, it should always migrate to the same relative position.

a. 7 bands identical with wild type

b. 7 bands identical with wild type

c. The largest and smallest bands would be expected to disappear. Two new intermediate bands would appear, unless they happened to comigrate with one of the other five wild-type bands.

d. One band would be larger than expected and one would be smaller than expected, compared with the wild type.

e. 7 bands identical with wild type

f. 7 bands

g. The largest wild-type band would be missing and an even larger new band would be present.

6. One approach would be to "knock out" wild-type function and then observe the phenotype. To do this, follow the one-step protocol described for Figure 13-4. Insert a selectable gene into the cloned gene of interest. Linearize the plasmid and use it for the transformation of yeast cells. Select for the selectable gene. Another possibility is site-directed mutagenesis.

7. After electrophoresis, Southern blot the gel and probe with radioactive copies of the cloned gene.

9. a. The cross is $B/b \cdot \text{RFLP}^1/\text{RFLP}^2 \times b/b \cdot \text{RFLP}^2/\text{RFLP}^2$, where B = bent tail, b = wild type, RFLP^1 = 1.7-kb *Hind*III fragment, and RFLP^2 = 3.8-kb *Hind*III fragment. The data suggest that the two markers are 40 m.u. apart (there is 40% recombination).

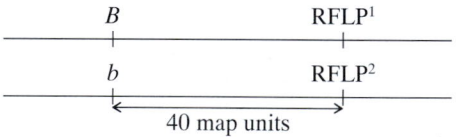

b. The wild-type progeny should be

60% "parental" $b\text{RFLP}^2/b\text{RFLP}^2$
40% "recombinant" $b\text{RFLP}^1/b\text{RFLP}^2$

10. a. 1: *A1, B2*, recombinant
2: *A2, B1*, recombinant
3: *A1, B1*, parental
4: *A2, B2*, parental

b. A ———— 30 ———— B   or   *A1*   *B1*
                                      ╫   ╫
                                      ╫   ╫
                                   A2   *B2*

12. a. Recall that yeast plasmid vectors can exist free in the cytoplasm, producing multiple copies that are distributed randomly to daughter cells, or they can be integrated into yeast chromosomes, in which case they replicate with the host chromosome and are distributed as any other yeast gene during mitosis and meiosis.

YP1 produces 100 percent *leu*⁺ progeny, suggesting that the *leu*⁺ gene is not integrated into a chromosome and that enough copies exist of the plasmid so that all progeny obtained at least one copy.

YP2 produces 50 percent *leu*⁺ progeny, which is indicative of orderly distribution and suggests that the *leu*⁺ gene is integrated into a yeast chromosome.

b. The "insert," which is not cut with the enzyme, is the *leu*⁺ gene inserted into the plasmid.

After digestion, electrophoresis, blotting, and probing, YP1 will produce one band that is somewhat heavier than that of the plasmid without an insert (that is, it migrates less distance than the plasmid during electrophoresis). YP2 will produce two bands that contain a part the integrated plasmid plus genomic DNA. The location of these bands relative to the plasmid cannot be predicted.

Chapter 14

1. a. The sequence is 2 1 3 (or 3 1 2).

b. YAC A spans at least a part of both 1 and 3. YAC B is within region 1. YAC C spans at least a part of regions 1 and 2. YAC D is contained within region 2. YAC E is contained within region 3. A diagram of these results is shown below. In the diagram, there is no way to know the exact location of the ends of each YAC.

```
|———— 2 ————+——— 1 ———+——— 3 ——|
                    ———————— A ————————
        —— D ——    —— B ——  —— E ——
            ———————— C ————————
```

3. The cross is

cys-1 RFLP-1ᴼ RFLP-2ᴼ × *cys-1*⁺ RFLP-1ᴹ RFLP-2ᴹ

A parental type will have the genotype of either strain and be in a majority, whereas a recombinant type will have a mixed genotype and be in a minority of the ascospores. Clearly, the first two ascospore types are parental, with the remaining being recombinant.

a. The *cys-1* locus is in this region of chromosome 5. If it were not in this region, linkage to either of the RFLP loci would not be observed.

b. The entire region is approximately 17 map units in length.

```
|—— 5 m.u. ——+———— 12 m.u. ——|
 RFLP-1      cys-1          RFLP-2
```

c. A number of strategies could be tried. Since this is an auxotrophic mutant, functional complementation can be attempted. Positional cloning and chromosome walking from the RFLPs are also very common strategies.

5. Remember that a gene is one small region of a long strand of DNA and that a cloned gene will contain the entire sequence of the gene under normal circumstances. If there are two cuts within a gene, three fragments will be produced, all of which will interact with the probe, as was seen with enzyme 1. Cuts external to a gene will produce one fragment that will interact with the probe, which was seen with enzyme 2. One cut within a gene will produce two fragments that will interact with the probe, as was seen with enzyme 3.

8. a. Rapidly amplified polymorphic DNAs are formed by being bracketed by two inverted copies of the primer sequence. In the following diagram, the primer is indicated by X's, and the amplified region appears in brackets. For convenience, the two amplified regions are placed on the same lengthy piece of DNA for strain 1.

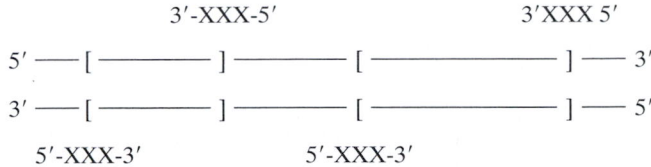

Strain 2 lacks one or two regions complementary to the primer.

b. Progeny 1 and 6 are identical with the strain 1 parent. Progeny 4 and 7 are identical with the strain 2 parent. Progeny 2 and 5 received the chromosome holding the upper band from the strain 1 parent and the chromosome holding the lower band from the strain 2 parent (resulting in no second band). Progeny 3 received the opposite: the chromosome holding the lower band from the strain 1 parent and the chromosome holding the upper band from the strain 2 parent (resulting in no second band). Therefore, bands 1 and 2 are unlinked.

c. Recall that a nonparental ditype has two types only, both of which are recombinant. Therefore, the tetrad would be composed of four progeny like progeny 2 and four progeny like progeny 3.

9. a. Gene *x* is located in region 5.

b. The location of gene *y* is region 8.

c. Both probes are able to hybridize with cosmid E because cosmid E is long enough to contain part of genes *x* and *y*.

11. a. DNA from each individual was obtained. It was restricted, electrophoresed, blotted, and then probed with the five probes. After each probing, an autoradiograph was produced.

b. The RFLP locus closest to the disease allele is 4°.

c. It appears that RFLP4 is the closest marker to the gene and could be used for positional cloning by chromosome walking. However, with only four offspring, the genetic distance between the gene and this marker could be quite large. The number of markers for each human chromosome is already large and increasing almost daily. If possible, it makes sense to analyze this family further (and as many other families with the same trait that can be found) to see if the gene can be further localized before the arduous task of "walking" is attempted.

Chapter 15

1. The petal will now be *W/w*, or blue, either in whole or in part, depending on the timing of the reversion.

3. Plate the cells on medium lacking proline. Nearly all colonies will come from revertants. The remainder will be second-site suppressors.

6. $1/10^{-6}$ per cell divisions

8. Stain pollen grains, which are haploid, from a homozygous *Wx* parent with iodine. Look for red pollen grains, indicating mutations to *wx*, under a microscope.

10. An X-linked disorder cannot be passed from father to son. Because the gene for hemophilia must have come from the mother, the nuclear power plant cannot be held responsible. It is possible that the achondroplastic gene mutation was caused by exposure to radiation.

13. a. Reddish all over

b. Reddish all over

c. Many small, red spots

d. A few large, red spots
e. Like c, but with fewer reddish patches
f. Like d, but with fewer reddish patches
g. Some large spots and many small spots

16. The mutants can be categorized as follows:

Mutant 1: An auxotrophic mutant
Mutant 2: A nonnutritional, temperature-sensitive mutant
Mutant 3: A leaky, auxotrophic mutant
Mutant 4: A leaky, nonnutritional, temperature-sensitive, mutant
Mutant 5: A nonnutritional, temperature-sensitive, auxotrophic mutant

18. **a.** This test is designed to identify the number of different genes that can mutate to a particular phenotype. In this problem, if the progeny of a given cross still express the twitcher phenotype, the mutations fail to complement and are considered alleles of the same gene; if the progeny are wild type, the mutations complement and the two strains carry mutant alleles of separate genes.

b. From the data,

1 and 5	fail to complement	gene A
2, 6, 8, and 10	fail to complement	gene B
3 and 4	fail to complement	gene C
7, 11, and 12	fail to complement	gene D
9	complements all others	gene E

There are five complementation groups (genes) identified by these data.

c. Mutant 1: $a^1/a^1 \cdot b^+/b^+ \cdot c^+/c^+ \cdot d^+/d^+ \cdot e^+/e^+$
Mutant 2: $a^+/a^+ \cdot b^2/b^2 \cdot c^+/c^+ \cdot d^+/d^+ \cdot e^+/e^+$
Mutant 5: $a^5/a^5 \cdot b^+/b^+ \cdot c^+/c^+ \cdot d^+/d^+ \cdot e^+/e^+$

$\frac{1}{2}$ hybrid: $a^+/a^1 \cdot b^+/b^2 \cdot c^+/c^+ \cdot d^+/d^+ \cdot e^+/e^+$
                      phenotype: wild type
                      1 and 2 are mutant for different genes

$\frac{1}{5}$ hybrid: $a^1/a^5 \cdot b^+/b^+ \cdot c^+/c^+ \cdot d^+/d^+ \cdot e^+/e^+$
                      phenotype: wiggles
                      1 and 5 are both mutant for gene A

21. The cross is $arg^r \times arg^+$, where arg^r is the revertant. However, it might also be $arg^- \cdot su \times arg^+ \cdot su^+$, where su^+ (suppressor) has no effect on the arg gene.

a. If the revertant is a precise reversal of the original change that produced the mutant allele, 100 percent of the progeny would be arginine independent.

b. If the suppressor mutation occurred on a different chromosome, then the cross is $arg^- ; su \times arg^+ ; su^+$. Independent assortment would lead to the following progeny:

1 arg^- ; su	arginine independent
1 arg^+ ; su^+	arginine independent
1 arg^- ; su^+	arginine dependent
1 arg^+ ; su	arginine independent

c. If the suppressor mutation occurred 10 map units from the arg locus, then the cross is $arg^- su \times arg^+ su^+$, but it is now necessary to write the diploid intermediate as $arg^- su/arg^+ su^+$.

The two parental types would appear 90 percent of the time, and the two recombinant types would appear 10 percent of the time. The progeny would be:

45% arg^- su	arginine independent
45% arg^+ su^+	arginine independent
5% arg^- su^+	arginine dependent
5% arg^+ su	arginine independent

Chapter 16

6. Depurination results in the loss of adenine or guanine from the nucleotide. Because the resulting apurinic site cannot specify a complementary base, replication is blocked. Under certain conditions, replication proceeds with a random insertion of a base opposite the apurine site. In three-fourths of these insertions, a mutation will result.

Deamination of cytosine yields uracil. If left unrepaired, uracil will be paired with adenine during replication, ultimately resulting in a transition mutation.

8-OxodG can pair with adenine, resulting in a transversion.

7. 5-Bromouracil is an analog of thymine. It undergoes tautomeric shifts at a higher frequency than does thymine and, therefore, is more likely to pair with G than is thymine during replication. At the next replication, this likelihood will lead to a GC pair rather than the original AT pair.

Ethyl methanesulfonate is an alkylating agent that produces O-6-ethylguanine. This alkylated guanine will mispair with thymine, which leads to an AT pair rather than a GC pair at the next replication.

10. Mismatch repair occurs if a mismatched nucleotide is inserted during replication. The new, incorrect base is removed, and the proper base is inserted. The enzymes required can distinguish between new and old strands because, in *E. coli,* the old strand is methylated.

Recombination repair occurs if lesions such as AP sites and UV photodimers block replication (there is a gap in the complementary strand). Recombination fills this gap with the corresponding segment from the sister DNA molecule, which is normal in both strands. This produces one DNA molecule with a gap across from a correct strand, which can then be filled by complementation, and one with a photodimer across from a correct strand.

11. Leaky mutants are mutants with an altered protein product that retains a low level of function. Enzyme activity may, for instance, be reduced rather than abolished by a mutation.

13. **a.** Because 5'-UAA-3' does not contain G or C, a transition to a GC pair in the DNA cannot result in 5'-UAA-3'. 5'-UGA-3' and 5'-UAG-3' have the DNA antisense-strand sequences of 3'-ACT-5' and 3'-ATC-5', respectively. A transition to either of these stop codons occurs from the nonmutant 3'-ATT-5'. However, a DNA sequence of 3'-ATT-5' results in an RNA sequence of UAA, itself a stop codon.

b. Yes. An example is 5'-UGG-3', which encodes Trp, changed to 5'-UAG-3'.

c. No. In the three stop codons, the only base that can be acted on is G (in UAG, for instance). Replacing the G with an A would result in 5'-UAA-3', a stop codon.

15. To understand these data, recall that half of the progeny should come from the wild-type parent.

a. A lack of revertants suggests either a deletion or an inversion within the gene.

b. *Prototroph A:* Because 100 percent of the progeny are prototrophic, a reversion at the original mutant site may have occurred.

Prototroph B: Half of the progeny are parental prototrophs, and the remaining prototrophs, 28 percent, are the result of the new mutation. Notice that 28 percent is approximately equal to the 22 percent auxotrophs. The suggestion is that an unlinked suppressor mutation occurred, yielding independent assortment with the *nic* mutant.

Prototroph C: There are 496 "revertant" prototrophs (the other 500 are parental prototrophs) and 4 auxotrophs. This result suggests that a suppressor mutation occurred in a site very close [100% (4 × 2)/1000 = 0.8 m.u.] to the original mutation.

Chapter 17

2. (a) 27 percent; (b) 36 percent.

5. A deletion occurred in the left arm of the large chromosome in nucleus 2, and *leu*$^+$, mating type *a*, and *un*$^+$ were lost.

8. The order is *b a c e d f*.

Allele	Band
b	1
a	2
c	3
e	4
d	5
f	6

10. **a.** Eighteen map units (m.u.) were either deleted or inverted. A large inversion would result in semisterility, whereas a large deletion would most likely be lethal. Thus, an inversion is more likely.

b. Normal

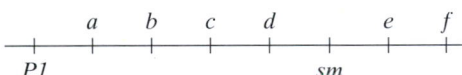

Inversion

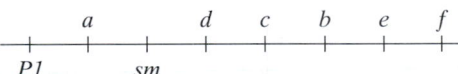

Alternatively, *sm* could be located external to the *e* locus on both the normal and the inversion chromosomes.

c. The semisterility is the result of crossing-over in the inverted region. All products of crossing-over would have both duplications and deletions.

13. The most likely explanation is that one or both break points were located within essential genes.

20. **a.** The inversion must involve *b* through *x*.

b. To obtain recombinant progeny when an inversion is involved, either a double crossover occurred within the inverted region or single crossovers occurred between *f* and the point of inversion, which occurred someplace between *f* and *y*.

22. **a.** and **b.** (**Note:** DCO = double recombinant.)

Class 1: parental
Class 2: parental
Class 3: DCO *y-cv* and *B-car*
Class 4: reciprocal of class 3
Class 5: DCO *cv-v* and *v-f*
Class 6: reciprocal of class 5
Class 7: DCO *cv-v* and *f-car*

Class 8: reciprocal of class 7
Class 9: DCO *v-cv* and *v-f*
Class 10: reciprocal of class 9
Class 11: This class is identical with the male parent's X chromosome and could not have come from the female parent. Thus, the male sperm must have donated it to the offspring. In *Drosophila*, sex is determined by the ratio of X chromosomes to the number of sets of autosomes. The ratio in males is 1X : 2A, where A stands for the autosomes contributed by one parent (the ratio in females is 2X : 2A). Thus, this class of males must have arisen from the union of an X-bearing sperm with an egg that was the product of nondisjunction for X and contained only autosomes. Because all nonparental progeny are double crossovers, there must be an inversion spanning the genes that were studied.

c. Class 11 should have only one sex chromosome, which can be checked cytologically.

32. **a.** The gray male 1 was crossed with a yellow female, yielding yellow females and gray males, which is reversed sex linkage. If the *e*$^+$ allele were translocated to the Y chromosome, the gray male would be X^{e}Y^{-e+}, or gray. When crossed with yellow females, the results would be:

$$X^eY^{-e+} \text{ gray males} \qquad X^eX^e \text{ yellow females}$$

b. The gray male 2 was crossed with a yellow female, yielding gray and yellow males and females in equal proportions. If the *e*$^+$ allele were translocated to an autosome, the progeny would be as follows, where "A" indicates an autosome involved in the translocation and the "/" separates male and female contributions:

P $\quad$ A^{e+}/A ; X^e / Y × A / A X^e/ X^e
F$_1$ $\quad$ A^{e+} / A ; X^e / X^e $\quad$ gray female
$\qquad$ A^{e+} / A ; X^e / Y $\quad$ gray male
$\qquad$ A / A ; X^e / X^e $\quad$ yellow female
$\qquad$ A / A ; X^e / Y $\quad$ yellow male

38. The percent degeneration seen in the progeny of the exceptional rat is roughly 50 percent larger than that seen in the progeny from the normal male. Semisterility is an important diagnostic for translocation heterozygotes. This could be verified by cytological observation of the meiotic cells from the exceptional male.

39. **Note:** CO = single recombinant.

a. Heterozygous reciprocal translocations lead to duplications and deletions. Therefore, in asci in which crossing-over occurs within the translocated region, each crossing-over event would lead to two white ascospores that abort and two viable dark ascospores. In asci in which no crossing-over occurs within the translocated region, the ascospores would be normal color. Alternate segregation is assumed.

No CO	CO
●	●
●	○
●	○
●	●

b. Heterozygous pericentric inversions lead to duplications and deletions. Therefore, in asci in which crossing-over occurs within the pericentric inversion, each crossing-over event would lead to two white ascospores that abort and two viable dark ascospores. In asci in which no crossing-over occurs within the pericentric inversion, the ascospores would be normal color. Alternate segregation is assumed.

No CO	CO
●	●
●	○
●	○
●	●

c. Heterozygous paracentric inversions result in an acentric fragment that has lost some genetic material (deletion) and a dicentric chromosome that has gained some genetic material (duplication) if crossing-over occurs within the paracentric inversion. Therefore, in asci in which crossing-over occurs within the paracentric inversion, each crossing-over event would lead to two white ascospores that abort and two viable dark ascospores. In asci in which no crossing-over occurs within the paracentric inversion, the ascospores would be normal color. Alternate segregation is assumed.

No CO	CO
●	●
●	○
●	○
●	●

40. Species B is probably the "parent" species. A paracentric inversion in this species would give rise to species D. Species E could then arise by a translocation of $z \, x \, y$ to $k \, l \, m$. Next, species A could result from a translocation of $a \, b \, c$ to $d \, e \, f$. Finally, species C could result from a pericentric inversion of $b \, c \, d \, e$.

Chapter 18

1.
Klinefelter syndrome	XXY male
Down syndrome	trisomy 21
Turner syndrome	XO female

3. a. 3, 3, 3, 3, 3/3, 3/3, 0, 0
   b. 7, 7, 7, 7, 8, 8, 6, 6

5. b

7. If there is no pairing between chromosomes from the same parent, then all pairs are A/a. The gametes are $1 \, A/A : 2 \, A/a : 1 \, a/a$. With selfing, the progeny are $\frac{1}{16} \, A/A/A/A$, $\frac{4}{16} \, A/A/A/a$, $\frac{6}{16} \, A/A/a/a$, $\frac{4}{16} \, A/a/a/a$, $\frac{1}{16} \, a/a/a/a$. If there is no pairing between chromosomes from different parents, then all pairs are B/B and b/b and all gametes are B/b. Thus 100% of the progeny are $B/B/b/b$.

11. b, f, and h, and sometimes c

12. Colorblindness appears to be an X-linked recessive disorder. That means that the woman with Turner syndrome had to have obtained her sole X from her mother. She did not obtain a sex chromosome from her father, which indicates that nondisjunction occurred in him. The nondisjunction could have occurred at either M_I or M_{II}.

If the colorblind patient had Klinefelter syndrome (XXY), then both X's must carry the allele for colorblindness. Therefore,

nondisjunction had to occur in the mother. Remember that during meiosis I, given no crossover between the gene and the centromere, allelic alternatives separate from each other. During meiosis II, identical alleles on sister chromatids separate. Therefore, the nondisjunctive event had to occur during meiosis II because both alleles are identical.

17. a. Loss of one X in the developing fetus after the two-celled stage.
   b. Nondisjunction leading to Klinefelter syndrome (XXY) followed by a nondisjunctive event in one cell for the Y chromosome after the two-celled stage, leading to XX and XXYY.
   c. Nondisjunction for X at the one-celled stage.
   d. Either fused XX and XY zygotes or fertilization of an egg and polar body by one sperm bearing an X and another bearing a Y, followed by fusion.
   e. Nondisjunction of X at the two-celled stage or later.

19. a. The extra chromosome must be from the mother. Because the chromosomes are identical, nondisjunction had to have occurred at M_{II}.
   b. The extra chromosome must be from the mother. Because the chromosomes are not identical, nondisjunction had to have occurred at M_I.
   c. The mother correctly contributed one chromosome, but the father did not contribute any chromosome 4. Therefore, nondisjunction occurred in the male at either meiotic division.
   d. One cell line lacks a maternal contribution, whereas the other has a double maternal contribution. Because the two lines are complementary, the best explanation is that nondisjunction occurred in the developing embryo in mitosis.
   e. Each cell line is normal, indicating that nondisjunction did not occur. The best explanation is that the second polar body was fertilized and was subsequently fused with the developing embryo.

20. a. The common progeny are $b \, e^+/b \, e$ and $b^+ \, e/b \, e$.
   b. The rare female could have come from crossing-over, which would have resulted in a gamete that was $b^+ \, e^+$. The rare female also could have come from nondisjunction that resulted in a gamete that was $b \, e^+/b^+ \, e$. Such a gamete might give rise to viable progeny.
   c. The female was a product of nondisjunction.

23. Radiation could have caused point mutations or induced recombination, but nondisjunction is a more likely explanation.

27. Aneuploidy is the result of nondisjunction. Therefore, any system that will detect nondisjunction will work. One of the easiest follows. Cross white-eyed females with red-eyed males and look for the reverse of X linkage in the progeny. Compare populations unexposed to any environmental pollutants with those exposed to different suspect agents.

Chapter 19

1. 3, 4, 6

5. First, notice that gene conversion has occurred. In the first cross, a_1 converted (1 : 3). In the second cross, a_3 converted. In the third cross, a_3 converted. Polarity obviously played a part. The results can be explained by the following map, in which hybrid DNA enters only from the left.

$$a_3 \text{ —— } a_1 \text{ —— } a_2$$

7. The ratios for a_1 and a_2 are both 3 : 1. There is no evidence of polarity, which indicates that gene conversion as part of recombination occurred. The best explanation is that two separate excision-repair events took place and, in both cases, the repair retained the mutant rather than the wild type.

8. a. and **b.** A heteroduplex that contains an unequal number of bases in the two strands has a larger distortion than does a simple mismatch. Therefore, the former would be more likely to be repaired. For such a case, both heteroduplex molecules are repaired (leading to 6 : 2 and 2 : 6) more often than one (leading to 5 : 3 or 3 : 5) or none (leading to 3 : 1 : 1 : 3). The preference in direction (that is, the addition rather than the subtraction of a base) is analogous to thymine dimer repair. In thymine dimer repair, the unpaired, bulged nucleotides are treated as correct and the strand with the thymine dimer is excised.

A mismatch more often than not escapes repair, leading to a 3 : 1 : 1 : 3 ascus.

Transition mutations would not cause as large a distortion of the helix, and each strand of the heteroduplex should have an equal chance of repair. This would lead to 4 : 4 (two repairs each in the opposite direction), 5 : 3 (one repair), 3 : 1 : 1 : 3 (no repairs or two repairs in opposite directions), and, less frequently, 6 : 2 (two repairs in the same direction).

c. Because excision repair excises the strand opposite the larger buckle (that is, opposite the frameshift mutation), the cis transition mutation also is retained. The nearby genes are converted because of the length of the excision repair.

9. **(a)** 6 : 2 = 31.25 percent; **(b)** 2 : 6 = 5 percent; **(c)** 3 : 1 : 1 : 3 = 11.25 percent; **(d)** 5 : 3 = 37.5 percent; **(e)** 3 : 5 = 15 percent.

Chapter 20

3. R plasmids are the main carriers of drug resistance. They acquire these genes by transposition of drug-resistance genes located between IR (inverted repeat) sequences. Once in a plasmid, the transposon carrying drug resistance can be transferred on conjugation if it stays in the R plasmid or it can insert into the host chromosome.

4. Boeke, Fink, and their co-workers demonstrated that transposition of the *Ty* element in yeast requires an RNA intermediate. They constructed a plasmid by using a *Ty* element that had a promoter that could be activated by galactose and an intron inserted into its coding region. First, the frequency of transposition was greatly increased by the addition of galactose, indicating that an increase in transcription (and production of RNA) was correlated to rates of transposition. More importantly, after transposition, they found that the newly transposed *Ty* DNA lacked the intron sequence. Because intron splicing takes place only in RNA processing, there must have been an RNA intermediate in the transposition event.

5. P elements are transposable elements found in *Drosophila*. Under certain conditions they are highly mobile and can be used to generate new mutations by random insertion and gene knockout. As such, they are a valuable tool to tag and then clone any number of genes. (See Problem 15 from Chapter 13 for a discussion on cloning by tagging.) P elements can also be manipulated and used to insert almost any DNA (or gene) into the *Drosophila* genome. P element–mediated gene transfer works by inserting the DNA of interest between the inverted repeats necessary for P element transposition and injecting this recombinant DNA along with helper intact P element DNA (to supply the transposase) into very early embryos and screening for (random) insertion among the injected fly's offspring.

7. The best explanation is that the mutation is due to an insertion of a transposable element.

8. a. and **b.** The soil bacterium *Agrobacterium tumefaciens* contains a large plasmid called the Ti (tumor-inducing) plasmid. When this bacterium infects a plant, a region of the Ti plasmid called the T-DNA is transferred and inserted randomly into the plant's genome. The T-DNA directs the synthesis of plant hormones that cause uncontrolled growth (a tumor) and also directs the plant's synthesis of compounds called opines. (These compounds cannot be metabolized by the plant but are used by the bacterium.)

When a piece of "normal" plant tissue is cultured with appropriate nutrients and growth hormones, cells are stimulated to divide in a disorganized manner, forming a mass of undifferentiated cells called a callus. These cells will differentiate only into shoots (or roots) if the levels of growth hormones are carefully adjusted. The T-DNA causes undifferentiated growth because it directs the unbalanced synthesis of these same hormones. The fact that some of the infected cultures produced shoots suggests that these cells "lost" the ability to overproduce these hormones. This would be consistent with the loss of the T-DNA (similar to the loss of other transposable elements that is observed in many species). Thus, the A graft would grow normally, and seeds produced by the graft would have no trace of the T-DNA. The fact that cells from the A graft grow like tumor cells when placed on synthetic medium suggests that the medium supplies the high levels of hormones necessary for undifferentiated growth even in the absence of T-DNA.

Chapter 21

3. Maternal inheritance of chloroplasts results in the green-white color variegation observed in *Mirabilis*.

Cross 1: variegated female $\times$ green male $\longrightarrow$ variegated progeny
Cross 2: green female $\times$ variegated male $\longrightarrow$ green progeny

5. Both yeast parents contribute mitochondria to the cytoplasm of the resulting diploid cell. Subsequent meiosis shows uniparental inheritance for mitochondria. Therefore, 4:0 and 0:4 asci will be seen.

7. The genetic determinants of R and S are cytoplasmic and are showing maternal inheritance.

9. Both yeast parents contribute mitochondria to the cytoplasm of the resulting diploid cell. Recombination of mtDNA can take place in these heteroplasmons, but cytoplasmic segregation results in tetrads that contain either one of the parental mtDNA genotypes or one of the reciprocal recombinant types.

$oli^R cap^R$	or	$oli^S cap^S$	or	$oli^R cap^S$	or	$oli^S cap^R$
$oli^R cap^R$		$oli^S cap^S$		$oli^R cap^S$		$oli^S cap^R$
$oli^R cap^R$		$oli^S cap^S$		$oli^R cap^S$		$oli^S cap^R$
$oli^R cap^R$		$oli^S cap^S$		$oli^R cap^S$		$oli^S cap^R$

11. If the mutation is in the chloroplast, reciprocal crosses will give different results; whereas, if it is in the nucleus and dominant, reciprocal crosses will give the same results.

13. After the initial hybridization, a series of backcrosses using pollen from plant B will result in the desired combination of cytoplasm A and nucleus B. With each cross, the female contributes all of the cytoplasm and half the nuclear contents, whereas the male contributes half the nuclear contents.

16. The suggestion is that the mutants are heteroplasmons and contain both wild-type (*sm-s*) and mutant (*sm-r*) organelle DNA. Subsequent mitotic divisions can lead to cytoplasmic segregation such that daughter cells contain only one organelle DNA type.

17. The red phenotype in the heterokaryon indicates that the red phenotype is caused by a cytoplasmic organelle allele.

21. **a.** and **b.** Each meiosis shows uniparental inheritance, suggesting cytoplasmic inheritance.

c. Because *ant*^R is probably mitochondrial and because petites have been shown to result from deletions in the mitochondrial genome, *ant*^R may be lost in some petites.

23. Some tetrads will show strain-1 type, some will show strain-2 type, and some will be recombinant.

24. **a.** No; in diploid budding, all the progeny receive one type of mtDNA. Most likely it is a plasmid or episome.

b. Three bands, one at 2μ, the others totaling 2μ.

26. First, prepare a restriction map of the mtDNA by using various restriction enzymes. Assume evolutionary conservation and, on a Southern blot, hybridize equivalent fragments from yeast or another organism in which the genes have already been identified.

Chapter 22

3. **a.** Dominant. The misexpression of FasL from one allele would be dominant to the normal expression of the wild-type FasL allele. In this case, each liver cell would signal its neighboring cells to undergo apoptosis.

b. No. The mutant would lead to excess cell death, not proliferation.

4. 1. Certain cancers are inherited as highly penetrant, simple Mendelian traits.
2. Most carcinogenic agents are also mutagenic.
3. Various oncogenes have been isolated from tumor viruses.
4. A number of genes that lead to the susceptibility of particular types of cancer have been mapped, isolated, and studied.
5. Dominant oncogenes have been isolated from tumor cells.
6. Certain cancers are highly correlated with specific chromosomal rearrangements.

6. Inhibition of apoptosis can contribute to tumor formation by allowing cells to have an unusually long lifetime in which to accumulate various mutations that lead to cancer. Additionally, the normal role of apoptosis in removing abnormal cells and, through p53, killing cells that have "damaged" DNA would be inhibited.

9. p53 detects and is activated by DNA damage. When activated, p53 activates p21, an inhibitor of the cyclin–CDK complex necessary for the progression of the cell cycle. If the DNA damage is repairable, this system will eventually deactivate p53 and allow cell division. However, if the damage is irreparable, p53 stays active and activates the apoptosis pathway, ultimately leading to cell death. For this reason, the "loss" of p53 is often associated with cancer.

Chapter 23

3. In humans, a single copy of the Y chromosome is sufficient to shift development toward the normal male phenotype. The extra copy of the X chromosome is simply inactivated. Both mechanisms seem to be all-or-none rather than to be based on concentration levels.

5. The concentration of *Sxl* is crucial for female development and dispensable for male development. The dominant *Sxl*^M male-lethal mutations may not actually kill all males but may simply produce an excessive amount of gene product so that only females (fertile XX and XY) result. The reversions may eliminate all gene product, resulting in XX (sterile) and XY males. The reversions would be recessive because, presumably, a single normal copy of the gene may produce enough gene product to "toggle the switch" in development to female.

7. **a.** The results suggest that ABa and ABp are not determined at this point in their development. Additionally, future determination and differentiation of these cells depend on their position within the developing organism.

b. Because an absence of EMS cells leads to a lack of determination and differentiation of AB cells, the EMS cells must be at least in part responsible for AB-cell development, either through direct contact or by the production of a diffusible substance.

c. Most descendants of the AB cells do not become muscle cells when P2 is present; all descendants of the AB cells become muscle cells when P2 is absent. Therefore, P2 must prevent some AB descendants from becoming muscle cells.

12. Proper *ftz* expression requires *Kr* in the fourth and fifth segments and *kni* in the fifth and sixth segments.

15. The anterior–posterior axis would be reversed.

Chapter 24

1. Selection, mutation, migration, inbreeding, and random genetic drift

2. $A1 = 0.57, A2 = 0.43$

4. 0.65

6. **a.** $p = [406 + 1/2(744)]/1482 = 0.52$, $q = [332 + 1/2(744)]/1482 = 0.48$. If the population is in Hardy-Weinberg equilibrium, the genotypes should be distributed as follows:

$$L^M/L^M = p^2(1482) = 401 \quad \text{Actual: } 406$$
$$L^M/L^N = 2pq(1482) = 740 \quad \text{Actual: } 744$$
$$L^N/L^N = q^2(1482) = 341 \quad \text{Actual: } 332$$

The population is in equilibrium.

b. If mating is random with respect to blood type, then the following frequency of matings should occur:

$L^M/L^M \times L^M/L^M = (p^2)(p^2)(741) = 54$ Actual: 58

$L^M/L^M \times L^M/L^N$ or $L^M/L^N \times L^M/L^M = (2)(p^2)(2pq)(741) = 200$
     Actual: 202

$L^M/L^N \times L^M/L^N = (2pq)(2pq)(741) = 185$ Actual: 190

$L^M/L^M \times L^N/L^N$ or $L^N/L^N \times L^M/L^M = (2)(p^2)(q^2)(741) = 92$
     Actual: 88

$L^M/L^N \times L^N/L^N$ or $L^N/L^N \times L^M/L^N = 2(2pq)(q^2)(741) = 170$
     Actual: 162

$L^N/L^N \times L^N/L^N = (q^2)(q^2)(741) = 39$ Actual: 41

The mating is random with respect to blood type.

7. a. and **b.**

Population	p	q	Equilibrium?
1	1.0	0.0	yes
2	0.5	0.5	no
3	0.0	1.0	yes
4	0.625	0.375	no
5	0.3775	0.625	no
6	0.5	0.5	yes
7	0.5	0.5	no
8	0.2	0.8	yes
9	0.8	0.2	yes
10	0.993	0.007	yes

c. 0.898

d. $p = 0.56$, $q = 0.44$

9. The frequency of a phenotype in a population is a function of the frequency of alleles that lead to that phenotype in the population. To determine dominance and recessiveness, do standard Mendelian crosses.

11. Wild-type alleles are usually dominant because most mutations result in lowered or eliminated function. To be dominant, the heterozygote has approximately the same phenotype as the dominant homozygote. This will typically be true when the wild-type allele produces a functional product and the mutant allele does not.

Chromosomal rearrangements are often dominant mutations because they can cause gross changes in gene regulation or even cause fusions of several gene products. Novel activities, overproduction of gene products, and so forth, are typical of dominant mutations.

14. Albinos may have been considered lucky and encouraged to breed at very high levels in comparison with nonalbinos. They may also have been encouraged to mate with each other. Alternatively, in the tribes with a very low frequency, albinos may have been considered very unlucky and destroyed at birth or prevented from marriage.

17. (a) 0.528; **(b)** 0.75.

18. 0.01

20. 6.5

Chapter 25

2. a. Shank length: $H^2 = 0.200$ $h^2 = 0.150$
Neck length: $H^2 = 0.600$ $h^2 = 0.010$
Fat content: $H^2 = 0.500$ $h^2 = 0.400$

b. Fat content would respond best to selection.

c. 8.9%

3. a. p(homozygous at 3 loci) $= 2(\frac{1}{2})^3 = \frac{2}{8}$

p(homozygotic at two loci) $= 3(\frac{1}{2})^3 = \frac{3}{8}$

p(homozygotic at three loci) $= (\frac{1}{2})^3 = \frac{1}{8}$

b. p(0 capital letters) $= \frac{1}{64}$ p(4 capital letters) $= \frac{15}{64}$

p(1 capital letter) $= \frac{3}{32}$ p(5 capital letters) $= \frac{3}{32}$

p(2 capital letters) $= \frac{15}{64}$ p(6 capital letters) $= \frac{1}{64}$

p(3 capital letters) $= \frac{10}{32}$

5. The population described would be distributed as follows:

three bristles	$\frac{19}{64}$
two bristles	$\frac{44}{64}$
one bristle	$\frac{1}{64}$

Note that the three-bristle class contains 7 different genotypes, the two-bristle class contains 19 different genotypes, and the one-bristle class contains only 1 genotype. It would be very difficult to determine the underlying genetic situation by doing controlled crosses and determining progeny frequencies.

7. Mean $= 4.7$
Variance $= 0.2619$
Standard deviation $= 0.5117$

9. a. H^2 has meaning only with respect to the population that was studied in the environment in which it was studied. Otherwise, it has no meaning.

b. Neither H^2 nor h^2 is a reliable measure that can be used to generalize from a particular sample to a "universe" of the human population. They certainly should not be used in social decision-making (as implied by the terms *eugenics* and *dysgenics*).

c. Again, H^2 and h^2 are not reliable measures, and they should not be used in any decision-making with regard to social problems.

11. First, define *alcoholism* in behavioral terms. Next, realize that all observations must be limited to the behavior that you used in the definition and that all conclusions from your observations are applicable to only that behavior. To do your data gathering, you must work with a population in which familiarity is distinguished from heritability. In practical terms, this population must consist of persons who are genetically close but who are found in all environments possible.

14. a. If you assume that individuals at the extreme of any spectrum are homozygous, then their offspring are more likely to be heterozygous than the original individuals. That is, they will be less extreme.

b. For Galton's data, regression is an estimate of heritability (h^2), assuming that there were few environmental differences between father and son.

Chapter 26

2. The three principals are: (1) organisms within a species vary from one another, (2) the variation is heritable, and (3) different types leave different numbers of offspring in future generations.

4. A population will not differentiate from other populations by local inbreeding if:

$$\mu 1/N$$

so

$$N1/\mu$$
$$N10^5$$

7. The mean fitness of population 1 [$p(A) = 0.8$, $p(B) = 0.9$] is 0.92. The mean fitness of population 2 [$p(A) = 0.2$, $p(B) = 0.2$] is 0.73. Because selection acts to increase the mean fitness, the frequencies of both A and B should increase in the next generation. There is a single adaptive peak at $A/A \cdot B/B$. By inspection, the fitness is lowest at $a/a \cdot b/b$ and highest at $A/A \cdot B/B$. The allelic frequency at the peak is 1.0 for both A and B.

10. The α and β gene families show remarkable amino acid sequence similarities (see Table 26-4 of the text). Within each gene family, sequence similarities are greater and, in some cases, member genes have identical intron-exon structure.

12. A geographical race is a population that is genetically distinguishable from other local populations but is capable of exchanging genes with those other local populations.

A species is a group of organisms that exchange genes within the group but cannot do so with other groups.

Populations that are geographically separated will diverge from each other as a consequence of a combination of unique mutations, selection, and genetic drift. For populations to diverge enough to become reproductively isolated, spatial separation sufficient to prevent any effective migration is usually necessary.

14.

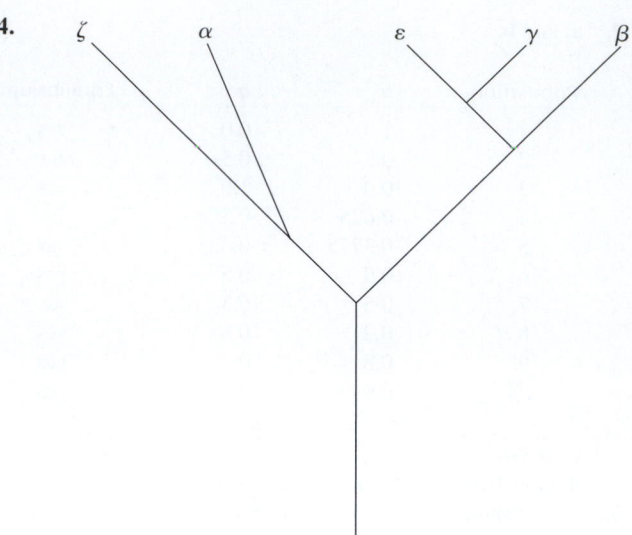

Index